MOTOR
Auto Body Repair

Third Edition

Robert Scharff
James E. Duffy

Delmar Publishers

an International Thomson Publishing company I**T**P®

Albany • Bonn • Boston • Cincinnati • Detroit • London • Madrid
Melbourne • Mexico City • New York • Pacific Grove • Paris • San Francisco
Singapore • Tokyo • Toronto • Washington

NOTICE TO THE READER

Publisher does not warrant or guarantee any of the products described herein or perform any independent analysis in connection with any of the product information contained herein. Publisher does not assume, and expressly disclaims, any obligation to obtain and include information other than that provided to it by the manufacturer.

The reader is expressly warned to consider and adopt all safety precautions that might be indicated by the activities herein and to avoid all potential hazards. By following the instructions contained herein, the reader willingly assumes all risks in connection with such instructions. The Publisher makes no representation or warranties of any kind, including but not limited to, the warranties of fitness for particular purpose or merchantability, nor are any such representations implied with respect to the material set forth herein, and the publisher takes no responsibility with respect to such material. The publisher shall not be liable for any special, consequential, or exemplary damages resulting, in whole or part, from the readers' use of, or reliance upon, this material.

Delmar Staff:

Publisher: Robert Lynch
Acquisitions Editor: Vernon Anthony
Developmental Editor: Catherine Wein
Project Editor: Christopher Chien
Production Coordinator: Toni Bolognino
Art and Design Coordinator: Cheri Plasse
Marketing Coordinator: Nicole Benson
Cover Design: Charles Cummings Art/Advertising, Inc.
Cover Image: Jeff Hinckley

COPYRIGHT © 1998

Delmar is a division of Thomson Learning. The Thomson Learning logo is a registered trademark used herein under license.

Printed in the United States of America
3 4 5 6 7 8 9 10 XXX 02 01 00 99

For more information, contact Delmar, 3 Columbia Circle, PO Box 15015, Albany, NY 12212-0515; or find us on the World Wide Web at http://www.delmar.com

International Division List

Japan:
Thomson Learning
Palaceside Building 5F
1-1-1 Hitotsubashi, Chiyoda-ku
Tokyo 100 0003 Japan
Tel: 813 5218 6544
Fax: 813 5218 6551

UK/Europe/Middle East:
Thomson Learning
Berkshire House
168-173 High Holborn
London
WC1V 7AA United Kingdom
Tel: 44 171 497 1422
Fax: 44 171 497 1426

Canada:
Nelson/Thomson Learning
1120 Birchmount Road
Scarborough, Ontario
Canada M1K 5G4
Tel: 416-752-9100
Fax: 416-752-8102

Australia/New Zealand:
Nelson/Thomson Learning
102 Dodds Street
South Melbourne, Victoria 3205
Australia
Tel: 61 39 685 4111
Fax: 61 39 685 4199

Latin America:
Thomson Learning
Seneca, 53
Colonia Polanco
11560 Mexico D.F. Mexico
Tel: 525-281-2906
Fax: 525-281-2656

Asia:
Thomson Learning
60 Albert Street, #15-01
Albert Complex
Singapore 189969
Tel: 65 336 6411
Fax: 65 336 7411

All rights reserved Thomson Learning 1998. The text of this publication, or any part thereof, may not be reproduced or transmitted in any form or by any means, electronics or mechanical, including photocopying, recording, storage in an information retrieval system, or otherwise, without prior permission of the publisher.

You can request permission to use material from this text through the following phone and fax numbers.
Phone: 1-800-730-2214; Fax 1-800-730-2215; or visit our Web site at http://www.thomsonrights.com

Library of Congress Cataloging-in-Publication Data:

Scharff, Robert.
    Motor auto body repair / Robert Scharff, James E. Duffy.—3rd ed/
        p. cm.
    Rev. ed. of: Motor auto body repair. 2nd ed. 1992.
    Includes index.
    ISBN 0-8273-6858-5
        1. Automobiles—Bodies—Maintenance and repair. I. Duffy, James
    E. II. Motor auto body repair. III. Title.
TL255.S33   1996
629.2'6'0288—dc21   96-47077
                                        CIP

TABLE OF CONTENTS

FOREWORD vi

PREFACE vii

ACKNOWLEDGMENTS ix

FEATURES OF THE TEXT xi

CHAPTER 1

Body/Paint Shop Work and Safety Procedures 1

1.1 Body and Paint Shop Operations 3
1.2 Shop Safety Practices 14
Review Questions 29

CHAPTER 2

Understanding Automobile Construction 32

2.1 Body Shapes and Parts 34
2.2 Construction Types 37
2.3 Conventional Body–Over–Frame Construction 39
2.4 Unitized Frame and Body Construction 44
2.5 Body Parts 57
Review Questions 62

CHAPTER 3

Body Shop Hand Tools 64

3.1 General Purpose Tools 64
3.2 Bodyworking Tools 79
3.3 Body Surfacing Tools 90
3.4 Hand Tool Safety 92
Review Questions 97

CHAPTER 4

Body Shop Power Tools 99

4.1 Air–Powered Tools 99
4.2 Electric–Powered Tools 117
4.3 Hydraulically Powered Shop Equipment 123
4.4 Power Jacks and Straightening Equipment 123
4.5 Hydraulic Tool Care 126
4.6 Hydraulic Lifts 127
Review Questions 130

CHAPTER 5

Compressed Air Supply Equipment 132

5.1 The Air Compressor 133
5.2 Air and Fluid Control Equipment 140
5.3 Compressor Accessories 144
5.4 Air System Maintenance 146
5.5 Air System Safety 147
Review Questions 148

CHAPTER 6

Auto Body Materials and Fasteners 150

6.1 Refinishing Materials 150
6.2 Fasteners 164
Review Questions 172

CHAPTER 7

Welding Equipment and Its Use 174

7.1 MIG Welding 181
7.2 MIG Welding Equipment 183
7.3 MIG Operation Methods 185
7.4 Basic Welding Techniques 193
7.5 MIG Welding Galvanized Metals and Aluminum 203
7.6 Testing the MIG Weld 205
7.7 MIG Weld Defects 205
7.8 Flux–Cored Arc Welding 205
7.9 TIG Welding 208
7.10 Resistance Spot Welding 208
7.11 Other Spot Welding Functions 218
7.12 Stud Spot Welding for Dent Removal 220
7.13 Mold Rivet Welding 220
7.14 Oxyacetylene Welding 220
7.15 Brazing 225
7.16 Soldering (Soft Brazing) 229
7.17 Plasma Arc Cutting 229
Review Questions 233

CHAPTER 8

Basic Auto Sheet Metal Work 235

8.1 Automotive Sheet Metal 235
8.2 Classifying Body Damage 241
8.3 Metal Straightening Techniques 247
8.4 Metalworking Techniques 249
8.5 Working Aluminum 261
Review Questions 263

CHAPTER 9

Minor Auto Body Repairs 265

9.1   Body Fillers 265
9.2   Applying Plastic Body Filler 272
9.3   Applying Lead Filler 281
9.4   Repairing Scratches 283
9.5   Repairing Nicks 285
9.6   Repairing Dings 286
9.7   Repairing Rust Damage 287
9.8   Repairing Small Rustouts 290
9.9   Repairing Large Rustouts 292
Review Questions 296

CHAPTER 10

Diagnosing Major Collision Damage 300

10.1   Impact and Its Effects on a Vehicle 304
10.2   Visually Determining Extent of Impact Damage 313
10.3   Measurement of Body Dimensions 315
10.4   Gauge Measuring Systems 317
10.5   Tram Gauges 319
10.6   Centering Gauges 326
10.7   Strut Centerline Gauge 332
10.8   Diagnosing Damage Using Gauge Measuring System 333
10.9   Universal Measuring Systems 336
10.10   Dedicated Bench and Fixture Measuring Systems 346
Review Questions 351

CHAPTER 11

Body Alignment 354

11.1   Body Alignment Basics 356
11.2   Straightening Equipment 357
11.3   Straightening and Realigning Techniques 368
11.4   Alignment Safety Considerations 368
11.5   Planning Collision Repair Procedures 369
11.6   Stress Relieving 385
11.7   Final Straightening Considerations 388
Review Questions 390

CHAPTER 12

Panel Replacement and Adjustment 393

12.1   Panel Removal 393
12.2   Installing New Panels 403
12.3   Structural Sectioning 410
12.4   Antirust Treatments 423
12.5   Door Panel Replacement 424
12.6   Panel Adjustments 427
12.7   Custom Body Panels 443
Review Questions 445

CHAPTER 13

Servicing Mechanical, Electrical, and Electronic Components 447

13.1   Power Train Construction 452
13.2   Suspension Systems 465
13.3   Steering Systems 471
13.4   Wheel Alignment 478
13.5   Brake Systems 488
13.6   Cooling Systems 496
13.7   Heater Operation 502
13.8   Air Conditioning and Heater Systems 503
13.9   Exhaust Systems 506
13.10   Emission Control Systems 508
13.11   Hose and Tubing Inspection 510
13.12   Checking Electrical Problems 511
13.13   Electronic System Service 529
Review Questions 535

CHAPTER 14

Repairing Auto Plastics 537

14.1   Types of Plastics 538
14.2   Plastic Repair 542
14.3   Chemical Adhesive Bonding Techniques 544
14.4   Plastic Welding 550
14.5   Hot–Air Plastic Welding 550
14.6   Airless Plastic Welding 553
14.7   Ultrasonic Plastic Welding 553
14.8   Plastic Welding Procedures 553
14.9   Repairing Vinyl 560
14.10   Ultrasonic Stud Welding 562
14.11   Reinforced Plastic Repairs 562
Review Questions 574

CHAPTER 15

Other Body Shop Repairs 576

15.1   Replacing Glass 576
15.2   Types of Glass 576
15.3   Removing Windshields and Rear Window Glass 578
15.4   Door Window Glass Service 591
15.5   Door and Trunk Locks 593
15.6   Locating Leaks 595
15.7   Headlights 598
15.8   Tail, Backing, and Stop Lights 600
15.9   Rattle Elimination 601
15.10   Adjusting or Replacing Bumpers 602
15.11   Restraint Systems 604
15.12   Installing Body Molding 613
15.13   Vinyl Roof Servicing 616
Review Questions 630

CHAPTER 16

Restoring Corrosion Protection 632

16.1   What is Corrosion? 633
16.2   Causes for Loss of Factory Protection 636
16.3   Anticorrosion Materials 640
16.4   Basic Surface Preparation 642
16.5   Corrosion Treatment Areas 643
16.6   Corrosion Protection Primers 643
16.7   Exposed Joints 648
16.8   Exposed Interior Surfaces 651
16.9   Exposed Exterior Surfaces 653

16.10   Exterior Accessories   654
16.11   Acid Rain Damage   654
Review Questions   656

CHAPTER 17

Vehicle Surface Preparation   658

17.1   Evaluation of Surface Condition   659
17.2   Sanding   660
17.3   Coated Abrasives (Sandpaper)   661
17.4   Methods of Sanding   666
17.5   Types of Sanding   672
17.6   Refinishing Surfaces   677
17.7   Painted Surface in Good Condition   677
17.8   Paint Work in Poor Condition   679
17.9   Bare Metal Substrate   683
17.10   Using Primers   686
17.11   Plastic Parts Preparation   689
17.12   Masking   691
Review Questions   699

CHAPTER 18

Refinishing Equipment and Its Use   702

18.1   Spray Guns   703
18.2   Spraying Techniques   710
18.3   Cleaning the Spray Gun   724
18.4   Spray Gun Troubleshooting   727
18.5   Other Spray Systems   734
18.6   Spray Booths   742
18.7   Spray Booth Maintenance   749
18.8   Drying Room   750
18.9   Other Paint Shop Equipment and Tools   753
18.10   Basic Paint Shop Materials   756
Review Questions   760

CHAPTER 19

Refinishing Procedures   762

19.1   Purpose of Refinishing   762
19.2   Types of Refinishing Repair   764
19.3   Applying Undercoats   765
19.4   Topcoats   768
19.5   Determining If the Auto Has Been Painted   772
19.6   Determining Type of Paint on Vehicle   773
19.7   Selecting Solvents (Reducers and Thinners)   774
19.8   Pre-Painting Review   778
19.9   Repainting Spray Methods   781
19.10   Spatter Finishes   788
19.11   Topcoats for Plastic Automotive Parts   788
Review Questions   797

CHAPTER 20

Color Matching and Custom Painting   799

20.1   Color Theory   800
20.2   Dimensions of Color   802
20.3   Plotting Color   802
20.4   Paint Color Matching   803

20.5   Analyzing Color   806
20.6   Tinting   808
20.7   Matching Solid Colors   810
20.8   Matching Metallic Finishes   810
20.9   Matching Basecoat/Clearcoat Finishes   813
20.10   Matching Pearl Luster Finishes   814
20.11   Matching Tri–Coat Finishes   815
20.12   Test Panels   817
20.13   Why a Color Mismatch?   818
20.14   Tinting Summary   818
20.15   Custom Painting   820
Review Questions   823

CHAPTER 21

Paint Problems and Final Detailing   824

21.1   Painting Problems   826
21.2   Removing Masking Materials   837
21.3   Final Detailing   838
21.4   Caring for New Finish   845
21.5   Decal Replacement   845
21.6   Final Detailing   847
Review Questions   849

CHAPTER 22

Repair Cost Estimating and Entrepreneurship   850

22.1   The Estimate   851
22.2   Estimating Bids   856
22.3   Computer Estimating   858
22.4   Estimating Sequence   864
22.5   Repairability of the Vehicle   868
22.6   Part Prices   869
22.7   Labor Costs   872
22.8   Refinishing Time   876
22.9   Estimate Total   878
22.10   Entrepreneurship   878
Review Questions   879

APPENDIX A

Auto Body Shop Terms   882

APPENDIX B

Abbreviations   903

APPENDIX C

Decimal and Metric Equivalents   905

APPENDIX D

Fluid and Air Nozzle Selection   906

APPENDIX E

Viscosity Conversion Chart   907

APPENDIX F

Chemical Substance Information Tables   908

INDEX   910

FOREWORD

Today is an exciting time for a young man or young woman entering the world of automotive body repair. Gone are the days when body repair meant simply banging dents out of a fender with hammer and dolly. Today's autobody technician is a skilled professional with expertise in many areas: computerized estimating, diagnostic equipment, sophisticated measuring systems, and shop management, to name a few. The successful collision repair specialist must be familiar with complex automotive systems as well as with the modern materials that go into today's automobiles. Most of all, the autobody technician must know how to access and use information, whether that information is about damage estimating, repair techniques, or making a profit.

Consider *MOTOR Auto Body Repair* as the first and most important piece of information for anyone contemplating a career in the collision repair industry. It begins by helping you to understand how the automobile is built; then it shows you what tools and equipment you will need to repair it; and it reviews the time-tested techniques used to restore a damaged vehicle to its pre-accident condition. It is your first step in acquiring the information you will need to become a specialist in this field.

I'm sure you realize by now that we live in the age of information. Information is the key to the future. Your future. Let me put it another way. Every successful collision repair specialist I know recognizes one elementary truth: in this business, as in any other business, *you must never stop learning*. And the best place to start is with a book like this one that raises the curtain on a stage that is brilliantly lit with high strength steel, structural glass, space age plastics, and an infinite variety of paint colors. It is a stage that invites you to play a part in an industry that rewards skill, knowledge, and hard work. We at Motor are proud to have had the leading role as information provider in the on–going drama of automotive repair since 1903. We sincerely hope that you will come and join us.

Kevin F. Carr
Vice–President & General Manager
Motor Publications

PREFACE

The late Robert Scharff took great pride in the success of MOTOR Auto Body Repair, and with good cause. This book has been used by thousands of students in hundreds of schools and colleges, and continues to this day to be a leader in its market. This edition brings a new author and a new era to this textbook, and both the author and the publisher hope that teachers and students will appreciate the improvements and updates that have been made. Most of all, we hope that the student hoping to become collision repair professionals will take from this book the information they need to get started the right way. Textbooks are just one part of a vocational-technical education; it is our sincere desire that this book adds something good to that education.

Writing is all about communication between the writer and the reader. This edition of *MOTOR Auto Body Repair* is the most "readable" edition ever published. If you have taught from prior editions, we believe that you will notice a real difference in ease of comprehension. We have worked hard to make the book easier to understand without diminishing its technical completeness or challenge. In particular, we think you will notice a much better handling of new technical vocabulary. Every effort has been made to help the student identify key new terms as they read, and to consistently define those terms clearly and for the beginner, upon first use.

Every textbook preface you have or will ever read in this field will suggest that the book you are reading reflects all of the many changes in the collision repair field. Readers then typically find that, of course, no textbook can possibly keep up to date with the trade. We don't pretend that this book is perfect, nor that as time passes teachers and students won't notice that products and procedures have changed. All textbooks must be augmented by manufacturer's manuals, I-CAR materials, and the teacher's knowledge of the state of the trade (as well as the student's own experiences as they read and compare the author's ideas to their actual experiences). What we do stand by is that this book will give the reader an absolutely sound introduction to the basics. We also have made every effort to update this book. Examples of this include:

- Artwork. This book has always had thousands of illustrations. This edition includes the most complete overhaul of the art since the book was published. We think you will find it to be the most current of all the collision repair books in print.
- New chapter coverage has been added on subjects including Auto Body Materials, Electronic Components, Color Matching, Custom Painting, Paint Problems, and Final Detailing. Hundreds of equipment and paint companies were contacted and their manuals and publications used to make this book current.
- New coverage of hand and power tools, basic measurement skills, service manuals, fasteners, materials, liquid mask, fine-line masking tapes, surface evaluation, sanding, scuffing, stripping, using paint mixing sticks, paint matching, waterbase paints, computerized estimating, entrepreneurship, and many other topics, is included.
- Increased coverage is now given to high-strength steels, aluminum, plastic and composite part repair are now given increased coverage reflecting the trend to use lighter materials in the construction of the modern vehicle.
- Greater emphasis on ASE certification I-CAR standards have been built in. This book is organized around NATEF competency tasks for collision repair, and lists of the tasks covered in each chapter appear at the beginning of the chapter.
- This book has a refined collection of chapter-end exercises, including ASE-style and essay questions and critical thinking problems. Math problems relevant to the collision repair student are now included in every chapter.
- Many new photographs were taken at a large collision repair shop and organized into photosequences. One is a sixteen-page, full-color overview of the complete repair procedure, and the others focus on procedures in detail.
- Every effort has been made to make this book correct. Collision repair teachers, I-CAR trainers, experts at MOTOR, and others were brought in to review this book and ensure its accuracy. We worked with all of their comments and criticisms.
- Safety is emphasized throughout the text. Safety cautions and warnings appear frequently, and we worked to make sure that all the illustrations represent safe practices.

- A completely new shop manual, written by an experienced automotive collision repair teacher, has been written for this edition. The job sheets in this shop manual are taken from actual exercises this teacher has performed in his own classes. They work.
- We have revised and expanded the instructor's guide for this edition.

We are anxious to know what you think of the effort to update and upgrade *MOTOR Auto Body Repair*. Send letters or call the publisher so I can hear what you think. We hope you will find this book a useful resource for many years to come.

James E. Duffy, "A Fellow Educator"

ACKNOWLEDGMENTS

Three-C Body Shop, Inc. was honored to play a role in creating the illustrations in the new edition of *MOTOR Auto Body Repair*, and to thus further the highly technical process of repair and training needed for the 21st century, and to help future technicians enter this challenging field.

Robert A. Juniper, President
Chris Sexton, Production Manager
Three-C Body Shop, Inc., Columbus, Ohio

REVIEWERS

John Brosda
Northern Alberta Institute of Technology
Edmonton, AB, CANADA

Kurt L. Carlson
Cape Cod Regional Technical High School
Harwich, MA

Michael Crandell
Carl Sandburg College
Galesburg, IL

Edward J. Curtis
Wyoming Technical Institute
Laramie, WY

Ronald D. Dohi
El Camino Community College
Torrance, CA

Peter Gall
Lakeshore Technical College
Cleveland, WI

Bruce Gamroth
Arizona Automotive Institute
Glendale, AZ

Michael Jund
Scott Community College
Bettendorf, IA

Keith Schieffer
Amarillo Technical Center
Amarillo, TX

Clifford J. Smith
Northwest Technical College
Wadena, MN

Steve White
Portland Community College
Portland, OR

CONTRIBUTING COMPANIES

Through the generosity of the following companies, technical information and art were supplied to enhance the book.

3M Automotive Trades Division
Accuspray Inc.
ALC Sandy Jet
American Engineered Components
Atlantic Pneumatic
Arn-Wood Company, Inc.
Audi
Babcox Publications
Badger Air Brush
BASF

ASE CERTIFIED

WE SUPPORT
VOLUNTARY TECHNICIAN
CERTIFICATION THROUGH

National Institute for
**AUTOMOTIVE
SERVICE
EXCELLENCE**

Bee Line Co.
Biddle Instruments
Binks Mfg. Company
Black & Decker (U.S.) Inc.
Blackhawk Automotive, Inc.
Bond Tite®, Division of U.S. Chemical and
    Plastics, Inc.
Brian R. White Co.
Bron Corp Mfg. Co.
Car-O-Liner
Carborundum Abrasives Co.
Chicago Pneumatic Tool Division
Century Mfg. Co.
Champion Pneumatic Machinery Co.
Chicago Pneumatic Tool Co.
Chief Automotive Systems, Inc.
Chrysler Corporation
Clements National Co.
Dataliner AB
Data Welder
Dedoes Industries, Inc.
Delta International Machinery Corp.
Dorman Products
DuPont Automotive Products
Dynatron/Bondo Corp.
Easco/K-D Tools
The Eastwood Co.
Elcometer
Equalizer Industries
Eurovac, Inc.
Fiberglass Evercoat Co., Inc.
Fitz and Fitz, Inc.
Florida Department of Vocational Education
Florida Pneumatic Mfg.
Ford Motor Company
Garmat USA, Inc.
General Motors Corporation, Service Technology
    Group
Goodson Shop Supplies
Graco, Inc.
Guy Chart
H&S Manufacturing
HAKO Minuteman, Inc.
Hankison International
Henning Hansen, Inc.
Herkules Equipment
American Honda Motor Company, Inc.
HTP America, Inc.
Hunter Engineering
Hutchins Mfg, Inc.
I-CAR
Isuzu
ITW DeVilbiss Automotive Refinishing Products
Kansas Jack, Inc.

Lenco, Inc.
Lincoln Automotive
Lisle Corporation
Lors Machinery, Inc.
Maaco Enterprises, Inc.
Majestic Tools
Marson Corporation
Martin-Senour Automotive Finishes
Mazda Motor of America
Mercedes-Benz
Mine Safety Appliance Co.
Mitchell International
Morgan Mfg
Mustang Monthly Magazine
National Automobile Dealers Used Car Guide Co.
National Institute for Automotive Service
    Excellence (ASE)
Nicator, Inc.
Nilfisk of America, Inc.
Nissan North America, Inc.
Norco Industries, Inc.
Norton
OTC Division of SPX
Paint Safe Products
Palnut Co.
Pittway Corp.
PPG Refinishing Industries
Proto Tools
Rotary Lift Corp.
S&G Tool Aid Corp.
Saab Cars, USA, Inc.
Seelye, Inc.
Sellstrom Manufacturing, Inc.
The Sherwin-Williams Company
S.M. Arnold
Snap-on-Tools Corporation
Spartan Plastics
Stanley Works
Steck Manufacturing Co., Inc.
Style-Line Corp.
Subaru of America, Inc.
Talsol, Inc.
Tech-Cor, Inc.
Thermal Devices, Inc.
Toyota Motor Corporation
Truman's, Inc.
TRW Fasteners Division
Unicorn Corp.
Urethane Supply Co., Inc.
Volkswagon of America, Inc.
Vulcan Materials Co.
Wedge Clamp
Willson Safety Products

FEATURES OF THE TEXT

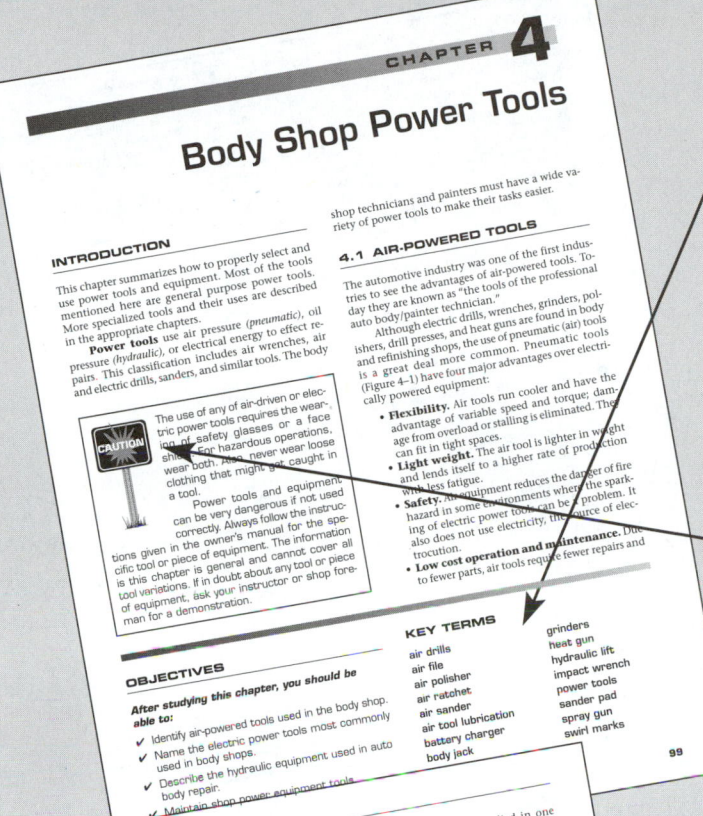

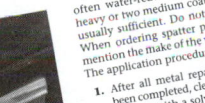

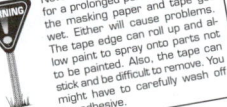

KEY TERMS

The "Key Terms" are the most important technical words you will learn in the chapter. These are listed at the beginning of each chapter and appear in bold print where they are first defined. For added study, you can write the key terms on a sheet of paper with its definition to make sure you can explain the terms. These terms are also given in the glossary at the back of the book.

CAUTIONS

"Cautions" summarize critical safety rules. They alert you to operations that could hurt you or someone else. They are not only covered in the safety chapter, but you will find them throughout the text where they apply. Read and remember all cautions. Your health is invaluable.

WARNINGS

"Warnings" provide important information to help prevent the kind of accidents that can damage parts or tools. They are common mistakes that should be avoided. They appear throughout the text where they apply to the instructions being given.

SHOP TALK

"Shop Talk" gives added information to help you complete a particular procedure successfully or to make a task easier. They are hints to help you work more efficiently and profitably.

CHAPTER SUMMARY

Each "Chapter Summary" gives a brief list of the most important information in the chapter. It will help you review and understand which points were the most important.

ASE TASK LISTS

The ASE Task Lists at the beginning of each chapter identify the tasks associated with the content of the chapter.

REVIEW QUESTIONS

"Review Questions" will help measure the skills and knowledge you learned in the chapter. To check your "brain power," different types of questions are given: ASE, Essay, Critical Thinking Problems, and Math Problems. The ASE questions will help you prepare to pass Auto Body Repair certification tests.

---

a continuous high frequency arc is best for punching through the nonconductive surface layer and for keeping the arc going while cutting. When cutting bare metal, a high frequency arc is needed only to

start the arc. Once the torch starts to cut, a direct current pilot arc is all that is needed to keep things going. The bare metal position gives the longest electrode and nozzle life.

Welding Equipment and Its Use   **233**

SUMMARY

- There are three basic methods of joining metal together in the automobile assembly:
  Mechanical (metal fastener) methods
  Chemical (adhesive fastening) methods
  Welding (molten metal) methods
- Welding is one method of repair in which heat is applied to the pieces of metal to fuse them together into the shape desired.
- Visible weld penetration is indicated by the height of the exposed surface of the weld on the back side. Full weld penetration is needed to assure maximum weld strength.
- MIG welding is recommended by all OEMs, not only for HSS and unibody repair, but for all structural collision repair.
- The resistance spot welder provides very fast, high-quality welds while maintaining the best control of temperature buildup in adjacent panels and structure.
- Always follow service manual recommendations when welding. This will assure structural integrity.
- During the welding process, either inert gas or active gas shields the weld from the atmosphere and prevents oxidation of the base metal.

- Flat welding means the pieces are parallel with the bench or shop floor.
- Horizontal welding has the pieces turned sideways. Gravity tends to pull the puddle into the bottom piece. When welding a horizontal joint, angle the gun upward to hold the weld puddle in place against the pull of gravity.
- Vertical welding has the pieces turned upright. Gravity tends to pull the puddle down the joint. When welding a vertical joint, the best procedure is usually to start the arc at the top of the joint and pull downward with a steady drag.
- Overhead welding has the workpieces turned upside down. The tack weld is exactly that: a tack—a relatively small, temporary MIG spot weld that is used instead of a clamp or sheet metal screw to tack and hold the fit-up in place while a permanent weld is made.
- In a continuous weld, an uninterrupted seam or bead is laid down in a slow, steady, ongoing movement.
- A plug weld is made in a drilled or punched hole through the outside piece (or pieces).
- If welding defects should occur, think of ways to change your procedures to correct them.

ASE-STYLE REVIEW QUESTIONS

1. Technician A uses a forward gun angle to achieve a deep penetration in the metal. Technician B uses the reverse gun angle to achieve a flat bead. Who is correct?
   A. Technician A
   B. Technician B
   C. Both A and B
   D. Neither A nor B

2. Technician A says that the main function of the gun nozzle is to provide gas protection. Technician B says that if the insulation in the gun nozzle area is bypassed the current will ignite the inert shielding gas. Who is correct?
   A. Technician A
   B. Technician B
   C. Both A and B
   D. Neither A or B

3. Welding current affects which of the following?
   A. Base metal penetration depth
   B. Arc stability
   C. Amount of weld spatter
   D. All of the above

4. When MIG welding, what happens if the tip-to-base metal distance is too long?
   A. The shield gas effect is reduced.
   B. The wire protruding from the end of the gun increases and becomes preheated.
   C. The melting speed of the wire increases.
   D. All of the above.

---

2   Chapter 1

  ASE TASK LIST
Job Skills covered in this chapter include:

PAINTING AND REFINISHING TEST (B2) TASK LIST

A. Surface Preparation
3. Inspect and identify substrate, type of finish, and surface condition; develop a plan for refinishing.
8. Mask trim, and protect other areas that will not be refinished.
9. Mix primer, primer-surfacer, or primer-sealer; spray onto surface of repaired area.
13. Clean area to be refinished using a proper cleaning solution.

C. Paint Mixing, Matching, and Applying
1. Determine type and color of paint already on vehicle.
4. Apply basecoat/clearcoat for spot and panel blending, and overall refinishing.

F. Safety Precautions and Miscellaneous
1. Identify and take necessary precautions with hazardous operations and materials according to EPA regulations.
2. Inspect personnel health and safety hazards according to OSHA guidelines.

4. Select proper respiratory protection system; inspect to insure proper fit and operation.

NONSTRUCTURAL ANALYSIS AND DAMAGE REPAIR TEST (B3) TASK LIST

A. Preparation
2. Lift, raise, and position vehicle to perform repairs.

C. Metal Finishing and Body Filling
1. Grind off paint from the damaged area of a body panel.

E. Welding and Cutting
5. Identify safety considerations: Eye protection, proper clothing, shock hazards, fumes, M.S.D.S., etc. before beginning any welding operation.
6. Apply knowledge of the proper procedures for safely handling gas cylinders.
9. Protect vehicle components, including computers and other electronic modules, from possible welding and cutting damage.

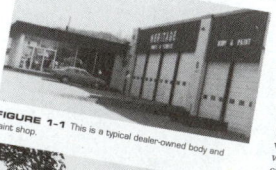

FIGURE 1-1 This is a typical dealer-owned body and paint shop.

FIGURE 1-2 An independent body shop is not affiliated with one make of vehicle or manufacturer. (Courtesy of Tech-Cor)

finish business. This tremendous volume of work requires a huge number of properly trained technicians and related personnel. The information in this text will help you enter this "in- demand" industry more successfully.

WHAT IS COLLISION REPAIR?

A **collision**, commonly called a "crash" or "wreck," is an impact caused by the vehicle hitting another vehicle or object. Since cars and light trucks can weigh well over a ton, metal parts are often crushed, bent, and torn. Plastic parts can be broken and deformed. The frame might even be forced out of alignment from the tremendous force of an auto crash.

A collision might be as minor as a "door ding" where someone accidentally opens the door and hits another car, or it might be severe enough to cause a **total loss**, where repairs would be more expensive than the cost of buying another car. In this case, the insurance company does not pay for repairs but gives the policy holder enough money to purchase a similar year, make, and model vehicle.

The direction and force of impact, type of body structure, and other factors are all important. For example, the damage may be the result of two

---

- The term "raising" means to work a dent outward or away from the body. The term "lowering" means to work a high spot or bump down or into the body.
- The metal that is pushed up is called a pressure area. Areas that are pushed down are called tension areas.
- Hammer-on-dolly repairs [are used] to smooth

Basic Auto Sheet Metal Work   **263**

small, shallow dents and bulges and to stretch metal so that it can return to its original shape.
- Hammer-off-dolly is used to straighten metal just before the final stage of straightening.
- A pull pin spot welder or nail gun fuses a metal pin to the dent area
- Aluminum is used for a variety of automotive panels, such as hoods and roof panels.

ASE STYLE REVIEW QUESTIONS

1. Which type of sheet metal is most often used in unibody construction?
   A. Hot-rolled
   B. Cold-rolled
   C. Both A and B
   D. Neither A nor B

2. Whenever a collapsed rolled buckle occurs, there are two other buckles adjacent to the collapsed roll. What type of buckles are these?
   A. Simple hinge buckles
   B. Collapsed hinge buckles
   C. Simple rolled buckles
   D. None of the above

3. The cross cut disc action is used to _____.
   A. Remove paint
   B. Smooth the filler
   C. Remove metal
   D. Both A and B

4. Which kind of spoon can be used instead of a hammer?
   A. Dinging spoon
   B. Slapping spoon
   C. Both A and B
   D. Neither A nor B

5. When filing a relatively flat area, Technician A holds the file at a 30-degree angle and pushes it straight. Technician B says that as long a stroke as possible should be taken. Who is correct?
   A. Technician A
   B. Technician B
   C. Both A and B
   D. Neither A or B

6. Technician A says that aluminum has a higher melting point than mild steel. Technician B says that martensitic steel loses strength when welded. Who is correct?
   A. Technician A
   B. Technician B
   C. Both A and B
   D. Neither A nor B

7. Technician A says that 80 percent of sheet metal damage is direct damage. Technician B says that bent metal is not necessarily buckled metal. Who is correct?
   A. Technician A
   B. Technician B
   C. Both A and B
   D. Neither A nor B

ESSAY QUESTIONS

1. In your own words, how do you use a body hammer to straighten metal?

2. Describe the hammer-on-dolly method.

3. Summarize the method for removing a gouge in metal.

4. List and explain five types of part loading.

CRITICAL THINKING PROBLEMS

1. An aluminum panel has a bulge in it after initial straightening. The bulge pops in and out with hand pressure. What must be done?

2. After shrinking a steel panel, the technician finds a flat area under tension and lower than the rest of the panel. What is wrong?

MATH PROBLEMS

1. If a fender is ¹⁄₁₆-inch (1.6 mm) thick and you grind away .031 inch (0.79 mm) of metal, how thick is the remaining panel?

2. If a 12-inch (305 mm) metal rod is heated and expands 2 percent in size, how long has it become?

PHOTO SEQUENCES

Several step-by-step photo sequences illustrate common Auto Body Repair procedures.

COLOR INSERT

The Color Insert takes you through the complete repair of a collision-damaged vehicle. It explains and shows the major steps needed to repair a modern vehicle. You are given a visual tour through a complete repair operation.

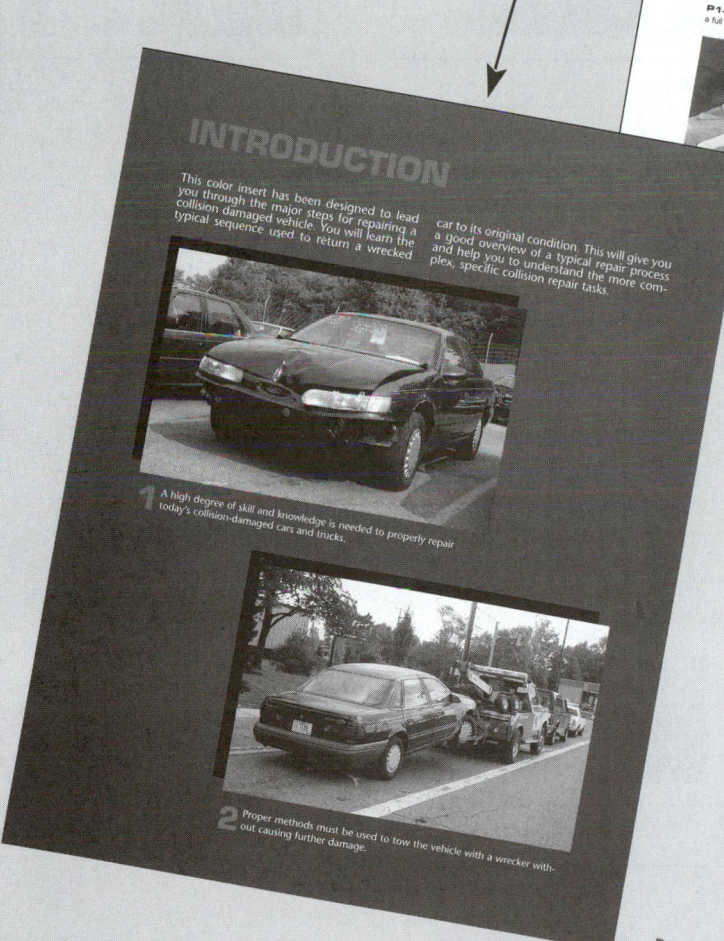

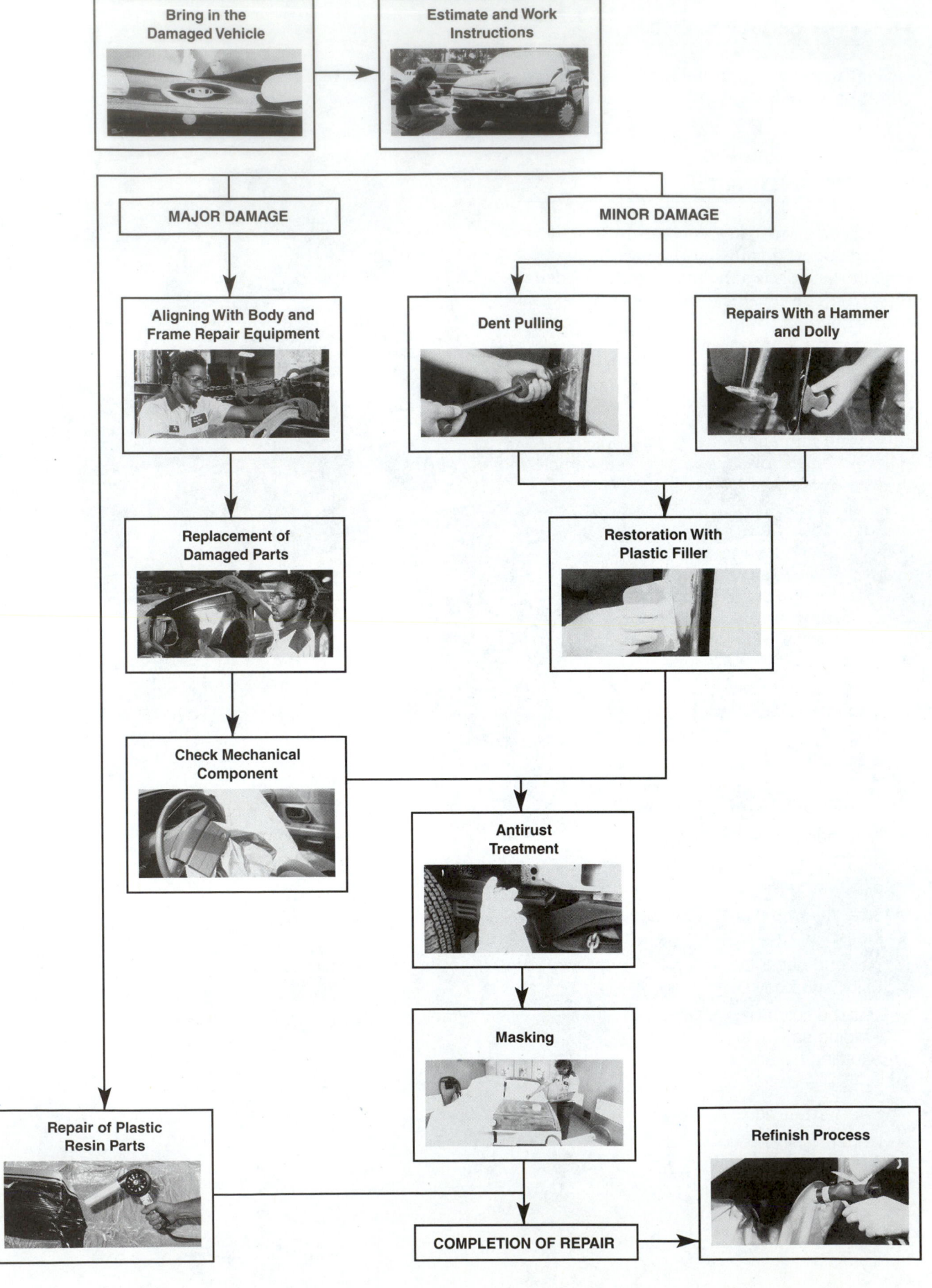

Bring in the Damaged Vehicle

Estimate and Work Instructions

MAJOR DAMAGE

MINOR DAMAGE

Aligning With Body and Frame Repair Equipment

Dent Pulling

Repairs With a Hammer and Dolly

Replacement of Damaged Parts

Restoration With Plastic Filler

Check Mechanical Component

Antirust Treatment

Masking

Repair of Plastic Resin Parts

Refinish Process

COMPLETION OF REPAIR

CHAPTER 1

Body and Paint Shop Work and Safety Procedures

INTRODUCTION

This chapter will introduce you to the challenging world of collision repair, also called auto body repair. It will summarize the major work areas in a typical shop and show how a wrecked vehicle moves through these areas when being repaired. You will learn about the different job positions and how they interact. This information will give you the background needed to more fully grasp later chapters and to become a collision repair professional.

The second section of the chapter details shop safety. It summarizes the types of hazards present in a body shop. You will learn how to best avoid injuries and how to work with hazardous materials and equip-

ment. You will learn that a body shop can be a safe and enjoyable place to work if proper procedures are followed. Conversely, it can be a very dangerous, even deadly place, if you fail to follow basic safety rules.

Of the nearly 100,000 body and paint shops in the United States, some 25,000 are owned by new-car dealers. Whether dealer-owned (Figure 1–1) or independent (Figure 1–2), no two are exactly alike. They vary in size, layout of work sections, location, number of employees, amount or type of equipment, or in work procedures. However, they all use similar methods and follow similar safety rules.

The collision repair industry is vast. Combined, U.S. body shops do an astounding annual volume of more than $10 billion in the collision repair and re-

OBJECTIVES

After studying this chapter, you should be able to:

✔ Summarize the basic steps needed to repair a car or truck damaged in an accident.

✔ Explain the major work areas of a body shop.

✔ Describe the types of positions or jobs available in the collision repair industry.

✔ List and describe the publications and associations that are available to paint and body shop employees.

✔ List the rules regarding personal safety, work area safety, tool and welding safety, environmental safety, fire safety, and the methods of handling hazardous wastes safely.

✔ Explain the "Right-to-Know" laws.

✔ Explain the requirements for ASE Body/Paint Technician Certification.

✔ Understand more fully certain ASE test questions.

KEY TERMS

ASE certification
asphyxiation
chemical burns
collision
combustibles
damage estimate
drying room
electrocution
EPA
explosions
frame rack
insurance adjuster
masking
measurement systems
metalworking
repair order
respirator
Right-to-Know laws
total loss

ASE TASK LIST

Job Skills covered in this chapter include:

PAINTING AND REFINISHING TEST (B2) TASK LIST

A. Surface Preparation

3. Inspect and identify substrate, type of finish, and surface condition; develop a plan for refinishing.
8. Mask trim, and protect other areas that will not be refinished.
9. Mix primer, primer-surfacer, or primer-sealer; spray onto surface of repaired area.
13. Clean area to be refinished using a proper cleaning solution.

C. Paint Mixing, Matching, and Applying

1. Determine type and color of paint already on vehicle.
6. Apply basecoat/clearcoat for spot and panel blending, and overall refinishing.

F. Safety Precautions and Miscellaneous

1. Identify and take necessary precautions with hazardous operations and materials according to EPA regulations.
2. Inspect personnel health and safety hazards according to OSHA guidelines.

4. Select proper respiratory protection system; inspect to insure proper fit and operation.

NONSTRUCTURAL ANALYSIS AND DAMAGE REPAIR TEST (B3) TASK LIST

A. Preparation

2. Lift, raise, and position vehicle to perform repairs.

C. Metal Finishing and Body Filling

1. Grind off paint from the damaged area of a body panel.

E. Welding and Cutting

5. Identify safety considerations: Eye protection, proper clothing, shock hazards, fumes, M.S.D.S., etc. before beginning any welding operation.
6. Apply knowledge of the proper procedures for safely handling gas cylinders.
9. Protect vehicle components, including computers and other electronic modules, from possible welding and cutting damage.

FIGURE 1-1 This is a typical dealer-owned body and paint shop.

FIGURE 1-2 An independent body shop is not affiliated with one make of vehicle or manufacturer. *(Courtesy of Tech-Cor)*

finish business. This tremendous volume of work requires a huge number of properly trained technicians and related personnel. The information in this text will help you enter this "in- demand" industry more successfully.

WHAT IS COLLISION REPAIR?

A **collision**, commonly called a *"crash"* or *"wreck,"* is an *impact* caused by the vehicle hitting another vehicle or object. Since cars and light trucks can weigh well over a ton, metal parts are often crushed, bent, and torn. Plastic parts can be broken and deformed. The frame might even be forced out of alignment from the tremendous force of an auto crash.

A collision might be as minor as a *"door ding"* where someone accidentally opens the door and hits another car, or it might be severe enough to cause a **total loss**, where repairs would be more expensive than the cost of buying another car. In this case, the insurance company does not pay for repairs but gives the policy holder enough money to purchase a similar year, make, and model vehicle.

The direction and force of impact, type of body structure, and other factors are all important. For example, the damage may be the result of two

FIGURE 1-3 Damage is often due to a collision with an object or another vehicle. *(Courtesy of Tech-Cor)*

vehicles hitting each other or one vehicle hitting a stationary object (Figure 1–3). As you will learn in later chapters, this information can be important when trying to repair the damage.

1.1 BODY AND PAINT SHOP OPERATIONS

There are two major work areas in every collision repair and refinishing shop: (1) **metalworking** (body work) and (2) *paint* (or refinishing work). Whatever the physical size of the overall shop, the work flow should be continuous, as on a factory assembly line. The vehicle should move through all areas efficiently, with each phase preparing the car for the next step.

Most body and paint shops can be classified by size. The shop layouts shown in Figure 1–4 are examples of some of the most necessary areas included in typical shops.

The repair method used for restoring a vehicle damaged in a collision is determined by many factor. Some of these include

1. Area damaged on vehicle
2. Extent or amount of damage
3. Type of painting/refinishing methods required
4. Repair costs
5. Value of vehicle

Remember that nearly any damaged automobile can be restored if the vehicle owner or insurance company is willing to pay for the repair. It is this cost that is a major consideration.

DAMAGE ESTIMATING

The vehicle is first brought into the shop where a **damage estimate** is prepared to calculate the cost of labor, parts, and materials for the repair. Then the decision is made whether the car has *major damage*

(vehicle requires body or structural repairs, Figure 1–5) or *minor damage* (vehicle requires only outer panel or cosmetic repairs, Figure 1–6).

Estimating involves analyzing damage and calculating how much it will cost to repair the vehicle. It is critical that the quote on the repair is not too high or too low. If too high, another shop with a lower bid will usually get the job. If too low, the shop's profits may not be enough to support the business. The shop could lose money!

In most shops, a well-trained *estimator* makes an appraisal of vehicle damage and determines the parts, materials, and labor needed to repair the vehicle to its original condition. This person must be well versed in how cars and trucks are made and be good with numbers, computers, and communicating with people.

Repair instructions are written down on the **repair order** (RO) and the repair operations are carried out according to those instructions. Once the repair order is received in the shop, the repair procedure follows a step-by-step pattern.

Manual estimating involves using an estimating sheet for writing out information about the vehicle, and using crash-estimating guides and collision damage manuals to make the repair estimate. *Crash estimating books* and *collision damage manuals* contain vehicle identification information, the price of new parts, time needed to install the parts, refinishing or painting data, and other information.

Computer estimating involves using electronic *hardware* (computers, printers, hard drives, CD-ROM drives) and *software* (floppy disks, computer programs, CDs) to speed up the estimating process. The estimator might use a handheld unit to input which parts must be replaced or repaired, saving time over writing the estimate out longhand. The office computer can streamline the estimating process by automating many of the necessary steps to arrive at a written estimate.

When making an estimate, the estimator must make sure no damage is overlooked. Many parts away from the inital impact area may be affected by a serious accident. For example, with a major impact, there can be damage to the chassis, engine, electrical system, and interior of the car.

The car may have to be measured to determine the extent of the damage (Figure 1–7). The estimate must also account for mechanical repairs or a wheel alignment (Figure 1–8).

The estimate is usually given to the customer, who submits it to the insurance company.

The **insurance adjuster** reviews the estimates and determines which one best reflects how the vehicle should be repaired and may inspect the wrecked car to ascertain that the repairs will be done cost effectively. The insurance company will give a check to the owner to cover the cost of the repair.

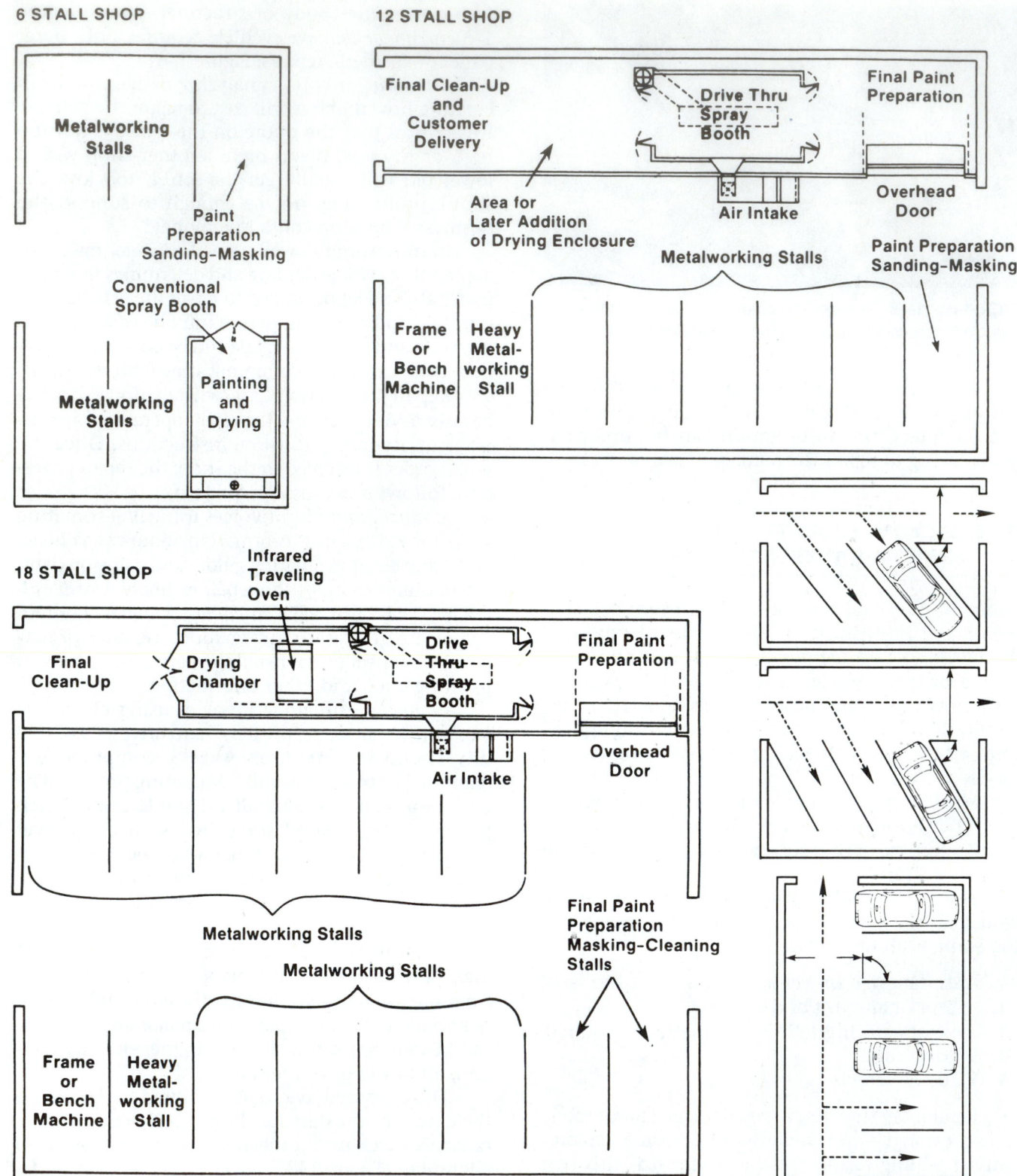

6 STALL SHOP

Metalworking Stalls

Paint Preparation Sanding-Masking

Conventional Spray Booth

Metalworking Stalls

Painting and Drying

12 STALL SHOP

Final Clean-Up and Customer Delivery

Area for Later Addition of Drying Enclosure

Drive Thru Spray Booth

Final Paint Preparation

Air Intake

Overhead Door

Metalworking Stalls

Paint Preparation Sanding-Masking

Frame or Bench Machine

Heavy Metal-working Stall

18 STALL SHOP

Final Clean-Up

Infrared Traveling Oven

Drying Chamber

Drive Thru Spray Booth

Final Paint Preparation

Air Intake

Overhead Door

Metalworking Stalls

Metalworking Stalls

Final Paint Preparation Masking-Cleaning Stalls

Frame or Bench Machine

Heavy Metal-working Stall

FIGURE 1-4 Study the layout of a small (6-stall) shop, and a large (18-stall) paint and body shop. Note the driveway widths but note parking should be used where it cannot be avoided.

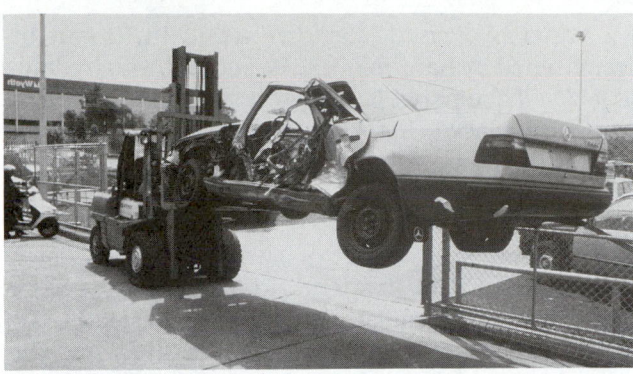

FIGURE 1-5 Major damage can require weeks to repair. If costs of parts and labor are more than the value of the car, it would be a total loss and sold for parts. *(Courtesy of Tech-Cor)*

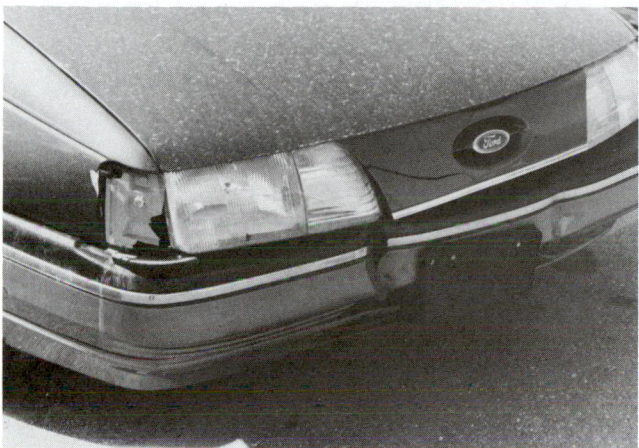

FIGURE 1-6 This minor front end damage would require parts replacement, bumper repair, and metal straightening before refinishing.

FIGURE 1-7 Accurate measurements must be taken to determine the extent and direction of major damage. *(Courtesy of Wedge Clamp)*

FIGURE 1-8 With today's unibody cars, body technicians must consider wheel alignment during major repairs. *(Courtesy of Bee Line, Inc.)*

FIGURE 1-9 The shop supervisor and body technician often discuss the repair order to determine the best methods of making repairs. *(Courtesy of Maaco Enterprises, Inc.)*

In addition to determining the cost of repairing and refinishing, the estimator must work with insurance companies' adjusters or appraisers. The *appraiser* is the insurance company's representative who estimates a vehicle's damage and authorizes payment to the shop. All estimates or *damage appraisals*, as they are sometimes called, must be in writing and signed by an insurance company's adjuster or the customer and the shop's estimator. The estimate then becomes a legal commitment among the parties involved as to work to be done, cost, and method of payment. More details on estimating and the importance of the estimator's function in a body/paint shop operation are given in Chapter 20.

Once the owner and the insurance company approve the repairs, the vehicle is turned over to the shop supervisor. The *shop supervisor* and sometimes a technician will then review the estimate to determine how to proceed with the repair, which may involve panel straightening or replacement (Figure 1–9).

Panel straightening involves using various hand tools and equipment to bend or shape the panel back into its original contour. Dollies, body hammers, plastic filler, and sanders are a few of the tools and materials used to fix metal panel damage. Similar techniques are needed with plastic panels.

Panel replacement involves removing a panel or body part that is too badly damaged to be fixed and installing a new one. You may have to unbolt and replace a fender, door, or spoiler, for example. With quarter panels and other welded body sections, you will have to cut off the damaged panel with power tools and then use a welder to install the new panel. This takes considerable skill.

WASHUP AREA

When a car is brought to the shop for repairs, the first step is washup. *Washup* involves a thorough cleaning of the vehicle with soap and water before beginning work. This is followed by wiping the body down with wax and grease remover. These steps will remove mud, dirt, wax, and water-soluble contaminants. These substances must be cleaned off before starting because they could contaminate the work area and the paint job. The car or truck must be completely dry before being moved to the metalworking area.

METALWORKING SHOP

The *metalworking area* is where the vehicle body structure is repaired (Figure 1–10). This damage can be the result of either a collision or deterioration. The repair tasks in this area of the shop are performed by body technicians or mechanics and their helpers.

In repairing any type of collision damage, the body technician must first study and diagnose the damage that has occurred and must use the information on the work order to determine what repairs are needed. The technician may need to consult with the estimator before proceeding. It is then up to the technician to decide how to accomplish the repairs outlined on the estimate.

Once the damage and repair methods are analyzed, the repairs must be completed. For example,

FIGURE 1-10 The shop metalworking area is the heart of most facilities. *(Courtesy of DuPont Automotive Products)*

if a panel is creased, torn, or caved in, it can be straightened by hammers, picks, spoons, and hydraulic jacks. If the panel is badly crushed and folded, it must be replaced. If the unibody or frame is damaged, it must be straightened or replaced according to factory specs (Figure 1–11). The labor time may be higher than the cost of the part.

Accurate adjustment of body assemblies, such as hoods, rear deck lids, and doors, may have to be made by the technician. If not done correctly, the assemblies are difficult to close and rattle when the car is driven over rough road. In addition, they are apt to leak excessive amounts of rain and dust. Such failures by the technician are bound to cause customer complaints.

Today, because of changing auto construction design, more and more body shops are offering complete collision services, such as:

1. Wheel alignment
2. Cooling system repairs
3. Electrical repairs
4. Suspension repairs
5. Air conditioning repairs
6. Other repairs

Many of these repairs are still done by so-called "auto specialty shops." However, the expanding scope of the body shop has made it necessary for the body technician to have some knowledge of these systems.

The *metalworking technician* must also be able to correct and repair minor defects, such as scratches, chips, dents, surface rust, and rustouts (Figure 1–12). More and more auto body shops are providing rustproofing services, either before rustout occurs or after repairing the problem. Except in large shops, methods of applying rustproofing are usually handled jointly by both the body and paint technicians.

Today's body technician must have a total knowledge of the tools used in the metalworking area. These include body repair hand tools, air and electric power tools, hydraulic body tools, and welding equipment; these are all covered in Chapters 6 through 15.

Sanding is done by the metalworking technician by using an abrasive coated paper to level and smooth a body surface being repaired. Coarse, rough sand-

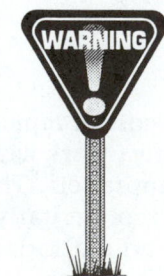

WARNING Always refer to a factory shop manual for details when working. It will give specifications or measurements, procedures, and other information needed to do quality repair work.

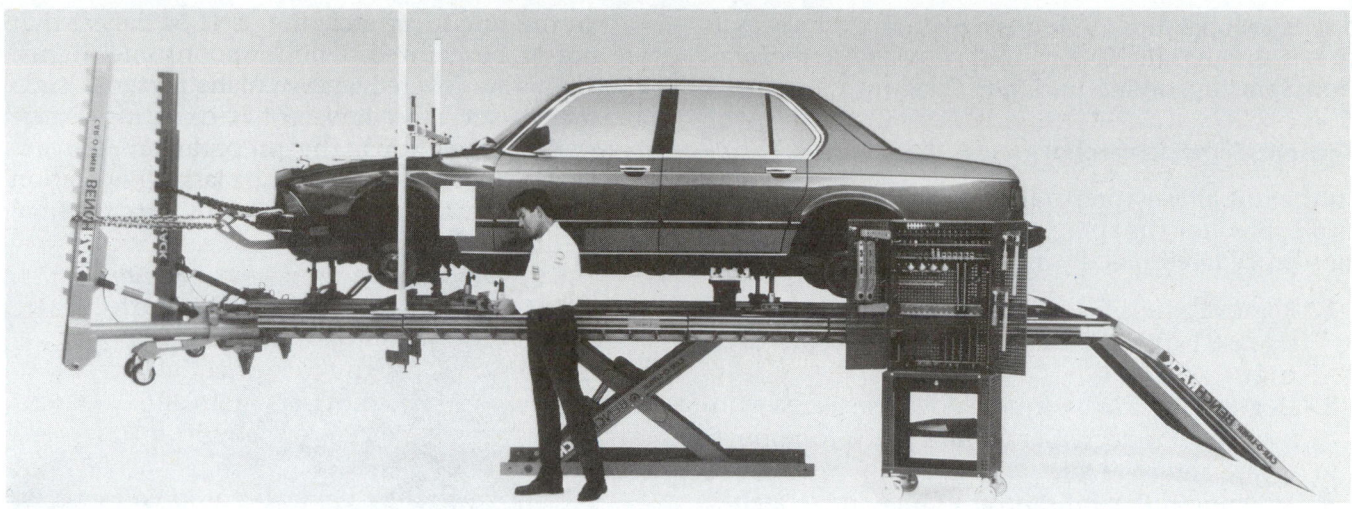

FIGURE 1–11 Hydraulic pulling equipment will exert tons of force to remove major structural damage to the unibody or frame. *(Courtesy of Car-O-Liner)*

FIGURE 1–12 Rust deterioration can be time-consuming to repair. *(Courtesy of Dobbs Publishing)*

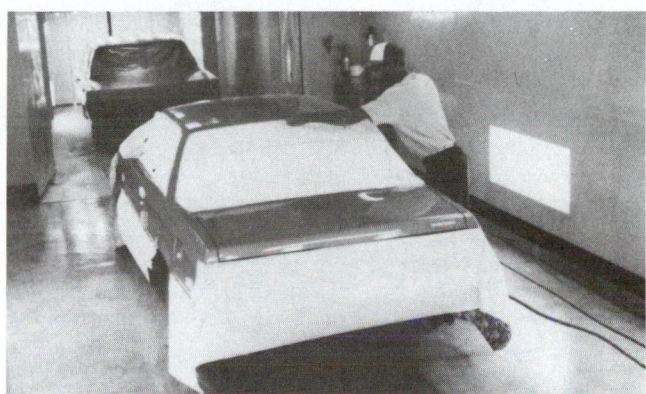

FIGURE 1–13 The paint shop must include a spray booth and a drying area to do quality work. *(Photo courtesy ITW DeVilbiss Automotive Refinishing Products)*

paper may be used to level plastic filler. Fine, smooth paper may be used to lightly scuff old paint so the new paint will stick.

Measurement systems allow the technician to check for frame or body misalignment resulting from a major collision. Various types of gauges and measuring devices can be used to compare known good body specs with the actual body measurements. The measurements will help determine what must be done to straighten any frame or body misalignment.

Once the extent and direction of frame misalignment are known, frame straightening equipment can be used to pull the frame back into alignment.

Frame straightening equipment (also called a **frame rack**) uses a large steel framework, pulling chains, and hydraulic power to pull or force the frame back into its original position. The vehicle frame or body is clamped down onto the frame straightening equipment so it cannot move. Clamps and chains are then fastened to the damaged portion of the vehicle. Then

tremendous pulling hydraulic force is applied to the chains to pull the frame or body in the opposite direction of deformation.

After pulling, more measurements are taken to determine if everything has returned to specs. Some frame rails and other body sections may have to be cut off and replaced instead of being repaired.

PAINT SHOP

Auto refinishing or painting is a very important part of the auto repair business. Not only do major collisions and wrecks and minor damage have to be painted, but also many automobiles are repainted to enhance their beauty. New- and used-car dealers repaint automobiles to attract buyers. Sometimes an owner gets tired of looking at the same old color. The finish on many automobiles may need attention because it has been neglected or damaged by weather conditions.

The *paint shop area* (Figure 1–13) is where the car is refinished by refinishing technicians or painters. Here, a series of operations is performed on the vehicle as it passes through the following stages: surface preparation, spraying, drying, and final detailing.

Paint Preparation

In the initial *prep* (preparation for painting), the car is prepared for the spraying operation. The preparation procedure generally includes the following steps:

1. Remove windshield wipers, emblems, nameplate (Figure 1–14), mirrors, and other small pieces of trim.
2. Degrease the car by wiping down each area with a solvent or degreaser (Figure 1–15) to remove grease and road tar.
3. Machine and hand-sand any chips and scratches as well as all areas to be painted (Figure 1–16).
4. Clean both the interior and exterior carefully (Figure 1–17A). All dust is removed with a blow gun and wiped with a tack rag (Figure 1–17B).
5. Inspect all surfaces to be sure that they are properly cleaned before moving the vehicle to the masking area.

In the **masking** area, the parts of the car that are not to be painted—windows, chrome, lights, vinyl top—are covered with masking material, such as paper, plastic, tape, or water-based coating (Figure 1–18). In large shops, the preparation jobs are handled by the sander, while the masking operation is performed by the masker. In small shops, the final

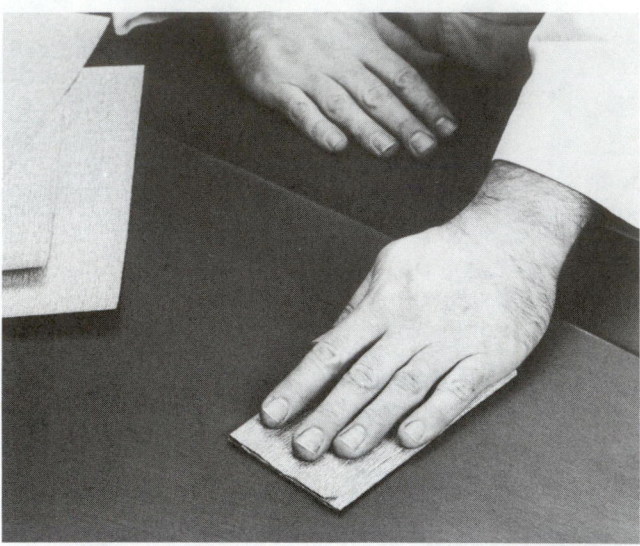

FIGURE 1-16 Sanding removes scratches and levels the surface. *(Courtesy of Carborundum Abrasives Co.)*

FIGURE 1-14 To prepare for refinishing, it is often necessary to remove emblems and nameplates. Removal is often easier than masking.

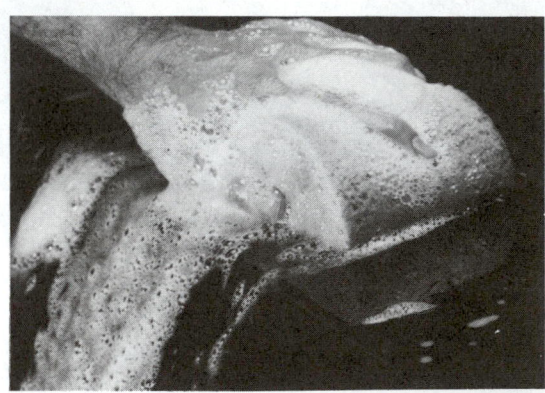

A

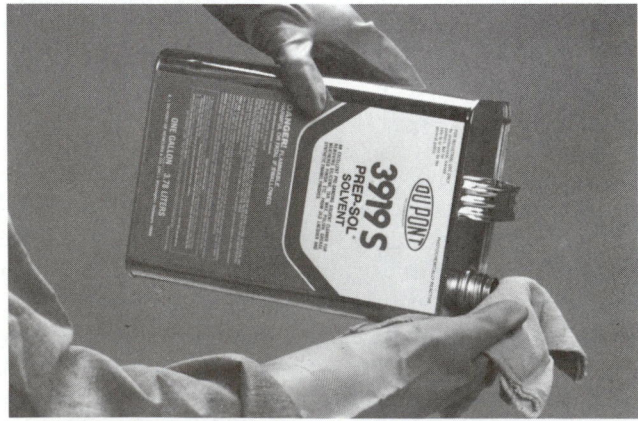

FIGURE 1-15 To keep contaminants out of new paint, clean all surfaces to be painted with an approved solvent. While still wet, wipe the surface dry with a clean cloth. This is often done before starting repairs. Always wear gloves to keep chemicals off your skin. *(Courtesy of DuPont Automotive Products)*

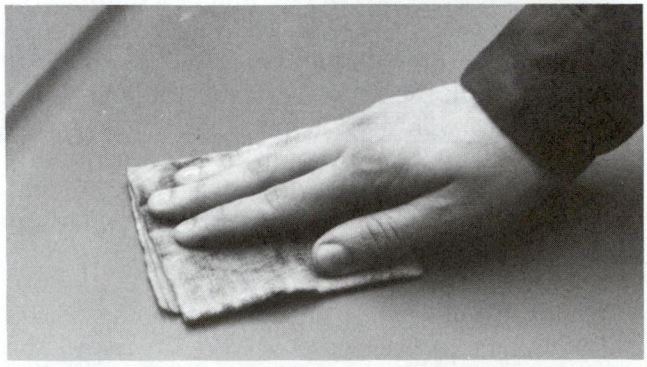

B

FIGURE 1-17 (A) After sanding, blow off all surfaces while wiping with a clean rag to remove sanding dust. (B) Finally, wipe the surface with a tack rag to remove any remaining dust and lint.

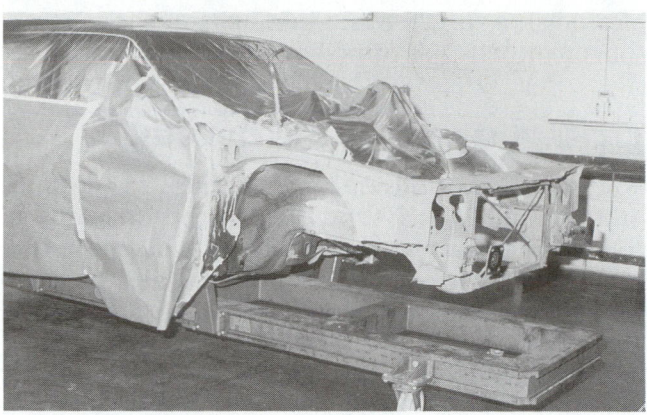

FIGURE 1-18 Masking-taped paper or plastic is used to protect areas not to be repainted. *(Courtesy of Tech-Cor)*

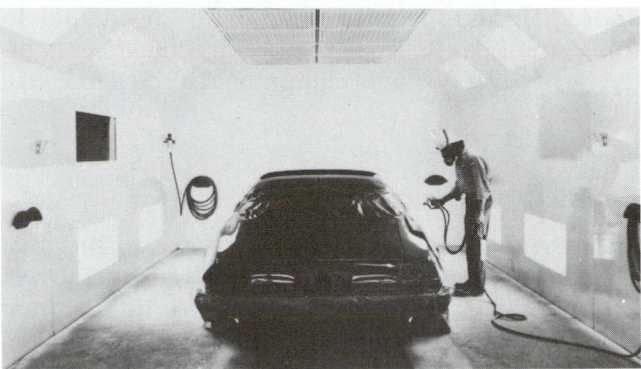

FIGURE 1-19 The spray booth is a clean area with forced air circulation and good lighting for painting vehicles. *(Photo courtesy ITW DeVilbiss Automotive Refinishing Products)*

prep jobs are usually done by the painter or a helper. Complete details of the preparation procedures are given in Chapter 17.

Spraying

The vehicle's paint or *finish* performs two basic functions—it beautifies and it protects. Can you imagine what a car body would look like without paint? For a day or so, it would be the drab, steel gray of bare sheet metal. Then, as rusting eats into the metal, the body would turn an ugly, reddish brown. This degeneration would continue until the body was solidly coated with rust.

The term *paint* generally refers to the visible topcoat. The most elementary painting system consists of a compatible primer and final topcoat over the *substrate* (body material). This process can vary considerably and is usually more complex, as you will learn later.

A car's finish consists of several coats of two or more different materials. The most basic finish consists of

1. Undercoat or primer coat
2. Topcoat (colorcoat or basecoat/clearcoat)

The *primer* or *undercoat* has to improve adhesion of the topcoat. It is often the first coat applied. Paint alone will not stick or adhere as well as a primer. If you apply a topcoat to bare substrate, the paint will peel, flake off, or look rough. This is why you must "sandwich" a primer undercoat between the substrate and the topcoat. Undercoats also prevent any chemicals from bleeding through and showing in the topcoats of paint.

The terms *topcoat* or *colorcoat* refer to the paint applied over the undercoat. This usually consists of several light coats of one or more paints. The topcoat is the "glamour coat" because it features the eye-catching color, color effects, and gloss.

Basecoat/clearcoat paint systems use a colorcoat applied over the primer with a second layer of

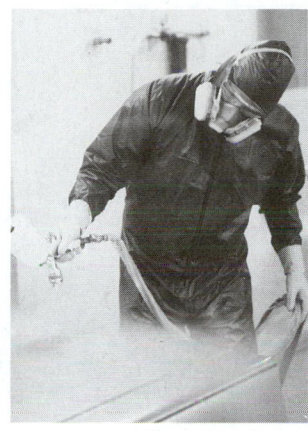

FIGURE 1-20 Finish topcoat must be applied evenly to produce a high gloss paint surface without imperfections. *(Courtesy of DuPont Automotive Products)*

clearcoat over the colorcoat. This is the most common paint system used today. The clear paint brings out the richness of the underlying color and also protects it. It makes the paint shine more than a single layer of color without a clearcoat.

Spraying involves using precision paint guns and a spray booth to apply these paint materials to the body. When the vehicle moves into the spray booth (Figure 1–19), the painter refinishes all or part of the body (Figure 1–20).

In addition to being able to apply the new finish properly, the painter or refinishing technician must have knowledge about paint products and how to mix and match them.

If the refinishing job looks good and the color matches well, the customer will usually be satisfied with all of the other repair work.

Note: Complete information on applying paint is given in Chapter 19.

Drying

In the **drying room**, most shops use drying equipment (Figure 1–21) to speed up the paint curing or drying. This is especially useful in drying enamel. In some small paint shops the drying operation is often done with portable equipment in the spray area (Figure 1–22).

Final Cleaning

After the finish is dried, the car moves to the final *assembly* or *detail area*. The masking is now removed. Compounding or polishing is done in the final cleanup area along with other tasks, such as polishing all chrome, cleaning glass and overspray, replacing removed items, vacuuming the interior, and

cleaning the vinyl top and tires. In large shops, these jobs are usually performed by the *detailer*.

OTHER SHOP PERSONNEL

Foremost in any collision repair business are the body and paint technicians and their helpers. In addition to part-time cleanup employees.who work one to two hours a day, there are other jobs that must be done as well. The personnel who handle these jobs include the following.

Shop Owner and Manager

The *shop owner* must be concerned with all phases of work done in the shop. In smaller shops, the owner and shop manager are one and the same person. In larger operations or in new-car dealerships, the owner might hire a shop manager to supervise the operation

FIGURE 1-21 A drive-through drying area will speed paint hardening so the vehicle can be released to the customer more quickly. *(Courtesy of Garmat Corp.)*

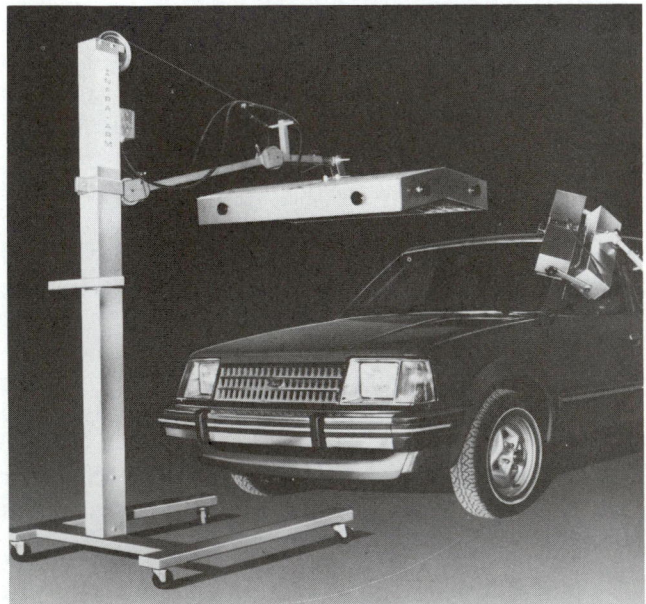

FIGURE 1-22 Portable drying equipment is good for spot repairs of small areas on the vehicle. *(Courtesy of Thermal Devices, Inc.)*

FIGURE 1-23 The parts manager is an important member of the body/paint shop team.

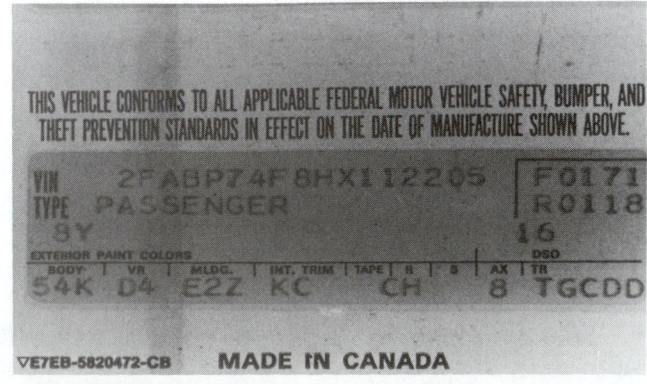

THIS VEHICLE CONFORMS TO ALL APPLICABLE FEDERAL MOTOR VEHICLE SAFETY, BUMPER, AND THEFT PREVENTION STANDARDS IN EFFECT ON THE DATE OF MANUFACTURE SHOWN ABOVE.

VIN	2FABP74F8HX112205	F0171
TYPE	PASSENGER	R0118
8Y		16

EXTERIOR PAINT COLORS

| BODY | VR | MLDG. | INT. TRIM | TAPE | R | S | AX | TR |
| 54K | D4 | E2Z | KC | | CH | | 8 | TGCDD |

VE7EB-5820472-CB   **MADE IN CANADA**

FIGURE 1-24 Note typical body code plate. Location of these tags can vary.

of a collision repair and refinishing shop. In all cases, the person in charge should understand all of the work done in the shop as well as its business operations.

Parts Manager

The *parts manager* (Figure 1–23) is in charge of ordering all parts (both new and salvaged), receiving all parts, and seeing that they are delivered to the ordering technician. In new-car dealerships, these tasks may be taken care of by the parts manager of the franchise.

It is important that the repair shop collect all of the necessary information about the vehicle in order to locate and order the correct parts. Accurate ordering of parts is vital not only to the job production, but also to the job cost and profit. For example, the price of a can of paint can vary as much as 200 percent from one color to another. The only way for the parts manager to order the correct color is for the painter to give all the information that appears on the body code plate (Figure 1–24). As described in Chapters 16 and 20, these tags are located on various parts of the vehicle.

In order for the parts manager to purchase the correct parts, it is sometimes necessary to know the name of the assembly plant that manufactured the original part. Many body parts require the same information. To get a wood grain overlay to match a certain vehicle, the parts department must know which assembly plant built the car. The plant code is provided on a body code plate somewhere on the vehicle.

The *VIN plate* and *VIN* are used to accurately identify the body style, model year, and engine, and to give other data about the vehicle. Since 1981 the VIN plate is riveted to the upper left corner of the instrument panel and is visible through the windshield (Figure 1–25). Check the service manual for

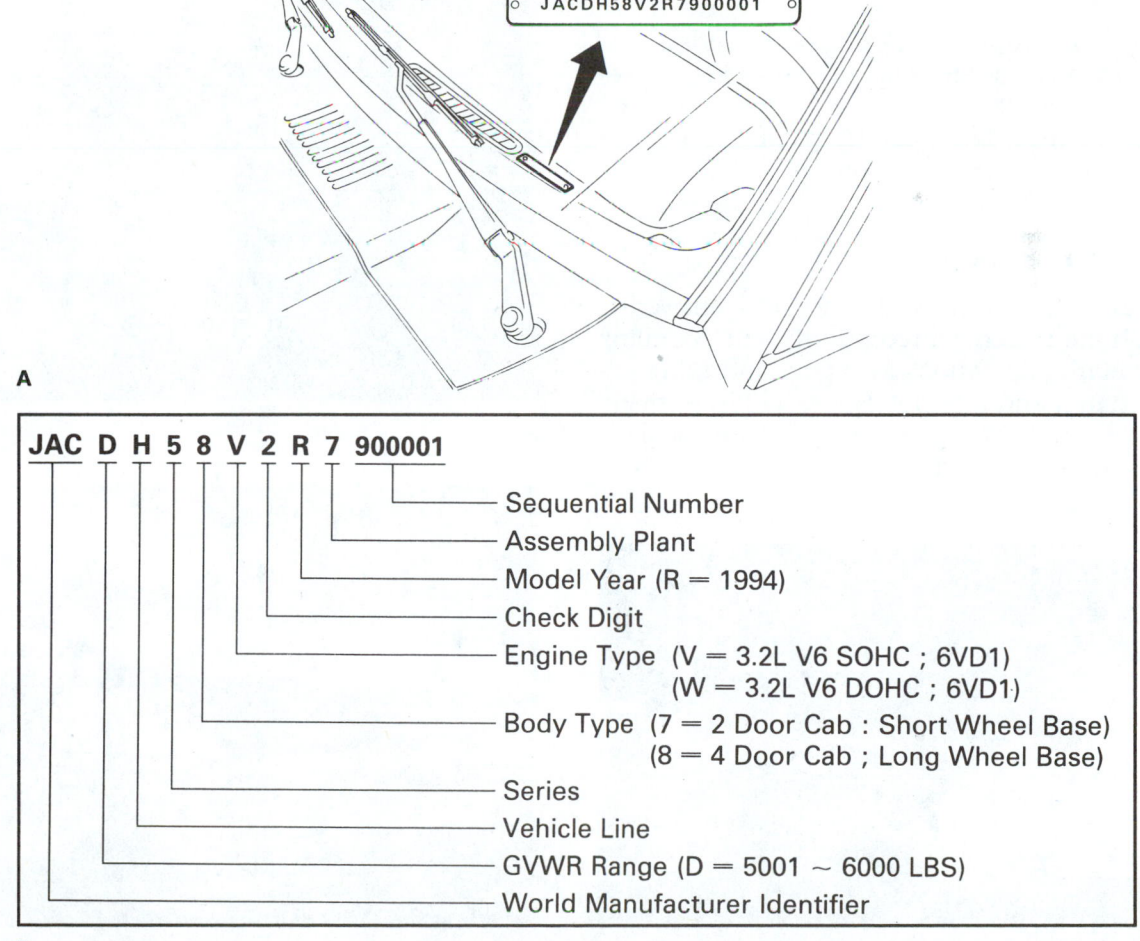

FIGURE 1–25 Study the use of the vehicle identification number. (A) The VIN is normally located on the dash and can be read through the windshield. (B) Note the meaning of the VIN for this specific car. *(Courtesy of Isuzu Motor Corp.)*

the location of the VIN, vehicle certification label, or body plate for vehicles made prior to 1981 and for foreign vehicles. Service manuals and crash-estimating guides also contain all of the necessary VIN decoding information.

The *body ID number* or *service part label* gives information about how the vehicle is equipped. It will give paint codes or numbers for ordering the right type and color paint, including lower and upper body colors if a two-tone paint. The body ID number will also give trim information. This number will be on the body ID plate on the door, console lid, or elsewhere on the body.

There are frequent "running changes" made by the vehicle manufacturers. Vehicles built in the same model year can have one of several individual part applications. The only way to keep on track with all of the changes is for the manufacturer to note either the build date when a different part is first installed or to use the VIN or body code plate number as reference.

Bookkeeper

The *bookkeeper* keeps the shop's record books, provides all invoices, writes checks, pays bills, makes bank deposits, checks bank statements, and takes care of tax payments. Many small shops hire an outside accountant to perform these jobs.

Office Manager

The duties of the *office manager* vary from answering the telephone to being a receptionist and handling such secretarial operations as typing and filing letters, estimates, and receipts (Figure 1–26). In many small shops, the office manager also acts as the parts manager and the bookkeeper.

FIGURE 1-26 Computers play an important role in the daily operations of the body and paint shop. The office manager must be able to operate a computer efficiently.

Tow Truck Operator

Larger shop firms generally have trained *tow truck operators* to operate its wrecker(s). Rather than own this expensive piece of equipment (Figure 1–27), many smaller shops depend on independent towing services or farm out such work to other repair garages.

FIGURE 1-27 Many body shops offer towing service.

A

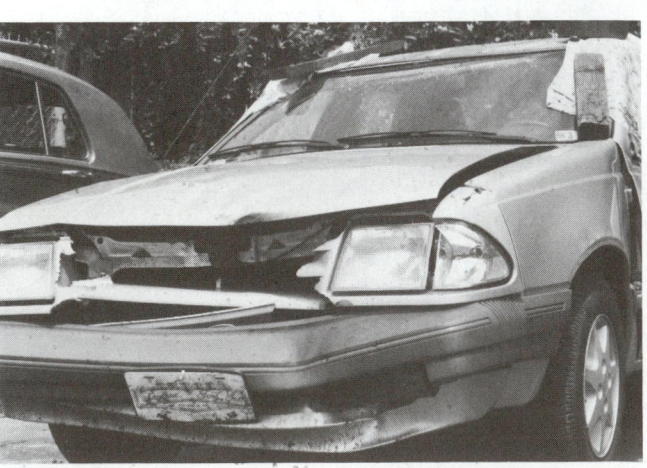

B

FIGURE 1-28 Collision damage can range from (A) a small dent to (B) major damage.

Towing is needed only with major damage or when the vehicle is not safe to drive after an accident (Figure 1–28). As soon as the vehicle is towed to the body shop, the estimator begins the work of analyzing damage (Figure 1–29).

OTHER CAREER OPPORTUNITIES

There are other career openings in the collision repair field which include

Insurance Adjuster or Appraiser
Vocational/Technical Instructor
Salvage Yard Technician
Dealership Parts Counterperson
Paint Company Representative
Auto Manufacturer Representative
Equipment Salesperson

To research these and other career opportunities, talk to your guidance counselor or visit your local library. There you can obtain more detailed information on the qualifications and training requirements needed for each position.

SHOP PUBLICATIONS

There are several publications that all body shop personnel should become familiar with. The most important of these to the body technician is the manufacturers' body service manuals (Figures 1–30 and 1–31). On the other hand, the crash-estimating guides are vital to the estimator (Figure 1–32). Replacement catalogs are very important to the parts manager. All shop personnel should be interested in trade publications (Figure 1–33).

All automobile companies publish yearly *service manuals* that describe the construction and repair of their vehicle makes and models. These manuals give important details on repair procedures and parts. Also called *shop manuals*, they give instructions, specifications, and illustrations for their specific cars and trucks. Service manuals have both mechanical and body repair information.

The *contents page* of a service manual lists the broad categories in the manual and gives page numbers. Each service manual section then concentrates on describing one area of repair.

Service manuals' abbreviations represent technical terms or words and save space. Each manufacturer uses slightly different abbreviations.

Aftermarket repair manuals are published by publishing companies (Mitchell Manuals, Motor Manuals, Chilton Manuals). They are not as detailed but can give enough information for most repairs.

Repair charts give diagrams that guide you through logical steps for making repairs. They can

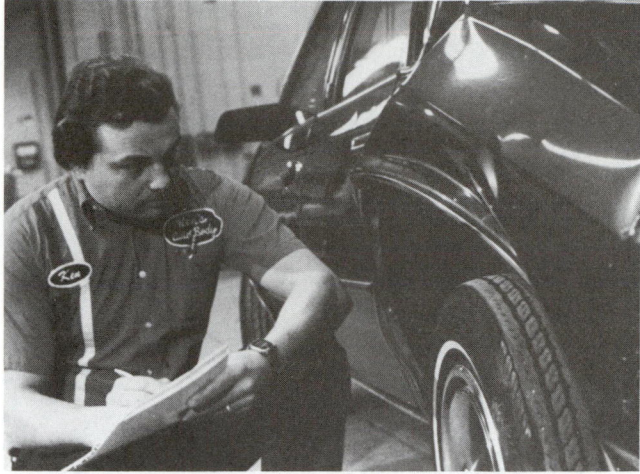

FIGURE 1–29 The estimator is making a written estimate. He or she must note all parts and repairs needed to restore the vehicle to pre-accident condition.

FIGURE 1–30 Part removal and installation instructions can be found in the factory body manuals and aftermarket manuals.

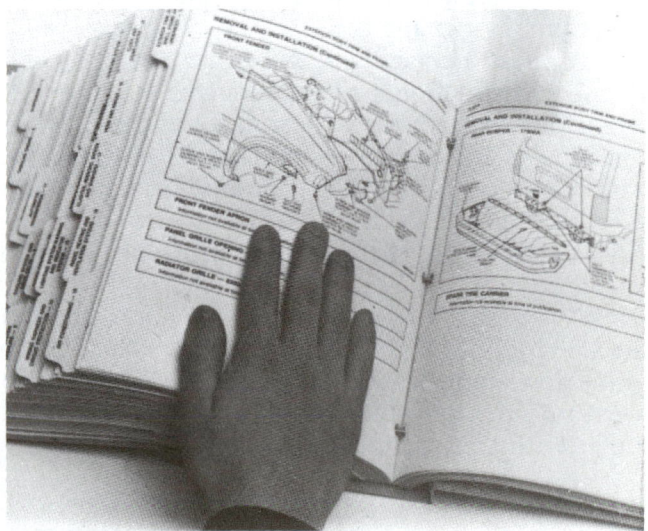

FIGURE 1–31 Typical page from a manufacturer's service manual shows an exploded view of body parts which can be helpful to a technician.

FIGURE 1-32 Estimating guides are used in the preparation of damage reports.

FIGURE 1-33 Trade publications will help keep you up-to-date on changes in technology that affect your work quality and speed.

vary in content and design. Most use arrows and icons (graphic symbols) that represent repair steps.

Diagnosis charts or *troubleshooting charts* give logical steps for finding the source of problems. Mechanical, body, electrical, and other types of troubleshooting charts are provided in service manuals. They give the most common sources of problems for the presented symptoms.

Paint reference charts in service manuals list comparable paints manufactured by different companies. This will help match the color of the new paint with the paint already on the vehicle.

Collision-estimating manuals or guides give information for calculating the cost of repairs. They list part numbers, prices, section illustrations, and other data to help the estimator. Electronic or computer-based estimating guides are also available and are discussed in later chapters.

A *vehicle dimension manual* gives the body and frame measurements of undamaged vehicles. Dimensions are given for every make and model car and truck. These known good dimensions can be compared with actual measurements taken of a wrecked car or truck and will show how badly the vehicle is damaged and what must be done to repair it.

Color matching manuals contain information needed for finishing panels so that the repair has the same appearance as the old finish. They have paint code information, color chips, blending and tinting data, tinting procedures, and other information.

1.2 SHOP SAFETY PRACTICES

The most important action taken in any body shop is accident prevention. Carelessness and the lack of safety habits cause accidents. Accidents have far-

reaching effects, not only on the victim, but also on the victim's family and society in general. More importantly, accidents can cause permanent disability or even death. Therefore, all shop employees must foster and develop a safety program to protect the health and welfare of those involved.

In the following chapters of this book, the text contains special notations labeled SHOP TALK, WARNING, and CAUTION. Each one has a specific purpose.

1. **SHOP TALK** gives added information that will help the technician to complete a particular procedure or make a task easier.
2. **WARNING** is given to prevent the technician from making an error that could damage the vehicle.
3. **CAUTION** reminds the technician to be especially careful of those areas where carelessness can cause personal injury.

Remember to read these special notes carefully. There are several kinds of accidents that must be prevented:

Asphyxiation refers to anything that prevents normal breathing. There are many mists, gases, and fumes in a body shop that can damage the lungs and affect the ability to breathe.

Chemical burns result when a corrosive chemical injures the skin or eyes. These can result from various chemicals in a body shop—paint removers, part chemicals, and so on.

Electrocution results when electricity passes through your body. Severe injury or death can result.

A *fire* is rapid oxidation of a flammable material, producing high temperatures. A burn from a fire can cause horrible, permanent scar tissue and death. There are numerous **combustibles** (paints, thinners, reducers, gasoline, dirty rags) in a body shop. Any can quickly cause a fire.

Explosions are air pressure waves that result from extremely rapid burning. For example, if you were to weld near a gas tank, the fumes in the tank could ignite and explode.

A *physical injury* is a general category that includes cuts, broken bones, strained backs, and similar injuries. To prevent these painful injuries, constantly think and evaluate every technique. Always think about what you are doing and try to do it better.

PERSONAL SAFETY AND HEALTH PROTECTION

The following are very important personal safety rules that must be heeded while working.

Air Passages and Lungs

Abrasive dusts, vapors from caustic solutions and solvents, and spray mists from undercoats and finishes can be harmful. They all present dangers to the air passages and lungs, especially for workers who are surrounded by them day in and day out.

Respirators are needed in the paint shop to keep airborne materials from being inhaled. They often must be used even when adequate ventilation is provided (Figure 1–34) if airborne chemicals are present (Figure 1–35).

A *dust* or *particle respirator* is basically a filter that fits over the nose and mouth to block small airborne particles (Figure 1–36). It is not designed to stop paint mists and fumes. A dust respirator is used to keep dangerous materials out of the lungs when sanding or grinding.

Sanding operations create dust that can cause bronchial irritation and possible long-term lung damage. Protection from this health hazard is often overlooked. Just because the sanding dust does not cause immediate symptoms does not mean that there will not be problems later in life. An approved dust respirator should be worn when sanding or grinding.

Follow the instructions provided with the dust respirator to ensure a proper fit. Bend or shape it so that air cannot leak around your face, nose, and mouth.

CAUTION
Remember that inexpensive dust masks or respirators do not protect against vapors and spray mists.

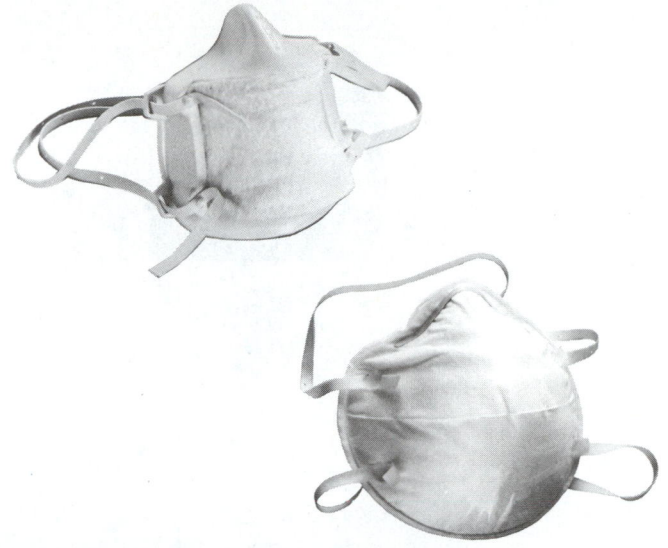
FIGURE 1-34 If you do not wear proper protection when power sanding and grinding, harmful paint dust and dirt can get into your lungs. Always wear a NIOSH-approved dust particle mask before grinding or sanding.

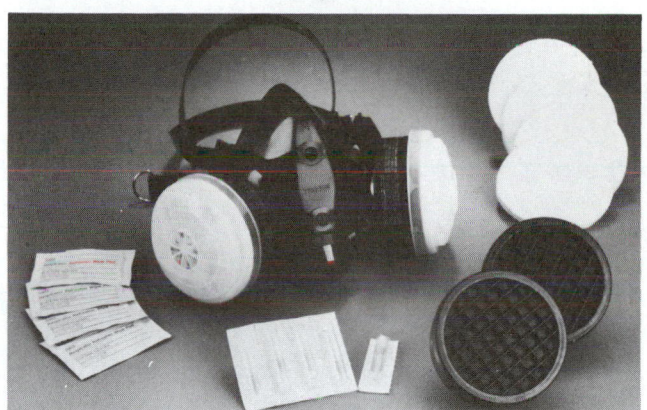
FIGURE 1-35 Cartridge-type respiratory masks will filter finer particles than fiber dust masks. *(Courtesy of Paint-Safe Products)*

FIGURE 1-36 This is a typical cartridge filter with a full facepiece *(Courtesy of Mine Safety Appliance Co.)*

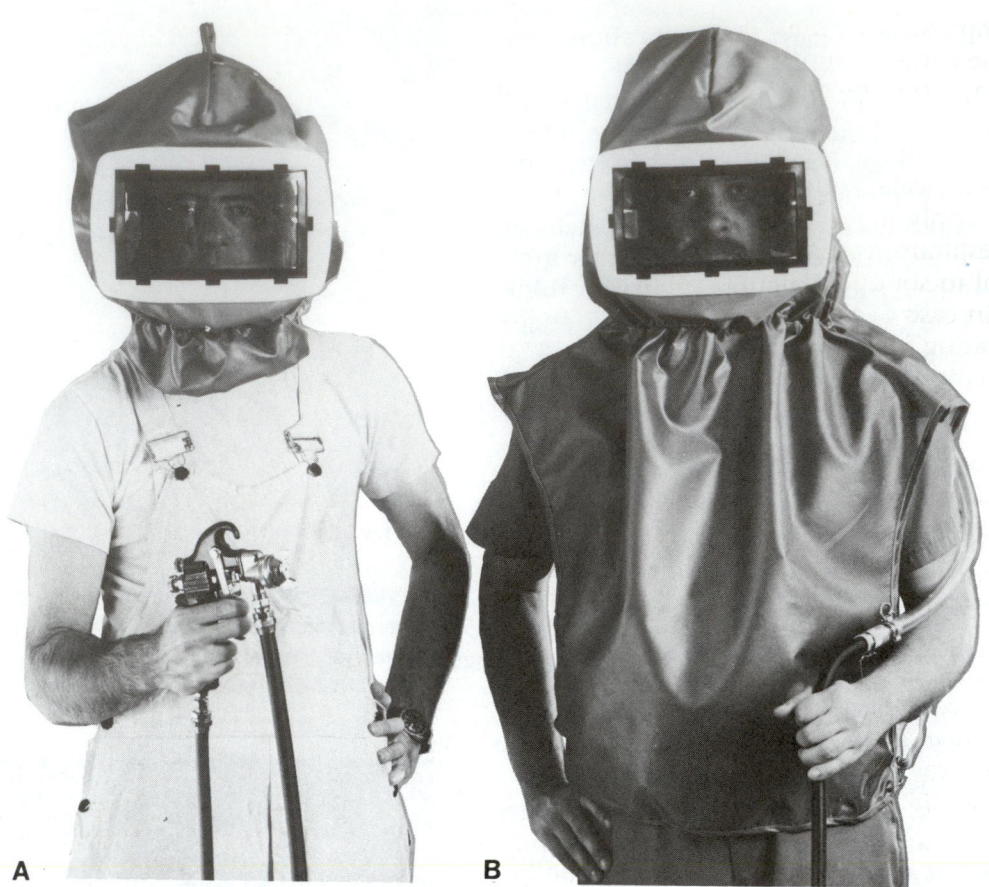

FIGURE 1–37 These are air-supplied respirators for painting (A) neck-length and (B) waist-length. They feed fresh air into the respirator so no paint fumes are inhaled. *(Courtesy of Binks Mfg. Co.)*

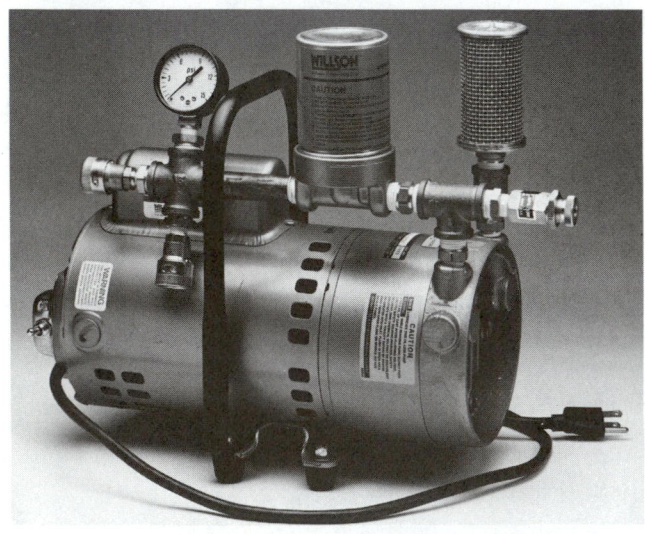

FIGURE 1–38 A typical oilless air pump that moves ambient air from a clean environment and supplies it to the air line respirators of the hood. The unit shown is suitable for two painters. *(Courtesy of Willson Safety Product, Inc.)*

A *cartridge filter respirator* protects against vapors and spray mists of nonactivated enamels (no hardener added), lacquers, and other nonisocyanate materials. Cartridge filter respirators should be used only in well-ventilated areas. They must not be used in environments containing less than 19.5 percent oxygen.

The cartridge filter respirator consists of a rubber face piece designed to conform to the face to form an airtight seal. It includes replaceable prefilters and cartridges that remove solvent and other vapors from the air. The respirator also has intake and exhaust valves, which ensure that all incoming air flows through the filters.

Cartridge filter respirators are available in several sizes. Purchase one that fits your face. To maintain the cartridge filter respirator, keep it clean and change the prefilters and cartridges as directed by the manufacturer.

An approved *air-supplied respirator* consists of a half mask, full face piece, hood or helmet, and an external air supply hose. Clean, breathable air is supplied through the hose from a separate air source (Figure 1–37). It provides protection from the dangers of inhaling paint with a hardener, isocyanate paint vapors and mists, and hazardous solvent vapors.

The fresh air supply respirator is comfortable and cool to wear and does not require fit testing. The fresh air supply respirator may include a self-contained oilless pump to supply air to either one hood or two half-mask respirators (Figure 1–38).

The pump's air inlet must be located in a clean air area. Some shops mount the pumps on an outside wall, away from the dust and dirt generated by shop operations. If shop compressed air is used, it must be filtered with a trap and carbon filter to remove oil, water, scale, and odor.

The air supply must have a valve to match air pressure to respirator equipment specs and an automatic control to sound an alarm or shut down the compressor in case of overheating and contamination. Overheating frequently causes carbon monoxide contamination of the air supply.

Respirator Testing and Maintenance

It is very important that an air-purifying cartridge filter respirator fits securely around your face. This will prevent contaminated air from entering your lungs. To check for respirator air leaks, a *fit test* should be done prior to using the respirator. Perform both negative and positive pressure checks.

To make a *negative pressure test*, place the palms of your hands over the cartridges and inhale. A good fit will be evident if the face piece collapses onto your face.

To perform a *positive pressure test*, cover up the exhalation valve and exhale. A proper fit is evident if the face piece billows out without air escaping from the mask. Another form of fit testing consists of exposing amyl acetate (banana oil) near the seal around the face. If no odor is detected, a proper fit is evident.

Replace the prefilters when it becomes difficult to breathe through the respirator. Replace the cartridge(s) at proper intervals, and always at the first sign of solvent odor. Regularly check the mask to make sure it does not have any cracks or deformities. Store the respirator in an airtight container.

Note that facial hair may prevent an airtight seal, presenting a health hazard. Therefore, refinishers with facial hair should use a fresh air supply respirator.

Follow detailed manufacturer's instructions to ensure proper maintenance and fit.

Head Protection

Be sure to tie long hair securely behind your head before beginning to work on a vehicle. The hair also must be protected against dust and sprays. To keep hair clean (and healthy) wear a cap at all times in the work area. Wear a protective painter's stretch hood in the spray booth.

A body technician should consider wearing a bump cap or hard hat when working beneath hoods or under the car.

Eye and Face Protection

Eye protection (safety glasses, goggles, or a face shield) is required at all times in most shops to comply with OSHA or insurance company requirements. Eye protection (Figure 1–39) should be worn when using grinders, disc sanders, power drills, or pneumatic chisels, and when removing shattered glass or working with paints and other chemicals. All of these activities carry the possibility of an eye injury from flying particles, chips, or chemicals.

Remember! When you are in the shop, there is always the possibility of flying objects, dust particles, or splashing liquids entering your eyes. Not only can this be painful, but it can also cause loss of sight.

A *welding helmet* or *welding goggles* (Figure 1–40) with the proper shade lens must be worn when welding. These will protect the eyes and face from flying

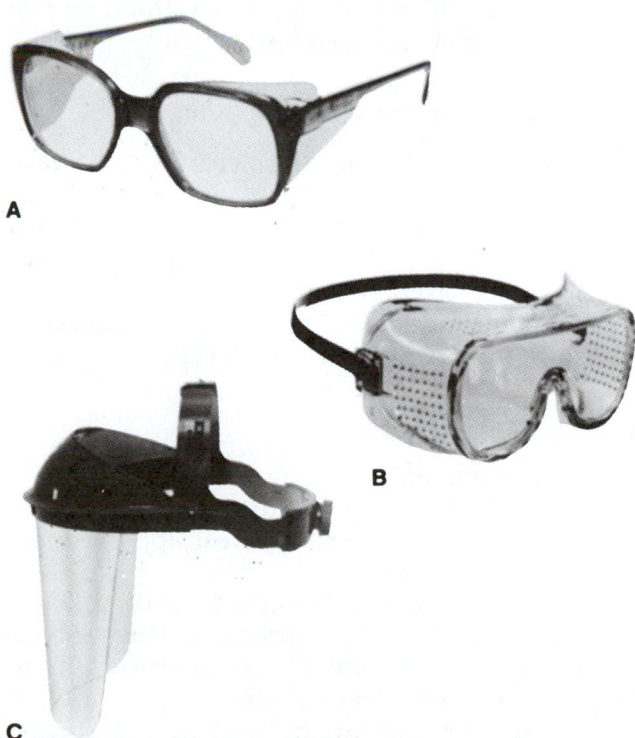

A

B

C

FIGURE 1–39 (A) Safety glasses, (B) goggles, and (C) face shield. These are "vital tools" that protect you from flying objects. *(Courtesy of Goodson Shop Supplies)*

CAUTION Your eyes are irreplaceable. Get in the habit of wearing safety goggles, glasses, or face shields in ALL work areas.

FIGURE 1-40 (A) A welding helmet is needed when electric welding to prevent eye and skin burns. (B) Goggles can be used when gas heating or cutting. *(Courtesy of Sellstrom Manufacturing, Inc.)*

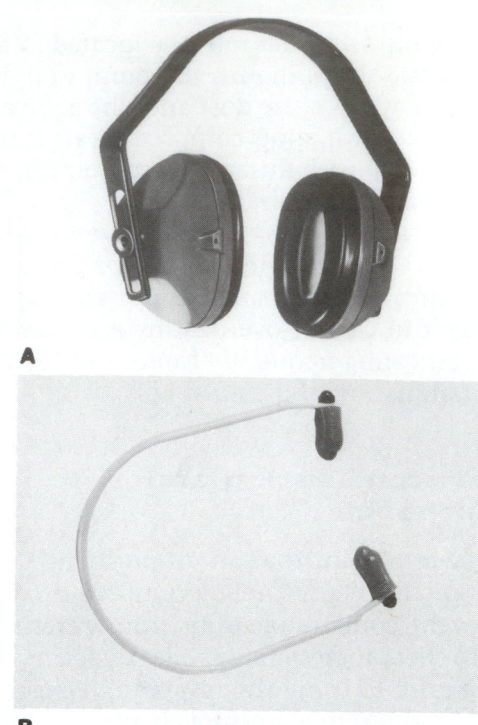

FIGURE 1-41 (A) Ear muffs or (B) ear plugs should be worn when grinding and doing other operations producing loud noise. *(Courtesy of Willson Safety Products, Inc.)*

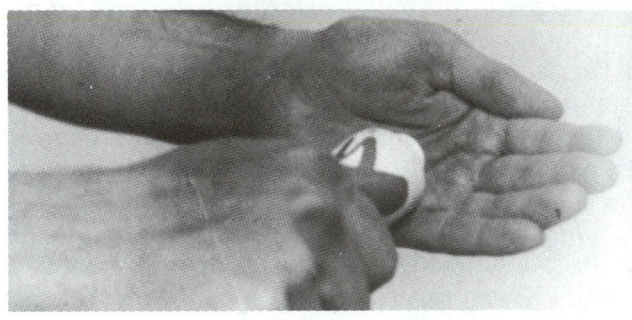

FIGURE 1-42 After working, use a little silicone-free moisturizing cream to soften your skin.

pieces of molten steel and from harmful light rays. Sunglasses are not adequate protection!

Ear Protection

The noise of panel beating, the piercing noise of sanding, the radio blaring full blast—it is impossible to hear anything else. It is enough to deafen a person, and that is exactly what it will do if proper precautions are not taken. When in metalworking areas, wear *ear plugs* or *muffs* (Figure 1–41) to protect the eardrums from damaging noise levels.

Body Protection

Loose clothing, unbuttoned shirt sleeves, dangling ties, jewelry, and shirts hanging out are very dangerous in a body shop. Instead, wear approved shop coveralls or jumpsuits.

A clean jumpsuit or lint-free coveralls should be worn when you are in the spray area. Dirty, solvent-soaked clothing will hold these chemicals against the skin, causing irritation or a rash.

Pants should always be long enough to cover the top of the shoes. This will prevent sparks from going down into the shoes, especially when you are using welding equipment. For added welding safety, welder's pants, leggings, or spats are often worn.

Upper body protection should include either a welder's jacket or apron (see Chapter 6).

Hand Protection

The harmful effects of liquids, undercoats, and finishes on the hands can be prevented very effectively by wearing work gloves. *Impervious gloves*, such as the nitrile latex type, should be used when working with solvents or two-pack primers and topcoats. These gloves offer special protection from the materials found in two-component systems. See the Material Safety Data Sheets (MSDS) for glove recommendations.

Thick, strong *work gloves* should be worn in the prep area to avoid cuts or abrasions. Sheet metal can cut your skin like a knife.

Always remember to wash your hands thoroughly before leaving the shop area. This provides protection from ingesting any harmful elements that may have been touched.

It is usually recommended that the hands be washed with a proper hand cleaner. At the end of a day's work, it is wise to oil the skin a little by applying a good skin cream (Figure 1–42). Do not use thinner as a hand cleaner.

Foot Protection

Wear *safety work shoes* that have metal toe inserts and nonslip soles. The inserts protect the toes from falling objects; the soles help prevent falls. In addition, good work shoes provide support and comfort for someone who is standing for a long time. Never wear gym shoes or dress shoes, as they do not provide adequate protection in a body shop.

When spraying, many technicians wear disposable shoe covers. In fact, disposable garments and hoods are becoming more commonly used by sprayers.

Day-by-day Personal Safety Guide

The following guidelines are designed to protect you while on the job:

- **Be informed!** Read the warnings on the product labels and in manufacturers' literature. If more information is desired, get copies of the Material Safety Data Sheets for specific products from the shop's office or from the material suppliers. *Material Safety Data Sheets* contain information on hazardous ingredients and protective measures that the technician should use.
- **Power sanding can be harmful!** When you are power sanding, dust and dirt fly into the air. These particles can get into your eyes, lungs, and scalp without proper protection. Safety glasses or goggles will protect the eyes. Do not wear contact lens when grinding, sanding, or handling solvents. Head covers provide scalp and hair protection. A NIOSH-approved dust particle mask should also be worn to prevent inhaling dust and particles. All masks should fit tightly to the skin.
- **Be careful when cleaning with compressed air!** When using a dust gun to clean doorjambs and other hard-to-reach places, eye protection and particle masks should always be worn.
- **Metal conditioning can hurt you!** Metal conditioners contain phosphoric acid. Breathing these chemicals or allowing them to come in direct contact with the skin, eyes, or clothing may cause irritation. The use of safety glasses (to prevent splashes into the eyes), coveralls, rubber gloves, and a NIOSH-approved organic vapor respirator is recommended when using these products. If the coveralls become soaked for any reason, make sure

they are changed right away or soak the spill with water to dilute the chemicals.

- **Mixing and handling paints can be dangerous!** Mixing and pouring of refinish materials should be done in a well-ventilated area.
- **Spraying body shop materials can damage your lungs!** Application of undercoats and topcoats requires the use of spraying equipment, which can be hazardous if not used properly. Static electricity is generated when using airless or electrostatic spraying methods. Special attention must be paid to grounding and bonding for this equipment to prevent explosions. Technicians should be fully protected when applying undercoat or topcoat products.
- **Store paint materials properly to prevent fumes or a fire!** All refinish products should be stored away from the actual work area. Paint kept in the work area should be limited to a one-day supply. Empty containers should be disposed of daily. All partially used containers should be kept securely closed and should be placed in proper metal (fire resistant) storage cabinets (Figure 1–43) at the end of the day.
- **Horseplay is unacceptable!** Proper conduct can also help prevent accidents. Horseplay is not fun when it sends someone to the hospital. Such things as air nozzle fights, creeper races, or practical jokes do not have any place in the shop.
- **Lifting improperly can cause long-term back problems!** When lifting and carrying objects, bend with the knees, not the back. Do not bend the waist when lifting; heavy objects should be lifted and moved with the proper equipment for the job.
- **Before leaving the shop, store toxic materials properly!** Solvents, chemicals, and

FIGURE 1–43 All flammable materials must be stored in a metal safety cabinet. *(Courtesy of The Sherwin-Williams Co.)*

other materials can contaminate clothing and wind up on the hands when you remove personal protective equipment or put away the refinishing tools.

1.3 GENERAL SHOP PROCEDURES

In addition to personal safety, the body/paint technicians must be aware of general shop safety procedures. The following are some of the rules and precautions that should be observed.

ENVIRONMENTAL CONTROLS

Persons working in body and/or paint shops are often exposed to dangerous levels of various gases, dusts, and vapors. Because of this exposure, control measures should be established for air contaminants and other hazardous substances.

- **Proper ventilation** is very important in areas where caustics, degreasers, undercoats, and finishes are used. The vapors from thinners used in most paints have a narcotic effect, and long-term exposure can eventually cause serious illnesses. In the shop and in the area where the vehicles are prepared, ventilation can be provided by means of an air-changing system, extraction floors, or central dust extraction. For the spray booth (Figure 1–44), adequate air replacement is necessary not only to promote evaporation and drying of the areas sprayed, but also to remove harmful mists and vapors. In addition to the ventilation in the spray area or paint booth, respirators should be worn.
- **Wear rubber or safety gloves** while handling paints and thinners to prevent chemical burns. If any of these materials get on the skin, wash the affected area promptly with soap and water.
- **Operate engines only in a well-ventilated area** to avoid the danger of carbon monoxide (CO) poisoning. CO poisoning from engine exhaust fumes can be fatal. Also, if the shop is equipped with a tailpipe exhaust system to remove CO from the garage, use it!

    Space heaters used in some shops can also be a serious source of CO. They should be periodically inspected to ensure they are adequately vented and have not become blocked.
- **Use a dustless sanding system** (Figure 1–45) to minimize exposure to toxic airborne dust particles created by sanding lead- or chrome-based automotive paints and primers. A dustless sanding system uses a blower or air pump to draw airborne dust into a storage container, much like

a vacuum cleaner. This action pulls airborne sanding dust through holes in a special sanding pad or through a shroud that surrounds the sanding pad. Some dustless system manufacturers claim that their machines can trap over 99 percent of the toxic dust created by sanding operations.

VEHICLE HANDLING IN THE SHOP

When handling a vehicle in the shop, keep the following safety rules in mind:

- Set the parking brake when you are working on the vehicle. If the car has an automatic transmission, set it in park. If the vehicle has a manual transmission, it should be in reverse (engine *off*) or neutral (engine *on*).
- If for some reason a work procedure requires you to work under a vehicle, use safety or *jack stands*. Never work under a car that is supported only by a hydraulic floor jack. The jack could be bumped and lower or fail and crush you to death!

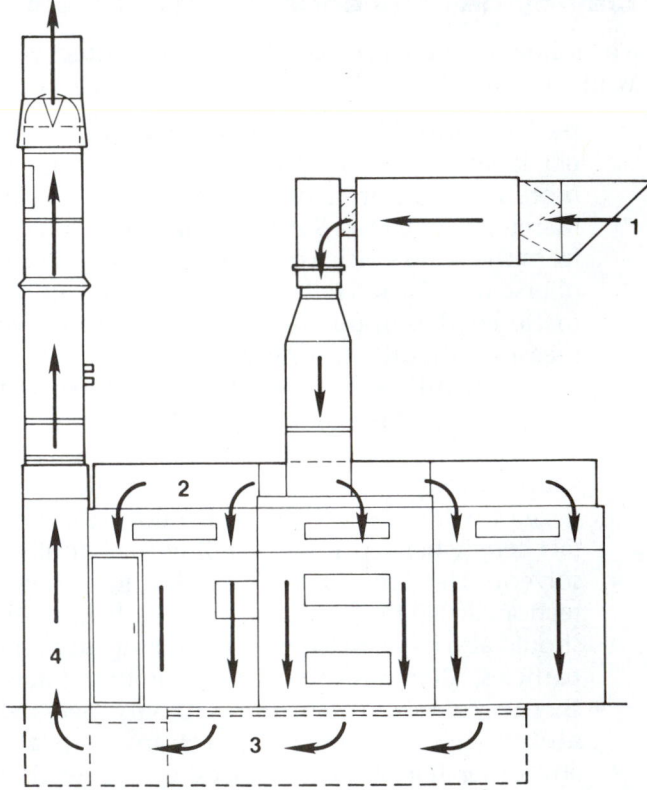

FIGURE 1–44 Study the air replacement unit for a downdraft spray booth: (1) the cycle begins with outside air being filtered and conditioned to proper temperature. Processed air then flows through (2) a ceiling plenum and filter system and down into the spray booth. As air flows downward, it passes around the vehicle and is drawn through (3) an opening in the floor that contains paint arrester filters. The (4) booth fan exhausts air from the pit, up a duct, and out the stack on the roof of the shop.

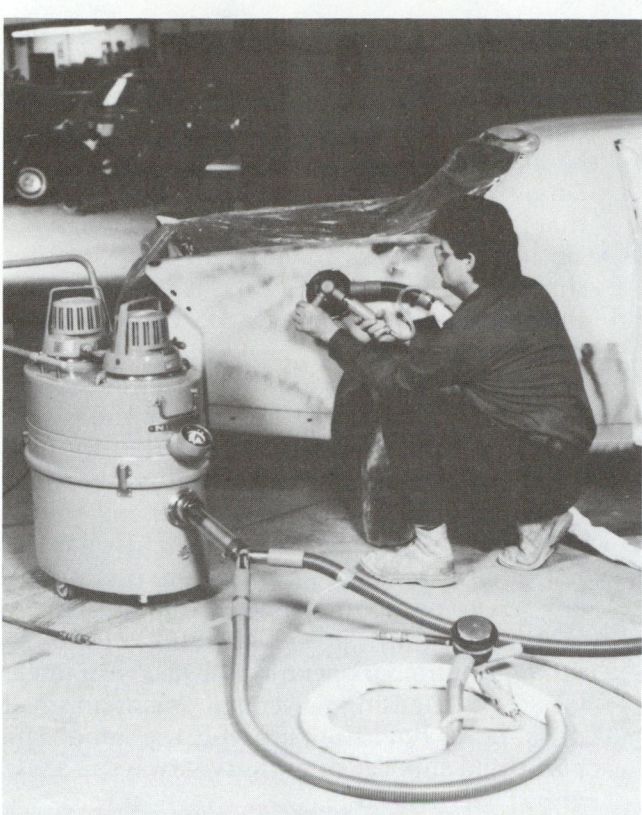

FIGURE 1-45 A portable vacuum system can be connected to a sander to help keep the work area clean. *(Courtesy of Nilfisk of America, Inc.)*

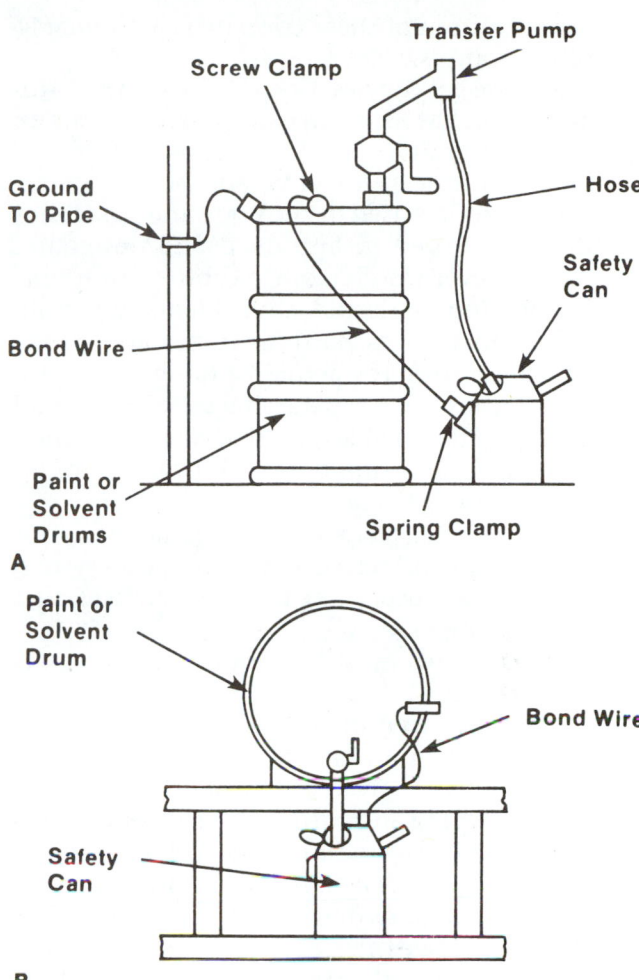

FIGURE 1-46 Two safe methods of moving flammable liquids from a drum to a portable safety can are shown.

- To prevent serious burns, avoid contact with hot metal parts, such as the radiator, exhaust manifold, tailpipe, catalytic converter, muffler, or parts being welded.
- Keep clothing and body parts away from moving parts when the engine is running, especially the radiator fan blades and belts.
- Be sure that the ignition switch is always in the off position, unless otherwise required for the procedure.
- When moving a vehicle around the shop, be sure to look in all directions and make certain that nothing is in the way.

HANDLING OF SOLVENT AND OTHER FLAMMABLE LIQUIDS

The body/paint technician will be working with various solvents to clean surfaces and equipment and to thin finishes. These solvents are extremely flammable. Fumes, in particular, can ignite explosively.

The following safety practices will help avoid fire and explosion:

- Do not light matches or smoke in the spraying and painting area. Make sure that your hands and clothing are free from solvent when lighting matches.
- All ignition sources should be carefully controlled and monitored to avoid any possible fire hazard where a high concentration of vapor from highly flammable liquids might at times be present.
- A *UL* (Underwriters Laboratories) approved drum transfer pump along with a drum vent should be used when transferring chemicals.
- Keep all solvent containers clearly labeled and closed, except when pouring.
- Handle all solvents or liquids with care to avoid spillage. Extra caution should also be used when transferring flammable materials from bulk storage. Remember to make sure the drum is grounded (Figure 1–46A) and that a bond wire connects the drum to a safety can (Figure 1–46B). Otherwise, static electricity can build up and create a spark that could cause an explosion.
- Discard or clean all empty solvent containers as prescribed by local regulations. Solvent fumes

in the bottom of these containers are prime ignition sources.

- Never use gasoline as a cleaning solvent. A tiny amount of gasoline can cause a tremendous explosion and fire.
- Paints, thinners, solvents, and other combustible materials used in the body and paint shop must be stored in approved and designated metal (never wood) storage cabinets or rooms. Storage rooms should have adequate ventilation, which takes harmful fumes and pollutants away from the actual working area. Many body shops use a separate facility for the bulk storage of flammable material. Never have more than one day's supply of paint outside of approved storage areas.
- Before spraying paint, be sure to remove any portable lamps and turn on the ventilation system.
- Spray areas must be free from hot surfaces, such as heat lamps.
- The spray area must be kept clean of combustible residue.
- The ventilation system must be left on while the paint is drying. Complete paint spraying safety precautions are given in Chapter 17.
- When welding and cutting, remember that very high heat and sparks can travel a long distance. Never weld or cut near paints, thinners, or other flammable liquids or materials. Cover open containers or move them to a safe area. Never cut or weld a container before checking what material was originally in that container.
- Never weld or grind near a battery. The battery charging operation produces hydrogen gas and a battery explosion can result.
- Fuel tanks should be drained and removed, if necessary, to repair panels next to them. It is easy to cut and rupture a tank with body tools. When welding or grinding near fuel filler lines, close them tightly and cover them with wet rags.
- When welding or cutting near car interiors, remove seats and floor mats, or cover them with a water-dampened cloth or a welding blanket. Always have a pail of water handy and a fire extinguisher nearby. Other welding safety tips are given in Chapter 6.

TOOL AND EQUIPMENT SAFETY

You must observe the following hand and power tool safety guidelines:

- Hand tools should always be clean and in working condition. Greasy, oily, or chipped hand tools can easily slip out of one's grasp, causing skinned knuckles or broken fingers.
- Check all hand tools for cracks, chips, burrs, broken teeth, or other dangerous conditions before using them. If any tools are defective, do not use them.
- Be careful when using sharp or pointed tools that can slip and cause injury. If a tool is supposed to be sharp, make sure it is sharp.
- Do not use hand tools for any job other than that for which they were specifically designed.
- Do not carry screwdrivers, punches, or other sharp hand tools in pockets. It is possible to injure oneself or damage the vehicle being worked on.
- When using an electric power tool, make sure that it is properly grounded. Check the wiring for cracks in the insulation, as well as for bare wires. Also, when using electrical power tools, never stand on a wet or damp floor.
- Do not operate a power tool without its guard(s).
- Disconnect electrical power before performing any service or maintenance on the tool.
- When doing any power grinding, chipping, sanding, or similar operation, always wear safety glasses. When using power equipment on small parts, never hold the part with the hand. The part could slip, causing injury. Always use vise grips pliers instead.
- Before plugging in any electric tool, make sure the switch is off to prevent serious injury. When you are not using an electric tool, turn it off.
- Do not attempt to use the tool beyond its stated capacity or for operations requiring more than the rated horsepower of the motor. Never use a tool for operations it was not designed for.
- Keep hands away from moving parts when the tool is under power. Never clear chips or debris when the tool is under power and never use your hands to clear chips. Use a brush or chip rake.
- Never overreach. Maintain a balanced stance and avoid slipping.
- Use utmost caution with compressed air. Pneumatic tools must be operated at the pressure recommended by their manufacturers. The downstream pressure of compressed air used for cleaning purposes must remain at a pressure level below 30 psi whenever the nozzle is dead-ended (Figure 1–47). Do not use compressed air to clean clothes. Even at low cleaning pressure, compressed air can cause dirt particles to become embedded in the skin, which can result in infection.
- Store all parts and tools properly by putting them away neatly where other workers will not trip over them. This will increase safety and reduce time wasted looking for a misplaced part or tool.
- When working with a hydraulic press, make sure that hydraulic pressure is applied in a safe manner. It is generally wise to stand to the side

when operating the press. Always wear safety glasses.

- If the shop has a hydraulic lift, be sure to read the instruction manual before using it. Check the pads to see that they are making proper contact with the frame. Then raise the vehicle about 6 inches and shake it to make sure it is well balanced on the lift. If there are any rattling or scraping sounds, it means that the vehicle is not locked in place properly. If this happens,

lower the lift and realign the pads to the vehicle. Test it again as previously described. Then, after lifting the vehicle to full height, put the safety catch on before working underneath the vehicle. Never permit anyone, either technician or customer, to remain in the car while it is being lifted.

- All bolts, nuts, lock rings, and other fastening components mentioned in the manufacturer's service manual are crucial to the safe operation of the car. Failure to use those specific items could cause extensive damage. Manufacturer's torque specifications must be followed.
- Do not risk injury through lack of knowledge; use shop tools or perform repair operations only after receiving proper instruction. Safety tips as well as operational procedures are given throughout this book.

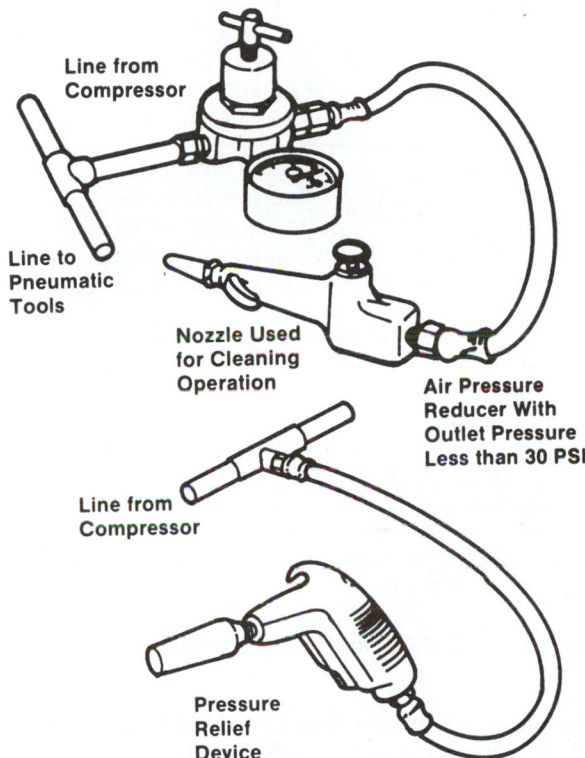

FIGURE 1-47 Air pressure reducing regulators are needed to keep the air pressure of tools at recommended levels for safe operation.

FIRE PROTECTION

During a fire, a few moments time can seem a lifetime if someone is being burned. Know the location of all fire extinguishers in the shop, so they can be reached quickly in an emergency.

To prevent an electrical fire where wires burn from excess current flow, disconnect the battery before working. This might be necessary before disconnecting any wires or if you suspect wire damage (Figure 1–48).

Every body shop requires fire extinguishers (Figure 1–49). Since fires are classified as Class A, B, C, and D type, there are different types of extinguishers specially designed for a particular class of fire. Table 1–1 gives the common classes of fire that are found in body shops and methods of containing them. Some extinguishers are capable of being used on more than one type of fire.

FIGURE 1-48 Disconnect the battery before working. During a collision, wires can be severed. If a hot wire touches the ground, an electrical fire could result. *(Courtesy of Tech-Cor)*

FIGURE 1-49 Body and paint work require plenty of fire extinguishers capable of fighting type A, B, and C fires.

TABLE 1-1: GUIDE TO EXTINGUISHER SELECTION

	Class of Fire	Typical Fuel Involved	Type of Extinguisher
Class **A** Fires (green)	**For Ordinary Combustibles** Put out a class A fire by lowering its temperature or by coating the burning combustibles.	Wood Paper Cloth Rubber Plastics Rubbish Upholstery	Water[*][1] Foam[*] Multipurpose dry chemical[4]
Class **B** Fires (red)	**For Flammable Liquids** Put out a class B fire by smothering it. Use an extinguisher that gives a blanketing, flame-interrupting effect; cover whole flaming liquid surface.	Gasoline Oil Grease Paint Lighter fluid	Foam[*] Carbon dixoide[5] Halogenated agent[6] Standard dry chemical[2] Purple K dry chemical[3] Multipurpose dry chemical[4]
Class **C** Fires (blue)	**For Electrical Equipment** Put out a class C fire by shutting off power as quickly as possible and by always using a nonconducting extinguishing agent to prevent electric shock.	Motors Appliances Wiring Fuse boxes Switchboards	Carbon dioxide[5] Halogenated agent[6] Standard dry chemical[2] Purple K dry chemical[3] Multipurpose dry chemical[4]
Class **D** Fires (yellow)	**For Combustible Metals** Put out a class D fire of metal chips, turnings, or shavings by smothering or coating with a specially designed extinguishing agent.	Aluminum Magnesium Potassium Sodium Titanium Zirconium	Dry powder extinguishers and agents only

*Cartridge-operated water, foam, and soda-acid types of extinguishers are no longer manufactured. These extinguishers should be removed from service when they become due for their next hydrostatic pressure test.

Notes:

(1) Freezes in low temperatures unless treated with antifreeze solution, usually weighs over 20 pounds, and is heavier than any other extinguisher mentioned.

(2) Also called ordinary or regular dry chemical. (sodium bicarbonate)

(3) Has the greatest initial fire-stopping power of the extinguishers mentioned for class B fires. Be sure to clean residue immediately after using the extinguisher so sprayed surfaces will not be damaged. (potassium bicarbonate)

(4) The only extinguishers that fight A, B, and C classes of fires. However, they should not be used on fires in liquefied fat or oil of appreciable depth. Be sure to clean residue immediately after using the extinguisher so sprayed surfaces will not be damaged. (ammonium phosphates)

(5) Use with caution in unventilated, confined spaces.

(6) May cause injury to the operator if the extinguishing agent (a gas) or the gases produced when the agent is applied to a fire are inhaled.

Operating instructions are imprinted on each extinguisher. However, during an emergency there might be no time to read the label. You must know how to use the fire extinguisher *before* any emergency.

A fire can be extinguished by depriving it of its essential ingredients, which are heat, fuel, and oxygen. Most extinguishers work by cooling the fire and removing the oxygen. If the fire extinguisher is going to be used effectively, it must be aimed at the base of the flame where the fuel is located. Fire extinguishers should be checked regularly and placed at strategic shop locations.

GOOD HOUSEKEEPING

Here are some simple good housekeeping precautions that should be followed in every shop:

- All surfaces should be kept clean, dry, and orderly. Any oil, coolant, or grease on the floor can cause slips that could result in serious injuries. To clean up oil, be sure to use a commercial oil absorbent.
- Keep all water off the floor; remember that water is a conductor of electricity. A serious shock hazard will result if a live wire hap-

pens to fall into a puddle in which a person is standing.

- Make sure that aisles and walkways are kept clean and wide enough for a safe clearance. Cluttered walking areas contain items waiting to cause accidents.
- There should be a list of emergency telephone numbers clearly posted next to the telephone. These numbers should include a doctor, hospital, and fire and police departments.
- The work area should have a first aid kit for treating minor injuries. This kit should include some sterile gauze, bandages, scissors, and other related items. Facilities for flushing the eyes should also be in or near the shop area.
- Make sure that so-called hazardous materials are not discharged through floor drains or other outlets leading to public waterways.
- Any dirty rags or other combustible material must be deposited in a metal container with a suitable metal cover. They should be removed to a safe place outside the building. Keep used paper towels and other paper products in a separate, covered container, which should be emptied every day.
- Customers and all nonemployees should never be allowed in any of the shop's work areas.

MANUFACTURER'S WARNINGS AND GOVERNMENT REGULATIONS

Right-to-Know laws give essential information and stipulations for safely working with hazardous materials. They started with OSHA's *Hazard Communication Standard*. This document was originally intended for chemical companies and manufacturers that require employees to handle potentially hazardous materials in the workshop. Since then the majority of states have enacted their own Right-to-Know laws. The federal courts have decided that these regulations should apply to all companies, including the auto collision repair and refinishing professions.

The general intent of the law is for employers to provide their employees with a safe working place as it relates to hazardous materials. Specifically, there are three areas of employer responsibility:

1. **Training and educating employees.** All employees must be trained about their rights under the legislation. They must learn the nature of the hazardous chemicals, the labeling of chemicals, and the information about each chemical. This information should be posted on Material Safety Data Sheets (MSDS). These sheets (Figure 1–50) detail product composition and precautionary information for all products that can present a

health or safety hazard. They are generally prepared by the product manufacturer. Employees must be familiarized with the general uses, characteristics, protective equipment, and accident or spill procedures associated with major groups of chemicals. This training must be given to employees annually and provided to new employees as part of their job orientation.

2. **Labeling and giving information about potentially hazardous chemicals.** All hazardous materials must be properly labeled, indicating what health, fire, or reactivity hazard they pose. Information about what protective equipment is needed when handling each chemical must be given. This safety data must be read and understood by the user before application.

3. **Record keeping.** Shops must maintain documentation on the hazardous chemicals in the workplace. They must provide proof of training programs, records of accidents and/or spill incidents, satisfaction of employee requests for specific chemical information via the MSDSs, and a general Right-to-Know compliance procedure manual used within the shop.

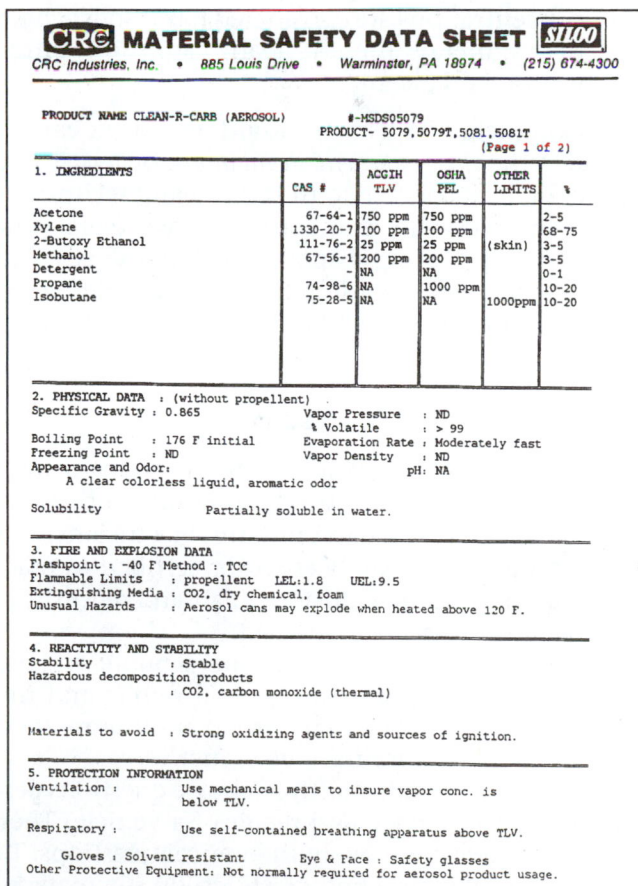

FIGURE 1–50 Material Safety Data Sheets will give information on hazardous materials. *(Courtesy of CRC Industries, Inc.)*

Hazardous waste as determined by the Environmental Protection Agency (**EPA**) can be a solid or liquid that can harm people and the environment. If the waste is on the EPA list of known harmful materials or has one or more of the following four known dangerous characteristics, it is considered hazardous.

1. **Ignitability.** The material or waste fails the ignitability test if it is a liquid with a flash point below 140 degrees Fahrenheit or a solid that can spontaneously ignite.
2. **Corrosiveness.** A material or waste is considered corrosive if it dissolves metals and other materials or burns the skin. It is an aqueous solution with a pH of 2 and below, or 12.5 and above. Acids have the lower value and alkalis have the higher value.
3. **Reactivity.** Any material that reacts violently with water or other materials or releases cyanide gas, hydrogen sulfide gas, or similar gases when exposed to low pH solutions (acid) is considered hazardous. This also includes material that generates toxic mists, fumes, vapors, and flammable gases.
4. **Toxicity.** Materials that leach one or more heavy metals in concentrations greater than 100 times primary drinking water's standard concentrations are considered hazardous. These heavy metals include lead, cadmium, chromium, and arsenic.

Complete EPA lists of hazardous wastes can be found in the *Code of Federal Regulations*. Materials and wastes of most concern to the body/paint technician are organic solvents that contain heavy metals, especially lead. During disposal, all hazardous waste must be handled according to the appropriate regulations (Figure 1–51 and 1–52).

ASE CERTIFICATION

ASE Certification is a testing program to help prove that you are a knowledgeable auto body technician. Just as doctors, accountants, electricians, and other professionals are licensed or certified to practice their profession, collision repair and refinishing technicians can also be certified.

Certification protects the general public and the professional. It assures the general public and the prospective employer that certain minimum standards of performance have been met.

ASE tests include multiple-choice questions pertaining to the service and repair of a vehicle. They do not include completely theoretical questions. To prepare for ASE auto body tests, study the material in this book carefully.

To help prepare for the Body Repair and/or Painting and Refinishing tests, some test questions at the end of the service chapters in this text are similar to those used by ASE. Table 1–2 gives a breakdown of the ASE tests.

Collision repair and refinishing technicians can get certified in one or more technical areas by taking and passing written certification tests. The National Institute for Automotive Service Excellence (ASE) offers a voluntary certification program that is recommended by the major vehicle manufacturers in the United States. The Body Repair and Painting and Re-

FIGURE 1–51 Many automotive service operations, including body shops, are hiring outside contractors to handle their hazardous waste materials. *(Courtesy of DuPont Automotive Products)*

FIGURE 1–52 Handle all hazardous wastes as per EPA regulations. *(Courtesy of DuPont Automotive Products)*

TABLE 1-2

TEST SPECIFICATIONS PAINTING AND REFINISHING

Content Area	Questions in Test	Percentage of Test
A. Surface Preparation	12	24%
B. Spray Gun Operation and Related Equipment	7	14%
C. Paint Mixing, Matching, and Applying	13	26%
D. Solving Paint Application Problems	8	16%
E. Finish Defects, Causes, and Cures	6	12%
F. Safety Precautions and Miscellaneous	4	8%
Total	**50**[1]	**100%**

NON-STRUCTURAL ANALYSIS AND DAMAGE REPAIR

Content Area	Questions in Test	Percentage of Test
A. Preparation	4	8.9%
B. Outer Body Panel Repairs, Replacements, and Adjustments	16	35.6%
C. Metal Finishing and Body Filling	6	13.3%
D. Moveable Glass and Hardware	4	8.9%
E. Welding and Cutting	8	17.8%
F. Plastic Repair	7	15.5%
Total	**45**[1]	**100.0%**

STRUCTURAL ANALYSIS AND DAMAGE REPAIR

Content Area	Questions in Test	Percentage of Test
A. Frame Inspection and Repair	11	24.4%
B. Unibody Inspection, Measurement, and Repair	18	40.0%
C. Fixed Glass	4	8.9%
D. Metal Welding and Cutting	12	26.7%
Total	**45**[1]	**100.0%**

MECHANICAL AND ELECTRICAL COMPONENTS

Content Area	Questions in Test	Percentage of Test
A. Suspension and Steering	12	26.6%
B. Electrical	8	17.8%
C. Brakes	5	11.1%
D. Heating and Air Conditioning	4	8.9%
E. Cooling Systems	4	8.9%
F. Drive Train	4	8.9%
G. Fuel, Intake, and Exhaust Systems	3	6.7%
H. Restraint Systems	5	11.1%
1. Active Restraint Systems	(1)	
2. Passive Restraint Systems	(1)	
3. Supplemental Restraint Systems (SRS)	(3)	
Total	**45**[1]	**100.0%**

FIGURE 1-53 An ASE body/paint technician's shoulder patch or emblem shows that you have passed the tests measuring your knowledge of proper repair methods.

FIGURE 1-54 Various trade associations work to improve training and communication among all those involved in the collision repair industry.

finishing tests contain 40 questions in each of the various areas.

Craftspeople who pass the written tests are awarded a certificate and a shoulder emblem for their work clothes.

Many employers now expect their collision repair and refinishing technicians to be certified. The certified technician is recognized as a professional by the public, employers, and peers (Figure 1–53). For this reason, the certified technician usually receives higher pay than one who is not certified.

For further information on the ASE certification program write to ASE, 1305 Dulles Technology Drive, Herndon, VA 22071-3145.

In addition to ASE certification, a professional body/paint technician or the shopowner should consider membership in the various trade associations. The growth and influence of these trade associations within the industry has led to increased communication, training, and the sharing of knowledge and ideas. The Inter-Industry Conference on Auto Collision Repair (I-CAR), Society of Collision Repair Specialists (SCRS), the Automotive Service Association (ASA), and the International Autobody Congress and Exposition (NACE) are several examples. They promote professional and consumer education and awareness (Figure 1–54).

SHOP TALK

It is important for body shop personnel to have good communication and cooperation. A wise old saying goes "A chain is only as strong as its weakest link." This applies to the smooth operation of a body shop. If anyone, from the estimator to the final get-ready, does not do the job right, everyone suffers. Customers will not return to the shop and everyone's paycheck will go down.

SUMMARY

- A collision, commonly called a "crash" or "wreck," is an impact caused by a vehicle hitting another vehicle or object.

- The repair method used for restoring a vehicle damaged in a collision is determined by many factors.

- The vehicle is first brought into the shop where a damage estimate is prepared to calculate the cost of labor, materials, and parts for the repair.

- Estimating involves analyzing damage and calculating how much it will cost to repair the vehicle.

- Panel straightening involves using various hand tools and equipment to bend or shape the panel back into its original contour.

- Panel replacement involves removing and installing a new panel or body part.

- The metalworking area is where the vehicle body structure is repaired. Accurate adjustment of body assemblies, such as hoods, rear deck lids, and doors, might have to be made by the technician.

- Measurement systems allow the technician to check for frame or body misalignment resulting from a major collision.

■ Frame straightening equipmentuses a large steel framework, clamps, pulling chains, and hydraulic power to pull or force the vehicle structure back into its original position.

■ The paint shop area is where the car is refinished by refinishing technicians or painters.

■ ASE Certification is a testing program to help prove that you are a knowledgeable auto body technician.

■ The most important program in any body shop is accident prevention. Carelessness and the lack of safety habits cause accidents and injuries.

ASE-STYLE REVIEW QUESTIONS

1. Which of the following presents dangers to the air passages and lungs of the technician?

   A. Dust
   B. Vapors from caustic solutions and solvents
   C. Spray mists from undercoats and finishes
   D. All of the above

2. Which of the following respirators covers the entire head and neck area?

   A. Cartridge filter respirator
   B. Dust respirator
   C. Air-supplied respirator
   D. None of the above

3. Technician A and Technician B are both in the practice of spraying continuously for extended periods of time. Technician A changes the prefilters when it becomes difficult to breath through the respirators. Technician B performs a fit test prior to using the respirator. Who is correct?

   A. Technician A
   B. Technician B
   C. Both A and B
   D. Neither A or B

4. Which respirator is used to protect against dust from sanding and grinding?

   A. Hood respirator
   B. Organic vapor-type respirator
   C. Air-supplied respirator
   D. None of the above

5. Eye protection should be worn when using

   A. Grinders
   B. Disc sanders
   C. Pneumatic chisels
   D. All of the above

6. Which of the following should not be worn in a body/paint shop?

   A. Jumpsuit
   B. Loose clothing
   C. Cap
   D. Both B and C

7. By what means can ventilation be achieved in the body/paint shop?

   A. Extraction floors
   B. Central dust extraction
   C. Air-changing system
   D. All of the above

8. What is the maximum amount of paint that should be left outside of approved storage areas?

   A. 10 quarts
   B. 20 quarts
   C. One day's supply
   D. A half day's supply

9. Technician A discards all empty solvent containers. Technician B uses gasoline to clean them out. Who is correct?

   A. Technician A
   B. Technician B
   C. Both A and B
   D. Neither A or B

10. Which type of extinguisher can be used on all classes of fires?

    A. Water
    B. Foam
    C. Multipurpose dry chemical
    D. None of the above

11. What is another name for the cartridge filter respirator?

    A. Dust mask
    B. Organic vapor-type respirator
    C. Air-supplied respirator
    D. Hood respirator

12. Technician A will operate a tool beyond its stated capacity only for a limited time. Technician B will not operate a power tool without its guards. Who is correct?

    A. Technician A
    B. Technician B
    C. Both A and B
    D. Neither A or B

13. Technician A wears a dust respirator when spray painting, while Technician B wears a cartridge respirator. Who is correct?

    A. Technician A
    B. Technician B
    C. Both A and B
    D. Neither A nor B

14. Technician A, when transferring solvents into smaller containers, labels them for proper identification later. Technician B has built a separate wooden storage facility for solvents. Who is correct?

    A. Technician A
    B. Technician B
    C. Both A and B
    D. Neither A or B

15. In attempting to find a qualified hauler to remove hazardous waste, Technician A consults local paint suppliers. Technician B checks with the regional EPA office. Who is correct?

    A. Technician A
    B. Technician B
    C. Both A and B
    D. Neither A nor B

ESSAY QUESTIONS

1. What happens to a vehicle during a collision?

2. Explain the difference between a "door ding" and a "total loss."

3. What is estimating?

4. Describe the differences between panel straightening and panel replacement.

5. What is the VIN and how is it used?

CRITICAL THINKING PROBLEMS

1. How would you determine if a car is a total loss?

2. A fire breaks out in the shop and a co-worker's clothes are on fire. What should you do?

3. What can you do to make your shop a safer place to work?

MATH PROBLEMS

1. A technician is paid $45 dollars per hour. The repair will take 11.5 hours. Parts for the repair will be $176.25 plus 7% tax. What will the total be for the estimate?

2. An older car is valued at $2,500. The parts for the repair will be $1,500. How many hours labor at $35 per hour can be completed before the car should be totaled?

P1-1 When sanding, wear a dust mask to protect your lungs from paint dust and abrasive dust.

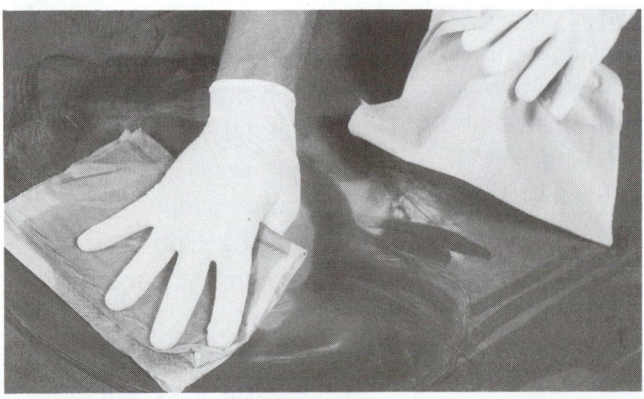

P1-2 When grinding, wear leather gloves, a respirator, and a full face mask. Grinding is more dangerous than sanding.

P1-3 When working with chemicals, wear plastic gloves to protect your skin from potentially harmful agents. To avoid inhaling fumes, wear an approved cartridge-type respirator. Read and follow the directions on the label of the container.

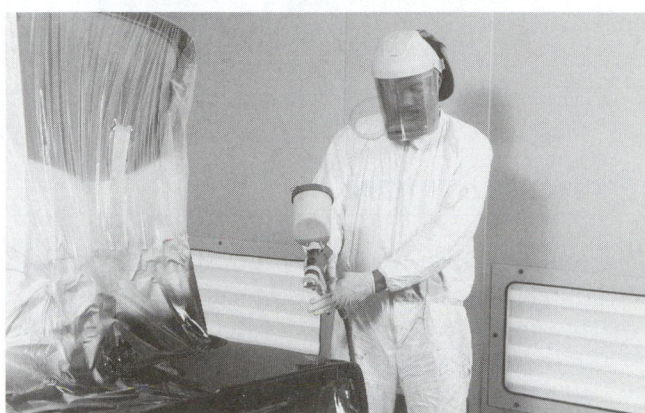

P1-4 When spraying modern primers and paints, wear an air-supplied respirator in a properly ventilated spray booth. Coveralls will protect your skin and help keep lint and dust out of the paint.

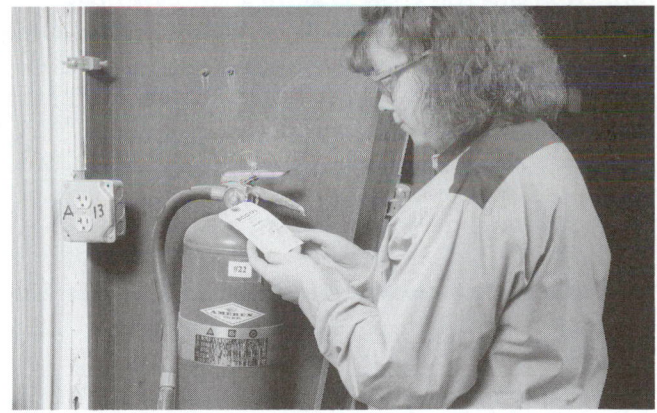

P1-5 Know the location and proper use of all fire extinguishers in your shop.

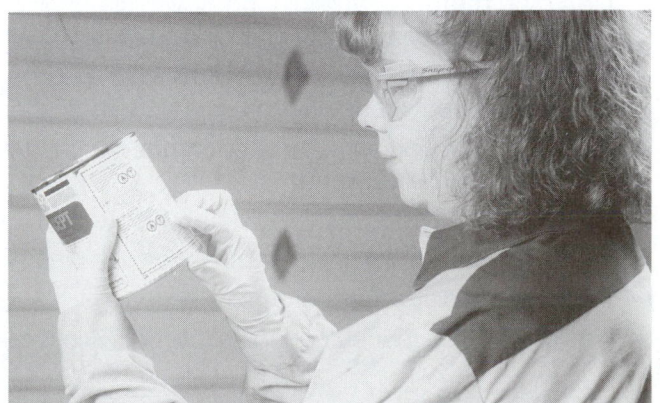

P1-6 Label directions will give safety precautions that must be followed.

CHAPTER 2

Understanding Automobile Construction

INTRODUCTION

The goal of collision repair is to restore the vehicle to its preaccident condition. To accurately repair a wrecked car or truck, you must fully understand how the vehicle is designed and constructed. You must be capable of accurately identifying all damaged components and the repair options available. You must know the materials used in the vehicle and how these materials may affect the repair process.

To fully comprehend the challenges faced by the technician, consider the radical changes in the industry over the years (Figure 2–1). With passenger cars, minivans, and many small pickup trucks, body-over-frame construction that served the industry for over sixty years has given way to unitized construction.

The *frame* is usually a high strength metal structure used to support other parts of the vehicle. It holds the engine, transmission, suspension, body, and other parts in position. The frame can be separate from the body or the body can be welded together to form the frame.

The vehicle *body* is a steel, aluminum, fiberglass, plastic, or composite skin forming the outside of the car. The body is normally painted to give the vehicle its attractive, shiny, color appearance.

Body-over-frame vehicles have separate body and chassis parts bolted to the frame. The engine and other major assemblies are mounted on the frame. This type of frame consists of two side rails connected by a series of cross members. Body-over-frame construction is still being used on full-size pickup trucks, some small pickups, and most full-size vans. Some larger luxury cars still use traditional coil spring, body-over-frame setups.

Unibody construction uses body parts welded and bolted together to form an integral frame. No separate heavy gauge steel frame under the body is needed. Unibody construction is a totally different

OBJECTIVES

After studying this chapter, you should be able to:

✔ Name the general body shapes
✔ Describe the general evolution of vehicle body design from early body-over-frame to present-day semiunitized, and unitized construction.
✔ List the major design characteristics of modern body-over-frame and modern unibody construction and how they affect repair procedures.
✔ Identify the major structural components, sections, and assemblies of body-over-frame vehicles.
✔ Identify the major structural components, sections, and assemblies.
✔ Identify the important parts of motor vehicles.
✔ Read a VIN plate number.
✔ Answer vehicle construction-related ASE test questions.

KEY TERMS

anticorrosion materials
body-over-frame
center section
cowl
dash panel
deck lid
floor pan
frame rails
front section
instrument panel
pillars
quarter panels
radiator core support
rear section
shock towers

ASE TASK LIST

Job Skills covered in this chapter include:

PAINTING AND REFINISHING TEST (B2) TASK LIST

A. Surface Preparation

7. Identify type of metal and apply suitable metal treatment or primer.

C. Paint Mixing, Matching, and Applying

9. Identify the types of rigid, semirigid, or flexible plastic parts to be refinished; determine the proper materials and refinishing procedures.

NONSTRUCTURAL ANALYSIS AND DAMAGE REPAIR TEST (B3) TASK LIST

A. Preparation

5. Remove undamaged, nonstructural body panels and components that may interfere with or be damaged during repair.

B. Outer Body Panel Repairs, Replacements, and Adjustments

3. Determine the extent of damage to aluminum body panels; repair, weld, or replace in accordance with manufacturers' specifications.
4. Remove, replace, and align hood, hood hinges, and hood latch/lock.
5. Remove, replace, and align deck lid, lid hinges, and lid latch/lock.
6. Remove and replace doors, tailgates, hatches, lift gates, latch/lock assemblies, and hinges.
7. Remove, replace, and align bumpers, reinforcements, guards, isolators, and mounting hardware.

8. Check and adjust clearances of front fenders, header, and other panels.
9. Check door hinge condition; check door frames for proper fit; check and adjust door clearances.
17. Diagnose and repair water leaks, dust leaks, wind noise, squeaks, and rattles.

D. Moveable Glass and Hardware

5. Inspect, repair, and install convertible top and related mechanisms.

E. Welding and Cutting

1. Identify weldable and non-weldable materials used in vehicle construction.

STRUCTURAL ANALYSIS AND DAMAGE REPAIR TEST (B4) TASK LIST

A. Frame Inspection and Repair

8. Remove and replace damaged frame horns, side rails, cross members, and front or rear sections.
10. Repair or replace weakened or cracked frame members in accordance with vehicle manufacturers'/industry standards.

C. Stationary Glass

1. Remove and replace front and rear stationary glass (heated and non-heated) in accordance with manufacturers' recommendations.

A
B

FIGURE 2-1 Great changes have occurred in automotive design since the early years of the automobile: (A) 1897 horseless carriage *(Courtesy of Oldsmobile)* and (B) experimental sports car. *(Photograph provided courtesy of Mercedes-Benz North America, Inc.)*

concept in vehicle design that requires new assembly techniques, new materials, and a completely different approach to repairs. In unibody designs, heavy gauge, cold-rolled steels have been replaced with lighter, thinner, high-strength steel alloys or aluminum alloy. This requires new handling, straightening, and welding techniques.

The ever increasing use of plastics, modular glass, and advanced paint systems is also changing the face of vehicle repair. On unibody vehicles, certain mechanical systems, such as suspension and steering, rely on the proper positioning of unibody components for alignment and smooth operation.

Space frame construction is similar to unibody construction. The main difference is that the metal structure is covered with an outer skin of plastic or fiberglass panels. These are attached with mechanical fasteners or adhesives. This type of construction requires technicians to be more careful when inspecting for collision damage because the plastic panels can hide damage to the metal structure. Greater care must also be taken with corrosion protection for this same reason.

2.1 BODY SHAPES AND PARTS

Various methods of classifying vehicles exist—by type of engine, fuel system, drive line, and so forth. The classifications most common to consumers are body shape, seat arrangement, number of doors, and so on. Six basic body shapes are used today:

1. **Sedan.** A vehicle with front and back seats that accommodates four to six persons is classified as either a two- or four-door sedan (Figure 2–2).
2. **Hardtop.** A vehicle with front and back seats, a hardtop is generally characterized by a lack of door or "B" pillars that extend to the roof (Figure 2–3). It can also be classified as either a two- or four-door hardtop.
3. **Convertible top.** After an absence from the domestic market for several years, the convertible top made a comeback in 1985. Today's convertible top vehicle has a vinyl or cloth roof that can be raised or lowered (Figure 2–4). Like a hardtop, a convertible top has no door pillars and, depending on the make, can be purchased with

2-DOOR SEDAN

FIGURE 2–2 Note typical sedan body shapes.

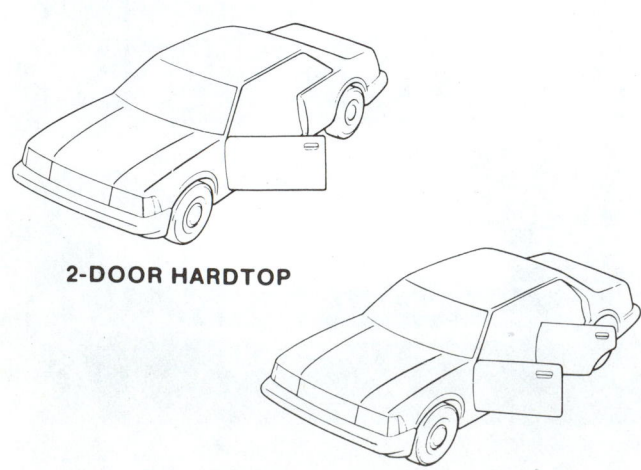

2-DOOR HARDTOP

FIGURE 2–3 These are typical hardtop body styles.

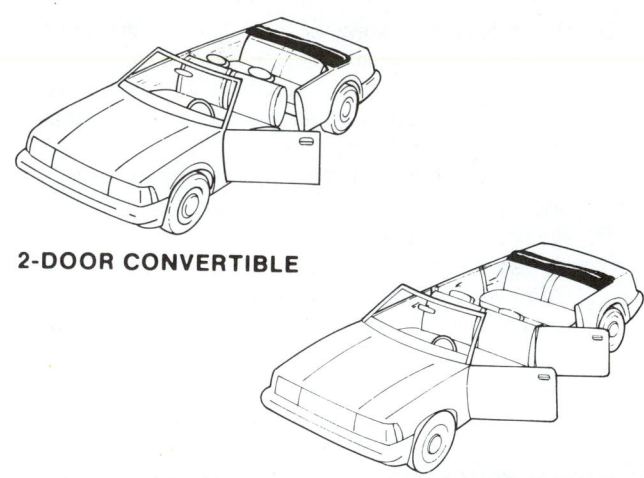

2-DOOR CONVERTIBLE

FIGURE 2–4 Compare convertible body shapes.

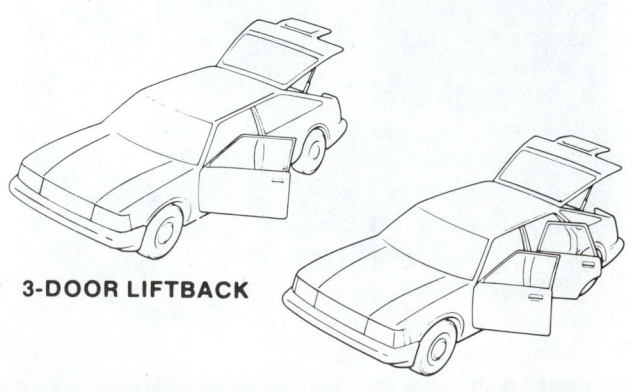

3-DOOR LIFTBACK

FIGURE 2–5 These are 3- and 4-door liftback bodies.

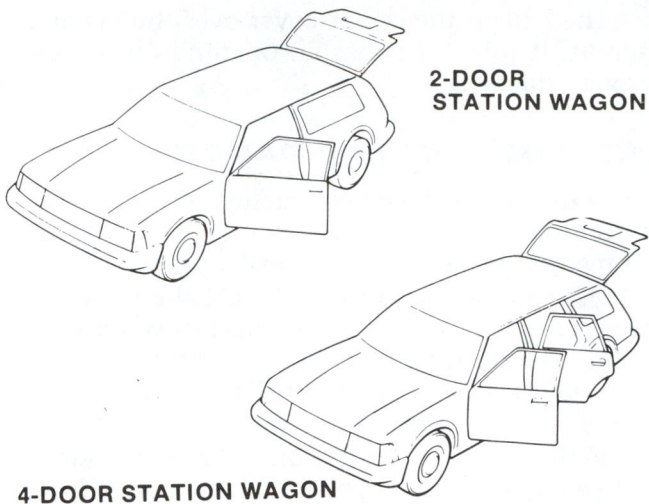

2-DOOR
STATION WAGON

4-DOOR STATION WAGON

FIGURE 2-6 Note typical station wagon body shapes.

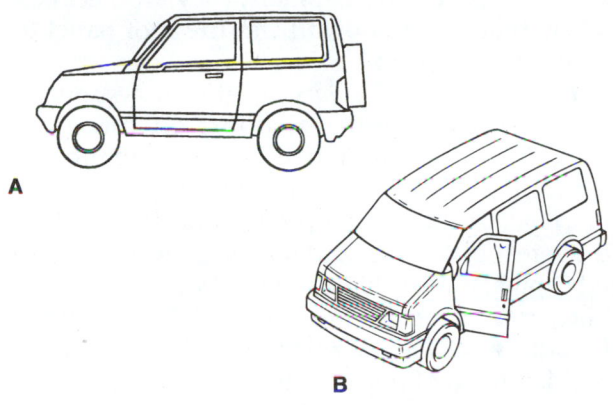

A

B

FIGURE 2-7 Compare (A) sports vehicle *(Courtesy of McGard, Inc.)* and (B) minivan body shapes.

or without a back window. It is available in two- and four-door models.

4. **Liftback or hatchback.** The distinguishing feature of this vehicle is its rear luggage compartment, which is an extension of the passenger compartment. Access to the luggage compartment is gained through an upward opening hatch-type door (Figure 2–5). The vehicle comes in three- and five-door models.

5. **Station wagon.** A station wagon (Figure 2–6) is characterized by its roof, which extends straight back for the length of the vehicle, allowing a spacious interior luggage compartment in the rear. The rear door, which can be opened in various ways depending on the model, provides access to the luggage compartment. Station wagons come in two- and four-door models and have space for up to nine passengers.

6. **Sport or multipurpose vehicles.** This new classification of vehicles covers a range of body

designs (Figure 2–7). They are available in two-wheel drive, four-wheel drive (4x4), or all-wheel drive. Pickup truck body designs are available with standard cab designs, with extended (larger) cab areas (some have added seats in back of the front seat), and some with open or closed pick-up spaces. While sport utilities appeal to the outdoor enthusiast who wants both road and off-road applications, van designs are considered sport vehicles, but for family use.

Although body types are sometimes classified according to these various descriptions, the body's strength depends on the type of vehicle and its body structure. Factors, such as door size, the presence or absence of a center pillar, front body pillar, quarter panels, roof panels, and so forth, greatly affect how much or how little the impact from a collision is absorbed. For vehicles with large luggage areas, such as vans or liftbacks, the structural members and reinforcements designed into the body measurably affect overall torsional rigidity.

MAJOR BODY SECTIONS

For simplicity and to help communication in auto body repair, a vehicle is commonly divided into *three body sections*—front, center, and rear. You should understand how these sections are constructed so that you can properly repair them.

The **front section**, also called *nose section*, includes everything between the front bumper and the firewall. The bumper, grille, frame rails, front suspension parts, and usually the engine are a few of the items included in the front section of a vehicle.

The nickname *"front clip"* or *"doghouse"* is sometimes used to refer to the front body section. It is often purchased and cut off from a wreck in one piece from an *automotive recycler* or salvage yard. The empty engine compartment forms the "doghouse."

The **center section** or *midsection* typically includes the body parts that form the passenger compartment. A few parts in this section include the floor pan, roof panel, cowl, doors, door pillars, glass, and related parts. A slang name for the center section is *"greenhouse"* because it is surrounded by glass.

The **rear section**, *tail section*, or *rear clip* commonly consists of the rear quarter panels, trunk or rear floor pan, rear frame rails, trunk or deck lid, rear bumper, and related parts. It is often sectioned or cut off of a salvaged vehicle to repair severe rear impact damage.

When discussing collision repair, body shop personnel often refer to these sections of the vehicle. It simplifies communication because everyone knows which parts are included in each section.

PANEL AND ASSEMBLY NOMENCLATURE

A *panel* is a stamped steel or molded plastic sheet that forms a body part. Various panels are used in a vehicle. Usually, the name of the panel is self-explanatory. When panels are joined into a large part, it is called an *assembly*.

The *vehicle left-side* is the driver's or steering wheel side on vehicles and trucks built for American roads. The *vehicle right-side* is the passenger side or the side opposite the steering wheel. Remember that vehicles built for other countries will often have the steering wheel on the right-side because they drive on the other side of the road.

Another way to determine the right- and left-sides of a vehicle is to stand behind the vehicle. Your right hand would be on the right side and left hand would be on the left side of the vehicle. Panels and parts are often called out as left- or right-side.

FRONT SECTION PARTS

The **frame rails** are the box frame members extending out near the bottom of the front section. They are usually the strongest part of a unibody.

The **cowl** is the body parts at the rear of the front section, right in front of the windshield. This includes the top cowl panel and side cowl panels.

The *front fender aprons* are inner panels that surround the wheels and tires to keep out road debris. They often bolt or weld to the frame rails and cowl.

The **shock towers** or *strut towers* are reinforced body areas for holding the upper parts of the suspension system. The coil springs and strut or shock absorbers fit up into the shock towers. They are normally formed as part of the inner fender apron.

The **radiator core support** is the framework around the front of the body structure for holding the cooling system radiator and related parts. It often fastens to the frame rails and inner fender aprons.

The *hood* is a hinged panel for accessing the engine compartment (front-engine vehicle) or trunk area (rear-engine vehicle). *Hood hinges*, bolted to the hood and cowl panel, allow the hood to swing open. The hood is normally made of two or more panels welded or bonded together to prevent flexing and vibration. Some hoods also hinge at the radiator support.

The **dash panel**, sometimes termed *firewall* or *front bulkhead*, is the panel dividing the front section and the center, the passenger compartment section. It normally is welded in place.

The *front fenders* extend from the front doors to the front bumper. They cover the front suspension and inner aprons. They normally bolt into place around their perimeter.

The *bumper assembly* bolts to the front frame horns or rails to absorb minor impacts.

The *grille* is the center cover over the radiator support. It sometimes has an opening for airflow through the radiator.

CENTER SECTION PARTS

The **floor pan** is the main structural section in the bottom of the passenger compartment. It is often stamped as one large piece of steel.

With front-wheel-drive vehicles, the floor pan can be relatively flat. With rear-wheel-drive vehicles, a *tunnel* is formed in the floor pan for the transmission and drive shaft. The drive shaft needs room to extend back to the rear axle assembly.

Pillars are vertical body members that hold the roof panel in place and protect in case of a rollover accident.

The *front pillars* extend up next to the edges of the windshield. They must be strong to protect the passengers. Also termed A-pillars, they are steel box members that extend down from the roof panel to the main body section.

Center pillars or *B-pillars* are the roof supports between the front and rear doors on four-door vehicles. They help strengthen the roof and provide a mounting point for the rear door hinges.

Rear pillars extend up from the quarter panels to hold the rear of the roof and rear window glass. Also called C-pillars, their shape can vary with body style.

Rocker panels or *door sills* are strong beams that fit at the bottom of the door openings. They normally are welded to the floor pan and to the pillars, kick panels, or quarter panels. The *kick panels* are small panels between the front pillars and rocker panels.

The *rear shelf* or *package tray* is a thin panel behind the rear seat and in front of the back glass. It often has openings for the rear stereo speakers. The *rear bulkhead panel* separates the passenger compartment from the rear trunk area.

The *doors* are complex assemblies made up of an outer skin, inner door frame, door panel, window regulator, glass, and related parts. *Door hinges* are bolted or welded between the pillars and door frame. The *window regulator* is a gear mechanism that allows you to raise and lower the door glass.

Side impact beams are metal bars or corrugated panels that bolt or weld inside the door assemblies to protect the passengers. Primarily, they prevent the door from opening upon impact. They also help keep anything from intruding into the passenger area. When made of ultra-high-strength steel, side impact beams should not be repaired.

The *roof panel* is a large multipiece panel that fits over the passenger compartment. It is normally welded to the pillars. Sometimes it includes a sunroof or removable top pieces, termed T-tops.

The *dash assembly*, sometimes termed **instrument panel**, is the assembly including the soft dash pad, instrument cluster, radio, heater and air-conditioning controls, vents, and similar parts. It can be damaged in a collision by human contact.

REAR SECTION PARTS

The *rear frame rails* are strong boxed structures that give strength to the rear of the vehicle.

The *trunk floor panel* is a stamped steel part that forms the bottom of the rear storage compartment. Quite often the spare tire fits down into this stamped panel. It often welds to the rear rails, inner wheel houses, and lower rear panel.

The **deck lid** or *trunk lid* is a hinged panel over the rear storage compartment. A *rear hatch* is a larger panel and glass assembly hinged for more access to the rear of the vehicle.

The **quarter panels** are the large, side body sections that extend from the side doors back to the rear bumper. They are welded in place and form a vital part of the rear body structure.

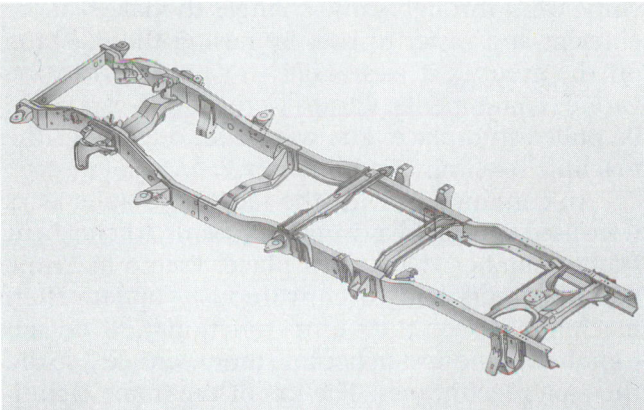

FIGURE 2-8 Typical conventional perimeter frame construction is common on trucks and luxury cars. It is easy to see why this type is called "body-over-frame" construction since the attractive body mounts over the frame. *(Courtesy of Chrysler)*

The *lower rear panel* fits behind the rear bumper and between the quarter panels.

Rear shock towers hold the top of the rear suspension. The *inner wheelhousings* surround the rear wheels and weld to the quarter panels.

The *upper rear body panel* is the area between the back glass and trunklid.

ANTICORROSION AND SOUND DEADENING MATERIALS

Anticorrosion materials are used to prevent rusting of metal parts. Various types of anticorrosion materials are available (weld-through primer, sealers, rubberized undercoating, etc.). When performing repairs, you must restore all corrosion protection.

Sound deadening materials are used to help quiet the passenger compartment. They are insulation materials that prevent engine and road noise from entering the passenger area.

2.2 CONSTRUCTION TYPES

Mentioned briefly, passenger cars of today use one of three types of construction:

- Conventional body-over-frame (Figure 2–8)
- Unitized or unibody (Figure 2–9)
- Spaceframe

Through the 1960s and early 1970s, American automobiles were manufactured in pretty much the same way with similar characteristics. These were

- Body/frame construction
- Rear drive
- Independent front suspension
- Symmetrical design

In 1974 a variety of events took place that rocked the foundations of the automobile industry. First of

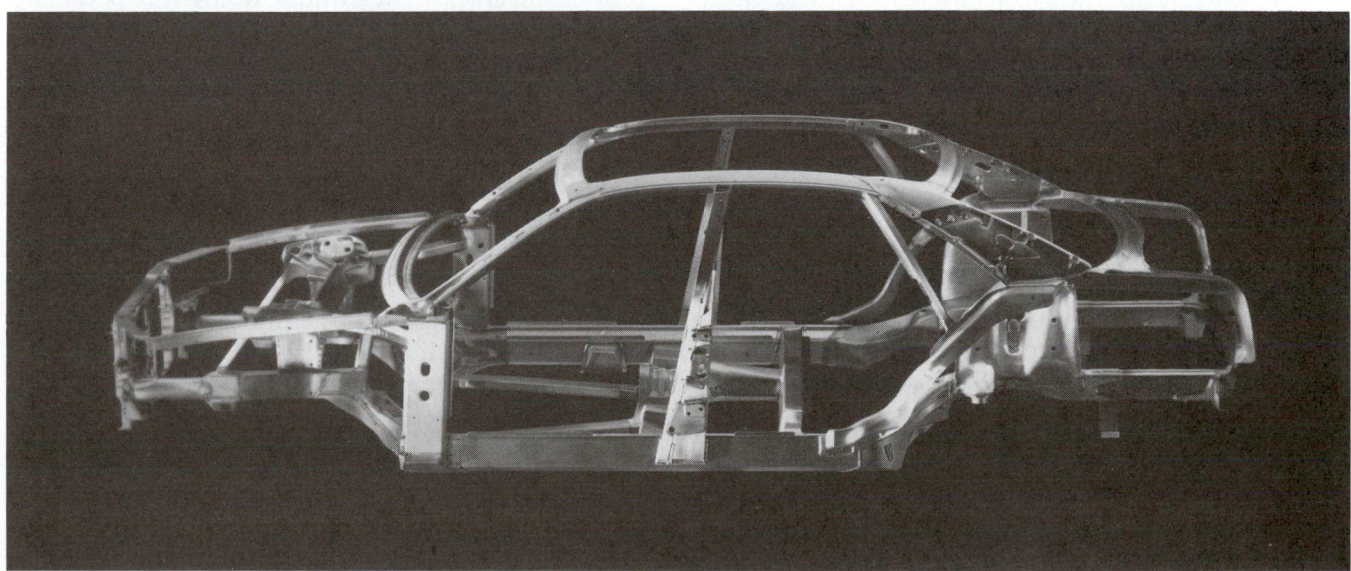

FIGURE 2-9 Unibody construction has body parts serve as the rigid frame structure. *(Courtesy of Audi)*

all, the government placed very strict fuel economy and emission control laws and standards on the manufacturers. The economy standard was known as CAFE for Corporate Average Fuel Economy, and the emission control standard was set by the Environmental Protection Agency. This meant that American automotive manufacturers had to start designing more efficient methods of combustion and emission control systems for engines.

As a result of a startling revelation by the media of poor safety records and operating conditions, there came a public demand for safe as well as clean-running vehicles. As if these were not enough problems for the manufacturers to deal with, the Arab oil embargo occurred as well. The price of gasoline escalated; consumers then demanded increased fuel efficiency. Foreign car makers, who had always manufactured smaller, lighter, more fuel-efficient vehicles, captured an increasing share of the domestic new-car market. American automakers were forced to produce smaller, more efficient cars. This resulted in the development of the unibody cars on the roads today.

The five construction areas where domestic automobiles have changed since the mid-1970s are:

1. Body/frame construction
2. Weight (average fleet)
3. Metal composition
4. Suspension/steering
5. Engine location/drive

In 1977 most cars used a perimeter type frame. They averaged around 4,500 pounds (2,038 kg), used 18-gauge mild steel, and were still conventional in design. Body weight began to decrease and thinner gauge metal was used. Also, the first American-made transverse engine, front-wheel drive, strut suspension car was introduced.

By 1981 unibodies were used in almost half the American-made cars. Fleet average weight decreased 600 pounds (272 kg); 22-gauge high-strength steel was used in construction; the steering changed to rack and pinion; the suspension changed to MacPherson strut type; and the drive changed from rear- to front-wheel drive.

At the present time, most unibodies are constructed of 24-gauge high-strength steel, have a fleet average weight 900 pounds (407 kg) less than in 1980, and feature MacPherson struts, rack-and-pinion steering, and front-wheel drive. Today, 95 to 97 percent of all passenger cars on American roads are unibodies.

As construction of vehicles has changed over the years, so has the collision repair profession. In the early days of automobiles, there were no specialized shops for automobile collision repair. When a car was brought in for repair, the damaged part was usually removed and replaced with a new one that was either forged from steel or cut from wood.

This method was expensive and time-consuming. Many times there was a long wait for parts. Most of the early body/frame technicians were carpenters or blacksmiths.

Because automobile bodies and frames became more complex in design, it became more practical to repair instead of replace. The early repair procedure involved placing the bent member over a massive structure and forcing it back into shape.

With more cars being manufactured, business got to such a point that repair people began to specialize as either a body or a frame technician. As repair technicians became more experienced, their techniques improved. Pushing from within the body compartment became accepted as a body alignment method. Frame technicians also used internal pushing as a means of frame alignment.

As procedures were refined, the stationary frame machine evolved. This machine was a more sophisticated version of the railroad iron and mechanical jack the blacksmith used. This stationary frame machine went through some changes to make it more efficient and easier to use. By raising the machine off the ground, it was easier to gain access to the various components. Ramps were built so cars could be pulled into place. Pits were later dug under the machine to eliminate the ramps.

In conjunction with the racks, hydraulic jacks were used to push heavy upright beams, forcing bent frame members back into place. Frame machine manufacturers had to continuously update their machines to keep their units functional. As the automobile frame design became more complex, so did the repair techniques. The job of the frame technician became a job for a highly skilled technician.

When Citroën introduced its unitized body in 1934, a whole new set of collision repair problems occurred. Since there was no frame to apply pushing pressure against, the technique of internal body and frame pushing was of little value. There was not enough material in any one place to push against.

The basic repair technique of pushing out damaged sections changed to that of pulling out damaged sections. Out of necessity, the portable body and frame puller was developed. It soon became accepted on a worldwide basis.

The manufacturers of stationary frame equipment again had to make modifications. The push technique was changed to a pull technique by adding adjustable pull towers (Figure 2–10). These units remained functional but became more massive, complicated, and expensive.

The portable body and fender puller and the updated stationary frame machine remained in use for a number of years. These machines became outdated when automobile engineers again changed the basic frame and body design. When the second generation perimeter frame and unitized bolt-on stub frame be-

FIGURE 2-10 Here are pulling towers at work on a modern aligner. They will exert tons of force to pull damage out of a frame or unibody. (Courtesy of Wedge Clamp)

came popular, a new repair system was needed. This system had to be flexible enough to repair both unitized and frame-constructed vehicles. These systems are the ones described in Chapters 9 and 10.

From the body technician's standpoint, it is important to know which type of construction is used. Repair work is different for each type of construction. For example, the mechanical components of a vehicle's steering and suspension, cooling system, drivetrain, and electrical systems are usually serviced by auto technicians. But with unibodies, because the mechanical components (Figure 2–11) are attached directly to the underbody, precision and skill are required. Since the body shop must return the vehicle to the customer completely repaired, the modern body technician requires much more knowledge than the counterpart of the era prior to the advent of the unibody.

2.3 CONVENTIONAL BODY-OVER-FRAME CONSTRUCTION

In the conventional body-over-frame construction, the frame is the vehicle's foundation. The body and all major parts of a vehicle are attached to the frame.

It must provide the support and strength needed by the assemblies and parts attached to it. The frame must also be strong enough to keep the other parts of the car in alignment should a collision occur. To the body technician, the frame can be considered the most important part of the vehicle.

The conventional frame is an independent, separate component because it is not welded to any of the major units of the body shell. The body is generally bolted to the frame. Large specially designed rubber "biscuits," or mounts, are placed between the frame and body structure to reduce noise and vibration from entering the passenger compartment (Figure 2–12). Quite often, two layers of rubber are used in the mounting pads to provide a smoother ride.

Today the strong steel frame side members of the modern conventional design are normally made of U-shaped channel sections or box-shaped sections. Cross members of the same material reinforce the frame and provide support for the wheels, engine, and suspension systems. Various brackets, braces, and openings are provided to permit installation of the many parts that make up the automotive chassis. The various cross members, brackets, and braces are welded, riveted, or bolted to the frame side rails.

Most conventional frames are wide at the center and narrow at the front and rear. The narrow front

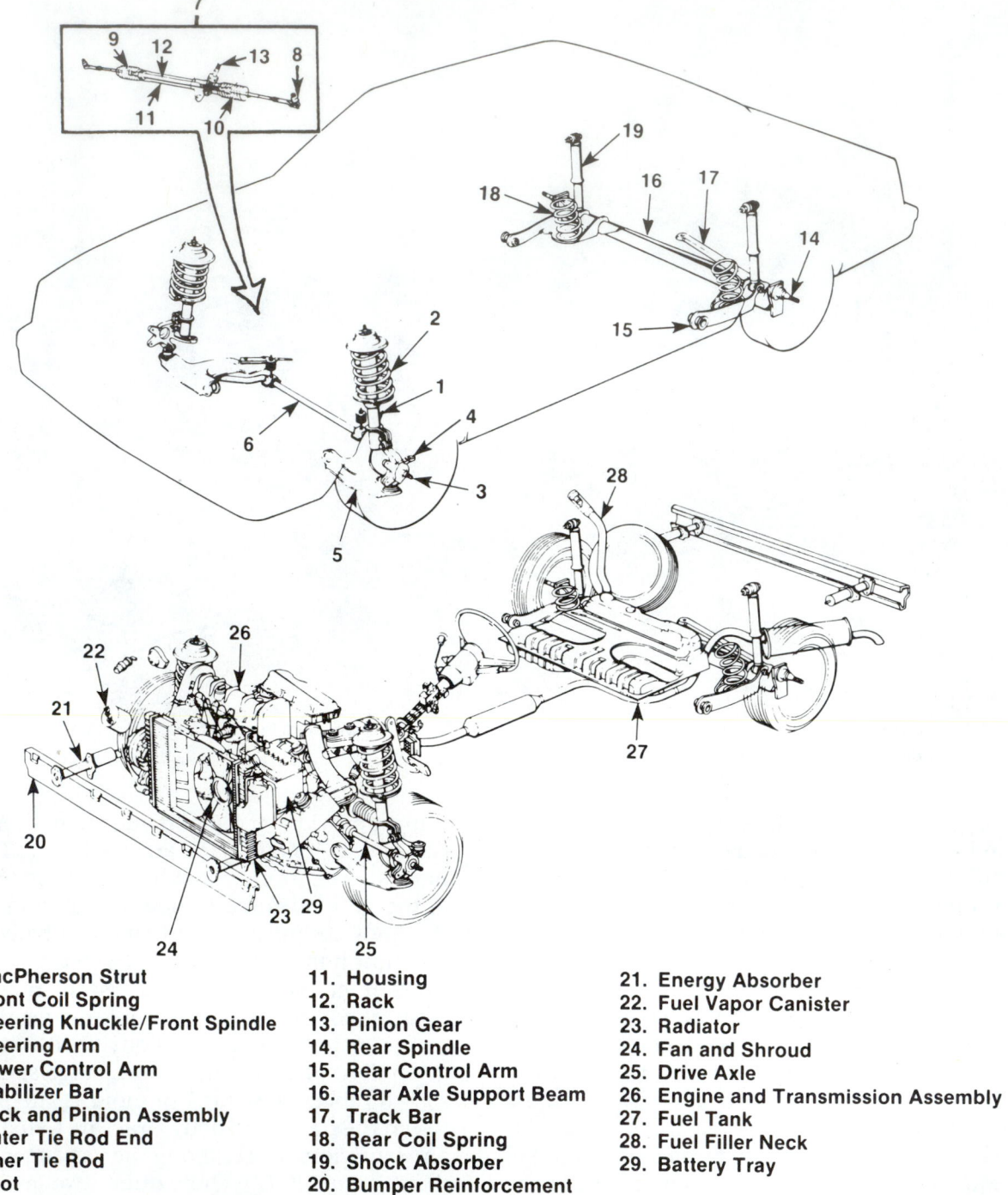

FIGURE 2-11 Study the mechanical components of a unibody car.

1. MacPherson Strut
2. Front Coil Spring
3. Steering Knuckle/Front Spindle
4. Steering Arm
5. Lower Control Arm
6. Stabilizer Bar
7. Rack and Pinion Assembly
8. Outer Tie Rod End
9. Inner Tie Rod
10. Boot
11. Housing
12. Rack
13. Pinion Gear
14. Rear Spindle
15. Rear Control Arm
16. Rear Axle Support Beam
17. Track Bar
18. Rear Coil Spring
19. Shock Absorber
20. Bumper Reinforcement
21. Energy Absorber
22. Fuel Vapor Canister
23. Radiator
24. Fan and Shroud
25. Drive Axle
26. Engine and Transmission Assembly
27. Fuel Tank
28. Fuel Filler Neck
29. Battery Tray

construction enables the vehicle to make a shorter turn. A wide frame at the rear provides better support of the body. Other characteristics of frame type vehicles are

- Load-induced vibrations that are transferred to the body via the frame, thus resulting in a smooth ride.
- Rubber mountings between the body and frame that insulate it from vibrations, providing a quiet interior.

- High amounts of energy are absorbed during a collision.
- Undersurfaces of the body are protected over rough roads.
- Suspension and power train parts can be quickly assembled on the basic frame.
- The heavy frame made of thick sheet metal is approximately $3/64$ to $1/8$ inch (1.2 to 3.1 mm).
- The vehicle profile is generally high off the ground.

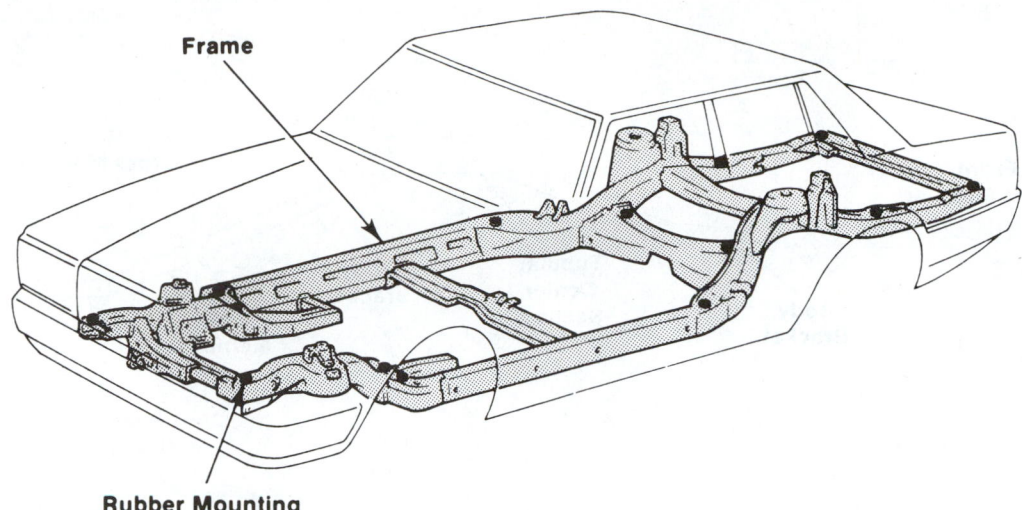

Frame

Rubber Mounting

FIGURE 2-12 Rubber "biscuits" or mounts fit between frame and body to reduce noise and vibration. *(Courtesy of Toyota Motor Corp.)*

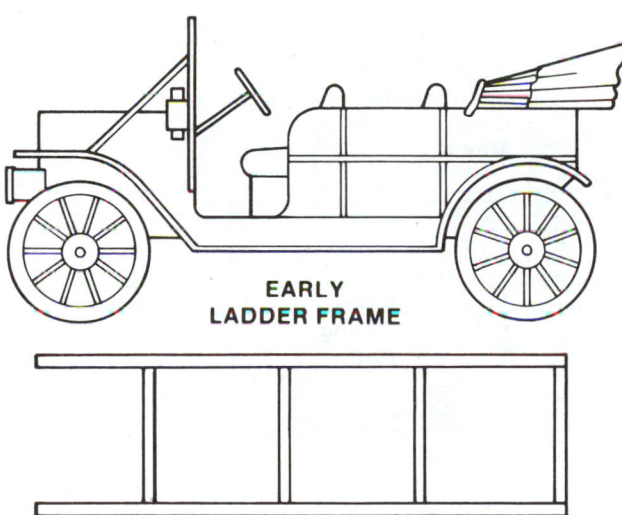

EARLY LADDER FRAME

FIGURE 2-13 An old-fashioned ladder frame actually looked something like a simple step ladder.

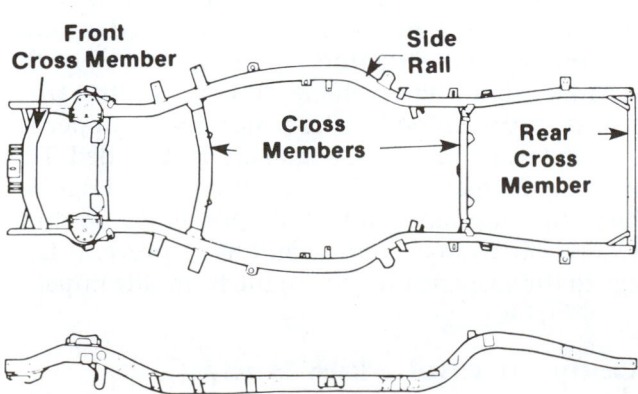

Front Cross Member   **Side Rail**

Cross Members   **Rear Cross Member**

FIGURE 2-14 A modern ladder frame is used in some small trucks. Note part names.

CONVENTIONAL FRAME DESIGNS

While several conventional frame designs have been used by the auto industry, the three that the body technician may come across are the

1. Ladder frame
2. X-frame (or backbone)
3. Perimeter frame

The *ladder frame* consists of two side rails, not necessarily parallel, connected to each other by a series of cross members like a ladder. In fact, as shown in Figure 2–13, some of the early car frames were perfect ladder shapes. While the ladder frame design is no longer used for passenger vehicles because of the "hard" ride, it still can be found on some trucks because of its strength (Figure 2–14).

The *X-frame* (Figure 2–15) narrows in the center, giving the vehicle a rigid structure that is designed to withstand a high degree of twist. A heavy front cross member is used to support the upper and lower suspension control arms and coil springs. The X-frame has not been used since the late 1960s and can be considered obsolete.

A variation of the X-frame is the *backbone frame*, which has a single thick beam in the middle section. It can be found on some sport cars.

The *perimeter frame* (Figure 2–16) is similar in construction to the ladder frame. The full-length side rails support the body at its greatest width, which provides more protection to the passengers in the case of a side impact to the body. The areas behind the front wheels and in front of the rear wheels are stepped to form a torque box structure. In a head-on collision, the stepped areas absorb much of the energy. In a side impact collision, the passenger compartment is protected from collapse since the center

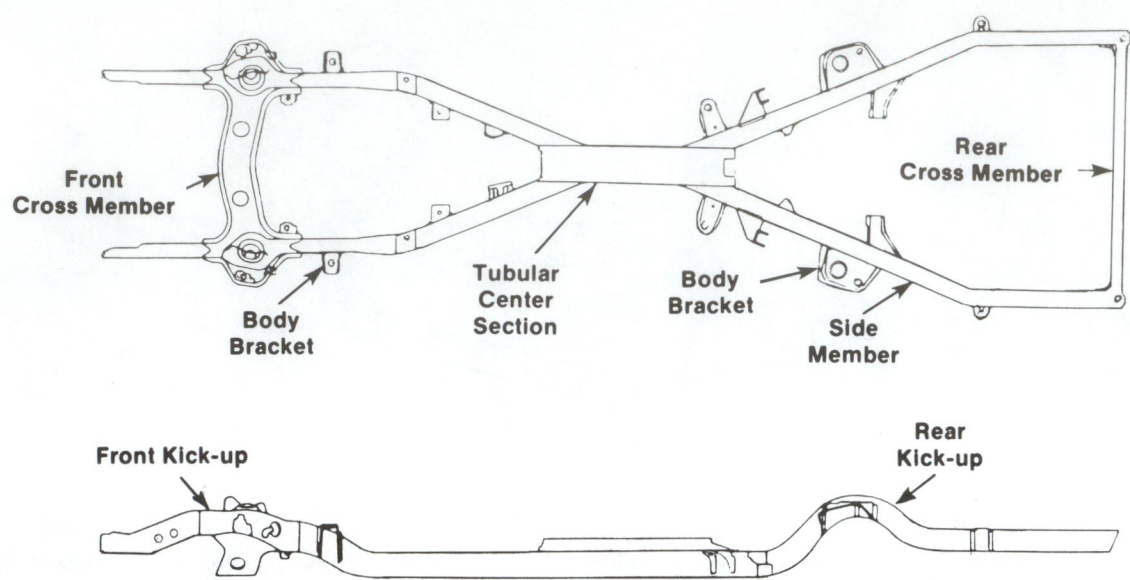

FIGURE 2-15 The X-frame is not very common.

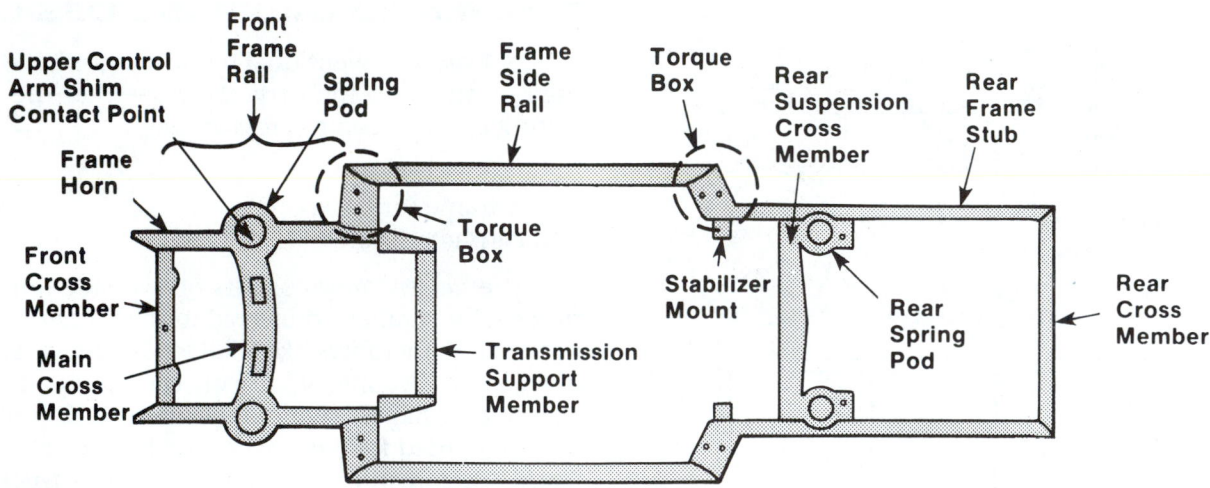

FIGURE 2-16 Study perimeter frame terminology.

frame rail is near the front floor side member. In rear end collisions, the rear cross members and kick-up absorb the shock. As for twisting and bending, strategic areas are reinforced with cross members.

The perimeter frame shown in Figure 2–17 features a center rail that passes very close to the inside of the front floor side member. Because of this, the passenger compartment floor can be made lower than in cars with other types of frames.

Most of the conventional frames used today are of the perimeter design and include the following body sections.

Conventional Front Body

The *front body section* is made up of the radiator support, front fender, and front fender apron. These components (Figure 2–18) are installed with bolts and form an easily disassembled structure. The radiator support is made up of the upper support, lower support, and left- and right-side supports welded together to form a single structure. The front fender of the separate frame type vehicle differs from the front fender of the unibody. The panels in the upper inside and rear ends of the fender are spot welded. This not only increases the fender's strength and rigidity, but also works along with the front fender apron to reduce vibration and noise and helps prevent damage to the suspension and engine from side impacts.

Conventional Main Body

The *main body* is made up of the dash panel, underbody, roof, and so on, to form the passenger and

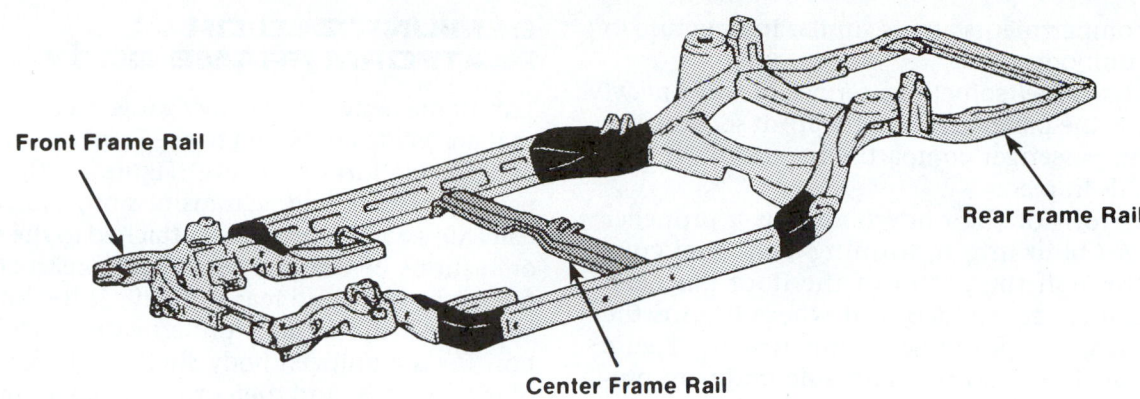

FIGURE 2-17 This perimeter frame features a center rail. Shaded portions are torque box structures. *(Courtesy of Toyota Motor Corp.)*

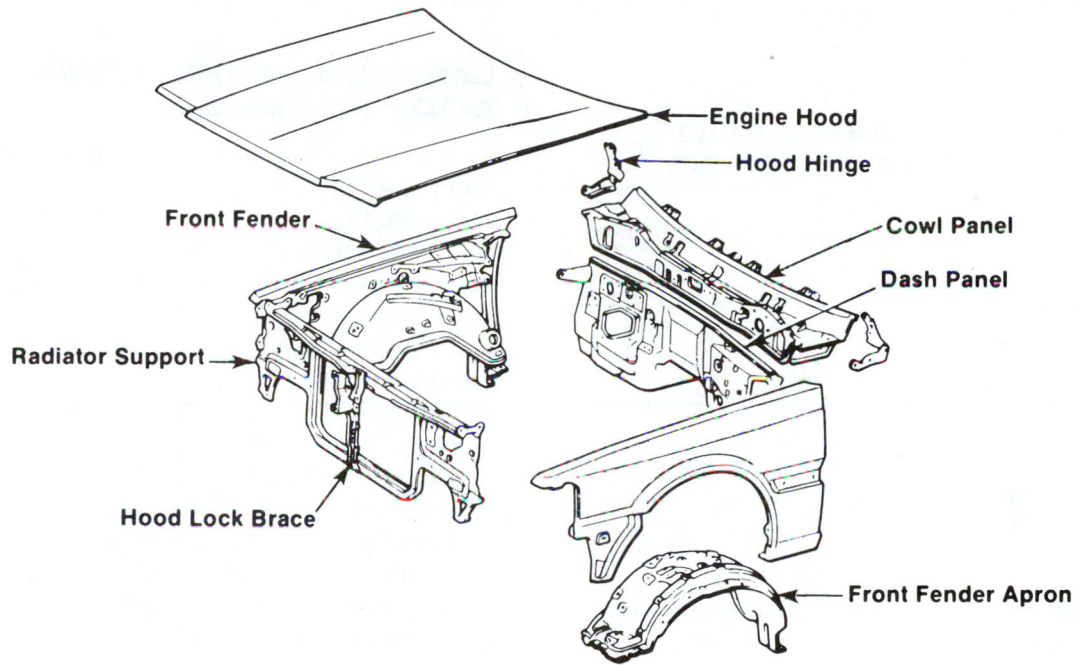

FIGURE 2-18 Note front body parts of a conventional vehicle. *(Courtesy of Toyota Motor Corp.)*

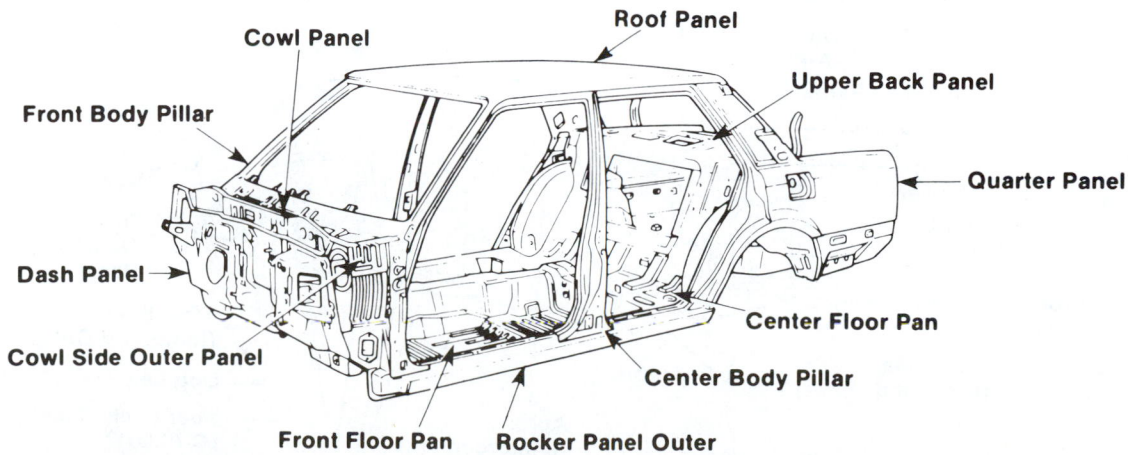

FIGURE 2-19 Carefully note the main body structural parts of this vehicle. *(Courtesy of Toyota Motor Corp.)*

luggage compartments and is similar in structure to that of a unibody.

The *dash panel*, sometimes termed *firewall* or *front bulkhead*, is the panel dividing the front section and the center, passenger compartment section. It normally welds in place.

The front of the underbody has a propeller shaft *tunnel* built into it, forming a channel cross section through the center of the floor pan, and cross members are welded to it where it joins the frame. Thus, the passenger compartment (Figure 2–19), roof side rail, door, and side body are protected from side impact collisions. In addition, the front, back, and left- and right-sides of the floor pan are made uneven in the stamping process, increasing the rigidity of the floor pan itself, which reduces vibration.

2.4 UNITIZED FRAME AND BODY CONSTRUCTION

Most newer models manufactured in the last few years in small to midsize classes (and even some full-size) are of the unitized or semiunitized body construction.

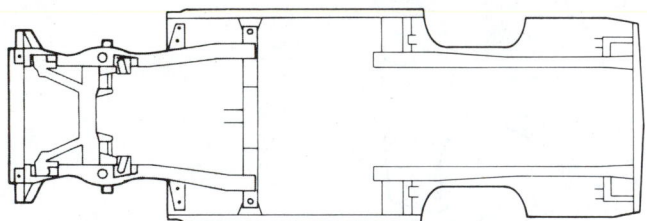

FIGURE 2-20 Bolt-on stub frame or carriage can be found on some vehicles and it simplifies repair since parts can easily be removed for replacement.

SEMIUNITIZED OR PLATFORM FRAME BODY

This frame design uses heavy gauge steel "stub" rails that are welded or bolted to the front and rear of the body or platform structure (Figure 2–20). The suspension system, engine, transmission, and, of course, rear axle assembly are either attached to the stub rails or sections of the platform-type construction. Between the front and rear stub rails is the underbody structure with no frame underneath to act as a support for the unitized body shell or platform. Many of the stub rails and their cross members are welded to both body shell and sheet metal components to form a single integral unit. Today bolt-on stub construction can be considered obsolete.

UNITIZED FRAME AND BODY ASSEMBLY

As previously mentioned, the unitized frame and body assembly has no separate frame (Figure 2–21). The unibody was a design concept used for the bodies of aircraft, and the eggshell is often cited as an example of this type of structure. Even when pressing hard on an eggshell, it is difficult to crush it. All the force or strength applied is not concentrated in one place but is dispersed effectively through the entire shell. In mechanics, this action is called a "stressed hull structure."

In a car body, there is no complete stressed hull structure. Generally, a body with a structure that integrates the frame and body to receive and hold outside forces is called a unibody. It is made by combining pieces of thin sheet metal pressed to form panels of various shapes and joined into an integrated structure by spot welding. This lightweight structure

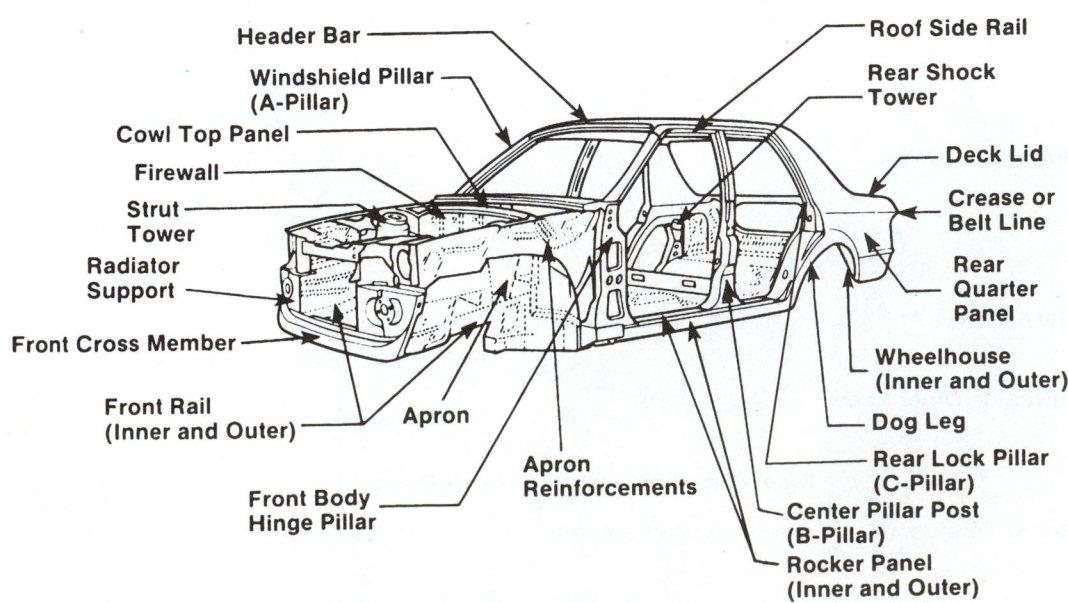

FIGURE 2-21 Memorize unibody terminology. *(Courtesy of Toyota Motor Corp.)*

is highly rigid to bending or twisting and has the following characteristics:

- The bulk taken by the frame can be used to make the car more compact.
- Vibration and noise from the drivetrain and suspension enter the floor pan and are amplified by the body, which acts as an acoustic chamber. This makes it necessary to add extra components to the body to suppress vibration and noise.
- Once deformed, special procedures are needed to restore it to its original shape.
- With the thin sheet metal body close to the road surface, adequate measures must be taken to prevent the deterioration in strength from corrosion. This is particularly important when dealing with reinforcing materials that make up the underbody.

The major advantage of unibody vehicles is that they tend to be more tightly constructed because the major parts are all welded together. This design characteristic that helps protect the occupants during a collision causes damage patterns that differ from those of frame-type vehicles. The stiffer sections used with unibody design tend to transmit and distribute impact energy throughout more of the vehicle, causing misalignment in areas remote from the impact point. Even sections that are buckled or torn loose might have passed along heavy force before deforming. Worse still, much of this remote damage can easily be overlooked in casual inspection but still be sufficient to cause handling or power train problems later.

Torque boxes are used in the design of some unitized frames/bodies (Figure 2–22).

The extra complexity and stiffness of the structure are especially critical in the front end, which houses not only the front suspension and steering linkage, but also the entire drivetrain—engine, transaxle, drive shafts, and constant velocity U-joints. To keep all these in proper alignment requires support, including that supplied by the front end sheet metal. Accordingly, a damaged unibody vehicle requires a more thorough damage analysis than a similar impact would require in a conventional frame/body car. If not, after a car is returned to its owner, it may later show unsafe handling qualities, water leaks, or a new family of strange noises in the power train.

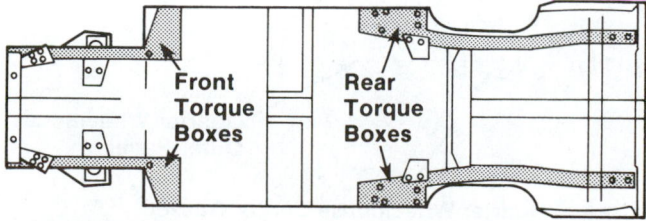

FIGURE 2-22 Torque boxes are used in some unibody designs.

The three basic unibody structures are a front-engine rear-drive (*FR*) vehicle, a front-engine front-drive (*FF*) vehicle, and a mid-engine rear-drive (*MR*) vehicle.

FR VEHICLE BODY STRUCTURE

The *front-engine rear-drive* has the engine mounted in the front and the driving wheels in the rear. The body of an FR vehicle is divided into three main sections: front body, passenger compartment (side body), and rear body. The engine, transmission, front suspension, and steering equipment are installed in the front body. The differential and the rear suspension are installed in the rear body. Since all impacts from the road surface are transmitted to the entire body through the front and rear wheels, the body must have high strength. This strength is supplied by the side and cross members welded to the floor.

FR vehicles are characterized as follows:

- Since the engine, transmission, and differential are in separate positions, weight can be distributed uniformly between the front and rear wheels, lightening the steering force.
- In an FR vehicle, the engine is placed longitudinally with respect to the vehicle. Most vehicles have a single suspension cross member placed laterally between the front side members at about the middle of the front body, which supports the engine.
- Since it is possible to remove and install the engine, propeller shaft, differential, and suspension independently, body restoration and repair workability are good.
- Since rear-wheel-drive equipment is necessary, a tunnel in the floor is necessary, which decreases interior space.
- Since the engine's output is transmitted to the rear wheels by the drive shaft and differential, the vehicle, vibration, and noise sources are widely distributed over the front and rear.

FR Front Body

The engine, suspension, and steering equipment are all mounted on the front fender apron and the front side member of the front body. This part is very important because it influences front-wheel alignment and the amount of vibration and noise that are transmitted into the passenger compartment. Therefore, it must be made with great accuracy and strength. With the exception of outer shell parts, such as the engine hood, front fenders, and front valance panel (installed with nuts and bolts), all other exterior parts are welded together, reducing body weight and increasing body strength (Figure 2–23).

FR Side Body

The side body (Figure 2–24) is joined to the front body and roof panel to form the passenger compartment. During travel, these panels distribute the loads from the underbody to the upper part of the vehicle and prevent bending of the left- and right-sides. The side body members also serve as door supports and maintain the integrity of the passenger compartment if the vehicle should overturn. Since the sides are weakened by large door openings, they are reinforced by joining the inner and outer panels, which forms a very strong boxed-type structure.

The basic arrangement and shapes of these members and floor pan will vary slightly depending on the size and shape of the suspension and the underbody (Figure 2–25).

- **Underbody front section** (Figure 2–26). Since the front side members and front cross members of the front underbody section directly affect front-wheel alignment, they are formed

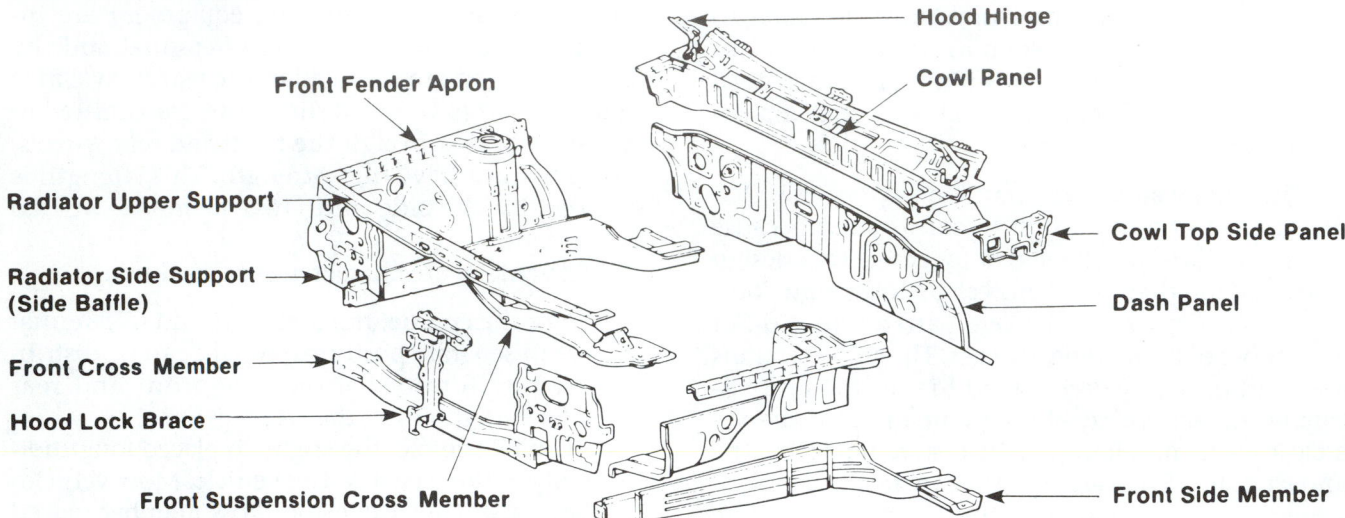

FIGURE 2-23 Front body structure components are listed for a typical FR vehicle. *(Courtesy of Toyota Motor Corp.)*

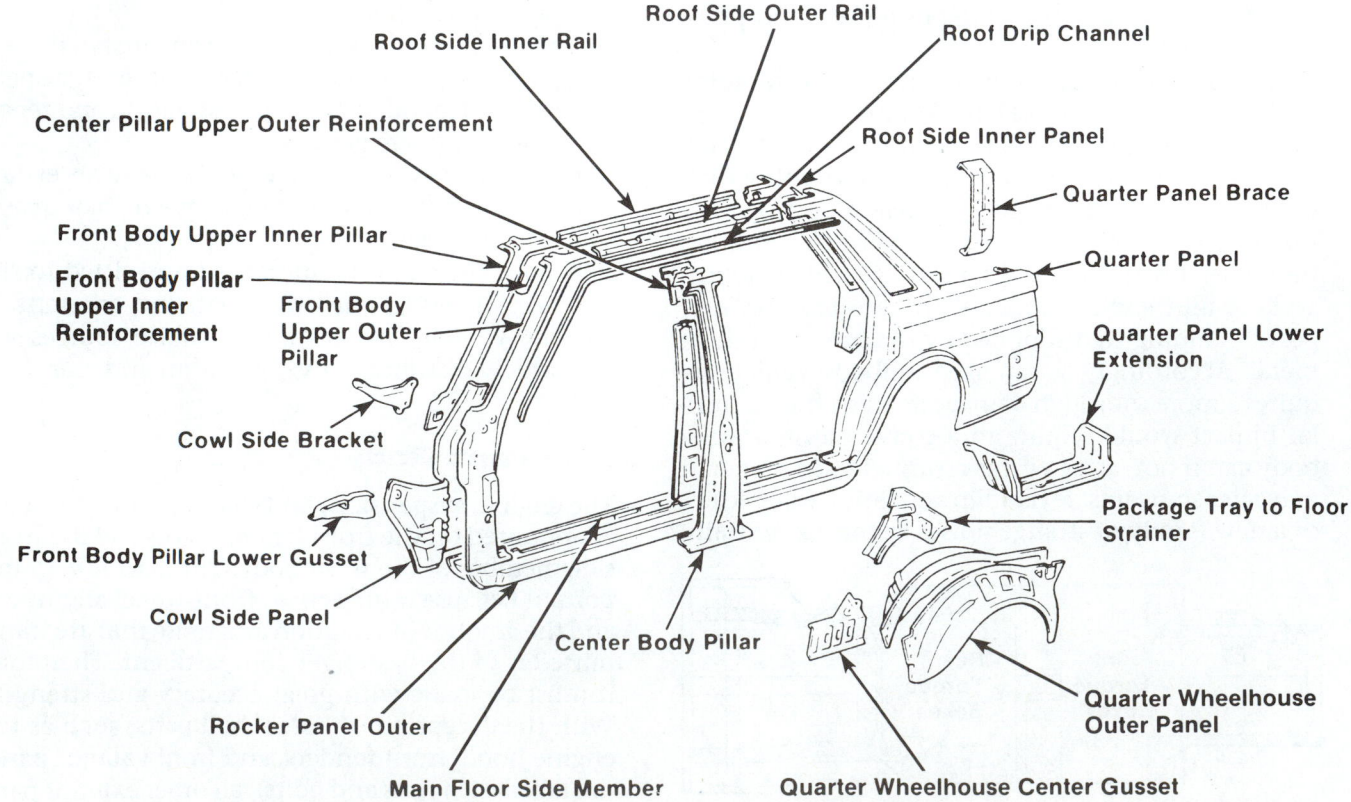

FIGURE 2-24 Study side body structure components of a typical FR vehicle. *(Courtesy of Toyota Motor Corp.)*

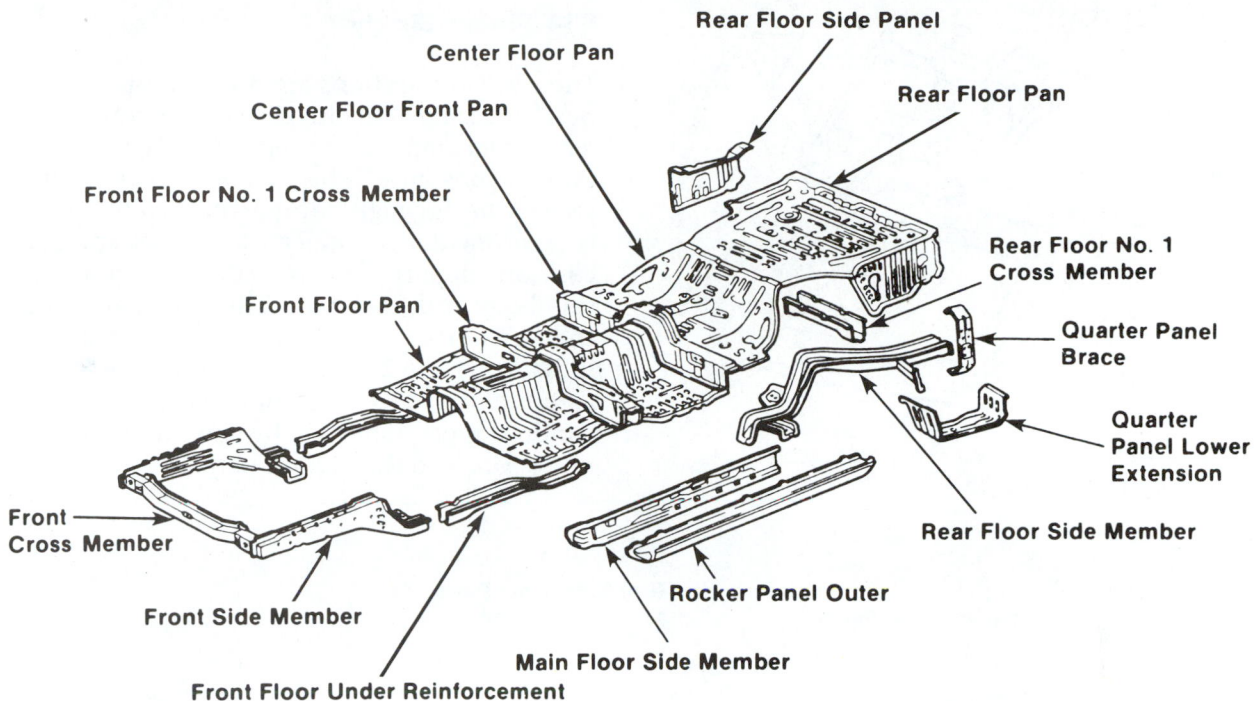

FIGURE 2-25 Note underbody structural components of a typical FR vehicle. *(Courtesy of Toyota Motor Corp.)*

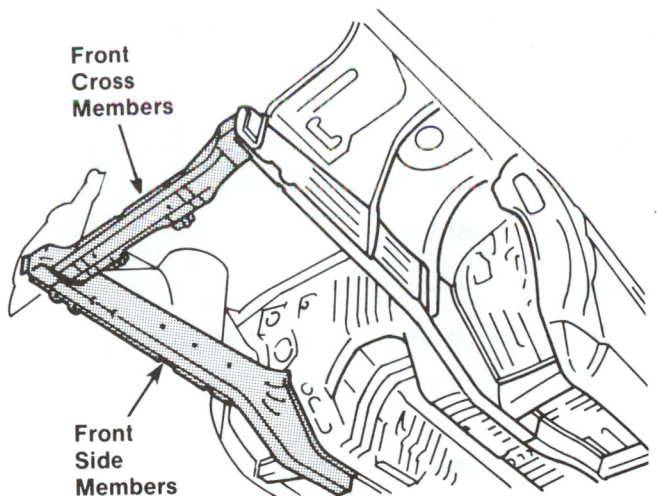

FIGURE 2-26 The underbody front section has strong side rails and cross members to support the engine and suspension system.

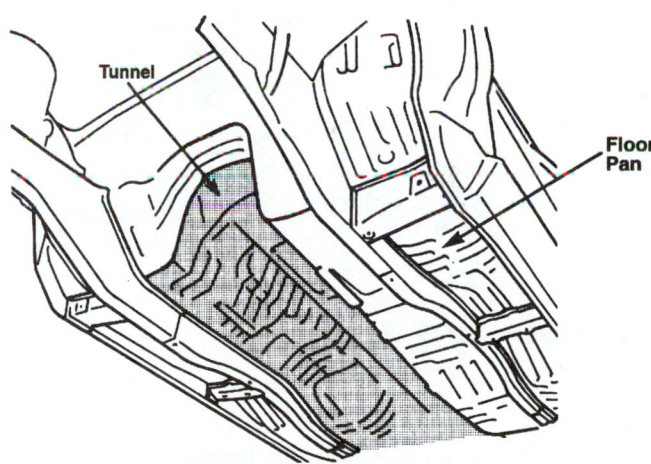

FIGURE 2-27 The underbody center section has a large floor pan and sometimes a tunnel for the transmission and drive shaft. *(Courtesy of Toyota Motor Corp.)*

into a boxed section. On some vehicles, they are made of high-strength steel. To prevent the collapse of the passenger compartment in a head-on collision, the front side members are made with a kick-up so that all the members will bend and absorb shock loads.

- **Underbody center section** (Figure 2–27). The center underbody section is mainly composed of the floor pan, cross member, and main floor side member. The center of the floor pan contains the propeller shaft tunnel, which prevents the floor from twisting. In addition, the

main floor side member and cross members below the front seats and in front of the rear seats strengthen the left- and right-sides and prevent the floor from folding in the event of a side collision.

- **Underbody rear section** (Figure 2–28). The rear side member of the underbody extends from under the rear seat to a point near the rear axle, where it forms a large kick-up and extends to the rear floor. This kick-up, like the front side members, is designed to absorb the energy of a rear end collision.

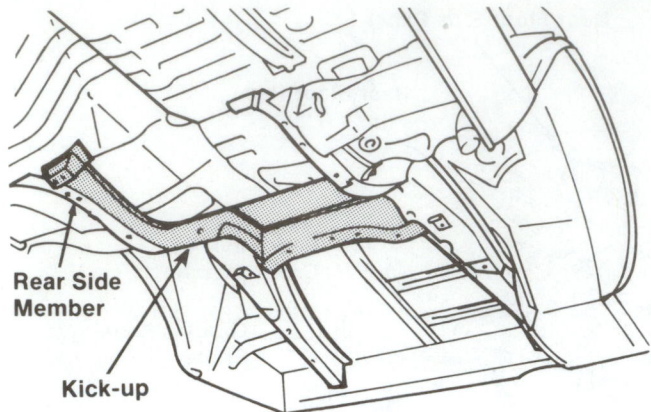

Rear Side Member

Kick-up

FIGURE 2-28 Note the underbody rear section. *[Courtesy of Toyota Motor Corp.]*

FR Rear Body

The rear body sections are divided into two categories. Sedans have the luggage compartment and the passenger compartment separated (Figure 2–29). Station wagons and liftbacks have no separation between the luggage compartment and passenger compartment. The upper back panel and rear seat cushion support brace in sedans are joined at the side body and floor pan. The back panel prevents the body from twisting. In station wagons and liftbacks (Figure 2–30), body rigidity is enhanced by adding enlarged roof side inner rear panels and a back window upper frame and by extending the roof side inner panels to the quarter panels.

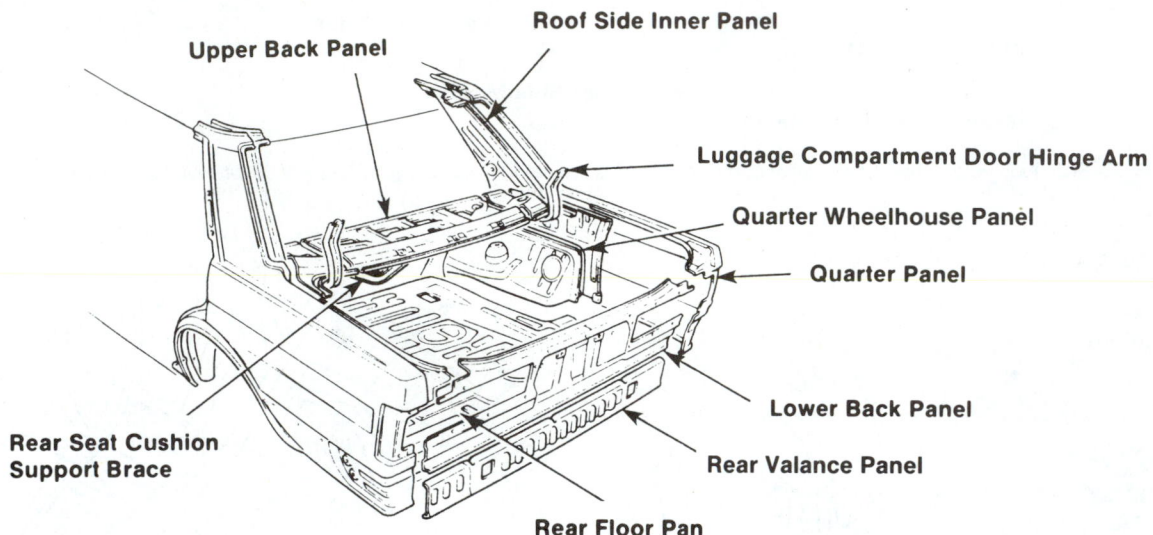

Upper Back Panel

Roof Side Inner Panel

Luggage Compartment Door Hinge Arm

Quarter Wheelhouse Panel

Quarter Panel

Lower Back Panel

Rear Valance Panel

Rear Floor Pan

Rear Seat Cushion Support Brace

FIGURE 2-29 Study the rear body structural components of a typical FR sedan. *[Courtesy of Toyota Motor Corp.]*

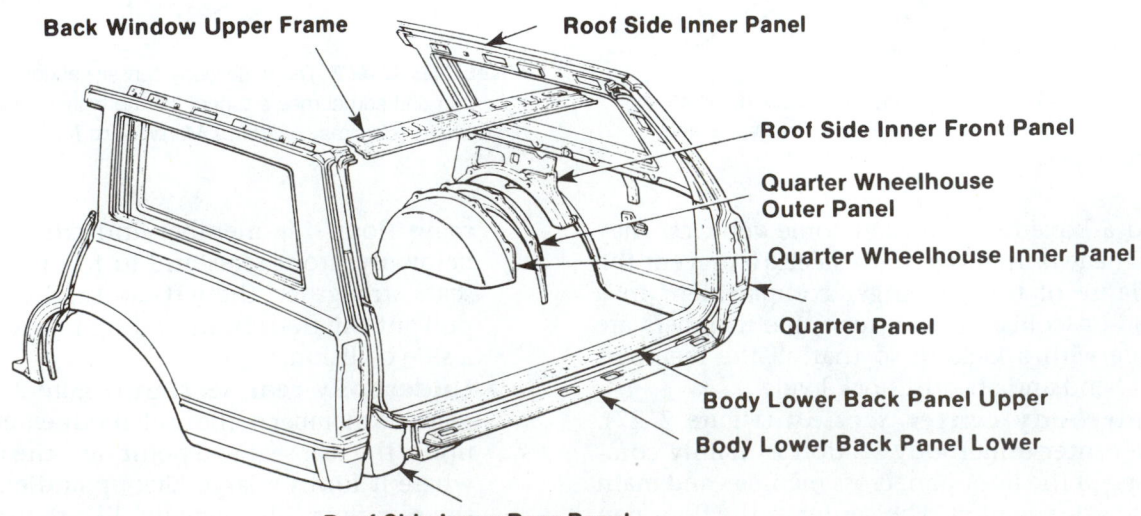

Back Window Upper Frame

Roof Side Inner Panel

Roof Side Inner Front Panel

Quarter Wheelhouse Outer Panel

Quarter Wheelhouse Inner Panel

Quarter Panel

Body Lower Back Panel Upper

Body Lower Back Panel Lower

Roof Side Inner Rear Panel

FIGURE 2-30 Rear body structural components are given for a typical FR station wagon. *[Courtesy of Toyota Motor Corp.]*

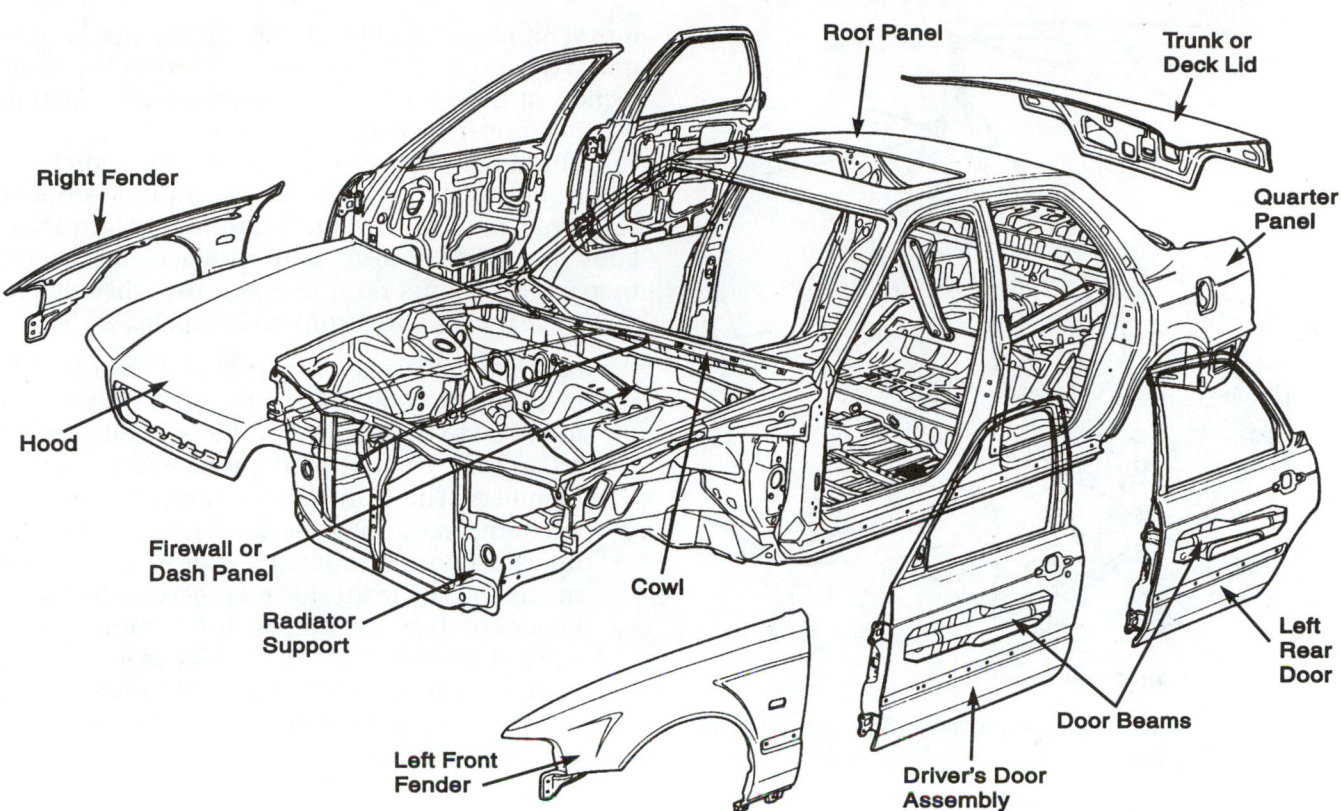

FIGURE 2-31 This is a typical FF vehicle main body. *(Courtesy of American Honda Motor Corp.)*

FF VEHICLE BODY STRUCTURE

In an FF passenger car (Figure 2–31), the engine is mounted in the front of the vehicle, and the engine drives the front wheels. It is also called a front-wheel-drive (FWD) vehicle. In the space ordinarily taken up by the rear axle, the passenger compartment can be enlarged and the rear suspension simplified. This results in substantial weight reduction. Since the engine, transaxle, front suspension, and steering equipment are all located in the front body section, the methods of reinforcement are much different from those used in the FR vehicles.

FF vehicles are characterized by the following:

- The transmission and differential are combined, and the propeller shaft is eliminated, providing a substantial weight reduction.
- Overall noise and vibration are reduced because they are confined to the front of the vehicle.
- Since the engine and transmission are located in the front, the load on the front suspension and tires is increased.
- The interior of the vehicle is larger because there is no need for a propeller shaft or rear-drive axle.

- Since the fuel tank can be placed under the center of the vehicle, the luggage compartment can be large and flat.
- Because of the location of the engine, there is a greater forward inertial weight in a head-on collision. Therefore, engine mounting components are reinforced accordingly.

The engine of an FR vehicle is mounted longitudinally. The engine of an FF vehicle can be mounted either longitudinally or transversely. Engine support methods differ between longitudinal mount FF vehicles and transverse mount FF vehicles.

1. *Longitudinally mounted FF engine supports.* The engine is supported by the front suspension members connected to the left and right front side members. The FF engine mounting is the same as the FR's engine mounting and is supported in the same manner (Figure 2–32).
2. *Transversely mounted FF engine supports.* The engine is supported at four points, the front and rear of the engine mounting center member, positioned longitudinally through the vehicle's center, and the left and right front side members (Figure 2–33).

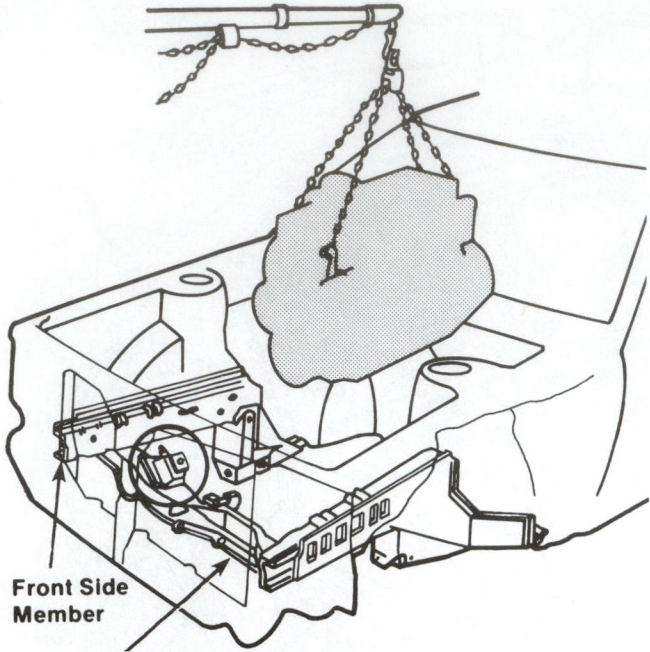

Front Side Member

Front Suspension Cross Member

FIGURE 2-32 Longitudinally mounted FF engine supports can be seen with the engine raised out of the vehicle.

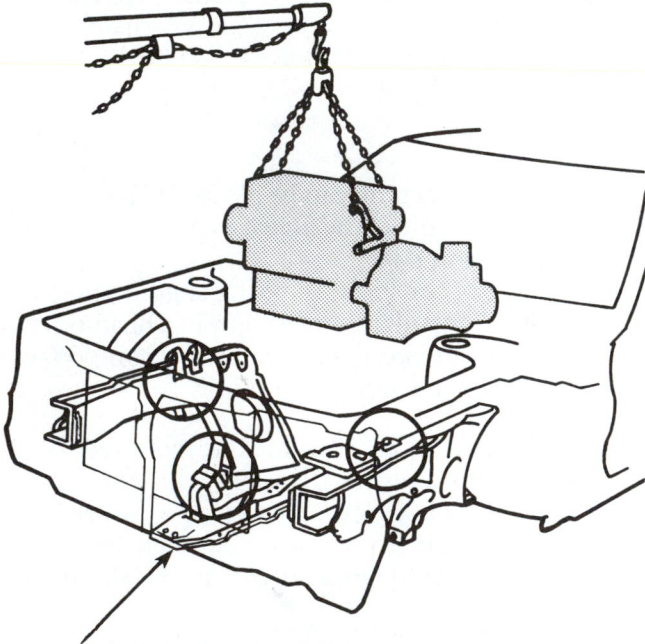

Engine Mounting Center Member

FIGURE 2-33 Transversely mounted FF engine supports are in different locations from longitudinal mounting.

FF Front Body

The front body components, consisting of the engine hood, front fenders, radiator upper support, radiator side support, front cross member, front side members, front fender apron, and dash panel, are stamped from thin sheet metal. A high-strength structure consisting of reinforced side members and

motor mounts capable of supporting the engine transaxle and suspension loads is used in the front section of the vehicle. Lightweight single structure plastic bumpers may be used on FF models.

The front suspension in FF and FR vehicles is almost identical. Both vehicles use an independent strut-type front suspension. The accuracy of the front body has a direct effect on front-wheel alignment; therefore, it is important to check the wheel alignment after performing front body repairs.

- *Longitudinal engine, front body.* The front body of a longitudinal FF (including 4WD) is nearly identical to that in an FR. The only differences are in their front fender aprons and front side members. The front fender apron is strengthened and reinforced by welding together the upper and lower front fender apron to cowl side members. The front side members of the FF are larger and heavier than their FR counterparts, since they must carry a heavier front vehicle load. A torque box is welded onto the rear end of the front side members with the suspension arms connected to it.
- *Transverse engine front body.* The lower dash panel and the front side members of FF vehicles with transversely mounted engines (Figure 2–34) are quite different from FR or FF vehicles with longitudinally mounted engines because the steering gear or rack is mounted in the lower portion of the dash panel. The steering linkage passes through a large opening in the rear portion of the front cross member, and the suspension arms are mounted to a structure that is directly below the opening.

FF Rear Body

The rear body section (Figure 2–35) consists of the back door panel, lower back panel, quarter panel, quarter wheelhouse outer panels, quarter wheelhouse inner panel, rear floor pan, and the rear floor side members. Since the entire power train is located in the front of FF vehicles, the fuel tank is located below the center floor, which allows the rear floor side member to be lower than in FR vehicles. The lower part of the rear floor side members is then connected to the rear suspension arm. An independent strut suspension is used to improve handling performance and driving stability. The rear suspension may also be a solid axle. Therefore, a rear end collision has a greater influence on rear-wheel alignment than it would in an FR vehicle. As a consequence, rear-wheel alignment should be checked whenever repairs are performed on a rear body section.

In the case of an FR vehicle, the front of the rear floor pan is joined to the end of the center floor pan

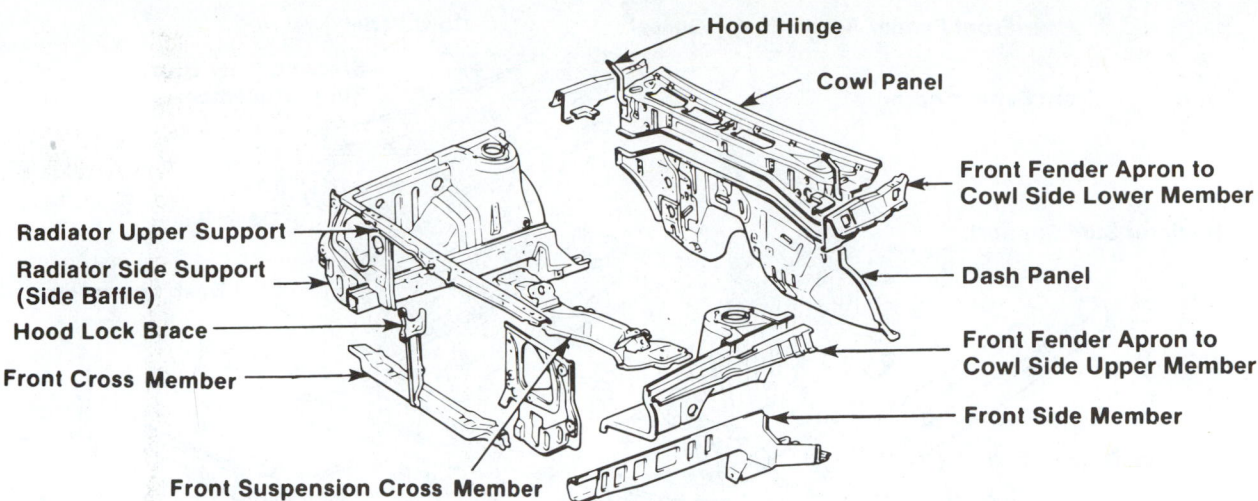

FIGURE 2-34 Study the front body structural components of a typical transversely mounted engine of an FF vehicle. *(Courtesy of Toyota Motor Corp.)*

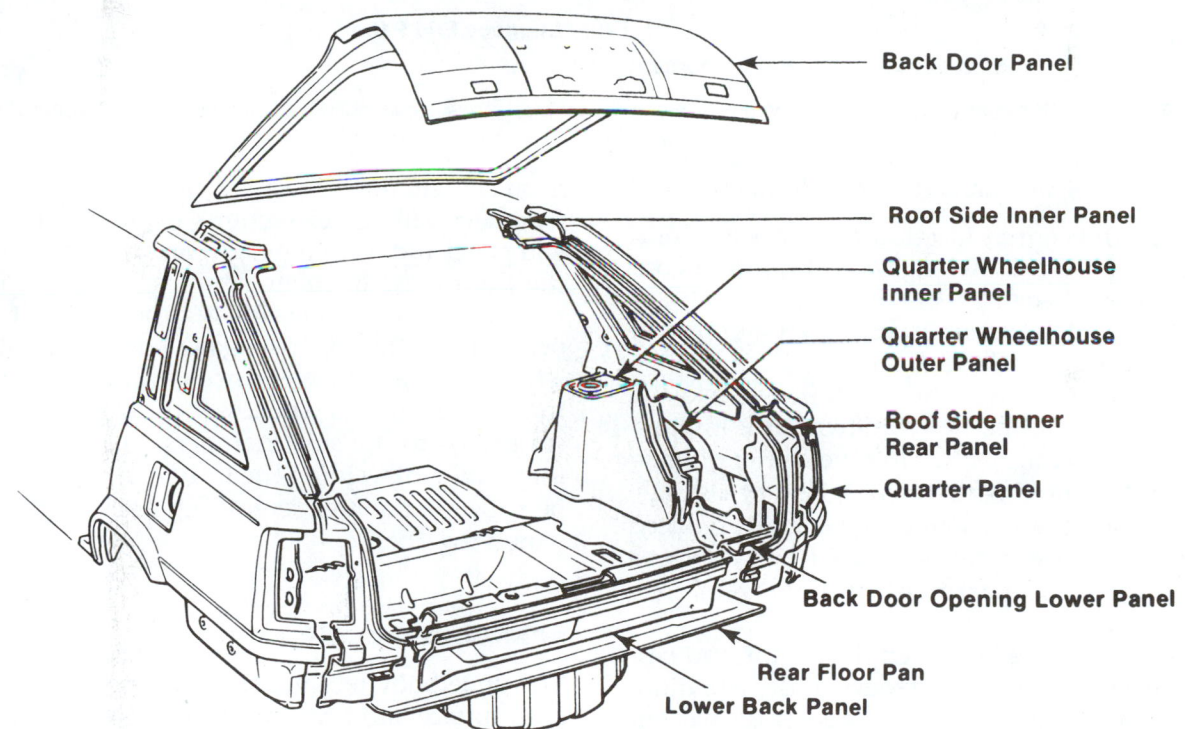

FIGURE 2-35 Note the rear body structural parts of a typical FF vehicle. *(Courtesy of Toyota Motor Corp.)*

with spot welds. However, in an FF vehicle, the center and rear floor pans are joined and reinforced with an interlocking structure. The rear body of a 4WD is similar to that of an FR vehicle.

MR VEHICLE BODY STRUCTURE

As previously mentioned, MR is the nomenclature derived from a mid-engine rear-drive vehicle, more commonly known as a mid-engine vehicle. The term *mid-engine* refers to the central positioning of the engine and power train between the passenger compartment and the rear axle.

Due to its unique engine placement, a mid-engine vehicle (Figure 2–36) has a lower profile and hence a lower center of gravity. Since this type of vehicle has the majority of its heavy components near the center of the vehicle, the strength of the center structure is higher. A high-strength box section that runs throughout the vehicle is used in the MR vehicle, resulting in further weight reductions.

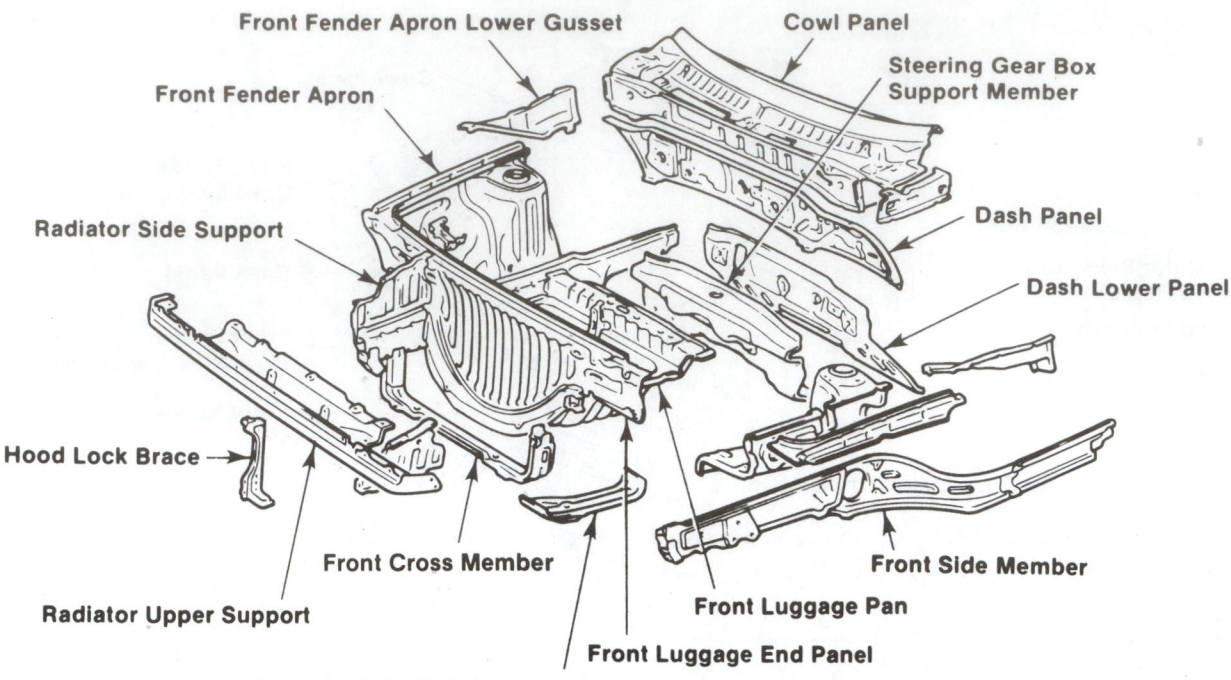

FIGURE 2-36 The front body structural components of a typical MR vehicle are identified. *(Courtesy of Toyota Motor Corp.)*

MR vehicles are characterized as follows:

- Due to the central location of the heavy components, such as the engine and transmission, the center of gravity is also concentrated toward the center of the vehicle, which gives it improved steering and handling.
- Since the engine is in the rear of the vehicle, the front hood can be sloped downward, improving aerodynamics, lowering the center of gravity, and improving the driver's field of vision.
- Engine access and cooling efficiency are reduced because the engine is located between the passenger compartment and the rear axle assembly.
- A barrier is placed between the engine and passenger compartment to reduce noise, vibration, and heat that might otherwise enter the passenger compartment.

MR Front Body

The front suspension, steering, radiator, and air condenser are mounted in the front body section (Figure 2–36). Since the engine and transaxle are located toward the rear body, the front body shape is low and sharp. The independent front suspension is supported by the front fender apron and front side members. Because of the engine's unique location, there is room for a front luggage compartment.

Various removable parts, such as the front fenders, hood, and front valance panel, are bolted on. The front fender apron, front cross member, and front side support are spot welded to the front side members. The upper sections of the front luggage end panel and front luggage pan are spot welded to the front cross member. The front luggage pan is spot welded to the steering gear box support member to form the front luggage compartment. The steering linkage passes through grommets in the front side members. The lower control arms are also connected to the side members. The body is reinforced by spot welding the front and rear side members together as well as joining each of the rocker panels together.

MR Underbody

The underbody receives the various loads from the road surface and distributes them to the side body, the various body pillars, and the roof. Many of the components (Figure 2–37) that make up the underbody are made of high-strength steel. In addition, the underbody is strengthened by raising the tunnel of the front floor pan.

MR Rear Body

The rear body consists of the quarter panels, luggage compartment door, engine hood, body lower back panel, rear floor pan, room partition panel, rear floor partition panel, and rear side members (Figure 2–38). The engine and rear luggage compartment are divided by the rear floor partition panel. The rear floor pan, room partition panel, and rear floor partition panel

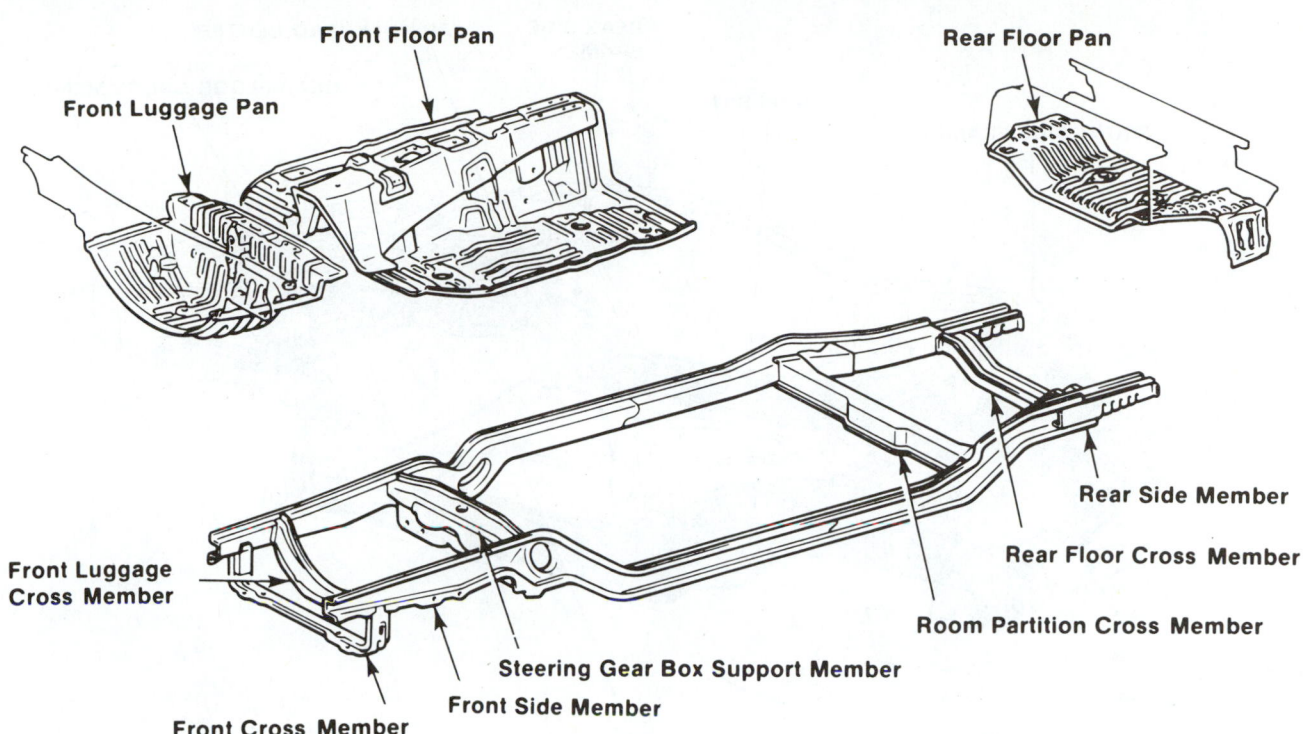

FIGURE 2-37 The underbody structural components are shown for a typical MR vehicle. *(Courtesy of Toyota Motor Corp.)*

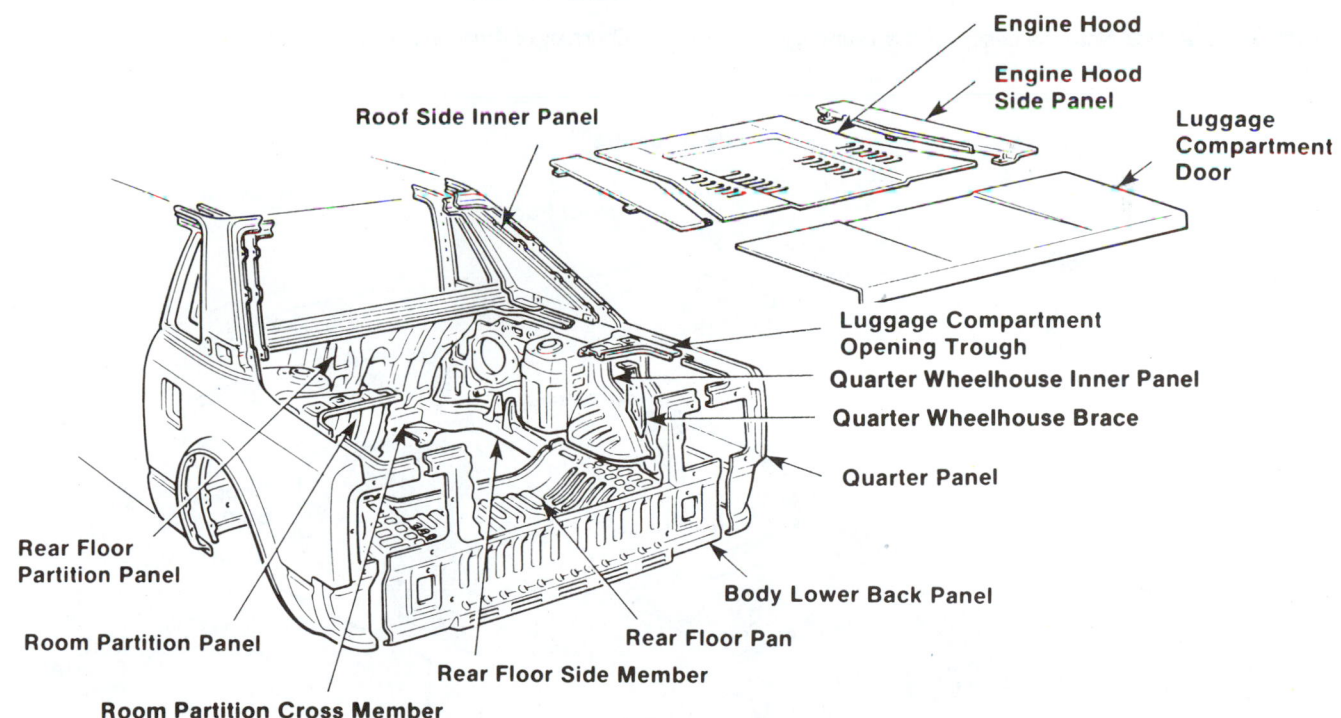

FIGURE 2-38 Study the rear body structural components of a typical MR vehicle. *(Courtesy of Toyota Motor Corp.)*

are reinforced with a deep bead structure and, together with the rear side members, form a rigid body.

The engine is positioned transversely and supported on engine mountings located at four points, on the left and right rear side members, the room partition cross member, and the rear floor cross member. Since the engine is mounted just behind the passenger compartment, the wall between the passenger compartment and the engine compartment is a three-layered structure to keep out noise, vibration, and heat. Also, since an independent strut suspension is used for the rear suspension, the body structure is made to maintain body accuracy for components that have an influence on rear-wheel align-

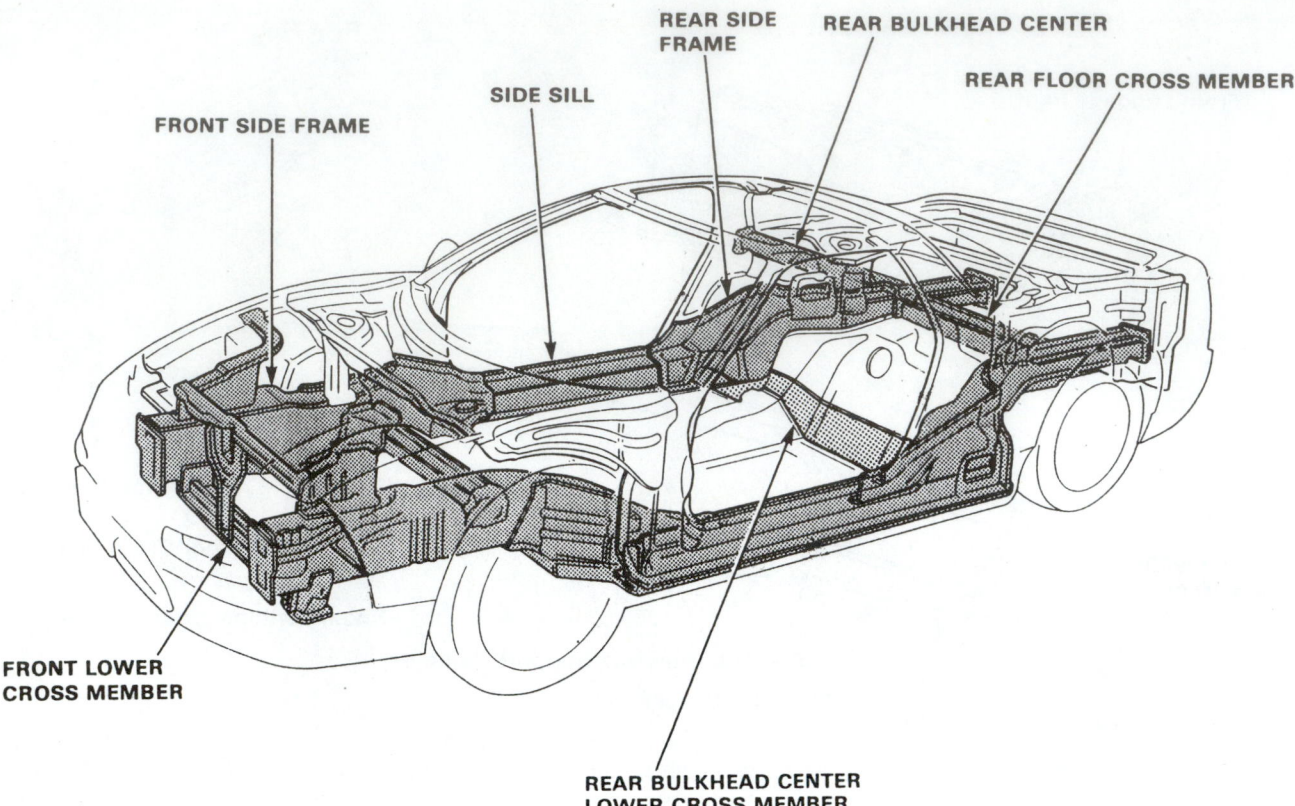

FRONT SIDE FRAME

SIDE SILL

REAR SIDE FRAME

REAR BULKHEAD CENTER

REAR FLOOR CROSS MEMBER

FRONT LOWER CROSS MEMBER

REAR BULKHEAD CENTER LOWER CROSS MEMBER

FIGURE 2-39 Note the design of this rear-engine sports car. *(Courtesy of American Honda Motor Co.)*

Body Structure

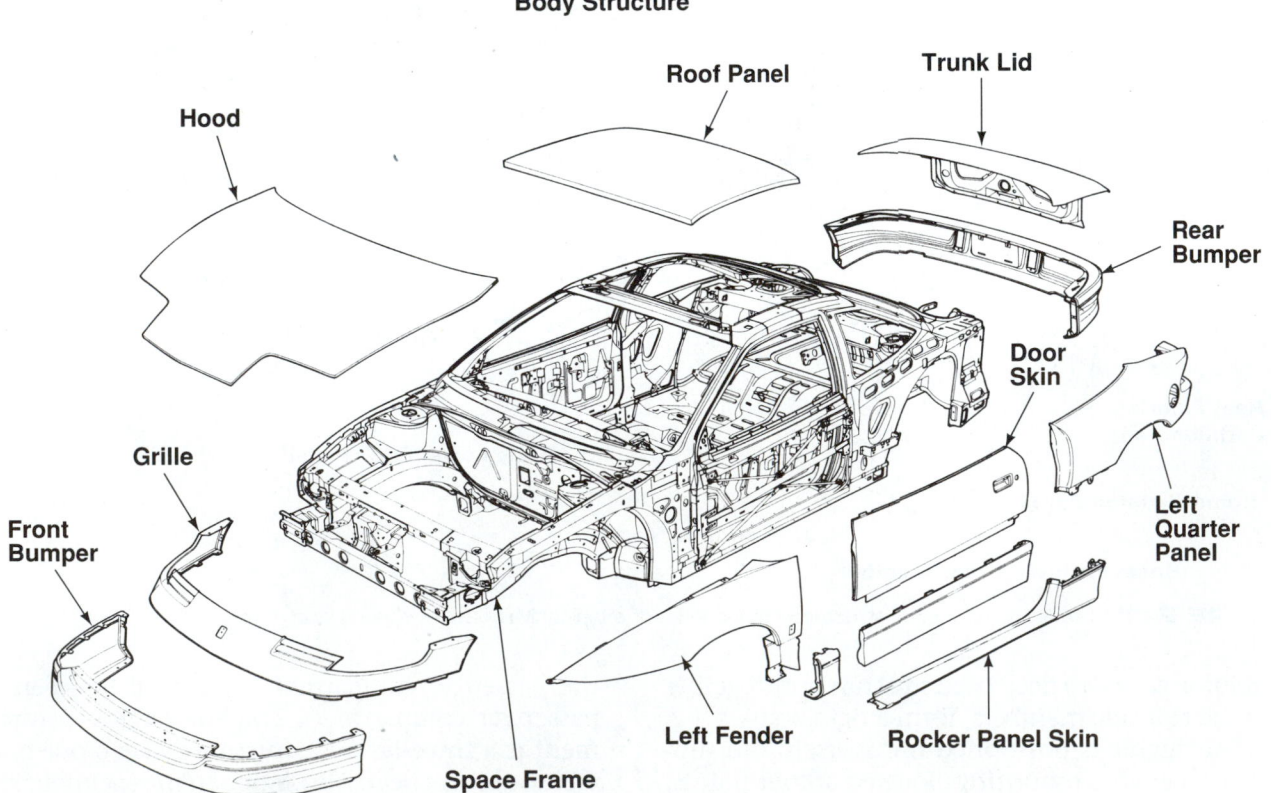

Hood

Roof Panel

Trunk Lid

Rear Bumper

Door Skin

Left Quarter Panel

Grille

Front Bumper

Left Fender

Space Frame

Rocker Panel Skin

FIGURE 2-40 The space frame has a metal frame structure covered with plastic or fiberglass body panels. This car's door panels are flexible and will not dent easily, so no more "door dings." *(Courtesy of Saturn Corp.)*

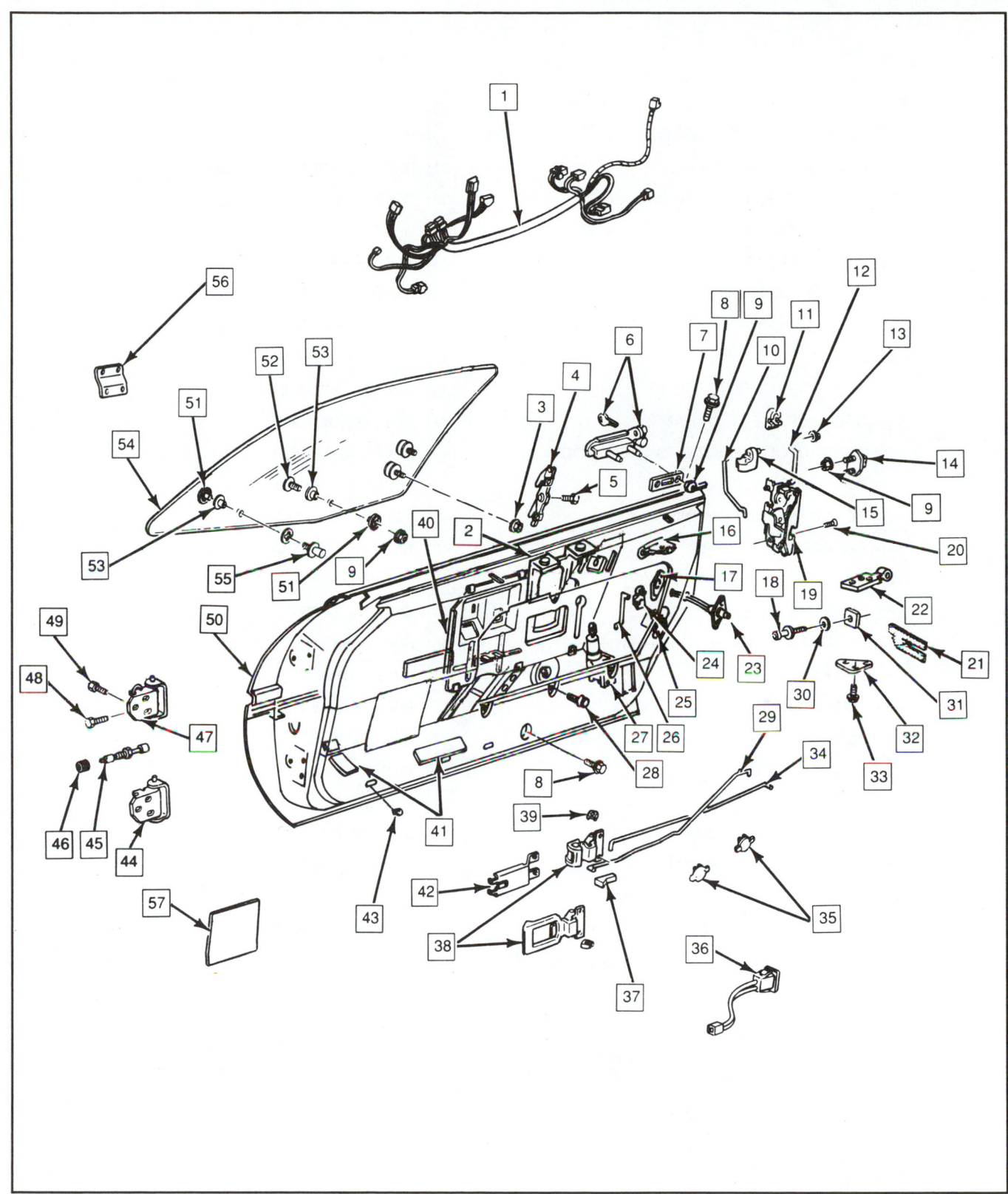

FIGURE 2-41 Study the parts of a typical door assembly. *(Courtesy of Chevrolet Motor Div.)*

1	HARNESS, POWER WINDOW		30	WASHER
2	REGULATOR, W/MOTOR		31	PLATE, LOCK STRIKER TAP
3	NUT, WINDOW GLASS W/WASHER		32	RETAINER, GUIDE PIN LOCATOR
4	GUIDE, STABILIZER		33	BOLT
5	BOLT, STABILIZER GUIDE		34	ROD, LOCK INSIDE HANDLE
6	HANDLE, OUTSIDE LOCK W/UNCODED KEY		35	GUIDES, LOCK ROD
7	BRACKET, DOOR HANDLE WIRING HARNESS		36	SWITCH, HATCH RELEASE
8	BOLT		37	KNOB, INSIDE LOCK
9	NUT		38	HANDLE, DOOR INSIDE
10	ROD, OUTSIDE LOCK HANDLE		39	CLIP, INSIDE HANDLE
11	CONNECTOR, REMOTE CONTROL ROD		40	PLATE, ACCESSORY MOUNTING
12	ROD, LOCK CYLINDER		41	SEAL, POLYURETHANE
13	RETAINER, LOCK CYLINDER ROD		42	INSERT, TRIM PANEL
14	PIN, GUIDE		43	NUT, NYLON (SEVERAL OTHER LOCATIONS)
15	CLIP, OUTSIDE LOCK HANDLE ROD		44	HINGE, LOWER
16	CUSHION, GLASS ANTI-RATTLE		45	SWITCH, JAMB
17	RETAINER, AJAR INDICATOR SWITCH		46	SEAL, JAMB SWITCH
18	STRIKER, LOCK		47	HINGE, UPPER
19	LOCK, SIDE		48	BOLT, HEX W/WASHER
20	SCREW, PAN HEAD TORX		49	BOLT, W/WASHER
21	SEAL, GUIDE PIN LOCATOR		50	PANEL, DOOR
22	LOCATOR, GUIDE PIN		51	NUT, WINDOW STOP
23	SWITCH, AJAR INDICATOR		52	STUD, WINDOW GLASS
24	CLIP, SIDE LOCK REMOTE CONTROL ROD		53	BUSHING, WINDOW STOP
25	BRACKET, WIRING LOWER REAR		54	GLASS, DOOR
26	ROD, SIDE LOCK REMOTE CONTROL		55	STUD, WINDOW GLASS
27	ACTUATOR, ELECTRIC LOCK W/BRACKET		56	BRACKET, POWER MIRROR WIRING HARNESS
28	SCREW, W/WASHER		57	PATCH, SOUND INSULATOR
29	ROD, LOCK REMOTE CONTROL			

FIGURE 2-41 Continued

ment, such as the rear floor side members and quarter wheelhousings.

Figure 2–39 shows the frame structure in a modern mid-engine sport car.

SPACE FRAME

Similar to a unibody, a *space frame* has a metal body structure covered with an outer skin of plastic or fiberglass panels (Figure 2–40). It is a relatively new type of vehicle construction. A space frame design is currently used on some vans and economy vehicles. Quite often, the roof and quarter panels are not welded to the structure as they are with traditional unibodies. Exterior body panels are attached with mechanical fasteners or adhesives.

After a collision, a space frame is more likely to have hidden damage because of the ability of plastic panels to hide more severe damage. Corrosion protection is also important since the plastic body panels may look good but the hidden metal frame structure may become deteriorated.

2.5 BODY PARTS

The body structures or sections are divided into small units called *assemblies*, which in turn are divided into even smaller units, called *components* or *parts*.

For example, *doors* are complex assemblies made up of an outer skin, inner door frame, door trim panel, window regulator, glass, and related parts (Figure 2–41). *Door hinges* bolt or weld between the pillars and door frame. The *window regulator* is a gear mechanism that allows you to raise and lower the door glass. The *door latch* engages the striker and allows you to lock and unlock the door. The *striker* is a post mounted on the body and it engages in the latch.

The door is formed by joining a high-strength inner and outer door panel that has inner access holes. The door opens and closes on hinges and is made air- and watertight by *weather-stripping*. The door reinforcement guard protects the passenger compartment in collisions or if the car should accidentally overturn.

Side impact beams are metal bars or corrugated panels that bolt or weld inside the door assemblies to protect the passengers. Primarily, they prevent the door from opening upon impact. They also help keep anything from intruding into the passenger area.

The *bumper assembly* bolts to the front frame horns or rails to absorb minor impacts. Its parts are shown in Figure 2–42.

Stationary parts, like the floor, roof, and quarter panels, are permanently welded or adhesive bonded

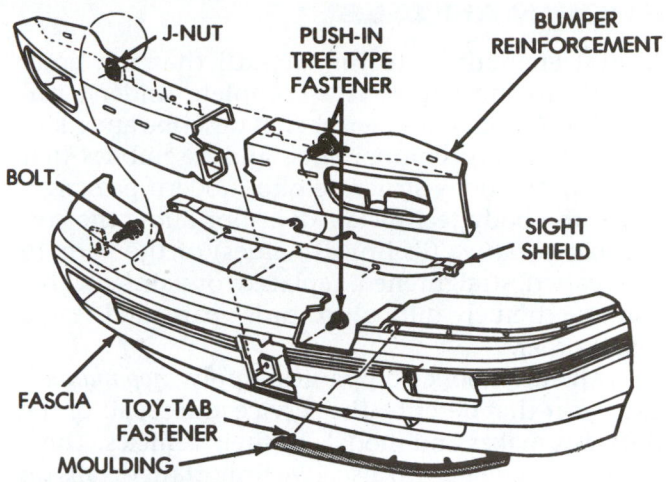

FIGURE 2–42 Here are the basic parts of a bumper assembly. *[Courtesy of Chrysler Corp.]*

into place. *Hinged parts*, like doors, hoods, and decklids, will swing out or up.

Fastened parts are held together with various fasteners (bolts, nuts, clips, etc.). Many parts, like the fenders, hood, and grille bolt into place. These bolted-on parts also add to the strength of the vehicle.

Welded parts are permanently joined by fusing the material so that it flows together and bonds when cooled. Both metal and plastic parts can be welded.

Press-fit or *snap-fit parts* use clips or an interference fit to hold parts together. This assembly method is becoming more common to reduce manufacturing costs.

Adhesive bonded parts use a high-strength epoxy or special glue to hold the parts together. Both metal and plastic parts can be joined with adhesive. *Structural adhesive* can also be used to bond parts together.

A *composite unibody* is made of specially formulated plastics and other materials, like carbon fiber, to form the vehicle. These parts are adhesive bonded to each other. The frame is made totally of plastics, keeping metal parts to a minimum. This cuts weight while increasing strength, rigidity, performance, and fuel economy. Although not mass produced, several manufacturers are experimenting with composite unibody construction.

Make sure you fully understand correct repair procedures and construction technology before working on a vehicle. It can be costly, dangerous, and even deadly if you do not understand how a car or truck is made and should be repaired.

SHOP MANUALS

A vital element of technical skill that the body technician must have is a complete understanding of commonly used terms that describe and identify parts, units, components, and assemblies that make up the body structure of a modern passenger car. If the body technician does not know the correct *nomenclature* (technical names) of the parts to be repaired, straightened, replaced, or painted, it becomes extremely difficult to order parts and read a repair order.

All automobile companies supply *shop manuals* each year that describe the service and repair of the different makes and models of their vehicles. These manuals (Figure 2–43) also give important details on body styles and parts.

Before using the manufacturer's shop manual or any type of manual, it is important to accurately identify the body style, model year, engine, and other pertinent details.

Check the shop manual for the location of the vehicle identification number (*VIN*), vehicle certifi-

cation label, or body number plate (Figure 2–44). In addition to collision repair or estimating guides, the shop manuals contain all necessary decoding information. Become familiar with each car maker's method of vehicle identification and the specific information it contains. It is wise to obtain all of the information possible on the vehicle being worked on (Figure 2–45).

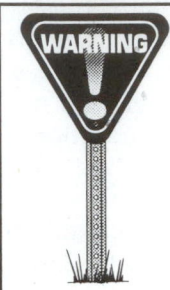

The procedures in this book are general. Always refer to a factory service manual when in doubt about any operation. It will give the specific procedures needed to do competent work on the specific make and model vehicle!

NOTE: Other chapters in this text give more information on using different types of manuals. Refer to the index if needed.

FIGURE 2-43 A service manual is vital when doing complex repair work. Always refer to one when in doubt.

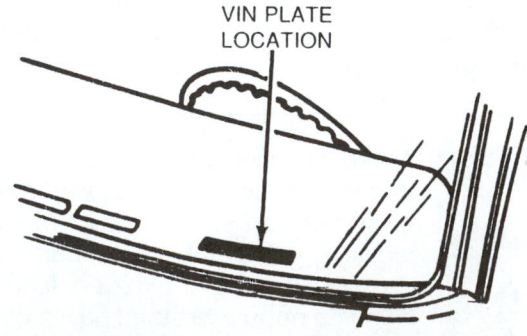

VIN PLATE
LOCATION

FIGURE 2-44 The typical VIN or vehicle identification number is normally located on the dash, behind the windshield. *(Courtesy of Chevrolet Motor Div.)*

CRASH TESTING

Automobile manufacturers are challenged by having to design vehicles that are light, *aerodynamic* (have low wind resistance), and yet strong and safe.

Computer-simulated crash testing is used before building a *prototype* or first real vehicle to determine how well the vehicle might survive a crash. It is critical that the passenger compartment is strong enough to help prevent injury to the driver and the passengers.

Certified crash tests are done using a real vehicle and sensor-equipped dummies that show how much impact the people would suffer during a collision. Computer readings from the sensors in the dummies give feedback about each crash test for body structure evaluation (Figure 2–46).

Crush zones are built into the frame or body to collapse and absorb some of the energy of a collision. The front and rear of the vehicle collapses while the passenger compartment tends to retain its shape. This helps reduce the amount of force transmitted to the occupants (Figure 2–47).

An *insurance rating system* uses a number scale to rate the vehicle's accident survivability. It indicates how well the vehicle will survive a crash and how much it will cost to repair it. A negative five is the worst rating (most damage) and positive five is the best (least damage). A zero rating is average.

SERVICE PARTS IDENTIFICATION LABEL

The Service Parts Identification Label provides identification of vehicle equipment to assist in servicing and determining replacement parts. Included on this label will be regular production options (RPO's) as well as standard and mandatory options. The label will be af-fixed to the inside of each passenger car vehicle at the assembly plant.

For additional information on the Service Parts Identification Label, see a GM Parts Catalog.

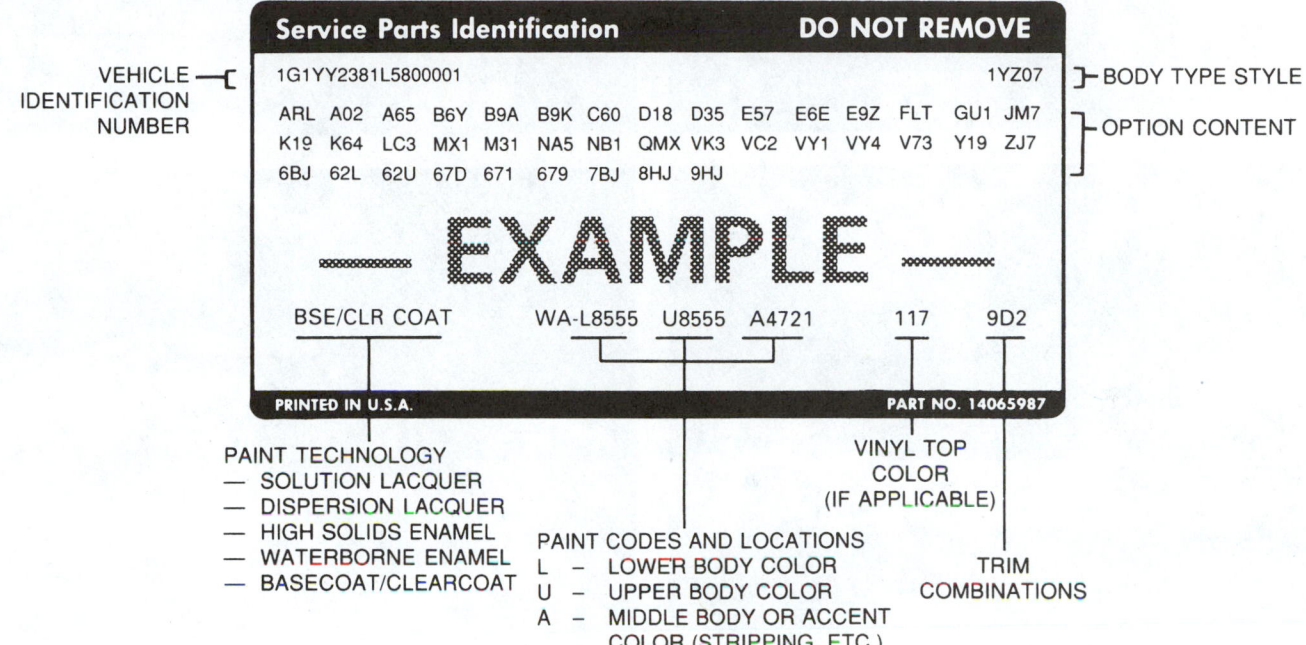

LABEL LOCATION

CORVETTE

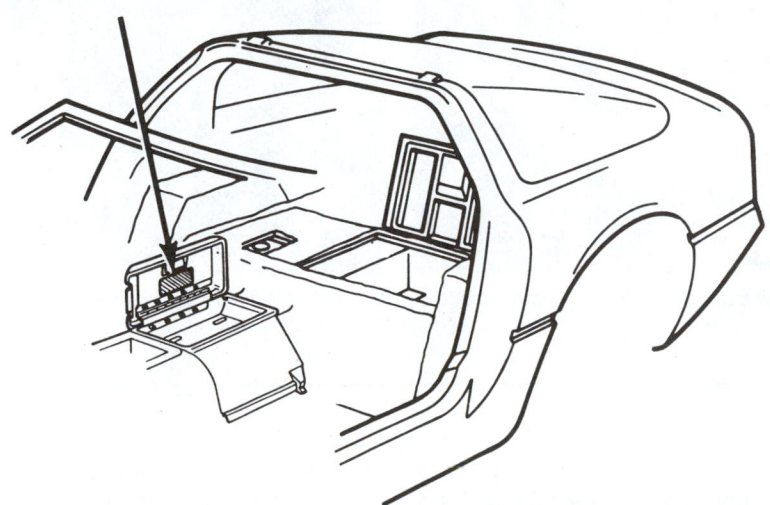

FIGURE 2-45 Here is an explanation of one service parts label. The service manual will explain the label for different makes and models. *(Courtesy of Chevrolet Motor Div.)*

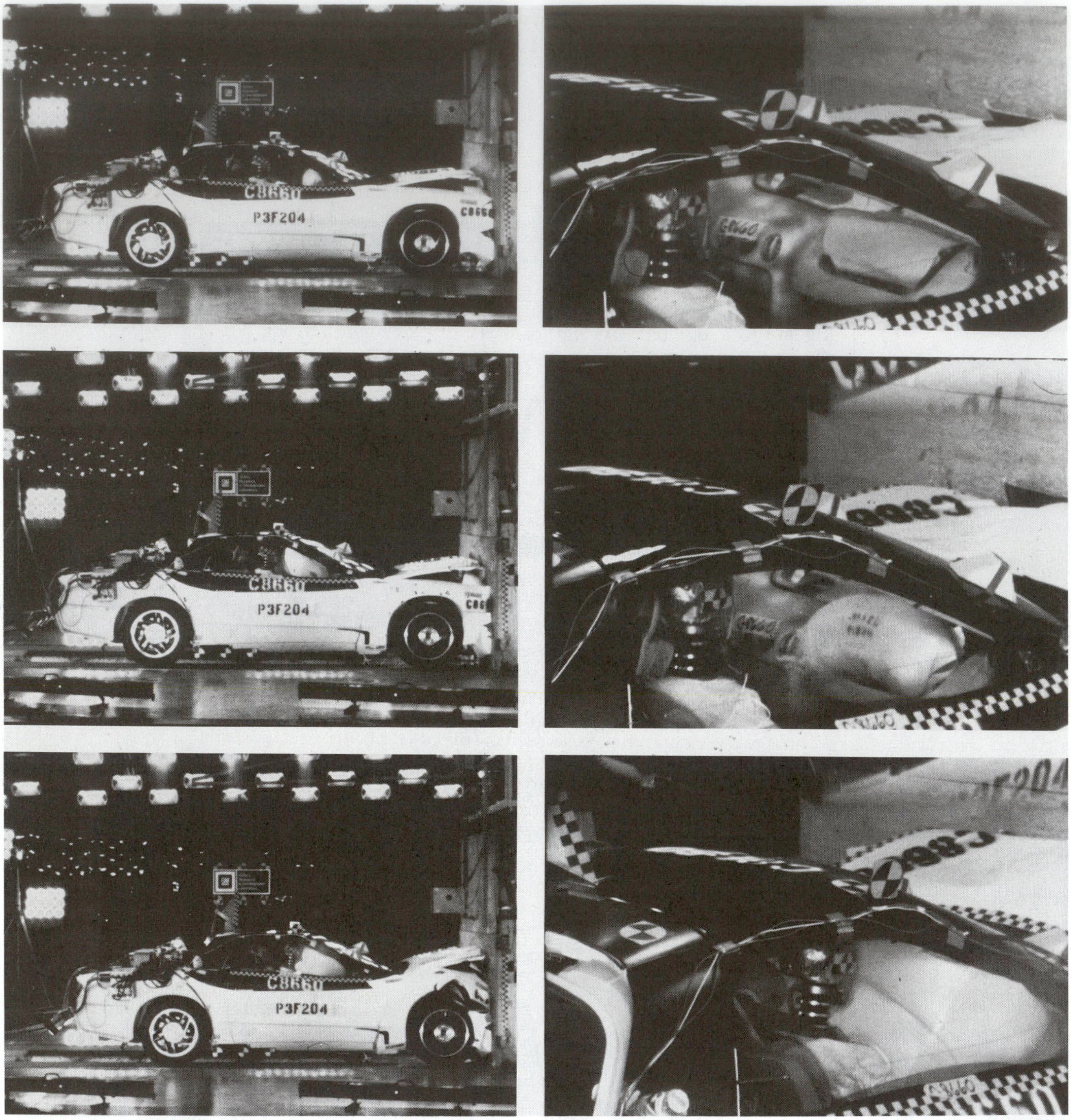

FIGURE 2-46 Crash tests analyze how well a vehicle and its passengers will survive an accident. *(Courtesy of Pontiac Motor Div.)*

SHOP TALK

If you have ever listened to a couple of veteran body technicians talking, you may have thought they were speaking in some kind of coded jargon. The conversation might have gone: "Well Joe, it bent the frame rails and shock towers from the frontal impact. We will have to mount it on the alignment rack to pull out the damage." If you don't know the basic parts of a vehicle, you will not be able to understand this kind of technical language or a shop manual.

A

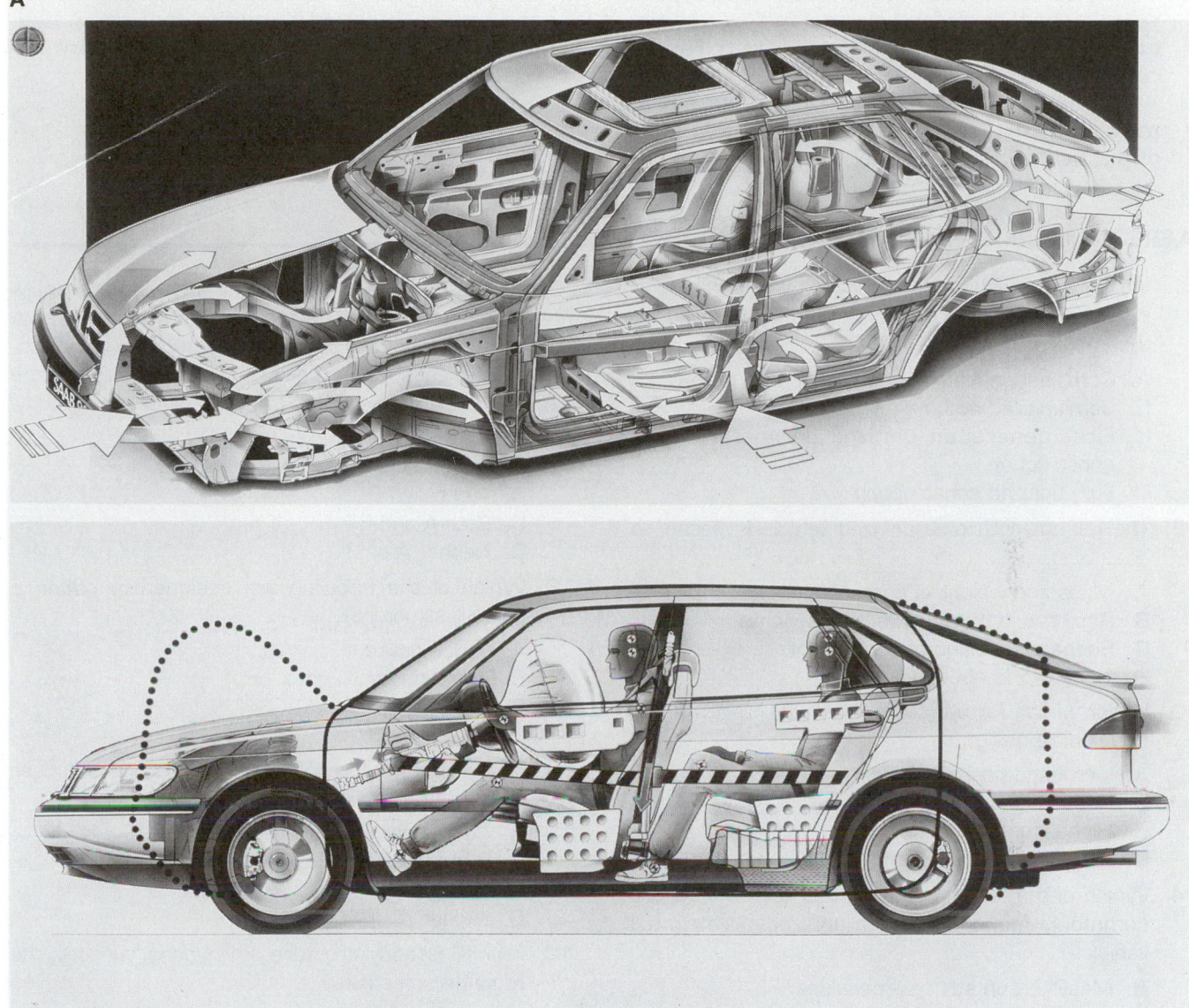

B

FIGURE 2-47 Crush zones help absorb the energy of a collision to protect the passengers. (A) Note how energy from front, side, and rear impacts flows through the body. (B) Study how the front and the rear of the vehicle collapse while the passenger compartment stays intact, providing a protective cage around the passengers. *(Courtesy of Saab Motor Co.)*

SUMMARY

- The goal of collision repair is to restore the vehicle to its preaccident condition.

- For simplicity and to help communication in auto body repair, a vehicle is commonly divided into three body sections—front, center, and rear.

- Anticorrosion materials are used to prevent rusting of metal parts. The five construction areas where domestic automobiles have changed since the mid-1970s are:

1. Body/frame construction
2. Weight (average fleet)
3. Metal composition
4. Suspension/steering
5. Engine location/drive

- In the conventional body-over-frame construction, the frame is the vehicle's foundation.

    While several conventional frame designs have been used by the auto industry, the three that the body technician may come across are the

1. Ladder frame
2. X-frame (or backbone)
3. Perimeter frame

■ Most newer models manufactured in the last few years in small to midsize classes (and even some full-size) are of the unitized or semiunitized body construction.

  Similar to a unibody, a space frame has a metal body structure covered with an outer skin of plastic or fiberglass panels.

■ Certified crash tests are done using a real vehicle and sensor-equipped dummies that show how much impact the people would suffer during a collision.

ASE-STYLE REVIEW QUESTIONS

1. Which type of vehicle construction uses a frame only in areas requiring extra support and a strong attachment point?
   A. Combination frame construction
   B. Semiunitized stub rail construction
   C. First generation unitized perimeter frame construction
   D. Fully unitized construction

2. The full strength of a unitized vehicle is based on _____.
   A. Mass and weight of components
   B. Rigidity and thickness of components
   C. Shape and design of components
   D. None of the above

3. Which of the following is not an advantage of unitized vehicle design?
   A. Increased passenger compartment safety
   B. Reduced vehicle weight
   C. Higher fuel efficiency
   D. Localized collision damage to components

4. Which of the following mechanical components are commonly found on newer unitized constructed vehicles?
   A. MacPherson strut suspensions
   B. Rack-and-pinion steering
   C. Front-wheel drive with CV joints
   D. All of the above

5. In front-engine, rear-wheel-drive unitized vehicles, the engine is mounted _____.
   A. Longitudinally
   B. Transversely
   C. Between the passenger compartment and rear axle
   D. Either A or B

6. Which of the following is not a use or characteristic of hot-rolled steels?
   A. Thicker components, such as frame legs and cross members
   B. A black oxidized surface appearance
   C. Highly accurate dimensional thickness
   D. Poorer workability than cold-rolled steels

7. Technician A gives a more thorough damage analysis to a unibody vehicle than to a conventional frame vehicle. Technician B says that the modern body technician needs much more knowledge than a technician of the era prior to the advent of the unibody. Who is correct?
   A. Technician A
   B. Technician B
   C. Both A and B
   D. Neither A or B

8. Which of the following are designed to stiffen a unibody structure?
   A. Torque boxes
   B. Frame horns
   C. Crush zones
   D. Stone deflectors

9. Which of the following frame designs are no longer used in automobile manufacturing?
   A. Perimeter
   B. Stub
   C. Hourglass
   D. Ladder

10. In an FF unibody structure, which panel supports the MacPherson struts?
    A. Front cross member
    B. Aprons
    C. Side rails
    D. Radiator support

11. Technician A checks the rear-wheel alignment whenever repairing a rear body section. Technician B says that a solid rear-axle equipped vehicle rarely suffers any damage in an accident. Who is correct?
    A. Technician A
    B. Technician B
    C. Both A and B
    D. Neither A or B

12. What type of body and frame is welded into one unit?
    A. Frame body
    B. Unibody
    C. Stub frame
    D. Nose frame

ESSAY QUESTIONS

1. What is the difference between body-over-frame construction and unibody construction?

2. Explain the terms *part* and *assembly*.

3. Describe the three body sections of a vehicle.

4. What are anticorrosion materials?

CRITICAL THINKING PROBLEMS

1. Why would you have to use different repair methods to repair a collision damaged full frame car and a unibody car?

2. From a repair standpoint, what are pros and cons of space frame construction?

MATH PROBLEMS

1. If a car weighs 4,000 pounds (1,816 kg) and each tire has the same amount of weight on it (equal weight distribution), how much weight is on each tire in pounds and kilograms?

2. In the previous question, if the tire has four square inches touching the road, how many pounds of weight would be pushing down per square inch?

CHAPTER 3

Body Shop Hand Tools

INTRODUCTION

Hand tools are general tools used by both auto mechanics and auto body technicians. They include wrenches, screwdrivers, pliers, and other common tools. Hand tools are needed to remove parts, fenders, doors, and similar assemblies. A full range of bodyworking hand tools are general purpose tools, metalworking tools, and body surfacing tools.

Hand tools are extensions to the human body. They allow you to do tasks otherwise impossible with your bare hands. By knowing how to select the right tool for the job, you will do higher quality work in less time. Hand tool knowledge is the sign of an experienced technician. Without the right tools, even the best body technician cannot do quality bodywork.

When purchasing tools, get high-quality tools from a reputable manufacturer. Most offer lifetime guarantees against failure. Never buy cheap, low-grade tools! Cheap tools slow down your work rate and efficiency because they are heavier, clumsier, and break more easily. You usually get what you pay for. Good tools will pay for themselves in a short period of time.

Figure 3–1 shows a typical set of body shop hand tools.

This chapter explains which hand tools an auto body repair technician must have and explains how they are used. This is a vital chapter that gives you the basics for understanding procedures given throughout this book and in service manuals.

3.1 GENERAL PURPOSE TOOLS

Many of the tools a body technician uses every day are common, general purpose hand tools: wrenches, screwdrivers, pliers, and so forth. An apprentice auto body repair technician will probably already have many of these in a tool collection. The less familiar tools are designed for specific types of industrial fasteners often encountered in bodywork.

WRENCHES

A complete collection of wrenches is indispensable for the auto body technician. A variety of auto body parts, accessories, and shop equipment utilizes common bolts and nuts. Fasteners can be standard SAE or metric size. A well-equipped auto body technician will have both metric and SAE wrenches in a variety of sizes and styles (Figure 3–2).

OBJECTIVES

After studying this chapter, you should be able to:

✔ List the most common hand tools used in auto body repair.

✔ Identify common body shop hand tools.

✔ Explain the importance of having a wide range of hand tools.

✔ Choose the correct tool for the job at hand.

✔ Use each tool correctly and in a safe manner.

✔ Answer ASE tool-related test questions.

KEY TERMS

adjustable wrench	Phillips screwdriver
Allen wrench	picking hammer
ball peen hammer	picks
body spoons	ratchet handle
box-end wrench	socket wrench
bumping hammer	spreaders
clip pullers	squeegee
combination pliers	standard screwdriver
dead blow hammer	surform file
dolly	Torx fastener
drive sizes	vise grips
finishing hammer	wrench size
needle nose pliers	

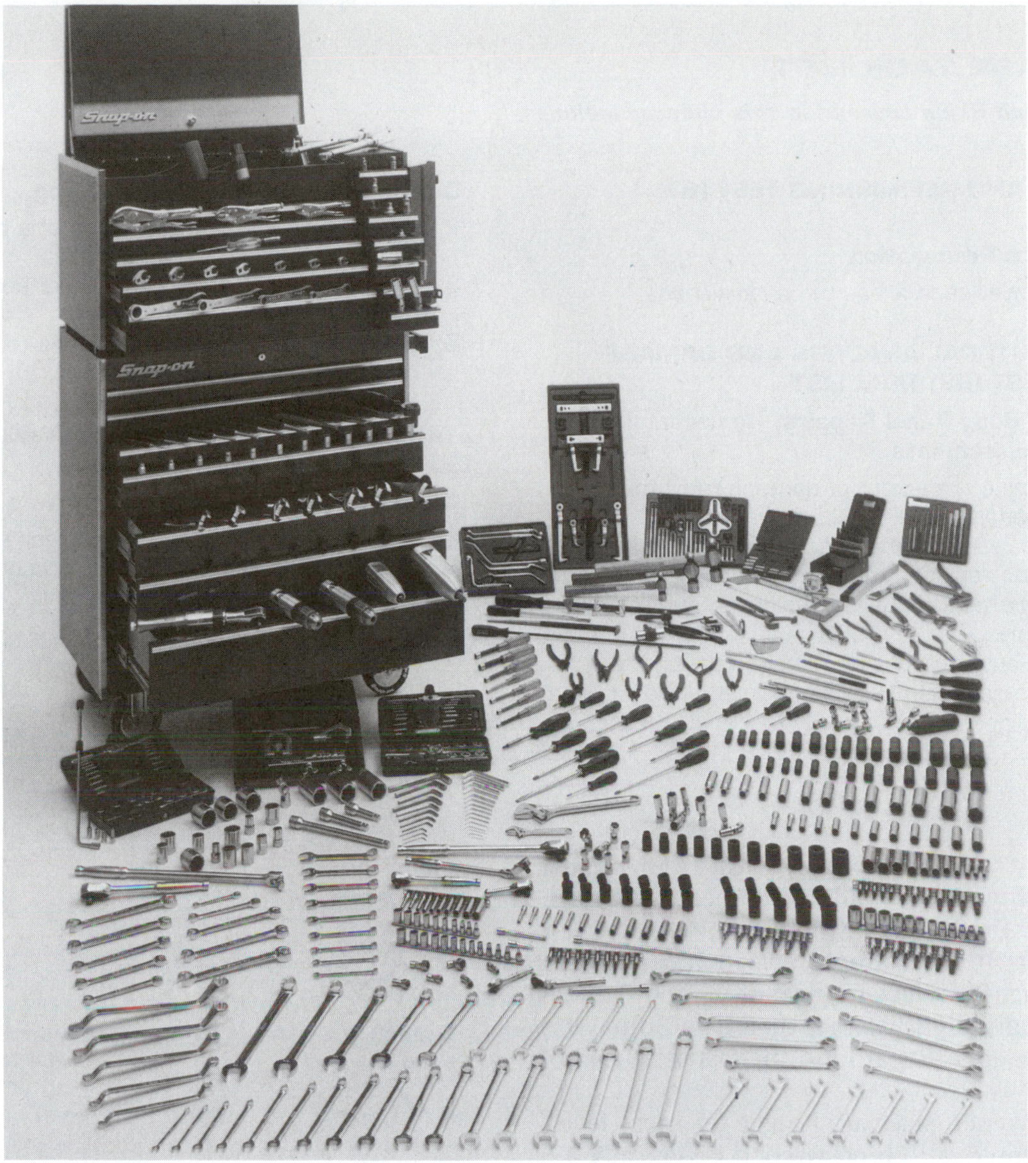

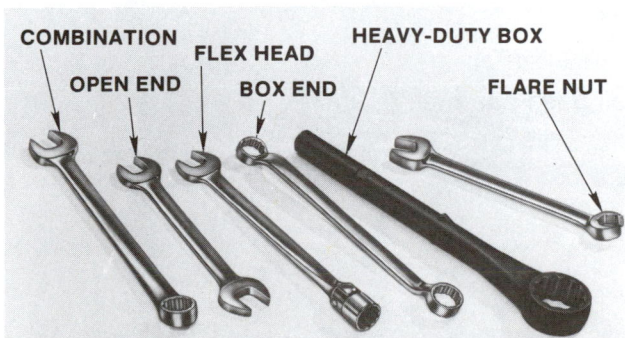

FIGURE 3–1 A professional set of hand tools is a big investment. Always purchase quality tools with a lifetime guarantee against failure. High-quality tools will save you time and effort and help you get more work done for more profit. *(Courtesy of Snap-On-Tools Company, Copyright Owner)*

FIGURE 3–2 Here is an assortment of wrenches. Study them! *(Courtesy of Snap-On-Tools Corporation)*

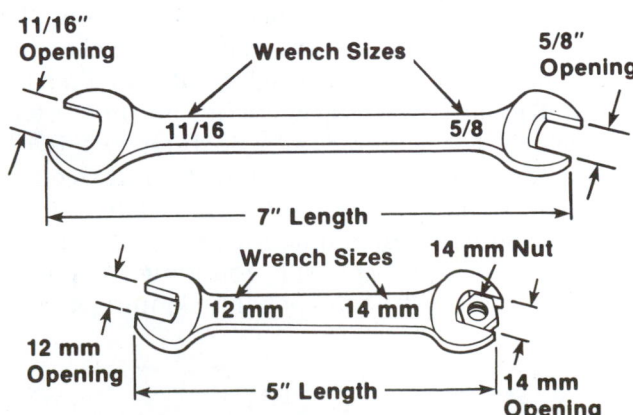

FIGURE 3–3 Wrench size is the size of the bolt or nut it will fit properly or the measurement across jaw openings.

ASE TASK LIST

Job Skills covered in this chapter include:

PAINTING AND REFINISHING TEST (B2) TASK LIST

A. Surface Preparation

5. Dry or wet sand areas to be refinished.

NONSTRUCTURAL ANALYSIS AND DAMAGE REPAIR TEST (B3) TASK LIST

B. Outer Body Panel Repairs, Replacements, and Adjustments

3. Determine the extent of damage to aluminum body panels; repair, weld, or replace in accordance with manufacturers' specifications.

7. Remove, replace, and align bumpers, reinforcements, guards, isolators, and mounting hardware.

10. Rough out contours of damaged panel to a surface condition suitable for metal finishing and body filling.

C. Metal Finishing and Body Filling

2. Pick and file the damaged area of a body panel to eliminate surface irregularities.

3. Prepare surface for application of body filler material.

5. Cold shrink stretched panel areas to proper contour.

STRUCTURAL ANALYSIS AND DAMAGE REPAIR TEST (B4) TASK LIST

C. Stationary Glass

1. Remove and replace front and rear stationary glass (heated and nonheated) in accordance with manufacturers' recommendations.

The word "wrench" means "twist." A *wrench* is a tool for twisting and/or holding bolt heads and nuts. The width of the jaw opening determines **wrench size** (Figure 3–3). For example, a $1/2$-inch wrench has a jaw opening (from face-to-face) of $1/2$ inch. The actual size is really slightly larger than its nominal size so that the wrench fits around the fastener head of equal size.

Larger wrench sizes are longer for more leverage. For example, a $1/4$-inch wrench is typically $4^1/2$ inches long. A $3/4$-inch wrench probably is 10 to 12 inches long. The extra length provides the user with more leverage to turn the larger size nut or bolt.

Most standard wrench sets include sizes from $7/16$ to 1 inch. Metric sets usually include 6- to 19-millimeter wrenches. Smaller and larger wrenches can be purchased but are rarely used in auto body repair.

Metric and SAE size wrenches are not interchangeable. For example, a $9/16$-inch wrench is $3/10$ millimeter larger than a 14-millimeter nut (see the conversion chart in Appendix C). If the $9/16$-inch wrench is used to turn or hold the 14-millimeter nut, the wrench will probably slip, rounding the points on the nut (Figure 3–4) and possibly skinning knuckles as well.

Open-End Wrenches

Every tool chest must have a set of open-end or combination wrenches (Figure 3–5). The jaws of the

Sharp Corners

Wrench is a snug fit.

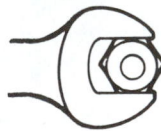

Round Corners

Wrench is too large for nut.

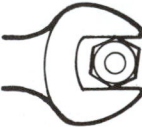

FIGURE 3–4 Use the right size wrench to avoid rounding off flats on a fastener.

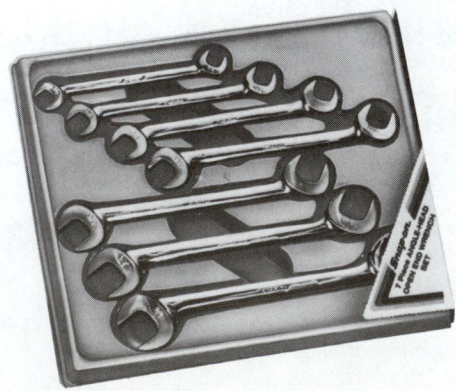

FIGURE 3–5 Note set of open-end wrenches. *(Courtesy of Snap-On Tools Corp.)*

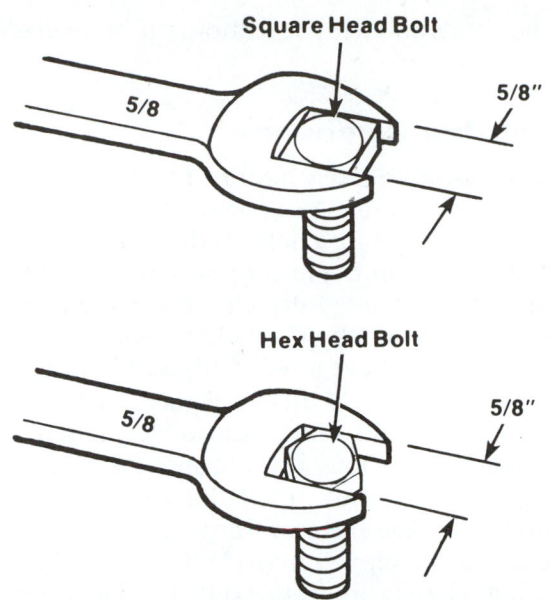

FIGURE 3–6 An open-end wrench grasps only two faces on a fastener and can slip off easily.

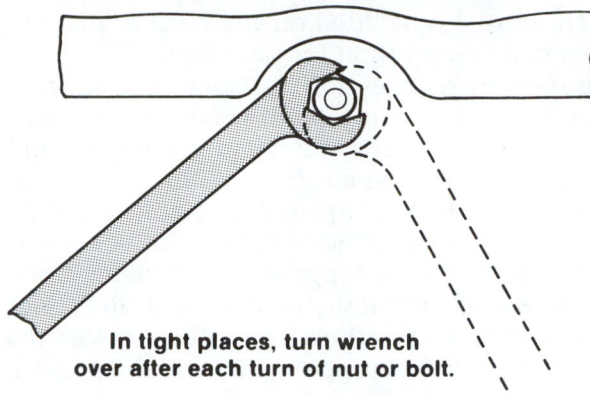

In tight places, turn wrench over after each turn of nut or bolt.

FIGURE 3–7 An offset wrench can increase the turning radius and can allow the wrench to fit over the fastener in an obstructed area.

open-end wrench will slide around bolts or nuts where there might be insufficient clearance above or on one side to accept a box wrench.

The open-end wrench fits both square head (four-cornered) or *hex head* (six-cornered) nuts. The disadvantage of using open-end wrenches is that only two faces of the nut are gripped by the jaws (Figure 3–6) and there is a greater tendency for it to slip off the bolt or nut. Rounded nuts and injured hands are too often the result. Because of this, the open-end wrench should be used for holding when there is not sufficient room for a box-end wrench to be used.

Open-end wrenches are often angled 15 to 80 degrees at both ends. The offset helps turn a bolt or nut that is recessed or is in a confined area. Flipping the wrench over after each turn maximizes the turning arc (Figure 3–7).

Box-End Wrenches

Figure 3–8 shows box-end wrenches in a variety of sizes, points, and offsets. The end of the **box-end wrench** is closed rather than open for better holding power. The jaws of the wrench fit completely around a bolt or nut, gripping each point on the fastener (Figure 3–9).

The box-end wrench is thus the safest to use. More force can be applied without slippage and rounding bolt or nut heads. The handle of many box-end wrenches is offset to provide hand clearance. Each end is usually a different size.

The box-end wrench does have limitations. There must be sufficient clearance for the jaws to fit over and around the head or nut. The box-end

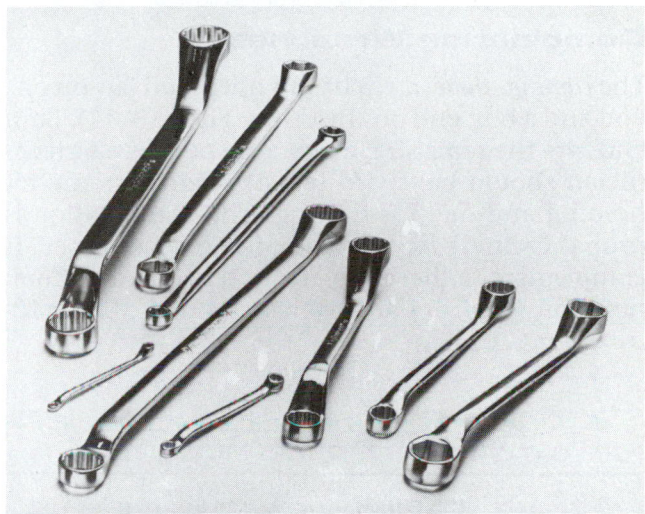

FIGURE 3–8 Here is an assortment of box-end wrenches. *(Courtesy of Snap-On Tools Corp.)*

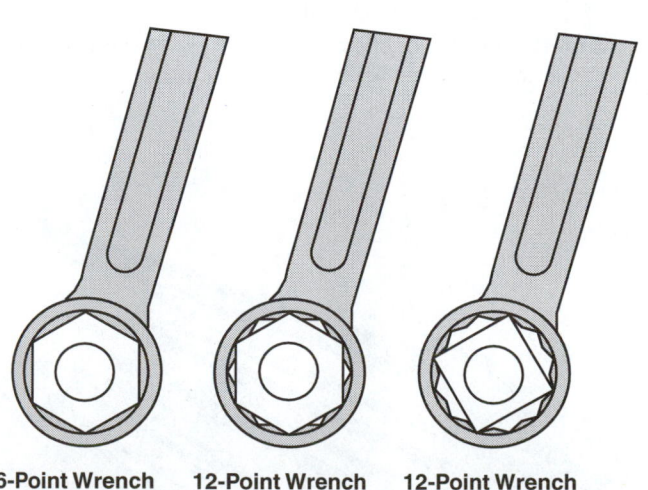

6-Point Wrench and 6-Point Nut   12-Point Wrench and 6-Point Nut   12-Point Wrench and 4-Point Nut

FIGURE 3–9 Note how the contact areas of a 6-point wrench are greater than those of a 12-point.

wrench must also be lifted off the head or nut and rotated to a new position for each pull.

Box-end wrenches are available in 6, 8, or 12 points (Figure 3–9). The 6-point wrench is the strongest because it completely surrounds the hex nut and brings force to bear on all six sides and points. The 12-point wrench also grips the six points but does not bear on the face surfaces of a hex nut, so there is a greater potential for slippage. The advantage of a 12-point wrench is that the wrench can grab the nut in twelve different positions. In confined spaces, the additional engagement points increase the possible turning radius. The 8-point box-end wrench is seldom used because it fits only square head nuts.

The handle of a box-end wrench is often offset 10 to 60 degrees (Figure 3–10), which allows recessed fasteners to be reached more easily.

Combination Wrenches

The *combination wrench* has an open-end jaw on one end and a box-end on the other (Figure 3–11). Both ends are the same size. Every auto body repair technician should have two sets of wrenches: one for holding and one for turning. The combination is probably the best choice for the second set. It complements either open-end or box-end sets. Combination wrenches are available with 6-, 8-, or 12-

point box ends and with or without offset open ends and handles.

Adjustable Wrenches

An **adjustable wrench** has one fixed jaw and one moveable jaw (Figure 3–12). The wrench opening can be adjusted by rotating a helical adjusting screw that is mated to teeth in the lower jaw. The jaw opening can be adjusted from fully closed ($1/2$ inch) to its maximum open width ($1 3/4$ inches). The auto body tool chest should have a set of adjustable wrenches in lengths from 4 to 24 inches. Figure 3–13 shows a set of 4-, 6-, 8-, and 10-inch adjustable wrenches.

Besides the obvious advantage of fitting various size bolt heads and nuts, the adjustable wrench has the same advantage and disadvantage of an open-end wrench. It can be slipped around bolts or nuts in tight places, but it also bears against only two faces. It, too, slips more easily than does the box-end wrench.

As the adjustable wrench wears with use, the adjusting screw will lose some of its holding power. The jaws develop a tendency to loosen as force is applied and then slip off the nut or bolt. Therefore, use an adjustable wrench only when a suitable box-end or open-end wrench is not available. Use it to

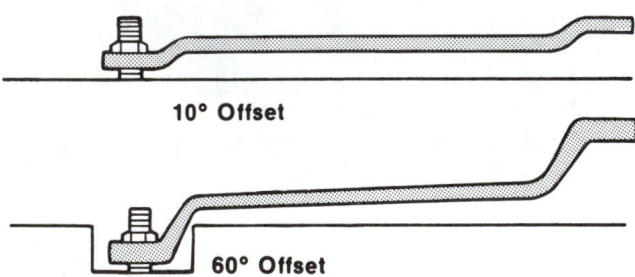

FIGURE 3–10 An offset box wrench will help reach down into an area without the handle hitting on a part.

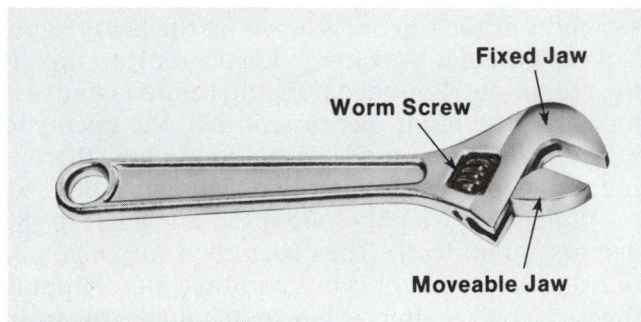

FIGURE 3–12 An adjustable wrench has a movable jaw for fitting different size fasteners. *[Courtesy of Stanely Works]*

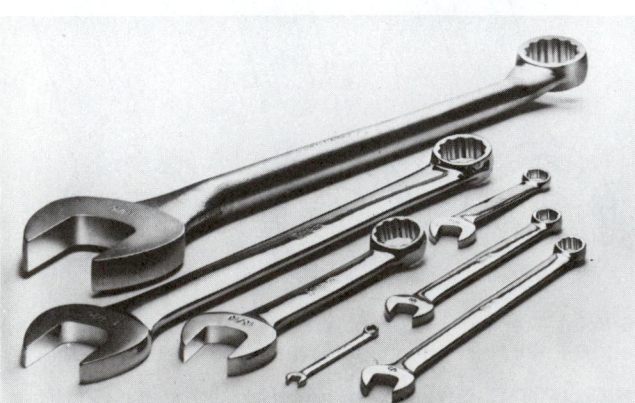

FIGURE 3–11 These are combination wrenches. Note the different head openings. *[Courtesy of Snap-On Tools Corp.]*

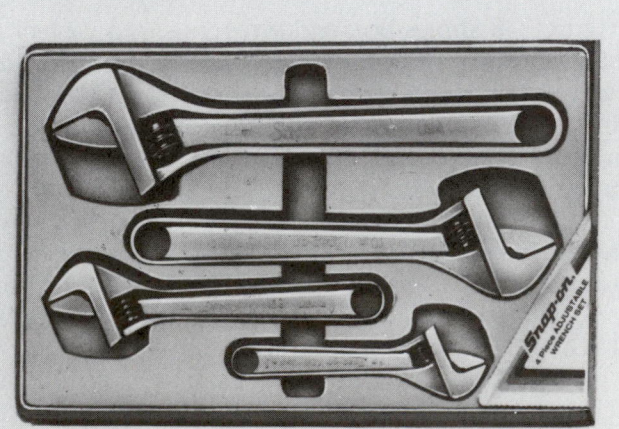

FIGURE 3–13 Here is a set of adjustable wrenches. *[Courtesy of Snap-On Tools Corp.]*

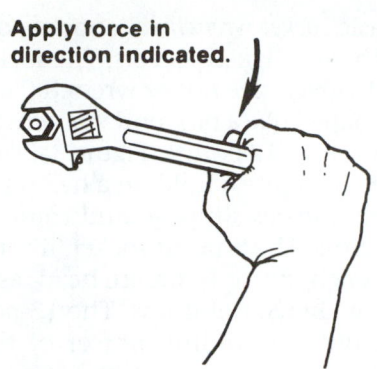

FIGURE 3-14 Pull on the wrench so that force bears against the fixed jaw.

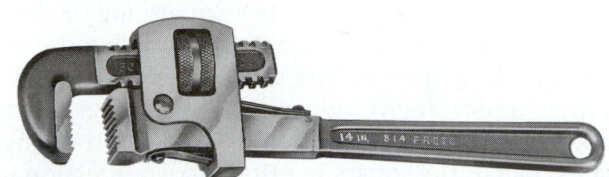

FIGURE 3-15 A pipe wrench will grasp and turn round objects with great force. However, it will usually mar the part surface as the jaws dig into it. *(Courtesy of Proto Tools)*

hold rather than turn when it must be used. Tighten it securely. Hold it flush, and pull the handle so that the force bears on the fixed jaw (Figure 3–14).

Pipe Wrenches

Another type of wrench that is occasionally used in the auto body shop is the *pipe wrench* (Figure 3–15). The pipe wrench gets its name from its most common use—turning pipes and pipe fittings. The advantage of the pipe wrench over other wrenches is that it will grab and turn round objects, such as pipes and studs.

Like the adjustable wrench, the pipe wrench opening is adjustable. The top or hook jaw is threaded through a stationary adjusting nut. Turn the adjusting nut to increase or decrease the jaw opening. Pipe wrenches are available in maximum openings of $^3/_8$ to 8 inches in various lengths. A 10-inch-long pipe wrench is probably adequate for most body shop applications. The best have replaceable teeth on the lower (heel) jaw.

Hex and Torx Wrenches

An **Allen wrench** is a hex, hexagon, or six-sided wrench. It will install or remove set screws. Set screws are often used on pulleys, gears, rear view mirrors, and knobs. A set of hex head wrenches, or Allen wrenches (Figure 3–16), should be in every tool box. Hex head sockets are also available for removing large setscrews (Figure 3–17).

The **Torx fastener** is a 6-point fastener that is easier to grip and drive without slippage. Sometimes

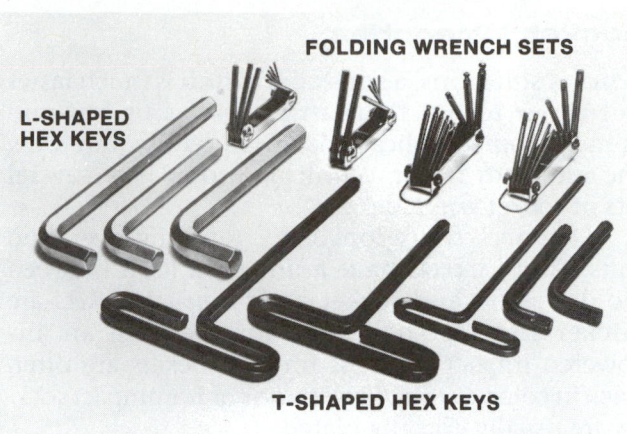

FOLDING WRENCH SETS

L-SHAPED HEX KEYS

T-SHAPED HEX KEYS

FIGURE 3-16 Note a typical set of Allen wrenches. *(Courtesy of Snap-On Tools Corp.)*

FIGURE 3-17 A hex socket set is much easier to use since the ratchet will rapidly turn the tool. *(Courtesy of Snap-On Tools Corp.)*

FIGURE 3-18 Note the shape of Torx socket heads. *(Courtesy of Lisle Corp.)*

called a "*starfastener,*" this type is used on most late model cars. On many vehicles, Torx fasteners are used in luggage racks, headlights, taillight assemblies, mirror mountings, door strikers, seat belts, and exterior trim. Torx wrenches or drivers are sold in sets of five or seven popular metric sizes (Figure 3–18).

Socket Wrenches

In many situations, a **socket wrench** is much faster and easier to use than an open-end or box-end wrench. Some applications absolutely require one. The auto body repair technician should have several sets of socket wrenches.

Deep sockets are longer for reaching over stud bolts. *Swivel sockets* have a universal joint between the drive end and socket body. *Impact sockets* are thicker and case-hardened for use with an air-powered impact wrench. Impact sockets are often black in color. *Conventional sockets* or nonimpact sockets are usually chrome-plated.

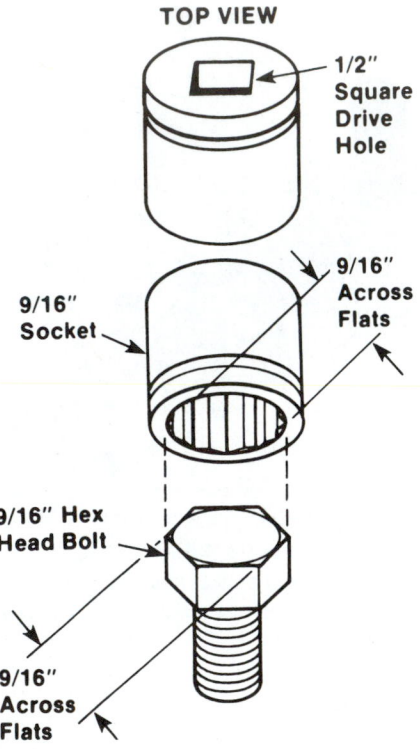

TOP VIEW

1/2" Square Drive Hole

9/16" Across Flats

9/16" Socket

9/16" Hex Head Bolt

9/16" Across Flats

FIGURE 3–19 Socket size is almost the same as bolt or nut size. It is the measurement across flats.

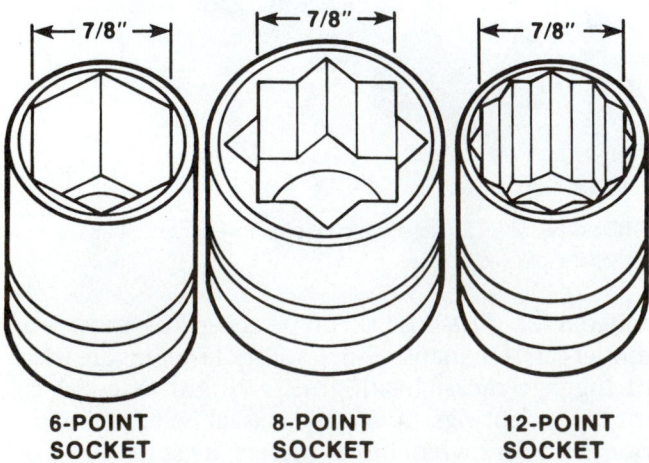

← 7/8" →   ← 7/8" →   ← 7/8" →

6-POINT SOCKET   **8-POINT SOCKET**   **12-POINT SOCKET**

FIGURE 3–20 Compare 6-, 8-, and 12-point sockets.

The basic *socket wrench set* consists of a handle and several barrel-shaped sockets. The socket fits over and around a given size nut or wrench (Figure 3–19). Inside it is shaped like a box-end wrench. Sockets are available in 6, 8, or 12 points (Figure 3–20). A 6-point socket gives the tightest hold on a hex nut. The face-to-face fit minimizes slippage and rounding of the fastener's points. The 8-point socket, like the 8-point box-end wrench, fits only square head fasteners and is thus limited in its usefulness. The 12-point socket does not have the holding power of the 6-point socket, but its numerous positions maximize the possible turning radius.

The socket is closed on one end. The closed end has a square hole that accepts a square lug or drive on the socket handle (Figure 3–21). One handle fits all the sockets in a set. The size of the lug ($^1/_4$ inch, $^3/_8$ inch, and $^1/_2$ inch) indicates the drive size of the socket wrench. On better-quality handles, a spring-loaded ball in the square lug fits into a depression in the socket. The ball holds the socket to the handle.

Socket Sizes. The size of the individual sockets depends on the drive size of the set as well as the size of the fastener head it fits. The socket size is slightly larger than the face-to-face dimension of the fastener. A $^1/_4$-inch set has sockets ranging from $^3/_{16}$ to $^1/_2$ inch or 4 to 13 mm. A $^3/_8$-inch drive set has sockets ranging in size from $^3/_8$ to $^3/_4$ inches or 9 to 25 mm. A good $^1/_2$-inch socket wrench set has sockets ranging in size from $^7/_{16}$ to $1^1/_4$ inches or 11 to 30 mm.

Both standard SAE and metric socket wrench sets are necessary for bodywork. A large percentage of the vehicles sold in the United States require metric tools.

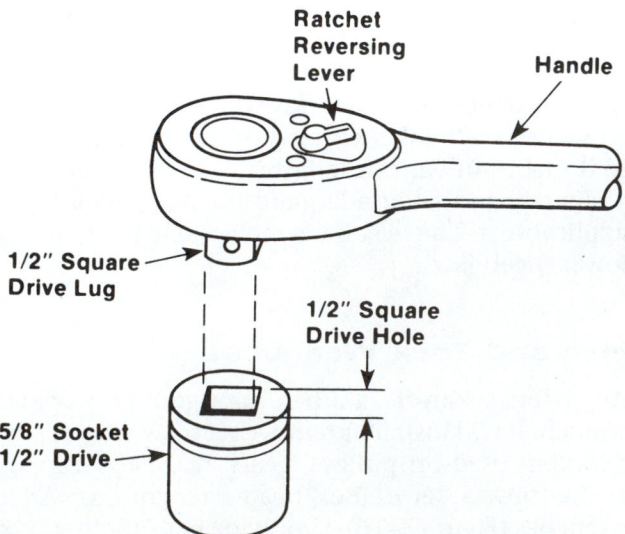

Ratchet Reversing Lever

Handle

1/2" Square Drive Lug

1/2" Square Drive Hole

5/8" Socket 1/2" Drive

FIGURE 3–21 Socket drive size is the measurement across the square hole for the drive lug on the handle. Make sure it is large and strong enough not to break from excessive torque.

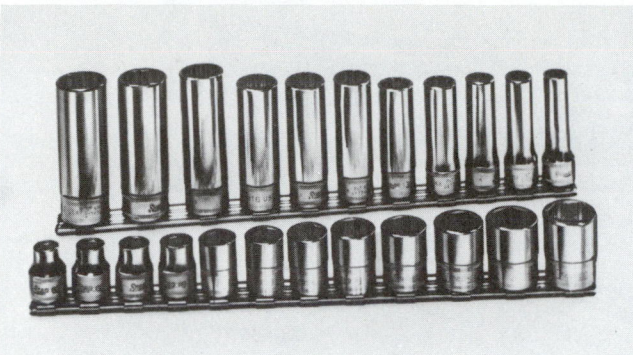

FIGURE 3-22 These are regular and deep-well sockets. *(Courtesy of Snap-On Tools Corp.)*

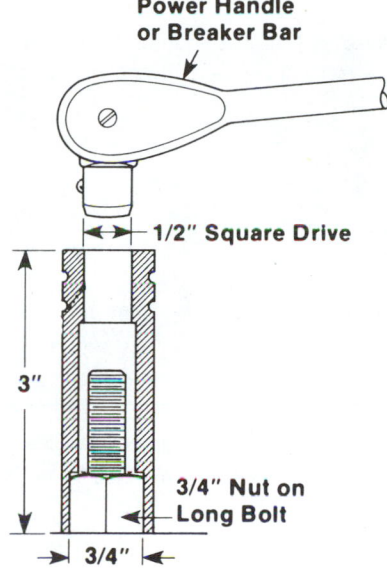

Power Handle or Breaker Bar

1/2" Square Drive

3"

3/4" Nut on Long Bolt

3/4"

CROSS-SECTION VIEW

FIGURE 3-23 A deep-well socket will slide down over studs to reach nuts.

Sockets are available not only in standard face-to-face diameters, but also in various lengths or bore depths (Figure 3–22). Normally, the larger the socket size, the deeper the well. Deep-well sockets (Figure 3–23) are made extra long to fit over bolt ends or studs to reach a nut. A spark plug socket is an example of a deep-well socket. Deep-well sockets are useful in a body shop for removing bumper bolts and also for reaching nuts or bolts in limited access areas. Deep-well sockets should not be used when a regular size socket will do the job. The longer socket develops more twisting torque and tends to slip off the fastener.

Drive Sizes. Socket wrench sets can be purchased in $1/4$-, $3/8$-, $1/2$-, $3/4$-, and 1-inch **drive sizes**. The small drive sizes are used for turning small fasteners on emblems and trim where little torque is required. The larger drive sizes with longer handles are used where greater torque is needed. A $3/4$- or 1-inch socket wrench is useful for truck and heavy equipment repair.

An auto body repair technician will need a set of $1/4$-, $3/8$-, and $1/2$-inch drive sockets (Figure 3–24). The $1/4$-inch drive set is a must for disassembly and removal of interior trim components. The $3/8$-inch

SHOP TALK

Always use the appropriate drive size for the job at hand. If the drive size is too small for the fastener, you can break the drive or socket. If the drive is too large, you will waste time trying to handle the clumsy tool.

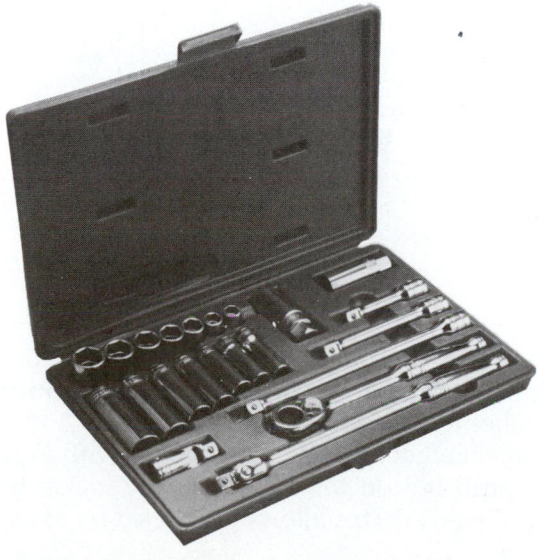
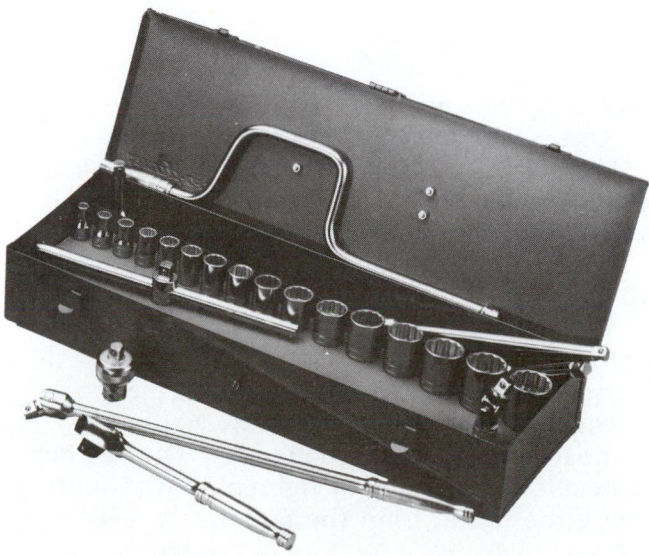

FIGURE 3-24 Note typical $3/8$- and $1/2$-inch socket wrench sets. *(Courtesy of Snap-On Tools Corp.)*

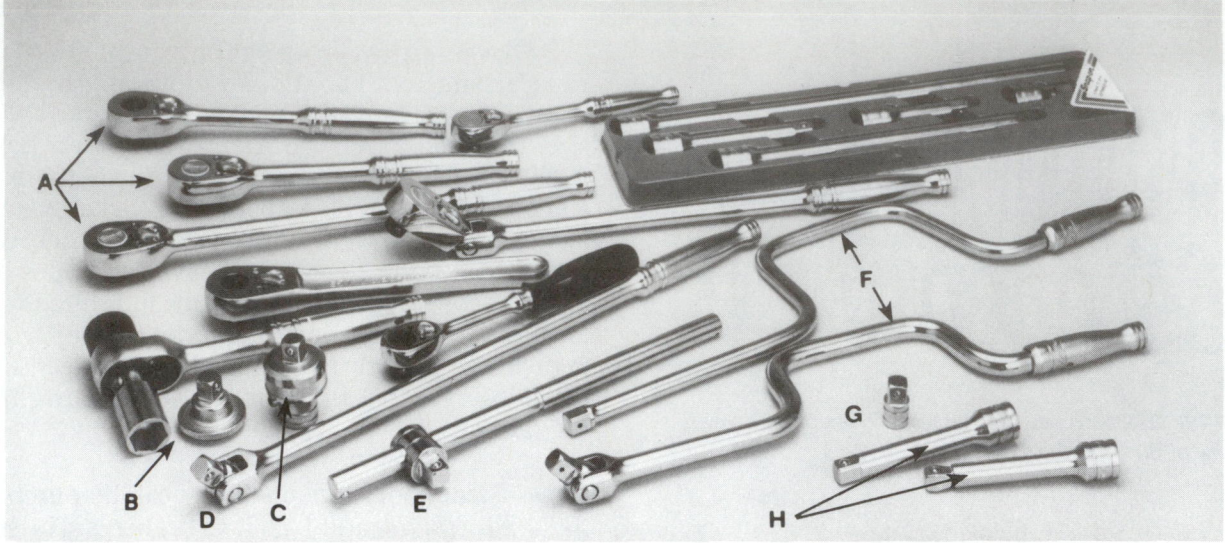

FIGURE 3–25 Study socket wrench accessories: (A) ratchet handles, (B) spinner, (C) ratchet adapter, (D) breaker bars, (E) sliding T-handle, (F) speed handles, (G) drive adapter, and (H) extension bars. *(Courtesy of Snap-On-Tools Corporation)*

drive sockets will fit almost all sheet metal bolts and nuts found on a vehicle. The sockets in a $^1/_2$-inch drive set are useful in removing exhaust systems, suspension parts, bumpers, and other related automotive parts commonly removed in body repair.

Socket Wrench Accessories

Figure 3–25 shows a number of socket wrench set accessories. Accessories multiply the usefulness of a socket wrench. A good socket wrench set has a variety of the following:

- Ratchet handles
- Spinner
- Ratchet adapter
- Breaker bars
- Sliding T-handle
- Speed handles
- Drive adapter
- Extension bars

Handles. Several different types of handles are available. One is the *breaker bar* or power handle (Figure 3–25D). Held at a 90-degree angle, the extra long handle provides the torque needed to loosen stubborn fasteners. After breaking the bolt loose, swing the handle straight out and you can turn the handle with your fingers to quickly remove the nut or bolt.

A *speed handle* (Figure 3–25F) is a bit-and-brace-type handle that can quickly turn off a nut or bolt. Its use requires sufficient clearance for turning the handle.

The **ratchet handle** (Figure 3–26) is probably the most commonly used handle. The ratchet handle allows removing or tightening without removing the socket from the fastener. A reversing

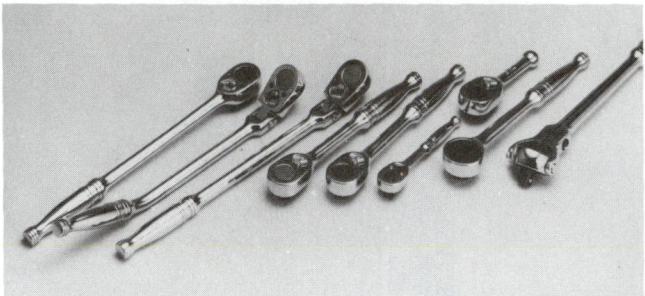

FIGURE 3–26 A variety of ratchet handles are needed in auto body repair. *(Courtesy of Snap-On-Tools Corporation)*

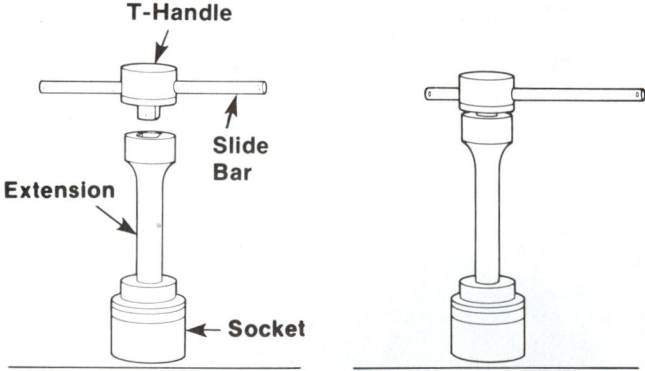

FIGURE 3–27 T-bars can be gripped with both hands or used as breaker bars.

lever allows the ratchet mechanism to slip (ratchet) in one direction and turn the socket in the other. The turning direction can be changed by turning the ratchet lever position.

Some ratchet handles are equipped with a quick release push button for unlocking the socket from the handle drive. This allows the socket to be easily removed from the handle. Ratchet handles are available not only with flexheads, but also with offset handles to help reach obstructed areas.

A *T-handle* or *slide bar* (Figure 3–27) is another handle that comes in handy when access is limited. The T-handle is similar to a long extension, but a slide bar fits in a hole in the upper end. The slide bar can be centered in the hole and gripped with both hands. The push-pull effort helps loosen stubborn fasteners with less likelihood of slippage. The slide bar can also be slid to one side and used as a breaker bar.

Many socket wrench sets contain sockets in two drive sizes, such as $1/4$- or $3/8$-inch and $1/2$-inch drives. An *adapter*, such as the one shown in Figure 3–25G, is often provided so that the larger drive handle can be used with the smaller drive sockets.

Most socket wrench sets also contain extensions and universal joints. The *extensions* (3- and 6-inch are common lengths, with up to 36-inch available) reach into otherwise inaccessible places. A universal joint (Figure 3–28) allows the work to be done at an angle to the fastener. With a universal joint adapter, one can reach around obstacles and use a socket wrench.

Flexockets. A *flexocket* is a combined socket and universal joint (Figure 3–29). The *universal joint* allows the handle to be held at an angle other than

90 degrees to the fastener. Flexockets are normally $3/8$-inch drive.

Screwdriver Attachments. Screwdriver attachments are also available for use with a socket wrench. Figure 3–30 shows a typical set and three specialty sockets:

- Hex driver
- Phillips driver
- Flat tip driver
- Clutch head driver
- Torx driver
- Three-wing socket
- Double square socket
- Torx socket

These socket wrench attachments are very handy when a fastener cannot be loosened with a regular screwdriver. The leverage that the ratchet handle provides is often just what it takes to break a stubborn screw loose.

Two other specialty fastener sockets are shown in Figures 3–31 and 3–32. Triple square and ribe fasteners are common on many European import vehicles.

SCREWDRIVERS

A variety of threaded fasteners used in the automotive industries are driven by a *screwdriver*. Some, like

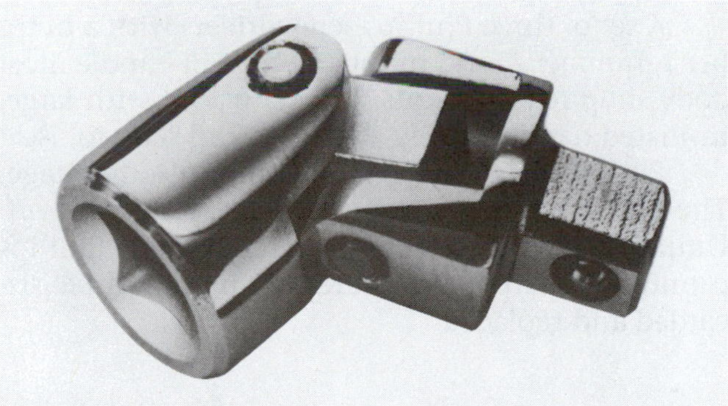

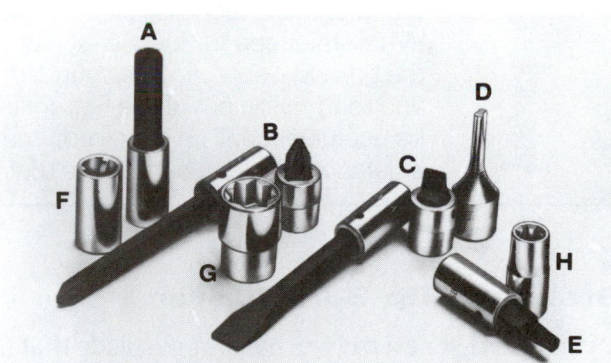

FIGURE 3–28 A universal joint will allow a handle or extension to be placed at an angle to drive around an obstruction. *(Courtesy of Mac Tools)*

FIGURE 3–30 Study a typical screwdriver attachment set: (A) hex driver, (B) Phillips driver, (C) flat tip driver, (D) clutch head driver, (E) Torx driver, (F) three-wing socket, (G) double square socket, and (H) Torx socket.

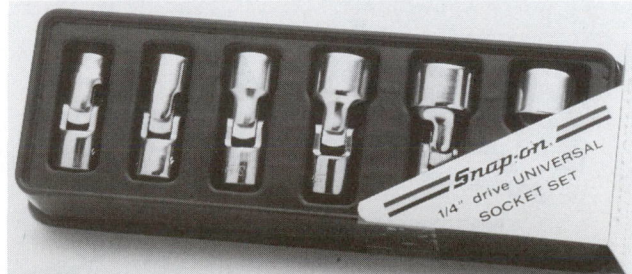

FIGURE 3–29 Flexockets have universal joints formed between socket and drive heads. *(Courtesy of Snap-On-Tools Company, Copyright Owner)*

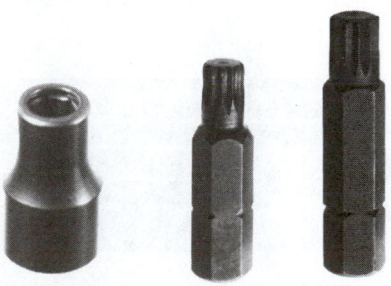

FIGURE 3–31 Triple square sockets are not that common.

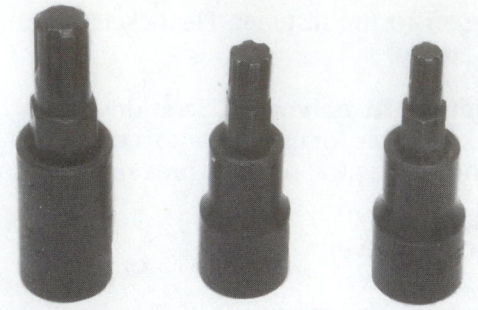

FIGURE 3-32 Ribe sockets may sometimes be needed.

the self-tapping sheet metal screw, are common. Others, like the Torx or clutch head are less common. Each fastener requires a specific kind of screwdriver. The well-equipped technician will have several sizes of each.

All screwdrivers, regardless of the type of fastener they are designed for, have several things in common. The size of the screwdriver is determined by the length of the shank or the blade. The size of the handle is important, too. The larger the handle diameter, the better grip it has and the more torque it will generate when turned.

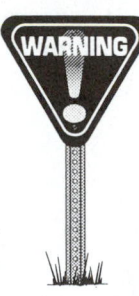

Do not use a screwdriver as a chisel, punch, or pry bar. Screwdrivers were not made to withstand blows or bending pressures. When misused in such a fashion, the tips will wear, become rounded, and tend to slip out of the fastener. Its usefulness will be impaired, and a defective tool is a dangerous tool.

Standard Tip Screwdriver

A **standard screwdriver** has a single blade that fits a screw with a slotted head (Figure 3–33). The blade and lengths should match the job. The blade tip width and thickness must fit the screw head perfectly.

A good set of standard tip screws will have five to seven screwdrivers from the 1½-inch stubby to a 10-inch driver with a heavy-duty ⁵⁄₁₆-inch blade.

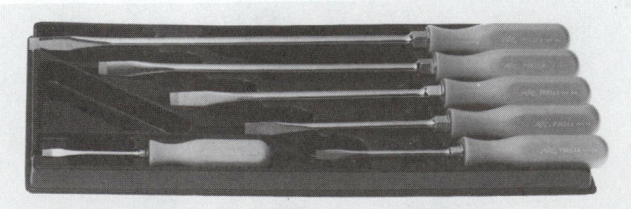

FIGURE 3-33 Several sizes of standard tip screwdrivers should be in your toolbox. *(Courtesy of Mac Tools)*

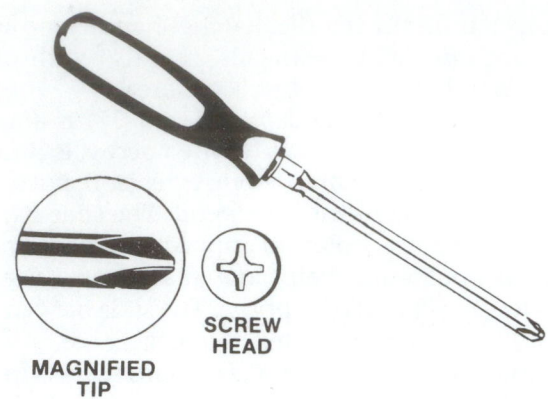

FIGURE 3-34 Purchase a good assortment of Phillips screwdrivers. *(Courtesy of Mac Tools)*

Phillips Screwdrivers

The tip of a **Phillips screwdriver** has four prongs that fit the four slots in the Phillips screw head (Figure 3–34). This type of fastener is often used in the automotive field. Not only does it look nicer than the slot head screw, but it also is easier to install and remove. The four surfaces enclose the screwdriver tip so that there is less likelihood that the screwdriver will slip off the fastener. Phillips screws, unlike the standard tip, can also be installed by automated machinery. This is the primary reason they are used on vehicles today.

A set of three Phillips screwdrivers with a number 1, number 2, and number 3 tip will handle most body shop requirements. Purchase a set with large, insulated handles. They are easier and safer to use.

Phillips screwdrivers have one disadvantage. The prongs of the tip tend to wear and round off. Unlike a standard tip screwdriver, a worn Phillips cannot be sharpened. Therefore, it should be discarded and replaced.

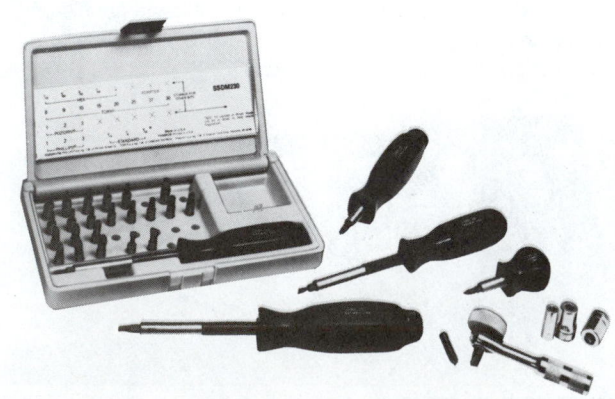

FIGURE 3-35 A magnet screwdriver and bit set is handy for holding small screws while working. *(Courtesy of Snap-On Tools Corp.)*

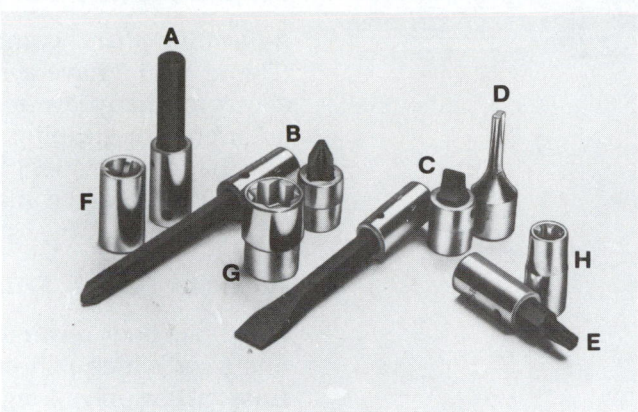

FIGURE 3-36 Typical screwdriver attachment set including (A) hex driver, (B) Phillips driver, (C) flat tip driver, (D) clutch head driver, (E) Torx® driver, (F) three-wing socket, (G) double square socket, and (H) Torx® socket. *(Courtesy of Snap-on-Tools Corporation)*

Specialty Screwdrivers

A number of specialty fasteners have been replacing slot and Phillips head screws. These new breeds of fasteners are designed to improve transfer of torque from screwdriver to fastener, slip less, result in less work fatigue, and offer some tamper resistance. Figure 3–35 shows a set of specialty screwdriver tips and assorted drivers. Most of these screwdriver bits (Figure 3–36) will prove useful in auto body repair. Three of the most often used screwdrivers are described below.

Clutch Head Screwdriver. There are two kinds of clutch head screws, the older G-style and the newer A-style. The G-style clutch type screwdriver tip (Figure 3–37) has an hourglass profile. It fits into a similarly shaped slot in a clutch head screw. Used most frequently by General Motors, this type of fastener system provides a more positive engagement with less slippage.

Pozidriv Screwdriver. This screwdriver is also like a Phillips but with a tip that is flatter and blunter (Figure 3–38). The square tip grips the screw head and slips less than a Phillips screwdriver. Less slippage results in less aggravation and fatigue and lengthens the life of the screwdriver as well.

Torx Screwdriver. The Torx fastener is becoming more and more common. It is used in a variety of industries, including the automotive industries. Many American-made automobiles use the star-shaped Torx fastener to secure headlight assemblies, mirrors, and luggage racks. Not only does the six-prong tip provide greater turning power and less slippage, but the Torx fastener also provides a measure of tamper resistance. The popularity of Torx fasteners makes having a complete set of Torx screwdrivers (Figure 3–39) a necessity for today's auto body repair technician.

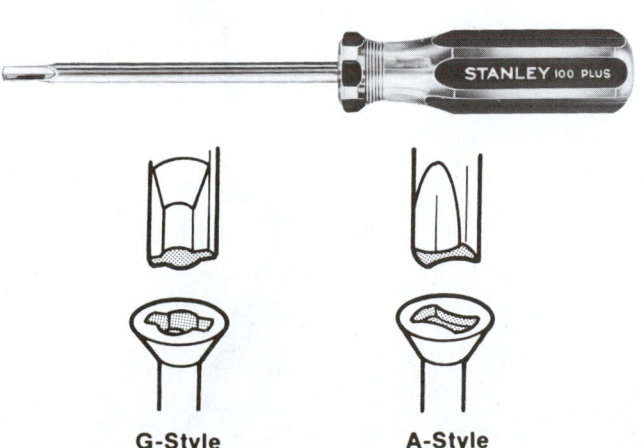

FIGURE 3-37 Note clutch screwdrivers. *(Courtesy of Stanley Works)*

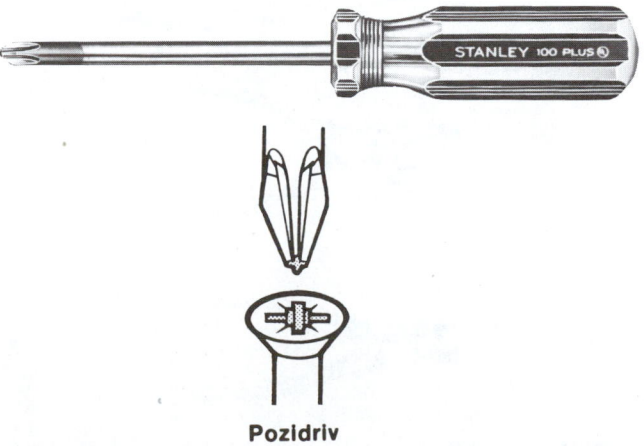

FIGURE 3-38 This is a Pozidriv screwdriver. *(Courtesy of Stanley Works)*

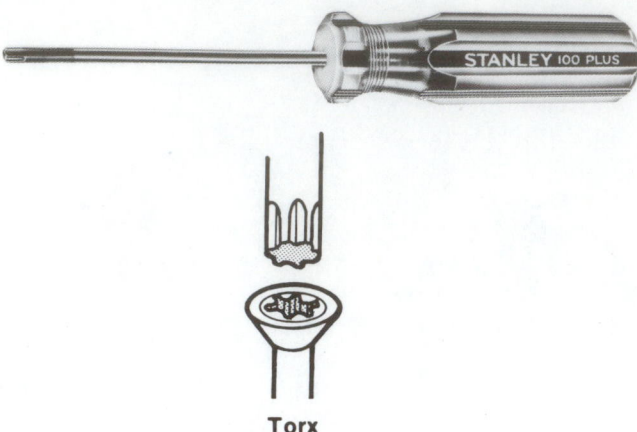

FIGURE 3-39 A Torx screwdriver has a strong head design. *(Courtesy of Stanley Works)*

PLIERS

Pliers are an all-around grabbing and holding tool for working with wires, clips, and pins. The auto body repair technician must own several types: standard pliers for common parts and wires, needle nose for the really small parts, and large adjustable pliers for heavy-duty work, including bending sheet metal.

Combination Pliers

Combination pliers (Figure 3–40) are the most common type of pliers. The jaws have both flat and curved surfaces for holding flat or round objects. Also called *slip-joint pliers*, combination pliers have two jaw-opening sides. One jaw can be moved up or down on a pin attached to the other jaw to change the opening.

Adjustable Pliers

Adjustable pliers, commonly called *channel-locks* (Figure 3–41), have a multiposition slip joint that allows for many jaw opening sizes. Adjustable pliers are useful for grasping objects of varying sizes. The long handle provides plenty of turning leverage. Adjustable pliers are available with flat or curved jaws.

Needle Nose Pliers

Every auto body repair technician should have at least one 6- or 8-inch pair of needle nose pliers. **Needle nose pliers** have long, tapered jaws (Figure 3–42). They are indispensable for grasping small parts or for reaching into tight spots. Many needle nose

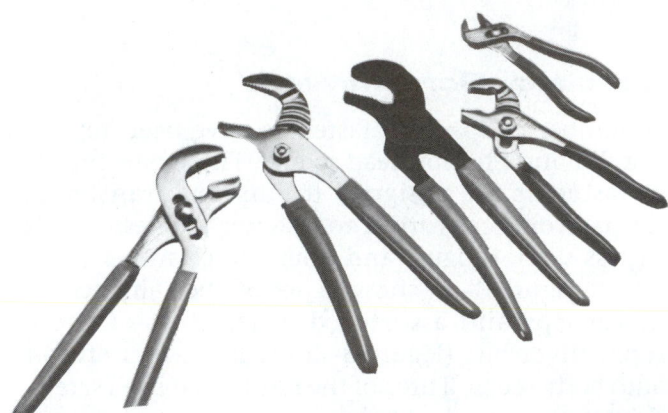

FIGURE 3-41 Adjustable pliers have strong holding power but should not be used in place of a wrench. They will damage bolt and nut heads. *(Courtesy of Snap-On Tools Corp.)*

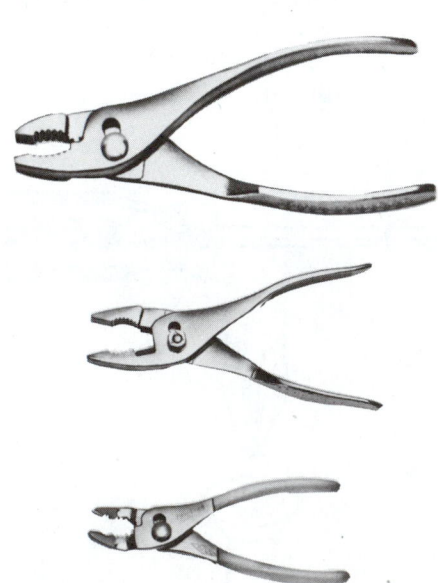

FIGURE 3-40 Slip joint combination pliers are commonly used to grasp and hold parts while working. *(Courtesy of Snap-On Tools Corp.)*

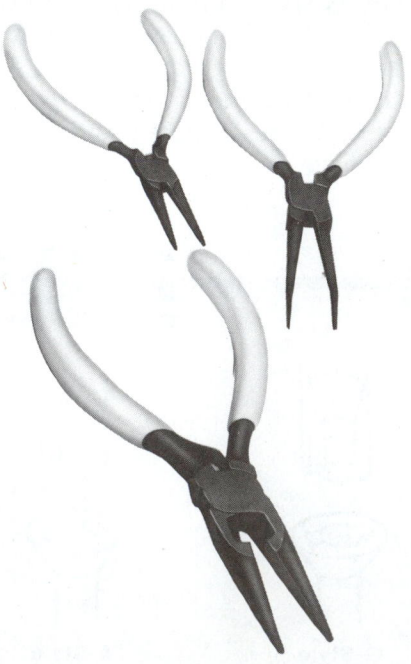

FIGURE 3-42 Needle nose pliers will reach into tight places and hold small parts. *(Courtesy of Snap-On Tools Corp.)*

pliers also have wire cutting edges and a wire stripper. These are very handy for electrical work (headlights, for example). A needle nose with a 90-degree bend in the jaws is also handy for reaching behind or around obstacles.

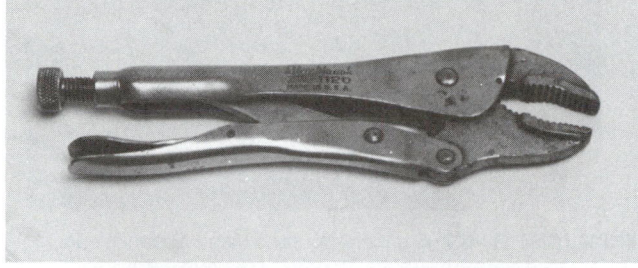

A

B

FIGURE 3-43 (A) Vise grips will lock onto a part, freeing your hands for other tasks. (B) These vise grips have large jaws for working sheet metal. *(Courtesy of Dobbs Publishing)*

Locking Pliers

Locking pliers or **vise grips** are similar to the standard kind except that they lock closed with a very tight grip (Figure 3–43). They are extremely useful for holding parts together. For example, several pairs of locking pliers will come in handy when holding a replacement panel in position for spot welding. They are also useful for getting a firm grip on a badly rounded fastener on which wrenches and sockets are no longer effective. Locking pliers come in several sizes and jaw configurations for use in many auto body jobs.

Of the ones illustrated in Figure 3–44, the C-clamp, welding, and duckbill types are among those frequently used. The C-clamp type is handy for reaching over and clamping pieces with flanges or beads. The welding vise grip has a special shaped jaw for gripping and aligning the weld joints in brazing or welding operations. The duckbill pliers have wide jaws for holding and bending sheet metal.

MISCELLANEOUS HAND TOOLS

A variety of other, miscellaneous hand tools will be useful from time to time. Many are inexpensive; most have a variety of uses. A few are very expensive and will probably be provided by the shop.

Tape Measure

A retractable *tape measure* or *steel rule* (Figure 3–45) is an essential measuring tool for every tool box. Its uses

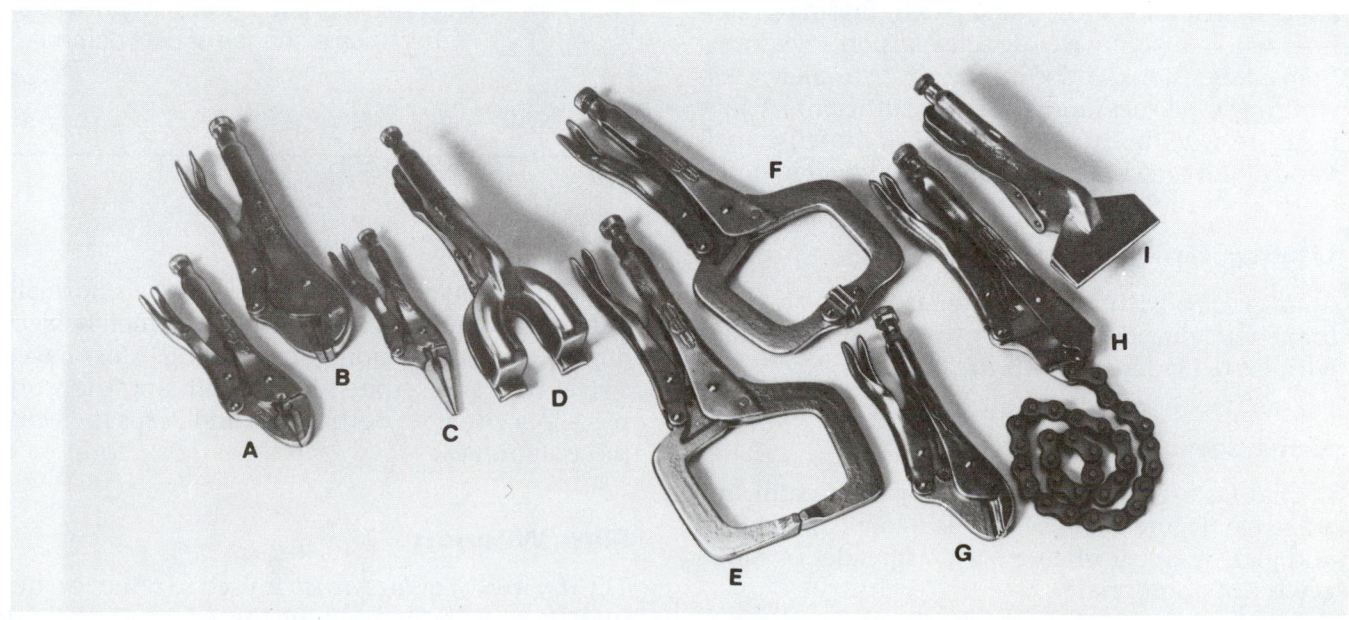

A. Wire Cutter	D. Welding Clamp	G. Hose Pinch Off
B. Standard	E. C-Clamp	H. Chain Wrench
C. Long Nose	F. C-Clamp With Swivel Pads	I. Duckbill

FIGURE 3-44 Study assorted vise grip locking pliers. *(Courtesy of Snap-On Tools Corp.)*

FIGURE 3-45 A retractable tape measure is essential for making large rough measurements during body work. *(Courtesy of Mac Tools)*

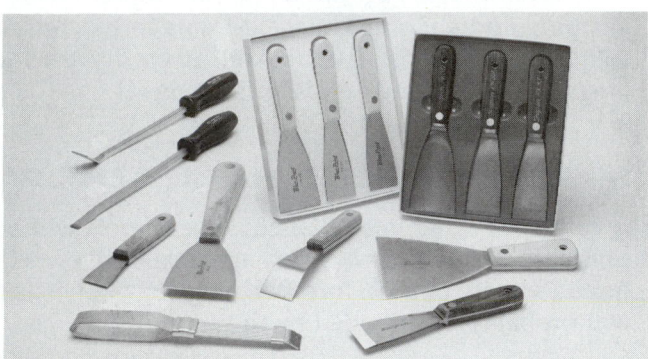

FIGURE 3-46 Note various size scrapers. *(Courtesy of Snap-On-Tools Company, Copyright Owner)*

in the auto body repair shop are many—measuring a repair area for a sheet metal patch, finding centerlines when applying woodgrain transfers, measuring frame dimensions, and so on. A 15-foot (5-meter) self-winding tape measure is usually sufficient. A tape measure with both metric and SAE dimensions is handy. Body measurements are often given in both.

Utility Knife

A *utility knife* with a retractable razor blade is useful for general purpose cutting or trimming. Most come with extra blades stored in the handle.

Scrapers

A *scraper* is used to remove old body filler, paint, and adhesive (Figure 3–46) as well as to apply body filler and glazing putty when a plastic spreader or squeegee is not appropriate.

Wire Brush

Wire brushes are often used for cleaning, such as cleaning the weld joints of welding flux. A wire brush

FIGURE 3-47 A bench-mounted vise will secure parts during disassembly, drilling, cutting, etc.

should be used sparingly on bare metal because it can leave scratches in the metal.

C-clamps

When a third hand is needed, a *C-clamp* is often the answer to hold parts in place while you are working. C-clamps come in various sizes and a variety will be very useful in a body shop.

WARNING

Clamping objects in a bench vise can damage them. Use soft lead or plastic jaw covers when there might be a chance of part damage.

Vise

A medium-duty *bench vise* (Figure 3–47) is normally a shop-provided tool used for holding metal objects during grinding, bending, or welding. Most have a swivel base and serrated jaws. Clamping the workpiece in a vise frees both hands and keeps the workpiece stationary.

Rim Wrench

The *rim wrench* or *lug wrench* is used to remove wheel lugs (nuts or bolts holding the wheel to the axle flange). A *four-way lug wrench* (Figure 3–48) has four hex-shaped wrenches, or sockets. Each lug socket is sized to fit one of the four popular sizes of wheel lug nuts available. Metric sizes are also available.

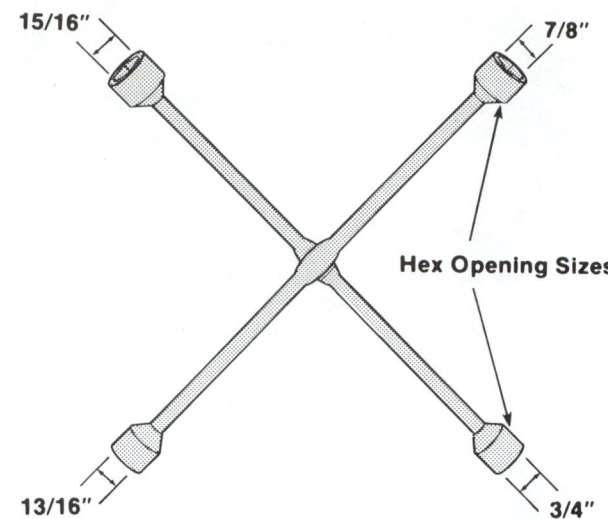

FIGURE 3-48 A four-way lug wrench will fit most lug nuts for wheel and tire service.

FIGURE 3-50 Heavy-duty tool chests on rollers will protect your tool investment and help keep the tools organized and ready for use. *(Courtesy of Snap-On-Tools Company, Copyright Owner)*

Tool Chest

A *tool box* stores and protects your tools. Most tool boxes consist of a large, bottom roll-around cabinet and an upper tool chest (Figure 3–50). The *upper chest* often holds commonly used tools at eye level. A chest of drawers organizes wrenches, screwdrivers, hammers, and so forth. A portable tray on top of the chest can often be lifted out so frequently used tools can be taken to the job.

The *lower cabinet* often stores heavier tools. The cabinet is on rollers so that it can be conveniently located in the work area. The box and chest of drawers can be locked to safeguard the tools.

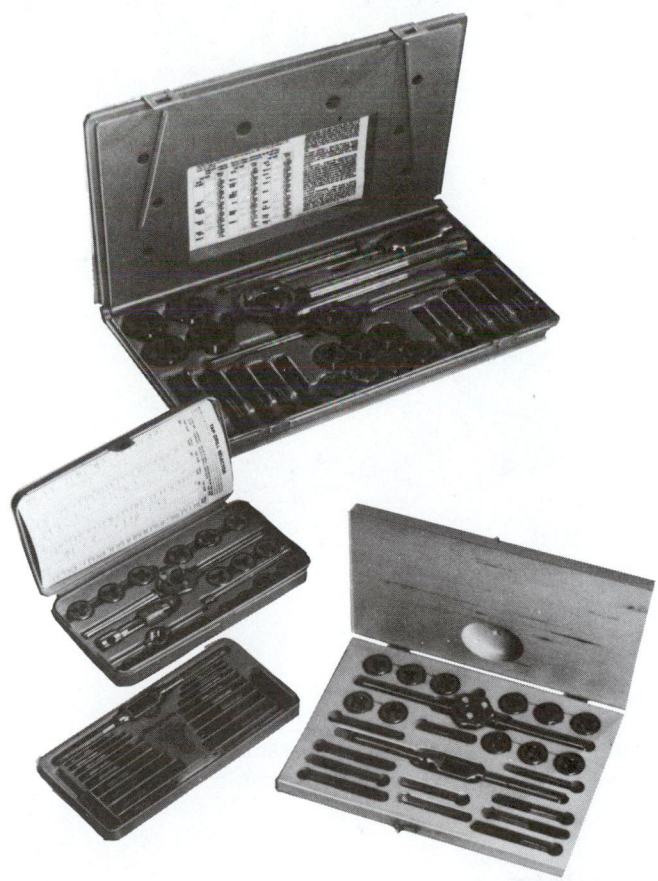

FIGURE 3-49 Tap and die sets are needed for thread repairs.

NEVER open more than one tool box drawer at the same time because the box can flip over. Death or injury could result! Close each drawer before opening the next one.

Tap and Die Set

A *tap* is a tool that cleans and rethreads holes. A *die* straightens damaged threads on bolts or studs. A tap and die set, such as the one shown in Figure 3–49, will perform most rethreading tasks in the body shop.

3.2 BODYWORKING TOOLS

Bodyworking tools include some familiar, general purpose metalworking tools as well as specialized tools used only in auto body repair. The following is a description of the most commonly used bodyworking tools. A typical set of bodyworking tools is shown in Figure 3–51.

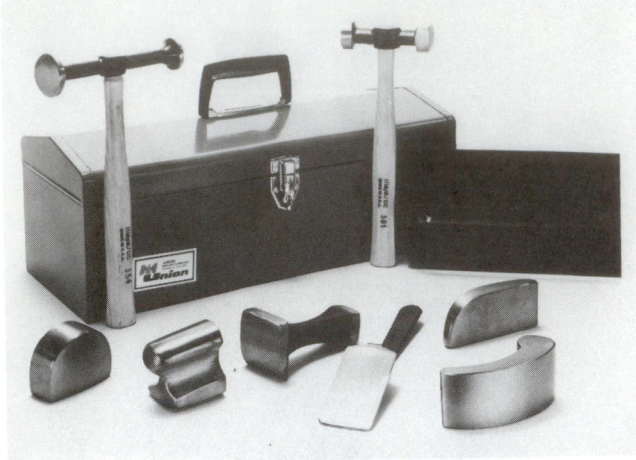

A

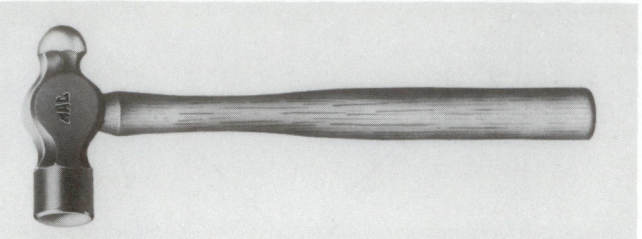

FIGURE 3-52 A ball peen hammer has a flat head and a rounded head. The flat head is for hammering on flat surfaces and the round head for concave surfaces. *(Courtesy of Mac Tools)*

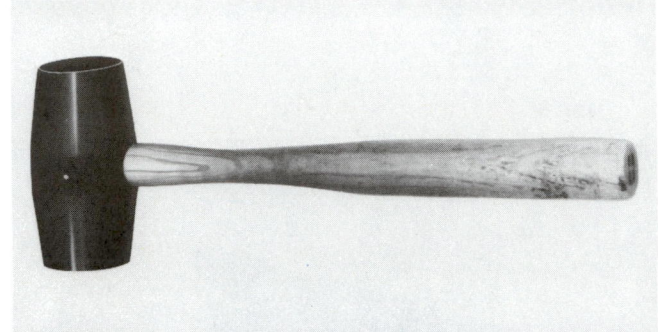

A

B

FIGURE 3-51 (A) Basic bodyworking set; (B) master bodyworker's bump set. *(Courtesy of Majestic Tools Mfg. Co.)*

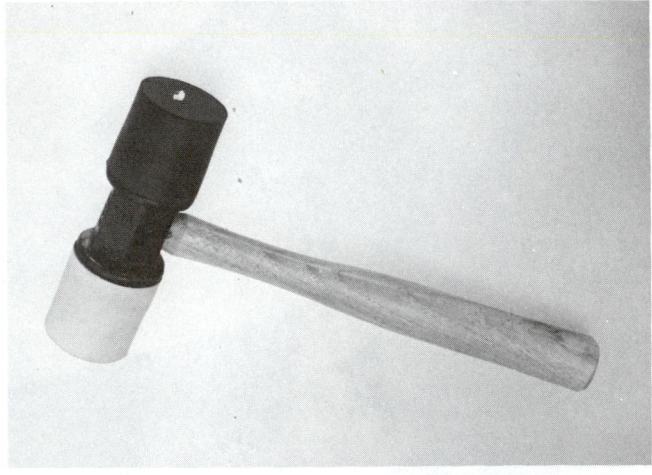

B

FIGURE 3-53 (A) Rubber mallet and (B) soft-faced hammer. *(Photo B courtesy of Majestic Tools Mfg. Co.)*

HAMMERS

A number of different hammers are useful in the body shop. Many are specially formed for a specific metal shaping operation.

Ball Peen Hammers

The **ball peen hammer** (Figure 3–52) is a useful, multipurpose tool for all kinds of work with sheet metal. Heavier than the body hammer, it is used for straightening bent underpinnings, smoothing heavy gauge parts, and roughly shaping body parts. It is sometimes used before work with a body hammer and dolly begins. Several ball peen hammers of different weights will see a lot of action in a body shop.

Mallets

The *rubber mallet* (Figure 3–53) gently bumps sheet metal without damaging the painted finish. It is often used with the suction cup on soft "cave-in" type dents. While you pull upward on the cup, the mallet is used to tap lightly all around the surrounding high spots. A popping sound occurs as the high spots drop and the low spot springs back to its original contour.

A steel hammer with rubber tips is another mallet useful in bodywork. The hammer shown in

Figure 3–53B has both hard and soft replaceable rubber heads. The *soft-faced hammer*, as it is sometimes called, is used to work chrome trim and other delicate parts without marring the finish.

A **dead blow hammer** has a metal face filled with lead shot (balls) to prevent rebounding. It will not bounce back up after striking. It is good for driving operations where part damage must be avoided.

Sledgehammer

A light **sledgehammer** (Figure 3–54) is an essential tool for the first stages of re-forming damaged thicker metal parts. Those with short handles can be used in tight places. The sledgehammer can be used to clear away damaged metal when replacing a panel.

BODY HAMMERS

Body hammers are the basic tools for working sheet metal back into shape. They come in many different designs. As shown in Figure 3–55, they have flat, square, rounded, or pointed heads. Each style is designed for a special purpose.

Picking Hammers

The **picking hammer** has a pointed tip on one end and usually a flat head on the other. It will take care of many small dents. The pointed end is used to raise low spots from the inside. A gentle tap in the center usually does it (Figure 3–56). The flat end is

for hammer-and-dolly work to remove high spots and ripples. Picking hammers come in a variety of shapes and sizes (Figure 3–57). Some have long picks for reaching behind body panels. Some have sharp pencil points; others have blunted bullet points. Select the head best suited for the job.

Bumping Hammers

Larger dents require the use of a **bumping hammer**. Bumping hammers can have round faces or square faces which are almost flat. The faces are large so that the force of the blows is spread over a large area. These hammers are used for initial straighten-

Wide Nose Peen

Picking and Dinging

Long Spot Pick

Spot Pick

Wide Nose Cross Peen

Cross Peen Shrinking

Shrinking

Long Picking

Shrinking Hammer

Wide Face Bumping

Reverse Curve, Light Bumping

Short-Curved Cross Peen

FIGURE 3–55 Study names for body hammers. *(Courtesy of Snap-On Tools Corp.)*

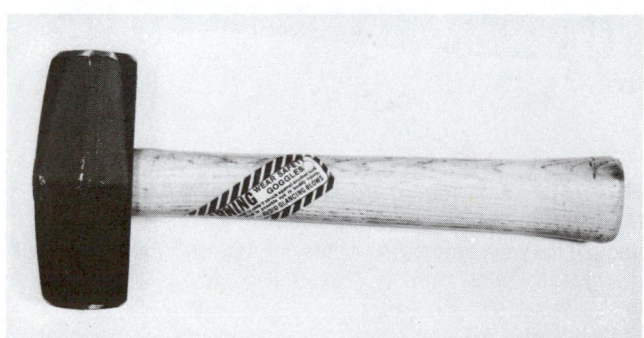

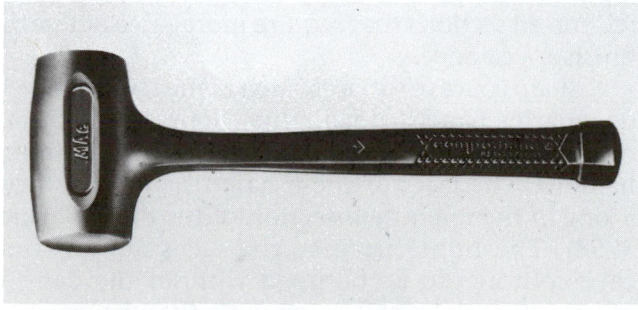

FIGURE 3–54 Light sledgehammers are for heavier driving tasks. *(Courtesy of Mac Tools)*

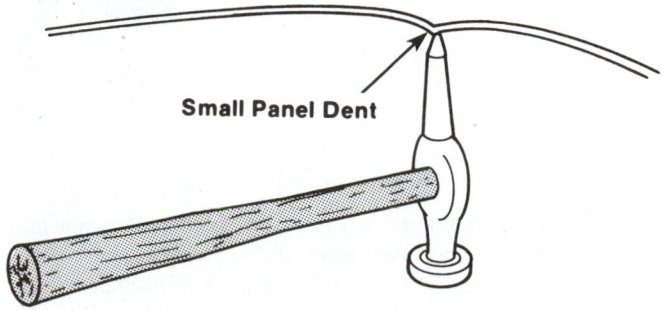

Small Panel Dent

FIGURE 3–56 Pick hammer's pointed end can be used to raise a dent in sheet metal.

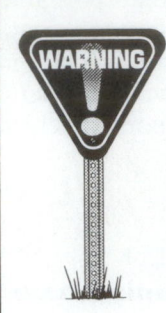

WARNING

Be careful when using the pick hammer. If swung forcefully, the pointed end can pierce the lighter sheet metals used in late model cars. Use the pick only on small dents and control impact force.

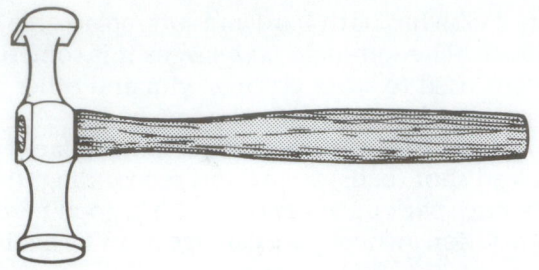

FIGURE 3-58 Bumping hammers are heavier for working larger dents.

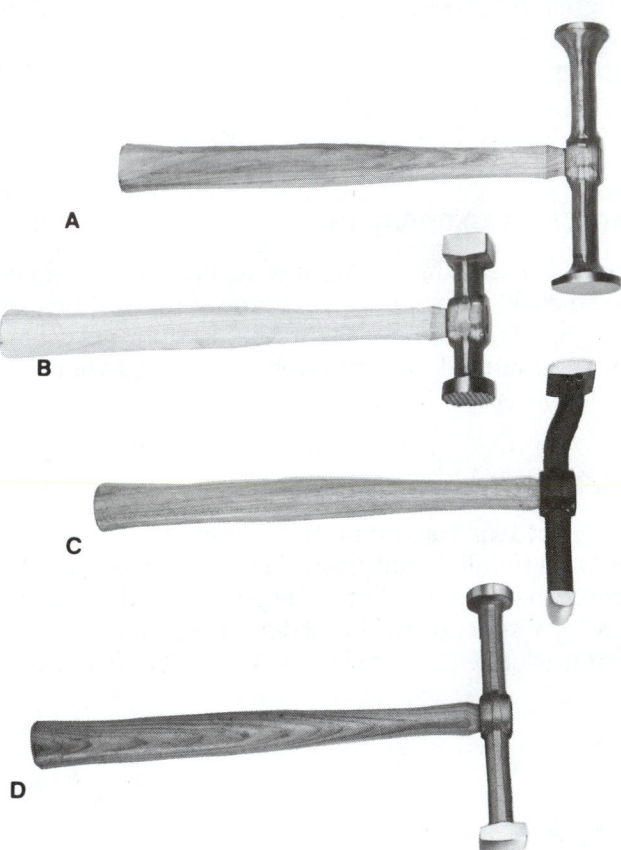

FIGURE 3-59 Here are several finishing hammers: (A) double round; (B) shrinking hammer; (C) offset bumping; and (D) dinging hammer. *(Courtesy of Majestic Tools Mfg. Co.)*

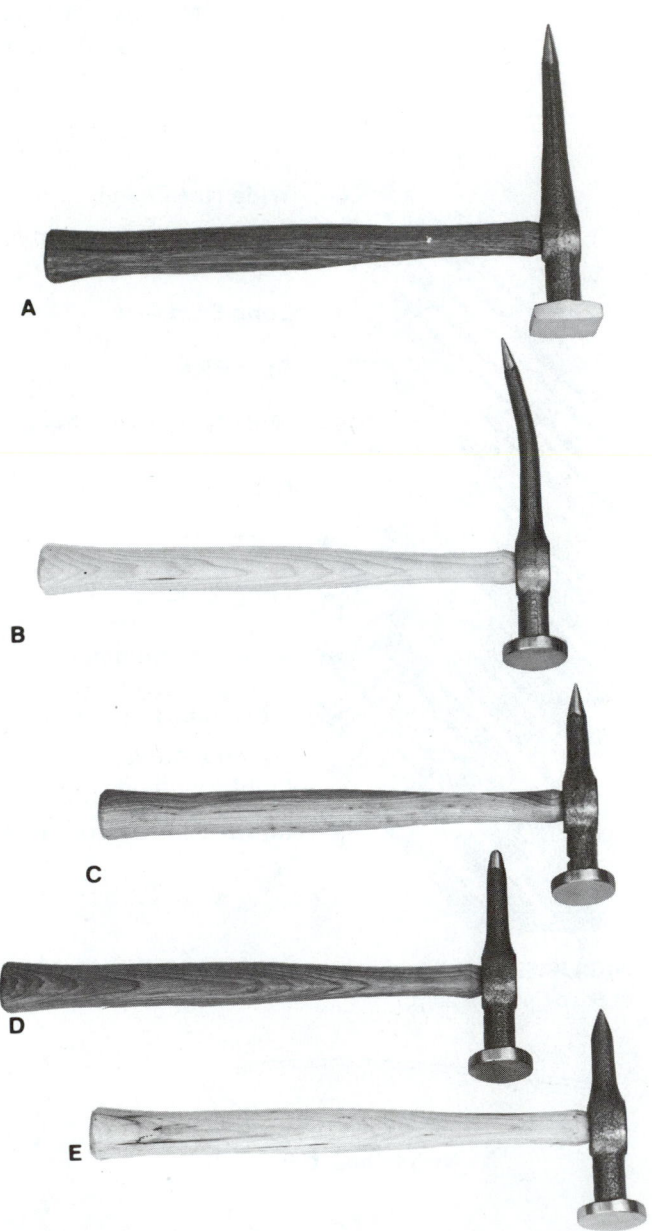

FIGURE 3-57 Note pick hammer names. (A) long pencil point; (B) long-curved pencil point; (C) short pencil point; (D) short bullet point; and (E) short chisel point. *(Courtesy of Majestic Tools Mfg. Co.)*

ing on dented panels or for working inner panels and reinforced sections that require more force but not a finish appearance.

Sharp concave surfaces, such as the reverse curves on quarter panels, headlights, doors, and so on, require the use of a reverse curve light bumping hammer. The faces of these hammers are crowned —one in the opposite direction of the other (Figure 3–58). The tight curve of the faces allows concave contours to be bumped without the danger of stretching the metal. Remember that the contour of the hammer must be smaller than the contour of the panel to avoid stretching the metal.

Finishing Hammers

After the bumping hammer is used to remove the dent, final contour is achieved with the **finishing hammer** (Figure 3–59). The faces on a finishing hammer are smaller than those of the heavier bumping hammer. The surface of the face is crowned to concentrate the force on top of the ridge or high spot.

A *shrinking hammer* (Figure 3–60) is a finishing hammer with a serrated or cross-grooved face. This hammer is used to shrink spots that have been stretched by excessive hammering.

DOLLIES

The **dolly** or *dolly block* is used like an small anvil while body damage is worked out. It is generally held on the backside of a panel being struck with a hammer. Together the hammer and dolly work high spots down and low spots up (Figure 3–61).

There are many different shapes of dollies (Figure 3–62). Each shape is intended for specific types of dents and body panel contours—high crowns, low crowns, flanges, and others. It is very important that the dolly fits the contour of the panel. If a flat dolly or one with a low crown is used on a high crown panel, additional dents will be the result.

A *general purpose* dolly has many contours. It can be used in most situations. A rail-type dolly is another commonly used dolly with many contours. Toe and heel dollies are used for bumping in tight places. The flat right angle edge is used for straightening flanges.

SPOONS

Body spoons (Figure 3–63) are another class of bodyworking tools used like a hammer or a dolly.

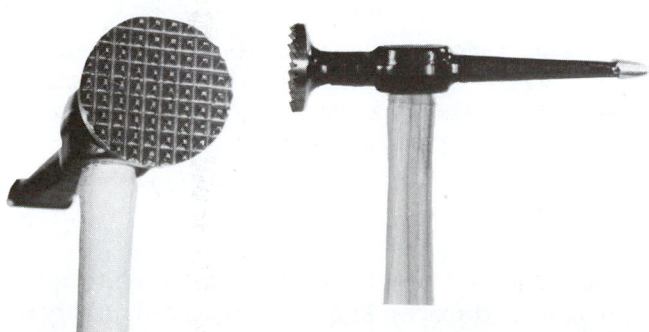

FIGURE 3-60 Note the serrated face of a shrinking hammer. *(Courtesy of Majestic Tools Mfg. Co.)*

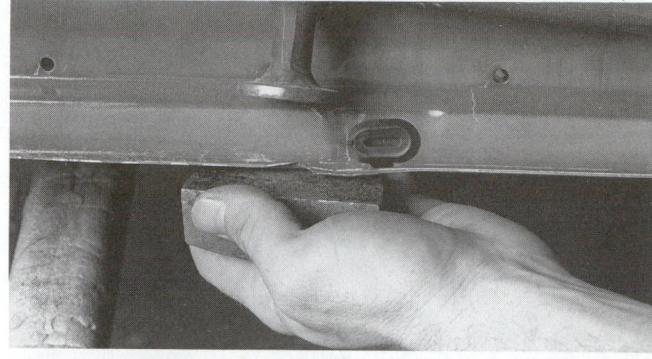

A

B

FIGURE 3-61 (A) Bumping a dent with a hammer and dolly is common. (B) Select the dolly and hammer shapes carefully to match the shape of body contour.

Universal Dolly

Fender Dolly

Toe Dolly

Heel Dolly

Comma Dolly

FIGURE 3-62 Study various dolly blocks.

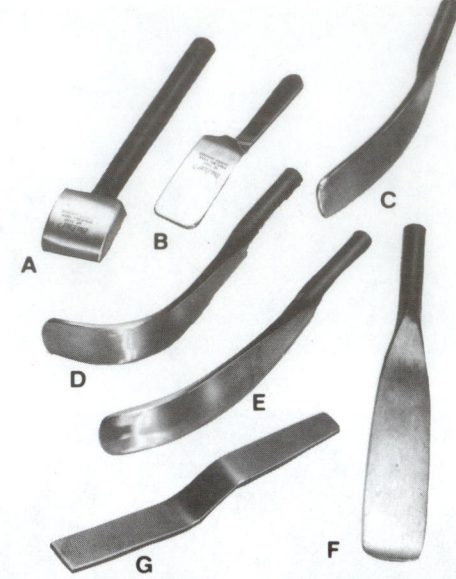

A. **Spoon Dolly**
B. **Light Dinging Spoon**
C. **Surfacing Spoon**
D. **Inside High Crown**
E. **Inside Medium Crown**
F. **Inside Heavy-Duty Spoon**
G. **Bumping File**

FIGURE 3-63 Note body spoon names. *(Courtesy of Snap-On Tools Corp.)*

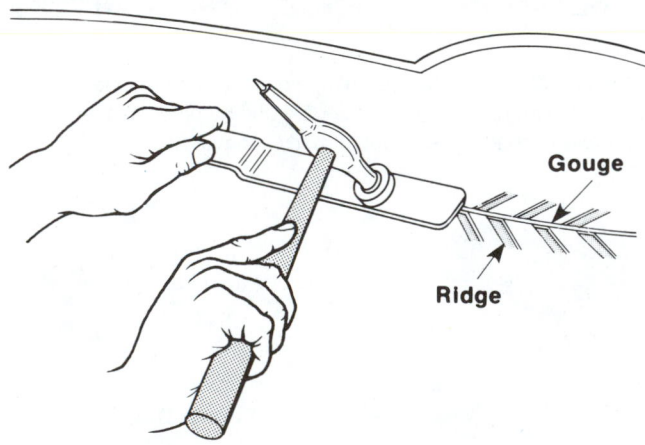

FIGURE 3-64 A ridge can be lowered by placing the spoon over a high spot and hitting it with a hammer. The spoon increases the surface area so that hammering will not produce dents.

They are available in a variety of shapes and sizes to match various panel shapes. The flat surfaces of a spoon distribute the striking force over a wide area (Figure 3–64). They are particularly useful on creases and ridges. A spoon dolly can be used as a dolly where the space behind a panel is limited. A dinging spoon is used with a hammer to work down ridges. Inside spoons can be used to pry up low places or can be struck with a hammer to drive up dents. Bumping files have serrated surfaces and are used to slap ridges or the underside of creases to bump the metal back to its original shape.

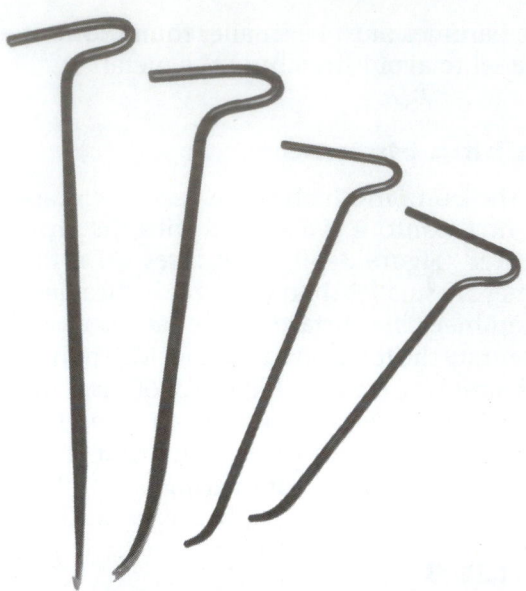

FIGURE 3-65 Several different body picks are shown. *(Courtesy of Snap-On Tools Corp.)*

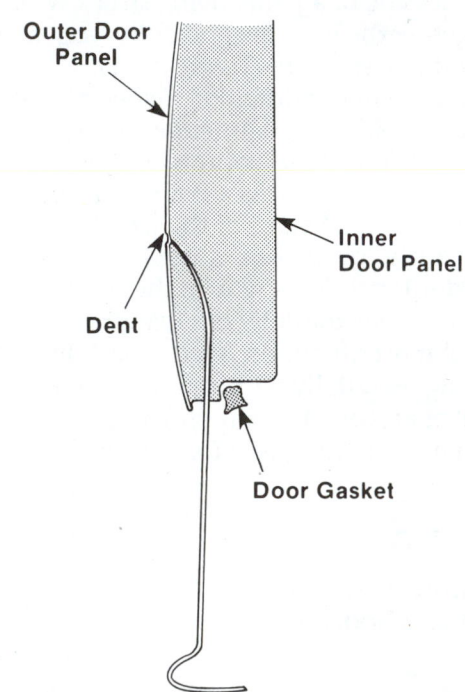

FIGURE 3-66 A pick can be inserted into obstructed areas, like inside a door, to help pry out small dents. This is the basis for paintless dent removal. If you carefully position and work the pick, you may not have to paint the panel.

PICKS

Picks (Figure 3–65), like spoons, are used to reach into confined spaces. The pick is used only to pry up low spots. They vary in length and shape, and most have a U-shaped end that serves as a handle. Picks are commonly used to raise low spots in doors, quarter panels, and other sealed body sections (Figure 3–66). Picks are often preferred to slide hammers and pull rods because they do not require drilling holes into the metal and subsequently welding them shut after the repair.

Picks are sometimes used during *paintless dent removal* (removing small body dings or dents without painting the panel).

NOTE: Straightening tools and techniques are discussed further in other text chapters. Refer to the index for additional information.

DENT PULLERS AND PULL RODS

Creases in sealed body panels that cannot be reached from the backside can be pulled out with a **dent puller** or *pull rod*. Technicians used to drill or punch holes in the crease for the puller or pull rod. Now, common practice is to weld a bracket or pull pin onto the surface instead of drilling. Either will give the rod something to grab and pull on.

A dent puller usually comes with a threaded tip and a hook tip (Figure 3–67). Either tip is inserted in the drilled hole or welded rod or bracket. Then the slide hammer is pulled back and struck against the handle. Tapping the slide hammer against the handle slowly pulls up the low spot (Figure 3–68).

A small dent or crease can be pulled up with a single pull rod (Figure 3–69). Three or four pull rods can be used simultaneously to pull up larger dents (Figure 3–70).

A body hammer can also be used with a pull rod. The high crown of a dent can be bumped down, while the low spot is pulled up (Figure 3–71). Simultaneous bumping and pulling return the panel to its original shape with less danger of stretching the metal.

It is important to weld closed the holes created by using dent pullers and pull rods. Simply patching the holes with body filler will not provide sufficient corrosion protection. Rust will result in a short time! You must restore all corrosion protection.

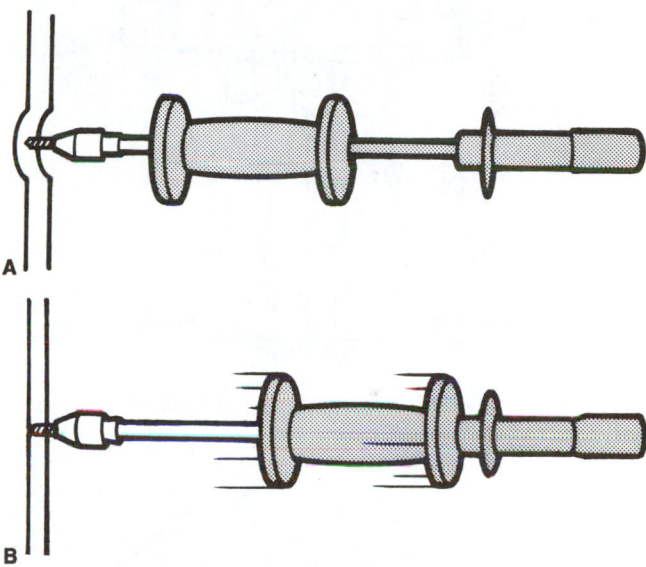

FIGURE 3-68 To pull a dent with a dent puller, (A) resistance weld a bracket or pull pin to the surface of the repair area and attach a dent puller with the appropriate tip. (B) Slide the hammer back to pull out the dent.

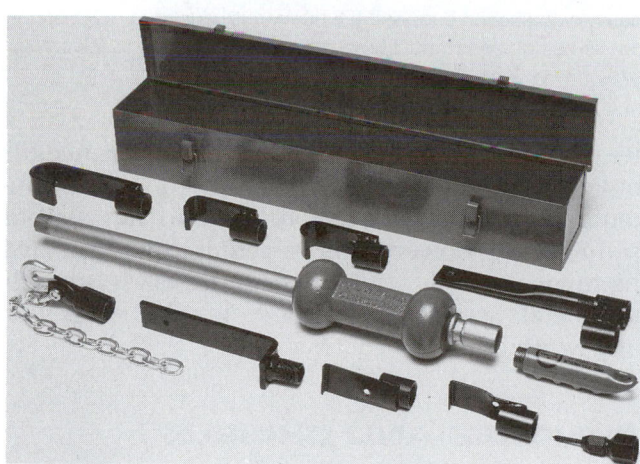

A

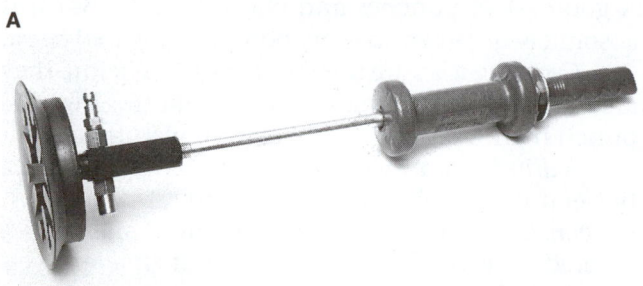

B

FIGURE 3-67 Dent puller. (A) Dent puller set has various attachments. (B) Puller equipped with suction cup head. It is attached to a compressed air source to form a powerful vacuum for popping out more stubborn dents. *(Courtesy of Morgan Manufacturing)*

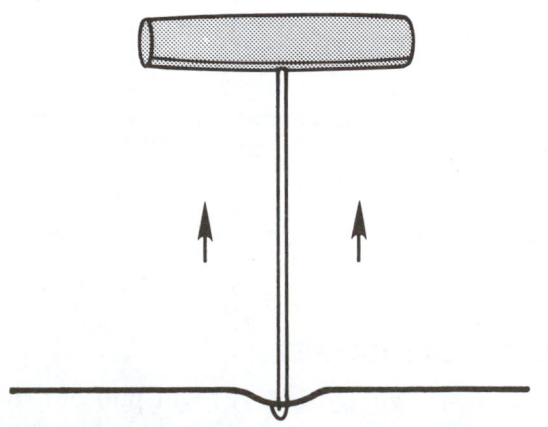

FIGURE 3-69 A small dent can sometimes be worked out with a pull rod. Drill a hole in the dent. Install the tool and pull. Weld the hole closed when done pulling.

SHOP TALK

Screw-in dent pullers and pull rods are not used much today. The need to weld the holes can create panel warpage and restoring corrosion protection can cause more repair time. Nail guns that weld pins onto the panel are now being used to avoid this problem.

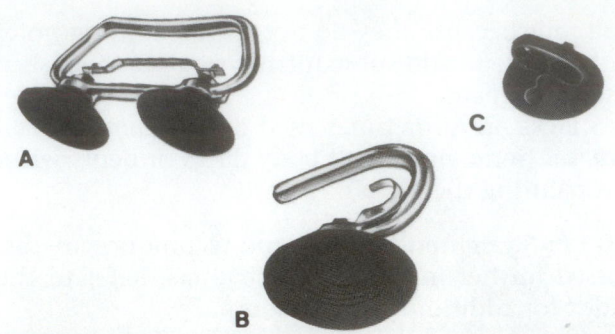

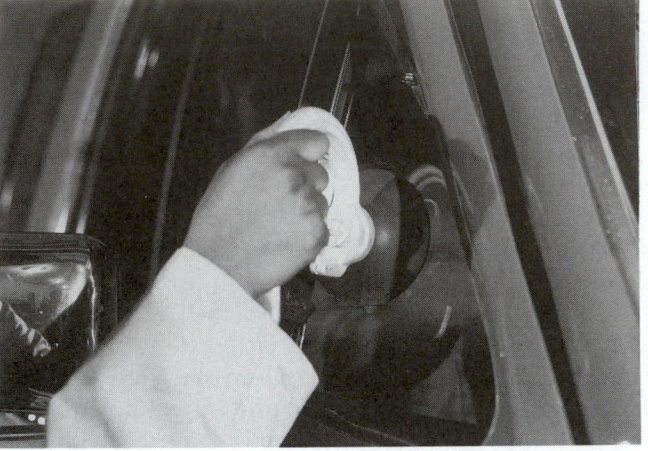

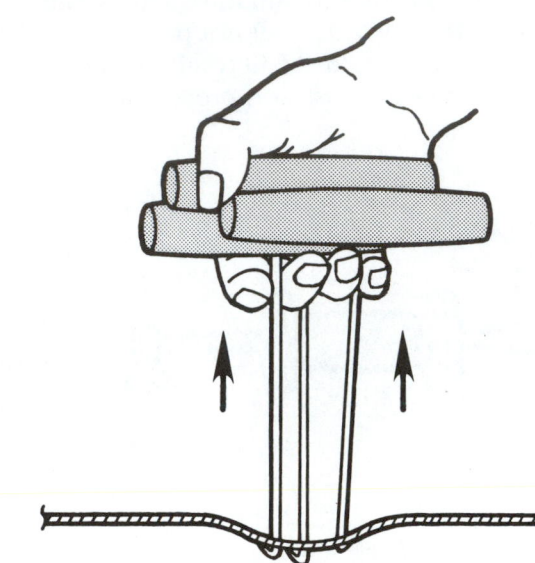

FIGURE 3-70 Several pull rods can be used to pull up larger dents.

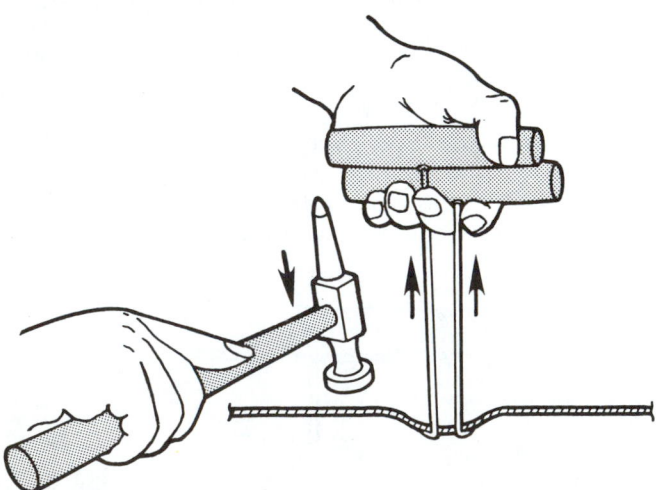

FIGURE 3-71 Bumping and pulling lowers crowns while raising dings.

SUCTION CUPS

The *suction cup* (Figure 3–72) will pull out shallow dents quickly if they are not locked in by a crease in the metal. Simply attach the suction cup to the cen-

FIGURE 3-72 Note suction cups: (A) two-finger cup; (B) heavy-duty suction cup, and (C) heavy-duty dual cup. (D) Suction cup is being used to hold wing glass during removal. *(Courtesy of Snap-On Tools Corp.)*

ter of the dent and pull. The dent might come right out with no damage to the paint and no refinishing required. It is an easy tool to use and can make a simple repair. However, once a dent is locked in, some hammer and dolly work will be necessary to smooth the metal. Even so, the suction cup method is usually worth a try.

PUNCHES AND CHISELS

A good set of punches and chisels (Figure 3–73) is absolutely necessary in every bodyworking tool chest.

Center punches are used to mark parts before they are removed and for marking a spot for drilling. The punch mark keeps the drill bit from wandering.

A *drift* or *starter punch* has a tapered point with a flat end used to drive out rivets, pins, and bolts. A *pin punch* is similar to the drift except its shaft is not tapered, so it can be used to drive out smaller rivets or bolts.

An *aligning punch* is a long tapered punch used to align body panels for welding and for starting bolts. For example, one might be used to align fender bolt holes and a bumper.

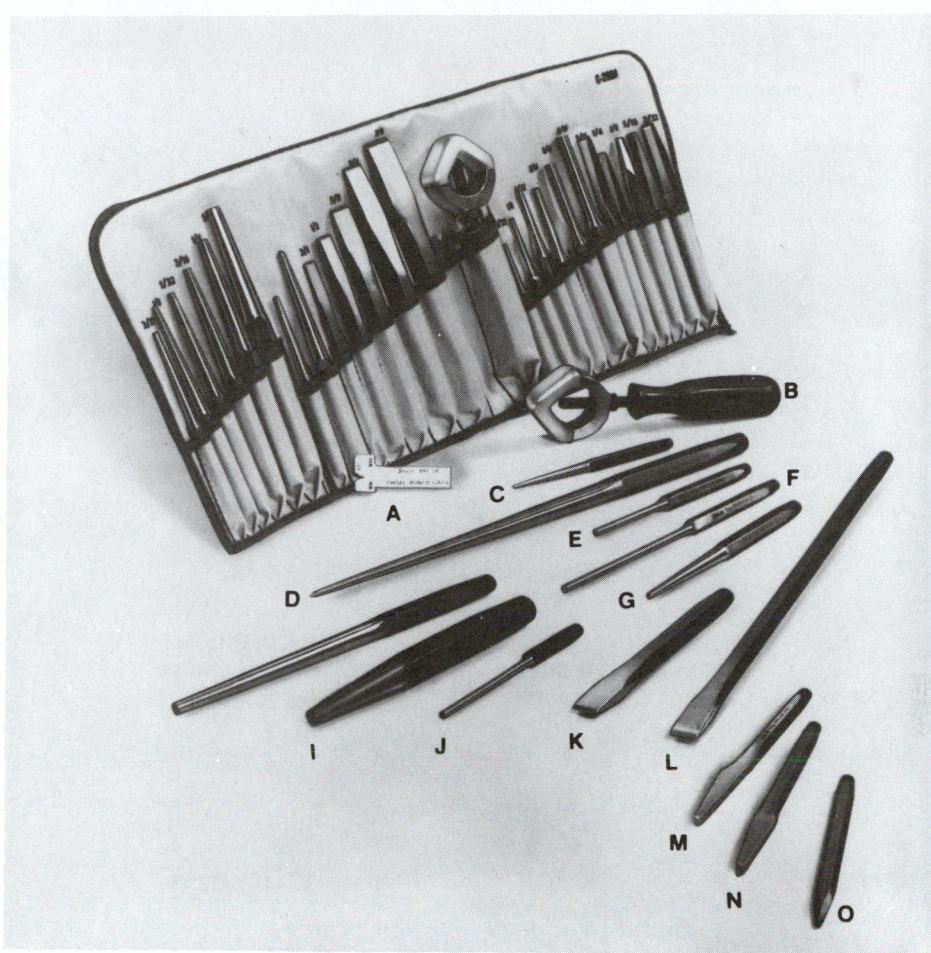

A. Chisel Gauge
B. Punch/Chisel Holder
C. Center Punch
D. Long Center Punch
E. Pin Punch

F. Long Pin Punch
G. Starter Punch
H. Long-Tapered Punch
I. Short-Tapered Punch
J. Roll Pin

K. Flat Chisel
L. Long Flat Chisel
M. Round Nose Cape Chisel
N. Cape Chisel
O. Diamond Point Chisel

FIGURE 3-73 Study names of punches and chisels. *(Courtesy of Snap-On Tools Corp.)*

A *chisel* is a steel bar with a hardened cutting edge for shearing steel. These chisels come in various sizes, and a set is necessary for both light- and heavy-duty work. The cold chisel is used to split frozen nuts, shear off rusted bolts, cut welds, and separate body and frame parts.

 **CAUTION** Keep the end of a chisel or punch ground properly. If the end is mushroomed or enlarged from hammering, grind it down. A mushroomed end could cut your hand and metal fragments could fly into your face.

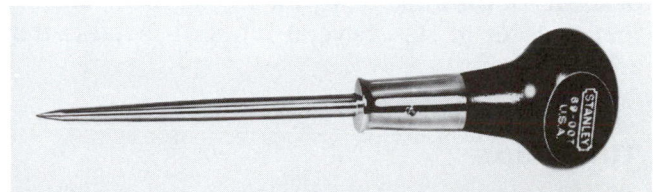

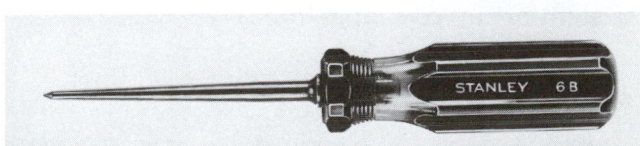

FIGURE 3-74 A scratch awl can be used to mark parts or penetrate plastic and thin sheet metal. *(Courtesy of Stanley Works)*

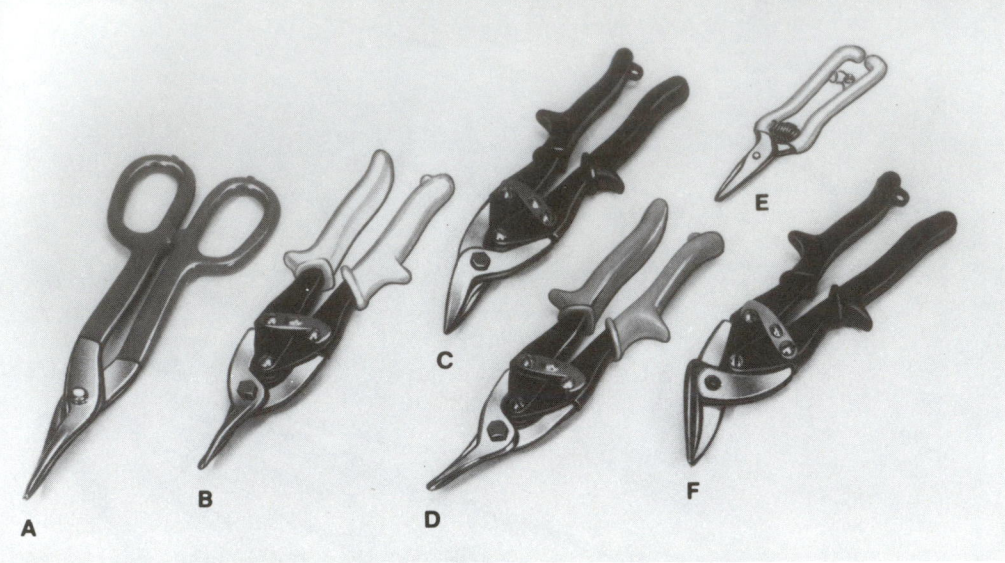

A. Tin Snips
B. Straight Cut Shears
C. Right Cut Shears

D. Left Cut Shears
E. Light-Duty Snips
F. Aviation Snips

FIGURE 3-75 Sheet metal cutting shears are identified. *(Courtesy of Snap-On Tools Corp.)*

SCRATCH AWL

A *scratch awl* (Figure 3–74) is very similar in appearance to an ice pick, but the pointed steel shank is heavier. A scratch awl is used to pierce holes in light gauge metal when a specific size hole is not required. It is also used to mark metal for cutting, drilling, or fastening. Keep the awl ground to a sharp point so it can be used effectively and safely in every job.

METAL CUTTING SHEARS

Most body repair technicians have at least one pair of shears or tin snips. *Snips* are used to trim panels or metal pieces to size. Several types of metal cutters are useful.

Tin Snips

Tin snips (Figure 3–75A) are perhaps the most common metal cutting tool. They can be used to cut straight or curved shapes in heavy steel.

Metal Cutters

Metal cutters (Figure 3–75B, C, and D), also called aviation snips (Figure 3–75F), are used to cut through hard metals, such as stainless steel. The narrow profile of the jaws allows the snips to slip between the cut metal. The jaws are serrated to cut through tough metal.

Panel Cutters

Panel cutters (Figure 3–76) are special snips used to cut through body sheet metal. These are used to make straight or curved cutouts in panels that require spot repair for rust or damage. They are designed to leave a clean, straight edge that can be welded easily.

RIVET GUN

Pop rivets are sometimes used to hold panels in place while repairs are made. They can be inserted into a blind hole through two pieces of metal and then drawn up with a riveting tool. This locks the pieces of metal together (Figure 3–77). There is no need to have access to the back of the rivets. They are used as temporary fasteners before the replacement sheet metal is welded. This prevents the extreme heat from distorting the metal or creating a safety hazard (such as around the gas tank).

A good rivet gun (Figure 3–78) does not cost much. The most commonly used rivets in bodywork

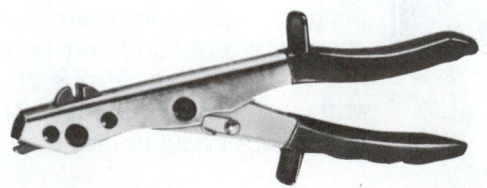

FIGURE 3-76 A panel cutter, or two-way nibbler, will cut thicker panels easily. *(Courtesy of S & G Tool Aid Corp.)*

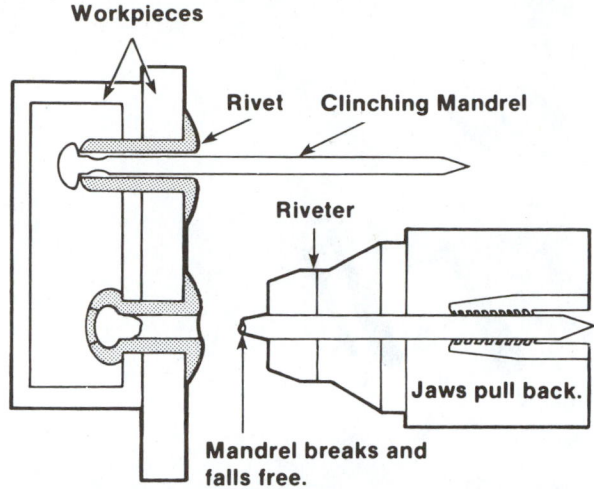

FIGURE 3–77 Note how rivets can hold two panels together.

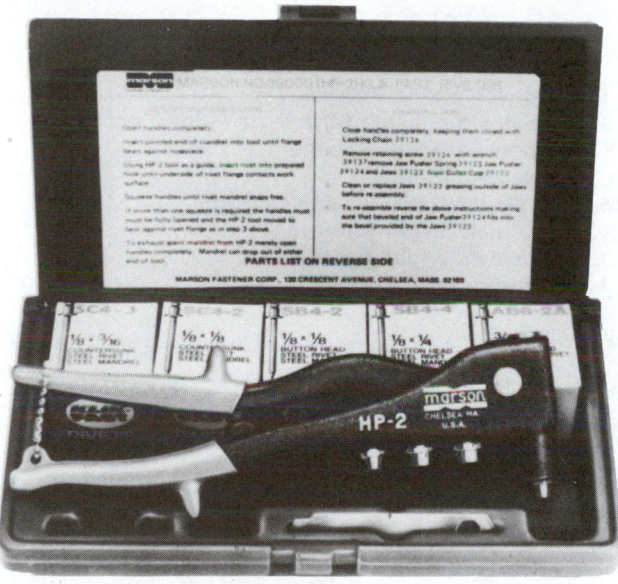

FIGURE 3–78 A hand riveter in kit form for blind rivets. *(Courtesy of Marson Corporation)*

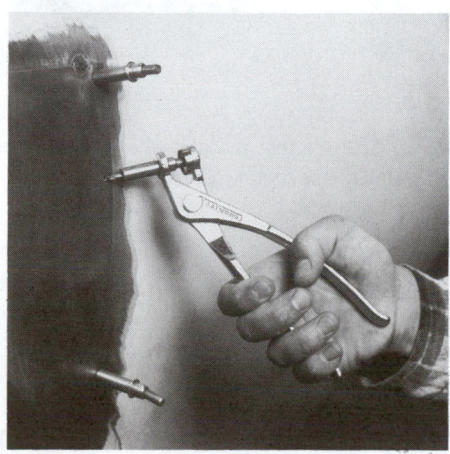

FIGURE 3–79 These spring-loaded rivets can be installed and removed easily. They are ideal for holding panels together while welding. *(Courtesy of Eastwood Co.)*

are $\frac{1}{8}$- and $\frac{3}{16}$-inch. A few others of assorted sizes might be needed for special jobs (Figure 3–79).

A heavy-duty riveter is used to rivet hard-to-reach places and heavier mechanical assemblies, such as a window glass regulator. It has long handles, a long nose, and sets $\frac{3}{16}$- to $\frac{1}{4}$-inch blind rivets.

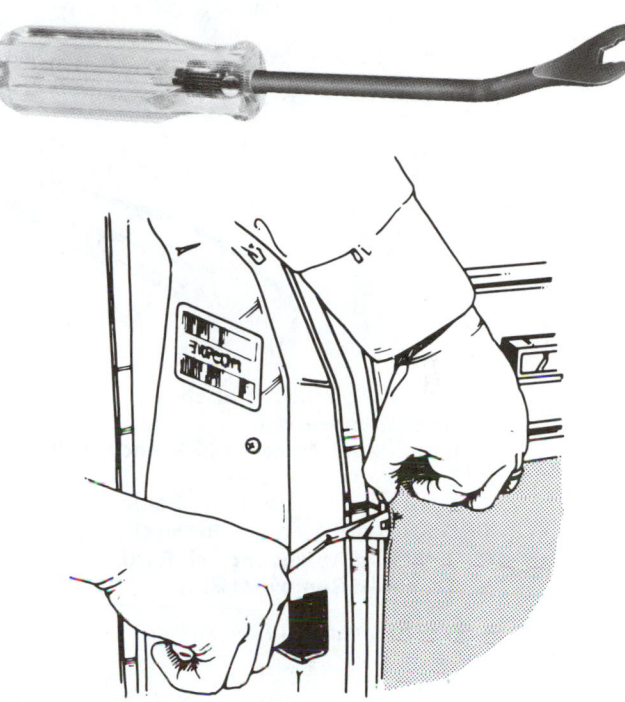

FIGURE 3–80 An upholstery tool will remove clips without damaging parts.

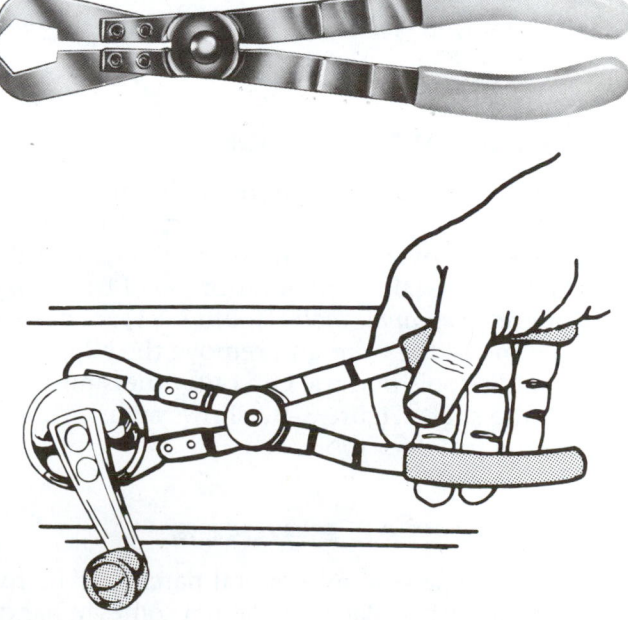

FIGURE 3–81 A door handle tool will reach behind the handle to remove spring clips. *(Courtesy of Lisle Corp.)*

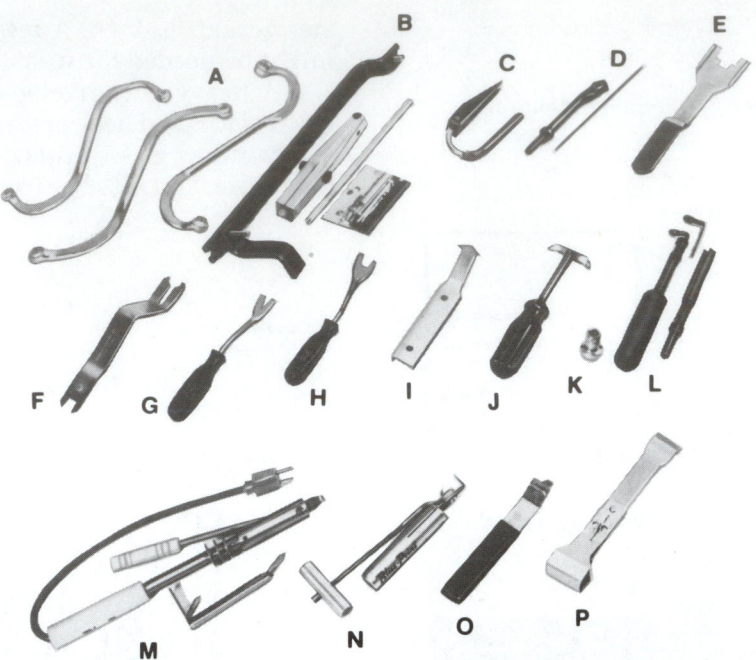

A. Door Hinge Bolt Wrenches
B. Door Removal Kit
C. Door Panel Remover (GM and Ford)
D. Door Panel Remover
E. Door Handle Tool (GM, Some Fords)
F. Door Handle Tool (Chrysler)
G. Trim Pad Remover (GM, Ford, Chrysler)
H. Trim Pad Remover (GM)

I. Window Molding Release Tool
J. Windshield Locking Strip Installation Tool
K. Window Sash Nut Spanner Socket
L. Windshield Remover
M. Hot-Tip Windshield Removing Kit
N. Windshield Wiper Removal Tool
O. Windshield Wiper Tool
P. All-Purpose Window Scraper

FIGURE 3–82 Study names of window and door tools. *(Courtesy of Snap-On Tools Corp.)*

UPHOLSTERY TOOLS

Any repair work that requires removing interior trim will be facilitated with an *upholstery tool* (Figure 3–80). This prong-shaped prying tool is used to slip under and pry up upholstery tacks, springs, clips, and other fasteners.

DOOR HANDLE TOOL

Interior door handles are often secured to the door panel by wire spring clips. These clips, shaped like horseshoes, fit over the handle shaft and hold the handle tightly against the interior panel trim. **Clip pullers** or *door handle tools* (Figure 3–81) are needed to reach inside the door and remove the clip. Some door handle tools pull the clip out; others push the clip off the shaft. Figure 3–82 shows an assortment of window and door tools.

SHEET METAL BRAKES

Many body repairs require metal patches to be riveted or welded into place. A tool that comes in handy for bending and breaking sheet metal is the *sheet metal brake*. One is shown in Figure 3–83. This sheet

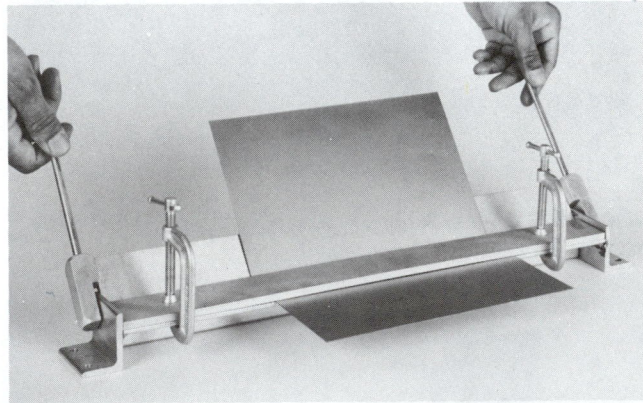

FIGURE 3–83 A sheet metal brake is used to bend, shape, and break sheet metal. *(Courtesy of Majestic Tools Mfg. Co.)*

metal brake bends sheet metal up to 20 gauge and sheet aluminum up to 16 gauge. Clean, smooth bends up to 90 degrees can be made with a brake.

3.3 BODY SURFACING TOOLS

A number of surfacing tools are used to give a repair its final shape and contour. Some are used to shape the repaired metal. Others are used to apply and shape plastic body filler and putty.

METAL FILES

After working a damaged panel back to its approximate original contour, a *metal file* is used to mark (scratch) the metal to find high and low spots. Two special files are necessary for most bodywork.

Reveal File

The *reveal file* (Figure 3–84) is a small file that is available in numerous shapes. Generally, it is curved to fit tightly crowned areas such as around windshields, wheel openings, and other panel edges. The reveal file is pulled, not pushed, when used. Pushing causes the file to chatter, resulting in nicks and an uneven surface.

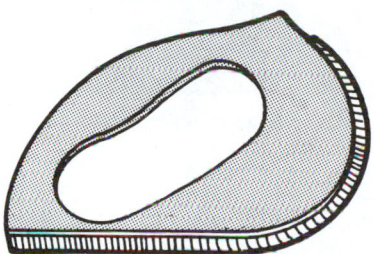

FIGURE 3-84 A reveal file is used to shape tight curves.

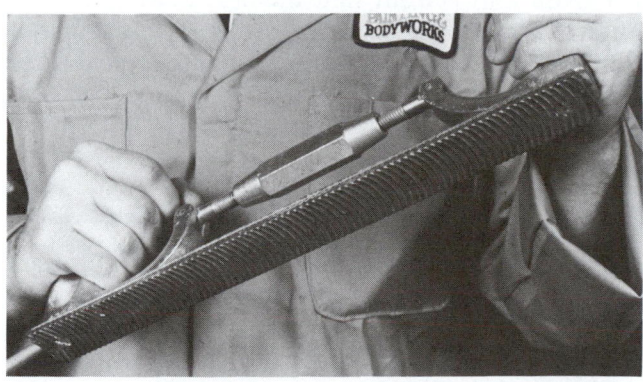

A

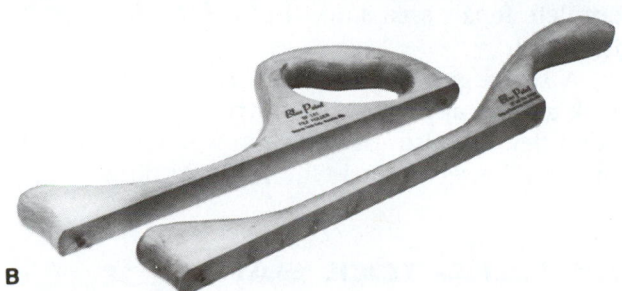

B

FIGURE 3-85 Shown are a body file (A) in a flexible holder and (B) rigid file holders. *(Photo B courtesy of Snap-On Tools Corp.)*

Body File

The body file is used for minor leveling of large surfaces. After a dent has been bumped or pulled back into shape, the body file will reveal any low or high spots that might require additional bumping. Keep in mind that it is possible to file through the thin metal used in some vehicles.

The blade of the body file is held in a holder. Figure 3–85A shows a flexible holder with a turnbuckle. The turnbuckle can be adjusted to flex the file. The flexible holder allows the shape of the file to fit the contour of the panel. Fixed file holders (Figure 3–85B) are also available for filing flat or slightly convex shapes.

Surform File

Body filler can be cut level and to rough contour with a **Surform file** (Figure 3–86). Commonly referred

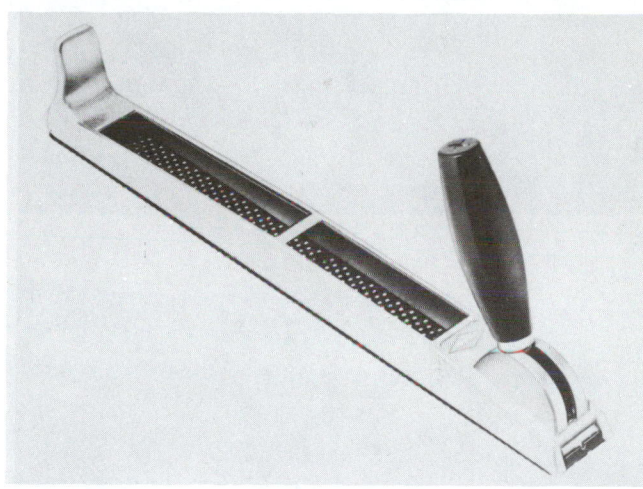

A

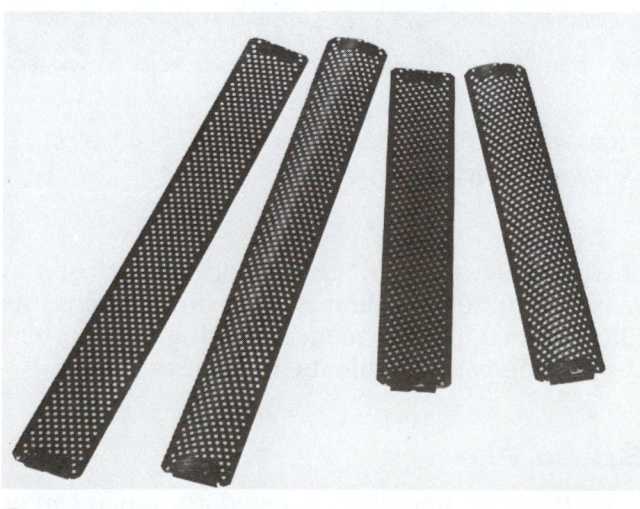

B

FIGURE 3-86 (A) Surform "cheese grater" and (B) replacement blades. *(Photo A courtesy of Bond-Tite® Division of U.S. Chemical and Plastics, Inc.)*

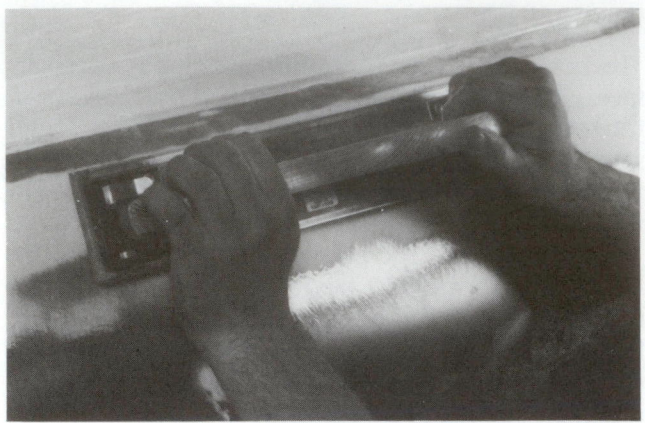

FIGURE 3-87 Speed files will cut to rough shape.

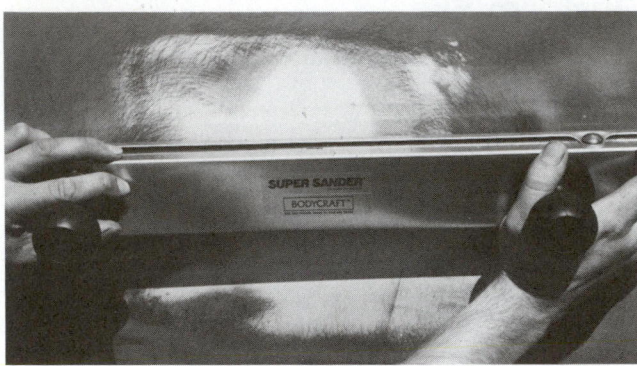

A

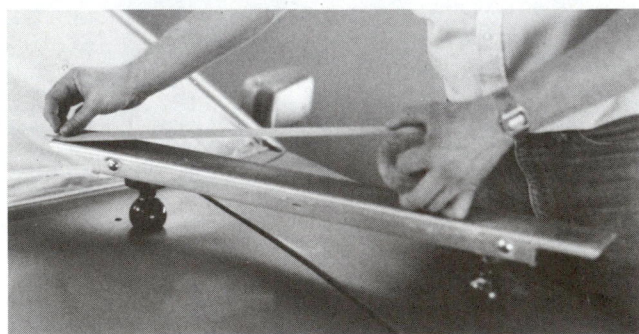

B

FIGURE 3-88 A flexible sander is handy on contours. *(Courtesy of Bodycraft Corp.)*

to as a "cheese grater," the surform file is used to shape body filler while it is semihard. Shaping the filler before it hardens reduces sanding and shortens the waiting period while the filler cures.

Speed File

Once the body filler has hardened, the repair can be shaped and leveled with a *speed file* (Figure 3–87). The speed file is a rigid wooden holder about 17 inches long and 2³/₄ inches wide. Also sometimes called a *"flatboy,"* the speed file allows a repair area

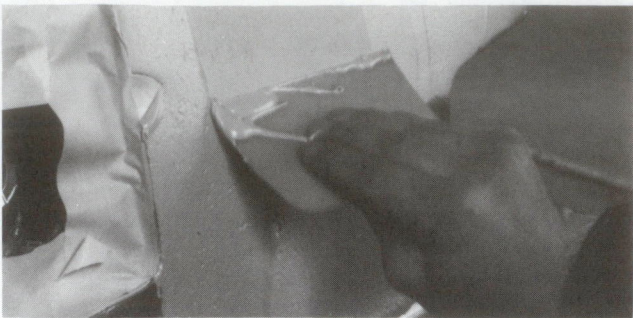

FIGURE 3-89 A body filler spreader will smoothly apply filler to save filing and sanding. *(Courtesy of DuPont Automotive Products)*

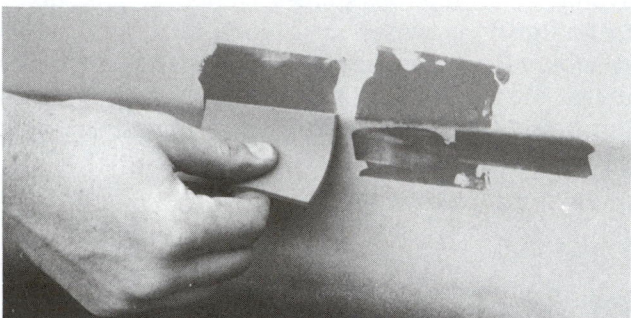

FIGURE 3-90 A rubber squeegee is good for applying spot putty.

to be sanded quickly with long, level strokes. This eliminates waves and uneven areas.

The lightweight aluminum sander shown in Figure 3–88A is designed to quickly level body filler. The extra long length helps avoid creating a wavy surface. The sander also flexes to match the panel contour. Adhesive-backed sandpaper is applied from a roll (Figure 3–88B). This particular sander can also be attached to a straight-line air sander.

SPREADERS AND SQUEEGEES

Spreaders and squeegees are two important tools used in auto body resurfacing. **Spreaders** are used to apply body filler and are made of rigid plastic and available in various sizes (Figure 3–89). Be sure to use one that is large enough to apply plastic filler over the complete repair area smoothly before the filler begins to set.

A **squeegee** (Figure 3–90) is a flexible rubber block used to apply glazing putty and light coats of body filler. It is also used to skim water and sanding grit from the repair area when wet sanding.

3.4 HAND TOOL SAFETY

Hand tool safety begins with purchasing quality tools. Quality tools may require a greater investment in money, but the dividends—safety and durability—are worth the expense. In the long run, they cost less than cheap tools. Many quality tools are warranted to ensure their quality and protect the investment.

The second step in safe tool usage is knowledge. Read the manufacturer's instructions and use the tool only for the tasks it was designed to do. Misuse of a tool causes accidents, as well as wears or weakens the tool, increasing the likelihood of slipping, chipping, or shattering.

The final step in tool safety is maintenance. Body shop tools must be kept clean, rust-free, sharp, and safely organized in a tool cabinet or chest. A tool should be maintained as close to the original condition as possible. Damaged and broken tools should never be used.

GENERAL PURPOSE TOOLS AND SAFETY

Proper use and care of tools will eliminate most accidents in the body shop. Follow these precautions when using general purpose hand tools.

Safety Rules for Pliers

The following basic safety rules should be applied to the use of pliers:

- Pliers should not be used for cutting hardened wire unless specifically manufactured for this purpose.
- Never expose pliers to excessive heat. This might draw the temper and ruin the tool (Figure 3–91).
- Always cut at right angles. Never rock the tool from side to side or bend the wire back and forth against the cutting blades.
- Do not bend heavy metal with light pliers. Needle nose pliers can be damaged if the tips are used to bend body metal, large wires, and so on. Use a sturdier tool.

- Never use pliers as a hammer (Figure 3–92) or use a hammer on the handles. They may crack or break, or the blades may be nicked by such abuse.
- Ordinary plastic-dipped handles are designed for comfort, not electrical insulation. Tools having high-dielectric insulation are available and are so identified. Do not confuse the two.
- Never extend the length of the handles to secure greater leverage. Use a larger pair of pliers or a bolt cutter.
- Pliers should not be used on nuts or bolts. A wrench will do a better job with less risk of damage to the fastener.
- Safety glasses should be worn when cutting wire or other such tasks to protect the eyes from being struck by the end of the object being cut.

Safety Rules for Wrenches

Here are a few basic rules that should be kept in mind when using wrenches:

- Keep wrenches clean and free of oil. Dirt and grit will prevent them from seating firmly around a nut. Oil and grease will make them slippery.
- Do not use a wrench to do the job of another tool. The job will not be done as well, and the wrench may be damaged or even broken. Using a wrench as a hammer or a pry bar or as anything but a wrench can be dangerous. Take the time to get the right tool.
- Never use a wrench opening that is too large for the fastener. This can spread the jaws of an open-end wrench and batter the points of a box or socket wrench. A wrench that has an opening that is too large can also spoil the points of the nut or bolt head. When selecting a wrench for proper fit, take special care to use inch wrenches on inch fasteners and metric wrenches on metric fasteners.

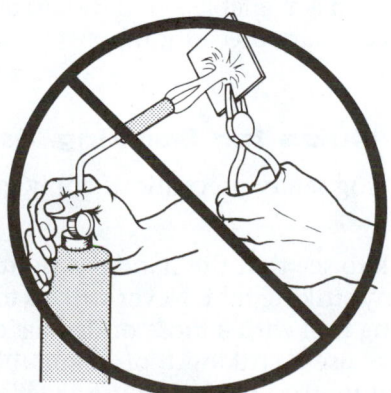

FIGURE 3–91 Do not subject pliers to heat.

FIGURE 3–92 Do not hammer with pliers.

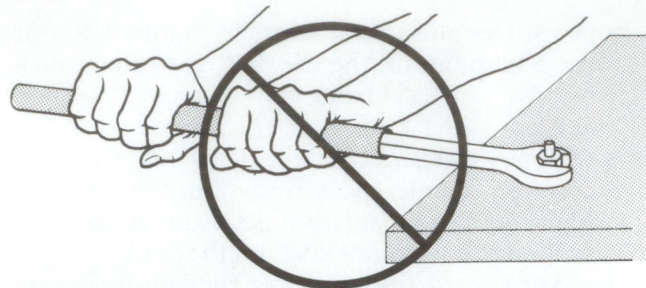

FIGURE 3-93 Do not use pipe extensions on wrench handles.

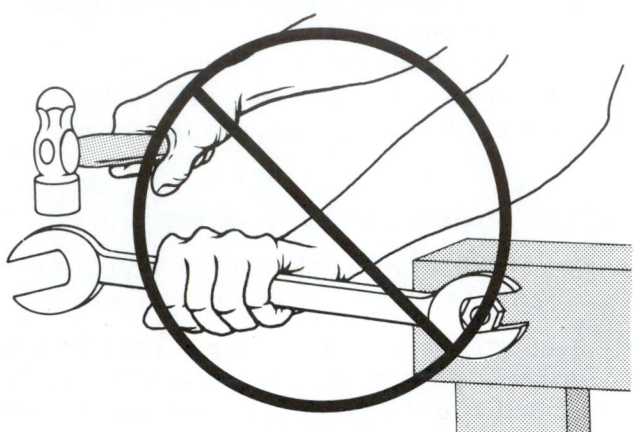

FIGURE 3-94 Do not hammer on wrenches.

- Match the wrench to the job, using box or socket wrenches for heavier jobs, open-end wrenches for medium-duty work, and adjustable wrenches for light-duty jobs and odd-sized nuts. A pipe wrench should be used only on pipes, never on a nut, because its teeth will damage the nut. Always use a straight, rather than offset, handle if conditions permit.
- Never push a wrench beyond its capacity. Quality wrenches are designed and sized to keep the leverage and intended load (torque) in safe balance. The use of an artificial extension (such as a pipe "cheater") on the handle of a wrench can break the wrench, spoil the work, and hurt the user (Figure 3–93). Instead, get a larger wrench or a different kind of wrench to do the job. The safest wrench is a box or socket type. Never hammer a wrench not made to be struck (Figure 3–94). (To free a frozen nut or bolt, use a striking face box wrench or a heavy-duty box or socket wrench. Never use an open-end wrench. And apply penetrating oil beforehand.)
- Determine which way a nut should be turned before trying to loosen it. Most nuts are turned counterclockwise for removal. This might seem obvious, but even experienced people have been observed straining at the wrench in the tightening direction when they wanted to loosen the nut.

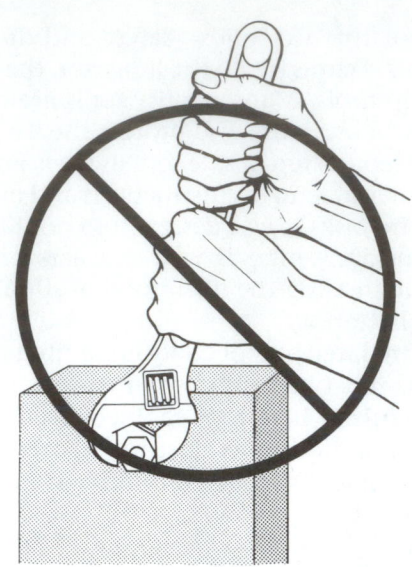

FIGURE 3-95 Do not use adjustable wrenches to tighten nuts or bolts.

- If possible, always pull on a wrench handle. Adjust your stance to prevent a fall if the tool slips. If it is ever necessary to push the wrench, do it with the palm of the hand, and hold the palm open. This will prevent smashed knuckles!
- Never expose a wrench to excessive heat. Direct flame can draw the temper from the metal, weakening and possibly warping it. This would make the tool unsafe to use.
- Place the tool in its correct position. For example, never cock or tilt an open-end wrench. Always be sure the nut or bolt head is fully seated in the jaw opening, for both safety and efficiency. A box or socket wrench should be used on tight fasteners. Adjustable wrenches should be tightly adjusted to the work and pulled so that the force is applied to the fixed jaw (Figure 3–95).
- Do not depend on plastic-dipped handles to insulate from electricity. Ordinary plastic-dipped handles are for comfort and a firmer grip. They are not intended for protection against electric shock. (Special high-dielectric-strength handle insulation is available, but it should only be used as a secondary precaution.)

Safety Rules for Striking Tools

The following safety precautions apply generally to striking tools:

- Check to see that the handle is tight before using any striking tool. Never strike a tool or use a striking tool with a loose or damaged handle.
- Always use a striking tool of suitable size and weight for the job. Do not use a tack hammer to drive a spike, or a sledge to drive a tack.

- Discard or re-dress any striking or struck tool if the face shows excessive wear, dents, chips, mushrooming, or improper re-dressing. Re-dressing involves using a grinding wheel to carefully re-shape the tool. This might be reshaping the tip or head of a chisel.
- Rest the face of the hammer on the work before striking to get the feel or aim. Then grasp the handle firmly with the hand near the extreme end of the handle.
- Strike squarely, but lightly, until the punch or tool to be driven is set. Make sure that the fingers of the other hand are out of the way before striking with force.
- A hammer blow should always be struck squarely with the hammer face parallel to the surface being struck (Figure 3–96). Always avoid glancing blows and over-and-under strikes.
- For striking another tool (cold chisel, punch, wedge, and so on) the face of the hammer should be proportionately larger than the head of the tool. For example, a $1/2$-inch cold chisel requires at least a 1-inch hammer face.

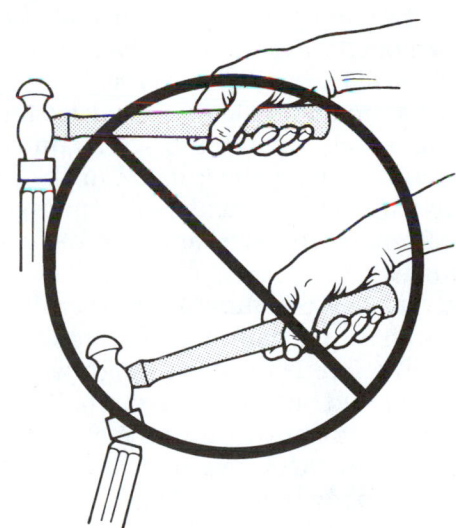

FIGURE 3–96 Strike squarely to prevent damage.

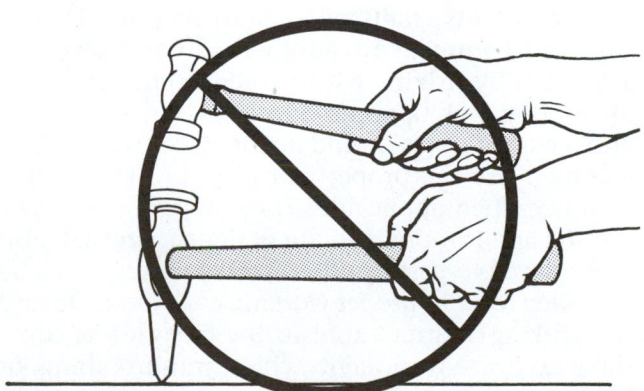

FIGURE 3–97 Do not hit hammers with hammers.

- Never use one hammer to strike another hammer (Figure 3–97).
- Replace or repair a damaged or loose hammer handle. A loose handle could allow the hammer head to fly off and injure someone. Replacement handles are available. Use a metal wedge driven into the end of the handle to secure it.
- Do not use the end of the handle of any striking tool for tamping or prying; it might split.

Safety Rules for Screwdrivers

The screwdriver is a very safe tool; it presents a hazard only when misused or when the user lacks the knowledge or skill to use it. Keep the following safety tips in mind:

- Do not hold the work in your hand while using a screwdriver; if the point slips, it can cause a bad cut. Hold the work in a vise, with a clamp, or on a solid surface. If that is impossible, follow this rule: Never get any part of your body in front of the screwdriver blade tip. That is a good safety rule for any sharp or pointed tool.
- All screwdrivers should have smooth, firm handles, and the blades should be kept in good condition. The screwdriver becomes useless and dangerous if the blade is broken.
- Do not use screwdrivers as chisels, hammers, can openers, crowbars, or for any purpose other than to turn screws (Figure 3–98).
- Never try to turn a screwdriver with a pair of pliers (Figure 3–99).
- Employ the right screwdriver for the job. That is, always use a screwdriver with a blade that fits the screw to be turned. The blade of the screwdriver should be seated squarely against the bottom of the screw slot.
- Screwdrivers with insulated handles and/or blades should be used for electrical repairs or

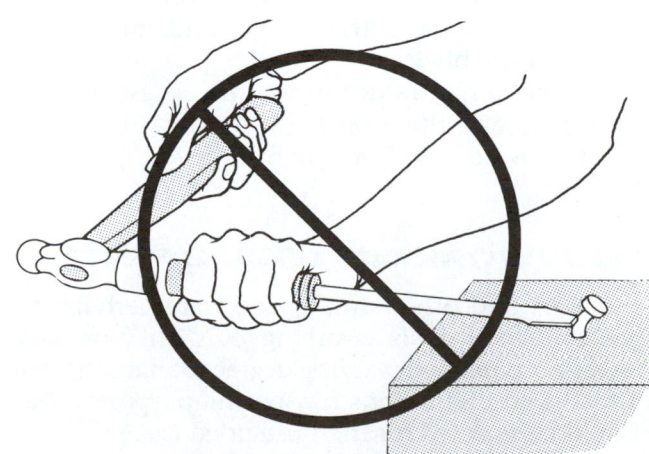

FIGURE 3–98 Do not use screwdrivers as chisels.

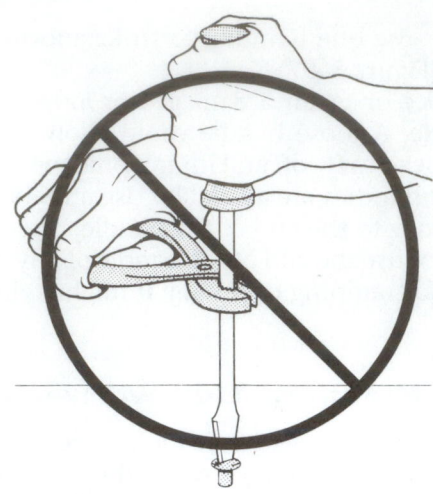

FIGURE 3-99 Do not force screwdrivers with vise grips or pliers.

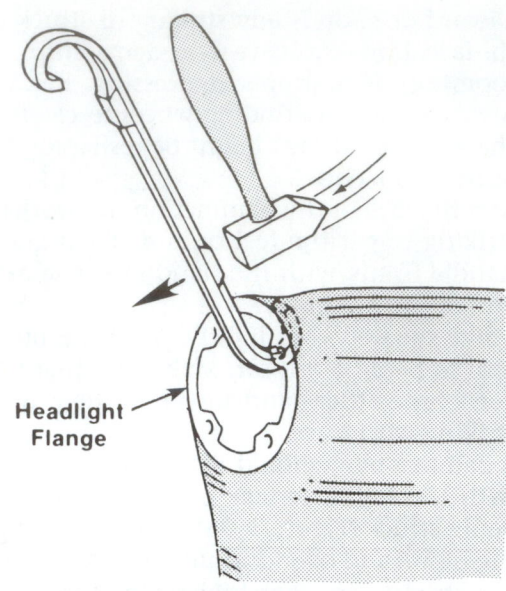

FIGURE 3-100 Driving out a dented fender with a double-end heavy-duty driver spoon.

installations. Be sure the battery is disconnected before attempting such work.

Safety Rules for Snips

Although snips are very safe tools, keep the following tips in mind:

- Learn to use snips properly. They should always be oiled and adjusted to permit ease of cutting and to produce a surface that is free from burrs. If the blades bind or are too far apart, the snips should be adjusted.
- Never use snips as screwdrivers, hammers, or pry bars. They break easily.
- Do not attempt to cut materials heavier than those that the snips are designed for. Never use tin snips to cut stainless steel or other hardened sheet metals. Such use will dent or nick the cutting edges of the blades.
- Never toss snips in a tool box where the cutting edges can come into contact with other tools. This dulls the cutting edges and might even break the blades.
- When snips are not in use, the safest thing to do is hang them on hooks or lay them on an uncrowded shelf or bench.

BODYWORKING TOOL SAFETY

Hammers and other striking tools are perhaps the most abused of all bodyworking tools. They are made in various types with varying degrees of hardness and different configurations for specific purposes. They should be selected for their intended use.

For example, do not use a body hammer to drive a light spoon that was not designed to be struck. Only

driving spoons (Figure 3–100) and spoons designed for spring hammering should be hit with a body hammer. Do not try to use an adjustable body file to file high-crown or concave surfaces. Overflexing the adjustable file will cause the file to break.

Body hammers are highly specialized tools used almost exclusively by automotive technicians for bumping and dinging sheet metal in the repair of automotive bodies and fenders. Drop-forged tools are made from selected steel and heat-treated to specific hardness.

Never use these hammers for other than bodywork. Always use a hammer of suitable size and weight for the job. Do not use a lightweight small hammer to pound out a heavy section. Do not use a heavy hammer to lightly finish metal.

Never strike cold chisels, punches, or other hard objects with a body hammer. The face can chip and possibly cause damage, not only to the hammer but also to the tool user. Eye protection, through the use of safety glasses or hood, is recommended.

Do not drive nails with a body hammer. Do not strike one hammer with another hammer. Never use a hammer by striking with its side or cheek. Do not use a body pick hammer as a punch. Cold chisels, punches, and rivets should not be struck with a body hammer. Use the proper ball peen or light sledge-hammer when applicable. They are made for their specific applications and are designed, dressed, and heat-treated to perform their tasks safely.

Keep tools in proper working condition. Discard any striking or struck tools if the face and its edges show excessive wear, dents, chips, mushrooming, or improper re-dressing. Mushroomed heads can chip and cause injury. Never use a hammer with a loose or

FIGURE 3-101 Always use fender and seat covers when needed. If a fender is not going to be painted, you do not want to scratch it. Also, to prevent stains, dirty clothes must never touch the vehicle's upholstery. *(Courtesy of Tech-Cor)*

You are only as good as your tools. Dirty, oily tools are a good indicator of a poor technician. A wise technician knows that quality, properly maintained tools will result in better and more work output, and safer working conditions. Never be a "grease monkey"! Keep your tools clean and organized.

Always use fender covers and seat covers when needed (Figure 3-101). It is very easy to scratch the existing finish during disassembly and reassembly. This makes for extra bodywork. It is easy also to stain seats if you have oil or grease on your work cloths. Always use covers to protect the vehicle from damage.

damaged handle. Keep hammer heads tightened on the handles. If only the handle is damaged, replace it with a new high-grade hickory or fiberglass handle.

Never re-dress hammers without proper re-dressing instructions. The tool must be returned to its original shape. Re-dressing and reshaping of tools having chipped, battered, or mushroomed heads is not recommended. Tool manufacturers cannot be held responsible for injury or damage caused by tools improperly used or from tools that have been abused or badly worn.

SUMMARY

- Hand tools are extensions to the human body. They allow you to do tasks otherwise impossible with your bare hands.
- When purchasing tools, get high-quality tools from a reputable manufacturer.
- A complete collection of wrenches is indispensable for the auto body technician.
- In many situations, a socket wrench is much faster and easier to use than an open-end or box-end wrench.
- The ratchet handle is probably the most commonly used handle. The ratchet handle allows removing or tightening without removing the socket from the fastener.
- A variety of threaded fasteners used in the automotive industries are driven by a screwdriver.

- Pliers are an all-around grabbing and holding tool for working with wires, clips, and pins.
- Body hammers are the basic tools for pounding sheet metal back into shape. They come in many different designs.
- The dolly or dolly block is used like a small anvil for working out body damage.
- Hand tool safety begins with purchasing quality tools. Quality tools might require a greater investment in money, but the dividends—safety and durability—are worth the expense.
- Proper use and proper care of tools will eliminate most accidents in the body shop.

ASE-STYLE REVIEW QUESTIONS

1. Which socket drive size should be used on large lug nuts?
   A. 1/4 inch
   B. 5/16 inch
   C. 3/8 inch
   D. 1/2 inch

2. A striking tool should be discarded or serviced when it is _____.
   A. Improperly re-dressed
   B. Mushroomed
   C. Chipped
   D. All of the above

3. Technician A says that a $\frac{1}{2}$-inch drive socket set is required to disassemble and remove interior trim components. Technician B says that a 1-inch drive socket set is useful for truck and heavy equipment repair. Who is correct?
   A. Technician A
   B. Technician B
   C. Both A and B
   D. Neither A or B

4. Technician A says that a wire brush scratches bare metal and thus should be used sparingly on it. Technician B uses a wire brush on weld joints to clean off welding flux. Who is correct?
   A. Technician A
   B. Technician B
   C. Both A and B
   D. Neither A or B

5. Which of the following wrenches provides the safest grip on a fastener?
   A. Open-end wrench
   B. Box-end wrench
   C. Adjustable wrench
   D. None of the above

6. Technician A sometimes uses a screwdriver as a chisel. Technician B sometimes uses a screwdriver as a pry bar. Who is correct?
   A. Technician A
   B. Technician B
   C. Both A and B
   D. Neither A nor B

7. Technician A says that a tap cuts external threads. Technician B says that a die cuts internal threads. Who is correct?

   A. Technician A
   B. Technician B
   C. Both A and B
   D. Neither A nor B

8. Which of the following wrenches are used with a ratchet handle?
   A. Combination
   B. Allen
   C. Socket
   D. Box

9. Which of the following screwdriver tips provides the most resistance to slipping out of a fastener?
   A. Standard
   B. Torx
   C. Phillips
   D. Slotted

10. Which of the following pliers is often used in electrical work?
    A. Adjustable pliers
    B. Needle nose pliers
    C. Retaining ring pliers
    D. Snap ring pliers

11. Which of the following hammers would be best used for driving gears or shafts?
    A. Brass
    B. Ball peen
    C. Plastic
    D. Pick

12. An extractor is used for removing broken _____.
    A. Seals
    B. Bushings
    C. Pistons
    D. Bolts

ESSAY QUESTIONS

1. Explain when you might use each drive size: $\frac{1}{4}$, $\frac{3}{8}$, and $\frac{1}{2}$ inch.

2. Describe the construction and typical organization of a typical tool chest.

3. Summarize the utilization of three types of punches.

4. What is the difference between a spreader and a squeegee?

CRITICAL THINKING PROBLEMS

1. What could happen if you used a body file excessively on an aluminum body panel?

2. A bolt is badly rusted and rounded off. How could you get it out?

MATH PROBLEMS

1. If a breaker bar is two feet long and you exert 125 lbs of force to the end of the handle, approximately how many pounds of force are exerted at the drive head?

CHAPTER 4

Body Shop Power Tools

INTRODUCTION

This chapter summarizes how to properly select and use power tools and equipment. Most of the tools mentioned here are general purpose power tools. More specialized tools and their uses are described in the appropriate chapters.

Power tools use air pressure *(pneumatic)*, oil pressure *(hydraulic)*, or electrical energy to effect repairs. This classification includes air wrenches, air and electric drills, sanders, and similar tools. The body

---

The use of any of air-driven or electric power tools requires the wearing of safety glasses or a face shield. For hazardous operations, wear both. Also, never wear loose clothing that might get caught in a tool.

Power tools and equipment can be very dangerous if not used correctly. Always follow the instructions given in the owner's manual for the specific tool or piece of equipment. The information is this chapter is general and cannot cover all tool variations. If in doubt about any tool or piece of equipment, ask your instructor or shop foreman for a demonstration.

---

shop technicians and painters must have a wide variety of power tools to make their tasks easier.

4.1 AIR-POWERED TOOLS

The automotive industry was one of the first industries to see the advantages of air-powered tools. Today they are known as "the tools of the professional auto body/painter technician."

Although electric drills, wrenches, grinders, polishers, drill presses, and heat guns are found in body and refinishing shops, the use of pneumatic (air) tools is a great deal more common. Pneumatic tools (Figure 4–1) have four major advantages over electrically powered equipment:

- **Flexibility.** Air tools run cooler and have the advantage of variable speed and torque; damage from overload or stalling is eliminated. They can fit in tight spaces.
- **Light weight.** The air tool is lighter in weight and lends itself to a higher rate of production with less fatigue.
- **Safety.** Air equipment reduces the danger of fire hazard in some environments where the sparking of electric power tools can be a problem. It also does not use electricity, the source of electrocution.
- **Low cost operation and maintenance.** Due to fewer parts, air tools require fewer repairs and

OBJECTIVES

After studying this chapter, you should be able to:

✔ Identify air-powered tools used in the body shop.
✔ Name the electric power tools most commonly used in body shops.
✔ Describe the hydraulic equipment used in auto body repair.
✔ Maintain shop power equipment tools.
✔ Answer ASE test questions relating to power tools.

KEY TERMS

air drills	grinders
air file	heat gun
air polisher	hydraulic lift
air ratchet	impact wrench
air sander	power tools
air tool lubrication	sander pad
battery charger	spray gun
body jack	swirl marks

ASE TASK LIST
Job Skills covered in this chapter include:

PAINTING AND REFINISHING TEST (B2) TASK LIST

A. Surface Preparation
3. Inspect and identify substrate, type of finish, and surface condition; develop a plan for refinishing.
4. Remove paint finish.
5. Dry or wet sand areas to be refinished.

B. Spray Gun Operation and Related Equipment
1. Inspect, clean, and determine condition and adequacy of spray guns and related equipment (air hoses, regulator, air lines, air source, and spray environment).
2. Check and adjust spray gun pressure for siphon-feed, pressure-feed, gravity-feed, HVLP (high volume, low pressure) and LVLP (low volume, low pressure) guns.

C. Paint Mixing, Matching, and Applying
8. Sand, buff, and polish finishes where necessary.
9. Identify the types of rigid, semirigid, or flexible plastic parts to be refinished; determine the proper materials and refinishing procedures.

NONSTRUCTURAL ANALYSIS AND DAMAGE REPAIR TEST (B3) TASK LIST

A. Preparation
1. Review damage report; analyze damage to determine appropriate methods for overall repair.
2. Lift, raise, and position vehicle to perform repairs.

B. Outer Body Panel Repairs, Replacements, and Adjustments
3. Determine the extent of damage to aluminum body panels; repair, weld, or replace in accordance with manufacturers' specifications.
7. Remove, replace, and align bumpers, reinforcements, guards, isolators, and mounting hardware.
10. Rough out contours of damaged panel to a surface condition suitable for metal finishing and body filling.

C. Metal Finishing and Body Filling
1. Grind off paint from the damaged area of a body panel.

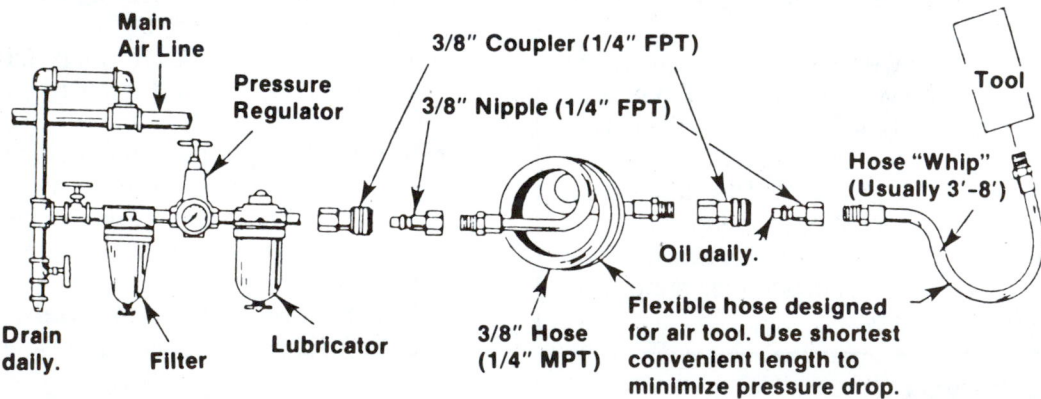

FIGURE 4-1 Study the parts of a typical body shop compressed air system to power air tools.

less preventive maintenance. The original cost of air-driven tools is usually less than the equivalent electric type.

There are pneumatic equivalents for nearly every electrically powered tool, from sanders to drills, grinders, impact wrenches, and screwdrivers. Furthermore, there are some pneumatic tools with no electrical equivalent, such as the chisel, ratchet wrench, grease gun, and various auto tools (Figure 4–2).

Hoists, lifts, and frame and panel straighteners can be used in conjunction with a compressed air

system. However, in most cases, these pieces of equipment are hydraulically operated. Described later in this chapter, hydraulic tools play a very important role in any body shop operation.

SPRAY GUNS

Spray guns are used to apply sealer, primer, paint, and other liquid finishing materials to the vehicle. Spray guns must atomize the liquid, often paint, so that it flows onto the body surface smoothly and evenly (Figure 4–3).

FIGURE 4-2 Common air-powered tools used in a body shop are: (A) ½-inch impact wrench with 2-inch extended anvil; (B) vertical air polisher (buffer); (C) ³/₈-inch ratchet wrench; (D) orbital sander; (E) finishing sander; (F) disc sander; (G) ³/₈-inch palm grip impact wrench; (H) angle polisher; (I) air chest; (J) straight-line sander; (K) ½-inch impact wrench; (L) ³/₈-inch angle head impact wrench; (M) ³/₈-inch drill.

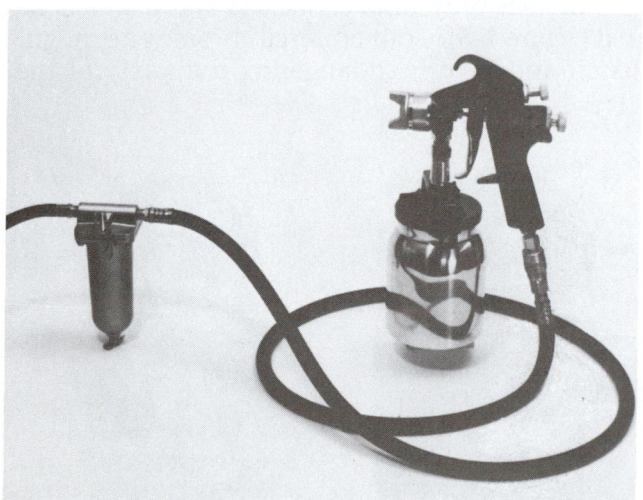

FIGURE 4-3 This is a conventional atomizing paint spray gun air system arrangement.

A spray gun must *atomize* the liquid, breaking it into a fine mist of droplets. This requires sufficient pressure and volume at the gun, which can be powered by air or electric energy, although air is more common.

Spray Gun Parts

To use, service, and troubleshoot a spray gun properly, you must understand the operation of its major parts.

The *gun body* holds the parts that meter air and liquid. The body holds the spray pattern adjustment valve, fluid control valve, air cap, fluid tip, trigger, and related parts (Figure 4–4).

The *spray gun cup* often fits onto the bottom of the body to hold the material to be sprayed. The top of the cup fits against a rubber seal to prevent leakage. The seal is mounted in a lid on the lower part of the gun body.

The spray gun's *fluid control valve* can be turned to adjust the amount of paint or other material emitted. It consists of a thumbscrew or knob, needle valve, and spring. Turning the knob affects how far the trigger pulls the needle valve open. The *fluid needle valve* seats in the fluid tip to prevent flow or can be pulled back to allow flow.

The spray gun's *air control valve*, or *spreader valve*, controls how much air flows out of the air cap side jets. It has an *air needle* that can be slid back and forth to open or close the air valve.

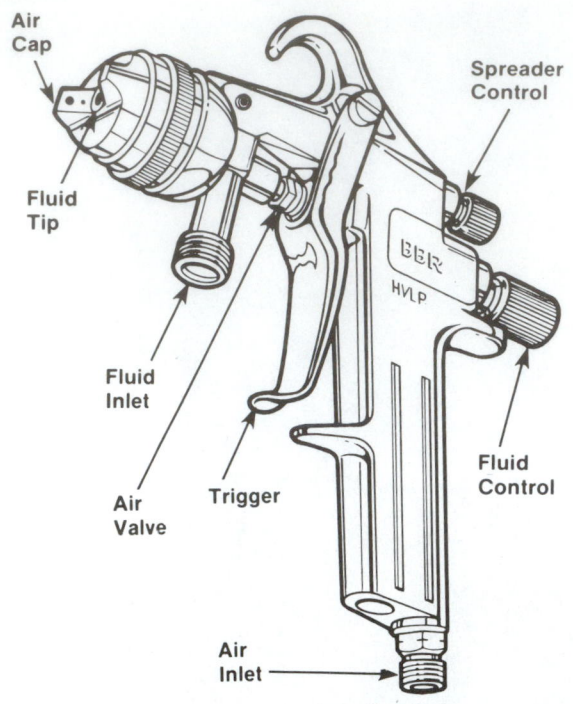

FIGURE 4-4 Study parts names of this modern, high volume, low pressure spray gun. *(Courtesy of Binks Mfg. Co.)*

The *spray gun trigger* can be pulled to open both the fluid and air valves. It uses lever action to pull back on the needle valves.

The *spray gun air cap* works with the air valve to control the spray pattern of the paint. It screws over the front of the gun head.

The *spray pattern* is the shape of the atomized spray when it hits the body. With little air flow out of the side jets on the cap, you would have a very round, concentrated spray pattern. As you adjust air flow up, the cap jets narrow and better atomize the paint flowing out of the gun.

The spray gun is probably the most used air-powered tool in the body/paint shop. It is used to do most of the refinishing work. Spray guns are also one of the most efficient of all pneumatic tools.

A conventional atomizing air spray gun is a precision tool using compressed air to atomize sprayable material. Replacing the conventional system in many body shops is the high volume, low pressure (HVLP) gun (Figure 4–4). Air and paint enter both types of guns through separate passages and are mixed and ejected at the air nozzle to provide a controlled spray pattern.

Pulling back slightly on the trigger opens the air valve to allow use of the gun as a blow gun. In this position the trigger does not actuate the fluid needle and no fluid flows. As the trigger is further retracted, it unseats the needle in the fluid nozzle and the gun begins to spray. The amount of paint leaving the gun is controlled by the pressure on the container, the viscosity of the paint, the size of the fluid orifice, and the fluid needle adjustment. In industrial finishing where pressure tanks or pumps are used, the fluid needle adjustment should normally be fully opened. In suction cup operation, the needle valve controls the flow of paint.

Complete details on the operation of the various types of spray guns and their use can be found in Chapter 17.

AIR-POWERED WRENCHES

Any job that involves threaded fasteners can be done faster and easier with air-powered wrenches. There are two basic types of wrenches: the impact and the ratchet.

Impact Wrenches

An **impact wrench** is a portable hand-held, reversible air-powered tool for rapid turning of bolts and nuts. When triggered, the output shaft spins freely over 2,000 rpm. The socket snaps over the square drive head and shaft. This depends, however, on the wrench's make and model. When the impact wrench meets resistance, a small spring-loaded hammer, which is situated near the drive head, strikes an anvil attached to the drive shaft onto which the socket is mounted. Thus each impact moves the socket around a little until torque equilibrium is reached, the fastener breaks, or the trigger is released.

When using an air impact wrench, it is important that only impact sockets and adapters be used (Figure 4–5). Conventional chrome ones might shatter and fly off, endangering the safety of the

FIGURE 4-5 An impact socket (left) is tougher than the normal socket (right). The impact type has thicker walls for a stronger 6-point design. Impact sockets are usually flat black color.

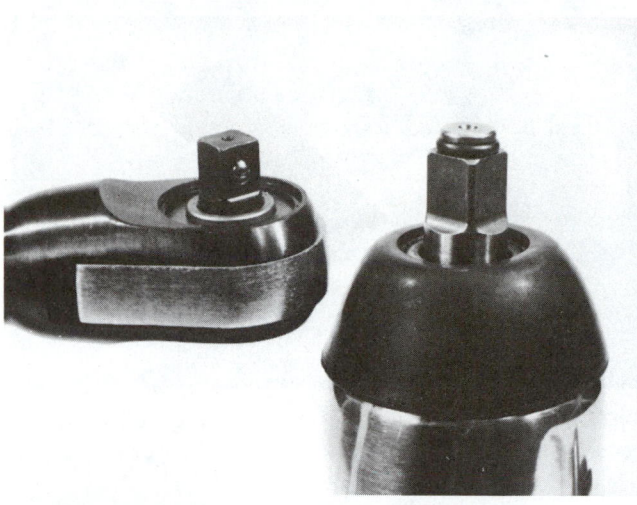

**Adjustable
Air Regulator**

FIGURE 4–7 The air regulator knob on a typical impact wrench can be adjusted to control speed and torque output.

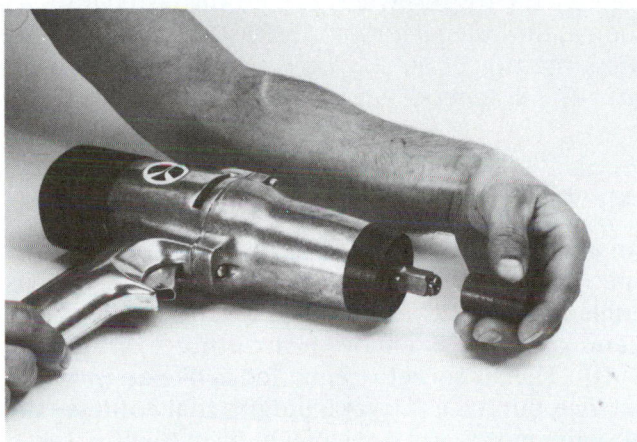

FIGURE 4–6 Two types of output shafts: (top left) detent ball and (top right) retaining or hog ring. (Bottom) Installing a retaining type on an impact wrench.

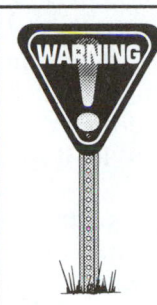

WARNING

Always use the simplest possible tool-to-socket hookup. Every extra connection, like extensions and universal joints, absorbs energy and reduces power. It will also cause undue wear on tools.

operator and others in the work area. Make certain that sockets and adapters are clearly marked or labeled for use with impact wrenches. Impact tools are usually flat black in color, not chrome. Impact sockets are sold in both SAE and metric system sizes.

To attach socket chucks and adapters to air impact wrenches, merely push them onto the output shaft as far as they will go. True power wrench output shafts come in two common variations (Figure 4–6), the detent ball and the retaining ring.

An adjustable air regulator is part of most pneumatic impact wrenches (Figure 4–7). Turning the air control knob allows you to adjust the tool's speed and torque. Input air pressure above the usual 90 to 125 psi (620 to 861 kPa) range can be fed to the inlet without excessive tool wear.

Air impact wrenches work equally well for tightening and loosening. Direction of the rotation is usually controlled by a switch or two-way trigger (Figure 4–8). Remember not to change the direction of rotation while the trigger switch is on.

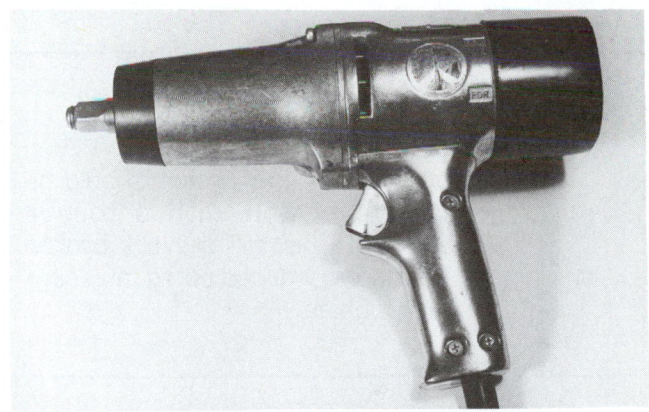

A

B

FIGURE 4–8 Impact wrenches can be triggered by (A) trigger switch or (B) butterfly throttle switch. The butterfly switch acts as a forward and reverse lever for this palm grip air impact wrench, permitting either a forward right-hand rotation or a reverse left-hand rotation.

FIGURE 4-9 The technician is loosening a nut using an extension socket on an impact wrench. *(Courtesy of Tech-Cor)*

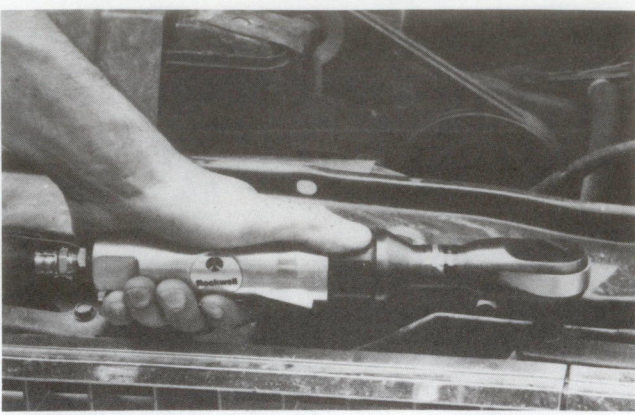

FIGURE 4-10 An air ratchet wrench is good for small fasteners with low torque and tight working quarters.

To remove conventional threaded fasteners, set the switch for left-hand rotation. Place the socket over the nut or fastener head. Exert forward pressure on the wrench while depressing the trigger switch. As soon as the nut or fastener becomes loosened, relax the forward pressure on the wrench to let it spin the nut or fastener free (Figure 4-9).

To install fasteners, normally set the switch for right-hand rotation. Start the nut on the stud or the bolt on the threads by hand to avoid cross threading. Place the socket over the nut or fastener head.

Depress the trigger switch to spin the fastener in. As soon as the fastener bottoms, release the trigger. Then, press the trigger quickly one or two more times to snug the fastener down.

Air Ratchet Wrenches

An **air ratchet** wrench, like the hand ratchet, has a special ability to work in hard-to-reach places. Its angle drive reaches in and spins fasteners where other hand or power wrenches just cannot work (Figure 4-10). The air socket wrench looks like an ordinary ratchet, but has a thicker handgrip that contains the air vane motor and drive mechanism (Figure 4-11).

WARNING

Do not use an air impact wrench to final torque fasteners. When important, final torque should be done by hand with a torque wrench. This will prevent broken fasteners, warped parts, and similar troubles.

A

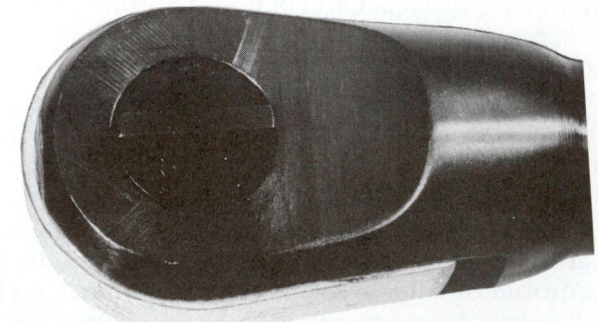

B

SHOP TALK

When using an impact wrench, keep the following pointers in mind:

- Always make sure the impact wrench socket is properly retained. To prevent injury, use an impact-type socket. If the impact wrench has a pin retainer, do not substitute a bent nail or piece of wire.

- If an air wrench fails to loosen a bolt in 3 to 5 seconds, use a breaker bar. Soak large rusted nuts with penetrating oil before using the wrench.

FIGURE 4-11 (A) Typical air ratchet wrench and (B) the forward and reversing lever is used to change the rotation of the tool. When the lever is set in the forward position, the tool will turn in a right-hand rotation to fasten nuts and bolts. When the lever is moved to the reverse position, the tool will run in reverse or left-hand rotation to unscrew nuts and bolts.

SHOP TALK

Actual torque on a fastener is directly related to joint hardness, tool speed, condition of socket, and the time the tool is allowed to impact.

After breaking loose a fastener by pulling the air ratchet handle, power can be used easily to spin the fastener out. When tightening, run in the bolt or nut under power and finish tightening it by hand-pulling. Remember that an air ratchet has little torque and hand tightening is important.

For all their torquing power, air impact wrenches have practically no recoil. Holding one is rather easy, but can be misleading. The flow of power through the socket is very strong. Therefore, the wrench should be held tightly.

Remember that there is no consistently reliable adjustment with any air socket or impact wrench. Where accurate preselected torque adjustments are required, a standard torque wrench should be used. The air regulator on air-powered wrenches can be employed to adjust torque to be slightly below the needed tightness of a known fastener.

AIR DRILLS

Air drills use shop air pressure to spin a drill bit. They can be adjusted to any speed and are more commonly used than electric drills. They are usually available in ¼-, ⅜-, and ½-inch sizes. Air drills are smaller and lighter than electric drills. This compactness makes them a great deal easier to use for most tasks (Figure 4–12) and especially for drilling operations in auto work (Figure 4–13).

Drill bits of different sizes fit into the chuck on drills for making holes in parts. They come in various sizes. The size is usually stamped on the upper part or shank of each bit.

A *key* is used to tighten the chuck. The *chuck* has movable jaws that close down and hold the bit. The key has a small gear that turns a gear on the chuck.

Never leave a key in a drill when not tightening. Unhook the air hose when installing a bit; otherwise, injury could result. This applies to a large drill press also.

To drill with an air tool into any material, the following general procedures should be kept in mind:

1. Accurately locate the position of the hole to be drilled. Mark the position distinctly with a center punch or an awl to provide a seat for the drill point and to keep it from "walking" away from the mark when pressure is applied.
2. Always know what is on the other side. Do not drill a wiring harness or through a trim panel.
3. Unless the workpiece is stationary or large, fasten it in a vise or clamp. Holding a small item in the hand might cause injury if it is suddenly seized by the bit and whirled out of grip. This is most likely to happen just before the bit breaks through the hole at the underside of the work.
4. Carefully center the drill bit in the jaws, while securely tightening the chuck. Avoid inserting the bit off-center because it will wobble and probably break when it spins. After centering, place the drill bit tip on the exact point to drill the hole; then start the motor by pulling the trigger switch. Never apply a spinning drill bit to the work.
5. Except when it is desirable to drill a hole at an angle, hold the drill perpendicular to the face of the work.
6. Align the drill bit and the axis of the drill in the direction of the hole. Apply pressure only along this line with no sidewise or bending pressure. Changing the direction of this pressure will distort the dimensions of the hole. It could

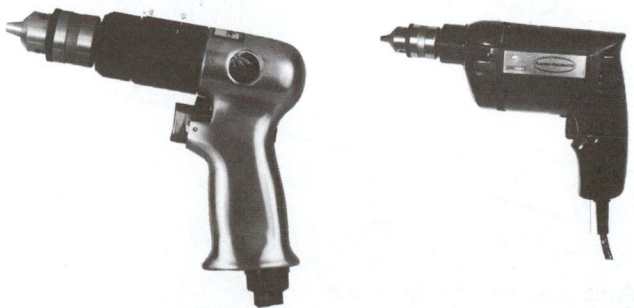

FIGURE 4–12 (Left) ⅜-inch air drill and (right) ⅜-inch electric drill. The air drill weighs 2½ pounds, while the electric type weighs 4½ pounds.

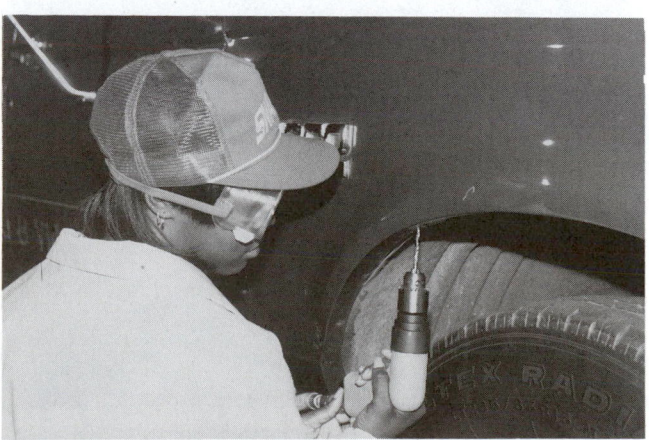

FIGURE 4–13 Note that the technician is wearing eye protection while using an air drill.

snap a small drill. To avoid stressing the drill bit, place only light pressure right above the housing on the drill body. If you push down with the handle, side loading could break the bit or affect the hole roundness.

7. Use just enough steady and even pressure to keep the drill cutting. Guide the drill; do not force it. Too much pressure can cause the bit to break or the tool to overheat. Too little pressure will keep the bit from cutting and dull its edges due to the excessive friction created by sliding over the surface.

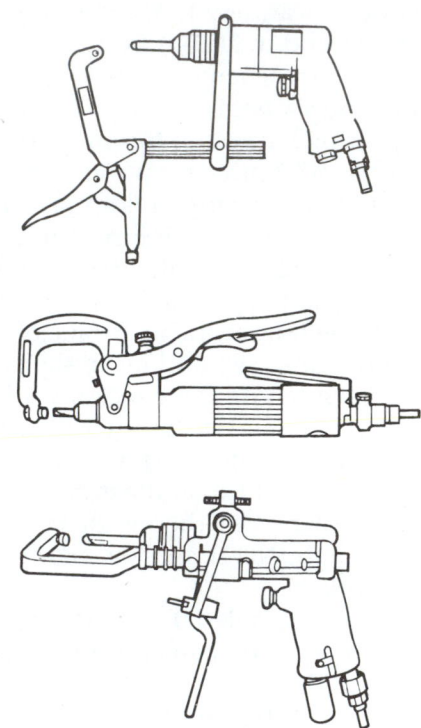

FIGURE 4–14 Several types of spot weld remover air drills are shown. They have plier attachments for helping to force the bit into spot weld for removal.

FIGURE 4–15 When drilling panels, clamping pliers are often used to hold parts in position. *[Courtesy of Dobbs Publishing]*

8. When drilling deep holes, especially with a twist drill, withdraw the drill several times to clear the cuttings. Keep the tool running when pulling the bit back out of a drilled hole. This will help prevent jamming.

9. Reduce the pressure on the drill just before the bit cuts through the work.

There are special *spot remover air drills* and/or attachments that are used for cutting out spot welds (Figure 4–14). When cutting out spot welds, the drill can be fastened to the weld area with a clamp attachment that makes the operation easier (Figure 4–15). The cutter will not deviate from the weld center during cutting.

There are two types of spot cutters available that can be mounted in an air drill for cutting out spot welds:

- **Drill type** (Figure 4–16A). This type does not damage the bottom panel, nor does it leave a nib in the bottom panel, so finishing is easy.
- **Hole saw type** (Figure 4–16B). The cutting depth of this type can be adjusted so that the bottom panel is not damaged. It is necessary to sand off the remaining portion of the weld.

When performing any air drill operation, keep the following maintenance and safety pointers in mind:

- Clean chuck jaws occasionally to prolong concentricity. Use sharp drills; they require far less effort and put less strain on the tool.
- To avoid the danger of breakage from high breakthrough torque, use properly sharpened twist drills and reamers, and select proper drill speed.
- Start drilling at low speed and gradually increase it. Avoid kickback by easing up at breakthrough.

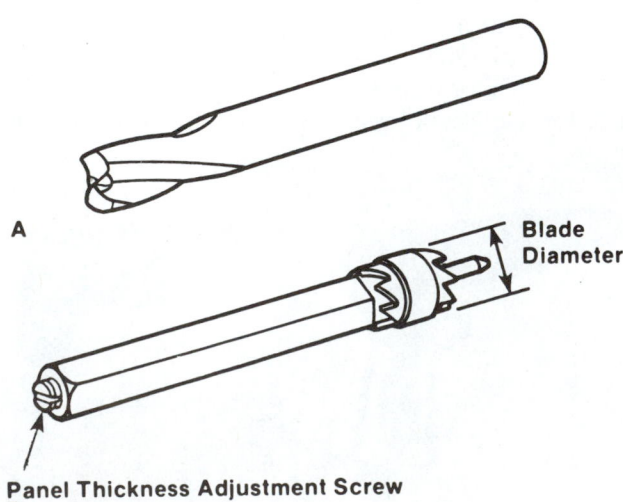

FIGURE 4–16 Note spot cutters: (A) drill type and (B) hole saw type.

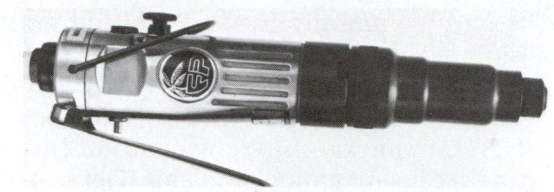

A

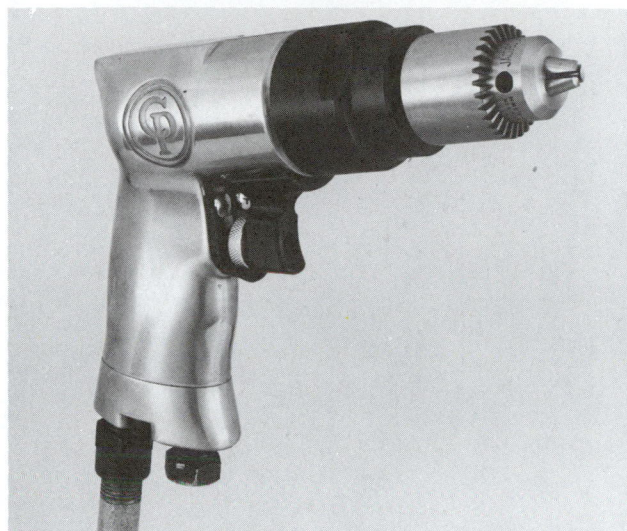

B

FIGURE 4-17 Air screwdriver handle types are shown: (A) straight and (B) pistol grip. *(Courtesy of Florida Pneumatic Mfg. Corp.)*

Motion of Abrasive Paper

Circular Motion (Single)

FIGURE 4-18 A disc sander and its motion are illustrated. *(Courtesy of Florida Pneumatic Mfg. Corp.)*

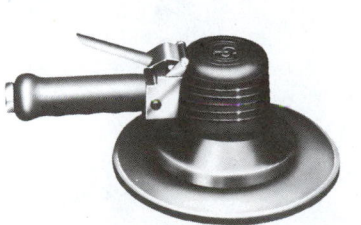 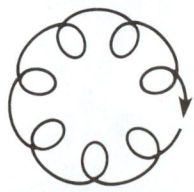

Motion of Abrasive Paper

Circular Motion (Double)

FIGURE 4-19 The dual- or double-action sander and its motion are demonstrated. *(Courtesy of Chicago Pneumatic Tool Division)*

AIR SCREWDRIVERS

Unlike electric screwdrivers, air screwdrivers run cool and will not burn out, even under constant use. They are designed to perform well in a wide variety of applications:

- Machine screws in tapped holes
- Self-tapping screws in plastic
- Sheet metal screws
- Self-drilling screws in metal sandwiches
- Fine-threaded screws in delicate assemblies
- Thread forming screws in blend die-cast holes
- Many more similar shop jobs

Pneumatic screwdrivers are available with straight- or pistol-grip handles (Figure 4–17).

AIR SANDERS

An **air sander** uses an abrasive action to smooth and shape body surfaces. Different coarseness sandpapers can be attached to the pad on the sander. *Coarse sandpaper* removes material more quickly. *Fine sandpaper* produces a smoother surface finish. Air sanders are one of the most commonly used air tools in auto body repair.

There are two basic types of air sanders: disc and orbital (finishing). Most rough sanding done in automotive work is done with a *disc sander* (Figure 4–18) or its counterpart, the *dual-action* (DA) orbital sander (Figure 4–19). The orbital sander oscillates while it is rotating and creates a buffing pattern rather than the swirls and scratches often caused by the disc sander.

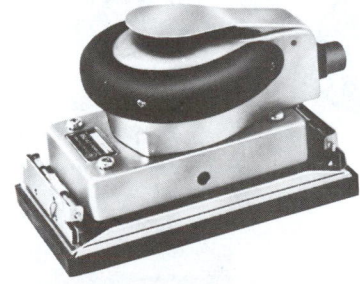 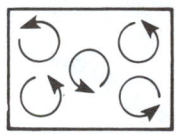

Motion of Abrasive Paper

Circular Motion (Single)

FIGURE 4-20 Study the pad or jitterbug sander and its motion. *(Courtesy of Chicago Pneumatic Tool Division)*

CAUTION

Wear respiratory protection any time a sander is being operated.

The *finishing orbital sanders*, also called "pad" or "jitterbug" sanders (Figure 4–20), are designed for fine finish sanding. It is possible to use a wider variety of abrasives with finish sanders. For the most part, the best work is done with comparatively fine grit abrasive paper. Finish sanders are also especially designed for hard-to-reach places and tight corners.

The **sander pad** is a soft mounting surface for the sandpaper. Disc adhesive, self-stick paper, or hook and latch paper holds the sandpaper onto the pad. *Disc adhesive* is special nonhardening glue that comes in a tube. It can be placed on the pad to adhere the paper to the sander. *Self-stick paper* already has a nonhardening glue on it.

Hook and latch paper is similar to Velcro®. One advantage is that you can take hook and latch paper off the tool without damage and reuse it.

An **air file** is a long, thin air sander for working large flat surfaces on panels. It is handy when you must true or flatten a large repair area. It will plane down filler so that a large area is flat. An air file is often used for rough shaping operations (Figure 4–21). It operates in either an orbital or straight line motion.

Details on operating sanders can be found in Chapter 18.

AIR GRINDERS

Grinders are used for fast removal of material. They are often used to smooth metal joints after welding and to remove paint and primer. They come in various sizes and shapes.

The most commonly used portable air grinder in collision repair and refinishing shops is the *disc-type grinder*. It is operated like the single-action disc sander (Figure 4–22). An air grinder should be used carefully. It can quickly thin down and cut through body panels, causing major problems (Figure 4–23).

FIGURE 4-22 A disc-type grinder is shown at work during major structural repair. *(Courtesy of Dobbs Publishing)*

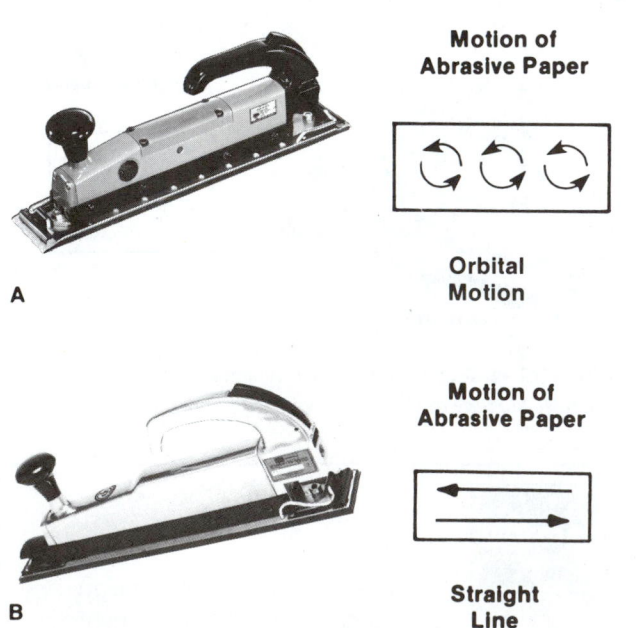

Motion of
Abrasive Paper

Orbital
Motion

Motion of
Abrasive Paper

Straight
Line

FIGURE 4-21 Here are two types of board sander: (A) orbital and its motion and (B) straight line and its motion. *(Photo A courtesy of Florida Pneumatic Mfg. Corp.; photo B courtesy of Chicago Pneumatic Tool Division)*

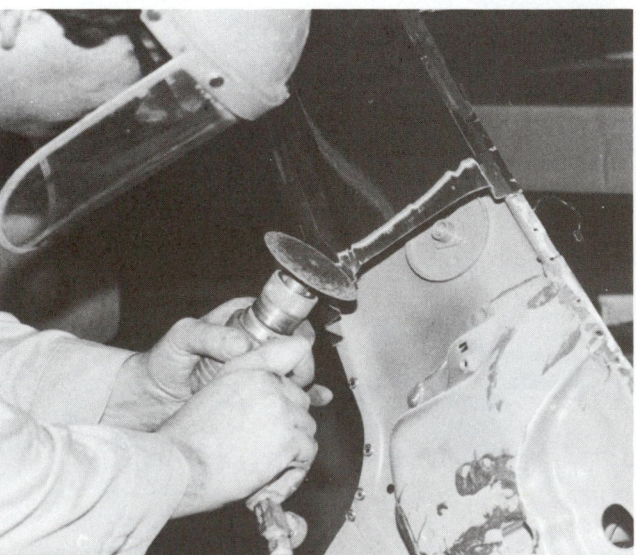

FIGURE 4-23 Take care not to catch projections, or the tool could fly into your body or face. Serious injury could result. Always wear face protection! *(Courtesy of Marson Corp.)*

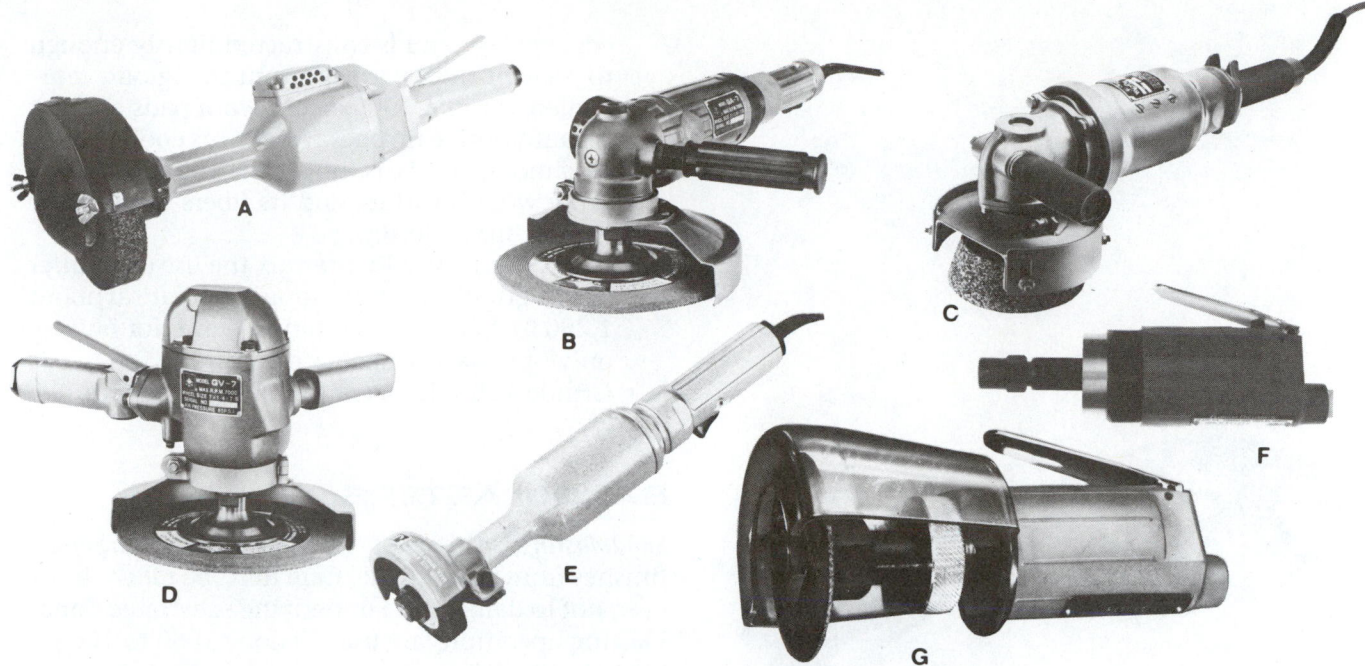

FIGURE 4-24 Study the various types of grinders found in a body shop: (A) horizontal grinder; (B) vertical grinder; (C) vertical grinder with cup wheel; (D) angle grinder; (E) small wheel grinder; (F) die grinder; and (G) cut-off grinder.

CAUTION

If an air grinder will cut metal, it will cut bone and flesh like "a hot knife going through butter." Wear leather gloves and a full face shield when grinding.

There are, of course, several other grinders used in the body shop. The more common ones include the following:

- *Horizontal grinder* (Figure 4–24A), used for heavy-duty grinding.
- *Vertical grinder* (Figure 4–24B), a larger version of the disc grinder. With a sanding pad, this grinder can be converted into a disc sander. Most vertical grinders can be used with both straight wheels and cup wheels (Figure 4–24C).
- *Angle grinder* (Figure 4–24D), used primarily for smoothing, deburring, and blending welds.
- *Small wheel grinder* (Figure 4–24E), used with cone wheels, wire brushes, or collet chucks and burrs, in addition to a straight grind.
- *Die grinder* (Figure 4–24F), used with mounted points and carbide burrs for a variety of applications, such as weld cleaning, deburring, blending, and smoothing. It is available in both straight and angle head designs.

- *Cut-off grinder* (Figure 4–24G), used to cut through muffler clamps and hangers with ease. It also slices through sheet metal and radiator hose clamps.

POLISHER/BUFFERS

An **air polisher** is used to smooth and shine painted surfaces by spinning a soft buffing pad. Polishing or buffing compound is rubbed over the paint with the polisher pad. This removes minor paint imperfections to increase *paint gloss* (shine). See Figure 4–25.

A *polishing pad* is a thick, cotton, synthetic cloth, lamb's wool, or foam cover that fits over the polisher's rubber backing plate. Sometimes the pad and backing plate are *integral* (one-piece). An integral pad would have a plastic backing with a foam pad. The pad could be smooth or wavy.

A *pad cleaning tool* is a metal star wheel and handle that will clean dried polishing compound out of the pad. The wheel is held onto the spinning pad to clean out and soften the cloth material. It will help prevent **swirl marks** (round or curved lines) in the paint caused by buffing. Note that a cleaning tool should not be used on foam pads because it damages the pad.

One of the most important considerations when operating a polisher/buffer is the selection of the proper buffing pad. Here are some points to consider when making a selection:

- Match the pad to the needs of the job. Low pile heights work best for the early stages of cutting

A

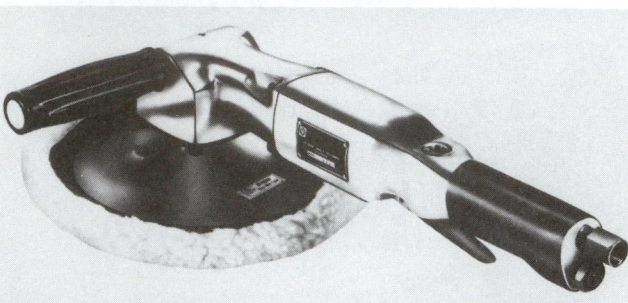

B

FIGURE 4-25 Two types of polisher/buffers are shown: (A) angle and (B) vertical. *(Photo A courtesy of Florida Pneumatic Mfg. Corp.; photo B courtesy of Chicago Pneumatic Tool Division)*

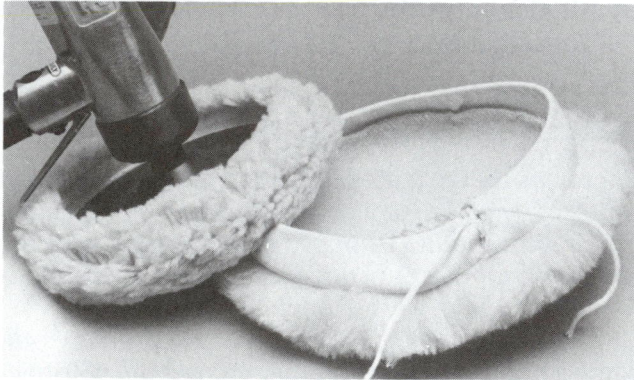

FIGURE 4-26 Typical polishing/buffing pads or bonnets fit over the tool. If kept clean and in good shape, they will make paint bright and shiny, with little or no buffing or swirl marks.

and compounding. High pile heights are better for light compounding and critical jobs. Thicker pads are good for touchups and blending (Figure 4-26) where raised body lines demand cushioning. For final finishing and waxing, use a clean lamb's wool bonnet or foam pad. These will run the coolest and offer the most polishing action. For further protection, consider using pads with rounded-up edges.

- Let the pad do the work. Using the design of a pad to its best advantage means changing pads at the various stages of the job.
- Be sure that the pad does not load up too fast, does not burn (a rounded-up edge helps prevent edge burns), and is constructed tightly enough to prevent wool particles from flying out.
- Remember that 100 percent wool pads are best for automotive finishes. Wool runs cooler, cushions more, and lasts longer than synthetics because wool breathes, and its fibers retain their natural spring longer.
- It is poor practice to intermix the use of a buffer with a grinder. A buffer should operate at about 1,200 to 2,200 rpm. Employ a buffer for buffing only or surface scratches can result.
- Grinders should operate at about 5,000 rpm.

SANDBLASTERS

Sandblasting is the most effective way to remove all finishes from any vehicle. Care must be taken, however, not to damage the underlying substrates. Sandblasting operations are usually done at 60 to 100 psi (414 to 690 kPa).

The use of plastic media blasting has replaced sand in many body shops. The process is very similar in principle to sandblasting. Instead of using hard silica sand, much softer reusable plastic particles are used at low blasting pressures of 20 to 40 psi (138 to 276 kPa). At the lower pressures, the plastic medium removes paint coatings without damaging the underlying substrates, including thin aluminum, steel, fiberglass, and even plastics.

Soda blasters are another new method of stripping paint. Baking soda is used, either wet or dry, under air pressure to strip away the old finish. As with plastic media blasting, soda can be safely used on almost any substrate.

As shown in Figure 4-27, there are two basic types of sandblasters:

- Standard sandblaster
- Captive sandblaster

The standard sandblaster is usually operated outdoors, while the captive units can be used indoors. The indoor sandblasters have a nozzle assembly that confines the blasting action. A vacuum in the machine sucks the abrasive and debris back into the machine. Plastic can be used in both types of blasters.

AIR CHISELS OR HAMMERS

Of all auto air tools, the air chisel or hammer (Figure 4-28) is one of the most useful, especially as a spot weld remover. Used with the accessories illustrated in Figure 4-29, this tool will perform the following operations:

- *Universal joint and tie rod tool.* It helps shake loose stubborn universal joints and tie rod ends.

A

FIGURE 4-28 A typical air chisel is at work removing a badly rusted panel. *(Courtesy of Dobbs Publishing)*

B

FIGURE 4-27 Two types of sandblasters are shown: (A) standard pressure and (B) captive. Captive can be used inside the shop since it prevents particles from flying all over.

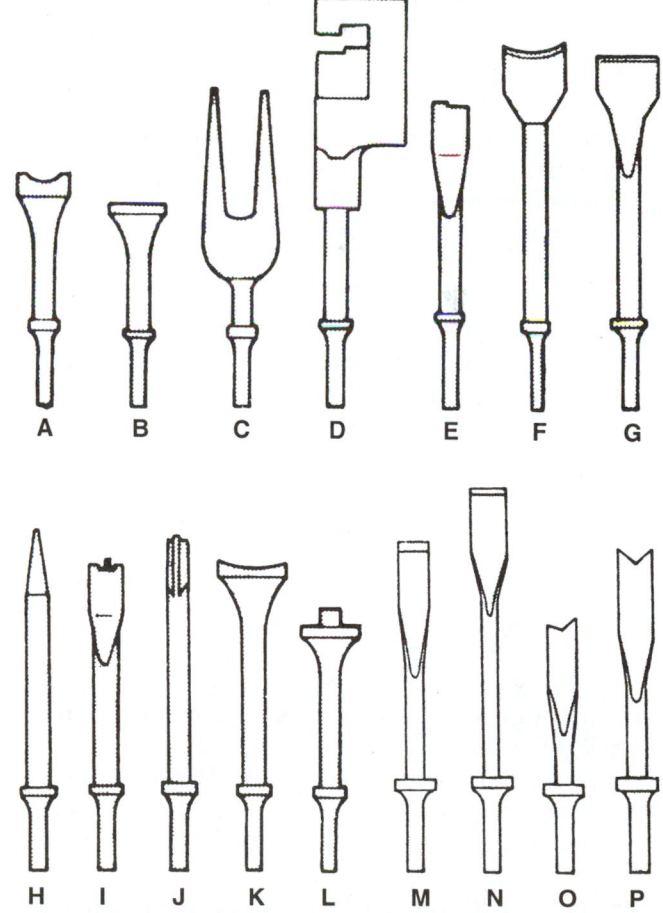

FIGURE 4-29 Typical air chisel accessories: (A) universal joint and tie rod tool; (B) smoothing hammer; (C) ball joint separator; (D) panel crimper; (E) shock absorber chisel; (F) tail pipe cutter; (G) scraper; (H) tapered punch; (I) edging tool; (J) rubber bushing splitter; (K) bushing remover; (L) bushing installer; (M) rivet cutter; (N) flat chisel; (O) sheet metal cutter; (P) spot weld breaker.

SHOP TALK

When using an air chisel or hammer, keep the following pointers in mind.

- Always use a chisel retainer when operating an air hammer.
- Position tool by starting slowly, then increasing power. Avoid running into hardware, frames, and so forth, with sheet metal cutting tools.
- Check chisel shanks periodically for peening, and grind a new chamfer when required.
- Do not let the chisels ride out of the air hammer.
- Keep cutters sharp.
- Wear thick gloves, a face shield, and hearing protection when power cutting metal.

- *Smoothing hammer.* A good accessory for reworking metal.
- *Ball joint separator.* The wedge action breaks apart frozen ball joints.
- *Panel crimper.* It forms a step in a panel where a damaged section has been removed. The filler panel will then fit flush, resulting in a strong, professional joint.
- *Shock absorber chisel.* Quick work of the roughest jobs is made without the usual bruised knuckles and lost time. It easily cracks frozen shock absorber nuts.
- *Tail pipe cutter.* The cutter slices through mufflers and tail pipes.
- *Scraper.* Removing undercoating in addition to other coverings is this accessory's function.
- *Tapered punch.* Driving frozen bolts, installing pins, and punching or aligning holes are some of the many uses for this accessory.
- *Edging tool (claw ripper).* It is used to slice through sheet metal, leaving a smooth edge.
- *Rubber bushing splitter.* Old bushings can be opened up for easy removal.
- *Bushing remover.* This accessory is designed to remove all types of bushings. The blunt edge pushes but does not cut.
- *Bushing Installer.* The installer drives all types of bushings to the correct depth. A pilot prevents the tool from sliding.

In addition to special chisels, there are the so-called standard types that can be used for cutting rivets, nut, and bolts, plus removing weld splatter and breaking spot welds.

A

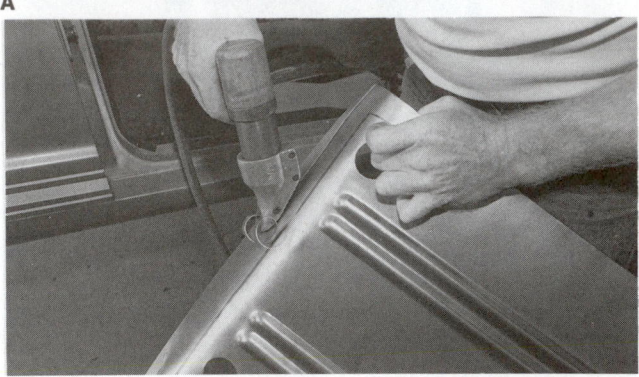

B

C

FIGURE 4–30 (A) An air punch will quickly drive holes in parts for riveting or welding. (B) Air shears will easily cut and trim sheet metal panel during installation. (C) The technician is using a nibbler to cut sheet metal while fabricating a rust repair patch out of sheet metal. *(Courtesy of Dobbs Publishing and Snap-On Tools Corp.)*

OTHER PNEUMATIC BODY SHOP TOOLS

There are several other air tools that can be found in some auto body repair shops (Figure 4–30). They include:

- *Needle scalers* (Figure 4–31A). They are used for derusting and cleaning of metals as well as for peening welded joints.

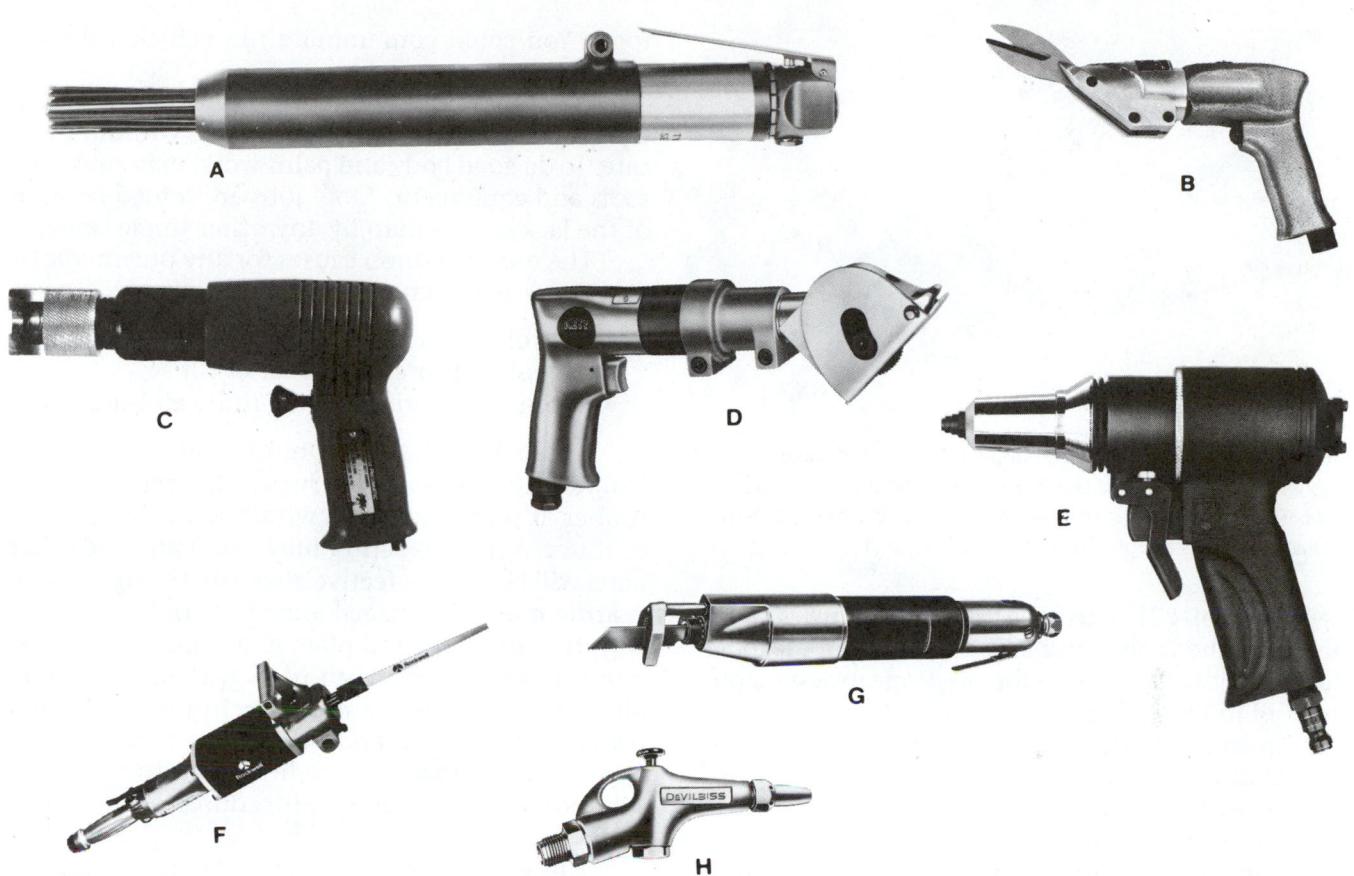

FIGURE 4-31 Other popular air tools found in auto body shops are: (A) needle scalers; (B) metal shears; (C) air nibbler; (D) panel saw; (E) power riveter; (F) air hacksaw; (G) reciprocating saw/file; and (H) air blow gun.

- *Metal shears* (Figure 4-31B). They cut, trim, outline, and shear plastic, tin, aluminum, and other metals up to 18-gauge rolled steel. Shears can be air- or electric-powered.
- *Nibbler* (Figure 4-31C). It can cut sheet metal up to 16-gauge in any configuration and also cuts holes as small as 1 inch in diameter. Nibblers can be air- or electric-powered.
- *Panel saw* (Figure 4-31D). It cuts mild steel (up to 16-gauge), plastic up to $3/8$-inch, and aluminum up to $1/4$-inch.
- *Power riveter* (Figure 4-31E). It sets rivets of up to $3/16$-inch steel or closed-end rivets. It provides an effective, high-strength fastening technique.
- *Air hacksaw* (Figure 4-31F). It has many metal cutting functions in any metal shop.
- *Reciprocating saw/file* (Figure 4-31G). It is a dual action tool that trims or shapes sheet metal and plastic.
- *Air blow gun* (Figure 4-31H). Possibly the smallest air tool in the shop, it is one of the most worthwhile. It blows away dust and dirt from any small hard-to-reach place.

There are some specialized air-powered auto mechanic tools that are found in commercial garages. The names of these tools usually imply their use; they include a radiator tester, cylinder hoist, brake tester, transmission flusher, engine cleaner, and spark plug cleaner. They are, however, seldom found in a typical auto body and refinishing shop.

PNEUMATIC TOOL MAINTENANCE

Air tools need little upkeep. However, you will have problems if basic maintenance is not performed. For example, moisture gathers in the air lines and is blown into tools during use. If a tool is stored with water in it, rust will form and the tool will wear out quickly.

Air tool lubrication should be done on all air tools. A few drops of special air tool oil or a mixture of oil and automatic transmission fluid should be used. Most air tool manufacturers recommend the use of special oil.

Squirt a couple of drops of oil into the air inlet or into special oil holes on the tool before and after

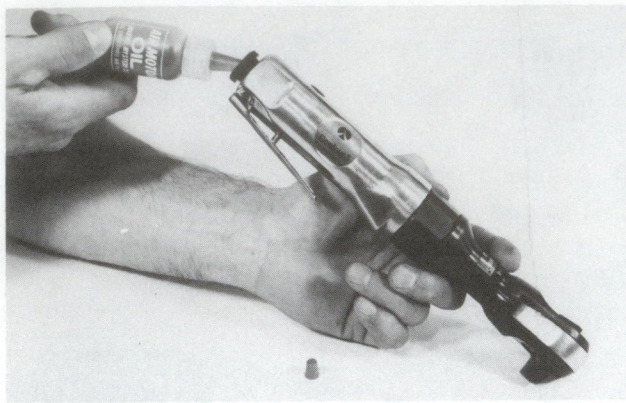

FIGURE 4-32 You must hand oil air-powered tools frequently. It is best to oil them before putting them away at night. Place a few drops of oil in them and then run the tool to circulate the oil through the tool.

use (Figure 4–32). This will prevent rapid wear and rusting of the vane motor and other parts in the tool. Run the tool after adding the oil. Wipe excess oil off the tool to keep it from the body parts.

An **in-line oiler** is an attachment that will automatically meter oil into air lines for air tools. It can be used on lines used exclusively for air tools but not for spray guns.

Remember not to use in-line oilers in the paint area. Never over-oil sanders, grinders, and other air tools. You could contaminate the vehicle's surface with oil.

It is important to remember that your tools and equipment cannot function correctly without good care. To do good body and paint work, maintain your tools and equipment. More jobs are ruined because of the lack of care than by any other single cause.

The most common causes for any pneumatic or air tool to malfunction are:

• Lack of proper lubrication
• Excessive air pressure or lack of it
• Excessive moisture or dirt in the air lines

All air tools have recommended air pressures (Table 4–1). If a tool is overworked, it will wear out sooner. If something goes wrong with the tool, fix it. If not, a chain reaction may occur and the other parts will become defective also. For example, if the gearing must be replaced and the tool is used anyway, the rotor and end plate may soon wear out as a result. A tool with worn parts will also use more air pressure. The air compressor, in turn, will then become overworked and put out air that is not as clean or dry and is shot right back into the tools. Air tool troubleshooting procedures are given in Table 4–2.

Full information on pneumatic air system operation is given in Chapter 5.

TABLE 4-1: AIR CONSUMPTION CHART†					
Tool	**Scfm***	**Psi***	**Tool**	**Scfm***	**Psi***
Air brush	1	10–50	Needle scaler	3–4	70–90
Air chisel	4	70–90	Nibbler	8	70–90
Air filter cleaner	3	70–90	Nutsetter	6–7.5	70–90
Air hammer	6–10	70–90	Paint sprayer	0.7–5	10–70
Blow gun	1–2.5	40–90	Panel saw	4–8	70–90
Brake tester	3.5	70–90	Pneumatic garage door	2	90–150
Burring tool	5	70–90	Polisher	2	70–90
Car washer	8.5	40–90	Ratchet wrench	4	70–90
Cut-off grinder	4–8	80–90	Riveter	4.5–5.5	70–90
Drill 3/8 inch	4–6	70–90	Sandblast gun	2.2–4	30–90
Engine cleaner	4–6	70–90	Sandblast gun/hopper	2–6	40–90
Grease gun	3–4	90–150	Sander, disc	4–6	60–80
Grinders, die	4–6	70–90	Sander, double action	6–8	60–80
Grinders, vertical	6–12	70–90	Sander, finish	6–8	60–80
Hacksaw	6–8	70–90	Sander, straight line	6–8	70–90
Hoist (1 ton)	1	70–90	Screwdriver	2–6	70–90
Hydraulic lift	5–7	90–150	Shears	5–8	70–90
Impact wrench 1/4 inch	1.4	70–90	Spark plug cleaner	5	70–90
Impact wrench 3/8 inch	3	70–90	Tire changer	1	90–150
Impact wrench 1/2 inch	4	70–90	Tire chuck	1.5	10–50
Material tank	1.8	10–50			

* SCFM: Standard cubic feet per minute
 PSI: Pounds per square inch
†Always check with the tool manufacturers for the actual air consumption of the tools being used. These figures are based on averages and should not be considered accurate for any particular make of tool.

TABLE 4-2: TROUBLESHOOTING AIR TOOLS

Problem	Probable Cause	Recommended Action
Air Drills		
Tool will not run or runs slowly, air flows slightly from exhaust, spindle turns freely.	Motor or throttle plugged with dirt	1. Check for dirt in air inlet. 2. Pour liberal amount of air tool oil in air inlet. 3. Operate trigger in short bursts. 4. Disconnect air supply, then turn empty and closed drill chuck by hand. Reconnect air supply. 5. If still not functional, tool should be checked by authorized service center.
Tool will not run, air flows freely from exhaust, spindle turns freely.	Rotor vanes stuck with dirt or varnish	1. Pour liberal amount of air tool oil in air inlet. 2. Operate trigger in short bursts. 3. Disconnect air supply, then turn empty and closed drill chuck by hand. Reconnect air supply. 4. If still not functional, tool should be checked by authorized service center.
Tool locked up, spindle will not turn.	Broken motor vane Gears broken or jammed by foreign object	Tool should be checked by authorized service center.
Tool will not shut off.	Throttle valve O-ring blown off seat	See parts list for part number and replace O-ring or send to authorized service center.
Air Hammers		
Tool will not run.	Cycling valve or throttle valve clogged with dirt or sludge Piston stuck in cylinder bore by rust or dirt	1. Pour liberal amount of air tool oil in air inlet (check for dirt). 2. Operate trigger in short bursts (chisel in place and against solid surface). 3. If not free, first disconnect air supply, then tap nose or barrel lightly with plastic mallet, reconnect air supply, and repeat above steps. 4. If still not free, disconnect air supply, insert a 6-inch piece of 3/8-inch diameter rod in nozzle and lightly tap to loosen piston in rearward direction. Reconnect air supply, and repeat Steps 1 and 2.
Chisel stuck in nozzle.	End of shank peened over	Tool should be sent to authorized service center.
Air Ratchets		
Motor runs, spindle doesn't turn or turns erratically.	Worn teeth on ratchet or pawl Weak or broken pawl pressure spring Weak drag springs fail to hold spindle while pawl advance for "another bite"	Replacement parts should be installed by authorized service center.
Motor will not run, ratchet head indexes crisply by hand.	Dirt or sludge in motor parts	1. Pour a liberal amount of air tool oil into air inlet. 2. Operate throttle in short bursts. 3. With socket engaged on bolt, alternately tighten and loosen bolt by hand. 4. If motor remains jammed, tool should be checked by authorized service center.

TABLE 4-2:   TROUBLESHOOTING AIR TOOLS (CONTINUED)

Problem	Probable Cause	Recommended Action
	Air Wrenches	
Tool runs slowly or not at all, air flows only slightly from exhaust.	Airflow blocked by accumulation of dirt Motor parts jammed with dirt particles Power regulator might have simply vibrated to closed position	1. Check air inlet strainer for blockage. 2. Pour liberal amount of air tool oil into air inlet. 3. Operate tool in short bursts—quickly reversing rotation back and forth. 4. Repeat as needed. 5. If this fails to improve performance, tool should be serviced at authorized service center.
Tool will not run, exhaust air flows freely.	One or more motor vanes stuck due to sludge or varnish buildup Motor jammed due to rust	1. Pour a liberal amount of air tool oil into air inlet. 2. Operate tool in short bursts of forward and reverse rotation. 3. Tap motor housing lightly with plastic mallet. 4. Disconnect air supply, then attempt to free motor by rotating drive shank manually. (Some clutches will not engage sufficiently for this operation.) 5. If tool remains jammed, it should be serviced by an authorized service center.
Sockets will not stay on.	Worn socket retainer ring or soft back-up ring	1. Wear safety glasses. 2. Disconnect air supply. 3. Using external retaining ring pliers, expand old retaining ring and remove OR if retaining ring pliers not available, clamp tool "lightly" in soft jaw vise. 4. Holding square drive with appropriate open-end wrench, pry old retainer ring out of groove with small screwdriver. 5. Always pry off ring away from body; it can be propelled outward at high velocity. 6. Replace backup O-ring and retainer ring with correct new parts. (See parts lists that accompanied tool.) 7. Place retaining ring on table, press tool shank into ring in a rocking motion. Snap into groove by hand.
Premature shank wear.	Use of chrome sockets or excessively worn sockets	Discontinue use of chrome sockets. Remember that chrome sockets have a hard surface and relatively soft core. Drive hole will become rounded, but still be very hard. Besides the danger of splitting, they will wear out wrench shanks prematurely.
Tool gradually losing power but still runs at full free speed.	Clutch parts worn, perhaps due to lack of lubricant. Engaging cam of clutch worn or sticking due to lack of lubricant.	*Oil Lubed* 1. Check for presence of clutch oil (where oil is specified for clutch) and, removing oil fill plug, tilt to drain all oil from clutch case. Refill with 30 weight SAE oil or that recommended by manufacturer, but only the amount specified. 2. Check for excess clutch oil. Clutch cases need only be 50 percent full. Overfilling can cause drag on high-speed clutch parts. A typical 1/2-inch oil-lubed wrench only requires 1/2 ounce of clutch oil.

TABLE 4-2:   TROUBLESHOOTING AIR TOOLS (CONTINUED)

Problem	Probable Cause	Recommended Action
Tool gradually losing power but still runs at full free speed (continued).	Clutch parts worn, perhaps due to lack of lubricant. Engaging cam of clutch worn or sticking due to lack of lubricant.	*Grease Lubed* Vibration and heat usually indicate insufficient grease in the clutch chamber. The average greasing interval is specified in parts list. Severe operating conditions might require more frequent lubrication. 1. Check for excess grease by rotating drive shank by hand. It should turn freely. Excess is usually expelled automatically. 2. If disassembly is required for greasing, it should be done carefully to maintain orientation of mating parts.
Tool will not shut off.	Throttle valve O-ring broken or out of position Throttle valve stem bent or jammed with dirt particles	1. Remove assembly and install new O-ring. 2. Lubricate with air tool oil and operate trigger briskly. If operation cannot be restored, tool should be checked by authorized service center.

4.2 ELECTRIC-POWERED TOOLS

As mentioned earlier in this chapter, shop tools such as sanders, polishers, impact tools (Figure 4–33), and drills can also be powered by electric motors. The most important electric-only tools are drill presses, bench grinders, vacuum cleaners, heat guns, and plastic welders. The two most popular automotive metal welding systems—MIG and spot—are electrically operated (Figure 4–34).

FIGURE 4-33 An electric-powered impact wrench is not common in a professional shop. *(Courtesy of Black & Decker, Inc.)*

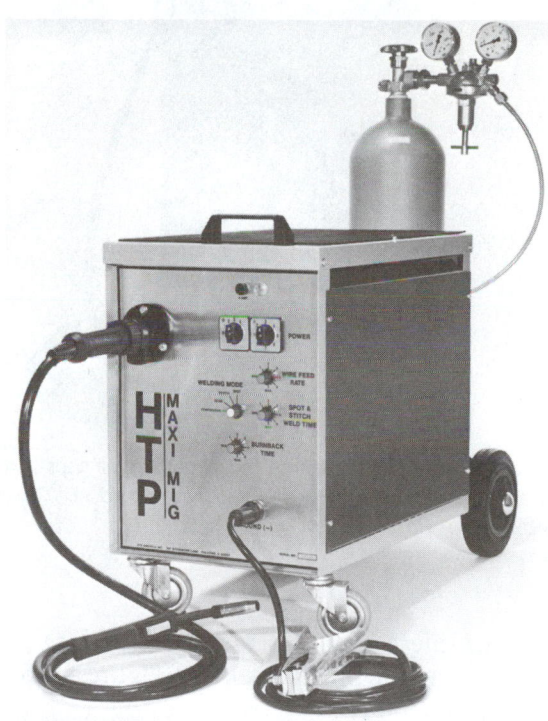

A

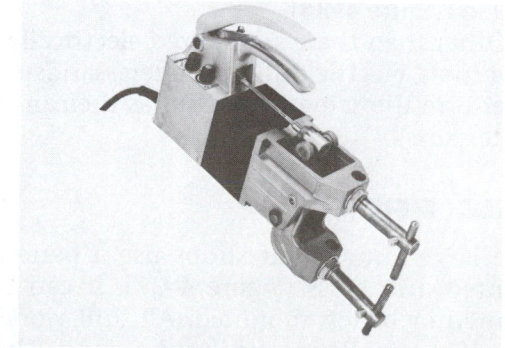

B

FIGURE 4-34 Study welder types: (A) MIG welder and (B) squeeze-type resistance spot welder.

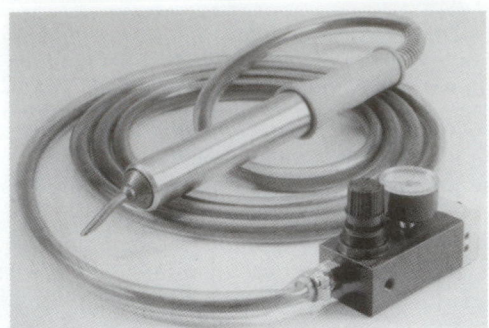

FIGURE 4–35 A typical hot-air plastic welder will join and repair many plastic parts. *(Courtesy of Seelye, Inc.)*

FIGURE 4–36 A plasma metal cutter will rapidly melt and cut metal panels for quick, accurate removal. *(Courtesy of Century Mfg. Co.)*

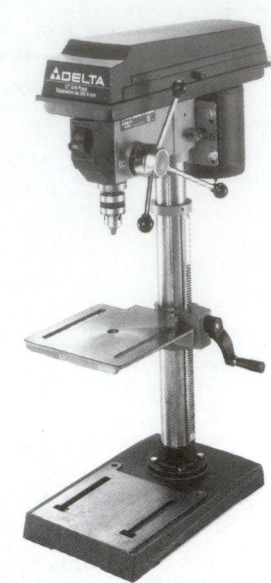

FIGURE 4–37 The bench type drill press is handy for drilling smaller, thicker parts. *(Courtesy of Delta International Machinery Corp.)*

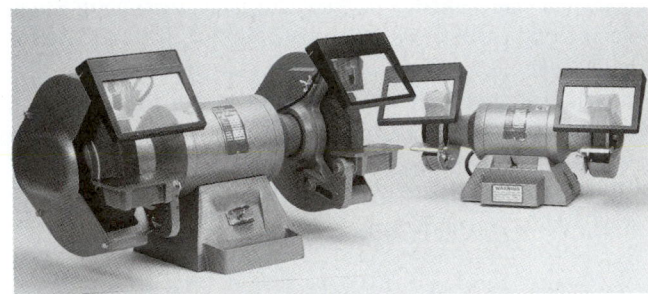

FIGURE 4–38 A bench grinder is often used to sharpen tools and dress smaller parts. *(Courtesy of Snap-on-Tools Company, Copyright Owner)*

Complete data on metal welders is given in Chapter 6 while plastic welders (Figure 4–35) are described in Chapter 13.

For metal cutting operations, the electric-powered plasma arc cutter is the one most recommended (Figure 4–36).

Other than these specialized electrically driven power tools, electric drills, polishers, sanders, and so on perform the same shop tasks as their pneumatic counterparts.

DRILL PRESS

Some large auto repair shops use a permanently mounted drill press (Figure 4–37). It can be floor mounted or bench mounted. All drill work is performed on a table attached to the stand. The worktable can be adjusted up or down. The drill speed is variable for various materials and thicknesses.

BENCH GRINDER

This electric power tool (Figure 4–38) is generally bolted to one of the shop's workbenches. A bench grinder is classified by wheel size, with 6- to 10-inch wheels the most common in auto repair shops. Three types of wheels are available with this bench tool:

- *Grinding wheel*, for a wide variety of grinding jobs from sharpening cutting tools to deburring.
- *Wire wheel brush*, used for general cleaning and buffing, removing rust and scale, paint removal, deburring, and so forth.
- *Buffing wheel*, for general purpose buffing, polishing, light cutting, and finish coloring operations.

When using a bench grinder (these safety rules also apply to portable air grinders), remember to:

- Always use a wheel with a rated speed equal to or greater than the grinder's.
- Always use a wheel guard.

A

B

FIGURE 4–39 (A) A portable body shop vacuum cleaner is sometimes used to capture sanding dust and other debris while you are working. (B) A hand-held vacuum cleaner can be used to do minor cleaning of the interior of the vehicle.

- Inspect wheels for wear or cracks before using them. Use the correct wheel for the job and mount it properly.
- Always wear safety goggles or a face shield and make sure the eye shields of the grinder operator are in position.
- Adjust the rest as needed whenever the gap between it and the grinding wheel exceeds $\frac{1}{3}$ inch.

VACUUM CLEANER

A must in every body and refinishing shop is a *vacuum cleaner* for removing dust and debris from the vehicle interior (Figure 4–39A). It should be one of the first tools used when a vehicle comes in for refinishing. An incoming vehicle should be completely washed and vacuumed before it is prepared for refinishing. This will greatly reduce the chance of dirt getting into the complete job. The vehicle should also be vacuumed after refinishing to make the customer happy.

There are two basic types of shop vacuum cleaners: the dry pickup-type and the wet/dry unit. For interior vehicle cleaning, the portable vacuum cleaner is popular (Figure 4–39B).

Vacuum tool attachments allow a vacuum cleaner to be attached to a power tool. They are available for sanders and other shop tools. For example, when on a sander, the vacuum cleaner will keep most of the sanding dust out of the shop. This is good for your lungs and the paint job.

POWER WASHERS

Power washers can be used in exterior car preparation, engine cleaning, undercarriage cleaning, shop degreasing and cleaning, and snow and salt removal

FIGURE 4–40 Typical gasoline powered pressure washer. *(Courtesy of Graco, Inc.)*

from vehicles. Figure 4–40 shows typical units. Before using a power washer, check the OSHA regulations.

To keep dust out of the refinishing shop, some power orbital sanders—both electric and air-powered—have a sanding dust pickup arrangement. This helps keep the shop cleaner. It also increases air cleanliness for safer working conditions.

Figure 4–41 shows a straight line sander equipped with its own vacuum system and a catching bag, while Figure 4–42 illustrates an adapter that connects to a central vacuum cleaner system. An air sander equipped with a vacuum pickup usually requires a special sandpaper.

FIGURE 4-41 A sander equipped with a dust collecting bag. *(Courtesy of Hutchins Mfg. Co.)*

HEAT GUN

Heat guns have many uses in the auto body shop where controlled heat is needed (Figure 4–43). They are used in almost all vinyl roof repairs as well as other plastic repairs. They can be used in some panel shrinking jobs as well as speeding up drying times. The overall use of heat guns can be found in several chapters of this book.

BATTERY CHARGER

A **battery charger** converts 120 volts AC into 13 to 15 volts DC for recharging drained batteries. One is shown in Figure 4–44.

To use a battery charger, connect red to positive and black to negative. The red lead on the charger goes to the positive terminal of the battery. The black lead goes to ground or the negative battery terminal.

After connecting the charger, adjust its settings as needed (12 volt battery, fast or slow charge, etc.). If a battery is low, it is best to slow charge the battery for several hours. A fast charge for a few minutes will not restore battery charge properly.

ELECTRIC POWER TOOL SAFETY

To protect the operator from electric shock, most power tools are built with an *external ground plug*. There is a wire that runs from the motor housing, through the power cord, to a third prong on the power plug.

FIGURE 4-42 This central vacuum system will pull sanding dust from the shop. If connected to the tailpipe, it will also keep deadly exhaust fumes out of the shop area. *(Courtesy of Eurovac, Inc.)*

FIGURE 4-43 A heat gun can be used to apply controlled heat to objects, as when removing stick-on decals. *(Courtesy of Black & Decker, Inc.)*

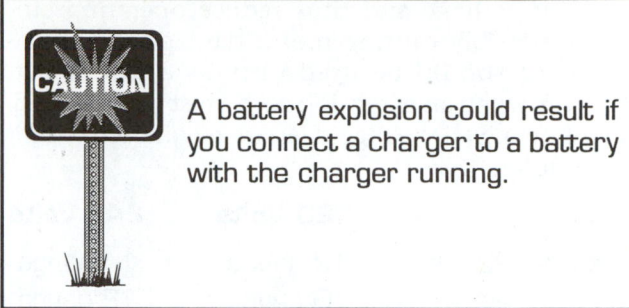

CAUTION

A battery explosion could result if you connect a charger to a battery with the charger running.

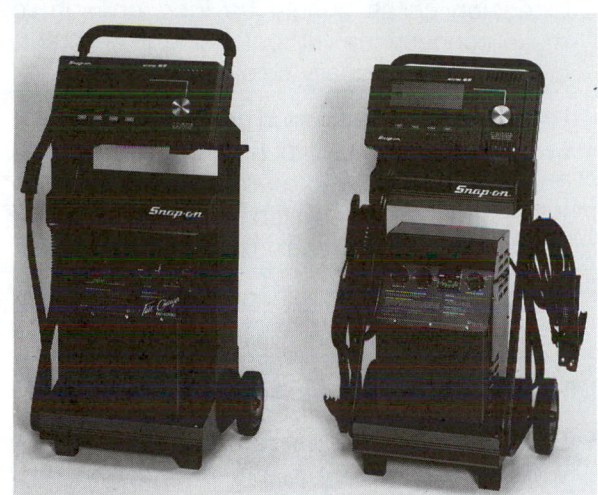

FIGURE 4-44 A battery charger is often needed to recharge dead batteries in a body shop. Connect red lead to battery positive and black lead to battery negative. *(Courtesy of Snap-On-Tools Company, Copyright Owner)*

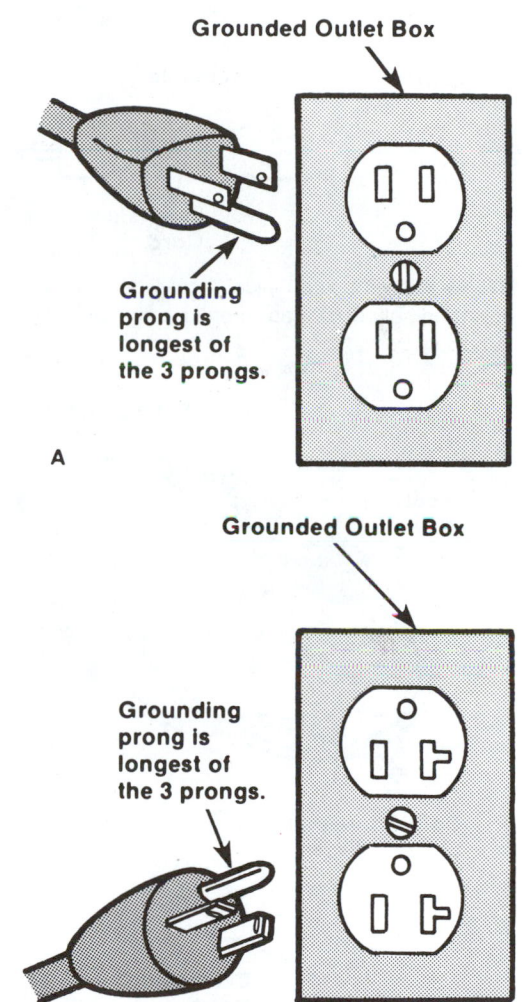

Grounded Outlet Box

Grounding prong is longest of the 3 prongs.

A

Grounded Outlet Box

Grounding prong is longest of the 3 prongs.

B

FIGURE 4-45 (A) Note the approved type of three-prong grounding plug and outlet box for 120 volts AC; (B) note the approved type of three-prong grounding plug and outlet box for 240 volts AC.

When this third prong is connected to a three-hole electrical outlet, the grounding wire will carry any current that leaks past the electrical insulation of the tool away from the operator and into the shop's wiring. In most modern electrical systems, the three-prong plug fits into a three-prong, grounded receptacle.

If the tool is operated at or less than 120 volts, it has a plug like that shown in Figure 4-45A. If it is for use on greater than 120 or but not more than 240 volts, it has a plug like that shown in Figure 4-45B. In either type, the green (or green and yellow) conductor in the tool cord is the grounding wire. This grounding wire should be connected to the longer, rounded prong of the plug and never to the shorter flat prongs.

Some of the new electric power tools are self-insulated and do not require grounding. These tools have only two prongs since they have a nonconducting plastic housing. In shop operations, never connect a three-prong to a two-prong adapter plug.

Extension Cords

If an extension cord is used, it should be kept as short as possible. Very long or undersized cords will reduce operating voltage and thus reduce operating efficiency, possibly causing motor damage. An extension cord should be used only as a last resort. However, when an extension cord must be employed, the following wire gauge sizes are recommended for different lengths:

Length	120 Volts	240 Volts
Less than 25 feet	12 gauge	14 gauge
25 to 50 feet	10 gauge	12 gauge
50 to 100 feet	8 gauge	10 gauge

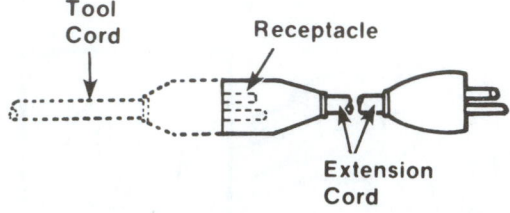

FIGURE 4-46 This is a typical three-wire extension cord with an approved connector cap to ensure continuity of the tool's grounding wire.

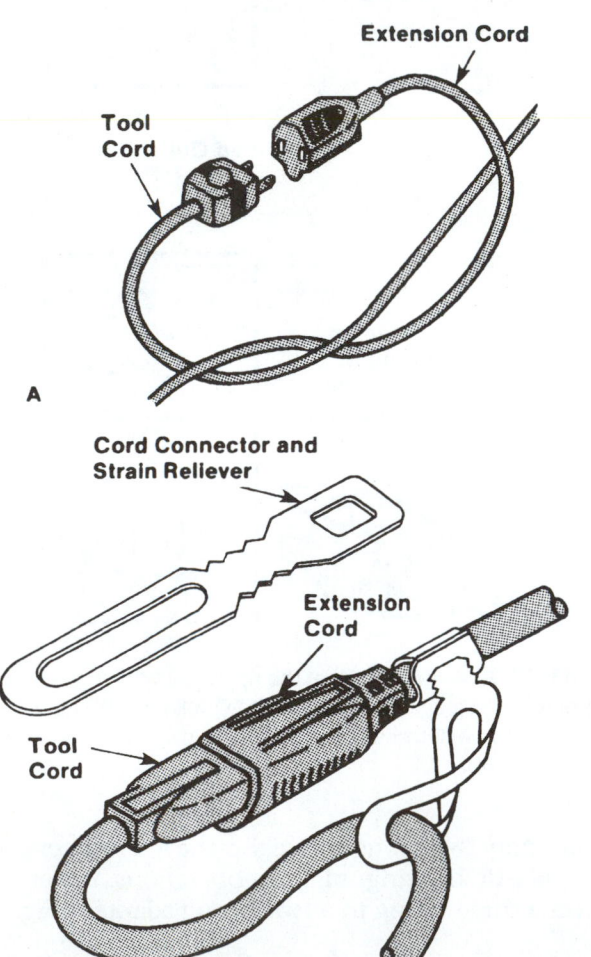

FIGURE 4-47 (A) A knot will prevent the extension cord from accidentally pulling apart from the tool cord during operation; (B) a cord connector will serve the same purpose effectively.

The smaller the gauge number, the heavier duty the cord. These are recommended minimum wire sizes.

Tools with three-prong, grounded plugs must only be used with three-wire grounded extension cords connected to properly grounded, three-wire receptacles (Figure 4-46).

Here are some safety tips to keep in mind when using extension cords:

- Always plug the cord of the tool into the extension cord before the extension cord is inserted into a convenience outlet. Always unplug the extension cord from the receptacle before the cord of the tool is unplugged from the extension cord.

- Extension cords should be long enough to make connections without being pulled taut, creating unnecessary strain and wear.

- Be sure that the extension cord does not come in contact with sharp objects. The cords should not be allowed to kink, nor should they be dipped in or splattered with oil, grease, or chemicals, or allowed to touch hot surfaces.

- Before using a cord, inspect it for loose or exposed wires and damaged insulation. If a cord is damaged, it must be replaced. This advice also applies to the tool's power cord.

- Extension cords should be checked frequently while in use to detect unusual heating. Any cable that feels more than comfortably warm to the bare hand, which is placed outside the insulation, should be checked immediately for overloading.

- See that the extension cord is in a position to prevent tripping or stumbling.

- To prevent the accidental separation of a tool cord from an extension cord during operation, make a knot as shown in Figure 4-47A, or use a cord connector as shown in Figure 4-47B.

Many insurance companies are now requiring automatic shut-off switches on electrical equipment to prevent it from inadvertently being left on.

CORDLESS TOOLS

As an alternative, cordless tools—cordless drills and sanders (Figure 4-48)—have made their way into

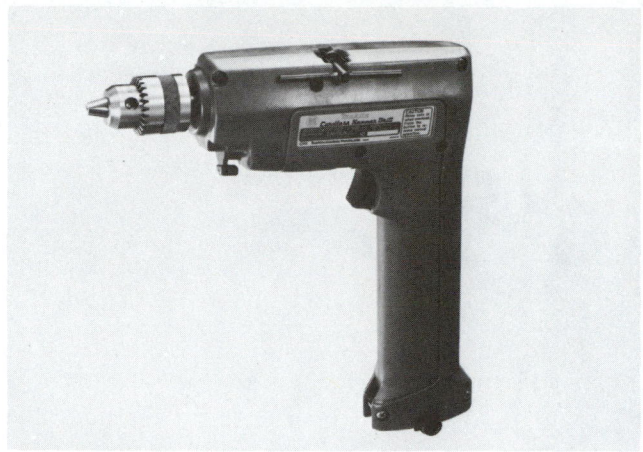

A

B

FIGURE 4-48 Cordless tools are handy for shorter tasks not requiring prolonged tool operation: (A) drill and (B) sander.

Never work under a vehicle supported only by a hydraulic jack. The jack could lower and kill you or someone else. Place jack stands under the car or truck before working (Figure 4–49).

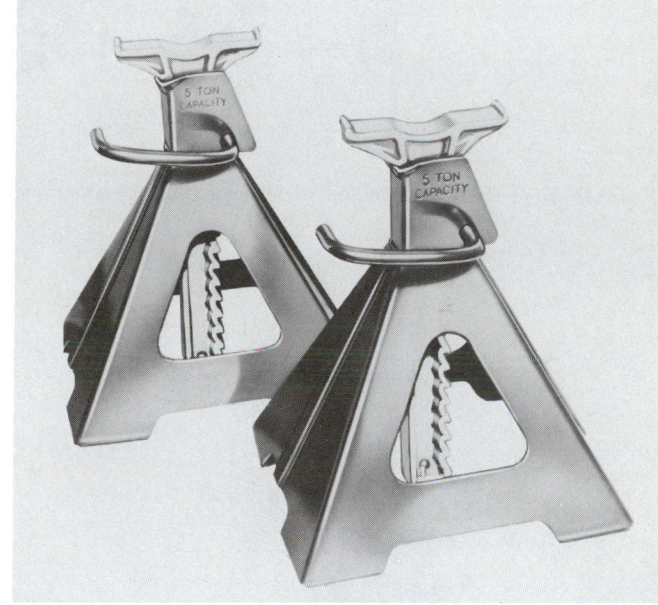

FIGURE 4-49 Jack stands must used before working under a car or truck. Note that the capacity is usually stamped on the jack. *(Courtesy of Lincoln, St. Louis)*

the auto body shop. These tools require no air hose or electric cord, but they require recharging. They are handy for smaller jobs where battery life is not a concern.

4.3 HYDRAULICALLY POWERED SHOP EQUIPMENT

Hydraulic body shop equipment uses an oil-like liquid, called hydraulic fluid, to develop the pressure necessary to operate it. This pressure is achieved manually by pumping on a handle or lever to build up the fluid pressure or by a small motor—either air or electrically driven—that provides the pressure needed to force the hydraulic fluid into the equipment's cylinder. The cylinder then causes the tool to operate when a button or a lever is turned.

Hydraulic power equipment is usually classified as "manual," "air over hydraulic," or "electric over hydraulic." Air or electric over hydraulic means that either an air-powered or an electric-powered motor is used to force the hydraulic fluid into the tool's cylinder.

4.4 POWER JACKS AND STRAIGHTENING EQUIPMENT

Hydraulic power equipment in the auto shop is used to operate various jacks. These range from body jacks to large frame/panel pulling and straightener units (Figure 4–50). There are small jack stands used for holding a vehicle after it has been raised in position by a floor jack.

The average shop has approximately a dozen jacks, either air, hydraulic, or a combination of both, depending on the preference of the technicians. (Manual jacks are practically obsolete except for use in confined spaces.) The most popular jacks are:

- *Hand* or *bottle jacks* (Figure 4–51). These tubular-shaped jacks are not specialized. Rather, they perform a variety of functions and range from $1^1/_2$ tons to 100 tons lifting capacity. They are useful when a service jack is too much.

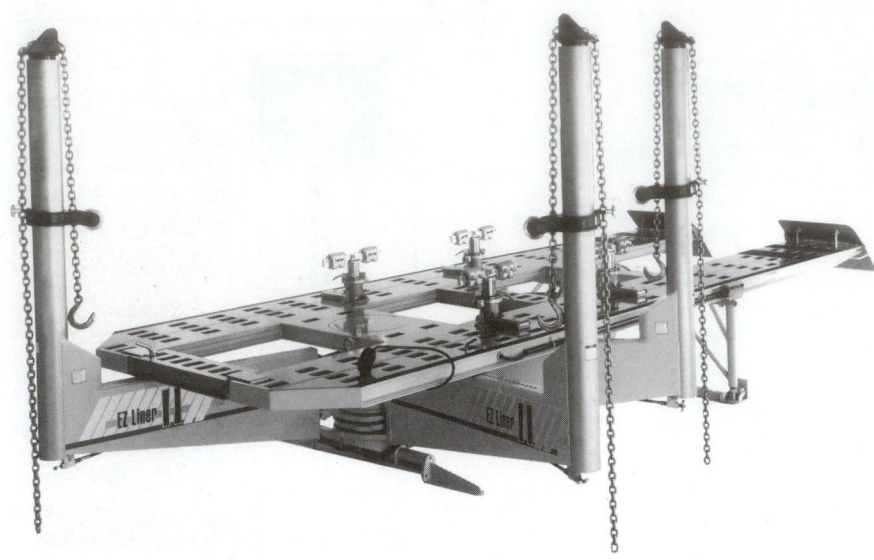

FIGURE 4–50 A frame/panel pulling and straightener unit, also called a frame rack. *(Courtesy of Chief Automotive Systems, Inc.)*

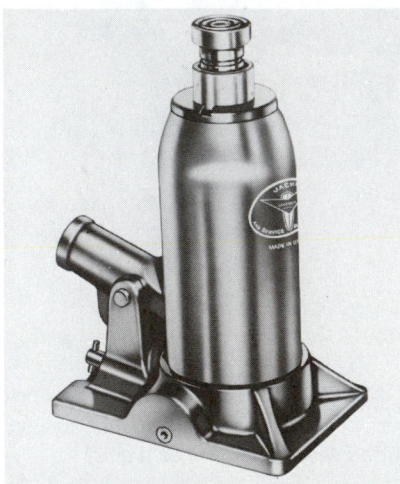

FIGURE 4–51 Even this small hand or bottle jack can exert tons of force. *(Courtesy of Lincoln, St. Louis)*

- *Service jacks* (Figure 4–52). These four-wheeled jacks with a pump handle are by far the most commonly used jack in the body shop. Ranging in lifting capacity from 1¹/₂ tons to 5 tons, these jacks are easily "dollied" around the shop and rolled under the car to lift a section of it, as opposed to the entire structure. Compact and portable, these jacks were developed for a variety of in-shop uses on full-size, intermediate, compact, or subcompact cars. They are also used for road service calls. Service jacks are recommended for all automotive, agricultural, and light truck repair facilities.
- *End lifts* (Figure 4–53). These jacks are either air-over-hydraulic or manual. As the name implies, they lift only a section of the vehicle by adher-

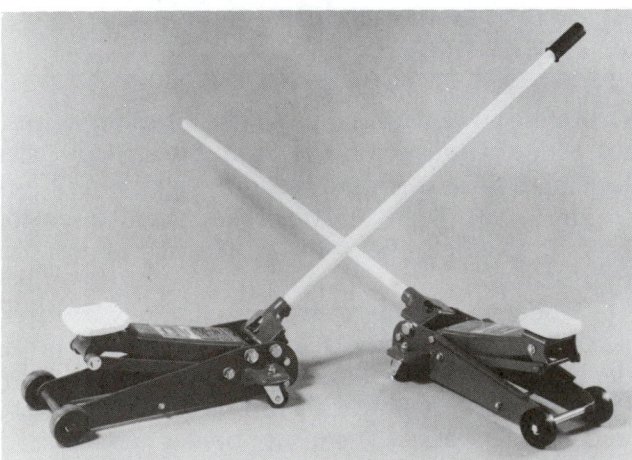

FIGURE 4–52 Service jacks must be placed under recommended lift point on the vehicle. If not, the underbody could be damaged. The service manual will give lift points for each make and model vehicle. *(Courtesy of Blackhawk Automotive, Inc.)*

FIGURE 4–53 A typical end lift jack is not common since new cars often have nonmetal bumpers. *(Courtesy of Lincoln, St. Louis)*

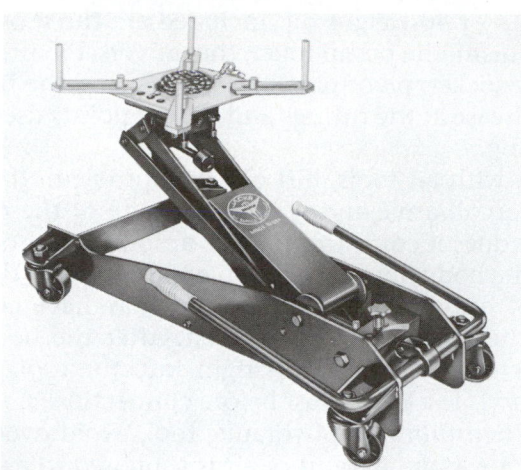

FIGURE 4-54 A transmission jack has a large saddle for fitting around and holding transmissions or other large assemblies, like differentials. *(Courtesy of Lincoln, St. Louis)*

ing to the bumper. They do not lift the sides of the vehicle. Lifting capabilities range from $1^1/_2$ to 7 tons.

- *Transmission jacks* (Figure 4–54). Often it is necessary to remove the transmission, engine, or drivetrain from a unibody before servicing a repair. This jack was developed specifically for this purpose. The lifting capacity ranges from $^1/_4$ to 1 ton and jacks are mechanical, air-over-hydraulic, or manual.

There are many styles of frame/panel straighteners on the market, but there are only two basic types: portable and stationary. Portable units (Figure 4–55) are less expensive to purchase, but they cannot make as many push and pull actions at one time as can the stationary units (Figure 4–56).

FIGURE 4-55 A portable frame/panel straightener unit can be moved around the shop to do minor frame straightening. *(Courtesy of Nicator, Inc.)*

Always use the rated tonnage of a jack for the tons specified. If a jack is rated for 2 tons, do not attempt to use it for a job requiring 20 tons. It is dangerous for both the body technician and the vehicle.

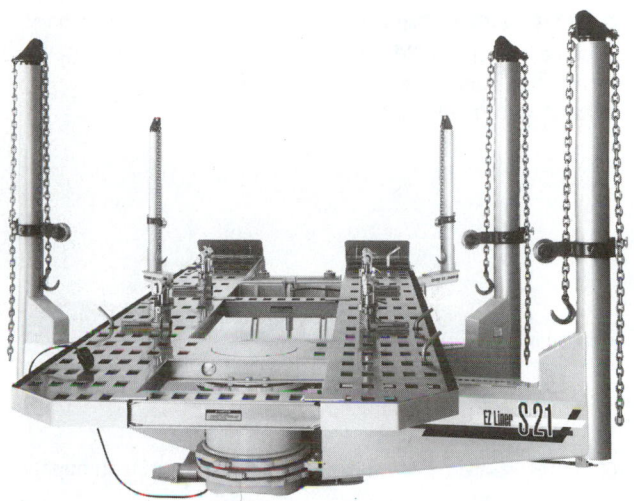

A

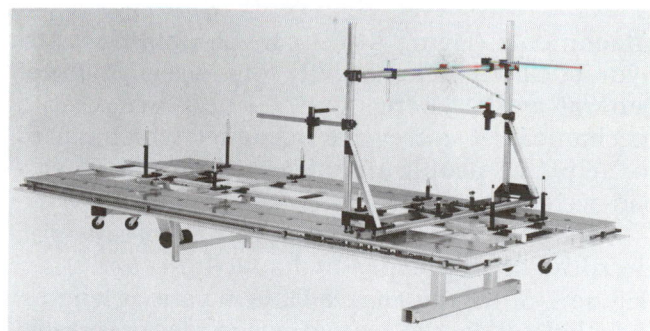

B

FIGURE 4-56 A stationary frame/panel straightener unit is a complex piece of machinery. It requires proper training for successful use. (A) Side view shows large towers with pull chains. (B) Front view shows rack with measuring gauge for checking direction and extent of damage. *(Courtesy of EZ-Liner)*

A **body jack** (also known as an auto body repair kit or portable power unit) is a portable hydraulic pump and ram for body and minor frame straightening operations. It can be used with frame/panel straighteners or can be used by itself. To perform the many different straightening operations involving pushing, pulling, or holding a panel to straighten or align metal, a large assortment of

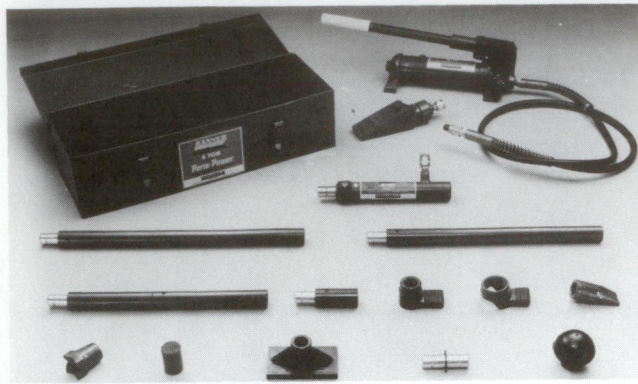

FIGURE 4-57 This is a body jack or body power repair kit, also called a portable power unit. Various attachments fit over rams for doing different repair tasks. *(Courtesy of Blackhawk Automotive, Inc.)*

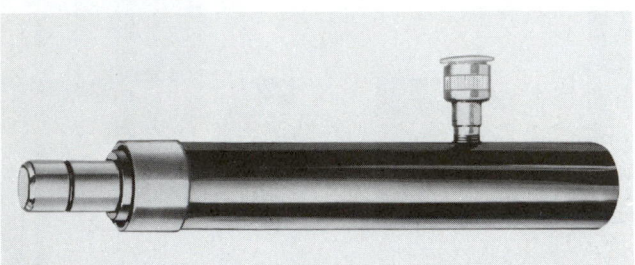

FIGURE 4-58 Even this small ram has a 10-ton capacity. *(Courtesy of Lincoln, St. Louis)*

attachments (Figure 4–57) can be obtained. The hydraulic body jack is usually sold in sets or kits for general work. There is also a bodywork set, a mechanical set, and even a rescue set, which is used to help free people trapped in a vehicle after a bad accident.

The basic *hydraulic jack* unit consists of a manual, electrical, or air pump, a hydraulic hose, and a ram (Figure 4–58). Rams are available in various lengths. Included under rams are two wedge-type or spreader rams used for getting into tight locations.

Body shop technicians must understand how to make hookups for correcting body or frame damage with power jacks. Complete information on using body jacks and frame/panel straighteners is given in Chapters 9 and 10.

4.5 HYDRAULIC TOOL CARE

Hydraulic tools always seem to be there when they are needed. Although they are often ignored, they also require their share of preventive maintenance to avoid failure at a critical moment.

Just because a hydraulic tool is filled with hydraulic fluid does not mean it has been lubricated properly. Moving parts should be lubricated regularly with 30- or 40-weight oil. Included are the moving mechanism, the pump roller, the universal joint, the handle socket, pivot pins, the wheels, and the bearings. Grease at the fittings and sliding points used in pumping.

As with air tools, dirt can be a problem. It will act as an abrasive and scratch the bore of the ram. Where does it come from? Look around the shop at all of the body filler or metal shavings. If the hydraulic tool has an air motor, the dirt may have come from the air hose sitting in dust. After the hose is connected, the dirt shoots right into the tool. The solution is to clean it just before connection.

When filling the hydraulic tool, avoid overfilling it. A certain amount of air is supposed to be left in the reservoir. If it is completely filled with fluid, too much vacuum is created and the fluid will not move out of the reservoir.

Following are major problems (Table 4–3) of hydraulic tools and equipment operation:

- *Spongy effect.* Air trapped in the hydraulic system easily compresses under pressure and causes sponginess. To bleed the system, place the pump at a higher elevation than the hose and ram. The objective is to "float" the air bubbles uphill and back to the reservoir where they belong. Close the valve and extend the unit as far as possible. Open the valve fully, allowing the oil and air to return to the reservoir. Repeat this procedure until the tool starts to extend on the first stroke of the handle. Usually two or three times will do the trick.

- *Tool will not extend all the way.* This is usually a sign of low hydraulic fluid. Fill to the mark on the dipstick. Do not overfill. The tool should be fully retracted to check the oil level. If the tool does not have a dipstick, refer to the manufacturer's manual for filling procedure and fluid level. The hydraulic unit needs the prescribed amount of air chambered in the reservoir because it works on a partial vacuum to avoid venting to the outside.

- *Tool will not retract.* Usually just too much oil and/or air in the system. Bleed the system and fill to the proper level. If that does not correct the problem, inspect for a bent plunger. If neither is the cause, the quick-coupler is probably damaged. Replace it.

- *Tool leaks under pressure.* Make sure that the release valve is fully closed. If it still leaks, there is probably dirt in the return. Check the ball valve. Flush the system with mineral spirits or kerosene. If the problem still exists, the valve is damaged and the unit should go to a repair facility.

- *Handle kickback.* Dirt is in the check valve; flush the system. A damaged check valve should be taken to a repair facility.

TABLE 4-3: TROUBLESHOOTING CHART

Problem	Reservoir Low on Hydraulic Fluid	Reservoir Over Full	Air in System	Bent Plunger	Release Valve Not Fully Closed	Dirt in Release Valve	Dirt in Check Valve	Loose Dirt or Air Bubble in Valve System	Damaged Quick-Coupler
Spongy effect			X						
Ram will not extend all the way	X	X							
Ram will not retract		X		X					X
Ram leaks down under pressure					X	X			
Handle kickback							X		
Works properly one time but not the next								X	

- *Works properly one time, but not the next.* There is loose dirt or an air bubble in the valve system. Flush and refill the system.

4.6 HYDRAULIC LIFTS

Another important piece of equipment is the **hydraulic lift** used to raise the whole vehicle in the air for easier working conditions. The traditional stationary in-the-ground unit was usually found only in service stations, muffler shops, transmission shops, and tire dealers for reasons such as oil and lube jobs, brake service, and other underbody repairs. Today, all body shops, because of unibody construction, are looking for ways to get the vehicle off the ground to estimate and repair. Technicians work better with cars on a lift. Damage reports are easier to write on a vehicle that has been up on a lift.

As shown in Figure 4–59, there are several ways to get the vehicle off the ground. Four-post and two-post hoists are neat and allow total movement under the vehicle, but they take up more space than a work stall in length and width. Side post hoists are great for estimating, but some access to the sides of

the vehicle is impaired. And lastly, old center post hoists make some areas under the vehicle hard to reach but take up less space.

The using of lifts is a much less physical and mental task on the body technicians, since they can work at a level that is comfortable for them. But lifts are a specialized piece of equipment and a technician using them needs to know exactly what is to be done, especially if there are any vehicle weight problems. It is important to note that the quality of today's lifts, as well as the number of safety devices on each model, makes them very safe to operate.

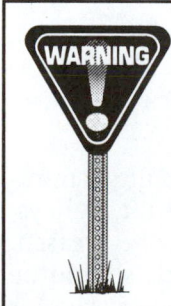

WARNING

If any jack or lift appears to be damaged, worn, or operates improperly, remove it from service until the necessary repairs are made.

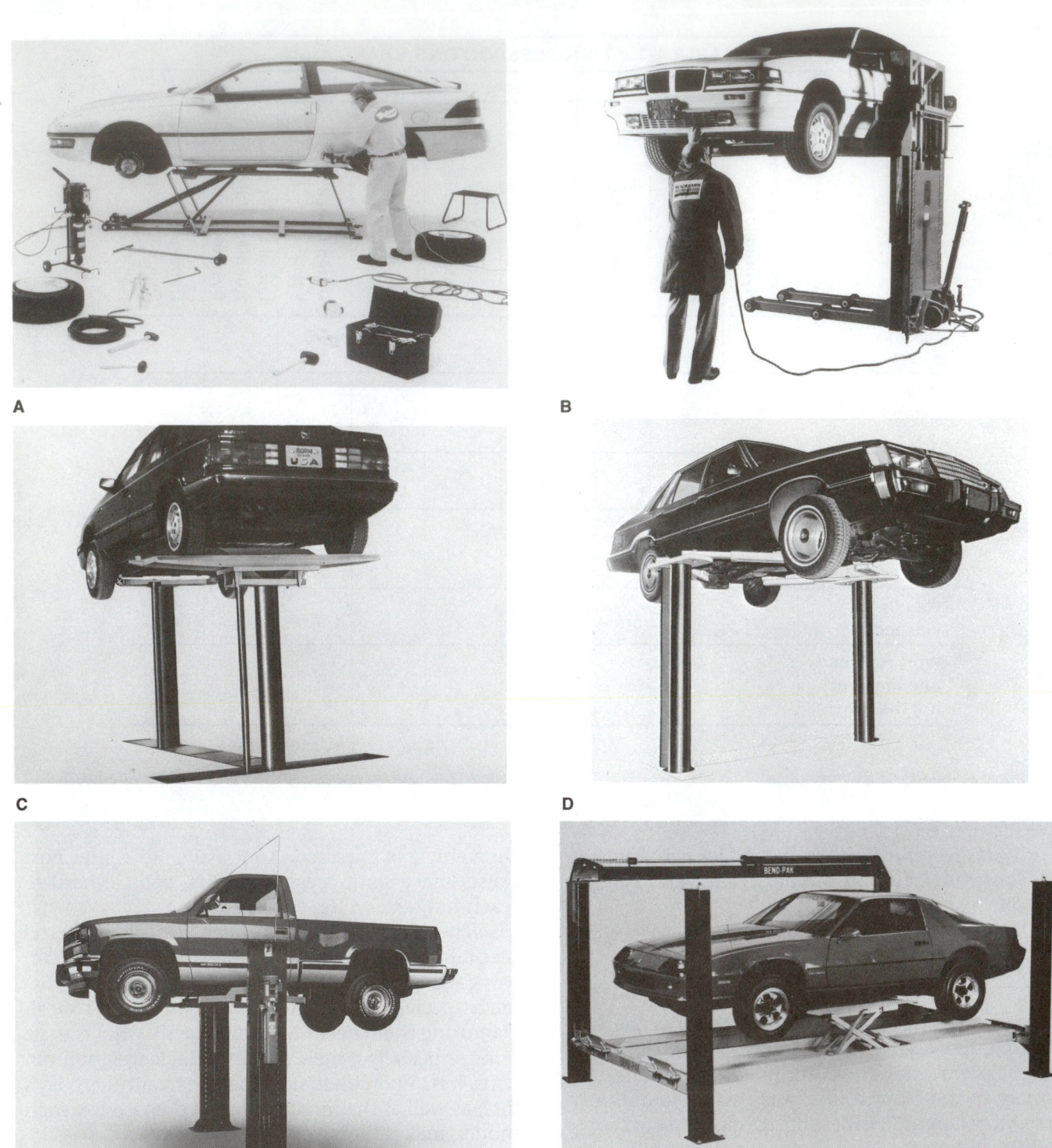

A

B

C

D

E

F

FIGURE 4-59 Several ways to get vehicles off the ground: (A) portable lift; (B) side lift; (C) in-ground lift—front to rear frame support; (D) in-ground lift—side to side support; (E) two-post above-ground hoist; (F) four-post above-ground hoist; (G) drive-through above-ground scissor lift; and (H) drive-on above-ground ramp plus a scissor lift. *(Courtesy of Rotary Lift Corp.)*

The maintenance on above-ground lifts is minimal, but important. Depending on the lift, pulleys, pivoting lift links, and wheels should be greased. Bearings, pins, and other moving parts should be oiled, and cables and chains should be checked for worn or frayed areas. Again, follow the manufacturer's directions.

General maintenance on lifts includes inspecting the lift pads and bumper cushions regularly and replacing them if necessary. It is advisable to appoint

G

H

FIGURE 4-59 Continued

a knowledgeable person to inspect the jacks and lifts daily, especially if the units are subjected to abnormal loads or shocks.

LIFT SAFETY

Raising a vehicle on a lift or a hoist requires specific care. Drive-on lifts are fairly safe. However, it is important to make sure that vehicles equipped with a catalytic converter have enough clearance between the hoist and the exhaust system components before driving the vehicle onto the ramps.

Adapters and hoist plates must be positioned correctly on twin-post- and rail-type lifts to prevent damage to the underbody of the vehicle. The catalytic converter, tie rod, rod bracket, and shock absorbers are some of the components that could be damaged if the adapters and hoist plates are incorrectly placed.

There are specific contact points to use where the weight of the vehicle is evenly supported by the adapters or hoist plates. The correct lifting points can be found in the vehicle's service manual.

SUMMARY

- Power tools use air pressure (pneumatic), oil pressure (hydraulic), or electrical energy to effect repairs.

- An impact wrench is a portable hand-held, reversible air-powered tool for rapid turning of bolts and nuts.

- Do not use an air impact wrench to final torque fasteners. Fastener or part failure could result. This could endanger the driver and passengers.

- Drill bits of different sizes fit into the chuck on drills for making holes in parts.

- There are special spot remover air drills and/or attachments that are used for cutting out spot welds.

- An air sander uses an abrasive action to smooth and shape body surfaces. Different coarseness sandpapers can be attached to the pad on the sander.

- Grinders are used for fast removal of material. They are often used to smooth metal joints after welding and to remove paint and primer.

- An air polisher is used to smooth and shine painted surfaces by spinning a soft buffing pad.

- Sandblasting is the most effective way to remove all finishes from any vehicle. The use of plastic media or soda blasting has replaced sand in many body shops.

- Air tools need little upkeep. However, you will have problems if basic maintenance is not performed.

- The most common causes for any pneumatic or air tool to malfunction are: lack of proper lubrication, excessive air pressure or lack of it, and excessive moisture or dirt in the air lines.

Here are some other lift safety tips that should always be kept in mind and followed when getting a vehicle up in the air:

- Never overload the lift. The manufacturer's rated capacity is shown on the nameplate affixed to the lift.

- Employees should stand to one side of a vehicle when directing it into position over a lift. Do not allow customers or bystanders to operate the lift or be in the lift area during its operation.

- Positioning of the vehicle and operation of the lift should be done only by trained and authorized personnel.

- Always keep the lift area free from obstructions, grease, oil, trash, and other debris.

- Operating controls are designed to close when released. Do not block open, or override them.

- Before driving a vehicle over a lift, position the arms and supports to provide unobstructed clearance. Do not hit or run over lift arms, adapters, or axle supports. This could damage the lift or the vehicle.

- Load the vehicle on the lift carefully. Check to make sure adapters or axle supports are in secure contact with the vehicle, per the manual instructions, before raising to the desired working height. Remember that unsecured loads can be dangerous.

- Make sure the vehicle's doors, hood, and trunk are closed prior to raising the vehicle. Never raise a car with passengers inside.

- Position the lift supports to contact at the vehicle manufacturer's recommended lifting points. Raise the lift until the supports contact the vehicle. Check supports for secure contact with the vehicle and raise the lift to the desired working height.

- After lifting a vehicle to the desired height, always lower the unit onto mechanical safeties.

- Note that with some vehicles, the removal (or installation) of components can cause a critical shift in the center of gravity and result in raised vehicle instability. Refer to the vehicle manufacturer's service manual for recommended procedures when vehicle components are removed.

- Make sure tool trays, stands, and so forth are removed from under the vehicle. Release locking devices as per instructions before attempting to lower the lift.

- Before removing the vehicle from the lift area, position the arms, adapters, or axle supports to assure that the vehicle or lift will not be damaged.

- Inspect the lift daily. Never operate it if it malfunctions or if it has broken or damaged parts. It should be removed from service and repaired immediately. A lift requires immediate attention if it

    Jerks or jumps when raised

    Slowly settles down after being raised

    Slowly rises when not in use

    Slowly rises when in use

    Comes down very slowly

    Blows oil out of the exhaust line

    Leaks oil at the packing gland

Repairs should be made with original equipment parts only.

ASE-STYLE REVIEW QUESTIONS

1. To locate the position of a hole to be drilled, Technician A uses an awl, Technician B says that the distinct mark left by the awl provides a seat for the drill point. Who is correct?

    A. Technician A
    B. Technician B
    C. Both A and B
    D. Neither A or B

2. Technician A uses a low pile height buffing pad for light compounding, touchups and blending. Technician B uses a screwdriver to clean dried compound off the pad. Who is correct?

    A. Technician A
    B. Technician B
    C. Both A and B
    D. Neither A or B

3. Which of the following operations can the air hammer perform?

   A. Bushing installer
   B. Bushing remover
   C. Shock absorber chisel
   D. All of the above

4. Which type of wheel is used for sharpening cutting tools?

   A. Grinding wheel
   B. Wire wheel brush
   C. Buffing wheel
   D. Cutting wheel

5. Which of the following is used to lift the entire automobile?

   A. Floor jack
   B. Hoist
   C. Hydraulic press
   D. All of the above

6. When working with a piece of hydraulic equipment that will not extend all the way, Technician A says a likely problem is low hydraulic fluid. Technician B says a likely problem is dirt in the check valve. Who is correct?

   A. Technician A
   B. Technician B
   C. Both A and B
   D. Neither A or B

7. What causes a pneumatic tool to fail?

   A. Lack of lubrication
   B. Excessive air pressure
   C. Lack of air pressure
   D. All of the above

8. Which air sander creates a buffing pattern?

   A. Disc sander
   B. Dual-action orbital sander
   C. Long board type sander
   D. All of the above

9. Which pad should be used on a polisher/buffer to do the early stages of cutting and compounding?

   A. $1/2$ to 1 inch pile height
   B. 1 to $1^1/4$ inch pile height
   C. $1^1/4$ to $1^1/2$ inch pile height
   D. $1^1/2$ to 2 inch pile height

10. Which type of sandblaster can be used indoors?

    A. Standard
    B. Captive
    C. Both A and B
    D. None of the above

11. What is the recommended air pressure for a disc sander?

    A. 60 to 80 psi
    B. 70 to 90 psi
    C. 40 to 90 psi
    D. 10 to 50 psi

12. Which of these tools is not used with an air hammer?

    A. Smoothing hammer
    B. Panel crimper
    C. Heat tip
    D. Tall pipe cutter

ESSAY QUESTIONS

1. Explain eight major parts of a typical paint spray gun.

2. How does an air impact tool operate?

3. Why should you use a torque wrench to final tighten fasteners and not an impact wrench?

4. Explain the differences between sandblasting and plastic media blasting.

CRITICAL THINKING PROBLEMS

1. If you have rough filed a large area of body filler and the material is hard, how would you continue rough straightening?

2. What can happen to air tools if they are not oiled periodically?

MATH PROBLEMS

1. If both air wrenches are made the same, how much stronger than a $1/4$-inch drive head is a $1/2$-inch drive head?

2. A small hydraulic jack has a pumping plunger that is one square inch in diameter. The ram piston has a diameter of 5 square inches. If you exert 500 pounds of pressure on the pumping plunger through the handle's lever arm, how much weight can you lift with the jack?

CHAPTER 5

Compressed Air Supply Equipment

INTRODUCTION

The **compressed air supply system** is designed to provide an adequate supply of clean, dry air at a predetermined pressure to insure efficient operation of all pneumatic equipment in the body shop. The system can vary in size from small portable units (Figure 5–1) to large in-shop installations (Figure 5–2).

The basic requirements for shop compressed air systems are the same for most body shops (Figure 5–3):

- An air compressor, sometimes referred to as an air pump, can be one compressor or a series of compressors. The compressor is the "heart" of the system. The power source is generally an electric motor. (Portable gasoline-driven compressors are available for work outside the shop.)
- A control or set of controls is necessary to regulate the operation of the compressor and motor.
- Air intake filters/silencers are designed to muffle intake noises as well as filter out dust and dirt.

- The air tank or receiver must be properly sized. It cannot be too small or it will cause the compressor to cycle too often, thus causing excessive load on the motor. It should not be too large

FIGURE 5-1 A portable piston compressor mounted on wheels may be adequate for small jobs and one air tool. *(Photo courtesy ITW DeVilbiss Automotive Refinishing Products)*

OBJECTIVES

After studying this chapter, you should be able to:

✔ Identify the various types of air compressors used in a body shop.
✔ Explain the operation of the various air compressors.
✔ Describe the function of air and fluid control equipment.
✔ Use the different accessory equipment of the air compressor.
✔ Maintain an air supply system properly.
✔ List the air system safety rules.
✔ Answer ASE test questions relating to compressed-air equipment.

KEY TERMS

adapter
aftercooler
air compressor
air line lubricator
air pressure regulator
air tank shutoff valve
air transformer
automatic unloader
compressed air supply
  system
compressor air tank
compressor drain
  valve

compressor oil plugs
coupling
cubic feet per minute
displacement
eccentric
gauges
intake air filter
pressure switch
psi
single-stage
  compressor
two-stage compressor

ASE TASK LIST

Job Skills covered in this chapter include:

PAINTING AND REFINISHING TEST (B2) TASK LIST

B. Spray Gun Operation and Related Equipment

1. Inspect, clean, and determine condition and adequacy of spray guns and related equipment (air hoses, regulator, air lines, air source, and spray environment).

MECHANICAL AND ELECTRICAL COMPONENTS TEST (B5) TASK LIST

A. Suspension and Steering

1. Identify suspension system fasteners which should not be reused.

FIGURE 5-2 The multicompressor setup from large body shops will handle the needs of numerous air tools simultaneously. *(Courtesy of Binks Mfg. Co.)*

because of space constraints and should not have unnecessary capacity.

- The *distribution system* includes the pipes and hoses or "arteries" that link the compressed air system. The hose and piping from the air receiver to distribution points requiring compressed air are included. The distribution system must consist of the proper sizes of hose or pipe, fittings, valves, air filters, oil and water extractors, regulators, gauges, lubricators, and other equipment for the effective operation of specific air tools and spraying equipment.

5.1 THE AIR COMPRESSOR

The **air compressor** is designed to raise the pressure of air from normal atmospheric to some higher pressure. This pressure is measured in pounds per

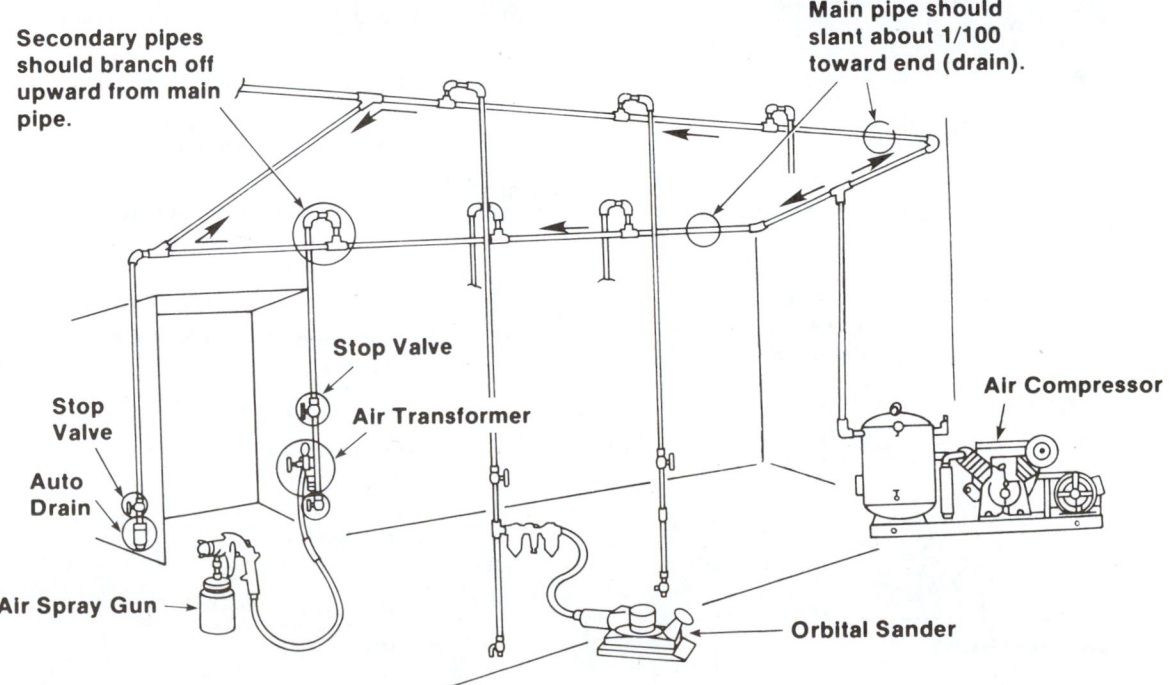

FIGURE 5-3 Study the typical piping arrangement found in a body/paint shop.

square inch (psi) or metric kilopascals (kPa). While normal atmospheric pressure is about 14.7 psi (101 kPa) at sea level, a compressor is capable of delivering air at pressures up to 200 psi or 1,378 kPa.

An air compressor is usually made up of an electric motor, air pump, and large air storage tank. The motor spins the air pump, which works like a small reciprocating piston engine. The air pump piston action pushes the air into a large, thick steel storage tank. The "heart" of any paint or body shop is its air compressor.

COMPRESSOR-TYPES

There are three basic types of air compressors: the diaphragm type, the piston type, and the rotary screw type.

Diaphragm-Type Compressor

The *diaphragm-type compressor* uses a flexible synthetic rubber membrane to produce a pumping action (Figure 5–4). This type of compressor is often used on very small compressors to power small air brushes

for doing custom painting and to provide air to respirators.

A durable diaphragm is stretched across the bore of a very shallow compression chamber. An **eccentric** (egg-shaped part), mounted on the motor shaft, acts on the diaphragm plate to pull the diaphragm up and down.

As the diaphragm is pulled down (Figure 5–5A), air is drawn into the small space above the diaphragm. When the diaphragm is thrust upward (Figure 5–5B), the air trapped in the compression chamber is squeezed and forced out into the delivery chamber and supply lines. Only a very small amount of air, in the 30 to 35 psi range, is compressed during each cycle. However, the pumping action is very rapid—in excess of 1,500 strokes per minute.

Piston-Type Compressor

The *piston-type air compressor* pump develops compressed air pressure through the action of a reciprocating piston. As shown in Figure 5–6, a piston, which is actuated by a crankshaft, moves up and down inside a cylinder. This is very much like a piston in an automobile engine.

On the downstroke (Figure 5–6A), air is drawn into the compression chamber through a one-way valve. On the upstroke (Figure 5–6B), as the air is compressed by the rising piston, a second one-way valve opens. The air is forced into a pressure tank or receiver. As more and more air is forced into the tank, the pressure inside the tank rises.

Piston compressor pumps are available in single or multiple cylinder and single- or two-stage models. Selection depends on the volume and pressure required.

When air is drawn from the atmosphere and compressed in a single stroke, the compressor is referred to as a **single-stage compressor.** Single-stage units normally are used in pressure ranges up to 125 psi (861 kPa) for intermittent service. Most single-

FIGURE 5–4 This diaphragm type compressor is designed to run a small touch-up spray gun or air brush.

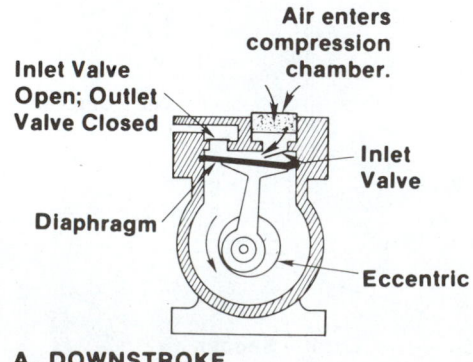

A. DOWNSTROKE

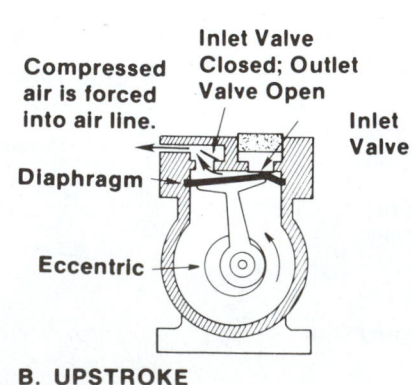

B. UPSTROKE

FIGURE 5–5 The operation of a diaphragm compressor is illustrated. As eccentric rotates, it pulls up and down on the flexible diaphragm. Valves then direct the flow in and out of the pump to produce pressure.

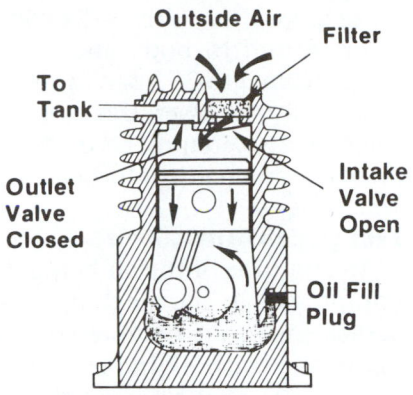

Outside Air
Filter
To Tank
Outlet Valve Closed
Intake Valve Open
Oil Fill Plug

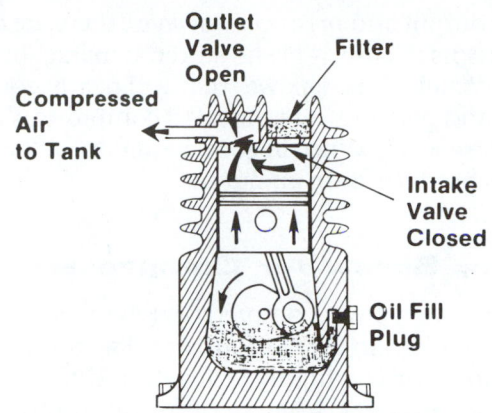

Outlet Valve Open
Filter
Compressed Air to Tank
Intake Valve Closed
Oil Fill Plug

A. DOWNSTROKE—AIR BEING DRAWN INTO COMPRESSION CHAMBER

B. UPSTROKE—AIR BEING COMPRESSED AND FORCED TO TANK

FIGURE 5-6 Study the operation of a piston-type compressor. As the piston slides down, an inlet reed allows outside air to flow into the cylinder. As the piston slides up, the inlet valve closes and the outlet valve opens; this forces air out of the cylinder under pressure.

FIGURE 5-7 Note the single-stage compressor pumps (left to right): single cylinder, angled V-cylinders, and two cylinders.

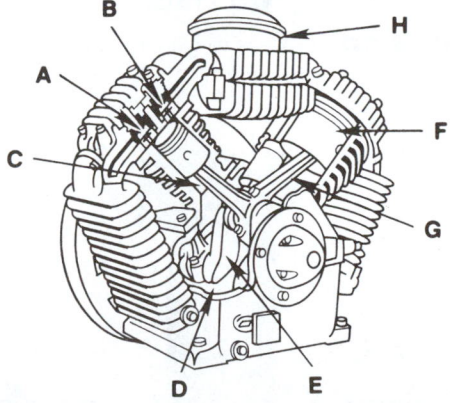

FIGURE 5-8 Study the principal parts of a four-cylinder, two-stage piston-type compressor: (A) intake and (B) exhaust valves; (C) second-stage piston; (D) crankcase; (E) crankshaft; (F) first-stage piston; (G) connecting rod assembly; and (H) air intake filter.

stage compressors are rated at *50 percent duty cycle* ($\frac{1}{2}$ the time on, $\frac{1}{2}$ the time off). They are available in single- or multicylinder compressors (Figure 5–7). The principal parts of a typical piston-type compressor are shown in Figure 5–8.

In a **two-stage compressor**, air is first compressed to an intermediate pressure and then further compressed to a higher pressure (Figure 5–9). Such a compressor has cylinders of unequal bore. The first stage of compression takes place in the large bore cylinder. In the second stage, the pressurized air passes through an intercooler. Air is then compressed for a second time to a higher pressure in the smaller bore cylinder (Figure 5–10).

Two-stage compressors are usually more efficient, run cooler, and deliver more air for the power consumed, particularly in the 100 to 200 psi (689 to 1,378 kPa) pressure range. This range of pressure is enough for most body or finishing applications.

Compressor oil plugs are provided for filling and changing air pump oil. The oil level in the compressor should be checked and changed periodically. Single weight, nondetergent oil or the oil recommended by the manufacturer is normally used.

In recent years, an oilless or oil-free piston compression system has been introduced that employs self-lubricating materials that do not require an oil lubricant. Until recently, most oilless compressors, like the diaphragm type, were considered compacts and limited in both output and pressure. However, there are oilless compressors (Figure 5–11) now on the market of up to approximately 5 horsepower that will nearly equal, in output and pressure, oil-lubricated compressors of the same horsepower. When in good condition, oilless compressors produce clean air output.

Rotary Screw Air Compressor

Rotary screw-type air compressors (Figure 5–12) have been a standard in other industries. The rotary screw air compressor is a highly efficient and dependable machine. However, because of an oil output problem, this type was never accepted by the automotive refinishing profession.

HOW COMPRESSORS ARE RATED

The following terms are used to measure the performance of a compressor.

- *Horsepower* (hp). Horsepower is the measure of the work. In a compressor, it is the capacity of the motor or engine that drives the compressor. Compressors found in body and paint shops usually range from 3 to 25 hp. As a general rule, the greater the horsepower, the more powerful the compressor. Also, in most cases, as the horsepower increases, so will the other compressor ratings that follow.

- **Cubic feet per minute (cfm).** Cubic feet per minute is the volume of the air being delivered by the compressor to the air tool. This spec is used as a measure of the compressor's capabilities. Compressors with higher cfm ratings provide more air through the hose to the tool, thus making higher cfm outfits more practical for larger jobs. Actually, compressors have two cfm ratings.

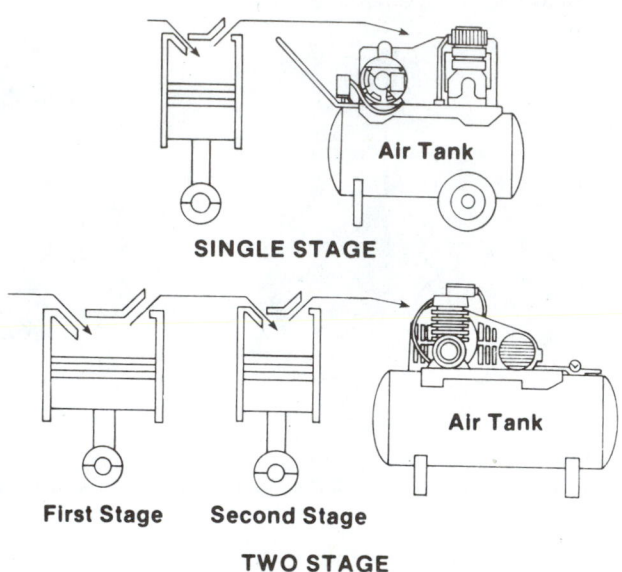

SINGLE STAGE

First Stage   Second Stage

TWO STAGE

FIGURE 5-10 Compare single- and two-stage compressors.

FIGURE 5-9 This is a typical two-stage compressor. *(Courtesy of Binks Mfg. Co.)*

FIGURE 5-11 Three mountings for compressors. From left to right: horizontal tank on wheels, stationary vertical tank, and stationary horizontal tank. *(Courtesy of Campell-Hausfeld)*

FIGURE 5-12 A typical rotary screw air compressor is shown. *(Courtesy of Champion Pneumatic Machinery Co., Inc.)*

- **Displacement cfm.** Displacement is the theoretical amount of air in cubic feet that the compressor can pump in one minute. It is a relatively simple matter to calculate the air displacement of a compressor if the piston diameter, length of stroke, and rpm are known. For example, the area of the piston multiplied by the length of the stroke and the shaft revolutions per minute equals the displacement volume. The formula for computing it is as follows:

$$\frac{\text{Area of piston x stroke x rpm x number of pistons}}{1,728}$$

$$= \text{Displacement in cfm}$$

The displacement of a two-stage compressor is always given for that of the first stage cylinder or cylinders only. This is because the second stage merely rehandles the same air the first stage draws in and cannot increase the amount of air discharged.

- *Free air cfm.* The free air rating is the actual amount of free air in cubic feet that the compressor can pump in one minute at working pressure. The free air delivery at working pressure, not the displacement or the horsepower, is the true rating of a compressor. It should be the primary cfm rating considered when selecting an air compressor.

  The *compressor's volumetric efficiency* is the ratio of free air delivery to the displacement rating, expressed in percent. For example, if a compressor unit for 100 pounds service has a displacement of 8 cfm and its volumetric efficiency is 75 percent, at this pressure the free air delivery will be: 8 cfm x 75 percent, or 6 cfm.

- **Pressure (psi).** Pounds per square inch (psi) is the measure of air pressure or force delivered by the compressor to the air tool. This is usually expressed as:

1. Normal or continuous working pressure
2. Maximum pressure

If the cfm of a compressor is too low for demand, the pump will not be able to keep up. The speed of the air tool will slow down as air pressure is consumed from the tank and you will have to wait while the compressor builds pressure again. Always purchase a compressor that will handle maximum air consumption.

A

B

FIGURE 5-13 Compare (A) horizontal- and (B) vertical-mounted piston compressors.

TANK SIZE

The **compressor air tank** is a heavy gauge steel tank for holding an extra supply of compressed air. Working pressure is not available until the tank pressure is above the required psi or kPa rating of the air tool. The compressor puts more into the tank than is required for application. The larger the tank, the longer a job can be done at the required pressure before a pause is needed to rebuild pressure in the tank.

FIGURE 5–14 HVLP spray guns are available in suction and gravity feed (shown) as well as pressure feed models and can operate with air compressors as small as 3 horsepower. *(Photo courtesy ITW DeVilbiss Automotive Refinishing Products)*

Air tanks or receivers usually have a cylindrical shape (Figure 5–12). The compressor motor and pump are usually mounted on top of it. Tanks can be purchased with either horizontal (Figure 5–13A) or vertical stationary mountings (Figure 5–13B) or can be mounted horizontally on wheels for portability (Figure 5–1).

An **air tank shutoff valve** is a hand valve that isolates the tank pressure from shop line pressure. It should be closed at night or when the compressor is not going to be used. If not closed, the compressor would run all night if a hose leaked or ruptured.

A **compressor drain valve** on the bottom of the tank allows you to drain water. The compression of the air tends to make moisture condense. This moisture must be drained periodically to prevent it from entering the air lines.

COMPRESSOR OUTFITS

There are two types of compressor outfits used in body shops: portable and stationary. A portable outfit is designed for easy movement and is equipped with handles, wheels, or casters and usually a small air receiver or pulsation chamber.

A stationary outfit is one that is permanently installed. It is usually equipped with a larger air receiver than the portable type. It might have a pressure switch or an automatic unloader as found on larger industrial units. Larger stationary models are generally equipped with a centrifugal pressure release.

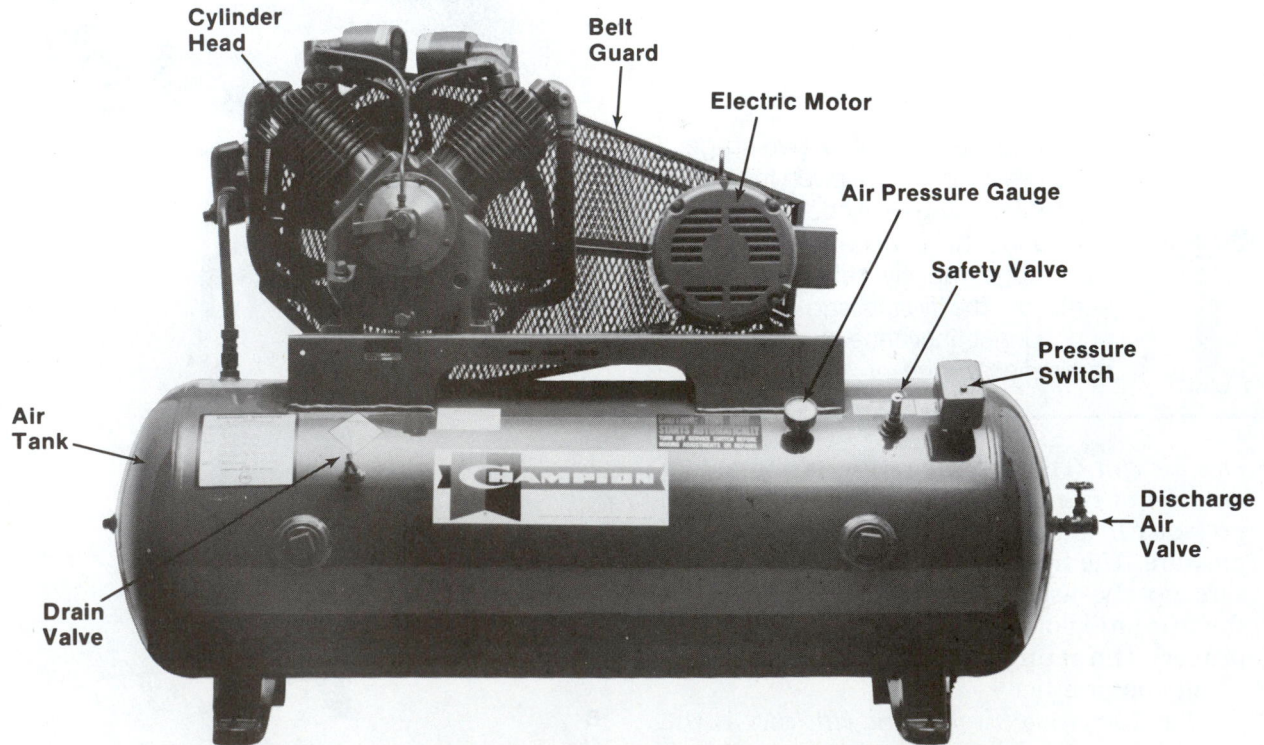

FIGURE 5–15 Study the parts of a stationary compressor. *(Courtesy of Champion Pneumatic Machinery Co., Inc.)*

SHOP TALK

Another source of compressed air found in many body repair/paint shops is the high volume, low pressure system's turbine generator (Figure 5–14). HVLP systems are fully described in Chapter 17.

Shown in Figure 5–15, the typical parts of a stationary compressor are:

- Air compressor pump (pressurizes air)
- Electric motor or gasoline engine (powers compressor)
- Air receiver or storage tank (holds compressed air)
- Check valve (prevents leakage of stored air)
- Pressure switch (automatically controls the air pressure)

Because of the importance of the system's safety controls, it is wise to know how they operate to pre-vent excessive pressure and electrical problems. The most common of these are:

- An **automatic unloader** is a device designed to maintain a supply of air within given pressure limits on compressors when it is not practical to start and stop the motor during operations. When the demand for air is relatively constant at a volume approaching the main capacity of the compressor, an unloader is recommended. When maximum pressure in the air receiver is reached, the unloader pilot valve (Figure 5–16A) opens to let air travel through a small tube to the unloader mechanism (Figure 5–16B). This holds open the intake valve on the compressor, allowing it to run idle. When pressure drops to a minimum setting, the spring loaded pilot automatically closes. Air to the unloader is shut off causing the intake valve to close. The compressor resumes normal operation. Maximum and minimum pressures can be varied by resetting the pressure adjusting screw on the pilot.

- A **pressure switch** is a pneumatically controlled electric switch for starting and stopping electric motors at preset minimum and maximum pressures. When system pressure reaches its maximum setting, the switch opens to cut current to the compressor motor. When pressure drops to a specific point, the switch closes to turn the compressor back on. The cycle is repeated to maintain a minimum and maximum pressure value in the system. Table 5–1 gives typical cut-in and cut-out times, which vary by compressor size, type, and cut-in/cut-out pressures.

- A *motor starter* is an electrical switch designed to provide overload protection or other necessary electrical control for starting motors of various types. The design of the switch varies with different motor sizes and current characteristics.

- *Overload protection* is usually provided on small units by fuses and on larger ones by thermal over-

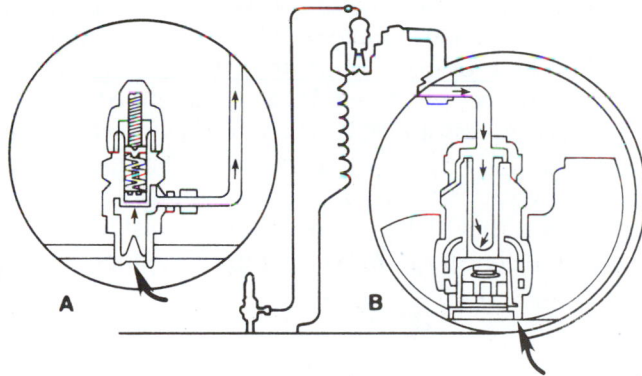

FIGURE 5–16 An automatic unloader is a spring-loaded valve that will release pressure from the compressor as needed. *[Photo courtesy ITW DeVilbiss Automotive Refinishing Products]*

Outfit	HP	Cut-In pounds pressure	Cut-out pounds pressure	Time to pump from Cut-In to Cut-out (In seconds)	Tank Size
Single Stage	1	80	100	83	30 gal.
Single Stage	2	80	100	69	60 gal.
Single Stage	3	80	100	51	60 gal.
Two Stage	1	140	175	284	60 gal.
Two Stage	3	140	175	115	80 gal.
Two Stage	5	140	175	75	80 gal.
Two Stage	10	140	175	56	120 gal.
Two Stage	15	140	175	42	120 gal.
Two Stage	20	140	175	36	200 gal.
Two Stage	25	140	175	30	200 gal.

TABLE 5-1: TYPICAL CUT-IN/CUT-OUT COMPRESSION TIMES

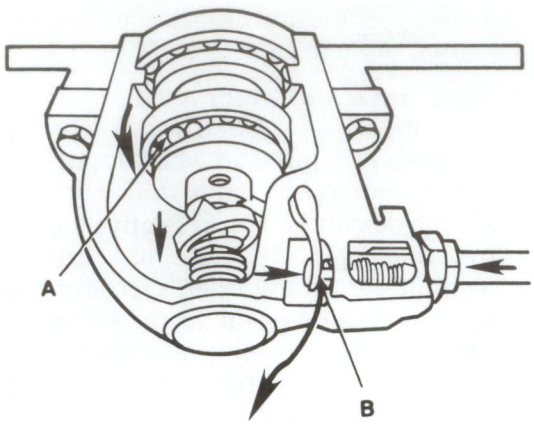

FIGURE 5-17 The centrifugal pressure release will also allow the compressor to run without building additional pressure. *[Photo courtesy ITW DeVilbiss Automotive Refinishing Products]*

load relays. Relays are recommended with time delay features so that circuits will not be opened by short duration surges that are not harmful to the motor. Overload protection should be employed on all compressor installations.

- A *centrifugal pressure release* is a device that allows the motor to start up and gain momentum before engaging the load of pumping air against pressure. When the compressor slows down to stop, rotating the crankshaft more slowly, steel balls (Figure 5–17A) move toward the center where they wedge against a cam surface forcing the cam outward. This opens a valve (Figure 5–17B), bleeding air from the line connecting to the check valve. With air pressure bled from the pump and aftercooler, the compressor can start up free of back pressure until it gets up speed. When normal speed is reached, the balls move out by centrifugal force, releasing the cam, closing the valve, and allowing air to again be pumped into the air receiver.
- A *fused disconnect switch* is a knife-type off/on switch containing the proper size fuse. This should be used at or near the compressor unit

⚠ WARNING

An electrician, who will be qualified to satisfy all electrical codes, should always make the electrical hookup of an air compressor. If something is wired incorrectly or connected wrong, severe part damage could result.

with the line going from the fused disconnect to the starter. Fuses should be the size recommended by the compressor's manufacturer.

5.2 AIR AND FLUID CONTROL EQUIPMENT

The control of the volume (amount), pressure, and cleanness of the air going to the pneumatic tools, especially spray guns, is of critical importance.

The **intake air filter** located on the compressor inlet is very important as all the air going into the compressor must pass through this filter. The filter element must be made of fine mesh or felt material to ensure that small particles of grit and abrasive dust do not pass into the cylinders. The intake filter must prevent excessive wear on cylinder walls, piston rings, and valves.

If possible, the compressor should be placed where it can receive an ample supply of clean, cool, dry air. If necessary, connect the air intake to the outside of the building. Distance between the intake and the compressor should be as short as possible for best efficiency. The outside intake should be protected from the elements with a hood or suitable weatherproof shield. The compressor air intake should not be located near steam outlets or other moisture-producing areas.

It must be remembered that raw air piped directly from a compressor is of little use to the body or refinishing shop. The air contains small but harmful quantities of water, oil, dirt, and other contaminants that will lessen the quality of the sprayed finish. And the air will likely vary in pressure during the job.

TABLE 5–2: MINIMUM PIPE SIZE RECOMMENDATIONS*			
Compressing Outfit		**Main Air Line**	
Size	Capacity	Length	Size
1-1/2 and 2 HP	6 to 9 CFM	Over 50 feet	3/4"
3 and 5 HP	12 to 20 CFM	Up to 200 feet	3/4"
		Over 200 feet	1"
5 to 10 HP	20 to 40 CFM	Up to 100 feet	3/4"
		Over 100 to 200 feet	1"
		Over 200 feet	1-1/4"
10 to 15 HP	40 to 60 CFM	Up to 100 feet	1"
		Over 100 to 200 feet	1-1/4"
		Over 200 feet	1-1/2"

*Piping should be as direct as possible. If a large number of fittings are used, large size pipe should be installed to help overcome excessive pressure drop.

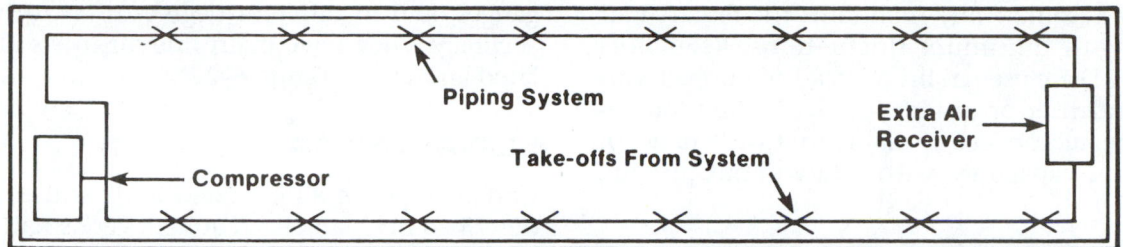

FIGURE 5-18 Some body/paint shop layouts require an extra air receiver. It provides more balanced pressure throughout the system.

DISTRIBUTION SYSTEM

The *air distribution system* carries compressed air from the compressor tank to various locations in the shop. The piping from the compressor to the tool input is often iron pipe. Table 5–2 shows the correct pipe size in relation to compressor size and air volume.

The compressor should be located as near as possible to operations requiring compressed air. This cuts down lengthy air lines that cause needless pressure drop.

A double loop or circle is accomplished by installing a tee in the line and then running a loop or circle in both directions back to the air tank. For this type of installation, it is recommended that an extra air tank (Figure 5–18) be installed at the far end to balance out peak loads. All piping should be installed so that it slopes toward the compressor air receiver or a drain leg should be installed at the end of each branch to provide for drainage of moisture from the main air line. This line should not run adjacent to steam or hot-water piping.

In the air distribution system, there should be a *shut-off valve* on the main line, close to the storage receiver tank. This valve is used to shut off the air at the air receiver. Keeping the air shut off at the storage tank overnight ensures a full tank of air when the shop is opened each day.

AIR TRANSFORMER

An **air transformer**, sometimes called a moisture *separator/regulator,* is a multipurpose device. It removes oil, dirt, and water from the compressed air. It filters and regulates the air. It also has a gauge that shows the regulated air pressure. The transformer may also provide multiple air outlets for spray guns, blow guns, air-operated tools, and so on.

Figure 5–19 illustrates typical air transformers. Some air transformers are equipped with a second gauge that indicates main line pressure.

Air transformers are used in all spray finishing operations since they require a supply of clean, dry, regulated air. They remove entrapped dirt, oil, and

FIGURE 5-19 Air transformers (air control units) are available with and without air regulators. *(Photo courtesy ITW DeVilbiss Automotive Refinishing Products)*

moisture by a series of baffles, centrifugal force, expansion chambers, impingement plates, and filters. They allow only clean, dry air to emerge from the outlets. The air regulating valve provides positive control, ensuring uniformly constant air pressure.

Gauges indicate regulated air pressure and, in some cases, main line pressure as well. Outlets with valves allow compressed air to be distributed where it is needed. The *drain valve* provides for elimination of sludge consisting of oil, dirt, and moisture. The air transformer should be installed at least 25 feet (7.6 meters) from the compressing unit.

AIR CONDENSER OR FILTER

An *air condenser* is basically a filter that is installed in the air line between the compressor and the point of use. It separates solid particles, such as oil, water, and dirt, out of the compressed air. No pressure regulation capability is supplied by this device. A typical air condenser is illustrated in Figure 5–20.

AIR PRESSURE REGULATOR

An **air pressure regulator** is a device for reducing the main line air pressure as it comes from the compressor. It automatically maintains the required air pressure with minimum fluctuations. Regulators (Figure 5–21) are used in lines already equipped with an air condenser or other type of air filtration device. Air regulators are available in a wide range of cfm and psi capacities, with and without pressure gauges, and in different degrees of sensitivity and accuracy. They have main line air inlets and regulated air outlets (Figure 5–22).

LUBRICATOR

Certain types of air-operated tools and equipment described in Chapter 4 require a very small amount of oil mixed in the air supply that powers them. An automatic **air line lubricator** (Figure 5–23) should

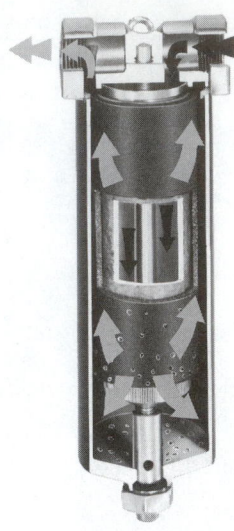

FIGURE 5-20 This air filter removes debris and moisture from air entering the system. *[Courtesy of Binks Manufacturing Co.]*

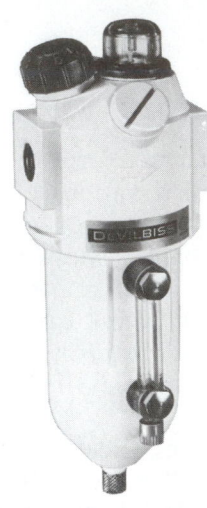

FIGURE 5-23 An air line lubricator will cause sanders, grinders, and similar air tools to last longer. It is not used with paint spray guns because the oil could contaminate the finish. *[Photo courtesy ITW DeVilbiss Automotive Refinishing Products]*

FIGURE 5-21 A simple air regulator can be adjusted to meet the needs of an air tool.

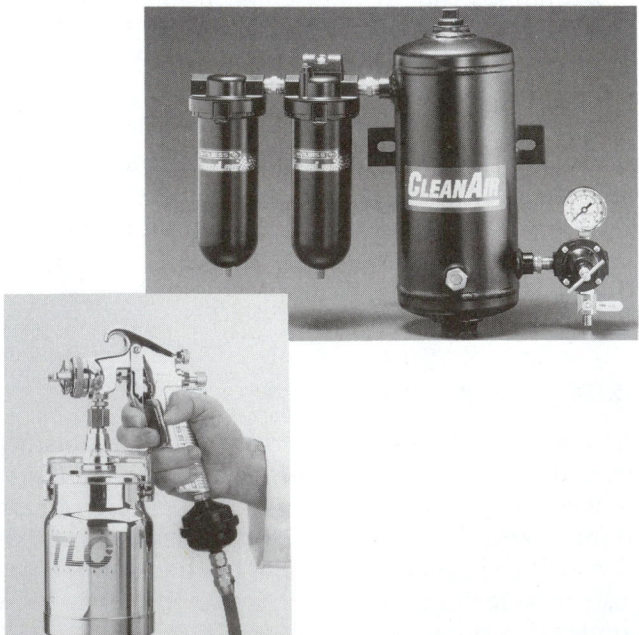

FIGURE 5-22 A filter/air regulator is designed for multiple air tools.

FIGURE 5-24 Air line filtration equipment ranges from the very basic gun-mounted disposable filter to a three-stage dessicant air drying system that removes water, dirt, oil, and water vapor from the air line. *[Photo courtesy ITW DeVilbiss Automotive Refinishing Products]*

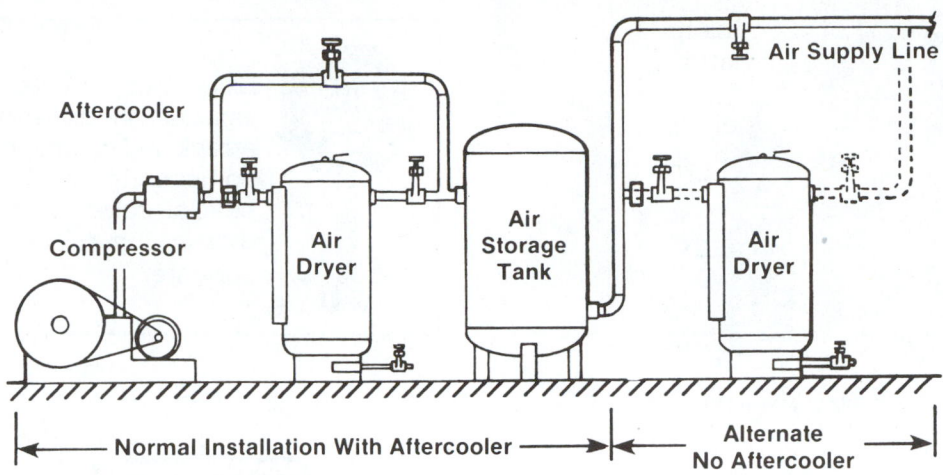

FIGURE 5-25 Note arrangement where dry air supply is required.

be installed on a leg or branch line furnishing air to pneumatic tools. (Never install a lubricator on a leg or branch air line used for paint spraying, since the oil supplied by it could damage the finish.) Lubricators are often combined with air filters and regulators in a single unit. Figure 5–24 shows a lubricator/ filter/regulator unit with a built-in sight glass for determining reserve oil level.

THERMAL CONDITIONING AND PURIFICATION EQUIPMENT

The air control devices already described in this chapter will remove contaminants from the compressed air most satisfactorily. However, there are some special problems concerning heat, dampness, and dirt that require special thermal conditioning and purification equipment. This equipment is usually installed between the compressor and the air

FIGURE 5-26 An air-cooled aftercooler for a system requiring drier air. It is installed between the two compressor stages. *(Courtesy of Hankison Corp.)*

storage receiver tank (Figure 5–25). It includes the following:

- **Aftercooler.** An aftercooler is used to reduce the temperature of compressed air. Heat, as well as some impurities, can be removed by installing an aftercooler in the system (Figure 5–26). Aftercoolers are very efficient in lowering air temperature and removing most of the oil and water. The residue of oil and water will be removed before it enters the air receiver. There are several different designs or types of aftercoolers available. The most common is the water-cooled "air tube" design in which air passes through small tubes. Recirculating water is directed back and forth across the tubes by means of baffles to reverse the direction of air flow. This cross-flow principle is accepted as the most efficient means of heat transfer.

- *Automatic dump trap.* An automatic dump trap, installed at the lowest point below the air receiver, will collect condensed moisture. It will open automatically to discharge a predetermined volume. Due to the air pressure behind the water, the trap opens and closes with a snap action that ensures proper seating. A small line strainer should be installed ahead of any automatic device to keep foreign particles from clogging the working parts.

- *Air dryers.* Good aftercoolers will remove the greatest percentage of water vapor. However, to remove any remaining residue, an air dryer (Figure 5–27) is often used. There are many designs of air dryers available. Among these are chemical, desiccant (drying agent), and refrigeration types. All dryers are designed to remove moisture from the compressed air supply so that no condensation will take place in the distribution system under normal working conditions.

FIGURE 5-27 Typical air dryers used in paint shops. *(Courtesy of Hankison Corp.)*

5.3 COMPRESSOR ACCESSORIES

The various types of hose used to carry compressed air and fluid to the spray gun and other power tools are important parts of the system. Improperly selected or maintained hose can create a number of problems.

HOSE TYPES

There are two types of hoses in a compressed air system: air hose and a fluid or material hose (Figure 5–28). The air hose in most compressed systems is usually covered in red rubber; in smaller, low pressure systems it might be covered with a black and orange braided fabric. The fluid or material hose is normally black or brown rubber.

The air hose is usually a simple braid-covered hose that consists of rubber tubing (1) reinforced and

SHOP TALK An air hose should never be used for solvent-based paints. Solvents will eat at the hose material and cause hole failure and rupture.

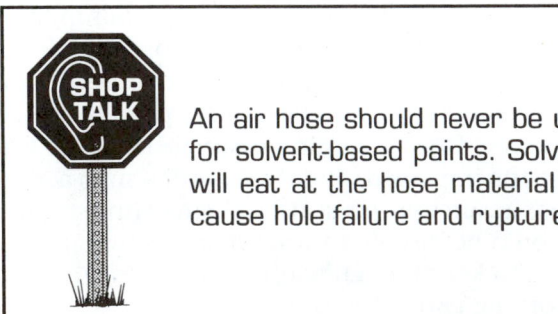

Fluid Hose

Air Hoses

FIGURE 5-28 Fluid and air hoses are made differently and should not be interchanged. *(Courtesy of Binks Mfg. Co.)*

SHOP TALK Since the solvents in some coatings used in refinishing would readily attack and destroy ordinary rubber compounds, fluid hose is lined with special solvent-resistant material that is impervious to all common solvents.

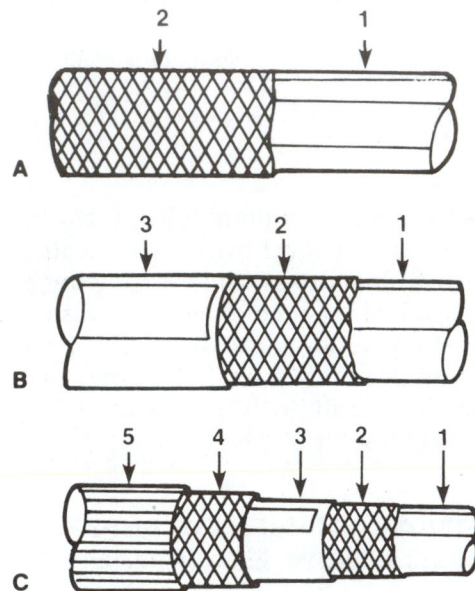

FIGURE 5-29 Study construction of hoses used in a compressed air system: (A) braid-covered hose; (B) single braid hose; and (C) double braid hose.

covered by a woven braid (2), as shown in Figure 5–29A. The single braid, rubber-covered hose (Figure 5–29B) consists of an inner tube (1), a braid (2), and an outside cover (3), all vulcanized into a single unit. The double braid hose, illustrated in Figure 5–29C, consists of an inner tube (1), a braid (2), a separator or friction layer (3), a second layer of braid (4), and an outer rubber cover (5), all vulcanized into one. Double braid hose has a higher working pressure than single braid hose.

HOSE SIZE

It is important to use the proper size and type of hose to deliver the air from the compressor and the material from its source to the air tools and guns. When air is compressed and must travel a long distance, its pressure begins to drop. However, for a distance of up to 100 feet, pressure drop can be kept minimal when the proper diameter hose and fittings are used.

TABLE 5–3: AIR PRESSURE DROP				
Size of Air Hose (ID)*	Air Pressure Drop			
	5-Foot Length	15-Foot Length	25-Foot Length	50-Foot Length
1/4 Inch	PSIG	PSIG	PSIG	PSIG
@ 40 PSIG†	0.4	7.5	10.5	16.0
@ 60 PSIG	4.5	9.5	13.0	20.5
@ 80 PSIG	5.5	11.5	16.0	25.0
5/16 Inch				
@ 40 PSIG	0.5	1.5	2.5	4.0
@ 60 PSIG	1.0	3.0	4.0	6.0
@ 80 PSIG	1.5	3.0	4.0	8.0
3/8 Inch				
@ 40 PSIG	1.0	1.0	2.0	3.5
@ 60 PSIG	1.5	2.0	3.0	5.0
@ 80 PSIG	2.5	3.0	4.0	6.0

* ID: Inner diameter

†PSIG: Pounds per square inch gauge

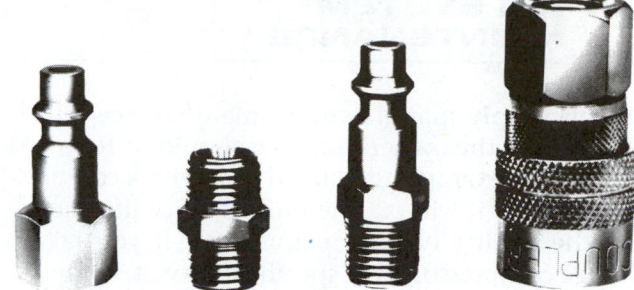

FIGURE 5–30 Note various hose ends. From left to right: female coupler, union, male coupler, quick-connect.

Employ only the hose constructed for compressed air use and with a rating of at least four times that of the maximum psi being used.

Table 5–3 indicates just how much pressure drop can be expected at different pressures with hoses of varying lengths and internal diameters. At low pressure and with short lengths of hose, this drop is not particularly significant. As the pressure is increased and the hose is lengthened, the pressure drop rapidly becomes very large and must be compensated for. Too often a tool is blamed for malfunctioning when the real cause is an inadequate supply of compressed air resulting from using too small an *inside diameter* (ID or distance measured across inside surfaces) hose.

MAINTENANCE OF HOSES

A hose will last a long time if it is properly cared for and maintained. Caution should be taken when it is dragged across the floor. It should never be pulled around sharp objects, run over by vehicles, kinked, or otherwise abused. Hose that ruptures in the middle of a job can ruin or delay the work.

The fluid hose can be cleaned using a hose cleaner, a device that forces a mixture of solvent and air through the fluid hose and spray gun, ridding them of paint residue. A valve stops the flow of solvent and allows air to dry the equipment. Clean the fluid hose internally with the proper solvent when the gun is cleaned.

The outside of both the air and the fluid hose should be wiped down with solvent at the end of every job. Wrap them into a large coil and hang up the hose during storage.

CONNECTORS

Connections are needed between the compressor, the ends of hoses, and the air tools. Of the many different types used, the most common are the threaded and quick-connect types. The screw-type fitting is usually connected and disconnected with a wrench. The quick-connect is readily attached and detached by hand by pulling back on its sleeve (Figure 5–30).

Both types of connections may use the compression ring system to mount the fittings to the air or fluid hose. They may also use pipe fittings.

To install a compression ring connection, slip the sleeve and the compression ring over the end of the hose. Hold the body of the connection in a vise, and push the hose into the body as far as it will go. Slide the compression ring up to the body and bring the sleeve over the ring and thread it on by hand. Tighten it with a wrench.

Most paint spray guns require either $^1/_4$- or $^5/_{16}$-inch (6.3 or 7.9 mm) hoses; air hoses for pneumatic tools usually have $^5/_{16}$- to $^3/_8$-inch (7.9 to 9.5 mm) inside diameters. A few of the air tools described in Chapter 4 require hoses of specified inside diameters, which the tool manufacturers usually supply.

ADAPTERS AND COUPLINGS

An **adapter** is a type of connection that is male on one end and female on the other. It is used to convert the connections on the hose and other equipment from one thread size to another. Adapters are available in a very wide variety of sizes and threads.

A **coupling** is a type of connection that is male on both ends. It is used to couple two pieces of hose or pipe together or to convert a female connection of one size thread to a male connection of another size thread.

5.4 AIR SYSTEM MAINTENANCE

The air supply manufacturer's maintenance schedule given in the owner's manual should be followed exactly. For example, if you fail to service a compressor properly, it will cut the unit's service life and affect the quality of your paint work. If you fail to change compressor oil at specific intervals, compressor parts will wear and leak. Excess oil can then enter the air lines and air tools. Some of this oil could get on the vehicle surface or into the spray equipment, ruining your paint work.

In general, all air systems require the following periodic maintenance.

DAILY MAINTENANCE

- Drain the air receiver and the moisture separator/regulator or air transformer. If the weather is humid, drain them several times a day. See Figure 5–31.
- Check the level of the oil in the crankcase. While it should be kept at full level, do not overfill. Overfilling causes excessive oil usage.
- SAE 10W-30, a multigrade oil, can be used as a substitute when SAE 10 or 20 weight oil is not

readily available. Multigrade oils do contain additives that can cause harmful carbon residue and varnish. Detergent-type oils are satisfactory if used before hard carbon deposits have developed. Before changing to a detergent-type oil, pistons, rings, valves, and cylinder heads should be cleaned since the detergent oil can loosen hard carbon deposits that can plug passages and damage cylinders and bearings.

WEEKLY MAINTENANCE

- Pull the ring on the safety valve and unseat it. If the valve is working properly, it will release air. Reseat the safety valve by pushing the stem down with your finger. If the valve sticks or fails to seat, repair or replace it immediately.
- Clean air filters. Felt and foam air filters should be washed in nonexplosive solvent, allowed to dry, and reinstalled. A dirty air filter decreases compressor efficiency and will increase oil usage.
- Clean or blow off fins on cylinders, heads, intercoolers, aftercoolers, and any other parts of the compressor or outfit that collect dust or dirt. A clean compressor runs cooler and provides longer service.
- Check the oil filter in the air line and change the filter element if necessary.

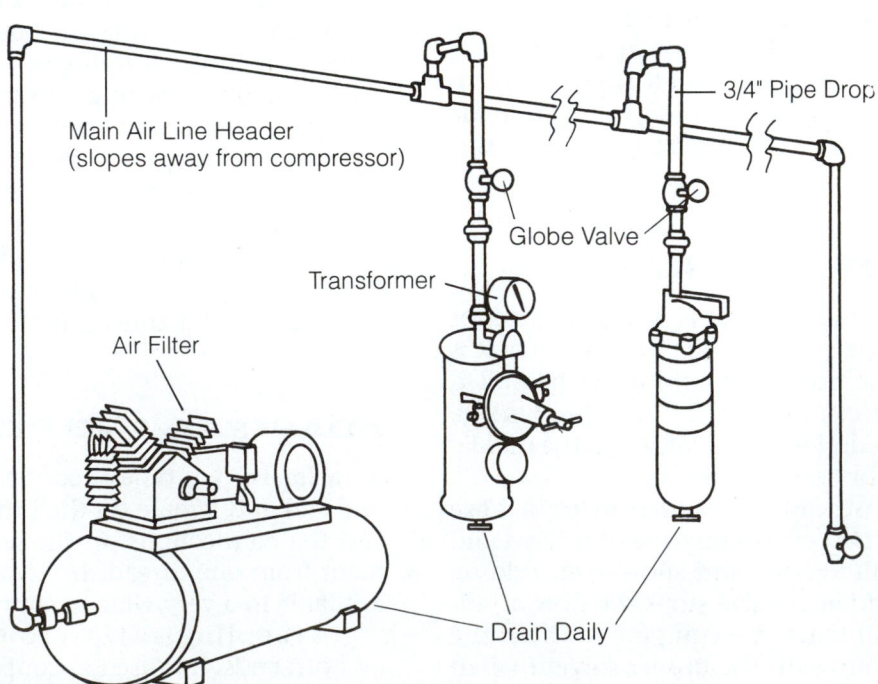

FIGURE 5-31 Note the drain valve locations on a typical compressed air system. *(Courtesy of PPG)*

MONTHLY MAINTENANCE

- Add or change the compressor crankcase oil. Under clean operating conditions, the oil should be changed at the end of 500 running hours or every six months, whichever occurs first. If operating conditions are not clean, change oil more frequently.
- Adjust the pressure switch cut-in and cut-out settings if needed.
- Check relief valve for exhausting of head pressure each time the motor stops.
- Tighten belts to prevent slippage. A heated motor pulley is a sign of loose belts. Overtightening of belts can cause motor overload or premature failure of motor and compressor bearings.
- Check and align a loose motor pulley or compressor flywheel. It will be necessary to remove the front section of the enclosed belt guard.
- Tighten all valve plugs and covers on the compressor head to ensure that each valve does not become loose and damage the valve or piston.
- Check for air leaks on the compressor outfit and air piping system.
- Check compressor pump-up time when the air receiver outlet valve is closed.
- Listen for unusual noises.
- Check and correct oil leaks.

5.5 AIR SYSTEM SAFETY

An air compressor system is a very safe arrangement to operate. Accidents seldom happen, but the few that do occur can usually be traced to human error.

- **Read the instructions.** Learn what each part of the compressor does by carefully reading the owner's manual that comes with the unit.

- **Inspect before each use.** Carefully check the hoses, fittings, air control equipment, and overall appearance of the compressor before each use. Never operate a damaged unit.
- **Use proper electrical outlets.** Electrical damage often results from using improperly grounded outlets. Use only a properly grounded outlet that will accept a three-prong plug.
- **Always run the compressor on a dry surface.** The compressor should be located where there is a circulation of clean, dry air. Avoid getting dust, dirt, and paint spray on the unit.
- **Be aware of starts and stops.** Most compressors start and stop automatically. Never attempt to service a unit that is connected to a power supply.
- **Keep hands away.** Fast moving parts will cause injury. Keep fingers away from the compressor while it is running. Do not wear loose clothing that will get caught in the moving parts. Unplug the compressor before working on it.
- **Keep the belt guard on.** Use all the safety devices available and keep them in operating condition. Also, remember that compressors become hot during operation. Exercise caution before touching the unit.
- **Release air slowly.** Fast moving air will stir dust and debris. Be safe! Release air slowly by using a pressure regulator to reduce pressure to that recommended for the tool.
- **Keep air hose untangled.** Keep the air, power, and extension cords away from sharp objects, chemical spills, oil spills, and wet floors. All of these can cause injury.
- **Depressurize the tank.** Be sure the pressure regulator gauge reads zero before removing the hose or changing the air tools. The quick release of high pressure air can cause injury.

SUMMARY

- The compressed air supply system is designed to provide an adequate supply of clean, dry air at a predetermined pressure to insure efficient operation of all pneumatic equipment in the body shop.
- The air compressor, the heart of the system, is designed to raise the pressure of air from normal atmospheric to some higher pressure. There are three basic types of air compressors: the diaphragm type, the piston type, and the rotary screw type.
- The piston-type air compressor pump develops compressed air pressure through the action of a reciprocating piston.

- Single-stage units normally are used in pressure ranges up to 125 psi (861 kPa) for intermittent service.
- Two-stage compressors are usually more efficient, run cooler, and deliver more air for the power consumed, particularly in the 100 to 200 psi (689 to 1,378 kPa) pressure range.
- Compressor oil plugs are provided for filling and changing air pump oil. Pounds per square inch (psi) is the measure of air pressure or force delivered by the compressor to the air tool.

■ The compressor air tank is a heavy gauge steel tank to hold an extra supply of compressed air. The typical parts of a stationary compressor are:

Air compressor pump (pressurizes air)

Electric motor or gasoline engine (power compressor)

Air receiver or storage tank (holds compressed air)

Check valve (prevents leakage of stored air)

Pressure switch (automatically controls the air pressure)

Air transformer (removes oil, dirt, and water from the compressed air)

ASE-STYLE REVIEW QUESTIONS

1. Which type is seldom used as the main type of compressor in a body shop?

   A. Reciprocating piston compressor
   B. Diaphragm compressor
   C. One-stage compressor
   D. Two-stage compressor

2. The piston compressor is _____.

   A. Less durable than the diaphragm compressor
   B. More durable than the diaphragm compressor
   C. One that requires lubrication
   D. Both B and C

3. This is the actual amount of free air in cubic feet that the compressor can pump in one minute at working pressure.

   A. Cfm
   B. Displacement cfm
   C. Free air cfm
   D. Psi

4. What does the displacement of a two-stage compressor equal?

   A. The sum of the two stages
   B. The difference between the stages
   C. That given for the first stage
   D. None of the above

5. Air tanks can have

   A. Horizontal mountings
   B. Vertical mountings
   C. Wheels
   D. All of the above

6. Which safety control is designed to maintain a supply of air within given pressure limits on gasoline and electrically driven compressors when it is not practical to start and stop the motor?

   A. Pressure switch
   B. Automatic unloader
   C. Centrifugal pressure release
   D. Overload protection

7. What is the recommended fuse rating for a fused disconnect switch?

   A. Equal to the current rating stamped on the motor
   B. Twice the current rating stamped on the motor
   C. Value recommended by manufacturer
   D. None of the above

8. The intake air filter must be made of _____.

   A. Fiberglass
   B. Fine mesh
   C. Felt material
   D. Both B and C

9. How far away from the compressor unit should an air transformer be installed?

   A. At least 5 feet (1.52 meters)
   B. At least 15 feet (4.56 meters)
   C. At least 25 feet (7.6 meters)
   D. At least 35 feet (10.6 meters)

10. Which of the following is often combined with a lubricator in a single unit?

   A. Thermal conditioning equipment
   B. Air condenser
   C. Air pressure regulator
   D. Both B and C

ESSAY QUESTIONS

1. List and explain the basic requirements for a compressed air supply system.

2. How is air pressure measured in an air supply system?

3. How does a piston-type compressor operate?

4. Explain cfm.

5. What is an air transformer?

6. What are some daily maintenance tasks on a compressed air supply system?

CRITICAL THINKING PROBLEMS

1. What will happen if the oil level is too low or dry in a piston-type compressor?

2. What could be some of the reasons that an air tool slows down and runs slowly?

MATH PROBLEMS

1. If a compressor piston has an area of 5 square inches, a 3-inch stroke, runs at 2,000 rpm, and has one piston, what is its theoretical cfm?

2. How much pressure is dropped at 60 psi (408 kPa) when using a 50-foot (15.2) length ¼-inch (6.3 mm) inside diameter air hose?

   A. 20.5
   B. 6.0
   C. 4.5
   D. 1.5

CHAPTER 6

Auto Body Materials and Fasteners

INTRODUCTION

When customers look at a car's paint job, they often see only a shiny, bright color. They seldom understand all of the technology involved in producing that long-lasting, tough, durable, high gloss finish. There is much hidden technology under the surface of the paint. A professional technician comprehends all of the "chemistry" and skill needed to do a good repair. This chapter will introduce you to the materials needed to do competent paint and body work.

Body shop materials include more than just refinishing or paint materials. They include the various fillers, primers, sealers, adhesives, sandpapers, and other compounds common to a body shop. It is critical that you understand their selection and use. This first part of the chapter will explain the purpose of these basic types of materials. The second section of the chapter summarizes fasteners. This informa-

tion will prepare you for later chapters that explain how to use materials and fasteners in more detail.

6.1 REFINISHING MATERIALS

A vehicle body is protected by a complete finishing system. All parts of the system work together to protect the vehicle from ultraviolet radiation, weathering, pollutants, and corrosion.

Refinishing materials is a general term referring to the products used to repaint a vehicle. Refinishing material chemistry has changed drastically in the past few years. New paints last longer but require more skill and safety measures for proper application.

The *substrate* is the metal, fiberglass, or plastic material used in the vehicle's construction. It will affect the selection of refinishing materials.

OBJECTIVES

After studying this chapter, you should be able to:

✔ Select the right materials for the job.

✔ Explain the basic purpose of common materials.

✔ Compare the use of similar shop materials.

✔ Summarize when to use different kinds of filler.

✔ Know how to select the right type of primer and paint.

✔ Understand the importance of using a complete paint system.

✔ Identify the various fasteners used in body construction.

✔ Remove and install bolts and nuts properly.

✔ Explain when specific fasteners are used in body construction.

✔ Explain bolt and nut torque values.

✔ Summarize the use of chemical fasteners.

✔ Answer ASE test questions relating to body shop materials and fasteners.

KEY TERMS

adhesives

bleeding

catalyst

compounding

enamel

fasteners

metal conditioner

paint

pearl paints

plastic filler

prep solvent

primer

primer-surfacer

sandpaper

sealers

waterbase paint

ASE TASK LIST

Job Skills covered in this chapter include:

PAINTING AND REFINISHING TEST (B2) TASK LIST

A. Surface Preparation

2. Remove dirt, road grime, and wax or other protective coatings from area to be refinished and adjacent vehicle surfaces.
3. Inspect and identify substrate, type of finish, and surface condition; develop a plan for refinishing.
4. Remove paint finish.
8. Mask trim, and protect other areas that will not be refinished.
9. Mix primer, primer-surfacer, or primer-sealer; spray onto surface of repaired area.
10. Apply two-component putty to minor surface imperfections.
11. Dry or wet sand area to which primer-surfacer and/or two-component putty have been applied.
13. Clean area to be refinished using a proper cleaning solution.
14. Remove, with a tack rag, any dust or lint particles from the area to be refinished.
15. Apply suitable paint sealer to the area being refinished when sealing is needed or desirable.
16. Remove imperfections from sealer.
18. Apply stone chip-resistant coating.
19. Restore corrosion resistant coatings, caulking, and seam sealers to repaired areas.

B. Spray Gun Operation and Related Equipment

1. Inspect, clean, and determine condition and adequacy of spray guns and related equipment (air hoses, regulator, air lines, air source, and spray environment).

C. Paint Mixing, Matching, and Applying

5. Apply single-stage topcoat for spot and panel blending, and overall refinishing.
6. Apply basecoat/clearcoat for spot and panel blending, and overall refinishing.
8. Sand, buff, and polish finishes where necessary.
12. Apply multistage (mica, pearl, etc.) coats for spot repair, panel blending, and overall refinishing.
13. Identify paint color formula and proper usage of mixing equipment and materials.

D. Solving Paint Application Problems

2. Identify blushing (milky or hazy formation); determine the cause(s), and correct the condition.
3. Identify contaminants in the painted surface; determine the source(s), and correct the condition.
8. Identify orange peel appearance of the refinished surface; determine the cause(s), and correct the condition.

9. Identify an overspray condition; determine the cause(s), and correct the condition.
12. Identify sandscratch swelling; determine the cause(s), and correct the condition.

F. Safety Precautions and Miscellaneous

1. Identify and take necessary precautions with hazardous operations and materials according to EPA regulations.

NONSTRUCTURAL ANALYSIS AND DAMAGE REPAIR TEST (B3) TASK LIST

A. Preparation

5. Remove undamaged, nonstructural body panels and components that may interfere with or be damaged during repair.
8. Remove dirt, grease, wax, and decals from areas to be repaired.
9. Remove corrosion protection, undercoatings, sealers, and other protective coatings as necessary to perform repairs.
10. Remove repairable plastics and other parts that are recommended for off-vehicle repair.

B. Outer Body Panel Repairs, Replacements, and Adjustments

2. Remove and replace bolted, bonded, and welded panels or panel assemblies.
7. Remove, replace, and align bumpers, reinforcements, guards, isolators, and mounting hardware.
12. Apply protective coatings and sealants to restore corrosion protection.

C. Metal Finishing and Body Filling

3. Prepare surface for application of body filler material.
6. Mix plastic filler.
7. Apply plastic body filler; rough shape during curing.
8. Sand cured plastic body filler to contour.

D. Moveable Glass and Hardware

5. Inspect, repair, and install convertible top and related mechanisms.

STRUCTURAL ANALYSIS AND DAMAGE REPAIR TEST (B4) TASK LIST

A. Frame Inspection and Repair

9. Restore corrosion protection to repaired or replaced frame areas.

The term **paint** generally refers to the visible topcoat. The most elementary painting system consists of a compatible primer and final topcoat over the substrate. This process can vary considerably and is usually more complex, as you will learn later.

UNDERCOATS AND TOPCOATS

A basic finish consists of several coats of two or more different materials. The most basic finish consists of:

1. Undercoat or primer coat.
2. Topcoat (colorcoat or basecoat/clearcoat).

The **primer** or *undercoat* improves adhesion of the topcoat. It is often the first coat applied. Paint alone will not stick or adhere as well as a primer. If you apply a topcoat to bare substrate, the paint will peel, flake off, or look rough. This is why you must "sandwich" a primer undercoat between the substrate and the topcoat. Undercoats also prevent any chemicals from bleeding through and showing in the topcoats of paint.

The terms *topcoat* or *colorcoat* refer to the paint applied over the undercoat. It is usually several light coats of one or more paints. The topcoat is the "glamour coat" because it features the eye-catching color, color effects, and gloss.

Basecoat/clearcoat paint systems use a colorcoat applied over the primer with a second layer of clearcoat over the colorcoat. This is the most common paint system used today. The clear paint brings out the richness of the underlying color and also protects it. It makes the paint shine more than a single layer of color not using a clearcoat.

PAINT TYPES

There are three general types of paint:

1. Lacquer paints
2. Enamel paints
3. Waterbase paints

As you will learn, there are variations within these categories. It is important that you know what type of finishes manufacturers use because there are slightly different methods required for refinishing them.

Lacquer is an older paint that dries quickly because of solvent evaporation. Lacquers have been phased out and replaced by the more durable enamel and waterbased paints by both OEMs and body shops.

Lacquer topcoats usually must be compounded or rubbed with a compound or polish to bring out their gloss. Some basecoat/clearcoat finish systems have a lacquer basecoat to be clearcoated with an enamel topcoat. This helps eliminate the need for compounding or polishing to bring out the gloss.

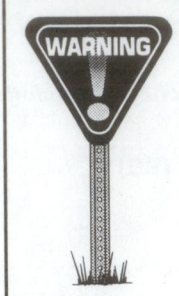

WARNING

Never spray lacquer over the top of enamel. The enamel can lift and cause problems. You can spray enamel over an old lacquer finish, however.

Enamel finishes dry with a gloss and do not require rubbing or polishing because the enamel dries through a chemical change rather than by solvent evaporation.

Since enamels generally dry slower, there is more of a chance for dirt and dust to stick in the finish. While there is usually a slight amount of surface roughness (orange peel) in an enamel film, too much will cause a lower gloss.

Two-stage paints consist of two distinct layers of paint: basecoat and clearcoat. *Basecoat/clearcoat enamel* is now the most common system used to repaint cars and trucks. First, a layer of color is applied over the undercoat of primer. Next, while the basecoat is still wet, a second coat of clear is sprayed over the color basecoat.

Acrylic enamel and *acrylic urethane enamel* are two specific types of enamel paint that are commonly used in the industry. Acrylic urethanes are slightly harder than plain acrylics. Each is available in a variety of colors.

It can be difficult to tell the difference between lacquer and enamel, especially if the vehicle has been refinished. As you become more familiar with automotive refinishing, it will be easier to distinguish between them. This will be discussed later.

Waterbase paint, as implied, uses water to carry the pigment. It dries through evaporation of the water. Some manufacturers are starting to used waterbase paints on new vehicles. This helps satisfy stricter emission regulations in some geographic areas. The basecoat of color is waterbase. Then, an enamel topcoat is applied over the waterbase paint to protect it from the environment.

Waterbase primers have been used for years as a fix for lifting problems. Waterbase primers serve as an excellent barrier coat when there are paint incompatibility problems.

OEM FINISHES

Today's passenger car and light truck *OEM finishes* (factory paint jobs) are either "thermo-setting" acrylic enamel (paint is oven-baked and hardened at the factory), new high solids basecoat/clearcoat enamels, or sometimes waterbase, low emission paints. Common enamel finishes are baked in huge ovens to shorten the drying times and cure the paint. This is done before installing the interior and other non-metal parts.

Vehicle manufacturers use several different types of finish materials, coating processes, and application processes. Each type of finish requires different planning and repair steps.

The most common types of OEM coating processes include:

1. Single stage
2. Two-stage (basecoat/clearcoat)
3. Three-stage paint (tri-coat)
4. Multistage

These will be explained fully in later chapters of this book.

CONTENTS OF PAINT

A paint's *chemical content* includes the following:

1. Pigments
2. Binders
3. Solvents
4. Additives

Each of these ingredients has a specific function within the paint formula.

Paint Pigments

The *pigments* are fine powders that impart color, opacity, durability, and other characteristics to the primer or paint. It is a nonvolatile film-forming ingredient. The main purpose of the pigment is to hide everything under the paint.

The size and shape of the pigment particles are also important. Pigment particle size affects hiding ability and appearance. In addition, pigment shape affects strength. Pigment particles may be nearly spherical, rod-, or platelike. Rod-shaped particles, for example, reinforce paint film like iron bars in concrete.

Medium size reflective pigment particles, such as mica, are added to **pearl paints** to give the paint a luster or shine that tends to change color with viewing angle. As you will learn later, pearlescent paints are now common and are the most difficult to match when repainting.

Large reflective pigment flakes are added to *metallic paints*. The size, shape, color, and material in the flakes can vary. Often called *metal flakes*, they can be made of tiny but visible bits of metal or polyester. When light strikes the flakes, they reflect the light at different angles, making them look like tiny glittering stars inside the paint.

If this is new to you, start looking at paint jobs more closely. See if you can tell the difference between a solid color, a pearl, and a metallic.

Paint solids are the nonliquid contents of the paint or primer. *High solids paints* are needed to reduce air pollution or emissions when painting. *Low solids paints* are less desirable because they contain more solvents that pollute the air.

Paint Binders

The *paint binder* is the ingredient in a paint that holds the pigment particles together. It is the backbone, or film-former, of the paint. The binder helps the paint stick to the surface. Various materials are used in the binder. The binder type determines the kind of paint—lacquer or enamel—because it contains the drying mechanism.

It is generally made of a natural resin (such as rosin), drying oils (like linseed or cottonseed), or a manufactured plastic and is usually modified with plasticizers and catalysts. They improve such properties as durability, adhesion, corrosion resistance, mar resistance, and flexibility.

Paint Solvents

The *paint solvent* is the liquid solution that carries the pigment and binder so it can be sprayed. Thinners and reducers are composed of one or several solvents and provide a transfer medium. Solvents are the volatile part of a paint in non-water-based systems. They are used to reduce a paint for spraying and give the paint its flow characteristics, evaporating as the paint dries.

Most solvents are made from crude oil or petroleum. However, waterbase paints are increasing in use to meet strict pollution regulations. The solvent reduces or thins the binder and transfers the pigment and binder through the spray gun to the surface being painted.

When used with lacquer, the solvent is called a *thinner*. When used with enamel, the solvent is called a *reducer*. This is an important distinction. The respective products are so labeled.

It is important to remember that you "thin" lacquer and "reduce" enamel. Thinning and reducing are needed to make the mixture the right thickness or *viscosity* to flow smoothly onto the surface.

Waterbase paints come *premixed* (ready to spray) and they are not normally reduced. In an emergency, distilled water can be added to make a thinner, more liquid solution.

When using waterborne materials, the water used for equipment cleaning:

1. Contains hazardous materials and should be disposed of as hazardous waste.
2. Cannot be poured down the drain for disposal.
3. Must not be combined with other waste solvents, such as reducers or thinners. Keep a separate container for storing water wastes.

Due to clean air regulations, some solvents are no longer being used. To meet these regulations, traditional solvents are being replaced by water or other solvents. Check local ordinances.

Paint Additives

Paint additives are ingredients added to modify the performance and characteristics of the paint. Additives are used to:

1. Speed up or slow down the drying process.
2. Lower the freezing point of a paint.
3. Prevent the paint from foaming when shaken.
4. Control settling of metallics and pigments.
5. Make the paint more flexible when dry.
6. Increase gloss or shine.
7. Decrease gloss or shine.

DRYING AND CURING

Drying is the process of changing a coat of paint or other material from a liquid to a solid state. Drying is due to the evaporation of solvent (Figure 6–1), a chemical reaction of the binding medium (Figure 6–2), or a combination of these causes.

The term *"drying"* refers to a paint material that evaporates its solvent to harden. The term *"curing"* refers to a chemical action in the paint or other material itself that causes hardening.

Flash is the first stage of drying where some of the solvents evaporate, which dulls the surface from a high gloss to a normal gloss.

A *retarder* is a slow evaporating thinner or reducer used to retard or slow drying. It is a slow-drying solvent often used in very warm weather. If a paint dries too quickly, problems like blushing can result.

An *accelerator* is a fast evaporating thinner or reducer for speeding the drying time. It is needed in very cold weather to make the paint dry in a reasonable amount of time. The term "accelerator" also refers to a hardener or catalyst in most paint systems.

A general rule to follow in selecting the proper thinner or reducer is: The faster the shop drying conditions, the slower drying the thinner or reducer should be. In hot, dry weather, use a slow-dry thinner or reducer. In cold, wet weather, use a fast-dry thinner or reducer.

A **catalyst** is a substance that causes or speeds up a chemical reaction. When mixed with another substance, it speeds the reaction but does not change itself. Catalysts are used with many types of materials—paints, putties, fillers, fiberglass, and plastics.

A *paint catalyst*, *drier*, or *hardener* is an additive used to make enamel paints cure quickly. Seldom is an enamel paint used alone. A hardener is almost always used by professional painters. The hardener speeds curing and makes the paint more durable.

Isocyanate resin is a principal ingredient in some urethane hardeners. Because this ingredient has toxic effects, you must always wear a NIOSH-approved respirator. Usually, a positive pressure or fresh-air-supplied respirator is required when spraying an isocyanate product.

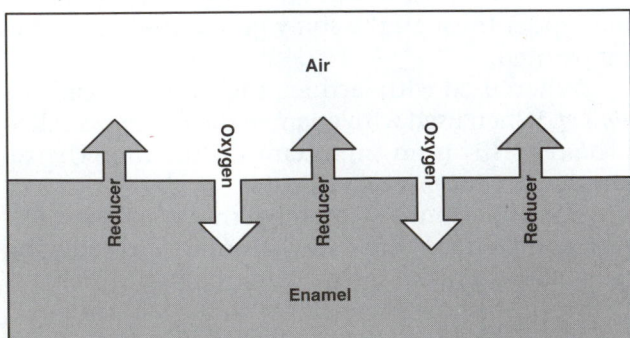

FIGURE 6–1 Automotive paints or finishes either dry or cure. Enamel dries by evaporation of its reducer or solvent first, then by oxidation. Resin reacts with oxygen in the air to dry. Heat speeds this operation. *(Courtesy of PPG Industries)*

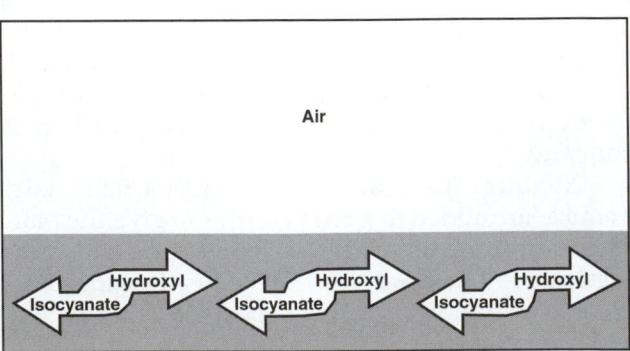

FIGURE 6–2 Urethane and polyurethane enamels cure by molecular crosslinking. Oxygen does not have to be present. This type of paint will cure inside an enclosed container, like your spray gun cup if you forget to clean it! *(Courtesy of PPG Industries)*

The hardener is added to the paint right before it is sprayed. When an enamel catalyst is used, the paint can be wet sanded and compounded (polished) the next day. If you make a mistake (paint run, dirt in paint, etc.), you can fix the problem after the short curing time. The hardener will make the enamel cure in just a few hours. Also the car can be released to the customer sooner with less chance of paint damage.

PRIMERS AND SEALERS

Primers come in many variations—primer, primer-sealer, primer-surfacer, primer-filler, etc. It is important to understand the functions of subcoating or undercoat materials. You must follow manufacturer's instructions. Deviation from these directions will result in unsatisfactory work.

A plain primer is a thin undercoat designed to provide good adhesion for the topcoat. Primers are generally used by automobile manufacturers rather than paint and body shops. However, primers can be used when the surface is very smooth and there is no potential problem with bleeding. If properly applied, primer does not require sanding.

Primers are usually enamel or epoxy-based products because they provide better adhesion and corrosion resistance than older lacquers.

A *self-etching primer* has acid in it to treat bare metal so that the primer will adhere properly. Some primer-sealers and primer-surfacers also have an etching material in them.

Sealers

Bleeding or *bleedthrough* is a problem where colors in the undercoat or old paint chemically seep into the new topcoats. This can discolor the new paint.

A *paint sealer* is an innercoat between the topcoat and the primer or old finish to prevent bleeding. Sealers differ from primer-sealers in that they cannot be used as primers. Sealers are sprayed over a primer or primer-surfacer, or a sanded old finish. Sealers do not normally need sanding but some are sandable.

Sealers are sometimes used when a sharp color difference is visible after sanding. They are also used to prevent sand scratch swelling problems.

A *primer-sealer* is an undercoat that improves adhesion of the topcoat and also seals old painted surfaces that have been sanded. It will solve two potential problems (adhesion and bleed) with one application.

Primer-sealers provide the same protection as primers—adhesion and corrosion resistance. But they also have the ability to seal over a sanded old finish to provide uniform color hold out, which prevents old color showing through new color.

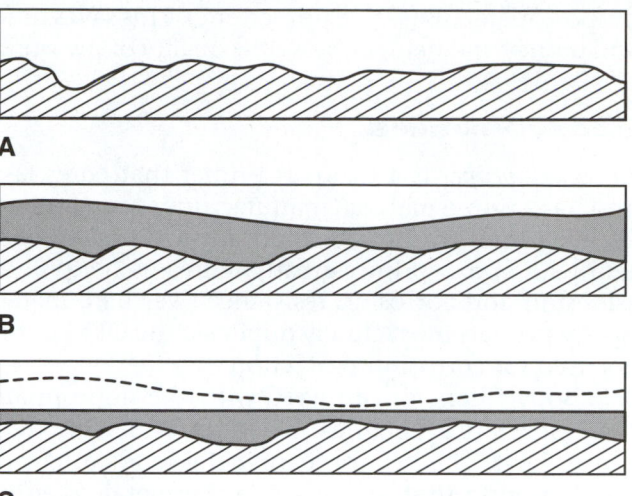

FIGURE 6-3 Primer-surfacer is the "workhorse" of the auto body repair industry. (A) Magnified view of a cross-section shows that the surface is slightly rough. This might be due to sanding or the texture of the plastic filler. (B) Primer-surfacer has been sprayed over the surface. It has a high solids content that flows and fills indentations. (C) Sanding the primer-surfacer will quickly level and smooth the surface. This readies the surface for topcoats. *(Courtesy of PPG Industries)*

Primer-Surfacers

A **primer-surfacer** is a high solids primer that fills small imperfections and usually must be sanded. It is often used after a filler to help smooth the surface. Primer-surfacers are the workhorses among the undercoats in refinish applications. They are used to build up and level featheredged areas or rough surfaces and to provide a smooth base for topcoats (Figure 6–3).

A good primer-surfacer should have the following characteristics:

1. Adhesion
2. Rust and corrosion resistance
3. Buildup
4. Sanding ease
5. Color hold out
6. Quick drying speed

Strong adhesion is the first prerequisite of a primer. All automotive topcoat colors require the use of a primer or primer-surfacer as a first coat over bare substrate. A good primer-surfacer should be ready to sand in as short a period as 30 minutes.

A *primer-filler* is a very thick form of primer-surfacer. It is sometimes used when a very pitted or rough surface must be filled and smoothed quickly. It might be used on a solid but badly rusted and pitted body panel, for example.

The industry trend is to use combination materials (primer-sealers or primer-surfacers) over single purpose materials (sealer, primer, etc.). This saves time and money and helps improve the quality of the work.

Epoxy Primers

An *epoxy primer* is a two-part primer that cures fast and hard. Some material manufacturers recommend epoxy primer prior to the application of body fillers. Using an epoxy primer greatly increases body filler adhesion and corrosion resistance over bare metal. Epoxy primers most closely duplicate the OEM primers used for corrosion protection.

Body fillers, once mixed with the appropriate catalyst, start a chemical reaction, which in turn causes heat. The heat on bare metal tends to create condensation that may corrode the metal. Eventually the plastic body filler cracks and loosens, leaving a corroded area. An epoxy primer protects against moisture entrapment caused by condensation and can result in a longer lasting repair.

COMPLETE PAINT SYSTEMS

Remember to always use a complete paint system. A *paint system* means all materials (primers, catalysts, paints) are compatible and manufactured by the same company. They are designed to work properly with each other. If you mix materials from different manufacturers, you can run into problems. The chemical contents of the different systems may not work well together (Figure 6–4).

OTHER PAINT MATERIALS

A **prep solvent** or wax and grease remover is a fast-drying solvent often used to chemically clean a vehicle. It will remove wax, oil, grease, and other debris that could contaminate and ruin the paint job.

A *flattener* is an agent added to paint to lower gloss or shine. It can be added to any color gloss paint to make it a *flat* (dull) color. For example, some factory and custom hoods are painted flat black for a high performance look. A flattening agent would be used in this instance. It can also be used where reflection off a high gloss paint could affect the driver's vision.

A *fisheye eliminator* is a paint additive that helps smooth the paint when small craters or holes in the paint film are a problem. Contaminants (usually silicone) make the paint flow away from small debris in the paint film. Fisheye eliminator is an oil-based material that makes the paint flow over the top of the contaminate. It should be used only when absolutely necessary. However, always have fisheye eliminator on hand for emergencies.

A *flex agent* is an additive that allows primers and paints to flex or bend without cracking. It is

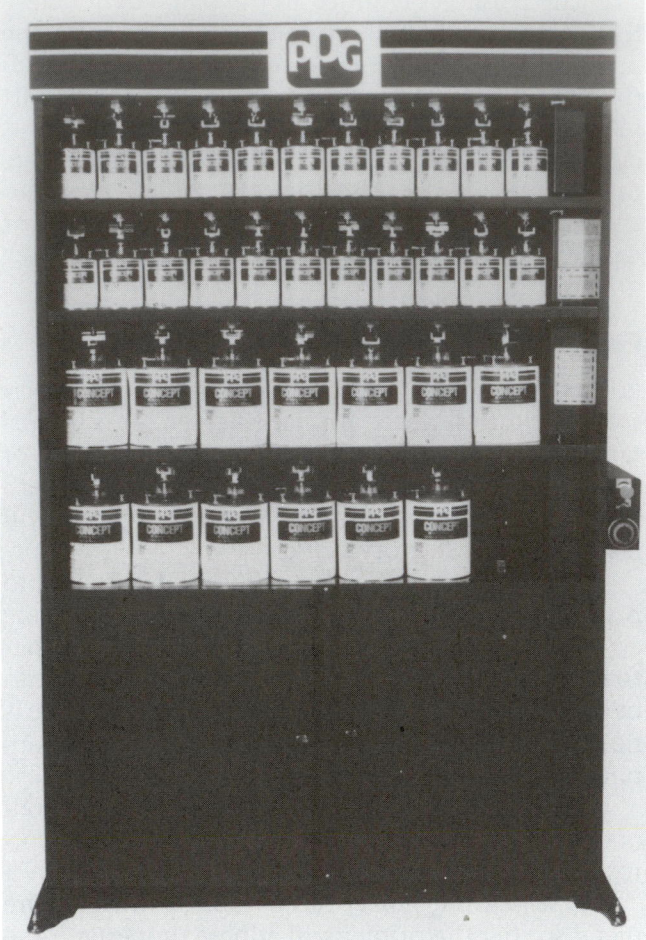

FIGURE 6–4 When painting, always use a complete paint system produced by one manufacturer. Then you are sure that all the chemicals used in the products will work together without problems. *(Courtesy of PPG Industries)*

commonly added to paints being applied to plastic bumper covers. Also called an *elastomer,* it is a manufactured compound with flexible and elastic properties that can be added to primers and paints.

Antichip coating, also called *gravel guard, chip guard,* or *vinyl coating,* is a rubberized material used along a vehicle's lower panels and on the front edge of hoods and fenders. It is designed to be flexible or rubbery to resist chips from rocks and other debris flying up off the tires. Antichip coatings are usually applied with a special spray gun, like the one in Figure 6–5.

Many manufacturers are using special chip resistant coatings in areas that are exposed to stones and gravel. These coatings are generally between the E-coat primer and the topcoats. Some chip resistant coatings are clear and can be applied over the topcoat. If a vehicle has chip resistant coatings, they must be replaced during the refinishing process.

Rubberized undercoat is a synthetic-based rubber material applied as a corrosion or rust preventive

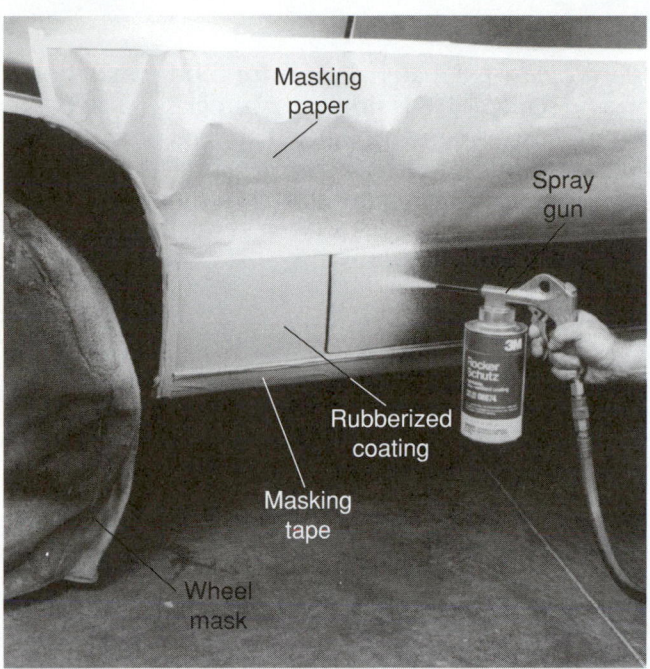

FIGURE 6–5 Antichip coating is rubberized material often sprayed along lower panels of a vehicle to help resist chips. *(Courtesy of Bond-Tite®, Division of U.S. Chemical and Plastics, Inc.)*

layer. It can be applied using a production gun or a spray can.

A **metal conditioner** is phosphoric acid used to etch bare sheet metal before priming. It is a chemical cleaner that removes rust and corrosion from bare metal and helps prevent further rusting.

Remember the following about metal conditioners:

1. Acid cleans the metal.
2. It dissolves light surface rust.
3. It etches metal, improving adhesion.
4. It needs to be completely neutralized with water after applying.
5. It may have to be diluted; follow product directions.
6. It is always followed by conversion coating.
7. It is necessary to wear rubber gloves, a respirator, and eye protection.

A *conversion coating* is a special metal conditioner or primer used on galvanized steel, uncoated steel, and aluminum to prevent rust. It is applied after acid etching or metal conditioning.

Corrosion is a chemical reaction of air, moisture, or corrosive materials on a metal surface. Corrosion of steel is usually referred to as *rusting* or oxidation.

Paint stripper is a powerful chemical that dissolves paint for fast removal of an old finish. If the old paint is cracking or peeling, you may have to use a chemical stripper. It is applied over the old paint. After it soaks into and lifts the paint, a plastic scraper is used to remove the softened paint.

SHOP TALK

Chemical paint strippers are not environmentally friendly. Check with local and national regulations regarding usage of chemical strippers. Plastic media and soda blasting are alternate, less polluting means of paint removal.

A *tack cloth* can be used to remove dust and lint from the surface right before painting. It is a cheesecloth treated with nondrying varnish to make it tacky. A tack cloth must be wiped gently over the surface to keep the varnish from contaminating the paint.

PLASTIC BODY FILLERS

A *filler* is any material used to fill or level a damaged area. There are several types of filler. You should understand their differences.

Body filler or **plastic filler** is a heavy-bodied plastic material that cures very hard for filling small dents in metal. It is a compound of resin and plastic used to fill dents on car bodies (Figure 6–6).

Body fillers come in cans and in plastic bags. When in a plastic bag, a dispenser is used to force the filler onto your mixing board. This keeps the filler clean. A *mixing board* is the surface (metal, glass, or plastic) used for mixing the filler and its hardener.

Light body filler is formulated for easy sanding and fast repairs. It is used as a very thin top coat of filler for final leveling and can be spread thinly over large surfaces for block or air tool sanding.

FIGURE 6–6 Plastic filler comes in cans, buckets, and plastic bags. A dispenser will save time and keep the filler uncontaminated. *(Courtesy of Bond-Tite®, Division of U.S. Chemical and Plastics, Inc.)*

FIGURE 6-7 Cream hardeners are added to the filler to make it cure. Note that hardeners can differ and can deteriorate with age. Always use the type recommended by the filler manufacturer. *(Courtesy of Bond-Tite®, Division of U.S. Chemical and Plastics, Inc.)*

Fiberglass body filler has fiberglass material added to the plastic filler and is used for rust repair or where strength is important. It can be applied on both metal and fiberglass substrates. Because fiberglass-reinforced filler is very difficult to sand, it is usually used under a conventional, lightweight plastic filler.

Short-strand fiberglass filler has tiny particles of fiberglass in it. It works and sands like a conventional filler, but is much stronger. *Long-strand fiberglass filler* has long strands of fiberglass for even more strength. It is commonly used for repairing holes in metal or fiberglass bodies.

Cream hardeners are used to cure body fillers and usually come in tubes. Once the hardening cream is mixed in, the plastic filler will heat and harden (Figure 6–7). Too much cream hardener can also cause problems with adhesion and pinholing.

Fiberglass

Fiberglass resin is another form of plastic body repair material. It is a thick resin liquid that comes in a can. The fiberglass resin must be mixed with its own special type of hardener to cure. If you accidentally use cream hardener in fiberglass resin, it will not harden and you will have a mess to clean up.

Fiberglass mat is a series of long fiberglass strands irregularly distributed to form a patch. It is used to strengthen and form a shape for the resin liquid.

Fiberglass cloth is made by weaving the fiberglass strands into a stitched pattern (Figure 6–8).

Fiberglass tape will stick to the repair surface to help form the body shape and is a fast method of applying fiberglass for repairs. Resin can be applied over the tape.

Glazing Putty

Glazing putty is a material made for filling small holes or sand scratches. It is similar to primer-surfacer but it has more solid content. Putty is applied over the undercoat of primer-sealer or primer-surfacer to correct small surface imperfections. The purpose of glazing

A

B

FIGURE 6-8 Fiberglass resin and mat or cloth is often used to repair fiberglass body panels and parts. It is sometimes used for rust repair. (A) Clear resin cures to a hard, brittle solid. A special catalyst is needed. Do not use cream hardener. (B) Fiberglass cloth, mat, or tape is used with resin. Cut pieces to size, coat them with resin, and apply them over the repair area. *(Courtesy of Bond-Tite®, Division of U.S. Chemical and Plastics, Inc.)*

putties is to fill imperfections that cannot be filled with a primer-surfacer.

Spot putty is the same as glazing putty except it has even more solids. Spot putty is recommended for scratches or nicks up to $^1/_{16}$ inch (1.5 mm) deep. It should not be used to fill large surface depressions. For larger depressions, use plastic filler or catalyzed putty.

One-part putty often comes in a tube. A rubber spreader is used to work the putty into small holes in the primer. After fully curing, the putty is sanded smooth. Only the small pinholes or scratches remain filled with the putty.

Two-part putty comes with its own hardener for rapid curing. This is the main advantage of two-part putty, as it cures much more quickly. If you paint over partially cured putty, it can shrink and cause problems.

WARNING A common mistake is to try to use putty like a filler. Putty is very weak and will not adhere to metal or filler. Only use putty over the top of primer to fill small pinholes or sand scratches. Don't be a "putty builder" or your work will be very weak.

MASKING MATERIALS

Masking materials are used to cover and protect body parts from paint overspray. *Overspray* is unwanted paint spray mist floating around from a spray gun. It can stick to glass and body parts and take considerable time to clean off.

Masking paper is special paper designed to be used to cover body parts that are not to be painted. It comes in a roll and, when mounted on a masking machine, masking tape is automatically applied to one edge of the paper, which speeds the work (Figure 6–9).

Masking plastic is used just like paper to cover and protect parts from overspray. It also comes in rolls and can cover large body areas more easily than paper.

Wheel masks are preshaped plastic or cloth covers for the vehicle's wheels and tires. Preshaped plastic antenna, headlamp, and mirror covers are also available.

Masking tape is used to hold masking paper or plastic in position. It is a high-tack, easy-to-tear tape. It comes in rolls of varying widths, $^3/_4$-inch (19 mm) wide being the most common. Refer to Figure 6–10. To save time and improve quality of work, always purchase quality masking tape.

Fine line masking tape is a very thin, smooth surface plastic masking tape. Also termed *flush masking tape*, it can be used to produce a better *paint part edge* (edge where old paint and new paint meet). When the fine line tape is removed, the edge of the new paint will be straighter and smoother than if conventional masking tape is used.

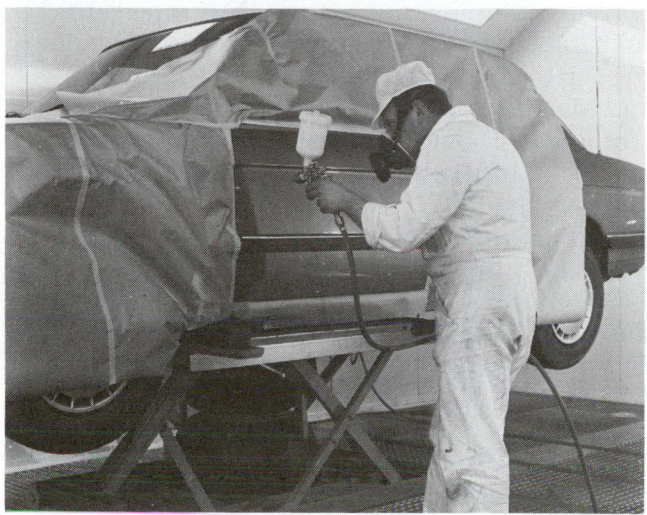

FIGURE 6–9 Masking is needed to keep overspray off panels and parts not to be painted.

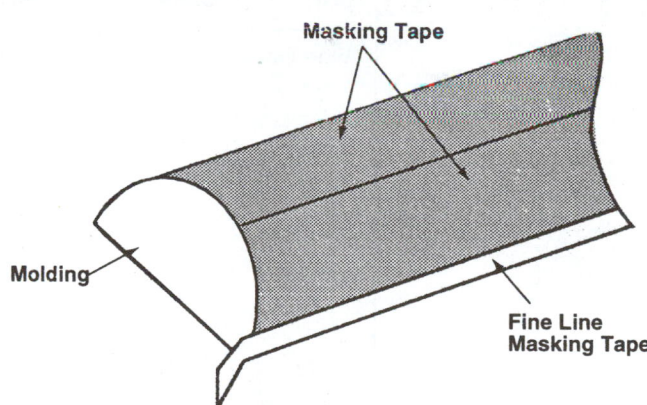

1. **Apply Masking Tape to Within 1/8" (3mm) of Edge.**

2. **Apply Masking Paper.**

3. **Mask to Edge of Part with 1/4" (6mm) Fine Line Tape.**

4. **Leave a Tail or Handle.**

5. **Remove Fine Line Tape While Paint is Still Wet.**

FIGURE 6–10 Masking tapes come in various widths, $^3/_4$-inch wide being the most common. Masking tape has a paper backing. Fine line tape is made of thin plastic for making a smoother paint edge. Note the example of how fine line tape is used along rubber molding to produce a good paint edge. Paper masking tape is then used to hold the paper in place next to the fine line tape. *[Courtesy of I-CAR]*

Fine line tape can help a painter mask flush mounted parts. It may be better to mask these parts than to remove them. Door handles, side trim, mirror mounts, and moldings are a few examples. Fine line tape is also used where two different paint colors come together, as when painting stripes or two-tone finishes.

Duct tape is a thick tape with a plastic body that is sometimes used to protect parts from damage when grinding, sanding, or blasting. Duct tape is thicker than masking tape and provides more protection for the surface under the tape.

Masking liquid, also called *masking coating*, is usually a water-based sprayable material for keeping overspray off body parts. Some are solvent-based. Masking liquids come in large, ready-to-spray containers or drums. These materials are sprayed on and form a paint-proof coating over the vehicle.

Some masking coatings are tacky and used only during priming and painting. They form a film that can be applied when the vehicle enters the shop. Others dry to a hard, dull finish.

Masking coatings can be removed when the vehicle is ready to be returned to the owner. They wash off with soap and water. Local regulations may require that liquid masking residue be captured in a floor drain trap and not put into the sewer system.

ABRASIVES

An *abrasive* is any material, such as sand, crushed steel grit, aluminum oxide, silicon carbide, or crushed slag, used for cleaning, sanding, smoothing, or material removal. Many types of abrasives are used by the auto body technician.

Grit refers to a measure of the size of particles on sandpaper or discs. A *coarse sandpaper* has large grit; *fine sandpaper* has smaller grit.

Grit Ratings

A *grit numbering system* denotes how coarse or fine the abrasive is. For example, 16 grit would be one of the coarsest and 1500 grit would be one of the finest. The grit number is printed on the back of the paper or disc. Look at Figure 6–11.

Very coarse grit of 16 to 24 is generally used for fast material removal. It will quickly remove paint down to bare metal. This grit is commonly used on grinding discs and air files for rapid cutting.

GRIT	ALUMINUM OXIDE	SILICON CARBIDE	AUTO BODY USE
Ultra Fine		1500 1200 800	Color-Coat Wet Sanding.
Very Fine		600	Color-Coat Wet Sanding Before Polishing.
	400 320 280 240	400 320 280 240	Sanding Primer-surfacer and Paint Before Painting. Wet or Dry Sanding Paper
	220	220	Dry Sanding Top Coat. Also Available for Wet Sanding.
Fine	180 150	180 150	Final Sanding of Bare Metal and Smoothing Old Paint. Dry Paper.
Medium	120 100 80	120 100 80	Smoothing Paint and Plastic Filler. Dry Paper.
Coarse	60 50 40 36	60 50 40 36	Rough-Sanding of Filler and Plastics. Dry Paper.
Very Coarse	24 16	24 16	Coarse Sanding or Grinding to Remove Paint. Dry Paper.

FIGURE 6–11 Note how sandpaper is made and classified by coarseness.

A *coarse grit* of 36 to 60 is basically used for rough sanding and smoothing operations. This coarseness might be used to get the general shape of a large plastic filler area.

Medium grit of 80 to 120 is often used for sanding plastic filler high spots and for sanding of old paint.

Fine grit of 150 to 180 is normally used to sand bare metal and for smoothing existing painted surfaces. This is also used for final sanding of plastic filler and to featheredge paint.

Very fine grit ranges from 220 to about 2000 and is used for numerous final smoothing operations. Larger grits of 220 to 320 are for sanding primer-surfacers and old paint. Finer grits of 400 to 2000 are for colorcoat sanding and sanding before polishing or buffing. Very fine grits are usually wet sandpaper, which is required to keep the paper from becoming clogged or filled with paint.

When starting your work, use the coarsest grit practical. This will remove and smooth the area quickly. Then gradually go to finer paper to achieve the desired surface smoothness. This will be detailed in later chapters.

Open and Closed Coat Grits

Sandpaper and discs come in either open coat or closed coat types of grit.

With an *open coat*, the resin that bonds the grit to the paper touches only the bottom of the grit. About 50 to 70 percent of its surface is covered by grit materials. Open coat grit will not clog as quickly.

With a *closed coat*, the resin completely covers the grit. This bonds the grit to the paper or disc more securely. About 90 percent of the surface is covered by grit materials. Closed coat will clog faster than open coat.

Grinding Discs

Grinding discs are round, very coarse abrasives used for initial removal of paint, plastic, and metal (weld joints). Some are very thick and do not require a backing plate. Others are thinner and require a *disc backing plate* mounted on the grinder spindle. They are used for material removal operations, with 24 grit being the most common.

Grinding disc size is measured across its outside diameter. The most common grinding disc sizes are 7 and 9 inch. The hole in the center of the disc must match the shaft on the grinder or sander.

Sandpapers

Sandpaper is a heavy paper coated with an abrasive grit. It is the most commonly used abrasive in

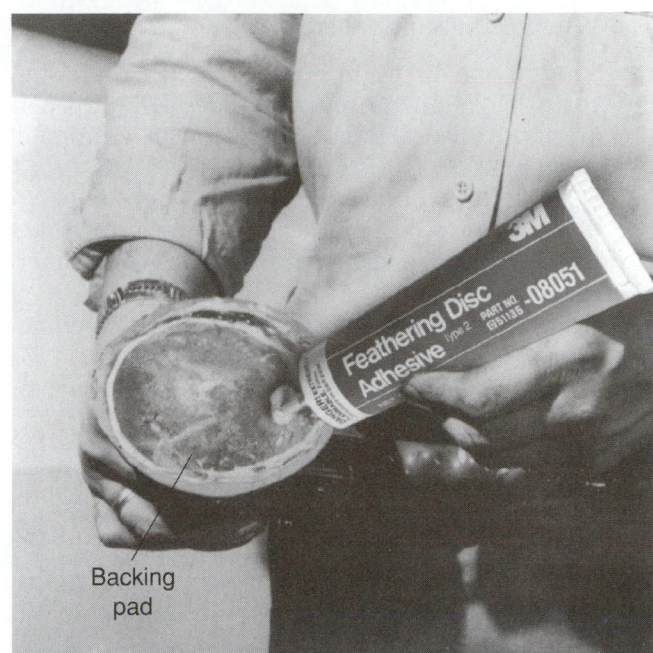

Backing pad

FIGURE 6-12 Disc adhesive is sometimes used to hold sandpaper onto the air tool's pad. Use only a small amount. Spread it over the pad surface with the back of the sandpaper. *(Courtesy of 3M Automotive Trades Division)*

auto body repair. There are several kinds, shapes, and grits of sandpaper.

Sanding discs are round and are normally used on air-powered orbital sanders. They might have self-stick coating or require the use of disc adhesive to hold the sandpaper onto the tool pad (Figure 6-12).

Sanding sheets are rectangular and can be cut to fit sanding blocks. Long sheets are also available for use on air files.

Dry sandpaper is designed to be used without water. Its resin is usually an animal glue, which is not water resistant and will dissolve when wet, ruining the sandpaper.

Dry sandpaper is often used for coarse to medium grit sanding tasks, like shaping and smoothing plastic filler. One example is 80-grit dry sandpaper, which is often used on plastic filler. It quickly cuts the filler down but does not leave deep sand scratches in the paint surrounding the filler.

Wet sandpaper, as implied, can be used with water for flushing away sanding debris that would otherwise clog fine grits. Wet sandpaper comes in finer grits from about 220 to 2000 for final smoothing operations before and after painting.

Wet sandpaper is commonly used to block sand paint before compounding or buffing. Wet sanding will knock down any imperfections (color sanding) in the paint film. Buffing or compounding is then needed to make the paint shiny again.

When using sandpapers:

1. Sand in one direction only; this is usually along the line of sight. If several grades of sandpaper are used on one area, cross sand to eliminate scratches.
2. Use the finest abrasive possible to do the job.
3. Start with as fine a grade as possible. If too fine, go to a coarser grade; then work back to the finer grade.
4. Adjust one grade finer for hand versus machine sanding.
5. Support the abrasive with a block or pad to avoid finger marks and crowning.
6. Adjust two grades finer when wet sanding, due to faster cutting abrasives.
7. Choose one manufacturer's line so you learn its cutting characteristics.
8. Follow the manufacturer's recommendations for use.

Scuff Pads

Scuff pads are tough synthetic pads used to clean and lightly scratch the surface of paints. Being like sponges, they are handy for scuffing irregular surfaces, like door jams, around the inside of the hood and deck lids, and other obstructed areas. They clean and lightly scuff these areas so the paint will stick. Scuff pads are also used to lightly scuff exterior surfaces prior to blending. This light roughening allows the paint and clearcoat to adhere without showing any scratches.

Compounds

Compounding involves using an abrasive paste material to smooth and bring out the gloss of the applied topcoat. It can be applied by hand or with a polishing wheel on an air tool. A compound has a fine volcanic pumice or dust-like grit in a water-soluble paste. When rubbed on a painted surface, a thin layer of paint is removed. It will remove the very top layer of old, weathered paint, leaving a new fresh surface of paint.

A *hand compound* is designed to be applied by hand with a rag or cloth. *Machine compound* is formulated to be applied with an electric or air polisher. It will not cut as fast and will not break down with the extra friction and heat of machine application.

Rubbing compound is the coarsest type of hand compound. It will rapidly remove paint but will leave visible scratch marks. Rubbing compound is designed for hand application, not machine application. It is often used on small areas to treat imperfections in the paint surface.

Polishing compound or *machine glaze* is a fine grit compound designed for machine application. A polisher is used to carefully run the compound over the cured or dried paint. Polishing compound is often used after a rubbing compound or after wet (color) sanding. It will make the paint shiny and smooth.

Hand glazes are for final smoothing and shining of the paint. They are the last process used to produce a professional finish. They are applied by hand, using a circular motion, like a wax.

Besides these, other compounds come in various formulations. Read the label on the compound to learn about its use.

A painter will generally use

1. Rubbing compounds to remove surface imperfections in paint
2. Machine glazes to restore paint gloss after wet or color sanding
3. Hand glazes to remove swirl marks after machine buffing

ADHESIVES

Adhesives are special glues designed to bond parts to one another. Various types are available.

Weatherstrip adhesive is designed to hold rubber seals and similar parts in place. Weatherstrip adhesive dries to a hard rubber-type consistency. This makes it ideal for holding door seals, trunk seals, and other seals onto the body (Figure 6–13).

Plastic adhesive or *emblem adhesive* is designed to hold hard plastic and metal parts. It is used to install

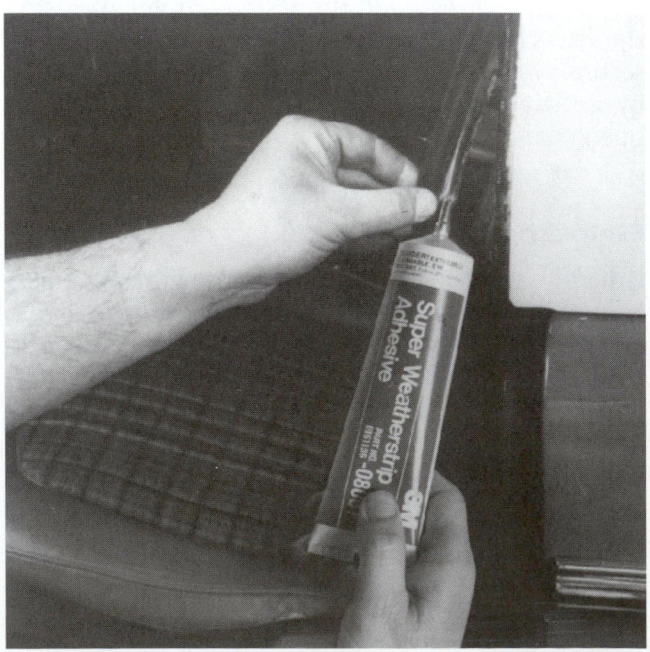

FIGURE 6–13 Weatherstrip adhesive is commonly used to install rubber seals around doors and trunks. Black color adhesive is best on black rubber seals. *(Courtesy of 3M Automotive Trades Division)*

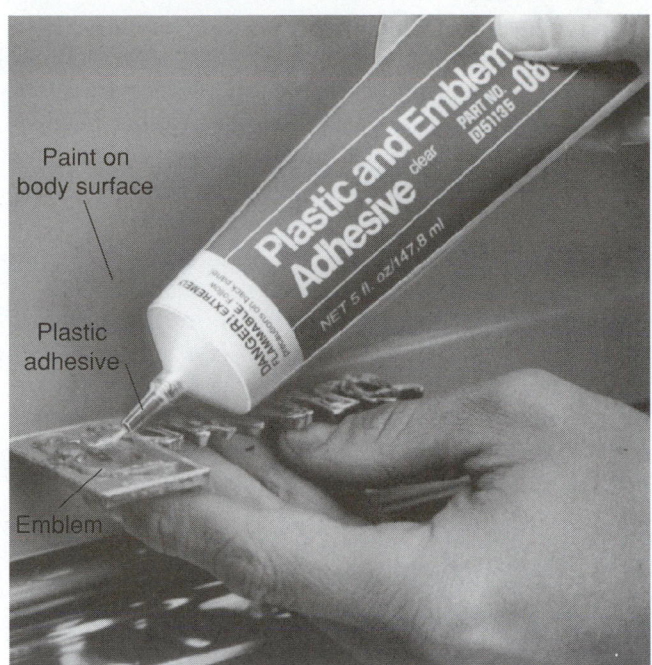

FIGURE 6-14 Plastic adhesive dries harder than weatherstrip adhesive. It is often used on trim pieces that bond directly to paint. *(Courtesy of 3M Automotive Trades Division)*

FIGURE 6-16 A release agent will dissolve adhesive so you can remove glued-on parts easily. *(Courtesy of 3M Automotive Trades Division)*

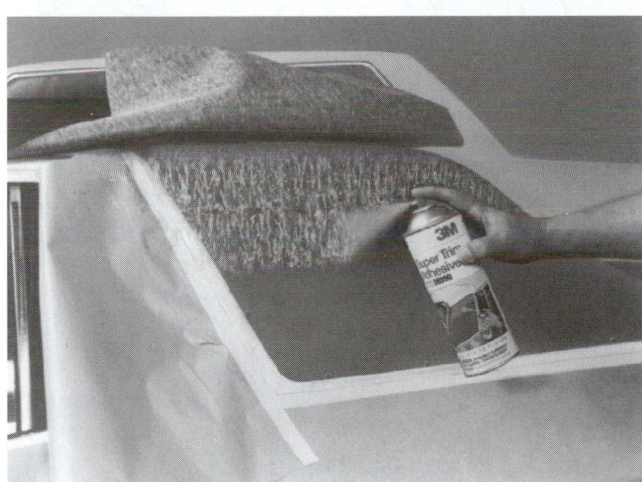

FIGURE 6-15 Vinyl adhesive is commonly used to bond the vinyl roof to the body. The spray can allows for a smooth coating of the adhesive so the roof covering will lay flat and smooth. *(Courtesy of 3M Automotive Trades Division)*

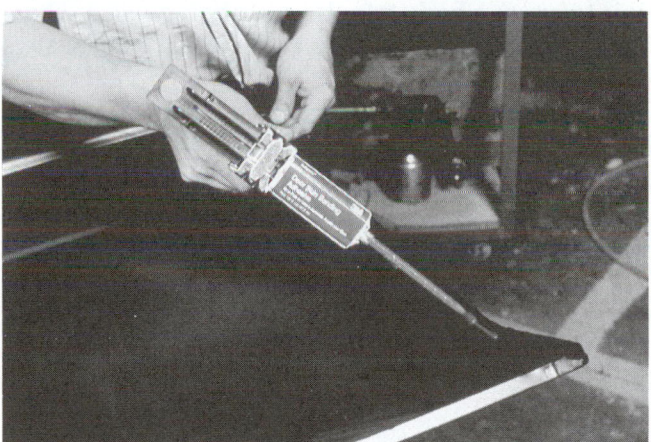

FIGURE 6-17 Numerous types of two-part epoxies are used to help bond parts. Make sure you use the type recommended by the manufacturer. *(Courtesy of 3M Automotive Trades Division)*

various types of emblems and trim pieces onto painted surfaces. See (Figure 6–14).

Vinyl adhesive is designed to bond a vinyl top to the vehicle body. This type of adhesive is often used when installing or repairing vinyl tops, interior roof liners (headliners), and similar parts (Figure 6–15).

An *adhesive release agent* is a chemical that dissolves most types of adhesives. It is used when you want to remove a part without damage. The agent is sprayed onto the adhesive to soften it so the part can be lifted off easily (Figure 6–16).

Epoxies

An *epoxy* is a two-part glue used to hold various parts together. The two ingredients are mixed together in equal parts. This makes the mixture cure through a chemical reaction. Always use the type of epoxy suggested by the auto manufacturer (Figure 6–17).

SEALERS

Sealers are used to prevent water and air leaks between parts. They are flexible to prevent cracking and come in several variations (Figure 6–18).

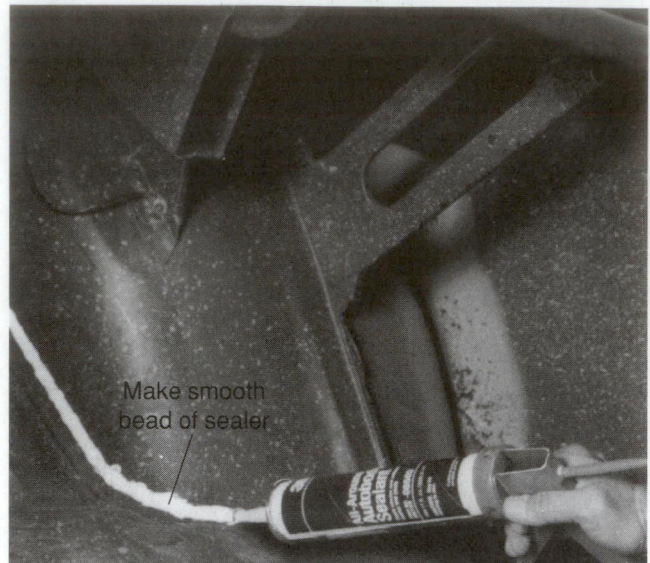

FIGURE 6–18 Various types of sealers are used to prevent leakage between body parts. The technician is using a caulking gun to seal a newly installed trunk floor panel. *(Courtesy of 3M Automotive Trades Division)*

Seam sealers are designed to make a leakproof joint between body panels. It is often needed where two panels butt or overlap each other. Seam sealers come in different forms and each is applied differently. Read the directions.

Tube sealers are applied directly from the tube or by using a caulking gun. They squirt out like toothpaste and cure in a few hours.

Apply primer before applying seam sealer. Seam sealers are paintable but may need to be reprimed if the product directions specify. Silicone sealers are not paintable and should not be used in auto body repair. Follow instructions on the product for finishing sealers.

Ribbon sealers come in strip form and are applied by hand. They are a thick sealer that must be worked onto the parts with your fingers.

NOTE: The proper use or application of these materials will be explained later in this book. Refer to the index if you need more information now.

6.2 FASTENERS

Fasteners include the thousands of bolts, nuts, screws, clips, and adhesives that literally hold a vehicle together.

As an auto body technician, you will constantly use fasteners when removing and installing body parts. This makes it important for you to be able to identify and use fasteners properly.

Remember that each fastener is engineered for a specific application. Always replace fasteners with

exactly the same type that was removed from the original equipment manufacture (*OEM*) assembly. Never try to re-engineer the vehicle. Keep in mind that using an incorrect fastener or a fastener of inferior quality can result in failure and possible injury to the vehicle occupants.

BOLTS

A *bolt* is a metal shaft with a head on one end and threads on the other. A *cap screw* is a term that describes a high-strength bolt. Bolts and cap screws are usually named after the body part they hold, such as fender bolt or hood hinge bolt. Their shape and head drive configurations also help name them.

Bolt Terminology

To work with bolts properly, you must understand basic bolt terminology. See Figure 6–19.

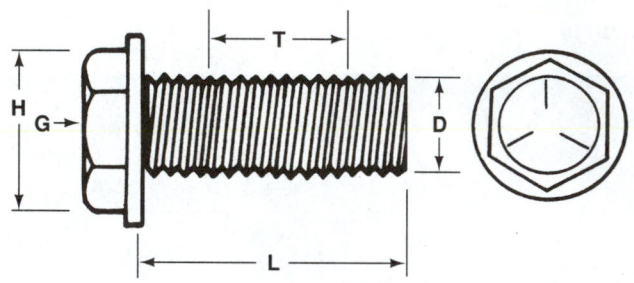

H - Head
G - Grade Marking (Bolt Strength)
L - Length (Inches)
T - Thread Pitch (Thread/Inch)
D - Nominal Diameter (Inches)

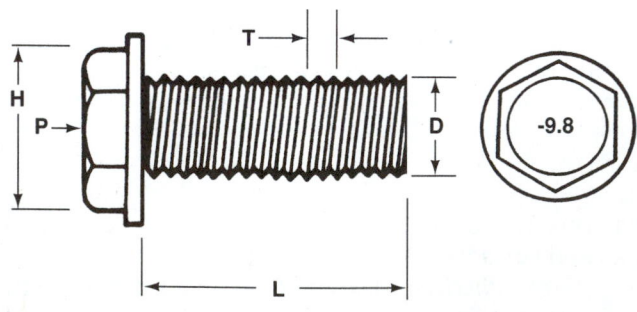

H - Head
P - Property Class (Bolt Strength)
L - Length (Millimeters)
T - Thread Pitch (Millimeters)
D - Nominal Diameter (Millimeters)

FIGURE 6–19 Bolt measurements are needed when you are working. Study each dimension of both English and metric bolts. *(Courtesy of Ford Motor Company)*

The *bolt head* is the top and is used to torque or tighten the bolt. A socket or wrench fits over the head, which enables the bolt to be tightened. Some English and metric sockets are very close in size. It is important not to use metric sizes for English bolts, or English sizes for metric bolts as the heads can be rounded and damaged.

Bolt length is measured from the end of the threads to the bottom of the bolt head. It is *not* the total length including the bolt head.

Bolt diameter, sometimes termed *bolt size*, is measured around the outside of the threads. For example, a $1/2$-inch bolt has a thread diameter of $1/2$ inch while its head or wrench size would be $3/4$ inch.

Bolt head size is the distance measured across the flats of the bolt head. In *USC* (United States Customary), head size is given as fractions, just like wrench size. A few common sizes are $7/16$, $1/2$, and $9/16$ inch. In the metric system, 3-, 4-, and 5-millimeter head sizes are typical (Figure 6–20).

Bolt thread pitch is a measurement of thread coarseness. Bolts and nuts can have coarse, fine, and metric threads. Bolt threads can be measured with a *thread pitch gauge.*

A number of terms have been used over the years to identify the various types of threads. The terms used in the automotive trade—the United States Standard (USS), the American National Standard (ANS), and the Society of Automotive Engineers Standard (SAE)—have all been replaced by the Unified National Series.

The two common metric threads are coarse and fine and can be identified by the letters SI (Système International or International System of Units) and ISO (International Standards Organization).

Do NOT accidentally interchange thread types or damage will result. It is easy to mistake metric threads for English threads. If the two are forced together, either the bolt or the part threads will be ruined.

Bolts and nuts are also available in right- and left-hand threads. *Right-hand threads* must be turned clockwise to tighten. Less common *left-hand threads* must be rotated in a counterclockwise direction to tighten the fastener. Left-hand threads may be denoted by notches or the letter "L" stamped on them.

When ordering bolts, it is necessary to designate the bolt diameter, thread pitch, and length. An example would be $1/4$-inch–20 × 1 inch.

Bolt Strengths or Grades

Bolt strength indicates the amount of torque or tightening force that should be applied. Bolts are made from different materials having various degrees of hardness. Softer or harder metal can be used to achieve different hardnesses and strengths for use in different situations.

Bolt grade markings are lines or numbers on the top of the head to identify bolt hardness and strength. The hardness or strength of metric bolts is indicated by using a property class indicator on the head of the bolt.

Bolt strength markings are given as lines (Figure 6–21). The number of lines on the head of the bolt is related to the strength. As the number of lines increases so does the strength.

Metric bolt strength markings are given as numbers. The higher the number is, the stronger the bolt. These markings apply to both bolts and nuts (Figure 6–22).

Common English (U.S. Customary) Head Sizes	Common Metric Head Sizes
Wrench Size (inches)*	Wrench Size (millimeters)*
3/8	9
7/16	10
1/2	11
9/16	12
5/8	13
11/16	14
3/4	15
13/16	16
7/8	17
15/16	18
1	19
1-1/16	20
1-1/8	21
1-3/16	22
1-1/4	23
1-5/16	24
1-3/8	26
7/16	27
1-1/2	29
	30
	32

* The wrench sizes given in this chart are not equivalents, but are standard head sizes found in both inches and millimeters.

FIGURE 6-20 These are common bolt head and wrench sizes. Never use a standard wrench on metric bolts and vise versa. This will round off the bolt head.

SAE	⬡	⬡⟨-⟩	⬡⟨+⟩	⬡⟨✳⟩	⬡⟨✳⟩
DEFINITION	No lines: unmarked indeterminate quality SAE Grades 0-1-2	3 Lines: common commercial quality Automotive & AN Bolts SAE Grade 5	4 Lines: medium commercial quality Automotive & AN Bolts SAE Grade 6	5 Lines: rarely used SAE Grade 7	6 Lines: best commercial quality NAS & Aircraft Screws SAE Grade 8
MATERIAL	Low Carbon Steel	Med. Carbon Steel Tempered	Med. Carbon Steel Quenched & Tempered	Med. Carbon Alloy Steel	Med. Carbon Alloy Steel Quenched & Tempered
TENSILE STRENGTH	65,000 psi	120,000 psi	140,000 psi	140,000 psi	150,000 psi

FIGURE 6-21 Bolt tensile strengths are denoted on the heads of bolts. Slash marks (SAE) or numbers (metric) are used. Always replace bolts with an equal or a higher rating to prevent failure.

Grade	Identification	Class	Identification
Hex Nut Grade 5	3 Dots	Hex Nut Property Class 9	Arabic 9
Hex Nut Grade 8	6 Dots	Hex Nut Property Class 10	Arabic 10
Increasing dots represent increasing strength		Can also have blue finish or paint dab on hex flat. Increasing numbers represent increasing strength.	

FIGURE 6-22 Quality nut strengths are also denoted. More dots or a high number means more strength.

Tensile strength is the amount of pressure per square inch the bolt can withstand just before breaking when it is pulled apart. The harder or stronger the bolt, the greater the tensile strength.

Never replace a high-grade bolt with a bolt having a lower grade marking as the weaker bolt could break. If in the steering or suspension, this could seriously endanger the passengers of the vehicle.

Bolt Torque

Bolt torque is a measurement of the turning force applied when installing a fastener. It is critical that bolts and nuts are torqued or tightened properly. Overtightening will stretch and possibly break the bolt. Undertightening may allow the bolt or nut to loosen and fall out.

Torque specifications are tightening values for the specific bolt or nut. They are given by the manufacturer. Discussed in the tool chapter, a torque wrench must be used to measure torque values.

If you cannot find the factory torque specification for a bolt, you can use a *general bolt torque chart*. It will give a general torque value for the size and grade of bolt. Such a chart is shown in Figure 6–23. Normally the bolt threads should be lubricated to get accurate results. Refer to the chart to see if the threads should be lubricated or dry.

A *tightening sequence*, or *torque pattern*, assures that parts are clamped down evenly by several bolts or nuts. Tighten fasteners in a crisscross pattern, so as to pull the part down evenly, preventing warpage. This is commonly recommended for wheels, as shown in Figure 6–24.

Tighten the fastener in steps, to about half torque, three-fourths torque, and then full torque, at least twice.

Be careful when tightening bolts and nuts with air wrenches. It is easy to stretch or break a bolt instantly. The air wrench can spin the bolt or nut so fast that it can hammer the fastener past its yield point. This can strip threads or snap off the bolt.

When torque is not critical, do not use the air impact to run the nut full speed onto the bolt. Instead, run it up slowly until it contacts the work. Then mark the socket and watch how far it turns. Smaller air-powered speed wrenches do not produce the severe force of impact wrenches and are much safer to use.

Metric Standard						SAE Standard / Foot Pounds				
Grade of Bolt	5D	.8G	10K	12K		Grade of Bolt	SAE 1&2	SAE 5	SAE 6	SAE 8	
Min. Tensile Strength	71,160 P.S.I	113,800 P.S.I	142,200 P.S.I	170,679 P.S.I		Min. Ten Strength	64,000 P.S.I	105,000 P.S.I	133,000 P.S.I	150,000 P.S.I	
Grade Markings on Head	5D	8G	10K	12K	Size of Socket on Wrench Opening	Markings on Head	●	◬	✛	✳	Size of Socket or Wrench Opening
Metric		Foot Pounds			Metric	U.S. Standard		Foot Pounds			U.S. Regular
Bolt Dia.	U.S. Dec Equiv.				Bolt Head	Bolt Dia.					Bolt Head	Nut
6 mm	.2362	5	G	8	10	10 mm	1/4	5	7	10	10.5	3/8	7/16
8 mm	.3150	10	16	22	27	14 mm	5/16	9	14	19	22	1/2	9/16
10 mm	.3937	19	31	40	49	17 mm	3/8	15	25	34	37	9/16	5/8
12 mm	.4720	34	54	70	86	19 mm	7/16	24	40	55	60	5/8	3/4
14 mm	.5512	55	89	117	137	22 mm	1/2	37	60	85	92	3/4	13/16
16 mm	.6299	83	132	175	208	24 mm	9/16	53	88	120	132	7/8	7/8
18 mm	.709	111	182	236	283	27 mm	5/8	74	120	167	180	15/16	1.
22 mm	.8661	182	284	394	464	32 mm	3/4	120	200	280	296	1-1/8	1-1/8

FIGURE 6-23 If factory specs are not available, use this general bolt torque chart. It gives different values for each bolt tensile strength rating.

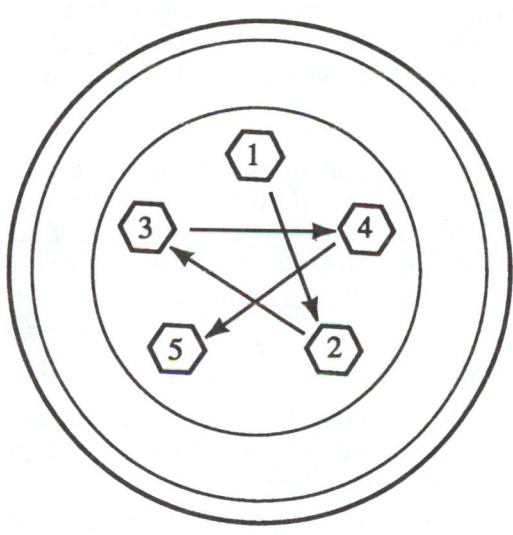

Lug Nut Torque Specifications

FIGURE 6-24 When tightening several bolts that hold one part, a wheel for example, always use a crisscross pattern. This will prevent part warpage and problems.

NUTS

A *nut* uses internal (inside) threads and an odd-shaped head that often fits a wrench. When tightened onto a bolt, a strong clamping force holds the parts together. Many different nuts are used by the automotive industry. Several are shown in Figure 6–25.

Castellated or *slotted nuts* are grooved on top so that a safety wire or cotter pin can be installed into a hole in the bolt. This helps prevent the nut from working loose. For example, castellated nuts are used with the studs that hold wheel bearings in position. Slotted nuts are also used on steering and suspension parts for safety.

Self-locking nuts produce a friction or force fit when threaded onto a bolt or stud. The top of the nut can be crimped inward. Some have a plastic insert that produces a friction fit to keep the nut from loosening. Locking nuts may need to be replaced after removal. Front-wheel-drive spindles sometimes use self-locking nuts.

Jam nuts are thin nuts used to help hold larger, conventional nuts in place. The jam nut is tightened down against the other nut to prevent loosening.

Wing nuts have two extended arms for turning the nut by hand. They are used when a part must be removed frequently for service or maintenance. Air cleaners sometimes use wing nuts.

Acorn nuts are closed on one end for appearance and to keep water and debris off the threads. They can be used when they are visible and looks are important.

Special types of nuts are used to hold parts onto the vehicle. Sometimes a washer is formed onto the nut. Termed *body nuts*, the flange on the nut helps distribute the clamping force of the thin body panel or trim piece to prevent warpage. Look at Figure 6–26.

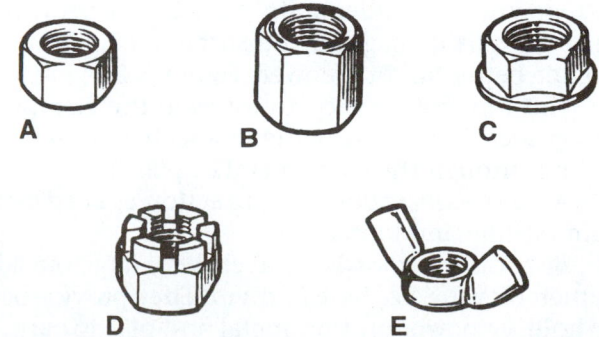

FIGURE 6-25 Memorize nut types: [A] hex nut, [B] high or deep nut, [C] flange nut, [D] castle or slotted nut, [E] wing nut. *(Courtesy of Dorman)*

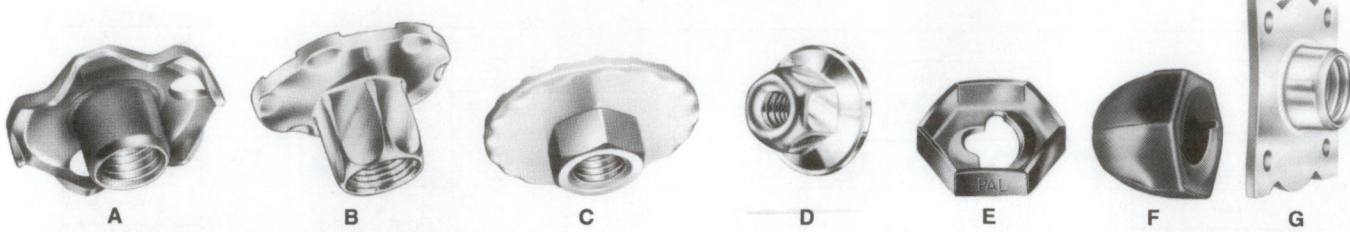

FIGURE 6-26 Body nuts are specially designed for specific holding applications: (A) counter boring nut, (B) hex barrel nut, (C) washer nut, (D) washer nut with rubber gasket, (E) pal or locknut, (F) acorn nut, (G) plate nut. *(Courtesy of TRW)*

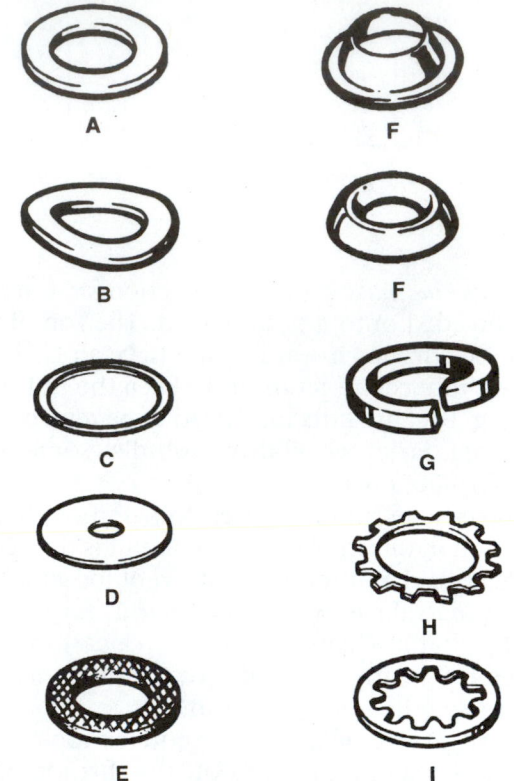

FIGURE 6-27 Study washer types: (A) plain flat washer, (B) wave or spring washer, (C) spacer washer, (D) fender washer, (E) fiber washer, (F) finishing washers, (G) split lock washer, (H) external lock washer, (I) internal lock washer.

WASHERS

Washers are used under bolts, nuts, and other parts. They prevent damage to the surfaces of parts and provide better holding power (Figure 6–27).

Flat washers are used to increase the clamping surface area. They prevent the smaller bolt head from pulling through the sheet metal or plastic.

A *wave washer* adds a spring action to keep parts from rattling and loosening.

Body or *fender washers* have a very large outside diameter for the size hole in them. They provide better holding power on thin metal and plastic parts.

Copper or *brass washers* are used to prevent fluid leakage, as on brake line fittings. *Spacer washers* come in specific thicknesses to allow for adjustment of

parts. *Fiber washers* will prevent vibration or leakage but cannot be tightened very much.

Finishing washers have a curved shape for a pleasing appearance. They are used on interior pieces.

Split lock washers are used under nuts to prevent loosening by vibration. The ends of these spring-hardened washers dig into both the nut and the work to prevent rotation. The lock washers should be placed next to the bolt head or nut.

Shakeproof or teeth lock washers have teeth or bent lugs that grip both the work and the nut. Several designs, shapes, and sizes are available. An external type has teeth on the outside, and an internal type has teeth around the inside. Lock washers are extremely hard and tend to break under severe pressure.

A rule of thumb is if the part did not come with a lock washer, do not add one. If the bolt or nut has a lock washer, use one. The manufacturer would not use one if it did not have a purpose.

SCREWS

Screws are often used to hold nonstructural parts on the vehicle. Trim pieces, interior panels, etc., are often secured by screws. Refer to Figure 6–28.

Machine screws are threaded their full length and are relatively weak. They come in various configurations and will accept a nut.

Set screws frequently have an internal drive head for an Allen wrench and are used to hold parts onto shafts.

Sheet metal screws have pointed or tapered tips; they thread into sheet metal for light holding tasks.

Self-tapping screws have a pointed tip that helps cut new threads in parts.

Trim screws have a washer attached to them, which improves appearance and helps keep the trim from shifting.

Headlight aiming screws have a special plastic adapter mounted on them. The adapter fits into the headlight assembly. Different design variations are needed for different makes and models of vehicles.

NONTHREADED FASTENERS

As implied, *nonthreaded fasteners* do not use threads. They include keys, snap rings, pins, clips, adhesives,

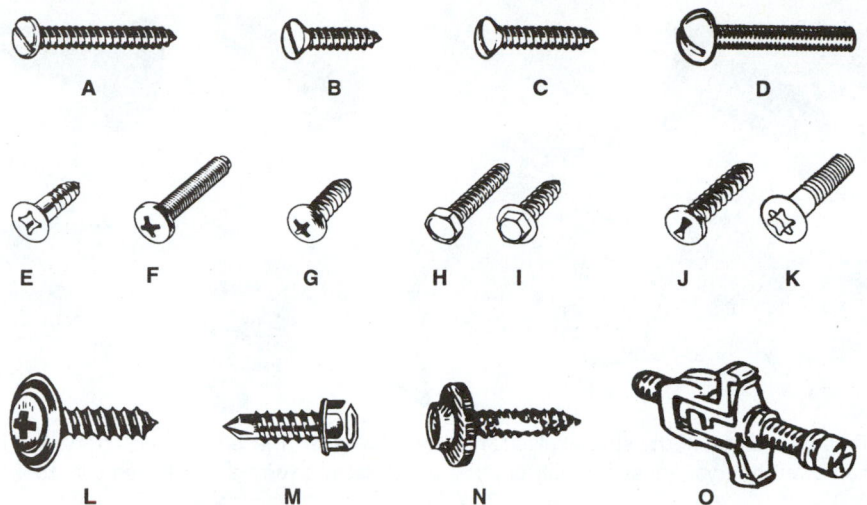

FIGURE 6-28 Study common screw names: (A) pan head sheet metal screw, (B) flat head sheet metal screw, (C) oval head, (D) round head, (E) Phillips head screw, (F) threaded screw, (G) oval head sheet metal screw, (H) hex or nutdriver screw, (I) hex screw with flange or integral washer, (J) clutch head, (K) Torx head, (L) trim screw, (M) self-taping screw, (N) body screw, (O) headlight aiming screw. *(Courtesy of Dorman)*

etc. Various *keys* and *pins* are used by equipment manufacturers to retain parts in alignment. It is important to be able to identify these keys and pins to order replacements.

Square keys and *Woodruff keys* are used to prevent hand wheels, gears, cams, and pulleys from turning on their shafts. These keys are strong enough to carry heavy loads if they are fitted and seated properly. See Figure 6–29.

Round *taper pins* have a larger diameter on one end than the other. They are used to locate and

position matching parts. They can also be used to secure small pulleys and gears to shafts (Figure 6–30).

Dowel pins have the same general diameter their full length. They are used to position and align the parts of an assembly. One end of a dowel pin is chamfered, and it is usually 0.001 to 0.002 inch greater in diameter than the size of its hole. When

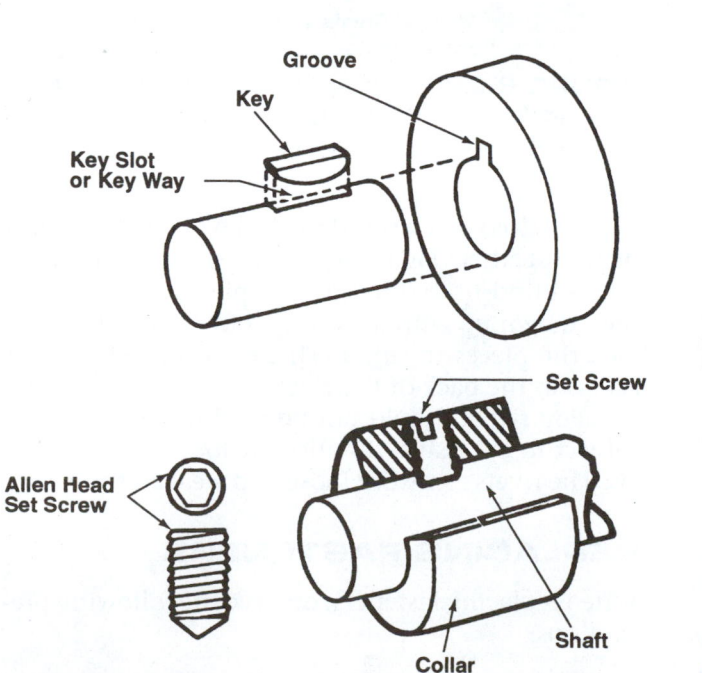

FIGURE 6-29 Keys and set screws are both used to align parts on shafts: (A) key and keyway, (B) set screw application. *(Courtesy of Florida Dept. Of Vocational Education)*

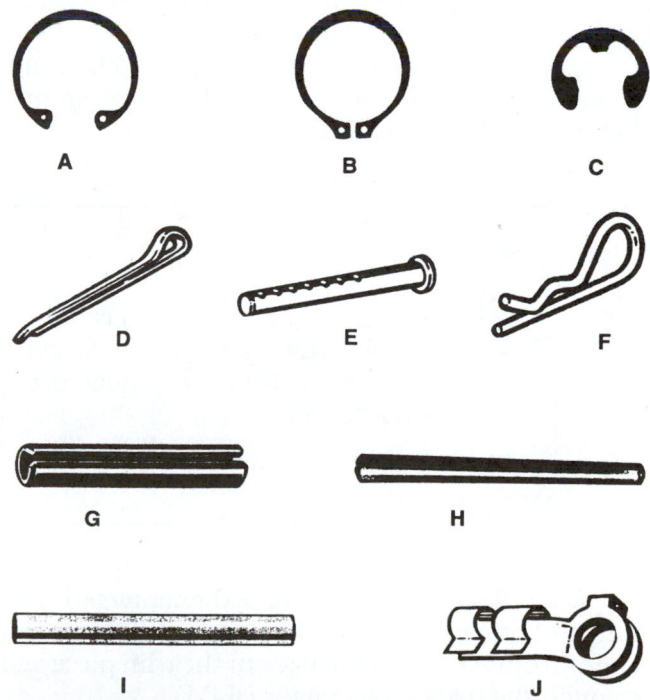

FIGURE 6-30 Learn these nonthreaded fastener names: (A) internal snap or retaining ring, (B) external snap ring, (C) E-clip or snap ring, (D) cotter pin, (E) clevis pin, (F) hitch pin, (G) split rollpin, (H) tapered pin, (I) straight dowel pin, (J) linkage clip. *(Courtesy of Dorman)*

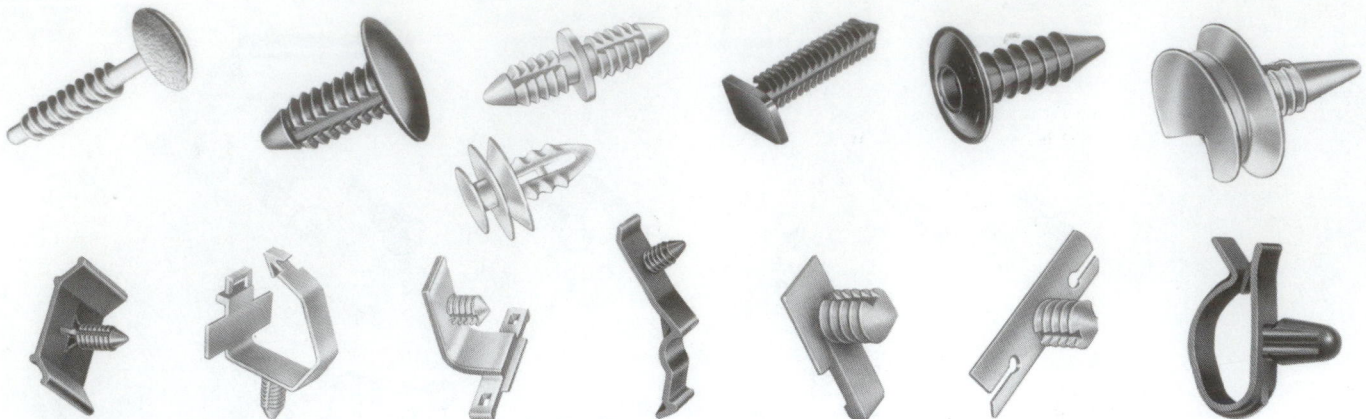

FIGURE 6-31 These are a few of the special plastic retainers available. These types are often used in interiors. They quickly press into a hole. To remove them, you must carefully pry next to the retainer with a flat, forked trim tool. *(Courtesy of TRW)*

replacing a dowel pin, be sure that it is the same size as the old one (Figure 6–30).

Cotter pins help prevent bolts and nuts from loosening or they fit through pins to hold parts together. They are also used as stops and holders on shafts and rods. All cotter pins are used for safety and should never be reused (Figure 6–30).

The cotter pin should fit into the hole with very little side play. If it is too long, cut off the extra length. After insertion, bend it over in a smooth curve with needle nose pliers. Sharply angled bends invite breakage. Final bending of the prongs can be done with a soft-faced mallet.

Snap rings are nonthreaded fasteners that install in a groove machined into a part. They are used to hold parts on shafts (Figure 6–30).

Special snap ring pliers are designed to flex and install or remove snap rings. They have special tips that will hold the snap ring.

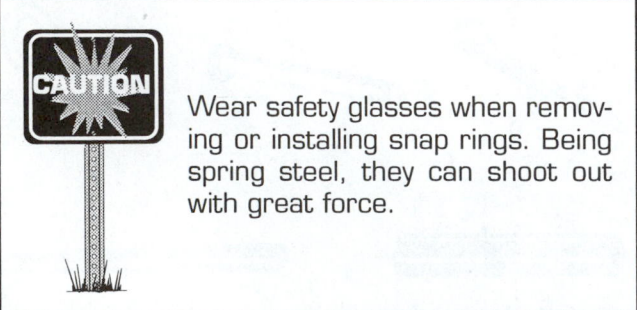

Wear safety glasses when removing or installing snap rings. Being spring steel, they can shoot out with great force.

Body clips are specially shaped retainers to hold trim and other body pieces requiring little strength. The clip often fits into the back of the trim piece and through the body panel (Figure 6–31).

Push-in clips are usually made of plastic and they force fit into holes in body panels. Push-in clips are used to hold interior door trim panels, for example. They install easily, but can be difficult to remove.

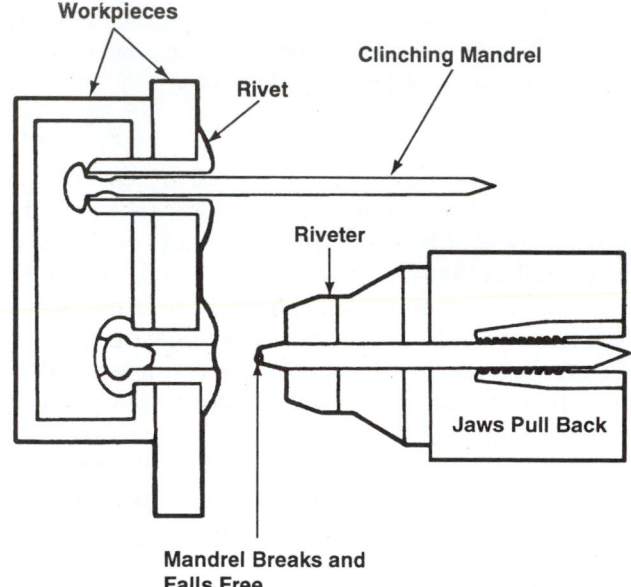

FIGURE 6-32 Pop rivets provide a quick way of holding sheet metal parts or panels. They are handy when the backside of the panel is not accessible.

Pop rivets can be used to hold two pieces of sheet metal together (Figure 6–32). They can be inserted into a blind hole through two pieces of metal and then drawn up with a riveting tool or gun. This will lock the pieces together. There is no need to have access to the back of the rivets.

Pop rivets should not be used in areas that are subject to excessive vibration or for structural panels. The rivets can work loose and weaken the repair.

REPLACING FASTENERS

When replacing fasteners, observe the following precautions:

1. Always use the same number of fasteners.
2. Use the same diameter, length, pitch, and type of fasteners.

3. Observe the OEM's recommendations given in the service manual for tightening sequence, tightening steps (increments), and torque values.
4. Always replace a used cotter pin.
5. Replace stretched fasteners or fasteners with any signs of damage.
6. Use the correct washers and pins as specified by the OEM.
7. Always replace "one-time" fasteners. They can be found in suspension and steering assemblies.

HOSE CLAMPS

Hose clamps are used to hold radiator hoses, heater hoses, and other hoses onto their fittings. See Figure 6–33.

A *spring hose clamp* is made of spring steel with barbs on each end. Squeezing the ends opens and expands the clamp. Special hose clamp pliers should be used to remove or install round wire type clamps. The pliers have a deep groove that will keep the clamp from slipping out of the jaws. Conventional pliers will work fine on spring strap clamps.

A *worm hose clamp* uses a screw that engages a slotted band. Turning the screw reduces or enlarges clamp diameter. This is the most common replacement type of hose clamp.

ACRYLIC RESIN ADHESIVE AND EPOXY

Adhesives provide an alternate means of bonding parts together. The two types of adhesives most often used with automotive metals are acrylic resin and epoxy.

Acrylic resin adhesive is a two-part, somewhat flexible adhesive (liquid and powder) used on small surface areas. It is waterproof and not affected by gasoline or water.

Follow the manufacturer's directions for mixing materials. Drying and setting times are controlled by the amounts mixed. For example, three parts of

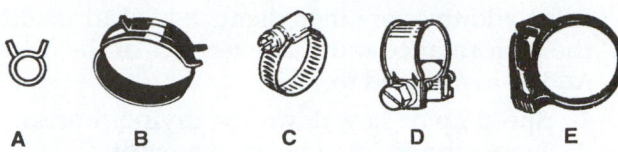

FIGURE 6–33 Note hose clamp types: (A) wire spring hose clamp, (B) wire strap hose clamp, (C) worm hose clamp, (D) screw-nut hose clamp, (E) plastic hose clamp. *(Courtesy of Dorman)*

powder to one part of liquid will set in about five minutes at 70 degrees Fahrenheit. Changing these proportions will allow for faster or slower drying and setting times.

While acrylic resin provides a very strong bond, it sets too fast for large area work and is generally used for heavy-duty repairs on smaller objects. Acetone can be used as a cleaning solvent for acrylic resin adhesive.

Epoxy is a two-part bonding agent that dries harder than adhesive. It comes in two separate containers, usually tubes. One contains the epoxy resin and the other contains a hardener. Epoxy does not shrink when it hardens and is waterproof and heat-resistant at moderate temperatures.

Read the instructions for the proper quantity to use. If both resin and hardener are not in proper proportions, the bond might fail. Some epoxy tubes automatically dispense the correct amount of resin and hardener.

Once mixed, the epoxy remains in a workable condition for only a brief time. Therefore, try to mix only as much as is required and use it as quickly as possible. Clamp the work while the glue cures, which can take several hours. Do not apply epoxy in low temperatures (below 50 degrees Fahrenheit or 10 degrees Celsius), because it will not harden. Once an epoxy is applied, it is difficult to remove it from a surface.

NOTE: Adhesives are discussed in more detail in other chapters. Refer to the text index if needed.

SUMMARY

- Body shop materials include more than just refinishing or paint materials. They include the various fillers, primers, sealers, adhesives, sandpapers, and other compounds common to a body shop.
- Refinishing materials is a general term referring to the products used to repaint the vehicle.
- The term *paint* generally refers to the visible topcoat.
- The primer or undercoat has to improve adhesion of the topcoat. It is often the first coat applied. Paint alone will not stick or adhere as well as a primer.

- Basecoat/clearcoat paint systems use a colorcoat applied over the primer with a second layer of clearcoat over the colorcoat.
- Lacquer is an older paint that dries quickly because of solvent evaporation.
- Enamel finishes dry with a gloss and do not require rubbing or polishing.
- A paint's chemical content includes the following:
  1. Pigments
  2. Binders
  3. Solvents
  4. Additives

- Paint additives are ingredients added to modify the performance and characteristics of the paint. Additives are used to:
  1. Speed up or slow down the drying process.
  2. Lower the freezing point of a paint.
  3. Prevent the paint from foaming when shaken.
  4. Control settling of metallics and pigments.
  5. Make the paint more flexible when dry.
  6. Increase gloss or shine.
  7. Decrease gloss or shine.
- A catalyst is a substance that causes or speeds up a chemical reaction.
- A primer-sealer is an undercoat that improves adhesion of the topcoat and also seals old painted surfaces that have been sanded.
- A primer-surfacer is a high-solids primer that fills small imperfections and usually must be sanded.
- A paint system means all materials (primers, catalysts, paints) are compatible and manufactured by the same company.
- A prep solvent or wax and grease remover is a fast-drying solvent often used to chemically clean a vehicle.
- A metal conditioner is a phosphoric acid used to etch bare sheet metal before priming.

- Body filler or plastic filler is a heavy-bodied plastic material that cures very hard for filling small dents in metal.
- Glazing putty is a material made for filling small holes or sand scratches.
- Masking materials are used to cover and protect body parts from paint overspray.
- Grit refers to a measure of the size of particles on sandpaper or discs. A coarse sandpaper would have large grit. A fine sandpaper would have smaller grit.
- Compounding involves using an abrasive paste material to smooth and bring out the gloss of the applied topcoat.
- Adhesives are special glues designed to bond parts to one another.
- Sealers are used to prevent water and air leaks between parts.
- Fasteners include the thousands of bolts, nuts, screws, clips, and adhesives that literally hold a vehicle together.
- Bolt torque is a measurement of the turning force applied when installing a fastener. It is critical that bolts and nuts are torqued or tightened properly.

ASE-STYLE REVIEW QUESTIONS

1. Which of the following terms refers to a factory applied finish?
   A. Lacquer
   B. Enamel
   C. OEM paint
   D. ASE paint

2. This is an undercoat that improves adhesion of the topcoat and seals old painted surfaces that have been sanded.
   A. Primer
   B. Primer-sealer
   C. Primer-filler
   D. Primer-surfacer

3. This would be the strongest bolt.
   A. Three head markings
   B. Two head markings
   C. No head markings
   D. One head marking

4. When installing a wheel on an older car, no service manual can be found for getting a factory torque specification. Technician A says to use an impact wrench on the medium setting. Technician B says that a crisscross pattern should be followed when tightening the wheel nuts. Who is correct?
   A. Technician A
   B. Technician B
   C. Both A and B
   D. Neither A or B

ESSAY QUESTIONS

1. What are two functions of a vehicle's paint or finish?
2. List and explain the three basic types of automotive paints.
3. What is the difference between thinner, reducer, and premix?
4. What is a catalyst?

5. What is corrosion or rust?
6. What is masking liquid or masking coating?
7. In detail, explain the sandpaper grit numbering system and how it is used.
8. What general sequence should be used when tightening a series of bolts or nuts?

9. Define the terms *undercoat* and *topcoat*.

10. Name the four ingredients in a paint.

11. Define the term *cap screw*.

12. What is bolt thread pitch and how is it measured?

13. If you turn right-hand threads clockwise, what will happen?

CRITICAL THINKING PROBLEMS

1. If lug bolts on a wheel are improperly tightened, what might happen?

2. When repairing a wrecked car, what are some of the fasteners that you might have to torque while working?

MATH PROBLEMS

1. If a paint is supposed to be reduced one part paint to one part thinner, what is the percentage of reduction?

2. If a bolt torque spec is 100 pounds of torque, what is the torque value in newton-meters?

CHAPTER 7

Welding Equipment and Its Use

INTRODUCTION

With major collision repair work, many of the panels on a vehicle must be replaced and welded into place. As you will learn, this requires considerable skill and care. The structural integrity of the vehicle is dependent upon how well you weld and install panels.

If not already trained, you may want to consider taking an MIG welding course in school or through another agency. Ask your instructor or guidance counselor for more information on welding courses in your school or area. Welding is an essential skill if you plan on becoming a master auto body technician. It is a good idea to become I-CAR trained and pass their Automotive MIG Welding Qualification Test to show your welding skills are competent.

JOINING METALS

There are three basic methods of joining metal together in the automobile assembly:

- Mechanical (metal fastener) methods (Figure 7–1)
- Chemical (adhesive fastening) methods

OBJECTIVES

After studying this chapter, you should be able to:

✔ Identify the three classes of welding.
✔ Explain how to use an MIG welding machine.
✔ Name the six basic welding techniques employed with MIG equipment.
✔ Describe differences between MIG electrode wires.
✔ Determine where and how to use resistance spot welding.
✔ Identify oxyacetylene welding equipment and techniques.
✔ Explain general brazing and soldering techniques used in a body shop.
✔ Describe plasma arc cutting of body panels.
✔ Explain plasma cutting techniques.
✔ List safety procedures important in each welding operation.
✔ Pass ASE Certification test questions on welding and cutting.
✔ Explain the information needed to pass I-CAR's Welding Qualification Test.

KEY TERMS

brazing	nondestructive check
burn mark	overhead welding
burn-through	oxidizing flame
carburizing flame	plasma arc cutting
continuous weld	plug weld
DC reverse polarity	pressure welding
destructive check	spot weld
flat welding	stitch weld
fusion welding	tack weld
heat crayons	vertical welding
horizontal welding	weld face
insert	weld legs
joint fit-up	weld penetration
lap spot weld	weld root
MIG welding	weld throat
neutral flame	welding filter lens

 **ASE TASK LIST**

Job Skills covered in this chapter include:

PAINTING AND REFINISHING TEST (B2) TASK LIST

A. Surface Preparation

1. Remove, assess, and store trim and moldings.
7. Identify type of metal and apply suitable metal treatment or primer.

NONSTRUCTURAL ANALYSIS AND DAMAGE REPAIR TEST (B3) TASK LIST

A. Preparation

3. Remove outside trim and moldings as necessary; store reusable parts.
4. Remove damaged or undamaged inside trim and moldings as necessary; store reusable parts.
7. Protect panels and parts adjacent to repair area, to prevent damage during repair.
10. Remove repairable plastics and other parts that are recommended for off-vehicle repair.

B. Outer Body Panel Repairs, Replacements, and Adjustments

3. Determine the extent of damage to aluminum body panels; repair, weld, or replace in accordance with manufacturers' specifications.
7. Remove, replace, and align bumpers, reinforcements, guards, isolators, and mounting hardware.
11. Weld cracked or torn steel body panels; reweld broken welds; replace molding studs.
13. Remove damaged sections of steel body panels; weld in replacements in accordance with manufacturers'/industry specifications.

C. Metal Finishing and Body Filling

4. Heat shrink stretched panel areas to proper contour.

E. Welding and Cutting

1. Identify weldable and nonweldable materials used in vehicle construction.
2. Understand the limitations of welding and cutting high-strength steels and other metals.
3. Determine correct welding process [GMAW (MIG), compression/resistance spot, oxyacetylene, GTAW (TIG)], electrode, wire type, diameter, and gas to be used in specific welding situations.
4. "Tune" the MIG welder by adjusting for the proper electrode stickout, voltage, polarity, flow rate, and wire speed required for the material being welded.
5. Identify safety considerations: Eye protection, proper clothing,, shock hazards, fumes, M.S.D.S., etc. before beginning any welding operation.

6. Apply knowledge of the proper procedures for safely handling gas cylinders.
7. Insure proper work clamp (ground) location.
8. Use the proper gun-to-joint angle, and direction of gun travel, for welds being made in all positions.
9. Protect vehicle components, including computers and other electronic modules, from possible welding and cutting damage.
10. Clean the metal to be welded; assure good metal fit-up; apply weld-through primer.
11. Perform the correct joint type (butt, lap, etc.) for the weld being made.
12. Determine the correct type of weld (continuous, stitch/pulse, tack, plug, spot, etc.) for each specific welding operation.
13. Identify the causes of spits and sputters, burn-through, lack of penetration, cracks in metal, porosity, incomplete fusion, excessive spatter, distortion, waviness of bead, and failure of wire to feed; make necessary adjustments.
14. Identify proper cutting process for different materials and locations in accordance with manufacturers' recommendations.

STRUCTURAL ANALYSIS AND DAMAGE REPAIR TEST (B4) TASK LIST

A. Frame Inspection and Repair

8. Remove and replace damaged frame horns, side rails, cross members, and front or rear sections.
10. Repair or replace weakened or cracked frame members in accordance with vehicle manufacturers'/industry standards.

B. Unibody Inspection, Measurement, and Repair

15. Determine the extent of damage to structural steel body panels; repair, weld, or replace in accordance with vehicle manufacturers' specifications.
16. Remove damaged sections of structural steel body panels, and weld in replacements in accordance with vehicle manufacturers' specifications.

D. Metal Welding and Cutting

1. Identify weldable and nonweldable materials used in vehicle construction.
2. Understand the limitations of welding and cutting high-strength steels and other metals.
3. Determine correct welding process [GMAW (MIG), compression/resistance spot, GTAW(TIG)], electrode, wire type, diameter, and gas to be used in specific welding situations.

ASE TASK LIST (continued)

Job Skills covered in this chapter include:

4. "Tune" the MIG welder by adjusting for proper electrode stickout, voltage, polarity, flow rate, and wire speed required for the material being welded.
5. Identify safety considerations: eye protection, proper clothing, shock hazards, fumes, etc. before beginning any welding operation.
6. Understand the proper procedures for safely handling gas cylinders.
7. Insure proper work clamp (ground) location.
8. Use the proper gun-to-joint angle, and the direction of gun travel, for welds being made in all positions.
9. Protect vehicle components, including computers and other electronic modules, from possible damage from welding and cutting operations.

10. Clean the metal to be welded; assure good metal fit-up; apply weld-through primer.
11. Perform the correct type of joint (butt, lap, etc.) for the weld being made.
12. Determine the correct type of weld (continuous, stitch/pulse, tack, plug, spot, etc.) for each specific welding operation.
13. Identify the causes of spits and sputters, burn-through, lack of penetration, cracks in metal, porosity, incomplete fusion, excessive spatter, distortion, waviness of bead, and failure of wire to feed; make necessary adjustments.
14. Identify the proper cutting process (abrasive, plasma arc, oxyacetylene) for different materials and locations in accordance with manufacturers' recommendations.

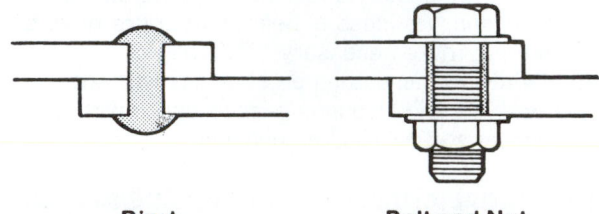

Rivet   **Bolt and Nut**

FIGURE 7-1 Mechanical joining methods use threaded or nonthreaded fasteners, which are heavier and less dependable than a welded joint.

- Welding (molten metal) methods (Figure 7–2)

Welding is a method of repair in which heat is applied to pieces of metal to fuse them together into the shape desired. Welding can be divided into three main categories:

- **Pressure welding.** The metal is heated to a softened state by electrodes. Pressure is applied, and the metal is joined. Of the various types of pressure welding, electric resistance welding (spot welding) is an indispensable method used in automobile manufacturing and to a lesser degree in repair operations.
- **Fusion welding.** Pieces of metal are heated to the melting point, joined together (usually with a filler rod), and allowed to cool.
- *Braze welding.* Metal with a melting point lower than the base metal to be joined is melted over the joint of the pieces being welded (without fusing pieces of base metal). Braze welding is classified as either soft or hard brazing, depending on the temperature at which the brazing material melts. *Soft brazing* is done with brazing

FIGURE 7-2 Welding is a fast, strong method of joining metal parts.

material that melts at temperatures below 850 degrees Fahrenheit (455 degrees Celsius). *Hard brazing* is done with brazing materials that melt at temperatures above 850 degrees Fahrenheit (455 degrees Celsius).

Shown in Table 7–1, there are distinct welding methods within each respective category. Many of these methods can be used in the auto body shop. Gas metal arc welding is the preferred method.

WELD TERMINOLOGY

The **weld root** is the part of the joint where the wire electrode is directed. The **weld face** is the exposed surface of the weld on the side that has been welded.

TABLE 7-1: WELDING METHODS

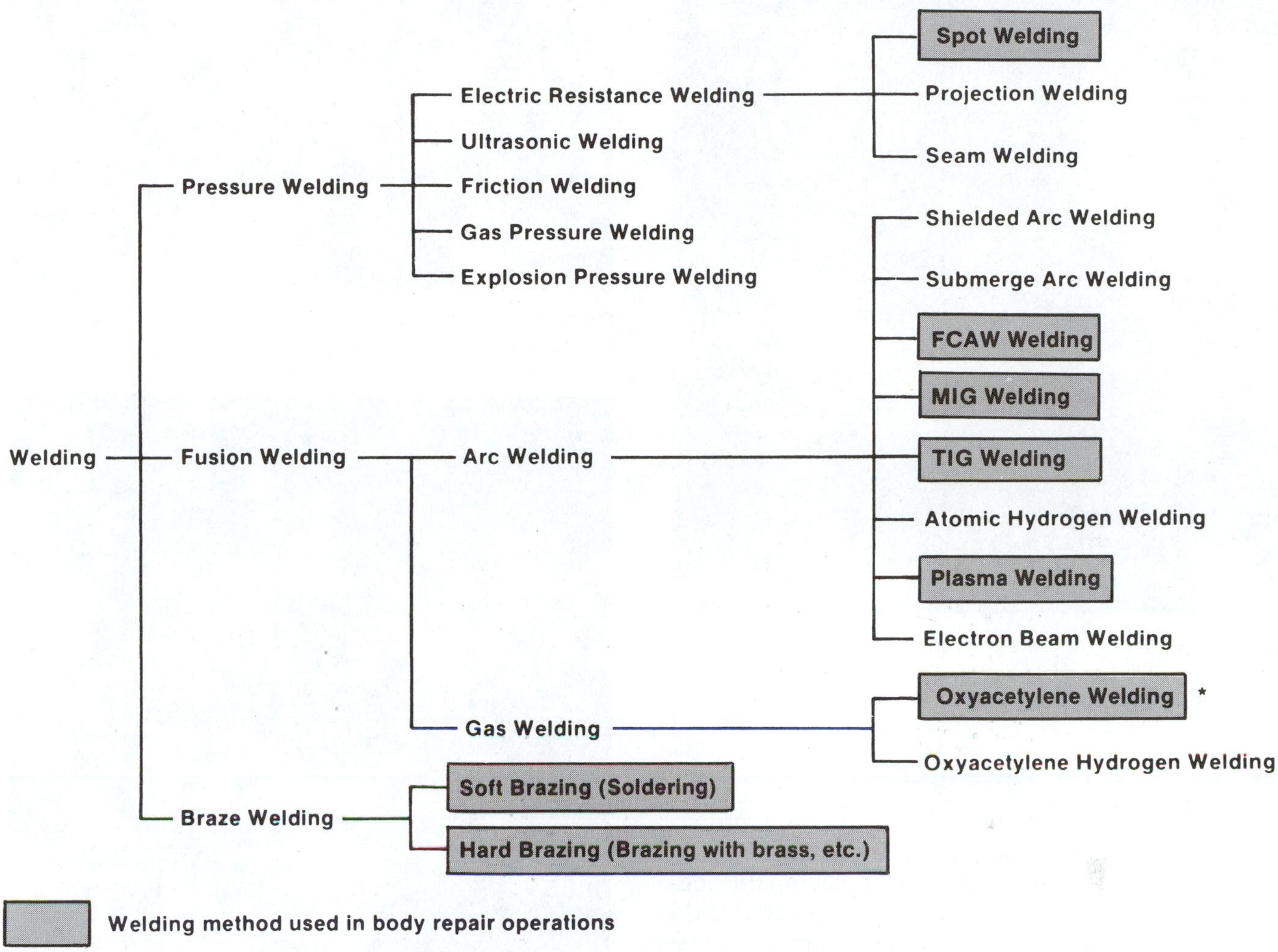

```
Welding ─┬─ Pressure Welding ─┬─ Electric Resistance Welding ─┬─ Spot Welding
         │                     ├─ Ultrasonic Welding            ├─ Projection Welding
         │                     ├─ Friction Welding              └─ Seam Welding
         │                     ├─ Gas Pressure Welding
         │                     └─ Explosion Pressure Welding
         │
         ├─ Fusion Welding ─┬─ Arc Welding ─┬─ Shielded Arc Welding
         │                  │                ├─ Submerge Arc Welding
         │                  │                ├─ FCAW Welding
         │                  │                ├─ MIG Welding
         │                  │                ├─ TIG Welding
         │                  │                ├─ Atomic Hydrogen Welding
         │                  │                ├─ Plasma Welding
         │                  │                └─ Electron Beam Welding
         │                  │
         │                  └─ Gas Welding ─┬─ Oxyacetylene Welding  *
         │                                   └─ Oxyacetylene Hydrogen Welding
         │
         └─ Braze Welding ─┬─ Soft Brazing (Soldering)
                            └─ Hard Brazing (Brazing with brass, etc.)
```

▭ Welding method used in body repair operations

*Not recommended for HSS (High Strength Steel) or unibody vehicles

Visible **weld penetration** is indicated by the height of the exposed surface of the weld on the back side. Full weld penetration is needed to assure maximum weld strength.

A **burn mark** on the back of a weld is an indication of good weld penetration. **Burn-through** results from penetrating too much into the lower base metal which burns a hole through the back side of the metal.

Fillet weld parts include the following. The **weld legs** are the width and height of the weld bead. The **weld throat** refers to the depth of the triangular cross section of the weld.

Joint fit-up refers to holding work pieces tightly together, in alignment, to prepare for welding. It is critical to the replacement of body parts!

WELDING CHARACTERISTICS

Joint welding is indispensable in the restoration of collision-damaged vehicles. The characteristics of welding can be summarized as follows:

- Since the shape of welding joints is limitless, it is the perfect method for joining a vehicle structure, while still maintaining body integrity.
- Weight can be reduced (no fasteners are necessary).
- Air and water tightness are excellent.
- Production efficiency is very high.
- Strength of a welded joint is greatly influenced by the level of skill of the operator.
- Surrounding panels will warp if too much heat is used.

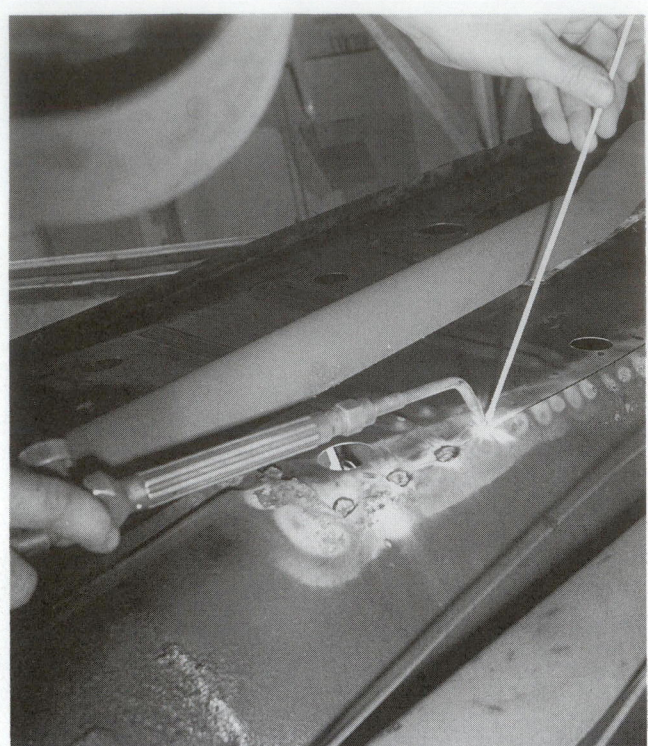

FIGURE 7-3 The once popular oxyacetylene welding process is no longer recommended for automobile work. *(Courtesy of Dobbs Publishing)*

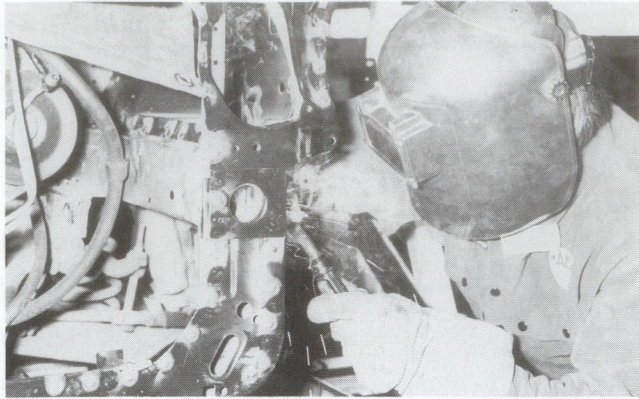

FIGURE 7-4 MIG is the number one welding method in auto body repair work. *(Courtesy of Tech-Cor)*

FIGURE 7-5 Conventional stick electrode or shielded arc welding spreads too much heat into the part for modern thin metal panels.

WELDING IN THE AUTO BODY SHOP

New welding techniques and equipment have entered the auto body repair picture, replacing the one-time popular arc and oxyacetylene processes (Figure 7–3). New steel alloys used in today's cars cannot be welded properly by these two processes. Presently gas metal arc welding (GMAW)—better known as metal inert gas (MIG) welding—offers more advantages than other methods for welding high-strength steels (HSS) and high-strength, low-alloy (HSLA) steel component parts used in modern cars. Most of the applications of HSS and HSLA steels are confined to body structures, reinforcement gussets, brackets, and supports, rather than large panels or outer skin panels.

The advantages of MIG welding (Figure 7–4) over conventional stick electrode arc welding (Figure 7–5) are so numerous that manufacturers now recommend it almost exclusively. MIG welding is recommended by all OEMs, not only for HSS and unibody repair, but for *all* structural collision repair. This recommendation extends also to independent collision repair shops.

Here are some of the advantages of MIG welding.

• MIG welding is easier to learn than arc or gas welding. The typical welder can learn to use MIG

welding equipment with proper training. Moreover, experience shows that even an average MIG welder can produce higher quality welds faster and more consistently than a highly skilled welder using older stick electrode welds.

• MIG welding produces 100 percent fusion in the parent metals. This means MIG welds can be dressed or ground down flush with the surface (for cosmetic reasons) without loss of strength.

• Low current can be used for thin metals. This prevents heat damage to adjacent areas that can cause strength loss and warping.

• The arc is smooth and the weld puddle small, so it is easily controlled (Figure 7–6). This ensures maximum metal deposit with minimum splatter.

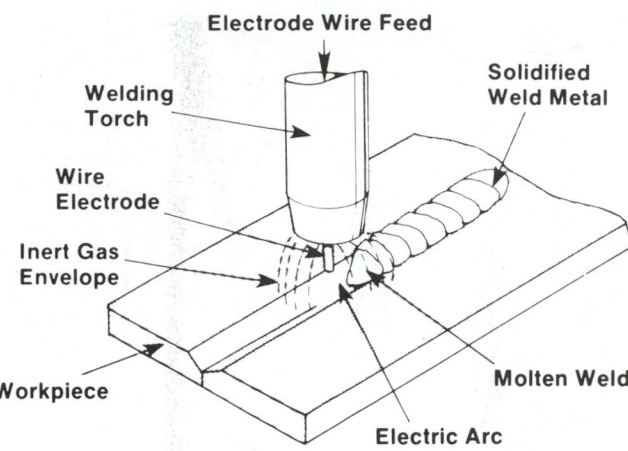

FIGURE 7-6 Study the basic MIG welding process.

- MIG welding is more tolerant of gaps and misfits. Several gaps can be spot welded immediately (no slag to remove) by making several spots on top of each other. Therefore, the area can be easily refinished.
- Almost all steels can be welded with one common type of weld wire. What is in the machine is generally right for any job.
- Metals of different thicknesses can be welded with the same diameter of wire. Again, what is in the machine is right for almost any job.
- The MIG welder can control the temperature of the weld and the time the weld takes place.
- With MIG welding, the small area to be welded is heated for a short period of time, thereby reducing metal fatigue, warpage, and distortion of the panel. Vertical and/or overhead welding is possible because the metal is molten for a very short time.

Portable resistance spot welding is also recommended today for some repairs (Figure 7–7). This type of equipment is used to form spot weld attachments like the production welds. To use this kind of spot

welding equipment, you must install the proper extensions and electrodes on the welder to provide access to the area being welded. The clamping force on a squeeze-type resistance spot welder must be properly adjusted. On some equipment, the amperage, current flow, and timing are all made with one adjustment. After the adjustments are made, the spot welder is positioned over the panels being joined, making sure the electrodes are directly opposite to each other. The trigger is squeezed and the spot weld takes place.

The resistance spot welder, which probably requires the least skill to operate, provides very fast, high-quality welds while maintaining the best control of temperature buildup in adjacent panels and structure.

When reference is made to resistance type spot welders, it generally describes the type of welding that requires the actual weld to take place on both sides of all panels at the same time. It normally does not mean the type of spot weld that welds panels together from the same side at the same time. Opposite side spot welding is a structural weld.

Be sure to consult the car manufacturer's recommendations in the vehicle's service manual before welding. When replacing body panels, all the new welds should be similar in size to the original factory welds. Except when spot welding, the number of replacement welds should be the same as the original number of welds in production. Strength and durability requirements differ depending on the location of the part that is to be welded to the body.

The manufacturer decides what is the most appropriate welding method (Figure 7–8) by first determining the intended use, the physical characteristics, and the location of the part.

FIGURE 7-7 The technician is using a portable squeeze-type resistance spot welding gun to install the left rear quarter panel to the lower back panel. *(Courtesy of Lors Machinery, Inc.)*

WARNING Always follow service manual recommendations when welding. This will assure structural integrity.

It is essential that appropriate welding methods, which do not reduce the original strength and durability of the body, are used when making repairs. This is accomplished if the following basic points are observed:

- Try to use either spot welding or MIG/MAG (metal inert gas/metal active gas) welding.
- Do *not* braze any body components other than those brazed at the factory.
- Do *not* use an oxyacetylene torch for welding late model auto bodies.

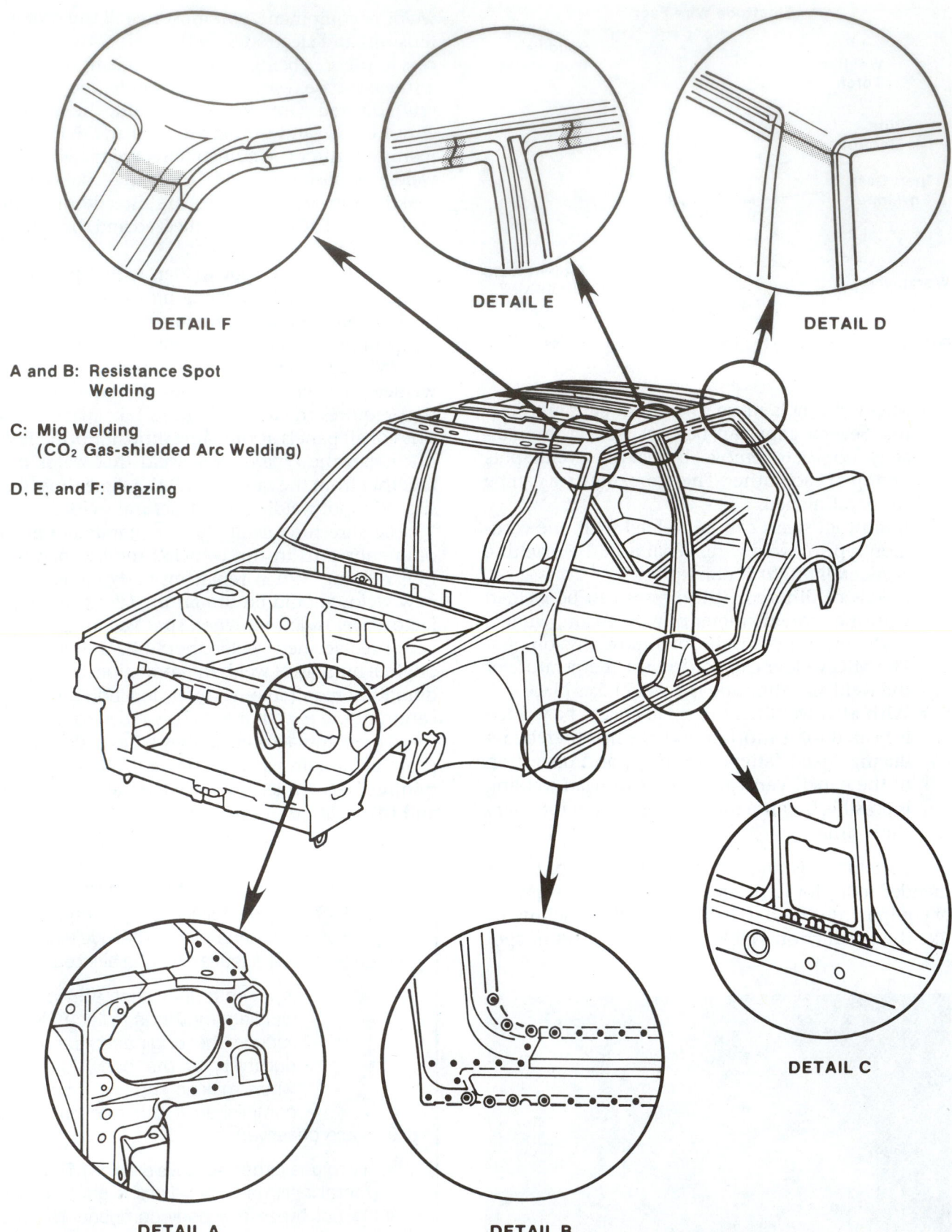

DETAIL F

DETAIL E

DETAIL D

A and B: Resistance Spot
Welding

C: Mig Welding
(CO₂ Gas-shielded Arc Welding)

D, E, and F: Brazing

DETAIL C

DETAIL A

DETAIL B

FIGURE 7-8 Compare the welding methods used in vehicle production. *(Courtesy of Nissan Motor Corp.)*

SHOP TALK

Regardless of the type of welding, you must properly clean the surface before starting the weld. Remove all surface materials back to the bare metal (Figure 7–9). When dirt, rust, sealers, paint, or other materials are left in the area of the weld, they will burn during the application of heat. The ash or oxidized material can become a part of the weld. The dirt and foreign material cause the weld to be weakened. If the weld fails, it could endanger the occupants of the car or truck.

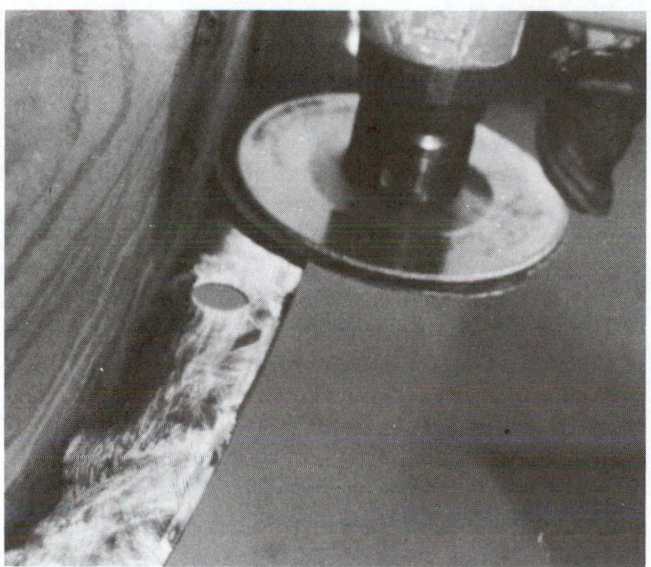

FIGURE 7–9 Make sure the surfaces to be welded are completely free of rust and scale contaminants loosened by grinding, sanding, or sandblasting.

7.1 MIG WELDING

MIG welding became popular in body shops when auto manufacturers began using thin-gauge, high-strength, low-alloy (HSLA) steels. Car makers insisted that the only correct way to weld HSLA and other thin-gauge steel was with MIG (or the similar gas metal arc welding [GMAW] system). And once the MIG welder was in place, it was easy to see that it provided clean, fast welds for all applications. Welding a rear quarter panel with an oxyacetylene welder averages about 4 hours. An MIG welder can do the same job in about 40 minutes.

MIG welding is not limited to body repairs alone. It is also ideal for exhaust work, repairing mechanical supports, installing trailer hitches and truck bumpers, and any other welds that would be done with either an arc or gas welder. In addition, it is possible to weld aluminum castings, like cracked transmission cases, cylinder heads, and intake manifolds.

MIG PRINCIPLES AND CHARACTERISTICS

MIG welding uses a welding wire that is fed automatically at a constant speed as an electrode. A short arc is generated between the base metal and the wire. The resulting heat from the arc melts the welding wire and joins the base metals together. Since the wire is fed automatically at a constant rate, this method is also called *semiautomatic arc welding*.

During the welding process (Figure 7–10), either *inert gas* or active gas shields the weld from the atmosphere and prevents oxidation of the base metal. The type of inert or active gas used depends on the base material to be welded. For most steel welds, *carbon dioxide* (CO_2) is used as the shield gas (Figure

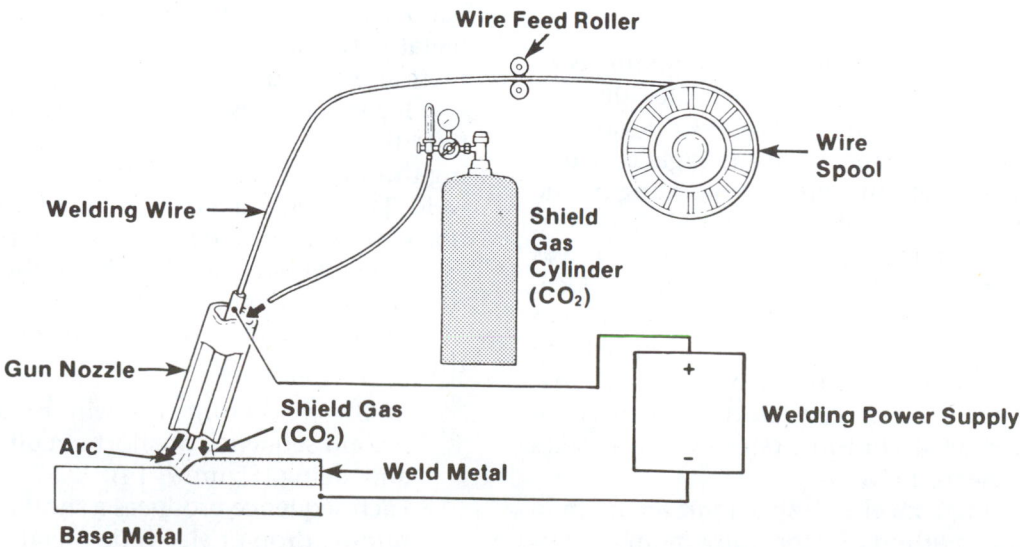

FIGURE 7–10 Study principles of MIG welding, especially hook-up and major welder parts. *(Courtesy of Toyota Motor Corp.)*

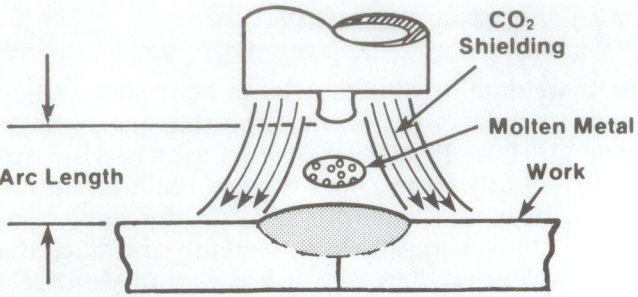

FIGURE 7-11 Carbon dioxide (CO_2) protects molten metal from contamination by the atmosphere.

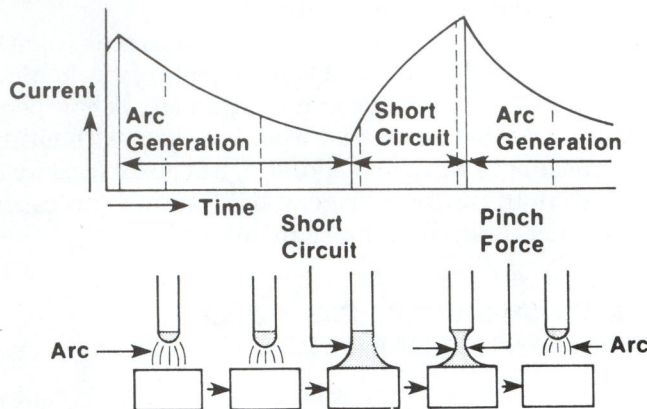

FIGURE 7-12 Graph and illustrations show how the short circuit arc method operates.

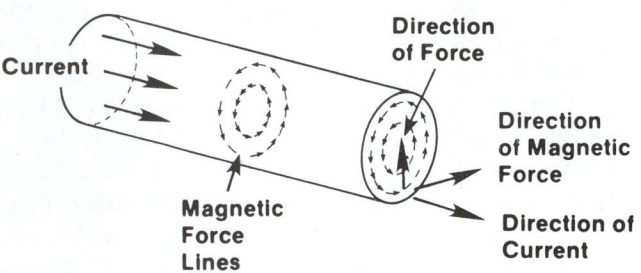

FIGURE 7-13 Study typical pinch force and how it is formed.

MIG welding is sometimes called carbon dioxide arc welding. Actually, MIG (metal inert gas) welding uses a fully inert gas, such as argon or helium, as a shield gas. Since carbon dioxide gas is not a completely inert gas, it is more accurately called MAG (metal active gas) welding. Although most auto body shop welding is done with carbon dioxide gas as the shield gas, the term MIG is used to describe all gas metal arc welding processes. In fact, many welders on the market can use carbon dioxide (a semiactive gas) or argon (an inert gas) by simply changing the gas cylinder.

7–11). Another common shielding gas mixture is 75 percent argon and 25 percent carbon dioxide. This latter mixture is usually referred to as C-25 gas.

With aluminum, either pure argon gas or a mixture of argon and helium is used, depending on the alloy and the thickness of the material. It is even possible to weld stainless steel by using argon gas with a little oxygen (between 4 and 5 percent) added.

MIG *flux core wire* has its own flux contained in a tubular electrode and does not require a shielding gas. As with stick welding, the flux forms slag that must be chipped off. Flux core electrode wire is not convenient for most collision repair work. It takes more time to clean the weld.

MIG welding uses the short circuit arc method that is a unique method of depositing molten drops of metal onto the base metal. Welding of thin sheet metal for automobiles can cause welding strain, blow holes, and warped panels. To prevent these problems, it is necessary to limit the amount of heat near the weld. The *short circuit arc method* uses very thin welding rods, a low current, and low voltage. By using this technique the amount of heat introduced into the panels is kept to a minimum and penetration of the base metal is quite shallow.

As shown in Figure 7–12, the end of the wire is melted by the heat of the arc and forms into a drop. The drop then comes in contact with the base metal and creates a short circuit. When this happens, a large current flows through the metal and the shorted portion is torn away by the pinch force or burnback, which re-establishes the arc. The bare wire electrode is fed continuously into the weld puddle at a controlled, constant rate, where it short circuits, and the arc goes out. While the arc is out, the puddle flattens and cools; but the wire continues to feed, shorting to the workpiece again. This heating and cooling happens on an average of 100 times a second. The metal is transferred to the workpiece with each of these short circuits.

If current is flowing through a cylindrical-shaped fluid (in this case molten metal) or current is flowing through an arc, the current is pulled toward the weld. This works as a constricting force in the direction of the center of the cylinder. This action is known as the pinch effect, and the size of the force is called the *pinch force* (Figure 7–13).

In summary, the MIG welding process works like this:

- At the weld point, the wire undergoes a split-second sequence of short circuiting, burnback, and arcing (Figure 7–14).
- Each sequence produces a short arc transfer of a minute drop of electrode metal from the tip of the wire to the weld puddle.

- A gas curtain or shield surrounds the wire electrode. This gas shield prevents contamination from the atmosphere and helps stabilize the arc.
- The continuously fed electrode wire contacts the work and sets up a short circuit, and resistance heats the wire and the weld site.
- As the heating continues, the wire begins to melt and thin out or neck down.
- Increasing resistance in the neck accelerates the heating in this area.
- The molten neck burns through, depositing a puddle on the workpiece and starting the arc.
- The arc tends to flatten the puddle and burn back the electrode.

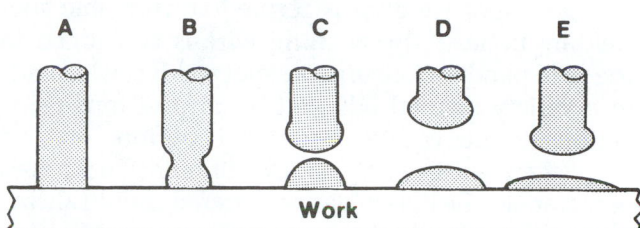

FIGURE 7-14 Here is the typical action of welding wire as it burns back from work during MIG welding.

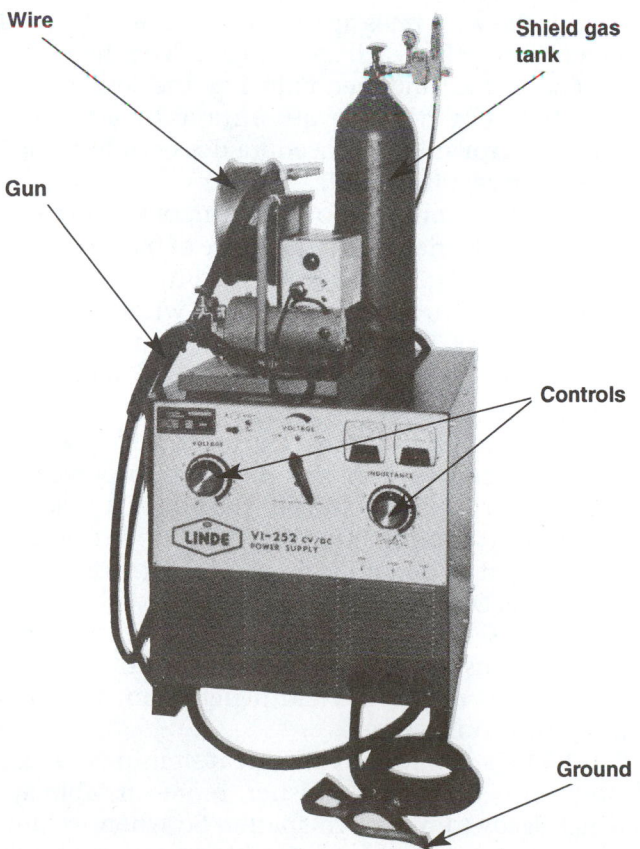

FIGURE 7-15 Note basic components of an MIG setup.

- With the arc gap at its widest, it cools, allowing the wire feed to move the electrode closer to the work.
- The short end starts to heat up again, enough to further flatten the puddle but not enough to keep the electrode from recontacting the workpiece. This extinguishes the arc, re-establishes the short circuit, and restarts the process.
- This complete cycle occurs automatically at a frequency ranging from 50 to 200 cycles a second.

7.2 MIG WELDING EQUIPMENT

Most MIG welding equipment for collision repair work is considered semiautomatic. This means that the machine's operation is automatic, but the gun is hand controlled. Before starting to weld, the operator sets

- Voltage for the arc
- Wire speed
- Shielding gas flow rate

Then the operator has complete freedom to concentrate entirely on the weld site, the molten puddle, and whatever welding technique that is used.

Regardless of the type of MIG equipment used, it will comprise the following basic components (Figure 7–15):

- Supply of shielding gas with a flow regulator to protect the molten weld pool from contamination
- Wire/feed control to feed the wire at the required speed
- Spool of electrode wire of a specified type and diameter
- MIG type of welder machine connected to an electrical power supply
- Work cable and clamp assembly
- Welding gun and cable assembly that the welder holds to direct the wire to the weld area

Fine diameter welding wires are used in collision repair and typically range from .025 inch (0.397 mm) through .030 inch (0.793 mm). This is roughly equivalent in diameter to ultrafine leads in today's mechanical pencils. A wire that is becoming more commonly used today is .025 inch (0.397 mm). Once a specialty wire, it is now stocked by most wire manufacturers. These small diameter wires can be used at low currents (10 to 20 amps) and voltages (120 volts), thus greatly reducing heat input to the base material. The welding wire must carry a minimum specification of AWS-ER70-6.

Because of the power demands in this process, it is necessary to use a constant potential, constant

FIGURE 7-16 This is a self-contained MIG unit capable of producing quality welds. *(Courtesy of Snap-On Tools Corp.)*

FIGURE 7-17 Look over the typical control panel of an MIG welder unit.

SHOP TALK

I-CAR worked with the American Welding Society to develop industry standards for auto body welding. You may want to consider enrolling in an I-CAR welding class to upgrade your skills.

voltage power source (Figure 7–16). The controls are a voltage adjustment and wire feed speed adjustment. Some optional controls available on this type of equipment (Figure 7–17) are a spot control and pulse control.

MIG spot welding is termed consumable spot welding because the welding wire is consumed in the weld puddle. Consumable spot welds can be made in a variety of methods and in all positions using various nozzles equipped with this option.

When you are spot welding different thicknesses of materials, the lighter gauge material should always be spotted to the heavy material.

Spot welding usually requires greater heat to the weld than continuous or pulse welding. It is best to use sample materials when setting the controls for spot welding. To check a spot weld, pull the two pieces apart. A good weld will tear a small hole out of the bottom piece. If the weld pulls apart easily, increase the weld time or heat. After each spot is complete, the trigger must be released and then pulled for the next spot.

MIG spot has the advantage of an easily grindable crown. The procedure does not leave any depression requiring a fill.

The pulse control allows continuous seam welding on the material with less chance of burn-through or distortion. This is accomplished by starting and stopping the wire for preset times without releasing the trigger. The weld "on" time and weld "off" time can be set for the operator's preference and the metal thickness.

The burnback control on most MIG gives an adjustable burnback of the electrode to prevent it from sticking in the puddle at the end of a weld.

In MIG welding the polarity of the power source is important in determining the penetration to the workpiece. DC power sources used for MIG welding typically use DC reverse polarity. **DC reverse polarity** means the wire (electrode) is positive and the workpiece is negative. Weld penetration is greatest using this connection.

Weld penetration is also greatest using CO_2 gas. However, CO_2 gives a harsher, more unstable arc, which leads to increased spatter. So when welding on thin materials, it is preferable to use argon/CO_2 (Figure 7–18).

| 75% Argon 25% CO₂ | 50% Argon 50% CO₂ | CO₂ |

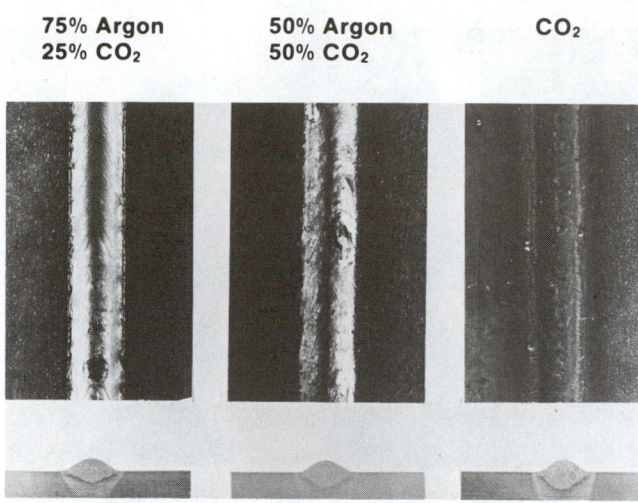

FIGURE 7-18 Note how gas affects weld penetration.

Voltage adjustment and wire feed speed must be set according to the diameter of the wire being used and metal thickness. It should be noted that when setting these parameters, the manufacturer's recommendations should be followed to reach approximate settings. When rough parameters are selected, change only one variable at a time until the machine is fine tuned for an optimum welding condition. MIG welders can be tuned in using both visual and audio signals.

WELDING LENS

A **welding filter lens,** sometimes called *filter plate,* is a shaded glass welding helmet insert for protecting your eyes from ultraviolet burns. The lenses are graded with numbers, from 4 to 12. The higher the number, the darker the filter. The American Welding Society (AWS) recommends grade 9 or 10 for MIG welding steel.

Note that there are *self-darkening filter lenses* available that instantly turn dark when the arc is struck. There is no need to move the face shield up and down.

FIGURE 7-19 Check the manufacturer's manual before hooking up equipment.

7.3 MIG OPERATION METHODS

To match MIG welding power to the available input voltage, follow the procedure prescribed on the machine or in the manufacturer's manual (Figure 7–19).

Handle the cylinder of shielding gas with care. It might be pressurized to more than 2,000 pounds per square inch (13,800 kPa). Chain or strap the cylinder

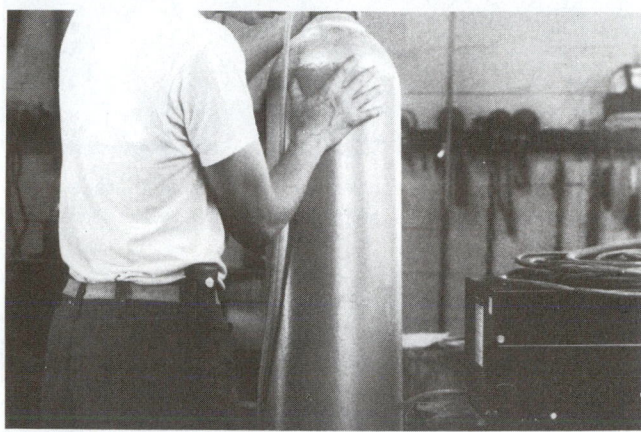

A

B

FIGURE 7-20 (A) Install the shielding gas cylinder with care and (B) chain or strap it in place.

FIGURE 7-21 Do not crossthread or strip fit when installing a regulator on a cylinder.

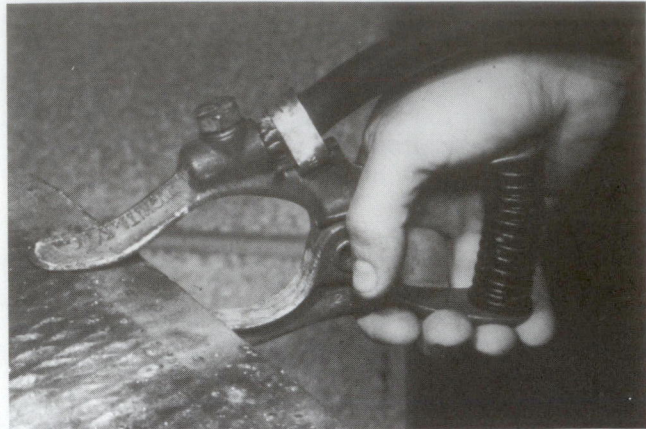

FIGURE 7-22 Attach the clamp to a clean metal surface to prevent electrical resistance that can affect weld current.

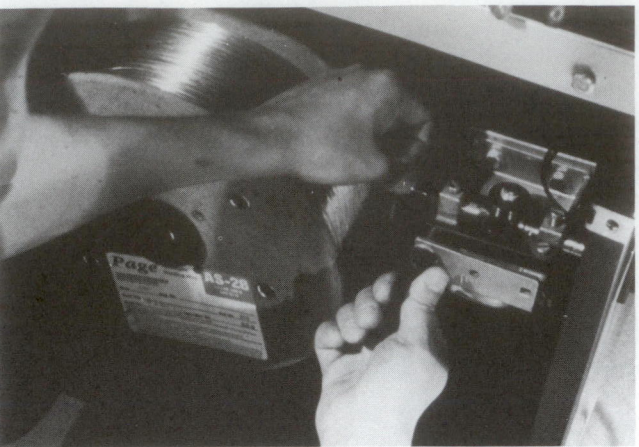

FIGURE 7-24 Make sure the drive roll grooves, wire guides, cable liner, and gun contact tube correspond with the size of the wire being used.

FIGURE 7-23 Make welder set-up and adjustments following the manufacturer's directions. *(Courtesy of I-CAR)*

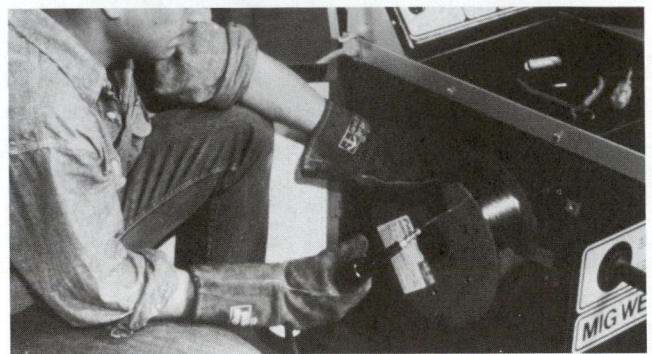

FIGURE 7-25 Tension should be light enough so the wire slips at the rollers when it is stopped at the nozzle of the gun.

to a support sturdy enough to hold it securely to the MIG machine (Figure 7–20). Install the regulator, making sure to observe the recommended safety precautions (Figure 7–21).

When the clamp is attached to clean metal on the vehicle (Figure 7–22) near the weld site, it completes the welding circuit from the machine to the work and back to the machine. This clamp is not referred to as a ground cable or ground clamp. The ground connection is for safety purposes and is usually made from the machine's case to the building ground through the third wire in the electric input cable.

Consult the manufacturer's manual as to the specific procedure for assembling, installing, and adjusting the wire feeder components (Figure 7–23). In general, the adjustment of the wire feeder can be done as follows:

- Mount the wire. Feed the wire manually for about 12 inches (305 mm), making sure that it travels freely through the gun assembly.
- A correct setting on the drive rollers will assure just enough pressure on the wire to pull it off

the wire spool and through the gun/cable assembly (Figure 7–24). The tension must be set so that the wire will slip at the rollers when the wire is stopped at the nozzle, but tight enough to withstand a 30 to 40 degree deflection. If too much pressure is applied, the wire will be deformed, creating a spiral effect through the liner and erratic feed.
- Stopping the wire at the tip with this much pressure will also cause the wire to bird-nest between the rollers and cable entrance. The tension on the wire spool spindle should also be set so that the wire can be pulled off easily but just tight enough to stop the spool from free wheeling when the trigger is released (Figure 7–25).

The proper handling of any welding equipment is an essential ingredient in successful welding. When tuning the MIG welder for any given welding job, you have to deal with a number of parameters, meaning values that are variable: input voltage to the welding equipment, welding current, arc voltage, tip-to-base metal distance, torch angle, welding direction, shield gas flow volume, welding speed, and

**TABLE 7-2: RELATIONSHIP BETWEEN WIRE DIAMETER,
PANEL THICKNESS, AND WELDING CURRENT**

Wire Diameter	Panel Thickness						
	1/64"	1/32"	Less Than 3/64"	3/64"	1/16"	3/32"	1/8"
1/64"	20–30A	30–40A	40–50A	50–60A	—	—	—
1/32"	—	—	40–50A	50–60A	60–90A	100–120A	—
More Than 1/32"	—	—	—	—	60–90A	100–120A	120–150A

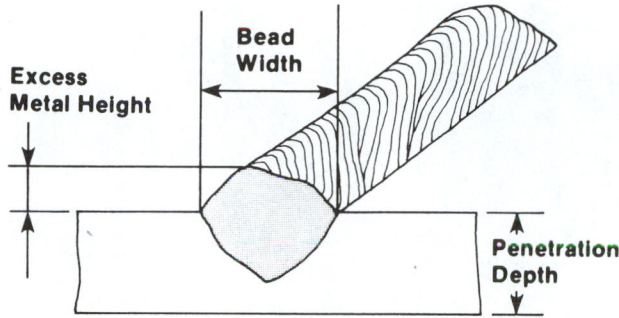

FIGURE 7-26 These are the three most important variables for a good weld: penetration depth, excess metal height, and bead width. *(Courtesy of Toyota Motor Corp.)*

wire speed. Most manufacturers of MIG welders provide tables that show the variable control parameters that apply to their machines.

MIG WELDING CURRENT

Welding current affects the base metal penetration depth (Figure 7–26), the speed at which the wire is melted, arc stability, and the amount of weld spatter. As the electrical current is increased, the penetration depth, excess metal height, and bead width also increase (Table 7–2).

MIG ARC VOLTAGE

Good welding results depend on a proper arc length. The length of the arc is determined by the arc voltage. When the arc voltage is set properly, a continuous light hissing or cracking sound is emitted from the welding area.

When the arc voltage is high, the arc length increases, the penetration is shallow, and the bead is wide and flat.

When the arc voltage is low, the arc length decreases, penetration is deep, and the bead is narrow and dome shaped (Figure 7–27).

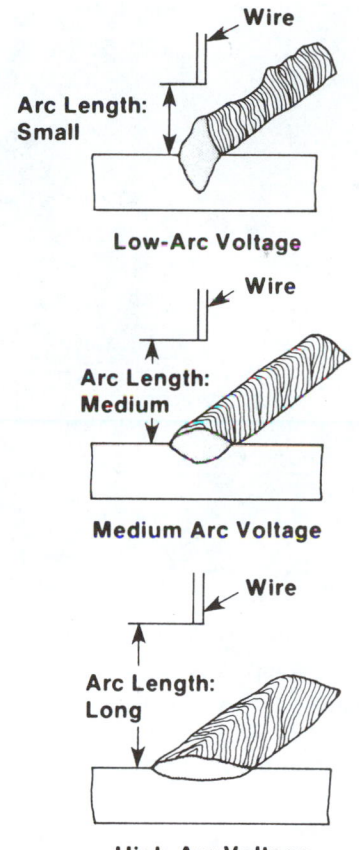

FIGURE 7-27 Note how arc voltage affects the bead shape.

Since the length of the arc depends on the amount of voltage, voltage that is too high will result in an overly long arc and an increase in the amount of weld spatter. A sputtering sound and no arc means that the voltage is too low.

MIG TIP-TO-BASE METAL DISTANCE

The tip-to-base distance (Figure 7–28) is also an important factor in obtaining good welding results. The

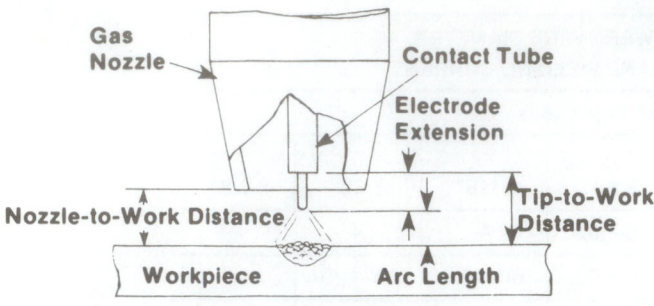

FIGURE 7-28 Tip-to-base metal distance should be kept constant when welding.

standard MIG welding tip-to-base distance is approximately $1/4$ to $5/8$ inches (6.3 to 40 mm).

If the tip-to-base metal distance is too long, the length of wire protruding from the end of the gun increases and becomes preheated, which increases the melting speed of the wire. Also, the shield gas effect will be reduced if the tip-to-base metal distance is too long.

If the tip-to-base metal distance is too short, it becomes difficult to see the progress of the weld because it will be hidden behind the tip of the gun.

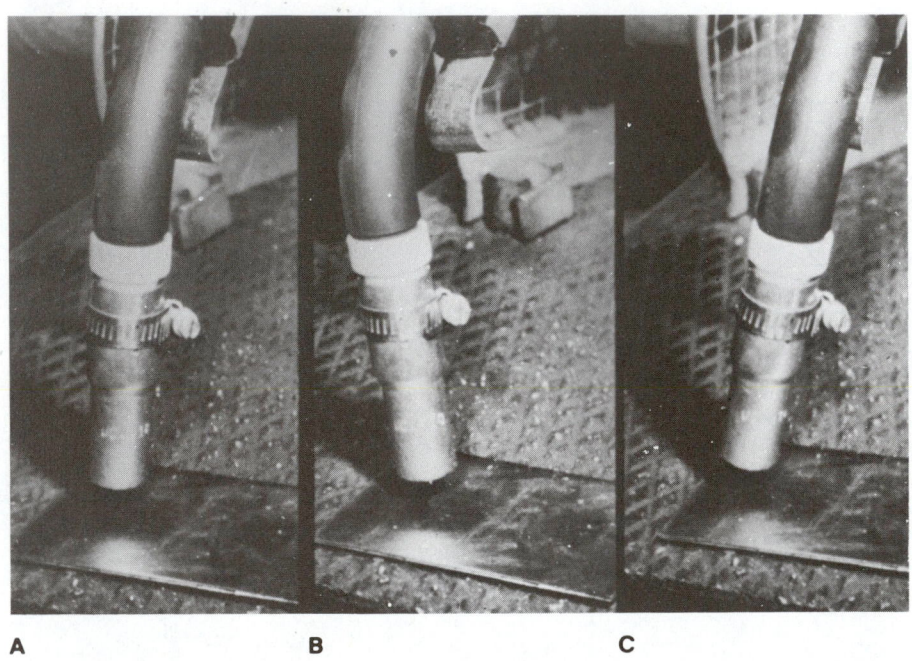

A   B   C

FIGURE 7-29 (A) Hold the torch at a transverse angle of 90 degrees directly over the center of the joint. Find the longitudinal angle by experimentation. (B) Trailing or dragging the torch angle—where the torch is pointing opposite or reverse to the direction of travel—should be tried first at about 10 degrees perpendicular. (C) Leading or pushing the torch angle—where torch is pointing forward in the direction of travel—should be used at the same angle.

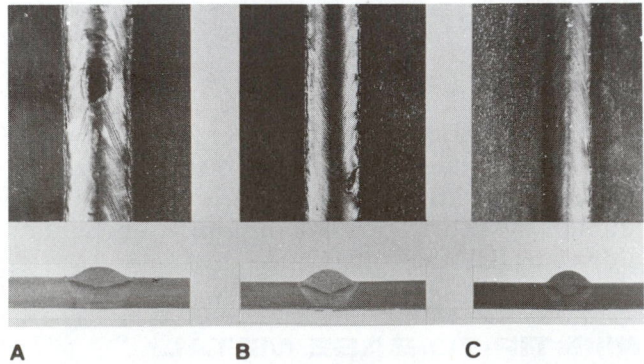

A   B   C

FIGURE 7-30 Penetration and weld pattern for three travel methods: (A) forehand; (B) transverse angle of 90 degrees; (C) backhand (reverse).

MIG GUN ANGLE AND WELDING

There are two methods: the forward or forehand method and the reverse or backhand method (Figure 7–29). With the forward method, the penetration depth is shallow and the bead is flat. With the reverse method, the penetration is deep and a large amount of metal is deposited (Figure 7–30). The gun angle for both methods should be between 10 and 30 degrees.

MIG SHIELD GAS FLOW VOLUME

Precise gas flow is essential to a good weld. If the volume of gas is too high, it will flow in eddies and reduce the shield effect. If there is not enough gas, the shield effect will also be reduced. Adjustment is made in accordance with the distance between the nozzle and the base metal, the welding current, welding speed, and welding environment (nearby air currents). The standard flow volume is approximately $1^3/_8$ to $1^1/_2$ cubic inches (0.022 to 0.024 liters) per minute or 15 to 25 cubic feet (420 to 700 cubic liters) per hour.

MIG WELDING SPEED

If you weld at a rapid pace, the penetration depth and bead width decreases, and the bead is dome shaped. If the speed is increased even faster, *undercutting* (weld surface level lower than base metal) can occur. Welding at too low a speed can cause burnthrough holes. Ordinarily, welding speed is determined by base metal panel thickness and/or voltage of the welding machine (Table 7–3).

MIG WIRE SPEED

An even, high-pitched buzzing sound indicates the correct wire-to-heat ratio producing a temperature in the 9,000 degrees Fahrenheit (4,986 Celsius) range. Visual signs of the correct setting occur when a steady reflected light starts to fade in intensity as the arc is shortened and wire speed is increased.

If the wire speed is too slow (Figure 7–31), a hiss and a plop sound will be heard as the wire melts away from the puddle and deposits the molten gob back. The visual signal will be a much brighter reflected light.

Too much wire speed will choke the arc. More wire is being deposited than the heat and puddle can absorb. The result is spitting and sputtering as the wire melts into tiny balls of molten metal that fly away from the weld. The visual signal is a strobe light arc effect.

Before this critical ratio can be obtained, a thorough understanding of what is happening to produce these signals is essential.

When the trigger is first activated, a solid steel wire makes its initial contact with a solid steel plate. Prior to contact, the wire has been charged with

TABLE 7-3: WELDING SPEED

Panel Thickness	Welding Speed (in./min.)
1/32"	41-11/32–45-9/32
More Than 1/32"	39-3/8
3/64"	35-7/16–39-3/8
1/16"	31-1/2–33-15/32

current and the gas flow has been started (Figure 7–32). The first contact produces tiny sparks of oxide being burned off the wire and base metal.

Immediately after the oxide sparks, tiny molten balls are produced as the wire melts prior to having a molten puddle that will absorb them. Once the heat creates the puddle, the balls stop. A consistent transfer and sound with only oxide sparks are present as they burn off the wire and base metal during the weld process.

In slow motion, after the arc transfer has been started, an on-off action occurs. Every time the metal is deposited a plop is heard. When it pulls away, a hiss is heard. Speeded up to 200 plops and hisses per second, it creates a smooth buzz, like the sound of bacon frying in a pan.

When welding overhead, the danger of having too large a puddle and ball are obvious. The ball is pulled by gravity down onto the contact tip or into the gas nozzle where it can create serious problems. Therefore, overhead welding should always be done with a higher wire speed with the arc and ball kept tiny and close together. Pressing the gas nozzle against the work insures that the wire is not moved out of the puddle. If it is moved out, the balls are produced by melting wire until a new puddle is formed to absorb them.

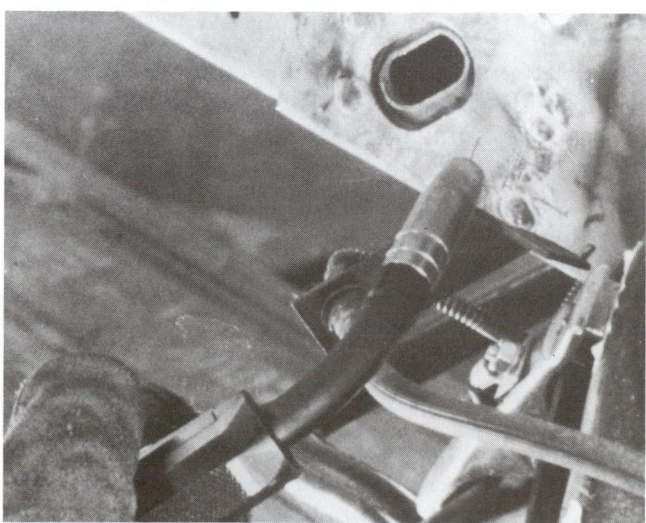

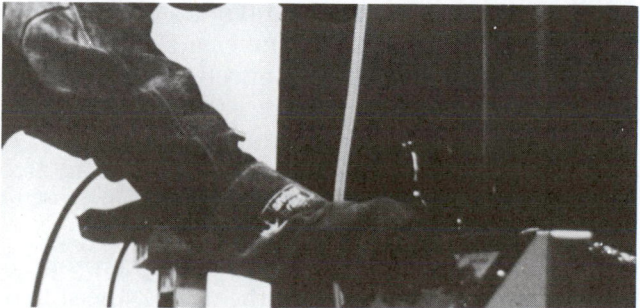

FIGURE 7-31 If the wire speed is too slow, increase the dial setting on the welder.

FIGURE 7-32 Prior to contact, the wire has been charged and the gas flow started.

TABLE 7-4: ADJUSTMENTS IN WELDING PARAMETERS AND TECHNIQUES

Welding Variables to Change	Desired Changes							
	Penetration		Deposition Rate		Bead Size		Bead Width	
	Increase	Decrease	Increase	Decrease	Increase	Decrease	Increase	Decrease
Current and Wire Feed Speed	Increase	Decrease	Increase	Decrease	Increase	Decrease	No effect	No effect
Voltage	Little effect	Little effect	No effect	No effect	No effect	No effect	Increase	Decrease
Travel Speed	Little effect	Little effect	No effect	No effect	Decrease	Increase	Increase	Decrease
Stickout	Decrease	Increase	Increase	Decrease	Increase	Decrease	Decrease	Increase
Wire Diameter	Decrease	Increase	Decrease	Increase	No effect	No effect	No effect	No effect
Shield Gas Percent CO_2	Increase	Decrease	No effect	No effect	No effect	No effect	Increase	Decrease
Torch Angle	Backhand to 25°	Forehand	No effect	No effect	No effect	No effect	Backhand	Forehand

Normal buildup of oxide sparks in the gas nozzle area must be carefully removed before they fall inside and short out the nozzle. Balls caused by too slow a wire speed must also be removed before a short is formed.

As a summary, Table 7–4 outlines the various effects of several welding parameters and the changes necessary to alter a variety of weld characteristics.

MIG GUN NOZZLE ADJUSTMENT

The guns used on automotive MIG welders serve two main functions:

1. To provide proper gas protection
2. To feed the wire into the arc at the right speed

If the insulation is bypassed by accidentally grounding the body of the gun, the power intended for the wire is transferred to the gas nozzle, causing the nozzle to burn up. Welding on dirty or rusty material can cause heavy bombardment into the nozzle and will require immediate cleaning if proper welding performance is to be achieved. To successfully weld on a poor, rusty surface, slow the wire speed. Set the burnback control to its maximum and tap the trigger floating the ball on and off the material.

Of the four main components in an MIG welder, the nozzle area is the most crucial. The wire feed delivery is second. A clogged or damaged liner will cause erratic wire speed and produce molten balls that, in turn, will short out the gas nozzle.

To summarize the basic adjustment procedure of the gas nozzle, proceed as follows:

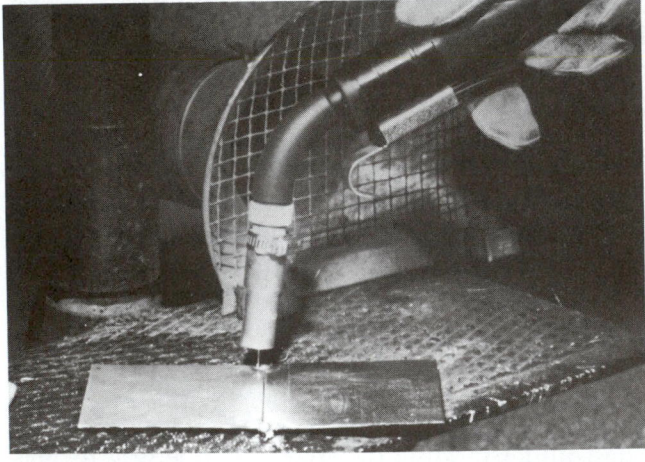

FIGURE 7–33 An activating gun switch starts the wire feed.

- *Arc generation.* Position the tip of the gun near the base metal. When the gun switch is activated (Figure 7–33), the wire is fed at the same time as the shield gas. Bring the end of the wire in contact with the base metal and create an arc. If the distance between the tip and the base metal is shortened a little, it will be easy to generate an arc (Figure 7–34). If the end of the wire forms a large ball, it will be difficult to generate an arc, so quickly cut off the end of the wire with a pair of wire cutters (Figure 7–35).
- *Spatter treatment.* Remove weld spatter promptly. If it adheres to the end of the nozzle, the shield gas will not flow properly and a poor weld will result. Antispatter compounds are available that

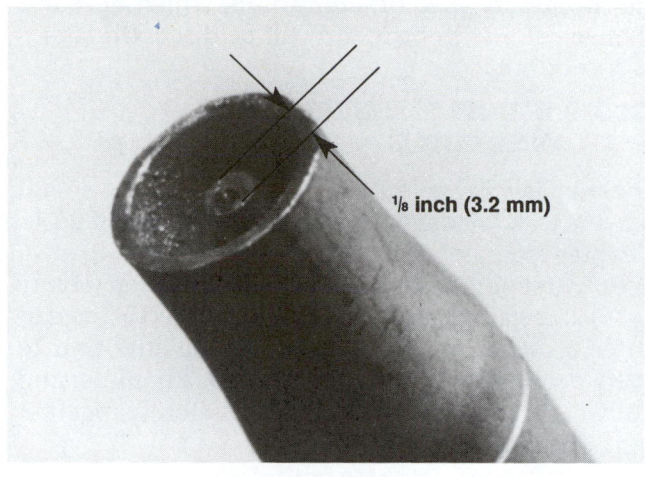

A

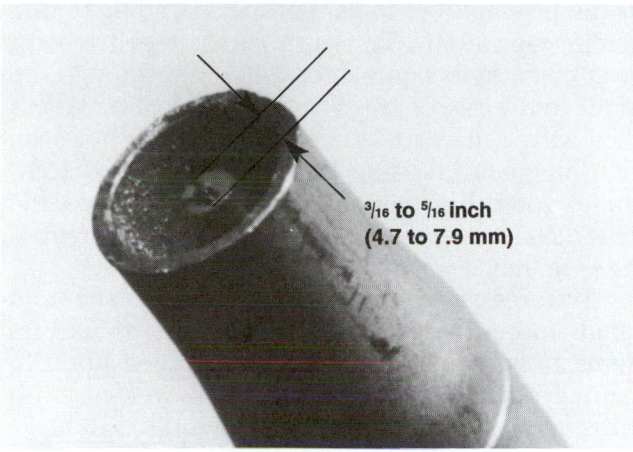

B

FIGURE 7-34 (A) The contact tip should be flush with the nozzle to ⅛ inch (3.2 mm) inside it. (B) The wire should extend from ³/₁₆ to ⁵/₁₆ inch (4.7 to 7.9 mm) beyond the contact tip.

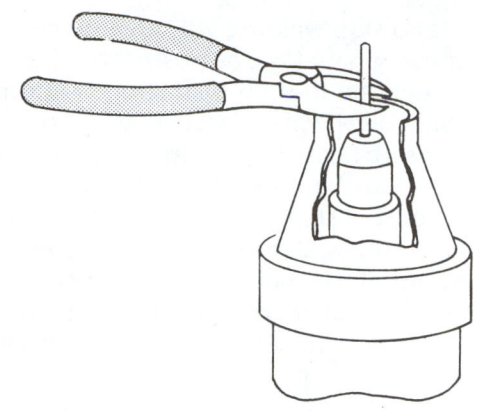

FIGURE 7-35 Cut off the end of the wire with side cutters if too long or damaged.

FIGURE 7-36 Spray antispatter compound on the tip to protect it from molten metal particles.

reduce the amount of spatter that adheres to the nozzle (Figure 7–36). Weld spatter on the tip will prevent the wire from moving freely. If the wire feed switch is turned on and the wire is not able to move freely through the tip, the wire will become twisted inside the welder. Use a suitable tool, such as a file, to remove spatter from the tip and then check to see that the wire comes out smoothly.

- *Contact tip conditions.* To ensure a stable arc, the tip should be replaced if it has become worn. For a good current flow and stable arc, keep the tip properly tightened (Figure 7–37).

HEAT BUILD-UP PREVENTION

Too much heat during welding distorts and weakens the metal. Always make sure you do not allow excess heat to transfer into any area of a panel. Use stitch or skip welding methods described earlier.

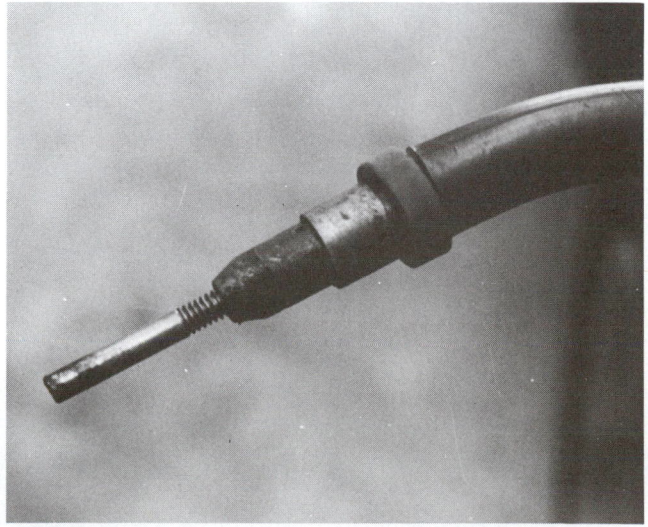

FIGURE 7-37 Check tip condition periodically and replace it when needed.

Stitch and skip welding will prevent costly and time consuming panel warpage. Another method of preventing heat build-up is heat sink compound.

Heat sink compound is a paste that can be applied to parts to absorb heat and prevent warpage. It comes in a can and can be applied and reused. The heat sink compound is sticky and can be placed on the panel next to the weld. Heat will flow into the compound and out of the metal to prevent heat damage.

Heat crayons or thermal paint can also be used to warn you when a panel is becoming too hot. They

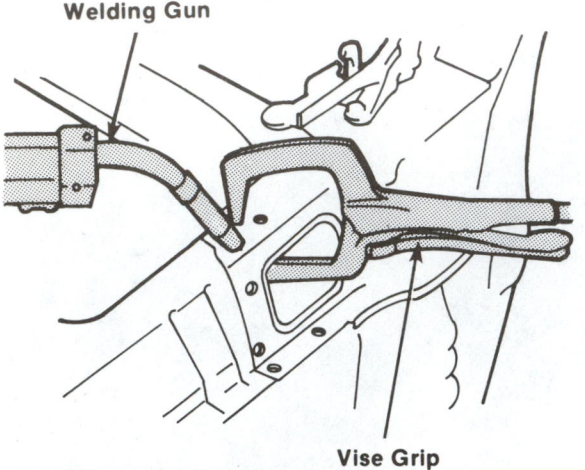

FIGURE 7-38 Note clamping for the welding of a front fender apron. *(Courtesy of Toyota Motor Corp.)*

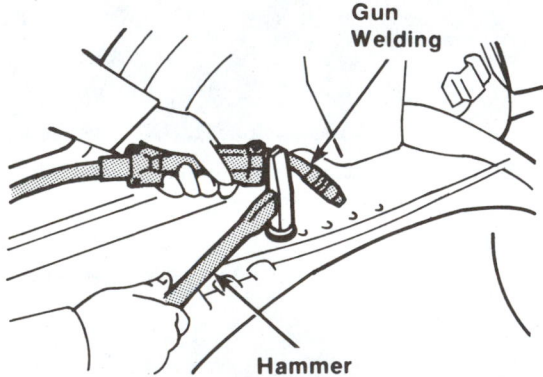

FIGURE 7-39 If needed, use hammer blows to move the panels together for a tighter joint before welding. *(Courtesy of Toyota Motor Corp.)*

are commonly used on aluminum, which does not change color with heat and will be discussed later.

CLAMPING TOOLS FOR WELDING

Locking jaw (vise) pliers, C-clamps, sheet metal screws, tack welds, or special clamps, described in Chapter 3, are necessary tools for good welding practices. Anybody can clamp panels together (Figure 7-38), but clamping panels together correctly to guarantee a sound weld will require close attention to every detail. As shown in Figure 7-39, a hammer and dolly can often be used to fit panels closely together in places that cannot be clamped.

Many times clamping both sides of a panel is not possible. In these cases, sheet metal screws or pop rivets can be employed to gain proper clamping during welding operations. To clamp panels together with sheet metal screws, punch or drill holes through the panel. In the case of plug welding, every other hole is filled with a sheet metal screw. The empty holes are then plug welded using proper plug welding techniques. After the original holes are plug welded, the sheet metal screws are removed and the holes left from the sheet metal screws are then plug welded.

Fixtures can also be used in some cases to hold panels to be welded in proper alignment. Fixtures alone, however, should not be depended upon to maintain tight clamping force at the welded joint. Some additional clamping will be required to make sure that panels are tightly clamped together and not just held in proper alignment.

WELDING POSITION

In collision repair, the *welding position* is usually dictated by the location of the weld in the structure of the car. Both the heat and wire speed parameters can be affected by the welding position (Figure 7-40).

Flat welding means the pieces are parallel with the bench or shop floor. Flat welding is generally easier and faster and allows for the best penetration (Figure 7-41). When welding a member that is off

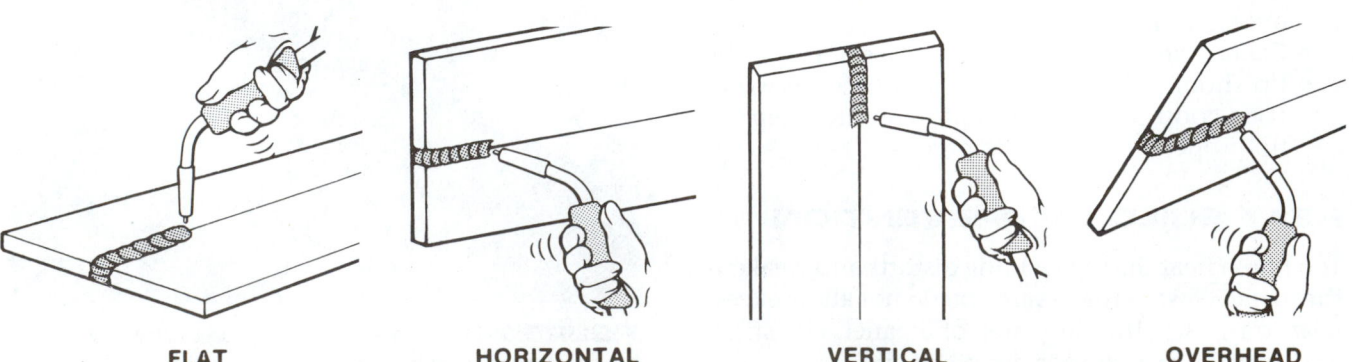

FLAT   HORIZONTAL   VERTICAL   OVERHEAD

FIGURE 7-40 Note the four basic welding positions.

FIGURE 7-41 The flat welding position is easiest to do because gravity pulls the puddle into the joint.

FIGURE 7-43 The vertical welding position makes the puddle flow along the joint.

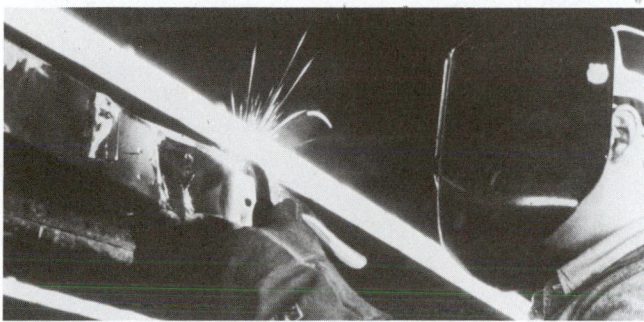

FIGURE 7-42 The horizontal welding position tends to make the puddle flow to one side of the joint.

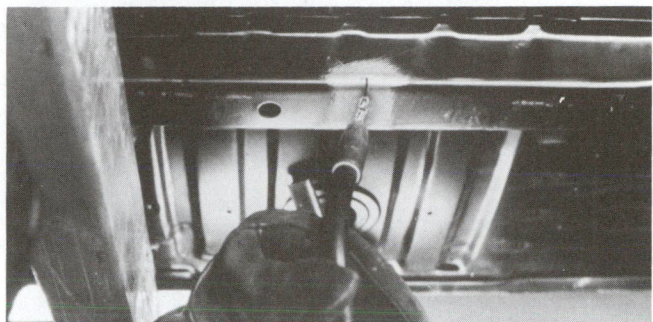

FIGURE 7-44 The overhead welding position tends to make the puddle fall back out of the joint.

the car, try to place it so that it can be welded in the flat position.

Horizontal welding has the pieces turned sideways. Gravity tends to pull the puddle into the bottom piece. When welding a horizontal joint (Figure 7–42), angle the gun upward to hold the weld puddle in place against the pull of gravity.

Vertical welding has the pieces turned upright. Gravity tends to pull the puddle down the joint. When welding a vertical joint (Figure 7–43), the best procedure is usually to start the arc at the top of the joint and pull downward with a steady drag.

Overhead welding has the workpieces turned upside down. Overhead welding is the most difficult. In this position (Figure 7–44), the danger of having too large a puddle is obvious; some of the molten metal can fall down into the nozzle, where it can create problems. So always do overhead welding at a lower voltage, while keeping the arc as short as possible and the weld puddle as small as possible. Press the nozzle against the work to ensure that wire is not moved away from the puddle. It is best to pull the gun along the joint with a steady drag.

7.4 BASIC WELDING TECHNIQUES

As shown in Figure 7–45, there are six basic welding techniques employed with MIG equipment.

- **Tack weld.** The tack weld is exactly that: a tack—a relatively small, temporary MIG spot weld that is used instead of a clamp or sheet metal screw to tack and hold the fit in place while proceeding to make a permanent weld. And like the clamp or sheet metal screw, a tack weld is always a temporary device. The distance between each tack weld is determined by the thickness of the panel. Ordinarily, a length of 15 to 30 times the thickness of the panel is appropriate (Figure 7–46). Temporary welds are very important in maintaining proper panel alignment and must be done accurately.

- **Continuous weld.** In a continuous weld, an uninterrupted seam or bead is laid down in a slow, steady, ongoing movement. Support the gun securely so it does not wobble. Use the forward method, moving the torch continuously at a constant speed, looking frequently at the welding bead. The gun should be inclined between 10 and 15 degrees to obtain the best bead shape, welding line, and shield effect (Figure 7–47). Maintain proper tip-to-base metal distance and correct gun angle. If the weld is not progressing well, the problem might be that the wire length is too long. If this is the case, penetration of the metal will not be adequate. For proper penetration and a better weld, bring the gun closer to the base metal. If the gun handling is smooth and even, the bead will be of consistent height and width, with a uniform, closely spaced ripple.

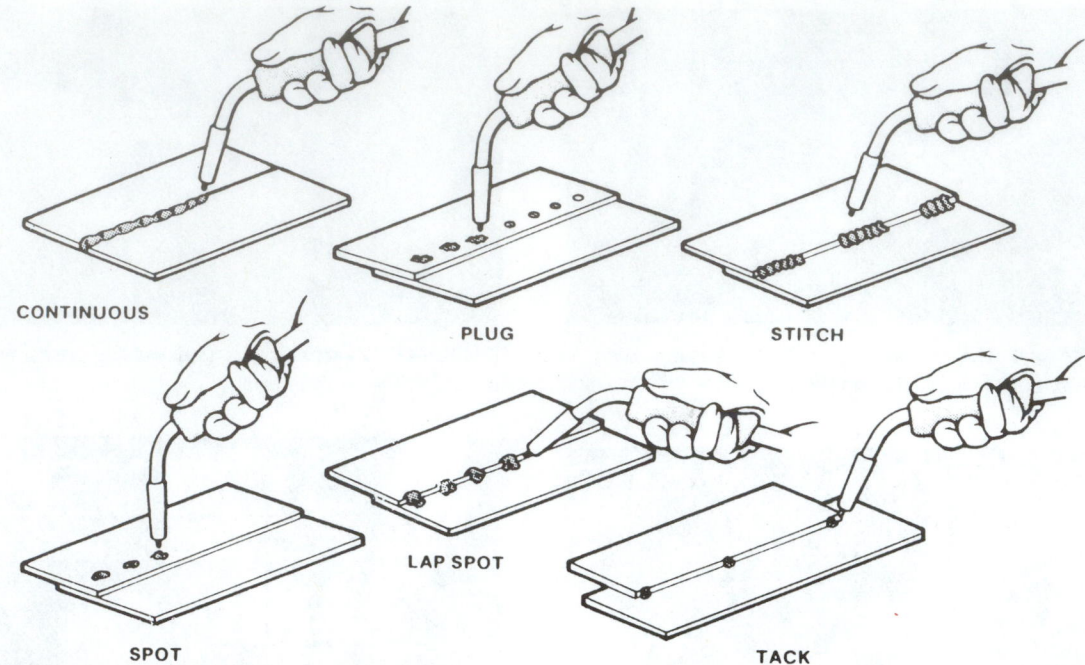

FIGURE 7-45 Memorize the basic welding techniques.

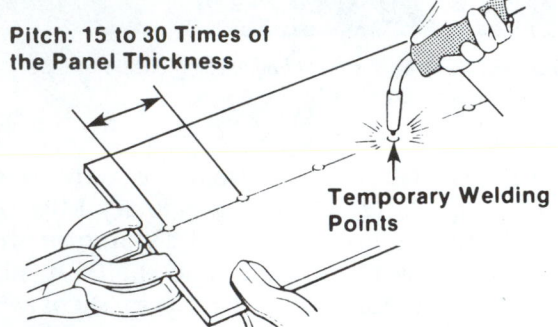

FIGURE 7-46 Temporary or tack welding is commonly used to hold parts in place before the final continuous weld. *(Courtesy of Toyota Motor Corp.)*

- **Plug weld.** A plug weld is made in a drilled or punched hole through the outside piece (or pieces). The arc is directed through the hole to penetrate the inside piece. The hole is then filled with molten metal.
- **Spot weld.** In an MIG spot weld, the arc is directed to penetrate both pieces of metal, while triggering a timed impulse of wire feed.
- **Lap spot weld.** In the MIG lap spot technique, the arc is directed to penetrate the bottom piece and the puddle is allowed to flow into the edge of the top piece.
- **Stitch weld.** A stitch weld is a series of connecting or overlapping MIG spot welds, creating a continuous seam.

BASIC WELDING METHODS

Each type of joint can be welded by several different techniques. The technique used depends mainly on the given welding situation:

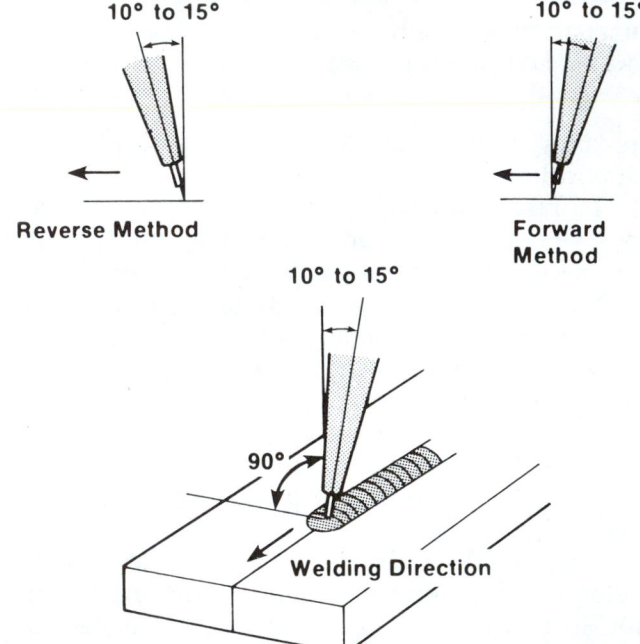

FIGURE 7-47 Continuous welding, if done properly, produces a joint stronger than the base metal. *(Courtesy of Toyota Motor Corp.)*

The thickness or thinness of the metal

The condition of the metal

The amount of gap, if any, between the pieces to be welded

The welding position

For example, the butt joint can be welded with the continuous technique or the stitch technique. And it can be tack welded at various points along the joint to hold the parts in place while completing the joint with a permanent continuous weld or a stitch weld. Lap and flange joints can be made using all six welding techniques.

Making Butt Welds

Butt welds are formed by fitting two edges of adjacent panels together and welding along the mating or butting edges of the panels.

In butt welding, especially on thin panels, it is wise not to weld more than $^3/_4$ inch (19 mm) at one time. Closely watch the melting of the panel, welding wire, and the continuity of the bead. Be sure the end of the wire does not wander away from the butted portion of the panels. If the weld is to be long, it is a good idea to tack weld the panels in several locations (stitch weld) to prevent panel warpage (Figure 7–48). Figure 7–49 shows how to generate an arc a short distance ahead of the point where the weld ends and then immediately move the gun to the point where the bead should begin. The bead width and height should be uniform at this time.

Weld in a sequence that allows an area to cool before the next area is welded (Figure 7–50). To fill the spaces between intermittently placed beads, first grind the beads along the surface of the panel. Then fill the space with metal (Figure 7–51). If weld metal is placed without grinding the surface of the beads, blowholes can be produced.

When welding thin panels that are $^1/_{32}$ inch (0.79 mm) or less, an intermittent or stitch welding technique is a *must* to prevent burn-through. The combination of the proper gun angle and correct cycling techniques will enable you to achieve a satisfactory weld bead (Figure 7–52). The reverse welding method (Figure 7–29) can be used for moving the gun because it is easier to aim at the bead.

Figure 7–53 shows a typical butt welding procedure for installing a replacement panel. If the desired results are not obtained, the cause can be that the distance between the tip of the gun and the base metal might be too great. Weld penetration decreases

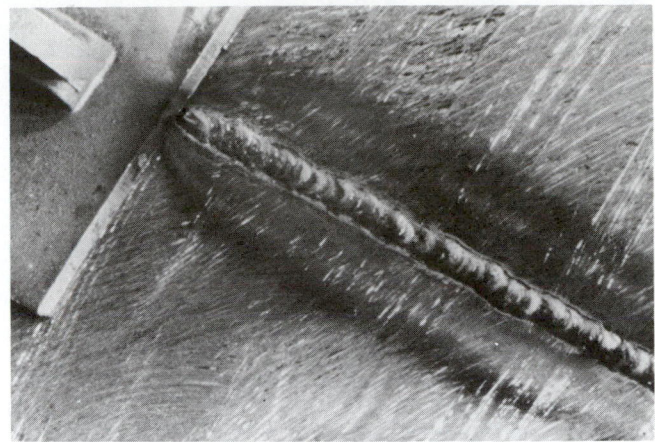

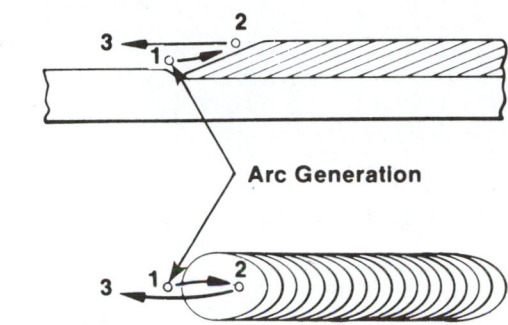

FIGURE 7-49 If gun handling is smooth and even, the bead will be of a consistent height and width, with a uniform, closely spaced ripple.

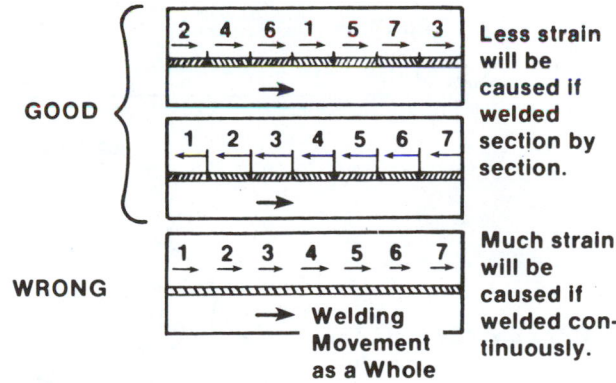

FIGURE 7-50 Compare the right and wrong welding sequence. *[Courtesy of Nissan Motor Corp.]*

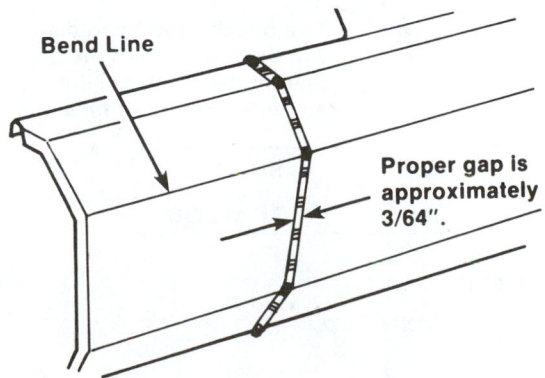

FIGURE 7-48 Tack weld of panels will also help prevent warpage. *[Courtesy of Nissan Motor Corp.]*

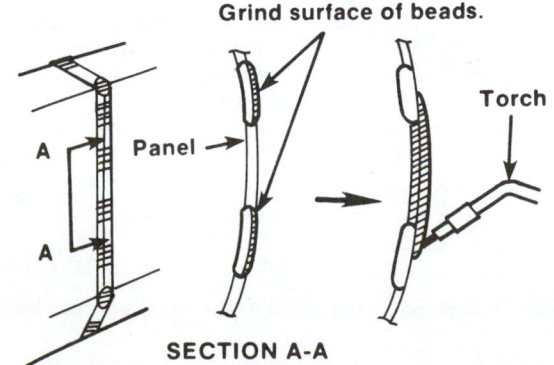

FIGURE 7-51 Filling the space between intermittently spaced beads will finish this weld.

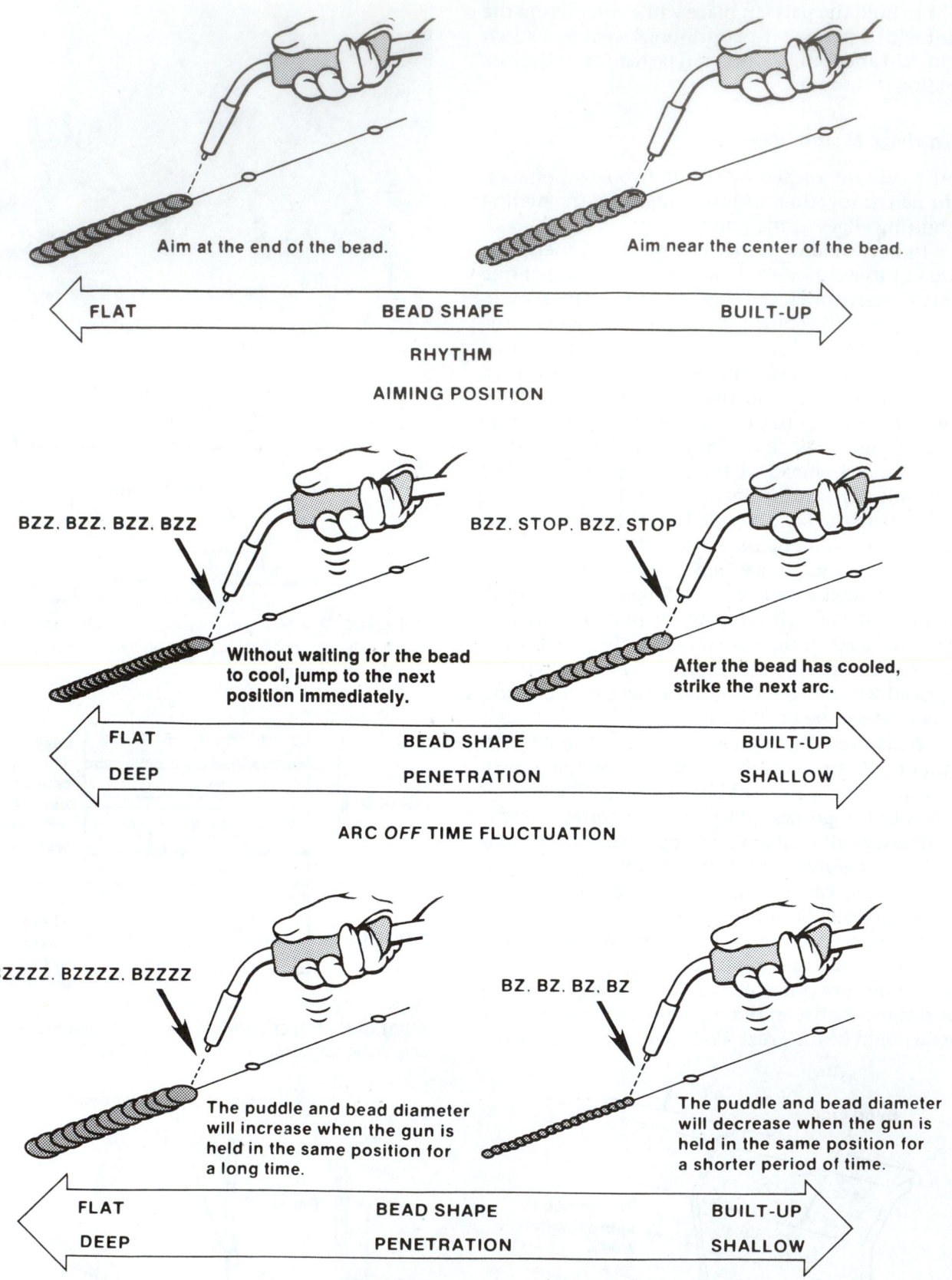

Aim at the end of the bead.

Aim near the center of the bead.

FLAT   BEAD SHAPE   BUILT-UP

RHYTHM

AIMING POSITION

BZZ. BZZ. BZZ. BZZ

BZZ. STOP. BZZ. STOP

Without waiting for the bead to cool, jump to the next position immediately.

After the bead has cooled, strike the next arc.

FLAT   BEAD SHAPE   BUILT-UP
DEEP   PENETRATION   SHALLOW

ARC *OFF* TIME FLUCTUATION

BZZZZ. BZZZZ. BZZZZ

BZ. BZ. BZ. BZ

The puddle and bead diameter will increase when the gun is held in the same position for a long time.

The puddle and bead diameter will decrease when the gun is held in the same position for a shorter period of time.

FLAT   BEAD SHAPE   BUILT-UP
DEEP   PENETRATION   SHALLOW

ARC *ON* TIME FLUCTUATION

FIGURE 7-52 Study the steps in achieving a proper bead. *(Courtesy of Toyota Motor Corp.)*

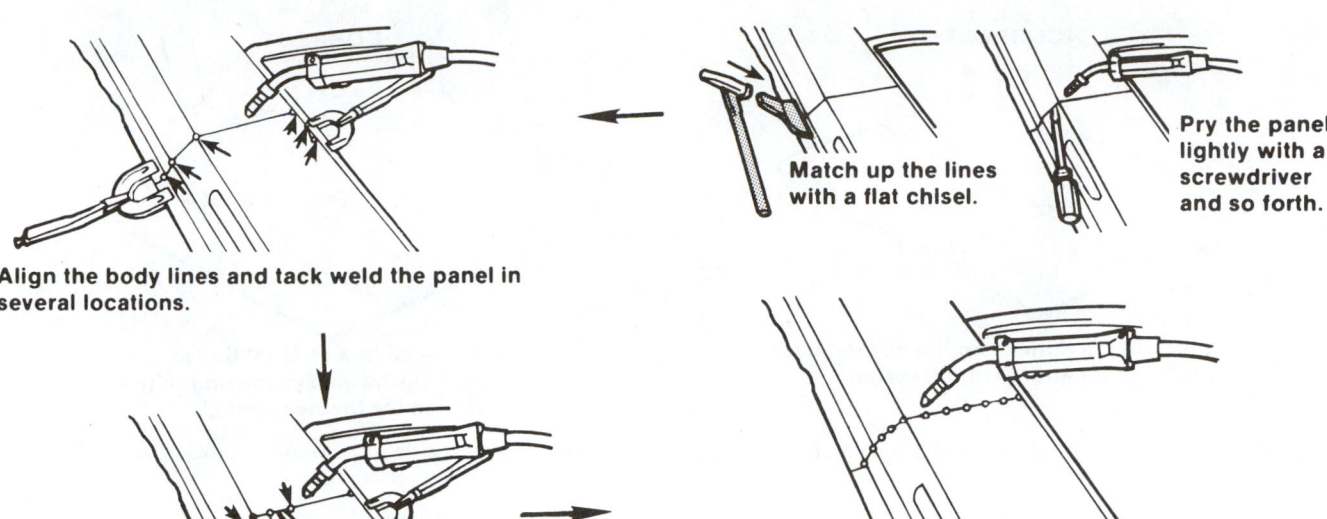

Align the body lines and tack weld the panel in several locations.

Match up the lines with a flat chisel.

Pry the panel lightly with a screwdriver and so forth.

Match up the level differences in the panel surfaces and tack weld the panel in place.

Do NOT weld continuously from one point to another. Use an interrupted (stitch type) weld.

FIGURE 7-53 Note the procedure for butt welding sectional areas. *(Courtesy of Toyota Motor Corp.)*

INCORRECT

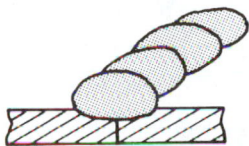

Insufficient penetration. Weld strength is poor and the panel could separate when the panel is finished with a grinder.

INCORRECT

There is good penetration but finish grinding will be both difficult and time consuming.

CORRECT

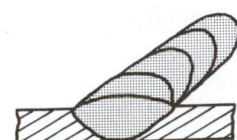

Good Penetration and Easy to Grind

FIGURE 7-54 Compare bead cutaway shapes.

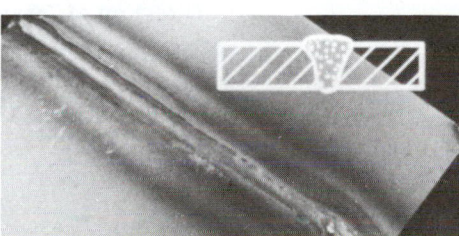

CORRECT

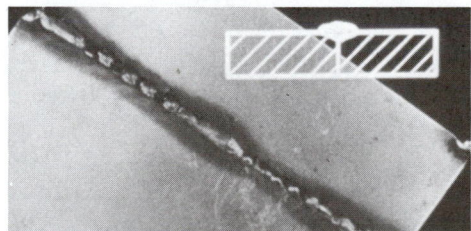

TOO FAST

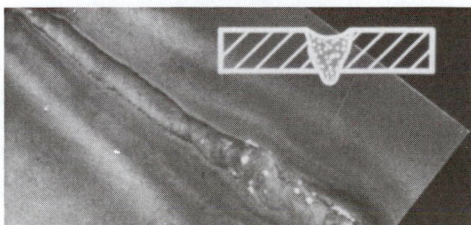

TOO SLOW

FIGURE 7-55 Gun movement speed affects the bead shape.

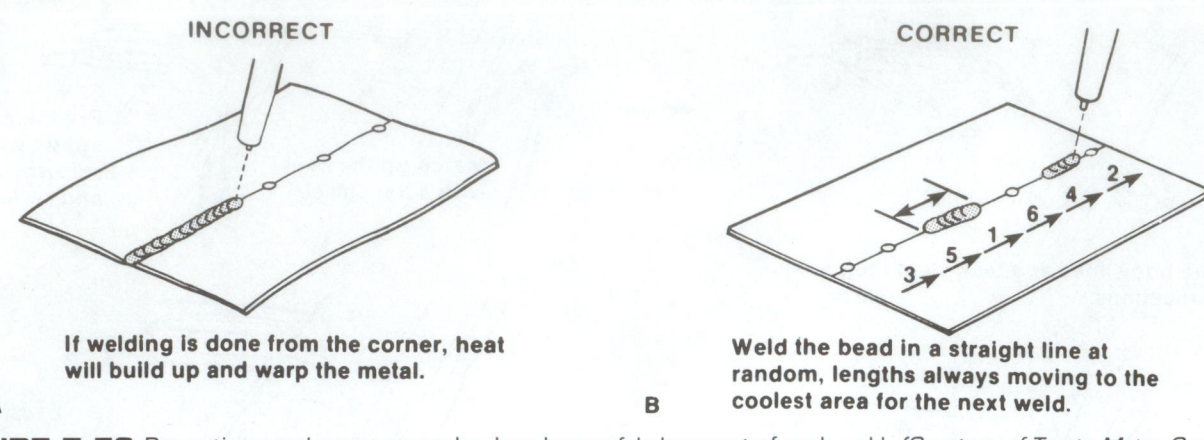

FIGURE 7-56 Preventing panel warpage can be done by careful placement of each weld. *(Courtesy of Toyota Motor Corp.)*

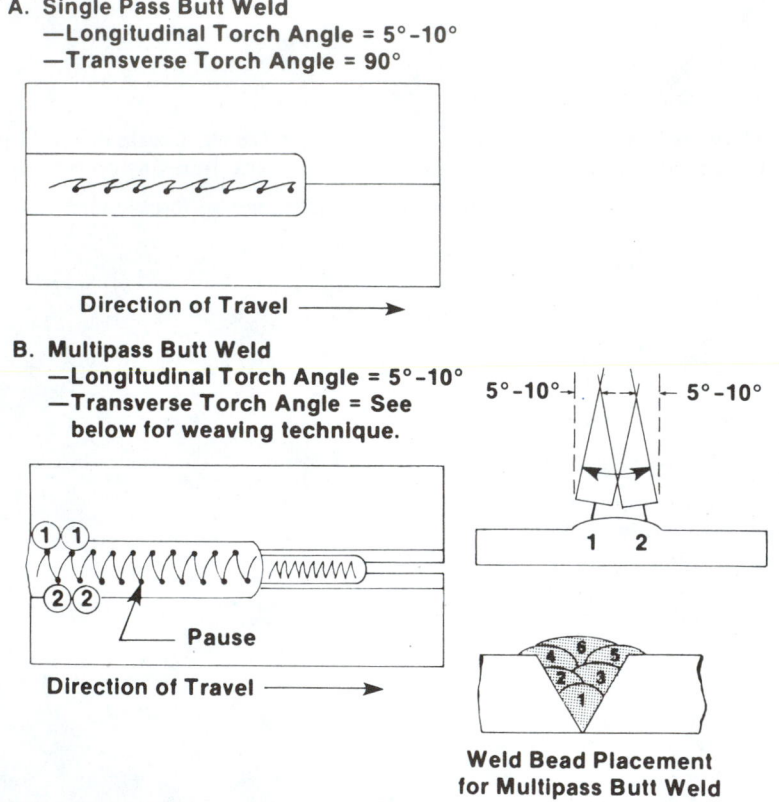

FIGURE 7-57 Study torch movement for flat position butt welding.

as the distance between the tip and the base metal increases. Try holding the tip of the gun at several distances away from the base metal until the proper distance gives the desired results (Figure 7–54).

Moving the gun too fast or too slowly (Figure 7–55) will give poor welding results (even if speed of wire feed is constant). A gun speed that is too slow will cause melt-through. Conversely, a gun speed that is too fast will cause shallow penetration and poor weld strength.

Even if a proper bead is formed during butt welding, panel warpage can result if the weld is started at or near the edge of the metal (Figure 7–56A). Therefore, to prevent warpage, disperse the heat into the

base metal by starting the weld in the center of the panel. Frequently change the location of the weld area (Figure 7–56B). The thinner the panel thickness, the shorter the bead length.

When welding a butt joint, be sure the weld penetrates all the way through to the backside of the joint. Where the metal thickness at a butt joint is $1/16$ inch (1.59 mm) or more, a gap should be left to assure full penetration. If it is not practical to leave a gap, grind a V-groove in the joint (Figure 7–57) so the weld can penetrate to the backside.

MIG *butt welds* are often used to make two joints when sectioning frame rails, rocker panels, and door pillars. For butt welds keep a gap between the two pieces the thickness of one piece. This helps weld penetration and prevents expansion and contraction problems. Also, hold the gun at 90 degrees to the joint.

An **insert**, or *backing strip,* made of the same metal as the base metal can be placed behind the weld. The backing helps proper fit, helps align the joint, and gives the joint the same strength and rigidity as the original structure.

Making Lap and Flange Welds

Lap and flange welds (Figure 7–58) are made with identical techniques. They are formed by welding or fusing two surfaces to be joined at the edge of the top one of two overlapping surfaces. This is similar to butt welds except only the top surface has an edge. Lap and flange welds should be made only in repairs where they replace original factory lap or flange welds, or where outer panels and not structural panels are involved. These welds should not be used to join more than two thicknesses of material together.

The same technique used for temperature control in butt welding should be followed for lap and flange welding. Welds should never be made continuously but should be sequenced to allow for natural cooling and to prevent temperature buildup in the welding area.

Making Plug Welds

The plug weld is the body shop alternative to the OEM resistance spot welds made at the factory because it can be used anywhere in the body structure that the factory used a resistance spot weld. Its use is not restricted. It has ample strength for welding load bearing structural members. It can also be used on cosmetic body skins and other thin-gauge sheet metal.

Plug welding (Figure 7–59) is a form of spot welding—spot welding through a hole. A plug weld is formed by drilling or punching (Figure 7–60) a hole in the outer panel being joined (Figure 7–61). The materials should be tightly clamped together. Holding the torch at right angles to the surface (Figure 7–62), aim the electrode wire in the hole, and trigger the arc while moving the gun in a circular motion around the hole (Figure 7–63). The puddle fills the hole and solidifies.

When plug welding, try to duplicate the number and the nugget size of the original factory spot welds. The hole that is punched or drilled should not be larger in diameter than the factory weld nugget. Drill or punch $^3/_{16}$- to $^3/_8$-inch (5–9mm) holes in top piece or pieces.

Start around the edge of the hole; then fill in the hole. A $^5/_{16}$-inch (8 mm) hole works well for most

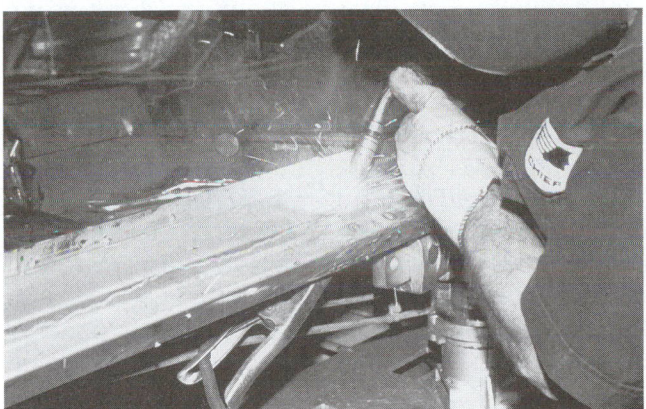

FIGURE 7-58 Welding takes great skill, knowledge, and patience. *(Courtesy of Tech-Cor)*

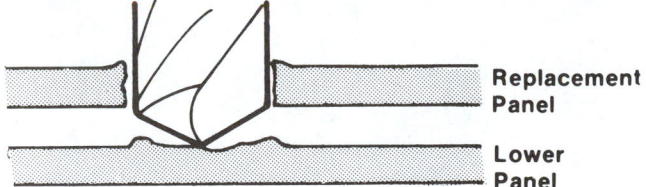

FIGURE 7-60 Plug welds are formed by drilling or punching a hole in the outer panel that is being joined.

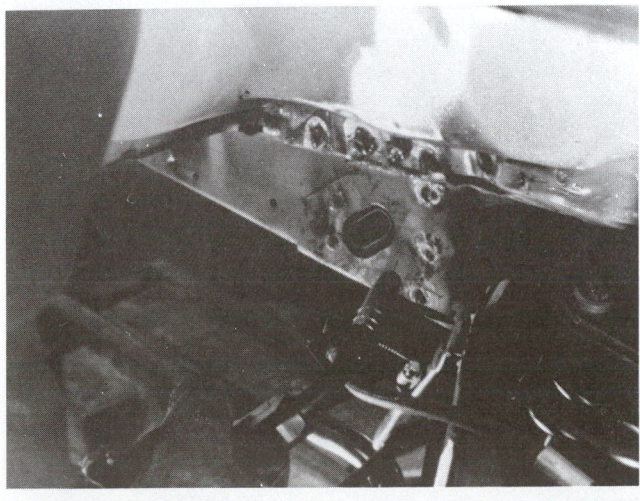

FIGURE 7-59 Plug welding is similar to spot welding.

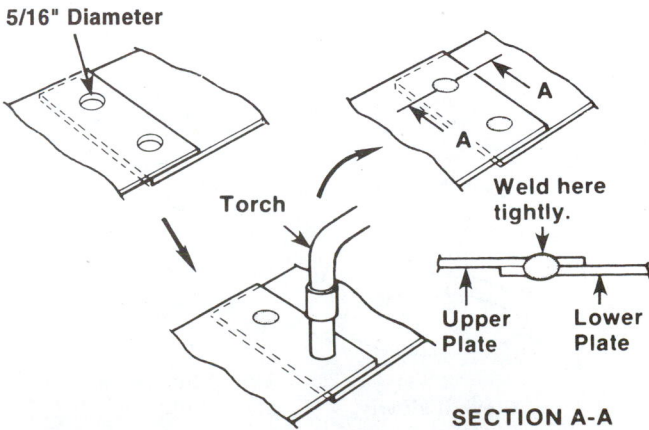

FIGURE 7-61 Study the steps in making a plug weld. *(Courtesy of Nissan Motor Corp.)*

collision repair. A ³/₁₆-inch (5 mm) hole is better with very thin metals (24 gauge and lighter), and a ³/₈-inch (9 mm) hole is better with heavier metal (14 gauge and heavier).

When plug welds are used to join three or more panels together, holes are punched or drilled in every piece except the bottom piece. The holes are made progressively smaller from the top down. This is done to get better fusion of each layer to the adjacent one.

Typical hole size combinations are as follows. With three layers of metal, use ⁵/₁₆-inch (8 mm) and ³/₈-inch (9 mm) holes. With four layers, use ¹/₄-inch (6 mm), ⁵/₁₆-inch (8 mm) and ³/₈-inch (9 mm) diameter holes.

Where MIG plug welds are used to replace factory spot welds:

1. Follow manufacturer's recommendations for number, size, and location of plug welds.
2. If this information is not available, duplicate the number, size, and location of original factory welds.

FIGURE 7-62 The technician is making a plug weld while installing the panel.

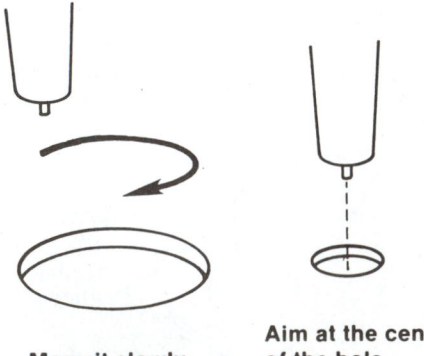

Move it slowly.

Aim at the center of the hole.

FIGURE 7-63 The gun movement is circular to fill the plug weld.

SHOP TALK Intermittent welding leads to the generation of oxide film on the surface and this causes blowholes. If this occurs, remove the oxide film with a wire brush (Figure 7–64).

Proper welding wire length is an important factor in obtaining a good plug weld. If the length of the wire protruding out of the end of the gun is too long, the wire will not melt properly, causing inferior weld penetration. The weld will improve if the gun is held closer to the base metal. Be sure the weld penetrates into the lower panel. Round dome-shaped protrusions on the underside of the metal are good indicators of proper weld penetration.

The area welded should be allowed to cool naturally before any adjacent welds are made. Areas around the weld should not be force cooled using water or air. It is important that they be allowed to cool naturally. Slow, natural cooling without using water or air will minimize any panel distortion and keep the strength designed into the panels.

Plug welds can also be used to join more than two panels together. A hole is punched in every panel except the lower panel (Figure 7–65). The diameter of the plug weld hole in each panel being joined should be smaller than the diameter of the plug weld hole on top. Likewise, if panels of different thicknesses are being joined, a larger hole is punched in the thinner panel to assure that the thicker panel is melted into the weld first. When welding panels of

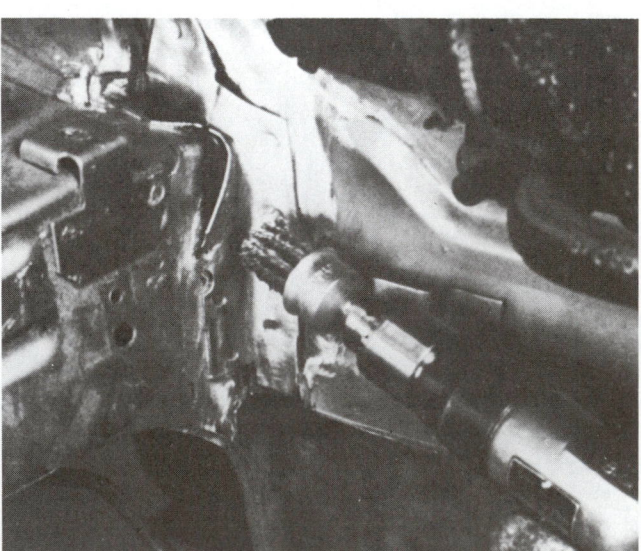

FIGURE 7-64 Remove oxide film with a wire brush or grinder.

SHOP TALK

Considerations important to high-quality plug welds are proper weld time, current flow, temperature, adjustments, clamping pressure, and filler rod type.

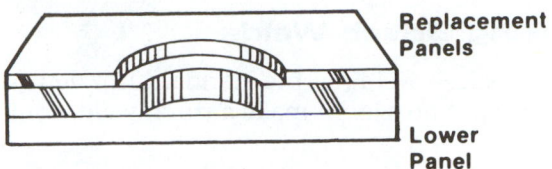

Replacement Panels

Lower Panel

FIGURE 7-65 Use a smaller size hole on a lower panel when plug welding.

FIGURE 7-66 Spot welding can be accomplished with most MIG machines.

different thicknesses using the plug weld method, the thinner panel should be on top.

A plug weld using an MIG welder can be accomplished in a minimum amount of time, creating less temperature buildup in adjacent panels. While adjacent welds should not be made immediately, the area being welded will cool in a very short period of time.

Making Spot Welds

Most MIG machines that are designed for collision repair work have built-in timers that shut off the wire feed and welding arc after the time required to weld one spot (Figures 7–66 and 7–67). Some MIG equipment also has a burnback time setting. It can be adjusted to prevent the wire from sticking in the puddle. The setting of these timers depends on the thickness of the workpiece. This information can usually be found in the machine's owner's manual.

For MIG spot welding, a special welding nozzle (Figure 7–68) must replace the standard nozzle. Once in place and with the spot timing, welding heat, and backburn time set for the given situation, the spot nozzle is held against the weld site and the gun triggered. For a very brief period of time, the timed pulses of wire feed and welding current are activated, during which the arc melts through the outer layer and penetrates the inner layer (Figure 7–69). After this, the automatic shutoff goes into action and no matter how long the trigger is squeezed, nothing will happen. The trigger must be released and then squeezed again to obtain the next spot pulse.

Because of varying conditions, the quality of an MIG spot weld is difficult to determine. On load bearing members, therefore, MIG plug welding is the preferred method.

The MIG lap spot technique is a popular one for the quick, effective welding of lap joints and flanges on thin-gauge nonstructural sheets and skins (Figure 7–70). Here again the spot timer is set, but this time the spot nozzle is positioned over the edge of the outer sheet at an angle slightly off 90 degrees. This will allow contact with both pieces of metal at the same time. The arc melts into the edge and penetrates the lower sheet.

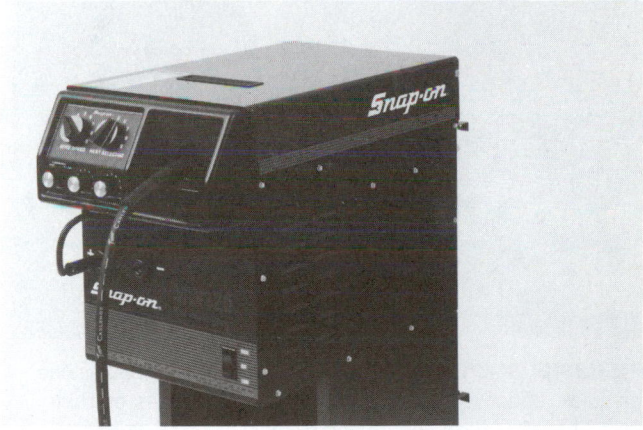

FIGURE 7-67 Note the controls on a typical body shop MIG machine. *(Courtesy of Snap-On-Tools Corporation)*

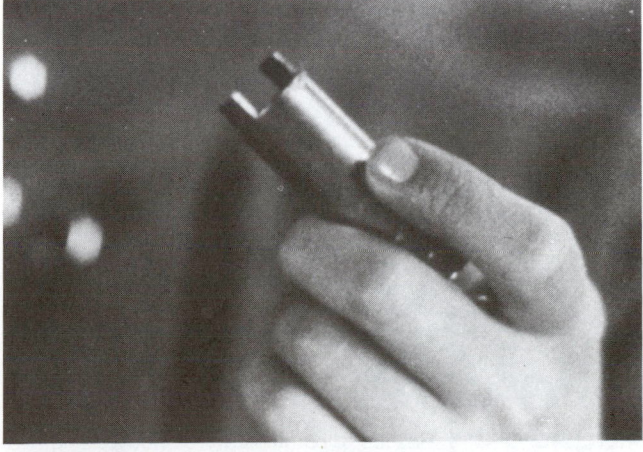

FIGURE 7-68 This is a special spot welding nozzle.

Making Stitch Welds

In MIG stitch welding, the standard nozzle is used, not the spot nozzle. To make a stitch weld, combine spot welding with the continuous welding technique (Figure 7–71). To do this, set the automatic timer—either a shutoff or pulsed interval timer—depending on the MIG machine (Figure 7–72). The spot weld pulses and shutoffs occur with automatic regularity: weld-stop-weld-stop-weld-stop as long as the trigger is held in.

The arc-off period allows the last spot to cool slightly and start to solidify before the next spot is deposited. This intermittent technique means less distortion and less melt-through or burn-through. These characteristics make the stitch weld preferable to the continuous weld for working thinner-gauge cosmetic panels.

The intermittent cooling and solidifying of the stitch weld also makes it preferable to continuous welding on vertical joints where distortion is a problem (Figure 7–73). The welder does not have to contend with a continuous weld puddle that gravity is trying to pull down the joint ahead of the arc. Stitch welding is also preferable in the overhead position

FIGURE 7-69 During one brief, timed pulse of wire feed and welding current, the arc melts through the outer layer and penetrates the inner layer.

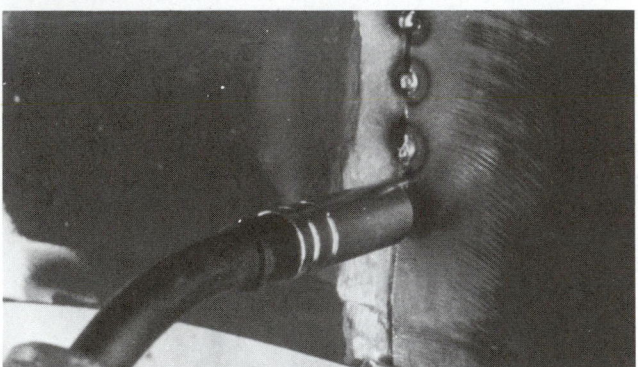

FIGURE 7-70 The lap spot technique is a popular one for quick, effective welding of lap joints and flanges on thin-gauge nonstructural sheets and skins.

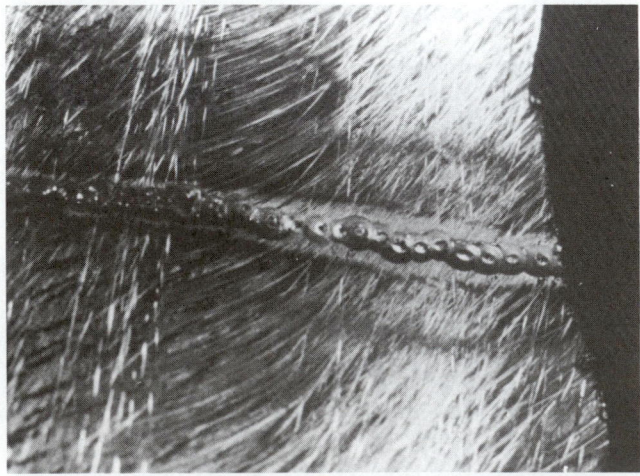

FIGURE 7-71 Results of combining the spot welding process with continuous welding gun technique.

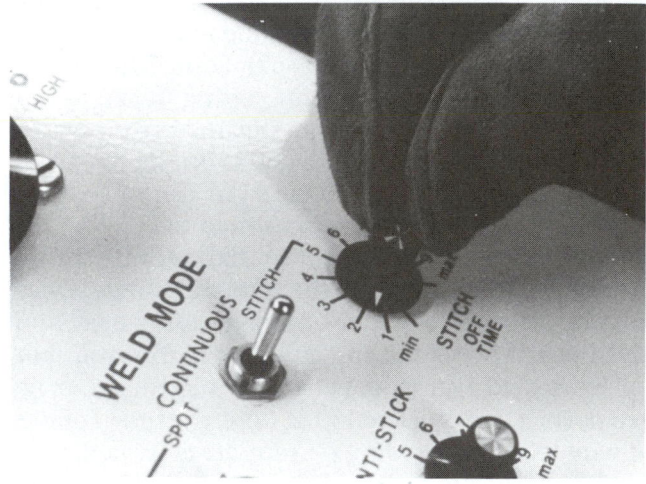

FIGURE 7-72 The technician is setting the spot-off, or interval, timer.

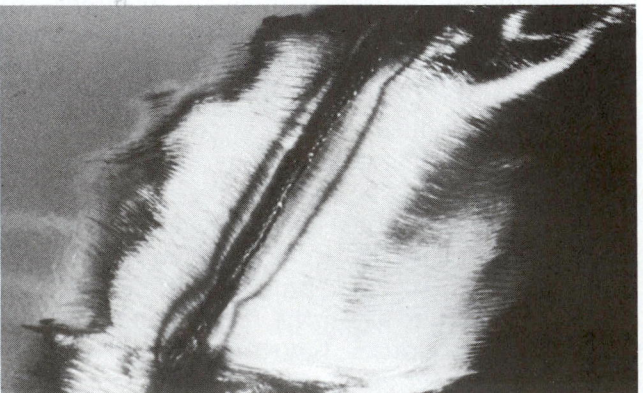

FIGURE 7-73 Intermittent cooling and solidifying of a stitch weld makes it preferable to continuous welding on vertical joints where distortion is a problem.

FIGURE 7-74 Stitch welding is also common in the overhead position.

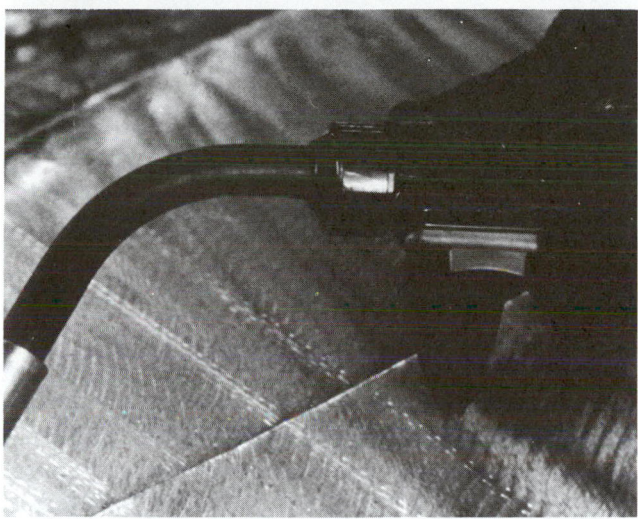

FIGURE 7-75 A hand triggering welding gun will help control the heat on difficult welds.

(Figure 7–74). There is virtually no weld puddle for gravity to pull.

If the MIG machine does not have the automatic stitch modes, the spot and stitch welds can be made manually. The operator merely has to be capable of triggering the gun on-off, on-off, on-off—the same way the automatic system does (Figure 7–75).

7.5 MIG WELDING GALVANIZED METALS AND ALUMINUM

When MIG welding galvanized or zinc-metallized steels, also called zinc-coated steels, do *not* remove the zinc. If zinc is ground away, the thickness of the metal is reduced and so is its strength. And when a zinc-free area around the weld site is created, it is an inviting target for corrosion.

FIGURE 7-76 When welding galvanized or zinc-coated steels, use a slower gun travel speed.

With galvanized or zinc-coated steels (Figure 7–76), use a slower gun travel speed than when welding uncoated steels. This is because the zinc vapors tend to rise into the arc zone and interfere with arc stability. A slower travel speed allows the zinc to burn off at the front of the weld pool. How much to reduce the gun travel speed will depend on the thickness of the zinc coating, the joint type, and the welding position. Experience is the best teacher with these variable conditions.

Since there is slightly less weld penetration with galvanized or zinc-coated steels than with uncoated steels, a slightly under gap in square edge butt welds is needed. To prevent burn-through or excessive penetration of the wider gap, the welding gun should be handled with a side-to-side weaving motion.

It must be remembered that there is more spatter when welding galvanized or zinc-coated steels than with uncoated steels. Therefore, it is a good idea to apply antispatter compound inside the gun nozzle and to clean the nozzle frequently.

Always wear a welding respirator when fusing galvanized metals. The fumes can cause serious lung or respirator illness.

WELDING ALUMINUM

Several vehicles now have body, frame, and chassis parts made of aluminum. Whole bodies made of aluminum are now available. As a result, the need for information on welding aluminum is growing.

Aluminum is light and relatively strong. It is naturally corrosion resistant. Aluminum forms its own corrosion barrier of aluminum oxide when exposed to air. A disadvantage of aluminum is its high cost.

There are some major differences to keep in mind when working with aluminum as opposed to steel, particularly when it comes to welding. Pure aluminum is lightweight and useful more for its ability to be formed than for its strength. When used on vehicles, it is alloyed with other elements and heat-treated for additional strength.

Concerning welding, pound for pound, aluminum is the best conductor of electricity. It conducts heat three times faster than steel. Aluminum becomes stronger when welded and is not brittle in extreme cold. It is also easily recyclable. Aluminum conducts heat faster than steel, and also spreads the heat faster. Therefore, it requires special attention when welding.

Aluminum looks similar to magnesium which, if welded, could start a flash fire. To make sure the part is aluminum, brush the part with a stainless steel brush. Aluminum turns shiny; magnesium turns dull gray.

WARNING

If the part is found to be magnesium, do NOT weld it. It can start to burn with tremendous heat.

When welding aluminum, be sure to protect wire harnesses and electronics from potential damage caused by spreading heat. Aluminum takes more voltage and amperage for the same thickness of material.

Use the following guidelines when MIG welding aluminum:

1. Match the wire to the aluminum alloy.
2. Set the wire speed faster than with steel.
3. Hold the gun closer to vertical. Tilt it only about 5 to 10 degrees from the vertical in the direction of the weld.
4. Use only the forward welding method. Always push; never pull. When making a vertical weld, start at the bottom and work up.
5. Set the tension of the wire drive roller lower to prevent twisting. But do not lower the tension too much or the wire speed will not be constant.

6. Use about 50 percent more shielding gas.
7. Because there tends to be more spatter, use an antispatter compound to control buildup at the end of the nozzle and contact tip. Only apply the compound to the nozzle and clean off any excess to keep it out of the weld puddle. Antispatter compound must be kept off all welding surfaces because it will contaminate the weld.
8. Shop squeeze-type resistance spot welders do not have enough amperage for aluminum. Do not use resistance spot welders.
9. Always use skip and stitch welding techniques to prevent heat warping. Set wire speed slightly faster. Hold gun closer to vertical compared to steel.
10. Use 100 percent argon for the shield gas.

An MIG gun for welding aluminum can be either the standard equipment or self-contained with a motor in the handle and aluminum wire spool mounted on top. The nozzle is straight, not tapered in at the end.

Aluminum electrode wire is classified by series, according to the metal or metals the aluminum is alloyed with, and whether the aluminum is heat-treated or not. The series are set up by the Aluminum Association, not the AWS. The number does not indicate the strength of the electrode.

Special cleaning instructions are needed for aluminum. Remove all aluminum oxide with a stainless steel brush before welding. The metal might look clean, but aluminum oxide always needs to be brushed off. Clean the metal until shiny.

Never use a brush and sanding discs on aluminum after it has been used on steel. If already used on steel, iron powder will remain on the surface of the aluminum and contaminate the weld.

The typical procedure for MIG welding aluminum is as follows:

1. Clean the weld area completely, both front and back. Use wax and grease remover and a clean rag.
2. If the pieces to be welded are coated with a paint film, sand a strip about $^3/_4$-inch (19 mm) wide to the bare metal, using a disc sander and a No. 80 disc. Do not press too hard or the sander will heat up and peel off aluminum particles, clogging the paper.
3. Clean the metal until shiny with a stainless steel wire brush.
4. Load 0.030 aluminum wire into the welder. Trigger it to extend about an inch (25 mm) beyond the nozzle.
5. Set the voltage and wire speed according to the instructions supplied with the welding machine. Remember that the wire speed must be faster

than for steel. Make a practice weld on scrap pieces.

6. Position the two pieces together and lay a bead along the entire joint. The distance between the contact tip and the weld should be ⁵/₁₆ to ⁹/₁₆ inch (8 to 14 mm).

7. If the arc is too large, turn down the voltage and increase the wire speed. The bead should be uniform on top, with even penetration on the backside.

The high heat conduction of aluminum means that the technician must protect against warpage. There are two methods for doing this: stitch welding and center out welding, which were explained earlier.

Parts made of aluminum are usually 1¹/₂ to 2 times as thick as steel parts. When damaged, aluminum feels harder or stiffer to the touch because of work hardening.

7.6 TESTING THE MIG WELD

Repair welds should be tested from time to time on every job. This can be done simply with test panels. Before welding on a vehicle, make some welds on pieces of scrap sheet metal like the panels that are going to be installed on the vehicle. If the proper settings on the MIG welder are obtained on the test pieces, the quality of the weld on the car can be assured. To check the quality of the weld, try to break it apart as illustrated in Figure 7–77.

7.7 MIG WELD DEFECTS

Defects in MIG welds and their causes are summarized in Table 7–5. Proper welding techniques assure good welding results. If welding defects should occur, think of ways to change your procedures to correct the defect.

When making any MIG repairs, the materials and panels must be similar enough to allow mixing when they are welded together. The melting and flowing of metals can be accomplished by many methods, depending upon the materials being joined. The combinations of cleanliness of the welded area, the mixing of proper metals, and the right heat application will result in a good MIG weld.

A *welding problem* causes a weak or cosmetically poor joint that reduces quality. Some common weld problems include:

1. *Weld Porosity* (holes in the weld)
2. *Weld Cracks* (cracks on the top or inside the weld bead)
3. *Weld Distortion* (uneven weld bead)

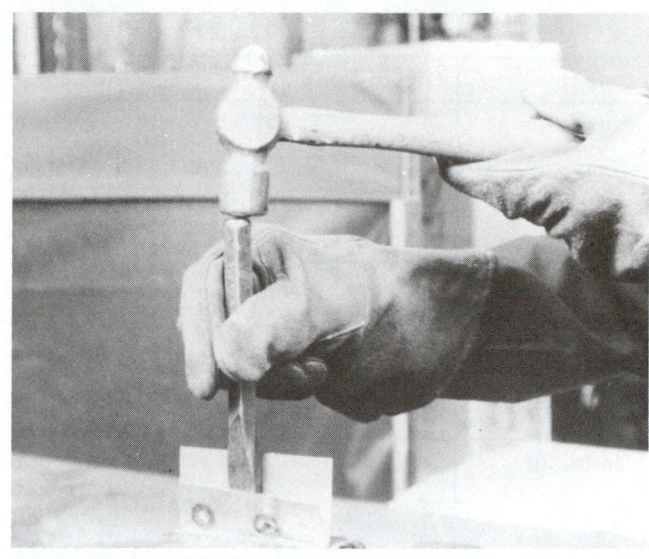

FIGURE 7-77 The technician is checking the quality of an MIG weld.

4. *Weld Spatter* (drops of electrode on and around weld bead)
5. *Weld Undercut* (groove melted along either side of weld and left unfilled)
6. *Weld Overlap* (excess weld metal mounted on top and either side of weld bead)
7. *Too little penetration* (weld bead sitting on top of base metal)
8. *Too much penetration* (burn-through beneath lower base metal)

7.8 FLUX-CORED ARC WELDING

Flux-cored arc welding (FCAW) is an electric arc welding process that uses a tubular wire with flux inside. With the development of 0.030 self-shielded flux-cored wire, the flux-cored welding process has proven to be valuable for work on high-strength steel (coated or uncoated). The FCAW process uses the same type of constant potential power source as MIG. It also uses the electrode feed system, contact tube, electrode conduit, welding gun, and many other pieces of equipment that are used in MIG. Nevertheless, the process itself differs somewhat from MIG.

There is no external shielding gas in FCAW. As the flux within the wire melts in the heat of the arc, the created gases shield the weld puddle, stabilize the arc, help control penetration, and reduce porosity. The melted flux also mixes with the impurities on the metal surface and brings them to the top of the weld where they solidify as slag. The slag can then be chipped or brushed away.

TABLE 7-5: WELDING PRECAUTIONS

Defect	Defect Condition	Remarks	Main Causes
Pores/Pits	Pit Pore	There is a hole made when gas is trapped in the weld metal.	1. There is rust or dirt on the base metal. 2. There is rust or moisture adhering to the wire. 3. Improper shielding action (the nozzle is blocked or wind or the gas flow volume is low). 4. Weld is cooling off too fast. 5. Arc length is too long. 6. Wrong wire is elected. 7. Gas is sealed improperly. 8. Weld joint surface is not clean.
Undercut		Undercut is a condition where the overmelted base metal has made grooves or an indentation. The base metal's section is made smaller and, therefore, the weld zone's strength is severely lowered.	1. Arc length is too long. 2. Gun angle is improper. 3. Welding speed is too fast. 4. Current is too large. 5. Torch feed is too fast. 6. Torch angle is tilted.
Improper Fusion		This is an unfused condition between weld metal and base metal or between deposited metals.	1. Check torch feed operation. 2. Is voltage lowered? 3. Weld area is not clean.
Overlap		Overlap is apt to occur in fillet weld rather than in butt weld. Overlap causes stress concentration and results in premature corrosion.	1. Welding speed is too slow. 2. Arc length is too short. 3. Torch feed is too slow. 4. Current is too low.
Insufficient Penetration		This is a condition in which there is insufficient deposition made under the panel.	1. Welding current is too low. 2. Arch length is too long. 3. The end of the wire is not aligned with the butted portion of the panels. 4. Groove face is to small.
Excess Weld Spatter		Excess weld spatter occurs as speckles and bumps along either side of the weld bead.	1. Arc length is too long. 2. Rust is on the base metal. 3. Gun angle is too severe.
Spatter (short throat)		Spatter is prone to occur in fillet welds.	1. Current is too great. 2. Wrong wire is selected.

TABLE 7-5: WELDING PRECAUTIONS (CONTINUED)

Defect	Defect Condition	Remarks	Main Causes
Vertical Crack		Cracks usually occur on top surface only.	1. There are stains on welded surface (paint, oil, rust).
The Bead Is Not Uniform.		This is a condition in which the weld bead is misshapen and uneven rather than streamlined and even.	1. The contact tip hole is worn or deformed and the wire is oscillating as it comes out of the tip. 2. The gun is not steady during welding.
Burn Through		Burn through is the condition of holes in the weld bead.	1. The welding current is too high. 2. The gap between the metal is too wide. 3. The speed of the gun is too slow. 4. The gun-to-base metal distance is too short.

Two very important advantages of the FCAW process over MIG are its ability to tolerate surface impurities (thus requiring less precleaning) and to stabilize the arc. Other beneficial characteristics of the process include the following:

- High deposition rate.
- Efficient electrode metal use.
- Requires little edge preparation.
- Welds in any position.
- Welds a wide range of metal thicknesses with one size of electrode.
- Produces high-quality welds.
- Weld puddle is easily controlled and its surface appearance is smooth and uniform even with minimal operator skill.
- Produces a weld with less porosity than MIG when welding galvanized steels.

While the FCAW process has a number of advantages over MIG, it has the following drawbacks:

- FCAW wires are more expensive than MIG hard wires. However, the cost is quickly recovered through higher productivity.
- The flux from the wire changes to slag as it cools. Until it does cool, the slag is sharp and hot and should be considered an eye and skin hazard. Once it cools, this slag must be removed prior to the application of fillers, seam sealers, primers, or paint.
- Spatter is worse when using flux-cored wires. Use nozzle gel and keep the nozzle scraped clean. Spatter buildup in the gun nozzle can jam the wire in the contact tip; it can also fall off during

welding and mix with the molten puddle, diminishing the quality of the weld.
- Excessive tension on the drive rollers or using the incorrect style of drive rollers can collapse the tubular wire. Check the owner's manual for flux-cored wire requirements.
- Only ferrous metals can be welded.

If a machine is used for both MIG and FCAW, the welder must have polarity switching capabilities. FCAW with .030- or .035-inch (0.58, 0.76, or 0.89 mm) wire uses straight polarity while .023-, .030-, and .035-inch (0.58, 0.76, or 0.89 mm) hard wires for MIG use DC reverse polarity. Many of the gas metal arc welding machines sold over the past few years were originally designed to run DC reverse. Without going inside the machine to change polarities, which is difficult and time consuming, this type of machine will not run DC straight polarity. Check the owner's manual for polarity reversing capabilities.

1. FCAW wires are more expensive per pound than hard wires for GMAW.
2. The .030-inch (0.58 mm) self-shielded cored wire contains fluoride compounds. **Use adequate ventilation.**
3. The flux in the core of the wire changes to slag upon cooling. This slag must be removed prior to the application of fillers, seam sealers, primers, or paint.
4. In addition, the slag is sharp and hot until it cools, so it must be considered an eye and skin hazard.

5. Spatter is worse when using cored wires. Use nozzle gel and keep the nozzle scraped clean. Spatter buildup in the gun nozzle can jam the wire in the contact tip or fall off during welding, mixing with the molten puddle and contributing to a poor quality weld.

6. Wire feed problems for FCAW are similar to those encountered with GMAW, but with one important difference. Because cored wires are not solid, excessive tension on the drive rolls or the incorrect style of drive rolls may collapse the tubular wire which leads to feeding problems. Again, check with the owner's manual for correct drive rolls and tension requirements for flux-cored wire.

7.9 TIG WELDING

Tungsten inert gas (TIG) welding, another form of GMAW, uses a nozzle-fed shielding gas and a hand-held filler rod. It has somewhat limited use in body shop repair applications. In a general auto repair or engine rebuilding, however, it does things that make it a valuable tool.

MIG welders lay down weld beads at the average of 25 inches (635 mm) per minute. TIG welding is much slower, with weld speeds ranging between 5 and 10 inches (127 to 254 mm) per minute. However, this slower speed gives much more control, and the end result is the best-looking weld obtainable. A TIG unit can be used to repair cracks in aluminum cylinder heads and reconstruct combustion chambers and other automotive components that need to be welded.

Like MIG (Figure 7–78), TIG welders use an inert gas, such as argon or helium, to surround the weld

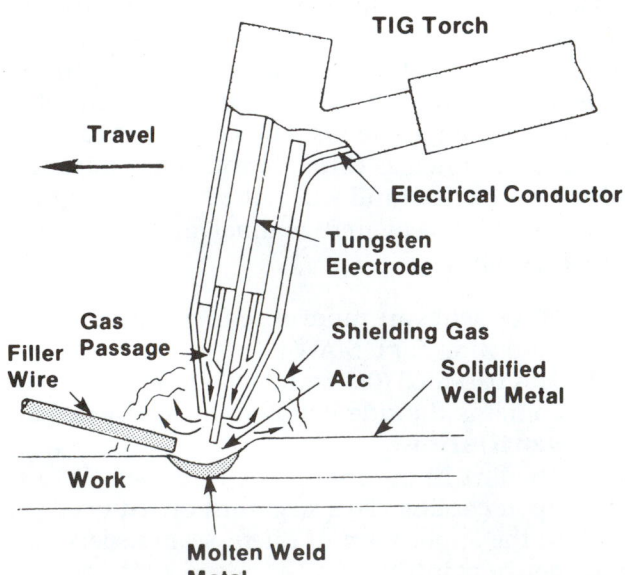

FIGURE 7-78 Study the principles of the TIG process. If filler metal is required, it is fed into the pool from a separate filler rod.

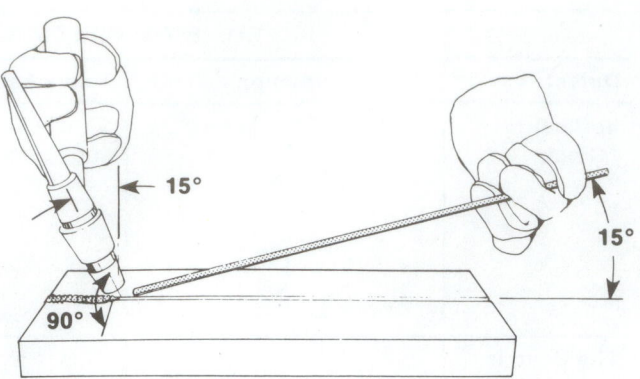

FIGURE 7-79 Proper position of torch and filler rod for manual TIG welding.

area and prevent oxygen and nitrogen in the atmosphere from contaminating the weld. But instead of having a wire feed welding electrode like MIG units, TIG machines use a tungsten electrode with a very high melting point (about 6,900 degrees Fahrenheit) to strike an arc between the welding gun and the work.

Since the tungsten electrode has such a high melting point, it is not consumed during the welding process. A filler rod must be used for welding thicker materials (Figure 7–79).

7.10 RESISTANCE SPOT WELDING

Resistance spot welding is the most important welding process used by automobile manufacturers. It is used on their assembly lines to make many of the OEM welds on unibody cars (Figure 7–80). It is estimated that between 90 to 95 percent of all factory welds in a unibody structure are spot welds. In this country, it is also widely used in the automotive aftermarket for sunroof installations and vehicle conversions, including recreational vehicles (RVs) and stretch limousines.

Since resistance spot welding is now specified by a growing number of automobile manufacturers for repair welding their vehicles, the repair specialist must know how to use a resistance spot welding gun.

The squeeze-type resistance spot welder (Figure 7–81) is ideal for repair welding many of the unibody's thin-gauge sections that require good weld strength and no distortion. Typical applications include roofs, window and door openings, rocker panels, and many exterior panels (Figure 7–82). Due to the strength requirements of unibody repairs, it is often important that a squeeze-type resistance spot welder be used and that the repair specialist know how to set it up, make test welds, and use it.

Resistance spot welding has several advantages:

- It reduces welding costs.
- No consumable filler wire, rod, or gas is required.
- It's clean with no smoke or fumes.

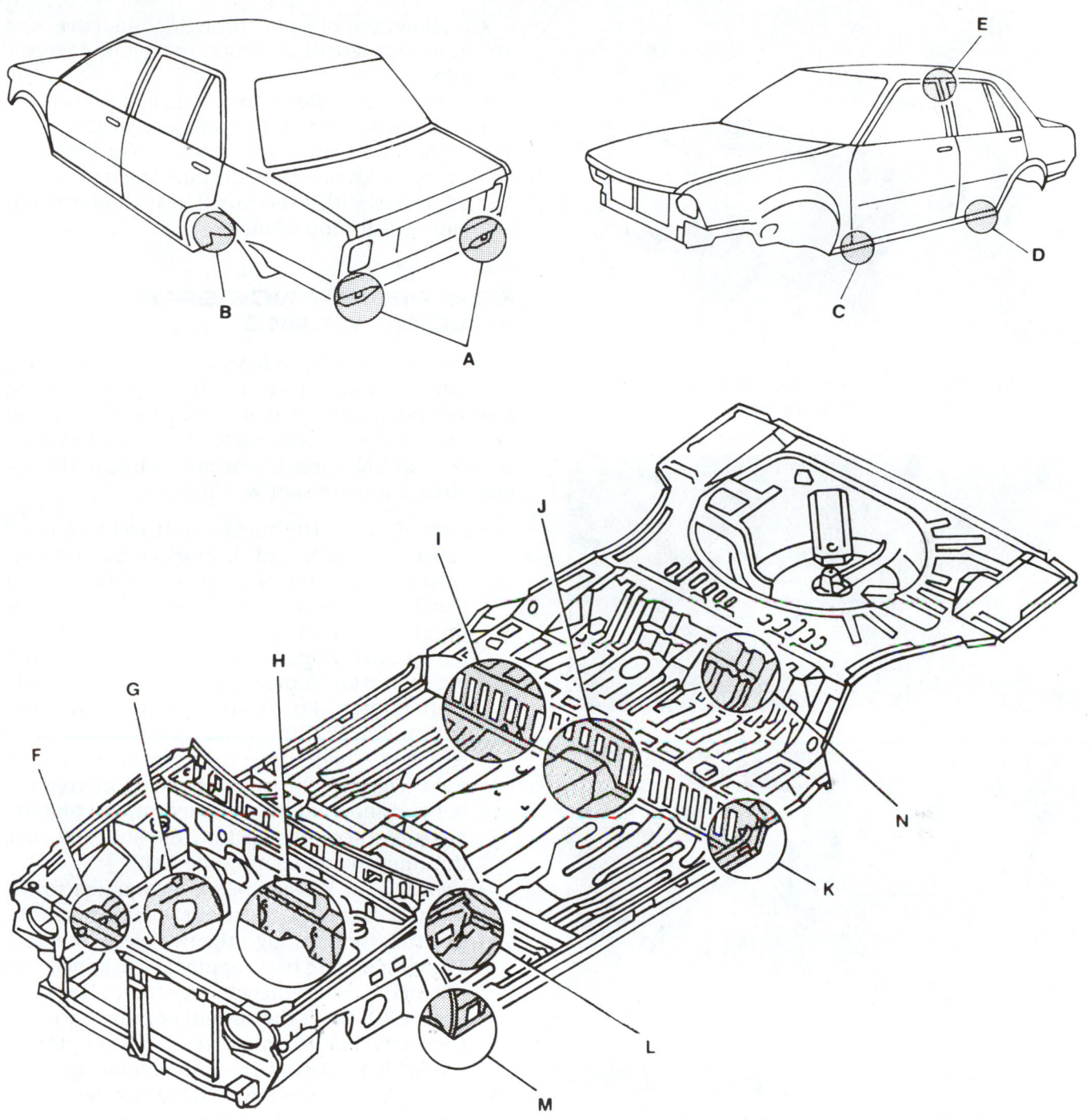

Construction	Location
Suspension mounting	G
Steering gear mounting	H
Fuel tank mounting	N
Engine. transmission mounting	F

Construction	Location
Belt anchor	E
Jack-up point	C, D
Major construction portions	A, I, J, K, L, M

FIGURE 7-80 This illustration gives important locations on the body for production spot welding. Refer to a specific service manual for the vehicle being repaired. *(Courtesy of Nissan Motor Corp.)*

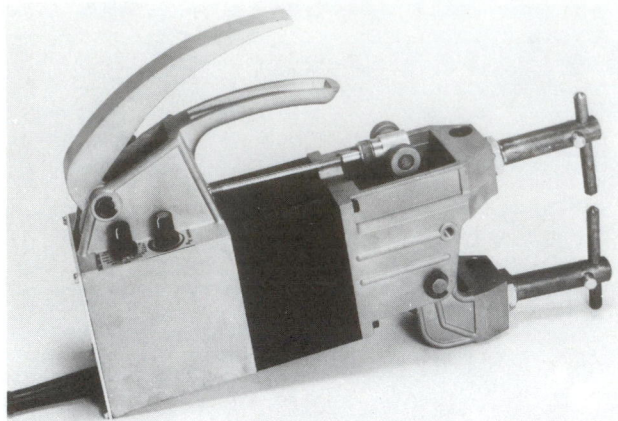

FIGURE 7-81 This is a squeeze-type resistance spot welder. *(Courtesy of Lors Machinery, Inc.)*

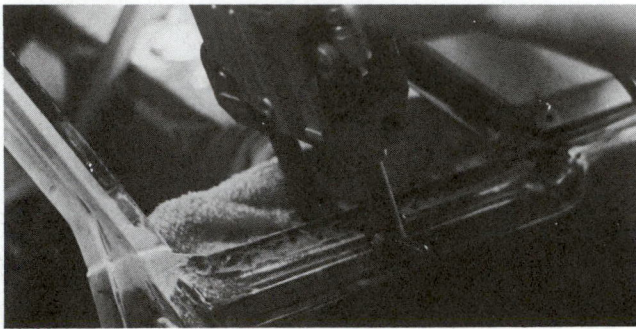

A

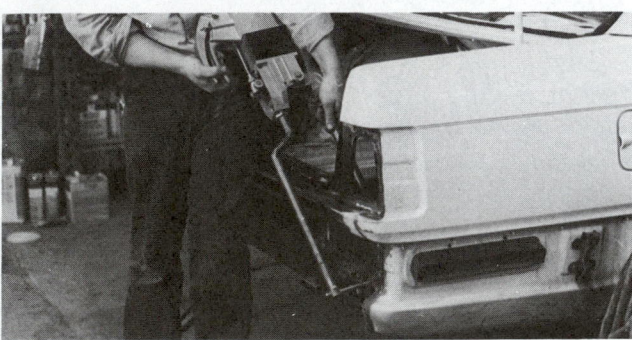

B

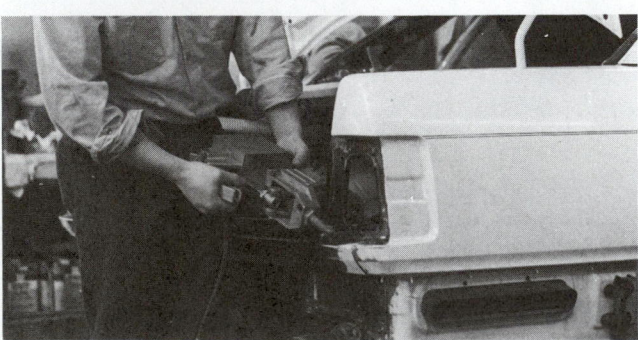

C

FIGURE 7-82 Note the typical applications of a squeeze-type resistance spot welder: (A) welding the side quarter window pinch weld flange; (B) welding the right quarter panel to the rear side member, trunk floor pan, and lower pack panel; and (C) welding the left rear quarter panel to the rear valance. *(Courtesy of Lors Machinery, Inc.)*

- It allows use of weld through conductive zinc primers to restore corrosion protection to repair joints.
- It duplicates OEM factory weld appearance.
- It eliminates need for grinding of welds.
- It's fast; weld times of one second or less make strong welds on HSS and HSLA steels as well as mild steels with a very small heated zone, eliminating distortion of metal.

HOW RESISTANCE SPOT WELDING WORKS

Resistance spot welding relies on the resistance heat generated by low-voltage electric current flowing through two pieces of metal held together, under pressure, by the squeeze force of the welding electrodes. Thus, the three important factors in the operation of resistance spot welding are

- *Pressurization.* The mechanical welding bond between two pieces of sheet metal is directly related to the amount of force exerted on the sheet metal by the welding tips. As the tips squeeze the sheet metal together, an electrical current flows from the tips through the base metal causing the metal to melt and fuse together. Weld spatter (internal or external) is the result of low pressure on the tip or excessive electrical current flow. A high tip pressure causes a small spot weld (Figure 7–83) and a reduced mechanical bond of the weld. In other words, the high tip pressure forces the tip into the softened area, thinning and weakening the weld.
- *Current flow.* When pressure is applied to the metal, a high electric current flows through the electrodes and through the two pieces of metal. The temperature rises rapidly at the joined portion of the metal where the resistance is greatest (Figure 7–84A). If the current continues to flow, the metal melts and fuses together (Figure 7–84B). If the electrical current becomes too great or the pressure too low, internal spatter will result. However, if the current is decreased or the pressure is increased, weld spatter will be held

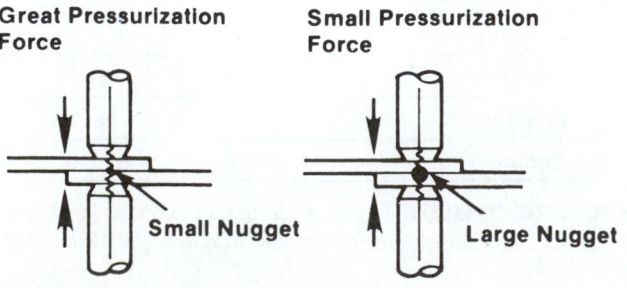

FIGURE 7-83 Electrode (tip) pressure affects the spot weld.

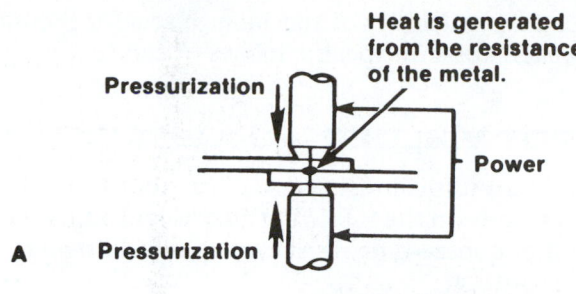

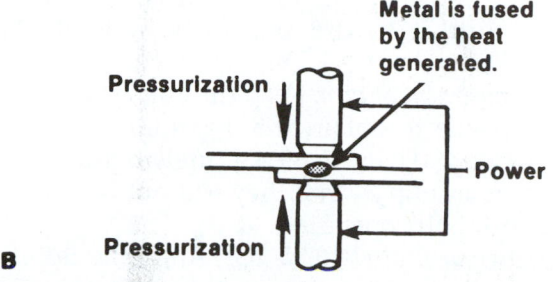

FIGURE 7–84 Note how electrical current (amperage) forms a weld.

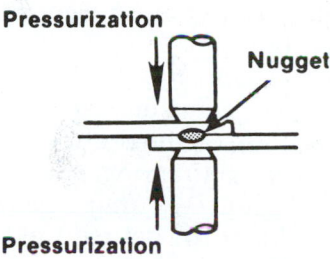

FIGURE 7–85 Note electrical current (holding) flow time.

to a minimum. As can be seen, there is a mutual relationship between the electrical current and the pressure applied to the spot weld.

- *Holding.* If the current flow is stopped, the melted portion begins to cool and forms a round flat bead of solidified metal (nugget) (Figure 7–85). This structure becomes very dense due to the pressurization force, and its subsequent mechanical bonding is excellent. Pressurization time is very important. Do not use less time than specified in the operator's manual.

RESISTANCE SPOT WELDING COMPONENTS

The components of a resistance spot welder (Figure 7–86) are the welding transformer, the welder control, and the welding gun with interchangeable arm sets.

The transformer converts low-amperage, 240-volt shop line current to high secondary amperage, low-voltage (2 to 5 volts) welding current, safe from electrical shock. The welder transformer can either be built into the welding gun or mounted remotely and connected to the gun by means of cables.

A built-in transformer is electrically more efficient since there is little or no loss of welding current between the transformer and the gun. A remote transformer must be larger and draw more shop line current to compensate for power losses through the long cables connecting it to the gun.

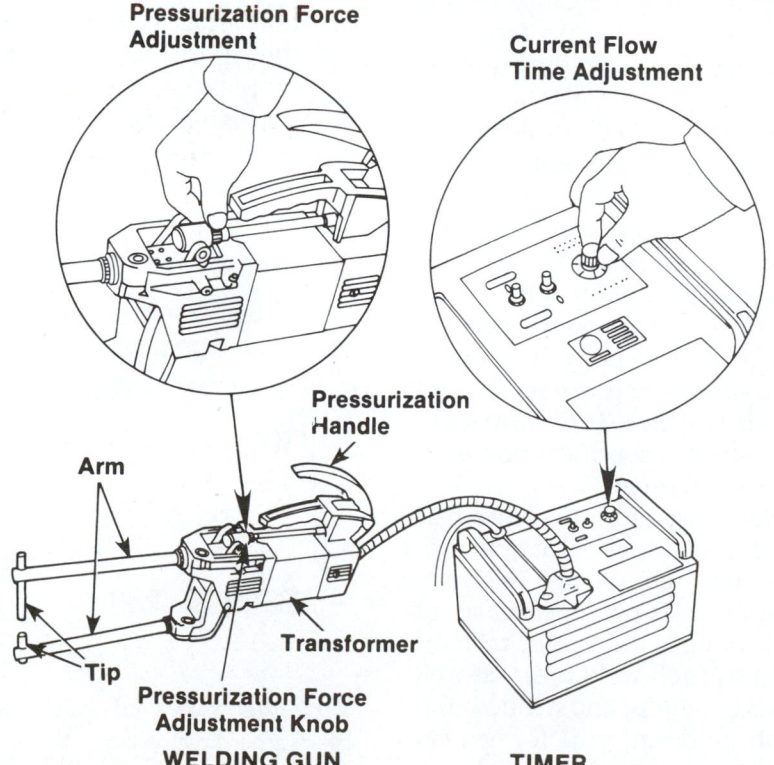

FIGURE 7–86 Study the components of a resistance spot welding system. *(Courtesy of Toyota Motor Corp.)*

Remember that this high weld current will decrease when long reach or wide gap arm sets are used. A high weld current output can be adjusted to a lower intensity by use of the welder control.

The welder control adjusts the transformer's weld current output and permits precise adjustment of the weld time during which the welding current is switched on and allowed to flow through the metal being welded and then switched off. It is desirable to have a range of timing adjustment from approximately $1/6$ of a second to 1 second (10 to 60 cycles) for typical collision repair welding applications. A repeatable accuracy of at least $1/10$ of a second is desirable for consistent weld quality.

The welder control should be capable of providing a full range of adjustment of the welding current. Weld current settings vary, depending upon the thickness of the steel to be welded and the length and gap of the arm sets needed to reach into the area being welded. It might be necessary to decrease weld current when welding with short reach arm sets, or increase weld current when using long reach or wide gap arm sets.

Some manufacturers of resistance spot welders designed for unibody repair work offer additional control features that compensate for small amounts of surface scale or slight rust on the metal. Such features permit the repair specialist to determine when a poor weld condition exists.

The welding gun applies the squeeze force and delivers the welding current through the welder arms to the metal being welded. Most resistance spot welders are designed with a force multiplying mechanism to produce the high electrode force required for consistent weld quality. These force multiplying mechanisms can be spring or pneumatically assisted. Squeeze-type resistance welders that do not use a force multiplying mechanism and rely solely on the operator's manual grip for pressure are not recommended for repair welding unibody structures.

The majority of welding guns in auto body shops should have a maximum capacity of up to two times $5/64$-inch-thick (1.98 mm) steel when equipped with short reach arm sets of 5 inches (127 mm) or less. Capacity with long reach or wide gap arm sets should be at least two times $1/32$-inch-thick (0.79 mm) steel. These capacities comply with the specifications listed in most factory body repair manuals.

Resistance spot welders used for unibody repair welding are available with a full range of interchangeable arm sets. Standard arm sets (Figure 7–87) are designed to reach difficult areas on most makes of cars, such as wheel well flanges, drip rails, taillight openings, and other tight pinch weld areas, as well as floor pan sections, rocker panels, and window and door openings. Repair shops doing work for new car dealers should check the factory repair manuals and look for availability of special arm sets for the hard-to-reach areas on specific makes of cars.

SPOT WELDER ADJUSTMENTS

To obtain sufficient strength at the spot-welded portions, perform the following checks and adjustments on the squeeze-type resistance spot welding gun before starting:

- *Arm selection.* It is important to select the arm according to the area to be welded (Figure 7–88).
- *Adjustment of arm.* Keep the gun arm as short as possible to obtain the maximum pressure for welding (Figure 7–89). Securely tighten the gun arm and tip so that they will not become loose during the operation.
- *Alignment of electrode tips.* Align the upper and lower electrode tips on the same axis (Figure 7–90). Poor alignment of the tips causes insufficient pressurizing, and this results in insufficient current density and insufficient strength at the welded portions.
- *Diameter of electrode tip.* The diameter of the spot weld decreases as the diameter of the electrode tip increases. Also, if the electrode tip is too small, the spot weld will not increase in size. The tip diameter (Figure 7–91) must be properly controlled to obtain the desired welding strength. Before starting operation, make sure that the tip diameter (D) is kept the proper size, and file it cleanly to remove burnt or foreign matter from the surface of the tip. As the amount of dirt on the tip increases, the resistance at the tip also increases, which reduces the current flow through the base metal that in turn reduces weld penetration resulting in an inferior weld. If the

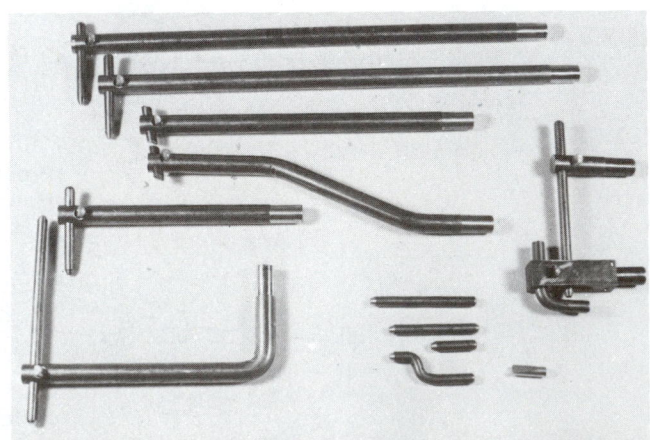

FIGURE 7-87 Various accessory arms are needed to reach around the panels to be welded. *(Courtesy of Henning Hansen, Inc.)*

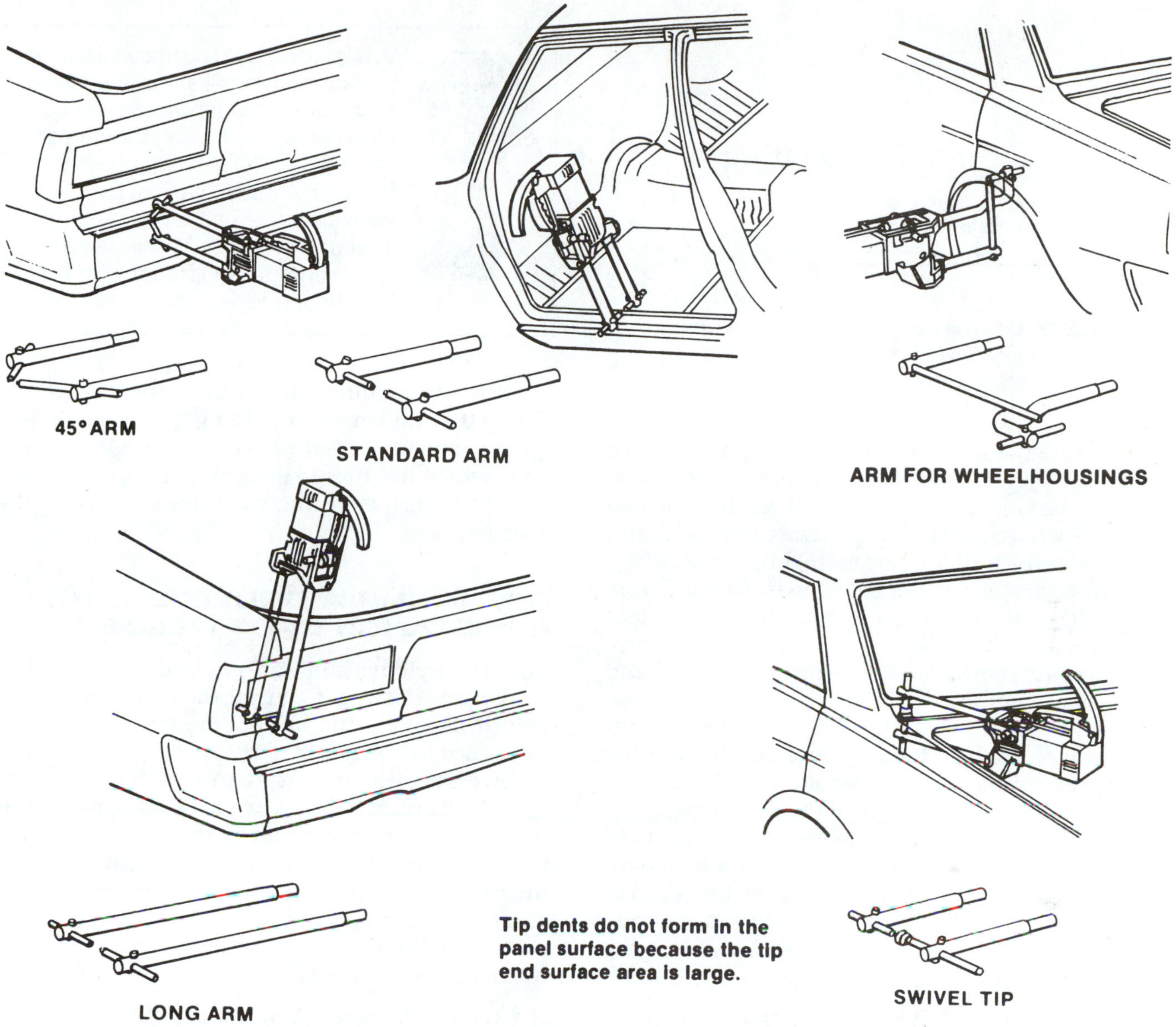

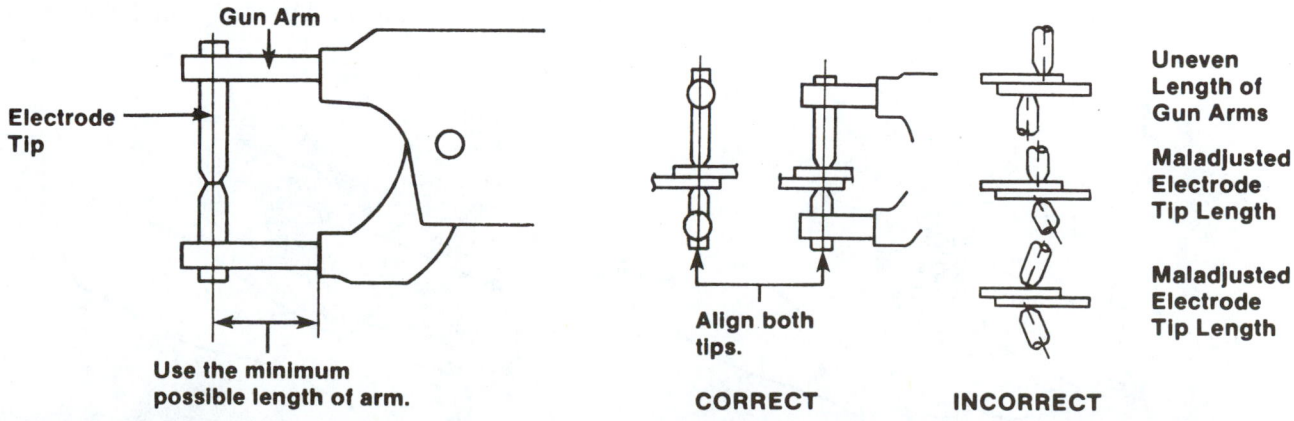

45° ARM

STANDARD ARM

ARM FOR WHEELHOUSINGS

LONG ARM

Tip dents do not form in the panel surface because the tip end surface area is large.

SWIVEL TIP

FIGURE 7-88 Select the proper type of arm for the job. *(Courtesy of Toyota Motor Corp.)*

Gun Arm

Electrode Tip

Use the minimum possible length of arm.

FIGURE 7-89 Adjust gun arms for proper alignment. *(Courtesy of Nissan Motor Corp.)*

Align both tips.

CORRECT   **INCORRECT**

Uneven Length of Gun Arms

Maladjusted Electrode Tip Length

Maladjusted Electrode Tip Length

FIGURE 7-90 Study correct and incorrect alignment of electrode tips. *(Courtesy of Nissan Motor Corp.)*

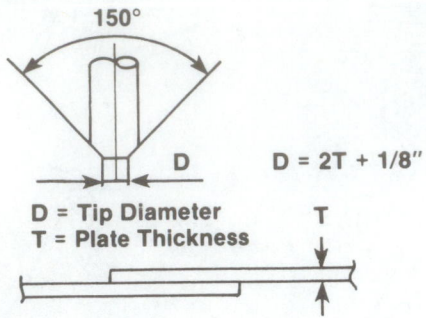

FIGURE 7-91 Use the formula to determine the tip diameter: two times the plate thickness plus 1/8 inch or 3.2 mm. *[Courtesy of Nissan Motor Corp.]*

While spot welding galvanized or zinc-coated steel panels used in auto bodies, offset the drop in current density by raising the current value 10 to 20 percent above that for ordinary steel panels. Since the current value cannot be adjusted in spot welders ordinarily used for body repairs, lengthen the current flow time a little.

tips are used continuously over a long period of time, they will not dissipate heat properly and will become red hot. This will result in premature tip wear that also increases resistance and causes the welding current to drop drastically. If necessary, let the tips cool down after five or six welds. If the tips are worn, use a tip dressing tool to reshape the tips (Figure 7–92).

- *Electrical current flow time.* Current flow time also has a relationship to the formation of a spot weld. When the electrical current flow time increases, the heat that is generated increases the spot weld diameter and penetration. The amount of heat that is dissipated at the weld increases as the current flow time increases. Since the weld temperature will not rise after a certain amount of time, even if the current flows longer than that time, the spot weld size will not increase. However, tip pressure marks and heat warping might occur.

The pressurization force and welding current of many spot welders cannot be adjusted and the current value might be low. However, welding strength can be assured by lengthening the current flow time (letting low current flow for a long time).

The best welding results can be obtained by adjusting the arm length or welding time according to the thickness of the panels. While the welder instruction manual has these values listed inside, it is best to test the quality of the weld using the methods described later.

OPERATING A SQUEEZE-TYPE RESISTANCE SPOT WELDER

Hold the welding gun and position it so that the welder arm electrodes contact the body parts to be welded. Then use the squeeze mechanism to apply weld force to both sides of the metal being welded. As force is applied and maintained on the metal, the force mechanism initiates an electrical signal to the welder control that switches on the flow of weld current for a preset time and then switches it off. Since the weld time is usually less than one second, the entire process is very fast.

Other important operational considerations when using a squeeze-type resistance spot welder are:

- *Clearance between welding surfaces.* Any clearance between the surfaces to be welded causes poor current flow (Figure 7–93). Even if welding can be made without removing such a gap, the welded area would become smaller, resulting in

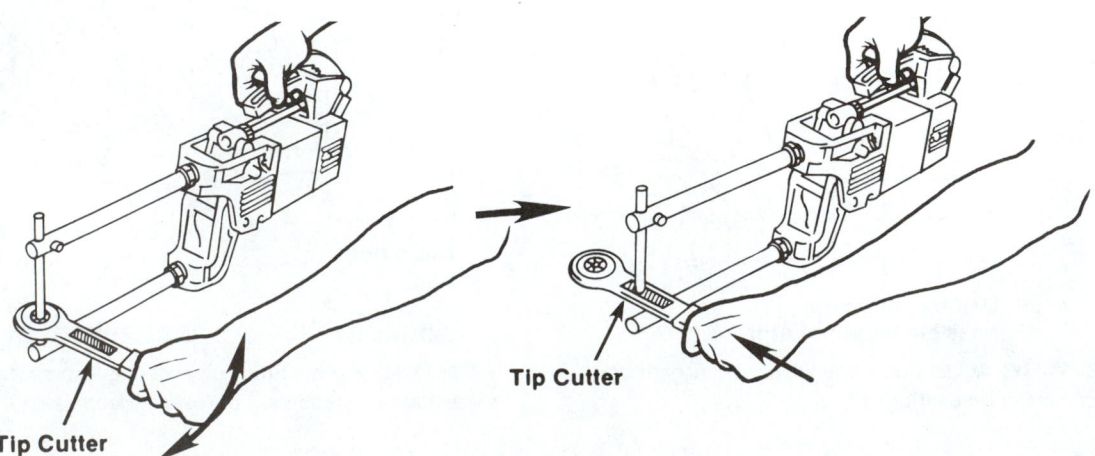

FIGURE 7-92 This special tool will reshape the ends of tips. *[Courtesy of Toyota Motor Corp.]*

CORRECT   INCORRECT   INCORRECT

FIGURE 7-93 Compare the correct and incorrect clearance between welding surfaces.

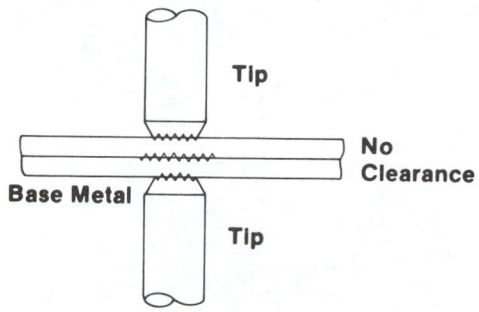

FIGURE 7-94 The condition of base metal surfaces is critical to weld quality.

insufficient strength. Flatten the two surfaces to remove the gap and clamp them tightly with a clamp before welding.

- *Metal surface to be welded.* Paint film, rust, dust, or any other contamination on the metal surfaces to be welded cause insufficient current flow and poor results. Remove such foreign matter from the surfaces to be welded (Figure 7–94).

- *Corrosion.* Coat the surfaces to be welded with an anticorrosion agent (see Chapter 16) that has higher conductivity. It is important to apply the agent uniformly even to the end face of the panel (Figure 7–95).

- *Performance of spot welding operations.* When performing spot welding operations, be sure to

Use the direct welding method. For the portions to which direct welding cannot be applied, use plug welding by MIG welding.

Apply electrodes at a right angle to the panel (Figure 7–96A). If the electrodes are not applied at right angles, the current density will be low, resulting in insufficient welding strength.

For the portion where three or more metal sheets are overlapping, spot welding should be done twice (Figure 7–96B).

- *Number of points of spot welding.* The capacity of spot welding machines available in a repair shop generally is smaller than that of welding machines at the factory. The number of points of spot welding should be increased accordingly by 30 percent in a service shop compared to spot welding in the factory (Figure 7–97).

- *Minimum welding pitch.* The strength of individual spot welds is determined by the *spot weld pitch* (the distance between spot welds) and *edge distance* (the distance of the spots from the panel edge). The bond between the panels becomes stronger as the weld pitch is shortened. However, over a certain point, the metal becomes saturated and further shortening of the pitch will not increase the strength of the bond because the current will flow to the spots that have previously been welded. This reactive current diversion increases as the number of spot welds increases, and the diverted current does not raise the temperature at the welds (Figure 7–98). The distance of the weld pitch must be beyond the area influenced by the reactive current diversion. In general, the values given in Table 7–6 should be observed.

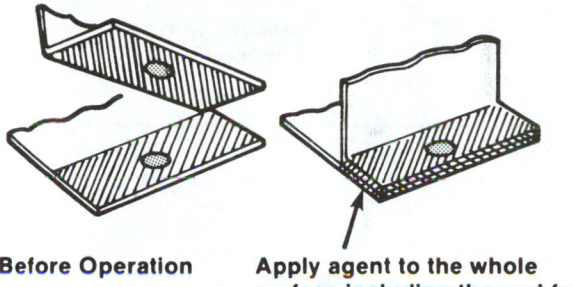

Before Operation   Apply agent to the whole surface including the end face.

FIGURE 7-95 Apply an anticorrosion agent to metal surfaces requiring protection from rust. The service manual will give locations of the areas needing an agent.

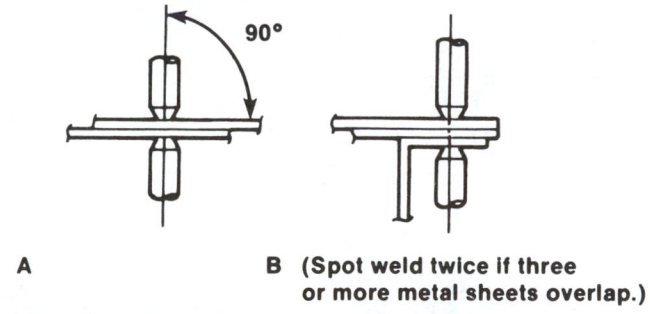

A

B (Spot weld twice if three or more metal sheets overlap.)

FIGURE 7-96 Precautions in performing spot welds. (A) Make sure tips are parallel and at right angles to the panels. (B) Spot weld twice for large total panel thickness.

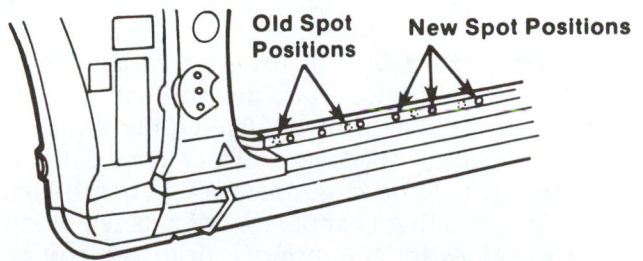

Old Spot Positions   New Spot Positions

FIGURE 7-97 The number of points to spot weld will also be given in the vehicle's service manual.

TABLE 7-6: SPOT WELDING POSITION

Panel Thickness	Pitch S	Edge Distance P
1/64"	7/16" or more	13/64" or more
1/32"	9/16" or more	13/64" or more
Less Than 3/64"	11/16" or more	1/4" or more
3/64"	7/8" or more	9/32" or more
1/16"	1-9/64" or more	5/16" or more

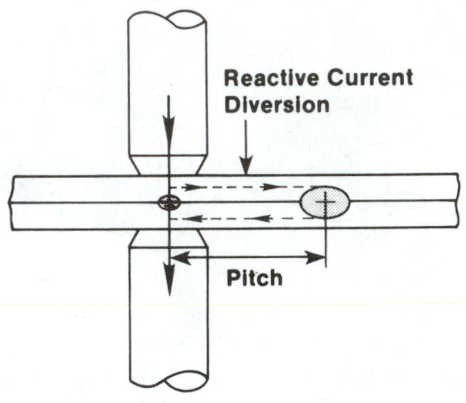

FIGURE 7-98 Minimum welding pitch is spec for how far apart the welds should be.

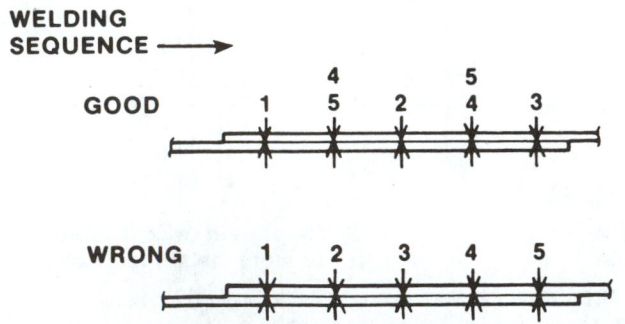

FIGURE 7-99 Memorize the proper welding sequence. [Courtesy of Nissan Motor Corp.]

- *Position of welding spot from edge and end of panel.* The edge distance is also determined by the position of the welding tip. Even if the spot welds are normal, the welds will not have sufficient strength if the edge distance is insufficient. When welding near the end of a panel, observe the values for the distance from the end of a panel given in Table 7–7. If the distance is too small, it results in insufficient strength and also in a strained panel.

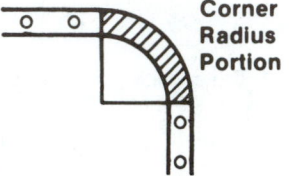

FIGURE 7-100 Often you do not weld corners; just weld right up to them or as directed in the manual. [Courtesy of Nissan Motor Corp.]

- *Spotting sequence.* Do not spot continuously in one direction only. This method provides weak welding due to the shunt effect of the current (Figure 7–99). If the welding tips become hot and change their color, stop welding and allow them to cool.
- *Welding corners.* Do not weld the corner radius portion (Figure 7–100). Welding this portion results in concentration of stress that leads to cracks. The following locations require special consideration:

Upper corner of front and center pillars

Front upper portion of the quarter panel

Corner portion of front and rear windows

INSPECTION OF SPOT WELDS

Spot welds are inspected either by outward appearance (visual inspection) or destructive testing. Destructive testing is used to measure the strength of a weld, and a visual inspection is used to judge the quality of the outward appearance.

Appearance Inspection

Check the finish of the weld visually and by touching. The items to check are:

- *Spot position.* The spot weld position should be in the center of the flange with no tip holes and

TABLE 7-7: POSITION OF WELDING SPOT FROM THE END OF PANEL

Thickness (t)	Minimum pitch (ℓ)
1/64″	7/16″ or over
1/32″	7/16″ or over
Less than 3/64″	15/32″ or over
3/64″	9/16″ or over
1/16″	5/8″ or over
5/64″	11/16″ or over

have no spot welds overriding the edge. As a rule, an old spot position should be avoided.

- *Number of spots.* There should be 1.3 or more times the number made by the manufacturers. (For example, 1.3 times 4 original factory spot welds equals roughly 5 new repair spot welds.)
- *Pitch.* It should be a little shorter than that of the manufacturer and spots should be uniformly spaced. The minimum pitch should be at a distance where reactive current diversion will not occur.
- *Dents (tip bruises).* There should be no dents on the surfaces that exceed half the thickness of the panel.
- *Pinholes.* There should be no pinholes that are large enough to see.
- *Spatter.* A glove should not catch on the surface when rubbed across it.

Destructive Testing

Most destructive tests require the use of sophisticated equipment, a requirement that most body shops are unable to meet. For this reason, simpler methods described here have been developed for general use in body shops.

- **Destructive test.** A test piece of the same metal as the welded piece and with the same panel thickness is made and welded in the positions shown in Figure 7–101. Force is then applied in the direction of the arrow and the spots are separated. It is then judged by how cleanly the weld has broken whether it is satisfactory or not. If the weld pulls out cleanly, like a cork from a bottle, the weld is judged to be good. It should be noted that since the weld performance cannot be exactly duplicated by this test, the results should only serve as a reference.
- **Nondestructive test.** To confirm a spot weld after it has been made, use a chisel and hammer and proceed as follows:

Insert the tip of a chisel between the welded plates (Figure 7–102) and tap the end of the chisel until a clearance of $1/8$ to $5/32$ inches (3.2 to 3.97 mm) (when the plate thickness is approximately $1/32$ inch (0.79 mm)) is formed between the plates. If the welded portions remain normal, it indicates that the welding has been done prop-

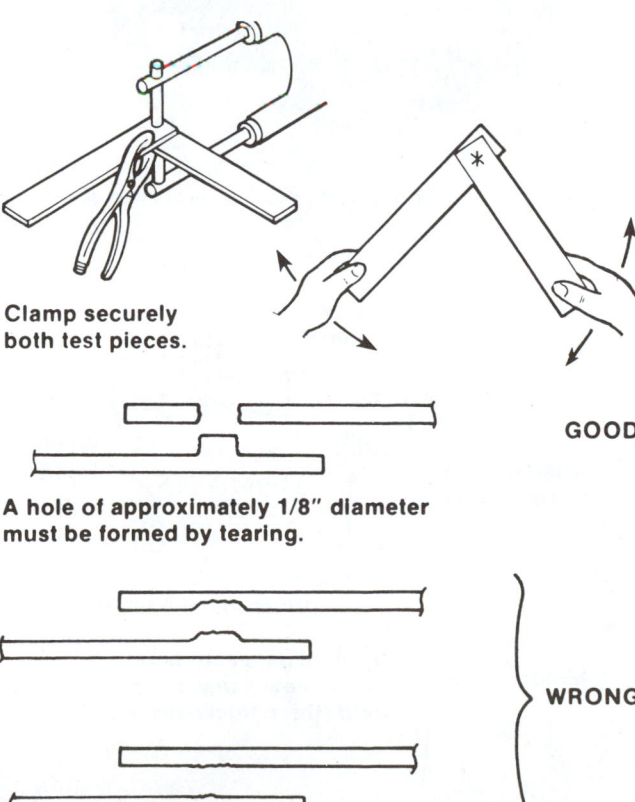

Clamp securely both test pieces.

GOOD

A hole of approximately 1/8″ diameter must be formed by tearing.

WRONG

FIGURE 7–101 When performing a destructive check, force the test weld apart and inspect it. *[Courtesy of Nissan Motor Corp.]*

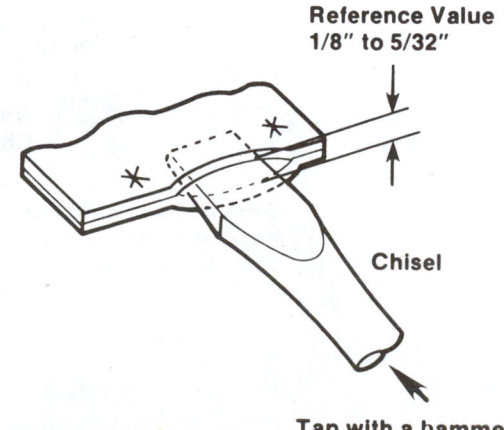

Reference Value 1/8″ to 5/32″

Chisel

Tap with a hammer.

FIGURE 7–102 Perform a destructive check on metal identical to that being welded on the vehicle. *[Courtesy of Nissan Motor Corp.]*

erly. This clearance varies with the location of the welded spots, length of the flange, plate thickness, welding pitch, and other factors. Note that the values given here are only reference values.

If the thickness of the plates is not equal, the clearance between the plates must be limited to $1/16$ to $5/64$ inches (1.58 to 1.98 mm). Note that further opening of the plates can become a destructive test.

Be sure to repair the deformed portion of the panel after inspection.

7.11 OTHER SPOT WELDING FUNCTIONS

While the squeeze-type welding gun is the most used in the repair shop, there are other types of guns used with spot welding equipment. With the proper gun attachment, the spot welder can be used as a panel spotter, stud welder, spot shrinker, and mold rivet welder.

PANEL SPOTTING

At one time spot welding equipment was called a *panel spotter* (Figure 7–103). When operating a panel spotter, the two electrode guns are placed on the nonstructural replacement panel. Figure 7–104 shows how both lap and flange joints can be made with a spliced panel or full panel installation.

After the adjustments are made following the manufacturer's directions, push both electrodes against the panel and apply moderate pressure to close any gaps. Press the weld button on the switch handle and hold it down until the welding cycle stops automatically. Release the weld button and move the electrodes to the next welding location.

Here are some other panel spotter operational tips:

- Thoroughly clean the surfaces along the weld seam. If a new replacement has been primed, strip off this coat on both sides of the panel and along the weld seam with a coarse abrasive paper. If the panel has a rust preventative film coating instead of a primer, merely wipe off both sides of the weld seam with a clean rag and solvent.
- Use vise grips on all flange joint and drip rail applications (Figure 7–105) to bring the parts closely together. Weld near the vise grip jaws where the fit-up is tight.

FIGURE 7-103 This is a typical panel spotter. *(Courtesy of Lenco Inc.)*

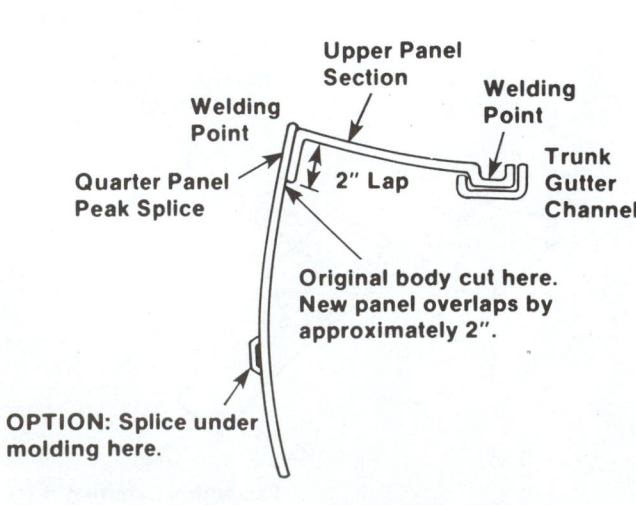

LAP JOINT

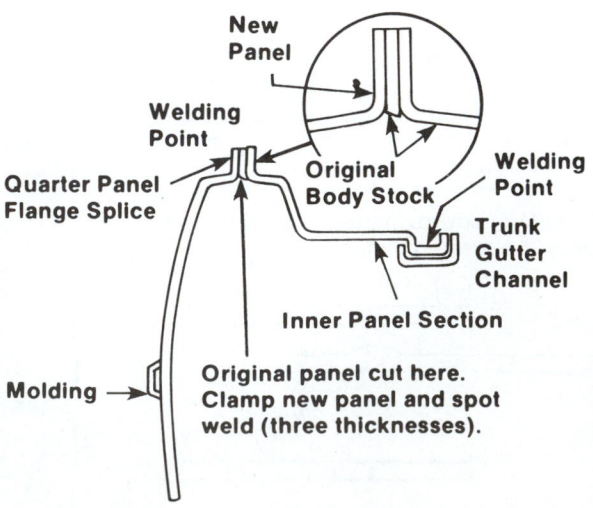

FLANGE JOINT

FIGURE 7-104 Note the method of panel spotting lap and flange joints.

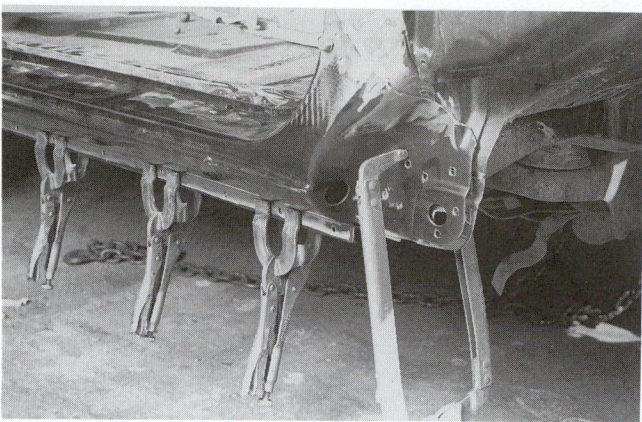

FIGURE 7-105 The technician is using vise grips to hold the joint together.

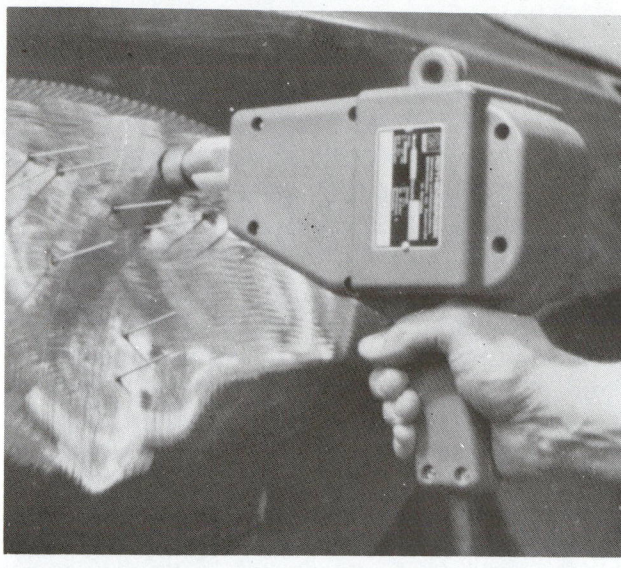

FIGURE 7-106 Note a typical stud or pin welder for removing dents. *(Courtesy of Henning Hansen, Inc.)*

- A few sheet metal screws can be used on lap joints to position the panel for spot welding. Make sure that the paint has been removed from the joints.
- On long splice jobs, start in the middle of the panel and spot weld in one direction. For example, go from the middle of the panel to the door post. Start again in the middle and complete the panel welds to the tail light area. This is an additional aid in eliminating distortion.
- Removal of burrs on the newly cut panel insures getting good metal-to-metal contact when body pressure is applied to the electrodes. Burrs and dents cause an air space between mating parts and prevent positive metal contact.

The twin electrodes of the panel spotter often permit spot welding in spaces where the squeeze-type has difficulty operating. In addition, the panel spotter can be converted to a squeeze-type spot welder

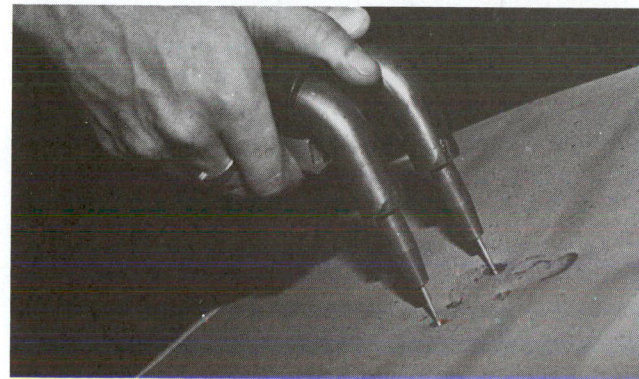

FIGURE 7-107 The panel spotter will weld on spot studs for pulling the dent. This allows pulling dents without drilling holes in the panel. *(Courtesy of Lenco Inc.)*

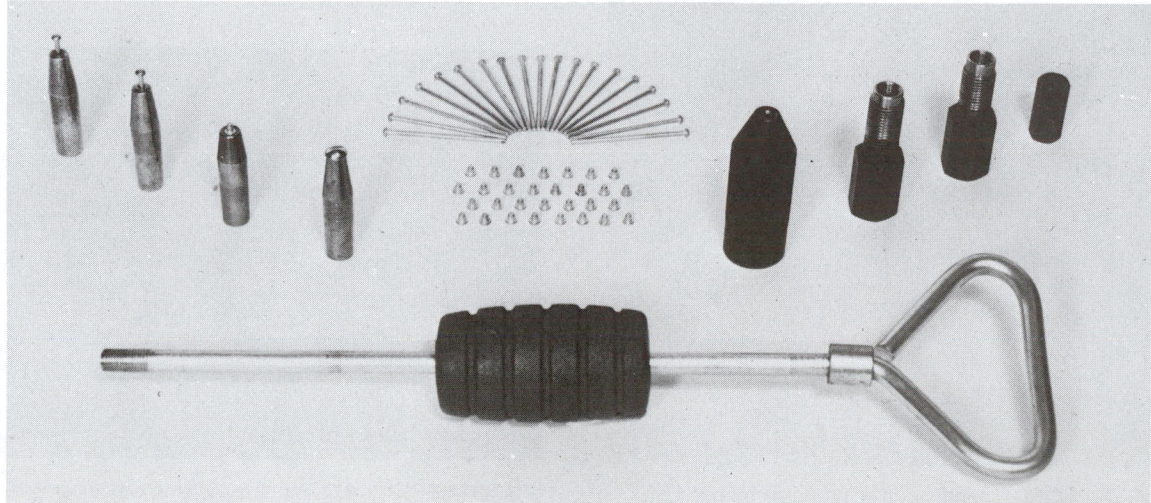

FIGURE 7-108 After welding on studs over the dent, a slide hammer and attachments can be used to pull out the dent. *(Courtesy of Lenco Inc.)*

with a gun attachment. However, this arrangement should be used only on nonstructural parts, never on structural parts.

7.12 STUD SPOT WELDING FOR DENT REMOVAL

Studs used in dent removal can be resistance welded with a special stud welder (Figure 7–106) or a panel spotter equipped with stud welding attachments (Figure 7–107). With either method, a stud pulling kit (Figure 7–108) containing all the necessary items (including a slide hammer) is a must for dent removal.

To remove a dent properly with either a stud or stud spot welder, a good quality stud is necessary. The stud should offer the necessary combination of pull strength and tensile strength, while remaining extremely flexible. The flexibility allows the stud to be bent out of the way when working on adjacent studs, then bent back when required. The importance of this stud is to minimize the heat required and, therefore, maintain the flexibility of the steel when being applied and removed. Complete details on using stud or panel welding for dent removal can be found in Chapter 8.

7.13 MOLD RIVET WELDING

Although many decorative strips are applied with adhesive, moldings are still applied with mold rivets and clips. For example, chrome strips on rocker pan-els and window and vinyl roof moldings are often held this way.

When patching or refinishing areas that are susceptible to moisture, salt, or high humidity, a technician is usually apprehensive about drilling holes exposing inner panels. Mold rivet welding with a stud or spot welder is a logical solution. As shown in Figure 7–109, one electrode has the mold rivet welding tip, while the other has the ground tip. No holes are made; rivets can be relocated or replaced without exposing vulnerable areas to outside elements. This one-step operation achieves a factory replica and is ideal for placing rivets on new skins. If rivets need to be removed or relocated, they require very little grinding.

7.14 OXYACETYLENE WELDING

Oxyacetylene welding is a type of fusion welding. Acetylene and oxygen are mixed in a chamber, ignited at the tip, and used as a high-temperature heat source (approximately 5,400 degrees Fahrenheit or 2,984 degrees Celsius) to melt and join the welding rod and base metal together (Figure 7–110).

Since it is difficult to concentrate the heat in one area, the heat affects the surrounding areas and reduces the strength of steel panels. Because of this problem, auto makers do not recommend the use of oxyacetylene in repairs of damaged vehicles. Although oxyacetylene is in disfavor with most automobiled manufacturers—with good reason—it has some use in the body shop. The oxyacetylene flame is still used to repair other damaged auto bodies, and for some heat shrinking operations, brazing, soldering, surface cleaning, and cutting of nonstructural parts. Oxyacetylene should not be used to cut structural parts of any vehicle unless special care is taken.

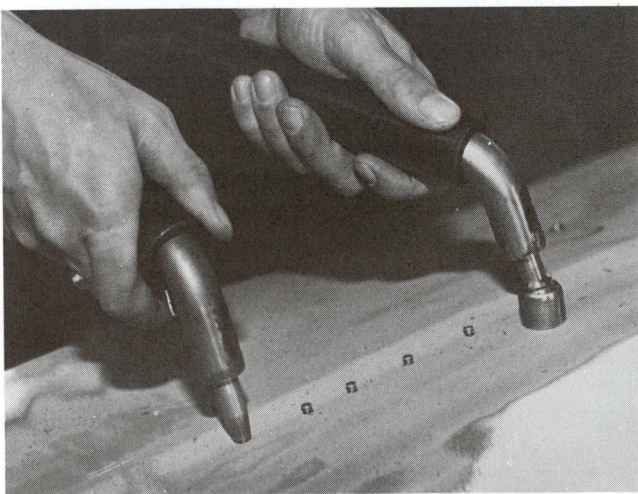

FIGURE 7-109 A panel spotter can also be used to install molding rivets. *(Courtesy of Lenco Inc.)*

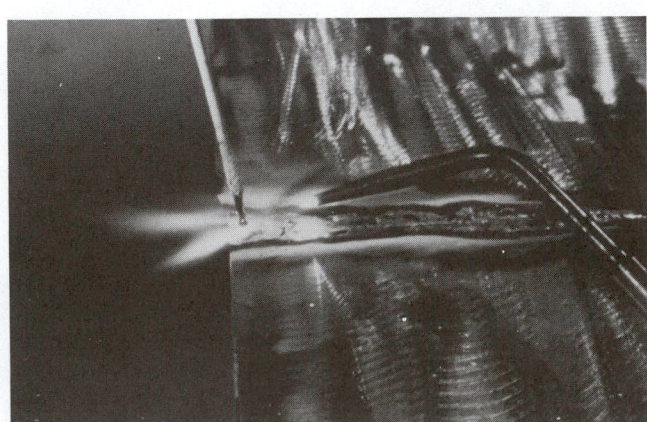

FIGURE 7-110 An oxyacetylene welder is seldom used to weld body panels because of the heat warpage that results.

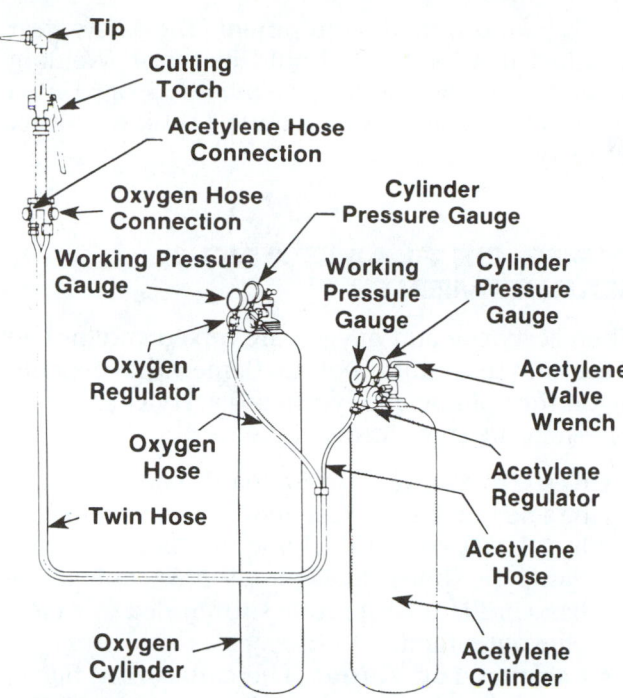

FIGURE 7-111 Study the parts of an oxyacetylene welding and cutting outfit.

WELDING AND CUTTING EQUIPMENT

In general, an oxyacetylene welding and cutting outfit (Figure 7–111) consists of the following:

- *Steel tanks (cylinders)* filled with

  Oxygen

  Acetylene

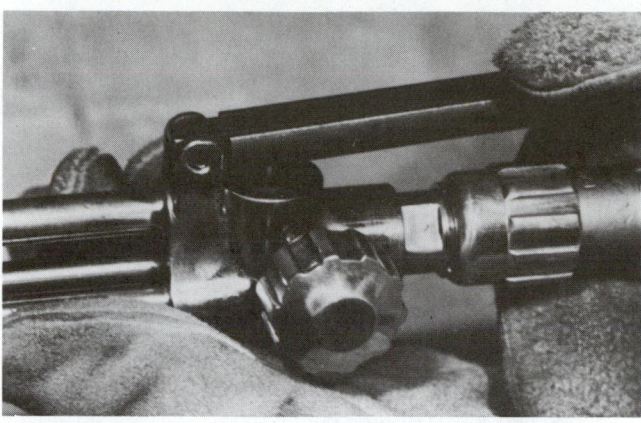

FIGURE 7-112 It is important to be familiar with cutting torch adjustments.

- *Regulators,* which reduce the pressure coming from the tanks to the desired level and maintain a constant flow rate of

  Oxygen pressure: 15 to 100 psi (103 to 689 kPa)

  Acetylene: 3 to 12 psi (21 to 83 kPa)

- *Hoses* from the regulators and cylinders connect the oxygen and acetylene to the torch.
- *Torch.* The torch body mixes the oxygen and acetylene from the tanks in the proper proportions and produces a heating flame capable of melting steel. There are two main types of torches:

  Welding torch

  Cutting torch

The low-pressure torch is generally used for acetylene welding. This torch can be used at an extremely

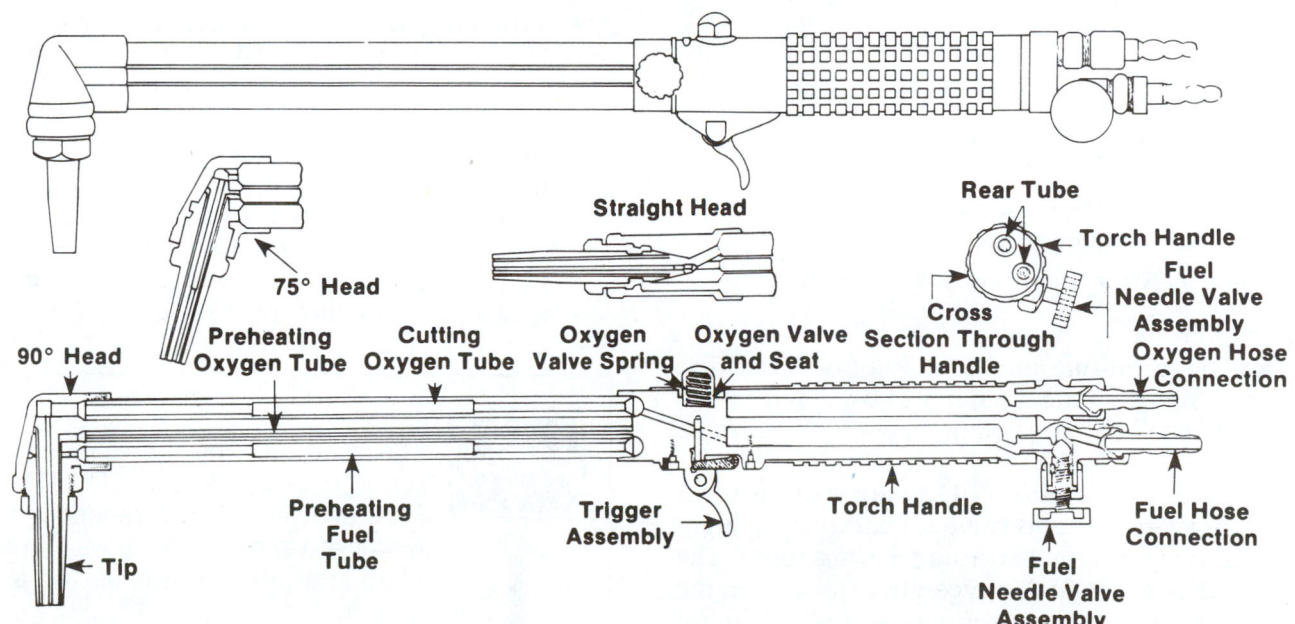

FIGURE 7-113 Study the parts of a typical cutting torch.

FIGURE 7–114 By squeezing the handle on a spark lighter, you can light the flame on the torch.

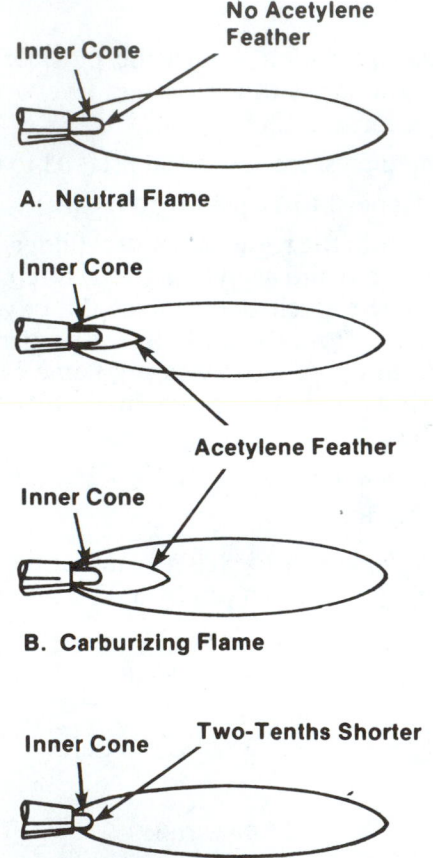

FIGURE 7–115 Compare types of cutting flames.

low acetylene pressure and has an injector nozzle. The gases are mixed by the discharge of oxygen from the center nozzle. The important operation control is shown in Figure 7–112.

As shown in Figure 7–113, the cutting torch has an oxygen tube and valve for conducting high-pressure oxygen attached to a welding torch. The flame outlet has a small oxygen hole located in the center of the tip that is surrounded by holes arranged in a spherical pattern. The outer holes are used for preheating.

To round out the equipment, the safety gear described in Chapter 1 should be worn. Welding should be done with either a number 4, 5, or 6 tinted filter shade. A spark lighter (Figure 7–114) is another necessity.

TYPES OF FLAME AND ADJUSTMENT

When acetylene and oxygen are mixed and burned in the air, the condition of the flame varies depending on the volume of oxygen and acetylene.

There are three forms of flame:

- **Neutral flame.** The standard flame is said to be a neutral flame. Acetylene and oxygen mixed in a 1 to 1 ratio by volume produces a neutral flame. As shown in Figure 7–115A, this flame has a brilliant white cone surrounded by a clear blue outer flame.
- **Carburizing flame.** The carburizing flame, also called a surplus or reduction flame, is obtained by mixing slightly more acetylene than oxygen. Figure 7–115B shows that this flame differs from the neutral flame in that it has three parts. The cone and the outer flames are the same as the neutral flame, but between them there is an intermediate light-colored acetylene cone enveloping the cone. The length of the acetylene cone varies according to the amount of surplus acetylene in the gas mixture. For a double surplus flame, the oxygen-acetylene mixing ratio is about 1 to 1.4 (by volume). A carburizing flame is used for welding aluminum, nickel, and other alloys.
- **Oxidizing flame.** The oxidizing flame is obtained by mixing slightly more oxygen than acetylene. The oxidizing flame (Figure 7–115C) resembles the neutral flame in appearance, but the acetylene cone is shorter and its color is a little more violet compared to the neutral flame. The outer flame is shorter and fuzzy at the end. Ordinarily, this flame oxidizes melted metal, so it is not used in the welding of mild steel, but it is used in the welding of brass and bronze.

The acetylene line pressure must never exceed 15 psi (103 kPa). Free acetylene has a tendency to dissociate at pressure above 15 psi (103 kPa) and could cause an explosion.

WELDING TORCH FLAME ADJUSTMENT

As stated in the overview of welding, oxyacetylene welding is not used for welding modern auto bodies, but it is used for brazing certain nonstructural panels at factory-brazed seams. When using a welding torch, proceed as follows:

1. Attach the appropriate tip to the end of the torch. Use the standard tip for sheet metal (each torch manufacturer has a different system for measuring the size of the tip orifice).
2. Set the oxygen and acetylene regulators at the proper pressure:

   Oxygen = 8 to 25 psi (55 to 172 kPa)

   Acetylene = 3 to 8 psi (21 to 55 kPa)
3. Open the acetylene valve about half a turn and ignite the gas. Continue to open the valve until the black smoke disappears and a reddish yellow flame appears. Slowly open the oxygen valve until a blue flame with a yellowish white cone appears. Further open the oxygen valve until the center cone becomes sharp and well defined. This type of flame is called a neutral flame and is used for welding mild steel (other than automobile bodies).

   If acetylene is added to the flame or oxygen is removed from the flame, a carburizing flame will result.

   If oxygen is added to the flame or acetylene is removed from the flame, an oxidizing flame will result.

GAS CUTTING TORCH FLAME ADJUSTMENT

The cutting torch is sometimes used in collision repair shops to rough cut damaged panels. Gas cutting torch flame adjustment and cutting procedures are as follows:

1. Adjust the oxygen and acetylene valves for a preheating neutral flame.
2. Open the preheating oxygen valve slowly until an oxidizing flame appears. This makes it difficult for melted metal to remain on the surface of the cut panel, allowing for clean edges.
3. *Thick panel cutting method.* Heat a portion of the base metal until it is red hot. Just before it melts, open the high-pressure oxygen valve and cut the panel. Advance the torch forward while making sure the panel is melting and being cut apart. This method is widely used for thick panels when there are several pieces overlapped together or for a side member, even when there is an internal reinforcement.

4. *Thin panel cutting method.* Heat a small spot on the base metal until it is red hot. Just before it melts, open the high-pressure oxygen valve and incline the torch to cut the panel. When cutting thin material, incline the tip of the torch so that the cut will be clean and fast (this prevents unwanted panel warpage).

 **SHOP TALK** As soon as the cutting operation is completed, quickly turn off the high-pressure oxygen flow used for cutting and pull the torch away from the base metal. This action prevents sparks from entering the tip and igniting the oxygen-acetylene mixture in the torch handle. In extreme cases the ensuing fire could melt the torch handle.

CUTTING HSS FOR SALVAGE PURPOSES

Salvage components must be cut with a grinding wheel disc, an air chisel and/or metal cutting saw, or with a plasma cutter. If the use of a gas torch is necessary when cutting HSS sheet metal components for salvage purposes or cutting a body structure for a front/rear clip, factory engineers advise the following approach:

- Cut the metal structure at least 2 inches (51 mm) away from the desired cut line. Sheet metal within the heat-affected area will lose strength when subjected to the high heat levels of a torch.
- After torch cutting, use a grinding wheel disc, an air chisel, or a metal saw to make the final cut at the originally intended dimension line. HSS damage will then be cut out of the salvaged part.

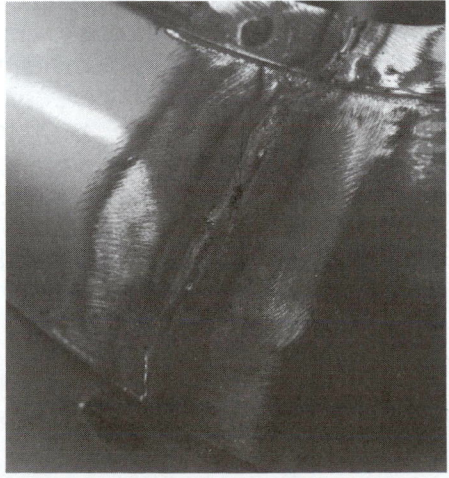

FIGURE 7-116 Discoloration of HSS heat-affected steel will weaken the metal and cause distortion.

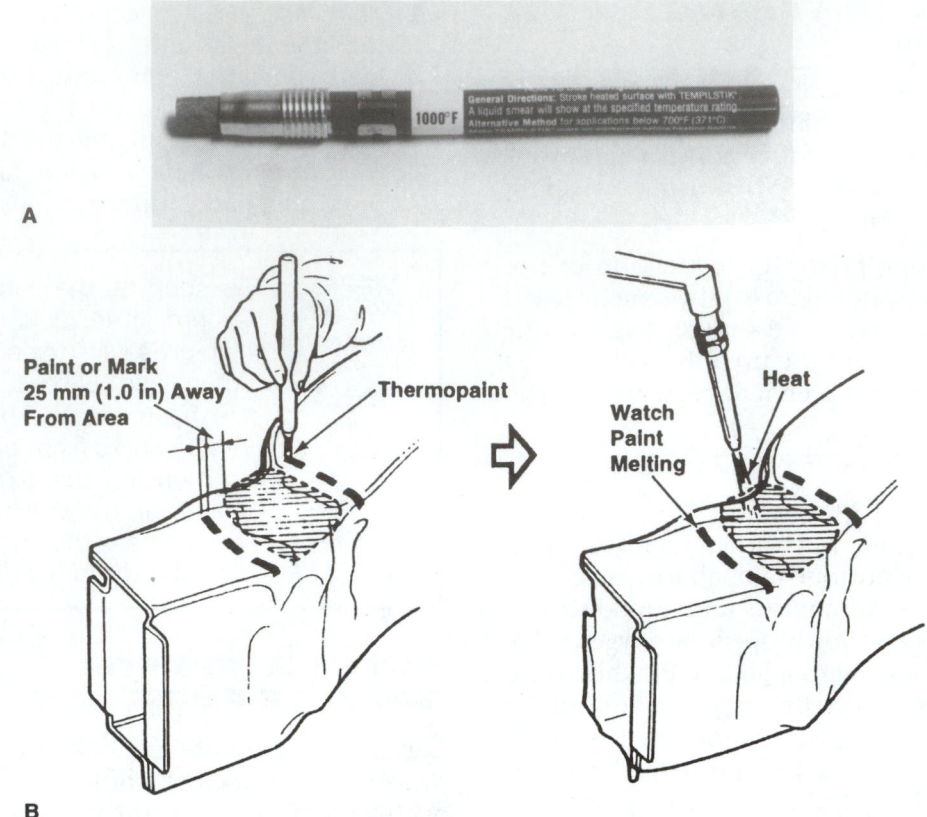

A

B

FIGURE 7-117 (A) A typical heat indicating crayon will melt at a specific temperature and help prevent overheating of parts while welding, cutting, or heating. (B) A thermal crayon or paint will melt at a predetermined temperature; this is especially helpful when welding aluminum, which does not change color with heat. *(Courtesy of American Honda Co., Inc.)*

As stated previously, oxyacetylene equipment should not be used on HSS components for welding or cutting. Vehicle manufacturer's engineers stress this point. There is just too much heat buildup that can reduce structural strength. However, in some instances an oxyacetylene torch can be used to heat HSS components or parts ("hot working"), provided the critical 1,400 degrees Fahrenheit temperature is not exceeded. (Check the manufacturer's shop manual on this point because some say 1,000 degrees Fahrenheit (538 degrees Celsius) is the critical temperature.)

High-strength steels should be exposed to high temperatures from an oxyacetylene torch for only a very short period of time. Three minutes is the recommended maximum time span for exposing HSS to a 1,400-degree Fahrenheit (760 degrees Celsius) temperature to reduce the amount of scaling that normally takes place on the metal surface. High-temperature exposure causes discoloration as shown in Figure 7–116.

To determine and control temperatures of high-strength steel parts and components being "heat worked" with oxyacetylene equipment, it is necessary to use a temperature indicating crayon (Figure 7–117).

Heat Crayons

With steel, the use of heat is avoided whenever possible to prevent reducing the strength of the metal. With aluminum, heat must be used to restore flexibility caused by work hardening. If not, it will crack when straightening force is applied.

Before straightening, heat is often applied to the damaged area of the aluminum. It is easy to apply too much heat since aluminum does not change color with high temperatures. It also melts at a relatively low 1,220 degrees Fahrenheit (660 degrees Celsius). Careful heat control is very important.

Heat crayons or *thermal paint* can be used to determine the temperature of the aluminum or other metal being heated. They will melt at a specific temperature and warn you to prevent overheating.

The crayon or paint is applied next to the aluminum area to be heated. The mark will begin to melt when the crayon's or paint's melting point is almost reached. The melting will let you know that you are about to reach the melting point of the aluminum.

The metal should be marked closely adjacent to the area being worked with a crayon rated no more than 1,400 degrees Fahrenheit (760 degrees Celsius). Using such a crayon will indicate to the welder whether or not an excessive amount of heat is being applied. Thus metal temperatures can be controlled within safe levels and HSS damage easily prevented.

CLEANING WITH A TORCH

It is important before starting any weld that the surfaces to be joined must be thoroughly clean. The weld site must be completely free of any foreign material that might contaminate the weld. The finished weld is quite likely to be brittle, porous, and of poor integrity.

To remove heavy undercoating, rustproofing, tars, caulking, sealants, road dirt, and primers, first use a scraper to get off the loose material. Then use a scraper and an oxyacetylene torch. Then, if needed, use a wire brush and the torch, using a carburizing flame (Figure 7–118). In any event, keep the torch at

a very low, controlled heat to prevent part damage. Use just enough heat to get the job done.

FLAME ABNORMALITIES

When changes occur during gas welding, such as overheating of the flame outlet, adhesion of spatter, or fluctuations in the gas adjustment pressure, the result will be variations in the flame and weld. Therefore, you must always be aware of the condition of the flame. Flame abnormalities, their causes, and remedies are described in Table 7–8.

7.15 BRAZING

Brazing is applied only to places for sealing. This is a method of welding in which a nonferrous metal,

TABLE 7-8: FLAME ABNORMALITIES AND REMEDIES

Symptom	Cause	Remedy
Flame Fluctuations	1. Moisture in the gas, condensation in hose. 2. Insufficient acetylene supply.	1. Remove the moisture from the hose. 2. Adjust the acetylene pressure and have the tank refilled.
Explosive Sound While Lighting the Torch	1. Oxygen or acetylene pressure is incorrect. 2. Removal of mixed-in gases are incomplete. 3. The tip orifice is too enlarged. 4. The tip orifice is dirty.	1. Adjust the pressure. 2. Remove the air from inside the torch. 3. Replace the tip. 4. Clean the orifice in the tip.
Flame Cut Off	1. Oxygen pressure is too high. 2. The flame outlet is clogged.	1. Adjust the oxygen pressure. 2. Clean the tip.
Popping Noises During Operation	1. The tip is overheated. 2. The tip is clogged. 3. The gas pressure adjustment is incorrect. 4. Metal deposited on the tip.	1. Cool the flame outlet (white letting a little oxygen flow). 2. Clean the tip. 3. Adjust the gas pressure. 4. Clean the tip.
Reversed Oxygen Flow (Oxygen is flowing into the path of the acetylene.)	1. The tip is clogged. 2. Oxygen pressure is too high. 3. Torch is defective. (The tip or valve is loose.) 4. There is contact with the tip and the deposit metal.	1. Clean the tip. 2. Adjust the oxygen pressure. 3. Repair or replace the torch. 4. Clean the orifice.
Backfire (There is a whistling noise and the torch handle grip gets hot. Flame is sucked into the torch.)	1. The tip is clogged or dirty. 2. Oxygen pressure is too low. 3. The tip is overheated. 4. The tip orifice is enlarged or deformed. 5. A spark from the base metal enters the torch, causing an ignition of gas inside the torch. 6. Amount of acetylene flowing through the torch is too low.	1. Clean the tip. 2. Adjust the oxygen pressure. 3. Cool the tip with water (letting a little oxygen flow). 4. Replace the tip. 5. Immediately shut off both torch valves. Let torch cool down. Then relight the torch. 6. Readjust the flow rate.

whose melting point (temperature) is lower than that of the base metal, is melted without melting the base metal (Figure 7–119). Brass brazing is frequently applied to automotive bodies.

Brazing is similar to joining two objects with adhesives; melted brass sufficiently spreads between the base metals to form a strong bond. Braze joint strength is less than that of the base metal, but the same as the melted brass. Therefore, never use brazing as a structural joint unless recommended by the vehicle manufacturer.

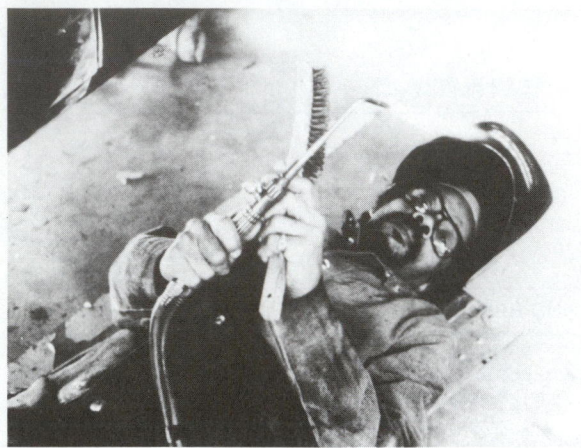

FIGURE 7-118 Cleaning of metal is sometimes done with a gas torch and wire brush.

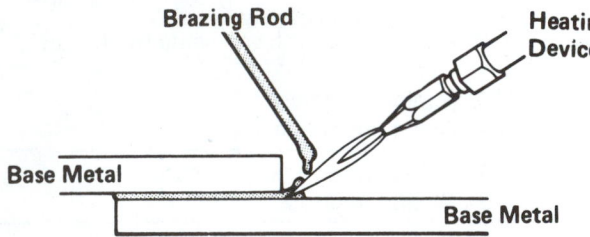

FIGURE 7-119 The brazing principle involves heating the base metal until the molten metal sticks to its surface. The base metals do not become molten as in welding. *(Courtesy of Toyota Motor Corp.)*

There are two types of brazing:

Soft brazing (soldering)

Hard brazing (brass or nickel)

Ordinarily, the term *brazing* refers to hard brazing. The basic characteristics of brazing are:

- The pieces of base metal are joined together at a relatively low temperature where the base metal does not melt. Therefore, there is a lower risk of distortion and stress in the base metal.
- Because the base metal does not melt, it is possible to join otherwise incompatible metals.
- Brazing metal has excellent flow characteristics; it penetrates well into narrow gaps and it is convenient for filling gaps in body seams.
- Since there is no penetration and the base metal is joined only at the surface, it has very low strength to resist repeated loads or impacts.
- Brazing is a relatively easy skill to master.

Automobile assembly plants sometimes use arc brazing to join the roof and quarter panels together (Figure 7–120). Arc brazing uses the same principles as MIG welding. However, argon is used with brazing metal instead of CO_2 or an argon/CO_2 mixture (Figure 7–121). Special brazing wire is also required. Since the amount of heat applied to the base metal is low, overheating is minimized. There is little distortion or warpage of the base metal. Compared to

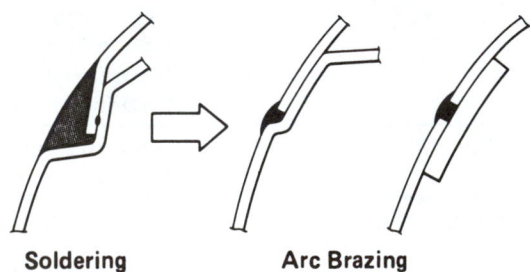

Soldering   Arc Brazing

FIGURE 7-120 Note the typical body construction using solder or arc brazing. *(Courtesy of Toyota Motor Corp.)*

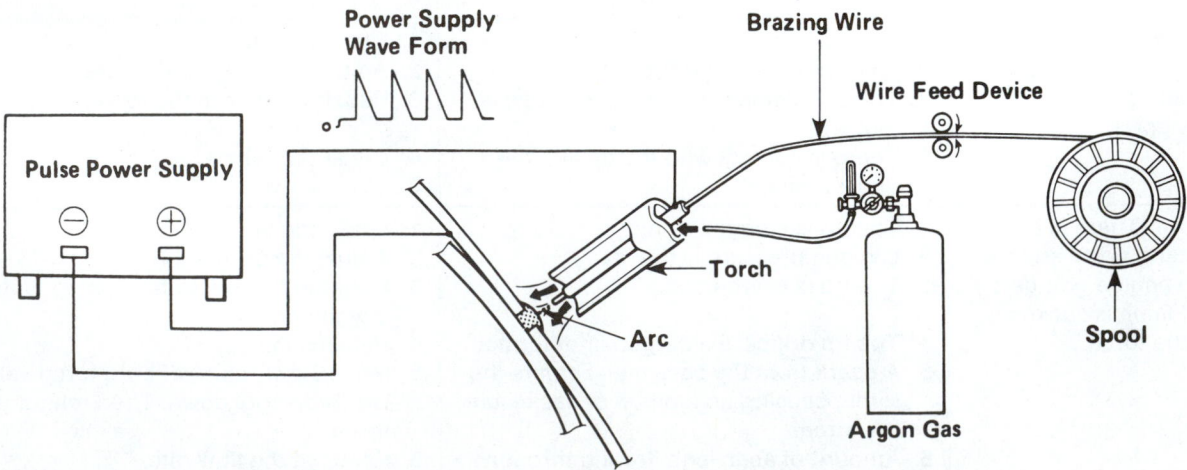

FIGURE 7-121 Study arc brazing principles. *(Courtesy of Toyota Motor Corp.)*

TABLE 7-9	
Types of Brazing Materials	**Main Ingredients**
Brass brazing metal	Copper, Zinc
Silver brazing metal	Silver, Copper
Phosphor copper brazing metal	Copper, Phosphorus
Aluminum brazing metal	Aluminum, Silicon
Nickel brazing metal	Nickel, Chrome

flame brazing, arc brazing shortens both the time for making the weld and finishing. Also, there is no danger of lead poisoning.

In the body shop, the brazing equipment is usually about the same as oxyacetylene welding. For brazing, an oxyacetylene torch, brass filler rods, flux welding goggles, gloves, and a torch lighter are needed. While the oxyacetylene torch can be used in soft brazing (soldering), it is best to use one designed for soldering.

To have brazing material with good qualities, such as flow characteristics, melting temperature, and compatibility with base metal and strength, it is made of two or more metals that form an *alloy* (Table 7–9). Copper and zinc are the main ingredients of the brazing rods used on auto bodies.

INTERACTION OF FLUX AND BRAZING RODS

Generally the surfaces of metals exposed to the atmosphere are covered with an oxidized film, which, if heat is applied, thickens. *Flux* not only removes this oxidized film, but prevents the metal surface from reoxidizing. It also increases the bond between the base metal and the brazing material.

If a brazing material is melted over a surface that has an oxidized film and foreign matter adhering to it, the brazing material will not adequately bond to the base metal. Surface tension will cause the brazing material to ball up and not stick to the base metal (Figure 7–122A).

The oxidized film can be removed by applying flux to the surface of the base metal and then heating it until it becomes liquid (Figure 7–122B). After the oxidation has been removed, the brazing material will adhere to the base metal and the flux will prevent further oxidation.

BRAZING JOINT STRENGTH

Since the strength of the brazing material is lower than that of the base metal, the shape of the joint and the clearance of the joint are extremely important. Figure

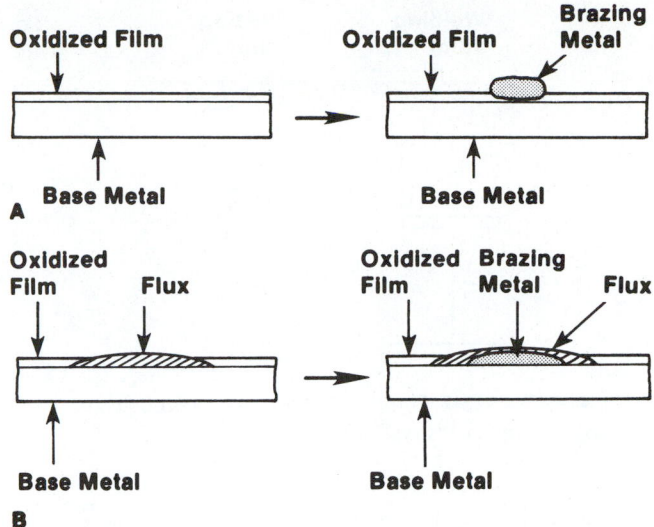

FIGURE 7-122 (A) Compare when flux is not used and (B) when flux is used. *(Courtesy of Toyota Motor Corp.)*

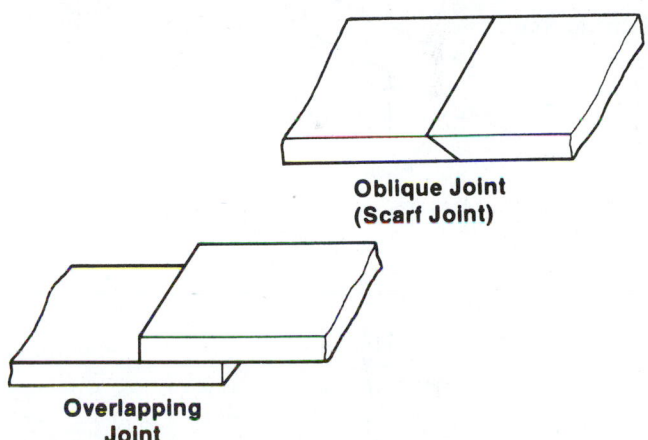

FIGURE 7-123 These are two basic brazing joints. *(Courtesy of Toyota Motor Corp.)*

7–123 shows a basic brazing joint. Joint strength is dependent on the surface area of the pieces to be joined. Therefore, make the joint overlap as wide as possible.

Even when the items being joined are of the same material, the brazed surface area must be larger than that of a welded joint (Figure 7–124). As a general rule, the overlapping portion must be *three* or more times wider than the panel thickness.

BRAZING OPERATIONS

General brazing procedure is as follows:

1. *Cleaning the base metal.* If there is oxidation, oil, paint, or dirt on the surface of the base metal, clean the surfaces before brazing. These contaminants, if allowed to remain on the surface, can cause eventual joint failures. Even though flux acts to remove oxidized film and most

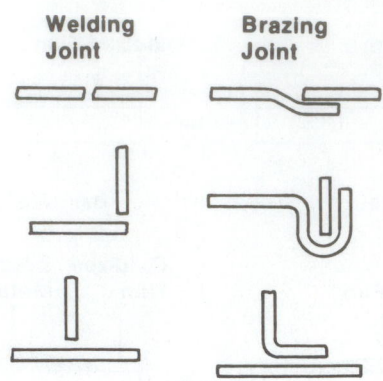

FIGURE 7-124 Compare welded and brazed joints. *(Courtesy of Toyota Motor Corp.)*

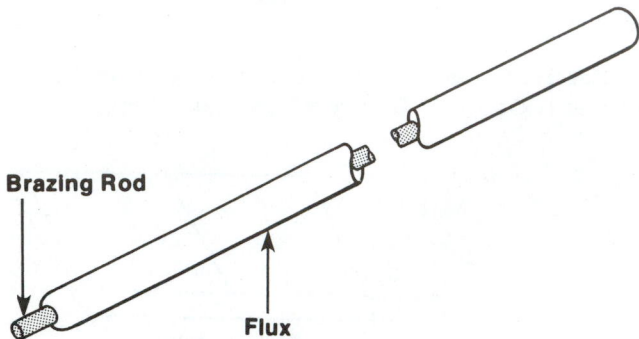

FIGURE 7-125 This is a brazing rod with flux.

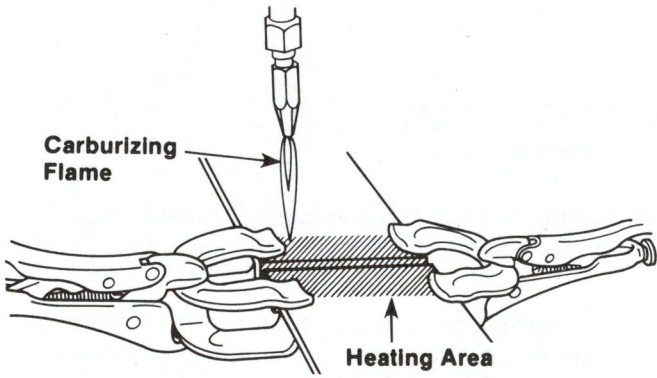

FIGURE 7-126 Use a carburizing flame to heat the base metal. *(Courtesy of Toyota Motor Corp.)*

contaminants, it is not strong enough to completely remove everything. Therefore, first clean the surface mechanically with a wire brush.

2. *Flux application.* After the base metal is thoroughly cleaned, apply flux uniformly to the brazing surface. (If a brazing rod with flux in it is used (Figure 7–125), this operation is not necessary.)

3. *Base metal heating.* Heat the joining area of the base metal to a uniform temperature capable of accepting the brazing material (Figure 7–126). Adjust the torch flame so that it is a slight carburizing flame. By watching the melting flux, you can estimate the proper temperature for the brazing material.

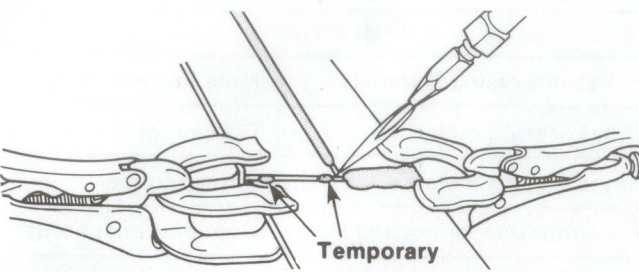

FIGURE 7-127 Once base metal is heated, brazing metal can be applied to surfaces. *(Courtesy of Toyota Motor Corp.)*

4. *Base metal brazing operation.* When the base metal has reached the proper temperature, melt the brazing material onto the base metal (Figure 7–127). Let the braze metal flow out naturally. Stop heating the area when the brazing material has flowed into the gaps of the base metal.

Other points to consider are:

- Since brazing material flows easily over a heated surface, it is important to remember to heat the entire joining area to a uniform temperature.
- Do not melt the brazing material prior to heating the base metal because the brazing material will not adhere to the base metal.
- If the surface temperature of the base metal becomes too high, the flux will not clean the base metal, resulting in a poor brazing bond and inferior joint strength.

The following additional precautions should be taken when brazing:

- Brazing temperature must be higher than the melting point of brass by 50 to 190 degrees Fahrenheit (10 to 89 degrees Celsius).
- The size of the torch tip must be slightly larger than the thickness of the panel.
- Preheat the panel to deposit brazing filler metal more efficiently.
- Secure the panel to prevent the base metal from moving and the brazing zone from breaking.
- Evenly heat the portion to be welded without melting the base metal.
- Control the heat by tilting the torch more horizontally (flatter to surface) or by removing the flame and allowing the area to cool briefly.
- The brazing time must be as short as possible (to prevent weld strength from lowering).
- Avoid brazing the same place again.

TREATMENT AFTER BRAZING

Once the brazed portion has cooled down sufficiently, rinse off the remaining flux sediment with water. Scrub the surface with a stiff wire brush. Baked and blackened flux can be removed with a sander or a sharp-pointed tool. If the remaining flux sediment is not adequately removed, the paint will not adhere properly. Corrosion and cracks might form in the joint.

7.16 SOLDERING (SOFT BRAZING)

Soldering is not used to reinforce the panel joints. It is used only for final finishing, such as in leveling the panel surface and correcting the surface of the welded joints. Because soldering functions by "capillary phenomenon," it has outstanding sealing ability.

Before attempting to solder a joint, remove paint, rust, oil, and other foreign substances.

SOLDERING PROCEDURE

After the surface has been thoroughly cleaned, proceed as follows:

1. Heat the portion to be soldered. Wipe it with a cloth after heating.
2. Stir solder paste well, and apply it with a brush. Apply it to an area 1 to $1^1/_2$ inches (25.4 to 38 mm) larger than the built-up area.
3. Heat it from a distance.
4. Wipe the solder paste from the center to the outside.
5. Make sure the soldered portion is silver gray. If it is bluish, it is due to overheating. If any spot is not soldered, reapply the paste for soldering.

When soldering, keep the following points in mind:

- It is desirable to use a special torch for soldering. If a gas welding torch is used, the oxygen and acetylene gas pressures must be 4.3 to 5.0 psi (29.7 to 34 kPa).
- The solder must contain at least 13 percent zinc.
- Maintain the appropriate temperature.
- Move the torch so that the flame evenly heats the entire portion to be soldered (without heating a single spot only).
- When the solder begins to melt, remove the flame and start finishing with a spatula.
- When additional solder is required, the previously built-up solder must be reheated.

7.17 PLASMA ARC CUTTING

Plasma arc cutting creates an intensely hot air stream, which melts and removes metal, over a very small area. Extremely clean cuts are possible with plasma arc cutting. Because of the tight focus of the

heat, there is no warpage, even when cutting thin sheet metal.

Plasma arc cutting is replacing oxyacetylene as the best way to cut metals. It cuts damaged metal effectively and quickly but does not destroy the properties of the base metal. The old method of flame cutting just does not work that well anymore.

In plasma arc cutting, compressed air is often used for both shielding and cutting. As a shielding gas, air covers the outside area of the torch nozzle, cooling the area so the torch does not overheat.

Air also becomes the cutting gas. It swirls around the electrode as it heads toward the nozzle opening. The swirling action helps to constrict and narrow the gas. When the machine is turned on, a pilot arc is formed between the nozzle and the inner electrode. When the cutting gas reaches this pilot arc, it is super heated—up to 60,000 degrees Fahrenheit (33,315 degrees Celsius).

Figure 7–128A shows that there are two areas for gas flow. In air plasma arc cutting, compressed air is used for both shielding and cutting. As a shield gas, air shields the outside area of the torch nozzle, cooling the area so the torch does not overheat. Air also becomes the cutting gas. The air swirls around the electrode as it heads toward the nozzle opening. The swirling action helps constrict and narrow the gas. When the machine is turned on, a pilot arc is formed between the nozzle and the inner electrode (Figure 7–128B). When the cutting gas reaches this pilot arc, it is super heated—up to 60,000 degrees Fahrenheit.

The gas is now so hot it ionizes and becomes capable of carrying an electrical current (ionized gas is actually the plasma). The small, narrow opening of the nozzle accelerates the expanding plasma toward the workpiece. When the workpiece is close enough, the arc crosses the gap, with the electrical current being carried by the plasma (Figure 7–128C). This is the cutting arc.

The extreme heat and force of the cutting arc melt a narrow path through the metal. This serves to dissipate the metal into gas and tiny particles. The force of the plasma literally blows away the metal particles, leaving a clean cut.

A 10- to 15-amp plasma arc cutter is generally adequate for mild steel up to $^3/_{16}$ inch (5 mm) thick; a 30-amp unit can cut metal up to $^1/_4$ inch (6 mm) thick; and a 60-amp unit will slice through metal up to $^1/_2$ inch (13 mm) thick.

Controls are usually quite simple. Plasma arc cutters made specifically for thinner metals might only have an on/off switch and a ready light. More elaborate equipment can include a built-in air compressor, variable output control, on-board coolant, and other features.

On some units, a switch is provided that allows you to alter the current mode depending on the

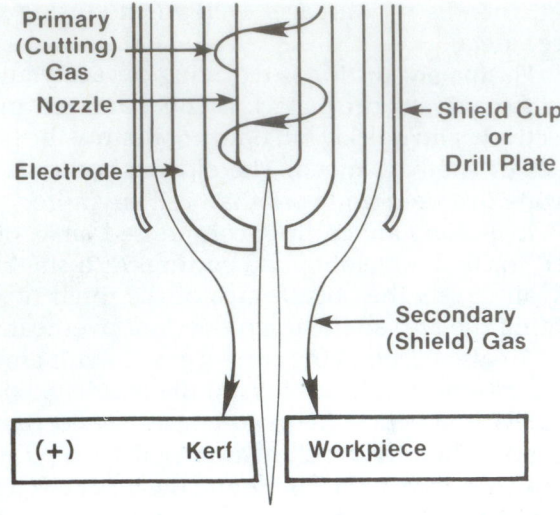

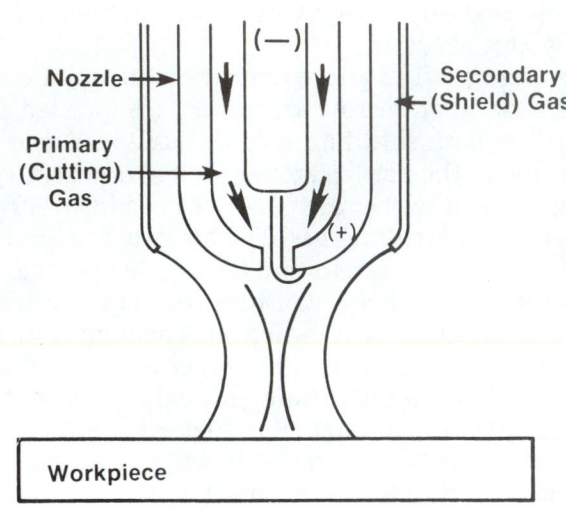

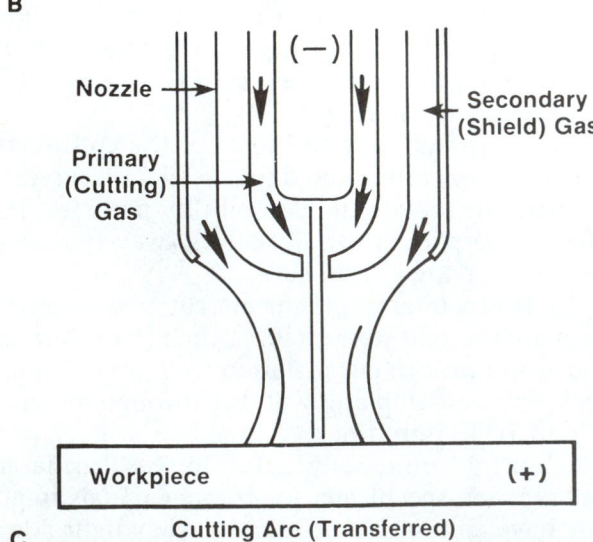

FIGURE 7-128 Note a typical plasma arc cutting setup. (A) Basic parts involved. (B) Pilot arc. (C) Cutting arc.

surface being cut. When cutting painted or rusty metal, a continuous high-frequency arc is best.

Two critical parts of the torch are the cutting nozzle and the electrode. These are the only consumables (besides air) in plasma arc cutting. If either the nozzle or the electrode is worn or damaged, the quality of the cut will be affected. They wear somewhat with each cut. Moisture in the air supply, cutting thick materials, or poor technique will make them fail more quickly. Keep a supply of electrodes and nozzles on hand and replace them when needed.

Today's plasma arc cutters do an excellent job using clean, dry compressed air. The air can be supplied through an external or built-in air compressor or by using a cylinder of compressed air. Cylinders of air can be expensive, while shop air is almost free. To reduce contaminants, use a regulator with a filter.

Also, check the air pressure regularly. Using the wrong pressure can reduce the quality of the cuts, damage parts, and decrease the cutting capacity of the machine.

OPERATING A PLASMA ARC CUTTER

To operate a typical plasma arc cutter (Figure 7–129), proceed as follows:

1. Connect the unit to a clean, dry source of compressed air with a minimum line pressure of 60 psi (413 kPa) at the air connection.
2. Connect the torch and ground clamp to the unit. After plugging the machine in, connect the ground clamp to a clean metal surface on the vehicle. The clamp should be as close as possible to the area to be cut. Various types of clamps are shown in Figure 7–130.
3. Move the cutting nozzle into contact with an electrically conductive part of the work. This must be done to satisfy the work safety circuit.
4. Hold the plasma torch so that the cutting nozzle is perpendicular to the work surface (Figure 7–131). Push the plasma torch down. This will force the cutting nozzle down until it comes in contact with the electrode. Then the plasma arc will start. Release downward force on the torch to let the cutting nozzle return to its normal position. While keeping the cutting nozzle in light contact with the work, drag the gun lightly across the work surface.
5. Move the plasma torch in the direction the metal is to be cut. The speed of the cut will depend on the thickness of the metal. If the torch is moved too fast, it will not cut all the way through. If moved too slowly, it will put too much heat into the workpiece and might also extinguish the plasma arc.

Other pointers that should be remembered when using a plasma arc cutter are:

1. When piercing materials ⅛ inch (3 mm) thick or more, angle the torch at 45 degrees until the plasma arc pierces the material. This will allow the stream of sparks to shoot off away from the gas diffuser.

   If the torch is held perpendicular to the work when piercing heavy gauge material, the sparks will shoot back up at the gas diffuser. The molten metal will collect on the diffuser.

This might plug the air holes and shorten the life of the diffuser.

2. Torch cooling is important to extend the life of the electrode and nozzle. At the end of a cut, the air continues to flow for several seconds. This prevents the nozzle and electrode from

CAUTION When angling the torch, be aware that the sparks can shoot as far as 20 feet (6 meters) away. Be sure that there are no combustibles or other workers in the area.

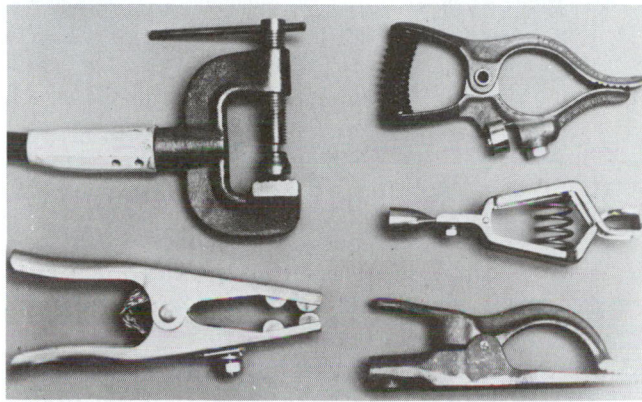

FIGURE 7-130 Study the various types of clamps.

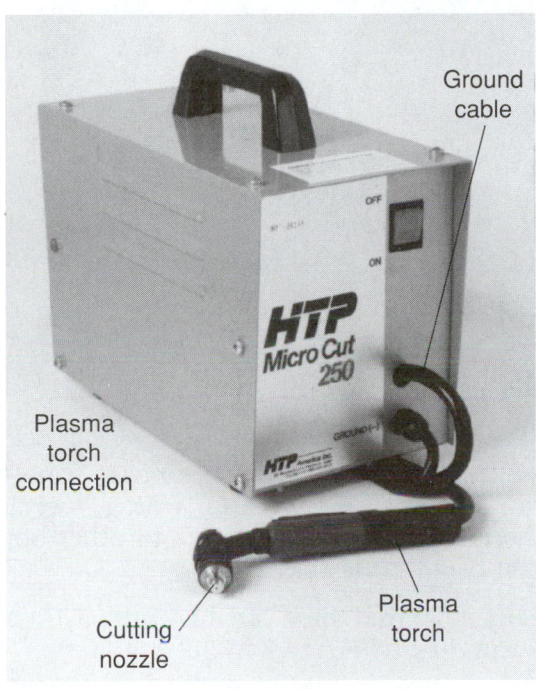

A

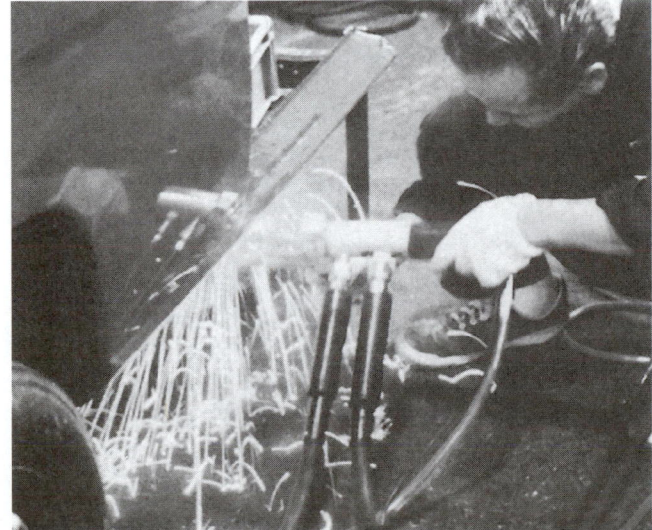

B

FIGURE 7-129 A typical plasma arc cutter will make rapid smooth cuts in metal. (A) Modern plasma arc cutter. (B) Technician using plasma arc cutter. *(Courtesy of HTP America, Inc.)*

FIGURE 7-131 When using a plasma arc cutter, be careful of molten metal spray on the backside of the cut. It could ignite and burn the interior parts of the vehicle. *(Courtesy of Century Mfg. Co.)*

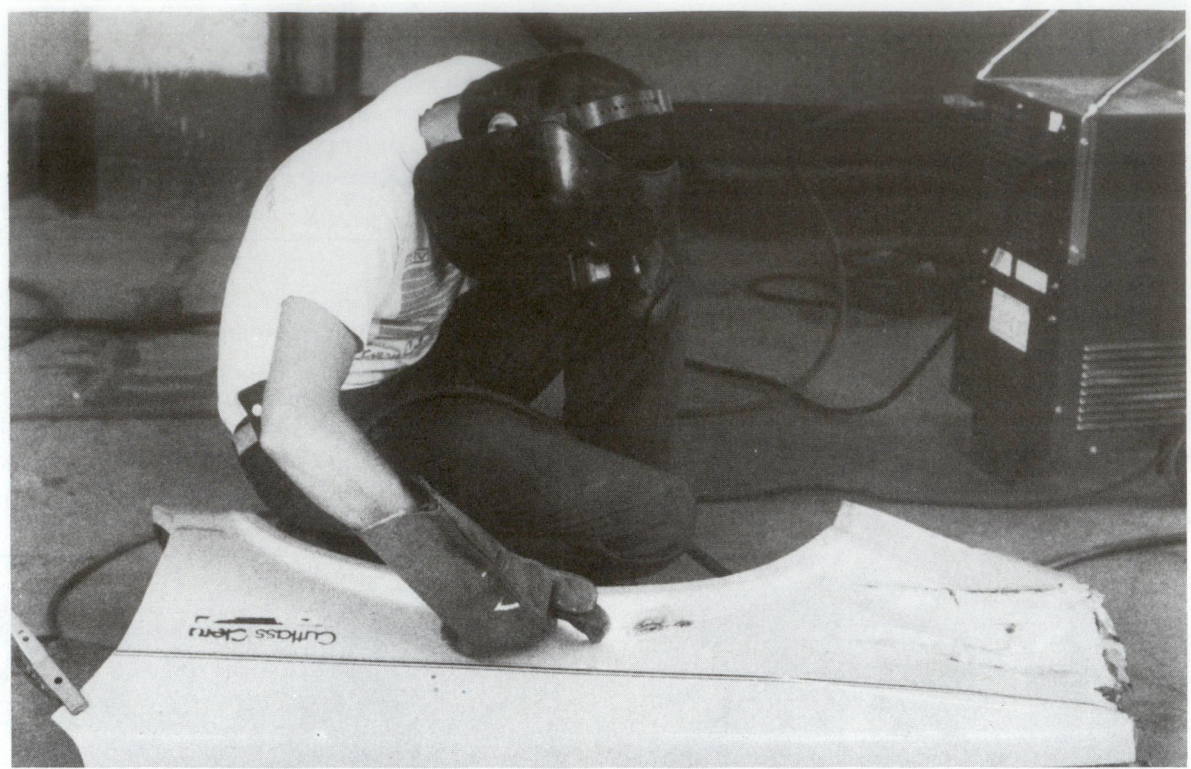

FIGURE 7-132 Making a quality cut takes practice. *(Courtesy of Century Mfg. Co.)*

overheating. Some equipment suppliers also recommend idling the unit for a couple of minutes after the cut is made.

3. When making long straight cuts, use a metal straightedge as a guide. Simply clamp it to the work to be cut. For elaborate cuts, make a template out of thin sheet metal or wood and guide the tip along that edge.

4. When cutting 1/4-inch (6 mm) materials, start the cut at the edge of the material.

5. When making rust repairs on cosmetic panels, it is possible to piece the new metal over the rusted area and then cut the patch panel at the same time that the rust is cut out. This process also works when splicing in a quarter panel.

6. Be aware of the fact that the sparks from the arc can damage painted surfaces and can also pit glass. Use a welding blanket to protect these surfaces.

SHOP TALK

Air ordinarily will not conduct electricity. But during very high voltage, the air molecules ionize and become electrically conductive. The air becomes super heated and forms a path along which voltage can easily flow.

7. Make sure there is nothing behind the panel that can be damaged. Check for wiring, fuel lines, sound deadening materials, and other objects that could cause a fire.

Remember that these variables will have a bearing on cut quality (Figure 7–132).

Travel speed. The thicker the material, the slower the speed. Travel is faster for thin material.

Parts wear. The tip and electrode will erode with use. The more wear, the poorer the quality of the cut.

Air quality. Moist or oil-contaminated air will contribute to a poor quality weld.

PLASMA AIR CUTTER

Some equipment (Figure 7–129) has a built-in safety protection system to protect the operator. This type of system cuts output power automatically if the safety cup is removed from the torch, if the tip and electrode are accidentally short circuited because of insufficient air pressure, or if the duty cycle is exceeded. The open circuit voltage of plasma cutting equipment can be very high (in the range of 250 to 300 volts), so insulated torches and internally connected terminals are also essential.

On some units, a switch is provided that allows you to alter the current mode when cutting bare or painted metal. When cutting painted or rusty metal, a continuous high frequency arc is best for punching through the nonconductive surface layer and for keeping the arc going while cutting. When cutting bare metal, a high frequency arc is needed only to start the arc. Once the torch starts to cut, a direct current pilot arc is all that is needed to keep things going. The bare metal position gives the longest electrode and nozzle life.

SUMMARY

- There are three basic methods of joining metal together in the automobile assembly:

  Mechanical (metal fastener) methods

  Chemical (adhesive fastening) methods

  Welding (molten metal) methods

  Welding is one method of repair in which heat is applied to the pieces of metal to fuse them together into the shape desired.

- Visible weld penetration is indicated by the height of the exposed surface of the weld on the back side. Full weld penetration is needed to assure maximum weld strength.

- MIG welding is recommended by all OEMs, not only for HSS and unibody repair, but for all structural collision repair.

- The resistance spot welder provides very fast, high-quality welds while maintaining the best control of temperature buildup in adjacent panels and structure.

- Always follow service manual recommendations when welding. This will assure structural integrity.

- During the welding process, either inert gas or active gas shields the weld from the atmosphere and prevents oxidation of the base metal.

- Flat welding means the pieces are parallel with the bench or shop floor.

- Horizontal welding has the pieces turned sideways. Gravity tends to pull the puddle into the bottom piece. When welding a horizontal joint, angle the gun upward to hold the weld puddle in place against the pull of gravity.

- Vertical welding has the pieces turned upright. Gravity tends to pull the puddle down the joint. When welding a vertical joint, the best procedure is usually to start the arc at the top of the joint and pull downward with a steady drag.

- Overhead welding has the workpieces turned upside down. The tack weld is exactly that: a tack—a relatively small, temporary MIG spot weld that is used instead of a clamp or sheet metal screw to tack and hold the fit-up in place while a permanent weld is made.

- In a continuous weld, an uninterrupted seam or bead is laid down in a slow, steady, ongoing movement.

- A plug weld is made in a drilled or punched hole through the outside piece (or pieces).

- If welding defects should occur, think of ways to change your procedures to correct them.

ASE-STYLE REVIEW QUESTIONS

1. Technician A uses a forward gun angle to achieve a deep penetration in the metal. Technician B uses the reverse gun angle to achieve a flat bead. Who is correct?

   A. Technician A
   B. Technician B
   C. Both A and B
   D. Neither A nor B

2. Technician A says that the main function of the gun nozzle is to provide gas protection. Technician B says that if the insulation in the gun nozzle area is bypassed the current will ignite the inert shielding gas. Who is correct?

   A. Technician A
   B. Technician B
   C. Both A and B
   D. Neither A or B

3. Welding current affects which of the following?

   A. Base metal penetration depth
   B. Arc stability
   C. Amount of weld spatter
   D. All of the above

4. When MIG welding, what happens if the tip-to-base metal distance is too long?

   A. The shield gas effect is reduced.
   B. The wire protruding from the end of the gun increases and becomes preheated.
   C. The melting speed of the wire increases.
   D. All of the above.

5. Technician A starts a butt weld in the center of the metal. Technician B says that it is wise not to weld more than ³/₄-inch (19 mm) at one time. Who is correct?

   A. Technician A
   B. Technician B
   C. Both A and B
   D. Neither A or B

6. Which of the following welds is the body shop alternative to the OEM resistance spot welds made at the factory?

   A. Spot
   B. Plug
   C. Stitch
   D. All of the above

7. What determines the length of a tack weld?

   A. Operator preference
   B. Thickness of the panel
   C. Type of base metal being welded
   D. Type of shielding gas being used

8. When using a resistance welder, Technician A installs a larger diameter electrode tip to increase the diameter of the spot weld. Technician B says that when the tips are worn, a tip dressing tool can be used to reshape the tips. Who is correct?

   A. Technician A
   B. Technician B
   C. Both A and B
   D. Neither A or B

9. Which of the following statements concerning plasma arc cutting is incorrect?

   A. The plasma arc process cuts mangled metal effectively.
   B. Plasma cutting is an extension of the TIG process.
   C. The nozzle must come in contact with an electrically conductive part of the work before the arc can start.

   D. When piercing material that is more than ¹/₈-inch (3.1 mm) thick, hold the torch perpendicular to the work.

10. The typical acetylene pressure for oxyacetylene welding is

    A. 15 to 100 psi (103 to 689 kPa)
    B. 3 to 12 psi (21 to 83 kPa)
    C. 3 to 25 psi (21 to 173 kPa)
    D. 30 to 120 psi (207 to 827 kPa)

11. Mixing slightly more acetylene than oxygen will obtain what type of flame?

    A. Neutral
    B. Standard
    C. Carburizing
    D. Oxidizing

12. Which of the following is *not* characteristic of brazing?

    A. Relatively high strength
    B. Can join parts of varying thickness
    C. Greater risk of distortion in the base metal
    D. Can join otherwise incompatible metals

13. When operating an MIG welder, which of the following indicates the correct wire-to-heat ratio?

    A. An even, high-pitched buzzing sound
    B. A steady, reflected light
    C. Both A and B
    D. Neither A nor B

14. Technician A says that all steels can be MIG welded with one common type of weld wire. Technician B says that metals of different thicknesses can be MIG welded with the same diameter wire. Who is correct?

    A. Technician A
    B. Technician B
    C. Both A and B
    D. Neither A nor B

ESSAY QUESTIONS

1. Summarize the MIG process.

2. Describe basic guidelines when MIG welding aluminum.

3. Describe plasma arc cutting.

CRITICAL THINKING PROBLEMS

1. If undercutting occurs while MIG welding, what should you do?

2. What can be done to prevent heat build-up during welding?

MATH PROBLEMS

1. When setting up a typical plasma arc cutter, the air pressure gauge shows only 20 psi (138 kPa). How much should this pressure be changed?

2. During a butt weld on a thin panel, the technician has welded ¹/₄ inch (6.4 mm). How much further can he or she go before stopping to allow cooling?

CHAPTER 8

Basic Auto Sheet Metal Work

INTRODUCTION

To do quality sheet metal repairs, you must first know how to return the sheet metal to its original shape. You can then use a thin layer of plastic filler to smooth the surface above the panel.

Metalworking skill is probably the most important craft a body technician can bring to a shop. It is also probably one of the most neglected skills. An untrained worker can spend more time shaping and sculpting an overthick layer of plastic filler than would need to be spent properly reworking the damaged metal. Not only does this waste valuable shop time and materials, repair quality also suffers.

This chapter will introduce you to basic metalworking methods. It will explain how to analyze minor damage to sheet metal before showing you how to repair the damage. Good metalworking skills are critical to your success as an auto body technician (Figure 8–1).

8.1 AUTOMOTIVE SHEET METAL

There are two types of sheet metal used in automobile construction—hot-rolled and cold-rolled.

Hot-rolled sheet metal is made by rolling at temperatures exceeding 1,472 degrees Fahrenheit (800 degrees Celsius). It has a standard manufacturing thickness range of $1/16$ to $5/16$ inch (1.6 to 8 mm). It is used for comparatively thick parts, such as frames and cross members.

OBJECTIVES

After studying this chapter, you should be able to:

✔ Describe different types of metals used in vehicle construction.
✔ Explain the strength ratings of metals.
✔ Summarize the deformation effects of impacts on steel.
✔ Use a hammer and dolly to straighten metal.
✔ Explain how to bump dents with spoons.
✔ List the steps for shrinking metal.
✔ Prepare a surface for filler.
✔ Mix filler and cream hardener properly.
✔ Apply filler correctly.
✔ Use recommended methods for shaping filler.
✔ List common mistakes made when using filler and spot putty.
✔ Pass ASE Certification tests relating to sheet metal work.

KEY TERMS

cleavage
cold-rolled sheet metal
compression
deformation
direct damage
gouge
hammer-off-dolly
hammer-on-dolly
high-strength steel
hot-rolled sheet metal
indirect damage
kinking
lowering
paintless dent removal

peel
pressure area
pull pin spot welder
raising
rough-out
shear
shrinking metal
spring-back
stretched metal
tensile strength
tension
tension area
work hardening
yield point
yield strength

ASE TASK LIST

Job Skills covered in this chapter include:

PAINTING AND REFINISHING TEST (B2) TASK LIST

A. Surface Preparation

5. Dry or wet sand areas to be refinished.
6. Featheredge areas to be refinished.

NONSTRUCTURAL ANALYSIS AND DAMAGE REPAIR TEST (B3) TASK LIST

B. Outer Body Panel Repairs, Replacements, and Adjustments

1. Determine the extent of the direct and indirect damage and the direction of impact; plan the methods and order of repair.
7. Remove, replace, and align bumpers, reinforcements, guards, isolators, and mounting hardware.
10. Rough out contours of damaged panel to a surface condition suitable for metal finishing and body filling.
13. Remove damaged sections of steel body panels; weld in replacements in accordance with manufacturers'/industry specifications.

C. Metal Finishing and Body Filling

1. Grind off paint from the damaged area of a body panel.
2. Pick and file the damaged area of a body panel to eliminate surface irregularities.
3. Prepare surface for application of body filler material.

5. Cold shrink stretched panel areas to proper contour.
6. Mix plastic filler.
7. Apply plastic body filler; rough shape during curing.
8. Sand cured plastic body filler to contour.

D. Moveable Glass and Hardware

3. Inspect, repair or replace, and adjust removable, manually operated roof panels and hardware.

STRUCTURAL ANALYSIS AND DAMAGE REPAIR TEST (B4) TASK LIST

B. Unibody Inspection, Measurement, and Repair

13. Use proper cold stress relief methods.
14. Remove folds, curves, creases, and dents, using power tools and hand tools, to restore damaged areas to proper contours and dimensions.
15. Determine the extent of damage to structural steel body panels; repair, weld or replace in accordance with vehicle manufacturers' specifications.

C. Stationary Glass

1. Remove and replace front and rear stationary glass (heated and nonheated) in accordance with manufacturers' recommendations.

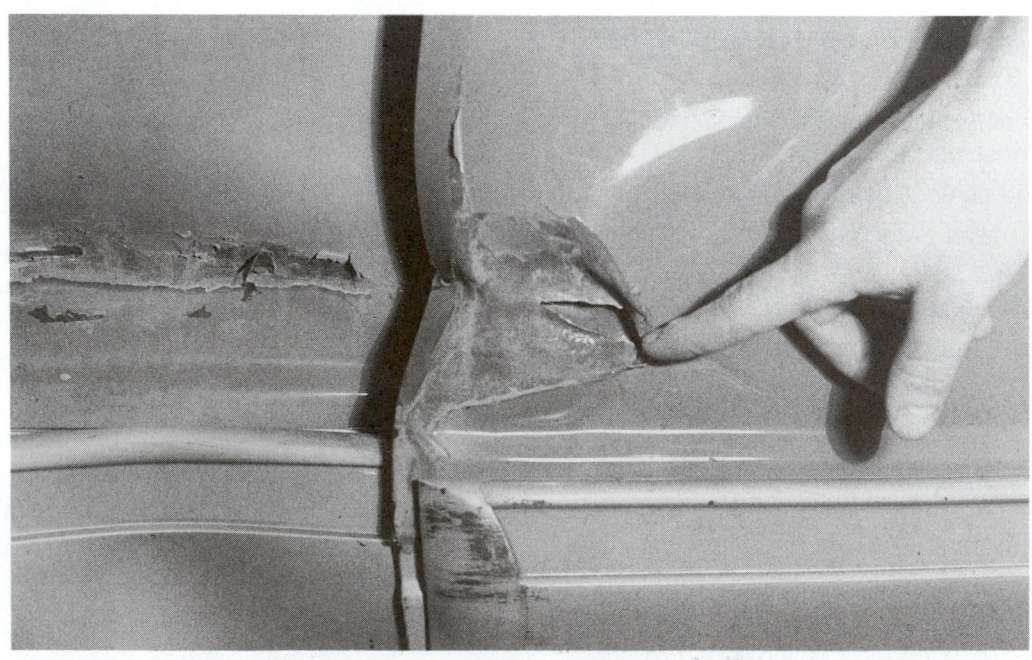

FIGURE 8-1 Damage to sheet metal, like this, can be difficult to repair if you are not properly trained.

Cold-rolled sheet metal is hot-rolled sheet metal that has been acid rinsed, cold rolled thin, then annealed. It has a dependable thickness accuracy, surface quality, and better workability than hot-rolled steel. Most unibodies are made from cold-rolled steel.

LOW-CARBON STEELS

Low-carbon, or *mild steel,* has a low level of carbon, is relatively soft and easy to work. A small amount of the sheet metal used in collision repair today can be low-carbon or mild steel (MS). It can be safely welded, heat shrunk, and cold worked without seriously affecting its strength. Mild steel has a yield strength of up to 30,000 psi (207,000 kPa).

Because MS is easily deformed and relatively heavy, vehicle manufacturers have begun using high-strength steels in load carrying parts of the vehicle.

HIGH-STRENGTH STEELS

High-strength steel (HSS) is stronger than low-carbon or mild steel because of heat treatment. Most new vehicles contain HSS in their structural components. It has a yield strength of up to 60,000 psi (413,400 kPa). HSS experiences an increase in stress, exceeding this yield strength, when deformed during a collision.

The same properties that give strength offer some unique challenges. When high-strength steel is deformed on impact, it is more difficult to restore than mild steel.

Types of High-Strength Steel

Many types of steel are generally classified as high-strength steel. Before explaining their differences, it is important to understand the definition of strength.

There are two types of strength, and both relate to the ability of the metal to resist permanent deformation.

- **Yield strength,** or yield stress, is measured as the minimum force per unit of area that causes the material to begin to permanently change its shape.
- **Tensile strength,** or tensile stress, is measured as the maximum force per unit of area that causes a complete fracture or break in the material.

Both stress and strength are expressed in pounds per square inch or kilograms per square millimeter. Steel strength can be increased by a variety of manufacturing processes that include heat treatment, cold rolling, and chemical additives.

High-strength sheet metal is divided into the following three types according to its strengthening process:

- *High-strength, low-alloy (HSLA) steel,* or rephosphorized steel, is produced by adding phosphorus to mild steel to upgrade its strength level. It has working characteristics that are similar to those of mild steel and was developed in recent years to provide better tensile strength to the exterior panels of auto bodies.
- *High-tensile-strength steel (HSS),* or Si-Mn solid solution hardened steel, contains increased amounts of silicon, manganese, and carbon to give it a higher tensile strength. It has been used in the past for suspension-related components, frames, and so on. It is used mainly for door side guards, bumper reinforcements, structural members on unibody vehicles, and so on.
- *Ultrahigh-strength steel (UHSS),* or dual phase steel, is made by quenching the steel on a continuous annealing line or in a hotstrip mill. This steel has a two-phase microstructure (quenched martensitic structure and ferritic structure). Dual phase steel has good formability for HSS. Martensitic steel is the best known of UHSS.

High-Tensile-Strength Steel

High-tensile-strength steel (HSS) is stronger than low-carbon or mild steel because of heat treatment. Most new vehicles contain HSS in body structural components. Conventional heating and welding methods will not adversely affect the strength level of HSS.

The material will experience an increase in stress, exceeding the yield strength, as it is deformed during a collision. When heat is applied to an HSS component to assist in straightening, the stresses resulting from the collision are decreased, thereby restoring the strength to a lower or normal level. If the collision stresses exceed the tensile strength, the material will tear or fracture.

Oxyacetylene can be used to aid in restoring the component, heating a part being straightened, for example. However, extreme caution must be exercised

 **WARNING** Door guard beams and some bumper reinforcements cannot be straightened and should be replaced. Minor damage on door guard beams can be ignored if it does not interfere with door alignment or function. If the corrugations in the beam are dented or deformed, the door beam should be replaced.

when using oxyacetylene. Thermal crayons or paint should be applied around the area to be heated with an oxyacetylene torch to restrict these temperatures to 1,200 degrees Fahrenheit (649 degrees Celsius).

All new or used replacement parts should be MIG welded, using AWS-ER-70S-6 wire, which has the same strength level of HSS.

High-Strength, Low-Alloy Steel

High-strength, low-alloy steel (HSLA) is sometimes used for body structural components, such as front and rear rails, rocker panels, bumper face bars, bumper reinforcements, door hinges, and lock pillars. Its strength is mainly due to the addition of special chemical elements.

To avoid substantially reducing the ability of the structure to react to normal road loads or collision forces, never exceed the factory-recommended temperatures. A safe rule of thumb is to never heat it over 700 to 900 degrees Fahrenheit (377 to 482 degrees Celsius). Do not apply heat longer than three minutes. MIG welding is the most accepted practice for HSS and HSLA. Most automobile manufacturers do not recommend using oxyacetylene to weld either type.

Ultrahigh-Strength Steel

Ultrahigh-strength steels are alloy-free with tensile strength almost ten times that of typical mild steel. Many door beams and some bumper reinforcements are martensitic, or ultrahigh-strength steels. The unusually high strength properties are a result of a special grain or crystalline composition imparted during forming and fabricating.

Any reheating of ultrahigh-strength steel for repair destroys this unique composition and reduces strength to that typical of mild steel. In addition, these steels are so hard that they cannot be straightened cold with typical shop equipment. Therefore, a damaged martensitic, or ultrahigh-strength, steel member must not be repaired; it must be replaced. New parts should be installed by MIG plug welding.

PART LOADING

Loading refers to the type of force applied to a part to damage it. As shown in Figure 8–2, the basic types of loading are as follows:

1. **Tension** is a load that tries to pull parts straight apart.
2. **Compression** is a load that forces the parts straight into each other.
3. **Shear** is a load that pulls parts sideways.
4. **Cleavage** is a load that tries to force parts apart from an angle.

5. **Peel** is a load that pulls parts straight away from each other.

During a collision, one or more of these types of loads may damage the parts of the vehicle.

PROPERTIES OF STEEL SHEET METAL

To repair collision damage, you must also understand what property changes have taken place in the metal.

Deformation refers to the new, undesired bent shape the metal takes after an impact or collision. There are various ways to measure the strength of a metal. All relate to the metal's ability to resist deformation.

1. *Yield stress* is the amount of strain needed to permanently deform a test specimen. *Ultimate strength* is a measure of the load that breaks a specimen. The tensile strength of a metal can be determined by a tensile testing machine.
2. *Compressive strength* is the property of a material to resist being crushed.
3. *Shear strength* is a measure of how well a material can withstand forces acting to cut or slice it apart.
4. *Torsional strength* is the property of a material that withstands a twisting force.

Strength is expressed in pounds per square inch (psi) or kilograms per square millimeter (kPa).

Even though most types of steel look alike, there are differences in their chemical makeup and crystalline structure. These invisible differences can affect strength and sensitivity to heat. There is a variety of high-strength steels. All have unique properties that dictate the way in which they can be repaired.

When flat sheet steel is formed into a shape for a part, it takes on certain properties that harden it.

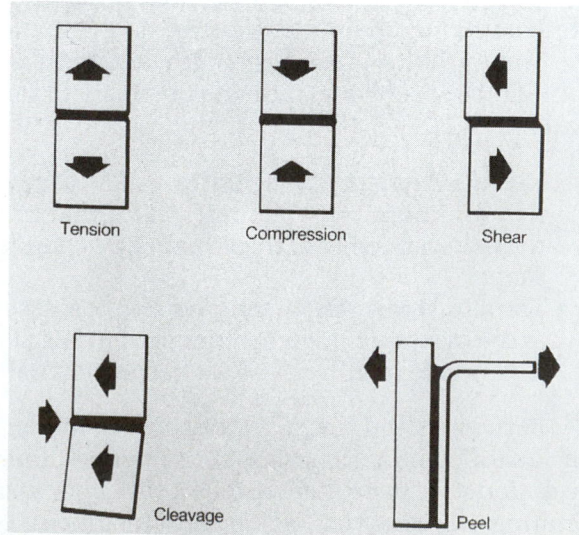

FIGURE 8–2 Note the types of loads that can be applied to parts during a collision. *(Courtesy of Tech-Cor)*

For example, a roof panel is relatively flat. If hit lightly in the center, the panel will usually bend and then pop back to its original shape. However, if you hit a panel that has a curved shape, the panel will hardly move. Although both are the same steel, the one that has been changed the most will be stronger and more resistant to bending.

The same is true for panels whose shape has been changed during a collision. The structure of the metal in the affected areas has changed, causing the metal to become harder and more resistant to corrective forces.

Physical Structure of Steel

The best way to define these properties is by first elaborating on the physical structure of sheet metal. Steel, just like all matter, is composed of atoms. These minute particles of matter are combined to form grains. Metal grain is so small that it can be seen only with the aid of a microscope. Grains are formed into patterns called *grain structures*.

In mild steels, the individual grains can withstand a considerable amount of change and movement before splitting or breaking occurs. To demonstrate this, take a welding rod and bend it back and forth several times. Notice that in the area of the bend the metal will become very hot. The heat is generated by the internal friction created as the individual grains move against each other in the area of the bend.

Effect of Impact Forces

The grain pattern of a metal will determine how it reacts to force. Sheet metal's resistance to change has three properties: elastic deformation, plastic deformation, and work hardening. All of these properties are related to the yield point.

Yield point is the amount of force that a piece of metal can resist without tearing or breaking.

Spring-back is the tendency for metal to return to its original shape after deformation. It will occur in any area that is still relatively smooth. Many such areas will spring back to shape if they are released by relieving the distortion in the buckled areas.

Plasticity is important to the collision repair technician because both stretching and permanent deformation take place in various areas of most damaged panels.

Elastic Deformation

Elastic deformation (Figure 8–3) is the ability of metal to stretch and return to its original shape. For example, take a piece of sheet metal and gently bend it to form a slight arc. When released, it will spring back to its original shape. This tendency to spring

back makes it necessary for you to recognize elastic deformation in damaged panels. This will enable you to plan your work and to take advantage of any tendency of the damaged metal to spring back.

Spring-back will be found in any area that is still relatively smooth, even though it has been carried out of position by buckles formed in adjoining areas. Many such areas will spring back to shape if they are released by relieving the distortion in the buckled areas that hold them out of place.

Plastic Deformation

Plastic deformation is the ability of metal to be bent or formed into different shapes. When metal is bent beyond its elastic limit, it will have a tendency to spring back. However, it will not spring all the way back to its original shape. This is because the grain structure has been changed (Figure 8–4).

A piece of sheet metal, if bent into a tight U-shape, will spring back when released but not to its original shape. This is because the grain structure has taken on a new set.

Elasticity and plasticity are illustrated in Figure 8–5. The graph shows the relationship between the size of the load and the elongation of sheet metal

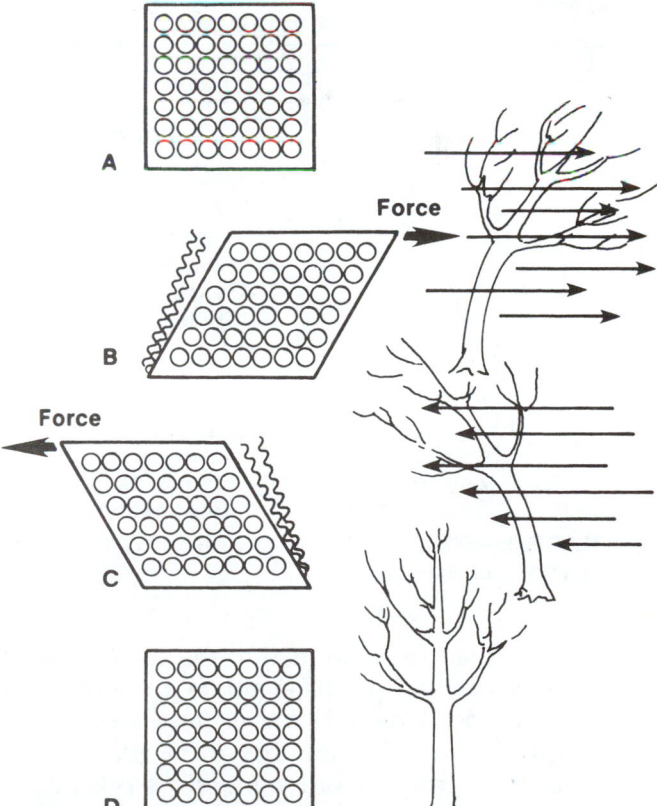

FIGURE 8–3 Elastic deformation. (A) Metal at rest. (B) Metal under pressure, metal bends like a tree in wind. (C) When force is released, metal rebounds. (D) Finally, metal returns to its original shape if not overbent, just like a tree bent in wind.

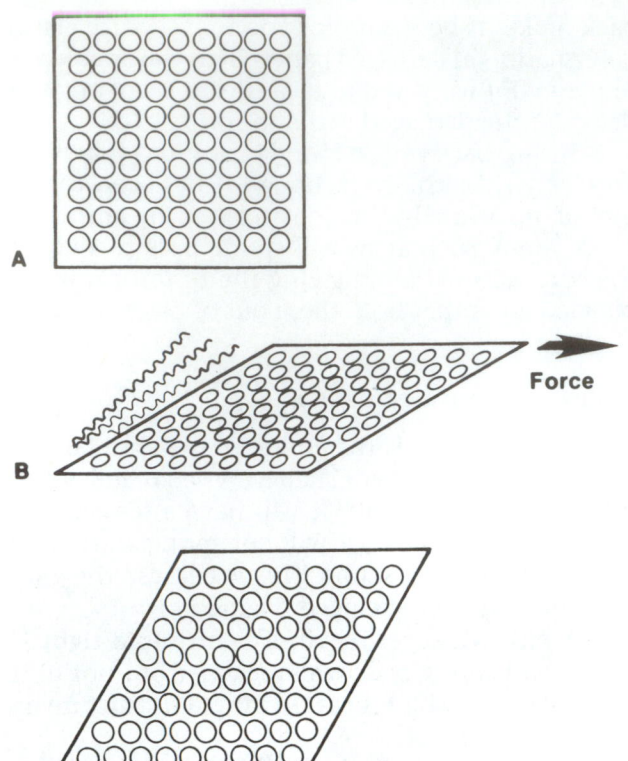

FIGURE 8-4 Plastic deformation occurs when grain structure is forced beyond elastic limit and takes on a new set. (A) Metal at rest. (B) Metal bent beyond elastic limit. (C) Metal takes on a new shape.

Point A in Figure 8–5 is called the elastic limit. If the load is lower than point A, deformation of the sheet metal will disappear when the load is removed. It will return to its original shape. This is called *elastic stress*. If the load exceeds point A, even if the load is removed, the deformation will remain. The panel will not return to its original shape. This is called *permanent plastic*, or *permanent stress*.

For example, if the load is removed at point P, the elongation of the panel will return to point E but permanent stress OE will remain. When a car is damaged in a collision, the stress sustained from the impact will remain unless it is taken out. This is a condition where an area with permanent stress is surrounded by a neighboring area with elastic stress that cannot be removed. When repairing body panels, first remove the permanent stress that is restricting the elastic stress. When that is done, the elastic stress will disappear naturally. The body panel will then return to its original shape by itself.

Work Hardening

Work hardening is the upper limit of plastic deformation, causing the metal to become very hard in the

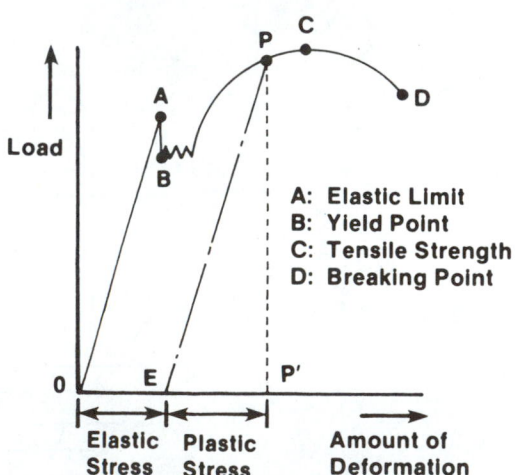

FIGURE 8-5 Study the graph showing load and deformation characteristics.

A: Elastic Limit
B: Yield Point
C: Tensile Strength
D: Breaking Point

when a tensile load is applied to the sheet metal. If the load is increased a little at a time, the elongation increases proportionally. However, if the load exceeds a certain limit, internal slipping of the grain pattern occurs. Even if the rate of load increase is kept constant, elongation will suddenly increase and the maximum load will be reached. After that, partial elongation will occur in one portion of the material and it will break.

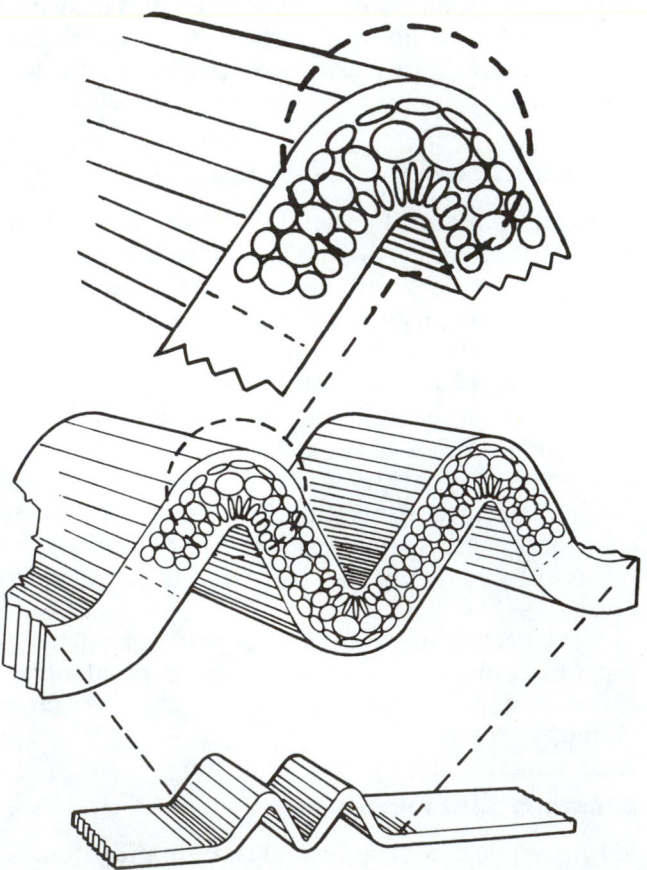

FIGURE 8-6 A bend, when severe, can result in work hardening.

bent area. For example, if a welding rod is bent back and forth several times, a fold or buckle will appear at the point of the bend. The plastic deformation has been so great at this point that the grain structure has been radically forced out of alignment, causing the metal to become very hard and stiff (Figure 8–6). This increased hardness is called work hardening.

The importance of understanding how metal stiffens, making it stronger in areas that are bent or worked, cannot be overemphasized. It is the basis of practically all damage repair.

Some work hardness will be found in any undamaged body panel. It is the result of the original forming process. The bending caused by a collision adds still more work hardening in the areas affected. Sometimes much more will be added by the cold working used by the body technician as the damaged area is straightened. If excessive work hardening is caused by working the metal improperly, the job will be made more difficult.

The untrained body technician can create as much damage during the repair as was done in the original impact. This is due to a lack of knowledge and skill. It is impossible to correct all the damage without creating some during straightening. With proper knowledge you can keep this problem to the absolute minimum.

8.2 CLASSIFYING BODY DAMAGE

The first step in auto body repair is analyzing the damaged area. A number of conditions that the body technician must recognize are present in any damaged panel. Each of the items listed below is a condition that occurs when metal is damaged by impact.

- Direct damage: tear, gouge, or scratch
- Indirect damage: buckle (fold or hinge in metal due to damage or tension) or pressures (unwanted force due to impact damage)
- Work hardening: normal and impact created

DIRECT DAMAGE

Direct damage is simple, easy to find, visible damage, such as a gouge, a tear, or a scratch. It is the damaged portion of the panel that came in direct contact with the object that caused the impact (Figure 8–7).

Direct damage is usually about 80 percent of the total damage. Direct damage repair at the point of impact is limited. Metal used in today's cars is often too thin to be reworked. Straightening is time consuming and usually not practical on areas containing direct damage. Direct damage usually requires some plastic filler or, on rare occasions, lead, after all

indirect damage has been handled. Direct damage varies from job to job.

INDIRECT DAMAGE

Indirect damage is caused by the shock of collision forces traveling through the body and inertial forces acting on the rest of the unibody. Indirect damage can be more difficult to completely identify and analyze. It may be found anywhere on the vehicle. Indirect damage represents, on the average, 10 to 20 percent of the overall damage.

BASIC METALWORKING METHODS

A few basic techniques will handle most bodywork with either direct or indirect damage.

Before doing anything else, take time to carefully perform an overall visual inspection and try to determine the direction of impact and the areas of indirect damage.

As previously stated, work hardening occurs whenever metal bends. It happens when the metal is first formed to shape by the manufacturer and also when it is damaged. To understand this, visualize a flat sheet of metal formed into a body panel, such as a fender. The flat sheet is placed in the stamping press. The edges are clamped securely. Then the center is stretched into the press until the once flat metal takes the shape of the press. The metal was relatively soft before it went into the press. Now it is quite hard. It has been work hardened by rearranging the grain structure. The areas that are still relatively flat are softer.

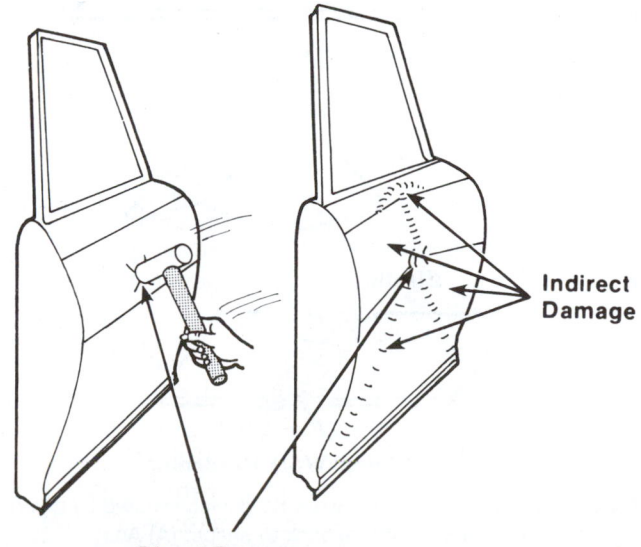

Indirect Damage

Direct Damage

FIGURE 8-7 Direct damage is the result of impact by another object. Indirect damage results from movement of metal, which affects other areas.

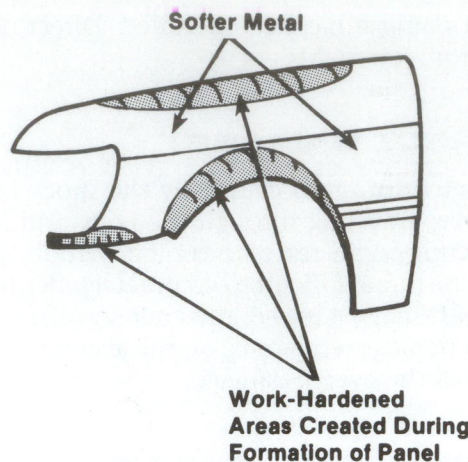

FIGURE 8-8 Factory-formed, work-hardened areas are often built into parts.

In the fender, illustrated in Figure 8–8, there are soft areas (unshaded) and hard areas (shaded). The shaded areas (crowns and ridges) are harder to damage. However, when damaged, they are more difficult to straighten. On the other hand, the flat metal will damage easily during straightening operations. The correct straightening techniques must be used to avoid damaging the undamaged metal.

Work hardening is in all sheet metal panels of a car to varying degrees. It is important to know where the metal was the hardest and softest before it was damaged.

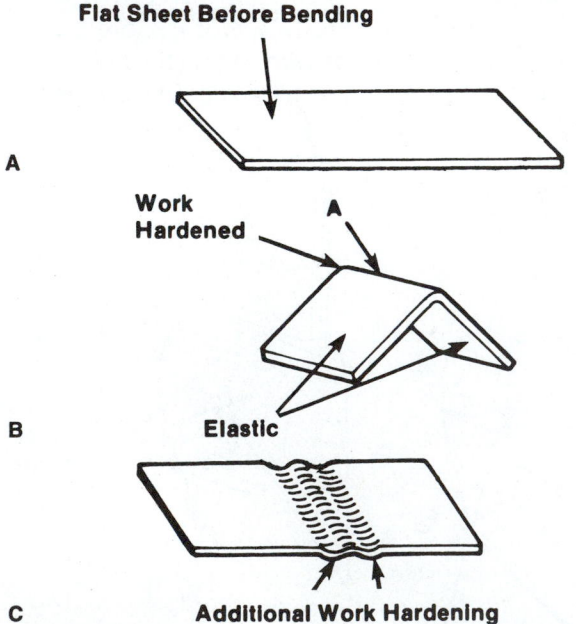

FIGURE 8-9 Additional work hardening created by trying to bend work-hardened metal back to shape. (A) An undamaged piece of sheet metal. (B) The sheet metal has been severely bent, exceeding elastic limit. (C) If you try to bend metal straight, the work-hardened areas in the bend will not straighten while unhardened areas on each side of the bend will bend. The result is more damage to the panel.

To demonstrate how work hardening affects the repair process, imagine a piece of steel about 12 inches (25 mm) long and 6 inches (12.5 mm) wide (Figure 8–9). One can bend this metal strip slightly and it will return to its original shape. If bent past a certain limit (elastic limit), the metal takes a set called a buckle. The metal surrounding the new bend returns to a straight condition. But at the point of the bend, work hardening has set in. If an attempt is made to bend the metal back to its original shape, two additional buckles (work hardened) are created adjacent to the original bend because the bend will not open up. It is too hard.

Buckles caused by impact create additional work hardening in an automotive sheet metal panel. Remember! A buckle is created only when the metal is bent to a point where it will not return to its original shape. Bent metal is not necessarily buckled metal. When metal is bent but returns to its original shape later, it is not buckled metal.

Work-hardened buckles are represented in Figure 8–10. The other areas are in the damage but are only bent metal, not buckled. It is important to recognize these areas as they play an important role in determining a sound repair procedure.

BUCKLES

Buckles are a result of bending metal past its elastic limit. If bent beyond this limit, metal will not return to its original shape. New work hardening has occurred and a new shape is formed. The buckles in indirect damage are classified as follows:

Simple hinge
Collapsed hinge
Simple rolled
Collapsed rolled

It is easier to think of them as two types (a hinge and a roll), each of which can be simple or collapsed.

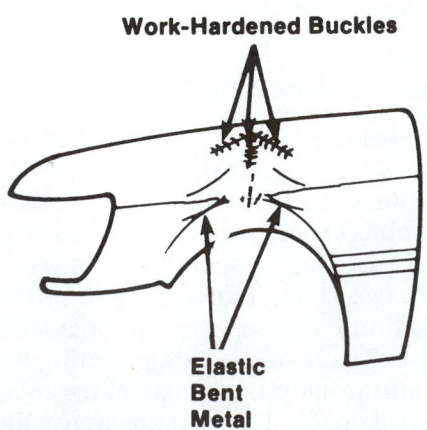

FIGURE 8-10 Always remember to locate the work-hardened and elastic areas in a typical dent.

Simple Hinge Buckle

The *simple hinge buckle* bends like a hinge equally along its entire length. The buckle usually causes little stretching or shrinking. If straightened incorrectly, however, it will cause considerable trouble. When bending is severe, it should be pulled out rather than pushed out.

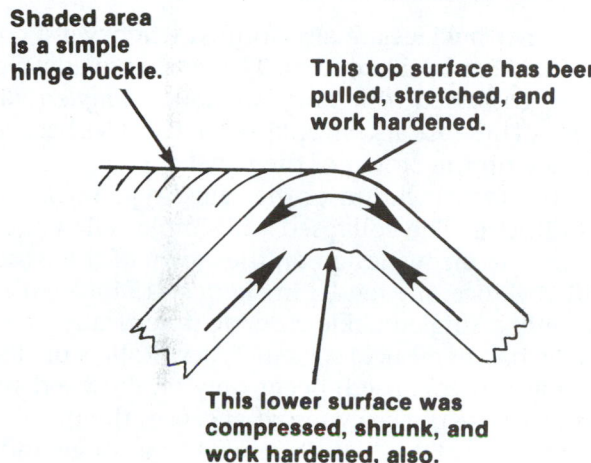

Shaded area is a simple hinge buckle.

This top surface has been pulled, stretched, and work hardened.

This lower surface was compressed, shrunk, and work hardened, also.

FIGURE 8-11 Analyze this hinge buckle.

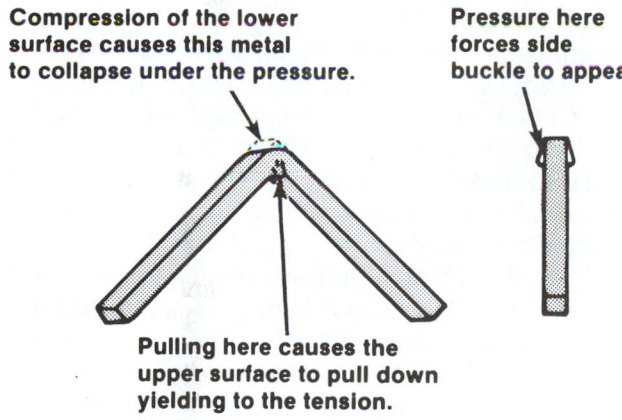

Compression of the lower surface causes this metal to collapse under the pressure.

Pressure here forces side buckle to appear.

Pulling here causes the upper surface to pull down yielding to the tension.

FIGURE 8-12 Study this collapsed box section.

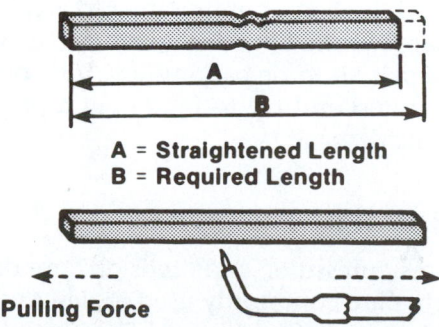

A = Straightened Length
B = Required Length

Pulling Force

FIGURE 8-13 Heat is applied to a work-hardened area to soften the metal and allow the rail to be pulled back to its original length.

Figure 8-11 is a drawing of a simple hinge buckle. It shows how the top surface stretches and how the bottom surface pushes together. This causes shrinkage on the underside of the metal. If there is stretching on the top and shrinkage on the bottom, it stands to reason that somewhere in the middle there is an area that is unchanged.

The correct repair procedures will result in straightening the metal and leaving the overall dimension the same. If corrected properly, the buckle reverses exactly as it was created. If not, additional damage is created to the adjacent metal and to the buckle. A simple hinge buckle always forms a "straight line" buckle.

The description of the simple hinge buckle refers to solid sheet metal. The basic rule of metal bending applies to box sections as well as solid metal. There is a difference, however, in the results of bending (Figure 8-12). The box section has no strength in its center and, as a result, the top is pulled downward instead of stretching. Little or no stretching occurs. The lower sheet metal is under pressure from both sides and buckles easily. If not straightened with care, the top surface will buckle also. Severe overall shrinkage is the result. Unlike the buckle in the solid metal, the box section collapses on both sides, along with the bottom (and the top, when incorrectly straightened).

Collapsed Hinge Buckle

Sheet metal gives very little resistance to pressure forces exerted on its end. It has enormous resistance to a pulling force along its length. To demonstrate this, take a sheet of metal about 1 inch (25 mm) x 8 inches (203 mm) in size. Try to push something with it. It takes very little effort to bend the metal. Now secure the metal to a bench or table and pull. The metal strip will pull hundreds of pounds of weight without stretching or creating any damage. This simply means that the top metal in the hinge buckle (Figure 8-12) is relatively undamaged compared with the bottom metal. The pressure side of the buckle has shrunk severely, far more than if the box section had been solid metal. This is called a *collapsed hinge buckle* or a *collapsed box section*.

When the box section is straightened, further collapsing of the top surface could easily occur. Special effort with heat and pulling equipment is required to prevent it. Figure 8-13 shows the results of improper and proper corrective procedures.

In the unibody cars of today there are quite a number of complete boxed sections. Boxed structural rails, rocker panels, windshield pillars, center pillars, and roof rails are just a few. Some boxed sections like the door assembly are quite large. Any metal that is bent to form an angle is considered a box section.

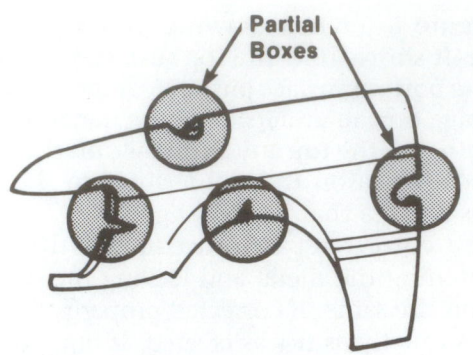

FIGURE 8-14 Note the partial boxed areas on this fender.

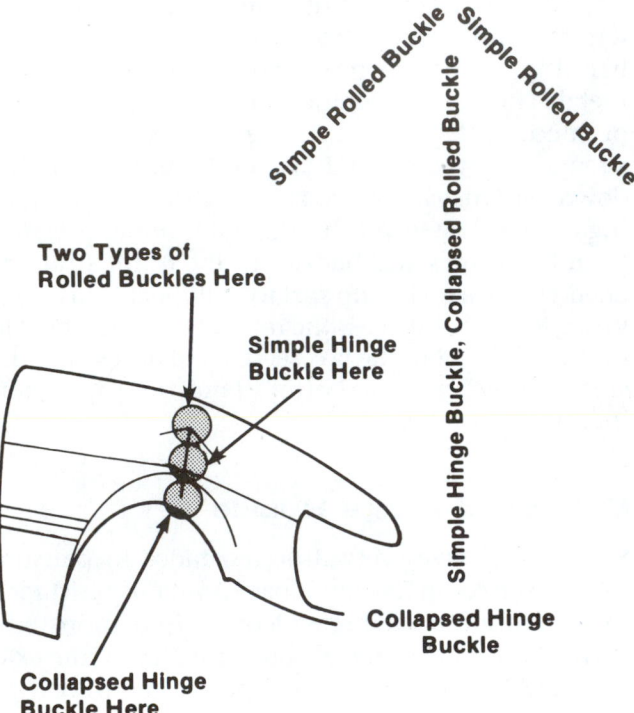

FIGURE 8-15 A combination of buckles commonly occurs in a dented fender like this one.

Late model cars have a great number of ridges and flanges in them. These are all areas where work hardening is built in and they are considered partial boxed sections (Figure 8–14). Entire fenders can be thought of as partial boxed sections. As did the complete boxed section in Figure 8–12, the partial box also collapses. Straightening improperly will have the same results as with the complete box—overall shortage in dimensions.

Collapsed Rolled Buckle

When a hinge buckle crosses a panel, it not only shrinks any and all boxed or partially boxed sections, but it also shrinks any crowned surfaces it crosses. When this happens, a new buckle is formed. This buckle tries to turn the panel inside out and rolls along,

increasing in length as it does. This "increasing in length" is the characteristic of the buckle. It is called a *collapsed rolled buckle*. Hinge type buckles (collapsed or simple) increase in depth, but not in length. Any buckles that are on crowned surfaces will shrink the metal. The collapsed rolled buckle is no exception. The shrinkage depends on the severity of the impact.

Simple Rolled Buckle

Two other buckles are also formed whenever a collapsed rolled buckle occurs. These two are adjacent to the collapsed roll. They are called *simple rolled buckles*. These are also shrinking type buckles because they are on the crown of the panel.

The identification of the rolled type buckles is not difficult. The collapsed and simple rolled buckles form an arrow design on the crown of the panel. At first glance, the fender in Figure 8–15 looks like it has only a single buckle crossing it vertically. Actually, it has five buckles, four types. Rolled buckles (simple and collapsed) occur only on crowned surfaces because the crown is what causes them.

If the metal is flat, it bends like a hinge and a simple hinge buckle occurs. If the panel is crowned, the buckle crossing it tends to roll as it travels deep into the metal. This happens because of the folding action on the surface, and because the metal shrinks within itself. Later if the metal is laid flat (no folds), the shrinkage will still be there, within the metal.

Collapsed rolled buckles always occur crosswise to all crowns they are in. The shrinkage they create is also in that direction. Simple rolled buckles shrink the metal as do collapsed rolled buckles, but in a different direction.

You should learn the four different kinds of buckles that are in the indirect damage area. The examples given here have been on only a few specific panels. Recognize the bulges that give clues to possible shrinkage somewhere. Be able to identify the buckles at a glance. Practice by examining many dents. Ask "Where is the direct damage?" and "What is the resulting indirect damage?" Find the work-hardened areas. Find the bent, but undamaged areas. Look for the collapsed rolled buckle. Examine damage found in the repair shop or parking lot. Take note of the buckles found and try to figure out a procedure to correct the damage.

PRESSURE FORCES

The terms "pressure" and "tension" are descriptive words that are commonly used to describe the conditions in metal after damage. These conditions are often described as high spots and low spots. A *high spot* or *bump* is an area that sticks up higher than the surrounding surface; a *low spot* or *dent* is just the opposite. It is recessed below the surrounding surface. Minor low and high spots in sheet metal can often be fixed with a body hammer.

The term **raising** means to work a dent outward or away from the body. The term **lowering** means to work a high spot or bump down or into the body.

It is important to understand their meanings. There are pressures and tensions (stresses and strains) within the metal before any damage occurs. All crowns, for example, are under pressure. Metal that is forced up has a new pressure applied to it. The pressure is being held there by the work-hardened buckles. If they were to suddenly disappear, the metal would return to its original shape. Consider the metal as being without pressure or tension before it was damaged when evaluating the changes that have taken place and the corrections that must be made.

The metal that is pushed up is called a **pressure area.** An area that is pushed down is called a **tension area.** The drawing in Figure 8–16 shows a typical cross section of a crowned panel that is damaged. The movement of the metal is obvious.

All panels are crowned to some degree. A panel with a large crown is called high crowned. A nearly flat panel is called low crowned. There are three types of outer panels:

Single crown
Combination crown
Double crown

The important thing to remember is that both the repair procedure and the application of tools are determined by whether the area is under tension or pressure. A hammer should never strike in a tension area; the dolly should never strike the underside of a pressure area. Power hookups are determined by the direction of pressure forces. No plastic filling can be done when there are pressure areas present because the filler could pop off from movement of the pressure area.

Single Crown Panels

Figure 8–17 shows a drawing of a single crowned panel. This panel is flat in one direction (left to right) and crowned in another (90 degrees or crosswise).

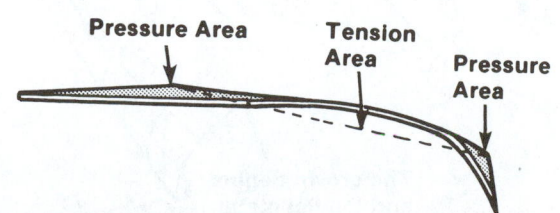

FIGURE 8–16 High and low tension areas often occur in a damaged combination type panel.

The damage shown in the drawing has tension in one direction and pressure in another. Look at the damage from the flat side (side view of drawing); it appears similar to that in Figure 8–18. Look at it from the end view; it appears similar to that of Figure 8–19.

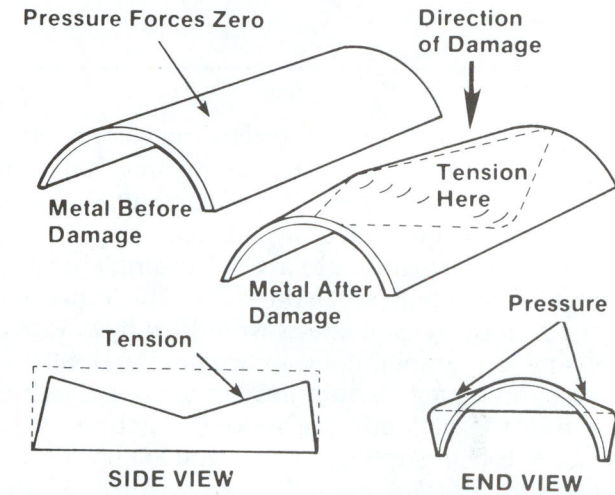

FIGURE 8–17 Study the high and low spots created in a crowned panel.

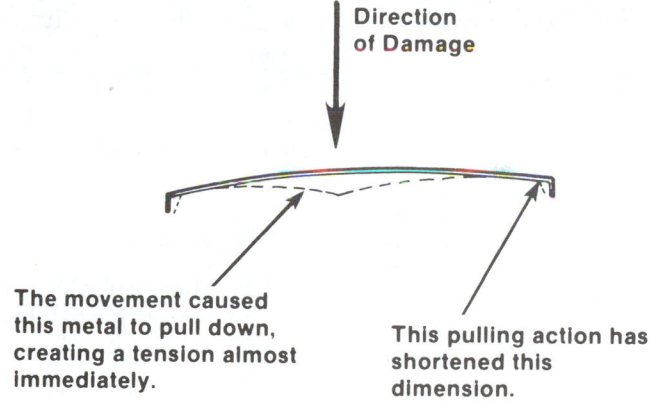

FIGURE 8–18 The tension area in a damaged low-crown area has affected the dimension at the right.

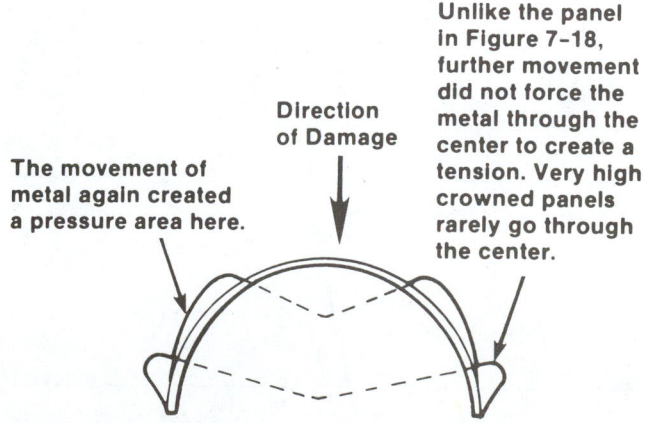

FIGURE 8–19 Study the results of pressure areas in a damaged high-crown panel.

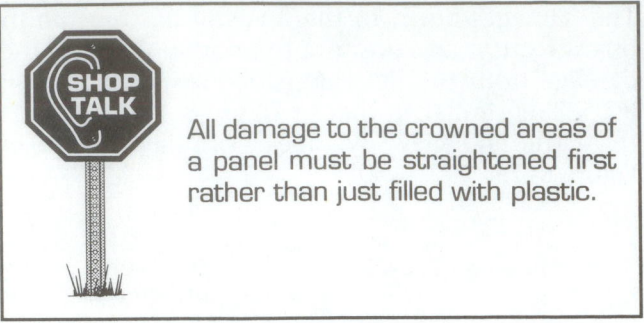

SHOP TALK

All damage to the crowned areas of a panel must be straightened first rather than just filled with plastic.

This means there is a three-dimensional effect on all pressure and tension areas. An area that is low is accompanied by a high (or pressure) area adjacent to it. This is true on all crowned areas.

Parking lot dings are a good example of pressure and tension areas (Figure 8–20). The impact creates a shallow tension area surrounded by a ridge or pressure area. Sometimes you can correct a small door ding by using a pick from inside the door. Carefully position the pick and push while watching where pressure is being applied. When you are on the center of the ding, use enough pushing force to move the ding out level with the panel. You may not even have to repaint the door.

With deeper dings having more serious tension areas, you may have to use a hammer and dolly to work the ding. The ridge around the ding must be taped down level with the panel and the low area must be pushed out. The area might then need to be leveled with a plastic body filler and refinished.

Combination Crown Panels

The shifting of the pressure areas on a combination panel is shown in Figure 8–21. The direction of damage shown is from above and almost straight down, yet there are two collapsed rolled buckles (P to BF and P to BC) that are of different lengths. This is because the crown is stronger than the flatter area and resisted the pressure forces more.

During the damage, the same force was applied to both sides of the arrow (at P). Yet, the metal to the left has a greater area of damage. The significance here is that if the novice body technician merely "drives" the metal upward to correct it, further damage would be caused at the flatter area of the panel. The flat area would yield to corrective forces (blows), and the greatest resistance area (P to BC) would remain intact. The correction here is to unroll the buckle (P to BC). This is the key to opening up the metal in the flatter area. It is the area of greatest resistance.

If a crowned panel has a shrunken area on it (caused by either welding, improper hammer and dolly techniques, or the results of a buckle in the crown), the level of that area will be lower than the normal level yet have no pressure areas present. Whenever a low area on a crown is not accompanied by a nearby pressure area, the low area is shrunken and can be corrected only by stretching. To attempt straightening by picking up the low area will only lower the adjacent metal, as shown in Figure 8–22.

A damaged crown panel will always have some pressure areas present somewhere, unless it is damaged from underneath. In this case, the metal will be pulled inward creating the opposite condition from those described in Figures 8–18 and 8–19.

Understanding these principles will help you determine correct repair procedures. For example, when welding on a concave surface (panel sectioning, perhaps), is the metal going to sink inward

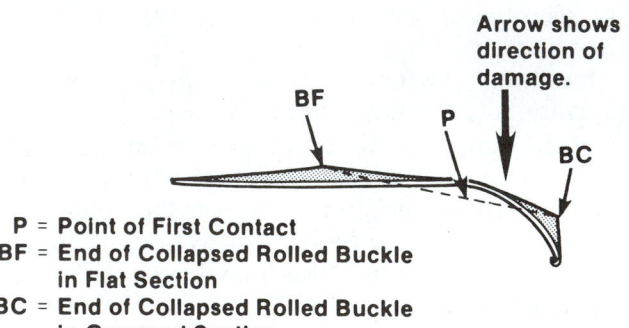

Arrow shows direction of damage.

BF
P
BC

P = Point of First Contact
BF = End of Collapsed Rolled Buckle in Flat Section
BC = End of Collapsed Rolled Buckle in Crowned Section

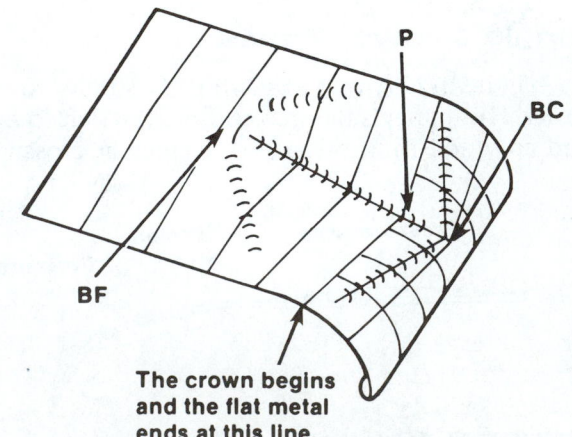

P
BC
BF

The crown begins and the flat metal ends at this line.

FIGURE 8-21 Work-hardened areas often result in a damaged combination panel.

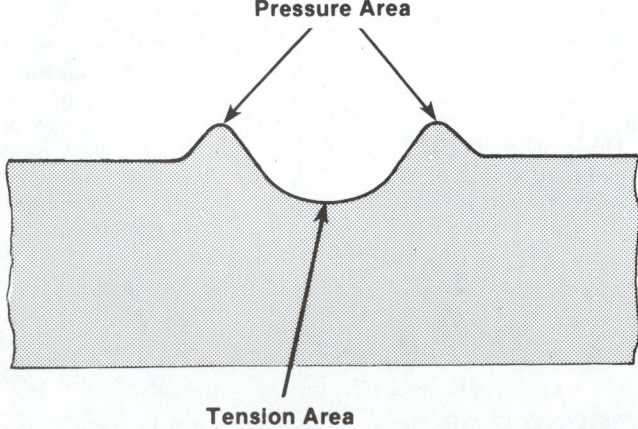

Pressure Area

Tension Area

FIGURE 8-20 A simple door ding or small dent easily shows pressure and tension areas in damage.

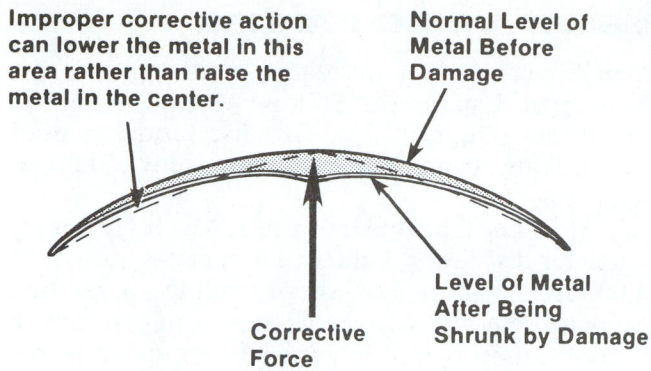

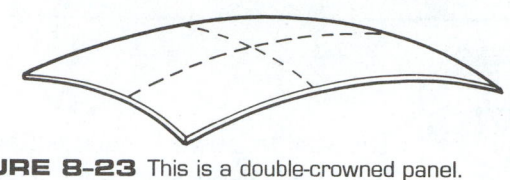

FIGURE 8-23 This is a double-crowned panel.

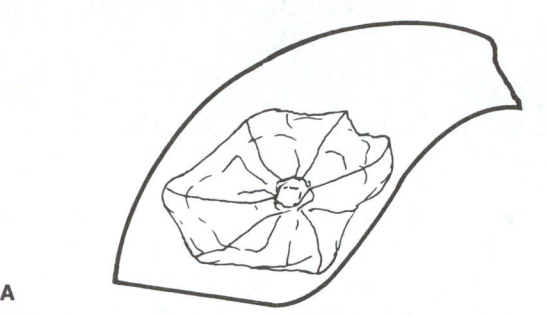

A

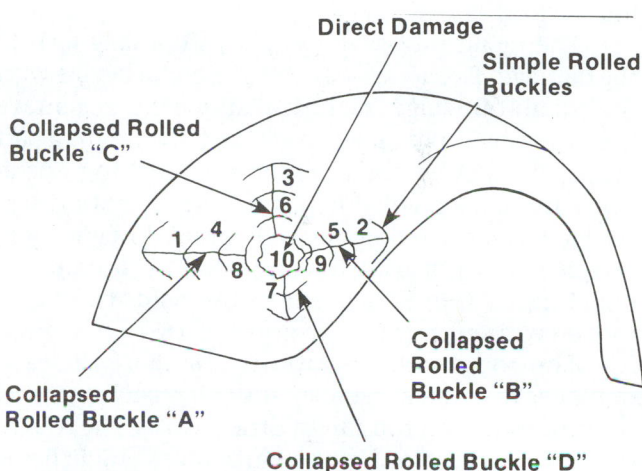

B

FIGURE 8-24 (A) Collapsed rolled buckles normally occur in a double-crowned panel. (B) Study the procedure for bumping out a collapsed rolled buckle.

FIGURE 8-22 A shrunken panel must be stretched to return it to its original contour.

or rise because of the weld shrinkage? The answer is that the metal will rise. A high area will be created. This problem can be corrected by using the hammer-on-dolly technique to lower the metal. The novice usually feels that stretching will raise metal. On a crowned panel, it will.

Double Crowns

For simplicity, the various buckles have been discussed as they occur on panels that are curved or crowned in one direction and flat in another direction. Most panels are reasonably close to this type of construction. Some, however, are crowned in both directions (Figure 8–23). These are called *double crowns.*

Rolled buckles occur on crowned surfaces and roll (or travel) toward the nearest flat area. In the case of a panel that has a double-crowned surface, the rolled buckles will travel normally in all directions from the point of impact. The collapsed rolled buckles spread out from the impact point, like spokes of a wagon wheel, the hub being the initial point of impact. The damage shown in Figure 8–24A is typical of an impact on that type of panel.

DETERMINING THE DIRECTION OF DAMAGE

All the information given so far will help determine the direction of damage. To fix collision damage, you must apply force in the reverse of how the damage occurred. Visual inspection can usually tell what happened. However, sometimes it becomes complicated when there are overlapping conditions.

Collapsed rolled buckles always move away from the point of first contact. When two or three of them are present, it becomes easier. Where they all converge (like spokes of a wheel to the hub) is the point of first contact. It is also usually approximately 90 degrees straight out from the collapsed rolled buckle (Figure 8–24B).

It helps to visualize the accident happening in slow motion. Then, by reversing the accident,

studying each buckle, and unfolding each work-hardened section of metal, you can determine how the repair operations should be formed.

Of course, it is not possible to always reverse the damage conditions in the order that they occurred. Buckles cannot be unfolded simultaneously. Shrunken metal that is work hardened cannot be softened in reverse. Tears in the metal will not reweld themselves. But before the damaged area can be repaired, you must know how the damage occurred, recognize the conditions existing in the metal, and use the correct repair tools and procedures.

8.3 METAL STRAIGHTENING TECHNIQUES

Analysis and theory will tell you what is wrong. Next you must have the basic skills to repair the damage. After this, you must know how to put these things together to produce the overall results required of a

SHOP TALK

The way to achieve success as a body technician is to concentrate on the prevention of damage.

good body technician. You must develop a good procedure for repair. A good procedure can save a great deal of technician-created damage so that the entire repair time is reduced to a minimum for higher profits.

The repair procedure begins with a diagnosis of the damage. The actual work on the metal begins with the rough-out stage. **Rough-out** means to remove the most obvious damage to get back the original part shape. It must be done properly if finishing operations are to succeed. When finishing operations are started too soon, it becomes difficult to do a good job. Rough-out is not usually completed on a damaged panel until 80 to 90 percent of the paint can be removed without turning the grinding disc on its edge.

The rough-out operations change with each damage, with each vehicle, and with each location of the damage on the car. In other words, the rough-out is very important to the particular vehicle being worked on. The analysis must be good for each damage and set of circumstances. Proper analysis in the rough-out stage can mean the difference between making money on the job or taking a loss. Poor rough-out always costs the body technician money in time lost. Typical of this is the situation in which the technician hits up all low areas and beats in all high ones thinking that eventually the metal will become straight.

A body technician with a clear understanding of damage analysis knows metal is not straightened that way. Rolled buckles (simple and collapsed) require special handling, as do hinge-type buckles. Flat low-crowned metal must be protected more than straightened. Hitting up a low-crowned flat area to straighten a work-hardened buckle could create severe damage to the (up to now) undamaged metal in the flat area. Every buckle has a definite method of correction. The basic rules always apply. If an area must be stretched, it might be done with pull clamps, solder plates in combination with a hydraulic jack, hammer and dolly, or several other means.

The rest of the chapter is devoted to explaining some of the common skills utilized by the technician from the rough-out stage of repair up to the plastic filling stage.

USING A DISC SANDER

Many sheet metal repairs require removing the paint finish first. Usually this is done with a disc sander or grinder (Figure 8–25). The disc sander is used throughout the repair for removing paint and metal (Figure 8–26).

Most body technicians use a 7-inch (177 mm) diameter disc and a sander that operates at least at 4,000 rpm or more. Low-speed sanding can be used for paint removal. Grit sizes range from #16 to #60. Size 24 is most commonly used for removing paints and #24 or #36 for grinding away metal. Higher grits are used for removing file marks and polishing metal.

Two types of backing pads are used. An inflexible pad is used for removing metal. A softer back-up pad, which allows the disc to roll and give with the metal, should be used when removing paint or polishing metal.

There are two different ways to use a grinder. One, called *buffing* (little overlap of grinding marks), is used to remove paint and smooth the filler. The other is called *cross cut* (overlap of grinding marks), and it is used to remove metal (Figure 8–27).

FIGURE 8-25 The technician is grinding off paint to repair for metal straightening. *(Courtesy of Tech-Cor)*

FIGURE 8-26 Paint has been ground off roughed-out dented area.

When using the grinder, only the top 1½ to 2 inches (38 to 51 mm) should contact the surface. Do not use excessive pressure. The weight of the grinder should just about be enough.

On vertical surfaces, use pressure equal to the weight of the grinder. The grinder should be held so the back of the disc is raised 10 to 20 degrees off the metal. It is sometimes difficult to use the round sanding disc in a sharp reverse crown area. The edge of the disc will cause a deep groove to be cut in the metal. This can be avoided by cutting the edge of

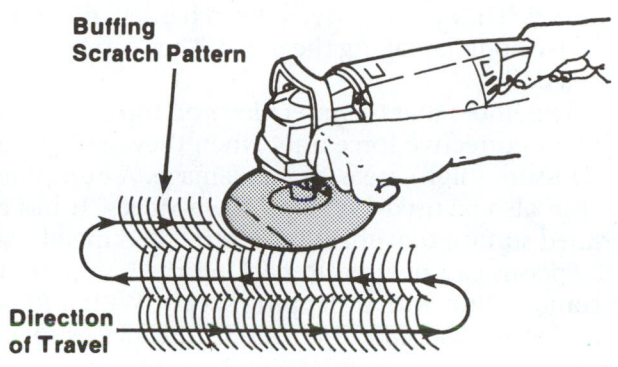

A

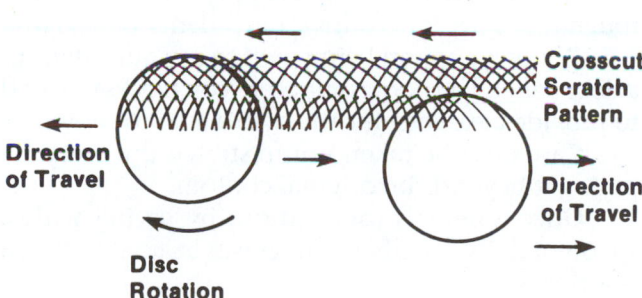

B

FIGURE 8-27 Two grinding actions: (A) buffing method and (B) crosscutting method.

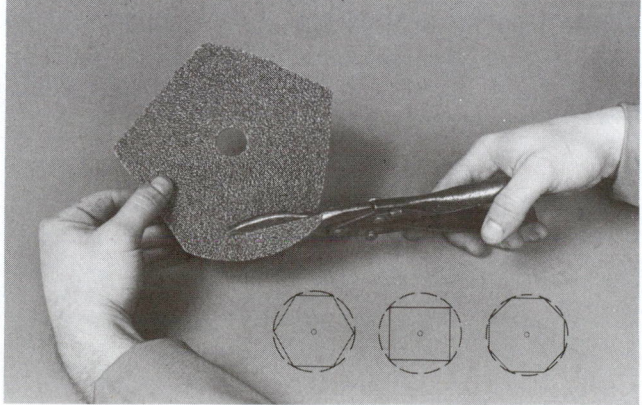

FIGURE 8-28 Round discs can be cut into star disc shapes for better cutting action.

the disc into points, commonly called a *star disc* (Figure 8–28).

8.4 METALWORKING TECHNIQUES

The buckles and creases in a dented panel can be unlocked in a variety of ways. On panels where the backside is accessible, hammers and dollies or spoons are used for the initial roughing out. On areas where the backside of the panel is difficult to reach, slide hammers, picks, and welded studs can be used to reverse the damage.

Damage in an exterior panel can be locked in by indirect damage to structural panels or inner reinforcements. Before metalworking techniques are used, structural damage must be repaired by pushing or pulling the damaged understructure into alignment. Usually when this is necessary, it is best to maintain the hydraulic push or pull until the exterior panels have been straightened using one of the techniques described here.

BUMPING DENTS

The body repair hammer is designed to strike the sheet metal and rebound off the surface. It is not designed to be driven down, as in driving a nail. A driving action would create additional damage in the sheet metal.

The secret of metal straightening is to hit the right spot at the right time with the right amount of force. When using a body hammer, swing in a circular motion at your wrist. Do not swing the hammer with your whole arm and shoulder. Hit the part squarely and let the hammer rebound off the metal. Space each blow ³/₈ to ½ inch (9.5 to 13 mm) apart until the damaged metal is level. The hammer should be held as shown in Figure 8–29.

The face of the hammer must fit the contour of the panel. Use a flat face on flat or low-crown panels. Use a convex-shaped or high-crown face when bumping inside curves.

Heavy body hammers should be used for roughing out the damage. *Finishing*, or *dinging, hammers*

Before bumping any sheet metal, make sure that both sides of the panel are clean of road tars, mud, undercoating, and so on. This will ensure that the tools come in direct contact with the metal.

should be used for final shaping. The secret to finish hammering is light, rapid taps. It is also important to hit squarely. Hitting with the edge of the hammer will put additional nicks in the metal.

BUMPING DENTS WITH SPOONS

Spoons can be used in a number of ways. Spoons can be used to pry out dents. Certain kinds can be struck with a hammer to drive out dents. In hard-to-reach areas, a spoon can be used as a dolly. Some are even designed to be used in place of a hammer.

Spring hammering is commonly done with a hammer and a *dinging spoon*. The dinging spoon is lightweight and has a low crown. When used, it is held firmly against the high ridge or crease. The spoon is then struck with a ball peen or bumping hammer (Figure 8–30). The force of the blow is distributed by the spoon over a large area of the crease or ridge. This reduces the likelihood of stretching the metal.

Always keep firm pressure on the spoon when spring hammering. It must never be allowed to bounce. Part of the corrective force is the pressure of the spoon. Begin at the ends of a ridge (hinge buckle) and work toward the high point on the ridge, alternating from side to side.

The effect of hammering is increased when pressure is kept on the underside of the panel. This can be done with a dolly, spoon, or hydraulic jack.

Slapping spoons are sometimes used instead of hammers. They can be driven down harder and more often without damaging the panel. They can be used with a dolly.

Remember that hammer blows on top of a panel can be a corrective force only when they are placed on pressure (high) areas of the damage. A bumping file can also be used to "slap" down ridges. It has a serrated surface that shrinks the stretched metal.

Spoons can be used to back up the hammer or in combination with a slapping spoon. With a long body spoon, you can often reach into places inaccessible to a hammer or dolly. Pressure can be applied to tension areas with the spoon while high areas are bumped down (Figure 8–31A).

Spoons can also be used to pry metal up in the rough-out stage or to drive deep dents out. Figure 8–31B shows a spoon being used to pry out a dent in a door panel. The door is supported on blocks of wood to provide clearance for the door panel to move.

Care must be taken not to stretch the metal by prying it beyond the original contour.

Once a dent is roughed out by prying with a spoon or dolly, a body hammer can be used to finish the area.

BUMPING DENTS WITH DOLLIES

The dolly block is also used throughout the sheet metal repair process (Figure 8–32). In the rough-out phase, it is used as an impact tool. The underside of

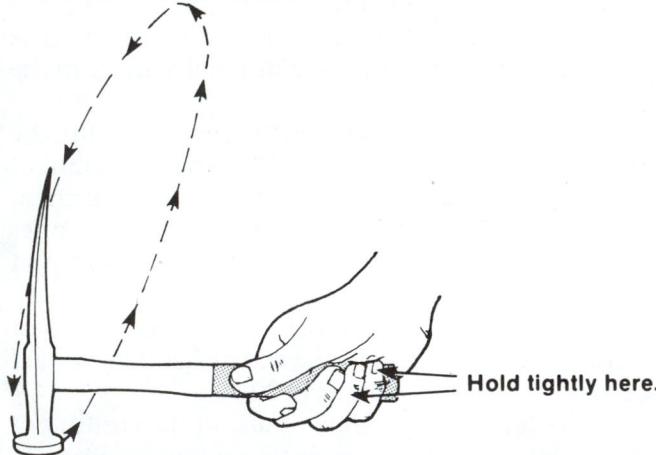

FIGURE 8-29 Hold the hammer with the third and fourth fingers and swing with a circular motion.

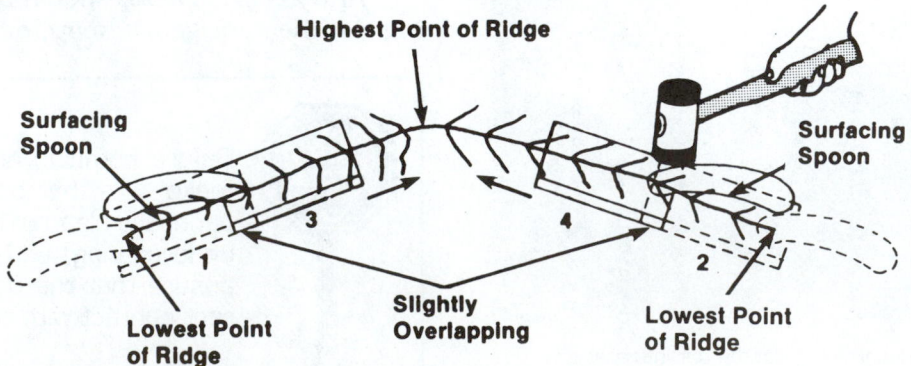

FIGURE 8-30 Note the method for spring hammering a ridge.

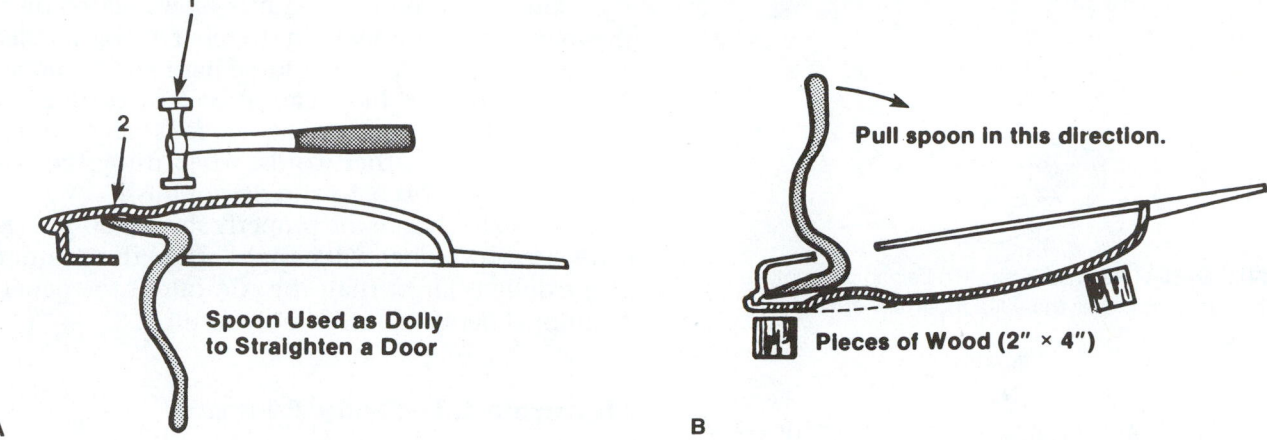

FIGURE 8-31 (A) Using a spoon as a dolly. (1) Use a hammer to work metal from the front (2) while holding a spoon at the rear of the panel. (B) Using a spoon to pry out a dent in a door panel.

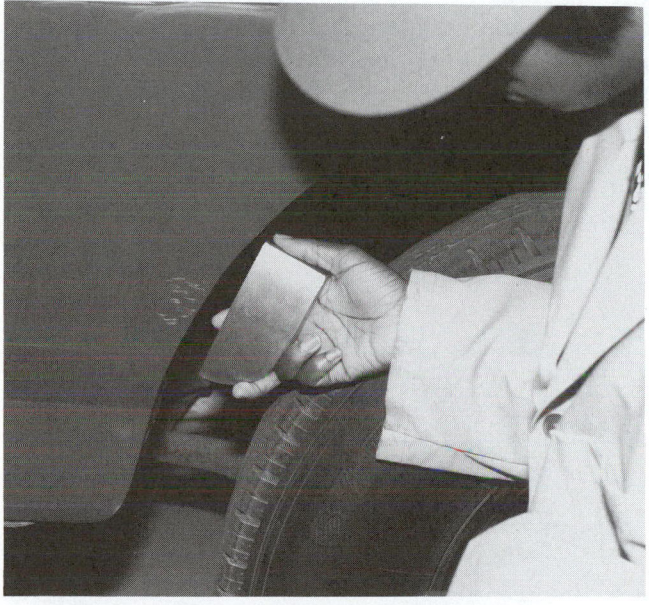

FIGURE 8-32 (A) Select a dolly the correct shape to fit the rear of the panel. (B) While holding the dolly behind the panel, use careful hammer blows to straighten.

the metal can be hit with the dolly to raise low areas and to unroll buckles. The dolly is also used as a backing block for the hammer. When it is used this way, there are two techniques:

Hammer-on-dolly
Hammer-off-dolly

Hammer-On-Dolly Method

Hammer-on-dolly repairs are used to smooth small, shallow dents and bulges and to stretch metal so that it can return to its original shape. This occurs usually on crowns and occasionally on flat panels. To flatten a bulge, place a dolly against the backside of the panel directly behind the bulge. Use a hammer from the front side to strike the damaged panel over the dolly.

There will be a slight rebound as the hammer hits the dolly. By increasing how hard you press on the heavy dolly, you will increase the flattening action on the panel (Figure 8–33).

Carefully monitor or watch how much the panel stretches as it is hammered. If not checked, the panel

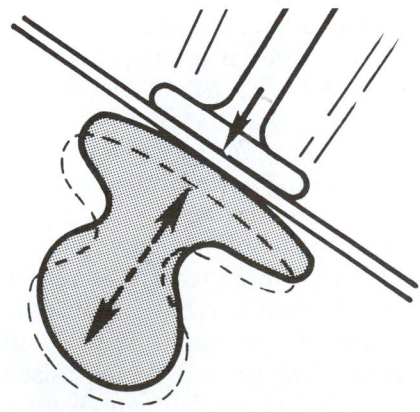

FIGURE 8-33 Rebound of the dolly forces the metal up.

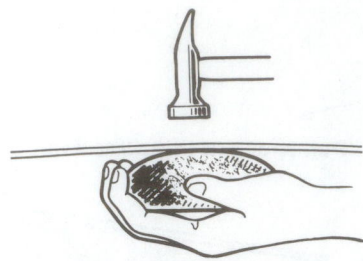

FIGURE 8–34 Hammer-on-dolly repairing is done by hitting the panel right over the dolly. *[Courtesy of American Honda Motor Co.]*

can overstretch and elongate too much. This could cause large warps in it. Since a great deal of body repair experience is needed to accurately judge how much a panel will stretch when hammered, an inexperienced technician should proceed slowly. Work the panel a little; then step back and inspect the area around and in the repair area.

You want to work out low spots so that they are almost level with the rest of the undamaged panel (Figure 8–34). You do not want to form bumps on the panel. A slightly concave (sunken) area should remain around the dent for applying a thin layer of plastic filler.

In the hammer-on-dolly action, there are two actions:

- When the hammer strikes the metal
- When the dolly rebounds upward and strikes the underside of the metal.

Any shrunken metal on a crown can be stretched up to its normal level faster and easier by the hammer-on-dolly than by any other method. The technique of on-dollying is used, provided that there is access to the underside of the panel. If not, a sliding hammer might be needed. Figure 8–35 shows a drawing of a typical low area (in this case a collapsed rolled buckle).

This panel was damaged and has been roughed out but is still low in the area of the collapsed rolled buckle (circled). Because this panel shows no evidence of a pressure area adjacent to the buckle, the panel is shrunken at A as shown in the drawing. The only way A can be raised to the original level B is by stretching. The stretching has to be along the line of the rolled buckle. Picking up the low area will not correct the condition. The hammer-on-dolly method should be used here on line A.

Tension is kept on the dolly at all times. The greater the tension, the greater the rebound action and the faster it returns. When hammering fast, the tension is increased on the dolly to ensure a quick rebound. When low metal is being raised, each rebound action should be deliberately driven up by the hand action.

Note that improper hammer blows placed on a crown will shrink rather than stretch it. All blows that are designed to stretch should be hard and accurate. An inaccurate hard blow can also cause damage to the panel. Light hammer blows are for straightening, not stretching. In other words, when using the on-dolly technique, hit hard and do not miss!

Be sure to choose the properly shaped dolly. Figure 8–36 shows what will happen when the contour of the dolly is larger than the contour of the panel. Additional dents will inevitably result.

Hammer-Off-Dolly Method

Hammer-off-dolly is used to straighten metal just before the final stage of straightening. In this procedure, hold the dolly under the lowest area, and hit the high area with your hammer (Figure 8–37). The dolly, like the hammer, is designed to correct damage, and must strike only in tension areas to be effective as an impact tool (when used under the panel in the normal way).

The hammer-off-dolly method is used mostly on flat or low-crown panels. These panels are soft in

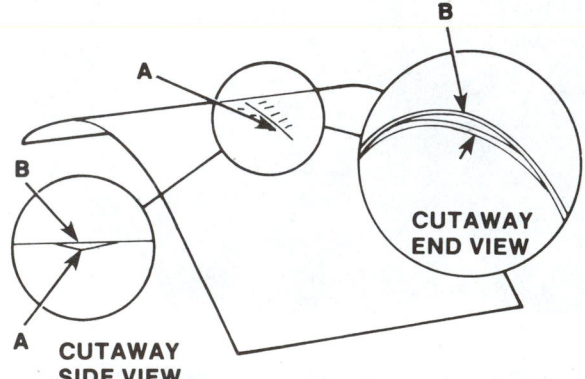

A = Metal Below Normal Level
B = Normal Level

FIGURE 8–35 The shrunken area must be stretched with the hammer-on-dolly method.

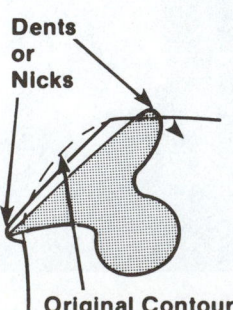

FIGURE 8–36 Using a dolly whose contour does not fit the contour of the panel will result in additional damage to the panel.

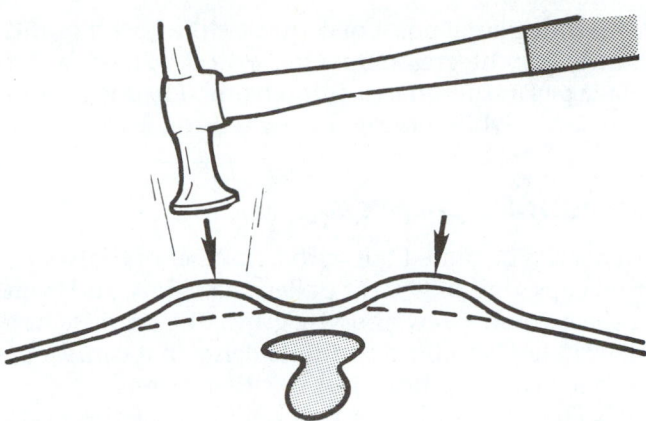

FIGURE 8-37 The hammer-off-dolly position is done by hitting the metal to one side of the dolly.

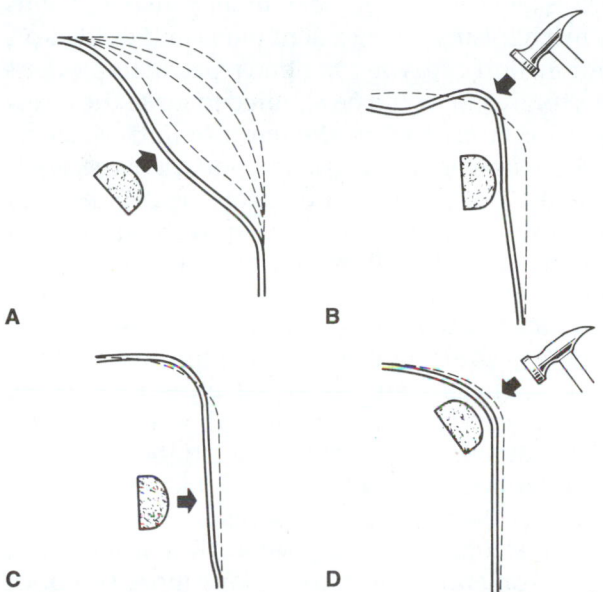

FIGURE 8-38 Study the basic steps for forming a bent panel. (A) Use blows from the dolly to remove a large dent. (B) Use the off-dolly method to further work the dent. (C) Blows from the dolly might shape a large area. (D) The on-dolly method will work smaller areas to the original contour. *(Courtesy of American Honda Motor Co.)*

comparison with crowned panels. Sometimes the dolly is directly under the hammer but does not actually strike it, as shown in Figure 8–38.

UNLOCKING DENTS WITH A HAMMER AND DOLLY

A minor dent (Figure 8–39A) is often straightened by using hammer and dolly to "roll out" the metal in the reverse order it happened. For example, the point of impact (POI) was the first area touched by the impact. As the metal is pushed in, a channel is gradually formed on either side of the point of impact. This channel (a collapsed rolled buckle) is usually deepest next to the POI, decreasing in size toward the outward end of the channel.

At the same time as the channels are being pushed in, ridges (simple rolled buckles) are being formed around the outer area of the dent. They also have the greatest degree of bend in their center, gradually decreasing in size toward each end. Both the ridges and channels contain work hardening, the amount depending on the degree of bend.

To remove the dent, roll out the damage from the outside, working toward the center in a reverse order of damage. Hold the dolly tightly under the channel at the outer end where there is least damage (Figure 8–39B). A flat-faced dinging hammer might be used to direct light to medium blows at the outer ends of the ridge closest to the dolly (off-dolly blows). The force from the hammer will gradually lower the ends of the ridges. Your hand or arm pressure on the heavy dolly will force the end of the channel upward. The same procedure is then repeated on the other end of the channel and the adjacent ridges.

The off-dolly method is gradually worked toward the center or the greatest degree of bend in the ridges and channels. As the pressure is released in the ridges and channels, the surrounding elastic metal tends to move back to its original position. The dolly can also be used as a driving tool to work the channel upward (Figure 8–39C). However, if the

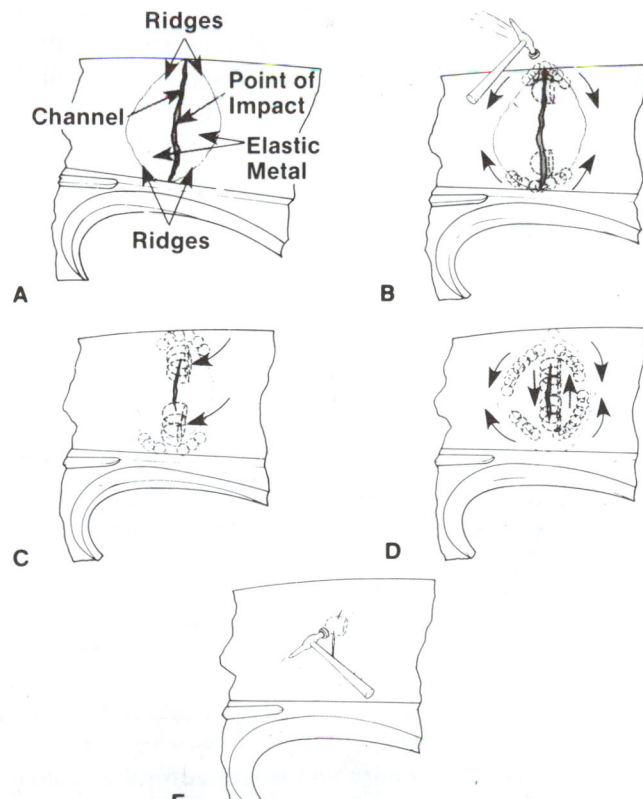

FIGURE 8-39 Study the steps in repairing a dented panel with a hammer and dolly.

dolly does not move when the channel is hit upward, there is still too much pressure on either or both the ridges and/or the channel. More dollying must be done to relieve the tension (Figure 8–39D).

Once the area has been brought back to its basic shape, use the light on-dolly method to smooth and level the area (Figure 8–39E). It is then ready for either the metal finishing or filling procedure.

PICKING DENTS

There are several methods of picking up metal with the use of a pointed (not necessarily sharp) tool. Pick hammers, long picking tools, the edge of a dolly, and even a scratch awl can be used. When picking up a small dent with a striking tool, it is better to use several light blows rather than one or two hard blows. After an area has been picked up, a file or grinder can be used to level the damaged area if needed. Figure 8–40 shows a low spot being raised with a pick.

Long picking tools can also be used to pry up metal in areas that cannot be reached with a dolly or spoon. A car door is a good example (Figure 8–41). A pick can sometimes be inserted through a drainage hole or a hole drilled behind the door gasket, which eliminates the need to remove the inside door trim or to drill holes in the outer panel for pulling the dent. Picks are used during **paintless dent removal** (removing small body dings or dents without painting the panel).

When prying with a pick, be careful not to stretch the metal by exerting too much pressure. Start with the original point of contact or the lowest point. Slowly pry the crease up. On larger dents, use a flat blade pick rather than a pointed one. Tap down pressure areas, while prying up low tension areas.

PULLING DENTS

Dents can be pulled out with a number of tools: suction cups, pull rods, dent pullers, nail guns, and even a sheet metal screw and vise grips. The purpose of a dent puller is to lift out simple dents that cannot be reached easily or lifted out by other means.

The dent puller is probably one of the body technician's most used tools. One reason is the growing complexity of automobile body construction and corrosion protection. Access to the inside of many panels is blocked by welded-in inner panels and window mechanisms. Using a dent puller and welded-on nail or suction cup, you can often repair a simple dent in less time than would be required to make the disassembly necessary to start the repair from the inside.

A *suction cup* can be used to pull out large, shallow dents. Wet the area and install the cup. If hand held, pull straight out on the cup's handle. If mounted on a slide hammer, use a quick blow to pop the dent out.

A *vacuum suction cup* uses a remote power source (separate vacuum pump or air compressor airflow) to produce negative pressure (vacuum) in the cup. This increases the pulling power because the cup will be forced against the panel tightly. Larger, deeper dents can be pulled with a vacuum suction cup. Figure 8–42 shows a pneumatic dent puller.

A slide hammer equipped with a sheet metal screw is sometimes used to remove more stubborn dents and creases. If a nail gun (stud welder) is not

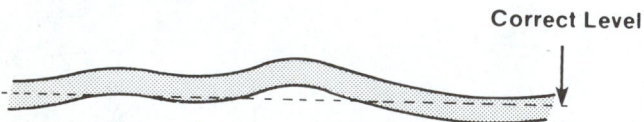

A. The metal is low and irregular.

B. The stretched metal is raised above the normal level.

C. The metal is filed hard and is leveled to the desired amount.

FIGURE 8–40 Note how to raise a dent with a pick hammer.

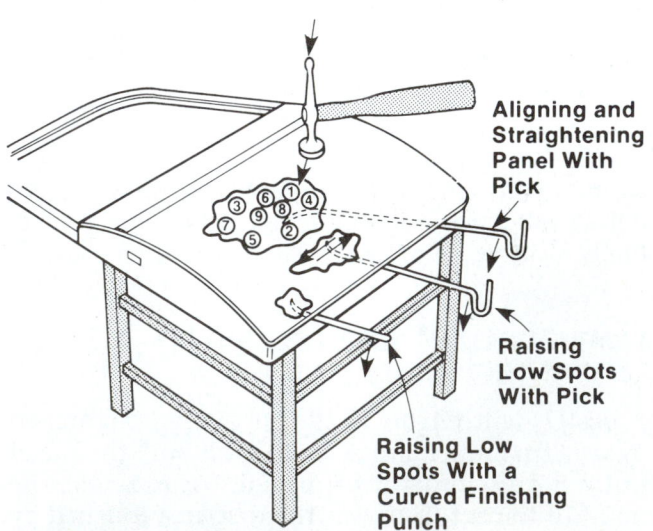

FIGURE 8–41 Paintless dent removal is commonly done on small door dings. Reach in behind the dent with a long pick. Carefully apply controlled pressure to the dent to push it out.

available, drill or punch holes one inch (25 mm) apart along the length of the initial crease. Punching is preferred to drilling because punching leaves more metal for the screw tip to grab. The series of holes will weaken the area in the crease and allow for its removal.

Beginning at the point of initial impact, thread the screw tip on the dent puller into the first hole. Hold the handle of the dent puller in one hand and slide the weight straight back against the handle (Figure 8–43).

Gradually work the crease out. Pull the first hole slightly. Then go on to the next hole. Work from the lowest spot to the crease ends in both directions. After pulling each hole out slightly, repeat the process. Work the surface as close to the finished contour as possible without pulling the metal beyond the original surface. Tap high spots down with a body hammer to relieve pressure in ridge areas. Weld the holes closed and restore corrosion protection on the

inside of the panel. Low spots can then be filled with body filler.

Shallow, small dents in enclosed panels can be pulled out with one or more pull rods. Again, a body hammer should be used to tap high spots down during pulling.

SHOP TALK

Avoid making holes in panels during straightening. Holes weaken the panel and require welding as well as restoration of the corrosion protection. It is senseless to pierce a large number of holes in the surface of a panel that could have been straightened by other means. The result is often an unsatisfactory job. More time may have been required to do it wrong than would have been required to do it right.

Pulling With Studs

The most advanced and common way to pull dents is with a nail gun and small metal studs welded to the dent. The stud might be a washer, a pull tab, or a pin. Regardless of the system used, pulling with spot-welded studs avoids drilling or punching through the metal and undercoating, potentially a technique that will result in corrosion damage.

Figure 8–44 shows a complete set of stud pulling equipment.

A **pull pin spot welder** or *nail gun* fuses a metal pin to the dent area (Figure 8–45A). Fusing the pin to the panel takes only a fraction of a second. The pin or pins can then be pulled with either a dent puller (Figure 8–45B) or a power jack (Figure 8–45C).

FIGURE 8-42 To remove a dent with a dent puller, weld a bracket or a pull pin to the surface of the panel. Attach the appropriate tip to the dent puller, insert it in the bracket or pull pin, and slide back on the hammer. Use soft blows to slowly pull out the dent.

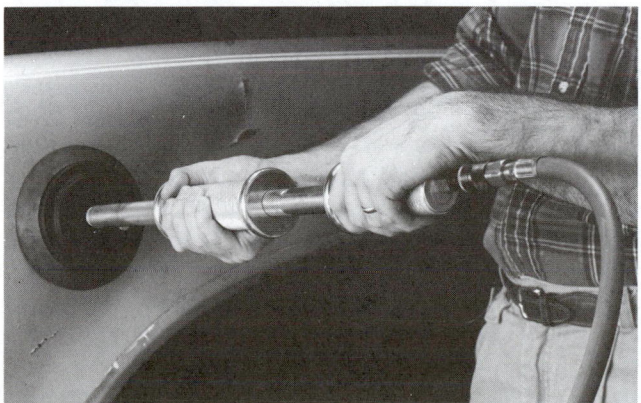

FIGURE 8-43 A pneumatic or suction cup dent puller will pop out large, shallow dents easily. *(Courtesy of Atlantic Pneumatic, Inc.)*

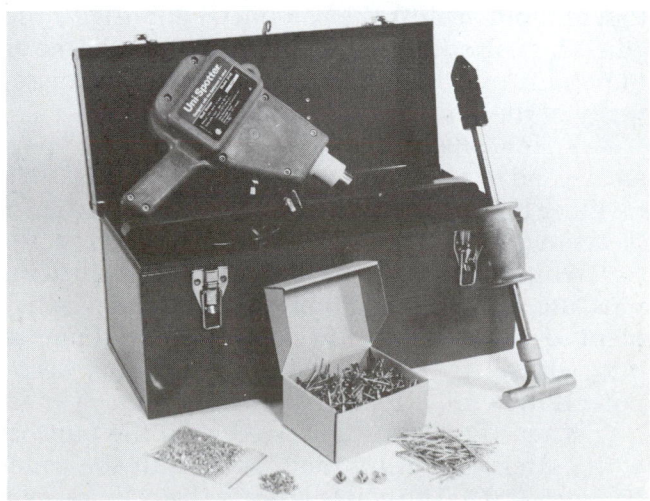

FIGURE 8-44 This is a welded stud dent removal kit. *(Courtesy of H & S Manufacturing Co. Ltd.)*

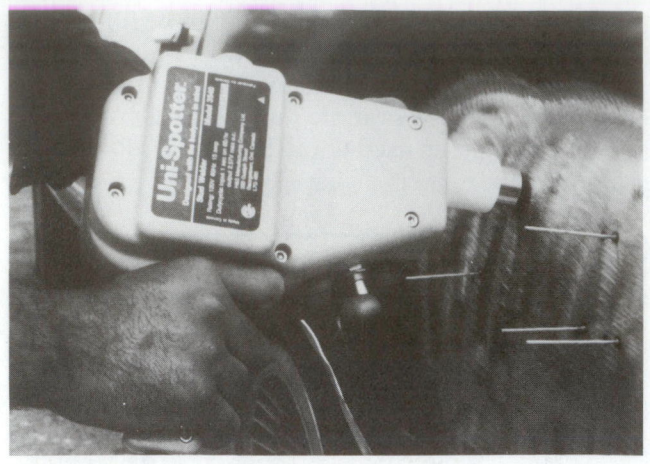

A

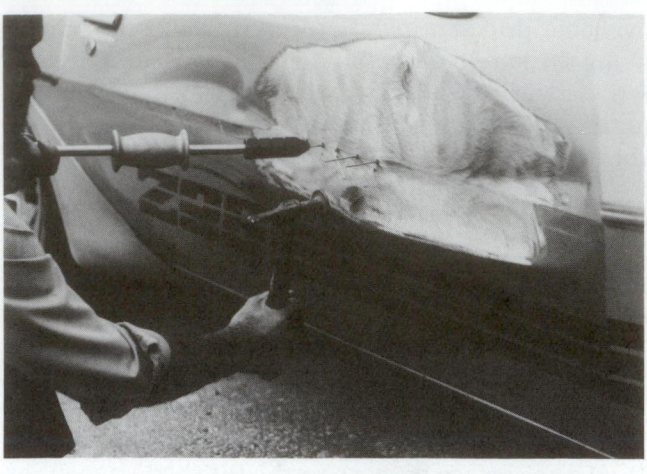

B

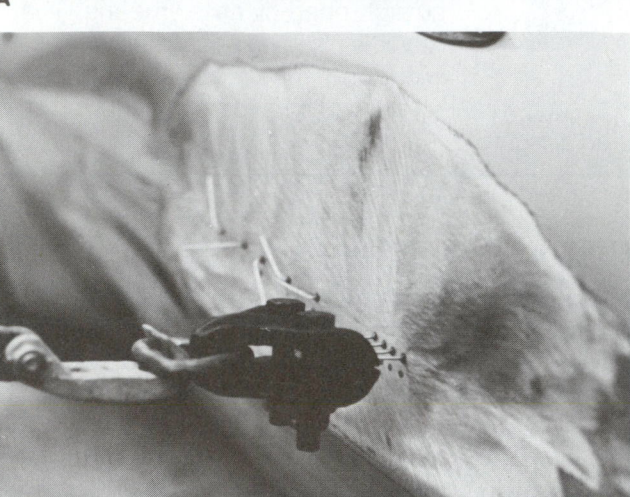

C

D

FIGURE 8–45 (A) A pull stud is being welded to a dented panel. (B) A slide hammer can then be attached to the welded-on stud for pulling the dent. (C) Power equipment can also be attached to several studs simultaneously for pulling tougher dents. (D) Grind off the welded studs after pulling. *(Courtesy of H & S Manufacturing Co. Ltd.)*

Grind off the area to be pulled to expose the bare metal. Place a pin in the spot welder. Press the tool and pin against the bare metal and trigger the gun. This will spot weld the pin to the panel. One or more pins may be needed depending upon the severity of the dent.

A dent puller is handy if only one or two pins are needed to straighten the panel. Tighten the slide hammer tip around the pin. Use moderate hammer blows to force the pin and panel outward.

With larger dents requiring several welded-on pins, use a power jack or frame straightening equipment to pull the pins. You can grasp several pins at once with a large clamp, and use hydraulic power to pull on the pins to remove the larger dent.

When the dent is corrected, snip the pins off with cutters and grind them flush with the panel (Figure 8–45D). The whole process is very quick with no damage to the panel and its corrosion protection. It is especially useful in pulling out small door,

quarter panel, fender, and other panel dents where the backside of the part is inaccessible.

SHRINKING DENTS

Shrinking metal is needed to remove strain or tension on a damaged, stretched sheet metal area. During impact, the metal can be stretched. When pulled or hammered straight, the area can still have tension or strain on it because the stretched metal no longer fits in the same area. The metal will tend to pop in and out when you try to final straighten it.

If a strained area is filled with plastic filler, road vibrations can cause the panel to make a popping or flapping noise. After prolonged movement of the strained area, the filler can crack or fall off. Eventually, you will be required to spend extra time correcting the work that should have been done properly in the first place.

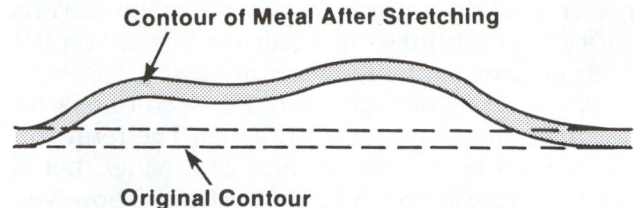

Contour of Metal After Stretching

Original Contour

FIGURE 8-46 Stretched metal must be shrunk to relieve pressure.

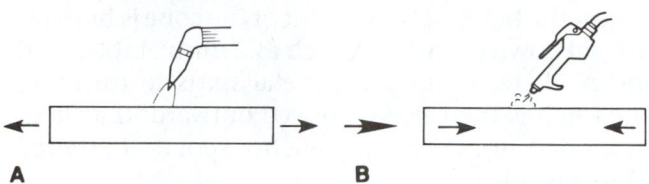

FIGURE 8-47 (A) Heat causes metal to expand. (B) Cooling causes metal to contract.

STRETCHED METAL

Stretched metal has been forced thinner in thickness and larger in surface area by impact. When metal is severely damaged in a collision, it is often stretched in the badly buckled areas. These same areas are also sometimes stretched slightly during the straightening process. Most of the stretched metal will be found along ridges, channels, and buckles in the direct damage area. When there are stretched areas of metal, it is impossible to correctly straighten the area back to its original contour. The stretched areas can be compared to a bulge on a tire. There is no place for the area to fit within the correct panel contour.

When an area is stretched, the grains of metal are moved farther away from each other. The metal is thinned and work hardened. Shrinking is needed to bring the molecules back to their original position and to restore the metal to its proper contour and thickness.

Before shrinking, dolly the damaged area back as close to its original shape as possible. Then you can accurately determine whether or not there is stretched metal in the damaged area. It will usually pop in and out if stretched. If it is stretched, you must shrink the metal (Figure 8–46).

Principles of Shrinking

Figure 8–47 shows that a steel bar, with both ends free to expand or contract, will expand when heated and contract to its original length when cooled.

If the same steel bar is heated while it is blocked or restricted at both ends, then cooled, its size will decrease. This is accomplished by the following:

• When heated, the steel bar tries to expand (Figure 8–48A), but since it is prevented from

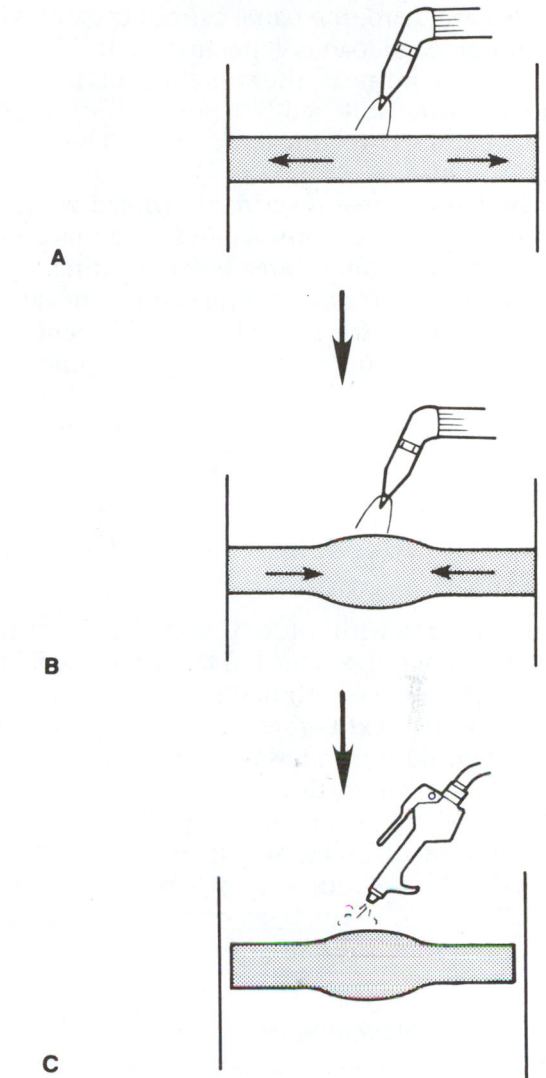

FIGURE 8-48 (A) Shrinkage occurs when expansion forces are restricted by panel rigidity. (B) This causes heat-softened metal to expand and thicken. (C) When metal cools, the panel contracts and, due to an increased area of hot spot, shrinks to an area smaller than its original size.

expanding at both ends, a strong compression load is generated inside the bar.
• When the temperature is increased even more, and the steel becomes red hot and soft, the compression load concentrates in the red-hot area and is relieved as the diameter of the red-hot area increases (Figure 8–48B).
• If the steel bar is suddenly cooled down, the steel contracts and the length of the bar is shortened (Figure 8–48C).

The above-stated principle of shrinking steel also applies to the shrinking of a warped area in a piece of sheet metal. A small spot in the center of the warped area is heated to a dull red. When the temperature rises, the heated area of the steel panel swells and attempts to expand outward toward the edges of the heated circle (circumference). Since the surrounding area is cool and hard, the panel cannot expand, so a strong compression load is generated.

If heating continues, the stretching of the metal is centered in the soft, red-hot portion, pressing it out. This causes it to thicken, thus relieving the compression load.

If the red-hot area is suddenly cooled while in this state, the steel will contract and the surface area will shrink to less than its area before heating.

A variety of pieces of welding equipment can be used to heat metal for shrinking. Attachments are available for spot and MIG welding equipment to transform them into shrinking equipment. Nail guns can also be used. The most commonly used tool, however, is the oxyacetylene torch with a #1 or #2 tip.

Shrinking Operation With a Gas Torch

To shrink an area with a torch, a small spot of the stretched area or bulge is heated to a cherry red. The "shrink" is placed in the highest spot of the stretched area, then in the next highest spot, and so on. This is repeated until the area has been shrunk back to its proper position (Figure 8–49).

The size of the shrink or hot spot is determined by the amount of excess metal in the area to be shrunk. The shrinks can be anywhere in size from a silver dollar down to the head of a thumbtack. The larger the hot spot, the harder the heat is to control.

An average-sized shrink is usually about the size of a quarter. Small shrinks should always be used on flat panels because panels tend to warp easily.

A very small hot spot would be used to take an oil can size bulge out of a flat panel. The term "oil can" is used to describe an area of a panel that is stretched very slightly. It can be pushed in; however, as soon as the pressure is released, the area will pop back out again, just as the bottom of an oil can does.

A neutral flame and a #1 or #2 tip are often used to heat the hot spots. The point of the cone is brought straight down to within 1/8 inch (3.2 mm) of the metal and held steady until the metal starts to turn red. The torch is then slowly moved outward in a circular motion until the complete hot spot is cherry red (Figure 8–50).

As the heat from the torch enters the small spot in the panel, the heated metal expands. The cooler metal surrounding the hot spot resists the expansion forces. As the temperature increases, the heated metal becomes softer. This soft metal piles up and forms a bulge in the hot spot (Figure 8–51).

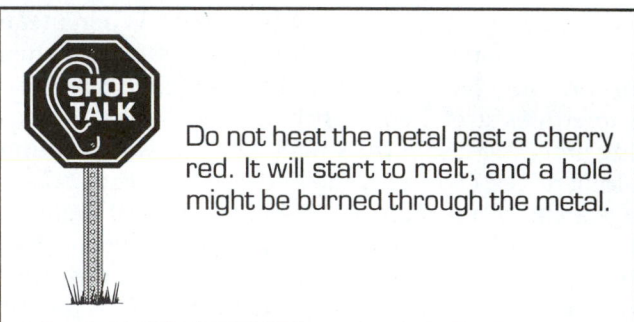

SHOP TALK

Do not heat the metal past a cherry red. It will start to melt, and a hole might be burned through the metal.

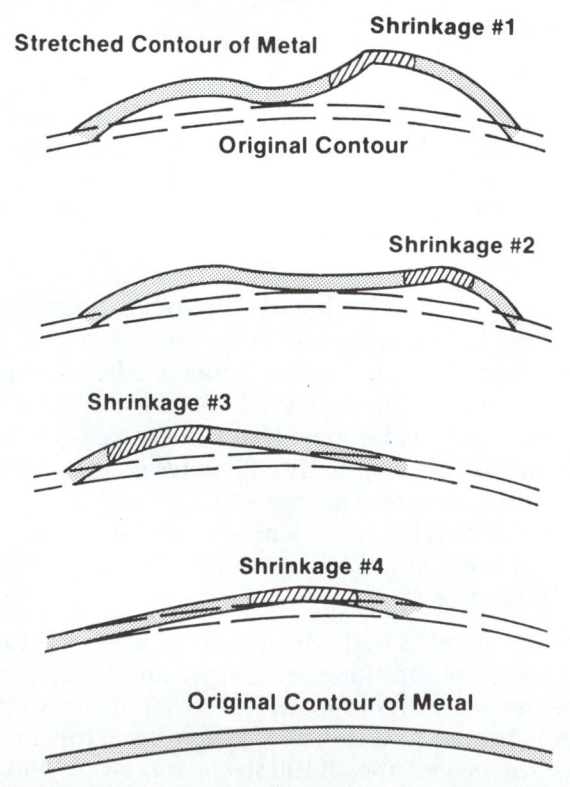

FIGURE 8-49 Shrinking stretched metal usually requires more than one hot spot. Always heat the highest spot.

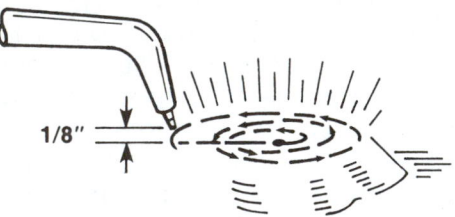

FIGURE 8-50 Keep the flame cone 1/8 inch (3.2 mm) from the metal and move the torch in a circular motion from the center out.

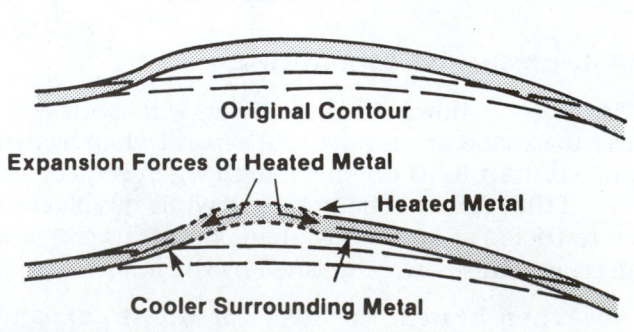

FIGURE 8-51 Heat-softened metal will usually bulge up.

The metal usually bulges up instead of down because the top of the metal is heated first. When it starts to bulge, the rest of the metal in the hot spot follows.

After the spot has been heated, the first hammer blow should be directed on the center of the heated spot. This will compress the excess metal into the "soft" spot created by heating the metal. The subsequent hammer blows should be directed around the perimeter of the hot spot while supporting the back side of the metal with the dolly (Figure 8–52). This is done to smooth the metal to control its movement caused by the contracting forces. One must be careful not to strike the metal too hard, as it can cause additional excess metal. This entire procedure must be accomplished within 30 seconds or the effect will be lost.

Once the redness has disappeared and the area has been dollied smooth, the shrink can be cooled with a wet rag or sponge. When this is done, a greater degree of contraction occurs, and a slight amount of distortion could result. Any warpage should be straightened before the next shrink is attempted.

The technician must be cautious not to quench the metal prematurely as it can cause some serious

SHOP TALK

Never use hard on-dolly blows to level the area. It will result in restretching the area of metal.

damage, such as overshrinking and hardening the metal. The metal may simulate a tempered effect if it is quenched prematurely, making it difficult to work smooth. This condition becomes more severe if a series of shrink spots are made in a small area. It should be understood that quenching is not necessary to accomplish the shrinking of metal. The open hammer blow on the hot spot and the contracting forces of the heat are responsible for reducing the excess metal. The amount of shrinking that occurs can be controlled much more readily by not quenching.

It is very hard to determine accurately the amount that each hot spot will shrink. One shrink can remove far more excess metal in one area than the same size would in another. It is not uncommon to find that an area has been overshrunk when it has completely cooled. When the area has been overshrunk, the metal in the area of the last shrink is usually collapsed or pulled flat. Sometimes the metal surrounding the shrink area can even be pulled out of the proper contour.

Overshrinking is corrected by using hard hammer-on-dolly blows to stretch the last shrink. The last shrunken area is usually the direct cause of overshrinking.

KINKING

Kinking involves using a hammer and dolly to create pleats, or kinks, in the stretched area to shrink the area's surface area. Instead of using heat to shrink the metal, kinking is another way to deal with stretched metal (Figure 8–53). Kinking the metal will lower the area slightly below the rest of the panel. The low spot should be filled with plastic body filler and filed and sanded level with the panel.

SHRINKING A GOUGE

A **gouge** is caused by a focused impact that forces a sharp dent or crease into a panel. A gouge causes the metal to be stretched. Gouges must be shrunk to their original size to properly repair the damage. Simply picking up the low area would distort the panel. Filling the gouge with filler without restoring the panel's

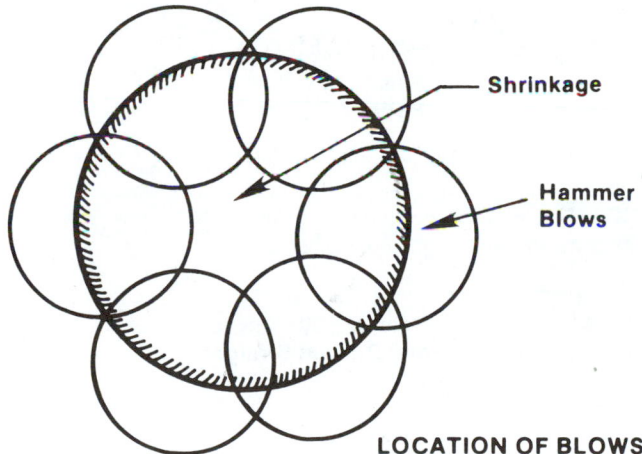

FIGURE 8-52 Hammer around the hot bulge to shrink it.

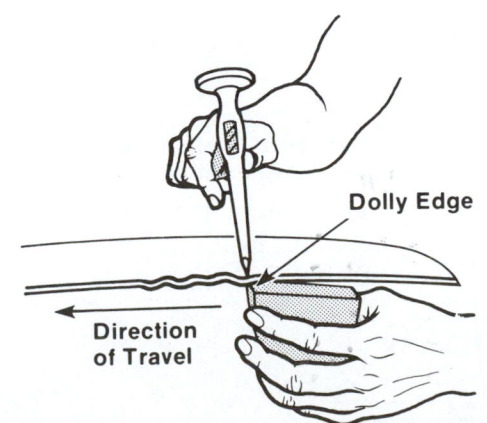

FIGURE 8-53 Shrinking metal by kinking the high spot can be done only when heat shrinking is not practical.

original contour will leave tension in the panel that could cause the filler to crack or pop off.

Follow this procedure for shrinking gouges:

1. Heat the lowest point of the gouge with a gas torch until the metal is a dull red (Figure 8–54).
2. Use a dolly to hammer up the hot spot. This will increase the tension on the soft spot, forcing it to swell and return to its original position.
3. While the metal is still hot, hold the dolly directly under the groove and tap down the ridges that will have developed on either side of the groove. This will not only drive down the ridges but will also bump up the gouged metal.
4. If the gouge is a long one, this process will have to be repeated several times to raise the whole length of the gouge. Heat only as much of the gouge as can be worked before the metal cools.

FILING THE REPAIR AREA

When the damaged area has been bumped and pulled as level and smooth as possible, the body file must be used to locate any remaining high and low spots (Figure 8–55).

Begin filing in the undamaged area on one side and progress across the damaged area to the undamaged metal on the opposite side. This way the correct plane can be maintained from undamaged metal to damaged area.

The file should be pushed forward by the handle for the cutting stroke. Downward pressure and direction is controlled by the hand holding the front of the file. As long a stroke as possible should be taken. On the return stroke, the file is pulled back over the metal by the handle. When filing relatively flat areas, the file is either held at a 30-degree angle and pushed straight or held straight and pushed at a 30-degree angle (Figure 8–56). On a crowned panel, the file is either held straight and pushed straight along the flattened crown of the panel, or held straight with the length of the flattest crown and pushed to one side at a 30-degree or less angle (Figure 8–57).

The scratch pattern created by the file identifies any low spots. The technician then "picks" up the low spots, bumps down the high spots, and files the

SHOP TALK

Be careful when filing or grinding high spots in metal panels. The metal on new vehicles is very thin. You can easily thin the metal too much or even form a hole in the panel.

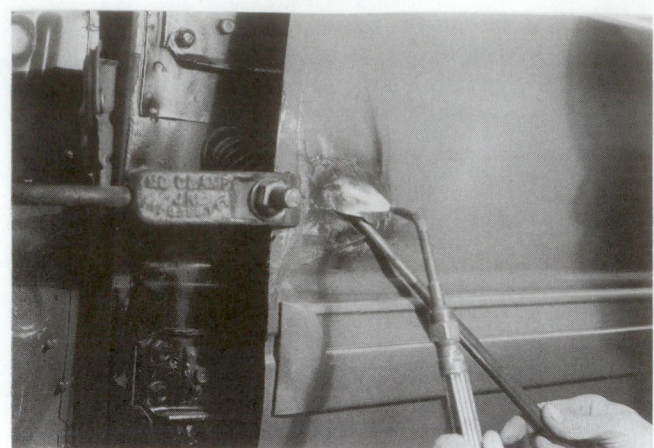

FIGURE 8-54 Controlled heat is being used in combination with pulling force to repair a dent and tear in a door panel.

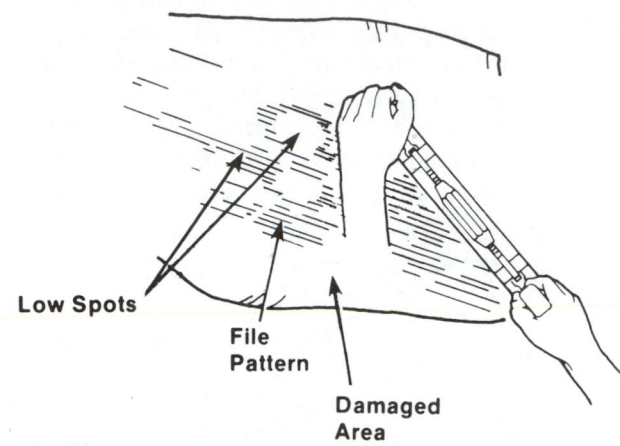

Low Spots

File Pattern

Damaged Area

FIGURE 8-55 Scratch the surface of the repair area with a file to reveal high and low spots.

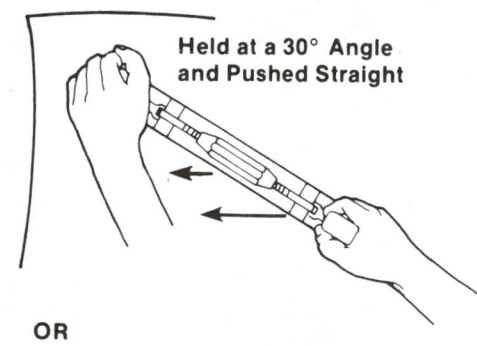

Held at a 30° Angle and Pushed Straight

OR

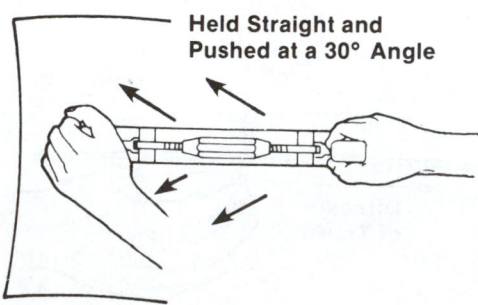

Held Straight and Pushed at a 30° Angle

FIGURE 8-56 Push the file at a 30-degree angle on flat or low-crown panels.

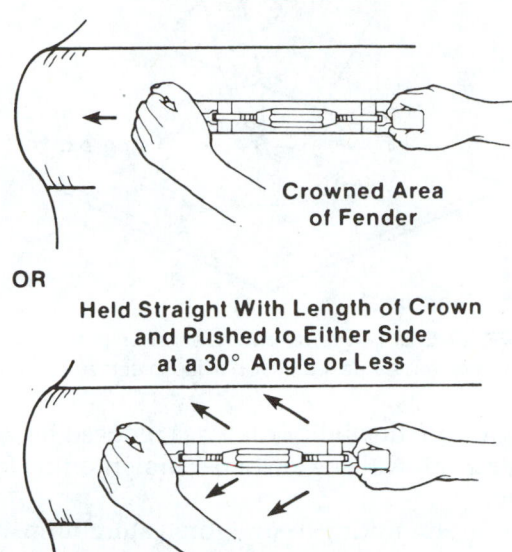

Held Straight With Length of Crown
and Pushed Straight

Crowned Area
of Fender

OR

Held Straight With Length of Crown
and Pushed to Either Side
at a 30° Angle or Less

FIGURE 8-57 Note the procedure for filing a crowned panel.

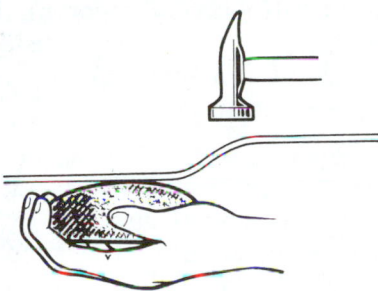

FIGURE 8-58 The hammer-off-dolly method is preferred when straightening aluminum. *(Courtesy of American Honda Motor Co.)*

area with the body file. This process is repeated until all the low spots disappear and the area has been filed relatively smooth. This will ready the repair area for plastic filler.

8.5 WORKING ALUMINUM

Aluminum is used for a variety of automotive panels, such as hoods and roof panels. Pure aluminum is lightweight and useful more for its ability to be formed than for its strength. When used on vehicles, it is alloyed with other elements and heat treated for more strength.

Aluminum's natural resistance to corrosion is a special advantage. When first exposed to air, a thin oxide film forms on the surface. This aluminum oxide serves as self-protection and prevents further corrosion. These properties vary depending on the type of alloy and whether or not it is heat treated.

The repair of aluminum panels requires much more care than the working of steel panels. Aluminum is much softer than steel, yet it is more difficult to shape once it becomes work hardened. It also melts at a lower temperature and distorts readily when heated.

It is important to keep in mind that aluminum body and frame parts are usually $1\frac{1}{2}$ to 2 times as thick as steel parts. When damaged, aluminum feels harder or stiffer to the touch because of work hardening. These characteristics must be taken into consideration when working damaged aluminum panels.

HAMMER AND DOLLY WORKING ALUMINUM

Straightening with a hammer and dolly is basically the same with aluminum as for steel, with the following exceptions:

- The hammer-off-dolly method is generally recommended for aluminum panel straightening (Figure 8-58). Because aluminum is less ductile than steel, it does not readily bend back to its original shape after being buckled by an impact. Therefore, aluminum does not respond well to off-dolly hammering. Care must also be exercised not to create additional damage when attempting to lower ridges with hammer and dolly blows.
- Aluminum alloys bend too quickly when the panel is sandwiched between the hammer and dolly, as with the hammer-on-dolly method. When it is necessary to hammer-on-dolly, hammering too hard or too much can stretch the panel. Use many light strokes rather than a few heavy blows.
- Shrinking hammers for working steel should not be used since they can cause cracking. Separate sets of tools should be used on steel and aluminum.

PICKING ALUMINUM

Raising small dents with a pick hammer or pry bar is an excellent way to repair aluminum panels.

SHOP TALK Do not apply the filler or putty to the bare surface of aluminum. Always apply an epoxy primer first. Also never use lead filler. Lead reduces aluminum's resistance to corrosion.

However, be careful not to raise the panel too far, stretching the soft aluminum.

SPRING HAMMERING ALUMINUM

Spring hammering with hammer and spoon is an excellent way to unlock stresses in high-pressure areas. The spoon distributes the force of the blow over a wider area, minimizing the possibility of creating additional dents in the unyielding buckle.

FILING ALUMINUM

Because aluminum is soft, reduce hand pressure on the body file. Use a file with rounded edges to avoid scratching and gouging the metal.

GRINDING ALUMINUM

Grinding must be done very carefully on aluminum panels. A coarse grit disc on a high-speed grinder can quickly burn through the soft metal. The heat from the grinding operation can also quickly warp the panel.

A #36 grit open coat disc can be used, but grind carefully in order to remove only paint and primer, and not the metal. Make two or three passes. Then quench the area with a wet rag to cool the metal and minimize heat gain. Grinding small areas and feathering small areas and featheredging should be done with a dual-action sander or an electric polish machine that rotates less than 2,500 rpm. Use #80 or #100 grit paper and a soft, flexible backing pad.

HEAT SHRINKING ALUMINUM

There is one major difference with straightening aluminum by heat shrinkage. With steel, use heat only when the metal is stretched and cannot be straightened by other means. With aluminum, heat must be

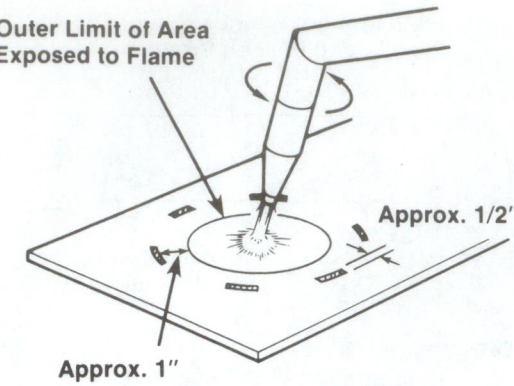

FIGURE 8-59 Use temperature sensitive paint or crayons to prevent excessive heat buildup in soft aluminum.

used to restore flexibility that was reduced by work hardening or it will crack when a straightening force is applied.

Before attempting to straighten aluminum, heat the damaged metal with a torch. It is easy to apply too much heat because aluminum does not change color with high temperatures. Aluminum melts at a low 1,220 degrees Fahrenheit (660 degrees Celsius), so careful heat control is very important. Use a temperature sensitive paint or a heat sensitive crayon made to change color at about 750 degrees Fahrenheit (417 degrees Celsius).

- Apply temperature sensitive paint or crayon in a circular pattern around the area that will be exposed to the flame (Figure 8–59).
- Heat the area, keeping the flame moving constantly.
- Stop heating when the paint or crayon color changes. The surface temperature at the center of the heated area will be between 750 and 800 degrees Fahrenheit, a safe margin from aluminum's melting point. A lack of caution will result in a melted panel. Also the shrink spot must be quenched very slowly to avoid distorting the panel by excessive contraction.

SUMMARY

- To do quality sheet metal repairs, you must first know how to return the sheet metal to its original shape.

- Metalworking skills are probably the most important craft a body technician can bring to a shop.

- High-strength steel (HSS) is stronger than low-carbon or mild steel because of heat treatment. Most new vehicles contain HSS in their structural components.

- To repair collision damage, you must also understand what property changes have taken place in the metal.

- Yield point is the amount of force that a piece of metal can resist without tearing or breaking.

- Spring-back is the tendency for metal to return to its original shape after deformation.

- Work hardening is the upper limit of plastic deformation, causing the metal to become very hard in the bent area.

- Direct damage is simple, easy-to-find, visible damage, such as a gouge, a tear, or a scratch.

- Indirect damage is caused by the shock of collision forces traveling through the body and inertial forces acting on the rest of the unibody.

- The term "raising" means to work a dent outward or away from the body. The term "lowering" means to work a high spot or bump down or into the body.

- The metal that is pushed up is called a pressure area. Areas that are pushed down are called tension areas.

- Hammer-on-dolly repairs are used to smooth small, shallow dents and bulges and to stretch metal so that it can return to its original shape.

- Hammer-off-dolly is used to straighten metal just before the final stage of straightening.

- A pull pin spot welder or nail gun fuses a metal pin to the dent area

- Aluminum is used for a variety of automotive panels, such as hoods and roof panels.

ASE STYLE REVIEW QUESTIONS

1. Which type of sheet metal is most often used in unibody construction?
   A. Hot-rolled
   B. Cold-rolled
   C. Both A and B
   D. Neither A nor B

2. Whenever a collapsed rolled buckle occurs, there are two other buckles adjacent to the collapsed roll. What type of buckles are these?
   A. Simple hinge buckles
   B. Collapsed hinge buckles
   C. Simple rolled buckles
   D. None of the above

3. The cross cut disc action is used to _____.
   A. Remove paint
   B. Smooth the filler
   C. Remove metal
   D. Both A and B

4. Which kind of spoon can be used instead of a hammer?
   A. Dinging spoon
   B. Slapping spoon
   C. Both A and B
   D. Neither A nor B

5. When filing a relatively flat area, Technician A holds the file at a 30-degree angle and pushes it straight. Technician B says that as long a stroke as possible should be taken. Who is correct?
   A. Technician A
   B. Technician B
   C. Both A and B
   D. Neither A or B

6. Technician A says that aluminum has a higher melting point than mild steel. Technician B says that martensitic steel loses strength when welded. Who is correct?
   A. Technician A
   B. Technician B
   C. Both A and B
   D. Neither A nor B

7. Technician A says that 80 percent of sheet metal damage is direct damage. Technician B says that bent metal is not necessarily buckled metal. Who is correct?
   A. Technician A
   B. Technician B
   C. Both A and B
   D. Neither A nor B

ESSAY QUESTIONS

1. In your own words, how do you use a body hammer to straighten metal?

2. Describe the hammer-on-dolly method.

3. Summarize the method for removing a gouge in metal.

4. List and explain five types of part loading.

CRITICAL THINKING PROBLEMS

1. An aluminum panel has a bulge in it after initial straightening. The bulge pops in and out with hand pressure. What must be done?

2. After shrinking a steel panel, the technician finds a flat area under tension and lower than the rest of the panel. What is wrong?

MATH PROBLEMS

1. If a fender is $1/16$-inch (1.6 mm) thick and you grind away .031 inch (0.79 mm) of metal, how thick is the remaining panel?

2. If a 12-inch (305 mm) metal rod is heated and expands 2 percent in size, how long has it become?

PHOTO SEQUENCE

DENT REMOVAL

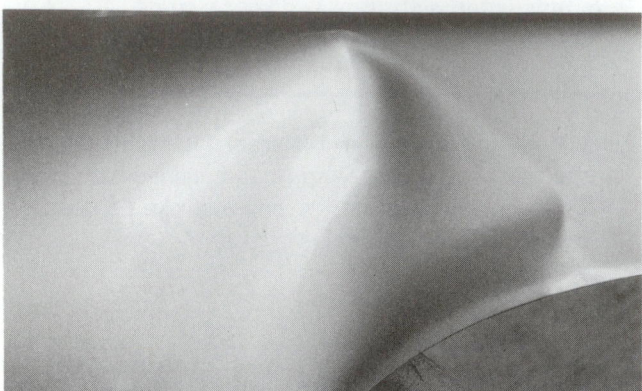

P8-1 A dent in a contoured part forms a crease. Proper metalworking methods are needed to remove the dent efficiently and without further stretching the metal.

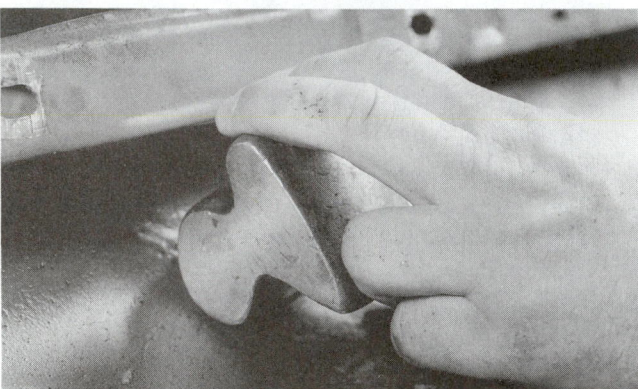

P8-2 Using a properly shaped dolly, start working the dent from the ends of the crease, not from the middle. Flatten the curve at the ends of the dent so that the metal will not be stretched as the center is moved back out.

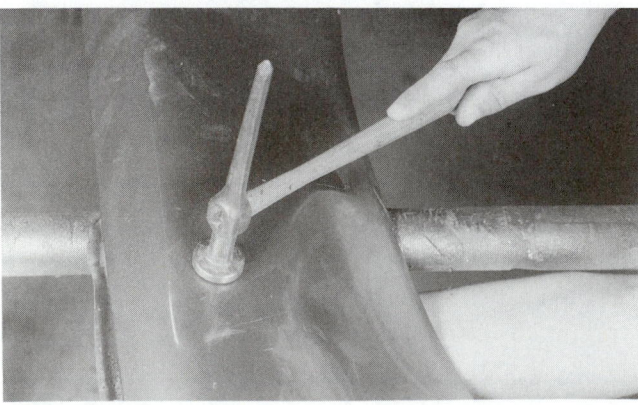

P8-3 Next, move the center of the crease out part way. You will need the edge of a dolly with a larger contour to match the larger contour of the center area of the crease.

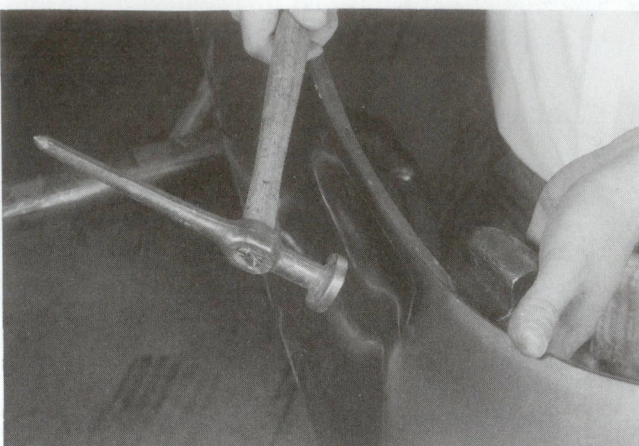

P8-4 Then, go back and work the ends. Try to remove the damage either as it occurred, or all at once.

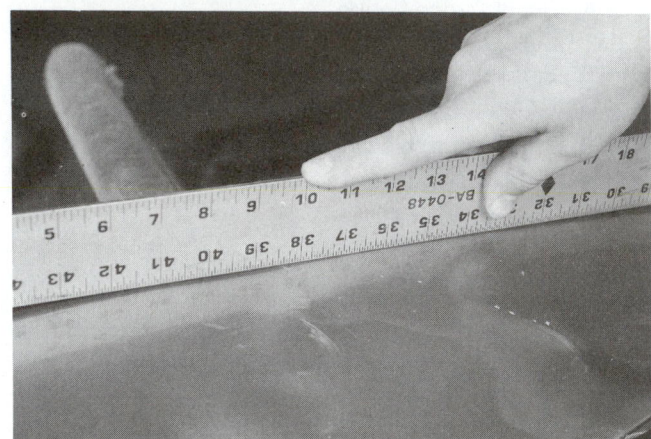

P8-5 Use a straightedge to check your progress.

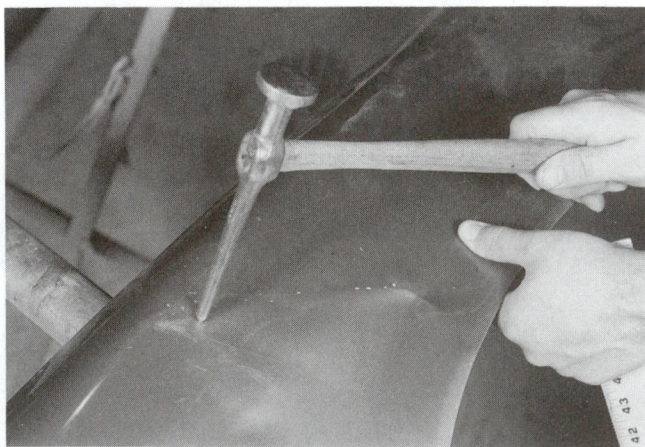

P8-6 Light hammer blows will help to lower any surfaces that have been raised too high. The repair area should be within 1/8 inch (3.1 mm) of level so it can be filled with plastic filler.

CHAPTER 9

Minor Auto Body Repairs

INTRODUCTION

The metalworking techniques, discussed in Chapter 8, are fundamental to any repair job. Damaged metal must be restored to its original contour before filling with plastic. This is done by unfolding buckles, relieving stresses in work-hardened ridges, shrinking stretched metal, and stretching shrunken areas. Repair quality depends on sound metalworking techniques!

Plastic body filler is the finishing touch for most sheet metal repairs. Restoring bent and stretched metal to its exact original shape and dimension can be very time-consuming and almost impossible in many instances (Figure 9–1). Remaining surface irregularities can be quickly and easily filled and smoothed with a thin coat of plastic filler. It is critical that you prepare the surface and apply fillers properly. If not, the filler could crack or pop off or the topcoat of paint may be adversely affected.

Plastic body filler is also used to fill scratches, dings, and pitted areas. The filler is then filed and sanded level with the panel. Spot putty and/or primer-surfacer is applied to fill any remaining surface imperfections.

A summary of minor metalworking procedures is given in Table 9–1.

9.1 BODY FILLERS

Body filler or **plastic filler** is a heavy-bodied plastic material that cures very hard for filling small dents in metal. It is a compound of resin and plastic used to fill dents on car bodies.

Only after the damaged panel has been bumped, pulled, pried, and dinged to within at least $1/8$ inch (3.1 mm) of the original contour can filler be applied. Then you can fill, shape, and smooth the repair area with a thin layer of plastic body filler (Figure 9–2).

Most auto body repairs require some application of plastic body filler (Figure 9–3), which is a fast, inexpensive way to restore the final contour of a damaged panel.

Unfortunately, some body shops fail to properly do sheet metal repairs and simply hide the damage under a thick layer of filler. Body fillers were never meant to replace proper metalworking techniques.

Before any filler is applied, all holes, cracks, and joint gaps must be welded. Some body fillers are

OBJECTIVES

After studying this chapter, you should be able to:

✔ List the different types of body fillers and glazes.

✔ Choose the correct plastic body filler for a particular repair job.

✔ Identify the correct way to mix filler and hardener.

✔ Describe how to use special sanding aids.

✔ Explain how to repair scratches, nicks, dings, and surface rust with plastic filler and glazing putty.

✔ Repair rustout damage properly.

✔ Understand the basics of lead filling.

✔ Answer ASE test questions pertaining to minor body repairs.

KEY TERMS

featheredging
fiberglass body filler
hardener
hardener kneading
light body filler
long-strand fiberglass filler
metal conditioner
mixing board

one-part glazing putty
plastic filler
polyester glazing putty
rustout
short-strand fiberglass filler
spoiled hardener
surface rust

ASE TASK LIST

Job Skills covered in this chapter include:

PAINTING AND REFINISHING TEST (B2) TASK LIST

A. Surface Preparation

4. Remove paint finish.
6. Featheredge areas to be refinished.
9. Mix primer, primer-surfacer, or primer-sealer; spray onto surface of repaired area.
10. Apply two-component putty to minor surface imperfections.
11. Dry or wet sand area to which primer-surfacer and/or two-component putty have been applied.
12. Remove dust from area to be refinished, including cracks or moldings of adjacent areas.
13. Clean area to be refinished using a proper cleaning solution.
14. Remove, with a tack rag, any dust or lint particles from the area to be refinished.
15. Apply suitable paint sealer to the area being refinished when sealing is needed or desirable.
16. Remove imperfections from sealer.

C. Paint Mixing, Matching, and Applying

2. Shake, stir, reduce, catalyze, and strain paint according to manufacturer's recommendations.
6. Apply basecoat/clearcoat for spot and panel blending, and overall refinishing.
8. Sand, buff, and polish finishes where necessary.

D. Solving Paint Application Problems

1. Identify blistering (raising of the paint surface); determine the cause(s), and correct the condition.
6. Identify lifting (surface distortion or shriveling) while the topcoat is being applied; determine the cause(s), and correct the condition.

E. Finish Defects, Causes, and Cures

1. Identify poor adhesion; determine the cause(s), and correct the condition.
4. Identify blistering in the paint surface; determine the cause(s), and correct the condition.

NONSTRUCTURAL ANALYSIS AND DAMAGE REPAIR TEST TASK LIST

B. Outer Body Panel Repairs, Replacements, and Adjustments

2. Remove and replace bolted, bonded, and welded panels or panel assemblies.

C. Metal Finishing and Body Filling

2. Pick and file the damaged area of a body panel to eliminate surface irregularities.
6. Mix plastic filler.
7. Apply plastic body filler; rough shape during curing.
8. Sand cured plastic body filler to contour.

A   B

FIGURE 9-1 You must determine if panel repairs are major or minor. (A) How would you repair damage on the door and quarter panel? (B) The door required replacement of the whole skin. The quarter panel was repaired with metal straightening and filler with plastic.

hydroscopic, which means they absorb moisture when exposed to humid conditions. Unless filled with a waterproof pigment, these fillers will absorb moisture through holes or cracks in the metal.

Body fillers and putties can also be used to repair minor surface defects, such as dings, stone chips, surface rust, and rustouts. Be aware, though, that plastic body fillers have limitations. Large panels, such

TABLE 9-1: MINOR DAMAGE REPAIR PROCEDURE

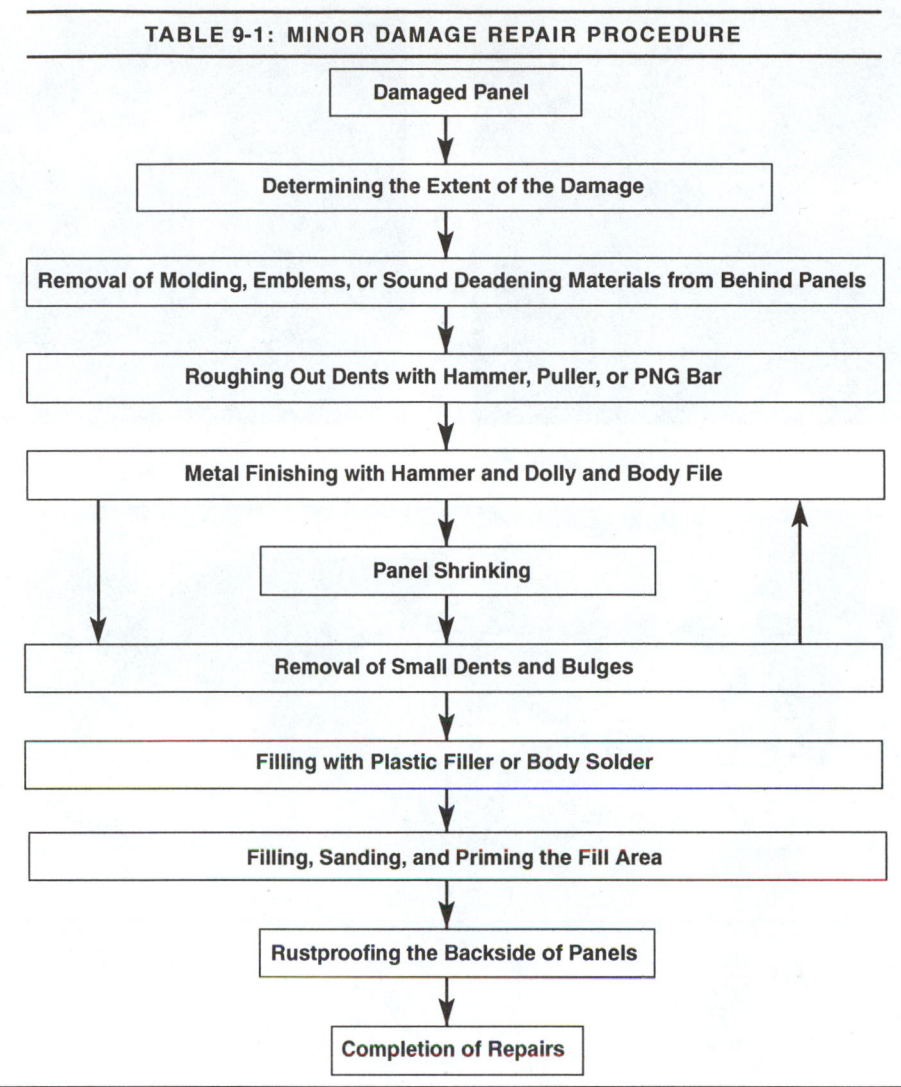

Damaged Panel

↓

Determining the Extent of the Damage

↓

Removal of Molding, Emblems, or Sound Deadening Materials from Behind Panels

↓

Roughing Out Dents with Hammer, Puller, or PNG Bar

↓

Metal Finishing with Hammer and Dolly and Body File

↓

Panel Shrinking

↓

Removal of Small Dents and Bulges

↓

Filling with Plastic Filler or Body Solder

↓

Filling, Sanding, and Priming the Fill Area

↓

Rustproofing the Backside of Panels

↓

Completion of Repairs

as hoods, deck lids, and door panels, tend to vibrate violently under normal road conditions. Vibrations can crack and dislodge filler that is applied over an area that is too large or applied too thickly.

Care must also be taken when applying filler to semistructural panels in unibody frames. Panels, such as quarter panels and roofs, absorb road shocks and torque flexing. Excessive fillers applied in these areas can be popped off by stresses in the panels. Plastic filler should also be used sparingly on rocker panels, lower rear wheel openings, and other areas subject to flying stones and rock chips.

GENERAL DESCRIPTION

Plastic body filler is very similar to paint in composition. Both are made of resins, pigments, and solvents. Most plastic body fillers have a polyester resin that acts as a binder. When the filler is applied and the solvents evaporate, the binders hold the pigments together in a tough, durable film. The basic pigment or filler in conventional fillers is talc.

Talc, also used in baby powders, absorbs moisture. That is good for the baby but bad for the car if proper steps are not taken to shield the filler from moisture. If holes in the metal or cracks in the paint expose the filler to the atmosphere, the talc in the filler absorbs moisture, which attacks the metal substrate and forms rust. The rust destroys the filler-to-metal bond, causing the filler to fall off. Waterproof fillers are available. Fiberglass strands or metal particles are used instead of talc as pigments.

Like some paints, plastic fillers harden by chemical action. Hardening, or curing, produces a molecular structure that will not shrink or soften. The chemical reaction is set off by oxygen. If the container of plastic filler is open and left exposed to the oxygen in the atmosphere, it will slowly harden.

To speed up the drying process, a chemical catalyst is provided by the manufacturer. The catalyst, in

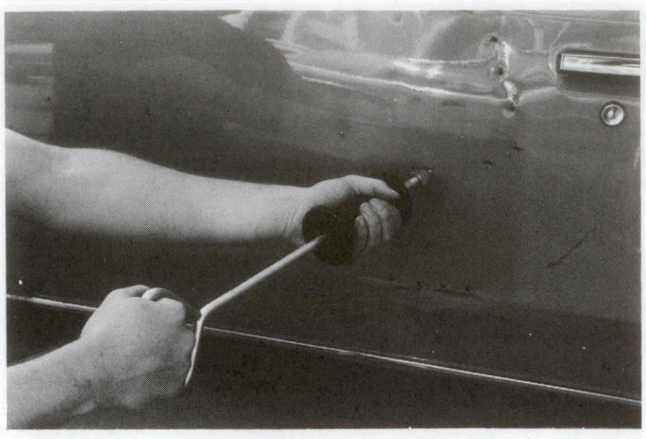

A

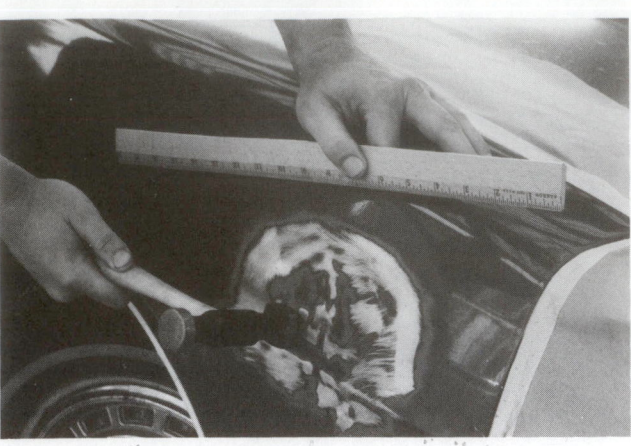

B

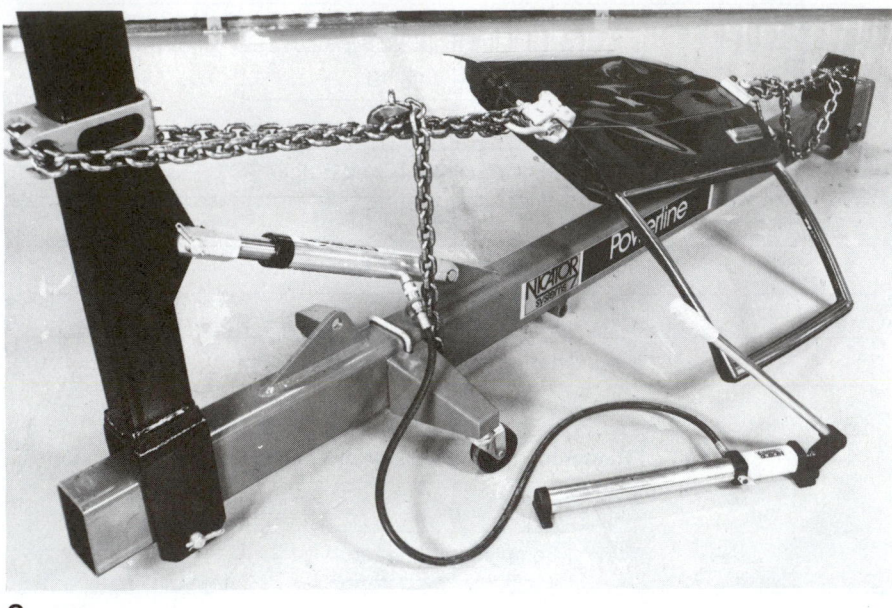

C

FIGURE 9-2 Always use proper metalworking techniques before filling with plastic: (A) pulling, (B) bumping, and (C) stretching. *(Photo C courtesy of Nicator, Inc.)*

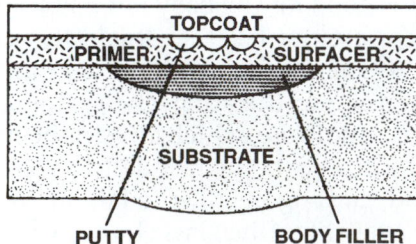

FIGURE 9-3 Study the basic sequence of a minor repair. Plastic filler goes over the metal first. Then the primer or primer-surfacer, followed by the spot putty before spraying on the paint or topcoat. *(Courtesy of DuPont Automotive Products)*

liquid or cream form, is called **hardener** (Figure 9–4). Hardener is basically a chemical compound called peroxide. The oxygen in the peroxide drastically speeds up the curing process. As Table 9–2

TABLE 9-2: EFFECT OF TEMPERATURE ON WORKING TIME	
Temperature	**Working Times**
100°F	3 to 4 minutes
85°F	4 to 5 minutes
77°F	6 to 7 minutes
70°F	8 to 9 minutes

shows, the filler will become too stiff to work in just a few minutes after adding hardener, depending on the ambient air temperature.

Curing fillers may also produce a *waxy coating*, or *paraffins*, on the surface. The purpose of the paraffins in the filler is to form a film that prevents oxygen absorption from the atmosphere. The paraffins are suspended in the filler solvent and are carried to

As the filler cures and hardens, the chemical reaction produces a tremendous amount of heat. For this reason, unused filler should not be discarded in trash cans containing solvent-wet paper or cloths. A serious shop fire could result!

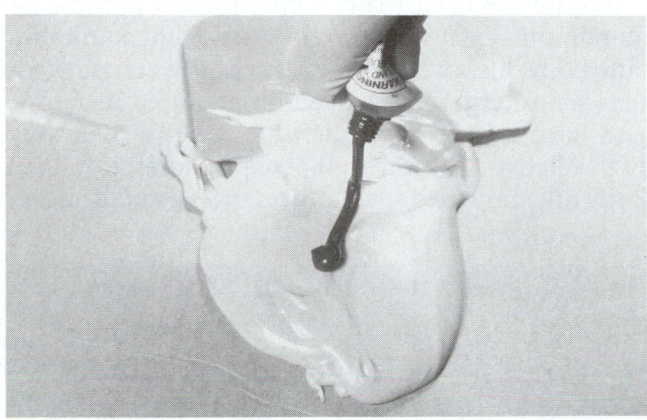

FIGURE 9-4 A chemical catalyst called hardener causes plastic filler to harden very quickly.

the surface when the solvents evaporate. The paraffins must be either removed with a wax and grease remover before being sanded or else filed off with a surform cheese grater.

TYPES OF BODY FILLERS

During the first fifty years of auto body repair, blemishes in sheet metal panels were corrected by applying lead filler. *Lead filler* or solder is an alloy of lead and tin. A welding torch was used to soften the solder and bond it to the body sheet metal.

Before World War II, automobiles were made with heavy gauge steel panels that were unaffected by the heat used in the "tinning" operation. But changes began to take place in automotive construction in the late 1940s and early 1950s. In the economic boom following World War II, Americans began demanding larger and fancier cars. Manufacturers responded with vehicles made with thinner, larger, and more complex body panels. The thinner metals, however, made the old lead repair methods almost obsolete. The heat required for the lead filler warped the thin panels, and hammer-and-dolly work stretched metals too thin for filing. There was a real need for an inexpensive, time-saving substitute.

In the early 1950s, epoxy-based fillers were developed. Usually mixed with aluminum powder, epoxy fillers cured very slowly and did not harden at all if applied too thickly.

In the middle 1950s, the first polyester resin-based body fillers were developed. These fillers were made from the same resin used to make fiberglass boats and required mixing with a liquid hardener and accelerator. Since the fiberglass resin is very brittle when cured and depends on cloth or matte for flexibility, the early polyester body fillers were also very brittle and hard.

The first successful filler was named "Bondo," and many body technicians still refer to plastic body fillers as Bondo. The early fillers were composed of approximately 40 percent polyester resin and 60 percent talc (by weight).

When the filler and hardener were properly mixed together, the filler dried very hard. Early fillers were difficult to file and had to be leveled with a grinder, resulting in choking clouds of dust that blanketed the shop. Low-dust, straight-line air fillers had not been developed yet.

As the technology developed, body fillers became softer, easier to apply, and easier to shape. Fillers soon appeared in black, red, gray, white, and yellow. Cream hardeners in contrasting colors—red, white, green, and blue—were also developed to provide a mixing reference for the various colored polyester fillers. The softer fillers could also be grated while still semicured, thus reducing the amount of sanding required. Note that the addition of color does not affect the working characteristics of the filler.

Conventional body fillers have over thirty years of development backing them today. The premium fillers use very fine grain talc to provide superior workability, sandability, and featheredging. High-quality resins ensure excellent adhesion and quick curing properties. Most plastic fillers can be grated in 10 to 15 minutes.

Fiberglass Fillers

Fiberglass body filler has fiberglass material added to the plastic filler. It is used for rust repair or where strength is important. It can be used on both metal and fiberglass substrates. Because fiberglass-reinforced filler is very difficult to sand, it is usually used under a conventional, lightweight plastic filler.

As thinner-gauge sheet metal replaced the heavy gauge steel, rust became a problem, especially in areas of the country where road salts are used in winter. A product was needed to repair rustouts. Because talc-filled body fillers absorb moisture readily, the available fillers did not provide long-lasting protection when used to repair rustouts.

To meet this demand for a waterproof filler, fiberglass-reinforced fillers were developed. Fiberglass fillers use fiberglass strands rather than talc as a bulking agent. These fillers are more flexible and stronger than conventional fillers. Because they are also waterproof, they can be used to bridge holes, tears, and rustouts.

Fiberglass fillers are available in two basic forms. One is formulated with short strands of fiberglass; the other is made with long strands.

Short-strand fiberglass filler has tiny particles of fiberglass in it. It works and sands like a conventional filler, but is much stronger. **Long-strand fiberglass filler** has long strands of fiberglass for even more strength. It is commonly used to repair larger holes in metal or fiberglass bodies.

Short-strand fiberglass fillers are generally used to repair small holes, approximately 1 to 1½ inches (25 to 38 mm) in diameter. When used to repair larger holes, a fiberglass cloth or screen should be used as a back support. The short-strand fillers can be sanded and finished as any conventional filler.

Long-strand fiberglass products are designed to fill holes larger than 1½ inches (38 mm) in size. The longer strands interlock and provide a much stronger patch. The long-strand filler might also be used with fiberglass cloths or mattes to bridge even larger rustouts. The long-strand fillers, however, are used only as a base. Smoother fillers, either short strand fiberglass fillers or conventional fillers, must be used for the final fill. Chopped fiberglass fibers are also available to be added to fillers to increase their strength.

Aluminum Fillers

Aluminum fillers actually contain metal particles of aluminum. The first premixed, 100 percent aluminum auto body fillers were introduced in 1965. These products are waterproof, use a red-tinted liquid hardener, and have a fairly good shelf life. Due to their very high relative cost, the 100 percent aluminum-filled body fillers are used sparingly on special applications,

A

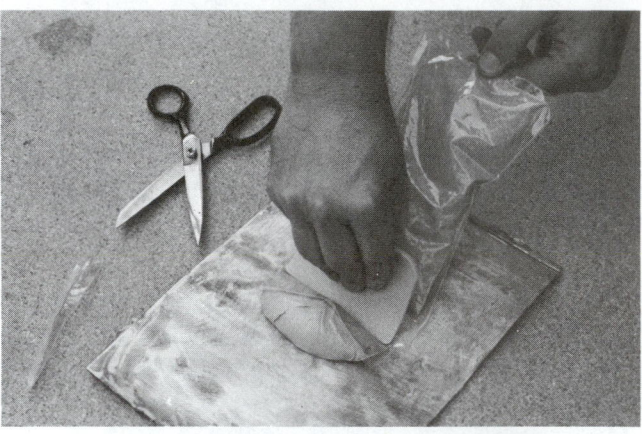

B

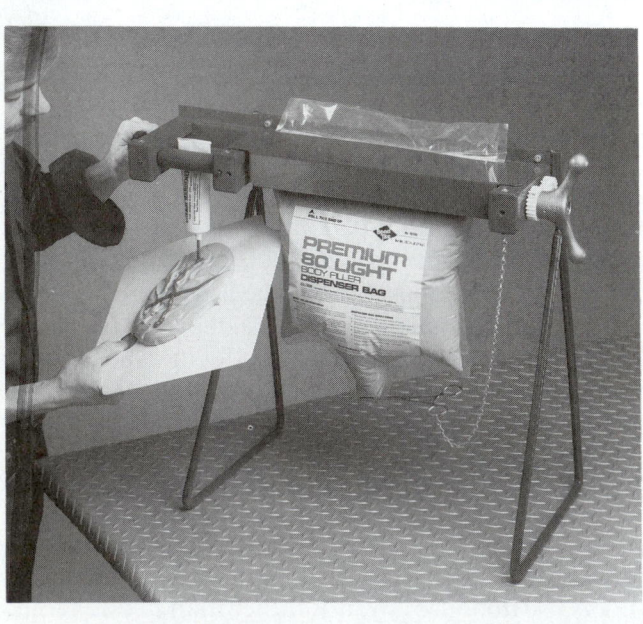

C

D

FIGURE 9-5 Plastic fillers come in different packaging. (A) Gallon can. (B) Plastic bag. (C) Dispenser for plastic and hardener. (D) Pneumatic 5-gallon can. (*Courtesy of Dynatron/Bondo Corp. and Bond-Tite®, Division of U.S. Chemical and Plastics, Inc.*)

such as restoring antique cars. Today there are several similar 100 percent aluminum products available. Metal fillers are nonshrinking, waterproof, and very smooth. Metal fillers have the look of lead but are easier to work. When cured, they are harder than talc or fiberglass-filled plastic fillers.

Lightweight Fillers

Light body filler is formulated for easy sanding and fast repairs (Figure 9–5). It is used as a very thin topcoat of filler for final leveling. It can be spread thinly over large surfaces for block or air tool sanding. Lightweight fillers were formulated by replacing about 50 percent of the talc in the filler with tiny glass spheres. The resulting higher resin content dramatically improved the filing and sanding characteristics of the filler as well as improved the filler's adhesion and water resistance.

Most lightweight fillers are homogenous. The glass bubbles remain suspended in the resin and do not settle to the bottom of the can. This homogenous composition allows lightweight fillers to be packaged in plastic bags or cans and dispensed with rollers or compressed air or squeezed out with a plastic spreader. The plastic bags keep filler fresh and eliminate much of the wasted filler sometimes associated with canned fillers.

Lightweight fillers represent more than 80 percent of the total filler used in body shops across the nation.

Premium Fillers

Filler manufacturers in the mid-1980s have taken advantage of new technology to produce premium-quality fillers. Premium fillers have superior performance qualities that go beyond the capabilities of conventional lightweight fillers. Premium fillers are moist and creamy. They spread easily yet will not sag on vertical surfaces and dry tack-free without pinholing.

Spot and Glazing Putties

Because body fillers usually have tiny pinholes and sand scratches, spot putty is used as a quick way of fixing them.

One-part glazing putties are applied directly out of a tube and cure slowly. They were developed to fill minor surface imperfections and produce a smoother surface. They are being phased out for quicker curing two-part putties.

Polyester Glazing Putty

Old-style one-part putties caused problems with the new basecoat/clearcoat paint systems. The rich

solvents and multicoats required for these "trick" paint jobs caused the pigment to "bleed" and stain the finish on light colors, usually after several days of exposure to sunlight. The widespread use of basecoat/clearcoat products and other multicoat systems has made this staining problem a more frequent occurrence.

Glazing putties should be used only to fill very shallow sand scratches and pinholes (Figure 9–6). Maximum filling depth of most putties is only $1/32$ inch (.794 mm).

Although glazing putties featheredge very nicely, they do not develop the hardness of a body filler. When coated with primer or paint, putties absorb paint solvents and swell. Sufficient time must be allowed for the putty to fully cure again before finish sanding of the finish coats. If sanded too soon, sand scratches will appear in the finish as the putty dries completely and shrinks below the sanded surface.

Conventional one-part glazing putty does not dry below its surface very quickly (Figure 9–7). It takes several hours, sometimes days, for one-part putty to dry all the way through. Even though the putty may sand normally, it still may not be completely dry at the bottom of the scratch or pit. If paint is applied over the partially cured putty, the surface may sink or shrink, as well as bleed, and cause a flaw in the paint or topcoat.

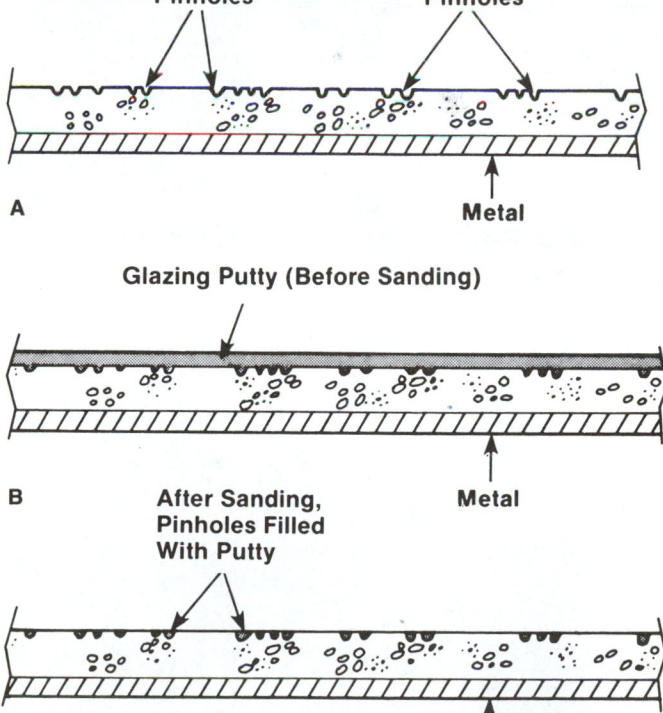

FIGURE 9–6 Glazing putty fills pinholes in primer and filler. (A) Pinholes can remain after sanding and priming. (B) Apply thin layer of putty over pinholes. (C) Sand off putty until flush with primer. Putty will remain and fill small holes or pits.

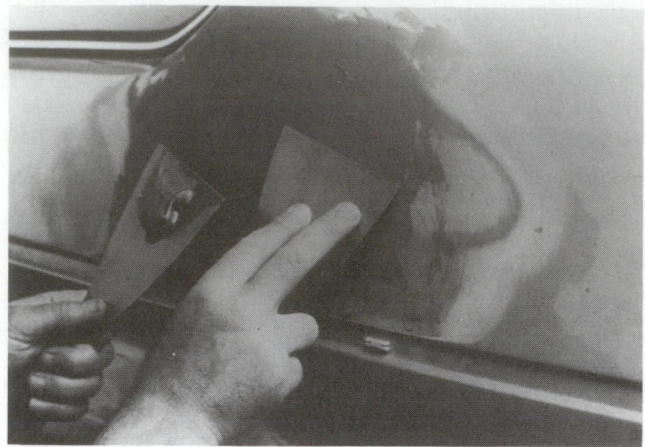

FIGURE 9-7 Polyester putty is gaining in popularity over one-part putty because it dries much more quickly, reducing the chance of bleeding and shrinking.

FIGURE 9-8 Sprayable polyester primer-filler. *(Courtesy of Marson Corporation)*

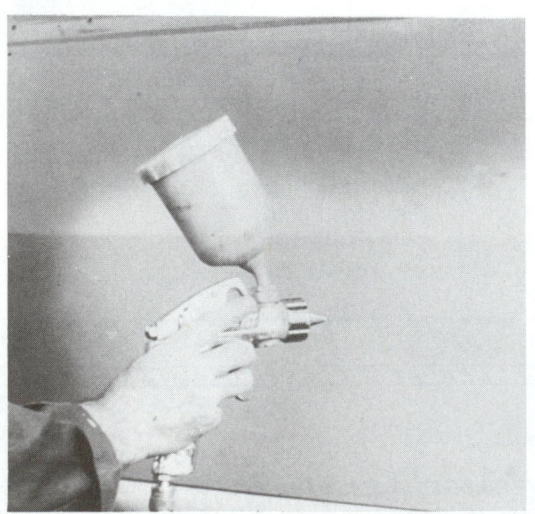

FIGURE 9-9 Sprayable polyester primer-filler is good for large surface areas with minor flaws. *(Courtesy of America Sikkens, Inc.)*

To solve the these problems, body filler manufacturers have developed a fine-grained, catalyzed **polyester glazing putty** or *two-part spot putty*. Like plastic filler, a hardener or catalyst must be mixed with the putty to initiate and speed curing.

Polyester glazing putty does not shrink, has excellent dimensional stability, and resists solvent penetration (the cause of bleed-through). When applied over traditional body fillers, polyester glazing putties effectively solve the bleed-through problem.

Bleed-through problems can also be avoided by using a sprayable polyester filler (Figure 9–8). This type of primer-surfacer has polyester resins and talc fillers and must be catalyzed with a liquid hardener. Sprayed from a conventional or gravity feed gun (Figure 9–9), polyester primers fill minor imperfections and seal both fillers and old paint finishes.

9.2 APPLYING PLASTIC BODY FILLER

Body fillers are designed to cover up the minor depressions that cannot be removed by metal straightening alone. As discussed, there are several types of filler for specific tasks. Table 9–3 summarizes the ingredients, characteristics, and applications of the currently available body fillers and putties. Study it carefully so you can use the product best suited for the job.

Taking shortcuts when mixing and applying fillers might save time at first, but it affects the quality of the repair. Fillers that are improperly used will eventually crack, lose adhesion, permit rust to form, and fall off the panel. Sooner or later, the repair will have to be redone. The loss in time, money, and reputation is hard to calculate.

PREPARING THE SURFACE FOR FILLING

One of the most important steps in applying plastic body fillers is surface preparation. Begin by washing the repair area to remove dirt and grime. Then clean the area with wax and grease remover to eliminate wax, road tar, and grease. Be sure to use a remover that will take away the silicones often present in car waxes. Also, wash brazed or soldered joints with soda water to neutralize the acids in the flux. Do not grind these areas before neutralizing the acids. Grinding simply drives the acids deeper into the metal.

Grind the area to remove the old paint (Figure 9–10). Remove the paint for 3 or 4 inches (76 to 101 mm) around the area to be filled. If filler overlaps any of the existing finish, the paint film will absorb solvents from the new primer and paint, destroying the adhesion of the filler. The filler will lift, cracking

TABLE 9-3: COMPARING FILLERS AND PUTTIES

Filler	Composition	Characteristics	Application
Conventional Fillers			
Heavyweight Fillers	Polyester resins and talc particles	Smooth sanding; fine feather-edging; nonsagging; less pinholing than lightweight fillers	Dents, dings, and gouges in metal panels
Lightweight Fillers	Microsphere glass bubbles; fine grain talc; polyester resins	Spreads easily; nonshrinking; homogenous; no settling	Dings, dents, and gouges in metal panels
Premium Fillers	Microspheres; talc; polyester resins; special chemical additives	Sands fast and easy; spreads creamy and moist; spreads smooth without pinholes; dries tack-free; will not sag	Dings, dents, and gouges in metal panels
Fiberglass-Reinforced Fillers			
Short Strand	Small fiberglass strands; polyester resins	Waterproof; stronger than regular fillers	Fills small rustouts and holes. Used with fiberglass cloth to bridge larger rustouts.
Long Strand	Long fiberglass strands; polyester resins	Waterproof; stronger than short strand fiberglass fillers; bridges small holes without matte or cloth	Cracked or shattered fiberglass. Repairing rustouts, holes, and tears.
Specialty Fillers			
Aluminum Filler	Aluminum flakes and powders; polyester resins	Waterproof; spreads smoothly; high level of quality and durability	Restoring classic and exotic vehicles
Finishing Filler/ Polyester Putty	High-resin content; fine talc particles; microsphere glass bubbles	Ultra-smooth and creamy; tack-free; nonshrinking; eliminates need for air dry type glazing putty	Fills pinholes and sand scratches in metal, filler, fiberglass, and old finishes.
Sprayable Filler/ Polyester Primer-Surfacer	High-viscosity polyester resins; talc particles; liquid hardener	Virtually nonshrinking; prevents bleed-through; eliminates primer/glazing/ primer procedure	Fills file marks, sand scratches, mildly cracked or crazed paint films, and pinholes. Seals fillers and old finishes against bleed-through.

SHOP TALK

A common mistake for the beginner is to apply filler over paint. This is easy to do but is a serious mistake that will affect repair quality. Remember that plastic filler is designed to be applied to the body substrate, usually metal, not to paint.

the paint and allowing moisture to seep under the filler. Rust will then form on the metal.

Use a #24 or #36 grit grinding disc to remove the paint. This coarse grit removes paint and surface rust quickly and also etches the metal to provide better adhesion. You can also blast off the paint to prevent any removal of metal.

If applying filler over a metal patch, avoid hammering down the excess weld bead. Grind it level with the surface. As shown in Figure 9–11, hammering the weld distorts the metal, creates stress in the panel, and increases the area to be filled.

Use metal straightening methods to straighten the metal as much as possible to match its original contour. If the panel is flat, use a long straightedge to gauge the panel straightness. Hold it over the repair area and look for high and low spots. You must return the panel to within ⅛ inch (3.1 mm) of its original shape.

After grinding away the finish from the repair area, blow away the sanding dust with compressed air and wipe the surface with a tack rag to remove any remaining dust particles.

FIGURE 9-10 Removing old paint finish around repair area. You cannot apply plastic filler to paint.

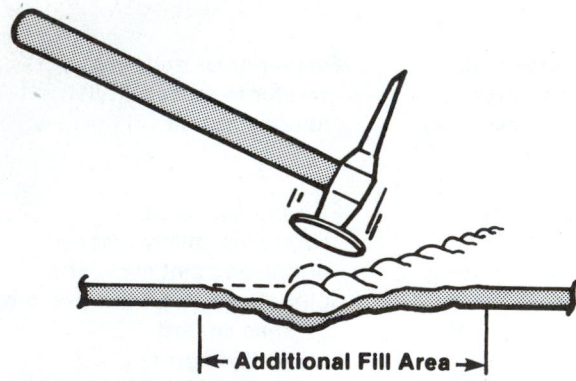

FIGURE 9-11 Beating weld bead down excessively will increase fill area and may cause strain in the panel. Grind it flush with the panel surface.

← Additional Fill Area →

CAUTION

Wear eye protection and a respirator when blowing off dust. It is very easy from something to fly into your eye. The dust can also contain materials that are harmful if inhaled.

MIXING THE FILLER

Body fillers come in cans and in plastic bags. When in a plastic bag, hand pressure or a dispenser can be used to force the filler onto your mixing board. This keeps the filler perfectly clean. A **mixing board** is the flat surface (metal, glass, or plastic) that is used for mixing the filler and its hardener.

Mix the can of filler to a uniform and smooth consistency free of lumps (Figure 9–12). Using a paint shaker will save time if the filler has been on the shelf for a while.

If the body filler is not stirred up thoroughly to a smooth and uniform consistency before use, the filler in the upper portion of the can will be too thin and that in the lower portion of the can will be too thick and very coarse and grainy.

Improper and inadequate plastic filler mixing can result in:

- runs and sags during application
- slow and poor curing
- a gummy condition when sanded
- poor featheredge
- a very tacky surface
- blistering and lifting
- poor adhesion
- rampant pinholing
- poor color holdout

KNEADING THE HARDENER

Loosen the cap of the cream hardener tube to prevent the hardener from being air bound. **Hardener**

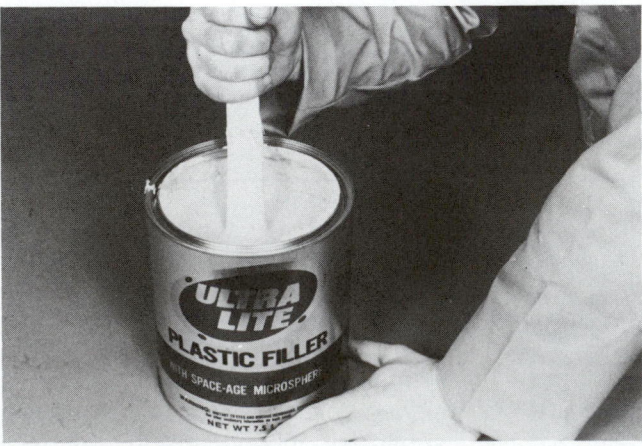

FIGURE 9-12 Mix filler thoroughly before using. Thick components can settle on bottom. If not mixed, drying and finish problems can result.

FIGURE 9-13 Knead hardener before use by squeezing it back and forth with your fingers. This will mix it up in the tube.

kneading is done by thoroughly squeezing its contents back and forth inside the tube. This will assure a smooth paste-like consistency (like toothpaste) when squeezed out (Figure 9–13).

If you do not knead the hardener, the result can be the same problems as just listed for poor filler mixing.

If the hardener is kneaded thoroughly and remains thin and watery, you have **spoiled hardener**. Hardener can spoil if frozen or if stored too long. It should not be used because it has broken down chemically.

MIXING FILLER AND HARDENER

Numerous problems can occur from improper *catalyzing* (mixing) of *cream hardener* (filler catalyst) and filler. Before catalyzing, make sure the materials (filler and hardener) to be used are compatible. They should be manufactured by the same company and be recommended for each other.

The following tips will help eliminate problems relating to mixing filler and its catalyst. Open the can of filler without bending its lid. Using a clean putty knife or spreader, remove the desired amount of filler. Place the filler on a smooth, clean filler mixing board. You can use sheet metal, glass, or hard plastic (Figure 9–14). Mixing boards with a handle are also available.

FIGURE 9-14 Mix filler and hardener together on nonporous mixing board.

Cardboard should not be used as a filler mixing board. It is porous and contains waxes for waterproofing. These waxes will be dissolved in the mixed filler and cause poor bonding. Cardboard also absorbs some of the chemicals in the filler and hardener, changing the filler's curing quality slightly. In addition, cardboard fibers can stick in the filler and ruin the finish.

Add hardener according to the proportion indicated on the can, usually 10 percent hardener (one part hardener for each ten parts filler). Too little hardener will result in a soft, gummy filler that will not adhere properly to the metal. It will also not sand or featheredge cleanly. Too much hardener will produce excessive gases, resulting in pinholing (Figure 9–15).

A general rule is to use a one-inch (25 mm) bead of hardener for each golf ball-sized glob of filler. If the filler is as big as a baseball, use about a 6-inch (152 mm) bead of hardener. However, always refer to the manufacturer's instructions for exact mixing proportions.

Filler overcatalyzation results when you use too much hardener for the amount of filler. This must be avoided because a reverse curing action may occur, causing poor adhesion and poor sanding properties, in addition to paint color bleed-through and pinholing.

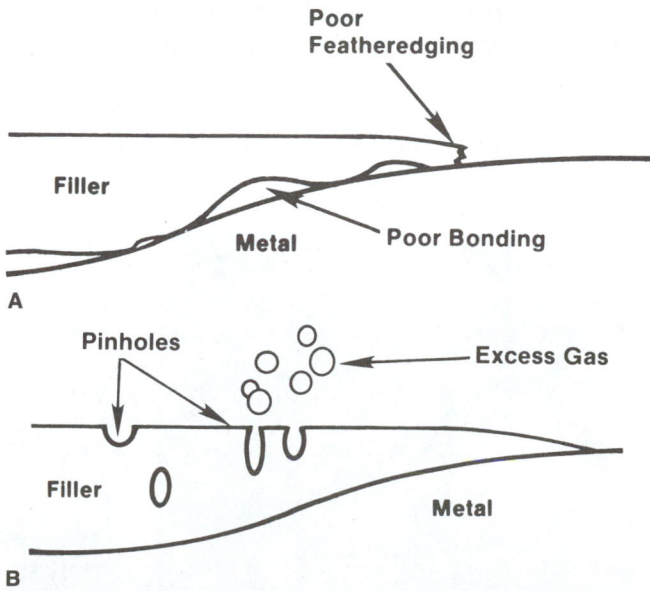

FIGURE 9-15 (A) Too little hardener often causes poor featheredging and weak bonding. (B) Too much hardener often results in pinholes. The plastic may also harden on the mixing board before you have time to apply it smoothly on the repair.

Filler undercatalyzation is caused by not using enough hardener, which causes the filler to not cure properly, resulting in tacky surfaces and poor adhesion properties.

With a clean putty knife or spreader, use a scraping motion (back and forth) to mix the filler and hardener together thoroughly and achieve a uniform color (Figure 9–16). Scrape filler off both sides of the spreader and mix it in. Every few back and forth strokes, scrape the filler into the center of the mixing board by circling inward.

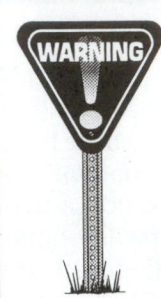

Do NOT stir the filler. Stirring whips air into the filler, causing air pockets and pinholes when it is applied.

If the filler and hardener are not thoroughly mixed to a uniform color, soft spots will form in the cured filler. The result is an uneven cure, poor adhesion, lifting, and blistering.

Reinstall the cover on the can of filler right away. This will keep out dust and dirt and also help prevent liquids in the filler from evaporating.

Always use clean tools when removing the filler from the can and mixing the filler and hardener together. Do not redip the knife, spreader, or mixed filler into the can. This causes the whole can of filler to harden with time. Hard lumps might form in the can and/or in the applied filler. It will cause problems the next time you try to use it.

Use different spreaders to mix and to apply the filler. A small amount of unmixed filler will always

The most common mistake of the apprentice is to use too much cream hardener! The plastic filler will set up or harden in a couple of minutes or before you have time to spread it on the body. Tons of plastic filler have been wasted because of hardener overuse. Using too much cream hardener can also cause problems with adhesion and pinholing.

remain on the mixing spreader. If any is accidentally applied, you will have soft spots in the cured filler and the paint finish may peel.

APPLYING FILLER

Apply the mixed filler promptly to a thoroughly clean and well-sanded surface. A first tight, thin application is recommended. Press firmly to force filler into sand scratches to maximize the bond (Figure 9–17). It is important to use the appropriate size spatula, or it will be difficult to apply a smooth layer of filler to the repair area. Rough filler takes extra time to sand off. Also, make sure you move the spatula over the repair area to match the shape of the part (Figure 9–18).

Always use a clean plastic spreader or putty knife to apply the filler. Use a different spreader to mix the filler. A small amount of unmixed (uncatalyzed) filler will always remain on the spreader and if any of it is transferred to the metal, soft spots will be present in the cured filler, and the paint finish will peel.

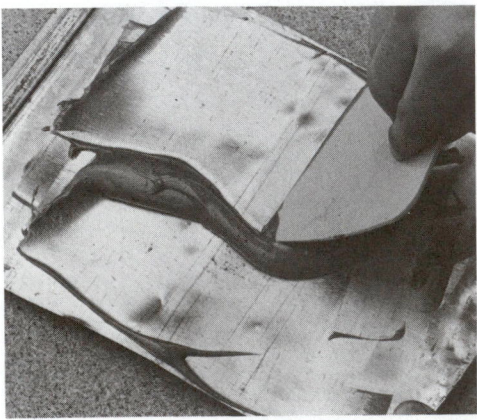
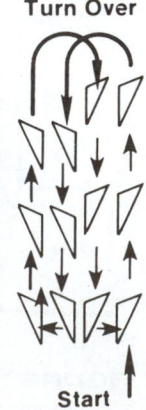

Turn Over

Start

FIGURE 9-16 Mix with a scraping motion. Do not use a stirring motion or air bubbles and pinholes can form in the filler.

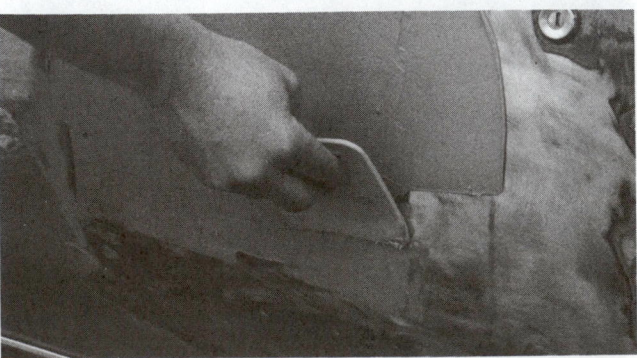

FIGURE 9-17 Applying filler with a plastic spreader. Lay down a thin coat first that roughly matches the shape of the panel.

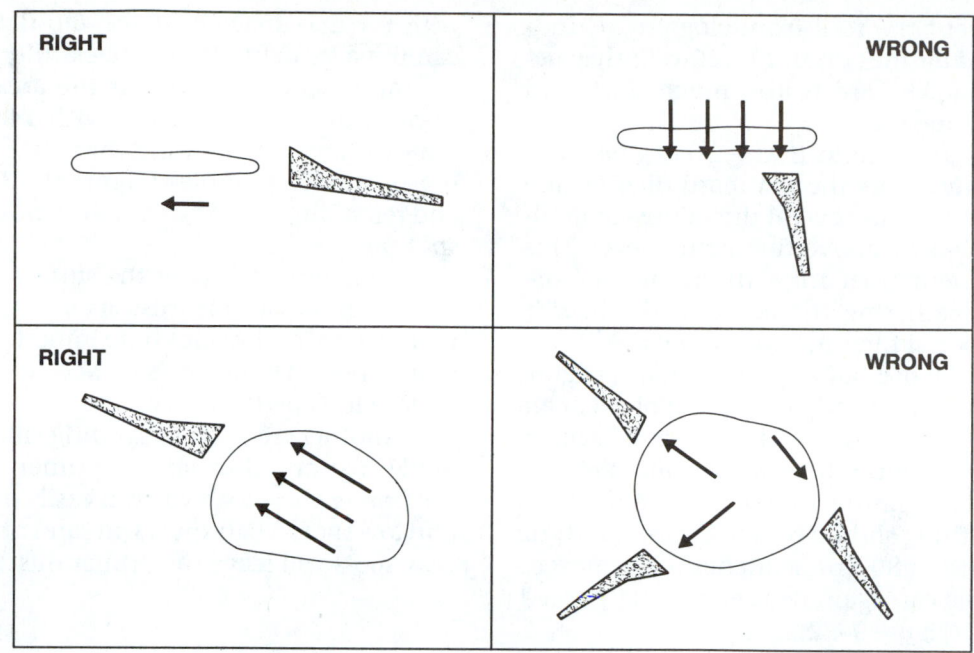

FIGURE 9-18 Note right and wrong ways of moving a spreader or spatula over repair areas when applying plastic filler. The size of the spreader should match the size of the repair.

When the filler is fully cured, you can apply additional coats as needed to build up the repaired area to the proper contour (Figure 9–19). Allow each application to set before applying the next coat of filler. Conventional body fillers should be built up slightly so that the waxy film that curing produces on the surface of the filler can be removed with a surform grater.

Applying the mixed filler thickly without first applying a thin, tight application causes poor bonding and pinholing. Wiping over the repaired area with

A. Thin skim coat

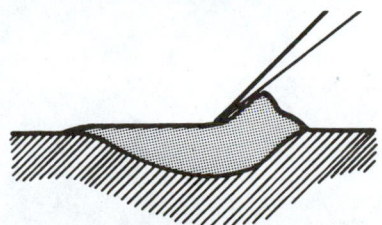

B. Fill slightly above panel.

FIGURE 9-19 Build filler thickness in several coats. Never apply a thick coat or cracks can develop from heat expansion during curing.

solvents before applying the mixed filler also causes pinholes and poor adhesion.

Avoid use of filler in cold temperatures. When the filler, shop, and/or body panel temperatures are cold, the body filler will not cure properly. This results in the filler being too soft and having a tacky surface and poor sanding properties. Tremendous pinholes could be created. Filler should be stored at room temperature. If needed, a heat lamp can be used to warm cold surfaces.

Avoid moisture on the repair area in conditions of high humidity in the shop. In winter, moisture will form on metal surfaces when a cold car is brought inside. In summer, condensation might form on the metal due to water evaporating from the floor. Again, use a heat lamp to warm and dry damp surfaces before applying filler. If the repair area is not first warmed to remove moisture accumulation on the surface, poor adhesion, poor featheredging, pinholing, and lifting when recoating with refinishing materials could result.

SHAPING THE FILLER

Allow the filler to cure to a semihard consistency. This usually takes 15 to 20 minutes. If a firm white track is left when the filler is scratched with a fingernail, it is ready to be filed. Filing is perhaps the most important factor in achieving a quality surface and controlling material cost and labor. The *surform* or *cheese grater file* is used to cut the excess filler to size quickly. Its long length produces an even, level surface. The teeth in the file are open enough to prevent the tool from clogging. Grinders, sanders, and air files do not level well; they become loaded quickly, create too much dust, and waste a lot of sandpaper.

To use the grater, hold it at a 30-degree angle and pull it lightly across the semihard filler (Figure 9–20). Work the filler in several directions. Stop filing when it is slightly above the desired level. This will leave sufficient filler for sanding out the file marks and for feathering the edges. If the filler is undercut, additional filler must be applied.

After grating, sand out all the file marks (Figure 9–21) with very coarse sandpaper. Use a block or air file on large, flat surfaces. Use a disc orbital sander on smaller areas. Then, follow with a finer #80 grit sandpaper until all grating scratches are removed.

Final smoothing should be done using #180 grit paper until all the #80 grit scratches are removed. The DA or air file can again be used or a long speed file can be used (Figure 9–22).

Be careful not to oversand. This results in the filled area being below the desired level, which, makes it necessary to apply more filler. Oversanding is a common mistake for the novice body technician. Always sand a little and then check your work. You want to slowly cut the filler down until flush with the undamaged surface. On flat surfaces, you can use a straightedge to check filler straightness.

After sanding, blow off the area with a high-pressure air gun and wipe it with a tack cloth. This removes any fine sanding dust that might be hiding in surface pinholes (Figure 9–23). These holes and remaining sanding scratches must be filled with spot putty.

Run your hand over the surface often to check for evenness. Do not trust eyeing for accuracy. Remember, paint does not hide imperfections; it highlights them. Do not be satisfied until the repaired surface feels perfectly even.

Another trick when sanding filler is to apply a guide coat or a thin mist of primer. By sanding off the primer guidecoat, you can easily detect filler high and low spots. High spots will sand off more quickly. Low spots will leave the primer mist intact.

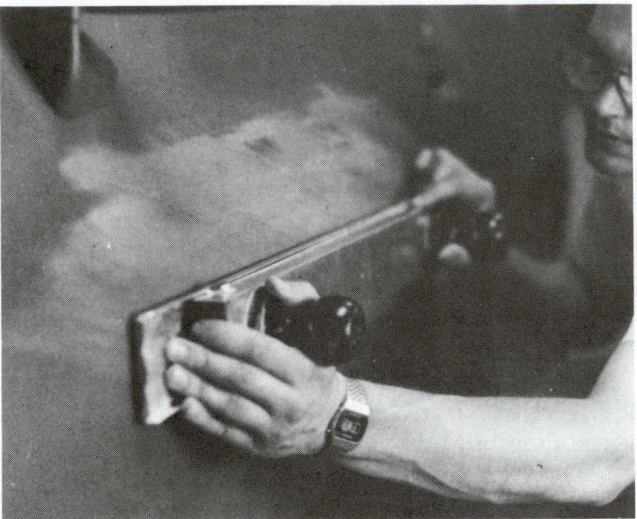

A

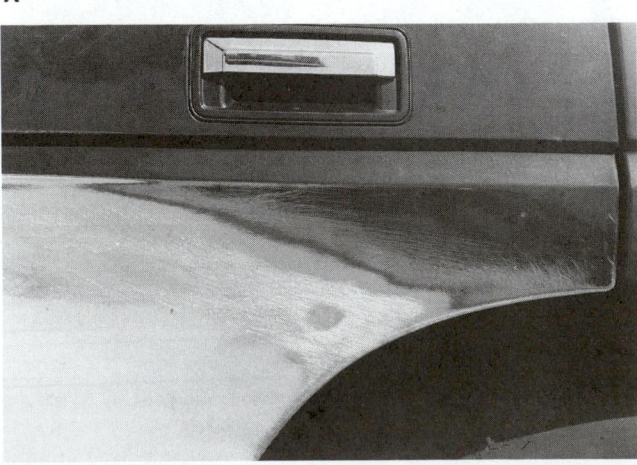

B

FIGURE 9-21 (A) Sand away file marks with progressively finer sandpaper. Use a block or long sanding board on large flat surfaces. (B) Close-up shows how the area has been sanded and feathered around the plastic filler.

A

Surform Moving Direction

30° to 40°

Surforming should be done in many directions.

B

FIGURE 9-20 (A) When filler is partially hard, file it to a rough shape with a super coarse, cheese grater file. (B) Hold the grater at a 30-degree angle and pull it across the filler. Knock off high spots but leave enough filler for sanding out deep file marks.

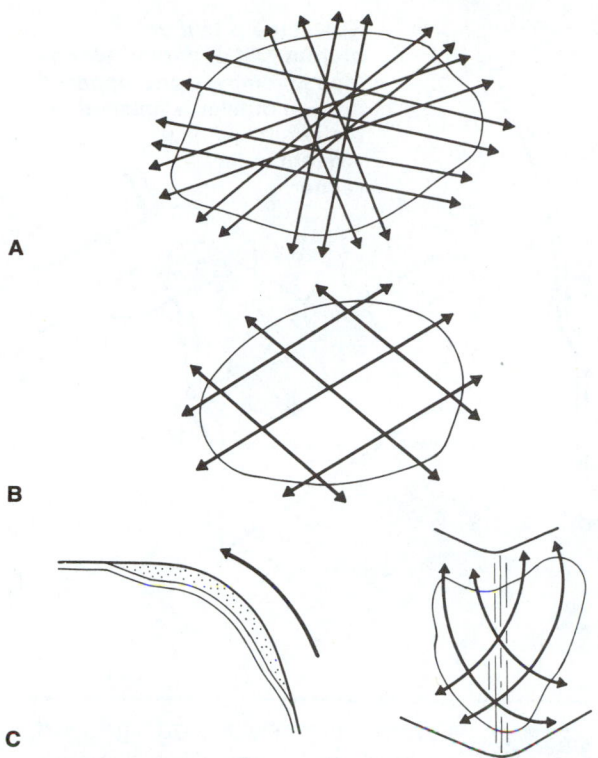

A

B

C

FIGURE 9-22 Basic sanding methods. (A) On a flat surface, move the sanding block in all directions to cut the area down evenly. (B) On a gently curved surface, move the sanding block in specific directions across or at an angle to the curve. (C) On a sharply curved part, move the sanding block with a rolling action to move over the top of the curve at an angle. *(Courtesy of American Honda Motor Co.)*

When satisfied with the smoothness of the filler surface, clean the area with a tack cloth. A tack cloth picks up bits of filler dust that normal cleaning leaves behind. Remember that the tiniest particle will mar or ruin the paint job.

After final sanding, spray the repair area with primer or primer-surfacer and recheck your work. The primer will help show surface imperfections that were not visible after sanding. Rework the repair area if needed until it is perfectly smooth.

APPLYING GLAZING PUTTY

Once the primer is dry, small pinholes and scratches can be filled with glazing putty. If using a polyester putty, mix the putty and hardener according to the manufacturer's instructions.

Place a small amount of putty onto a clean rubber squeegee. Apply a thin covering over any pits or other imperfections in the filler. Use single strokes and a fast scraping motion. Use a minimum number of strokes when applying putties. They skim over or dry very fast. Repeated passes of the spreader may pull the putty away from the filler.

Allow the putty to dry completely before sanding it smooth with #240 grit sandpaper. Sanding the putty before the solvents in it have completely evaporated results in subsequent sand scratches in the finish. Sand the putty only until flush with the surface of the primer, being careful not to cut through the

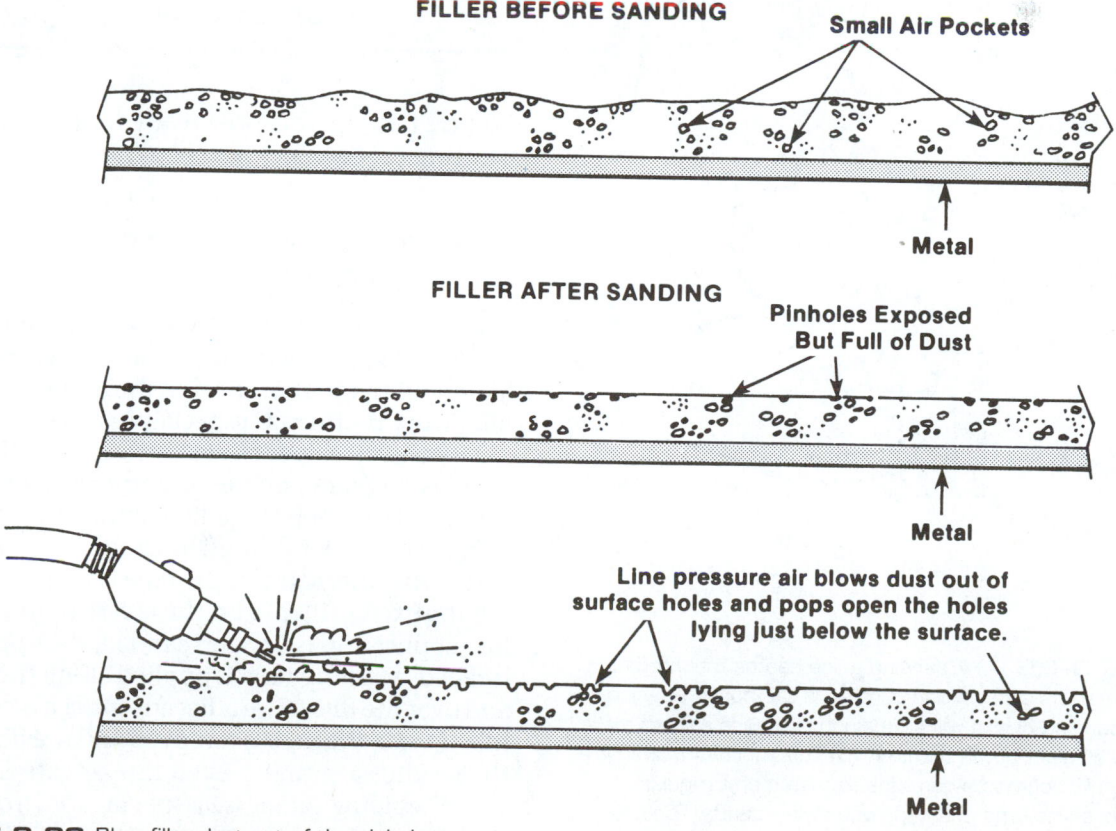

FILLER BEFORE SANDING

Small Air Pockets

Metal

FILLER AFTER SANDING

Pinholes Exposed
But Full of Dust

Metal

Line pressure air blows dust out of surface holes and pops open the holes lying just below the surface.

Metal

FIGURE 9-23 Blow filler dust out of the pinholes.

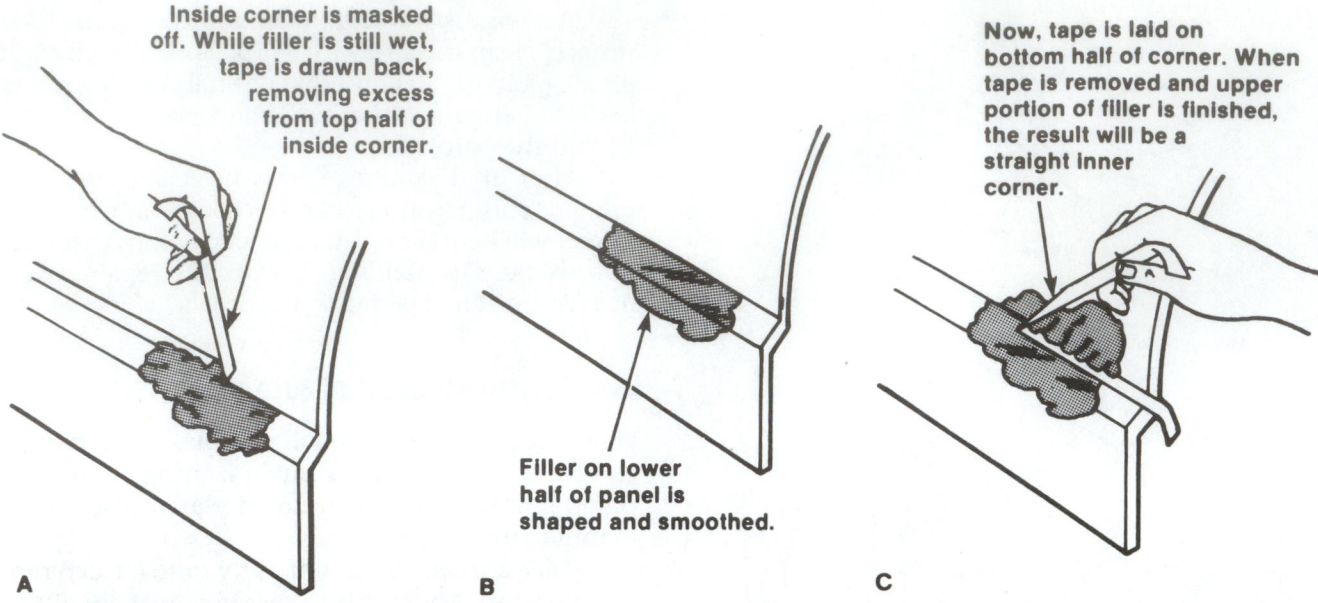

Inside corner is masked off. While filler is still wet, tape is drawn back, removing excess from top half of inside corner.

Filler on lower half of panel is shaped and smoothed.

Now, tape is laid on bottom half of corner. When tape is removed and upper portion of filler is finished, the result will be a straight inner corner.

A   B   C

FIGURE 9–24 Note the steps for filling angular body lines with tape.

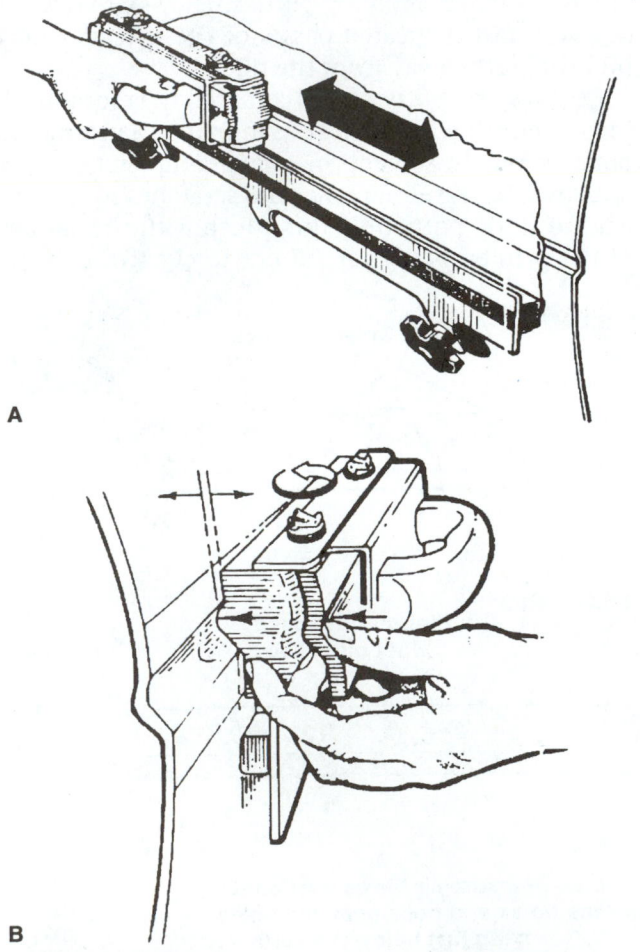

A

B

FIGURE 9–25 (A) A sanding guide can be mounted on a panel to sand the straight line contour in the panel. Specially designed sanding blocks will slide along the guide rails to make a perfectly straight cut in the filler. (B) This tool has a shapable sanding block. Segments can slide to match an irregular shape. This saves time and improves repair quality. *(Courtesy of Style-Line Systems)*

When properly mixed, applied, sanded, and primed, a quality body filler will have no imperfections. Excessive use of glazing putties is usually an indication of a lack of expertise on the part of the body technician. Remember that glazing putty is designed only for tiny pits and pinholes, and should not be used as a filler.

primer. The repair is now ready for the priming and painting processes.

APPLYING FILLER TO BODY LINES

Many cars and trucks today have sharp body lines in doors, quarter panels, hoods, and other body parts. Maintaining the sharpness of these lines when doing filler work is difficult, especially in recessed areas.

One way to get straight, clean lines by hand is to fill each plane, angle, or corner separately. To do this, masking tape is applied along one edge (Figure 9–24). Then filler is applied to the adjacent surface. Before the filler sets up, the tape is pulled off, removing the excess filler from the crease or line. After the first application is dry and sanded, the opposite edge is taped. Masking tape is applied along the body line and over the filler. The adjacent surface is coated with filler. When the tape is removed and the filler sanded, the result is a straight, even line or corner.

A *sanding guide* is a special tool for sanding straight, special contour lines on panels (Figure 9–25).

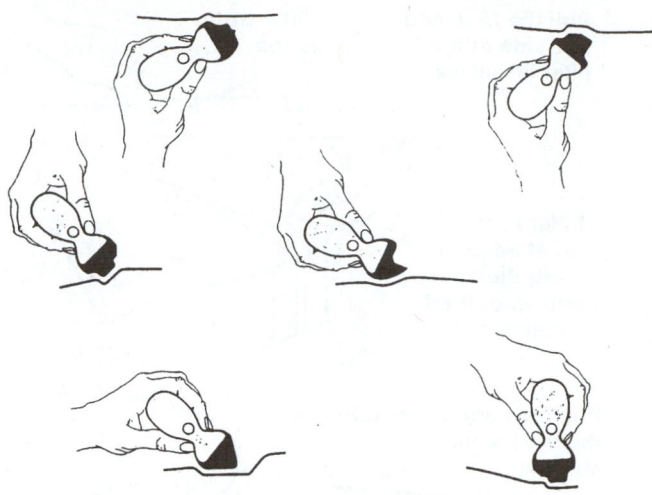

FIGURE 9-26 Fixed irregular-shaped sanding blocks can be selected to match existing contours. *(Courtesy of Style-Line Systems)*

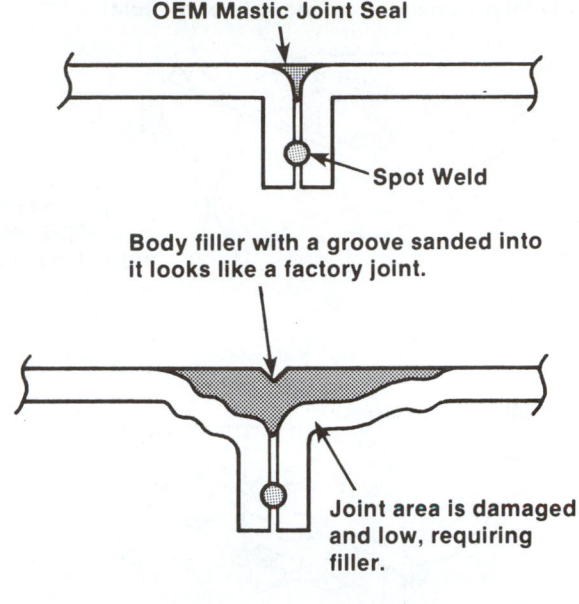

FIGURE 9-27 This body joint was accidentally covered with filler, which cracked.

The tool framework mounts on the panel. It has long slides that hold the sanding block for making perfectly straight cuts in the filler.

The sanding block is designed in segments so that it can be shaped to match the irregular shape of the edge. After fitting the block over a good section of the panel and moving its segments to match the contour, the segments are locked in place. The block is then mounted in the guide and moved over the filler. This lets you cut a straight, contoured edge in the panel quickly and easily.

Fixed-shaped sanding blocks are also available to help sand filler in contours (Figure 9–26).

APPLYING FILLER TO PANEL JOINTS

Many panels on unibody vehicles have joints that are factory finished with a flexible mastic that allows the panel to flex and move. Often both halves of the body joint suffer damage and require filling. Many untrained people make the mistake of covering the damaged joint with body filler.

Figure 9–27 shows what happens when the filler is subjected to the twisting action of the panels under normal road conditions. A crack develops that allows moisture to seep under the filler. This causes rust to form on the metal surface and eventually results in the failure of the filler-to-metal bond and a weakened sheet metal joint.

The original flexibility of the joint can be preserved by taping its alternate sides. As shown in Figure 9–28, tape is applied to one panel. Filler is then applied to the other panel and the tape is pulled up, removing the excess filler. Both panels are filled in

the same way. After the filler in both panels is cured and shaped, a sealer is forced into the joint.

9.3 APPLYING LEAD FILLER

Most body shops use plastic body fillers exclusively in dent repair. Those shops that do use lead filler, or *body solder* as it is sometimes called, use it only when restoring antique and classic automobiles or doing custom work. Some shops use lead filler to fill door edges and welded seams, but generally lead work is done only on request from the customer. It is a specialized skill that few body technicians have.

Lead filler is an alloy of lead and tin. Most lead solder used in body repair is 30 percent tin and 70 percent lead and is often called 30/70 solder. At approximately 360 degrees Fahrenheit (182 degrees Celsius), 30/70 body solder becomes soft or plastic. It becomes liquid at approximately 490 degrees Fahrenheit (255 degrees Celsius). Within this heat range, body solder is plastic enough to be worked and shaped.

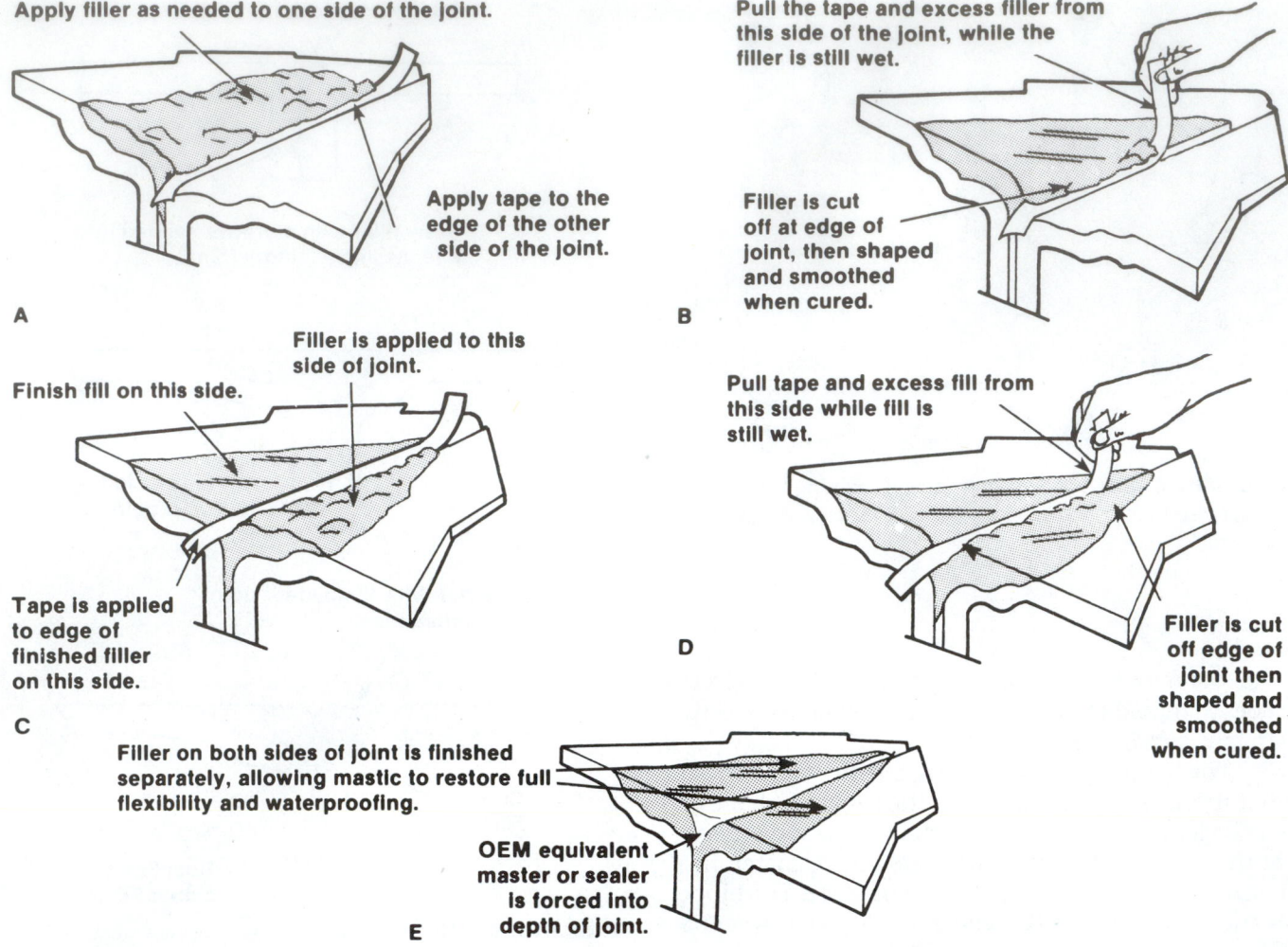

Apply filler as needed to one side of the joint.

Apply tape to the edge of the other side of the joint.

A

Pull the tape and excess filler from this side of the joint, while the filler is still wet.

Filler is cut off at edge of joint, then shaped and smoothed when cured.

B

Finish fill on this side.

Filler is applied to this side of joint.

Tape is applied to edge of finished filler on this side.

C

Pull tape and excess fill from this side while fill is still wet.

Filler is cut off edge of joint then shaped and smoothed when cured.

D

Filler on both sides of joint is finished separately, allowing mastic to restore full flexibility and waterproofing.

OEM equivalent master or sealer is forced into depth of joint.

E

FIGURE 9-28 Study how to create a flexible joint in a filler-coated panel.

An oxyacetylene welding torch or a specially designed soldering torch is used to heat and soften the solder. Use a medium size welding tip. The acetylene pressure is set at 4 to 5 psi (28 to 34 kPa). Adjust the torch for a carburizing flame. The low heat and wide flame are adequate to heat the solder and the repair area without overstressing most mild steel sheet metal panels.

After bumping the damaged sheet metal back to the original contour, grind the metal bare and clean it with a metal conditioner. Then heat the repair area and brush on a tinning flux. Tinning fluxes are used to clean microscopic rust particles from bare metal. This prevents rust from forming and promotes adhesion of the solder to the sheet metal. A variety of fluxes are available.

After applying the flux, heat the metal with the torch and rub the solder bar over the hot metal. This will deposit a thin layer of solder over the repair area. While the solder is still plastic, wipe the area with a clean shop cloth to spread the solder over the bare metal and remove any impurities. The tinned area will be silvery white.

Next, fill the area with solder. Do this by heating one inch (25 mm) of the solder rod or bar and the adjacent metal (Figure 9–29A). Press the heat-softened bar against the hot metal. This will deposit the solder on the sheet metal. Do this as much as is necessary to fill the repair area.

Now shape and smooth the solder. Heat it by moving the welding torch with a carburizing flame back and forth over the solder until it begins to sag. Then spread and smooth it over the repair area with a soldering paddle (Figure 9–29B). A flat paddle is

> **CAUTION**
> Lead and flux are toxic materials. Wear rubber gloves and an approved respirator when working with these substances. They can cause serious health problems if inhaled or absorbed into the skin.

used on flat and convex surfaces; a curved paddle is used on concave surfaces. The paddle must be clean and properly waxed so that the solder will stick to it. The solder level should be slightly above the panel.

While the metal is still hot, quench the area with cool water. Quenching will cool the metal so that it

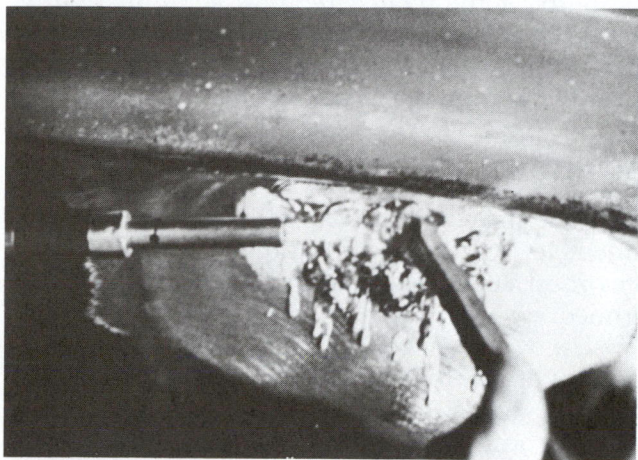

A

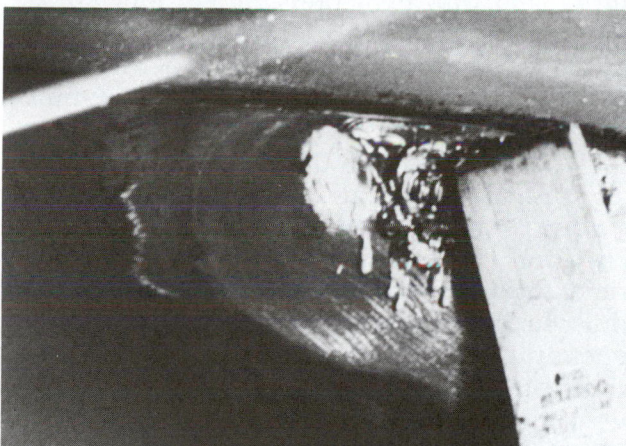

B

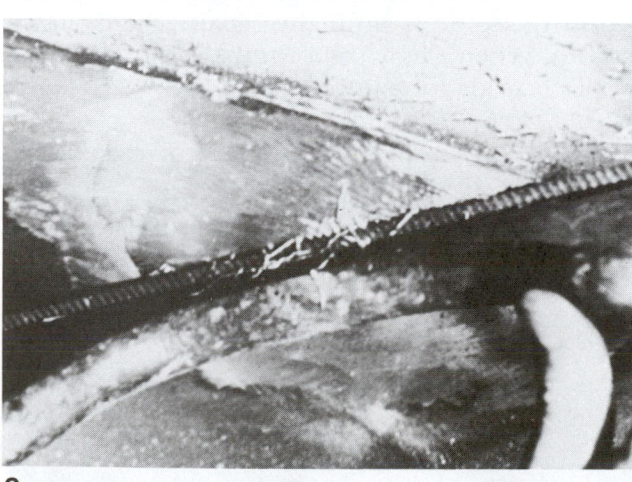

C

FIGURE 9-29 Apply lead filler when (A) melting lead solder, (B) smoothing solder with a wooden paddle, and (C) filing the solder.

can be filed and will also relieve the heat stress in the panel.

Use a body file to level the solder even with the panel (Figure 9–29C). Use a dull file if one is available. A sharp file will cut the solder too quickly. File with long strokes, working from the edge across the middle in every direction. Sand the solder to the final contour with #80 or #100 grit sandpaper and a speed file. Prep the metal before priming.

9.4 REPAIRING SCRATCHES

Most vehicles brought to the auto body shop for damage repair and/or refinishing have a variety of minor imperfections in the paint. Some defects, such as chalking or scuff marks, can be removed with rubbing compound. The abrasive compound removes the damaged surface paint and brings out the luster in the paint beneath without repainting. Compounding is discussed at length in later chapters.

Other defects, such as scratches, are too deep to buff out with rubbing compound. If the scratch penetrates through the primer and exposes the metal underneath, the scratch must be sanded, primed, and painted. On the other hand, shallow scratches that are too deep to be buffed out, but do not reach the metal underneath the paint, can be filled with putty or a polyester primer-surfacer before painting. An example of such a scratch is shown in Figure 9–30.

PREPARING THE SURFACE

Wash and clean the repair area with a wax and grease removing solvent. Then lightly sand the scratched areas. Use a sanding block when sanding large areas. A normal size sheet of #240 or finer grit paper folded up is fine for small areas. Many technicians like to wet sand the area.

A light sanding will rough up the finish coat so the glazing putty and new primer will adhere to the

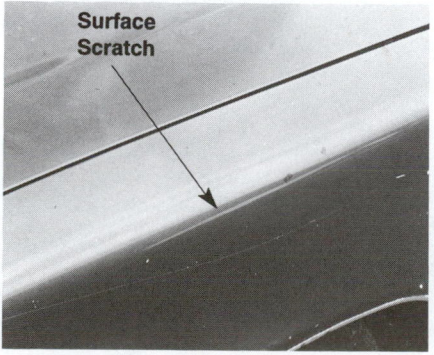

FIGURE 9-30 The shallow scratch resulted from vandalism.

old finish. Do not press hard on the sandpaper. Excessive pressure can result in low spots or a wavy surface that will require additional filling and sanding. After the rough sanding is complete, clean the sanded area with compressed air or a soft cotton cloth and wipe with a tack cloth.

APPLYING GLAZING PUTTY

If the scratch is part of a larger panel being refinished, try filling the repair area with primer-surfacer. Some sprayable polyester primers will provide up to 15 mils of fill. Another solution is to fill the scratch with glazing putty.

Following the instructions on the container of putty, apply some on the edge of a clean rubber contour squeegee. A little bit of putty goes a long way. When filling only scratches, do not squeeze too much putty onto the spreader.

Using moderate pressure, spread putty over the repair area with a rubber squeegee (Figure 9–31). Use a fast scraping motion and apply in one direction only. Do not pass the squeegee over the same area more than once. Multiple passes can pull the putty away from the body and out of the scratch.

Allow the putty to dry completely. Drying time varies with the putty's thickness, but it usually takes between 20 and 60 minutes. For best results, allow the one-part putty to dry overnight before sanding. Sanding the putty before it completely cures will result in sand scratches in the finish.

SANDING THE GLAZING

After the putty dries, sand the repair area with #240 or finer grit sandpaper. Wet sand to prevent the putty from clogging the paper and creating more scratches

in the finish. Use a sanding block to avoid making low spots with finger pressure.

When sanding, rub the palm of your hand over the puttied area to feel for high spots on the surface. After finishing, rinse the sludge away and wipe the surface dry. Clean the repair area with a tack cloth.

Inspect the scratch for low spots and voids in the putty. If the scratch requires additional putty, repeat the above procedure. When the surface of the previously scratched area is free of imperfections, it is ready for priming and refinishing.

PRIMING THE GLAZING

Once the scratched surface is filled with putty and sanded level with the surrounding panel, the repair area must be sanded to a smooth finish and then primed.

Use water, a sanding block, and #400 grit or finer sandpaper to finish sanding the puttied scratch (Figure 9–32). Wet sanding prevents the paper from clogging and creating additional scratches. The sanding block helps avoid creating low spots in the finish. Sand with light pressure and long strokes across the face of the repair. Do not concentrate or bear down on any one spot; doing so will almost guarantee a low spot. Low spots created by finish sanding must be filled with additional glazing putty.

When satisfied with the smoothness of the repair, rinse away sanding sludge and wipe the surface dry. Clean the area with a tack cloth. When it is dry and free of dust, spray a medium coat of primer-surfacer over the entire repair area. Allow the primer-surfacer to dry thoroughly. Drying times vary depending on the product being used. A-type primer-surfacer can be sanded in as little as 30 minutes while a high build primer-surfacer may require up to 4 hours dry time before sanding. Then wet sand the primer with #600 grit or finer sandpaper. Repeat this process until the repair area is glassy smooth. When glassy smoothness is achieved, the surface is ready for painting.

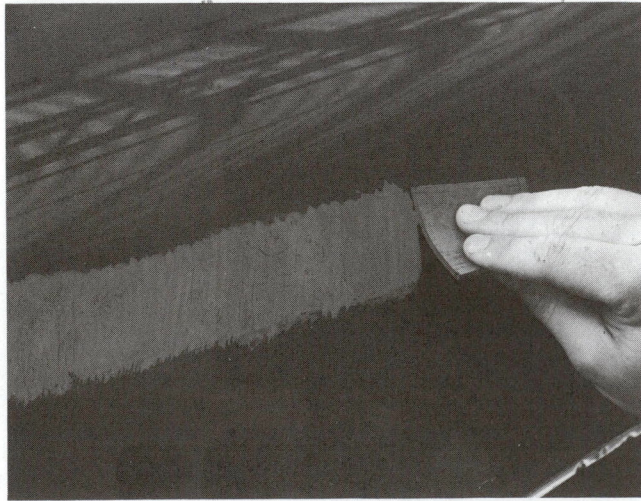

FIGURE 9-31 After wet sanding the area, apply glazing putty to the scratch.

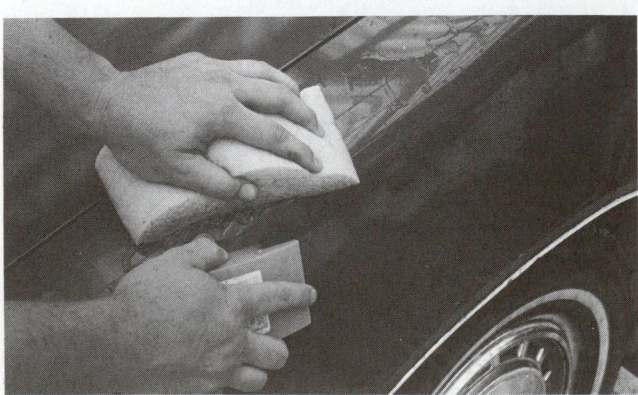

FIGURE 9-32 Wet sand putty until it is flush with the existing surface.

9.5 REPAIRING NICKS

Minor bumps and scrapes often leave nicks and scratches in a car's finish. A stone thrown up by a passing vehicle can chip the paint, exposing the sheet metal beneath. Side swipe collisions result in scrapes and gouges. Any time bare metal is exposed to the air, the metal must be primed before the new paint finish is applied. A large nick is shown in Figure 9–33.

FEATHEREDGING THE OLD FINISH

First, clean and dewax the repair area with an approved solvent. Wipe the area down with a clean towel or rag and sand the ragged edges of the chipped paint to a smooth, feathered surface. This is commonly referred to as featheredging. **Featheredging** tapers the edges of the paint so that it gradually blends in with the metal surface.

Featheredging chips and nicks is quickly done with a #180 grit disc and a disc orbital (DA) sander. In tight spots, use a sanding block (Figure 9–34). Sand the edges of the old finish to a fine taper. When the sanded area is smooth to the touch, switch to #240 grit or finer sandpaper. Sand any remaining scratches.

APPLYING PRIMER AND PUTTY

Never leave bare metal surfaces exposed to air. Moisture in the air quickly encourages rust to form. The slightest film of rust will prevent the primer and paint from properly adhering to the metal. Subsequent lifting and blistering will eventually ruin the paint. The area will have to be sanded down and refinished again. Priming the bare metal areas with a zinc chromate base primer inhibits rust formation and ensures good bonding of the finish paint.

Blow away any sanding dust and wipe the area with a tack cloth. Apply a coat of primer-surfacer to build up the area and fill any uneven featheredging. After the primer-surfacer has dried, apply a mist coat of another color or contrasting color guide coat. Block sand the area to identify low spots. Apply two-part glazing putty to fill the low spots or pits (Figure 9–35) if needed.

APPLYING FINISHING PRIMER

Final sanding and priming are necessary to achieve a super-smooth surface. Wet sand with #400 grit or finer sandpaper and a sanding block. Sand in long, straight strokes to avoid creating low spots. When sanding curved surfaces, sand very lightly holding the paper with the palm of your hand or use a flexible sander.

Clean, dry, and wipe the sanded surface with a tack cloth. Then, spray the repair area with primer-

FIGURE 9-33 There is a nick in the finish.

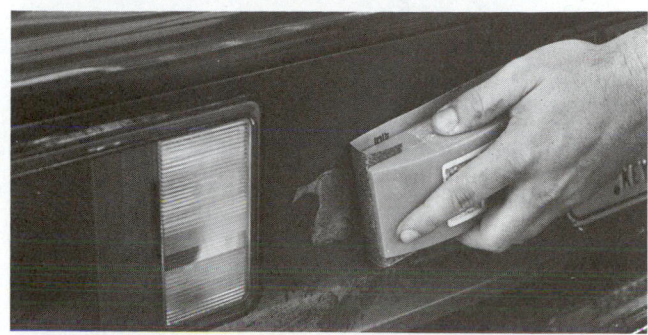

FIGURE 9-34 Wet sand the area around the nick. Featheredge it if needed.

FIGURE 9-35 After priming, apply glazing putty over the nick. After drying, sand the putty flush with the panel surface. Then finish.

surfacer. Completely cover the puttied area and several inches of the old finish around it. Allow the primer-surfacer to flash (surface dry) for an adequate time; then sand lightly with water and #240 or finer grit sandpaper.

Clean and prime once or twice more as needed. Between coats, wet sand lightly with #400 or #600 grit sandpaper to achieve an extremely smooth surface. The surface is now ready to be painted.

9.6 REPAIRING DINGS

One kind of surface imperfection that sometimes requires minor metalwork in addition to filling is a ding. Dings are small dents often caused by carelessly opened doors (Figure 9–36). When a panel is struck by the edge of another car door, the impact creates a shallow depression in the metal. This small tension area is also usually accompanied by a pressure ridge surrounding the ding (Figure 9–37).

PREPARING THE SURFACE

To repair a deep ding, first wash and dewax the surface. Whenever possible the ding should be removed with a body pick, hammering and dollying, or by whatever means may be available. The repair may still be visible after painting if this is not done. Then grind the finish from the repair area, using a lightweight air grinder. As shown in Figure 9–38, use up and down buffing strokes to remove the paint. Press the top edge of the disc against the metal and move the grinder up and down. Avoid removing too much metal which thins the panel. On smaller dings, a body file can be used to remove the finish over the ding and show the low spot (Figure 9–39). A die grinder with a cone bit can also be used to remove the paint on smaller dings. To remove paint from inside the ding, you could sandblast or plastic media blast the ding.

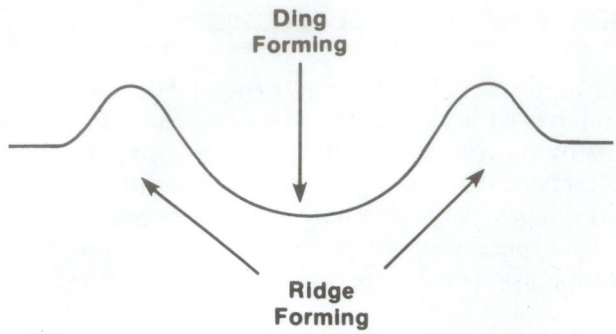

FIGURE 9-37 A minute ridge surrounds a ding.

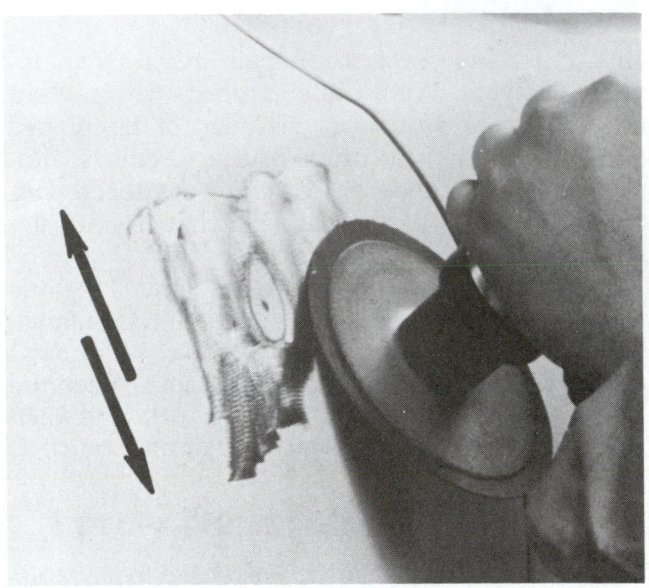

FIGURE 9-38 Buffing strokes with a grinder will remove the ridge around a ding. You must also remove paint around the repair area.

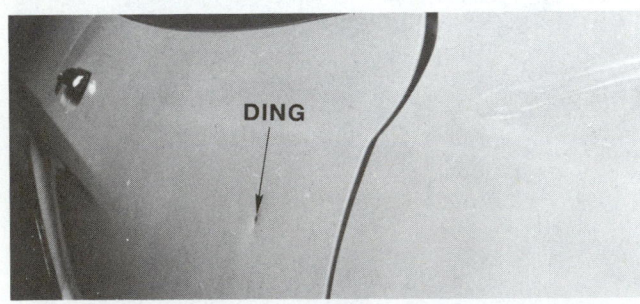

A

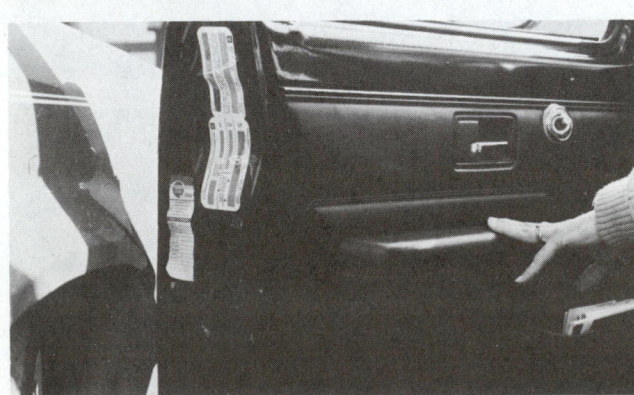

B

FIGURE 9-36 Minor bumps from sharp objects, like a door edge, cause dings. (A) Deep door ding. (B) Parking lots and careless people cause most door dings.

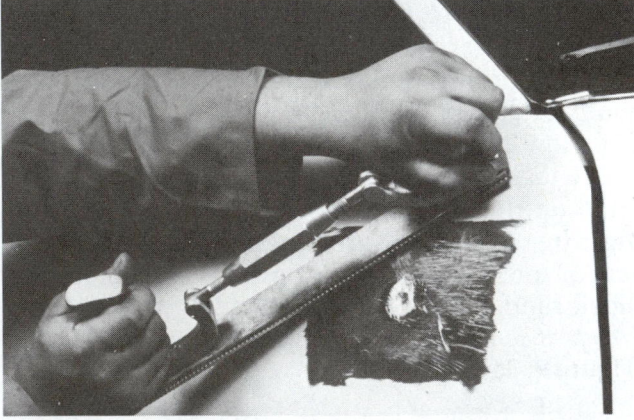

FIGURE 9-39 File the bare metal to reveal low spots. Low spots will not be scratched by the file.

APPLYING THE FILLER

After cleaning the panel, apply a skim coat of body filler (Figure 9–40). Allow the plastic to harden. Block sand the plastic smooth with #80 grit paper (Figure 9–41). Run your hand over the sanded plastic to feel

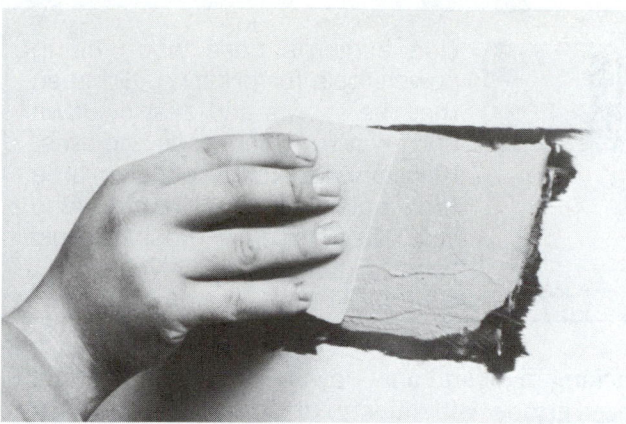

FIGURE 9-40 Apply a skim coat of filler. Do not apply filler over paint or problems will result.

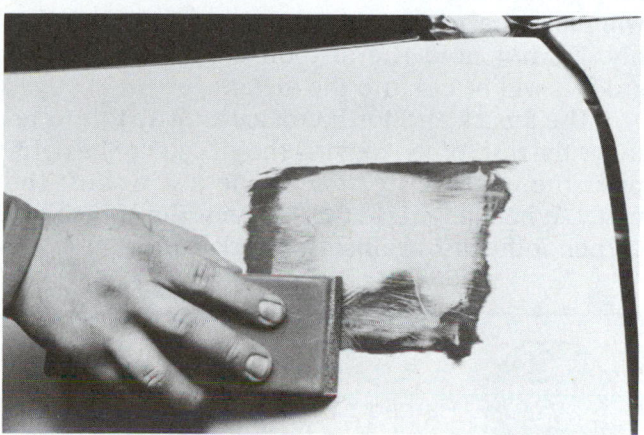

FIGURE 9-41 Block sand the filler until it is smooth. Go slowly and check your work. It is easy to oversand and sink the repair area too low in the panel.

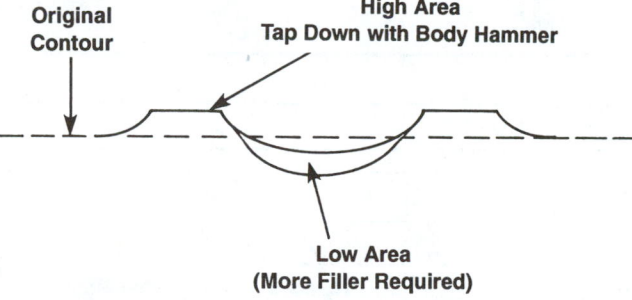

FIGURE 9-42 Feel for high and low spots carefully. Move your hand or a clean rag over the surface quickly in different directions. This will let you detect bumps and dips in the surface.

for high and low spots. High metal areas might need to be tapped down. Low spots will need another application of plastic (Figure 9–42).

Once the ding has been properly filled and leveled, the surrounding paint edges must be featheredged. A dual-action sander or a block sander can be used to featheredge the area, depending on the size of the repair.

CAUTION When working with plastic filler, it is advisable to wear gloves (neoprene or surgical) to keep the material from contacting your skin.

9.7 REPAIRING RUST DAMAGE

In addition to minor collision damage, the body repair technician must recognize and repair corrosion damage created by rusting sheet metal.

Rust is produced by a chemical reaction known as *oxidation*. Oxidation occurs when metal is exposed to moisture and air. The oxygen in the air, water, or other chemicals combines with the steel molecules to form iron oxide.

Iron oxide is the reddish brown compound commonly referred to as rust. By turning the metal into flaky or powdery iron oxide, rust will eat completely through a sheet metal panel if not treated.

If a crack in the car's finish allows moisture and air to seep under the paint film, the sheet metal beneath will begin to rust (Figure 9–43). The same is true if a chip or nick exposes sheet metal to the air and no immediate action is taken to repair it. Left alone to do its work, rust will soon form on the metal surface and begin to eat pits into the sheet metal. These pits harbor rust and prevent sanding the surface rust away. If simply painted over, rust in the pits will eventually bubble up and break through the new paint.

Corrosion damage on a vehicle panel takes two forms: surface rust and rustouts. **Surface rust** in its early stage can simply leave a reddish coating on the metal surface. Given time, the rust will eat pits into the surface. Eventually the pitting will develop into rust holes, or **rustouts**.

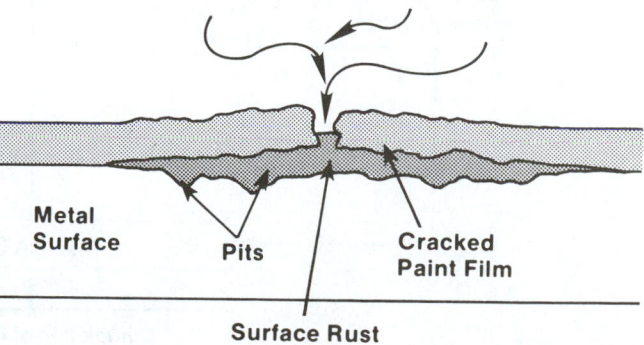

FIGURE 9-43 Note the structure of surface rust.

Both types of rust damage require different repair procedures. These are outlined in Table 9–4 and are explained in the rest of this chapter.

PREPARING SURFACE RUST

Repairing an area affected by light surface rust can be as simple as grinding the rust film away and chemically neutralizing the area with metal conditioner. However, if the metal is pitted, additional steps are required. The surface rust shown in Figure 9–44 requires minor metalworking and filling.

Prepare the defective area for sanding by first washing with a mild detergent. Then clean with a wax and grease remover solvent. Apply masking tape to nearby trim before grinding away paint and rust. The tape will protect the trim from sparks, chips, and accidental contact with the sanding disc.

The lightweight air grinder shown in Figure 9–45 is ideal for doing minor metalwork. Used with a rigid

Use extreme care when using power tools for grinding. Grind so that the sparks and dust fly down and away from your face and eyes. Always wear safety goggles or a face shield when grinding. Also wear an air filtering mask to avoid breathing paint dust.

backing plate and a #24 grit sanding disc, this high-speed grinder will quickly cut through paint and rust.

With the disc spinning, hold the grinder against the work surface at a 10-degree angle (Figure 9–46). Do not hold the disc flat on the surface because this will make the grinder skip and bounce uncontrollably. Do not hold the disc on edge or unwanted grooves will be cut into the metal.

Use a back and forth crosscutting action to remove the rust. When moving the grinder to the right, press the upper left corner of the disc against the metal. When moving to the left, press the upper right corner of the disc against the work surface.

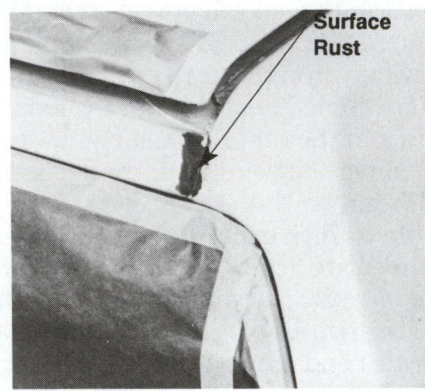

FIGURE 9–44 A surface rust formation usually shows up first as bubbles in the old paint.

Remember! Sandblasting or plastic media blasting can be used in place of disc and die grinding. Blasting will not remove and thin the metal like grinding.

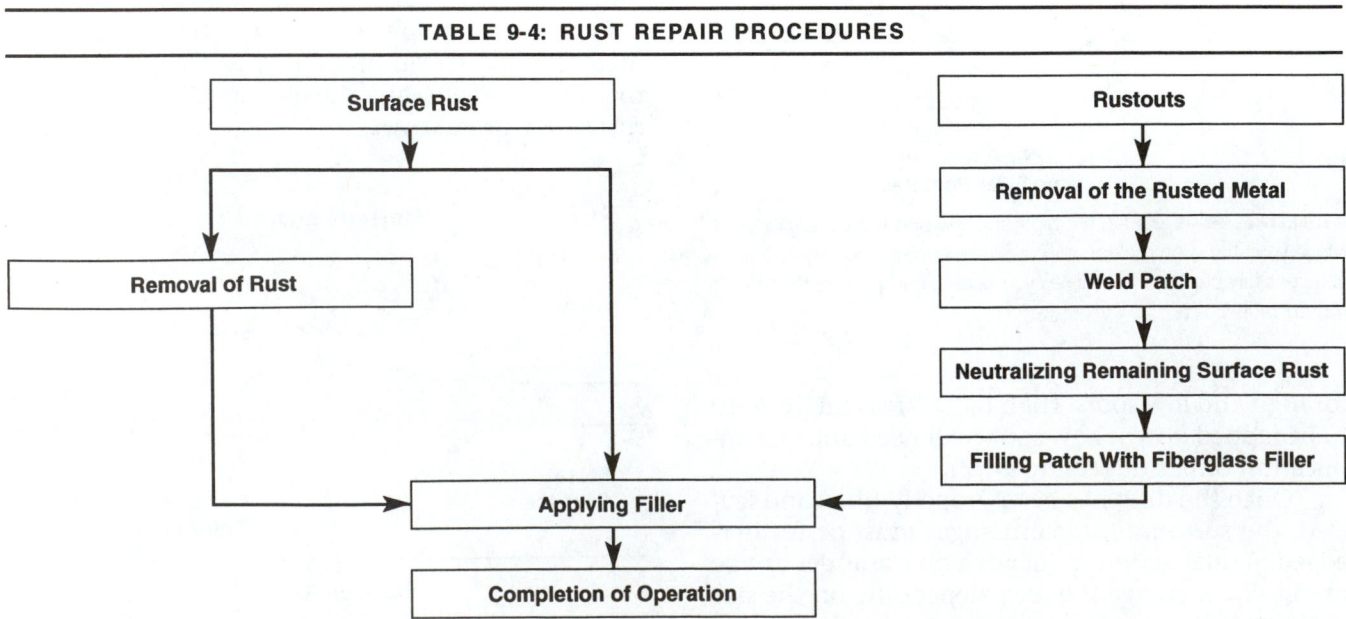

TABLE 9-4: RUST REPAIR PROCEDURES

Surface Rust	Rustouts
Removal of Rust	Removal of the Rusted Metal
	Weld Patch
	Neutralizing Remaining Surface Rust
	Filling Patch With Fiberglass Filler
Applying Filler	
Completion of Operation	

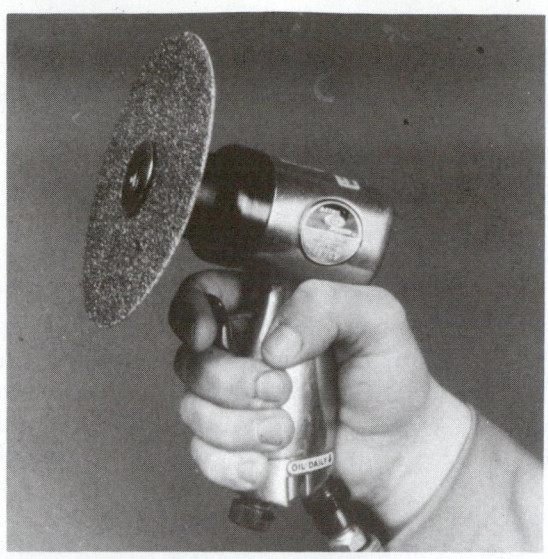

FIGURE 9-45 A lightweight air grinder is ideal for removing paint without damage to the panel.

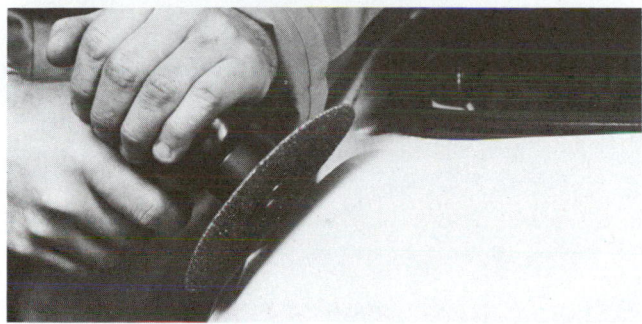

FIGURE 9-46 Hold the grinder at a 10-degree angle to the surface.

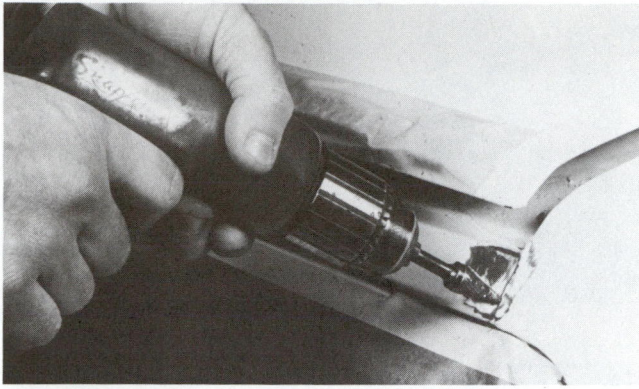

FIGURE 9-47 Grind or blast rust from the pits.

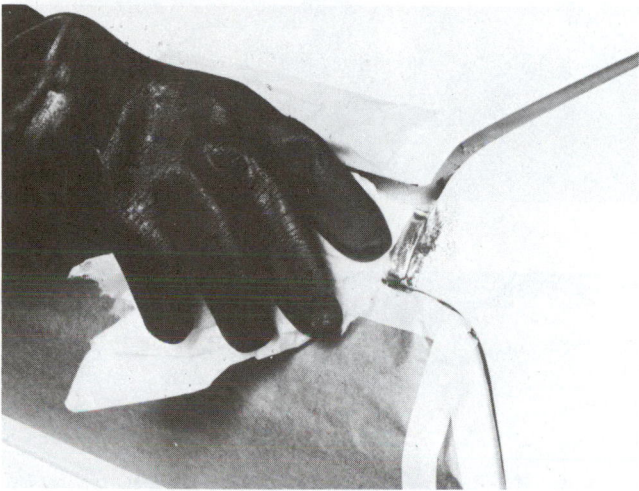

FIGURE 9-48 Clean bare metal with metal conditioner.

After removing surface rust with the grinder, use a die grinder attachment to remove rust from the pits, panel edges, and other hard-to-reach places (Figure 9–47).

Clean the bare metal with metal conditioner (Figure 9–48). **Metal conditioner** is an acid compound that neutralizes microscopic rust particles. The acid also etches the metal surface to improve the bond between the metal and the primer-surfacer. Metal conditioner is bottled in concentrated form and must be diluted with water before use. Always wear rubber gloves and safety glasses when handling conditioners. Follow the manufacturer's instructions carefully when diluting the conditioners.

APPLYING FILLER TO SURFACE RUST

Even though it is not advisable to metal condition a bare steel surface over which plastic is to be applied, there are times when an exception is advised. When a surface has become pitted with rust, the following alternative repair may be made. Grind the paint from the surface until the pitted area and approximately a two-inch area around it are free of paint and surface coatings. Apply metal conditioner to the pitted area and scrub with Scotch-Brite™. Dry it and neutralize it as normal. Use a sander with #80 grit sandpaper to abrade the bare surface where the metal conditioner was applied. This will neutralize the rust in the metal pores, which would otherwise remain active and begin to spread under the plastic filler in a short time.

Mix waterproof plastic filler and hardener together. Remember not to stir the plastic. Stirring results in air bubbles in the hardened filler. Use a back-and-forth wiping action.

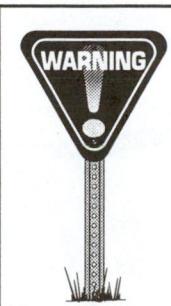

WARNING

Never apply plastic filler directly to a conditioned metal surface. The plastic will not adhere to the metal and could crack when in service.

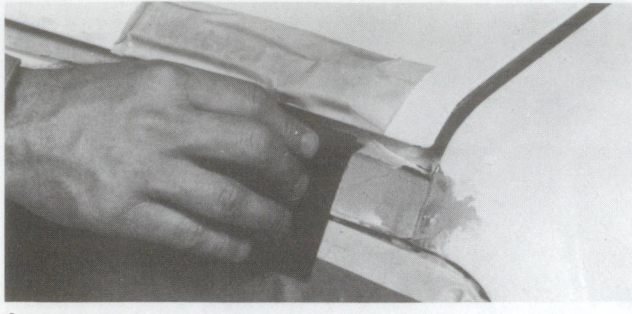

A

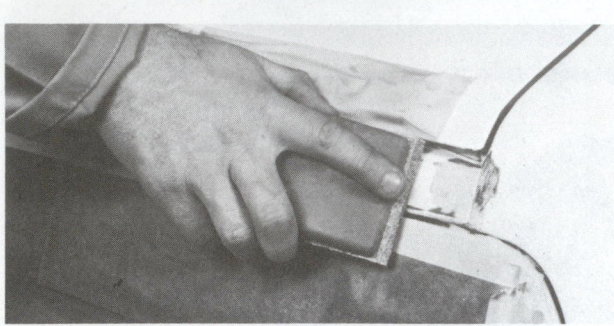

B

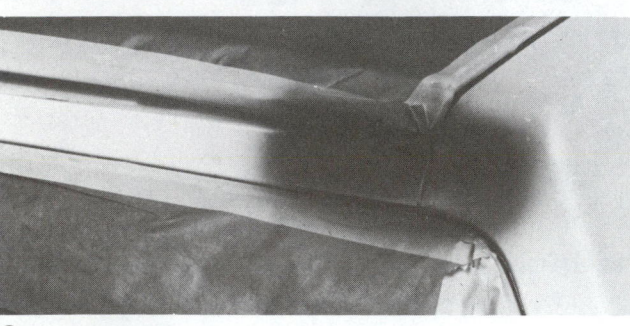

C

FIGURE 9–49 Cover the pitted area with filler; sand and prime before finishing.

Scoop up some filler onto the edge of the spreader and smear a thin skim coat over the pitted area. Apply moderate pressure to force the plastic filler into the pits (Figure 9–49A). Allow the filler to harden. Block sand with #80 grit sandpaper (Figure 9–49B) until the filler is level with the panel surface.

Use compressed air to blow filler dust from any still visible pits and pinholes. If necessary, apply spot

CAUTION

Always wear a dust mask while working body fillers. Inhaling filler dust can be harmful.

putty to any pits. Allow the filler or putty to dry before sanding and priming. Follow the sanding and priming (Figure 9–49C) procedures already described in this chapter.

9.8 REPAIRING SMALL RUSTOUTS

Rust can form on either side of the metal panel. Rust that is present on the backside of a panel might go unnoticed until the paint begins to bubble and lift. By this time, the rust has eaten completely through the panel. Spots of surface rust, as shown on the fender in Figure 9–50, might be a sign that

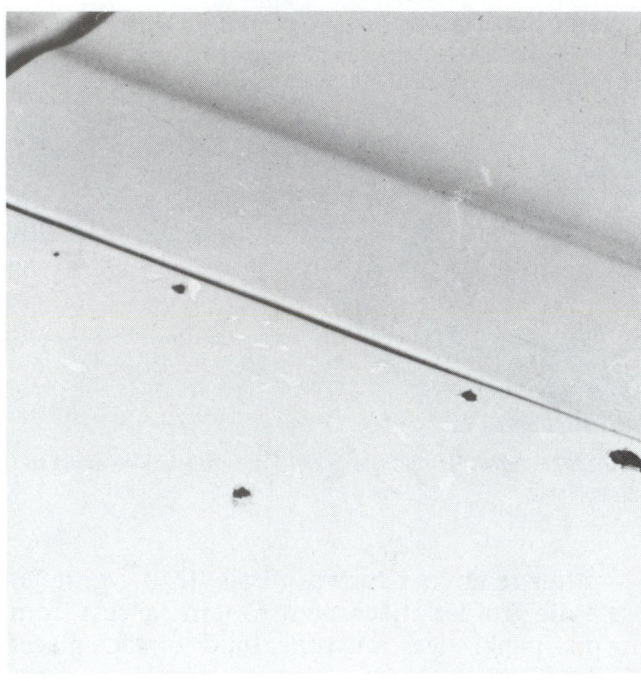

FIGURE 9–50 Small spots or bumps on the paint surface often indicate rustout from the inside out.

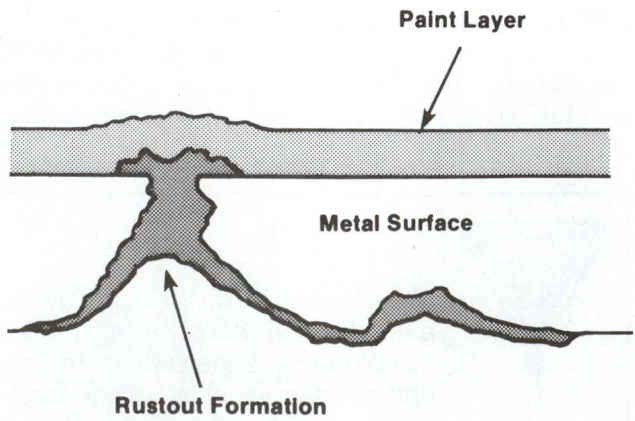

FIGURE 9–51 Study rustout formation.

rust has eaten through from the backside of the panel (Figure 9–51). When the surface rust and paint are ground or blasted away, small holes called rustouts will be uncovered.

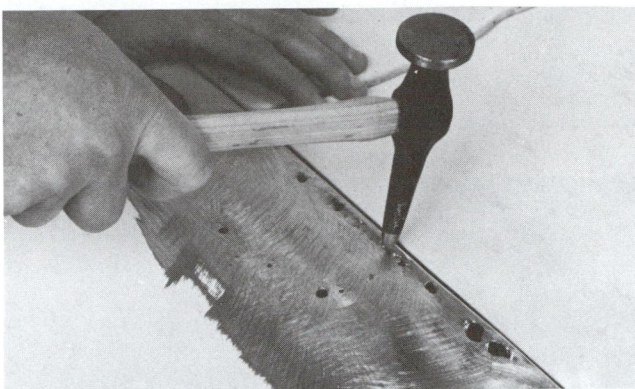

FIGURE 9-52 Light hammer blows show rust-weakened, thinned areas. These areas should be repaired with a metal patch.

FIGURE 9-53 Fill rustouts with waterproof filler. Conventional plastic filler will absorb moisture and rust out again in a short period of time.

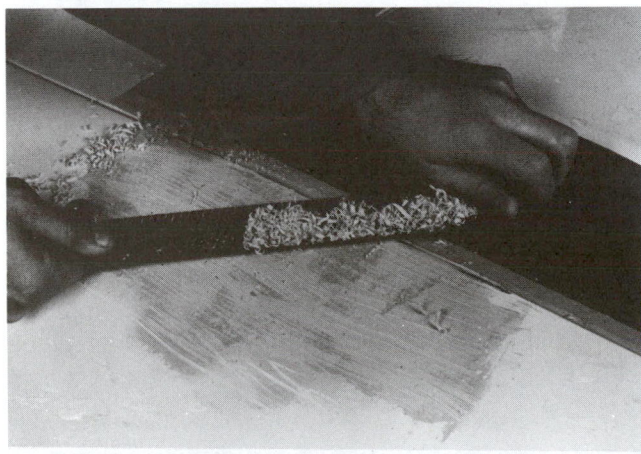

FIGURE 9-54 To save time and effort, always use a grater-type file to shape filler while it is still in a semisolid state. Once hardened, fiberglass plastic filler can be tough to sand.

PREPARING THE SURFACE

After blasting or grinding the area with the air grinder and a #24 to #36 grit rigid disc, use a pick hammer to find thin spots in the metal (Figure 9–52). They will dent easily because they are very thin.

If the backside of the panel is accessible, remove accumulated dirt and undercoating. A wire brush, scrapers, or blaster can be used to remove the rust on the backside of the repair area.

If it is difficult to remove, you might want to apply one of the commercially available rust deactivators. The chemical reacts with the rust to form a hard, black polymer coating over the rust. This coating seals out air and moisture, preventing any further oxidation.

It is best to weld in a metal patch to the back of the rustout hole. This will make the repair as strong as the original panel. If you simply fill a rust hole with fiberglass, it will reduce structural integrity.

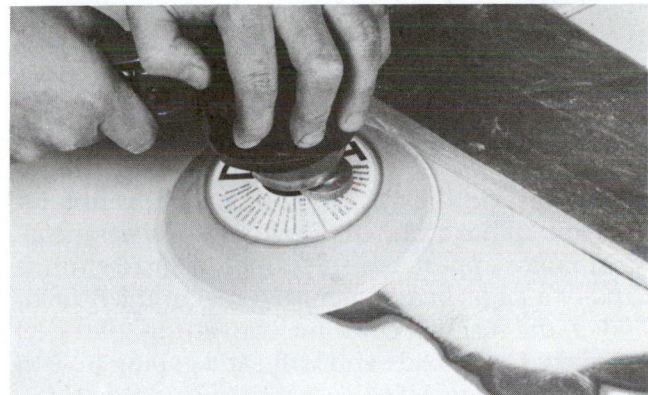

A

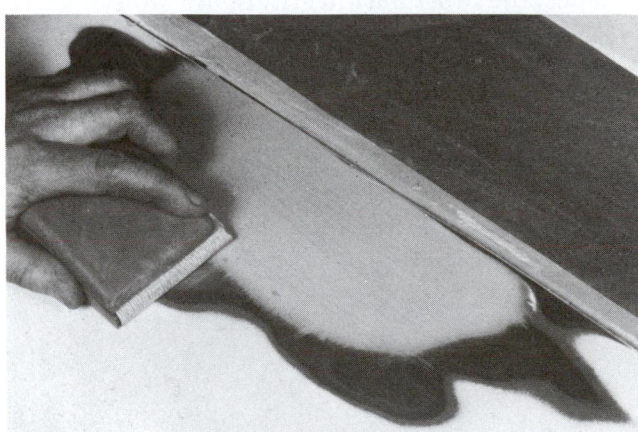

B

FIGURE 9-55 Sanding the filler to a featheredge. You want the filler to slowly blend into the existing shape of the panel. (A) DA is the common way of featheredging. (B) Block sanding should be used for most final featheredging because it is more controllable.

APPLYING FILLER TO RUSTOUT

After welding a metal patch around the rust hole, cover the metal patch with a waterproof, fiberglass reinforced filler (Figure 9–53). Use the plastic spreader to force the filler into the weld and patch area.

Remember! Regular plastic fillers with a talc bulking agent absorb moisture and are not suitable for filling rustouts.

After the waterproof filler hardens, sand the filler smooth and wipe the dust away with a clean cloth.

Cover the rustout patch with a layer of regular plastic filler. After the plastic turns rubbery, but before it hardens, knock off the high spots with a cheese grater (Figure 9–54). Hold the cheese grater at a 10- to 20-degree angle and pull it across the repair area. If the repaired area is small, it might not be necessary to use a cheese grater.

After the plastic completely hardens, sand it down level with the panel surface (Figure 9–55A and B). Featheredge the repair area into the surrounding panel. The sanded repair is now ready to be primed and puttied.

9.9 REPAIRING LARGE RUSTOUTS

Usually rust on the underside of a panel is not noticed until it has attacked a large area of sheet metal. What might appear as a small spot of surface rust is actually a large area of damaged metal. If left unattended, the rust will continue converting solid steel into flaky ferric oxide and will eat a gaping hole in the metal (Figure 9–56).

Not only is the rustout unattractive, but rustouts also affect the structural integrity of the vehicle. Structural members with this kind of damage should either be replaced or the rusted area cut away and new metal welded in place.

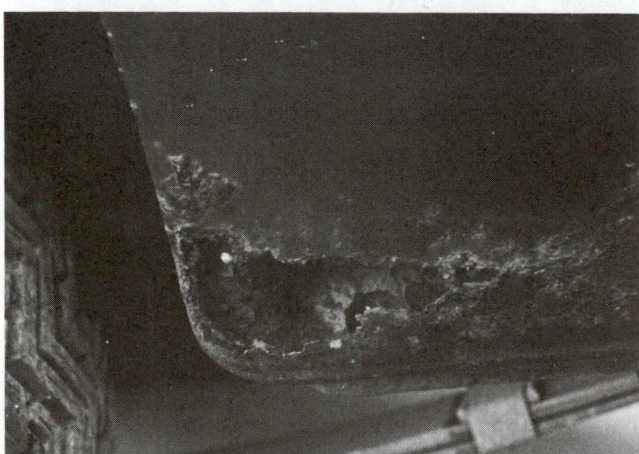

FIGURE 9-56 This is a large rustout in the bottom of a rear quarter panel.

Applying excessive pressure with the grinder against weak, rusted sheet metal can burn a hole right through the metal. The grinder could then catch on the sharp metal and kick back dangerously at you. Use light pressure when grinding and hold the tool securely.

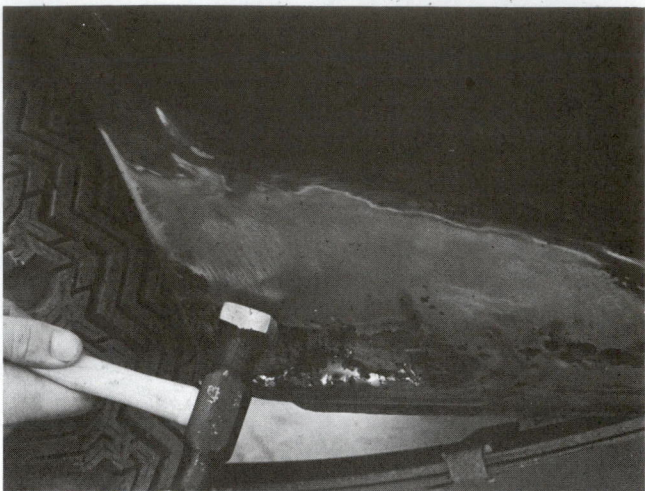

FIGURE 9-57 Light hammering will show thin rusted areas that should be cut out.

REMOVING RUST

Before sanding away paint and rust from the repair area, cover nearby trim pieces with masking tape. Masking protects the trim from grinding sparks, chips, and accidental bumps of the disc.

Use #24 or #36 grit disc and a lightweight air grinder to remove paint and rust from the rustout. Exercise extreme care when grinding paint and rust from a rustout area. Rust has often deteriorated a panel until it is very thin and weak. If possible, it best to blast the rust off the panel to avoid thinning the metal.

Remove as much rust as possible. Cut away rust-softened metal around the rustout. Then make a repair panel to fit the rustout and weld in place (Figure 9–57).

APPLYING RUST DEACTIVATOR

After cleaning and sanding the repair area, apply metal conditioner or a rust deactivator to neutralize rust remaining in pitted areas and around the edges of the rustout. Always wear rubber gloves when handling rust deactivator and metal conditioners. Shake the container well before using and follow instructions carefully.

Apply the rust deactivator to the backside of the panel, too. With a wire brush or scraper, first clean dirt and undercoating from the backside of the repair area. Then apply a rust deactivator. Be sure to carefully follow the manufacturer's instructions. A black coating will soon develop over the rust. For maximum protection, this coating should have a solid black color. Wait 1 to 2 hours between coats. If the color is splotchy and uneven, apply additional coats. Two or three thin coats neutralize the rust better than one thick coat. If excess rust deactivator runs over the finish paint, wipe it off immediately. Use a cloth dampened with mineral spirits.

You may also want to blast the area to remove rust from pits. This may eliminate the need to use a rust converter.

PATCHING AND FILLING THE RUSTOUT

After the repair area is clean, a metal patch should be made of the same metal as the original panel (Figure 9–58). It should be welded well above or beyond the rusted area being repaired. The heat generated by welding could cause the weld area to weaken and rust out prematurely. Another alternative is to use adhesives to bond the repair patch. This may be advisable particularly if the repair area is on a nonstructural part. The bond surface should be reinforced with a few resistance spot welds. The rustout patch

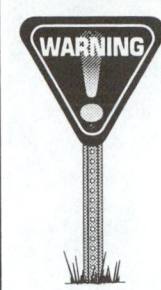

> **WARNING**
>
> A rustout in a structural panel should be repaired by replacing the part or repairing it by welding in a metal patch. The repair of the structural part is critical to the integrity of the vehicle. If you try to repair a large rustout hole with only fiberglass, the part will be weaker than when new.

area can then be filled with waterproof fiberglass-reinforced filler. Use filler with long fiberglass strands to give maximum strength to the repair. Make sure that the filler is mixed thoroughly before applying.

PATCHING THE RUSTOUT

Apply a waterproof filler over the welded repair patch. Do not use conventional filler because it is not waterproof. Even the best welds can have tiny flaws that allow moisture entry through the weld. For this reason, always use waterproof filler as the first coat over a repair area (Figure 9–59).

COMPLETING THE RUSTOUT REPAIR

With one coat of waterproof filler over the metal patch, mix enough conventional filler to fill the

FIGURE 9-59 Apply a fiberglass patch over the repair panel to seal out the moisture. Fiberglass cloth and fiberglass-reinforced plastic filler are now used in most situations. They are less messy and easier to apply than fiberglass resin.

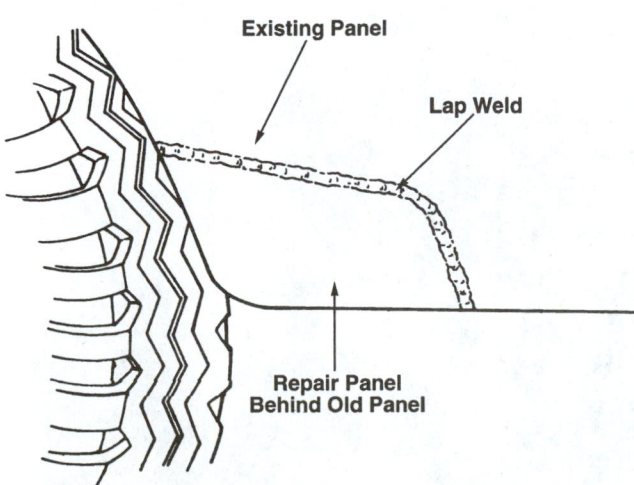

FIGURE 9-58 Cut out all thinned rusted areas. Then weld in repair panel.

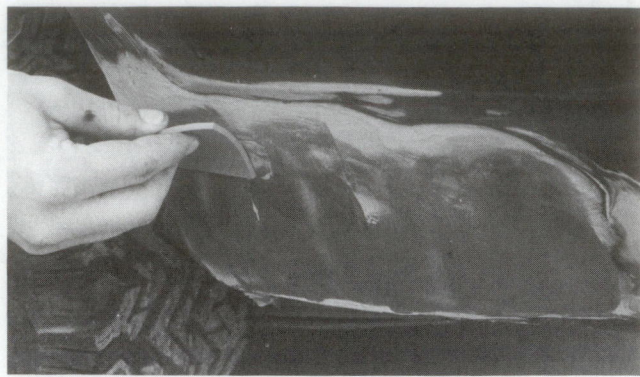

FIGURE 9-60 Apply short-strand fiberglass plastic over long-strand to speed final shaping.

depressed area. Using a larger plastic spreader, apply a coat of filler over the repair area. Press hard on the spreader to force out air bubbles. Spread from the edges to the center. Fill the depression and other low spots. A smoother application can be achieved by placing a piece of plastic film over the filler and smoothing the filler under the plastic (Figure 9–60).

SANDING AND PRIMING THE REPAIR

After the filler has dried, it can be sanded and shaped. Sand the repair area with #36 or #40 grit sandpaper and a sanding block to within $1/16$ inch (1.6 mm) of the finish level. Switch to #80 grit sandpaper and sand to the finish level.

While wearing clean plastic gloves, run the palm of your hand over the finish repeatedly to locate high and low spots. Avoid oversanding; this creates low spots (undercut) that will need additional filler. Feather the edges level with the surrounding metal surfaces.

When the filler has been sanded to the desired level and smoothness, clean the area and dry it with an unsoiled cloth. Wipe the repair area with a tack cloth. Blow with compressed air to remove any remaining dust particles. Spray with primer. After the primer dries, coat the filled area with polyester glazing putty. Apply with a clear rubber squeegee using single, smooth strokes and allow to dry.

Using #80 and progressing to #150 or #180 grit sandpaper and long, even strokes with the sanding block, sand the glazed repair to a smooth surface. If pinholes or voids still remain, fill with additional putty and sand.

Lightly wet sand with #400 grit sandpaper to polish sandpaper scratches from the repaired rustout. Clean with a tack cloth and prime the repair area once more. The repaired rustout is ready to be painted.

More information on rust and corrosion protection can be found in Chapter 15.

Make sure that the labor for making and installing the metal patch is worth it. You may be able to purchase a replacement panel and install it for a lower total cost.

FABRICATING A METAL PATCH

The best way to repair a large rustout is to make and weld in a metal patch. This makes the repair as strong as the original panel and the repair will last longer.

When installing a small metal patch in an antique car door, it is preferable to blast the area to be repaired after making sure that the area around the repair is sound. To check for badly thinned metal, push on areas of the panel with your thumb. If the panel flexes easily, it is rusted too thin.

After determining the amount of area damaged, grind or blast off the paint and primer (Figure 9–61). Mark around the area to be replaced. Then cut it out with a cutoff tool, power cutter, or plasma arc cutter (Figures 9–62 and 9–63).

Use the cutout piece of metal as a template to make a new repair piece. Use a scribe to mark the new metal piece to be cut (Figure 9–64). Make sure the repair piece is the same thickness and type of metal as the original. Cut along the scribed line (Figure 9–65). If the area is accessible from the rear, you might want to make the patch larger than the

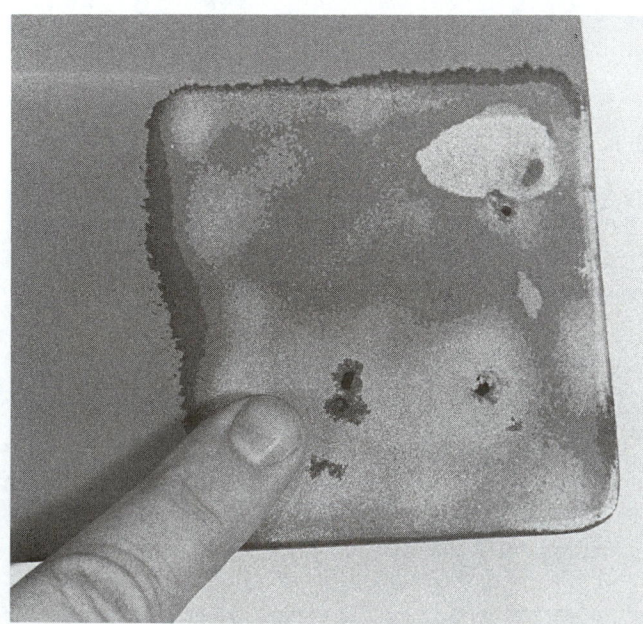

FIGURE 9-61 Plastic media blasting shows minor rustout in this collector car door.

FIGURE 9-62 Mark and cut out the rusted area.

FIGURE 9-63 Once cut out, treat any internal rust with a rust converter solution.

FIGURE 9-64 Use the old, cut-out piece to scribe mark the shape of the replacement sheet metal.

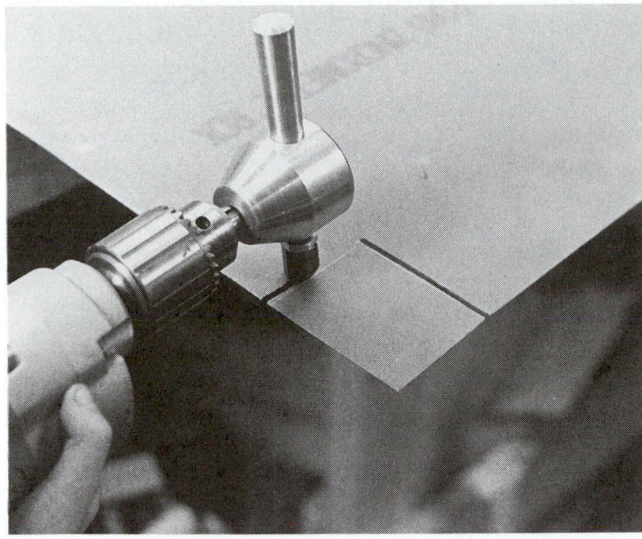

FIGURE 9-65 Cut the sheet metal patch along the scribe marks. If you can fit it in place from the rear, you may want to make the patch larger than the hole for making a lap weld.

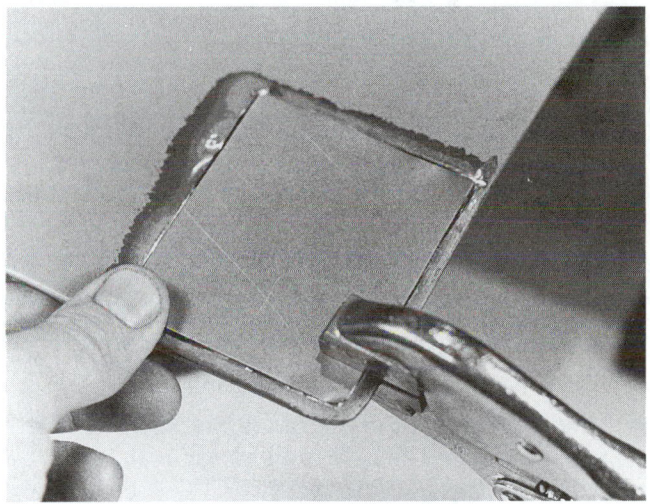

FIGURE 9-66 Fit and secure the piece in place for welding.

hole. You can then fit the patch in from the back and make a lap weld.

Fit and clamp the repair piece into place in the hole. Use clamps or vise grip pliers to secure the repair piece in position (Figure 9–66).

MIG weld the metal patch. Start by spot welding the patch in the corners; then stitch weld it to prevent warpage. Finish by continuous welding along each edge to form a leakproof weld joint. You might want to use *heat sink compound* to help prevent panel warpage (Figure 9–67).

Grind the weld down (Figure 9–68). If applicable, apply undercoating or anticorrosion compound to the inside of the panel to protect from further rusting. Apply filler over the metal patch as summarized earlier.

FIGURE 9-67 Using a heat sink compound to prevent panel warpage, MIG spot or tack weld the patch in place. Finish with a continuous bead all the way around the patch.

FIGURE 9-68 Grind weld flush with the panel. Then fill the area with a waterproof fiberglass plastic.

SUMMARY

- Plastic body filler is the finishing touch for most sheet metal repairs.

- Body filler or plastic filler is a heavy-bodied plastic material that cures very hard for filling small dents in metal.

- As the filler cures and hardens, the chemical reaction produces a tremendous amount of heat. For this reason, unused filler should not be discarded in hazardous materials containers that contain solvent-wet paper or cloths. It should be discarded in a hazardous waste container that has been specifically reserved for filler.

- Fiberglass body filler has fiberglass material added to the plastic filler. It is used for rust repair or where strength is important.

- Light body filler is formulated for easy sanding and fast repairs

- Body filler manufacturers have developed a fine-grained, catalyzed polyester glazing putty or two-part spot putty. Like plastic filler, a hardener or catalyst must be mixed with the putty to initiate and speed curing.

- Hardener kneading is done by thoroughly squeezing the contents back and forth inside the tube.

- Add hardener according to the proportions indicated on the can, usually 10 percent hardener (one part hardener for each ten parts filler).

- The most common mistake of the apprentice is to use too much cream hardener!

- The surform or cheese grater file is used to cut the excess filler to size quickly.

- A sanding guide is a special tool for sanding straight, special contour lines on panels.

- Featheredging tapers the edges of the paint so that it gradually blends in with the metal surface.

- Final sanding and priming are necessary to achieve a super-smooth surface. Wet sand with #400 grit or finer sandpaper and a sanding block.

- Surface rust in its early stage can simply leave a reddish coating on the metal surface. Given time, the rust will eat pits into the surface. Eventually the pitting will develop into rust holes, or rustouts.

ASE STYLE REVIEW QUESTIONS

1. Fiberglass fillers are

   A. Waterproof
   B. Available in three basic forms
   C. Never used to bridge rustouts
   D. All of the above

2. Lightweight fillers have improved

   A. Filing characteristics
   B. Sanding characteristics
   C. Water resistance
   D. All of the above

3. If too little hardener is used, the filler
   A. Will not adhere to the metal
   B. Will be subject to rampant pinholing
   C. Will be easier to handle
   D. None of the above

4. The first step in repairing nicks is _____.
   A. Featheredging
   B. Dewaxing
   C. Both A and B
   D. None of the above

5. Once a ding has been properly filled and leveled, Technician A uses a dual action sander to featheredge the surrounding paint edges. Technician B says to proceed slowly, because it is easy to oversand and sink the repair area too low in the panel. Who is correct?
   A. Technician A
   B. Technician B
   C. Both A and B
   D. Neither A or B

6. Metal conditioner is _____.
   A. An acid compound
   B. Used to etch the metal surface
   C. Used to neutralize rust
   D. All of the above

7. Technician A grinds a brazed joint before neutralizing the acids in the flux. Technician B says that soda water is used to neutralize these acids. Who is correct?
   A. Technician A
   B. Technician B
   C. Both A and B
   D. Neither A or B

8. Technician A uses the same putty knife to mix and apply plastic body filler. To save time, Technician B uses a piece of clean cardboard as a mixing board because it is clean and can be thrown away after being used. Who is correct?
   A. Technician A
   B. Technician B
   C. Both A and B
   D. Neither A or B

ESSAY QUESTIONS

1. How and why do you remove paint before filling with plastic?

2. How do you make and install a metal patch for a rustout?

3. What is a metal conditioner?

4. Describe how to mix and apply plastic filler.

CRITICAL THINKING PROBLEMS

1. If you use spot putty to fill a large dent on bare metal, will the repair be sound?

2. Describe some of the problems that can result if body filler is not stirred up thoroughly to a smooth and uniform consistency.

MATH PROBLEMS

1. If a panel is .031 inch (0.79 mm) thick, and you grind off .024 inch (0.61 mm) of metal, how thick is the remaining panel?

2. If maximum filler thickness is $3/4$ inch (19 mm), and you have already applied a thickness of .5 inch (12.7 mm), how much more thickness is allowable?

PHOTO SEQUENCE

USING PLASTIC FILLER

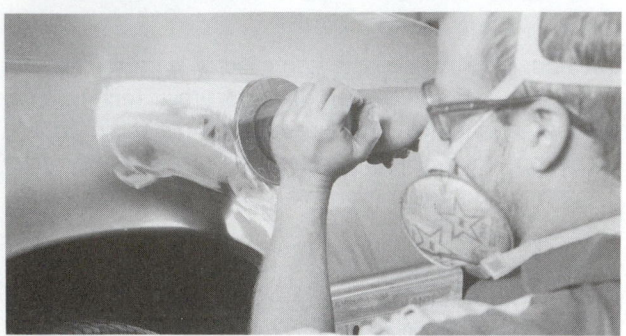

P9-1-1 Plastic filler should not be placed over existing paint. To prepare the surface for adhesion, grind or sand off the paint in the area to be filled. Grind the paint carefully without cutting too deeply into the metal. Too much grinding will weaken today's thin metal panels.

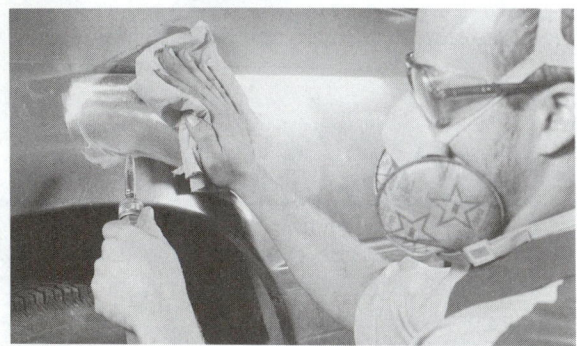

P9-1-2 Blow and wipe off the area to remove debris. If dust is left on the repair area, the filler may not stick or adhere to the body. Vibration and flexing of the body could cause the filler to crack or pop off.

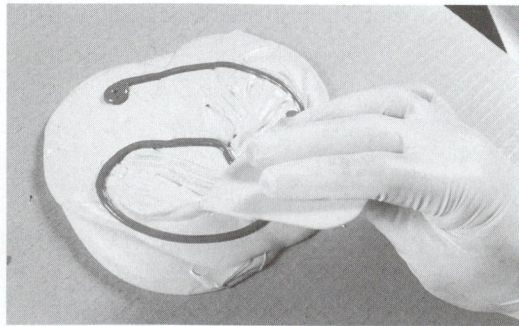

P9-1-3 Mix quality plastic filler with catalyst or hardener. Use a one-inch bead of hardener with a golf-ball size scoop of filler. Wipe the ingredients back and forth to mix them properly and to avoid stirring in air bubbles.

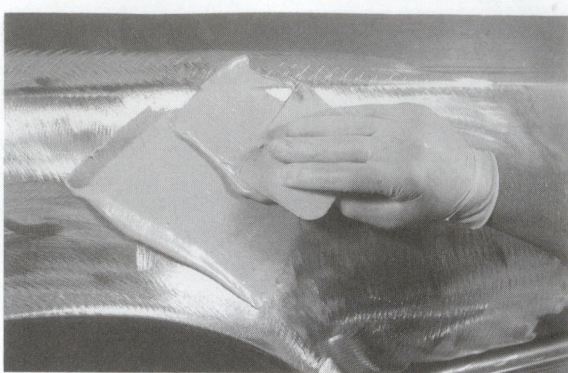

P9-1-4 Place the filler on a spreader. Wipe the filler across the repair area. Use hand motions to make the filler match the general shape of the part contour. Try to make the filler slightly higher than the surrounding body surface.

P9-1-5 Allow the filler to cure for a few minutes or until it has the consistency of paste. Then use a surform or "cheese grater" file to roughly shape the filler. Do not let the file cut grooves below the desired surface level; use the file to cut off any high spots of filler.

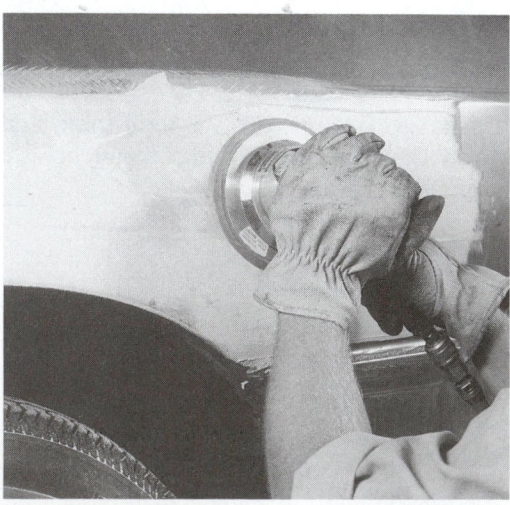

P9-1-6 Now, use either a dual-action sander or a hand block with medium grit sandpaper to sand the area. Knock down any remaining high spots as you featheredge or taper the outer edge of the filler into the undamaged body surface.

USING
SPOT PUTTY

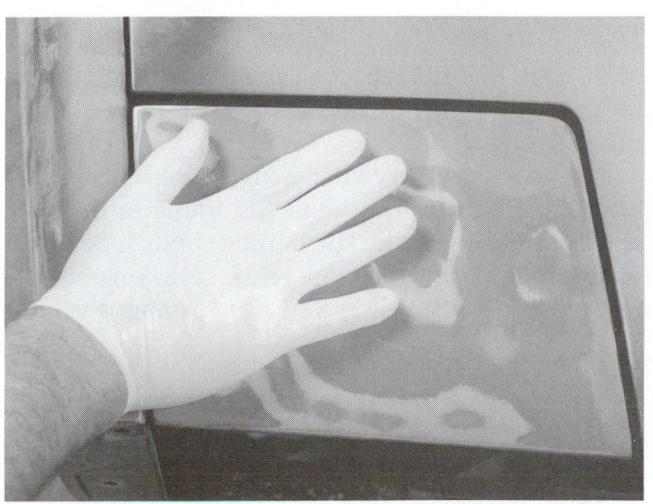

P9-2-1 Carefully inspect the repair area for low spots and small imperfections such as air bubble holes.

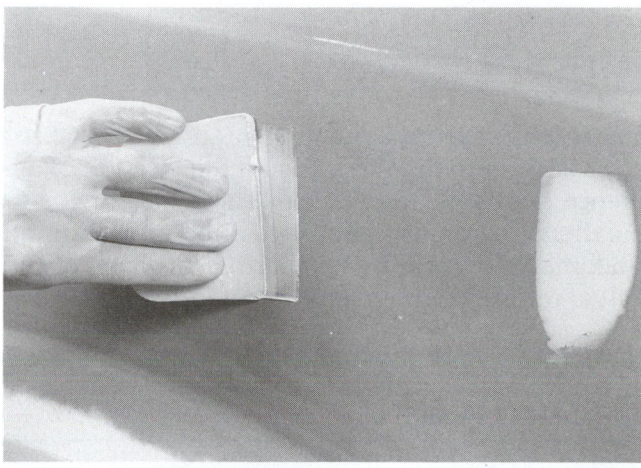

P9-2-3 Use a rubber squeegee to apply spot putty. Work the squeegee across low spots or pinholes to ensure that they are completely filled.

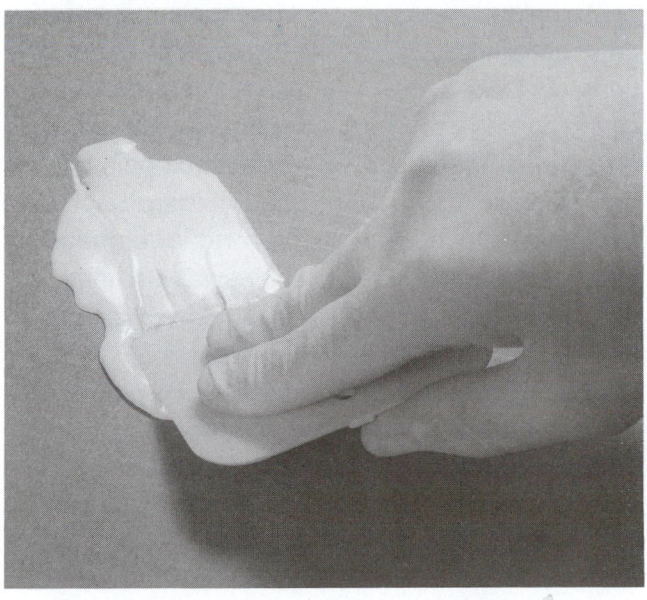

P9-2-2 Mix the correct proportions of polyester spot putty and catalyst.

P9-2-4 Use a finer grit sandpaper to sand the spot putty level with the body surface. A dual-action sander works well if it is handled carefully.

CHAPTER 10

Diagnosing Major Collision Damage

INTRODUCTION

Vehicle measurement involves using specialized tools and equipment to measure the location of reference points on the vehicle. These measurements are then compared to published dimensions from an undamaged vehicle. By comparing known good and actual measurements, you can determine the extent of damage. The difference in the two measurements indicates the direction and amount of frame or body misalignment.

When a car or truck is in a high-speed crash, powerful impact forces can bend the frame or unibody structure of the vehicle. The frame or body is designed to absorb some of the energy of the collision and protect its occupants. When a heavily damaged vehicle

OBJECTIVES

After studying this chapter, you should be able to:

✔ Explain how impact forces are transmitted through both frame and unibody construction vehicles.

✔ Describe how to visually determine the extent of impact damage.

✔ List the various types and variations of body measuring tools.

✔ Analyze damage by measuring body dimensions.

✔ Analyze impact damage to mechanical parts of the vehicle.

✔ Explain the importance of the datum plane and centerline concepts as related to unibody repair.

✔ Interpret body dimension information and locate key reference points on a vehicle, using body dimension manuals.

✔ Discuss the use of tram bars, self-centering gauges, and strut tower gauges.

✔ Diagnose various types of body damage, including twist, mash, sag, and sidesway.

✔ Locate and measure key points using a tape measure, tram bar, and self-centering gauges, when given a damaged vehicle and a body specification manual.

✔ Answer ASE test questions pertaining to vehicle measurement and damage analysis.

KEY TERMS

beam splitters
body center marks
body dimensions
body dimensions chart
center plane
cone concept
control points
crush zones
damage analysis form
datum plane
dedicated bench
diamond damage
fixtures
height dimensions
laser
laser guides
laser power unit
lateral dimensions

left-to-right symmetry check
mash damage
measurement gauges
mechanical measuring systems
pivot measure system
primary damage
reference points
sag damage
secondary damage
sidesway damage
strut tower gauge
symmetrical
three zero plane sections
twist damage
vehicle measurement

ASE TASK LIST

Job Skills covered in this chapter include:

PAINTING AND REFINISHING TEST (B2) TASK LIST

A. Surface Preparation

15. Apply suitable paint sealer to the area being refinished when sealing is needed or desirable.

NONSTRUCTURAL ANALYSIS AND DAMAGE REPAIR TEST (B3) TASK LIST

A. Preparation

5. Remove undamaged, nonstructural body panels and components that may interfere with or be damaged during repair.

B. Outer Body Panel Repairs, Replacements, and Adjustments

7. Remove, replace, and align bumpers, reinforcements, guards, isolators, and mounting hardware.
8. Check and adjust clearances of front fenders, header, and other panels.
13. Remove damaged sections of steel body panels; weld in replacements in accordance with manufacturers'/industry specifications.
17. Diagnose and repair water leaks, dust leaks, wind noise, squeaks, and rattles.

D. Moveable Glass and Hardware

2. Repair or replace electrically driven power sun roofs and related controls.
3. Inspect, repair or replace, and adjust removable, manually operated roof panels and hardware.

STRUCTURAL ANALYSIS AND DAMAGE REPAIR TEST (B4) TASK LIST

A. Frame Inspection and Repair

1. Diagnose structural damage using tram and self-centering gauges in accordance with industry specifications.
2. Straighten and align mash (collapse) damage.
3. Straighten and align sag damage.
4. Straighten and align sidesway damage.
5. Straighten and align twist damage.
6. Straighten and align kickup damage.
7. Straighten and align diamond frame damage.
8. Remove and replace damaged frame horns, side rails, cross members, and front or rear sections.

STRUCTURAL ANALYSIS AND DAMAGE REPAIR TEST (B4) TASK LIST

A. Frame Inspection and Repair

10. Repair or replace weakened or cracked frame members in accordance with vehicle manufacturers'/industry standards.

11. Diagnose misaligned or damaged steering, suspension, and power train components that can cause vibration, steering, and wheel alignment problems; align or replace steering and suspension components in accordance with vehicle manufacturers' recommendations.

B. Unibody Inspection, Measurement, and Repair

1. Recognize that measuring, dimensioning, and tolerance limits in unibody vehicles are critical to repair of these vehicles; recognize that suspension/steering mounting points and engine/power train attaching points are critical to vehicle handling, performance, and safety.
2. Diagnose misaligned or damaged steering, suspension, and power train components that can cause vibration, steering, and wheel alignment problems; realign or replace steering and suspension components in accordance with vehicle manufacturers' recommendations.
3. Diagnose and analyze unibody vehicle damage using mechanical equipment (tram and self-centering gauges).
4. Determine the proper locations of all suspension, steering, and power train component attaching points on the body.
5. Diagnose and measure unibody vehicles using a dedicated (fixture) measuring system.
6. Diagnose and measure unibody vehicles using a universal measuring system (mechanical, electronic, laster, sonar).
7. Determine the extent of direct and indirect damage, and the direction of impact; plan the methods and sequence of repair.
8. Attach proper anchoring devices.
9. Straighten and align center unibody section, cowl, bulkhead, roof, roof rails, pillars, floor, windshield, back glass openings, door openings, rocker panels, and floor pans.
10. Straighten and align rear unibody section quarter panels, wheelhouse assemblies, and rear body panel (including rails, suspension, and power train mounting points).
11. Straighten and align front unibody sections (aprons, strut towers, upper and lower rails, steering, suspension, and power train mounting points).
12. Recognize the limitations of applying heat to high-strength steel structural components; use proper stress relief methods for high-strength steel; weld in accordance with vehicle manufacturers' recommendations.
14. Remove folds, curves, creases, and dents, using power tools and hand tools, to restore damaged areas to proper contours and dimensions.

ASE TASK LIST (continued)

Job Skills covered in this chapter include:

15. Determine the extent of damage to structural steel body panels; repair, weld, or replace in accordance with vehicle manufacturers' specifications.
16. Remove damaged sections of structural steel body panels, and weld in replacements in accordance with vehicle manufacturers' specifications.

C. Stationary Glass
1. Remove and replace front and rear stationary glass (heated and nonheated) in accordance with manufacturers' recommendations.

TABLE 10-1

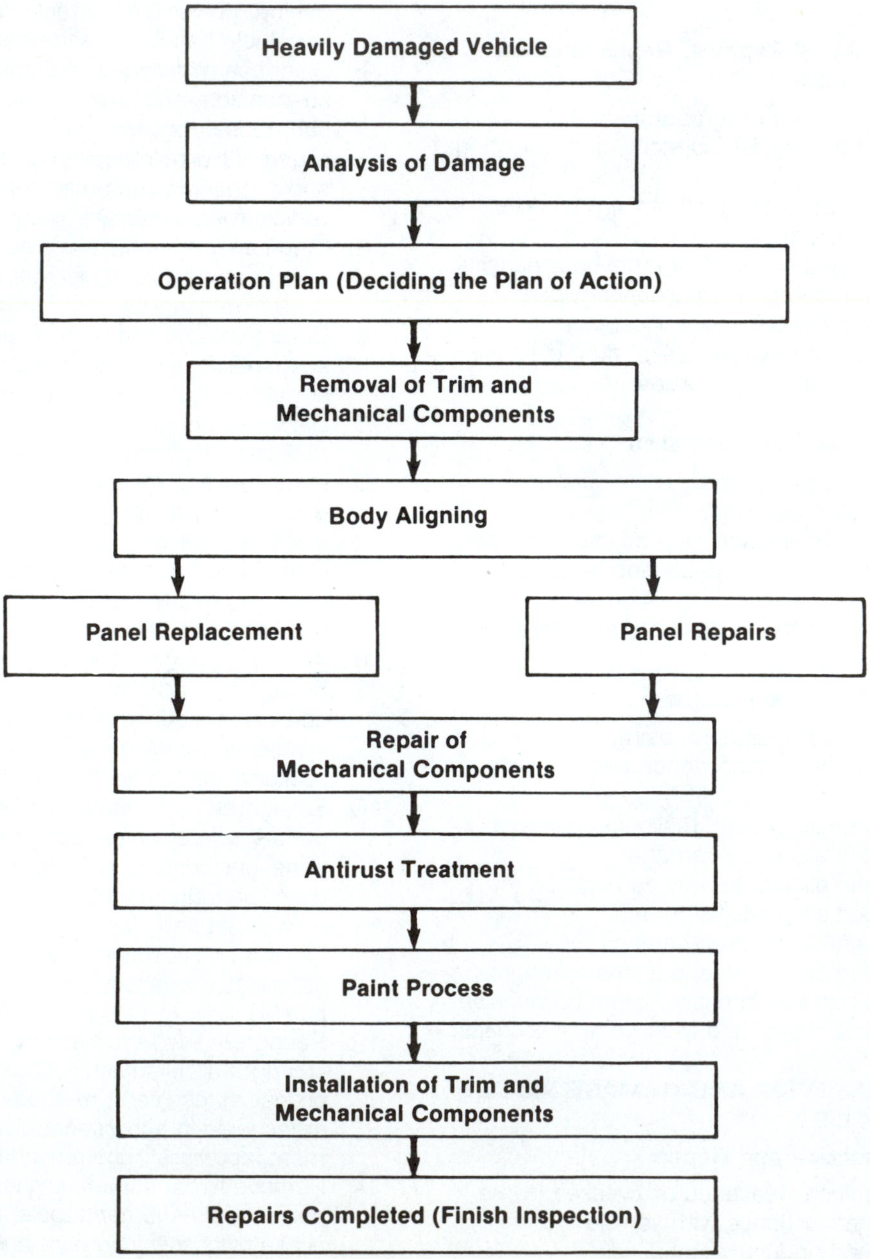

```
┌─────────────────────────────────────┐
│      Heavily Damaged Vehicle        │
└─────────────────────────────────────┘
                 ↓
┌─────────────────────────────────────┐
│         Analysis of Damage          │
└─────────────────────────────────────┘
                 ↓
┌─────────────────────────────────────┐
│ Operation Plan (Deciding the Plan   │
│            of Action)               │
└─────────────────────────────────────┘
                 ↓
┌─────────────────────────────────────┐
│       Removal of Trim and           │
│      Mechanical Components          │
└─────────────────────────────────────┘
                 ↓
┌─────────────────────────────────────┐
│            Body Aligning            │
└─────────────────────────────────────┘
        ↓                    ↓
┌──────────────────┐  ┌──────────────────┐
│ Panel Replacement│  │   Panel Repairs  │
└──────────────────┘  └──────────────────┘
        ↓                    ↓
┌─────────────────────────────────────┐
│           Repair of                 │
│      Mechanical Components          │
└─────────────────────────────────────┘
                 ↓
┌─────────────────────────────────────┐
│         Antirust Treatment          │
└─────────────────────────────────────┘
                 ↓
┌─────────────────────────────────────┐
│           Paint Process             │
└─────────────────────────────────────┘
                 ↓
┌─────────────────────────────────────┐
│      Installation of Trim and       │
│      Mechanical Components          │
└─────────────────────────────────────┘
                 ↓
┌─────────────────────────────────────┐
│ Repairs Completed (Finish Inspection)│
└─────────────────────────────────────┘
```

is brought to the shop, the extent of the damage must be carefully evaluated. Sometimes measurements are needed to help the estimator calculate the costs of the repairs.

After studying damage measurements, frame straightening equipment is used to pull the frame or body back into alignment. Straightening procedures are explained in the next chapter.

Table 10–1 illustrates the major collision repair processes after the estimate for a heavily damaged vehicle has been made.

To repair any vehicle properly, you must accurately diagnose the collision damage. The severity and extent of damage to parts must be analyzed. Once the total damage has been determined, a plan can be made for repair.

A complete and accurate damage diagnosis cannot be overstressed. Any inaccurately diagnosed damage on a vehicle will be uncovered during repair. When this happens, the repair method or procedure must be changed. The finished product will be less than satisfactory, resulting in the need for further repairs. Therefore, the best person for the body technician to talk to is the person who prepares the estimate.

Physical damage is rarely missed during an inspection by a competent estimator and body technician. However, the effects of the damage on unrelated systems and damage occurring next to the impacted part are sometimes overlooked. A visual inspection alone is generally inadequate with modern vehicles. Accident damage should be assessed carefully by measurements with the proper tools and equipment.

The following is a basic diagnostic procedure:

1. Know the vehicle construction type.
2. Visually locate the point of impact.
3. Visually determine the direction and force of the impact; once determined, check for possible damage.
4. Determine if the damage is confined to the body or if it involves functional parts (wheels, suspension, engine, and so on).
5. Systematically inspect damage to the components along the path of the impact. Find the point where there is no longer any evidence of damage. For example, pillar damage can be determined by checking the door fitting conditions.
6. Measure the major components (Figure 10–1). Check **body dimensions** (known correct body measurements of an undamaged vehicle) by comparing the actual measurements with the values in the repair manual or body dimensions chart.
7. Check for suspension and overall body damage with the proper equipment.

Vehicle damage conditions are diagnosed from the procedures given in Table 10–2.

Before starting damage evaluation, keep the following safety pointers in mind:

- Once the car is in the shop, check for broken glass edges and jagged metal. Edges of broken glass should be masked with tape and labeled "DANGER." Sharp jagged metal edges can be taped, but it is better to grind them down with a portable power grinder or a file.
- If fluids, such as lubricants or transmission fluid, are leaking from the vehicle, wipe them up to reduce the possibility of someone slipping on the floor.
- Remove the gas tank before welding or cutting is begun on the vehicle. Never just drain the tank because remaining fumes are explosive.
- Disconnect the battery to open the electrical system circuit. This will avoid the possibility of a charge igniting flammable vapors. It also protects the electrical system.
- Make the damage diagnosis in a well-lit shop. If the damage involves functional or mechanical parts, a detailed inspection of the underbody, using a lift or a bench is required.
- Other safety measures when diagnosing repair work would include any measures usually

FIGURE 10–1 Measuring is the most important step in major body repair work. Special measuring tools like this one are often required. *(Courtesy of Wedge Clamp Corp.)*

TABLE 10-2: FACTORS TO CONSIDER IN THE DIAGNOSIS OF COLLISION DAMAGE

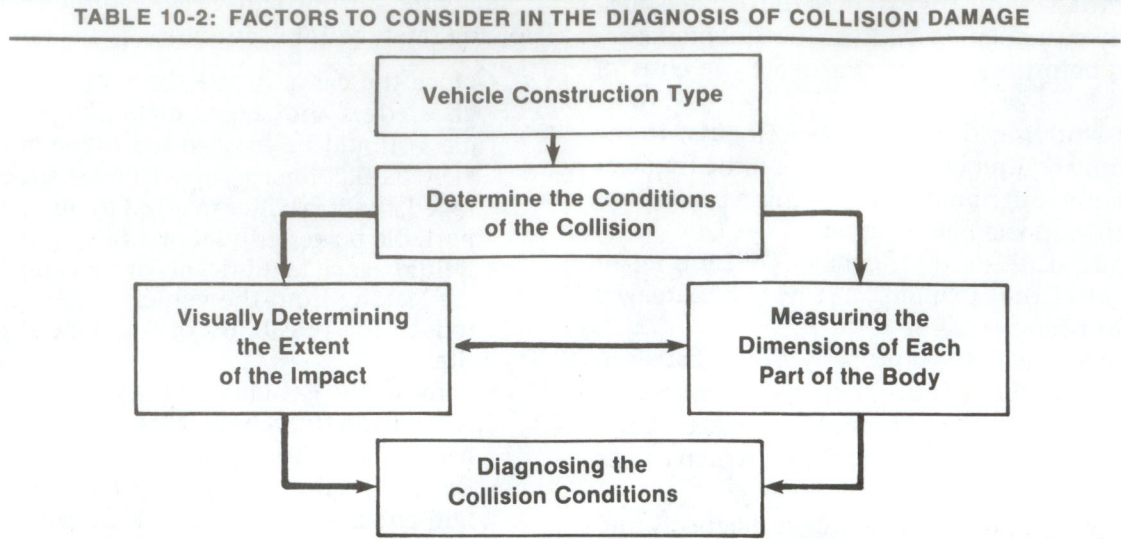

followed while in the body shop as detailed in Chapter 1. Being conscious of what one is doing is the most important precaution to remember.

10.1 IMPACT AND ITS EFFECTS ON A VEHICLE

The body of a vehicle is designed to withstand the shocks of normal driving and to provide safety for the occupants in the event of a collision. Special consideration is given to designing the body so that it will collapse and absorb the maximum amount of energy in a severe collision, while protecting its occupants. For this purpose, the front body and rear body are to some extent made to deform easily, forming a structure that absorbs impact energy.

During a head-on collision with a barrier at 30 miles per hour (mph) or 48 kilometers per hour (km/h) the engine compartment compacts by about 30 to 40 percent of its length. However, the passenger compartment is only compacted 1 to 2 percent of its length.

When diagnosing collision damage, you must compare known good measurements with the ones on the vehicle being repaired. With minor damage, this might be as simple as a visual inspection. With major damage, this will involve using complex measuring equipment (Figure 10–2).

To correctly analyze damage on a unibody vehicle, the entire structure must be considered. To do this, it is necessary to be able to take proper measurements to locate damage. It will also help to plan where to pull.

Measurement gauges are special tools used to check specific frame and body points. They allow you to quickly measure the direction and extent of vehicle damage.

Specific measurement points or locations on the frame or body are given for making measurements. They might be holes, specific bolts, nuts, panel edges, or other locations on the vehicle. To repair a badly damaged vehicle, you must restore these reference points to their factory dimensions while reference points in the undamaged area remain in their correct locations.

Therefore, the collision repair technician must work with the whole vehicle. This is done by measuring and recording dimensional changes. The most widely accepted method of checking body dimensions is to use the charts supplied in the body dimension manuals.

When the collision damage has been identified using the proper identification and analysis procedures, anyone skilled in the mechanics of collision damage repair is capable of repairing the car or truck.

The terms "control point" and "reference point" have different meanings. The **control points** used in manufacturing are not necessarily the same as the reference points the collision repair technician uses to measure the vehicle. **Reference points** refer to the points, bolts, holes, etc., used to give unibody and frame dimensions in body specification manuals. The distance between reference points can be measured with either a tram bar or a tape measure.

DETERMINING THE CONDITIONS OF THE COLLISION

The extent of the damage differs depending on the conditions at the time of the accident. To put it another way, the damage can be partly determined by understanding how the collision occurred. To understand the circumstances of the collision, it would be necessary to contact persons directly involved or eyewitnesses. Such a task would undoubtedly be a waste of time. However, it is possible for the person

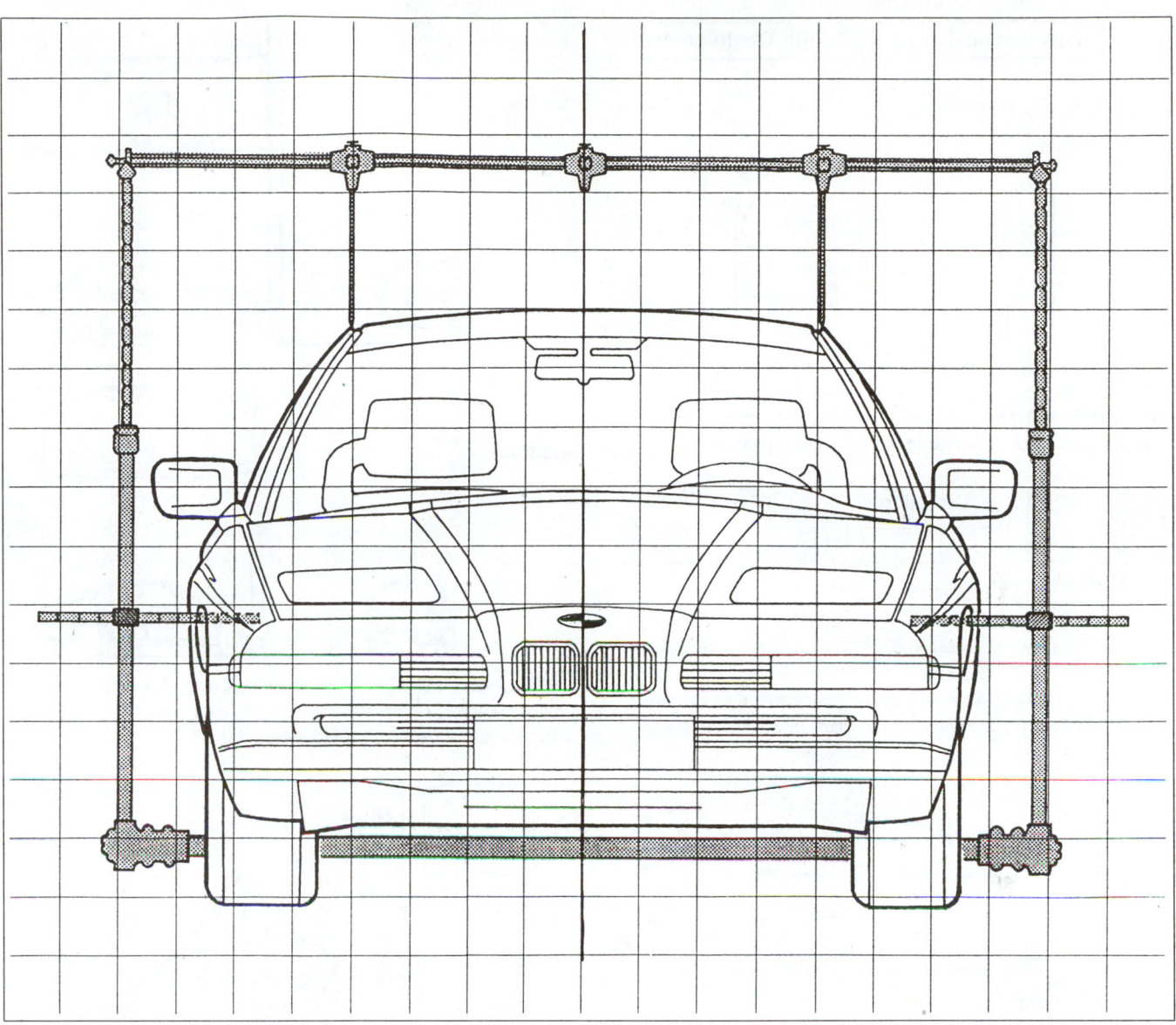

FIGURE 10-2 During major body or frame damage, you may have to use measuring equipment to compare known good dimensions on the vehicle with those on the damaged car. This will let you know what must be done to straighten the damage. *(Courtesy of Wedge Clamp Corp.)*

responsible for making the estimate to get a direct response from the customer. This method of damage assessment is sometimes necessary in order to estimate the cost of the repair. Therefore, the body technician should talk to the estimator to help analyze the methods of repair needed.

The body technician should know the following items:

- The size, shape, position, and speed of the vehicles involved in the collision
- Speed of the vehicle at the time of the collision
- Angle and direction of the vehicle at the time of the impact
- The number of passengers and their positions at the time of the impact

A good body/frame or structural technician can usually determine what actually happened in the collision to cause the damage. Because of the predictable nature of a driver's reactions before a collision, certain types of damage almost invariably occur in a rather predictable pattern and sequence.

If a driver's first reaction (Figure 10–3) is to turn away from the danger, the vehicle will be forced to take the hit on the side. If the driver's reaction is to slam on the brakes (Figure 10–4), the direction of impact will be frontal. A frontal collision where the point of impact is high on the vehicle could cause the cowl and roof to move rearward and the rear of the vehicle to move downward. Or, if the point of impact is low at the front, the inertia of the body mass could cause the rear of the vehicle to distort

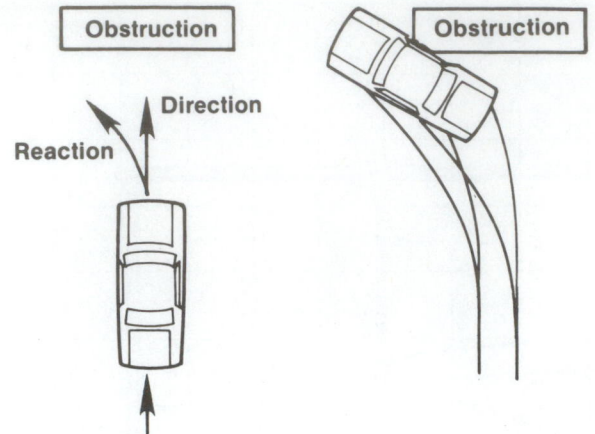

FIGURE 10-3 A driver's first reaction is to turn away from danger, forcing the hit to a side, causing sidesway.

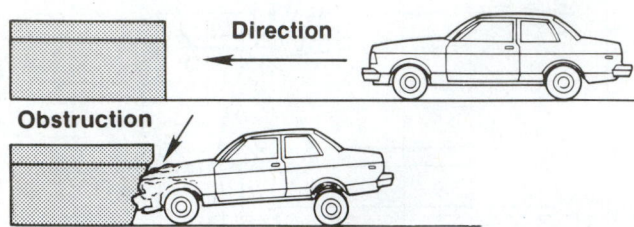

FIGURE 10-4 A driver's second reaction is to slam on the brakes, forcing the front end to drive down, causing sag.

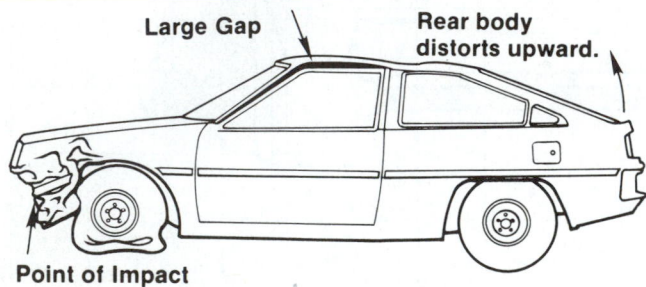

FIGURE 10-5 A hard front impact (primary damage) often causes secondary damage (buckles in the roof, for example).

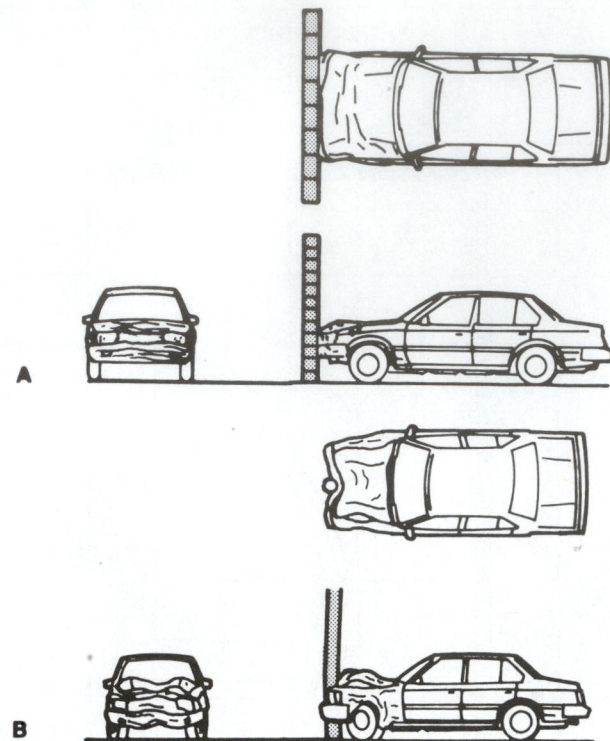

FIGURE 10-6 (A) This is an example of a large impact surface area (a brick wall) and (B) an example of a small impact surface area (a telephone pole).

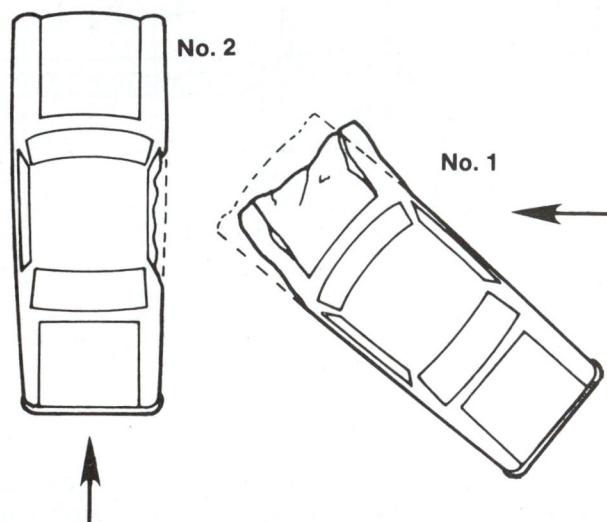

FIGURE 10-7 Note a typical broadside collision. If car No. 2 is moving, car No. 1 will have its front end mashed back and to one side.

upward, forcing the roof forward. This would leave an excessively large opening between the front upper part of the door and the roofline (Figure 10–5).

Given vehicles with similar weights traveling at about the same speed, vehicle damage will vary significantly depending on what is struck, for example, a telephone pole or a wall. If the impact is spread over a larger area, such as a wall, the damage will be spread over a wide body surface area (Figure 10–6A).

Conversely, the smaller the area of impact, such as a telephone pole, the greater the severity of the damage in a smaller area. In this example (Figure 10–6B), the bumper, hood, radiator, and so forth have been severely deformed. The engine has been pushed back and the effect of the collision has extended as far as the rear suspension.

Another consideration is when one car hits another while moving (Figure 10–7). If car Number 1 drives into the side of car Number 2 while Number 2 is moving, the motion of the first car will drive the front end of the car back. Simultaneously, the motion of car Number 2 will also drag that same front end to the side. There is only one collision but the damage is in two directions.

On the other hand, there might be two collisions in only one direction. This is a fairly common occurrence in freeway pile-ups. A car that collides with another car and then leaves the road to hit a pole or guard rail ends up with two completely separate types of damage.

There are many other variables and possible combinations of damage. It is important to determine what actually happened before an accurate diagnosis can be made. Get as many facts as possible. Combine them with physical measurements and centerline gauge readings to determine exactly the collision repair procedure that should be taken.

A little extra time spent evaluating damage can save many hours in the overall repair time. "Think time" saves "work time" and increases profits.

INFLUENCE OF IMPACT ON A BODY-OVER-FRAME VEHICLE

Figure 10–8 illustrates a body with a perimeter frame with its built-in collapsible sections. The circled areas indicate the softer sections of the frame designed to absorb the major impact of a collision.

The body is attached to the frame by rubber mounts. The *rubber body mounts* reduce the effects of road shocks traveling from the frame to the body. This quiets the ride in the passenger compartment. In the event of a large impact or collision, the bolts of the rubber mounts might bend, resulting in a gap between the frame and the body. Also, depending on the magnitude and direction of impact, the frame might experience damage while the body does not.

Frame deformation can be broken down into five categories:

- **Sidesway damage.** Collision impacts that occur from the side often cause a *sidesway damage* or side bending frame damage condition (Figure 10–9). Sidesway usually occurs in the front or rear of the vehicle. Generally, it is possible to spot sidesway by noting if there are buckles on the inside of one rail and buckles on the outside of the opposite side rail (Figure 10–10).

  Sidesway can be recognized by abnormalities, such as a gap at the door on the long side (Figure 10–11) and wrinkles on the short side. Look for impact damage obvious from the side,

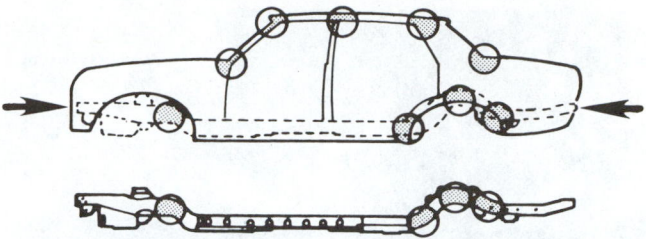

FIGURE 10-8 These are typical perimeter frame collapsible sections.

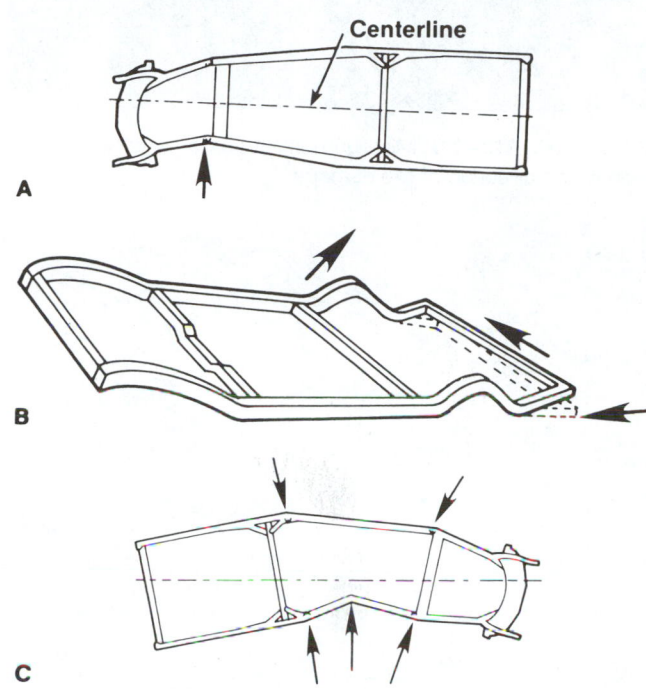

FIGURE 10-9 Study various sidesway damages: (A) sidesway at the front of the frame caused by a front end collision; (B) rear sidesway; and (C) double sidesway on the frame's outer section.

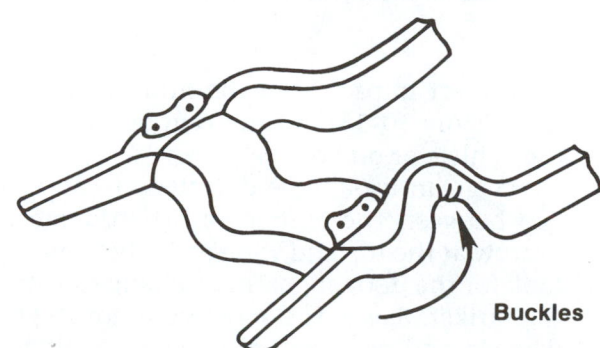

FIGURE 10-10 A good clue to frame misalignment is buckles in the crush zones of the frame rails.

such as the hood and deck lid do not fit into the proper opening.

- **Sag damage.** *Sag damage* is a condition where one area, often the cowl area, is lower than normal (Figure 10–12). The structure has a swayback appearance. Sag damage generally is caused

FIGURE 10-11 Misalignment of doors gives clues to the extent and direction of the damage.

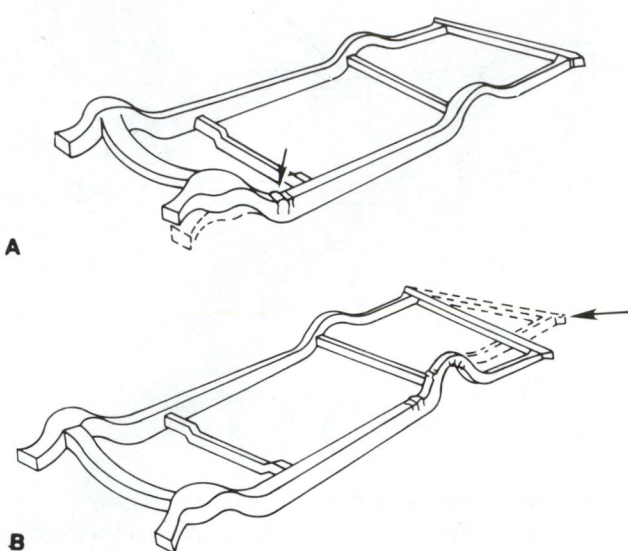

FIGURE 10-12 (A) Note the sag condition on the left front frame section and (B) rear end sag.

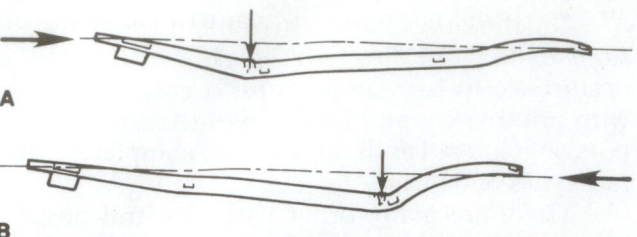

FIGURE 10-13 (A) Side rail sag resulted from a front end collision; (B) side rail sag resulted from a rear end collision.

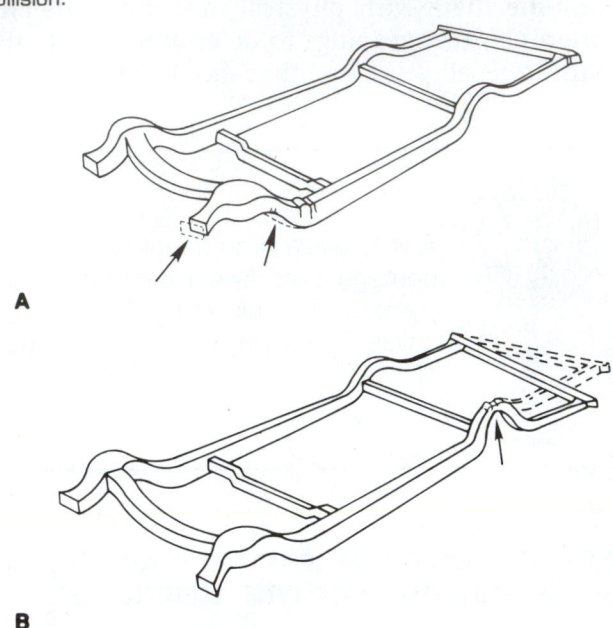

FIGURE 10-14 (A) Note the mash damage on the left front side rail; (B) mash damage on the left rear side rail.

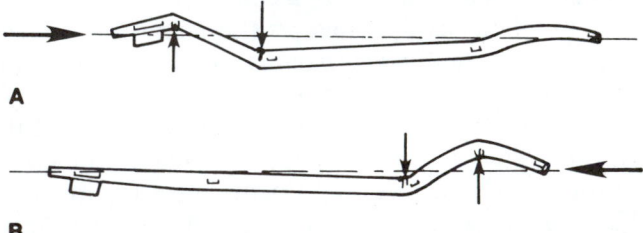

FIGURE 10-15 (A) A frame mashed and buckled from a front end collision. (B) A frame mashed from a rear end collision.

by a direct impact from the front or from the rear (Figure 10–13). It can occur on one side of the vehicle or on both sides.

Sag can usually be detected visually by a gap between the fender and the door being narrow at the top and wide at the bottom. Also look for the door appearing to hang too low at the striker. Sag is the most common type of damage and occurs in most vehicles that are involved in an accident. Enough sag can be present in the frame to prevent body panel alignment even though wrinkles or kinks are not visible in the frame itself.

• **Mash damage.** *Mash damage* is present when any section or frame member of the car is shorter than factory specifications (Figure 10–14). Mash is usually limited to forward of the cowl or rearward of the rear window. Doors might fit well and appear to be undisturbed. Wrinkles and

severe distortion will be found in fenders, hood, and possibly frame rails or horns. The frame will rise upward at the top of the wheel arch causing the spring housing to collapse (Figure 10–15). With mash damage, there is very little vertical displacement of the bumper. The damage results from direct front or rear collisions.

• **Diamond damage.** *Diamond damage* is a condition where one side of the car has been moved to the rear or front causing the frame and/or body to be out of square (Figure 10–16). The resulting shape will be a figure similar to a parallelogram and is caused by a hard impact on a corner or off-center from the front or rear. Diamond damage affects the entire frame, not just the side rails. Visual indications are hood and trunk lid misalignment. Buckles might appear in the quarter panel near the rear wheelhousing or at the roof to quarter panel joint. Wrinkles and buckles probably will appear in the passenger compartment and/or trunk floor. There usually will be some mash and sag combined with the diamond.

- **Twist damage.** *Twist damage* (Figure 10–17) is a condition where one corner of the car is higher than normal; the opposite corner might be lower than normal. Twist can happen when a car hits a curb or median strip at high speed. It is also common in rear corner impacts and rollovers.

    A careful inspection reveals no apparent damage to the sheet metal. However, the real damage is hidden underneath. One corner of the car has been driven upward by the impact. Most likely, the adjacent corner is twisted downward. If one corner of the car is sagging close to the ground as though a spring is weak, the car should be checked for twist.

Diamond damage (Figure 10–18) usually occurs when the vehicle is struck off-center. However, a frame will rarely experience deformation involving the whole frame.

The most frequent order of occurrence of damage is:

1. Sidesway
2. Sag
3. Mash
4. Diamond
5. Twist

As described in Chapter 11, the most important rule in body/frame alignment is REVERSE DIRECTION AND SEQUENCE. This means, to correct collision damage on a conventional vehicle, the pulling or pushing of the damaged area must be done in the opposite direction of impact. The repair must also be made in the reverse sequence that it happened.

Unfortunately, most accidents result in a mix of one or more of these damage problems. Sidesway and sag frequency occur almost simultaneously. Some damage affects the frame's cross members, especially the front member. In a rollover accident, for example, the front cross member on which the motor mounts are attached is often forced out of shape because of the engine's weight. This will result in a sag

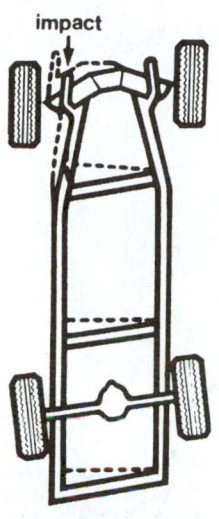

diamond

FIGURE 10-16 A diamond condition affecting the entire frame alignment resulted from a hard impact from the front, but only on one side. *(Courtesy of Guy-Chart)*

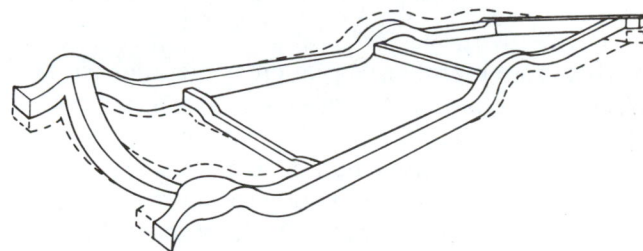

FIGURE 10-17 Twist conditions affect the entire frame alignment.

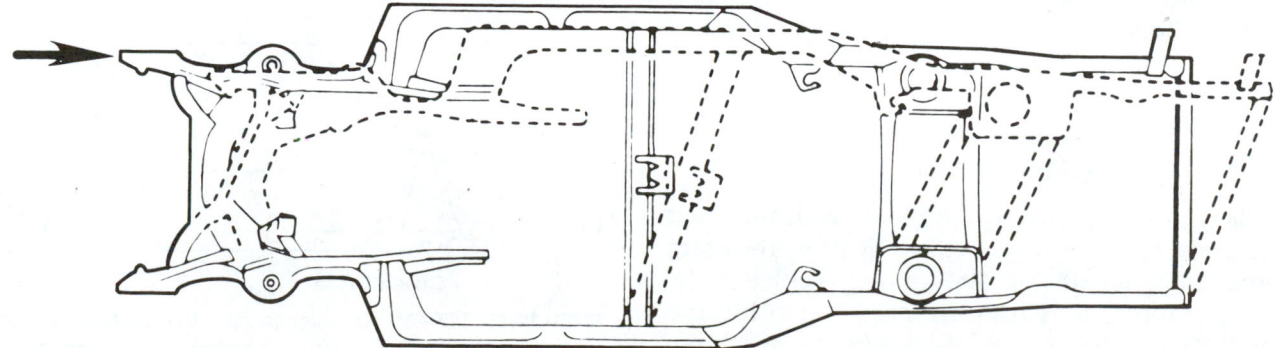

FIGURE 10-18 Study diamond conditions. Solid lines are the undamaged frame and dotted lines represent the damaged frame.

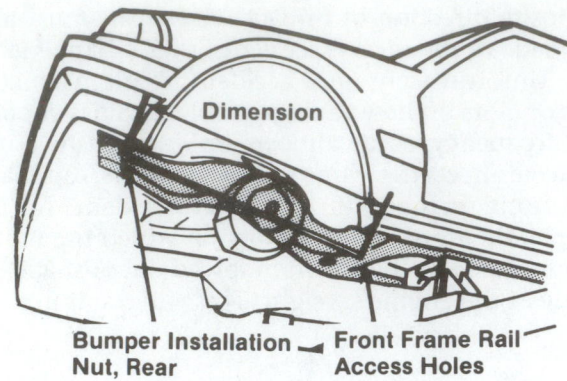

FIGURE 10-19 Measurements of undersurface dimensions are often needed.

increasing area of the unibody. This characteristic spreads the force until it is completely dissipated. Visualize the point of impact as the tip of the cone.

The centerline of the cone will point in the direction of impact. The depth and spread of the cone indicate the direction and area that the collision force traveled through the unibody. The tip of the cone is the **primary damage** area.

Since unibodies are structured entirely from the joining of pieces of thin sheet metal, the shock of a collision is absorbed by a large portion of the body shell. The effects of the impact shock wave as it travels through the body structure (Figure 10–21) is called **secondary damage**. This damage is toward the

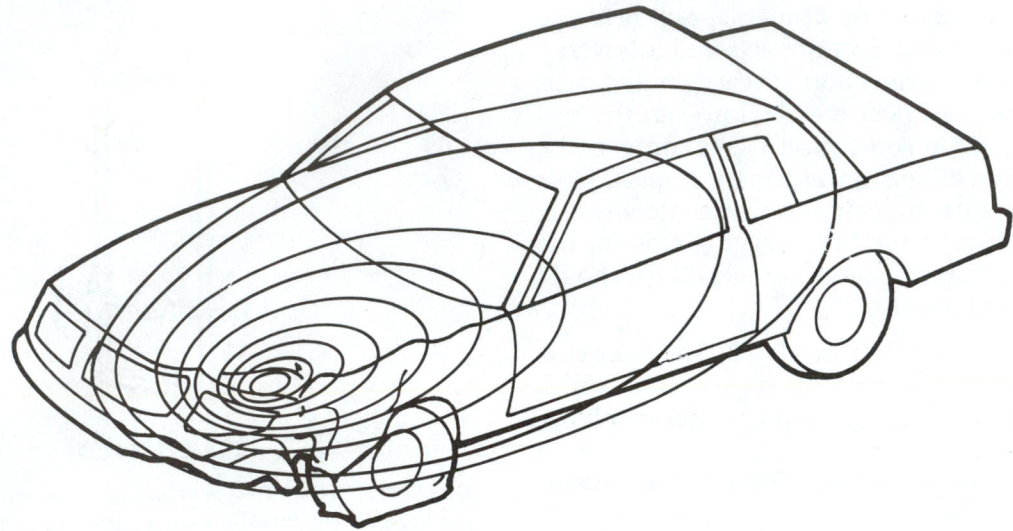

FIGURE 10-20 The best way to describe an impact effect on unibodies is the cone concept. Forces move out in a cone shape away from the point of impact.

of this cross member. While cross member damage is rather rare, it must be corrected since its alignment can affect the handling of the vehicle.

A deformed frame can be inspected by comparing the space between the body rocker panel and the front and back of the frame. Also compare the space between the front fender and the front and back of the wheel hub (Figure 10–19). To inspect front frame deformation, compare the left and right measurements from the rear hole for the front bumper to a point on the front frame rail.

IMPACT EFFECT ON UNIBODY VEHICLES

The damage that occurs to a unibody car as the result of an impact can best be described by using the **cone concept** (Figure 10–20). The unibody vehicle is designed to absorb a collision impact. When hit, the body folds and collapses as it absorbs energy. As the force penetrates the structure, it is absorbed by an ever

inner structure of the unibody or toward the opposite end or side of the vehicle (Figure 10–22).

To provide some control on secondary damage distortion and to give a much safer compartment for passengers, a unibody vehicle is designed with crush zones or areas at the front and rear (Figure 10–23).

Crush zones are engineered to collapse in a predetermined fashion to protect the vehicle's passengers and to localize damage. The effects of the

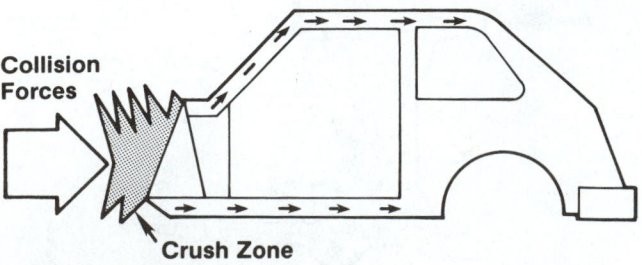

FIGURE 10-21 Collision energy often dissipates around the passenger compartment through the components. *(Courtesy of Babcox Publications)*

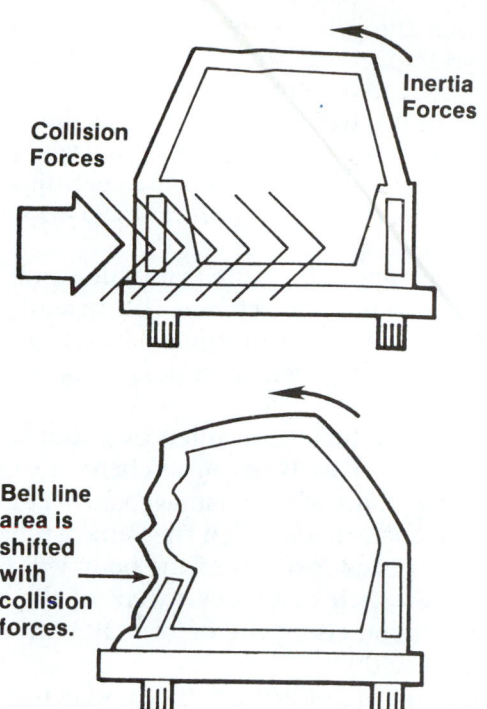

FIGURE 10-22 The roof shifted toward the side of impact because of weight/mass inertia. *(Courtesy of Babcox Publications)*

impact shock wave to the body structure are reduced. In other words, front impact shocks are absorbed by the front body and crush zones (Figure 10–24). Rear shocks are absorbed by the rear body and crush zones (Figure 10–25). Side shocks will be absorbed by the rocker panel, roof side frame, center pillar, and door.

Impact damages on unibody vehicles can be described in the following ways.

Frontal unibody damage results from a head-on collision with another object or vehicle. The impact of a collision depends upon the vehicle's weight, speed, area of impact, and the source of impact. In the case of a minor impact, the bumper is pushed back, bending the front side members, bumper stay or bracket, front fender, radiator support, radiator upper support, and hood lock brace (Figure 10–26). If the impact is further increased, the front fender will contact the front door. The hood hinge will bend up to the cowl top. The front side members may also buckle into the front suspension cross member, causing it to bend. If the shock is great enough, the front fender apron and front body pillar (particularly the front door hinge upper area) will be bent, which will cause the front door to drop down. In addition, the front side members will buckle and the front suspension member will

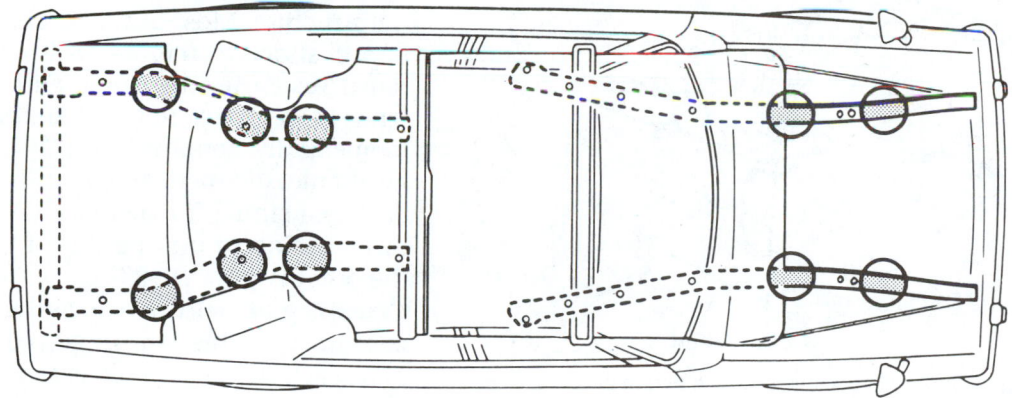

FIGURE 10-23 These are typical unibody impact absorbing areas.

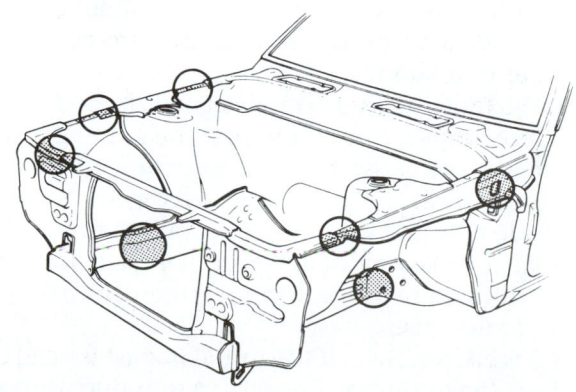

FIGURE 10-24 Here are common locations for unibody front crush zones. They should be inspected when analyzing damage. *(Courtesy of Toyota Motor Corp.)*

bend. The dash panel and front floor pan may also bend to absorb the shock.

If a frontal impact is received at an angle, the attachment point of the front side member becomes a turning axis. Lateral as well as vertical bending occurs. Since the left and right front side members are connected together through the front cross member, the shock from the impact is sent from the point of impact to the front side member of the opposite side of the vehicle (Figure 10–27).

Rear unibody damage occurs when the vehicle is moving backward and hits something or is hit by another vehicle from behind. When the impact is comparatively small, the rear bumper, the back panel, trunk lid, and floor pan will be deformed. The quarter panels will also bulge out. If the impact is severe enough, the quarter panels will collapse to the base of the roof panel. On four-door vehicles, the center body pillar might bend. Impact energy is absorbed by the deformation of the above parts and by the deformation of the kick-up of the rear side member.

Side unibody damage will cause the door, front section, center body pillar, and even the floor to deform. When the front fender or quarter panel receives a large perpendicular impact, the shock wave extends to the opposite side of the vehicle.

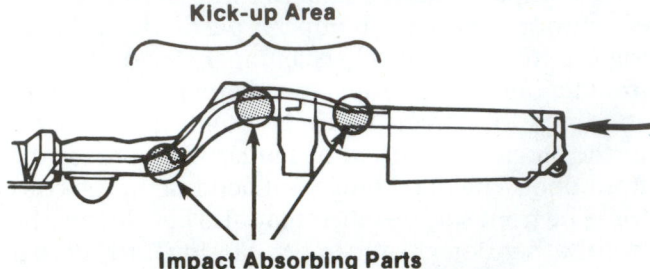

FIGURE 10-25 The rear side member impact absorbing areas should also be inspected for crumples.

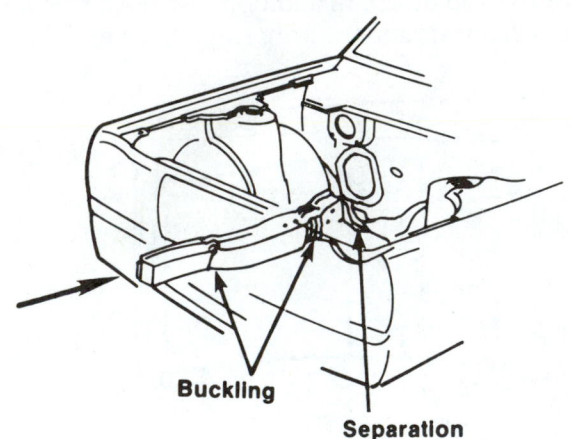

FIGURE 10-26 Buckling and separation action in a unibody vehicle indicate major damage. Cracked undercoating is also an indicator of more serious damage. *(Courtesy of Toyota Motor Corp.)*

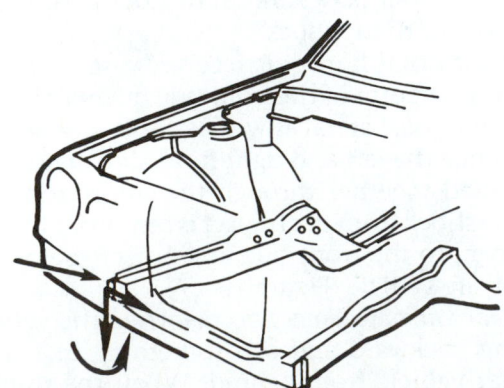

FIGURE 10-27 Both lateral and vertical bending movement of a unibody vehicle often happen during a collision. *(Courtesy of Toyota Motor Corp.)*

When the central area of the front fender receives an impact, the front wheel is pushed in. The shock wave extends from the front suspension cross member to the front side member. If the impact is severe, the suspension parts are damaged and the front wheel alignment and wheelbase may be changed. The steering gear or rack can also be damaged by side impacts.

Top impacts can result from falling objects or from a rollover of the vehicle. This type of damage not only involves the roof panel but also the roof side rail, quarter panels, and possibly the windows as well.

When a vehicle has rolled over and the body pillars and roof panels have been bent, the opposite ends of the pillars will be damaged as well. Depending on the manner in which the vehicle rolled over, the front or back sections of the body will be damaged too. In such cases, the extent of the damage can be determined by the deformation around the windows and doors.

The typical collision damage sequence (Figure 10–28) on a unibody structure is as follows:

- **Bending.** In the first microseconds of impact, a shock wave attempts to shorten the structure, causing a lateral or vertical bending in the central structure. Most of the forces that broadcast impact shock to remote areas occur at this instant. Since the structure is stiff and springy, it tends to snap back to its original shape—at least momentarily. Bending is usually indicated by the height measurement being out of tolerance. This damage—similar to sag in a conventional structure—can occur on one side of the car and not the other (Figure 10–29).

- **Crushing or collapsing.** As the collision event continues, visible crushing occurs at the point of impact. Impact energy is absorbed in the deforming structure (helping protect the passenger compartment). Remote areas might buckle, tear, or pull loose. Crush damage, which is similar to mash on BOF (body over frame) vehicles, is indicated by the length measurement being out of tolerance.

- **Widening.** In a well-designed unibody structure, impact forces reaching the passenger compartment cause the side structure to bow out away from the passengers (never in) distorting side rails and door openings. Widening is similar to sidesway damage in BOF vehicles and is indicated by the width measurement being out of tolerance.

- **Twisting.** Even if the initial impact is dead center, the secondary impact can introduce torsional loads that cause a general twisting of the structure. Unibody structural twisting, like twisting of a conventional vehicle frame, is usually the

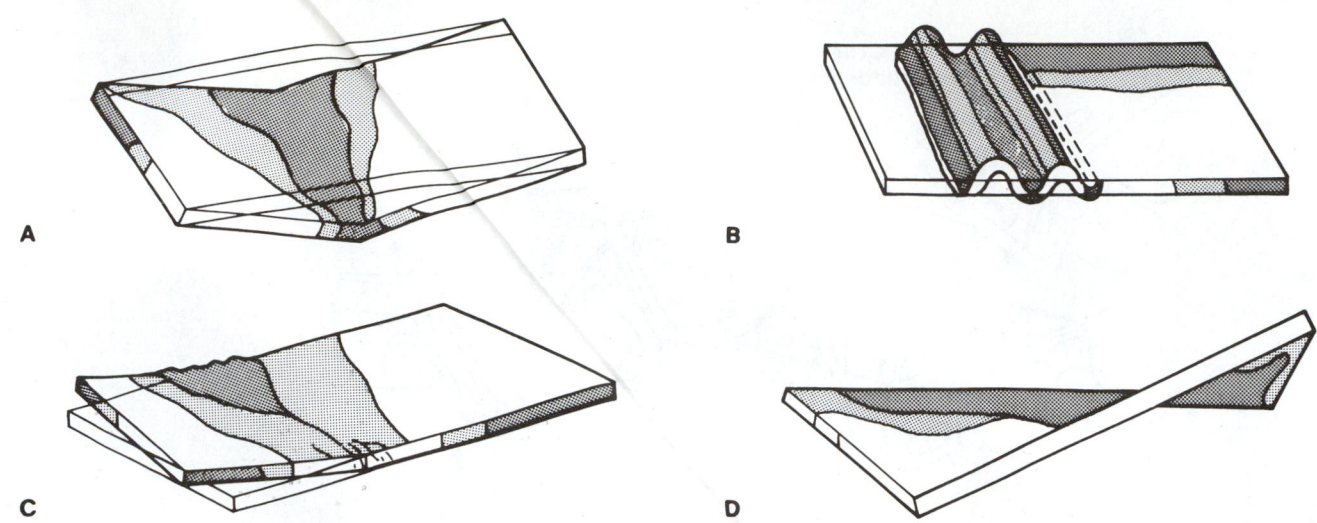

FIGURE 10-28 Study the types of unibody collision damage: (A) bending; (B) crushing or collapsing; (C) widening; and (D) twisting. *(Courtesy of Blackhawk Automotive Inc.)*

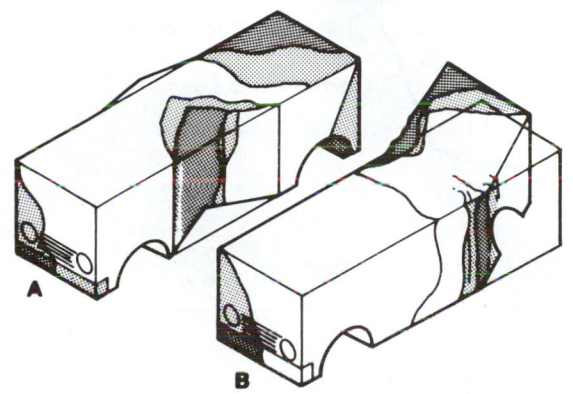

FIGURE 10-29 If a vehicle is viewed as a rectangular box, one can see how force moves through the structure: (A) center bending and (B) rear bending. *(Courtesy of Blackhawk Automotive Inc.)*

FIGURE 10-30 Visual inspection must be done carefully and repair notes made since procedures for pulling damage can be complex. *(Courtesy of Mitchell Manuals)*

last collision event. It is indicated by combinations of height and width measurements being out of tolerance.

There is a great similarity between the types of damage that can occur on body-over-frame and unibody vehicles. Unibody damage is often more complex. Note that a severe collision will not normally cause diamond damage on unitized cars. Also like conventional aligning, pulling secondary damage (last-in) so that it is corrected first (first-out) is the best way to correct damage to a unibody car. Secondary damage is identified by accurate measurement.

10.2 VISUALLY DETERMINING EXTENT OF IMPACT DAMAGE

Damaged parts show signs of structure deformations or fractures in most cases. When making a visual inspection, stand back from the vehicle to get an overall view. Estimate the size and direction of the impact (place where impact was received). Estimate how the impact was propagated and the damage sustained (Figure 10–30).

Also investigate whether or not there is any twisting, bending, or slanting of the vehicle overall. Look over the entire vehicle to try to determine where the damage occurred and whether or not all the damage was the result of the same collision.

Remember that impact force can flow through the vehicle and may damage many parts besides those at the point of impact. An impact force can pass easily through the strong portions of the body, finally ending up in the weak portions, damaging them also. Therefore, in searching for damage, inspection must be made along the path of propagation of the impact through the weak portions of the body.

You must look for the presence of strain, panel joint misalignment, cracks in and peeling of the paint film, cracked undercoat and sealer, and so on.

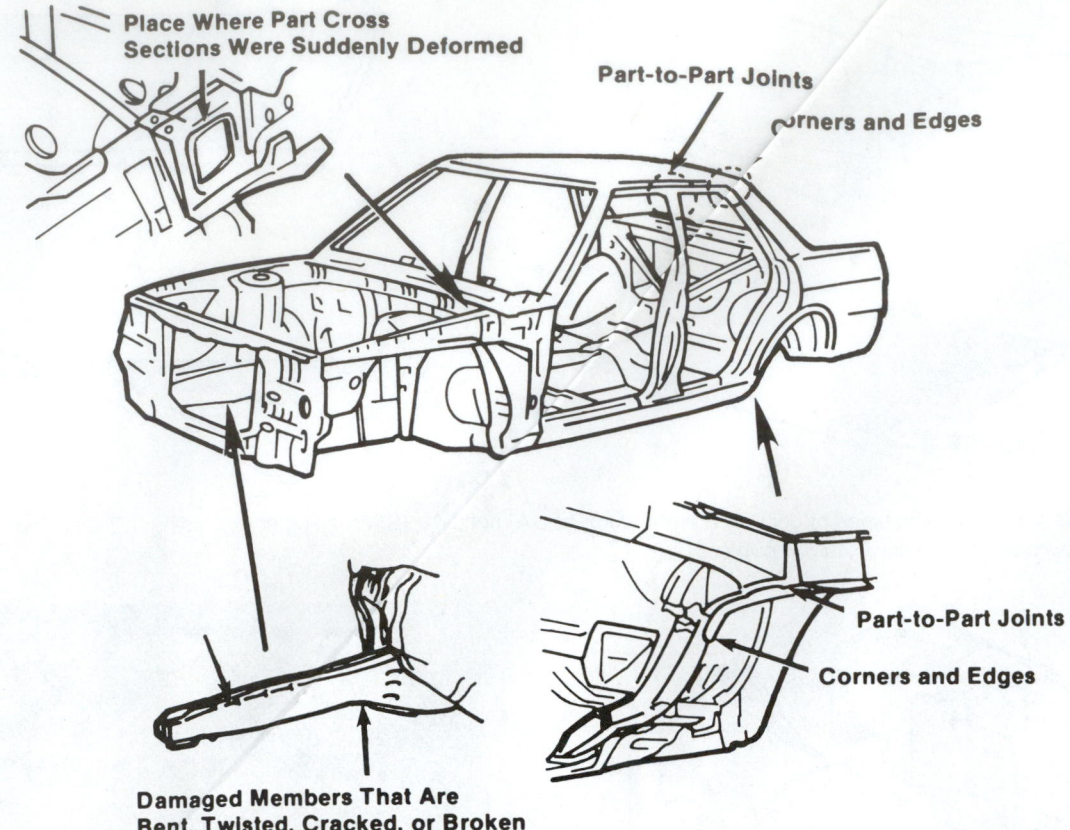

Place Where Part Cross Sections Were Suddenly Deformed

Part-to-Part Joints

Corners and Edges

Part-to-Part Joints

Corners and Edges

Damaged Members That Are Bent, Twisted, Cracked, or Broken

FIGURE 10–31 Study some of the parts that show damage easily. *(Courtesy of Toyota Motor Corp.)*

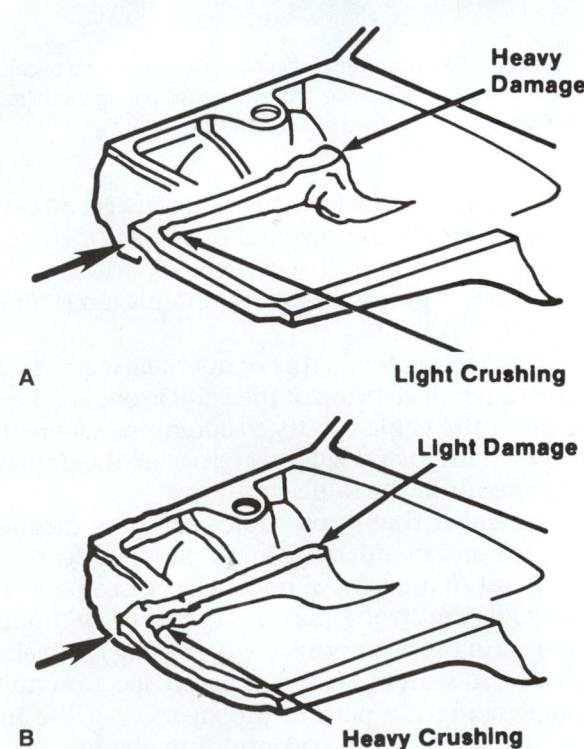

Heavy Damage

Light Crushing

A

Light Damage

Heavy Crushing

B

FIGURE 10–32 The same impact force can cause different types of damage depending upon the body structure of the vehicle. *(Courtesy of Toyota Motor Corp.)*

Damage can be detected easily by finding these types of symptoms (Figure 10–31):

- Areas where the cross sections of the components were suddenly deformed
- Parts that are broken or missing
- Gaps in strengthening materials, such as reinforcements or patches
- Part-to-part joints shifted
- Corners and edges of components misaligned

When surveying the extent of damage to the frame components, such as side members, it is easier to locate the damage on the concave side of the component. The concave side might appear as a sharp dent or kink rather than a minor bulge that would appear on the opposite side of the member.

A body is designed so that the energy received during impact travels along a predetermined path. Energy flow starts at the point of impact and flows through the structure until all the energy has been dissipated (Figure 10–32). Therefore, the evidence of damage will usually be greater near the point of impact because the extent of damage is reduced as the energy is dissipated into the adjacent structure. However, in some cases, the energy is passed through the impact point (with little evidence of damage showing) and is propagated to a point that is deep within the body.

INSPECTING CLEARANCE AND FIT OF EACH PART

Checking the door alignment makes it easy to determine if the body pillar has been damaged. Simply open and close the door, observing its alignment and action (Figure 10–33).

In the event of a front end collision, it is important to check the clearances and level differences between the rear doors and quarter panels and rocker panels. Another good method is to compare the clearances on the left and right sides of the vehicle.

Hinges on a vehicle wear over a period of time, and doors tend to drop down. This is especially true of the door on the driver's side, which is opened and closed quite frequently. Inspection should be made with the vehicle on a level shop floor or drive-on rack. If you place the car on a lift or stands, the fit of the doors can be affected by the flexibility of the body.

INSPECTING FOR INERTIA DAMAGE

In the case of heavy objects, such as engines mounted on rubber mounts, inertia becomes a powerful force during a collision. Inspect for damage to the mounts or to surrounding parts and panels. During a collision, the powerful impact usually causes the body and frame to become misaligned, damaging the body isolator mountings.

Inspect all mounting hardware for signs of inertia damage. With a unibody vehicle, look for a dent in the roof. On pickup trucks, inspect for an out-of-parallel condition between the cab and the bed.

INSPECTING FOR DAMAGE FROM PASSENGERS AND LUGGAGE

Passengers and luggage can cause secondary damage to the vehicle as a result of inertia during a collision. The damage will vary depending on the position of the passengers and the severity of the impact. Parts with a high-damage frequency are the instrument panel, steering wheel, steering column, and seat backs. Luggage in the trunk has also been known to cause damage to the body quarter panels.

10.3 MEASUREMENT OF BODY DIMENSIONS

Measuring is critical to the success of any major repair collision job regardless of the type of body structure. But with unibody vehicles, measurements are vital to successful repair because the steering and suspension are mounted to the body structure. In addition, some of the suspension geometry is built into

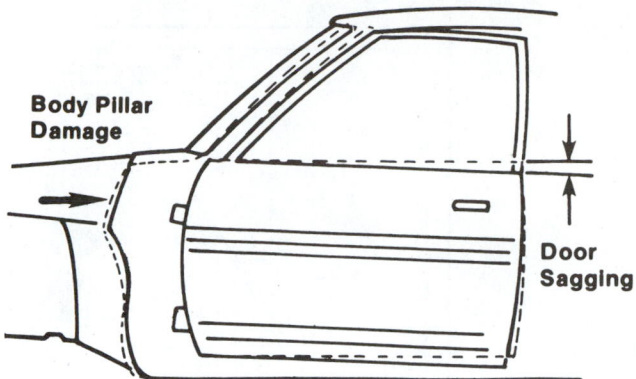

FIGURE 10–33 Inspection of door alignment will give valuable information about damage.

the body structure. As a result, the angles of caster and camber often have a fixed (nonadjustable) value. Body damage often seriously affects suspension geometry. The rack-and-pinion control box for the steering assembly is also mounted to a panel, resulting in a fixed relationship to the steering arms. The mechanical components, engine, transmission, and differential are all mounted directly to body members or to cradles supported by body members (panels or integral rails).

A distortion of any of these measuring points will change steering or suspension geometry, or misalign mechanical components. This can result in improper steering and handling, vibration and noise in the drivetrain, and excessive wear of tie rod ends, tires, rack-and-pinion assemblies, universal joints, or other drive or steering components. To maintain proper steering, handling, and drivability, repair tolerances must be held to within a maximum value of less than $1/8$ inch, or 3 millimeters.

BODY DIMENSIONS CHARTS

Accurate damage assessment can be made at specific points on the body using a **body dimensions chart**. The body dimensions chart gives measurement points and measurement specifications for a specific type of vehicle. You would need to find the chart for the specific make and model vehicle being repaired. The chart information will enable you to use measurement tools to compare the damaged vehicle to known good measurements (Figure 10–34).

In the body dimensions chart, measurements are based on the diagonal line measuring method (Figure 10–35). Engine compartment and body dimensional data should be compared to the chart and recorded.

Measurement points and tolerances are determined by inspection of the damaged area. Normally, in front end collisions that cause slight amounts of door sag, the damage does not extend beyond the center of the vehicle, so measurement in the rear section is not necessary. In a situation where a large

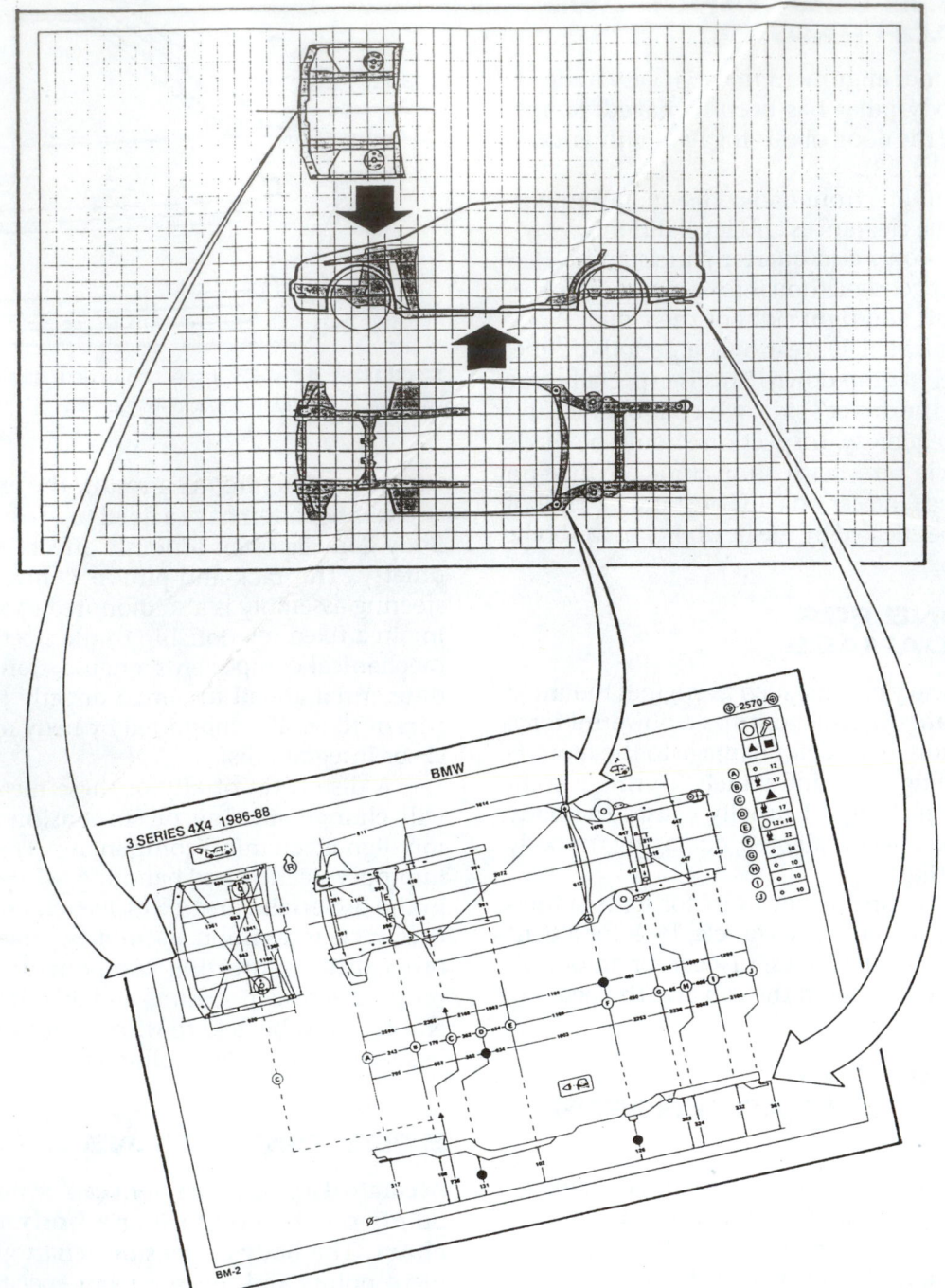

FIGURE 10-34 Compare spec values for known good measurements with your measurements at specific points on the vehicle for comparison. If yours are different, you know which direction and how far to power straighten the frame or body structure. *(Courtesy of Wedge Clamp)*

impact has occurred, many measurements must be taken to assure proper alignment procedures. However, taking and recording too many measurements may cause unnecessary confusion.

VEHICLE MEASURING BASICS

In unibody construction, each section should be checked for diagonal squareness by comparing diagonal lengths. Length and width should also be compared. The center section should be used as a base when reading structural alignment. All measurements and alignment readings should be taken relative to the center section.

Start measuring in the center or middle section. If it is not square, then move to the undamaged end of the vehicle to find three correctly positioned reference points.

Keep in mind that to accurately measure a vehicle, you must start with at least three reference points you know are right. The way to do this is to check the squareness of the vehicle. If the vehicle is not symmetrical, refer to the dimensions chart for correct measurements.

MEASUREMENT IMPORTANCE

In the entire repair process of both conventional and unibody cars, it is not possible to overemphasize the importance of measuring. A vehicle cannot be satisfactorily repaired unless all of the major manufacturing control points in the damaged area are returned to the manufacturer's specifications. To achieve this, the body technician must:

> Measure accurately
> Measure often
> Recheck all measurements

Because of the importance of measuring, many kinds of equipment have been developed and marketed by automotive equipment manufacturers strictly for the purpose of providing the capability to measure quickly and accurately. While there are a number of styles of measuring equipment that can

be found in body shops, most of it can be divided into five basic systems:

> Gauge measuring system
> Universal measuring system
> Dedicated fixture system
> Universal/Laser
> Computer/Electronic

10.4 GAUGE MEASURING SYSTEMS

The tram gauge, the centering gauge, and the MacPherson strut centerline gauge can be used separately or in conjunction with one another. The *tram gauges* are scaled rods used for measurement, while the *centering gauges* are metal rods used to check for misalignment. Supported by suspension system strut tower domes, the centerline gauge allows visual alignment of the critical control points of unibody vehicles. The tram centering and strut centerline gauges are available as a unit (Figure 10–36) or as separate diagnostic tools.

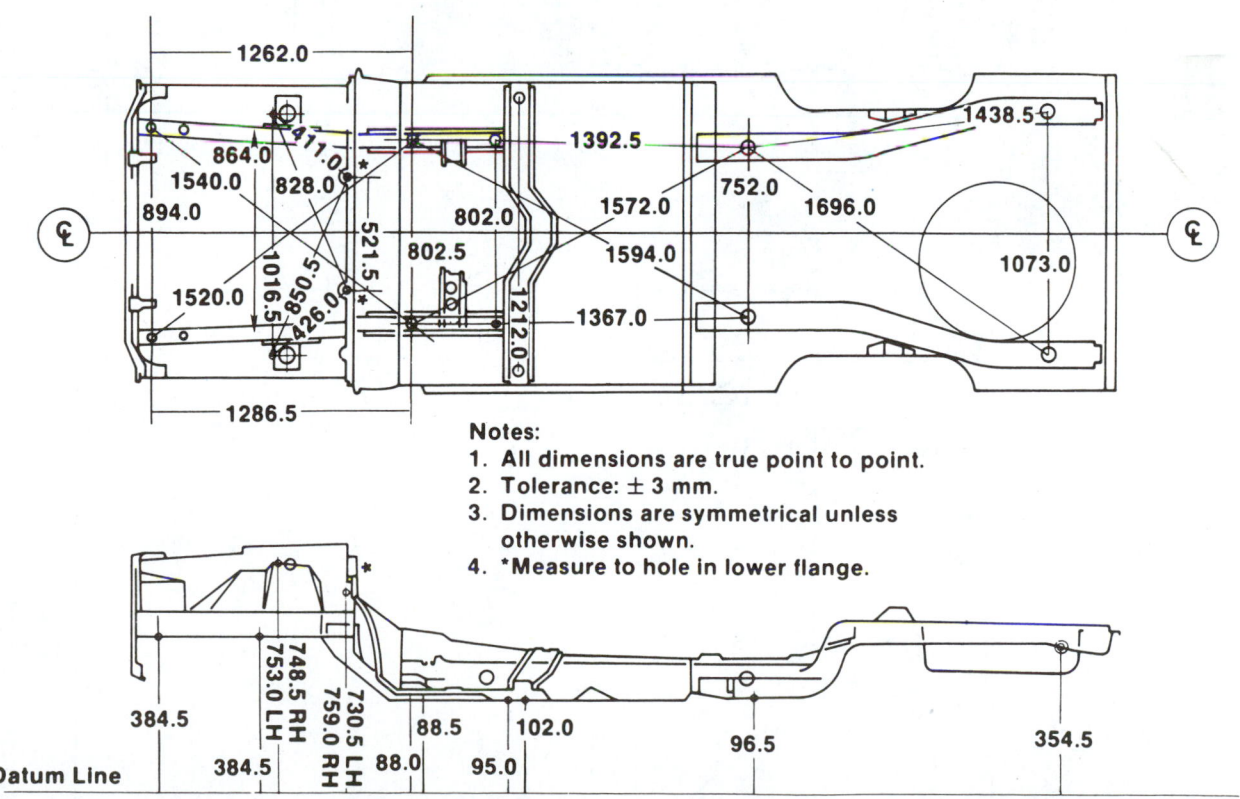

FIGURE 10–35 Read through the underbody/frame dimensions and specifications as taken from a typical automobile manufacturer's manual. (A) One auto manufacturer type. (B) Equipment manufacturer example. Find points on the drawing that you want to measure. Write the spec down in an analysis chart. Then measure any possible damage areas and compare them to good, unhurt measurements. Try using the metric conversion chart in this book's appendix to find U.S. equivalents. *(Courtesy of Ford Motor Co.)*

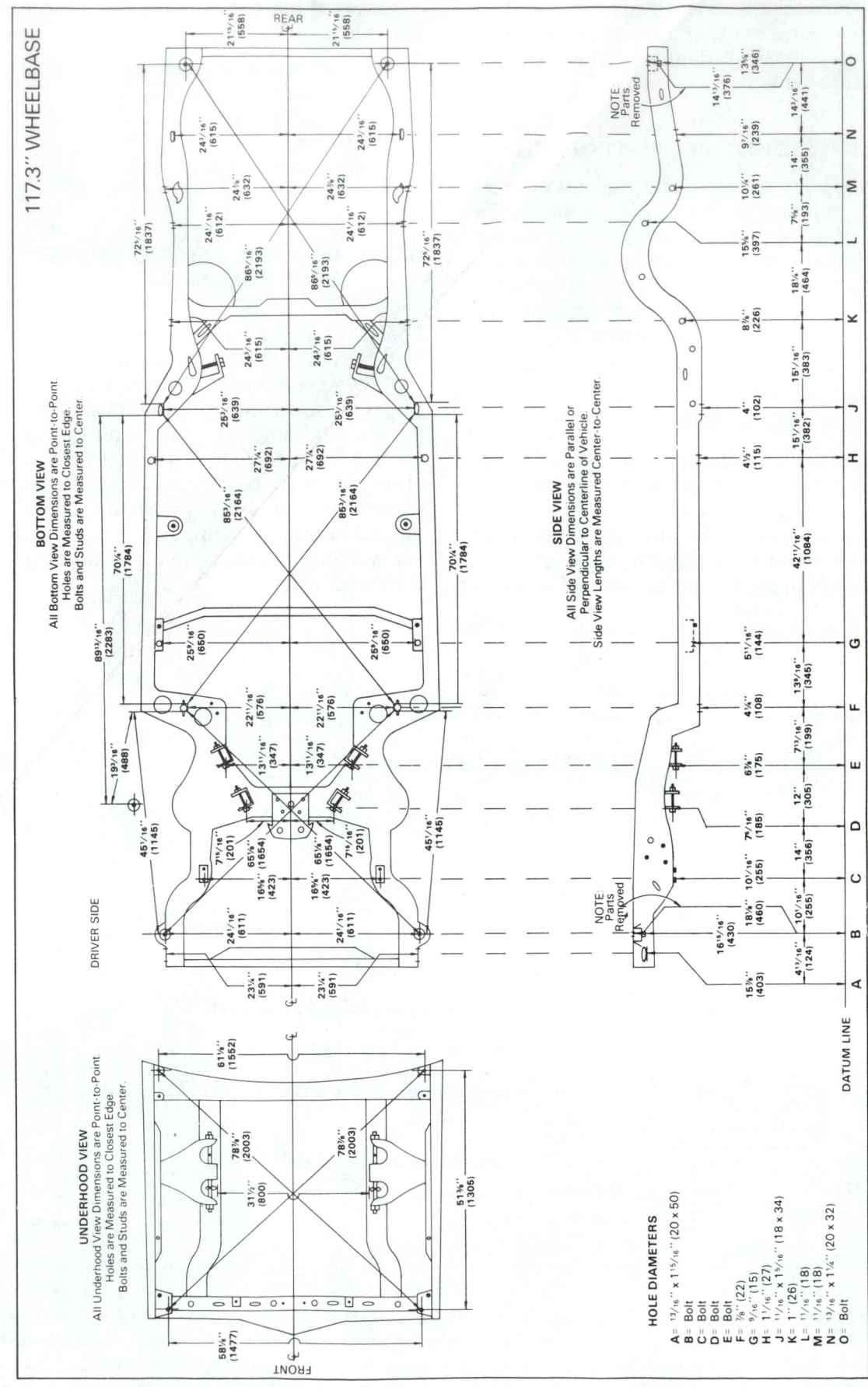

FIGURE 10-35 (Continued)

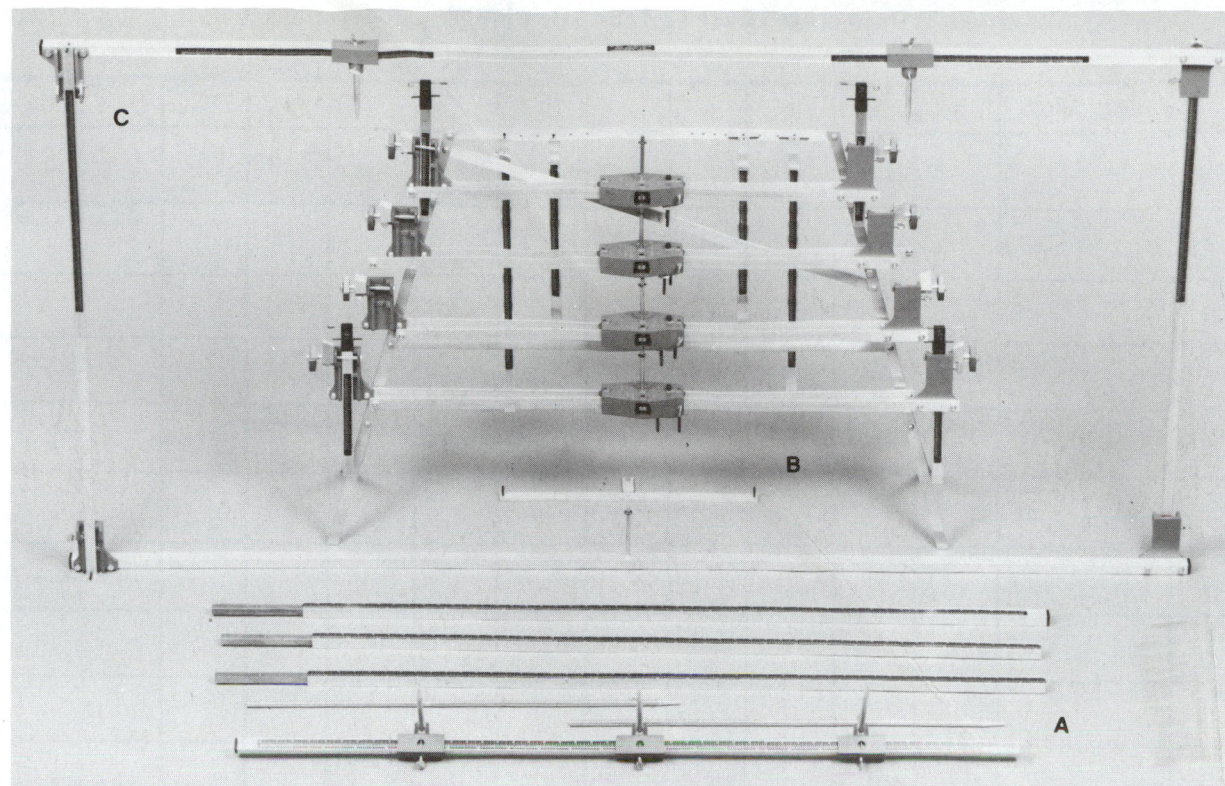

FIGURE 10-36 Here is a typical gauge set: (A) tram gauges; (B) centering gauges; and (C) strut centerline gauge. *(Courtesy of Arn-Wood Co., Inc.)*

Another gauge similar to the tram type is the tracking gauge. This gauge is used to check alignment of the front and rear wheels. If the front and rear wheels are not in alignment, the vehicle will not handle properly.

10.5 TRAM GAUGES

The tram gauge (Figure 10–37) measures one dimension at a time. Each dimension must be recorded and must be cross-checked from two additional control points—at least one being a diagonal measurement. The best areas to select for tram gauge measurements are the attachment points for suspension and mechanical components, since these are critical to alignment. Throughout the repair operation, critical control points must be measured (and recorded) repeatedly with the tram gauge in order to monitor progress and to prevent overpulling.

Since these control point tram measurements must be taken and written down several times in a repair operation, a method of tabulation must be devised. One of the ways to accomplish this is to use a data or tabulation chart similar to the one shown in Figure 10–38.

To use this data chart or a similar measurement sheet, the manufacturer's specifications taken from the service manual (Figure 10–39) are written down in the first column. The A-B-C designations are the actual measuring point dimensions; the 1-2-3 designations are the readings taken at measurement Step 1, measurement Step 2, and so on. As each step of a restoration repair is made, the measurements should be recorded, including those dimensions that have just been corrected. This measurement data chart tells the body technician at a glance if the job has succeeded in restoring the vehicle to its original state.

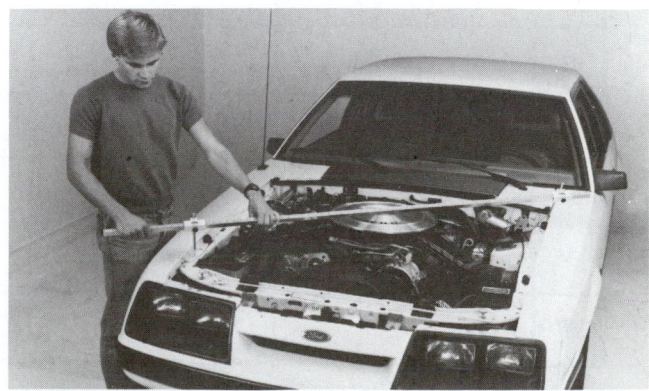

FIGURE 10-37 A tram gauge is simply a ridged tape measure with pointers. Place the pointer on the bolt or in the holes to measure between the reference points.

TRAM GAUGE DATA MEASUREMENT CHART

	Mfg. Spec.	1	2	3	4	5	6	7	8	9	10	11	12
A													
B													
C													
D													
E													
F													
G													
H													
I													
J													
K													
L													
M													
N													
O													
P													
Q													
R													
S													
T													
U													
V													
W													

FIGURE 10-38 To save time, make up and use a tabulation chart. Look up and write down spec values. Then take the chart to the vehicle so you can record actual measurements before and during frame/body power straightening.

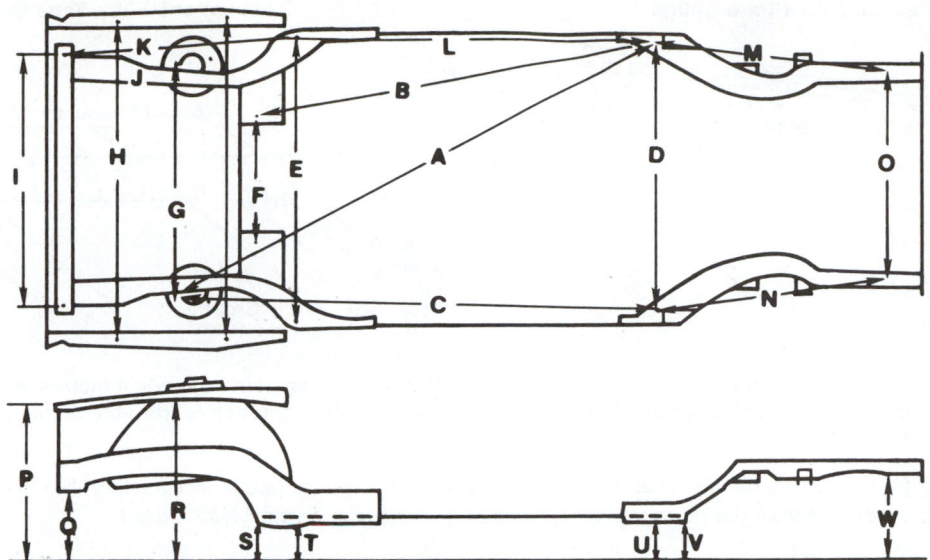

FIGURE 10-39 Many body technicians find it is easier to substitute letters in place of numbers when making up a tabulation chart. If the manual drawing has only letters, you need to refer to the accompanying chart giving number values for each letter. *(Courtesy of Chrysler Corp.)*

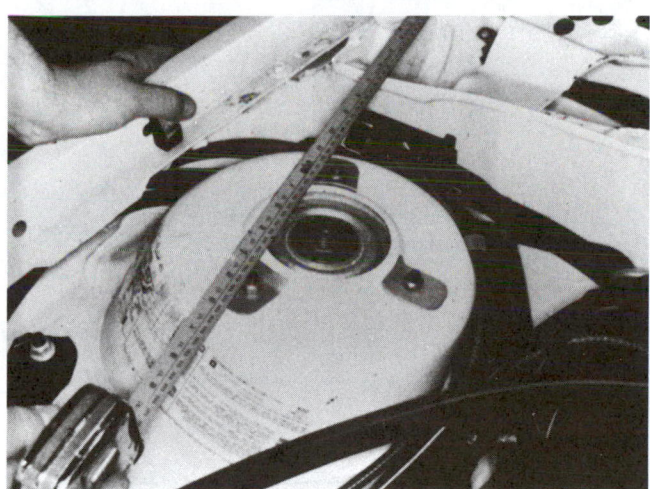

FIGURE 10-40 Quick measurement of reference points can be taken with a measuring tape. By grinding the tip sharp, you will get better readings in holes and on parts.

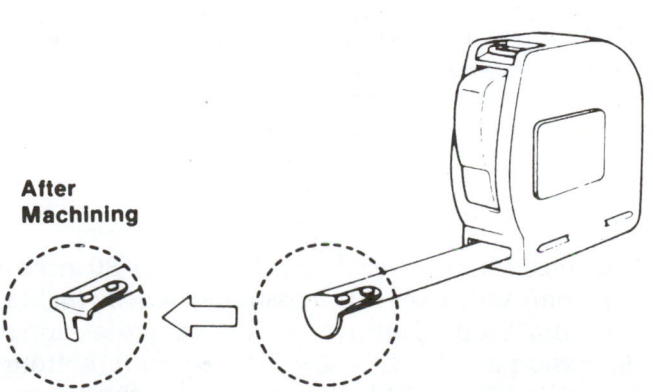

FIGURE 10-41 As shown above, machining tape measure's tip gives more accurate measurements.

The tram gauge may have a scale superimposed on it. However, since almost all manufacturer specifications list measurements in metric, use a steel tape with both fractional inches and metric scales to set the tram gauge up. The tape can also be used to take quick measurements between control points (Figure 10–40). Be sure that the tape has been checked for accuracy.

Accurate measurements can be taken if the tip of a tape measure is machined as in Figure 10–41. The pointed tip will allow you to insert the measurement tape fully into the control measurement hole.

Most reference points are actually holes in the vehicle structure, and dimensions are center-to-center distances (Figure 10–42). Control point holes are frequently larger in diameter than the tram gauge tip. To measure accurately with the tram gauge (when holes are same diameter), measure like-edge to like-edge. A few gauge manufacturer's spec books give the measurement based on the gauge's bar length. Always check the method used for specification measurements.

When the holes are not the same size, find the center-to-center measurement. They will usually be

Measurement With a Tram Gauge

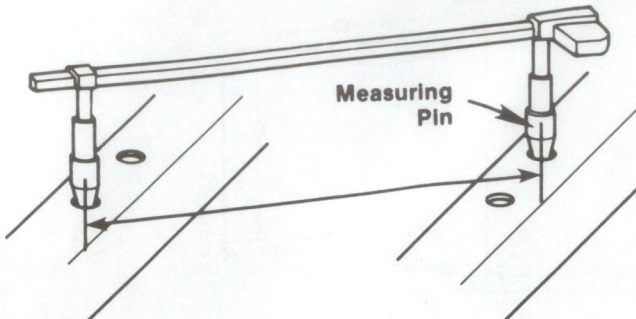

If the measuring pin is inserted securely into the measuring hole, the hole center distance can be measured.

A

Measurement With a Tape Measure

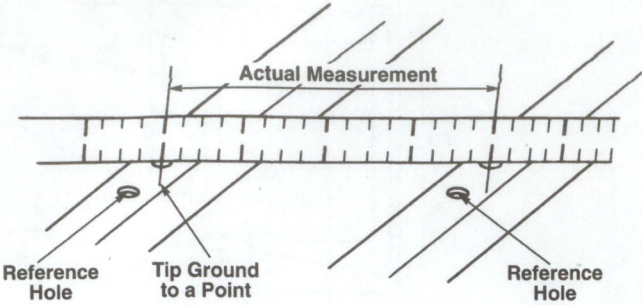

If the reference point is made 4 inches, measurement is easier. Subtract 4 inches from the measured dimension.

B

FIGURE 10-42 (A) Measuring the distance between hole centers with a tram gauge. (B) You can also make quick accurate measurements with a tape measure if the holes are the same size. *(Courtesy of Toyota Motor Corp.)*

In a Situation Where the Hole Diameters Are the Same

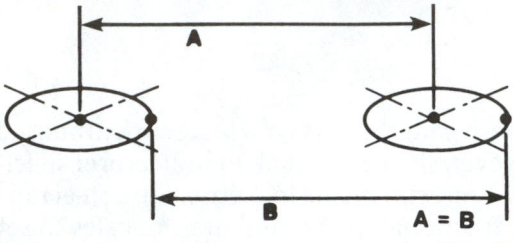

A

In a Situation Where the Hole Diameters Are Different

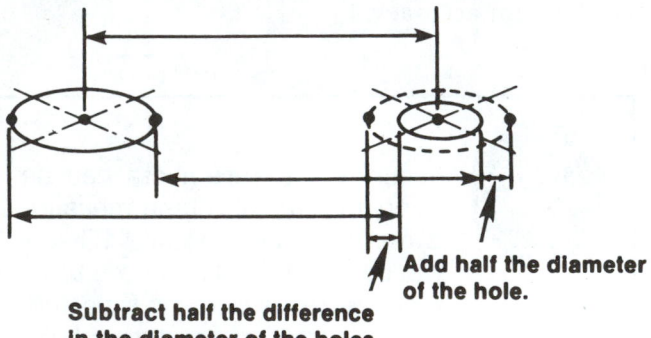

Add half the diameter of the hole.

Subtract half the difference in the diameter of the holes.

B

Tracking Gauge

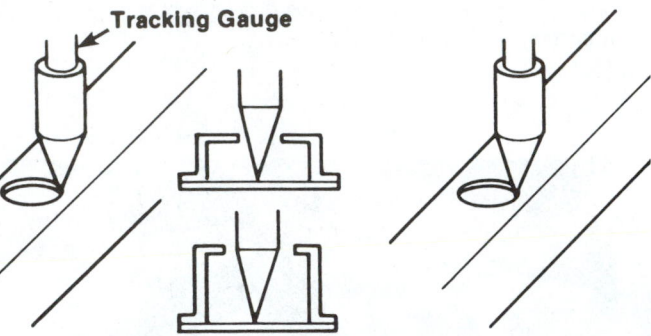

The measuring pin hits the bottom of the hole, or the measuring hole is too large.

C

Tape Measure

Hook onto the measuring hole.

D

FIGURE 10-43 (A) Note two ways of measuring when the holes are the same diameter. (B) If the holes are different diameters, measure from the centers if possible. If measuring them from the edges with a tape measure, subtract half of the hole size difference if going from a small hole to a large hole. Add half the diameter difference if going from a large hole to a small hole. (C) Pin on tram gauge fully insert and contact hole or inaccurate readings will result. (D) With a helper to hold the other end of the tape, you can also eyeball centers by reading tape as shown. Subtract the starting point on the tape (since not at zero or end tip of tape measure). *(Courtesy of Toyota Motor Corp.)*

the same type of hole: round, square, oblong, and so on. In this case, measure inside edge to inside edge, then outside edge to outside edge (Figure 10–43). Add the results of the two measurements and divide by 2.

For example, two round holes, one being $1/2$ inch (13 mm) in diameter, the other $1^1/2$ inches (38 mm)

in diameter, have an inside measurement of 30 inches (762 mm) and an outside measurement of 32 inches (813 mm) (Top Figure 10–44). The center-to-center dimension is 30" + 32" ÷ 2 = 31" (787 mm) (Bottom Figure 10–44). The 31 inches (787 mm) is the dimension for the tram gauge.

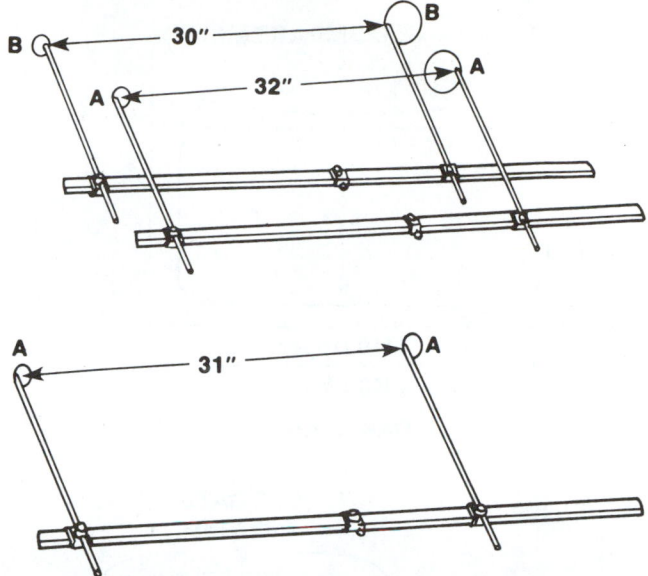

FIGURE 10–44 (Top Illustration) Two round holes, one being ¹/₂ inch (13 mm) in diameter, the other 1¹/₂ inches (38 mm) in diameter, have an inside measurement of 30 inches (762 mm) or an outside measurement of 32 inches (813 mm). (Bottom) Center-to-center dimension is 30" + 32" ÷ 2 = 31" (787 mm). The 31 inches (787 mm) is the dimension for the tram gauge.

In using a tram gauge for measuring, the manufacturer's specifications for the vehicle are needed. They will enable you to accurately assess the damage and restore the vehicle structure to factory dimensions.

If the manufacturer's specifications are not available, use an undamaged vehicle of the same make, year, model, and body style as a source for correct dimensions. Frequently, if only one side of a vehicle is damaged, it is possible to take measurements on the undamaged side. You can then apply them to the damaged side for a comparison measurement.

UPPER BODY DIMENSIONING

Upper body damage can also be determined with tracking trams and a steel measuring tape. Their use is basically the same as when doing an underbody evaluation. Manufacturers furnish specifications on the most important upper body reference points (Figure 10–45).

Measurement of the Front Body

In the case of a damaged vehicle that needs the hood edge and front side member replaced, it is reasonable to take measurements along with the repair. Even if only the front right side of the body received the impact, the left side will usually be damaged also. Therefore, the extent of deformation must be checked before remeasuring.

Figure 10–46 shows the typical front body reference points, which can be checked against the manufacturer's body dimensions diagram. You will need to measure across these points to analyze damage. If any measurements are shorter or longer than specs, you need to use frame straightening equipment to pull or push the body parts back into alignment.

When checking front end dimensioning, the best areas to select for the tram gauge measurements are the attachment points for suspension and mechanical components. These are critical to proper alignment. Each dimension should be checked from two additional reference points, with at least one reference point being a diagonal measurement. The longer the dimension, the more accurate the measurement. For example, a measurement from a lower cowl to the front engine mount cradle is better than from a lower cowl area to another lower cowl area. The longer dimension takes in a larger area of the vehicle. The use of two or more measurements from each reference point assures greater accuracy. It helps identify the extent and direction of any panel damage.

Measurement of the Body Side Panel

Any deformation of the body side structure can often be found by irregularities in the door when it is opened and closed. Depending on where the deformation is located, attention should be given to possible water and air leakage. It is important that accurate measurements be taken. The tracking tram gauge is primarily used to measure the body side panel (Figure 10–47).

Symmetrical means that the dimensions on the right side of the vehicle are equal to the dimensions on the left side of the vehicle. If the vehicle is asymmetrical, these dimensions are not the same.

A **left-to-right symmetry check** compares measurements on the undamaged side of the vehicle to the damaged side. Warping can generally be detected if the left-to-right symmetry is different in each side. Measure diagonal lines, as shown in Figure 10–48A. Use this measuring method if the data on the engine compartment and underbody is missing, if there is no data available in the body dimensions chart, or if the vehicle has been severely damaged in a rollover.

The diagonal line measurement method is not adequate when inspecting damage to both sides of the vehicle or in the case of twisting. The left-to-right difference in the diagonal lines cannot be measured (Figure 10–48B). If deformation is the same on the left and the right, a difference will not be apparent (Figure 10–48C).

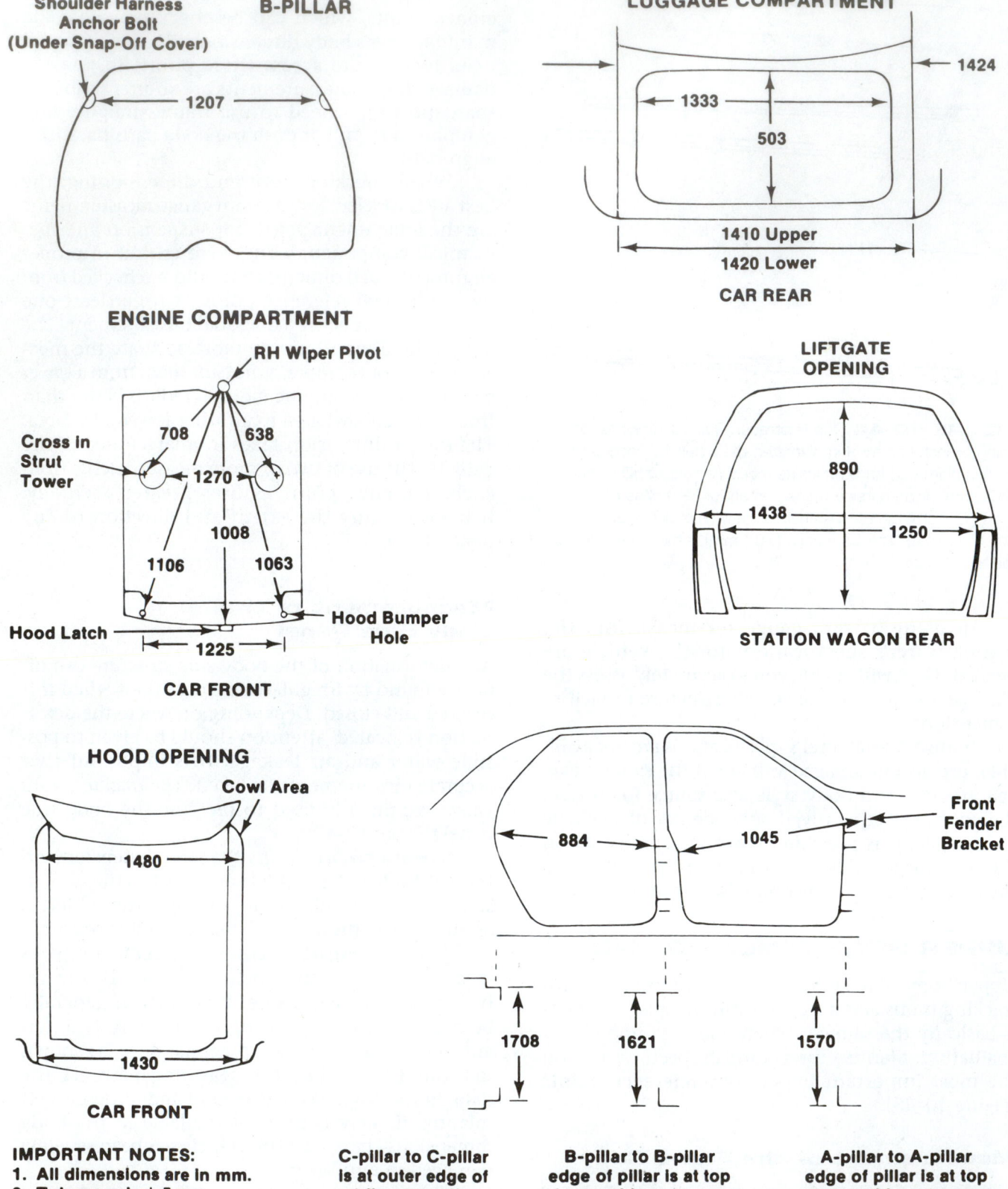

IMPORTANT NOTES:
1. All dimensions are in mm.
2. Tolerance is ± 5 mm.
3. All dimensions are true length.

C-pillar to C-pillar is at outer edge of striker mounting surface.

B-pillar to B-pillar edge of pillar is at top of upper hinge to same location on opposite side.

A-pillar to A-pillar edge of pillar is at top of upper hinge to same location on opposite side.

FIGURE 10-45 Read this example of upper body dimensions and specifications from a typical manufacturer's manual. Many body dimension charts are in metric measure or both U.S. and metric. *(Courtesy of Ford Motor Corp.)*

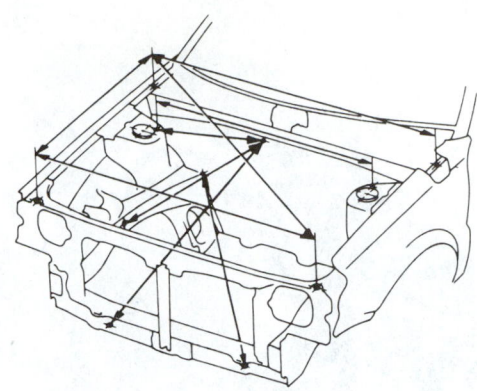

FIGURE 10-46 Front body measurement points are similar in all charts and manuals, but always use the correct one for the specific vehicle.

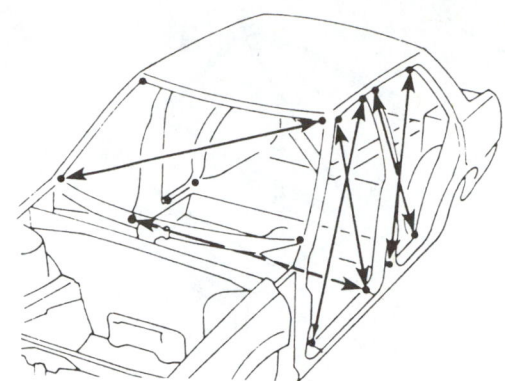

FIGURE 10-47 These are typical body side panel measurement points. Diagonal measurements across the door and windshield openings tell an important story about major center body damage.

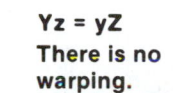

Yz = yZ
There is no
A   warping.

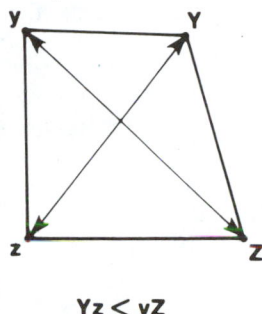

Yz < yZ
Deflection
B   on the left

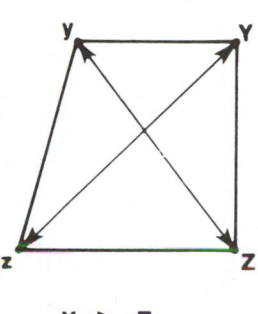

Yz > yZ
Deflection
C   on the right

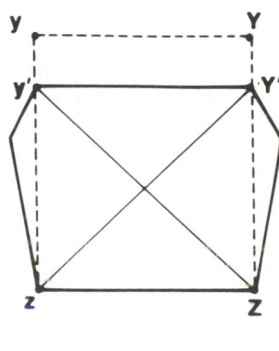

D   Y'z = yZ

FIGURE 10-48 Carefully study the diagonal line measurement method. (A) No straightening needed or same as specs. (B) Damage to left so pull to right for repair. (C) Door opening or other section has been damaged and pushed right. (D) With this damage, diagonal measures would be the same and might even be within specs. Other measurement points are needed to pull out damage.

In Figure 10–49, the measurement and comparison of the left and right lengths between yz and YZ will give an even better indication of damage conditions (this method should be used in conjunction with the diagonal line measurement method). It can be applied where there are parts that are symmetrical on the left and right sides.

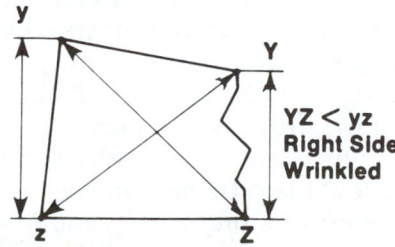

YZ < yz
Right Side
Wrinkled

In the diagonal line measurements,
YZ < yz, and it is judged that
there is deflection on the left.

FIGURE 10-49 Measurement and comparison of the left and right lengths between yz and YZ will give an even better indication of damage conditions. This method should be used in conjunction with the diagonal line measurement. This method can be applied where there are parts that are symmetrical on the left and right sides. *[Courtesy of Toyota Motor Corp.]*

Measurement of the Rear Body

Any deformation of the rear body can be roughly estimated by appearance and irregularities evident when the trunk lid is opened and closed. The trunk lid might rub or catch on the body. Since this can cause water leakage and paint damage, measurements are needed to check for the probable cause (Figure 10–50).

Furthermore, any wrinkle in the rear floor is usually due to buckling of the rear side member. Measure the rear body together with the underbody. In this way, the straightening work can be performed effectively.

When using a tram gauge, be sure to keep the following pointers in mind:

- Measurements are made to fixed points on the vehicle, such as bolts, plugs, or holes.
- A point-to-point measure is the direct actual measurement between two points (Figure 10–51).
- The tram bar should be parallel to the car body (Figure 10–52). This might require the pointers on the tram bar to be set at different lengths.
- Some body dimensions manuals show dimensions in bar length. Other dimension books show

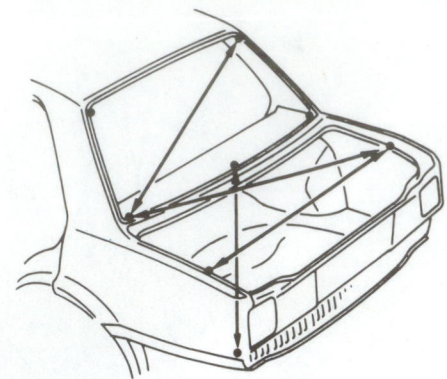

FIGURE 10-50 Typical rear body measurement points can be checked like other parts already discussed.

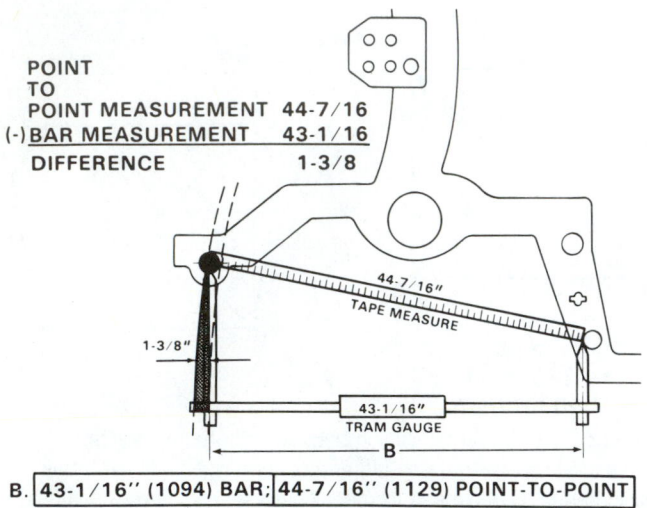

POINT
TO
POINT MEASUREMENT 44-7/16
(-) BAR MEASUREMENT   43-1/16
DIFFERENCE   1-3/8

44-7/16"
TAPE MEASURE

1-3/8"

43-1/16"
TRAM GAUGE

B

B. | 43-1/16" (1094) BAR; | 44-7/16" (1129) POINT-TO-POINT |

FIGURE 10-51 Point-to-point measurement is direct actual measurement between two points. If you hold the tram at an angle as shown, you get an incorrect reading. Refer to spec chart footnotes for getting correct values and procedures. *(Courtesy of Blackhawk Automotive Inc.)*

FIGURE 10-52 By changing the tram pointer's lengths, you can measure over an obstruction yet still keep the bar parallel to the car's body. *(Courtesy of Blackhawk Automotive Inc.)*

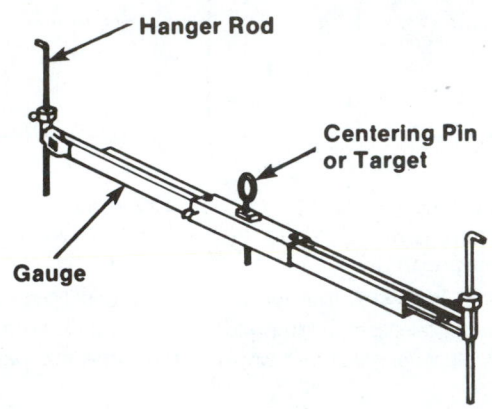

Hanger Rod

Centering Pin or Target

Gauge

FIGURE 10-53 The typical self-centering gauge has a center pin or target for viewing the centerline of the vehicle. Hangers are provided for suspending the tool from the underbody of the vehicle.

dimensions in point-to-point lengths. Some manuals use both. Keep this in mind when reading specs and measuring.

- Make all measurements on the damaged vehicle at the points specified in the body manual. The amount of damage can usually be determined by subtracting the actual measurement from the specified measurement.

10.6 CENTERING GAUGES

While self-centering gauges (Figure 10–53) are closely related to the tram gauges, they do not measure; they show alignment or misalignment by projecting points on the vehicle's structure into the technician's line of sight. They are installed at various control areas on the vehicle.

Self-centering gauges have two sliding horizontal bars that remain parallel as they move inward and outward. This action permits adjustment to any width for installation on various areas of the vehicle. After the gauges are hung on the car (usually three or four sets), the horizontal bar will be parallel to the portion of the structure to which it is attached.

Place one centering gauge at the front of the vehicle, one at the extreme rear, one just behind the front wheels (front torque box), and one forward of the rear wheels (rear torque box) (Figure 10–54). The two gauges in the center are usually considered to be the baseline gauges.

When inspecting for collision damage, first hang centering gauges from two places where there is no visible damage, usually the center section of the body or frame. Then hang two more gauges where there is obvious damage (Figure 10–55). Then look or sight along the gauges and analyze how the self-centering gauges line up. Check for parallel misalignment of the gauges or misalignment of the centering pins. This will help determine the direction and extent of major body or frame damage.

Centering gauges are equipped with center pins or sights that always remain in the center of the gauge regardless of the width of the horizontal bars. This allows the body technician to read the centerline throughout the length of the vehicle.

Each self-centering gauge accommodates two vertical scales—one on the left side, one on the right. These scales can be adjusted vertically. This assures

SHOP TALK

There should not be any deformation at the point the gauge is installed. There are many instances where the alignment of the gauge holes has been deformed by previous collisions or other causes. Do not use deformed holes unless they can be repaired satisfactorily (Figure 10–56).

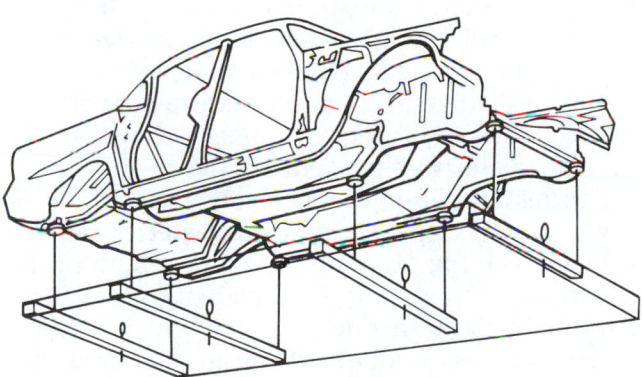

FIGURE 10–54 Typical starting locations for centering gauges. A least two gauges must be hung from undamaged areas on the vehicle. This will let you sight down and see any misalignment of the gauges in the crushed area. *(Courtesy of Blackhawk Automotive Inc.)*

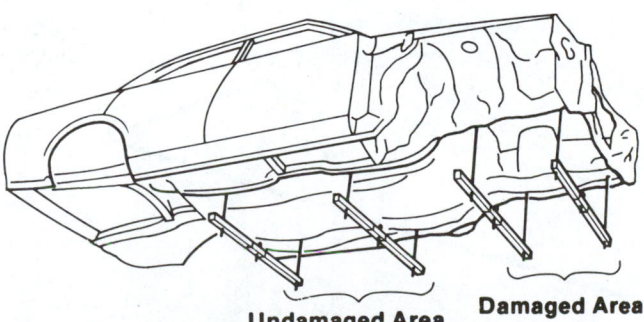

Undamaged Area **Damaged Area**

FIGURE 10–55 Placement of centering gauges to check a damaged area. For example, if the center section is undamaged, you can view and analyze damage to the front body/frame structure for straightening.

that the horizontal bars accurately reflect the true positions of the parts to which they are attached (Figure 10–57). Once hung in specific locations, these gauges generally remain on the car throughout the entire repair operation, unless one or all of them interfere with straightening or with tram gauge measurements.

Front Floor Under Reinforcement Reference Hole

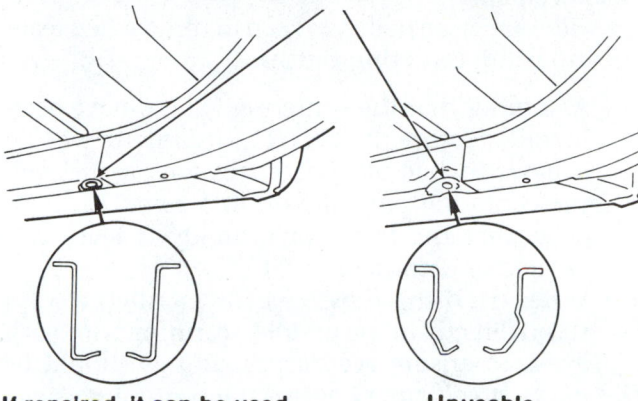

If repaired, it can be used.   **Unusable**

FIGURE 10–56 Never use body holes that are deformed or damaged. Fix or reshape them before measuring. *(Courtesy of Toyota Motor Corp.)*

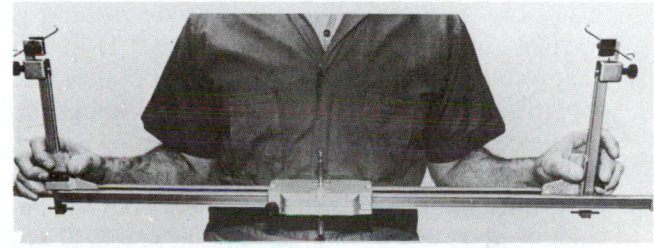

FIGURE 10–57 Many centering gauges have adjustable, calibrated arms or pointers.

Centering Pin

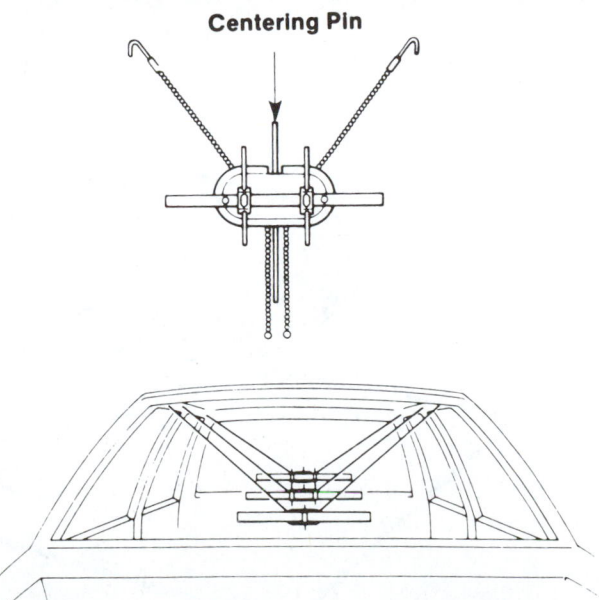

FIGURE 10–58 These centering gauges are being used to check body pillar damage. *(Courtesy of Toyota Motor Corp.)*

Special centering gauges are available that can be used to check such items as body pillar damage (Figure 10–58). The same system of alignment is employed when using centering gauges to check underbody damage.

The centering gauge reading is the visual alignment of parallel bars and pins. The final objective is to achieve a level and centered structural alignment in the vehicle.

Here are some points to keep in mind when reading and using centering gauges.

- Assuming that the centering gauges have been installed properly, correct alignment will be achieved when all the gauges are parallel and the centering pins line up in a row. This indicates that the frame or unibody is level, not twisted or deformed.
- When sighting crossbars for parallel, always stand directly in the middle, scanning with both eyes. To ensure accuracy, readings should be made at the outer edge of the centering gauge, not in the middle.
- The farther one stands from the centering gauges while reading, the more accurate the reading will be. Standing close changes the line of sight to the front gauges so that an accurate reading is nearly impossible.
- Centering gauges should always be set at the same height or plane. Different heights will change the angle of sight and give a false reading.
- It is sometimes beneficial to sight over one gauge and under another. Going to the end of the vehicle, opposite the damage, to make readings will sometimes result in a more accurate reading. This is true because you are able to read the base gauges before sighting into the damaged area. With

practice and a certain amount of experimentation, you can improve the damage analysis.

- The sighting of centerline pins must be done with one eye. Since the center section is always the base for gauging, the line of sight must always project through the pins of the base gauges. Observing pins in other sections of the frame will then reveal how much they are out of alignment.
- Never attach the centering gauges to any movable parts, such as control arms or springs.

Self-centering alignment gauges are used to read three major elements of collision damage: datum, center, and zero planes. Critical measurements—the fourth major element of analysis—are handled with tape measure and tram gauge.

DIMENSIONAL REFERENCES

Two major dimensional references are indicated in all body dimension manuals: the datum plane and centerline.

A *datum line*, or **datum plane**, is an imaginary flat surface parallel to the underbody of the vehicle at some fixed distance from the underbody. It is the plane from which all vertical or **height dimensions** are taken by the vehicle manufacturer (Figure 10–59). It is also the plane that is used to measure the vehicle during repair. The datum is normally shown on dimension charts from the vehicle's side view.

Using this line of reference, centering gauges can be strategically suspended under the vehicle from side to side at varying distances along the length of the chassis frame or unibody. First, place the base gauges at the main platform: one across the vehicle beneath the rear seat (rear torque box) and another under the

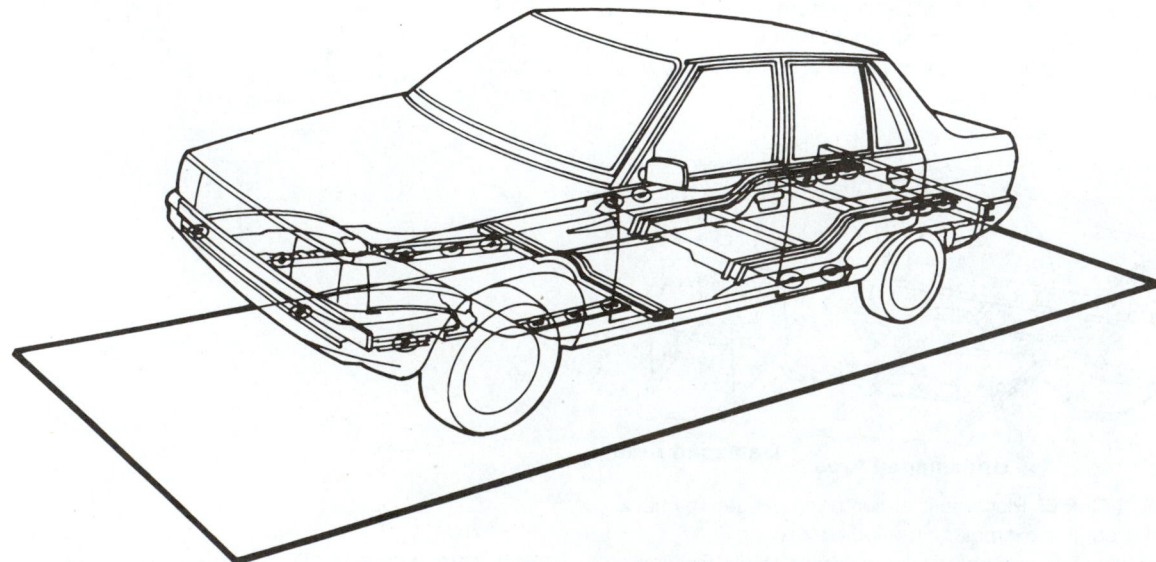

FIGURE 10-59 The datum plane is a reference line for vertical body dimension. *(Courtesy of Blackhawk Automotive Inc.)*

cowl area (front torque box). Add two more gauges before and after the base gauges; one located at the front cross member and a second at the rear cross member. Additional gauging of the front cross member area and/or strut tower completes the picture.

To read datum, all gauges must be on the same plane, as indicated by the spec sheets. After hanging all four gauges, read across the top to determine if datum is correct. If all four gauges are level at the top (Figure 10–60A), the vehicle is on datum. If they are not level, the vehicle is off datum (Figure 10–60B).

Since the datum line is an imaginary plane, datum heights can be raised or lowered to facilitate gauge readings. If the datum height is changed at one gauge location, all gauges must be adjusted an equal amount to maintain accuracy.

While datum readings are usually obtained from centering gauges, there are individual gauges available for measuring datum heights. These datum gauges are usually held in position by magnetic holders. Remember that the dimensions that allow the vehicle to be level with the road are measured from the datum plane.

THE CENTER PLANE

The **center plane**, or *center line*, divides the vehicle into two equal halves: the passenger side and the driver's side. Note how the center plane or

line cuts up through the middle of the vehicle (Figure 10–61).

The center line is shown on dimension charts in either the bottom or top views. It can be found on some vehicles in the form of body center marks.

Body center marks are often stamped into the sheet metal in both the upper and lower body areas of the vehicle. They can save time when taking measurements.

All width or **lateral dimensions** of symmetrical vehicles are measured from the center. The measurement from the centerline to a specific point on the right side will be exactly the same as the measurement from the centerline to the same point on the left side. One side of the structure should be a perfect mirror image of the other.

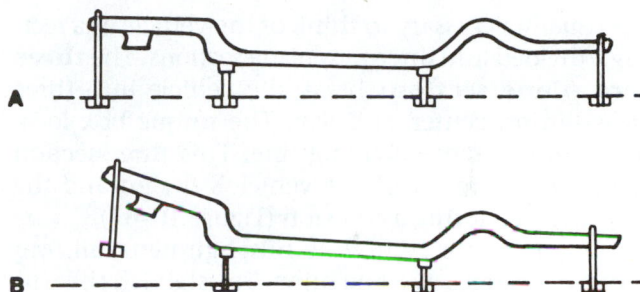

FIGURE 10–60 (A) Datum is correct and (B) datum is off at front. The front centering gauge would be too high when you sighted down the other three gauges.

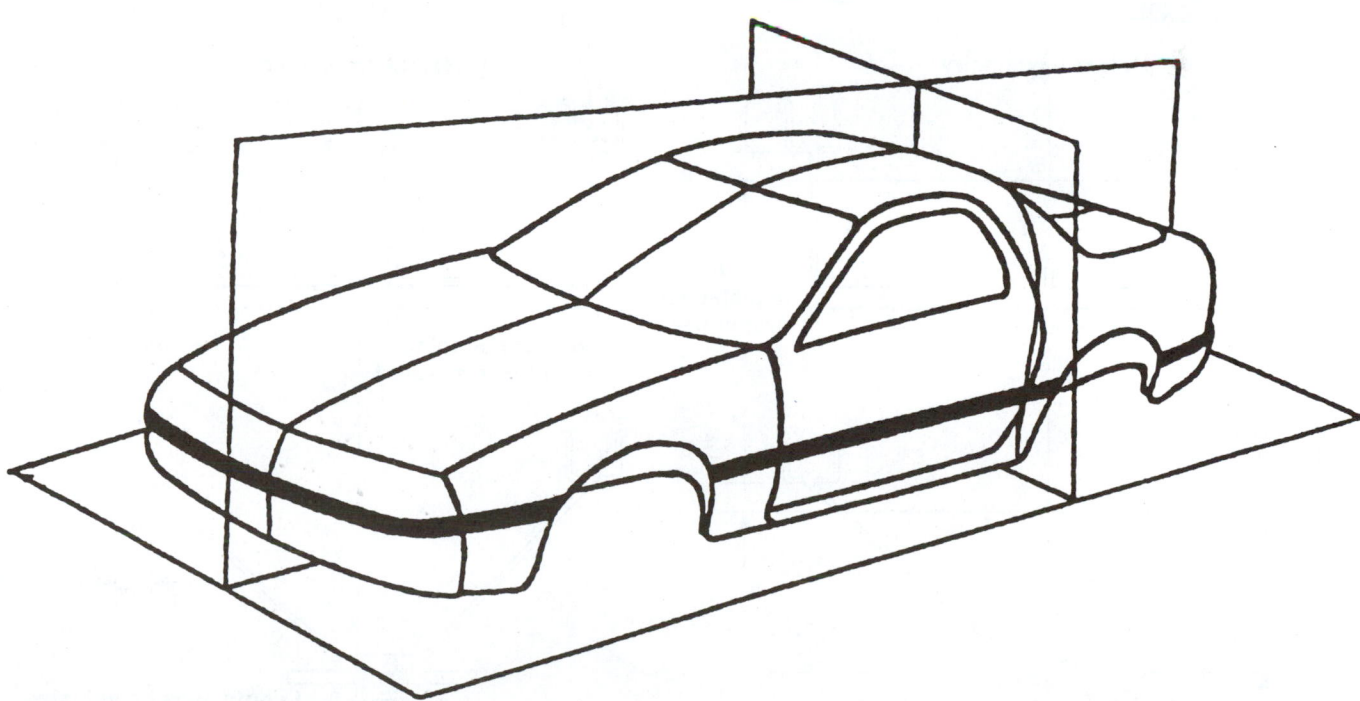

FIGURE 10–61 The center plane or centerline of a vehicle allows for horizontal measurements while the datum plane allows for vertical measurements. *(Courtesy of Chief Automotive Systems, Inc.)*

SHOP TALK

Most vehicles are built symmetrically. But if the vehicle is not symmetrical (asymmetrical or has unequal measurements on each side), the self-centering gauges will not align and will not indicate a true center reference. In such a case, a centering gauge that compensates for the asymmetry of the underbody can be used in conjunction with a body dimensions chart that has a built-in compensation factor (Figure 10–62).

ZERO PLANES

It is usually necessary to think of the vehicle as a rectangle divided into three zero plane sections. The **three zero plane sections** break the vehicle into three areas—front, center, and rear. The torque box location is used as the dividing line. This three-section principle is a result of the vehicle's design and the way it reacts during a collision (Figure 10–63).

To check for centerline misalignment, all four centering gauges must be hung. To establish the true centerline, the center pin on the #2 gauge must be lined up with the center pin on the #3 gauge. Then the center pins of #1 and #4 can be read relative to the centerline of this base.

Of course, other damage conditions will affect a centerline reading. If a vehicle has a sway condition or an out-of-level condition, the centerline reading will be affected. Further inspection by gauging or measuring might be necessary to determine the presence or absence of these problems.

The controlling points of any car underbody are the front cross member, the cross member at the cowl, the cross member at the rear door, and the rear cross member (Figure 10–64). The center section or the area between the cowl and rear door cross member is the portion used when doing a major straightening operation.

Level in a zero plane means the condition in which all areas of the vehicle are parallel to one another. Level refers to parallel conditions in the vehicle structure only and has nothing to do with any outside reference, such as the floor. Check for an out-of-level condition in the front or rear sections. When the #2 and #3 base gauges are hung, the center section is read for level. When these base gauges are parallel, no twist can exist. However, the front or rear sections could still be out of level. To check for this condition, hang #1 and #4 gauges and read relative to the nearest base gauges. If #1 hangs parallel to #2, the front section is level, relative to the base. If #4 hangs parallel to #3, the rear section is level, relative to the base. If an out-of-level condition exists in the front, #1 will not hang parallel to #2. This same type of reading should be done in the rear section.

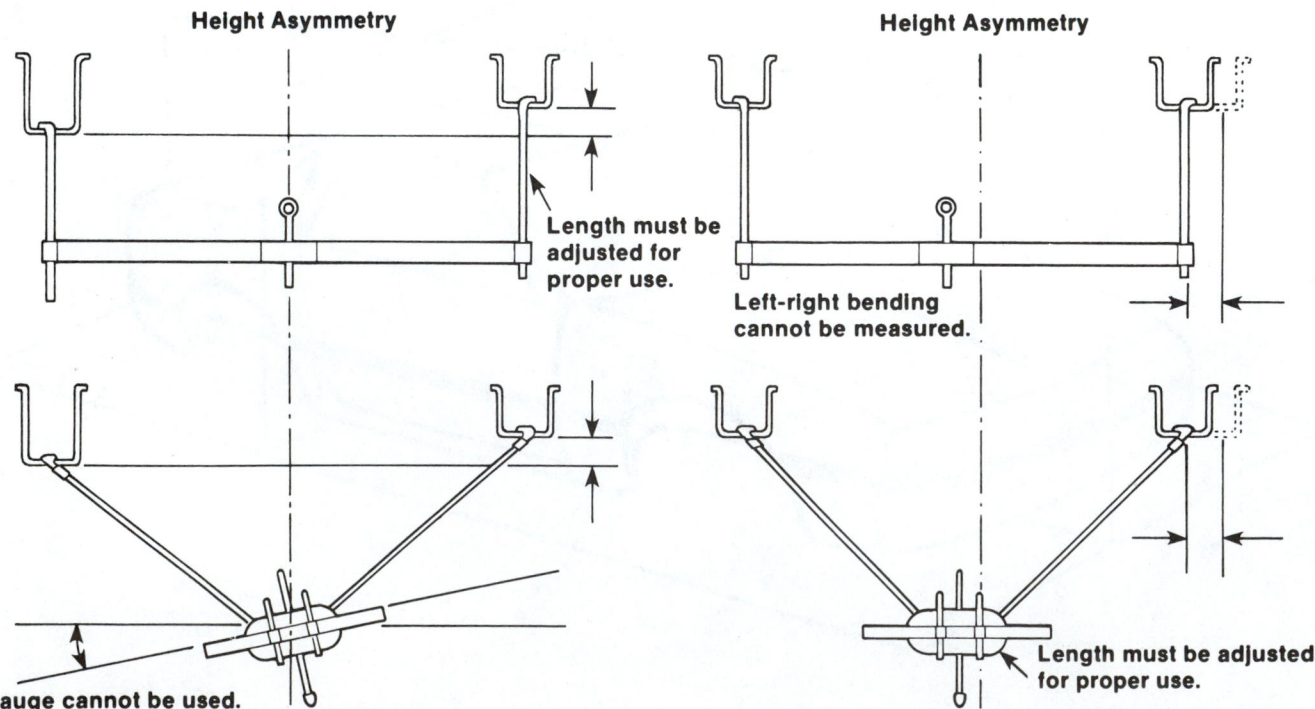

Height Asymmetry

Length must be adjusted for proper use.

Gauge cannot be used.

Height Asymmetry

Left-right bending cannot be measured.

Length must be adjusted for proper use.

FIGURE 10–62 Note how these gauges are arranged when measuring asymmetric installation points. *(Courtesy of Toyota Motor Corp.)*

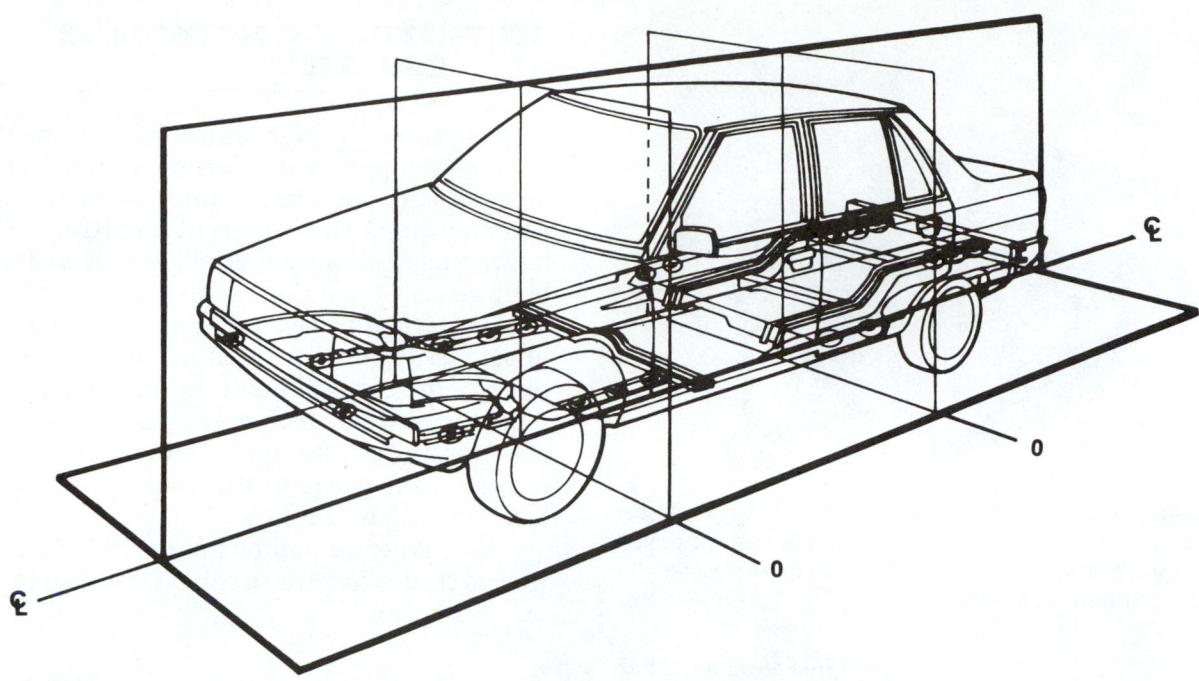

FIGURE 10-63 Three vehicle planes are combined in this illustration. This allows for 3-D or three-dimensional measurements. *(Courtesy of Blackhawk Automotive Inc.)*

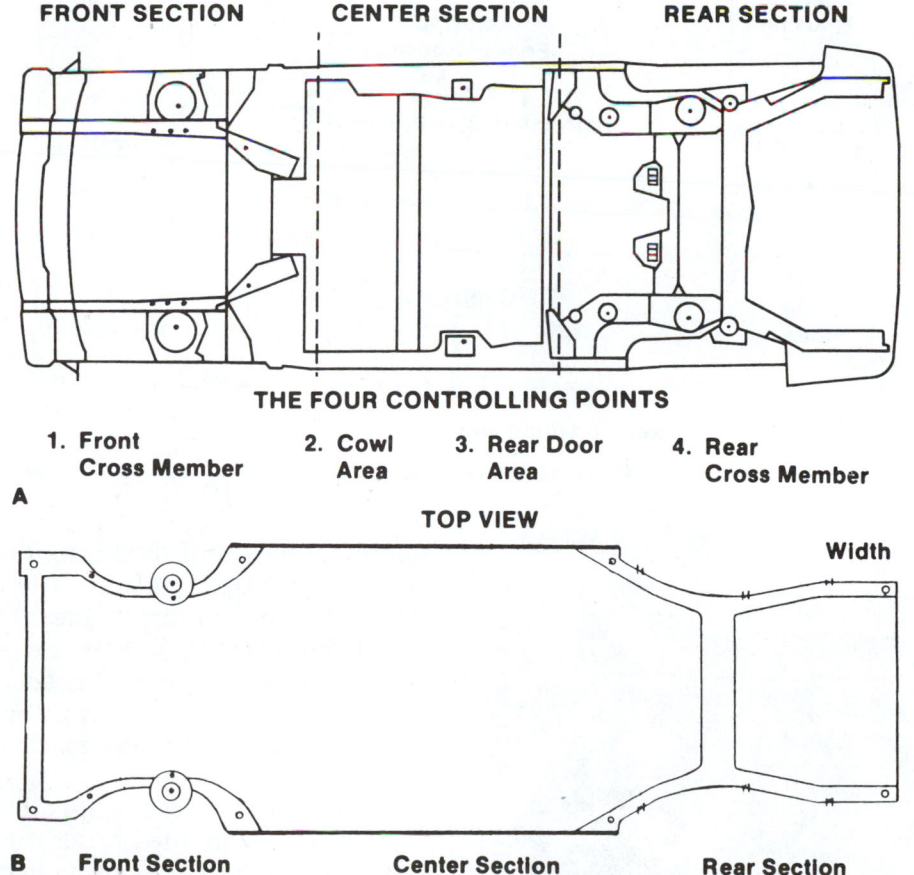

FRONT SECTION   **CENTER SECTION**   **REAR SECTION**

THE FOUR CONTROLLING POINTS

1. **Front Cross Member**   2. **Cowl Area**   3. **Rear Door Area**   4. **Rear Cross Member**

A

TOP VIEW

Width

B   **Front Section**   **Center Section**   **Rear Section**

FIGURE 10-64 During major repairs, the vehicle is normally divided into three zones (sections) and four controlling areas. Compare (A) a conventional frame vehicle and (B) a unibody car.

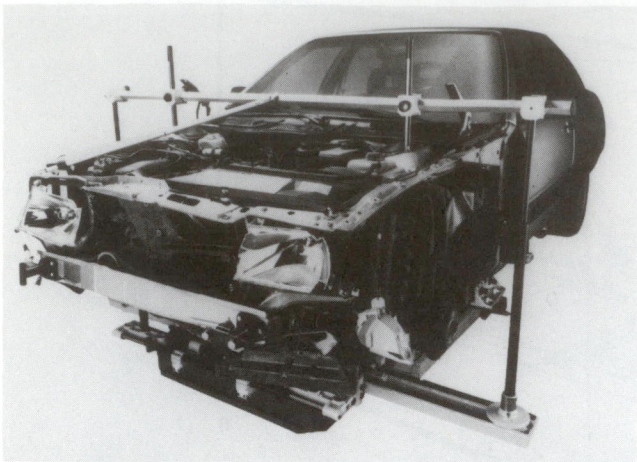

FIGURE 10-65 A typical strut centerline gauge in place. The top of the strut towers is a common measurement point since they are often moved out of alignment during frontal impact. *(Courtesy of Car-O-Liner)*

10.7 STRUT CENTERLINE GAUGE

A **strut tower gauge** shows misalignment of the strut tower/upper body parts in relation to the centerline plane and datum line plane. It is usually mounted on the strut towers (Figure 10–65). The strut tower gauge allows visual alignment of the upper body area.

The strut tower gauge (Figure 10–66) features an upper and lower horizontal bar, each with a center pin. The upper bar is usually calibrated from the center out. Pointers, which are positioned in an adjustable housing on the upper horizontal bar, are used to mount the gauge to the strut tower/upper body locations (Figure 10–67).

Two types of pointers are provided: cone and reverse cone. The reverse cone is notched to provide

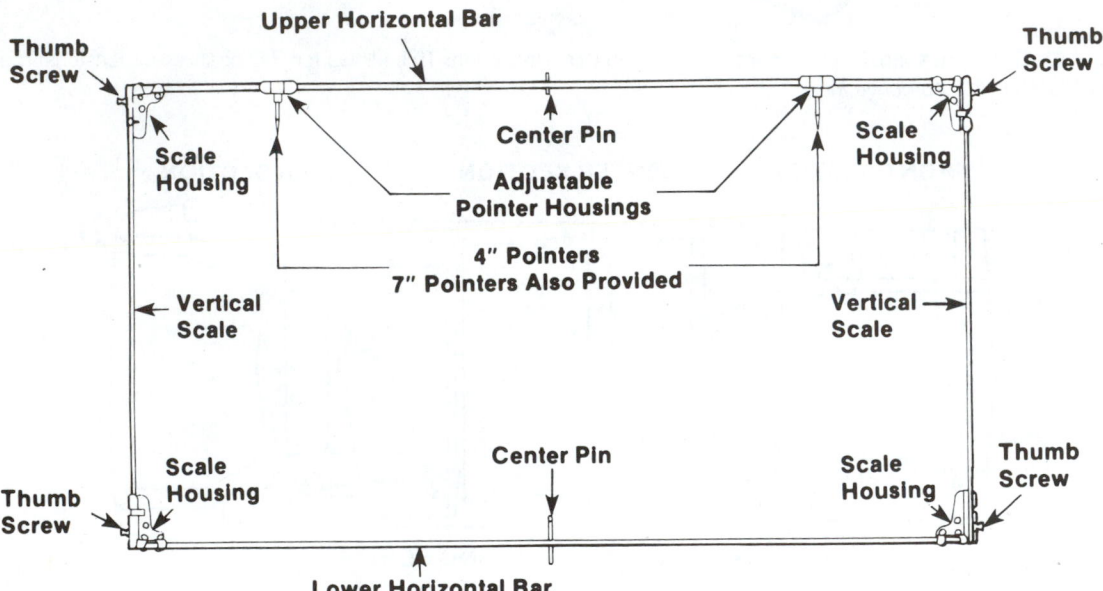

FIGURE 10-66 Study parts of a typical centerline strut tower/upper body gauge. *(Courtesy of Chief Automotive Systems, Inc.)*

FIGURE 10-67 Adjust and mount the upper horizontal bar between two strut towers. Pointers should contact a specific part, often a bolt head. *(Courtesy of Chief Automotive Systems, Inc.)*

additional means of mounting on the vehicle, for example, on ridged surfaces. The pointers are usually held in the housing by means of thumbscrews.

Different length pointers are provided for situations when more length is needed to position the gauge (Figure 10–68). When using different length pointers to mount the gauge, remember that they change the scale reading.

The vertical scales that link the upper and lower horizontal bars are used to set the lower bar at the datum height of the mounting parts (Figure 10–69). The scales fasten in housings at the ends of the horizontal bars. Height adjustments are made at the housings of the upper horizontal bar.

Using dimensions charts, the technician can adjust the strut tower gauge to the correct dimensions

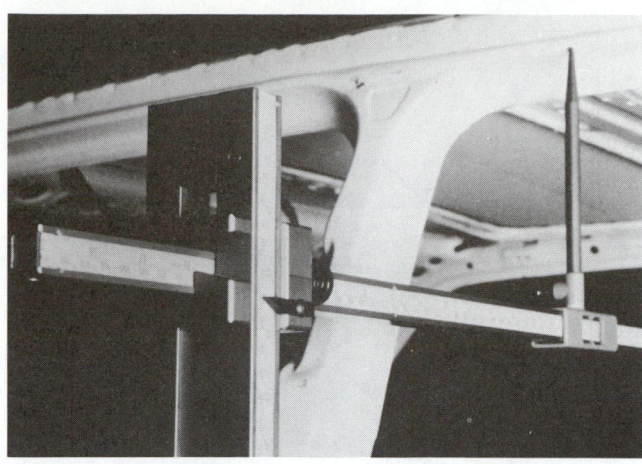

FIGURE 10–68 Longer pointers can be used to measure roof structures by extension of the bar through the passenger compartment. Project pointers upward from pointer housings to contact reference points on the roof. *(Courtesy of Chief Automotive Systems, Inc.)*

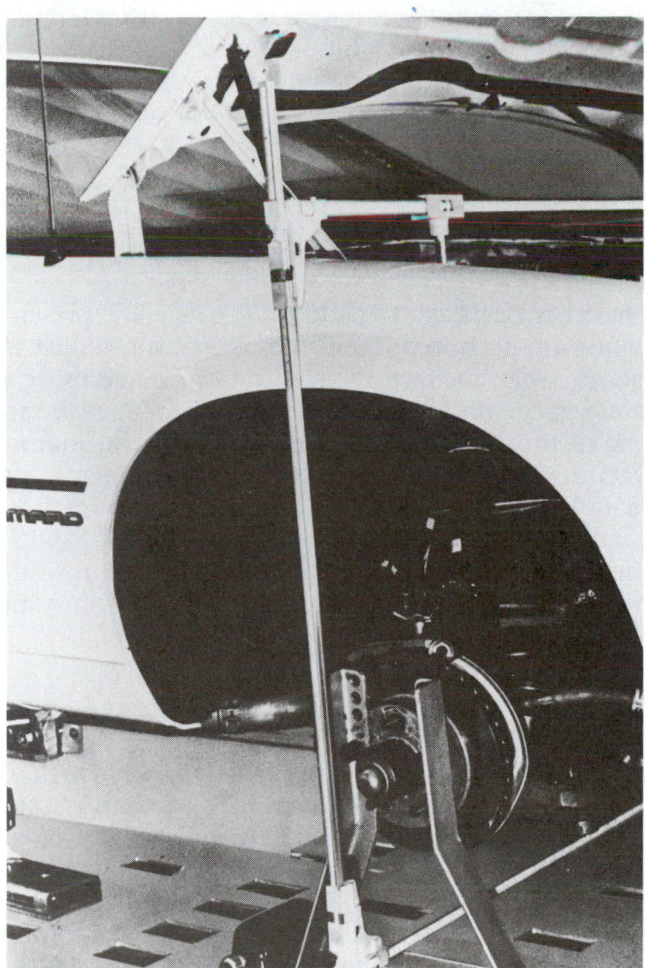

FIGURE 10–69 The vertical side bar of a strut tower/upper body assembly must be mounted and secured as directed in the owner's manual for the specific equipment and manufacturer. *(Courtesy of Chief Automotive Systems, Inc.)*

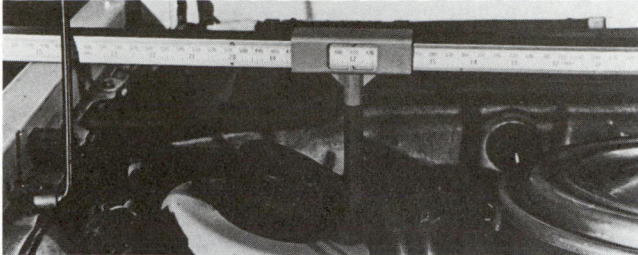

FIGURE 10–70 Check the strut for misalignments. This is done by comparing actual measurements to known good ones. *(Courtesy of Chief Automotive Systems, Inc.)*

using the vertical scales that link the upper and lower horizontal bars to set the lower bar at the datum plane. With the lower bar set to align properly, the upper pointers should be located at reference points on the strut towers. If not, the strut towers are damaged and pushed out of alignment. This would tell the technician that straightening is needed so that the front suspension and the wheels can be aligned properly.

The strut tower upper body gauge is used most often to detect misalignment of the strut towers (Figure 10–70). However, it can also be used to detect misalignment of a radiator support, center pillar, cowl, quarter panel, and so on.

10.8 DIAGNOSING DAMAGE USING GAUGE MEASURING SYSTEMS

As previously mentioned, the most common rule in body/frame alignment is: "Reverse direction and sequence." To correct collision damage pull or push the damaged area in the opposite direction of impact, and since the repair must be made in the reverse sequence, the damage must also be measured in the reverse sequence.

SHOP TALK

When measuring damage, keep in mind that a vehicle is similar to a building. If the foundation is not square and level, the rest of the structure will be uneven also. The vehicle's foundation, which is the center section of the vehicle, is measured for twist and diamond first. These two measurements will tell the collision repair technician if the foundation is square and level. The remaining measurements use the foundation as a reference.

If a true twist exists, the gauges would read like this.

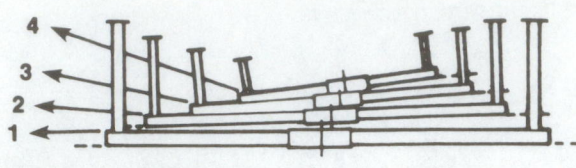

Twisting

FIGURE 10-71 If self-centering gauges read like this, the body or frame has twist damage.

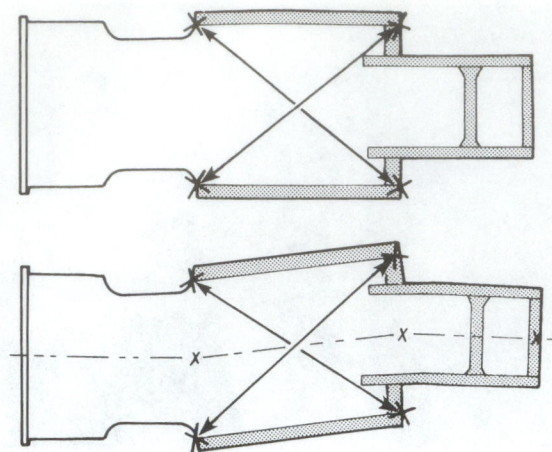

FIGURE 10-72 Unequal diagonal measurements show diamonding.

MEASURING TWIST DAMAGE

Twist damage is a condition where one corner of the car is higher than normal; the opposite corner may be lower than normal.

The first damage condition to look for and measure is twist. Twist is the last damage condition to occur to the vehicle and exists throughout the entire vehicle.

To check for twist, sight down your properly hung self-centering gauges. Twist will show up when the gauges are not parallel.

Twist can be checked only in the center section; otherwise, additional misalignment in the front or rear might give an inaccurate reading of twist.

To check for twist, two base gauges must be hung. These base gauges are also referred to as the #2 (front center) and #3 (rear center) gauges. The #2 should be hung as far forward of the center section, up to the cowl, as possible. The #3 is hung as far rearward of the center section, toward the rear kickup, as possible. The #2 gauge is then read relative to #3.

If the gauges are parallel, no twist exists. If the gauges are not parallel, then a twist may exist. Remember that a true twist must exist throughout the entire structure. To check for a true twist versus an out-of-level condition in the center section, hang another gauge. Go to the undamaged section of the car and hang either the #1 (front) or #4 (rear) gauge. These gauges will be read relative to the nearest base gauge. The #1 will be read relative to #2 and the #4 will be read relative to #3.

If the front or rear gauge reads parallel to the nearest base gauge, true twist cannot exist, and there is an out-of-level condition in the center section. If a true twist exists, the gauges will read like those shown in Figure 10-71.

MEASURING DIAMOND DAMAGE

Diamonding is a condition in which one rail or rocker is pushed either forward or rearward of the opposite rail or rocker. This condition will often be found in conventional frames. Rarely do you find a diamond condition in unibody vehicles, but it is possible.

The check for diamonding is simple. Using the tram gauge, measure from the front corner of one rail or rocker to the rear corner of the opposite side. If one measurement is longer than the other (Figure 10-72), a diamond condition exists. If one measurement is one inch (25 mm) longer than the other, the condition is referred to as "one inch diamond" or "25 millimeter diamond."

MEASURING MASHING (CRUSHING)

Mash can be measured with a tram gauge. It is present when any section or frame member of the vehicle is shorter than the factory specification. When using a tram gauge on a mash-damaged vehicle, be sure to make the measurement specified on the manufacturer's specification sheets or in the dimensions data book. The amount of mash is determined by subtracting the actual measurement from the specified measurement. The proper methods of measuring various impacts with a tram gauge are shown in Figure 10-73C, D, and E.

MEASURING SAG DAMAGE

Sag is a condition where the cowl or another area of the vehicle is lower than normal. Sag can also occur at the front cross member. The ends of the cross member will be closer than normal and the center will be too low.

Three centering gauges are used to check for a sag condition. One gauge is placed at the front cross member, the next one at the cowl area (torque box), and the third one at the rear door area (torque box). The gauges are on-center and parallel with each other, but the front frame gauge is lower than the others

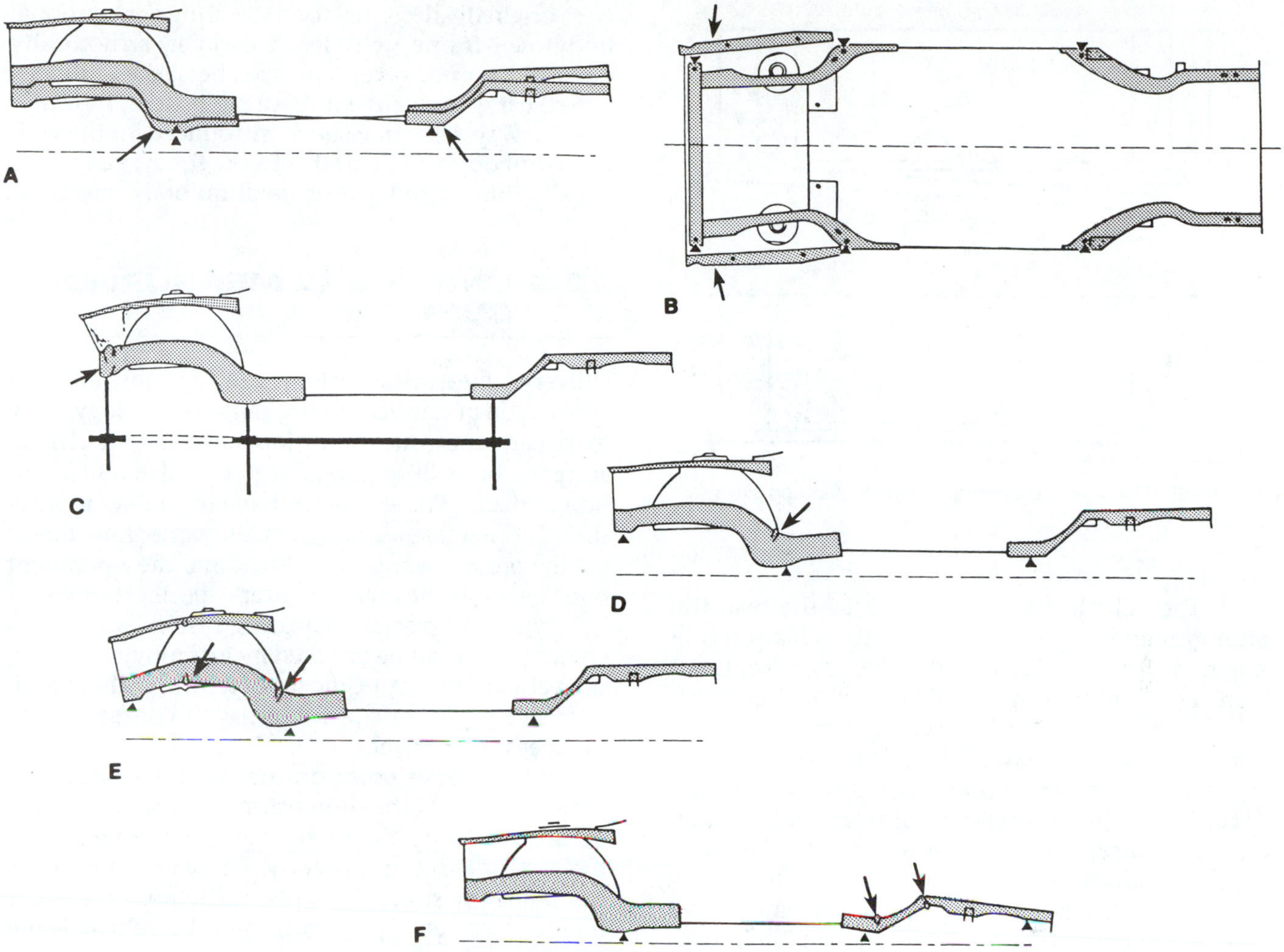

FIGURE 10-73 Study variations of unibody distortion under impact force. (A) High front impact with secondary damage to the rear of the assembly; (B) right front corner impact; (C) direct front impact; (D) low front impact; (E) high front impact; and (F) high rear impact.

FIGURE 10-74 Cowl area sag is shown by the gauge pointer being lower in undamaged areas.

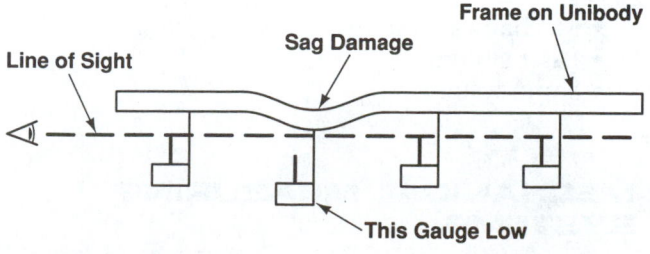

FIGURE 10-75 A side view gives a better idea of sag.

(Figure 10-74); this indicates a sag condition at the front cross member area.

In Figure 10-75 observe the relationship of the centering gauges to each other. Note that the parallel bar in the cowl area is about 2 inches lower than the other bars. This means there is 2 inches of sag.

MEASURING SIDESWAY DAMAGE

Sidesway is present when the front, center, or rear portion of the vehicle is pushed out of alignment by a side impact. Three centering gauges are used to check for sidesway.

If the vehicle was hit in the front, base gauges #2 (cowl area) and #3 (rear door area) are hung. The sighting gauge #1 is located in the front cross member area. If gauge #1 does not line up with the other two, front sidesway is present (Figure 10-76).

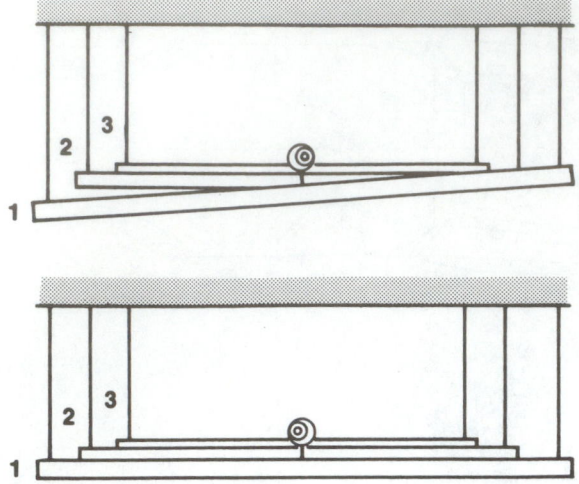

FIGURE 10-76 Self-centering gauges showing sidesway damage.

If the vehicle has been struck in the rear, the misalignment would appear on the self-centering gauges in a way similar to front sidesway, except that the rear pin or bull's-eye would be out of alignment.

A center hit on a vehicle causes a misalignment known as *double sidesway*. It results from a severe impact in the center section, but it affects the entire vehicle. The dimensions of both front and rear sections must be checked during the pulling of double sidesway damage.

While the self-centering and datum gauges give a total picture of frame and body damage, their functions can be adapted into a so-called frame gauge. By viewing body damage with a frame gauge arrangement, it is possible to measure the amount of frame or body damage the vehicle has incurred.

An undamaged frame would give a frame gauge indication as shown in Figure 10-77. The horizontal bars are parallel to each other, indicating that the frame is level. The targets are centered within each other indicating a perfect centerline. The gauges reveal horizontal and vertical alignment for certain body and frame damages.

Measuring gauges—tram, self-centering, datum, and frame—have been in use for many years and

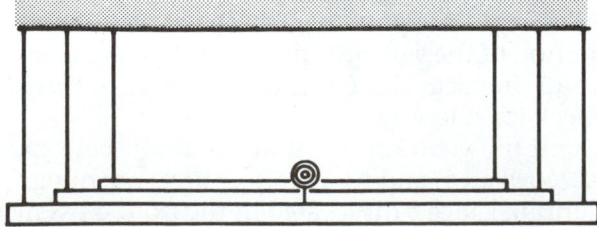

FIGURE 10-77 When the body/frame is fixed by power straightening, no deviation should be seen in the centering circles or pointers of gauges.

were originally designed for measuring conventional body-over-frame vehicles. However, structurally damaged unibody vehicles can be successfully repaired using the gauge measuring system. In recent years new systems have been introduced for use with both unibody and BOF structures. Today, gauges are usually limited to light or medium body damage.

10.9 UNIVERSAL MEASURING SYSTEMS

Universal measuring systems are the most efficient application of tram/centering gauge technology. They make parts of the measuring job much easier and more accurate, but still require a degree of skill and attention to detail. These systems have the ability to measure all the reference points at the same time. But to get the proper measurement reading, the equipment must be set to the manufacturer's specifications.

With a universal measuring system, all the reference points can be checked by just moving around the vehicle. You can quickly determine where each reference point on the vehicle is in comparison to the measuring system.

If a reference point on the vehicle is not in the same position as the dimension chart says it should be, the reference point on the vehicle is wrong. When the system is set up properly, you can monitor the key points by simply looking at the pointers. If the pointers are out of position, then the vehicle is not dimensionally correct. A reference point that is out of position must be brought back to preaccident specifications.

Before beginning any universal measuring operations, be sure to:

1. Remove detachable damaged body parts, both mechanical and sheet metal body panels.
2. If the damage is severe, perform rough straightening to the center section or foundation of the vehicle.
3. If the mechanical parts are left in the vehicle and an overhang condition exists, this must be compensated for.

Universal measuring systems fall into three groups:

- Mechanical systems
- Laser systems
- Sonic systems

MECHANICAL MEASURING SYSTEMS

With most mechanical measuring systems, several mechanical pointers are attached to a precision

FIGURE 10-78 Universal measuring offers the advantages of measuring a third dimension for better collision damage analysis. This type of tool mounts to the vehicle, which allows measurements to be taken at any time during repair. *(Courtesy of Chief Automotive Systems, Inc.)*

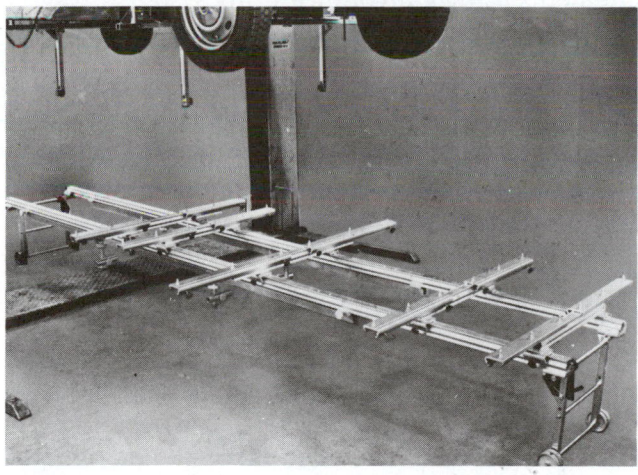

FIGURE 10-79 A free-standing-type universal bridge mounts next to but not on vehicle. *(Courtesy of Nicator, Inc.)*

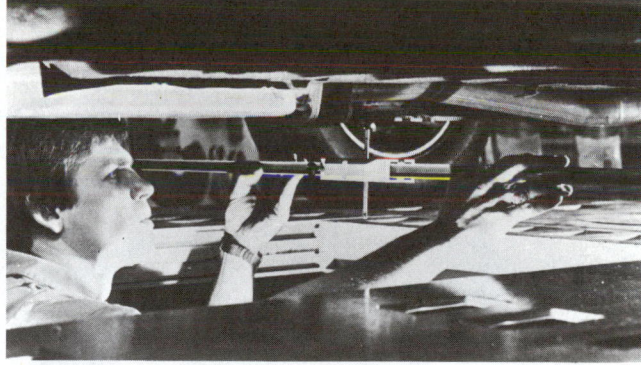

FIGURE 10-80 With modern universal systems, gauge adjustments are simple and readings are easy to make. *(Courtesy of Chief Automotive Systems, Inc.)*

measurement bridge (Figure 10–78). Free-standing-type bridges are available (Figure 10–79). The pointers on the measuring system are positioned on the bridge or framework according to the vehicle's correct factory specifications for horizontal and vertical dimensions. This allows simultaneous observation of a number of reference points on the damaged vehicle.

Care must be taken to ensure that the measurement bridge is not stressed or damaged during the repair process. The accuracy of this system is dependent upon the location and precision of the pointers on the measurement.

The advantage of a universal system over a tram gauge is that readings are instantaneous; the pointers either align with the reference points or they do not (Figure 10–80).

In practice, a universal measurement system offers the technician the advantage of being able to visually inspect all the reference points by just walking around the car and looking at the pointers. It can be quickly determined where each reference point is in comparison to where it should be (Figure 10–81). If the reference point is not aligned with its pointer, the reference point on the car is wrong and must be straightened.

Universal measuring systems vary from the complex units shown in Figure 10–82A to simple ones like the unit shown in Figure 10–82B. The latter is actually a tram/centering gauge system that is fastened to the vehicle. Since equipment designs vary, you must read the owner's manual for each piece of equipment to learn the specific procedures for proper use.

Most mechanical universal measuring systems work on both the unitized and conventional frame vehicle. They measure the lower and upper body reference points of a vehicle as identified in dimensions manuals and make comparison measurements of components from one side of a vehicle to the other. They measure all three dimensions of the vehicle: length, width, and height.

Figures 10–83 and 10–84 show how measurements are made in a typical mechanical measuring system.

A universal mechanical measuring system assesses a damaged vehicle by showing how far components are out of alignment. It also remains on the vehicle to guide the technician and verify that components are back in their proper places when the repair is complete. As you use frame straightening equipment to pull or push the structure into alignment, you can watch the pointers to check your work.

A **pivot measure system** uses rotating rods and pointers to measure vehicle damage (Figure 10–85, page 341). Its use is similar to the ones just described. The main difference is how the pointers are mounted and positioned for measuring.

Basically, the poles and pointers are mounted and calibrated from vertical and horizontal reference posts. Then the bars can be used to measure reference points in the vehicle.

Figure 10–86 (page 341) shows how this system is used to measure various points on the front of a car. Figure 10–87 (page 342) shows other sample measurements using a pivot measuring system.

When using any measuring system, record keeping is critical. A **damage analysis form** (Figure 10–88, page 343) is handy because it helps you organize specs, actual measurements, and differences in good and actual measurements. The form will help you more quickly and accurately determine what must be done to straighten the frame or unibody.

FIGURE 10–81 Although the equipment looks complex at first glance, it is easy to use once you know the basics and study the user's manual. *(Courtesy of Chief Automotive Systems, Inc.)*

A

B

FIGURE 10–82 (A) This type of equipment mounts under the vehicle on its own framework. (B) This type hangs under the vehicle. *(Courtesy of Chief Automotive Systems, Inc.)*

LASER MEASURING SYSTEMS

The laser measuring system uses a beam of light and targets mounted on the vehicle to measure damage. It is an extremely accurate system when properly installed and used. The word **laser** stands for light amplification by stimulated emission of radiation. A laser is good for measuring because it shoots out a perfectly straight beam of light.

Body shop laser measuring systems operate in basically the same way. The laser source is aimed at a target that is either hung or attached to the car. Some systems even use parts of the vehicle as targets.

Measurements are taken by observing the laser beam on the target. Some targets (Figure 10–89, page 344) are clear, allowing the laser beam to shine through. Several clear targets can be used with one light source. Mirrored targets (Figure 10–90, page 344), on the other hand, are capable of reflecting the laser beam to additional targets. (In some laser systems these targets are called laser guides.) Using combinations of transparent and mirrored targets, it

INTRODUCTION

This color insert has been designed to lead you through the major steps for repairing a collision damaged vehicle. You will learn the typical sequence used to return a wrecked car to its original condition. This will give you a good overview of a typical repair process and help you to understand the more complex, specific collision repair tasks.

1 A high degree of skill and knowledge is needed to properly repair today's collision-damaged cars and trucks.

2 Proper methods must be used to tow the vehicle with a wrecker without causing further damage.

3 The estimator must study the damage carefully to determine the cost of repairs.

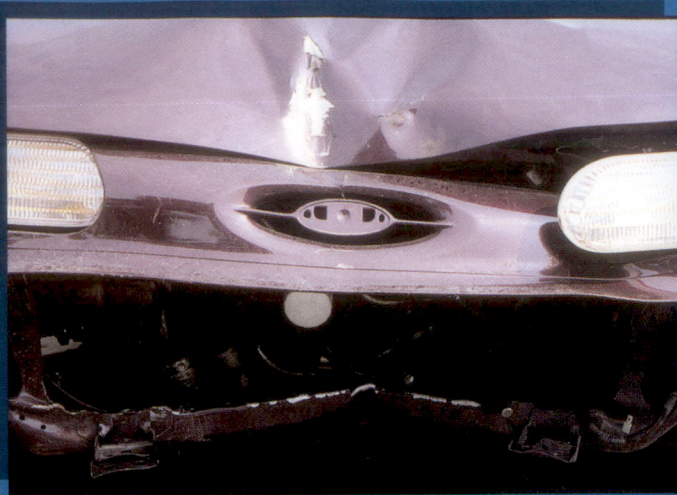

4 Direct damage occurs at the point of impact on the vehicle. It is the most obvious damage.

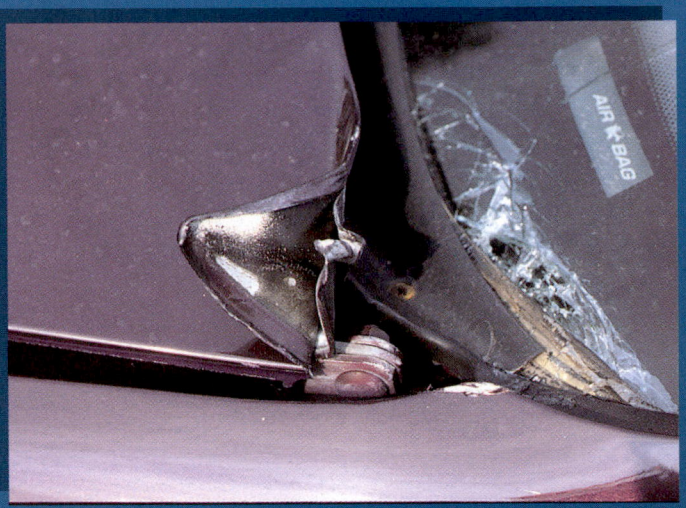

5 Indirect damage is damage at points other than the area of impact. The estimator must find all direct and indirect damage.

6 Estimators commonly use an electronic storage unit equipped with a bar code reader to input data from the parts list quickly.

7 The vehicle interior is often damaged in a major collision. It should be included in the estimator's inspection. Here, air bag and sensor replacement costs must be added to the repair expenses.

8 Damage to the engine, engine mounts, pulleys, frame rails, and other parts is more visible from underneath the vehicle.

9 Once the estimator has listed all the damaged parts, the data can be downloaded into a personal computer in the business office.

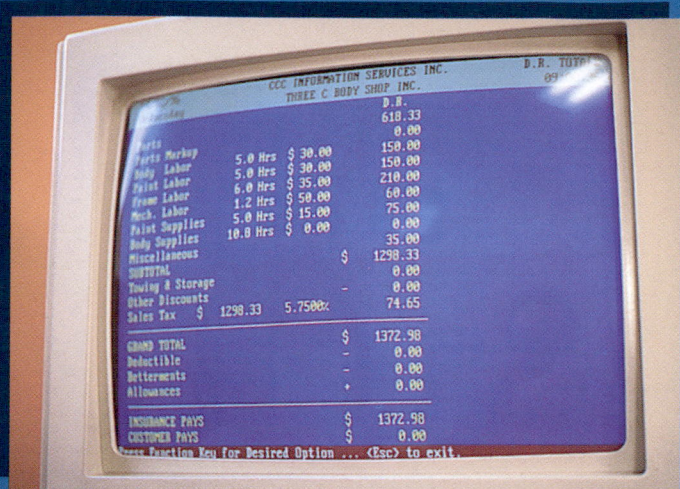

10 The computer helps tabulate labor rates and part costs for repairs.

11 Once the repair costs have been tabulated, the estimator gives a written estimate to the customer or the insurance company adjuster.

12 Before starting repairs, wash the vehicle thoroughly to remove road dirt. Keeping the shop clean is important to refinishing quality.

13 The body shop can be a safe and enjoyable place to work if safety rules are followed. If not, it can be a very dangerous place!

14 Always know the location and use of fire extinguishers in your shop. A few seconds' time can mean a lifetime during a fire.

15 Remove damaged outer body parts to gain access to hidden parts that require straightening or replacement.

16 With the most severely damaged parts out of the way, repair of the unibody damage can begin.

17 Here the technician is attaching a sensor to the damaged vehicle to allow the laser to measure the vehicle and assess the damage.

18 A measuring system shows where straightening is needed. Compare known good measurements with those taken from the damaged vehicle to determine how to proceed.

19 Frame straightening equipment designs vary but basic procedures are similar. The vehicle must be anchored securely before pulling by using devices such as pinch weld clamps (shown here).

20 After the vehicle has been anchored, attach pulling clamps and chains to the area to be straightened.

21 Apply pulling power carefully to remove damage in the reverse order of which it occurred. Use a small amount of overpull before stopping so parts will flex back closer to their original shape.

22 Controlled heat may be needed to relieve stress from damaged parts when pulling.

23 Measure constantly when pulling out damage. This will help you pull in the right direction with the right amount of power.

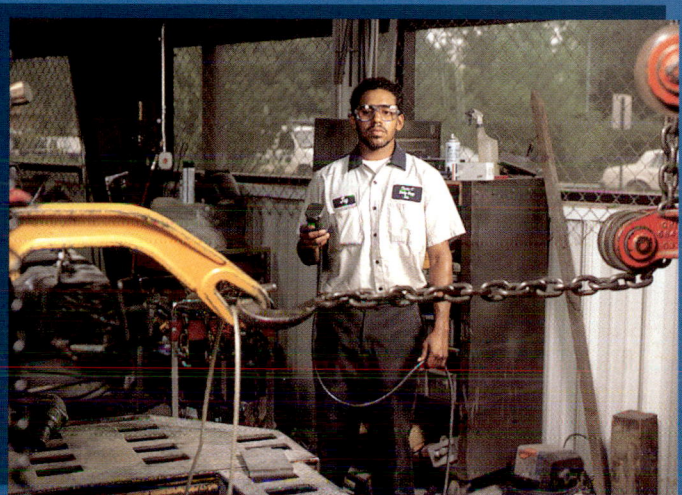

24 To prevent injury, stand to one side of the pulling chains. They are under tremendous tension and could fly off with deadly force.

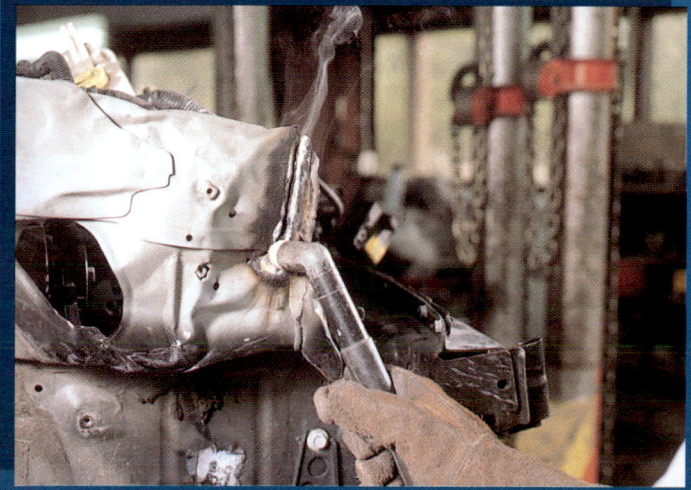

25 After the frame or unibody is straightened, the damaged parts that require replacement can be cut off. Here the technician uses a plasma arc cutter to slice a part off for replacement.

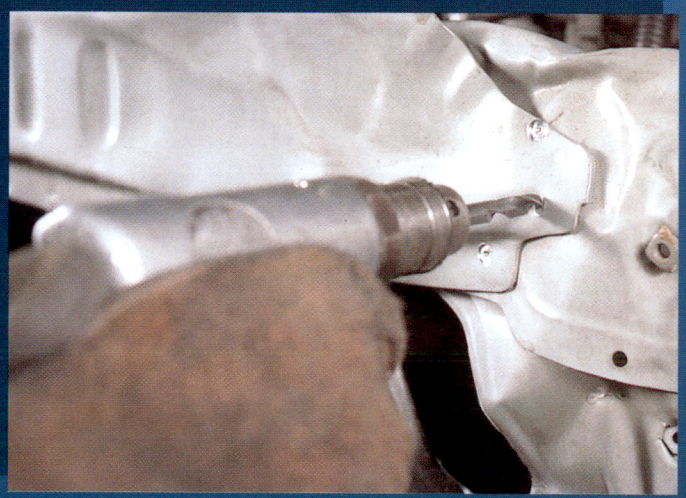

26 Many parts are attached with spot welds. A spot weld remover, a specialized drill, is often used to cut out each spot weld for part removal.

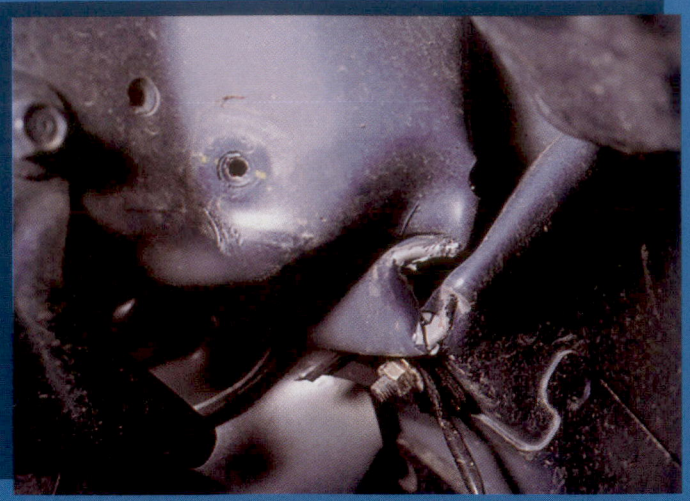

27 Frame rails are often damaged in frontal collisions. They often have crush zones to absorb impact forces. Proper procedures must be used during frame rail replacement.

28 Manufacturers will recommend the proper locations for either sectioning or complete replacement of their frame rails. Always follow manufacturer's directions.

29 Mechanical parts such as motor mounts must sometimes be disconnected or removed to repair major structural damage. This is often easier to do after the appropriate body parts are removed.

30 Proper welding skills are needed to do major collision damage repair. Successfully completing a welding class can open up new career opportunities.

31 After tack welding the part, measure its location to determine if any adjustments are needed before making the final weld.

32 After replacing the frame rail, other parts can be replaced as needed. Here, the apron is being fitted into place. Clamping pliers are often used to hold parts while welding.

33 When replacing damaged parts, start with the innermost layers of the body structure and work outwards. Make sure that each part is welded into its proper position before attaching the next part.

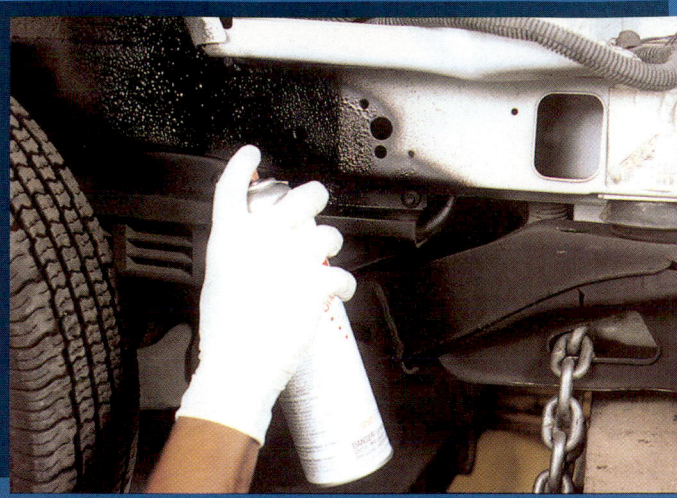

34 Corrosion protection must be restored to all parts during repairs. Various products are available to prevent corrosion or rusting of parts.

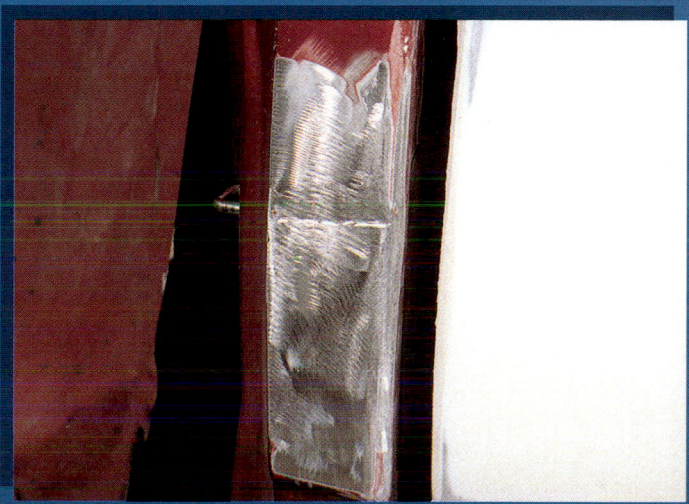

35 Many panels will have minor damage that can be repaired. The cost of labor and materials must be weighed against the cost of new or salvaged parts.

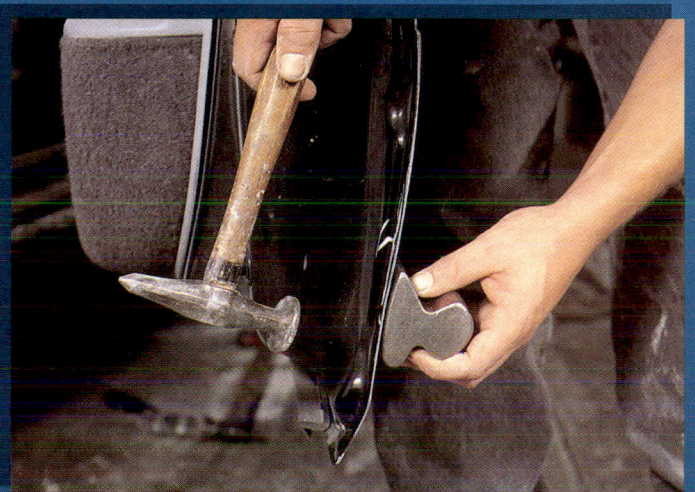

36 Hammers and dollies are used to work out minor sheet metal damage. A dolly is used on the back side of the panel to raise low spots, while controlled hammer blows lower high spots to level out the panel surface.

37 For damage that cannot be accessed from behind, a nail or pin gun is helpful. The gun welds metal pull pins onto the damaged panel.

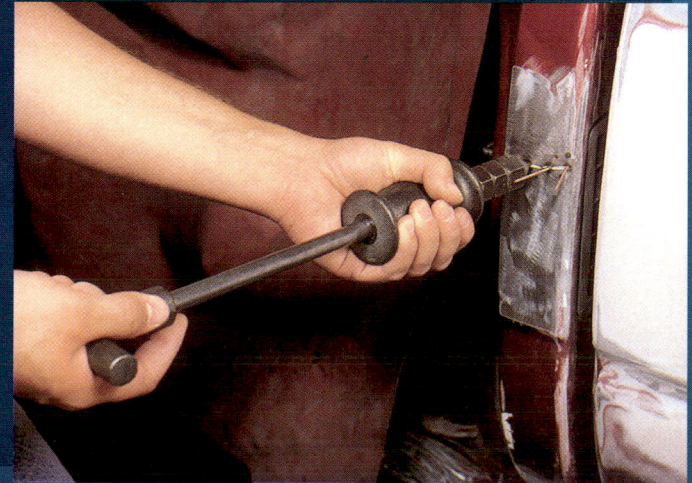

38 After welding the pins into the dent, attach a puller to each pin and pull out the low spot. The metal must be straightened to within 1/8 inch of level before plastic filler can be used.

39 After pulling out the damage, cut off the pins and grind them until they are flush with the panel surface.

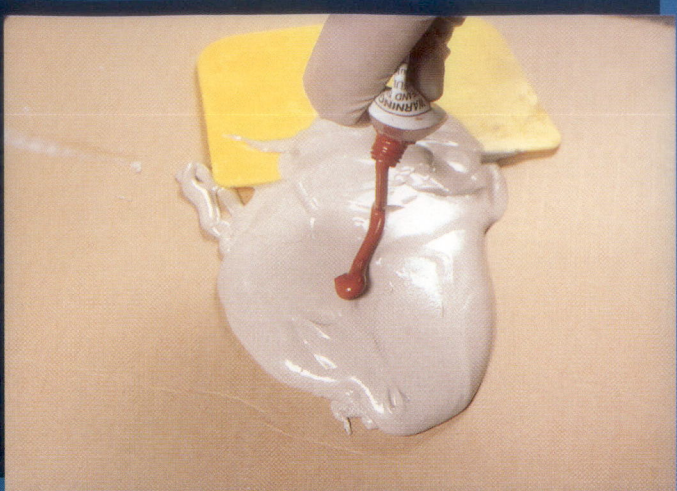

40 Mix plastic filler and hardener in correct proportions. Generally, a "golfball-size glob" of filler requires a 1" (25mm) bead of hardener. A "baseball-size glob" requires a 6" (152mm) bead of hardener.

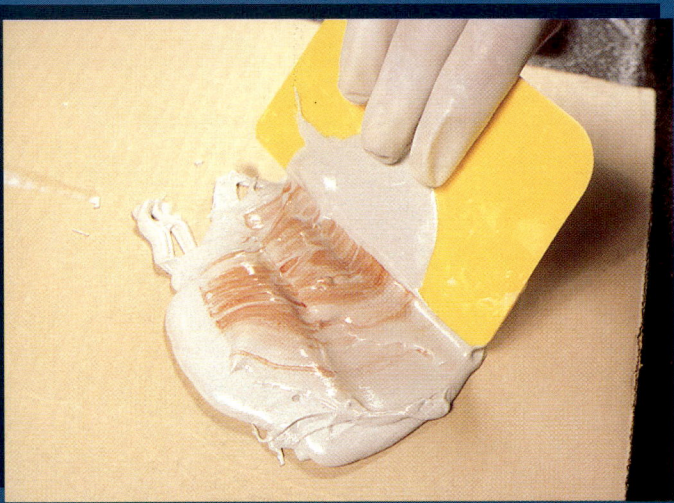

41 Do not stir the hardener into the plastic filler; this can cause air bubbles and pin holes. Instead, wipe the two back and forth as shown to mix the ingredients completely.

42 Apply the plastic filler to the clean body surface right away. It will start to harden in a few minutes. Try to spread it out to match the body shape to reduce sanding time later.

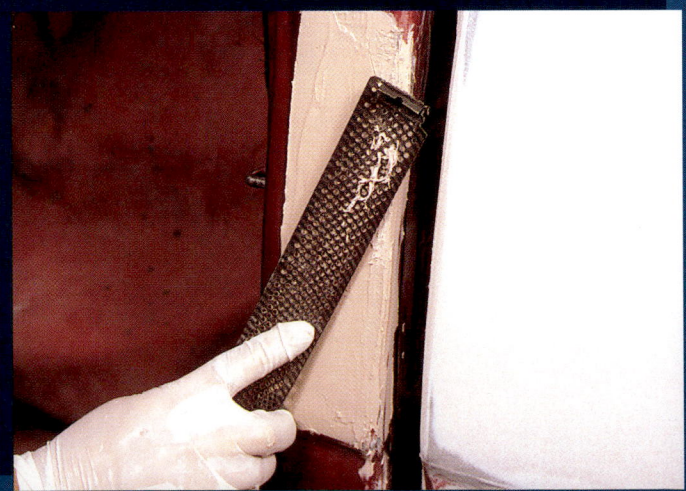

43 When the filler starts to solidify, rough out the shape with a "cheese grater" file. File until the filler is contoured properly. The filler should stick up above the panel surface a little so it can be sanded smooth.

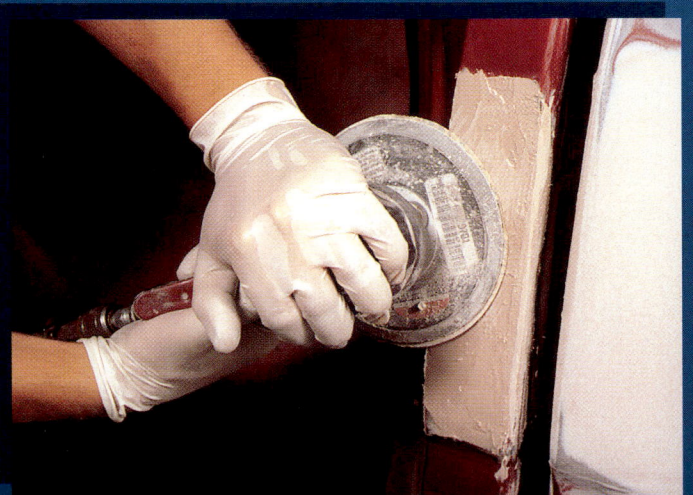

44 After grating the filler, use an air sander and medium grit sandpaper to cut the filler down more. Featheredge the filler and old paint for a smooth transition into the panel.

45 Before priming, mask parts that might get covered with overspray.

46 Reduce, mix, and filter primer-surfacer properly. Use one paint system from the same manufacturer for best results. Materials from different manufacturers may not be totally compatible.

47 Spray test patterns to ensure that the spray gun is adjusted properly before actually priming the vehicle.

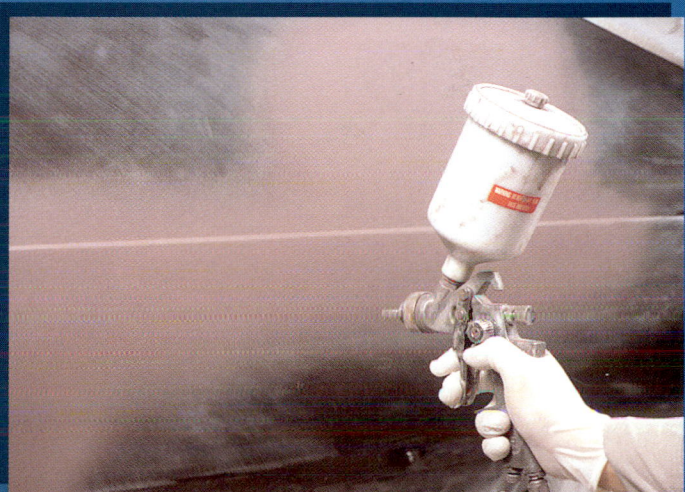

48 Apply one or two coats of self etch primer-surfacer to the repair area. Allow each coat to flash properly before applying the next coat.

49 Some technicians spray a light "guide coat" of a different color primer over the repair to use as a guide when wet sanding the surface smooth.

50 Wet sand the guide coat. It will sand off easily if the surface is level. If you cut through the guide coat, this indicates a high spot that needs more sanding; if it does not sand off, this indicates a low spot.

51 After doing body work and priming repair areas, remove old masking materials. These collect and hold dust and debris that could ruin the topcoat of paint.

52 After all structural parts are welded into place and minor damage has been straightened, start installing the parts held by fasteners. Be careful not to hit and dent parts during installation.

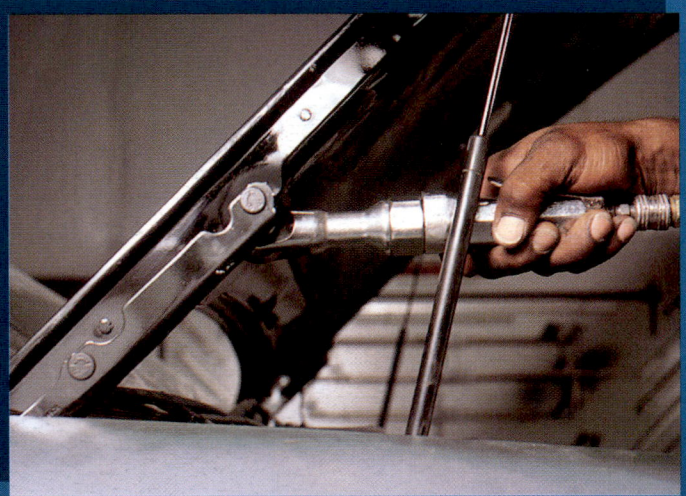

53 The hood and fenders must be aligned properly. The hood should be centered between the fenders and have the proper gap at the rear. Slotted holes in the hood hinges allow for adjustment.

54 When first closing the hood, make sure it does not hit the fenders or cowl. It is easy to damage parts if the hood is out of adjustment.

55 Before final masking and moving the vehicle into the paint booth, use a blow gun to clean the vehicle of dust and debris . Blow out all gaps between parts that could hold dust and dirt.

56 Spray-on masking material is handy when masking a large area, like underhood components. It is waterbased and can be washed off with soap and water after painting.

57 The vehicle can now be moved into a clean paint booth for final masking and painting.

58 Special wheel masks will save time and protect tires and wheels from overspray.

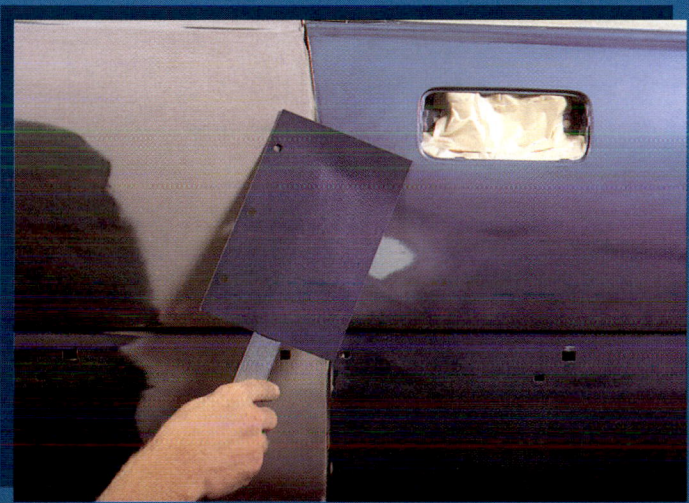

59 To match existing paint, especially pearl paints, always make a test panel or spray-down panel. This helps determine tinting and how many coats of clear are needed to match the color correctly.

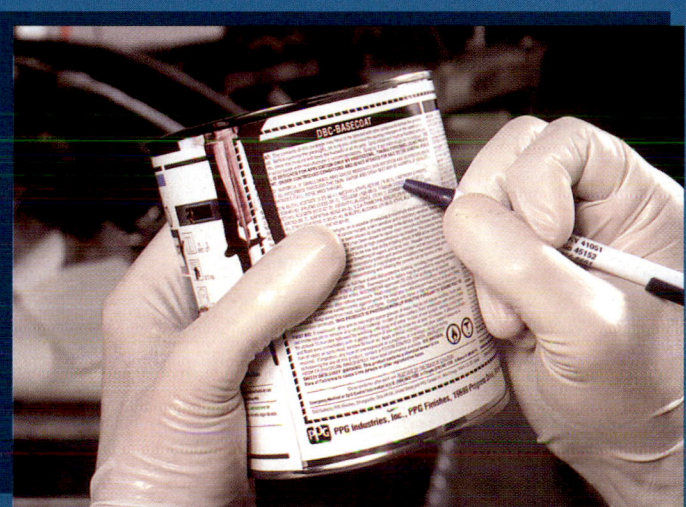

60 Mix paint according to label directions. The label gives ratios for reducer, catalyst, and paint.

61 The Touch Mix measuring system is used to measure paint viscosity.

62 Mixing sticks are another common way to mix paints. First, find the correct stick for the paint material. Then pour in the paint, reducer, and hardener until each is level with the appropriate mark on the stick.

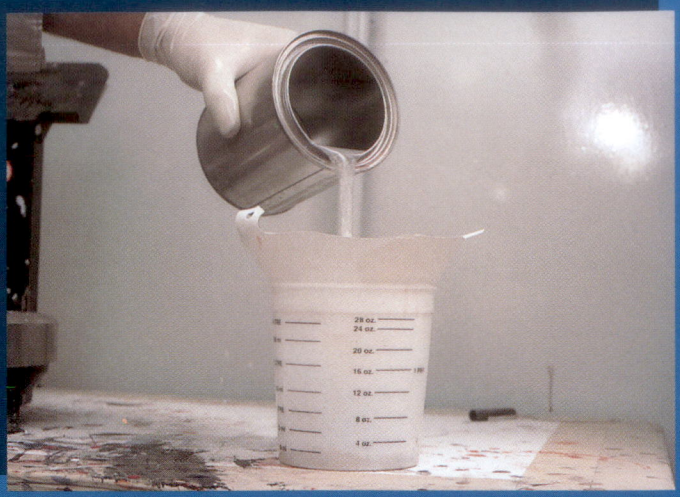

63 A graduated container can also be used to measure out paint, reducer, and hardener.

64 Filter everything that goes into the spray gun cup. One piece of dirt or cured paint can ruin a paint job.

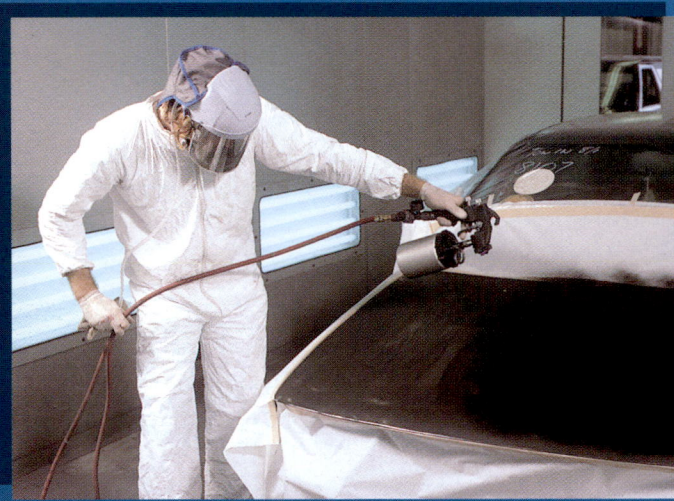

65 Apply a sealant to all new and repaired parts before painting. Be sure to wear protective clothing and equipment, including a safety shield and an air supplied respirator.

66 Wipe with a tack cloth after every step.

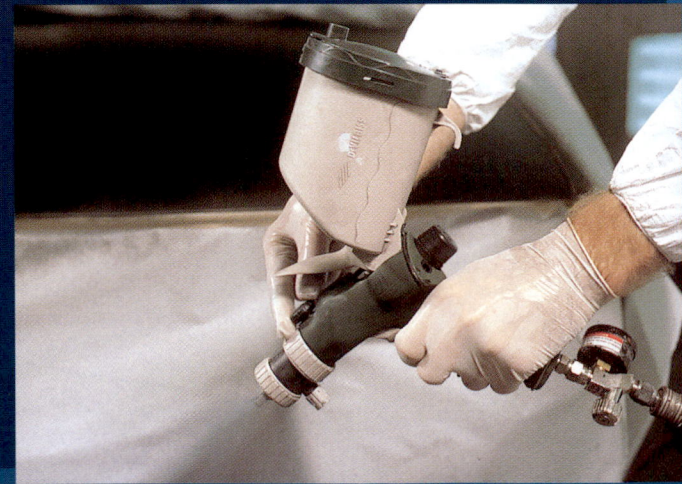

67 Adjust the spray gun pressure and spray pattern. Test the gun on an old part or piece of masking paper before spraying the vehicle.

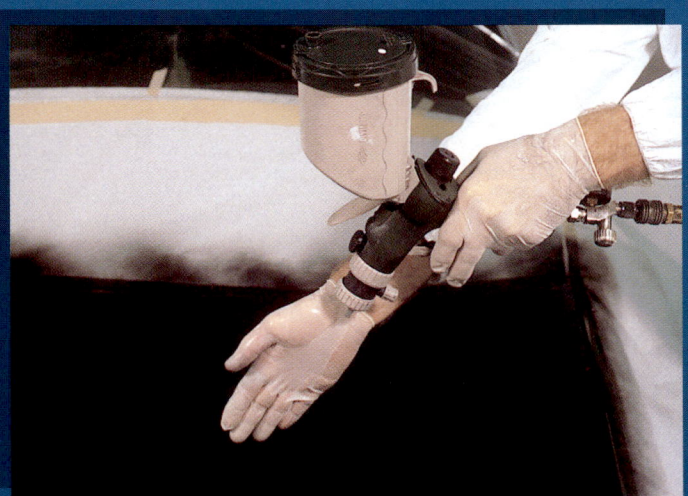

68 When painting the vehicle, keep the gun a correct distance from the surface. Also keep the gun parallel with the surface. Use the same application techniques as used on the test panel.

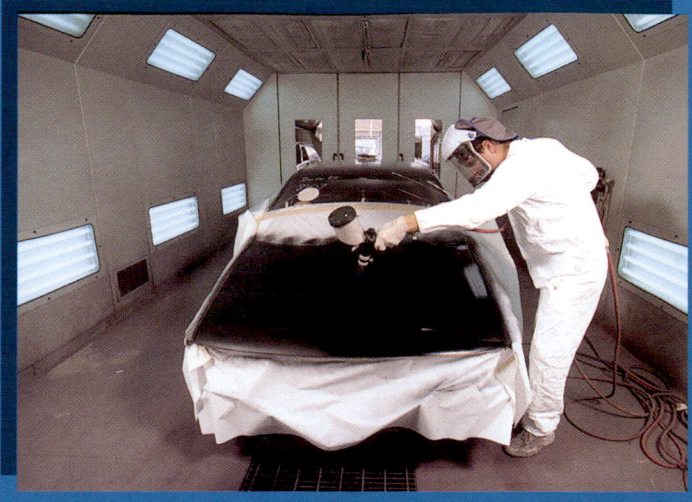

69 When painting, move the spray gun as evenly as possible. Make straight line paint strokes that overlap evenly. If either the gun speed or distance from the surface changes, paint problems will result.

70 For spot repairs with basecoat/clearcoat paint, base or color only the damaged area. Blend new paint into existing paint so the repair is not noticeable. It is usually best to clearcoat the entire panel.

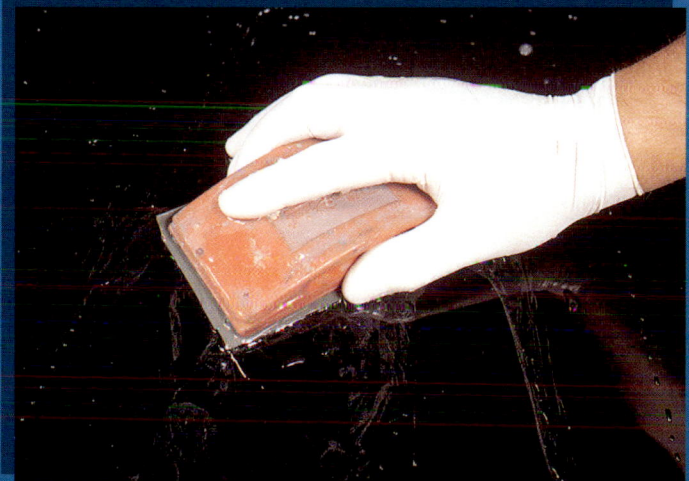

71 After the paint cures overnight or has baked, wet sand any imperfections with a sanding block and ultra-fine sandpaper. Be careful not to cut through the paint or repainting will be required.

72 After wet sanding, compound the area to return the paint gloss. When machine buffing, try not to cut through the paint. Some technicians apply masking tape to sharp edges to avoid cutting through the paint.

73 A final cleanup of the interior and exterior of the vehicle is important. The customer's first impression of your work should be pleasing.

74 Delivery of the vehicle to the customer. It should look and drive as well as it did before the collision.

PAINT FAULTS

Dust in finish

Runs

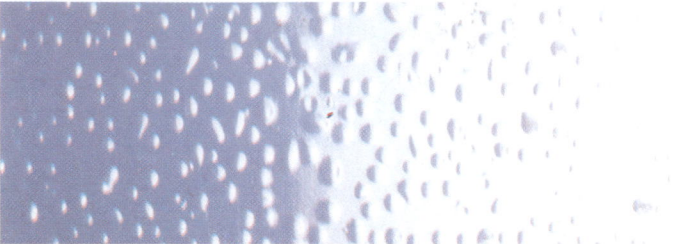

Blistering

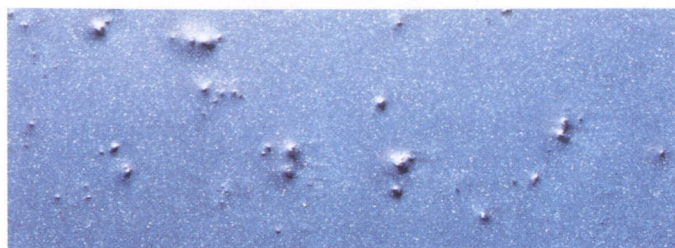

Mottling

Sand scratch swelling

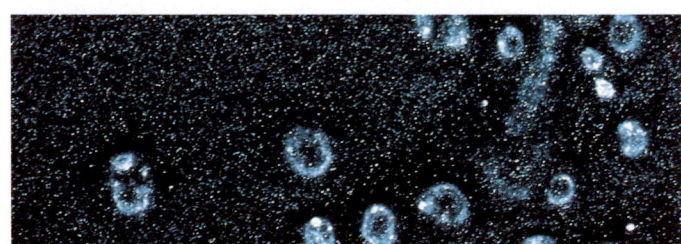

Fisheyes

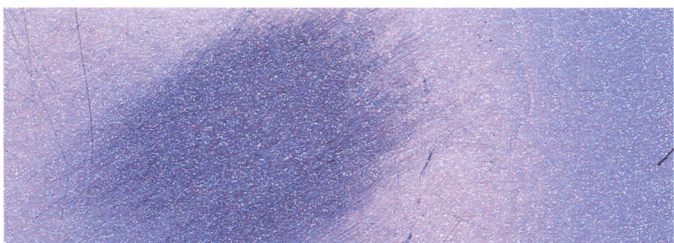

Cold cracking

Burned through

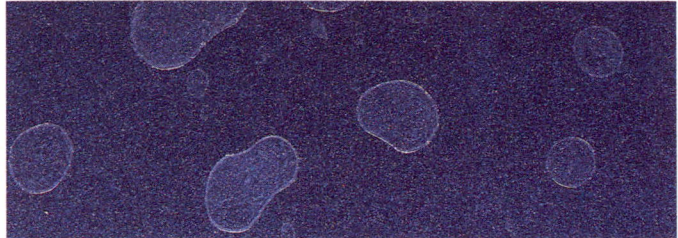

Water spotting

Metallic sag

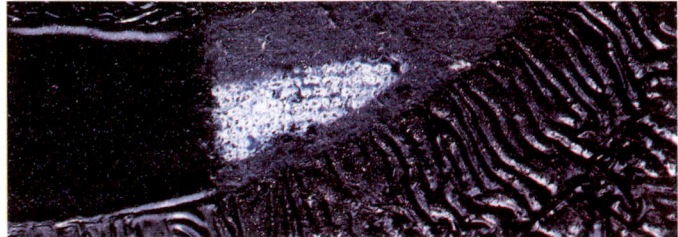

Lifting

Chemical spotting

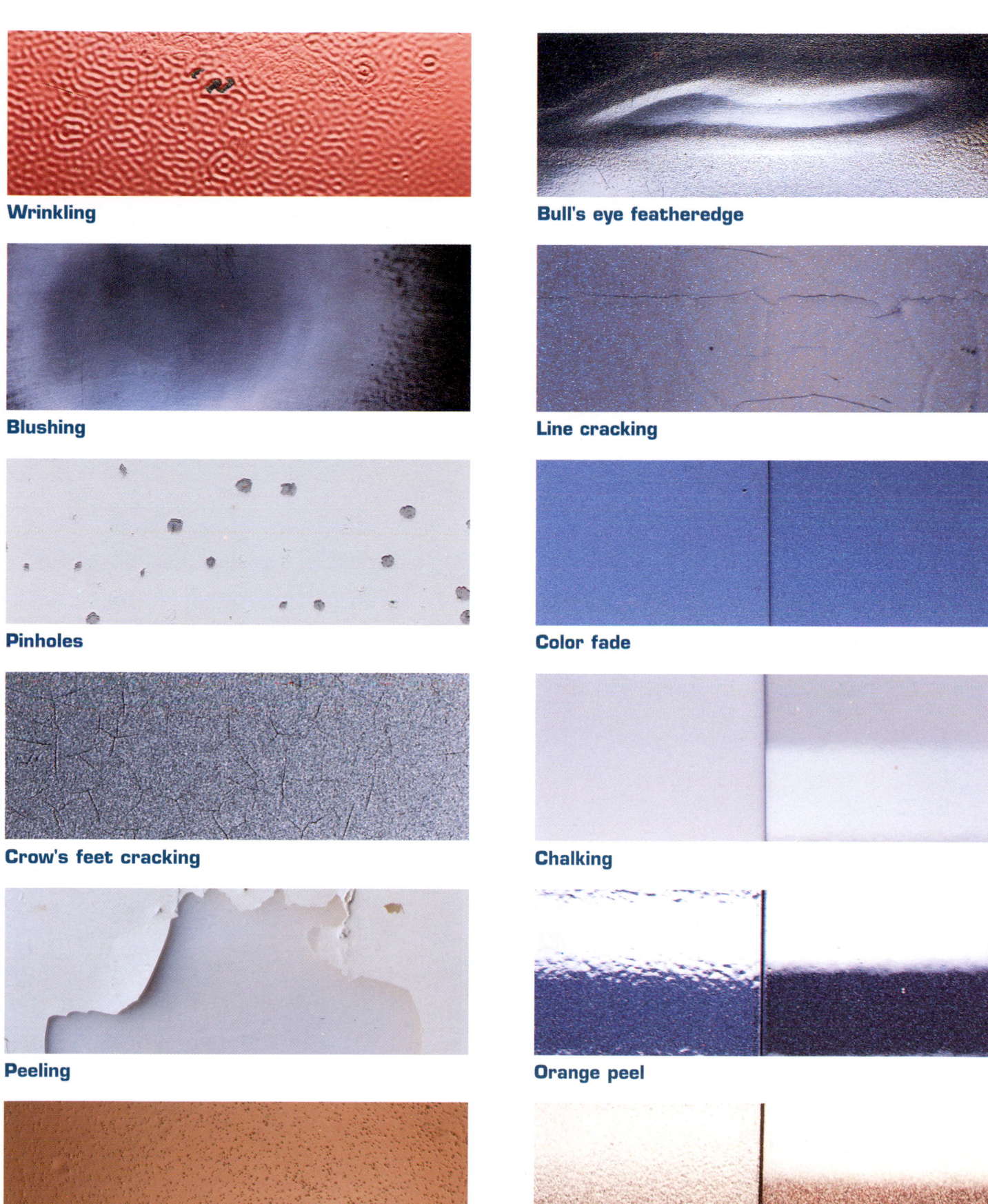

Wrinkling

Bull's eye featheredge

Blushing

Line cracking

Pinholes

Color fade

Crow's feet cracking

Chalking

Peeling

Orange peel

Seedy

Dulling

TINTING

BASE COLOR	ALUMINUM LETDOWN	MASS TONE	WHITE LETDOWN	BASE COLOR	ALUMINUM LETDOWN	MASS TONE	WHITE LETDOWN
1. Yellow Gold				15. Indo Orange			
2. Lt. Chrome Yellow	*Not to be used with Aluminum Letdown.*			16. Moly Orange (Red Shade)	*Not to be used with Aluminum Letdown.*		
3. Oxide Yellow				17. Red Oxide			
4. Indo Yellow				18. Transparent Red Oxide			
5. Transparent Yellow Oxide				19. Deep Violet			
6. Rich Brown				20. Quindo Violet			
7. Black				21. Magenta Maroon			
8. Strong Black				22. Phthalo Green (Yellow Shade)			
9. Organic Orange (Light)				23. Phthalo Green			
10. Oxide Red				24. Scarlet Red	*Not to be used with Aluminum Letdown.*		
11. Permanent Red				25. Perrindo Maroon			
12. Organic Scarlet				26. Phthalo Blue (Medium)			
13. Phthalo Blue (Green Shade)				27. Phthalo Green			
14. Permanent Blue				28. Phthalo Green (Yellow)			

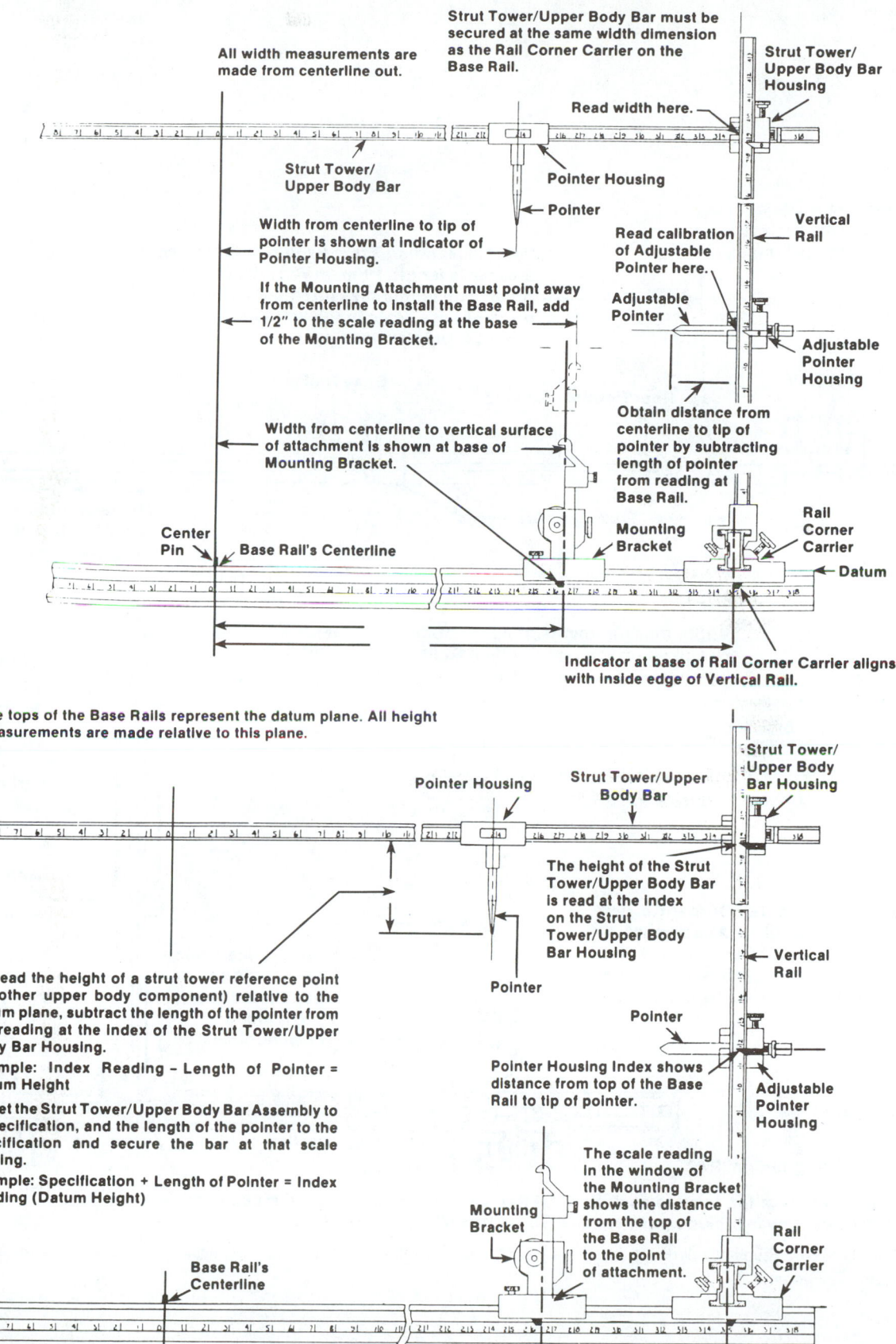

All width measurements are made from centerline out.

Strut Tower/Upper Body Bar must be secured at the same width dimension as the Rail Corner Carrier on the Base Rail.

Strut Tower/Upper Body Bar Housing

Read width here.

Strut Tower/Upper Body Bar

Pointer Housing

Pointer

Width from centerline to tip of pointer is shown at Indicator of Pointer Housing.

If the Mounting Attachment must point away from centerline to install the Base Rail, add 1/2" to the scale reading at the base of the Mounting Bracket.

Read calibration of Adjustable Pointer here.

Vertical Rail

Adjustable Pointer

Adjustable Pointer Housing

Width from centerline to vertical surface of attachment is shown at base of Mounting Bracket.

Obtain distance from centerline to tip of pointer by subtracting length of pointer from reading at Base Rail.

Center Pin

Base Rail's Centerline

Mounting Bracket

Rail Corner Carrier

Datum

Indicator at base of Rail Corner Carrier aligns with inside edge of Vertical Rail.

The tops of the Base Rails represent the datum plane. All height measurements are made relative to this plane.

Pointer Housing

Strut Tower/Upper Body Bar

Strut Tower/Upper Body Bar Housing

The height of the Strut Tower/Upper Body Bar is read at the Index on the Strut Tower/Upper Body Bar Housing

Pointer

Vertical Rail

To read the height of a strut tower reference point (or other upper body component) relative to the datum plane, subtract the length of the pointer from the reading at the Index of the Strut Tower/Upper Body Bar Housing.

Example: Index Reading – Length of Pointer = Datum Height

To set the Strut Tower/Upper Body Bar Assembly to a specification, and the length of the pointer to the specification and secure the bar at that scale reading.

Example: Specification + Length of Pointer = Index Reading (Datum Height)

Pointer

Pointer Housing Index shows distance from top of the Base Rail to tip of pointer.

Adjustable Pointer Housing

The scale reading in the window of the Mounting Bracket shows the distance from the top of the Base Rail to the point of attachment.

Mounting Bracket

Rail Corner Carrier

Base Rail's Centerline

FIGURE 10-83 Study how widths and heights are measured on a typical mechanical universal system. *(Courtesy of Chief Automotive Systems, Inc.)*

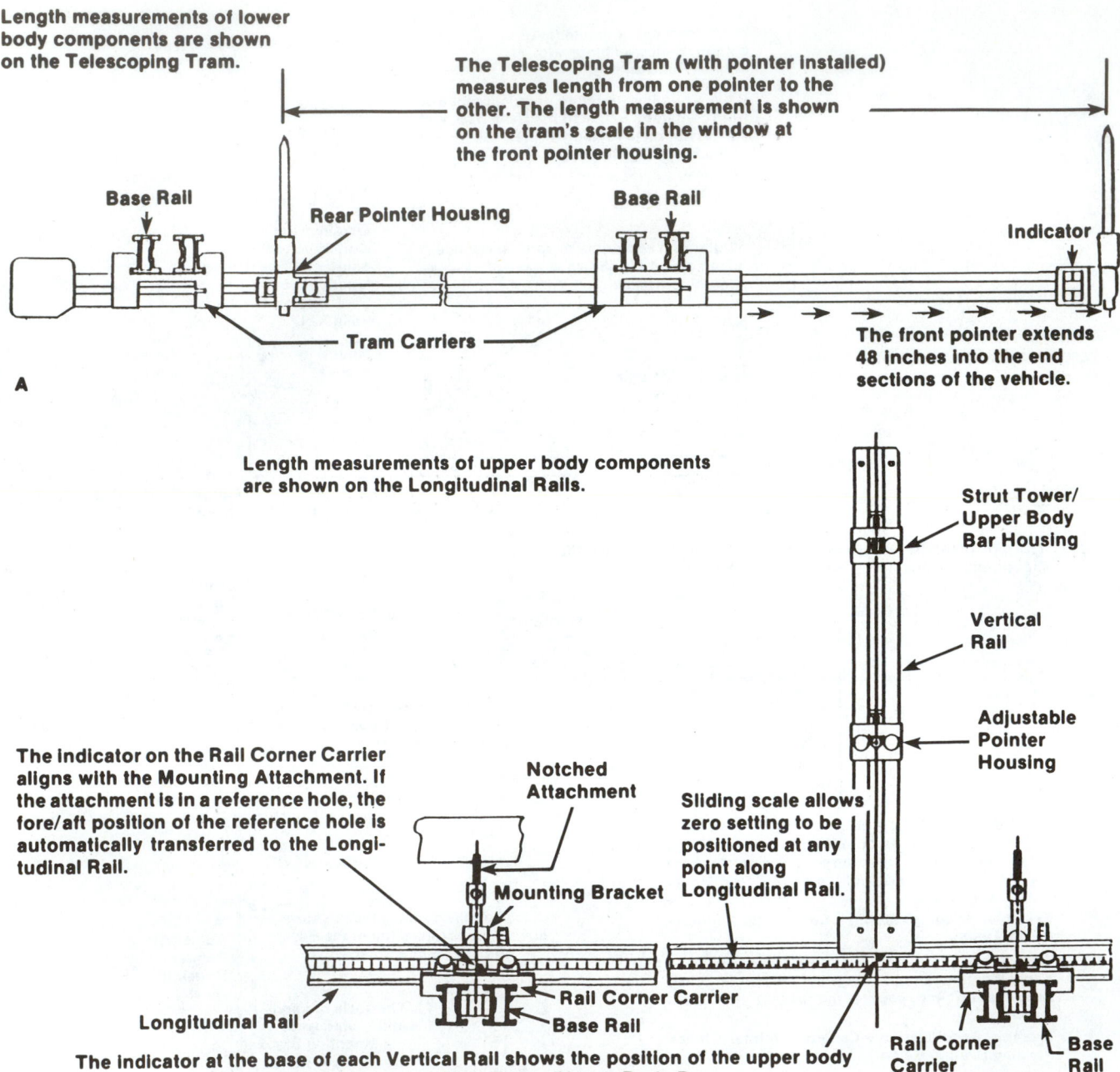

Length measurements of lower body components are shown on the Telescoping Tram.

The Telescoping Tram (with pointer installed) measures length from one pointer to the other. The length measurement is shown on the tram's scale in the window at the front pointer housing.

Base Rail

Rear Pointer Housing

Base Rail

Indicator

Tram Carriers

The front pointer extends 48 inches into the end sections of the vehicle.

A

Length measurements of upper body components are shown on the Longitudinal Rails.

Strut Tower/ Upper Body Bar Housing

Vertical Rail

Adjustable Pointer Housing

The indicator on the Rail Corner Carrier aligns with the Mounting Attachment. If the attachment is in a reference hole, the fore/aft position of the reference hole is automatically transferred to the Longitudinal Rail.

Notched Attachment

Mounting Bracket

Sliding scale allows zero setting to be positioned at any point along Longitudinal Rail.

Rail Corner Carrier

Base Rail

Longitudinal Rail

Rail Corner Carrier

Base Rail

The indicator at the base of each Vertical Rail shows the position of the upper body indicators. These include pointers and Strut Tower/Upper Body Bar.

B

FIGURE 10-84 Study how lengths are measured (A) using a telescoping tram and (B) using longitudinal rails. *(Courtesy of Chief Automotive Systems, Inc.)*

FIGURE 10-85 This type of measuring system uses long rods that pivot to different angles to measure any location on the vehicle. *(Courtesy of Wedge Clamp)*

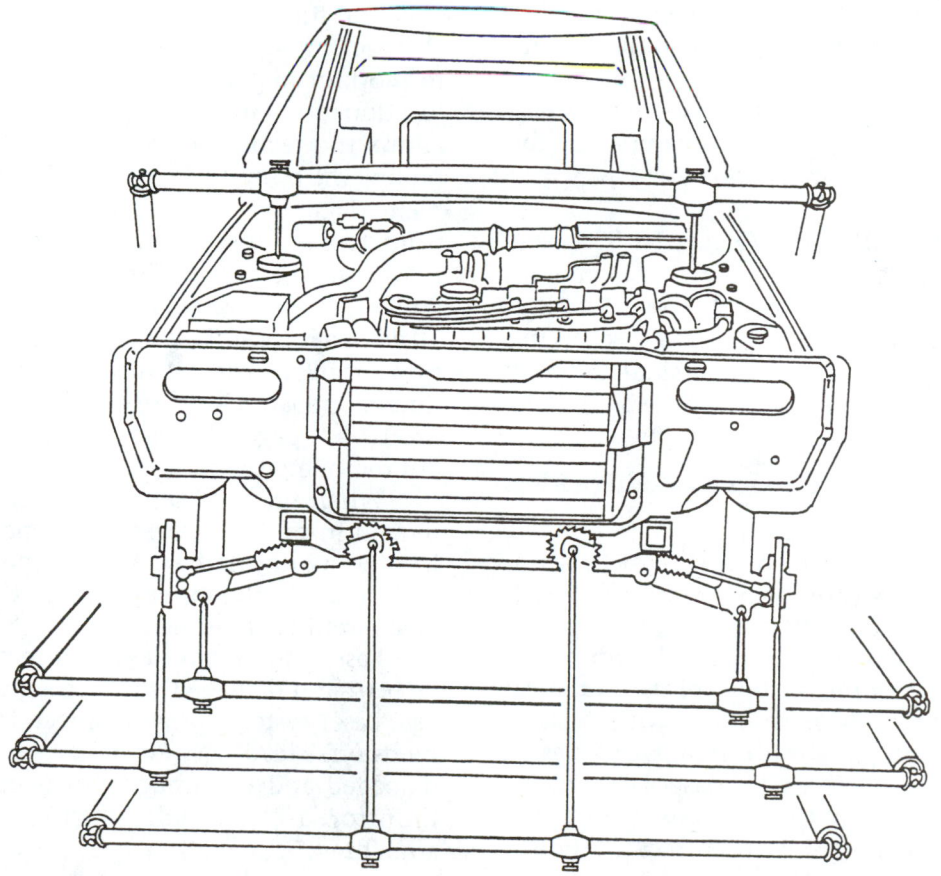

FIGURE 10-86 Note how the pivot system can be positioned to measure different points. *(Courtesy of Wedge Clamp)*

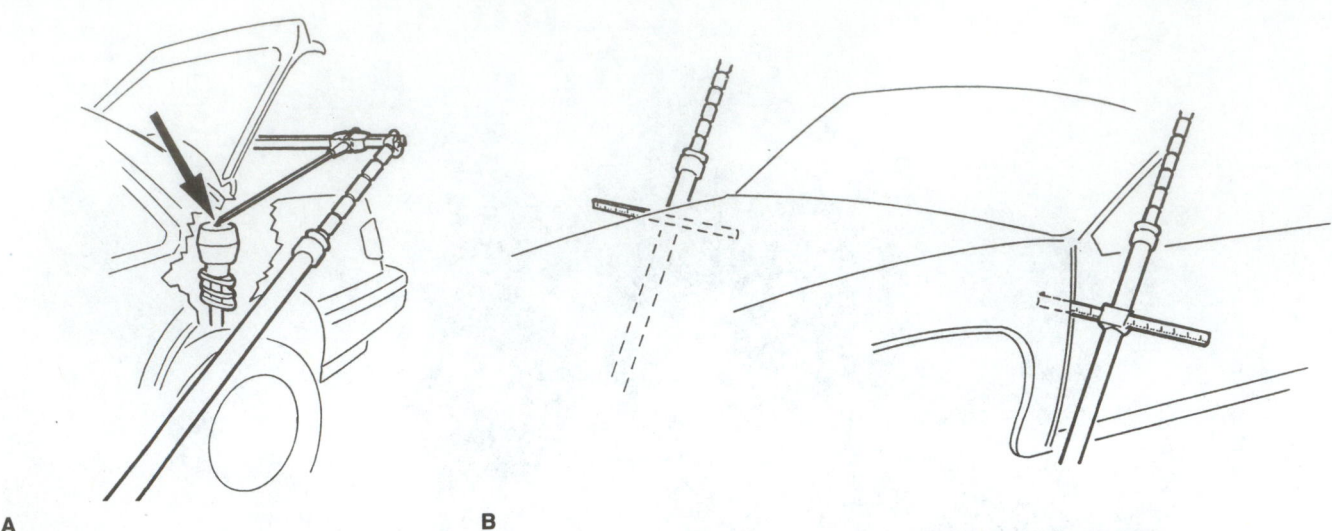

FIGURE 10-87 (A) Measurement is being taken at the rear strut towers, a hard-to-reach area, with a pivot-type system. (B) Measurement at one side of the cowl by extending the pointers through door gaps. *(Courtesy of Wedge Clamp)*

is often possible to simultaneously measure several dimensions on a vehicle using a single laser source.

Some laser systems use up to three laser guns to give length/width/height coordinates anywhere on the vehicle—decks, cowls, door openings, hinges, posts, and rooflines (Figure 10–91, page 344). With such a three-dimensional system, a single gun can be used to make measurements in conjunction with other measuring devices, such as a metal tape (Figure 10–92, page 344).

Laser measuring systems are made up of both optical and mechanical parts. For example, in the clear target system, the optical parts are:

- **Laser power unit** or gun (Figure 10–93A, page 345), which emits a safe, low-powered laser beam
- **Beam splitters** (Figure 10–93B, page 345) that project beams at a precise right angle so height, width, and length can be measured simultaneously
- **Laser guides** (Figure 10–93C, page 345) to deflect the laser beam at exactly 90 degrees

The mechanical items on laser measuring devices include:

- *Calibrated bars* that attach to the car or act as support devices for the laser gun itself
- Measurement data *specification sheets*
- *Transparent scales* that create a datum plane, which brings the measuring points on the car to a level that can be illuminated with the laser beam. Scales are hung from fixtures or holes on the underside of the car. The laser beam passes through the center of the scale target area when the measuring point is in its correct position.

After the laser measuring system is set up and the transparent scales hung under the vehicle (in accordance with data sheets), measuring can be started. First, use two sets of undamaged measuring points under the vehicle to adjust the lasers and guides (Figure 10–94, page 345). Calibration is completed when the laser beam passes through the two target scales on the undamaged area of the vehicle.

Some laser measuring systems will permit you to monitor upper body information, such as pillar locations for windshields. Upper body information allows the technician to have more control when pulling the body back into location for proper alignment (Figure 10–95, page 345). It assures that the front door just installed will fit or that the windshield to be installed is snug. Some laser systems also offer an integral four-wheel alignment capability (Figure 10–96, page 345). This is beneficial because suspension problems can be measured and corrected for unibody repair. Using an accessory as shown in Figure 10–97 (page 345), struts can be measured without removing them.

When properly set up, most laser systems can remain in position during the repair operation unless the mounting hardware or the laser targets interfere with the operation of the pulling and straightening equipment.

Laser equipment has fewer mechanical parts to be bent and damaged while straightening the damage. Laser systems provide direct, instantaneous dimensional readings. Reference points in both the damaged and undamaged areas of the car can be monitored continually during the pulling and straightening operation.

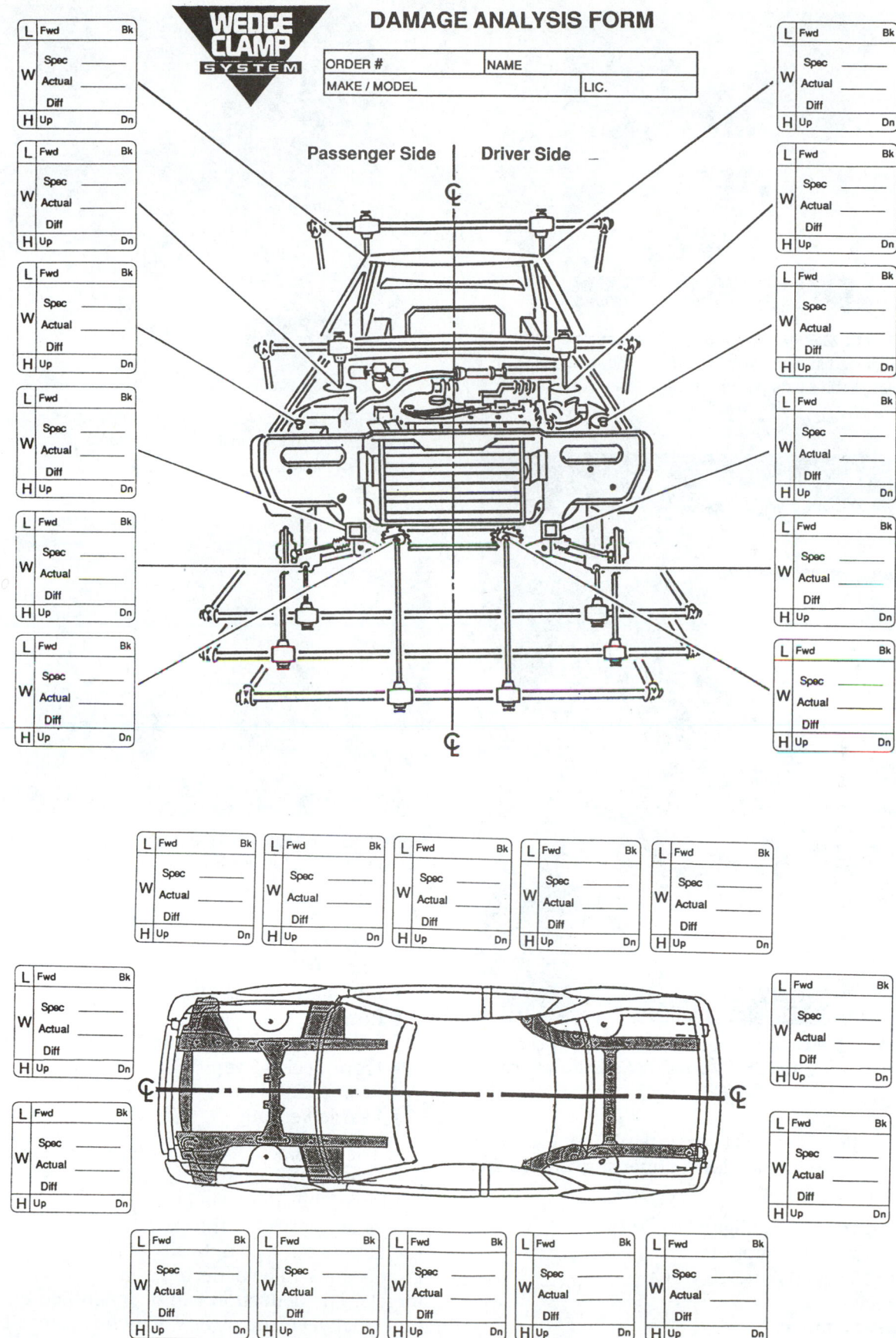

WEDGE CLAMP SYSTEM

DAMAGE ANALYSIS FORM

| ORDER # | | NAME | |
| MAKE / MODEL | | LIC. | |

Passenger Side   Driver Side

FIGURE 10-88 A damage analysis form will simplify the frame/body straightening task. Write down specs from the manual. Then measure and record values taken from the vehicle. By subtracting the two measurements, you can determine the direction and extent of damage. *(Courtesy of Wedge Clamp)*

FIGURE 10-89 These are transparent scales or sighting units used in some laser systems. They are mounted or positioned at reference points on the vehicle. *(Courtesy of Nicator, Inc.)*

FIGURE 10-91 A three-dimensional laser system permits length/width/height measurements. *(Courtesy of Kansas Jack, Inc.)*

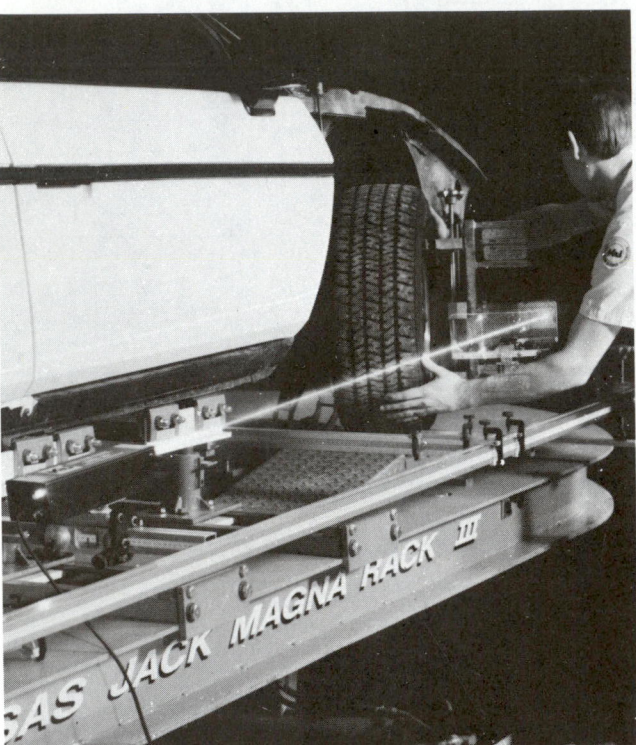

FIGURE 10-90 Mirrors allow one laser to reflect to several points on the vehicle. *(Courtesy of Kansas Jack, Inc.)*

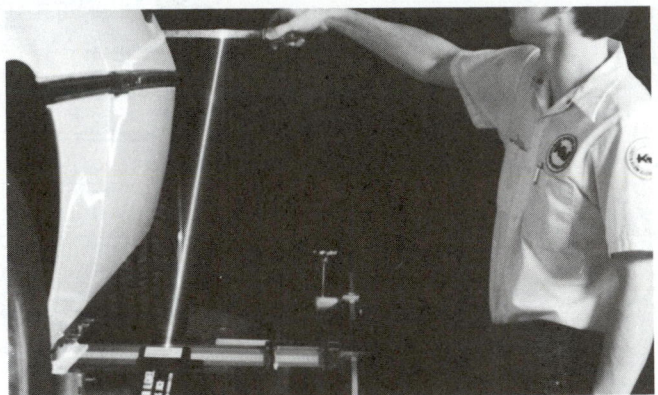

FIGURE 10-92 Laser is bright light. Never let a laser hit you straight in the eye. *(Courtesy of Kansas Jack, Inc.)*

- Door hinges
- Door strikers
- B-pillars
- Rockers
- Rear wheel offset (Figure 10–98D)
- Front deck lid gap
- Rear deck lid gap
- Rear lower rails

In addition, a typical laser system (Figure 10–98A) can make approximately forty dimension checks, including the following:

- Lower front rails (Figure 10–98B)
- Upper radiator support
- Strut towers (Figure 10–98C)
- Wheels
- Cross members
- Fender gaps
- Cowl

The accuracy of laser system measurement depends upon the accuracy of the targets and the calibration of the laser systems to the targets. When laser beams are projected through one target to another, the targets must be optically perfect; otherwise, the laser beam will be deflected as it passes through the targets. The deflection will be magnified by the distance between the targets, which can result in serious error. Laser targets that become scratched or warped should be discarded.

A　　　　　　　　**B**　　　　　　　　**C**

FIGURE 10-93 Note optical parts of one type of laser system: (A) laser gun; (B) beam splitter; and (C) laser guide. *(Courtesy of Nicator, Inc.)*

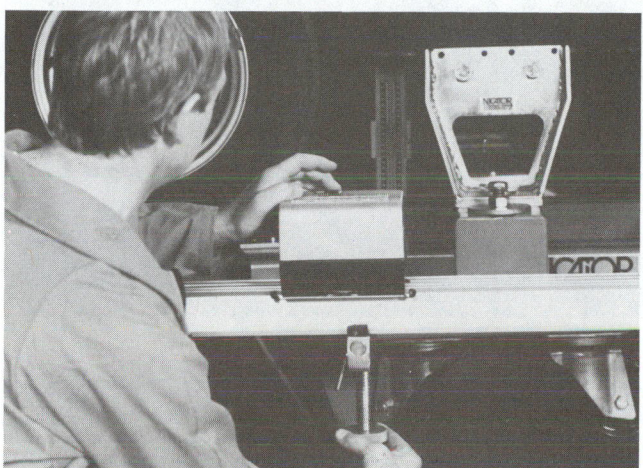

FIGURE 10-94 When properly set up, laser produces perfectly straight beams of light that will let you accurately check body/frame alignment values. *(Courtesy of Nicator, Inc.)*

FIGURE 10-96 This laser measuring system is measuring four-wheel alignment. *(Courtesy of Kansas Jack, Inc.)*

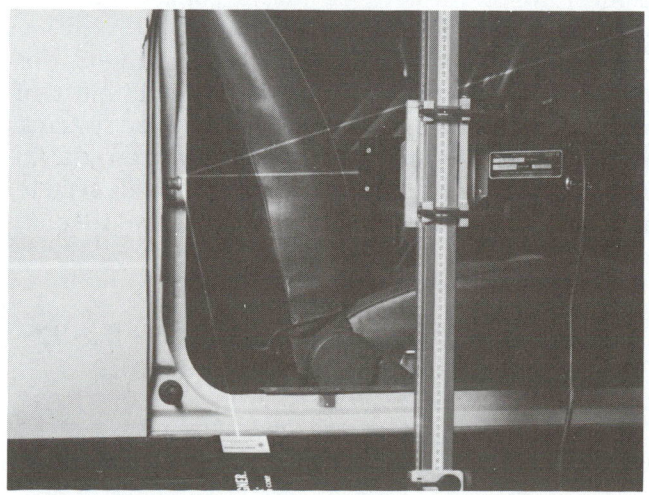

FIGURE 10-95 Here a three-dimensional check is being made of a side body point. *(Courtesy of Kansas Jack, Inc.)*

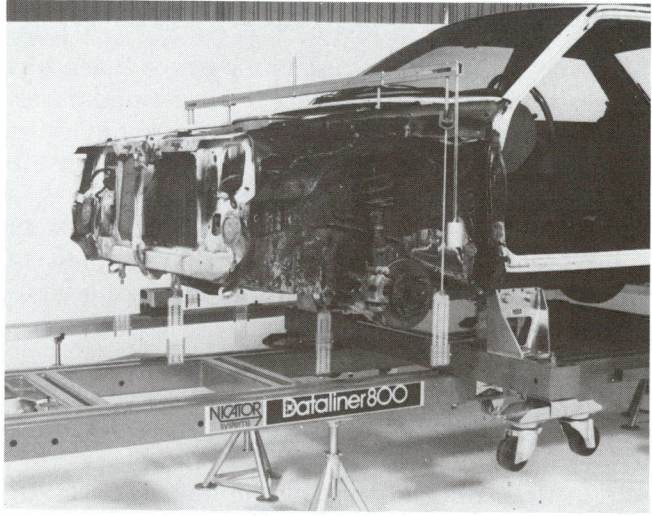

FIGURE 10-97 A laser measuring system is being used in conjunction with a strut gauge balanced on two bolts with a counterweight. *(Courtesy of Nicator, Inc.)*

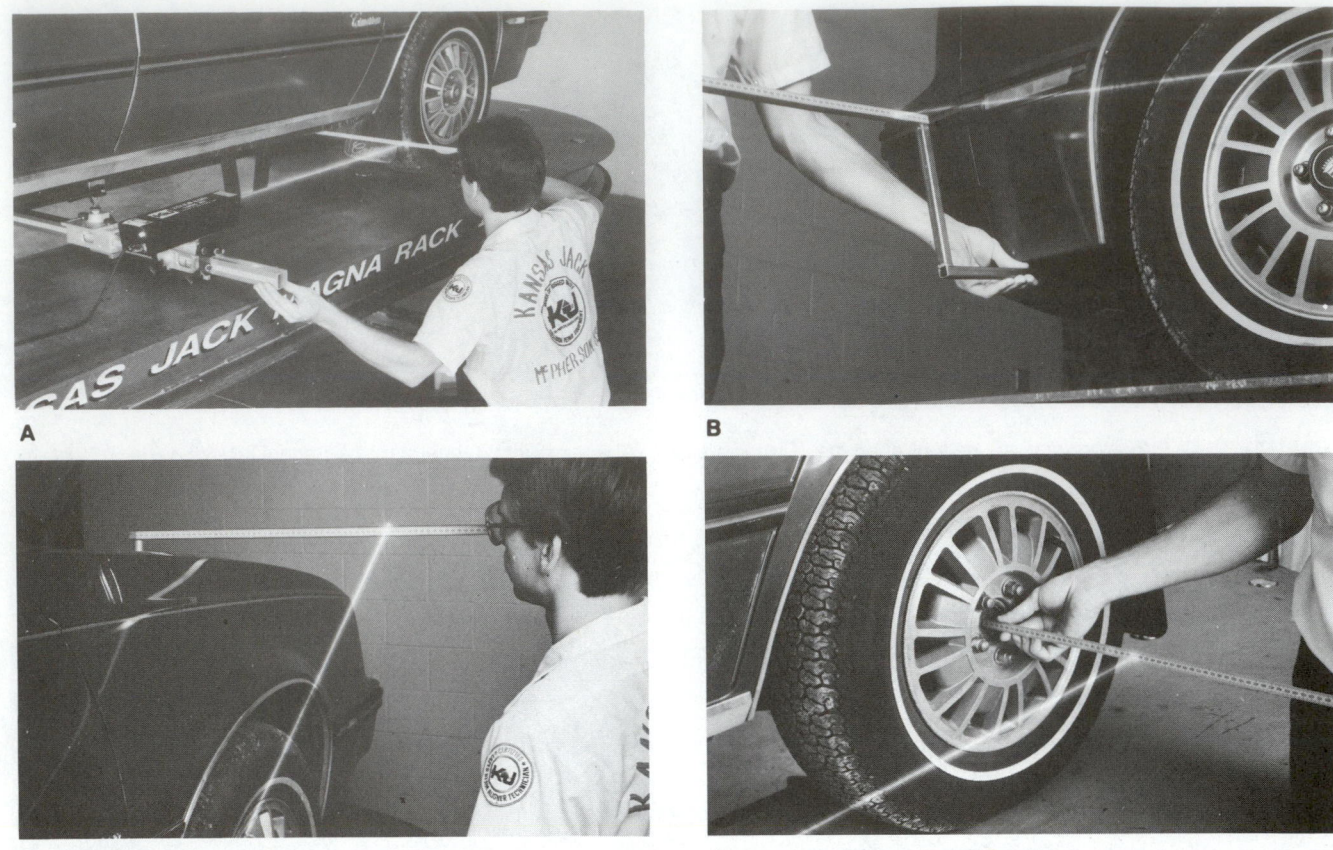

FIGURE 10-98 Checking for structural damage using a laser system is quick and accurate: (A) setting up and aligning carriage and laser; (B) checking lower front rails; (C) another method of checking strut towers; and (D) checking rear wheel offset. *(Courtesy of Kansas Jack, Inc.)*

SONIC MEASURING SYSTEMS

Sonic measuring systems use sound waves as a means of measuring vehicle dimensions. Since sound travels at a constant speed, the time it takes for sound waves to travel between different reference points on the vehicle can be measured quickly and accurately.

The setup and use of a sonic measuring system is similar to a laser system. However, in this relatively new system, transmitting-sensing units are used to make measurements.

MEASURING SYSTEM VARIATIONS

While the important vehicle dimensions can be found in body dimensions manuals, most measuring equipment manufacturers have specific dimension charts for their equipment. These charts, one for each vehicle model manufactured, serve as guides to use before and during the repair.

The dimensions chart usually illustrates two views of the vehicle underbody. Some charts also give under hood and upper body dimensions. The latter is most important with the plastic paneled mill and drill pad dimensions.

Many equipment manufacturer's dimensions charts are intended only for that specific piece of equipment. Because of this variation between systems, it would be difficult to explain how each manufacturer's system measures a vehicle. You will need to read the equipment owner's manual for these details.

10.10 DEDICATED BENCH AND FIXTURE MEASURING SYSTEMS

The *dedicated bench* and *fixture system* acts a "go–no-go" gauge. It is a completely different type of measuring method. Instead of taking actual measurements, dedicated fixtures are used to check body or frame alignment (Figure 10–99).

The **dedicated bench** consists of a strong, flat work surface to which fixtures are attached.

FIGURE 10-99 A dedicated bench is often used by large new-car dealerships. *(Courtesy of Blackhawk Automotive Inc.)*

Fixtures are thick metal parts that bolt between the vehicle and the bench to check alignment. They are designed from the vehicle manufacturer's drawings. The fixtures allow the technician to physically check mountings or other key locations of the underbody (Figure 10–100).

If the fixtures fit the vehicle properly, the technician knows that the underbody and strut towers are in perfect alignment. All that is necessary is to straighten the vehicle until the reference points match the fixtures. No other underbody measurements are usually required.

The dedicated bench and fixture measuring system requires a specific set of fixtures for each family of body styles. If the collision repair shop does not own the required fixtures, they can be rented. The bench has built-in reference positions, and the fixtures are positioned to the specific references according to instructions supplied by the manufacturer. At least three fixtures must be positioned on the bench before the vehicle is mounted. Generally these are the torque-box fixtures. However, the damage to a specific vehicle will dictate which fixtures should be used.

Figure 10–101 shows four of the more common types of fixtures.

- *Bolt-on fixtures* are used when the attachment is required for suspension or bumper mountings. The studs or bolts normally used to attach these parts to the car are also used to attach the fixtures. Depending on the damage, the fixtures can be either bolted to the car first and lined up with the bench during the repair or attached to the bench first and lined up with the car during repair.

- *Pin-type fixtures* are used most often to mate with manufacturing control holes in the underbody.

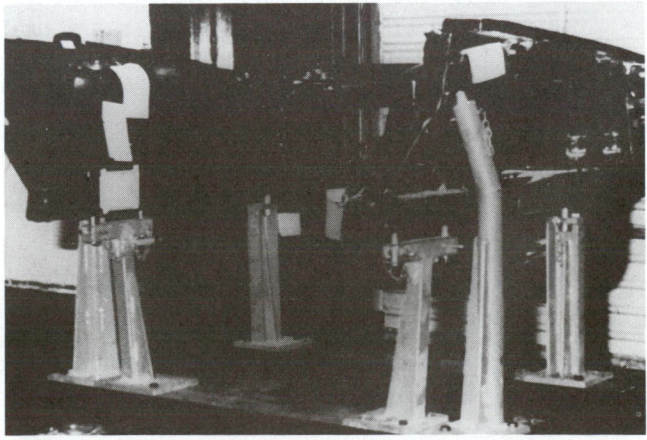

FIGURE 10-100 Fixtures must be bolted or clamped in place. *(Courtesy of Blackhawk Automotive Inc.)*

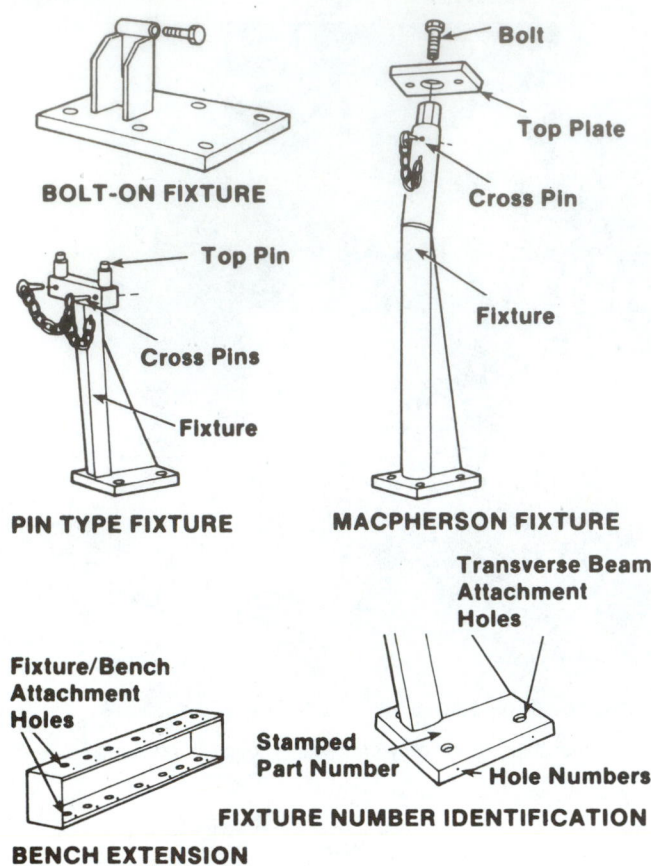

BOLT-ON FIXTURE

PIN TYPE FIXTURE

MACPHERSON FIXTURE

BENCH EXTENSION

FIXTURE NUMBER IDENTIFICATION

FIGURE 10–101 Study typical fixtures and bench extensions. *(Courtesy of Blackhawk Automotive Inc.)*

They can also be used to mate with suspension mounting holes. These fixtures have the advantage that they can be left in place if overpulling is necessary.

- *Strut fixtures* are used in the same manner as a pin-type fixture; a typical MacPherson fixture consists of a bottom plate assembly, a sliding shaft with a cross hole and cross pin, and a bolted-on top plate.

- *Bench extensions* are included with certain fixture sets where the length of the car requires that fixtures be positioned beyond the bench surface. The extensions are always used at the rear of the car. Each extension is drilled with seven holes on the top and seven holes on the bottom. These holes are directly in line with each other and are numbered beginning with 0 in the center to 3 at either end. The extension is always used between the transverse beam and the bench. The positions are shown on the data fixture diagram.

Most fixtures share a common attachment method to the transverse beams. The base plate for each fixture is drilled with four, eight, or twelve holes and is marked with a part number and also the hole number location on the transverse beam. The part

number may denote whether the fixture is for the right or left side of the vehicle.

Stamped at the rear of each base plate are the numbers that correspond to the numbers on the transverse beam. Always make sure these numbers are on the same side as the numbers on the beam. In cases where there are more than four holes on the base plate, minor length differences in some models require two or three sets of four holes. They are stamped in the base of the fixture and noted on the data sheet.

A typical fixture consists of 14 to 25 units that can be used individually or together. Many are designed so that they can be used either with mechanical parts in place or removed. One set of fixtures can be used to measure several models or body styles within a given car family. Each set is shipped in a color-coded container determined by the car manufacturer. In some areas of the country, fixtures are available on a rental basis.

FIGURE 10–102 New parts can often by held in place using fixtures. *(Courtesy of Blackhawk Automotive Inc.)*

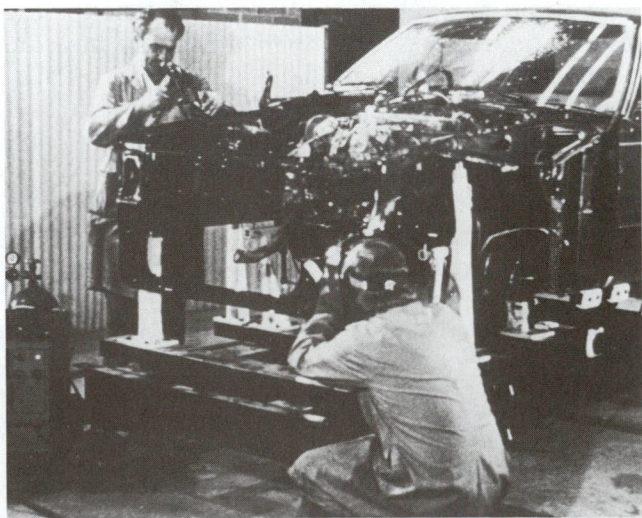

FIGURE 10–103 Once bolted in place and held with fixtures, they can be MIG welded to the vehicle. *(Courtesy of Blackhawk Automotive Inc.)*

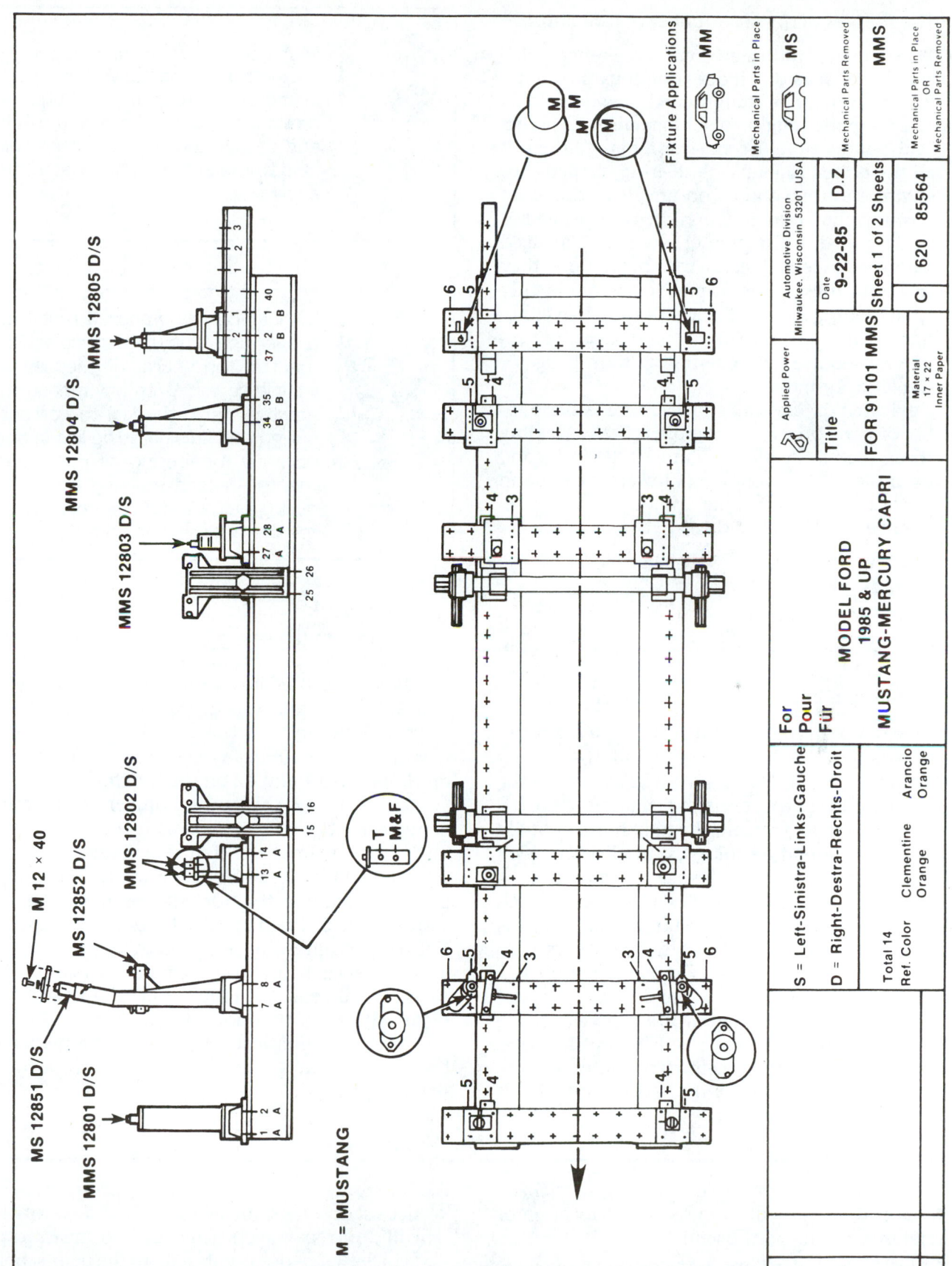

FIGURE 10-104 A dedicated bench diagram will give specific instructions for proper use. *(Courtesy of Blackhawk Automotive Inc.)*

The fixtures perform a number of functions:

- They visually indicate where a reference point should be located. If it does not line up with the fixture, it must be straightened.
- They provide gauging of all the reference points at the same time. No measuring is required. If all the points line up, the steering, suspension, engine mounts, and so on are all in the exact position they were in when the car was made.
- As described in Chapter 11, once the damaged parts are properly lined up with the fixtures, they can hold these parts in position while further straightening is done (Figure 10–102). This eliminates the "pull-measure" "pull-measure" sequence required with centering/tram gauge or universal systems.
- Dedicated fixtures can allow accurate subassembly of parts on a bench before actually welding those pieces together (Figure 10–103). A good example of this would be a lower box rail and tower assembly on some unibody cars. The sequence would go like this:

Position and hold the lower rail pieces on the fixtures.

Weld them together.

Position and hold the strut tower pieces on top of the rail.

Weld them in their correct positions.

One of the most common errors in bench setup is misreading of the diagram (Figure 10–104). Along the bottom edge of the diagram is a list of each of the fixtures included in the set: an explanation of the letters D for right and S for left; the total number of fixtures in the set and their color; the models on which the fixture set can be used; the part number (title) of the fixture set; and an explanation of the terms, such as MM for mechanical parts, MS for mechanical parts removed, and MMS for fixtures that can be used either with the mechanical parts in place or removed.

The upper illustration in the diagram is a side view and the lower illustration is a top view of the bench with transverse beams and fixtures in place. The front of the bench is to the left of the diagram. Each fixture is numbered with a part number that is

Manufacturers of all measuring systems are constantly furnishing informational updates and bulletins on their products. Be sure to read and study them because they will help to make the repair procedure easier.

It must be remembered that the proper use of any measuring equipment is the secret of successful vehicle repair. With any measurement system, the key to correct pulling and straightening lies in accurately monitoring all measurements—before starting the pull, during pulling, and immediately after the pull has been made.

also stamped in the base plate of the fixture. In addition, it has numbers corresponding with those on the bench holes and on the A or B holes of the transverse beams. (For example, the first transverse beam in the diagram shown indicates holes 1A and 2A. This means that the A holes in the transverse beam should be lined up with holes #1 and #2 on the bench.)

The top view shows the number on the transverse beam that corresponds with the hole pattern in the base plate of the fixture. These numbers are also stamped in the rear edge of the fixture bottom plate. For example, the front fixtures in the diagram shown are to be lined up with holes #4 and #5 in the transverse beam. Also shown on the fixture diagram are any special instructions pertaining to specific fixtures, car models, or double-ended pins and base plates with more than four holes.

More information on the use of a dedicated bench and fixtures in bodyshop work can be found in Chapter 11.

SUMMARY

- Vehicle measurement involves using specialized tools and equipment to measure the location of reference points on the vehicle.
- Check body dimensions (known good body measurements of the undamaged vehicle) by comparing the actual measurements with the values in the repair manual or body dimensions chart.

- The control points used in manufacturing are not necessarily the same as the reference points the collision repair technician uses to measure the vehicle. Reference points refer to the points, bolts, holes, etc., used to give unibody and frame dimensions in body specification manuals.

- Collision impacts that occur from the side often cause a sidesway damage or side bending frame damage condition

- Sag damage is a condition where one area, often the cowl area, is lower than normal.

- Mash damage is present when any section or frame member of the car is shorter than factory specifications.

- Diamond damage is a condition where one side of the car has been moved to the rear or front causing the frame and/or body to be out of square.

- Twist damage is a condition where one corner of the car is higher than normal; the opposite corner might be lower than normal.

- The most important rule in body/frame alignment is REVERSE DIRECTION AND SEQUENCE. This means, to correct collision damage on a conventional vehicle, the pulling or pushing of the damaged area must be done in the opposite direction of impact.

- Crush zones are engineered to collapse in a predetermined fashion to protect the vehicle's passengers and to localize damage.

- The body dimensions chart gives measurement points and measurement specifications for a specific type of vehicle.

- Tram gauges are scaled rods used for measurement, while centering gauges use metal rods to check for misalignment.

- Symmetrical means that the dimensions on the right side of the vehicle are equal to the dimensions on the left side of the vehicle.

- A datum line, or datum plane, is an imaginary flat surface parallel to the underbody of the vehicle at some fixed distance from the underbody.

- The center plane, or center line, divides the vehicle into two equal halves: the passenger side and the driver's side.

ASE-STYLE REVIEW QUESTIONS

1. Which type of damage occurs most often in a body-over-frame vehicle?

   A. Sidesway
   B. Sag
   C. Mash
   D. Diamond

2. In a unibody structure, which of the following occurs last in the typical collision damage sequence?

   A. Bending
   B. Widening
   C. Twisting
   D. Crushing

3. Technician A says that the tolerance of critical manufacturing dimensions must be held to within a maximum value of 5 millimeters. Technician B will check for door sag in the rear section of a vehicle after a front end collision. Who is correct?

   A. Technician A
   B. Technician B
   C. Both A and B
   D. Neither A or B

4. To accurately measure a vehicle, how many correct dimensions are required as a starting point?

   A. At least two
   B. At least three
   C. At least four
   D. One

5. Technician A says that any wrinkle in the rear floor is usually due to buckling of the rear side member. Technician B says that self-centering gauges are used to measure in a manner closely related to the tram gauges. Who is correct?

   A. Technician A
   B. Technician B
   C. Both A and B
   D. Neither A nor B

6. When proceeding through the measurement of the damage, Technician A measures the center section of the vehicle for twist and diamond first. Technician B measures for mash with a strut tower upper body gauge. Who is correct?

   A. Technician A
   B. Technician B
   C. Both A and B
   D. Neither A nor B

7. Technician A says that the larger the area over which the impact is spread, the greater severity of damage in a smaller area. Technician B says that a visual inspection is sufficient to accurately assess impact damage. Who is correct?

   A. Technician A
   B. Technician B
   C. Both A and B
   D. Neither A or B

8. Technician A says that a laser guide deflects a laser beam at exactly 90 degrees. Technician B says that

beam splitters project the laser beams at a precise right angle so height, width and length can be measured simultaneously. Who is correct?

A. Technician A
B. Technician B
C. Both A and B
D. Neither A or B

9. Technician A leaves the laser system in position during the repair operation. Technician B discards a laser target when it is scratched or warped. Who is correct?

A. Technician A
B. Technician B
C. Both A and B
D. Neither A nor B

10. How many gauges are usually hung to check for centerline misalignment?

A. 2
B. 3
C. 4
D. 5

11. Technician A frequently uses tram gauges and centering gauges in conjunction with one another. Technician B stands as close as possible to centering gauges in order to read them accurately. Who is correct?

A. Technician A
B. Technician B
C. Both A and B
D. Neither A nor B

ESSAY QUESTIONS

1. Explain the difference between primary and secondary damage.

2. Give seven steps in a basic collision damage diagnosis procedure.

3. What is diamond damage?

4. Explain the cone concept of damage.

5. What is a left-to-right symmetry check?

6. Before beginning any universal measuring operations, what three things should you do?

CRITICAL THINKING PROBLEMS

1. If you see an accident, and car no. 1 is braking hard and hits a stationary car no. 2 center broadside, what damage would you expect to each vehicle?

2. During an inspection, you find the right front frame rail crush zone badly collapsed and pushed to the left. What was the direction of impact?

MATH PROBLEMS

1. When measuring damage, you find that two round holes, one being $1/2$ inch in diameter, the other $1\frac{1}{2}$ inches in diameter, have an inside measurement of 32 inches and an outside measurement of 34 inches. What is this dimension?

2. A body dimension is 35.2 inches (890 mm). Your measurements show a reading of 37.1 inches (942 mm). If the tolerance is $1/16$ inch, what is the mini-

mum amount you must straighten the body to be within tolerances?

3. Use Figure 10–35 to find six spec measurements for rear frame rails on a specific make vehicle. Then use the metric conversion chart in the Appendix of this textbook to find the equivalent of U.S. values for these specs.

MEASUREMENT

P10-1 Visual inspection of vehicle damage will help you determine where and how to measure. For example, if the front wheels are obviously out of alignment, you would know to measure for front frame rail and strut tower misalignment.

P10-2 The technician refers to the computer or a dimensions manual to find known good measurements.

P10-3 Use an undamaged portion of the vehicle to establish a reference plane for measuring the direction and the extent of the damage.

P10-4 Measure the distance between factory-recommended reference points on the body structure. By comparing actual measurements with specifications, you can determine what must be done to straighten the unibody or the frame.

Body Alignment

INTRODUCTION

Vehicles with major damage must often have their frame or body structures straightened. Vehicle straightening involves using high-powered hydraulic or pneumatic equipment, mechanical clamps, and chains to bring the frame or body structure back into its original shape (Figure 11–1). At the same time, measuring equipment is commonly used to monitor and direct the straightening operation.

Body aligning or pulling is often thought to be a rough and tough physical operation. Actually, it is a relatively easy step-by-step task if proper equipment and methods are used. An important requirement when straightening is accuracy. For example, wheel alignment is directly affected by body/frame alignment. If it is not properly straightened, you will not be able to align the wheels of the vehicle.

Improper straightening techniques are costly and time-consuming mistakes. Accurate vehicle alignment affects safety, repair time, repair quality, and the confidence of your customer. This chapter will summarize the most important methods for realigning a vehicle with major damage.

FIGURE 11-1 Modern body aligning equipment makes body straightening a fairly straightforward task if basic rules are followed. *(Courtesy of Car-O-Liner Co.)*

OBJECTIVES

After studying this chapter, you should be able to:

✔ List the types of straightening equipment and explain how they are used.

✔ Describe the basic straightening and aligning techniques.

✔ Identify safety considerations for using alignment equipment.

✔ Plan and execute collision repair procedures.

✔ Identify signs of stress/deformation on a unibody vehicle and make the necessary repairs.

✔ Determine if a repair or replacement can be done before, during, or after pulling.

✔ Answer ASE test questions pertaining to body alignment procedures.

KEY TERMS

anchor chains
anchor clamps
anchor pots
anchor rails
anchoring equipment
bench system
bench-rack system
chain tighteners
composite force
cross tube
hydraulic rams
in-floor straightening
   systems

multiple pull method
operating manual
original state
overpulling damage
pulling
pulling equipment
pulling posts
single pull method
straightening system
   accessories
stress concentrators
traction
vector system

ASE TASK LIST
Job Skills covered in this chapter include:

PAINTING AND REFINISHING TEST (B2) TASK LIST

A. Surface Preparation
1. Remove, assess, and store trim and moldings.

E. Finish Defects, Causes, and Cures
3. Check for rust spots (corrosion); determine the cause(s), and correct the condition.

NON-STRUCTURAL ANALYSIS AND DAMAGE REPAIR TEST (B3) TASK LIST

A. Preparation
3. Remove outside trim and moldings as necessary; store reusable parts.
4. Remove damaged or undamaged inside trim and moldings as necessary; store reusable parts.
5. Remove undamaged, non-structural body panels and components that may interfere with or be damaged during repair.

B. Outer Body Panel Repairs, Replacements, and Adjustments
1. Determine the extent of the direct and indirect damage and the direction of impact; plan the methods and order of repair.
8. Check and adjust clearances of front fenders, header, and other panels.

D. Moveable Glass and Hardware
2. Repair or replace electrically-driven power sun roofs and related controls.
3. Inspect, repair or replace, and adjust removable, manually-operated roof panels and hardware.

STRUCTURAL ANALYSIS AND DAMAGE REPAIR TEST (B4) TASK LIST

A. Frame Inspection and Repair
2. Straighten and align mash (collapse) damage.
3. Straighten and align sag damage.
4. Straighten and align sidesway damage.
5. Straighten and align twist damage.
6. Straighten and align kickup damage.
7. Straighten and align diamond frame damage.
8. Remove and replace damaged frame horns, side rails, cross members, and front or rear sections.
10. Repair of replace weakened or cracked frame members in accordance with vehicle manufacturers'/industry standards.

B. Unibody Inspection, Measurement, and Repair
3. Diagnose and analyze unibody vehicle damage using mechanical equipment (tram and self-centering gauges).
4. Determine the proper locations of all suspension, steering, and power train component attaching points on the body.
5. Diagnose and measure unibody vehicles using a dedicated (fixture) measuring system.
6. Diagnose and measure unibody vehicles using a universal measuring system (mechanical, electronic, laser, sonar).
7. Determine the extent of direct and indirect damage, and the direction of impact; plan the methods and sequence of repair.
8. Attach proper anchoring devices.
9. Straighten and align center unibody section, cowl, bulkhead, roof, roof rails, pillars, floor, windshield, back glass openings, door openings, rocker panels, and floor pans.
10. Straighten and align rear unibody section quarter panels, wheelhouse assemblies, and rear body panel (including rails, suspension, and power train mounting points).
11. Straighten and align front unibody sections (aprons, strut towers, upper and lower rails, steering, suspension, and power train mounting points).
12. Recognize the limitations of applying heat to high-strength steel structural components; use proper stress relief methods for high-strength steel; weld in accordance with vehicle manufacturers' recommendations.
14. Remove folds, curves, creases, and dents, using power tools and hand tools, to restore damaged areas to proper contours and dimensions.
15. Determine the extent of damage to structural steel body panels; repair, weld, or replace in accordance with vehicle manufacturers' specifications.
16. Remove damaged sections of structural steel body panels, and weld in replacements in accordance with vehicle manufacturers' specifications.

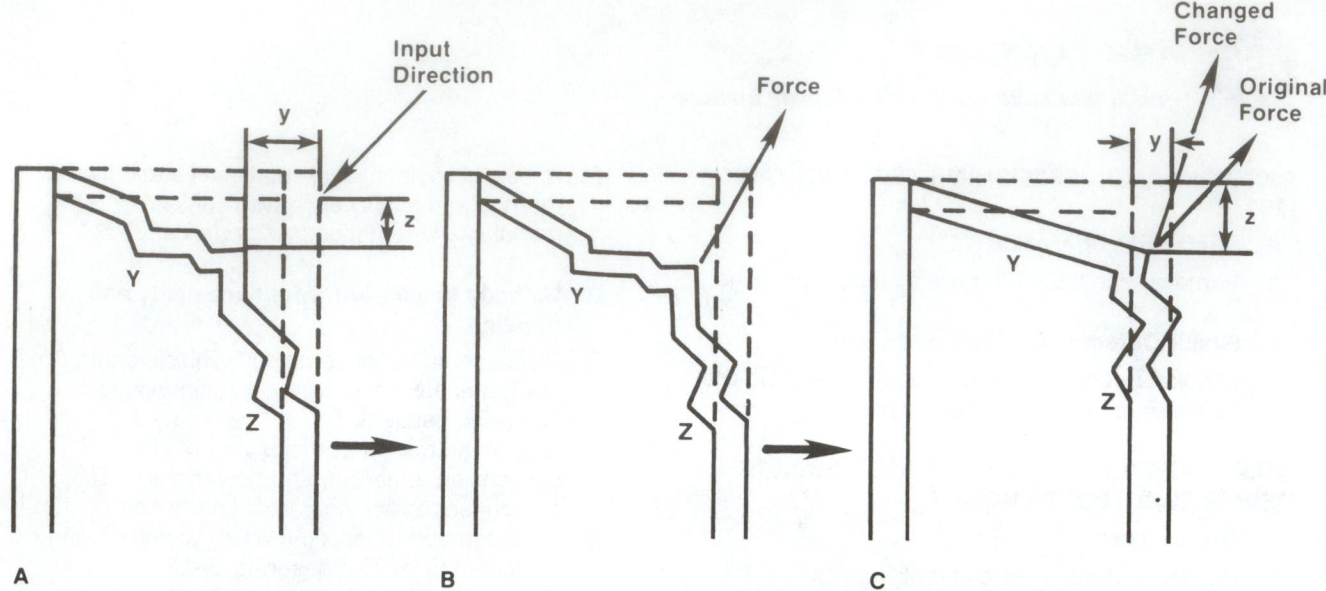

FIGURE 11-2 Study basic pulling direction to correct damage: (A) Input was in the direction of the back arrow causing damage in directions Y and Z; (B) apply force in the direction opposite to the input force; and (C) if a difference in the degree of repair between Y and Z occurs, change the pulling direction accordingly. *(Courtesy of Toyota Motor Corp.)*

11.1 BODY ALIGNMENT BASICS

ALIGNMENT BASICS

The term **pulling** refers to using alignment equipment to stretch the damaged metal back to its original shape. The vehicle is secured and held stationary by the equipment. Clamps and chains are then attached to the damaged area. When the hydraulic system is activated, the chains slowly pull out the damage. Measurements are made at body/frame reference points while pulling to return the vehicle to its original dimensions (Figure 11–2).

When realigning a vehicle, a pull force or **traction** should be applied in the opposite direction of the force of the impact. When determining the direction of a pull, basically you must set the equipment to pull perpendicular to the damage.

The **single pull method** only uses one pulling chain. This method works well with minor damage on one part. A small bend in a part can often be straightened with a single pull.

With major damage to several panels, a **multiple pull method** with several pulling directions and steps is needed. With major damage, body panels are often deformed into complex shapes with altered strengths in the damaged areas. To pull only in the opposite direction would not work because of the differences in the strength and recovery rates of each panel. With multiple pulls, force is applied to all pulling chains at one time. This helps to prevent metal from tearing and to more uniformly move damage back into position.

Use the method that works best for the given situation. Since applying force in only one place will not always work, you often have to exert pulling force on many places at the same time. For convenience, the term "direction opposite to input" will be used to describe the effective pulling direction.

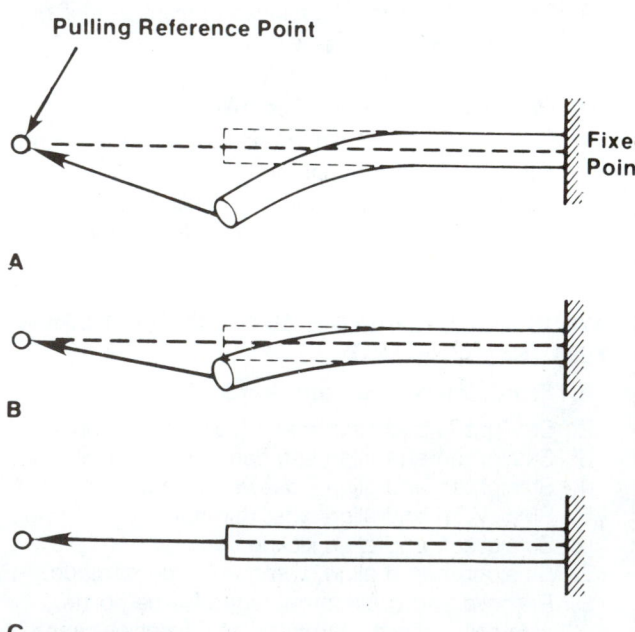

FIGURE 11-3 (A and B). Think of the condition wanted after repairs are completed. Set a reference point along an imaginary line extending along the desired axis from which to exert force, and pull from that point. (C) When force is applied and the bend is repaired, the part will be straightened. *(Courtesy of Toyota Motor Corp.)*

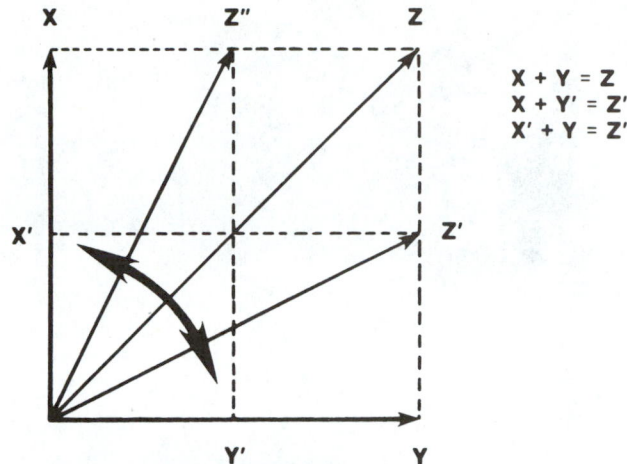

$$X + Y = Z$$
$$X + Y' = Z'$$
$$X' + Y = Z''$$

FIGURE 11-4 If the pulling force is divided between two directions (X, Y), the composite force direction (Z) will change freely with adjustments to the force in the two directions. *(Courtesy of Toyota Motor Corp.)*

To alter the direction while pulling, divide the pulling force into two or more directions. This will allow you to change the direction of the **composite force** (force of all pulls combined).

With *push-pull force,* the straightening equipment is used to pull out damage while a portable ram pushes out damage at the same time. This reduces the total force needed and restores the parts with less effort and repair time.

The pulling and straightening process must remove both direct and indirect damage. It must return all of the damaged metal back to preaccident dimensions. To do this, the equipment must reverse the direction and sequence in which the damage occurred. The damage that occurred last during the collision should be pulled out first (Figure 11-3).

Use the method that works the best for the given situation (Figure 11-4). Since applying force in only one place will usually not result in proper repairs, it is recommended to exert pulling force on many places at the same time.

11.2 STRAIGHTENING EQUIPMENT

Straightening equipment is used to apply tremendous force to move the frame or body structure back into alignment. Straightening equipment includes anchoring equipment, pulling equipment, and other accessories.

The **anchoring equipment** holds the vehicle stationary while pulling and measuring. Anchoring can be done by fastening the frame or unibody of the vehicle to anchors in the shop floor or to the straightening equipment rack, frame, or bench. The

objective of the anchoring system is to hold the vehicle solidly in place while pulling forces are applied. It must also distribute pulling forces throughout the vehicle.

The **pulling equipment** uses hydraulic power to force the body structure or frame back into position. There are many different types of pulling equipment available. Regardless of their design or operating features, each system uses the same basic pulling theory and is used in a similar manner.

Hydraulic rams use oil pressure from a pump to produce a powerful linear motion. When you electrically activate the system, oil is forced into the ram cylinder. The ram is then pushed outward with tremendous force. This pulls on the chain attached to the vehicle to remove the damage. The rams can be mounted in or on the pulling towers or posts, or between the vehicle and anchoring system.

Pulling posts or towers are strong steel members used to hold the pulling chains and hydraulic rams. Depending upon equipment design, they can be positioned at whatever location is needed to make the pull. They push against the rack or bench as the pull is made. This eliminates the need for separate anchoring to keep the pulling equipment from sliding under the rack or bench as the pull is made.

Pins or hardened bolts lock the tower to the rack. The tower can be rotated sideways for different pulling angles. Some will also tilt into an angle for even more pulling flexibility. Make sure all lock pins are securely in their holes before applying pulling power.

You should be familiar with a variety of anchoring and pulling systems and their general operation.

On a partial or full frame type vehicle, if the lower structure (frame) is restored to its proper alignment, generally the suspension and power train will be in proper alignment. However, with a unibody vehicle, the frame is the entire unibody structure. For this reason, pulling on the underbody structure will not straighten the upper structure. In addition, body-on-frame structures are tolerant of the trial-and-error pulling procedure. Because of the unibody's

WARNING The amount of pulling pressure required to remove damage should not be too high. If the pulling equipment is straining during the pulling process, something is wrong. If this happens, stop pulling! Release tension, and reevaluate the setup to find the problem. If too much pressure is applied, parts or equipment can be damaged and serious injuries could result.

thin-gauge structure, improper pulling can badly damage the vehicle. For unibody repair, the alignment equipment must also show the amount of misalignment at each reference point and the direction of the misalignment.

You may have to make some general measurements with a tape measure or tram gauge. These would include diagonal measurements to check for diamond and length measurements to check for mash. Try to get as good an idea as possible of where the damage begins and ends. Use all the dimension data available, including body/frame dimensions books, the vehicle manufacturer's manuals, or by checking against an undamaged car. Remember that these are general measurements and do not have to be made as accurately as when straightening the vehicle.

It is necessary to have at least THREE REFERENCE POINTS on the undamaged part of the vehicle that can be used to set the car up properly. These three locations will then form the datum plane on which all of the other measurements will be based. If there are more than three locations that are undamaged, they can also be used for setup.

In a situation where there are not three undamaged reference points, such as a severe side impact, it might be necessary to do some rough straightening of the underbody until three points can be secured.

There are several different types of frame straightening equipment on the market that are suitable for both body-on-frame and unibody collision repair work. The most popular systems are discussed below.

IN-FLOOR SYSTEMS

Two types of in-floor systems available are the anchor-pot system (Figure 11–5A) and the modular rail frame system (Figure 11–5B).

In-floor straightening systems have anchor pots or rails cemented or mounted in the shop floor. Some use **anchor pots** or small steel cups in various locations in the shop floor. Others use a system of steel **anchor rails** in the floor so that an infinite number of pulling and holding locations can be used. Both systems must be balanced in both direction and force of the pull (Figure 11–6).

To provide the pulling force for straightening, the anchor-pot system can use hydraulic rams and pulling posts or towers. The floor grid system generally uses hydraulic rams to provide the pulling force.

An in-floor system is ideal for a small body shop. After the rams and the other power accessories have been neatly stored away, the area can be used for other body shop purposes. In-floor systems also provide single or multiple pulls, and positive anchoring without sacrificing shop space.

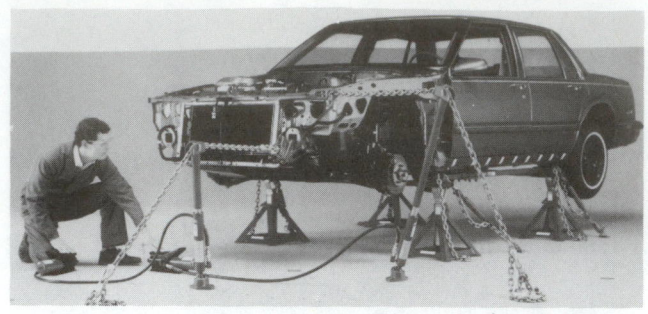

A

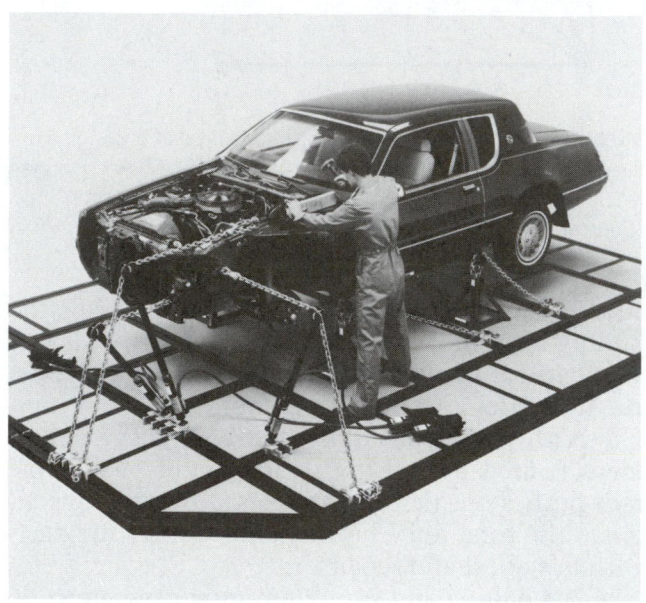

B

FIGURE 11–5 These are typical in-floor systems: (A) the anchor-pot system and (B) the modular rail frame. *(Courtesy of Blackhawk Automotive Inc.)*

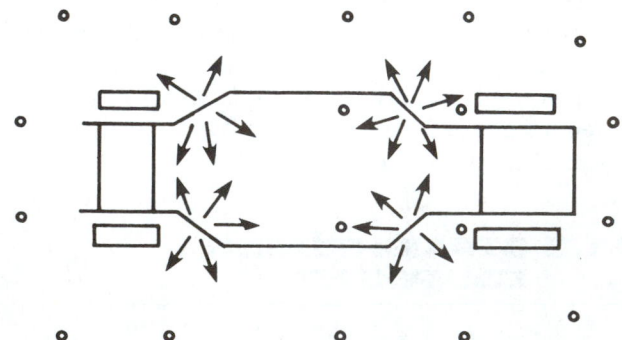

FIGURE 11–6 The anchoring setup must balance in the direction and force of the pull. If not, the vehicle could be pulled sideways by the pulling force.

Anchor clamps are bolted to specific points on the vehicle (unibody pinchwelds, for example) to allow the attachment of anchor chains. They distribute pulling force to prevent metal tearing (Figure 11–7).

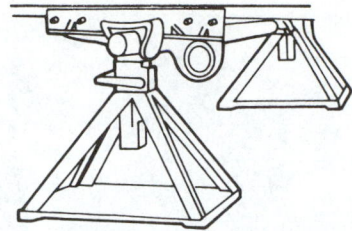

FIGURE 11-7 Anchoring clamps often bolt around a thick metal flange (rocker pinch weld) on the bottom of the body structure. *(Courtesy of Blackhawk Automotive Inc.)*

The vehicle is usually supported using **cross tube anchor clamps** which link both sides of the vehicle. The cross tube anchor clamps are placed over the cross tube and tightened securely. Chains are then attached to the cross tube anchor clamps.

Anchor chains are attached from the floor anchors to the clamps attached to the vehicle. The anchor chains and clamps must hold the vehicle securely while straightening. **Chain tighteners** or shorteners are used to take slack out of the anchor chains.

The vehicle is usually supported on a car/truck stand using the cross tube anchor clamps. The cross tube anchor clamps are placed over the cross tube and tightened securely. Chains are attached to the cross tube anchor clamps and to the anchor pots or rails. Pull the chains tight for secure anchoring. Chain shorteners should be used to remove slack from the anchor chains (Figure 11-8).

To make further preparations for the pull, position the ram in the ram foot so it will exert force in the desired direction. Build the ram up to the desired height. Pull the chain tight and lock it in the chain head with the cross pin. Hook the chain onto the anchor. The anchor, ram foot, and attachment point on the vehicle should be in a straight line in the direction of pull. Hook the pump to the ram and connect

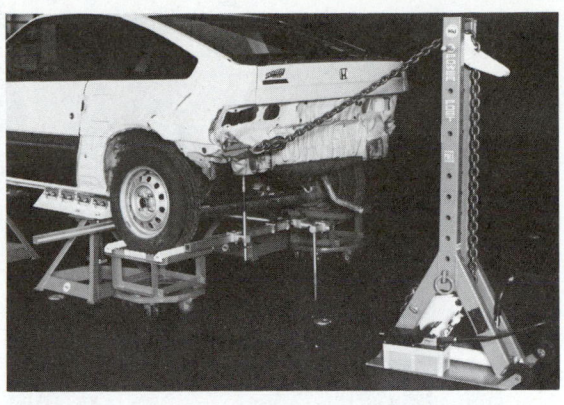

FIGURE 11-9 A power tower allows you to pull out damage once the vehicle is anchored securely. *(Courtesy of Kansas Jack, Inc.)*

an air hose to the pump. Engage the pump to take up slack in the chain, and pulling can be started.

It is possible with most in-floor systems to use a power tower or post to provide extra pulling force (Figure 11-9). A power tower or post is frequently used to supply pulling energy when straightening minor frame or unibody damage.

CHAINLESS ANCHORING SYSTEMS

A chainless anchoring system has a low profile to keep it close enough to the floor to load a vehicle quickly, yet high enough to allow easy access to the vehicle's underbody. The unit's adjustable length and width allows the body technician to set it easily to all vehicle sizes.

The following steps are required to set up a vehicle for pulling with a chainless anchoring system:

1. Raise and secure the vehicle and attach the underbody clamps (Figure 11-10A).
2. Insert the support tube through the underbody clamp and the base (Figure 11-10B). Lower the car onto the base.
3. Secure the system to the in-floor anchors by positioning the lock arm. Hammer the wedges in place to lock the vehicle to the system (Figure 11-10C).
4. Repeat the operation on the opposite end of the vehicle (Figure 11-10D).

To clamp a box or full perimeter frame, spacers may be available for the pinch weld clamps. They allow the clamps to have a wider jaw opening for clamping around a thick steel box frame.

Once the vehicle is secured, it is locked to stay. It is possible to pull from any angle, 360 degrees around the vehicle. A power tower or post is often used as a pulling force (Figure 11-11).

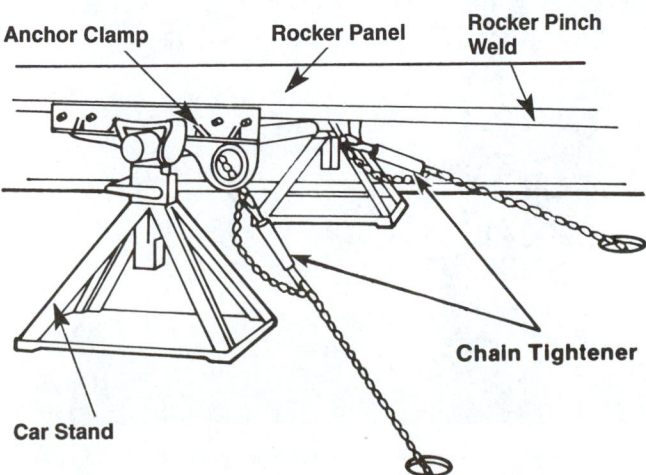

Anchor Clamp   Rocker Panel   Rocker Pinch Weld

Chain Tightener

Car Stand

FIGURE 11-8 Anchor chains extend from body clamps to floor pots with this setup.

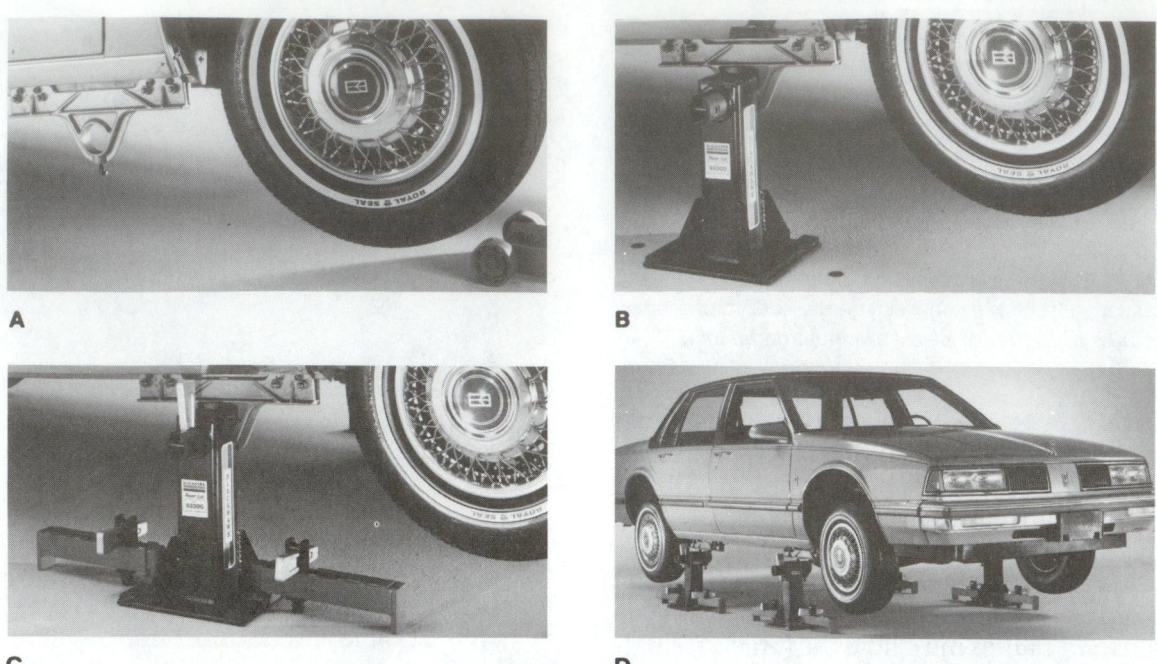

FIGURE 11-10 Steps to install a vehicle on the stands of a chainless anchoring system. (A) Raise the car and install a clamp on the body flange. (B) Install a special stand onto the clamp. (C) Anchor the stand to the floor holding system. (D) Repeat on the other corners of the vehicle. *(Courtesy of Blackhawk Automotive Inc.)*

FIGURE 11-11 A body and frame puller may seem complex and confusing at first. However, if you methodically analyze the directions of the pulls, it is fairly easy to use. *(Courtesy of Chief Automotive)*

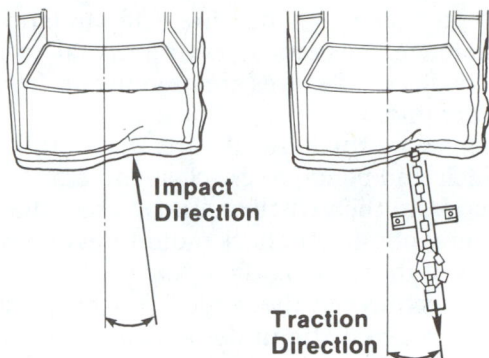

FIGURE 11-12 A portable body and frame puller can be quickly set so that the traction direction is opposite to the damage input direction. *(Courtesy of Nissan Motor Co.)*

PORTABLE BODY AND FRAME PULLERS

Portable body and frame pullers have a hydraulic pressure system installed between the removable main frame and the mast. This type of system is designed to extract damage by means of chains and clamps. It is often used for minor damages.

Since they are easily movable, you can quickly set the traction direction opposite to the damage input (Figure 11–12). Many units of this type, however, are able to pull only in one direction. These units are more dangerous to use than rack or floor systems.

RACK SYSTEMS

Most *rack systems* have thick steel ramps to which pulling towers and anchoring clamps are attached. This usually gives the technician infinite positioning to pull from any angle and height, 360 degrees around the vehicle. It is possible to make up-pulls or down-pulls (Figure 11–13). In fact, pulls can be made with the rack positioned at full rack height or flush to the floor (Figure 11–14). Most racks tilt hydraulically so that vehicles can either be driven on or pulled into position with the optional power winch, shown in Figure 11–15. Most rack systems also provide an excellent measuring system (Figure 11–16).

BENCH SYSTEMS

A **bench system** is generally a portable or stationary steel table for straightening severe vehicle damage. Some benches tilt. Others have drive-on ramps, like a rack, that can be raised up or down as needed. Alignment benches are available in fixed (Figure 11–17) and movable types.

The *dedicated bench* is an older system that consists of a strong, flat work surface to which measuring fixtures are attached. The dedicated bench acts a

A

B

FIGURE 11-13 Many racks can make (A) pulls up or (B) pulls down. *(Courtesy of Kansas Jack, Inc.)*

FIGURE 11-14 Some racks can be positioned at full rack height or flush to the floor. *(Courtesy of Kansas Jack, Inc.)*

FIGURE 11-15 Many racks tilt hydraulically so that vehicles can either be driven on or pulled into position with a power winch, shown here. *(Courtesy of Kansas Jack, Inc.)*

FIGURE 11-16 Universal measuring systems can be used on most racks. They are needed to monitor direction and amount of pull to avoid overpulling damage. *(Courtesy of Kansas Jack, Inc.)*

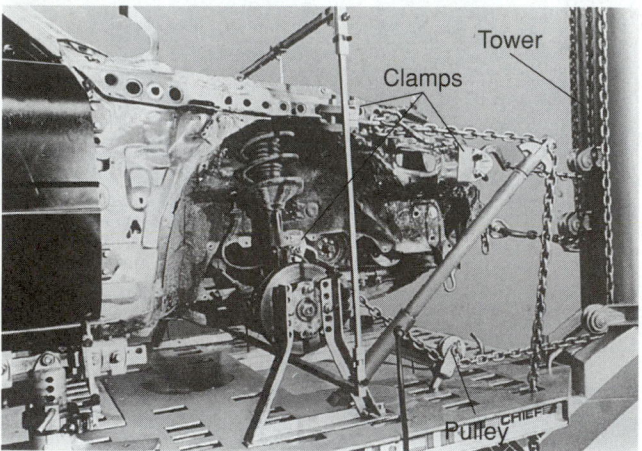

FIGURE 11-17 When the vehicle is securely anchored, a pulling post or tower can be used to apply corrective force to damaged areas through pulling chains. You must constantly measure to determine the pulling direction and force needed.

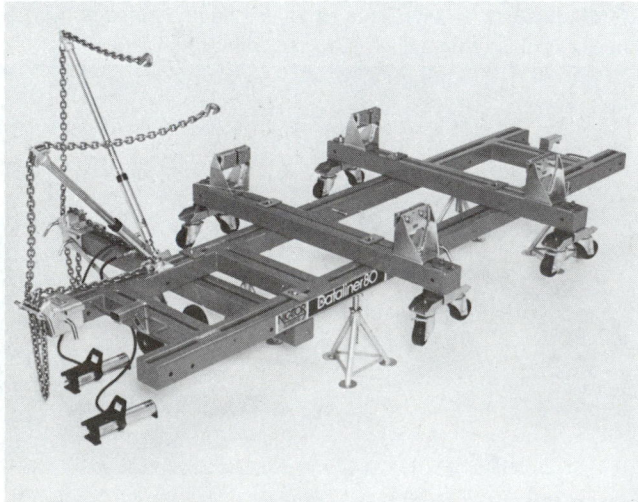

FIGURE 11-18 This is a portable or movable type alignment bench. *(Courtesy of Nicator, Inc.)*

"go–no-go" gauge to measure and straighten vehicle damage. Instead of taking actual measurements, dedicated fixtures are used to check body or frame alignment.

Fixtures are thick metal parts that bolt between the vehicle and bench to check alignment. They are designed from the vehicle manufacturer's drawings. The fixtures physically check mountings or other key locations of the underbody.

If the fixtures fit the vehicle properly, the technician knows that the underbody and strut towers are in perfect alignment. All that is necessary is to straighten the vehicle until the reference points match the fixtures. No other underbody measurements are usually required.

A **bench-rack system** is a hybrid machine that allows quick loading like a rack and has other features of a bench. The table often tilts like a rack for quick loading of the vehicle. It also provides the accuracy and convenience of a bench once the vehicle is in place. This is the most modern type of bench in use today.

Some bench-rack systems have a drive-on ramp with a rolling dolly system. This allows for obstructions under the vehicle and easier access to repair areas and reference points. The ramp may also be used as stands while the vehicle is installed on its anchoring systems.

EQUIPMENT SETUP

To set up an alignment machine, such as shown in Figure 11–18, proceed as follows:

1. Mount the vehicle by anchoring it to the machine. Using a service jack, first lift the end of the vehicle that has the longest overhang in relation to the axles. Roll the transport beams under the car. Adjust the chassis brackets and attach them to the pinch weld of the sill. Tighten the bracket bolts (Figure 11–19A).

   NOTE: Do not work under the vehicle unless it is on jack stands!

2. Perform the corresponding operation for the other end of the car. Check that all chassis brackets are fixed tight. As the transport beam and the main frame are both fitted with wheels, the vehicle can now roll to the workplace (Figure 11–19B).

3. Place the jack on the jack attachment under the vehicle. When the frame is raised to meet the transport beams, attach the beam nearest the jack to the frame (Figure 11–19C). Continue raising the frame until a suitable working height has been attained. Then place the axle stand under the frame. Perform the corresponding operation at

A

B

C

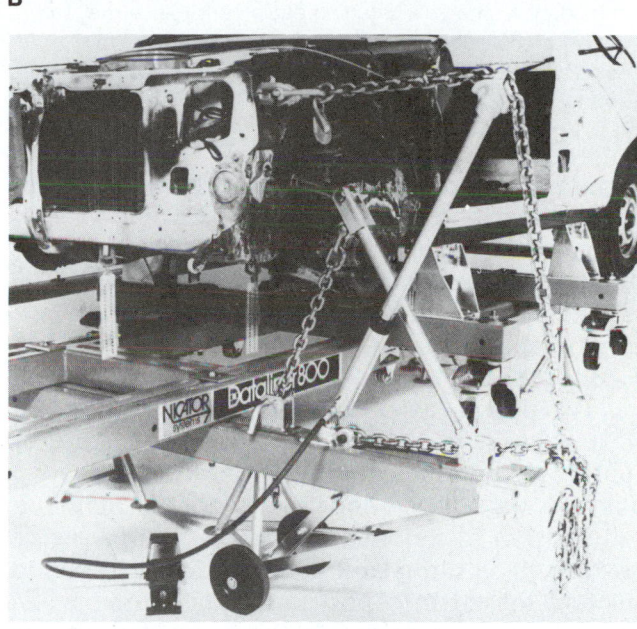

D

FIGURE 11-19 Set-up procedures for alignment on a typical portable bench. (A) Raise the car and mount it on anchor mechanism. (B) Move the car to the repair area. (C) Anchor it on stationary stands. (D) Pull out the damage. *(Courtesy of Nicator, Inc. and Car-O-Liner)*

the other end of the vehicle. Place the remaining bolts in position and tighten them.

4. When making a pull, the aligners or rams are mounted on an aligning beam with wheels. Then they can be rolled into the desired position and bolted in place. The direction of the beam determines the direction of the pull. If, for example, pulling sideways, position the aligning beam at right angles to the main frame (Figure 11-19D).

With a *scissor lift system* or with four small air jacks, the four anchor clamps can be installed simultaneously. This reduces repair time because you do not have to lift one end of the car at a time, as with some older systems.

A *hand-held controller* is often provided for operating the lift. When activated, a large hydraulic ram operates the scissor arms to raise or lower the vehicle.

STRAIGHTENING ACCESSORIES

Mentioned briefly, **straightening system accessories** include the various chains, clamps, hooks, adapters, straps, and stands needed to mount various makes and models of vehicles (Figure 11-20).

Various adapters are often supplied with frame straightening equipment. Some are holding devices to clamp parts in place so that they do not move while pulling other parts (Figure 11-21). For example, large diameter stationary adapters bolt to the side of

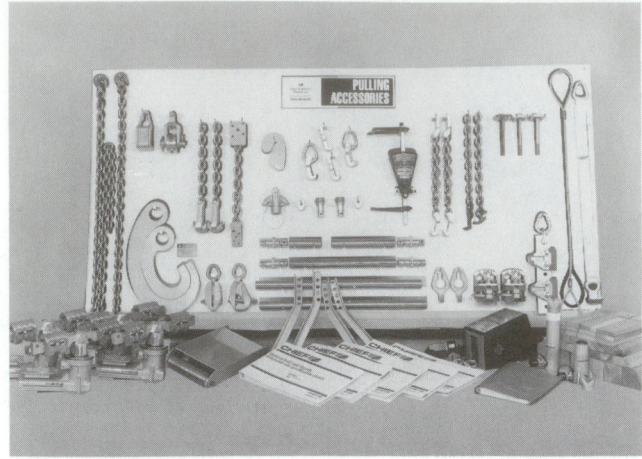

FIGURE 11-20 Various accessories are provided with modern frame straightening equipment. Since they vary in design and purpose, refer to the owner's manual for specific directions before use. *(Courtesy of Chief Automotive Systems, Inc.)*

the rack for making down pulls or changing the direction of pulls. Also, adapters are available for widening pinch weld clamps so that they can be mounted on a conventional frame.

Refer to the equipment owner's manual to learn about adapter installation and use. Procedures vary (Figure 11–22).

Two popular accessories that are often used with most alignment systems are the engine stand or holder and a portable pulling or pushing arm.

A *restraint bar* can be used to hold or maintain a dimension in an opening during pulling (Figure 11–23A). It an adjustable steel bar that can be slid out and locked into position. Many times when pulling, you need to retain the movement of one area while straightening another (Figure 11–23B).

A door aligner is a special bar designed to flex a door and its hinges for correcting alignment (Figure 11–23C). It is a special bar that snaps onto the door lock and striker. By pulling up or down on the handle, you can adjust minor door misalignment quickly.

When it is necessary to remove the engine or transmission mounts, the *engine holder* can be used to support the engine. It rests on the inner fenders and is adjustable in width. An adjustable chain hook is used to hold the chain attached to the engine. This allows you to remove engine and transmission mounts or the cradle for repairs.

With a *pulling/pushing arm*, the unit can pivot completely around the end of the bench from the center position on a special flange. In the other positions, it can reach the end and one side of the bench. The unit can also be used anywhere along the side of the bench by hooking the inner clamp on the outer flange on the opposite side.

A

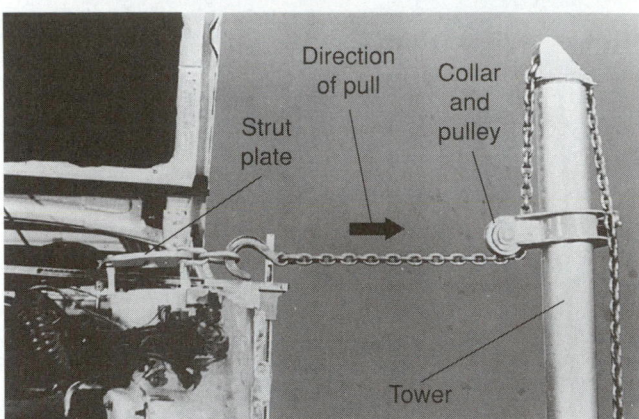

B

C

FIGURE 11-21 Study examples of frame rack accessories. (A) the device is being used to push inward on the apron while pulling to help remove buckle. (B) A strut adapter is being used to pull MacPherson strut tower sideways. Note how the collar and pulley on the tower are set to the same height for this pull. (C) Nylon pull straps are handy when it is difficult to install pulling clamps. *(Courtesy of Car-O-Liner, Chief Automotive, Bee Line Industries)*

FIGURE 11-22 Modern pulling equipment is very versatile. Here it is being used to remove an engine from a vehicle. *(Courtesy of Kansas Jack, Inc.)*

Portable hydraulic rams are possibly the most versatile of all aligning tools (Figure 11–24) since they can be used to push, spread, clamp, pull, and stretch (Figure 11–25). Figure 11–26 summarizes the various applications of portable hydraulic equipment.

Some pulling towers are designed so you can pull and push the damage at the same time. A portable power cylinder can be mounted between the damage to be pushed and the tower. It will push damage inward while pulling chains force the damage outward.

An **operating manual** or user's manual and other publications are provided to give detailed information on using the exact type of straightening equipment properly. You must always refer to these materials when working (Figure 11–27, page 368). They give anchoring and pulling instructions, the accessories needed for each vehicle, and other essential information.

It is beyond the scope of this text to detail the use of every type of body-frame alignment equipment. Always refer to the manufacturer's instructions when in doubt!

Always return all adapters and equipment to their proper storage locations after use.

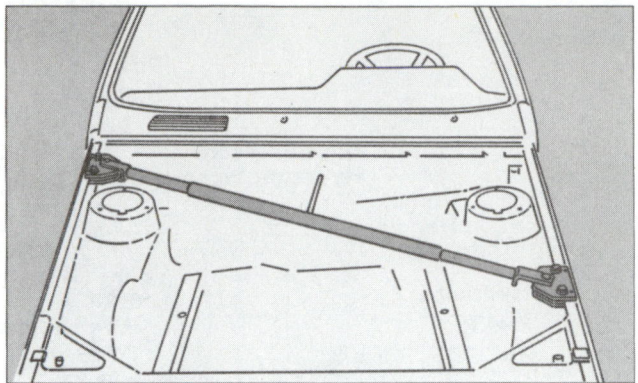

A

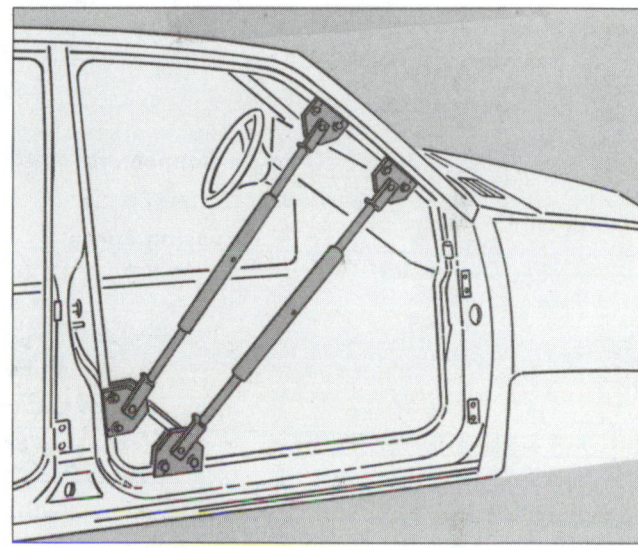

B

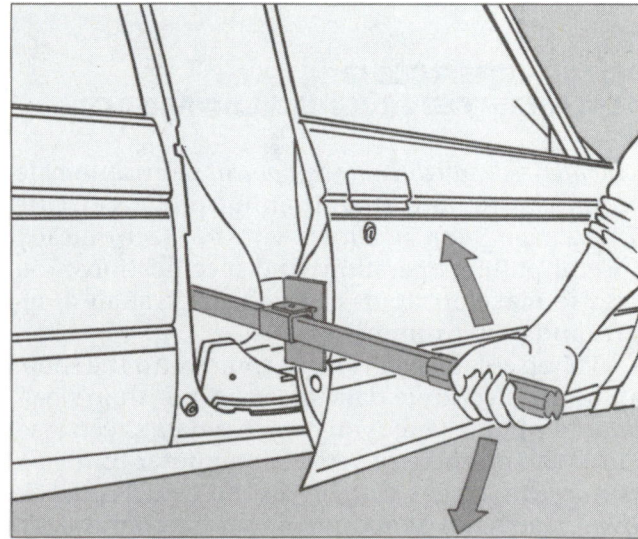

C

FIGURE 11-23 A few frame rack accessories. (A) Single holding restraint is secured across fender aprons to hold good dimension while pulling. (B) Two restraints are being used to secure the door opening dimension. (C) A door aligner fits into the door lock and catch so you can flex the door up or down for minor adjustment. *(Courtesy of Wedge Clamp)*

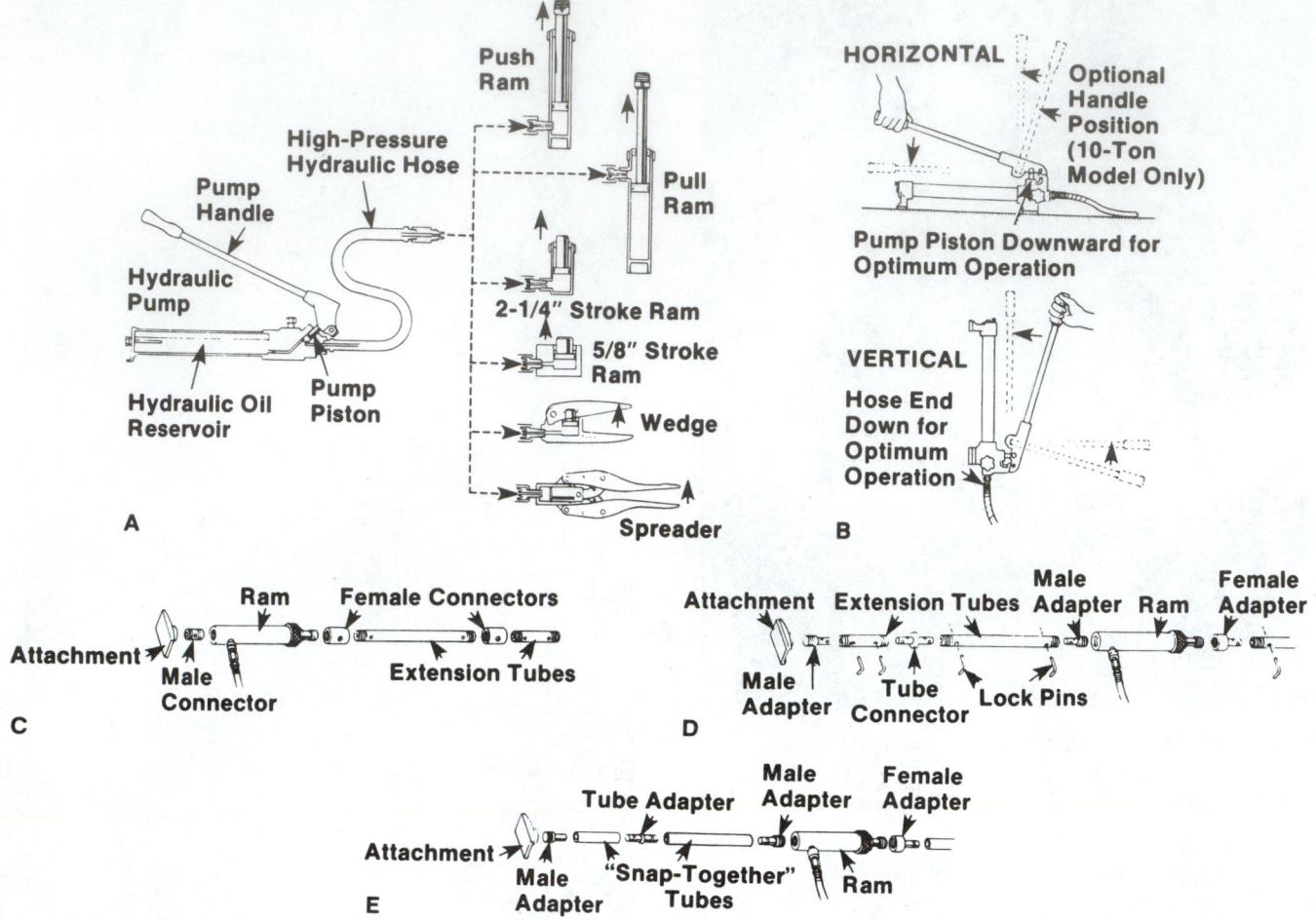

FIGURE 11-24 Study major components of a portable hydraulic ram system: (A) the heart of the system—portable pump, high-pressure hydraulic hose and hydraulic ram, wedge, or spreader; (B) positioning and operation of the pump; (C) threaded connection; (D) quick-fitting connection; and (E) snap-together connection. The latter should not be used for pulling. *(Courtesy of Norco Industries, Inc.)*

COMPUTERIZED STRAIGHTENING EQUIPMENT

Computerized straightening equipment helps automate the measuring and straightening process (Figure 11–28, page 368). It allows you to electronically monitor pulling operations and specifications. You can also make printouts of the damage analysis reports and repair summaries.

When a damaged vehicle comes into the shop, call up its electronic data sheet on the shop's personal computer. Once you have the data sheet, load it into the portable or remote computer terminal. With most systems, data on several vehicles can be downloaded and stored in the mobile terminal at one time.

The portable unit is then taken back out into the shop and attached to the frame straightening machine (Figure 11–29, page 368). A measurement bridge is mounted on the bench rack. A measurement slide-arm unit mounts and slides on the bridge to take measurements of the vehicle. A graduated scale is burned in the bridge and measurement arm so the computer can tell how far each is moved.

To calibrate the machine, a programming procedure must first be used. This can vary with the type and make of equipment. To calibrate the system for the setup, measurements are first taken of undamaged points on the vehicle. To take a measurement with this electronic system, the reference arm and pointer are installed at a specific reference hole or point. Then, by pushing a button on the machine, the point is stored in memory in the electronic terminal. This calibrates the unit to this vehicle. This is comparable to mounting centering gauges on undamaged control points.

When the measuring arm is moved to other reference points, the computer system can tell either that the point is within specs or the amount and direction of movement of the point away from specifications. As you straighten the vehicle, the arm can quickly be moved into the new location to measure its movement. Some electronic terms help show you the direction and amount of movement needed to pull a reference point back into specifications.

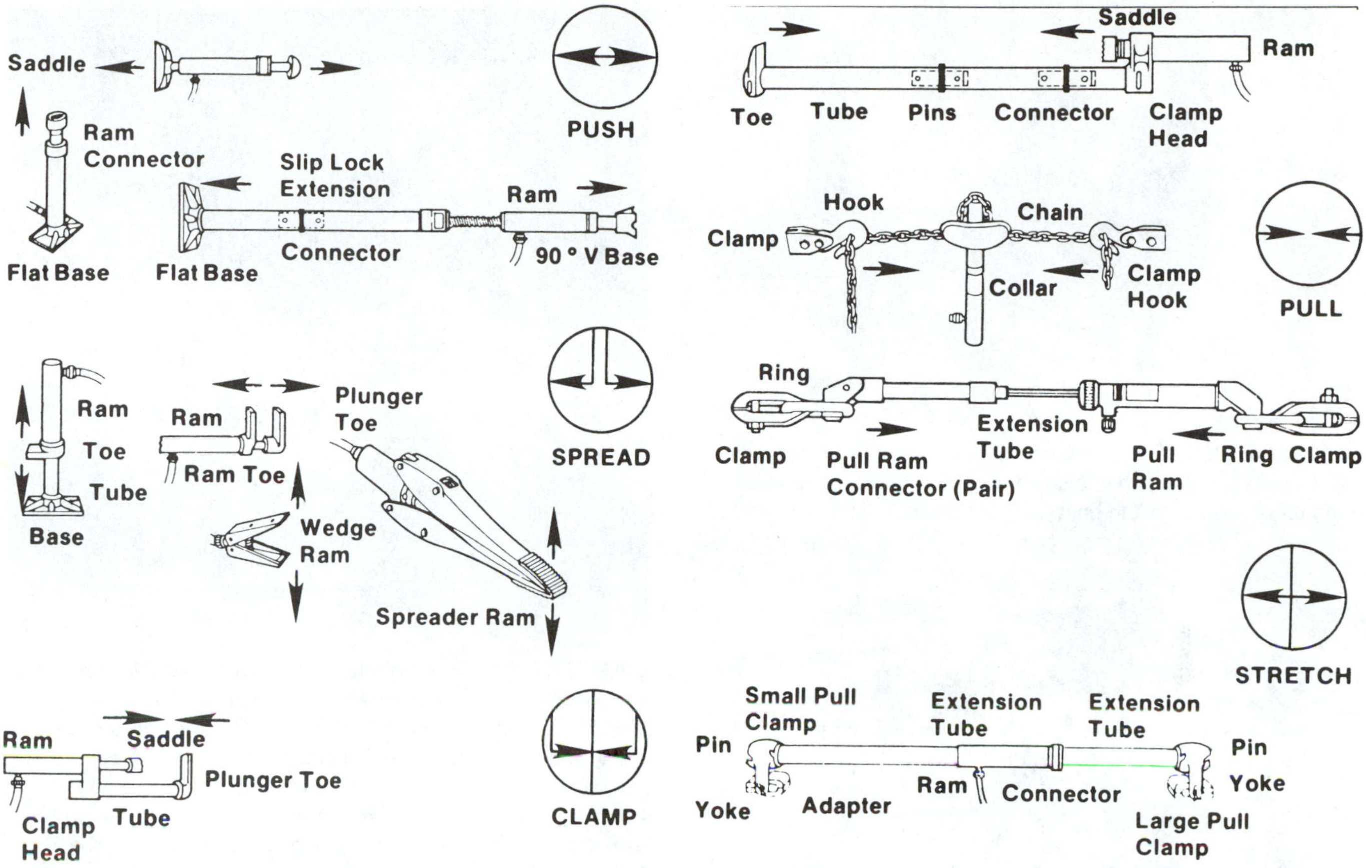

Saddle ← ⇄ →

PUSH

Saddle →
Ram

Ram Connector

Saddle ←

Slip Lock
Extension
Connector

Ram →

Toe   Tube   Pins   Connector   Clamp
Head

Flat Base   Flat Base   90° V Base

Hook   Chain

Clamp   Clamp
Hook

Collar

PULL

Ram
Toe
Tube

Base

Ram
Ram Toe

Plunger
Toe

Wedge
Ram

Spreader Ram

SPREAD

Ring

Clamp   Pull Ram
Connector (Pair)

Extension
Tube

Pull
Ram   Ring Clamp

STRETCH

Ram   Saddle

Plunger Toe

Clamp   Tube
Head

CLAMP

Small Pull
Clamp
Pin

Yoke   Adapter

Extension
Tube

Extension
Tube
Pin

Ram   Connector   Yoke

Large Pull
Clamp

FIGURE 11-25 Note a basic ram system setup. The standard push, spread, clamp, pull, and stretch symbols tell the technician how each setup will work. *(Courtesy of Blackhawk Automotive Inc.)*

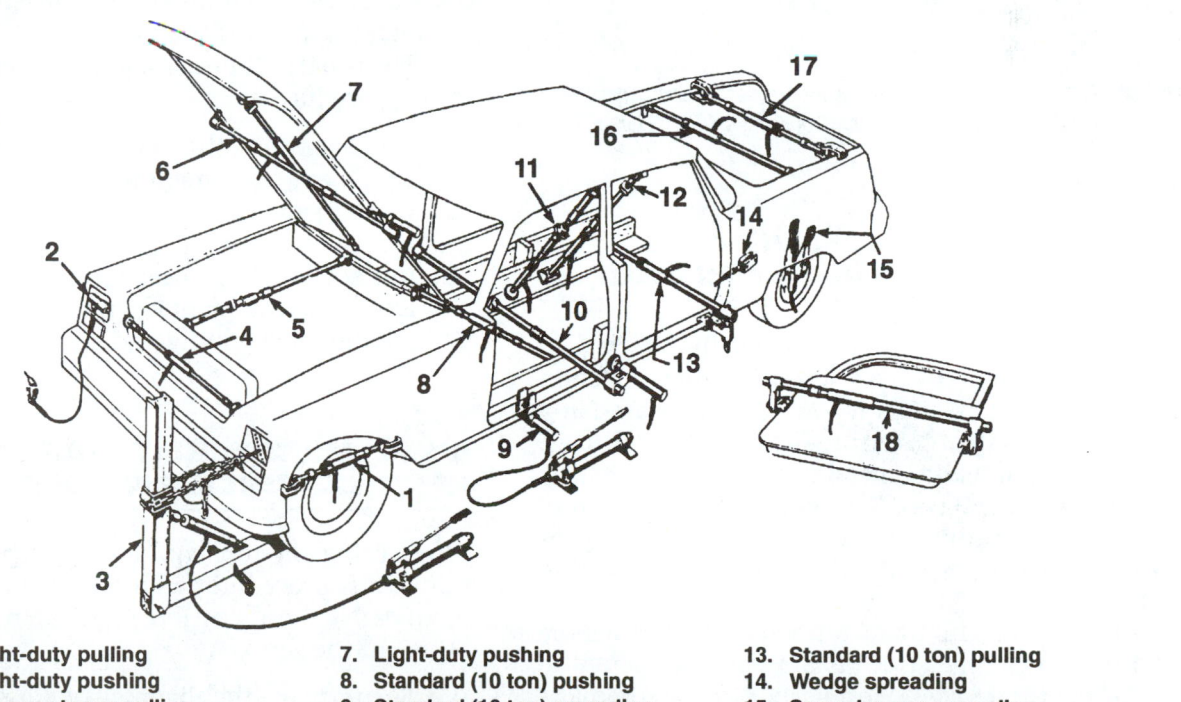

1. Light-duty pulling	7. Light-duty pushing	13. Standard (10 ton) pulling
2. Light-duty pushing	8. Standard (10 ton) pushing	14. Wedge spreading
3. Dozer or tower pulling	9. Standard (10 ton) spreading	15. Spreader ram spreading
4. Light-duty spreading	10. Standard (10 ton) clamping	16. Light-duty pushing
5. Standard (10 ton) pushing	11. Standard (10 ton) pushing	17. Light-duty pulling
6. Light-duty clamping	12. Standard (10 ton) pushing	18. Standard (10 ton) pulling

FIGURE 11-26 This summary shows various body straightening operations that can be accomplished with a portable ram. *(Courtesy of Blackhawk Automotive Inc.)*

FIGURE 11-27 Since equipment designs and methods of proper use vary, refer to the owner's or operating manual and other literature if in doubt. *(Courtesy of Car-O-Liner)*

FIGURE 11-28 This rack-bench system is equipped with electronic measuring equipment. *(Courtesy of Car-O-Liner)*

FIGURE 11-29 An office computer is used to download vehicle body dimensions into a small terminal. The terminal then stores dimensions of specific vehicles and can be taken out to the frame straightening equipment for connection to an electronic measurement arm. By moving the arm to specific reference points, the terminal will tell you the direction and extent of the damage on an alpha-numeric readout of the terminal. *(Courtesy of Car-O-Liner)*

11.3 STRAIGHTENING AND REALIGNING TECHNIQUES

The body-on-frame vehicle can usually be straightened and realigned with a series of single-direction pulls. Single, hard pulls in one direction are fairly effective for straightening full frame vehicles. Overpulling or tearing metal was not a very big issue when frame metal was $\frac{1}{8}$- to $\frac{1}{4}$-inch (3.2 to 6.3 mm) thick. However, this is seldom the case when repairing modern cars, especially in unibody construction with its thin 24-gauge steel.

Remember also that a unibody vehicle is a more complex structure and has a greater tendency to spread collision forces. Most unibody repairs demand multiple pulls, which sometimes means four or more pulling points and directions during a single straightening and alignment setup. The equipment must provide clamps that will prevent further damage of the structure during the pull. A single, hard pull in one direction on a unibody vehicle will usually tear the metal before it is straight.

The usual sequence for a total structure realignment procedure is:

1. Understanding the safety considerations of the alignment equipment used.
2. Damage analysis (this is fully covered in Chapter 10).
3. Initial clamping with alignment checks.
4. Executing the planned pulling sequences with additional clamping and alignment checks.

11.4 ALIGNMENT SAFETY CONSIDERATIONS

When using aligning equipment, inadequate attention to any procedure can result in vehicle or equipment damage and possible physical injury to you or others in the shop.

- Be sure to use the alignment equipment correctly according to the instruction manual prepared by the manufacturer.
- Never allow unskilled or improperly trained personnel to operate aligning equipment without supervision.

- Make sure the rocker panel pinch welds and chassis clamp teeth are tight. As you pull, check the clamps to make sure they are not slipping.
- Always anchor the vehicle securely before making a pull. Check that the chassis clamps and anchor bolts are tightened.
- Always use the size and grade (alloy) chain recommended for pulling and anchoring. Use only the chain and bolt grades supplied with the aligning equipment.
- Drawing chains must be positively attached to the vehicle and/or anchoring locations so that they will not come off during the pulling operation. Avoid placing chains around sharp corners.
- Before powerful side pulls are executed, apply counter supports to prevent pulling the vehicle off the straightening equipment.
- Never use a service jack for supporting the vehicle while working on or under it.
- Always use car stands for supporting the vehicle. Use only the stands recommended for the aligning equipment.
- A pull clamp can slip and cause a sheet metal tear. Prevent bodily harm and material damage by always using safety wires. Watch them closely while pulling.
- Never stand in line with a chain or clamp. Chain breakage, clamp slippage, or sheet metal tearing could cause injury or damage. Remember it can be dangerous to work inside the vehicle at the same time pulls are being made outside.
- If equipment does not have automatic chain locks, cover pulling chains with a heavy blanket. This will keep the chain from being thrown across the shop if it should break.
- Wear leather gloves to prevent hand injuries.
- Wear a hard hat or bump cap when under the vehicle.

Before doing any pulling work, protect the body and externally attached parts as follows:

- When welding or plasma arc cutting, cover glass, seats, instruments, and carpet with a heat-resistant material. Also disconnect the battery and computer modules to protect them from welding or cutting current.
- When removing external parts (moldings and trim) attached to the body, apply cloth or protection tape to the body to prevent scratching.
- If the painted surface on an undamaged part is accidentally scratched, make a note to repair it. Even a small flaw in the painted surface might cause corrosion and an upset customer.

PART REMOVAL

As a general rule, only remove the parts that prevent you from getting to the area of the vehicle being

FIGURE 11–30 It may be necessary to remove the engine and the transaxle or transmission before making pulls and repairs. *(Courtesy of Tech-Cor)*

repaired (Figure 11–30). At one time, you often had to remove the suspension and driveline completely from a unibody vehicle before putting it on the straightening machine. With most of the current straightening systems and such accessories as the engine holder, this is no longer necessary. Major straightening operations can be done with major mechanical parts intact.

Depending on the construction of the vehicle and the location and degree of damage, there will be cases where it will be more convenient to remove bolt-on parts before proceeding with the repair. Carefully analyze the vehicle and the damage to determine what must be removed. It is sometimes best to remove bolt-on parts before putting the vehicle on the rack. You might have better access to the fasteners.

Take the time to carefully study the locations of engine and transmission mounts, suspension mounts, and whether or not these parts themselves are damaged.

11.5 PLANNING COLLISION REPAIR PROCEDURES

When planning the pulling process, you or the technician should

1. Determine direction of the pulls.
2. Find out how to repair the damage in the reverse (first-in, last-out) sequence to which it occurred during the collision.
3. Plan the pulling sequence with the pulls in the opposite direction from those that caused the damage.
4. Find the correct attachment points of the pulling clamps.
5. Estimate the number of pulls required to correct the damage.
6. Determine which parts must be removed to make the pulls.

Many times it may be best to draw the repair plan prior to actually pulling the vehicle. This drawing should show OEM and actual dimensions, anchoring, and pulling locations.

The easiest way to determine where to pull from is to picture the damage being removed by pulling with your bare hands. The pulling process will work in the exact same manner.

Before attempting any repair work, determine exactly the collision procedure that should be taken. A little extra time spent on such an analysis and operational plan can save hours of work.

As a general rule, vehicle straightening is needed whenever the damage involves the suspension, steering, or power train mounting points or major damage to the center section of the vehicle. This, of course, would include situations such as a side collision where the suspension parts and their mountings are not damaged directly, but because of deformation in the center section of the vehicle's structure, the whole body is out of alignment.

Determine whether a particular collision meets this rule either by eye, where there is obvious damage, or by making some general measurements with a tape measure or tram gauge. These would include diagonal measurements to check for diamond and length measurements to check for mash. Try to get as good an idea as possible of where the damage begins and ends. Use all the dimension data available, including body/frame dimensions books or vehicle manufacturer's manuals, or by checking against an undamaged vehicle.

When it is determined how far the damage traveled in the unibody structure and it is fully identified, the damaged area can be pulled and straightened. The corrected reference points provide a larger guide for subsequent pulls.

In planning the repair (pulling) sequence, remember the two basic guides to assure that misalignment and damage will be corrected with minimum metalworking and without further damage:

- Repair the damage in the reverse (first-in, last-out) sequence to which it occurred during the collision.
- Plan the pulling sequence with the pulls in the opposite direction from those that caused the damage.

INITIAL CLAMPING WITH ALIGNMENT CHECKS

There are two pulling systems:

1. *Single-pull systems.* A single-pull system (Figure 11–31) is capable of making a single, very directional pull on the damaged area of a vehicle. These systems are effective on primary damage on frame-type cars. They do not, however, have the ability to hold an undamaged or a corrected reference dimension while pulling. Other devices must be used to provide the hold capability when making these single, very directional pulls (Figure 11–32).

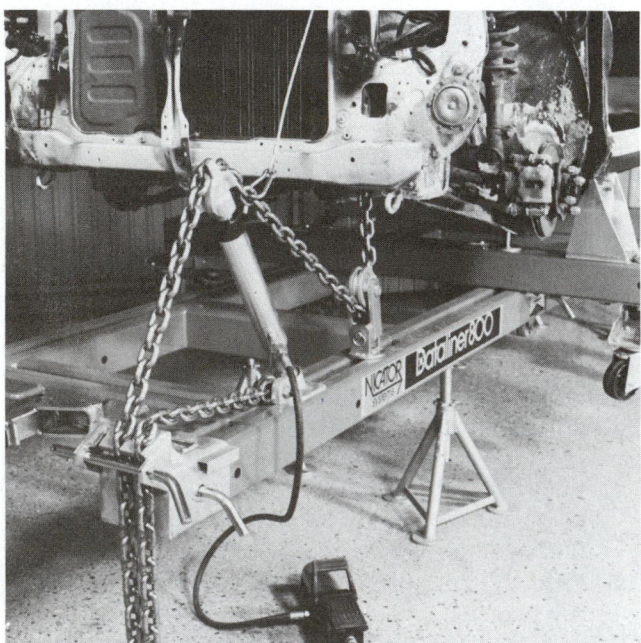

FIGURE 11-31 Single pull uses only one ram. It is OK when complex damage has not occurred. *(Courtesy of Nicator, Inc.)*

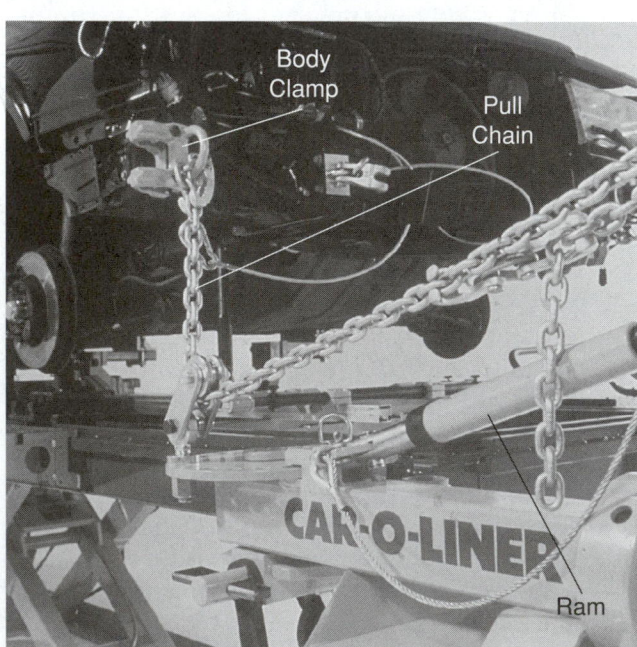

FIGURE 11-32 This is a single down pull to minor damage to front frame rail. Note how the pulley is used to change the direction of the pull to move the rail downward. *(Courtesy of Car-O-Liner)*

2. *Multiple-pull systems.* Multiple-pull systems allow hold and pull or bi-directional pull capability that is often required in correcting secondary damage in unibody cars. Multiple-pull systems, when used, provide the ability to exert a great deal of control over any pulling task. This improves the precision with which a pull can be made. Multiple-pull systems also eliminate the need for disconnecting and moving the power posts.

The multiple-pull approach (Figure 11–33) accomplishes these objectives:

- The exact desired direction of pull can easily be achieved from three or four points at one time. This gives the control needed in the repair of modern unibody construction.
- The use of multiple pull points reduces the amount of force required at any single point. This reduces the risk of tearing lightweight metals. Due to the design of today's vehicles, there simply is not enough strength available in any one place to transmit sufficient force to complete a repair. Again, as in the anchoring system, the pull load must be distributed through several attaching points (Figure 11–34).

A unibody car should be anchored by attaching pinch weld clamps and cross bar as shown in Figure 11–35. To hold the vehicle down and remove slack, pass the chain under the hold down.

When anchoring the car in preparation for pulling, attempt "overanchoring" or "overclamping." An extra anchor point or two takes very little time and it improves safety (Figure 11–36). However, there are many cases where a clamp cannot be fastened to the exact area of deformation. If the section is to be replaced, a piece of steel can be temporarily welded to the section.

A frame-type vehicle can be anchored by placing a suitable plug hook in the fixture holes located on the bottom of the frame rail. Blocking should be used to keep the hook in line with the frame rail. If a hard pull is to be made, it is advisable to weld a washer around the hole as a reinforcement. Make an identical hookup on both sides of the vehicle (Figure 11–37).

It is necessary to set the pulling clamp so that the line extending along the path of the pulling force passes through the middle of the teeth of the clamp. If this is not done, rotational force will act on the clamp to pull it off, further damaging the section (Figure 11–38).

Move the pulling towers into position. Attach the tower chain to the vehicle. Make sure the chains are not twisted. Repeat this on the other towers. If damage has pushed one area of the vehicle up and back, then you need to position the tower and chains to pull down and out.

Collars on the towers allow you to raise or lower pulling points. Take the slack out of the chain and you are ready to pull.

When hooking up to make a pull on a unibody vehicle, consider these pointers:

- The unitized body has made multiple anchoring a must (Figure 11–39). At least four anchors are required, one on each of the body clamps. Depending on the vehicle construction, additional anchoring might be required.
- Always look for the possibility of more than one hookup for both damage correction and restraints. Multiple pulls and/or restraints allow

FIGURE 11-33 Multiple pull is often needed with today's auto designs. You apply pulling tension in more than one location and direction to remove damage as it occurred. *(Courtesy of Nicator, Inc.)*

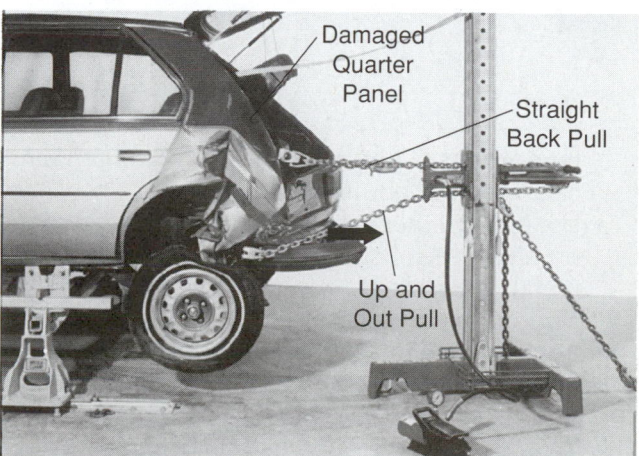

FIGURE 11-34 Here multiple pull is being used to remove rear damage. One pull is up and out and the other is straight out. *(Courtesy of Wedge Clamp International, Inc.)*

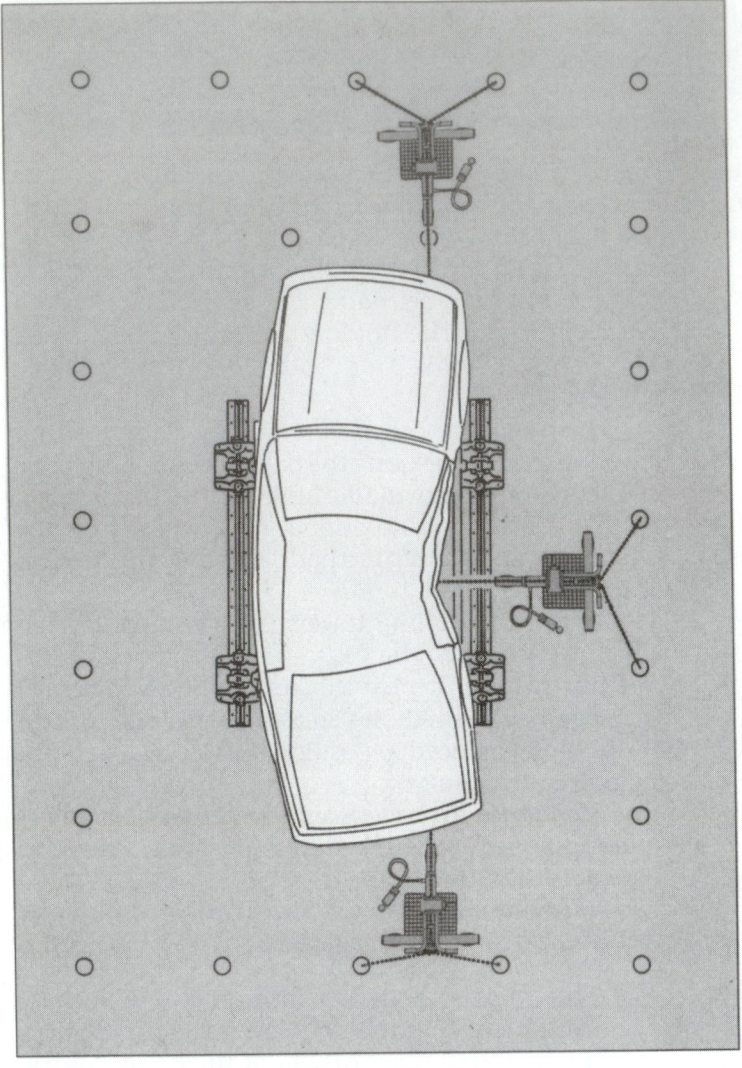

A

B

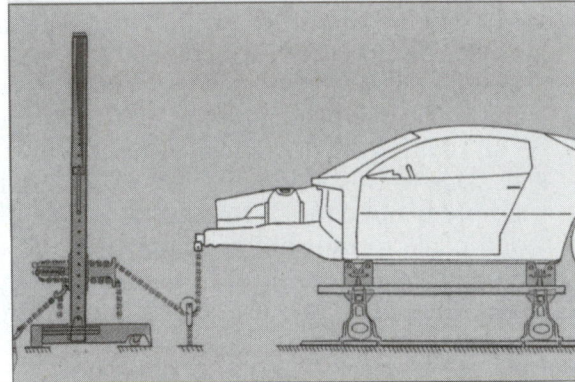

C

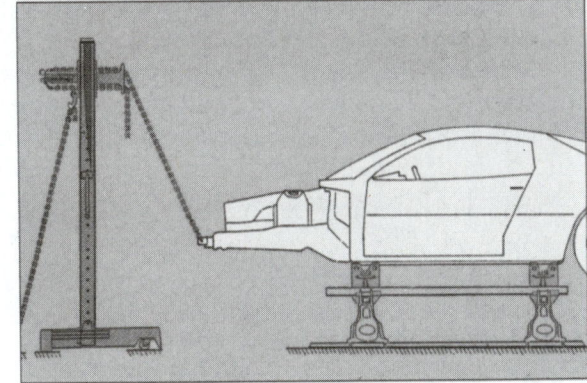

D

FIGURE 11-35 Note examples of portable towers being used to fix damage. (A) Three towers are pulling out side impact damage. Two stretch vehicle lengthwise while other pulls sideways. (B) Tower is pulling up on roof panel. (C) Tower is being used to make downward pull on front rail. (D) Tower is pulling up on rail. *(Courtesy of Wedge Clamp International, Inc.)*

twice the pull potential with less damage being caused at the points of attachment.

- Use multiple hookups on structural members and on sheet metal sections to be worked. Today's metals will shift, shrink, and stretch quite readily. This is why an incorrect (too localized) pull can cause more damage than it removes.
- Always install additional security chain or chains to a substantial member on the vehicle chassis.
- Treat each damaged area as individually as possible since the cars of today are manufactured for isolated collapse upon impact.

- Carefully observe the "last-in, first-out" rule in areas of primary as well as secondary damage. This principle can be occasionally violated for initial pulls, but nearly always holds true in the fine-tuning phases of unibody alignment.
- Use imagination in utilizing available clamps for multiple hookups, including the shaping of straps and other attaching devices.
- After placing a small amount of pressure on the pulling chains, the anchor clamps will be seated, and they should be retightened to prevent slippage.

FIGURE 11-36 Anchor clamps are bolted around pinch welds on the underbody to keep a unibody vehicle from moving while pulling. *(Courtesy of Car-O-Liner)*

Up and Out Pull

Straight Out Pull

FIGURE 11-37 Straightening a full frame vehicle is similar to straightening a unibody vehicle. However, more pulling force must be applied to the frame rails and less to the body structure. *(Courtesy of Wedge Clamp International, Inc.)*

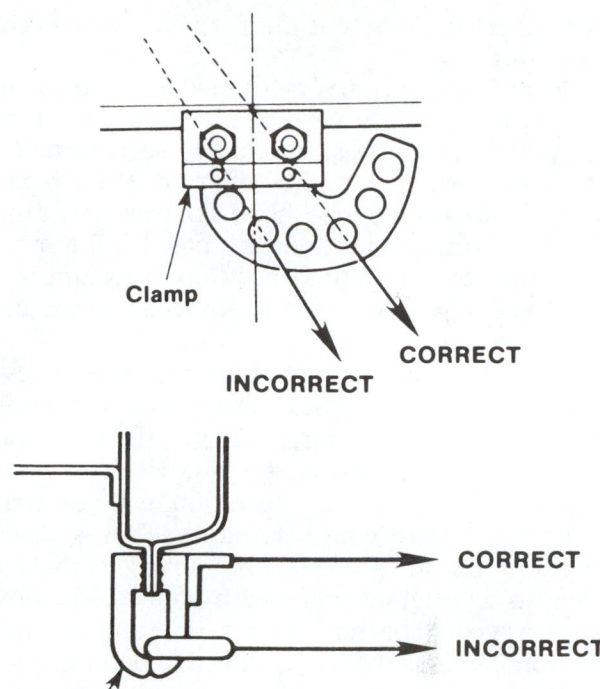

Clamp

CORRECT

INCORRECT

CORRECT

INCORRECT

Clamp

FIGURE 11-38 Note the right and wrong ways to set clamps. *(Courtesy of Toyota Motor Corp.)*

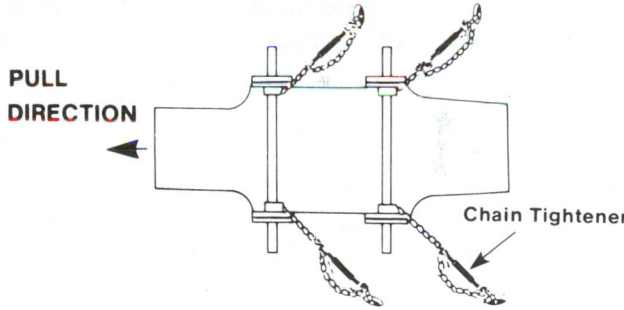

PULL DIRECTION

Chain Tightener

FIGURE 11-39 This drawing shows four-point anchoring. *(Courtesy of Blackhawk Automotive Inc.)*

EXECUTING A PLANNED PULLING SEQUENCE

The progress toward alignment should be monitored during the pull. Since the body (sheet metal) has elasticity, the structure will partially return to its original damaged condition (to a certain extent) even if the body is pulled back to the prescribed dimensions. Therefore, estimate the amount of return in advance and make allowance (overpull) for it during the pulling operations.

The pull procedure or sequence simply consists of solving a variety of small problems rolled into one. Find your first problem, and begin working on it. Then move to the next problem and so on.

Because of the power of the rams, the metal will begin moving as soon as the chain slack is taken up.

Always check your dimensions frequently to prevent overpulling.

Several attempts may be needed to get the damaged area to remain in the proper position. You may have to pull and release tension to see where the panel moves with the tension released. Then, repull and release to slowly move the part or panel to within specs. Each time this is done the panel will move a little closer to the desired position. Shocking the metal in an adjacent area will help relieve stress and keep it moving as needed.

Make the pulls a little at a time; relieve the stress; take a measurement. Typically, work from the center section outward, achieving first length, then sidesway removal, and finally correct height.

Approach the pulling operation as if you were going to do it with your bare hands. That is, determine how the metal should be moved to mold it back into its original shape with your hands. How many areas could be moved at one time and in which directions? This is the key to effective pulling.

There are a number of setups for pulling or pushing. The pulling arrangement with the **vector system** is determined by a simple triangle. The setup shown in Figure 11–40 is used for pulling up and out. The ram, the base unit, and the chain form a triangle. As the ram is extended, one side of the triangle becomes longer. This causes the ram to swing

to the right because the chain is locked to the ram. As it swings to its new position, the damaged vehicle is pulled upward and over. This simple procedure is based on the principle of vectors. Tremendous forces can be exerted in a carefully planned direction by using this principle.

Figure 11–41 shows a triangular arrangement that will provide more of a straight outward pull. Note that the ram is placed at an angle to the right of true vertical. As force is applied, the ram will swing to the right pulling the damaged car sections with it. It is important that the ram is at the proper height. This can be controlled by adding the proper length of tubing to arrive at the correct height before the hookup is finished.

By setting up the equipment as designed, the vector system will provide strong pull capabilities and commonly allow 10 inches (254 mm) of chain travel in one continuous pull (Figure 11–42). As higher, longer pulls are needed, the vector system automatically trades power for motion and typically allows as much as over 3 feet (0.9 m) of chain travel in one

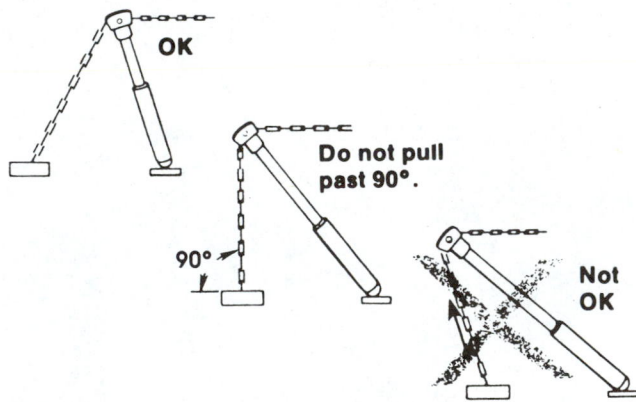

FIGURE 11–41 Study the right and wrong triangular arrangement for a straighter outward pull. *(Courtesy of Blackhawk Automotive Inc.)*

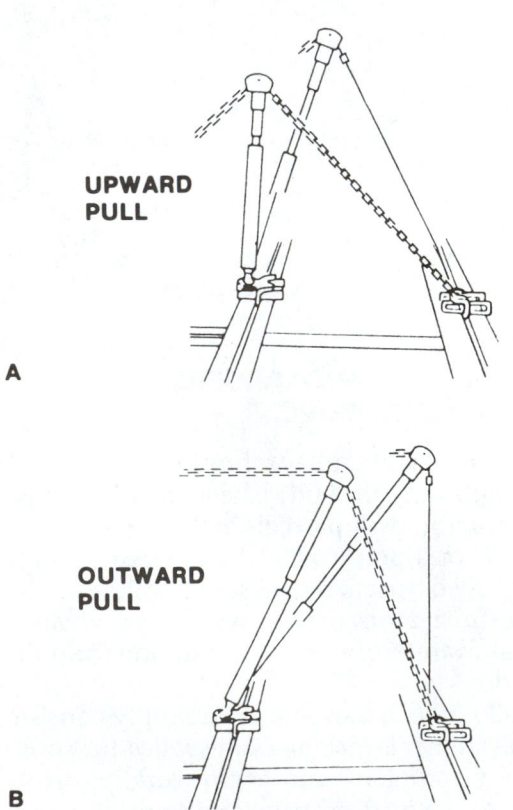

FIGURE 11-40 Note the arrangement for different pulling directions. (A) upward and (B) outward. *(Courtesy of Blackhawk Automotive Inc.)*

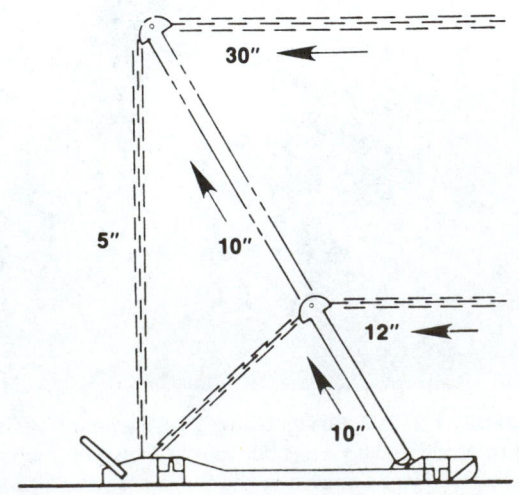

FIGURE 11-42 A vector-type pull for varying triangle between chain and ram. *(Courtesy of Blackhawk Automotive Inc.)*

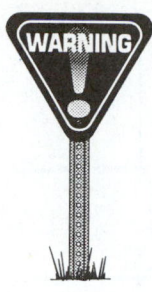

WARNING At no time should pulling continue if the chain between the ram and the anchor goes beyond perpendicular. If this condition should occur, the possibility of overloading a chain could result because of the added stress placed on the anchored end of the chain. To avoid this condition, be sure that the chain lock head is not placed behind the chain anchor.

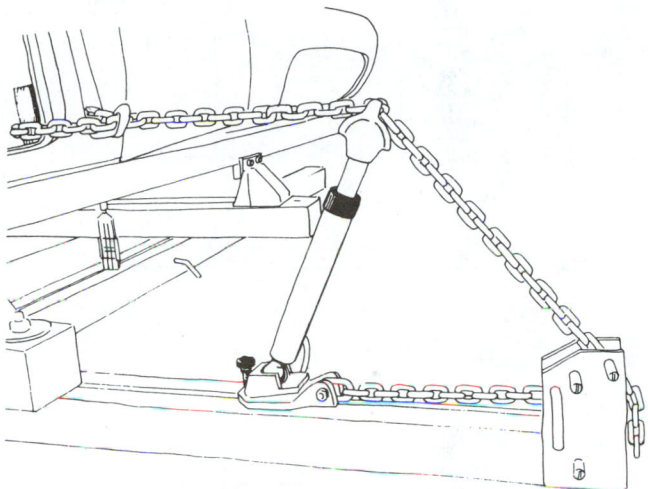

FIGURE 11-43 A lower structure pull is being done with a short ram mounted on a rack framework. *(Courtesy of Nicator, Inc.)*

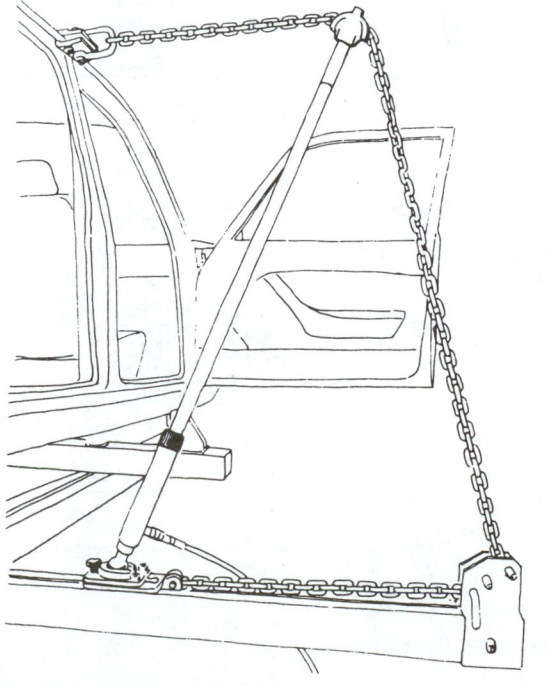

FIGURE 11-44 Typical roof rail height pulls can be done by adding extensions sleeves to the ram. *(Courtesy of Nicator, Inc.)*

continuous pull. The system can easily pull up, out, or down. These pulls are accomplished by controlling the angle of the ram and by adjusting the length of the tubing used with the ram.

In a typical pull setup at frame rail height, the power ram is set so that the angle between the ram and the pulling chain is equal to the angle between the ram and the anchor (Figure 11–43). With approximately 10 tons (9 metric tons) of force and 10 inches (254 mm) of pulling chain travel available, there is ample power and movement for the lower tough structure pulls.

As the pulls move higher, part of the available power is converted to longer pull chain travel (Figure 11–44). For example, at roof rail height 10 tons (9 metric tons) of force are not needed to position a top section. However, more than 10 inches (254 mm) of continuous pull chain travel might be needed to get it into proper alignment with one setup. In a typical roof rail height pull, approximately three feet (0.9 m) of continuous pull is usually available automatically. This saves set-up time and speeds the repair procedure.

There are other basic single-pull setups when using a ram:

- For a high pull, more tubing is required. For an out and down pull (Figure 11–45A) less tubing is needed. Another way to make a down pull is to attach a chain between the vehicle and floor anchors. By pulling on the chain bridge, the structure is forced down (Figure 11–45B).
- A horizontal pull on a rail (Figure 11–45C) can be accomplished by placing the ram at about a 45-degree angle.
- By adding tubing to the ram, a straight out pull on the cowl can be accomplished (Figure 11–45D).
- To pull straight out at the roofline, use the ram with extension tubes, as shown in Figure 11–45E.
- Upward pulls are very easy to set up (Figure 11–45F). In most cases, the ram is in a vertical position. This pull setup will produce an upward and slightly outward pull.
- The same type of setup can be used at roof height by adding extensions to the ram (Figure 11–45G).
- Although pushing is not used to the extent it once was in collision damage repair, the capability to push is still important (Figure 11–45H). The vector system provides push capability from any angle around the vehicle by means of a simple triangular setup.
- It is also possible to push from underneath the car at whatever angle is needed (Figure 11–46A). This push setup can be used to effectively remove sag at the cowl area (Figure 11–46B).
- In most situations, more than one pull will be needed to effectively repair the vehicle for a

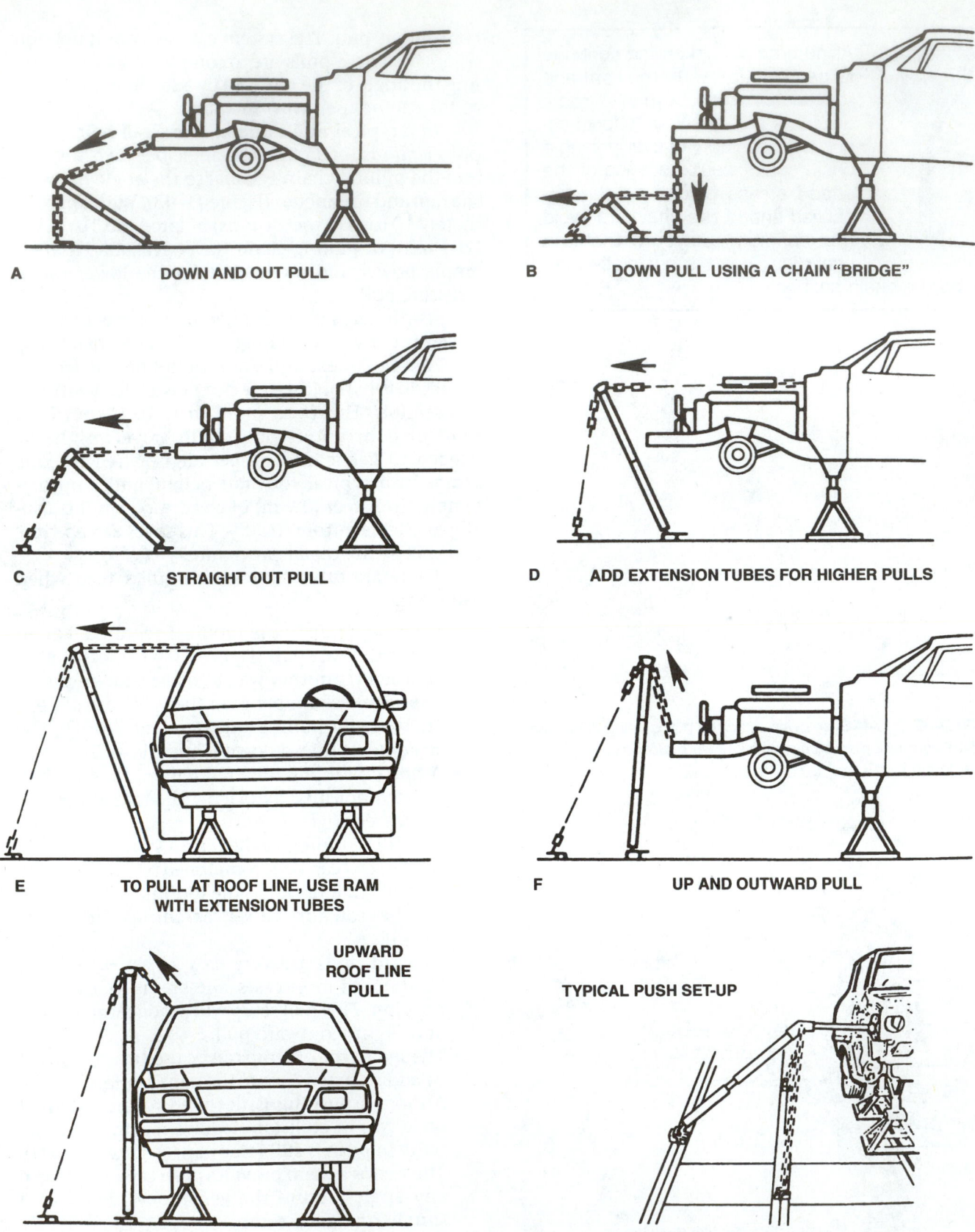

A   DOWN AND OUT PULL

B   DOWN PULL USING A CHAIN "BRIDGE"

C   STRAIGHT OUT PULL

D   ADD EXTENSION TUBES FOR HIGHER PULLS

E   TO PULL AT ROOF LINE, USE RAM WITH EXTENSION TUBES

F   UP AND OUTWARD PULL

UPWARD ROOF LINE PULL

TYPICAL PUSH SET-UP

G

H

FIGURE 11-45 Study basic ram single pull setups. *(Courtesy of Blackhawk Automotive Inc.)*

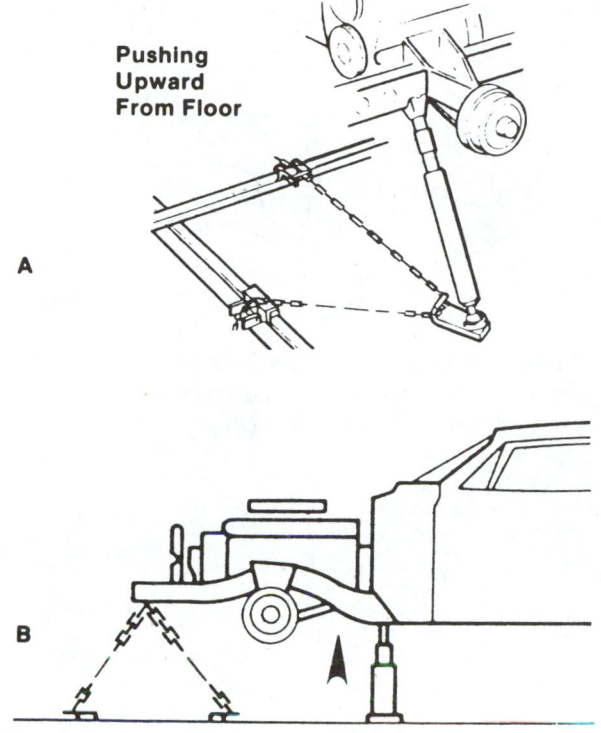

FIGURE 11-46 These are common pushing ram setups. (A) A ram anchored by chains will push up and sideways on the part. (B) A ram will push straight up. *(Courtesy of Blackhawk Automotive Inc.)*

variety of reasons. Some important multiple pull setups will be discussed later in this chapter.

Due to the high-strength, heat-sensitive, characteristics of the unibody structure, it is usually best not to attempt to make an alignment or straightening pull in one step. Instead, use a sequence that consists of a pull, hold the pull, more pull, hold, and so on. This allows more time for working the metal, and gives the metal time to relax, enabling you to check the progress of alignment (for clamping, repair or reattachment by welding, and so on).

Start the hydraulics moving, slowly and carefully. Watch the movement closely. Is it doing what it is supposed to do? If it is on the right track, keep going. If it is not, determine why, make the angle or direction adjustment, and try again.

Relieve the stressed or locked-up metal by hammering as described in Chapter 8. Pull the damaged metal to tension; then loosen it by hammering. Increase the tension and loosen it again. If you are not sure if the metal is locked, try freeing it with hammering.

Remember that a thin-gauge unibody structure can still be overstressed and damaged where it is held by the pulling clamp. It might be a good idea to distribute the load over more of the structure with additional clamps (Figure 11-47).

The repair of a bent closed cross-sectional structure such as a side member is done by clamping the surface of the bent-in side and pulling. The pulling direction should be such that force is applied in the direction of an imaginary straight line extending through the original position of the part. The minor dented portion of the part can often be repaired by welding studs and pulling them with a sliding hammer (see Chapter 9).

If some of the buckles are folded so tightly that they threaten to tear, it may be necessary to use a little heat. However, use heat carefully and only on the corners and double panels. Heat on a low spot in the side of a frame rail or box section will only drive the damage deeper. Use heat carefully and as a means of releasing locked metal and not as a means to soften an area. Although a torch is not recommended on HSS metals, it can sometimes be used with care as described in Chapter 7.

By bringing the damaged metal back into shape slowly and carefully, a first-class solid, safe repair is easily achieved. Although there will be exceptions, a good rule of thumb to follow is to achieve proper length, width, and height, in that order.

Overpulling

Overpulling is done by pulling the damage slightly beyond its original dimension. If done in a controlled way, the metal will flex back slightly when tension

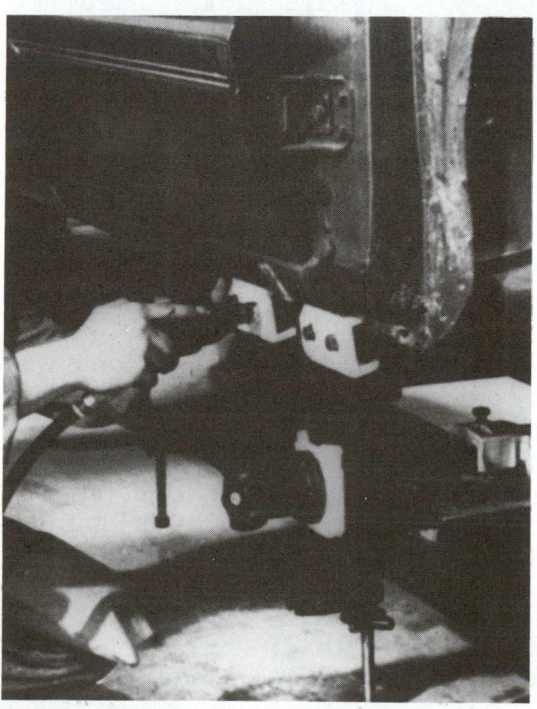

FIGURE 11-47 Attach additional clamps where needed to prevent metal tearing. *(Courtesy of Blackhawk Automotive Inc.)*

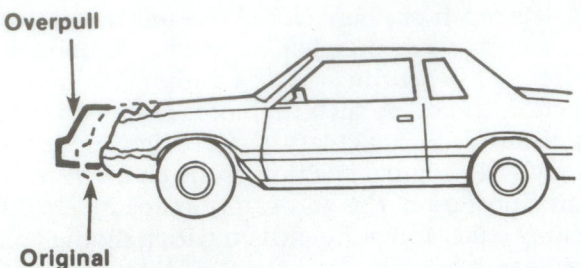

FIGURE 11-48 Slight overpulling is needed so that metal can spring back to its original dimensions. If overpulling is excessive, expensive and time-consuming part replacement will often be needed.

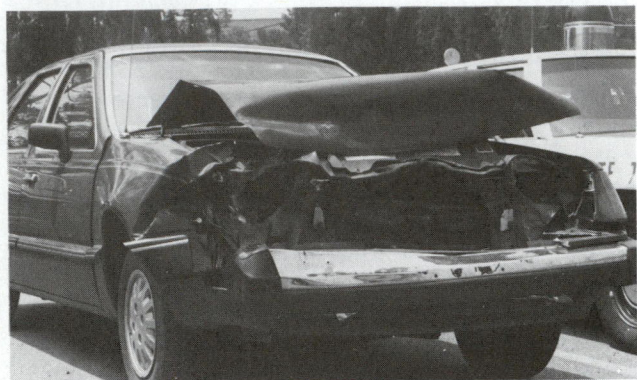

FIGURE 11-49 An example of front end damage that would require R&R of bumper, hood, fenders, and related parts before pulling.

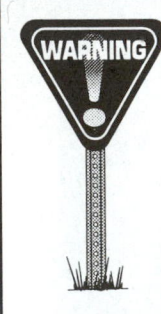

WARNING !

Remember that it is possible to pull a piece of string in a straight line, but there is no way to push it straight. It is difficult to shrink or compress any damaged metal pulled or stretched beyond its critical reference dimension. In most instances, the only way the overpulled panel can be repaired is by replacement.

is released. The body/frame reference points will then line up properly. If done too much, overpulling might not be a correctable error (Figure 11-48).

Overpulling damage results from failing to measure accurately and often. To prevent overpull damage, measure the progress when pulling the damaged area.

STRAIGHTENING FRONT END DAMAGE

The general repair method for front end damage is best covered by going through a typical example (Figure 11-49). It is important to begin the repair by

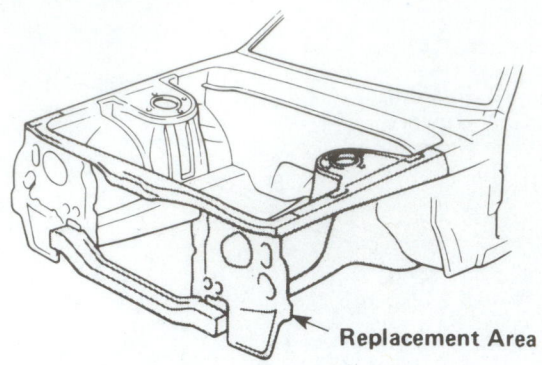

FIGURE 11-50 Restoring the front fender apron will often require replacement of a badly damaged radiator support. *(Courtesy of Toyota Motor Corp.)*

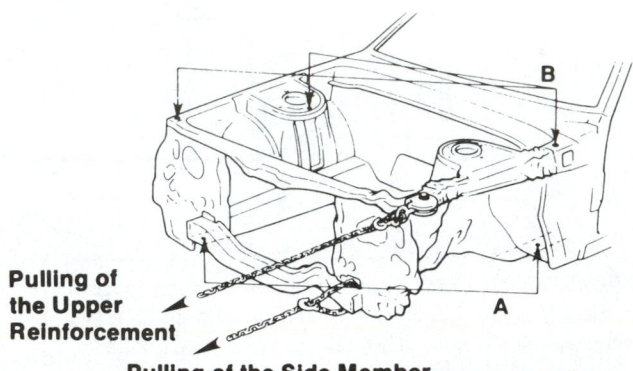

FIGURE 11-51 Constantly check the front dimensions as you pull out damage. Make sure the direction and amount of pull is correct. *(Courtesy of Toyota Motor Corp.)*

restoring the front fender apron and side member to their predamaged condition and to repair the support structure on the replacement side (Figure 11-50).

First, pull the side member on the replacement side in the direction opposite to the impact direction. Then repair the fender apron and side member on the repair side. At the same time, repair the front fender apron and side member installation areas on the replacement side. There are many cases where the entire fender apron or side member on the repair side is deflected left or right only. Since there is practically no warping in the lengthwise direction, repairs involve measuring the diagonal dimensions A and B, as shown in Figure 11-51. This measurement would be compared to the frame/body manual to determine the extent of the damage. Correct this distance while keeping an eye on the repair condition. The operation can be done efficiently if the fender apron upper reinforcement is pulled at the same time as the side member. Nylon pull straps are sometimes used on double pull hookups (Figure 11-52).

If there is severe bending damage, it might be best to separate the front cross member and radiator upper support and repair or replace them separately. Grip the inside broken face of the side member, and while pulling it forward, pull the bent piece from the inside or push it from the outside (Figure 11–53). After repairing the bent portion, match the dimensions to specs.

To repair the replacement side front fender apron and side member area, the main repairs are near the dash and cowl panels. If the impact was severe, the damage will extend into the front body (A) pillar (the door will fit poorly in this case). Simply gripping the front edge of the side member of the fender apron and pulling will not repair the major damage to the front body (A) pillar or the dash panel (Figure 11–54A). In this case, cut the fender apron and side member near the installation area, clamp near the major panel damage, and pull (keep an eye on door fit conditions). Good results can be obtained using this method. Also, at the same time that the pillar is being pulled forward, pushing can be done from the interior side with a power ram (Figure 11–54B).

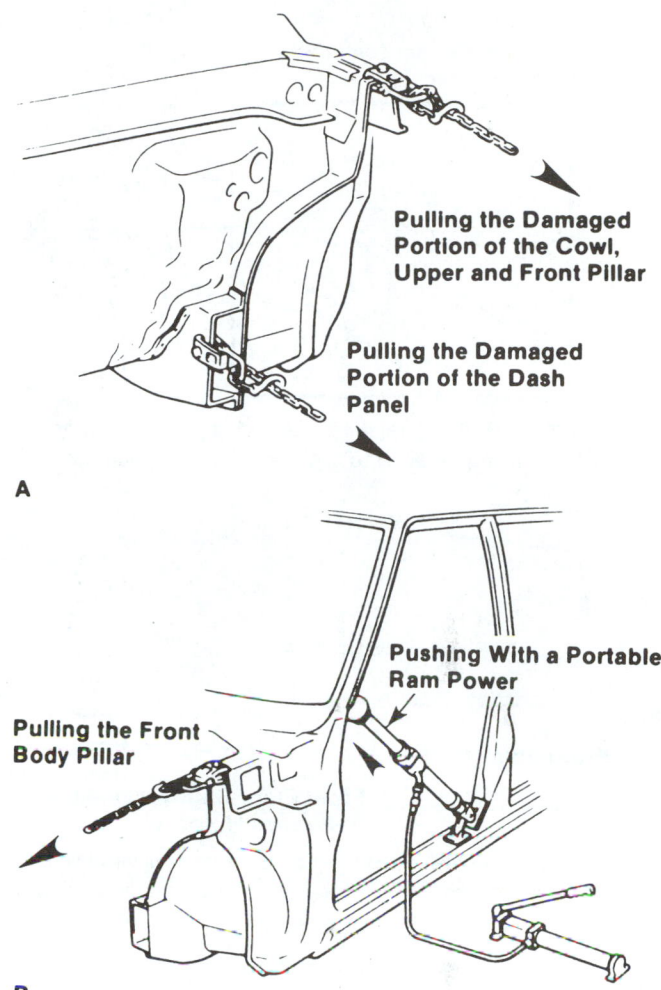

Pulling the Damaged Portion of the Cowl, Upper and Front Pillar

Pulling the Damaged Portion of the Dash Panel

A

Pushing With a Portable Ram Power

Pulling the Front Body Pillar

B

FIGURE 11-54 If major damage extends to the cage around the passenger compartment, this type of pulling might be needed. (A) Pulling on the cowl and rail at the same time. (B) While pulling from the front, you may also need to use a portable power unit to spread the door opening. *(Courtesy of Toyota Motor Corp.)*

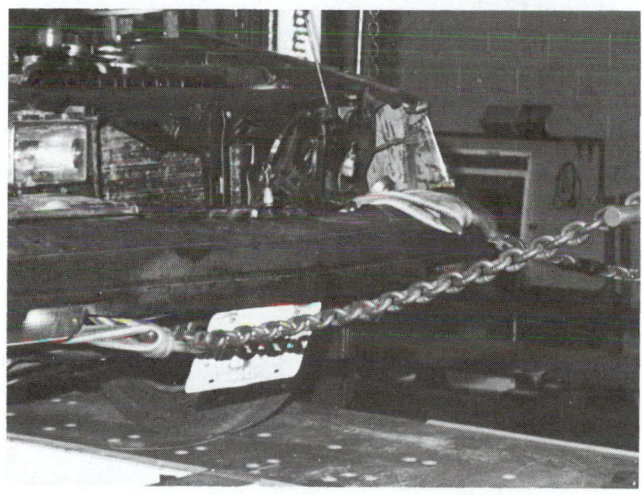

FIGURE 11-52 Use of nylon pull straps is handy and timesaving. *(Courtesy of Bee Line Co.)*

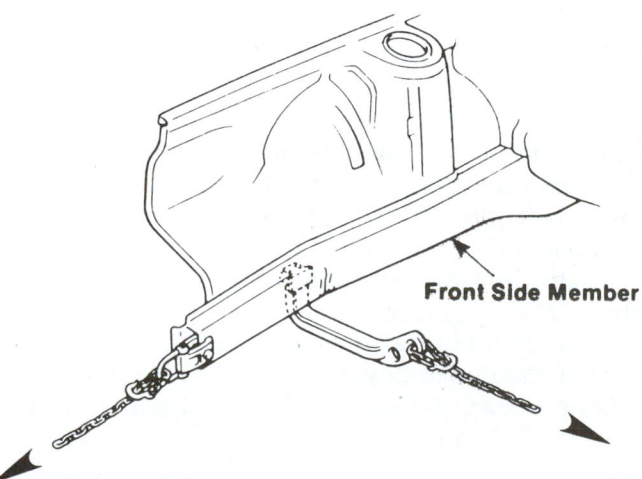

Front Side Member

FIGURE 11-53 Minor front side member damage often requires pulling from the front and side at the same time. *(Courtesy of Toyota Motor Corp.)*

During body aligning, confirm the degree of restoration by measuring critical dimensions. Reference holes in the underbody front floor and rear of the front fender installation hole are the standard reference points. Therefore, it is important to confirm that the damage does not extend to these areas during initial diagnosis (Figure 11–55). If the impact to the front side member structure, which is used particularly for FR vehicles, is severe, there is a tendency for it to take the shape shown in Figure 11–56. The height of the standard measuring point might be distorted, so use caution during repairs of this nature. Further, the front side member used in vehicles has a reference point in the rear that has a tendency to be deflected upward when damaged (Figure 11–57).

To correct lateral bending damage of the front caused by side impact, frame straightening equipment is necessary. The clamping point (Figure 11–58)

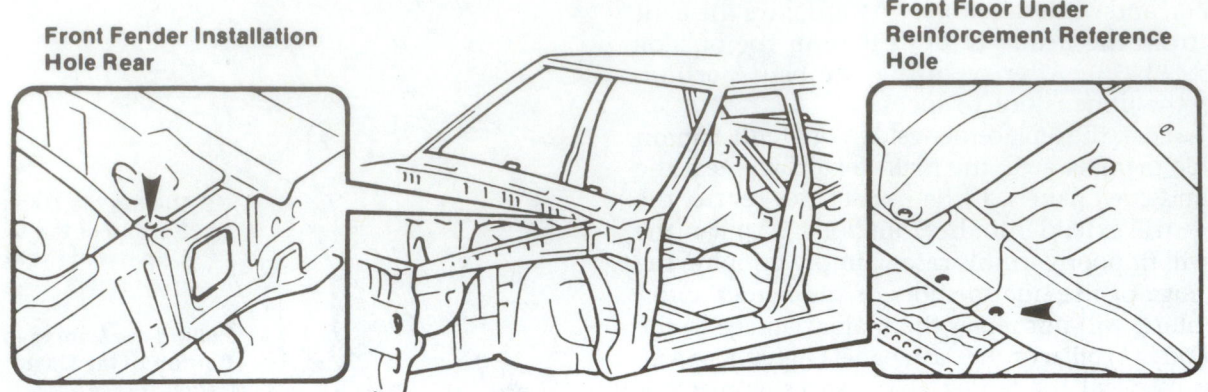

FIGURE 11-55 Standard measuring points will vary from vehicle to vehicle. Refer to the specs for the exact make and model car or truck being repaired to get the right dimensions or measurements. *(Courtesy of Toyota Motor Corp.)*

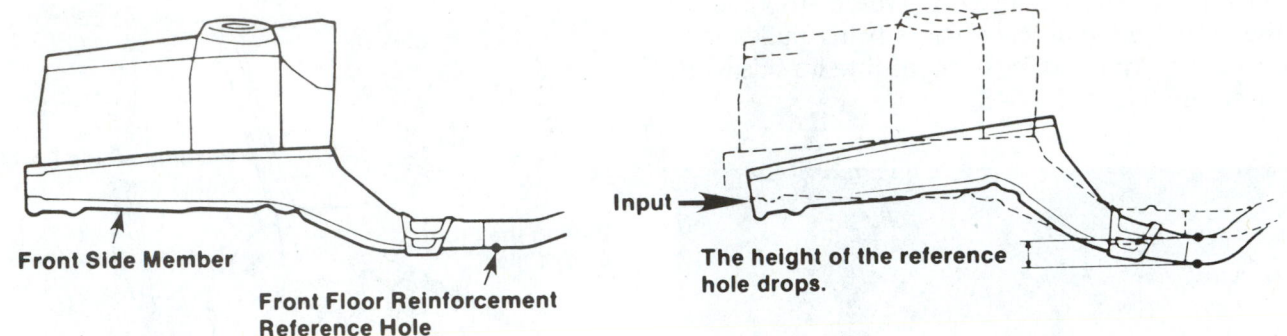

FIGURE 11-56 With front side member downward impact, damage may extend into the floor and affect its reference points. This is a front-engine, rear-wheel-drive vehicle example. *(Courtesy of Toyota Motor Corp.)*

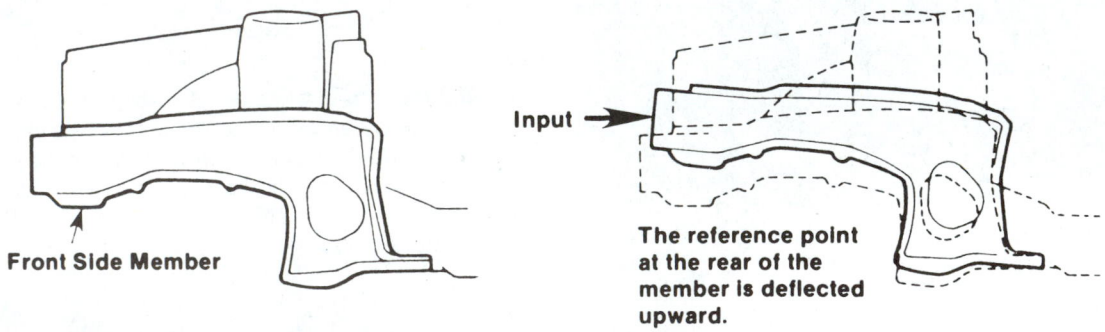

FIGURE 11-57 Note typical front-engine, front-wheel-drive vehicle front side member damage. *Courtesy of Toyota Motor Corp.)*

receiving the greatest force is point B, which must be clamped securely. If point C is not secured, point A cannot be pulled.

STRAIGHTENING REAR DAMAGE

Since panel construction of the rear body is not as strong as the front body, damage can be more complex and more extensive. In most cases, the bumper is impacted during rear end collisions (Figure 11–59). The impact force will usually radiate through the ends of the rear side members or nearby panels and cause damage to the kick-up area. Next, the wheel housings will deform, causing the entire quarter panel to

move forward, which will cause clearance problems between other components. If the impact is severe enough, it will affect the roof, door panels, and center body pillar.

Attach clamps or hooks to the rear portion of the rear side member, rear floor pan, or quarter panel and pull while measuring the dimensions of each part of the underbody. Determine the degree of repairs necessary by the conditions of panel fit and clearances (Figure 11–60).

NOTE: Do not clamp and pull a quarter panel that has little or no strain on it with major rear damage. When the rear side member is pushed into the wheel housing or there are clearance problems at the rear

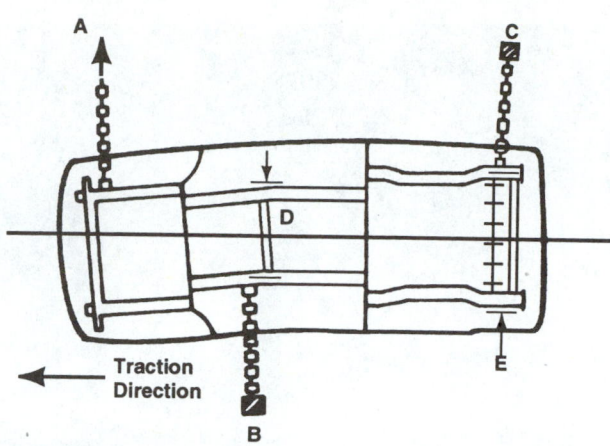

FIGURE 11-58 Here is a common setup for correcting lateral bending damage. B is the point of impact which pulls against A and C. Blocking devices at points D and E prevent part damage and help pull all damaged areas together. *(Courtesy of Nissan Motor Corp.)*

FIGURE 11-60 Note typical rear damage repair pulling setup. *(Courtesy of Kansas Jack, Inc.)*

FIGURE 11-59 Typical rear end damage has ruined bumper, quarter panel, and related parts. Pulling of the right, rear frame rail would be needed.

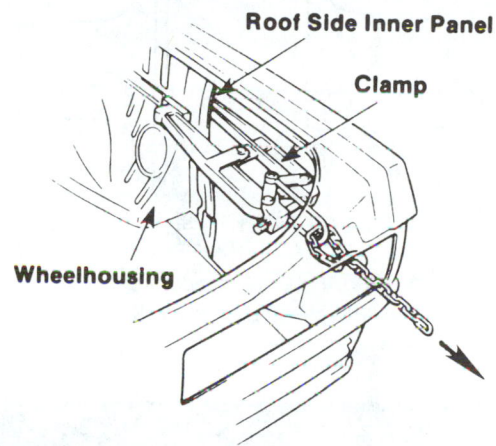

FIGURE 11-61 Here the pulling chain is reaching through the lens opening to grasp and pull on the inner panel. *(Courtesy of Toyota Motor Corp.)*

door, pulling on the quarter panel should not be done. Relieve the stress in the quarter panel by pulling on the side member only. If the wheel housing or the roof side inner panel is clamped and pulled along with the rear side member, the clearances with the door panel can be maintained properly (Figure 11–61).

Where there is rear structure twist from a front hit, the rear lower structure should be clamped to the straightening equipment. The preliminary pulling will restore some of the lower alignment points. Clamps can be repositioned to preserve the alignment. As you move forward with subsequent pulls, the alignment and number of anchoring points will, of course, move right along with them.

Install upper structure alignment fixtures as soon as upper damage is reduced enough to permit their

positioning. Also, sections too damaged for repair and requiring replacement can be cut away at this time.

STRAIGHTENING SIDE DAMAGE

If there is a severe impact to the center of the rocker panel, the floor pan will deform and the entire body will take on a curved shape like a banana. To align this type of damage, use a method similar to straightening a piece of bent wire. The two ends of the body are pulled apart and the caved-in side is pulled outward (three-way pulling, Figure 11–62).

Anchoring unitized body vehicles for side pulls can be very difficult due to the limited chain hookup areas. Figure 11–63 shows an anchoring method. Keep in mind that when making a side pull on the end of a vehicle, the center section can be anchored by passing a chain around the pinch weld clamp and hooking it to the edge of the bench or rack

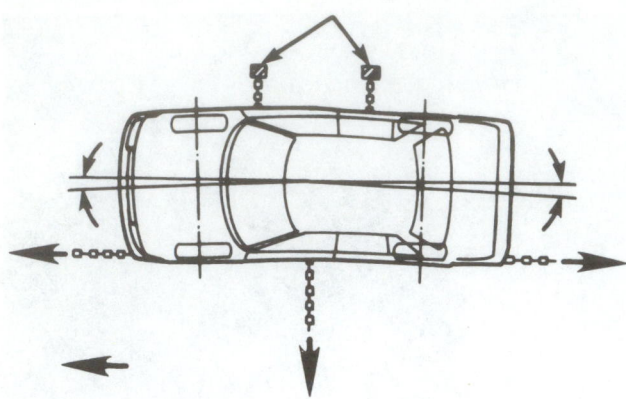

FIGURE 11-62 For inside damage repairs, it may be necessary to pull in three directions. *(Courtesy of Nissan Motor Corp.)*

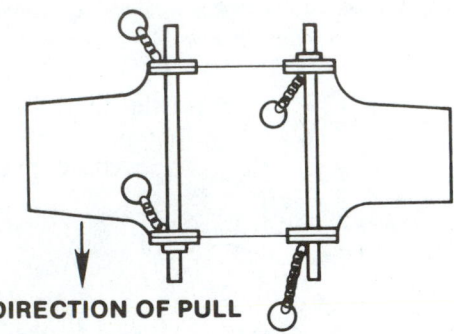

DIRECTION OF PULL

FIGURE 11-63 Anchoring a side pull in front of a vehicle like this will keep the body from swiveling on its anchors. *(Courtesy of Blackhawk Automotive Inc.)*

FIGURE 11-64 Double clamps are being used to anchor this unibody car. *(Courtesy of Bee Line Co.)*

A

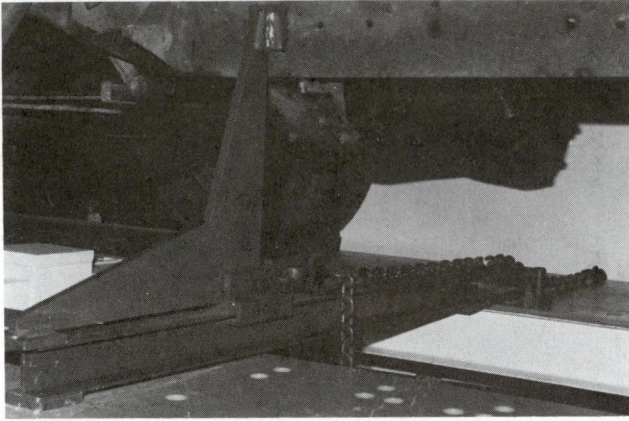

B

FIGURE 11-65 Use of a portable beam and knee is common with full frame vehicles. *(Courtesy of Bee Line Co.)*

(Figure 11–64A). Tension/anchoring can be applied by attaching the pull chain to the pinch weld clamp (Figure 11–64B).

The portable beam and knee can be used as a side anchor with either inside or outside contact (Figure 11–65). By attaching the pull chain to the portable beam and knee, it can be used as a pulling or anchoring attachment.

It is advisable to make an end-to-end stretch pull whenever pulling outward on the center section of a vehicle. If pulling high on the body, it is necessary to tie the vehicle down on the opposite side (Figure 11–66). Pulling outward on the center section of the vehicle can also be done with the portable beam and knee attached to the pulling tower or ram (Figure 11–67). The chain roller can be in the lowest position on the power tower.

• **Upward movement.** Pulling upward can be done by locking two power towers across from each other and connecting the pull chains as shown (highest roller position) in Figure 11–68. Suitable sheet metal clamps are attached to the tower pull chains. Direction of the lift can be controlled by opening or closing pull ram valves. The vehicle must be securely tied down to the frame rack.

Raising the vehicle can be done by positioning towers as shown in Figure 11–69. Wrap nylon straps or chain around the bumper or isolators. You could also pass tower pull chains between the underbody and stabilizers.

- **Sag correction.** Blocking under the low area and pulling down on the high end will correct sag (Figure 11–70A). The vehicle must also be tied down to the equipment at the opposite end. Anchoring the high portion of the vehicle to the rack with chains and pushing up at the low spot will also correct the datum line (Figure 11–70B). When using the pulley and base for

the downward pull, the tower pull chain must be in the lowest position.

Sag can also occur at the front frame cross member. The ends of the cross member will be closer than normal and the center will be too low. This condition can be corrected by using three hydraulic rams and two chains plus an anchoring rail or pots. The correcting hookup for sag on both sides of the cross member is shown in Figure 11–71A. Figure 11–71B shows the hookup for sag at one side. Check the repair with a tram gauge and compare the measurements to the specifications in the body manual or chart (Figure 11–71C).

- **Twist correction.** Position and lock a ram or tower on the side of the platform or bench next to the low side of the vehicle. With the tower chain in the highest position, route the chain

FIGURE 11-66 The technician is making an end-to-end stretch pull. *[Courtesy of Bee Line Co.]*

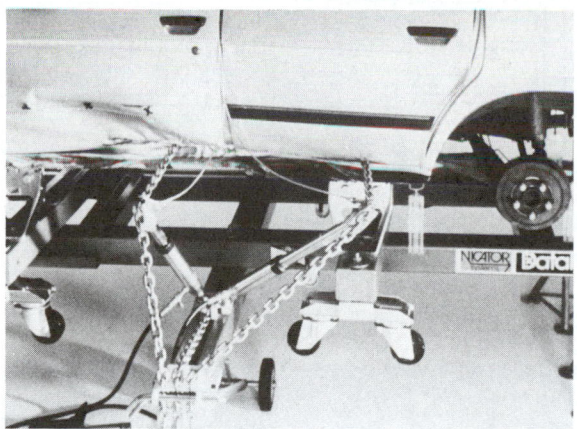

FIGURE 11-67 A frame rack is being used to pull outward on the center section. *[Courtesy of Nicator, Inc.]*

FIGURE 11-68 Making an upward pull to repair front impact damage. *[Courtesy of Bee Line Co.]*

FIGURE 11-69 The vehicle can be raised easily with rack and straps for mounting and part removal. *[Courtesy of Bee Line Co.]*

A

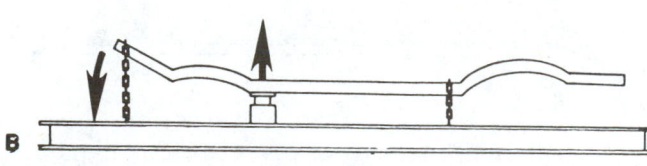

B

FIGURE 11-70 Correcting sag damage can be done like this. *[Courtesy of Bee Line Co.]*

under the lower frame rail end and over the high rail end. Attach a chain hook to the outside edge of the platform bed. Make an identical hookup at the opposite end of the vehicle or tie down and block under the center section of the vehicle. Apply pressure to the pull chain.

An alternate method of correcting twist condition can be made by pulling down on the high side as described previously and blocking or lifting under the low side. The center section of vehicle should be blocked and tied down (Figure 11–72).

- **Diamond correction.** Place a pulling tower or ram on each end of the frame rack on opposite sides. Adjust chain height and attach to the vehicle as described for end pull corrections. Block or anchor one side of the vehicle to prevent side movement of the vehicle (Figure 11–73). Activate the pull ram.
- **Strut pull.** To pull a MacPherson strut tower into position, attach the pull plate to the vehicle and connect a tower pull chain (Figure 11–74). Most MacPherson adapter plates can also be used

FIGURE 11-72 A setup is being used to fix twist damage. *(Courtesy of Bee Line Co.)*

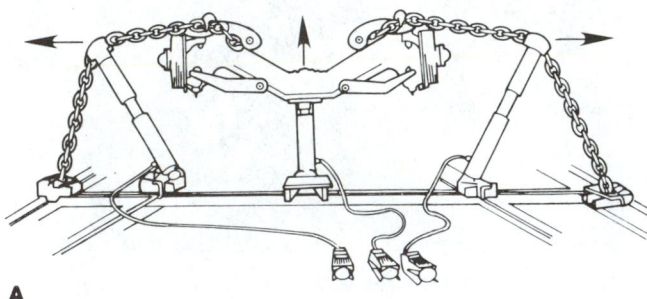

A

Shim Contact Point

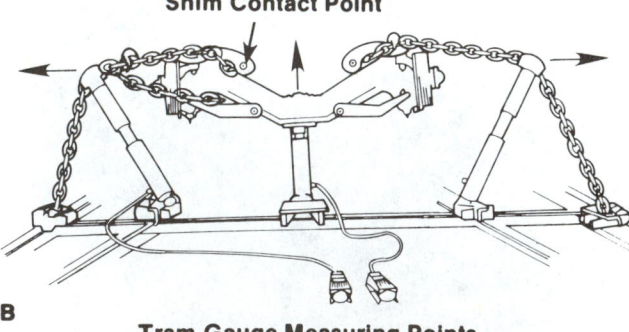

B

Tram Gauge Measuring Points

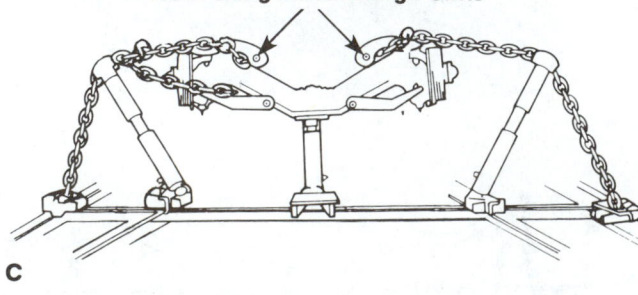

C

FIGURE 11-71 Note methods of correcting cross member sag. *(Courtesy of Blackhawk Automotive Inc.)*

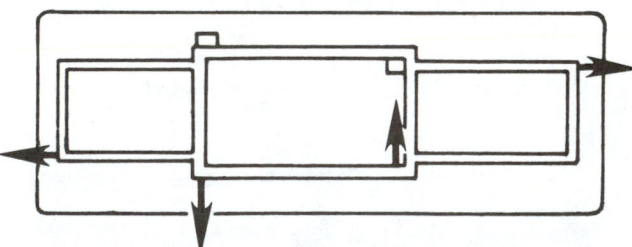

FIGURE 11-73 Correct diamond damage by pulling in these directions.

FIGURE 11-74 Strut tower pull hookup often requires a plate that bolts over the top of the tower. *(Courtesy of Bee Line Co.)*

SHOP TALK By analyzing all critical reference points, you can determine multiple pull setup and methods. Pulling setup can seem confusing at first; so take your time (Figure 11–75).

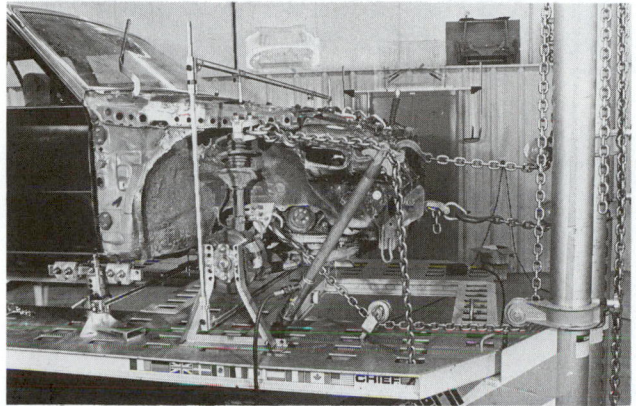

FIGURE 11–75 By analyzing all critical reference points, you can determine multiple pull setup and methods. *(Courtesy of Chief Automotive)*

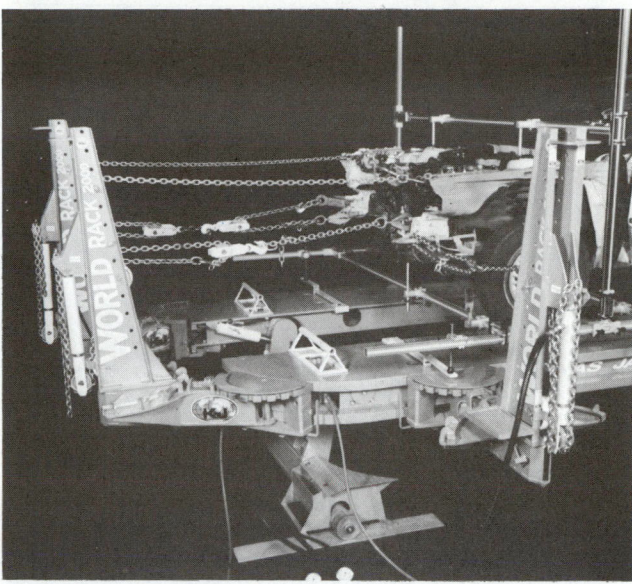

FIGURE 11–76 Correcting multiple damage conditions can be complex. Take your time and proceed slowly while monitoring reference points closely. *(Courtesy of Kansas Jack, Inc.)*

on frame ends, hinge mounts, or similar locations. If both towers are repositioned left or right, they can be positioned by mounting adapter plates to both towers and installing a strap to make the pull. After the pull is made, a dimension check should be performed using a strut measuring gauge.

Correcting multiple damage conditions can be done by making any of the individual hookups in combination with each other (Figure 11–76). Other pull techniques are shown in the color section.

11.6 STRESS RELIEVING

Stress relieving uses hammer blows, and sometimes carefully controlled heat, to help return damaged metal to its original shape and state. It is important to note that shape and state do not necessarily mean the same thing. Something can be manipulated back into its original shape and still be far from its original state. There are two separate problems in the pulling procedure:

- Restoring the vehicle to its original shape.
- Relieving all the stress in the metal accumulated when its shape was distorted in the accident. This is called **original state.**

Original state means back to the original form. Metal has an elastic property because it remembers its original state. The metal will be "comfortable" once it is returned to that condition. The job is to remove all the stress caused by the accident.

Unbent metal (Figure 11–77A) contains layers of grain or molecules, all in a relatively relaxed state. As a piece of metal is bent (Figure 11–77B), these grains become slightly distorted, introducing stress. If a piece of metal is flexible enough, once pressure is released the grain will return to its original state. If the metal is bent too far, as in a collision, the grain on the outside of the bend is severely distorted by tension, while the grain of the inside is distorted by compression (Figure 11–77C). Because a large amount of stress is present, it will remain in this shape. If a pull moves these grains back into shape without affecting their state, the piece of metal will look like that shown in Figure 11–77D.

With the use of controlled heat (or no heat) and hammering, the grain can be revived and relaxed back to its original state (Figure 11–78).

Stress is defined in metallurgical books as the internal resistance a material offers to being deformed when subjected to a specific load (force). In the collision repair industry, stress can be defined as the internal resistance a material offers to corrective techniques. This resistance or stress can be caused by

- OEM stamping
- Deformation
- Overheating
- Improper welding techniques
- Undesirable stress concentrations

- Metal is unbent.
- Grain (molecules) is in a relaxed, comfortable state.

A

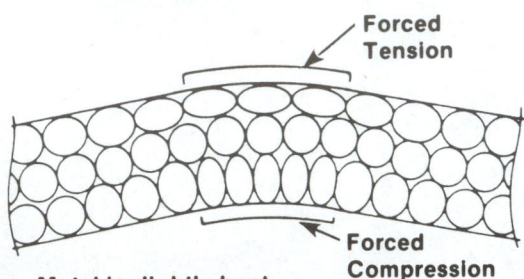

Forced Tension

Forced Compression

- Metal is slightly bent.
- Grain is forced to expand on outside of bend and compress on inside.
- If metal is flexible, grain will return to comfortable state.

B

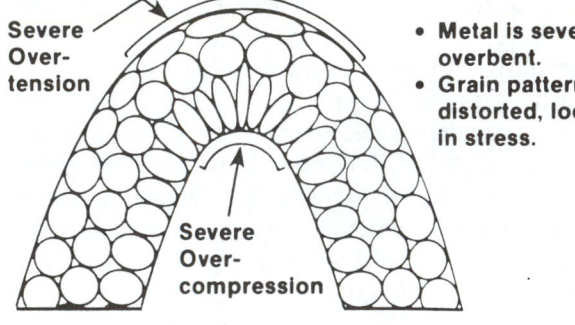

Severe Over-tension

Severe Over-compression

- Metal is severely overbent.
- Grain pattern is distorted, locking in stress.

C

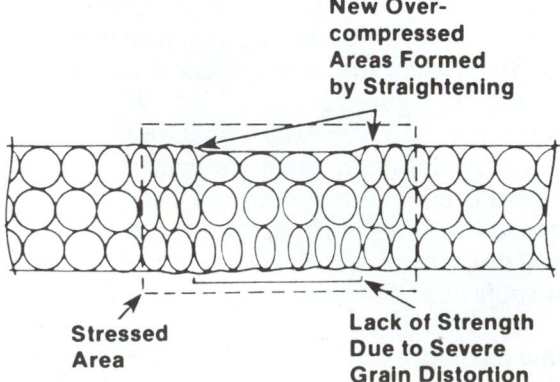

New Over-compressed Areas Formed by Straightening

Stressed Area

Lack of Strength Due to Severe Grain Distortion

D

FIGURE 11-77 Condition of metal is important when straightening: (A) unbent, (B) slightly bent, (C) overbent, and (D) straightened (without relieving stress).

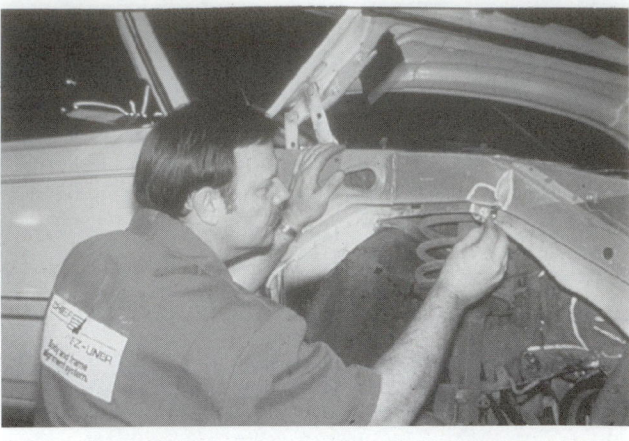

A

B

C

FIGURE 11-78 Stress relieving is important to repair quality. (A) Mark all areas that suffer from bad buckling and stress to aid in analyzing them during pull. (B) Hammer blows on stress areas with pulling tension applied will help free them. (C) If absolutely necessary, use heat and hammer blows to stress relieve. Keep in mind that too much heat will weaken metal. *(Courtesy of Chief Automotive)*

Signs of stress/deformation (Figure 11–79) on unibody cars are

- Misaligned door, hood, trunk, and roof openings
- Dents and buckles in aprons and rails
- Misaligned suspension and motor mounts
- Damaged floor pans and rack-and-pinion mounts
- Cracked paint and undercoating
- Pulled or broken spot welds
- Split seams and seam sealer

Since the damaged area will frequently give greater resistance to alignment than adjacent areas, additional holds for clamping are sometimes required. One example is a damaged rail with greater resistance to tension than where it usually attaches to the front cowl/bulkhead. By using a clamp, movement at the front cowl is prevented and corrective forces are applied directly to the rail. During the pulling operation, all critical reference points must be measured to monitor direction and prevent overpulling.

Generally, if the strain generated by the force of an impact does not lead to buckling, effective pulling force during aligning will alleviate the problem. Use common sense while hammering to remove the strain in a panel or member. Overdoing it will lead to time-consuming surface smoothing operations later on. Use particular caution when pulling large panels such as roof panels that are easily strained

when pulled. For example, while applying a backward pulling, use a spoon to press on the backside of the strained area (Figure 11–80).

Once the damage has been analyzed and the angle and direction of the pull decided upon, tension is applied and spring hammering is used to relieve the stress. Spring hammering is usually done with a spoon or a block of wood to distribute the force of the blow over a large area. This releases the tension and allows the elasticity of the metal to return to its original size and shape. Spring hammering should also be applied to areas adjacent to the major damage.

A dolly or large wood block and hammer will work out a lot of stress. Most of the stress relieving will be "cold work." Not much heat will be used. But if heat is needed, control the heat carefully to prevent part damage and warpage.

Heating will usually result in a certain amount of oxidation or scaling and can also result in decarburization. Scaling represents a loss of metal and mars the surface finish. Decarburization results in a soft surface and can seriously affect the fatigue life. The amount of scaling is largely determined by the time and temperature of the heating operation. The scale will always be heavier on the backside of the heated piece than on the side exposed to the flame. The exposed side is protected from oxidation by the burning gas until the torch has been removed. But the backside is subjected to oxidation as soon as the proper temperature is reached. Each reheating of the same area causes more scaling.

If the damage requires the use of heat, follow the vehicle manufacturer's recommendations exactly. When using heat on a unibody rail, for example, heat only the corners of the rail. Never attempt to cool the heated area by using water or compressed air. Allow it to cool naturally. Rapid cooling can cause the metal to become hard and brittle.

As explained in Chapter 7, the best way to monitor heat applications is with a heat crayon. Stroke or

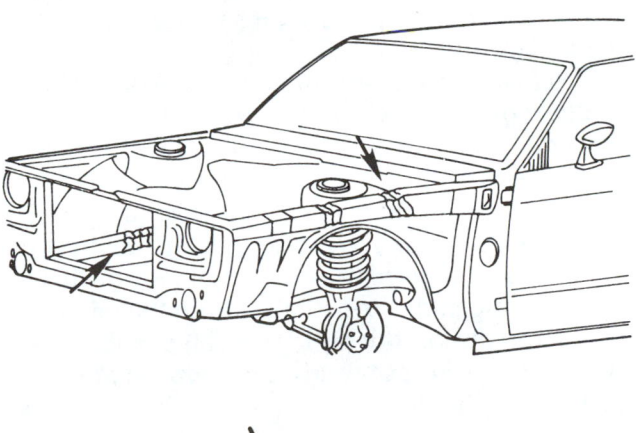

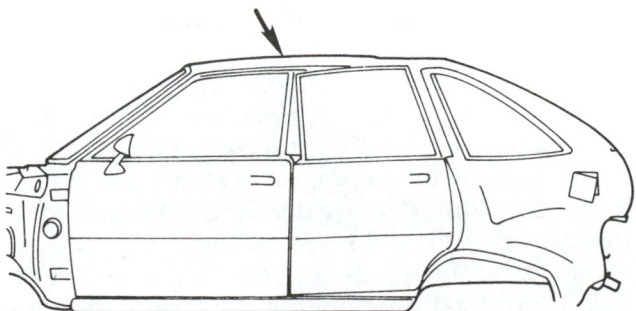

FIGURE 11-79 Signs of stress/deformation on unibody cars are cracked sealer or undercoating, bulges, buckles, and misaligned panels.

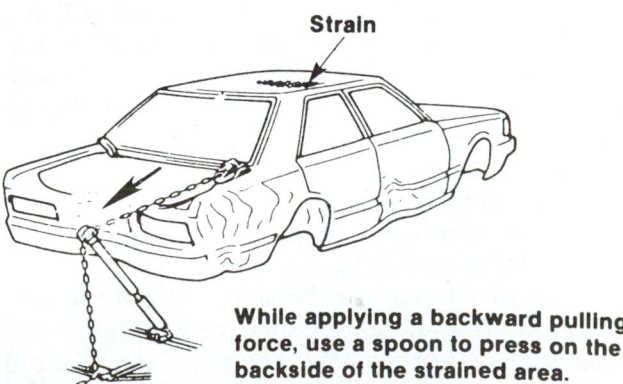

Strain

While applying a backward pulling force, use a spoon to press on the backside of the strained area.

FIGURE 11-80 Panel strain repair might be needed in large panels a great distance from the point of impact. *(Courtesy of Toyota Motor Corp.)*

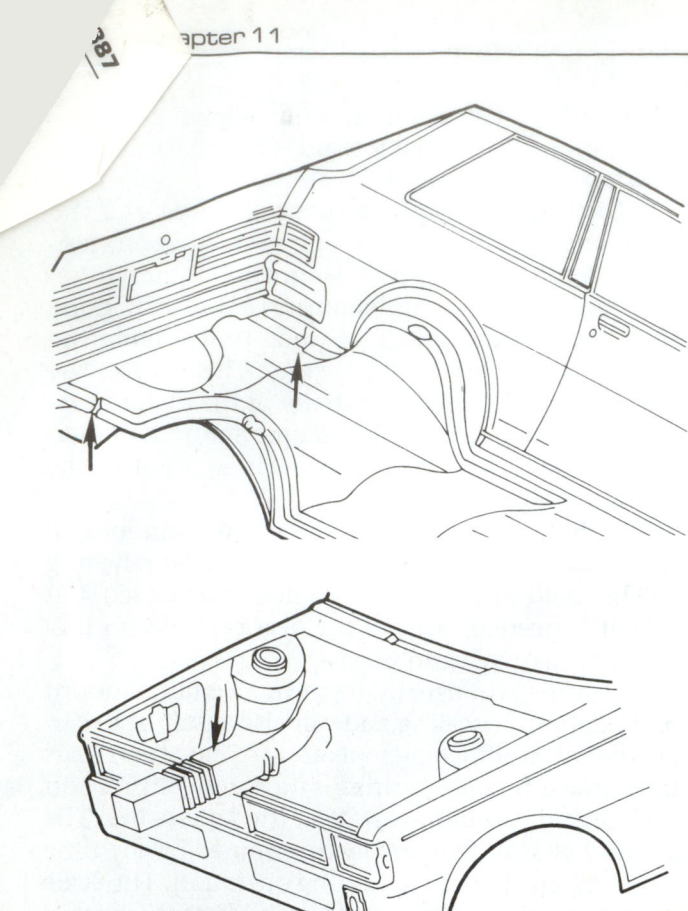

FIGURE 11–81 Stress concentrators are used to absorb impact energy to protect the passenger compartment.

mark the cold piece with the crayon. When the stated temperature has been reached, the crayon mark will liquefy. Heat crayons are quite precise and far more accurate than the common body shop technique of watching for specific color change. There is a plus-minus one percent factor with a heat crayon.

Certain conditions or possible defects of a metal structure can reduce its strength. These conditions are called stress concentrators (Figure 11–81). Stress concentrators, as the name implies, result in a localized concentration of stress as a load is applied.

Stress concentrators are designed into unibody vehicles to control and absorb collision forces, minimize structural damage, and increase occupant protection. They also make damage more predictable and allow for easier damage detection when analyzing damage or estimating.

Do not remove designed stress concentrators. Follow the car manufacturer's recommendations for straightening or replacement of parts that have designated stress concentrators.

A quality repair can be achieved only if function, durability, and appearance have been restored. When stress is not removed, the following possibilities can occur:

- Fatigue caused by loading and unloading of suspension and steering components.

- In the event of a second similar collision, less force is required to cause the same or greater damage and could endanger the occupants of the vehicle.
- The vehicle can dimensionally distort causing handling problems.

11.7 FINAL STRAIGHTENING CONSIDERATIONS

The structural qualities of vehicle construction (especially heat sensitive, thin-gauge unibody panels) bring up questions about repairing or replacing severely damaged sections that cannot be restored by the pulling operations. The basic questions are:

- When should the repair or replacement be undertaken in relation to the pull—before, after, or during?
- Which procedures and tools should be used to perform the repair or replacement?

The basic rule of thumb here is to make any and all repairs when part alignment permits. In other words, during the pulling, when the torn edges become aligned or sheared spot welds move back and line up is when they should be welded.

In the beginning, apply the pulling force intermittently while checking the movement of the panel and confirming that the force is working effectively on the damaged area. If no effect can be seen, consider changing the pulling direction or the area being pulled.

Remember that pulling generates forces opposite in direction to those that caused the initial damage. Accordingly, it can take as much force to straighten the panel as it did to damage it. Buckled sections often have high tension strength to resist straightening while others do not. As a result, one panel may pull out and into alignment before another. For this reason, it is imperative to visually check the progress of alignment during the pull. Make all possible repairs (welding of torn spot welds, for example) at the moment when the alignment is correct. You do not want to accidentally pull one panel past its alignment point while straightening another panel.

Use *blocking* or *holding devices* when pulling across an area that has an opening. This would include a roof panel with a sun roof opening, a door opening, a wheel opening, and so on. Use a turnbuckle or holding device to apply light tension across the opening. This will keep the opening from growing larger from the pulling force. Failure to block or hold openings during pulling can be a serious mistake that will ruin the panel.

The repair of a bent boxed part, such as a side member, is done by clamping the surface of the bent-in side and pulling. The pulling direction should be applied in an imaginary straight line extending through the original position of the part (Figure 11–82).

In repairing unibody structures, only use manufacturer methods to salvage members containing damage such as tears, fractures, or buckles. In a modern, well-designed vehicle, some members, such as rails, are designed to provide "controlled damage" in a collision. This prevents or postpones the damage of more critical areas, such as those that surround the passenger compartment.

A reinforcing patch that overstrengthens such a section might prevent "controlled damage" and defeat the design purpose of the section. Consequently, as mentioned earlier, when a fractured, torn, or buckled area cannot be repaired without patching, the only acceptable alternative is to replace the entire section.

Once the parts are welded into place, restore corrosion protection. Follow the vehicle manufacturer's recommendations for sealing, caulking, and restoring corrosion protection to all the structural parts. Described in later chapters, this is extremely important on a unitized body to assure lasting strength and quality of the repair.

Once the repair is completed, including all straightening and welding operations, the alignment procedures are ready for a final check. If the vehicle was repaired with fixtures, the technician knows that the underbody of the car is within specifications. If the car was repaired without fixtures, final measurements should be made and compared to the body/frame dimensions book. All measurements must be checked against factory specifications to ensure quality work. This is time consuming, but it must be done.

Begin the final checks by slowly walking around the vehicle looking for obvious signs of misalignment. Large gaps between the roof lines and doors (Figure 11–83) are a good sign that small amounts of damage are still present. If fixtures or jigs were used, go back and check all the fixture reference points (Figure 11–84). Do they still line up without having to force or hammer the fixture? Go back over the repair order or estimate to be sure everything was done. Remember that it is much easier to set up and make additional pulls now rather than to wait until more steps of the damage repair procedure have been completed.

FIGURE 11–83 Check for abnormal or irregular gaps around doors and other panels to help check straightening.

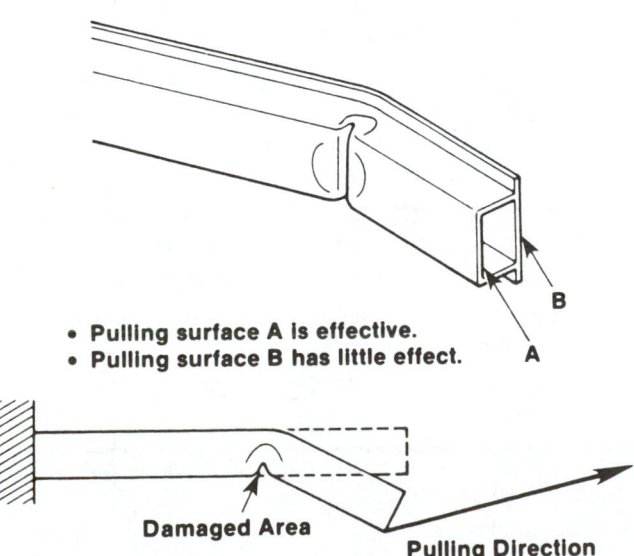

- **Pulling surface A is effective.**
- **Pulling surface B has little effect.**

Damaged Area

Pulling Direction

FIGURE 11–82 Note the point of clamping and the direction of pull to straighten a box section like this one.

FIGURE 11–84 When all fixtures fit into their reference points, straightening operations may be complete.

Among other items that should be carefully inspected are

- the alignment between the door and rocker sill. This should be a straight and narrow gap.
- the general alignment of all the upper body areas. Everything should look as though nothing was ever out of alignment.
- the opening and closing of the doors and lids. Do they feel tight and secure when latched? They should close smoothly and open easily.

After the final alignment inspection, the vehicle can often be left on the frame rack for replacement of many components. This will allow you to make any minor straightening needed for the final fit of new or replacement parts.

SHOP TALK Straightening equipment and procedures can be very complex. As an apprentice, do not be afraid to ask coworkers questions. It is better to ask a silly question than to make a major mistake that could cost you or the shop thousands of dollars or cause harm to a person from an incorrect repair.

SUMMARY

- Vehicle straightening involves using high-powered hydraulic or pneumatic equipment, mechanical clamps, and chains to bring the frame or body structure back into its original shape.

- With major damage to several panels, a multiple pull method with several pulling directions and steps is needed.

- Hydraulic rams use oil pressure from a pump to produce a powerful linear motion.

- The anchoring equipment holds the vehicle stationary while pulling and measuring.

- Straightening equipment is used to apply tremendous force to move the frame or body structure back into alignment.

- It is necessary to have at least THREE REFERENCE POINTS on the undamaged part of the vehicle that can be used to set the car up properly.

- Straightening system accessories include the various chains, clamps, hooks, adapters, straps, and stands needed to mount various makes and models of vehicles.

- Computerized straightening equipment helps automate the measuring and straightening process.

- When using aligning equipment, inadequate attention to any procedure can result in vehicle or equipment damage and possible physical injury to you or others in the shop.

- When planning the pulling process, you or the technician should

1. Determine direction of the pulls.
2. Find out how to repair the damage in the reverse (first-in, last-out) sequence to which it occurred during the collision.
3. Plan the pulling sequence with the pulls in the opposite direction from those that caused the damage.
4. Find correct attachment points of the pulling clamps.
5. Estimate the number of pulls required to correct the damage.
6. Determine which parts must be removed to make the pulls.

- Typically, work from the center section outward, achieving first length, then sidesway removal, and finally correct height.

- Use heat carefully and as a means of releasing locked-up metal and not as a means to soften an area. Although a torch is not recommended on HSS metals, it may be used following manufactuer's recommendations.

- Overpulling is done by pulling the damage slightly beyond its original dimension. If done in a controlled way, the metal will flex back slightly when tension is released.

- Stress relieving uses hammer blows, and sometimes carefully controlled heat, to help return damaged metal to its original shape and state.

- Once the parts are welded into place, restore corrosion protection.

ASE-STYLE REVIEW QUESTIONS

1. Technician A says that in-floor systems provide fast hookup and positive anchoring without sacrificing space. Technician B says that some in-floor systems can be stored by hanging the components on the wall. Who is correct?

   A. Technician A
   B. Technician B
   C. Both A and B
   D. Neither A or B

2. Technician A says a rack system should be used whenever the damage involves the suspension, steering, or power train mounting points. Technician B says that it is necessary to have at least three reference points on the undamaged part of the car that can be used to set the vehicle up properly on the frame straightening equipment. Who is correct?

   A. Technician A
   B. Technician B
   C. Both A and B
   D. Neither A nor B

3. Technician A always removes the suspension and driveline completely from a unibody vehicle before putting it on the frame straightening equipment. Technician B says that single-pull systems cannot hold an undamaged or a corrected reference dimension while pulling to correct other damaged areas of the car. Who is correct?

   A. Technician A
   B. Technician B
   C. Both A and B
   D. Neither A nor B

4. Which should the technician work toward achieving first?

   A. Sidesway removal
   B. Length
   C. Height
   D. It depends on the situation.

5. When raising a vehicle during an upward pull, Technician A wraps chains around the bumper or isolators. Technician B uses this technique to help in anchoring the vehicle to the frame machine. Who is correct?

   A. Technician A
   B. Technician B
   C. Both A and B
   D. Neither A or B

6. When a vehicle has been hit from the side, a _____ hit often results.

   A. Banana
   B. Sway
   C. Sag
   D. Accordion

7. Which of the following is true?

   A. Cracked paint and undercoating is a sign of stress.
   B. Most of the stress relieving will be "cold work."
   C. The best way to monitor heat applications is with a heat crayon.
   D. All of the above

8. When should the repair or replacement of severely damaged sections that cannot be restored by the pulling operation take place?

   A. Before pulling
   B. During pulling
   C. After pulling
   D. Repairs during pulling and replacements after pulling

ESSAY QUESTIONS

1. Explain the usual sequence for a total structure realignment.

2. Describe six things to remember when planning a pull.

3. Before doing any pulling work, list three steps to protect the body and externally attached parts.

CRITICAL THINKING PROBLEMS

1. A unibody car has been driven over a cement barrier, badly scraping and damaging its underbody. There are only two unbent flanges for installing clamps. What must be done?

2. During final inspection, you find a door badly out of alignment. How should you proceed?

MATH PROBLEMS

1. If the manufacturer states that a 100 psi (689 kPa) reading on a pressure gauge equals one ton (0.9 metric ton) of pulling power, what would 63 psi (425 kPa) equal?

2. If a pulling chain is rated at 2,100 pounds (945 kg) and it should not be strained over 50 percent of its rating, what is the maximum force that should be applied to the chain?

PHOTO SEQUENCE

UNIBODY/FRAME STRAIGHTENING

P11-1 Remove any parts that are in the way of clamping and pulling operations.

P11-2 Anchor the vehicle so it will not move when pulling force is applied. Pinch weld clamps are often tightened down around the lower rocker panel flange to secure the vehicle.

P11-3 Attach pulling chains at the locations of damage. Generally, you want to pull out damage in the reverse order of which it was formed by the collision impact forces.

P11-4 Stand to one side of the chains as force is applied. Do not exceed the recommended equipment pressure.

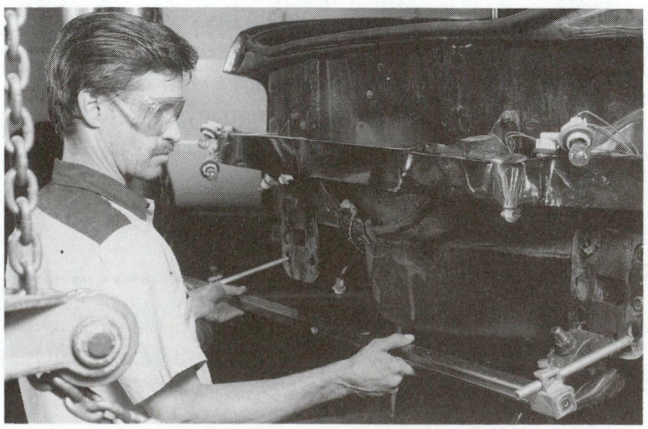

P11-5 Stop periodically to measure progress. Remove the pulling force and measure the area being pulled. Keep pulling until all reference points are within specifications.

CHAPTER 12

Panel Replacement and Adjustment

INTRODUCTION

Panel replacement involves removing and installing a new panel or body part. You might have to unbolt and replace a fender, door, or spoiler. With quarter panels and other welded body sections, you will have to cut off the damaged panel with power tools and then use a welder to install the new panel. This takes considerable skill.

A collision-damaged vehicle can require a variety of repair operations. Repair steps will depend on the nature and location of the damage. Panels with minor damage can often be straightened and filled with plastic. However, quite often the damage is too great and part replacement is the only logical answer.

Bent structural panels may have to be pulled and realigned using hydraulic equipment. Some panels, however, might be so badly damaged that replacement is the only practical and effective procedure for cost effective repair.

Table 12–1 outlines the general procedure for replacing both bolted and welded panels.

12.1 PANEL REMOVAL

When starting work, refer to the estimate to get guidance on how to begin. The estimator will have determined which parts need to be repaired and which should be replaced. You would use this information and shop manuals to remove and replace parts efficiently (Figure 12–1).

OBJECTIVES

After studying this chapter, you should be able to:

✔ List parts of the vehicle that are considered structural.

✔ List the steps necessary for replacing a part along factory seams.

✔ Describe how spot welds are separated.

✔ Explain how new body panels can be positioned on a vehicle body.

✔ List the steps for welding new body panels in place.

✔ Describe how to install foam panel fillers.

✔ Section rails, rocker panels, A- and B-pillars, floor pans, and trunk floors.

✔ Identify the principal methods of corrosion protection.

✔ Answer ASE test questions pertaining to panel replacement.

KEY TERMS

A-pillars
B-pillars
closed section
door skin
foam fillers
full body sectioning
inserts
open hat channel
panel replacement
part R&R
recycled assemblies

replacement panels
sectioning partial
spot weld cutter
structural adhesives
structural panels
test weld
three-way tailgate
weld-bond adhesives
weld-through primer
window regulator

 **ASE TASK LIST**

Job Skills covered in this chapter include:

PAINTING AND REFINISHING TEST (B2) TASK LIST

A. Surface Preparation

15. Apply suitable paint sealer to the area being refinished when sealing is needed or desirable.
16. Remove imperfections from sealer.
19. Restore corrosion resistant coatings, caulking, and seam sealers to repaired areas.

E. Finish Defects, Causes, and Cures

3. Check for rust spots (corrosion); determine the cause(s), and correct the condition.

NONSTRUCTURAL ANALYSIS AND DAMAGE REPAIR TEST (B3) TASK LIST

A. Preparation

9. Remove corrosion protection, undercoatings, sealers, and other protective coatings as necessary to perform repairs.

B. Outer Body Panel Repairs, Replacements, and Adjustments

2. Remove and replace bolted, bonded, and welded panels or panel assemblies.
4. Remove, replace, and align hood, hood hinges, and hood latch/lock.
5. Remove, replace, and align deck lid, lid hinges, and lid latch/lock.
6. Remove and replace doors, tailgates, hatches, lift gates, latch/lock assemblies, and hinges.
8. Check and adjust clearances of front fenders, header, and other panels.
9. Check door hinge condition; check door frames for proper fit; check and adjust door clearances.
12. Apply protective coatings and sealants to restore corrosion protection.
13. Remove damaged sections of steel body panels; weld in replacements in accordance with manufacturers'/industry specifications.

14. Repair or replace door skins in accordance with vehicle manufacturers' specifications; inspect intrusion beams.
15. Replace or repair plastic panels in accordance with manufacturers'/industry specifications.
16. Restore sealers, mastic, sound deadeners, and foam fillers.

D. Moveable Glass and Hardware

1. Inspect, adjust, repair, or replace window regulators, run channels, glass (including electrically-heated glass), power mechanisms, and related controls.

STRUCTURAL ANALYSIS AND DAMAGE REPAIR TEST (B4) TASK LIST

A. Frame Inspection and Repair

8. Remove and replace damaged frame horns, side rails, cross members, and front or rear sections.
10. Repair or replace weakened or cracked frame members in accordance with vehicle manufacturers'/industry standards.

B. Unibody Inspection, Measurement, and Repair

16. Remove damaged sections of structural steel body panels, and weld in replacements in accordance with vehicle manufacturers' specifications.
17. Restore corrosion protection to repaired or replaced unibody structural areas.

C. Stationary Glass

1. Remove and replace front and rear stationary glass (heated and nonheated) in accordance with manufacturers' recommendations.

The estimate is an important reference tool for doing repairs. It must be followed. The insurance company and estimator have both determined which parts must be repaired. If you fail to follow the estimate, the insurance company may not pay for your work.

The estimate is also used to order new parts. You might want to make sure all ordered parts have arrived. Compare new parts on hand with the parts list. If anything is missing, have the parts person order them. This will save time and prevent your work area from being tied up while waiting for parts.

WHERE TO START REPAIRS

Generally, start removing large, external badly damaged parts first. For example, if the front end was hit hard, you might remove the hood first. This will give you more room to access rear fender bolts. It will also allow more light into the front for finding and removing hidden bolts in the frontal area. Use this kind of logic to remove parts efficiently.

If in doubt about how to remove a part, refer to the vehicle's service manual. Factory service manuals normally have a body repair section. The *body*

TABLE 12-1: TYPICAL PANEL REPLACEMENT PROCEDURE

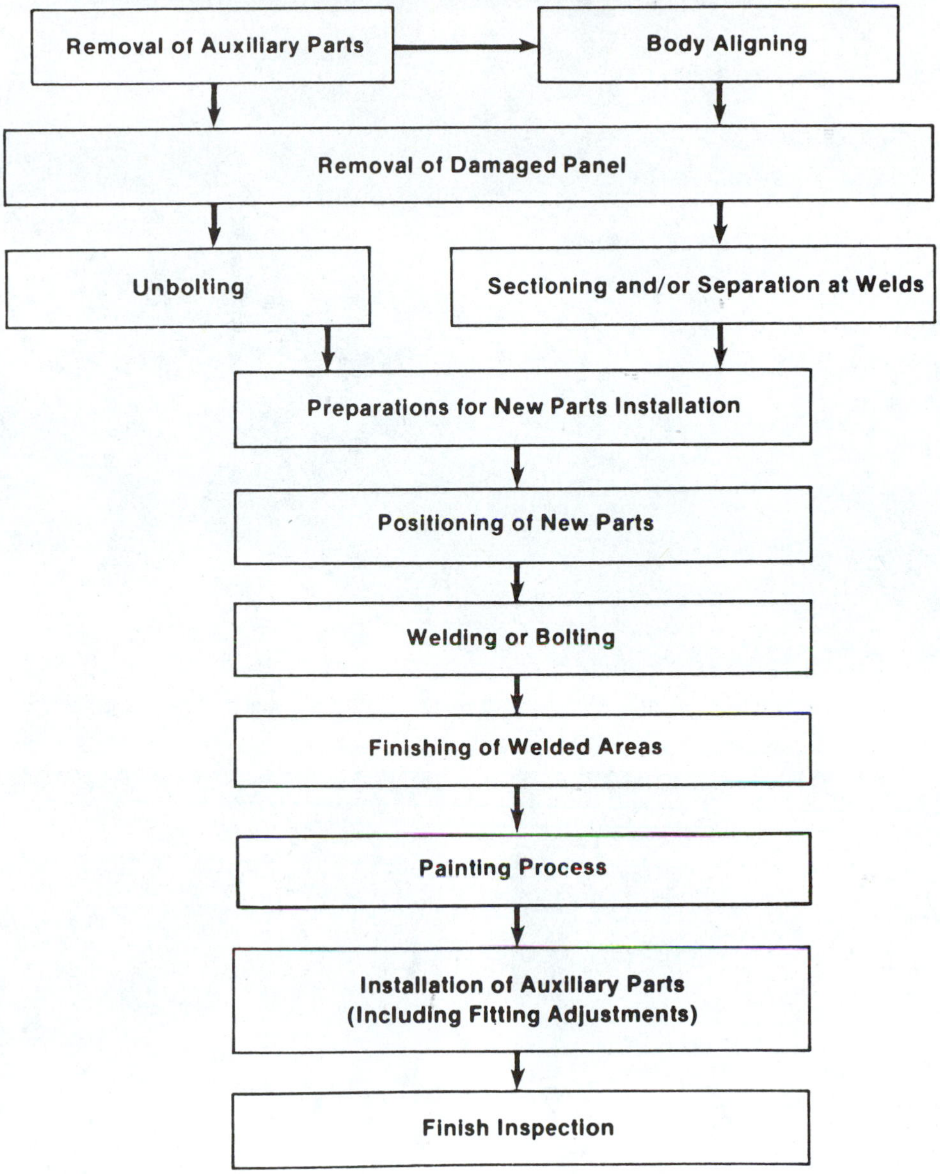

```
┌─────────────────────────┐        ┌─────────────────────────┐
│ Removal of Auxiliary    │───────▶│    Body Aligning        │
│        Parts            │        │                         │
└───────────┬─────────────┘        └───────────┬─────────────┘
            │                                  │
            ▼                                  ▼
┌─────────────────────────────────────────────────────────────┐
│              Removal of Damaged Panel                       │
└───────────┬─────────────────────────────────┬───────────────┘
            │                                  │
            ▼                                  ▼
┌─────────────────────────┐        ┌─────────────────────────┐
│        Unbolting        │        │ Sectioning and/or       │
│                         │        │ Separation at Welds     │
└─────────────────────────┘        └───────────┬─────────────┘
                                               │
                                               ▼
                         ┌─────────────────────────────────────┐
                         │ Preparations for New Parts          │
                         │         Installation                │
                         └─────────────────┬───────────────────┘
                                           ▼
                         ┌─────────────────────────────────────┐
                         │     Positioning of New Parts        │
                         └─────────────────┬───────────────────┘
                                           ▼
                         ┌─────────────────────────────────────┐
                         │       Welding or Bolting            │
                         └─────────────────┬───────────────────┘
                                           ▼
                         ┌─────────────────────────────────────┐
                         │     Finishing of Welded Areas       │
                         └─────────────────┬───────────────────┘
                                           ▼
                         ┌─────────────────────────────────────┐
                         │        Painting Process             │
                         └─────────────────┬───────────────────┘
                                           ▼
                         ┌─────────────────────────────────────┐
                         │   Installation of Auxiliary Parts   │
                         │    (Including Fitting Adjustments)  │
                         └─────────────────┬───────────────────┘
                                           ▼
                         ┌─────────────────────────────────────┐
                         │       Finish Inspection             │
                         └─────────────────────────────────────┘
```

FIGURE 12-1 When starting work on a vehicle with structural panel or part damage, refer to the estimate. It will give information stating what should be done to fix the car or truck. *(Courtesy of Tech-Cor)*

repair section of the manual explains and illustrates how parts are serviced. The manual will give step-by-step instructions for the specific make and model vehicle. It will give bolt locations, torque values, removal sequences, and other important information.

To give you an idea of the types of things to consider when starting a repair, refer to Figure 12–2.

Now refer to Figure 12–3. It shows a collision-damaged door and quarter panel. The collision created a buckle that work hardened in the flange and body line areas of the quarter panel. How should it be repaired?

Although the quarter panel could have been repaired, the customer and insurance company recommended replacement. The door panel was so badly damaged that it could not be repaired. A new door skin was installed on the door frame. To replace the

FIGURE 12-2 This vehicle had major damage. Which parts were replaced and which were repaired?

Work-
Hardened
Areas

FIGURE 12-3 Note the damage to the door and quarter panel. Since the quarter panel had major stress, it was replaced rather than repaired.

FIGURE 12-4 The damaged quarter panel has been cut and removed.

FIGURE 12-5 The new quarter panel has been welded in place.

quarter panel, section cuts were made in the rear pillars and rocker panel (Figure 12–4). Spot welds were drilled out in the factory seams. The new panel, cut to fit at the pillars, was welded in place (Figure 12–5).

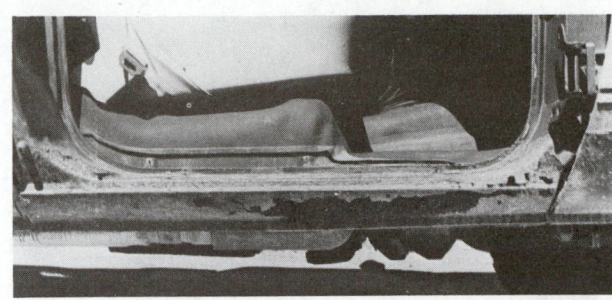

A

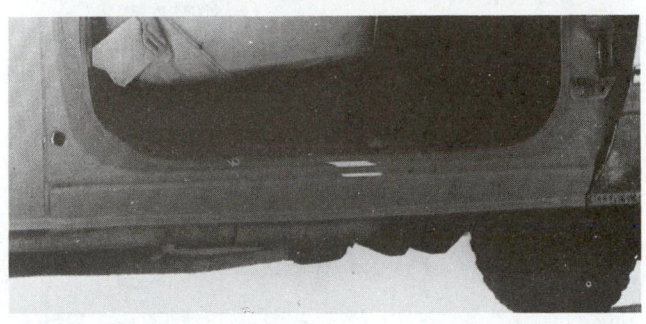

B

FIGURE 12-6 (A) Rocker panel on older car is badly rusted and in need of replacement. (B) Replacement rocker panel has been installed.

FIGURE 12-7 Occasionally, you may have to fabricate repair pieces out of sheets of metal. Make sure the type and thickness of metal sheet is the same as the original part.

Panel replacement is often the only permanent remedy for corrosion damage. Figure 12–6A shows a rusted out rocker panel and cab corner on a truck. The repair was made by cutting the rusty metal away and welding new partial panels in place (Figure 12–6B).

Sectioning involves cutting the part in a location other than a factory seam. This may or may not be a factory-recommended practice. Special care must be taken when sectioning a part to make sure it will not jeopardize structural integrity.

Some manufacturers do not allow resectioning of structural panels. Others approve of sectioning only if proper procedures established by the manufacturer are followed. All manufacturers stress: Do not section areas that might reduce passenger protection, drivability of the car, or where critical dimensions can be affected.

Partial replacement panels are often designed to replace only a section or area of a large panel. They are ideal for body areas commonly subject to rustout and are available from a number of aftermarket parts manufacturers, local salvage yards, or the original equipment manufacturer.

Fabricated panels are hand-made repair parts to fix small problems (gouge or rusted holes for example) in panels when a new or partial panel is not available or practical. When making a fabricated panel, use the same metal type and thickness found on the vehicle. Cut out the damaged section of the part. Then, use it as a template to make the new part

(Figure 12–7). Usually the fabricated part is made larger than the cut out section so a lap joint can be formed. This produces a strong joint for welding in the repaired section of metal.

HOW PARTS ARE FASTENED

Fastener variations can make repair more challenging. Parts can be held by screws, bolts, nuts, metal or plastic clips, adhesives, and other methods. To efficiently replace parts, you must carefully study part construction. Inspect parts closely to find out how they are held on the vehicle. This allows you to make logical decisions on the order in which parts should be removed and the methods needed.

The methods of fastening parts to cars and trucks has changed in the past few years. Many parts that were held with bolts and screws in the past now snap fit into place. Plastic retainers now hold these parts onto the vehicle. This was done to save time during vehicle manufacturing. The part is simply pressed or popped into position on the assembly line.

Keep in mind that on-the-job experience is the only way to become competent and fast at body **part R&R** (part removal and replacement). Sometimes you must remove one part at a time. In other instances, it is better to remove several parts as an assembly. This chapter will give you the background information to make this learning process easier.

Figure 12–8 shows the mechanical fasteners used to attach the front fender and splash shield on a Ford Mustang.

STRUCTURAL PANELS

In modern unitized construction, all the **structural panels**, from the radiator support to the rear end panel, are welded together to make a one-piece structure

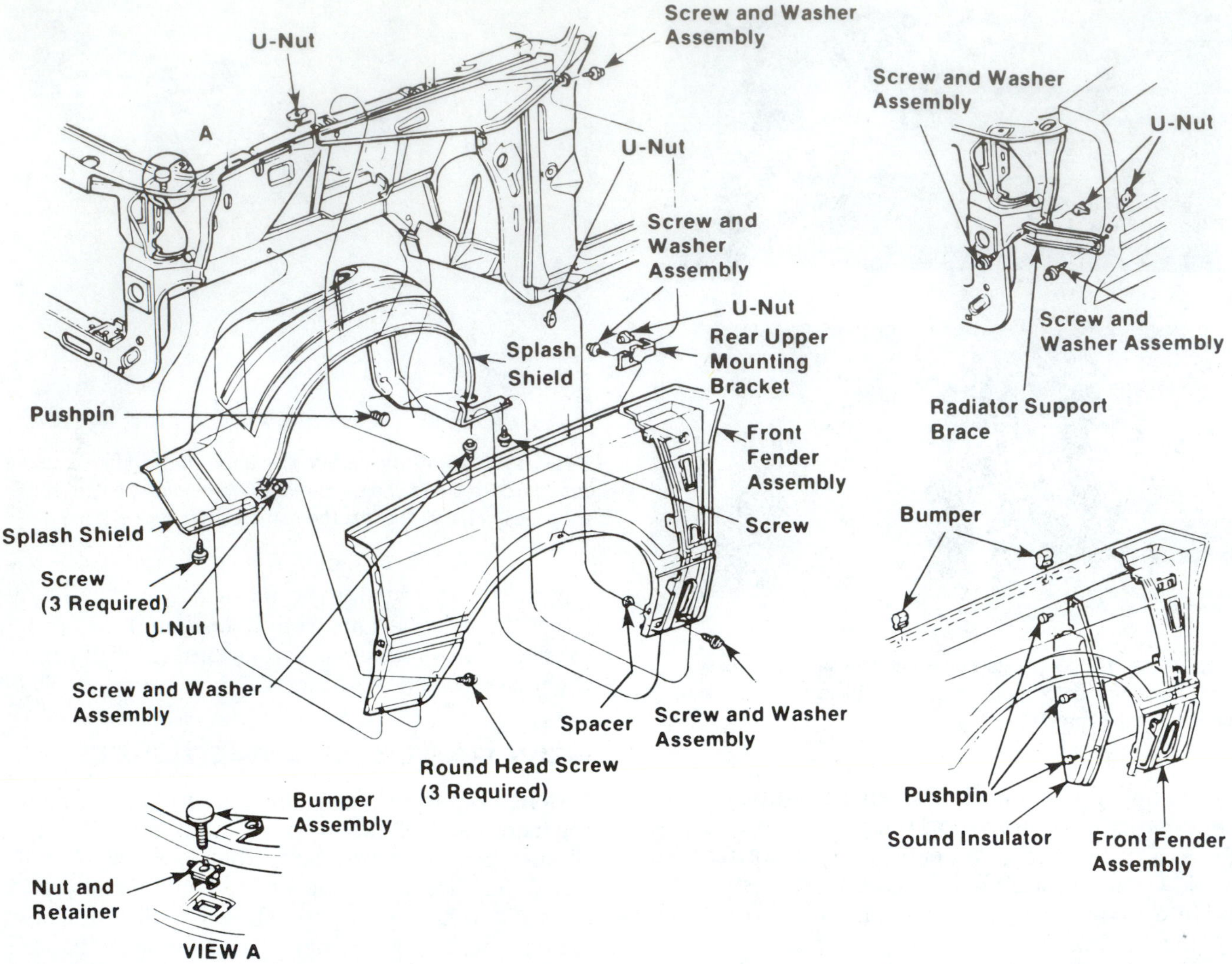

U-Nut

Screw and Washer Assembly

A

Screw and Washer Assembly

U-Nut

U-Nut

Screw and Washer Assembly

U-Nut

Splash Shield

Rear Upper Mounting Bracket

Screw and Washer Assembly

Radiator Support Brace

Front Fender Assembly

Screw

Pushpin

Splash Shield

Bumper

Screw (3 Required)

U-Nut

Screw and Washer Assembly

Spacer

Screw and Washer Assembly

Round Head Screw (3 Required)

Pushpin

Sound Insulator

Front Fender Assembly

Bumper Assembly

Nut and Retainer

VIEW A

FIGURE 12-8 Note typical fasteners holding the front fender assembly and splash shield.

(Figure 12–9). Some examples of structural panels in unibody construction are the radiator supports, the inner fender aprons, the floor pan, the rocker panels, the engine compartment side rails, upper reinforcements, lower body rails in the rear, inner fender wells, and the luggage compartment floor.

The integrity of the whole vehicle is dependent on the interconnection of all the individual structural panels. The individual panels are joined together at flanges or mating surfaces usually formed at the edges of the panels during factory production.

The structural panels provide the foundation to which all the mechanical components are mounted, and all outer panels are attached to them. Therefore, all appearance fits and suspension alignments are determined by the accuracy of the positioning of the welded structural panels. Welded panels cannot be shimmed to correct sloppy fit-up procedures. Struc-

It is required to pull the damaged structure back into factory specified alignment before removing the old damaged parts.

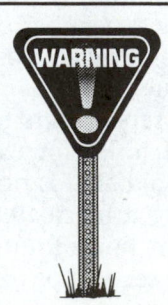

It is very important to always follow the manufacturer's recommendations when servicing structural panels. This is especially true concerning sectioning.

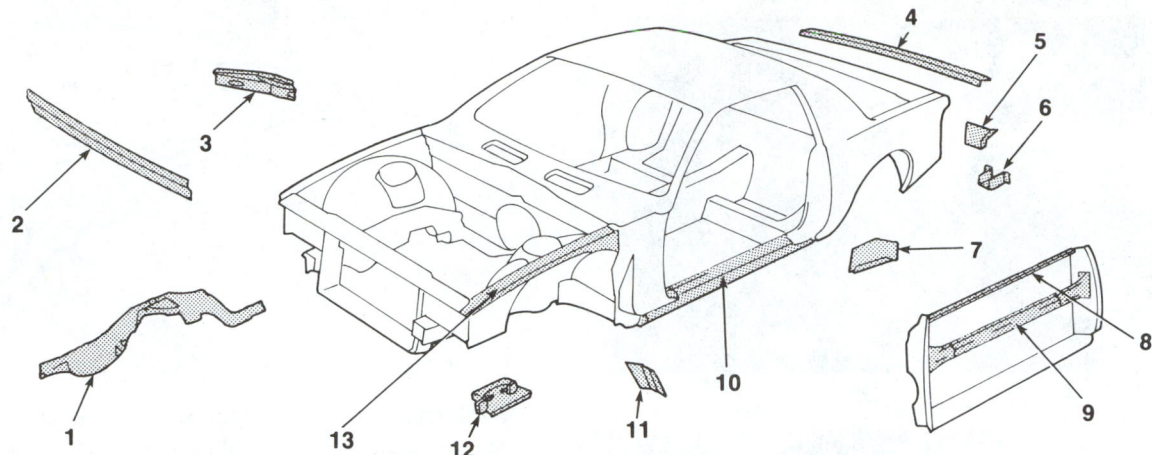

1. Engine Compartment Outer Side Rail
2. Windshield Lower Reinforcement With Extension Support
3. Body Lock Pillar Brace to Wheelhouse
4. Rear End Panel Bumper Retainer
5. Rear End Panel Reinforcement at Bumper Support
6. Compartment Panel Rail Reinforcement at Bumper Mount
7. Inner Compartment Panel Rail Extension
8. Door Inner Panel Reinforcement at Belt
9. Door Outer Panel Bar, Stiffener, and Reinforcement
10. Outer Rocker Panel
11. Engine Compartment Side Rail Reinforcement at Lower Dash
12. Engine Compartment Side Rail Reinforcement at Stabilizer Bar
13. Outer and Lower Engine Compartment Upper Rail Panels

FIGURE 12-9 Unibody construction often uses thin, high-strength steel panels.

tural panels must be accurately positioned BEFORE final welding.

Another important area is ultra-high-strength steel panels, such as bumper reinforcements and door intrusion beams. These panels must be replaced when damaged. Under no condition can heat be applied to straighten ultra-high-strength steel panels.

REMOVING STRUCTURAL PANELS

Structural body panels are often joined together in the factory by spot welding. Therefore, removing panels involves mainly the separation of spot welds. Spot welds can be drilled out, blown out with a plasma torch, chiseled out, or ground out with a high-speed grinding wheel. The best method for removing a spot-welded panel is determined by the number and arrangement of mating panels and the accessibility of the weld.

FINDING SPOT WELDS

It is usually necessary to remove the paint film, undercoat, sealer, or other coatings covering the joint area to find the locations of spot welds. To do this, remove the paint using a DA sander with medium grit paper or use a scuff wheel in a grinder. A coarse wire wheel or brush attached to a drill can also be used to remove paint over spot welds (Figure 12–10).

Scrape off thick portions of undercoating or wax sealer before trying to remove the paint.

Try to avoid using an oxyacetylene or propane torch to remove paint because they could overheat the metal. If you do use a torch, however, do not burn through the paint film so that the sheet metal panel begins to turn color. Heat the area only enough to soften the paint and then brush or scrape it off.

NOTE: It is not necessary to remove paint from areas where the spot welds are visible through the paint film.

In areas where the spot weld positions are not visible after the paint is removed, drive a chisel between the panels as shown in Figure 12–11. Doing so will cause the outline of the spot welds to appear.

SEPARATING SPOT WELDS

After the spot welds have been located, the welds can be drilled out and removed, using a **spot weld cutter** (Figure 12–12). Two types of cutting bits can be used: a drill type or a hole saw type.

Table 12–2 shows when each type should be used to drill out spot welds. Be careful not to cut into the lower panel. Also, be sure to cut the plugs out precisely to avoid creating an excessively large hole.

Drilling out the numerous spot welds in a panel can be tedious. To make the job easier, use a spot removing drill with an integral clamping mechanism (Figure 12–13). Hand pressure and lever action forces

A

B

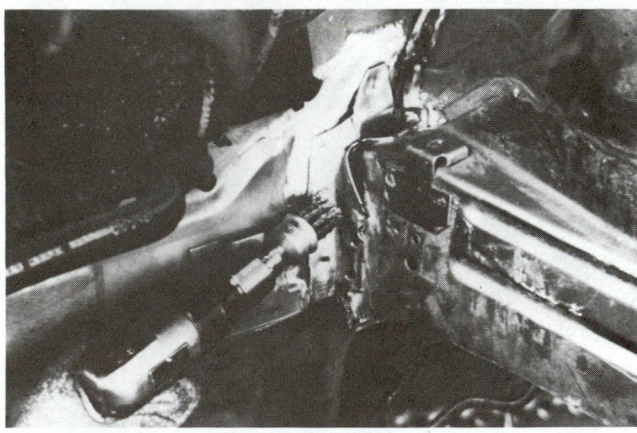

C

FIGURE 12-10 Always clean the part flanges to expose spot welds and to prepare for welding. (A) To determine the spot weld locations, use a torch with care to prevent warpage. (B) After softening the material with heat, scrape off the excess. (C) After scraping, final clean the flange with a power brush.

the special rounded bit into the weld for faster cutting action.

A plasma arc torch is seldom recommended for removal of spot welds, although it can be used (Figure 12–14). The plasma torch will quickly blow a hole in all the thicknesses of metal at the same time. Obviously, the use of a plasma torch does not preserve the integrity of the underlying panels and the less heat used the better with today's metals.

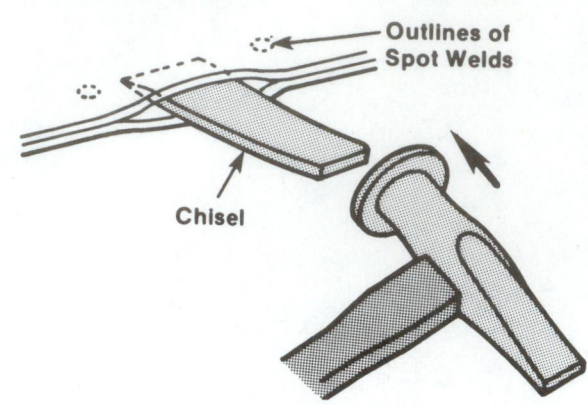

FIGURE 12-11 You can determine a spot weld location with a chisel. By spreading the flange slightly, spot weld dimples will form. *(Courtesy of Toyota Motor Corp.)*

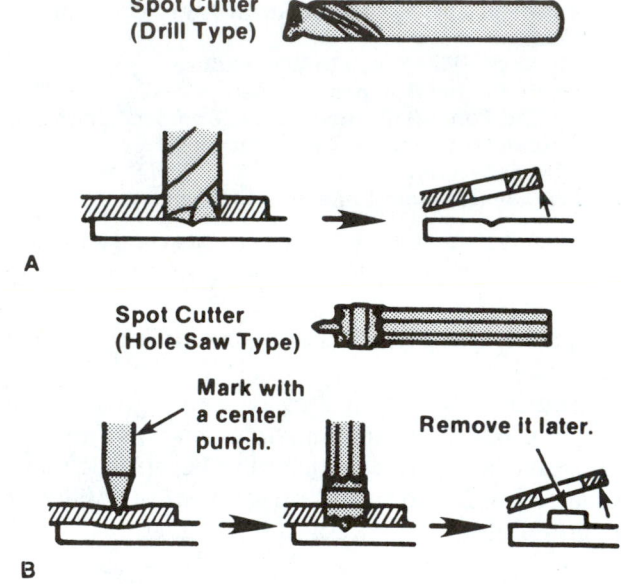

FIGURE 12-12 Note types of spot weld cutters: (A) drill type and (B) hole saw type. *(Courtesy of Toyota Motor Corp.)*

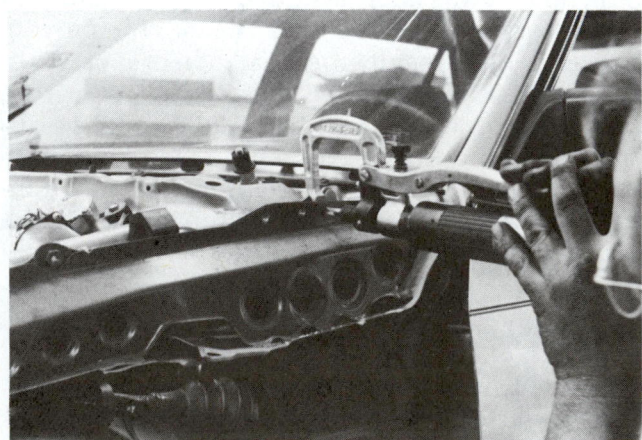

FIGURE 12-13 The spot weld removing tool has a clamp mechanism for forcing the cutter into the weld and holding the cutter in place.

TABLE 12-2: SEPARATION OF SPOT WELDS

Type			Application Method	Characteristics
Spot Cutter	Drill Type	Small	Places where the replacement panel is between other panels and welding cannot be done from the backside Places where the replacement panel is on top and the weld is small	The separation can be accomplished without damaging the bottom panel. Since the nugget is not left in the bottom panel, finishing is easy.
		Large	When the replacement panel is on top When the panel is thick (places where nuggets are large) Places where the weld shape is destroyed	
	Hole Saw Type		When the replacement panel is on top	Separation can be accomplished without damaging the bottom panel. Since only the circumference of the nugget is cut, it is necessary to remove the nugget remaining in the bottom panel after the panels are separated.
	Drill		When the replacement panel is on bottom When the replacement panel is between and welding can be done from the backside (Select a drill diameter that is appropriate for the panel thickness and the weld diameter.)	Lower cost Recently, a labor saving spot weld removing tool has been developed that is easy to use and has a built-in attaching clamp.

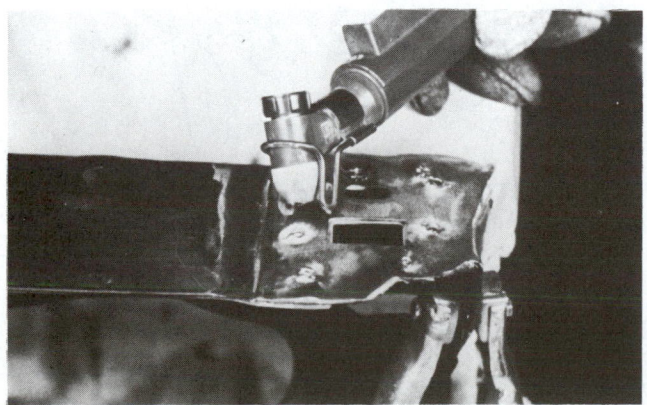

FIGURE 12-14 Spot welds can also be removed with a plasma torch. Make sure nothing behind the cut is flammable.

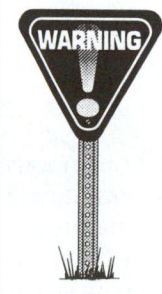

WARNING When using a plasma torch, remember that tremendous heat is blown through the cut parts. This could burn wires, undercoating, and other parts. It could also start a serious fire that could cause severe vehicle damage or even injury.

A high-speed grinding wheel can also be used to separate spot-welded panels (Figure 12–15). Use this technique only when the weld is not accessible with a drill, the replacement panel is on top, or a plug weld (from a previous repair) is too large to be drilled out.

CAUTION

Serious cuts can result from the sharp metal left from drilling out spot welds. Never run your hand over a drilled hole. This could cause a painful accident. It is also advisable to wear heavy leather gloves in case the drill or cutting tool slips.

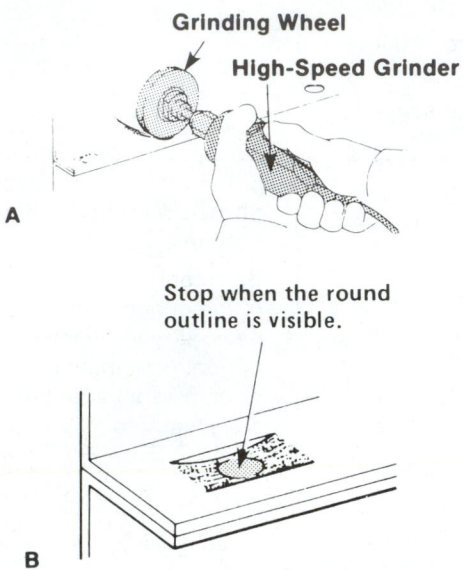

FIGURE 12-15 To use a grinder to remove spot welds, carefully remove the metal over the weld. (A) Hold the cutter securely and keep it from walking sideways. (B) Form a small pocket over the weld nugget. *(Courtesy of Toyota Motor Corp.)*

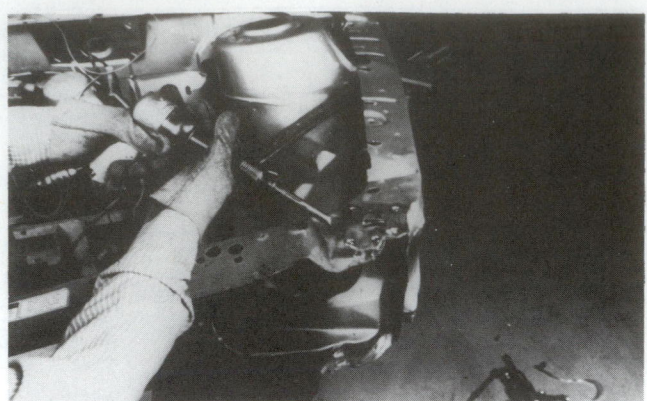

FIGURE 12-16 An air chisel will quickly spread and separate panels after spot welds are cut out.

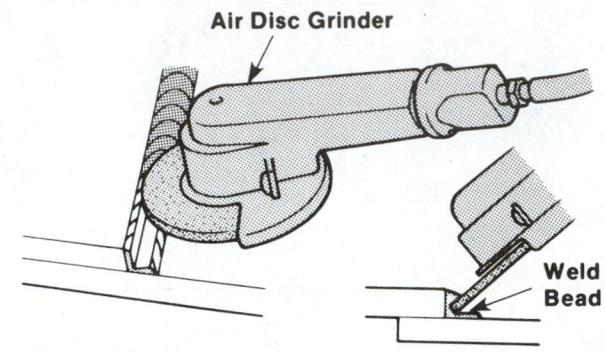

FIGURE 12-17 A continuous weld can be removed with a disc grinder. Again, guide the cutter carefully to prevent damage to the lower panel or the panel to remain on the vehicle. *(Courtesy of Toyota Motor Corp.)*

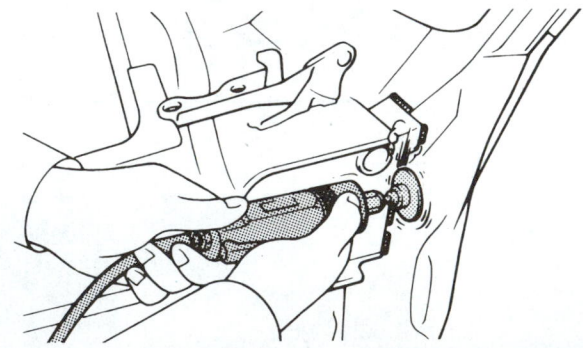

FIGURE 12-18 A high-speed grinder will quickly remove a continuous weld. *(Courtesy of Toyota Motor Corp.)*

After the spot welds have been drilled out, blown out, or ground down, drive a chisel between the panels to separate them (Figure 12–16). Be careful not to cut or bend the undamaged panel.

SEPARATING CONTINUOUS WELDS

In some vehicles, panels are joined by continuous MIG welding. Since the welding bead is long, use a grinding wheel or high-speed grinder to separate the panels. As shown in Figures 12–17 and 12–18, cut through the weld without cutting into or through the panels. Hold the grinding wheel at a 45-degree angle to the lap joint. After grinding through the weld, use a hammer and chisel to separate the panels.

SEPARATING BRAZED AREAS

Brazing is used at the ends of outer panels or at the joints of the roof and body pillars to improve finish quality and the body seal. Generally, separation of brazed areas can be done by grinding the metal out from the joint. With soft brazing, it can also be done by melting the brazing metal with an oxyacetylene or propane torch or by grinding. Grinding is preferred because less heat stress is involved.

Note that in areas where arc brazing is used, the fusion temperature of the brazing metal is higher than with ordinary brazing. Melting the brazing

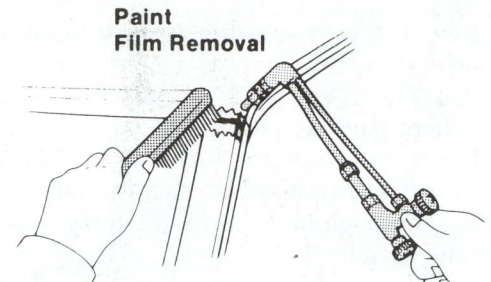

Paint Film Removal

FIGURE 12–19 Controlled heat from a torch will remove paint from the brazed area. *(Courtesy of Toyota Motor Corp.)*

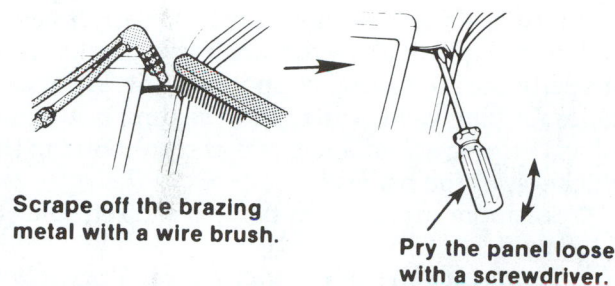

Scrape off the brazing metal with a wire brush.

Pry the panel loose with a screwdriver.

FIGURE 12–20 Use a light prying action to separate the brazed joints. *(Courtesy of Toyota Motor Corp.)*

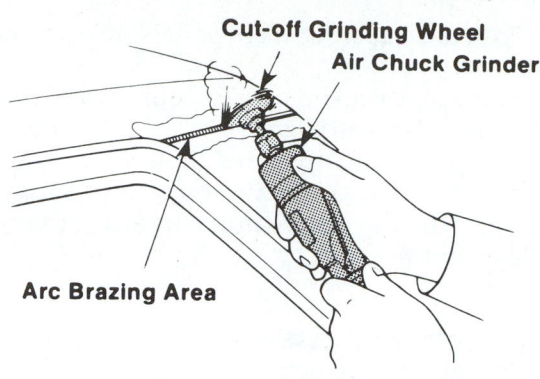

Cut-off Grinding Wheel
Air Chuck Grinder

Arc Brazing Area

A. Separation of arc brazed areas

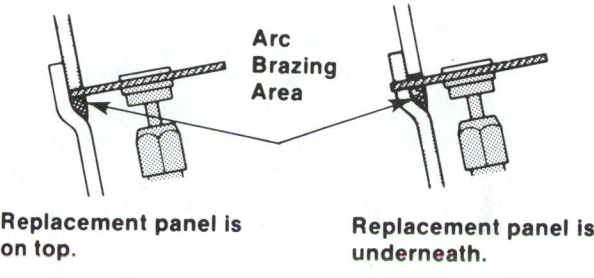

Arc Brazing Area

Replacement panel is on top.

Replacement panel is underneath.

B. Cutting depth

FIGURE 12–21 Note the basic method to separating panels connected by arc brazing. *(Courtesy of Toyota Motor Corp.)*

metal would result in damaging the panels underneath. Therefore, areas that are arc brazed are normally separated by grinding.

Ordinary soft brazing can be distinguished from arc brazing by the color of the brazing metal. Ordinary brazed areas are the color of brass, but arc-brazed areas are a reddish copper color.

First, soften the paint with an oxyacetylene torch and remove it with a wire brush or scraper. Then, heat the brazing metal until it starts to melt and puddle and quickly brush it off (Figure 12–19). Be careful not to overheat the surrounding sheet metal. Drive a chisel between the panels to separate them (Figure 12–20). Keep the panels separated until the brazing metal cools and hardens.

If, after removal of the paint, it is determined that the brazed joint is arc brazed, use a high-speed grinder and grinding wheel to cut through the brazing (Figure 12–21). If replacing the top panel, do not cut through the panel below it. After grinding

through the brazing, separate the lapped panels with a chisel and hammer.

12.2 INSTALLING NEW PANELS

As stated earlier, exterior sheet metal panels are attached with either fasteners or welds. The fastener method of installing panels is simple and fast. It is a matter of bolting the new panel in place and adjusting the fit.

Inspect and measure the adjacent or adjoining panels for proper alignment before tightening the bolts. If necessary, straighten any adjacent panel that is out of alignment. One indication of misalignment or damaged inner structure is that the bolt holes do not line up with the bolt holes in the new

SHOP TALK

It is easier to separate the brazed areas after all the other welded parts have been separated.

SHOP TALK

When replacing a bolt-on panel, start all bolts or fasteners into their holes before tightening any of them. If this is not done, it may be impossible to shift the part and start some of the bolts.

panel. Tapered gaps between panels also indicate misalignment.

Welding replacement panels requires much more preparation and care in alignment. The following procedure is typical of many panel replacement operations. Always refer to the appropriate body repair manual provided by the manufacturer for the type and placement of welds.

PREPARING FOR PANEL WELDING

After removal of the damaged panels, prepare the vehicle for installation of the new panels. To do this, follow these steps:

1. Grind off the welding marks from the spot welding areas. Use a wire brush to remove dirt, rust, paint, sealers, zinc coatings, and so on from the joint surfaces. Do not grind the flanges of structural panels. Grinding will remove metal, thinning the section and weakening the joint. Also, remove paint and undercoating from the back sides of the panel joining surfaces on parts that will be spot welded during installation.

2. Smooth the dents and bumps in the mating flanges with a hammer and dolly (Figure 12–22).

3. Apply **weld-through primer** to areas where the base metal is exposed after the paint film and rust have been removed from the joining surfaces. It is very important to apply the antirust primer to joining surfaces or to areas where painting cannot be done in later processes.

To make sure that the MIG machine is correctly adjusted for the specific joint being welded, always

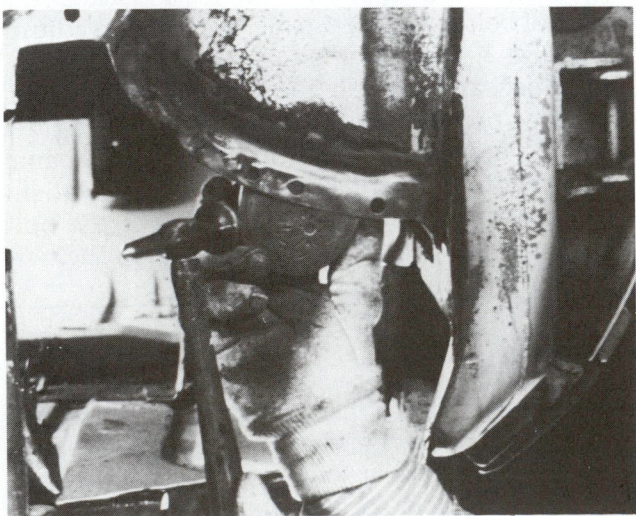

FIGURE 12-22 After removing the damaged panel, straighten the remaining panel flange with a hammer and dolly to prepare it to accept a new part and weld.

do a test weld. A **test weld** is done on scrap pieces of metal the same thickness and type as the parts to be repaired. This is especially important on a closed section where the backside of the weld cannot be checked. A test weld is the only way to ensure that the welding techniques and machine adjustments will restore the original strength, integrity, and alignment of the panel.

REPLACEMENT PANEL PREPARATION

Since new parts are coated with a primer, this coating must be removed from the mating flanges before welding. This will allow the welding current to flow properly during resistance spot welding. Also, drill holes for plug welds where spot welding is not possible. Use plug hole diameters that correspond to the thickness of the panels.

For more information on this subject, refer to Chapter 7.

To prepare the new panel for welding, follow these steps:

1. Use a disc sander to remove the paint from both sides of the spot welding area. Do not grind into the steel panel. Do not heat the panel so that it turns blue or begins to warp.

2. Make holes for plug welding with a punch and drill. Always refer to the body repair manual for each type of vehicle to determine the number of holes needed. Generally, you will have to make more holes than used on the factory assembly line on nonstructural parts. Be sure to make plug welding holes of the proper diameter. If the size of the welding holes is too large or too small for the thickness of the panel, either the metal will melt through or the weld will be weak. Space the holes evenly.

3. Apply weld-through primer to the welding surfaces where the paint film was removed. Apply the weld-through primer carefully so that it does not ooze out or run from the joining surfaces. If the primer does squeeze out, it will have a detrimental effect on painting, necessitating extra work. Weld-through primer is in an aerosol spray can or brushable material. Remove any excess with a solvent-soaked rag.

4. If the new panel is sectioned to overlap any of the existing panels, rough cut the new panel to size using an air saw, cut-off grinding wheel, or similar tool. The edges should overlap the portion of the panel remaining in the sectioning area on the body by $3/4$ to 1 inch (19 to 25 mm). If the overlap portion is too large, it will make matching the position of the panel more difficult during temporary installation.

POSITIONING NEW PANELS

Aligning new parts with the existing body is a very important step in body repairing. Improperly aligned panels will affect both the appearance and the drivability of the repaired vehicle.

Basically, there are two methods of positioning body panels. With major damage, use dimension measuring instruments to determine the correct part position. With minor body damage, you can often visually find the correct panel position by the relationship between the new part and the surrounding panels.

The dimensional accuracy of the engine compartment, fender aprons, front side members, rear side members, and similar rear structural parts has a direct effect on wheel alignment and driving characteristics. Therefore, when replacing structural panels in unibody vehicles, accurately measure part position. Whether structural or cosmetic panels, the emphasis is on proper fit. You must often use both methods together to assure accuracy and the fit necessary for a high-quality vehicle repair.

POSITIONING BY DIMENSIONAL MEASUREMENT METHODS

When installing major structural parts (Figure 12–23), measurement should be done throughout the repair process. All straightening must be done before replacing panels. Otherwise, proper alignment of the new panels will be impossible. As panels are fit into place and welded, they should be measured again.

When using a bench system for panel replacement at a factory joint, place the fixtures or gauges on the bench in their correct locations and tie them to the bench. Then place the new panel in position on the fixtures. See how it lines up with the good panels on the vehicle. Make any necessary adjustments. Clamp the panel in place, and weld it to the mating panels.

The illustrations in Figures 12–24 through 12–34 demonstrate the basic methods for installing front end

Failure to measure properly before welding new panels into place can be very embarrassing and expensive. If you ever make the mistake of welding a part in place out of alignment, you will probably have to cut off the new part and reweld it in place correctly. This will waste time, money, and your reputation!

parts after a major frontal impact. In the procedure being illustrated, a vehicle is being fitted with a front fender apron assembly, a front cross member, and a radiator support.

1. Match the assembly reference marks on the installation areas of the front fender apron and the side member. Fasten them in place with vise grips. Parts that have no assembly reference marks should be installed in the same location as the old parts.
2. Match the length dimensions by setting the tracking gauge at the reference values, adjusting the length dimensions so they match those values. Temporarily install the front cross member. Use a hammer and block of wood to shift the parts in the desired direction (Figure 12–24).
3. If the length dimensions match the reference values, temporarily install the front floor reinforcement by tack welding one spot. Choose a spot weld location in an area where it will be easy to remove if necessary. Scribe a positioning line at the end of the part that is not welded and drill a small hole. Fasten the parts together with a sheet metal screw. Scribe a line on the apron installation area but do not weld the panels together yet.
4. Use a centering gauge to match the height of the new components to the components on the

FIGURE 12-23 As you install a new structural part, use a measuring system to align it precisely on the vehicle.

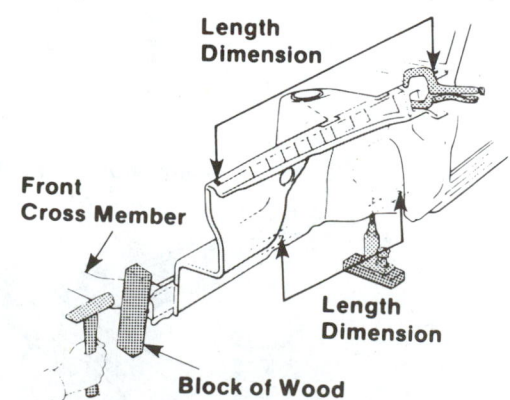

FIGURE 12-24 When making length adjustments, use a measuring system to compare the part location with the correct dimensions. Use a block of wood and light hammer blows to move the part as needed. *(Courtesy of Toyota Motor Corp.)*

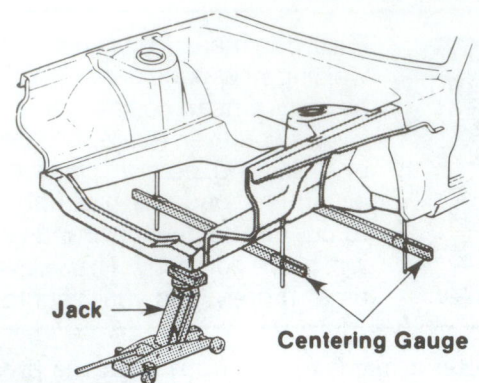

FIGURE 12-25 Here is an example of checking height adjustments with centering gauges. *(Courtesy of Toyota Motor Corp.)*

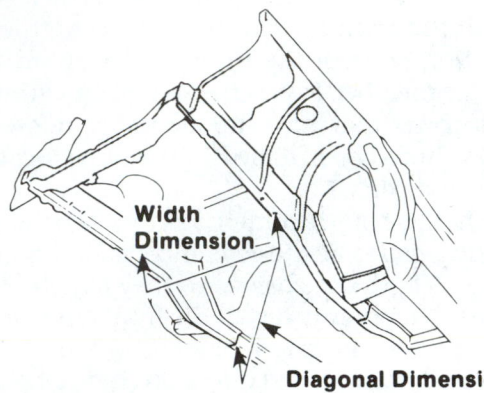

FIGURE 12-26 Diagonal measurement is made from an undamaged reference point to a point in a part being installed. Width dimension is checked from the undamaged to the damaged side. *(Courtesy of Toyota Motor Corp.)*

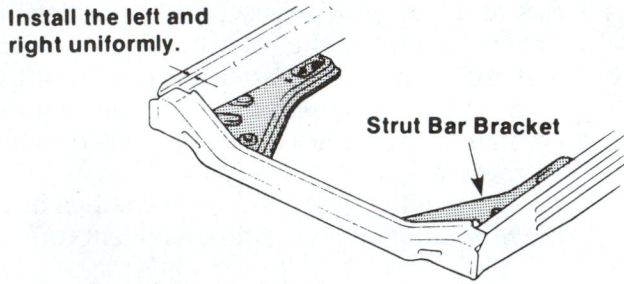

FIGURE 12-27 The correct position of a strut bar bracket is critical to the action of the suspension system. *(Courtesy of Toyota Motor Corp.)*

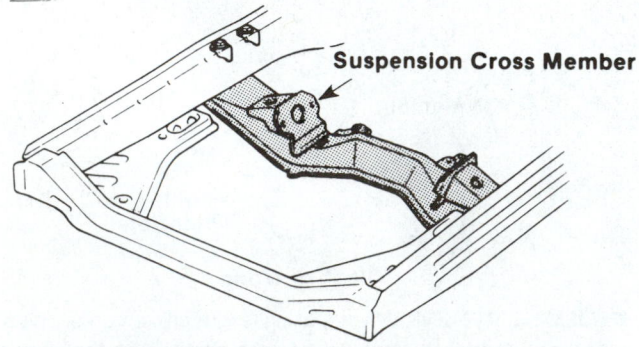

FIGURE 12-28 Attaching the suspension cross member is also critical because it supports the lower ends of the strut towers and the engine. *(Courtesy of Toyota Motor Corp.)*

opposite side of the vehicle (Figure 12–25). Support the new parts with a hydraulic jack so that the height does not change.

5. Match the diagonal and width dimensions (Figure 12–26). Then move the side member back and forth to match the dimensions.

6. Confirm the height dimensions again.

7. Position the front cross member (Figure 12–27). The strut bar bracket can be installed with a fixture (jig). Install the cross member so that both the left and right ends are uniform.

8. Once the dimensions of the side member match the reference dimensions, secure the member in place. The suspension cross member may be installed with fixtures (jigs) (Figure 12–28). Use plug welds at several locations to fasten the side member, the under reinforcement, and the side member to the front cross member.

9. Confirm that the apron upper length has not changed by checking at the scribed line.

10. Match the diagonal dimensions between the fender rear installation hole and the spring support hole or fender front installation hole (Figure 12–29). It is also a good idea to match the spring support dimension from side to side at this time.

11. Verify the width dimension of the spring support and the front of the fender installation hole and fasten them together. If the width dimension does not match the reference value, make a small adjustment, being careful of changes in the diagonal dimensions. Temporarily install and fasten the radiator upper support and the radiator support (Figure 12–30).

12. Match the radiator support width dimensions (Figure 12–31). Set the tracking gauge to the reference value and adjust the support so that the dimensions match that on the gauge. Lightly fasten it with vise grips.

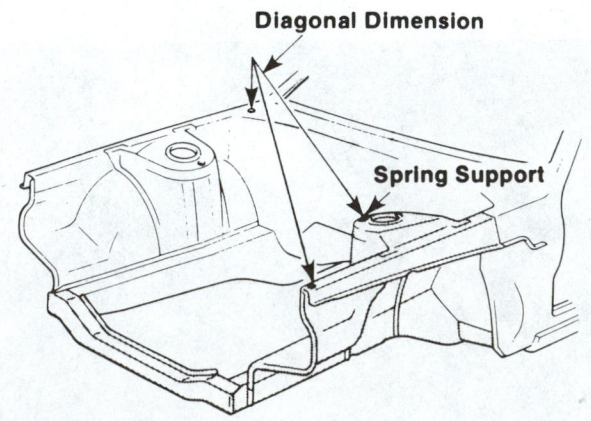

FIGURE 12-29 After partial welding, you may need to adjust the fender apron dimensions at the top. Check them continuously during the welding process. *(Courtesy of Toyota Motor Corp.)*

13. Match the diagonal dimensions for the radiator support (Figure 12–32). See that the diagonal dimensions of the supports match. Verify their height and ensure that the left and right sides are installed in the same manner.

14. Visually verify the left-right balance. Stand back and visually compare the new parts with the existing ones.

15. Temporarily install the front fender and inspect it for proper fit with the door. If the clearance is not correct, it may be that the fender apron or the side member height is off on both the left and right sides. Temporarily install the hood to check its alignment with the cowl and fenders.

16. Verify the overall dimensions once more before welding (Figures 12–33 and 12–34).

When using the tram and centering gauge method of component positioning, it is important to remember that measurement points for the new parts should be the same as the opposite side of the vehicle. If the dimensions do not match, the reference points must be verified and changed if necessary.

Complete information on the use of fixtures and adjusting of panels is given in Chapters 10 and 11.

Positioning by Visual Inspection

Nonstructural outer panels can sometimes be visually aligned with adjacent panels without the precise measurements necessary in replacing structural panels. This is true of both mechanically fastened panels and welded panels. The emphasis here is on appearance. Body lines must be flush and aligned, and gaps between panels must be even, not tapered.

For example, when installing a hood, it is impractical to measure it during installation. You simply want it to fit in its opening properly. If the cowl, fenders, and radiator support are installed correctly,

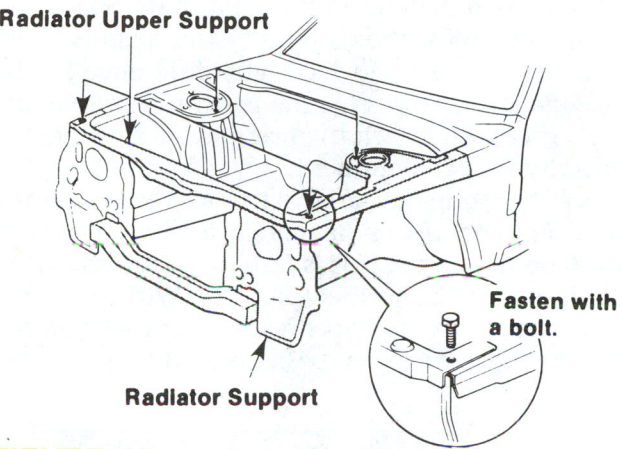

FIGURE 12–30 Check the fender apron width dimensions as you install the radiator support. *(Courtesy of Toyota Motor Corp.)*

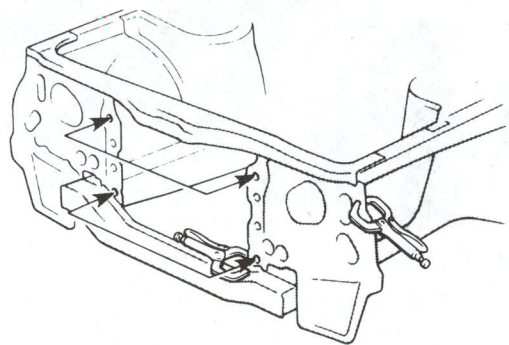

FIGURE 12–31 Match the width dimensions to make sure the front is aligned properly. *(Courtesy of Toyota Motor Corp.)*

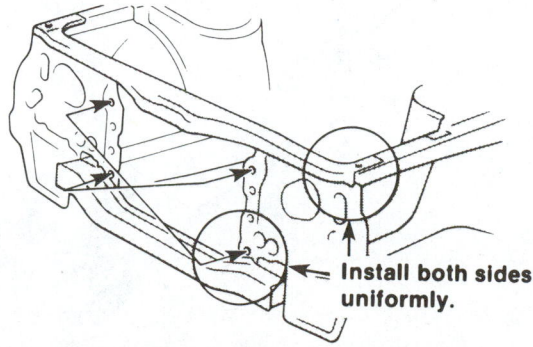

FIGURE 12–32 Also match the diagonal dimensions. *(Courtesy of Toyota Motor Corp.)*

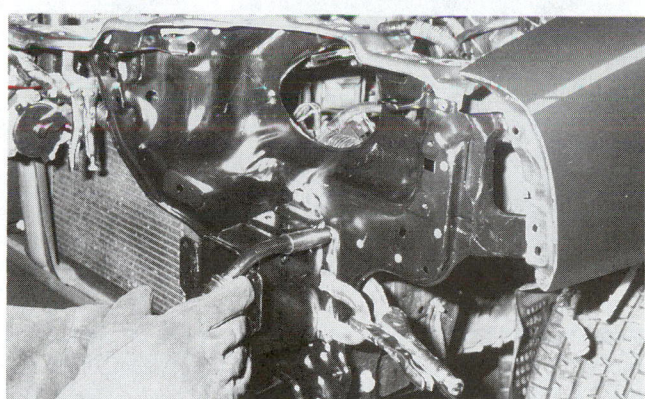

FIGURE 12–33 With the radiator support perfectly aligned and clamped, you can start the final MIG welding. *(Courtesy of Tech-Cor)*

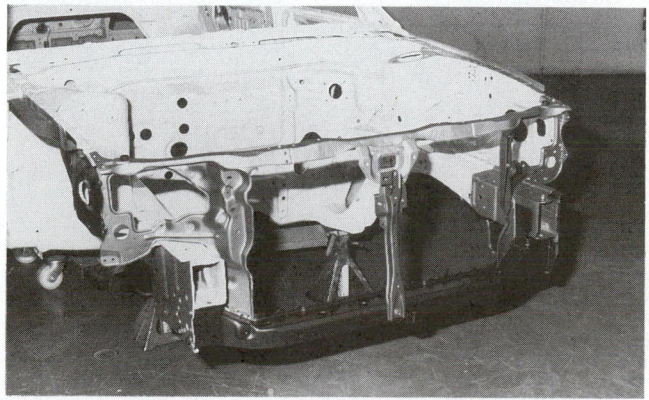

FIGURE 12–34 These painted parts show some typical parts that must be replaced after major frontal impact.

simply center the hood in its opening by aligning it
with the cowl. Then install the fenders and align
them with the hood. You must also make sure the
hood latch and safety catch engage properly and the
hood hinges operate smoothly.

FITTING A QUARTER PANEL

Figures 12–35 through 12–40 demonstrate the basic
steps for installing a rear quarter panel. The quarter
panel is too damaged for metal straightening (Figure

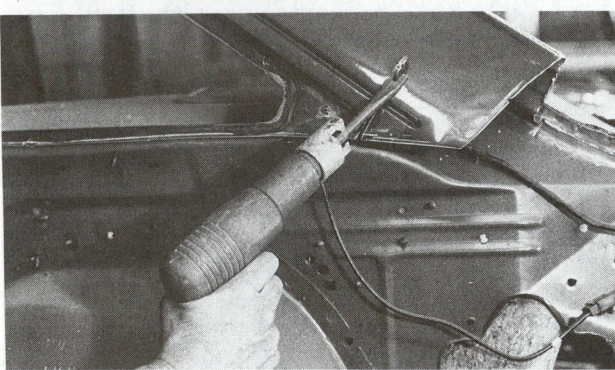

FIGURE 12-35 Study major steps for removing a
quarter panel. (A) Damage is too severe for straightening.
(B) Cut off damaged quarter panel.

12–35A). First, rough out the old quarter panel. Then
section the pillar at the manufacturer's recommended
point (Figure 12–35B). Also cut out the spot welds
holding the panel around the rest of its perimeter.
Remove the damaged quarter panel (Figure 12–36).

Next, the new quarter panel is carefully aligned
with adjacent body parts and secured with spot welds.
Temporarily install the quarter panel and fasten it at
several points with vise grips or another holding tool.
Check that the panel end and flange match.

Carefully adjust the fit with the surrounding
panels. Adjust the panel so that the clearance with
the door and body lines match each other (Figure
12–37). Then install the trunk lid in its correct position and adjust the clearances and heights. Confirm that there is no left-right difference in the
diagonal dimensions for the rear window opening
(Figure 12–38). Match the rear glass to the opening
to verify proper alignment.

After fitting the panel to the door and the trunk
lid, drill some small holes and fasten the panel with
self-tapping screws. If it is fastened with vise grips,
the fit cannot be verified properly. With the trunk
lid aligned with the upper rear body panel, you can
align the quarter panel to the trunk lid.

FIGURE 12-36 Be careful not to damage the parts
under the panel while cutting.

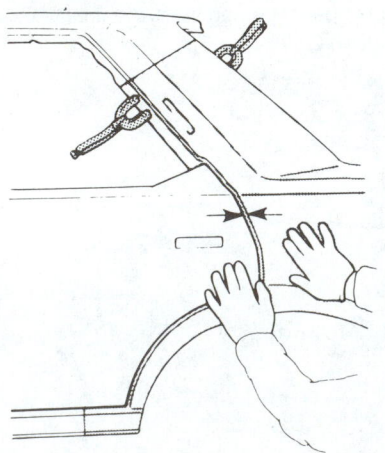

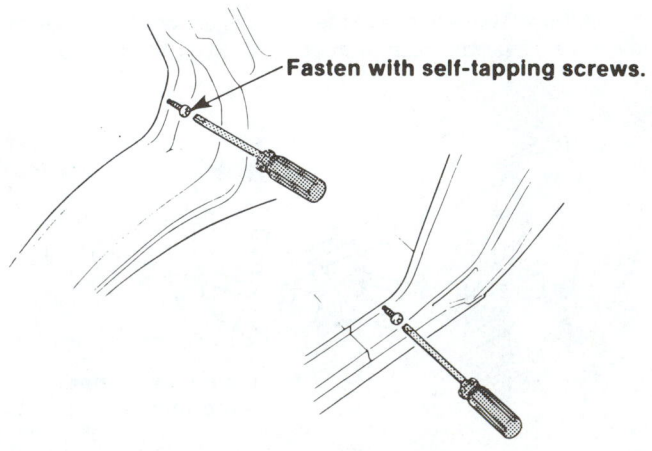

FIGURE 12-37 (A) Use clamps to position the quarter panel and check its alignment with the door panel. (B) Small screws
can be used to secure the position of the panel. *(Courtesy of Toyota Motor Corp.)*

Adjust the body line and panel overlap to match the lower back panel and the rear valance panel. Install the rear combination lamp and fit the panel to the lamp assembly (Figure 12–39).

When the clearance, body line, and height differences of each part have been adjusted, visually check for overall twisting or bending.

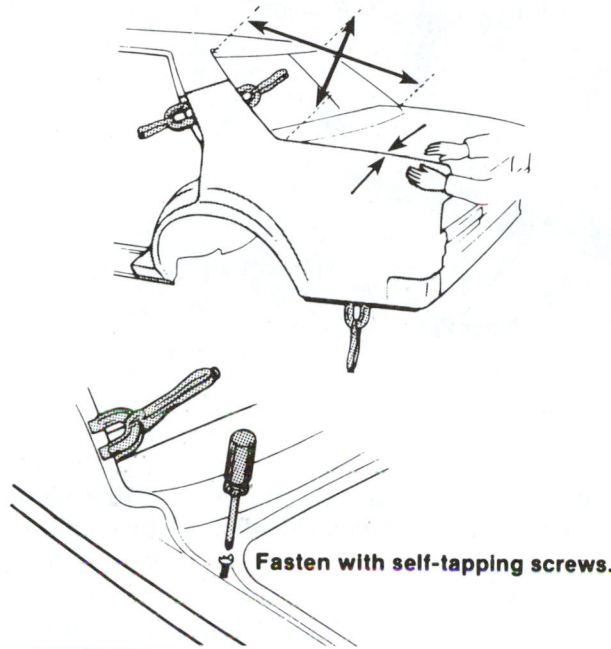

Fasten with self-tapping screws.

FIGURE 12-38 With the front of the panel secured, adjust the alignment to the rear window and deck lid. Install other sheet metal screws as needed to hold the panel precisely. *(Courtesy of Toyota Motor Corp.)*

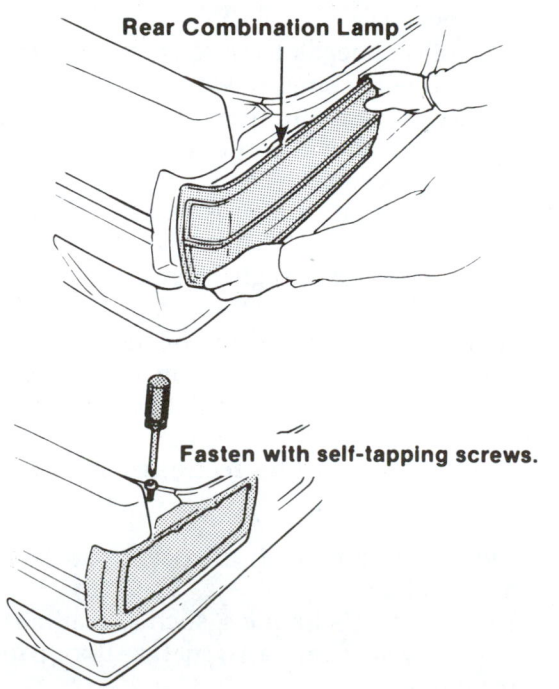

Rear Combination Lamp

Fasten with self-tapping screws.

FIGURE 12-39 With the quarter panel dimensionally correct, you can install the lower back panel and check the fit of the lens. *(Courtesy of Toyota Motor Corp.)*

CUTTING THE OVERLAPPING PANEL

After the panel is properly positioned, cut the overlapping portion of the joining area with an air saw or cut-off grinding wheel. Be precise when making cuts in the sectioning area. If a gap opens up at the cut or if the panels overlap, welding will be difficult. Therefore, after matching the panel fit at every point, it is important that cutting be done accurately.

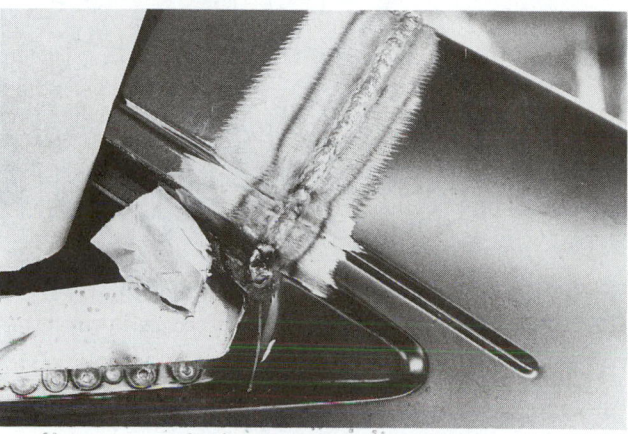

A

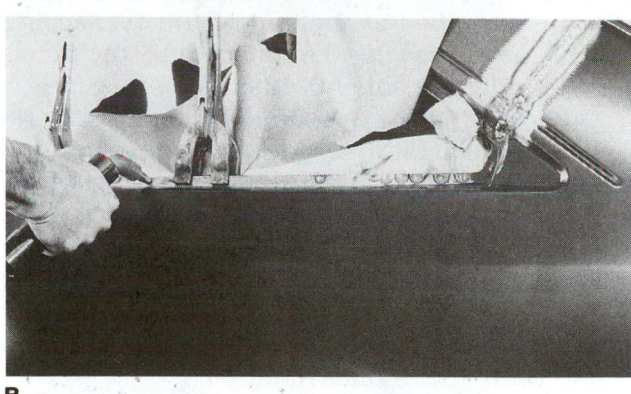

B

C

FIGURE 12-40 (A) When you are sure the quarter panel fits properly, final weld it to the vehicle. Note the continuous weld at the pillar and the spot welds above the window. (B) The technician is using a spot welder on the quarter panel upper flange. (C) Grind welds down flush.

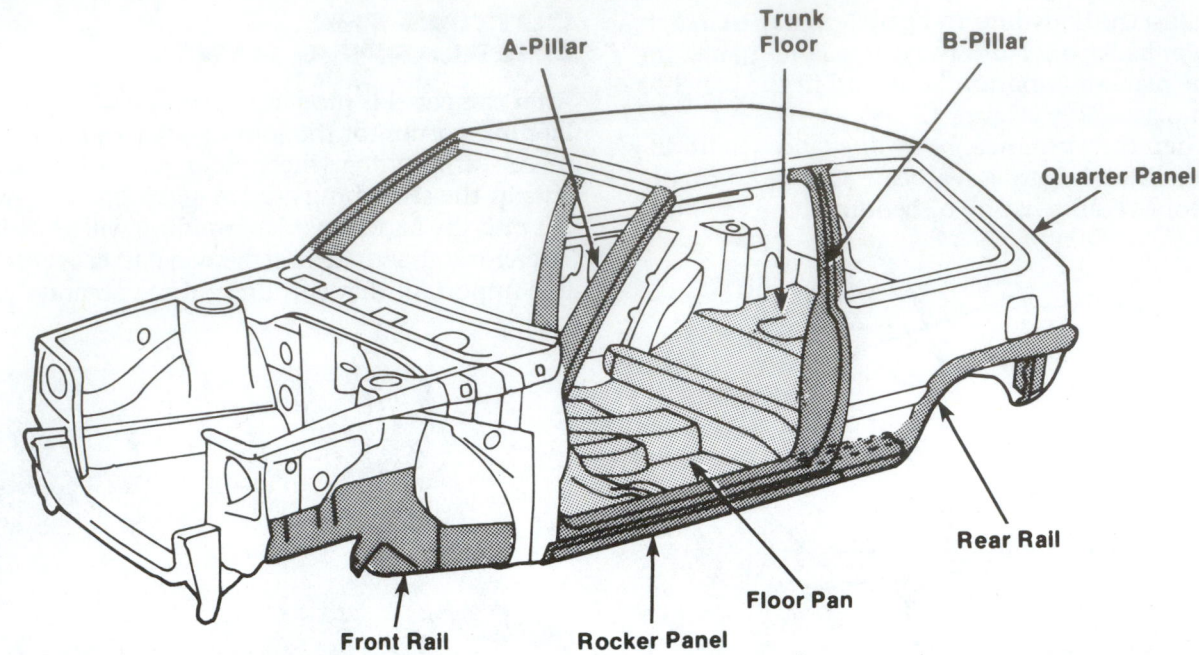

FIGURE 12-41 These are some other areas commonly sectioned on this body type.

If there is sufficient overlapping, both panels can be cut simultaneously. If the overlapping is small, a line can be scribed at the end of the overlapping panel. When cut along the scribed line, the panels should fit together snugly with little or no gap.

After cutting the overlap to fit, remove the replacement panel. Clean off any metal chips and other foreign material from the inside of the panel before proceeding. Drill or punch holes for any plug welds and apply a weld-through primer.

Apply body sealer around the inside perimeter of the quarter panel. Install the panel and other parts with self-tapping screws in the same screw holes as before. Verify the fit once more.

WELDING QUARTER PANELS

Once the dimensions and position of the new part are correct, weld the quarter panel in place. Typically, a continuous weld is used across the pillar (Figure 12–40A). Spot welds are often used along the edge near the window (Figure 12–40B). After welding, grind down the weld until flush with the panel surface (Figure 12–40C).

Welding operations for new panels are explained in detail in Chapter 7. The installation of fiberglass (SMC) body panels is discussed in Chapter 14.

12.3 STRUCTURAL SECTIONING

As mentioned, sectioning involves cutting and replacing panels at locations other than factory seams. When body parts need to be replaced, replacing them at factory seams is the logical first choice. However, this is impractical when many seams have to be separated in undamaged areas. In some repairs, sectioning of parts, such as rails, pillars, and rocker panels, may be required to make their repair economically feasible.

Remember that sectioning requires precision as well as strict adherence to recommended procedures. Always check the body repair manual for the manufacturer's procedure for the specific sectioning location.

As a collision repair technician, never forget that sectioning finally comes down to sound judgment. It is the technician who must assure the quality of the repair through the proper application of tested and proven procedures.

With structural parts, it is best to treat them all like high-strength steel. Then you do not have to identify every piece of high-strength steel, and you can use a MIG welder on all repairs.

Figure 12–41 shows some of the parts that are often sectioned: rocker panels, quarter panels, floor pan, front rails, rear rails, trunk floor, A-pillars, and B-pillars.

Unibody parts to be sectioned involve these types of construction:

1. Closed sections, such as rocker panels and A- and B-pillars
2. Hat or open U-channels, such as rear rails
3. Single layer or flat parts, such as floor pans and trunk floors

The closed-type sections are the most critical because they provide the principal strength in the unibody structure. They possess much greater strength per pound of material than other types of sections.

OTHER CAUTION AREAS

There are other areas to stay away from when making cuts. Stay away from holes in a part. Do not cut through any inner reinforcements, meaning double layers in the metal. Careless cutting through a closed section with inner reinforcements may make it impossible to restore the area to preaccident strength.

Stay away from anchor points, such as those for the suspension, seat belts in the floor, and shoulder belt D-ring anchor points. For example, when sectioning a B-pillar, make an offset cut around the D-ring area to avoid disturbing the anchor reinforcement.

When deciding where to section, look for an area with a uniform cross section. Check the body repair manuals and bulletins provided by the vehicle manufacturers. Much of this literature provides specific instructions on how and where to section.

BASIC TYPES OF SECTIONING JOINTS

There are three basic types of sectioning joints. They are the

1. Lap joint
2. Offset butt joint
3. Butt joint with insert

One of these joints, or a combination of these joints, will be used for all sectioning procedures. The type of joint used for a specific repair will depend upon the location and design of the structural part.

A butt joint with insert is used mainly on closed sections, such as rocker panels, A- and B-pillars, and rails (Figure 12–42). **Inserts** make it easy to fit and align the joints correctly. They also help make the welding process easier and more structurally sound.

Another basic joint is an *offset butt joint* without an insert (Figure 12–43). This type is also known as a *staggered butt joint*. The staggered butt joint is used on A- and B-pillars and front rails.

The third type is a lap joint, which is used on rear rails, floor pans, trunk floors, and B-pillars (Figure 12–44).

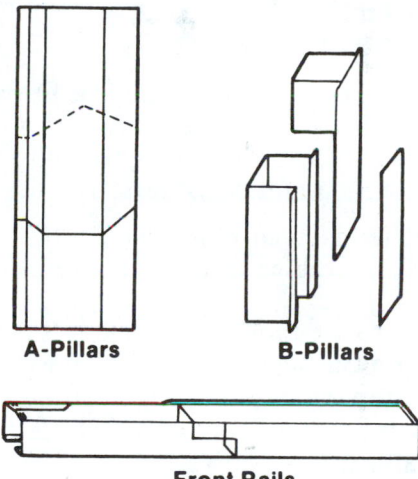

A-Pillars   **B-Pillars**

Front Rails

FIGURE 12-43 An offset butt joint without an insert is sometime used on these areas.

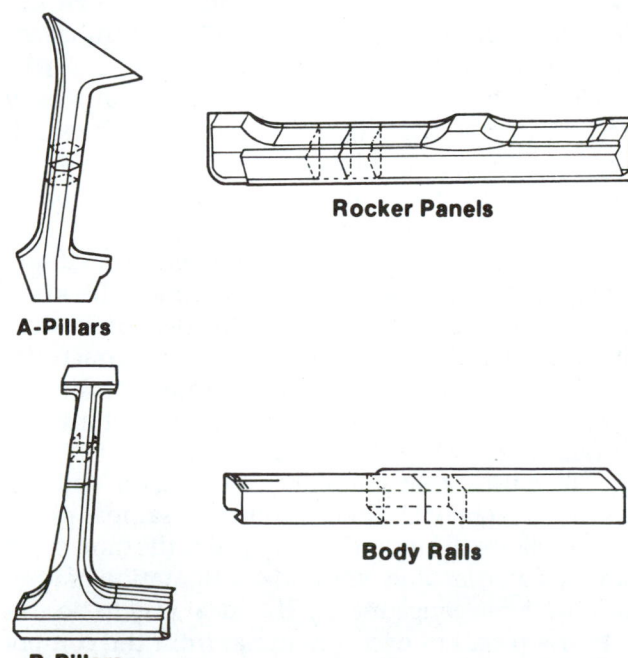

Rocker Panels

A-Pillars

Body Rails

B-Pillars

FIGURE 12-42 A butt joint with insert may be used on pillars, rocker panels, and rails. You need to refer to the manual or use common sense to determine if this type of joint works best.

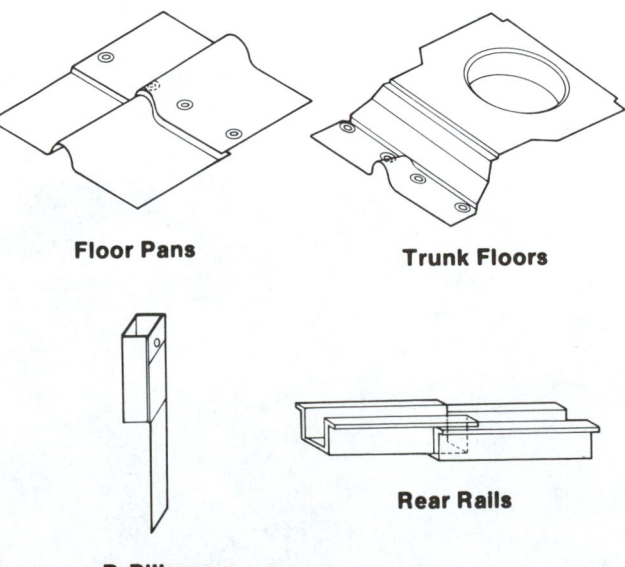

Floor Pans   **Trunk Floors**

B-Pillars   **Rear Rails**

FIGURE 12-44 Overlap joints are common on floor and trunk pans, and sometimes pillars and rails.

The configuration and make-up of the component being sectioned may call for a combination of joint types. Sectioning a B-pillar, for example, may require the use of an offset cut with a butt joint in the outside piece and a lap joint in the inside piece.

PREPARING TO SECTION

When preparing to section and replace a panel or structural member of a damaged vehicle, certain steps must be taken to ensure the quality of the repair. The first of these is the sectioning of the replacement part by the recycler or technician. Specific instructions must be provided to the recycler as to the placement of the section and the method of sectioning to be used. Other important considerations are the welding techniques used and the cleanliness of the joint metal.

USING RECYCLED OR SALVAGED PARTS

Recycled assemblies are undamaged parts from another damaged vehicle that are used for repairs. The use of recycled assemblies in collision repair makes sense for a number of reasons:

1. Fewer welds need to be made when using recycled assemblies compared to new, separate parts.
2. Less factory corrosion protection is disturbed.
3. More measuring is required when welding separate new parts and attaching them to the vehicle.
4. There is an abundance of recyclable assemblies available in most areas.

When using recycled parts, tell the recycler exactly where to make the cuts. Have the required part removed with a metal saw if possible. If the recycler uses a cutting torch, specify that at least 2 inches (51 mm) of extra length is to be left on the part to ensure that the heat dispersion from the cut does not invade the joint area. Instruct the recycler to make the cut so that reinforcing pieces that are welded inside the component are not cut through.

When a recycled or salvaged part is received, examine it for corrosion. If it has a lot of rust on it, do not use it. Ask for another one. Before installing a recycled part (Figure 12–45), check it for possible damage and make sure it is dimensionally accurate.

Remember! Using quality replacement parts is a must to achieve a quality repair. If the recycled part is almost rusted through, the repair will be inferior and may even endanger the passengers of the repaired vehicle.

Careful joint preparation is another necessity for doing a proper job of structural sectioning used parts. Before starting to weld, be sure to thoroughly clean the surfaces to be joined. Use a scraper to remove thick materials (heavy undercoatings, rustproofing, tars, caulking and sealants, and road dirt). Be sure to remove rustproofing, lead, plastic filler, and other contaminants from the inside of structurally closed sections when preparing them for welding. Then use an scuff wheel or a sander to remove thinner, less flammable primers, and paint. Do the finish cleaning with a wire brush.

The surfaces to be welded must be completely free of rust and scale. This is best removed by sanding or sand (media) blasting until there is a clean metal welding surface. In some cases, it is possible to do this with a power wire brush.

The weld site must be completely free of any foreign material that might contaminate the weld. Improper cleaning can result in a brittle, porous weld of poor integrity. In addition, you must attach the welder's cable clamp to a clean surface to have a trouble-free welding circuit. Application of a weld-through primer is also recommended.

To ensure that complete penetration and full fusion are achieved, do test welds on sample pieces that duplicate the intended workpiece: the same types of welds on the same type and configuration of joint and the same gauge metal. The ideal way to do this is to use pieces of excess material from the components being joined on the car, such as the scrap cut off to make the fit-up. While making the test welds, adjust the MIG machine to suit the given situation.

After completing the test welds, check them for strength as described in Chapter 7.

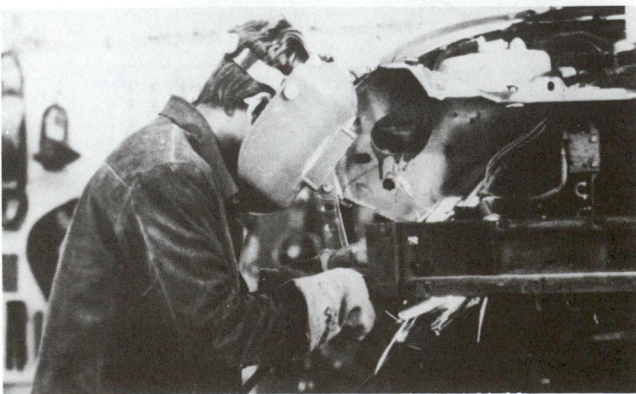

FIGURE 12-45 MIG welding is a must when sectioning and installing recycled parts to prevent warpage and for strong joints.

SECTIONING BODY RAILS

Virtually all front and rear rails are closed sections, but the closures are of two distinct types. One is called a **closed section.** It comes from the factory or the recycler with all four sides intact. Sometimes it is referred to as a *box section*. The other type comes as an **open hat channel** and is closed on the fourth side by being joined to some other component in the body structure (Figure 12–46).

The butt joint with insert is commonly used for repairing a closed section rail (Figure 12–47). Most rear rails, plus various makes of front rails, are of the hat channel type. Some of the hat channel closures are vertical, such as a front rail joined to a side apron. Some of them are horizontal, a rear rail joined to a trunk floor.

In most cases, when sectioning the open hat channel type of rail, the procedure is a lap joint with plug welds in the overlap areas and a continuous lap weld along the edge of the overlap (Figure 12–48).

SECTIONING ROCKER PANELS

Rocker panels are constructed differently depending on the make and model of the vehicle (Figure 12–49). The rocker panel might contain reinforcements. The reinforcements might be intermittent or continuous. Before starting work, you should know how the rocker panel is made.

Depending on the nature of the damage, the rocker panel can be replaced with the B-pillar or without it (Figure 12–50). To section and repair the rocker panel, a straight-cut butt joint with an insert can be used. The outside piece of the rocker panel can also be cut and the repair piece installed with overlap joints. Generally, the butt joint with insert is used when installing a recycled rocker panel with a B-pillar attached and when installing a recycled quarter panel.

To do a butt joint with insert, cut straight across the panel. An insert is fashioned out of one or more pieces cut from the excess length on the repair panel or from the end of the damaged panel. The insert should typically be 6 to 12 inches (152 to 305 mm) long and should be cut lengthwise into two- to four-pieces, depending on rocker panel configuration (Figure 12–51). Remove the pinch weld flange so

An insert should be twice the width of the cross section. For example, if the widest dimension is 2 inches (51mm), the insert should be 4 inches (102mm) long. For specific recommendations, refer to OEM Repair Manuals.

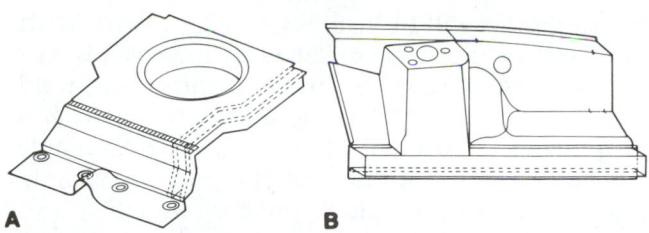

FIGURE 12–46 These are typical hat channels: (A) front rail and (B) rear rail.

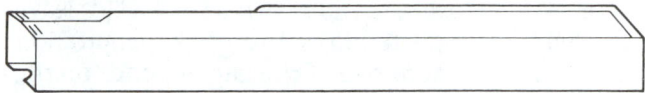

FIGURE 12–47 Note rear rail with butt joint and insert.

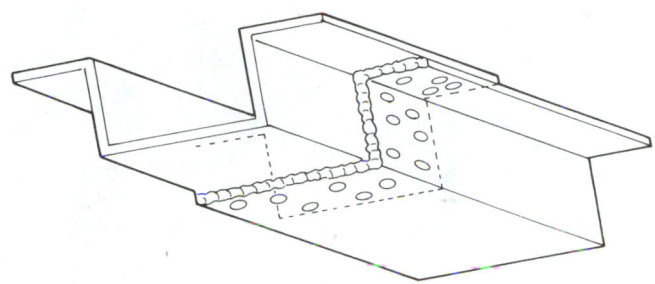

FIGURE 12–48 Joining open rails can be done with plug welds and continuous weld for strong repair.

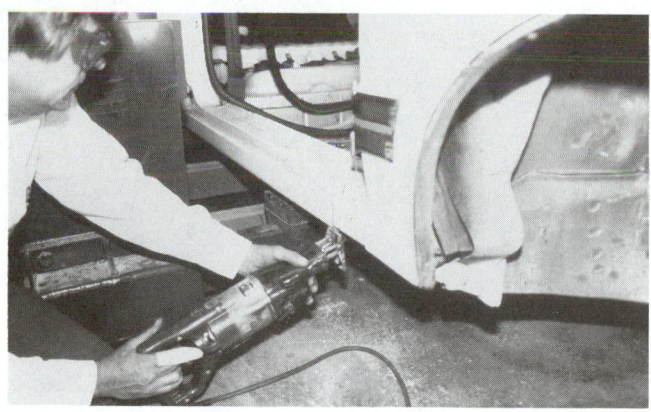

FIGURE 12–49 After determining the internal structure of the rocker panel and how it will be repaired, cut it off. A power saw works well on a rocker panel.

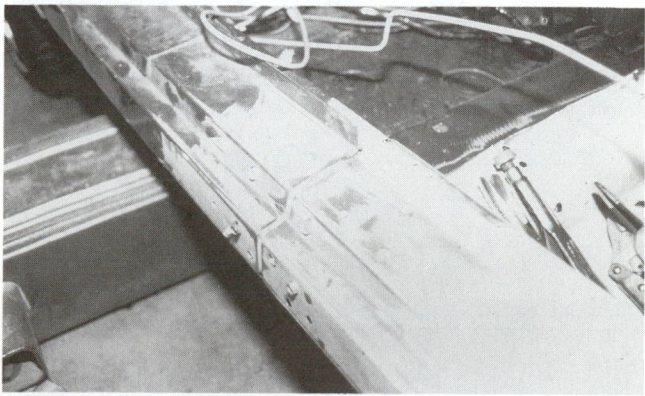

FIGURE 12–50 A properly welded rocker panel will be as strong as a new one. Its installation is critical to vehicle integrity.

that the insert will fit inside the rocker panel. With the insert in place, secure it with plug welds. For structural sectioning, 5/16-inch (8 mm) plug weld holes are typical to achieve an adequate nugget and acceptable weld strength. This 5/16-inch (8 mm) hole requires a circular motion of the gun to properly fuse the edge of the hole to the base metal.

When installing an insert in a closed section, whether it is a rocker panel, A- or B-pillar, or body rail, make sure the closing weld fully penetrates the insert. When closing the job with a butt weld, leave a gap wide enough to allow thorough penetration into the insert. The width of the gap depends on the thickness of the metal, but ideally it should not be less than 1/16 inch (1.59 mm) nor more than 1/8 inch (3.2 mm).

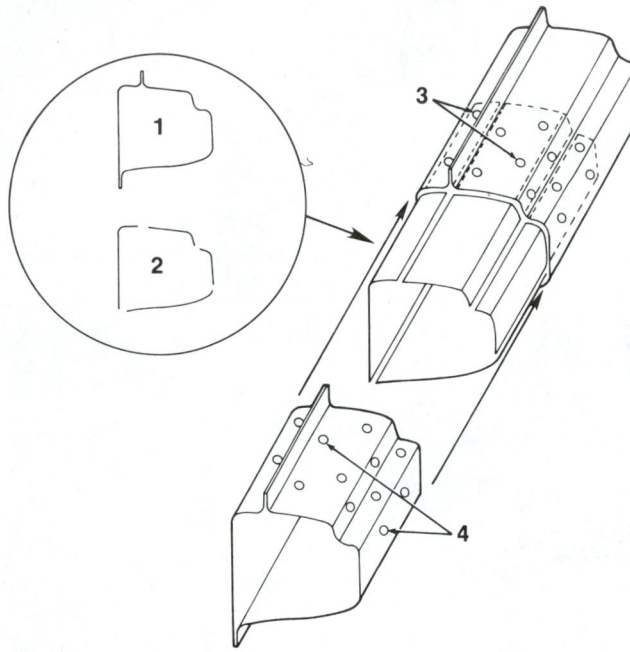

1. **Cross Section of Rocker Panel Insert Material Before Cutting Lengthwise**
2. **Insert Cut Lengthwise into Sections**
3. **Insert Inside Rocker Panel, Secured With Plug Welds or Sheet Metal Screws**
4. **5/16" Holes for Plug Welds**

FIGURE 12-51 If practical, cut an insert to fit a rocker panel. It can be made from an old section of rocker panel or from an unused new section. By removing the flange from the insert, you can slide it inside the rocker to provide strong repair.

FIGURE 12-52 Burrs in a joint will weaken the weld. Grind them off before fitting the parts.

Be careful to remove the burrs from the cut edges before welding. Otherwise, the weld metal tends to travel around and up under the burr. This can create a flawed weld, resulting in cracks and weakening of the joint (Figure 12–52).

In general, use the overlap procedure on a rocker panel when installing only the outer rocker or a portion of it. Leave the inner piece intact and cut only the outer piece. One way to make an overlap joint is to make the cut in the front door opening and allow for an overlap there when measuring. When making this cut, stay several inches away from the base of the B-pillar to avoid cutting any reinforcement underneath it (Figures 12–53 and 12–54).

1. Cut around the bases of the B- and C-pillars, leaving overlap areas around each (Figure 12–55).
2. Cut out the new outer rocker panel so that it overlaps around the bases of the pillars and the original piece of the outer rocker still affixed to the car.

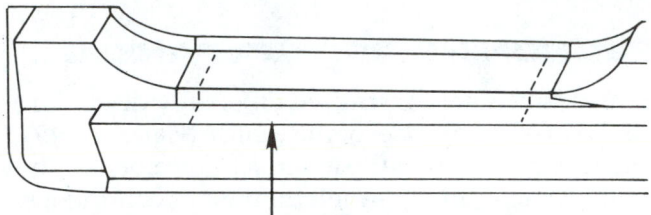

Stay Several Inches Away

FIGURE 12-53 Overlapping the outer rocker panel section will work fine if the ends are not damaged. This will avoid complex cutting and welding around the pillars.

FIGURE 12-54 A technician is installing only the outer half of a rocker panel to fix a minor dent. *(Courtesy of Chief Automotive)*

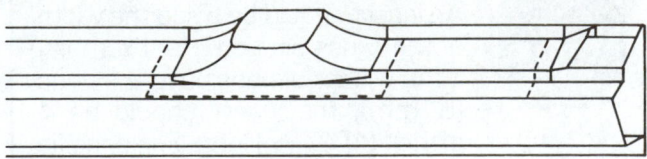

Leave overlap areas.

FIGURE 12-55 An overlap joint has been formed around the pillar.

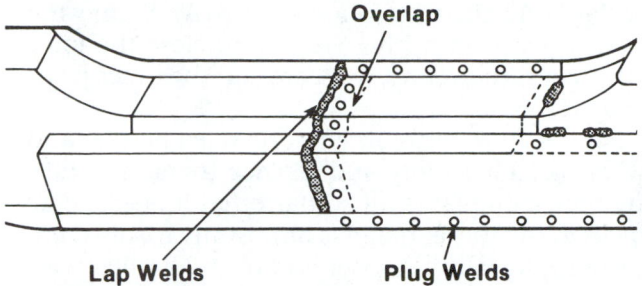

FIGURE 12-56 Plug welds and a lap weld will secure the rocker panel.

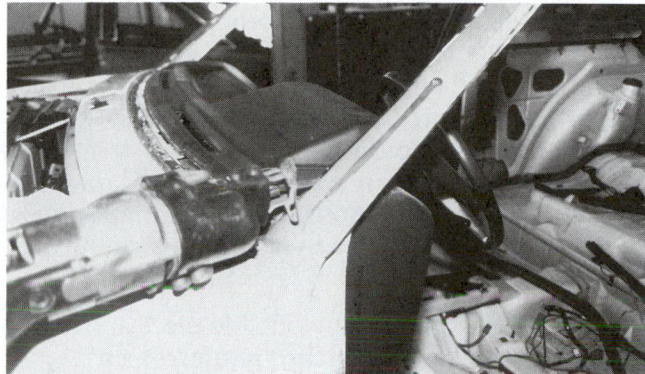

FIGURE 12-57 Normally, cut A-pillars near the center. Use a power saw to make a clean cut. Refer to the service manual for directions since construction and procedures vary. *(Courtesy of Tech-Cor)*

FIGURE 12-58 Form the recommended joint in the pillar and then clamp the parts. Measure the position of the panels before welding. *(Courtesy of Tech-Cor)*

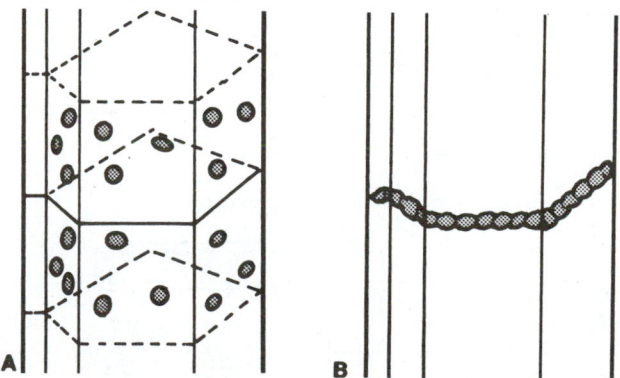

FIGURE 12-59 Note the welding butt joint with insert in an A-pillar with (A) plug welds and (B) butt welds.

3. In the pinch weld flanges, use plug welds to replace the factory spot welds.
4. Plug weld the overlaps around the B- and C-pillars, using approximately the same spacing as in the pinch weld flanges.
5. Then lap weld the edges with about a 30-percent intermittent seam; about $1/2$ inch (12.7 mm) of weld in every $1^1/2$ inches (38 mm) of overlap edge.
6. Put plug welds in the overlap area in the door opening. Lap weld around the edges to close the joint (Figure 12–56).

Depending on the nature of the hit, you may need to make the overlap cut in the rear door opening and cut out and overlap around the bases of the A- and B-pillars. Use this same basic technique to replace the entire outer rocker. In this version, cut around the bases of all three pillars and overlap all three bases in the same way as before.

SECTIONING A-PILLARS

The front pillars or **A-pillars** extend up next to the edges of the windshield. They must be strong to protect the passengers. They are steel box members that extend down from the roof panel to the main body section.

A-pillars can be either two-piece or three-piece components. They can be reinforced at the upper end or the lower end, or both. However, they are not usually reinforced in the middle. Therefore, A-pillars should be cut near the middle to avoid cutting through any reinforcing pieces (Figure 12–57). It is also the easiest place to work.

To section an A-pillar, use a straight-cut butt joint with an insert or an offset butt joint without an insert. The butt joint with insert repair is made in the same manner as already described for the rocker panel. The A-pillar insert should be 4 to 6 inches (102 to 152 mm) in length. After cutting the insert lengthwise and removing any flanges, tap the pieces into place. Secure the insert in place with plug welds. Then close all around the pillar with a continuous butt weld (Figure 12–58 and 12–59).

To make the offset butt joint, cut the inner piece of the pillar at a different point than the other piece was cut, creating an offset. Whenever possible, try to make the cuts between the factory spot welds, so that it will not be difficult to drill them out. Make the cuts no closer to each other than 2 to 4 inches (50 to 100 mm). Butt the sections together and continuous-weld them all around.

SECTIONING B-PILLARS

Center pillars or **B-pillars** are the roof supports between the front and rear doors on four-door vehicles.

They help strengthen the roof and provide a mounting point for the rear door hinges.

For sectioning B-pillars, two types of joints can be used: the butt joint with insert and a combination of offset cut and overlap. The butt joint with insert is usually easier to align and fit when the B-pillar is a relatively simple two-piece cross section without a lot of internal reinforcing members. The insert provides additional strength (Figure 12–60).

Be sure to cut below the seat belt D-ring mount low enough to avoid cutting through the D-ring anchor reinforcement. The majority of B-pillars have them. In the case of the B-pillar, use a channel insert in only the outside piece of the pillar. The D-ring anchor reinforcement welded to the inside piece prevents the installation of an insert there.

Begin by overlapping the new inside piece on the existing one, rather than butting them together,

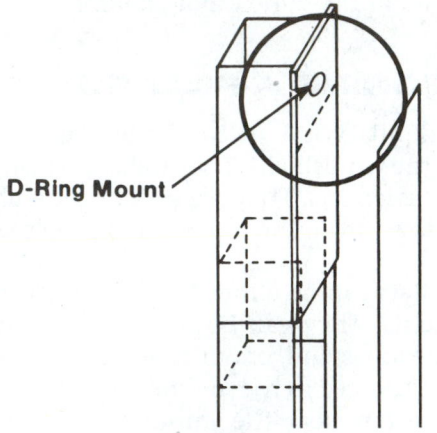

D-Ring Mount

FIGURE 12-60 A two-piece B-pillar requires different repair methods. Stay away from reinforced areas like around the D-ring.

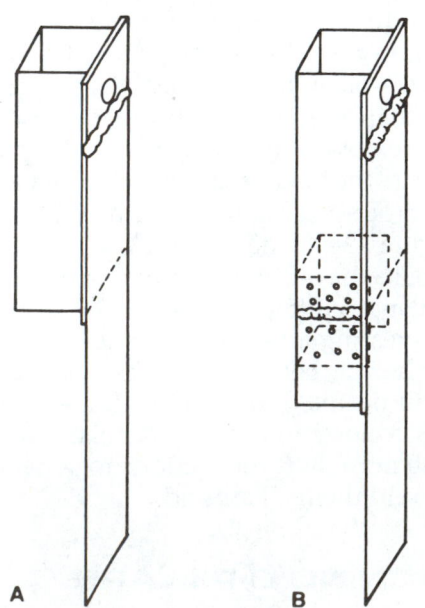

A   B

FIGURE 12-61 (A) Lap weld the inner panel; (B) plug and butt weld the outer panel.

and lap weld the edge (Figure 12–61A). Secure the insert in place with plug welds and close the joint with a continuous butt weld around the outer pillar (Figure 12–61B).

You may want to obtain a recycled B-pillar and rocker panel assembly and replace them as a unit. Any time a B-pillar is hit so hard that it needs to be replaced, the rocker panel is almost invariably damaged also. Install the upper end of the B-pillar with either of the two approved types of joints and make a butt joint with insert in the rocker panel in the manner already shown.

If the main damage is in the rear door opening, make the butt joint with insert in the front door opening. Install the other end of the rocker in its entirety. If the main damage is in the front door opening, reverse the procedure.

The combination offset and overlap joint (Figure 12–62) is used more often when installing new parts and when working with separate inside and outside pieces.

1. Cut a butt joint in the outside piece above the level of the D-ring anchor reinforcement.
2. Make an overlap cut in the inside piece below the D-ring anchor reinforcement.
3. Install the inside piece first with the new segment overlapping the existing segment.
4. Lap weld the edge (Figure 12–63A).
5. Put the outside pieces in place; make plug welds in the flanges. Close the section with a continuous weld at the butt joint (Figure 12–63B).

It is usually best to use the offset and overlap joint on a B-pillar with three or more pieces in its cross section. This design would make it difficult to install an insert. In fact, sometimes the offset and overlap procedure is mandatory, because it is not possible to install an insert.

REPLACING FOAM FILLERS

Some manufacturers place foam inside the panels. **Foam fillers** are used to add rigidity and strength

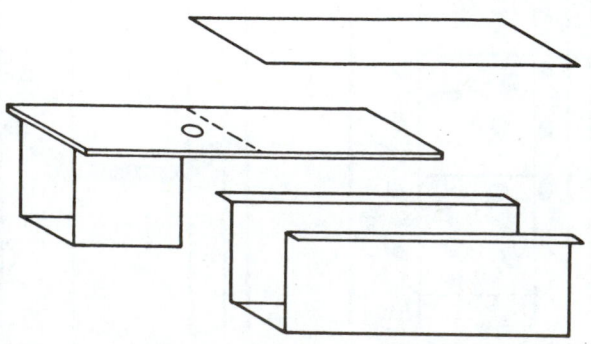

FIGURE 12-62 This is a combination offset and overlap joint.

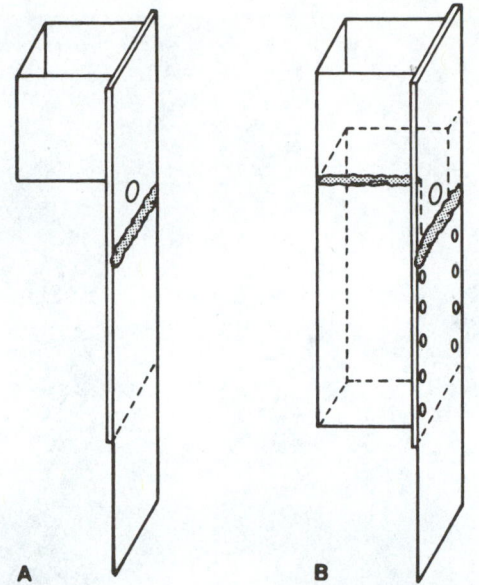

FIGURE 12–63 Here is an example of creating a combination offset and overlap joint: (A) lap welding inside and (B) plug and lap welding outside.

to structural parts. They also reduce noise and vibrations. Cutting and welding will damage the foam. Replacing the foam fillers must be part of the repair procedure.

Some vehicle manufacturers are using urethane foam in A- and B-pillars and other locations. The manufacturer may or may not consider the foam filler to be structural. The use and location of foam fillers are different from vehicle to vehicle. Follow the manufacturer's recommendations for replacing or sectioning foam-filled panels.

Some OEM replacement parts come with the foam already in the part. When the parts come without foam filler, or foam filler needs to be replaced, a product designed specifically for this application must be used to fill the panel.

When sectioning foam-filled A-pillars, the foam filler is removed in the repair area. It is then replaced after all welding is completed.

Some manufacturers have specific recommendations for the type of replacement foam filler

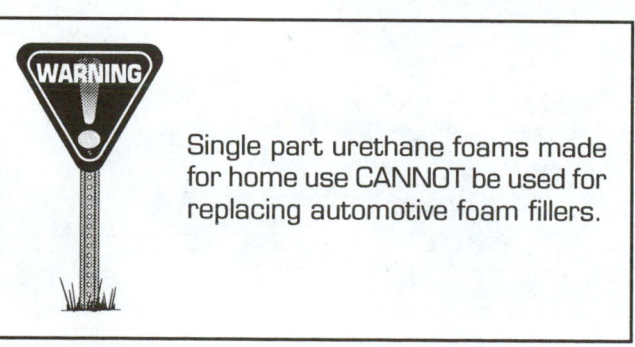

WARNING

Single part urethane foams made for home use CANNOT be used for replacing automotive foam fillers.

needed. The repair usually calls for a foam filler that, when cured, does not change in volume due to differences in temperature and humidity. There may also be specific requirements for foam density given in ounces per cubic inch (grams per cubic centimeter).

SECTIONING FLOOR PANS

When sectioning a floor pan, do not cut through any reinforcements, such as seat belt anchors. Always make sure the rear section overlaps the front section. You want the edge of the bottom piece, under the car, to point rearward. This type of overlap prevents road debris and water splash from entering the joint between the parts (Figure 12–64).

1. Join all floor pan sections with an overlap.
2. Plug weld the overlap, putting the plugs in from the topside, downward (Figure 12–65A).
3. Caulk the top, forward edge with a recommended, flexible body caulk.
4. On the bottom side, lap weld the floor pan edge with a continuous bead.
5. Cover the lap weld with a primer, a seam sealer, and a topcoat (Figure 12–65B). The primer helps the sealer hold better, and the topcoat completes the protection. This assures that there will be no deadly carbon monoxide intrusion from any leaking engine exhaust system through the joint and into the passenger compartment.

SECTIONING TRUNK FLOORS

When sectioning a trunk floor, in general, follow the basic procedures just described for the floor pan with some variations.

1. There is generally some kind of a cross member under the trunk floor near the rear suspension.

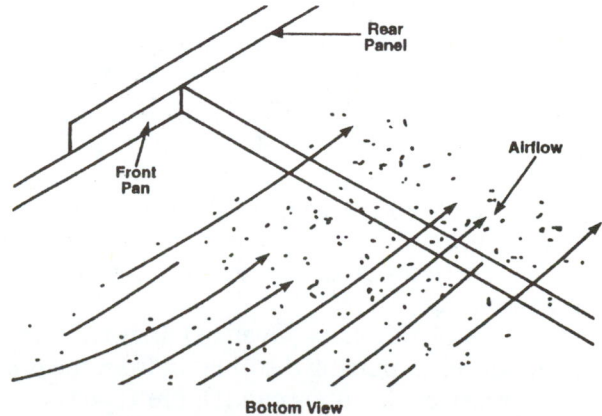

FIGURE 12–64 The rear section of a floor pan should overlap to shield the joint from windstream.

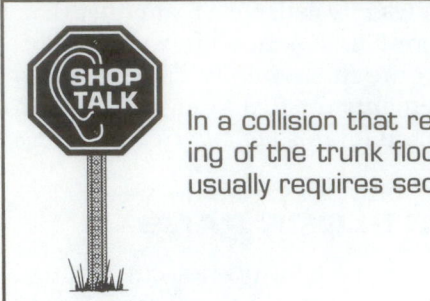

SHOP TALK

In a collision that requires sectioning of the trunk floor, the rear rail usually requires sectioning.

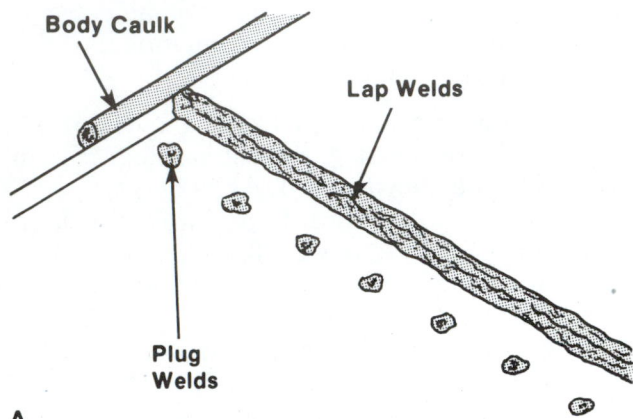

A

Body Caulk

Lap Welds

Plug Welds

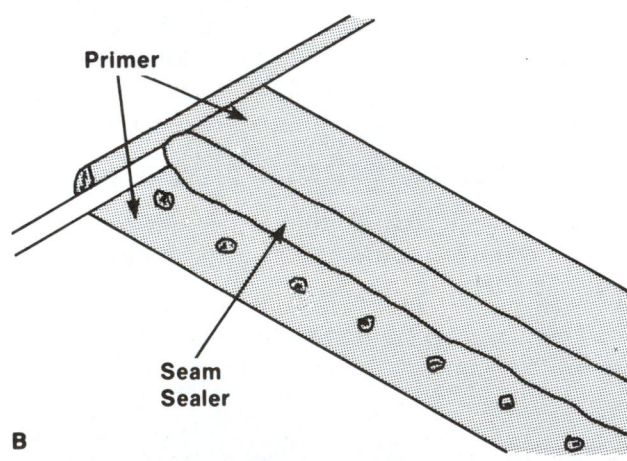

B

Primer

Seam Sealer

FIGURE 12-65 Note the basic method of installing floor pans. (A) Plug weld from top and lap weld bottom edge; (B) seal and prime bottom edge.

Whenever possible (Figure 12–66), section the trunk floor above the cross member's rear flange. Section the rail just rearward of the cross member.

2. Plug weld the trunk floor overlap joint to the cross member, again putting the plugs in from the top, downward, as in a floor pan (Figure 12–67).

3. Caulk the top, forward edge just like a floor pan seam (Figure 12–68).

4. On the bottom side, a lap weld is not always necessary because of the strength provided by

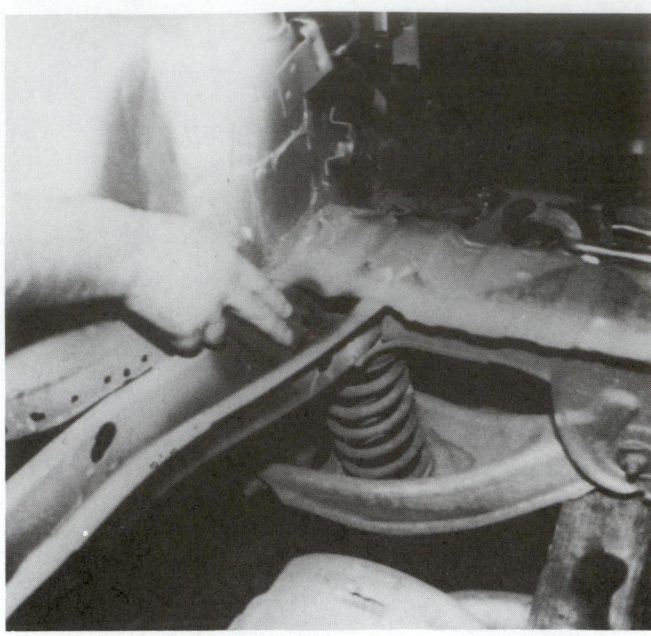

FIGURE 12-66 You may need to section the trunk floor above the cross member so members can be straightened.

FIGURE 12-67 Plug weld the trunk floor to members, making sure it laps on top of the frontal area and is welded properly.

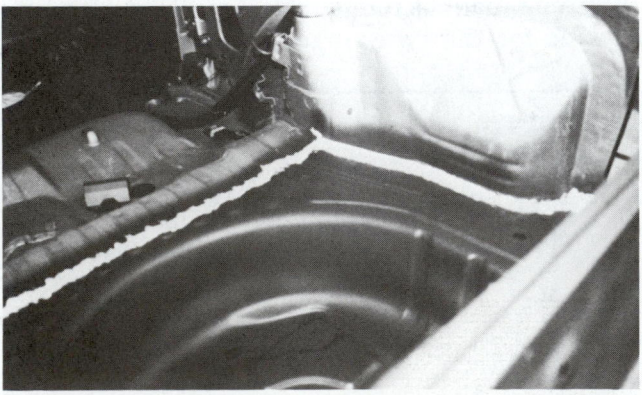

FIGURE 12-68 Caulk the inside seams to keep out water and poison gas from the exhaust system.

the cross member. However, on cars where the trunk floor section is not above a cross member, the lower edge must be lap welded.

In both cases, cover the bottom side seam with a primer, a seam sealer, and a topcoat. With the trunk floor, sealing against poison carbon monoxide gas is critical because of the proximity of the tail pipe.

SECTIONING RAILS

Certain structural components have crush zones, or buckling points, designed into them for absorbing the impact energy in a collision. This is particularly true of the front and rear rails because they take the brunt of the impact in most collisions. Crush zones are in all front and rear rails (Figure 12–69).

You can often identify crush zones by their appearance. Some are in the form of convoluted or crinkled areas. Others have dents or dimples, and others have holes or slots so the rail will collapse at these points. Crush zones are ahead of the front suspension and behind the rear suspension.

Avoid cutting near crush zones. Sectioning procedures can change the designed collapsibility if improperly located. If a rail has suffered major damage, it will be buckled in the crush zone, so the crush zone will usually be easy to locate. Where only moderate damage has occurred, be very careful. The hit might not have used up the entire crush zone. So be aware of other potential areas where designed-in collapse might occur.

Vehicle and part manufacturers often give recommendations about how to install frame rail replacements (Figure 12–70). Refer to these instructions. They will give the critical information about were and how to cut (Figure 12–71). They will give the type of weld joint that will work best with the specific rail (Figure 12–72). These factor instructions will also tell you the best way to secure and align the new rail section (Figure 12–73).

Recent testing has determined that lap welding of front section frame rails and rocker panels can yield a tighter fitting section and superior corrosion protection when compared to inserts (Figure 12–74). The following example involves a vehicle that has sustained damage requiring the left rail to be sectioned.

FIGURE 12-69 Here you can see what happens to a crush zone on a member. The member should be replaced. *(Courtesy of Tech-Cor)*

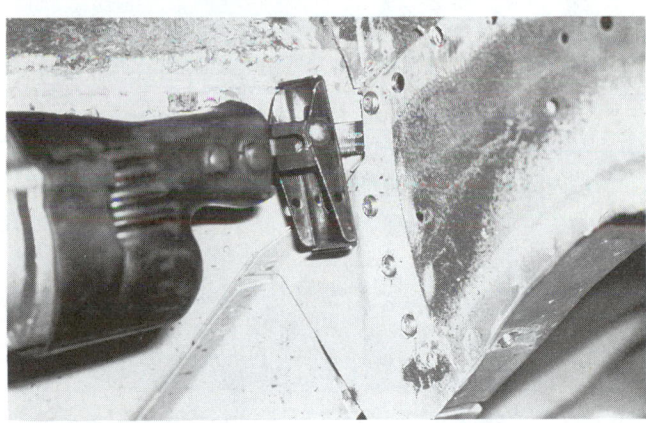

FIGURE 12-71 Cut the member at the manufacturer's recommended location. *(Courtesy of Tech-Cor)*

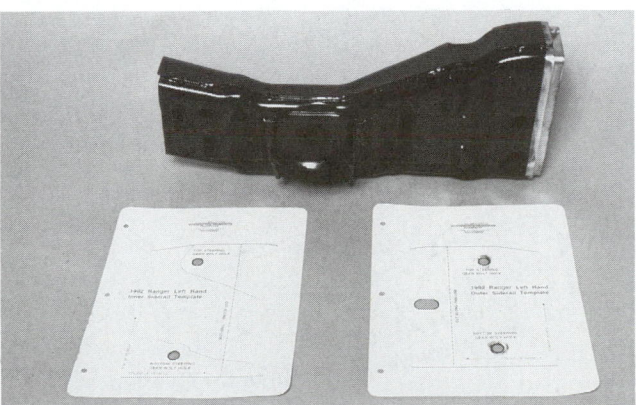

FIGURE 12-70 Templates may be available for installing replacement members. These will simplify the process and assure quality repair. *(Courtesy of Tech-Cor)*

FIGURE 12-72 Note how the member and apron have been sectioned to remove badly deformed metal. *(Courtesy of Tech-Cor)*

1. Locate and drill out the factory spot welds that attach the upper rail to the cowl at the base of the windshield. A propane torch, scraper, and wire brush may be needed to remove sealant or caulk from the spot weld areas. Set the torch on low heat to avoid burning the seam sealer.

2. Remove the two hidden spot welds that secure the upper rail to the rear outer flange of the strut tower. They are normally visible through a hole at the rear portion of the upper rail (Figure 12–75).

3. Remove the spot welds that attach the strut tower to the rail extension panel at the base of the strut tower. Remove any seam sealer covering these welds inside the engine compartment. The under surface is often coated with sealer and sound-deadening material.

4. The lower rail sectioning is often done forward of the center of the strut tower. The sectioning procedure uses a staggered cut of the inner and outer lower rail, with a lap joint at both cut lines. There is usually an inner rail reinforcement located in the area of the section. Because of this

reinforcement, this part of the rail is ideal for sectioning.

5. There are spot welds that attach the reinforcement to the inside of the lower rail. These must be removed before any cuts are made. They are visible on the wheelhouse side of the rail.

6. The location of the cut on the engine side is usually about 12 inches (305 mm) from the cowl (Figure 12–76). This positions the cut line near the end of the inner reinforcement. Refer to the service manual to get an exact dimension for making this cut.

7. The outer rail cut line (wheelhouse side) is typically made 3 to 5 inches (76 to 127 mm) rearward of the engine side cut line (Figure 12–77).

8. To achieve the correct overlap, carefully split the corners on the original structure at the exposed

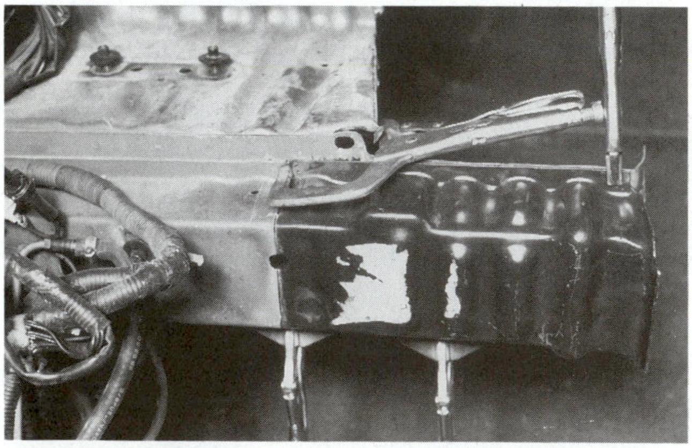

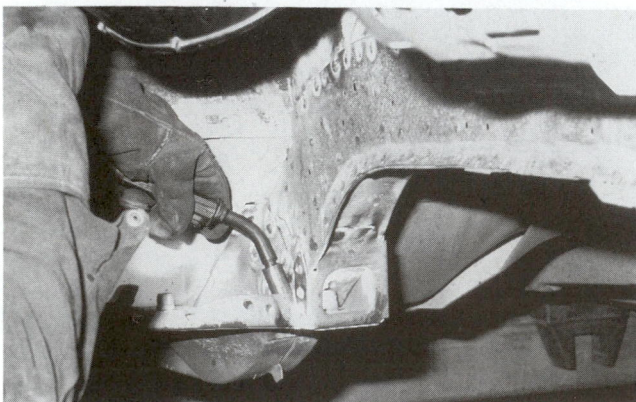

FIGURE 12-73 Clamp the new section of the member in place and measure it for accuracy before welding. *(Courtesy of Tech-Cor)*

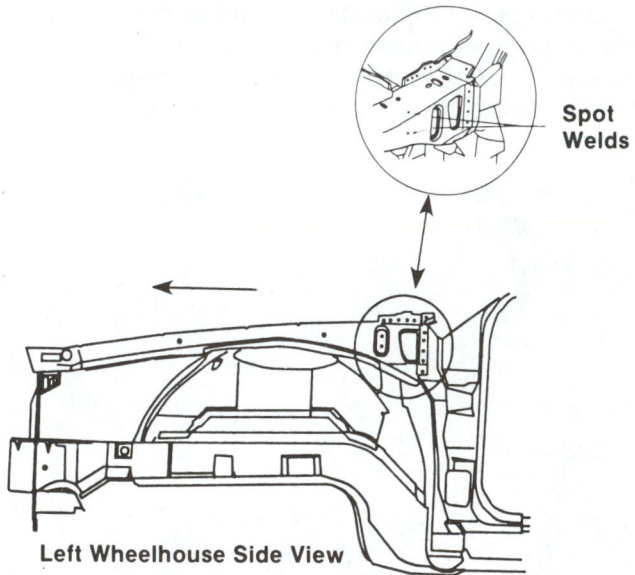

FIGURE 12-75 Spot welds that secure the upper rail to the strut tower can sometimes be seen through a hole at the rear of the upper rail.

FIGURE 12-74 MIG weld the member to the vehicle while double-checking measurements. *(Courtesy of Tech-Cor)*

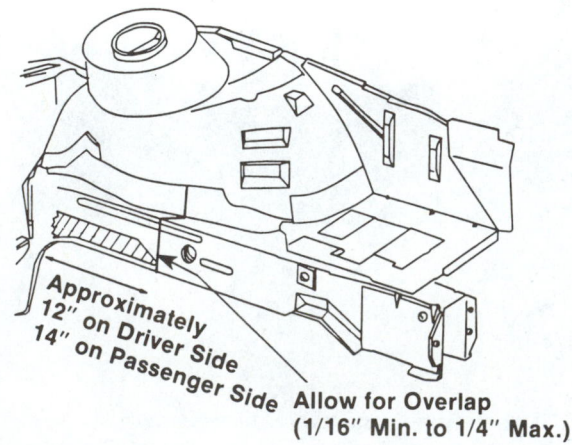

FIGURE 12-76 This sectioning cut on the engine side is typical.

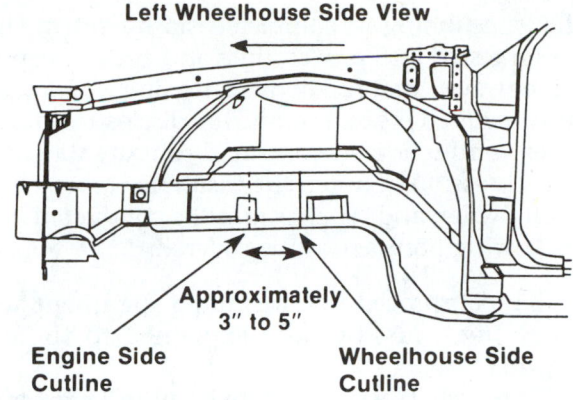

Left Wheelhouse Side View

Approximately 3" to 5"

Engine Side Cutline

Wheelhouse Side Cutline

FIGURE 12-77 This sectioning cut on the outer rail is also typical. Refer to the manufacturer's directions for the exact location.

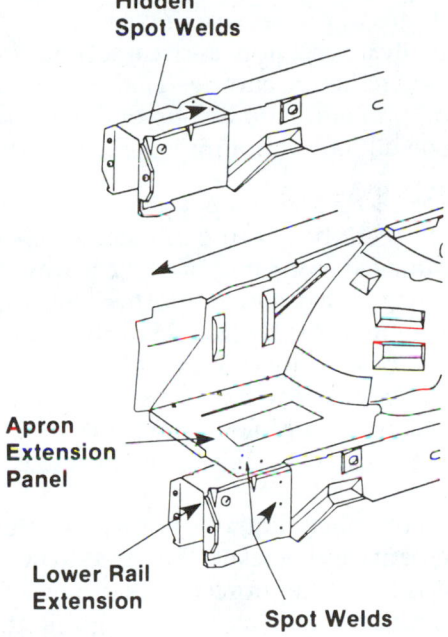

Hidden Spot Welds

Apron Extension Panel

Lower Rail Extension

Spot Welds

Right Engine Side View

FIGURE 12-78 Fold the apron extension panel up to expose the spot welds.

FIGURE 12-79 Check the dimensional accuracy after the new assembly is clamped into place.

end. These splits should not exceed $\frac{1}{4}$ inch (6.4 mm). Any part of the splits that is exposed after fit-up must be welded closed.

9. It is very important that the replacement structure is positioned over the original structure. This will allow the application of the corrosion protection to be more effective. The open portion of the joint will face the open end of the rail.

10. Separate the opposite side lower rail extension from the lower rail by first drilling the spot welds that secure the radiator support and apron extension panel. Next, carefully fold the apron extension panel upward to expose the other spot welds that attach the extension to the rail (Figure 12–78).

11. After all the spot welds have been removed and the offset cuts made, the damaged assembly can be removed from the vehicle.

12. Preparation of the used assembly (for example, spot weld removal and lower rail offset cut) is identical to that of the damaged assembly. It is very important to inspect, measure, and, if necessary, straighten the used assembly to the proper dimensions prior to installing it on the vehicle. Remember to add to the length measurement on the replacement rail to allow for the overlap.

13. All adjoining flanges and weld areas must be cleaned using a propane torch and wire brush. Do not grind or burn off any galvanized coatings. After cleaning and before welding, a weld-through primer must be applied to all bare metal mating surfaces.

14. After the used assembly is installed, measuring equipment can be used to check the correct position. Then the assembly can be clamped in place (Figure 12–79).

15. After checking that all dimensions are within tolerance, the assembly can be welded. Continuous welds should be made in alternating segments of $\frac{1}{2}$ to $\frac{3}{4}$ inch (13 to 19 mm) to minimize distortion. Be sure to completely close all seams and do not leave any gaps.

16. Corrosion protection, including refinishing of the replacement pieces, should be completed as detailed in Chapter 16.

FULL BODY SECTIONING

One of the most drastic repairs that can be performed is full body sectioning. **Full body sectioning** is replacing the entire rear section of a collision-damaged vehicle with the rear section of a salvaged vehicle. It may be more economical than trying to rebuild the damaged vehicle using new parts. Full body sectioning requires the highest quality workmanship possible (Figure 12–80).

Jigs are often used to help locate and guide the cut when sectioning. They are used on any sectioning procedure where an offset butt joint is used. Precise cuts are essential when sectioning (Figure 12–81).

Full body sectioning procedures require sectioning the two A-pillars, two rocker panels, and the floor pan. When this procedure is properly performed, sectioned vehicles have been shown to be as strong and serviceable as an undamaged vehicle (Figure 12–82).

Full body sectioning is complicated by antilock brake systems. The replacement vehicle must be so equipped, or the ABS system parts must be retrofitted.

Location and function of body computers may change, even within the same production year of a given vehicle. Check the locations of these computers on both the damaged vehicle and the salvaged section.

When the individual components are properly sectioned, aligned, and welded using the proper techniques and procedures, full body sectioning is a suitable and satisfactory procedure. Vehicles repaired by full body sectioning are completely crash-worthy. This has been tested and proved time and again. Keep in mind, however, that full body sectioning is not a frequently required procedure, and full disclosure should be given to the car owner before repairs are started.

A discussion between the insurance representative, car owner, and repairer must be conducted and the following points must be covered:

- All repair procedures, including alignment and welding, must be fully explained to the car owner.
- The recycled sections—both body and mechanical—must be of like kind and quality. Always verify that all VIN code identifications and EPA emission control requirements are met and that all suspension, braking, and steering components are in proper working order.
- Carefully inspect front and rear sections for proper alignment before cutting. If either is out of alignment, proper fit and line-up of the section joints will be difficult if not impossible to achieve.

Figure 12–83 shows a popular compact that was fully body sectioned in a commercial repair shop. The undamaged front half of one car was joined to the undamaged rear half of another. Butt joints and inserts were used in the middle of the A-pillars and in the two rocker panels. An overlap joint was used in the floor pan. The rocker panel and floor pan cuts were made in the middle of the front door opening to avoid any brackets or reinforcements in the A- and B-pillars.

But remember that the floor pan might have reinforcements and brackets that need to be removed before sectioning. Reinforcements can be left on the replacement rear half to aid in alignment. Proper corrosion protection must be restored when replacing brackets and reinforcements.

FIGURE 12-80 Full body sectioning joins undamaged front and rear halves of the vehicle. This is a complex repair. *(Courtesy of Tech-Cor)*

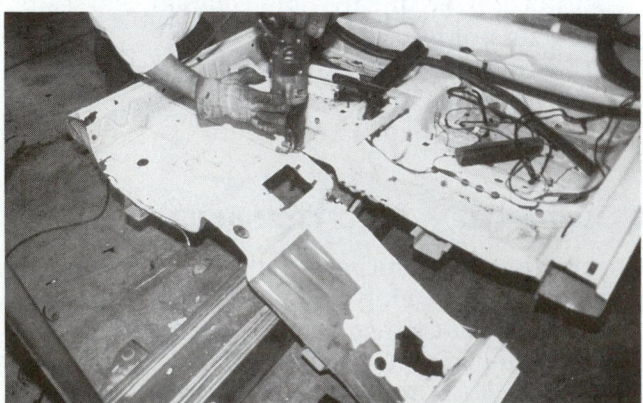

FIGURE 12-81 Accurate cuts must be made on both vehicle sections around the floor, rocker panels, and pillars. *(Courtesy of Tech-Cor)*

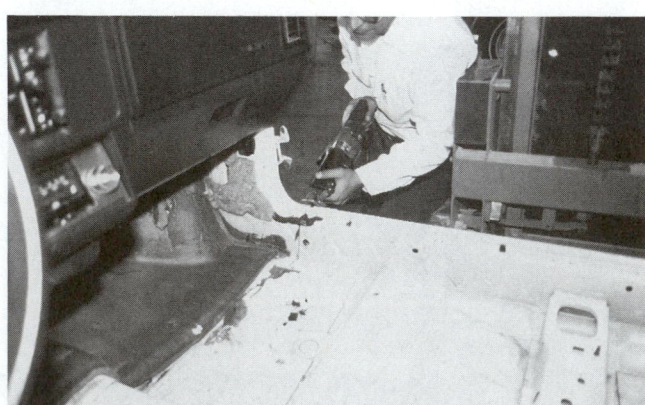

FIGURE 12-82 When cutting, make sure you do not sever fuel lines, wiring, and other parts or a deadly fire could result. *(Courtesy of Tech-Cor)*

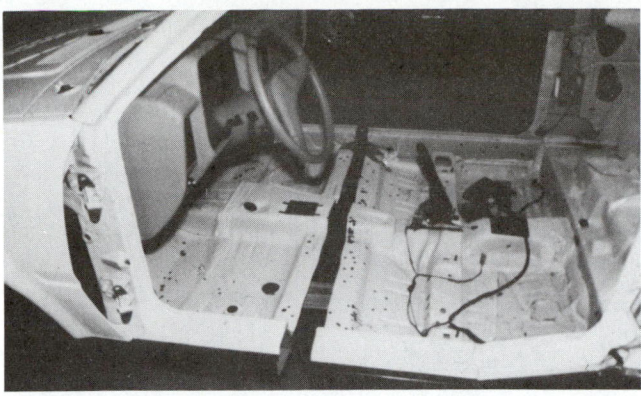

A

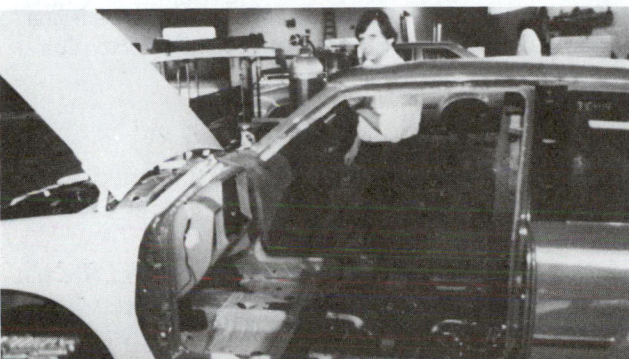

B

FIGURE 12-83 (A) After accurate section cuts, two halves can be "married" or fit together. (B) Jigs may have to be fabricated to hold the sections when measuring and welding.

JOINING THE FULL BODY SECTIONS

After the front and rear sections have been trimmed to fit, drilled for plug welds, and prepared with weld-through primer, follow these steps to join the sections.

1. Install the rocker and pillar inserts. Clamp them in place with sheet metal screws.
2. Place the A-pillar inserts in the upper or lower portion of the windshield pillar, depending on the angle and contour of the windshield.
3. Fit the two halves together by first joining the rocker panels and then the A-pillars. Clamp the rocker and pillar flanges to prevent the sections from pulling apart.
4. Check the windshield and door openings for proper dimension, using a tram gauge or a steel rule. If possible, install the doors and windshield to verify proper alignment.
5. When proper alignment is achieved, secure overlapping areas with sheet metal screws to pull the seam areas together and hold the sections together during welding.

FIGURE 12-84 If done properly, a section car or truck will be as good as new. This same principle is used to make stretch limousines. *(Courtesy of Tech-Cor)*

6. Using centerline gauges and a tram gauge, double-check vehicle dimensions and section alignment before welding the sections together.
7. Weld sections together using techniques already described in this chapter for joining rocker panels, A-pillars, and the floor pan (Figure 12–84).

Remember to check the OEM Repair Manuals for specifications.

12.4 ANTIRUST TREATMENTS

The application of antirust agents is necessary not only before welding, but also before and after the painting process. Welded panel joints are treated with weld-through primer before they are joined together. The weld joints must also be sealed with body sealer before finishing undercoating or an antirust treatment must be applied to the joints after finishing to seal out moisture and prevent rust formation.

FIGURE 12-85 A floor jack and a block of wood covered with shop rag can be used to support the door while removing the hinge bolts. *(Courtesy of Mitchell International)*

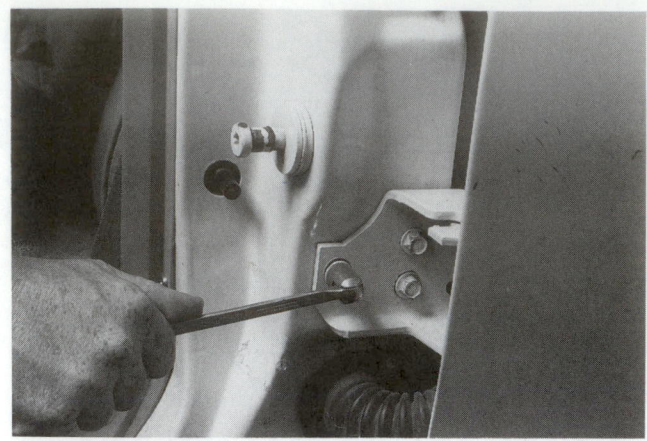

FIGURE 12-86 When loosening door hinge bolts, make sure the door is held so it does not fall off the jack.

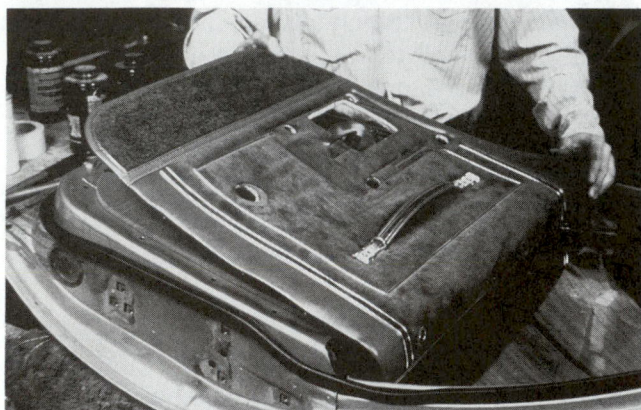

FIGURE 12-87 Door panel trim can be removed before or after door removal. *(Courtesy of 3M Automotive Trades Div.)*

Rustproofing and corrosion protection techniques are discussed fully in Chapter 15.

12.5 DOOR PANEL REPLACEMENT

Like other damaged panels, a door can be bumped or pulled into shape, or replaced. The decision is determined by the amount of time required to repair the door versus the cost of a replacement door.

The outer panel or **door skin** wraps around the door frame and is clinched to the pinch weld flange. The skin is secured to the frame either with plug or spot welds or with adhesives.

REPLACING WELDED DOOR SKINS

1. Before removing the door, check to see if the hinges are sprung. Check the alignment of the door with respect to its opening.

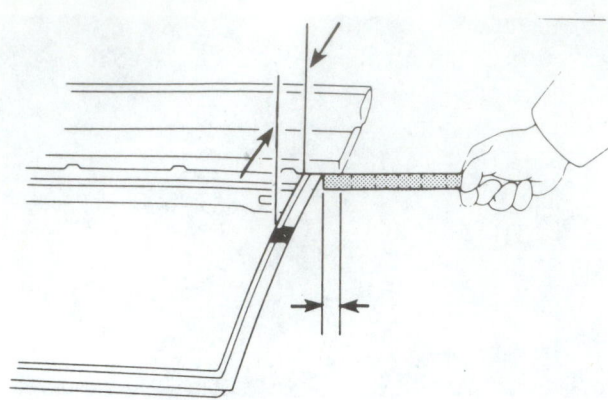

FIGURE 12-88 Measure the location of the old door panel on its frame. Write down the horizontal and vertical positions. Here a piece of tape has been placed on the frame for measurement reference. *(Courtesy of Toyota Motor Corp.)*

2. Observe how the door skin is fastened to the door so you can tell how much interior hardware must be removed.
3. Remove the trim panel and disconnect the battery to isolate all door power accessories.
4. To prevent loss, place parts inside the vehicle. This also applies to the door glass.
5. To remove some of the damage and possibly straighten or align the inner door frame, use a hydraulic or body jack (Figure 12–85).
6. Remove the door glass to prevent breakage. Now remove the door from the vehicle and move it to a suitable work area (Figure 12–86).
7. Remove all hardware from the door (Figure 12–87).
8. Using an oxyacetylene gas torch and wire brush, remove paint from the spot welds in the panel hem. Using a drill and spot cutter or grinder, remove the spot welds.
9. As a reference, apply tape to the door frame. Measure the distance between the lower line of the tape and the outer panel edge. Also measure the distance between the front or rear edge of the outer panel and the door frame (Figure 12–88).
10. Use a plasma cutter or cut-off grinder to remove the brazed portion from the outer panel door frame connections (Figure 12–89).
11. The quickest way to remove an exterior door panel is to grind off the edge of the hem flange (Figure 12–90). Only grind off enough metal so that the panel can be separated from the inner flange. Do not grind into the inner panel. Do not use a welding torch or power chisel to separate the panels. The inner panel can become distorted or be accidentally cut.
12. Separate the reinforcing strip on the top of the panel (if installed or used).

FIGURE 12-89 Cut off the old door panel without damaging the frame. *(Courtesy of 3M Automotive Trades Div.)*

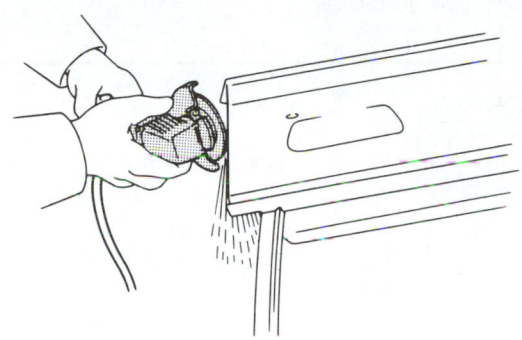

A

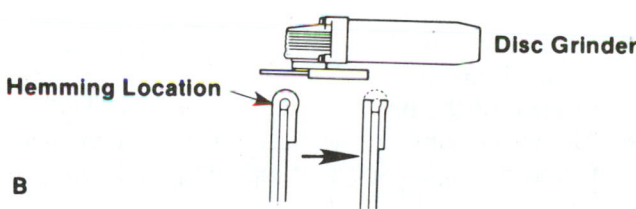

B

FIGURE 12-90 (A) Grind off door hemming flange; (B) Note the hem flange cross section before and after grinding. *(Courtesy of Toyota Motor Corp.)*

13. Using a hammer and chisel, loosen the two panels. Use a pair of tin snips to cut around any spot welds that could not be drilled or ground off (Figure 12–91). When the exterior panel moves freely, remove the panel. Use vise grips or pliers to remove what remains of the inner hem flange. Any remaining spot welds or brazing should be ground off using a disc grinder.

14. With the exterior panel removed, examine carefully the inner panel and frame construction for damage. If necessary, straighten or repair any remaining inner door damage at this time (Figure 12–92). Remove dents on the inner flange with a hammer and dolly. Media blast away any rust.

15. Apply weld-through primer to any areas to be spot welded. Cover other bare metal areas with a rust resistant primer or other manufacturer recommended rust treatment.

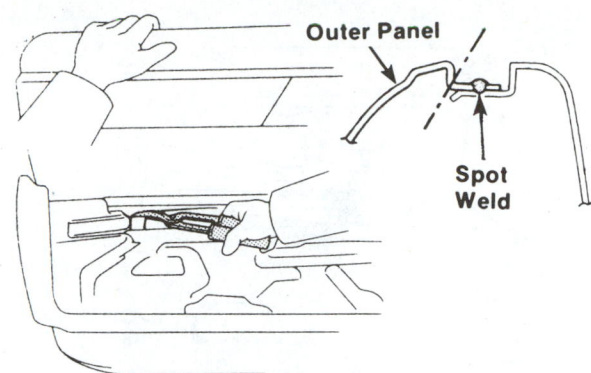

FIGURE 12-91 Cut around spot welds as needed. *(Courtesy of Toyota Motor Corp.)*

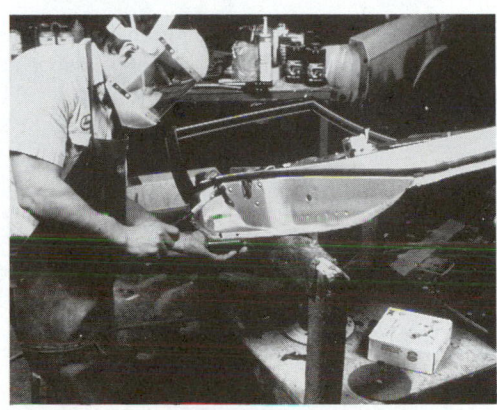

FIGURE 12-92 Make sure you wear proper safety gear when grinding and using other power tools. *(Courtesy of 3M Automotive Trades Div.)*

16. Prepare the new panel for installation. Drill holes for any plug welds. Using a sander, remove the paint from the weld and braze locations. Apply weld-through primer to the bare metal seam areas and prime any other bare metal areas.

17. Some outer door panels are accompanied by a silencer pad that must be glued to the outer panel. To do this, clean the outer panel with alcohol or suggested solvent. Heat the outer panel and silencer pad with a heat lamp. Then, glue the silencer pad to the outer panel.

18. Before installing the new panel, apply body sealer to the backside of the new panel. Apply the sealer evenly $3/8$ inch (9.5 mm) from the flange in a $1/8$-inch (3.2 mm) thick bead.

19. Using vise grips, attach the new outer panel to the door. Align it properly, using the dimensions determined in Step 2. Braze the outer panel where required.

20. Use a hammer and dolly to bend the outer panel flange (Figure 12–93). Cover the dolly face with cloth tape to avoid scarring the panel. Bend the hem gradually in three steps. Be careful not to tap the panel edge; that would throw the panel out of alignment. Do not create bulges or creases in the body lines of the outer panel.

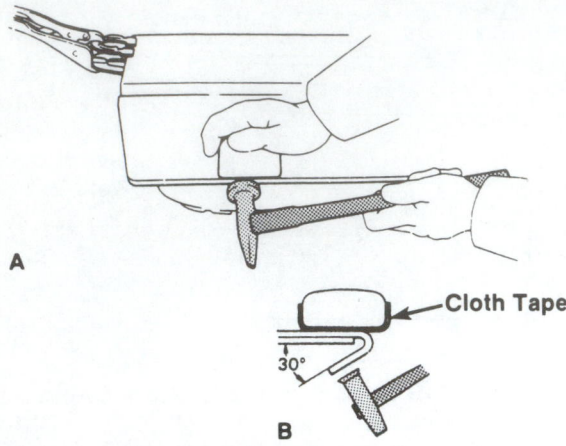

FIGURE 12-93 Carefully straighten the door frame flange after old panel removal. (A) Dollying panel flange and (B) cross section. *(Courtesy of Toyota Motor Corp.)*

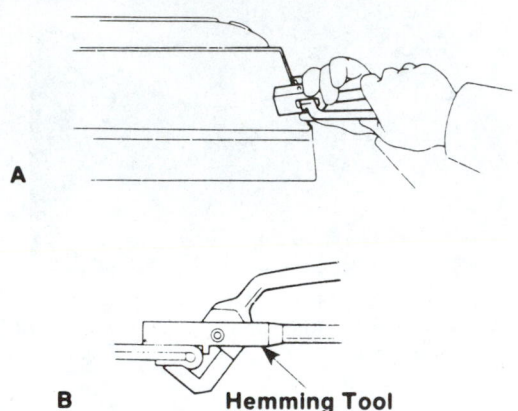

FIGURE 12-94 (A) Special tools are used when forming the panel hem and (B) cross section. *(Courtesy of Toyota Motor Corp.)*

21. After working the flange within 30 degrees of the inner panel, use a hemming tool to finish the hem (Figure 12–94). Again, finish the hem in three steps, being careful not to deform the panel.
22. Test fit the door in the body opening. Look for correct gaps and make sure the door is flush.
23. Weld the plug or spot weld locations of the glass opening and tack weld the hemming edge of the outer panel flange (Figure 12–95).
24. Apply body sealer to the hemming edge of the flange. Apply antirust agents on the inside of the spot welds, plug welds, and brazed areas.
25. Drill holes into the new panel for moldings, trim, and so forth.
26. At this point, you have two possible methods to choose from: 1) continue to prepare the panel for refinishing, or 2) place the door on the vehicle to check its alignment prior to final welding, prepare the door for refinishing, and then reinstall the door.

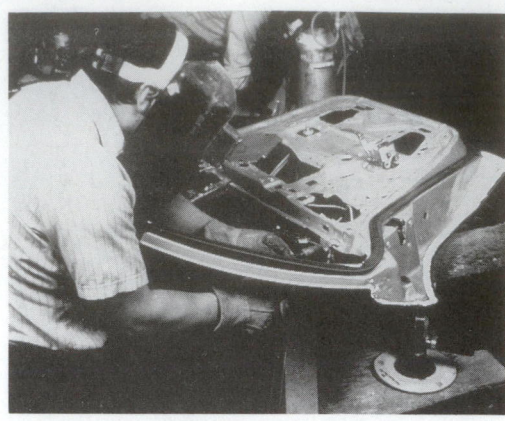

FIGURE 12-95 Measure the final weld door panel in place. Then reinstall the window and other parts. *(Courtesy of 3M Automotive Trades Div.)*

 **CAUTION** Wear leather gloves when working with door skins and other sharp sheet metal panels. Serious cuts can result from the jagged metal edges.

27. Install the door glass before refinishing the door. This will prevent overspray from getting on the interior of the door and avoid paint damage.
28. Align the door with all adjacent panels and check for proper closure or latching and panel gaps.

REPLACING PANELS WITH ADHESIVES

Some vehicle manufacturers use structural adhesives along certain weld seams. These two-part epoxy adhesives are sometimes called **weld-bond adhesives** because spot welds are placed through the adhesive.

Weld-bond adhesives are used to add strength and rigidity to the vehicle body. They also improve corrosion protection in weld seams and help control noise and vibrations.

Parts most commonly weld-bonded are

1. A- and B-pillars
2. Rocker panels
3. Roof panels
4. Rear quarter panels
5. SMC door panels

If adhesives are disturbed by repairs, they must be replaced. Follow recommendations in the body repair manual.

Some manufacturers use **structural adhesives** in place of welds. One example is around the wheel openings and sail panel reinforcement. This is a different type of adhesive than the weld-bond adhesive. Check the body repair manual for information on the use of structural adhesives.

To replace and repair fiberglass plastic door skins and other parts, see Chapter 14.

REPLACING BONDED DOOR PANELS

Bonded door panels are held to their frame with structural adhesive. The service of bonded door panels is similar to servicing other panels held in place with adhesives.

To begin removal of the damaged panel, mark the location of the old panel on the door frame. This will allow you to install the new panel in the right location on the frame. Cut out the center of the door panel. Use power shears or a cut-off wheel. With the center section of the door panel cut out, you will have better access to the inside hem flange. You can also inspect the door frame for damage. If the frame is damaged, you should consider purchasing a new or recycled door assembly.

Use a heat gun to soften the adhesive that holds the door panel to the door frame. As you heat the adhesive, it will soften so you can wedge a putty knife or chisel into the joint. Be careful not to damage the door frame when separating the door panel. Keep working around the flange with heat and your tool until the complete panel is removed.

Clean off the remaining old adhesive with your putty knife. A very thin layer of adhesive can remain on the frame as long as it is not thick enough to affect new panel installation.

Apply a liberal amount of new adhesive in a continuous bead around the frame flange. Place the new panel down into place. Align your reference marks so that the new panel is positioned properly. Use clamping pliers and light pressure to secure the panel in place as the adhesive cures. Allow the adhesive to cure the suggested time before moving the door.

12.6 PANEL ADJUSTMENTS

Manufacturers provide for adjustments of mechanically fastened panels in several ways. One of the most frequent is slotted or oversized holes. Because the openings are larger than the bolts, the panels can be moved up or down, forward or rearward, and in or out. Washers are provided under the bolt heads to provide a sufficient bearing surface on which the bolt is to be tightened.

Another common fastener that allows for adjustments is a caged plate (Figure 12–96). The *caged plate* is like a large nut with a heavy steel plate

threaded to accept two or more bolts. The plate is housed in a cage of thin sheet metal spot welded to the supporting panel (Figure 12–97). The cage is larger than the plate, so that the plate can be moved around, but the cage prevents the plate from falling away from the panel. Oversized holes in the panel allow the panel to be adjusted in any direction.

Caged plates are often used in doors and door pillars (Figure 12–98). In this application, the cluster of oversized bolt holes would weaken the panel without the reinforcement of the steel backing plate.

A shim is another means of making an adjustment. A *shim* is a thin U-shaped piece of metal for making part adjustments. By loosening a bolt, a shim can be slipped under the bolt head and around the bolt. When retightened, the position of the attached panel is changed. Shimming body panels was once a very common operation. But with the welded panels of today's unibody construction, there are few body panels that can be shimmed. They can sometimes be found on full frame vehicles.

Another means of adjusting body parts (hoods, trunk lids, and hatchback lids) is the adjustable stop.

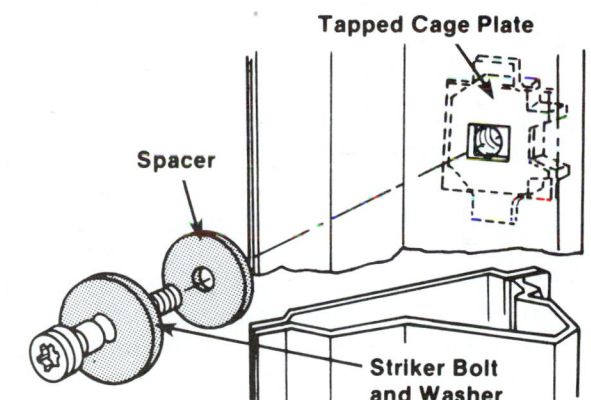

FIGURE 12-96 A caged plate permits striker adjustments and provides a strong mounting point.

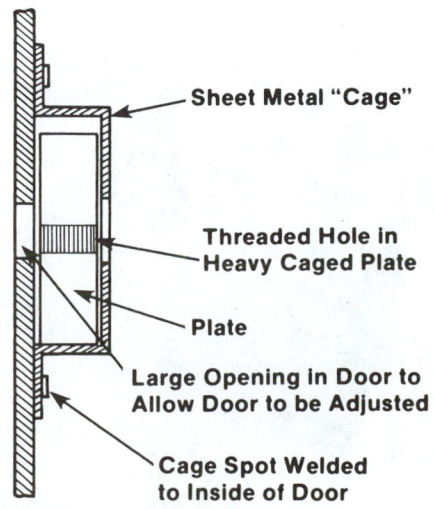

FIGURE 12-97 Study the caged plate cross section.

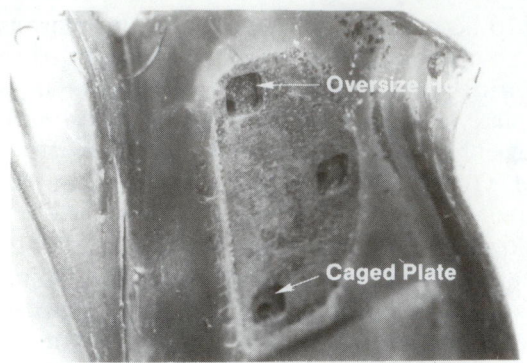

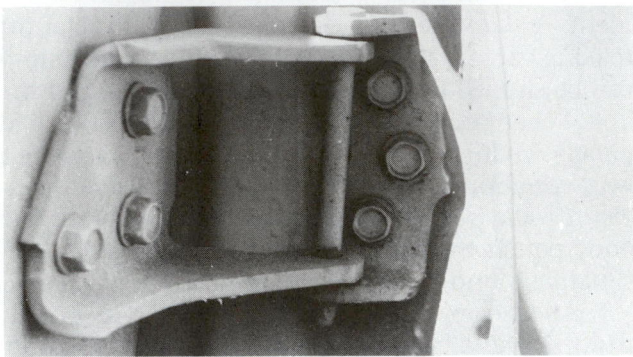

FIGURE 12-98 Caged plate connections allow door adjustments.

The *adjustable stop* is a threaded bolt with a rubber cap. It is usually attached to a body panel by a fixed nut in a clip. A locknut secures the stop in position. The adjustable stop may be completely rubber with a spiral groove to allow height adjustment. Adjustable stops are usually found under the hood. These stops bear against the hood to prevent it from vibrating and rattling on its hinges and to control the height of the hood. The rubber cap also prevents the stop from chipping the paint and denting the hood.

ADJUSTING DOORS

Door adjustment is needed so the doors will close easily, not rattle, and not leak water and dust. This section will describe various door adjustments and the servicing of the door glass.

Types of Doors

Two basic types of doors are used on today's vehicles: framed and hardtop (Figure 12–99). The *framed door* (sedan) uses an upper door frame structure that surrounds the glass. The *hardtop door* does not use a frame structure around the door glass. Both door types normally use one-piece glass. However, one new model car uses two-piece door glass.

The glass mechanism on framed doors is rather simple because the upper frame serves as a guide and support for the door glass. Hardtop doors do not have this advantage, and glass generally rests against the top door opening. A soft rubber gasket is used in the door opening to protect the glass when it is fully raised or closed. The glass must have some means of support and height control when it is lowered or raised.

The hardtop glass must also be properly tilted to make contact with the upper gasket. If it does not, the glass will leak water and dust. If it is tilted too far

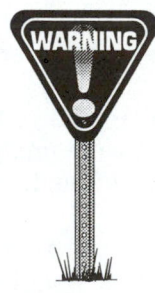

WARNING Hardtop door glass is prone to damage if it is slammed against the roof drip rail when the door is closed. Make sure the door glass is down or rough adjusted before closing the door. Never slam it until all adjustments are completed.

A

B

FIGURE 12-99 Note the differences in (A) a framed door and (B) a hardtop door.

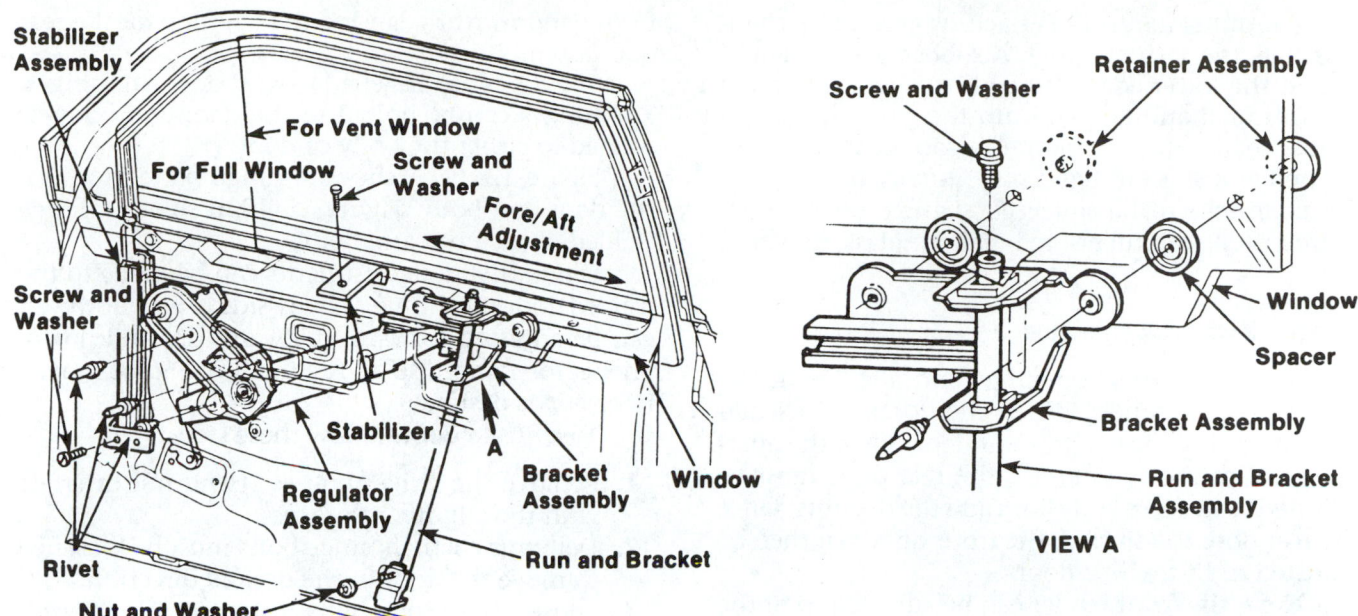

FIGURE 12-100 Study the construction of the inside of a door.

in, the door will be hard to close and the gasket can be damaged. By far the framed door requires the least amount of adjustments for the door glass.

Various methods are used to attach the door glass to the window lifting mechanism:

- Bolt-through method
- Adhesive or bonding method
- Sash channel method

The bolt-through method utilizes bolts or rivets that have plastic or rubber gaskets to prevent direct contact with the glass. The fasteners pass through the window and secure it to the lift channel or bracket. Rubber spacers separate the glass from the bracket and fasteners (Figure 12–100). The bolts are used to attach the lift channel or brackets to the glass. The bolts are inserted through the glass.

The use of adhesives is another method of securing the lower lift bracket to the glass. Usually a U-channel, with insulator stays, is used to prevent the glass from contacting the metal channel (Figure 12–101).

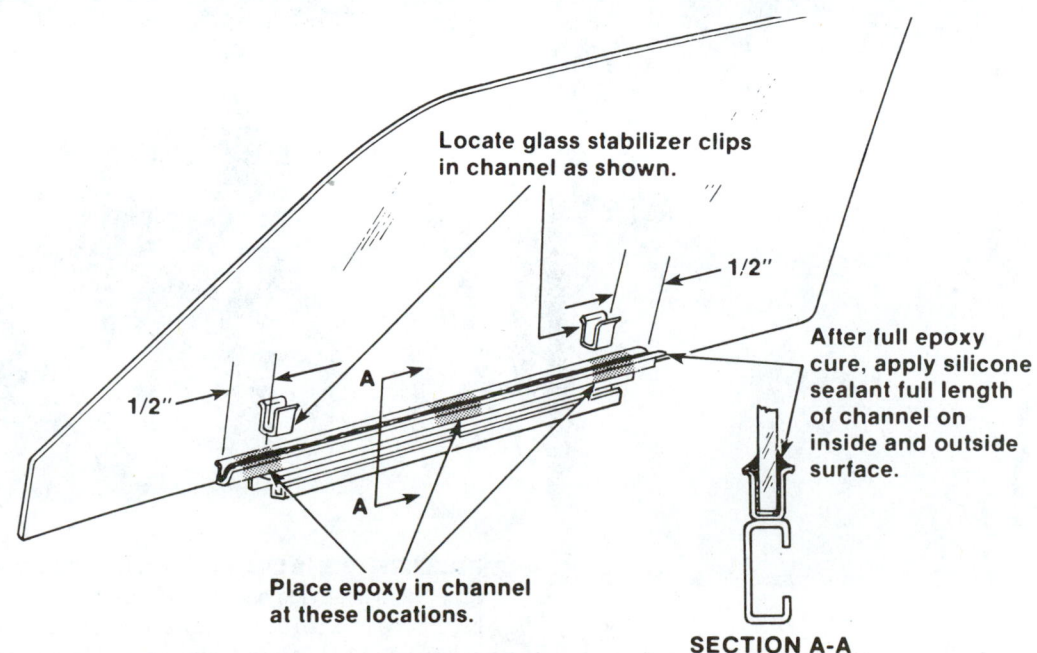

FIGURE 12-101 This glass is adhesive-bonded glass in a bracket.

The oldest method of attaching glass to lift channels uses the sash channel. A rubber seal or tape is put on the lower edge of the glass. Then, a channel is positioned and tapped onto the glass by using a rubber mallet. If the channel is too loose, tape can be used as a shim to tighten it. Usually the edges of the channel can be squeezed slightly for an even tighter fit. Be careful not to break the glass, however.

Door Adjustments

Doors must fit their openings and align with the adjacent body panels. When the doors on a sedan need adjusting, start at the rear door. Since the quarter panel cannot be moved, the rear door must be adjusted to fit these body lines and the opening. Once the rear door is adjusted, the front door can then be adjusted to fit the rear door.

Next, the front fender can be adjusted to fit the door. On hardtop models, the windows can then be adjusted to fit the weatherstripping. The windows are usually adjusted starting with the front and working toward the back. The front is adjusted to fit the front door pillar, and the front window is then adjusted to it. The rear door window is adjusted to the front window rear edge and the opening for the rear door assembly.

The doors are attached to the body with hinges. The hinges can be bolted to the door and body or welded to either the body or door.

Figure 12–102A shows a hinge bolted to both the door and body. Figure 12–102B shows a hinge welded to the door and body.

Obviously, no adjustments can be made to the welded door side hinge. A body side hinge, however, can be adjusted forward, rearward, up, and down. The use of shims also allows the hinge to be moved in or out as desired.

To adjust a door, follow these steps:

1. Remove the striker bolt so it will not interfere with the alignment process.
2. Determine which hinge bolts must be loosened to move the door in the desired direction.
3. Loosen the hinge bolts just enough to permit movement of the door with a padded pry bar or jack and wooden block. Figure 12–103 shows a specially designed pry bar being used to adjust a

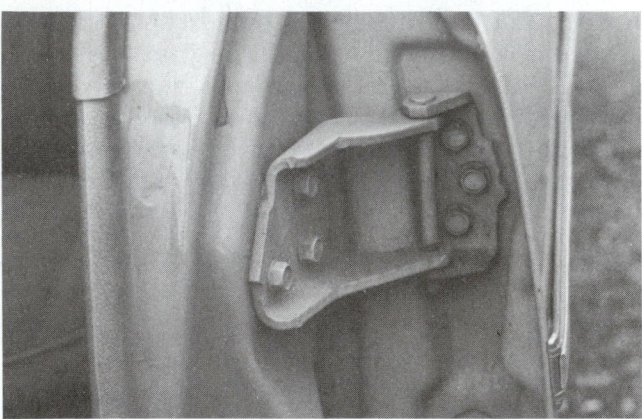

A

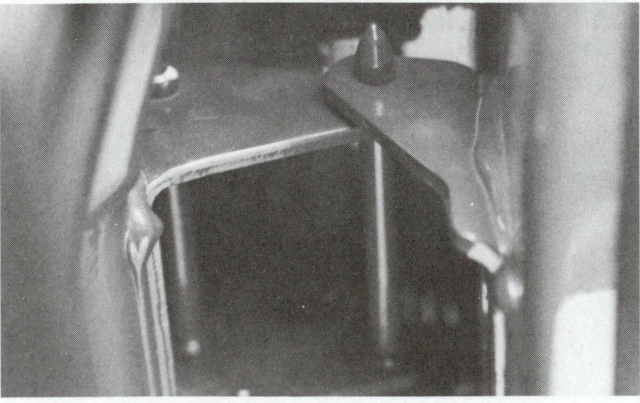

B

FIGURE 12-102 Compare (A) the bolted door hinge and (B) the welded door hinge.

A

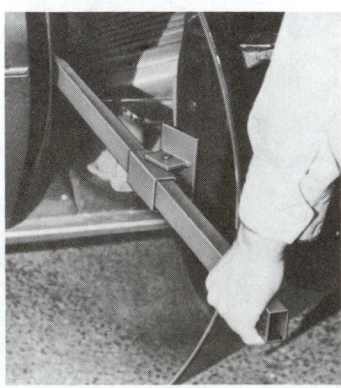

B

FIGURE 12-103 (A) A handy holding mechanism for a door has been made out of two scissor jacks. (B) An adjusting bar makes door adjustments easier. You can pry the door up or down to make minor adjustments. Do not overflex the door mounting hardware or serious damage could result.

door. The end of the bar hooks over the striker bar and a U-shaped bracket engages the latch. On some vehicles, a special wrench must be used to loosen and tighten the bolts.

4. Move the door as needed. Tighten the hinge bolts. Then check the door fit to be sure there is no bind or interference with the adjacent panel.

5. Repeat the operation until the desired fit is obtained.

6. Install the striker bolt and adjust it so the door closes smoothly and flush with rear door or quarter panel. Check that the door is in the fully latched position and not in the safety latch position.

7. On all hardtop models, the door and quarter glass must be checked to assure proper alignment to the roof rail and weatherstripping.

Worn door hinges will have play that allows up and down movement of the rear of the door. If the hinge pins are worn out, you should replace the hinges. Some hinges use bushings around the hinge pins. When these bushings are worn out, replace them. This will retighten the pin in the hinges and also readjust the door to a certain extent. Make sure replacement hinge bushings are available.

If the hinges are to be removed, scribe a line around the hinge to mark its position on the body and door. This will simplify reinstallation and positioning of the new hinge. You might have to loosen the fender at the rear bottom edge to reach the hinge bolts.

In-and-out adjustments are also very important.

The door must fit the opening and be aligned in and out to fit the body panels. The door must also provide a good seal between the weatherstripping and the body opening. The weatherstripping must be compressed sufficiently in the opening to prevent water, dust, drafts, and wind noises from entering the automobile.

Care must be taken when adjusting the in-and-out movement of the door. If the door is moved out on the top hinge, it will not only affect the top of the door but it will also move the opposite bottom corner in. If the bottom of the door is moved in on the hinge, it will move the top opposite corner out. But if the door is moved in or out equally on both hinges, it will only affect the front of the door because the amount of adjustment decreases toward the back of the door. The center door post, striker bolt, and lock will determine the position of the door at the location. The front leading edge of the door should always be slightly in on the front edge from the rear of the other panel (usually the front fender). This will help stop wind noises at the leading edge of the door panel. If the front edge is sticking out, wind noise will annoy the car owner.

Another type of hinge used in some compact models is the welded-on type that has no adjustment provisions (Figure 12–104). A pin is provided to remove the door for servicing of the hinges. The half of the hinge that is to be installed on the door is predrilled to permit a bolt-on installation with tapped caged plates and bolts. The half of the hinge on the hinge pillar must be rewelded on the pillar when it

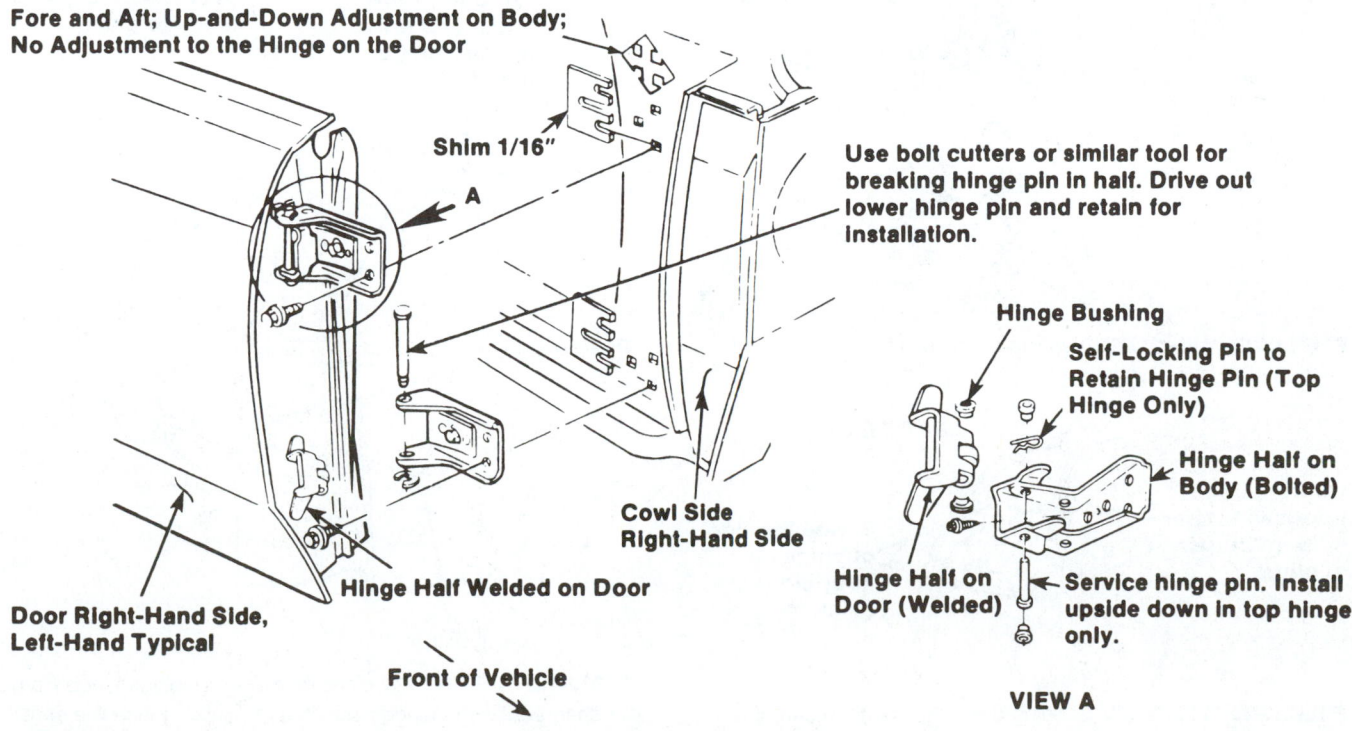

Fore and Aft; Up-and-Down Adjustment on Body; No Adjustment to the Hinge on the Door

Shim 1/16″

A

Use bolt cutters or similar tool for breaking hinge pin in half. Drive out lower hinge pin and retain for installation.

Cowl Side Right-Hand Side

Hinge Half Welded on Door

Door Right-Hand Side, Left-Hand Typical

Front of Vehicle

Hinge Bushing

Self-Locking Pin to Retain Hinge Pin (Top Hinge Only)

Hinge Half on Body (Bolted)

Hinge Half on Door (Welded)

Service hinge pin. Install upside down in top hinge only.

VIEW A

FIGURE 12–104 Study adjustments to a welded door hinge.

SHOP TALK

The striker plate is not adjusted properly if the door rises or it is forced down when the door is closed. The striker should slide and engage smoothly into the latch when the door is closed. The striker can be moved up and down, in and out, and back and forth.

is replaced.

When removing the door hinge pins, use a special spring compressing tool similar to that shown in Figure 12–105. The spring must be seated properly in the tool before compressing it. Otherwise, the spring could slip and cause damage or personal injury. The pin is then removed in each hinge and the door can then be removed from the vehicle.

To replace the welded door side hinge, first scribe the outline of the hinge on the door. Then center punch the spot welds and drill a $^1/_8$-inch (3.2 mm) pilot hole completely through the welds (Figure 12–106). This is then drilled out to $^1/_2$ inch (12 mm), but only deep enough to penetrate the hinge base to

release the hinge from the panel. A chisel is then driven between the hinge and the base to break it free from the panel.

The new part is installed on the door by drilling recommended size holes into the attaching holes. The holes will provide a slight amount of adjustment on the door assembly since the bolts are often smaller than the holes.

To remove the body side hinge, scribe the hinge position as shown in Figure 12–107. Measure $1^3/_4$ inches (44 mm) from the forward edge and scribe a mark above and below the hinge. Drill a slight depression in the $1^3/_4$ inch (44 mm) scribe mark, using a $^1/_8$-inch drill bit. Do not drill completely through the pillar. Drill another shallow $^1/_8$-inch (3.2 mm) hole at the corners of the tabs to mark the hinge position. Then use a cutting torch to cut the tabs holding the hinge together. The door sill plate and carpet should be removed or covered with an asbestos sheet to protect them from the hot slag of the cutting operation.

The welds holding the separated hinge tabs are then twisted or rotated to break them with a suitable

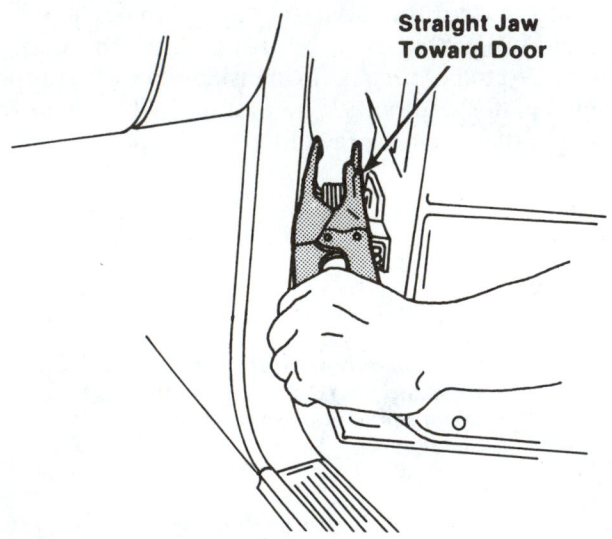

FIGURE 12-105 A spring compressing tool might be needed with some hinges.

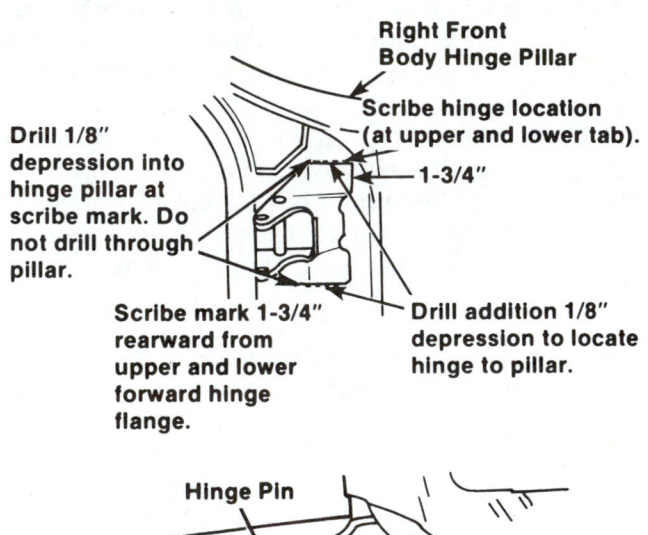

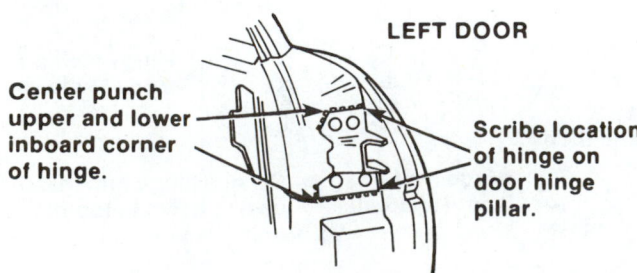

FIGURE 12-106 Scribing the hinge location can help rough adjust a door when reinstalling it.

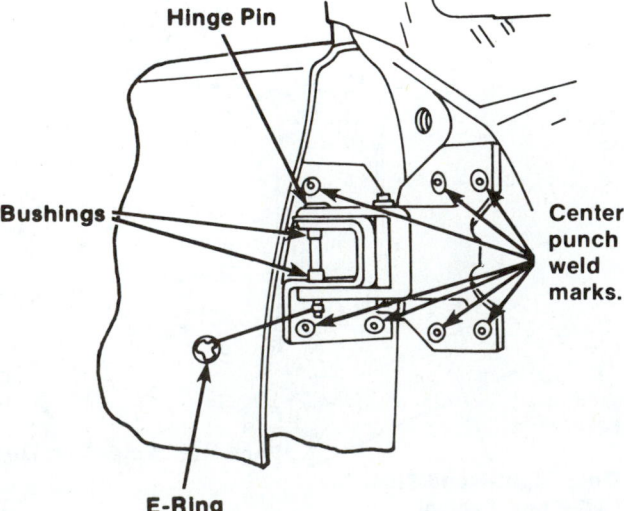

FIGURE 12-107 Before removing a welded hinge from the door, scribe the hinge position and center punch the spot welds.

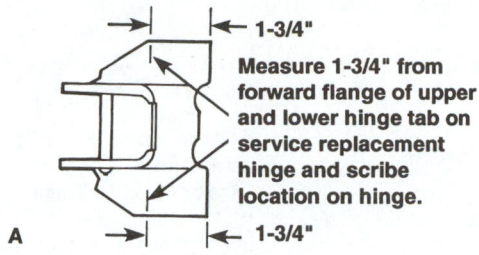

Measure 1-3/4" from forward flange of upper and lower hinge tab on service replacement hinge and scribe location on hinge.

A

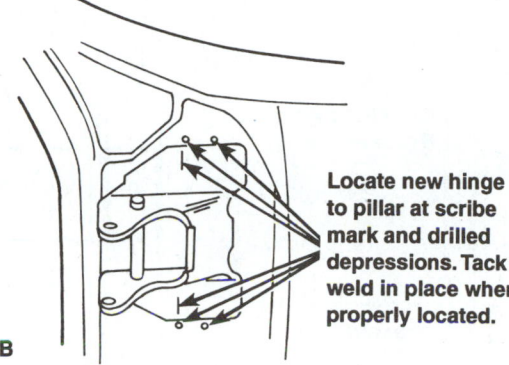

Locate new hinge to pillar at scribe mark and drilled depressions. Tack weld in place when properly located.

B

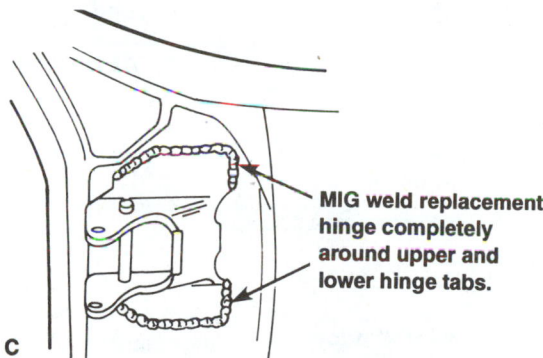

MIG weld replacement hinge completely around upper and lower hinge tabs.

C

FIGURE 12-108 (A) Measure the hinge location; (B) transfer these measurements to the new hinge; and (C) weld the hinge tabs.

tool such as grip-type pliers. Once the tabs are removed, the pillar is then ground smooth and prepared to receive the new part.

To install the new hinge strap, the measurements, as shown in Figure 12–108A, must be transferred to the new part. The new part is then lined up to the scribed marks on the hinge pillar as shown in Figure 12–108B. It is tack welded carefully in place and then the door is rehung and pins installed to check the fit of the door to the opening and surrounding panels. If it fits properly, the door is removed and the hinge is then welded completely around the upper and lower hinge tabs (Figure 12–108C). The area is cleaned properly and a paintable sealer is applied around the perimeter of the hinge. The area is then refinished to the proper color before the door is reinstalled.

Door Window Alignment and Service

On some vehicle doors, channels are used to control the forward and rearward window adjustments. Some vehicle windows utilize adjustable guide rollers in a movable channel. Still others have a center lift guide and are adjustable forward and rearward as well as in and out by tilting the glass. These adjustments are controlled at the bracket that is attached to the lower sash channel or where the guide attaches to the inner door panel.

If the window binds or is stiff, check the channel or add lubricant to the glass runs or guide channels. A door window that tips forward (or rearward) and that binds can be caused by improper adjustment of the lower sash brackets, a loose channel or cam roller, or a channel that is out of adjustment.

On some sedan doors, a full or partial length rubber glass run or channel is used (Figure 12–109). If the channels are too tight or lack proper lubricant, the glass or rubber will bind. To free up the glass, a dry silicone spray should be applied to the glass run.

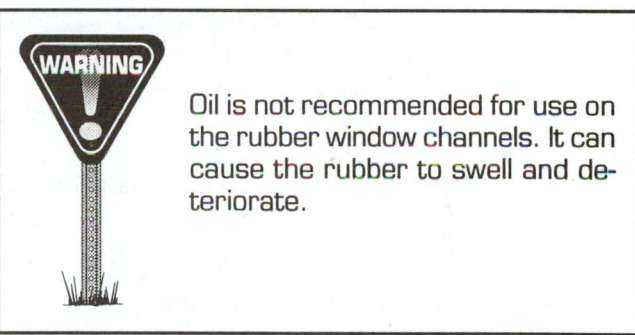

WARNING

Oil is not recommended for use on the rubber window channels. It can cause the rubber to swell and deteriorate.

Vehicle doors that use a full trim panel sometimes have a set of brackets at the top of the door. The trim panel is attached to these brackets. If they are set too far inward, the window glass will bind. Other items such as antirattle slides or other devices can cause the window glass to bind when raised. If not set correctly, they will cause binding.

If the door glass has to be adjusted to align with the edge of the quarter glass, be sure to check for proper quarter glass adjustment. Some of the quarter glasses are movable and some are stationary. To adjust or remove a quarter glass, some of the interior trim, such as the rear seat cushion, back rest, inner trim panel, and water shield, might have to be removed to gain access to the attaching mechanism. On some types of stationary or swing-out glass, it is not necessary to remove the seats. The stationary glass can be adjusted by loosening the retainer bolts or screws and shifting the glass to align it with the door glass. The manufacturer's specifications or procedures should be consulted for specific details.

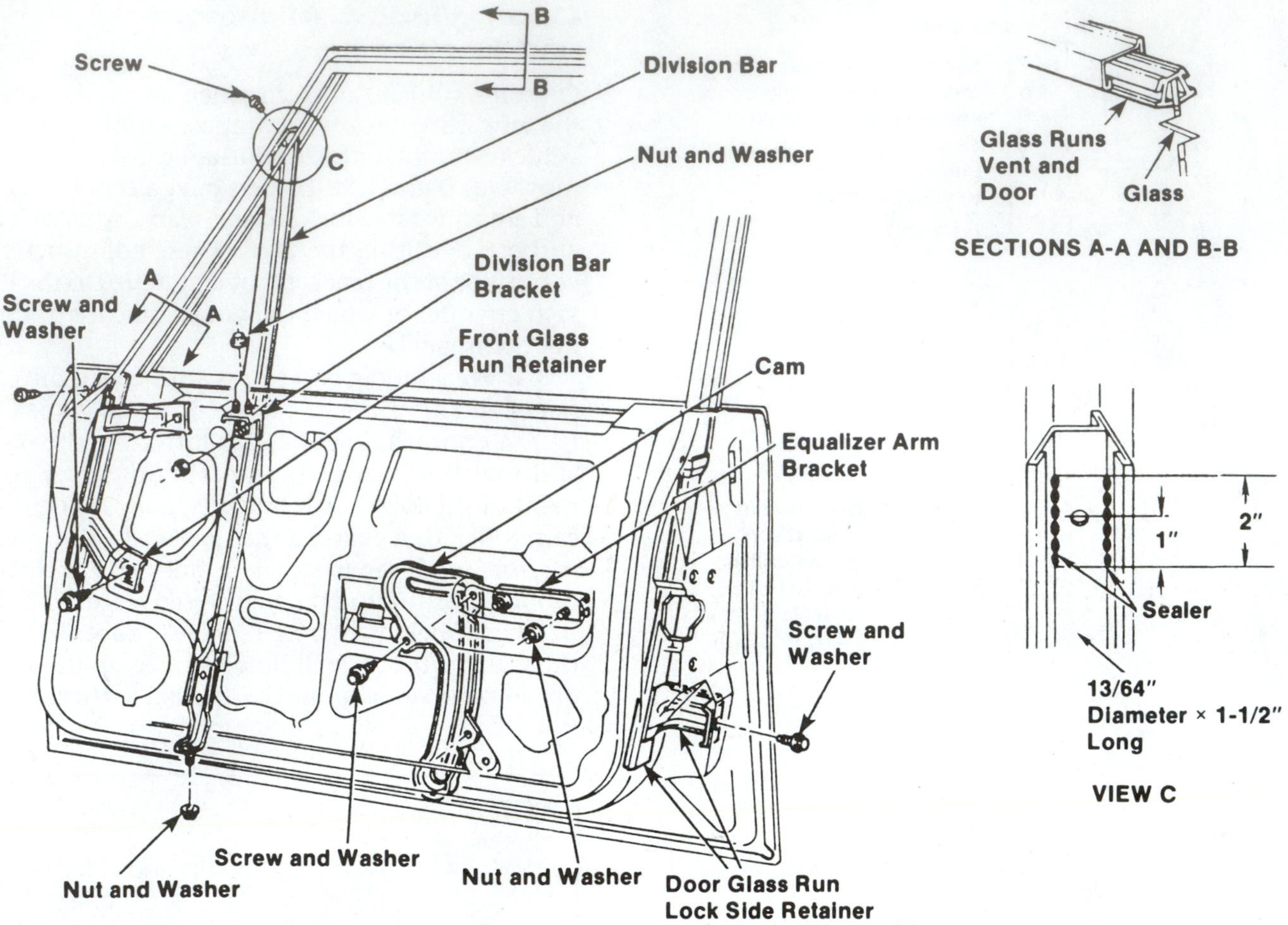

Screw — Division Bar

Nut and Washer

Screw and Washer

A — A

Division Bar Bracket

Front Glass Run Retainer

Cam

Equalizer Arm Bracket

Screw and Washer

Nut and Washer

Screw and Washer

Nut and Washer

Door Glass Run Lock Side Retainer

Glass Runs Vent and Door

Glass

SECTIONS A-A AND B-B

2"

1"

Sealer

13/64" Diameter × 1-1/2" Long

VIEW C

FIGURE 12-109 Glass runs guide glass as it slides open and closed. They are sometimes adjustable.

Operation of the movable quarter glass is the same as that for door glass. Stops are used to control the up and down movements. The up-stop controls the up and forward movement of the glass. This stop must be set to obtain the proper spacing between the door and quarter glass. Some vehicles use two up-stops—one for the front and one for the rear sections. Just like the door glass, the quarter glass is provided with adjustments for inward and outward tilt of the glass. Consult the manufacturer's service manual or make a careful inspection of what is necessary to remove the quarter glass.

To replace the vent window, it must be aligned as shown in Figure 12–110. An awl or dowel can be used to align the holes in the door frame and vent window. Then reinstall the mounting screws and torque them to the manufacturer's specifications. Replace the door glass run in the division bar.

Manual and Power Regulators

The **window regulator** is a gear mechanism that allows you to raise and lower the door glass. Regulators can be manual or powered electrically. Both types of regulators are very similar, the only

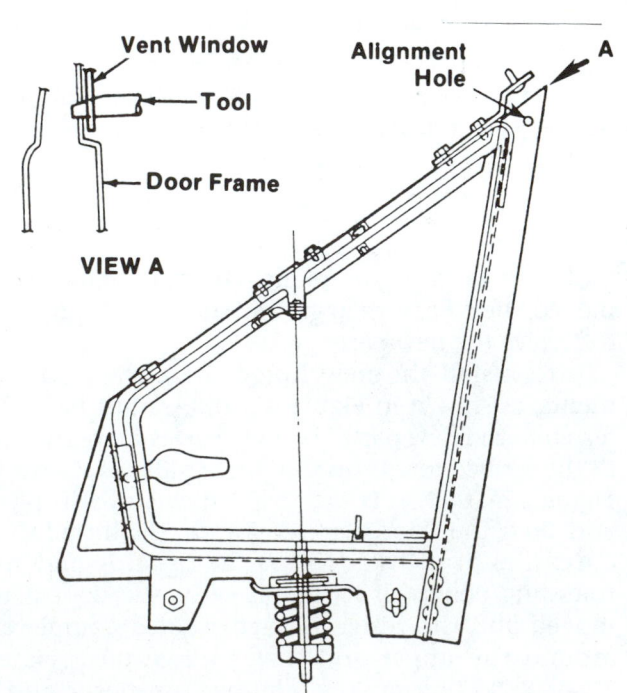

Vent Window

Tool

Door Frame

VIEW A

Alignment Hole

A

FIGURE 12-110 Vent window alignment is needed on some vehicles, like pickup trucks.

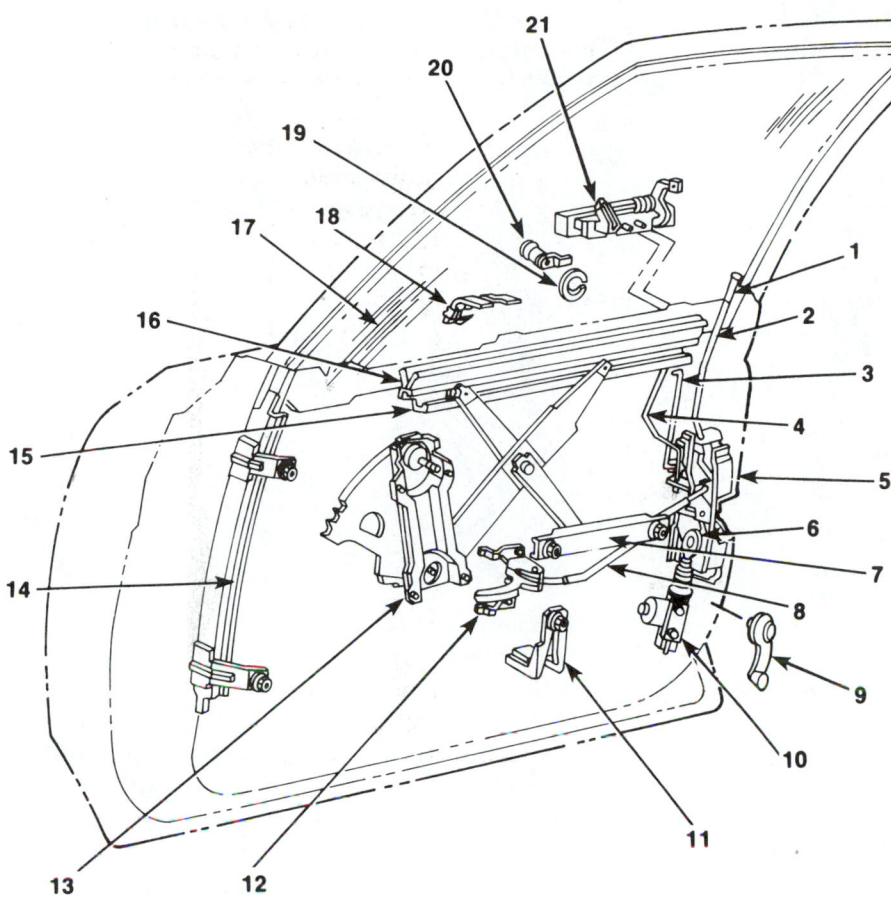

1. **Inside Locking Rod Knob**
2. **Inside Locking Rod**
3. **Lock Cylinder to Lock Connecting Rod**
4. **Outside Handle to Lock Connecting Rod**
5. **Door Lock**
6. **Inside Locking Rod to Electric Actuator Connecting Rod**
7. **Inner Panel Cam**
8. **Inside Remote Handle to Lock Connecting Rod**
9. **Manual Window Regulator Handle**
10. **Power Door Lock Actuator**
11. **Down-Travel Stop**
12. **Inside Remote Handle**
13. **Manual Window Regulator**
14. **Glass Run Channel Retainer**
15. **Lower Sash Channel Cam**
16. **Lower Sash Channel**
17. **Window Glass**
18. **Lock Cylinder Retainer**
19. **Lock Cylinder Gasket**
20. **Lock Cylinder Assembly**
21. **Outside Handle Assembly**

FIGURE 12–111 Study the X-design window regulator.

difference being the handle crank mechanism on manual regulators and electromotor-driven gear mechanism on powered regulators. The lift arms are the same for both types.

One or two lift arms can be used depending on the make of the vehicle. If two lift arms are used, it is usually referred to as an *X-type regulator* (Figure 12–111). The X design uses an auxiliary arm that is mounted into a cam or stabilizer channel that is adjustable. The cam adjustments allow the glass to be tilted or rocked so that it can be raised in a parallel position.

The regulator and its associated parts are sometimes riveted to the door structure in lieu of being bolted. In this case, drill out the rivets in accordance with good shop practice and reinstall the necessary parts using the appropriate rivet gun and rivets.

For those window regulators that are spot welded to the inner door panel, use a spot welder cutter to drill out the welds. If necessary, use a chisel between the regulator and inner panel to separate the two structures. Generally, the replacement regulator is reinstalled with bolts.

If the regulator is to be removed without removing the glass, secure the glass in an up position to prevent it from dropping inside the door shell. Heavy cloth tape or a wedge can be used for this purpose. Always consult the manufacturer's manual for the proper removal and installation procedures for the regulator.

Power regulators require additional installation care because of the use of counterbalance springs in their design (Figure 12–112). On some vehicle models, the counterbalance spring must be released before servicing the regulator motor or other associated parts.

- Move the window to position the regulator under an access hole in the inner door panel. Clamp the regulator to the panel. Use a C-clamp.
- Using a special tool, release the spring tension. After servicing the motor or part, the spring can be reinstalled in its original position. Removing the spring avoids possible hand injury and damage to the door.

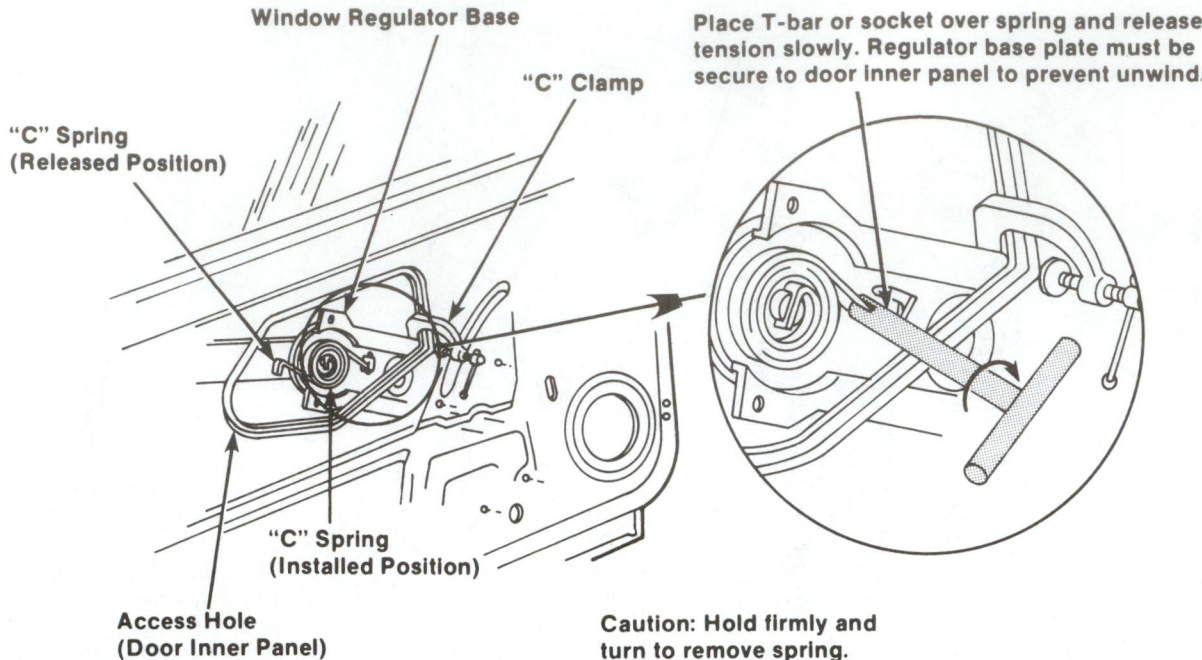

Window Regulator Base

"C" Clamp

"C" Spring
(Released Position)

Place T-bar or socket over spring and release
tension slowly. Regulator base plate must be
secure to door inner panel to prevent unwind.

"C" Spring
(Installed Position)

Access Hole
(Door Inner Panel)

Caution: Hold firmly and
turn to remove spring.

FIGURE 12-112 A special tool may be needed to free the counterbalance spring with some window regulators.

- Secure the regulator to the inner door panel by drilling a hole through the regulator gear and back plate. Insert a screw or bolt through the hole. Use a nut to lock the assemblies together. Be sure to remove the screw or bolt after servicing the regulator.

FENDER ADJUSTMENTS

Fenders are bolted to the radiator core support, the inner fender panel in the engine compartment, and the cowl behind the door and under the car. When these bolts are loosened the fender can be moved for adjustment (Figure 12–113).

The curvature of the fender must match the shape of the front door edge. Sometimes, a mounting bolt is provided at the center rear of the fender. It can be tightened when you have the correct curvature. If not, you will need to adjust the position of the upper and lower rear mounting holes (up and down) so the fender matches the door.

The fender-to-door alignment can be made by shimming the two large bolts attaching the fender

to the cowl. The top bolt is usually in the door pillar. The bottom bolt is either in the hinge post or under the car in the rocker panel. By shimming the top bolt, the upper fender can be moved out. Shimming the lower bolt will move the lower portion out. If

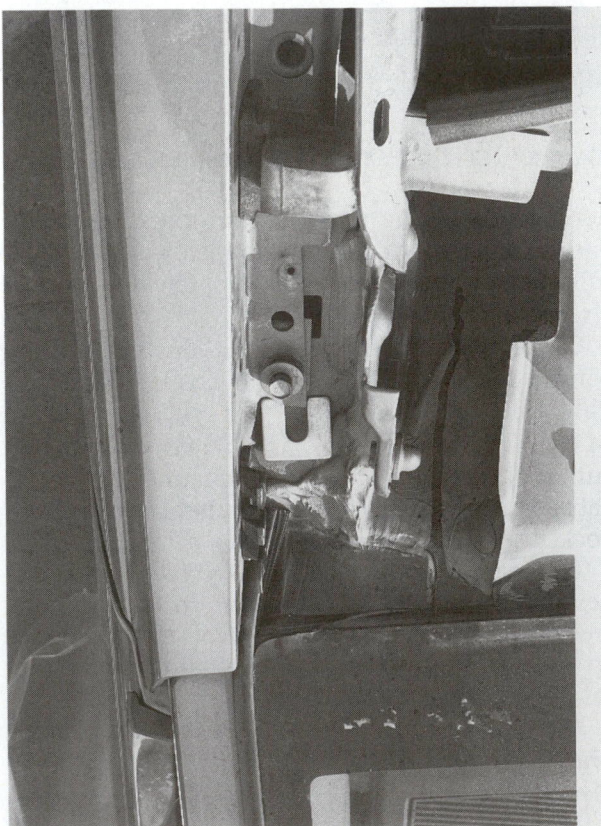

FIGURE 12-113 Minor adjustments of fenders can be made by using metal shims. *(Courtesy of Mitchell International).*

SHOP TALK The screws holding the splash guard to the fender might also have to be loosened for fender adjustment.

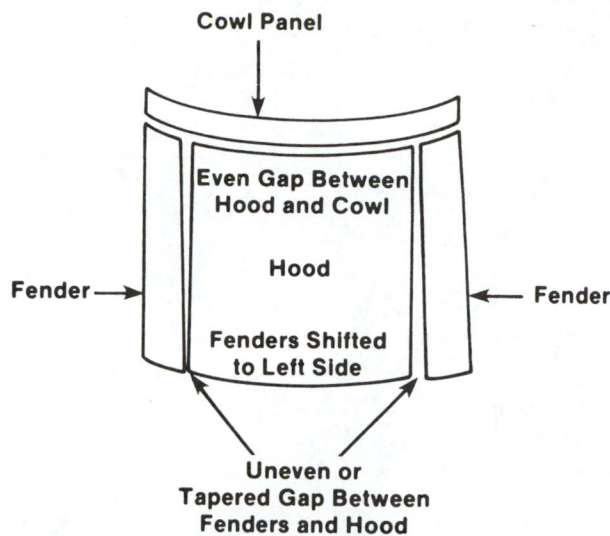

FIGURE 12-114 Note the uneven hood gaps around the hood. Fenders would have to be adjusted for equal gap.

FIGURE 12-115 Double-check all gaps around fenders, doors, and hood before releasing the vehicle.

 Make sure there is enough gap at the back edge of the hood to clear the cowl panel (Figure 12-114). Before opening the hood, check the clearance at the back of the hood. A common mistake is to have the hood hit the cowl, chipping paint or denting the cowl (Figure 12-115).

the fender is in too far and not flush with the door, the protruding door edge will cause noisy wind turbulence when the vehicle is in motion.

These adjustments allow the fender, hood, and door to be properly aligned. Often the fender and hood adjustments must be made simultaneously to achieve a pleasing result. The gap between fender

FIGURE 12-116 When removing the hood, place a towel or fender covers under the hood to prevent it from damaging paint on the cowl. *(Courtesy of Mitchell International)*

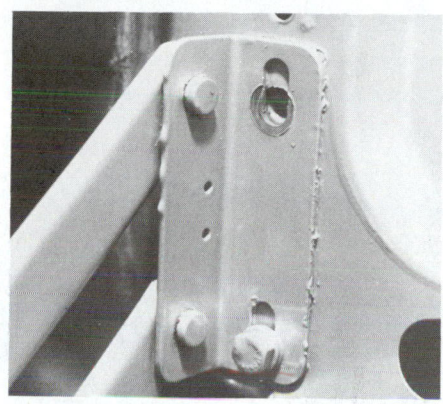

FIGURE 12-117 If the hood is to be reinstalled, scribe mark its location to simplify reassembly. Marks of hinges and bolts can also be used to realign parts for a rough, initial adjustment.

and hood should be no more than $3/16$ inch (4.8 mm). The door-to-fender gap should be no more than $3/16$ inch (4.8 mm) also. The front of the fender and the hood should be aligned as well. The result will be even spacing all around the fenders and hood.

Complete alignment can be achieved on many fenders without using any shims. Shims should be used only if alignment cannot be achieved without them.

HOOD ADJUSTMENTS

The hood is the largest adjustable panel on most vehicles. It can be adjusted at the hinges, at the adjustable stops, and at the hood latch. The adjustments allow the hood to be moved up and down, forward and rearward to align it with the fenders and cowl. The hood should align with the cowl and fenders with an equal gap of approximately $5/32$ inch (4 mm) between them. The front edge of the hood should also be even with the front edge of the fender.

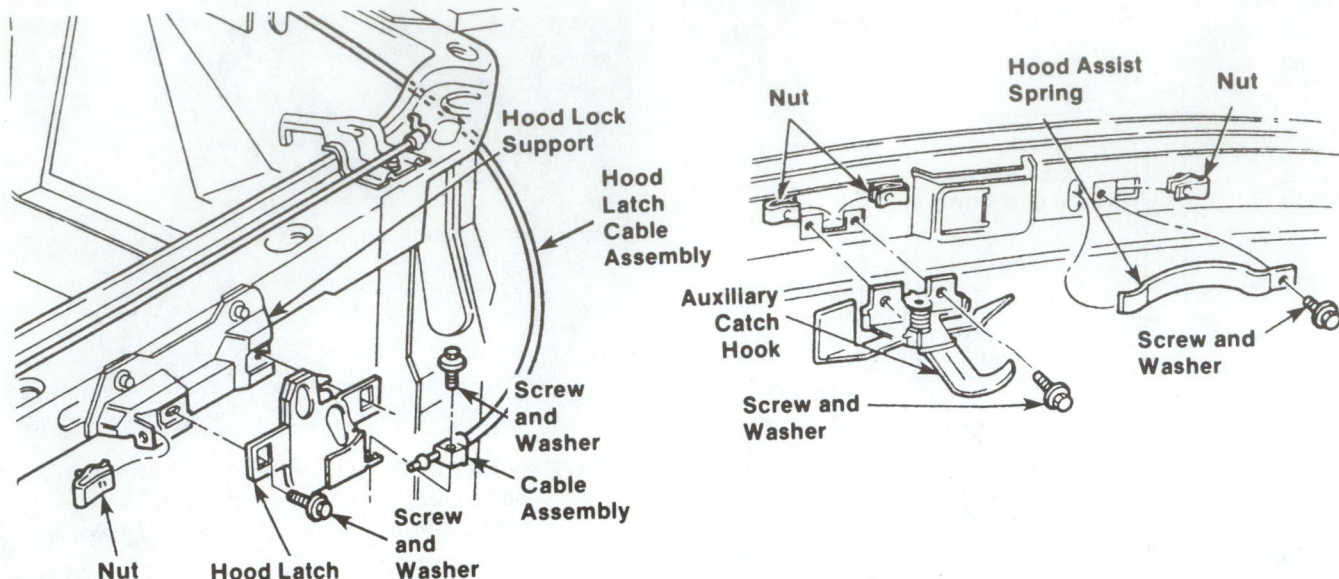

FIGURE 12-118 Note the mounting and construction of hood latches.

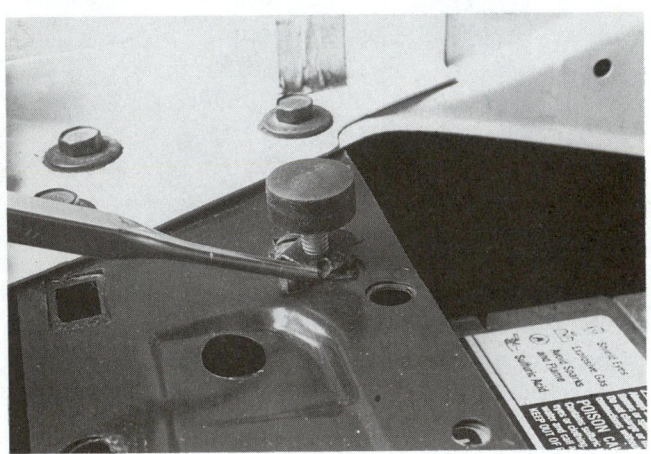

FIGURE 12-119 Rubber stops are often used to adjust the height of the hood. They can be screwed up and down to contact the hood and keep it from vibrating. The hood latch must be locked while stops are being set.

Many technicians remove the hood latch or leave it off while adjusting the position of the hood. If in place, the latch could kick the hood sideways and give a false impression of proper alignment. After positioning the hood to the cowl and fenders, install and adjust the latch.

Hood hinges bolted to either the cowl or the inner fender (Figure 12–116) hold the rear of the hood to the vehicle. The holes in the hinges are slotted to allow the hinges to be adjusted (Figure 12–117). They can be raised or lowered on the cowl or fender and moved forward or backward.

The front of the hood is held in place by a *hood latch* (Figure 12–118). The latch is used to secure the front of the hood so that it latches tightly and aligns with the fenders. Slotted holes are usually provided on the latch for alignment purposes. A safety catch

keeps the hood from flying open while driving if the latch should accidentally open.

Most vehicles have adjustable *hood bumper stops* mounted on the radiator support or along the inner edges of the fenders. These bumper stops provide up and down adjustment points to set the hood height in relation to the fenders. They also prevent hood flutter when driving. They determine the position of the hood when it is closed (Figure 12–119).

The *hood safety catch* prevents the hood from flying off should it accidentally open when the vehicle is in motion. It is simply a spring-loaded hook or lever that engages the hood.

Hood-to-Hinge Adjustment

To adjust a hood, slightly loosen the bolts attaching the hood to the hinges. Keep them tight enough to hold the hood during adjustment but loose enough to allow you to shift the hood.

Close the hood and line it up properly. Shift it by hand until the gap around all sides of the hood are equal. Carefully raise the hood far enough for another technician to tighten the bolts. The front of the hood must align with the front of the fenders and any panel in front of the hood. There must be sufficient clearance between the hood and cowl to allow the hood to be raised without rubbing the cowl.

If you cannot obtain the right clearance between the fenders and hood, the fenders may be out of adjustment.

Hood Height Adjustments

To correct the alignment of the hood up and down at the rear, slightly loosen the bolts holding the

SHOP TALK In order to lower the hood, it may be necessary to turn down the adjustable stops under the rear corners of the hood. If the hood is lowered without lowering the stops, the hood might bend and crease. Additional bodywork will be required to fix the damage.

hinges to the fenders or cowl. Then slowly close the hood and raise or lower its back side as necessary. When the back of the hood is level with the adjacent fenders and cowl, slowly raise the hood and tighten the bolts.

Once the rear of the hood is adjusted to the correct height, the adjustable stops must be checked. The rear stops must be adjusted to bear against the hood. This eliminates hood movement and rattle. The front stops control the height of the front of the hood. Turn the stops in or out until the front of the hood is even with the top of the fenders. Be sure to retighten the locknut on the stop after adjustment.

Hood Latch Adjustments

After making height and position adjustments, test the hood for proper latching. Slowly lower the hood and make sure it engages the latch in the center. If the hood must be slammed excessively hard to engage the latch, the latch should be raised. If the hood

does not contact the front stoppers when latched, the latch should be lowered. To adjust the hood latch, do the following:

1. Remove the hood latch assembly from the radiator support and lower the hood.
2. Check that all the gaps around the hood are properly aligned.
3. Reinstall the hood latch and lower the hood until it engages or contacts the first latch (auxiliary latch or safety catch).
4. Attempt to raise the hood. If it does open, adjust the safety catch so that it engages. Sometimes the hook can be shifted or bent until the auxiliary latch catches.
5. Lower the hood slowly. Check to see if it shifts to one side or the other when it is locked. The striker bar bolted to the hood should be centered in the "U" of the latch. When the hood is latched, it should be even with the surrounding sheet metal and fit tightly.
6. Loosen the hood latch just enough to maintain a tight fit but with enough give for you to move the latch.
7. Move the latch from side to side to align it with the hood latch hook. Move the latch up or down as required to obtain a flush fit between the top of the hood and the fenders when an upward pressure is applied to the front of the hood.
8. Tighten the hood latch attaching hardware.
9. Open the hood and double-check its action.
10. Close the hood. Make sure it is still at the same height as the fenders. If necessary, again adjust the bumper stops to eliminate any looseness

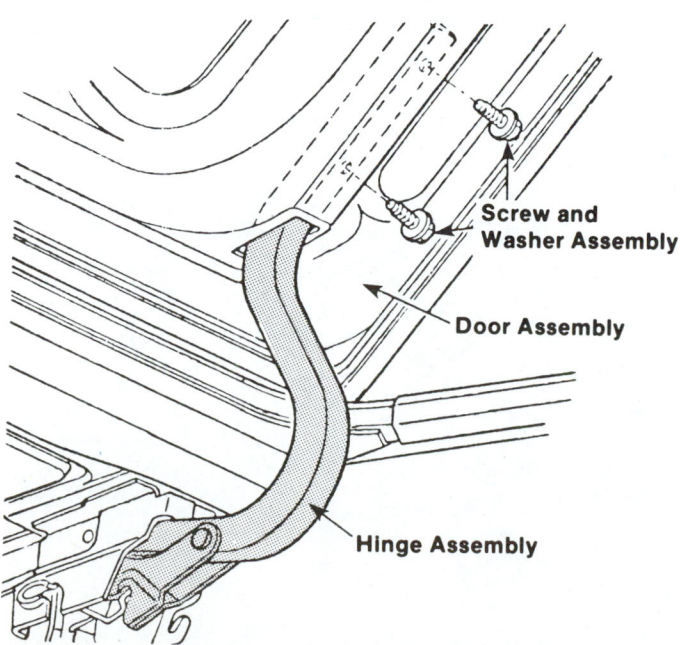

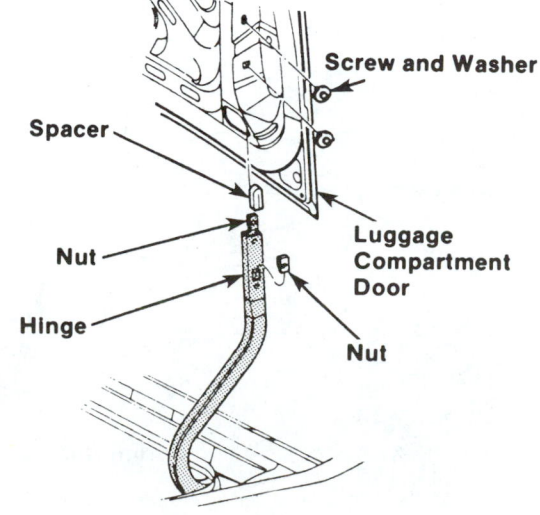

FIGURE 12-120 Service of trunk hinges is similar to hood hinges.

at the front of the hood and ensure a good, tight fit.

11. Tighten the attaching hardware on the bumper stops.

12. Check to see that the side bumper stops (if any) are in place and in good condition.

13. Make sure that the safety catch is working properly.

TRUNK LID ADJUSTMENTS

The trunk lid is very similar to the hood in construction. Two hinges (Figure 12–120) connect the lid to the rear body panel. The trailing edge is secured by a locking latch. The trunk lid seals the trunk area from dust and water. Weatherstripping is used to provide the proper seal. For the seal to be effective, the trunk

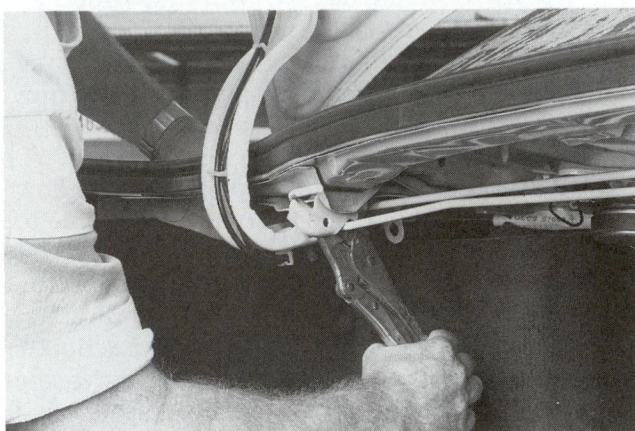

FIGURE 12-121 Some trunk lids have torsion bars. *(Courtesy of Mitchell International)*

lid must contact all of the weatherstripping when the lid is closed.

The lid must be evenly spaced between the adjacent panels. Slotted holes in the hinges and/or caged plates in the lid allow the trunk lid to be moved forward, rearward, and side to side. To adjust the lid forward or backward, slightly loosen the attaching hardware on both hinges. Close and adjust the lid as required. Then raise the lid carefully and tighten the attaching hardware.

In some cases, it might be necessary to use shims between the bolts and the trunk lid to raise or lower the front edges. If the front edge must be raised, place the shim(s) between the hinge and the lid in the front bolt area. To lower the front edge of the lid, place the shim(s) at the back of the hinge.

On some vehicles, the trunk lid has hinge assemblies that utilize torque rods to counterbalance the weight of the lid. This arrangement makes the trunk lid easier to raise and holds it in the up position. The torque rods can be tightened or loosened by moving the torque rod end to a different hole or slot (Figure 12–121). Using a pipe inserted over the end of the torque rod is one way to safely move the rod to a new position.

Figure 12–122 shows a typical trunk lid lock assembly. The lock assembly is usually in the trunk lid and the striker plate is bolted to the rear body panel. On some cars, this position is reversed. The trunk lid latch and striker can usually be adjusted up, down, and sideways to properly engage, align, and tightly hold the trunk lid. The necessary adjustments are similar to those given for the hood.

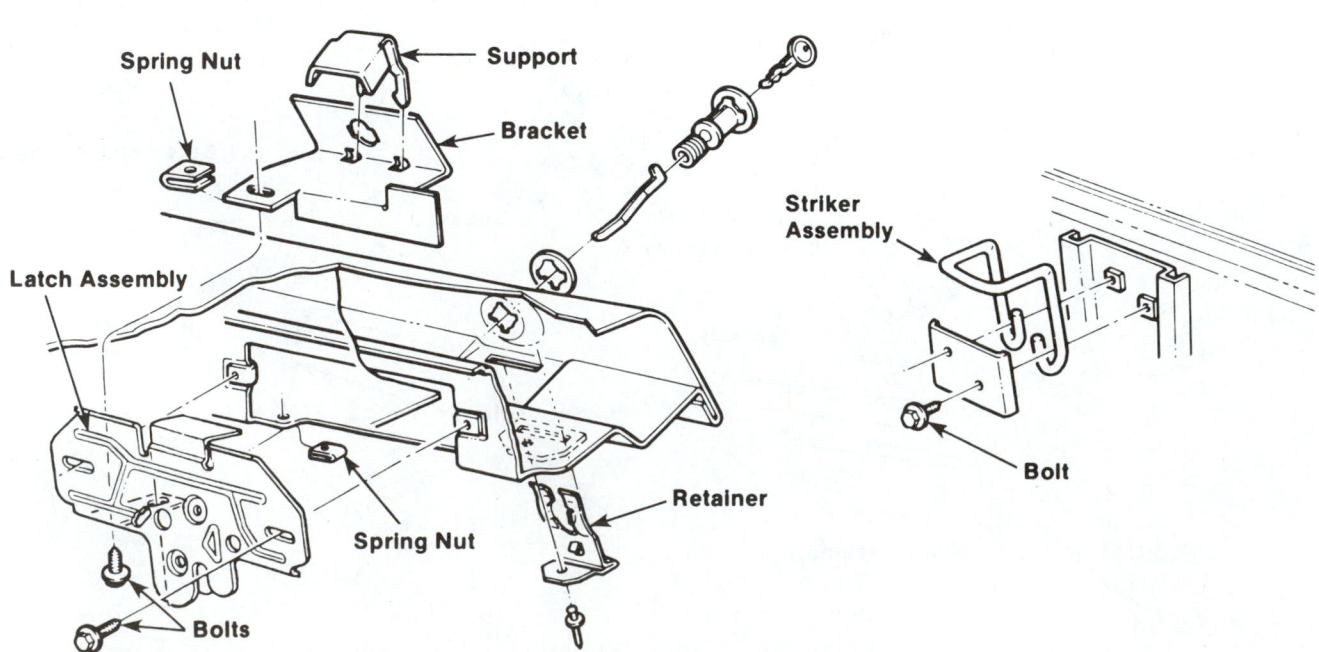

FIGURE 12-122 Study the construction of a typical trunk lock mechanism.

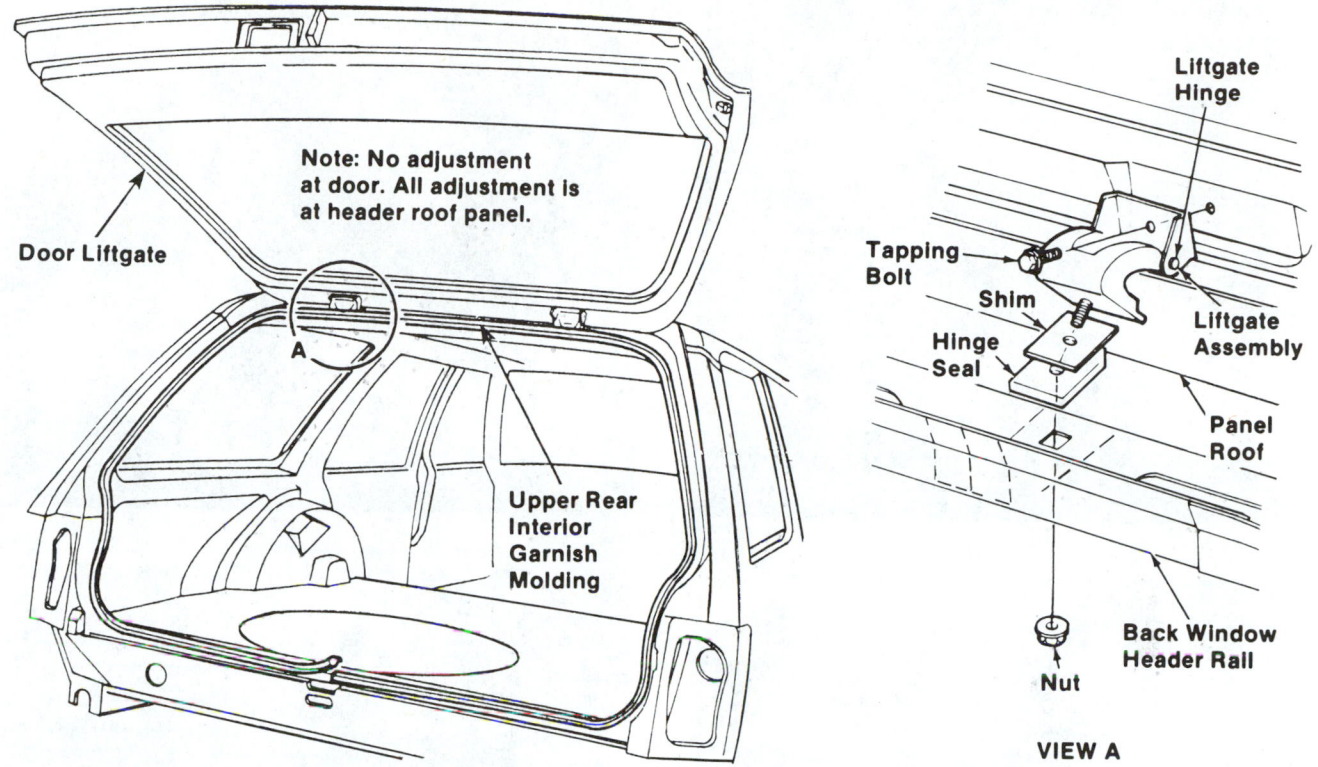

FIGURE 12-123 Note the basic method for adjusting a hatchback lid.

Vehicles with hatchback-type trunk lids are usually difficult to align mainly because of their size. Many lids of this type are nearly horizontal in design, which makes them more prone to water and dust leaks. Some models use adjustable hinges. Others use welded hinges. The hatchback types also use gas-filled door lift or shock assemblies, or springs, one at each upper corner of the lid. Some play may be available in the door lift support brackets to allow adjustment of the hatchback trunk lid (Figure 12–123).

STATION WAGON TAILGATE ADJUSTMENTS

Most late model, full size station wagons have a three-way tailgate. (Compact cars are the exception; most of these have a hatchback-style rear gate.) The **three-way tailgate** has a unique hinge and locking arrangement that allows the tailgate to be operated as a tailgate with the glass fully down or as a door. Figure 12–124 shows the hinges and locks in a typical three-way tailgate.

The lock system on the three-way tailgate performs the following functions:

- Allows the tailgate to be opened and closed as a door with the glass up or down. It also prevents the accidental operation of the upper left lock, which allows the gate to be opened as a gate.
- Allows the tailgate to be opened and closed as a tailgate with the glass down. At the same time,

it prevents the accidental operation of the lower right lock, which allows the gate to be opened as a door.

Before doing any station wagon tailgate alignment, closely examine the area to determine where the misalignment exists. It might be necessary to remove the right upper striker bolt and adjust the tailgate as a regular door. If the right lower striker forces the tailgate up or down and out of alignment, readjust the striker bolt to provide a smooth open/close action.

The left hinges can be adjusted to position the tailgate in the vehicle body opening and for flush alignment with the adjacent body panels. They are also adjustable for smooth lock operation.

Closely examine the hinges to determine what adjustments are available. On some vehicles, the lower left hinge provides up and down as well as in and out adjustments for the tailgate. Some vehicles provide adjustments on the body side of the hinge. Others allow adjustment to the tailgate side of the hinge.

After adjusting the left hinges, it is necessary to reinstall and adjust the strikers. If all adjustments have been done correctly, the tailgate should operate smoothly and fit properly.

The following is a typical procedure for adjusting a three-way tailgate.

1. The lower left hinge assembly at the body attachment is adjustable up or down and laterally. To

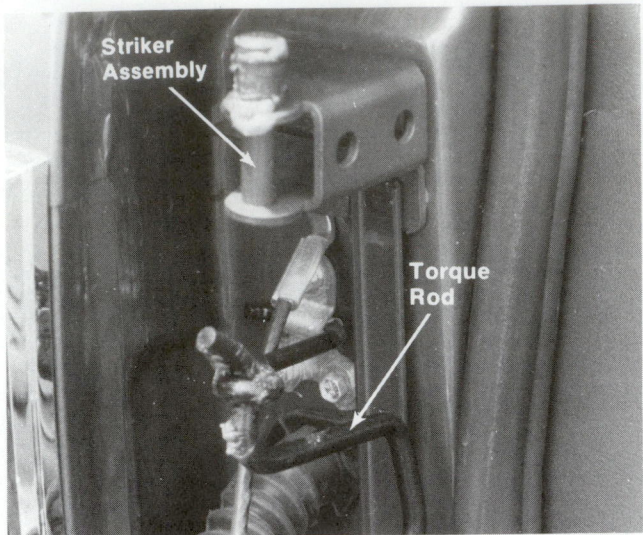

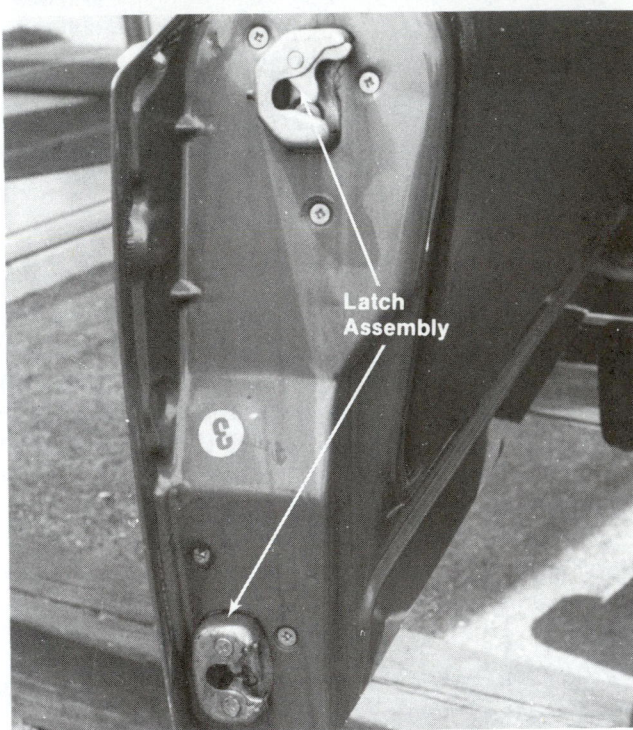

FIGURE 12-124 A three-way tailgate can be a challenge to adjust.

gain access to the lower left hinge-to-body attachments, remove the left quarter trim. Loosen the hinge-to-body attachments and adjust the hinge up, down, laterally, or rotate as required.

Rotating the hinge slightly will raise or lower the right side of the gate. This can be accomplished by loosening the lower hinge-to-body attachments. Then, with the tailgate open as a door, support the right side of the tailgate in the desired position and tighten the hinge attachments. If this adjustment is performed or if the tailgate is moved sideways, it may be necessary to also adjust the left upper hinge striker assembly.

If the lower left hinge is adjusted upward or downward, clearance between the upper left lock frame on the tailgate and the hinge lock striker on the body pillar should be checked. Check service manual instructions for clearance specifications.

2. The lower left hinge assembly at the gate attachment is adjustable forward or rearward. This adjustment is primarily for flush alignment of the tailgate to the adjacent panels.

The lower left hinge-to-tailgate attaching nuts are located inside the tailgate. To loosen nuts for the adjustment of the tailgate on the hinge, remove the tailgate trim panel to gain

SHOP TALK Prior to performing any adjustments, the position of the hinge lock, or striker, should be scribed or marked to ease realignment.

access to the hinge-to-tailgate attachments. Adjust the tailgate on the hinge as required, then tighten the nuts and replace the previously removed parts.

3. The upper left hinge lock assembly is adjustable forward and rearward. This adjustment is available to provide a flush alignment of the tailgate outer panel with adjacent body panels in the area of the upper left lock.

   Prior to adjusting the upper left lock, mark the position of the lock on the tailgate. Then loosen the hinge-to-gate attaching nuts. Adjust the lock as required and replace all previously removed parts.

   After any adjustment of the upper left lock, synchronization of the lock system should be checked. Consult the appropriate manufacturer's service manual for proper synchronization procedures.

4. The upper left striker assembly is usually adjustable up, down, and laterally. The up or down adjustment is to provide adequate clearance between the bottom of the lock frame (on the tailgate) and the top of the hinge pin striker plate.

   To check the clearance, open the gate as a door and measure the distance between the upper surface of the upper left hinge pin and striker plate and the lower surface of the upper left lock frame.

   To make any necessary adjustment, remove the left quarter trim to gain access to the attachments. Loosen the attaching nuts and reposition the striker hinge assembly as required.

   This adjustment is available to provide proper engagement of the hinge pin and lock striker with the lock. It is not intended as a means of raising or lowering the left or right side of the gate.

   To find the correct adjustment, open the tailgate as a gate; then, while closing the gate, carefully observe how the striker pin engages in the slot in the bottom surface of the lock. The striker pin should enter into the slot with side pressure.

5. The right upper and lower lock striker assemblies are adjustable forward or rearward, up or

down, and laterally by using spacers. The upper and lower right strikers should be removed prior to performing any other hinge or lock adjustment.

To adjust the upper or lower right strikers, open the tailgate as a door and remove the striker. Check the alignment of the tailgate in the body opening. The tailgate should be aligned with the left upper hinge lock prior to adjustment of the strikers. Install the striker slightly more than finger tight. Then carefully close the gate to allow the striker to self-align. Carefully open the gate and tighten the striker.

Operate the tailgate both as a door and a gate and check for flush alignment of outer panels in the area of the striker. If any further minor adjustment is required, mark the position of the striker on the body pillar, loosen the striker, and make the required adjustments from the marked position and tighten the striker.

Do not use the right upper and lower striker to align the right side of the gate up or down in the body opening. This causes part wear and improper opening-closing action.

6. When the tailgate is aligned properly, check the glass for a good seal with the body. Glass adjustment is similar to methods described for hardtop-style door glass.

12.7 CUSTOM BODY PANELS

Body customizing is a growing trend. Many car owners, when faced with the need to replace damaged panels, are spending the extra money to give their cars a performance face-lift. A number of companies are marketing spoiler and air dam kits that can turn any street car into an "Indy impression."

Figure 12–125 shows what can be done to a plain economy car to give it a sporty look. The accessory panels not only improve the appearance of the car but also improve the aerodynamic performance during high-speed driving (Figure 12–126).

Air dams restrict air passing under the car body (minimizing undercar turbulence and reducing

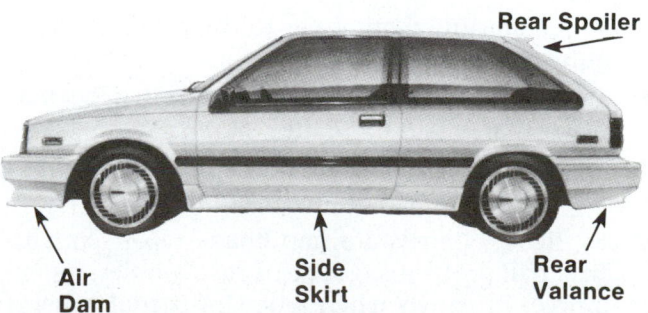

FIGURE 12–125 Spoilers and an air dam are common.

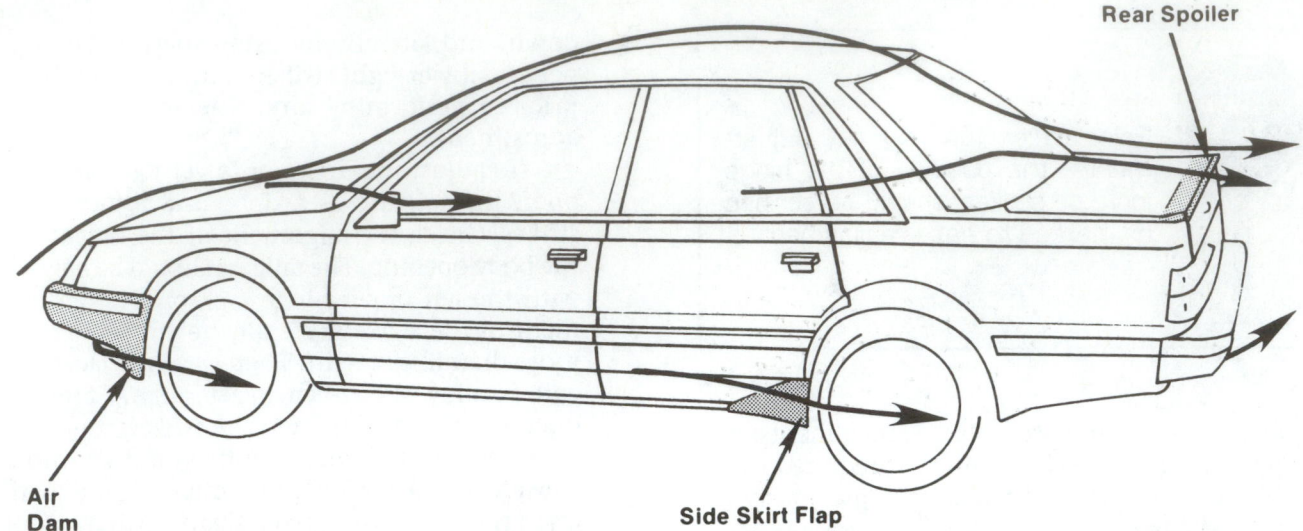

FIGURE 12-126 The aerodynamic performance of a vehicle with air dam, side skirt, and rear spoiler is improved.

resistance to airflow) and prevent the front wheels from lifting. Side skirts channel airflow away from the rear wheels, thus reducing turbulence and resistance to airflow. The rear spoiler alters the airstream at the rear body end to again reduce resistance to airflow and to prevent the rear wheels from lifting.

The spoilers, side skirts, air dams, and other parts are often made from a molded polyurethane plastic. The product usually comes with a silicone release that must be removed with a wax and grease remover. Then the panel must be scuff sanded with #240 grit sandpaper followed by #320 grit paper to improve adhesion. The painting of plastic panels is covered in detail in Chapter 20.

Most add-on body panels utilize original fasteners. You may have to drill additional mounting holes for the fasteners provided by the kit manufacturer. Double-sided adhesive tape is sometimes used to keep the plastic panels fitting tightly against the original sheet metal panels. The tape application area should be dewaxed and scuff sanded to ensure a positive bond between the car and new panel.

The key to successful add-ons is proper alignment, careful positioning of the new fasteners, and preparation of the mating panel surfaces.

1. Remove the lower body panel from under the front bumper.
2. Replace the new air dam under the front bumper and align the air dam with the wheel wells. Make sure that the front top lip of the air dam is located inside the front panel.
3. Clamp the corners of the air dam to the wheel well with vice grips.
4. Transfer the front body panel mounting holes with a scribe onto the air dam.
5. Use a scribe to transfer the mounting holes in the ends of the air dam to the wheel well.
6. Using recommended size drill bit, drill the right amount of holes through the sheet metal and air dam.
7. Loosely bolt the air dam in position and check it for correct alignment.
8. Without overtightening them and causing part distortion, snug down all fasteners securely.

SUMMARY

- Panel replacement involves removing and installing a new panel or body part.
- When starting work, refer to the estimate to get guidance on where to begin.
- Sectioning involves cutting the part in a location other than a factory seam.
- Partial replacement panels are often designed to replace only a section or area of a large panel.
- Fabricated panels are handmade repair parts to fix small problems (gouge or rusted holes, for example) in panels when a new or partial panel is not available or practical.

- In modern unitized construction, all the structural panels, from the radiator support to the rear end panel, are welded together to make a one-piece frame structure.
- Always obtain OEM Repair Manuals for specific instructions for sectioning.
- It is usually necessary to remove the paint film, undercoat, sealer, or other coatings covering the joint area to find the locations of spot welds.
- After the spot welds have been located, the welds can be drilled out and removed, using a spot weld cutter.

- Apply weld-through primer to areas where the base metal is exposed after the paint film and rust have been removed from the joining surfaces.

- A test weld is done on scrap pieces of metal the same thickness and type as the parts to be repaired.

- Unibody parts to be sectioned involve these types of construction:

  1. Closed sections, such as rocker panels and A- and B-pillars
  2. Hat or open U-channels, such as rear rails
  3. Single-layer or flat parts, such as floor pans and trunk floors

- There are three basic types of sectioning joints. They are the

  1. Lap joint
  2. Offset butt joint
  3. Butt joint with insert

- Recycled assemblies are undamaged parts from another damaged vehicle that are used for repairs.

- The front pillars or A-pillars extend up next to the edges of the windshield.

- Center pillars or B-pillars are the roof supports between the front and rear doors on four-door vehicles.

- Foam fillers are used to add rigidity and strength to structural parts. They also reduce noise and vibrations.

- Full body sectioning is replacing the entire rear section of a collision-damaged vehicle with the rear section of a salvaged vehicle.

- The outer panel or door skin wraps around the door frame and is clinched to the pinch weld flange.

- Some manufacturers use structural adhesives in place of welds.

- Door adjustment is needed so the doors will close easily, not rattle, and not leak water and dust.

- Worn door hinges will have play that allows up and down movement of the rear of the door.

- Oil is not recommended for use on the rubber window channels. It can cause the rubber to swell and deteriorate.

- The window regulator is a gear mechanism that allows you to raise and lower the door glass.

ASE-STYLE REVIEW QUESTIONS

1. Technician A sometimes positions nonstructural outer panels visually, without making precise measurements. Technician B says that when positioning a hood one of the final checks is to make sure the safety catch is functioning correctly. Who is correct?

   A. Technician A
   B. Technician B
   C. Both A and B
   D. Neither A or B

2. Rocker panels and pillars are known as what type of sections?

   A. Open surface
   B. Crush zone
   C. Closed
   D. Compound

3. Where should A-pillars be cut when sectioning?

   A. At the upper end
   B. Near the middle
   C. At the lower end
   D. Both A and C

4. Which of the following statements concerning replacement at factory seams is incorrect?

   A. Replacing panels at factory seams is common in unibody repair.
   B. Damaged rails and panels should not be returned to factory specifications until after they have been removed from the vehicle.

   C. It is easy to destroy more of the factory welds than is necessary when doing the job.
   D. All of the above

5. What is the best method for separating spot welds?

   A. Blowing them out with a plasma torch
   B. Cutting them out with a spot cutter
   C. Drilling them out
   D. Grinding them down with a high-speed grinding wheel

6. What color is metal that has been arc-brazed?

   A. Black
   B. Reddish copper
   C. Silver
   D. Brass-colored

7. When preparing a vehicle for the installation of replacement body panels, Technician A grinds the flanges of the structural panels. Technician B says that weld-through primer must be applied to areas where bare metal is exposed on a joining surface. Who is correct?

   A. Technician A
   B. Technician B
   C. Both A and B
   D. Neither A or B

8. When a replacement body panel overlaps any existing panels, the overlap portion should be _____.

   A. $3/4$ to 1 inch
   B. 1 to $1\frac{1}{2}$ inches
   C. At least 2 inches
   D. No more than $1/2$ inch

9. When positioning a nonstructural outer body panel visually, Technician A makes sure that the gaps between the panels are tapered. Technician B makes sure that the gaps between the panels are even. Who is correct?

   A. Technician A
   B. Technician B
   C. Both A and B
   D. Neither A nor B

10. Which of the following should be avoided when making sectioning cuts?

    A. Structural member mounts
    B. Compound member mounts
    C. Dimensional reference holes
    D. All of the above

11. When separating an exterior door panel from the inner panel, Technician A uses a power chisel. Technician B uses vise grips to remove the remains of the inner hem flange. Who is correct?

    A. Technician A
    B. Technician B
    C. Both A and B
    D. Neither A or B

ESSAY QUESTIONS

1. How would you use the estimate when starting work on a wrecked car?

2. Describe how to clean a typical part before removing its spot welds.

3. Why should you be careful when using a plasma arc torch to cut parts?

4. Why does the use of recycled assemblies in collision repair make sense?

5. Summarize how you fully section a car.

6. How do you adjust a door?

CRITICAL THINKING PROBLEMS

1. You are working on a late model Porsche. You are not quite sure how to section a damaged front rail. What should you do?

2. After welding in a new quarter panel, you find that you cannot make the trunk lid fit properly in its opening. What are your options to fix this problem?

MATH PROBLEMS

1. If factory specs say that panel overlap should be $3/4$ inch, and you only have .25 inch overlap, how much more is needed?

2. If a recycled panel has been cut 1.3 inches longer than what is needed for proper overlap, and the total length is 2 feet, 1.5 inches, what should the part's final length be?

3. A foreign service manual gives a 12.7 mm spec for a drilled hole; what size English or U.S. drill bit should be used?

CHAPTER 13

Servicing Mechanical, Electrical, and Electronic Components

INTRODUCTION

Mechanical repairs include tasks like replacing a damaged water pump, radiator, or engine bracket. Mechanical components like these are often damaged in a major collision. Many mechanical parts are easy to replace and the work can be done by the auto body technician. However, other mechanical repairs may require special skills and tools. In this case, the vehicle would be sent to a professional mechanic.

Electrical repairs include tasks like repairing severed wiring, replacing engine sensors, and scanning for computer or wiring problems. During a collision, the impact on the vehicle body and the resulting metal deformation can easily crush wires and electrical components. For this reason, today's auto body

OBJECTIVES

After studying this chapter, you should be able to:

✔ Explain the procedure for removing a power train from a unibody vehicle.

✔ Describe how suspension and steering systems work.

✔ List the elements of proper wheel alignment.

✔ Recognize the typical problems caused by improper suspension system servicing.

✔ Perform the diagnosis and servicing of a power steering system.

✔ List the various driveline variations.

✔ Name the service procedures for the major parts of a cooling system, including the radiator, fan and fan clutches, and belts and hoses.

✔ Describe the service procedures for an air-conditioning system.

✔ Perform the diagnosis and servicing of an emission control system.

✔ Describe the test procedures used to repair electrical and electronic systems.

✔ Answer ASE test questions relating to mechanical and electrical systems.

KEY TERMS

A/C high side
A/C low side
active suspension
actuator
axial runout
brake pads
brake rotors
camber
caster
computer
computer self-diagnostics
computer system
control arms
curb height
current
CV-axles
dog tracking
fault code
jounce
master cylinder
motor mounts
multimeter
oil dipstick
parallelogram steering system

R-134a
rack-and-pinion steering
radial runout
radiator
rear axle assembly
rebound
resistance
scan tool
scrub radius
sensor
shock absorbers
steering axis inclination
thrust angle
toe
transaxle
turning radius angle
voltage
wheel alignment
wheel puller
wiring diagram
wiring harness

ASE TASK LIST

Job Skills covered in this chapter include:

NONSTRUCTURAL ANALYSIS AND DAMAGE REPAIR TEST (B3) TASK LIST

A. Preparation

6. Remove all vehicle mechanical and electrical components that may interfere with or be damaged during repair.
11. Identify potential safety and environmental concerns associated with vehicle components and systems, i.e. ABS, air bags (SRS), refrigerants, coolants, etc.

STRUCTURAL ANALYSIS AND DAMAGE REPAIR TEST (B4) TASK LIST

A. Frame Inspection and Repair

11. Diagnose misaligned or damaged steering, suspension, and power train components that can cause vibration, steering, and wheel alignment problems; align or replace steering and suspension components in accordance with vehicle manufacturers' recommendations.

B. Unibody Inspection, Measurement, and Repair

2. Diagnose misaligned or damaged steering, suspension, and power train components that can cause vibration, steering, and wheel alignment problems; realign or replace steering and suspension components in accordance with vehicle manufacturers' recommendations.

C. Stationary Glass

1. Remove and replace front and rear stationary glass (heated and nonheated) in accordance with manufacturers' recommendations.

MECHANICAL AND ELECTRICAL COMPONENTS TEST (B5) TASK LIST

A. Suspension and Steering

1. Identify suspension system fasteners that should not be reused.
2. Inspect and replace rack-and-pinion steering gear, inner tie rod ends, and bellows boots.
3. Remove and replace power steering pump, belts, hoses, and fittings; inspect pump mounts.
4. Remove and replace power steering gear (non-rack-and-pinion type).
5. Remove and replace power rack-and-pinion steering gear; inspect and replace mounting bushings and brackets; ensure proper mounting location.
6. Inspect and adjust (where applicable) steering linkage geometry (attitude/parallelism).
7. Inspect and replace pitman arm.
8. Inspect and replace replay (center link/intermediate) rod.

9. Remove and replace idler arm and mountings.
10. Remove and replace tie rod sleeves, clamps, and tie rod ends.
11. Remove and replace steering linkage damper.
12. Remove and replace upper and lower control arms.
13. Remove and replace upper and lower ball joints on short and long arm suspension systems.
14. Remove and replace steering knuckle/spindle/hub assemblies.
15. Remove and replace coil springs and spring insulators (silencers).
16. Inspect, replace, and adjust front suspension system torsion bars and inspect mounts.
17. Inspect and replace MacPherson strut cartridge or assembly, upper bearing, and mount.
18. Inspect, remove, and replace rear suspension system transverse links, control arms, stabilizer bars, bushings, and mounts.
19. Inspect, remove, and replace rear suspension system leaf spring(s), leaf spring insulators (silencers), shackles, brackets, bushings, and mounts.
20. Inspect rear axle assembly for damage and misalignment.
21. Inspect and replace shock absorbers, air shock absorbers, load-leveling devices, air springs, and associated lines and fittings.
22. Diagnose, inspect, adjust, or replace components of electronically controlled suspension systems.
23. Measure vehicle ride height; determine needed repairs.
24. Remove, replace, and align front and rear subframes.
25. Diagnose steering column damage, looseness, and binding problems (including tilt mechanisms); determine needed repairs.
26. Diagnose manual and power steering gear (non-rack-and-pinion type) noises, binding, uneven turning effort, looseness, hard steering, and lubricant leakage problems; determine needed repairs.
27. Diagnose manual and power rack-and-pinion steering gear noises, vibration, looseness, and hard steering problems; ensure proper mounting location.
28. Inspect and replace steering shaft U-joint(s), flexible coupling(s), collapsible columns, and steering wheels.
29. Diagnose front and rear suspension system noises and body sway problems; determine needed repairs.
30. Diagnose MacPherson strut suspension system noises and body sway problems; determine needed repairs.

ASE TASK LIST

Job Skills covered in this chapter include:

31. Diagnose vehicle wandering, pulling, hard steering, bump steering, memory steering, torque steering, and steering return problems; determine needed repairs.
32. Check and adjust front and rear wheel camber on suspension systems with camber adjustments.
33. Check front and rear wheel camber on non-adjustable suspension system; determine needed repairs.
34. Check and adjust caster on suspension systems with caster adjustments.
35. Check caster on nonadjustable suspension systems; determine needed repairs.
36. Check and adjust front wheel toe; center steering wheel if necessary.
37. Identify toe-out-on-turns (turning radius) related problems; determine needed repairs.
38. Identify SAI (steering axis inclination)/KPI (king pin inclination) related problems; determine needed repairs.
39. Check rear wheel toe; determine needed repairs.
40. Identify thrust angle related problems; determine needed repairs.
41. Check for front wheel setback; determine needed repairs.
42. Diagnose tire wear patterns; check and adjust air pressure.
43. Diagnose wheel/tire vibration, shimmy, and tramp (wheel hop) problems; determine needed repairs.
44. Measure wheel, tire, axle, and hub runout; determine needed repairs.
45. Diagnose tire pull (lead) problems; determine corrective actions.
46. Check wheels for dents, cracks, mounting surface damage, and worn lug holes.

B. Electrical

1. Check voltages in electrical wiring circuits with a DMM (digital multimeter); determine repair procedure.
2. Check continuity and resistance in electrical wiring circuits and components with a DMM (digital multimeter); determine repair procedure.
3. Check electrical circuits, wiring, and connectors; repair according to manufacturers' specifications.
4. Inspect, test, and replace fusible links, circuit breakers, and fuses.
5. Inspect, clean, and replace battery, battery cables, connectors, and clamps.
6. Perform slow/fast battery charge in accordance with manufacturers' recommendations.

7. Identify programmable electrical/electronic components; record data for reprogramming before disconnecting battery.
8. Remove and replace alternator, alternator drive belts, pulleys, and fans; inspect and adjust alignment.
9. Remove and replace headlights, parking/taillights, stoplights, flashers, turn signals, and backup lights; check operation; aim headlights as necessary.
10. Check operation of retractable headlight assembly.
11. Remove and replace motors, switches, relays, connectors, and wires of retractable headlight assembly circuits.
12. Inspect, test, and repair or replace switches, relays, bulbs, sockets, connectors, and wires of all light circuits including four-wire taillight systems.
13. Remove and replace horn(s); check operation.
14. Check operation of windshield wiper/washer system.
15. Check operation of power side windows and power tail-gate window.
16. Remove and replace power seat, motors, linkages, cables, etc.; check operation.
17. Remove and replace components of electric door and hatch/trunk lock; check operation.
18. Remove and replace components of keyless lock/unlock devices and alarm systems; check operation.
19. Remove and replace electrical components of sunroof or convertible top; check operation.
20. Check operation of electrically heated mirrors, windshields, backlights, panels, etc.; repair as necessary.
21. Remove and replace components of power antenna circuits; check operation.

C. Brakes

1. Inspect brake lines and fittings for leaks, dents, kinks, rust, cracks, or wear; tighten loose fittings and supports; replace brake lines (double flare and ISO types), hoses, fittings, and supports.
2. Inspect flexible brake hoses for leaks, kinks, cracks, bulging, or wear; remove and replace hoses; tighten loose fittings and supports.
3. Select, handle, store, and install brake fluids (including silicone fluids).
4. Bleed (manual, pressure, vacuum, or surge) and/or flush hydraulic brake system in accordance with manufacturers' procedures.
5. Pressure test brake hydraulic system; determine needed repairs.

6. Adjust brake shoes; reinstall brake drums or drum/hub assemblies and wheel bearings.
7. Reinstall wheels and torque lug nuts; make final checks and adjustments.
8. Remove and replace caliper assembly.
9. Clean and inspect caliper mountings and slides for wear and damage.
10. Check parking brake system operation.
11. Identify and replace ABS wheel speed sensor components according to manufacturers' specifications.
12. Depressurize ABS hydraulic system according to manufacturers' specifications.

D. Heating and Air Conditioning

1. Identify type of refrigerant and refrigerant oil.
2. Recover refrigerant from A/C system.
3. Evacuate recharge A/C system with refrigerant (liquid or vapor); perform leak test.
4. Inspect and correct oil level in A/C system.
5. Inspect, adjust, and replace A/C compressor drive belts; check pulley alignment.
6. Remove and replace A/C compressor; inspect, repair, or replace A/C compressor mountings.
7. Inspect, repair, or replace A/C system mufflers, hoses, lines, fittings, and seals.
8. Inspect A/C condenser for air flow restrictions; clean and straighten fins.
9. Inspect, test, and replace A/C system condenser and mountings.
10. Inspect and replace receiver/drier or accumulator/drier.
11. Inspect and replace evaporator.
12. Inspect and repair evaporator housing water drain.
13. Inspect, test, repair, or replace heating, ventilating, and A/C vacuum components.
14. Inspect and repair A/C component wiring according to manufacturers' specifications.
15. Inspect, test, and repair heating, ventilating, and A/C ducts, doors, hoses, and outlets.
16. Performance test heating and A/C system.

E. Engine Cooling Systems

1. Inspect and replace engine cooling and heater system hoses and belts.

2. Inspect, remove, and replace radiator, pressure cap, coolant recovery system, and water pump.
3. Remove and replace thermostat, by-pass, and housing.
4. Recover, flush, and refill bleed system with proper coolant and level of protection; leak test system.
5. Remove and replace fan, fan pulley (both electrical and mechanical), and fan shroud.
6. Inspect, remove, and replace auxiliary oil coolers.
7. Inspect, remove, and replace electric fan sensors and wiring.

F. Drivetrain

1. Remove and replace power train assembly; inspect, replace, and align power train mounts.
2. Remove, replace, and adjust cables or linkages for throttle valve (TV), kickdown, and accelerator pedal.
3. Remove and replace electronic sensors, wires, and connectors.
4. Remove, replace, and adjust mechanical or hydraulic shift or clutch linkage as required.
5. Remove and replace front and/or rear drive axle assembly.
6. Remove, inspect, and replace front-drive half shafts and axle constant velocity (CV) joints.
7. Inspect, remove, and replace front and rear drive shafts and universal joints.

G. Fuel, Intake and Exhaust Systems

1. Inspect, remove, and replace exhaust manifold, exhaust pipes, mufflers, converters, resonators, tail pipes, and heat shields.
2. Inspect, remove, and replace fuel tank, fuel tank filter, fuel cap, fuel filler hose, quarter-to-body seal, and inertia switch; inspect and replace fuel lines and hoses; check fuel for contaminants.
3. Inspect, remove, and replace engine components of air intake systems for collision damage.
4. Inspect, remove, and replace canister, filter, and vent, and purge lines of fuel vapor control systems.

technician must have the basic skills needed to work with and repair electrical and electronic components.

In many cases, major driveline parts such as engines, transmissions, and drive axles are mounted directly to structural unibody panels (Figure 13–1). In other cases, these parts are mounted to support-

ing cross members, subframe assemblies, or cradle assemblies that are mounted to the unibody structural panels.

Today's body shop technician must sometimes remove, install, and diagnose problems with mechanical systems. Also, most modern mechanical

A

B

C

FIGURE 13-1 Three photos show different views of how a car is made. (A) The car body provides an attractive exterior. It can be made of steel, fiberglass, plastics, and composite materials. (B) The body has been sectioned away to show the engine and other mechanical/electrical assemblies under the body. (C) This view shows the power train for the same car. It is a front-engine, rear-wheel-drive configuration. *(Courtesy of Ford Motor Co.)*

systems are monitored or controlled by on-board computers. This makes electrical and electronic knowledge important to a modern collision repair technician. This chapter will give you the basic information needed to work with the mechanical and electrical systems of present-day vehicles.

13.1 POWER TRAIN CONSTRUCTION

The *power train* comprises all the parts that produce and transfer power to the drive wheels, including the engine, transmission or transaxle, drive axle, and

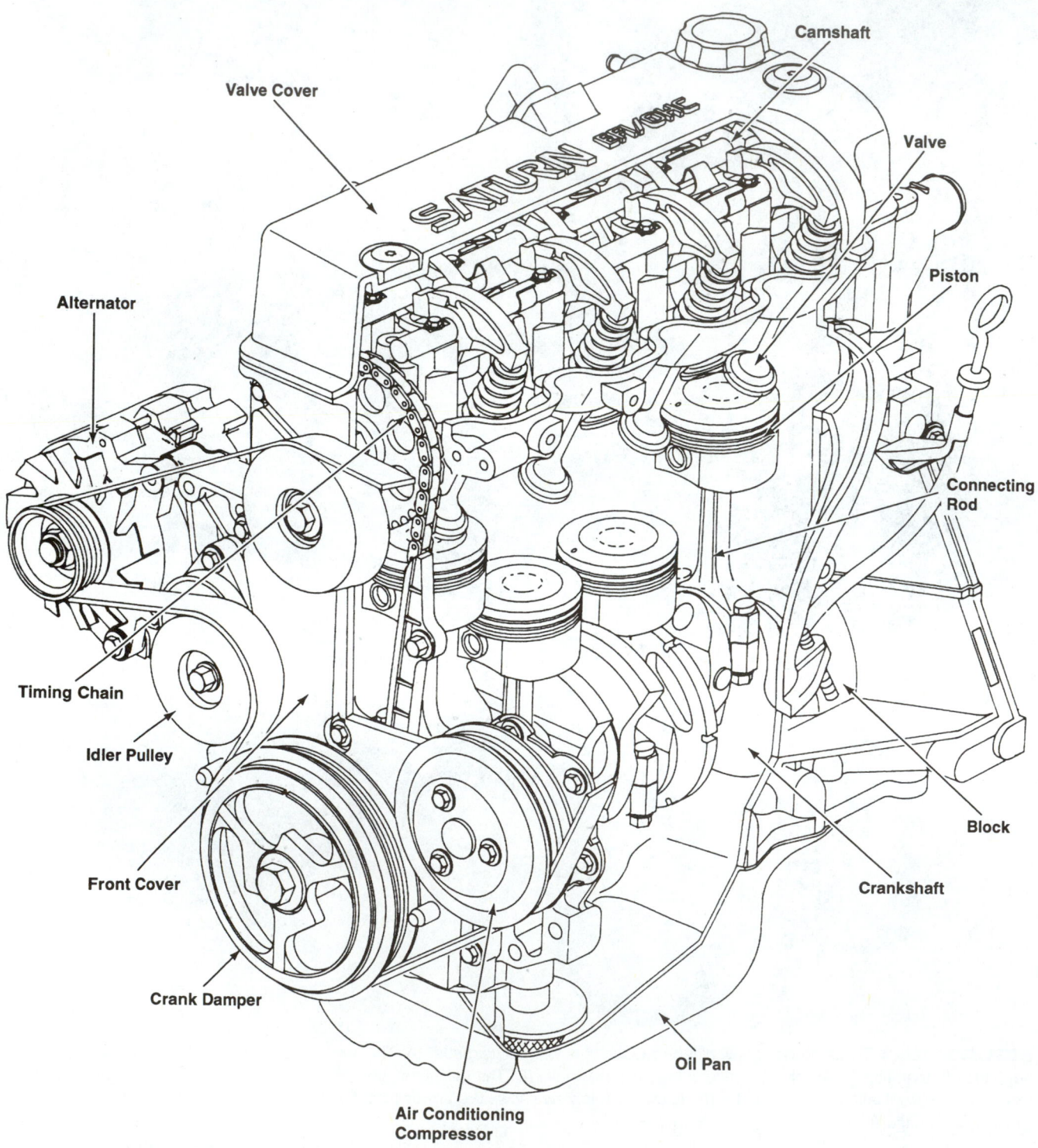

FIGURE 13-2 Review fundamental engine parts. *(Courtesy of General Motors Corp.)*

WARNING

When diagnosing and repairing mechanical parts, always refer to the service manual for the specific vehicle. Never attempt to work with mechanical components without the aid of a service manual. It gives the detailed procedures and specifications for doing competent work.

other related parts. The *drivetrain* typically includes everything that sends power to the drive wheels except the engine.

ENGINE

The *engine* provides energy to move the vehicle and power all accessories (Figure 13–2). Most cars and trucks use gasoline engines while a few use diesel engines. The basic parts of a typical internal combustion, piston engine include the following.

The *block* is the foundation of the engine; all the other engine parts are either housed in or attached to the block. A *cylinder* is a round hole bored (machined) in the block that guides piston movement.

The *piston* transfers the energy of *combustion* (burning of an air-fuel mixture) to the connecting rod. *Rings* are circular seals installed around the top sides of the piston. They keep combustion pressure and oil from leaking between the piston and the cylinder wall (cylinder surface). A *connecting rod* is a link that attaches the piston to the crankshaft.

The *crankshaft* changes reciprocating (up and down) motion of the piston and rod into more useful rotary (spinning) motion. Power to turn the driving wheel comes from the rear of the crank and accessories are driven off the front.

A *cylinder head* covers and seals the top of the cylinder. It contains valves, rocker arms, and sometimes the camshaft. The *combustion chamber* is a small enclosed area between the top of the piston and bottom of the cylinder head. The burning of the air-fuel mixture occurs in the combustion chamber.

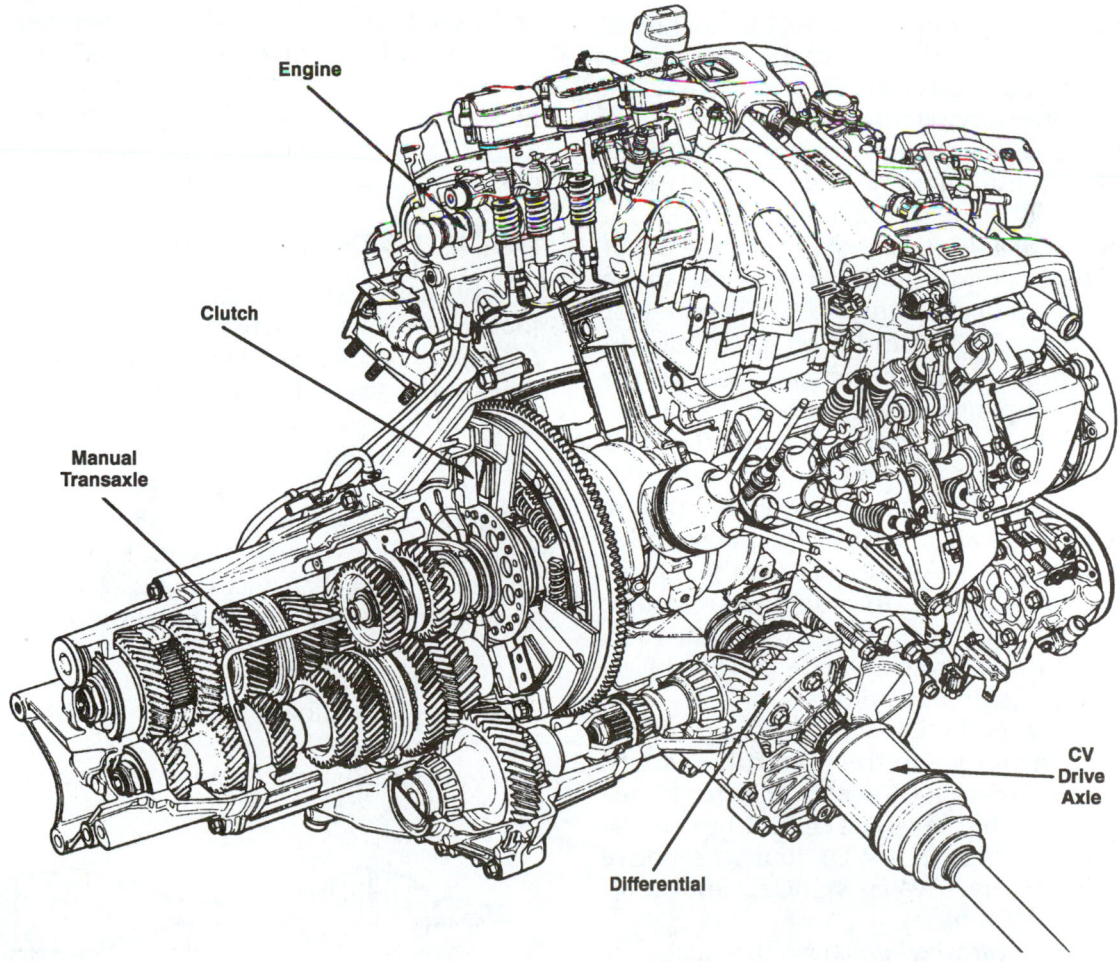

Engine

Clutch

Manual
Transaxle

Differential

CV
Drive
Axle

FIGURE 13–3 The transaxle or transmission bolts to the rear of a rear-wheel-drive engine. The clutch or torque converter fits between the engine and the transmission. The differential is part of the transaxle. With the rear-drive transmission, the differential is in the rear axle assembly. With front-wheel drive, CV axles transfer power to drive the wheel and tires. *(Courtesy of Honda Motor Co.)*

Valves are flow control devices that open to allow the air and fuel mixture into and exhaust out of the combustion chamber. *Valve springs* hold the valves closed when they do not need to be open. They also return the valve train parts to the at-rest position.

The *camshaft* controls the operation of valves. It can be located in the block or the cylinder head. A *lifter* is a cylindrical-shaped part that rides on the camshaft lobes and transfers motion to the push rods. The *push rods* are hollow tubes that transfer motion from the lifters to rocker arms. The *rocker arms* are levers that transfer camshaft action from the push rods to the valves.

The *flywheel* is a heavy metal disc used to help keep the crankshaft turning smoothly. It also connects engine power to the transmission. A larger gear on the outside of the flywheel engages the starting motor when cranking the engine for starting.

In a collision, the oil filter, oil pan, and related parts are sometimes damaged. They are made of thin metal and can be crushed and ruptured easily.

DRIVETRAIN CONSTRUCTION

The drivetrain uses engine power to turn the drive wheels. It includes everything after the engine—the clutch, transmission, drive shaft, and drive axles. Drivetrain designs vary (Figure 13–3).

Some cars use a manual transmission (hand shifted). Others use an automatic transmission (shifts gear automatically using internal oil pressure).

The *transmission* is an assembly with a series of gears for increasing torque to the drive wheels so the car can accelerate properly. It provides high power for acceleration in lower gears and good gas mileage in higher gears. With an automatic transmission, a *torque converter* (fluid coupling) is used in place of a clutch.

A **transaxle** is a transmission and differential combined into a single housing or case. Both automatic and manual transaxles are available (Figure 13–3). After collision repairs, check the transmission or transaxle fluid level before test driving the vehicle.

A *clutch* is a device used to couple and uncouple engine power to a manual transmission or transaxle. It uses a friction disc, pressure plate, flywheel face, and release bearing for activation.

Front-wheel-drive (FWD) vehicles use a transaxle to transfer engine torque to the front drive wheels. *Constant velocity axles* or **CV-axles** transfer torque from the transaxle to the wheel hubs. They can be found on rear-wheel-drive (RWD), four-wheel-drive (4WD), all-wheel-drive (AWD) vehicles, and front-wheel-drive (FWD) vehicles.

Front-engine, *rear-wheel-drive* (RWD) vehicles use a conventional transmission, drive shaft, and rear axle assembly to transfer power to the rear drive wheels.

A *drive shaft* is a long tube that transfers power from the transmission to the rear axle assembly. It

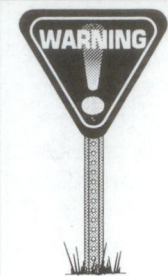

WARNING! When starting an engine before or after repairs, check the oil level with the **oil dipstick**. Also, always look under the vehicle for oil leakage. If you find an oil leak, shut the engine off right away. Find and fix the source of the oil leak.

has *universal joints* or U-joints at both ends that allow flexibility of the suspension while maintaining the driving force.

The **rear axle assembly** is the housing that contains the ring gear, pinion gear, differential assembly, and axles. Rear suspension springs attach to the housing.

A *differential assembly* is a unit within the drive axle assembly. It uses gears to allow different amounts of torque (turning force) to be applied to each drive wheel while the vehicle is making a turn (Figure 13–3).

Constant velocity (CV) joints have overcome the design limitations of conventional universal joints and have become common to unibody drivetrains. They can be found on some rear-wheel-drive (RWD), four-wheel-drive (4WD), all-wheel-drive (AWD) vehicles, and on all front-wheel-drive (FWD) vehicles. They eliminate the vibration problem typical of older cross- and roller-type U-joints.

FRONT-WHEEL DRIVE

In a typical FWD application (Figure 13–4) two CV joints are used on each half shaft, or a total of four CV joints. Two outboard joints are installed near the wheels and two inboard joints near the transaxle. The outboard joints are usually fixed and the inner ones are generally plunging types.

Front-wheel-drive half shafts can be solid or tubular, of equal or unequal length (Figure 13–5A), or with or without damper weights. Equal length shafts (Figure 13–5B) help reduce torque steer (the tendency to steer to one side as engine power is applied). In these applications, an intermediate shaft links the transaxle to one of the half shafts. This intermediate shaft uses an ordinary universal joint to couple to the transaxle. At the outer end is a support bracket and bearing assembly. Be sure to inspect these drive-

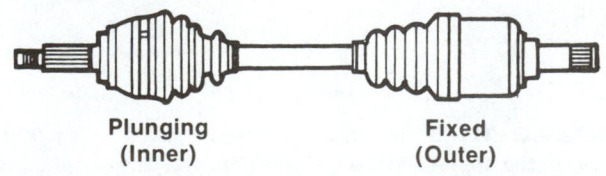

Plunging   Fixed
(Inner)   (Outer)

FIGURE 13-4 Note inner and outer CV joints on a front-wheel-drive half shaft.

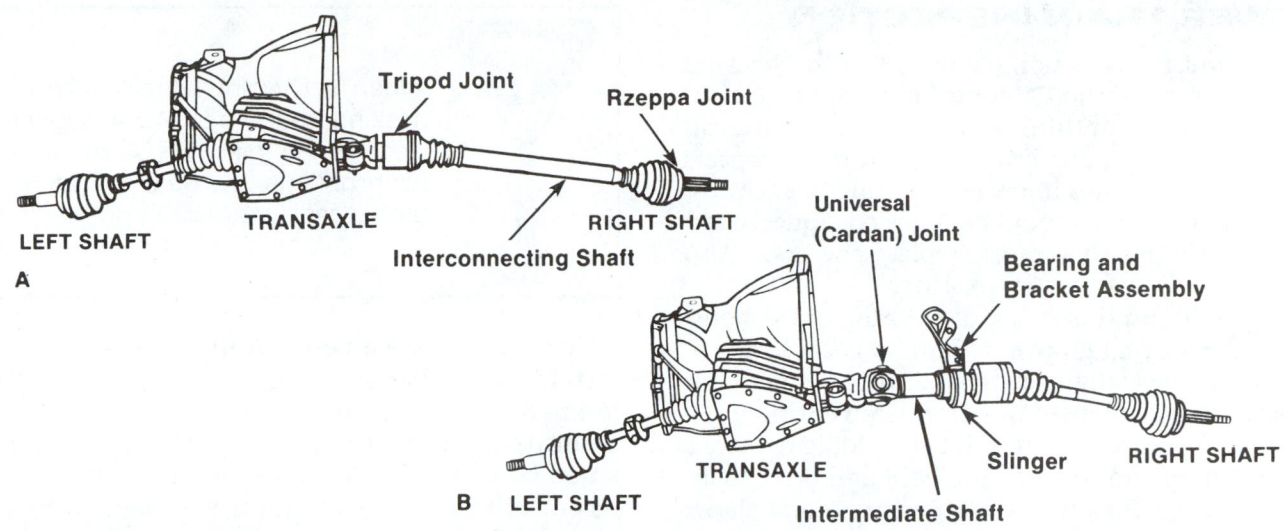

FIGURE 13-5 This car has (A) unequal and (B) equal length FWD half shafts.

train components because loose bearings and/or brackets vibrate.

Because the half shafts on a front-wheel-drive vehicle turn at roughly one-third the speed of the drive shaft in a rear-wheel-drive vehicle, half shaft balance and runout are not very important. The small damper weight that is sometimes attached to one half shaft serves to dampen harmonic vibrations in the drivetrain and to stabilize the shaft as it spins, not to balance it.

REAR-WHEEL DRIVE

There are two basic types of CV joint applications found on RWD unibody vehicles: independent rear suspension (Figure 13–6A) and solid axle housing (Figure 13–6B). In RWD with independent rear suspension (IRS), CV joints can be found at both ends of the axle shafts (for a total of four).

In propeller shaft applications, the CV joint still needs a protective boot, but it is a different design from the boots used on front-wheel-drive or rear-wheel-drive axle half shafts. Instead of wrapping around the outside of the joint, it is mounted within the outer housing. This protects the boot but also makes inspection more difficult.

CV-jointed propeller shafts revolve at high speeds and require balancing if removed for service. Remember that many RWD vehicles still use the conventional U-joint drive. Check the service manual whether a CV- or U-joint system is used.

FOUR-WHEEL DRIVE

There are several four-wheel-drive drivetrain configurations (Figure 13–7). CV joints and/or universal joints are used on front and/or rear axle shafts as well as on the front and rear drive shafts. On the typical 4WD vehicle, a *transfer case* is used to send power to front and rear axles. It is mounted to the side, underneath, or to the back of the transmission. A chain or gear drive within the case receives the power flow from the transmission and transfers it to two separate drive shafts leading to the front and rear axles.

A selector switch or shifter located in the driving compartment controls the transfer case so that power is directed to the axles as the driver desires. Power can be directed to all four wheels (4WD), to only two wheels—normally the rear wheels (2WD), or to no wheels (neutral). On many vehicles, the driver is also given the option of a "low" 4WD range for extra traction in especially rough conditions, such as deep snow or mud.

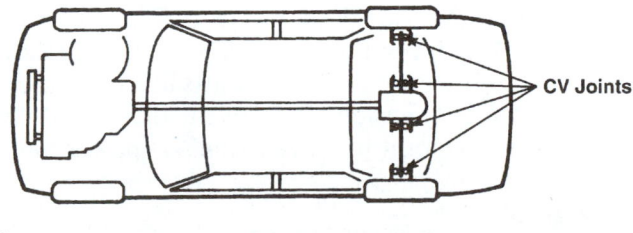

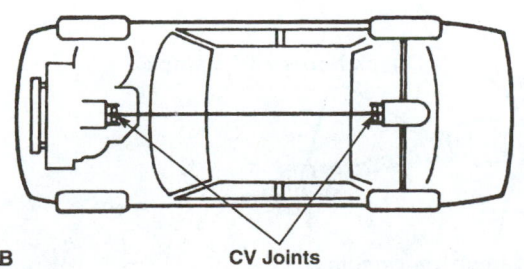

FIGURE 13-6 Compare (A) rear-wheel drive with independent rear suspension CV joint positions; (B) CV joint applications on the propeller shaft of a RWD vehicle.

POWER TRAIN INSPECTION

Begin damage inspection (Figure 13–8) by checking the condition of the CV-joint boots. Splits, cracks, tears, punctures, or thin spots caused by rubbing call for immediate boot replacement. If the boot appears rotted, this indicates improper greasing or excessive heat, and the boot should be replaced. Squeeze test all boots. If any air escapes, replace the boot. Also replace any boots that are missing.

Keep in mind that any discoloring of the housing at the bearing grooves is normal because all CV joints are especially heat treated. If the inner boot appears to be collapsed or deformed, venting it (allowing air to enter) might solve the problem. Place a round-tipped rod between the boot and drive shaft. This equalizes the outside and inside air and allows the boot to return to its normal shape.

Make sure that all boot clamps are tight. Missing or loose clamps should be replaced. If the boot appears loose, slide it back and inspect the grease inside for possible contamination. A milky or foamy appearance indicates water contamination. A gritty

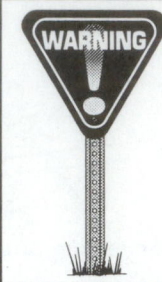

WARNING

When a CV-joint boot is torn or missing, there is often damage or wear in the joint. Check the joint for problems any time a boot requires replacement.

feeling when rubbed between the fingers indicates dirt. In either case, as a minimal service before the joint can be dried, the old grease will have to be cleaned out, the joint repacked with fresh grease, and a new boot installed. However, in many cases a water- or dirt-contaminated joint will have to be replaced rather than regreased.

The drive shafts should be checked for signs of contact against the chassis or rubbing. Rubbing can be a symptom of a weak or broken spring, engine mount, or chassis misalignment.

On front-wheel-drive transaxles with equal-length half shafts, inspect the intermediate shaft U-joint, bearing, and support bracket for looseness by rocking the wheel back and forth and watching for any movement.

Various drivetrain and suspension problems can be confused with symptoms produced by a bad CV joint. The following list of symptoms should help guide the technician to a proper diagnosis:

- **A popping or clicking noise when turning.** This signals a worn or damaged outer joint. The condition can be aggravated by putting the car in reverse and backing in a circle. If the noise gets louder, the outer joint(s) should be replaced.
- **A "clunk" when accelerating, decelerating, or when putting the transaxle into drive.** This kind of noise can come from excessive play in the inner joint of FWD applications, either inner or outer joints in a RWD independent suspension, or from the drive shaft CV joints or U-joint in a RWD or 4WD power train.

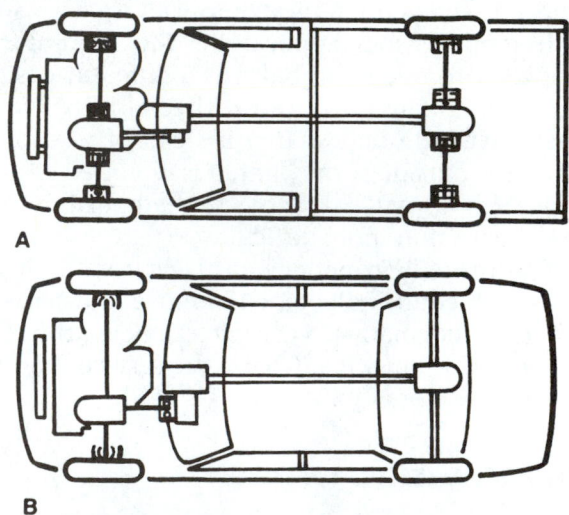

FIGURE 13-7 Study the differences between (A) four-wheel-drive system with independent front and rear suspension; (B) four-wheel-drive system with solid beam rear axle.

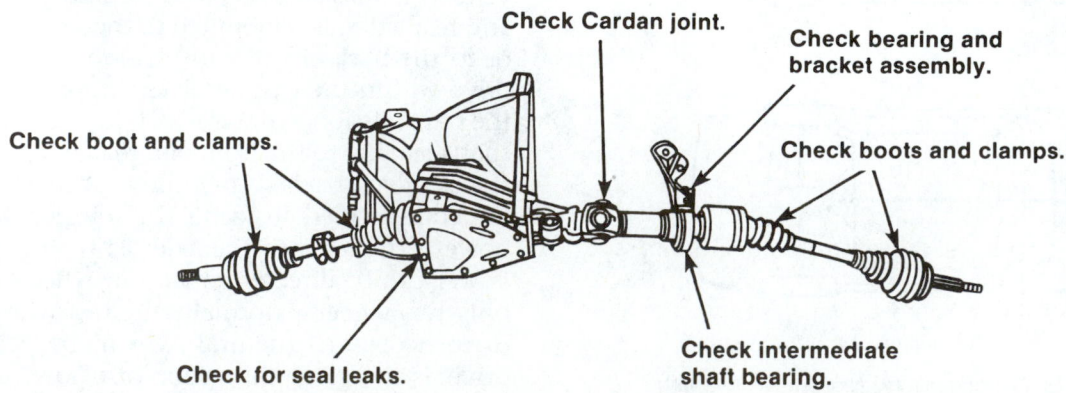

Check Cardan joint.

Check bearing and bracket assembly.

Check boot and clamps.

Check boots and clamps.

Check for seal leaks.

Check intermediate shaft bearing.

FIGURE 13-8 These are typical inspection points for FWD vehicles. They can be damaged during a collision.

Be warned, though, that the same kind of noise can also be produced by excessive backlash in the differential gears and transmission.

- **A humming or growling noise.** This is sometimes due to inadequate lubrication in either the inner or outer CV joint. It is more often due to worn or damaged wheel bearings, a bad intermediate shaft bearing on equal-length half shaft transaxles, or worn shaft bearings within the transmission.

- **A shudder or vibration when accelerating.** There is excessive play in either the inboard or outboard joints but more likely the inboard plunge joint. These kinds of vibrations can also be caused by a bad intermediate shaft bearing on transaxles with equal-length half shafts. On FWD vehicles with transverse mounted engines, this kind of vibration can also be caused by loose or deteriorated engine/transaxle mounts. Be sure to inspect the rubber bushings in the upper torque strap on these engines to rule out this possibility. Note, however, that shudder could also be inherent to the vehicle itself.

- A vibration that increases with speed is rarely due to CV-joint problems or FWD half shaft imbalance. An out-of-balance tire or wheel, an out-of-round tire or wheel, or a bent rim are the more likely causes. It is possible that a bent half shaft as the result of collision or towing damage could cause a vibration, as could a missing damper weight.

POWER TRAIN REPAIRS

Many mechanical parts such as engine mounts and transmission supports are through-bolted. The position of these mountings must be maintained parallel to each other to allow for the correct movement of the mechanical parts. When these mechanical mountings are not in proper alignment, free movement of parts may be restricted.

For example, misalignment in transmission linkages can easily cause erratic transmission performance. Proper drive shaft angles must be maintained to prevent vibration and chatter of the drive shaft and universal joints.

Motor mounts prevent minor engine vibrations and noise from being transferred into the body. Misalignment of these motor mounts can cause vibrations to be transferred directly to the passenger compartment. In order to provide the necessary structural support mountings for mechanical parts, special fasteners are frequently used.

At times it is desirable to completely remove the drivetrain from the unibody to make repairs to the body (Figure 13–9). Removal of the drivetrain allows ready access to structural unibody panels for repair or replacement.

In some cases, the time taken to remove the drivetrain pays off in considerable time savings in the repair or replacement of body panels. Repairs of damaged mechanical parts can sometimes become easier and faster after the piece is removed from the car. The decision whether to remove the drivetrain or to work around it in the car can be made by the repair technician or estimator (Figure 13–10).

When more time is saved in the repair of adjacent panels than is necessary to remove and reinstall the drivetrain, the drivetrain should be removed. When no time savings will result, the repairs should be made with the mechanical components in the car.

Some engines must be removed out the top of the engine compartment. Others must be removed from the bottom. Some should be separated from the transmission; others should be removed together. Since procedures vary, remember to refer to the service manual when in doubt.

With the car on the ground, start disconnecting wires, hoses, and cables that prevent engine removal (Figure 13–11).

FIGURE 13-9 The vehicle power train must sometimes be removed to repair major structural damage to the body or frame. *(Courtesy of Tech-Cor)*

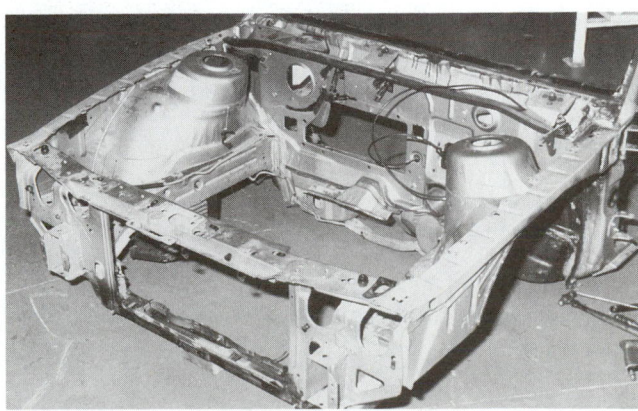

FIGURE 13-10 With the power train removed, you can use frame straightening equipment to bring the structure back into alignment. *(Courtesy of Tech-Cor)*

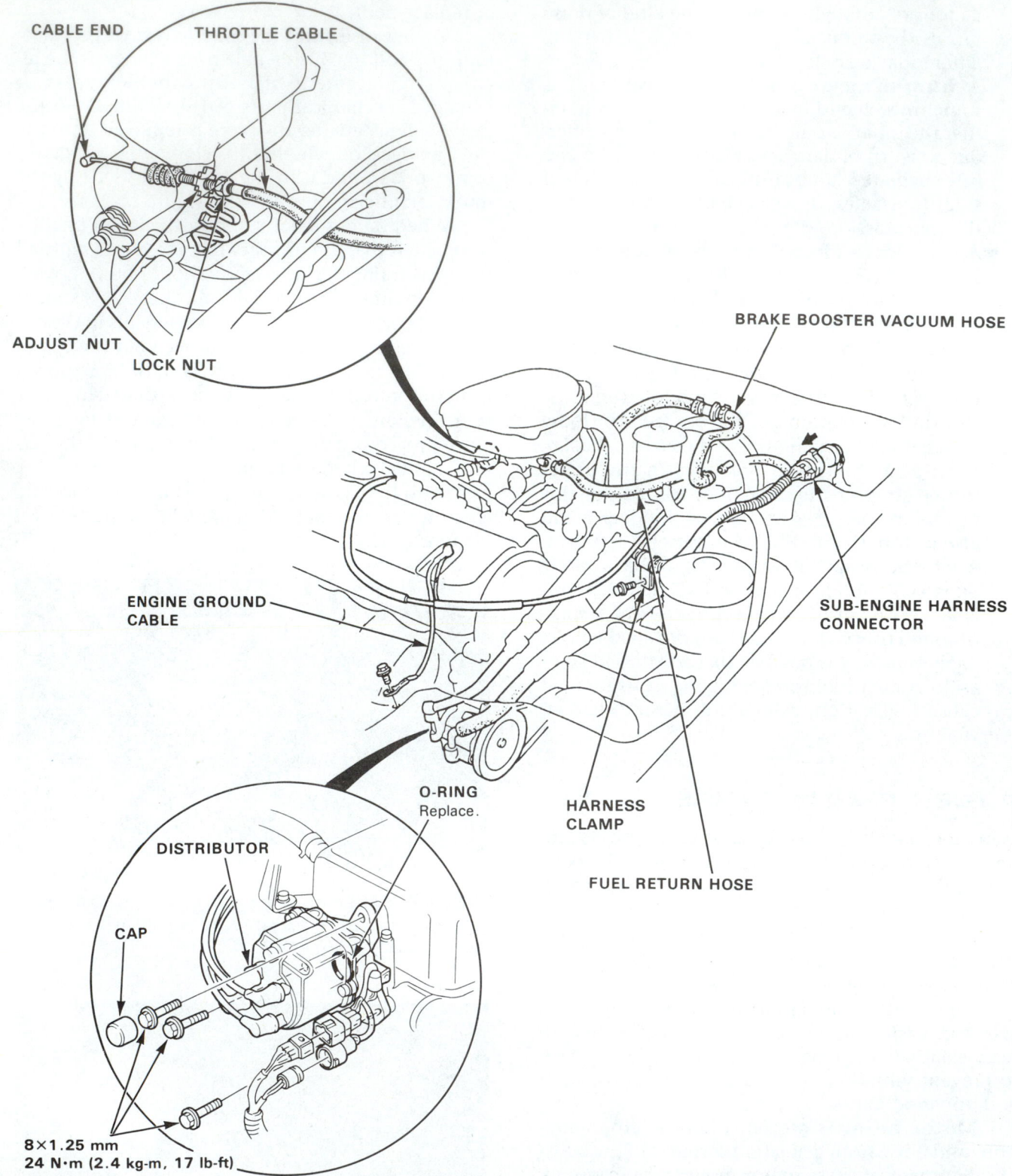

CABLE END

THROTTLE CABLE

ADJUST NUT

LOCK NUT

BRAKE BOOSTER VACUUM HOSE

ENGINE GROUND
CABLE

SUB-ENGINE HARNESS
CONNECTOR

O-RING
Replace.

HARNESS
CLAMP

DISTRIBUTOR

CAP

FUEL RETURN HOSE

8×1.25 mm
24 N·m (2.4 kg-m, 17 lb-ft)

FIGURE 13-11 If engine removal is needed, begin by disconnecting cables, hoses, and wires that attach between the engine-transmission assembly and the body. *(Courtesy of Honda Motor Co.)*

Make a *masking tape label* to identify where each part goes before removal. Print the name of the part on the masking tape to simplify reassembly (Figure 13–12). Mark the same code letter or number on both sides of what has been disconnected. Once all parts and connections have been identified, you can feel assured that everything will work properly when put back together.

When removing the drivetrain from a unibody vehicle, refer to the Service Manual for specific instructions or proceed as follows:

1. Disconnect both battery cables from the battery and the body ground from the battery tray (Figure 13–13). The cables will remain attached to the engine.
2. Remove the air cleaner to aid visibility and to increase the working area.
3. Drain the cooling system. If provided, open the *radiator drain cock* on the bottom radiator tank. The bottom radiator hose can also be disconnected at either end.
4. Disconnect the vacuum hoses connected from the body to the engine and transmission (Figure 13–14). Also disconnect the engine electrical harnesses. If possible, disconnect the main engine harness at the bulkhead connector on the firewall.
5. Disconnect the throttle body linkage and the transmission or transaxle linkage (Figure 13–15).
6. With a manual transaxle, disconnect the clutch cable or linkage.
7. Disconnect the speedometer cable. It is often fastened to the side of the transmission case (Figure 13–16). Disconnect the transaxle or transmission cooling lines at the radiator.
8. Disconnect the heater hoses (Figure 13–17). If they are connected to the heater core, be careful not to damage the heater core by twisting and pulling too hard.
9. If needed, disconnect the power steering pump lines where accessible (Figure 13–18). It may be easier to remove the pump from the engine.
10. Remove the air-conditioning (A/C) compressor. If the A/C system is not damaged and a recharge is not required, remove the A/C compressor from its mounting bracket and leave it with the body. If there are open lines, they should be plugged.

FIGURE 13–13 Always disconnect the battery cables before starting any work on the car. Impact damage could sever or short wires and possibly start an electrical fire. *(Courtesy of Tech-Cor)*

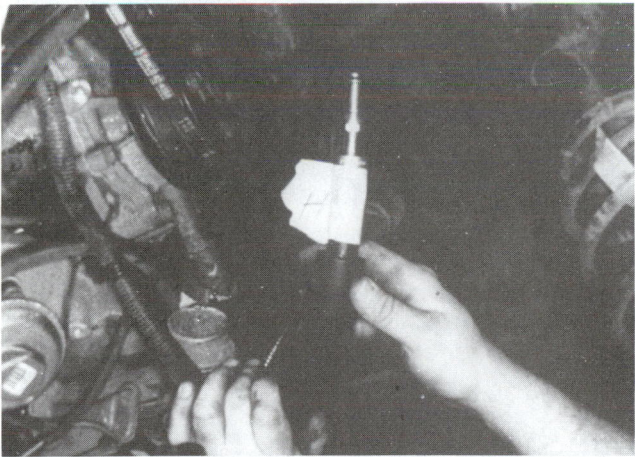

FIGURE 13–12 Label hoses, wires, and cables as they are disconnected to simplify and speed reassembly. This can save you time.

When disconnecting lines (fuel, power steering, transmission, air conditioning, etc.), use two wrenches when needed. One is used to hold the fitting while the other turns the flare nut. This will prevent you from twisting, kinking, and damaging the metal line.

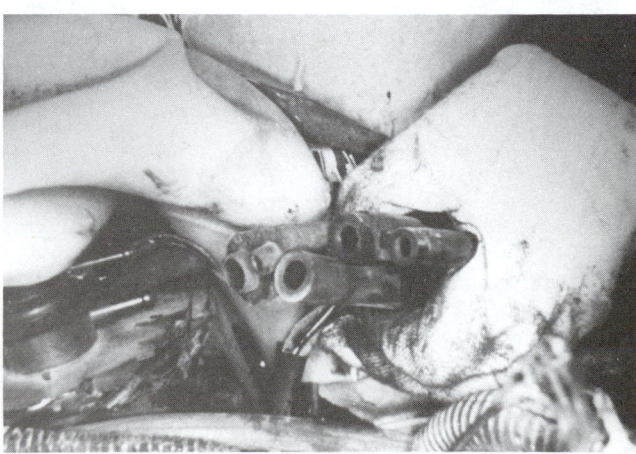

FIGURE 13–14 Disconnect all vacuum hoses. Again, make sure you label them as needed.

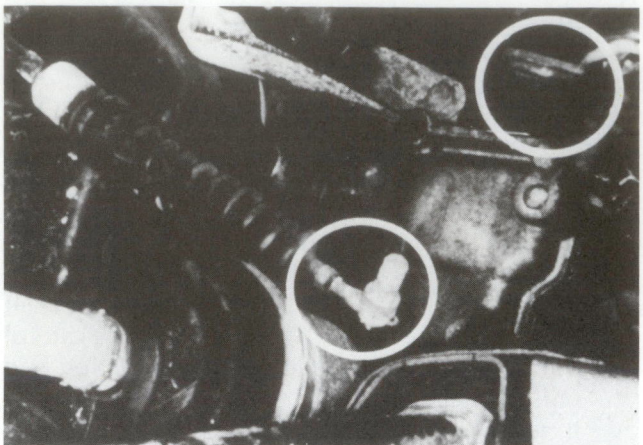

FIGURE 13-15 Disconnect the shift linkage cable(s) from the transaxle if needed.

FIGURE 13-16 Do not forget to disconnect the speedometer cable from the transmission or transaxle before removing the power train from the vehicle.

FIGURE 13-17 You will need to remove heater hoses at the engine or on the firewall.

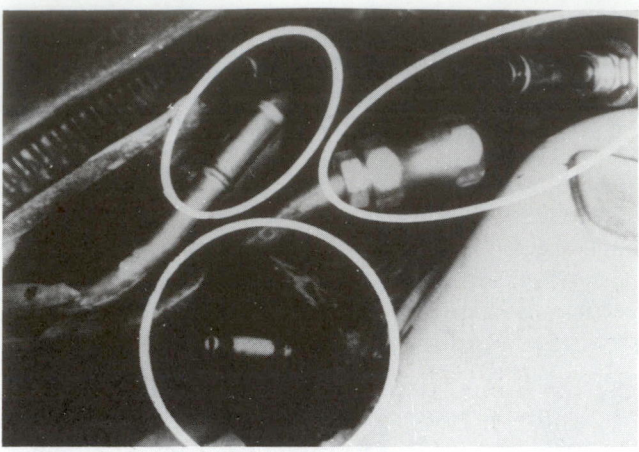

FIGURE 13-18 Disconnect the power steering system.

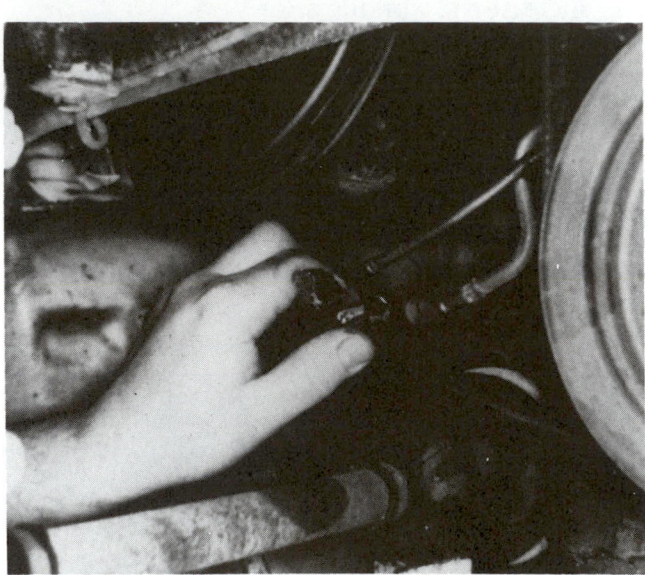

FIGURE 13-19 Disconnect and plug fuel flex lines near the fuel pump or engine.

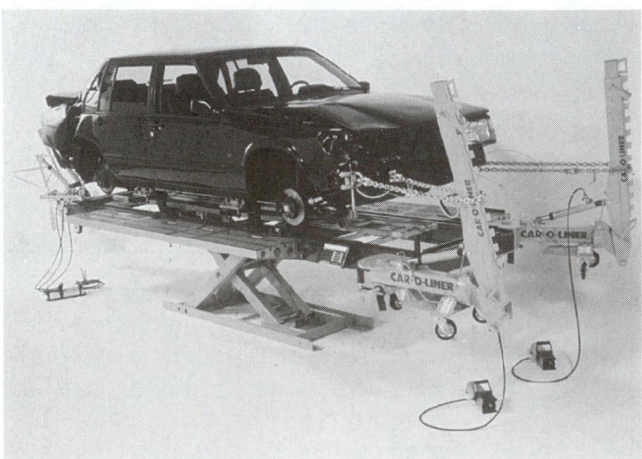

FIGURE 13-20 Modern frame straightening equipment will let you raise the vehicle for part removal and straightening. *(Courtesy of Car-O-Liner)*

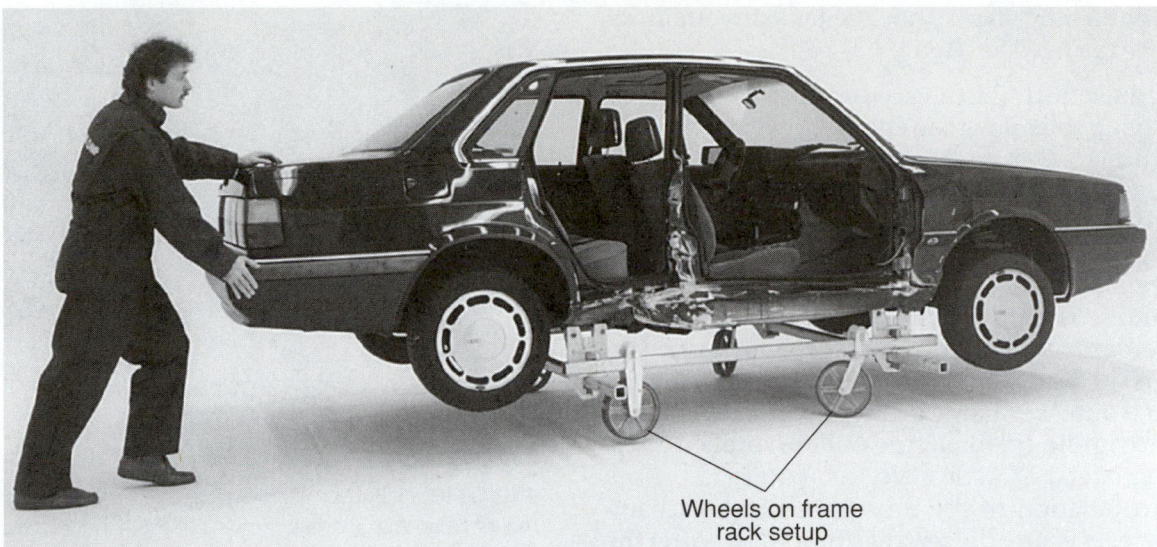

FIGURE 13-21 This vehicle has been mounted on pinch weld clamps and dolly wheels for moving the vehicle around the shop for removal of mechanical parts. *(Courtesy of Wedge Clamp International, Inc.)*

Remove the radiator fan and shroud for additional clearance if necessary. Double-check for any individual wires attached to the engine and traveling to the vehicle body.

11. Disconnect the fuel lines between the engine and body (Figure 13–19). Plug the fuel lines to prevent leakage.

12. To get the vehicle up in the air, use the frame straightening equipment if possible (Figure 13–20). Most racks and bench-racks allow for quick raising of the vehicle with built-in jacks or a scissor lift inside the rack. The vehicle can be placed on stands or a mechanical safety catch must be engaged on a scissor lift before working. Dolly wheels can also be used to move the vehicle to other areas in the shop for part removal (Figure 13–21).

If using older equipment, you might have to use a floor jack under the car and place jack stands in position on the cross tubes before removing some parts.

13. Once the vehicle is in the air and on jack stands, remove other parts as needed (Figure 13–22). This might include the CV-axles, exhaust system, steering linkage, suspension system, brake calipers, etc. (Figure 13–23). This prevents the need to bleed the brake system on reassembly; however, the caliper assembly must be fastened to the car and not allowed to hang by the hose.

14. Disconnect the exhaust pipe at the coupling behind the engine (Figure 13–24).

15. If required, remove the three upper strut tower mounting bolts from each side (Figure 13–25).

FIGURE 13-22 Make sure the vehicle is on jack stands or supported safely before working under it.

FIGURE 13-23 Brake calipers may have to be removed so you can free the CV-axles.

Remove other parts that prevent drivetrain removal as needed (Figure 13–26).

Double-check that everything is disconnected before trying to remove the engine/drivetrain assembly (Figure 13–27). As always, support the vehicle with jack stands. Care must be taken when separating the drivetrain from the body. Remember to keep checking all sides for wires or hoses that might not have been disconnected.

Remove the drivetrain (Figure 13–28) and repair the damaged unibody panels as described in Chapters 11 and 12. While the body is being repaired, an auto mechanic can service damaged parts of the drivetrain (bent engine pulleys, broken alternator, damaged CV-axles, etc.) as required.

Reinstallation of the drivetrain can be accomplished by reversing the removal procedure. After the unibody structure has been accurately repaired, the cradle can be quickly and correctly positioned by using the line-up holes located at the right front and right rear cradle mounting points. An incorrectly

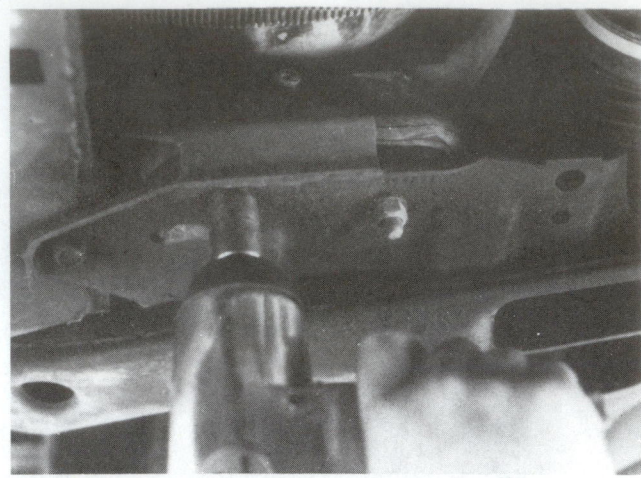

FIGURE 13–26 Here a technician is removing the motor mount bolts that prevent removal of the drivetrain from the vehicle.

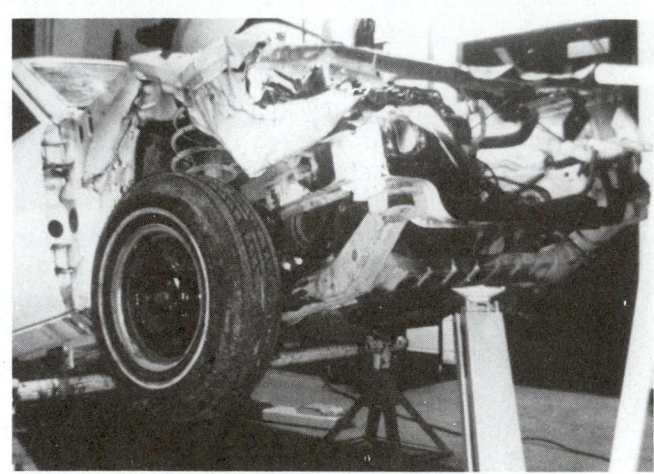

FIGURE 13–27 You may need to remove the engine from the top or bottom of the engine compartment. Front-wheel drivetrains normally come out the bottom. If vehicle is front-engine, rear-wheel-drive, the engine normally comes out the top of the engine compartment.

FIGURE 13–24 An impact wrench will help you to disconnect the exhaust system.

FIGURE 13–25 An air ratchet will save time on smaller fasteners, like on the upper strut tower.

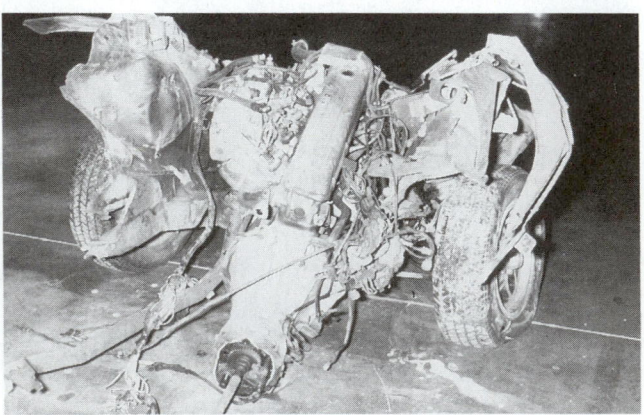

FIGURE 13–28 Once the drivetrain has been removed, replace all damaged parts while the assembly is out of the vehicle. It is easier to service at this time. *(Courtesy of Tech-Cor)*

positioned cradle can give the customer a wheel alignment problem.

It must be remembered that procedures mentioned in this chapter are general. The technician must be aware of the specifics of removal for each particular body style. These are found in the OEM Service Manuals. An example would be a body style without a full cradle. On that type of body, it is necessary to tie the suspension system to the engine and drivetrain assembly to help prevent the CV joints from separating from their assembly.

On a vehicle with a rear-wheel drive, the drivetrain removal procedure for bottom removal would be almost the same. With the car on the ground, disconnect everything you can reach from the top. Raise the vehicle off the ground to disconnect the remaining pieces from underneath (Figures 13–29 and 13–30)

1. Remove the drive shaft at the rear of the car by disconnecting the U-bolts (Figure 13–31). Make sure to mark the U-bolts for identification when reassembling.

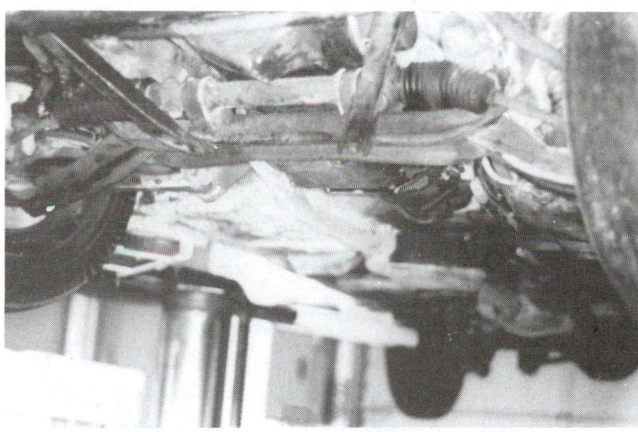

FIGURE 13–29 With the vehicle raised, make sure all components under the engine and transmission are disconnected.

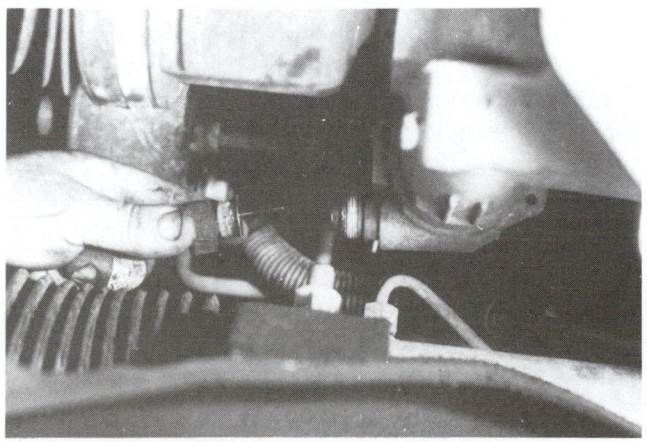

FIGURE 13–30 Remove transmission shift and speedometer cables. Various methods can secure them to linkage arms and the case.

2. Remove the bolts that secure the small cross member that helps support the transmission from the body (Figure 13–32). Remove the motor mounts (Figure 13–33). If the car has a manual transmission, remove the clutch linkage and return spring.

3. Disconnect the transmission cable, the wiring for the back-up lights, and the ground strap. Also

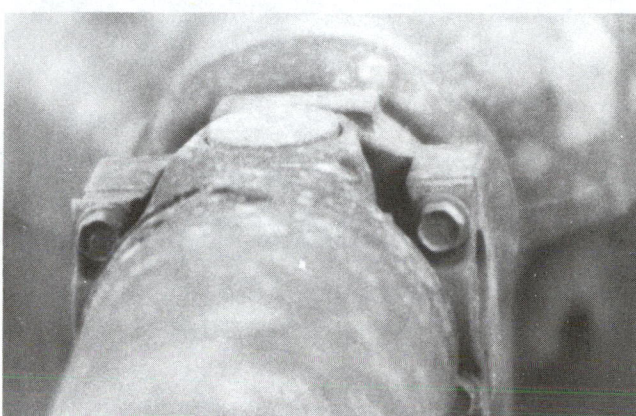

FIGURE 13–31 Scribe mark parts before the drive shaft is removed. Marks must be realigned during reassembly to prevent vibration.

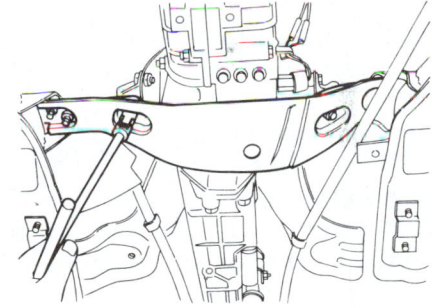

FIGURE 13–32 Remove the cross member that supports the transmission. *(Courtesy of Subaru Motor Co.)*

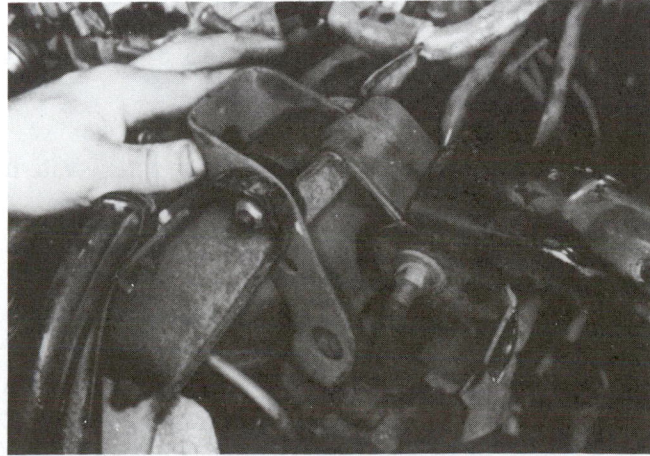

FIGURE 13–33 Make sure all motor mounts are disconnected before trying to raise the engine.

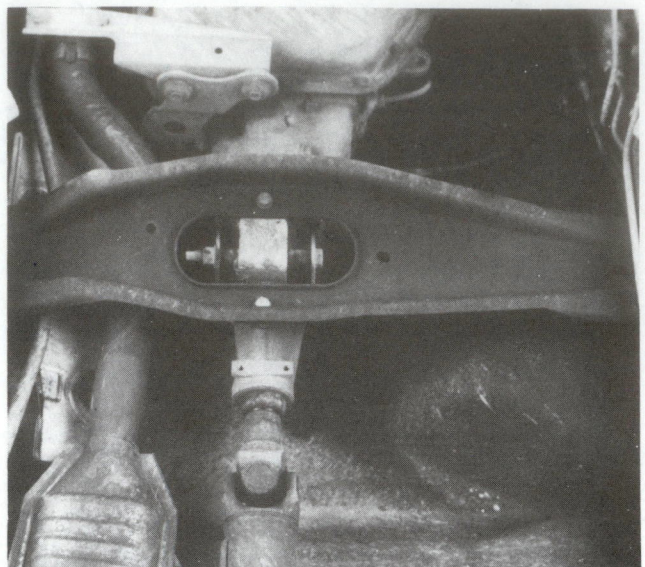

FIGURE 13-34 Here you can see a rear motor or transmission mount on a rear-wheel-drive vehicle.

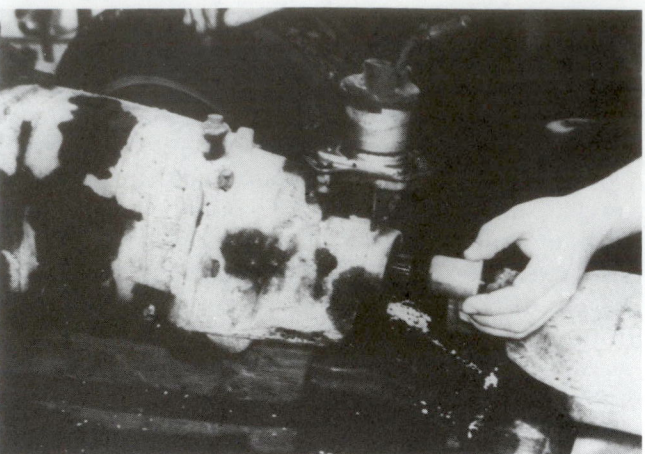

FIGURE 13-35 Plug transmission to prevent loss of fluid.

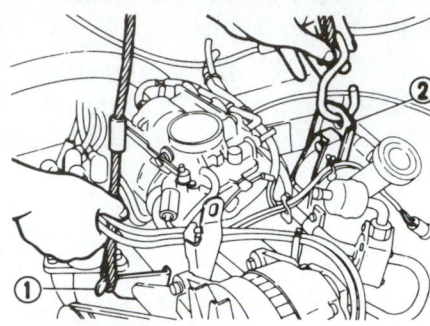

1　Front engine hanger
2　Rear engine hanger

FIGURE 13-36 When lifting the engine, keep hands and feet out from under the heavy assembly at all times. Connect the chain or cable to recommended lift points. (Courtesy of Subaru Motor Co.)

disconnect the transmission rear mount (Figure 13–34). All of these are attached to the transmission.

4. Disconnect the drive shaft at the transmission and be sure to protect the spline or gear that sticks out from the end of the transmission. Use a plastic plug or a rag (Figure 13–35). Disconnect the exhaust system at the catalytic converter.

5. Disconnect the steering rack from the steering column at the rack.

6. Disconnect the shocks from the top of the towers or at the bottom of the shocks. When disconnecting a shock at the bottom, it might be necessary to remove the wheel to have access to the through-bolt that attaches the shock to the rest of the suspension system.

7. Remove the brake lines at the wheel and support the disc brake caliper.

8. Connect the lift chain and hoist to recommended lift points on the engine and transmission (Figure 13–36).

A vehicle's suspension and steering system (Figure 13–37) performs three basic functions:

- It acts as the overall connection between the wheels and the vehicle body.
- It damps and controls the ride; that is, it acts to partially absorb road shock and sway.
- It provides directional control of the vehicle.

Up to a few decades ago, suspension and steering systems were not the concern of the body technician. Today, however, quality body shop repairs include the return of steering and suspension system parts to their original factory location.

CAUTION

An engine/transaxle or transmission assembly is very heavy. If dropped, it can easily chop off toes and fingers or crush bones. Keep your hands and feet out from under the engine assembly while moving it.

When starting to remove any drivetrain from any unibody, make the actual separation very slowly, while constantly walking around the entire vehicle to make sure everything is clear and disconnected.

SHOP TALK

Wire the struts together inboard after removing the engine. This will prevent damage to the CV-joints and rubber boots. Also, place plastic bags over the CV-joints for protection.

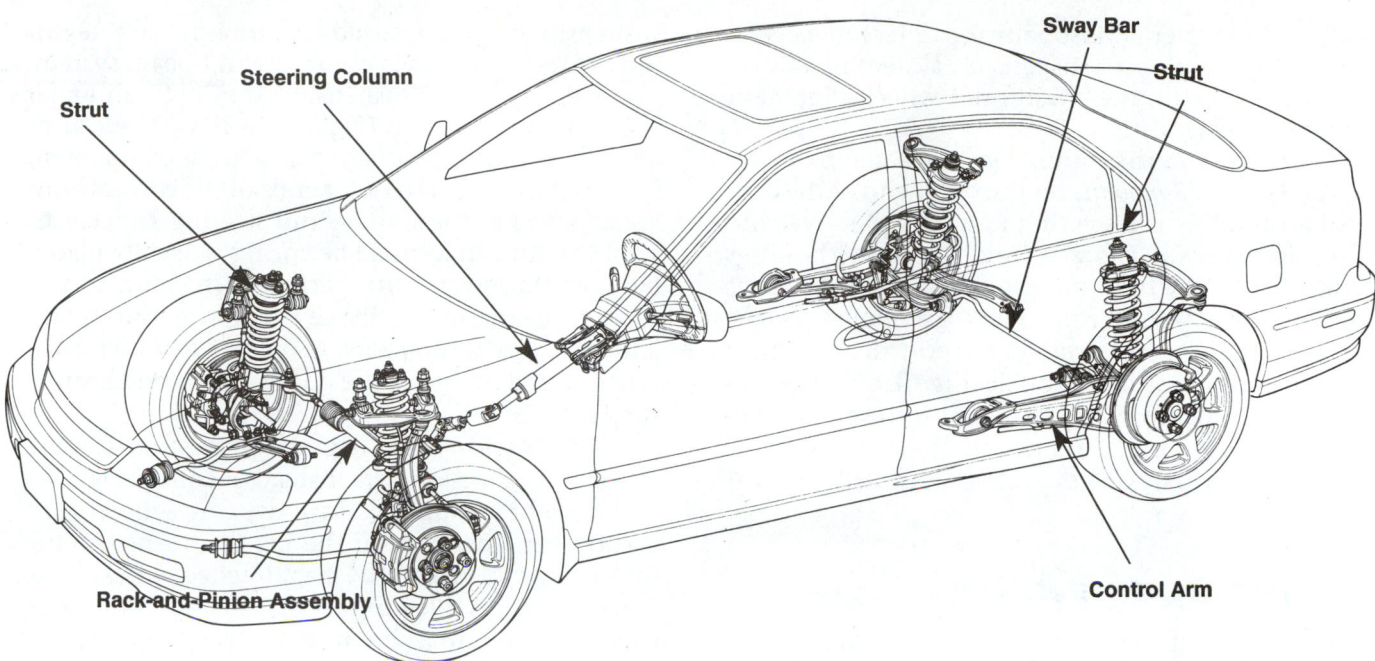

FIGURE 13-37 Here you can see major parts of the steering and suspension systems. *(Courtesy of Honda Motor Co.)*

In unibody construction, this is especially important because body panels provide the critical mounting positions for the suspension and steering systems. For example, the inner fender skirts in conventional body frame construction prevent dirt and splash from entering the engine compartment. In unibody construction, they are called strut towers and also provide the upper mounting controls for the strut suspension system.

In some cases, the steering gear of a unibody car is bolted directly to a body panel, like a cowl (Figure 13–38). When this is done, the body panel must hold the steering gear in its correct location. The unibody structure must maintain the proper relationship of the steering and suspension parts to each other.

Rear suspension components are also mounted directly to the unibody structural panels. The proper relationship of the rear suspension parts to each other and to front suspension parts depends on the position of the unibody structural panels.

All vehicles—not only those of unibody construction—require an interrelated suspension and body design. The difference in today's vehicles is that safe, efficient collision repairs demand that both body and mechanical parts be corrected together.

Collision repairs on body-over-frame vehicles, where the mechanicals and suspension are connected only to the frame, demand that the frame be restored to within certain tolerances. From there a wheel alignment shop can fine-tune the various suspension angles that affect road handling and safety. Frame vehicles have plenty of suspension adjustments built into their design. The system of repair requires the frame shop to straighten the frame, the body shop to hang sheet

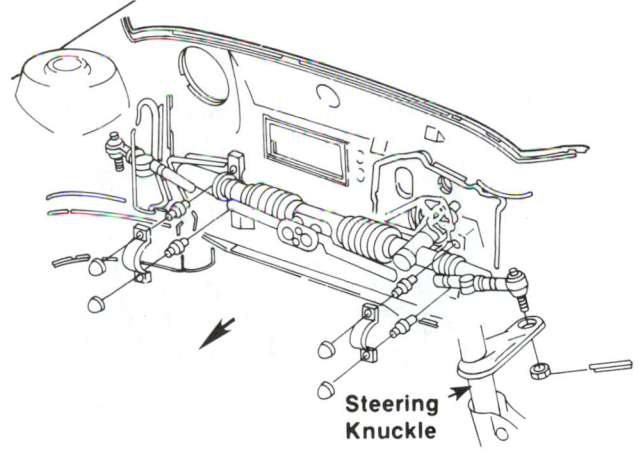

FIGURE 13-38 Note how this rack-and-pinion assembly bolts to the body. It may require removal for major straightening or replacement.

metal and straighten panels, and then the wheel alignment shop to adjust the suspension angles.

While the mechanical technician might make the repair to the suspension and steering system, it is a must for the body repair technician to know and understand what is involved and why it is important.

13.2 SUSPENSION SYSTEMS

Proper collision repairs of suspension systems and supporting unibody structural panels must restore the ability of these panels to support the high dynamic loads experienced during operation of the suspension system. Most body repair technicians focus their attention to suspension system repairs on the ability to restore traditional wheel alignment angles to specification.

Control arms mount on the frame to swivel up and down. *Ball joints* on the outer end of the control arms allow the steering knuckles to swivel and turn. The *steering knuckles* hold the wheel bearings and wheels. The hubs mount on the wheel bearings to hold the wheels or rims. The *wheels* hold the tires.

Suspension system springs support the weight of the car and allow suspension flexing. **Shock absorbers** are dampening devices that absorb spring oscillations (bouncing) to smooth the vehicle's ride quality. They may be gas-, oil-, or air-filled.

FRONT SUSPENSION

There are basic types of suspension systems used in passenger cars and light-duty trucks: coil spring, leaf spring, torsion bar, strut, and modified strut (Figure 13–39). Most frame bodies use either the coil spring, leaf spring, or torsion bar system. The strut

suspension is widely used in unibody cars. Light-duty trucks sometimes use the twin I-beam system.

The coil spring suspension uses both an upper and lower control arm (Figure 13–39A). These arms are attached with pivots to a structural component, such as the frame. The outer ends of the control arms are attached to the spindle and steering knuckle assembly with ball joints. The spring is usually placed between the lower control arm and the frame. Some types place the spring above the upper control arm and others use torsion bars. A separate shock absorber is connected to one of the control arms and a structural member.

The torsion bar suspension system (Figure 13–39B) uses torsion bars instead of coil springs. Vehicle weight is supported by a twisting action of the torsion bar. The front of the bar is attached to the lower control arm and the rear attached to the frame. Torsion bars installed in this manner are commonly called "longitudinal" because they run lengthwise in the vehicle.

The most commonly used front suspension system for unibody vehicles is the MacPherson (conventional) strut suspension (Figure 13–39C) and the

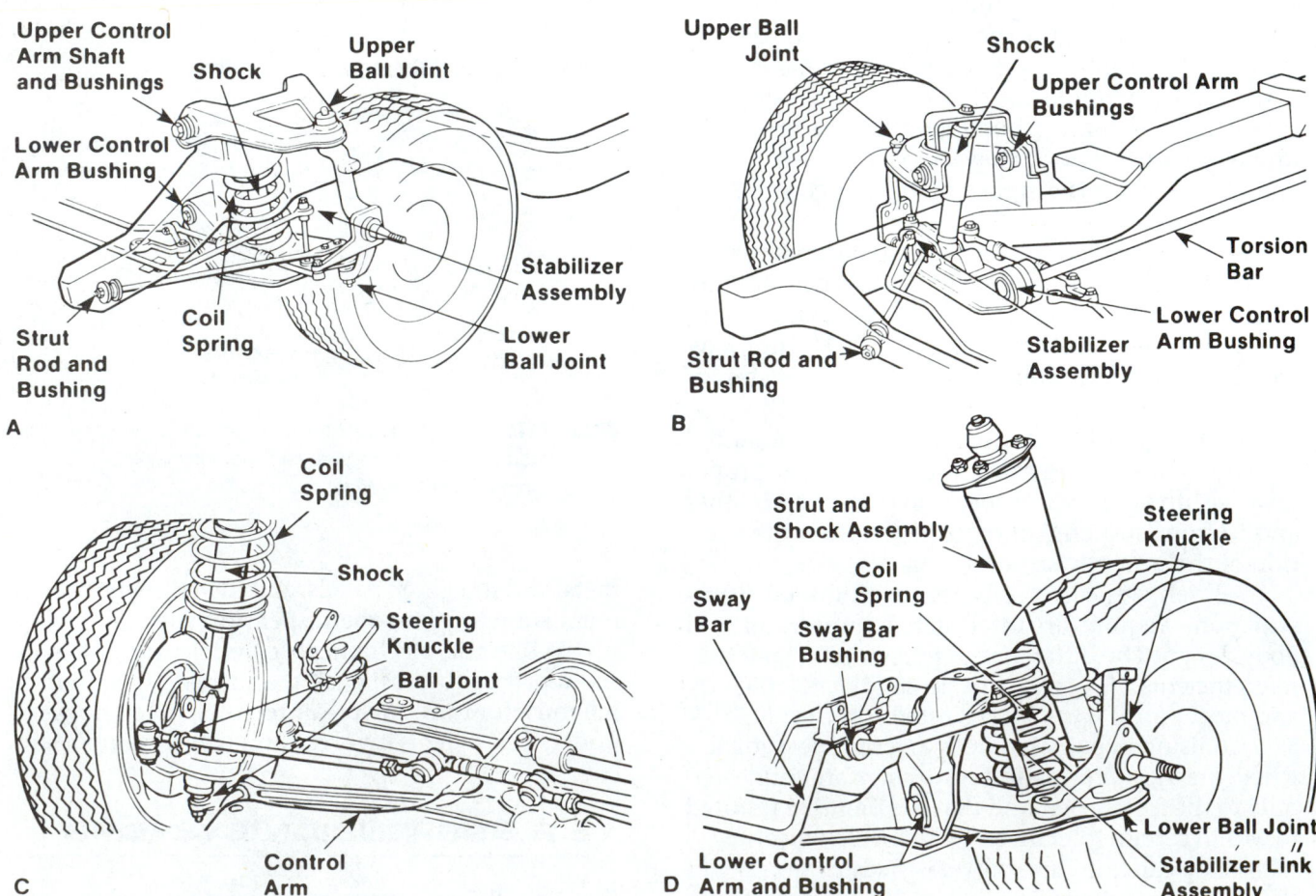

FIGURE 13-39 Study front suspension systems: (A) conventional coil spring system; (B) conventional torsion bar system; (C) conventional strut system; and (D) modified strut.

modified version (Figure 13–39D). The design and operation of a strut suspension system are simple compared with the more familiar parallel arm suspension system.

Like the parallel arm suspension, the MacPherson strut suspension has a lower control arm and spring. The strut replaces the shock absorber and the upper control arm (Figure 13–40). The strut suspension system uses a coil spring that is part of the strut assembly. In some cases, the coil spring is placed between the lower control arm and the unibody structure. In either case, the loads generated by the strut suspension are transferred directly to the unibody structure through the spring mounting.

The twin I-beam front suspension was developed to combine independent front wheel action with the strength and dependability of the mono beam axle. Twin I-beam axles allow each front wheel to absorb

A coil spring has deadly force when compressed! If this force is accidentally released, the spring can shoot out, possibly killing or seriously injuring someone. Use extreme caution when removing coil springs (Figure 13–41).

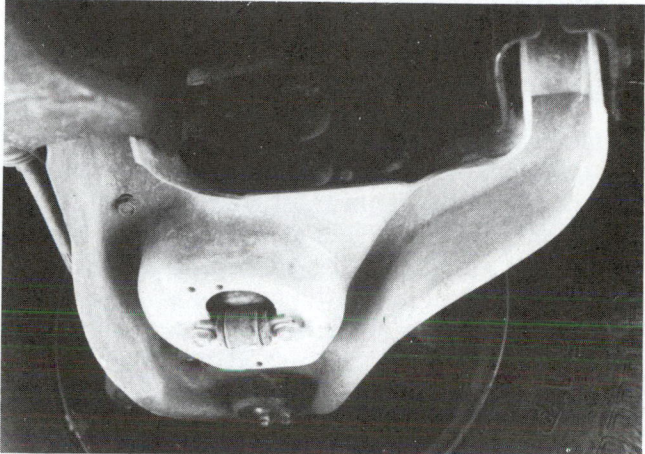

FIGURE 13–42 The lower control arm normally holds the ball joint, shock end, and bushings.

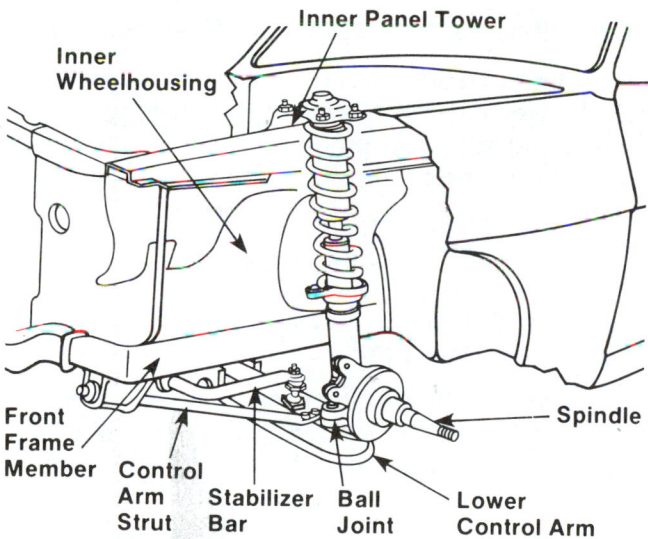

FIGURE 13–40 Note major components of a MacPherson suspension system.

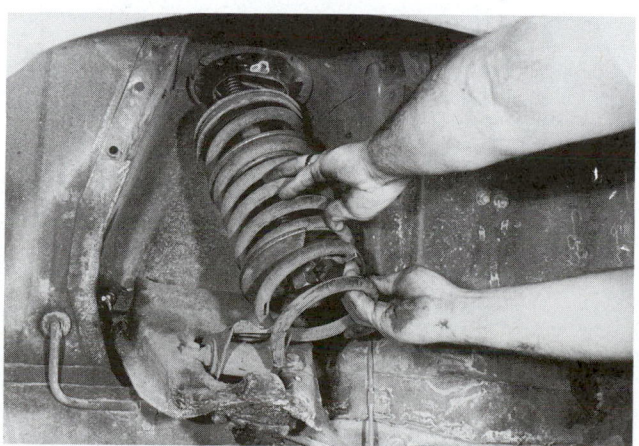

FIGURE 13–41 A spring compressor is needed to remove coil springs. Be extremely careful when doing this because a compressed spring has tremendous stored energy.

bumps and road irregularities independently, while providing sturdy, simple construction. The outer ends of the I-beams are attached to the spindle and to the radius arms. The inner ends are attached to a pivot bracket fastened to the frame near the opposite side of the vehicle.

The radius arms permit the I-beam to move up and down, stabilize any front to rear movement of the I-beam, and help maintain the proper caster setting. The spindle is mounted to the I-beam by a spindle bolt (kingpin). There are no ball joints.

Many unibody vehicles have lower control arms attached directly to the front body side rails. The use of a progressive coil spring (a spring that becomes increasingly stiff as it is compressed) provides good riding qualities on normal roads and sturdiness off the road.

Other unibody vehicles attach lower control arms to suspension subframes or cross members that are bolted directly to the front body side rails (Figure 13–42).

The suspension system mounting locations in the unibody structural panels (Figure 13–43) must be able to withstand the load transferred by this vertical movement. In the case of strut suspensions, the vertical load is transferred directly from the upper strut mount to the top unibody panel.

Front suspension ball joints (Figure 13–44) are used to connect the spindle to the upper and lower control arms. They provide a pivot for the wheel to turn and also allow for vertical movement of the control arms as the vehicle moves over irregularities in the road. While holding the front wheel in position, each ball joint performs a specific function. One joint supports vehicle weight, while the other functions as a steering dampener or resistance joint. But remember, both are subject to the stresses of road shock (up and down movement), braking, and cornering forces.

REAR SUSPENSION

Generally, rear suspensions require no special service. Broken or worn parts should be replaced. Remember that rear wheels, just like front wheels, are affected by road shock, acceleration, and braking forces. Control arm or leaf spring bushings are constantly flexing. In addition, bushings keep the rear wheels in line with the front wheels and when worn can upset the settings of the entire suspension and driveshaft systems.

FIGURE 13-43 Steering and suspension parts are often damaged in a collision, especially when the wheels hit objects. *(Courtesy of Tech-Cor)*

Loose, worn, or broken attaching parts will allow the rear wheels to shift, causing premature tire wear as well as short U-joint service life. A metallic jingling sound when driving over small bumps or unusual tracking (sometimes called **dog tracking**) also indicate the need for inspection. Usually a visual inspection is enough to determine repair requirements.

The coil spring and leaf spring nonindependent rear suspensions (Figure 13–45) are the most common today on rear-drive vehicles. The solid axle design will exhibit some of the same teeter-totter characteristics as noted with solid axle front systems. However, the effect is not nearly as dramatic since the rear wheels do not pivot.

Figure 13–46 shows various rear wheel suspension system variations.

Table 13–1 gives a diagnosis of suspension problems.

COMPUTERIZED SUSPENSION

Computerized suspension systems use sensors and an electronic control unit to adjust height, leveling of the car, and ride firmness in some models. Several systems are in current use:

1. Electronic air suspension (EAS)
2. Electronic level control (ELC)
3. Variable damping shock (VDS)
4. Springless electronic suspension (SES)
5. Computer-controlled suspension (CCS)

Some of these systems are relatively new, and repairing them will become more common as more vehicles come equipped with them. Anyone involved in the collision repair process needs a good basic knowledge of how these systems work. The best source of information about these systems is the service manual.

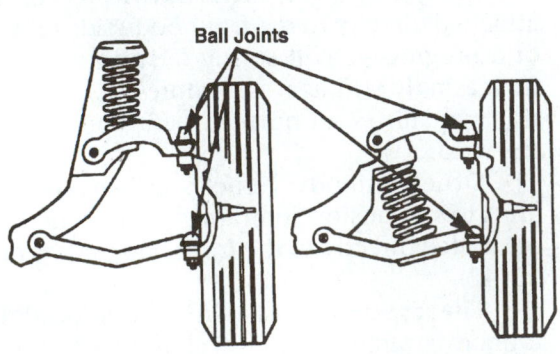

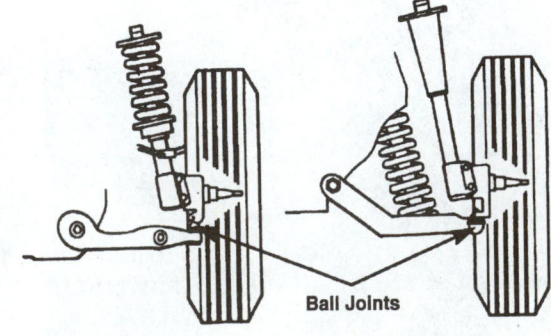

| Coil Spring or Torsion Bar Mounted on Upper Control Arm | Coil Spring or Torsion Bar Mounted on Lower Control Arm | MacPherson Strut | Coil Spring Mounted on Lower Control Arm with Modified Strut |

FIGURE 13-44 Study ball joint location variations.

TABLE 13-1: SUSPENSION PROBLEM DIAGNOSIS

Check	Noise	Instability	Pulls to One Side	Excessive Steering Play	Hard Steering	Shimmy
			Problem			
Tires/Wheels	Road or tire noise	Low or uneven air pressure; radials mixed with belted bias ply tires	Low or uneven air pressure; mismatched tire sizes	Low or uneven air pressure	Low or uneven air pressure	Wheel out of balance or uneven tire wear or overworn tires; radials mixed with belted bias ply tires
Shock Dampers (Struts/ Absorbers)	Loose or worn mounts or bushings	Loose or worn mounts or bushings; worn or damaged struts or shock absorbers	Loose or worn mounts or bushings	—	Loose or worn mounts or bushings on strut assemblies	Worn or damaged struts or shock absorbers
Strut Rods	Loose or worn mounts or bushings	Loose or worn mounts or bushings	Loose or worn mounts or bushings	—	—	Loose or worn mounts or bushings
Springs	Worn or damaged	Worn or damaged	Worn or damaged, especially rear	—	Worn or damaged	—
Control Arms	Steering knuckle control arm stop; worn or damaged mounts or bushings	Worn or damaged mounts or bushings	Worn or damaged mounts or bushings	—	Worn or damaged mounts or bushings	Worn or damaged mounts or bushings
Steering System	Component wear or damage	Component wear or damage	Component wear or damage	Component wear or damage	Component wear or damage	Component wear or damage
Alignment	—	Front and rear, especially caster	Front, camber and caster	Front	Front, especially caster	Front, especially caster
Wheel Bearings	On turns or speed changes: front-wheel bearings	Loose or worn (front and rear)	Loose or worn (front and rear)	Loose or worn (front)	—	Loose or worn (front and rear)
Brake System	—	—	On braking	—	On braking	—
Other	Clunk on speed changes: transaxle; click on turns: CV joints; ball joint lubrication	—	—	—	Ball joint lubrication	Loose or worn friction ball joints

Keep in mind that the computer suspension systems are vulnerable to certain problems, any one of which can upset the ability to maintain the desired distance between chassis and road. A plugged or leaky air hose, a bad solenoid, a dead compressor, or a misadjusted or faulty height sensor, for example, can interfere with the suspension's ability to level itself.

Because of this, check the operation of the system after a collision to make sure it is functioning correctly.

To understand the basic operation of one type of electronically controlled suspension system, look at Figure 13–47.

An *air spring* replaces the coil spring used in a conventional independent suspension system. The

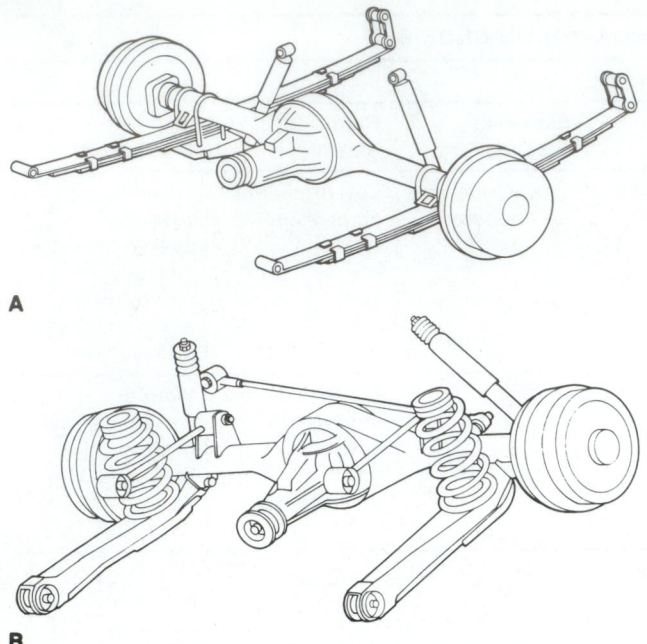

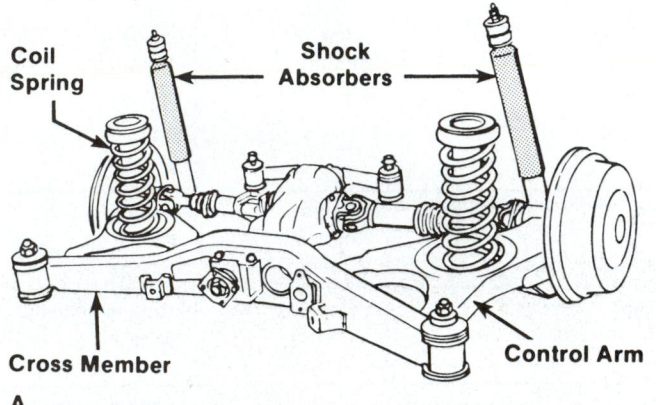

FIGURE 13-45 These are two of the more popular rear suspension systems: (A) leaf spring nonindependent and (B) coil spring nonindependent systems.

computer controls the car's height by either telling a battery driven air compressor to pump air into an air spring or telling a valve to let air out. Three height sensors, two in the front and one in the rear, tell the computer if the car is too high or too low. When the car reaches the right height, the sensors send a trim signal. The pump is shut off, the air spring valve is closed, and the air is trapped in the spring.

An **active suspension** system uses computer control to move the suspension arms up and down with road irregularities. They have the ability to adapt themselves to changing road and driving conditions. Instead of conventional springs or air springs to support the vehicle, double-acting hydraulic cylinders (called *actuators*) are mounted at each wheel. Each actuator maintains a sort of hydraulic equilibrium with the others to carry the vehicle's weight, while maintaining the desired body altitude. At the same time, each actuator serves as its own shock absorber, eliminating the need for yet another traditional suspension component.

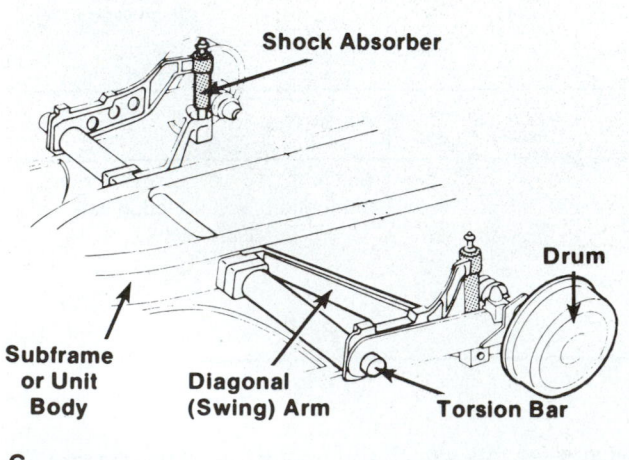

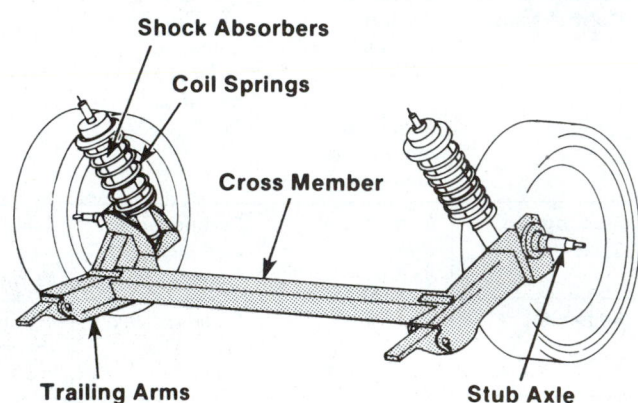

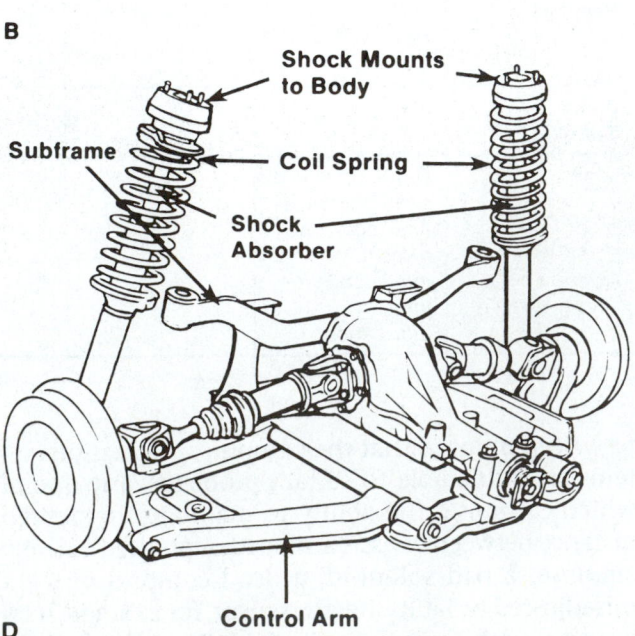

FIGURE 13-46 Compare rear suspension systems on front-wheel-drive vehicles: (A) independent rear suspension; (B) independent rear axle suspension; (C) swing arm rear suspension; and (D) strut rear suspension.

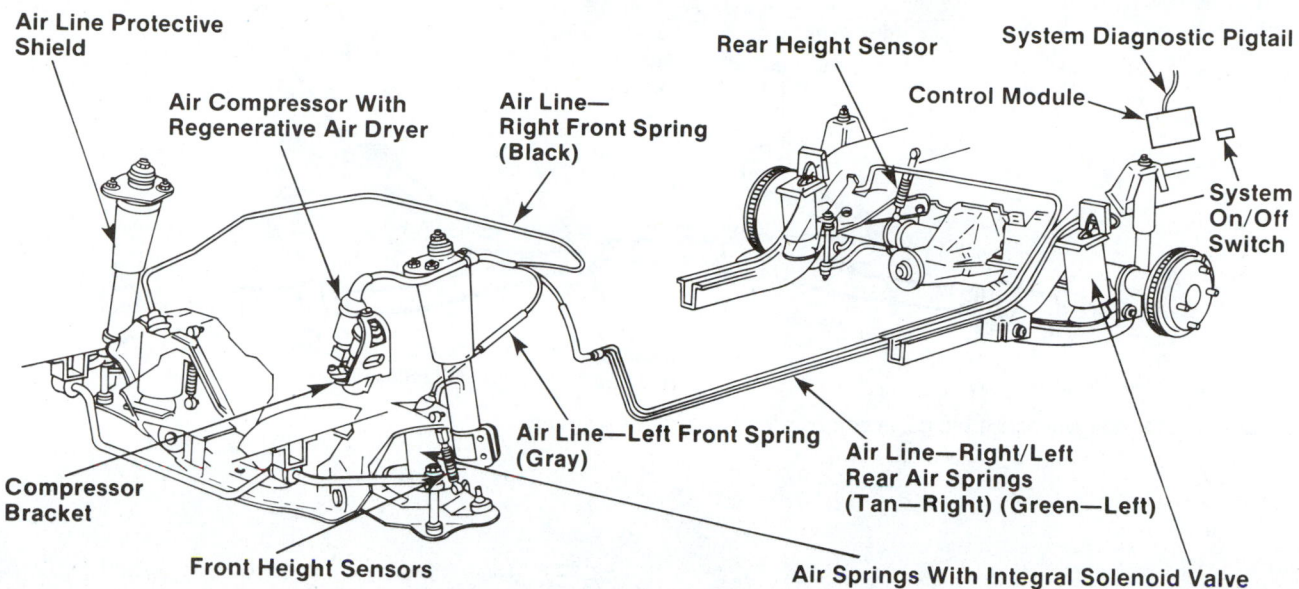

Air Line Protective Shield

Air Compressor With Regenerative Air Dryer

Air Line—Right Front Spring (Black)

Rear Height Sensor

System Diagnostic Pigtail

Control Module

System On/Off Switch

Compressor Bracket

Front Height Sensors

Air Line—Left Front Spring (Gray)

Air Line—Right/Left Rear Air Springs (Tan—Right) (Green—Left)

Air Springs With Integral Solenoid Valve

The system consists of an electric air compressor with regenerative air dryer, electronic height sensors, eight quick connect air fittings, four air springs with integral solenoids, four one-piece air lines connecting each spring to the compressor, and a control module with a single chip microcomputer.

FIGURE 13-47 Study parts of a typical electronic air suspension system.

In other words, each hydraulic actuator acts as both a spring (with variable-rate damping characteristics) and a variable-rate shock absorber. This is accomplished in an active suspension system by varying the hydraulic pressure within each cylinder and the rate at which it increases or decreases. By bleeding or adding hydraulic pressure from the individual actuators, each wheel can react independently to changing road conditions.

A steering angle sensor is used to signal the computer when the vehicle is turning. To monitor body motions, a roll sensor and lateral acceleration and G-sensors are used. The computer also monitors hydraulic pressure within the system and the speed of the pump motor.

Once it has all the necessary inputs, the computer can then regulate the ebb and flow of hydraulic pressure within each individual actuator according to any number of variables and its own built-in program. Bonus features are, for example, leaning into turns or even raising a flat tire on command to change the tire without using a separate jack.

13.3 STEERING SYSTEMS

The *steering system* transfers steering wheel motion through gears and linkage rods to swivel the front wheels. When you turn the *steering wheel*, a *steering shaft* extends down through the *steering column* and rotates the steering gearbox.

The *steering gearbox*, either a worm or rack-and-pinion type, changes the wheel rotation into side

movement for turning the wheels. A series of *linkage rods* connect the steering gearbox with the steering knuckles.

- The standard or **parallelogram steering system** (Figure 13–48) is the most common type on conventional frame cars. A *pitman arm* attaches the steering box to the linkage rods. Steering action is relayed via the *center link*, again attached by either ball sockets or bushings. The *idler arm* supports the center link at the opposite end, holding the system parallel and transmitting horizontal steering action. If up and down movement is excessive, toe change might exceed manufacturer's limits thereby creating premature and rapid tire wear.

  Tie-rod ends attach the linkage to the steering knuckles. They are the final wearable pivots of the system. Looseness in the ball sockets causes steering play. If you can wiggle the tires sideways during the inspection (car off the ground, hand pressure only), the tie-rod ends are worn or damaged and should be replaced.

- The **rack-and-pinion steering system** (Figure 13–49) is fast becoming a standard system for unibody vehicles. It gets its name from the *pinion gear* attached to the steering column and the *rack gear* in the steering gear housing. This rack gear is moved right to left within the housing by the rotation of the pinion gear. The ends of the steering rack are attached to the front wheel spindles by tie rods. In unibody construction, the rack-and-pinion steering gear assembly on some cars is mounted to the cowl panel.

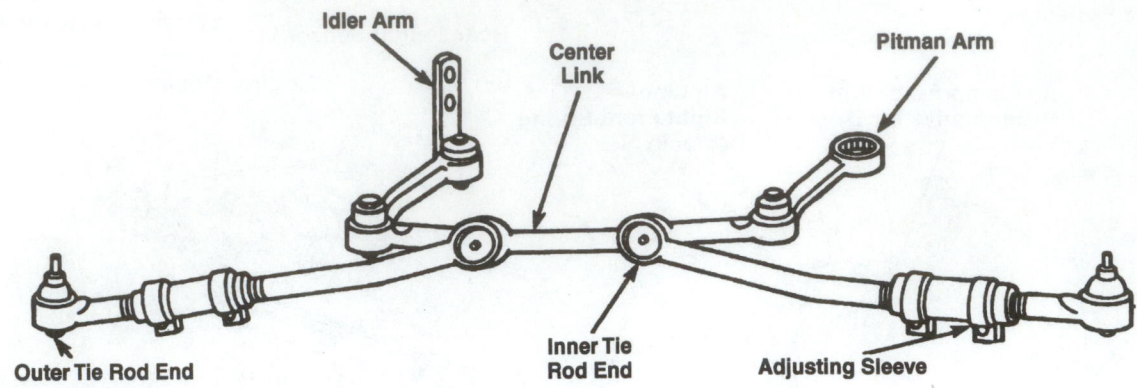

FIGURE 13-48 Memorize the parts of a parallelogram steering system.

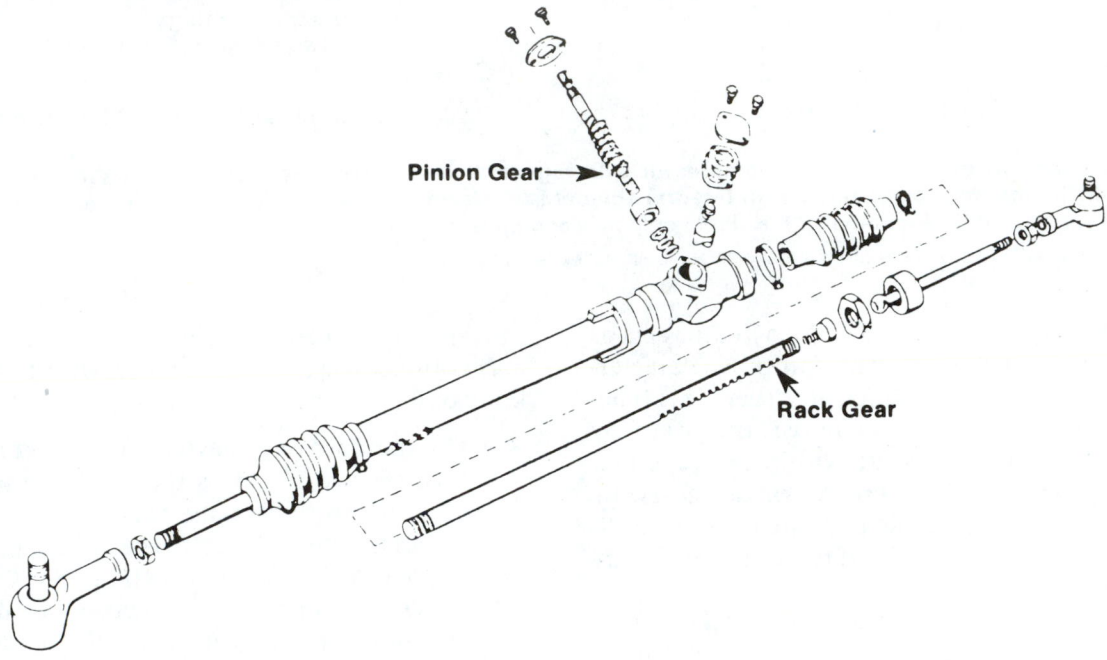

FIGURE 13-49 The rack-and-pinion steering system is the most common type today.

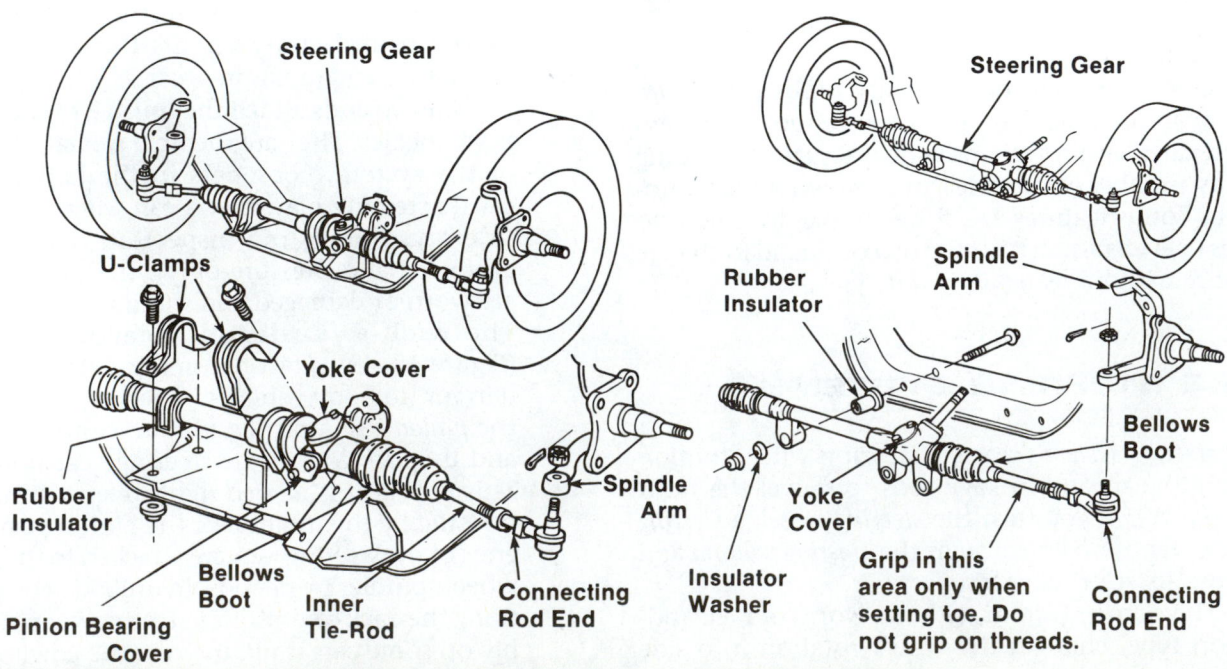

FIGURE 13-50 Here are typical attaching methods for manual rack-and-pinion assembly.

SHOP TALK

Remember that rack-and-pinion units must be mounted on a level plane. The closer the mounting brackets are to the middle of the rack, the more critical this measurement becomes (Figure 13–50). Misalignment of the rack and pinion will cause changes in the steering geometry during jounce/rebound. This condition cannot be corrected by changing the length of the tie rods.

In other cases, the rack-and-pinion steering gear is mounted to the front suspension cross member or the engine cradle assembly. The rack-and-pinion steering gear must be mounted securely because any movement will cause the car to wander as it travels down the road.

Rack-and-pinion has fewer friction points than a traditional steering system, so more energy and movement from road forces get through to the steering wheel. This gives the driver a more positive feel of the road. A power-assisted rack-and-pinion system responds faster to input changes. It also steers more easily without a boost than hydraulic systems.

STEERING INSPECTION

To check a rack-and-pinion system, begin by raising the car and taking the weight off the front suspension. Visually inspect the steering system for any physical damage. Check the boots for leaks, inspect the tie rods, and examine the mounting points for any distortion. Inspect the tie-rod ends. Grab the tie rod near the tire and try pushing it up and down. Any vertical looseness indicates damage or wear.

Check the inner tie rod socket by squeezing the bellows until the socket can be felt. With the other hand, push and pull on the tire. Looseness in the socket indicates damage or wear. Take a front tire in each hand and see if they can be moved back and forth in opposite directions. If excessive movement is noted, wear or damage is likely. Observe the rack-and-pinion at the same time. Any movement might indicate a problem.

POWER STEERING

The *power steering* unit is designed to reduce the amount of effort required to turn the steering wheel (Figure 13–51). It also reduces driver fatigue on long drives and makes it easier to steer the vehicle at slow road speeds, particularly during parking.

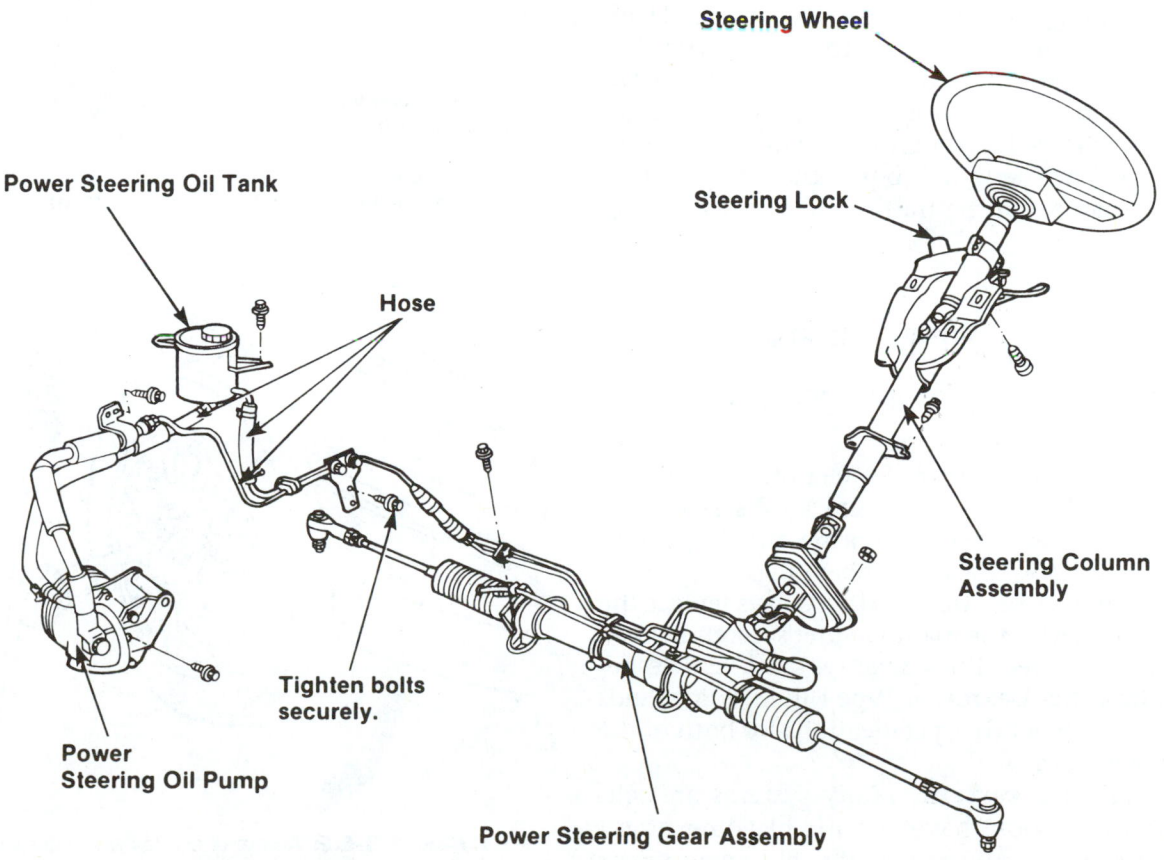

FIGURE 13–51 A power steering system often uses a pump to force hydraulic oil through the system to ease steering wheel effort.

Power steering can be broken down into two design arrangements: hydraulic and electrically controlled.

In the conventional arrangement, *hydraulic power* (fluid under pressure) is used to assist the driver. With the electric-type assist, a motor and electronic controls provide power assistance in steering.

Power steering hoses carry the oil to and from the pump. A *hydraulic piston* on the steering linkage or in the gearbox helps turn the wheels. *Hydraulic valves* control power assist.

In an electronically controlled power steering arrangement, an electric/electronic rack-and-pinion unit replaces the hydraulic pump, hoses, and fluid associated with conventional power steering systems. The design features a DC motor armature with a hollow shaft to allow passage of the rack through it. The outboard housing and rack are designed so that the rotary motion of the armature can be transferred to linear movement of the rack through a ball nut with thrust bearings. The armature is mechanically connected to the ball nut through an internal/external spline arrangement.

With a *computer-assisted steering* system, *sensors* provide feedback for the computer. The *computer* or electronic control unit can then precisely control power assist as variables change. Mechanical steering is still provided with an electrical failure.

With *electronically controlled power steering*, the conventional power steering parts are replaced with electronic controls and an electric motor. The electric motor is mounted in the rack assembly. A DC motor armature with a hollow shaft is used to allow passage of the rack through it. The outboard housing and rack are designed so that the rotary motion of the armature can be transferred to linear movement of the rack.

POWER STEERING SERVICE

Here are some power steering service tips that should be kept in mind:

- **Protect the system.** Protect the system from invasion by dirt and moisture. If the system must be open, be sure to plug or tie off all openings with a plastic sheet or rubber plugs.
- **Use recommended fluid.** Always replace the lost fluid with the manufacturer's recommended type to protect the warranty. Most vehicles require either Dexron or Type F fluid. Some fluids claim to meet the specifications for both of the above types.
- **Bleed the system.** Many systems are self-bleeding. Some have specific bleeding procedures in the service manual to eliminate air. Air in the system can cause noise, vibration, and erratic performance.

- **Check the hose routing.** Check the hose routing when reassembling power steering systems. Always route and hang the same as in the factory installation. Avoid contact with other parts. Especially watch rubbing against moving parts.

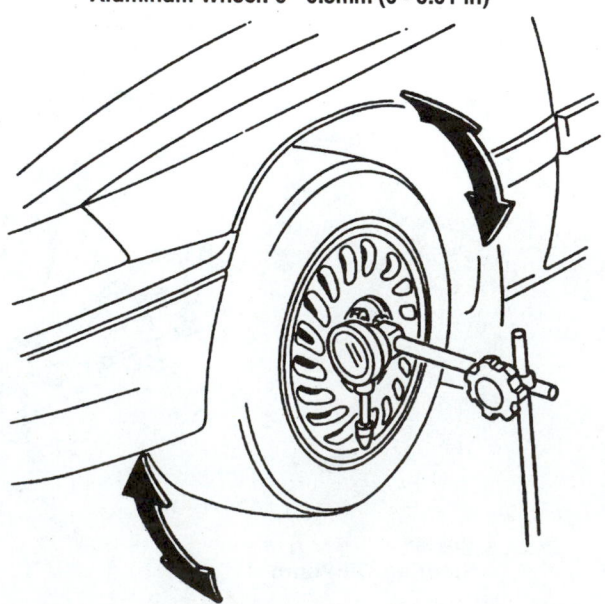

Front and Rear Wheel Axial Runout

Standard: Aluminum Wheel: 0 - 0.3 mm (0 - 0.01 in)

Front and Rear Wheel Radial Runout

Standard: Aluminum Wheel: 0 - 0.3mm (0 - 0.01 in)

FIGURE 13-52 Always check for wheel damage when needed. Check both axial and radial runout. Specs given are typical, but refer to the service manual for a specific vehicle. *(Courtesy of Honda Motor Co.)*

TABLE 13-2: STEERING PROBLEM DIAGNOSIS

Check	Problem					
	Noise	Instability	Pulls to One Side	Excessive Steering Play	Hard Steering	Shimmy
Tires/Wheels	Road/tire noise	Low/uneven tire pressure; radial tire lead	Low/uneven tire pressure; radial tire lead	Low/uneven tire pressure	Low/uneven tire pressure	Unbalanced wheel; uneven tire wear; over-worn tires
Tie-rods	Squeal in turns: worn ends	—	Incorrect toe: tie rod length	Worn ends	Worn ends	Worn ends
Mounts/ Bushings	Parallelogram steering: steering gear mounting bolts, linkage connections; rack & pinion steering: rack mounts	Idler arm bushing	—	Parallelogram steering: steering gear mounting bolts, linkage connections; rack & pinion steering: rack mounts	Parallelogram steering: steering gear mounting bolts, linkage connections; rack & pinion steering: rack mounts	Parallelogram steering: steering gear mounting bolts linkage connections; rack & pinion steering: rack mounts
Steering Linkage Components	Bent/damaged steering rack	Incorrect center link/ rack height	Incorrect center link/ rack height	Worn idler arm, center link, or pitman arm studs; worn/damaged rack	Idler Arm binding	Worn idler arm, center link, or pitman arm studs
Steering Gear	Improper yoke adjustment on rack & pinion steering	—	—	Improper yoke adjustment on rack & pinion steering; worn steering gear/ incorrect gear adjustment on parallelogram steering; loose or worn steering shaft coupling	Parallelogram steering: low steering gear lubricant, incorrect adjustment; rack & pinion: bent rack, improper yoke adjustment	—
Power Steering	—	—	—	—	Fluid leaks, loose/worn/ glazed steering belt, weak pump, low fluid level	—
Alignment	—	—	Unequal caster/camber	—	Excessive positive caster, excessive scrub radius (incorrect camber and/or SAI)	Incorrect caster

• **Check for leaks.** After making repairs, always check for fluid leaks before releasing the vehicle to the customer.

To perform a diagnostic check of possible steering problems, see Table 13–2.

WHEEL SERVICE

During a collision, wheels are often bent and damaged, which results in serious vibration when driven after repairs. Always check for wheel damage when making repairs.

Wheel runout is caused by a bent, damaged rim. **Radial runout** causes the diameter of the wheel to change as it is rotated. **Axial runout** causes the wheel to wobble sideways as it rotates. A dial indicator can be used to quickly check for wheel runout (Figure 13–52).

When installing wheels, especially lightweight aluminum ones, use a torque wrench to tighten lug

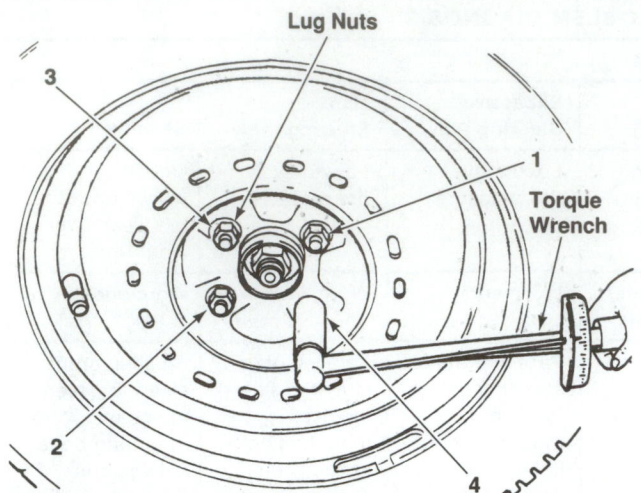

FIGURE 13-53 Always use a torque wrench to properly tighten lug nuts to specs. This is critical with today's lightweight aluminum "mag wheels." *(Courtesy of Chrysler Corp.)*

nuts. This will prevent you from warping the hub and wheel and causing runout vibration (Figure 13–53).

FOUR-WHEEL STEERING

The industry is now offering four-wheel independent steering systems, where the rear wheels also help to turn the car. This is done by either electrical or mechanical means. Because these systems are going to be more common in the future, the repair industry will have to adapt to them (Figure 13–54).

The idea behind four-wheel steering is that a vehicle requires less driver input for any steering maneuver if all four wheels are steering the vehicle. As with two-wheel steering vehicles, grip holds the four wheels on the road. However, when the driver turns the wheel slightly, all four wheels react to the steering input, causing slip angles to form at all four wheels. The entire vehicle moves in one direction

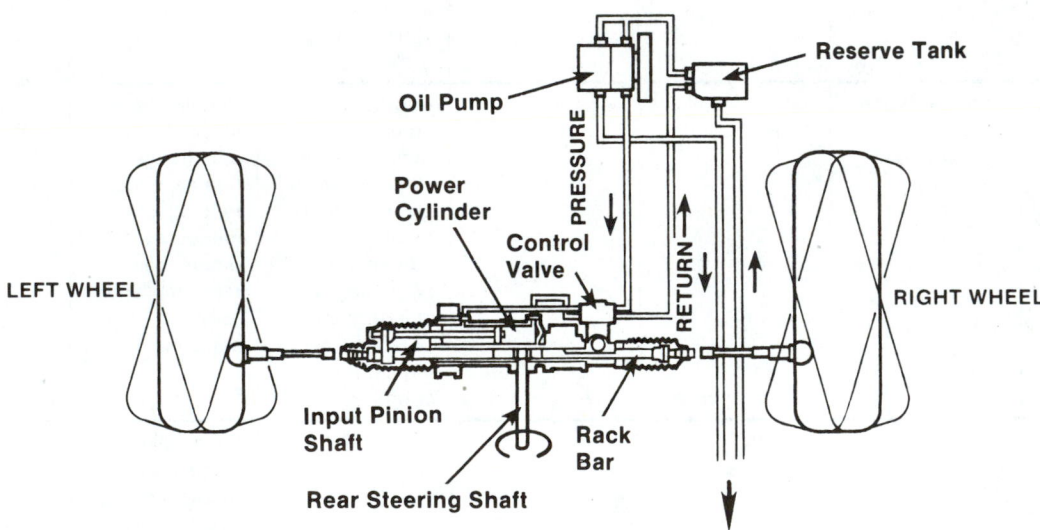

FRONT STEERING SYSTEM

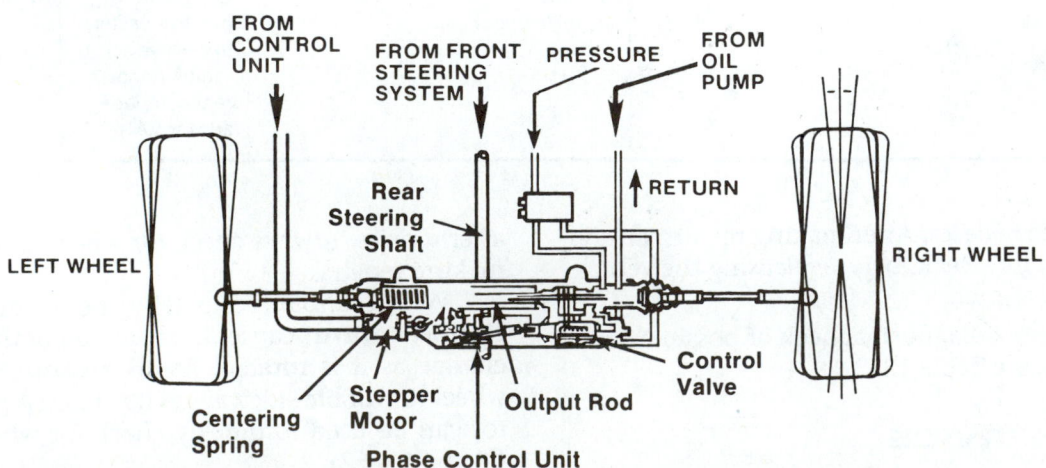

FIGURE 13-54 The four-wheel steering system normally operates in relation to speed and steering wheel movement.

rather than the rear half attempting to catch up to the front half (Figure 13–55). There will also be less yaw when the wheels are turned back to a straight-ahead position. The vehicle responds more quickly to steering input because rear-end lag is eliminated.

Currently there are three types of four-wheel steering systems: mechanical, hydraulic, and electro-hydraulic designs. Since each system is unique in its construction and repair needs, the service manual for the vehicle must be followed for proper diagnosis, repair, and alignment of a four-wheel system.

COLLAPSIBLE STEERING COLUMNS

To reduce the chance of injury, automotive engineers have designed *collapsible steering columns* that crush when hit by the driver's body during a collision. There are several designs.

Lower steering sections can be linked by two or more universal joints. These joints allow the sections to fold (Figure 13–56).

Methods used to lock the shaft to the tube include plastic inserts or steel balls held in a plastic retainer that allow the shaft to "roll" forward inside the tube.

There are also collapsing steel mesh or accordion-pleated devices that give way under pressure. During damage analysis, check the steering column for evidence of collapse. Although the car can be steered with a collapsed column that has been pulled back, the collapsed portion must be replaced. All service manuals provide explicit instructions for doing this.

The steering wheel is often held on by a large nut and a press-fit. A **wheel puller** is used to remove a steering wheel (Figure 13–57). This is a common task in a body shop since steering wheels are frequently damaged. Use hardened bolts to hold the puller into the steering wheel. Then tighten down the large bolt in the center of the puller to force the steering wheel off its shaft.

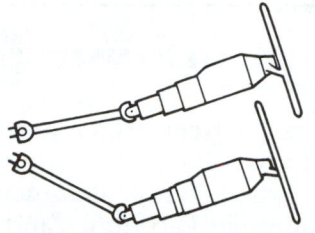

PLATE SEPARATES

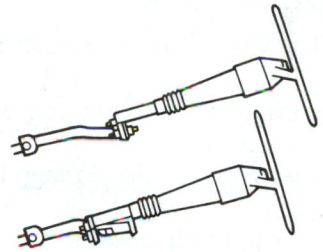

U-JOINTS PIVOT TO ABSORB CRASH IMPACT

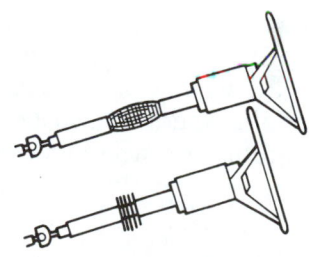

MESH COMPRESSES

FIGURE 13–56 Note common types of collapsible steering columns.

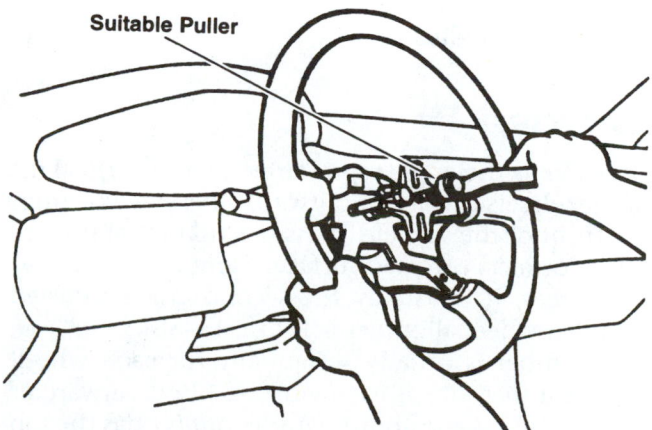

FIGURE 13–57 A wheel puller is needed to remove the steering wheel from its shaft. *(Courtesy of Mazda Motor Co.)*

		Steering Wheel Turned Quickly		Steering Wheel Turned Slowly	
		Phase 1	Phase 2	Phase 1	Phase 2
HIGH	SPEED				
MEDIUM	SPEED				
SLOW	SPEED				

FIGURE 13–55 Note rear-wheel positions with four-wheel steering. Wheel angles and scale are exaggerated.

SHOP TALK

Radial tires cause problems when they have defective belts, unusual wear patterns, uneven air pressure, or are mismatched. These tire problems can cause the technician to misdiagnose steering and alignment problems.

13.4 WHEEL ALIGNMENT

In collision repair, **wheel alignment** involves adjusting the vehicle's tires so that they roll properly over the road surface. Wheel alignment is essential to safety, handling, fuel economy, and tire life.

Following a collision, a vehicle would require an alignment if:

1. There is damage to any steering and suspension parts.
2. There is damage to any steering or suspension mounting locations.
3. There was engine cradle damage or a position change.
4. Suspension or steering parts were removed for access to body parts.

Wheel alignment is done to fine-tune body-frame adjustments. The job of the collision repair technician is to make sure that everything can be fine-tuned and the wheels can be aligned properly.

The proper alignment of a suspension/steering system centers around the accuracy of seven control angles:

1. Camber
2. Caster
3. Steering axis inclination (SAI)
4. Scrub radius
5. Toe (in and out)
6. Thrust line
7. Turning radius

CAMBER

Camber is the angle represented by the vertical tilt of the wheels inward or outward when viewed from the front of the vehicle. It assures that all of the tire tread contacts the road surface. Camber is measured in degrees. It is usually the second angle adjusted during a wheel alignment (Figure 13–58).

Camber is usually set equally for each wheel. Equal camber means each wheel is tilted outward or inward the same amount. *Positive camber* has the top of the wheel tilted out, when viewed from the front. The outer edge of the tire tread contacts the road.

Negative camber has the top of the wheel tilted inward when viewed from the front. The inner tire tread contacts the road surface. Note how camber changes when turning.

Camber is controlled by the control arms and their pivots. It is affected by worn or loose ball joints, control arm bushings, and wheel bearings. Anything that changes chassis height will also affect camber (Figure 13–59).

Camber is adjustable on most vehicles. Some manufacturers prefer to include a camber adjustment at the spindle assembly. Camber adjustments are also provided on some strut suspension systems at the top mounting position of the strut. Remember that camber adjustment also changes SAI or the included angle.

Very little adjustment will be required if the strut tower and lower control arm positions are in their proper places. If you find serious camber error and suspension mounts have not been damaged, it is an indication of bent suspension parts. In this case, diagnostic angle and dimensional checks should be made to the suspension parts. Damaged parts must be replaced.

CASTER

Caster is the angle of the steering axis of a wheel from true vertical, as viewed from the side of the vehicle. It is a directional stability adjustment. Caster is measured in degrees (Figure 13–60).

Caster has little effect on tire wear. Caster affects where the tires touch the road compared to an imaginary centerline drawn through the spindle support. Caster is the first angle adjusted during an alignment.

Positive caster tilts the tops of the steering knuckles toward the rear of the vehicle. It aids in keeping the vehicle's wheels traveling in a straight line. The wheels resist turning and tend to return to the straight-ahead position.

Negative caster tilts the tops of the steering knuckles toward the front of the vehicle. Negative caster makes the wheels easier to turn. However, it produces

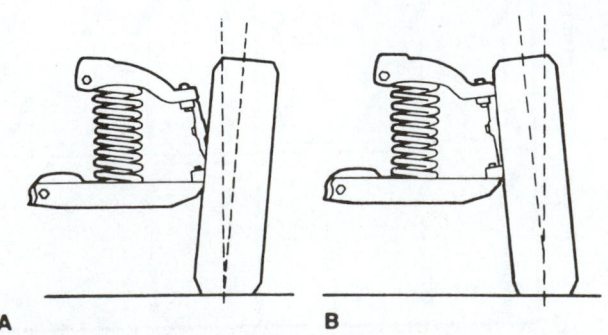

A   B

FIGURE 13–58 Compare types of camber: (A) positive and (B) negative.

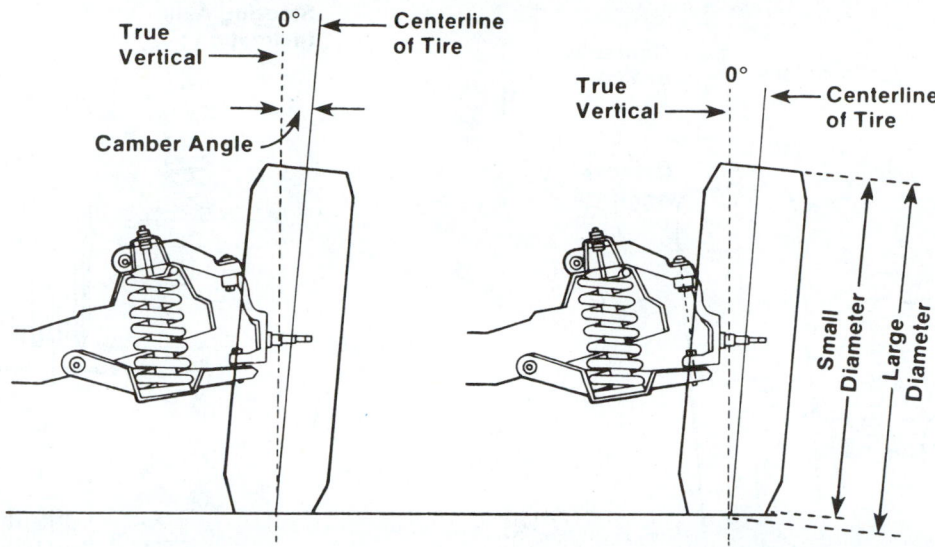

FIGURE 13-59 When camber is out of specifications, it can create changes in the diameter of the tire from inside to outside.

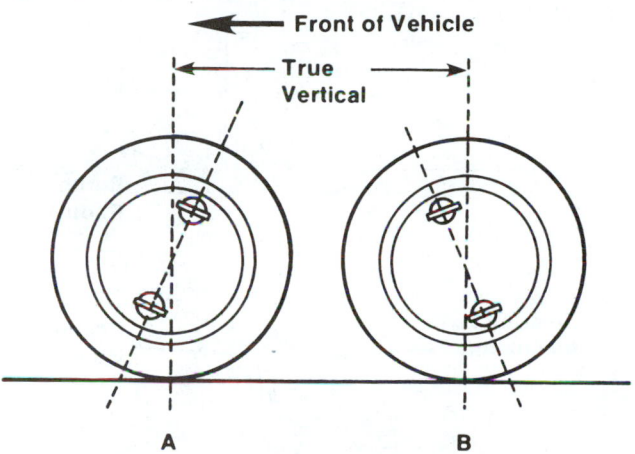

FIGURE 13-60 Study two types of caster: (A) positive and (B) negative.

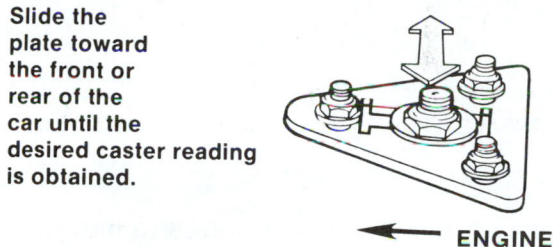

Slide the plate toward the front or rear of the car until the desired caster reading is obtained.

ENGINE

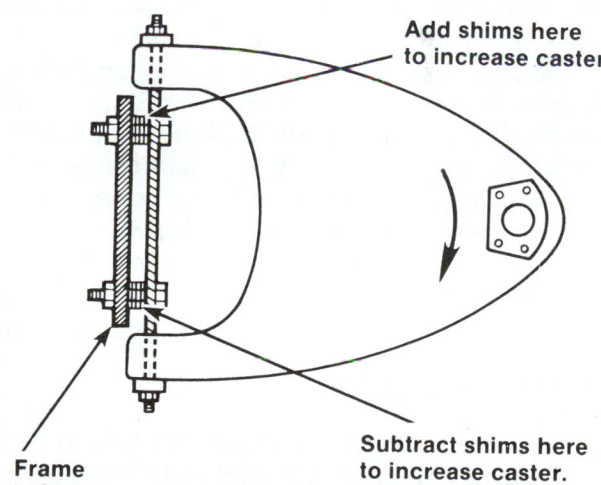

Add shims here to increase caster.

Subtract shims here to increase caster.

Frame

FIGURE 13-61 On some struts, the caster can be adjusted by sliding a plate. With older suspension systems, it can be changed by adding or subtracting shims.

less directional stability. The wheels tend to follow imperfections in the road surface.

Caster is designed to provide steering stability. The caster angle for each wheel should be almost equal. Unequal caster angles will cause the vehicle to steer toward the side with less caster. Too much negative caster can cause the vehicle to have sensitive steering at high speeds. The vehicle might wander as a result of too much negative caster.

Several factors can adversely affect caster. The most common problem is worn or loose strut rod bushings and control arm bushings. Caster adjustments are not provided on some strut suspension systems. Where they are provided, they can be made at the top or bottom mount of a strut suspension.

Caster is measured in degrees from true vertical. Specifications for caster are given in degrees positive or negative. Typically, more positive caster is used with power steering. More negative caster is used with manual steering to reduce steering effort. Also, a ve-

hicle pulls to the side with the least amount of caster (Figure 13–61).

STEERING AXIS INCLINATION

Steering axis inclination is the inward tilt of the steering axis at the top (Figure 13–62). It also contributes to directional stability. Because the steering

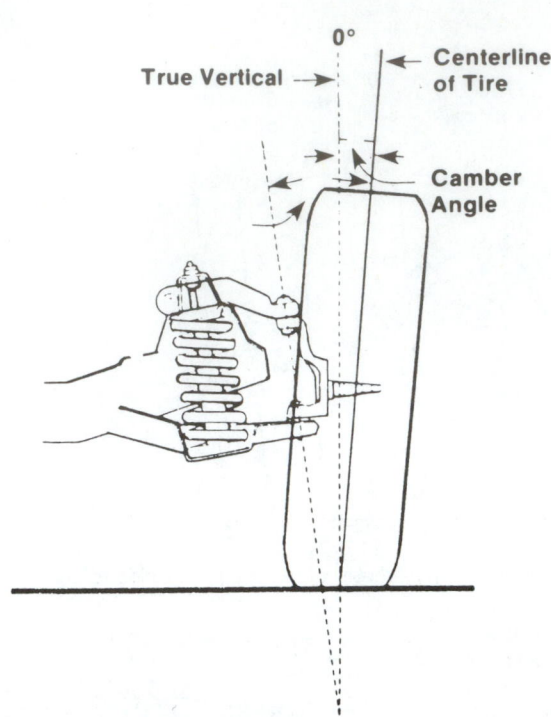

FIGURE 13-62 This illustrates steering axis inclination.

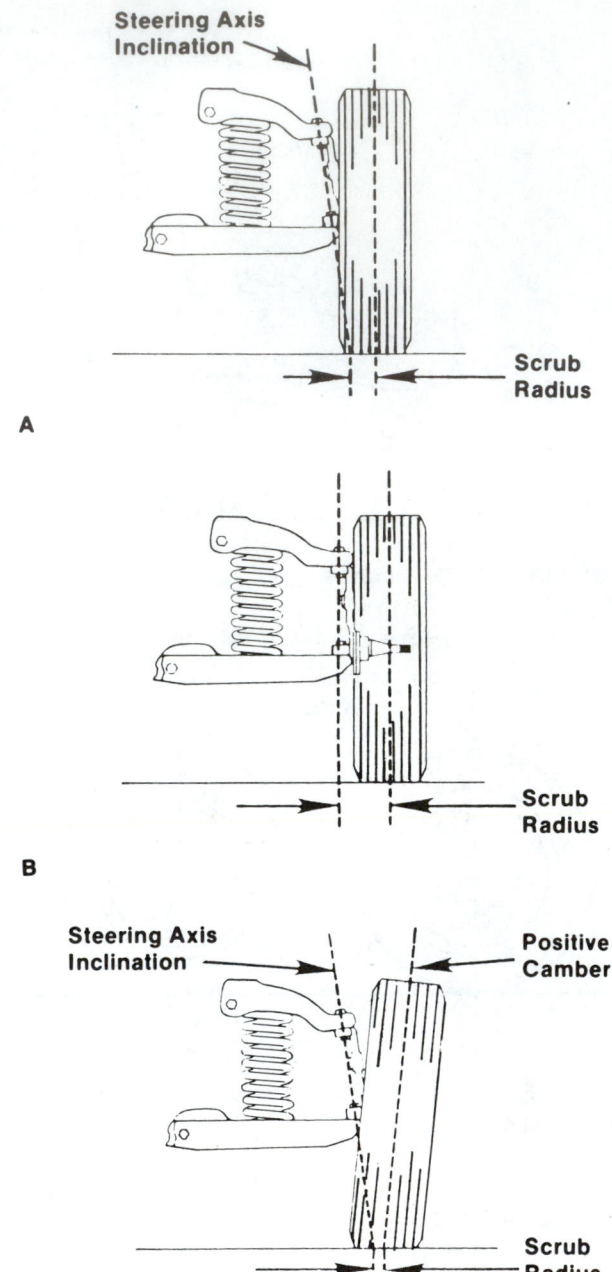

FIGURE 13-63 Study steering scrub radius.

axis is inclined, the spindle is forced to move in an arc downward as the wheel is turned. This action causes the vehicle to rise as the wheel is turned in either direction, so the weight of the car forces the wheels back to the straight-ahead position.

Steering axis inclination is not generally considered a tire wear factor unless there is an extreme change. The amount of inclination is preset and should not change unless there is damage to either the spindle support arm, strut tower, lower control arm, or the lower control arm mounting.

Camber and steering axis inclination are sometimes measured together as the "included angle." The amount of tilt is measured in degrees from vertical.

SCRUB RADIUS

The importance of steering axis inclination to steering ease and stability centers around the reduction of scrub radius. **Scrub radius** is the distance between the centerline of the ball joints and the centerline of the tire at the point where the tire contacts the road surface (Figure 13–63A). When the ball joint centerline (pivot point) is inboard of the point of tire contact, the tire does not pivot where it touches the road. Instead, it has to move forward and backward to compensate as the driver turns the steering wheel. Steering effort is greatly increased as the tires scrub against the road during turns.

If the control arm assembly were designed with no steering axis inclination, scrub radius would be quite large, as can be seen in Figure 13–63B.

Both positive camber and steering axis inclination combine to reduce scrub radius to a minimum (Figure 13–63C).

TOE

Toe is the difference in the distance between the front and rear of the left- and right-hand wheels. Toe can be measured in inches, millimeters or degrees, depending upon the equipment used (Figure 13–64). Toe should be the last wheel alignment adjustment made.

Toe adjustment is critical to tire wear. If properly adjusted, toe makes the wheels roll in the same

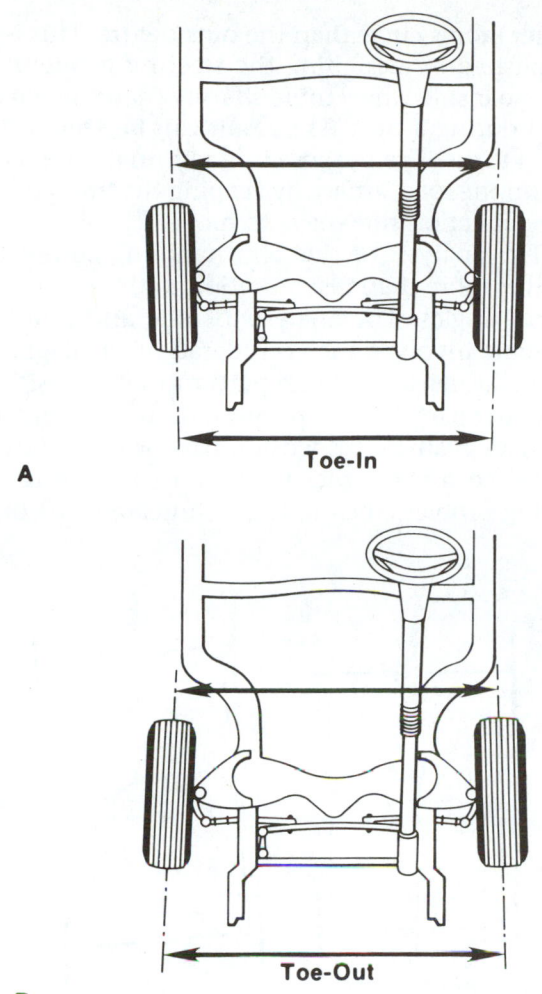

A

Toe-In

B

Toe-Out

FIGURE 13-64 Compare toe conditions: (A) Toe-in is the amount that wheels are closer together at the extreme front of the tires than they are at the extreme rear. (B) Toe-out is just the opposite of toe-in, with a greater measurement in front than in the rear.

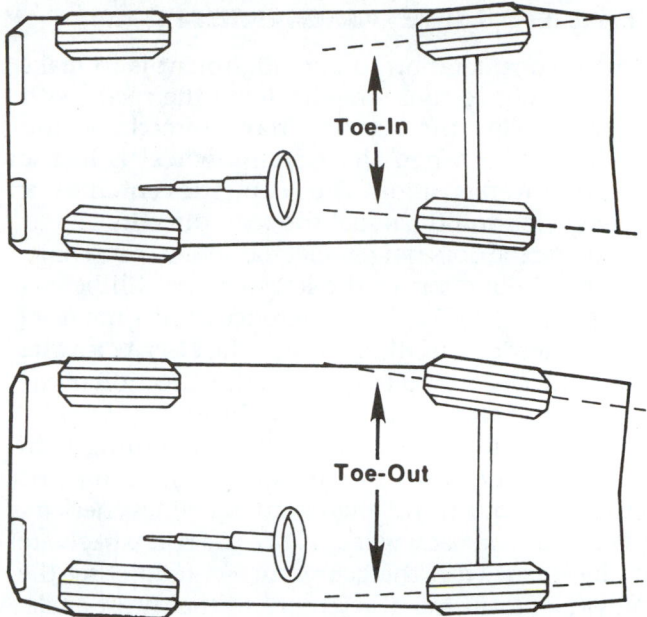

Toe-In

Toe-Out

FIGURE 13-65 These are typical rear toe conditions.

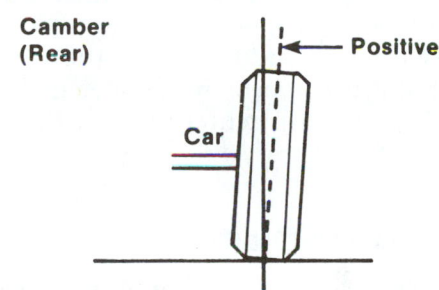

Camber (Rear)

Positive

Car

FIGURE 13-66 Rear camber and the position of the rear wheel should be checked with rear damage.

direction. If toe is not correct, the misaligned wheels will scuff or drag the tires sideways, causing rapid tire wear.

Remember that *excessive toe* (in or out) will cause a sawtooth edge on the tire tread from dragging the tire sideways.

Toe-in results when the front of the wheels are set closer than the rear. The wheels point in at the front. *Toe-out* is just the opposite. It has the front of the wheels farther apart than at the rear. The wheels point out at the front.

Toe is a very critical tire wearing angle. Wheels that do not track straight ahead have to drag as they travel forward.

Rear-wheel-drive vehicles are often adjusted to have toe-in at the front wheels. Toe-in is needed to compensate for tire rolling resistance, play in the steering system, and suspension system action. The tires tend to toe-out while driving. By setting the wheels

for a small toe-in of about $1/16$ inch (1.5 mm), the tires will roll straight ahead over the road surface.

Most front-wheel-drive vehicles need to have their front wheels set for a slight toe-out. The front wheels pull and propel the vehicle. As a result, they are forced forward by drivetrain torque. This tries to make the wheels point inward while driving. Front-wheel drive toe-out of $1/16$ inch (1.5 mm) is typical.

Rear toe condition (Figure 13–65) refers to the angle of the rear wheel in or out at the front of the wheel as viewed from the top. It might or might not be adjustable depending on the design of the car. However, it has an important effect on the handling of the car. Some cars with independent rear suspensions also have at-rest toe settings to compensate for play in the rear suspension.

Rear camber (Figure 13–66) refers to the position of a rear wheel in or out at the top as viewed from the rear of the rear wheel. It might or might not be adjustable depending on the design of the car. However, it has an important effect on the handling of the car.

THRUST LINE ALIGNMENT

A main consideration in any alignment is to make sure the vehicle runs straight down the road. With proper *tracking,* the rear tires travel directly behind the front tires when the steering wheel is in the straight-ahead position. The geometric centerline of the vehicle should parallel the road direction.

If rear toe does not parallel the vehicle centerline, a "thrust" direction to the left or right will be created (Figure 13–67). This difference of rear toe from the geometric centerline is called the **thrust angle.** The vehicle will tend to travel in the direction of the thrust line, rather than straight ahead.

To correct this problem, begin by setting individual rear toe equal in reference to the geometric centerline. Four-wheel alignment machines check individual toe on each wheel. Once the rear wheels are in alignment with the geometric centerline, set the individual front toe in reference to the thrust angle. Following this procedure assures that the steering wheel will be straight ahead for straight-ahead travel. If you set the front toe to the vehicle geometric centerline, ignoring the rear toe angle, a cocked steering wheel will result. The direction of the front wheels would be trying to compensate for differences between the vehicle geometric centerline and thrust angle (Figure 13–68).

TURNING RADIUS

Turning radius angle or cornering angle is the amount of toe-out on turns (Figure 13–69). As a car goes around a corner, the inside tire must travel in a smaller radius circle than the outside tire. This is accomplished by designing the steering geometry to turn the inside wheel more sharply than the outside wheel during a turn. The result can be seen as toe-out on turns. The purpose is to eliminate tire scrubbing on the road surface by keeping the tires pointed in the direction they have to move.

The analysis of ride and handling complaints involves more than just attention to the accuracy of control angles. The analysis of ride and handling problems involves the consideration of diagnostic angles. Suspension system parts must be considered as moving parts in the operation of the car. Diagnostic angles evaluate suspension parts as they move.

When a car comes in with a collision-caused steering problem, before the technician starts disas-

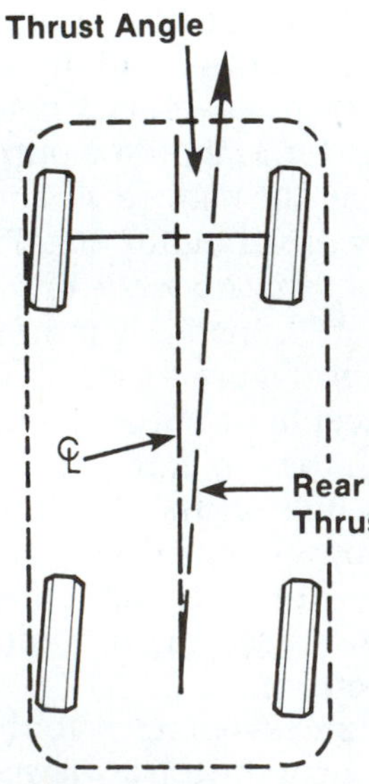

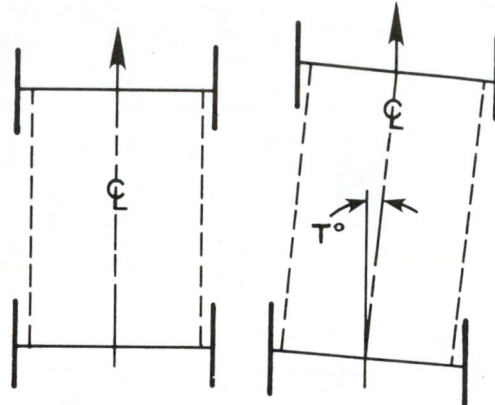

FIGURE 13-68 With proper tracking, alignment of the rear wheels is parallel with vehicle centerline and front wheels. When the rear wheels are not set parallel to the centerline and to the front wheels, the car will "dog track."

FIGURE 13-67 If the rear axle or wheels are not in alignment, it will affect thrust line alignment and tracking of the vehicle down the road.

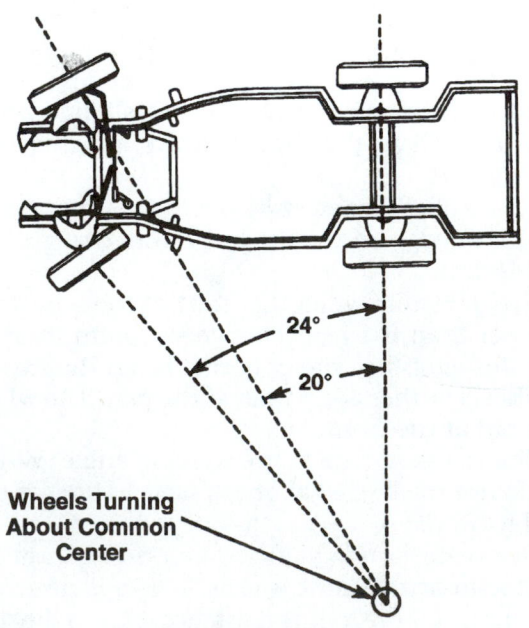

Typical turning radius

FIGURE 13-69 This is a typical turning radius.

sembling the car, the first diagnostic check should be a visual inspection of the entire vehicle for anything obvious: bent wheels, obvious misalignment of the cradle, and the wheels to the wheel opening. If there is not a thing obviously wrong with the car, make the following diagnostic checks without disassembling the vehicle.

One of the very useful diagnostic checks that can be made with a minimum of equipment is a jounce-rebound toe-in change check, which can help determine the condition of the suspension system.

- **Jounce** is the motion caused by a wheel going over a bump and compressing the spring. During jounce, the wheel moves up toward the chassis. Jounce can be simulated for in-shop testing

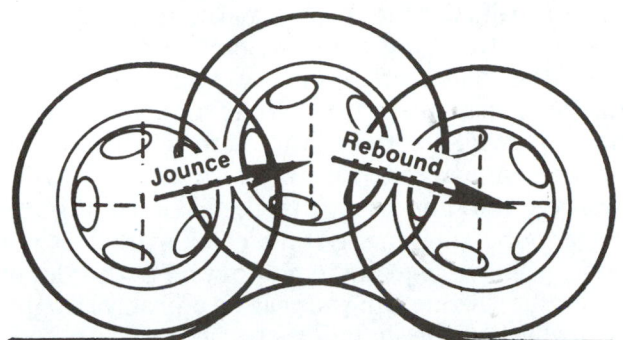

FIGURE 13-70 Jounce and rebound refer to the reactions of the suspension to an irregular road surface.

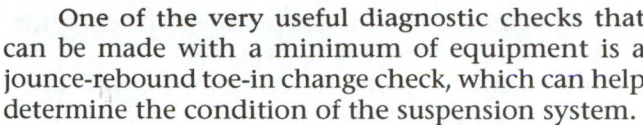

FIGURE 13-71 A simple gauge can be used in camber checks.

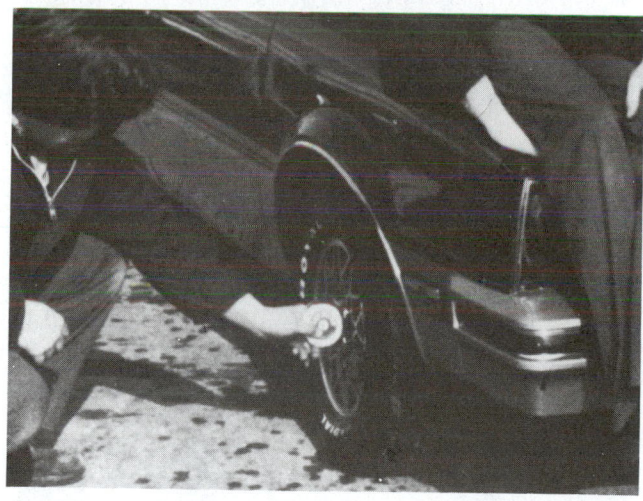

A

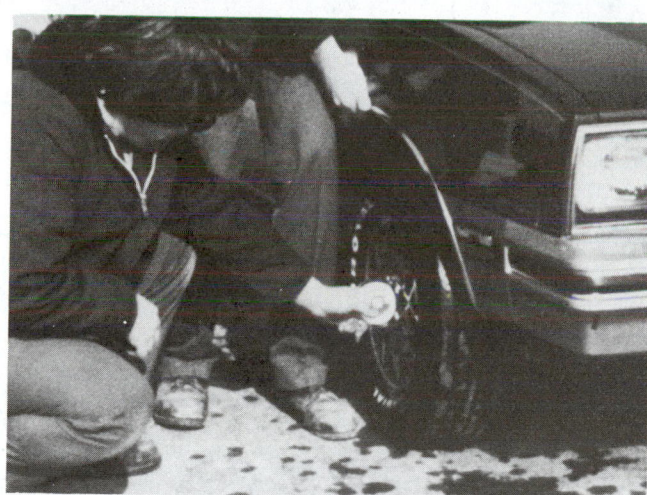

B

C

D

FIGURE 13-72 Jounce and rebound camber checks involve rotating the wheel and taking measurements.

by pushing down on the bumper. The car must be jounced equally on both sides.

• **Rebound** is the motion caused by a wheel going into a dip or returning from a jounce and extending the spring. During rebound, the wheel moves down away from the chassis. Rebound can be simulated for in-shop testing by lifting

up on the fender. The car must be lifted equally on both sides.

This jounce-rebound check (Figure 13–70) will determine if there is some misalignment to the rack-and-pinion gear. For a quick check, unlock the steering wheel and see if it moves during the jounce and/or rebound.

For a more careful check, employ a pointer and a piece of chalk. Use the chalk to make a reference mark on the tire tread and place the pointer on the same line as the chalk mark. Jounce and rebound the suspension system a few times, while someone watches the chalk mark and the pointer. If the chalk mark on the wheel moves unequally in and out on both sides of the car, the chances are that there is a steering arm or gear out of alignment. If the chalk mark does not move, or moves equally in and out on both sides of the car, the steering arm and gear are probably all right. Each wheel or side should be checked.

The next diagnostic check is for cornering angle. The cornering angle check evaluates the proper relationship of the two front wheels as they are turned through a steering arc. To measure cornering angle, one wheel is turned on a turn plate or protractor a given amount; the amount of rotation of the opposite wheel is measured in a similar manner. The results are compared right to left to determine if the two front wheels are rotating through the same arc.

During a cornering angle check, the left front wheel should be turned out 20 degrees. Then the right wheel rotation is measured. The readings will usually not match those of the right wheel. The right wheel should turn in the same amount or about 2 degrees less. The difference accounts for the turning radius difference between the inside and outside wheels during cornering.

The process is repeated with the right wheel; the right wheel is turned out 20 degrees. The movement of the left wheel is measured on the protractor or turn plate. The left wheel should turn in the same amount or about 2 degrees less.

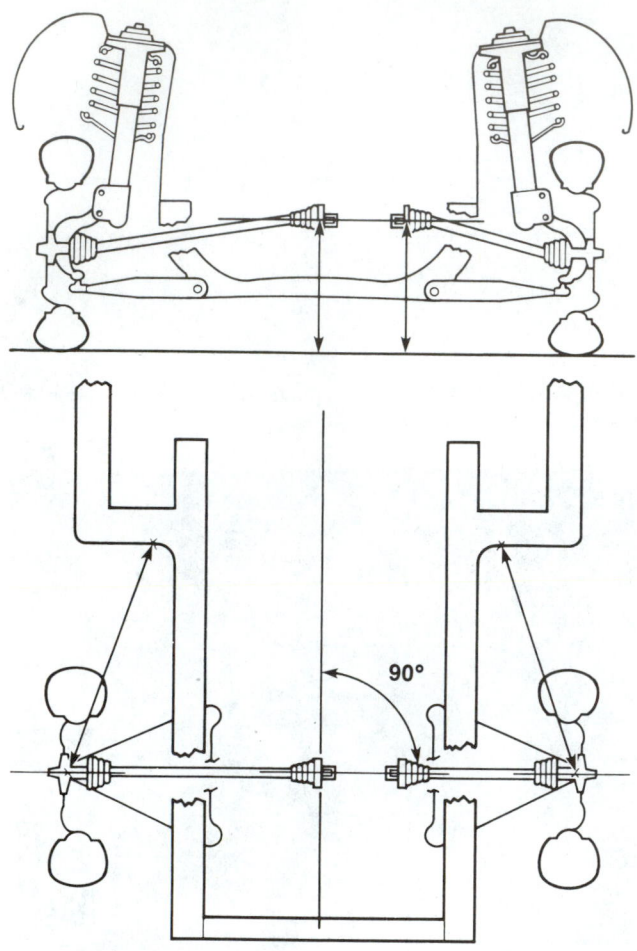

FIGURE 13-73 Drive shafts are a part of the steering and suspension systems. Their positioning can affect those systems. If the drive shaft system parts are bent or damaged, they can cause handling problems.

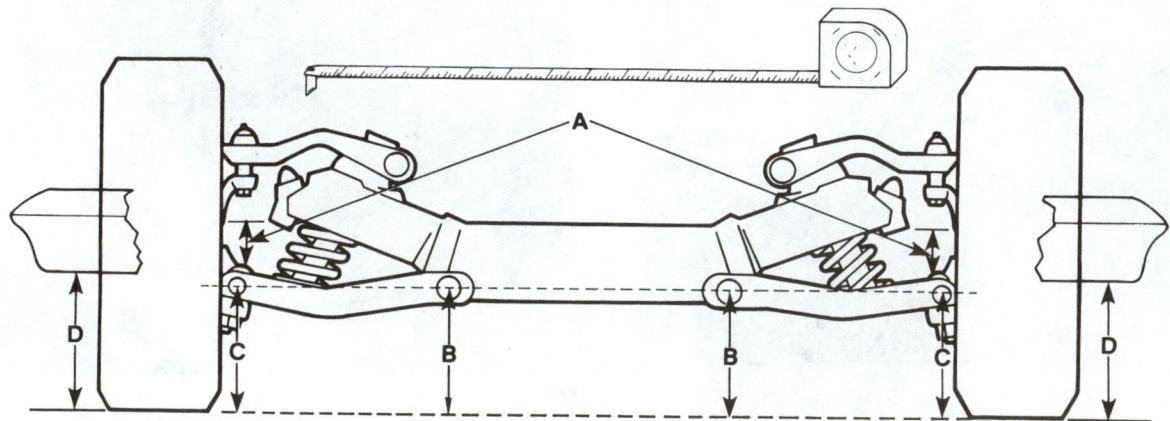

FIGURE 13-74 Note typical locations for measuring curb or ride height.

By design, a vehicle might use a different turning radius from one side to the other. If in doubt, refer to the manufacturer's specifications. If these measurements do not repeat within 2 degrees, damage to the steering arms or gear is indicated. Cornering angle measurements are especially useful in determining whether improper toe conditions are caused by poor wheel alignment or damaged suspension components.

CAMBER CHECKS

Some camber checks can be made to diagnose the condition of a strut and can be measured easily with a camber gauge (Figure 13–71). One is called a jounce-rebound camber measurement (Figure 13–72A) and can be made by loading the suspension in a similar fashion to jounce-rebound toe change and measuring the camber angle from an individual wheel.

The suspension is then unloaded as in the jounce-rebound toe check and a second camber reading of the same wheel is made (Figure 13–72B). The two readings are compared; these readings should not differ more than 2 degrees on a MacPherson strut-type suspension. In most cases, the readings will be the same.

The jounce-rebound camber change will tell the technician if the strut is bent either inboard or outboard. Check each wheel individually before deciding if one strut is bad based on the readings. If the readings differ between wheels more than 2 degrees, a bent strut is indicated.

A swing camber measurement is made by turning the front wheel in a given amount and performing a jounce-rebound camber check (Figure 13–72C).

The front wheel is then turned out the same amount and the camber angle is measured again

(Figure 13–72D). If the camber angle change differs more than 3 degrees from left wheel to right wheel, it is likely that either the strut is bent forward or rearward from its normal position or the caster angle is incorrect. As a further test for a bent strut, perform a jounce-rebound check while the wheels are turned in and while they are turned out. Check each wheel and compare the readings. These diagnostic angles are especially helpful in determining the cause of vehicle handling and tracking problems.

ENGINE CRADLE POSITION

Proper positioning of the engine cradle can affect the steering angles. Since the cradle provides the lower pivot point, movement of the cradle will cause a camber change. Both wheels will show an equal camber change, one side negative and one side positive. It will also cause an SAI but not an included angle change. Make sure the cradle's position is within the specifications given in the service manual.

The positioning of the drive shaft can also affect the steering and suspension systems. If any of these parts are bent, it can cause a shimmy or handling problems. If there is any doubt about the positioning of the drive shafts, measure them as shown in Figure 13–73.

CURB HEIGHT

For proper alignment, each of the front and rear wheels must carry the same amount of weight. The car is designed to ride at a specific height, sometimes referred to as **curb height** (Figure 13–74). Curb height specs are published in the service manuals and some of the alignment spec books.

If the vehicle leans to one side or seems to be lower on one side than on the other, something is wrong. Either the front or rear suspension on that side of the vehicle can cause the condition.

To isolate the height problem, place a jack in the center of the main cross member in the front of the vehicle (Figure 13–75). Raise the vehicle several inches, and look at the rear of the car. If the rear of the car looks level, the problem is in the front suspension on the side that shows the lean. If the rear suspension is not level, the problem is the rear suspension on the low side.

WHEEL ALIGNMENT PROCEDURE

Before making any adjustment affecting caster, camber, or toe-in, the following checks should be made to ensure correct alignment readings and adjustments.

• Make sure the vehicle is sitting on a level surface (side to side and front to rear).

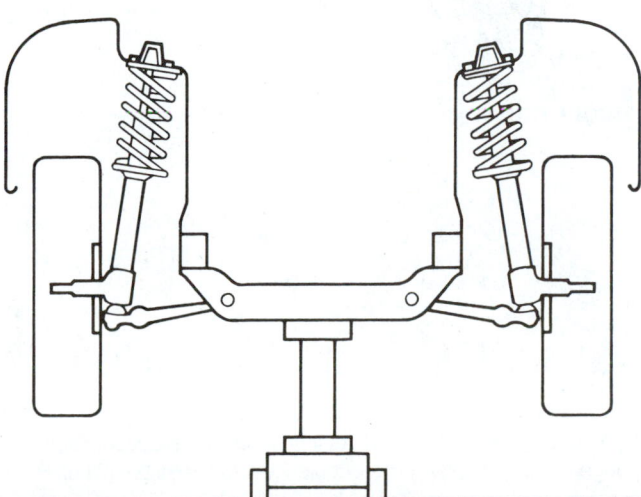

FIGURE 13–75 Raise the vehicle by jacking in the very center of the main cross member. You can then check height at both sides of the rear of the vehicle to isolate the cause of irregular curb height measurements.

A

B

C

D

FIGURE 13-76 An alignment gauge that makes it easy to check wheel alignment while the vehicle is still on the repair bench: (A) measure passenger side to drive side at the front and rear of the gauge for toe; (B) read the bubble level for camber; (C) check the rear suspension; and (D) read the camber with the car on a level surface. *(Courtesy of Steck Mfg. Co. Inc.)*

A

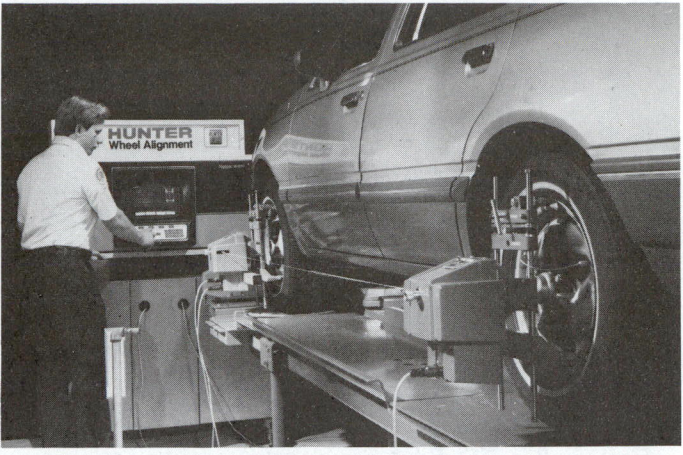

B

FIGURE 13-77 Sophisticated wheel alignment, once found only in alignment specialty shops, has now found its way into body shops because of unibody construction. (A) Typical four-wheel alignment system; (B) computerized system. Four-wheel and computerized alignments are becoming a "must" today. *(Photo A courtesy of Bee Line Co.; photo B courtesy of Hunter Engineering Co.)*

SHOP TALK Caster and camber angles are measured with gauges (Figures 13–76 and 13–77) available from specialty tool manufacturers. They must be used as directed to get proper measurements.

- Rotate the tires if needed. (Check the tires for similar size, tread design, depth, and construction).
- Make sure all tires are inflated to recommended pressure.
- Inspect for worn or bent parts and replace. Much of this should be checked during body/frame correction.
- Check and adjust wheel bearings if necessary. (Spin tires; check for looseness or unusual noises.)
- Check for unbalanced loading (proper chassis height). This should be checked after body/frame correction.
- Check for loose ball joints, tie-rod ends, steering relay rods, control arms, and stabilizer bar attachments.
- Check for run-out of wheels and tires.
- Check for defective shock absorbers.
- Consider excess loads, such as tool boxes.
- Consider the condition and type of equipment being used to check alignment and follow the manufacturer's instructions.

The adjustment order—caster, camber, toe—is recommended regardless of the make of car or its type of suspension. Methods of adjustment vary from vehicle to vehicle and, in some cases, from year to year of the same make car. Refer to the manufacturer's service manual for details. A typical alignment procedure could be considered as follows:

1. Obtain manufacturer's specifications.
2. Camber. Tilt of wheel inward and outward.
3. Caster. Forward or rearward tilt of steering axis.

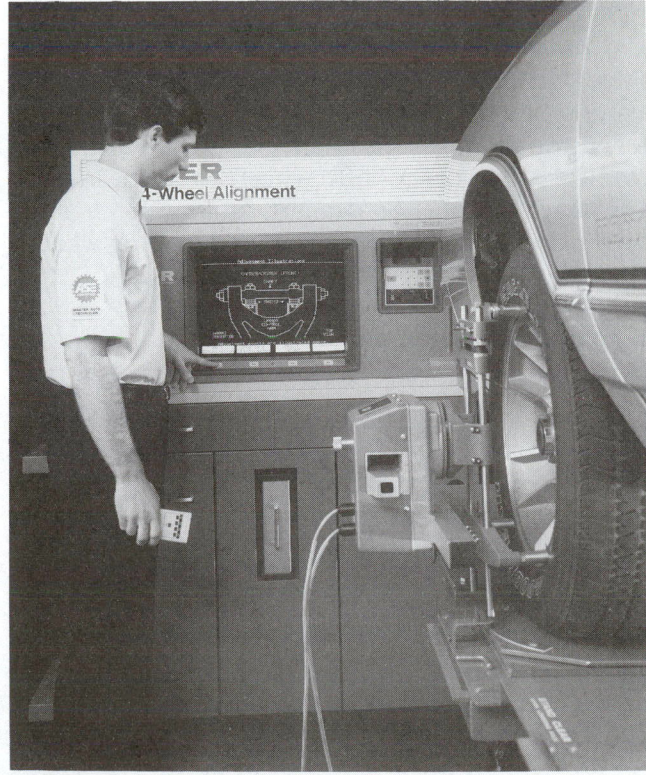

FIGURE 13-78 Modern wheel alignment equipment will help guide you through procedures. *(Courtesy of Hunter Engineering Co.)*

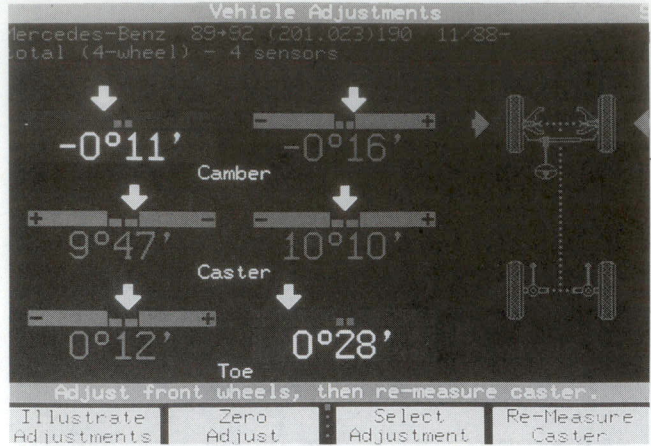

FIGURE 13-79 The computer screen of this wheel alignment machine gives specs and instructions for adjustment. *(Courtesy of Hunter Engineering Co.)*

4. Steering axis inclination. Inward tilt of steering axis at the top.

5. Turning radius. Wheel angles while turning.

6. Toe. Difference in distance between the front and rear of the tire.

Many of today's wheel alignment machines are computerized (Figure 13–78). They will give you exact specs, tell where adjustments are needed, and may even show a picture of what is wrong. Modern equipment saves time and requires less training (Figure 13–79).

Today, most vehicles use a four-wheel alignment. Table 13–2 summarizes a typical steering problem diagnosis. But, it is important to remember that the typical customer judges the quality of a wheel alignment by the position of the "fifth wheel"—the one in his or her hands. It must be straight. Make sure all alignments end with a properly centered steering wheel.

13.5 BRAKE SYSTEMS

The *brake system* uses hydraulic pressure to slow or stop wheel rotation with brake pedal application (Figure 13–80).

The *brake pedal* transfers the driver's foot pressure into the master cylinder. The **master cylinder** develops hydraulic pressure (oil pressure) for the system (Figure 13–81).

Brake lines and *hoses* carry fluid out to the wheel cylinders. The *wheel cylinders* use hydraulic pressure to push the brake pads or shoes outward.

The **brake pads** or **shoes** have friction lining for rubbing on the brake rotor or drum. The **brake rotors** or **drums** provide heavy metal friction surfaces bolted between the hub and wheel. A *caliper* holds the piston(s) and brake pads on disc brakes.

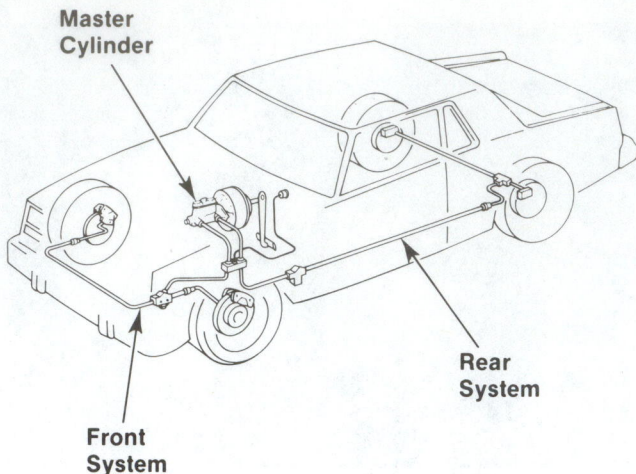

A

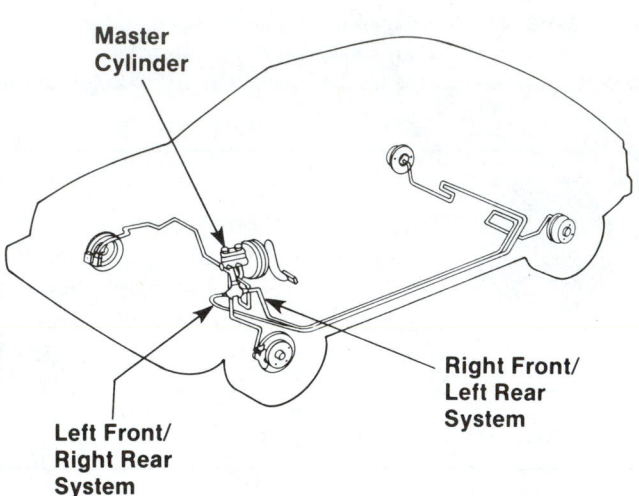

B

FIGURE 13-80 Compare (A) front/rear split hydraulic system; (B) dual diagonal split hydraulic system.

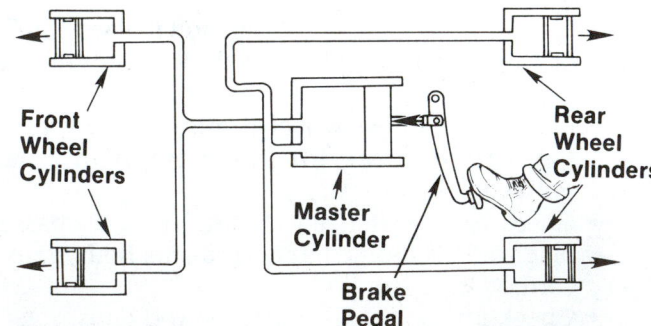

FIGURE 13-81 This simple automotive hydraulic system diagram shows how a brake system operates.

Power brakes is a standard hydraulic brake system with a vacuum, hydraulic, or electric assist. A booster unit is added to help apply the master cylinder and brakes.

In today's front-wheel-drive automobiles, the front brakes are now doing as much as 80 percent of the work in stopping the car. Thus, the conventional front/rear split hydraulic system has given way to a dual diagonal split system. This combines a front brake with its opposite rear brake. This system allows straight stopping and provides 50 percent of the braking capacity in case of failure in either of the two hydraulic systems.

Two basic types of hydraulic brakes are used in unibody vehicles. They are drum brakes and disc brakes.

DRUM BRAKES

A drum brake assembly consists of a cast-iron drum bolted to the vehicle axle. A fixed *brake backing plate* holds the shoes and other components—wheel cylinders, automatic adjusters, linkages, and so on (Figure 13–82). Additionally, there might be some extra hardware for parking brakes.

The brake shoes are surfaced with frictional linings, which contact the inside of the drum when the brakes are applied. The shoes are forced outward, against the action of the return springs, by pistons or wheel cylinders that are actuated by hydraulic pressure. As the drum rubs against the shoes, the energy of the moving drum is transformed

SHOP TALK

Keep the brake system closed to the atmosphere as much as possible. This will keep moisture from entering the system. Moisture is readily absorbed by the fluid and causes sludge and corrosion to form. In time, this will cause partial or complete loss of brake effect. Moisture also lowers the boiling point of the fluid. Boiling results in vapor, which has the same effect as air in the system. Moisture can enter the system in several ways: leaving the system open, improper storage of fluid, and incorrect bleeding equipment or techniques.

into heat, and this heat energy is passed into the atmosphere.

When the brake shoe is engaged, the frictional drag acting around its circumference tends to rotate it about its hinge point, the brake anchor. If the rotation of the drum corresponds to an outward rotation

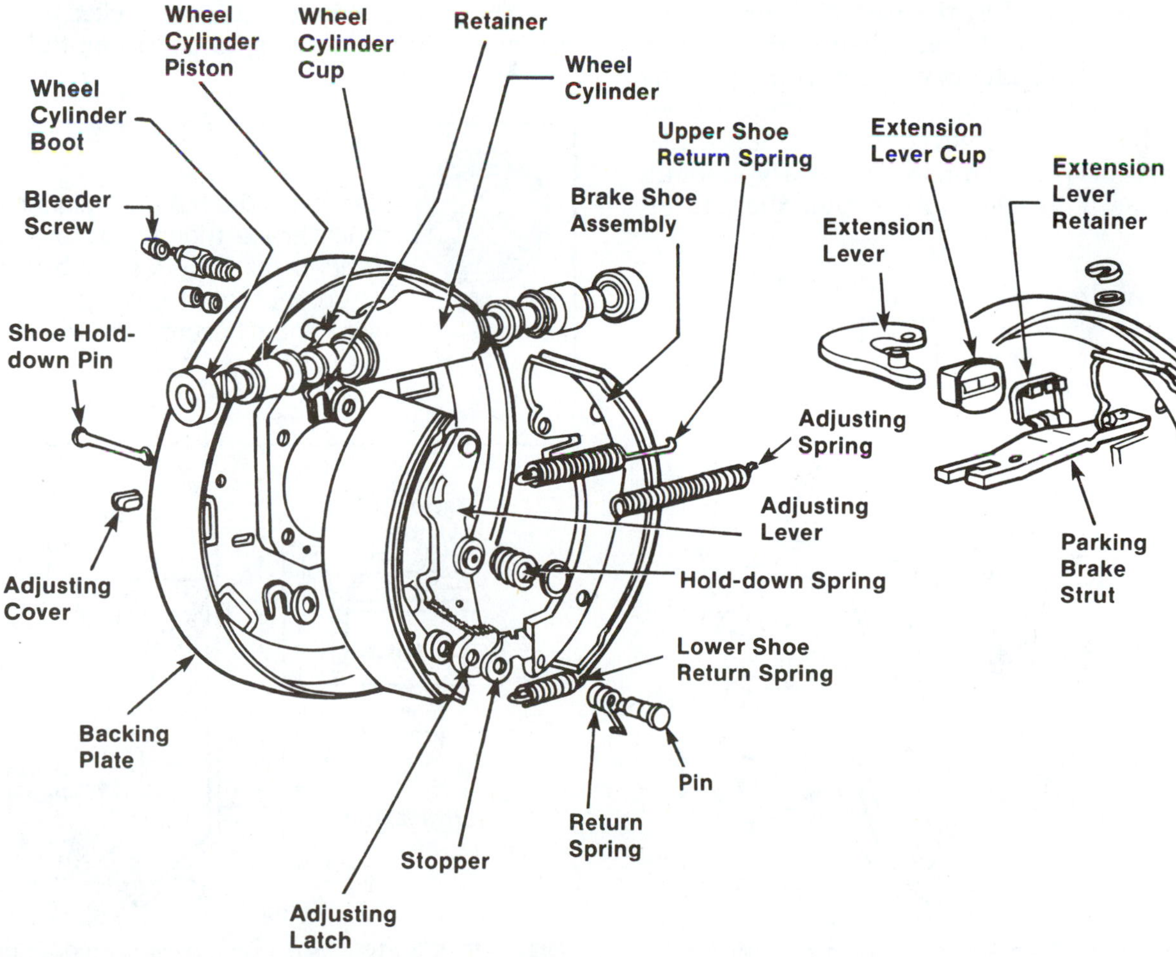

FIGURE 13–82 Study the parts of a rear drum brake assembly.

of the shoe, the drag will pull the shoe tighter against the inside of the drum, and the shoe will be self-energizing.

DISC BRAKES

Disc brakes resemble the brakes on a bicycle: the friction elements are in the form of pads, which are squeezed or clamped about the edge of a rotating wheel. With automotive disc brakes, this wheel is a separate unit, called a *rotor*, inboard of the vehicle wheel (Figure 13–83). The rotor is made of cast iron and, since the pads clamp against both sides of it, both sides are machined smooth. Usually the two surfaces are separated by a finned center section for better cooling. The pads are attached to metal shoes, which are actuated by pistons, like drum brakes. The pistons are contained within a *caliper assembly*, a housing that wraps around the edge of the rotor. The caliper is kept from rotating by bolts holding it to the car's suspension, usually the steering knuckle (front wheels) or just knuckle (rear wheels).

The caliper is a housing containing the pistons and related seals, springs, and boots, as well as the cylinders and fluid passages necessary to force the friction linings or pads against the rotor. The caliper resembles a hand in the way it wraps around the edge of the rotor. It is attached to the steering knuckle. Some models employ light spring pressure to keep the pads close against the rotor; in other caliper designs this is achieved by a unique seal that pushes out the piston for the necessary amount, then retracts it just enough to pull the pad off the rotor.

Unlike shoes in a drum brake, the pads act perpendicular to the rotation of the disc when the brakes are applied. This effect is different from that produced in a brake drum, where frictional drag actually pulls the shoe into the drum. Disc brakes are nonenergized and require more force for the same braking effort. For this reason, they are ordinarily used with a power brake unit.

Actual work on hydraulic brake system parts— master cylinder, combination valve, wheel cylinders, brake shoe, and caliper assemblies—is best left to brake specialists. However, disassembly, reassembly, and replacement of some damaged components are skills of the service technician.

MASTER CYLINDER

The *master cylinder* is the heart of the hydraulic system (Figure 13–84). It is located in the engine compartment, usually on the driver's side, and is connected to the brake pedal by a special rod. The master cylinder initiates braking when the brake pedal is depressed by pushing out a piston inside the cylinder, exerting pressure that is transferred through the system. To protect against total failure of the system, all cars are now required to have two hydraulic systems.

The master cylinder must be checked before the car is put back in service. To check the fluid level,

SHOP TALK

Never use the master cylinder and other brake mounts as a pulling attachment. If a pull must be made to correct cowl damage, use a plate and different bolts to anchor the pull.

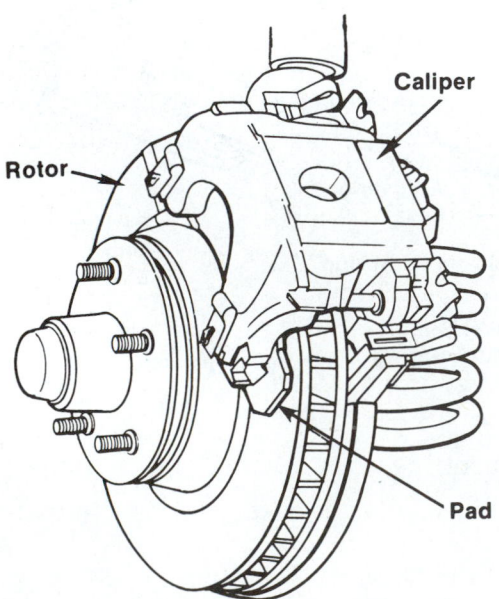

FIGURE 13–83 A typical disc brake has a caliper that clamps around the rotor.

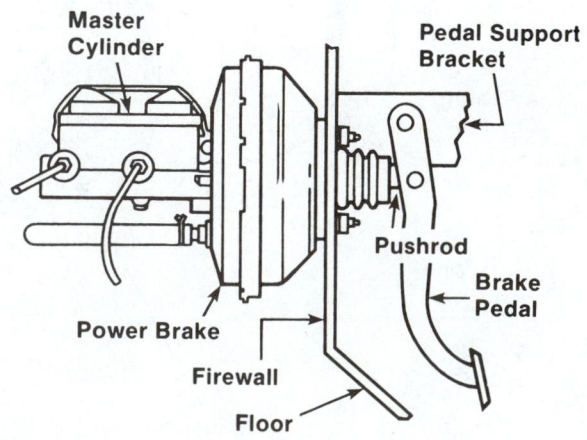

FIGURE 13–84 The master cylinder normally bolts to the firewall.

clean all dirt and grease off the unit, then simply pop the wire bracket (or whatever locking device is on the top) and remove the lid. The level should not be more than ¼ inch (6.3 mm) below the top of the reservoir. If the level is below this, check the brake line connections and refill the reservoir with fluid. If the system is leaking anywhere but at the brake line connections, the master cylinder should be replaced or rebuilt by a brake specialist.

BRAKE FLUID

When brake fluid absorbs moisture, it drastically reduces the boiling point of the fluid. This effect is even more pronounced in high-temperature fluids that are used in heavy-duty and disc brake service. Typical vehicles in the field for eighteen months accumulate 2 to 3 percent water in the brake fluid. Absorption of a mere 3 percent of moisture reduces the boiling point of these fluids by 25 percent.

This effect bears a bit of investigation. Given a system contaminated with water dispersed throughout the brake fluid, the system acts properly when cold. But after some heavy braking, the fluid in the wheel cylinder or caliper heats up and the contaminated fluid vaporizes. This vapor is now a gas in the system and behaves just like air or any other gas, creating a spongy pedal or, in the extreme case, no pedal at all. The danger of entrapped moisture is that the symptoms do not show up until moments of heavy braking, when the brakes are needed most.

To prevent contamination, the following precautions must be strictly observed when handling brake fluid:

- Keep the master cylinder tightly covered.
- Always recap it immediately after filling.
- Use the smallest possible can of fluid, and use it all if possible. If, for example, there is a choice of using two small cans or a portion of a large can, use the two small ones.
- Tightly cap the fluid container after use.

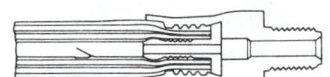

Torn inner lining restricts
flow, acts as valve.

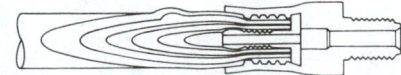

Fitting leakage seeps
out or forms bubble.

FIGURE 13-85 These are examples of internal defects in brake hoses.

- If using a pressure brake bleeder, keep its fluid reservoir tightly closed, just like the master cylinder.
- If any fluid has become contaminated, throw it out.
- Do not reuse old brake fluid.
- Do not reuse an old brake fluid container since it is not possible to know what else might have been in the can.
- Do not transfer brake fluid from its original container to anything other than a container specifically designed to hold brake fluid, such as a pressure bleeder.

BRAKE LINES

The brake lines carry fluid pressure between the master cylinder and wheel cylinders, and related parts. They are generally the major brake component that a body technician must repair.

When making a collision inspection, check the brake lines for chafing, crimps, loose or missing tube clips, kinks, dents, and leakage. Leaks are evidenced by fluid seepage at the connections or stains around hose ends (Figure 13–85). Blockages are not so readily apparent, but they are just as detrimental to the function of the braking system, often acting as a check valve to prevent proper release of the brakes. During a brake application the pressure forces the fluid past the obstruction, but when the pressure is relaxed, the fluid does not readily flow back past the blockage and the brakes drag. Brake lines are usually steel, except where they have to flex—between the chassis and the front wheels, and the chassis and the rear axle. At these places flexible hoses are used.

When replacing damaged brake lines, use the same type of material as OEM. This includes stainless steel, armor plate tubing, or ribbed hose. Local availability might be limited on special types, but it is important to try to match the factory materials. Never use a weaker material to make a brake line or catastrophic brake failure may result.

When cutting tubing to length, it is important to duplicate the factory flare (Figure 13–86). Most cars use a double flare connection, so check for details carefully. Do not use compression fittings in brake line repairs. Replace all supporting clamps removed during the repair. Support springs prevent kinking and serve a very important role. Be sure to replace them just as they were, and install new ones if damaged. Always replace brake lines in the original routing to avoid later damage to the lines.

Remember that you are repairing the car, not re-engineering it. A change in routing may result in rubbing or chafing of brake lines as the suspension moves. Most brake hoses have a male fitting on one end and a female fitting on the other. Disconnect the female end first, remove the clip or jam nut holding it down; then unscrew the male end. Install the new hose by connecting the male end. If a copper gasket was used, replace it with a new one. When the male end is tight, connect the female end. Tighten it in such a way as to keep the hose from touching any part of the chassis or suspension. Check for interference during suspension deflection and rebound and turning of the front wheels.

BLEEDING

To remove or replace a brake component, follow the instructions in the service manual. Remember that any time the brake system is open, it must be bled. Keep the system open for as short a time as possible to prevent moisture from entering and causing sludge and corrosion.

Bleeding removes air from the brake system. Air is lighter than liquid and it seeks high points in the hydraulic system. Bleeder screws are provided at each of these collecting points: calipers, wheel cylinders, and on some master cylinders. Bleeding involves opening up these screws in a specific order to let the trapped air escape. Fluid is added to the master cylinder to replace whatever is lost in bleeding.

Bench Bleeding the Master Cylinder

When the master cylinder is removed for rebuilding or replacement, bench bleeding is necessary to ensure that air does not remain in the cylinder when it is reinstalled. Mount the cylinder in a vise with the bore angled slightly downward (Figure 13–87).

Attach two short brake lines or purge tubes to the outlet ports so they curl back into the reservoirs with the ends below the fluid level. Stroke the piston back and forth. This pumps air out of the cylinder and into the reservoir. Do this until only clear brake fluid comes out of the tubes. The same method will work using threaded plugs instead of purge tubes.

Bleeding the System

Whether the pressure or manual bleeding technique is used, both methods of bleeding follow a common sequence for opening the bleeder screws. Begin at the master cylinder if it has bleeder screws. Move to the combination valve if it has bleeder screws. Next, bleed the wheel cylinders or calipers. Start with the wheel located farthest from the master cylinder. Work back from there.

The bleeding sequence at the wheels is different for dual front/rear systems than for dual diagonal systems. In addition, each manufacturer might have a preferred sequence for any given model design. Check the service manual for each car.

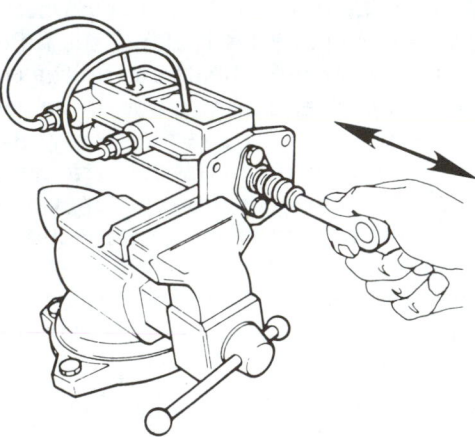

FIGURE 13–87 Before installing the new master cylinder, bleed it. Fill it with fluid and install bleeder hoses from the outlets to the reservoirs. Then, pump the piston back and forth to remove air.

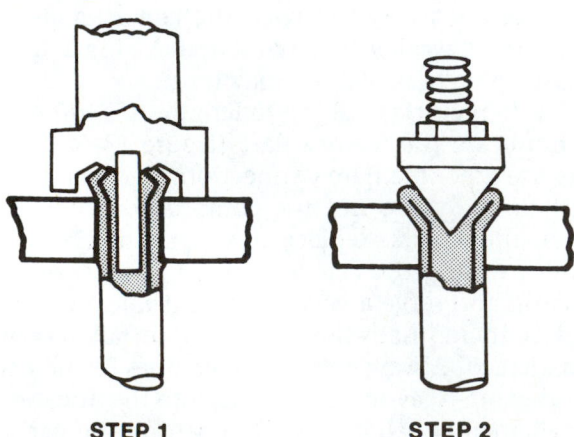

STEP 1　　　　STEP 2

FIGURE 13–86 Brake lines normally require double flare. With step 1, form the steel tubing inward with a special tool. With step 2, use a pointed tool to double lap the tubing in on itself.

FIGURE 13–88 Refer to the service manual for servicing modern antilock brakes. They can be complex. *(Courtesy of Tech-Cor)*

Some four-piston calipers have two bleeder screws. In this case, bleed the lower one first. On diagonal systems, bleed one system at a time. Do one front disc brake first, then the diagonally connected rear drum.

Always check the master cylinder first. If the brake fluid falls below the level of the intake ports, air will get into the system. Refill the reservoir and pump the brake pedal slowly a number of times. Oftentimes this will purge it of all unwanted air. If this does not work, bleed the system.

When bleeding modern antilock brake systems, refer to the service manual. It will give the detailed instructions needed to do good work (Figure 13–88)

Manual Bleeding

Manual bleeding should be done only if a pressure bleeder is not available. Begin at the master cylinder. Clean the cover before removing it and then the diaphragm gasket. Fill the reservoir to $1/4$ inch (6.3 mm) from the top. Apply pressure to the brake pedal slowly and with a smooth action. Open the bleeder screw on the first wheel in the sequence. Drain the aerated fluid through the bleeder hose into a jar partially filled with clean brake fluid (Figure 13–89).

Keep up a pedal pressure while the bleeder screw is open. When the pedal bottoms out, close the screw and release the pedal. If all the air is not yet purged and air bubbles can be seen in the fluid, repeat the process. When only clear fluid with no bubbles appears, go on to the next wheel in the sequence.

While bleeding the brakes, watch the fluid level in the reservoir. About every six pedal applications more fluid will have to be added so it does not fall below the level of the intake port. If it does, more air will enter into the system.

Pressure Bleeding

This procedure is the recommended method for ridding the hydraulic system of air. It is the most efficient, requiring only one person to perform it.

The pressure unit used in this process is a tank divided into two sections by a flexible diaphragm (Figure 13–90). Pressurized air comes into the bottom chamber, compresses the fluid in the top, and brings it up to the desired pressure. The fluid then goes into the hydraulic system through a hose attached to a master cylinder adapter cap. In using a pressure bleeder unit, make sure to use the correct adapter for the particular master cylinder.

Bring the pressure unit up to a level of 15 to 20 psi (103 to 138 kPa). Make sure the master cylinder cover is clean so that no loose particles of dirt fall into the reservoir. Remove the gasket and clean the gasket seat. Fill the reservoir and attach the adapter

cap and hose. Check the coupling sleeve and make sure it is fully engaged before opening the fluid supply valve.

Follow the sequence for bleeding as recommended by the service manual. Allow the aerated fluid to flow out of the bleeder screws through a short bleeder hose into a jar. Once completed, close the supply valve of the pressure unit. Wrap the coupling sleeve in a rag to prevent brake fluid from dripping onto the car finish. (Flush off brake fluid immediately with water or it will ruin the paint.) Undo the coupler. Disconnect the unit and check to determine that the brake fluid level in the reservoir is not more than $1/4$ inch (6.3 mm) from the top. Replace the gasket and cover and check the vehicle's brakes.

The recommended method of pressure bleeding a brake system is with a vacuum-type bleeder. This technique withdraws the fluid from the system rather than pumping it, which has certain advantages:

- It will not create foaming.
- It does not activate the pressure differential valve.
- There is less chance of fluid contamination.

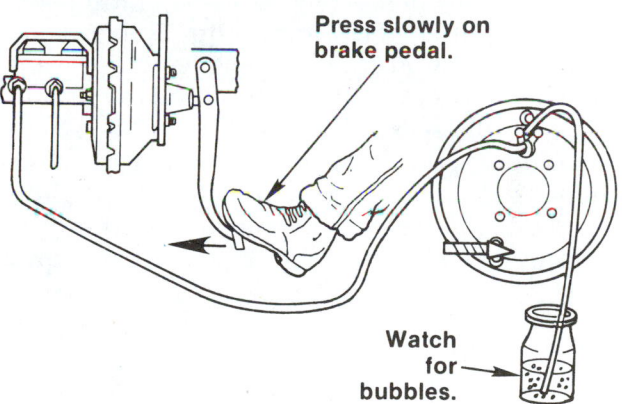

FIGURE 13–89 To manually bleed brakes, open the wheel cylinder bleeder screw. Connect a hose from the screw to a container of brake fluid. By pumping the brake pedal, you will force air from the system. Keep the reservoir full.

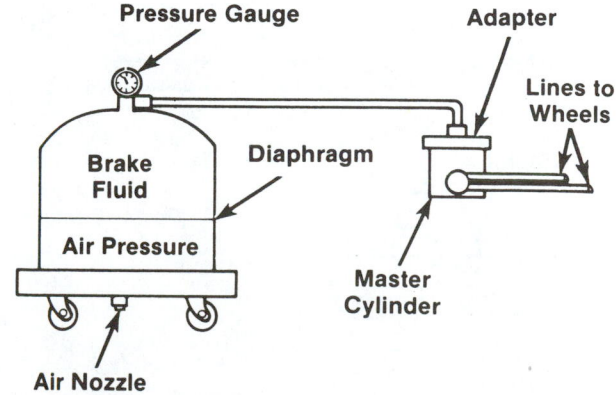

FIGURE 13–90 Pressure bleeding is needed with some brake systems.

Check the connection pattern of the system's wheels. Some cars have the two front and two back wheels connected; some are crossed diagonally. Others combine the front two with one rear, and the rear two with one front wheel. Always check the service manual.

FINAL CHECK OF BRAKES

Late-model cars sometimes have air left in the master cylinder bore above the outlet ports. As a final check, raise the rear of the car so the bore angles downward. Take off the reservoir cover and tap the pedal lightly a number of times. Very small bubbles should come up through the fluid to the top of the reservoir. When a spurt of fluid appears and the stream of bubbles ceases, the vehicle is ready to be road tested.

POWER BRAKES

Power brakes are nothing more than a standard hydraulic brake system with a vacuum assist or booster unit between the pedal and the master cylinder to help activate the brakes (Figure 13–91). Most power brake units consist of a piston, control valves, and a vacuum connection from the engine intake manifold. When the foot is off, the brake and the vacuum unit are in the released position, and the vacuum system intake port is closed. However, a special atmospheric port remains open to allow air to pass from one side of the vacuum piston to the other. This maintains equal pressure on the piston and keeps it

SHOP TALK

Check if the booster is working by pushing the pedal down, hold it down, and start the engine. Note that when the car first starts, the brake pedal will tend to go down. This is normal and is related to the vacuum buildup.

in the off position. When the brake is hit, the atmospheric port closes and the vacuum port opens. Vacuum from the engine then withdraws the unit's piston forward against the master cylinder operating rod and actuates the brakes.

When a unibody car is involved in a collision, the power brake booster should be carefully inspected. Pay particular attention to vacuum hoses, check valves, fasteners, and the master cylinder itself. Replace all damaged pieces. This system builds up to 2,500 psi (17,250 kPa) to make the car stop—do not take any chances with it.

ANTILOCK BRAKES

Modern antilock or antiskid brake systems can be thought of as electronic/hydraulic "pumping" of the brakes for straight-line stopping under panic conditions. That is, this system is another control arrangement that is used in conjunction with a basic hydraulic braking operation. During hard braking

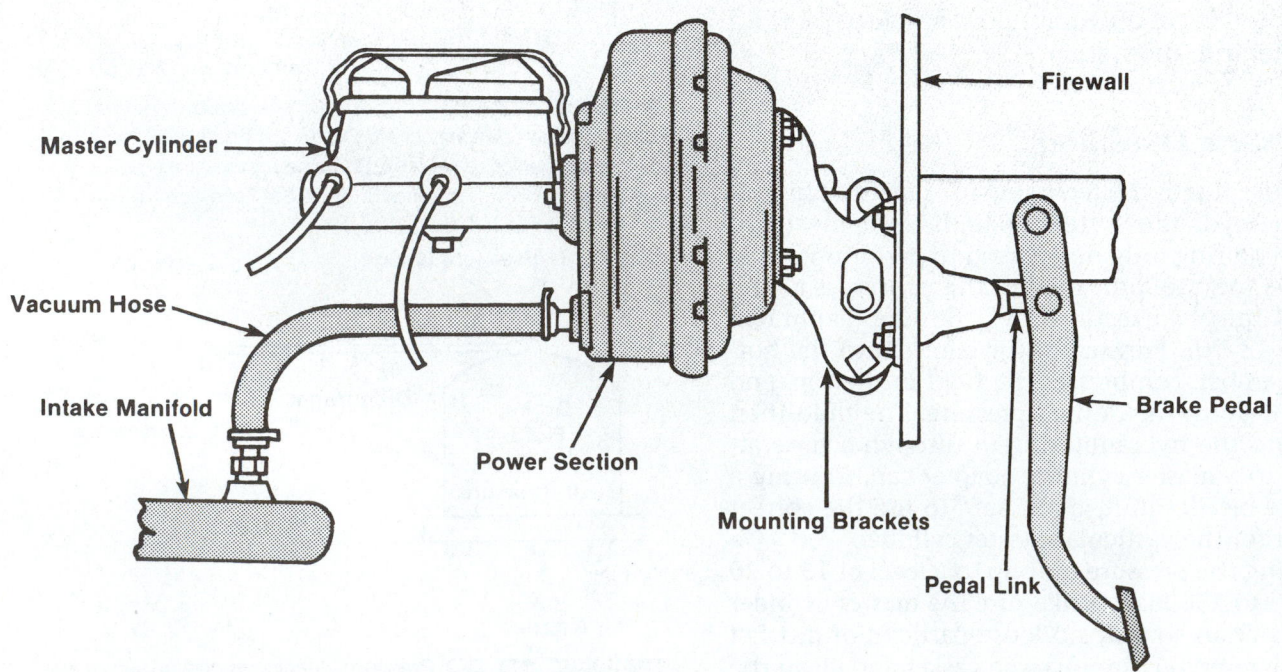

FIGURE 13-91 The brake booster normally bolts between the master cylinder and the firewall.

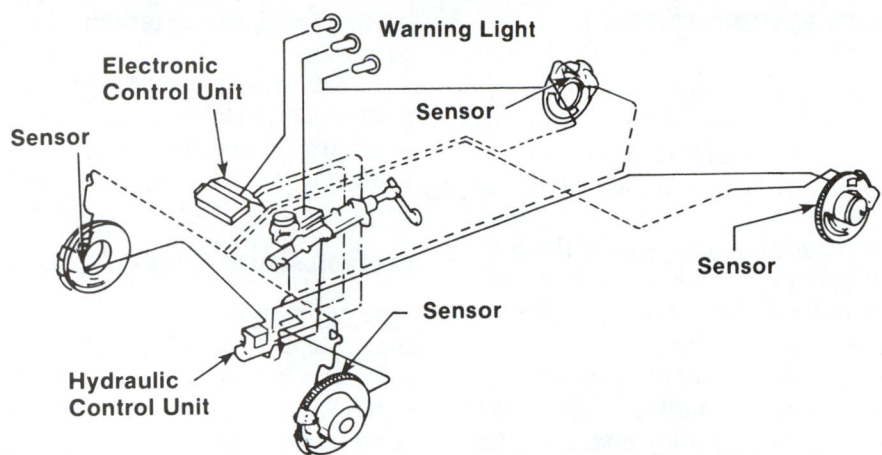

FIGURE 13-92 Study the basic parts of an ABS four-channel or four-sensor system.

conditions with a conventional hydraulic system, it is possible for the wheels of a vehicle to lock, resulting in reduced steering as well as braking. On vehicles equipped with the antilock brake system (ABS), however, an electronic sensor constantly monitors wheel rotation (Figure 13–92). If one or more of the wheels begins to lock, the system opens and closes solenoid valves, cycling up to 10 times per second. This applies and releases the brakes rapidly and repeatedly, so that the front wheels alternately steer and brake. This makes it possible for vehicles equipped with the antilock brake system to avoid skidding under conditions that might cause vehicles not so equipped to handle differently.

The antilock or antiskid brake system has a controller that senses rotation at each of the wheels through wheel sensors. It can apply the antilock brake system to each of the front wheels independently, to the rear wheels as a pair, or to any combination of these three, as the need arises. Since there are several antilock or antiskid systems, check the service manuals for diagnosis and service procedures.

Antilock brake service is similar to conventional brakes. However, electronic parts are added to operate the system. Most ABS brakes have self-diagnosis. The computer will output a trouble code if an electrical-electronic malfunction develops. You can refer to charts in the service manual to see what each number code means. This is will tell which part might be at fault.

For example, if a trouble code indicates a problem with one of the wheel speed sensors, check that it is adjusted properly and undamaged. You may also need to test the sensor and its wiring.

PARKING BRAKES

The *parking* or *emergency brake* uses a steel cable to physically apply the brake shoes or pads. The rear wheel brakes act to hold the car stationary for parking. Although shown in Figure 13–93, parking brakes are not actually a part of the hydraulic brake service system. They are actuated mechanically rather than by hydraulic pressure.

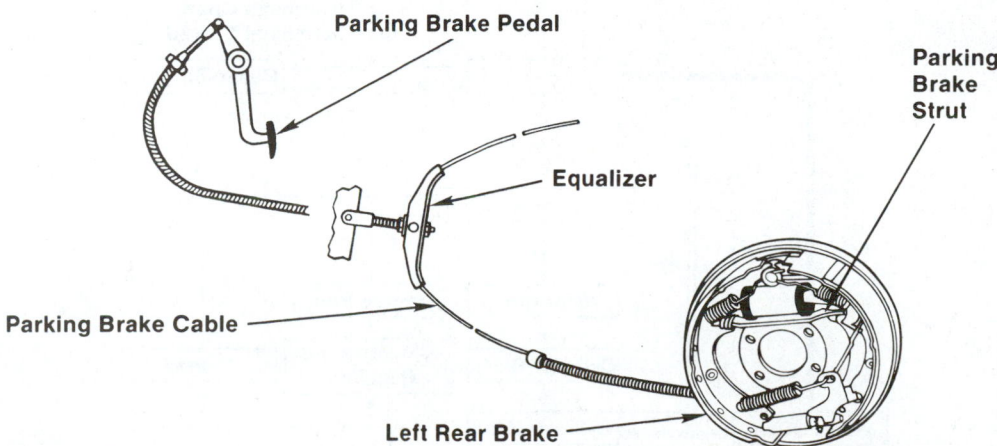

FIGURE 13-93 The emergency or parking brake is simply a mechanical system of cables and levers that apply shoes or pads.

13.6 COOLING SYSTEMS

A *cooling system* maintains the correct engine operating temperature. It is often damaged in a collision and must be restored to its preaccident condition. Shown in Figure 13–94, the basic parts of a cooling system are:

The **radiator** transfers coolant heat to the outside air. The radiator *pressure cap* prevents the coolant from boiling. A *radiator fan* draws outside air through the radiator to remove heat.

The *water pump* circulates coolant through the inside of the engine, hoses, and radiator. The *water jackets* are passages in the engine for coolant. The *thermostat* regulates coolant flow and system operating temperature.

A *heater system* uses coolant heat and a heater core (small radiator under dash) to warm the passenger compartment. The *automatic transmission cooler* uses the radiator to reduce transmission fluid temperature.

Antifreeze is used to prevent freeze-up in cold weather and to lubricate moving parts. Antifreeze also prevents engine overheating. A *coolant recovery system* stores an extra supply of coolant for the system.

COOLANT SERVICE

One of the most frequently missed areas of the cooling system is the strength of the antifreeze. A common idea is that the stronger the concentration, the better. This is not so. Pure water transfers heat better than pure antifreeze, but it does not protect the system from freezing or corrosion.

In addition, water has a boiling point of only 212 degrees Fahrenheit. Pure antifreeze has a higher boiling point (330 degrees Fahrenheit) than pure water. But, due to its lack of heat transferability, it

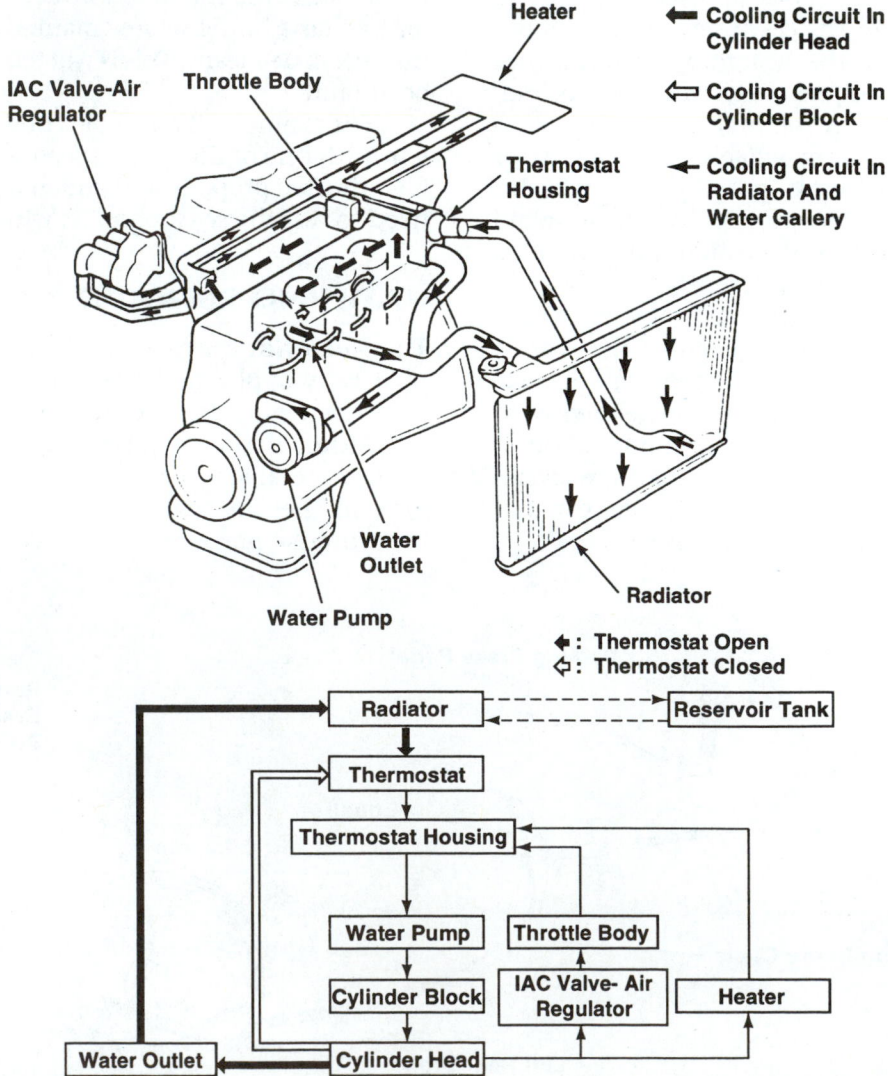

FIGURE 13–94 Study cooling system components and flow.

WARNING

Do not let antifreeze, brake fluid, and other chemicals drip on painted surfaces. They can discolor or damage paint!

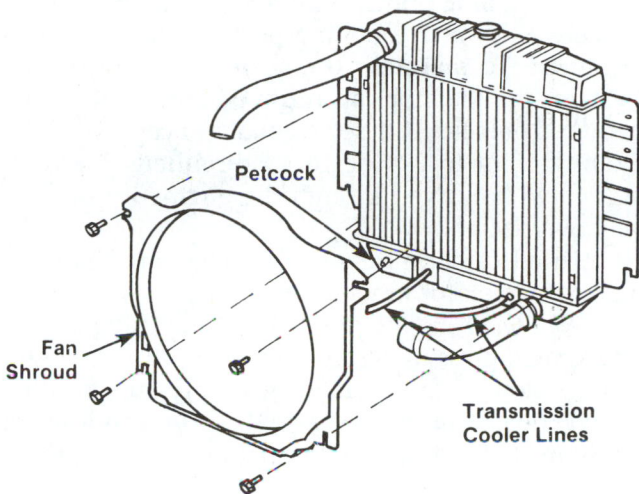

Petcock

Fan
Shroud

Transmission
Cooler Lines

FIGURE 13-95 Transmission cooler lines may go into the radiator.

can cause an engine to overheat. In addition, pure antifreeze offers no corrosion protection, since the anticorrosion chemical must be mixed with water to activate it.

The ideal antifreeze-to-water ratio is 50/50. This ratio provides freezing protection to –34 degrees Fahrenheit, while increasing the coolant's boiling point to 224 degrees Fahrenheit.

An *antifreeze tester*, commonly called a *hydrometer*, is used to determine the freeze-up protection of the coolant mixture. Pull a sample of the vehicle's coolant solution into the tester. Then read the lowest temperature the coolant will withstand without freezing. Add more coolant if needed.

Some systems, such as mid-engine cars and vans, and those having dual heaters, can require up to three to four gallons of antifreeze. Check the specifications. Measure the level of installed protection with an antifreeze tester, following the manufacturer's directions.

Coolant Leaks

A low coolant level reduces cooling capacity. In addition to obvious leaks resulting in squirting steam and water, coolant leaks may also be internal. Internal leaks

can result from a blown head or intake manifold gasket or warped cylinder heads.

Another type of internal leak occurs when the automatic transmission fluid (ATF) cooler tank leaks from the inside into the radiator (Figure 13–95). A sure sign of this is a thick, pink solution in the radiator caused by the mixing of coolant and transmission fluid. Due to the impact forces in a collision, this area should be a high priority in any post-repair inspection. The loss of ATF due to a fractured radiator or ATF-to-radiator line connection causes two problems:

- Eventual loss of hydraulic pressure needed to operate the transmission
- Loss of lubricant needed to protect the internal mechanical and friction surfaces

Coolant Recovery Bottle. This bottle is normally plastic and can be easily damaged. Check for cracks or abrasions in the bottle and make sure the hose leading to the radiator is connected and in good shape. These plastic tanks are normally not repaired. If cracked or distorted, replace with new components.

Refilling a System. Before making any replacements in the cooling system, drain the coolant from the system and dispose of it properly (check local regulations). Never reuse the old fluid. When refilling, make sure that the proper coolant is used in vehicles with aluminum engines or radiators. Some warranties will not be honored if the coolant recommended by the manufacturer is not used.

After installation of the engine coolant, bleed the cooling system to ensure proper coating. Always follow the manufacturer's recommendations. On some vehicles there are bleed valves; on others, the thermostat must be taken out to bleed the system; and on others, the upper radiator hose must be removed. Check the directions in the service manual.

RADIATOR CONSTRUCTION

The coolant flows from the engine to tubes located inside the fins of the radiator where the airflow cools it. If these tubes become plugged, either by being bent or through maintenance neglect, the flow of coolant through the radiator is reduced and overheating can result. This condition is more noticeable at highway speeds and/or with heavier loads. If the vehicle is not air conditioned, plugged areas of the radiator can be identified by cold spots felt on the front of the radiator after the vehicle is warmed up.

Chances are that a collision-damaged vehicle will have damaged or bent areas in the radiator (Figure 13–96), even though leaks might not be present. So always check that area carefully.

The *radiator cap pressure rating* is stamped on the cap. If the cooling system is disassembled during repair or parts are replaced, a pressure test should be performed.

A *cooling system leak test* is performed by installing the tester on the radiator neck. Pump the tester handle until its gauge equals the cap pressure rating. A loss of pressure or coolant means there is a leak (Figure 13–97).

Crushed radiator fins can be straightened with a special tool designed for that purpose and tubes that are not too badly mangled can be soldered. But if large hunks of cooling fins have been pulled loose or if multiple tubes have been crushed or ruptured, a new core is recommended. Time is money and if the cost to repair core damage begins to approach the cost of a new core, most shops will opt for the new core. Besides, a new core offers greater reliability, especially if the radiator is showing its age.

Due to the aerodynamic designs of some vehicles, the radiator ends up being the low point in the cooling system. Because the radiator is lower than other areas, the cooling system might not be able to be completely filled after it is drained. This will cause overheating problems due to the cooling

system being only partially full. To eliminate this problem, some vehicles have a separate filler neck that is higher than the radiator. If the vehicle is not so equipped, jacking up the front of the vehicle will place the fill point higher than the rest of the system. This allows the system to be completely filled.

RADIATOR CAP OPERATION

Most cooling systems operate under a pressure of 15 psi. This is because increased pressure on a liquid raises its boiling point. With a 50/50 mixture of antifreeze and water, a boiling point of 263 degrees is achieved. To maintain the correct pressure in the system, the radiator cap must be able to hold the required pressure. A defective radiator cap lowers the boiling point to 224 degrees Fahrenheit. The coolant could boil, even though the engine is not actually overheated.

Dried calcium deposits in a radiator cap can make the radiator inoperative.

A *radiator cap pressure test* is done using a cooling system pressure tester (Figure 13–98). The tester gauge should stop increasing its pressure reading when the cap rating is reached. If the cap leaks or does not open at its spec temperature, replace it!

THERMOSTAT OPERATION

The *thermostat* (Figure 13–99) is the engine's temperature control. It keeps the coolant in the engine until the engine reaches its peak operating temperature. Once the proper operating temperature (usually 180 to 195 degrees Fahrenheit) has been reached, the thermostat opens up, allowing the coolant to flow to the radiator for cooling.

FIGURE 13-96 The radiator is often damaged in a frontal collision.

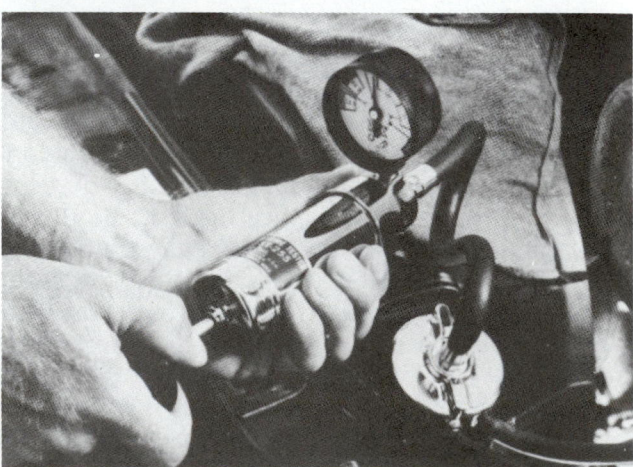

FIGURE 13-97 To check for coolant leaks, install the tester on the radiator filler neck. Pump until the gauge reads at pressure cap rating. Then look for leakage or loss of coolant.

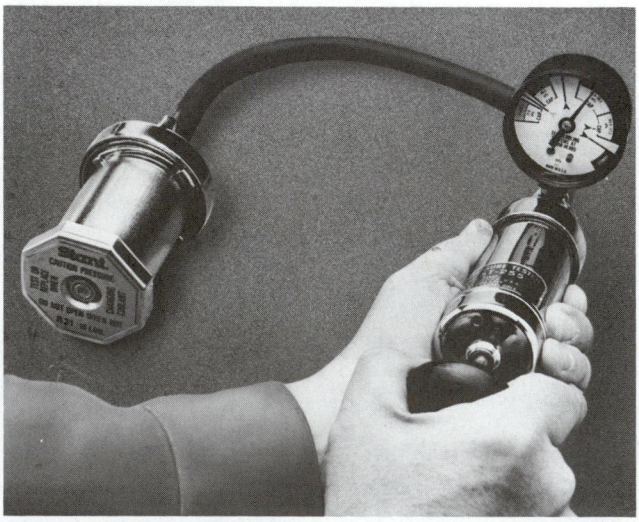

FIGURE 13-98 Test the radiator cap if needed. It should hold spec pressure without leakage.

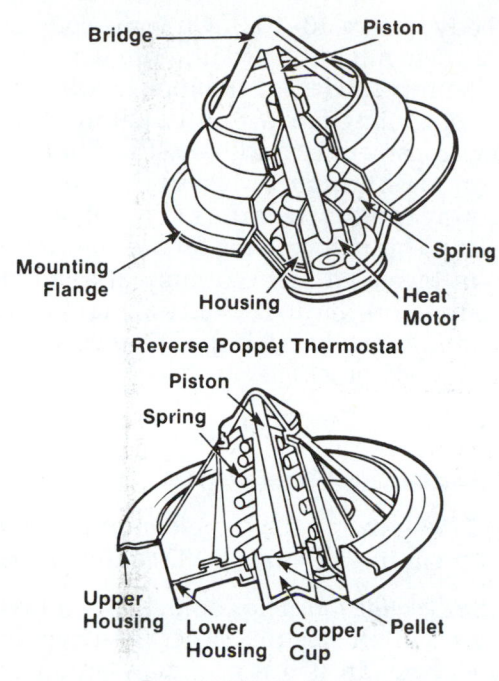

FIGURE 13-99 Note two types of thermostats: (A) reverse poppet and (B) balanced sleeve.

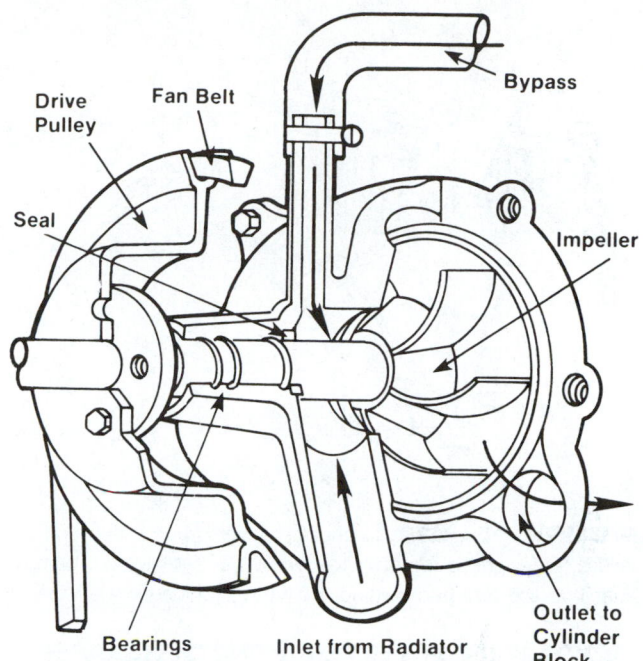

FIGURE 13-100 Note basic parts of water pump.

A *stuck thermostat* can cause the engine to overheat or run too cool. A quick way to tell if this is happening is to feel the upper radiator hose. If it is cold but the engine is hot, the thermostat is probably stuck closed. On the other hand, if the thermostat is stuck open, the engine will take a long time or possibly fail to reach proper operating temperature.

WATER PUMP OPERATION

The water pump circulates the coolant throughout the cooling system with internal blades called *impellers* (Figure 13–100). They push the fluid through the system. If they are loose or corroded, the coolant will not circulate properly and will cause overheating. This is more noticeable at highway speeds. Check for coolant movement with the cap removed and the engine warm enough for the thermostat to be open. If the coolant does not flow, replace the pump.

Radiator System Belts and Hoses

While working on the cooling system during collision repair, check that the hoses are in good condition and are securely clamped (Figure 13–101). The lower radiator hose routes coolant from the radiator to the water pump. The turning water pump draws coolant, creating a low-pressure area in the lower hose. A coiled spring inside the lower hose keeps it from collapsing. The lower hose should not show signs of collapse.

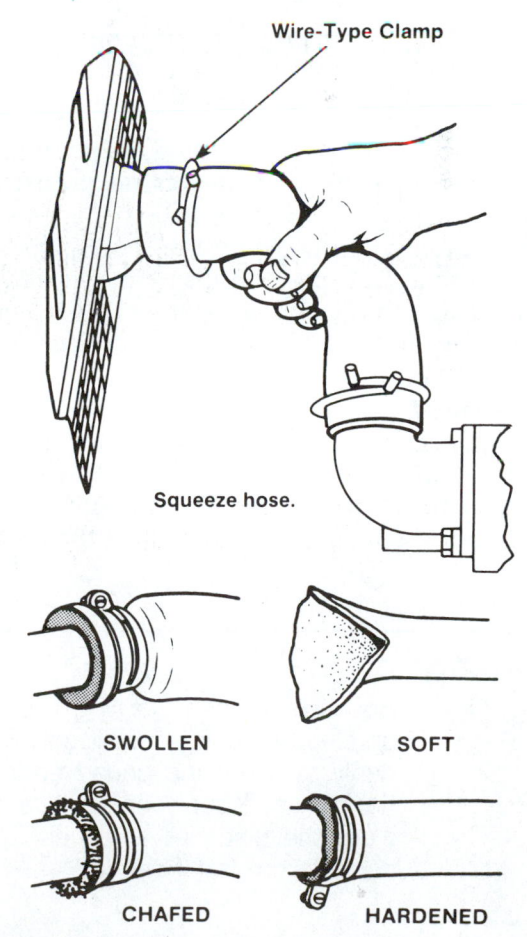

FIGURE 13-101 Always check hoses for damage or deterioration.

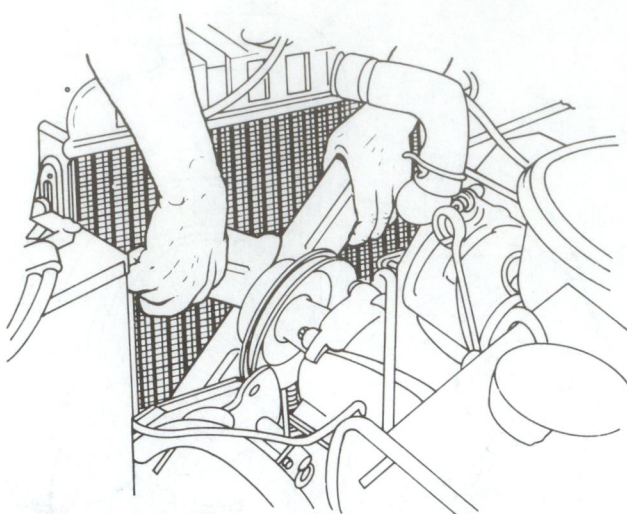

FIGURE 13-102 To check for water pump damage or failure, try wiggling fan or pulley sideways. If it wiggles, pump bushings are bad (fan shroud not shown for better clarity).

COOLING SYSTEM PROBLEMS

Although similar, front- and rear-drive vehicles have some important cooling system differences.

RWD Cooling Problems

Certain cooling system problems are limited to RWD vehicles. This is due to the layout of the engine and cooling fans.

- *Belt tension.* On vehicles where the cooling fan is driven by belts from the engine, proper tension must be maintained or belt slippage will occur. When this happens, the cooling fan does not turn at full speed, resulting in reduced airflow through the radiator at idle. This is a problem only at idle. At highway speeds, the airflow is sufficient to maintain cooling.
- *Fans and fan clutches.* On vehicles with electric fans, check for loose electrical connections and bare, burnt, or cut wires. Check that the fan blades turn with no interference and that the fan mounts are not rubbing against the radiator

Be careful when working around an electric cooling fan. It can suddenly turn on and cause serious hand injury. Keep your hands away from the blade at all times. Disconnect the fan if needed.

or body (Figure 13–102). On some cooling systems there might be two or more fans.

Some vehicles are equipped with a special type of cooling fan with a clutch built in. Fan clutches can either be filled with a fluid or use a thermostat spring that allows the fan to slip at highway speeds when the fan is not needed for airflow. This reduces drag on the engine, resulting in increased fuel economy. If a fan clutch does not work due to its fluid leaking or its spring becoming defective, fan speed is reduced at idle, causing overheating.

FWD Cooling Problems

Here are some of the more common problems that can occur in a unibody FWD cooling system.

- *Electric cooling fan.* The cooling fan on FWD vehicles is often electric, rather than belt driven. An electric fan is less of a draw on an engine than a belt-driven fan, resulting in better fuel economy.

  Most unibody vehicles have one single-speed cooling fan (Figure 13–103). Some vehicles are equipped with two fans or one two-speed fan. These are designed to increase airflow through the radiator. If engine overheating is present at idle, check to make sure that the fan or fans are operating correctly.
- *Cooling fan and air conditioners.* When the air conditioner is turned on, the cooling fan should also come on, regardless of coolant temperature. This prevents the engine from overheating due to the increased load placed on it from the air-conditioning.

REPLACING A RADIATOR

There are two basic types of radiator design. They are distinguished from one another by the direction of the coolant flow and location of the two tanks. In the *downflow radiator* (Figure 13–104), the coolant flows from the top tank downward to the bottom tank. In the *crossflow radiator* (Figure 13–105), the tanks are located at either side and the coolant flows across the radiator core from tank to tank.

When replacing a radiator, first measure it upright with the filler neck opening facing up. Determine whether it is a downflow or crossflow type. When measuring either a crossflow or downflow radiator, measure only the core to determine the size (Figure 13–104). The core is the central part of the radiator (between the tanks and mounting brackets) and consists of parallel rows of tubes and fins.

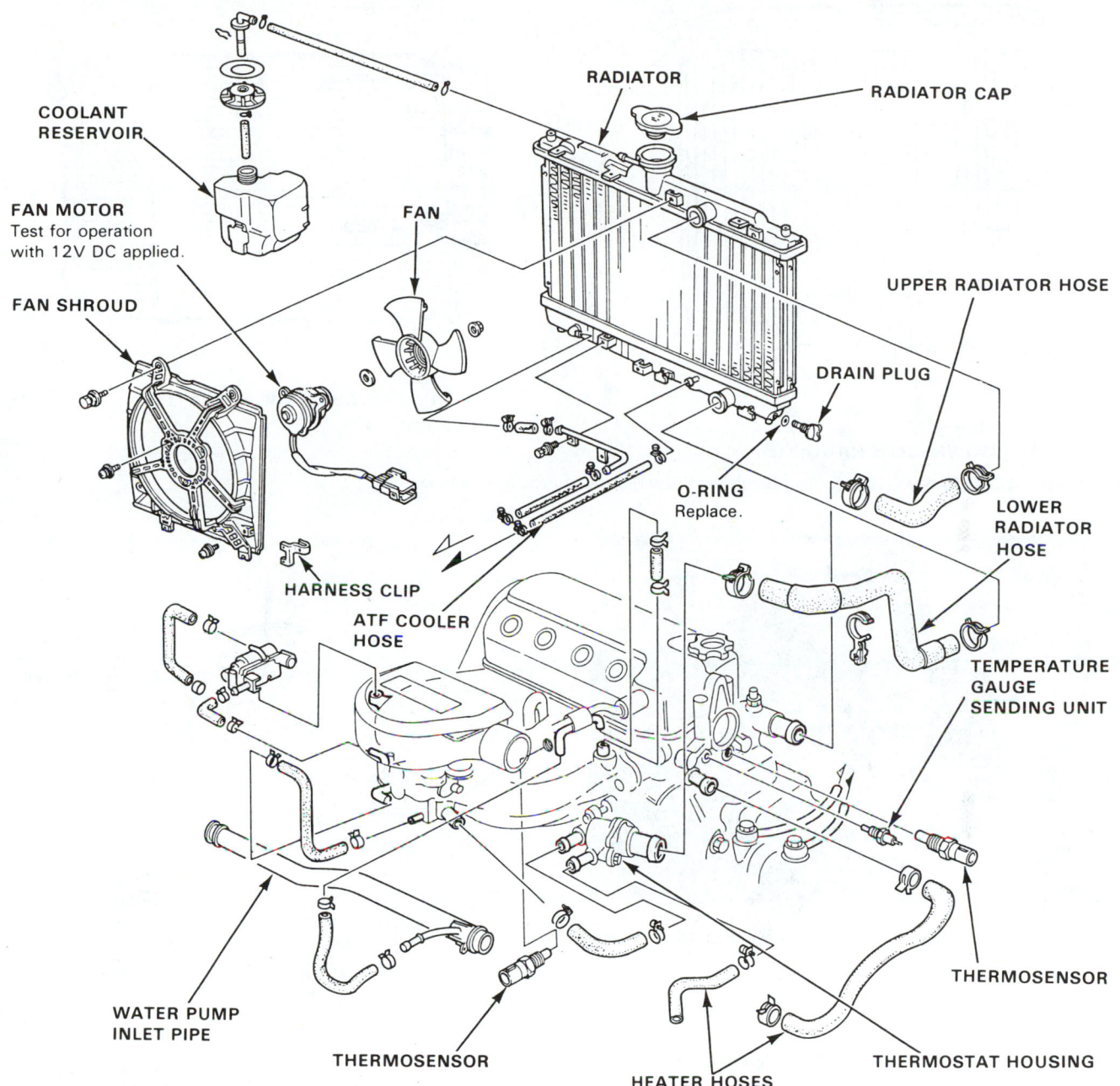

FIGURE 13-103 Study parts of cooling system and electric cooling fan. They are often damaged in frontal collision. *(Courtesy of Honda Motor Co.)*

Measure the height from the top edge of the core to the bottom edge, and from side to side to determine the width. To measure the thickness of the core, simply take a piece of straight wire, for example, and insert it through the core until the end becomes flush with the other side. Mark the other end of the wire with the thumb and forefinger (at the point where it is flush with the core), withdraw the wire, measure it, and obtain the core thickness.

Replacing radiators (Figure 13-105) on vehicles involved in collisions requires a special check of tolerances, making certain the mounting does not bring the surface of the radiator too close to the fan. This could create a situation in which hitting a bump could result in the radiator hitting the fan. Again, use care in reconnecting the transmission lines, avoiding a connection at an angle where the threads could be stripped. Follow the manufacturer's torque specifications.

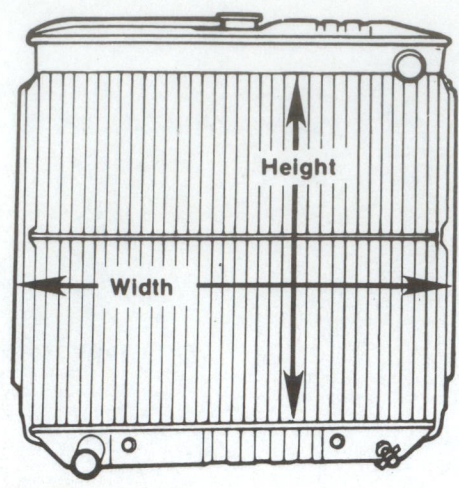

DOWNFLOW RADIATOR

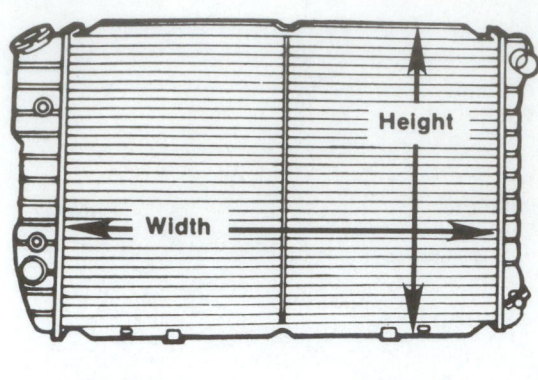

CROSSFLOW RADIATOR

FIGURE 13-104 Measuring a radiator like this will allow you to order a new one.

FIGURE 13-105 Be very careful when replacing a new radiator. It can be damaged easily.

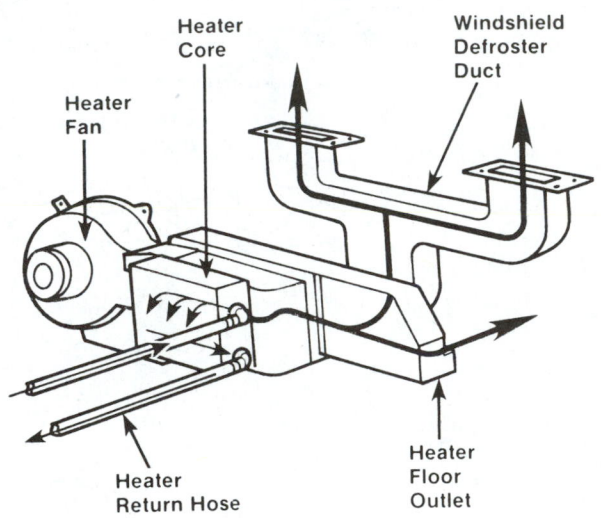

FIGURE 13-106 The heater core is the heart of the heater mechanism.

13.7 HEATER OPERATION

The heater is a comfort control item especially in colder climates. Actually, the heater core could be considered a miniature version of the radiator. That is, as hot coolant flows through the heater core, a fan blows air over the tubes, warming it and delivering it to the passenger compartment. The blower fan, located in the heater housing, forces air through the heater core and into the passenger compartment.

The air heating distributor system is a duct system. Outside air enters the system through a grille, usually located directly in front of the windshield, and goes into a plenium chamber where rain, snow, and some dirt is separated from it. The air from the plenium is directed through the car's heater core, through the air conditioner evaporator, or into a duct that runs across the fire wall of the car. Outlets in

the duct direct the airflow into the passenger compartment (Figure 13-106).

Doors inside the system either recirculate the air inside the compartment or circulate outside air, according to the control settings. They also route the air inside the system through or around the heater or air conditioner, or to the windshield defroster.

Some cars have an additional distributor system that brings air into the vehicle through intakes located in the engine compartment and carries it through ducts to side vents located in the sidewall of the passenger compartment near the front seat passenger's feet. The airflow through these vents can be controlled mechanically, electronically, or by a vacuum system.

The heating system is controlled by a number of cables, valves, and switches. Be sure the cables are correctly reconnected if you disconnected any of them during repair. Follow the manual for trouble-shooting. Also keep in mind that the heating system may have collected some dust during repair. Be sure to run it during cleanup so that the customer is not caught in a cloud of sanding dust after picking up the car.

13.8 AIR-CONDITIONING AND HEATER SYSTEMS

Proper handling of the air-conditioning (A/C) system during collision repair is both one of the most important and one of the least understood aspects of working with mechanical components. Many needless repairs are caused unknowingly by service technicians who do not understand the importance of following some very strict rules for working with air-conditioning systems. What compounds the problem is that malfunctions often occur several months after the collision repair work is completed, so the customer is unaware of who caused the problem.

A/C OPERATION

An *air-conditioning system* is designed to cool the passenger compartment. System designs vary. For example: some air-conditioning systems use an accumulator while others use a receiver/drier (Figure 13–107).

Receiver/driers and *accumulators* serve the same basic purposes. They use a desiccant bag to remove moisture from the system. The difference is their location. The accumulator is between the evaporator and the compressor. The receiver/drier is between the condenser and the expansion device. They act as storage tanks.

Air-conditioning systems are divided into two sides, high and low. The dividing points are the compressor and the expansion device.

The **A/C high-side** contains high pressure/high temperature refrigerant. Its hoses feel hot to the touch. High-side hoses are generally smaller in diameter than the low-side (Figure 13–108).

The **A/C low-side** contains low pressure/low temperature refrigerant. Its hoses feel cold to the touch. Low-side hoses are generally larger in diameter than the high-side.

In the basic air-conditioner system, the heat is absorbed and transferred in the following six steps (Figure 13–108).

1. Refrigerant leaves the compressor as a high-pressure, high-temperature vapor.
2. By removing heat via the condenser, the vapor becomes a high-pressure, lower-temperature liquid.
3. Moisture and contaminants are removed by the receiver/drier, where the cleaned refrigerant is stored until it is needed.
4. The expansion valve converts the high-pressure liquid into a low-pressure liquid by controlling its flow into the evaporator.
5. Heat is absorbed from the air inside the passenger compartment by the low-pressure, low-

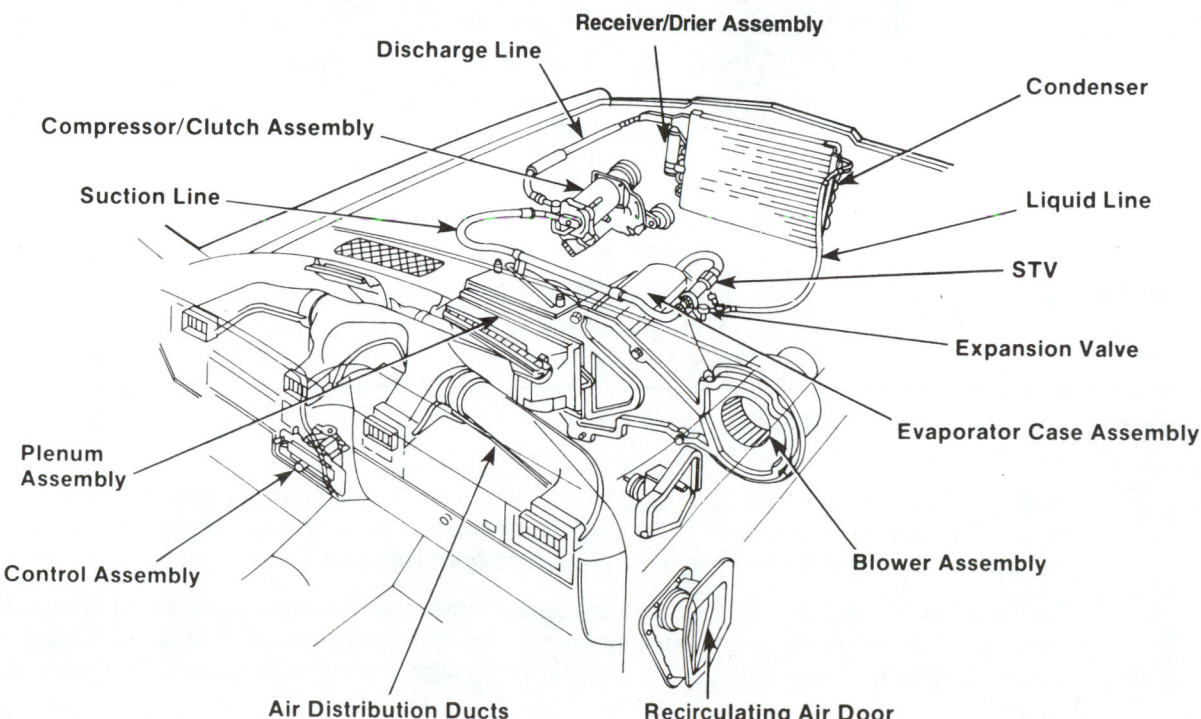

FIGURE 13-107 Study the parts of a typical air-conditioning system.

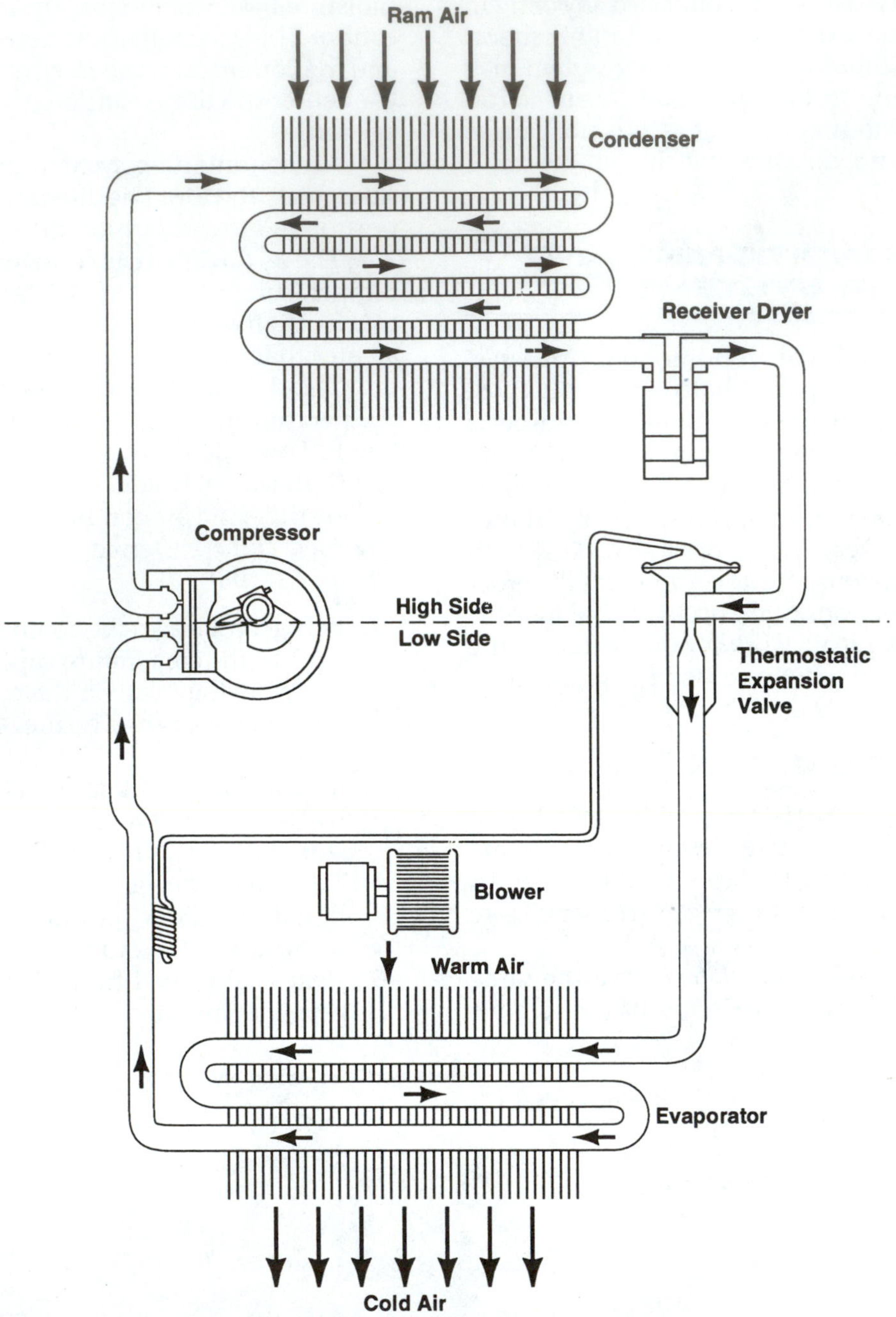

Ram Air

Condenser

Receiver Dryer

Compressor

High Side

Low Side

Thermostatic
Expansion
Valve

Blower

Warm Air

Evaporator

Cold Air

FIGURE 13-108 Note refrigerant flow cycle. *(Courtesy of Mitchell International, Inc.)*

WARNING

R-134a is NOT compatible with R-12. If you install the wrong kind of refrigerant, the A/C system will not work properly.

Also, R-134a oils are not compatible with R-12 oils. This requires separate service equipment. To avoid a mistake, R-134a uses metric quick-connect service ports. The high-side port is larger, so the same charging hoses cannot be used.

Mixing of R-12 and R-134a, even in trace amounts, can be fatal to a system. This mistake can cause damage to seals, bearings, compressor reed valves, and pistons. Mixing refrigerants can also cause desiccants used in R-12 systems to break down and form harmful acids.

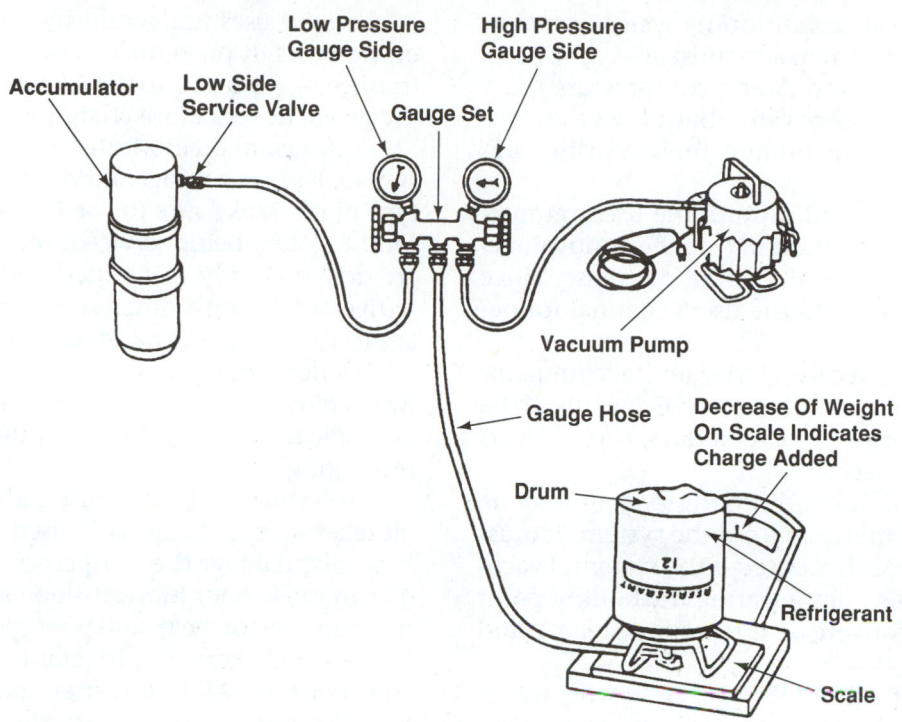

FIGURE 13-109 This setup will allow (A) evacuation and (B) recharging.

temperature refrigerant, causing the liquid to vaporize.

6. The refrigerant returns to the compressor as a low-pressure, higher-temperature vapor.

Due to their possible depleting effect on the ozone layer, chlorofluorocarbon (CFC) and hydrochlorofluorocarbon (HCFC) are being phased out. This includes R-12 and other refrigerants used in air-conditioning and refrigeration systems.

Environmental agencies call for a gradual phaseout of most ozone-depleting substances. They state that R-12 systems can be serviced using recovered and recycled refrigerant. After filtering, this R-12 can then be used again in another vehicle. This reuse is designed to extend the supply of refrigerant.

R-134a is the present replacement for *R-12*. It is less harmful to the ozone layer. New vehicles are being designed to run on this new refrigerant. The compressor and other parts are designed to be used with R-134a.

A/C HANDLING TIPS

Although most A/C repairs and service work are done in specialty air-conditioning shops, here are pointers that the collision repair technician must keep in mind:

Most A/C failures are caused by moisture entering the system. Moisture interacting with refrigerant causes sludge and hydrochloric acid to form. This

will attack the delicate parts of the compressor. The resulting acid will also eat away at the aluminum components of the evaporator and condenser.

When removing or opening up the A/C unit, seal all openings. These can be synthetic rubber, tight-fitting caps, plugs, or plastic wraps. Use sturdy rubber bands or wire ties to hold plastic wraps in place securely.

If an A/C system has been open to the atmosphere more than a few hours, do the following:

1. Change the oil.
2. Flush each component separately with nitrogen gas before charging.
3. Replace the receiver/drier or accumulator.
4. During evacuation, hold the system at high vacuum for a minimum of 30 minutes to pull out air and moisture.
5. Recharge without leaking refrigerant (Figure 13–109).

 **CAUTION** Wear hand and face protection when working on an air-conditioning system. When refrigerant escapes from the system or if you touch the supply tank, it can cause severe frostbite burns.

Discharging an air-conditioning system removes refrigerant from the system and must always be done before parts are removed. Some compressors use a special back seating service valve that allows the compressor to be removed without completely discharging the system.

A *recovery system* will capture the used refrigerant and keep it from contaminating the atmosphere. Most will also filter the refrigerant for reuse. Since equipment varies, refer to the user's manual for detailed procedures.

Manufacturer's receiver/drier and accumulator replacement recommendations vary. Generally, if the system has been open for several days, the receiver/drier should be replaced.

Evacuating an air-conditioning system is done to remove air and moisture from the system. It must be done any time air has entered the system. Evacuating is done by vacuum lowering the boiling point of the moisture, converting it to vapor (steam), and removing it.

Before *charging* (filling) the system with refrigerant, determine the amount and type of refrigerant used. This information is found in the service manual and on the label on the radiator support or compressor. Do not mix different types of refrigerants. Charging can be done with a gauge set or with a charging station.

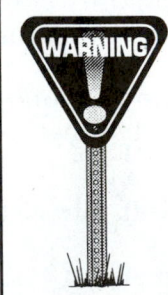

WARNING The release of R-12 into the atmosphere is prohibited by current environmental regulations. Never vent the refrigerant into open air. Use a recovery/recycling machine (Figure 13-110).

Purging uses refrigerant to push air and dirt out of the hoses. It prevents air and other contaminants from being pushed into the A/C system. Always purge the gauge hoses before charging.

Refrigerant oil lubricates moving parts in the A/C system. Use only refrigerant oil. Do not use any other type of oil. Make sure to use the type recommended for the system being serviced. For example, some oils are designed only to be used with specific types of refrigerant. Using a different type can result in damage to the compressor and seals and other parts.

General rules are to add the amount of oil that was removed during discharge. There are adapters available to use refrigerant pressure to add oil during recharging.

Too much oil can cause reduced cooling. The oil takes up space normally used by the refrigerant. It can also damage the compressor and seals. Too little oil can cause poor lubrication of the system, premature compressor wear, and poor system performance.

An open can of refrigerant oil can collect dirt and moisture. Adding contaminated refrigerant oil to the system can cause corrosion, which can result in the failure of the compressor and other parts.

There are several ways to find refrigerant leaks:

1. Electronic leak detector
2. Refrigerant cans with dye
3. Soap and water solution in spray bottle

A sight glass is used to check the amount of refrigerant in a system. It is located on the receiver/drier or in-line.

Readings of the sight glass are interpreted as follows:

1. Clear—completely full or completely empty
2. Oil streaks—no refrigerant
3. Foam or constant bubbles—low refrigerant charge
4. Clouded—desiccant being circulated through system

Electronic leak detectors are battery-operated instruments that use an audio sound to announce the presence of a gas leak. They are designed to detect different types of gases.

When checking for refrigerant leaks, always check along the bottom of the hoses, fitting, seals, and other possible leakage points. This is because the refrigerant is heavier than air and is easier to detect below these parts.

13.9 EXHAUST SYSTEMS

The *exhaust system* collects and discharges exhaust gases caused by the combustion of the air/fuel mixture within the engine. It also quiets the noise of the

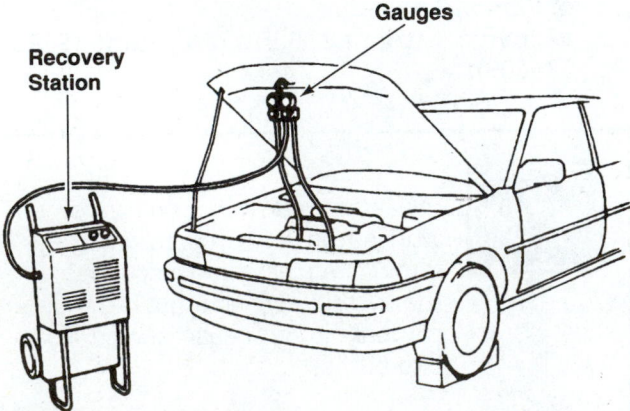

FIGURE 13-110 A recovery system is needed to keep refrigerant from entering and harming the atmosphere.

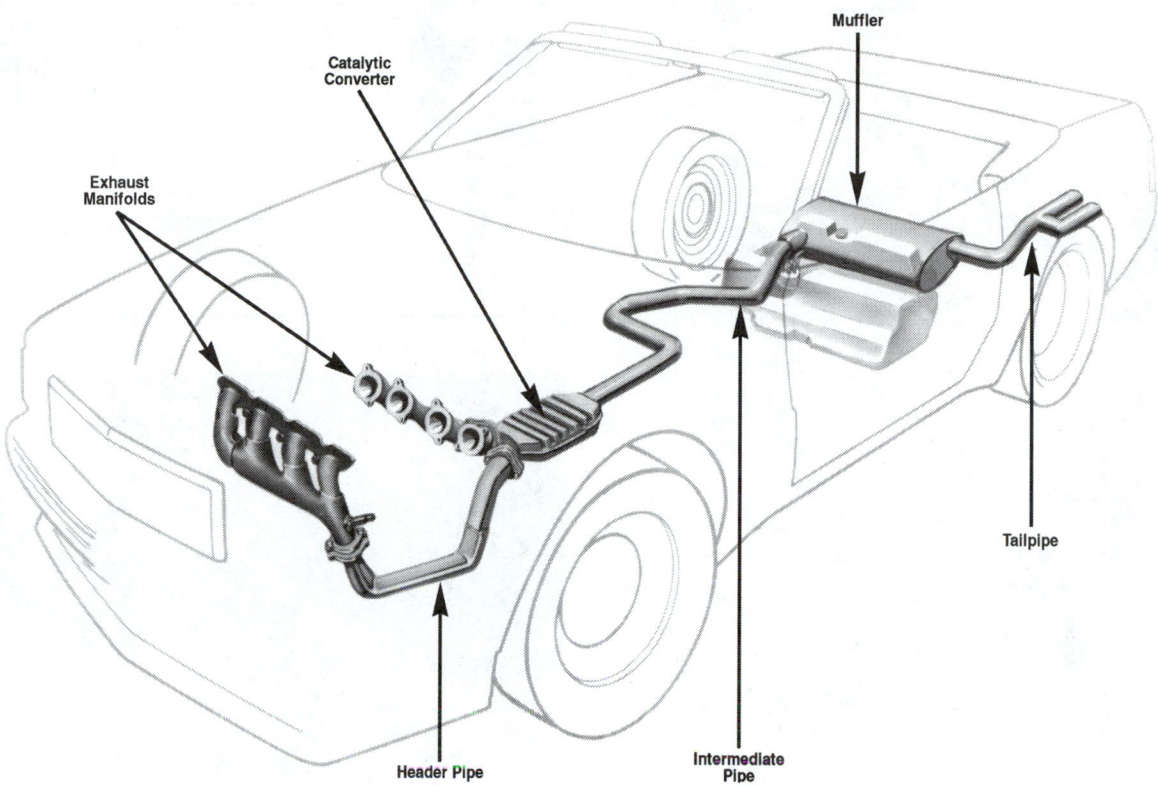

Muffler

Catalytic
Converter

Exhaust
Manifolds

Tailpipe

Header Pipe

Intermediate
Pipe

FIGURE 13-111 Note parts of a typical exhaust system.

running engine. The major parts of an exhaust system are shown in Figure 13–111.

The *header pipe* is steel tubing that carries exhaust gases from the engine's exhaust manifold to the catalytic converter. The *catalytic converter* is a thermal reactor for burning and chemically changing exhaust byproducts into harmless gases (Figure 13–112). The *intermediate pipe* is tubing that is sometimes used between the header pipe and catalytic converter or muffler.

A *muffler* is a metal chamber for dampening pressure pulsations to reduce exhaust noise. The *tail pipe* is a tube that carries exhaust gas from the muffler to the rear of the vehicle.

EXHAUST SYSTEM SERVICE

The exhaust system can also be damaged during a collision, requiring partial replacement. Its parts may need removal during major collision repairs.

Because of constant changes in recommended catalytic converter servicing and installation requirements, check with the vehicle manufacturer for the latest data regarding replacement.

To check the exhaust system's condition, grab the tail pipe (when cool). Try to move it up and down and side to side. There should be only slight movement in any direction.

To check further, you must start the engine. Once you have started the engine, feel around every joint of the exhaust system for leaks. If you find one, try tightening the clamp. If this does not stop the leak, it must be repaired.

If needed, raise the vehicle. Check the clamps and hangers that fasten the exhaust system to the underbody. Also, probe all rusted areas in the system with the blade of an old screwdriver. If the blade sinks through the metal at any point, that part is badly rusted. You can also tap on parts with a hammer or mallet (Figure 13–113). A ringing sound indicates that the metal is good. A dull thud indicates the thinned metal of a badly corroded part.

If louder than normal, there is usually a large leak or an internally rusted muffler.

When inspecting or working on the exhaust system, remember that its parts get very hot when the engine is running. Contact with them could cause a severe burn.

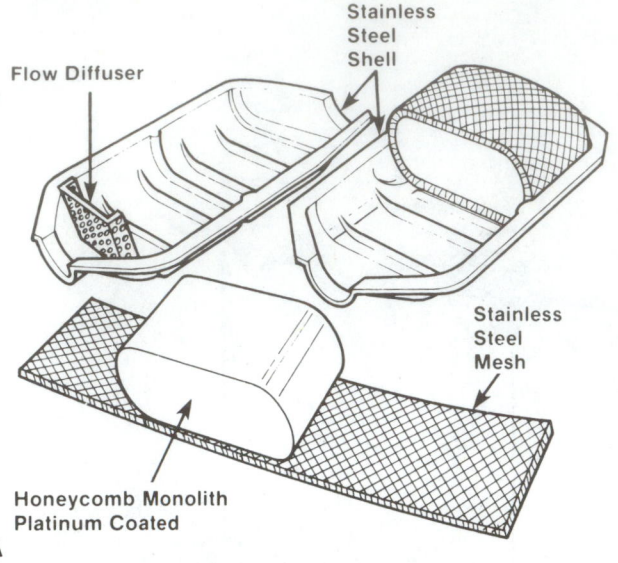

Flow Diffuser

Stainless Steel Shell

Stainless Steel Mesh

Honeycomb Monolith Platinum Coated

A

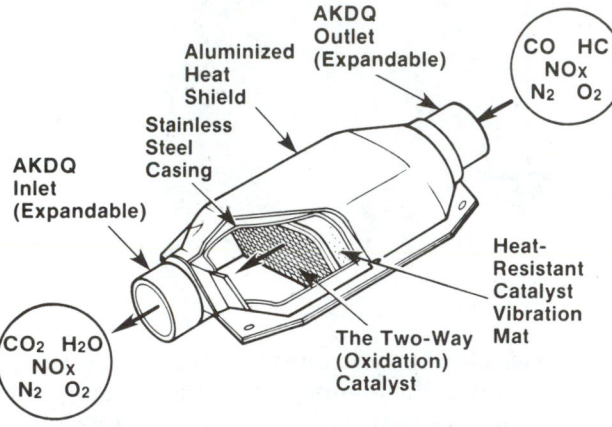

AKDQ Outlet (Expandable)

Aluminized Heat Shield

Stainless Steel Casing

AKDQ Inlet (Expandable)

CO HC NO_x N_2 O_2

CO_2 H_2O NO_x N_2 O_2

Heat-Resistant Catalyst Vibration Mat

The Two-Way (Oxidation) Catalyst

B

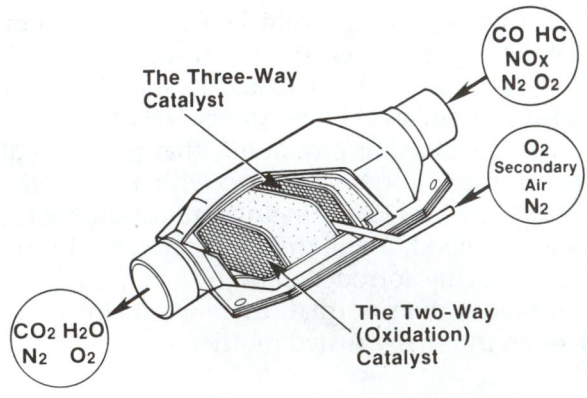

The Three-Way Catalyst

CO HC NO_x N_2 O_2

O_2 Secondary Air N_2

CO_2 H_2O N_2 O_2

The Two-Way (Oxidation) Catalyst

C

FIGURE 13-112 Compare types of catalytic converters: (A) single-bed, (B) dual-bed, and (C) three-way.

There is only one way to repair faulty exhaust system parts: replace them. It might not be necessary to take off all the exhaust system parts. You can usually separate each part and replace it individually.

Make sure that the work area is adequately ventilated before you start the engine. Either work outside or connect the vehicle to a tailpipe exhaust system to remove carbon monoxide gas from the garage.

FIGURE 13-113 Tap on the old muffler to see if it is rusted thin. If rusted, it will dent easily.

Because of constant changes in EPA catalytic converter servicing and installation requirements, check with the CC manufacturer or the EPA for the latest data regarding replacement, if damaged in a collision.

13.10 EMISSION CONTROL SYSTEMS

Emission control systems are used to prevent potentially toxic chemicals from entering our atmosphere. The most common of these are the exhaust gas recirculation (EGR), catalytic converter, air injection, and positive crankcase ventilation (PCV) systems.

Many times emission control systems are damaged in a collision and must be serviced as part of the collision repair. The *Clean Air Act*, which is a federal law, makes the repair technician responsible for the emission control systems. The law requires service technicians to restore emission control systems to their original design. The law prescribes penalties for repair shops and technicians who alter emission control systems or fail to restore them to their original design.

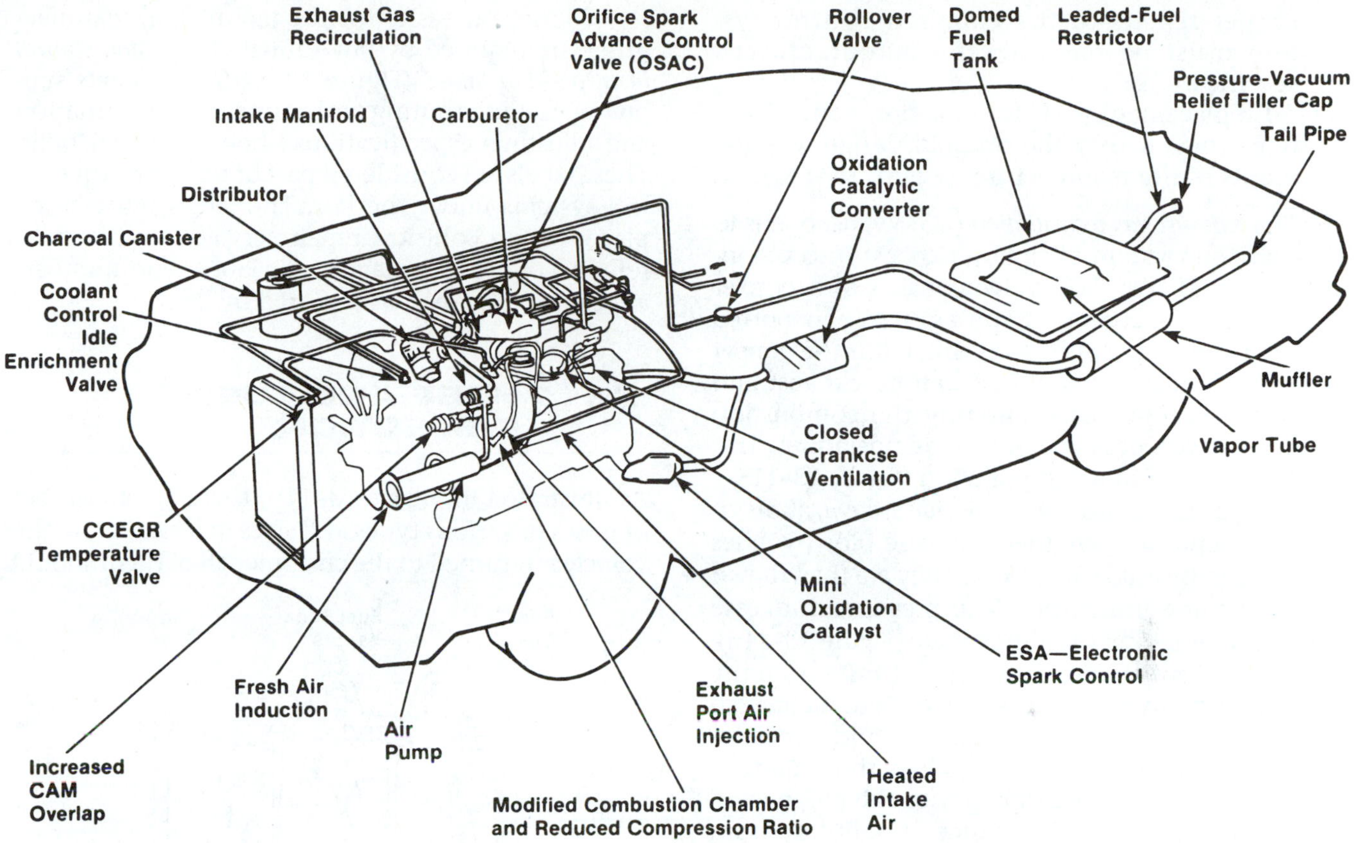

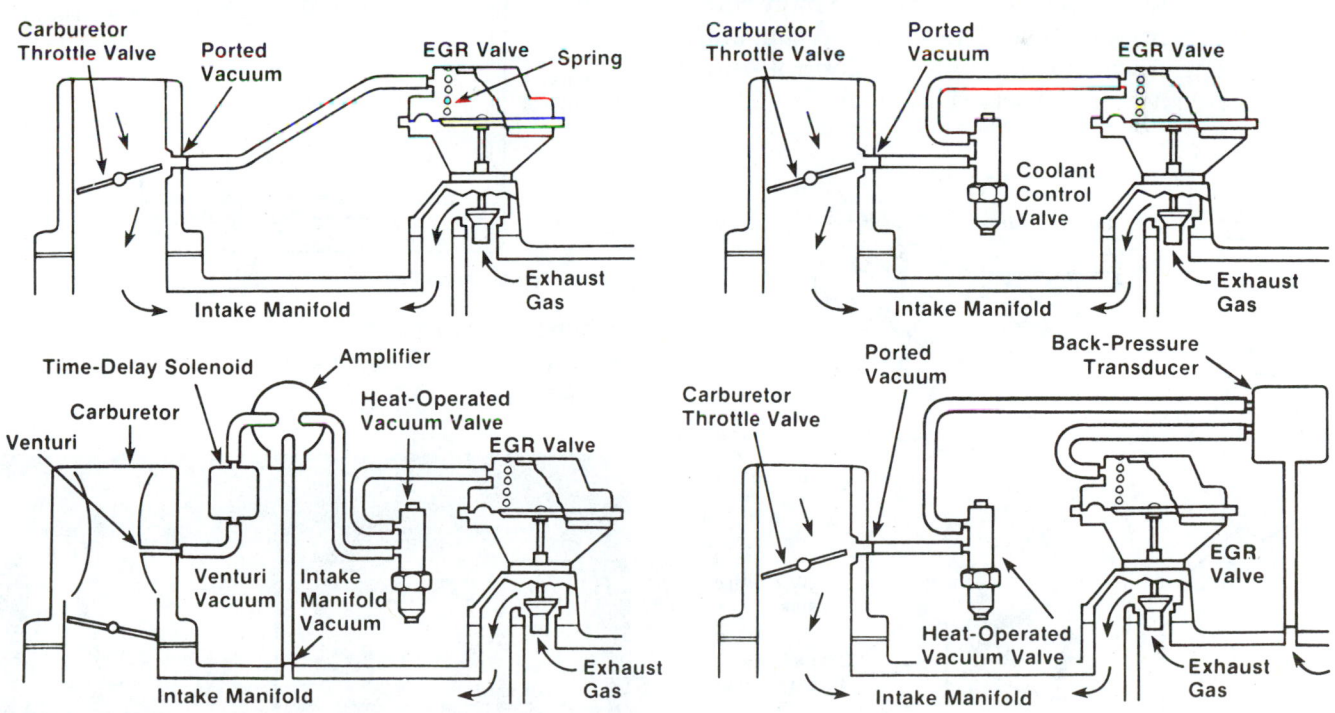

FIGURE 13-114 Study emissions control systems.

FIGURE 13-115 These are EGR variations.

- Damaged parts must be replaced with good parts. Eliminating damaged parts to avoid replacement parts is against the law.

- Using damaged parts that prevent proper operation of the emission control system is also against the law.

- Proper repairs to the emission control system must be made to the manufacturer's specifications.
- All replacement parts for emission control systems must satisfy the original design requirements of the manufacturer (Figure 13–114).

The *exhaust gas recirculation* (EGR) valve opens to allow engine vacuum to siphon exhaust into the intake manifold. The EGR valve consists of a poppet and a vacuum-actuated diaphragm. When ported vacuum is applied to the diaphragm, it lifts the poppet off its seat. Intake vacuum then siphons exhaust into the engine. The exhaust entering the combustion chambers lowers peak combustion temperatures. This reduces oxides of nitrogen pollution (Figure 13–115).

The *positive crankcase ventilation system*, abbreviated PCV, channels engine crankcase blowby gases into the engine intake manifold. They are then drawn into the engine and burned. This prevents crankcase fumes from entering the atmosphere (Figure 13–116).

The *fuel evaporative system* pulls fumes from the gas tank and other fuel system parts into a charcoal canister. The *charcoal canister* absorbs and stores vaporized fuel (Figure 13–117). When the engine is started, these vapors are drawn into the engine and burned. This prevents this source of pollution from entering our atmosphere.

Some emission control systems have more controls than mentioned here. The service, repair, and replacement procedures for emission control components are given in the service manual.

To make it easier for the technician, manufacturers are required by law to install *emission control identification labels* (Figure 13–118) and labels supplying vacuum routing and connection information and adjustment specifications when the cars are built. These labels are considered part of the emission control systems under the law. The labels must be replaced when collision repair services require their removal or cause damage to the labels. Part numbers appear on the labels as required by law.

13.11 HOSE AND TUBING INSPECTION

As illustrated in Figure 13–119, there are a number of hoses of various types and sizes in a car. Before the vehicle is returned to the customer all of them should

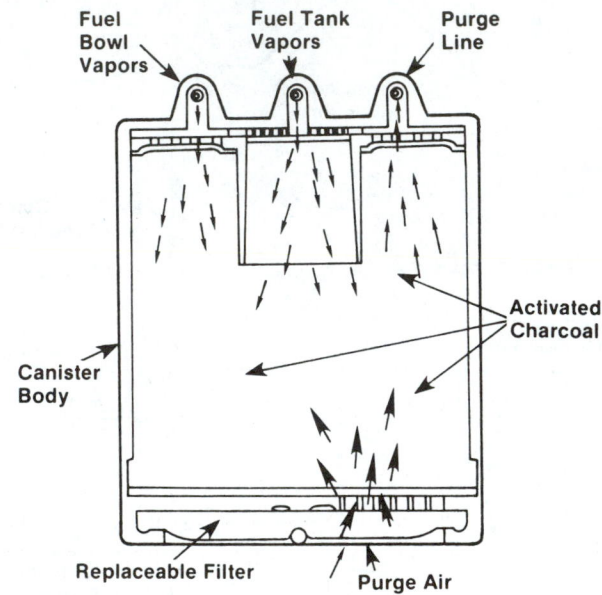

FIGURE 13-117 The charcoal canister captures and stores fuel system fumes so they can be burned in the engine upon startup.

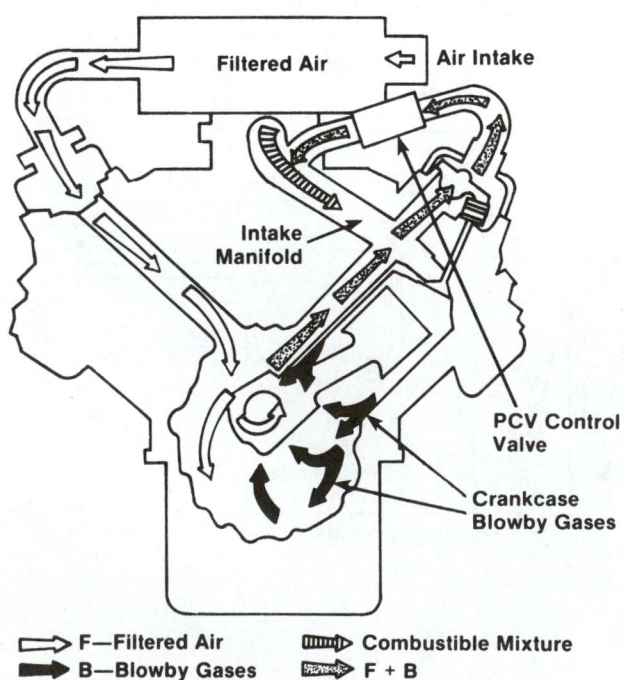

FIGURE 13-116 The PCV system prevents crankcase fumes from entering the atmosphere.

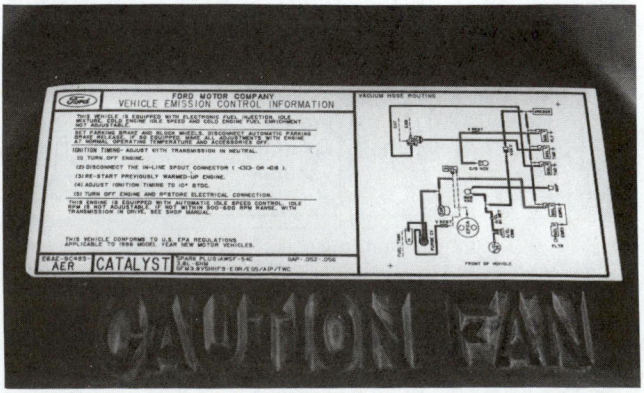

FIGURE 13-118 A typical emission control identification label gives information about equipment on a specific vehicle.

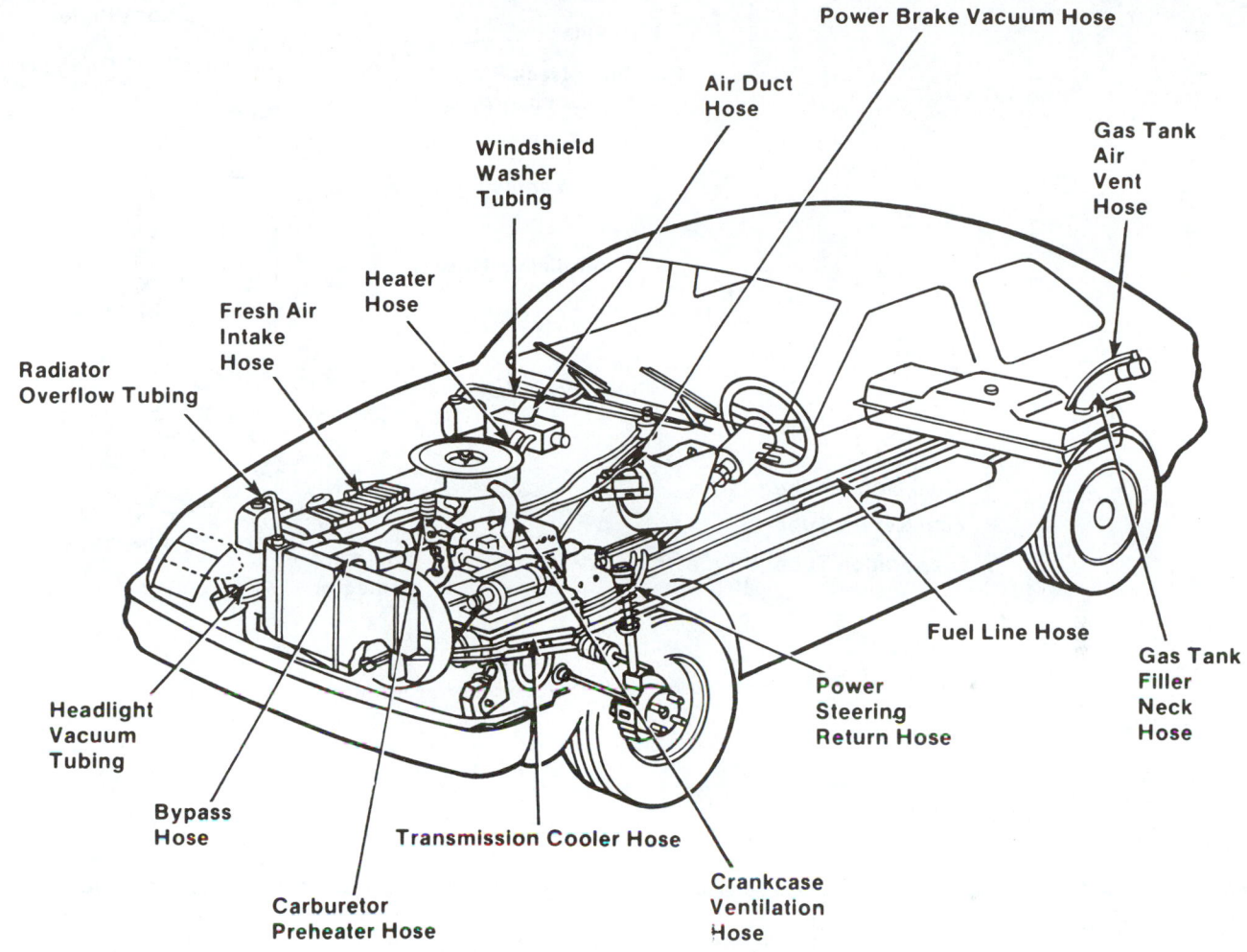

FIGURE 13-119 These hoses should be checked carefully after a collision. *(Courtesy of Dana Corporation)*

be checked. If any hoses, tubing, or clamps appear damaged, they should be replaced. This is especially true of the fuel line system (Figure 13–120).

Hose clamps might not seem like an important item, but OEM-type ring clamps are best discarded rather than reused. The ring-type clamps can lose tension and might not hold the hose securely if reused. Worm screw clamps are the preferred replacement.

13.12 CHECKING ELECTRICAL PROBLEMS

An often overlooked area of collision repair is the electrical system. The modern automobile is threaded with literally miles of wires. Most of these are bundled together in harnesses. These harnesses route the wires from the battery to all electrical body parts—dome lights, headlights, electric door locks, remote control side mirrors, sensors, computer, and so on.

A typical example of a vehicle's wiring harnesses is shown in Figure 13–121. These harnesses snake along body parts such as windshield pillars, rocker panels, doors, quarter panels, and roof panels, among others. Damage in these areas often cuts or abrades the insulation protecting the wires, and a short or open circuit is the result. Collision forces can also pull wires from their connections, and corrosion damage can loosen ground wire connections, again breaking electrical circuits.

Before any collision-damaged vehicle is returned to the owner, every electrical component should be operated to verify that it works and that it stops working when turned off. If either condition does exist, the problem must be traced to its source and the faulty wires or loose connections repaired. Oftentimes, the problem is simply a blown fuse or a loose connection.

For example, when replacing the outer door skin, the door, of course, is removed from the vehicle and any electrical components are disconnected from the related wiring harness. If the connectors are not properly connected when the door is replaced, the windows, lock, or exterior mirror will not be operable. Always double-check connections. Make sure harnesses are secured in their clips and that ground wires are tightly secure. Typical ground connections are shown in Figure 13–122.

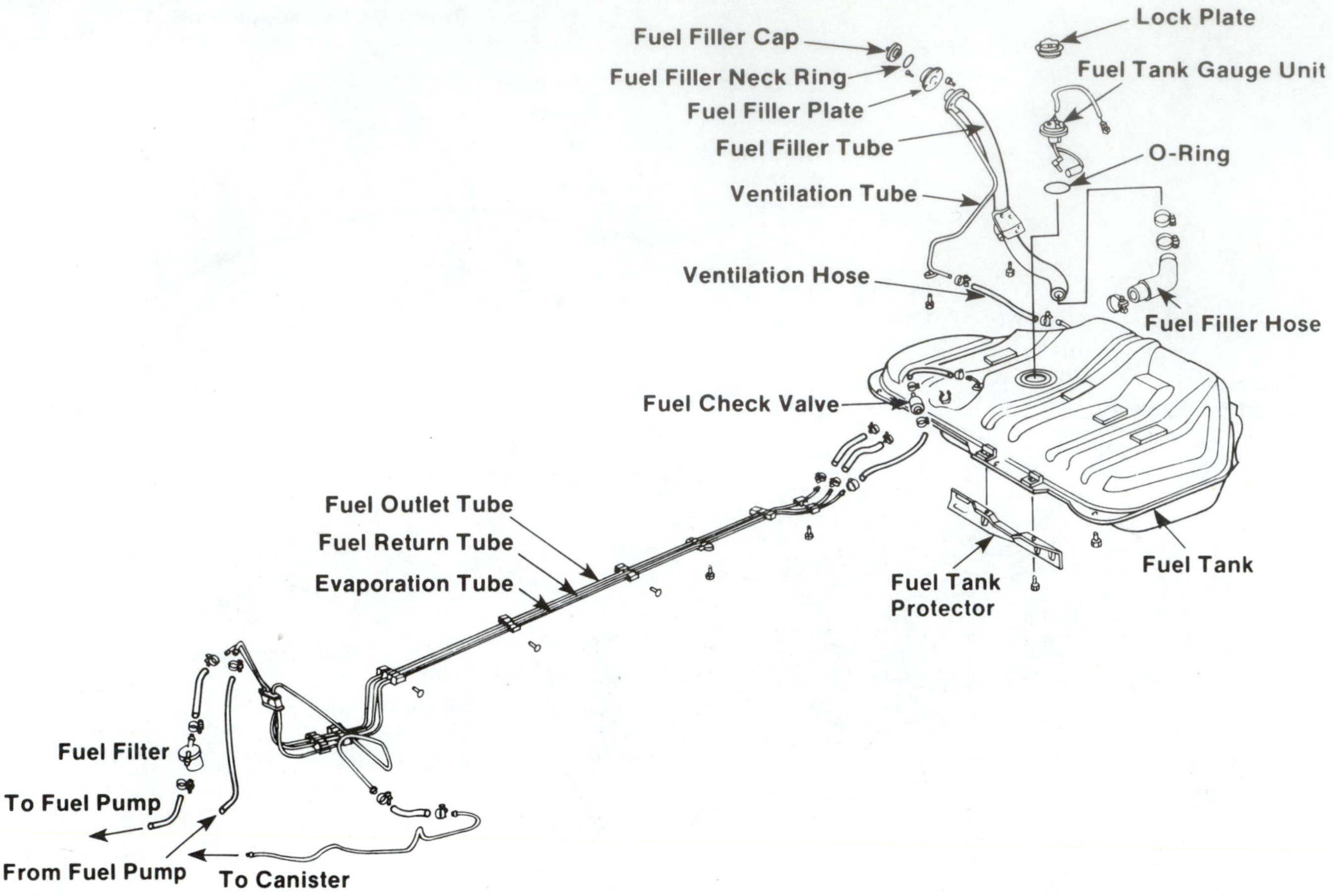

Lock Plate

Fuel Filler Cap

Fuel Tank Gauge Unit

Fuel Filler Neck Ring

Fuel Filler Plate

O-Ring

Fuel Filler Tube

Ventilation Tube

Ventilation Hose

Fuel Filler Hose

Fuel Check Valve

Fuel Outlet Tube

Fuel Return Tube

Evaporation Tube

Fuel Tank
Protector

Fuel Tank

Fuel Filter

To Fuel Pump

From Fuel Pump   To Canister

FIGURE 13-120 The fuel line tubing should be visually checked for any leak or damage and looseness of clamps. *(Courtesy of Nissan Motor Corp.)*

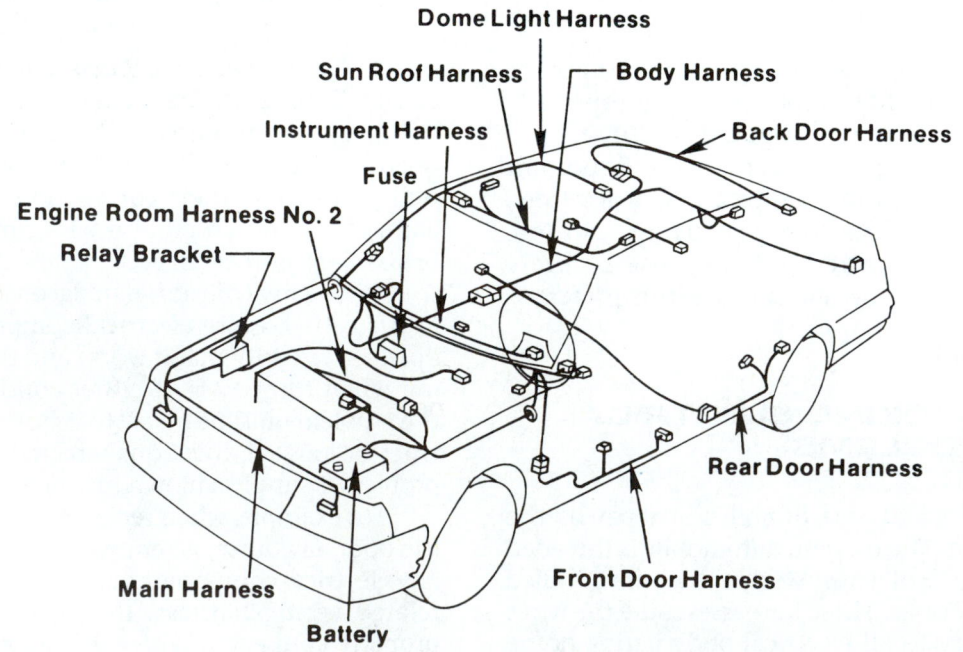

Dome Light Harness

Sun Roof Harness

Body Harness

Instrument Harness

Back Door Harness

Fuse

Engine Room Harness No. 2

Relay Bracket

Rear Door Harness

Main Harness

Front Door Harness

Battery

FIGURE 13-121 The electrical harness must be checked for any kinking, missing clamps, and loosened terminals.

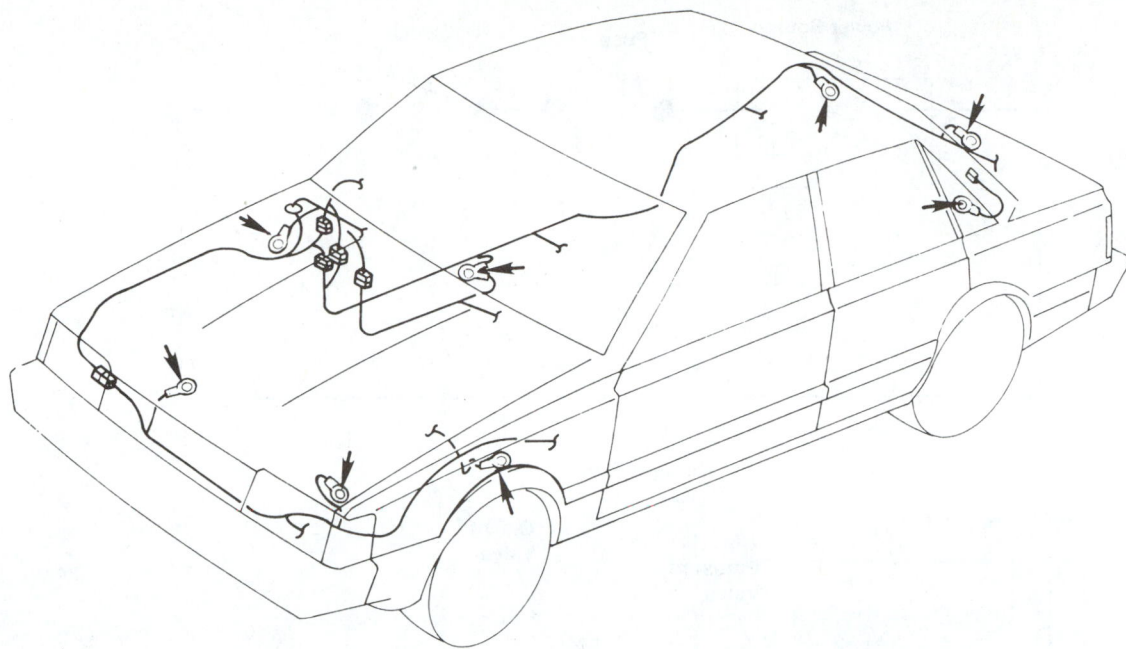

FIGURE 13-122 Bad grounding points can cause electrical malfunctions.

Most body shops send major electrical problems to a garage for repair—particularly damage to starter and ignition systems. Testing the major components in these systems often requires expensive diagnostic equipment. But there are many minor problems in the lighting and accessory circuits that can be quickly diagnosed and repaired in the body shop. Minor electrical troubleshooting requires only basic knowledge of electrical theory and a few simple diagnostic tools.

ELECTRICAL TERMINOLOGY

Various vehicle systems are controlled by a series of electrical controls and devices. To understand electricity, you must become familiar with three electrical terms: current, voltage, and resistance.

Current is the movement of (electrons) electricity through a wire or circuit. It is measured in amperes or amps using an ammeter. The common electrical symbol for current is "A" or "I."

Voltage is the pressure that pushes the electricity through the wire or circuit. The *power source* generates the voltage which causes current flow. Voltage is measured in volts using a voltmeter. The symbol for voltage is "E" or "V."

Resistance is a restriction or obstacle to current flow. It tries to stop the current caused by the applied voltage. Circuit or part resistance is measured in ohms using an ohmmeter. The symbol for resistance is "R" or "Ω."

A *conductor* carries current to the parts of a circuit. *Hot wires* connect the battery positive to the components of each circuit. *Insulation* stops current flow and keeps the current in the metal wire conductor. The body structure provides the *ground* conductor back to the battery negative cable.

ELECTRIC CIRCUITS

An *electric circuit* contains a power source, conductors, and a load. Some resistance is designed into a circuit in the form of a load. The *load* is the part of a circuit that converts electrical energy into another form of energy (light, movement, heat, etc.), as in Figure 13–123. Other parts are added to this simple circuit to protect it from damage and to do more tasks.

A *series circuit* has only one conductor path or leg for current through the circuit. Current must flow through the wires and components one after the other. If any part of the circuit is *opened* (disconnected), all of the series circuit stops working.

A *parallel circuit* has two or more legs or paths for current. Current can flow through either leg independently. One path can be *closed* (electrically connected) and the other opened, and the closed path will still operate.

A *series-parallel circuit* has both series and parallel branches in it. It has characteristics of both circuit types.

All the circuits in a vehicle can be classified and tested using the rules of these three circuits.

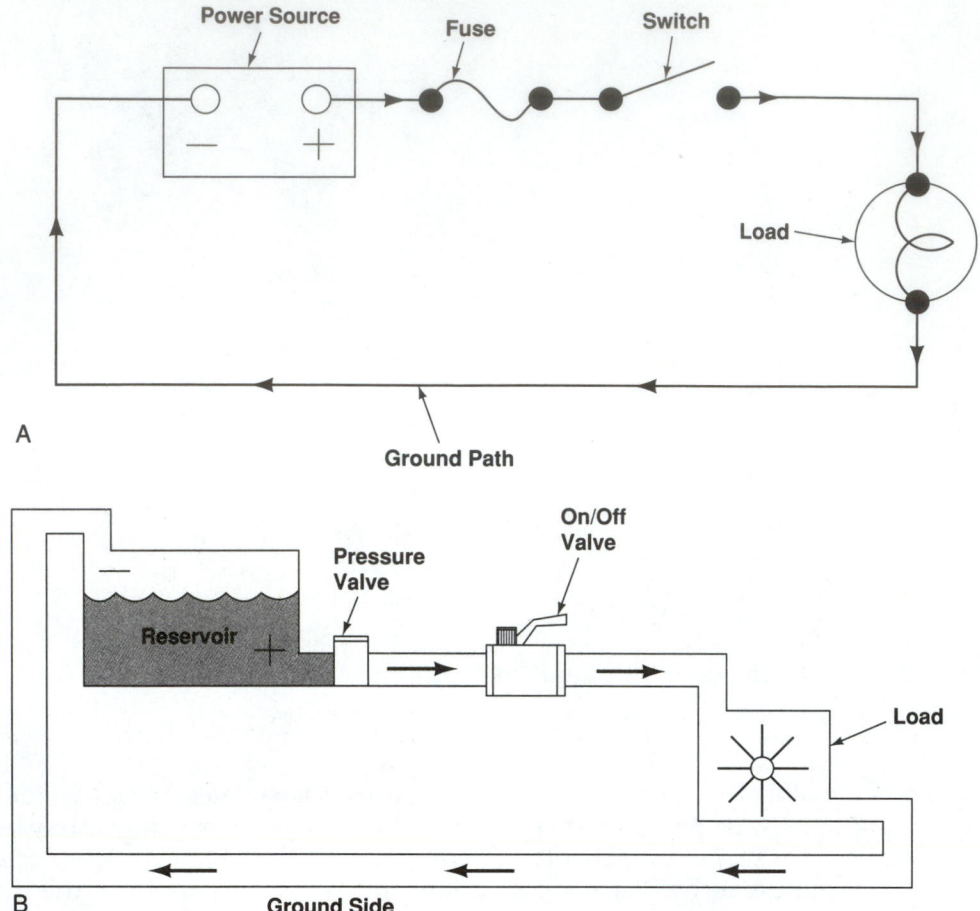

FIGURE 13-123 A hydraulic circuit will help you understand the operation of an electric circuit.

OPEN CIRCUITS

An *open circuit* is disconnected and does not have a complete electrical path. Most electrical problems are a result of a break in the wiring circuit.

Most electrical problems encountered in the auto body repair shop are the result of a break in the circuit. When the continuity is broken, the circuit is said to be open. Power does not flow to the component. Too often this problem is not discovered until the owner has picked up his or her vehicle and (days or weeks later) attempts to turn on the wipers, lock the doors, use the map light, or operate some other electrical device. These problems should be discovered and repaired in the body shop.

SHORT CIRCUITS

A *short circuit* has an unwanted path for current. A collision will sometimes create a short circuit. There are several kinds of short circuits.

A *dead short* to ground is usually a bare conductor touching directly against the vehicle frame. This type of short always opens a circuit breaker, blows a fuse, or pops a fusible link.

With an *intermittent short*, the shorted wire only touches ground and shorts momentarily when the vehicle bounces heavily or jars. A flickering dash lamp is such as example.

A *cross circuit short* occurs when two hot wires come in contact. This is usually caused by abrasion of the protective plastic coverings or by an overload that causes the coverings to melt. Such a short can cause more than one component to operate on a single switch. For example, when the lights are turned on, the windshield wipers operate. This is because the shorted wires are sharing current. Supplying current to one (by actuating a switch) supplies current to both.

The last type of short circuit is called a *high resistance short* to ground. The circuit is not broken but contact is present between the hot wire and the ground. A high resistance ground might not blow a fuse until the circuit is loaded to full capacity. Or it might not blow a fuse at all, but will slowly drain the battery.

BATTERY SAFEGUARDS

The battery is a power plant. Therefore, it must be treated with respect as well as protected. Because of the potential electrical problems incurred in a collision, always disconnect the battery ground cable as soon as a vehicle is received at the shop.

There are several precautions that must be taken when recharging a battery. The following precautions protect the battery, vehicle, and technician:

- Always disconnect the battery ground wire before charging a battery on the vehicle. Charging the battery with it connected might damage electrical components on the vehicle. The transistors and microcircuits on many cars are very sensitive to current levels. The high-amperage chargers can burn out many expensive parts: regulator, alternator, power transistors, computerized control modules, and so on.
- Never disconnect the battery with the ignition switch in the on position. The resulting voltage "spike" will destroy many microcircuits in today's electronic systems.
- Check the battery carefully before charging. Look for low water level, cracked case, and so forth. Add distilled water or electrolyte if needed. Use a hydrometer to check the state of charge

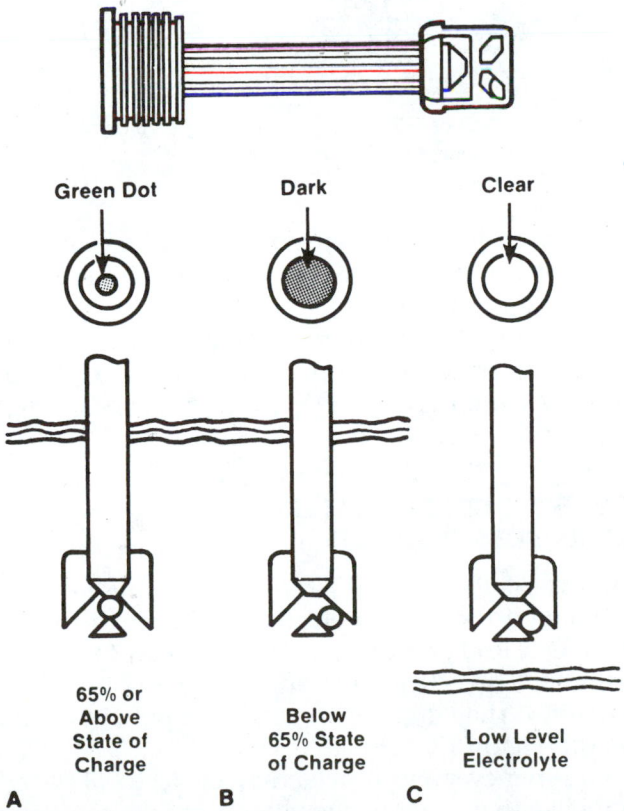

| Green Dot | Dark | Clear |

| A | B | C |
| 65% or Above State of Charge | Below 65% State of Charge | Low Level Electrolyte |

FIGURE 13-124 Most batteries have a built-in hydrometer that will indicate the state of charge and condition.

TABLE 13-3: SPECIFIC GRAVITY VS STATE OF CHARGE AT 80°F		
Specific Gravity	State of Charge	Open Circuit Cell Voltage
1.260	100%	2.10
1.230	75%	2.07
1.200	50%	2.04
1.170	25%	2.01
1.110	0	1.95

(specific gravity) or read the built-in hydrometer (Figure 13–124).

Table 13–3 shows the state of charge that corresponds to specific gravity reading.

- Follow the manufacturer's instructions carefully. Do not overheat the battery by charging too long. A standard battery charging guide is usually given in the owner's manual. Test the specific gravity once an hour. Once no change is noticed in readings, disconnect the charger. Overcharging the battery will destroy the active material on the plates.
- Never charge the battery near any welding operations, open flames, or other heat source. Do not smoke near the charging battery. The battery gives off very flammable hydrogen gas while charging. If the gas ignites, the battery will explode.
- Ensure protection from the battery acid. Wear eye protection when handling the battery and immediately wash off acid that splashes on clothing or skin. If battery acid does get in an eye, hold the eye open and flush with water at room temperature. See a physician immediately. Battery acid can cause blindness.

Although it is a common practice in some repair shops, avoid jump starting whenever possible. The discharged battery can explode or create voltage spikes, which can damage electronic components. This is true of both the car you are trying to start and the car providing the jump.

In an emergency, *jumper cables* may be needed to start the engine for bringing the vehicle into the shop. If the car must be jump started, connect the red jumber cable to the positive terminal of the dead battery. Connect the other end to the good battery positive terminal. Next, connect the black jumper cable to the negative terminal of the dead battery. Connect the other end of the black cable to the running car's chassis ground or to the good battery's negative terminal, as shown in Figure 13–125. If using another car for the jump, make sure the two vehicles are not touching each other.

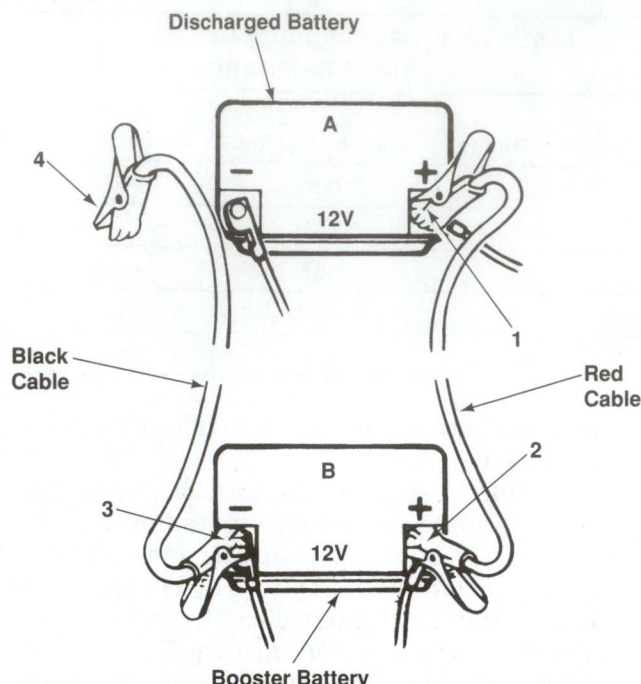

FIGURE 13-125 To prevent a battery explosion, connect the jumper cable carefully to prevent sparks around a dead battery. Connect in the number sequence shown. *(Courtesy of Volkswagen of America)*

FIGURE 13-126 An accurate voltmeter will let you check the general state of the charge on modern sealed batteries. A reading of 12.6 volts normally shows a fully charged battery.

FIGURE 13-127 During a collision, wires can be smashed and shorted. *(Courtesy of Tech-Cor)*

In addition, consider the following to avoid damage from voltage spikes in the electronic circuits of a vehicle with a dead battery:

- Make sure every electrical device in that car, including the dome light, is turned off before connecting the batteries.
- Only after the hookups are properly made, turn the key in the dead car to get it started.
- Once the dead car is running, remove all jumper connections before turning on any electrical devices.

SHOP TALK

Since most new batteries are sealed and you cannot check specific gravity, use a voltmeter to check state of charge. A fully charged 12-volt battery should show about 12.6 volts. Anything below this value shows battery drain, requiring a charge or possibly a new battery (Figure 13–126).

ELECTRICAL DIAGNOSTIC EQUIPMENT

Locating an electrical fault is not possible without using *diagnostic tools* (meters, test lights, jumper wires,

etc.). Keep in mind that today's delicate electronic systems can be damaged if the wrong methods and equipment are used.

During a collision, wiring damage is common. Wires can be severed, torn apart, or abraded (Figure 13–127). You may be required to find and fix these damaged wires.

TEST LIGHTS AND JUMPER WIRES

An externally powered *test light* is often used to determine if there is current flowing through the circuit. One lead of the test lamp is connected to a good ground. The other lead connects to a point in the circuit. If the lamp lights, current is present at that point (Figure 13–128).

Whenever a technician has an electrical system problem that does not directly concern the computer, a test light is handy.

Connect the light to the voltage source and to ground. If the test light does not glow, there is an

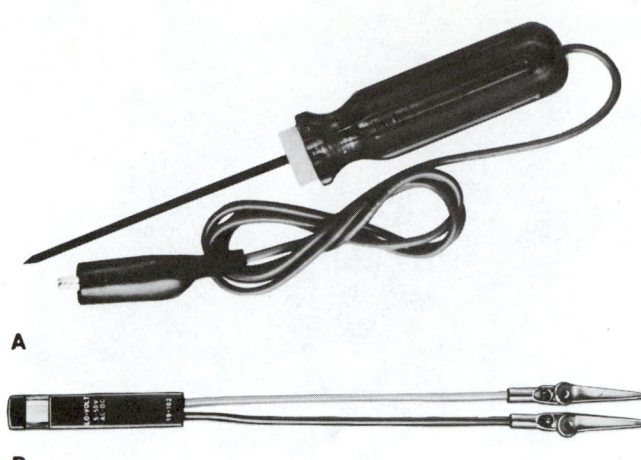

A

B

FIGURE 13-128 Compare test lights: (A) handle of probe light and (B) twin lead tester. *(Courtesy of Vaco Products Division)*

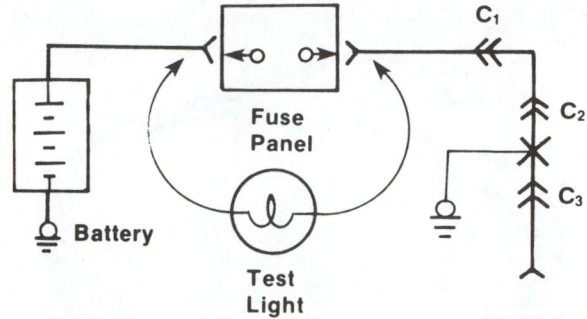

FIGURE 13-130 Short circuit check with test light; the light will glow as long as the short remains.

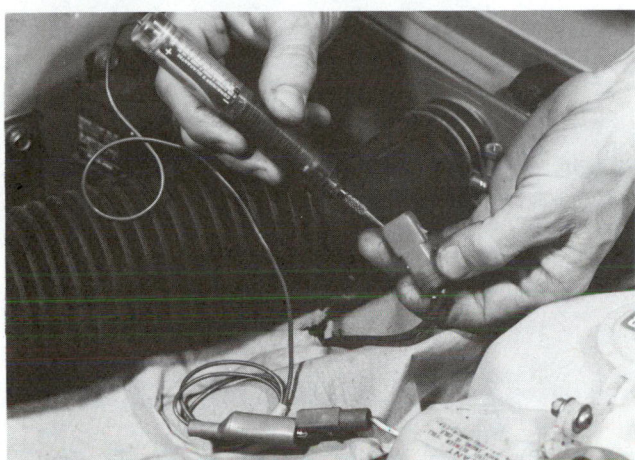

FIGURE 13-131 A test light will quickly check the continuity of an electrical harness. With a nonpowered light, you must connect the source of voltage to the other end of the wire. *(Courtesy of Easco/K-D Tools)*

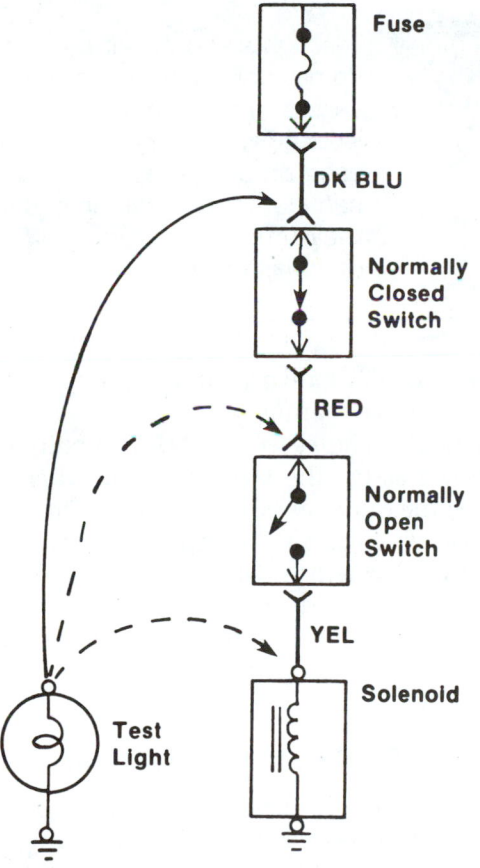

FIGURE 13-129 A test light will quickly check for voltage or power at different locations in the circuit to isolate any areas of trouble.

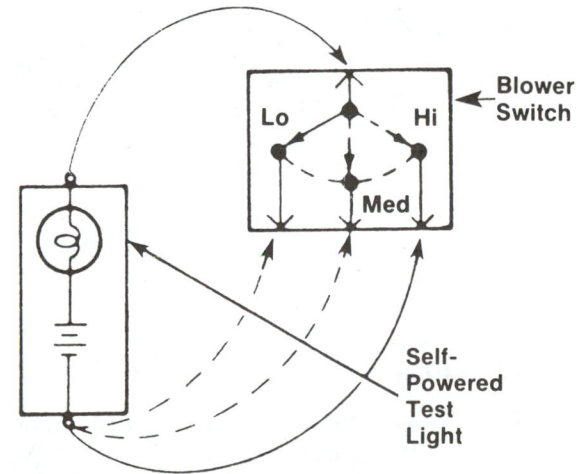

FIGURE 13-132 A self-powered test light does not need an external power source to make continuity checks.

open circuit somewhere. If the light is on, but the part does not work, the part is probably bad (Figure 13–129).

Figure 13–130 shows how to test for a short with a test light. Figures 13–131 and 13–132 demonstrate how to check *wire continuity* (if the wire is broken or has complete path) with test lights.

Jumper wires are used to temporarily bypass circuits or components for testing. They consist of a length of wire with an alligator clip at each end. They can be used to test circuit breakers, relays, lights, and other components (Figures 13–133 and 13–134).

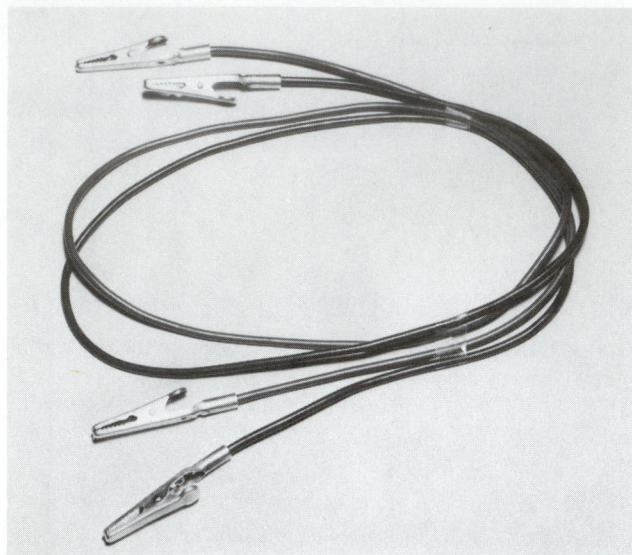

FIGURE 13-133 Jumper wires provide another way to make quick checks of circuits and parts. *(Courtesy of Vaco Products Division)*

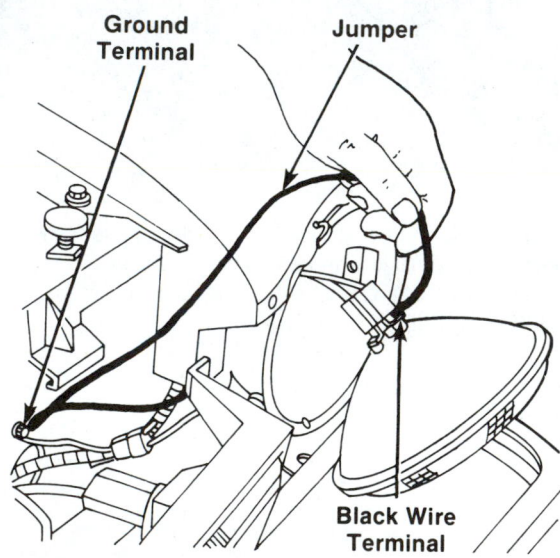

Ground Terminal

Jumper

Black Wire Terminal

FIGURE 13-134 Here jumper wires are being used to check for a bad ground wire. If the bulb starts working with a jumper, you have found high resistance.

Multimeters

A *multimeter* is a voltmeter, ohmmeter, and ammeter combined into one case. Also called a *VOM* (Volt-Ohm-Ammeter), it can be used to measure actual electrical values for comparison to known good values (Figure 13–135).

A *digital multimeter (DVOM)* has a number readout for the test value. This type is recommended by auto manufacturers because it will not damage delicate electronic components. A high-impedance (10 mega-ohm input) DVOM is recommended to avoid damaging sensitive components. Digital readouts give the precise measurement needed for proper

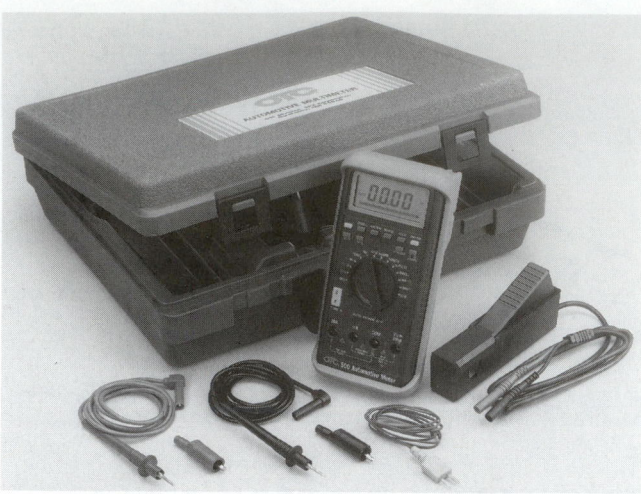

FIGURE 13-135 The multimeter is an important tool to today's technician. *(Courtesy of OTC Division of SPX)*

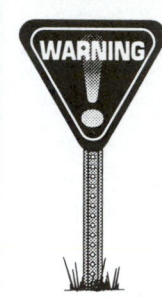

WARNING Test lights should not be used randomly, only when specified in the service manual procedures or on isolated wires. They are not to be used on solid state circuits. They can damage electronic parts if their *impedance* (internal resistance) is too low.

diagnosis. DVOMs are used in conjunction with the vehicle's service manual.

An *analog multimeter (AVOM)* has a pointer needle that moves across the face of a scale when making electrical measurements. Use of an AVOM can damage sensitive electronic components. They should only be used when testing all electrical, NOT electronic, circuits. They help show a fluctuating or changing reading, like from an intermittent (changing) problem.

MEASURING RESISTANCE

When measuring resistance, always disconnect the circuit from the power source. The multimeter must never be connected to a circuit in which current is flowing. Doing so can damage the meter.

Use the service manual to determine the normal resistance of the part being checked. For example, a computer system sensor may have a normal resistance reading of 45 to 55 ohms. You would select the "200" range on the multimeter because the resistance reading of 44 to 55 ohms will be 200 ohms or less.

Always refer to the multimeter owner's manual before using the multimeter for diagnostic purposes. You must make sure that you understand how to use the multimeter before performing diagnostics. All multimeters are fairly similar, but there can be differences between makes.

To measure resistance:

1. Connect the multimeter test leads to opposite ends of the circuit or wire being tested.
2. Set the range selector switch on the highest range position. Turn the multimeter on.
3. Reduce the range setting until the meter shows a reading near the middle of its scale. Some DVOMs have an *auto ranging* function that adjusts the settings automatically.

MEASURING VOLTAGE

The multimeter allows you to select either alternating current voltage (ACV) or direct current voltage (DCV). The ACV is selected when measuring alternating current voltage. *AC current* is the type of current that is found in your home wiring.

DCV is selected when measuring direct current voltage. *DC current* is what is normally measured in the automobile. Signals from sensors can be AC or DC.

When using the multimeter to measure voltage, the selection of the range scale is very important. If the voltage reading will be 12 to 14 volts, select the 20 V range on the multimeter. This range is selected because the voltage reading will be less than 20 volts.

Consider the taillight circuit in Figure 13–136A for an example of meter use. If the lights do not work, check the voltage available to the fuse. If approximately 12 volts are not displayed on the meter, a short exists between the battery and the fuse box. If 12 volts are available, check the voltage between the switch and the first lamp (Figure 13–136B). If voltage is not available, bridge the switch with a lead wire. If the light comes on or the voltmeter reads 12 volts, the switch is defective. If no voltage shows on

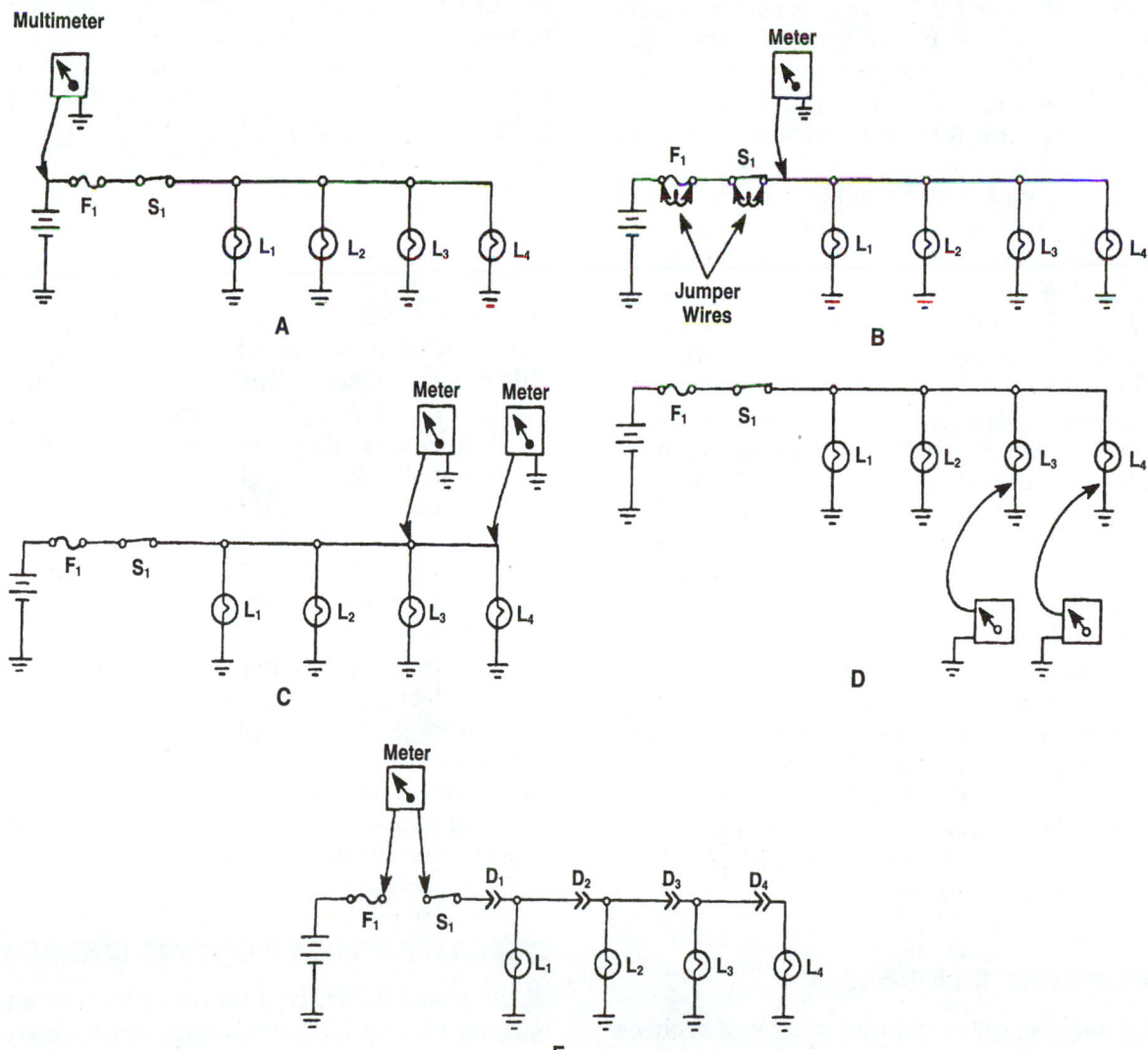

FIGURE 13-136 Checks that can be made with a multimeter to determine taillight circuit problems. This type of analysis also applies to other types of circuits.

the gauge, bridge the fuse box with the jumper wire. The short might be either in the box or in the wiring between the fuse box and the switch.

Assume that after repairing the shortage in the same circuit, only the first two lights come on. Then check the voltage available to the third and fourth light (Figure 13–136C). If the meter reads 0 volts, the wire between the two lights is probably broken or disconnected. If 12 volts register on the meter, the problem must be narrowed down to one of the following:

- Defective ground connection
- Defective light socket
- Burned out bulbs

Try new bulbs in each of the sockets. If that does not solve the problem, test the ground wires of the third and fourth lights with the AVR multimeter set to read voltage (Figure 13–136D). Connect the leads to the ground side terminal of the socket and to a suitable ground. If the meter reads 12 volts, this indicates a defective ground. If 0 volts is read, the sockets are probably bad. Verify that no continuity exists in the bulbs by setting the multimeter to read ohms. Connect the leads to either side of a socket. If the ohmmeter reads "infinity," no continuity exists and the sockets are defective.

If the problem with the taillights is narrowed down to a fuse that blows whenever the lights are turned on, the lights are drawing too much current. This indicates a high-resistance short somewhere in the circuit. To determine where the short is, connect the multimeter in series with the circuit (Figure 13–136E). Then, disconnect the hot wire from the first light. If current registers on the ammeter, a shortage somewhere between the switch and the first taillight is grounding the circuit.

If no current flow registers, remove the light bulbs and replace them one at a time. If the circuit is designed to draw 5 amperes, each bulb should draw 1.25 amperes ($5 \div 4 = 1.25$). Replace the first bulb. The ammeter should read 1.25 amperes. When the second lamp is added, the reading should be 2.5 (1.25 + 1.25). Add the third lamp to the circuit, and the reading should be 3.75 amperes; add the fourth lamp, and the reading should be 5 amperes. If a higher than normal reading is achieved at any light, the short drawing the additional current lies between that light and the preceding one. This type of analysis will let you find any circuit problem.

MEASURING CURRENT

Current or amperage is sometimes measured to check the consumption of power by a load. For example, current draw is often measured when checking the condition of a starting motor.

Modern ammeters have an *inductive pickup* that slips over the wire or cable to measure current. With older ammeters, the circuit had to be disconnected and the meter connected in series to measure current.

A high current draw indicates a low resistance, like from a dragging or partially shorted motor. A low current draw means there is a high resistance in the circuit, like from a bad connection or dirty motor brushes.

ELECTRIC COMPONENTS

A *switch* is used to turn a circuit on or off manually (by hand). When the switch is closed (on), the circuit is complete (fully connected) and will operate. Various types can be found in today's vehicles. A bad switch will often be open in both the closed and open positions.

An *inertia switch* is designed to shut the electric fuel pump off after a collision. It must be reset for the engine to restart after a collision. A button on the side of the inertia switch must be pressed to reclose the fuel pump circuit.

A *solenoid* is an electromagnet with a movable core or plunger. When energized, the plunger is pulled into the magnetic field to produce motion. Solenoids are used in many applications: door locks, engine idle speed, emission control systems, etc. A bad solenoid can develop winding opens, shorts, or a high-resistance problem.

A *relay* is a remote control switch. A small switch can be used to energize the relay. The relay coil then acts upon a movable arm to close the relay contacts. This allows a small switch to control high current going to a load. A relay is commonly used with electric motors since they draw heavy current. The service manual will often give relay locations for the specific make and model vehicle.

A bad relay will often have burned points that prevent current flow to the load. They can also develop coil opens and shorts that keep the points from closing.

Motors use permanent and electromagnets to convert electrical energy into a rotation motion for doing work. Some examples are the electric starting motor for the engine and stepper motors for computer control of parts.

Faulty motors can have worn bushings, brushes which decrease efficiency. They can also have winding shorts and opens that prevent motor operation.

CIRCUIT PROTECTIVE DEVICES

Circuit protection devices prevent excess current from burning wires and components. With an overload or short, too much current tries to flow. Without a fuse or breaker, the wiring in the circuit would heat up. The insulation would melt and a fire could result.

FUSES

Fuses burn an internal circuit with excess current to protect a circuit from further damage. They are normally wired between the power source and the rest of the circuit. There are three types of fuses in automotive use: cartridge, blade, and ceramic.

The *cartridge fuse* is found on most older domestic vehicles and a few imports. It is composed of a strip of metal enclosed in a glass or transparent plastic tube. To check the fuse, look for a break in the internal wire or metal strip. Discoloration of the glass cover or glue bubbling around the metal end caps is an indication of overheating.

Late-model domestic vehicles and many imports use *blade* or *spade fuses*. To check the fuse, pull it from the fuse panel and look at the element through the transparent plastic housing. Look for internal breaks and discoloration.

The *ceramic fuse* is used on many European imports. The core is a ceramic insulator with a conductive metal strip along one side. To check this fuse, look for a break in the contact strip on the outside of the fuse.

All fuse types can be checked with a circuit tester or multimeter. A *blown fuse* will have infinite resistance.

Fuse ratings are the current at which the fuse will blow. Fuse ratings are often printed on the fuse. Always replace a fuse with one of the same amp rating. If not, part damage can result from excess current flow.

A *fuse box* holds the various circuit fuses, breakers, and flasher units for the turn and emergency lights. It is often under the instrument panel, behind a panel in the foot well, or in the engine compartment.

Never permanently bypass a fuse or circuit breaker with a jumper wire. Do it only for test purposes.

FUSE LINKS

Fuse links or *fusible links* are smaller diameter wire spliced into the larger circuit wiring for over-current protection. Fuse links are normally found in the engine compartment near the battery. They are often installed in the positive battery lead that powers the ignition switch and other circuits that are live with the key off.

Fuse link wire is covered with a special insulation that bubbles when it overheats. This indicates that the fuse link has melted. If the insulation appears good, pull lightly on the wire. If the fuse link stretches, the wire has burned in half. When it is hard to determine if the fuse link is burned out, perform a continuity check.

When replacing fuse links, first cut the protected wire where it is connected to the fuse link. Then solder a new fuse link of the same rating in place. Since the insulation on manufacturers' fuse links is flameproof, never fabricate a fuse link from ordinary wire because the insulation may not be flameproof.

A number of new electrical systems use maxifuses in place of traditional fuse links. Maxi-fuses look and operate like two-prong blade fuses, except they are much larger and can handle more current. Maxi-fuses are located in their own underhood fuse block.

Maxi-fuses are easier to inspect and replace than fuse links. To check a maxi-fuse, look at the fuse element through the transparent colored plastic side housing. If there is a break in the element, the maxi-fuse has blown. To replace it, pull it from its fuse box or panel. Always replace a blown maxi-fuse with a new one having the same ampere rating.

CIRCUIT BREAKERS

Circuit breakers heat up and open with excess current to protect the circuit. They do not suffer internal damage like a fuse. Many circuits are protected by circuit breakers. They can be fuse panel mounted or in-line. Like fuses, they are rated in amperes.

Each circuit breaker conducts current through an arm made of two types of metal bonded together (known as a *bimetal arm*). If the arm starts to carry too much current, it heats up. As one metal expands farther than the other, the arm bends, opening the contacts and breaking the current flow.

In the *cycling breaker*, the bimetal arm will begin to cool once the current to it is stopped. Once it returns to its original shape, the contacts close and power is restored. If the current is still too high, this cycle of breaking the circuit will be repeated.

When replacing fuses and circuit breakers, install one with the same amp rating. A higher rated unit could cause an electrical fire.

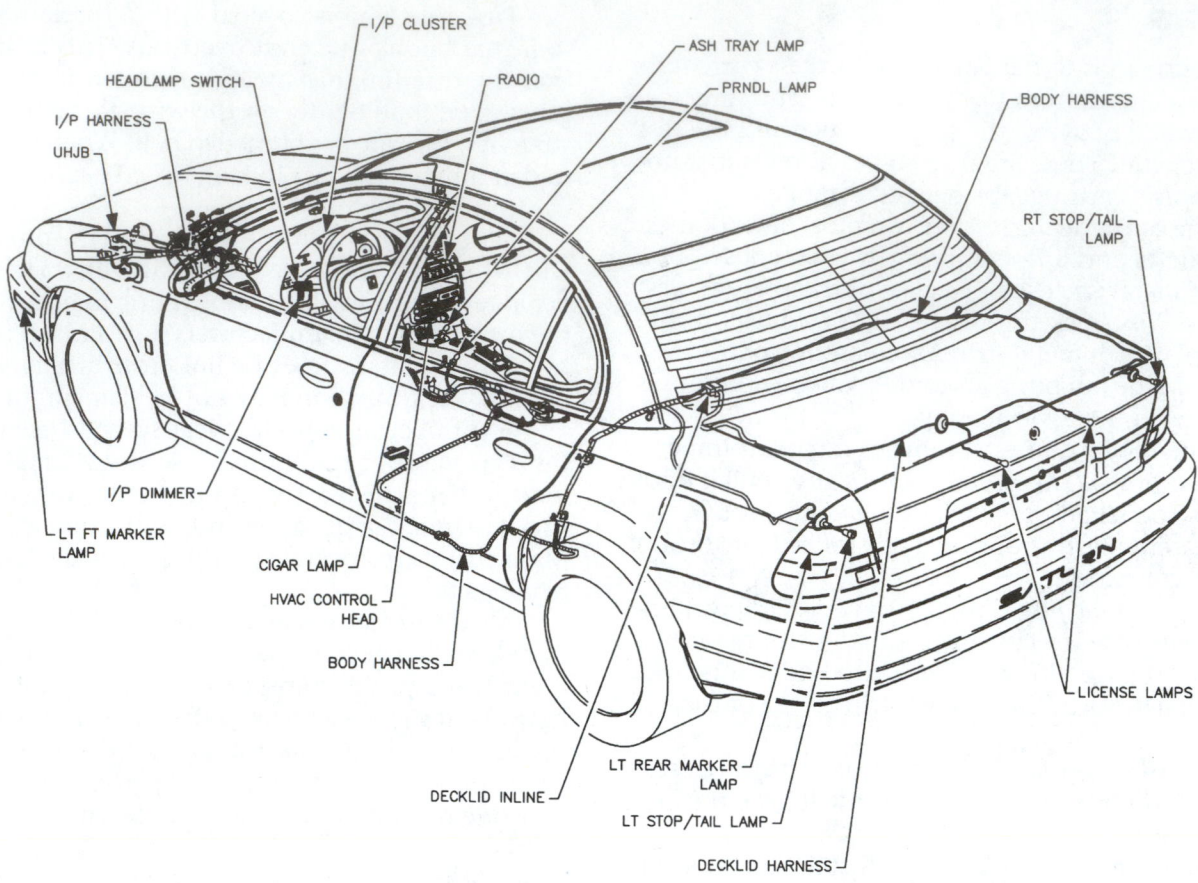

FIGURE 13-137 Study typical automotive lighting systems. (*Courtesy of Saturn Corporation © 1995*)

Cycling circuit breakers are generally used in circuits that are prone to occasional overloads. These include power windows, in which a jammed or sticking window can overwork the motor. In this situation, a circuit breaker prevents the motor from burning out.

Noncycling breakers open and must be manually reset to close the circuit. In automotive work, two types of noncycling circuit breakers are used. One is reset by removing the power from the circuit. The other type is reset by depressing a reset button.

LIGHTING AND OTHER ELECTRIC CIRCUITS

Automotive lighting systems have become increasingly more sophisticated. Headlights and taillights have grown into multiple-light systems. Indicator lights on the dashboard commonly warn of failure of the charging system, seat belts, brake systems, parking brakes, door latches, directional lights, and computer system (Figure 13–137).

Headlights have both a high and a low beam. Two circuits feed out to the bulbs. A switch in the turn signal cluster often operates the high and low beams. A few trucks still have a floor-mounted switch.

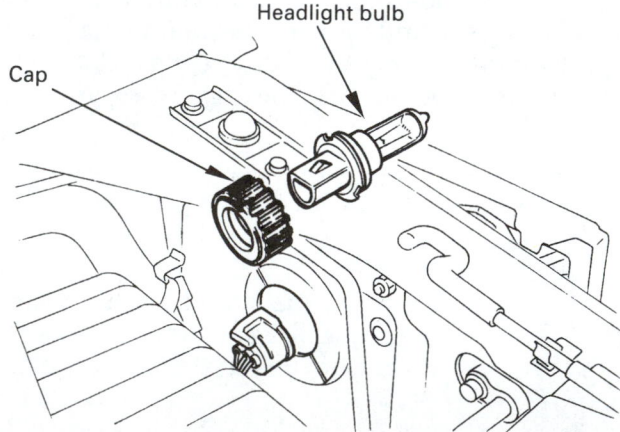

FIGURE 13-138 Most modern headlight bulbs are small cartridge-type bulbs that snap in from the rear. (*Courtesy of Isuzu Motor Co.*)

Always check the operation of both high and low beams after repairs. Many late-model cars use a cartridge-type headlight bulb (Figure 13–138). The headlights should be aimed or adjusted after a frontal collision.

Headlight aimers are available for adjusting the headlights to shine at the right height and side

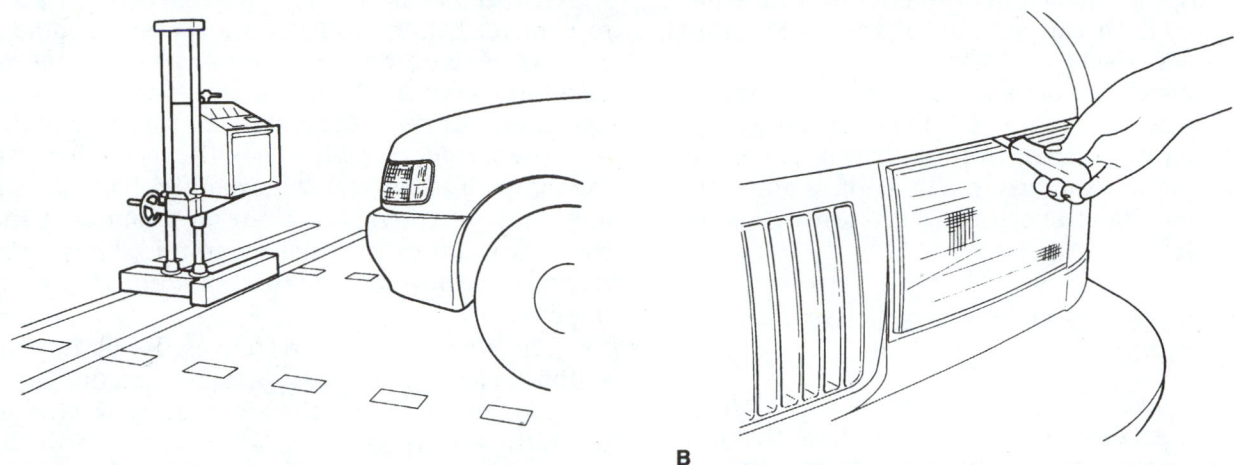

A   **B**

FIGURE 13-139 Always aim headlights after frontal repairs. (A) Use a headlight aimer to check the direction of headlight beams. (B) Turn the screws on the headlight mounting to adjust them as needed. *(Courtesy of Isuzu Motor Co.)*

direction in front of the vehicle (Figure 13–139A). Headlight adjustment screws can be turned to tilt the headlight lens up, down, right, or left (Figure 13–139B).

The turn, brake, and running lights use bulbs that fit into removable sockets. The service manual for the specific make vehicle will give directions for accessing and changing bulbs. Quite often, you must reach in behind the bulb and turn. This allows the bulb and socket to pop out for service (Figure 13–140).

Before releasing any vehicle to the customer, ensure that all light bulbs are working properly. Turn all lights on and walk around the car. If any bulbs are burned out or not functional, find out why. You may have to install a new bulb or fix wiring problems.

A few of the other circuits in an automobile include power seats, windows, door locks, mirrors, and cruise control. Other electrical devices include radios, cassette players, speaker systems, chimes, buzzers,

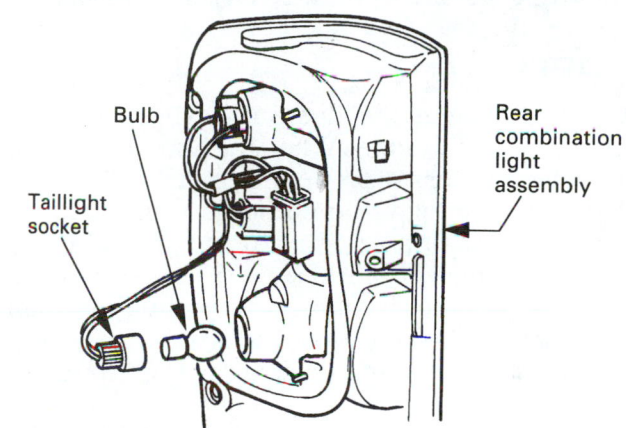

FIGURE 13-140 Note typical mounting of tail and brake light bulbs. They should be checked before releasing vehicle. The crushing jar of collision impact can fracture bulb elements. *(Courtesy of Isuzu Motor Co.)*

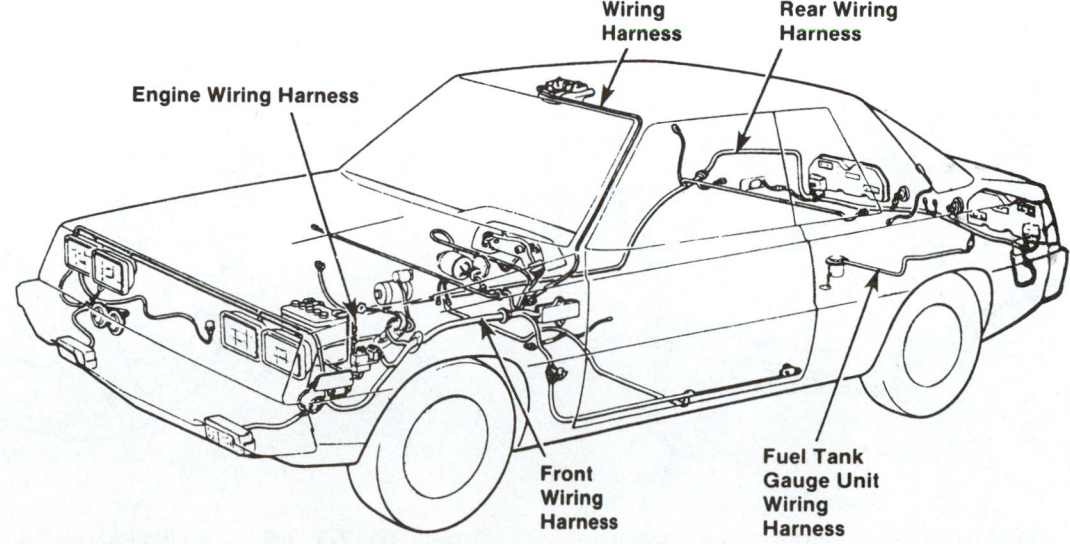

FIGURE 13-141 Be sure the wiring harness groupings are secure in their clips or holds.

graphic displays, analog instruments, and computer commands. Each can be fixed if the basic testing methods described earlier are followed.

Faulty grounds are another common source of circuit problems (Figure 13–141). A bad ground may be keeping current from returning to the battery, causing a dead circuit. The shop manual will give ground locations for circuit testing with a multimeter.

WINDSHIELD WIPERS AND WASHERS

A typical windshield wiper operates on a small single- or multi-speed electric motor. A switch on the steering wheel assembly or dashboard activates the motor. The spray washer generally has its own motor, plastic container, or reservoir and pump, which forces liquid through tubing to nozzles which spray the liquid washer on the windshield (Figure 13–142).

HORN

Most horn systems use the steering wheel switch and relay for horn sounding. When the horn button, ring, or padded unit is depressed, electricity flows from the battery through a horn lead into an electromagnetic coil in the horn relay to the ground (Figure 13–143). A small flow of electric current through the coil energizes the electromagnet, pulling a movable arm. Electrical contacts on the arm touch, closing the primary circuit and causing the horn to sound.

STARTING AND CHARGING SYSTEMS

The *starting system* has a large electric motor that turns the engine flywheel. This spins or "cranks" the crankshaft until the engine starts and runs on its own power.

The *ignition switch* in the steering column is used to connect battery voltage to a starter solenoid or relay. Other ignition switch terminals are connected to other electrical circuits. A *starter solenoid*, when energized, connects the battery and starting motor.

The *starting motor* is a large DC motor for rotating the engine flywheel. It normally bolts to the rear, lower side of an engine. A few are mounted inside the engine, under the intake manifold. The *flywheel ring gear* meshes with the starter-mounted gear during cranking.

The *charging system* recharges the battery and supplies electrical energy when the engine is running. An alternator or belt-driven DC generator produces this electricity.

A *voltage regulator*, usually mounted on the alternator, controls alternator output. *Charging system voltage* is typically 13 to 15 volts.

To quickly check the condition of a charging system, connect a voltmeter across the battery. With the engine off, you will read battery voltage. It should be above 12.6 volts. If not, the battery needs charging or is defective. When you start the engine with all electrical accessories on (lights, radio, etc.), the voltage must stay above battery voltage. If not, there is something wrong with the charging system.

No-start Engine Problem

If an engine cranks but fails to start, check for "spark" and "fuel." Both are needed for an engine to operate.

To CHECK FOR SPARK, pull off one of the spark plug wires. Install an old spark plug into the wire and lay the spark plug on engine ground. When you crank the engine, a bright spark should jump across

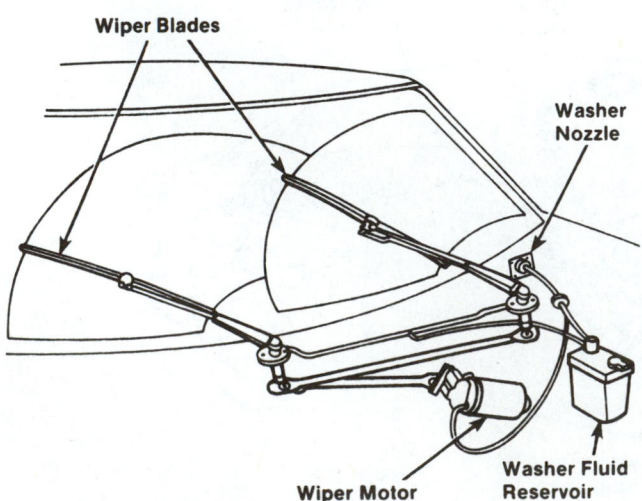

FIGURE 13–142 This is a typical configuration of windshield wipers and washers.

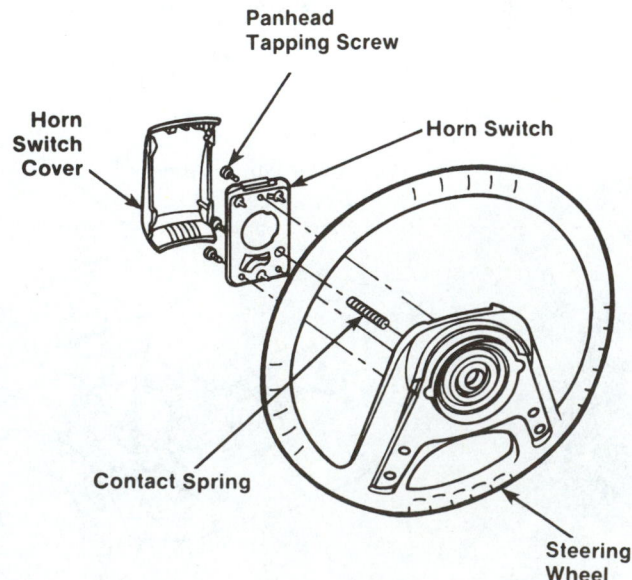

FIGURE 13–143 The horn switch often uses a spring-loaded plunger to make contact between rotating parts.

the spark plug gap. If not, something is wrong with the ignition system (blown fuse, damaged wires, crushed components, etc.).

If you have spark, CHECK FOR FUEL. On fuel-injected engines, this can often be done by installing a pressure gauge on the engine's fuel rail. A special test fitting is usually provided for a pressure gauge. With the engine cranking or the key on, the gauge should read within specs. If not, something is keeping the electric fuel pump(s) from working normally. Check for a clogged fuel filter, blow pump fuse, or wiring problem.

With older carbureted engines, you can look inside the carburetor to check for fuel. With the engine and air cleaner off, move the throttle opened and closed. This should make fuel squirt into the carburetor and engine. If not, something is preventing normal fuel flow from the tank to the engine.

Battery Drain Problem

A *battery drain* is a problem that causes current to flow out of the battery when everything should be off. For example, a low amperage short from a severed hot wire can gradually drain the vehicle's battery. Too often this happens and the body shop remedy is only to recharge the battery. Charging the battery can get the vehicle back to the owner, but this solution is only temporary. The owner will experience another dead battery in a day or two.

To find a battery drain, connect an ammeter between one of the battery cables and the battery. Turn everything off. Disconnect the fuse for the clock, and the meter should show little or no current draw. If you still have a drain, keep removing fuses until the meter shows zero. This will tell you which circuit has the short. Pinpoint test this circuit until the

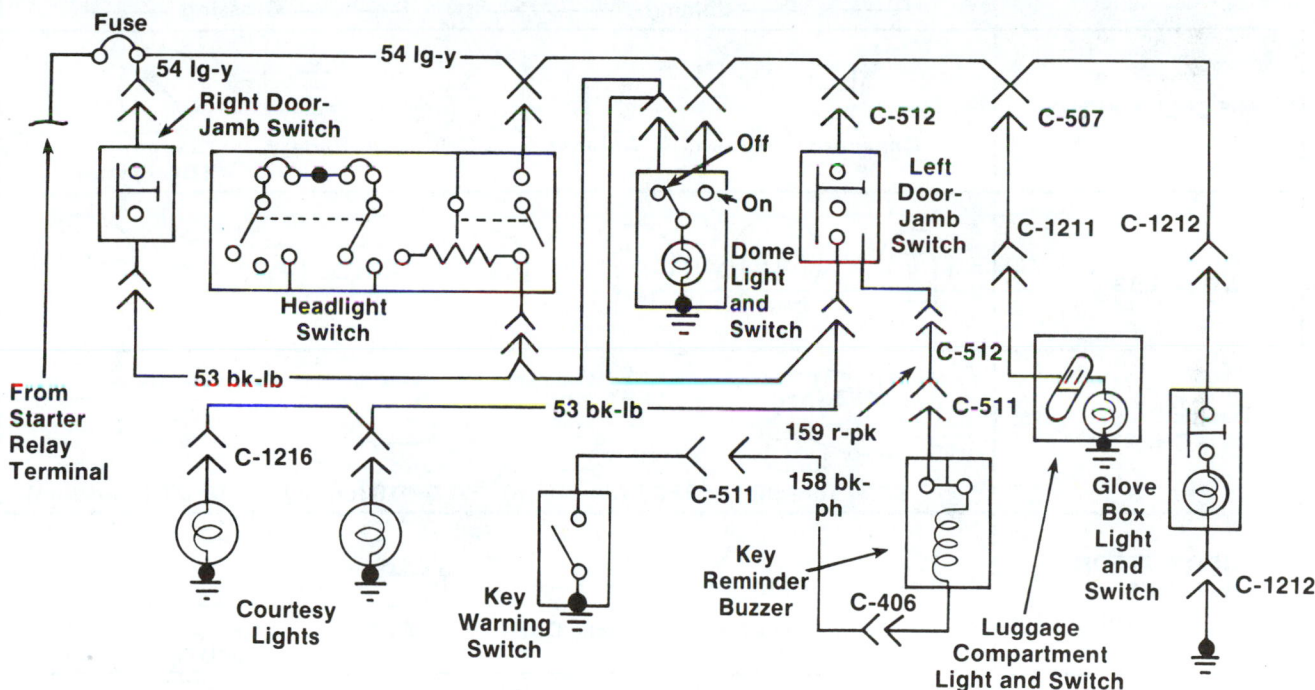

FIGURE 13-144 Note the symbols used in this interior light wiring diagram.

TABLE 13-4: COMMON ELECTRICAL ABBREVIATIONS			
A	Ampere	POS	positive
ac	alternating current	PRES	pressure
ACC	accessory	SOL	solenoid
BAT	battery	SPDT	single-pole double-throw
C/B	circuit breaker	SPST	single-pole single-throw
dc	direct current	TEMP	temperature
DPDT	double-pole double-throw	TOG	toggle (switch)
MOM	momentary	V	Volt
MOT	motor	W	Watt
(n)	none	–	negative
NC	(nc) normally closed	Ω	Ohm
NEG	negative	+	positive
NO	(no) normally open	±	plus or minus
PB	push button	%	percent

CIRCUIT BREAKERS	Circuit Breaker	Automatic Resetting	Manual Resetting		
RESISTORS	Fixed	Variable	Thermistor		
CONNECTORS	Male/Female	Polarized	Nonpolarized	Bulkhead	Female Male
WIRES	Joining			Crossing	
MISCELLANEOUS	Capacitor	Ground	Antenna	Cell	Battery	Gauge (Designate)
LIGHTS AND BULBS						
TOGGLE SWITCHES	SPST (nc) SPDT (on-on)	SPST (no) SPDT (on-off-on)	DPDT (on-on)	DPDT (on-off-on)	
PUSH BUTTON SWITCHES (MOMENTARY)	Normally Closed	Normally Open	SPDT	DPDT	
PRESSURE ACTUATED SWITCHES	Open On Fall	Open On Rise	SPDT	DPDT	
MISCELLANEOUS SWITCHES	Mercury	Wiper	Rotary	Slide	
FUSES	Fuse	In-Line	Fusible Link		

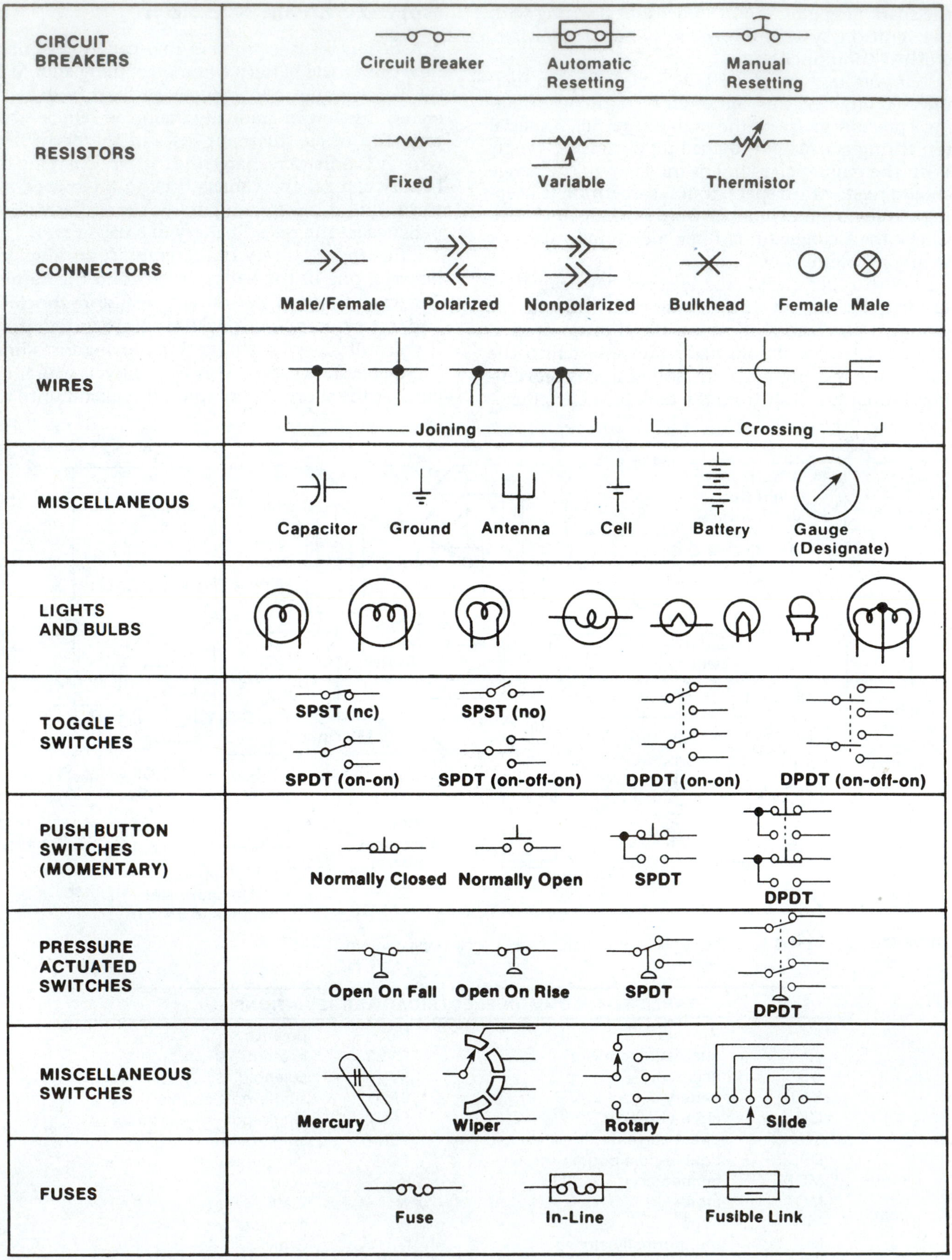

FIGURE 13-145 Memorize these common automobile electrical symbols.

shorted wire is found and repaired. When corrected, the ammeter should show only a tiny current flow to the clock.

WIRING DIAGRAMS, SCHEMATICS, AND DIAGNOSTIC FLOW CHARTS

To determine and isolate electrical problems, it is often necessary to trace through the electrical circuit using a **wiring diagram.** It is a graphic representation of all parts and wires in the circuit (Figure 13–144).

Abbreviations are used on wiring diagrams so that more information can be given. The service manual will have a chart explaining each abbreviation, number, and symbol on the diagram (Table 13–4).

Electrical symbols are graphic representations of electrical-electronic components. Symbol charts can be found at the front of the wiring diagram section of the service manual (Figure 13–145).

The symbol used to identify a part is either a universal symbol or an auto manufacturer's symbol. Since manufacturers use different symbol designs for some parts, follow the symbol chart for the specific wiring diagram being used.

Most wires on wiring diagrams will be identified by their insulation color. *Wire color coding* allows you to find a specific wire in a harness or in a connector. The color code abbreviation chart can also be found at the front of the wiring diagram section of the service manual. Different auto makers use different color-coding abbreviations on their wiring diagrams. Typical color codes are given in Table 13–5. The first letter in a combination of letters usually indicates the base color. The second letter usually

refers to the stripe color (if any). Tracing a circuit through a vehicle is basically a matter of following the colored wires.

Many wiring diagrams found in the service manuals also have circuits numbered. *Circuit numbering* is used to specify exactly which part of the circuit the service manual is referring to.

A wiring diagram is more like a book than a picture. You cannot understand a wiring diagram just by glancing at it. Like a book, you must read the diagram carefully all the way through for a complete understanding.

A **wiring harness** has several wires enclosed in a protective covering. A vehicle has several wiring harnesses, usually named after their location in the vehicle. The service manual will give illustrations with code numbers for locating parts and connections.

A *troubleshooting diagnostic flow chart* is a chart that gives diagnosis procedures that are read from the top to the bottom in a specific sequence. One is shown in Figure 13–146. The most efficient way to locate an electrical fault is to follow the manufacturer's diagnostic steps. These step-by-step procedures are used hand-in-hand with the wiring diagrams. If, however, troubleshooting flowcharts are not available, begin at the battery and systematically trace the circuit from

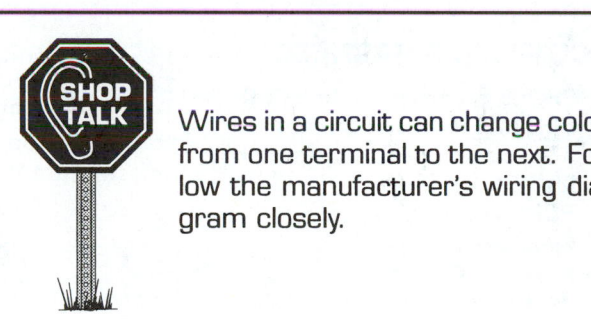

Wires in a circuit can change color from one terminal to the next. Follow the manufacturer's wiring diagram closely.

TABLE 13-5: COMMON WIRE COLOR CODES			
Color	**Abbreviations**		
Aluminum	AL		
Black	BLK	BK	B
Blue (Dark)	BLU DK	DB	DK BLU
Blue (Light)	BLU LT	LB	LT BLU
Brown	BRN	BR	BN
Glazed	GLZ	GL	
Gray	GRA	GR	G
Green (Dark)	GRN DK	DG	DK GRN
Green (Light)	GRN LT	LG	LT GRN
Maroon	MAR	M	
Natural	NAT	N	
Orange	ORN	O	ORG
Pink	PNK	PK	P
Purple	PPL	PR	
Red	RED	R	RD
Tan	TAN	T	TN
Violet	VLT	V	
White	WHT	W	WH
Yellow	YEL	Y	YL

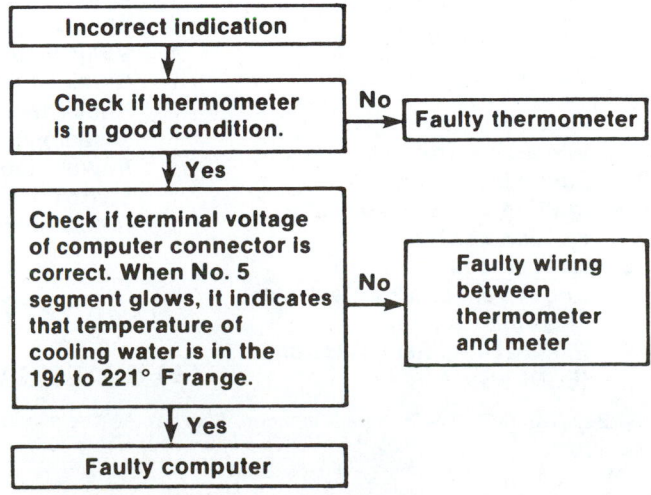

FIGURE 13–146 A manufacturer's troubleshooting flow chart for diagnosing a faulty temperature gauge. It is used from the top of the flow chart to the bottom.

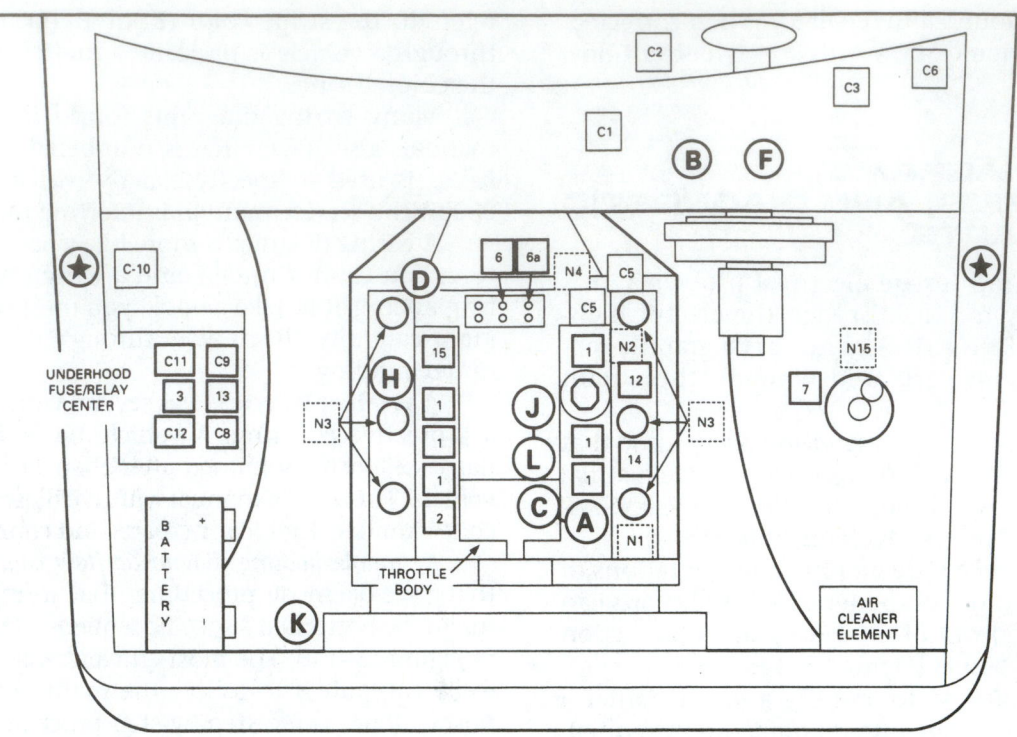

COMPUTER HARNESS

C1 Engine control module (ECM)
C2 DLC diagnostic connector
C3 "Check Engine" Malfunction indicator lamp
C5 ECM harness ground
C6 Fuse panel
C8 ECM main relay
C9 Fuel pump fuse 15A
C10 Injector resistor
C11 Oxygen 10A heater fuse
C12 30A ECM main fusible link

ECM CONTROLLED COMPONENTS

1 Fuel injector
2 Idle air control
3 Fuel pump relay
6 Ignition control module ignition control
6a Ignition coils
7 Knock sensor Module under charcoal canister
12 Exhaust Gas Recirculation (EGR) VSV

13 Air conditioning relay
14 Evaporate emission canister purge VSV
15 Induction air control plate system VSV

ECM INFORMATION SENSORS

A Manifold Absolute Pressure
B Heated oxygen sensor
C Throttle position sensor
D Engine coolant temperature
F Vehicle speed sensor
H Crank angle sensor
J Knock sensor (under intake assembly)
K Power steering pressure switch (in-line)
L Intake air temperature

⬡ **EGR VALVE**

★ **CHASSIS GROUNDS**

EMISSION COMPONENTS (NOT ECM CONTROLLED)

N1 Crankcase vent valve (PCV)
N2 Exhaust gas recirculation valve back pressure transducer
N3 Spark plugs
N4 Fuel rail test fitting (for fuel pressure test)
N15 Fuel vapor canister

* The ECM is located behind the console in the lower dash.

FIGURE 13-147 A part location diagram like this one is handy when you cannot find a part on a vehicle. *(Courtesy of Isuzu Motor Co.)*

there. Do not jump back and forth in the circuit or between circuits. Be patient and narrow down possible trouble spots by a process of elimination.

The service manual may also give a *part location diagram* for finding electrical parts (harnesses, sensors, switches, computers, etc.). It is helpful when you are having trouble finding something (Figure 13–147).

13.13 ELECTRONIC SYSTEM SERVICE

Many electronic diagnostic and repair procedures are not that difficult. If the technician goes back to the basics of electrical circuits and follows a few simple rules, he or she can find and fix a lot of electronic problems. Knowledge of electronics is necessary for the technician to perform the job (Figure 13–148).

The use of the word "electronics" in this book refers to on-board computers and other "blackbox"-type items, while electrical systems as just described means wiring and electrical components, such as alternators, lights, heater motors, and so on.

ELECTRONIC DISPLAYS

Electronic instrument displays are becoming more and more popular. The technology has become less expensive and more reliable. And many customers like the high tech, state-of-the-art image of electronic displays. But, in collision repair, electronic displays call for some special cautions. These complex and expensive parts must be handled carefully to avoid damage.

Most dashboard gauges are driven by some type of electrical signal. These gauges can be either analog

FIGURE 13–148 Today's auto body technician should have a strong understanding of electricity and electronics.

or digital. A *digital display* uses numbers instead of a needle or graphic symbol. In an *analog display,* an indicator moves in front of a fixed scale to give variable readout. The indicator is often a needle, but it can also be a liquid crystal or graphic display. An example is a speedometer in which the speed is shown by a set of vertical bars that light up or dim as the speed changes (Figure 13–149).

The advantage of analog displays is that they show relative change better than digital displays. They are useful when the driver must see something quickly, and the exact amount of change is not important. For example, an analog tachometer shows the rise and fall of the engine speed better for shifting than a digital display. Here the driver does not have to know exactly how many rpm's the engine is running. The most important thing is how fast the engine is reaching the red line on the gauge.

A digital display is better for showing exact data such as miles or operating hours. Many speedometer/odometer combinations are examples of both analog (speed) and digital (distance). The choice of display types is a matter of designer and buyer preferences.

An *analog signal* is continuously variable. An analog current is like the water flowing from a faucet, which is gradually turned up and down. Sometimes it flows a lot, sometimes only a little, and sometimes not at all. As an example, a temperature sensor causes the current to change as the temperature changes. As the temperature rises, the resistance decreases. This causes an increase in the circuit current. As the sensor cools, the current decreases.

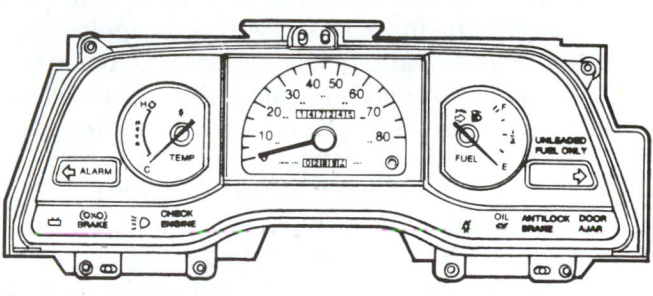

A

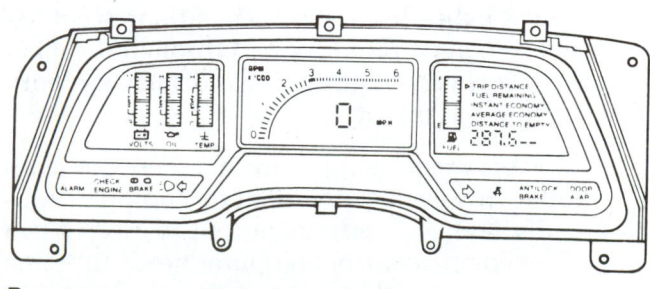

B

FIGURE 13–149 (A) An analog display has gauge needles that sweep around. (B) A digital display has number displays and sometimes analog indicators that move or slide across to show vehicle speed, for example.

The changing current is used to drive a gauge. The higher the temperature and pressure, the more current flows in the gauge circuit. The current creates a magnetic field that moves the pointer. In a temperature gauge, the higher the current (temperature), the greater the magnetic field and the more the pointer moves. These are called magnetic gauges and are used widely.

A *digital signal* has only two states—either on or off. If a switch is turned on and off many times, the number of pulses can be counted. For example, a sensor can be made to turn on and off each time a wheel moves a certain distance. The number of pulses that are counted in a given period of time allows the computer to display the speed. The pulses can also be used by the computer to change the odometer reading. This is the principle of computer system operation.

Following are the types of electronic displays used today:

- **Light-emitting diode (LED).** These are used as either single indicator lights, or they can be grouped to show a set of letters or numbers. LED displays are commonly red, yellow, or green. But, LED displays use more power than other displays. They can also be hard to see in bright light.
- **Liquid crystal diode (LCD).** These have become very popular for many uses, including watches, calculators, and dash gauges. They are made of sandwiches of special glass and liquid. That is where the term "liquid" comes from. A separate light source is required to make the display work. The display has wires on the glass. When there is no voltage, light cannot pass through the fluid. When voltage is applied, the light passes through the segment. LCDs do not like cold temperatures, and the action of the display slows down in cold weather. These displays are also very delicate, and must be handled with care. Any rough handling or force on the display can damage it.
- **Vacuum fluorescent diode (VFD).** These displays use glass tubes filled with argon or neon gas. The segments of the display are little fluorescent lights, like the ones in a fluorescent fixture. When current is passed through the tubes they glow very brightly. These displays are both durable and bright.

All gauges require input from a sensor or sending units. However, with modern computer-controlled displays, the sensor's output is used in two ways. The engine control computer needs the same information as the electronic display, so the information passes through the computer first. It then travels to the gauge. As an example, compare the temperature sensor on the vehicle of ten years ago with a modern vehicle. On the ten-year-old car, the tem-

perature gauge was connected directly to a sensor that checked the engine temperature. A rise in temperature resulted in increased current in the gauge circuit. This caused the pointer and magnetic gauge to move, showing the temperature to the driver on an analog scale.

On the modern vehicle the system works identically, with one very important exception. The information from the sensor is first fed through the vehicle's engine control computer. The computer uses the information to manage a variety of systems, including air/fuel ratio, spark timing, and switching of emission control system components. In addition, the computer uses the information to operate the temperature gauge—digital, analog, or just a temperature warning lamp (a form of digital display).

SELF-DIAGNOSTIC DISPLAYS

Most of the electronic displays today have some sort of built-in diagnostics. These differ with the vehicle make and year. However, in recent years there has been a trend toward more onboard checks. Each time the key is turned to the on position, the system does a self-check. The self-check ensures that all the bulbs, fuses, and electronic modules are working.

If the self-check finds a problem, it may store a code for later servicing. It may also instruct the computer to turn on a trouble light to show that service is needed. For example, some systems run an instrument cluster self-test on start-up. It goes through a series of tests, called *prove out,* checking for function of the display, connectors to the sensors, and the condition of the LCD displays themselves. During prove out, all illuminated parts of the display are briefly lighted. Passing all parts of this test tells that the display is working properly. If portions of the display do not light, there is a problem. The technician should check to see that all of the connectors are properly reinstalled to the back of a replacement display.

AIR BAGS

The *air bag system* uses impact sensors, the vehicle's onboard computer, an inflation module, and a nylon balloon in the steering column to protect the driver during a head-on collision. It can be expensive to service since the bag and all sensors often require replacement after inflation.

For details of air bag system operation and service, refer to Chapter 15.

COMPUTER SYSTEMS

Almost all vehicle systems are now controlled by computer. This includes the fuel, ignition, charging,

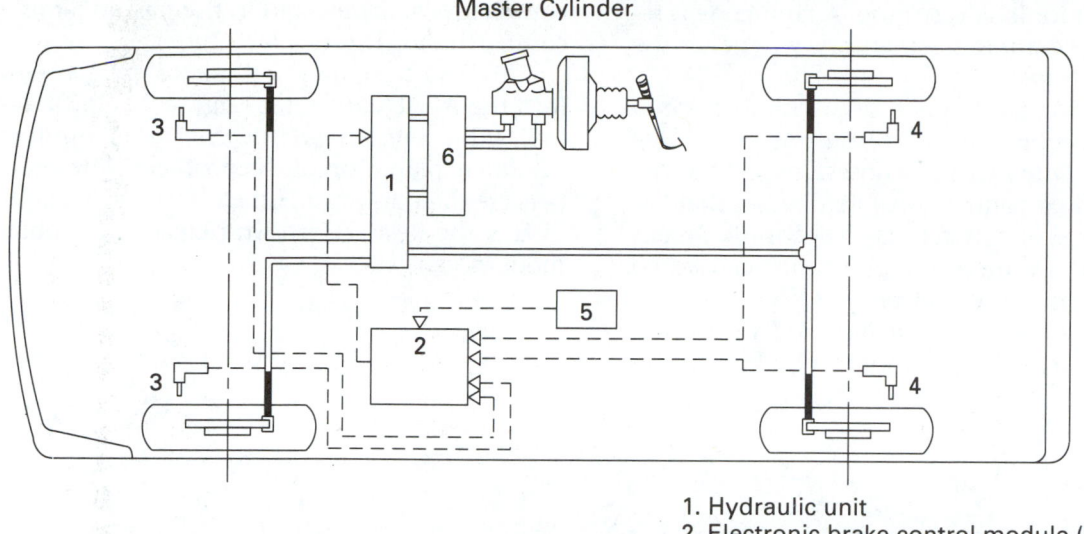

Master Cylinder

1. Hydraulic unit
2. Electronic brake control module (EBCM)
3. Front wheel speed sensor
4. Rear wheel speed sensor
5. G-sensor
6. Proportioning and bypass (P&B) valve
   (Refer to "Hydraulic Brakes"(Sec. 5A).)

– – – – – – – Electric

========= Hydraulic

FIGURE 13-150 Study the parts of this computer system for controlling the antilock brake system. *(Courtesy of Isuzu Motor Co.)*

suspension, brake, climate control, air bag, and other systems. Figure 13–150 shows one type of computer system for antilock brakes.

A basic **computer system** consists of:

1. Sensors (input devices)
2. Actuators (output devices)
3. Computer (electronic control unit)

The **sensors** are devices that convert a condition (temperature, pressure, part movement, etc.) into an electrical signal. They send an electrical input signal back to the computer. Once the computer analyzes the sensor data, it produces a preprogrammed output that is sent to system actuators.

Actuators are devices (solenoids or servo-motors, for example) that move when responding to electrical signals from the computer. In this way, a computer system can react to sensor inputs and then act upon these conditions by operating the motors or solenoids.

The **computer** is a complex electronic circuit that produces a known electrical output after analyzing electrical inputs. Today's vehicles can have one or more computers that monitor and control the operation of electrical systems.

Self-diagnostics

Computer self-diagnostics means the computer system can detect its own problems. Most electronic systems have built-in, self-diagnostic capabilities. Examples would be ABS braking systems and air bag systems.

If a problem exists, the computer will turn on a warning light in the instrument panel. This tells you to use self-diagnostics to find the source of the trouble. The computer performs a self-check each time the ignition key is turned to the on position (Figure 13–151).

The self-check makes sure that all of the bulbs, fuses, sensors, and electronic control modules are working. If the self-check finds a problem, it may store a **fault code** (code representing specific circuit or part

DIAGNOSTIC CODE DISPLAY

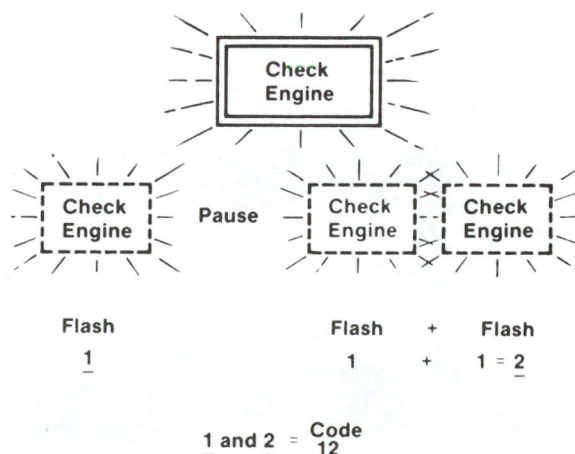

FIGURE 13-151 The check engine light will signal any computer trouble codes. Code 12 is a signal that the computer's diagnostic program is working properly. Note the Morse-type code of on-off signals or flashes. This code will vary with the make and model of the vehicle.

with a problem) for later servicing. A fault code is recorded into the computer's memory whenever the system malfunctions.

Before reading fault codes, do a visual check to make sure the problem is not a result of wear, loose connections, or faulty vacuum hoses. Inspect all wire and vacuum-hosed connections. Remember that the low level electrical signals in today's electronic circuits cannot tolerate the increased resistance caused by corrosion in connector contacts.

An *assembly line diagnostic link (ALDL)* connector is provided for reading fault or trouble codes. This

connector can be located in the passenger or engine compartment (Figure 13–152).

To read computer fault codes, plug a **scan tool** into the ALDL connector (Figure 13–153). The scan tool will then convert the computer number code into an explanation of the problem. The code numbers are then displayed on the tester's display screen. This is the best way to find electrical problems on modern vehicles.

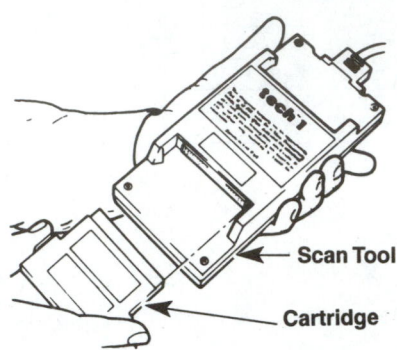

FIGURE 13-154 A scan tool will normally have different cartridges for different years and makes of vehicles. Some scan tools will actually tell you what to do while troubleshooting. *(Courtesy of Isuzu Motor Co.)*

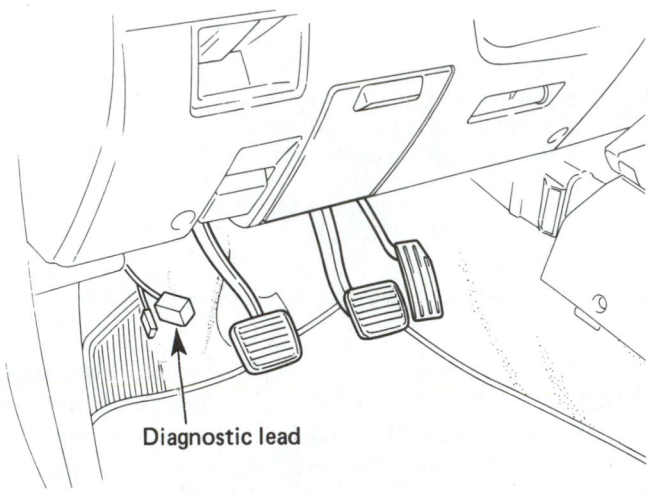

FIGURE 13-152 The diagnostic lead may be under the dash or in the engine compartment. Refer to the service manual for its location and use. *(Courtesy of Isuzu Motor Co.)*

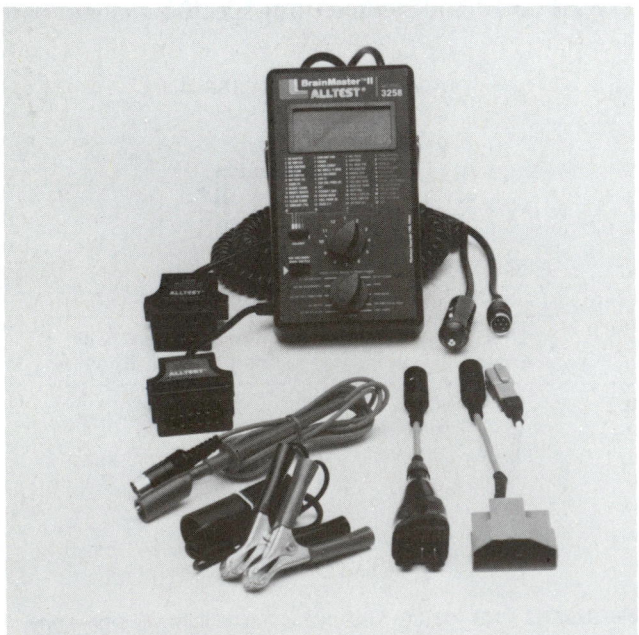

FIGURE 13-153 Scan tools can be used to more quickly access and analyze computer system trouble codes. *(Courtesy of All-Test, Inc.)*

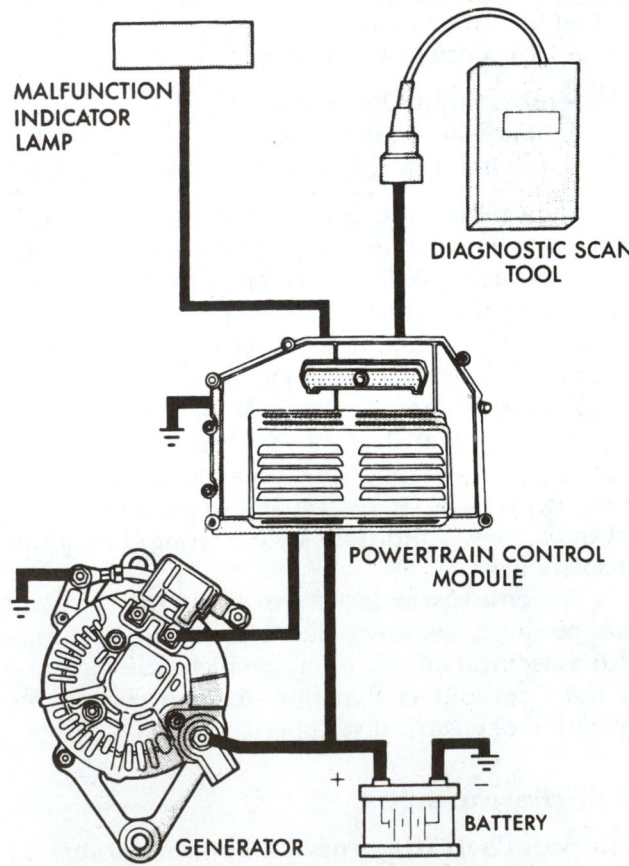

FIGURE 13-155 The trouble code chart from a specific service manual will explain what each code number means and what should be done to correct the problem. *(Courtesy of Dodge Trucks)*

Since fault codes vary from model to model, the service manual and scanner owner's manual must be consulted. Different cartridges must usually be installed in the scan tool to match the year, make, and model of the vehicle (Figure 13–154).

Connect the scan tool to the vehicle's diagnostic connector. It will then be able to communicate with the vehicle's onboard computer to analyze possible problems. (See Figure 13–155).

If you get a fault code, the scan tool may be able to describe the problem or it may only give a fault code number. If you only get a number, refer to the service manual fault code chart. It will explain the number code and describe which parts might be causing the trouble (Figure 13–156).

If you do not have a scan tool, you can use *computer self-diagnosis* to find circuit problems. By jumping across specific terminals on the ALDL connector or turning the ignition key on and off as directed, the computer will flash a Morse-type code indicating the problem. You can note these flashes and compare them to manual details to find the problem.

The computer itself is a highly reliable electronic device. Trouble is more often caused by damaged wiring, connectors, or sensors. The computer is one of the last parts to suspect with computer system malfunctions.

Computers can be found in just about any location on the vehicle. These include

1. Under the dash
2. Under the seats
3. Under the hood
4. Behind kick panels
5. In the trunk

A *PROM* is a programmable read-only memory chip mounted in the computer. If you have to replace a computer, you must install the PROM from the old computer into the new one. This will assure that the computer is programmed for the equipment on the vehicle being serviced.

Disconnecting the battery can erase stored electrical/electronic system fault codes on some vehicles. Sometimes you can pull and reinstall the computer fuse to erase fault codes. Always refer to the service manual for specific information about computer system service.

Protecting Electronic Systems

The last thing a technician wants to do when a vehicle comes into the shop for collision repair is create problems. This is especially true when it comes to electrical systems and electronic components. There are proper ways to protect automotive electrical systems and electronic components during storage and repair.

1. Disconnect the battery cables (negative cable first) before doing any kind of welding. To avoid the possibility of explosion, completely remove the battery when welding under the hood or on the front end. The ground connection must be clean and tight. Position the ground clamp as close as possible to the work area to avoid current seeking its own ground.

Diagnostic Trouble Code	DRB Scan Tool Display	Description of Diagnostic Trouble Code
11*	No Crank Reference Signal at PCM	No crank reference signal detected during engine cranking.
15**	No Vehicle Speed Sensor Signal	No vehicle distance (speed) sensor signal detected during road load conditions.
34*	Speed Control Solenoid Circuits	An open or shorted condition detected in the Speed Control vacuum or vent solenoid circuits.
	or	
	Speed Control Switch Always Low	Speed Control switch input below the minimum acceptable voltage.
	or	
	Speed Control Switch Always High	Speed Control switch input above the maximum acceptable voltage.
55*	N/A	Completion of fault code display on Check Engine lamp.

* Check Engine Lamp will not illuminate at all times if this Diagnostic Trouble Code was recorded. Cycle Ignition key as described in manual and observe code flashed by Check Engine Lamp.

**Check Engine Lamp will illuminate during engine operation if this Diagnostic Trouble Code was recorded.

FIGURE 13–156 The Service Manual Fault Code Chart explains the fault codes you see on the scan tool.

2. Whenever disconnecting or removing the battery on a computer-controlled vehicle, remember that the memory for radio station selection, seat position, climate control setting, and any other "driver programmable" options is erased. When delivering the vehicle, advise the customer to reprogram these settings. Or better yet, record them before beginning and reprogram them yourself before delivery.

3. Static electricity can cause problems. Avoid it by grounding yourself before handling any displays. One way is to touch a good ground with one hand before handling the display with the other hand. Another way is to use a grounding strap that attaches to the wrist and then to the vehicle.

4. Avoid touching bare metal contacts. Oils from your skin can cause corrosion and poor contacts.

5. Be careful about the placement of welding cables. Keep the electrical path as short as possible by placing the ground clamp near the point of welding. Also, do not let the welding cables run close to electronic displays or computers.

6. Take care when handling electronic displays and gauges. Never press on the gauge face because this could damage it.

7. If a fault code indicates a problem with the oxygen sensor, extra caution is required. The oxygen sensor wire carries a very low voltage and must be isolated from other wires. If not, nearby wires could add more induced voltage. This gives false data to the computer and can result in a drivability problem. Some manufacturers use a foam sleeve around the oxygen sensor wire to keep it separate and insulated from other wires.

8. The sensor wires that connect to the computer should never be rerouted. The resulting problem might be impossible to find. When replacing sensor wiring, always check the service manual and follow the routing instructions.

9. Remove any computer that could be affected by welding, hammering, grinding, sanding, or metal straightening. Be sure to protect the removed computer and its connectors by wrapping them in plastic antistatic bags to shield them from moisture and dust.

10. Be careful not to damage wiring when welding, hammering, or grinding.

11. Be careful not to damage connectors and terminals when removing electronic components. Some may require special tools to remove them.

12. Always route wiring in the same place it was originally. If not, electronic crossover from the current-carrying wires can affect the sensing and control circuits. Reuse or replace all electrical shielding for the same reason.

SUMMARY

- Mechanical repairs include tasks like replacing a damaged water pump, radiator, or engine bracket.

- Electrical repairs include tasks like repairing severed wiring, replacing engine sensors, and scanning for computer or wiring problems.

- The power train includes all of the parts that produce and transfer power to the drive wheels.

- The engine provides energy to move the vehicle and power all accessories.

- When starting an engine before or after repairs, check the oil level with the oil dipstick.

- When a CV-joint boot is torn or missing, there is often damage or wear in the joint. When more time is saved in the repair of adjacent panels than is necessary to remove and reinstall the drivetrain, the drivetrain should be removed.

- Always refer to the OEM Service Manual.

- An engine-transaxle or transmission assembly is very heavy. If dropped, it can easily chop off toes and fingers or crush bones. A coil spring has deadly force when compressed!

- Visually inspect the steering system for any physical damage. Check the boots for leaks, inspect the tie rods, and examine the mounting points for any distortion.

- In collision repair, wheel alignment involves adjusting the vehicle's tires so that they roll properly over the road surface.

- Camber is the angle represented by the vertical tilt of the wheels inward or outward when viewed from the front of the vehicle.

- Caster is the angle of the steering axis of a wheel from true vertical, as viewed from the side of the vehicle.

- Steering axis inclination is the inward tilt of the steering axis at the top.

- Toe is the difference in the distance between the front and rear of the left- and right-hand wheels.

- With proper tracking, the rear tires travel directly behind the front tires when the steering wheel is in the straight-ahead position. The brake system uses hydraulic pressure to slow or stop wheel rotation with brake pedal application.

- R-134a is the present replacement for R-12. It is less harmful to the ozone layer. A recovery system will capture the used refrigerant and keep it from contaminating the atmosphere.

- Emission control systems are used to prevent potentially toxic chemicals from entering our atmosphere. The most common of these are the exhaust gas recirculation (EGR), catalytic converter, air injection, and positive crankcase ventilation (PCV) systems.

- Current is the movement of electricity (electrons) through a wire or circuit. It is measured in amperes or amps using an ammeter. The common electrical symbol for current is "A" or "I."

- Voltage is the pressure that pushes the electricity through the wire or circuit.

- Resistance is a restriction or obstacle to current flow. It tries to stop the current caused by the applied voltage.

- An open circuit is disconnected and does not have a complete electrical path.

- A short circuit has an unwanted path for current. A collision will sometimes create a short circuit.

- If an engine cranks but fails to start, check for "spark" and "fuel."

ASE-STYLE REVIEW QUESTIONS

1. To begin a cornering angle check, Technician A turns the left front wheel out 20 degrees. Technician B says that at this point the readings on the right wheel should read about 2 degrees more than the left. Who is correct?
   A. Technician A
   B. Technician B
   C. Both A and B
   D. Neither A or B

2. In unibody construction, what provides the critical mounting positions for the suspension and steering systems?
   A. Body panels
   B. Drivetrain
   C. Upper and lower control arms
   D. Cradle assembly mounting biscuits

3. After examining a collapsed steering column Technician A attempts to repair the collapsed portion. Technician B says that to remove the steering wheel you may need to use a wheel puller. Who is correct?
   A. Technician A
   B. Technician B
   C. Both A and B
   D. Neither A or B

4. What is the definition of camber?
   A. The forward or backward tilt of the steering axis
   B. The distance between the centerline of the ball joints and the centerline of the tire at the point where the tire contacts the road surface
   C. The amount of toe-out present on turns
   D. The inward or outward tilt of the tire as measured at the top

5. To determine whether an improper toe condition is caused by poor wheel alignment or damaged suspension components, Technician A makes a cornering angle check; Technician B makes a jounce/rebound check. Who is correct?
   A. Technician A
   B. Technician B
   C. Both A and B
   D. Neither A nor B

6. Which of the following statements concerning brake systems is incorrect?
   A. All cars are now required to have two hydraulic systems.
   B. The major brake component that a technician must repair is the brake line.
   C. A hose that is blistered does not necessarily have to be replaced.
   D. The wheel cylinder converts hydraulic pressure to mechanical force.

7. Which emission control subsystem is responsible for channeling blowby gases into the fuel intake area?
   A. Engine control
   B. Positive crankcase ventilation
   C. Evaporative
   D. Exhaust gas recirculation

8. After recovering the refrigerant from an A/C system, Technician A uses plastic wrap with wire ties to keep moisture out of the components of the system. Technician B replaces the receiver-drier if the system has been open to the atmosphere for several days. Who is correct?
   A. Technician A
   B. Technician B
   C. Both A and B
   D. Neither A or B

ESSAY QUESTIONS

1. Describe the three functions of a vehicle's suspension system.

2. How does an electronically controlled steering system work?

3. List five power steering service tips that should be kept in mind.

4. What is the A/C system low side?

5. Explain the difference between voltage, current, and resistance.

6. Summarize the three major parts of a computer system.

CRITICAL THINKING PROBLEMS

1. A steering wheel has been badly bent by the driver's body flying forward without a seat belt. What should be done?

2. You must replace the air-conditioning system compressor. How should you proceed?

MATH PROBLEMS

1. A worn ball joint moves up and down $1/32$ inch (0.8 mm). The ball joint moves $1/64$ inch (0.4 mm). What, if anything, should be done? Give measurements.

2. The front wheels of a car are toed in $1/16$ inch (1.5 mm). Specs call for $1/16$-inch (1.5 mm) toe out. How much adjustment is needed?

CHAPTER 14

Repairing Auto Plastics

INTRODUCTION

The term **plastics** refers to a wide range of materials synthetically compounded from crude oil, coal, natural gas, and other natural substances. Unlike metals, plastics do not occur in nature and must be manufactured. Because plastic is much lighter in weight than sheet metal, it has become an important part of today's vehicles. Today, more and more plastic is being used in automobile manufacturing.

Plastic parts include bumpers, fender extensions, fascias, fender aprons, grille openings, stone shields, instrument panels, trim panels, fuel lines, door panels, quarter panels, and engine parts. Fuel saving and weight reduction programs by auto makers have made plastic parts more common (Figure 14–1).

Many of the new reinforced plastics are almost as strong and rigid as steel. Some are even more stable dimensionally. Plastic parts are also extremely corrosion resistant. Tests are being made on plastic engine blocks and plastic frame parts. Plastic suppliers are projecting increased use of plastic in floor pans, windows, steering shafts, springs, wheels, bearings, and other mechanical components.

This increasing use of plastic has resulted in new approaches to collision repair (Figure 14–2, page 540). Many plastic parts can be repaired more economically than they can be replaced, especially if the part does not have to be removed. Cuts, cracks, gouges, tears, and punctures are all repairable. When necessary, some plastics can also be re-formed back to their original shape after distortion. Since parts are not always avail-

OBJECTIVES

After studying this chapter, you should be able to:

✔ Identify and explain the difference between the two types of plastics used in automotive production.
✔ Identify unknown plastics.
✔ Repair minor cuts and cracks in plastics using adhesives.
✔ Repair gouges, tears, and punctures in plastics by means of a chemical bonding process.
✔ Set up and operate a typical welding torch.
✔ Explain the keys to good plastics welding.
✔ Describe the proper plastics welding repair sequence.
✔ Explain the safety precautions used when working with fiberglass.
✔ Explain how fiberglass is used in adhesives to reinforce the damaged surface.
✔ Make SMC and RRIM repairs.
✔ Answer ASE test questions relating to repairing plastics.

KEY TERMS

adhesion promoter
airless plastic welding
backing strip
carcinogens
composite plastics
fiber-reinforced
   plastics
melt-flow plastic
   welding
mill and drill pads
molded core
plastic flame treating
plastic flexibility test
plastic memory
plastic speed welding
plastic stitch-tamp
   welding

plastic welding
plastics
reinforced plastic
sheet molded
   compounds
tack welds
thermoplastics
thermosetting
   plastics
two-part adhesive
   systems
ultrasonic plastic
   welding
vinyl

ASE TASK LIST

Job Skills covered in this chapter include:

PAINTING AND REFINISHING TEST (B2) TASK LIST

A. Surface Preparation

5. Dry or wet sand areas to be refinished.
9. Mix primer, primer-surfacer, or primer-sealer; spray onto surface of repaired area.
11. Dry or wet sand area to which primer-surfacer and/or two-component putty have been applied.
12. Remove dust from area to be refinished, including cracks or moldings of adjacent areas.
13. Clean area to be refinished using a proper cleaning solution.

C. Paint Mixing, Matching, and Applying

9. Identify the types of rigid, semi-rigid, or flexible plastic parts to be refinished; determine the proper materials and refinishing procedures.
10. Refinish rigid, semi-rigid, or flexible plastic parts.
11. Clean, condition, or refinish vinyl (e.g., upholstery, dashes, and tops).

D. Solving Paint Application Problems

2. Identify blushing (milky or hazy formation); determine the cause(s), and correct the condition.

NONSTRUCTURAL ANALYSIS AND DAMAGE REPAIR TEST (B3) TASK LIST

A. Preparation

10. Remove repairable plastics and other parts that are recommended for off-vehicle repair.

B. Outer Body Panel Repairs, Replacements, and Adjustments

15. Replace or repair plastic panels in accordance with manufacturers'/industry specifications.
16. Restore sealers, mastic, sound deadeners, and foam fillers.

F. Plastic Repair

1. Identify the types of plastic(s); determine repairability.
2. Identify the proper plastic repair procedure; clean and prepare the surfaces of plastic parts in accordance with manufacturers'/industry guidelines.
3. Repair rigid plastic parts by welding.
4. Repair rigid plastic parts with urethane or epoxy adhesives, with and without reinforcement materials.
5. Repair flexible plastic parts by welding.
6. Repair flexible plastic parts with urethane or epoxy adhesives, with and without reinforcing materials.
7. Repair holes and cuts in rigid and flexible plastic parts using backing materials and adhesives.
8. Retexture plastic parts.
9. Repair vinyl-clad urethane foam parts.
10. Reshape and shrink flexible exterior plastic parts.
11. Remove damaged areas from rigid exterior SMC (sheet molded compound); repair with partial panel installation.
12. Repair deep gouges, holes, and cracks in SMC (sheet molded compound) panels.
13. Replace bonded SMC (sheet molded compound)-type body panels; straighten or align panel supports.
14. Replace bonded non-SMC-type plastic body panels; straighten or align panel supports.
15. Prepare repaired areas for refinishing.

---

able, this means less downtime for the vehicle and more profits for you and your shop.

14.1 TYPES OF PLASTICS

Two general types of plastics are used in automotive construction: thermoplastics and thermosetting plastics.

Thermoplastics can be repeatedly softened and reshaped by heating, with no change in their chemical makeup. They soften or melt when heated and harden when cooled. Thermoplastics are weldable with a plastic welder or they can be adhesively repaired.

Thermosetting plastics undergo a chemical change by the action of heating, a catalyst, or ultraviolet light. They are hardened into a permanent shape that cannot be altered by reapplying heat or catalysts. Thermosets are usually repaired with flexible parts repair materials. In general, chemical adhesive bonding is used to repair thermosetting plastics, and welding is used for thermoplastics.

Figure 14–3 explains more fully the effects of heat on the two types of plastics.

Table 14–1 (see page 541) shows some of the more common plastics with their full chemical name, common name, and where on the vehicle they might be found. Their designation as to thermosetting or thermoplastic is also included.

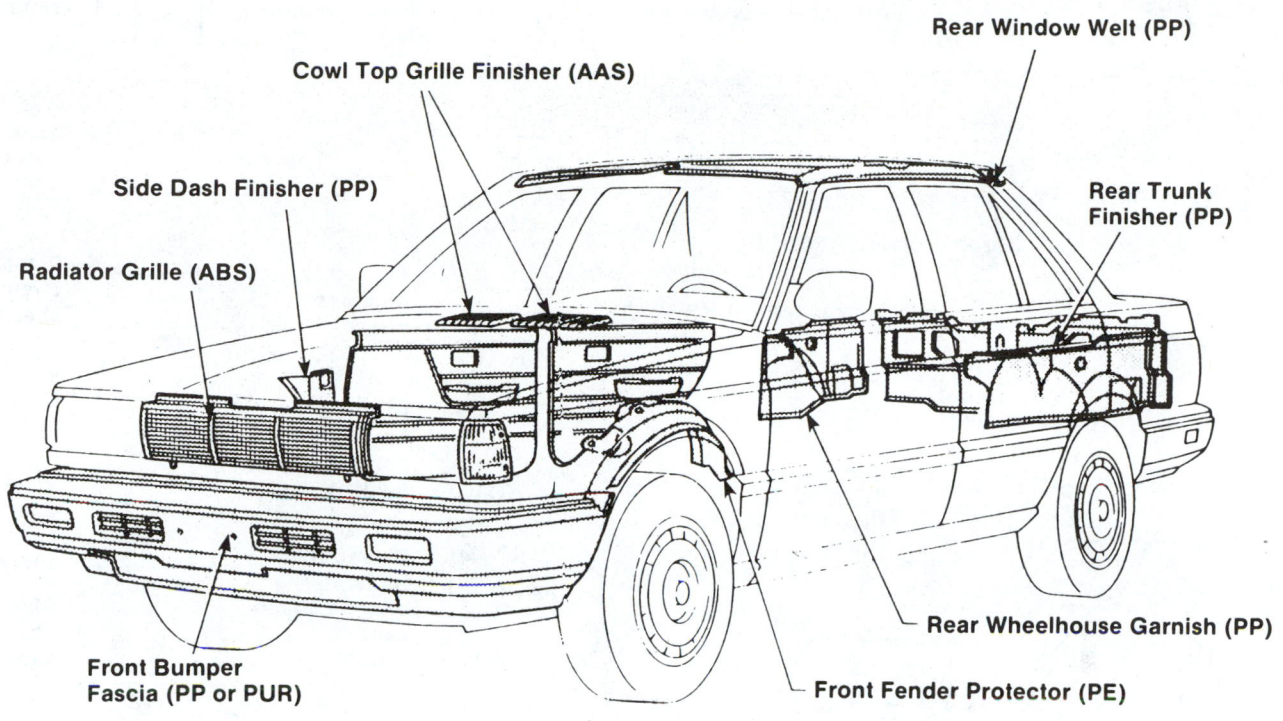

Rear Window Welt (PP)

Cowl Top Grille Finisher (AAS)

Side Dash Finisher (PP)

Rear Trunk
Finisher (PP)

Radiator Grille (ABS)

Rear Wheelhouse Garnish (PP)

Front Bumper
Fascia (PP or PUR)

Front Fender Protector (PE)

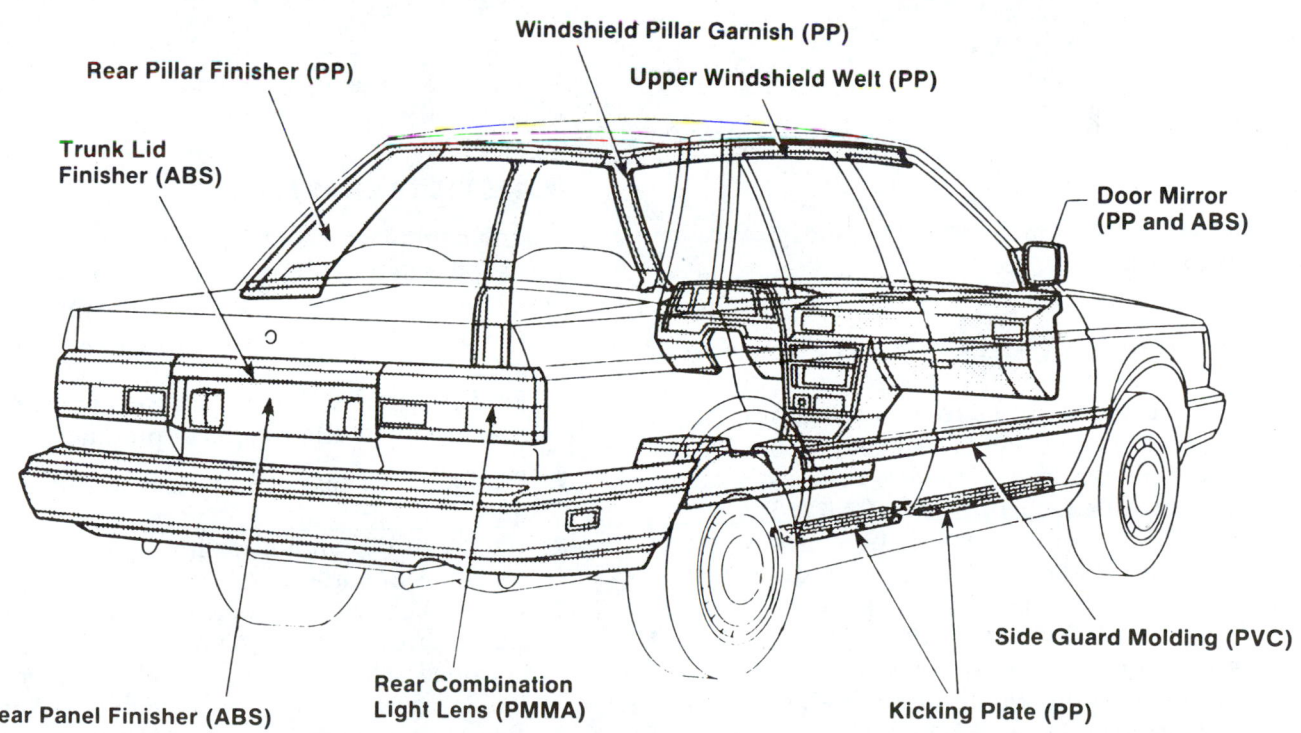

Windshield Pillar Garnish (PP)

Rear Pillar Finisher (PP)

Upper Windshield Welt (PP)

Trunk Lid
Finisher (ABS)

Door Mirror
(PP and ABS)

Side Guard Molding (PVC)

Rear Panel Finisher (ABS)

Rear Combination
Light Lens (PMMA)

Kicking Plate (PP)

FIGURE 14–1 Study the locations of plastic parts on the basic auto.

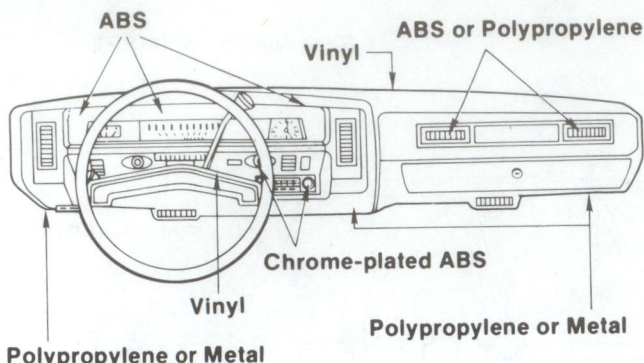

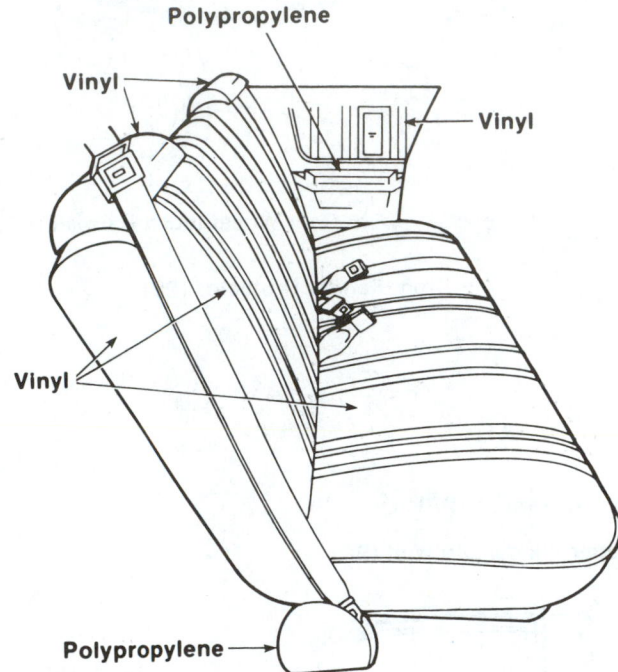

FIGURE 14-2 Note common interior uses of plastic in today's automobile.

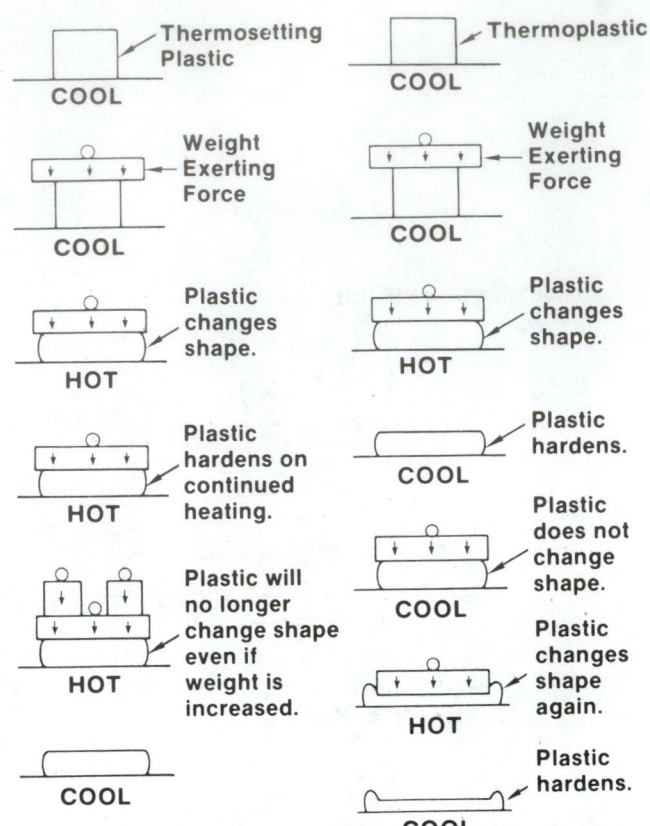

FIGURE 14-3 Here are the effects of heat on thermosetting plastic and thermoplastic.

Table 14-2 (see page 543)contains additional information on types of automotive plastics.

PLASTICS SAFETY

Working with plastics and fiberglass requires you to think about safety at all times. The resin and related ingredients can irritate your skin and stomach lining. The curing agent or hardener can produce harmful vapors.

Read and understand the following safety points before using any of these types of products:

1. Read all label instructions and warnings carefully.
2. When cutting, sanding, or grinding plastics, dust control is important.
3. Wear rubber gloves when working with fiberglass resin or hardener. A long-sleeved shirt with buttoned collar and cuffs is helpful to prevent sanding dust from getting on your skin. Disposable paint suits will keep dust away from clothes.
4. A protective skin cream should be used on any exposed areas of the body.
5. If the resin or hardener comes in contact with the skin, wash with borax soap and hot water or alcohol.

Composite plastics, or *hybrids*, are blends of different plastics and other ingredients designed to achieve specific performance characteristics.

A good example of this change is the use of fiber-reinforced composite plastic panels, commonly known as **sheet molded compounds** (SMC). The reason for using SMC is simple. It is light, corrosion proof, dent resistant, and relatively easy to repair compared to more traditional materials.

The use of SMC and other **fiber-reinforced plastics** (FRP) is not new. They have been used in various applications on automobiles for years. The use of large external body panels of reinforced plastic is not unusual either. What is new is that now, unlike the external panels on earlier vehicles, these panels are bonded to a metal space frame using structural adhesives, adding overall structural rigidity to the vehicle. SMC repair will be detailed later.

**TABLE 14-1: IDENTIFICATION SYMBOL, CHEMICAL NAME, TRADE NAME, AND
DESIGN APPLICATIONS OF MOST COMMONLY USED PLASTICS**

Symbol	Chemical Name	Common Name	Design Applications	Thermosetting or Thermoplastic
AAS	Acrylonitrile-styrene	Acrylic Rubber	—	Thermosplastic
ABS	Acrylonitrile-butadiene-styrene	ABS, Cycolac, Abson, Kralastic, Lustran, Absafil, Dylel	Body panels, dash panels, grilles, headlamp doors	Thermoplastic
ABS/MAT	Hard ABS reinforced with fiberglass	—	Body panels	Thermosetting
ABS/PVC	ABS/Polyvinyl chloride	ABS Vinyl	—	Thermoplastic
EP	Epoxy	Epon, EPO, Epotuf, Araldite	Fiberglass body panels	Thermosetting
EPDM	Ethylene-propylene-diene-monomer	EPDM, Nordel	Bumper impact strips, body panels	Thermosetting
PA	Polyamide	Nylon, Capron, Zytel, Rilsan, Minlon, Vydyne	Exterior finish trim panels	Thermosetting
PC	Polycarbonate	Lexan, Merlon	Grilles, instrument panels, lenses	Thermoplastic
PPO	Polyphenylene oxide	Noryl, Olefo	Chromed plastic parts, grilles, headlamp doors, bezels, ornaments	Thermosetting
PE	Polyethylene	Dylan, Fortiflex, Marlex, Alathon, Hi-fax, Hosalen, Paxon	Inner fender panels, interior trim panels, valances, spoilers	Thermoplastic
PP	Polypropylene	Profax, Olefo, Marlex, Olemer, Aydel, Dypro	Interior moldings, interior trim panels, inner fenders, radiator shrouds, dash panels, bumper covers	Thermoplastic
PS	Polystyrene	Lustrex, Dylene, Styron, Fostacryl, Duraton	—	Thermoplastic
PUR	Polyurethane	Castethane, Bayflex	Bumper covers, front and rear body panels, filler panels	Thermosetting
PVC	Polyvinyl chloride	Geon, Vinylete, Pliovic	Interior trim, soft filler panels	Thermoplastic
RIM	"Reaction injection molded" polyurethane	—	Bumper covers	Thermosetting
R RIM	Reinforced RIM-polyurethane	—	Exterior body panels	Thermosetting
SAN	Styrene-acrylonitrite	Lustran, Tyril, Fostacryl	Interior trim panels	Thermosetting

TABLE 14-1: IDENTIFICATION SYMBOL, CHEMICAL NAME, TRADE NAME, AND DESIGN APPLICATIONS OF MOST COMMONLY USED PLASTICS (CONTINUED)

Symbol	Chemical Name	Common Name	Applications	Thermosetting or Thermoplastic
TPR	Thermoplastic rubber	—	Valance panels	Thermosetting
TPUR	Polyurethane	Pellethane, Estane, Roylar, Texin	Bumper covers, gravel deflectors, filler panels, soft bezels	Thermoplastic
UP	Polyester	SMC, Premi-glas, Selection Vibrin-mat	Fiberglass body panels	Thermosetting

6. Safety glasses are always a necessity.
7. Always work in a well-ventilated area.
8. Wear an approved respirator to avoid inhaling sanding dust and resin vapors.

PLASTICS IDENTIFICATION

There are several ways to identify unknown plastics. One way to identify plastics is by the *international symbols*, or *ISO codes*, that are molded into the parts. Many manufacturers are using these symbols. The symbol or abbreviation is formed in an oval on the backside of the part. One problem is that you usually have to remove the part to read the symbol.

If the part is not identified by a symbol, the body repair manual will give information about plastic types used on the vehicle. Body manuals often name the types of plastic used in a particular application.

The *burn test*, which is no longer recommended, involved using a flame and the resulting smoke to determine the type of plastic. However, the test was not always reliable. Many parts are now being manufactured from composite plastics that use more than one ingredient, and the burn test would be no help in such cases. It is also environmentally unsound to burn plastics, which can produce **carcinogens** (cancer-causing agents).

A reliable means of identifying an unknown plastic is to make a weld rod adhesion test or a trial-and-error weld on a hidden or damaged area of the part. Try several different filler rods until one sticks. Most suppliers offer only a few types of plastic filler rods; the range of possibilities is somewhat limited. The rods are color coded. Once you find a rod that works, the base material is identified (Figure 14–4).

Another way to help identify a plastic part is the **plastic flexibility test**. To do a plastic flexibility test, use your hands to flex and bend the part and compare it to the flexibility of samples of plastic. Use the repair material that most closely matches the characteristics of the part's base material.

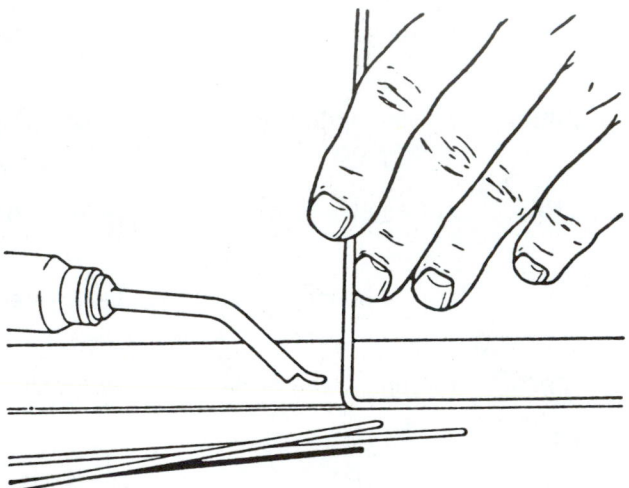

FIGURE 14-4 A trial-and-error weld will help identify the type of plastic and the best repair method.

14.2 PLASTIC REPAIR

Plastic repair, like any other kind of body repair work, begins with the estimation process. It must be determined if the part should be repaired or replaced. A minor crack, tear, gouge, or hole in a nose fascia or large panel that is difficult to replace, costly, or not readily available probably indicates a repair should be considered. Extensive damage to the same component or damage to a fender extension or plastic trim item that is cheap and easy to replace would dictate replacement. In short, it is up to the repair person or estimator to decide if it makes more sense to repair a plastic part than to replace it.

If repair is the answer, it must be determined if the part needs to be removed from the vehicle. The entire damaged area must be accessible to do a quality repair. If it is not accessible, the part must be removed. Keep in mind that the part will also have to be refinished.

TABLE 14-2: HANDLING PRECAUTIONS FOR PLASTICS

Code	Material Name	Heat Resisting Temperature* °F	Resistance To Alcohol or Gasoline	Notes
AAS	Acrylonitrile Acrylic Rubber Styrene Resin	176	Alcohol is harmless if applied only for short time in small amounts (example, quick wiping to remove grease).	Avoid gasoline and organic or aromatic solvents.
ABS	Acrylonitrile Butadiene Styrene Resin	176	Alcohol is harmless if applied only for short time in small amounts (example, quick wiping to remove grease).	Avoid gasoline and organic or aromatic solvents.
AES	Acrylonitrile Ethylene Rubber Styrene Resin	176	Alcohol is harmless if applied only for short time in small amounts (example, quick wiping to remove grease).	Avoid gasoline and organic or aromatic solvents.
EPDM	Ethylene Propylene Rubber	212	Alcohol is harmless. Gasoline is harmless if applied only for short time in small amounts.	Most solvents are harmless, but avoid dipping in gasoline, solvents, etc.
PA	Polyamide (Nylon)	176	Alcohol and gasoline are harmless.	Avoid battery acid.
PC	Polycarbonate	248	Alcohol is harmless.	Avoid gasoline, brake fluid, wax, wax removers, and organic solvents.
PE	Polyethylene	176	Alcohol and gasoline are harmless.	Most solvents are harmless.
POM	Polyoxymethylene (Polyacetal)	212	Alcohol and gasoline are harmless.	Most solvents are harmless.
PP	Polypropylene	176	Alcohol and gasoline are harmless.	Most solvents are harmless.
PPO	Modified Polyphenylene Oxide	212	Alcohol is harmless.	Gasoline is harmless if applied only for quick wiping to remove grease.
PS	Polystyrene	140	Alcohol and gasoline are harmless if applied only for short time in small amounts.	Avoid dipping or immersing in alcohol, gasoline, solvents, etc.
PUR	Thermosetting Polyurethane	176	Alcohol is harmless if applied only for very short time in small amounts (example, quick wiping to remove grease).	Avoid dipping or immersing in alcohol, gasoline, solvents. etc.
PVC	Polyvinylchloride (Vinyl)	176	Alcohol and gasoline are harmless if applied only for short time in small amounts (example, quick wiping to remove grease).	Avoid dipping or immersing in alcohol, gasoline, solvents. etc.
SAN	Styrene Acrylonitrile Resin	176	Alcohol is harmless if applied only for short time in small amounts (example, quick wiping to remove grease).	Avoid dipping or immersing in alcohol, gasoline, solvents, etc.

TABLE 14-2: HANDLING PRECAUTIONS FOR PLASTICS (CONTINUED)

TPO	Thermoplastic Polyolefin	176	Alcohol is harmless. Gasoline is harmless is applied only for short time in small amounts.	Most solvents are harmless, but avoid dipping in gasoline, solvents, etc.
TPUR	Thermoplastic Polyurethane	140	Alcohol is harmless if applied only for very short time in small amounts (example, quick wiping to remove grease).	Avoid dipping or immersing in alcohol, gasoline, solvents, etc.

*Temperature higher than listed here could result in material deformation during repair.
NOTE: When repairing metal body parts that adjoin plastic body parts (by brazing, flame cutting, welding, painting, etc.), consideration **must** be given to the properties of the plastic.
Chart courtesy of Toyota Motor Corporation

As mentioned earlier, there are two methods of repairing plastics:

- Use of chemical adhesives
- By plastic welding

Table 14–3 indicates the best repair systems for the plastics most often used by the automotive industry.

TABLE 14-3: PLASTIC PARTS REPAIR SYSTEMS

KEY

AR	Adhesive repair	S	Anerobic (instant) adhesive
FGR	Fiberglass repair	PC	Patching compound
HAW	Hot-air welding	AW	Airless welding

ISO Code	Name	Repair System
ABS	Acrylonitrile-butadiene-styrene (hard)	HAW, S FGR, AW
ABS/PVC	ABS/Vinyl (soft)	PC, AW
EPI II or TPO	Ethylene propylene	AR, AW
PA	Nylon	S, FGR, AW
PC	Lexan	S, FGR, AW
PE	Polyethylene	HAW, AW
PP	Polypropylene	HAW, AW
PPO	Noryl	FGR, AW
PS	Polystyrene	S
PUR, RIM, or RRIM	Thermoset polyurethane	AR, AW
PVC	Polyvinyl chloride	PC, AW
SAN	Styrene acrylonitrile	HAW, AW
TPR	Thermoplastic rubber	AR, AW
TPUR	Thermoplastic polyurethane	AR, AW
UP	Polyester (Fiberglass)	FGR

14.3 CHEMICAL ADHESIVE BONDING TECHNIQUES

Adhesive repair systems are of two types: cyanoacrylate (CA) and two-part. Two-part is the most commonly used.

Cyanoacrylates, or *CAs*, are one-part, fast-curing adhesives used to help repair rigid and flexible plastics. They are often used as a filler or to tack parts together before applying the final repair material. CAs are sometimes known as "super glues." They can be a valuable tool for the repair of plastic parts. CAs set up very quickly.

Although one-part, an activating agent can be used to accelerate the bonding process of a CA. Care must be used not to apply too much activating agent. If too much is used, the product foams, causing a weaker bond.

CAs do not work equally well on all plastics. There is no hard and fast rule. When CAs are used, be sure to use products from reliable suppliers and follow the manufacturer's guidelines for using them.

Two-part adhesive systems consist of a base resin and a hardener (catalyst). The resin comes in one container and the hardener in another. When mixed, the adhesive cures into a plastic material similar to the base material in the part. Two-part adhesive systems are an acceptable alternative to welding for many plastic repairs. They are also stronger than CAs.

Not all plastics can be welded, while adhesives can be used in all but a few instances. If adhesive repair is chosen, you must first identify the type of plastic. A good way to do this is the plastic flexibility test described earlier.

REPAIRS OF MINOR CUTS AND CRACKS

Adhesives are usually used for repairing minor cuts and cracks in plastic parts (Table 14-4). First, wash

TABLE 14-4: ADHESIVE REPAIR SYSTEM

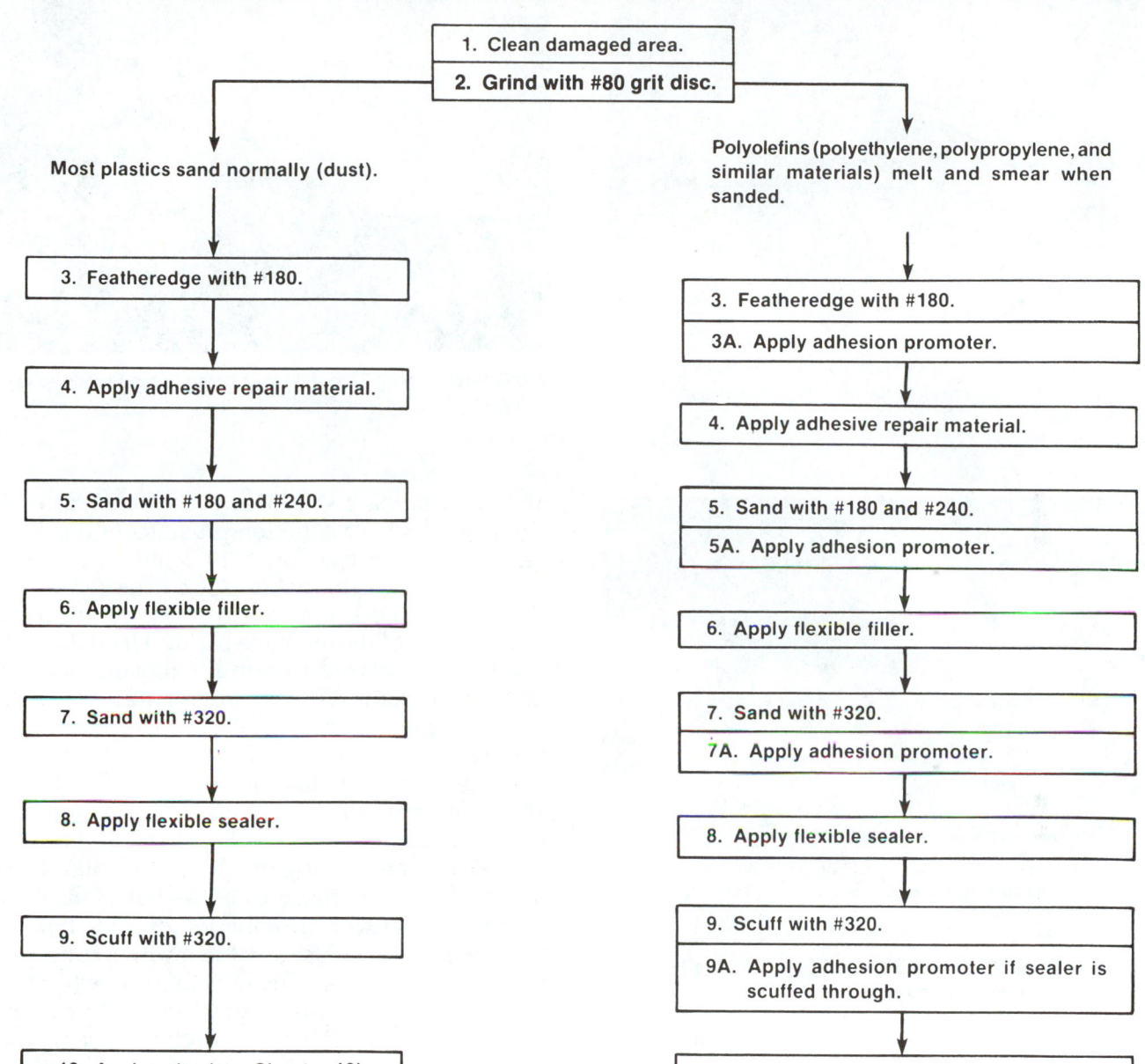

1. Clean damaged area.
2. **Grind with #80 grit disc.**

Most plastics sand normally (dust).

3. Featheredge with #180.

4. Apply adhesive repair material.

5. Sand with #180 and #240.

6. Apply flexible filler.

7. Sand with #320.

8. Apply flexible sealer.

9. Scuff with #320.

10. Apply color (see Chapter 19).

Polyolefins (polyethylene, polypropylene, and similar materials) melt and smear when sanded.

3. Featheredge with #180.
3A. Apply adhesion promoter.

4. Apply adhesive repair material.

5. Sand with #180 and #240.
5A. Apply adhesion promoter.

6. Apply flexible filler.

7. Sand with #320.
7A. Apply adhesion promoter.

8. Apply flexible sealer.

9. Scuff with #320.
9A. Apply adhesion promoter if sealer is scuffed through.

10. Apply color (see Chapter 19).

the area thoroughly with soap and hot water. Second, wipe or wash the repair area clean with water and a plastic cleaner. The surfaces must be free of wax, dust, or grease. Allow the part(s) to warm to 70 degrees Fahrenheit (21 degrees Celsius) before applying adhesives.

After cleaning, prepare the crack with an adhesive kit. The kit should have two elements: an accelerator and an adhesive. Spray one side of the crack with the accelerator, as shown in Figure 14–5. Then apply the adhesive to the same side of the crack.

Carefully position the two sides of the cut or crack in their original position, and quickly press them together with firm pressure. Hold for a full minute to achieve good bond strength. Then allow the repair to cure for 3 to 12 hours for maximum strength, or according to the instructions on the label. Note the precautions and instructions on the container for the adhesive used.

If the original paint was not damaged and the repair was properly positioned, painting may not be required. Where painting is required, special procedures are needed as described in Chapter 20.

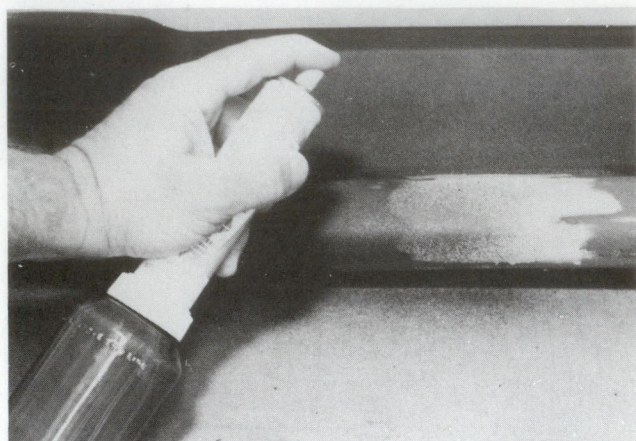

FIGURE 14-5 Spray the crack with accelerator to prepare it for the adhesive. *(Courtesy of Urethane Supply Company, Inc.)*

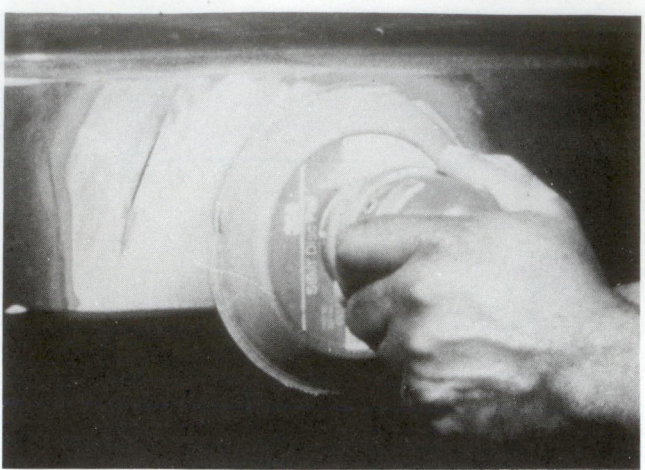

FIGURE 14-7 Use a finer grit disc to featheredge the paint.

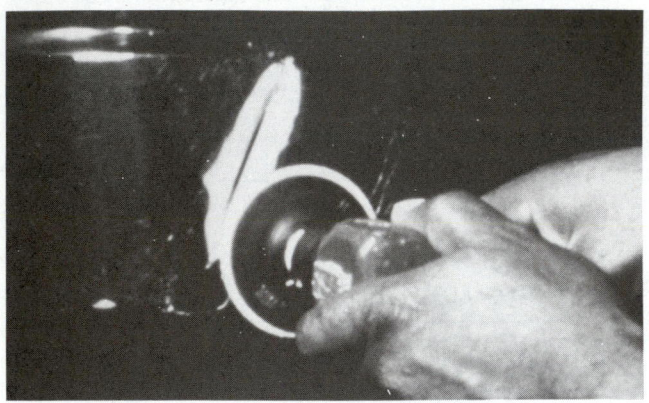

FIGURE 14-6 Bevel the damaged area with a 3-inch grinding disc to strengthen the repair.

REPAIRS OF GOUGES, TEARS, AND PUNCTURES

The procedure for repairs of gouges, tears, and punctures is somewhat more involved than the previous one but requires no special skills or tools.

First, wash the area thoroughly with soap and hot water. Then clean around the damaged area thoroughly with a wax-, grease-, and silicone-removing solvent applied with a water-dampened cloth. Then wipe the area dry.

To prepare for the structural adhesive, bevel the edges of the hole back about $1/4$ to $3/8$ inch (6.4 to 9.5 mm). The technician in Figure 14–6 is using a small grinding disc, with medium grit. Use a slow speed when grinding (2,000 rpm or less). In this repair, the beveling has left a coarse surface for good adhesion. In any repair, the mating surfaces should be scuffed to improve adhesion.

If the sanded area has a greasy appearance or smears when sanded, apply a coat of adhesion promoter to the surface. Apply more adhesion promoter

after each sanding step. However, some plastic repair materials have adhesion promoter in them. Refer to material instructions if in doubt.

Featheredge the paint around the repair area (Figure 14–7). Use a finer grit disc. Remove the paint, but very little of the urethane plastic. Blend the paint edges into the plastic. Continue removing paint until there is a paint-free band around the hole—about 1 to $1^1/2$ inches (25 to 38 mm) wide. The repair material must not overlap the painted surface.

Carefully wipe off all paint and urethane dust. The repair area must be absolutely clean for proper bond strength.

If recommended by the product manufacturer, flame treat the beveled area of the hole. This flame treatment improves the adhesion of some types of structural adhesive. Use any torch with a controlled flame, and develop a 1-inch (25 mm) cone tip. Very carefully direct the flame onto the beveled area. Keep moving the flame all the time until the area is properly heated. Be extremely careful to accomplish this without warping the urethane and without burning the paint.

The next step is to apply auto backing tape to the repair area. An aluminum foil with a strong adhesive on one side and a moisture-proof backing is recommended. Clean the inner surface of the repair area with silicone and wax remover. Then install the tape. Cover the hole completely, with about a 1-inch (25 mm) adhesion surface around the edges.

Then, before applying the structural repair adhesive material, the back of the opening should be thoroughly cleaned and taped with aluminum auto body repair tape to provide support for the repair. For best results slightly dish the aluminum tape so that the repair materials overlap the repaired area on the backside. It is often possible to install the tape without loosening or removing any parts from the

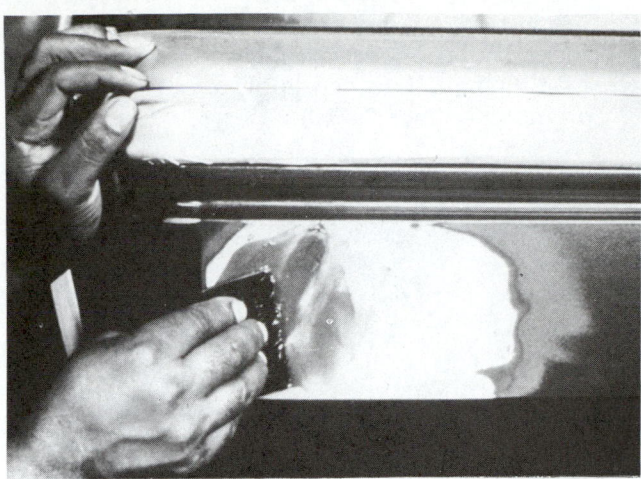

FIGURE 14-8 Apply structural adhesive with a squeegee to match the shape of the part.

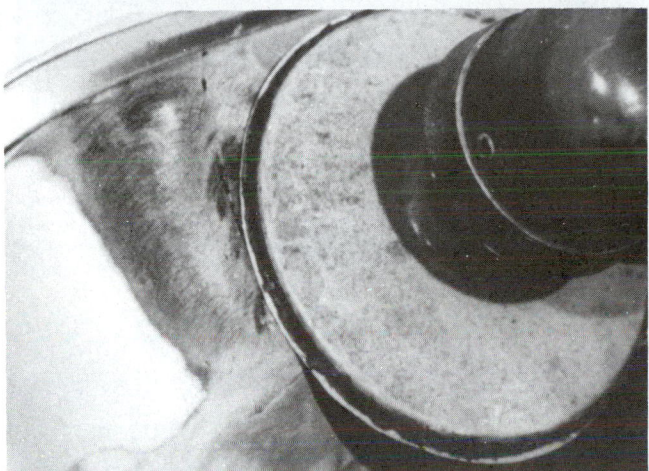

FIGURE 14-9 Grind down high spots on the adhesive as you would with conventional filler.

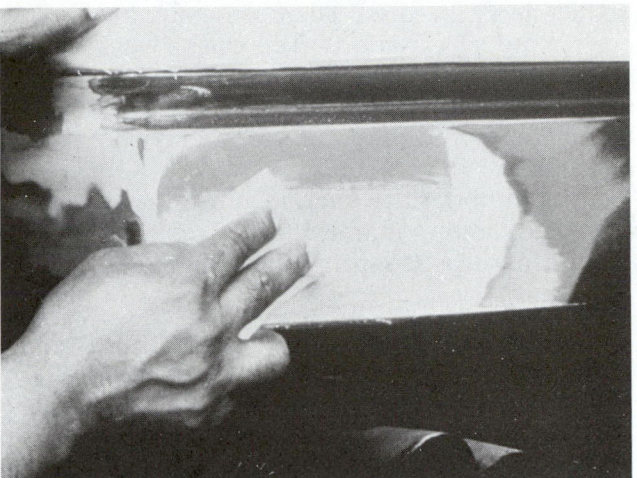

FIGURE 14-10 Apply a second coat of adhesive if needed.

car. For repairs in enclosed areas, partial disassembly may sometimes be necessary.

Some technicians like to back the repair with cloth rather than tape. The cloth will remain in place and reinforce the repair. Saturate the entire cloth on both sides with the adhesive. This will make the cloth bond to the back of the plastic properly and also seal the cloth.

Prepare the repair adhesive material on a clean, flat, nonporous surface such as metal or glass, as directed by the manufacturer. Most adhesive compounds come in two tubes. Squeeze out equal amounts of the repair mix. Then, with an even paddling motion to reduce air bubbles, completely mix the two components until a uniform color and consistency is achieved.

Paddle the structural adhesive into the hole, using a squeegee or plastic spreader (Figure 14-8). This must be done carefully and swiftly, as the structural adhesive material will begin to set in about 2 to 3 minutes. Two applications of the adhesive are usually required. The first application is used to fill the bottom of the hole. It is not necessary to worry about contour at this time. In the first application of patch material, try to fill the greater part of the hole's volume. Then cure for about an hour at room temperature, or 20 minutes with a heat lamp or gun at 190 to 200 degrees Fahrenheit (88 to 93 degrees Celsius) if the manufacturer's directions allow heat curing.

Before the final application of the adhesive, use a fine grit disc to grind down the high spots of the first application (Figure 14-9). Wipe the dust from the repair area.

After the first application is ground and wiped clean, mix the second application of the adhesive, squeezing the components together as before for about 2 minutes. Then apply the second adhesive mixture, paddling it into an overfill contour of the area (Figure 14-10). A flexible squeegee or spatula is useful in approximating the panel contours.

When the adhesive repair material has dried, establish a rough contour to the surrounding area with a #80 grit abrasive on a sanding block. You can also feather-sand the area using a disc sander with a #180 sandpaper followed by a #240 sandpaper to achieve an accurate level with the surface of the part. Check the repaired surface to see if there are any low areas, pits, or pinholes in it. If there are, additional material can be spread over them to fill the holes and raise the area as needed.

Final feathering and finish sanding can be done with a disc sander and a #320 grit disc. When the final sanding is completed, the area is cleaned to remove all dust and loose material. The plastic surface is then ready to be painted.

USE THE RIGHT ADHESIVE

When working with an adhesive system, use the manufacturer's categories to decide on a repair product and procedure. There are many plastic and repair material variations. The car or truck manufacturer's manual is the most accurate source of information. The service manual will recommend products and procedures for the exact type of plastic in the part.

It is important to keep in mind that there are differences among manufacturers' repair materials. When using plastic repair adhesives, remember the following:

1. Mixing product lines is not acceptable. Choose a product line and use it for the entire repair.
2. Most product lines have two or more adhesives designed for different types of plastic.
3. The product line usually includes an adhesion promoter, a filler product, and a flexible coating agent. Use each as directed.
4. Some product lines are formulated for a specific base material. For example, one manufacturer offers individual products for use on each type of plastic (TPO, urethanes, or Xenoy, for example), regardless of plastic flexibility.

A product line might use a single flexible filler for all plastics, or there may be two or more flexible fillers designed for different types of plastic.

An **adhesion promoter** is a chemical that treats the surface of the plastic so the repair material will bond properly. Some plastics (TPO, PP, and E/P) require an adhesion promoter. There is a simple test to perform that indicates whether or not the plastic will require an adhesion promoter. Lightly sand a hidden spot on the piece using a high-speed grinder and #36 grit sandpaper.

If the material gives off dust, it can be repaired with a standard structural adhesive system. If the material melts and smears or has a greasy or waxy look, then you must use an adhesion promoter. Many plastic fillers and adhesives contain an adhesion promoter. Check their labels.

FLEXIBLE PART REPAIR

Here is a typical way to use a two-part epoxy adhesive to repair a flexible part, a bumper cover in this example:

1. Clean the entire cover with soap and hot water. Wipe or blow-dry. Then clean the surface with a good plastic cleaner.
2. V-groove the damaged area. Then grind about a $1\frac{1}{2}$-inch (38 mm) taper around the damage for good adhesion and repair strength (Figure 14–11).
3. Use a sander with #180 grit sandpaper to feather-edge the paint around the damaged area. Then blow off the dust. Depending on the extent of the damage, the backside may need reinforcement. To do this, use steps 4 through 6.

4. To reinforce the repair area, sand and clean the backside of the cover with plastic cleaner (Figure 14–12). Then, if needed, apply a coat of adhesion promoter (Figure 14–13).

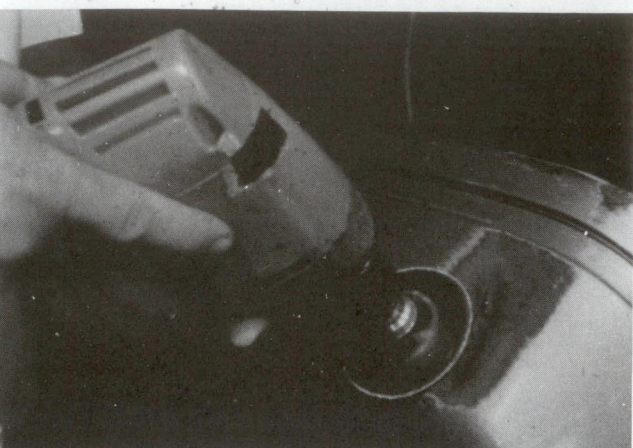

FIGURE 14–11. Grind and scuff around the damage for good adhesion.

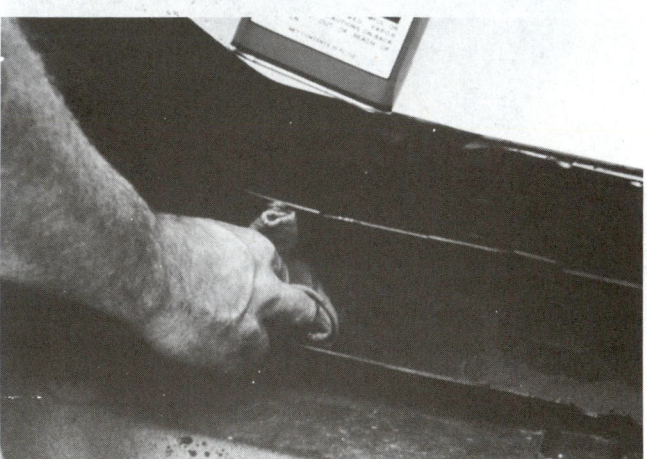

FIGURE 14–12 If needed, clean the back side with a plastic cleaner.

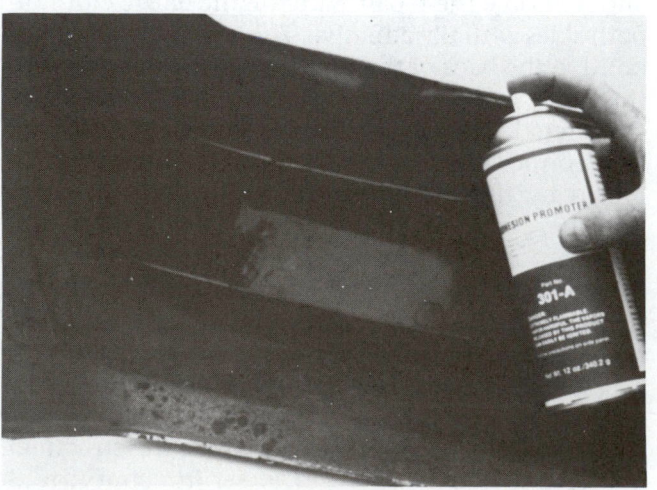

FIGURE 14–13 Apply the adhesion promoter to the back side.

5. Dispense equal amounts of both parts of the flexible epoxy adhesive. Mix them to a uniform color. Apply the material to a piece of fiberglass cloth using a plastic squeegee (Figure 14–14).

6. Attach the plastic-saturated cloth to the backside of the bumper cover. Fill in the weave with additional adhesive material.

7. With the backside reinforcement in place, apply a coat of adhesion promoter to the sanded repair area on the front side. Let the adhesion promoter dry completely (Figures 14–15 and 14–16).

8. Fill in the area with adhesive material. Shape the adhesive with your spreader to match the shape of the part (Figure 14–17). Allow it to cure properly.

9. Rough grind the repair area with #80 grit sandpaper, then sand with #180 grit, followed by smoother #240 grit.

10. If additional adhesive material is needed to fill in a low spot or pinholes, be sure to apply a coat of adhesion promoter again.

TWO-PART ADHESIVE DASH REPAIR

Some two-part adhesive products are made for dash pad repair. A typical repair procedure is as follows:

1. Thoroughly clean the part.
2. Sand or grind away the broken or loose vinyl covering to get to the foam beneath. A V-groove need not be used.
3. Apply the adhesive material according to the manufacturer's directions. Make sure the recommended cure times are followed.
4. After curing, sand or grind to contour. Apply a skim coat of adhesive if necessary. Allow the adhesive to cure. Then block sand with medium grit sandpaper, followed by fine sandpaper.
5. Apply a sealer. Then spray retexture as much area as needed to hide the repair.
6. Refinish the whole part according to paint system recommendations.

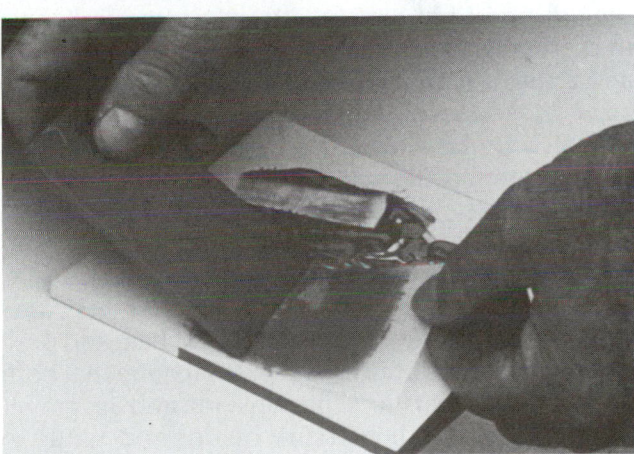

FIGURE 14–14 Apply the flexible epoxy adhesive to a piece of fiberglass cloth. The cloth should be cut to size to fit the repair area.

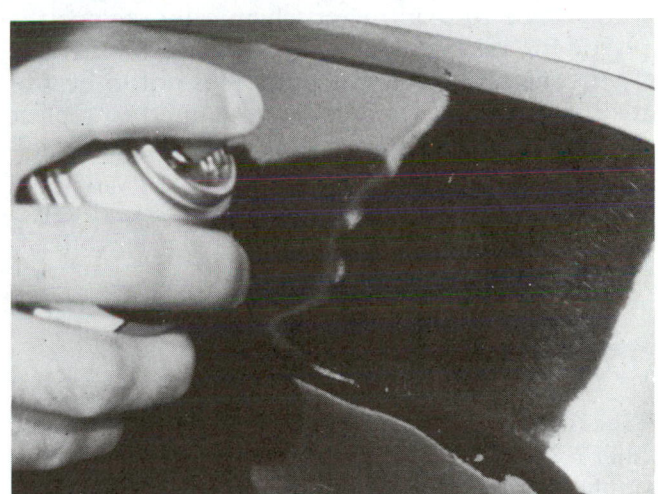

FIGURE 14–16 Apply the adhesion promoter to the front side.

FIGURE 14–15 Finish the back side reinforcement before filling the front.

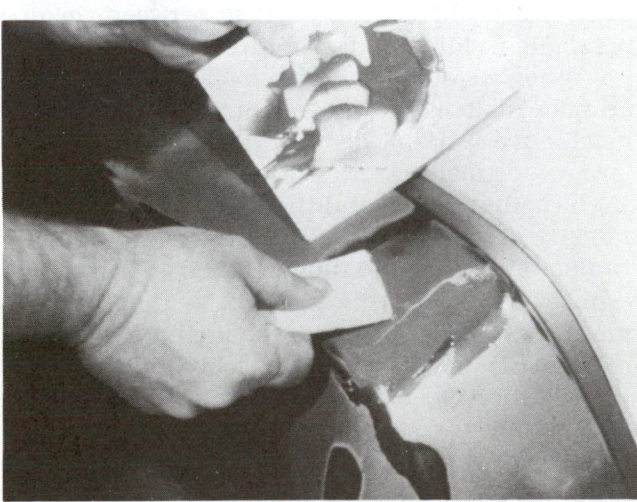

FIGURE 14–17 Fill in the groove with adhesive material.

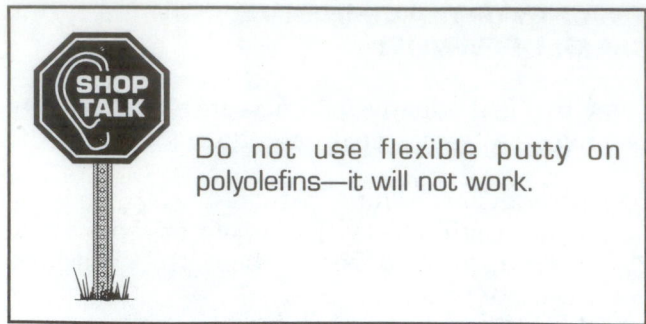

Do not use flexible putty on polyolefins—it will not work.

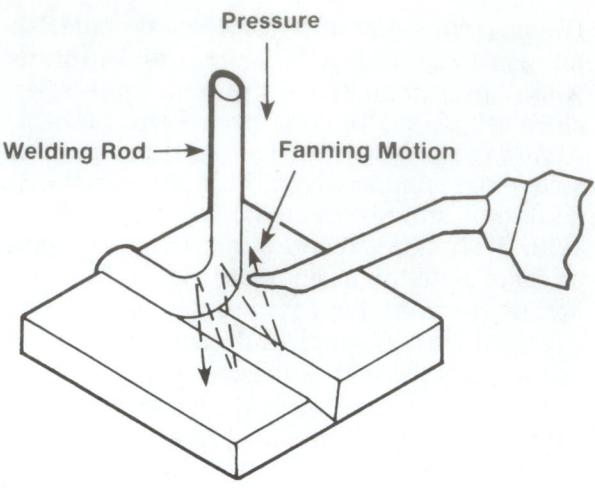

FIGURE 14–18 Successful plastic welding requires the proper combination of heat and pressure.

Be aware that some dash pad damage might be the result of age and exposure to sunlight. In these cases, repairs will not work. To check for this condition, press on the pad to flex the vinyl covering near the damage. If more cracks appear, it is beyond repair and should be replaced.

PLASTIC FLAME TREATING

Mentioned briefly, **plastic flame treating** uses a torch flame to oxidize and chemically prepare the plastic for adhesion. It is sometimes recommended by the manufacturer. The purpose is not to heat or melt the surface. Flame treating is used to replace the action of adhesion promoters. Immediately following flame treatment, complete the repair using methods described earlier for a two-part adhesive. Do not touch and contaminate the plastic while using the adhesive.

14.4 PLASTIC WELDING

Plastic welding uses heat and sometimes a plastic filler rod to join or repair plastic parts. The welding of plastics is not unlike the welding of metals. Both methods use a heat source, welding rod, and similar techniques (butt joints, lap joints, etc.). Joints are prepared in much the same manner, and evaluated for strength. There are differences between welding metal and welding plastics, however.

When welding metal, the rod and base material are made molten and puddled into a joint. And while metals have a sharply defined melting point, plastics have a wide melting range between the temperature at which they soften and the temperature at which they char or burn. Also, unlike metals, plastics are poor conductors of heat and thus are difficult to heat uniformly. The decomposition time at welding temperature is shorter than the time required to completely soften many plastics for fusion welding. The result is that a plastic welder must work within a much smaller temperature range than a metal welder.

Because a plastic welding rod does not become completely molten and appears much the same before and after welding, a plastic weld might appear incomplete to the technician who is used to welding metal. The explanation is simple; since only the outer surface of the rod has become molten while the inner core has remained hard, the welder is able to exert pressure on the rod to force it into the joint and create a permanent bond. When heat is taken away, the rod reverts to its original form. Thus, even though a strong and permanent bond has been obtained between the rod and base material, the appearance of the rod is much the same as before the weld was made, except for molten flow patterns on either side of the bead.

When welding plastics, the materials are fused together by the proper combination of heat and pressure (Figure 14–18). Successful welds require that both pressure and heat be kept constant and in proper balance. Too much pressure on the rod tends to stretch the bead; too much heat will char, melt, or distort the plastic. With practice, plastic welding can be mastered as completely as metal welding.

14.5 HOT-AIR PLASTIC WELDING

Hot-air plastic welding uses a tool with an electric heating element to produce hot air (450 to 650 degrees Fahrenheit or 232 to 345 degrees Celsius) that blows through a nozzle and onto the plastic (Figure 14–19). The air supply comes from either the shop's air compressor or a self-contained portable compressor that comes with the welding unit (Figure 14–20).

Most hot-air welders use a tip working pressure of around 3 psi (21 kPa). Air pressure regulators reduce the air pressure first to around 50 psi (345 kPa), and then to the working pressure of about 3 psi (21 kPa).

The torch is used in conjunction with the welding rod, which is normally $3/16$ inch (5 mm) in diameter. The plastic welding rod must be made of the same material as the plastic being repaired. This will ensure proper strength, hardness, and flexibility of the repair.

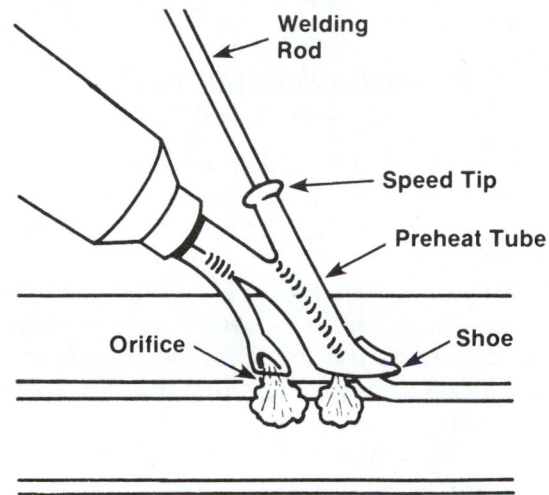

FIGURE 14–19 This plastic welder is equipped with a high-speed tip. [Courtesy of I-CAR]

One of the problems with hot-air welding is that the plastic welding rod is often thicker than the panel to be welded. This can cause the panel to overheat before the rod has melted. Using a smaller diameter rod can often correct such warpage problems.

Three types of welding tips are available for use with most hot-air plastic welding torches:

Tacking tips are shaped to tack weld broken sections of plastic together before welding (Figure 14–21). If necessary, tack welds can be easily pulled apart for realigning.

Round tips are used to make short welds to weld small holes, to weld in hard-to-reach places, and to weld sharp corners (Figure 14–22).

Speed tips hold, feed, and automatically preheat the plastic welding rod (Figure 14–23). This design feeds the rod into the base material, thus allowing for faster welding speeds. They are used for long, fairly straight welds.

Some hot-air welder manufacturers have developed specialized welding tips and rods to meet specific needs. Check the product catalog for more information.

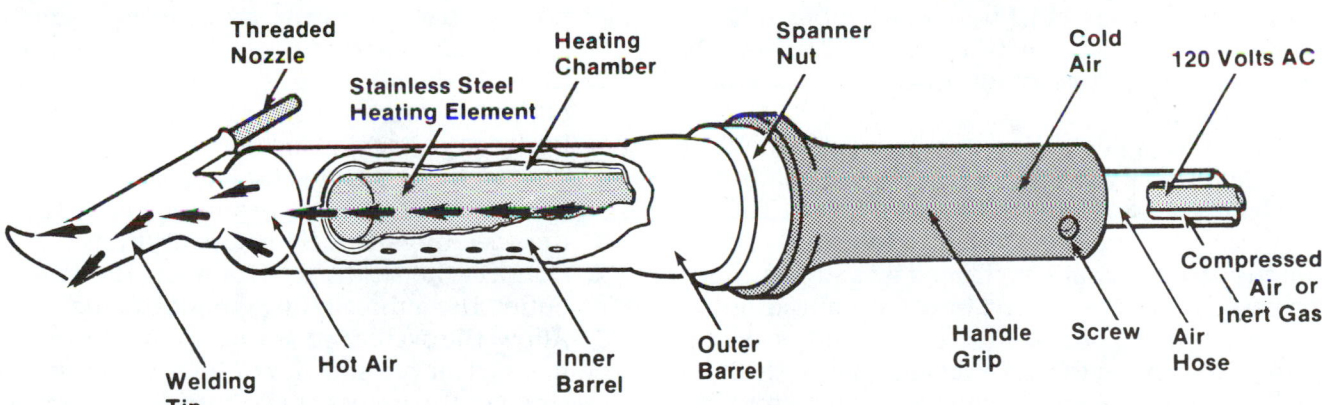

FIGURE 14–20 Study the parts of a typical hot-air welder.

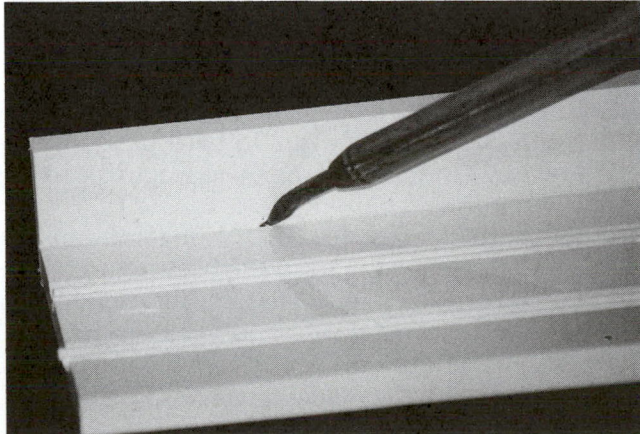

FIGURE 14–21 A tacking welding tip can be used to make small welds to hold parts before final welding. [Courtesy of Seelye Inc.]

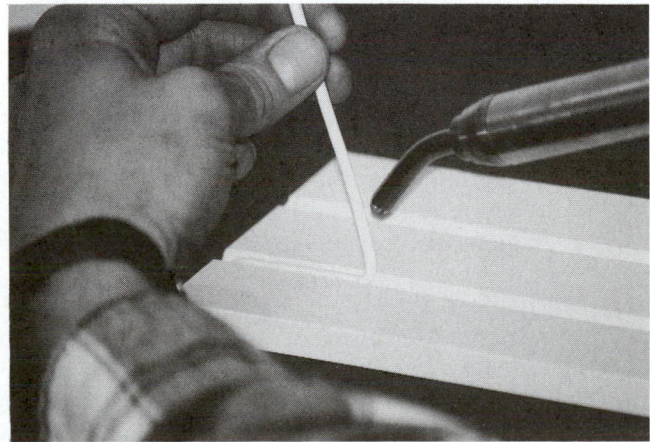

FIGURE 14–22 Round tips are used for short welds, small holes, hard-to-reach places, and sharp corners. [Courtesy of Seelye Inc.]

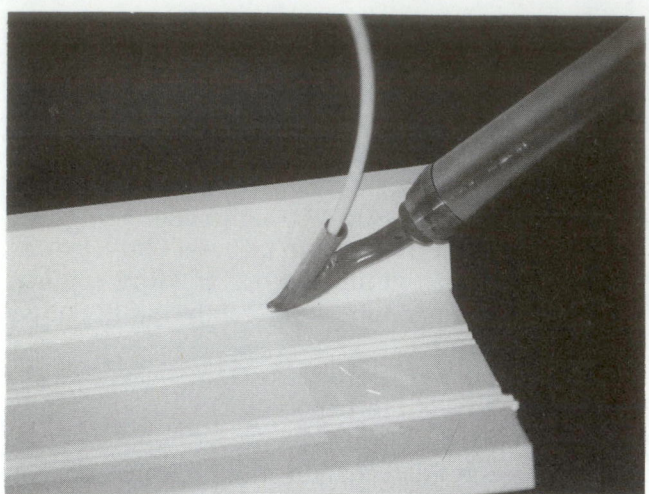

FIGURE 14-23 A speed welding tip will work fastest when space allows. *(Courtesy of Seelye Inc.)*

HIGH-SPEED WELDING

High-speed welding incorporates the basic methods utilized in hand welding except a specially designed and patented high-speed tip is used (Figure 14–19), which enables the welder to produce more uniform welds and work at a much higher rate of speed. As with hand welding, constant heat and pressure must be maintained.

The increased efficiency of high-speed welding is made possible through preheating of both the rod and base material before the point of fusion. The rod is preheated as it passes through a tube in the speed tip; the base material is preheated by a stream of hot air passing through a vent in the tip ahead of the fusion point. A pointed shoe on the end of the tip applies pressure on the rod, thus eliminating the need for the operator to apply pressure. At the same time the shoe smooths out the rod, creating a more uniform appearance in the finished weld.

In high-speed welding, the conventional two-hand method is replaced by a faster and more uniform one-hand operation. Once started, the rod is fed automatically into the preheating tube as the welding torch is pulled along the joint. High-speed tips are designed to provide the constant balance of heat and pressure necessary for a satisfactory weld. The average welding speed is about 40 inches (1,000 mm) per minute.

High-speed welding does have its advantages. Because increased speeds must be maintained to achieve the best possible weld, the high-speed welding torch is not suited for small, intricate work. Also, when the operator is new to this technique, the position in which the welder is held might seem clumsy and difficult. However, experience will en-

able the operator to successfully make all welds that can be made with a hand welder, including butt welds, V-welds, corner welds, and lap joint welds. Speed welds can be made on circular as well as flat work. In addition, inside welds on tanks can be speed welded, provided the working space is large enough to manipulate the torch.

PLASTIC WELDER SETUP, SHUTDOWN, AND SERVICING

No two hot-air plastic welders are exactly alike. For specific instructions, always refer to the owner's manual and other material provided by the welder manufacturer.

Some manufacturers advise against using their welder on plastic thinner than $1/8$ inch (3 mm) because of distortion. It is sometimes acceptable to weld thin plastics if they are supported from underneath while welding.

To set up a typical hot-air welder, proceed as follows:

1. Close the pressure regulator valve by turning it counterclockwise until loose. This will prevent possible damage to the gauge from a sudden surge of air pressure.
2. Connect the air pressure regulator to a supply of either compressed air or inert gas. If inert gas is used, a pressure-reducing valve is needed.
3. Turn on the air supply. Starting pressure depends upon the wattage of the heating element. Check the operating manual for specifications.
4. Connect the welder to a common 120-volt AC outlet. Use a three-prong grounded plug.
5. Allow the welder to warm up at the recommended air pressure. Air or inert gas must flow through the welder at all times, from warm-up to cool-down. This will prevent burnout of the heating element and damage to the gun.
6. Select the proper tip. While wearing work gloves insert the tip into the torch with pliers to avoid touching the barrel while hot.
7. After the tip has been installed, the temperature will increase slightly due to back pressure. Allow 2 to 3 minutes for the tip to reach operating temperature.
8. Check the temperature by holding a thermometer $1/4$ inch (6 mm) from the hot air end of the torch. For most thermoplastics, the temperature should be in the 450 to 650 degrees Fahrenheit (232 to 343 degrees Celsius) range. Information supplied with the welder usually includes a chart of welding temperatures.
9. If the temperature is too high to weld the material, increase the air pressure slightly until the

temperature goes down. If the temperature is too low, decrease the air pressure slightly until the temperature rises. When increasing and decreasing the air pressure, allow at least 1 to 3 minutes for the temperature to stabilize.

10. Remember, the element can become overheated by too little air pressure. When decreasing air pressure, never allow the round nut that holds the barrel to the handle of the welder to become too hot to the touch. This indicates overheating.

11. A partially clogged dirt screen in the regulator can also cause overheating. Watch for these symptoms.

12. When you are done welding, disconnect the electrical cord. However, allow the air to flow through the welder for a few minutes or until the barrel is cool to the touch. Then disconnect the air supply.

14.6 AIRLESS PLASTIC WELDING

Airless plastic welding uses an electric heating element to melt a smaller $1/8$-inch (3 mm) diameter rod with no external air supply. Airless welding with a smaller rod helps eliminate two troublesome problems: panel warpage and excess rod buildup (Figure 14–24).

When setting up an airless welder, set the temperature dial at the appropriate setting. This will depend upon the specific plastic being worked on. It will normally take about 3 minutes for the welder to fully warm up.

Make sure the rod is the same material as the damaged plastic or the weld will be unsuccessful. Many

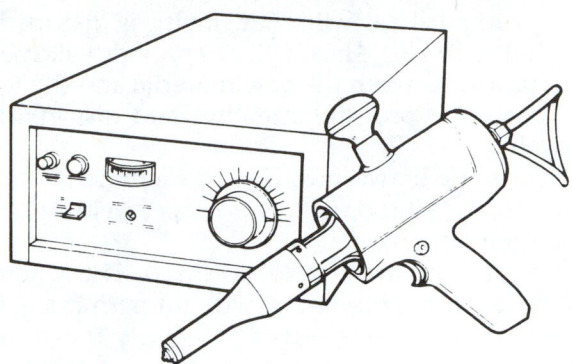

FIGURE 14-25 Here is a typical hand-held ultrasonic welder and power supply.

airless welder manufacturers provide rod application charts. When the correct rod has been chosen, it is good practice to run a small piece through the welder to clean out the tip before beginning.

14.7 ULTRASONIC PLASTIC WELDING

Ultrasonic plastic welding relies on high-frequency vibratory energy to produce plastic bonding without melting the base material. Hand-held systems are available in 20 and 40 kHz frequencies. They are equally adept at welding large parts and tight, hard-to-reach areas. Welding time is controlled by the power supply (Figure 14–25).

Most commonly used injection-molded plastics can be ultrasonically welded without the use of solvents, heat, or adhesives. Ultrasonic weldability depends on the plastic's melting temperature, elasticity, impact resistance, coefficient of friction, and thermal conductivity.

Ultrasonic welding is seldom used in the body shop. However, you should know that this process exists and is used occasionally.

14.8 PLASTIC WELDING PROCEDURES

The basic methods for hot-air and airless welding are very similar. To make a good plastic weld with either procedure, keep the following factors in mind:

1. Plastic welding rods are frequently *color coded* to indicate their material. Unfortunately, the coding is not uniform among manufacturers. It is important to use the reference information provided. If the rod is not compatible with the base material, the weld will not hold.

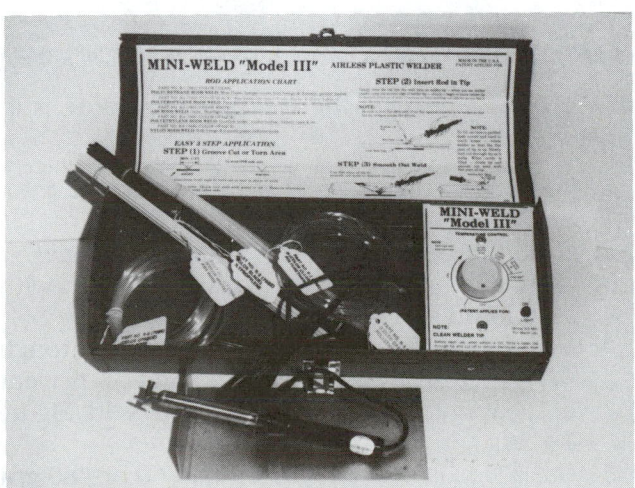

FIGURE 14-24 This is a typical airless welder.

2. Too much heat will char, melt, or distort the plastic. Too little heat will not provide weld penetration between the base material and the rod.
3. Too much pressure stretches and distorts the weld.
4. The angle between rod and base material must be correct. If too shallow, a proper weld will not be achieved.
5. Use the a correct welding speed. If the torch movement is too fast, it will not permit a good weld. If the tool is moved too slowly, it can char the plastic.

The basic repair sequence is generally the same for both plastic welding processes:

1. Prepare the damaged area.
2. Align the damaged area.
3. Make the weld.
4. Allow it to cool.
5. Sand. If the repair area has pinholes or voids, bevel the edges of the defective area. Add another weld bead and resand.
6. Apply a topcoat.

GENERAL WELDING TECHNIQUES

Welding plastic is not difficult when done in a careful and thorough manner. The following guidelines cannot be stressed enough:

- The welding rod must be compatible with the base material in order for the strength, hardness, and flexibility of the repair to be the same as the part.
- Always test a welding rod for compatibility with the base material. To do this, melt the rod onto a hidden side of the damaged part, let the rod cool, then try to pull it from the part. If the rod is compatible, it will adhere.
- Pay close attention to the temperature setting of the welder; it must be correct for the type of plastic being welded.
- Never use oxygen or other flammable gases with a plastic welder.
- Never use a plastic welder, heat gun, or similar tool in wet or damp areas. Remember: electric shock can kill.
- Become proficient at horizontal welds before attempting the more difficult vertical and overhead types.
- Make welds as large as they have to be. The greater the surface area of a weld, the stronger the bond.
- Before beginning an airless weld, run a small piece of the welding rod through the welder to clean out the torch tip.

- Consult a supplier for the brands of tools and materials that best fit the shop's needs. Always read and follow the manufacturer's instructions carefully.

HOT-AIR WELDING PROCEDURE

The typical hot-air plastic welding procedure is as follows:

1. Set the welder to the proper temperature (if a temperature adjustment is provided).
2. Wash and clean the part with soap and hot water. Allow it to dry. Then clean the part with a plastic cleaner. Do not use conventional prep solvents or wax and grease removers. To remove silicone materials, use a conventional cleaner first, making sure to completely remove all residue.
3. V-groove the damaged area.
4. Bevel the part $1/4$ inch (6 mm) on each side of the damaged area.
5. Tack weld or tape the break line with aluminum body tape.
6. Select the welding rod and welding tip best suited to the type of plastic and the damage.
7. Make the weld. The plastic weld should penetrate 75 percent through the base material. Allow it to cool and cure for about 30 minutes.
8. Grind or sand the weld to the proper contour and shape.

When welding plastic, single- or double-V butt welds produce the strongest joints. When using a round or V-shaped welding rod, prepare the area by slowly grinding, sanding, or shaving the adjoining surfaces to produce a single- or double-V. Wipe any dust or shavings from the joint with a clean, dry rag. Do not use cleaning solvents because they can soften the plastic edges and cause poor welds.

AIRLESS WELDING PROCEDURE

The typical airless plastic welding procedure is as follows:

1. Wash the damaged part with soap and hot water and wipe or blow-dry.
2. Clean the damaged part with plastic cleaner.
3. Align and tape the broken or split sections with aluminum body tape.
4. V-groove at least 50 percent of the way through the panel for a two-sided weld and 75 percent of the way through for a one-sided weld (Figure 14–26).
5. Use a slow speed grinder and a #60 or #80 grit disc to remove the paint from around the damaged area. Blow dust free.

6. After setting the welder to the proper temperature, slowly feed the rod into the melt tube.

7. Apply light pressure to the rod to slowly force it out into the grooved area.

8. As the rod melts, start to move the torch tip very slowly in the direction of the intended weld. Overlap the edges of the groove with melted plastic while progressing forward.

9. After completing the weld, use the flat "shoe" part of the torch tip to smooth it out.

When welding plastic, single- or double-V butt welds (Figure 14–27) produce the strongest joints; lap fillet welds are also good. When using a round- or V-shaped filler rod, the damaged area is prepared by slowly grinding, sanding, or shaving the adjoining surfaces with a sharp knife to produce a single- or double-V. For flat ribbon filler rods, V-grooving is not necessary. Wipe any dust or shavings from the joint with a clean, dry rag. The use of cleaning solvents is not generally recommended because they can soften the edges and cause poor welds.

TACK WELDING

On long tears where backup is difficult, small **tack welds** can be made to hold the two sides in place before doing the permanent weld (Figure 14–28). For larger areas, a patch can be made from a piece of plastic and tacked in place.

To tack weld, proceed as follows:

1. Hold the damaged area in alignment with clamps or aluminum body tape.

2. Using a tacking welding tip, fuse the two sides to form a thin hinge weld along the root of the crack. This is especially useful for long cracks because it allows for easy adjustment and alignment of the edges.

3. Start tacking by drawing the point of the welding tip along the joint. Press the tip in firmly, making sure to contact both sides of the crack. Draw the tip smoothly and evenly along the line of the crack. No welding rod is used when tacking.

4. The point of the tip will fuse both sides in a thin line at the root of the crack. The fused parts will hold the sides in alignment. Then you can fuse the entire length of the crack.

HAND WELDING PROCEDURE

Hand welding involves several different stages, each of which must be mastered for a quality job to be performed.

Starting Hand Weld

Prepare the rod for welding by cutting the end at approximately a 60-degree angle. When starting a weld, the tip of the welder should be held about $1/4$ to $1/2$ inch above and parallel to the base material. The filler rod is held at a right angle to the work as shown in Figure 14–29, with the cut end of the rod positioned at the beginning of the weld.

Direct the hot air from the tip alternately at the rod and the base, but concentrating more on the rod. Always keep the filler rod in line with the V while

One-sided Weld   **Two-sided Weld**

FIGURE 14-26 Proper V-grooves are critical to weld strength.

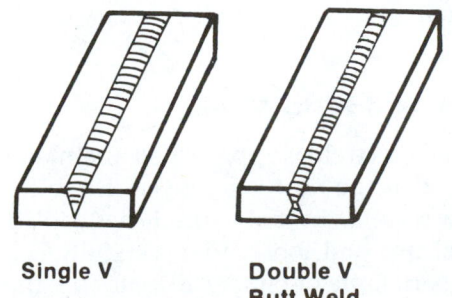

Single V   **Double V Butt Weld**

FIGURE 14-27 Single- and double-V butt welds produce strong joints.

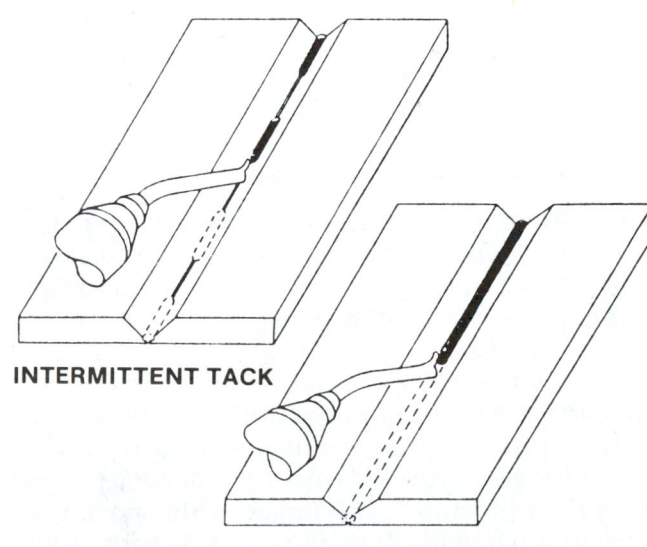

INTERMITTENT TACK

CONTINUOUS TACK

FIGURE 14-28 Two methods of making a tack weld are intermittent and shallow continuous.

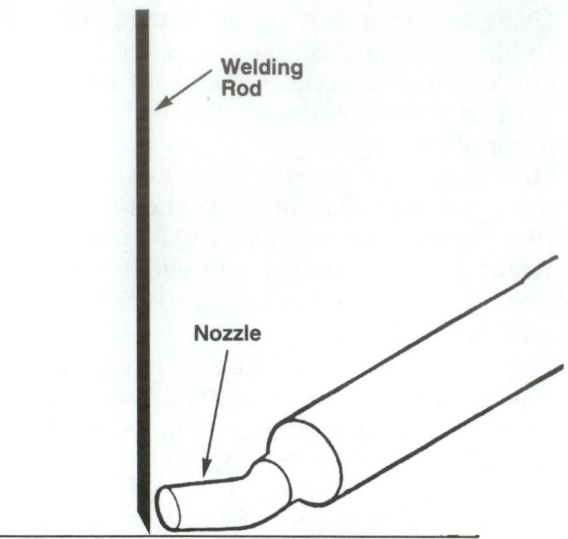

FIGURE 14-29 Keep the nozzle parallel to the base material and the rod at a right angle to the surface.

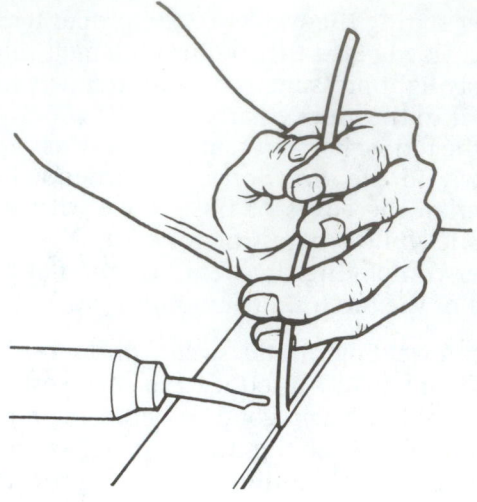

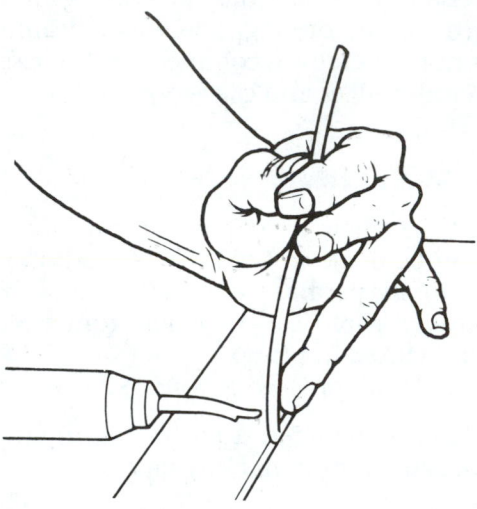

FIGURE 14-30 Continuously apply pressure on the rod while repositioning your fingers. This can be done by applying pressure with the third and fourth fingers while moving the thumb and first finger up the rod. Another way is to hold down the rod in the weld with the third and fourth finger while repositioning the thumb and first finger.

pressing it into the seam. Light pressure (about 3 psi) is sufficient for achieving a good bond. Once the rod begins to stick to the plastic, start to move the torch and use the heat to control the flow. Be careful not to melt or char the base plastic or to overheat the rod. As the welding continues, a small bead should form ahead of the rod along the entire weld joint. A good start is essential because this is where most weld failures begin. For this reason, starting points on multiple-bead welds should be staggered whenever possible.

Continuing Hand Weld

Once the weld has been started, the torch should continue to fan from rod to base material. But because the rod now has less bulk, a greater amount of heat must be directed at the base material. Experience will help develop the proper technique.

Hand Feeding the Rod

Throughout the welding process the rod is gradually being used up, making it necessary for the welder to renew his or her grip on the rod. Unless this is done carefully, the release of pressure might cause the rod to lift away from the weld and allow air to become trapped under the weld and weaken it. To prevent this, the welder must develop the skill of continuously applying pressure on the rod while repositioning the fingers. This can be done by applying pressure with the third and fourth fingers while moving the thumb and first finger up the rod. Another way is to hold down the rod in the weld with the third and fourth fingers while repositioning the thumb and first finger. The rod is cool enough to do this because only

the bottom of it is heated. However, care should be observed in touching new welds or aiming the torch near the fingers. Both methods of grip repositioning are shown in Figure 14–30.

Finishing Hand Weld

As the end of a weld is approached, maintain pressure on the rod as the heat is removed. Hold the rod still for a few seconds to make sure it has cooled enough so it will not pull loose, then carefully cut the rod with a sharp knife or clippers. Do not attempt to pull the rod from the joint. About 15 minutes cooling time is needed for rigid plastic and 30 minutes for thermoplastic polyurethane.

TABLE 14-5: PLASTIC WELDING TROUBLESHOOTING GUIDE

Problem	Cause	Remedy
Porous weld	1. Porous weld rod 2. Balance of heat on rod 3. Welding too fast 4. Rod too large 5. Improper starts or stops 6. Improper crossing of beads 7. Stretching rod	1. Inspect rod. 2. Use proper fanning motion. 3. Check welding temperature. 4. Weld beads in proper sequence. 5. Cut rod at angle, but cool before releasing. 6. Stagger starts and overlap splices 1/2 inch.
Poor penetration	1. Faulty preparation 2. Rod too large 3. Welding too fast 4. Not enough root gap	1. Use 60° bevel. 2. Use small rod at root. 3. Check for flow liners while welding. 4. Use tacking tip or leave 1/32-inch root gap and clamp pieces.
Scorching	1. Temperature too high 2. Welding too slowly 3. Uneven heating 4. Material too cold	1. Increase airflow. 2. Hold constant speed. 3. Use correct fanning motion. 4. Preheat material in cold weather.
Distortion	1. Overheating at joint 2. Welding too slowly 3. Rod too small 4. Improper sequence	1. Allow each bead to cool. 2. Weld at constant speed; use speed tip. 3. Use larger sized or triangular-shaped rod. 4. Offset pieces before welding. 5. Use double V or backup weld. 6. Backup weld with metal.
Warping	1. Shrinkage of material 2. Overheating 3. Faulty preparation 4. Faulty clamping of parts	1. Preheat material to relieve stress. 2. Weld rapidly—use backup weld. 3. Too much root gap 4. Clamp parts properly; back up to cool. 5. For multilayer welds, allow time for each bead to cool.
Poor Appearance	1. Uneven pressure 2. Excessive stretching 3. Uneven heating	1. Practice starting, stopping, and finger manipulation on rod. 2. Hold rod at proper angle. 3. Use slow, uniform fanning motion, heating both rod and material (for speed welding: use only moderate pressure, constant speed, keep shoe free of residue).
Stress cracking	1. Improper welding temperature 2. Undue stress or weld 3. Chemical attack 4. Rod and base material not same composition 5. Oxidation or degradation of weld	1. Use recommended welding temperature. 2. Allow for expansion and contraction. 3. Stay within known chemical resistance and working temperatures of material. 4. Use similar materials and inert gas for welding. 5. Refer to recommended application.
Poor fusion	1. Faulty preparation 2. Improper welding techniques 3. Wrong speed 4. Improper choice of rod 5. Wrong temperature	1. Clean materials before welding. 2. Keep pressure and fanning motion constant. 3. Take more time by welding at lower temperatures. 4. Use small rod at root and large rods at top—practice proper sequence. 5. Preheat materials when necessary. 6. Clamp parts securely.

Rough Grinding the Weld

The welded area can be smoothed by grinding with #36 grit emery or sandpaper. A 7- or 9-inch disc on a low-speed electric grinder will smooth large weld beads. Excess plastic can be removed with a sharp knife before grinding. Care must be taken not to overheat the weld area because it will soften. To speed up the work without damaging the weld, periodic cooling with water is necessary.

Checking Hand Weld

After rough grinding, the weld should be checked visually for defects. Any voids or cracks will make it unacceptable. Bending should not produce any cracks because a good weld is as strong as the part itself. Table 14–5 shows some typical welding defects, their causes, and corrections.

Sanding Hand Weld

The weld area can be finish sanded by using #220 grit sandpaper followed by a #320 grit. Either a belt or orbital sander may be used, plus hand sanding as required. If refinishing is to be done, follow the procedure designed specifically for plastics.

SPEED WELDING PROCEDURE

As mentioned, **plastic speed welding** uses a specially designed tip to produce a more uniform weld at a higher rate of speed. You must preheat both the rod and base material. The rod is preheated as it passes through a tube in the speed tip. The base material is preheated by a stream of hot air passing through a vent in the tip.

A pointed shoe on the end of the tip applies pressure on the rod. You do not have to apply pressure to the rod. The shoe smoothes out the rod, creating a more uniform appearance in the finished weld. On panel work, speed welding is commonly used.

With plastic speed welding, the conventional two-hand method is replaced by a faster and more uniform one-hand operation. Once started, the rod is fed automatically into the preheat tube as the welding torch is pulled along the joint. Speed tips are designed to provide the constant balance of heat and pressure. The average welding speed is about 40 inches per minute (1,016 mm per minute).

Following are some techniques essential for quality speed welding:

1. Hold the speed torch like a dagger. Bring the tip over the starting point a full 3 inches from the base material. You do not want the hot air to affect the part.
2. Cut the welding rod at a 60-degree angle. Insert it into the preheat tube. Immediately place the pointed shoe end of the tip on the base material at the starting point.
3. Hold the torch perpendicular to the base material. Push the rod through until it stops against the base material at the starting point. If necessary, lift the torch slightly to allow the rod to pass under the shoe (Figure 14–31).
4. Keep a slight pressure on the rod and only the weight of the torch on the shoe. Then pull the torch slowly toward you to start the speed weld (Figure 14–32).
5. In the first 1 to 2 inches (25–50 mm) of travel, push it into the preheat tube with slight pressure.
6. Once started, swing the torch to a 45-degree angle. The rod will now feed without a need for pressure. As the torch moves along, inspect the quality of the weld (Figure 14–33).

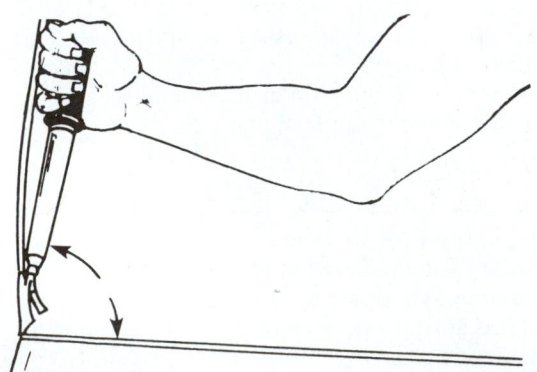

FIGURE 14-31 Start a speed weld by hold the torch perpendicular to the base material. Push the rod through until it stops against the base material at the starting point. If necessary, lift the torch slightly to allow the rod to pass under the shoe.

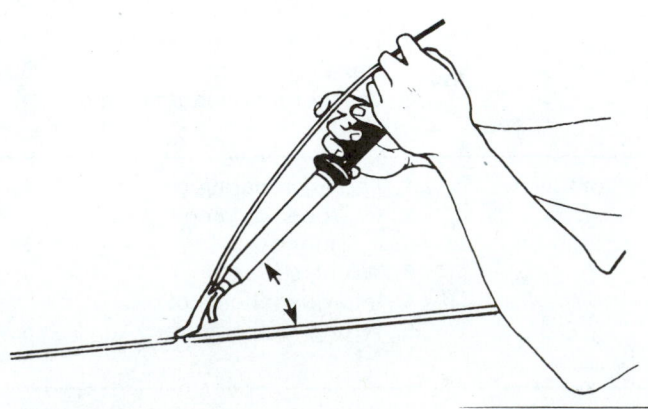

FIGURE 14-32 Continue a speed weld by pulling the tool over the parts to form a good joint.

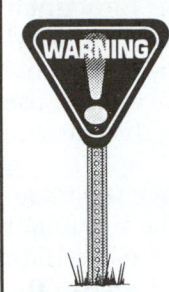

WARNING

Once the speed weld is started, do NOT stop. If you must pause for any reason, pull the speed tip off the rod immediately. If this is not done, the rod will melt into the preheat tube.

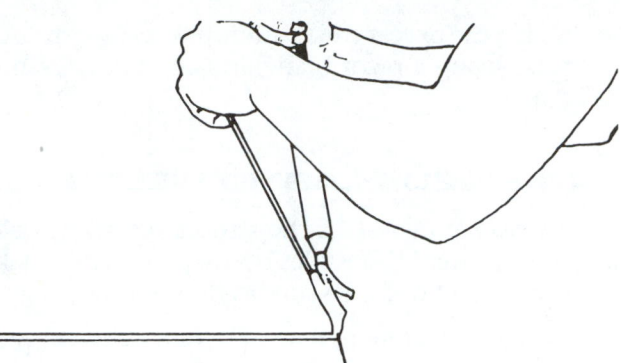

FIGURE 14-33 Finish a speed weld and then check weld strength.

The angle between the torch and the base material helps determine the *speed welding rate*. For this reason, hold the torch at a 90-degree angle when starting the weld. Hold the torch at 45 degrees after you begin welding.

If the welding rate is too fast, bring the torch back to the 90-degree angle temporarily to slow it down. Then gradually move to the desired angle for proper welding speed.

Remember, once started, speed welding must be maintained at a fairly constant rate. The torch cannot be held still. To stop welding before the rod is used up, bring the torch back past the 90-degree angle and cut off the rod at the end of the shoe.

A good speed weld in a V-groove will have a slightly higher crown and more uniformity than the normal hand weld. It should appear smooth and shiny, with a slight bead on each side. For best results and faster welding speed, clean the shoe on the speed tip with a wire brush to remove any residue that might create drag on the rod.

AIRLESS MELT-FLOW PLASTIC WELDING

Melt-flow plastic welding is the most commonly used airless welding method. It can be utilized for both single-sided and two-sided repairs.

A typical melt-flow procedure is as follows:

1. With the welding rod in the preheat tube, place the flat shoe part of the tip in the V-groove.
2. Hold it in place until the rod begins to melt and flow out around the shoe.
3. A small amount of force is needed to feed the rod through the preheat tube. The rod will not feed itself and care should be used not to feed it too fast.
4. Move the shoe slowly. Crisscross the groove until it is filled with melted plastic.
5. Work the melted plastic well into the base material, especially toward the top of the V-groove.
6. Complete a weld length of about 1 inch (25 mm) at a time. This will allow smoothing of the weld before the plastic cools.

PLASTIC STITCH-TAMP WELDING

Plastic stitch-tamp welding is used primarily on hard plastics, like ABS and nylon, to ensure a good base and rod mix.

After completing the weld using the melt-flow procedure, remove the rod. Turn the shoe over and slowly move the pointed end of the tip into the weld area to bond the rod and base material together. Stitch-tamp the entire length of the weld. After stitch-tamping, use the flat shoe part of the tip to smooth out the weld area.

SINGLE-SIDED PLASTIC WELDS

Single-sided plastic welds are used when the part cannot be removed from the vehicle. To make a single-sided weld, proceed as follows:

1. Set the temperature dial on the welder for the plastic being welded. Allow it to warm up to the proper temperature.
2. Clean the part by washing with soap and hot water, followed by a good plastic cleaner.
3. Align the break using aluminum body tape.
4. V-groove the damaged area 75 percent of the way through the base material. Angle or bevel back the torn edges of the damage at least $1/4$ inch (6 mm) on each side of the damaged area. Use a die grinder or similar tool.
5. Clean the preheat tube and insert the rod. Begin the weld by placing the shoe over the V-groove and feeding the rod through. Move the tip slowly for good melt-in and heat penetration.
6. When the entire V-groove has been filled, turn the shoe over and use the tip to stitch-tamp the rod and base material together into a good mix along the length of the weld.
7. Resmooth the weld area using the flat shoe part of the tip, again working slowly. Then cool with a damp sponge or cloth.

8. Shape the excess weld buildup to a smooth contour, using a razor blade and/or abrasive sandpaper.

TWO-SIDED PLASTIC WELDS

A *two-sided plastic weld* is the strongest type of weld because you weld both sides of the part. When making a two-sided weld, be sure to do the following:

1. Allow the welder to heat up. Then clean the preheat tube.
2. Clean the part with soap and hot water and plastic cleaner.
3. Align the front of the break with aluminum body tape, smoothing it out with a stiff squeegee or spreader.
4. V-groove 50 percent of the way through the backside of the panel.
5. Weld the backside of the panel using the melt-flow method. Move slowly enough to achieve good melt-in.
6. When finished, smooth the weld with the shoe.
7. Quick-cool the weld with a damp sponge or cloth.
8. Remove the tape from the front of the piece. V-groove deep enough so that the first weld is penetrated by the second V-groove.
9. Weld the seam, filling the groove completely.
10. Use a razor blade or slow speed grinder to reshape the contour.

14.9 REPAIRING VINYL

Vinyl is a soft, flexible, thin plastic material often applied over a foam filler. Vinyl over foam construction is commonly used on interior parts for safety. Common vinyl parts are the dash pads, arm rests, inner door trim, seat covers, and exterior roof covering. Dash pads or padded instrument panels are expensive and time-consuming to replace. Therefore, they are perfect candidates for repair (Figure 14–34).

Most dash pads are made of vinyl-clad urethane foam to protect people during a collision. Surface dents in foam dash pads, arm rests, and other padded interior parts are common in collision repair. These dents can often be repaired by applying heat as follows:

1. Soak the dent with a damp sponge or cloth for about half a minute. Leave the dented area moist.
2. Using a heat gun, heat the area around the dent. Hold the gun 10 to 12 inches (254 to 305 mm) from the surface. Keep it moving in a circular motion at all times, working from the outside in.

3. Heat the area to around 130 degrees Fahrenheit (54 degrees Celsius). Do not overheat the vinyl because it will blister. Keep heating until the area is too uncomfortable to touch. If available, use a digital thermometer to meter the surface temperature.
4. Using gloves, massage the pad. Force the material toward the center of the dent. The area might have to be reheated and massaged more than once. In some cases, heat alone might repair the damage.
5. When the dent has been removed, cool the area quickly with a damp sponge or cloth.
6. Apply vinyl treatment or preservative to the part.

WELDING VINYL-FOAM CUTS

To plastic weld a cut in vinyl-foam parts, a dash pad for example, proceed in the following manner:

1. Set up the welder for the proper temperature for welding urethane.
2. Wash the pad with soap and hot water, dry, and clean with plastic cleaner.
3. If the damaged area is brittle, warm it with a heat gun. If there are curled or jagged edges, cut them away.
4. V-groove at least $\frac{1}{4}$ inch (6 mm) into the foam padding. Bevel the edges as much as possible. Rough up the area for about $\frac{1}{4}$ inch (6 mm) around the V-groove.
5. Turn the welder so the shoe is facing up. Feed the rod slowly through the welder. Start the weld at the bottom of the groove. Completely fill it with melted plastic until it is flush with the surface.
6. Smooth out the excess rod buildup. Feather it out over the beveled edges for at least $\frac{1}{4}$ inch (6 mm) on each side of the groove.
7. Cool the weld and grind away any remaining excess rod buildup. Rough up the vinyl for about

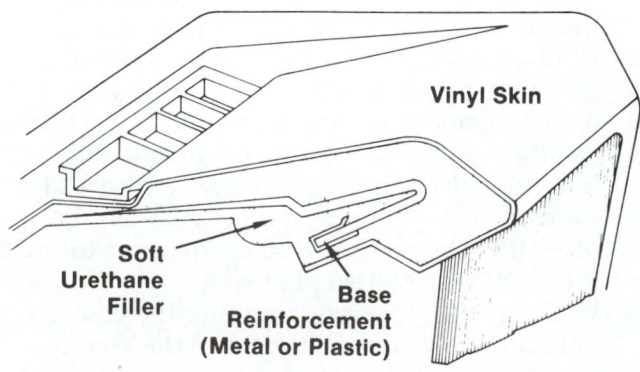

FIGURE 14–34 Most dash pads are made of vinyl over a foam material and metal frame.

2 inches (50 mm) beyond the weld on each side for good filler adhesion.

8. Use a flexible filler material designed for use on vinyl to get the desired contour. Allow the filler to cure.

9. Contour sand to remove the tacky glaze. If any filler is accidentally sanded through, apply a skim coat and resand.

10. Spray on a vinyl retexture material to match the grain of the existing vinyl. Blend it out to a break or contour line on the cover. This will help hide the repair.

11. Mask and paint the entire part the correct color vinyl paint. For more information on vinyl painting, refer to the text index.

SPRAYING VINYL PAINTS

Vinyl repair paints are usually ready for spraying as packaged. Since application properties cannot be controlled with thinners or other additives, air pressure is an important factor.

When siphon feed guns are used to apply vinyl paints the normal air pressure range is between 40 and 50 psi (275 and 345 kPa) at the gun. With pressure systems, the air pressure is reduced to 30 to 40 psi (207 and 275 kPa). Overspray can be controlled by decreasing the air pressure.

There is no retarder for vinyl paints. If blushing occurs, allow the initial coat to set up and reapply the color in a much lighter coat.

Vinyl and soft ABS plastics should be thoroughly cleansed with vinyl cleaner and allowed to dry. Then treat the surface with vinyl prep. *Vinyl prep* is a chemical solution that opens the pores of the vinyl material. Just after it is applied, wipe off the vinyl prep. The surfaces are then ready for color and/or clearcoating.

USING HEAT TO RESHAPE PLASTICS

Many bent, stretched, or deformed plastic parts, such as flexible bumper covers and vinyl-clad foam interior parts, can often be straightened with heat. This is because of **plastic memory**, which means the piece wants to keep or return to its original molded shape. If it is bent or deformed slightly, it will return to its original shape if heat is applied (Figure 14–35).

To reshape a distorted bumper cover, use the following procedure:

1. Thoroughly wash the cover with soap and hot water.

2. Clean with plastic cleaner. Carefully remove all road tar, oil, grease, and undercoating.

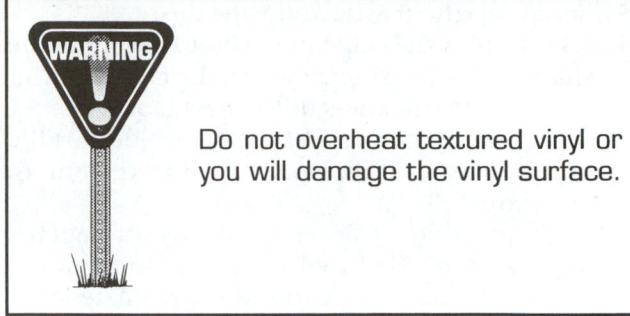

WARNING

Do not overheat textured vinyl or you will damage the vinyl surface.

3. Dampen the repair area with a water-soaked rag or sponge.

4. Apply heat directly to the distorted area. Use a concentrated heat source, such as a heat lamp or high-temperature heat gun. When the opposite side of the cover becomes uncomfortable to the touch, it has been heated enough.

5. Use a paint paddle, squeegee, or wood block to help reshape the piece if necessary.

6. Quick-cool the area by applying cold water with a sponge or rag.

TAB REPLACEMENT FOR PLASTIC PARTS

When a bumper cover or similar part has been torn away from its mounting screws, the mounting tabs will often be broken or torn away. Mounting tabs must be repaired with a two-sided weld to provide enough weld strength.

The following is the procedure for rebuilding a mounting tab that has been torn off:

1. Begin by cleaning the piece as described above.

2. Bevel back the torn edges of the mounting tab at least $1/4$ inch (6 mm) on both sides.

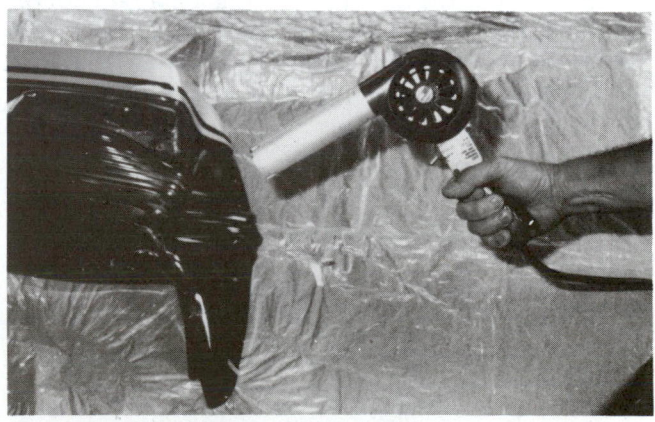

FIGURE 14–35 A heat gun can be used to straighten some plastic parts and speed the adhesive curing process. *(Courtesy of Tech-Cor)*

3. Rough up the plastic and wipe dust free.
4. Use aluminum body tape to build a form in the shape of the missing tab. Turn the tape edges up to form the thickness of the new tab.
5. Set the temperature dial on the welder for the type of plastic being repaired. Allow the unit to warm up.
6. Begin the weld. Push the rod slowly through the preheat tube. Slightly overfill the form, working the melted plastic into the base material.
7. Smooth and shape the weld. Quick-cool the weld area.
8. Remove the tape. V-groove along the tear line on the other side about halfway through the piece.
9. Weld the groove and quick-cool. Finish the weld to the desired contour using a slow-speed grinder with a #60- or #80-grit disc.

14.10 ULTRASONIC STUD WELDING

Ultrasonic stud welding uses high-frequency movement and friction to generate heat that bonds the plastic parts together. It can be used to join plastic parts at a single point or at numerous locations. In many applications, a continuous weld is not required. The welding cycle is short, almost always less than half a second.

Ultrasonic stud welding is made along the circumference of the stud. Its strength is a function of the stud diameter and the depth of the weld. Maximum tensile strength is achieved when the depth of the weld equals half the diameter of the stud.

Figure 14–36 shows the basic stud weld joint before, during, and after welding.

The radial interference (dimension A) must be uniform and should generally be 0.008 to 0.012 inch (0.20 to 0.30 mm) for studs having a diameter of 0.5 inch (12.7 mm) or less. The hole should be a sufficient distance from the edge to prevent breakout; a minimum of 0.125 inch (3.2 mm) is recommended.

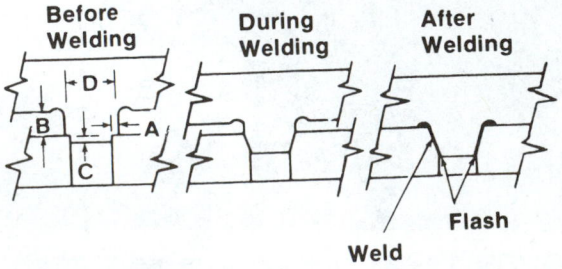

FIGURE 14-36 Study basic ultrasonic stud welding joints.

In the joint, the recess can be on the end of the stud or in the mouth of the hole, as shown in the examples. With the latter, a small chamfer can be used for rapid alignment. To reduce stress concentration, a good-sized fillet radius should be incorporated at the base of the stud. Recessing the fillet below the surface allows flush contact of the parts.

A variation can be used where appearance is important or an uninterrupted surface is required. The stud is welded into a boss, whose outside diameter can be no less than twice the stud diameter. When welding into a blind hole, it might be necessary to provide an outlet for air.

14.11 REINFORCED PLASTIC REPAIR

Reinforced plastic—including sheet-molded compound (SMC), fiber-reinforced plastic (FRP), and reinforced reaction-injection-molded polyurethane (RRIM)—parts are being used in many unibody vehicles (Figure 14–37). They often provide a durable plastic skin over a steel unibody.

Table 14–6 provides an overview of reinforced plastic repair materials.

The damage that generally occurs in reinforced plastic panels includes:

1. One-sided damage, such as a scratch or gouge.
2. Punctures and fractures.
3. Panel separation, where the panel pulls away from the metal space frame.
4. Severe damage, which requires full or partial panel replacement.
5. Minor bends and distortions of the space frame, which can be repaired by pulling and straightening.
6. Severe kinks and bends to the space frame, which require replacement of that piece along factory seams or by sectioning.

Combinations of these types of damage often occur on a single vehicle. Depending on the location and amount of damage, there are four different types of reinforced plastic repairs. These are

1. Single-sided repair
2. Two-sided repair
3. Panel sectioning
4. Full panel replacement

To select a repair method, thorough examination of the vehicle is required (Figure 14–38). Examine all affected reinforced plastic panels. First, check the entire panel for signs of damage. Also check all panel seams for adhesive bond failure. Examine the back of the panel to determine the extent of the damage.

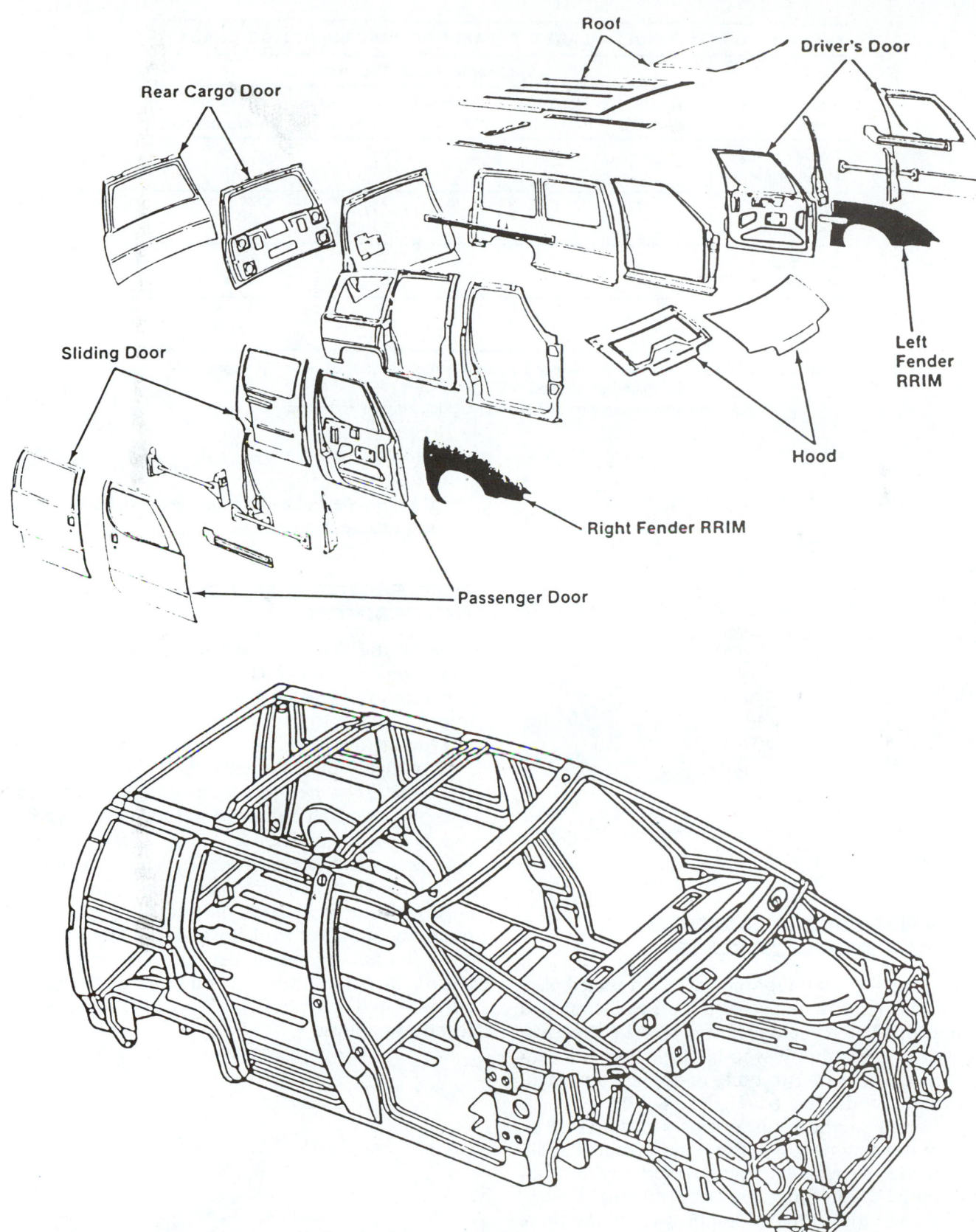

FIGURE 14-37 SMC panels are often bonded to a steel space frame.

TABLE 14-6: REINFORCED PLASTIC REPAIR MATERIAL SELECTION CHART

Type of Repair	Applicable Repair Product				
	Panel Adhesive	Patching Adhesive	Structural Filler	Cosmetic Filler	Glass Fiber Reinforcement
Panel Replacement	X				
Panel Sectioning	X		X_1	X_1	X
One-Sided Repairs				X_1	
Two-Sided Repairs	X_2	X_2	X	X	X

Notes: 1. Some panel adhesives can also be used as structural and cosmetic fillers, depending on sanding characteristics.
2. Panel adhesives can also be used as patching adhesives, but not vice versa.

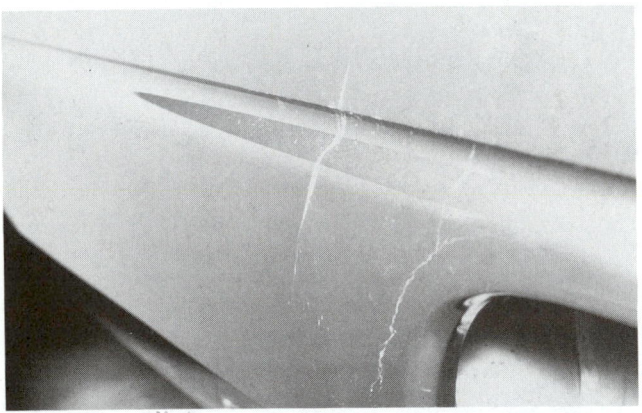

FIGURE 14-38 Inspect the extent and location of SMC damage to determine appropriate repair methods. *(Courtesy of Tech-Cor)*

REINFORCED PLASTICS REPAIR APPLICATORS

Most of the tools used for repairs of reinforced plastics should already be available in a well-equipped body shop. The *reinforced plastic adhesive applicator* allows two-part adhesives to be dispensed at a constant rate. There are two types of applicators: pneumatic and hand-operated (Figure 14–39).

The pneumatic applicator uses compressed air to force the materials out. The hand-operated applicator works like a caulking gun. Hand-applied pressure is used to force material out of the tubes.

To use either type of applicator, follow these simple rules:

- Follow the manufacturer's instructions.
- Check for proper product flow.
- Check for consistent mix of the two-component product.
- When changing cartridges, run a new test bead.

- If saving part of a cartridge, leave the static mixing nozzle in place.

REINFORCED PLASTIC ADHESIVES

Many of the materials that are used for reinforced plastic repair are two-part adhesive products. Two-part adhesive means a base material and a hardener must be mixed to cure the adhesive. Each must be mixed together in the proper ratio. Both parts must be thoroughly mixed together before use.

Work life or *open time* is the time when it is still possible to work the adhesive and still have the adhesive set up for a good bond. This work life time will be provided by the manufacturer. The cure time of some adhesives used in reinforced plastic repair can be shortened with the application of heat. Temperature and humidity can affect work life and cure time.

After mixing, remember that each product has a work life or open time. If you move or disturb the adhesive as it starts to harden, you will adversely affect its durability.

FIGURE 14-39 A dual cartridge applicator will dispense equal parts of two-part adhesive quickly. It can be either hand operated (like this one) or pneumatic.

SHOP TALK

Do not use fillers designed for sheet metal on reinforced plastic. The repair will be weak and will crack and fail quickly. This would be an embarrassing mistake!

REINFORCED PLASTIC FILLERS AND GLASS CLOTH

Two filler products are specifically formulated for use on reinforced plastic. They are cosmetic filler and structural filler.

Cosmetic filler is typically a two-part epoxy or polyester filler used to cover up minor imperfections.

Structural filler is used to fill the larger gaps in the panel structure while maintaining strength. Structural fillers add to the structural rigidity of the part.

All two-part products will shrink to some degree. The use of heat will help to speed the drying time and will eliminate some of the shrinkage.

Check with the product manufacturer for temperatures and dry times. If the product is not properly heated to a full cure, shrinkage will occur as the product cures with time. The "rule of thumb" is to heat the material to a surface temperature higher than any temperatures that the vehicle will be subject to when it is on the road. If it is a black vehicle sitting in the sun in mid-summer in Florida, this could be about 170 degrees Fahrenheit (77 degrees Celsius) or more.

Check with the product manufacturer for recommendations for heat curing. Generally, 200 to 250 degrees Fahrenheit (93 to 121 degrees Celsius) for 20 to 40 minutes should do it. Remember that at lower temperatures, the product will have to be heated longer. Also, if there is high humidity, the cure process will take longer.

There are several different types of glass cloth available. Rovings and mattings are not appropriate for reinforced plastic repair. Choose unidirectional cloth, woven glass cloth, or nylon screening. The cloth weave must be loose enough to allow the adhesive to fully saturate the cloth, leaving no air space around the weave.

SINGLE-SIDED REPAIRS OF REINFORCED PLASTIC (FIBERGLASS)

Single-sided damage is surface damage that does not penetrate or fracture the back of the panel. Damage might pass all the way through a panel, but no pieces of the panel have broken away. If the break is clean and all of the reinforcing fibers have stayed in place, then a single-sided repair is adequate.

For a single-sided repair, you must bevel deep to penetrate the fibers in the panel (Figure 14–40A). The broken fibers must come into contact with the adhesive.

The following is a typical single-sided repair procedure for reinforced plastic:

1. Clean the repair area with soap and hot water.
2. Clean again using mild wax and grease remover.
3. Remove any paint from the surrounding area by sanding with #80 grit sandpaper.
4. Scuff sand the area surrounding the damage.
5. Bevel the damage to provide an adequate area for bonding.
6. Mix two-part filler according to the manufacturer's instructions.
7. Apply the filler and cure as recommended.

Once the filler has been sanded, apply additional coats as required and resand. The product manufacturer will provide grit recommendations.

TWO-SIDED REPAIRS OF REINFORCED PLASTIC (FIBERGLASS)

A two-sided repair is normally needed on damage that passes all the way through the panel (Figure 14–40B). This would include damage to the reinforcing fibers.

A **backing strip** or *backing patch* is bonded to the back of the repair area to restore the reinforced plastic's strength. The patch also forms a foundation for forming the exterior surface to match the original contour of the panel.

To make a two-sided repair in a reinforced plastic or fiberglass panel, proceed as follows:

1. Clean the surface surrounding the damage with a good wax and grease remover. Use a #36 grinding disc to remove all paint and primer at least 3 inches (76 mm) beyond the repair area.

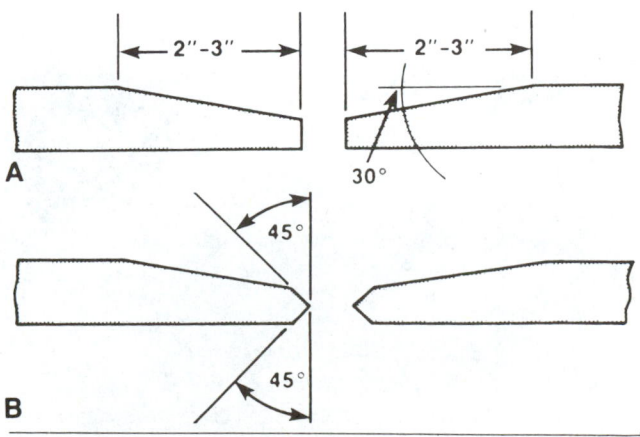

FIGURE 14–40 Beveling the inside and outside of the repair area permits better adhesion.

2. Grind, file, or use a hacksaw to remove all cracked or splintered material away from the hole on both the inside and outside of the repair area.

3. Remove any dirt, sound deadener, and the like from the inner surface of the repair area. Clean with reducer, lacquer thinner, or a similar solvent.

4. Scuff around the hole with #80 grit sandpaper to provide a good bonding surface.

5. Bevel the inside and outside edge of the repair area about 30 degrees to permit better patch adhesion.

6. Clean the repair area thoroughly.

7. Cut several pieces of fiberglass cloth large enough to cover the hole and the scuffed area. The exact number of pieces will depend on the thickness of the original panel.

8. Prepare a mixture of resin and hardener. Follow the label recommendations.

9. Using a small paintbrush, saturate at least two layers of the fiberglass cloth with the activated resin mix (Figure 14–41).

10. Apply the material to the inside or back surface of the repair area. Make sure the cloth fully contacts the scuffed area surrounding the hole.

11. Saturate three more layers of cloth with the mix. Apply it to the outside surface. These layers must also contact the inner layers and the scuffed outside repair area (Figure 14–42).

12. With all of the layers of cloth in place, form a saucerlike depression in them. This is needed to increase the depth of the repair material. Use a squeegee to work out any air bubbles.

13. Clean all tools with a lacquer thinner immediately after use.

14. Let the saturated cloth patch become tacky. An infrared heat lamp can be used to speed up the process. If one is used, keep it 12 to 15 inches (305 to 381 mm) away from the surface. Do not overheat the repair area because too much heat will cause distortion.

15. With #50 grit sandpaper, disc sand the patch slightly below the contour of the panel (Figure 14–43).

16. Prepare more resin and hardener mix. Use a plastic spreader to fill the depression in the repair area. You need a sufficient layer of material for grinding down smooth and flush.

17. Allow the patch to harden. Again, a heat lamp can be used to speed the curing process.

18. When the patch is fully hardened, sand the excess material down to the basic contour (Figure 14–44). Use #80 grit sandpaper and a sanding block. Finish sand with #120 or finer grit sandpaper.

There are several ways to hold the patch in place from the front side of the panel. One method is to use a pull rod. Drill a hole in the middle of the patch. Insert the end of the rod. Position the patch, and pull it snug with the pull rod. Apply heat using a heat gun on the front, or have someone else apply the heat to the back of the patch.

This same two-sided repair can be made by attaching sheet metal to the backside of the panel with sheet metal screws (Figure 14–45). Sand the sheet metal and both sides of the part to provide good adhesion. Before fastening the sheet metal, apply resin and hardener mix to both sides of the rim of the hole. Follow the procedures described earlier for the remainder of the repair.

When the inner side of the hole is not accessible, apply a fiberglass patch to the outer side only. After the usual cleaning and sanding operations, apply several additional layers of fiberglass cloth to the outer side of the hole. Before it dries, make a saucer-

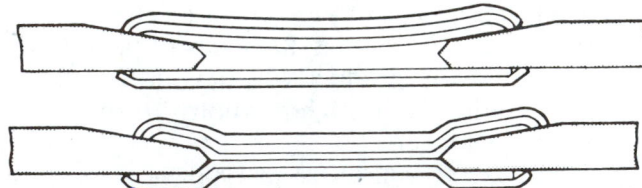

FIGURE 14–42 The inside and outside layers of saturated glass cloth should contact each other.

Original Panel Level

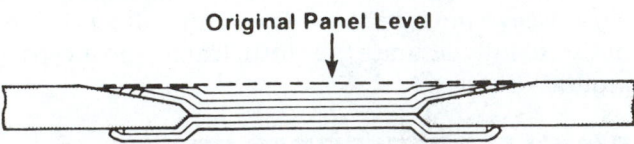

FIGURE 14–43 Sand the patch slightly below the contour of the panel.

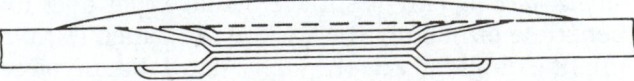

FIGURE 14–44 Featheredge sand the patching material into the surrounding panel as you would with conventional body filler.

FIGURE 14–41 Applying resin to the mat strips will produce a patch.

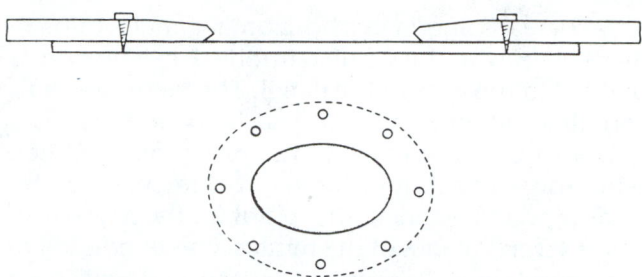

FIGURE 14-45 Sheet metal can be attached to the back side of a damaged panel for supporting repair material.

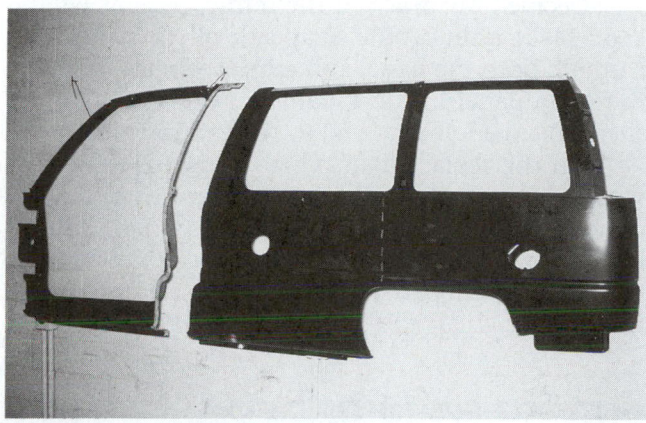

A

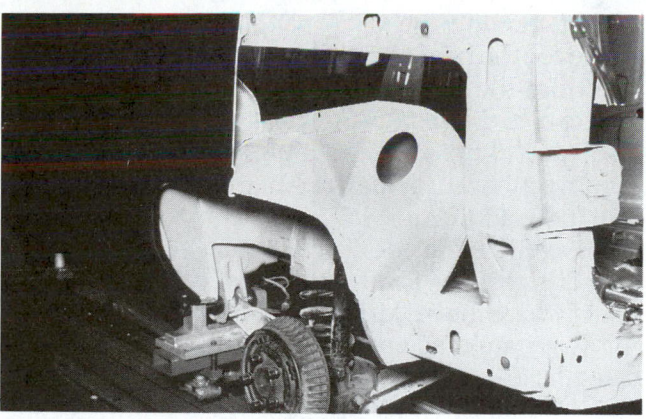

B

FIGURE 14-46 Sectioning procedures of SMC parts must be analyzed carefully. (A) New SMC parts. (B) Old parts must be sectioned off in an appropriate location that clears internal steel frame members. *(Courtesy of Tech-Cor)*

like depression in the cloth to provide greater depth for the repair material.

SECTIONING REINFORCED PLASTIC PANEL

Proper sectioning requires that you understand what areas are most appropriate for sectioning (Figure 14-46A). You must also know how to avoid problems with horizontal bracing, rivets, and concealed parts (Figure 14-46B). The replacement panel used

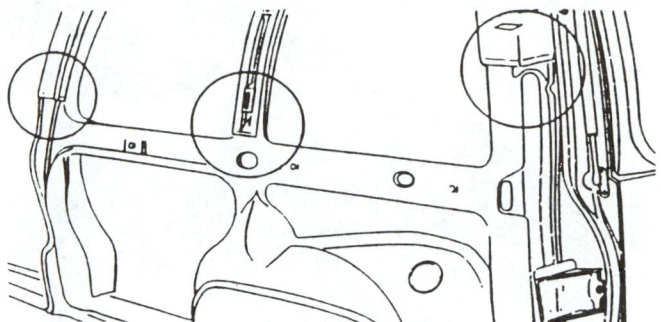

FIGURE 14-47 Sectioning locations are often given in the service manual and are typical of nonplastic parts.

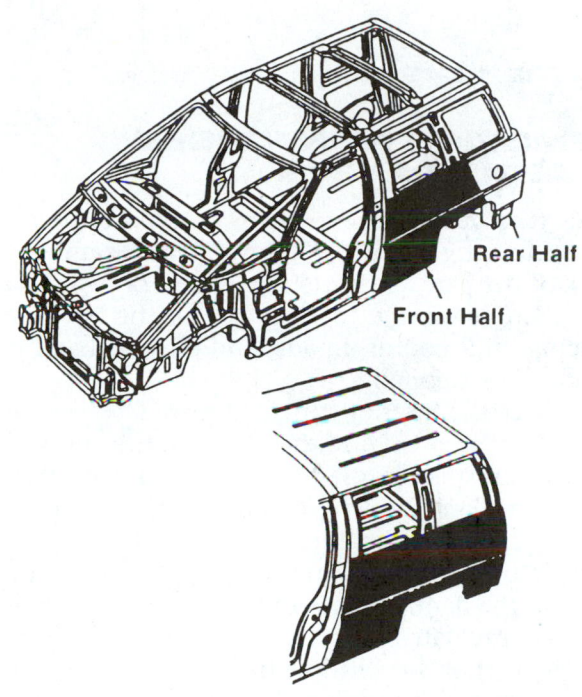

Rear Half

Front Half

FIGURE 14-48 Depending upon the extent of the damage, you may want to install a complete panel or section it down to replace only the damaged area.

will depend on the amount and location of the damage (Figure 14-47).

Using the left rear quarter panel as an example, there are three possibilities. The entire panel can be ordered, or just a front or rear half (Figure 14-48).

Remember that reinforced plastic is a very forgiving and workable material. Just because quarter panels come split at the wheel well, the sectioning point does not have to be located there. With proper backing strips to reinforce the joints, sectioning can be done almost anywhere.

The **mill and drill pads** are used to help the factory hold panels in place while the adhesive cures. These mill and drill pads will also help you to hold, align, and level replacement panels. If a panel is to be sectioned, it should be done between mill and drill pad locations (Figure 14-49).

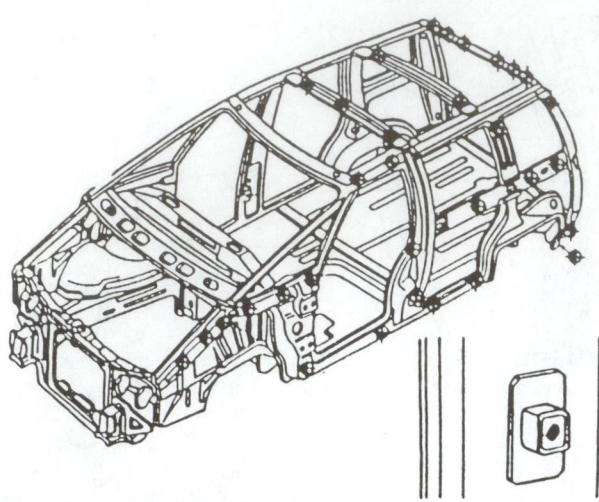

FIGURE 14-49 Mill and drill pad locations.

REMOVING REINFORCED PLASTIC PANELS

First, remove the interior trim to expose the horizontal bracing and mill and drill pads. Examine the back of the panel to gauge the extent of panel damage. Also determine the location of the horizontal bracing, mill and drill pads, and electrical and mechanical components.

Once the interior trim is removed, a "window" can be cut. Controlling the depth of the cut is very important. Space frame components, as well as electrical lines and heating/cooling elements, may be located behind the panels. When cutting the window, know what is behind the panel being cut into, or limit the depth of the cut to $\frac{1}{4}$ inch (6.35 mm) to avoid doing damage.

Now that the window has been cut, the rest of the panel can be removed from the space frame. This can be done using heat and a putty knife, or by carefully using an air chisel. Choose a flat chisel, beveled on one side only. Be careful not to damage the space frame.

If the door surround panels are to be left attached to the vehicle, the air chisel method may not be the best choice for separating the seam between two panels. Use the heat and putty knife method to separate the seams to avoid doing damage to the door surround pieces.

The adhesive bond must be broken to remove the remaining quarter panel. Using approximately 400 degrees Fahrenheit (204 degrees Celsius), heat the outboard section of the panel. Concentrate the heat in the area of the bond. Heat a span of about 5 to 7 inches (127 to 178 mm). After a short time, the adhesive will soften. Then the panel can be removed using a panel cutter, chisel, or pry bar. More than one person may be needed to assist in the heating and prying operations. Apply the heat and pry at the same time.

There is sometimes a horizontal panel reinforcement. It can be made of reinforced plastic and is bonded to the back of the panel. This reinforcement contains the mill and drill pad bolts and acts as a spacer to hold the panel in its correct place. When replacing a panel, use this reinforcement as a sectioning point and anchoring point for the new panel. Leave several inches of the reinforcement bonded to the space frame. Determine the sectioning location. Make the cut, using extra care to avoid cutting through the reinforcement. Cut through the outer panel only at the sectioning location.

Decide how much of the reinforcement will be kept. Make a cut on the scrap side of the panel. Cut through both the panel and reinforcement. Remove the scrap panel in the usual way. Then remove the strip of panel left attached to the reinforcement.

On the replacement panel, mark off the corresponding piece of panel reinforcement that must be removed to fit the panel. Be cautious when estimating how much reinforcement must be removed from the replacement panel. Leave enough reinforcement to enable the part to be trimmed to fit.

FITTING REINFORCED PLASTIC PANELS

After removing the scrap panel, prepare the space frame for the new panel. First, remove the old adhesive from the space frame. In some cases, the adhesive can be peeled away from the frame. You may also have to use heat and a putty knife or a sander. Remove all of the adhesive since the heat will break down the adhesive.

Bevel the outside edges of the existing panel to a 20-degree taper. Sand and clean the backside of the panel where the backing strips will be attached. Backing strips are made using scrap material that duplicates the original panel contour as closely as possible. They should extend about 2 inches (50 mm) beyond either side of the sectioning location. Clean the backing strips. Remove the paint from those places where adhesive will be applied.

Measure the replacement panel for fit. Trim the panel to size. Check the fit again. Leave a $\frac{1}{2}$-inch (13 mm) gap between the existing panel and the replacement panel. When proper fit has been established,

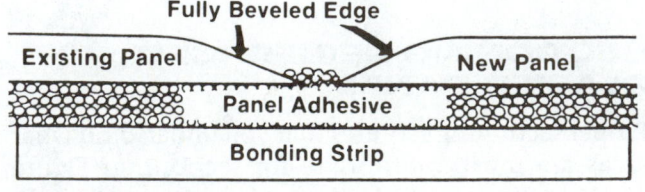

FIGURE 14-50 Note the basic method for fitting a sectioned SMC panel.

the new panel can be prepared for the adhesive. Sand or grind bevels into the panel where they mate to the existing panel (Figure 14–50).

Bevel the mating edges of the new panel to a shallow taper, just like on the existing panel. Make sure to bevel all the way through the panel. Do not leave a shoulder.

Apply the adhesive material in a continuous bead all the way around the panel. Make sure the bead is $3/8$ to $1/2$ inch (9.5 to 13 mm) in diameter. Check for horizontal bracing, and apply a bead of adhesive to correspond with it. Then fit the panel onto the vehicle and clamp it into place. Install the mill and drill pad nuts and tighten securely.

The work life will be recommended by the adhesive manufacturer. This time should be followed to ensure that the proper fit has been achieved.

USING MOLDED CORES

A **molded core** is a curved body repair part made by applying plastic repair material over a part and then removing the cured material. Naturally, holes are much more difficult to repair in a curved portion of a reinforced plastic panel than those on a flat surface. Basically, the only solution (short of purchasing a new panel section) is to use the molded core method of replacement. This is often the quickest and cheapest way to repair a curved surface.

While this relates to a rear fender section, the principle can be applied to any type of curved reinforced plastic panel.

1. Locate an undamaged panel on another vehicle that matches the damaged one. It will be used as a model or pattern. A new or used vehicle can be used since it will not be harmed. You can also use a new or undamaged used panel removed from a vehicle (Figure 14–51).

2. On the model vehicle, mask off an area slightly larger than the damaged area. Apply additional masking paper and tape to the surrounding area, especially on the low side of the panel. This will prevent any resin from getting on the finish.

3. Coat the area with paste floor wax. Leave a wet coat of wax all over the surface. A piece of waxed paper can be substituted for the coat of wax (Figure 14–52). Make sure the waxed paper is taped firmly in place.

4. Cut several pieces of thin fiberglass mat in sizes larger than the area to be repaired.

5. Mix the fiberglass resin and hardener following the label instructions.

6. Starting from one corner of the mold area, place pieces of fiberglass mat on the waxed area so each edge overlaps the next one; use just one layer of matting (Figure 14–53).

7. Apply the resin/hardener to the matting with a paintbrush. Force the mixture into the curved surfaces and around corners with the tips of the bristles.

8. Use the smaller pieces of matting along the edges and on difficult curves. Additional resin/hard-

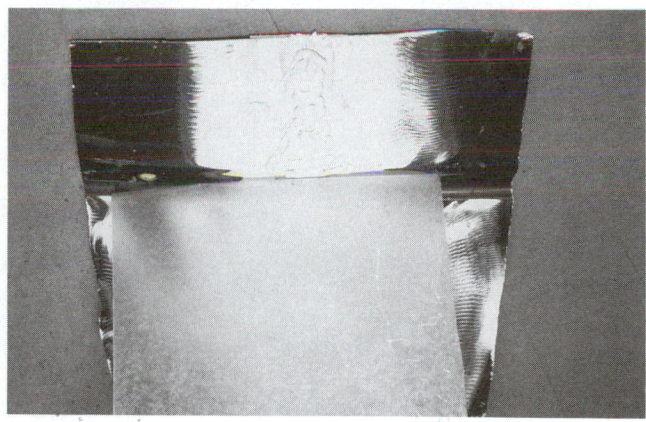

FIGURE 14-52 Wax or wax paper can be used as a release agent for the molded core repair piece. *(Courtesy of Tech-Cor)*

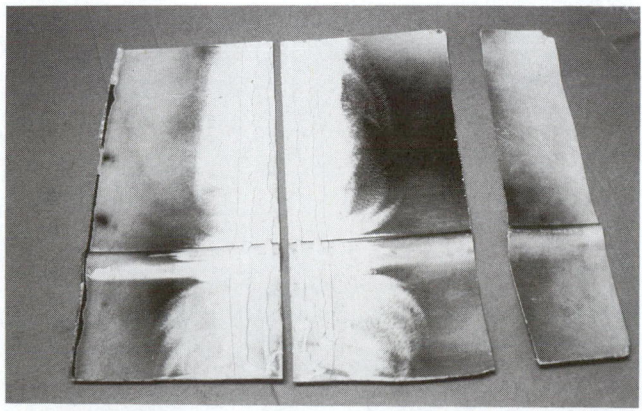

FIGURE 14-51 An old or undamaged panel can be used to make a molded core that will match the shape of the damaged panel. *(Courtesy of Tech-Cor)*

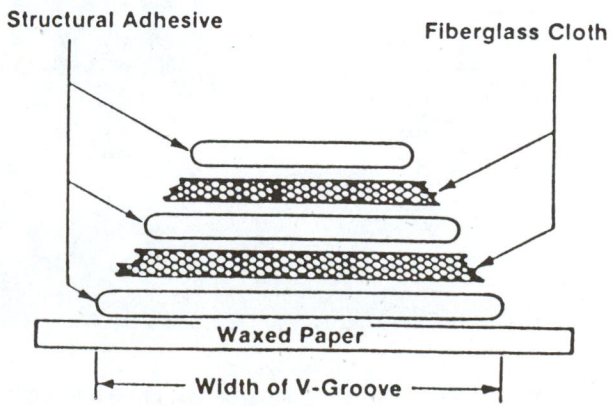

FIGURE 14-53 Note the method for making a filler patch.

ener can be applied if needed, brushing in one direction only to force the material into the indentations. In all cases, use only one layer of matting.

9. After matting has been applied to the entire waxed area, allow the molded core to cure a minimum of one hour.

10. Once the molded core has hardened, gently work the piece loose from the model vehicle. The core should be an exact reproduction of this section of the panel.

11. Remove the wax or waxed paper from the model vehicle. Then polish this section of the panel.

12. Since the molded core is generally a little larger than the original panel, place it under the damaged panel and align. If necessary, trim down the edges of the core and the damaged panel slightly where needed for better alignment. The edges of the damaged panel and core must also be cleaned.

13. Using fiberglass adhesive, cement the molded core in place on the inside of the panel. Allow the core and panel to cure (Figure 14–54).

14. Grind back the original damaged edges to a taper or bevel, maintaining the desired contour.

15. Lay a fiberglass mat, soaked in resin/hardener, on the taper or bevel and over the entire core (Figure 14–55). Once the mat has hardened, level it with a coat of fiberglass filler. Then prepare it for painting (Figure 14–56).

In some instances it might not be possible to place the core on the inside of the damaged panel. In this case, the damaged portion must be cut out to the exact size of the core. After the panel has been trimmed and its edges beveled, tabs must be installed to support the core from the inside. These tabs can be made from pieces of the panel or from fiberglass strips saturated in resin/hardener.

After cleaning and sanding the inside sections, attach the tabs to the inside edge of the panel and bond with fiberglass adhesive. Clamping pliers can be used to hold the tabs in place. Taper the edge of the opening and place the core on the tabs. Fasten the core to the tabs with fiberglass adhesive. Grind down any high spots so that layers of fiberglass mat can be added.

Place the saturated mats over the core, extending about $1\frac{1}{2}$ to 2 inches (38 to 50 mm) beyond the damaged area in all directions. Work each layer with a spatula or squeegee to remove all air pockets. Additional resin/hardener can be added with a paintbrush to secure the layers. Allow sufficient curing time. Then sand the surface level. For a smooth surface,

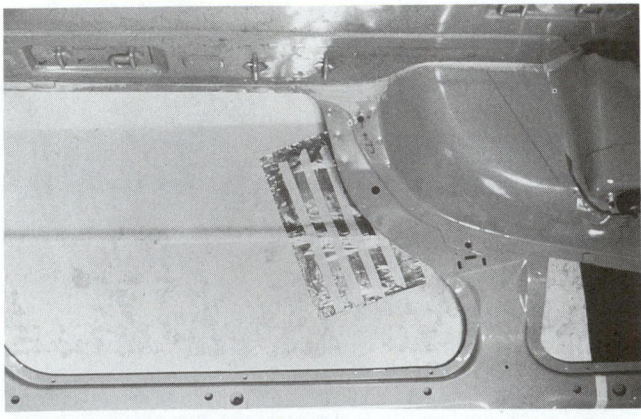

A

B

FIGURE 14-54 (A) If space permits, place the patch or backing over the inside of the panel. (B) Heat can be used to speed curing. *(Courtesy of Tech-Cor)*

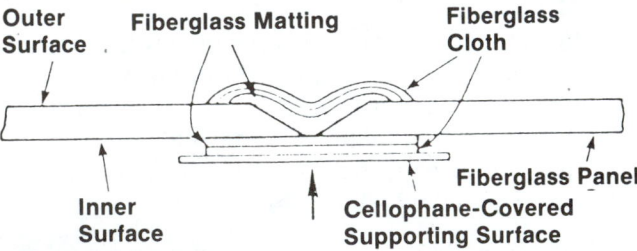

FIGURE 14-55 The repair area has been covered with fiberglass cloth.

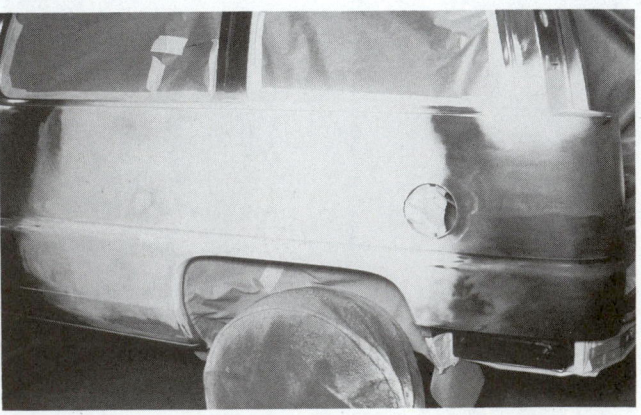

FIGURE 14-56 Featheredge the repair area and prime. *(Courtesy of Tech-Cor)*

FIGURE 14-57 Paint the repair area and buff as needed. *(Courtesy of Tech-Cor)*

use fiberglass filler to finish the job before painting (Figure 14–57).

Figure 14–58 summarizes making a repair with a molded core.

SMC DOOR SKIN REPLACEMENT

A door skin replacement is a straightforward repair because most are made of an inner and an outer piece of SMC bonded together. The exterior of the door might be repaired using a single-sided or two-sided repair, or the door skin might be replaced. Outer door skins are available as service parts.

The outer panel of the door usually overlaps the inner panel slightly. This forms a little lip around the door. Grind away this lip to expose the joint between the outer and inner panels. Use caution to avoid damaging the inner panel that must be saved.

There are two methods of separating the door pieces. They are:

1. Use heat and a putty knife.
   - Remove the lip of the panel with an air grinder.
   - Apply heat to the edge of the panels.
   - Force the putty knife between the panels, separating the adhesive bond.
2. Use an air chisel (Figure 14–59).
   - Force the chisel between the panels.
   - Do not damage the inner panel.
   - If the chisel begins to cut through the panel, remove it and try cutting from the outer direction.

The mating edges of the inner panel are now cleaned of loose adhesive or SMC or fiberglass parts.

Usually there is an inner UHSS reinforcement that runs the length of both front doors and the sliding cargo door. There will also be some inner SMC reinforcements attached directly to the SMC outer door skin itself. Look for any inner reinforcements

by removing the inner door skin. The leading and trailing edges inside the doors are bonded with a metal reinforcement to which the intrusion beams are bolted. Look at these areas carefully for damage if there has been an impact to the intrusion beam.

REPAIRING RRIM

Reinforced reaction injection-molded (RRIM) is a two-part polyurethane composite plastic. Part A is the isocyanate. Part B contains the reinforced fibers, resins, and a catalyst. The two parts are first mixed in a special mixing chamber, then injected into a mold. RRIM parts are becoming more common in fenders and bumper covers.

Since RRIM is a thermosetting plastic, heat (100 to 140 degrees Fahrenheit or 38 to 60 degrees Celsius) is applied to the mold to cure the material. The molded product is made to be stiff yet flexible. It can absorb minor impacts without damage. This makes RRIM an ideal material for exposed areas.

Gouges and punctures can be repaired using a structural adhesive. If the damage is a puncture that extends through the panel, a backing patch is required. To make a typical backing patch repair, proceed as follows:

1. Clean the damaged area thoroughly using the plastic cleaner recommended by the manufacturer and a clean cloth. Wipe dry.
2. Remove any paint film in and around the damage with an orbital sander and #180 grit disc.
3. Using a #50 grit disc, enlarge the damaged area, tapering out the damage for about 1 inch (25 mm). Wipe or blow away any loose particles.
4. Clean the backside of the damaged area with the plastic cleaner.
5. Use a #50 grit disc to scuff sand the area. Extend the area to about $1\frac{1}{2}$ inches (38 mm) beyond the damage. Align the front of the panel with body tape, if necessary.
6. Cut a piece of fiberglass cloth to cover the damaged area and the part of the panel that has been scuff sanded.
7. Mix the adhesive according to the manufacturer's recommendations. Apply a layer of adhesive to the backside of the panel about $\frac{1}{8}$ inch (3 mm) thick.
8. Place the fiberglass patch into position on the adhesive. Cover it with a sheet of waxed paper. Use a roller to force the adhesive into the fibers of the patch.
9. Remove the waxed paper and add another layer of adhesive. Work out the adhesive to just beyond the edges of the patch. Allow the adhesive to cure following the manufacturer's recommendation.

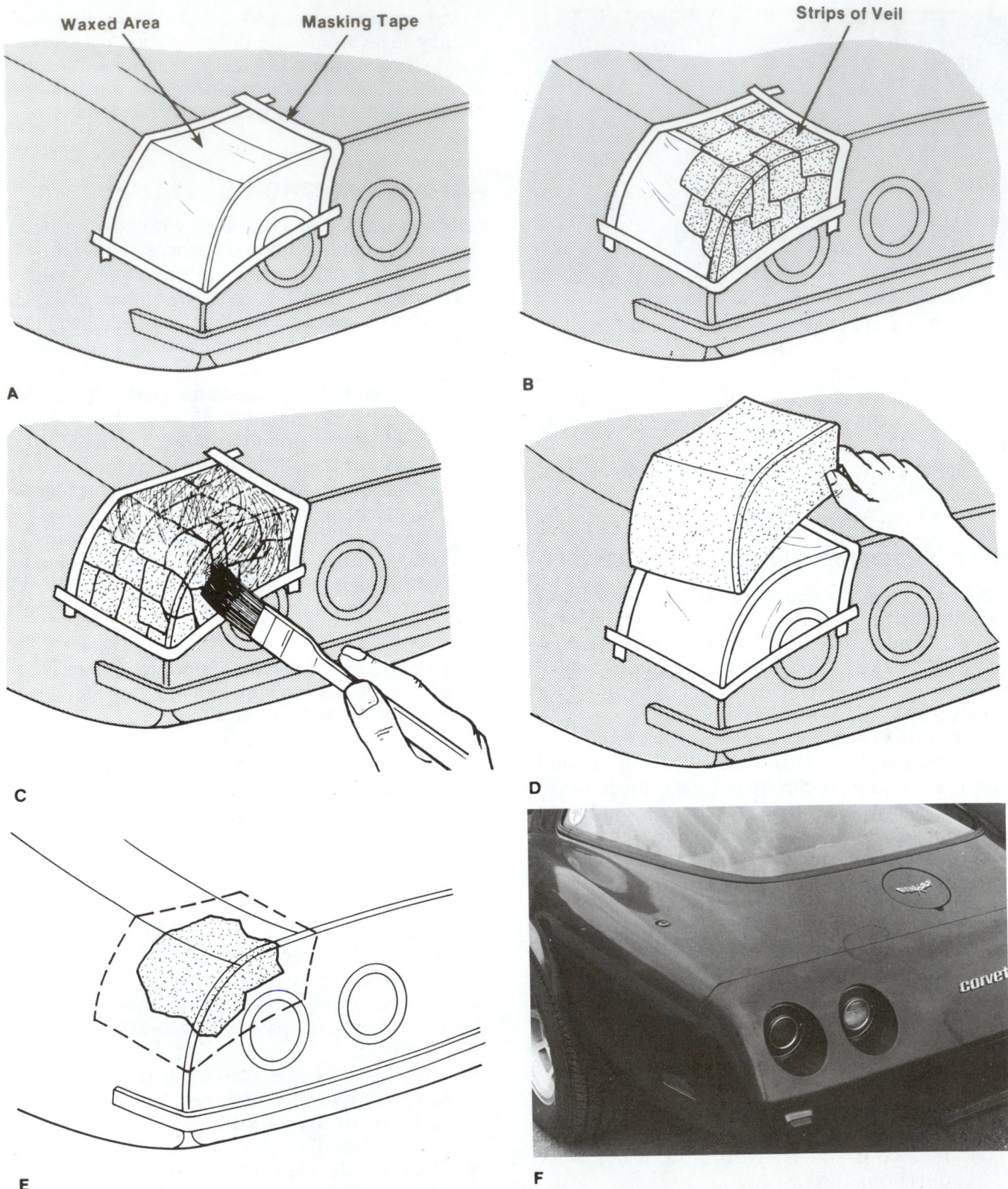

FIGURE 14-58 Study this summary of the steps in making a fiberglass core: (A) coat the area being used as a mold model with paste floor wax or a piece of waxed paper; (B) place pieces of fiberglass veil over the waxed or waxed paper surface; (C) apply resin/hardener to the veil material; (D) remove the mold core from the model; (E) cement the core piece in place; (F) the completed job. *(Courtesy of Unican Corp.)*

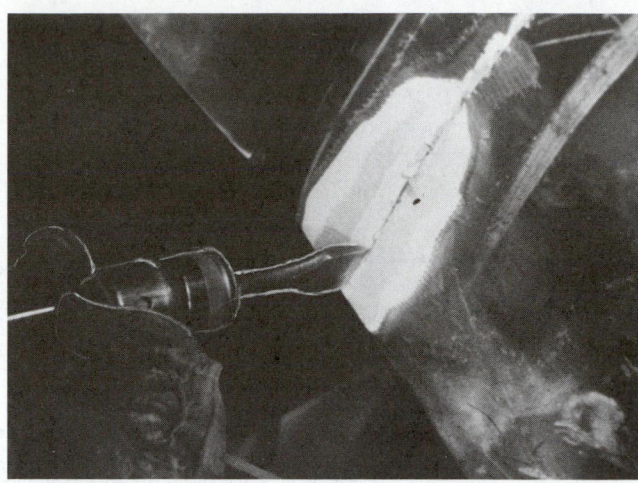

FIGURE 14-59 Using an air chisel to separate the panel from the body.

10. Now move to the front side of the panel. Apply a layer of adhesive, completely covering the

Plastic welding is not the repair choice for RRIM. Damage is normally repaired using an adhesive.

damaged area. Build it up to slightly higher than the surrounding contour. Allow it to cure.

11. Apply heat to help speed the cure of the patch.

12. Contour the adhesive to the adjoining surface by block sanding using #220 grit paper.

13. Finish by feathering with an orbital sander and a #320 disc.

14. Follow the recommendations given in Chapter 16 for priming and painting.

SUMMARY

- The term "plastics" refers to a wide range of materials synthetically compounded from crude oil, coal, natural gas, and other natural substances.

- Plastic parts include bumpers, fender extensions, fascias, fender aprons, grille openings, stone shields, instrument panels, trim panels, fuel lines, door panels, quarter panels, and engine parts.

- Thermoplastics can be repeatedly softened and reshaped by heating, with no change in their chemical makeup.

- Thermosetting plastics undergo a chemical change by the action of heating, a catalyst, or ultraviolet light.

- Composite plastics, or "hybrids," are blends of different plastics and other ingredients designed to achieve specific performance characteristics.

- Working with plastics and fiberglass requires you to think about safety at all times. The resin and related ingredients can irritate your skin and stomach lining. The curing agent or hardener can produce harmful vapors. It is also environmentally unsound to burn plastics, which can produce carcinogens (cancer-causing agents).

- Prepare the repair adhesive material as directed by the manufacturer. When working with an adhesive system, use the manufacturer's categories to decide on a repair product and procedure.

- Plastic welding uses heat and sometimes a plastic filler rod to join or repair plastic parts.

- Hot-air plastic welding uses a tool with an electric heating element to produce hot air (450 to 650 degrees Fahrenheit or 232 to 345 degrees Celsius), which blows through a nozzle and onto the plastic.

- High-speed welding incorporates the basic methods utilized in hand welding except a specially designed and patented high-speed tip is used. Airless plastic welding uses an electric heating element to melt a smaller $1/8$-inch (3 mm) diameter rod with no external air supply.

- The basic repair sequence is generally the same for both plastic welding processes:
  1. Prepare the damaged area.
  2. Align the damaged area.
  3. Make the weld.
  4. Allow it to cool.
  5. Sand.
  6. Apply a topcoat.

- On long tears where backup is difficult, small tack welds can be made to hold the two sides in place before doing the permanent weld.

- Vinyl is a soft, flexible, thin plastic material often applied over a foam filler.

- Reinforced plastic—including sheet-molded compound (SMC), fiber-reinforced plastic (FRP), and reinforced reaction injection-molded polyurethane (RRIM)—parts are being used in many unibody vehicles.

- Work life or open time is the time when it is still possible to work the adhesive and still have the adhesive set up for a good bond.

- Cosmetic filler is typically a two-part epoxy or polyester filler used to cover up minor imperfections.
- Structural filler is used to fill the larger gaps in the panel structure while maintaining strength.

Structural fillers add to the structural rigidity of the part.

- A backing strip or backing patch is bonded to the rear of the repair area to restore the reinforced plastic's strength.

ASE-STYLE REVIEW QUESTIONS

1. When tack welding, Technician A does not use filler rod. Technician B says that the thin, fused hinge weld produced, will hold the parts in alignment. Who is correct?

    A. Technician A
    B. Technician B
    C. Both A and B
    D. Neither A or B

2. In high-speed welding, what applies pressure on the welding rod?

    A. Operator
    B. Pointed shoe
    C. Preheat tube
    D. None of the above

3. Which type of welding tips is ideal for working in hard-to-reach places?

    A. Round
    B. Speed
    C. Both A and B
    D. Neither A nor B

4. The recommended welding rod size for airless welding is _____.

    A. $3/16$-inch diameter
    B. $1/32$-inch diameter
    C. $1/16$-inch diameter
    D. $1/8$-inch diameter

5. Ultrasonic weldability depends on the plastic's _____.

    A. Melting temperature
    B. Elasticity
    C. Impact resistance
    D. All of the above

6. A good speed weld in a V-joint will have _____.

    A. A smooth and shiny appearance
    B. More uniformity than a normal hand weld
    C. A slightly higher crown than a normal hand weld
    D. All of the above

7. Which of the following statements is incorrect?

    A. CAs work equally well on all automotive plastics.
    B. PP and TPO are examples of plastics that require an adhesion promoter as part of the repair process.
    C. The best way to identify a plastic for adhesion bonding is by using the flexibility test.
    D. Both A and B

8. When Technician A grinds the base material, it melts and smears, so he or she uses an adhesion promoter to make the repair. Technician B says that if during sanding the raw plastic is again exposed there is no need for further applications of adhesion promoter. Who is correct?

    A. Technician A
    B. Technician B
    C. Both A and B
    D. Neither A or B

9. Technician A uses a pneumatic adhesive applicator when repairing a reinforced plastic. Technician B says that the pneumatic applicator uses hydraulic fluid to force the materials out. Who is correct?

    A. Technician A
    B. Technician B
    C. Both A and B
    D. Neither A or B

10. Which of the following statements concerning minor cut and crack adhesive repairs is incorrect?

    A. Allow the parts to warm to 70 degrees Fahrenheit before applying the adhesive.
    B. The adhesive is applied after the accelerator.
    C. Both sides of the cut or crack must be sprayed with accelerator.
    D. For maximum strength, allow 3 to 12 hours of curing time.

11. Which of the following is the correct repair method for RRIM?

    A. Adhesive
    B. Welding
    C. Both A and B
    D. Neither A nor B

ESSAY QUESTIONS

1. Summarize the basic procedures for making a good plastic weld.

2. How do you make a typical hot-air plastic weld?

3. Summarize how to make a two-sided plastic weld.

4. List the procedures for making a plastic weld in a cut in a vinyl-foam part.

CRITICAL THINKING

1. How would you determine if it is better to repair an SMC panel or replace it?

2. You have a tear in a plastic part. How can you tell what type of plastic it is?

MATH PROBLEMS

1. A technician made a bevel $1/8$ in wide while specs call for a 6.4 mm wide bevel. How incorrect is this groove?

CHAPTER 15

Other Body Shop Repairs

INTRODUCTION

This chapter summarizes how to service the many parts not covered in the other chapters of this textbook. The replacement of windshields, molding, trim, bumpers, and similar parts is described. Many of these operations are done in the body shop. Therefore, this information will be very useful to you when on the job in the future.

There are a wide variety of methods used to service the parts discussed in this chapter. Always refer to a manufacturer's service manual when in doubt. It will give the details needed to do quality work.

15.1 REPLACING GLASS

Today's vehicles are built with a lot of glass for greater visibility (Figure 15–1). Frequently this glass is broken out or cracked as a result of a collision, flying gravel, or vandalism. Glass is sometimes considered a structural component. It is important for the body shop technician to be familiar with the various techniques to remove and install vehicle glass. Glass must also be removed from areas of major damage before the damage can be straightened.

15.2 TYPES OF GLASS

There are two types of glass used in today's vehicles: laminated and tempered. Both are considered safety glass. They may or may not be tinted.

Laminated plate glass consists of two thin sheets of glass with a thin layer of clear plastic (vinyl) between them. It is used to make all windshields. Some glass manufacturers have increased the thickness of the plastic material for greater strength. When this type of glass is broken, the plastic material will tend to hold the shattered glass in place and prevent it from causing injury (Figure 15–2). The plastic or vinyl material is usually clear to provide an unobstructed view from all angles.

Tempered glass is used for side and rear windows but rarely for windshields. It is a single piece of heat-treated glass, which has more resistance to impact than regular glass of the same thickness. The strength of tempered glass results from the high

OBJECTIVES

After studying this chapter, you should be able to:

✔ Describe windshield glass replacement procedures.
✔ Describe how to replace a bumper.
✔ Locate and correct wind and water leaks.
✔ Install body accessories such as moldings.
✔ Explain the types of restraint systems.
✔ Explain how to replace and repair vinyl roofs.
✔ Remove and install moldings.
✔ Answer ASE test questions relating to glass, trim, and other service operations.

KEY TERMS

active restraint
air bag
air bag controller
air bag igniter
air bag module
air leaks
antilacerative glass
arming sensor
clock spring
dash assembly
electronic
   stethoscope
full cutout method
garnish moldings

grain mold die
headliner
impact sensors
inertia sensors
instrument cluster
laminated plate glass
partial cutout method
passive restraint
propellant charge
reveal moldings
sensor arrow
tempered glass
water leaks

ASE TASK LIST

Job Skills covered in this chapter include:

PAINTING AND REFINISHING TEST (B2) TASK LIST

A. Surface Preparation

1. Remove, assess, and store trim and moldings.

NONSTRUCTURAL ANALYSIS AND DAMAGE REPAIR TEST (B3) TASK LIST

A. Preparation

3. Remove outside trim and moldings as necessary; store reusable parts.
4. Remove damaged or undamaged inside trim and moldings as necessary; store reusable parts.

B. Outer Body Panel Repairs, Replacements, and Adjustments

7. Remove, replace, and align bumpers, reinforcements, guards, isolators, and mounting hardware.
17. Diagnose and repair water leaks, dust leaks, wind noise, squeaks, and rattles.
18. Install interior and exterior trim and moldings.

STRUCTURAL ANALYSIS AND DAMAGE REPAIR TEST (B4) TASK LIST

C. Stationary Glass

2. Remove and replace side modular glass in accordance with manufacturers' recommendations.
3. Determine when and how to use the partial cutout or the full cutout method of repair when replacing a windshield.

MECHANICAL AND ELECTRICAL COMPONENTS TEST (B5) TASK LIST

A. Restraint Systems

1. Active Restraint Systems

1. Inspect, remove, and replace seatbelt and shoulder harness assembly and components in accordance with manufacturers' recommendations.
2. Inspect restraint system mounting areas for damage; repair in accordance with manufacturers' recommendations.
3. Verify proper operation of seatbelt in accordance with manufacturers' recommendations.

2. Passive Restraint Systems

1. Remove and replace seatbelt and shoulder harness assembly and components in accordance with manufacturers' recommendations.
2. Inspect restraint system mounting areas for damage; repair as necessary.
3. Verify proper operation of seatbelt in accordance with manufacturers' recommendations.
4. Remove, inspect, and replace track and drive assembly, lap retractor, torso retractor, inboard bucklelap retractor, and knee bolster (blocker) in accordance with manufacturers' recommendations.

3. Supplemental Restraint Systems (SRS)

1. Disarm airbag system in accordance with manufacturers' procedures.
2. Inspect and replace sensors and wiring in accordance with manufacturers' procedures; ensure proper sensor orientation.
3. Inspect, replace, and dispose of deployed airbag modules in accordance with manufacturers' procedures.
4. Verify that system is armed and operational in accordance with manufacturers' procedures.
5. Inspect, remove, replace, and dispose of non-deployed airbag in accordance with manufacturers' procedures.
6. Use fault codes to diagnose and repair airbag system.

compression of its surfaces. This high compression is induced by rapidly heating the glass. The high temperature softens the glass. The glass is then cooled rapidly by blowing air on both flat surfaces. The resulting rapid contraction adds compressive stress to the surface, which strengthens it. Because of this compression, tempered glass cannot be cut, drilled, or ground after the tempering process.

When tempered glass is broken, the pieces are small and have a granular texture. The shattered glass has an interlocked structure to it that obstructs visibility. This is one reason tempered glass is not used in windshields. Another reason is that this type of glass does not give readily and can cause more severe head injuries in a collision. Tempered glass will also shatter if previously damaged or stressed. For example, a chipped edge or a stone striking it can weaken the glass so that it suddenly shatters some time after the initial defect is incurred.

Zone-tempered glass is a better choice for windshields. Zone-tempered glass has a lesser degree of tempering in the area directly in front of the driver; this prevents small cracks from developing in the event of breakage.

FIGURE 15-1 Laminated plate glass consists of two thin sheets of glass with a thin layer of clear plastic between them. It is used to make all windshields. Tempered glass is used for side and rear window glass but rarely for windshields. *(Courtesy of Buick Motor Division of GM)*

FIGURE 15-2 Lamination kept this windshield glass intact upon impact. Tempered glass would have shattered into small pieces.

Tinted glass can be laminated or tempered glass. Tinted laminated glass often contains a light green or blue shade in the vinyl material to filter out most of the sun's glare. This type of windshield glass is helpful in reducing eye strain and driver tension and fatigue, and prevents fading of the interior furnishings.

Some windshields are shaded to reduce the sun's glare. This type uses only a dark band or section across the top part of the windshield. Tinted or shaded glass is usually recommended if the vehicle is to be equipped with air-conditioning.

Glass can also be tinted by adding small quantities of metal powder to the other normal ingredients to give it a particular color. The addition of cobalt gives glass a blue tint; iron gives glass a reddish tint.

Safety glass can also be fitted with a defrost circuit or antenna or additional plastic laminates. Defrost glass is common in rear windows and sometimes in the windshield. Before heat treatment, metal powder, which conducts electricity, is printed on the glass surface in the form of heating wires. The metal

powder is baked on the surface during the tempering process.

An antenna wire for radio reception is either placed between the layers of laminated windshield glass or printed on the surface of the rear window glass. Some windows have antenna wires and heating wires side by side. This type of glass is used in the rear window and front windshield.

Antilacerative glass is similar to conventional multilayered glass, but it has one or more additional layers of plastic affixed to the passenger compartment side of the glass. This glass is used in the front windshield only and is for added protection against shattering and cuts during impact.

15.3 REMOVING WINDSHIELD AND REAR WINDOW GLASS

Windshields and rear windows are usually secured in place by rubber weatherstripping or by an adhesive. Generally, moldings are used on the interior and exterior of the body around the glass opening.

Interior moldings are called **garnish moldings** and exterior moldings are called **reveal moldings.** The replacement of the windshield and the rear window follow almost identical procedures, varying slightly for different makes of vehicles.

It is important to remember that one of the first steps in glass service is to protect the vehicle's interior. Place seat covers over the seats and fender covers over the dash as needed. To keep glass bits or slivers out of the defrost ducts, cover them with duct tape.

REMOVAL AND REPLACEMENT OF WINDSHIELD GLASS

Replacement of windshield glass involves two different methods based on the materials used: gasket

FIGURE 15-3 This windshield glass is secured with a rubber gasket.

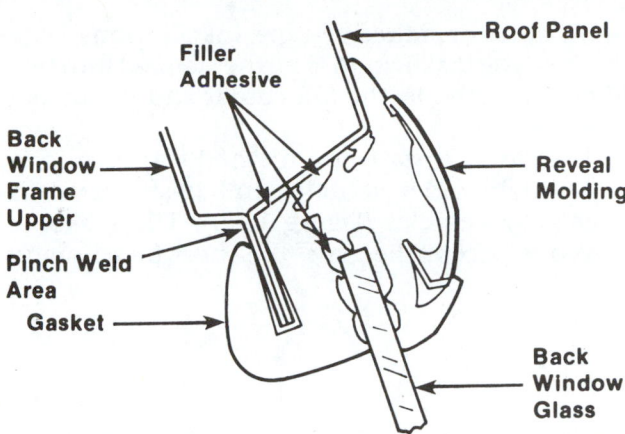

FIGURE 15-4 Cross section shows how the rubber gasket secures the glass in the body structure.

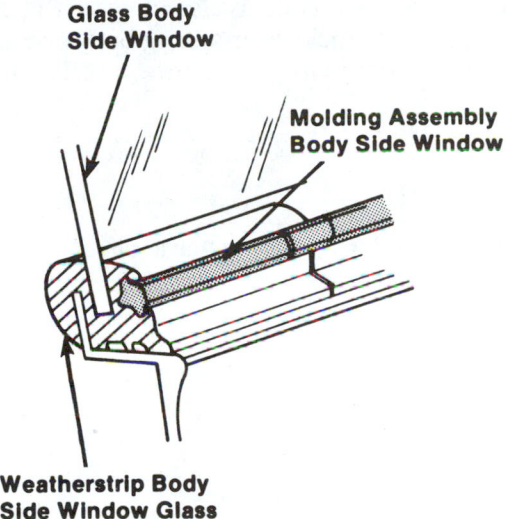

FIGURE 15-5 Note the construction of a locking strip gasket.

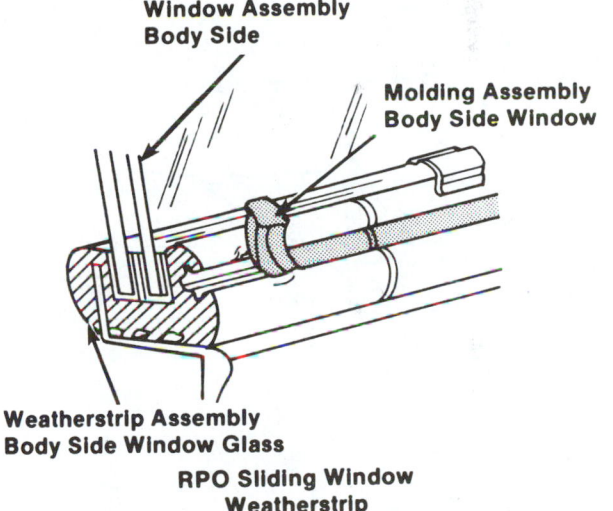

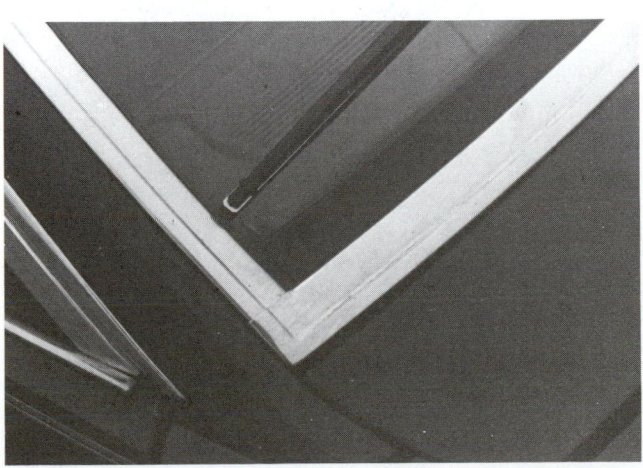

FIGURE 15-6 A cross section of bonded windshield shows design features.

FIGURE 15-7 This typical adhesive-bonded windshield is very common.

installations or adhesive-type installations. The adhesive-type installation is further refined into two additional methods: the full cutout and the partial cutout method.

The gasket installation method was more predominant in older vehicles and is still used in present-day vehicles (Figure 15–3). The gasket is grooved to accept the glass, the sheet metal pinch weld flange, and sometimes the exterior reveal molding (Figure 15–4).

Locking Strip Gaskets for Glass

The windshields and rear glass on many older-model vehicles are held in place by a locking strip of rubber (Figure 15–5). The locking strip fits into a groove between the glass and the pinch weld flange. The strip forces the gasket tightly against the glass and the flange. This locking strip must be removed BEFORE the glass can be removed from the opening. Once removed, the glass can be replaced without removing the gasket from the pinch weld flange.

The adhesive-type installation, as the name implies, uses an adhesive material to secure the glass in place (Figure 15–6). The use of adhesive permits the windshield to be mounted flush with the roof panel, decreasing wind drag and noise (Figure 15–7). Adhesive-bonded windshields also increase the overall rigidity of the vehicle, minimizing body twist and helping keep the glass in place during a collision.

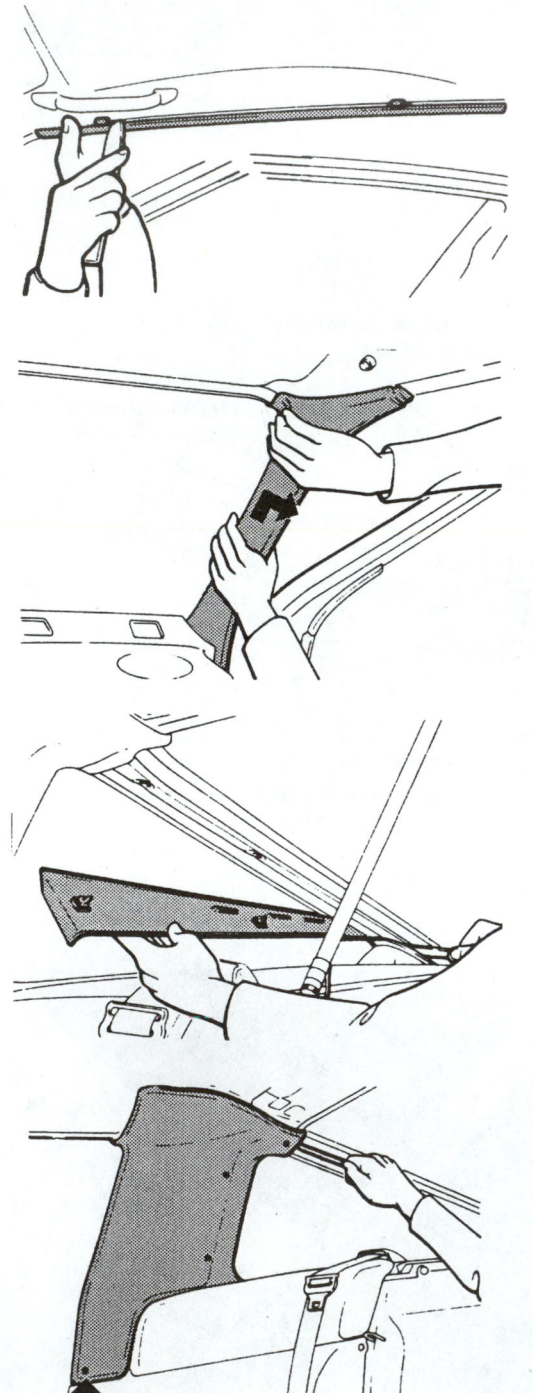

FIGURE 15–8 Interior garnish molding must sometimes be removed during glass service. *(Courtesy of Chrysler Corporation)*

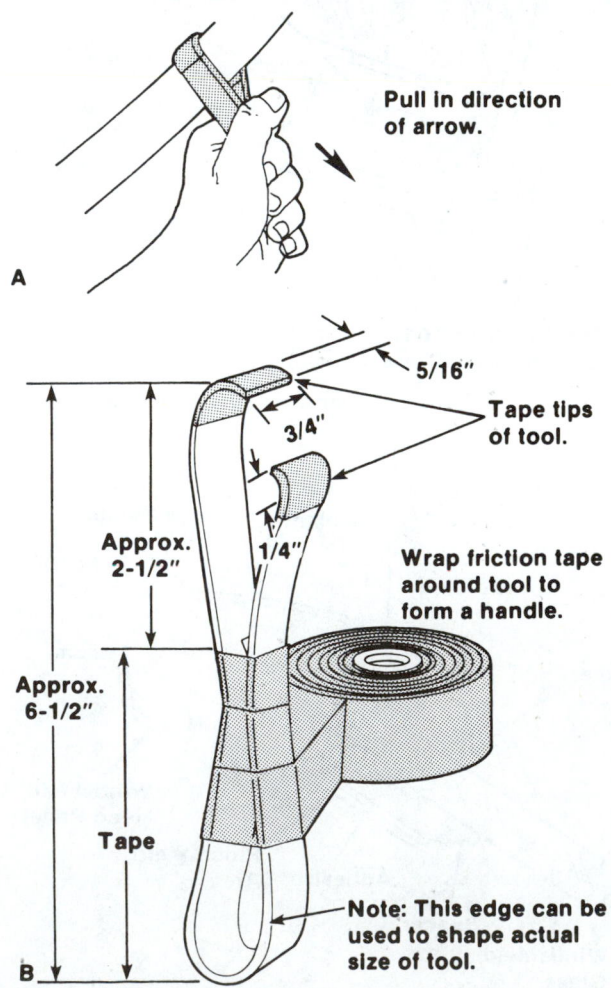

FIGURE 15–9 If needed, a spring clip removal tool can be made in your shop.

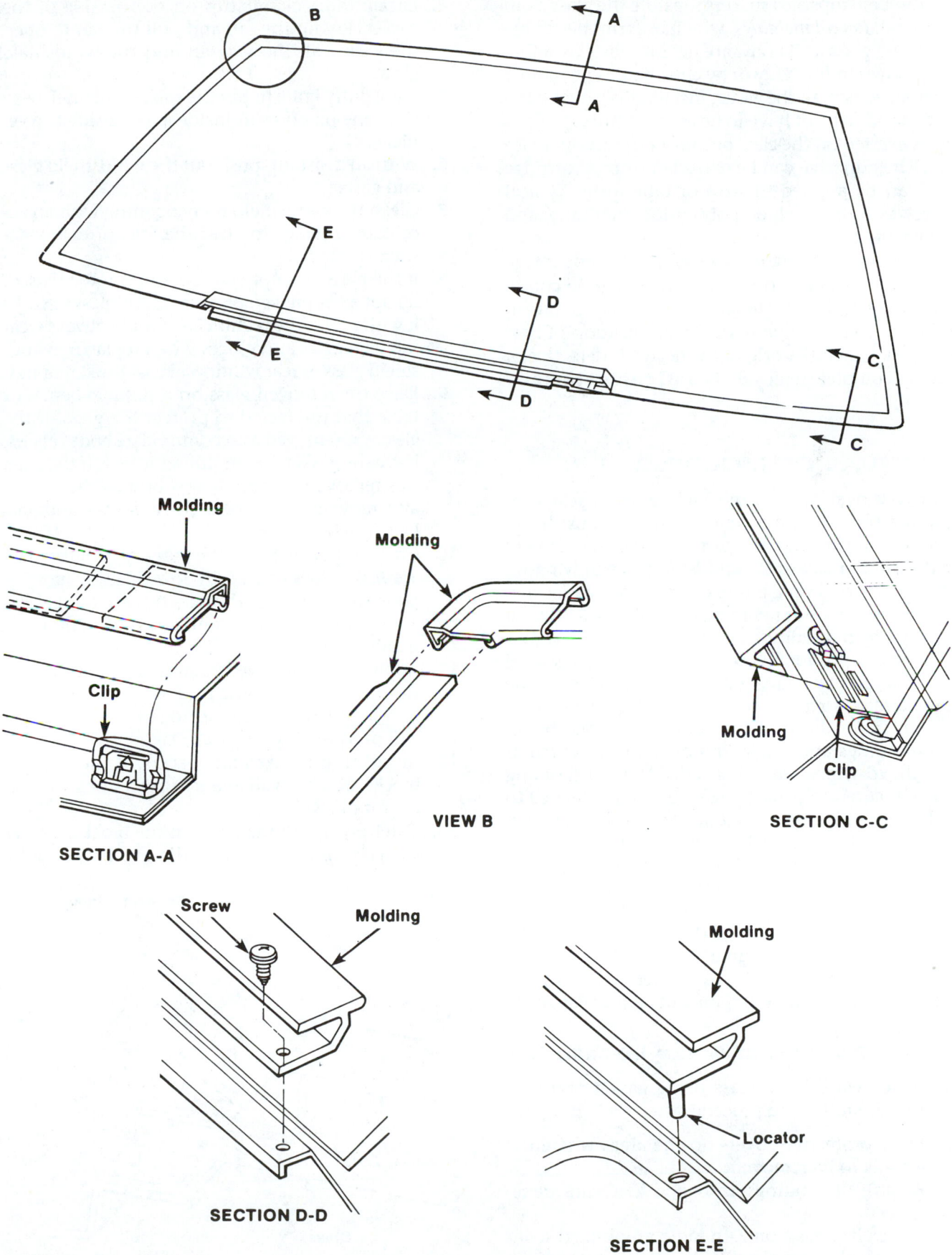

FIGURE 15-10 Study the construction of the windshield reveal molding.

Rubber stops and spacers separate the glass from the metal. *Reveal moldings*, which are trim pieces between the glass and body, are held in place by adhesive grooves in the body or by clips. This type is more advantageous than the gasket installation because the pinch welds do not have to be as exact since no pressure is exerted on the glass outside of the reveal moldings. The adhesive can be extruded from a cartridge or it can be applied in strip or tape form. Typical adhesives in use include polysulfide, urethane, and butyl rubber.

The **partial cutout method** takes advantage of the fact that most of the adhesive in good condition and of sufficient thickness is allowed to remain and be utilized as a base for the application of new adhesive. When the original adhesive is defective or requires complete removal, the **full cutout method** is used.

REMOVAL OF MOLDING

Before removal of the windshield glass or rear window, the interior and exterior moldings must be removed. In most cases, the garnish moldings are used on the interior face of the windshield or rear window. They consist of several pieces or strips that are secured in place by screws or retaining clips (Figure 15–8). All of the garnish moldings, as well as the rear view mirror (if possible), should be removed first with a special tool. If unavailable, a suitable remover can be made from banding strap steel as shown in Figure 15–9.

On the exterior of the vehicle, remove the reveal moldings and other trim or hardware (such as windshield wiper arms) if needed. Reveal molding can be secured in place by clips that are attached to the body opening by welded-on studs, bolts, or screws (Figure 15–10). A projection on the clip engages the flange on the reveal molding, thereby retaining the molding between the clip and body metal.

To disengage or remove the molding from the retaining clips, a special tool (Figure 15–11) must be used. Reveal moldings can also be anchored in the adhesive material. Exercise care when removing the reveal moldings so that they are not bent or damaged.

Gasket Method of Glass Installation

To replace windshield glass using gasket material, perform the following procedures:

1. Place protective covers on the areas where the glass is to be removed.
2. Be sure all moldings, trim, and hardware are removed.
3. If the glass has a built-in radio antenna, disconnect the antenna lead at the lower center of the windshield and tape the lead to the glass.

4. Locate the locking strip on the outside of the gasket. Pry up the tab and pull the tab to open the gasket all the way around the windshield glass.
5. Use a putty knife to pry the rubber channel away from the pinch weld inside and outside the vehicle.
6. With an assistant, push out the windshield glass and gasket.
7. Clean the windshield body opening with an acceptable solvent to clear the area of dirt or residual sealant.
8. If the glass was not cracked and is to be reused, do not exert uneven pressure to the glass or strike it with tools. The technician should always wear safety goggles and gloves when replacing windshield glass or rear window glass—broken or not.
9. Place the removed glass on a suitable bench or table that is covered to protect the glass. If the glass was removed to accommodate body repairs, leave the gasket and moldings intact. If the glass was removed because it was broken, be sure to also remove the associated moldings and gaskets from the glass.
10. Cracks that develop in the outer edge of the glass are sometimes caused by low or high spots or poor spot welds in the pinch weld flange. Examine the pinch weld and correct the problem if applicable.
11. Apply a double layer of masking tape around the outside edge of the glass with a $1/4$-inch (6.4 mm) overlap onto the inside of the glass. This will prevent chipping or breaking the glass.
12. Install stop blocks and spacers. If the original blocks are not available, cut pieces of used gasket for blocks.
13. Carefully install the glass on the blocks. Center the glass and then check the gap between the

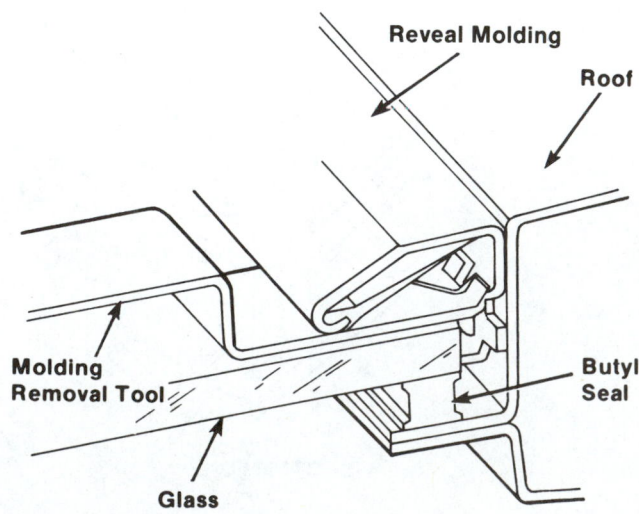

FIGURE 15-11 A clip removal tool will reach behind the molding to engage the clip.

glass and the pinch weld. The gap should be even around the entire pinch weld. Remove the masking tape around the edges of the glass.

14. Apply a bead of approved sealer in the glass channel and install the gasket on the glass.

15. Insert a cord (vinyl or nylon) in the pinch weld groove of the gasket. Start at the top of the glass. The cord ends should meet in the lower center of the glass. Tape the ends of the cord to the inside surface of the glass (Figure 15–12). Squirt a soapy solution in the pinch weld groove to ease installation.

16. Apply factory recommended sealer to the base of the gasket.

17. With the aid of an assistant, install the glass and gasket assembly in the body opening and center it. Slip the bottom groove over the pinch weld.

18. Very slowly pull the cord ends so that the gasket slips over the pinch weld flange (Figure 15–13). Work the bottom section of the glass in first, then do the sides, and finally the top section (Figure 15–14). Be sure to work the sections evenly because the glass might crack if the cord end is pulled from one side only.

19. Extrude a small bead of sealer around the body side of the gasket.

20. Remove excess sealer with a suitable solvent that will not harm the paint.

21. Install the reveal and garnish moldings.

22. Check the windshield for water leaks using a low pressure stream of water. Start at the bottom and work your way up each side slowly. Do the top last to help isolate the location of the leak.

23. Place a soapy solution in the locking strip groove and, with a tool designed for the job, replace the locking strip. The wedge-shaped tool shown in Figure 15–15 spreads the groove and feeds the strip into the opening. The soapy solution lubricates the groove and makes it easier to slide the tool through the rubber groove.

It is critical that the sheer strength and tensile strength of the adhesive is within specs. For this reason, most technicians use the windshield adhesive

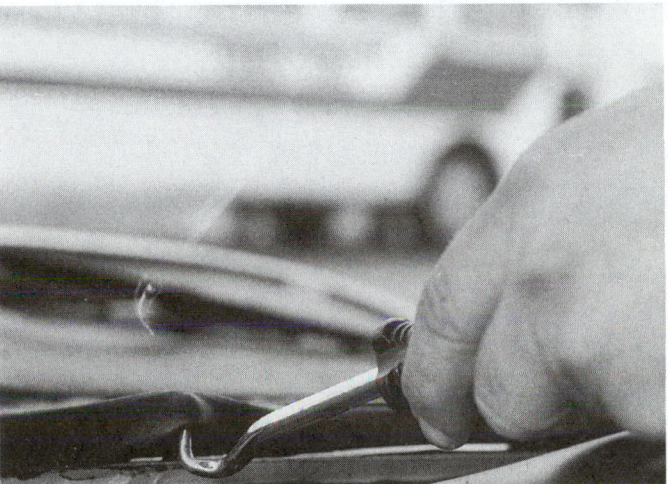

FIGURE 15–14 A hooked tool can be used to work any stuck lip of rubber gasket over the lip for complete installation. *(Courtesy of Mustang Monthly Magazine)*

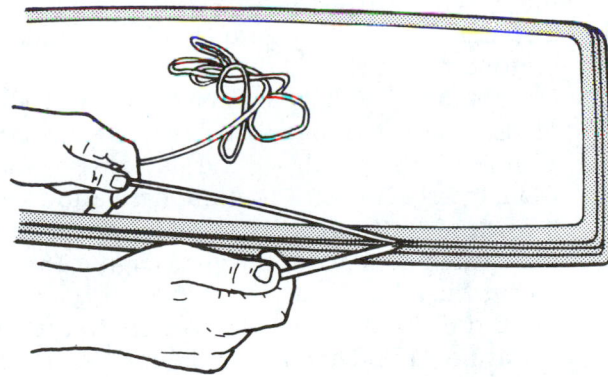

FIGURE 15–12 Cord in pinch weld opening can be used to work rubber gasket and windshield in position during installation.

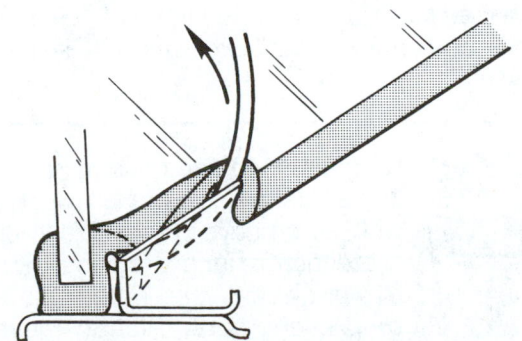

FIGURE 15–13 Pull the cord to slip the gasket lip over the pinch weld flange.

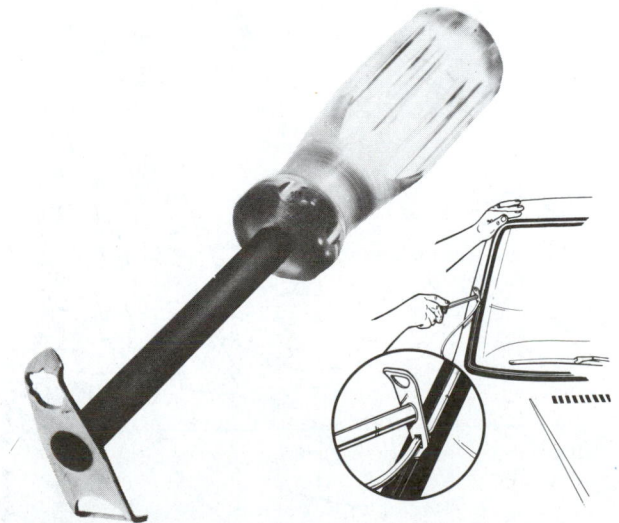

FIGURE 15–15 (A) Windshield locking strip tool and (B) installing a locking strip. *(Courtesy of Lisle Corp.)*

recommended by the manufacturer. Then there is no doubt that the adhesive is strong enough to secure the windshield properly.

Full Cut-out Method of Installing Glass

This method involves the complete removal of all the old adhesive sealer. To remove the glass, the adhesive must be cut first. Several devices are available to do this: a steel wire (piano wire), a hot knife, a pneumatic knife, an electric knife, or a cold, fine sharp knife. Each device has its own advantages and disadvantages. The pneumatic knife with a thin steel blade is preferred by many technicians.

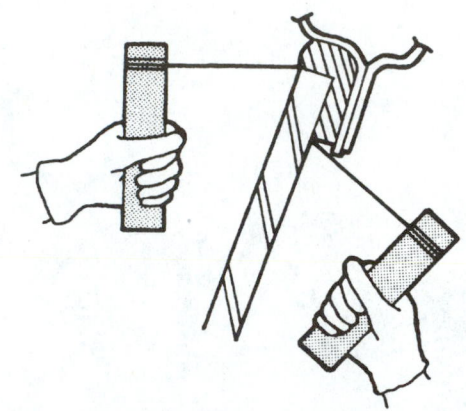

FIGURE 15-16 You can cut through windshield adhesive with a piano wire. Be extremely careful not to cut your hand!

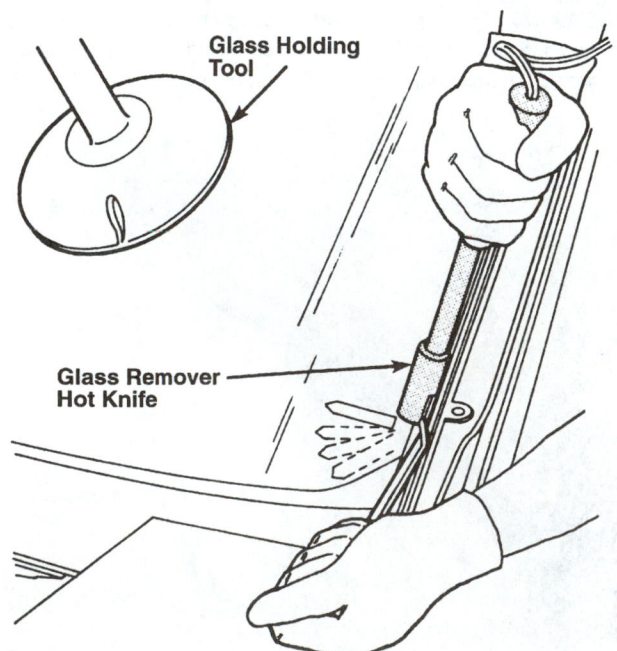

Glass Holding Tool

Glass Remover Hot Knife

FIGURE 15-17 Windshield adhesive can also be cut with a hot knife.

A 3-foot (914 mm) length of single strand steel music wire (smallest diameter available) is the safest to use to prevent glass breakage. The hot or cold knife can crack the glass in the areas where the reveal molding clips are very close to the glass.

Prior to removing the glass, check that you have the correct replacement glass. Be sure to remove all the reveal and garnish moldings and other accessories such as wiper arms, rear view mirrors, and so on. Also place protective covers inside as well as outside the vehicle in the general area of the glass to be replaced. If a window defogger or windshield antenna is installed, be sure to disconnect the appropriate electrical leads. Tape the defogger leads to the inside of the glass. Tape the windshield antenna leads to the outside of the glass.

The following procedure for replacing an adhesive-bonded windshield uses both butyl rubber tape and a urethane sealant.

1. If a steel wire is used to remove the glass, soften the adhesive by using a heat gun. Cut excessive adhesive from the glass edge to the pinch weld with a sharp knife. Attach one end of the wire to a wooden handle. Force the other end of the wire through the adhesive and under the bottom of the glass. Attach this end to a wooden handle also. With one technician inside the vehicle, work the wire back and forth to cut through the sealant (Figure 15–16). Cut out the bottom, the sides, and finally the top.

2. If a hot knife, such as the one shown in Figure 15–17, is used to remove the glass, cut excessive adhesive from the edge of the glass to the pinch weld. Insert a hot knife in the adhesive and keep it as close to the glass as possible. Cut around the entire perimeter of the glass (Figure 15–18). To cut the adhesive at the corners of the glass, move the handle of the tool as close to the corner as possible. Then rotate the tool to cut the adhesive seal. Be careful not to twist the blade of the knife because it will break. Use wedges (wooden, plastic, and so on) if the adhesive tends to reseal itself after being cut.

3. If a cold knife, as shown in Figure 15–19, is used to remove the glass, cut excessive adhesive from

CAUTION Remember to use caution and follow all safety rules when working with automotive glass. Wear eye protection to guard against flying bits of glass. Wear leather gloves to prevent cuts. Plastic gloves should be worn to keep adhesives off your skin.

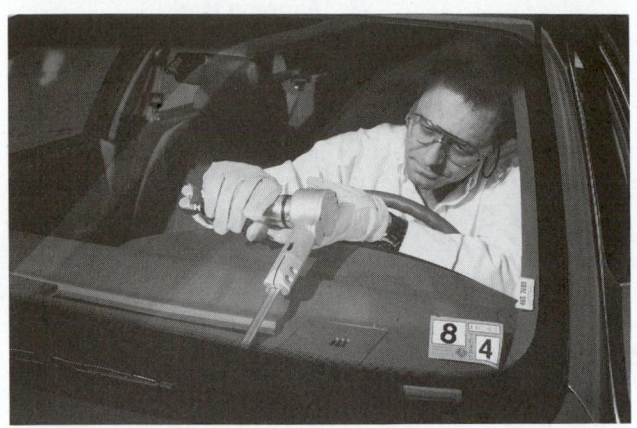

FIGURE 15-18 This technician is using a power knife to cut out a windshield for replacement. *(Courtesy of Equalizer Industries, Inc.)*

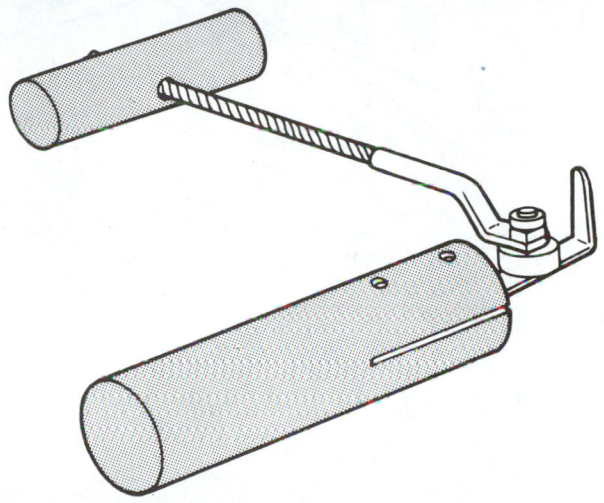

FIGURE 15-19 A cold knife is more difficult to use than a hot knife. The stationary handle is for guiding the knife. The pivoting handle is for pulling the knife in the desired direction.

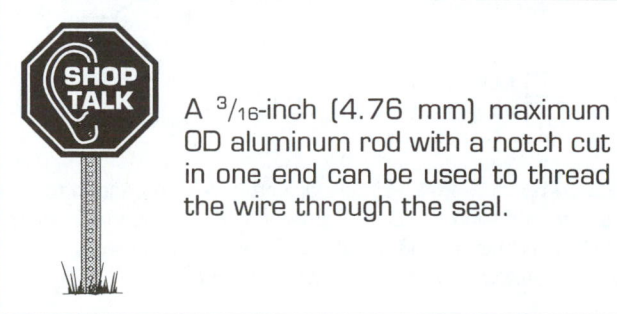

A ³/₁₆-inch (4.76 mm) maximum OD aluminum rod with a notch cut in one end can be used to thread the wire through the seal.

the edge of the glass to the pinch weld (Figure 15–20). Soften the adhesive by using a heat gun. Insert the knife and pull it carefully through the sealant (Figure 15–21). Tip the knife slightly so that the forward edge of the blade scrapes along the glass surface. Cut around the entire perimeter of the glass. Sharpen the knife blade as required.

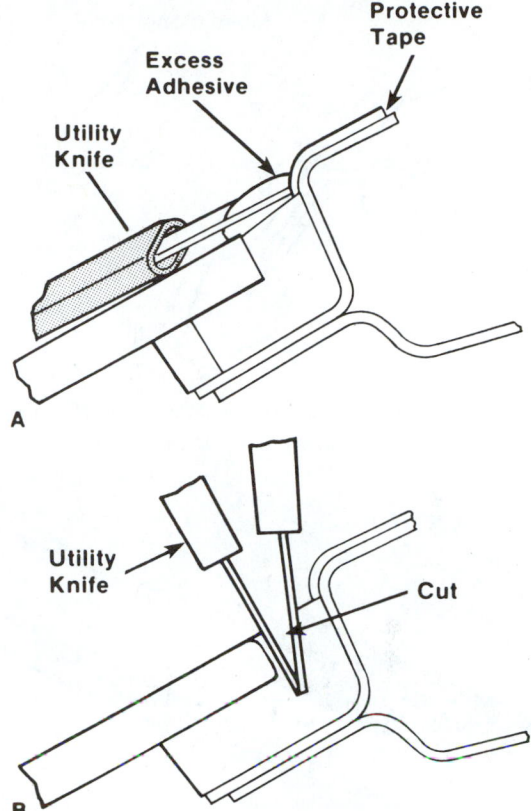

FIGURE 15-20 Cut the excess adhesive from the pinch weld flange.

Figure 15–22 shows a cold knife designed to be driven with an air hammer. This knife even cuts through tough urethane sealants with ease.

4. When the adhesive has been cut, remove the glass and place it in a safe area if it is to be reused. If the glass has been damaged, remove as required and discard. Be sure to wear safety goggles and gloves when handling glass.

5. Position replacement windshield into opening. Align for uniform fit and adjust setting blocks (spacers) as needed. To allow for sufficient bonding of urethane, make sure there is a minimum of ¹/₄ inch (6.4 mm) of glass, in addition to the space that will be taken up by the butyl tape around the entire perimeter of the glass. Mark the position with a crayon or by applying masking tape to the windshield and car body (Figure 15–23). Slit the tape at the edge of the glass. Remove windshield.

6. Remove the remaining adhesive from the body opening using a putty knife or scraper.

7. Inspect all reveal molding clips. Replace all broken or rusted clips; if bent, straighten them.

8. Check the pinch weld flange for rust. Remove any with a wire wheel or #50 grit sanding disc. Treat the bare metal with a metal conditioner, and prime the areas with a urethane primer.

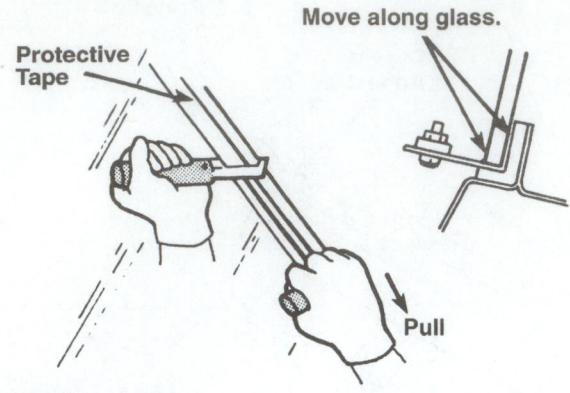

FIGURE 15-21 Pull the cold knife carefully through the adhesive without binding on the body or windshield.

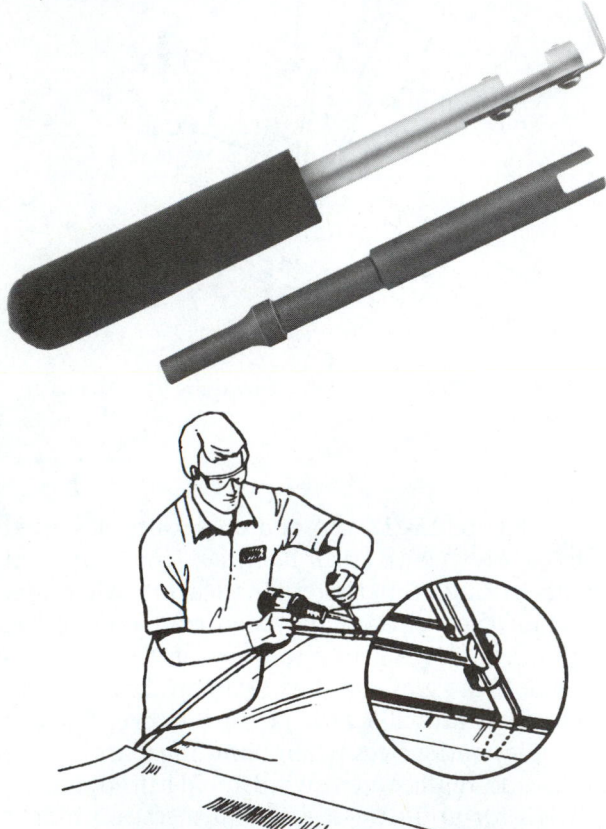

FIGURE 15-22 A cold knife can be driven with an air hammer if done carefully. This method should only be used when the windshield is ruined and is going to be replaced. The chances of breakage are too great otherwise. *(Courtesy of Lisle Corp.)*

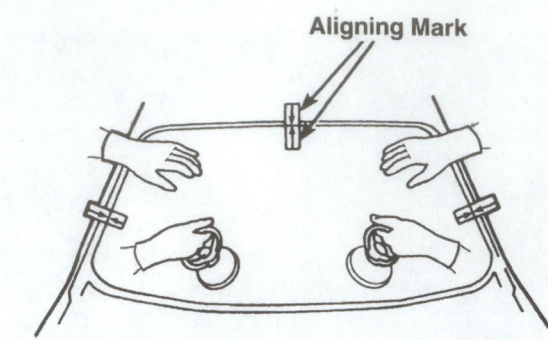

FIGURE 15-23 Mark glass position with masking tape. This will simplify installation.

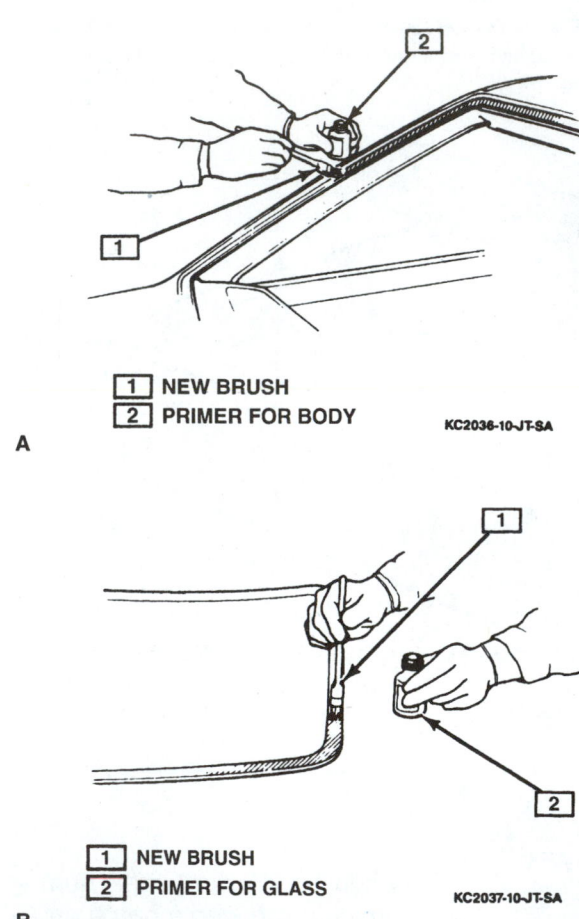

1 NEW BRUSH
2 PRIMER FOR BODY

KC2036-10-JT-SA

A

1 NEW BRUSH
2 PRIMER FOR GLASS

KC2037-10-JT-SA

B

FIGURE 15-24 Primers are often recommended to increase adhesion. (A) Body primer is being applied to inner lip. (B) Glass primer should be applied to area that accepts adhesive. *(Courtesy of General Motors Corp.)*

9. Clean the inside surface of the glass thoroughly with a recommended glass cleaner and wipe dry with a clean, lint-free cloth or towel. Note that glass cleaners containing ammonia could contaminate surfaces to be painted. Use a recommended product line so that all materials are compatible.

   Apply a uniform ¹/₂-inch (12.7 mm) wide coat of urethane primer to the inside edge of the glass (Figure 15–24). Allow primer to dry 1

SHOP TALK

If too much sealant is applied, excessive squeeze-out will occur. Taking time to do the job correctly now will minimize cleanup time later (Figure 15–28).

to 10 minutes (see manufacturer's instructions for suggested drying time).

10. Ensure the glass supports or spacers are in place (Figure 15–25). Install new ones if necessary. Cement the flat rubber spacers in place using just enough cement to attach them. The spacers should provide equal support around the perimeter of the glass and those on the sides will keep the glass from shifting left or right.

11. If replacing butyl ribbon adhesive, apply the appropriate size of rectangular adhesive ribbon sealer to the inside edge of the pinch weld (Figure 15–26). Start in the center bottom of the window opening to help avoid leakage. Do not stretch the strip of sealer. Cut the ends at a 45-degree angle and butt together.

12. Apply a bead of urethane sealant around the glass or the perimeter of the pinch weld flange as shown in Figure 15–27. Cut the cartridge nozzle at a 45-degree angle with an opening to achieve a bead size slightly larger than the ribbon sealer. Apply the sealant directly behind the ribbon sealer dam on the pinch weld. (Do not apply sealant on antenna lead wires.)

13. With the help of an assistant and suction cups, carefully position the glass in the body opening using the masking tape as a guide. Be careful not to smear the adhesive when positioning the glass. Lay the glass in the body opening and press firmly to properly seal the installation.

14. Paddle squeeze-out around the edge of the glass and remove any excess urethane. If necessary, paddle additional sealant between the glass and the car body to fill voids. Remove masking tape and protective coverings.

15. If the glass has an imbedded antenna that uses a butyl strip, put additional adhesive at the ends of the strip to get a watertight seal.

16. Water check the installation with a soft water spray. Do not use a direct water spray on the

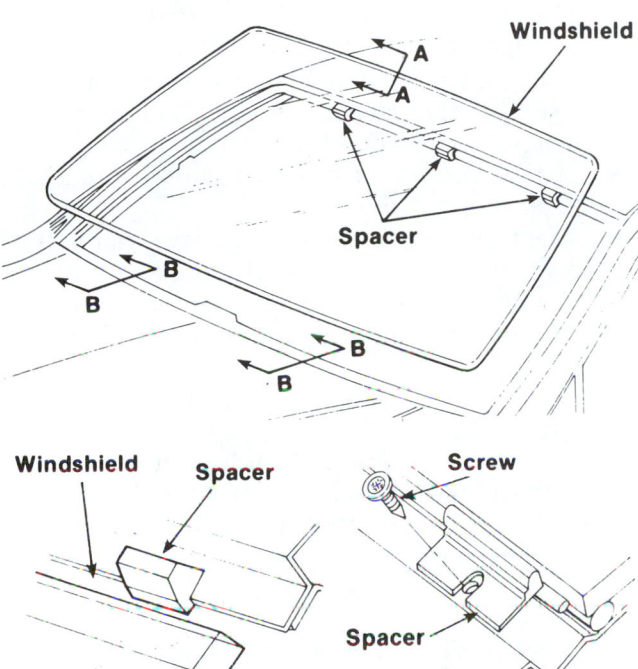

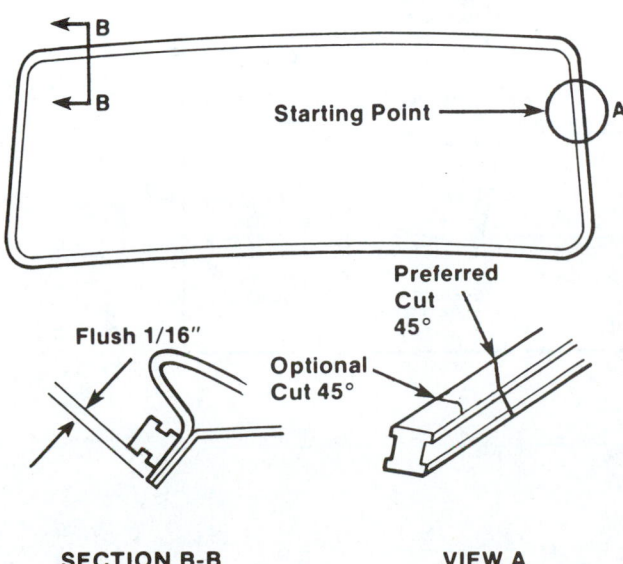

FIGURE 15–25 When used, these are typical spacer positions for a windshield.

FIGURE 15–26 Applying ribbon sealer as directed by manufacturer.

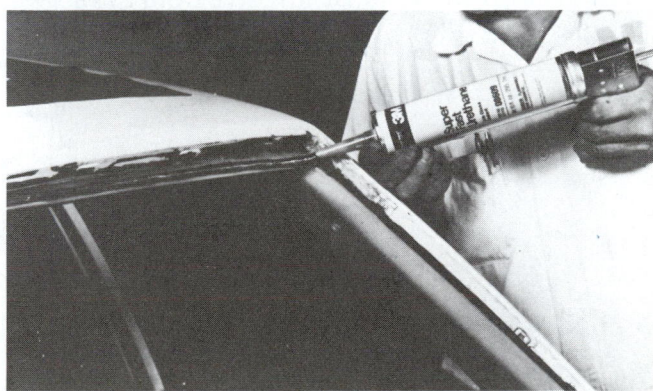

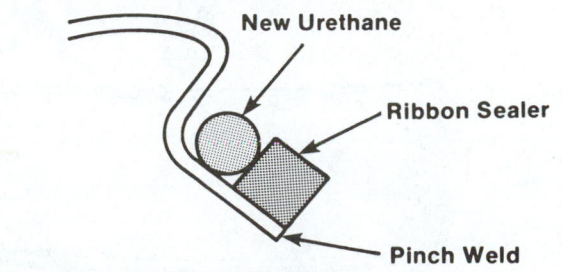

FIGURE 15–27 Applying urethane adhesive following service manual directions. *(Courtesy of I-CAR)*

fresh adhesive. Let water flow over the edges of the glass. If a leak is found, apply additional sealant at the leak point.

17. Install all necessary trim parts and attach the antenna lead and/or defogger lead.

18. Allow the adhesive to cure the manufacturer's recommended time, typically 6 to 8 hours or more at room temperature, before the vehicle is returned to its owner. However, since the ribbon sealer will hold the windshield in place while the urethane is curing, the car can be moved immediately.

Partial Cut-out Method of Installing Glass

If the partial cut-out method is to be used to install the glass, thoroughly inspect the remaining adhesive first before attempting the procedure. There must be sufficient adhesive remaining in the pinch weld to give adequate clearance between the body and the glass. This remaining adhesive must be tightly bonded to the pinch weld so that a good base exists for the new adhesive to be added.

Also check for rust under the adhesive. If it exists, the adhesive must be removed and the pinch weld ground down and refinished. Extensive rusting will require the technician to resort to the full cut-out method of installing glass.

If the original glass is to be reinstalled, remove all traces of adhesive from the glass and clean the areas with either denatured alcohol or approved solvent to clean any residual adhesive from the edge of the glass.

To replace a windshield using the partial cut-out method, do the following:

1. Place protective coverings on the vehicle to prevent damage to the paint or interior.

2. Remove windshield wiper arms, trim, antenna, and so on to expose the entire perimeter of the glass.

3. Using a utility knife, make a cut into the existing urethane sealant around the entire perimeter of glass. Cut as close to the edge of the glass as possible.

4. Using a cut-out knife or piano wire, cut out the glass, keeping your tool as close to the edge as possible. Remove the windshield. Trim any high spots on the urethane bed to assure a flat surface. The remaining adhesive should be approximately $3/32$ inch (2.38 mm) thick.

5. Inspect the reveal molding clips for damage. Replace any clips if necessary.

6. Select the proper type of adhesive that will be compatible with the adhesive used on the body pinch welds. Refer to manufacturer recommendations on the type and amount of adhesive to use.

7. Replace lower glass supports or spacers where applicable.

8. With the help of an assistant, position the glass in the body opening. Ensure that the gap is equal on both sides and that there is ample clearance on the top. Lower or raise the lower supports or spacers as required to get the correct placement of the glass.

9. Apply two pieces of masking tape from the bottom portion of the glass to the body about 6 to 8 inches (152 to 203 mm) in from the corner.

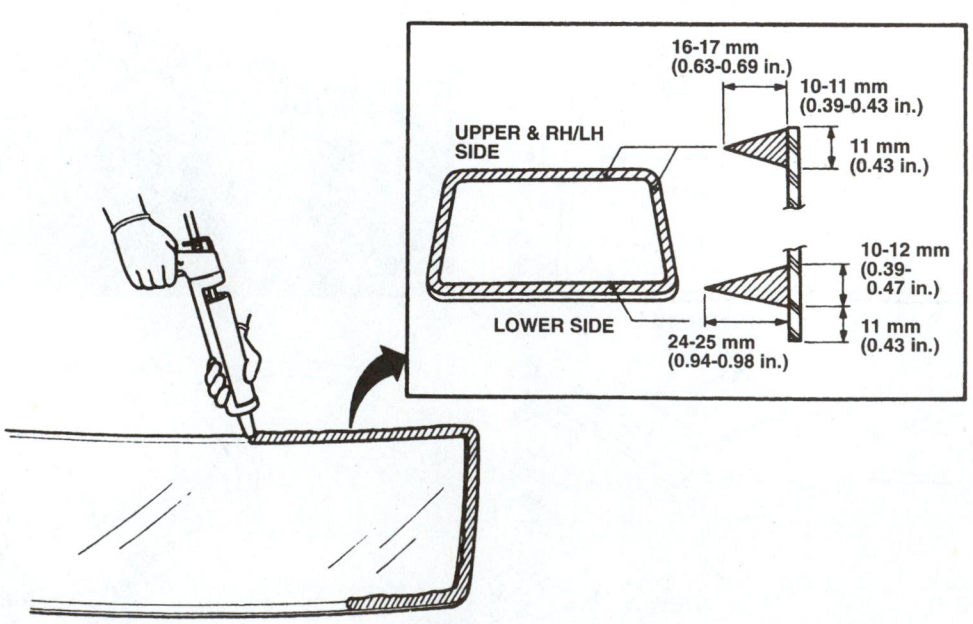

FIGURE 15-28 Here is a typical service manual illustration giving specifics for applying glass adhesive. *(Courtesy of General Motors Corp.)*

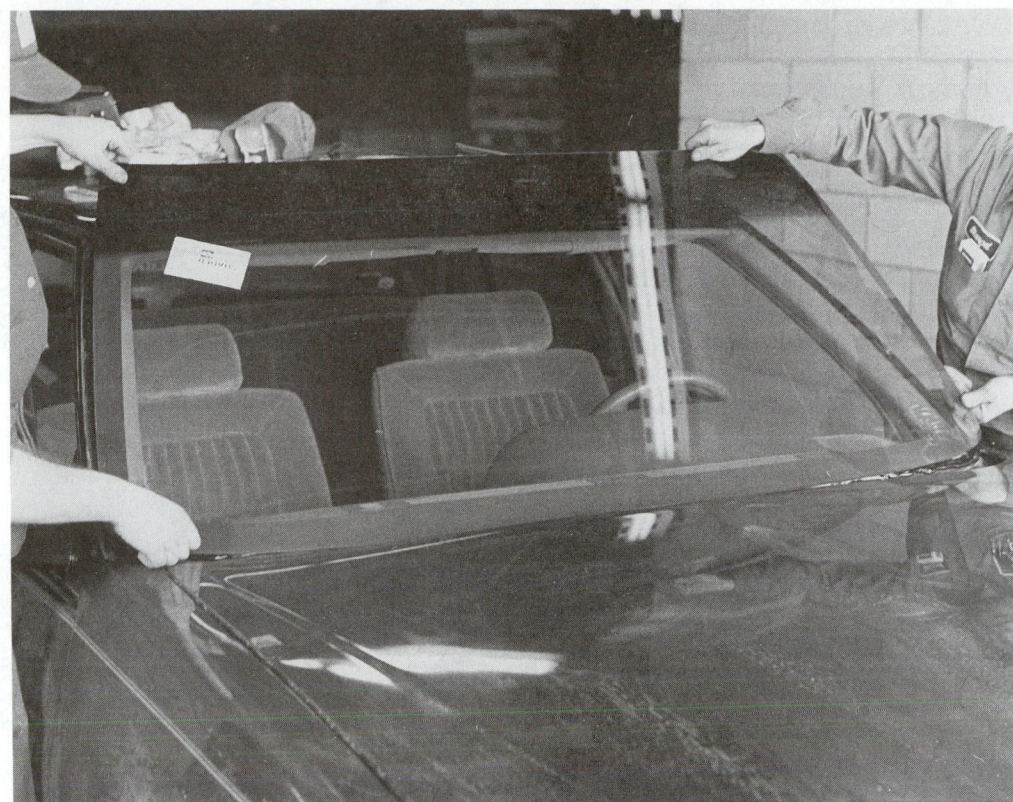

FIGURE 15–29 Slowly lower the windshield into place without smearing and breaking the bead of adhesive.

Repeat this procedure at the top of the glass. Use a razor blade or knife to cut the masking tape strips and remove the glass. The tape strips will help align the glass when reinstalling it.

10. Using a clean, dry, lint-free cloth, clean the surface of the urethane sealer remaining on the pinch weld. Replace the butyl tape strip in the antenna area with a new piece of butyl tape.

11. Clean the inside surface of the windshield thoroughly with an ammonia-free, noncontaminating glass cleaner. Wipe dry with a clean, lint-free cloth or towel. Apply a uniform 1/2-inch (12.7 mm) wide coat of urethane primer to the inside edge of the glass. Allow primer to dry 3 to 5 minutes. Also apply adhesive primer to the existing or remaining adhesive.

12. If the windshield contains an antenna, place a piece of butyl tape about 8 inches (203 mm) from the antenna pigtail. Do not use urethane or primer near the pigtail because it will interfere with radio reception.

13. Apply the manufacturer-recommended, new adhesive directly over the existing adhesive.

14. Apply masking tape about 1/4 inch (6.4 mm) from the outer edge of the inside of the glass on the top and both sides. This will aid the cleanup process when the glass is installed. Apply a smooth bead of the adhesive around the outer end of the glass or to the pinch weld.

15. With the help of an assistant, install the glass into the body opening. Place the glass on the lower supports or spacers with the masking tape strips properly aligned. (See Steps 5 and 6 above.)

16. Open the vehicle front doors. Place one hand inside the opening and gently set the glass in position. Use suction cups to control glass movement. An alternate method is to rest the glass on the lower supports or spacers. Then one technician can go inside the vehicle and help position the glass.

17. Firmly press the glass in place to set the adhesive material (Figure 15–29).

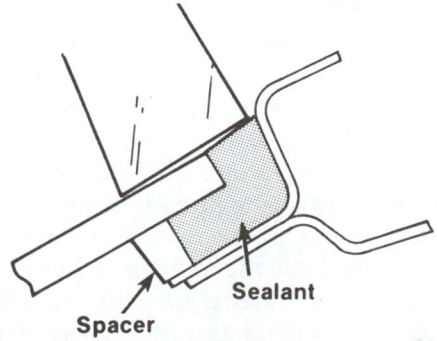

Spacer

Sealant

FIGURE 15–30 Smooth the glass adhesive with a putty knife.

18. If adhesive was placed in the pinch weld, a dark line in the glass will indicate a sealed area. The dark line should go completely around the glass. Any light spots that appear will indicate improper sealing.
19. If cartridge-type adhesive was used, the adhesive can be smoothed out along the edge of the glass (Figure 15–30).
20. Water test the installation using a fine water spray. Do not use a direct flow of water on the fresh adhesive. Correct leaks by adding additional adhesive in the applicable areas.
21. Install necessary trim and moldings and connect the antenna and/or defogger pigtails as applicable.
22. Clean excess adhesive from the glass area or body.
23. Allow the adhesive to cure for 6 to 8 hours before moving the vehicle.

WINDSHIELD WIPER SERVICE

The windshield wipers must often be serviced. The wiper arms may be held in place by spring clips or nuts. With spring clips, you must normally lift up on the end of the spring under the arm. This will free the arm for removal. A special tool is also available for spring tension mounted wiper arm removal. If a nut is used, a cover over the nut must be pivoted upward or popped off. Then you can remove the retaining nut and wiper arm.

When installing windshield wiper arms, make sure they are adjusted properly. Specs are usually given for the arms in the down position (Figure 15–31A). By engaging the arm into different teeth on the drive shaft, you can change the adjustment. Operate the wipers to make sure they are adjusted correctly and do not sweep too far one way or the other.

If the wiper blades must be replaced, disengage them from the wiper arm (Figure 15–31B). A small spring on the end of the blade must often be engaged into the arm.

REMOVING AND REPLACING REAR WINDOWS

The removal and replacement procedures for rear windows and many stationary side windows follow the same methods rather closely. Methods can vary slightly for different vehicle makes. However, many of the operations that are applied to one make of vehicle can readily be applied to others.

On some vehicles, the rear windows are removed toward the inside of the vehicle rather than to the outside. This procedure requires the removal of certain interior trim parts and accessories and the use of a special knife. For this type of glass removal, consult the manufacturer's service manual for specific procedures.

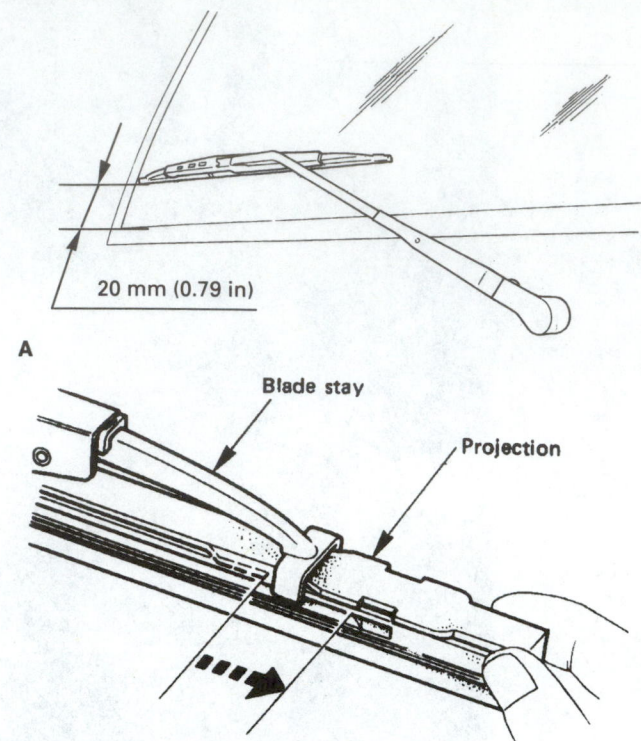

FIGURE 15–31 (A) When installing windshield wiper arms, make sure they are adjusted properly. (B) Most wiper refills are installed by compressing the small spring on the end of the arm. *(Courtesy of Isuzu Motor Corp.)*

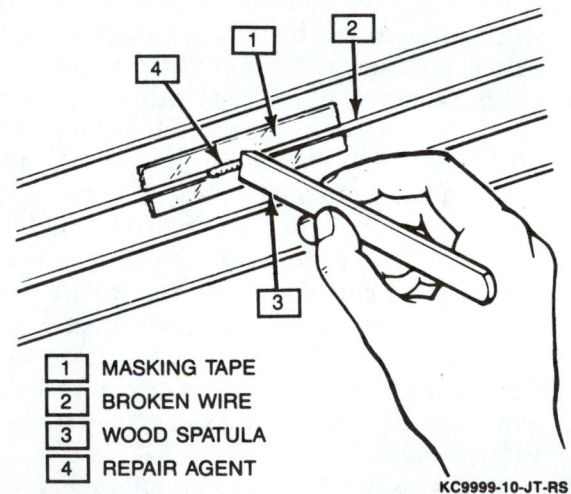

1	MASKING TAPE
2	BROKEN WIRE
3	WOOD SPATULA
4	REPAIR AGENT

KC9999-10-JT-RS

FIGURE 15–32 Some glass defrost elements can be repaired using a special agent that conducts current. Masking tape is used as a guide for applying the agent. *(Courtesy of General Motors Corp.)*

Some heating elements built onto rear glass can be repaired when damaged. A special electrically conductive adhesive is used to bridge the gap in any breaks in the heating element (Figure 15–32).

You can reattach broken antenna and heating element wires by soldering. Refer to the service manual for details.

15.4 DOOR WINDOW GLASS SERVICE

Removing or servicing door glass methods will vary. As explained in Chapter 12, door window glass is secured in a channel with either bolts, rivets, or adhesive. Doors on sedans are basically the same as are different makes of hardtop doors. Some hardtop doors require the removal of the upper window stops, the lower lift brackets or bolts, the front or rear glass run channel (if required), the upper glass stabilizers, and many other parts. If the glass is to be reinstalled, be sure to store it in a safe place.

Door glass requires servicing when it is broken or must be removed for other body or door repairs. The glass may also have to be removed to replace a broken channel assembly. On some vehicle makes, in order to remove the door glass, it may be necessary to remove the door trim panel, the water shield, the lower window stop, and the hardware securing the glass channel bracket to the glass channel. The forward edge of the glass assembly must be tilted up

to remove it from the door. At this point the channel assembly can be removed from the glass.

Mark the position of the channel on the glass. The sash channel can then be unbolted or rivets drilled out. If the glass is glued into the channel, follow this procedure:

1. Remove the channel from the glass by applying heat from a welding torch with a #2 or #3 tip along the full bottom length of the channel. Slowly pass the tip back and forth for 60 to 90 seconds, then grip the channel with pliers and pull loose. If the channel does not separate easily, repeat the heating operation.

2. Clean the replacement glass. If the original glass is to be used, scrape all traces of adhesive off with a sharp-bladed tool. If the original channel is to be reused, clamp it in a vise and burn out the remaining adhesive with a welding torch. While still hot, use a wire brush to remove adhesive traces from the channel. After it has cooled, remove the remaining adhesive from the

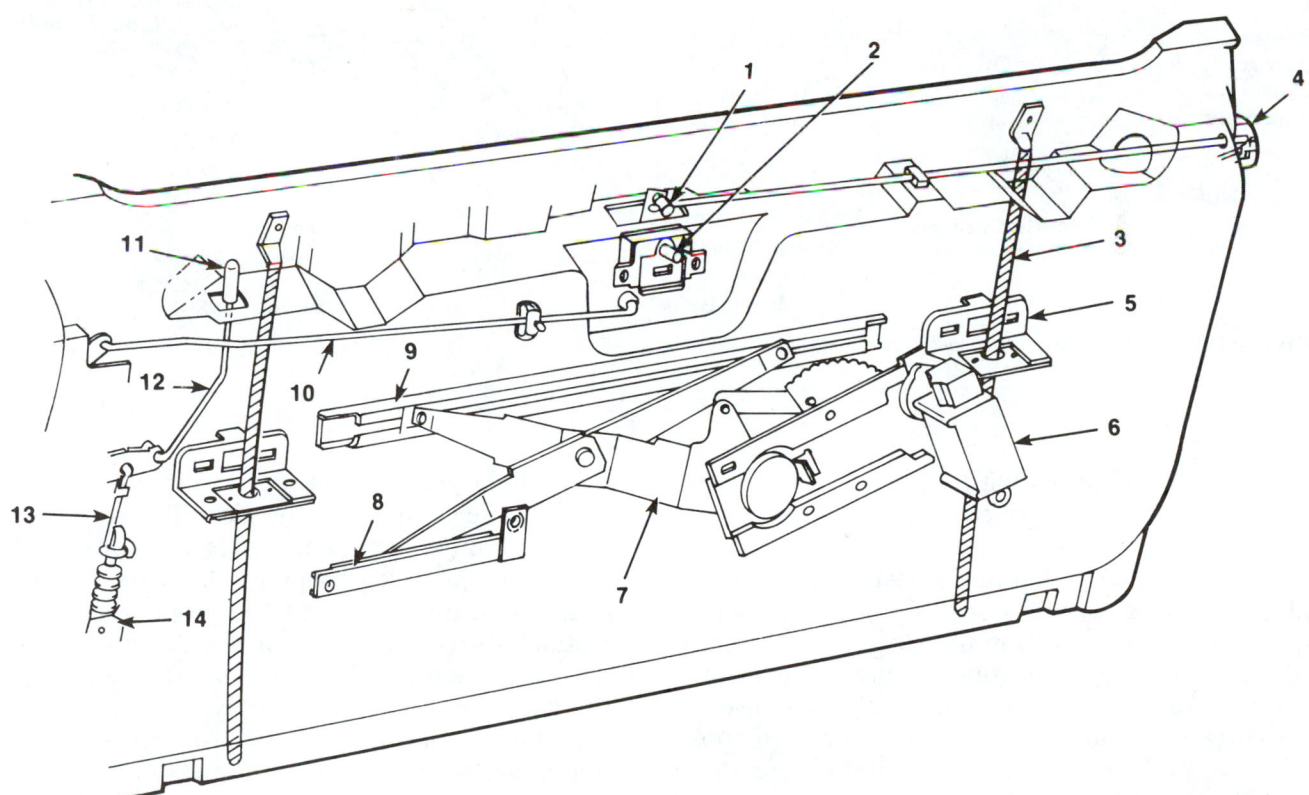

1. **Left Upper Remote Synchronization Lock Rod**
2. **Remote Control**
3. **Guide Tube**
4. **Left Upper Hinge Lock**
5. **Guide Plate Assembly**
6. **Window Regulator Motor**
7. **Tailgate Window Regulator**
8. **Tailgate Inner Panel Cam**
9. **Tailgate Glass Regulator Cam**
10. **Right Upper Remote Locking Rod**
11. **Knob Door Inside Locking**
12. **Tailgate Inside Locking to Lock Rod**
13. **Rod Tailgate Lock to Power Actuator**
14. **Electric Lock Power Actuator**

FIGURE 15-33 Note the parts of a typical tailgate glass regulator.

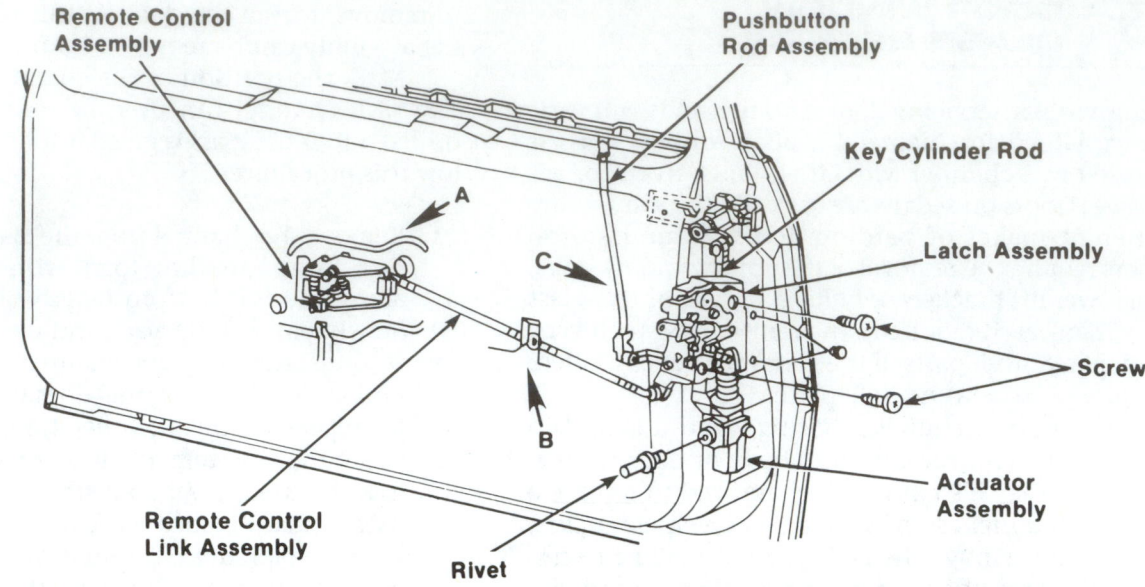

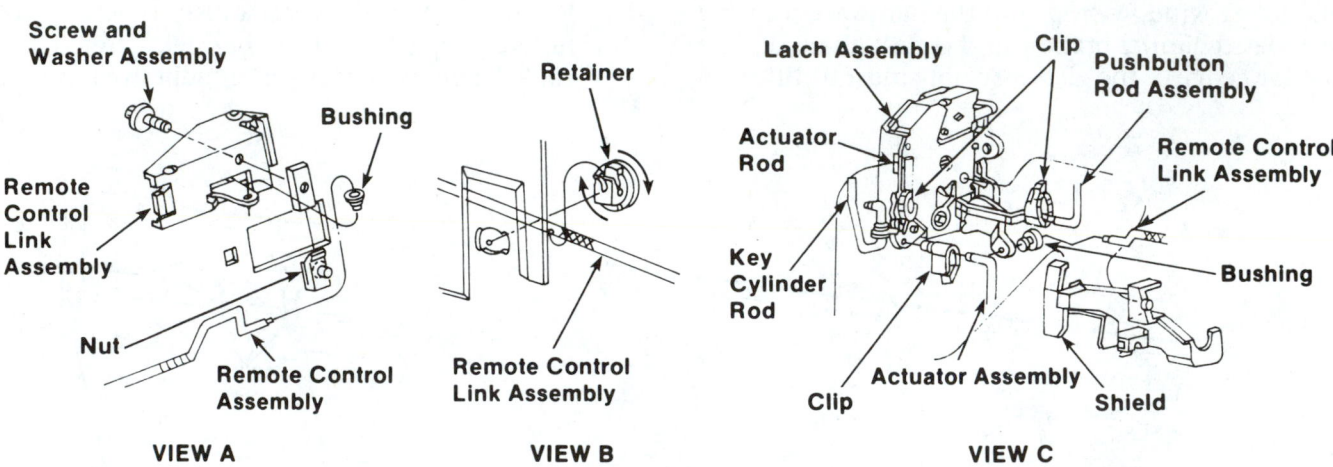

FIGURE 15-34 Study door lock assemblies.

glass and channel with lacquer thinner. Complete the cleaning operation with water.

When the sash channel is clean and dry, the replacement glass can be glued in place. If using new glass, transfer the position markings from the original glass to the replacement glass. The vehicle manufacturer's service manual can also be consulted for correct channel position. Then apply an epoxy adhesive to the channel. Position the spacers about $1/2$ inch (12.7 mm) from the ends of the channel and replace the glass in the sash. Tape the channel to the glass using cloth-backed body tape, and allow the adhesive to cure for 1 hour minimum prior to reinstallation into the car.

To reinstall glass in a bolted or riveted channel, a strip of tape of the proper thickness should be applied to the bottom of the glass. Rest the top part of the glass on a piece of soft wood or carpeting. Then

position the channel on the glass and use light blows to force the channel on the glass. If possible, use a rubber hammer. If the channel is loose on the glass, use a thicker piece of tape to close the gap in the channel a little to provide the proper width. Then reattach the channel bolts or rivets and spacers.

After the channel has been installed on the glass, the glass and channel assembly can be positioned into the door and secured with the necessary attaching hardware. Install the lower glass stop and adjust it if necessary. Finally, all the hardware and trim panels can be reinstalled.

To remove quarter glass (on rear doors on some vehicle makes), it might be necessary to remove some interior items such as the rear seat, the window regulator handle, the trim panel(s), and the water shield. While supporting the glass, remove the lower frame-to-glass attaching screws and then lift out the glass from the panel. Then remove the spacers from the

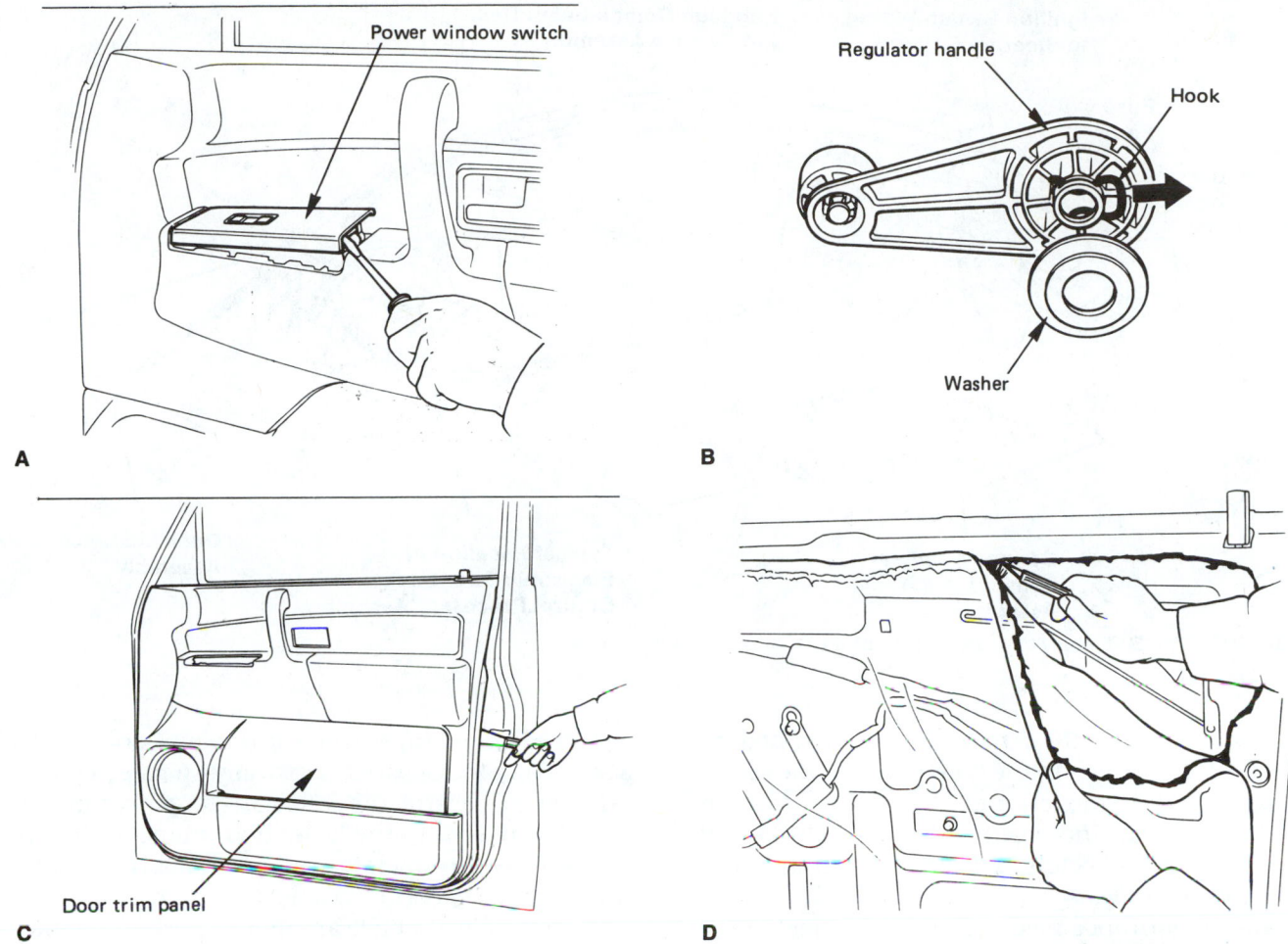

FIGURE 15-35 (A) To remove the inner door panel, first remove any snap-on parts, like this window switch. (B) The door handle can be secured with a snap ring or screws. A hook can be used to pull out the snap ring. (C) After removing any screws around the perimeter of the panel, use a wide tool to pry out the clips. (D) Try not to tear the dust cover under the door panel. It keeps dirt, noise, and moisture out of the passenger compartment. *(Courtesy of Isuzu Motor Corp.)*

openings in the glass. To install quarter glass, reverse the procedure.

NOTE! Window moldings are sometimes bent and ruined during removal. If needed, order new parts to replace all damaged ones.

TAILGATE GLASS SERVICE

Tailgate glass is generally secured to a regulator with screws or bolts. Figure 15–33 is a typical example of a tailgate window and regulator assembly. Removal or replacement of tailgate glass is similar to that presented for door glass. If in doubt, refer to the factory service manual for details.

To remove tailgate glass, run the glass to the down position (manually or electrically) and remove the inner trim panel cover. Run the glass to the up position and out of the tailgate until the glass channel screws are visible and accessible. Remove the glass channel screws to disconnect the glass from the channel assembly. Slide the glass out of the belt opening.

Then remove the attaching hardware from inside the tailgate. Finally, remove the slides from the glass in four places. The installation procedure is a reverse of the removal procedure.

15.5 DOOR AND TRUNK LOCKS

Door lock assemblies usually consist of the outside door handle and linkage, the inside door lock mechanism, and the inside locking rod (Figure 15–34). Various types of exterior door handles are in use and usually depend on the vehicle model. Among those in use are the push-button and the lift-handle types. On the push-button type, the button directly contacts the lock lever that releases the mechanism to open the door. Most exterior door handles operate a lock mechanism by using one or more rods. The handles can be replaced by raising the window and removing the interior trim and panel and water shield to gain access to the attaching hardware and linkage (Figure 15–35).

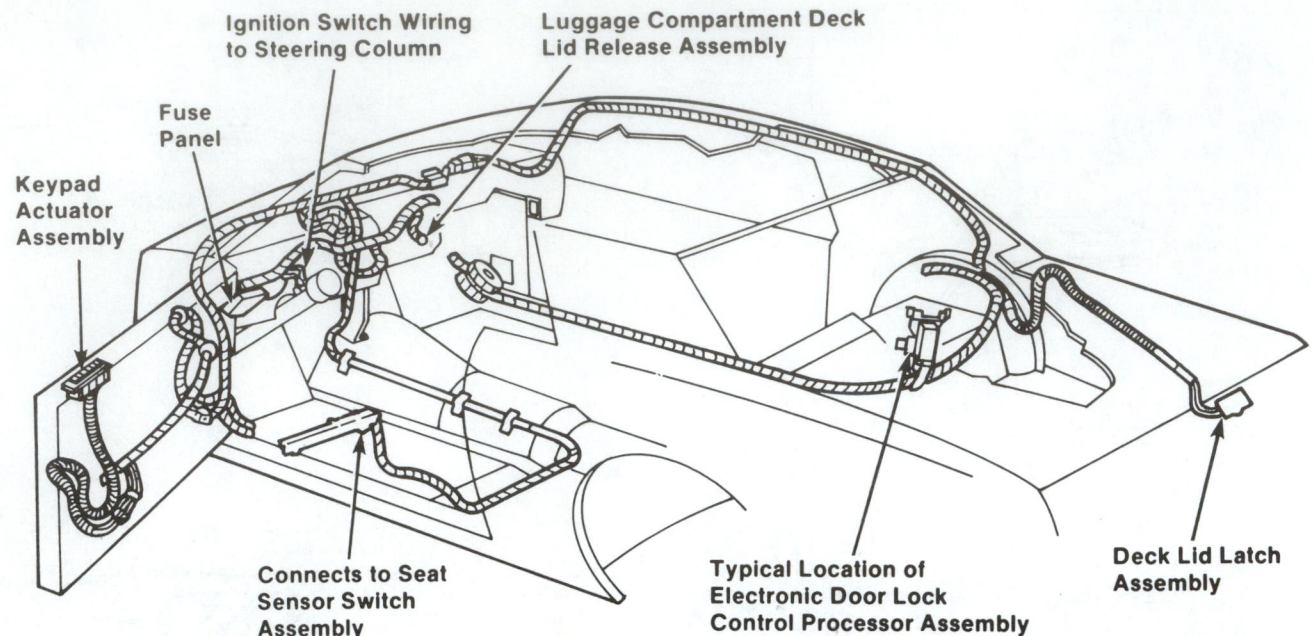

Ignition Switch Wiring
to Steering Column

Luggage Compartment Deck
Lid Release Assembly

Fuse
Panel

Keypad
Actuator
Assembly

Connects to Seat
Sensor Switch
Assembly

Typical Location of
Electronic Door Lock
Control Processor Assembly

Deck Lid Latch
Assembly

FIGURE 15-36 Study the major parts of a typical keyless entry system.

Some exterior door handles are attached by screws, bolts, or rivets to the door panel. Some causes of malfunctioning exterior door handles are: worn bushings; bent or incorrectly adjusted rods; lack of lubrication on the handle, linkage, or latch; and worn or damaged latches.

Inside door lock mechanisms are generally the pull-handle type. The mechanisms are connected to the lock by one or more rods and are accessible by removing the interior trim panel and water shield. The mechanism is attached to the door handle with screws, bolts, or rivets, depending on the vehicle make. The rods are usually long and must be supported to prevent them from engaging or getting caught in other components within the door panel. Clips or bushings are used to secure the rods in place.

Trunk or hatch doors usually do not have exterior or interior door handle mechanisms. They operate with a key (or dash switch on powered units) and lock mechanism. Some are adjustable, while others are riveted in place.

The lock cylinder on door and trunk lids is usually held in the panel with a retainer. A sealing gasket is also used on the cylinder to protect the paint. The retainer must be disengaged to remove the lock cylinder. A pointed tool or screwdriver can be used to accomplish this. On some vehicles the cylinder can be removed once the linkage is disconnected. The cylinder is either directly connected to the latch or connected by a rod. The lock is usually lubricated with dry graphite for ease of operation.

Optional equipment on some vehicles includes a feature whereby all the latches on the vehicle can be locked at one time from the driver's door control panel. The control panel electrically operates a power cylinder to control the latches. The power unit is bolted or riveted to the inner door frame. Failure of a power unit to operate could be caused by faulty wiring or connections or a defective power unit.

Some vehicles have an interior push rod or button to lock the doors. On this type, the doors cannot be opened from the inside once they are locked. This is a safety feature that is especially useful when small children are placed in the rear seat area.

Keyless door entry systems are becoming a more common feature on late-model vehicles. The system consists of two main components:

- Multibutton keypad on the outside panel of the driver's door
- Electronic microprocessor/relay module

A typical system (Figure 15-36) performs the following functions:

- It unlocks the driver's door. A keypad code is programmed into the system at the factory. The factory-programmed code is permanently recorded on the owner's warranty card (usually located inside the luggage compartment deck lid) and on a separate code card.

Some systems also allow owners to select and program their own personal code (a birth date or part of a social security number, for example) by pressing a specified sequence of keypad buttons (refer to procedure in the owner's guide). When either the factory-programmed

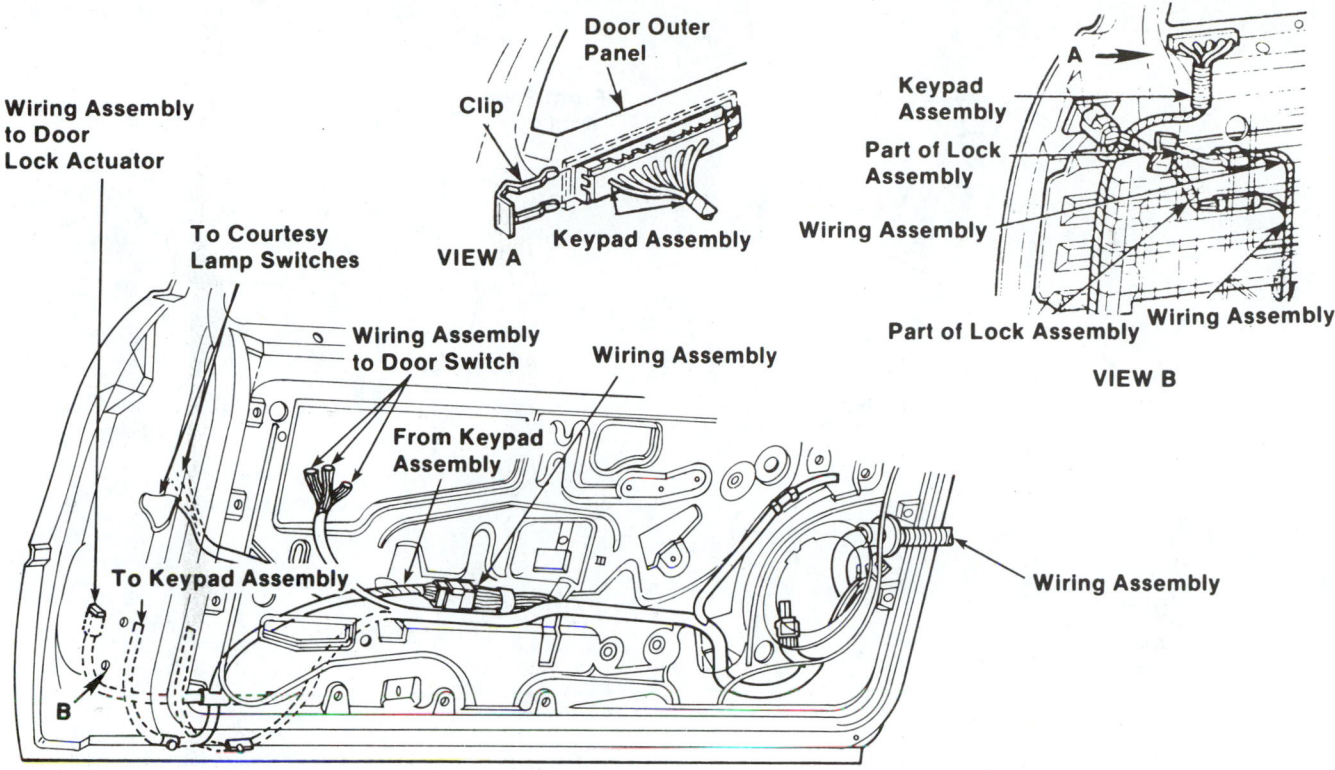

FIGURE 15-37 This wiring harness is for the keypad entry system. If the door is badly damaged, inspect the wiring closely.

code or the owner's code is entered, the driver's door unlocks.

- It unlocks the other doors of the vehicle if the keypad button is pressed within a certain time limit of the driver's door unlocking.

- It turns on the interior lamps and the illuminated keyhole on the driver's door. The lamps are turned on by pressing any keypad button or lifting the door handle.

- It unlocks the luggage compartment deck lid when the keypad button is pressed within a certain time limit after the driver's door is unlocked. It locks all the doors automatically when:

The driver's seat is occupied.

All the doors are fully closed.

The ignition switch is turned to run.

The transmission selector passes through the R position.

- It locks all the doors from outside the vehicle when the keypad buttons are pressed at the same time.

Collision repair or a malfunction in the keyless entry system might require the removal of related components from a vehicle. Figure 15–37 shows a typical wiring harness. The keypad can be secured by clips or screws. The microprocessor/relay module is located either under the dash or on the package tray in the luggage compartment. It is secured with screws or bolts. After reassembly of the keyless entry system components, test the system to ensure that it is fully operational.

15.6 LOCATING LEAKS

Water leaks are noticed when moisture or rain enters the passenger compartment and collects on the carpeting. **Air leaks** normally cause a whistling or hissing noise in the passenger compartment during driving. They are the cause of frequent customer complaints when a vehicle is brought to a body shop. Such problems are often difficult to locate. Water leaks frequently occur at panel joints and glass-to-metal joints due to cracked or insufficient sealer (Figure 15–38A). Dust and water leaks also occur at doors, windows, trunk lids, and windshields whenever the weatherstripping becomes damaged or loose or when the doors or window glass are improperly adjusted (Figure 15–38B).

Wind noises are high-frequency sounds heard during driving. They are heard mainly around the door when the window is closed. This is generally due to loose, worn, or improperly applied weatherstripping or misaligned doors, which allow air to leak

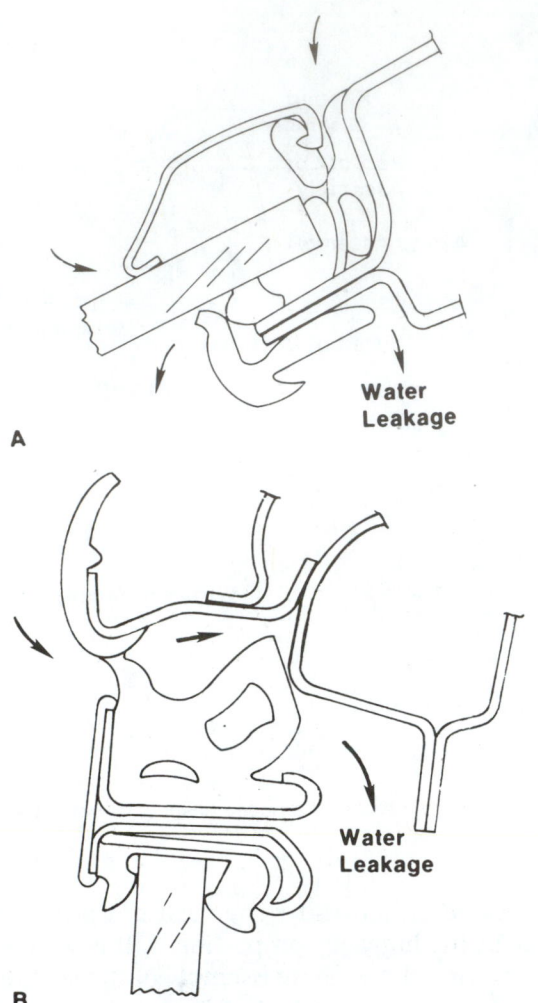

A

B

FIGURE 15-38 These are common locations for water leakage around the (A) windshield glass and (B) door weatherstripping.

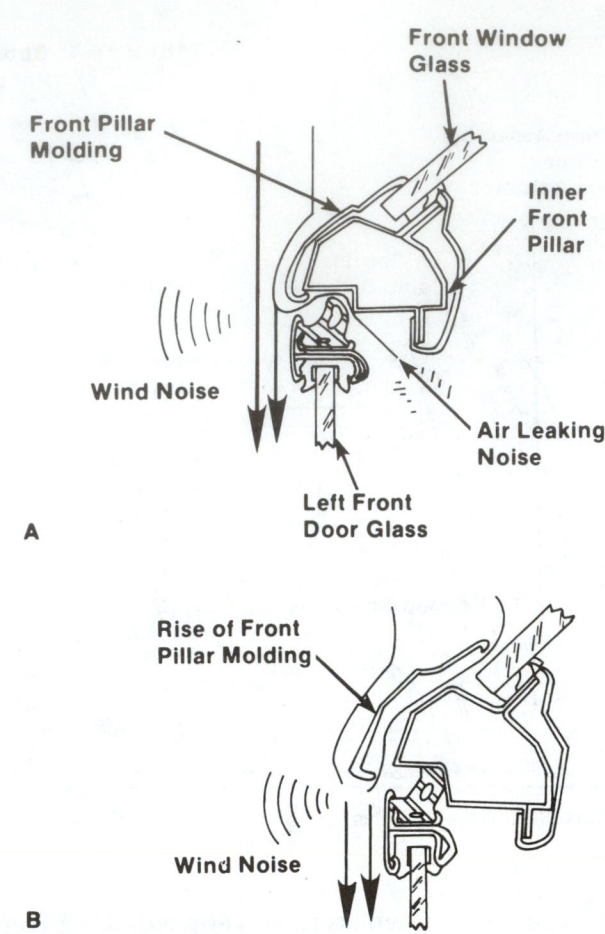

A

B

FIGURE 15-39 Wind noise can be caused by (A) loose fitting weatherstripping and (B) loose molding.

into the passenger compartment (Figure 15–39A). Wind noise is also produced when the wind hits a projection (Figure 15–39B). This disturbance produces an eddy or swirl behind the object, creating a noise (the principle of flute and bugle sounds).

A loose body molding, a poorly aligned front fender, or an improperly adjusted hood are just some examples of causes of wind noises. A troubleshooting chart on how to identify and solve wind noises is given in Table 15–1.

The principal methods used to locate air and water leaks are listed below:

- Spraying water on the vehicle
- Driving the vehicle over very dusty terrain
- Directing a strong beam of light on the vehicle and checking for light leakage between the panels
- Using a listening device

Before making an actual leak test, remove all applicable interior trim from the general area of the reported leak. The spot where dust or water enters the vehicle might be some distance from the actual leak. Therefore, remove all trim, seats, or floor mats from areas that are suspected as possible sources of the leak. Entrance dust is usually noticed as a pointed shaft of dust or silt at the point of entrance. These points should be sealed with an appropriate sealing compound and then rechecked to verify that the leak is sealed.

LEAK CHECKS USING WATER

After all the applicable trim has been removed, place one person inside the vehicle with all the doors and windows closed. Then spray the vehicle with a low pressure stream of water in the suspected area of the leak (Figure 15–40). The person inside the vehicle should act as an observer to locate just where the water enters.

The hose should be used as is, or press the end of it lightly with the thumb, according to the condition of the panel joint. Water should be sprayed for more than 10 minutes.

Another way to discover water leaks around a windshield or back light is to apply a soapy solution

TABLE 15-1: ELIMINATING NOISE LEAKS

Sources	Causes	Corrections
Weatherstrip	Imperfect adhesion to contact surface and improper contact of lip due to separation, breakage, crush, and hardening	Repair or replace weatherstrip.
Door Sash and Related Parts	1. Improper weatherstrip contact due to a bent door sash 2. Gap caused by corner piece improperly installed. 3. Gap caused by corner sash badly finished. 4. Separation and breakage of the rubber on the door glass run	1. Repair. 2. Install properly. 3. Repair with body sealer and masking tape. 4. Repair.
Door Assembly	Improper weatherstrip contact due to improper fitting door.	Correct door fit.
Door Glass	Gap caused due to ill-fitting door glass	Align door glass.
Body	Improper body finishing on contact surface for door weatherstrip (uneven panel joint, sealer installed improperly, and spot welding splash)	Repair contact surface.
Drip Molding	Rise and separation of molding	Repair or replace.
Front Pillar	Rise and separation of molding	Repair or replace.
Waist Molding	Door glass gap due to rise of molding and deformation of rubber seal	Repair.

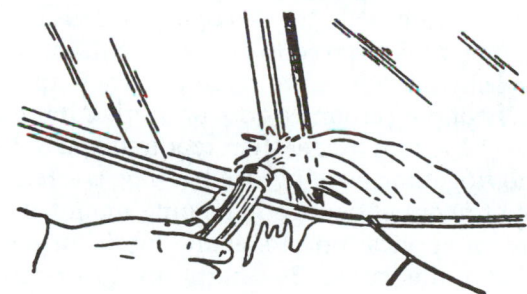

FIGURE 15-40 Water spray from a hose should be directed over potential leakage points on the body. If water starts dripping inside the vehicle, you have found the source of trouble. *(Courtesy of Chevrolet Motor Division)*

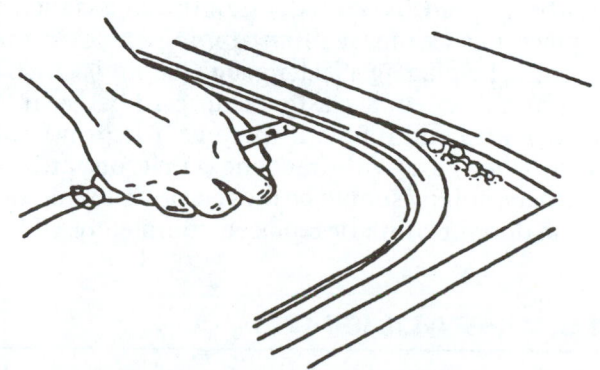

FIGURE 15-41 You can also use soapy water and compressed air to find leaks. A bubble will form inside parts at the point of leakage. *(Courtesy of Chevrolet Motor Division)*

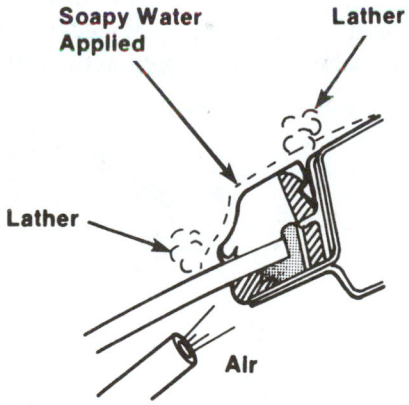

FIGURE 15-42 This cutaway view shows how air pressure will form a bubble when a soapy solution is used to find leaks.

around the outside edge of the window. Then, from inside the vehicle, apply compressed air to the window to panel joint (Figure 15–41). Any gap in the sealant will result in bubbling of the soapy solution (Figure 15–42).

LEAK CHECKS USING LIGHT

Simple leaks can often be located by moving a strong light source around the vehicle, while an observer remains inside the vehicle. This method is useful only if the leakage course is in a straight path. If the path is circuitous, the light beam will not pass through the turns and curves.

LEAK CHECKS USING A LISTENING DEVICE

A stethoscope (doctor's type of listening device) with the metal probe removed or a piece of vacuum hose can also help locate air leaks. While having someone drive the vehicle, move the hollow tube around potential leakage points. The sound of any air leakage will become very loud when you move the hose past the leak.

Special vacuum leak detectors that listen for the high pitch of an air or vacuum leak are also available. They will emit a warning sound when a test probe moves near an air leak.

REPAIRING LEAKS

Plugs and grommets are used in floor pans, dash panels, and trunk floors of a vehicle to keep dust and water from the interior. These items should be carefully checked to ensure they are in good condition.

Vehicle windshields and rear windows usually develop water leaks that can be repaired without removing the glass. A majority of leaks occur at the top of the windshield or top and bottom of the rear window. On a station wagon, the leaks usually occur on the rear side windows. If several leaks are detected,

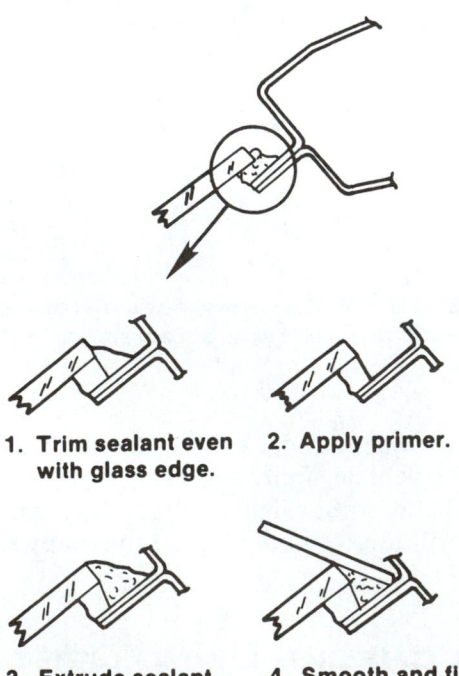

1. **Trim sealant even with glass edge.**
2. **Apply primer.**
3. **Extrude sealant.**
4. **Smooth and fill with putty knife.**

FIGURE 15-43 Here is the basic method for sealing a leaky windshield.

it is better to seal all around the area rather than at each leak point.

To repair an adhesive glass water leak, first clean the leak area and blow the area dry. Then trim off the surplus adhesive that extends beyond the edge of the glass (Figure 15-43). After removing the surplus adhesive, dry the area using compressed air. It is advisable to use a solvent to clean the area of oil or grease that might be present. Prime the repair area with a urethane primer sealer. Allow the primer to dry according to the manufacturer's instructions (approximately 5 minutes).

Then apply the windshield sealer along the cleaned area and use a putty knife to smooth it out. The sealant should be applied and spread so that it is even with the top edge of the glass and tapered back to the molding clip area. Be sure the sealant is worked into any existing crevices. While the sealant is still soft, water check the area again. Use a very soft stream or spray of water so as not to disturb the sealant. If no leaks are detected, reinstall trim and remove any surplus sealant on the glass or vehicle.

Doors and windows are sealed against wind and water with rubber gaskets called weatherstripping. Weatherstripping usually fits over a pinch weld flange or inside a channel. The rubber gaskets can be glued on, held with screws or clips, or simply held securely by the design of the gasket (Figure 15-44).

The weatherstripping in Figure 15-45 is held by a retainer that is screwed to the vehicle frame before the gasket is placed in the retainer. When applying weatherstripping around a door or trunk, cut the strip $1/2$ inch (12.7 mm) longer than required and butt the cut ends together. A sponge rubber plug is often used to hold the cut ends together. Some manufacturers require an application of silicone lubricant jelly to the base of the gasket. Be careful not to stretch the weatherstripping during installation. Pulling the strip too tight will result in an improper seal.

When the weatherstripping on doors and trunk lids becomes loose, damaged, or deteriorated, dust and water leakage results. On most vehicles the weatherstripping used on doors or trunks is cemented in place. Check the weatherstripping for correct positioning by placing a feeler gauge (about $1/64$ inch or 0.39 mm thick) or a plastic credit card between the weatherstrip and frame. If there is little or no resistance when you withdraw the gauge or card, the weatherstripping should be moved closer to the edges of the door or trunk or replaced completely.

15.7 HEADLIGHTS

If the vehicle was involved in a front-end collision, the headlights should be adjusted after needed parts

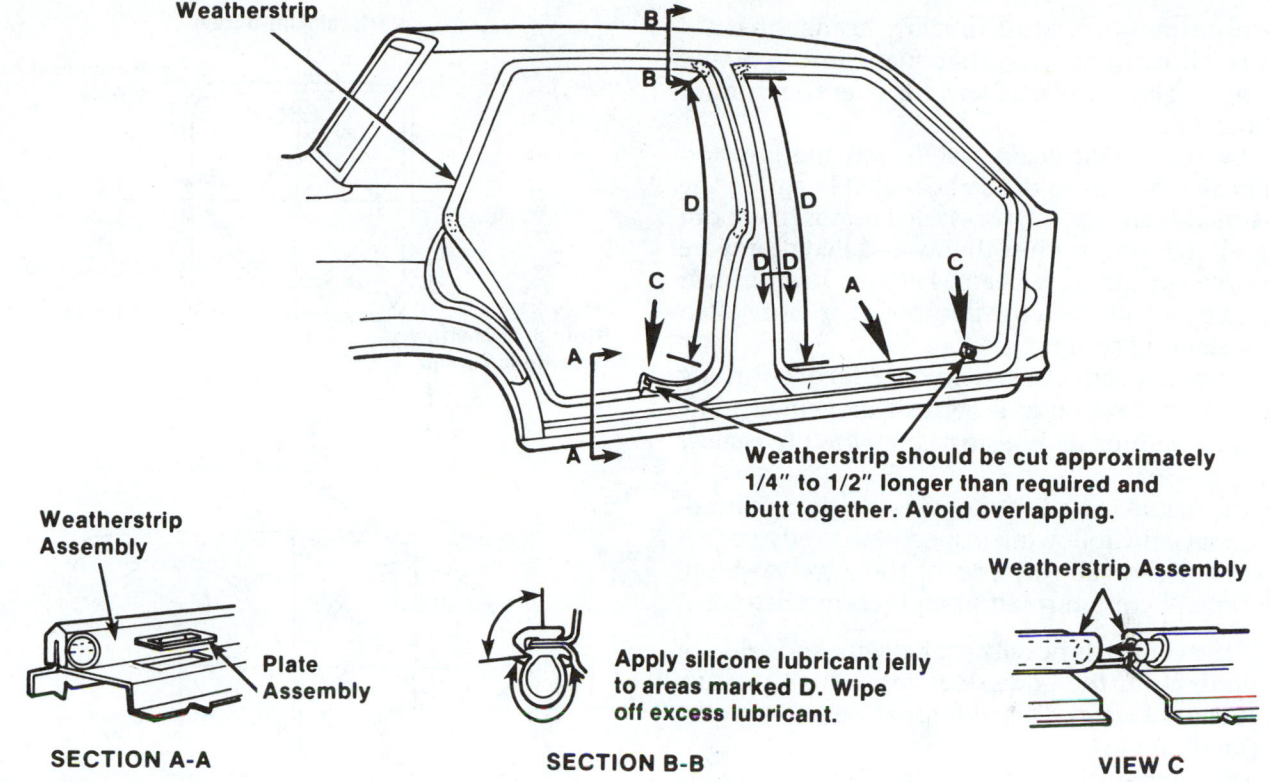

Weatherstrip should be cut approximately 1/4" to 1/2" longer than required and butt together. Avoid overlapping.

Apply silicone lubricant jelly to areas marked D. Wipe off excess lubricant.

SECTION A-A

SECTION B-B

VIEW C

FIGURE 15-44 Note typical door gasket installation.

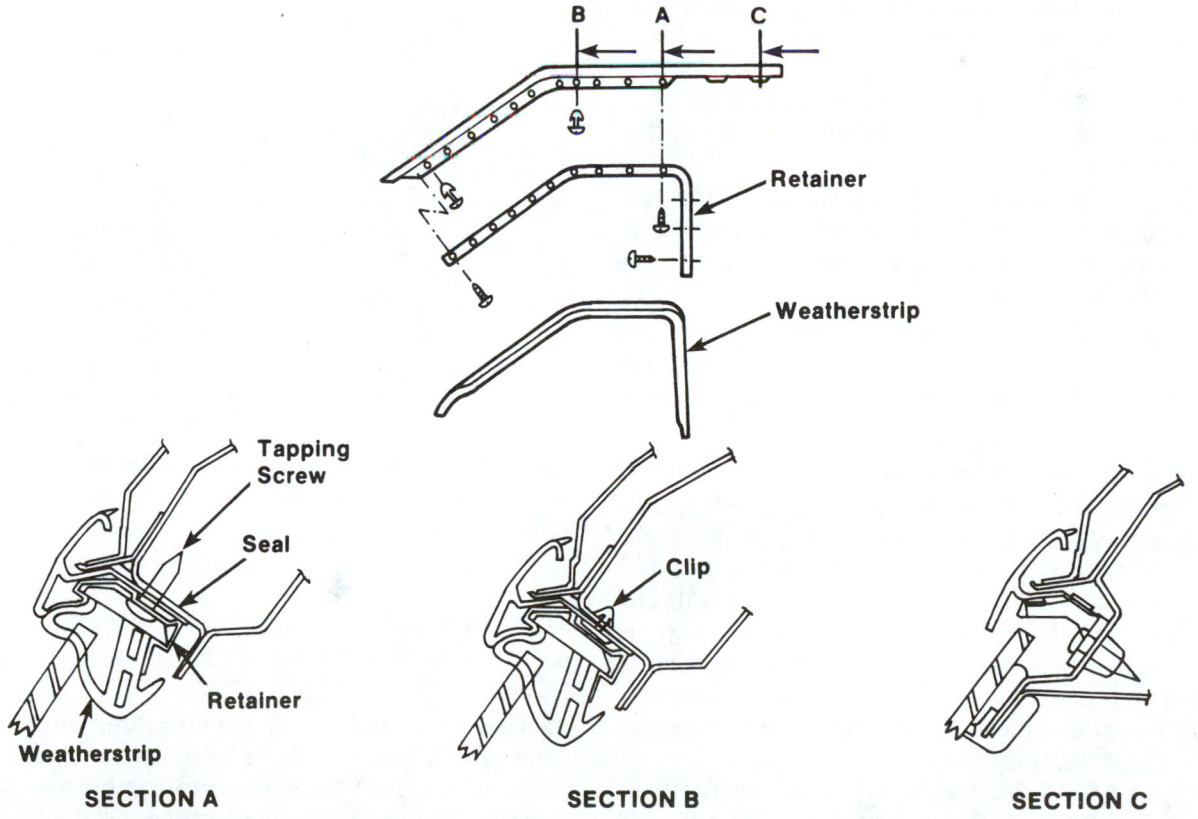

SECTION A

SECTION B

SECTION C

FIGURE 15-45 This weatherstripping is held by a retainer.

have been replaced. Most modern headlight bulbs are a small cartridge-type that snaps into a plastic housing. A glass or plastic lens fits over the front of the housing.

The two-sealed beam system has the low and high beams built into the same sealed beam. In the four-sealed beam system, two-sealed beams have both low and high beams; the other two-sealed beams are used only on the high beam. High or low beam is selected by a switch on the floorboard or more often on the steering column.

A typical headlight system consists of adjusting screws for horizontal and vertical movement. The headlight bulb or its housing is retained by screws (Figure 15–46).

Before making adjustments to a vehicle's headlights, make the following inspections to ensure that the vehicle is level. Any one of the adverse conditions listed here can result in an incorrect setting.

- If the vehicle is heavily coated with snow, ice, or mud, clean the underside with a high-pressure stream of water. The additional weight can alter the riding height.
- Ensure that the gas tank is half full. Half a tank of gas is the only load that should be present on the vehicle.
- Check the condition of the springs or shock absorbers. Worn or broken suspension components will affect the setting.
- Inflate all tires to the recommended air pressure levels. (Take into consideration cold or hot tire conditions.)
- If collision damage requires straightening the frame, make sure that the wheel alignment and rear axle tracking path are correct before adjusting the headlights.
- After placing the vehicle in position for the headlight test, bounce the vehicle by pushing down on the bumper or front fenders to settle the suspension.

Normally, the body shop will have and use a headlight alignment unit to make the necessary and correct adjustments. Once the vehicle is properly interfaced with the alignment unit, the horizontal and/or vertical adjustment screws on the headlight are adjusted for the proper reading or indication on the unit. If no headlight alignment equipment is available, an alternate method is available whereby an alignment setup can be laid out on the floor and wall of the auto body shop.

Figure 15–47 illustrates such a layout that can be used. The high and low beams should be adjusted until they appear as shown in Figure 15–48. The preferred method to align the headlights, however, is to use the headlight alignment unit.

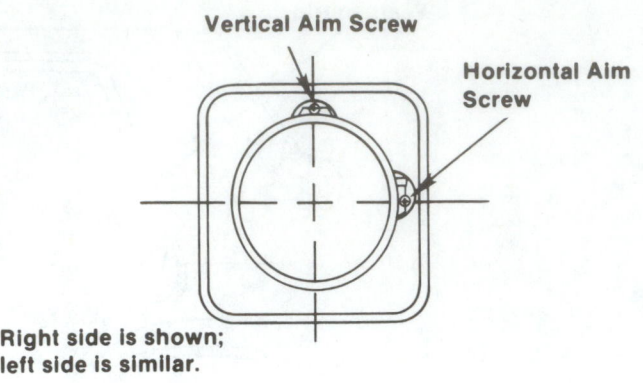

Right side is shown; left side is similar.

A

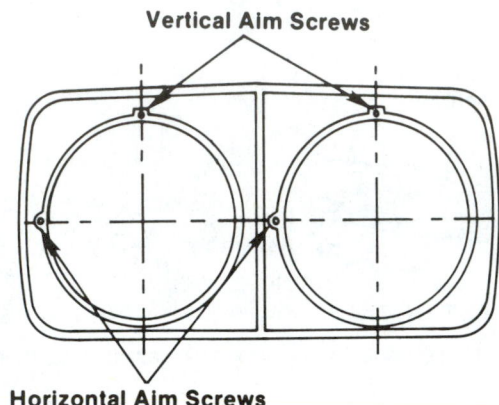

B

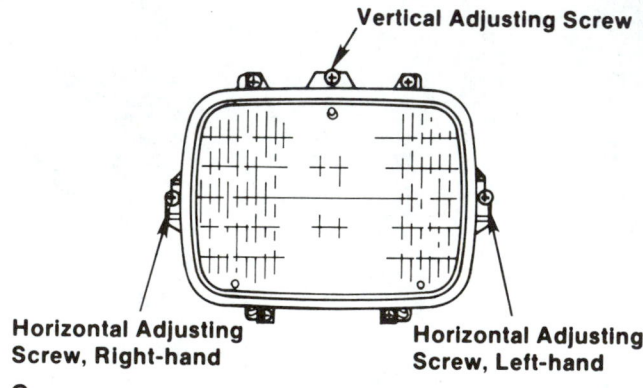

C

FIGURE 15-46 Headlight bulb adjusting screws are provided for moving the beam direction right or left and up or down.

15.8 TAIL, BACKUP, AND STOP LIGHTS

Failure of a tail, backup, stop, or directional light on the vehicle can usually be attributed to a faulty bulb. Oftentimes moisture gets into the bulb socket and causes corrosion of the electrical contacts and the bulbs. Corrosive conditions can be repaired by using sandpaper on the affected areas. For severe cases, replace the socket and/or bulb. After any repair, always

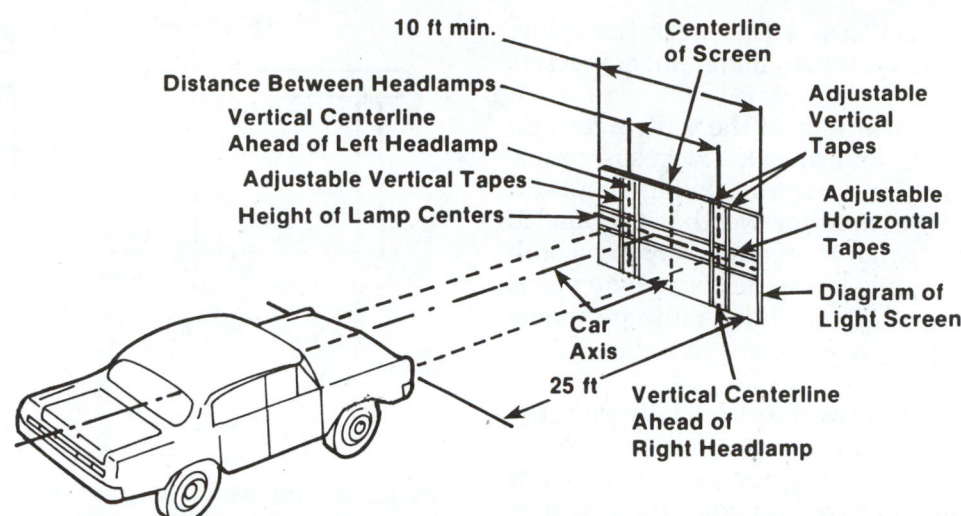

FIGURE 15-47 This is a headlight alignment setup. Beams should be set to hit the wall at a prescribed direction.

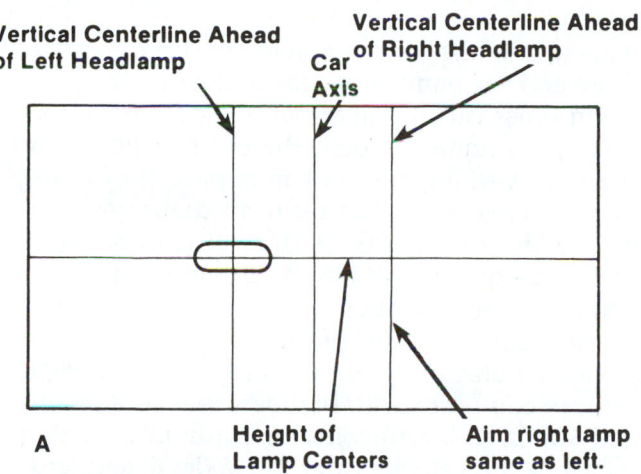

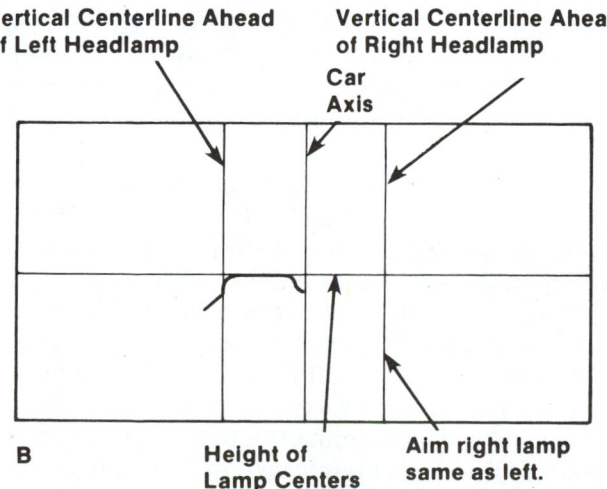

FIGURE 15-48 (A) Correct upper headlight beam alignment; (B) correct lower headlight beam alignment.

attempt to waterproof the assembly to prevent future problems. Sometimes, *dielectric grease* (special electrically conductive grease) is recommended in connectors and sockets. This grease should always be replaced if connectors and sockets are cleaned.

These types of lights are also contained within a lens- or bezel-type assembly, usually amber or red in color. Cracked or broken assemblies are easily replaced; they are secured by attaching hardware readily accessible to the body technician.

15.9 RATTLE ELIMINATION

Rattles and squeaks are sometimes caused by sheet metal that is loose or rubbing adjacent parts. They are also caused by loose bolts and screws, and improperly adjusted doors, hoods, or body panels. Other rather simple things such as a broken or loose exhaust mount or an improperly secured jack or tire, or articles in the trunk can also be the cause of rattle.

Oftentimes, a noise will be pinpointed by the customer to be in a certain area of the vehicle when in fact it might be caused by something in another area of the vehicle. This is caused by the sound traveling through the body. Usually a thorough investigation and a test drive of the vehicle is recommended so that the rattle or noise can be located.

Most rattle or noise repairs involve readjustment or replacement of parts, tightening loose attaching hardware, and welding broken parts. Check the hood for proper alignment at the front and the back. If the paint is chipped off or scratched on one end, it is probably hitting another edge or rubbing against it. Check the hood latch pin for looseness and proper fit as well as the rubber hood bumpers. If the back of the hood flutters, readjust the hood so that it fits properly at the back seal. Also check the grille, wheelhousing, trim moldings, and bumper brackets for tightness.

Many areas on the body of the vehicle can also cause rattles, noises, and squeaks. Most susceptible areas are the dash, doors, steering column, and seat tracks. It is also possible for weatherstripping to squeak, especially when it becomes very dry. Lubrication should be applied to applicable moving parts such as door hinges. All attaching hardware should be checked for tightness, especially in the area of the suspected noise source.

An **electronic stethoscope** is handy when trying to find rattles and other mechanical noises. Some have alligator clips and long test leads that can be connected to different components. You can then drive the vehicle while listening to the noise through the tester. When you connect the alligator clip to a part and the noise is the loudest, you have found the source of the rattle (Figure 15–49).

15.10 ADJUSTING OR REPLACING BUMPERS

Bumpers are designed to protect the front and rear of a vehicle from damage during a low-speed collision. Most modern bumper covers are made of flexible plastic. Some bumpers are made of heavy gauge spring steel plated with bright chromium metal. Many trucks are still equipped with chrome bumpers. Other vehicles, however, can have aluminum bumpers. These are much lighter than the steel bumpers and maintain a bright finish without chrome plating. Painted steel bumpers are popular on many trucks.

Bumpers on many late-model cars are covered with urethane or other plastics. The use of urethane, polypropylene, or other plastic allows the bumper to be shaped to blend with the body contour. Plastic bumper covers can also be painted to match the body finish color. Underneath the plastic covers might be a steel or aluminum face bar or reinforcement bar or a thick energy-absorbing pad made of high-density foam rubber or plastic. The repair of flexible bumper covers, as well as other plastic bumper parts such as bumper strips soft nose parts, gravel deflectors, and filler panels is explained in detail in Chapter 14.

Chrome bumpers that are severely damaged or that have chipped chrome plating are usually sent to a specialty shop that specializes in repairing and rechroming bumpers. The damaged bumper is exchanged for a refurbished bumper. Painted bumpers that are not too severely bent can be pulled and bumped back to shape, using common shop procedures. The damaged area can then be ground down

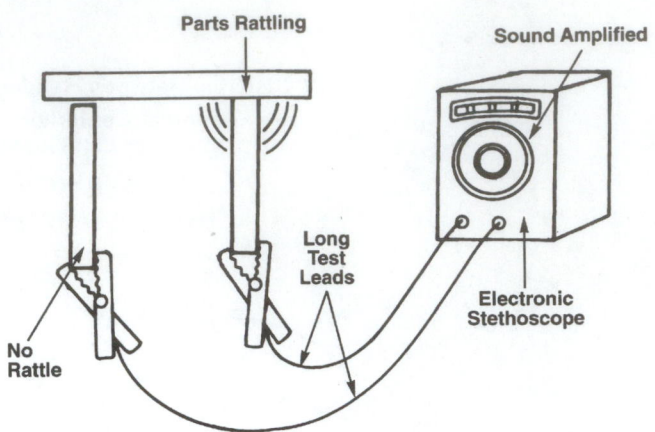

FIGURE 15–49 An electronic stethoscope is handy for finding rattles and other mechanical noises. Long alligator clips can be attached to parts. While the vehicle is driven, you can sit and listen to which part is causing the noise.

to bare metal, filler can be applied to surface irregularities, and the bumper primed and painted.

On older cars, bumpers were rigidly bolted to the vehicle's frame. At best, the old bumpers only resisted the bending forces of an impact; they transferred the energy shock directly to the frame and, thus, to the occupants of the vehicle. Modern bumpers are designed to absorb the energy of a low-speed impact minimizing the shock to both vehicle frame and the occupants of the vehicle. A few years ago, car manufacturers were required by federal regulations to equip cars with bumpers that could withstand 5 mph (8 km/h) collisions without incurring damage to the vehicle. Later, the federal standards were relaxed to 2.5 mph (4 km/h).

In order to comply with the federal regulations, manufacturers fitted their bumpers with energy absorbers. Most energy absorbers are mounted between the bumper face bar or bumper reinforcement and the frame (Figure 15–50).

There are many types of energy absorbers. The most common is similar to a shock absorber. The typical bumper shock is a cylinder filled with hydraulic fluid. Upon impact, a piston filled with inert gas is forced into the cylinder. Under pressure, the hydraulic fluid flows into the piston through a small opening. The controlled flow of fluid absorbs the energy of the impact. Fluid also displaces a floating piston within the piston tube, which compresses the inert gas. When the force of the impact is relieved, the pressure of the compressed gas forces the hydraulic fluid out of the piston tube and back into the cylinder. This action forces the bumper back to its original position.

Another energy absorber design is shown in Figure 15–51. Upon impact, fluid flows from a reservoir through a metering valve into an outer cylin-

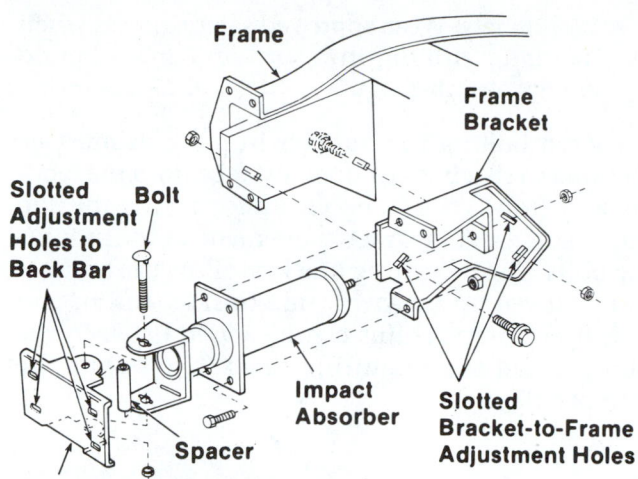

FIGURE 15-50 A bumper shock absorber bolts between the body or frame and the bumper.

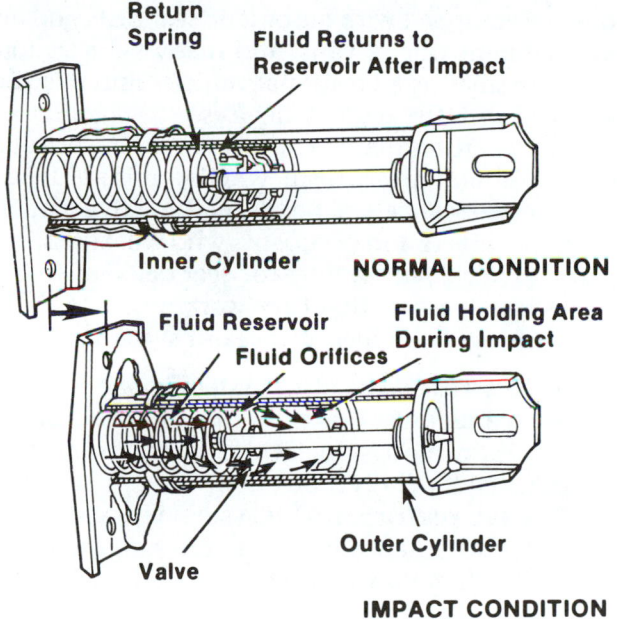

FIGURE 15-51 Note the construction of spring-loaded bumper shocks.

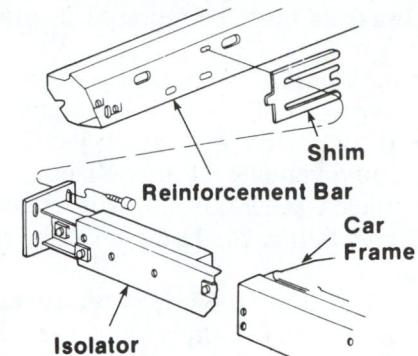

FIGURE 15-52 This is a typical bonded isolator.

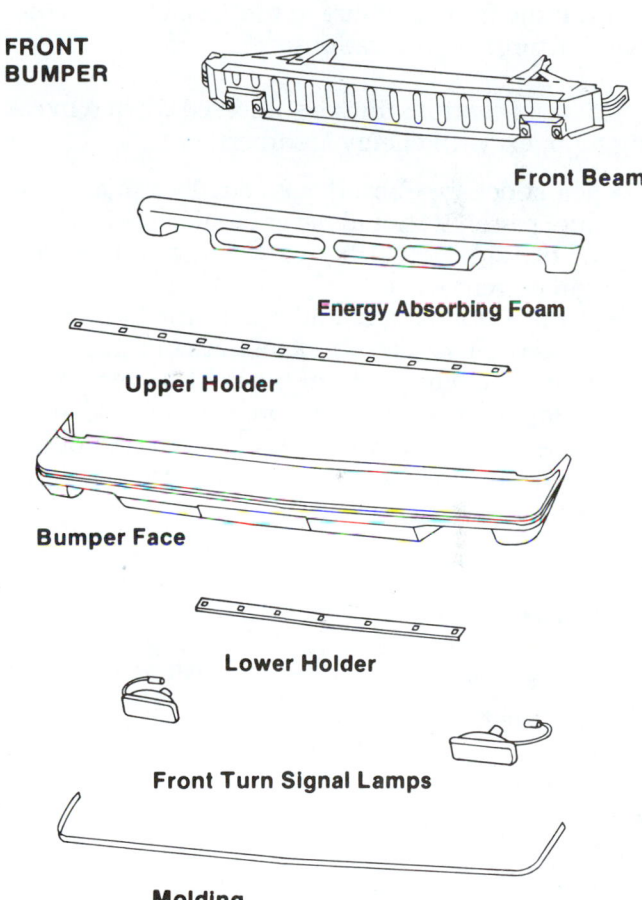

FIGURE 15-53 This plastic-covered bumper has an energy absorbing foam pad.

der. When impact forces are relieved, a spring in the absorber returns the bumper to its original position.

Another type of bumper energy absorber is called an isolator (Figure 15–52). It works in principle like a motor mount. A rubber pad is sandwiched between the isolator and the frame. Upon impact, the isolator moves with the force, stretching the rubber pad. The give in the rubber absorbs the energy of the impact. When the force is relieved, the rubber will retract to its original shape (unless it is torn from its base by the impact) and return the bumper to its normal position.

Another type of energy absorber is found on many light imports and sport-model vehicles. Instead

of shock absorbers mounted between the frame and the face bar or reinforcement bar, a thick urethane foam pad is sandwiched between an impact bar and a plastic face bar or cover (Figure 15–53). The pad is designed to give and rebound to its original shape in a 2.5 miles per hour collision.

On some vehicles the impact bar is attached to the frame with energy absorbing bolts (Figure 15–54). The bolts and brackets are designed to deform during a collision in order to absorb some of the impact force. The brackets must be replaced in most collision repairs.

Replacing a bumper is basically a matter of removing the correct bolts. This job is made easier if the bumper is supported by a scissors jack lift (Figure 15–55). On some vehicles, stone deflectors, parking lights, windshield washer hoses, and other items must be disconnected before the bumper can be removed from the car.

Figures 15–56 and 15–57 show typical urethane-covered front and rear bumpers and the related assembly hardware. Bumpers with energy absorbers should be unbolted from the absorber brackets. Fixed bumpers should be removed by unbolting the brackets from the frame. Be sure to torque all bolts to the manufacturer's specifications.

Several cautions must be observed when removing bumpers with energy absorbers.

- The shock-type absorber is actually a small pressure vessel. It should never be subjected to heat or bending. If cutting or welding near an absorber, remove it.
- If the absorber is bound due to the impact, relieve the gas pressure before attempting to remove the bumper from the vehicle. Secure the bumper with a chain to prevent its sudden release and drill a hole into the front end of the piston tube to vent the pressure. Then remove the bumper and absorber.

- Work safely. Wear approved safety glasses when handling, drilling into, or removing a bound energy absorber.

After bolting the bumper in place, it must be adjusted so that it is an equal distance from the fenders and front grille. The clearance across the top must be even. Adjustments are made at the mounting bolts. The mounting brackets allow the bumper to be moved up or down, side-to-side, and in and out. If necessary, shims can be added between the bumper and the mounting bracket to adjust the bumper alignment.

15.11 RESTRAINT SYSTEMS

The National Highway Traffic Safety Administration (NHTSA) requires that all new cars sold in the United States be equipped with automatic seat belts and air bags. All light trucks, vans, and many vehicles formerly classified as recreational vehicles must also be equipped with these safety devices.

The assertion that lives are saved by the use of a passive restraint mechanism, whether it is an air bag or motorized seat belt, depends on the active participation by drivers and occupants who will still be required to buckle up. Automatic seat belts must also be checked to assure that they work.

There are two types of restraint systems:

- **Active restraint.** This system is one that the occupants must make an effort to use. For example, in most vehicles the seat belts must be fastened for crash protection.
- **Passive restraint.** This system is one that operates automatically. No action is required to make it functional. Two types are automatic seat belts and air bags.

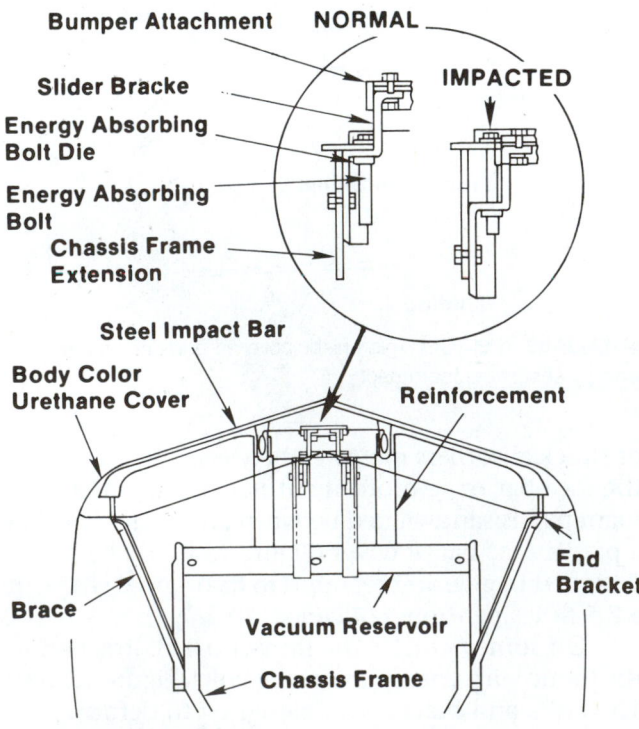

FIGURE 15–54 This urethane bumper has energy absorbing bolts.

FIGURE 15–55 This technician is supporting the bumper with a scissors jack. *(Courtesy of Fitz & Fitz, Inc.)*

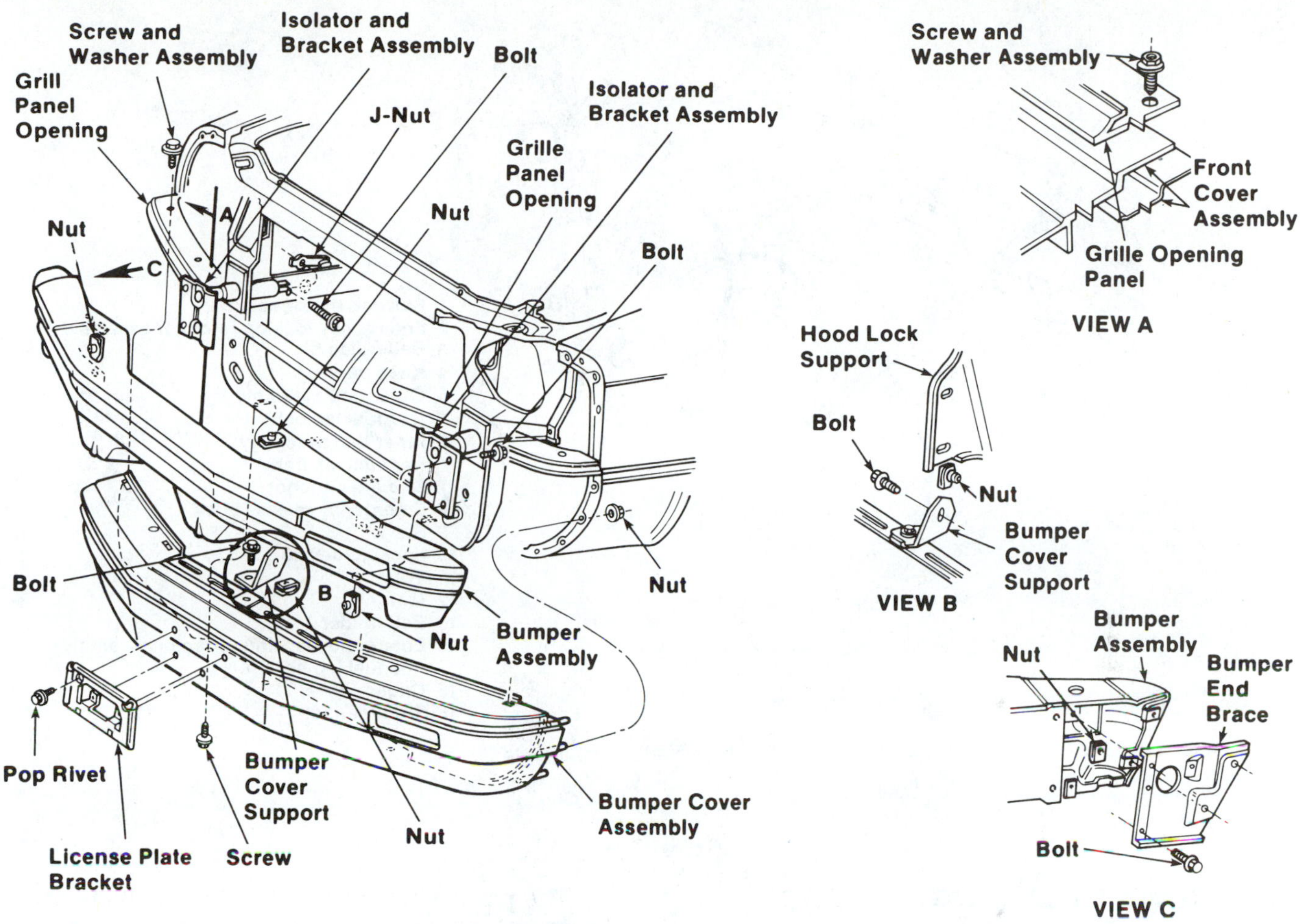

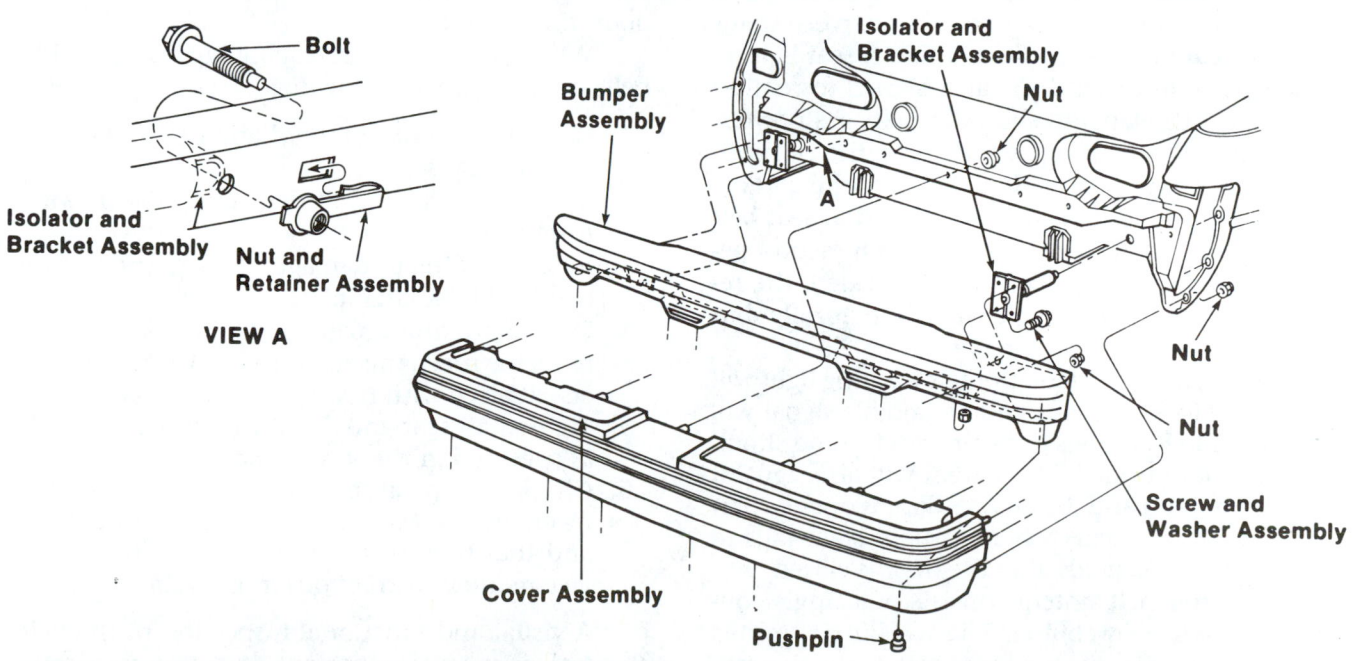

FIGURE 15-56 Study typical front bumper attachment methods.

FIGURE 15-57 Note typical rear bumper attachment methods.

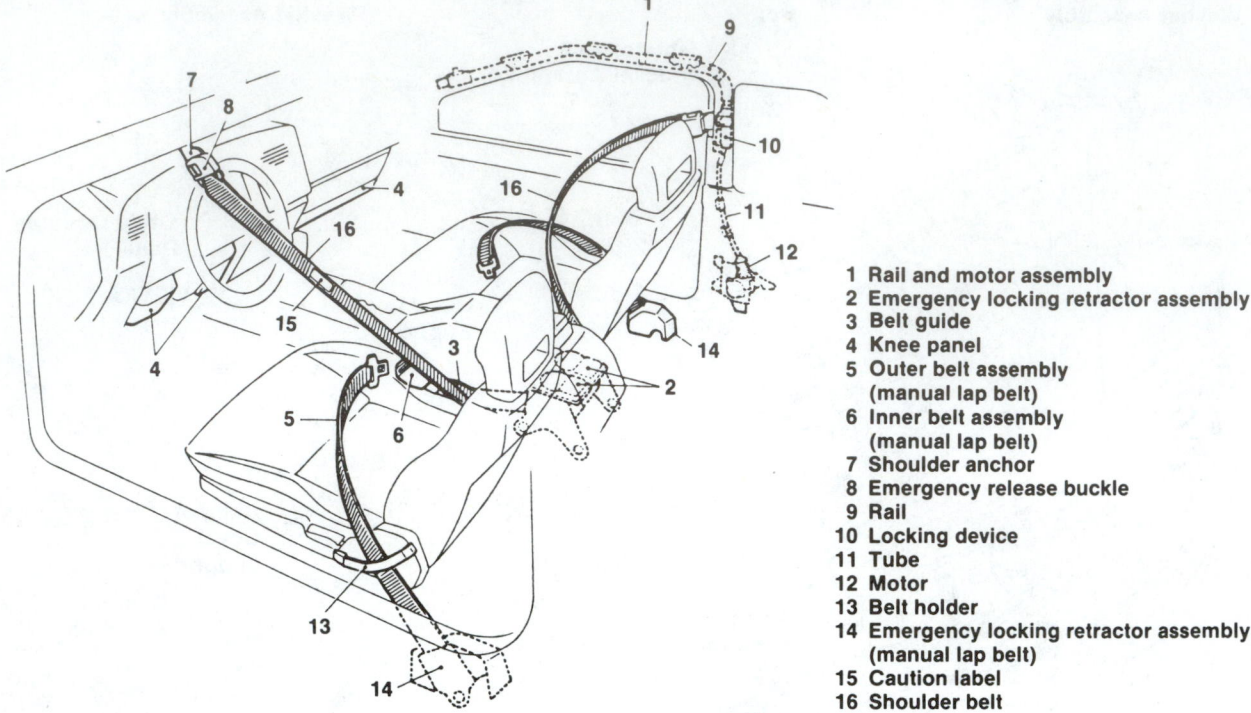

1 Rail and motor assembly
2 Emergency locking retractor assembly
3 Belt guide
4 Knee panel
5 Outer belt assembly
   (manual lap belt)
6 Inner belt assembly
   (manual lap belt)
7 Shoulder anchor
8 Emergency release buckle
9 Rail
10 Locking device
11 Tube
12 Motor
13 Belt holder
14 Emergency locking retractor assembly
   (manual lap belt)
15 Caution label
16 Shoulder belt

FIGURE 15-58 Study the parts of a passive seat belt system.

LAP AND SHOULDER BELTS

Most modern vehicles have a passive restraint system, while most late-model sedans were built with both passive and active systems. On active systems the front seat belt incorporates a fasten seat belt reminder light and sound signal designed to remind the driver if the lap and shoulder belts are not fastened when the ignition is turned to the on position. If the driver's seat belt is buckled, the audible signal will not sound; however, the fasten seat belt reminder light will stay on for a 4- to 8-second period. If the driver's seat belt is not buckled, the reminder light and sound signal will automatically shut off after a 4- to 8-second interval.

On the passive system, the belt warning light will glow for 60 to 90 seconds and an audible signal will sound for 4 to 8 seconds if the driver's lap and shoulder belt is not buckled. The system will also signal if the ignition is on and the driver's door is opened or if a system failure occurs wherein the system fails to deactivate the solenoids after the door is closed.

The active belt system consists of a single continuous length of webbing. The webbing is routed from the anchor (at the rocker panel), through a self-locking latch plate (at the buckle), around the guide assembly (at the top of the center pillar), and into a single retractor in the lower area of the center pillar.

The passive system for all vehicles differs from the active in that two retractors are used—one for the seat belt and a second for the shoulder belt (Figure 15–58). Both retractors are located behind the front door trim panels.

When servicing or replacing lap and shoulder belts, remember the following:

1. Do not intermix standard and deluxe belts on front or rear seats.
2. Keep sharp edges and damaging objects away from belts.
3. Avoid bending or damaging any portion of the belt buckle or latch plate.
4. Do not attempt repairs on lap or shoulder belt retractor mechanisms or lap belt retractor covers. Replace with new replacement parts.
5. Tighten all seat and shoulder belt anchor bolts as specified in the service manual.
6. Check the operation of the automatic seat belt mechanisms. Make sure they are not damaged and that they work properly. Refer to the service manual instructions if in doubt.

A visual and functional inspection of the belts themselves is very important to assure maximum protection for vehicle occupants. The following inspection checklist provides a typical, detailed seat belt inspection.

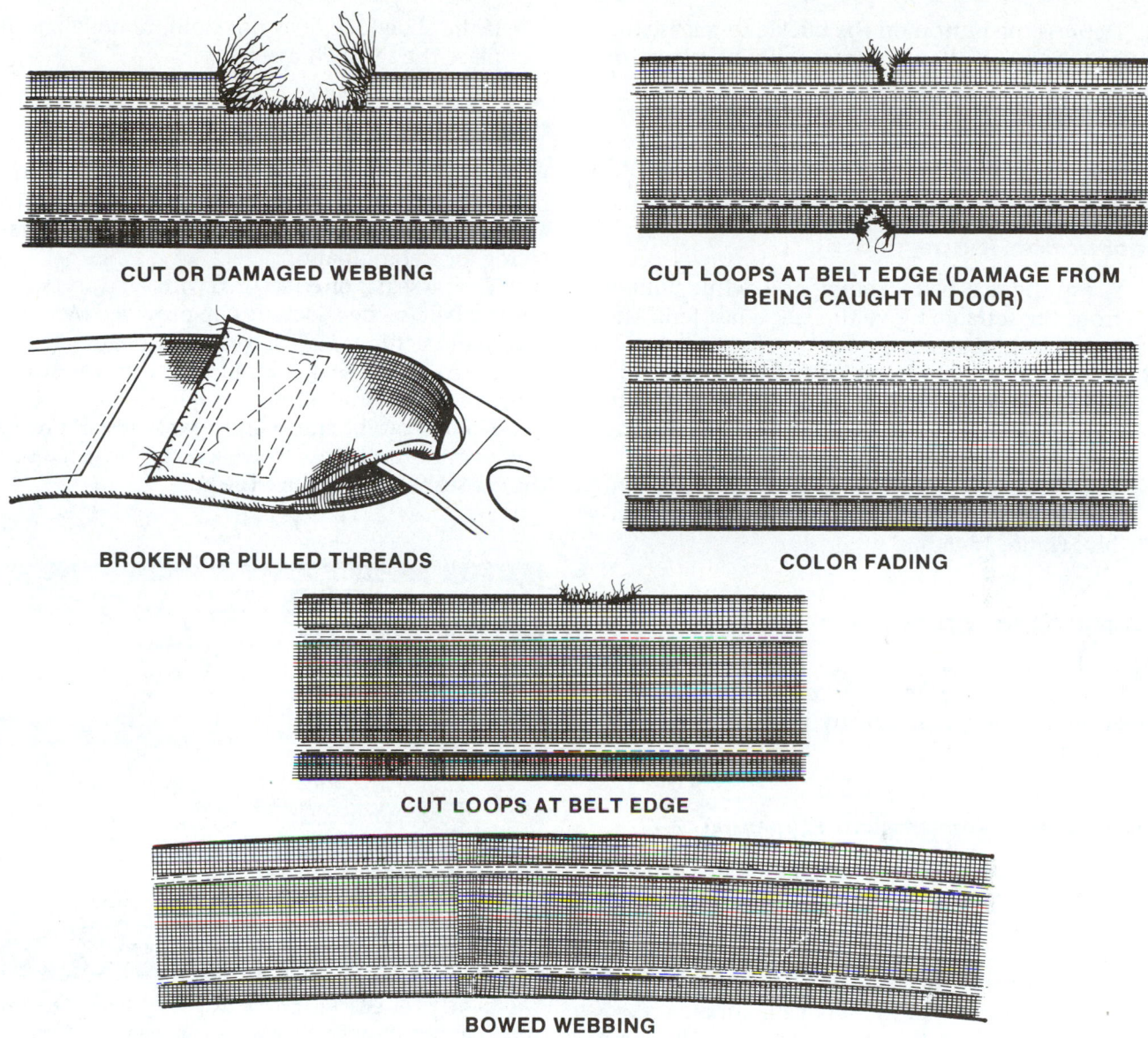

CUT OR DAMAGED WEBBING

CUT LOOPS AT BELT EDGE (DAMAGE FROM BEING CAUGHT IN DOOR)

BROKEN OR PULLED THREADS

COLOR FADING

CUT LOOPS AT BELT EDGE

BOWED WEBBING

FIGURE 15-59 Always inspect seat belts for these kinds of webbing defects.

Front Seat Webbing Inspection

1. Check for twisted webbing due to improper alignment when connecting the buckle.
2. Fully extend the webbing from the retractor. Inspect the webbing and replace with a new assembly if the following conditions are noted (Figure 15–59):
   - Cut or damaged webbing
   - Broken or pulled threads
   - Cut loops at belt edge
   - Color fading as a result of exposure to sun or chemical agents
   - Bowed webbing
3. If the webbing cannot be pulled out of the retractor or will not retract to the stowed position,

check for the following conditions and clean or correct as necessary:
   - Dirty webbing coated with gum, syrup, grease, or other foreign material
   - Twisted webbing
   - Retractor or loop on B-pillar out of position

Buckle Inspection

1. Insert the tongue of the seat belt into the buckle until a click is heard. Pull back on the webbing quickly to assure that the buckle is latched properly.
2. Replace the seat belt assembly if the buckle will not latch.

3. Depress the button on the buckle to release the belt. The belt should release with a pressure of approximately 2 pounds.

4. Replace the seat belt assembly if the buckle cover is cracked, the push button is loose, or the pressure required to release the buckle is too great.

Retractor Inspection

1. Grasp the seat belt webbing and, while pulling from the retractor, give the belt a fast jerk. The belt should lock up.

2. Drive the vehicle in an open area away from other vehicles at a speed of approximately 5 to 15 mph (8 to 24 km/h) and quickly apply the foot brake. The belt should lock up.

3. If the retractor does not lock up under these conditions, remove and replace the seat belt assembly (Figure 15–60).

Anchorage Inspection

Check the seat belt anchorage for signs of movement or deformation. Replace if necessary. Position the replacement anchorage exactly the same as in the original installation.

Rear Seat Restraint System

The method of removal and installation of a rear seat restraint system will be obvious upon inspection. Check the position of the factory installed lap belt and single loop belt anchors and reinstall the anchor plates in the same position as shown. Torque the bolts as specified in the service manual.

Some models have a rear center seat belt. These belts do not have a retractor. In addition to checking the webbing and anchorage, the adjustable slide locking of the belt must be checked.

- Fasten the tongue to the buckle and adjust by pulling the webbing end at a right angle to the connector and buckle.
- Release the webbing and pull upward on the connector and buckle.

SHOP TALK Do not bleach or dye the belt webbing (clean with a mild soap solution and water). Chemicals could deteriorate and weaken the belt fabric.

- If the slide lock does not hold, remove and replace the seat belt assembly.

CHILD SEAT

A *child seat* is a small, supplemental seat designed to protect small children from collision injuries. There are three types of child car seats: rear facing, forward facing, or a combination.

The car seat is often secured with a lap and shoulder seat belt (using a locking clip provided by the car seat manufacturer) or just a lap belt (Figure 15–61). An integral child seat is build into the rear seat by some vehicle manufacturers.

No one can be sure a child car seat will prevent injury in an accident. However, the proper use of the car seat should reduce the risk.

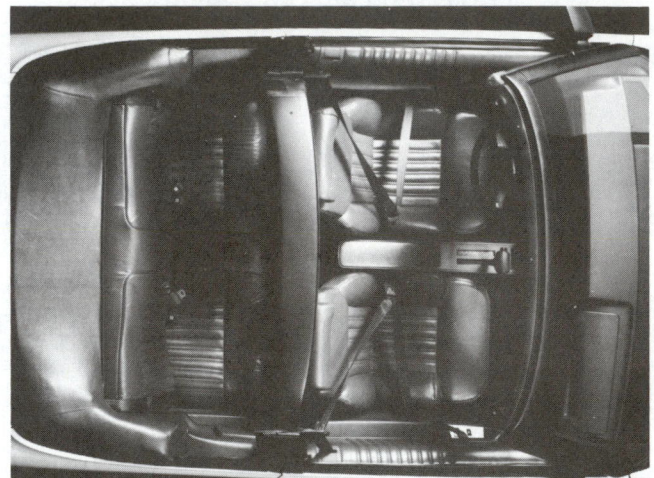

FIGURE 15–60 Remember that seat belts are critical to the safety of the vehicle's driver and passengers. Always service them carefully and torque all fasteners to specs.

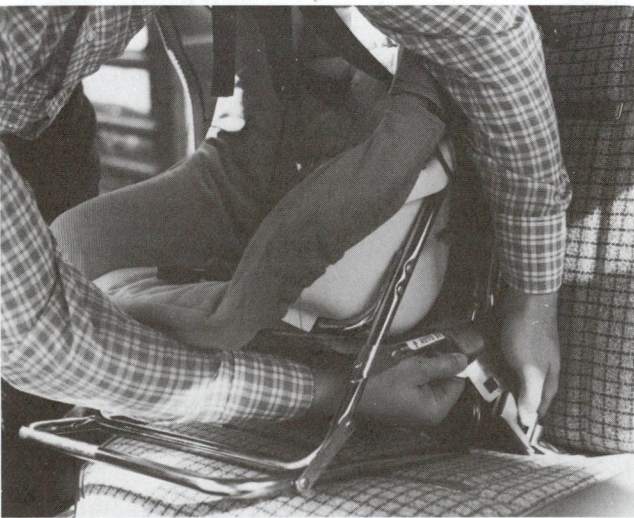

FIGURE 15–61 Many child car seats are not suitable for use in a front seat equipped with an air bag.

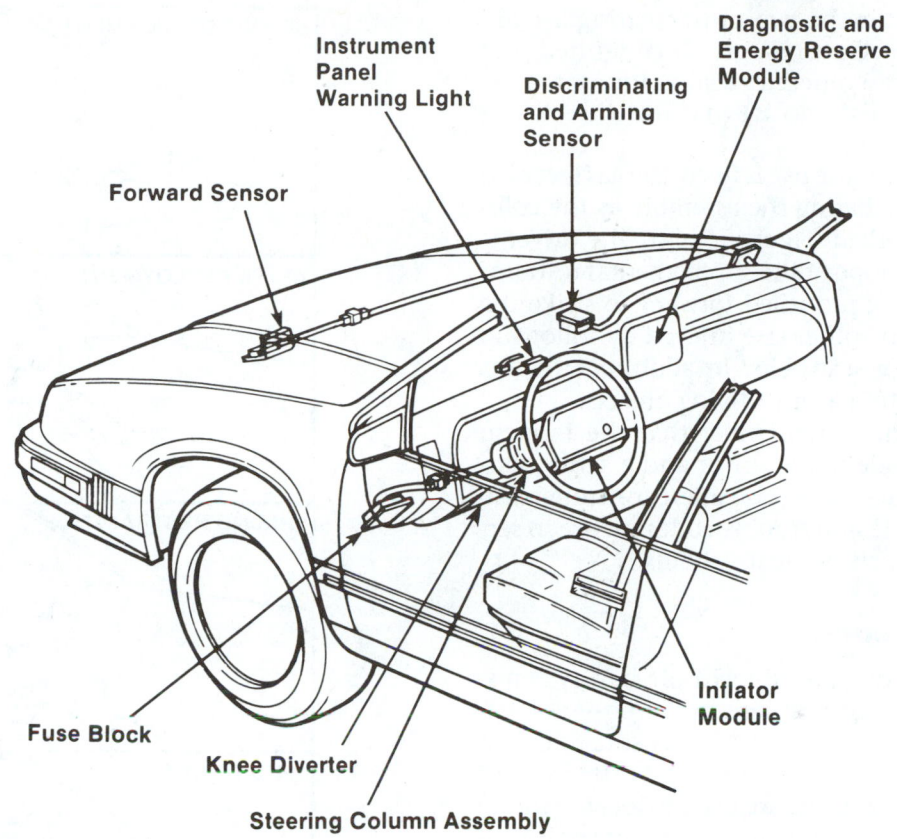

Instrument
Panel
Warning Light

Diagnostic and
Energy Reserve
Module

Discriminating
and Arming
Sensor

Forward Sensor

Inflator
Module

Fuse Block

Knee Diverter

Steering Column Assembly

FIGURE 15-62 Study air bag system components.

AIR BAG SYSTEMS

An *air bag system* automatically deploys a large nylon bag during frontal collisions (Figure 15–62). One or more air bags can be used. The *driver's side air bag* deploys from the steering wheel center pad. The *passenger side air bag* deploys from behind a small door in the right side of the dash. *Side impact air bags* may be used in the door panels or seats to protect against side impact injury.

While the location and design of the air bag system varies from manufacturer to manufacturer, all air bag systems have similar parts. These include the following:

1. *Air bag module* (inflator mechanism and nylon bag that expands to protect driver or passenger during collision)
2. *Air bag system sensors* (inertia sensors that signal computer of collision)
3. *Air bag control unit* (computer that operates system and detect faults)
4. *Air bag harness* (wiring and connectors that link system parts)
5. *Air bag warning lamp* (dash bulb that warns of system problem)

The **air bag module** is composed of the nylon bag and an igniter-inflator mechanism enclosed in a metal or plastic housing. All air bag module

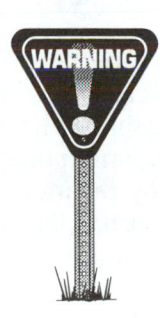

Aftermarket child seats may not be suitable for use in the front seat with an air bag system. Some seats locate the child too close to the air bag. Also, NEVER place a child seat in the front seat facing rearward in a vehicle equipped with airbags. Injury or death to the child could result from bag deployment. Warn your customers of this danger!

components are packaged in a single container mounted in the center of the steering wheel pad or dash. This entire assembly must be serviced as one unit when repair of the air bag system is required.

The **air bag** is a strong nylon bag attached to the metal frame of the module. Vent holes or a porous fabric in the bag allow for rapid deflation after deployment.

Inflation of the air bag is caused by an explosive release of gas. In order for the rapid expansion to occur, a chemical reaction must be started. This is done by the **air bag igniter**. The igniter is a two-pin bridge device that is activated when it receives a signal from the air bag monitor in the form of an electrical current. When the electrical current is applied, it arcs across the two pins, creating a spark that ignites a propellant charge. Once ignited, the **propellant charge** generates a large amount of gas extremely rapidly. This gas is what fills the air bag (Figure 15–63).

Almost as soon as the bag is filled, the gas is cooled and vented, thus deflating the assembly as the collision energy is absorbed (Figure 15–64). The driver is cradled in the envelope of the supplemental restraint bag instead of being propelled forward to strike the steering wheel or be otherwise injured by follow-up inertia energy from seat belts. In addition, there is some facial protection against flying objects.

It is important to remember that the tandem action of at least one main sensor and a safing sensor will activate the system. The microcontroller also provides failure data and trouble codes for use in servicing various aspects of most systems.

Air Bag Sensors

Two or more sensors are used in air bag systems: impact sensors and arming sensors.

Impact sensors are the first sensors to detect a collision because they are mounted at the front of the vehicle. Impact sensors are usually located in the engine compartment, while the safing sensor is usually located in the passenger compartment.

The *safing* or **arming sensor** ensures that the particular collision is severe enough to require that the air bag be deployed.

Both impact and arming sensors are inertia sensors. **Inertia sensors** detect a rapid deceleration to produce an electrical signal. Some air bag sensors have a small metal ball held in place by a permanent magnet. The sensor ball is thrown forward by the inertia of the collision. It then touches two electrical terminals that close the sensor circuit to the computer.

Another air bag sensor design uses a weight attached to a coil spring. During impact, the weight is thrown forward. This overcomes spring tension and

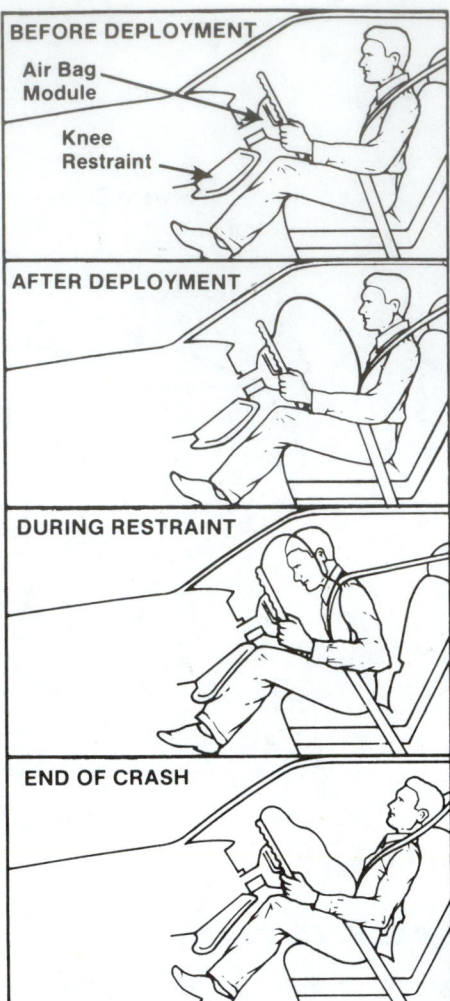

FIGURE 15-63 Note the crash sequence of a typical air bag system.

closes the sensor contact. It also closes the circuit to signal the computer of a possible collision.

Both the impact sensor and a safing sensor must close at the same time for air bag inflation. They work together to provide a fail-safe system to prevent accidental air bag deployment. When both an impact sensor and a safing sensor close, the diagnostic control module sends a signal to the igniter, which starts a chemical reaction to inflate the bag.

Passenger side air bags are very similar in design to the driver's unit. The actual amount of gas required to inflate the bag is much greater because the bag must span the extra distance between the occupant and the dashboard at the passenger seating location. The steering wheel and column make up this difference on the driver's side. Side air bags are smaller and require less propellant.

Even with the battery disconnected, the reserve module can fire the air bag. If you are working near the bag, serious injury could result.

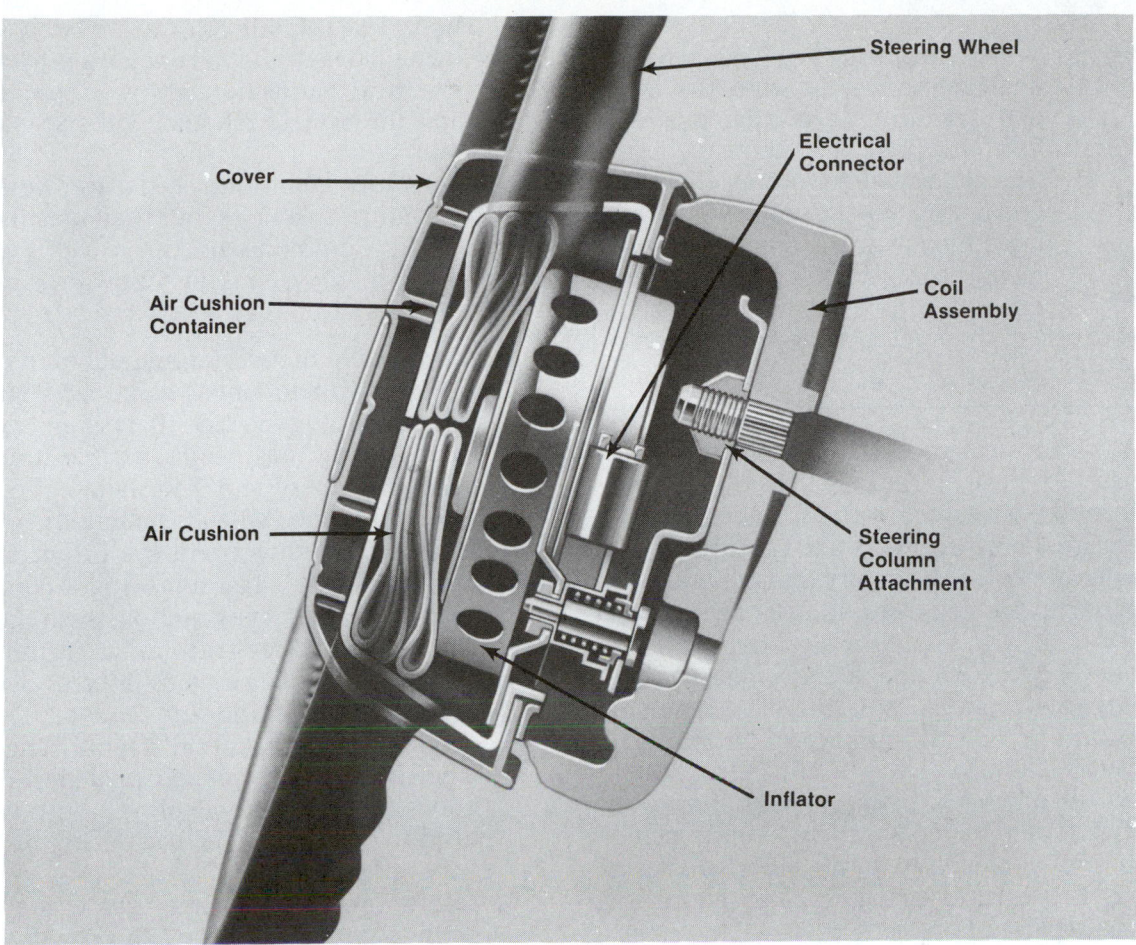

FIGURE 15-64 This cutaway shows the internal parts of an air bag in the steering wheel.

It is important to remember that the tandem action of at least one main sensor and a safing sensor will activate the system.

Air Bag Controller

The **air bag controller** analyzes inputs from the sensor to determine if bag deployment is needed. If at least one impact sensor and the arming sensor are closed, it sends current to the air bag module. This fires or deploys the air bag. The electronic control unit also provides failure data and trouble codes for use in servicing various aspects of most systems.

 **WARNING** Air bag systems may be equipped with an *energy reserve module* that allows the air bag to deploy in the event of a power failure. It must be removed from the system or allowed to discharge for a period of time ranging from a few seconds to 30 minutes after disconnecting the battery.

Air Bag System Servicing

Before servicing a vehicle equipped with an air bag, the system must be *disarmed* (all sources of electricity for igniter disconnected). Procedures for disarming the air bag system vary.

Manufacturers may specify removal of the system fuse or disconnection of the module. *Always refer to the service manual for exact procedures for disarming the system.* This will help prevent electrical system damage and accidental deployment of the new air bag. Always disconnect the negative battery cable.

After deployment, use a shop vacuum to clean the passenger compartment. Residual powder, which is an eye and skin irritant, can be present. Vacuum the dash, vents, seats, carpet, and other surfaces contaminated with this powder (Figure 15–65).

Air bag system parts replacement after a deployment will vary. Always consult the specific manufacturer's recommendations on parts replacement. Some manufacturers recommend replacement of all sensors and sometimes the electronic control unit during deployed air bag service.

When replacing air bag system sensors, double-check that the system is disarmed before removing any sensor. The service manual will give sensor

When carrying a live (undeployed) air bag module, be sure the bag and trim cover are pointed away from your body. This will help reduce the chances of serious injury if the bag accidentally inflates. When laying a module down on a work surface, make sure the bag and trim cover are face up to minimize a "launch effect" of the module if the bag suddenly inflates.

locations. Make sure you have the correct replacement sensor. During installation, check that the **sensor arrow** (directional arrow stamped on sensor) is facing forward. If a sensor is installed backwards, the air bag will not deploy during the next accident.

To remove the deployed air bag, remove the small screws from the rear of the steering wheel. You can then lift out the module and disconnect its wires. Wear safety glasses and a respirator while removing the deployed bag. This will protect you from the residual powder.

Inspect all parts for damage. Parts that have visible damage should be replaced. This would include the steering wheel, steering column, and related parts. Damage to the electrical wiring may also require wiring harness replacement.

Obtain the correct replacement parts from the manufacturer. Always refer to the service manual for exact procedures. System designs vary.

Do not carry any system parts by the wire harness or pigtails. Follow the manufacturer's policies if any part is dropped or shows visible signs of damage, or following a deployment.

Do not attempt to repair any parts or apply any electrical power to any part unless specified by the manufacturer.

The air bag module must be replaced following a deployment. Often the coil or **clock spring**, which is the electrical connection between the steering column and the air bag module, will also have to be replaced.

Remember! There is no need to fear working with air bag equipped vehicles, but they must be treated with respect. Common sense can go a long way when working on these systems. By following some simple rules, air bags can be safely serviced.

1. *Always* have the service manual on hand when working with an air bag equipped vehicle.
2. When servicing a vehicle that has an undeployed air bag, follow the manufacturer's instructions for disarming the system. You should also have the system disarmed when installing a new air bag.
3. Wear rubber gloves and eye protection when servicing the air bag following a deployment. In case of skin or eye irritation, wash thoroughly with water and seek medical attention.
4. Disarm the air bag system prior to performing any welding operations.
5. Keep your arms out of the steering wheel spokes when working on an air bag. It can shatter bones if accidentally deployed. Also keep your head to one side of the bag during installation (Figure 15–66).
6. Follow the manufacturer's guidelines on force drying paint on vehicles equipped with an air bag.
7. Air bag disposal procedures vary depending on whether or not the air bag has been deployed.
8. If the air bag module is defective or the vehicle is to be scrapped, the air bag should be manually deployed using the procedures described in the service manual. Do not dispose of an undeployed air bag.
9. On air bag modules that cannot be manually deployed, the disposal procedure is to ship it back to the manufacturer using the packaging that

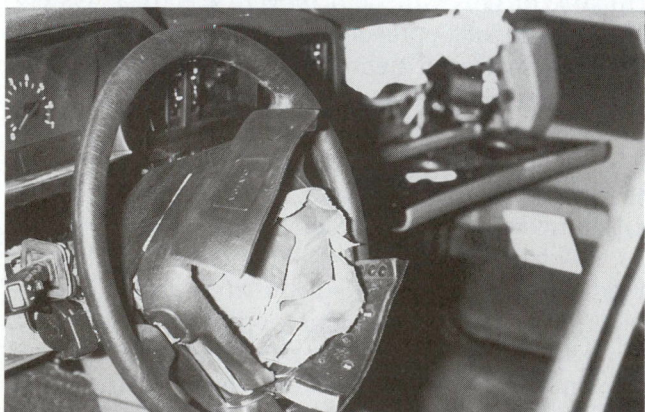

FIGURE 15–65 When an air bag is deployed, use a vacuum to clean up all the dust. Wear a respirator and eye protection since dust can be an irritant. *(Courtesy of Tech-Cor)*

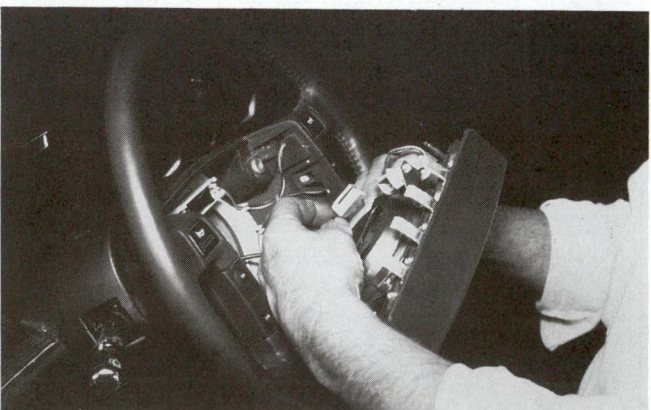

FIGURE 15–66 Make sure the system is disabled when installing a new air bag. If the bag were to deploy, serious injury could result. *(Courtesy of Tech-Cor)*

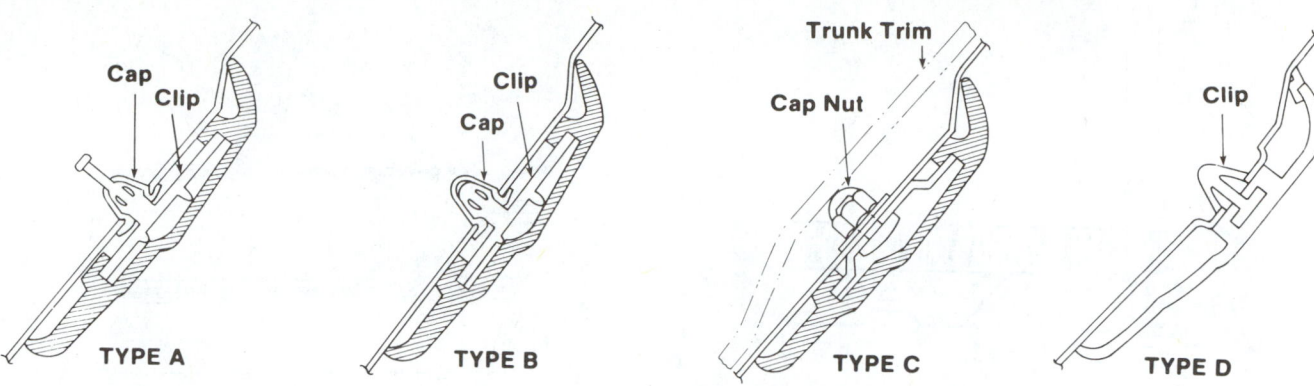

FIGURE 15-67 Here are a few molding attachment methods.

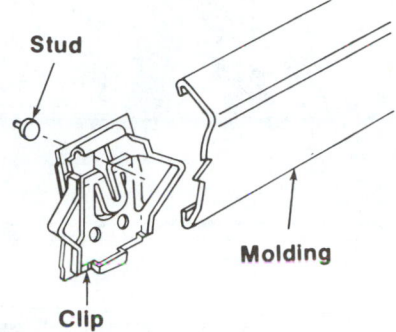

FIGURE 15-68 This is a weld stud molding attachment.

FIGURE 15-69 A power knife will simplify and speed molding and emblem removal. *(Courtesy of Equalizer Industries, Inc.)*

the replacement module came in. By using the replacement part's packaging, all of the needed warning labels are already on the package.

The air bag system performs a self check every time the ignition is turned to the on position. During the self check the air bag dash lamp indicator will light steady or blink. When the self check is completed the lamp should go off. If the lamp stays lit, there is a system fault present.

Make sure a final check is made for codes or accident information using the approved scan tool. Carefully recheck the wire and harness routing before releasing the car.

A final inspection of the job should include checking that the sensors are properly torqued to their mounting fixtures, with the arrows on them facing forward. Ensure that all the fuses are correctly rated and replaced.

15.12 INSTALLING BODY MOLDING

Every vehicle has a variety of moldings. Moldings enhance the appearance of a vehicle by hiding panel joints, framing windshield or back lights, and accenting body lines. They also help to weatherproof by channeling wind and water away from windows and doors. Moldings often must be replaced due to

collision damage, or they can be added as a custom accessory.

A variety of fasteners is used to secure moldings to a vehicle. The clips and bolts shown in Figure 15–67 are examples. The clips shown in A and B must be removed with a special clip puller. Removing fasteners C and D requires removing the interior trim. Type D is removed by compressing the clip and pushing it through the hole.

One of the most common molding attachment methods is a stud welded to the body. A clip fits over the stud and the molding slides over the clip (Figure 15–68). If a weld stud is bent or broken off, replace it with an oval head blind rivet, or weld a new stud in place with a stud welder equipped with a special rivet electrode.

Other methods of attachment include adhesives and rivets. Body side moldings installed as add-ons by dealers, trim shops, and body shops are installed by one of these two methods. Adhesive moldings, when properly applied, are permanent. Rivet-on moldings, while also permanent, might create buckles in large, low-crown panels and require drilling through the corrosion protection on the body. Instructions for both types are given here.

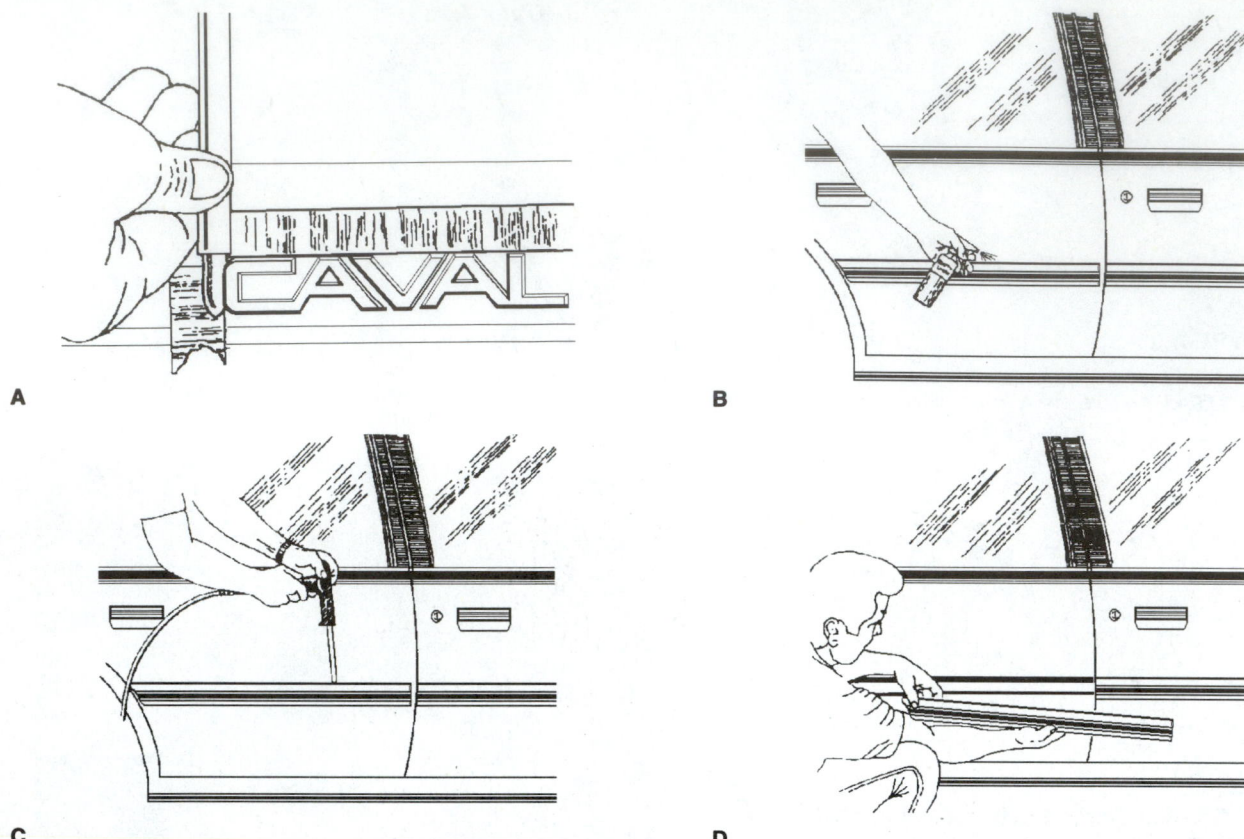

A

B

C

D

FIGURE 15-70 (A) Apply masking tape to protect paint before cutting. Adjust blade length to match width of emblem or molding. (B) Apply soapy solution to prevent blade overheating. (C) Hold tool securely as you cut between body and emblem. (D) Before installing new molding, repair the area as needed and clean properly to provide good adhesion. *(Courtesy of Equalizer Industries, Inc.)*

☐ : Double-faced adhesive tape

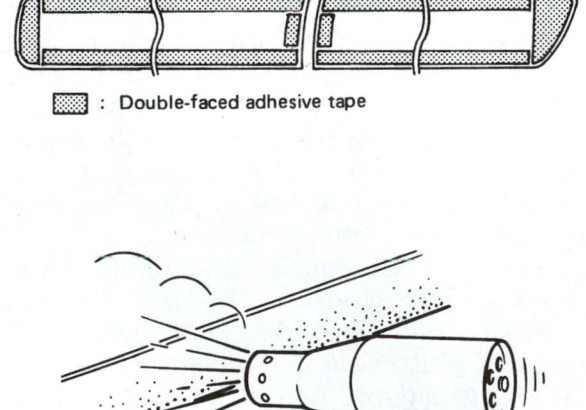

Heat gun

FIGURE 15-71 A heat gun is sometimes recommended to remove emblems and moldings. It will soften adhesive so the part can be pulled off easily. Do not apply too much heat! *(Courtesy of Nissan Motor Co.)*

REMOVING ADHESIVE-HELD MOLDINGS

A molding tool is the best way to remove adhesive-held moldings (Figure 15–69). It has a thin blade that will cut through the adhesive without causing part damage.

To cut off an old molding, apply masking tape around the molding to protect the paint (Figure 15–70A). Then adjust the blade depth to match the molding size. Apply a soapy solution to the molding to reduce friction and cool the cutter blade (Figure 15–70B). Slip the molding tool blade between the molding and body panel (Figure 15–70C). While holding the tool square and firmly, begin cutting slowly from one end to the other (Figure 15–70D).

A heat gun is sometimes recommended to soften the adhesive before molding removal (Figure 15–71). Be careful not to apply too much heat or you could blister the paint or ruin the molding.

INSTALLING ADHESIVE BODY SIDE MOLDINGS

1. Park the car on a level surface. It should be at room temperature to ensure proper adhesion.

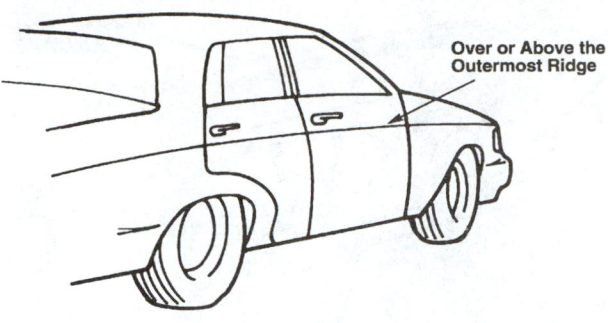

A

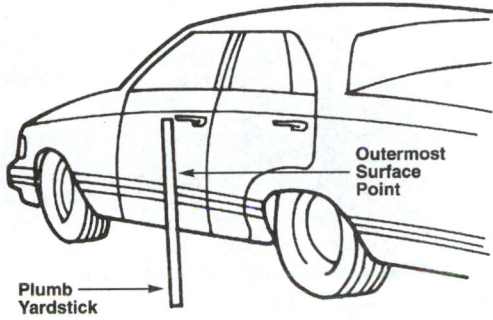

B

FIGURE 15-72 Apply exterior molding (A) along the outermost ridge or (B) the outermost surface point.

FIGURE 15-73 Clean the body with a wax and grease remover before installing adhesive-held molding or emblems.

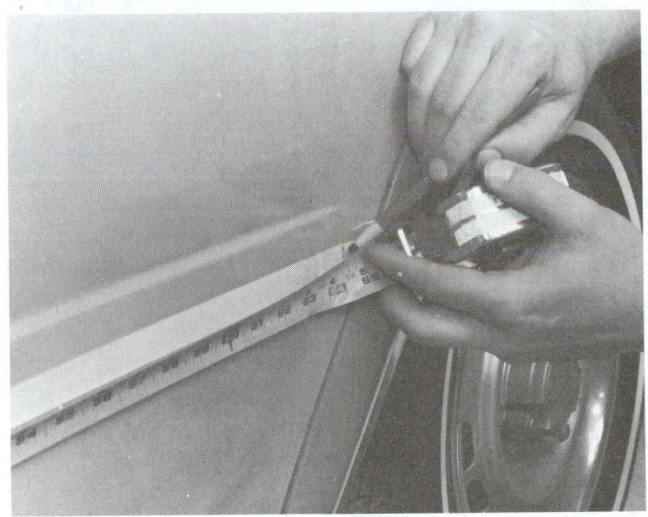

FIGURE 15-74 Apply masking tape as a straightedge and cut molding to length.

2. Select the area to which the molding will be applied. For greatest protection, it should be applied to the outermost surface of the vehicle. If the car has an outermost ridge, install the molding $\frac{1}{8}$ inch (3.1 mm) above or $\frac{1}{8}$ inch (3.1 mm) below the ridge, but not on the ridge itself. If the panel does not have a prominent ridge, select the outermost surface of the body contour (figure 15–72).

3. After determining the best location for the molding, thoroughly clean the area with water and detergent. Then use a clean rag wetted with enamel reducer or a wax and grease remover to remove waxes and silicones (Figure 15–73). Use a clean cloth for each side of the vehicle.

4. If the body molding will not be aligned above or below a body ridge that can be used as a guide, mark the correct height of the molding with a steel rule and a soft lead pencil or a china marker. Mark the height at the rear and front of the car and at each door gap. Then stretch a piece of masking tape from front to rear connecting the marks. Keep the tape taut and sight along its length to ensure a straight line. Magnetic plastic tape can also be used as a straightedge.

5. The next step is cutting the molding to length. Allow about $\frac{1}{8}$ inch (3.2 mm) clearance between the molding and the edge of the fender. Cut the molding to size (Figure 15–74). Repeat this procedure for the rear quarter panel piece of molding. When measuring for door pieces, leave $\frac{1}{8}$ inch (3.2 mm) clearance at both ends of the molding. To enhance the molding's appearance and to prevent binding when the door is opened, cut the ends of the molding at a 45-degree angle, using a single edge razor blade (Figure 15–75).

6. Peel 6 inches of the protective backing paper from the cut end of the fender molding. Do not touch or dirty the exposed adhesive after removing the backing. Begin installing the fender molding $\frac{1}{8}$ inch (3.2 mm) from the fender rear edge. Align the molding with the

FIGURE 15-75 Cut a 45-degree angle on the molding ends.

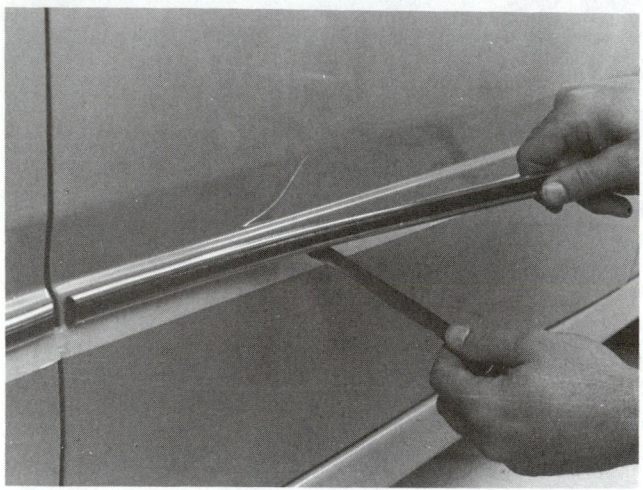

FIGURE 15-76 Adhere molding to body. Pull off backing strip as you guide molding into place against body.

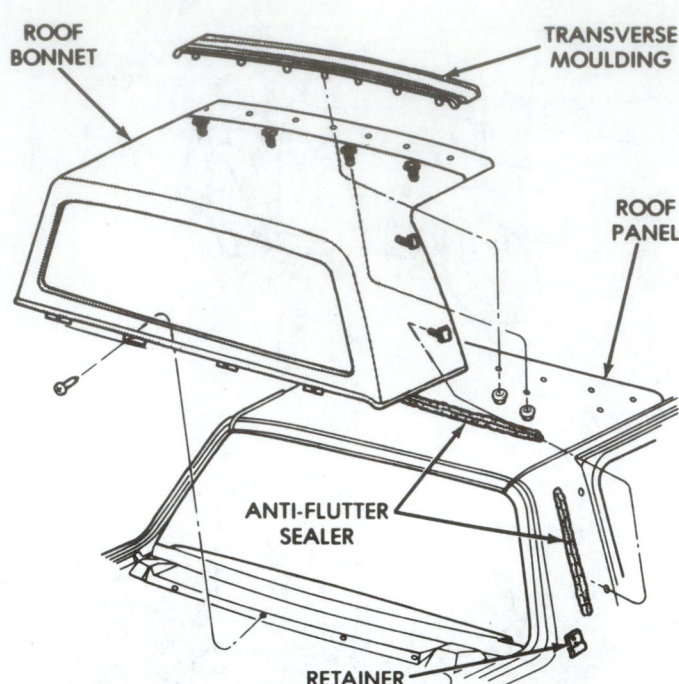

FIGURE 15-77 Vinyl-covered roofs are commonly found on luxury cars.

top edge of the tape and lightly press against the panel (Figure 15–76). Progressively remove the backing and press the molding against the surface and along the edge of the tape. Do not attempt to reposition a piece after it is applied. After the whole length of molding is applied, press along the entire length with the heel of the hand or with a roller.

Repeat this process with the door moldings and the rear quarter panel molding.

INSTALLING RIVET-ON BODY SIDE MOLDINGS

Before applying rivet-on moldings, park the vehicle on a level surface and clean the sides of the vehicle where the molding will be applied. Then determine the best location for the molding, following the procedure described in installing adhesive moldings. After applying masking tape as a guide, hold the

molding in position and mark the desired length. Be sure to leave a $1/2$-inch (12.7 mm) clearance on both ends for caps and spears.

Cut the molding to length. Use a sharp pair of metal cutters or a razor blade to cut the ends square. Duplicate each piece for installation on the opposite side of the vehicle.

Slide the molding out of the track, and hold the track in place. Some moldings have an adhesive backing to assist in positioning the track. With the track in place, use a $9/64$-inch (3.6 mm) drill bit, which is ($1/64$ inch (0.39 mm) larger than the rivet diameter, to drill through holes in the track. Also drill holes $3/4$ inch (19.1 mm) from each end of the molding. Install the rivets with a rivet gun.

Replace the molding in the track and install the end caps and spears. If necessary, tap the caps with a rubber mallet to seat them. Be careful not to scratch the finish.

15.13 VINYL ROOF SERVICING

Roof panels on luxury cars are commonly covered with a vinyl-coated fabric (Figure 15–77). Full vinyl roofs and landau roofs can sometimes be repaired or they might have to be completely replaced. Procedures vary depending on the roof design, attachment method, and trim molding design. Most vinyl covers are glued directly to the roof panel. Others are padded with foam rubber.

Vinyl covers are adhered to the roof panels with adhesive. A number of mechanical fasteners are also used to secure the vinyl cover to the edges of the roof: clips over weld-on studs, reveal moldings, finishing lace, drive nails, screws, drip scalp moldings, weatherstrip retainers, or finishing plugs.

Some replacement covers come with a preapplied adhesive backing. Foam rubber pads are applied following the basic procedure described for vinyl top installation.

Vinyl top repairs fall into three categories. One is complete replacement of the vinyl material. This can be necessary due to direct damage to the roof or due to weathering of the material. If not properly maintained, the vinyl material will discolor, blister, and peel. Partial removal of a vinyl roof is sometimes necessary to repair damage to the rear quarter panel or windshield A-pillar. Minor repairs, such as cuts and abrasions, can also be made with a patching compound or rebonded with a soldering iron.

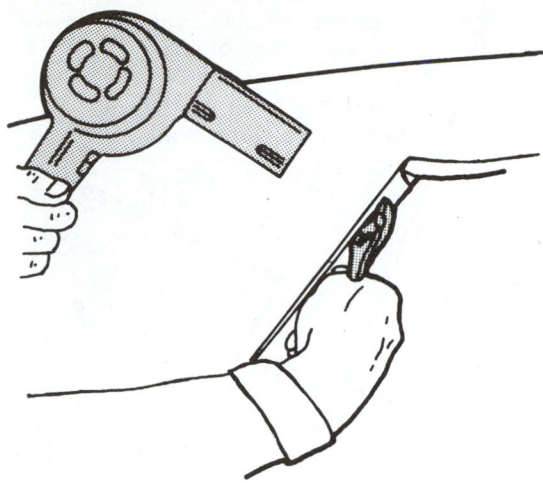

FIGURE 15-78 Use heat to soften the glue if necessary for vinyl top removal.

VINYL TOP REPLACEMENT

When a vinyl top has been damaged beyond repair, replace it following this procedure:

1. Unpack the new cover. Spread it out to flatten wrinkles. Controlled heat from a heat gun or lamp will help soften and flatten wrinkles.
2. Remove the trim moldings. If necessary, remove the quarter inner trim to gain access to any ornament-retaining nuts. Some vehicles also require removal of interior garnish molding to gain access to weatherstrip-retaining screws. Be very careful when removing drive nails. Avoid stretching the holes so that the nails can be reinstalled in the same holes. Refer to the service manual for details if in doubt.
3. If needed, use a heat gun to soften the cement to start removal of the vinyl top material.
4. Use a pair of pliers to grasp the vinyl edge and peel the cover from the roof (Figure 15–78). Continue using the heat gun to soften the cement as you pull off the vinyl (Figure 15–79). Carefully observe how the vinyl is fitted to the roof panel. The new cover must be installed identically.
5. After the vinyl has been removed, remove the old cement from the roof using an xylol-based adhesive remover. The roof must be smooth and level before the new fabric is applied, or the vinyl will be lumpy.
6. Mask around the roof to avoid getting cement on the glass and the paint. Masking is especially important if using a spray adhesive.
7. Lay the new roofing material over the roof. Verify that it fits properly.
8. Remove the fabric. Use chalk to mark the centerline of the roof. Measure carefully so that

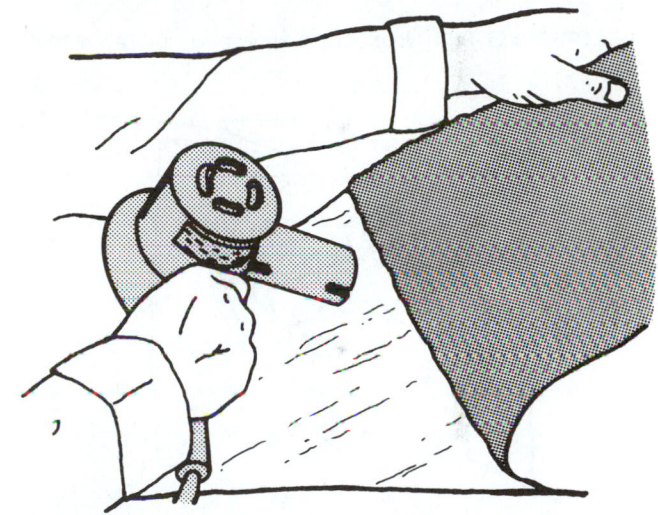

FIGURE 15-79 Continue heating as you pull off the old material.

the line is accurate. Then fold the fabric in half and mark the centerline on the cloth side of the vinyl.
9. Place the vinyl on the roof and align the centerlines. Fold one half of the vinyl back to the centerline.
10. Apply a vinyl roof cement to one half of the fabric. Do not apply adhesive to areas that cover the upper quarter panel, roof sides, or around the door, windshield, and rear windows.
11. Starting at the centerline, also apply vinyl roofing cement to the exposed half of the roof top. Remember not apply adhesive to sides or around windows. Be sure to carefully follow the adhesive manufacturer's instructions. Some adhesives must be brushed or rolled on. Others are designed to be sprayed on. The cement must be applied evenly. There should be no skips, voids, globs, or pools of adhesive.

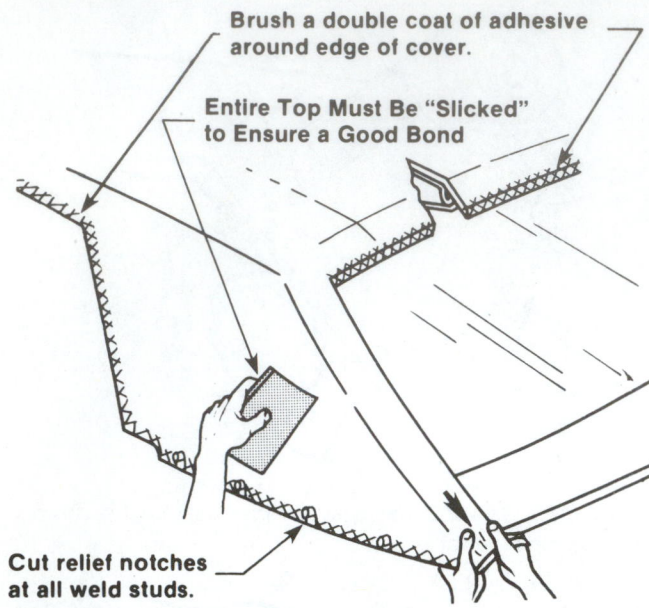

Brush a double coat of adhesive around edge of cover.

Entire Top Must Be "Slicked" to Ensure a Good Bond

Cut relief notches at all weld studs.

FIGURE 15-80 Note how to apply new vinyl to the roof panel.

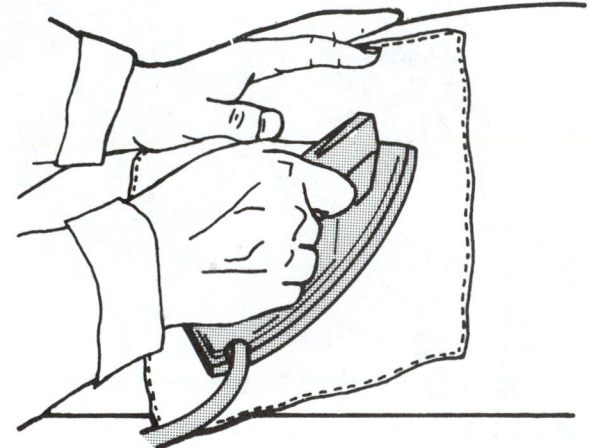

FIGURE 15-81 Smooth out any wrinkles with an iron.

12. After allowing for the manufacturer's suggested tack time, carefully roll the vinyl out over the cement, smoothing out any wrinkles or air bubbles. Be careful to keep the vinyl aligned with the roof. A stiff plastic squeegee can be used to work out air bubbles (Figure 15–80). To work out wrinkles, use a heat gun to soften the vinyl and adhesive. Hold the gun 1 to 2 inches (25 to 50 mm) from the vinyl. Rotate the gun in a circular motion. Apply pressure with the squeegee to smooth out the wrinkles.

13. Apply adhesive to the remaining half of the roof and cover as described in Steps 10 through 12. Smooth the cover over the roof and stretch out all wrinkles.

14. Apply adhesive to the body and underside of the cover in the area below the rear window (if applicable). Position the cover and remove any wrinkles.

15. Apply adhesive to the roof sides and door openings and to the underside of the cover. Make angle cuts at the corners as necessary. Then stretch the cover and cement in place. Pull the material taut, using fabric roof cover pliers.

16. On vehicles with frenched backlights, carefully apply adhesive around the backlight on padding and the plastic panel. Be sure to use a spatula-type tool to fold the roof cover back under the rear window opening panel.

17. Trim the roof cover at the windshield and rear window openings. Leave $1/2$ inch (12.7 mm) of material at the openings for cover attachment in the window opening recess.

18. Apply adhesive to the rear window opening and the underside of the roof cover. Make angle cuts as necessary at the corners and cement the material in the rear window opening recess. Trim the cover around the molding retainer weld studs.

19. Apply adhesive to the windshield opening and the underside of the roof cover. Make angle cuts as necessary at the corners and cement the material in the windshield opening recess. Trim the cover around the molding retainer weld studs.

20. At each roof rear quarter, locate and punch holes for the roof side ornament on models so equipped. Install any side ornaments with the studs entering the punched holes. Then, from inside the vehicle, install the retaining nuts on the studs. Apply a silicone sealant around the nuts.

21. Reinstall interior trim panels, garnish moldings, weatherstrips, and exterior moldings.

22. Clean the glass, moldings, top, and surrounding area.

PARTIAL VINYL TOP REMOVAL

If partial removal of the vinyl roof is necessary because of damaged roof metal, follow this procedure. Remove trim from the affected area and pull the vinyl material up. Peel it away from the area, following techniques discussed earlier. Place shop rags in the fold of the material to prevent it from creasing.

If the repair area requires welding or heat shrinking, make sure that the material is pulled far enough away so that it will be unaffected by the heat. If necessary, place an asbestos dam between the area and the vinyl to absorb the heat. After the area has been repaired and refinished, apply cement to the area and smooth the vinyl back into place. Pull it tight and work out any wrinkles. Clean up excess cement, and reapply the trim.

After the repair has cured for several days, wrinkles can appear in the vinyl. Use an electric household iron pressed over a damp cloth to soften the material and smooth the wrinkles out (Figure 15–81). A heat gun and a stiff squeegee can also be used. If the wrinkles cannot be removed with one of these methods, pull the fabric up. Stretch and reglue the area.

PATCHING A VINYL ROOF

A torn vinyl roof can sometimes be repaired with a vinyl patching compound and a grain mold die. The **grain mold die** is made from a special die material or from body filler. The die is used to impress the vinyl texture in the patching compound.

To make the die, clean a 6- to 8-inch (152 to 203 mm) square portion of the undamaged vinyl roof with a vinyl cleaner. Then spray the area with a silicone mold release. Apply a properly catalyzed amount of die material or body filler to the area. Cover the material with a scrap piece of vinyl, place a block of wood on top of the vinyl. Weight it down to force the material into the grain pattern. Allow

the die to harden. Peel the die up, trim the edges, and spray with a silicone mold release.

After the grain mold die has been made, patch the torn vinyl following this procedure:

1. If the area is torn, trim off any loose fabric and frayed edges (Figure 15–82). If the damaged area is larger, cement the loose material back into place with trim cement and allow to dry (Figure 15–83). Cut out at least $^1/_{16}$ inch (1.6 mm) of material on both sides of the tear line to allow for the vinyl patching compound.

2. Clean the area with a vinyl cleaner by spraying the cleaner on a cloth and wiping the area. Do not spray the cleaner directly onto the repair area.

3. Apply the vinyl patching compound to the repair area with a suitable applicator trowel. Use a trowel or putty knife to work the compound under any loose edges of the vinyl. Apply the first layer to the metal/vinyl edge to make a root fillet (Figure 15–84). Wipe off any excess compound from the vinyl. Because most vinyl compounds have a thin viscosity, several coats will be necessary to completely fill the void. Be sure to follow the manufacturer's instructions for curing time between coats. Normally, a heat gun is used to speed the curing time. Hold the gun 1 inch (25 mm) above the compound and move it in a circular motion to distribute the heat evenly. Curing takes about 20 seconds. Press down on the roof material around the repair during the heat curing process to keep the fabric flat.

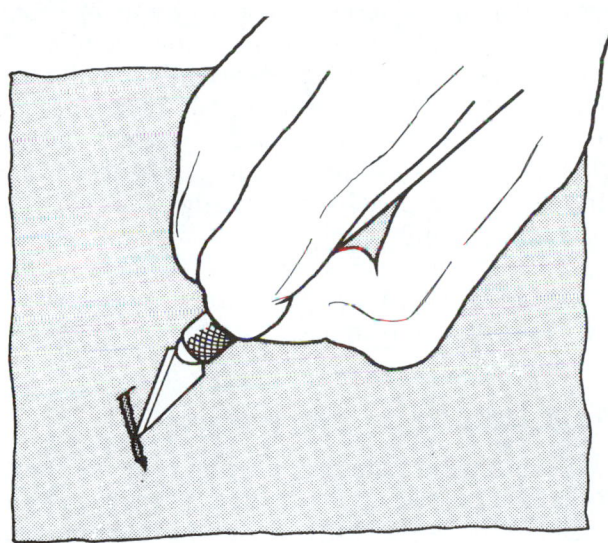

FIGURE 15–82 Trim off all frayed edges.

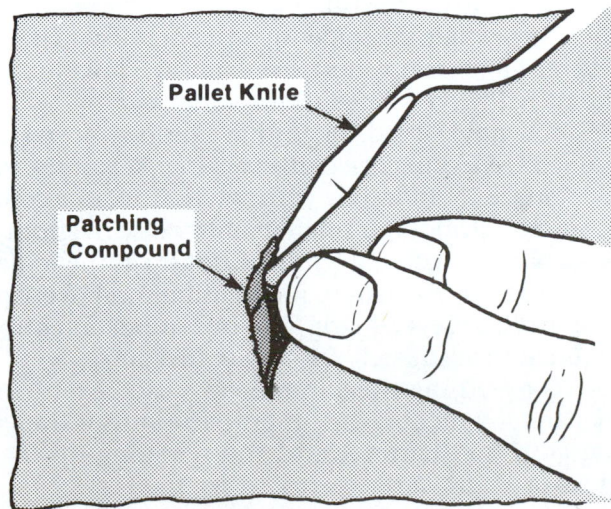

FIGURE 15–83 Glue the loose material.

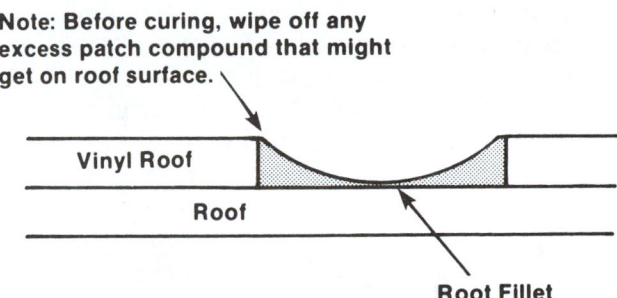

Note: Before curing, wipe off any excess patch compound that might get on roof surface.

Vinyl Roof

Roof

Root Fillet

FIGURE 15–84 Fill the void up to the vinyl edges.

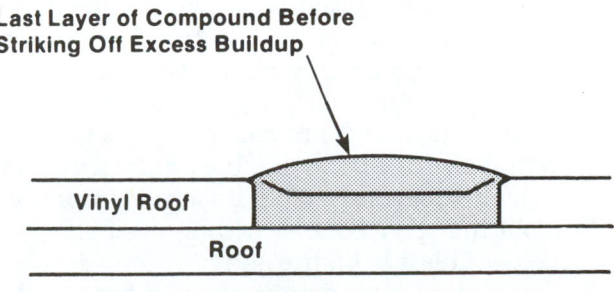

Last Layer of Compound Before Striking Off Excess Buildup

Vinyl Roof

Roof

Do not cure with heat gun at this stage.

FIGURE 15–85 Build up the last layer until it is above the existing vinyl material.

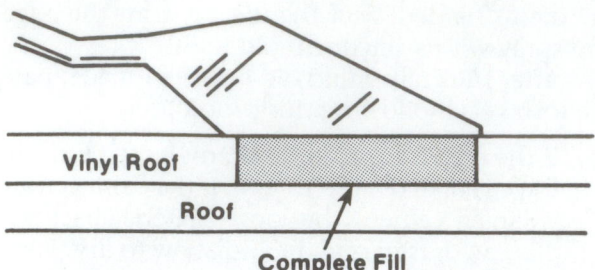

FIGURE 15-86 Level the compound with the vinyl material.

Before making any vinyl repair, thoroughly clean the vinyl roof with a vinyl cleaner. Use a scrub brush to remove all dirt imbedded in the graining. Repeat the cleaning operation as often as required, particularly on white vinyl tops.

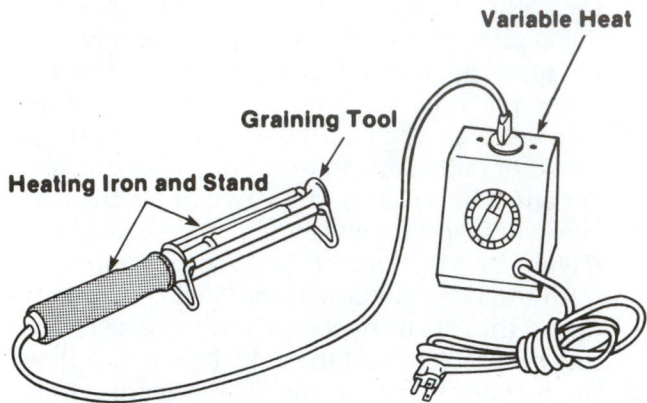

FIGURE 15-87 A heating iron is used to make minor repairs in vinyl.

4. Build up the repair area until the compound is slightly above the level of the vinyl (Figure 15–85). Usually three layers are adequate to complete the job, but larger holes might require up to four or five layers. Before the last application hardens, draw the side edge of the trowel tool over the face of the patch area to strike off the excess buildup to get a level surface (Figure 15–86). Remove any excess compound that might smear on the vinyl roof material adjacent to the patch area; otherwise, it will cure to the roof material and cause unwanted buildup on the surrounding area. The joint is now ready for the final heat cure and regraining procedure.

5. Cure this last leveled layer by heating the patched area for about 30 seconds at 1-inch (25 mm) nozzle distance (always moving the gun nozzle in a circular motion) until the patched area develops a shiny glazed appearance. Have the grain pattern die ready in the other hand.

6. Remove heat and immediately press grain pattern die with force. Hold die on service area for at least 10 seconds to obtain a good impression.

7. Cool the serviced area with a moist sponge or cold air blast from the gun.

8. Spray vinyl prep conditioner on a clean cloth and lightly dab the area. Do not rub.

9. Spray the repaired area with a matching color vinyl paint. Complete instructions for refinishing vinyl materials are given in later chapters.

MINOR VINYL ROOF REPAIRS

A variety of minor vinyl roof repairs can be made. Most require special tools not usually found in a body shop. Many minor repairs require the use of a soldering iron connected to a variable transformer (Figure 15–87). The transformer allows the heat of the soldering iron to be set at 225 degrees Fahrenheit (107 degrees Celsius). Other tools needed for minor repairs are a hypodermic syringe available from any medical supply company, an abrasive sponge, and a rubber cement pick-up block available at most craft or art supply stores. With these tools, the following repairs can be made.

Vinyl Scuffs or Abrasions

Connect the soldering iron and set the transformer heat range to approximately 225 degrees Fahrenheit (107 degrees Celsius). Clean the soldering iron tip thoroughly with the abrasive pad. This must be done frequently while performing the service to avoid vinyl buildup on the tip.

Lightly slide the soldering tip over the scuff mark several times using short overlapping strokes until the frayed vinyl is fused to the surface.

The vinyl surface might have been removed by the scuff, exposing the cloth backing. Repair the vinyl surface by filling in with vinyl. This is accomplished by stripping a small quantity of vinyl from a piece of scrap material with the hot tip of the solder gun. Carefully fill in as required using short overlapping strokes. Several stripping operations might be required to adequately fill the scuffed area.

The graining effect in the vinyl can be restored by carefully etching the grain pattern into the vinyl with the sharp edge of the soldering tip. The gloss or shiny surface created by the service area can be removed with the dulling agent or by spraying the area with liquid vinyl. Several color coats may be required. The last coat should be a fog coat to minimize gloss.

This service can be applied to scuffs or abrasions in most areas of the vinyl top.

Minor Vinyl Surface Cuts

The edges of a surface cut that has not penetrated the cloth backing can be welded together with a soldering iron.

Connect the soldering gun and set the transformer heat range to approximately 225 degrees Fahrenheit (107 degrees Celsius).

Clean the soldering iron tip thoroughly with the abrasive pad. Lightly slide the soldering tip across the cut surface using very short strokes—1/4 inch (6.3 mm) or less—until the cut is covered with vinyl. Go over the cut lengthwise again with the soldering tip to smooth out the surface. The graining effect can be restored by carefully etching the grain pattern into the vinyl with the sharp edge of the soldering tip.

If required, the gloss or shine on the vinyl surface created by the service can be removed with the dulling agent and then spraying the area with liquid vinyl. Several color coats of liquid vinyl can be required. The last coat should be a fog coat to minimize gloss.

MAJOR CUTS IN VINYL

A major cut—one that has penetrated through both the vinyl and the cloth backing—can also be repaired with the soldering iron. Connect the soldering iron and set the transformer heat range to approximately 225 degrees Fahrenheit (107 degrees Celsius). If damage is more than 1 inch (25 mm) long, apply a light coat of adhesive (Figure 15–88). (It is not necessary to use cement if the cut is over a padded area.)

Start at the center of the cut and lightly slide the soldering tip across the cut surface (Figure 15–89), using strokes 1/4 inch (6.3 mm) or less until the cut is covered with vinyl. Go over the cut lengthwise again with the soldering tip to smooth out the surface and restore the grain as described.

VINYL BUBBLES

Generally, an air bubble can be eliminated by removing the air with a hypodermic needle. Insert the needle under the vinyl material as shown in Figure 15–90. Press the air out. It is necessary to activate the adhesive under the vinyl after the air has been expelled by applying heat with the gun until the vinyl surface is hot to the touch. Then work the material down with the fingers.

To avoid overheating and possible damage to the vinyl, hold the heat gun about 10 to 12 inches (250 to 304 mm) from the surface and constantly move the gun in a circular motion. In some cases

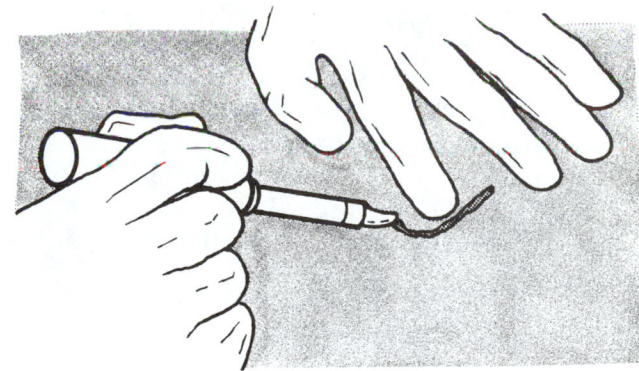

FIGURE 15-89 Solder the cut edges together.

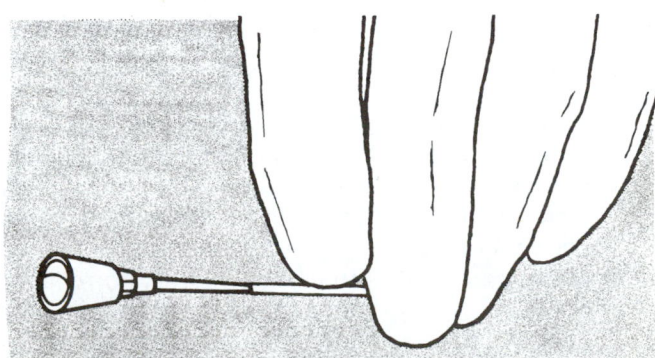

FIGURE 15-90 You can remove any air bubbles with a hypodermic needle. Draw out the air and then use heat to adhere the material.

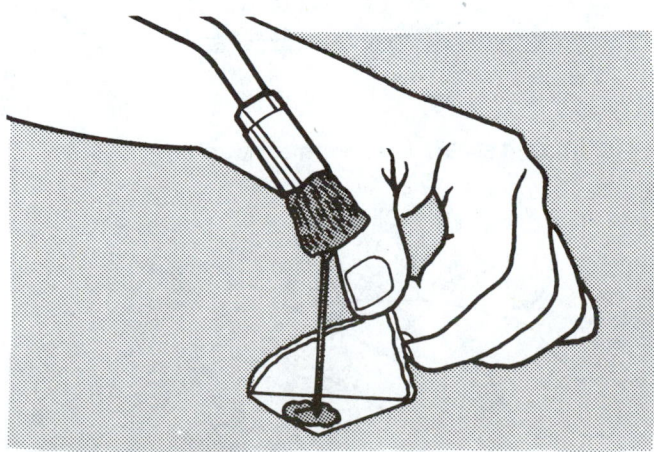

FIGURE 15-88 Glue the loose material using a recommended agent.

where a good bond cannot be obtained, insert a small amount of roof adhesive under the vinyl material with a syringe.

VINYL WRINKLES

Do not confuse a wrinkle with a bubble. Usually a wrinkle has small radial folds with slack material that cannot be displaced without rearranging the material. The following service procedure describes the correction of a wrinkle at the front corner:

1. Partially or completely remove any moldings or ornamentation in the immediate area of the wrinkle.
2. Pull the vinyl material free from the roof panel up to the bonded seam and along the side drip rail for approximately 10 inches (250 mm).
3. Clean the surface thoroughly to remove the old adhesive. Apply a thin film of adhesive both to the vinyl material and along the drip rail. Allow the adhesive to air dry for several seconds.
4. Grasp the material firmly and draw it tight until all wrinkles are removed. Fold the material under the drip rail flange.
5. Secure the front edge of the material with as many drive nails as needed to prevent the material from creeping. Apply sealer under the drive nail to prevent water leakage. Some manufacturers suggest securing the vinyl edge with the correct size pop rivets. Caution must be used when riveting the top. If the rivet head is exposed after installing the moldings, the vinyl roof must be replaced.
6. Install the moldings and weatherstrips.

VINYL LOOSENESS

Loose vinyl roof material usually occurs at padded areas of the vinyl roof on body styles so equipped. The following is a typical repair of loose material at the blind quarter area and generally applies to other areas such as around the back light opening:

1. Partially or completely remove moldings or ornamentation in the immediate area to be repaired.
2. Carefully peel the edge of the material free from the sheet metal.
3. Clean the metal surface thoroughly to remove the old adhesive. Apply a thin film of adhesive to both the vinyl and metal surface. Allow the adhesive to air dry for several seconds.
4. Grasp the material firmly and draw it tight. Secure the edge of the material with drive nails or sheet metal screws as needed to prevent the material from creeping.

5. Apply sealer over the drive nails or sheet metal screws to prevent water leaks. Install the moldings.

SEPARATED BONDED SEAMS IN VINYL

Connect the soldering iron and set the transformer heat range to approximately 225 degrees Fahrenheit (107 degrees Celsius).

Frequently clean the soldering iron tip thoroughly with the abrasive pad to avoid vinyl buildup on the tip. Start at one end of the separated seam and insert the tip of the soldering iron between the bonded seams (Figure 15–91). Note the position of the curved tip. Slowly move the tip along the underside of the seam. Immediately follow with the fingers to press the seams together.

Reverse the soldering iron tip (Figure 15–92) and again slowly move the tip along the edge of the seams.

This completes the service and, if performed correctly, should not require paint touch-up. Use a dulling agent or liquid vinyl to restore finish if necessary. Several coats might be required. The last coat should be a fog coat to minimize gloss.

DASH PANEL SERVICE

The **dash assembly**, sometimes termed *instrument panel*, is the assembly including the soft dash pad,

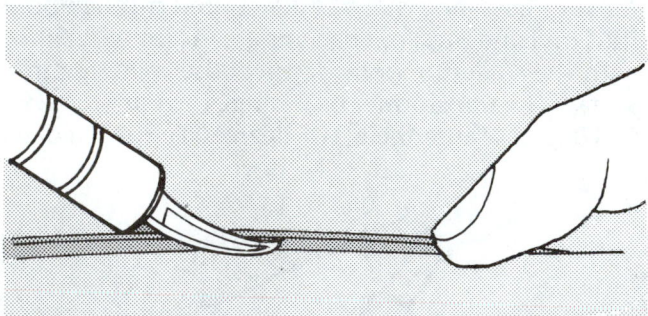

FIGURE 15-91 Soldering the underside of the seam.

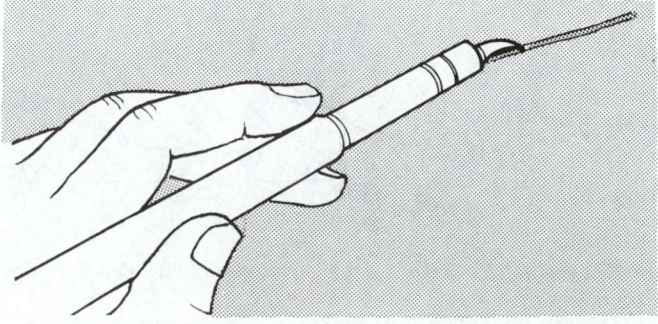

FIGURE 15-92 Soldering the outer edge of the seam.

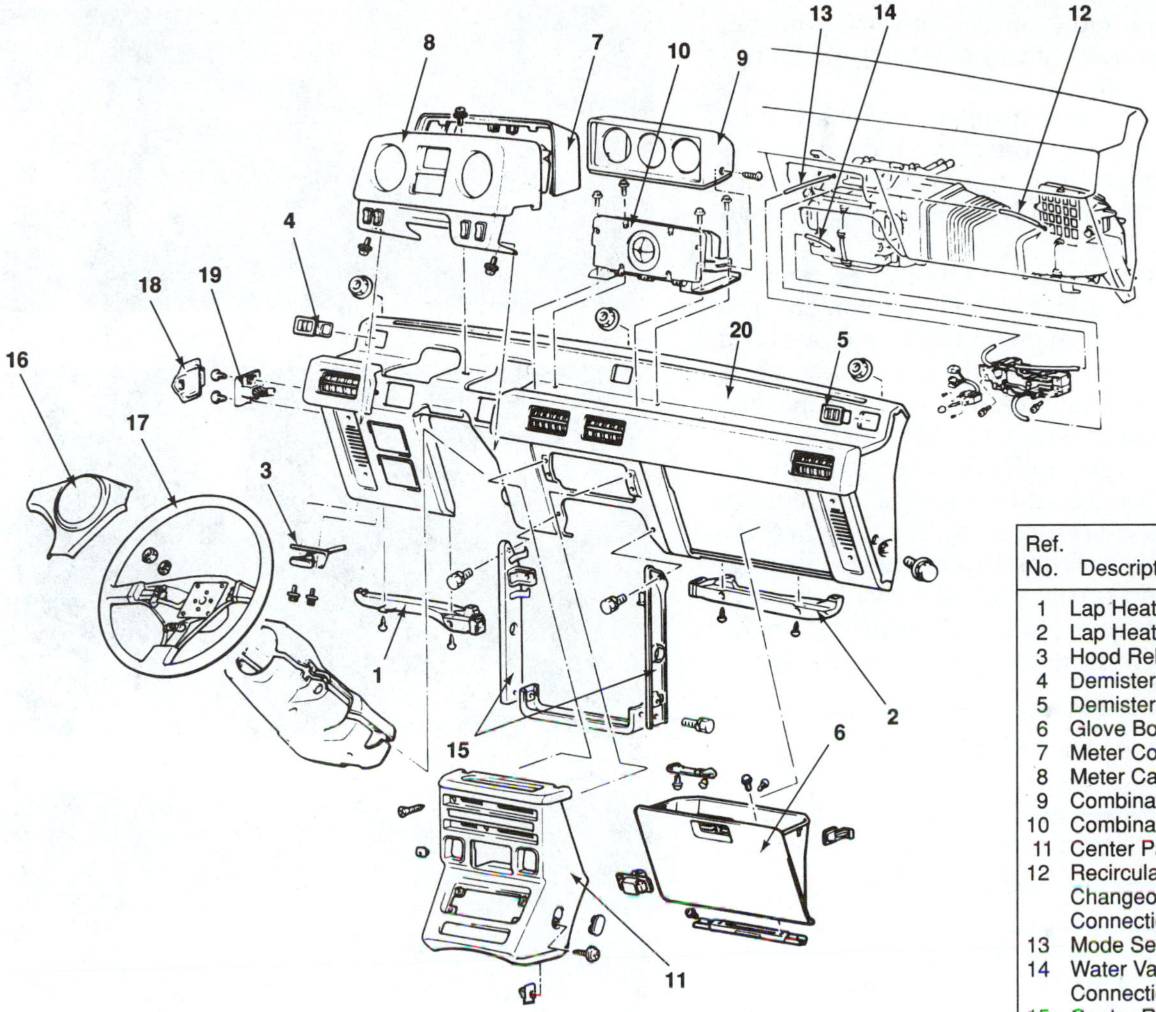

Ref. No.	Description
1	Lap Heater Duct
2	Lap Heater Duct
3	Hood Release Cable Bracket
4	Demister Grille
5	Demister Grille
6	Glove Box
7	Meter Cover
8	Meter Case
9	Combination Meter Pad
10	Combination Meter Case
11	Center Panel
12	Recirculation/Fresh Air Changeover Control Wire Connection
13	Mode Selection Control Wire
14	Water Valve Control Wire Connection
15	Center Reinforcement
16	Horn Pad
17	Steering Wheel
18	Fuse Box Cover
19	Fuse Box Assembly
20	Instrument Panel

FIGURE 15-93 Study the parts of the dash panel. They are often damaged when seat belts are not used. *(Courtesy of Chrysler Corp.)*

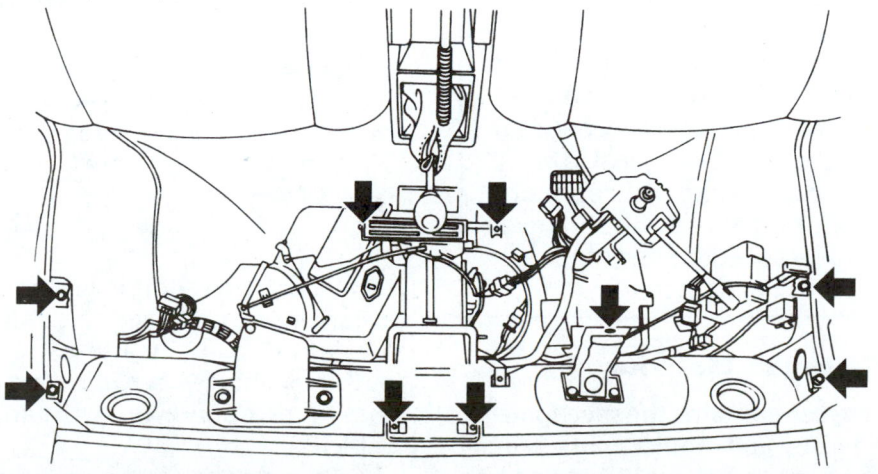

FIGURE 15-94 The service manual will give the locations of bolts and screws that hold the dash panel to the body structure. Some can be hard to find. *(Courtesy of Subaru of America)*

instrument cluster, radio, heater and A/C controls, vents, and similar parts. It can be damaged in a collision by human body parts.

When parts of the instrument panel are damaged in a collision, they must be removed and replaced. An exploded view of a typical instrument panel is shown in Figure 15–93. Study the relationship of the parts.

Many instrument panel parts can be replaced without unbolting the dash pad. The instrument cluster, vents, and many trim pieces can be removed and replaced with the main part of the dash intact. Vents often snap into place. A thin screwdriver can often be used to release and remove most vents.

Some of the screws and bolts that secure instrument panel parts can be difficult to find and remove. Some are along the bottom of the dash. Others are on the sides. A few fasteners can be inside openings in the instrument panel. You will have to remove parts to access these fasteners. Consult the service manual for exact procedures (See Figure 15–94).

WARNING
Keep your fingers off the inside of the instrument lens. Fingerprints on the inside of the lens can collect dust that cannot be wiped off after you reinstall the cluster.

With the lens removed, you can replace gauges and the speedometer head. Screws on the rear of the cluster normally hold each unit in place. See Figure 15–96B. Again, keep fingerprints off the faces of the gauges and speedometer. They will show after the installation.

WARNING
Make sure the speedometer reading of the new or replacement unit is the same as the old one (Figure 15–96C). You are breaking a federal law if you alter a speedometer reading. Check local laws for rules pertaining to reporting speedometer service. Never change a speedometer reading or you could be arrested.

Instrument Cluster Service

An **instrument cluster** contains the speedometer, gauges, indicating lights, and similar parts. It may require service when damaged in a collision or when parts are not working. Refer to the service manual for exact procedures.

To service an instrument cluster, first disconnect the battery. This will prevent the chance of an elec-

A

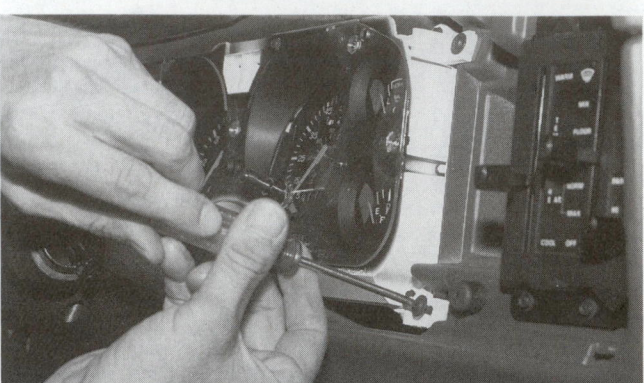

B

C

FIGURE 15-95 (A) Screws often secure the cover over the instrument cluster. (B) More screws hold the cluster to the dash. (C) Disconnect all wires and the speedometer before pulling the cluster all the way out. *(Courtesy of Mustang Monthly Magazine)*

trical fire if wires short to ground. Also note that static electricity can damage the chips in a digital dash. You might want to use a static strap to ground out static charges. A *static strap* is a soft metal strap that wraps around your wrist and is grounded to the vehicle.

Remove the screws that secure the instrument panel cover (Figure 15–95A), and remove the cover. Next, remove the screws that hold the cluster to the dash (Figure 15–95B). Pull the cluster out far enough to disconnect the wires and the speedometer cable.

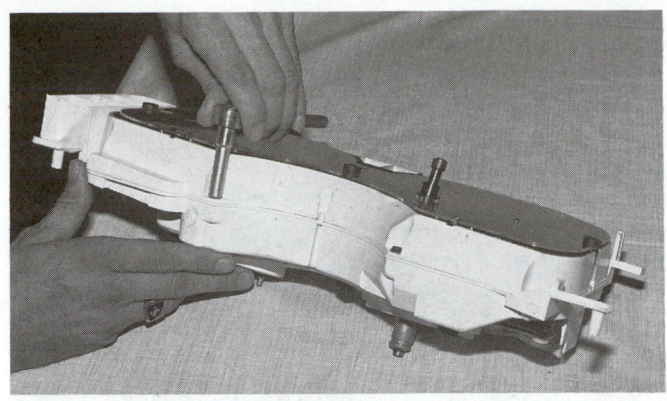

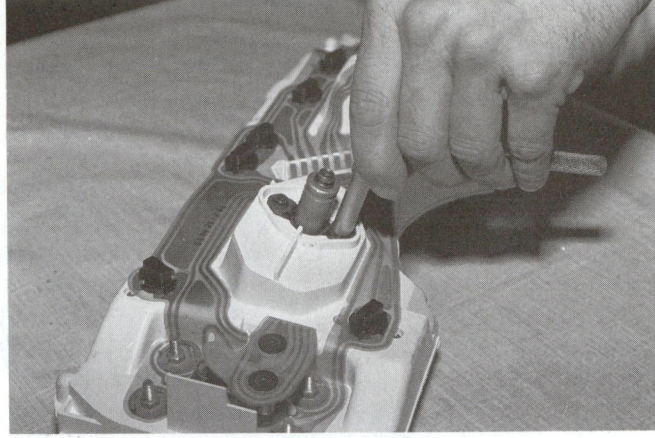

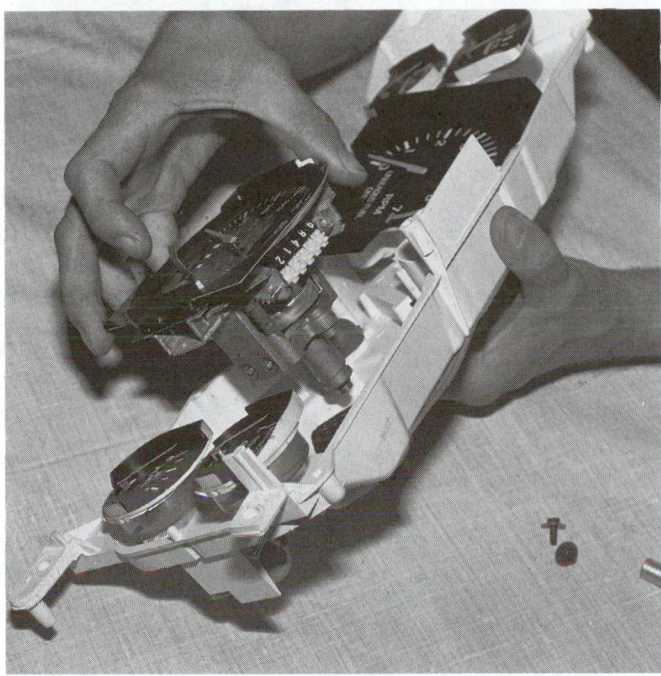

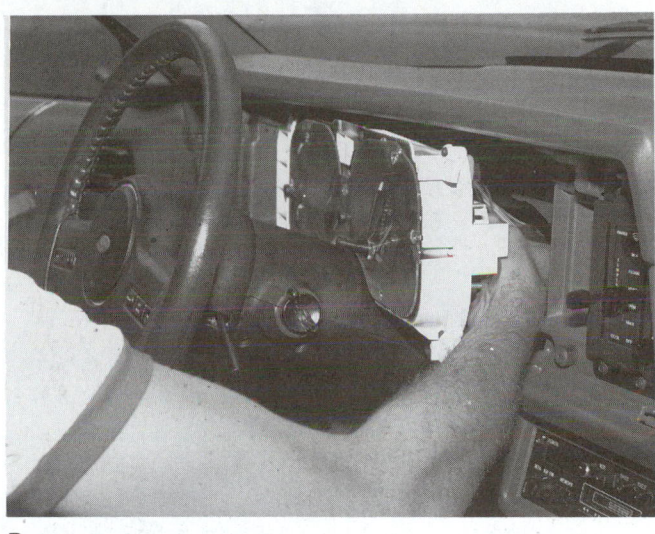

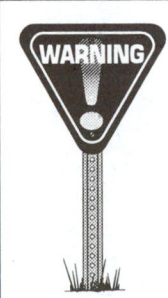

FIGURE 15-96 (A) To replace the speedometer head or the gauges, remove the small fasteners that hold the cluster together. (B) The speedometer head is often held by two fasteners on the rear of the cluster. (C) Make sure you do not change the speedometer reading when installing new parts. (D) Reconnect the wires and the speedometer cable before bolting the cluster into the dash. *(Courtesy of Mustang Monthly Magazine)*

Then you can lift the cluster out (Figure 15–95C). Bulbs can be replaced from the rear of the cluster.

To replace gauges or the speedometer, you must disassemble the cluster. To disassemble the instrument cluster, remove the small screws that hold the plastic lens plate over the housing (Figure 15–96A).

Install the instrument cluster parts in reverse order of removal. Remember to connect all wires and the speedometer cable, if used, to the cluster (Figure 15–96D). Check the operation of all dash lights and gauges after installation.

INTERIOR TRIM SERVICE

Various pieces of trim are used in the passenger compartment for appearance and safety. Most are held

> **WARNING**
>
> To prevent possible exposure to communicable diseases, do not come into direct contact with human blood in the passenger compartment. Wear plastic gloves and use seat covers to keep the blood from contacting your skin.

by snap-in clips or small screws. Sometimes screw heads are covered by small plastic plugs. Screws can also be hidden under protruding parts. Your service manual will give locations for the fasteners holding interior trim parts (Figure 15–97).

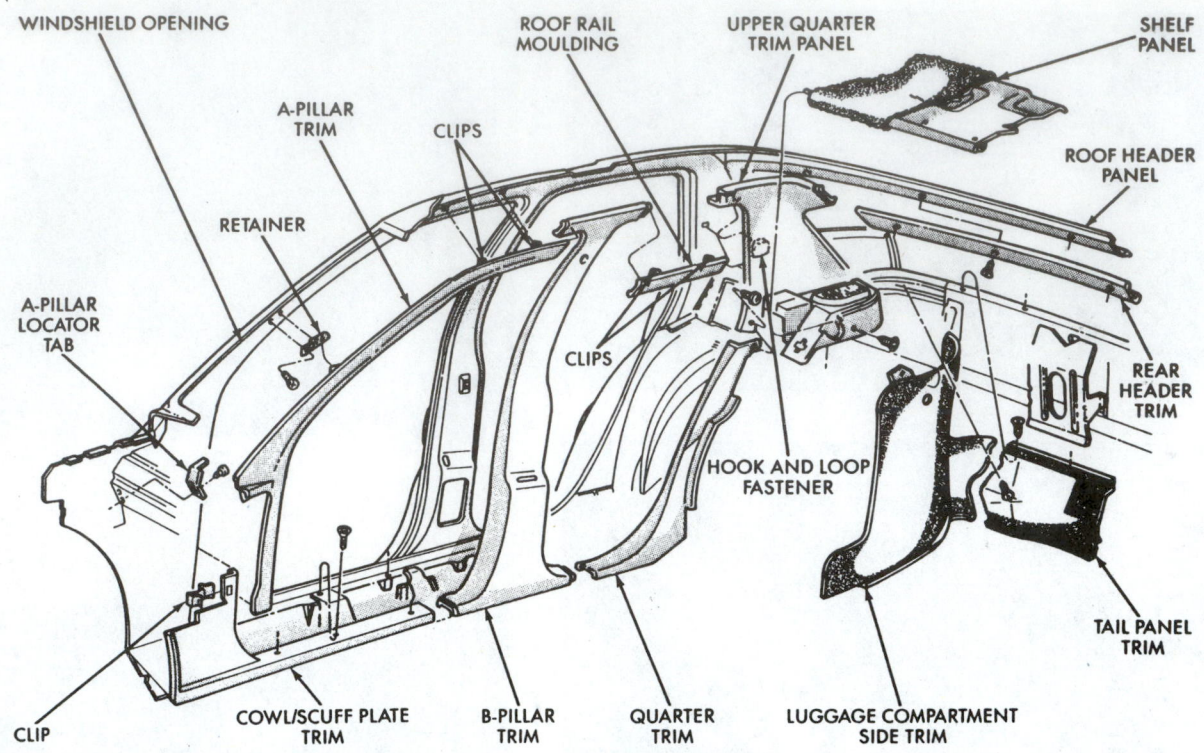

FIGURE 15-97 Note the interior trim pieces that sometimes must be serviced after a collision. *(Courtesy of Chrysler Corporation)*

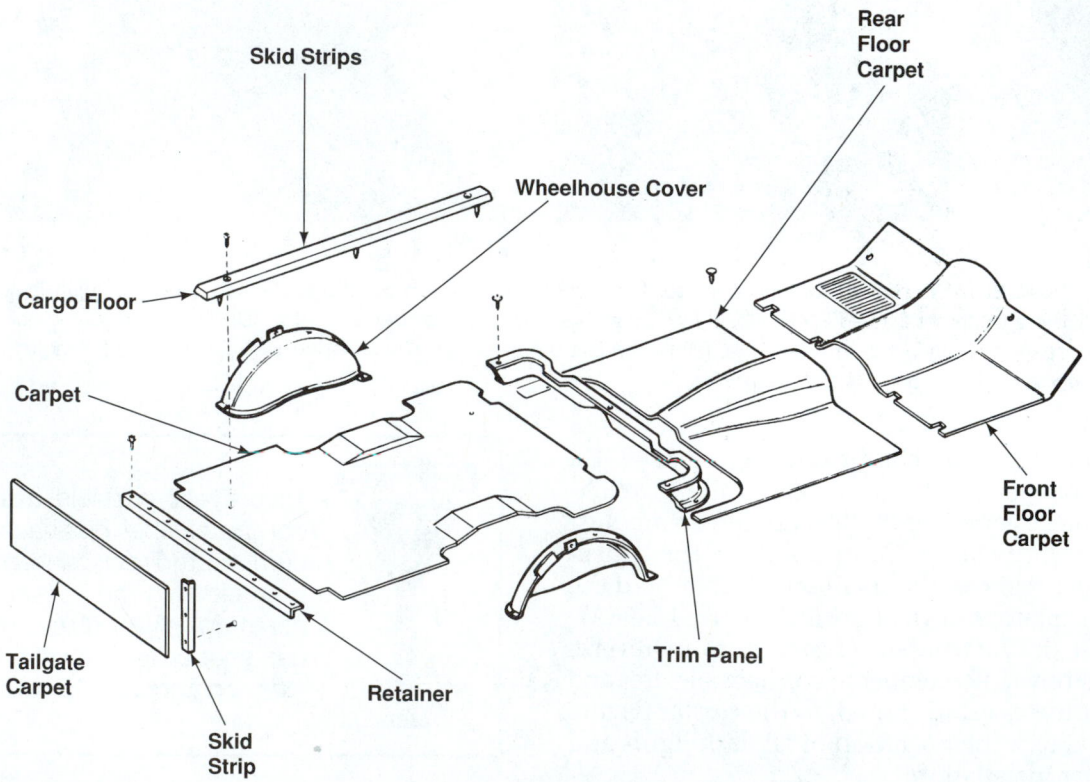

FIGURE 15-98 Carpeting often comes in several sections. Damaged sections can be replaced. *(Courtesy of Chrysler Corporation)*

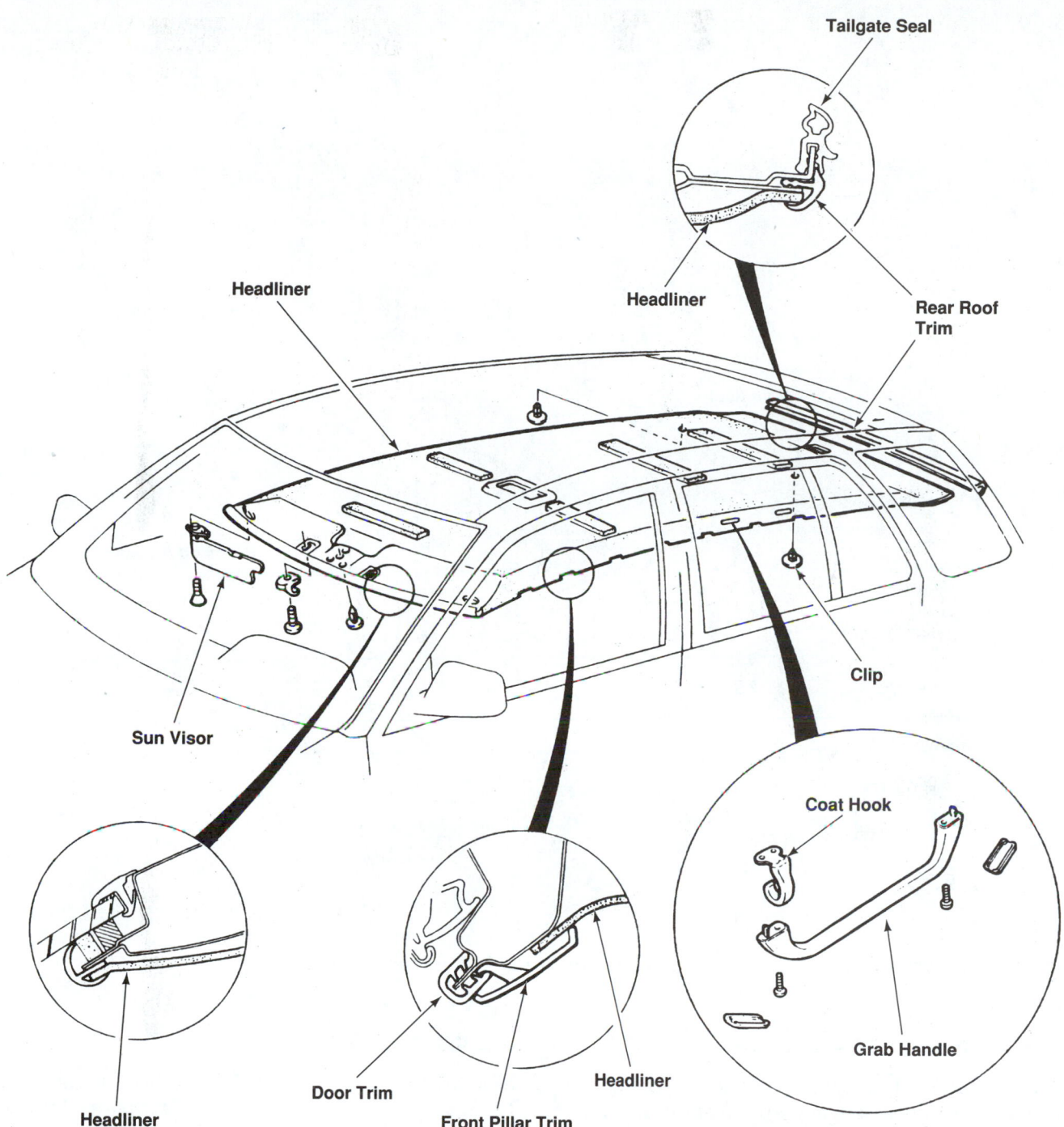

FIGURE 15-99 Procedures for replacing a headliner will vary. This is a typical design. *(Courtesy of American Honda Motor Co.)*

CARPETING SERVICE

Carpeting can be stained or torn during a collision. If it cannot be cleaned with a strong carpet cleaner or is torn, the carpeting must be replaced. A skilled upholsterer can sew small tears and holes in carpeting.

The major parts of interior carpeting are shown in Figure 15–98.

To replace carpeting, you must remove the seats, seat belt anchors, trim pieces, and other parts mounted over the carpeting. This might include the console, any electronic control units, and wiring harness bolted down to the carpet. Screws and clips hold down these parts and hold the carpet in place.

After removal, the new carpet is installed in the reverse order of removal. Make sure the new carpet is stretched out smoothly and is properly centered before installing any fasteners. An adhesive may be required between the carpet and the floor in some locations. Refer to the service manual for exact procedures.

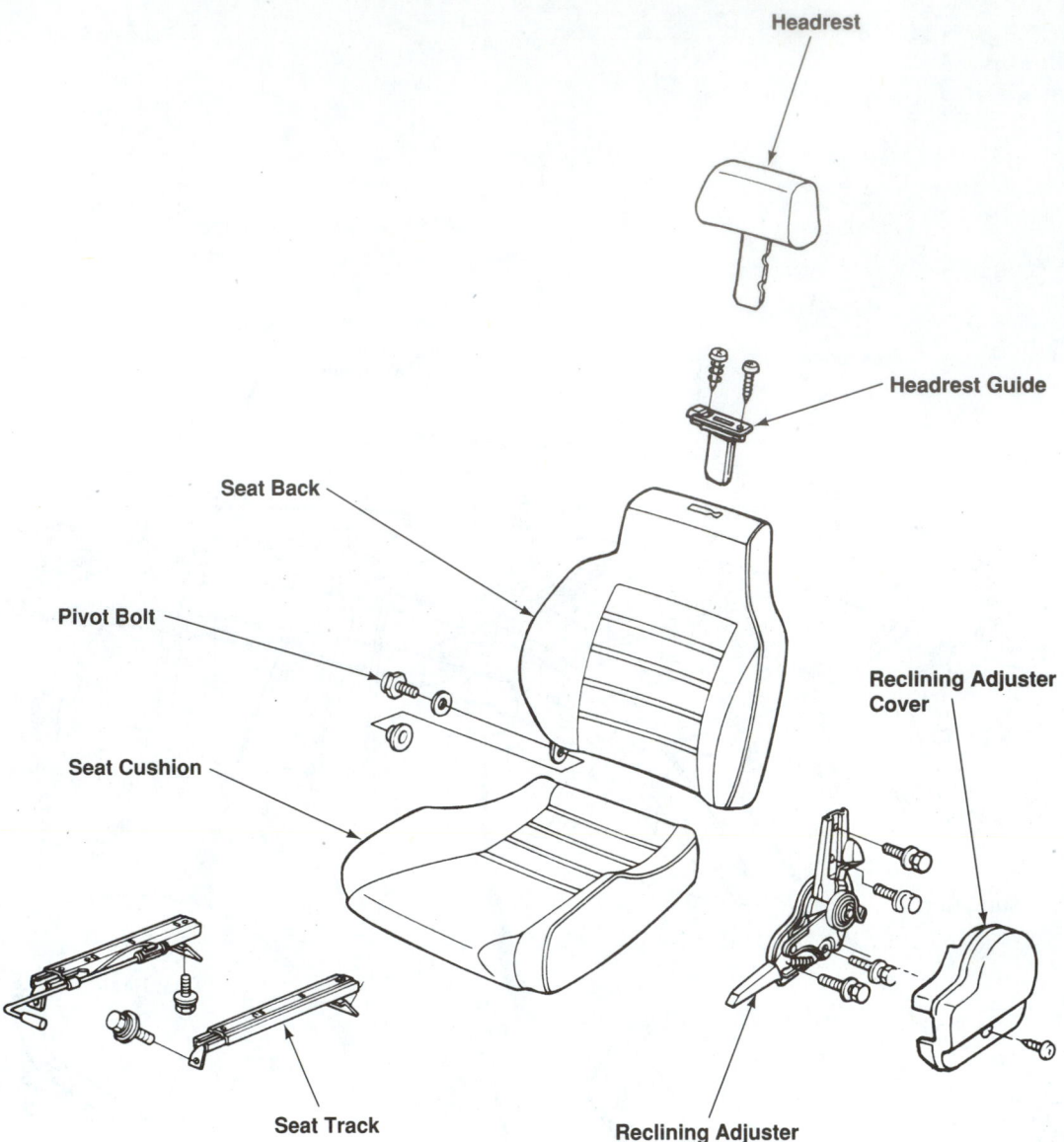

FIGURE 15-100 Note the basic parts of a seat. *(Courtesy of American Honda Motor Co.)*

HEADLINER SERVICE

A **headliner** is a cloth or vinyl cover over the inside of the roof in the passenger compartment. It can be torn or damaged during a collision. Some thick cloth or vinyl-covered foam headliners are bonded directly to the roof panel. Others are thin vinyl suspended by metal rods and bonded around the edges of the roof.

To service a headliner, first remove all of the trim pieces around the edges of the roof. Various screws and clips secure the trim pieces. You may also have to remove the sun visors, grab handles, and other parts for headliner service (Figure 15–99).

When installing a foam-backed headliner, be careful not to overbend and kink it. Center it in position. Then install it in reverse order of removal. Again, refer to the service manual if in doubt.

SEAT SERVICE

Seats are often damaged during a collision. They can be damaged by the inertia of the occupants, by side impact intrusion into the passenger compartment, or stained by blood. You may also have to remove seats for carpet replacement or floor panel repairs.

A *bucket seat* is a single seat for one person. A *bench seat* is a longer seat for several people. Both require similar methods during service.

The typical parts of a front seat include the following (Figure 15–100):

1. *Seat cushion* (bottom section of seat, which includes cover, padding, and frame)
2. *Seat back* (rear assembly that includes cover, padding, and metal frame)
3. *Headrest* (padded frame that fits into top of seat back)

4. *Headrest guide* (sleeve that accepts headrest post and mounts in seat back)
5. *Recliner adjuster* (hinge mechanism that allows adjustment of seat back to different angles)
6. *Seat track* (mechanical slide mechanism that allows seat to be adjusted forward or rearward)

SEAT REMOVAL

Four bolts normally secure the seat to the floor. To remove the front seat hold-down bolts, slide the seat fully backward. This will allow easier access to the front bolts. Then slide the seat forward to remove the two rear hold-down bolts.

After disconnecting any wiring and other parts attached to the seat, carefully lift it out of the vehicle.

A rear bench seat is often held in position by screws or spring-loaded clips. The screws are normally at the front, bottom of the seat. When removed, the seat can be pushed back and lifted up and out.

With spring clips, use your hands to force the seat down and back. This will free the hidden clips and allow you to lift out the bench seat.

When servicing seats, refer to the manufacturer's manual for details. Procedures vary. Improper seat installation could endanger the vehicle's passengers. Always use a torque wrench and factory specified torque values when tightening seat fasteners.

SEAT COVER SERVICE

The seat cover is a cloth, vinyl, or leather cover over the seat assembly. The cover may require replacement when damaged. With minor damage, an upholsterer can sometimes repair small holes and tears. You must disassemble the seat to replace the covers.

Hog rings and clips normally stretch and hold the seat cover over the seat frame and padding. They are located on the bottom of the seat cushion or rear of the seat back. Remove them and you can lift off the seat cover. The new cover can then be installed in reverse order of removal.

SUMMARY

- Today's vehicles are built with a lot of glass for greater visibility.
- Laminated plate glass consists of two thin sheets of glass with a thin layer of clear plastic between them.
- Tempered glass is used for side and rear window glass but rarely for windshields.
- Antilacerative glass is similar to conventional multilayered glass, but it has one or more additional layers of plastic affixed to the passenger compartment side of the glass.
- Interior moldings are called garnish moldings and exterior moldings are called reveal moldings.
- Replacement of windshield glass involves two different methods based on the materials used: gasket installations or adhesive-type installations.
- Most technicians use the windshield adhesive recommended by the manufacturer.
- Remember to use caution and follow all safety rules when working with automotive glass. Wear eye protection to guard against flying bits of glass. Wear leather gloves to prevent cuts. Plastic gloves should be worn to keep adhesives off your skin.
- Water leaks are noticed when moisture or rain enters the passenger compartment and collects on the carpeting.
- Air leaks normally cause a whistling or hissing noise in the passenger compartment during driving.
- The principal methods used to locate air and water leaks are:
  Spraying water on the vehicle
  Driving the vehicle over very dusty terrain
  Directing a strong beam of light on the vehicle and checking for light leakage between the panels
  Using a listening device
- If the vehicle was involved in a front-end collision, the headlights should be adjusted after needed parts have been replaced.
- An air bag system automatically deploys a large nylon bag during frontal collisions. Both impact and arming sensors are inertia sensors.
- Inertia sensors detect a rapid deceleration to produce an electrical signal.
- The air bag controller analyzes inputs from the sensor to determine if bag deployment is needed.
- Before servicing a vehicle equipped with an air bag, the system must be disarmed (all sources of electricity for igniter disconnected). Even with the battery disconnected, the reserve module can fire the air bag.

■ Most automakers recommend replacement of all sensors and sometimes the electronic control unit during deployed air bag service.

■ The dash assembly, sometimes termed instrument panel, is the assembly including the soft dash pad, instrument cluster, radio, heater and A/C controls, vents, and similar parts.

ASE-STYLE REVIEW QUESTIONS

1. Technician A attempts to repair lap and shoulder belt retractor mechanisms before replacing them. Technician B always tightens the anchor bolts as specified in the service manual. Who is correct?

   A. Technician A
   B. Technician B
   C. Both A and B
   D. Neither A or B

2. What is used to remove wrinkles from a newly repaired vinyl roof?

   A. Heat gun and squeegee
   B. Iron and damp cloth
   C. Both A and B
   D. Neither A nor B

3. Prior to patching a vinyl roof, Technician A cleans the repair area by spraying vinyl cleaner directly on the vinyl. Technician B frays the edges of the tear to provide extra bonding area for the patching compound. Who is correct?

   A. Technician A
   B. Technician B
   C. Both A and B
   D. Neither A or B

4. Tempered glass is never used for _____.

   A. Windshields
   B. Rear windows
   C. Side windows
   D. Both A and B

5. What type of glass fits the contours of a vehicle very closely?

   A. Modular
   B. Tempered
   C. Laminated
   D. Channel

6. When servicing an air bag system, Technician A probes the electrical connectors on the air bag module with a test light. Technician B always carries the module with the trim cover facing away from him or herself. Who is correct?

   A. Technician A
   B. Technician B
   C. Both A and B
   D. Neither A or B

7. Gasket glass installation is more common in _____.

   A. Older vehicles
   B. Newer vehicles
   C. Vehicles with modular glass
   D. None of the above

8. Which of the following items is used in the full cutout windshield replacement method?

   A. Butyl ribbon sealer
   B. Setting blocks
   C. Utility knife
   D. All of the above

9. In the partial cutout windshield replacement method, what serves as the base for the new adhesive?

   A. Butyl ribbon sealer
   B. Butyl tape
   C. Masking tape
   D. The old adhesive

10. Where do the majority of windshield leaks occur?

    A. Sides
    B. Top
    C. Bottom
    D. Corners

11. Which of the following should not be done before adjusting vehicle headlights?

    A. Inflate the tires to the recommended pressures.
    B. Clean off any snow, ice, or mud from the underside.
    C. Fill the gas tank.
    D. Check the condition of the suspension components.

ESSAY QUESTIONS

1. Describe the principal methods used to locate air and water leaks.

2. How do you find wind and water leaks?

3. Before making adjustments to a vehicle's headlights, what should you do to make sure the vehicle is level?

4. How do you use an electronic stethoscope?

5. What cautions must be observed when removing bumpers with energy absorbers?

6. Explain the difference between active and passive restraint systems.

7. Summarize the five major parts of an air bag system.

8. What happens when an air bag inflates?

9. After servicing an air bag system, what should you do to check for problems?

CRITICAL THINKING PROBLEMS

1. If you have your arm through the spoke of a steering wheel, and the air bag inflates, what could happen?

2. What would happen if you install an air bag sensor with its arrow facing to the rear?

MATH PROBLEMS

1. When adjusting headlights, you find a front curb height measurement to be 12.6 inches (320 mm). Specs call for a minimum ride height of 13.2 inches (335 mm). How much is the vehicle out of specs?

CHAPTER 16

Restoring Corrosion Protection

INTRODUCTION

Corrosion protection involves using various materials to protect steel body parts from rusting. When doing repairs, you must always use recommended methods of protecting repair areas from rust damage. Rustproofing is often a joint task or effort of both the collision repair and paint technicians.

With the recent developments in the automotive industry, the words "corrosion prevention" have taken on new meaning to body shop personnel:

• **Corrosion prevention** differs from rustproofing, undercoating, and sound deadening. Corrosion prevention implies a vehicle lifetime maintenance responsibility to the consumer. Rustproofing, undercoating, and sound deadening suggest a one-time application to the new vehicle by car dealerships and rustproofing franchises.

• Car manufacturers are including in their owner's manual instructions and recommendations for sheet metal repair or replacement. They suggest that the body shop should apply an anticorrosive material to the part repaired or replaced so that corrosion protection is restored.

• In affiliation with the Inter-Industry Conference on Auto Collision Repair (I-CAR), the insurance companies are promoting corrosion prevention repair to the body shops.

• The increased usage of replacement panels in the body shops requires widespread corrosion prevention treatment.

Possibly the major reason, however, is the advent of the unibody car. In unibody construction, the car's body panels are no longer cosmetic sheet metal. They now constitute the structural integrity of the vehicle. This means that rust is not just an "eyesore." The unibody car has more welded joints

OBJECTIVES

After studying this chapter, you should be able to:

✔ Define corrosion and describe the common factors involved in rust formation.

✔ Describe the anticorrosive materials used to prevent and retard rust formation.

✔ Explain the conditions and events that lead to corrosion on a vehicle body.

✔ Choose the correct anticorrosive application equipment for specific applications.

✔ Outline the correct corrosion treatment procedures for each of the four general corrosion treatment areas.

✔ List the four types of seam sealers and explain where each should be used.

✔ Answer ASE test questions about corrosion protection materials and methods.

KEY TERMS

acid rain
anticorrosion
   compound
anticorrosion
   compounds
antirust agents
body sealer
brushable seam
   sealers
corrosion prevention
corrosion protection
epoxy primers

galvanic corrosion
galvanizing
heavy-bodied sealers
industrial fallout
metal treating
rust
rust converters
solid seam sealers
thin-bodied sealers
undercoating
weld-through primer

ASE TASK LIST

Job Skills covered in this chapter include:

PAINTING AND REFINISHING TEST (B2) TASK LIST

A. Surface Preparation

4. Remove paint finish.
13. Clean area to be refinished using a proper cleaning solution.
19. Restore corrosion-resistant coatings, caulking, and seam sealers to repaired areas.

B. Spray Gun Operation and Related Equipment

1. Inspect, clean, and determine condition and adequacy of spray guns and related equipment (air hoses, regulator, air lines, air source, and spray environment).
3. Adjust spray gun using fluid and pattern control valves.

C. Paint Mixing, Matching, and Applying

8. Sand, buff, and polish finishes where necessary.

D. Solving Paint Application Problems

3. Identify contaminants in the painted surface; determine the source(s), and correct the condition.
9. Identify an overspray condition; determine the cause(s), and correct the condition.

E. Finish Defects, Causes, and Cures

1. Identify poor adhesion; determine the cause(s), and correct the condition.

NONSTRUCTURAL ANALYSIS AND DAMAGE REPAIR TEST (B3) TASK LIST

A. Preparation

9. Remove corrosion protection, undercoatings, sealers, and other protective coatings as necessary to perform repairs.

B. Outer Body Panel Repairs, Replacements, and Adjustments

1. Determine the extent of the direct and indirect damage and the direction of impact; plan the methods and order of repair.
12. Apply protective coatings and sealants to restore corrosion protection.
16. Restore sealers, mastic, sound deadeners, and foam fillers.

STRUCTURAL ANALYSIS AND DAMAGE REPAIR TEST (B4) TASK LIST

A. Frame Inspection and Repair

9. Restore corrosion protection to repaired or replaced frame areas.

B. Unibody Inspection, Measurement, and Repair

17. Restore corrosion protection to repaired or replaced unibody structural areas.

---

in critical structural areas where corrosion can do serious damage. It is an ever present danger to the unibody vehicle since rusting of structural panels and rails can affect the drivability of the car and the safety of its passengers.

If you fail to restore proper corrosion protection, it can endanger the driver and passengers of the vehicle. After prolonged service, rust could weaken the body structure. This could cause failure of the body where it supports the suspension components. The vehicle could become unstable and dangerous to drive because of this rust.

16.1 WHAT IS CORROSION?

Corrosion or **rust** is the oxidation and chemical change of metal. When it occurs on steel, it is the product of a complex chemical reaction with serious and costly consequences (Figure 16–1). In other words, the formula for rust in a car body is:

Iron + Oxygen + Electrolyte = Rust (Iron Oxide)

Chemical corrosion requires three elements (Figure 16–2):

- Exposed metal
- Oxygen
- Moisture (electrolyte)

There are three basic types of corrosion protection used on today's automobiles:

- Galvanizing or zinc coating
- Paint
- Anticorrosion compounds

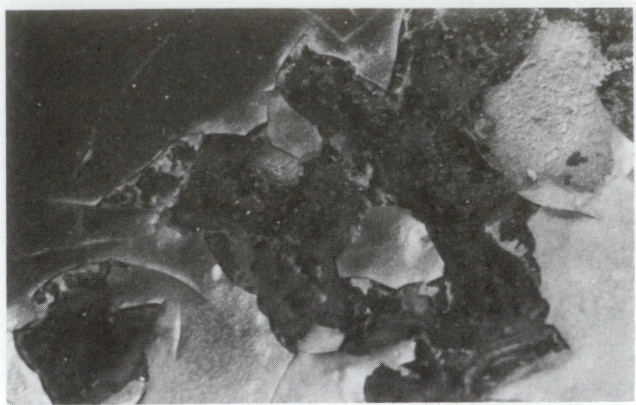

FIGURE 16-1 This close-up shows a car's number-one enemy—rust.

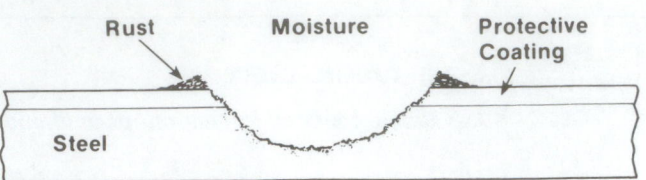

FIGURE 16-2 Breakdown in protective paint coating causes rapid rust corrosion formation.

Galvanizing is a process of coating steel with zinc (Figure 16–3). It is one of the principal methods of corrosion protection applied during the manufacturing process. On galvanized steel, the zinc forms a natural barrier between the steel and the atmosphere. As the zinc corrodes, a layer of zinc oxide will form on the surface. Unlike iron oxide, or rust, the zinc oxide adheres to the zinc coating tightly, forming a barrier between the zinc and the atmosphere.

When the surface of the vehicle's finish is damaged by a scratch or nick, the zinc coating undergoes corrosion, sacrificing itself to protect the iron under it. The resulting zinc oxide actually forms a protective coating and repairs the exposed area of the steel. Thus, zinc performs a two-fold protective process. First, it provides chemical, galvanic protection, and second, it forms a repair over the exposed steel with a layer of zinc oxide.

A paint system, such as those described in later chapters of this book, also provides a barrier between the atmosphere and the steel surface. When this barrier is in place (Figure 16–4), the moisture and impurities in the air cannot interact with the steel surface and the steel is protected from corrosion.

If the paint surface or barrier is broken by a stone chip or scratch, the steel in this area is no longer isolated from the moisture and impurities in the air. Corrosion will then take place in this region. Corrosion will spread between the paint and steel surface. If the adhesion of the paint to the steel is poor, large sections of the paint can be separated from the steel.

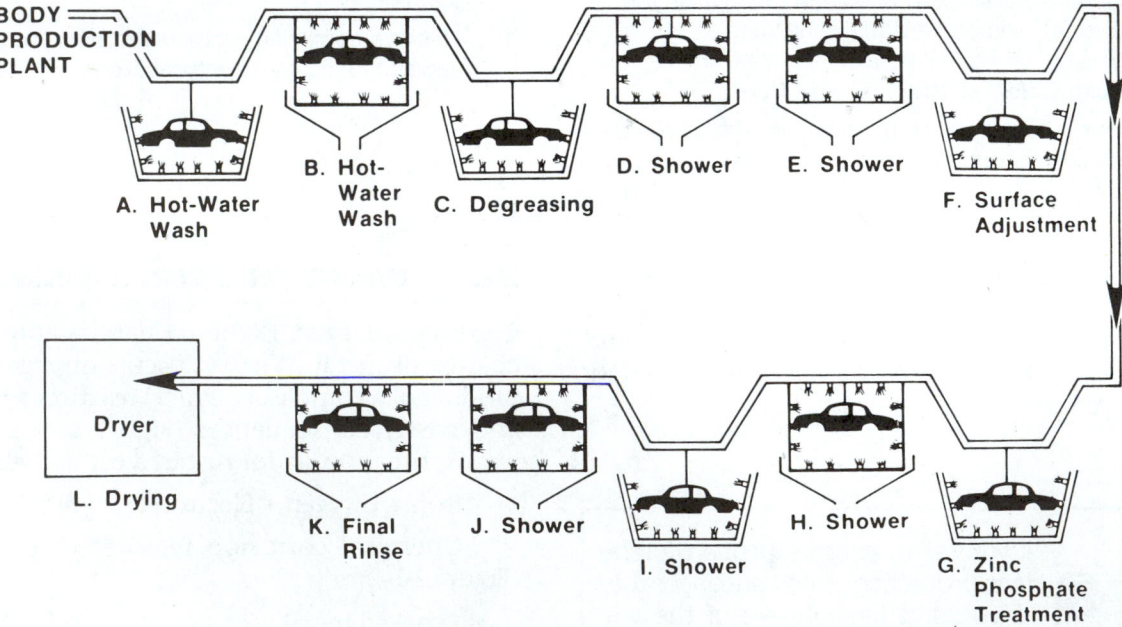

FIGURE 16-3 Zinc corrosion treatment utilizing the full dip method as done by a vehicle manufacturer's body production technician: (A) and (B) Metal chips, dirt, and other foreign particles are washed off with hot water at 100 to 125 degrees Fahrenheit (36 to 52 Celsius); (C) Press oil and anticorrosion oil are removed with a weak alkali degreasing agent; (D) and (E) The degreasing agent is washed off with water in two stages; (F) The nucleus of the zinc phosphate film is adhered to the panel surfaces; (G) The body is dipped into a tank of zinc phosphate for crystallization; (H), (I), and (J) The zinc phosphate liquid is washed off by water in three stages; (K) The body is given a final rinse to prevent blistering; (L) The body is dried at 212 to 300 degrees Fahrenheit (100 to 149 degrees Celsius). *Courtesy of Toyota Motor Corp.*

This will result in a large area of the steel being left unprotected. Severe rust in this region will quickly follow. If impurities are present between the paint and the steel, oxygen in the air can pass through the paint, reacting with the impurities and the steel to form rust. In this case, corrosion will take place on the steel surface and the protective paint barrier will

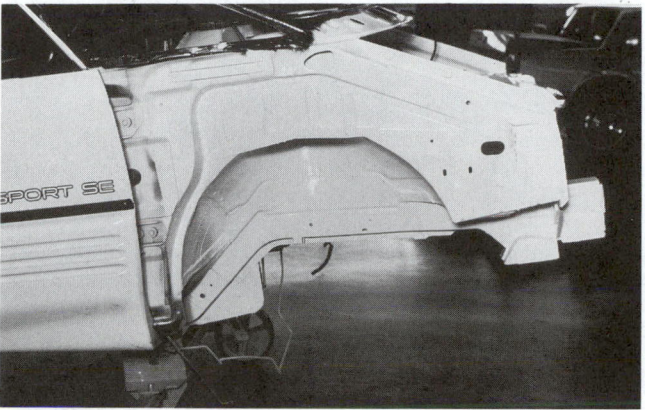

FIGURE 16-4 Paint is the first and most important barrier against corrosion on the vehicle. *(Courtesy of Tech-Cor)*

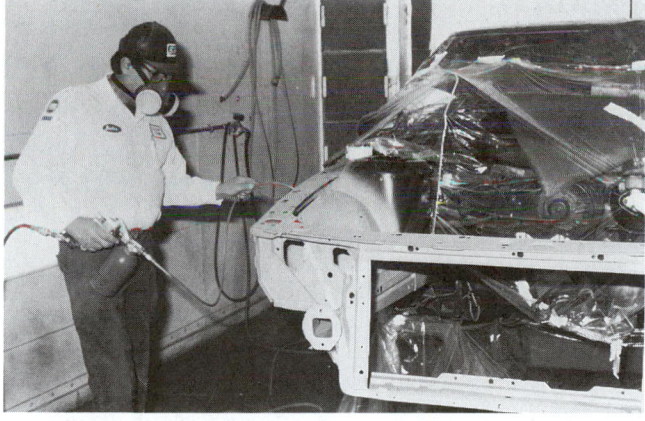

FIGURE 16-5 Anticorrosion material is being applied by auto body technician. *(Courtesy of Tech-Cor)*

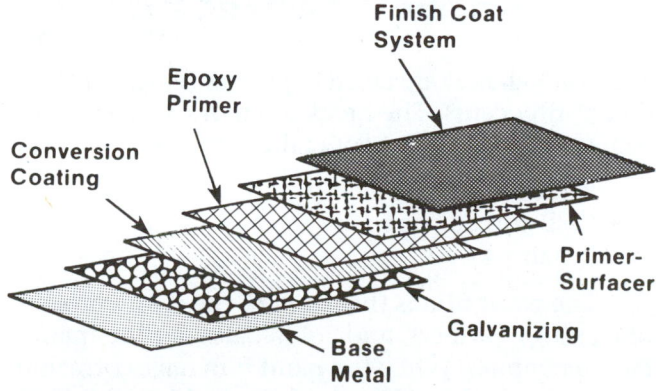

FIGURE 16-6 Note typical makeup of corrosion prevention material used by car manufacturers.

be destroyed. Paint, by itself, is only effective as long as the paint film remains intact.

Anticorrosion compounds are additional coatings applied over and under the paint film. Protective coatings can be applied either by the manufacturer or as an aftermarket process. The two most popular types of anticorrosion coatings are

- Petroleum-based compounds
- Wax-based compounds

Anticorrosion compounds are primarily used in enclosed body sections (Figure 16–5) and other rust-prone areas.

The auto manufacturers are increasing their corrosion protection measures all the time. New processes and methods, including the use of coated steels, zinc-rich primers, and more durable base coatings, have made it possible for modern cars to survive corrosive forces for longer periods than before.

The following is a typical new car finishing sequence (Figure 16–6) used by major auto manufacturers:

1. Use coated or galvanized steel (Figure 16–7).
2. Chemically clean and rinse.
3. Apply conversion coating.
4. Apply epoxy primer.
5. Bake primer.
6. Seam seal process.
7. Apply primer-surfacer.
8. Apply colorcoats.
9. Bake colorcoats.
10. Apply anticorrosion materials.

Because of these better procedures, corrosion protection warranties (Figure 16–8) of several years are common. With these dramatic improvements in the performance of OEM products, the repair industry must rise to the challenge of producing corrosion resistance in repaired areas that matches or exceeds the durability of the original product. Repair work that does not stand up will draw attention to itself next to the outstanding durability of many original finishes. It can also draw liability challenges where issues of vehicle safety are involved.

Remember that you, the body shop technician, are responsible for the quality and durability of the completed repairs. Remember that the customer is entitled to a car restored to the way it was before the damage occurred.

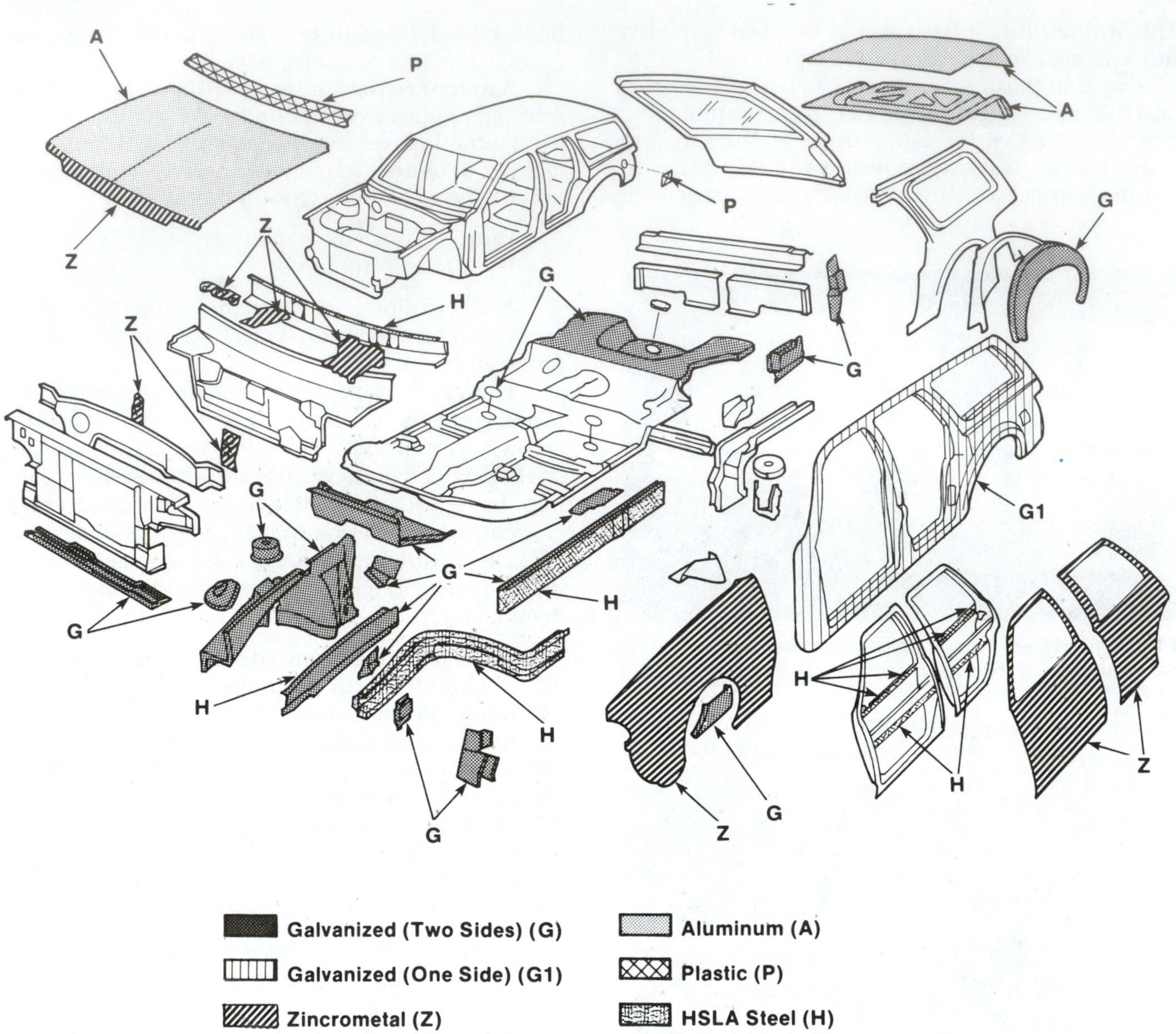

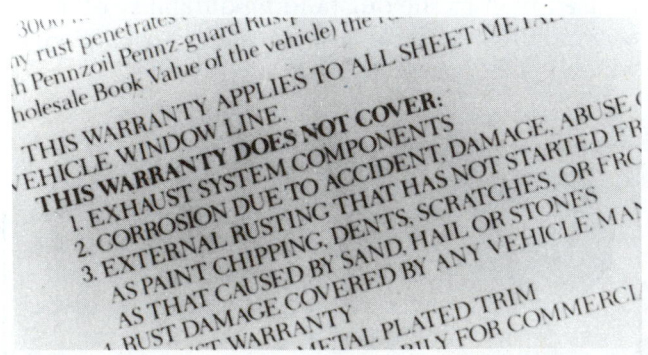

■	**Galvanized (Two Sides) (G)**	▦	**Aluminum (A)**
▥	**Galvanized (One Side) (G1)**	✕	**Plastic (P)**
▨	**Zincrometal (Z)**	▩	**HSLA Steel (H)**

FIGURE 16-7 This exploded view of a car body shows the parts and the types of coating used.

FIGURE 16-8 Proper methods are needed so you do not void the auto manufacturer's rust protection warranty.

16.2 CAUSES FOR LOSS OF FACTORY PROTECTION

Even with all the care taken to protect vehicles, breakdown still occurs. The breakdown of corrosion protection falls into three general categories:

- Paint film failure
- Collision damage
- Repair process

The paint film is the result of the entire process of coatings, primers, and colorcoats that the manufacturer applies. When the paint film fails, corrosion begins. Stone chips (Figure 16–9), moisture, and improper surface preparation can all lead to film failure.

FIGURE 16-9 Stone chips can break the paint film and can lead to rust spots.

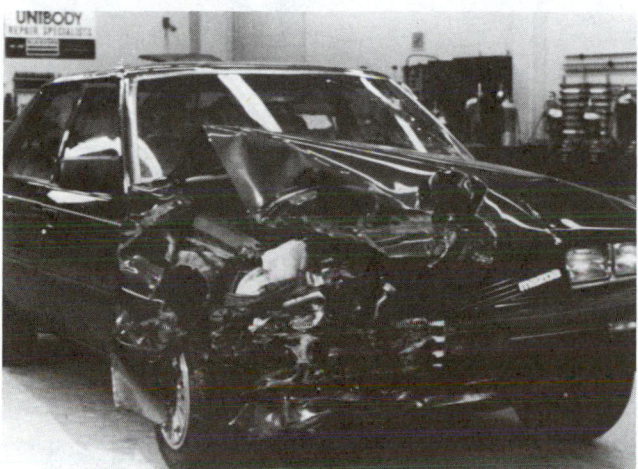

FIGURE 16-10 During a collision, corrosion protection is usually damaged.

FIGURE 16-11 Heat from welding and cutting operations destroys the factory corrosion protection.

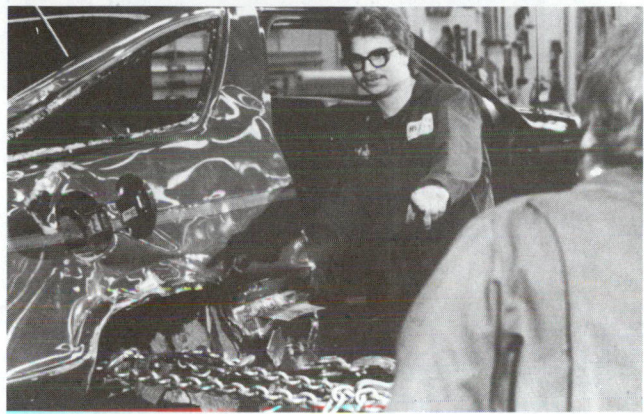

FIGURE 16-12 When clamping for body pulling, be sure that all hold areas are corrosion-treated after the repairs are made.

During a collision, the protective coatings present on a car are usually damaged (Figure 16–10). Corrosion protection can be damaged not just in the areas of direct impact, but also at indirect damage zones. Seams pull apart, caulking breaks loose, and paint can crack and chip. Locating and restoring the protection to all affected areas remains a key challenge.

Vehicle repair is possibly one of the major causes of protective coating damage. For example, repair procedures often require cutting body panels and seams either mechanically or with a plasma torch. Even minor straightening and stress relieving procedures can damage these protective coatings. Normal welding temperatures cause zinc to vaporize and be lost from the weld area (Figure 16–11). Abrasive operations during repair and refinishing can also leave areas unprotected. After all welding and repair work has been completed, these damage points need careful attention to eliminate contaminants. Then steps must be taken to keep the atmosphere off the metal by sealing all surfaces thoroughly.

Other precautions that should be taken to protect the factory corrosion protection are the following:

- Remove only the minimum amount of paint film from affected areas such as welded points.
- Be extremely careful not to scratch any part not being repaired. If there is an accidental scratch, take necessary remedial measures.
- When clamping or holding the affected panels during body repair work (Figure 16–12), clamping tools can cause scratches, which must be treated to avoid rusting.
- While grinding (Figure 16–13), cutting, or welding panels, place protective covers over adjacent painted surfaces and surrounding areas to protect them from the flame or sparks.
- Cover any opening of the body sills and similar area with masking tape to prevent metal chips from entering during the grinding, cutting, or welding operation.

FIGURE 16-13 Be sure all metal chips caused by grinding are cleaned up. If left on the surface, they can both cause and speed corrosion.

- Completely remove any metal chips from inside the body. Use a vacuum cleaner, not compressed air, to remove metal chips. If compressed air is used, metal chips can be blown out and accumulate in other corner areas.

There are also some environmental and atmospheric conditions that help influence the rate of corrosion:

- *Moisture.* As the water on the underside of the body increases, corrosion accelerates. Floor sections that have snow and ice trapped under the floor matting will not dry. Likewise, if holes at the bottom of the doors and side sills (Figure 16–14) are clogged or sealed shut, water will accumulate. Remember, water is one of the requirements for rust.
- *Relative humidity.* Corrosion will be accelerated in areas of high relative humidity, especially those areas where the temperatures stay above freezing and where atmospheric pollution exists and road salt is used.
- *Temperature.* A temperature increase will accelerate the rate of corrosion to those parts that are not well ventilated.

Acid rain is the term given to rain containing pollutants from the manufacturing and chemical industries. It causes discoloration and even destruction of the paint surface that can lead to corrosion damage. In Figure 16–15, the higher the pH number, the less chance of acid rain problems. Any condition below 6.0 pH is considered acid.

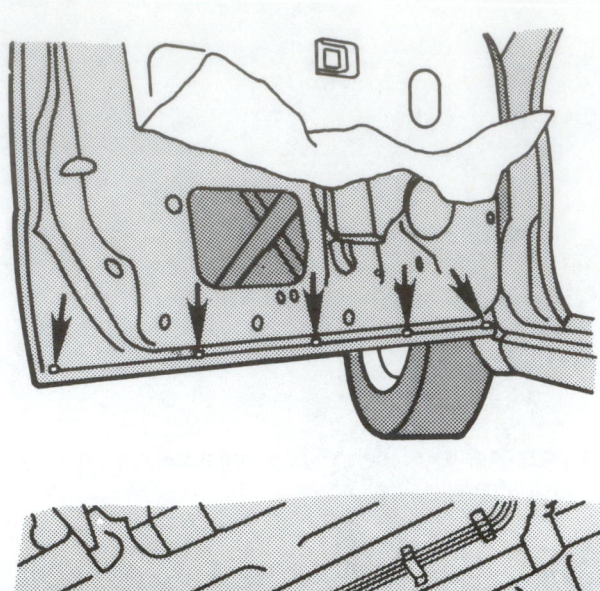

FIGURE 16-14 Keep drain holes open at the bottom of the doors, side sills, and so forth to avoid water accumulation inside panels. *[Courtesy of Nissan Motor Corp.]*

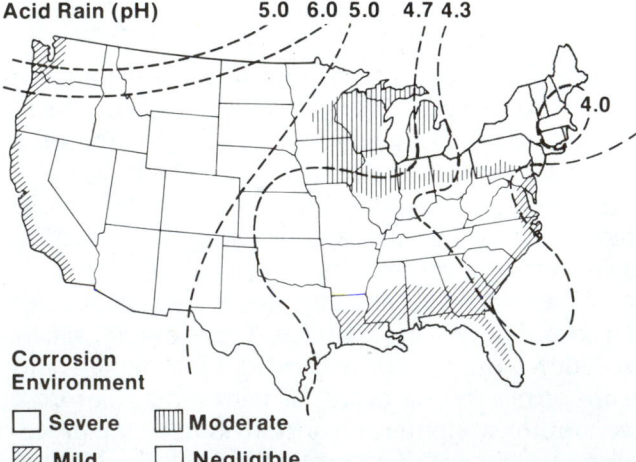

FIGURE 16-15 While corrosive environments are most severe in the northeast and along the southern seaboard, acid rain has become a factor in much of the country.

- *Air pollution.* Industrial pollution, the salty air of coastal areas, or the use of heavy road salt accelerate the corrosion process. Road salt will also accelerate the disintegration of paint surfaces. (Methods for combating acid rain damage are given later in this chapter.)

Another type of corrosion that must be considered when working on automobiles is known as galvanic corrosion. **Galvanic corrosion** occurs when two dissimilar metals are placed in contact with each other. The more chemically active of the two metals will corrode. As shown in Table 16–1, this is why zinc will sacrifice itself to protect steel. In the case of other metals, as mentioned later in the chapter, galvanic corrosion can cause problems.

Regardless of the cause, if corrosion prevention is not practiced, the cost to the body shop and insurer is "comebacks" or lost customers. Inadequate preparation that leaves dirt, grease, or acids on the metal will cause the loss of adhesion. Rust will start, a little at first, creating corrosive hot spots. Surface failure will progress quickly, spreading under the surface coatings, eating deeper and deeper into the metal.

Figure 16–16 illustrates some of the more common hot spots found on an automobile.

TABLE 16-1:RELATIVE ACTIVITY OF METAL	
Magnesium	Most Active
Aluminum	
Zinc	
Chromium	
Iron	
Cadmium	
Cobalt	
Nickel	
Tin	
Lead	
Copper	Least Active

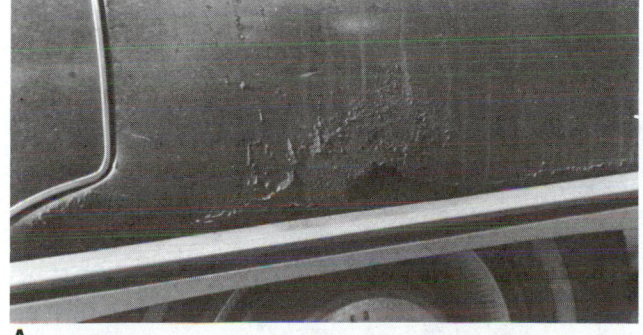

A

B

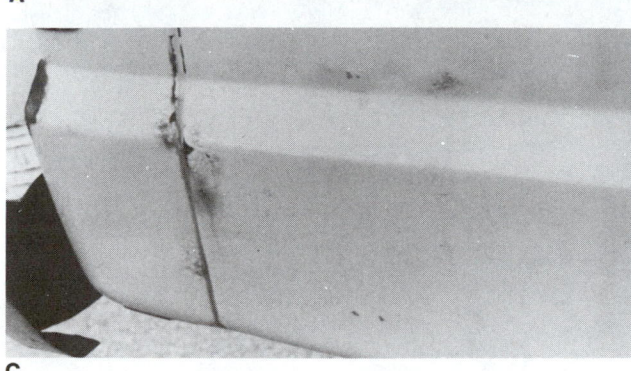

C

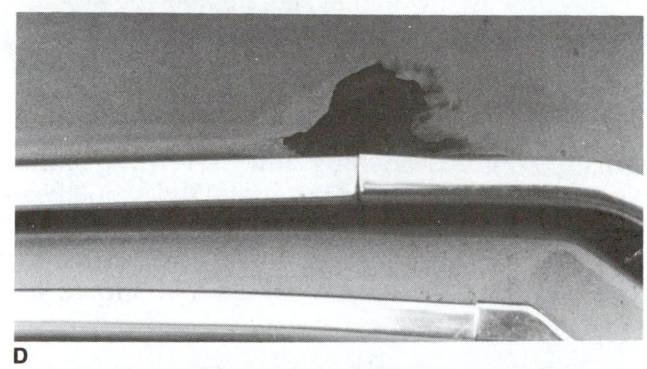

D

E

F

FIGURE 16-16 Note the common rust hot spots found on vehicles: (A) wheel housings, (B) rocker panels, (C) lower part of doors and fenders, (D) around roof drip rails, (E) front and rear pans, and (F) anywhere the paint film has been broken.

The body shop's interest in rust is two-fold:

1. The body technician must be able to repair rust damage.
2. The body technician must be able to provide treatment that will prevent rust from recurring.

Chapter 9 describes how to repair rust damage. This chapter is devoted to restoring corrosion prevention to damaged vehicles.

16.3 ANTICORROSION MATERIALS

The body and paint shop's efforts in protecting metal from rusting should focus on creating a clean, chemically neutral surface on the sheet metal. Then it should be on sealing the material under layers of paint. Under certain conditions, a wax- or petroleum-based anticorrosion compound is used to exclude air and moisture from the metal surface.

More and more new vehicles come off the assembly line today with anticorrosive materials that are available to the body shop. Being able to replace or install these materials properly is a very important skill for today's auto body technician.

Corrosion prevention has not always been a popular body shop operation. The original rustproofing was called **undercoating**, and it was an asphalt-based product. Applying this "tar" was sheer agony because it smelled bad and was messy. But, worst of all, it did not work very well. In time, the solvents used would evaporate and the asphalt would harden and crack. The moisture that causes oxidation would actually become trapped under the undercoating.

The asphalt undercoats did have benefits in terms of sound deadening and preventing stone marks under fenders. And it is useful today on fiberglass panels for the same reasons.

When selecting a modern anticorrosion material, there are several things that should be considered:

- The material should be thin enough to flow or penetrate pinch weld cracks. It must creep adequately to protect the exposed metal in areas adjacent to spot welds—a particularly tough rustproofing proposition.
- The material should have good adherence to both bare metal and painted surfaces. It should be highly resistant to water and commonly used cleaners or solvents. It must also resist cutting and damage from stones thrown up from the road. In other words, it should not only protect initially, it should also continue to protect. Material that does not retain some pliability and strength will not do the job.

- It is important to choose a material without solvents that have a lingering bad odor, which can be present in a car when it is delivered and make the best body work and paint seem bad.
- The product should be easy to clean up with ordinary and safe solvents.

Anticorrosion materials or agents can be divided into several broad categories:

- **Anticorrosion compound.** It can undercoat, sound deaden, and completely seal large surface areas from the destructive causes of rust and corrosion. It should be applied to the undercarriage and inside body panels so that it can penetrate into joints and body crevices to form a pliable, protective film.
- **Body sealer** or *sealant* (Figure 16–17). It prevents the penetration of water or mud into panel joints and serves the important role of preventing rust from forming between adjoining surfaces (Figure 16–18).
- **Weld-through primers** (Figure 16–19). Weld-through primer is used to provide anticorrosion protection to weld zones. This primer must be

A

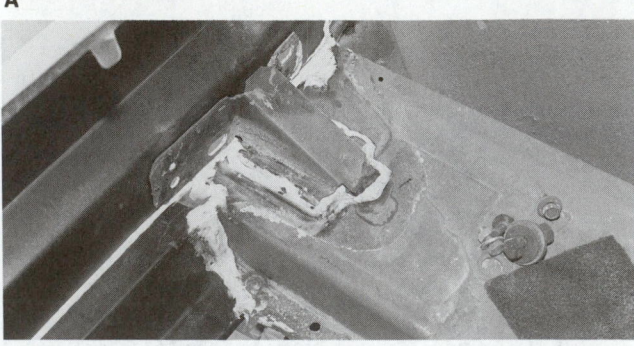

B

FIGURE 16–17 During repairs you must commonly (A) remove and (B) apply anticorrosion materials.

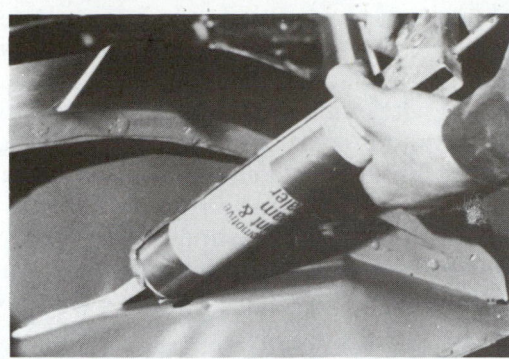

FIGURE 16-18 Applying a typical body sealer or sealant is often done with a cartridge-type dispenser.

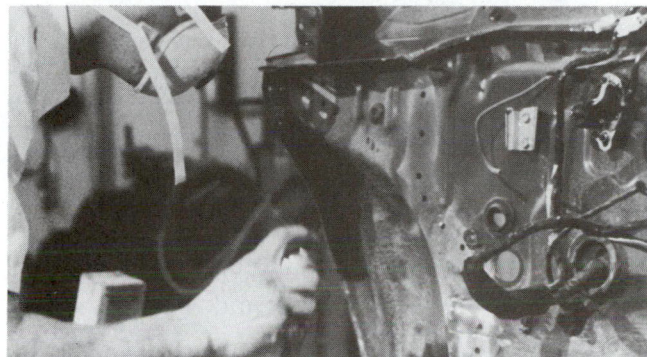

FIGURE 16-19 When applying a typical antirust agent, wear eye protection and a respirator mask.

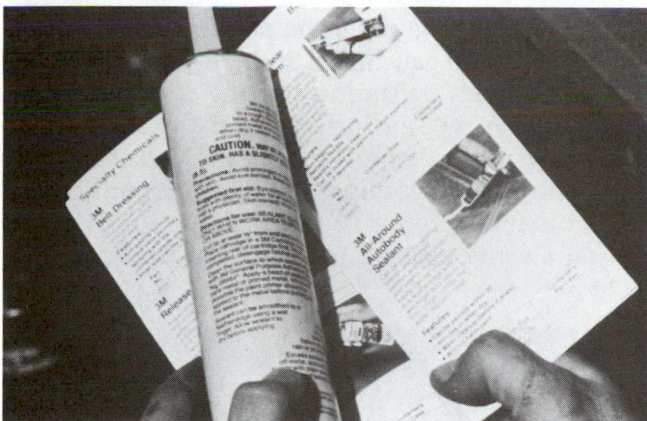

FIGURE 16-20 Always read the manufacturer's instructions and literature before using any chemical product.

applied to clean surfaces. Most weld-through primers have poor adhesion qualities. Do not overuse them, and always follow directions closely. Weld-through primer can be applied to galvanic mating surfaces where the coating was removed during repair. After welding, remove the excess primer.

- *Self-etching primers.* Self-etching primers etch the bare metal to improve paint adhesion and corrosion resistance, while providing the priming and filling properties of primer-surfacers. Self-etching

primers work best on lightly sanded surfaces where a slight-to-moderate amount of filling is required. They must be applied as directed by the manufacturer.

- Two-part **epoxy primers.** Two-part epoxy primers provide very strong base coating with good adhesion to bare metal. They are mixed together and cure very quickly, which helps prevent corrosion by more tightly bonding the coating over the metal.

- **Rust Converters** change ferrous (red) iron oxide to ferric (black/blue) iron oxide. Rust converters also contain some type of latex emulsion which seals the surface after the conversion is complete. These products offer an interesting alternative for areas that cannot be completely cleaned. However, some manufacturers do not recommend the use of rust converters.

SHOP TALK

Be sure to carefully read the manufacturer's instructions on the container (Figure 16–20) and follow them. Failure to follow directions can lead to failure of your repair!

New and better anticorrosion materials are constantly appearing on the automotive market. For example, a new line of anticorrosives is available that is effective for coating over existing rusted areas and retarding any further corrosion. This type of product can be helpful on repairs of older vehicles or repairs involving salvaged parts.

Be sure to check trade publications, I-CAR bulletins, manufacturer's representatives, and automotive suppliers for updated information on anticorrosion materials.

CORROSION PROTECTION SAFETY

As with other materials used in collision repair, the use of corrosion protection materials requires that you follow safety rules. The most basic rules are the following:

1. Wear gloves and avoid skin contact. Epoxy systems can create skin irritation.
2. If skin contact has occurred, wash hands with soap and hot water. Then apply a skin cream.
3. If adhesive accidentally contacts the eyes, wash immediately with clean water for 15 minutes. Then consult a physician.

4. Be sure to work in a well-ventilated area and wear a respirator. Spot welding in weld-bond joints can generate gases that can be harmful if inhaled.

16.4 BASIC SURFACE PREPARATION

Surface preparation is one of the most important steps in assuring long-term corrosion resistance of body panels and other metal parts. Without the proper surface (especially bare metal), the rest of the repair procedure and refinishing efforts will be futile. A common system generally consists of the following three-step process called **metal treating.**

- *Cleaning to remove contaminants.* Use a *wax and grease remover* to dissolve and float off oily, greasy film as well as other contaminants from the surface. Apply the remover with a clean, white cloth (Figure 16–21). Work small areas of no more than 2 to 3 square feet (0.6 to 0.9 square meters). Wet the surface liberally and keeping it wet, use a

second cloth to wipe the surface to remove the contaminants. Turn the cloth frequently while drying the surface (Figure 16–22).

- *Cleaning with metal conditioner.* A *metal conditioner* is a phosphoric acid used to etch bare sheet metal before priming. It is a chemical cleaner that removes rust and corrosion from bare metal and helps prevent further rusting.

  Remember the following about metal conditioners:

  1. Acid cleans the metal.
  2. It dissolves light surface rust.
  3. It etches metal, improving adhesion.
  4. It needs to be completely neutralized with water after applying.
  5. It may have to be diluted, following product directions.
  6. It is always followed by conversion coating.
  7. Wear rubber gloves and eye protection.

  Dilute the conditioner with water in a plastic bucket according to label instructions (Figure 16–23) and apply it to the metal. A spray bottle is often recommended (Figure 16–24). Then rinse with clear water and wipe dry with a clean cloth.

- *Applying conversion coatings.* The *conversion coating* forms a zinc phosphate coating that is chemically bonded to the metal. This layer makes an ideal surface for the primer and prevents rust from creeping under the paint. Use conversion coatings on galvanized and uncoated steels and aluminum. Be sure to use the correct product for each type of surface. Pour the appropriate conversion coating into a plastic bucket and mix with water according to the instructions on the container (Figure 16–25). Using a spray bottle, apply the coating to the metal surface. Then

FIGURE 16–21 Saturate a clean, white cloth with wax and grease remover.

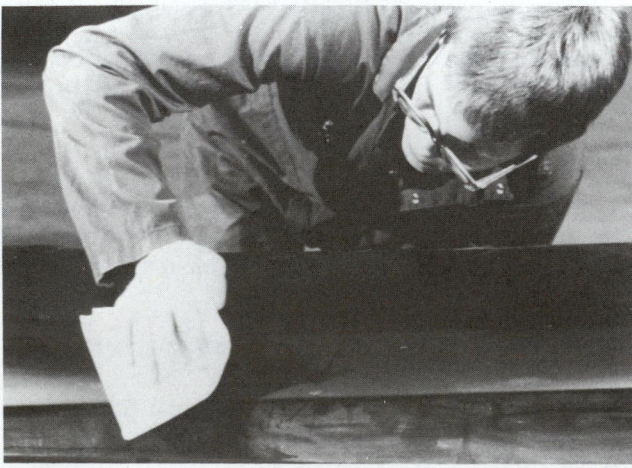

FIGURE 16–22 Use the saturated cloth to remove wax and grease to assure good adhesion of anticorrosion materials.

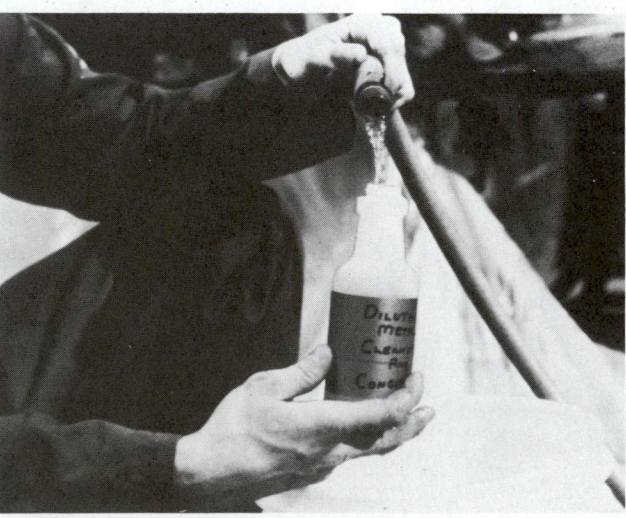

FIGURE 16–23 You must sometimes dilute a metal conditioner with clean water. Follow the label directions!

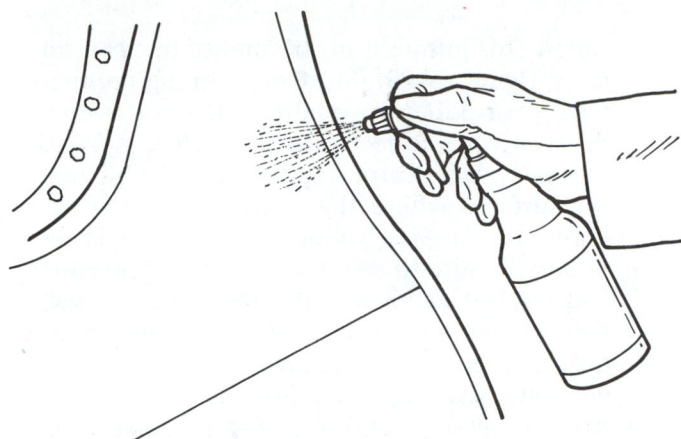

FIGURE 16-24 Applying a metal conditioner with a spray bottle is often the most efficient method.

FIGURE 16-25 Read application instructions very carefully on all products.

FIGURE 16-26 Flush off any excess conversion coating with clean water.

leave the conditioner on the surface 2 to 5 minutes. Work only as much area as can be coated and rinsed before the solution dries. Reapply if the surface dries before the rinsing. Flush the coating from the surface with clean water (Figure 16–26), or mop with a damp sponge or cloth that is rinsed occasionally in clean water. Wipe dry with a clean cloth and allow to air dry.

• *Applying self-etch primer.* Use a self-etch primer on bare metal to increase adhesion. This type of primer will actually eat into the metal to produce a strong bond. It will help prevent paint peeling and other problems.

Note that it is normal for metal to become rust-colored after a metal conditioner has been used on it.

Apply the self-etch primer in two thin, wet coats. Wet coats will help the material etch and bond with the metal. Avoid dry coats that will not adhere properly.

Note that most self-etching primers do not recommend the use of metal conditioners or conversion coatings before their application.

16.5 CORROSION TREATMENT AREAS

The corrosion treatment areas that must be considered when body repair work is done can be grouped in four categories. They are the following:

• *Enclosed interior surfaces,* which include body rails and rocker assemblies.
• *Exposed interior surfaces,* including floor pan, apron, and hood sections.
• *Exposed joints,* such as quarter-to-wheelhousing and quarter-to-trunk floor joints.
• *Exposed exterior surfaces,* such as fenders, quarter panels, and door skins.

The term *exposed* as used in this chapter refers to a panel surface that is accessible without having to remove a welded component.

16.6 CORROSION PROTECTION PRIMERS

Of all the areas to be protected during a repair job, the enclosed interior surfaces (Figure 16–27) are the most important. These include underbody structures such as front rails, rear rails, and rocker panels. The reason for the importance of these sections is that they represent the principal load-carrying members of the unibody car (Figure 16–28). Corrosion of these components can have a severe effect on the crashworthiness and durability of the vehicle.

Metal conditioners and conversion coatings are not recommended for use inside closed sections. The reason is that the chemicals and moisture might be difficult, if not impossible, to fully remove from the inside seams. Since stone chipping is not a problem here, the primer should develop adequate adhesion in these areas without conversion coating.

With this in mind, begin the process on closed sections with a thorough cleaning and degreasing. Because of the closed construction of these components, the cleaning must be done before the part is welded into the repair area.

After cleaning and degreasing, the enclosed metal surfaces must be protected with a primer. There are many different primers used by the auto trade. However, for corrosion protection, especially for enclosed interior surfaces, the three used most often are the following:

- *Two-part epoxy primers* are recommended by most automobile makers in place of the standard primer. When using a two-part primer, be sure to follow the manufacturer's instructions to the letter. Epoxy primer is sprayed on in the same way as other primers. However, it comes in two parts and must be mixed properly.

    Reduce the epoxy primer as recommended by the manufacturer. Use one or two wet coats to saturate the metal. This will help produce a strong bond. Care must be taken not to inhale the fumes because they can be very toxic. Use epoxy primer when a strong, durable undercoating is desired.
- *Applying weld-through primer.* Weld-through primer is used to provide anticorrosion protection to weld

zones. This primer must be applied to clean surfaces (Figure 16–29). Most weld-through primers have poor adhesion qualities. Do not overuse them. Always follow directions closely. Weld-through primer can be applied to galvanic mating surfaces where the coating was removed during repair. After welding, remove the excess primer. Be sure to select a weld-through primer and not just a galvanized spray coating, which may interfere with welding. Any overspray on the outside of panels should be removed before conventional priming and painting.
- *Applying self-etch primer.* Self-etch primers contain acids that will eat into the metal to produce

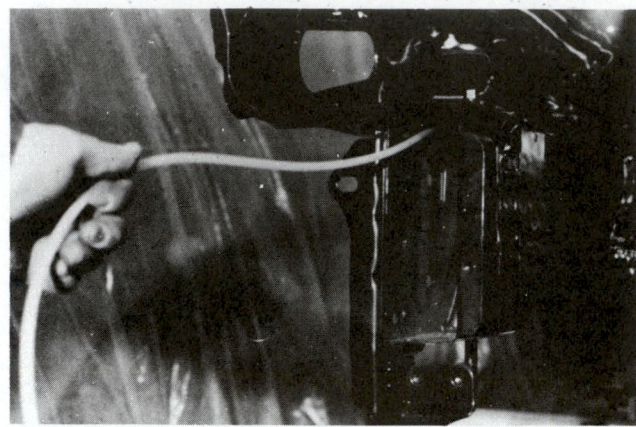

FIGURE 16-27 Enclosed surfaces such as this require anticorrosion attention.

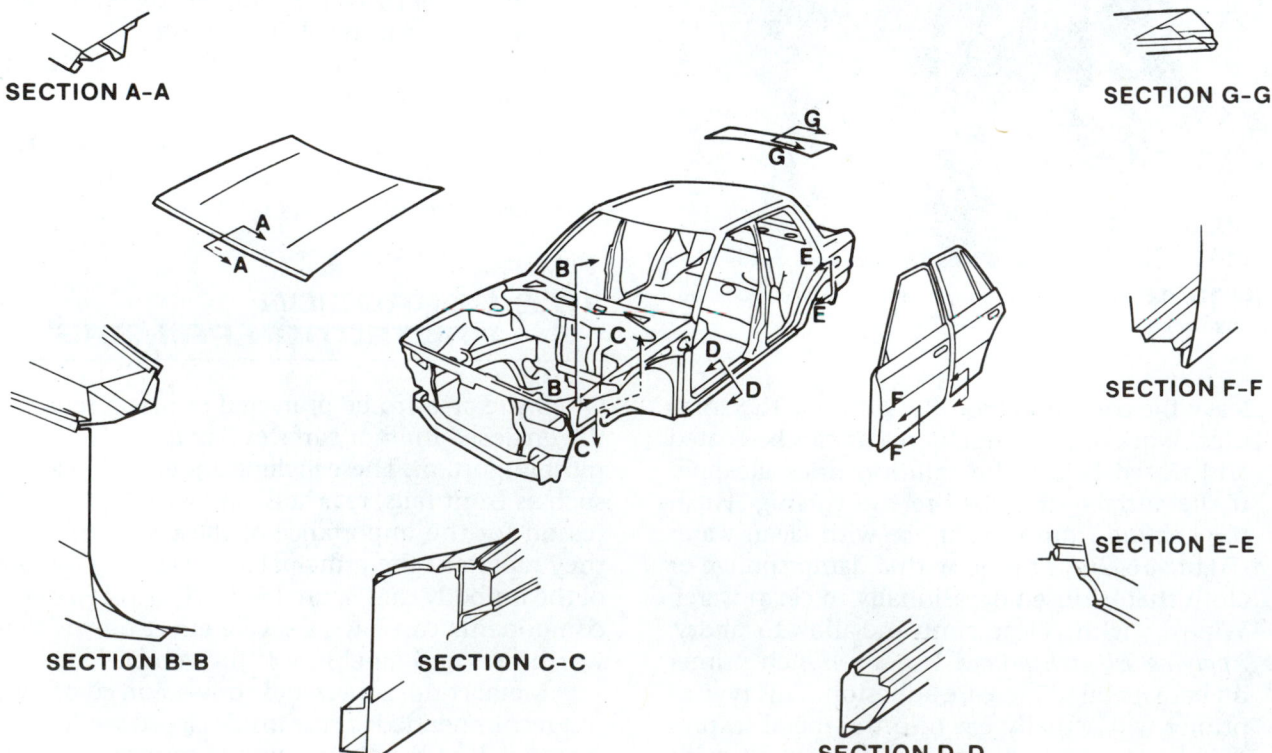

FIGURE 16-28 Note some enclosed interior surfaces that must be protected. *(Courtesy of Nissan Motor Corp.)*

SHOP TALK

Do not use lacquer-based primers since they do not provide enough adhesion under enclosed interior conditions.

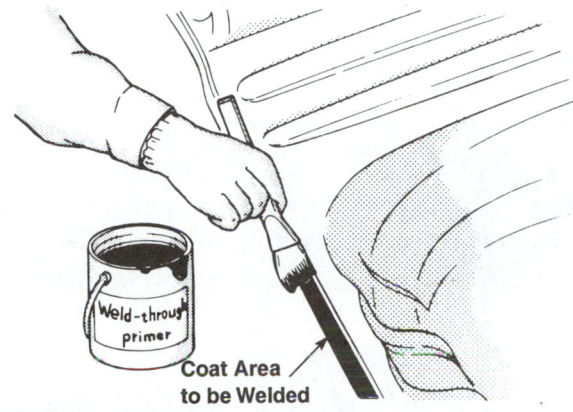

FIGURE 16-29 Brush weld-through primer on any flange that will be welded. If not treated, two bare, unprotected metal surfaces would be in contact. *(Courtesy of Nissan Motor Corp.)*

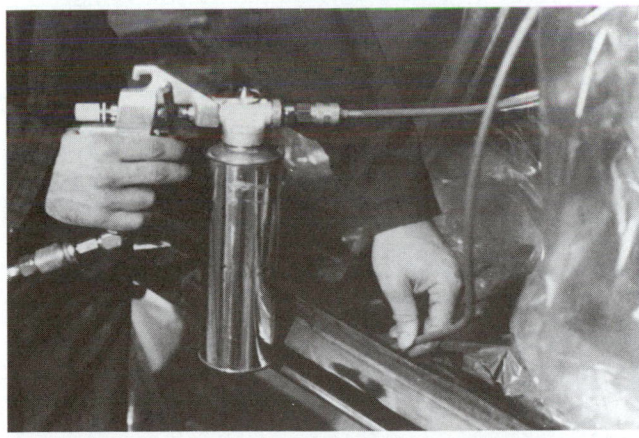

A

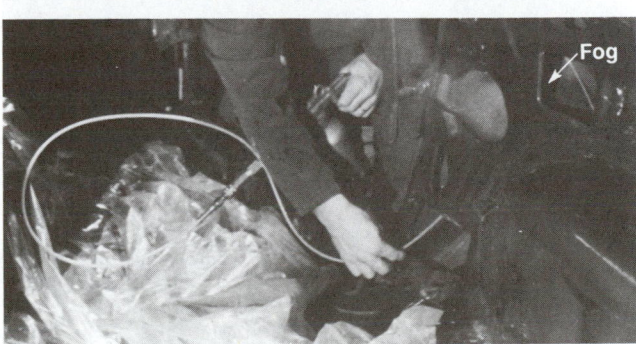

B

FIGURE 16-30 (A) A spray gun like this one is often used to apply material to enclosed areas. (B) Note "fog" coming out of the top hole. The fog mist helps coat all interior surfaces.

better adhesion. This type of primer should be used when there is a concern about paint peeling or lifting. Mix and spray self-etch primer according to the primer manufacturer's directions. Wear an approved respirator to prevent fumes from entering your throat and lungs.

The application of both the primer and anticorrosion materials to the inside of closed sections must be done only with the manufacturer's recommended equipment. This is normally the airless or pressure-feed type of spray gun, although some suction equipment is also recommended.

Aerosol or conventional spray gun equipment will not work in enclosed interior sections because you cannot spray the material directly on the surface. This work requires special wands to reach all the inside cavities and joints where the material must go.

The airless or pressure-feed spray gun uses compressed air behind the fluid to force the liquid through the wand and nozzle. The fluid is broken up into a very fine atomized state, sometimes called a *fog*. When the substance is sprayed inside the closed section in this atomized state, it spreads rapidly and evenly into all areas including tiny crevices (Figure 16–30).

To use a spray wand, insert it into the cavity (Figure 16–31) to the farthest point that the coat is needed. Begin the spray, and pull the wand out at an

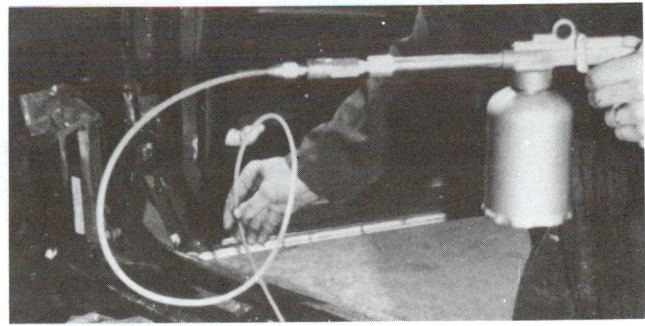

FIGURE 16-31 Be sure that the wand is inserted all the way into the cavity.

SHOP TALK

Once a week or as needed:

* Fill the gun or tank with cleaning solvent or mineral spirits.

* Spray cleaning solvent through all wands.

* Hang wands to drain overnight.

Failure to clean the gun and wands on a regular basis can result in clogging or total blockage of the wand, which will prevent proper function, work delays, and a time-consuming service of the tool.

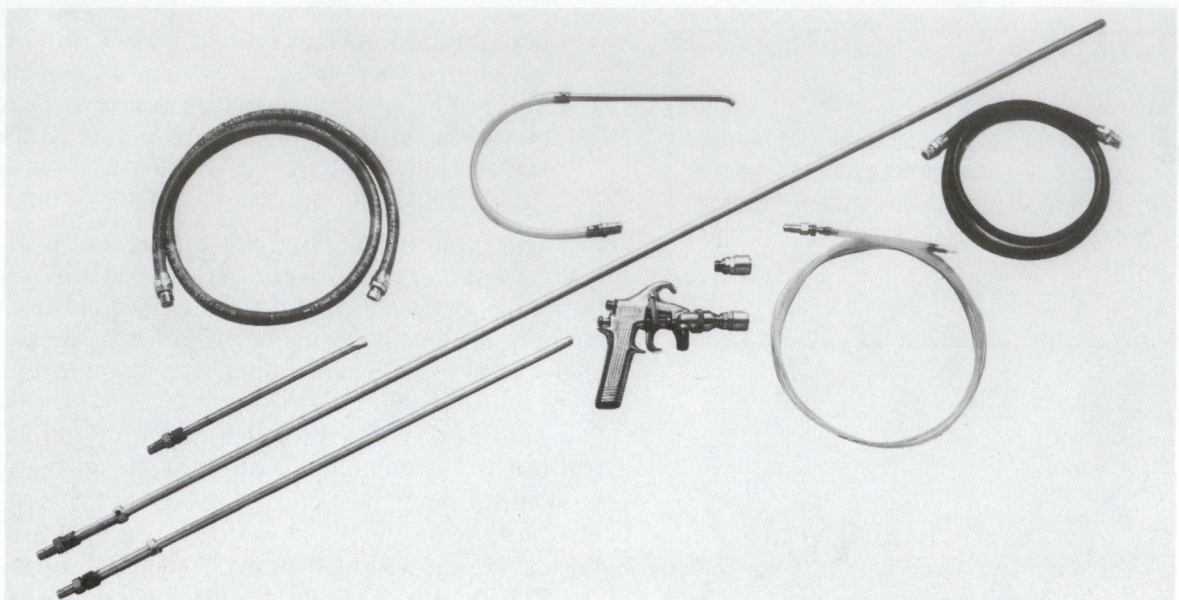

FIGURE 16-32 These are typical wands, connecting air hoses, and a typical spray gun used with an airless corrosion protection system. *(Courtesy of Binks Manufacturing Co.)*

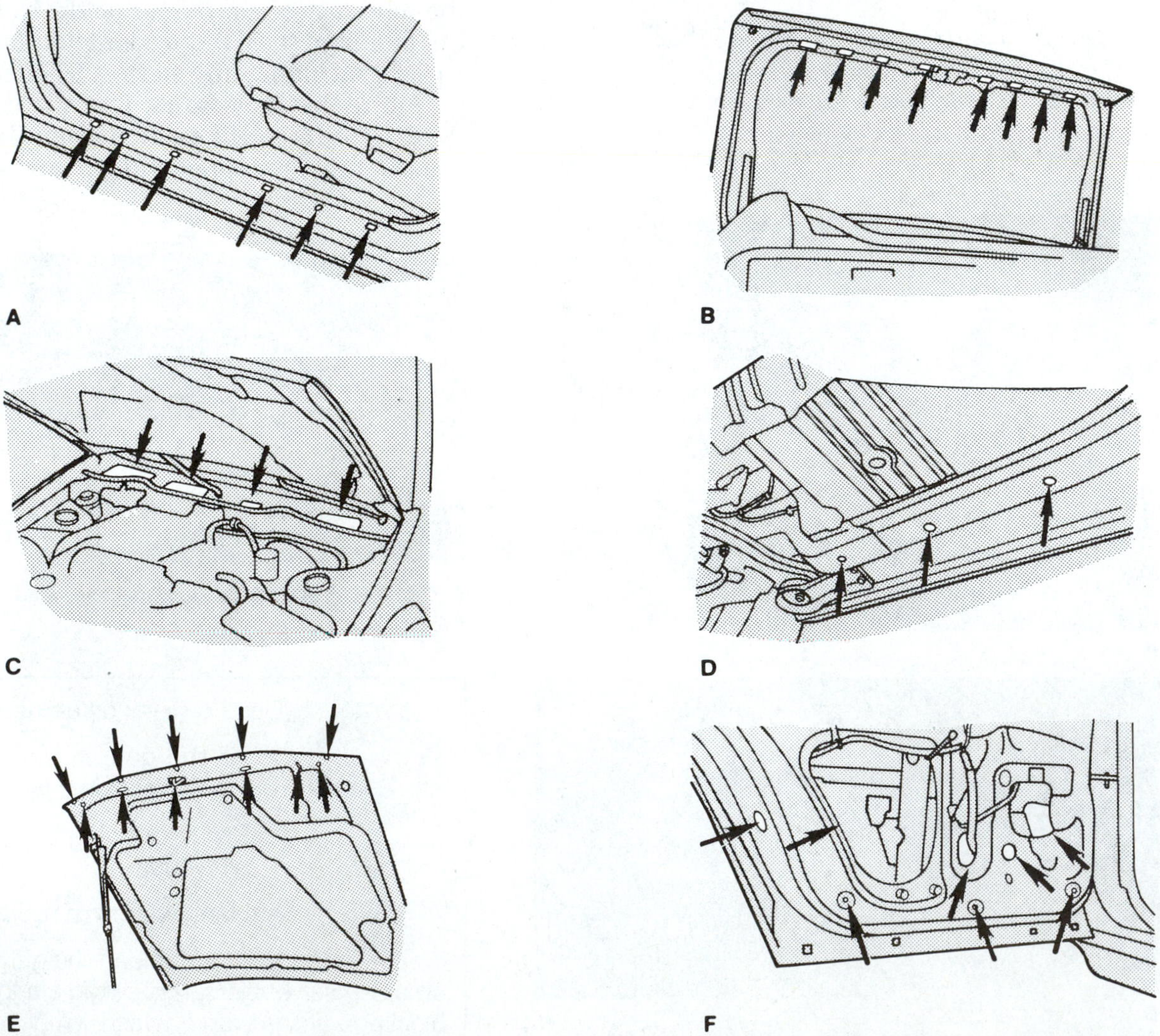

A

B

C

D

E

F

FIGURE 16-33 Inspect these typical service or access clip holes: (A) interior of body sill; (B) rear end of trunk lid; (C) inside cowl top; (D) inside member; (E) front end of hood; and (F) lower part of door. Keep the mechanical parts of the door free of compounds; windows should be in a closed position. *(Courtesy of Nissan Motor Corp.)*

even rate, coating the section of the cavity evenly as it moves along.

There are several wand styles available (Figure 16–32). Before spraying with a wand, be sure that the spray pattern is checked and corrected, and the pistol or barrel is filled with the desired primer or anticorrosion material.

The general corrosion restoration process for an enclosed interior surface is as follows:

1. Clean the enclosed interior surface with wax and grease remover.
2. Apply weld-through primer only to bare steel areas to be welded. Do not apply over paint, primer, or galvanized surfaces.
3. Apply only in the immediate weld area since this product has poor adhesion characteristics. After welding, thoroughly remove all welding residue and surplus primer from the joint area.
4. Research has shown that wire brushing is not the best way to clean a weld area. Wire brushing can leave scratches in the original primer that are not always filled by the new primer. The primer tends to "float" over the scratches, creating minute voids in which corrosion can start. A better way to clean the weld area is to use a plastic abrasive. Another way is to sandblast or plastic media blast with a captive blaster or with regular blasting equipment. Thorough cleaning after welding should again be stressed here.
5. After the area is thoroughly cleaned, apply a primer. Two-part or self-etching epoxy type is usually recommended for the inside area. Be sure to allow sufficient drying time according to the primer manufacturer's recommendations.
6. Apply anticorrosion compound according to manufacturer's direction. The material is applied using service or access clip holes (Figure 16–33) and drain holes. When the rustproofing material is dry, in approximately 1 hour, the water drain holes must be cleared.

To spray specific enclosed interior surfaces, special techniques might be required. These area considerations include the following:

- *Trunk*. Remove the spare tire, tools, floor mat, board, and padding on each side of the trunk.

WARNING

Refer to the vehicle manufacturer's recommendations before drilling holes in panels for applying corrosion protection materials. Holes in structural panels may weaken the vehicle and reduce its integrity.

The rear quarter panel behind the wheels is coated from inside the trunk using the flexible spray wand (Figure 16–34) to spray downward in the recess between the trunk and the quarter panel. Spray the back edge of the trunk, getting under the beads.

Spray the trunk lid by inserting the flexible wand into the existing holes, making sure the material reaches the edges.

When the spraying is completed, replace the padding, floor extensions, floor mats, tools, and tire. Wipe off overspray with a cloth dampened lightly with enamel reducer, solvent, or kerosene.

- *Doors*. Doors can be treated through their drain holes after the interior panel has been removed. A hole can also be drilled in the edge of each door, approximately 6 to 9 inches (152 to 228 mm) above the bottom. Center punch and drill the hole (Figure 16–35).

With the windows up, insert the wand into the drilled hole as far as possible. Slowly retract the wand, while spraying the length of the bottom third of the door (Figure 16–36). Just before the wand is withdrawn from the hole, point it

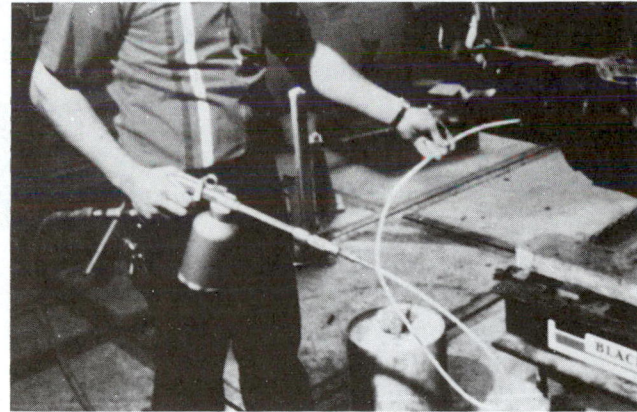

FIGURE 16–34 This technician is applying anticorrosion compound to the trunk's enclosed surfaces.

FIGURE 16–35 You may have to drill an access hole in the door.

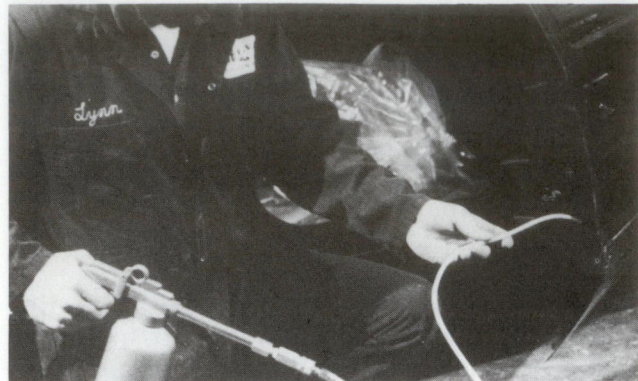

FIGURE 16-36 Move the wand so that all surfaces in the door are protected.

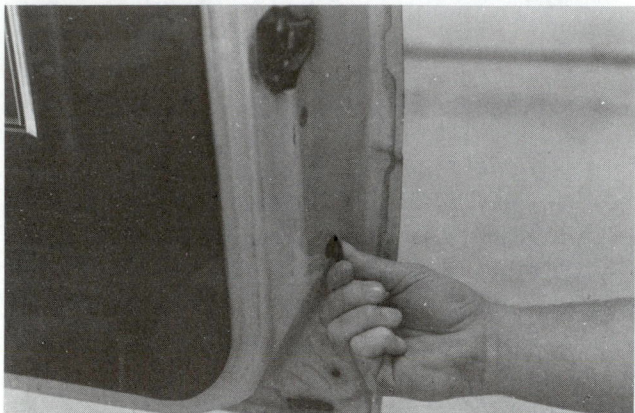

FIGURE 16-37 Plug all holes with plastic body plugs of the correct diameter.

down to assure direct coverage into the inside corner of the open door. Plug the hole with the right size plastic body plug (Figure 16–37).

- *Rear post and quarter panel.* The reverse side of the rear post and quarter panel can sometimes be sprayed from the trunk area. To ensure coverage, the front edge of the wheel well and the quarter panel area should be coated. You may have to spray through holes drilled in the rear post or by removing the ventilator cover. Insert the flexible spray in the drilled hole in the rear post. Spray while gradually withdrawing the wand. Plug the drilled hole with a plastic plug.

- *Front post (and four door center post).* Drill a hole in the front step plate at the center of the curve where it meets the front supports. Insert the flexible cone spray wand and spray thoroughly. Also spray the lower edge of the front fender and any boxed-in areas in the vicinity. Drill a hole in the center of the curve formed by the step plate and rear support. Spray with the flexible spray wand. If the car is a four door, the wand must reach the rear of this center post. If it does not, drill a hole from the rear of the center post.

- *Behind front fender.* Some cars might have a boxed cavity behind the front fender into which the flexible cone spray can be inserted either from under the hood or alongside the front door post.

- *Rocker panel.* Check to see if the rocker panels are boxed. If not, work from both ends. Before drilling any holes, check both ends of the rockers underneath for existing plugs. If satisfactorily located, these can be used to spray the entire length of the rocker with the flexible cone spray wand. If it is inconvenient or undesirable to do the rocker panel from above, drill a hole from below into the rocker panel at about the center and spray in both directions using the flexible cone spray wand. Be sure to spray on both sides of internal baffles, if present.

16.7 EXPOSED JOINTS

Body panel joints and seams require special attention. These areas are highly vulnerable to corrosion and must be protected correctly. This is because of the effect of welding on the metal as well as the tendency of water, snow, dirt, mud, and other contaminants to become trapped in the joint area. As a general rule, a body sealant must be applied over all the joints. The sealant must be applied so there are no gaps between the material and the panel surface (Figure 16–38).

The following are some factors needed when selecting and applying seam sealers:

- *Paintability.* All sealants must be paintable and have good adhesion to bare and primed metal. A seam sealer should be allowed to adequately dry before painting. The necessary drying time depends on the sealer itself, the thickness applied, and the temperature and humidity during the drying period. Normally, the lower the temperature is below 70 degrees Fahrenheit (21 degrees Celsuis), the longer the necessary drying time. The higher the humidity is above 50 percent relative humidity, the longer the necessary drying time.

- *Flexibility.* This is a critical issue with today's unibody automobiles. The sealer must be able to withstand the motion associated with the automobile. If not flexible, vibration could crack and damage the sealer.

- *Tooling sealants.* A finger wetted with solvent or water makes tooling easier and helps to keep the sealer from sticking to the finger of your plastic glove. This is a good application tip to help improve the finished seam's appearance. Brushable seam sealer should be tooled with a stiff bristle brush. It should be stroked in one direction only to help it match the original equipment appearance.

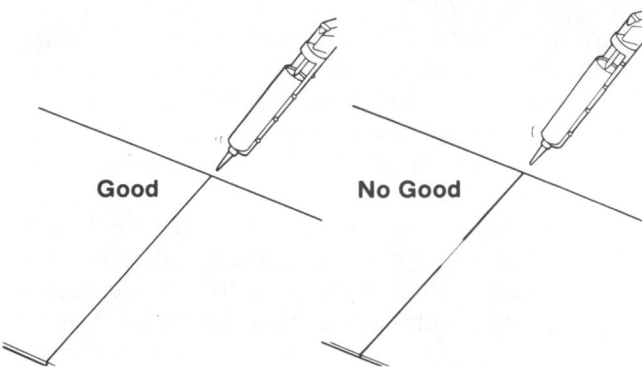

FIGURE 16-38 Proper application (left) of a sealant does not permit gaps in bead as shown at the right.

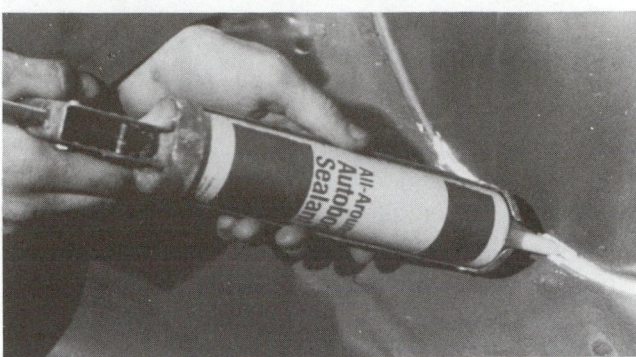

FIGURE 16-40 Heavy-bodied sealer is used for slightly larger gaps.

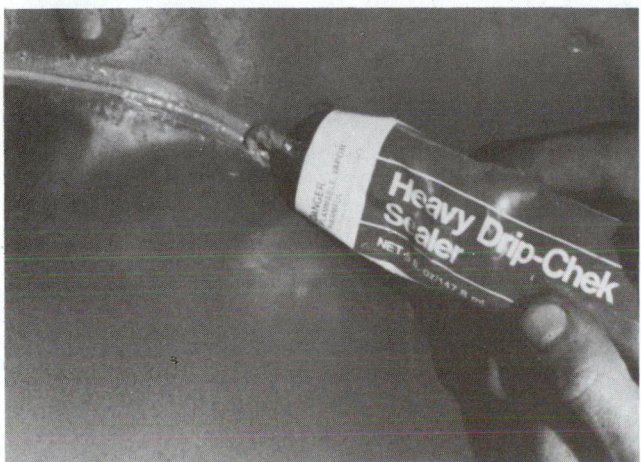

FIGURE 16-39 Thin-bodied sealer is sometimes recommended.

FIGURE 16-41 Brushable seam sealer is usually more durable but its appearance can be less attractive.

- *Silicone sealants.* These should not be used as a body seam sealer. They typically are not paintable, attract dust and dirt with time, and do not offer the adhesion of other types of sealants.

There are four types of seam sealers that are commonly used in auto corrosion protection work:

- **Thin-bodied sealers** are designed to fill seams under ¹⁄₈ inch (3.2 mm) wide. This sealer will shrink slightly to provide definition to the joint, while remaining flexible to resist vibration. Adhesion is good to both primed metal and bare metal surfaces. Since many of the seams are on a vertical surface, sag control is important so the sealer does not run out of the seam. Typically thin-bodied sealants carry the generic names of drip-check (Figure 16-39) or flow grade.
- **Heavy-bodied sealers** are used to fill seams from ¹⁄₈ to ¹⁄₄ (3.2 to 6.4 mm) inch wide (Figure 16-40). These sealers can be tooled to hide the seam or can be left in bead form. Shrinkage should be minimal, with good resistance to sagging and high flexibility to resist cracking in service. Heavy-bodied sealants are used on both coach joints and overlap seams. They are typically dispensed from cartridges. Some products are available in squeeze tubes.
- **Brushable seam sealers** are used on interior body seams where appearance is not important. These seams are normally hidden and not seen by the customer. Brushable sealers are designed to hold brush marks and to resist salt and automotive fluids such as gasoline, transmission fluid, and brake fluid. Any seams such as those under the hood and under the carriage that may be exposed to automotive fluids should have a brushable seam sealer. Applied with a brush, it normally has overlap seams (Figure 16-41).
- **Solid seam sealers** containing 100 percent solids are used to fill larger voids at panel joints or holes. This product comes in strip caulking form, designed to be pressed into place with your thumb (Figure 16-42).

Be sure to follow the manufacturer's instructions carefully for the use of these versatile products.

The application sequence is basically the same as it is for interior exposed panels with the addition

Some anticorrosion compounds can be harmful if placed in contact with human skin. Read the material directions. Wear plastic or rubber gloves when in doubt or when the material could be harmful.

of two more steps. After the welded areas are thoroughly cleaned and primed, seal all body panel joints with a seam sealer. Finish the joints by applying another coat of primer over the seam sealers, then topcoat with the same material used over the rest of the repaired area.

It is important to use a nozzle with a small hole in the end to apply the sealer. Then spread out the bead of sealer with a fingertip. If the nozzle hole is small, the finish can be kept neat (Figure 16–43A). If

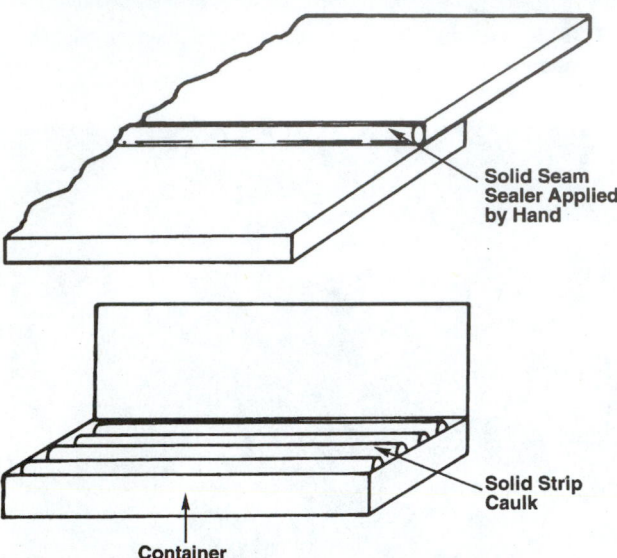

FIGURE 16–42 Solid seam sealer often comes in strips. After pulling a strip from its box, force the soft strip down into the joint. It will fill large gaps in panels and even holes.

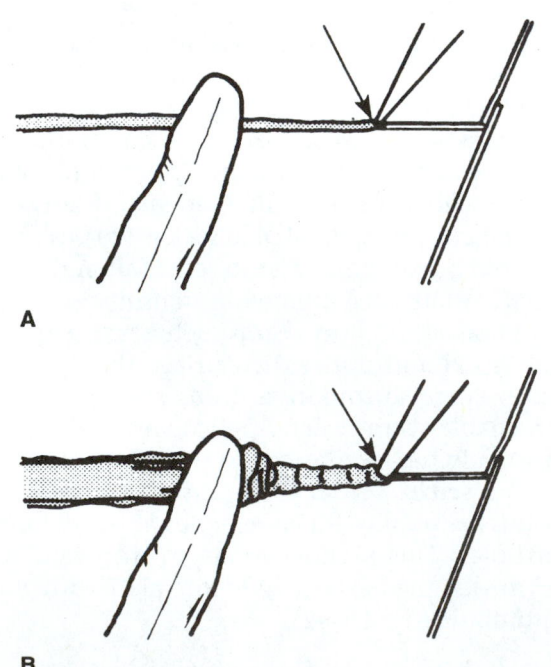

FIGURE 16–43 Use the proper nozzle size to make the repair more attractive. (A) A small hole in the nozzle resulted in little excess material. (B) The hole is too large and the repair looks sloppy. *(Courtesy of Toyota Motor Corp.)*

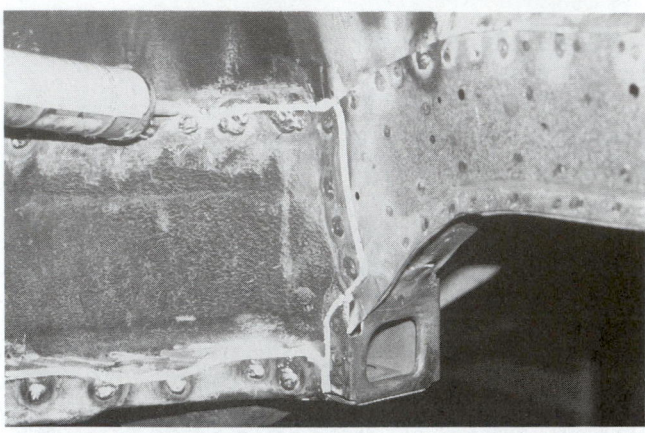

A

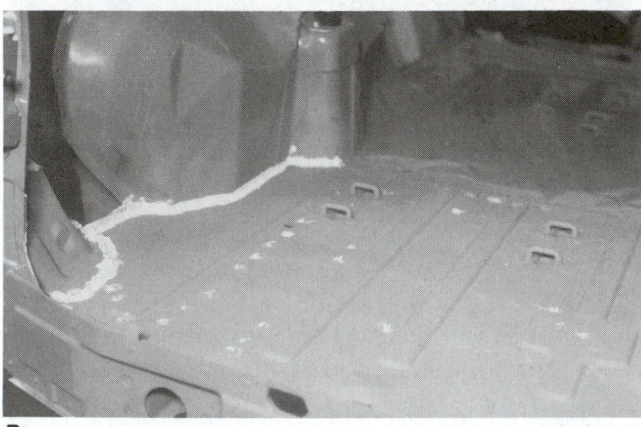

B

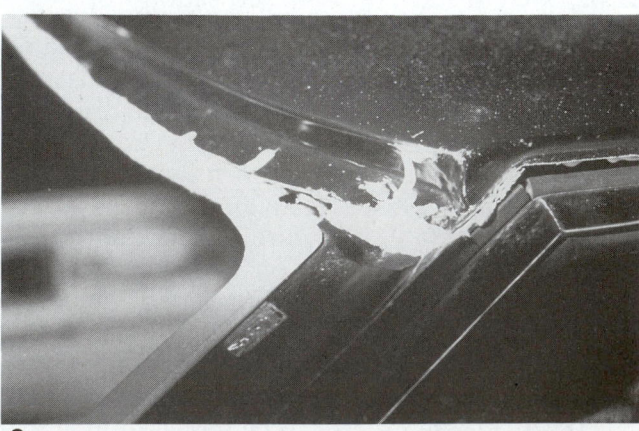

C

FIGURE 16–44 The service manual will usually give the details for application of seam sealer, and the locations that must be sealed. (A) Sealer in trunk area. (B) Sealer on frame rail. (C) Sealer on roof panel. *(Courtesy of Tech-Cor)*

SHOP TALK Do not use latex-based seam sealers designed for home use! These may draw moisture into the joint and cause rapid rust formation and joint failure. Use only recommended sealers.

SHOP TALK More than one type of seam sealer can be used on any given joint. For example, a brushable seam sealer could be used over a thin-bodied sealer. The material instructions and service manual will give details for proper application.

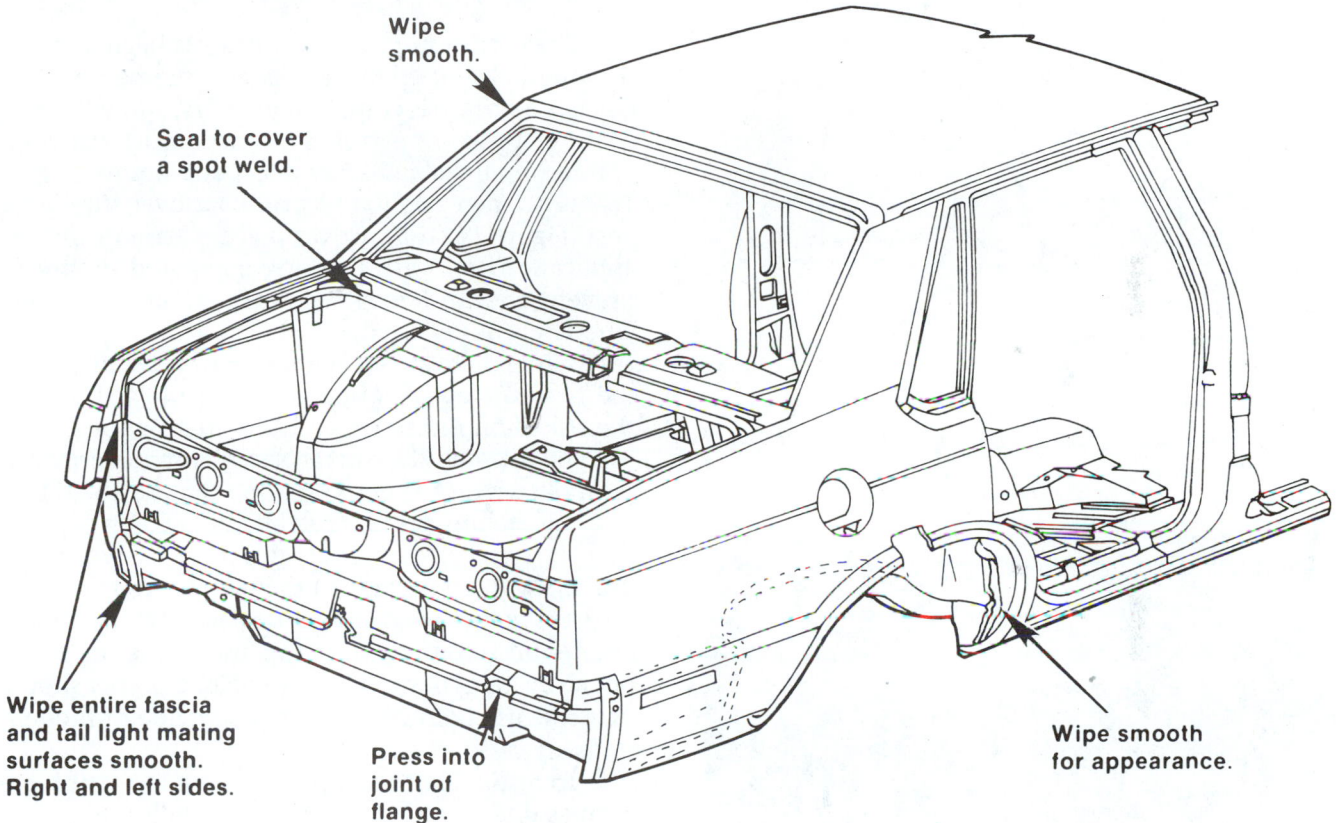

Wipe smooth.

Seal to cover a spot weld.

Wipe entire fascia and tail light mating surfaces smooth. Right and left sides.

Press into joint of flange.

Wipe smooth for appearance.

FIGURE 16–45 Service manual illustrations, like this one, will give valuable information for seam sealers and other anticorrosion materials.

the nozzle hole is too large, the sealer will spread too wide and may cause a poor-looking finish (Figure 16–43B).

Automotive masking tape can also be used to make a parting line for the sealer application area. In some situations, this makes the repair look more OEM.

When applying sealant, refer to the shop manual for the vehicle being repaired. Determine the sealer application area (Figure 16–44) or look at the other side of the vehicle to see where the sealer is applied (Figure 16–45).

To summarize the corrosion protection process for exposed joints and seams, proceed as follows:

1. Thoroughly clean the joint or seam.
2. Apply primer or primer-sealer.
3. Seal the joints with seam sealer.
4. Apply a second coat of primer or primer-sealer.
5. Finish with a colorcoat in a spray booth.

16.8 EXPOSED INTERIOR SURFACES

The bottom surfaces of the underbody and inside of the wheelhousing can be damaged by flying stones, causing rust to develop. These areas are given an undercoat treatment with a material such as shock-absorbing wax. Apply treatment from below the underbody.

Do not spray anticorrosion material into the passenger compartment. Metal conditioners and

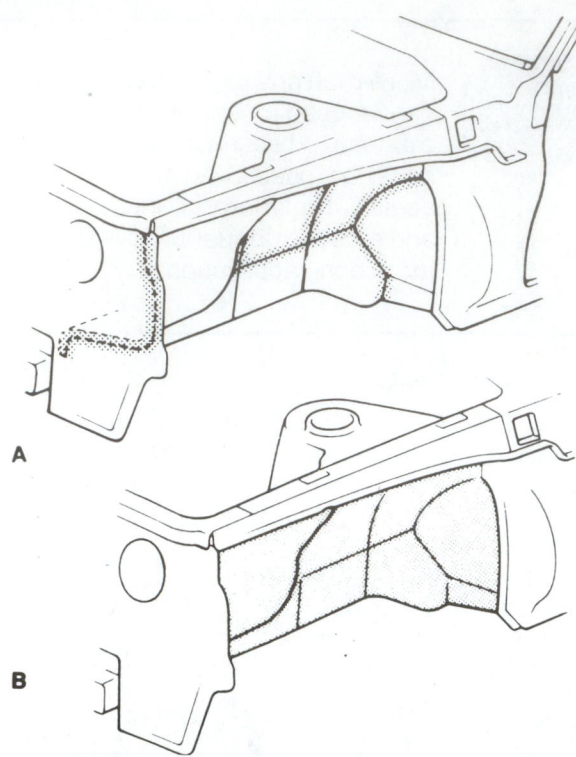

FIGURE 16-46 (A) Apply undercoat to all welded areas and panel joints; (B) then apply to the entire area. *(Courtesy of Toyota Motor Corp.)*

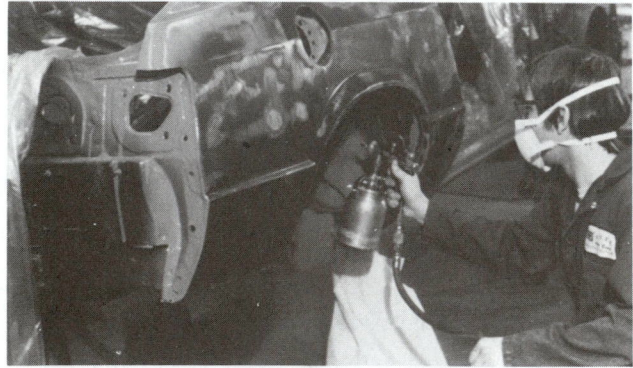

FIGURE 16-47 Exterior body panels require the most protection.

conversion coatings also are not recommended for interior surface protection. There are three reasons for this:

- These surfaces are not exposed to physical damage the way exterior surfaces are.
- These areas contain joints and seams that should not be contaminated with etching and conversion-coating chemicals and are generally difficult to rinse clean.
- They could cause harmful odors to remain in the passenger area.

The corrosion protection process begins with a thorough cleaning with a wax and grease remover. Once the surface is completely air dry, spray the first coating of wax- or petroleum-based undercoat compound on all welded areas and panel joints (Figure 16–46A). Then apply a second coat over the entire area (Figure 16–46B). Cover places surrounding the application area with masking paper and/or tape to prevent the undercoating from sticking to areas where it is not wanted.

While there are several different types available, the two-part epoxy primers most closely duplicate the baked-on electrode position coating, or E-coat, used by car manufacturers. Nearly any material can be applied over the epoxy primer (Figure 16–47).

Self-etching primers can also be used for exposed interior surfaces. However, common lacquer-based primers will not provide proper adhesion when used on bare metal, even if the area has been properly cleaned and conversion coating used. This point cannot be overemphasized. Lacquer-based primer should never be used directly on the bare metal of modern unibody cars.

To restore corrosion protection to specific areas such as under the hood, proceed as follows:

1. Lift the hood and spray the front fender or apron, between it and the wheel well. Be sure to apply the material right down to the fender beads. Use

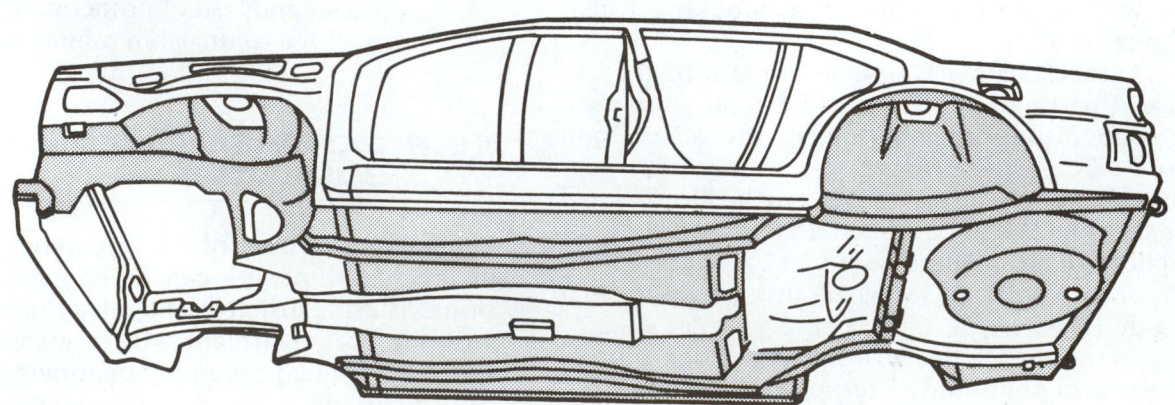

⊞ **Indicates Undercoated Portions**

FIGURE 16-48 A shop manual will give underbody panels that require protection. *(Courtesy of Nissan Motor Corp.)*

the flexible cone spray wand to reach all recessed areas.

2. Cover the large open spaces with a 45-degree flat spray wand.

3. Spray the leading edge and the side channels of the hood with the flexible cone spray wand.

4. Loosen or remove the battery and coat the battery tray and surrounding areas.

The headlight areas on some cars might be reached from under the hood. On other makes, the headlight areas can be reached from existing holes under the hood by means of the flexible spray wand. If not accessible from under the hood, the headlight areas can be sprayed from under the car, working forward in the front wheel well when the car is put on a lift. This can be a baffled area with a rubber edge that can be depressed to insert a flexible cone spray or flat spray wand. Choose the best method to assure complete coverage.

16.9 EXPOSED EXTERIOR SURFACES

Exterior surfaces are subjected to much greater exposure to chips and nicks than interior surfaces. Therefore, the use of etching and conversion coating agents is of critical importance on exterior surfaces. Conversion coating provides the kind of superior paint film adhesion that retards creeping rust from working its way under the paint when chips and nicks do occur.

Exposed exterior surfaces are of two types: cosmetic and underbody.

Anticorrosion procedures for exterior cosmetic surfaces are generally as follows:

1. Clean with a wax and grease remover.

2. Apply a metal conditioner. Rinse with water. Apply a conversion coating and allow to thoroughly air dry. Drying can be speeded with compressed air or a clean, white rag. Rinse with water.
   or
   Apply a self-etch primer that does not require the use of a metal conditioner. This can save time and effort.

3. Apply a primer-surfacer.

4. Apply a colorcoat or paint system.

If a lift is available, it makes underbody corrosion protection work easier (Figure 16–48). When corrosion-proofing the underbody, start by spraying the fenders and wheel wells, paying particular attention to the fender beads. On some cars, it will be necessary to remove the wheels to do an adequate spraying job.

Spray the remaining underbody and splash pans adjacent to the front and rear bumpers. Spray the underside of the floor pan, welded joints, frame, tank straps, and seams. Remove any loose debris or sound deadener, particularly around joints, before spraying. Loose sound-deadening materials or dirty surfaces will prevent the rustproofing material from reaching the metal and will create pockets in which rust will form.

Anticorrosion procedures for exterior underbody surfaces are generally as follows:

1. Clean with a wax and grease remover.

2. Apply a metal conditioner.

3. Rinse with water.

4. Apply a conversion coating.

5. Rinse with water.

6. Apply a recommended primer–self-etching primer.

7. Apply anticorrosion compound and sound-deadening materials to restore to factory specifications.

8. Most undercoat overspray can be removed with enamel reducer, solvent, or kerosene and by washing.

Care is needed when applying anticorrosion compounds. Keep the material away from parts that conduct heat, electrical parts, labels, identification numbers, and moving parts. Avoid applying corrosion protection materials to

1. Seat belt retractors and passive restraint guide rails

2. Hidden headlamp assemblies

3. Power window motors and cables

4. Exhaust system

5. Engine and accessories

6. Air filter

7. Air lift shock absorbers

8. Transmission parts

9. Shift linkages

10. Speedometer cables

11. Brake parts

12. Locks, key cylinders, and door latches

13. Power antennas

14. Theft prevention labels

15. Driveshaft

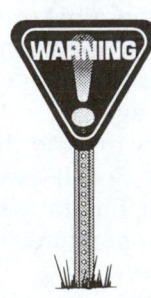 **WARNING** Make sure you mask or cover parts that you do not want to undercoat (exhaust system, brake rotors, fan belts, etc.). For example, if undercoat gets on the exhaust system, it could cause heavy smoke and possibly a fire upon engine operation—an embarrassing mistake.

16.10 EXTERIOR ACCESSORIES

To prevent corrosion, it is very important to install a barrier between dissimilar metal components such as aluminum bumpers and stainless and aluminum body trim. The plastic or rubber isolating pads accomplish this effectively. Mounting stainless and aluminum body trim must be done correctly to avoid galvanic corrosion. For example, when mounting trim requires drilling holes in a new or repaired panel, drill all holes before applying the primer, coating the inside edges of all holes completely.

When using a kit for replacement trim, be sure to use all parts supplied with the kit. If parts are not purchased as a kit, duplicate the original assembly exactly. Clearly, there is a great variety of body trim and accessories requiring many different application techniques. In all cases, be sure to follow the manufacturer's recommendations to avoid problems in making these repairs.

16.11 ACID RAIN DAMAGE

As mentioned earlier, air pollutants can damage an automotive finish. Since most of their damage is done to exterior, finished surfaces, they are a major concern of the refinisher.

Acid rain and other pollutants have generated a lot of controversy in recent years. There has been some confusion as to their causes and effects. Sulfur dioxide or nitrogen oxides create acid rain when released into the atmosphere. They combine with water and the ozone to create either sulfuric or nitric acid. It is estimated that the United States alone pumps out 30 million tons (27 metric tons) of sulfur dioxide and 25 million tons (22.5 metric tons) of nitrogen oxides yearly. More than two-thirds of the sulfur is emitted from power plants burning coal, oil, or gas. Iron and copper smelters, automobile exhaust, and natural sources like volcanoes, wetlands, and forest fires account for most of the remaining pollutants.

The standard for measuring acid rain is the pH scale. It runs from zero to 14, with 7 being neutral or equal to distilled water. A pH reading of 4 is ten times more acidic than a solution of acid and water with a pH of 5, and 100 times more acidic than a pH of 6. Once released into the ozone, these acids are readily dissolved into cloud droplets which, if low enough in pH, can cause significant paint damage.

The level of acid rain varies greatly around the country (see Figure 16–15). For example, South Carolina is reported to be one of the most acidic states in the nation. In Los Angeles, fog has been measured to have the acidic strength of lemon juice.

Rainfall in the northeastern states is extremely corrosive to car paints and finishes. For example, the average pH of rainfall in New Jersey is an acidic 4.3. Several manufacturers now have clauses in some of their new car warranties that exempt them from liabilities involving paint damage in high pH areas.

Acid rain damage generally occurs to the paint pigments, with lead-based pigments the most susceptible. Typically, the damage looks like water droplets that have dried on the paint and caused discoloration. Sometimes the damage appears as a white ring with a clear, dull center. Severe cases show pitting. Discoloration varies depending on the color. For example, acid rain damage to a yellow finish might appear as a white or dark brown spot. Medium blue may have a whitening look. A white finish may be a discolored pink, and a medium red may be purple.

Metallic finishes can be damaged because the acidic solution reacts with the aluminum particles and etches away the finish. A fresh finish is more easily damaged than an aged finish. Lacquers and uncatalyzed enamel finishes are most susceptible to damage, followed closely by catalyzed enamels.

Clear-coated finishes add a layer of protection against acid rain, so late-model vehicles with two- and three-coat finishes are less susceptible to damage. A clearcoat protects the paint pigments from discoloration, but it is still possible for acid rain to create a peripheral etch, or ring, on the clearcoat.

ACID RAIN REPAIRS

The procedure for restoring acid rain damage varies depending on the level and depth of the damage. The following steps outline repair procedures according to the level of damage as illustrated in Figure 16–49. When the problem has been corrected, stop at that stage. Remember that polishing or compounding removes part of the original finish and thereby reduces its overall life.

If the surface damage is like that shown in Figure 16–49A, proceed as follows:

1. Wash with soap and water.
2. Clean with wax and grease remover.

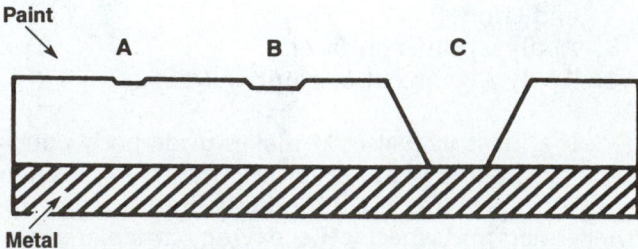

FIGURE 16–49 Different levels of acid rain damage will denote the repair methods required.

3. Neutralize the area by washing with a baking soda solution (1 tablespoon baking soda to 1 quart of water) and rinse thoroughly.

If the damage is embedded in the surface coat (Figure 16–49B), proceed as follows:

1. Follow cleaning and neutralizing steps already listed.
2. Hand polish problem area (inspect and continue if necessary).
3. Buff with polishing pad (inspect frequently and remove as little of the original finish as possible to cure the problem).
4. Use rubbing compound (inspect and continue if necessary).
5. Wet sand with #1500- or #2000-grit sandpaper and compound. If damage is still visible, repeat with #1200-grit. Do not use grits coarser than #1000.

If the damage is through to the undercoat (Figure 16–49C), proceed as follows:

1. Follow cleaning and neutralizing steps listed in Figure 16–49A.
2. Sand with #400- to #600-grit sandpaper.

3. Reclean and reneutralize prior to priming and repainting.

INDUSTRIAL FALLOUT SURFACE DAMAGE

Generally speaking, damage from **industrial fallout** is caused when small, airborne particles of iron fall on and stick to the vehicle's surface. The iron can eventually eat through the paint, causing the base metal to rust. Sometimes the damage is easier to feel than see. Sweeping a hand across the apparent damage will likely reveal a gritty or bumpy surface. Rust-colored spots might be visible, however, on light-colored vehicles.

The steps for repairing damage caused by industrial fallout are similar to those used when repairing acid rain damage, but with the following exception. After washing the car, treat the repair area with a "fallout remover," a chemical treatment product made especially for industrial fallout damage. Do not buff the damaged area before removing the fallout because buffing will drive the particles into the paint surface. If the particles break loose and become lodged in the buffing pad, deep gouges can occur.

SUMMARY

- Corrosion protection involves using various materials to protect steel body parts from rusting. When doing repairs, you must always use recommended methods of protecting repair areas from rust damage.

- If you fail to restore proper corrosion protection, it can endanger the driver and passengers of the vehicle.

- Corrosion or rust is the oxidation and chemical change of metal.

- Galvanizing is a process of coating steel with zinc. It is one of the principal methods of corrosion protection applied during the manufacturing process.

- During a collision, the protective coatings present on a car are usually damaged.

- Acid rain is the term given to rain containing pollutants from manufacturing and chemical industries. It causes discoloration and even destruction of the paint surface that could lead to corrosion damage.

- Weld-through primer is used to provide anticorrosion protection to weld zones.

- Self-etching primers etch the bare metal to improve paint adhesion and corrosion resistance, while providing the priming and filling properties of primer-surfacer.

- Two-part epoxy primers provide very strong base coating with good adhesion to bare metal.

- Rust converters change ferrous (red) iron oxide to ferric (black/blue) iron oxide.

- Be sure to carefully read the manufacturer's instructions on the container and follow them.

- Refer to the vehicle manufacturer's recommendations before drilling holes in panels for applying corrosion protection materials.

- Some anticorrosion compounds can be harmful if placed in contact with human skin.

- Generally speaking, damage from industrial fallout is caused when small airborne particles of iron fall on and stick to the vehicle's surface.

ASE-STYLE REVIEW QUESTIONS

1. Corrosion prevention is the phrase that is replacing _____.
   A. Rustproofing
   B. Undercoating
   C. Sound deadening
   D. All of the above
   E. Both A and B

2. Corrosion will be accelerated in areas _____.
   A. Of high relative humidity
   B. Where temperatures drop below freezing
   C. Both A and B
   D. None of the above

3. The higher the pH number rises above 6.0, the _____.
   A. Greater chance of acid rain
   B. Less likely chance of acid rain
   C. Both A and B
   D. None of the above

4. When two dissimilar metals are placed in contact with each other, the more chemically active will corrode, protecting the other metal in the process. This is called _____.
   A. Zinc coating
   B. Galvanic corrosion
   C. Both A and B
   D. None of the above

5. Technician A uses a conversion coating and then a metal conditioner. Technician B says that conversion coatings are usually mixed with water before using them. Who is correct?
   A. Technician A
   B. Technician B
   C. Both A and B
   D. Neither A or B

6. Technician A uses a conversion coating on inside closed sections. Technician B states that inside closed sections must be cleaned and degreased before being welded into place. Who is correct?
   A. Technician A
   B. Technician B
   C. Both A and B
   D. Neither A or B

7. This sealant is not paintable and attracts dust and dirt with time.
   A. Silicone sealant
   B. Tooling sealant
   C. Thin-bodied sealant
   D. Heavy-bodied sealant

8. Heavy-bodied sealers are used to fill seams from _____.
   A. $1/16$ to $1/8$ inch wide
   B. $1/8$ to $1/4$ inch wide
   C. $1/4$ to $1/2$ inch wide
   D. All of the above

9. Technician A uses conversion coating on aluminum. Technician B says that conversion coatings are applied using a spray bottle and rinsed off before drying. Who is correct?
   A. Technician A
   B. Technician B
   C. Both A and B
   D. Neither A or B

10. When restoring corrosion protection to an enclosed interior surface, Technician A applies the weld-through primer over existing paint. Technician B applies the weld-through primer over galvanized surfaces. Who is correct?
    A. Technician A
    B. Technician B
    C. Both A and B
    D. Neither A nor B

11. Any seams that might be exposed to automotive fluid should have a _____.
    A. Thin-bodied sealer
    B. Heavy-bodied sealer
    C. Brushable seam sealer
    D. Solid seam sealer

12. Undercoating compounds should never be applied to _____.
    A. The exhaust pipe or muffler.
    B. Suspension parts
    C. Drivetrain parts
    D. Brake drums
    E. Any of the above

ESSAY QUESTIONS

1. Summarize the corrosion process for exposed joints and seams.

2. Explain the four broad categories for anticorrosion materials.

3. Summarize four basic safety rules to follow when working with anticorrosion compounds.

4. What is weld-through primer?

5. Summarize the corrosion process for exposed joints and seams.

6. How would you repair acid rain damage embedded in the surface coat?

CRITICAL THINKING PROBLEMS

1. An untrained worker fails to apply anticorrosion materials after major structural repairs to a car's frontal frame rail and shock tower areas. The car will be driven on salty roads in winter months. What can happen after a few years of service?

2. A poor worker applies undercoating on a replaced floor pan. He failed to clean the area properly. What could happen?

MATH PROBLEMS

1. A can of sprayable undercoating will cover 10 square feet. You must spray panels that are 2 x 3 feet, 1 x 2.5 feet, and 1.1 x 1.1 feet. How many cans of undercoating will be needed?

2. To neutralize an area damaged by acid rain, you need 3 gallons of solution. If you must mix 1 tablespoon of baking soda to 1 quart of water, how many tablespoons will be needed?

CHAPTER 17

Vehicle Surface Preparation

INTRODUCTION

The life and appearance of a finish will depend upon the condition of the surface over which the paint is applied. In other words, proper surface preparation is the "foundation of a good paint job." Without it, there will be a weak base for the topcoat and the paint will fail or not look good.

The words **surface preparation** refer to getting the body surface clean, smooth, and ready for the application of the final colorcoats. To get a smooth, level surface often requires minor filling and sanding operations. Any painter knows that the colorcoat does little filling of rough areas and that the finished job is no smoother than the surface over which these materials are applied.

Refer to the index in the back of this book to find more information on the topics discussed in this chapter. Several earlier chapters gave the basics for fully comprehending the following information.

SHOP TALK

A dull primer surface may look smooth and ready for paint to the untrained eye. You must remember that the paint will actually work like a "magnifying glass" to exaggerate any scratches or irregularities. If you ever paint over a flawed surface, you will never do it again. It is time-consuming and frustrating to resand and repaint after making surface preparation mistakes.

OBJECTIVES

After studying this chapter, you should be able to:

✔ Determine whether or not the existing finish is defect-free and adheres soundly to the vehicle.

✔ Recognize surface defects that require additional surface preparation.

✔ Select the correct abrasive and sanding techniques for specific sanding operations.

✔ Prepare existing paint films and bare metal substrates for refinishing.

✔ Describe the three methods of removing a deteriorated paint film.

✔ Explain how conversion coating enhances primer adhesion to bare metal.

✔ Determine when to apply a primer, a primer-sealer, a primer-surfacer, or glazing putty.

✔ Prepare plastic parts for refinishing.

✔ Mask a car, panel, or spot repair for refinishing.

✔ Answer ASE test questions relating to vehicle surface preparation.

KEY TERMS

blasters
bullseye
coated abrasives
compounding
conversion coating
dry sanding
fine-line masking tape
guide coat
hand-rubbing
   compound
liquid masking
   material
machine compound
masking
masking covers
masking paper

masking tape
mil gauge
mils
orbital sander
overspray leak
paint edge
paint thickness
pot life
power grinding
reverse masking
sanding
scuff sanding
spot putty
surface preparation
wet sanding

ASE TASK LIST

Job Skills covered in this chapter include:

PAINTING AND REFINISHING TEST (B2) TASK LIST

A. Surface Preparation

2. Remove dirt, road grime, and wax or other protective coatings from area to be refinished and adjacent vehicle surfaces.
3. Inspect and identify substrate, type of finish, and surface condition; develop a plan for refinishing.
4. Remove paint finish.
5. Dry or wet sand areas to be refinished.
6. Featheredge areas to be refinished.
8. Mask trim, and protect other areas that will not be refinished.
11. Dry or wet sand area to which primer-surfacer and/or two-component putty have been applied.
12. Remove dust from area to be refinished, including cracks or moldings of adjacent areas.
13. Clean area to be refinished using a proper cleaning solution.

B. Spray Gun Operation and Related Equipment

1. Inspect, clean, and determine condition and adequacy of spray guns and related equipment (air hoses, regulator, air lines, air source, and spray environment).

C. Paint Mixing, Matching, and Applying

5. Apply single stage topcoat for spot and panel blending, and overall refinishing.
6. Apply basecoat/clearcoat for spot and panel blending, and overall refinishing.
8. Sand, buff, and polish finishes where necessary.
9. Identify the types of rigid, semirigid, or flexible plastic parts to be refinished; determine the proper materials and refinishing procedures.
10. Refinish rigid, semirigid, or flexible plastic parts.
11. Clean, condition, or refinish vinyl (e.g., upholstery, dashes, and tops).

D. Solving Paint Application Problems

1. Identify blistering (raising of the paint surface); determine the cause(s), and correct the condition.
3. Identify contaminants in the painted surface; determine the source(s), and correct the condition.
6. Identify lifting (surface distortion or shriveling) while the topcoat is being applied; determine the cause(s), and correct the condition.
9. Identify an overspray condition; determine the cause(s), and correct the condition.
12. Identify sandscratch swelling; determine the cause(s), and correct the condition.
15. Identify tape tracking; determine the cause(s), and correct the condition.

E. Finish Defects, Causes, and Cures

1. Identify poor adhesion; determine the cause(s), and correct the condition.
4. Identify blistering in the paint surface; determine the cause(s), and correct the condition.

F. Safety Precautions and Miscellaneous

5. Apply/remove decals, transfers, tapes, woodgrains, pinstripes (painted and taped), etc.

NONSTRUCTURAL ANALYSIS AND DAMAGE REPAIR TEST (B3) TASK LIST

A. Preparation

7. Protect panels and parts adjacent to repair area, to prevent damage during repair.
8. Remove dirt, grease, wax, and decals from areas to be repaired.

F. Plastic Repair

15. Prepare repaired areas for refinishing.

17.1 EVALUATION OF SURFACE CONDITION

Before painting, you must first identify the type of paint and overall condition of the existing paint system. Failure to identify defects at this stage can be very expensive to correct. It could even involve the complete removal of the repair and the original finish.

To evaluate the surface condition:

• Clean the areas to be inspected.
• Look carefully for any signs of paint film breakdown, such as checking, cracking, and blistering

(Figure 17–1). Horizontal surfaces usually show the greatest film deterioration. Careful inspection of the hood and trunk areas will give a good indication of the overall condition of the paint system.

• Note particularly the gloss level. Low gloss will often indicate surface irregularities caused by defects like checking or microblistering, which will need more thorough investigation with a magnifying glass.
• Any signs of disfigurement or discoloration of the paint film due to attack by industrial fallout or acid rain must be completely removed.
• Determine whether the old finish has good adhesion and that rust is not developing under the

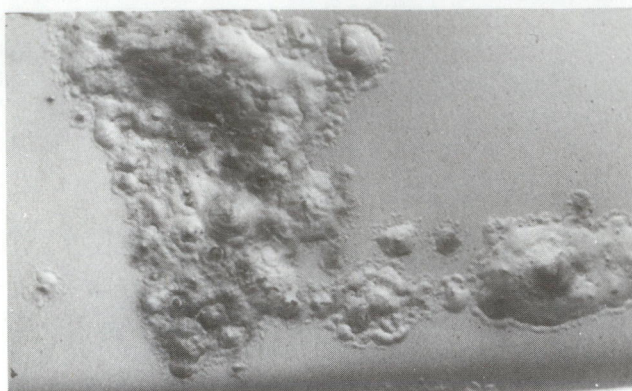

FIGURE 17-1 Carefully check the surface for signs of film breakdown such as blistering. This will let you know what should be done during surface preparation.

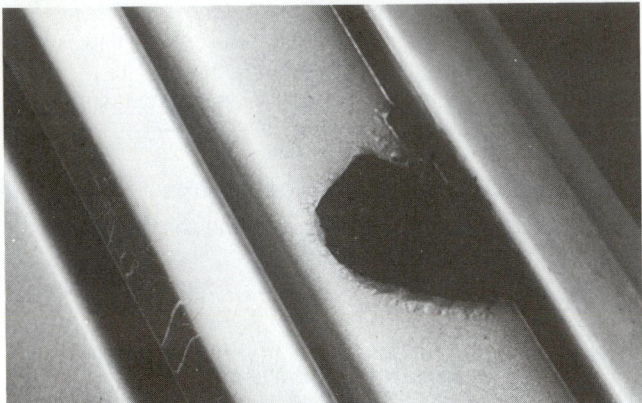

FIGURE 17-2 To check to see if rust has developed under a paint film, sand or poke through a small spot.

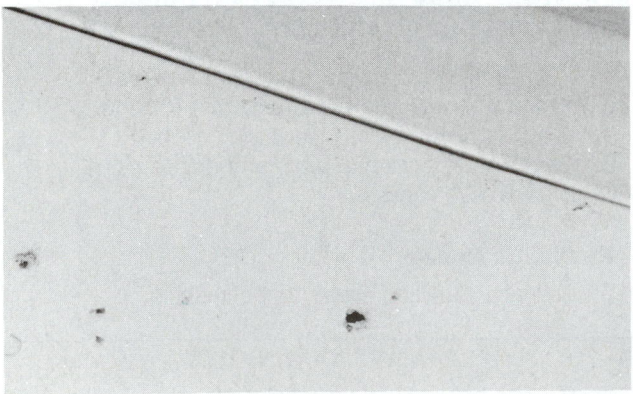

FIGURE 17-3 Small paint chips can easily be solved by proper sanding.

paint film. To test adhesion, sand through the finish (Figure 17–2), and featheredge a small spot. If the thin edge does not break or crumble, it is reasonable to assume that the old paint will stay on when the refinish color is applied over it. Developing rust can be detected by a roughness or pitting of the surface. The paint on those areas where either poor adhesion or rust is found must be removed to bare metal.

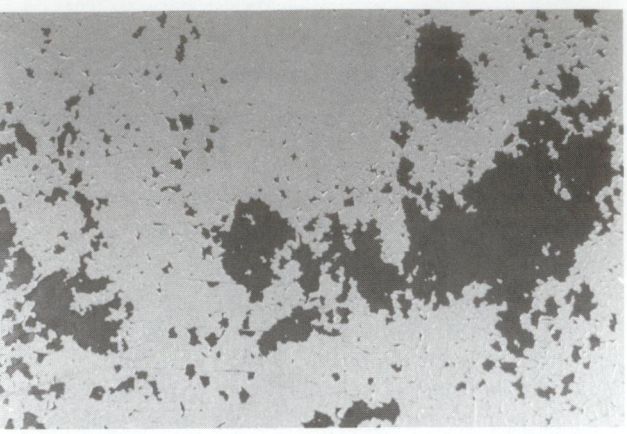

FIGURE 17-4 Cracking or peeling paint is a more serious problem indicating possibly poor adhesion of the old paint. *(Courtesy of Maaco Enterprises, Inc.)*

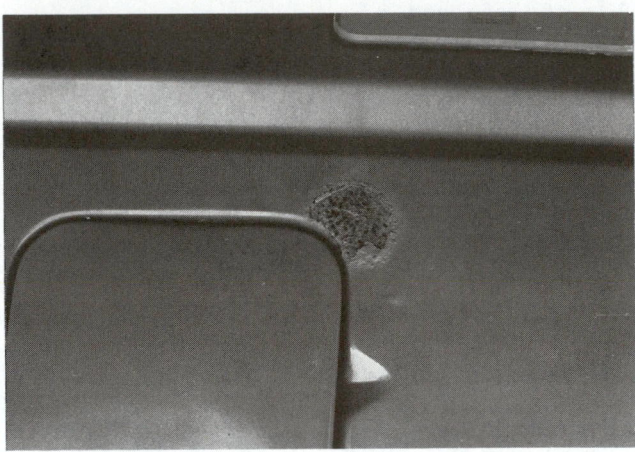

FIGURE 17-5 All rust must be removed by sanding or blasting before a new finish can be applied. If you paint over even the smallest bit of rust, rusting will continue and damage the new finish.

17.2 SANDING

Sanding uses an abrasive coated paper or plastic backing to level and smooth a body surface being repaired. Sanding with coarse, rough paper might be done to level plastic filler. Sanding with fine, smooth paper is often done to lightly scuff the old paint so the new paint will stick.

Sanding is one of the most important steps of surface preparation. In fact, this operation is a standard part of most surface preparation procedures. Sanding prepares the surface for painting in several ways:

- Chipped paint (Figure 17–3) is sanded to taper the sharp edges that would show up as ridges under the new finish.
- Cracking or peeling paint (Figure 17–4) and minor surface rust (Figure 17–5) must be removed before applying a fresh topcoat. If not, these

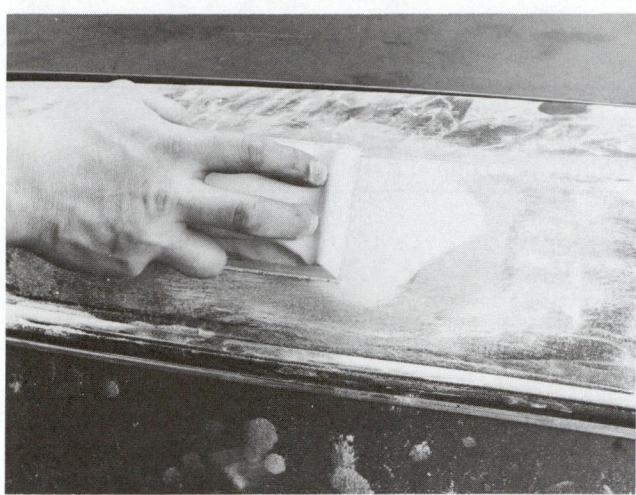

FIGURE 17-6 Primed and puttied areas must be smoothed and leveled by sanding.

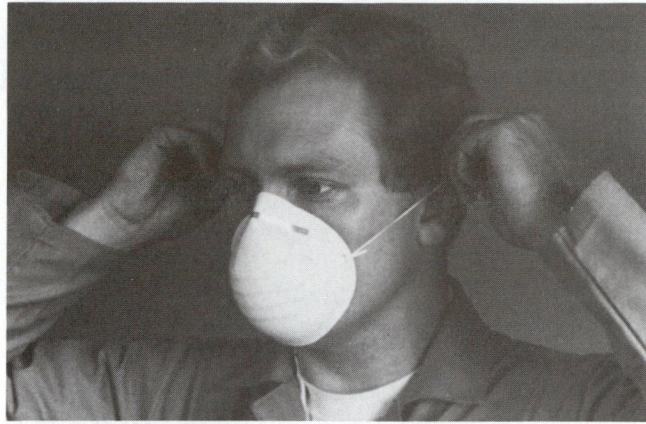

A

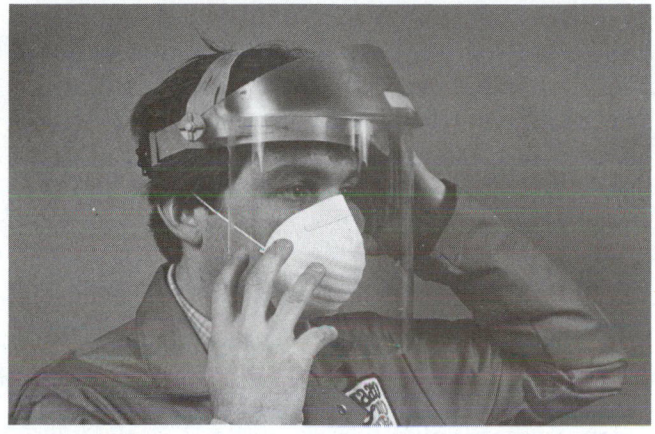

B

FIGURE 17-7 When just sanding, wear a dust respirator (A), but when sanding and grinding, wear both a dust respirator and face mask (B). *(Courtesy of Maaco Enterprises, Inc.)*

conditions will continue to deteriorate and will eventually ruin the new finish.

- Primed and puttied areas must be smoothed and leveled (Figure 17–6).
- The entire surface to be refinished must be scuff-sanded to improve adhesion of the new paint. **Scuff sanding** removes any trace of contaminants on the existing finish. A clean, scuffed surface is very important for proper bonding of the new topcoat.

Because **coated abrasives** (sandpaper) perform the actual cutting and leveling in the sanding operation, selecting the correct abrasive is critical to the quality of the finished work.

 Always wear recommended protective gear when sanding. Wear a respirator to protect your lungs from the paint dust. When power grinding, wear a face shield. When power sanding, wear safety glasses (Figure 17–7).

17.3 COATED ABRASIVES (SANDPAPER)

When modern coated abrasives (sandpaper) are made, a flexible or semirigid backing attaches to the abrasive grains, which are bonded by an adhesive. Hence, the most efficient results on a particular application depend on the selection and manufacturing of suitable combinations of available grains, adhesives, and backings. The automotive refinisher must then select and correctly use the proper sandpaper product for optimum productivity, material cost efficiency, and the best finish.

ABRASIVE TYPES

The abrasive grains used to manufacture sandpaper products used in automotive refinishing are selected on the basis of their hardness, toughness, resistance to grinding heat, fracture characteristics, and particle shape. The kind of grain a refinisher chooses depends on the purpose for which the coated abrasive is to be used. As for abrasive types, most body shops stock two: silicon carbide and aluminum oxide.

Silicon carbide sandpaper has a very sharp and fast-penetrating grain. It is customarily used (in paper sheet and disc form) for featheredging and dry sanding soft materials, such as old paint, fiberglass, and body putty. The major limitation of silicon carbide grain is that it tends to break down and dull rather readily when sanding hard surfaces.

Aluminum oxide sandpaper has an extremely tough, wedge-shaped grain that better resists fracturing and dulling. Traditionally it has been popular in coarse grits for grinding damaged metal, stripping old paint, and shaping plastic filler. Numerous tests have demonstrated the superior performance of aluminum oxide sanding sheets and discs over silicon carbide on today's modern paint systems. Aluminum oxide is also preferred for use with today's paint finishes, which are predominantly basecoat/clearcoat, have harder surfaces, and are applied in thinner layers than traditional paints.

The blocky shape of the aluminum oxide abrasive when compared to silicon carbide makes it not as likely to create deep scratches right through to the base material and so reduces the risk of overcutting. The greater durability of aluminum oxide versus silicon carbide enables the abrasive sheet or disc to better resist edge wear and dulling for longer effective life on these harder finishes.

A third type of abrasive, *zirconia alumina grain*, has a unique, self-sharpening characteristic that provides continuous new cutting points during the sanding operation for reduced labor and increased efficiency and longer effective life compared to traditional abrasives.

Zirconia alumina sandpaper has been developed through advanced technology. It continues to gain widespread preference in auto body repair shops. The fact that zirconia alumina products run cooler is also particularly important when removing OEM clearcoat finishes because of the extra heat generated when sanding these harder paint surfaces. A hot-running disc or sheet will load faster as the material being sanded softens and balls up in the abrasive. The self-sharpening action reduces the amount of sanding pressure required. Auto body professionals find that they can save money by using one grit finer and get a better finish. The net result is that zirconia alumina abrasive products are being recognized as the more cost-effective alternative to traditional aluminum oxide and silicon carbide for a growing number of auto body repair and refinish operations.

GRIT NUMBERING SYSTEM

The rough side of the sandpaper is called the *grit side*. Grit sizes vary from coarse to micro fine grades and are ordered by number (Table 17–1). The lower the number, the coarser the grit (Figure 17–8). For example, a #24 grit is used to remove old paint film, while a #320, #360, or #400 grit is used to sand the gloss off an old finish to be repainted. Very fine and ultra-fine abrasive papers are used primarily for colorcoat sanding. The so-called compounding papers, the #1200, #1500, and #2000 grits, are used to solve problems on basecoat/clearcoat paint surfaces such as those shown in Figure 17–9.

TABLE 17-1: TYPES OF GRIT AND NUMBERING SYSTEM

Grit	Aluminum Oxide	Silicon Carbide	Zirconia Alumina	Primary Use for Auto Body Repair
Micro fine	—	2000 1500 1250	—	Used for basecoat/clear coat paint system.
Ultra fine	—	800	—	Used for color-coat sanding.
Very fine	—	600	600	Used for color-coat sanding. Also for sanding the paint before polishing.
	400 320 280 240	400 320 280 240	400 — 280 240	Used for sanding primer-surfacer and old paint prior to painting.
	220	220	—	Used for sanding of topcoat.
Fine	180 150	180 150	180 150	Used for final sanding of bare metal and smoothing old paint.
Medium	120 100 80	120 100 80	— 100 80	Used for smoothing old paint and plastic filler.
Coarse	60 50 40 36	60 50 40 36	60 — 40 —	Used for rough sanding plastics filler.
Very coarse	24 16	24 16	24 —	Used on sander or grinder to remove paint.

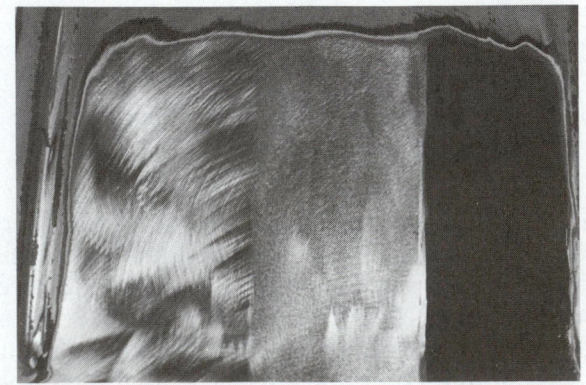

FIGURE 17-8 As the grit number increases, surface smoothness increases. *(Courtesy of Maaco Enterprises, Inc.)*

All domestic manufacturers conform to the same grading system for uniform consistency of standards. Differences in performance when using the same mineral, grit, bond, and backing from different manufacturers can be attributed to differences in manufacturing processes or quality, and/or operator methods.

Note that imported abrasives sometimes use a different grading system from those made in the United States. Refer to the manufacturers' information for comparison.

As shown in Figure 17–10, the abrasive papers are available in various sizes and shapes. The most common forms found in paint/body shops are sheet stock and discs. The sheet stock—usually 9 by 11 inches—can then be cut into smaller pieces. Sheets are also available in jitterbug and board or body file sizes.

The most common abrasive sanding disc sizes for disc and dual-action sanders are 5, 6, and 8 inches. Sandpaper disc grit sizes generally range from #40 to #400 grit.

Remember that European "P-grade" sandpaper grits are available. Their coarseness varies from conventional grit sizes. Refer to the sandpaper manufacturers' charts and information to convert to customary grit numbers if needed.

COATED ABRASIVE SURFACES

Coated abrasives are generally manufactured in two types of surface distributions:

- Closed coat abrasive paper
- Open coat abrasive paper

A *closed coat* product is one in which the surface grains completely cover the sanding side of the backing. An *open coat* product is one in which the abrasive grains are spaced to cover between 50 and 70 percent of the backing surface (Figure 17–11).

As for uses, open coat products are the popular choice on softer materials, such as old paint, body filler, and putty, plastic, and aluminum, where premature loading of the abrasive would otherwise be a problem. Closed coat products generally provide a finer finish and are most commonly used in wet sanding applications.

SANDPAPER ATTACHMENT

Sandpaper can be attached to sanders by

1. A tube of adhesive
2. Self-stick adhesive on the paper
3. Velcro or hook and latch

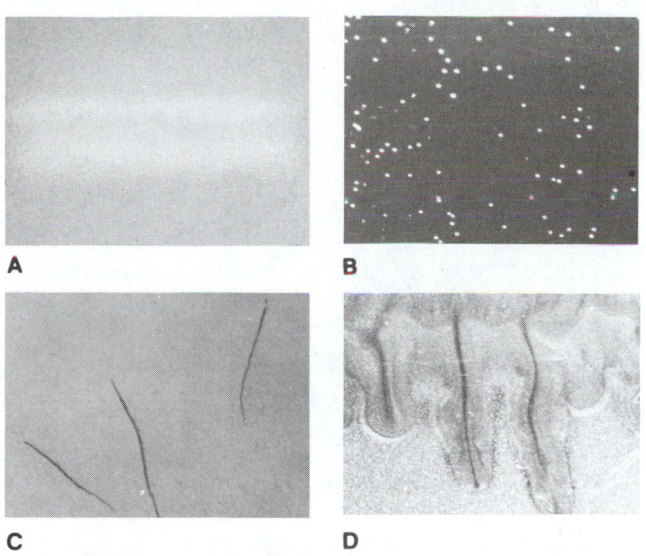

A   **B**   **C**   **D**

FIGURE 17-9 The micro fine grits are frequently used to remove such problems as: (A) orange peel; (B) dust nibs; (C) small surface scratches; and (D) paint sags.

FIGURE 17-10 Various sizes and shapes of abrasive papers and discs are needed for surface preparation.

FIGURE 17-11 Note (top) closed coat abrasive paper; (bottom) open coat abrasive paper. *(Courtesy of Norton Co.)*

To apply the sandpaper to the backing pad of a disc, orbital, or dual-action sander with tube adhesive, squeeze a few drops of adhesive on the backing pad. Spread the adhesive evenly on the pad (Figure 17–12A). Then press the disc on and off the pad to help make the adhesive sticky (Figure 17–12B).

With self-adhesive paper, simply pull a sheet off its backing and press it on the pad. Be sure to center the paper on the pad (Figure 17–13). With Velcro attachment, you must also center the paper on the backing pad. If not, the paper could fly off. To protect the self-stick adhesive from dust, always close and seal the box of unused paper.

Immediately after finishing the sanding operation, remove the used sandpaper from the backing pad. If it is not removed right away, the adhesive will harden and cause the disc to stick fast to the backing pad. Should this occur, use solvent on a rag to dissolve the adhesive and then remove the paper (Figure 17–14).

GRINDING DISCS

Grinding discs are used for rough jobs, such as grinding rust and paint. They are available in numbers of #16 to #50 grits and in diameters of 3 to 9 inches. A dual-action sander can be used to remove minor surface problems, but heavy surface damage must be removed with an air or electric grinder (Figure 17–15). The grinder disc is first assembled to the backing plate (Figure 17–16). Then the disc/plate assembly is attached to the grinder (Figure 17–17).

Some sandpaper discs are available with a center hole and are fastened to the sander in the same manner as the grinding abrasive disc. This manner of fastening is necessary in some wet sanding operations.

Although grinding discs are thicker and stronger than sandpaper discs, they are rather thin and easily bent. For this reason, the backing plate is necessary to provide stiffness for the revolving disc.

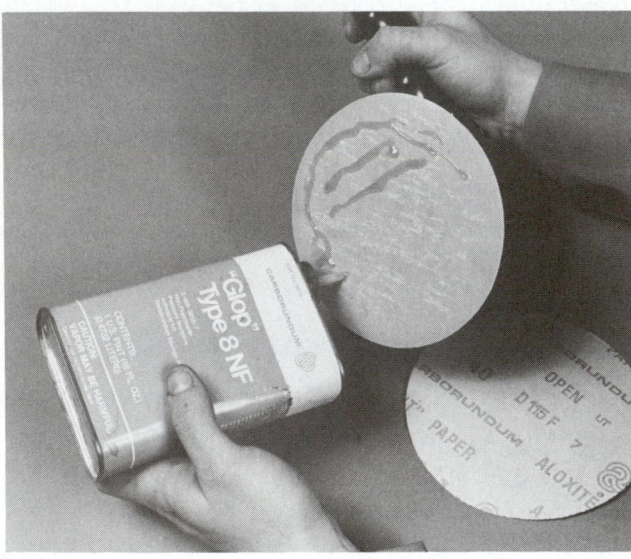

A

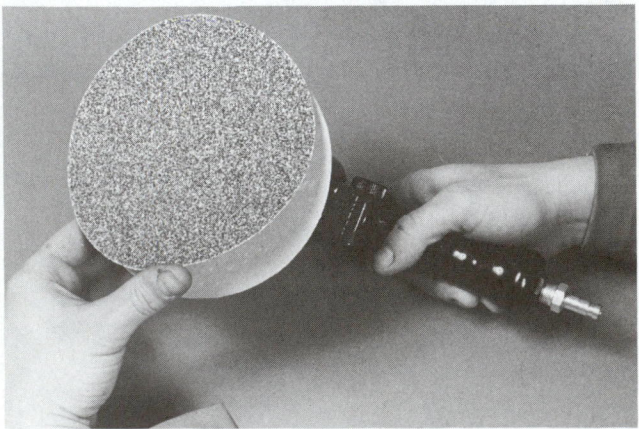

B

FIGURE 17-12 (A) Spread the adhesive on the pad, then (B) press the sandpaper disc in place. *(Courtesy of Carborundum Abrasives Co.)*

FIGURE 17-13 Center the paper on the pad so it does not fly off when sanding. *(Courtesy of Maaco Enterprises, Inc.)*

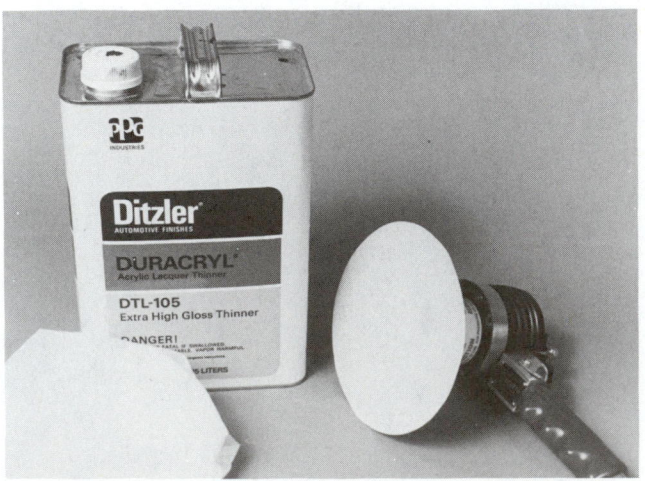

FIGURE 17-14 A solvent can be used to remove adhesive from the pad before storage.

Two types of back-up pads or plates are shown in Figure 17–18.

Safety Pointers When Grinding

In addition to the safety procedures given in Chapter 4, the following pointers must be remembered:

- When disc grinding, hold the grinder firmly at a low 5- to 10-degree angle to the work surface.
- Grind so that dust is directed away from the face and toward the floor.
- Be conscious of the grinder or polisher cord at all times to prevent entanglement.
- Do not grind or sand too close to trim, bumpers, or any projection that might snag or catch the grinding disc's edge.
- Never start or stop a disc grinder in contact with the work surface.
- Never "free run" a grinding disc or set a grinder down until it stops completely.
- Make certain the back-up pads are designed for the work, free of cuts or nicks at the edge or at the center hole. Make certain pads are seated on the shaft properly. Check for proper balance. Retainer nuts should not show excessive thread wear, must have at least three-thread contact, and should not cause damage to the grinding discs.

- Back-up pads for use with the self-adhesive type of discs must be dry, clean, and dust free. Avoid using pads with frayed, torn, dirty, or paper-contaminated surfaces. If necessary wipe the pad face with a clean, dry cloth. Do not immerse pad or clean pad face in solvent.

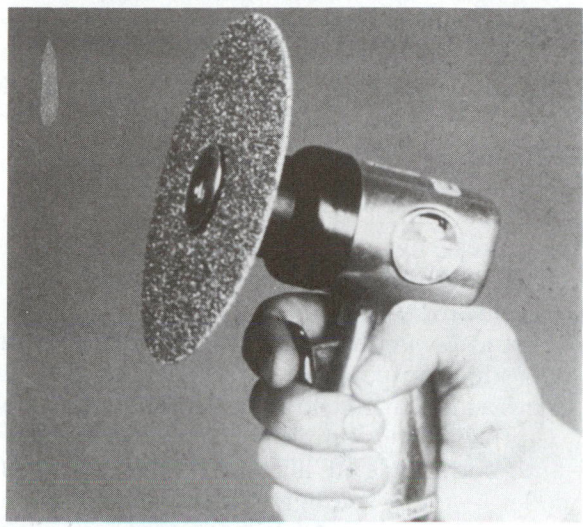

FIGURE 17-17 A grinder can be dangerous if misused. If the grinding disc contacts your skin, serious abrasion will result. *(Courtesy of Maaco Enterprises, Inc.)*

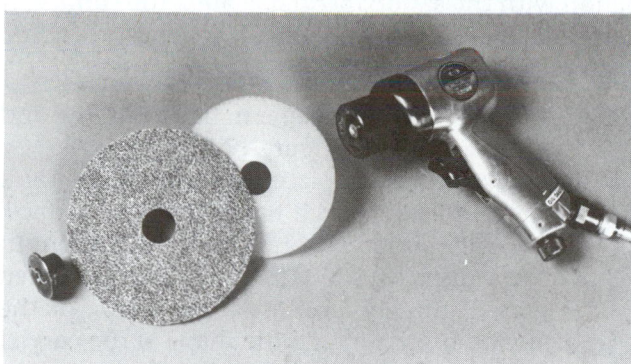

FIGURE 17-15 A grinder is used for fast material removal. Note how the disc attaches to the tool.

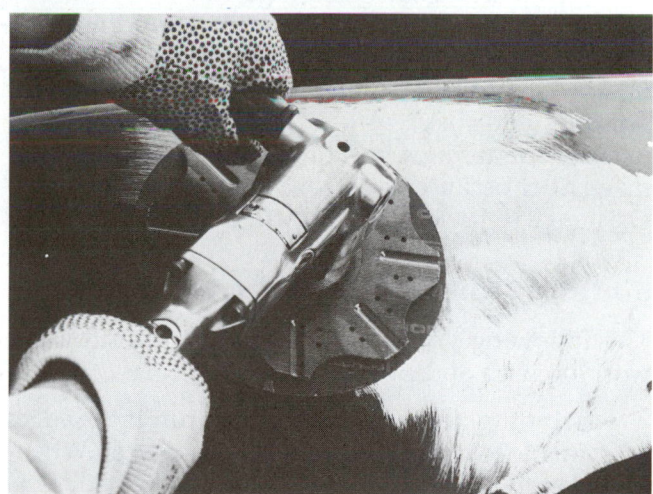

A

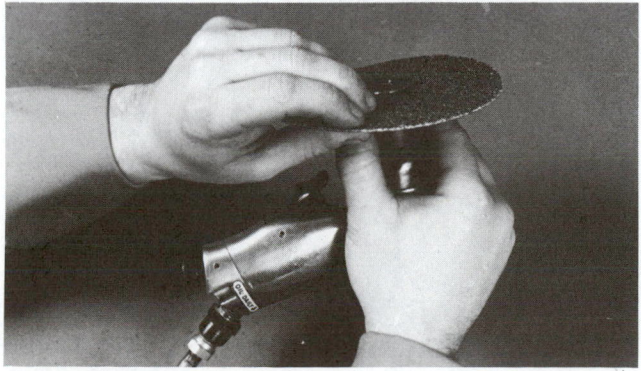

FIGURE 17-16 Make sure the nut that holds the parts on the grinder is tightened properly. *(Courtesy of Maaco Enterprises, Inc.)*

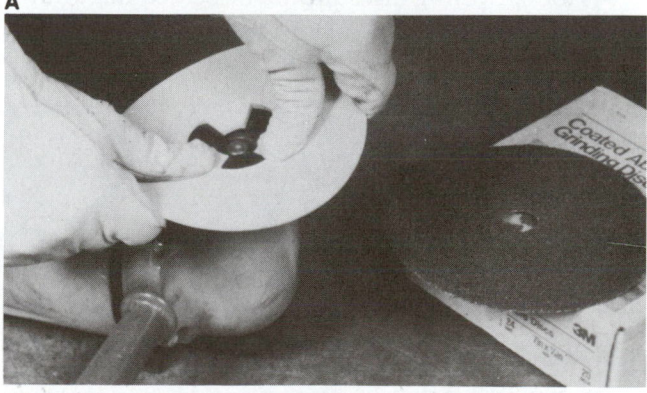

B
FIGURE 17-18 Note (A) the hard rubber backing plate and (B) a smaller fiber back-up plate.

WEIGHT OF PAPER

The proper selection of backings depends on the application involved. Paper-backed abrasive products used in automotive refinishing are designated under uniform standards by all manufacturers as A-, C-, D-, or E-weight.

A-weight paper is the lightest, most conformable paper backing available. It is popular for wet color sanding and dry finish sanding. The C- and D-weight paper products are progressively heavier, tougher, and less flexible. They are suitable for coarser sanding applications.

E-weight paper is being more widely used by refinishing personnel for paint stripping and shaping of filled areas, as it is more durable than the traditional D-weight paper backings once popular for these applications.

D- and E-weight papers are sometimes referred to as production papers because their construction produces a fast-cutting, long-lasting abrasive surface.

Cloth backings employed in products used by the auto body trade are likewise designated by a letter code. J-weight is a light, flexible cloth, popular for general clean-up and deburring, in sheet or handy roll form. X- and Y-weight cloths are heavier, with more rigid backings, often used in small disc form for tight-quarter coarse sanding.

Fiber backing is most common in grinding discs. This very tough, semirigid backing is best suited for heavy operator pressure applications, such as weld grinding and rust removal. The most suitable fiber backing for automotive application is 30-mil vulcanized fiber because of its extra durability and greater resistance to breakdown and edge chipping.

Safety Pointers With Abrasives

The following points must be kept in mind when working with abrasives:

- Grinding discs should never be run if the edges are nicked, torn, or show excessive wear. Whenever in doubt, do not use the product.
- Fiber grinding discs should be seated flat against a proper back-up pad and never overhang a pad by more than $1/4$ inch (6.4 mm).

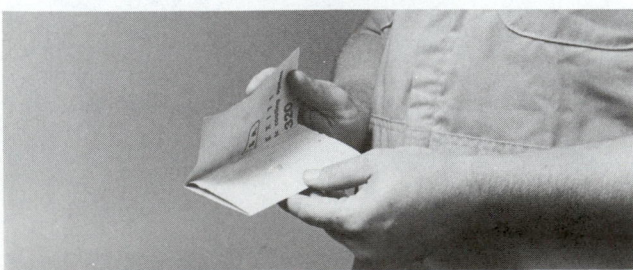

FIGURE 17-19 If you must use your hand to sand a curved surface, fold the sandpaper in thirds. *(Courtesy of Maaco Enterprises, Inc.)*

NOTE: old grinding discs should never be used as back-up pads.

- When paper discs are used on a slow-speed polisher, the recommended speed is 3,000 rpm or less.
- Curled discs generally indicate improper storage and should not be used until the shape is corrected. Storage of discs at 65 to 75 degrees Fahrenheit (18 to 40 degrees Celsius) will prevent excessive curling of abrasive products prior to usage.
- Ensure proper ventilation at all times when grinding or sanding and particularly avoid breathing dusts or fumes that are generated by "grinding aid" disc products. Refer to precautions on box labels, discs, or charts for detailed instructions.

17.4 METHODS OF SANDING

Refinishing sanding can be done by hand or by using power equipment. Most heavy sanding—such as removing the old finish—is done by power sanders. But some conditions, particularly the delicate operations, dictate hand sanding.

HAND SANDING

Hand sanding is a simple back and forth scrubbing action with the sandpaper flat against the surface. It can be achieved by following a general procedure such as this:

1. Cut the sheet of sandpaper in half crosswise and then fold in thirds (Figure 17-19).
2. On curved surfaces, place the paper in the palm of the hand and hold it flat against the surface. Apply even, moderate pressure along the length of the sandpaper using the palm and extended fingers (Figure 17-20). Make the shape of the palm and fingers match the shape of the curved surface. Sand back and forth with long, straight

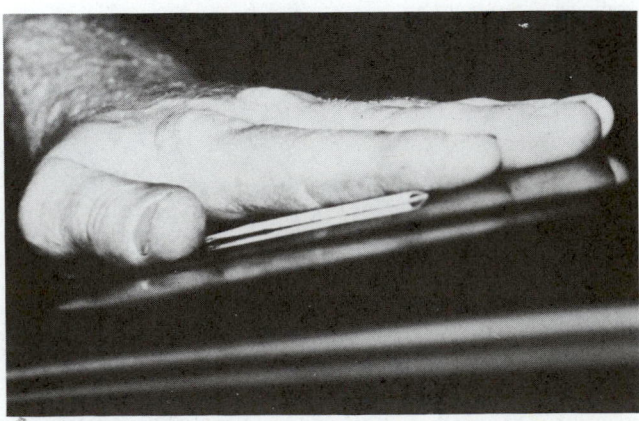

FIGURE 17-20 When sanding without a block, hold the paper like this and use hand pressure to match the shape of the body panel.

strokes. Avoid using only your hand to sand a flat surface. Use a sanding block. Your hand is not a flat surface; your fingers will be doing the sanding. This will result in uneven pressure being applied in the spaces between your fingers (Figure 17–21). Finger sanding should be avoided on flat surfaces.

3. Do not sand in a circular motion. This will create sand scratches that might be visible under the paint finish. To achieve the best results, always sand in the same direction as the body lines on the vehicle (Figure 17–22).

4. Be sure to carefully sand around unremoved parts, such as trim, moldings, door handles, radio antennas, and behind bumpers. Paint will not adhere properly to a smooth, unsanded surface (Figure 17–23).

5. On larger straight or flat surfaces, use a sanding block or pad for best results. To sand convex or concave panels (Figure 17–24), employ a flexible sponge-rubber backing pad. Use a sanding block (Figure 17–25) to sand level surfaces.

6. Hard-to-reach areas are easier to sand with a small abrasive pad similar to the one shown in Figure 17–26. If you fail to scuff these dirty, hidden areas, the paint can peel (Figure 17–27).

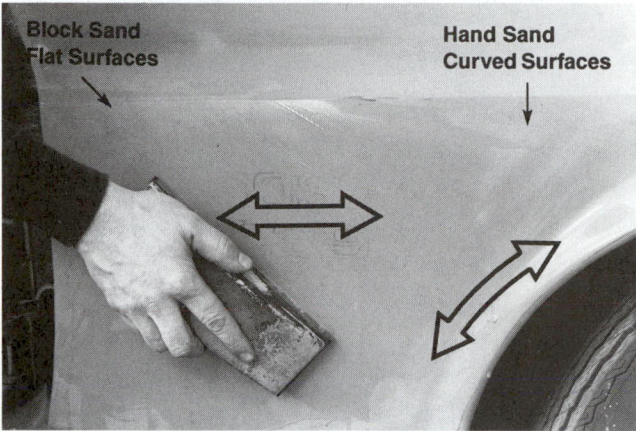

FIGURE 17-23 Sand around trim, handles, molding, and other similar items. When block sanding up against a curved surface, stop before the paper hits the curve, on this wheel well for example. Hand sand the curve of the wheel well.

FIGURE 17-21 Finger sanding will cause irregular surface. Use a sanding block on all flat surfaces! *(Courtesy of Maaco Enterprises, Inc.)*

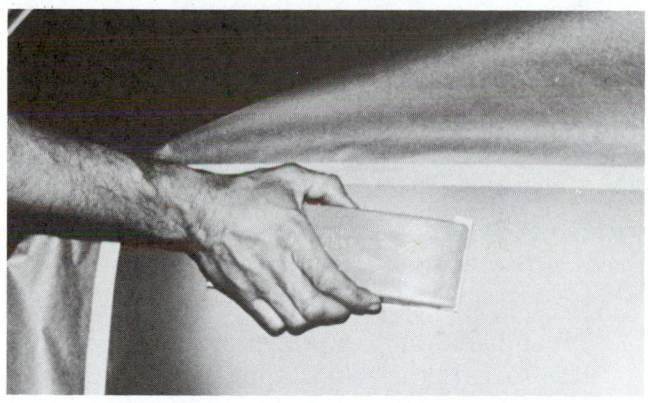

FIGURE 17-24 Use a flexible sponge rubber pad on convex and concave panels. *(Courtesy of Maaco Enterprises, Inc.)*

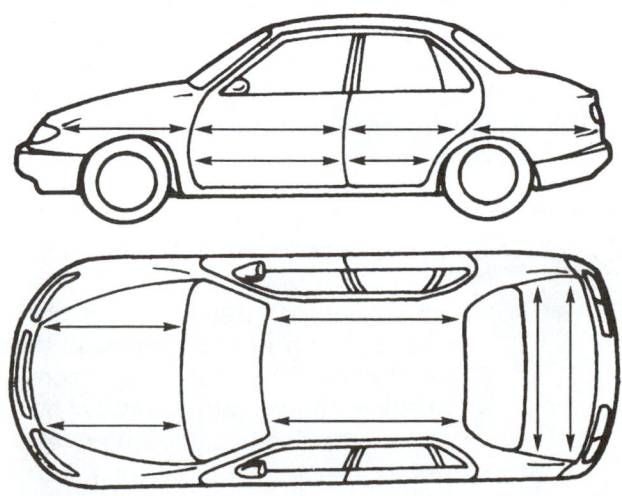

FIGURE 17-22 Sanding with the body lines will speed the work and help prevent paint runs on the sides of the vehicle.

FIGURE 17-25 Different size sanding blocks are available. Use a size appropriate for surface area. *(Courtesy of Maaco Enterprises, Inc.)*

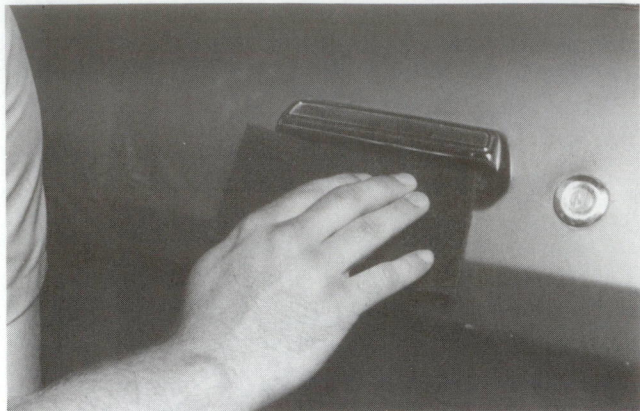

FIGURE 17-26 Use an abrasive pad in tight spots. It will scuff the good paint to provide good adhesion of new paint.

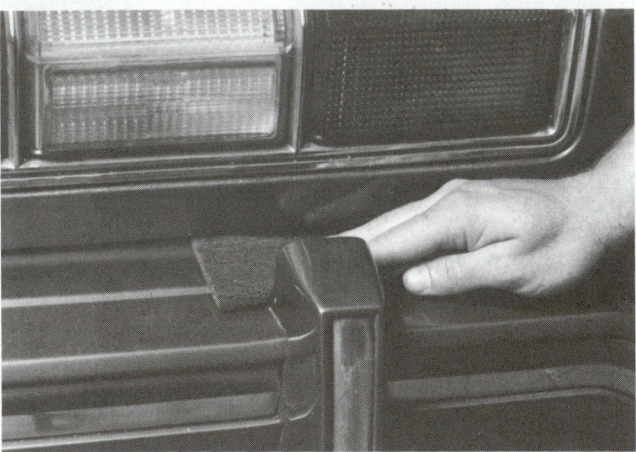

FIGURE 17-27 Scuff all hidden areas like on this bumper to provide a good, clean surface for painting.

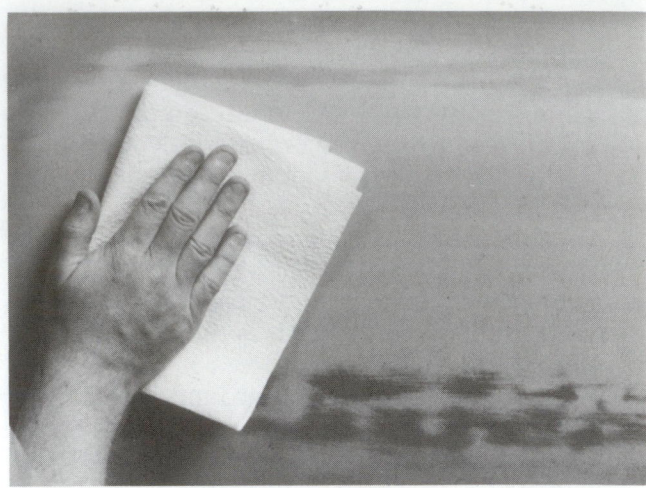

FIGURE 17-28 Avoid wiping the surface with your bare hands because skin oil can contaminate the surface. Wipe with a clean cloth to feel for rough spots that need more

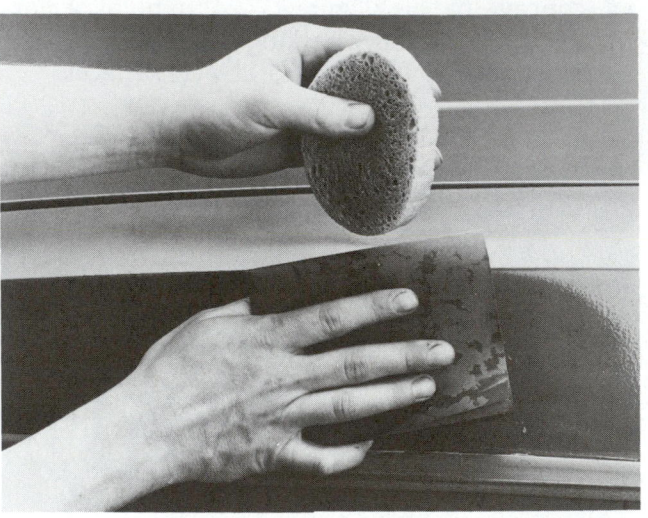

FIGURE 17-29 When wet sanding, water can be applied by a sponge or hose. *(Courtesy of Carborundum Abrasives Co.)*

7. When hand sanding primer or putty, make certain to sand the area until it feels smooth and level. Rub a clean cloth (Figure 17–28) over the surface to check for rough spots. The cloth will keep skin oil off the surface and also help magnify the feel of any roughness.

For hand sanding, you can use the dry or the wet sanding method:

- **Dry sanding.** This is basically the back-and-forth procedure just described. But one of the problems with it is that the paper tends to clog with paint or metal dust. Tapping the paper from time to time will remove some of the dust.

- **Wet sanding.** Wet sanding solves the problem of paper clogging. It is basically the same action as dry sanding except that water, a sponge, and a squeegee are used in addition to the sanding block. Sandpapers are available in dry, wet, or wet-or-dry abrasive types.

Most sandpaper manufacturers recommend that you soak wet sandpaper overnight before color sanding. This will soften the paper backing and help prevent tiny scratches in the paint.

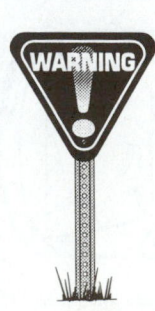

Avoid wet sanding plastic body filler! If not the waterproof type (fiberglass impregnated filler), the plastic filler can absorb and hold the moisture. The water can come back to "haunt you" when it shows up in the fresh paint when refinishing! Power sand the repair area level with dry paper and prime the area before wet sanding.

When wet sanding, dip the paper in the water or wet the surface with the sponge (Figure 17–29). Use plenty of water, employing short strokes and light pressure. Never allow the surface to dry during the wet sanding operation. Also do not allow paint residue to build up on the abrasive paper.

It is possible to tell how well the paper is cutting by the amount of drag felt as it moves across the surface. When the paper begins to slide over the surface too quickly and easily, it is no longer cutting. The grit has become filled with paint particles or sludge. Rinse the paper in water to remove the paint and sponge the surface to remove the remaining particles. Then the sandpaper will cut the surface again.

Check your work periodically by sponging off the surface and wiping it dry with a squeegee. This will remove all excess water, so that it is easier to evaluate the surface condition. It is usually wise to complete one panel or body section at a time. Then remove the sanding residues with the sponge and dry off with the squeegee before sanding the next panel.

Once the wet sanding operation is completed, be sure that all surfaces are dry. Blow out the seams and molding with a low-pressure stream of compressed air and tack-rag the entire surface.

A comparison of the advantages and disadvantages of wet and dry sanding is given in Table 17–2.

TABLE 17-2: COMPARISON OF WET AND DRY SANDING

Item	Wet Sanding	Dry Sanding
Work speed	Slower	Faster
Amount of sandpaper required	Less	More
Condition of finish	Very good	Final finish difficult
Workability	Normal	Good
Dust	Little	Much
Facilities required	Water drain necessary	Dust collector and exhaust necessary
Drying time	Necessary	Not necessary

POWER SANDING

As described in Chapter 4, there are four types of power sanders commonly used by the refinisher (Figure 17–30):

- Disc sander or grinder
- Orbital or jitterbug pad sander
- Dual-action (DA) sander
- Straight line or board sander

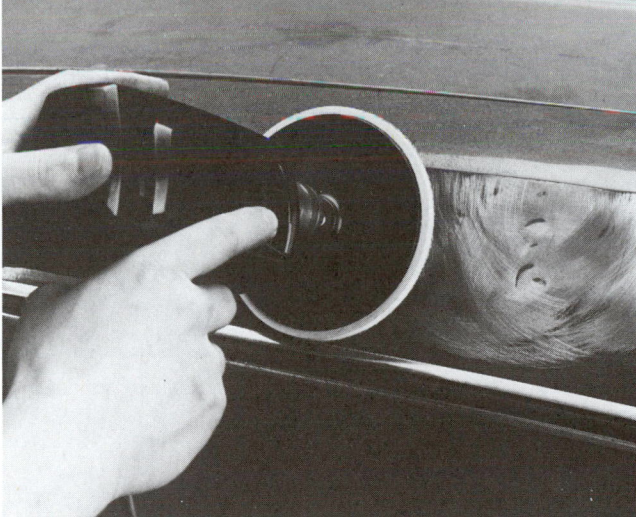

A

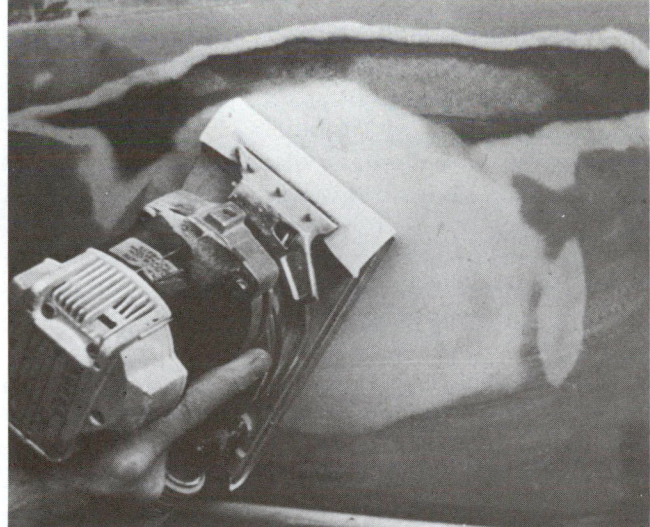

B

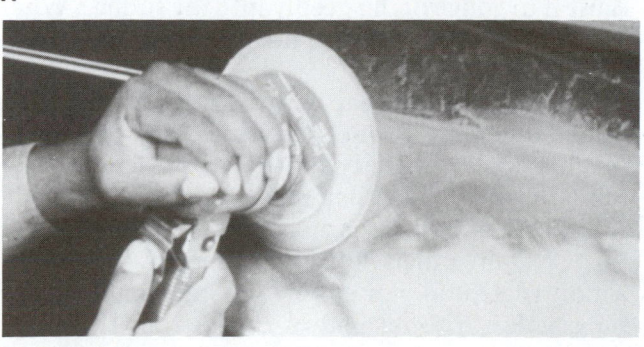

C

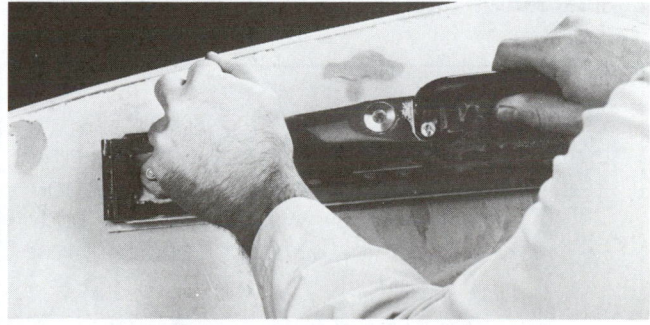

D

FIGURE 17–30 Study various types of sanders in operation: (A) disc sander; (B) jitterbug pad sander; (C) dual action sander; and (D) straight line sander or air file.

TABLE 17-3: USE OF SANDERS

Sander Type	Normal Area of Operation	Normal Use						
		Paint stripping	Feather-edging	Rough sanding of solder	Rough sanding of metal putty	Rough sanding of poly putty	Sanding of metal putty	Sanding of poly putty
Disc Sander		A	C	B	C	C	C	C
Dual Action Sander	Suitable for narrow areas	B	A	C	A	A	A	A
Orbital Sander		B	B	C	A	A	A	A
Straight Line Sander	Suitable for wide open spaces	B	C	C	A	B	A	B
Long Orbital Sander		B	C	C	A	B	A	B

NOTE: It is important that the correct type of sander and abrasive paper be used for each type of job. Also, always wear a mask or use some sort of dust arrester when using the sander.
A Preferred
B Acceptable
C Least preferred

All four types of sanders are powered by air or electricity.

In general, the type of power sander dictates sanding procedures (Table 17–3). Disc sanders or grinders, for example, have high-speed discs that turn from 2,000 to 6,000 rpm. They take circular discs from 5 to 9 inches (127 to 230 mm) in diameter and are used for such operations as grinding off an old finish. Heavier grinders (Figure 17–31) generally take a 9-inch (230 mm) diameter disc and—because of the obvious safety hazard involved—many have both a rear and side handle for better control.

Power grinding is done to quickly remove large amounts of old paint and other materials. An air grinder is one of the fastest methods to remove material.

When using a disc grinder, care must be taken to tilt it slightly so that only about 1 inch (25 mm) of the leading edge of the sanding disc contacts the surface (Figure 17–32). Never use the disc flat on the surface because it will twist the grinder and can even cause it to fly out of your grip. Also, when held flat it makes circular sand scratches, which are difficult to remove.

When grinding, position the grinding disc against the surface so that sparks fly away from your body. Wear leather gloves and a full face shield.

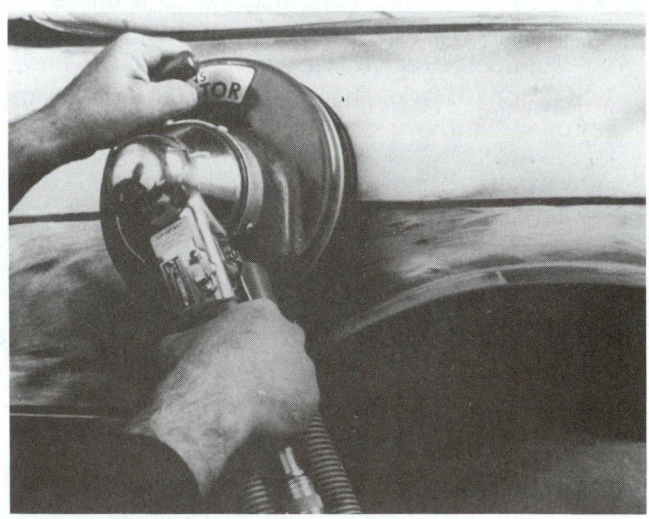
FIGURE 17-31 This is a heavy-duty sander/grinder in operation.

Never use a disc grinder at a sharp angle with just the edge of the disc in contact, because this will cause it to gouge or dig deeply into the surface. When a disc grinder is properly held, the sanding marks are nearly straight.

An **orbital sander** or *dual-action* sander moves in two directions at the same time. This produces a much smoother surface finish. A dual-action or DA sander is used to featheredge a repair area. It is the workhorse of body technicians.

Orbital sanders have an eccentric (off center) action that produces either a partly circular scrubbing action (orbital pad or dual-action type) or a straight back-and-forth reciprocating action (flat orbital or straight line type). Unlike the disc sander just

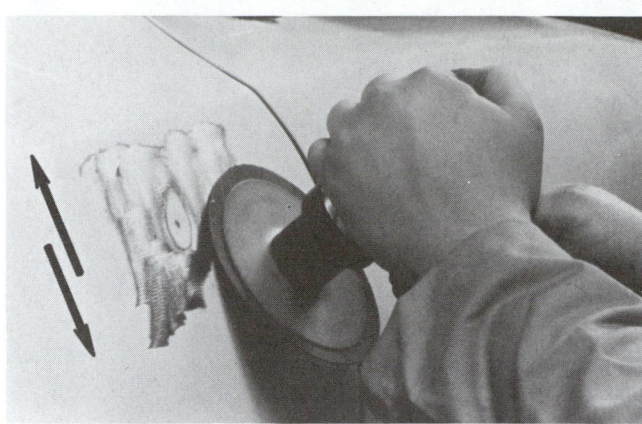

FIGURE 17–32 When using a disc grinder, be sure only the leading edge does the cutting.

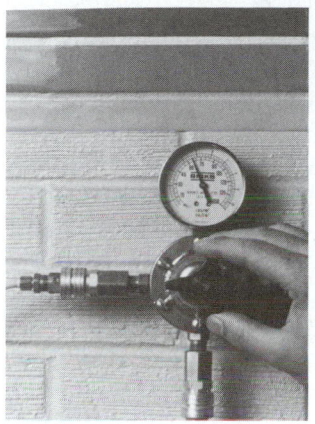

A

B

FIGURE 17–33 (A) Set the correct air pressure so tool runs at the appropriate speed. (B) On flat surfaces, do not tilt the sander or you will "dig a hole" in the surface. Place one hand on the handle and the other over the top of the sander to keep it square with the surface. *(Courtesy of Maaco Enterprises, Inc.)*

FIGURE 17–34 Avoid ornamental and chrome items. They can be damaged and will tear paper.

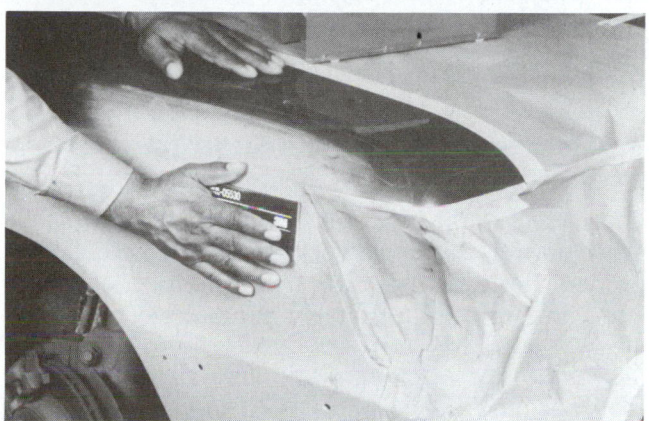

FIGURE 17–35 Masking before sanding will protect surfaces not to be painted.

FIGURE 17–36 Note how lower trim has been double masked for protection while grinding.

discussed, orbital sanders should be pressed flat so they will not leave surface scratches.

To operate an air sander, set the air pressure at the equipment manufacturer's specifications (typically about 70 psi or 476 kPa). If right-handed, hold the handle of the sander in the right hand, while using the left hand to apply light pressure and guide the tool (Figure 17–33).

To protect the chrome from damage, do not sand too close to the trim and moldings (Figure 17–34). Mask nearby trim, decals, glass, handles, and emblems (Figure 17–35) to prevent metal sparks from pitting these surfaces. In fact, it is a good idea to either double-tape (Figure 17–36) or remove all moldings and trim on the panel before sanding.

When using any mechanical sander—and particularly a disc grinder—keep it moving so that no deep scratches, gouges, or burn-throughs develop. And do not, except when sanding bare metal, power sand styling lines as this will quickly distort the styling edge.

When power sanding, replace the sandpaper when paint begins to cake or ball up (Figure 17–37). This paint buildup can scratch the surface and reduce the sanding action of the disc. Slowing down the speed of the sander will also help prevent paint buildup on the sanding disc and prolong sandpaper

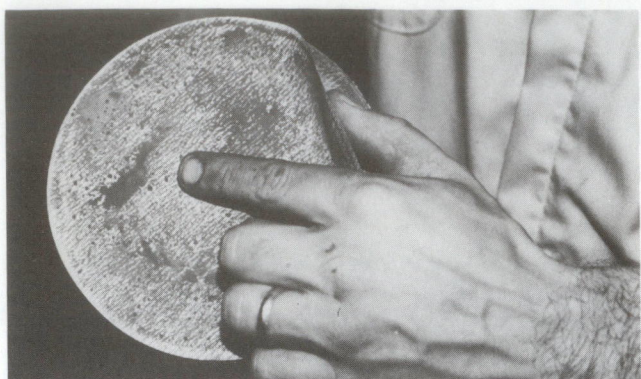

FIGURE 17-37 If paint cakes on paper excessively, go to a coarse paper before sanding with finer paper. *(Courtesy of Maaco Enterprises, Inc.)*

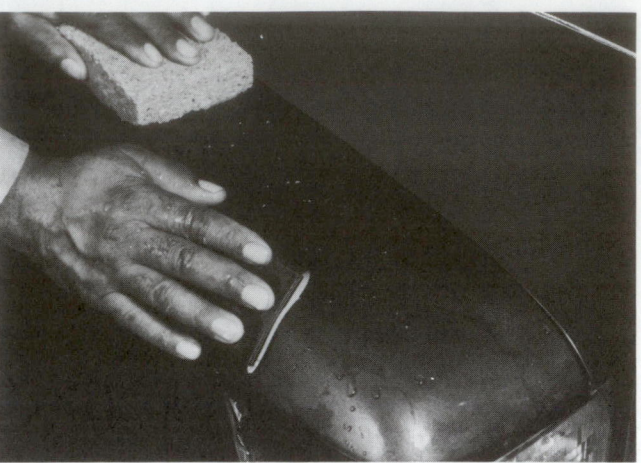

FIGURE 17-38 Lightly wet sand all surfaces to be refinished when the old finish is in good condition.

life. Generally, 6 to 8 sanding discs or pads will be required to featheredge the chips and scratches on the average automobile.

17.5 TYPES OF SANDING

There are several types of sanding that a refinisher must master. Some can be completed with power sanders alone, others with a combination of power and hand, and still others by hand alone.

BARE METAL SANDING

If the metal work has been done properly, little sanding of bare metal should be required. But once in a while the metal arrives in a very rough condition from coarse sanding in the metal shop. In such cases, it might be necessary to sand it with #50 grit to level out the burrs, nibs, and deep scratches. Remember that the smoother the bare metal, the easier the repair work will be.

THOROUGH SANDING

Use a thorough sanding procedure for three specific conditions:

- Where the old finish is rough or in poor shape
- To level and smooth primed areas
- To reduce paint mil thickness before refinishing.

Since the primer-surfacer is primarily intended to fill low spots and scratches, sanding must be done in a manner that will leave material in the low spots and cut away high spots. Block sanding is highly recommended for this purpose.

A **guide coat** is very helpful and assists in pointing out unlevel surfaces areas. Spray a very light coat of a different color sandable primer over the primer-surfacer. Block sand the area. If you cut through the

high spots and see unsanded low spots, more body work is needed to level the surface. If you sand off the guide coat without cutting through the primer-surfacer, the surface is ready for paint.

The sanding itself can be done mechanically or by hand with a sanding block. For the average hand wet sanding job, use #360 or #400 grit when applying the topcoat. Use #500 to #600 for basecoat/clearcoat systems.

LIGHT SANDING

Light sanding should be done on all areas where the old finish is in good condition. The purpose is to eliminate the gloss and to improve adhesion. Use an orbital or dual-action sander, or do it by hand (Figure 17–38), but NEVER use a disc grinder or sander.

If the new topcoat will be lacquer or enamel, use a #360 or #400 sandpaper. If it will be alkyd enamel, use a coarser #320 sandpaper. For basecoat/clearcoat finishes, proper surface preparation is critical. It is important to sand all surfaces to be refinished with #400 grit or finer paper. Sanding can be wet or dry.

COLOR SANDING

To achieve the smoothest finish and best results in acrylic lacquer work, wet sand (Figure 17–39) the next-to-the-last coat of color with #600 or #800 grit paper.

FEATHEREDGING

If a new coat of paint were to be applied right over the broken areas of the old finish, the broken film would be very noticeable through the topcoat (Figure 17–40). So the broken areas must be featheredged. That is, the sharp edge of the broken film must be tapered down by sanding (Figure 17–41). Then the bare metal

FIGURE 17-39 Use plenty of water when color sanding.

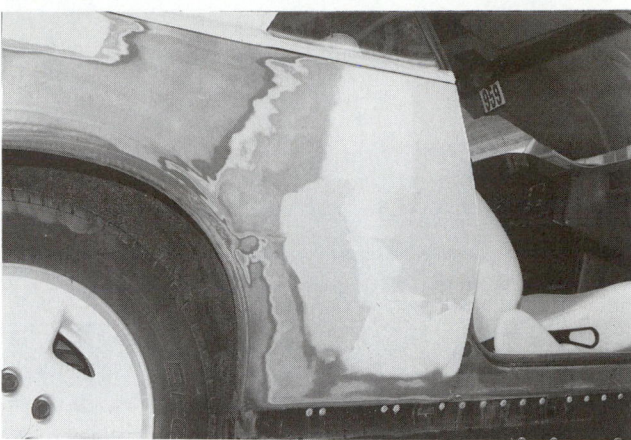

FIGURE 17-41 Featheredging tapers successive coats of paint and primer away from the metal to create a smooth surface.

FIGURE 17-40 A squeegee helps show the featheredge areas. A lip will be visible if more sanding is needed.

areas are filled with a primer-surfacer and the entire area is sanded smooth and level.

Featheredging by hand is usually a two-step procedure:

1. Cut down the edges of the broken areas with a coarse #80 to #220 sandpaper. Start with coarse paper. Use progressively finer paper as the area becomes level.
2. Complete the taper of the featheredge by hand with a sanding block and either a #360 or #400 grit sandpaper and water. This will produce a finely tapered edge and eliminate any coarse sandpaper scratches.

When featheredging with a power sander, an orbital or dual-action equipped with a flexible backing pad is recommended. Use a #80 grit for the rough cut, followed by a #220 or #400 sandpaper for the fine work. When featheredging a chip, start by positioning the sanding disc at a 5- to 10-degree angle from the work surface (Figure 17–42). Using the outer

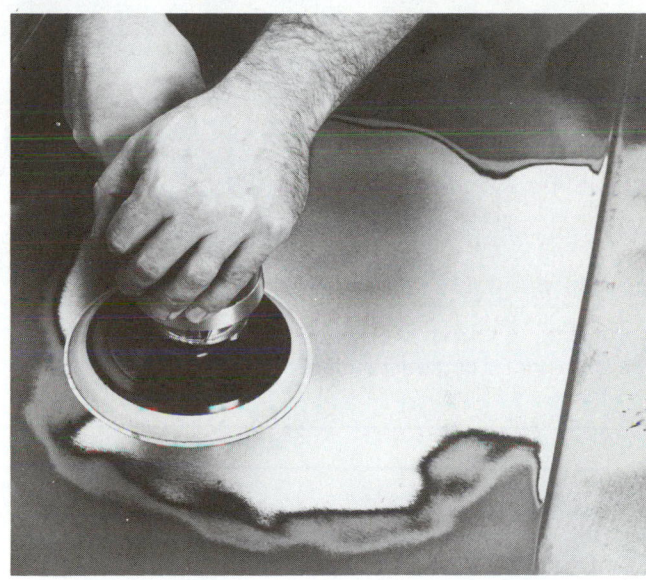

FIGURE 17-42 Featheredging requires correct coarseness of sandpaper. Too fine a paper will not cut fast enough on different materials and a lump will form on harder material. *(Courtesy of Maaco Enterprises, Inc.)*

edge, approximately 1 inch of the sanding disc, cut away the rough paint edges. Do not hold the sander at an angle greater than 10 degrees from the surface. Doing so will cut a deep gouge in the paint.

After initially leveling the rough paint edges, flatten the sander on the panel and finish tapering the paint layers by moving the sander back and forth in a crosscutting pattern (Figure 17–43). Start over the chipped area and work in an outward direction. Stop frequently and run a hand over the sanded area to feel for rough edges (Figure 17–44). When the surface feels smooth, and rings of old paint and primer color are visible, the featheredging is complete (Figure 17–45).

Certain localized peeling paint problems can be corrected using the featheredging technique.

FIGURE 17-43 Move the sander over the surface to cut the surface level. To featheredge even a small area, a much larger area must be sanded.

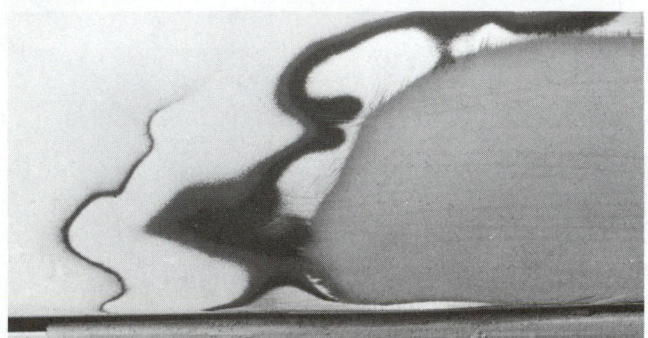

FIGURE 17-44 After being featheredged, the surface is ready for primer or primer-surfacer which will fill small sand scratches.

Slowing down the sander's speed helps produce a smooth edge on brittle paint. However, if the paint is extremely brittle, it will continue to chip away as sanding progresses. When this happens, move the sander several inches beyond the edge of the peeling paint and feather an edge in the undamaged finish. Once the layers of paint film have been

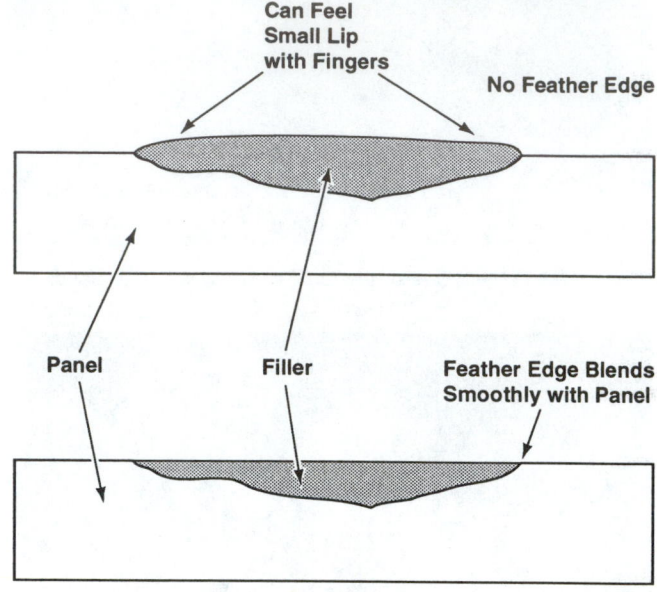

FIGURE 17-45 Featheredging if done properly will result in a level surface that blends outward without bumps or dips in surface. *(Courtesy of Maaco Enterprises, Inc.)*

	Body Filler	**Polyester Putty**	**Lacquer Putty**
Primary Use	Used to smooth out large depressions and fill in scratches.	Used to fill holes in body filler and sandpaper scratches in the metal.	Used to cover pinholes and small scratches after application of primer-surfacer, and to fill in small scratches in the old paint film.
Maximum Film Thickness per Application	Below 1/4" 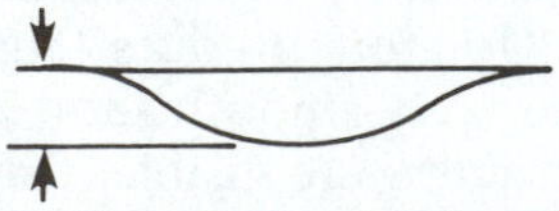	Below 1/8" 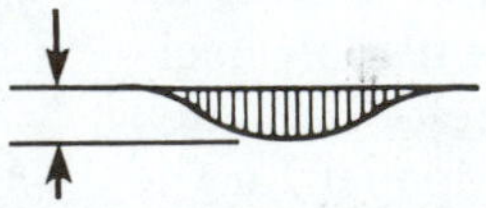	Below 1/16" 

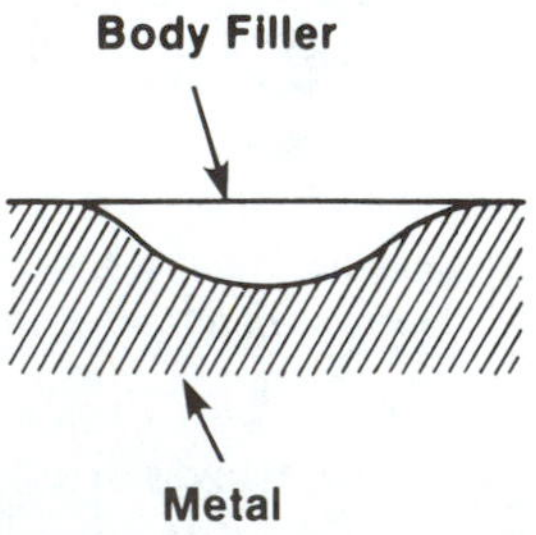

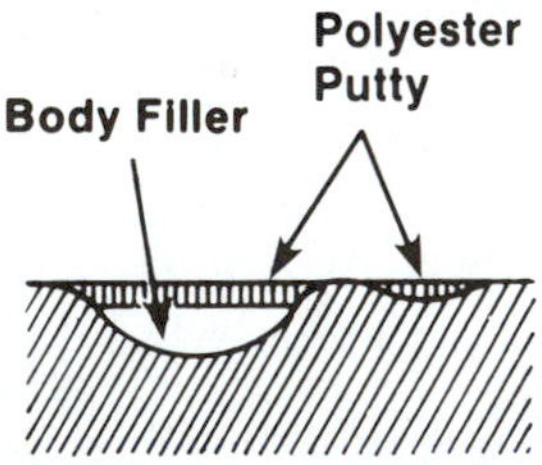

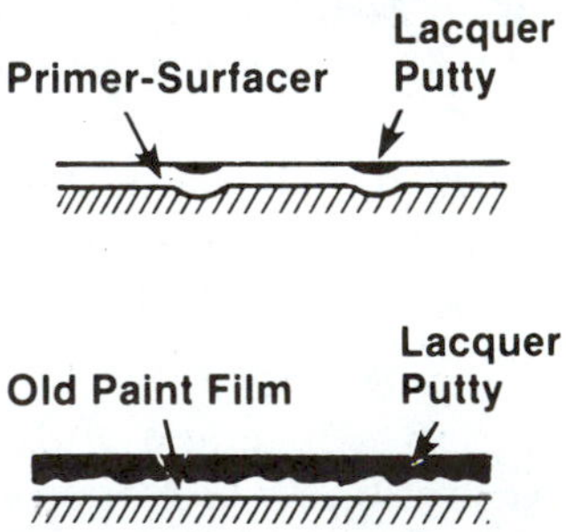

FIGURE 17-46 Compare use of filler and putty. Materials must not be applied beyond recommended thickness or problems will result.

successfully tapered, remove the damaged paint between the feathered edge and the original bare metal area using a buffing action.

When featheredging and sanding, make sure you use proper methods when applying filler and putty. Figure 17–46 shows typical recommendations for each. Spot putty is normally applied over the primer or primer-surfacer (Figure 17–47).

If the successive layers of paint are not properly tapered, a depression called a **"bullseye"** will show up under the new paint finish (Figure 17–48A). This condition can usually be corrected by extending each paint and primer ring farther from the bare metal. Do this until the depression can no longer be felt when a hand is run over the featheredged

area. Occasionally, when featheredging areas with several layers of paint, primer and putty might be necessary to fill the bullseye to the level of the existing film buildup (Figure 17–48B).

USING THE CORRECT GRIT

When sanding, make sure you are using the correct grit number for the job (Figure 17–49).

Very coarse grit of 16 to 24 is generally used for fast material removal. It will quickly remove paint and take it down to bare metal. This grit is commonly used on grinding discs and air files for rapid cutting.

A *coarse grit* of 36 to 60 is basically used for rough sanding and smoothing operations. This coarseness might be used to get the general shape of a large plastic filler area.

Medium grit of 80 to 120 is often used for sanding plastic filler high spots and for sanding off old paint.

Fine grit of 150 to 180 is normally used to sand bare metal and for smoothing existing painted surfaces. This is also used for final sanding of plastic filler and to featheredge paint.

Very fine grit ranges from 220 to about 2000 and is used for numerous final smoothing operations. Larger grits of 220 to 360 are for sanding primer-surfacers and old paint. Finer grits of 400 to 2000 are for colorcoat sanding, and sanding before polishing or buffing. Very fine grits are usually wet sandpaper to keep the paper from becoming clogged or filled with paint.

Generally, start your work with the coarsest grit appropriate for the task as described above. This will remove and smooth the work quickly. Then gradually go to finer paper to achieve the desired surface smoothness. This will be detailed in later chapters.

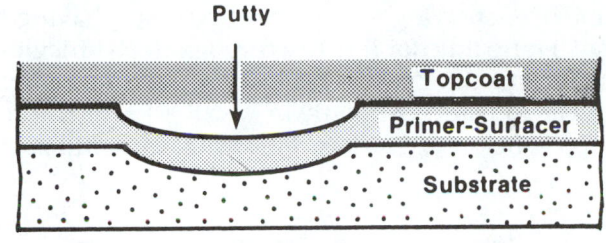

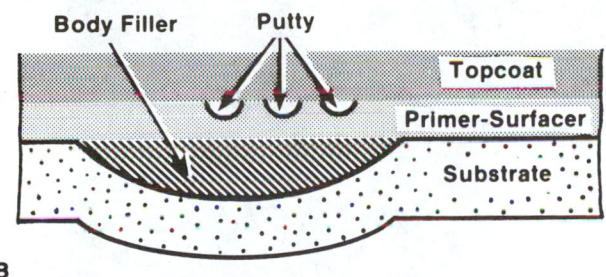

FIGURE 17-47 Remove brittle or damaged paint and featheredge to good areas of paint.

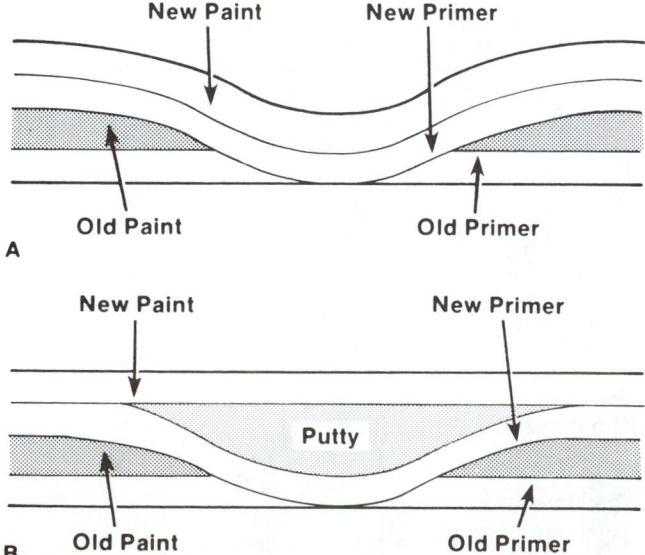

FIGURE 17-48 Note causes of "bullseye" and how to correct it.

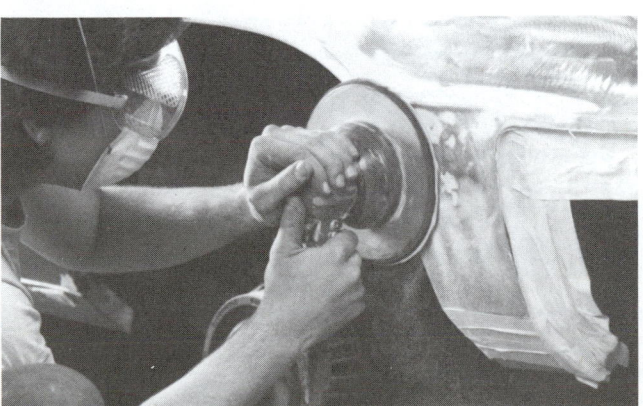

FIGURE 17-49 When grinding or sanding, use the grit number that corresponds to the amount of material that must be removed. If you use too coarse a grit for the job, it will require more work to remove deep scratches. Too fine a grit will clog, work slowly, and may not featheredge properly. *(Courtesy of DuPont Automotive Products)*

SCUFFING

Once all surface reconditioning is completed, the final sanding operation is to scuff the surface to remove nibs and dust specks on nonsanding primers, sealers, or where dirt shows up. It should be done with a very fine grit sandpaper such as #400.

Apply even, moderate pressure along the length of the sandpaper. Sand back and forth with long, straight strokes. Remember that scuffing the surface is only to improve adhesion of the new paint. Care should be taken not to cut into the film. When sanding large panels, one stroke back and forth in an overlapping pattern will be sufficient. More sanding than this is not only a waste of energy, but could possibly risk creating scratches that could show up under the new finish. Do not oversand.

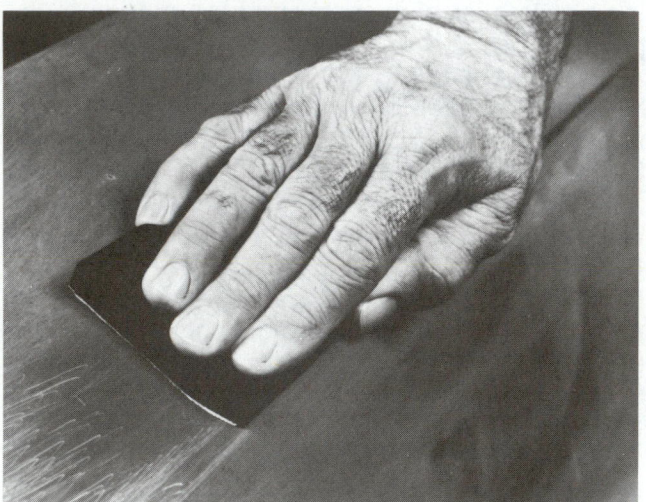

FIGURE 17-50 Any sanding operation is going to produce some scratches. The idea is too keep them very small. #800 grit or finer grit paper will keep them small enough that paint can fill them without being noticeable.

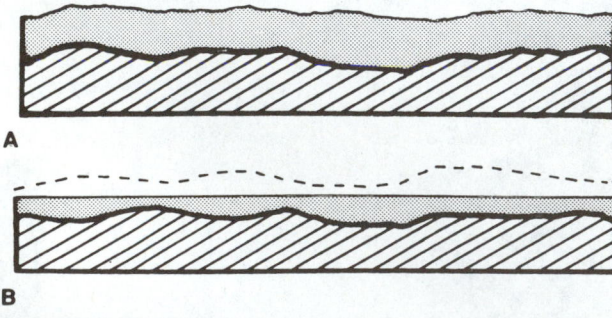

FIGURE 17-51 Study this enlarged cross-section of rough metal. (A) Some metal coated with primer-surfacer. The surface of the undercoat follows the approximate contours of the original metal when thoroughly dry. (B) Sanding levels off high spots, producing a flat, smooth surface. If the undercoat is not dry when sanded, further shrinking over deep fills produces an uneven surface and shows abrasive marks.

SAND SCRATCHES

The body technician can make it doubly hard for the painter if the metal is not properly finished (Figure 17–50). Because there are often little burrs or fins on the crests of the scratches, there can be uneven shrinkage in the surface coat (Figure 17–51A).

To eliminate sand scratches in metal, first sand with #220 or finer paper to round off the tops of these crests (Figure 17–51B). Do not worry about getting the metal too smooth. Sanded metal that looks and feels smooth will still have plenty of "tooth" for the surfacer (Figure 17–52).

Modern primer-surfacers will do a lot of filling in one coat (see Tables 17–4 and 17–5). The thicker the coat, the slower the drying, so spray two or three coats, allowing 5 to 15 minutes between them. This will save time over spraying a very heavy coat and having to wait a long time for it to dry through. It is difficult to tell when a thick coat is really dry. The surface might appear to be dry while there is still a lot of thinner trapped below the surface and shrinkage is still going

FIGURE 17-52 Note sand scratches in primer-surfacers enlarged 40 times. *(Courtesy of PPG Industries, Inc.)*

TABLE 17-4: FUNCTIONS OF UNDERCOATS				
Undercoat Function	**Primer**	**Primer-Surfacer**	**Primer-Sealer**	**Sealer**
Resists rust and corrosion	Yes	Yes	Yes	No
Makes topcoat adhere better	Yes	Yes	Yes	Yes
Fills scratches and nicks	No	Yes	No	No
Provides uniform hold out of the topcoat	No	No	Yes	Yes
Prevents show through of sand scratches	No	No	Yes	Yes

TABLE 17-5: SURFACES FOR UNDERCOATS

Undercoat Surface	Primer	Primer-Surfacer	Primer-Sealer	Sealer
Bare substrate (metal, fiber-glass, or plastic)	Yes	Yes	Yes	No
Sanded old finish	No	Yes	Yes	Yes

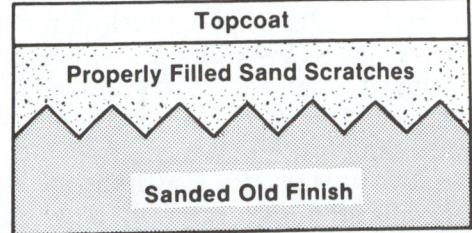

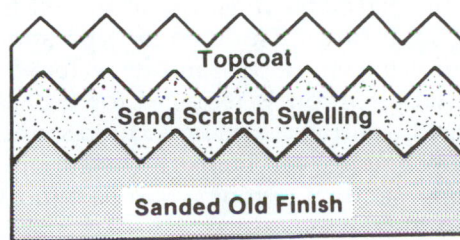

FIGURE 17-53 The best way to prevent sand scratches from showing is to use the proper grit sandpaper and proper sanding technique. In hand sanding always sand in a straight line, never circular. *(Courtesy of DuPont Automotive Products)*

on. Where the imperfections or scratches in the metal are unusually deep, the use of a glazing putty will save time in getting a smooth surface.

After the primer-surfacer has dried thoroughly, the next thing to consider is the sanding operation. The use of coarse sandpaper such as #220 or #240 will produce scratches in the primer-surfacer that will be hard to fill with the final finish coats. With the present-day surfacers, sanding is so easy it is not necessary to use paper coarser than #320 or #360. In order to get the smoothest finish, the use of #400 paper as a final sanding is recommended.

Where the undercoat is the heaviest, as in the deep scratches, the swelling will be the greatest. If the color is compounded and polished before all of the thinner has evaporated from the primer-surfacer, there will be further shrinkage at the point of deepest fill. Therefore, it is important to give finish coats plenty of drying time before sanding and polishing.

The danger of sandscratch swelling is greatest on the featheredge. The spraying of a light fog coat for the first colorcoat keeps the solvent content on the low side when it first comes in contact with the old featheredge finish.

It can be seen from Figure 17–53 that the shrinkage and swelling of undercoats is an important point to consider in the elimination of sand scratches. If the undercoat is not allowed to dry down to its final position before sanding or applying finish coats, scratches are likely to result.

17.6 REFINISHING SURFACES

It is unwise to apply any kind of finish to a surface that has not been prepared properly. Quality suffers; customer dissatisfaction is inevitable; and costs increase because the job usually has to be done over. A good beginning pays off in a savings of materials, time, and in a higher quality refinishing job.

Even if the original paint finish is in good condition, it should be thoroughly sanded or scuffed after washing to remove dead film and to smooth out imperfections. Sanding also reduces the existing paint thickness to avoid too much paint mil thickness after refinishing. If the surface is in poor condition, all the paint should be removed down to the bare metal. In this way, a good foundation is achieved.

17.7 PAINTED SURFACE IN GOOD CONDITION

It is simple to repaint over an existing paint film in good condition, providing it is stable and does not react to the solvent of the refinish paint. The procedure for surface preparation in good condition is as follows:

CLEANING THE VEHICLE

The vehicle should be washed to remove any mud, dirt, or other water-soluble contaminants before being brought into the shop (Figure 17–54). Hose down the car, sponge with detergent and water, then rinse thoroughly. Wash the top, front, deck, and sides; allow the vehicle to dry.

CLEANING WITH WAX AND GREASE REMOVER

Be sure there is no wax, grease, or other contaminants imbedded in the old finish.

Before the job is sanded, use a specially blended wax and grease remover or recommended solvent to thoroughly clean the surface. Be sure to thoroughly

678 Chapter 17

FIGURE 17-54 Wash the vehicle surface very carefully. The smallest contamination can ruin the paint job. *(Courtesy of Maaco Enterprises, Inc.)*

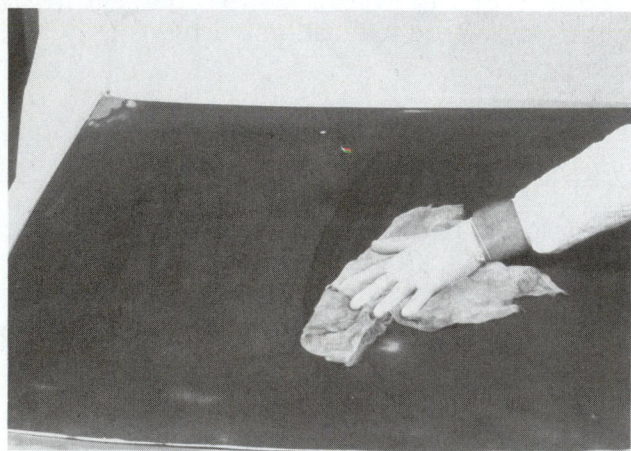

FIGURE 17-55 Applying wax and grease remover to all surfaces to be painted.

clean areas where a heavy wax buildup can be a problem, such as around trim, moldings, door handles, radio antennae, and behind the bumpers. Paint will not adhere properly to a waxy surface.

To apply the wax and grease remover (or silicone and wax remover as it is sometimes called) place the cleaner in a hand-pump type spray bottle. While wearing plastic gloves, spray the cleaner onto the body surface (Figure 17–55). While the surface is still wet, fold a clean cloth and wipe it dry. This will remove any impurities and deposit them on the clean rag or towel.

Work small areas (about three square feet or meters), wetting the surface liberally. Never attempt

When wiping body panels down, use a lint-free rag or towel. Special lint-free cleaning towels are available and should be used. They come in box dispensers or rolls. Avoid using rags with frayed edges that will deposit lint on the vehicle surface. When painting, "lint can be a nightmare!"

to clean too large an area. The solvent will dry before the surface can be wiped. Maximum effectiveness will be achieved by wiping the wax and grease remover while it is still wet. Always use new wiping cloths because laundering might not remove all oil or silicone residue.

To remove any last trace of moisture and dirt from seals and moldings, carefully blow out with compressed air at low pressure. Wax and silicone can penetrate beneath the surface. This contamination is not easily detectable. It is wise to assume that it is present, so always include some wax and grease cleaner or detergent in the sanding water.

Special attention is required for tar, gasoline, battery acid, antifreeze, and brake fluid stains. These can also penetrate well beneath the surface of old paint films and their residues must be removed during the sanding operation.

It is a common mistake to use synthetic reducers for cleaning vehicle surfaces because reducers can absorb into the paint film. Blistering or lifting can result from improper cleanup. Use only a recommended wax and grease remover.

REPAIRING FLAWS IN PAINTED SURFACES

Sand or grind off the rust and old paint in the damaged areas (Figure 17–56). If the grinding operation goes down to bare metal, it will be necessary to perform the appropriate metal conditioning. These steps are described later in this chapter.

Be sure that the repair of all dings, dents, and built-up areas have been made as described in Chapter 9. Many dings and dents are too deep to be filled by a primer-surfacer and/or putty. In such cases use a lightweight body filler. But when using a body filler, there are some precautions that should be kept in mind:

- *Do not* use body filler directly over a metal conditioner.

- *Do not* use too much hardener because it will cause pinholes.
- *Do not* return any unused mixture to the can.

The following is a review of the procedure for using body filler. For small dents, squeeze a 1¹/₂-inch (38 mm) ribbon of hardener on a mass of body filler about the size of a golf ball. Mix well with a putty knife or paint paddle (Figure 17–57). Mix only as much as can be handled properly because the mixture will harden. Apply immediately with a spreader or squeegee. Work to the contour of the surface. Then let it harden for 8 to 10 minutes. Use a grater to shape the repair, bridging the dent to prevent gouging. Sand as needed.

NOTE: For more information on body fillers types and use, refer to the textbook index. Several other chapters cover this subject.

If using a power sander, such as an orbital sander, use a #80 sandpaper for the rough cut followed by a #180 or #220 grit for the fine work. Next, feather the broken paint edges. This must be done so that a continuous smooth surface can be developed when filled with a primer-surfacer. Feather-edging can be done by hand with a sanding block or with an orbital sander.

Taper the broken edges first. If hand sanding, use a #220 grit sandpaper for the rough work. Then complete the job with a #240 or #320 paper to produce a fine-tapered edge and eliminate coarse sandpaper scratches.

To remove decals from a painted surface, a razor blade slipped under the edge of a decal will start a small area that can be pulled up and the whole decal peeled off. If the decal will not peel off, there are several other ways of removing it. One of these is disc grinding. Various companies sell rubber discs that attach to drills or disc grinders. These are designed to remove decals and plastic stripes without harming the

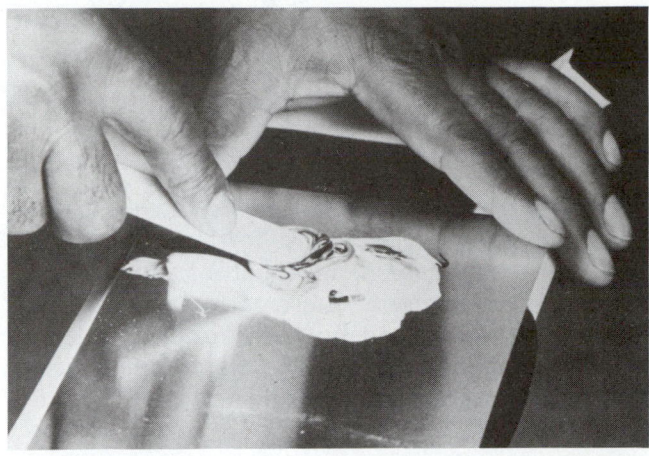

A

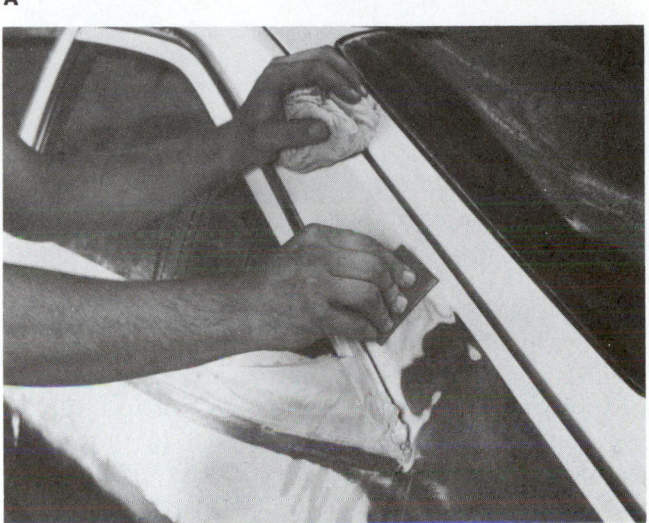

B

FIGURE 17–57 (A) Mix body filler and hardener in correct proportions. (B) Then apply to body to fill any small dents.

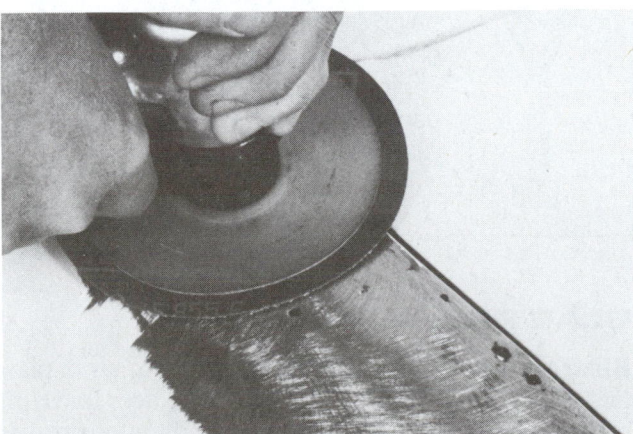

FIGURE 17–56 Grinding may discover hidden rust under old paint. Metal patching or part replacement would be needed. *(Courtesy of Maaco Enterprises, Inc.)*

finish. Another way to remove decals is with a heat gun. If one is available, use it to heat the decal and the surrounding surface to soften the adhesive, and then peel the decal off. There are also chemical decal removers on the market; however, these must be used with care since they may damage the surface on which the decal is applied.

17.8 PAINT WORK IN POOR CONDITION

Most forms of paint failure are progressive. These conditions cannot be stopped by any form of repairing. In fact, repairing will usually accelerate the deterioration of the original finish. If the old finish is badly weathered or scarred, it is not suitable for recoating. When this situation occurs, the old finish should be completely removed. This can be a labor-intensive and time-consuming process.

There are three common ways of stripping paint from metal surfaces:

- Sanding or grinding
- Media blasting
- Chemical stripping

With any of these methods, remove all trim, lamp surrounds, badges, and so forth attached to the area to be repaired. A chemical paint remover can be trapped and retained by these parts, or they can be accidentally damaged by the sander/grinder or blaster. In any case, corrosion is often found beneath exterior trim parts. This can only be dealt with if they are removed.

PAINT THICKNESS MEASUREMENT

Paint thickness is measured in **mils** or thousandths of an inch (hundredths of a millimeter). Original OEM paints are typically about 2 to 6 mils thick. With basecoat/clearcoats, the basecoat is approximately 1 to 2 mils thick. The clearcoat is about 2 to 4 mils thick. This is approximately the thickness of a piece of typing paper.

If a panel has been repainted, paint thickness will increase. If too much paint is already on the vehicle, it may have to be removed prior to refinishing. Paint buildup should be limited to no more than 12 mils. The OEM finish and one refinish usually equal just under 12 mils. Exceeding this paint thickness could cause cracking in the new finish. Chemical stripping, blasting, or sanding would be needed to remove the old paint buildup.

A **mil gauge** can be used to measure the thickness of the paint on the vehicle (Figure 17–58). This can be done before refinishing, after refinishing, and during other finishing operations.

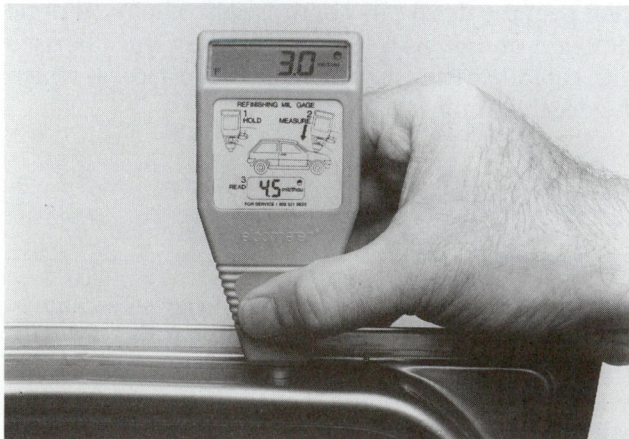

FIGURE 17-58 A mil gauge will measure the thickness of the old paint. If too thick, the paint must be removed or stripped before repainting. Thick paint will crack.

The *pencil mil gauge* paint measuring tools are spring-loaded and have markings for mil thickness. They have a magnet in them that is placed against the metal body or paint. It is pulled away from the body and a reading is taken.

The *electronic mil gauge* is similar but has a digital readout showing paint thickness. When placed and held on the surface, the tool will automatically register paint thickness. A scale on either tool will read in mils.

SANDING OR GRINDING OFF PAINT

Machine sanding or grinding is suitable for removing old finish from small flat areas and gently curved areas. Start with a #24 grit open-coated disc, and by holding the face of the disc at a slight angle to the surface, work forward and backward evenly over the area to get the bulk of the old finish down to the metal. Follow this with a #50 or #80 close-coated disc; go over the entire area and slightly out on the surrounding surface to clean up the work and eliminate the troughs or steps caused by the coarse disc. When using the grinder, care must be taken to prevent gouging or scarring the metal.

After all of the paint is removed with the grinder and the coarse grit disc, resand the area with the orbital or dual-action sander and #100 grit paper to remove the metal scratches. Then finish sanding the panel using #180 grit sandpaper. In this way most of the scratches created by the stripping operation will be eliminated. Remember that any metal that has been scratched with very coarse abrasive paper will require filling to the depth of the scratch plus the height of the burr.

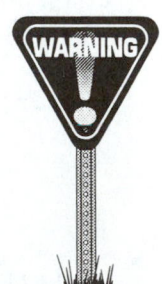

WARNING

Never use a grinder to remove paint from a very large surface area on a panel or from the whole vehicle. Heat from grinding will usually warp the panel and thin the metal too much.

BLASTING OFF PAINT

Blasters are air-powered tools for forcing sand, plastic beads, or another material onto surfaces for paint removal. For example, they are handy when trying to remove surface rust from body panels. They will blast out all of the rust without further thinning of the panel.

With today's thin-gauge, high-strength steel, blasting is often recommended over grinding to clean rust pockets. Grinding thins the metal and makes it weaker.

Blasting can be done with caution on nearly all types of body construction—even aluminum sheet. It leaves a clean, dry surface in an ideal condition for refinishing. It is a very fast method and has the further advantage of revealing rusted areas and places where hidden rusting can result in scaling after the job has been refinished. In addition, blasting makes hard-to-reach areas accessible to the technician. Also this method saves time when compared with sanding or grinding and chemical stripping.

A blaster concentrates the pressure and flow of air and sand. The technician can vary the blast volume, focusing the pattern on the spot at hand, rather than blasting in a wide pattern.

Blasters in the shop are one of two kinds: pressure or siphon. *Pressure blasters* are pressurized containers filled with abrasive material, such as silica sand or plastic beads. The material travels down one hose; the high velocity air comes down another hose. Both meet at a third hose and travel out toward the surface together at tremendous speed and force.

In a *siphon blaster*, compressed air draws the abrasive from the reservoir by producing a suction or vacuum at the blaster head. The abrasive accelerates and is shot out of the nozzle at the intended surface. Small bottle blasters are available for spot-type jobs (Figure 17–59).

The basic procedure in operating blasters is as follows:

1. Mask off the areas that will not be affected by the spot repair. For example, when spot-repairing a quarter panel, mask the wheel covers, glass, trim, and the top of the vehicle. Thick duct tape should be used for added protection where needed.

2. Put on the necessary safety gear. Wear gloves, eye protection, a helmet, and a respirator (Figure 17–60). A respirator must be worn because dust can build up in the lungs over an extended period of time.

3. Before blasting, check the manufacturer's instructions for proper blasting pressures, sand load procedures, and setup arrangements. When ready to blast, apply the abrasive material directly on the area to be blasted. Eventually, the area will turn a gray or white color. Blasting has textured the surface by opening the pores of the metal in these colored areas. This etched texture makes an excellent surface for primer adhesion. When the area shows no signs of brown rust, remove the pressure.

   Pressure should be applied by holding the nozzle the recommended distance from the area being repaired. It should hit the surface at a 20- to 30-degree angle. That way, the media will fly sideways and not back at you.

4. Watch the surface carefully. The blasting might reveal a rust hole. If you find major corrosion, blast out as much of the hole as possible. Blasting is designed to reveal weak spots like these. Before priming the rusted out area, cut out all corrosion-thinned metal and weld a metal patch on it.

5. After the paint has been removed, use an air blowgun (Figure 17–61) to remove the sand from all areas of the vehicle. If it is not removed, it could get into the paint. The abrasive could also get stuck in windshield wiper blades or window slots and scratch the windows.

FIGURE 17-59 Typical bottle type blaster is good for removing paint from small areas. *(Courtesy of A.L.C. Co.)*

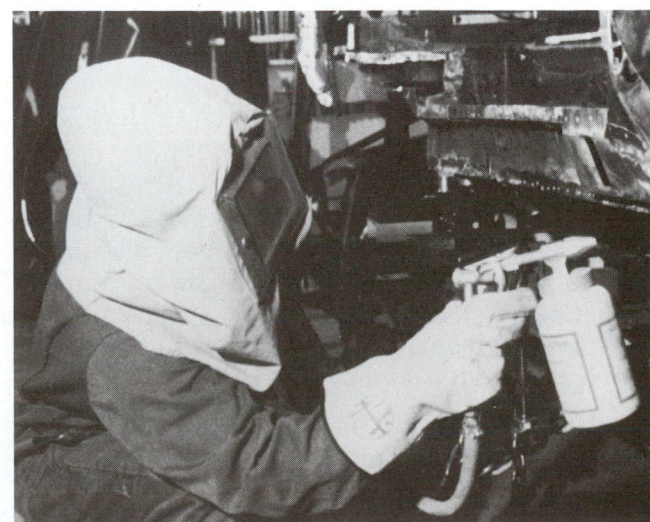

FIGURE 17-60 Wear safety gear to protect yourself from flying debris.

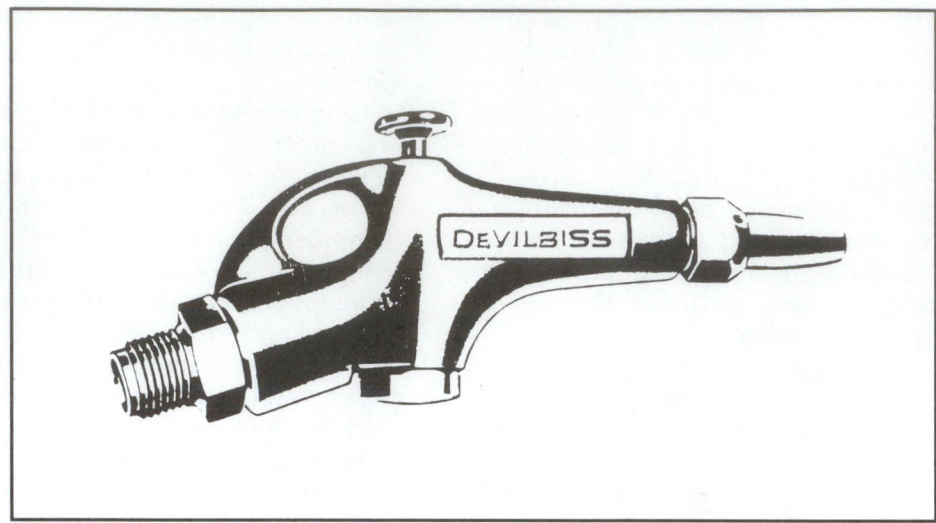

FIGURE 17-61 Typical blowgun. *(Courtesy of ITW DeVilbiss Automotive Refinishing Products)*

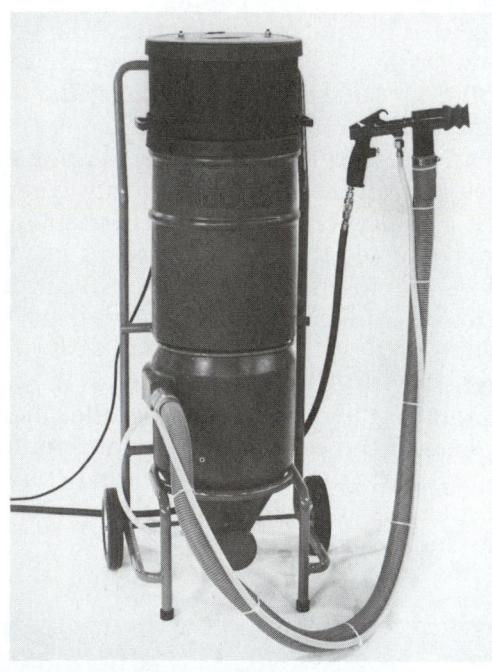

FIGURE 17-62 This is a typical dust-free "captive" sand blaster. *(Courtesy of Clements National Co.)*

FIGURE 17-63 Blasting with a "captive" sand blaster will keep debris from flying into the shop. *(Courtesy of Clements National Co.)*

6. It is advisable to prime coat the metal as soon as possible after any stripping process. Blasting requires that the job be primed almost immediately because the metal is in a raw state after this treatment. The bare metal will start rusting if allowed to stand overnight.

Until recently, all blasting work was done outdoors. Some of the newer blaster models provide dust-free, captive blasting (Figure 17–62). They contain a built-in vacuum and filtration system that cleans up and recycles the abrasive, while it separates paint, rust, and other debris.

To operate this newer type of blaster, hold the designed nozzle directly against the surface being treated (Figure 17–63). As the abrasive strikes the surface, it is sucked back by the vacuum, along with rust and debris. A rubber nozzle and stiff brush seal in the abrasive and debris to keep them from escaping. The rust and debris fall into an easy-to-empty pail, while the blasting abrasive is recycled and sent back into action.

CHEMICAL STRIPPING

A chemical paint remover can also be used for stripping large areas of paint if environmental regulations allow. It is an alternate method to blasting in those places that a power sander cannot reach. One advantage of chemical stripping is that there is no danger of the metal warping.

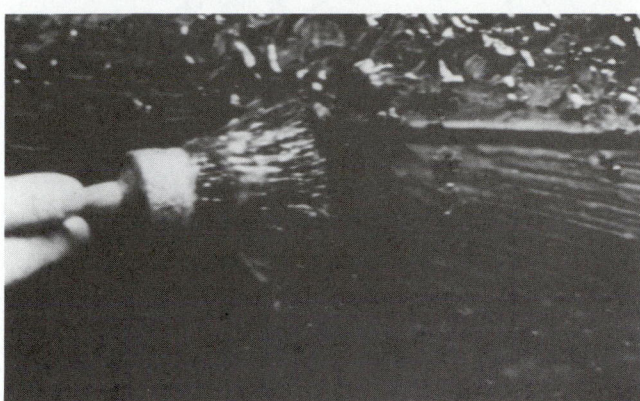

FIGURE 17-64 You can apply chemical paint remover with a brush. Put it on thick and let it soak into the old finish. *(Courtesy of America Sikkens, Inc.)*

Before applying paint remover, mask off the area to ensure that the remover does not get on any area that is not to be stripped. Use two or three thicknesses of masking tape to give adequate protection. Cover any crevices to prevent the paint remover from seeping to the undersurface of a panel. Slightly scoring the surface of the paint to be stripped will help the paint remover to penetrate more quickly.

Paint remover should be applied following the manufacturer's instructions. Pay attention to warnings regarding ventilation, smoking, and the use of protective clothing such as PVC or rubber gloves, long-sleeved shirts, and safety glasses or goggles. If remover comes in contact with the skin or eyes, it will cause irritation and burning.

To apply, brush on a heavy coat of paint remover in one direction only to the area being treated (Figure 17–64). Use a soft bristle brush, but do not brush the material out. Allow the paint remover to stand until the finish is softened.

Although paint remover is effective on most vehicle topcoats, some modern car undercoats can prove stubborn. If the finish resists the remover, more than one application might be needed.

Caution should be taken when removing the loosened paint coatings. Some paint removers are designed to be neutralized by water. Remove the dissolved paint with a squeegee or scraper (Figure 17–65). Be sure to rinse off any residue that remains on the body using cleaning solvent and steel wool. Follow this by wiping with a clean rag. This rinsing operation is essential. Many paint removers contain wax, which, if left on the surface, will prevent the refinish paint from adhering, drying, and hardening properly.

Rusting occurs very rapidly on metal that has been chemically stripped. In fact, any bare metal substrate should be treated immediately. But before selecting the type of metal treatment or conditioning system, first consider the types of rust. The least amount of rust might be considered *microscopic rust* (Figure

FIGURE 17-65 Removing softened paint with a plastic scraper to prevent scratches in the metal. *(Courtesy of America Sikkens, Inc.)*

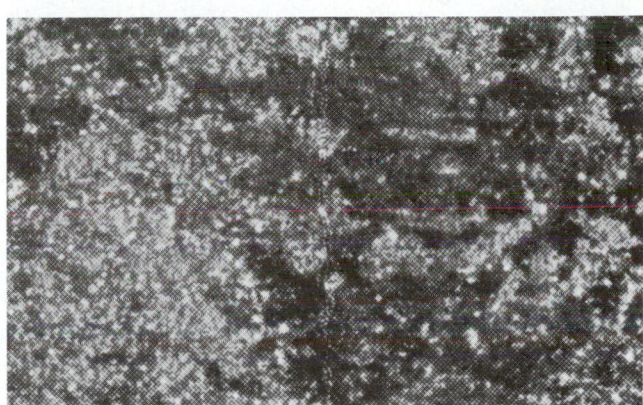

FIGURE 17-66 A rust-pitted sheet steel sanded clean to the naked eye. Under a microscope (enlarged 125 diameters) traces of rust still appear in the pits. This is why blasting is better to remove rust than sanding or grinding. *(Courtesy of PPG Industries, Inc.)*

17–66) that is not really visible to the eye but can be a hazard to the performance of a refinish job. The second type of rust might be called *flash rust* that usually develops when there is moisture or humidity present. The other types of rust are the types that are very visible and might even be large and scabby.

The decision about which metal-conditioning system to use depends on the type of rust and the type of substrate.

17.9 BARE METAL SUBSTRATE

Proper bare metal treatment is a critical step in every successful automotive painting operation. Yet it is often ignored or carried out in a haphazard manner. This can only result in poor adhesion, corrosion, and customer complaints. Though not noticeable when the original finish has been stripped down to bare metal, it is the single most important factor in original equipment finish life. Recognizing its importance, auto manufacturers devote more attention to this step than they do to priming and topcoating, using a seven-stage zinc phosphate metal treatment process to ensure adhesion of primers to the substrate.

Though the techniques and equipment used on the OEM level are not adaptable to body shops, some metal treatment products on the market today can enable you to simulate original equipment metal treatment.

Why is bare metal treatment so important? Water vapor penetrates all paint films. The fresher the paint and the more humid the weather, the further the vapor penetrates, sometimes reaching the bare metal. Once water droplets form under the paint film, pressure starts to build, causing bubbling, blistering, and loss of adhesion. Rust can also begin to form, further pushing the paint film away from the metal.

The only way to prevent this potential problem is to create such a strong bond between the primer or primer-surfacer and the metal that water vapor cannot penetrate down to the substrate. If the water vapor is not allowed to condense under the paint film, it will return to the surface of the finish and eventually evaporate.

This bond between a negatively charged metal car surface and a positively charged primer can be created electrochemically. All that is needed is the application of an acidic metal treatment system that contains both positive and negative parts. The negative parts are attracted to the metal, while the positive parts are attracted to the primer, forming a superior bond. This type of system is called a **conversion coating**.

A typical system consists of the following three-step process called metal treating:

- **Cleaning to remove contaminants.** Apply a wax and grease remover to the surface. While the surface is still wet, fold a second clean cloth and wipe dry. Work small areas, 2 to 3 square feet (0.6 to 1.8 square meters), wetting the surface liberally (Figure 17–67).
- **Cleaning with metal conditioner.** Mix the appropriate cleaner with water in a plastic bucket according to label instructions. Apply with a cloth, sponge, or spray bottle. If rust is present, work the surface with a stiff brush or abrasive plastic pad (Figure 17–68). Then while the surface is still wet, wipe it dry with a clean cloth.
- **Applying conversion coatings.** Pour the appropriate conversion coating into a plastic bucket (Figure 17–69). Using an abrasive pad, brush, or spray bottle, apply the coating to the metal surface. Then leave the conditioner on the

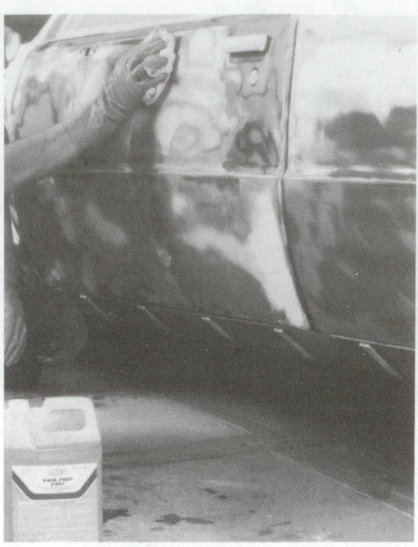

FIGURE 17–67 Wear rubber gloves when applying a wax and grease remover. *(Courtesy of DuPont Automotive Products)*

FIGURE 17–68 Apply metal conditioner so the primer will adhere to the metal properly. *(Courtesy of DuPont Automotive Products)*

FIGURE 17–69 Mixing a conversion coating according to label directions.

surface 2 to 5 minutes. Apply only to an area that can be coated and rinsed before the solution dries. If the surface dries before the rinsing, reapply. Flush the coating from the surface with cold water. Wipe dry with a clean cloth and allow to air dry completely. The desired primer or primer-surfacer can then be applied.

Another type of metal treatment system also can be applied to enhance adhesion and assure corrosion resistance. A wash-primer is a sprayable surface treatment that eliminates the need for a conversion coating. The vinyl resin in the wash primer provides corrosion resistance and the reducer contains phosphoric acid for strong bonding.

A wash-primer with phosphoric acid reducer not only cleans, it also etches the metal and assists the adhesion of the subsequent paint film. It helps prevent the occurrence of rust and also eases sanding marks. The wash-primer is applied by the following procedure:

1. Carefully read the manufacturer's directions and special instructions, which should be followed closely.
2. Pour the wash-primer into a container and add a special wash-primer thinner to achieve a sprayable viscosity. Do not use a metal container because the wash-primer reacts with metal. If applying a two-component wash-primer, the solution must be used within 8 hours after mixing the main and subagents.
3. Pour the mix solution or thinned solution into the spray bottle container and spray the metal immediately. Do not allow any wash-primer to get on any part of the vehicle that is not to be repainted. The gun air pressure and discharge pressure should be kept low and the gun held close to the surface. The area should be masked off to prevent other areas from coming into contact with the wash-primer.
4. Apply a thin coat. Too thick an application will result in paint peeling and blistering.
5. The wash-primer should not be allowed to dry on the metal. Should this happen a second application of the material will soften and dissolve the dried residues. After the wash-primer has been applied, wash well with plenty of clean water and dry thoroughly.
6. Wash out the spray gun immediately after spraying wash-primer. Wash-primer left in the spray gun or container will cause a chemical film to form on the metal, making the spray gun useless. Because of this, some manufacturers recommend that wash-primer be applied with an acid-resistant brush or sponge.

Typically, metal conditioning and priming/surfacing are considered separate surface preparation steps. Some new products, however, actually make it possible to combine these steps. Etching primer-fillers etch the bare metal to improve paint adhesion and corrosion resistance, while providing the priming and filling properties usually offered by primer-surfacers. Etching primer-fillers work best on lightly sanded surfaces where a slight-to-moderate amount of filling is required. They must be applied as directed by the manufacturer.

SPECIFIC METAL TREATMENTS

The preparation of the various metals used in automotive construction requires slightly different techniques. The more common bare metal procedures are as follows:

Steel-Body Metal (Including Blue Annealed) Preparation

1. Sand metal thoroughly. Remove all visible scale or rust.
2. Clean the surface with wax and grease remover and wipe dry.
3. Use any of the three bare metal treatments—conversion coating, wash-primer, or etching primer-filler—as previously described.
4. Apply an undercoat (primer or primer-surfacer). If the etching primer-filler is used, this step might not be necessary.
5. Once the undercoat refinish system is dry and sanded, wipe with a tack rag. The surface is ready for the colorcoat.

Galvanized or Other Zinc-coated Metal

1. Follow Steps 1 and 2 of steel-body preparation.
2. Use either a conversion coating or special zinc metal conditioner. Apply the latter according to manufacturer's directions. Never use a wash-primer since it will attack galvanized and other zinc surfaces and must not be allowed to come into contact with them.
3. Apply one wet double coat of epoxy primer. If filling is required, allow the epoxy primer to dry a minimum of 1 hour and then apply a primer-surfacer.
4. Sand the primer-surfacer after a 30-minute dry period. Once the undercoat system is completed (as mentioned later in this chapter), the surface is ready for the topcoat.

Anodized Aluminum or Untreated Aluminum and Oxidized Aluminum Preparation

1. Follow Steps 1, 2, and 3 of steel-body preparation.

2. Apply one wet double coat of epoxy primer or zinc chromate. If filling is required, allow the material a minimum of 1 hour to dry and then apply a primer-surfacer.

3. Sand the primer-surfacer after the recommended period of time for thorough drying. Once the undercoat system is completed, the surface is ready for the colorcoat.

Chromium Plating Preparation

Chromium is very hard. The primer and paint will not adhere to chrome very well. When painting is desired, prepare the surface by cleaning and sanding. Then proceed with the following system described for stainless steel preparation:

1. Clean the metal thoroughly with a wax and grease remover.

2. Sand metal thoroughly, using #320 wet or dry sandpaper.

3. Reclean with a wax and grease remover.

4. Apply any of the metal treatments described earlier in this chapter.

5. Spray two coats of primer-surfacer, allowing adequate drying time before dry sanding.

6. Blow out cracks; then use a tack rag on the entire surface. The final coat can now be applied.

Regardless of the cleaning procedure, once the metal is clean and prepared, it must not be contaminated by fingerprints, so clean cotton or plastic gloves should be worn when handling.

Sometimes painters rub their bare hands over an area to determine the effect of sanding without realizing that they are transferring oil from their hands to the surface. Oil comes from the skin and from shop tools. Even if the hands are freshly washed, a fine oily film will be left on the surface because there are not many people who have oil-free skin. Wiping off the surface with a good wax and grease remover, just before applying the finishing coat, is excellent insurance against peeling and/or blistering.

PREPARING METAL REPLACEMENT PARTS

Many car manufacturers and component suppliers protect panels with a primer. The function of this primer coat is to protect the metal against corrosion during storage and shipping. This coating does not usually provide a firm basis for a paint system.

If in doubt about the quality of the new part protective coating, check with the manufacturer of the part for the recommended finishing procedures. This will assure a quality, long lasting paint job.

Certain major auto manufacturers supply components in electro-coat primers that are an essential part of their warranty repair systems and should not be removed. They should be suitably prepared for the painting process.

It must be remembered that some replacement parts are provided from the manufacturer with various types of coatings. Some are protective coatings for shipping and others are primed for painting. Refer to published information to determine what must be done to prepare the new part for painting, as methods vary.

Quite often, the protective coating on new parts is not intended to serve as the primer. It is simply a protective coating and you will sometimes have to sand, blast, or chemically remove it completely. Then spray the part with self-etch primer and primer-surfacer to provide a smooth, solid base for the new paint. If not cleaned and sanded properly, the colorcoat will not stick properly to the new part.

If the part is already factory-primed and ready for paint, clean it with wax and grease remover. Then examine the part for imperfections such as drips or scratches. If drips or scratches are present, sand these imperfections until smooth. You may not have to remove the coating completely. Scuff sand the entire panel; then apply primer or primer-surfacer before painting.

Any bare metal replacement panels protected with grease should also be cleaned with a wax and grease solvent. They should be washed with liberal amounts of solvent. Change to a clean rag frequently. Then treat with a bare metal conditioner and flush down with water before drying the part.

17.10 USING PRIMERS

The decision to apply a primer, a primer-sealer, or a primer-surfacer by itself or combined with putty and/or a sealer depends on three factors. These are as follows:

- The condition of the substrate—smooth or rough, bare or painted
- The type of finish on the substrate—if painted
- The type of finish to be used for the topcoat

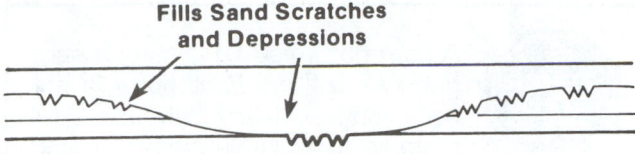

FIGURE 17-70 Note how undercoats protect bare metal.

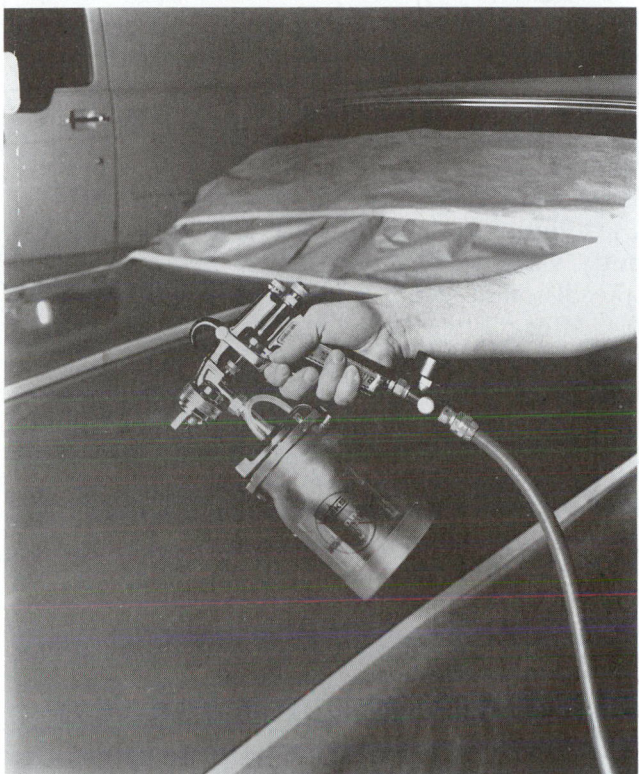

FIGURE 17-71 Apply the undercoat, in this case a primer-surfacer, in wet coats. Allow each coat to flash before applying the next coat.

These products are applied primarily to protect the bare metal against corrosion and to improve adhesion of the topcoats of paint (Figure 17-70). Due to their excellent filling and leveling qualities, they also fill minor sand scratches and level rough edges or depressions that remain after sanding. Reduce the undercoat chosen according to the manufacturer's instructions. Be careful to select the proper solvent for the weather conditions and mix the material thoroughly.

Generally, only 1 to 2 coats of primer or primer-sealer are required. Primer-surfacer, however, requires 3 to 4 coats for proper buildup.

Apply the first coat of undercoat (Figure 17-71). Allow this coat to flash dry, following the recommendations on the label for flash time.

Flash time is the time needed for a fresh coat of sprayed material to partially dry or cure. Flash time is needed to prevent the material from sagging, running,

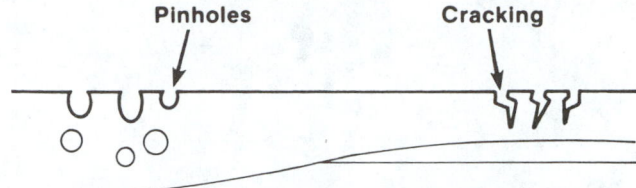

FIGURE 17-72 If primer is put on too thick in one coat, pinholes and cracking can result.

cracking, or experiencing other problems when another coat is to be applied. Check the paint manufacturer's specifications for recommended flash time.

Then apply the next coat or coats as medium wet coats for additional film buildup, with flash time between each application. When making a spot repair, extend the undercoat (primer material) several inches or millimeters around the first coat.

SHOP TALK

A common mistake is to apply primer materials too dry or too thick. Both can cause problems when painting. Spray the primer on in thin coats. Allow each coat to flash before applying another.

Allow the undercoat to dry thoroughly. Do not apply extra heavy coats to speed up the operation. Primer applied too thick will require more time to dry and can lead to cracking, crazing, pinholes, and poor holdout (Figure 17-72).

It is difficult to tell when a thick coat of primer-surfacer is really dry. The surface will appear dry while there is still a lot of solvent trapped below the surface. The lower layer of the primer-surfacer is still trying to dry and shrink. Follow manufacturer's guidelines to avoid problems!

If the primer-surfacer is sanded before all of the solvent has evaporated, the material in the scratches will continue to shrink down in the scratches. They will show up in the final finishing color topcoats as sanding scratches.

At the other extreme, thin dry coats of primer-surfacer can cause loss of adhesion, not only to the substrate, but also to the topcoat color. Always spray wet coats of primer-surfacer. Wait an hour or more before sanding the primer-surfacer.

After the undercoat is dry, block sand the area until it is smooth (Figure 17-73). For best results, use #320 grit sandpaper. If very fine scratches still appear, another coat of primer-surfacer might be all that is required to fill them.

After the primer dries, you can dry or wet sand the filler to check your work. Some technicians like

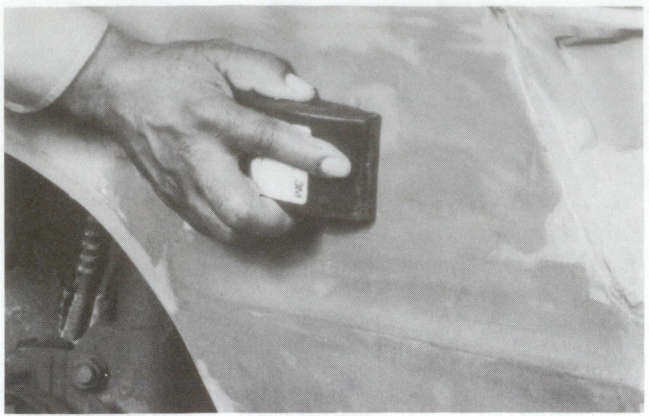

FIGURE 17-73 Block sand the undercoat or primer with fine sandpaper.

to apply a *guide coat,* which is a thin coat of a different color primer, over the full coat of primer-surfacer. Then, by sanding the area, you can easily find high and low spots. If the second color primer does not sand off, you have found a low spot. If it sands off too quickly, you have found a high spot.

Ideally, the second color primer should sand off at the same time. This shows that the surface is flat and ready for sealer, a colorcoat, and other operations.

PUTTY APPLICATIONS

Once the primer is dry, small pinholes and scratches can be filled with **spot putty** or *glazing putty* (Figure 17–74).

Older lacquer-based glazing putties are not recommended because they dry very slowly and can cause shrinkage problems. They have been replaced by polyester putties (two-part finishing fillers) and by polyester primer-fillers. Both products must be mixed with hardener before starting application and most can be applied to filler, metal, or old paint finishes. The use of these high viscosity, finely textured fillers eliminates the traditional primer/putty/primer process. Because they chemically harden, they cure quickly and can be primed and refinished without the worry of sandscratch swelling commonly associated with lacquer-based glazing putties.

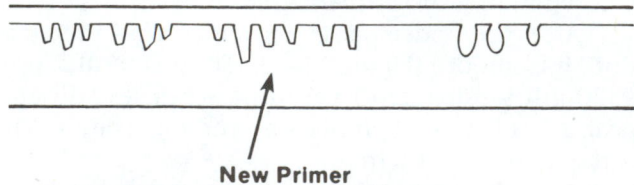

Thin coat fills scratches and pinholes.

New Primer

FIGURE 17-74 Glazing putty is like a primer, only it will fill minor surface imperfections like pinholes.

SHOP TALK

A common mistake is to use spot putty as a filler. Spot putty is not as strong as filler. Only use spot putty to fill small imperfections in the primer. Do not apply it to bare metal or painted surfaces. Most spot putties are designed to be applied over primer.

Place a small amount of properly mixed putty onto a clean rubber squeegee (Figure 17–75). Wipe a thin coat over the primer imperfections. Use single strokes and a fast scraping motion (Figure 17–76). Use a minimum number of strokes when applying putty, which dries very fast. Repeated passes of the spreader might pull the putty away from the primer.

After curing, sand the putty flush with the surrounding surface. Although wet sanding works well, it is not recommended by some putty manufacturers because moisture can soak into the putty. Refer to label instructions for details. Dry sand the putty if needed.

Allow the putty to air dry until it is hard. Test with a fingernail for hardness before sanding. If it is

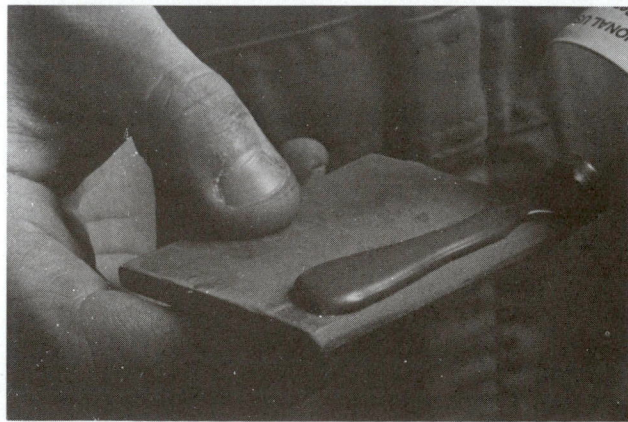

FIGURE 17-75 Spot putty is often applied with a soft rubber squeegee.

FIGURE 17-76 Apply putty to the surface to fill pinholes and other minor surface problems.

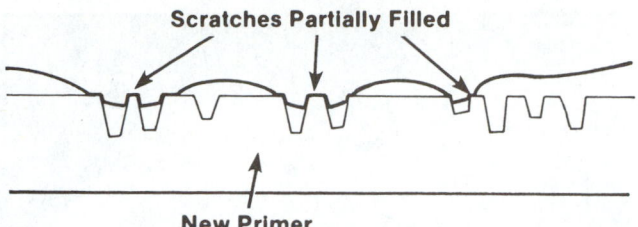

FIGURE 17-77 Putty shrinks as it cures; so do not sand putty until fully dry.

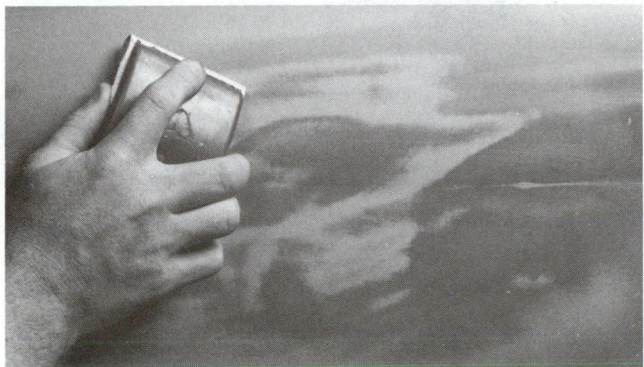

FIGURE 17-78 Sand the putty until level with rest of surface. This will quickly fill small imperfections.

FIGURE 17-79 Re-prime the puttied area and sand again if needed.

sanded too soon, the putty will continue to shrink, leaving part of the scratch unfilled (Figure 17-77). Once it hardens, the putty should be dry sanded with #220 grit paper or wet sanded (Figure 17-78). Wet sanding is carried out in the same manner as dry sanding; however, it requires special paper and plenty of water. After sanding the puttied area, clean the surface and then reprime (Figure 17-79). If the putty has been wet sanded, make sure to dry the surface thoroughly before applying primer-surfacer.

NOTE: Excessive use of glazing putties is usually an indication of a lack of skill and training. They should be used on small pinholes and other small surface problems in the primer.

COMPOUNDING

Compounding is sometimes done as a final smoothing step to remove light scratches, small dirt particles, and minor grinding or sanding marks before applying a final topcoat. Compounding can be done either by hand or machine. Rubbing and polishing compounds are available in various cutting strengths for both hand and machine as a final smoothing operation.

A **hand-rubbing compound** is usually coarser than a machine compound. It is used on small spot repairs, but it can be used on an entire car. It is applied with a damp rag to one small area at a time in a straight back-and-forth motion, not in circles. It is then buffed by machine or by hand.

Machine compound is made for use with a portable polisher or buffer (see Chapter 4). It is finer than hand-rubbing compound because the machine provides more power.

Details on both hand and machine compounds can be found in other chapters. Refer to the index if needed.

17.11 PLASTIC PARTS PREPARATION

Over 100 types of plastic are currently being used in the manufacture of vehicles, and approximately 40 need preparation before painting. The painter must be able to identify these plastic parts before refinishing them. Methods of identifying plastic and how to make plastic repairs are given in Chapter 14.

Plastic parts are usually considered either hard (rigid) or flexible (semirigid). Some flexible plastic auto body replacement parts come from the factory already primed, while others are delivered unprimed. If the parts are factory-primed, no additional priming is necessary. If they are not, both rigid and semirigid plastics might benefit from the use of a special plastic primer or primer-sealer to improve paint adhesion.

TPO, in particular, has an extremely slick, waxy surface that makes it difficult for the topcoat to form a strong bond to the substrate unless a primer is used.

PREPARATION OF FLEXIBLE PLASTIC

Prepare the surface of semirigid unpainted material as follows:

1. Clean the entire part with a recommended plastic cleaning solvent applied with a clean cloth. Wipe dry with another clean cloth before the solvent dries.
2. Featheredge the scuff or filler repair with #320 sandpaper, blow off dust, and tack wipe (Figure 17-80).

SHOP TALK Unpainted rigid plastic parts should also be solvent-cleaned to remove mold-release agents (typically silicones) before a primer or topcoat is applied.

3. Mix and apply four medium-dry coats of flexible primer-surfacer (Figure 17–81). Follow the manufacturer's instructions for specific mix ratios and additives.

4. Allow to dry at least 1 hour and block sand with #400 sandpaper. Sand the entire part to remove all gloss in preparation for color application.

When undercoats are modified with a flex additive, the possibility of mixture pot life exists.

Pot life refers to how long or much time you have while the material is still usable or sprayable. For example, a 3-hour pot life means you have 3 hours at a specific temperature (approximately 70 degrees Fahrenheit) before you should discard the material. Therefore, spray equipment should be emptied and flushed before pot-life time is reached.

Flex additives are needed for semirigid plastics because they expand, contract, and bend more easily than other substrates. The flex agent will keep the paint film flexible so it can accommodate the movement of the substrate without cracking.

SHOP TALK Use a fast-evaporating thinner or reducer as recommended to thin the primer-surfacer and do not apply excessively wet coats. Bare flexible plastic surface and/or flexible filler materials have a tendency to swell from thinner absorption, resulting in a visible or "highlighted" repair.

PREPARATION OF POLYPROPYLENE PLASTIC PARTS

The system for painting polypropylene parts involves the use of a special primer. Since polypropylene plastic is hard, it can be colorcoated after priming with conventional interior paint. To prepare the surface, proceed as follows:

1. Wash the part with an approved solvent. Follow the label directions.

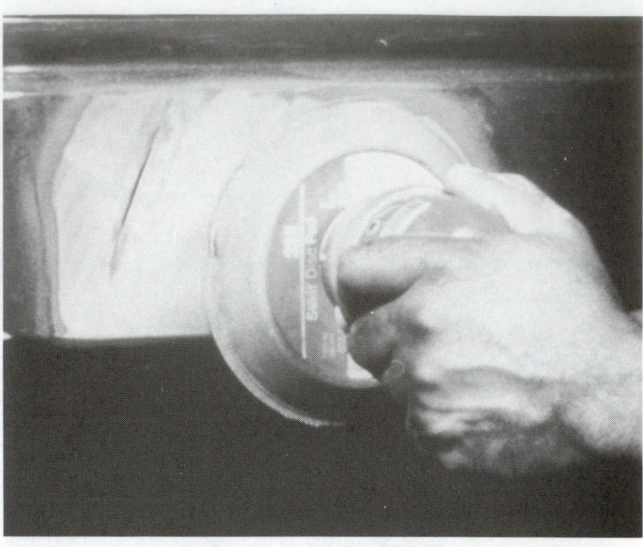

FIGURE 17–80 Using plastic repair agents is similar to using conventional body fillers. They must be mixed, applied, and sanded.

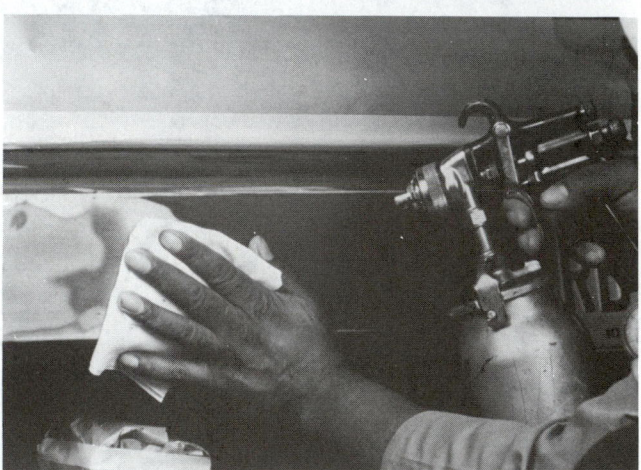

FIGURE 17–81 A flexible primer-surfacer must be used on flexible parts or cracking can result. Tack rag the surface before spraying.

2. Apply a thin, wet coat of polypropylene primer according to label directions. Wetness of primer is determined by observing gloss reflection of spray application in adequate lighting. Be sure primer application includes all edges. Allow primer to flash dry 1 minute minimum and 10 minutes maximum.

3. During the above flash time period (1 to 10 minutes), apply conventional interior finish color as required and allow to dry before installing the part. Application of color during the flash time range promotes the best adhesion of colorcoats.

PREPARATION OF RIGID PARTS

Exterior hard (rigid) parts should be treated as fiberglass when in doubt as to their makeup. In fact,

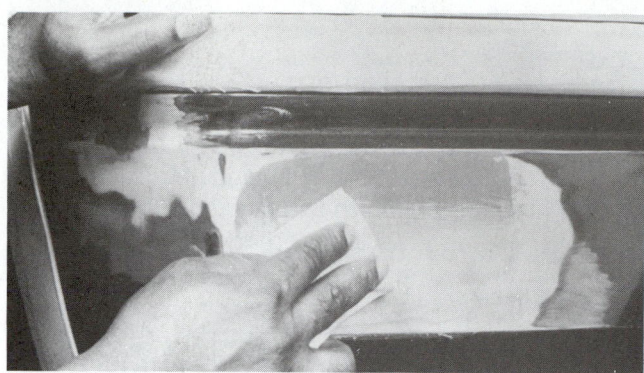

FIGURE 17-82 Plastic part repair agent is being applied over the damaged area.

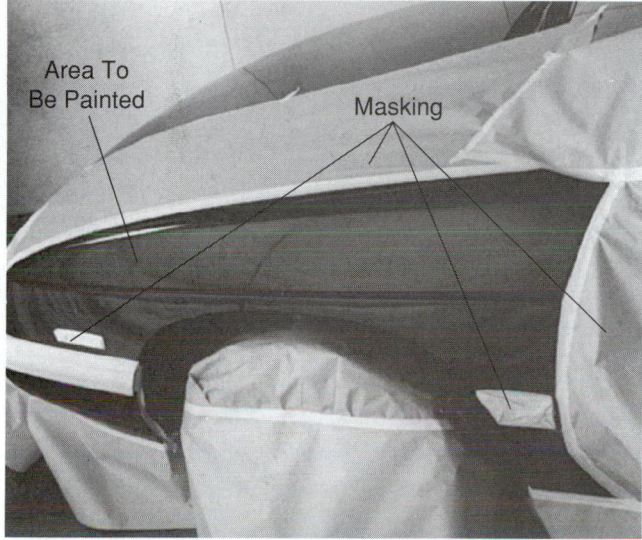

Area To Be Painted

Masking

FIGURE 17-83 This car has been masked for a panel or fender refinishing job. *(Courtesy of DuPont Automotive Products)*

fiberglass should be treated much the same—in preparation for a final coat—as body steel. It must be remembered that fiberglass parts do not require chemical conditioners. Replacement or new panels can contain contaminants on the surface due to the release agent used in the molds. Several common release agents are composed of silicone oils. These contaminants must be removed.

1. Newly molded parts should be washed with denatured alcohol used liberally on a clean cloth.
2. Thoroughly clean the surface with an approved material.
3. Sand the exposed fiberglass with #220 or #280 grit paper by hand or #80 to #120 grit with sander.
4. Reclean the surface and wipe dry with clean rags.

   Alternate Step: If there are joints to be filled or the sanding operation exposes air pockets or glass strands, glaze a coat of body filler over the entire surface (Figure 17–82). Allow to cure; then sand and reclean. Apply a single coat of sealer or a double coat of epoxy chromate primer.

5. Apply a primer-surfacer as directed on the label. Allow to dry and sand smooth with fine sandpaper to minimize sand scratches. Blow off with air and tack-rag the surface. If necessary apply another coat of sealer.

   Alternate Step: A synthetic primer-surfacer is also recommended if topcoats are to be enamel or acrylic enamel.
6. The surface is now ready for the colorcoats.

When refinishing previously painted fiberglass parts, care should be taken not to sand through the gel coat, and a sealer should be used. Fiberglass parts are extremely porous. The gel coat keeps topcoat solvents from being absorbed into the substrate.

17.12 MASKING

Masking keeps paint from contacting areas other than those to be refinished or painted (Figure 17–83). It is a very important step in the vehicle preparation process.

Masking has become even more important because of two-part paints. Once these paints dry, the overspray cannot be removed with a thinner or other solvent. It must be removed with a rubbing compound or by other time-consuming means.

There are several ways of masking the parts of a vehicle:

1. Masking paper and masking tape
2. Plastic sheeting and masking tape
3. Special shaped cloth or plastic covers (for wheels, antenna, rear view mirrors).
4. Liquid masking material

CAUTION

Traditional masking products may not work with waterborne paints because of their high water content. If using waterborne products, check with your supplier for masking papers and tapes made for use with them.

MASKING MATERIALS

Automotive **masking paper** is heat resistant so that it can be used safely in baking ovens. It also has good wet strength, freedom from loose fibers, and resistance to solvent penetration. Automotive paper comes in various widths, from 3 to 36 inches (76 to 914 mm).

Masking tape is very sticky paper tape designed to cover small parts and to also hold the

SHOP TALK

NEVER use newspaper for masking a vehicle since it does not meet any of the just-mentioned requirements. Newspaper also has the added disadvantage of containing printing inks that are soluble in some paint solvents. These inks can be transferred to the underlying finish, causing paint staining.

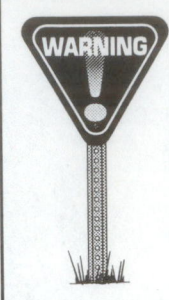
WARNING

Automotive masking tape should not be confused with tape bought in hardware or paint stores for home use. That kind of tape will not hold up to the demanding requirements of automotive refinishing.

FIGURE 17-84 Size and type of tape should match application. Narrow tapes are good for going around corners. Wide tapes are for long runs or for small parts. Duct tape is used for protection of parts when grinding, sanding, or blasting.

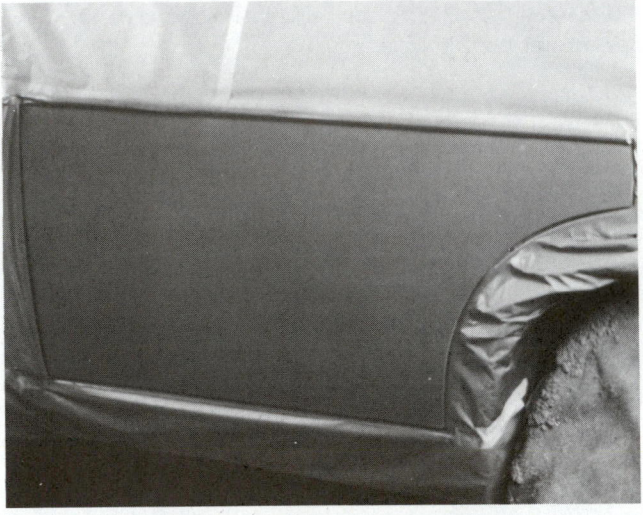

FIGURE 17-85 Make sure all surfaces not to be painted are masked and that no leaks are present where paint can flow through gaps in tape or paper.

masking paper in place. Automotive masking tape comes in various widths—from $1/16$ to 2 inches (2 to 51 mm). The most frequently used tapes are shown in Figure 17–84. Larger width tapes are used only occasionally since they are expensive and difficult to handle (Figure 17–85).

A *masking paper dispenser* automatically applies tape to one edge of the paper as the paper is pulled out (Figure 17–86). This saves time when masking a car or truck. The use of masking paper and tape-dispensing equipment makes it easy to tear off the exact amount needed. Some dispensers permit tape to adhere to both edges of the paper as it is rolled out.

FIGURE 17-86 This is a typical masking paper and tape dispenser.

SHOP TALK

It is interesting to note that the average size vehicle takes 2 to 2½ rolls of tape to be completely masked.

Masking covers are specially shaped cloth or plastic covers for masking specific parts (Figure 17–87). There are several types of masking covers available. They can save time when masking.

One is the *tire cover* that eliminates the need for masking off the tire (Figure 17–88). Others include a body cover, often made of disposable plastic (Figure

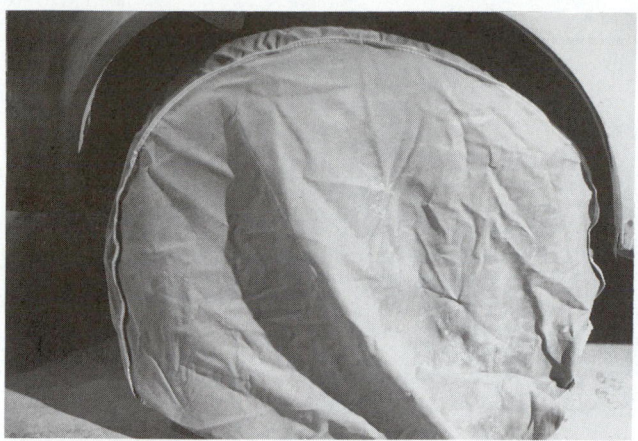

FIGURE 17–87 Special masking aids, like this tire cover, will speed masking operation. *(Photo courtesy ITW DeVilbiss Automotive Refinishing Products)*

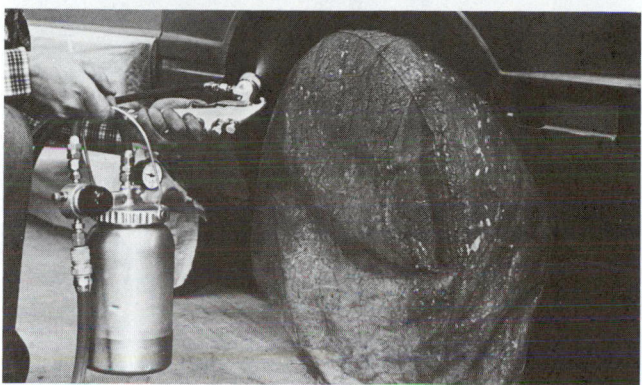

FIGURE 17–88 The technician is priming around the wheel well. The tire cover is protecting the wheel and tire from overspray.

FIGURE 17–89 A plastic car cover is handy for protection of a large surface area not to be painted. Note the door panel is the area to be painted. *(Courtesy of Fibre Glass Evercoat Co., Inc.)*

17–89). Covers are also available in a variety of sizes and shapes to mask headlights, taillights, antennas, mirrors, and so on. These various covers can be used as the situation demands.

HOW TO MASK

Before any types of masking materials are applied, the vehicle must be completely cleaned and all dust blown away. Masking tape will not stick to surfaces that are not clean and dry. It is important that the tape is pressed down firmly and that it adheres to the surface. Otherwise, paint will creep under it.

In the case of a two-tone finish where the color break is not hidden by a stripe or molding, use fine-line masking tape and press its edge down firmly.

There are no clear ground rules on when to mask and when to remove parts. This would include parts like trim, moldings, and door handles. The decision to remove or mask depends upon the design of the vehicle, and the expectations of the customer.

If a part can be removed easily, it is better to remove it than to try to mask around it. Also, if you cannot sand and clean right up to it, the part should

be removed. If it will be difficult to mask the part, you might save time by removing it from the vehicle. Each part will require an individual decision.

Removal of trim is more often necessary when using a basecoat/clearcoat system than with a single-stage finish. The film buildup can be greater with basecoat/clearcoats. This additional thickness makes the paint edge more likely to crack or chip.

It is wise to completely clean and detail the vehicle before masking and again after the refinishing job is completed. This is because masking over a dirty vehicle can cause a "dirty paint job!"

If the painting environment is cold and damp, the masking tape may not stick to the glass or chrome parts. Condensation on them can prevent the tape from sticking properly. Wipe off the parts before masking them. When masking doorjambs, be sure to cover both the door lock assembly and the striker bolt. They can become filled and clogged with paint if not masked.

Although masking tape is elastic, do not stretch the tape when making a straight line (Figure 17–90). Stretch masking tape only to cover curved surfaces. This is especially true when masking newly applied finishes that are still soft underneath. It is also wise to avoid stretching the tape because this can increase the degree of tape marking on the finish.

When applying masking tape, hold and peel the tape with one hand (Figure 17–91). Use your other hand to guide and secure the tape to the vehicle. This provides tight edges and allows you to change directions and go around corners.

To cut the tape easily, quickly tear upward against your thumbnail. This permits a clean cut of the tape without stretching (Figure 17–92).

Loop or overlap the inner tape edge to follow curves. The tape will stretch to conform to the curves.

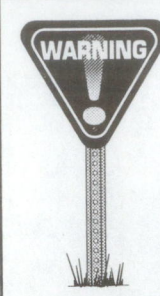 **WARNING**
Be careful that the masking tape does not overlap any area to be painted (Figure 17–93). After painting, you can remove paint from parts but you cannot add missing paint to the body.

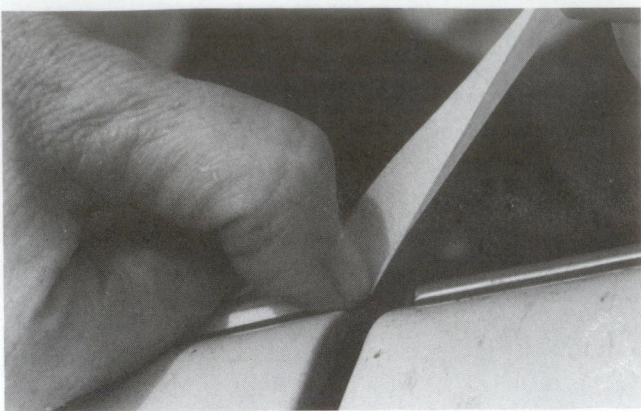

FIGURE 17-92 A thumb- or fingernail is a common way of cutting off masking tape.

FIGURE 17-90 Masking is often done on body contour lines. Fine line tape is applied first. Then, cover the body area with conventional tape and paper.

FIGURE 17-93 When masking the windshield, start at the bottom. The top layer will then overlap the bottom to help prevent leakage. Note how wide tape was wrapped or folded around the antenna.

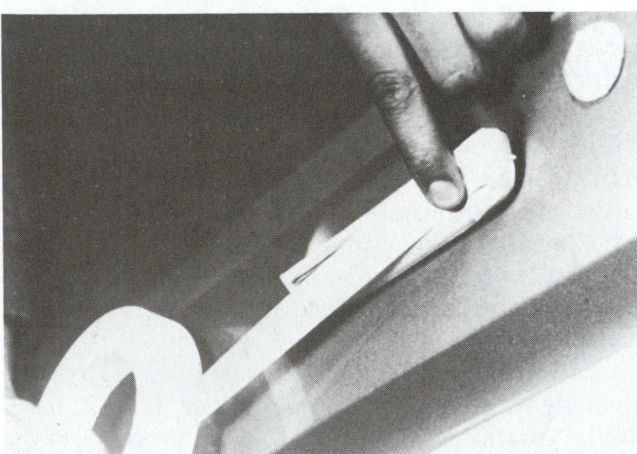

FIGURE 17-91 Several overlapping layers of tape are being used to mask this door handle.

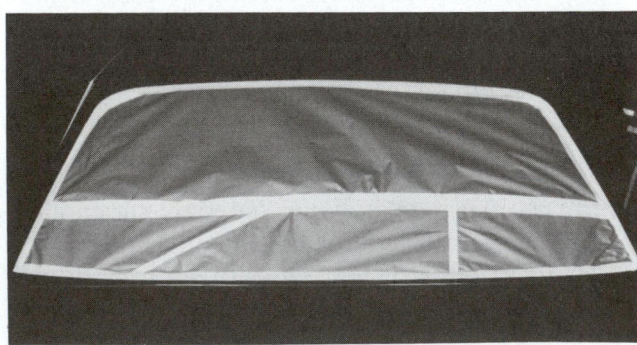

FIGURE 17-94 Mask the rear window by first running tape around the perimeter of the trim. Then, apply paper to the center of the glass.

Difficult areas such as wheels can be masked using this process, but more often covers are used to save time.

When masking large areas, such as bumpers, tape the paper to the middle of the bumper first. Then each side can be secured without the paper dragging on the floor and getting in the way.

Before masking glass areas, remove accessories such as wiper blades. The wiper shafts can be protected

in the same manner as radio antennas and door handles. Use tape or special covers.

To mask glass, first apply tape along the very top and along the edges of the moldings. Then, use two pieces of masking paper to cover the glass. Overlap the tape on the paper with the tape already on the molding (Figure 17–94). The top piece of paper

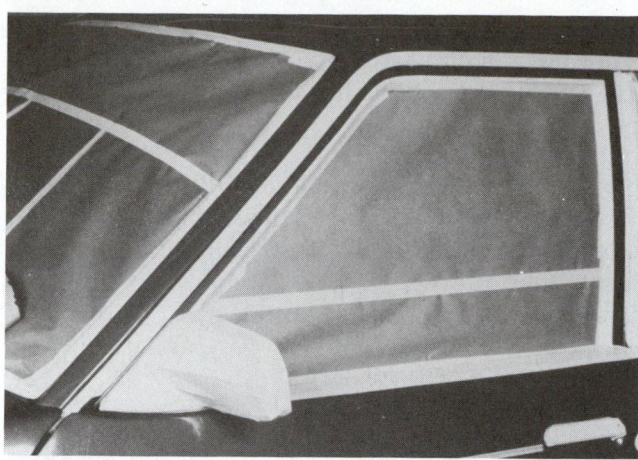

FIGURE 17-95 Masking the side window, door handles, and side mirror can be time consuming.

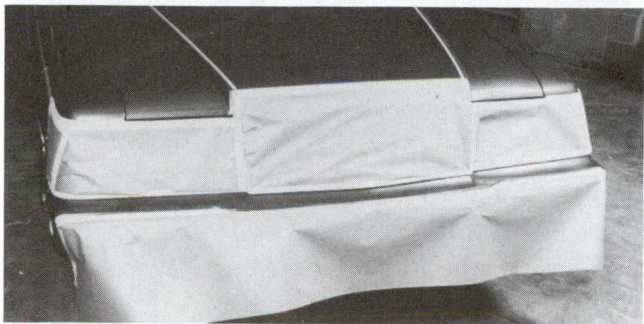

FIGURE 17-96 Mask grille and bumper to keep overspray off the lenses and chrome.

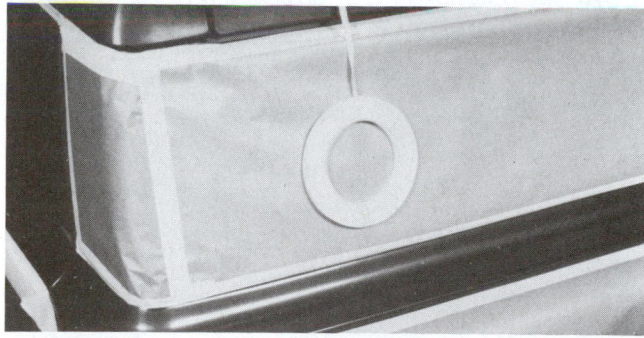

FIGURE 17-97 Mask taillights. With some designs, it may be easier to remove the lens before masking.

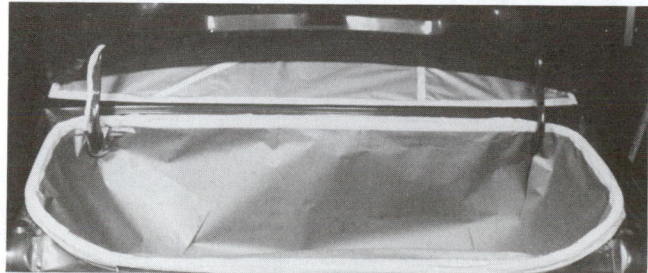

FIGURE 17-98 Mask a trunk by using large sheets of paper to form a protective cover over the weatherstripping and the floor area.

should overlap the bottom piece. If necessary, fold and tape any pleats in the paper to prevent dust and paint seepage.

Figures 17–95 through 17–99 give specific examples for proper masking of specific parts of a vehicle.

An **overspray leak** results when you do not seal the masking paper or plastic and overspray gets on unwanted surfaces. You would have to then take the time to use solvent or compound to remove the overspray problem.

Fine-line masking tape is a very thin, smooth surface plastic masking tape. Also termed *flush masking tape*, it can be used to produce a better **paint edge** (edge where old paint and new paint meet). When the fine-line tape is removed, the edge of the new paint will be straighter and smoother than if conventional masking tape were used (Figure 17–100).

Fine-line tape can be used to protect existing stripes from overspray. Also use fine-line tape for precise color separation in two-tone painting and for creating vivid, clean stripes. Its added flexibility

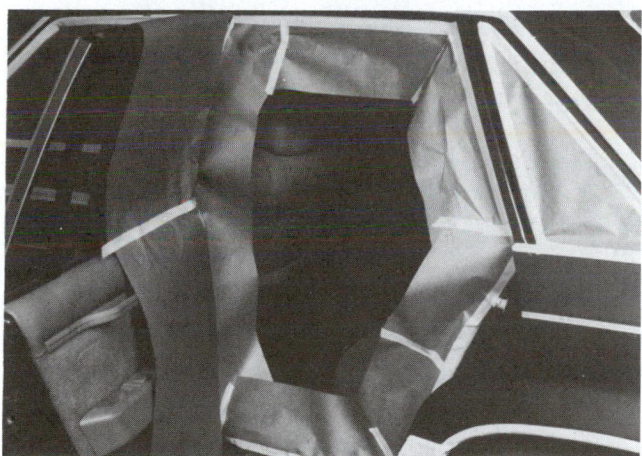

FIGURE 17-99 Masking doorjambs can be a challenge. Start by masking round edges of the door opening. Then apply paper carefully to the center hole in the paper to keep paint mist off interior parts.

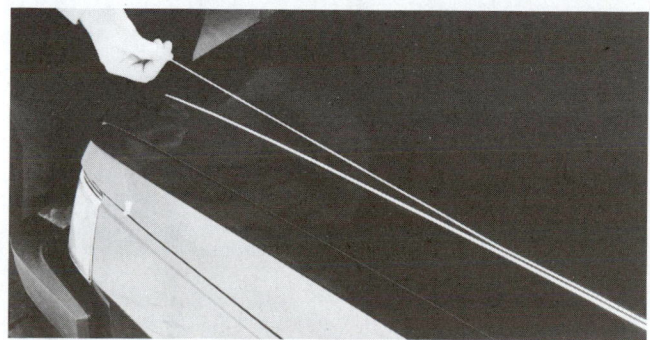

FIGURE 17-100 Fine-line tape can be used to paint stripes. Special tapes are available to create custom designs. *(Courtesy of Spartan Plastics, Inc.)*

makes painting of curved lines easier, with less re-working (Figure 17–101).

Fine-line tape is often used at the paint parting line (Figure 17–102). Then conventional tape is placed over half of the fine-line tape to secure the masking paper.

Back masking is done on the back of panels to prevent a visible parting line on the exterior of the part. As shown in Figure 17–103, apply the tape to the back of the panel, a door in this example. Close the door and push the tape back into the part gap. You can then paint the panels and an almost invisible parting line will be hidden on the inner edge of the panels.

Double masking uses two layers of masking paper to prevent bleed-through or finish-dulling from solvents. It is needed when spraying horizontal surfaces (hood, trunk, etc.) next to other horizontal surfaces. Overspray will tend to soak through the adjacent masked area and onto the old paint.

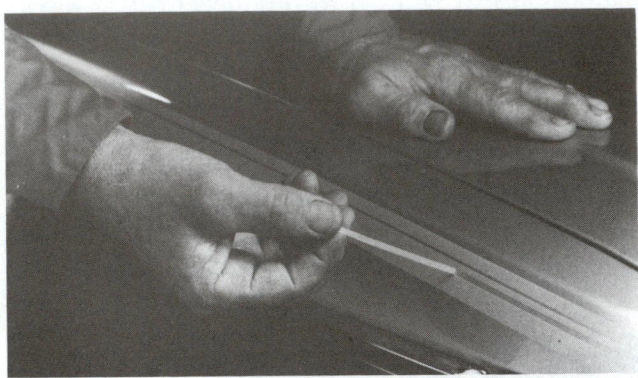

FIGURE 17-101 This technician is removing fine-line tape to produce painted stripes. *(Courtesy of 3M)*

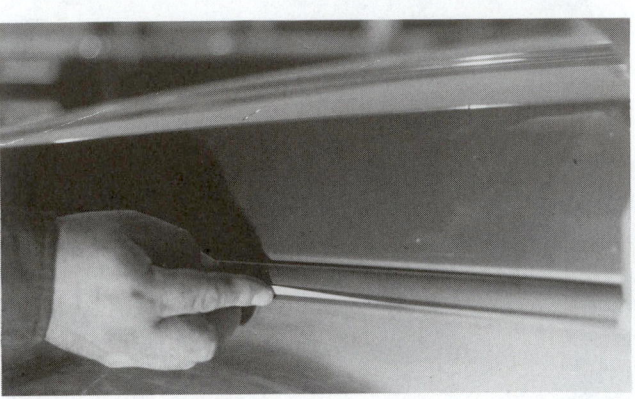

FIGURE 17-102 Fine-line masking tape can be used anywhere masking accuracy is critical.

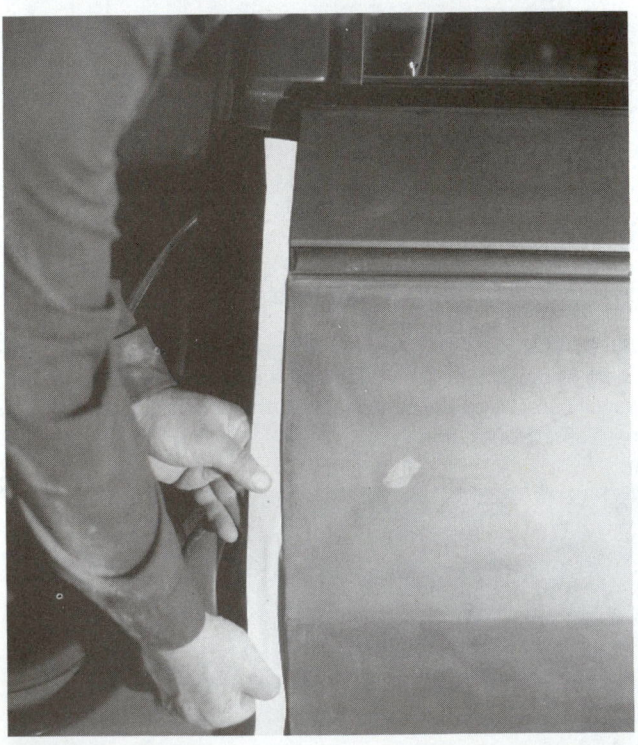

A

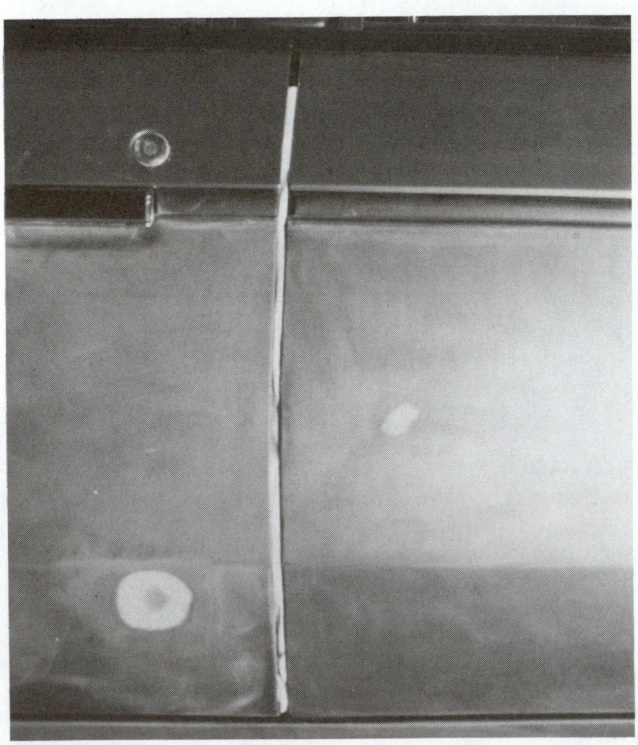

B

FIGURE 17-103 (A) Apply masking tape to the back of the door edge, with most of the tape sticking out. (B) Close the door and push the tape back into the crack between the doors. You can then paint both doors or a door and panel and a paint parting line will not be visible.

REVERSE MASKING METHOD A

REVERSE MASKING METHOD B

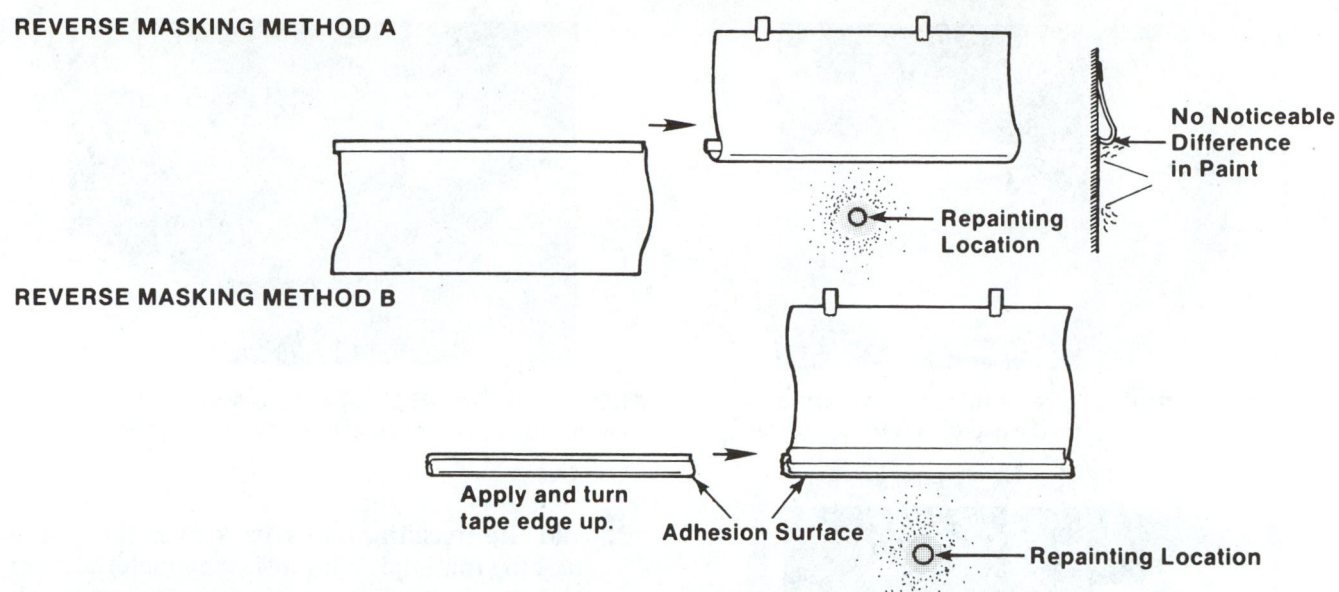

No Noticeable Difference in Paint

Repainting Location

Apply and turn tape edge up.

Adhesion Surface

Repainting Location

FIGURE 17-104 Note the procedure for making a reverse mask. *(Courtesy of Toyota Motor Co.)*

A **reverse masking** method requires you to fold the masking paper back and over the masking tape (Figure 17-104). It is often used during spot repairs to help blend the painted area and make it less noticeable. This also helps prevent bleed-through. The paper is taped on the inside and allowed to bellow slightly, which keeps it lifted a bit from the surface.

A door edge or body line next to a refinished panel can sometimes show a slight difference in an otherwise acceptable color match, particularly with metallics. This situation can be avoided by crossing the line and sanding the adjacent panel when preparing for refinishing. Then reverse mask the adjacent panel and refinish as desired.

If there is a slight difference after removing the masking paper, just paint across the line and blend in smoothly as with any spot repair. Reverse mask very carefully. Any overmasked or undermasked areas will lead to extra work after the vehicle is painted.

Overmasked areas mean that the painter must touch up the part of the vehicle that should have been painted. On the other hand, *undermasked* areas must be cleaned with solvent or polishing compound to remove overspray. Each would detract from the appearance of an otherwise good job.

LIQUID MASKING MATERIAL

Liquid masking material seals off the entire vehicle to protect undamaged panels and parts from paint overspray. It is the newest masking system available to painters. Liquid masking is used on areas where masking is necessary but difficult to apply, including wheel wells, headlights, grille, underbody chassis, and even the engine compartment.

Masking liquid, also called masking coating, is usually a water-based sprayable material for keeping overspray off body parts. Some are solvent-based. Masking liquid comes in a large, ready-to-spray container or drum. It is sprayed on and forms a paint-proof coating over the vehicle.

Some masking coatings are tacky and used only during priming and painting. They form a film which can be applied when the vehicle enters the shop. Others dry to a hard, dull finish.

Masking coatings can be removed when the vehicle is ready to return to the owner. It washes off with soap and water. Local regulations may require that liquid masking residue be captured in a floor drain trap, and not put into the sewer system.

To mask a vehicle using the liquid masking system, proceed as follows:

1. Partially mask the area to be painted by going around it with masking paper (Figure 17-105). Fold the paper over onto the area to be painted. Secure the paper with masking tape.
2. Apply the liquid masking material. Use a heavy, single overlapping coat. Apply the material to all surfaces *not* to be painted. This would include bumpers, grilles, doors, windshields, body panels, wheels, wheel wells, door jambs, and even the engine compartment (Figure 17-106).

An airless spray system is generally recommended for applying the masking material.

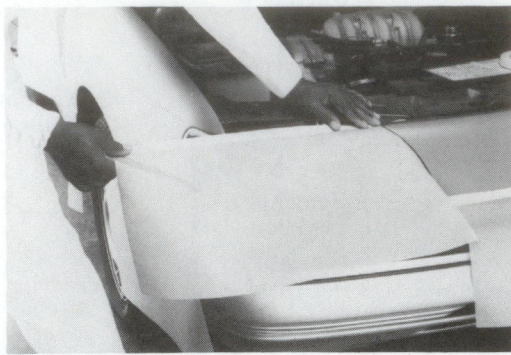

FIGURE 17-105 Use paper to mask right next to the area to be painted. Then, fold the paper back to expose the area to be painted.

FIGURE 17-106 Spray liquid masking material on areas not to be painted, like the engine compartment.

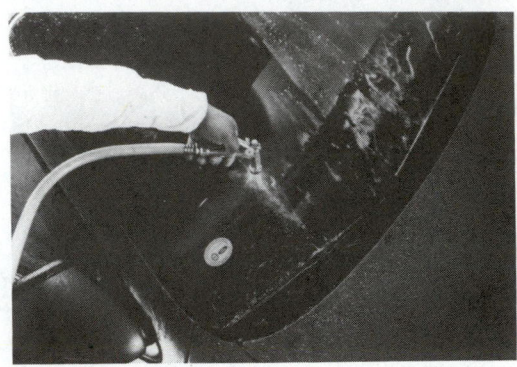

FIGURE 17-107 After painting, wash off liquid masking material.

3. Fold the masking paper back over the liquid masking material. Wipe away any material from the area to be painted with a damp sponge. Allow the surface to dry.
4. Prepare the surface. Then apply primer and paint according to the manufacturer's instructions.
5. Allow the paint to dry; then unmask the vehicle. Liquid masking may be used in both air dry or bake conditions.
6. After the paint is cured, wash off the dried liquid masking material with a garden hose or pressure wash (Figure 17–107).

SUMMARY

- The words "surface preparation" refer to getting the body surface clean, smooth, and ready for the application of the final colorcoats.

- A dull primer surface may look smooth and ready for paint to the untrained eye. You must remember that the paint will actually work like a "magnifying glass" to exaggerate any scratches or irregularities.

- Before painting, you must first identify the type of paint and overall condition of the existing paint system.

- Sanding uses an abrasive coated paper or plastic backing to level and smooth a body surface being repaired.

- Scuff sanding removes any trace of contaminants on the existing finish.

- Grit sizes vary from coarse to micro-fine grades and are ordered by number.

- Power grinding is done to quickly remove large amounts of old paint and other materials.

- An orbital sander or dual-action sander moves in two directions at the same time. This produces a much smoother surface finish.

- A dual-action or DA sander is used to feather-edge a repair area. It is the workhorse of body technicians.

- A guide coat is very helpful and assists in pointing out unlevel surfaces areas.

- When sanding, make sure you are using the correct grit number for the job.

- Be sure there is no wax, grease, or other contaminants imbedded in the old finish.

- Do not use body filler directly over a metal conditioner.

- Do not use too much hardener because it will cause pinholes.

- Do not return any unused mixture to the can.

- Paint thickness is measured in mils or thousandths of an inch (hundredths of a millimeter). Original OEM paints are typically about 2 to 6 mils thick.

- Blasters are air-powered tools for forcing sand, plastic beads, or other materials onto surfaces for paint removal. A chemical paint remover can also be used for stripping large areas of paint if environmental regulations allow.

- If the primer-surfacer is sanded before all of the solvent has evaporated, the material in the scratches will continue to shrink down in the scratches.

- After the undercoat is dry, block sand the area until it is smooth.

- Once the primer is dry, small pinholes and scratches can be filled with spot putty or glazing putty.

- Compounding is sometimes done as a final smoothing step to remove light scratches, small dirt particles, and minor grinding or sanding marks before applying a final topcoat.

- Pot life refers to how long it will take for the material to cure while still in the paint gun.

- Masking keeps paint from contacting areas other than those to be refinished or painted.

- Automotive masking paper is heat resistant so that it can be used safely in baking ovens.

- Masking tape is very sticky paper tape designed to cover small parts and to also hold the masking paper in place.

- Masking covers are specially shaped cloth or plastic covers for masking specific parts.

- Fine-line masking tape is a very thin, smooth surface plastic masking tape.

- Liquid masking material seals off the entire vehicle to protect undamaged panels and parts from paint overspray.

ASE-STYLE REVIEW QUESTIONS

1. Technician A uses a dual-action sander to remove heavy surface rust. Technician B uses a dual-action sander to featheredge repair areas. Who is correct?

   A. Technician A
   B. Technician B
   C. Both A and B
   D. Neither A or B

2. In preparation for topcoating, when the original paint surface is in good condition, Technician A will prepare the surface by washing it. Technician B will make sure the finish is stable and does not react to the solvent of the refinish paint. Who is correct?

   A. Technician A
   B. Technician B
   C. Both A and B
   D. Neither A or B

3. When filling a dent, Technician A first applies a metal conditioner and then the body filler. Technician B says that using too much hardener in the filler can cause pinholes. Who is correct?

   A. Technician A
   B. Technician B
   C. Both A and B
   D. Neither A or B

4. Which of the following methods is used to strip paint from the metal surfaces of a vehicle?

   A. Sanding
   B. Blasting
   C. Chemical stripping
   D. All of the above

5. Paint removers are _____.

   A. Neutralized and removed with water
   B. Removed with a squeegee or scraper
   C. Not to be used on fiberglass
   D. All of the above

6. Technician A applies a metal conditioner before applying a conversion coating. Technician B says that using certain wash-primers eliminates the need for conversion coating. Who is correct?

   A. Technician A
   B. Technician B
   C. Both A and B
   D. Neither A or B

7. A wash-primer with phosphoric acid _____.

   A. Cleans
   B. Etches the metal
   C. Assists in adhesion
   D. All of the above

8. Lacquer-based glazing putties are being partially replaced by _____.

   A. Polyester putties
   B. Finishing fillers
   C. All of the above
   D. None of the above

9. Technician A applies compound in a back-and-forth motion. Technician B uses a circular motion. Who is correct?

   A. Technician A
   B. Technician B
   C. Both A and B
   D. Neither A nor B

10. Newspaper _____.

    A. Has freedom from loose fibres
    B. Has resistance to solvent penetration
    C. Contains inks that are soluble in some paint solvents
    D. All of the above

ESSAY QUESTIONS

1. How do you evaluate the surface condition of a vehicle?

2. List some points that must be kept in mind when working with abrasives.

3. Describe the purpose of wet sanding.

4. Explain the use of a guide coat.

5. When using a body filler, what are some precautions that should be kept in mind?

6. Explain the use of fine-line masking tape.

7. How do you mask a vehicle using the liquid masking system?

CRITICAL THINKING PROBLEMS

1. If you are going to spray a two-part paint on a vehicle and there are leaks in the masking paper, what can be the result to the vehicle, to you, and to the shop?

2. If too coarse a sandpaper is used before painting, what would happen?

3. If measurements show a paint to be 24 mils thick, what does that tell you?

MATH PROBLEMS

1. If a paint measures 6 mils thick and paint buildup should be limited to no more than 12 mils, how much more paint could be applied to the body?

2. If a pint of material will cover 3 square feet (1.8 square meters), how much material would be needed for 23 square feet (6.9 square meters)?

USING PRIMER-SURFACER

P17–1 Reduce primer-surfacer according to label directions. Observe all safety precautions. Make sure the solvent is at the correct temperature and measure it accurately.

P17–2 Filter the primer-surfacer into the spray gun cup without spilling.

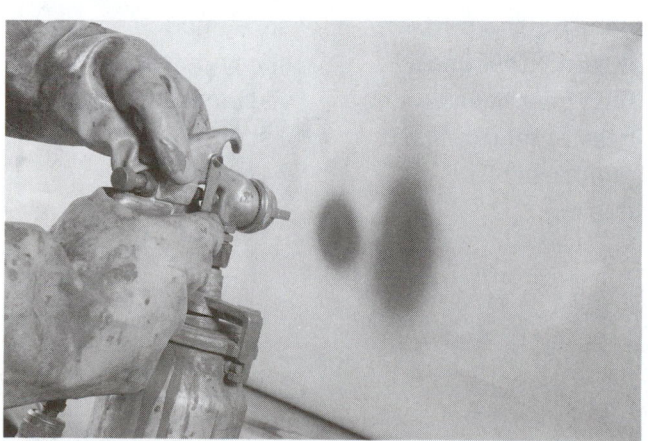

P17–3 Spray a test pattern on a test sheet attached to the spray booth wall. Make sure you are applying a smooth, wet layer of primer-surfacer.

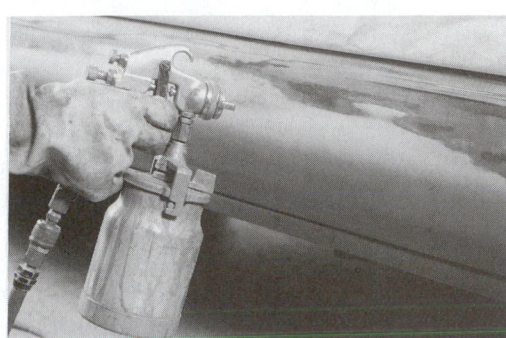

P17–4 Spray the repair area with one or two wet coats of primer-surfacer. Allow each coat to tack before applying the next. If you apply too many wet coats, the primer-surfacer can crack or check when it dries.

P17–5 On large repair areas, you may want to mist on a different color primer over primer-surfacers. This is called a guide coat and will help you find minor surface irregularities more easily.

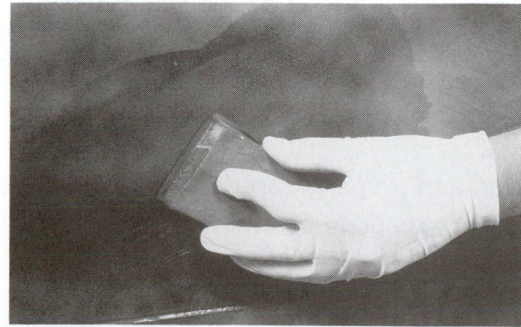

P17–6 Wet sand the guide coat. Any area where the guide coat does not sand off is a low area. If the guide coat sands off in a small area, that area is too high and requires further block sanding. If you sand off the entire guide coat without cutting through the primer-surfacer, you are ready for paint.

CHAPTER 18

Refinishing Equipment and Its Use

INTRODUCTION

This chapter summarizes the steps needed for preparing the finishing equipment and paint area for spraying the topcoat. There are a number of shop and equipment variables that affect the refinishing operation. These variables include the painting environment, as well as the painting equipment and their adjustments. They are ALL important. You must pay close attention to these variables because they can affect the quality of your work.

To do a good paint job, your shop and equipment must be in perfect condition! A dirty spray booth, a poorly maintained paint spray gun, contaminated air supply, and other avoidable situations will all ruin your work. Sloppy shop conditions will usually result in an inferior paint job. A professional painter will spend more time maintaining the shop

OBJECTIVES

After studying this chapter, you should be able to:

✔ Identify the spray painting equipment used in auto refinishing.
✔ Explain how a spray gun works.
✔ Identify the basic techniques of good spray painting and recognize variables that influence the quality of the spray finish.
✔ Adjust the spraying equipment to test and develop a good spray pattern.
✔ Implement the stroke technique procedure for single- and double-coat applications and recognize common errors made by apprentice refinishers.
✔ Identify the various types of spray coats.
✔ Determine when and how to make spot repairs.
✔ Clean and properly care for a spray gun.
✔ Identify situations for which airless, electrostatic, and HVLP spray systems or airbrushes are recommended.
✔ Explain the operation of spray booths and respirators.
✔ Answer ASE test questions about refinishing equipment.

KEY TERMS

agitation paddle
air brush
air cap
air valve
air vent hole
airless spraying equipment
arcing
atomization
electrostatic spraying
flash
flooding the pattern
fluid control knob
fluid needle valve
gravity feed system
gun stroke
heeling
high transfer efficiency
HVLP
improper coverage
improper overlap
overspray
pattern control valve
pressure cup spray guns
siphon spray guns
spray booth
spray gun
spray pattern test
viscometer
viscosity
wet coat

ASE TASK LIST

Job Skills covered in this chapter include:

PAINTING AND REFINISHING TEST (B2) TASK LIST

A. Surface Preparation

12. Remove dust from area to be refinished, including cracks or moldings of adjacent areas.
14. Remove, with a tack rag, any dust or lint particles from the area to be refinished.

B. Spray Gun Operation and Related Equipment

1. Inspect, clean, and determine condition and adequacy of spray guns and related equipment (air hoses, regulator, air lines, air source, and spray environment).
2. Check and adjust spray gun pressure for siphon-feed, pressure feed, gravity feed, HVLP (high volume, low pressure), and LVLP (low volume, low pressure) guns.
3. Adjust spray gun using fluid and pattern control valves.

C. Paint Mixing, Matching, and Applying

2. Shake, stir, reduce, catalyze, and strain paint according to manufacturer's recommendations.
3. Use appropriate spray technique (gun arc, gun angle, gun distance, gun speed, and spray pattern overlap) for finish being applied.
6. Apply basecoat/clearcoat for spot and panel blending, and overall refinishing.
8. Sand, buff, and polish finishes where necessary.

12. Apply multistage (mica, pearl, etc.) coats for spot repair, panel blending, and overall refinishing.
13. Identify paint color formula and proper usage of mixing equipment and materials.

D. Solving Paint Application Problems

4. Identify a dry spray appearance in the paint surface; determine the cause(s), and correct the condition.
7. Identify mottling or streaking in metallic and mica paint finishes; determine the cause(s), and correct the condition.
8. Identify orange peel appearance of the refinished surface; determine the cause(s), and correct the condition.
9. Identify an overspray condition; determine the cause(s), and correct the condition.
11. Identify sags and runs in the paint surface; determine the cause(s), and correct the condition.

E. Finish Defects, Causes, and Cures

1. Identify poor adhesion; determine the cause(s), and correct the condition.
13. Measure paint film thickness.

F. Safety Precautions and Miscellaneous

3. Inspect spray environment.

---

and his or her equipment than on any other single task. Spraying the vehicle only takes a very short amount of time by comparison.

Spray painting equipment for the auto refinishing shop consists of the following:

- Spray guns and cups (either suction, pressure, or gravity feed)
- Air compressor of adequate size
- Oil and water extractor and air regulator combination to filter air and regulate pressure
- Air hose of sufficient size (inside diameter) to convey air from the extractor to the spray gun without causing an excessive drop in pressure
- Spray booth or enclosure to ensure a healthy, safe, dust-free working area
- Separate drying area and force-drying equipment

NOTE: The selection and use of air compressors, air control equipment, and air hose connectors are thoroughly described in Chapter 5.

18.1 SPRAY GUNS

The **spray gun** (Figure 18–1) breaks the liquid primer or paint into a fine mist and forces it onto the surfaces of the vehicle. It is the key component in a refinishing system. It is a precision engineered and manufactured tool. Each type and size available is specifically designed to perform a certain number of tasks. Even though all spray guns have many parts and components in common, each gun type or size is suited for only a certain, defined range of jobs. As in most other areas of refinishing work, having the right tool for the job goes a long way toward getting a professional job done right in minimum time.

ATOMIZATION AND THE SPRAY GUN

A thorough understanding of atomization is the key to using a spray gun correctly. **Atomization** breaks paint into a spray of tiny, uniform droplets (Figure 18–2). When properly applied to the vehicle's surface, these droplets flow together to create an even film thickness with a mirrorlike gloss. Proper atomization is essential when working with today's basecoat/clearcoat finishes.

Atomization takes place in three basic stages (Figure 18–3):

- In the first stage, the paint siphoned from the fluid tip is immediately surrounded by air streaming from the annular ring. This turbulence begins the breakup of the paint.
- The second stage of atomization occurs when the paint stream is hit with jets of air from the containment holes. These air jets keep the paint stream from getting out of control and aid the breakup of the paint.
- In the third phase of atomization, the paint is struck by jets of air from the air cap horns. These air streams hit the paint from opposite sides, causing the paint to form into a fan-shaped spray.

PARTS OF A SPRAY GUN

The principal parts or components of a typical air spray gun, which are listed below, are illustrated in Figure 18–4. Most guns are equipped with a removable spray head unit containing the air cap, fluid tip, and fluid needle.

Air cap or nozzle
Fluid tip or nozzle
Fluid needle valve
Trigger
Fluid control (or spreader) knob
Air valve
Pattern (or fan adjustment) control knob
Gun body (or handle)

The **air cap** directs the compressed air into the material stream to atomize it and form the spray pattern. There are three types of orifices (holes) (Figure 18–5): the center orifice, the side orifices or ports,

and the auxiliary orifices. Each of the orifices has a different function.

The *center orifice* located at the nozzle tip creates a vacuum for the discharge of the paint. The *side orifices* determine the spray pattern by means of air pressure. The *auxiliary orifices* promote atomization of the paint. Figure 18–6 illustrates the relationship between the auxiliary orifices and the gun's performance.

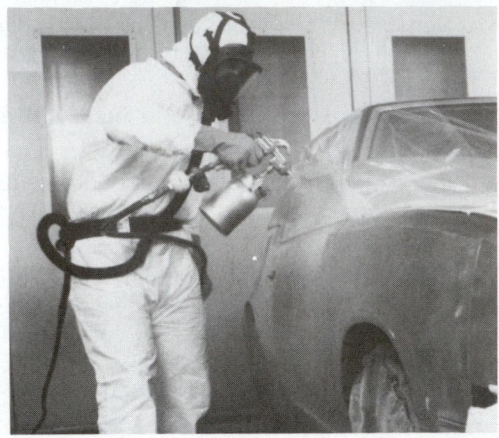

FIGURE 18-1 The proper use of a spray gun is a must in any refinishing paint shop. Note protective gear. *(Courtesy of Sherwin Williams Co.)*

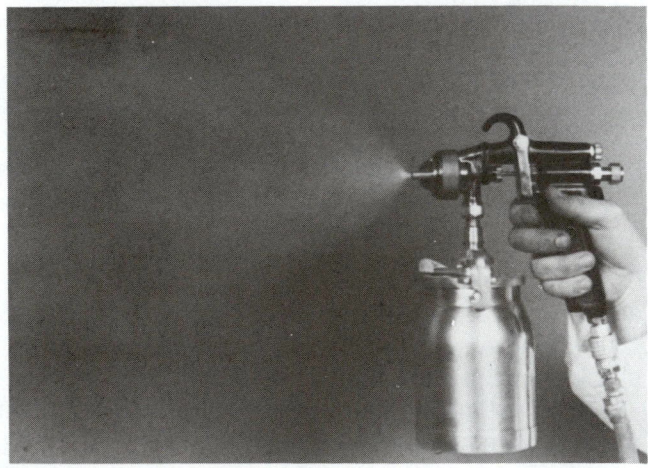

FIGURE 18-2 The spray gun breaks liquid into atomized spray for even deposit onto surface.

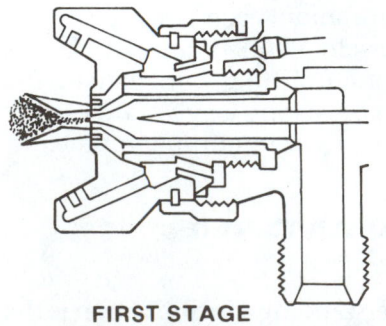

FIRST STAGE

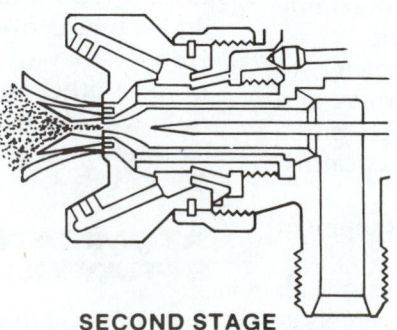

SECOND STAGE

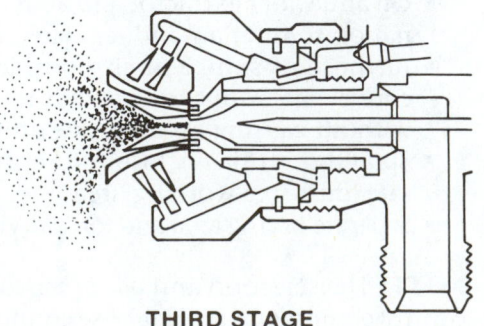

THIRD STAGE

FIGURE 18-3 Note three stages of atomization.

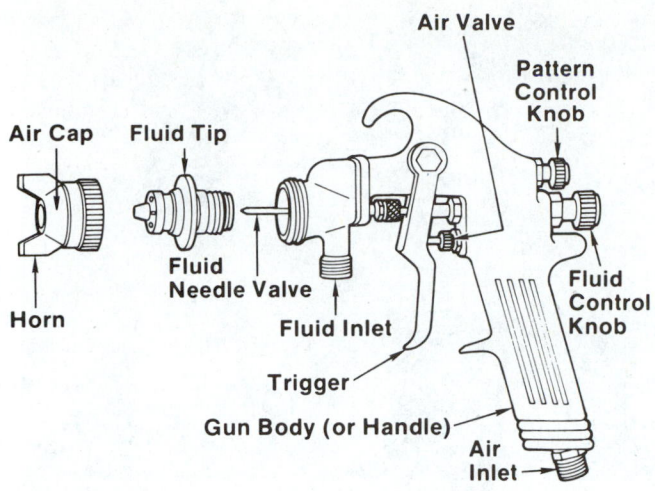

FIGURE 18-4 Study parts of a typical air spray gun.

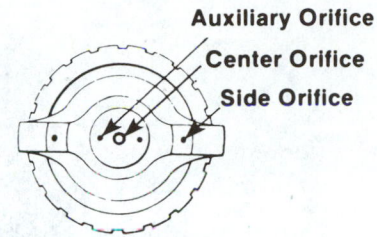

FIGURE 18-5 Memorize nomenclature of air orifices.

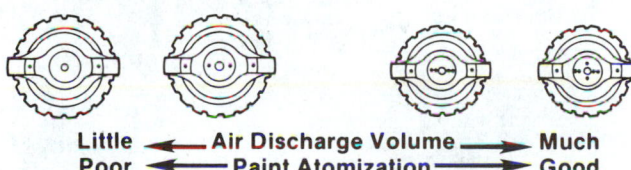

Little ← Air Discharge Volume → Much
Poor ← Paint Atomization → Good

FIGURE 18-6 Compare number of auxiliary holes and gun performance.

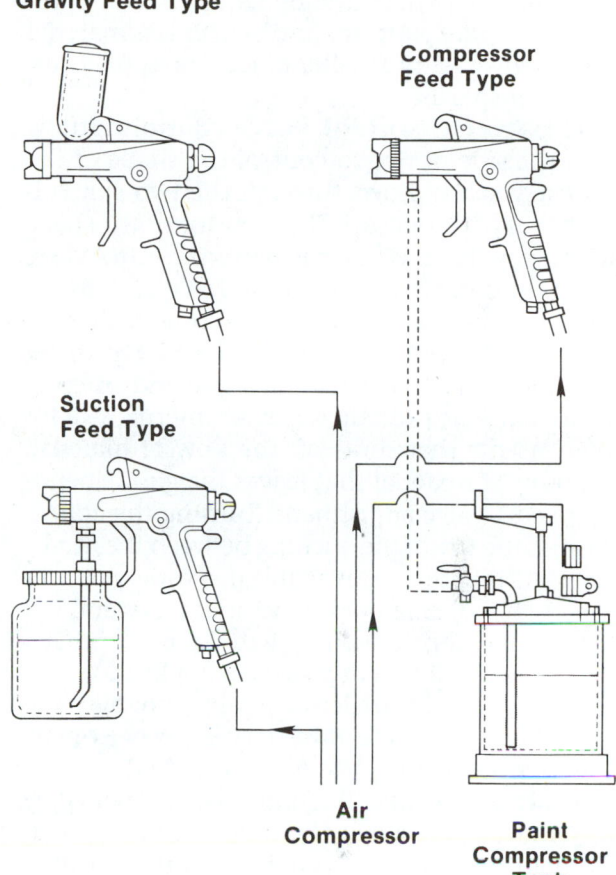

FIGURE 18-7 Study paint feed methods to air spray guns.

TABLE 18-1: TYPES OF AIR SPRAY GUNS

Type	Paint Feed Method	Advantages	Disadvantages
Suction Feed Type	Paint container is installed below the spray nozzle and paint is supplied by suction force alone.	Stable gun operation. Easy to refill container or make color changes.	Difficult to spray on horizontal surfaces and some variations occur in discharge volume due to variations in viscosity. Has a larger paint container than gravity feed type, but this causes quicker painter fatigue.
Pressure Type	Paint is pressurized by a compressed air tank or pump.	Large surfaces can be painted without stopping to refill container. A paint with a high viscosity can also be used.	Not suitable for small area painting. Color changes and gun cleaning take time.
Gravity Feed Type	As the paint cup is installed above the spray nozzle, paint is supplied by gravity and a suction force at the nozzle tip.	Because there is no change in paint viscosity, there is no variation in the injection volume. The position of the cut can be changed according to the configuration of the the painted item.	Because the cup is installed above the injection nozzle, it adversely affects gun stability. Cup capacity is small so not useful for painting larger surfaces.

Large orifices increase the ability to atomize more material for painting large objects with great speed. Fewer or smaller orifices usually require less air, produce smaller spray patterns, and deliver less material to conveniently paint smaller objects or apply coatings at lower speeds.

The **pattern control valve** controls airflow through the side orifice to control the shape of the paint mist. Air also flows through the two side orifices in horns of the air cap. This flow forms the shape of the spray pattern. When the pattern control valve is closed, the spray pattern is round. As the valve is opened, the spray becomes more oblong in shape.

The **fluid needle valve** and **fluid tip** meter the amount of material leaving the gun and entering the air stream. The fluid tip forms an internal seat for the fluid needle that shuts off the flow of material. The amount of material that leaves the gun depends on the needle valve adjustment. Turning the adjustment varies the size of the opening between the needle and its seat. Fluid tips are available in a variety of sizes to properly handle materials of various types and viscosities. Each will pass the required volume of material to the cap for different speeds of application.

The **fluid control knob** (valve) changes the distance the fluid needle valve moves away from its seat in the nozzle when the *trigger* is pulled.

The **air valve,** like the fluid valve, is opened by moving the trigger. When the trigger is pulled partway, the air valve opens. When it is pulled a little farther, the fluid valve opens. Conventional guns have an air needle valve. Newer high efficiency guns may use an adjustable air cap to control air flow.

TYPES OF AIR SPRAY GUNS

As pointed out in Table 18–1, there are three basic methods of paint supply or feed to the air spray gun (Figure 18–7):

- Suction (siphon) feed (Figure 18–8A)
- Pressure feed (Figure 18–8B)
- Gravity feed (Figure 18–8C)

Suction or **siphon spray guns** use air flow through the gun head to form a suction that pulls paint into the air stream. This is the most common type of gun in most areas. The paint material is often held in a 1-quart (0.94 liter) cup attached to the gun.

An **air vent hole** and hose on the siphon spray gun allow atmospheric pressure to enter the cup. This vent can become clogged with dry primer or paint. If the vent is plugged, paint will not flow out of the gun.

When the spray gun trigger is partially depressed, the air valve opens and air rushes through the gun. As the air passes through the openings in the air cap, a partial vacuum is created at the fluid tip (Figure 18–9A). Further squeezing of the trigger withdraws the fluid needle from the fluid tip. The vacuum sucks paint

A

B

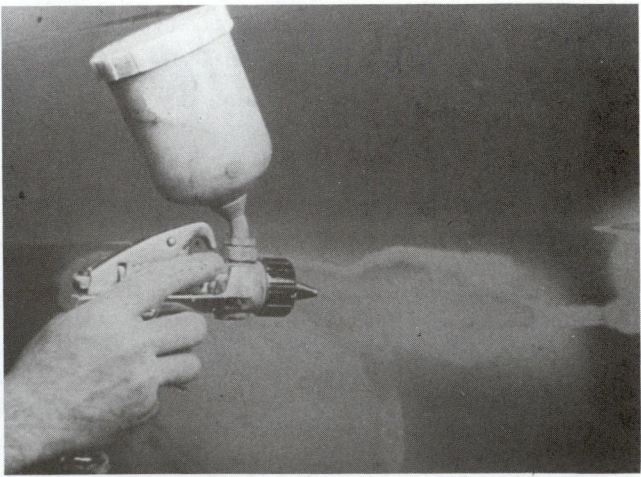

C

FIGURE 18-8 These are three common types of air spray guns: (A) suction, (B) pressure, and (C) gravity.

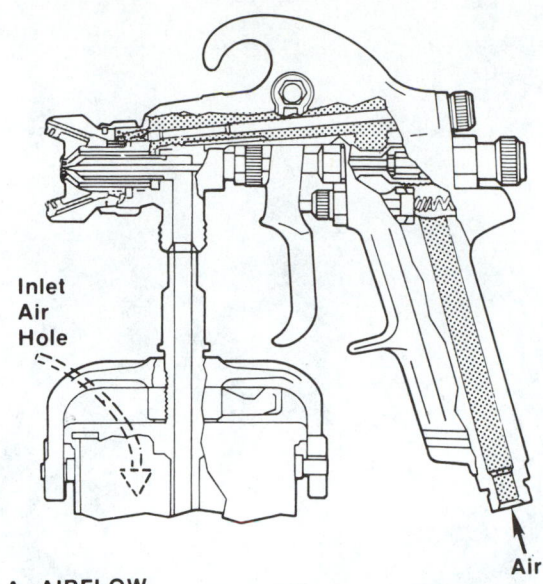

A. AIRFLOW

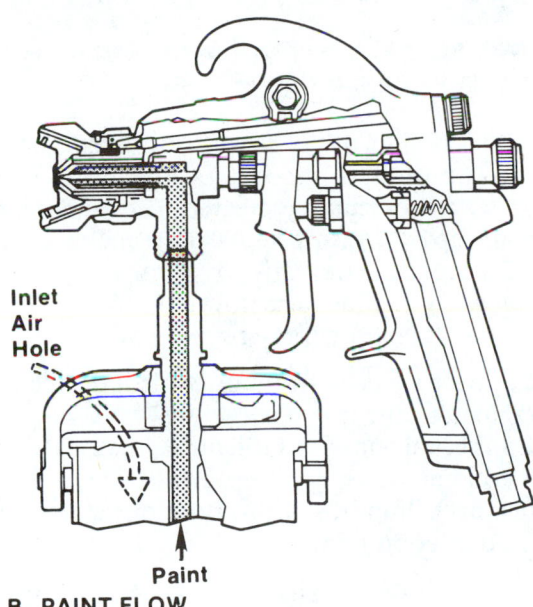

B. PAINT FLOW

FIGURE 18-9 Follow airflow and paint flow of pressure feed gun.

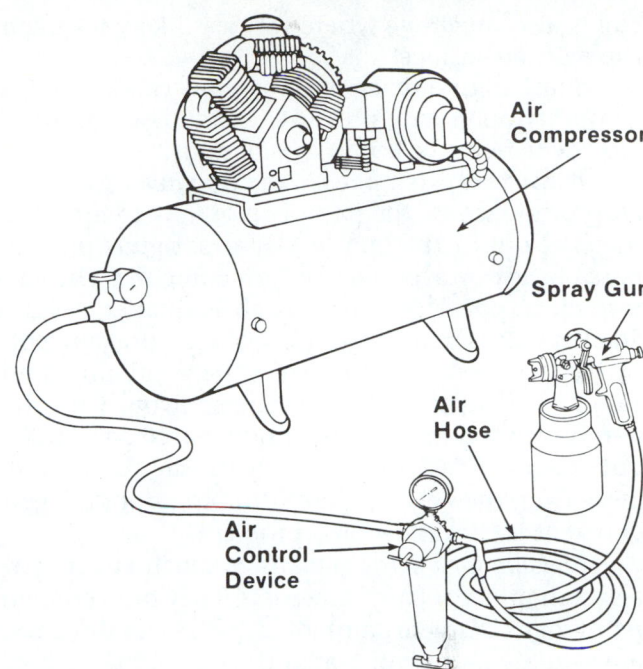

FIGURE 18-10 Study suction feed equipment hook-up.

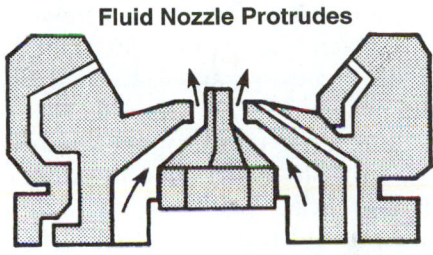

A. SUCTION AIR NOZZLE

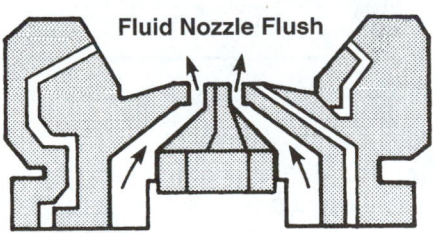

B. PRESSURE AIR NOZZLE

FIGURE 18-11 Compare difference between suction and pressure feed gun air nozzles.

from the cup, up the fluid inlet, and out through the open fluid tip. Air enters through the air hole and replaces the siphoned paint (Figure 18–9B). The inlet air vent holes in the cup lid must be open.

The suction feed equipment (Figure 18–10) is hooked up for operation as follows:

- Connect air line from the compressor outlet to the air control device or regulator inlet.
- Connect air hose leading from the air outlet on the air control device to the air inlet on the spray gun.
- After the material has been reduced to proper consistency, thoroughly mixed, and strained into the cup, attach the gun to the cup.

It is easy to identify a suction feed gun by its fluid tip that extends slightly beyond the face of the air cap, as shown in Figure 18–11A.

In the **gravity feed system**, the paint is supplied by gravity and the material is suction-forced at the nozzle tip. You can identify a gravity feed gun easily because the cup is on top of the gun, not under it. This high efficiency system is ideal for all spraying operations. The handling of the gun is the same as a suction feed gun. A gravity feed gun is easier to handle because of its better balance. The cup is also up above where it is less likely to touch the painted surface.

Some high efficiency spray guns can also use a combination gravity-pressure feed cup design. It offers the benefits of both types of guns.

Pressurized pot or **pressure cup spray guns** use air pressure inside the paint cup or tank to force the material out of the gun. Pressure pot guns provide possible advantages over siphon cup guns. They allow more paint speed through a smaller nozzle. Smaller paint streams atomize better. A pressure pot, by having a remote cup, makes the gun lighter and easier to handle. It also permits spraying with the gun horizontal for painting under flared parts without danger of "spitting." Also, with the cup or tank away from the vehicle, paint dripping from the gun cup vent is eliminated as a problem.

Pressure tank spray guns use a much larger storage container for paint materials. They hold enough paint for a complete paint job, which saves time. You are sure the paint will match throughout the whole job. This might help when spraying hard-to-match metallic or pearl paints, for example.

Remember that pressure cups have seals that must be kept clean and regularly inspected for damage. A loss of cup pressure affects the delivery of fluid to the spray gun.

Pressure cups also require some bleed-down time if the cup is initially over-pressurized. For example, if a cup is pressurized to 10 psi (69 kPa) and the painter desires only 6 psi (41 kPa), the cup maintains that 10 psi (69 kPa) pressure momentarily when the trigger is first pulled back. This could force out too much paint and cause problems.

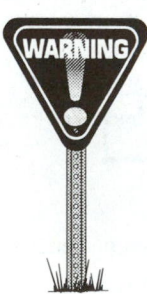

Always adjust line pressure to specs to prevent damage or rupture of the cup or tank. The specs will be given by the manufacturer.

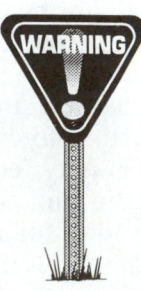

Pressure cups also hold pressure after being disconnected from the air source. This can be embarrassingly messy if you open the lid and paint blows all over you and the shop. Make sure you release the cup pressure before opening the lid.

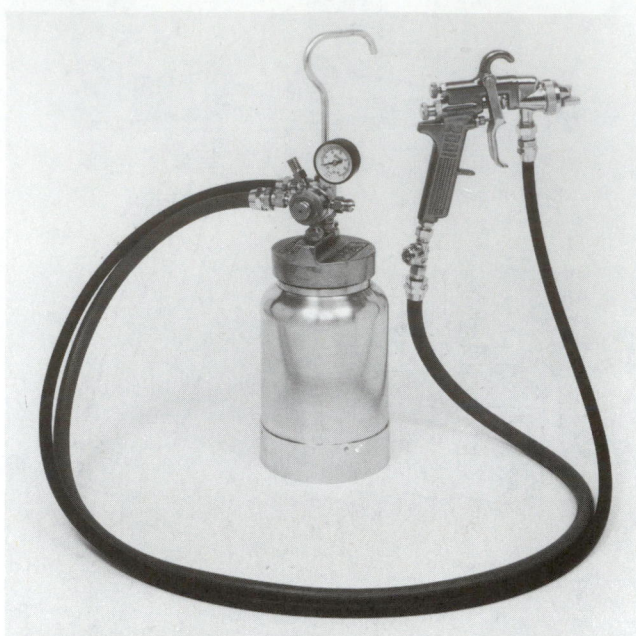

FIGURE 18-12 Air and fluid hoses connect gun and pressure cup. *(Courtesy of Binks Mfg. Co.)*

In the design of an air pressure feed gun, the fluid tip is flush with the face of the air cap (Figure 18–11B) and no vacuum is created. The fluid is forced to the air cap by pressure kept on the material in the system: a separate cup, tank, or pump.

Figure 18–12 illustrates how to hook up the regulated pressure cup for spraying:

- Connect air hose from air control device to air regulator on cup.
- Connect air hose or tank air regulator to air inlet on gun.
- Connect fluid hose from fluid outlet on cup to fluid inlet on gun.

Figure 18–13 illustrates how to hook up the equipment of the pressure tank spraying system:

- Connect regulated air hose from air control device on tank to air inlet on gun.
- Connect mainline air hose from main regulating device to air regulator inlet on tank.
- Connect fluid hose from fluid outlet on tank to fluid inlet on gun.

Paint pressure tanks are available in sizes from 2 quarts to 10 gallons (1.8 to 38 liters). They are available in dual, single, or nonregulated models (Figure 18–14). Dual air regulators control both material and atomization air pressure. Single models regulate material pressure only.

An advantage of a large tank pressure feed gun is that you will not run out of paint and have to refill the cup while doing a complete paint job. With the tank or cup away from the vehicle, there is also less chance of the cup dripping on the vehicle.

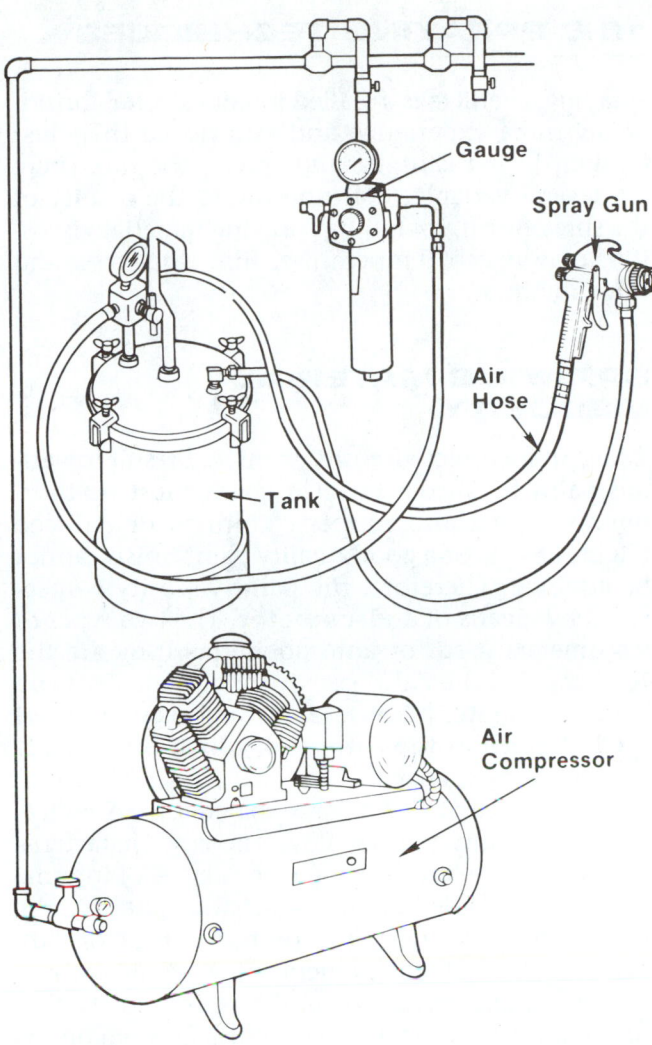

Gauge

Spray Gun

Air
Hose

Tank

Air
Compressor

FIGURE 18-13 Study hook-up for pressure tank.

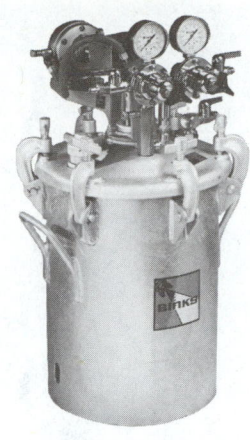

A

B

C

FIGURE 18-14 Compare pressure tanks (A) dual regulated, (B) single regulated, and (C) nonregulated. *(Courtesy of Binks Mfg. Co.)*

A disadvantage of the pressure system is clean-up. The tank and fluid lines must be cleaned and flushed right after use. This takes a little more time and solvent than conventional suction feed and HVLP guns.

Some tanks have an **agitation paddle** system to keep the pigments and solids thoroughly mixed at all times, assuring color uniformity. Some siphon gun cups also have an agitator system (Figure 18–15). These cups provide constant mixing of all automotive finishes and primers; they even keep metal flakes and metallics in total suspension and complete dispersion.

The main requirement of gravity feed is that the container be vented so that atmospheric air can replace the material as it is being sprayed. The gravity

FIGURE 18-15 Agitator type paint cup has paddle that moves up and down to mix paint. This is important with metallic paints that tend to settle in the cup. (Courtesy of Binks Mfg. Co.)

feed equipment is relatively inexpensive in initial cost. Viscosity and flow characteristics of the material directly affect rate of flow to the gun, as do hose size, hose length, and nozzle size. Flow is also affected by changes in the pressure head, which will vary with the vertical position of the gun and with the material in the container. A pressurized gravity feed gun eliminates all of these concerns.

The container can be of any convenient size with $1/2$ and 1 quart the most common in body shops. Its location should suit the material supply requirements of the gun.

Suction and gravity feed guns are a popular type of gun in auto refinishing shops for all types of work (spot, panel, and overall). The pressure feed gun is mainly used for overall painting, for spraying some heavier refinishing materials that are too heavy to be siphoned from a container, or where volume painting is required.

Gravity feed guns can be used for basecoat/clearcoat work and to spray undercoat refinishing materials, such as primers and sealers, as well as some lighter spray-on fillers. Modern gravity-pressure feed guns are as easy to use as older, less efficient designs.

SHOP TALK Some modern HVLP guns look like gravity feed guns but they are pressure-gravity feed. The cup is sealed and gun pressure is used to help move the sprayed material out of the cup and into the gun. This is an excellent design.

18.2 SPRAYING TECHNIQUES

Spraying a vehicle is a skilled job. It calls for considerably more experience and knowledge than just holding down the trigger and moving the gun. There are several variables contributing to the quality of the spray finish, including spraying material viscosity, spray booth temperature, film thickness, and spray method.

SPRAYING MATERIAL VISCOSITY

Using an incorrect viscosity paint will result in various paint finish defects. The paint must be thoroughly mixed and properly thinned or reduced (Figure 18–16) or a good-quality paint finish cannot be attained. Therefore, the paint viscosity is measured by means of a **viscometer**. The two types of viscometers used for automobile painting are the Ford cup and the Zahn cup. Although the Ford cup is very accurate, because of its high cost it is not used as much as the Zahn cup, which is less accurate but less expensive.

The **viscosity** of a spray material is an indication of its ability to resist flow. The flow characteristics of liquids relate directly to the degree of internal friction. Therefore, anything that will influence the internal friction (such as solvents, thinners, or temperature change) will influence flow. Similarly, it is the flow characteristics that determine how well a material will atomize, how well it will flow out on the work, and the type of equipment needed to move it.

When preparing material for spraying, thin to the proper viscosity according to the directions on the can or in the product manual, using the thinner

FIGURE 18-16 To achieve correct atomization, the paint must be reduced with the proper solvent. (Courtesy of Maaco Enterprises, Inc.)

FIGURE 18-17 Robots are used by most car manufacturers for the application of the finishing system. *(Courtesy of Sherwin-Williams Co.)*

or reducer best suited for the shop temperature and conditions. It can be demonstrated that at a given temperature, a 3-second difference in spraying viscosity will have a distinct influence on the flow of the material being sprayed. It can therefore be seen that exact reduction is essential if the painter is going to spray at the viscosity at which the paint will spray the easiest and the best results can be obtained. In auto factory operations (Figure 18-17), where new cars are sprayed, spraying viscosity is held within a tolerance of 1 second at a given temperature.

No method other than measurement of the thinner or reducer does the job adequately. Because appearance is affected by the temperature, the way the paint runs off the stirring paddle is not a reliable method of determining viscosity.

The amount of reduction should be the same regardless of temperature. At a higher temperature the viscosity of the reduced material is actually slightly lower, but this is offset by the faster evaporation of the thinner as it travels between the gun and the surface being painted. The result is that the paint reaches the surface at the correct viscosity. The re-

verse is true in a cold shop. The reduced paint is a little thicker, but evaporation in the air is less so that the paint reaches the surface being sprayed at the proper viscosity.

Various automotive finishes are manufactured to spray at ideal viscosities. Refer to the label on the material to find the recommended viscosity.

Generally, with a #2 Zahn cup, reduce:

- Very thin materials (wash primers, dyes, and stains) to 14 to 16 seconds.
- Thin materials (sealers, primers, zinc chromates, and acrylics) to 16 to 20 seconds.
- Medium materials (synthetic enamels, primer-surfacers, epoxies, urethanes, basecoat/clearcoat, etc.) to 19 to 30 seconds.

Ford Cup

The Ford cup (Figure 18-18) used for automobile painting comes in two sizes: #3 and #4. It has a cylindrical container, made of either aluminum or stainless steel. The bottom of the cup is conical-shaped with an orifice in the center. The #3 and #4 cups are distinguished by the diameter of this orifice. Ford cups are precision-made and care should be taken to prevent any damage or deformation of the inner surface of the cup or to the orifice. To measure the viscosity of a paint material with a Ford cup, proceed as follows:

1. Keep the temperature of the paint and Ford cup at room temperature.
2. Secure the cup with the set bolt and place the glass plate on top.
3. Place the level on top of the glass plate, and adjust the level of the frame with the level adjusting bolts. Then place a container below the cup.
4. While supporting the bottom of the cup with one hand, place a piece of thick rubber in be-

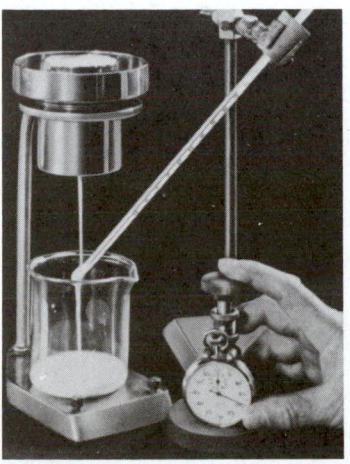

FIGURE 18-18 This is a Ford viscosity cup with a thermometer and stopwatch for testing paint viscosity. In practice, the thermometer is immersed in the paint. *(Courtesy of PPG Industries, Inc.)*

tween to prevent transmission of body heat to the orifice and pour in the paint, being careful that no air bubbles enter.

5. Slide the glass plate horizontally over the top of the cup to remove any excess paint and set it aside.

6. Release the rubber plate supporting the orifice and at the same time begin measuring the time of the continuous downward drain of the paint with a stopwatch. Measure until the continuous paint flow stops. This time is used as an indicator of the paint viscosity. For example, if it takes 15.4 seconds, the viscosity of the paint is said to be 15.4 seconds at 68 degrees Fahrenheit (20 degrees Celsius).

Zahn Cup

The #2 Zahn cup is very popular in auto refinishing shops. It is cylindrical in shape and has an orifice (hole) at the bottom. To determine viscosity with a #2 Zahn cup (Figure 18–19), proceed as follows:

1. Prepare the material to be tested. Mix, strain, and reduce as directed by the manufacturer.
2. Fill the cup by submerging it in the material.
3. Release the flow of the material and trigger the stopwatch. Keep eyes on the flow, not on the watch.
4. When the solid stream of material "breaks" (indicating air passing through the orifice), stop the watch.

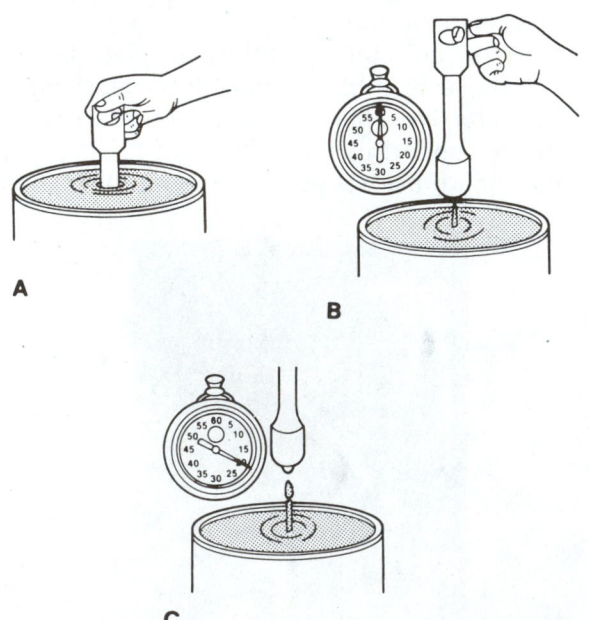

FIGURE 18–19 (A) Dip cup into paint until it is full. (B) Remove cup, and as it clears the surface of the paint, begin timing the flow of paint from the small hole in the bottom of the cup. (A stopwatch is preferred for this step.) (C) Stop the timer when the stream of paint breaks. *(Courtesy of DuPont Automoative Products)*

TABLE 18-2: SPRAYING VISCOSITIES USING THE #2 ZAHN CUP

Material	Reduction	Viscosity
Acrylic enamel	33-1/3%	19 seconds
Acrylic enamel	50%	18 seconds
Acrylic enamel with hardener	75%	16 seconds
Acrylic lacquer	150%	15 seconds
Basecoats	50%	18 to 21 seconds
Clearcoats	10%	18 to 22 seconds
Polyurethane enamel	per manufacturer's instructions	20 to 22 seconds
Primer-surfacer (2-part)	100–200%	18 to 25 seconds

5. The result is expressed in seconds. Table 18–2 gives typical desired results of the #2 Zahn cup.

PAINT MIXING STICKS

Graduated *paint mixing sticks* have conversion scales that allow you easily to convert ingredient percentages into part proportions. They are used by painters to help mix paints, solvents, catalysts, and other additives right before spraying. Detailed instructions for using paint mixing sticks are given in the next chapter.

TEMPERATURE

The temperature at which material is sprayed and dried has a great influence on the smoothness of the finish. This involves not only the air temperatures of the shop, but the temperature of the work as well. A job should be brought into the shop long enough ahead of spraying time to arrive at approximately the same temperature as the shop. Spraying warm paint on a cold surface or spraying cool material on a hot surface will completely upset flow characteristics. The rate of evaporation on a hot summer day is approximately 50 percent faster than it is on an average day with a shop temperature of 72 degrees Fahrenheit (22 degrees Celsius). Appropriate thinners or reducers should be used for warm and cold weather applications.

SHOP TALK

A stopwatch is necessary for measuring paint viscosity with either viscometer system. Most painters prefer a digital stopwatch to the standard type because it is easily read.

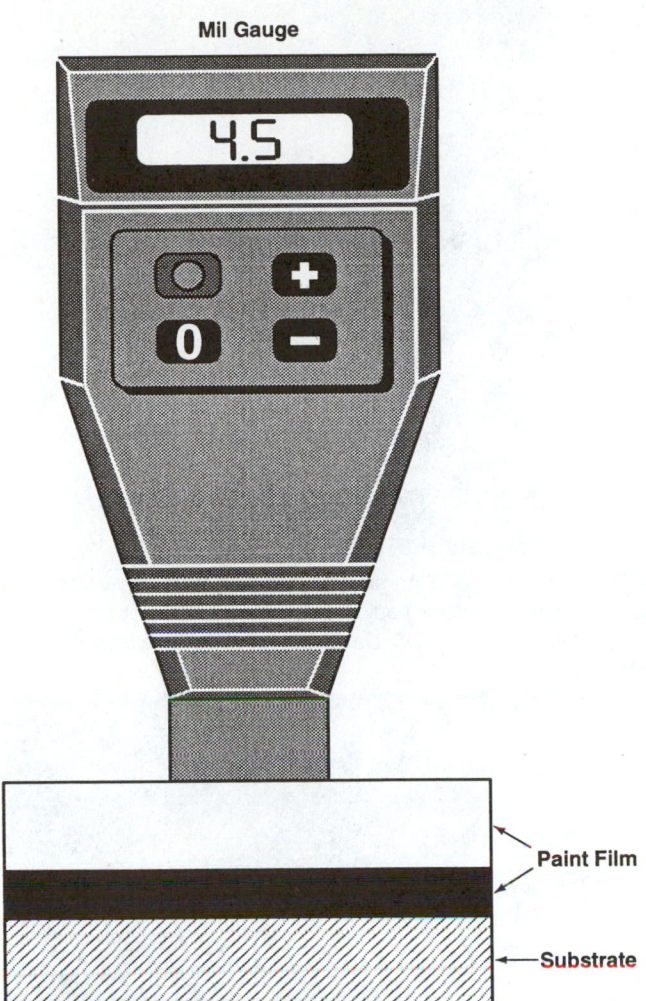

Mil Gauge

Paint Film

Substrate

FIGURE 18-20 Mil gauge will measure paint thickness. If too thick, old paint must be stripped before refinishing. *(Courtesy of Talsol Corporation.)*

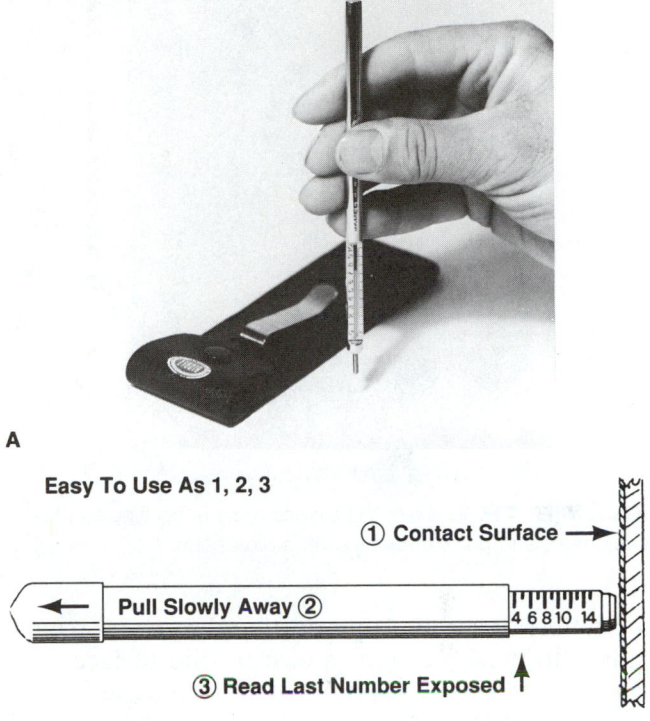

A

Easy To Use As 1, 2, 3

① **Contact Surface** →

← **Pull Slowly Away** ②

③ **Read Last Number Exposed** ↑

B

FIGURE 18-21 (A) A Tinsley gauge is a paint thickness gauge. (B) By pulling the gauge, the magnet will stick to the metal body. The gauge will read the paint thickness. Thicker paint will make the tool pull off the body with less pull. *(Courtesy of Biddle Instruments)*

FILM THICKNESS

Older acrylic lacquers dry by evaporation only. Alkyd and acrylic enamels dry by both evaporation and oxidation. Urethanes dry by evaporation and chemical cross-linking reaction.

The thicker the film applied, the longer the drying time. The difference in film thickness shows up plainly in primer-surfacers and enamels. A lacquer primer-surfacer that can be sanded in 30 minutes at 70 degrees will take over an hour if sprayed twice as heavily. Alkyd enamel of normal film thickness should dry tack free in 4 to 6 hours at 70 degrees and be hard enough for unmasking and handling in 16 hours. If sprayed twice the normal thickness, this will take 2 to 3 times longer. The thicker the film, the greater the depth of paint from which the thinner or reducer must work its way out. In enamels, the greater the distance the oxygen from the air must penetrate in order to dry or oxidize the finish, the longer the drying time will be. This process is complicated in thick coats by surface skins or crusts as the paint dries.

The technician should develop a technique so that the coat sprayed on a surface will remain wet long enough for proper flow-out, and no longer. Heavier coats are not necessary. They can produce sags, curtains, or wrinkles, as well as strongly influence metallic color when matching.

The amount of material sprayed on a surface with one stroke of the gun will depend on the width of the fan, the distance from the gun, the air pressure at the gun, the amount of reduction, the speed of the stroke, and the selection of thinner or reducer.

Discussed in previous chapters, many paint shops have a paint thickness measuring meter such as the one shown in Figure 18–20. This instrument is able to determine the thickness of the paint on a vehicle. It measures the thickness of paint and any body filler by sensing the distance between the paint surface and the metal body. It will give a digital readout of paint thickness.

Another popular and less expensive paint thickness gauge is the Tinsley gauge (Figure 18–21). This gauge consists of a special lightweight magnet attached to a spring and contained within a pencil-like tube. To take a measurement, the exploring head or magnet is placed on the surface and the body of the gauge is drawn away, thus extending the spring. The spring extension, the amount of which is ob-

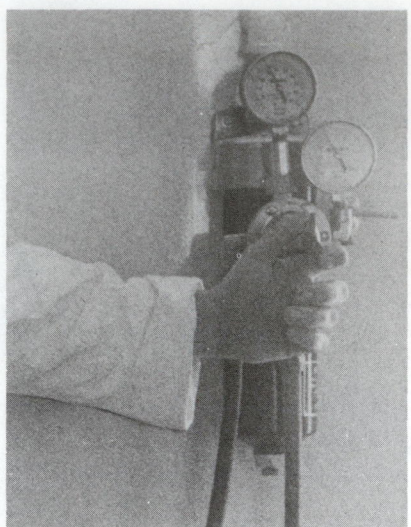

FIGURE 18-22 Set the air pressure according to the equipment and paint manufacturer's directions.

served on the scale, is proportional to the force required to detach the magnet from the surface. The reading is taken at the point when the magnet breaks away from the surface, and the thickness is read directly from the scale.

ADJUSTING THE SPRAY

A good paint spray pattern depends on the proper mixture of air and paint droplets. This is much like a fine-tuned engine that depends on the proper mixture of air and gasoline. The sprayed material should go on smoothly in a medium to wet coat without sagging or running. There are three basic adjustments that will give the proper spray pattern, degree of wetness, and air pressure for suction feed guns.

1. Adjust the pressure to the spray gun and paint manufacturer's recommendations (Figure 18–22). Air pressure, as described in Chapter 5, generally is set at the separator-regulator (or *transformer*). But due to friction as air passes from the regulator through the hose to the gun, pressure will be lost. The difference between the reading at the regulator and the reading at the gun will vary depending upon the length and diameter of the hose. For example, a 50-foot, $1/4$-inch diameter hose will yield a lower reading than a 15-foot, $5/16$-inch diameter hose. For this reason, pressure should be measured at the gun and all recommended pressures in this test are for readings at the gun.

   The surest method to measure this pressure drop is with an air gauge inserted between the hose coupler and the gun. Some guns are equipped with regulators that allow for checking and setting pressure at the gun (Figure 18–23),

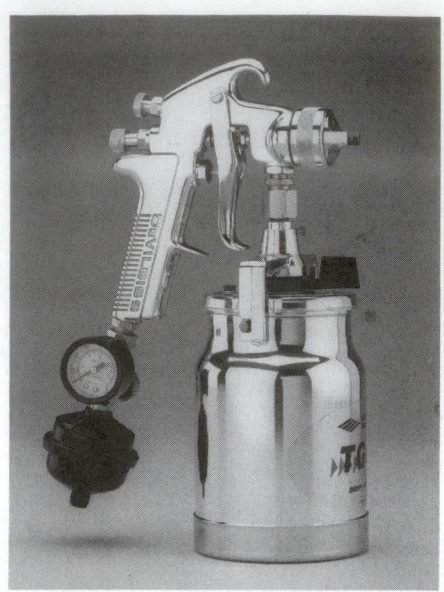

FIGURE 18-23 An air adjusting valve with a gauge will allow accurate pressure adjustment and readings at the spray gun. *(Photo courtesy ITW DeVilbiss Automotive Refinishing Products)*

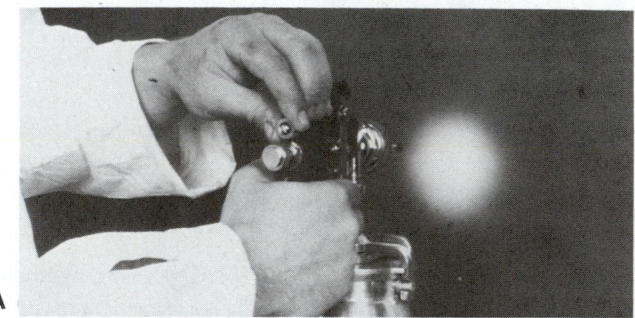

A

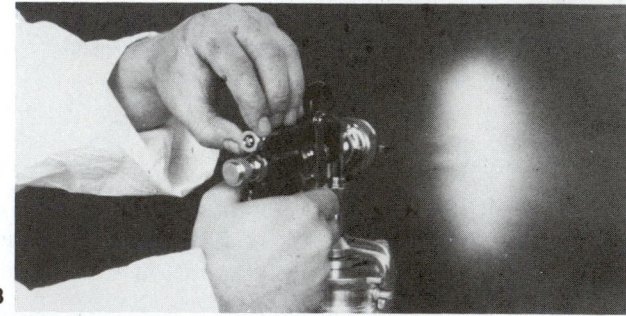

B

FIGURE 18-24 (A) Turn fan valve in to focus spray and (B) back it out to get desired fan pattern. *(Courtesy of Maaco Enterprises, Inc.)*

while others have optional accessories to do the same thing. Another method is to consult Table 18–3.

2. Set the size of the spray pattern using the fan adjustment or pattern control knob. To adjust the fan pattern, turn the pattern control knob all the way in to create a small, round pattern (Figure 18–24A). Then back out or unscrew the pattern control to produce a tall spray pattern (Figure 18–24B). Use narrower patterns for spot

TABLE 18-3: ESTIMATED AIR PRESSURES AT THE GUN

Pressure Reading (lbs.) at Gauge		Pressure at the Gun for Various Hose Lengths					
		5 feet	10 feet	15 feet	20 feet	25 feet	50 feet
1/4-Inch Hose	30	26	24	23	22	21	9
	40	34	32	31	29	27	17
	50	43	40	38	36	34	22
	60	51	48	46	43	41	29
	70	59	56	53	51	48	36
	80	68	64	61	58	55	43
	90	76	71	68	65	61	51
5/16-Inch Hose	30	29	28-1/2	28	27-1/2	27	23
	40	38	37	37	37	36	32
	50	48	47	46	46	45	40
	60	57	56	55	55	54	49
	70	66	65	64	63	63	57
	80	75	74	73	72	71	66
	90	84	83	82	81	80	74

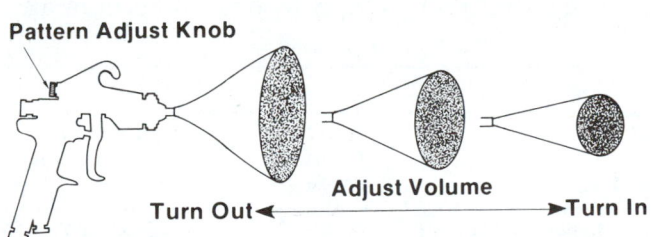

Pattern Adjust Knob

Turn Out ◄— Adjust Volume —► Turn In

FIGURE 18-25 Pattern width adjustment is done by turning adjustment knob as shown.

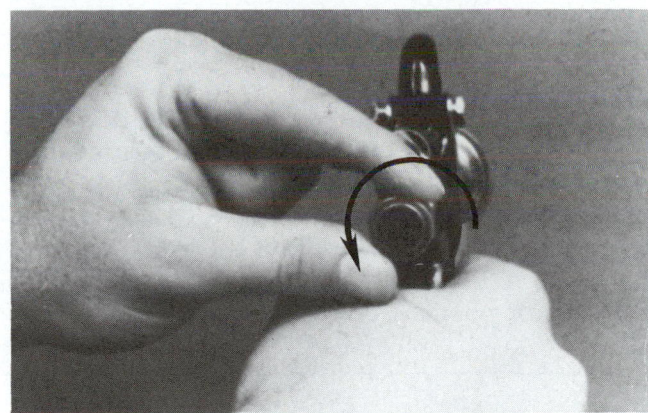

FIGURE 18-26 Fluid control valve is often set for near maximum flow by turning the knob counterclockwise the prescribed amount.

SHOP TALK

When making your spray gun pattern adjustments, tape a large sheet of masking paper to the wall of the spray booth. This will let you see how the spray pattern is forming on a test surface.

Figure 18–25 represents the adjustment of spray pattern from all the way in to all the way out.

repairs and wider patterns for panel repairs or overall painting.

3. Set the fluid control knob (Figure 18–26) to regulate the amount of paint according to the selected pattern size. Back the knob out to increase the paint flow and turn the knob in to decrease paint flow (Figure 18–27).

The optimum spraying pressure is the lowest needed to obtain proper atomization, emission rate, and fan width. A pressure that is too high results in excessive paint loss through overspray and poor flow, due to high solvent evaporation before the paint reaches the surface being sprayed.

SHOP TALK

With most paint technicians, the fluid control knob is set in the full open position. Typically, two or three threads are left showing on the adjusting valve screw. This will allow the experienced painter to refinish the vehicle in minimum time.

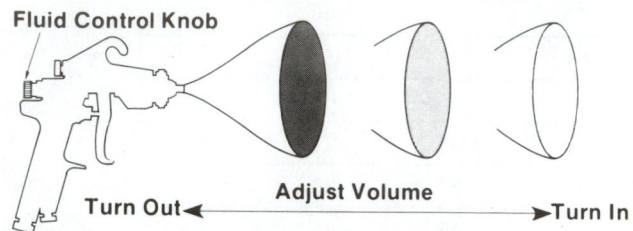

FIGURE 18-27 Study what happens by turning fluid control knob.

A pressure that is too low gives a paint film with poor drying characteristics, due to high solvent retention. Low gun pressure also makes the paint prone to bubbling and sagging. The recommended pounds of air pressure vary with the kind of material to be sprayed and the type of gun.

The typical pressure ranges for conventional and HVLP guns are given in Table 18–4. Remember that gun inlet pressures will vary with the manufacturer. Many low *volatile organic compound* (VOC) regulations require 10 psi (69 kPa) or less at the air cap.

Remember! Always refer to the gun's owner's manual for published pressures to get best results.

Always follow the spray gun and paint manufacturer's air pressure recommendations for the type of material to be sprayed.

Also note that each nozzle will require a different inlet pressure. Generally, the thicker the material being sprayed the higher the needed inlet pressure and the smaller the gun tip needed. If you change material thickness (to a VOC, high solid paint, for example), you may have to use a smaller nozzle and higher pressure to get a smooth paint film without orange peel.

HVLP guns will have the maximum inlet pressure stamped on the gun body or nozzle. Do not exceed this value to maintain the required 10 psi (69 kPa) at the nozzle.

BALANCING PRESSURE FEED GUN SYSTEM

To balance the pressure tank (Figure 18–28) or cup for spraying, the procedure is as follows:

1. After the paint is poured into the container (Figure 18–29), open the pattern control knob for maximum pattern size. Open the fluid control knob until the first thread is visible.
2. Shut off the atomization air to the gun. Set the fluid flow rate by adjusting the air pressure in the paint container. Use about 6 psi (41 kPa) for a remote cup and about 15 psi (103 kPa) for a 2-gallon or larger container. Adjust the fluid flow in either of the following ways:

   Remove the air cap, aim the gun into a clean container, and pull the trigger for 10 seconds. Measure the amount of material that flowed in that time and multiply by 6 (or 30 seconds and

Topcoats	HVLP Gun Presssure	Conventional Gun Pressure (psi)	Undercoats	HVLP Gun Pressure	Conventional Gun Pressure (psi)
Polyurethane enamel	18–20 20–30	50–55 (solids) 60–65 (metallic)	Lacquer primer-surfacers	15–18 16–20	25–30 (spot) 35–45 (panel)
Acrylic lacquer	12–18	20–45	Multipurpose primer-surfacers	16–20	30–40 as primer-surfacer
Acrylic enamel	18–20	50–60	Multipurpose primer-surfacers as nonsanding	17–20	35–40
Alkyd enamel	18–30	50–60	Nonsanding primer-sanders	18–20	45
Flexible finishes	14–28	35–40	Enamel primer-surfacers	18-20	45
Basecoat	14–16	30–35	Epoxy primer	18–20	45
Clearcoat	18–20	35–40	Zinc chromate primer	18–20	45
Sealers	HVLP Gun Presssure	Conventional Gun Pressure (psi)	Miscellaneous	HVLP Gun Pressure	Conventional Gun Pressure (psi)
Acrylic lacquer	12–16	25–30	Uniforming finishes	12–14	15–20
Universal sealer	14–18	35–45			
Bleederseal	14–16	35–40			

TABLE 18-4: TYPICAL AIR PRESSURE RANGES

NOTE: Spot repairs should be made at the low end of the air pressure range.

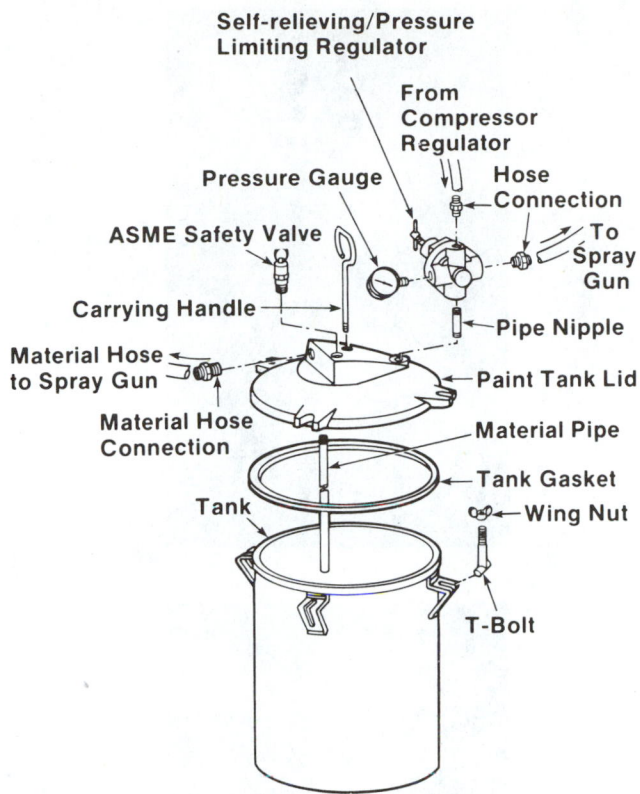

FIGURE 18-28 Study major parts of pressure tank.

FIGURE 18-29 Pour pre-filtered paint into tank without spilling.

multiply by 2). This is the fluid flow rate in ounces per minute. For standard refinishing, it should be about 14 to 16 ounces (413 to 472 mm) per minute. If the flow rate is less than this, increase the air pressure in the container and repeat. If it is faster than this, decrease the pressure slightly. When the flow rate is correct, reinstall the air cap.

OR

Pull the trigger and adjust the pressure on the paint container until the stream of paint discharging from the gun squirts about 3 to 4 feet (2.7 to 3.6 meters) before it starts to drop. This indicates a fluid flow of about 14 to 16 ounces (413 to 472 mm) per minute.

3. Turn on the atomization air to about 50 at the gun. Then spray a fast test pattern.

SHOP TALK Some gravity feed gun designs pressurize the cup to help paint flow into the gun. Their set-up, operation, and service is similar to conventional guns.

FIGURE 18-30 The best way to learn the effects of gun movement and gun adjustments is to experiment on a test surface. Old newspapers help recycle waste paper but should not be attached to vehicle because printing ink could contaminate surface. Never start out spraying the vehicle because the gun may not be working properly.

TESTING THE SPRAY PATTERN

A **spray pattern test** checks the operation of the spray gun on a piece of paper. Before attempting to paint the vehicle, it is very important to test the spray pattern (Figure 18–30).

Typically, hold the gun 8 to 10 inches (203 to 254 mm) away with a conventional gun. Hold the gun 6 to 8 inches (152 to 203 mm) away from the paper if using an HVLP gun. Pull the trigger all the way back and release it immediately. This burst of paint should leave a long, slender pattern on the test paper (Figures 18–31 and 18–32). Make a couple of

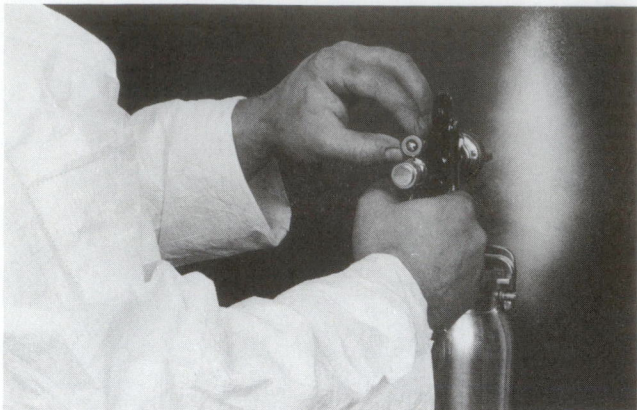

A

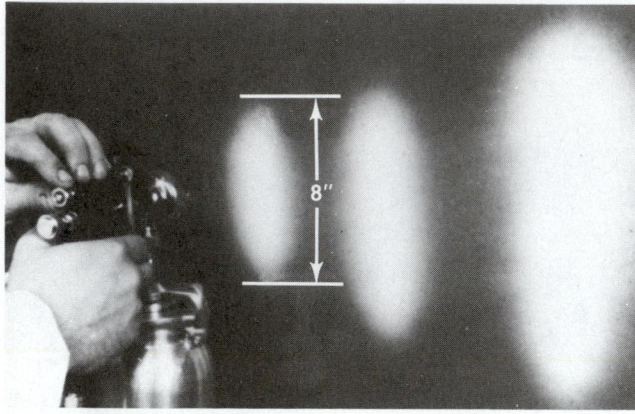

8"

B

FIGURE 18-31 Test patterns on sheet of masking paper to make sure everything is working properly.

SHOP TALK

If you recycle and use old newspaper as a spray pattern test sheet, do not attach it to the vehicle. The newspaper ink can bleed through the newspaper, contaminating the body surface. Attach the test sheet to a fabricated stand or holder or to the wall in the spray booth.

spray passes over the test sheet to make sure the gun is not spitting, leaking, or having other problems.

To narrow the pattern, adjust the air valve inward (clockwise). To widen the spray pattern, turn the air valve outward (counterclockwise).

A spray pattern test that is:

1. Heavy in the middle—could mean too little air flow.
2. Divided in the middle—indicates too much air flow.
3. Heavy at top or bottom—might be caused by a restriction at the fluid needle or air cap horn.

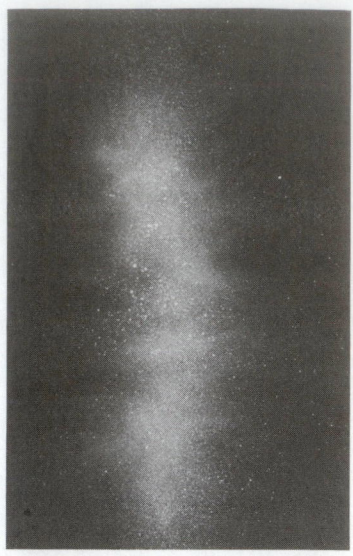

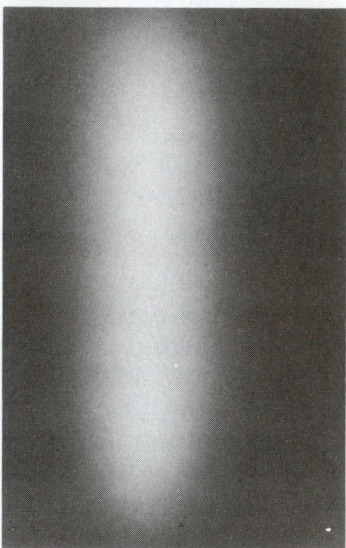

FIGURE 18-32 Inspect the texture of the spray pattern. (Top) Pattern is dry and coarse. (Bottom) Pattern looks good.

SHOP TALK

Some HVLP, high-efficiency guns have the fan adjustment at the cap nut. By turning the cap nut out, you widen the fan. By turning the cap nut in, you narrow the fan width. An air needle and knob is not needed with this gun design.

4. Leaning to one side—could mean that there is a restriction at the fluid needle or air cap horn.

If the pattern is heavy on one side or the top or bottom, try turning the air cap 180 degrees. If the pattern remains the same, clean or replace the fluid

SHOP TALK Remember that the objective of spraying on a test surface is two-fold. First, make sure all atomized paint particles are of uniform size. Second, make sure this size is fine enough to achieve proper flow-out (Figure 18–33).

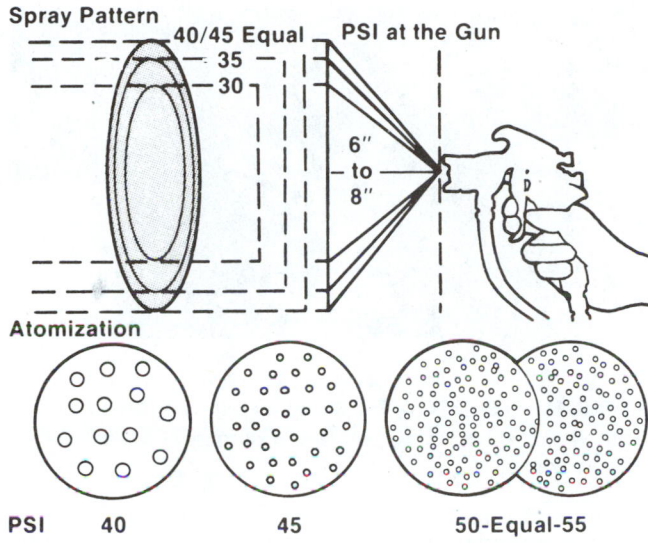

FIGURE 18–33 Note affects of gun pressure on spray pattern.

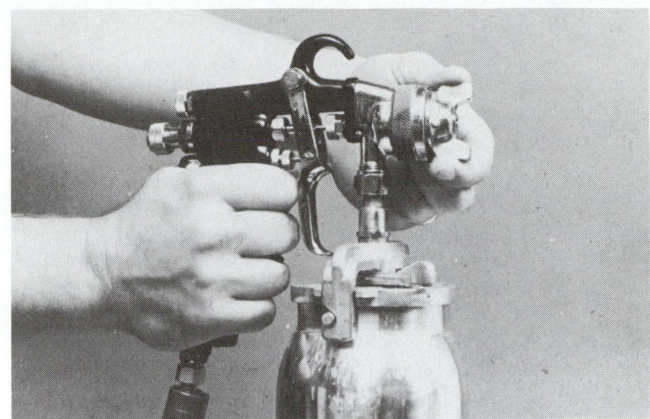

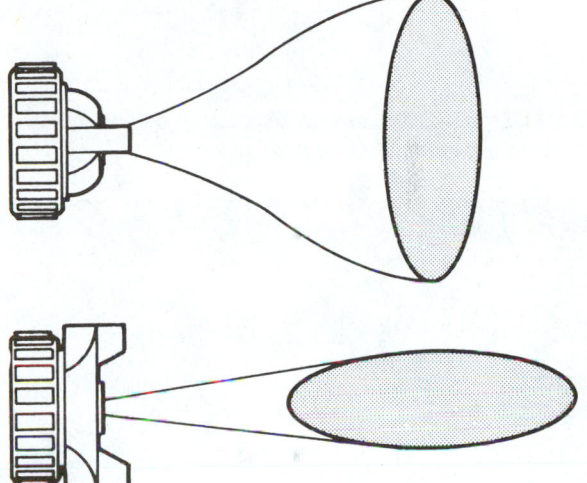

FIGURE 18–34 You can rotate air caps to change pattern. Usually, use the vertical pattern. Use horizontal when doing flood test.

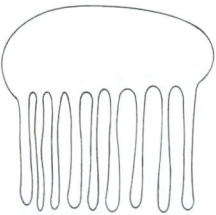

FIGURE 18–35 When flooding the test sheet, balanced spray pattern should result. Runs should be equal across pattern.

SHOP TALK Most spray pattern problems are usually caused by a clogged passage in the gun. Improper maintenance is usually the cause of a spray pattern problem.

needle and fluid nozzle. If the pattern rotates 180 degrees, then the problem is in the air cap horns.

Spraying primer-surfacer usually requires a smaller spray pattern. Turn in the pattern control knob until the spray pattern is 6 to 8 inches (152 to 203 mm) wide. For spot repair, the pattern should be about 5 to 6 inches (127 to 152 mm) from top to bottom.

If the paint droplets are coarse and large, close the fluid control knob about one half turn or increase the air pressure 5 psi (34 kPa). If the spray is too fine or too dry, either open the fluid control knob about one half turn or decrease the air pressure 5 psi (34 kPa).

Next, test the spray pattern for uniformity of paint distribution (Figure 18–34). Loosen the air cap retaining ring and rotate the air cap so that the horns are straight up and down. In this position, you will get a horizontal spray pattern instead of a vertical one.

Spray again on your test paper. However, hold down the trigger until the paint begins to run. This is known as **flooding the pattern**. Inspect the lengths of the runs. If ALL adjustments are correct, the runs will be almost equal in length (Figure 18–35).

The uneven runs in the split pattern shown in Figure 18–36 are a result of setting the spray pattern

FIGURE 18-36 Split pattern shows too much spray on each end of pattern. Air flow from sides of cap is insufficient.

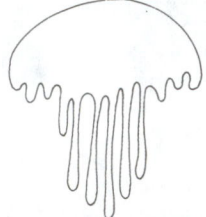

FIGURE 18-37 Heavy center pattern means too much air is being metered out of side orifices in cap.

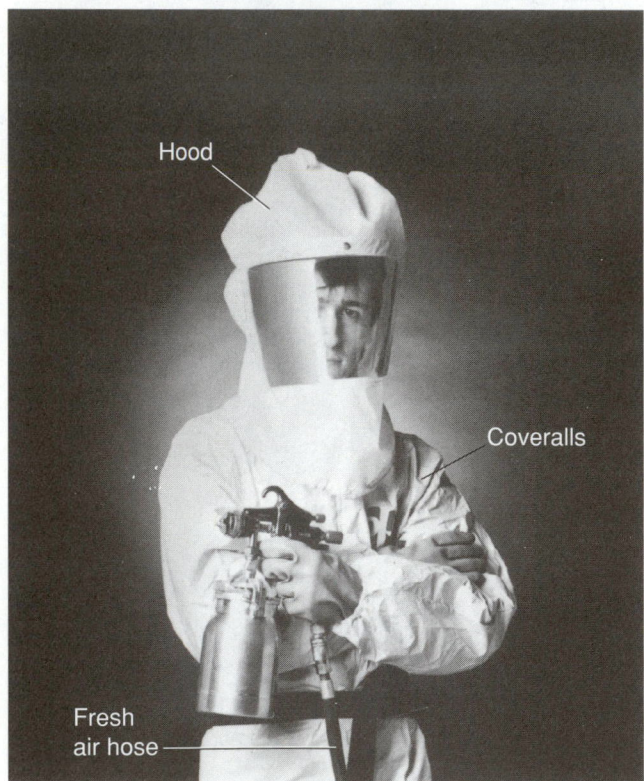

FIGURE 18-38 Be sure to wear an air respirator when spraying. *(Courtesy of PPG Industries, Inc.)*

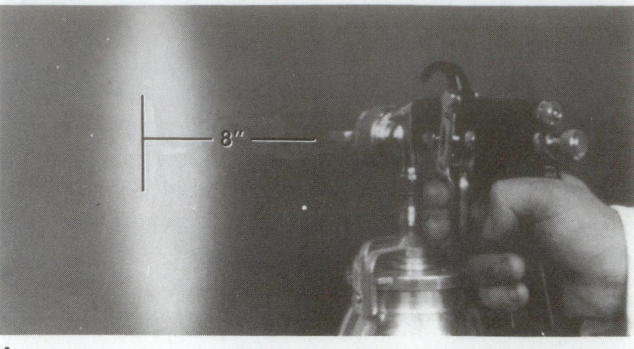

A

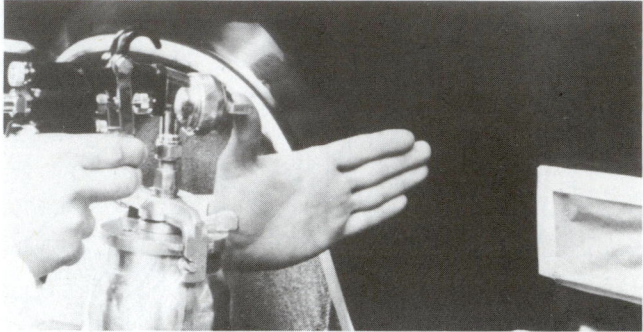

B

FIGURE 18-39 (A) Proper spray distance; (B) easy method of checking spray distance. *(Courtesy of Maaco Enterprises, Inc.)*

 Always wear a suitable air respirator when doing any spraying (Figure 18–38).

THE APPLICATION STROKE

Gun stroke refers to the hand movement used to move the gun while spraying. The proper stroke is important in obtaining a good refinishing job. To obtain a good stroke technique, proceed as follows:

1. Hold the spray gun at the proper distance from the surface—6 to 8 inches (152 to 203 mm) for lacquer, 8 to 10 inches (203 to 254 mm) for enamel (Figure 18–39). If the humidity is high, a shorter distance might be necessary. If the spraying is done from a shorter distance, the high velocity of the spraying air tends to ripple the wet film. If the distance is increased beyond that, there will be a greater percent of reducer evaporated, resulting in orange peel or dry film, and adversely affecting color where matching is

too wide or the air pressure too low. Turn the pattern control knob in ½ turn or raise the air pressure 5 psi (34 kPa). Alternate between these two adjustments until the runs are even in length.

If paint runs are longer in the middle than on the edges (Figure 18–37), too much paint is being discharged. Turn the fluid control knob in until the runs are even in length.

required. A lower evaporating thinner will permit more variation in the distance of the spray gun from the job but will produce runs if the gun gets too close (Figure 18–40). Excessive spraying distance also causes a loss in materials due to overspray.

2. Hold the gun level and perpendicular to the surface (Figure 18–41). If the spray gun is not kept at a right angle even at curves, an uneven paint film will result (Figure 18–42). On flat surfaces, such as the hood or roof, the gun should be pointed straight down (Figure 18–43).

3. The gun should be in motion before the trigger is pulled, and the trigger should be released before the gun motion stops. This technique gives a fade-in and fade-out effect, which prevents overloading where one series of strokes is joined to the next by overlapping the stroke ends.

4. Do not fan the gun if a uniform film is desired. The only time fanning is permissible is on a small spot repair area where the paint film at the edges of the spot should be thinner than the center portion.

5. Move the gun with a steady deliberate pass, about 1 foot (0.3 meter) per second. Moving the gun too fast will produce a thin film, while moving it too slowly will result in the paint running. The speed must be consistent or it will result in an uneven paint film. Never stop in one place or the sprayed coat will drip and run!

6. Release the trigger at the end of each pass. Then pull back the trigger when beginning the pass in the opposite direction. In other words: "trigger" the gun and turn off the gun at the end of each sweep. This avoids runs, minimizes overspray, and saves paint. Proper triggering involves four steps (Figure 18–44): (A) begin the stroke over the masking paper, triggering the gun halfway to release only air; (B) when the starting edge of the panel is reached, squeeze the trigger all the way to release the paint; (C) release the trigger halfway to stop the paint flow when directly over the finishing edge; and (D) continue the stroke several more inches before reversing the direction and repeating the sequence.

7. Difficult areas such as corners and edges should be sprayed first. Aim directly at the area so that half of the spray covers each side of the edge or corner. Hold the gun an inch or two (25 to 50 mm) closer than usual, or screw in the pattern control knob a few turns. Either technique will reduce the pattern size. If the gun is just held closer, the stroke will have to be faster to compensate for a normal amount of material being applied to a smaller area. After all of the edges and corners have been sprayed, the flat or nearly flat surfaces should be sprayed.

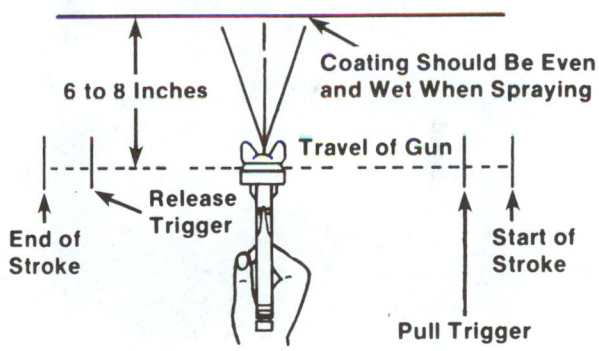

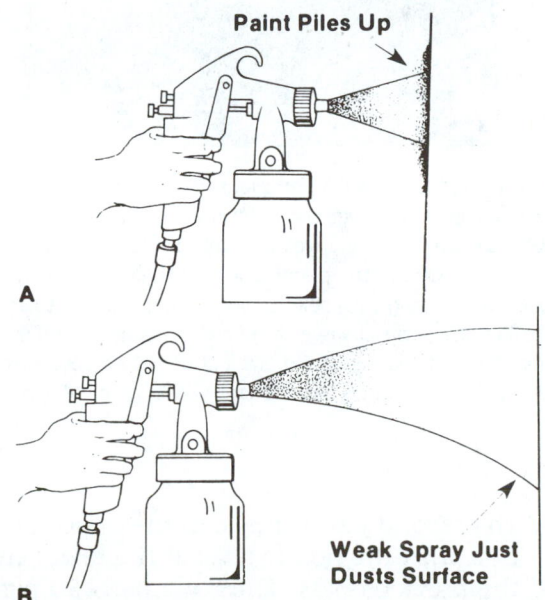

FIGURE 18–40 The correct gun-to-work distance is important. If the gun is too close. (A) the finish material piles up and causes runs and sags. When the gun is too far away, (B) material tends to dry into dust before it reaches the surface. Adjust distance accordingly.

FIGURE 18–41 (Top) Always try to hold the gun parallel with the surface being sprayed. (Bottom) Never fan the gun over a flat surface, or the surfaces to the right and left will have less paint on them.

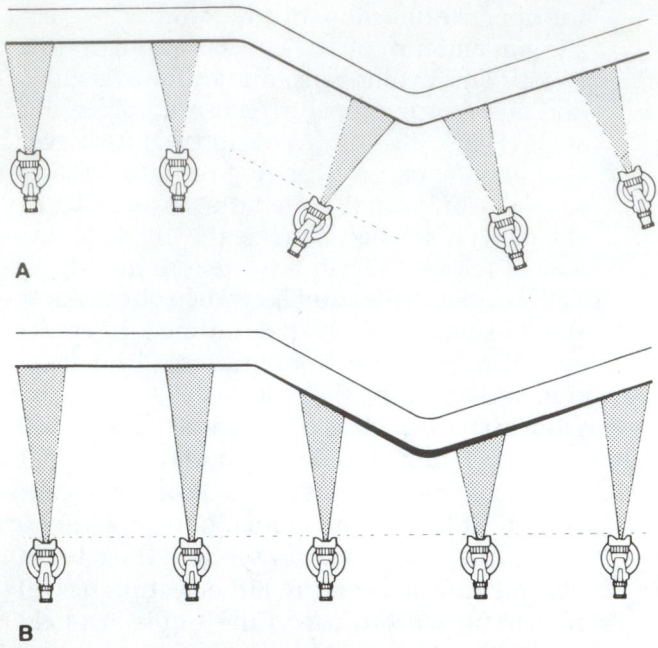

FIGURE 18-42 (A) Proper spray gun movement. (B) If the spray gun is not kept at a right angle even at curves in the body, an uneven paint film will result.

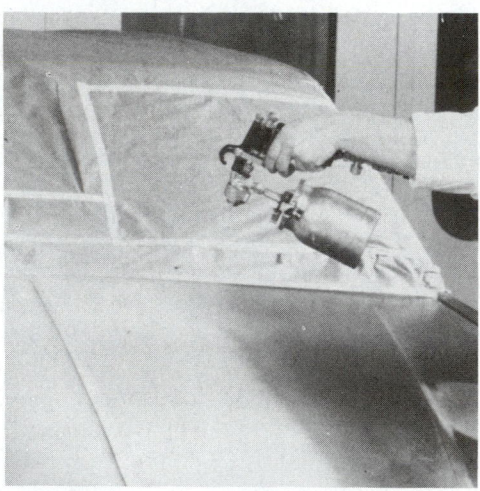

FIGURE 18-43 You must move the spray gun like a machine or robot to get a uniform deposit of paint on the vehicle. *(Courtesy of Maaco Enterprises, Inc.)*

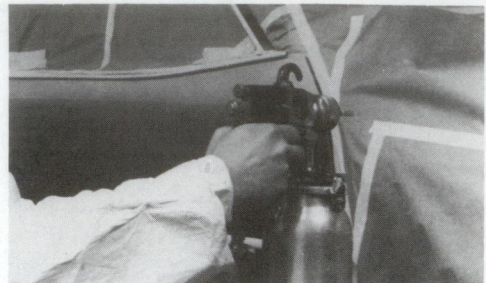

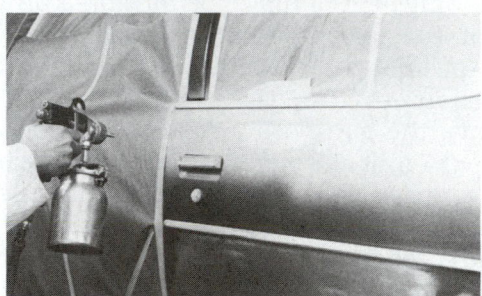

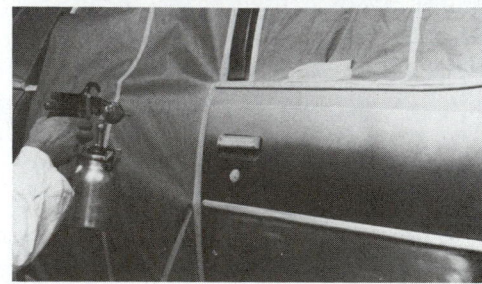

FIGURE 18-44 Proper triggering involves four steps: (A) begin the stroke over the masking paper and as the gun is moved, trigger halfway to release only air; (B) when reaching the starting edge of the panel, squeeze the trigger all the way to release the paint; (C) release the trigger halfway to stop the paint flow when directly over the finishing edge; and (D) continue the stroke several more inches before reversing the direction and repeating the sequence. *(Courtesy of Maaco Enterprises, Inc.)*

8. For painting very narrow surfaces, switch guns or caps with a smaller spray pattern to avoid having to readjust the full size pattern gun. The smaller pattern guns are easier to handle in critical areas. As an alternate, a full size gun can be used by reducing the air pressure and fluid delivery and triggering properly.

9. Generally, start at the top of an upright surface such as a door panel. The spray gun nozzle should be level with the top of the surface. This means that the upper half of the spray pattern will hit the masking.

10. The second pass is made in the opposite direction with the nozzle level at the lower edge of the previous pass. Thus one half (50 percent) of the pattern overlaps the previous pass and the other half is sprayed on the unpainted area (Figure 18–45).

11. Always blend into "the wet edge" of the previous section sprayed (Figure 18–46). Proper triggering

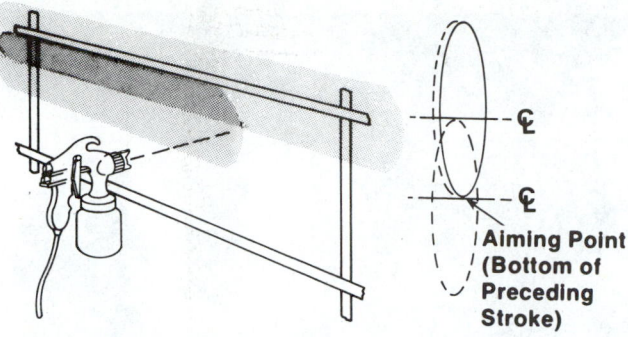

FIGURE 18-45 Overlap the strokes like this to get good, even paint coverage.

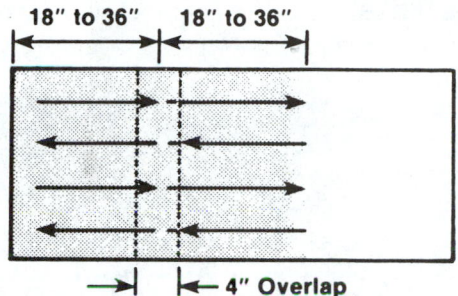

FIGURE 18-46 Always blend into the wet edge.

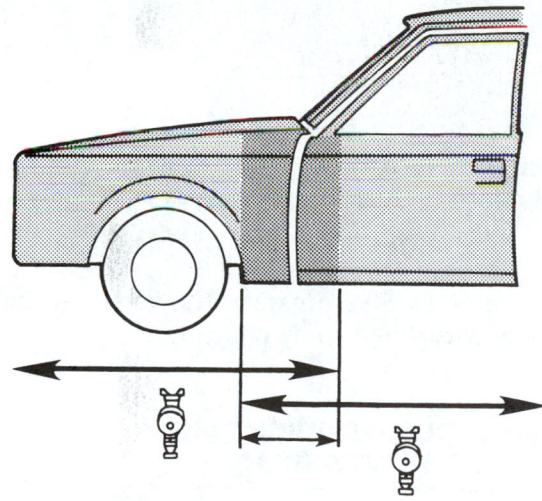

FIGURE 18-47 Remember that the gun overlap area is where runs occur easily in overall painting jobs. Move the gun to avoid runs in these areas.

technique at the area where the sections are joined will avoid the danger of a double coat at this point and the possibility of getting a sag (Figure 18–47).

12. Continue passes back and forth, triggering the gun at the end of each pass, and lowering each successive pass one half the top-to-bottom width of the spray gun pattern.

13. The last pass should be made with the lower half of the spray pattern below the surface being painted. If it is a door, the pattern would shoot off into the space below it.

14. The procedure just followed is called a *single coat*. For a *double coat*, repeat the single coat procedure immediately. Two or three single coats are normally required for enamel topcoats. Allow the first coat to set up (become *tacky* or partially dry or cured) before applying additional coats.

GUN HANDLING PROBLEMS

The inexperienced painter is prone to several spraying errors, including:

- **Heeling.** This occurs when the painter allows the gun to tilt (Figure 18–48). Because the gun is no longer perpendicular to the surface, the spray produces an uneven layer of paint, excessive overspray, dry spray, and orange peel.
- **Arcing.** This occurs when the gun is not moved parallel with the surface (Figure 18–49). At the outer edges of the arced stroke, the gun is farther away from the surface than at the middle of the stroke. The result is uneven film buildup, dry spray, excessive overspray, and orange peel.
- **Speed of stroke.** If the stroke is made too quickly, the paint will not cover the surface evenly (Figure 18–50). If the stroke is made too slowly, sags and runs will develop (Figure 18–51). The proper stroking speed is something that comes with experience.
- **Improper overlaps.** Improper overlapping results in uneven film thickness, contrasting color

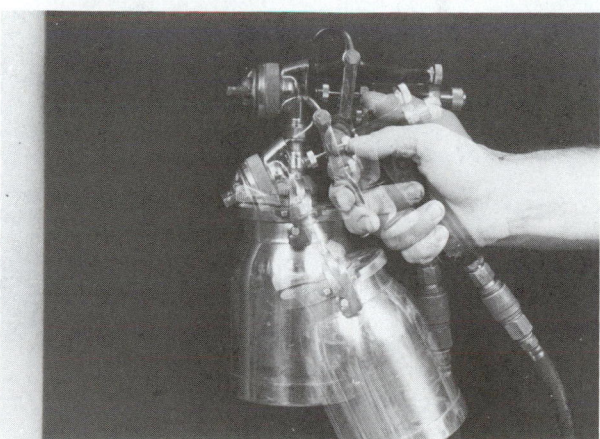

FIGURE 18-48 Heeling is a common gun handling error.

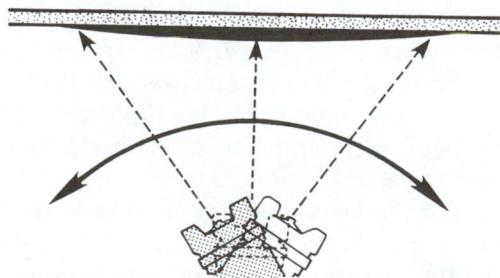

FIGURE 18-49 Another common gun handling error is arcing.

FIGURE 18-50 If gun movement is too fast, not enough paint will be deposited, resulting in weak coverage.

FIGURE 18-51 If gun movement is too slow, too much paint will cause runs and sags.

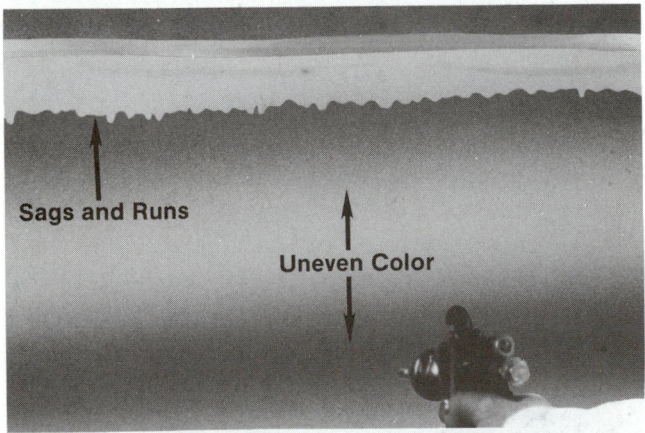

Sags and Runs

Uneven Color

FIGURE 18-52 Improper overlapping can cause problems.

FIGURE 18-53 This is an example of wasteful overspray. Only spray surfaces needing paint, not masking paper.

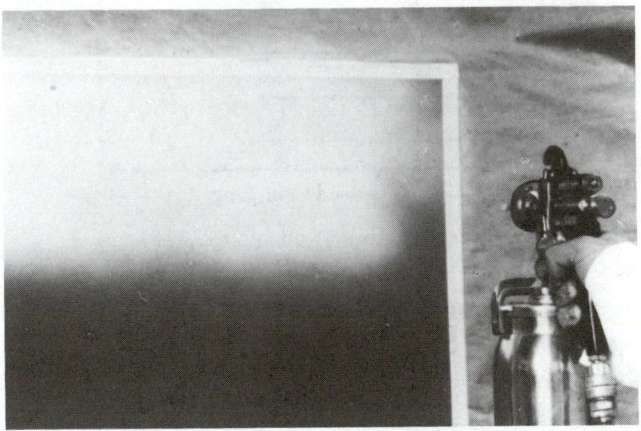

FIGURE 18-54 An example of improper coverage shows lack of coverage near top of masking paper.

hues, and sags and runs, as shown in Figure 18–52.

- **Wasteful overspray.** Failure to trigger the gun before and after each stroke results in wasteful overspray and excessive buildup of paint at the beginning and end of each stroke (Figure 18–53).
- **Improper coverage.** Triggering at the wrong time is another common error. Failure to trigger exactly over the edge of the panel results in uneven coverage and film thickness (Figure 18–54).

Table 18–5 summarizes the variables that control quality when spray painting.

18.3 CLEANING THE SPRAY GUN

Neglect and poor maintenance are responsible for the majority of spray gun difficulties. Proper care of a gun requires little time and effort. Thorough cleaning of the gun and accessory equipment IMMEDIATELY AFTER USE is critical. You must also lubricate bearing surfaces and packings at recommended intervals. Be careful when handling to avoid dropping and damaging the gun.

Even if you use a spray gun cleaning tank, you should periodically disassemble the gun for thorough service. To manually clean a gun, follow these general instructions.

To clean a suction feed gun, first loosen the cup from the gun (Figure 18–55A). With a gravity feed, remove the lid. Pour out any remaining, unused material into an approved container for proper disposal.

TABLE 18-5: SUMMARY OF VARIABLES CONTROLLING QUALITY IN SPRAY FINISHING

Atomization	1. Fluid viscosity 2. Air pressure 3. Fan pattern width 4. Fluid velocity or fluid pressure 5. Fluid flow rate 6. Distance of spray gun from work
Evaporation Stages	1. Between spray gun and part 2. From sprayed part
Evaporation Variables Between Spray Gun and Sprayed Part	1. Type of reducing thinner 2. Atomization pressure 3. Amount of thinner 4. Temperature in spray area 5. Degree of atomization

Evaporation Variables Affecting	1. Physical properties of solvents (i.e., fast or slow evaporation)
	2. Temperature
	  a. Fluid
	  b. Work
	  c. Air
	3. Exposed area of the surface sprayed

Evaporation Variables from the Sprayed Part	1. Surface temperature 2. Room air temperature 3. Air pressure velocity 4. Flash time between coats 5. Flash time after final coats 6. Physical properties of the solvents (i.e., fast or slow evaporation)

Operator Variables	1. Distance of spray gun from the work surface
	2. Stroking speed over the work surface
	3. Pattern overlap
	4. Spray gun attitude
	  a. Heeling
	  b. Arcing
	  c. Fanning
	5. Triggering

WARNING Avoid spraying solvent into the air when cleaning a spray gun. This pollutes the atmosphere. If a gun cleaning tank is not available, direct the flushing spray into a closed container of solvent. Then you will catch most of the solvent spray in the container and produce less air pollution.

Pour some gun cleaning solvent into the gun. Slosh it around to partially remove the paint film in the cup. Spray the solvent through the gun to remove most of the paint.

With a solvent soaked-rag, wipe the inside and outside of the cup, the gun body, air cap, and all external parts. Remove all traces of paint film.

Following the manufacturer's instructions, remove any parts that require further cleaning (air cap, nozzle, needles, vent tube, etc.). While wearing plastic or rubber gloves, wipe them clean with a solvent-soaked rag (Figure 18–55B). When blowing off a spray gun, use very low pressure (5 psi or 34 kPa).

SHOP TALK When cleaning a spray gun, it is best to use a recommended spray gun cleaning solvent. It will remove all deposits while not damaging gun parts. You should also use a special spray gun oil to lubricate parts as needed. Spray gun oil is formulated to not contaminate the paint and cause fisheyes like conventional oils.

If needed, clean small, hard-to-reach areas on the gun with a thin, soft bristle brush (Figure 18–55C). Wipe off residue with a clean rag soaked with solvent. Then pour 1 inch of clean solvent in the cup again. Spray the solvent through the gun to clean out the fluid passages.

With either type of gun, periodically remove the air cap and soak it in thinner or solvent. Clean out clogged holes with a soft item such as a round toothpick or a broom straw (Figure 18–55D).

To avoid gun damage, never use wires or nails to clean the precision-drilled openings. Clean the fluid tip with a gun brush and solvent (Figure 18–55E). With a clean rag soaked in thinner, wipe the outside of the gun to remove all traces of paint (Figure 18–55F).

Areas in the United States with air pollution problems require the use of an enclosed spray gun cleaning tank.

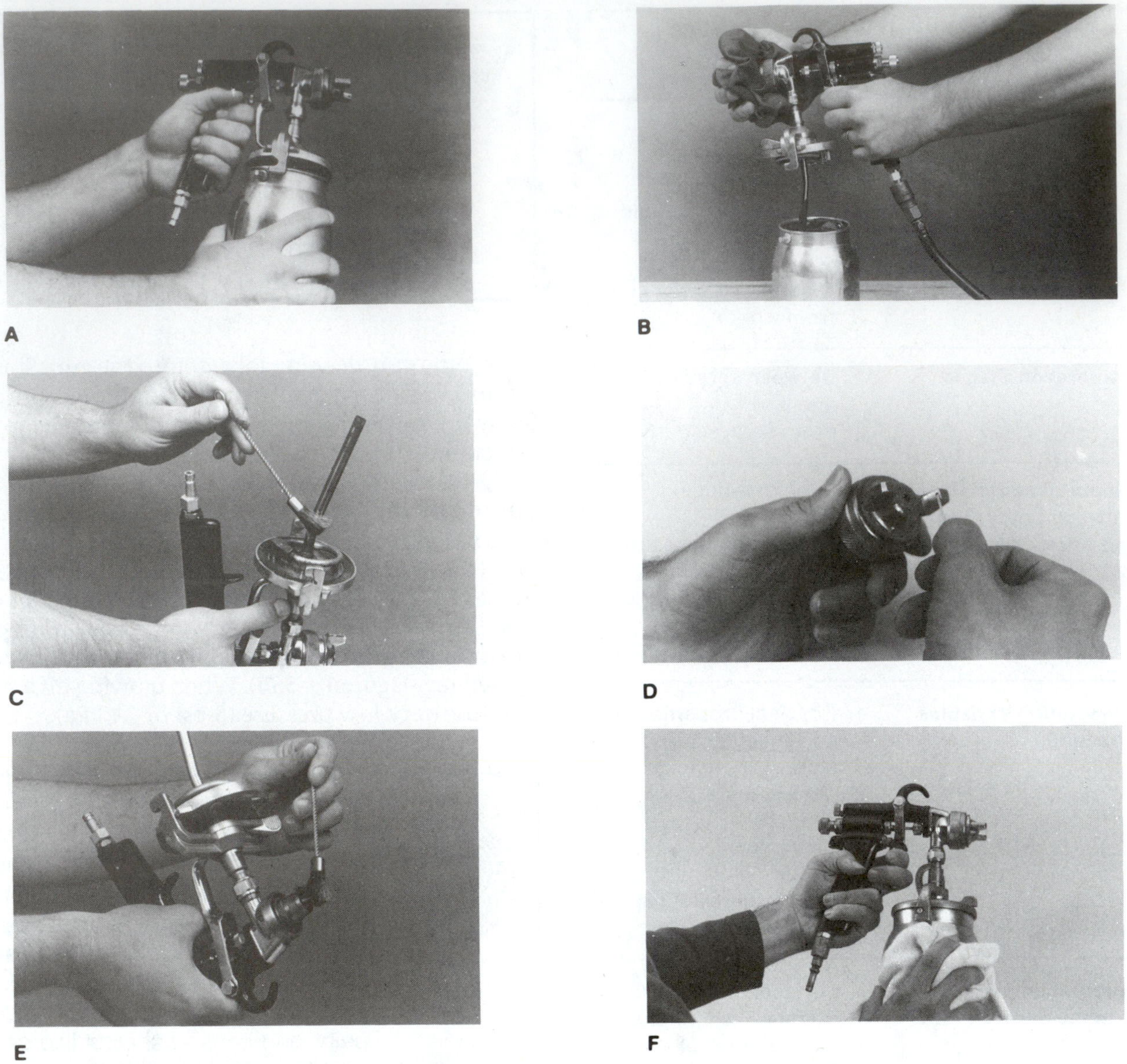

A

B

C

D

E

F

FIGURE 18-55 To clean a spray gun: (A) loosen cup; (B) discharge paint; (C) clean paint pipe; (D) clean clogged holes in air cap; (E) clean paint nozzle; and (F) clean cup cover. *(Courtesy of Maaco Enterprises, Inc.)*

WARNING

On pressure feed air guns, release tank or cup pressure first. Remove the cap or lid and pour out any remaining paint. Empty the contents into a suitable container for disposal. Partially fill the cup with a clean solvent. Spray solvent from the gun and repeat this process until clean solvent is flowing from the gun.

SPRAY GUN CLEANING TANK

A *spray gun cleaning tank*, also called a gun washer/recycler, is a pressurized container for flushing the gun and other tools with a cleaning solution. It is used by most modern body shops because it saves time and keeps the gun in a good, clean condition.

Paint-covered equipment (guns, cups, stirrers, and strainers) is placed in the larger tub of the gun washer/recycler (Figure 18–56). The lid is closed; then the pump recirculates the solvent into the upper portion of the tub. In less than 60 seconds, the equipment is clean and ready for use.

The automatic gun washer/recycler saves the body technician time. Compared with traditional manual cleaning methods, the gun washer/recycler machine saves 10 minutes on each color change. The cleaning system offers increased safety because solvent fumes are contained in the tank. The system is designed so that sludge from the cleaning action settles to the bottom for easy drainage and disposal

FIGURE 18-56 This is a typical automatic gun washer/recycler.

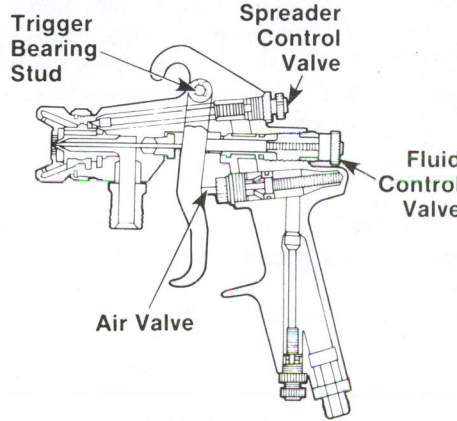

FIGURE 18-57 These are some of the parts to lubricate in an air spray gun.

with other shop wastes. Check the owner's manual for complete operational details and the proper solvents to use.

To use a gun cleaning tank, pour excess paint into a container for proper disposal. Remove any parts that might be damaged by the cleaning solvent. These parts would include the pressure gauge-regulator on HVLP guns and any plastic vent hoses. Place the cup into the tank. Then place the gun into the cleaning mechanism so that the trigger is engaged and the inlet tub is in place over the cleaning nozzle. Close the lid and turn on the machine. This will force solution through all passages in the gun. When cleaned, remove the gun and wipe it off with a clean rag.

Most spray gun manufacturers recommend lubricating the parts shown in Figure 18–57 with spray gun oil. Spray gun oil is compatible with paint and will not contaminate the gun.

Packings and springs plus needles and nozzles must periodically be replaced due to normal wear and tear. This should be done only in accordance with the manufacturer's instructions.

Avoid using conventional oil to lubricate a spray gun. Excess oil could overflow into the paint and air

SHOP TALK If the gun is not cleaned soon after use, the nozzle may clog, causing the gun to spit (eject pieces of dried paint) or form the wrong spray pattern. For enamel paints with additives, the enamel paint may harden right in the gun, an expensive mistake.

passages, mixing with the paint and resulting in a defective paint film. Conventional oil and paint do not mix; fisheyes in the fresh paint will result.

For best results in refinishing, use separate guns for topcoats and undercoats. Ideally, there should be at least three guns: one gun for spraying undercoats like primer-surfacers, another for colorcoats, and a third gun for spraying clearcoats and single-stage paints. If these guns are kept clean and in good working order, much time will be saved over trying to use one gun and having to adjust it each time that the operation is changed. Each gun will be preset to spray the specific type of material.

WARNING If you ever forget to clean a spray gun filled with a curing type of material (enamel with hardener or epoxy primer, for example), you will probably have to throw the gun in the trash and buy a new one! This is a costly mistake.

18.4 SPRAY GUN TROUBLESHOOTING

If the air spray gun is not adjusted, manipulated, and cleaned properly, it will apply a defective coating to the surface. Fortunately, defects from incorrect handling and improper cleaning can be tracked down and corrected without much difficulty.

NOTE: The most common spray gun application problems, with their possible causes and suggested remedies, are given in Chapter 20.

If not properly maintained, the air spray gun itself (Figure 18–58) can also create some problems. Table 18–6 contains the causes of and possible solutions to some of the more common spray gun difficulties.

Failure of the compressed air supply system to perform properly can cause the paint problems shown in Table 18–7.

TABLE 18-6: TROUBLESHOOTING AN AIR SPRAY GUN

Trouble	Possible Cause	Suggested Correction
Spray pattern top heavy or bottom heavy	1. Horn holes partially plugged (external mix).	1. Remove air cap and clean.
	2. Fluid tip clogged, damaged, or not installed properly.	2. Clean, replace, or reinstall fluid tip.
	3. Dirt on air cap seat or fluid tip seat.	3. Remove and clean seat.
Spray pattern heavy to right or to left	1. Air cap dirty or orifice partially clogged.	1. To determine where buildup occurs, rotate cap 180 degrees and test spray. If pattern shape stays in same position, the condition is caused by fluid buildup on fluid tip. If pattern changes with cap movement, the condition is in the air cap. Clean air cap, orifice, and fluid tip accordingly.
	2. Air cap damaged.	2. Replace air cap.
	3. Paint nozzle clogged or damaged.	3. Clean or replace paint nozzle.
	4. Too low a setting of the pattern control knob.	4. Adjust setting.
Spray pattern heavy at center	1. Atomizing pressure too low.	1. Increase pressure.
	2. Fluid of too great viscosity.	2. Thin fluid with suitable thinner.
	3. Fluid pressure too high for air cap's normal capacity (pressure feed).	3. Reduce fluid pressure.
	4. Caliber of paint nozzle enlarged due to wear.	4. Replace paint nozzle.
	5. Center hole enlarged.	5. Replace air cap and paint nozzle.
Spray pattern split	1. Not enough fluid.	1. Reduce air pressure or increase fluid flow.
	2. Air cap or fluid tip dirty.	2. Remove and clean.
	3. Air pressure too high.	3. Lower air pressure.
	4. Fluid viscosity too thin.	4. Thicken fluid viscosity.
Pinholes	1. Gun too close to surface.	1. Stroke 6 to 8 inches from surface.
	2. Fluid pressure too high.	2. Reduce pressure.
	3. Fluid too heavy.	3. Thin fluid with thinner.
Blushing or a whitish coat of lacquer.	1. Absorption of moisture.	1. Avoid spraying in damp, humid, or too cool weather.
	2. Too quick drying of lacquer.	2. Correct by adding retarder to lacquer.
Orange peel (surface looks like orange peel)	1. Too high or too low an atomization pressure.	1. Correct as needed.
	2. Gun too far or too close to work.	2. Stroke 6 to 8 inches from surface.
	3. Fluid not thinned.	3. Use proper thinning process.
	4. Improperly prepared surface.	4. Surface must be prepared.
	5. Gun stroke too rapid.	5. Take deliberate, slow stroke.
	6. Using wrong air cap.	6. Select correct air cap for the fluid and feed.
	7. Overspray striking a previously sprayed surface.	7. Select proper spraying procedure.
	8. Fluid not thoroughly dissolved.	8. Mix fluid thoroughly.
	9. Drafts (synthetics and lacquers).	9. Eliminate excessive drafts.
	10. Humidity too low (synthetics).	10. Raise humidity of room.

TABLE 18-6: TROUBLESHOOTING AN AIR SPRAY GUN (CONTINUED)

Trouble	Possible Cause	Suggested Correction
Excessive spray fog or overspray	1. Atomizing air pressure too high or fluid pressure too low.	1. Correct as needed.
	2. Spraying past surface of the product.	2. Release trigger when gun passes target.
	3. Wrong air cap or fluid tip.	3. Ascertain and use correct combination.
	4. Gun stroked too far from surface.	4. Stroke 6 to 8 inches from surface.
	5. Fluid thinned out too much.	5. Add correct amount of thinner.
No control over size of pattern	1. Air cap seal is damaged.	1. Check for damage, replace if necessary.
	2. Foreign particles are lodged under the seal.	2. Make sure surface that this sets on is clean.
Sags or runs	1. Dirty air cap and fluid tip.	1. Clean cap and fluid tip.
	2. Gun manipulated too close to surface.	2. Hold the gun 6 to 8 inches from surface.
	3. Not releasing trigger at end of stroke (when stroke does not go beyond object).	3. Release trigger after every stroke.
	4. Gun manipulated at wrong angle to surface.	4. Work gun at right angles to surface.
	5. Fluid piled on too heavy.	5. Learn to calculate depth of wet film of fluid.
	6. Fluid thinned out too much.	6. Add correct amount of fluid by measure.
	7. Fluid pressure too high.	7. Reduce fluid pressure with fluid control knob.
	8. Operation too slow.	8. Speed up movement of gun across surface.
	9. Improper atomization.	9. Check air and fluid flow; clean cap and fluid tip.
Streaks	1. Dirty or damaged air cap and/or fluid tip.	1. Same as for sags.
	2. Not overlapping strokes correctly or sufficiently.	2. Follow previous stroke accurately.
	3. Gun moved too fast across surface.	3. Take deliberate, slow strokes.
	4. Gun held at wrong angle to surface.	4. Same as for sags.
	5. Gun held too far from surface.	5. Stroke 6 to 8 inches from surface.
	6. Air pressure too high.	6. Use least air pressure necessary.
	7. Split spray.	7. Reduce air adjustment or change air cap and/or fluid tip.
	8. Pattern and fluid control not adjusted properly.	8. Readjust.
Gun sputters constantly	1. Connections, fittings, and seals loose or missing.	1. Tighten and/or replace as per owner's manual.
	2. Leaky connection on fluid tube or fluid needle packing (suction gun).	2. Tighten connections; lubricate packing.
	3. Lack of sufficient fluid in container.	3. Refill container with fluid.
	4. Tipping container at an acute angle.	4. If container must be tipped, change position of fluid tube and keep container full of fluid.
	5. Obstructed fluid passageway.	5. Remove fluid tip, needle, and fluid tube and clean.

Sputtering Spray

TABLE 18-6: TROUBLESHOOTING AN AIR SPRAY GUN (CONTINUED)

Trouble	Possible Cause	Suggested Correction
Gun sputters constantly (continued)	6. Fluid too heavy (suction feed).	6. Thin fluid.
	7. Clogged air vent in canister top (suction feed).	7. Clean.
	8. Dirty or damaged coupling nut on canister top (suction feed).	8. Clean or replace.
	9. Fluid pipe not tightened to pressure tank lid or pressure cup cover.	9. Tighten; check for defective threads.
	10. Strainer is clogged up.	10. Clean strainer.
	11. Packing nut is loose.	11. Make sure packing nut is tight.
	12. Fluid tip is loose.	12. Tighten fluid tip. Torque to manufacturer's specifications.
	13. O-ring on tip is worn or dirty.	13. Replace O-ring if necessary.
	14. Fluid hose from paint tank loose.	14. Tighten.
	15. Jam nut gasket installed improperly or jam nut loose.	15. Inspect and correctly install or tighten nut.
Uneven spray pattern	1. Damaged or clogged air cap.	1. Inspect air cap and clean or replace.
	2. Damaged or clogged fluid tip.	2. Inspect fluid tip and clean or replace.
Fluid leaks from spray gun **Nozzle Drip**	1. Fluid needle packing not too tight.	1. Loosen nut; lubricate packing.
	2. Fluid needle packing dry.	2. Lubricate needle and packing frequently.
	3. Foreign particle blocking fluid tip.	3. Remove tip and clean.
	4. Damaged fluid tip or fluid needle.	4. Replace both tip and needle.
	5. Wrong fluid needle size.	5. Replace fluid needle with correct size for fluid tip being used.
	6. Broken fluid needle spring.	6. Remove and replace.
Fluid leaks from packing nut **Packing Nut Leak**	1. Loose packing nut.	1. Tighten packing nut.
	2. Packing is worn out.	2. Replace packing.
	3. Dry packing.	3. Remove and soften packing with a few drops of light oil.
Fluid leaks through fluid tip when trigger is released	1. Foreign particles lodged in the fluid tip.	1. Clean out tip and strain paint.
	2. Fluid needle has paint stuck on it.	2. Remove all dried paint.
	3. Fluid needle is damaged.	3. Check for damage; replace if necessary.
	4. Fluid tip has been damaged.	4. Check for nicks; replace if necessary.
	5. Spring left off fluid needle.	5. Make sure spring is replaced on needle.
Excessive fluid	1. Not triggering the gun at each stroke.	1. It should be a habit to release trigger after every stroke.
	2. Gun at wrong angle to surface.	2. Hold gun at right angles to surface.
	3. Gun held too far from surface.	3. Stroke 6 to 8 inches from surface.
	4. Wrong air cap or fluid tip.	4. Use correct combination.
	5. Depositing fluid film of irregular thickness.	5. Learn to calculate depth of wet film of finish.
	6. Air pressure too high.	6. Use least amount of air necessary.
	7. Fluid pressure too high.	7. Reduce pressure.
	8. Fluid control knob not adjusted properly.	8. Readjust.

TABLE 18-6: TROUBLESHOOTING AN AIR SPRAY GUN (CONTINUED)

Trouble	Possible Cause	Suggested Correction
Fluid will not come from spray gun	1. Out of fluid. 2. Grit, dirt, paint skin, etc., blocking air gap, fluid tip, fluid needle, or strainer. 3. No air supply. 4. Internal mix cap using suction feed.	1. Add more spray fluid. 2. Clean spray gun thoroughly and strain spray fluid; always strain fluid before using it. 3. Check regulator. 4. Change cap or feed.
Fluid will not come from fluid tank or canister	1. Lack of proper air pressure in fluid tank or canister. 2. Air intake opening inside fluid tank or canister clogged by dried-up finish fluid. 3. Leaking gasket on fluid tank cover or canister top. 4. Gun not converted correctly between canister and fluid tank. 5. Blocked fluid hose. 6. Connections with regulator not correct.	1. Check for air leaks or leak of air entry; adjust air pressure for sufficient flow. 2. This is a common trouble; clean opening periodically. 3. Replace with new gasket. 4. Correct per owner's manual. 5. Clear. 6. Correct as per owner's manual.
Sprayed coat short of liquid material	1. Air pressure too high. 2. Fluid not reduced or thinned correctly. (Suction feed only) 3. Gun too far from work or out of adjustment.	1. Decrease air pressure. 2. Reduce or thin according to directions; use proper thinner or reducer. 3. Adjust distance to work; clean and adjust gun fluid and spray pattern controls.
Spotty, uneven pattern, slow to build	1. Inadequate fluid flow. 2. Low atomization air pressure. (Suction feed only) 3. Too fast gun motion.	1. Back fluid control knob to first thread. 2. Increase air pressure, rebalance gun. 3. Move at moderate pace.
Unable to get round spray	1. Pattern control knob not seating properly.	1. Clean or replace.
Dripping from fluid tip	1. Dry packing. 2. Sluggish needle. 3. Tight packing nut. 4. Spray head misaligned on type MBC guns causing needle to bind.	1. Lubricate packing. 2. Lubricate. 3. Adjust. 4. Tap all around spray head with wood and rawhide mallet and retighten locking bolt.
Excessive overspray	1. Too much atomization air pressure. 2. Gun too far from surface. 3. Improper stroking, i.e. arcing, moving too fast.	1. Reduce. 2. Check distance. 3. Move at moderate pace, parallel to work surface.
Excessive fog	1. Too much or quick drying thinner. 2. Too much atomization air pressure.	1. Remix. 2. Reduce.
Will not spray on pressure feed	1. Control knob on canister cover not open. 2. Canister is not sealing. 3. Spray fluid has not been strained. 4. Spray fluid in canister top threads.	1. Set this knob for pressure spraying. 2. Make sure canister is on tightly. 3. Always strain before using. 4. Clean threads and wipe with grease.

TABLE 18-6: TROUBLESHOOTING AN AIR SPRAY GUN (CONTINUED)

Trouble	Possible Cause	Suggested Correction
Will not spray on pressure feed (continued)	5. Gasket in canister top worn or left out. 6. No air supply. 7. Fluid too thick. 8. Clogged strainer.	5. Inspect and replace if necessary. 6. Check regulator. 7. Thin fluid with proper thinner. 8. Clean or replace strainer.
Will not spray on suction feed	1. Spray fluid is too thick. 2. Internal mix nozzle used. 3. Spray fluid has not been strained. 4. Hole in canister cover clogged. 5. Gasket in canister top worn or left out. 6. Plug or clogged strainer. 7. Fluid control knob adjusted incorrectly. 8. No air supply.	1. Thin fluid with thinner. 2. Install external mix nozzle. 3. Always strain before use. 4. Make sure this hole is open. 5. Inspect and replace if necessary. 6. Clean or replace strainer. 7. Correct adjustment. 8. Check regulator.
Air continues to flow through gun when trigger has been released (on nonbleeder guns only)	1. Air valve leaks. 2. Needle is binding. 3. Piston is sticking. 4. Packing nut too tight. 5. Control valve spring left out.	1. Remove valve, inspect for damage, clean valve, and replace if necessary. 2. Clean or straighten needle. 3. Clean piston, check O-ring, and replace if necessary. 4. Adjust packing nuts. 5. Make sure to replace this spring.
Air leak at canister gasket	1. Canister not sealing on canister cover.	1. Check gasket, clean threads, and tighten canister.
Leak at setscrew in canister top	1. Screw not tight. 2. Damaged threads on setscrew.	1. Clean threads and tighten screw. 2. Inspect and replace if necessary.
Leak between top of canister cover and gun body	1. Retainer nut is not tight enough. 2. Gasket or gasket seat damaged.	1. Check nut to make sure it is tight. 2. Inspect, clean, and replace if necessary.
Pressure Fluid Tank Problems		
Leaks air at the top of the tank lid	1. Gasket not seating properly or damaged. 2. Wing screws not tight enough. 3. Fittings leak. 4. Air pressure too high.	1. Drain off all of the air from fluid tank thus allowing the gasket to seat. Retighten wing nuts, and fill with air again. Lid will seat tightly. 2. Make sure all wing screws are tight. By following remedy #1 (above), wing screws can be pulled down even tighter. 3. Check all fittings and apply pipe dope if necessary. 4. Maximum 60 psi. Normal w.p. 25–30 psi.
No fluid comes through the spray gun	1. Not enough pressure in tank. 2. Out of fluid. 3. Fluid passages clogged.	1. Increase regulator setting until fluid flows; do not exceed 60 psi. 2. Check fluid supply. 3. Check tube, fittings, hose, and spray gun. Clean out fittings, hose, tube, and spray gun making sure all residual fluid is removed.

TABLE 18-7: TROUBLESHOOTING A COMPRESSED AIR SUPPLY

Fault	Result	Blistering	Nondrying	Poor Adhesion	Contamination	Poor Atomization	Poor Flow	Overloading	Sags	Popping	Slow Application	Off-shade Metallic	Uneven Application	Dry Spray	Dirt	Remedy
Oil/water not adequately condensed out.	Oil/water at spray gun	A	C	A	C											Ensure regular drainage of air receiver, separator, and transformer. Site transformers of adequate capacity in cool places. Lubricate compressors with recommended grade of mineral oil of good emulsifying properties.
Long air line; inadequate internal bore of air line; connectors, fittings, compressor, air transformers, and regulators of inadequate capacity.	Pressure drop					B	C	C	A	A	C	A				Ensure adequate air supply with 30 feet 5/16 inch (8mm) internal bore air line with appropriate fittings. NOTE: Reduction of viscosity to give improvement may produce other defects.
Inadequate compressor capacity. No pressure regulator. Regulator diaphragm broken.	Pressure fluctuation							A	A	A		A	A	A		Increase capacity. Use pressure regulator. Replace regulator diaphragm.
Compressed air intake filter breached. Transformer filter not properly maintained. Compressor sited in dusty area.	Dirt in compressed air														A	Repair air intake filter. Replace transformer filter. Clean dust and dirt from compressor site.

A Most likely failure to be associated with the fault
B Likely failure
C Failure less likely to be associated with the fault

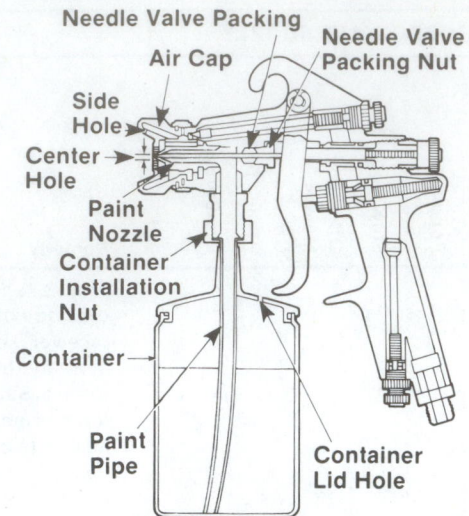

FIGURE 18-58 These are possible trouble spots of an air spray gun.

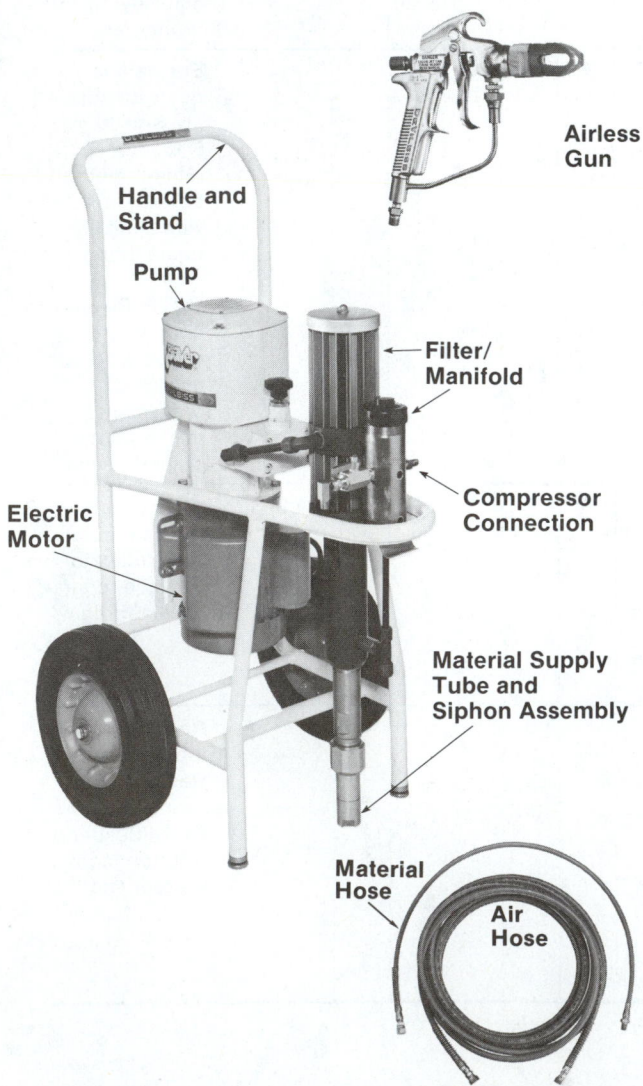

FIGURE 18-59 Note typical airless spray equipment. *[Photo courtesy ITW DeVilbiss Automotive Refinishing Products]*

18.5 OTHER SPRAY SYSTEMS

There are four other types of spray systems that can be found in some shops: airless spray gun, electrostatic system, HVLP system, and the airbrush. Operation is basically the same as the air spray system just described.

AIRLESS SPRAY GUN SYSTEM

Airless spraying equipment (Figure 18–59) uses hydraulic pressure rather than air pressure to atomize paint . With the airless spray method, pressure is applied directly to the paint, which is injected at high speed through small holes in the nozzle and formed into a mist. Unlike the air spray method, there is less mixing of air in the paint. Consequently, there is less mist dispersion. Also, since the paint is pressurized directly, less energy is used for atomization so that with the same amount of power, a degree of atomization is accomplished that is several times that for air spraying. In fact, the pressure developed in airless equipment ranges from 1,500 to 3,000 psi (10,350 to 20,500 kPa). Actual pressure depends on the pump ratio of the equipment.

The airless system reduces overspray and rebound to a minimum, and application of the finish is much faster than with conventional atomized air. Because of the higher pressures involved, the airless system can be used with paints and other materials that have a higher viscosity. However, this system of application can only be used where a fine finish is not required. It is often employed to apply the finishing coating in the truck fleet commercial vehicle refinishing business. It also has found a place for auto underbody and corrosion work. The so-called air-assisted airless system that uses some air to assist in the spraying operation tends to give a better finish.

Figure 18–60 shows a typical assembly of an airless system. The gun is connected to the pump with a single hose. When the gun is spraying, the pump delivers fluid under pressure adjusted by the air pressure to the pump. When the gun is not spraying, the fluid pressure and air pressure are balanced and the pump stops. The quality and economy of the finish is dependent upon operator skill, fluid preparation, and nozzle size. There are six ways that a painter can control the operation of this system. They are as follows:

- *Orifice size.* This determines the amount of paint sprayed through the gun. More paint will be applied through the gun with a larger orifice.
- *Paint viscosity.* This is controlled by the amount of reduction. Viscosity ranges can be from 24 to 36 seconds on a #2 Zahn cup (18 to 28 seconds on a #4 Ford cup).

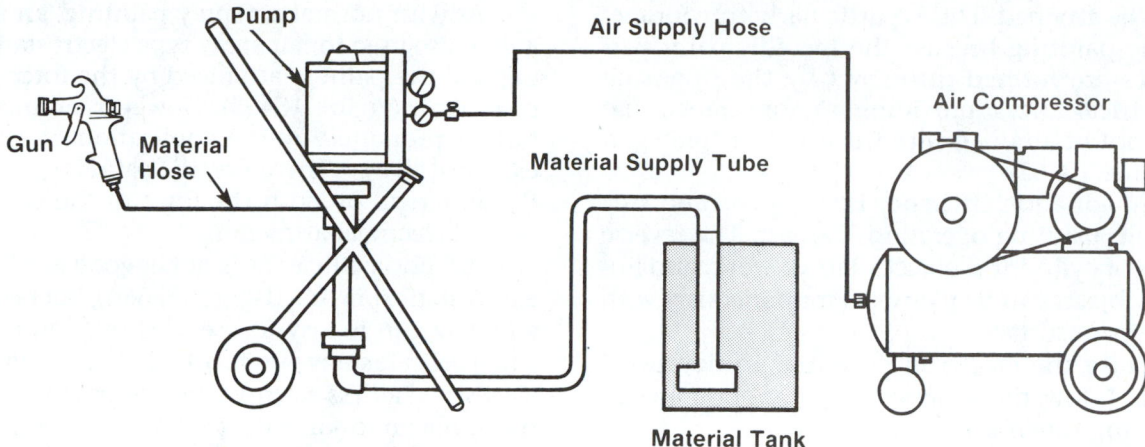

FIGURE 18-60 An airless spray equipment setup is fairly simple.

- *Speed of the reducer.* Generally, use the fastest reducer consistent with flow and sagging. Airless equipment sprays much wetter than conventional air-atomized equipment.
- *Speed of gun movement.* Because of the wetter spray with airless equipment, the painter will generally have to move faster than with conventional spray equipment.
- *Gun distance.* Because of wetter spray patterns, the gun distance to the work should be around 14 inches (356 mm).
- *Coating material.* Prepare the coating material and use the air pressure as recommended in the manufacturer's instruction manual.

The basic operating techniques of an airless spray gun are the same as those for conventional guns. The gun should be held perpendicular to and moved parallel with the surface in order to obtain a uniform coating of fluid. The wrist, elbow, and shoulder must all be used. Once the best working distance is determined, the spray gun should be moved across the work at this optimum distance throughout the stroke.

Some object shapes do not allow this practice, but it should be used whenever possible. The proper speed allows a full wet coat application with each stroke. If the desired film thickness cannot be obtained with a single stroke or pass because of sagging, then two or more coats can be applied with a flash-off period between each coat. The spray movement should be at a comfortable rate. If the spray gun movement is excessive in order to avoid flooding the work, then the fluid nozzle orifice is too large or the fluid pressure is too high. If the stroke speed is very slow in order to apply full wet coats, then the fluid pressure should be increased slightly or a larger tip is required.

ELECTROSTATIC SPRAYING SYSTEM

Electrostatic spraying utilizes the principle that positive (+) and negative (−) electrical charges each attract the other but oppose a like charge. Therefore, when paint particles are given a negative charge by a high-voltage generator (Figure 18–61), the particles oppose each other, causing them to become atomized.

On the other hand, because the adherend (the surface to which the paint adheres) is grounded, it is under a positive electrical charge. In this manner, when high voltage is applied between the adherend and the electrostatic painting equipment, an electrical field is formed and the air in the field allows the electricity to pass through easily. In other words, electrical passages are formed and the atomized paint passing through these passages is sent to and adheres to the object that is being painted.

Advantages and disadvantages of electrostatic painting are as follows:

- Because the paint particles are drawn to the adherend by electrical attraction, there is less paint loss compared to normal spray painting.
- Because atomization is promoted by opposing electrical forces, a very good-quality paint finish

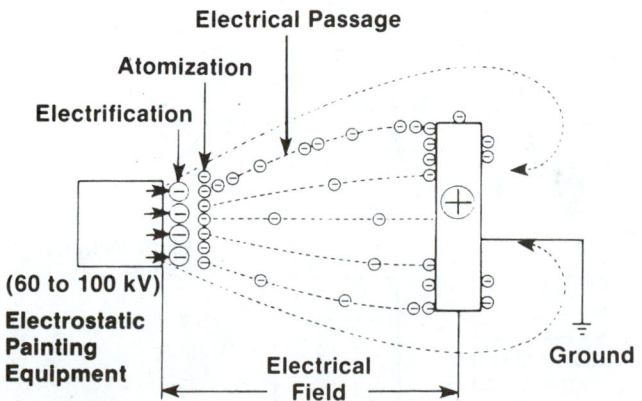

FIGURE 18-61 Note principle of electrostatic painting.

can be attained. This is particularly true for metallic painting because the metallic paint particles are formed into rows by the opposing electrical forces, providing an appearance that cannot be attained with the usual air spray gun (Figure 18–62).

- Paint adhesion efficiency is very good and, as a result, painting operations are fast. The reverse side of cylindrical objects, lattice work, and linear objects can be painted simultaneously with the front surface.
- Because the electrical potential in depressed areas is low, the adhesion is not as good, necessitating touch-up.
- Unless nonconductors such as plastic, glass, and rubber are made conductive, painting is not possible.

As for portable electrostatic painting equipment, there are both the air spray type (Figure 18–63) and the airless spray type (Figure 18–64).

As with normal air spray painting, an air spray gun is also used for air spray type electrostatic painting and the paint is atomized by the force of compressed air (Figure 18–65). However, atomization is further promoted by the application of a negative electrical charge. Therefore, the paint is sprayed onto the adherend by both the force of the compressed air and electrical attraction.

Adhesion efficiency is not as good as with airless electrostatic spraying (Figure 18–66A), but because the air spray gun is easy to use, this method is suitable when delicate spray gun manipulation is required. The air-assisted airless electrostatic equipment overcomes this problem to some degree (Figure 18–66B).

Like the normal airless spray method, airless electrostatic spraying utilizes high pressure to atomize the paint by injecting it through small holes in the nozzle, but it also gives the paint a negative electrical charge to further promote atomization. Paint is adhered by means of both injection pressure and electrical attraction. This method provides a very good adhesion efficiency and work is faster due to the large discharge volume. However, because compressed air is not used,

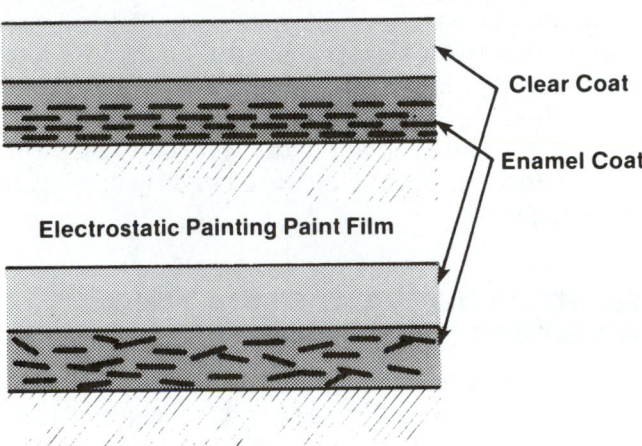

FIGURE 18-62 This is an illustration of an electrostatic painting film and a spray paint film.

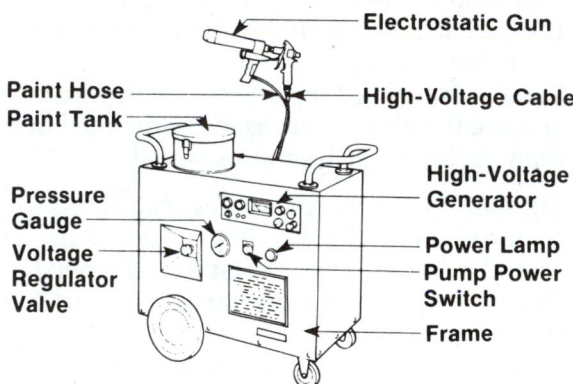

FIGURE 18-64 Study parts of airless type electrostatic painting equipment.

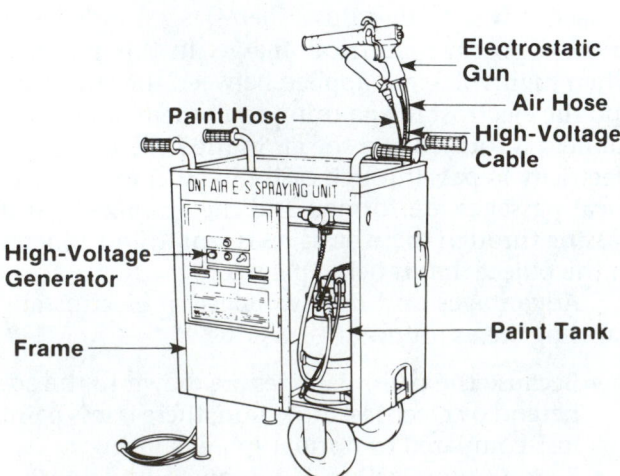

FIGURE 18-63 Note parts of air type electrostatic painting equipment.

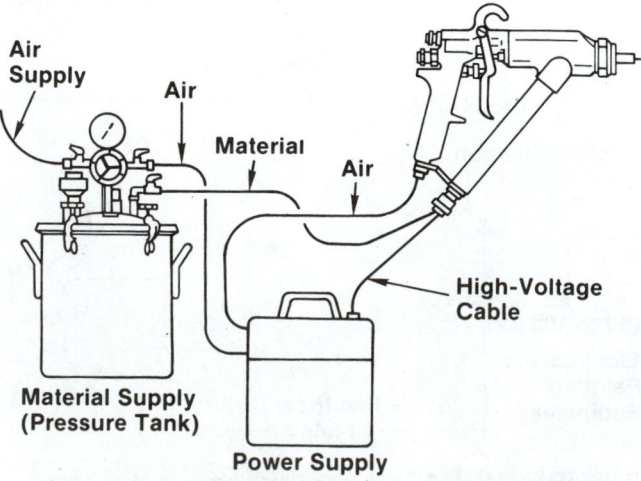

FIGURE 18-65 Note electrostatic equipment layout.

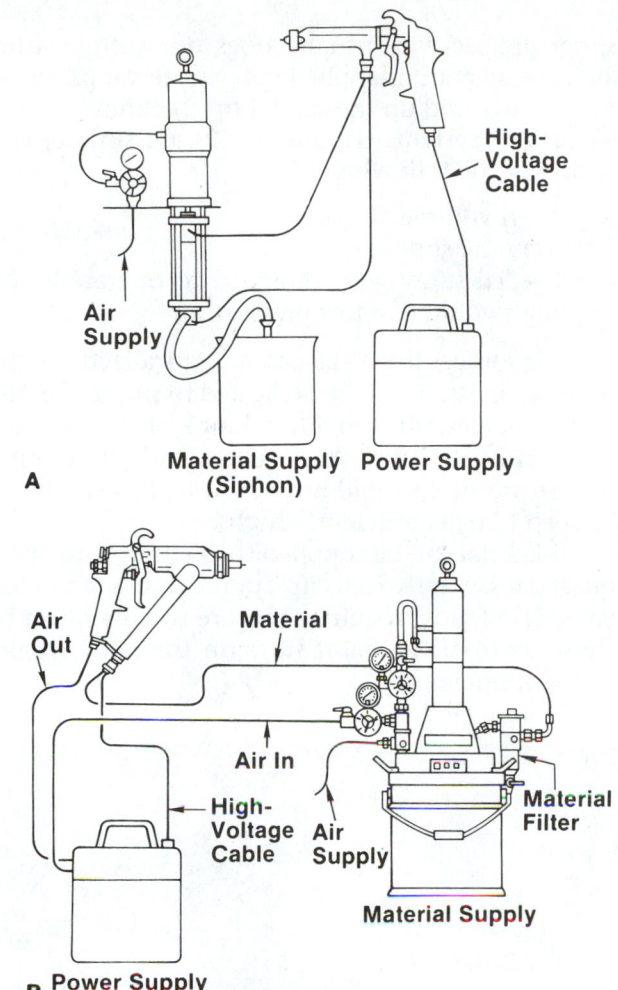

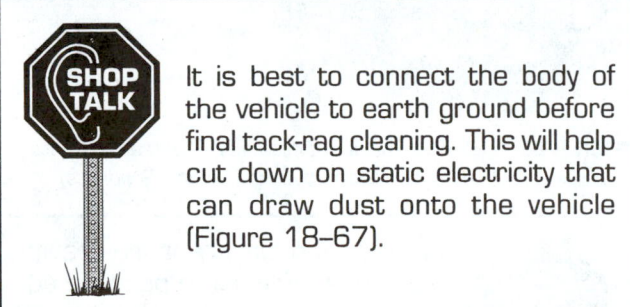

FIGURE 18-66 (A) Airless electrostatic equipment layout; (B) air-assisted airless electrostatic equipment.

SHOP TALK

It is best to connect the body of the vehicle to earth ground before final tack-rag cleaning. This will help cut down on static electricity that can draw dust onto the vehicle (Figure 18-67).

pass strict air emission or pollution standards. The most important way it differs from conventional spray systems is its high transfer efficiency.

High transfer efficiency means that more of the paint leaving the gun stays on the surface being painted. Less is wasted and enters the atmosphere as air pollution. This is the primary purpose of HVLP guns.

The high pressure of conventional sprays tends to blast the paint into small particles. In the process, it creates a fair amount of overspray. The transfer efficiency of high pressure systems suffers as a result of overspray, particle bounce, and blow back.

In contrast, HVLP relies on air delivered to the tip at 10 psi (69 kPa) or less to break the paint into small particles (Figure 18-68). As the material flows into the air stream, far less is lost in overspray, bounce, and blow back, hence the dramatic improvement in transfer efficiency. HVLP will work with any material that can be atomized by a spray gun. This includes two-component paints, urethanes, acrylics, epoxies, enamels, lacquers, stains, primers, and so on.

Some HVLP guns, especially retrofit guns, require lower than normal inlet pressure (12 to 30 psi or 82 to 207 kPa). Most completely redesigned HVLP guns use conventional inlet pressure to help

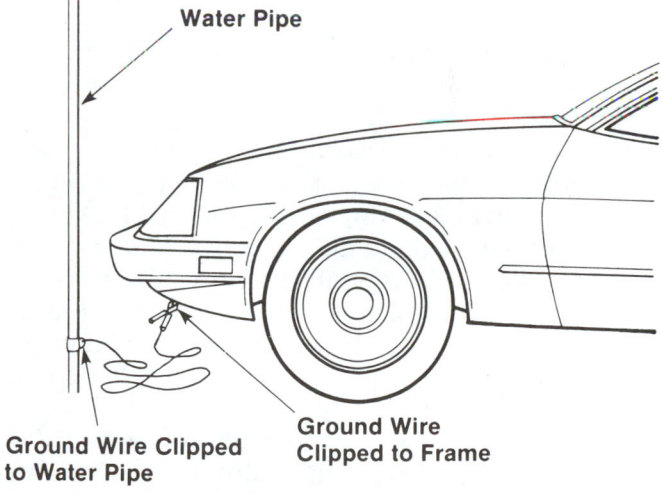

FIGURE 18-67 For safety, ground a vehicle in spray booth before spraying. *(Courtesy of Binks Mfg. Co.)*

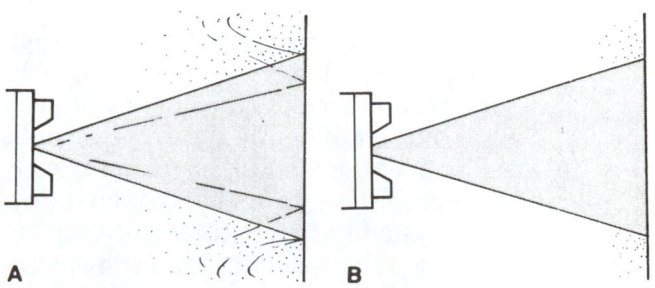

FIGURE 18-68 HVLP system has less overspray and waste than conventional spray equipment.

injection energy is not as strong and air spray prepainting of depressed areas like the underside of the hood and inner side of the doors is necessary.

HVLP SYSTEMS

The **HVLP** (high volume, low pressure) spray system (also known as the *high solids* system) uses a high volume of air, delivered at low nozzle pressure, to atomize paint into a pattern of low-speed particles. This type system is required by some geographic areas to atomization. These guns lower nozzle pressure to 10 psi (69 kPa) internally while allowing a high volume of air and paint to pass through the gun. This increases gun efficiency so that more paint is applied to the body surface and less is wasted as overspray. Unlike early gun designs, modern HVLP guns are easy to use and produce an excellent paint finish.

High transfer efficiency is attractive for several reasons. It reduces air pollution and reduces paint waste. In many states (California, for example), new laws require the use of spray equipment that is at least 65 percent transfer efficient. Low pressure spray (up to 10 psi or 69 kPa at the nozzle) and electrostatic spray methods have been approved by this legislation. Similar legislation has been passed in many geographic locations. The forecast is that high transfer efficiency will be a nationwide requirement to help reduce air pollution.

The purpose of this legislation is to protect the environment. However, there are other good reasons for HVLP. Higher transfer efficiency improves the quality of both the workplace and the finished product. Overspray not only makes painting work less desirable; it also reduces visibility, which contributes to mistakes and low productivity. Overspray is one of the main causes of paint booth maintenance, so cutting overspray cuts downtime. All paint spraying equipment can be affected by overspray, but the booth and its filters are affected the most.

To illustrate how much of a difference transfer efficiency makes in booth maintenance, consider that HVLP can be two to three times as efficient as conventional air spray. Depending on how it is used, conventional air spray is as little as 20 to 30 percent efficient. That means for every 3 gallons (11.4 liters) of paint sprayed, more than 2 gallons (7.6 liters) are wasted. With HVLP typically between 65 and 90 percent efficient, only 1 pint of paint would be wasted for every gallon applied. That is how a 3:1 difference in transfer efficiency becomes a 16:1 advantage in terms of overspray.

For example, if a conventional gun and a HVLP gun are used side by side to paint identical surface areas, the older gun will run out of paint while the HVLP gun may have enough paint to finish the job. This saves on the cost of paint and the time needed to refill the gun.

One of the most troublesome problems high transfer efficiency can solve is waste disposal. In air spray systems where overspray volume normally means using a water wash booth, the easy-to-handle dry filter media may now be sufficient, completely eliminating the hazardous waste that is often the by-product of these systems. High transfer efficiency can also make existing water wash filtration systems virtually maintenance-free, particularly when using the new sludge removal techniques. Conventional air

spray productivity usually does not suffer, either. Since more paint is applied per pass, fewer passes are needed to build up the same film thickness.

HVLP systems (Figure 18–69) are simple, consisting of the following:

- High volume air source
- Material supply system
- Special spray guns designed to operate with a high volume of low pressure air

Air sources for HVLP can be centralized, serving multiple guns, or can be dedicated to single-gun use. These sources will provide a range of delivery volumes and pressures. As a general rule, maximum nozzle pressure should be limited to 10 psi (69 kPa) to keep transfer efficiently high.

Material can be supplied through pressure pots, quart cups, gravity feed cups, or pressure-gravity feed cups. HVLP flow requirements are usually lower because more of the paint stays on the body (higher transfer efficiency).

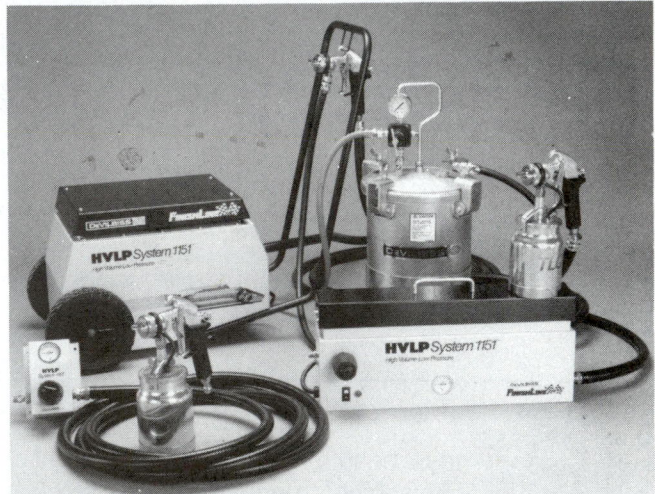

FIGURE 18–69 This is a typical HVLP system. *(Photo courtesy ITW DeVilbiss Automotive Refinishing Products)*

SHOP TALK

Gravity feed or pressure-gravity feed guns offer some possible advantages over suction feed guns besides high efficiency. They will use all of the material inside the cup for easier cleanup and less waste. Since the cup is on top of the gun, there is less chance of accidentally touching the wet paint with the cup when spraying hard-to-reach roof panels. There is also less chance of drips since the cup is tilted on the top of the gun. The gravity feed gun also has better balance and feel in your hand for easier control of hand-arm movements.

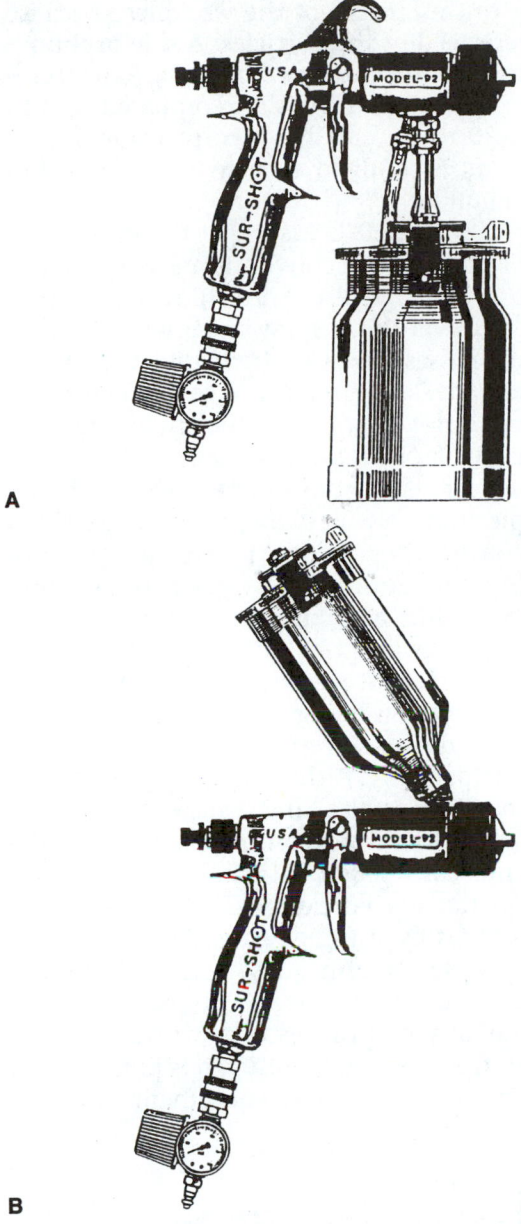

A

B

FIGURE 18-70 Typical HVLP spray guns use low pressure (10 psi at nozzle) to reduce overspray and reduce waste. (A) Suction feed HVLP gun. (B) Gravity feed HVLP gun. *(Courtesy of Bosco Mfg. Co.)*

HVLP guns are designed differently from conventional guns (Figure 18–70). They have no obstructions that could restrict air flow and increase the pressure drop inside the gun. In many designs, the airflow through the gun is continuous; the trigger opens the material flow valve. Special caps and tips are also used to assure proper atomization. If the HVLP gun requires heated air, an insulated handle may be used (Figure 18–71).

Other HVLP gun designs place the air valve in the handle, replacing the conventional air needle. The gun handle and body are larger to allow a high volume of air to flow to the tip. Some designs use only one needle valve to control paint flow. The air valve in the handle is right at the inlet fitting.

There are two basic air supply designs. One generates airflow from a turbine generator; the other converts 80 to 100 psi (552 to 690 kPa) shop air to the required 2 to 10 psi or 13.8 to 69 kPa inside the gun. Each of these approaches has advantages and disadvantages:

- *Turbine generators.* The turbine approach offers portability (Figure 18–72) that generally is not available when connecting to shop air lines. Existing shop air sources also may not provide sufficient volume. For example, in a body shop with its existing air compressor already working at full capacity, a turbine unit would work well, as it would not draw from the existing shop air. Due to friction, the turbine generates enough

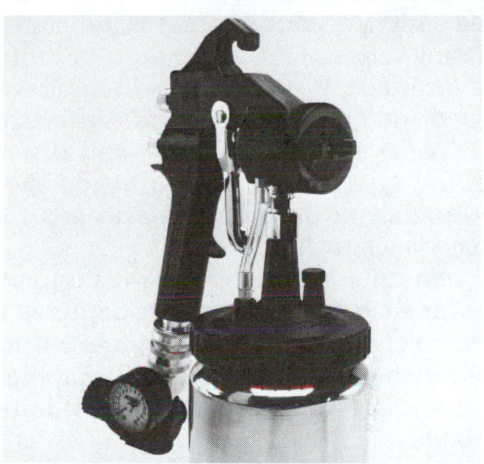

FIGURE 18-71 HVLP conversion kit is used to upgrade a compressed air system for low pressure operation. *(Courtesy of Accuspray)*

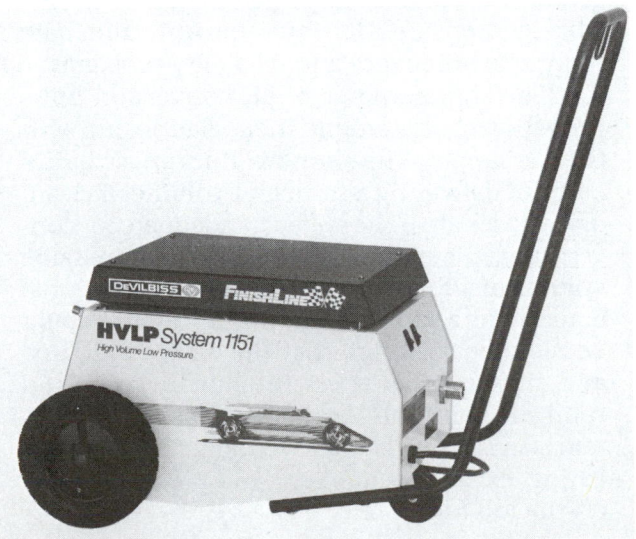

FIGURE 18-72 Small turbine generator produces pressure for system. *(Photo courtesy ITW DeVilbiss Automotive Refinishing Products)*

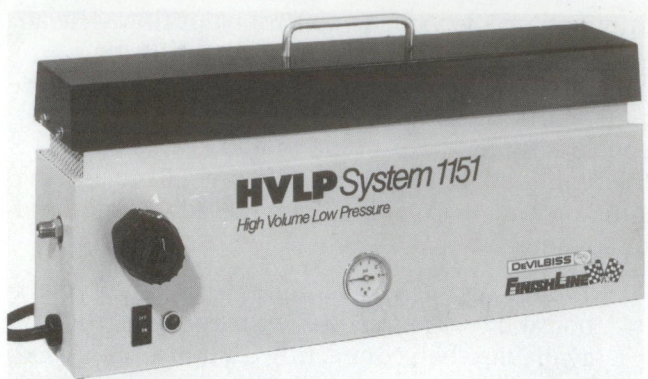

FIGURE 18-73 This is a typical air conversion unit. *(Photo courtesy ITW DeVilbiss Automotive Refinishing Products)*

heat to provide moderate air temperatures. However, the temperature of the air is not always controllable, depending on such factors as turbine design, distance from the gun, and the insulating values of the air conduit. Assuming that the air intake system and its filter are well designed and maintained, turbine generators provide relatively clean, oil-free, and dry air. For turbine systems to be used inside the spray booth, they must be of explosion-proof design, which is relatively costly.

- *Air conversion units.* A standard shop air compressor such as described in Chapter 4 and an air conversion unit provide more control over the variables that can affect HVLP application. Heat and pressure are the two most important variables.

When fitted with air heaters (Figure 18–73), air conversion units offer controlled air temperature. The heat can be varied or turned off completely. Furthermore, they can deliver a consistent 10 psi (69 kPa) or be regulated to provide somewhat less pressure. Air conversion units eliminate turbine maintenance and reliability problems.

The shortcomings of air conversion units primarily revolve around their relationship with the air supply. The shop air lines must be capable of delivering a sufficient volume of clean, dry, oil-free air. If the system is adequate for conventional air spray systems, however, the same volume of air will suffice with HVLP. Additional equipment and maintenance procedures could be required to assure that the air is clean, dry, and oil-free; otherwise, contaminated air could spoil the paint job. Again, if the air supply system is maintained for conventional air spray finishing, existing equipment and procedures will do the job for HVLP.

Except for a few subtle differences, the HVLP and conventional air spray gun operate in basically the same manner. For example, the HVLP gun should be held closer to the surface of the workpiece because of the lower speed of the particles. A rule of thumb would be to hold the gun 6 to 8 inches (15 to 20 cm) away when spraying with HVLP, compared to 8 to 10 inches (20 to 25 cm) for a conventional gun. Greater distances result in excessive dry spray and lack of film buildup.

Many first-time HVLP users get the impression that HVLP is slower than conventional air spray, but this is not always the case. Film thickness is often greater than conventional spray systems. This results in fewer total passes for the desired build. Sometimes the application is slower, but this is generally because the air source is not delivering sufficient air pressure. Remember that not all systems deliver their rated pressure under actual conditions. Another reason that some people think HVLP is ineffective is that it is quieter than what they are used to. The high air pressures used in conventional air spray systems cause them to sound like a leaking tire when airflow is present. The lower air pressures used in HVLP systems make them less noisy. This makes some people think that less work is being done.

Table 18–8 details some of the features of the six types of spray painting. Although at the present, airless, air-type electrostatic, and airless-type electrostatic are used primarily in commercial vehicles, the environmental consideration in some states is toward a 65 percent transfer efficiency. The only way this seems possible is by using electrostatic, air-assisted airless, and airless techniques. Spray equipment manufacturers are hard at work attempting to improve the quality of finishes obtained from these methods. In the foreseeable future, it is possible that electrostatic and airless spray equipment might be used on topcoat finishes.

AIRBRUSHES (TOUCH-UP SPRAY GUNS)

Touch-up spray guns are very small and are ideal for painting small repair areas. Often called an **airbrush** or *door jamb gun*, they have a tiny cup for holding a small amount of material. They operate like a conventional siphon gun.

Airbrushes range from simple types used for touch-up work (Figure 18–74) to complex and exacting tools used in custom finishing (Figure 18–75). The latter, of course, is generally found only in paint shops that do custom auto finishes.

It is important to select the correct airbrush for the type of work to be performed. Consider the size and type of the work to be done, the fineness of the line desired, and the fluids to be sprayed. Airbrushes used for custom auto finishing are generally in two categories: double-action and single-action types (Figure 18–76).

TABLE 18-8: COMPARISON OF VARIOUS SPRAYING SYSTEMS

	Conventional Air Spraying	Conventional Airless Spraying	Air-Assisted Airless Spraying	Air-Assisted Electrostatic Spraying	Airless Type Electrostatic Spraying	HVLP Spraying
Adhesion of Spray Efficiency	20 to 40%	50 to 60%	40 to 60%	60 to 70%	70 to 80%	65 to 90%
Quality of Finished Surface	Excellent	Poor	Fair to Good	Good	Fair	Good to Excellent
Work Environment (paint mist dispersion)	Poor	Good	Good	Excellent	Excellent	Excellent
Paint Speed	Slow	Fast	Fast	Very Fast	Very Fast	Slow
Paint of Depressed Areas	Excellent	Poor	Fair	Good	Fair	Excellent
Gun Handling (partial repainting and touch up)	Excellent	Fair	Fair	Fair	Good	Excellent

FIGURE 18-74 An airbrush can be used for simple touch-up jobs. *(Courtesy of Binks Mfg. Co.)*

FIGURE 18-75 The airbrush is commonly used by experienced, talented painters to do custom finishes. *(Courtesy of Badger Air Brush Co.)*

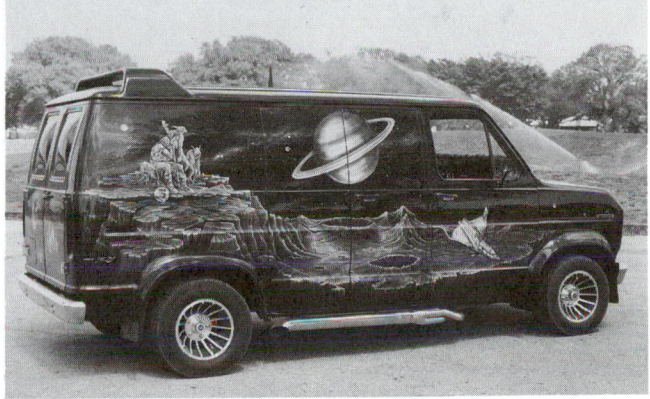

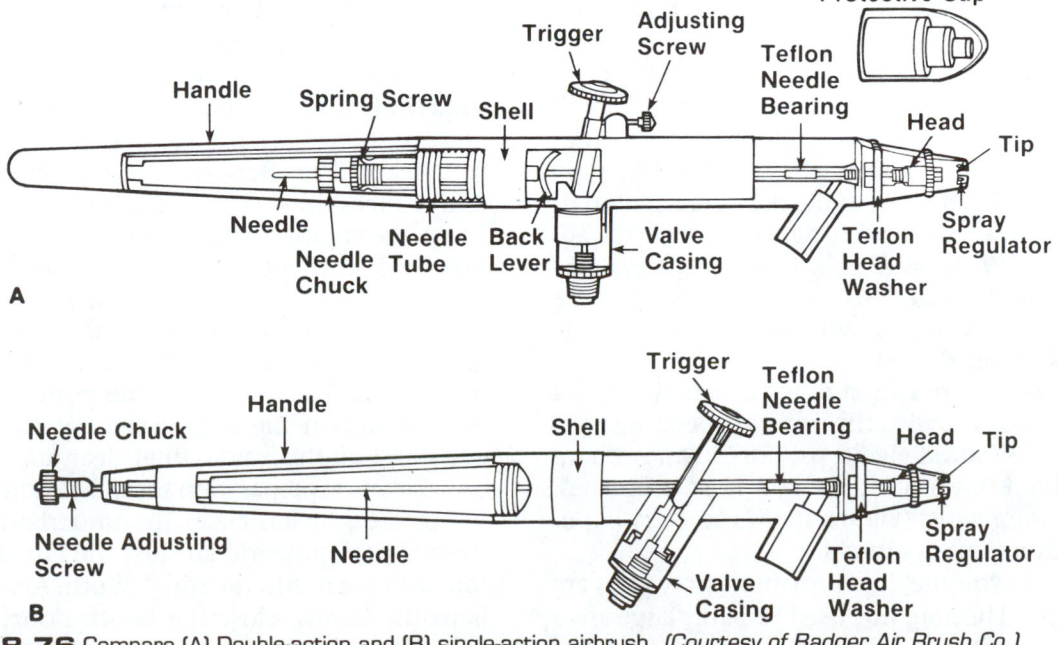

FIGURE 18-76 Compare (A) Double-action and (B) single-action airbrush. *(Courtesy of Badger Air Brush Co.)*

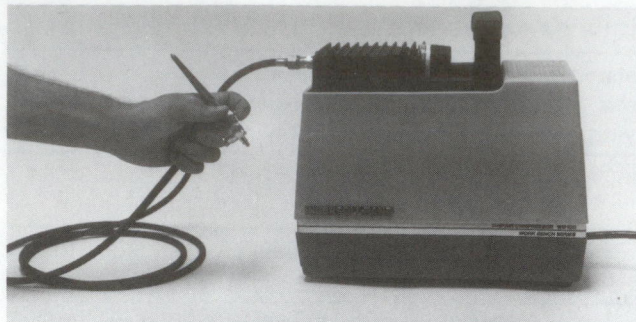

FIGURE 18-77 A typical airbrush is operated with a diaphragm type compressor.

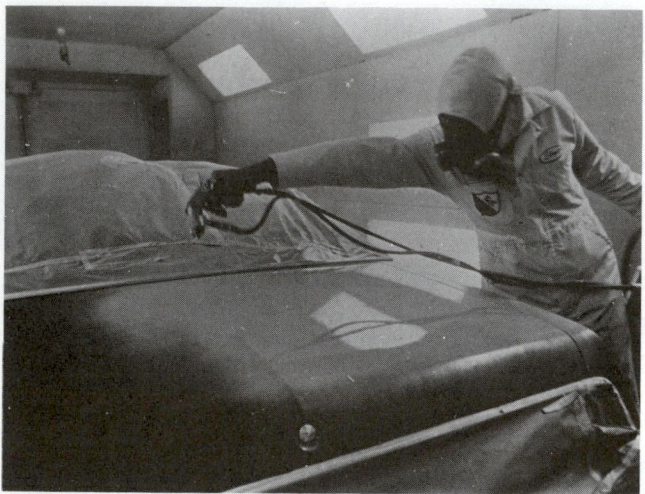

FIGURE 18-78 A well-designed spray booth will improve finish quality and safety. *(Courtesy of Sherwin-Williams Co.)*

The double-action brushes are the ones commonly found in most custom paint shops. They are available with a choice of tips to further increase their versatility. The double-action airbrush is usually recommended for projects that require very fine detailing. They produce a variable spray that works by depressing the finger-controlled front lever for air and pulling back on the same lever for the proper amount of color to be sprayed.

With single-action airbrushes, air is released by depressing the finger lever, while the amount of color desired is controlled by rotating the rear needle adjusting screw. While working, it is not possible to change the amount of color being sprayed because the operator must stop spraying to rotate the needle adjusting screw in the rear.

Airbrushes operate on a range of 5 to 50 psi (34 to 345 kPa) pressure, with the normal operating pressure being approximately 30 psi (207 kPa). An scfm rating of about 0.7 is sufficient for most airbrushes. Compact compressors (Figure 18–77) are very popular with custom auto painters.

LVLP (low volume, low pressure) spray guns are also available. They are not used to paint large areas because they are too slow. However, they can be handy for touch-up work on small surface areas.

18.6 SPRAY BOOTHS

A **spray booth** is designed to provide a clean, safe, well-lit enclosure for painting. It isolates the painting operation from dirt and dust to keep this debris out of the paint job. It also confines the volatile fumes created by spraying and removes them from the shop area.

Providing a clean, safe, well-illuminated enclosure for painting is the primary purpose of a spray booth (Figure 18–78). It isolates the painting operation from the dirt- and dust-producing activities and confines and exhausts the volatile fumes created by spraying automotive finishes.

Modern spray booths are scientifically designed to create the proper air movement, provide necessary lighting, and enclose the painting operation safely. In addition, their construction and performance must conform to federal, state, and even local safety codes, not to mention those of insurance underwriters. In most areas, automatically operated fire extinguishers are required because of the highly explosive nature of the refinishing materials.

The spray booth should be located as far as possible from the area where dust and dirt are prevalent. Therefore, it should be isolated from the mechanical and metalworking portions of the shop whenever possible. This can be accomplished with partitions, walls, or a separate building arrangement. A workbench should be handy in the spray booth for thinning the paint and filling the gun cup. Paint storage, however, should be outside the booth, but nearby.

When a spray booth must be located in the same room with metalworking stalls or other locations where there is excessive dust, the intake air can be drawn from the outdoors, utilizing an air replacement system. This arrangement greatly reduces the number of filter changes required in the booth doors and reduces the chances of ruined paint jobs.

If the volume of paint work is sufficient, a straight line work flow is recommended (Figure 18–79). Utilizing a drive-through type of spray booth, the layout is designed for maximum efficiency of manpower and equipment. Jobs are started in the metalworking stalls in the normal manner. From this point the work flows in a production line manner through each of the various stages all the way to final cleanup.

Cleaning preparation should be done outside the booth area. Steam clean the underbody of the vehicle thoroughly and air dust the entire vehicle before moving it into the spray booth. After the vehicle is in the booth, close the booth doors tightly and

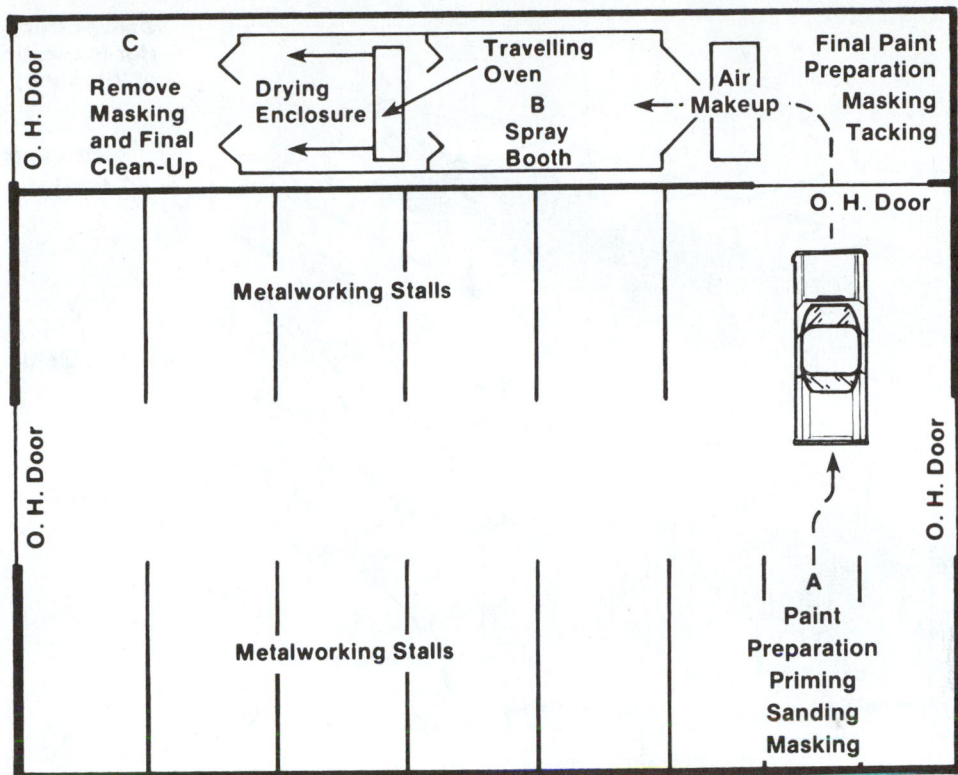

FIGURE 18-79 Typical body shop layout shows straight line work flow finishing operation. The important stops in such an arrangement are: (A) paint preparation, (B) spray booth, and (C) final clean-up.

tack-rag the entire vehicle again before proceeding with the painting operation.

All spray booth doors must be kept tightly closed during painting. If it becomes necessary to open the door, be sure the air supply is turned off. In fact, many spray booths are equipped with door switches that shut off the air supply when the doors are opened. The air compressor should be outside the booth with the air delivery pipes slanting back toward the compressor. The drive-through principle can be used in a one- or two-booth arrangement (Figure 18–80).

An *air makeup* or *air replacement system* is important because of the large volume of air exhausted from spray booths. This exhaust is sufficient to produce two or more complete changes every hour. Under such conditions in winter, the spray area can become cold and uncomfortable. Finish problems can arise because of spraying with cold materials on cold products in cold air. An air makeup system provides even temperatures and clean filtered air, and assures proper booth performance.

Sometimes paint shops employ an independent air replacement system specifically designed for the spray booth (Figure 18–81). This provides clean, dry, filtered air from the outside to the booth, heating the air in colder weather. Replacement air can be delivered to the general shop area or directly into the booth for a completely closed system.

There are four air makeup systems in use today:

- Regular flow booth
- Reverse flow booth
- Crossdraft booth
- Downdraft booth

Both the regular and reverse flow types of booths were once considered standards in spray booth construction. However, since the late 1970s, they have been replaced to a great degree by crossdraft and downdraft airflow types.

As shown in Figure 18–82, in the regular system, the airflow is from back to front. In the reverse flow process, the airflow is from front to back. The reverse flow type of booth generally has a solid back (Figure 18–83), while a regular flow usually is of the drive-through style. It is interesting to note that a good number of vehicles that are sprayed in a reverse type of booth are backed in.

The *downdraft spray booth* forces air from the ceiling down through exhaust vents in the floor. It is the most popular air movement system used today. The downward flow of air from the ceiling to the floor pit creates an envelope of air passing by the surface of the vehicle (Figure 18–84).

By taking clean, tempered air and directing it downward, the downdraft prevents contamination and overspray from settling on the freshly painted

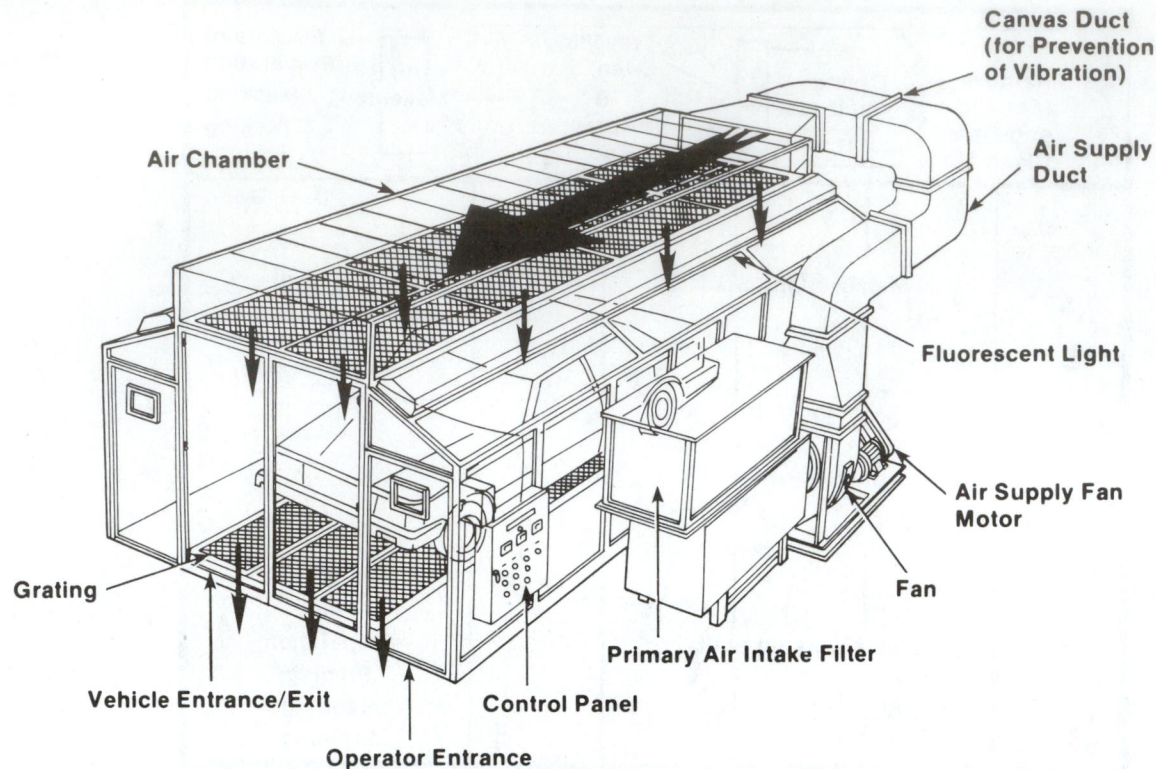

Canvas Duct
(for Prevention
of Vibration)

Air Chamber

Air Supply
Duct

Fluorescent Light

Air Supply Fan
Motor

Fan

Primary Air Intake Filter

Grating

Vehicle Entrance/Exit

Control Panel

Operator Entrance

ONE ROOM BOOTH

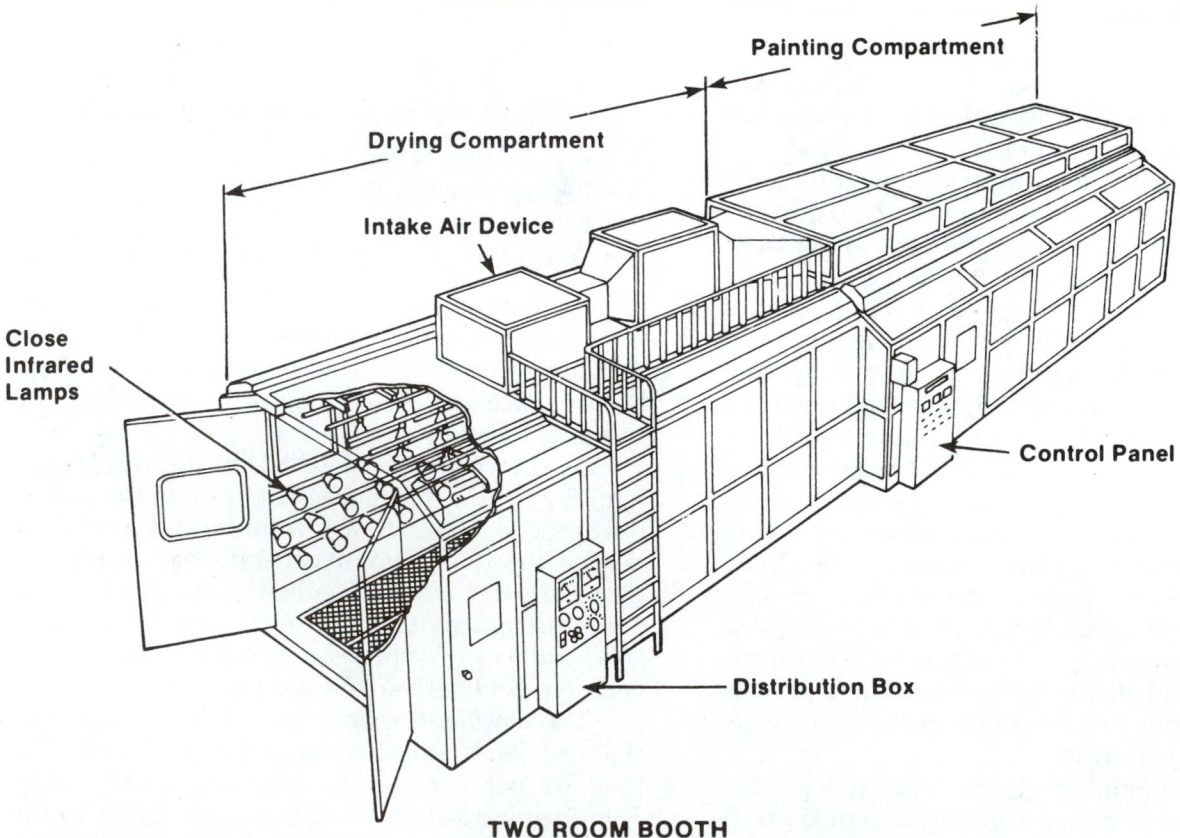

Painting Compartment

Drying Compartment

Intake Air Device

Close
Infrared
Lamps

Control Panel

Distribution Box

TWO ROOM BOOTH

FIGURE 18-80 Compare one- and two-room booths. *(Courtesy of Toyota Motor Corp.)*

surface of the vehicle. This air movement also helps to remove toxic vapors from the breathing zone of the painter, providing a safer working environment. Many shops also provide a supplied-air breathing system that ensures fresh outside air is provided to workers in the spray booths at all times. The downdraft booths as illustrated in Figure 18–85 are available in raised platform models and floor models and are usually of the drive-through type. Some more important features of this system are shown in Figure 18–86.

A *side draft* or *crossflow spray booth* moves air sideways over the car or truck. An air inlet in one wall pushes fresh air into the booth. A vent on the opposite wall removes the booth air.

Crossdraft systems are less expensive to install since they do not require a raised platform or a pit under the booth. The crossdraft-type booth provides a horizontal airflow and many of the advantages of the downdraft system. It is available in solid back and drive-through models.

Because of the many OSHA, state, and local regulations regarding spray booths, the use of solid concrete or cinder block types has been on a decline. While many are still in use, building a "do-it-yourself" spray booth is seldom done today.

AIR FILTRATION SYSTEMS

The most important safety feature of spray booths is the filtration system. Currently there are two common types in use (Figure 18–87): wet filtration systems and dry filtration systems.

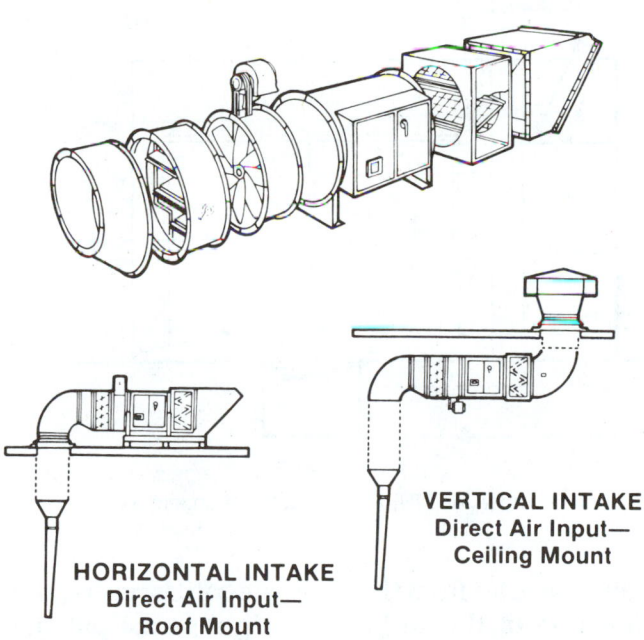

FIGURE 18–81 Air replacement is available in various configurations to fit building needs.

HORIZONTAL INTAKE
Direct Air Input—
Roof Mount

VERTICAL INTAKE
Direct Air Input—
Ceiling Mount

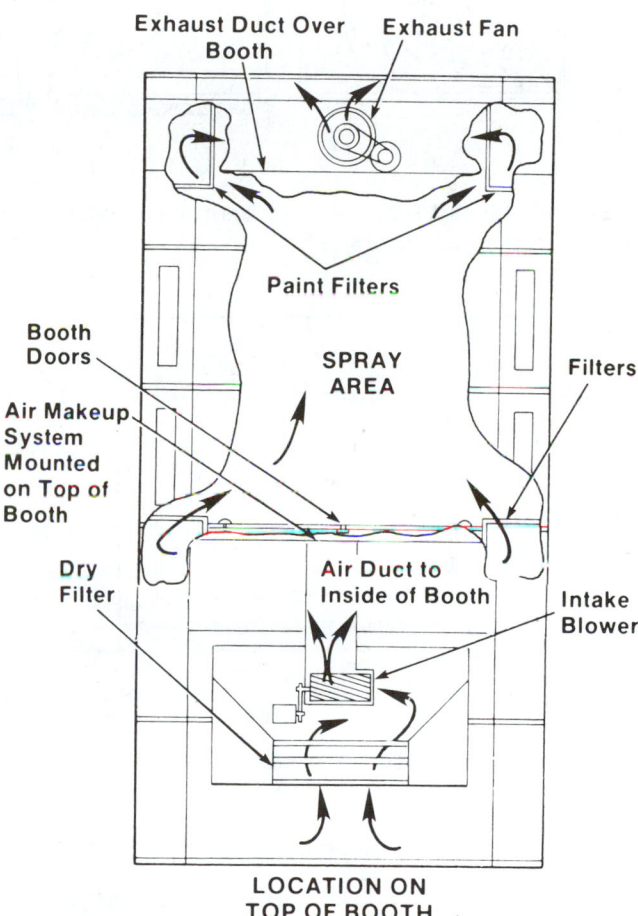

FIGURE 18–82 This is a typical air make-up system.

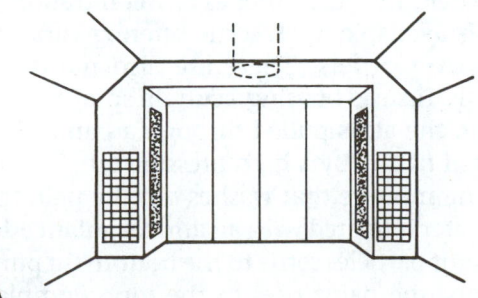

A

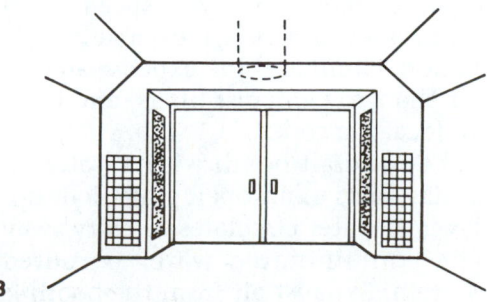

B

FIGURE 18–83 Note two designs of spray booths: (A) solid back and (B) drive-through. The solid design is generally found in small shops.

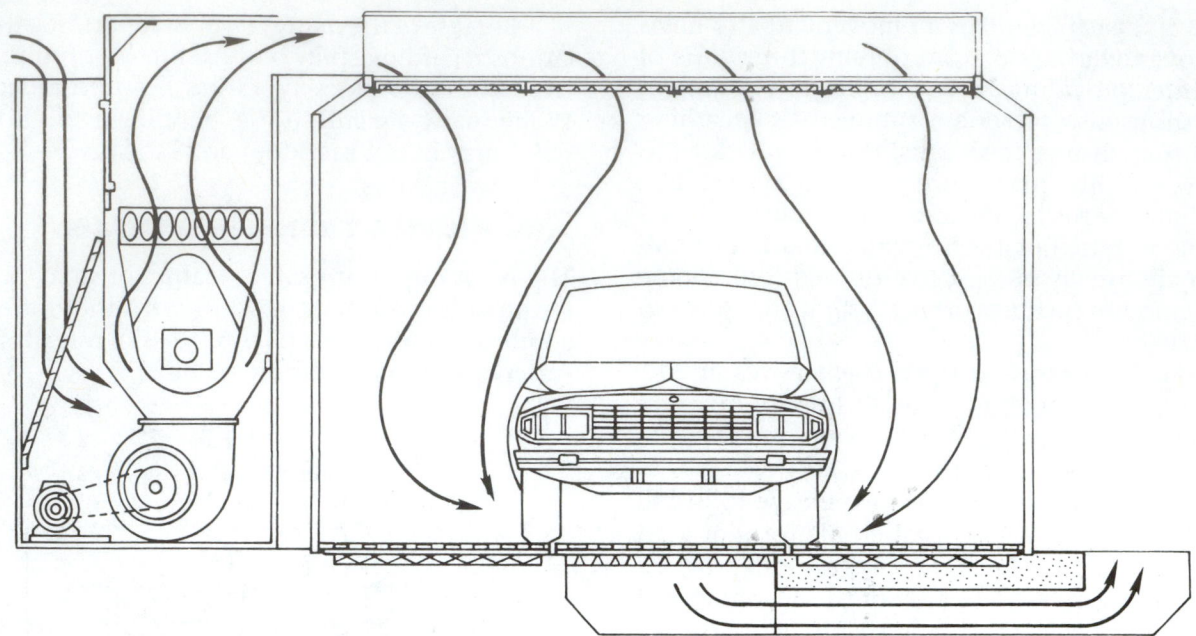

FIGURE 18-84 Follow airflow pattern of a downdraft spray booth. The location of the intake and exhaust system will depend on the system manufacturer.

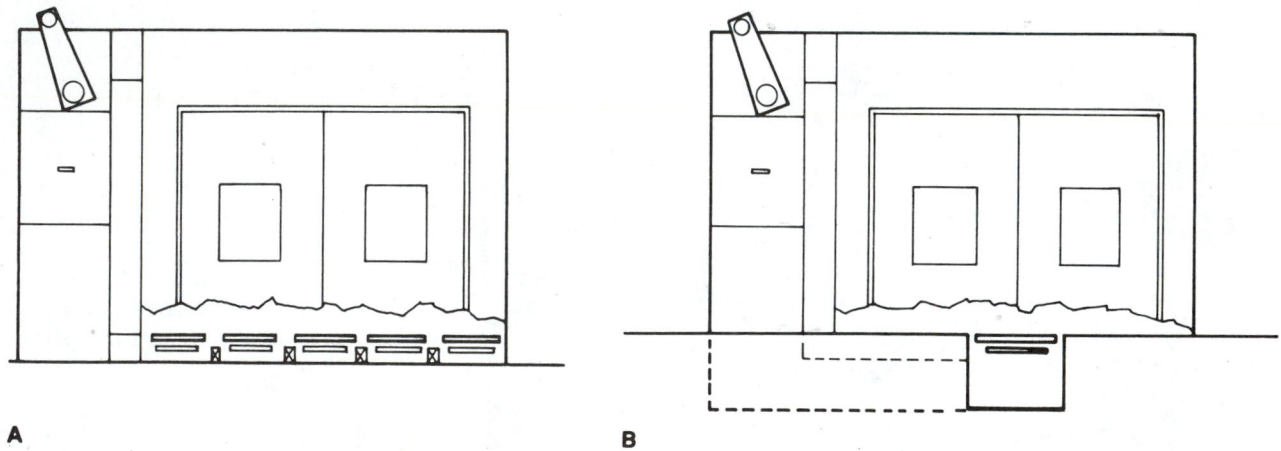

A   B

FIGURE 18-85 Note two models of downdraft booths: (A) raised platform model and (B) floor model with underfloor pit.

Wet Filtration Systems

Wet or *wash filtration* has a higher initial cost than a dry filter system. There can also be the additional cost of waste disposal. However, it does an excellent job of removing paint particles from exhaust air regardless of the paint viscosity or drying speed. It can handle a variety of spray materials, is capable of high volume production, eliminates the expense and inconvenience of changing exhaust filters, and is accepted by most local fire codes.

The typical downdraft booth with a water filtration system has ducts or an open grate floor under which a layer of water circulates to carry away overspray. The contaminated water is routed through the system. Exhaust air from the booth is purified by routing it through a water curtain wash system. A continuous spray mist of water scrubs the paint particles from the air, while baffles reverse the direction of the airflow to help separate out the particles by centrifugal action. The air that emerges is as clean or cleaner than that achieved by a quality dry filtration system (Figure 18–88).

There are wide varieties of wet filtration configurations available, with some offering various advantages over others. There are also pumpless wash systems. Instead of using a curtain spray wash to clean the air, the air is pulled through a pan of water and series of baffles by a high pressure fan. This creates a swirling mixture that washes out the paint particles. The water is treated with an anticoagulant additive so the paint particles settle to the bottom (in pump wash systems, the paint rises to the top). Pumpless wash systems are very sensitive to the water level and air pressure, so proper maintenance is very important.

The paint residue that collects in the water must be removed periodically and the water kept at the specified level. The rate at which water evaporates from the system will depend on temperature, humidity, the volume of usage, and the design of the system. It is necessary to add makeup water at least once a week unless the booth has an automatic water makeup feature. Water additives must be placed in the water to prevent the growth of bacterial/germ growth and an unpleasant smell.

A

B

C   D

FIGURE 18-86 Study important parts of a downdraft refinish system, which include: (A) spray booth itself; (B) air replacement unit; (C) ceiling plenum and filter system; and (D) spray booth itself, showing floor gratings and filter. *(Part A courtesy Binks Manufacturing Company; Parts B, C, and D photo courtesy ITW DeVilbiss Automotive Refinishing Products)*

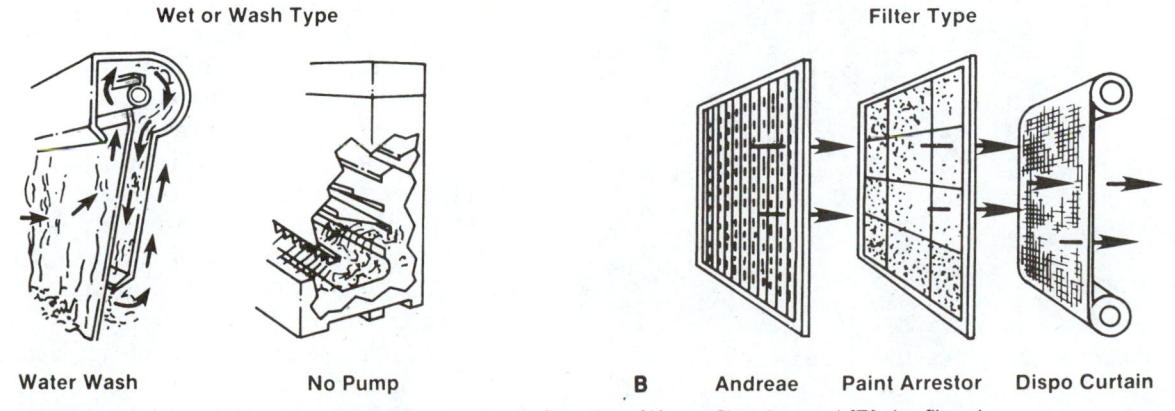

| Wet or Wash Type | | Filter Type | |

A   Water Wash   No Pump   B   Andreae   Paint Arrestor   Dispo Curtain

FIGURE 18-87 Here are different methods of spray booth filtration: (A) wet filtration and (B) dry filtration.

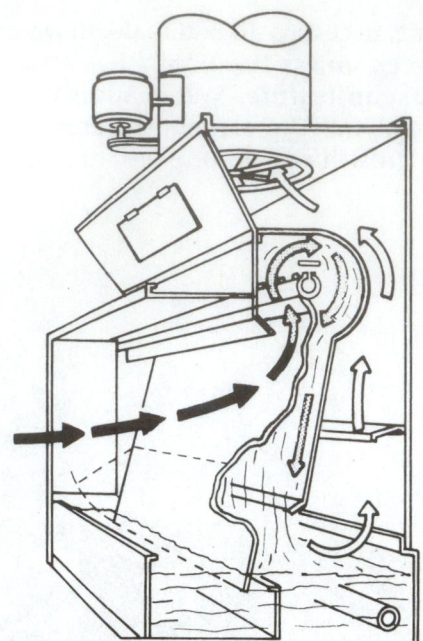

FIGURE 18-88 Air filtration system must remove contaminants from air circulating through spray booth.

FIGURE 18-89 Commercial paint waste pickup service is usually required to meet EPA regulations for hazardous waste disposal.

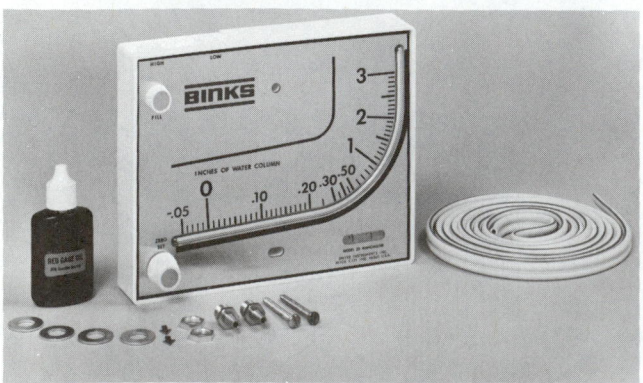

FIGURE 18-90 Manometer and its accessories are needed to control airflow through paint booth. *[Courtesy of Binks Mfg. Co.]*

The paint residue that is separated from the water hardens and can be disposed of by bagging it and, in most areas, sending it to a landfill (Figure 18–89). One innovative approach is to recycle it by mixing it with undercoating and spraying it under cars. The water is continuously recycled, so there is no reason to worry about water disposal. If the system has to be drained for some reason, the waste water must be disposed of in accordance with whatever local sewer restrictions might apply.

Dry Filtration Systems

Dry filtration systems come in various configurations and filter media (paper, cotton, fiberglass, polyester, and so on). What is notable is the efficiency of this system and how quickly the filters clog up. Dry filters work like a sieve. They mechanically filter out particles of paint and dirt by trapping the particles as air flows through the filter. Some are also coated with a tacky substance so particles will adhere to the surface of the fibers.

Most dry filtration systems can remove virtually 100 percent of the particulates that are large enough to cause a noticeable blemish in a paint job. Anything larger than about 14 microns (0.0005 inch, the smallest particle that can be seen by the naked eye) can leave a noticeable speck in the paint. Anything smaller than 14 microns is usually encapsulated in the paint and will not cause problems. Most of the filters that are used in the ceilings of downdraft booths or the doors of crossdraft booths today will stop anything larger than about 10 microns from getting through.

As a filter traps more particles, it becomes more dense and thus more efficient. But at the same time, it also offers increasingly greater resistance to the flow of air. Eventually the point is reached where the filter restricts airflow through the spray booth. Ideally, the filter should be changed before it reaches that point. It is something that has to be watched very closely.

The best way to judge a filter's condition is to measure its air resistance with a water column pressure differential gauge (manometer gauge). Some booths have built-in gauges; others do not (Figure 18–90). Comparing air pressure upstream of the filter to that which is downstream is a good indication of whether or not it is time to replace filters. The amount of restriction that is considered acceptable will vary according to filter construction and media, spray booth construction, and air volume.

Because the amount of restriction that is considered acceptable can vary so much from one type of filter to another, it is important to check with the filter supplier before using any replacement filters that are different from those originally supplied. The type of filter media used will also have a significant bearing on maintenance costs, filtration efficiency, and filter longevity. Filters are not a generic product; one type might be much better suited to a particular application than another. That is why spray booth manufacturers typically put such a high emphasis on filter selection.

There are still other considerations to keep in mind when selecting a filtration system. In some areas of the United States, for example, shops over a certain size are now facing an additional air filtration expense. Neither dry nor wet filtration can remove harmful chemicals and solvents such as isocyanates from the exhaust air, so additional exhaust filtration is now being required. To date, unfortunately, the only approved methods of treating exhaust air are with an after-burner system (which is very expensive) or with activated carbon filtration. Some booth manufacturers claim that water filtration can neutralize isocyanates, but this has not yet been approved in some areas.

18.7 SPRAY BOOTH MAINTENANCE

Regardless of the type of filtration that a paint shop employs, spray booth maintenance is a prime consideration, not only from the standpoint of cost and convenience, but also because it is essential to achieving quality paint jobs. The best air filtration system in the world will not be able to do its job if it is poorly maintained. The first task in learning how to avoid dirt is to understand where it comes from. Anything that is brought into the booth can bring dirt with it. Potential sources of dirt include the airborne particulates, the vehicle, the painter, the equipment and supplies, and even the paint.

Incoming air is a prime source of dirt. Dirt is generated by dirty filters, imbalanced air pressures, and open doors. Check the intake filters daily and change them as soon as the manometer indicates. When dust and dirt start to clog filters and restrict airflow, the velocity of air passing through the filters begins to climb. Increased velocity increases the likelihood of pulling dirt through the filters. Balance the input air pressure against the exhaust air to provide slightly positive pressure in the booth. This balance can change as filters load up and also differs from car to car. Therefore, check and adjust it with each new job.

Enter the booth only with the fans running. The positive pressure helps keep the dirt out. Once a vehicle is inside, keep traffic flow in and out of the booth to an absolute minimum. Also, make sure that the body shop doors are closed at all times during the painting operation. Opening and closing these doors can cause the booth balance to fluctuate, creating turbulence and dirt inside the booth.

The booth itself can be a main contributor to dirt problems through air leaks, poor housekeeping

habits, exhaust air, and floor coverings. There are recommended seals for door frames, light openings, and panel seams that must be installed properly and replaced periodically. Heavy usage and temperature extremes quickly destroy these seals. Use caulking as an inexpensive gap sealant to keep dirt out of the air stream.

When operating the spray booth, keep the following points in mind:

1. Follow the manufacturer's recommendations for the minimum velocity needed to exhaust spray vapors properly. If that recommendation is exceeded, turbulence cancels out the screening performed by the filters. If the velocity is too low, the air will not move fast enough to remove overspray and airborne dirt before it causes defects.

2. Paint arresters are a high-consumption item requiring frequent changing. Check filter resistance daily on the manometer. When paint accumulation builds up, velocity goes down, and air movement is too slow.

3. In a dry filtration system, the filters must be periodically inspected and replaced (Figure 18–91). And when they are replaced, the multistage filters designed for the booth should be used.

4. Be sure the water level in the wet filtration system is kept at its proper working level and that the correct water additive is used.

In order to get the best results from any type of spray booth, it is important to follow a good housekeeping program such as this:

- Periodically wash down the booth walls, floor, and any wall-mounted air controls to remove dust and paint particles. Many shops require that

FIGURE 18-91 Good air compressor maintenance includes changing filters, checking oil levels (weekly), and draining the storage tank daily! *(Photo courtesy ITW DeVilbiss Automotive Refinishing Products)*

floor and walls be wiped down after every job. Always pick up any scrap, rags, and so forth.

- The booth is no place to store parts, paint, trash cans, or work benches because dirt will accumulate on these things and will eventually land on the vehicle. Keep these items in a sealed, ventilated storage area.

- Be sure that all bodywork and most paint preparation procedures are done outside of the spray booth. Make certain no sanding or grinding operations are performed in or near the spray booth. The dust created will spread all over and ruin not only a present job but many future jobs.

- Water is most often used to contain dirt. It is cheap and effective at trapping the dirt. But it can splash on the car midway through the job or, in a heated booth, dry out before the paint job is finished. If water is sprayed on the floor to keep any stray dust down, eliminate all puddles to prevent splashes. Water can also rust the walls of the spray booth, resulting in premature deterioration.

- Roofing felt held to the floor with duct tape provides an inexpensive method of containing dirt. It attracts and holds lint, lasts longer than water, is not a hazardous waste, and does not deteriorate in the booth.

- Clay tiles look nice and provide an easy-to-clean surface, but are expensive and make it difficult to load a car in the booth on dollies.

- Concrete sealant provides a smooth, easily cleaned surface that is somewhat inexpensive. It should be noted that any slick surface treatment adds to the turbulence, while a textured surface tends to impede the turbulence or at least scrub the air. There are also strippable spray-on

SHOP TALK

Cotton is perhaps the greatest source of contamination and should never be worn in the booth. Lint-free paint suits, rubber form-fitting gloves, a dirt-free head cover, and the appropriate respirator should be worn inside the booth. Remain in the booth between the application of coats rather than risk dragging dirt back inside. If this is not possible, remove the protective suit inside the booth and leave it there. Upon returning, put the suit back on to contain the dirt collected outside the booth. (Anyone not wearing the proper attire should view the work through an observation window rather than risk contaminating the paint.)

coverings. When the overspray becomes too thick, strip and recoat.

The vehicle itself is often the greatest source of dirt in the spray booth. Dirt hides in cracks and crevices, behind bumpers, and in the engine compartment. Even a thoroughly cleaned vehicle collects dirt when left in the general sanding area before being brought into the booth. When the spray gun hits this dirt, it kicks it out of its hiding places and deposits it into the finish. That is why a good prep job is so important.

Spray guns and cups should always be kept spotless inside and out. Do not use those dirt-collecting cloth wheel covers. Spray guns, masking paper, paint cans, tape, wheel covers, air transformers, hoses, respirators, coveralls, tack rags, and various other supplies can all collect dirt if stored in a dirty environment. All of these items should be kept in a filtered, ventilated storage/mix room. If subjected to sanding dust, they will quickly ruin a paint finish.

Unbelievable as it might seem, dirt from compressed air lines often causes blemishes in paint jobs. Air transformers, with properly cleaned and regularly drained filters, keep the air clean and dry. Oil and water separators are absolutely necessary to eliminate dirt and contamination.

A buildup of overspray can collect on the air cap and turn into a kind of fuzz. Clean the gun frequently to prevent the fuzz from blowing off and ruining a paint job. Paint will set up in and on the gun. If the dried paint flakes, it will land on the job and cause a defect. Clean the gun inside and out after each job.

Improper viscosity can cause excessive overspray, increased booth maintenance, runs or sags, pebble-dry finishes, and color mismatches. Always mix the paint according to the manufacturer's recommendations and check the viscosity with a Zahn cup and stopwatch as described earlier in this chapter.

Paints are complex formulations. A combination of two or more brands of ingredients can result in unbalanced viscosity, poor adhesion, dry spray, mottling, low gloss, off-standard soft finish, and solvent pop. Until wrinkle finishes become popular, avoid this condition at all times by using the manufacturer's recommended products.

Oil the fan pulley and motor bearings of the spray booth regularly, if required. Always switch off the main fan power supply before oiling the fan. If the spray booth is not properly maintained, it can cause finish problems (Table 18–9).

18.8 DRYING ROOM

A dust-free drying room following the spray booth will speed up drying, turn out a cleaner job, and in-

TABLE 18-9: TROUBLESHOOTING SPRAY BOOTH PROBLEMS

Fault	Result	Dirty Job	Thin Coats	Poor Opacity	Sags	Overloading	Popping	Softness	Overspray	Uneven Application	Recoat Failure	Fire Hazard	Water Splashes
Dirty filters	Vacuum in booth (hot air drawn from oven) OR	C				B	A	A	C	C,D	A		
	Not pressurized (low air movement and dirty air drawn in from preparation area)	A	A*	C	B					C,D			
Breached or damaged filter	Turbulence	A							B	B,D			
	Over-pressurized				A†	A	A	A**	B	C,D	A		
Water level Low	Increased extraction					A	A	A		C,D	A		A‡
High	Restricted extraction		A*	C	B				C	C,D			A‡
Empty	Increased extraction with buildup of dry paint in reservoir	A										A	
Use of incorrect water additive, or incorrect use of water additive	Blocked water jets and filters. Formation of dry powder on anti-splash panels.	A										C	A
	E Bacterial/germ cultivation (unpleasant smell)												
	Corrosion of paint. Paint deposits difficult to remove											A	
Flatting paper, rags, masking paper, old cans, and so on in booth.	Dirt accumulation	A										A	
Spraying on walls of booth.	Poor light reflection									C,D			
Loose deposits of dirt, dry spray, rust, and so forth on booth walls.	Dirt in atmosphere	A											

A Most likely failure to be associated with the fault
B Likely failure
C Failure less likely to be associated with the fault
D Will affect color of metallics
E Health hazard

* Poor build
† In oven
** Cold air forced into oven
‡ Alkaline contamination

Note: 1. Use only lint-free overalls and head gear in the spray booth. (Use them only for this purpose.)
2. Clean and blow off prepared vehicles outside the spray booth, paying particular attention to the engine compartment. Do not exceed 40 psi (3 bars).
3. Repair damaged or ill-fitting spray booth doors promptly.
4. Affix clean sheet of paper daily to spray booth wall for testing gun.

crease the volume of refinishing work that can be handled. The drying rooms of more sophisticated paint shops have permanent infrared or sodium quartz units for the forced drying of paint, particularly enamels. These ovenlike units (Figure 18–92) can speed up the dry time of enamels as much as 75 percent. The use of forced drying on putty, primer, and sealer coats will reduce waiting time between operations and can also be used for fast drying spot and panel finish coats.

Infrared or sodium quartz drying equipment is available in portable panels for partial or sectional

FIGURE 18-92 Infrared drying units in this drying room will speed production so you can release the vehicle to the customer in less time. Temperature must be carefully controlled to prevent soft-part damage. *(Courtesy of Garmat, Inc.)*

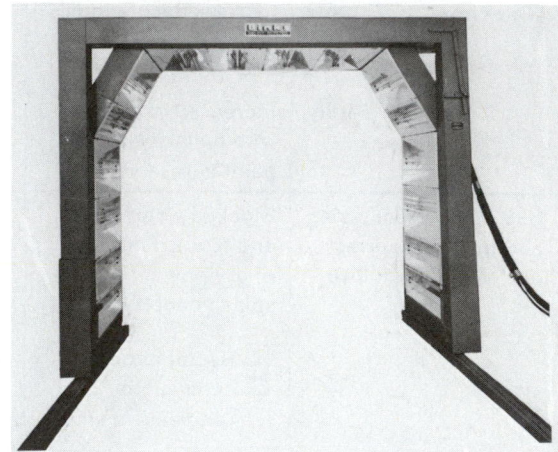

FIGURE 18-93 Compare typical near infrared drying equipment. The unit on the top has a movable top and is portable, while the one on the bottom moves over the vehicle on a track. *(Courtesy of Binks Mfg. Co.)*

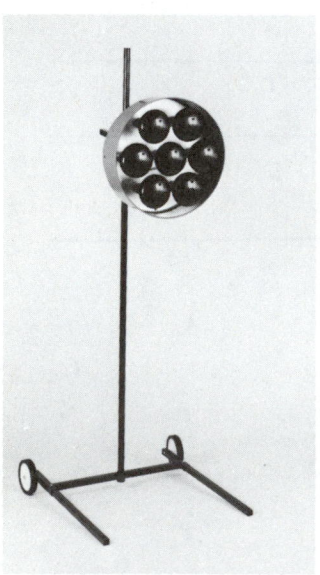

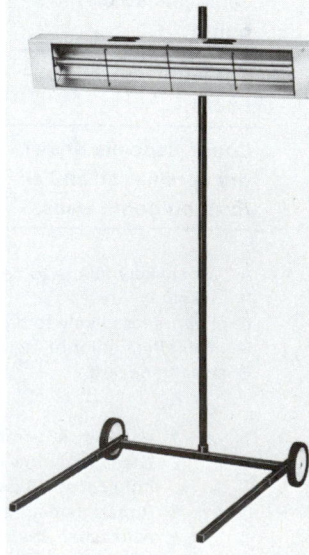

FIGURE 18-94 Here are examples of portable far drying equipment.

drying, or in large traveling ovens capable of moving automatically on track over the vehicle to dry a complete overall job. There are two types of infrared drying equipment:

- *Near drying equipment.* Because drying equipment uses lamps as the heat source, this type of equipment is easy to handle. The radiation angle can be varied easily. The construction, relocation, and assembly are simple, so it is the most common type used for automobiles. There are several shapes and sizes of this equipment, depending on what it is used for, but the most common types are illustrated in Figure 18–93.
- *Far drying equipment.* Far drying or sodium quartz equipment affects paint drying by means of heat radiated from a tubular or plate-type heater. The heat source is either gas or electricity. Far drying equipment also comes in various types and sizes, depending on its use (Figure 18–94).

Drying can best be accomplished in a separate drying chamber attached to the back of a downdraft system or conventional drive-through booth (Figure 18–95A) where the traveling oven is housed and operated. In this configuration, the highest production is achieved since both the painting and drying operations can be performed simultaneously.

Drying can also be performed directly in the spray booth after painting. A storage vestibule is used to store the traveling oven until it is needed (Figure 18–95B). After the vehicle is painted, the oven is rolled out of the vestibule and into the spray booth for the drying operation.

When using a drying room, certain precautions must be taken not to destroy the finish. Table 18–10 gives the common difficulties that can be caused in the drying room.

18.9 OTHER PAINT SHOP EQUIPMENT AND TOOLS

There are several pieces of paint shop equipment that can help the refinishing technician perform paint jobs better. These items include the following:

- *Wet sanding stand.* A wet sanding stand (Figure 18–96) is used for wet sanding individual components or small parts. These cabinets are made by individual paint shops with the size and installation location depending on shop requirements and conditions.
- *Paint hanger.* Paint hangers are used to suspend or secure individual components or small parts for spray painting. As with the wet sanding

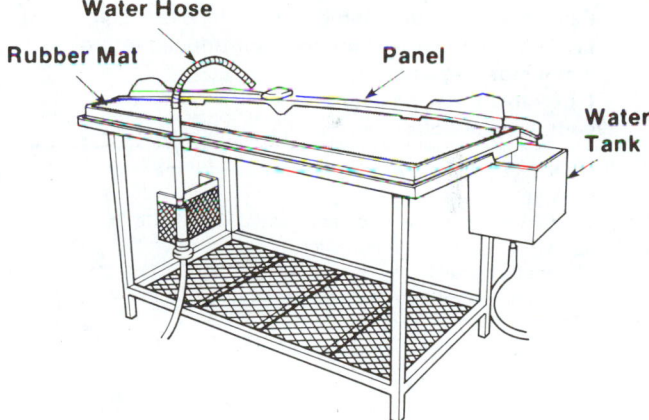

FIGURE 18-96 Typical wet sanding stand is good for smaller parts.

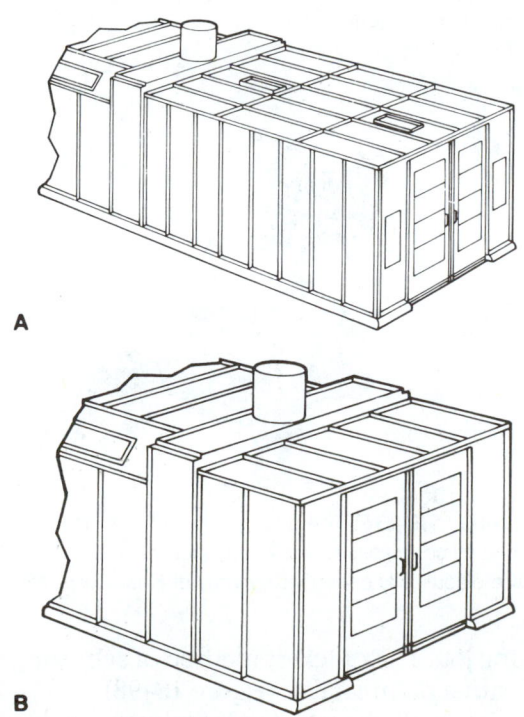

FIGURE 18-95 (A) Spray booth with drying chamber; (B) spray booth with storage vestibule.

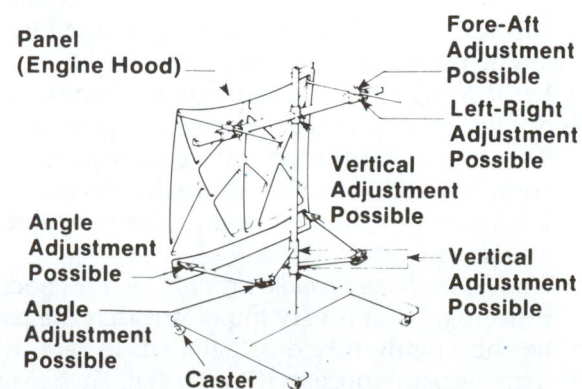

FIGURE 18-97 Typical paint shop hanger stand is good for larger panels and parts that must be painted on both sides. *(Courtesy of Herkules Equipment Corp.)*

TABLE 18-10: TROUBLESHOOTING DRYING ROOM PROBLEMS

Fault	Result	Popping	Softness	Dirty Job	Overspray	Impaired Durability	Polishing Impaired	Fire and Explosion Hazard	Loss of Gloss	Recoat Failure	Discolorate
Dirty filters	Diminished air velocity	A¹	A³			C	B			A	
	Diminished oven pressure		A⁴	C	B⁵	C	B				
	Spray booth/oven pressure imbalance			B							
Filters damaged or breached	High velocity jet streams and turbulence	A²	B²	A		C	C				
Thermostat probe not correctly sited in moving airstream and/or insufficiently sensitive.	Excessive high/low temperature modulation	A	A			C	C				
10% Bleed duct closed **10% Make-up filter clogged**	Foul oven Excessive fumes						B	A	A⁶		A⁷
Failure to remove deposits of rust, dust, and flaking paint from oven surfaces	Excessive dirt circulation				A						
Failure to clean unpainted areas on vehicles. Failure to clean masking or remask. Operators entering oven with dirty overalls	Unnecessary dirt introduced into oven				A						

A   Most likely failure to be associated with the fault
B   Likely failure
C   Failure less likely to be associated with the fault
D   Will affect color of metallics
E   Health hazard

¹Upper parts
²Local
³Lower parts
⁴Cold air drawn from booth

⁵Drawn from spray booth
⁶Microshrivel
⁷Chemical reaction

Note: Repair ill-fitting or damaged oven doors immediately.

stands, these are made by the individual shop in accordance with the shape of the item to be painted, the quantity required, and so on. Paint hangers keep the panel from dropping during painting. They must be made of a material that will withstand heat during paint drying. An example is shown in Figure 18–97.

- *Panel drying ovens.* These are small ovens used to dry test pieces. There are various types—from a very simple kind using infrared lamps to more complicated kinds with an electric heater, vent fan, and a timer for controlling the temperature and drying time.

- *Paint shakers and paddle agitators.* For a good refinishing job, it is very important that the paint be thoroughly mixed or agitated. In fact, with metallic paint topcoats it is essential. These paints contain metallic particles that are heavier than the paint itself and quickly settle to the bottom of the container. For this reason, metallic paint, as well as most other types, needs a proper mix-

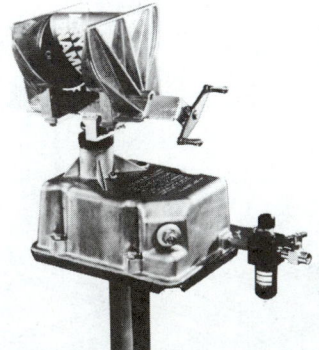

FIGURE 18-98 The paint shaker shown here is operated by compressed air. Paint shakers are also available that are electrically driven. *(Courtesy of Bron Corp. Mfg. Co.)*

ing job. The quickest method of achieving this is with a paint shaker (Figure 18–98).

Another type of paint mixer is the blade agitator (Figure 18–99). The blades of the agitator are dipped into the paint, and the paint can sealed by the agitator cover. The cover locks over

FIGURE 18-99 The blade agitator type mixer unit fits right on the top of the paint can. *(Courtesy of Dedoes Industries, Inc.)*

FIGURE 18-101 Weight type scales are used with most color mixing systems. Exact weight scales are available with standard or digital scales. *(Courtesy of DuPont Automotive Products)*

FIGURE 18-100 Most custom color mixing service centers use blade agitator mixers. Blades are built right into the shelf units. *(Courtesy of DuPont Automotive Products)*

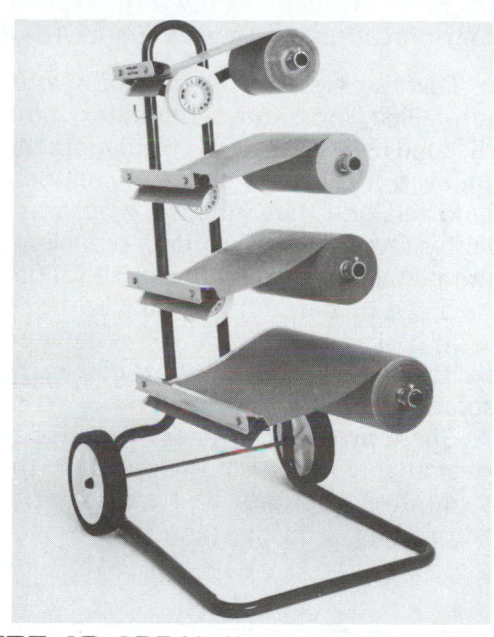

FIGURE 18-102 Masking paper and tape dispenser will save time. The tape is automatically applied to the edge of paper as it is rolled off. *(Courtesy of Marson Corp.)*

the can opening by spring action. These agitators usually come in 1 and 4 quart sizes and are the types of mixer used by color mixing centers (Figure 18–100).

- *Churning knives.* Churning knives are also used to stir paint. The handle tip is designed as a paint can lid opener. Some churning knives have a scale for measuring paint or hardening agents.
- *Color matching scales.* Color matching scales are used to match the paint with the original color or tone. There are volume-type and weight-type scales. The paint is matched against a color formula card. Use of these scales enables even a relatively inexperienced person to match paint. However, a final visual check is recommended before using the paint since the scales are not always 100 percent accurate. There are three types of color matching scales:

1. *Volume-type scale.* This type mixes the paint according to volume. The container for the paint to be mixed should have smooth walls and straight sides. To mix the paint, first set the color card into the scale, insert the ingredients as specified, and mix them.

2. *Mechanical weight-type scale.* This type (Figure 18–101) matches paint according to the weight of the ingredients. Weighing must be accurate, according to specifications on the formula card, in order to obtain a correct color mix. First, combine the ingredients and then mix thoroughly to obtain the desired color.

3. *Computer-type scale.* This type of scale allows very small quantities of paint to be mixed and its accuracy makes ingredients mix easily.

- *Masking paper dispenser.* A masking paper dispenser (Figure 18–102) allows dispensing of both masking paper and masking tape at the same time; as the paper is pulled, masking tape automatically adheres to the paper edge. Two or three sizes of roll paper can be set in the dispenser to help upgrade work efficiency.

- *Metal paint cabinet.* Paint cabinets are used for storage and stock control of paint, thinner, and putties. These cabinets should be selected for the amount of paint and thinner normally stored, conditions of the shop layout, and local fire codes.
- *Vehicle lift.* A portable vehicle lift, described in Chapter 4, is handy when preparing a car for refinishing (Figure 18–103).
- *Respirators.* Spray finishing creates a certain amount of overspray, hazardous vapors, and toxic fumes. This is true even under ideal conditions, and there is no way to avoid it entirely. Anyone who is around a spray finishing operation must consider wearing some type of respirator or breathing apparatus (Figure 18–104).

There are two good reasons for wearing a respirator. First, some sort of respiratory protection is dictated by OSHA/NIOSH regulations. And second, even if this were not true, common sense would tell one that inhaling overspray is not healthy. Overspray can contain particles of toxic paint pigments, harmful dust, and vapor fumes, which can be harmful to your health. Depending on design, a respirator can remove some or all of the dangerous elements from the air around a spray finishing operator.

There are three primary types of respirators available to protect the operator: the fresh air supplied respirator, the cartridge filter respirator, and the dust respirator. But, when there

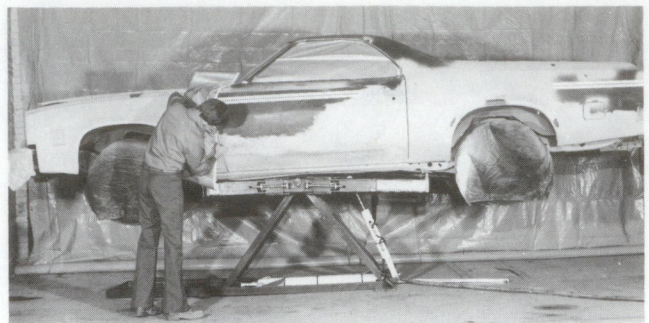

FIGURE 18-103 A portable vehicle lift is handy in refinishing preparation work. *(Courtesy of Herkules Equipment Corp.)*

FIGURE 18-104 Wear a respirator even when mixing paint.

is any doubt about the respirator to be worn, always use the air line hood respirator.

It is important to note that symptoms of inhalation overexposure might not appear until 4 to 8 hours after the exposure. Depending upon the severity of overexposure, symptoms might persist for 3 to 7 days.

Because high concentrations of vapor will irritate the eyes, goggles or sufficient eye protection should be worn at all times during application. If spray residue finds its way to the eyes, they should be flushed immediately with water for at least 15 minutes. If irritation persists, see a physician.

18.10 BASIC PAINT SHOP MATERIALS

Paint shop materials differ from refinishing equipment in one key way: paint shop refinishing materials are expendable, they are used up in the day-to-day

SHOP TALK

The right-to-know laws and personal protection equipment for the body shop painter were described in Chapter 1. Most of the materials handled in paint areas can generally be considered hazardous. Therefore, the paint technician must know:

- What the material is
- Hazardous properties of the material
- Operation of personal protection equipment needed for working with that material (Figure 18-105)
- Proper use of the protection equipment
- Protective equipment fits properly and is in working order

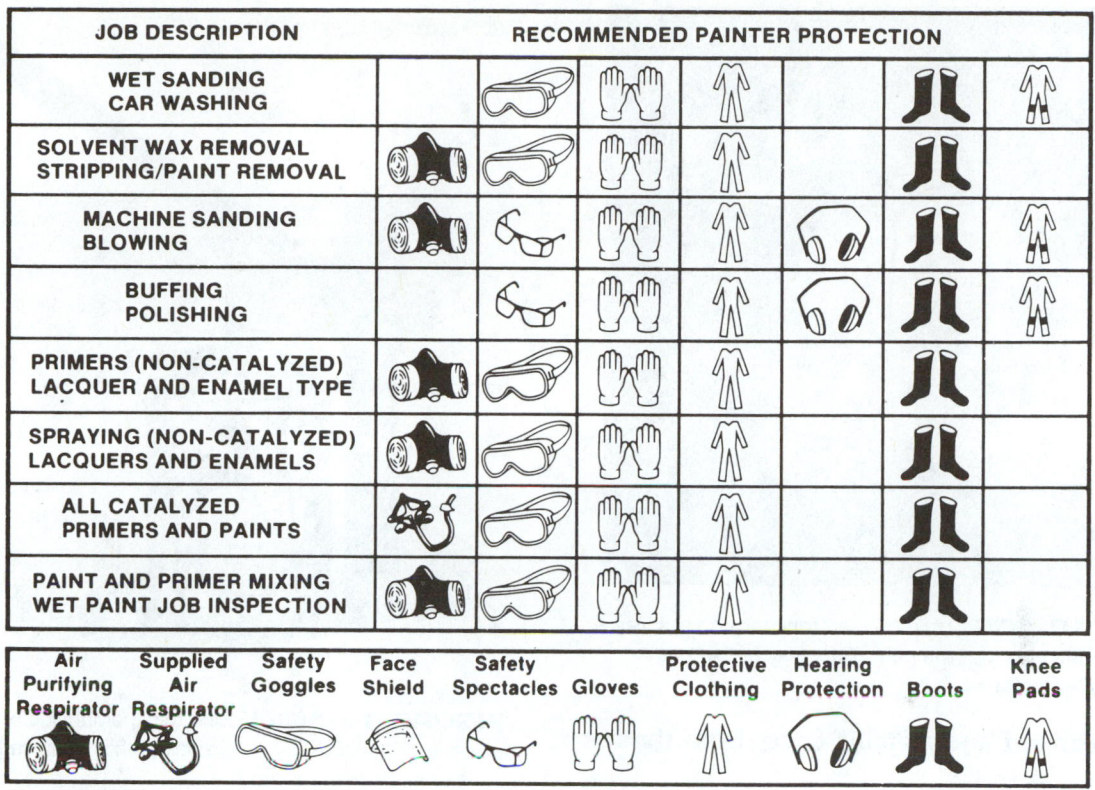

JOB DESCRIPTION	RECOMMENDED PAINTER PROTECTION						
WET SANDING CAR WASHING		Safety Goggles	Gloves	Protective Clothing		Boots	Knee Pads
SOLVENT WAX REMOVAL STRIPPING/PAINT REMOVAL	Air Purifying Respirator	Safety Goggles	Gloves	Protective Clothing		Boots	
MACHINE SANDING BLOWING	Air Purifying Respirator	Safety Spectacles	Gloves	Protective Clothing	Hearing Protection	Boots	Knee Pads
BUFFING POLISHING		Safety Spectacles	Gloves	Protective Clothing	Hearing Protection	Boots	Knee Pads
PRIMERS (NON-CATALYZED) LACQUER AND ENAMEL TYPE	Air Purifying Respirator	Safety Goggles	Gloves	Protective Clothing		Boots	
SPRAYING (NON-CATALYZED) LACQUERS AND ENAMELS	Air Purifying Respirator	Safety Goggles	Gloves	Protective Clothing		Boots	
ALL CATALYZED PRIMERS AND PAINTS	Supplied Air Respirator	Safety Goggles	Gloves	Protective Clothing		Boots	
PAINT AND PRIMER MIXING WET PAINT JOB INSPECTION	Air Purifying Respirator	Safety Spectacles	Gloves	Protective Clothing		Boots	

Air Purifying Respirator	Supplied Air Respirator	Safety Goggles	Face Shield	Safety Spectacles	Gloves	Protective Clothing	Hearing Protection	Boots	Knee Pads

FIGURE 18-105 Chart shows recommended personal protection equipment.

operation, whereas equipment is used over and over. Paint shop materials include:

- *Abrasive paper or sandpaper.* The rough side of paper is called the "grit side." Grit sizes vary from coarse to micro fine and are ordered by number (see Chapter 17). The lower the number, the coarser the grit.
- *Clean cloths or paper towels.* It is important that the areas to be painted are clean. Most paint shops provide clean cloths or special disposable paper wipes. Whichever is used, these simple tips will help:

Use clean, dry cloths folded into a pad.

When using a cleaning solvent, be sure to pour enough onto the pad being used to thoroughly wet the surface to be cleaned.

Do not wait for the solvent to dry. Wipe it dry with a second clean cloth.

Refold the cloth often to provide a clean section.

Change cloths often.

Once an area is clean, do not touch it with the hands, as it might affect adhesion.

- *Tack rags or cloths.* These are specially treated sticky cloths (varnish-coated cheese cloth) that are used in the make-ready operation to wipe the surface clean just before the paint is applied. They should be used on the area to be painted to remove all sanded particles, dirt, old paint chips, and so on. Often the painter will simply blow off the area with an air nozzle or will use an old rag to wipe off the area. These procedures will leave minute impurities on the surface that will detrimentally affect adhesion and performance of the product. A tack cloth will pick up fine particles that are invisible to the naked eye. Tack rags should be stored in an airtight container to conserve their tackiness.

- *Paint paddles.* Made of either wood, metal, or plastic, paint paddles are used to stir the paint material. If the paddle is wood, it is recommended that the end be tapered to a sharp edge like a chisel. This will make it easier to dislodge

SHOP TALK Cheesecloth strainers can add as much as a tablespoon of lint to the paint. The wrong mesh size can allow dirt to pass through the strainer that will later show up in the finish. Metallics require the coarsest mesh; clearcoats the finest. Clearcoats can develop small globs in the can that show up like dirt in the finish. Therefore, use only approved paper, or better yet, properly maintained metal strainers. Always strain the paint.

FIGURE 18–106 Strain thinned topcoat into a spray cup to keep debris out of the gun. *(Courtesy of DuPont Automotive Products)*

FIGURE 18–107 Device for applying lids will help prevent air leaks that will ruin paint while in storage. *(Courtesy of Dedoes Industries, Inc.)*

the pigment and metallic flakes from the bottom of the can.

- *Strainers.* Consisting of a cardboard funnel with cotton mesh, a strainer is used when pouring thinned topcoat and other materials into a spray cup (Figure 18–106). This is done to make sure it is free of any dirt or foreign material.

- *Containers.* Paint shop containers come in six common sizes and/or shapes and contain various materials:

Tubes that contain putty

Round cans such as gallon, quart, and pint containers for topcoats and some undercoats, including putty and body filler

Square cans for thinners, reducers, primer-surfacers, sealers, and clear topcoats

Pails that contain thinners, reducers, undercoats, and topcoats

Drums that contain thinners, reducers, and undercoats

Plastic containers that contain metal conditioner, body filler, polish, and buffing compounds

Lids on round containers of a gallon or less—called "friction lids"—should be carefully opened with a proper opener. After pouring whatever amount of material is needed from a round can, the lip should be wiped and the lid replaced tightly to form a good seal (Figure 18–107). A pouring spout made of masking tape will keep liquid from collecting in the rim, while a rubber

mallet is recommended for tapping the lid around the edge. Proper resealing of the can will keep air out and minimize the formation of the film on the top—called "skinning." It will also prevent the loss of solvents. Screw-top cans should be carefully wiped and closed tightly for the same reasons.

Where there is heavy usage of undercoats, thinners, and reducers, the material can be purchased in drums at a considerable savings. When lacquer undercoats are stocked in large containers, the solvents keep the solids properly mixed, so there is less waste of materials. Drums of undercoats should be fitted with a gate valve for pouring, while drums of thinners and reducers should be fitted with either a faucet or a pump.

SHOP TALK

Keep the pouring spouts of the stirring heads clean to avoid a buildup of paint residues around the spout that can ultimately affect the accuracy of pouring. Wiping the spout after every pour is the simplest method. Alternatively, application of masking tape around the spout that can be replaced at regular intervals, removing solidified residues, is fairly effective.

SHOP TALK Store clean empty containers upside down to prevent entry of dirt or other forms of contamination. Many shops now use can crushers to make container disposal easier (Figure 18–108).

FIGURE 18-108 A container crusher will save space in the trash container. *(Courtesy of Herkules Equipment Corp.)*

Plastic measuring cups are used for matching or thinning paint. Generally, their sizes range from 1 quart to 5 quarts and they are made of easy-to-use and easy-to-clean plastic.

• *Masking paper and tape.* As discussed in the previous chapter, masking paper is used to cover surrounding areas not to be painted so that paint mist does not settle there. It is necessary that the masking paper be capable of preventing solvent in the paint mist from seeping through to the surface of the object. Some paint shops use newspapers for this purpose, but this cannot be recommended because thin fibers from the paper come off and adhere to the painted surface, resulting in dirt. Also, solvent will seep through newspaper or even transfer newsprint onto the covered area.

Masking tape is used to stick the masking paper to the areas to be covered or it can be used by itself. Masking tape is made of different types of materials, such as paper, cloth, and vinyl, so that adhesion performance is assured regardless of the season or weather. The adhesive performance of masking tape does not change when heat is applied and will not leave traces of adhesive when removed. Also, it is easy to cut or tear off.

There are several types of paper masking tape depending on what it is used for, but they can be roughly classified into general masking tape used for air drying and heat-resistant tape used for baking enamel. The proper tape for the job must always be used. Full information on masking paper and tape and how it is used is given in Chapter 17.

SUMMARY

- To do a good paint job, your shop and equipment must be in perfect condition!

- The spray gun breaks the liquid primer or paint into a fine mist and forces it onto the surfaces of the vehicle.

- Pressurized pot or pressure cup spray guns use air pressure inside the paint cup or tank to force the material out of the gun.

- Suction or siphon spray guns use air flow through the gun head to form a suction that pulls paint into the air stream.

- Many low volatile organic compound (VOC) regulations require 10 psi (69 kPa) or less at the air cap.

- In the gravity feed system, the paint is supplied by gravity and the material is pressure or suction forced at the nozzle tip.

- Some high-efficiency spray guns can also use combination a gravity-pressure feed cup design. It offers the benefits of both types of guns.

- Graduated paint mixing sticks have conversion scales that allow you to easily convert ingredient percentages into part proportions.

- A pressure that is too high results in excessive paint loss through overspray and poor flow, due to high solvent evaporation before the paint reaches the surface being sprayed.

- A pressure that is too low gives a paint film with poor drying characteristics, due to high solvent retention. Low gun pressure also makes the paint prone to bubbling and sagging.

- A spray pattern test checks the operation of the spray gun on a piece of paper.

- Gun stroke refers to the hand movement used to move the gun while spraying.

- Thorough cleaning of gun and accessory equipment immediately after use is critical.

- Avoid spraying solvent into the air when cleaning a spray gun. This pollutes the atmosphere.

- A spray gun cleaning tank, also called a gun washer/recycler, is a pressurized container for flushing the gun and other tools with cleaning solution.

- Most spray gun manufacturers recommend lubricating the moving parts with spray gun oil.

- For best results in refinishing, use separate guns for topcoats and undercoats. Ideally, there should be at least three guns: one gun for spraying undercoats like primer-surfacers, another for colorcoats, and a third gun for spraying clearcoats.

If the air spray gun is not adjusted, manipulated, and cleaned properly, it will apply a defective coating to the surface.

- The HVLP (high volume, low pressure) spray system (also known as the "high solids" system) uses a high volume of air, delivered at low nozzle pressure, to atomize paint into a pattern of low-speed particles.

- High transfer efficiency means that more of the paint leaving the gun stays on the surface being painted.

- Touch-up spray guns are very small and are ideal for painting small repair areas. Often called an "air brush" or "door jamb gun," they have a tiny cup for holding a small amount of material.

- A spray booth is designed to provide a clean, safe, well-lit enclosure for painting.

ASE-STYLE REVIEW QUESTIONS

1. Which of the following promotes atomization of the paint?
   A. Center orifices
   B. Side orifices
   C. Ports
   D. None of the above

2. Which type of spray gun has the advantage of having a large cup capacity?
   A. Suction feed type
   B. Pressure type
   C. Air brush type
   D. Both A and B

3. A spraying pressure that is too high results in which of the following?
   A. Poor flow
   B. Poor drying characteristics
   C. Bubbling
   D. Sagging

4. What happens when testing the spray pattern for uniformity of paint distribution?
   A. A vertical spray pattern is used
   B. The trigger is pulled all the way back and released immediately.
   C. The pattern is flooded
   D. Both A and B

5. This spray system (also known as the *high solids* system) uses a high volume of air, delivered at low pressure, to atomize paint into a pattern of low-speed particles.
   A. LVLP
   B. HVLP
   C. HVHP
   D. LVHP

6. Which spray booth provides the safest environment for the technician?
   A. Regular flow booth
   B. Reverse flow booth
   C. Downdraft booth
   D. Crossdraft booth

7. Technician A monitors the manometer readings hourly. Technician B monitors the manometer readings daily. Who is correct?
   A. Technician A
   B. Technician B
   C. Both A and B
   D. Neither A nor B

8. Technician A uses the same spray gun for the topcoat that was used for the undercoat. Technician B uses a different spray gun. Who is correct?
   A. Technician A
   B. Technician B
   C. Both A and B
   D. Neither A nor B

ESSAY QUESTIONS

1. What happens if the air vent hole becomes plugged in a siphon cup?

2. Summarize some advantages of a pressure pot gun.

3. How do you determine viscosity with a #2 Zahn cup?

4. How do you do a spray pattern test?

5. Describe five problems often found during a spray pattern test.

6. What are some common gun handling problems of an inexperienced painter?

CRITICAL THINKING PROBLEMS

1. What will happen if you hold the spray gun too close when spraying?

2. When spraying, you have the paint viscosity too thick. What will be the result?

MATH PROBLEMS

1. An HVLP spray gun requires air delivered at 10 psi (69 kPa). Your gauge shows only 6.3 psi; what percentage is the pressure too low?

2. If you are holding a spray gun 4.2 inches (106 mm) away when spraying enamel and specs call for a distance of 8 inches (203 mm), how much should you move the gun?

CHAPTER 19

Refinishing Procedures

INTRODUCTION

From the customer's standpoint, the topcoat or colorcoat is the most important, if not the only factor determining the quality of the paint job. Many customers do not understand all of the work that goes under, and in, a paint job. Paint chemistry and spray equipment are much more complex than in the past. Proper material selection, skill, and knowledge are needed to produce a durable, long-lasting finish.

The expert refinisher takes special pride in producing a beautiful finish that matches both the color and texture of the original paint. It is your job to satisfy the customer with professional paint application.

19.1 PURPOSE OF REFINISHING

Since the customer sees only the topcoat and judges the quality of the refinisher's work on its appearance and its appearance alone, there is little appreciation for all the work done underneath the topcoat. As already pointed out in previous chapters, the cleaning, filling, and sanding of the substrate must be done painstakingly. A perfectly smooth surface must be readied before the topcoat is applied. Otherwise, any imperfection—even the smallest—will show in the topcoat.

Automobile finishes perform four very important functions:

- **Protection.** The automobile is constructed primarily of steel sheet metal. If this steel were left

OBJECTIVES

After studying this chapter, you should be able to:

✔ Explain the functions of the four types of undercoats.

✔ Name the types of topcoats.

✔ Discuss the advances made in refinishing by basecoat/clearcoat finishes.

✔ Explain the advantages of basecoat/clearcoat finishes.

✔ Describe the role of solvents and the variables that affect their spraying.

✔ Select and mix paint solvents.

✔ Determine the type of paint on a car and whether or not the car has been repainted.

✔ Identify the steps in applying various types of colorcoats.

✔ Apply basecoat/clearcoat systems.

✔ Describe the paint finishing systems applicable to plastic parts.

✔ Summarize methods for applying topcoats.

✔ Answer ASE test questions concerning refinishing procedures.

KEY TERMS

banding	paint flex agent
color book	paint mixing sticks
dust coat	paint stirring sticks
fast-drying solvent	paint strainer
force drying enamels	panel repair
full wet coat	pearl paints
induction time	percentage reduction
medium-drying solvent	retarder
medium wet coat	slow-drying solvent
metal flakes	spot repair
mist coat	stabilizer
mixing chart	tack cloth
OEM finishes	tack coat
orifice	topcoat
overall refinishing	undercoat
paint blending	

ASE TASK LIST

Job Skills covered in this chapter include:

PAINTING AND REFINISHING TEST (B2) TASK LIST

A. Surface Preparation

3. Inspect and identify substrate, type of finish, and surface condition; develop a plan for refinishing.
4. Remove paint finish.
9. Mix primer, primer-surfacer, or primer-sealer; spray onto surface of repaired area.
12. Remove dust from area to be refinished, including cracks or moldings of adjacent areas.
13. Clean area to be refinished using a proper cleaning solution.
14. Remove, with a tack rag, any dust or lint particles from the area to be refinished.
17. Prepare the repaired and adjacent areas for blending.

B. Spray Gun Operation and Related Equipment

1. Inspect, clean, and determine condition and adequacy of spray guns and related equipment (air hoses, regulator, air lines, air source, and spray environment).

C. Paint Mixing, Matching, and Applying

2. Shake, stir, reduce, catalyze, and strain paint according to manufacturer's recommendations.
3. Use appropriate spray technique (gun arc, gun angle, gun distance, gun speed, and spray pattern overlap) for finish being applied.

5. Apply single-stage topcoat for spot and panel blending, and overall refinishing.
6. Apply basecoat/clearcoat for spot and panel blending, and overall refinishing.
7. Check for color matching of all applied finishes.
12. Apply multistage (mica, pearl, etc.) coats for spot repair, panel blending, and overall refinishing.

D. Solving Paint Application Problems

1. Identify blistering (raising of the paint surface); determine the cause(s), and correct the condition.
7. Identify mottling or streaking in metallic and mica paint finishes; determine the cause(s), and correct the condition.

E. Finish Defects, Causes, and Cures

4. Identify blistering in the paint surface; determine the cause(s), and correct the condition.
13. Measure paint film thickness.

F. Safety Precautions and Miscellaneous

3. Inspect spray environment.

---

uncovered, the reaction of oxygen and moisture in the air would cause it to rust. Painting serves to prevent the occurrence of rust, thereby protecting the body.

- **Appearance improvement.** The shape of the body is made up of several types of surfaces and lines, such as elevated surfaces, flat planes, curved surfaces, straight and curved lines, and so forth. Therefore, another objective of painting is to improve the body appearance by giving it a three-dimensional color effect.
- **Increased value.** When comparing two vehicles of identical shape and performance capabilities, the one with the most beautiful paint finish will have a higher market value. Hence another objective of painting is to increase the resale value.
- **Color designation.** Still another reason to paint automobiles is to make them easily distinguishable by the application of certain colors

or markings. Examples are police and fire department vehicles.

To achieve this, the typical automotive finishing system consists of several coats of two or more different materials:

- Undercoat or primer coat(s)
- Topcoat (colorcoat or basecoat/clearcoat)

The **undercoat** provides a sound foundation for the topcoat and makes it adhere better (Figure 19–1). If topcoats are applied to bare substrates (metal, fiberglass, or plastic), they might peel or look rough. This is why the undercoat is sandwiched between the substrate and the topcoat. The undercoat also protects against rusting and fills scratches and other flaws in the metal or plastic.

The **topcoat** is the finish that is seen on the car. From an appearance standpoint, it is smooth, glossy, and eye-catching (Figure 19–2). Functionally,

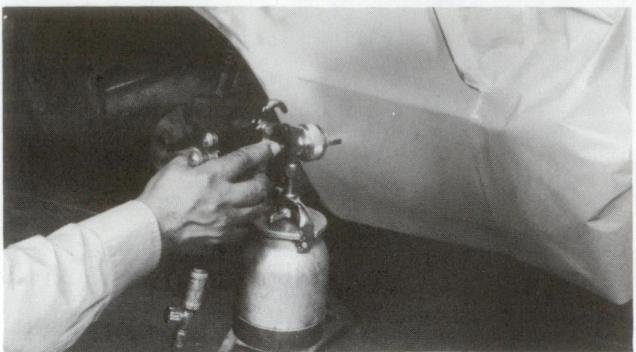

FIGURE 19-1 Undercoats of primer, sealer, or primer-surfacer are critical to the service life of the topcoat of paint. *(Courtesy of 3M)*

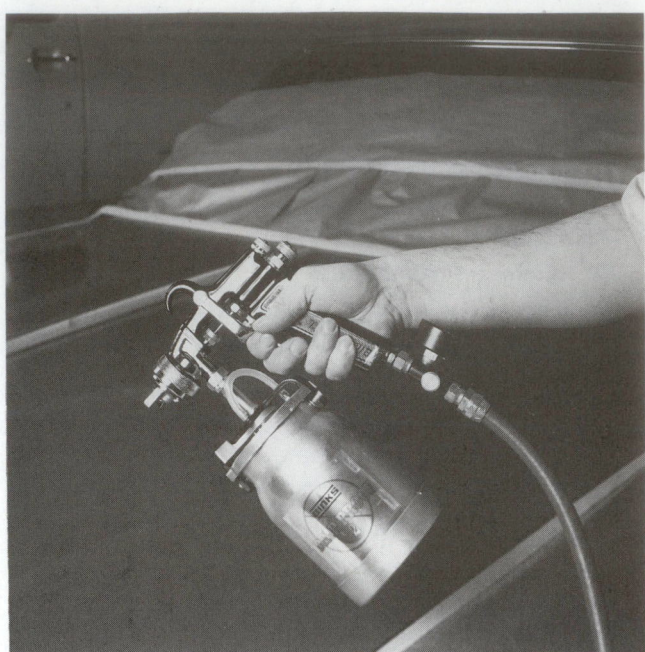

FIGURE 19-2 The topcoat produces the shiny, colorful finish that is seen by the customer. *(Courtesy of Binks Mfg. Company)*

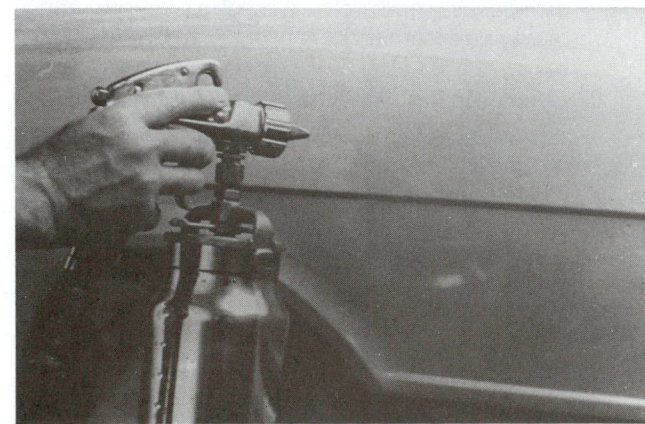

FIGURE 19-3 Spot refinishing repair generally involves minor body repair, featheredging, application of primer-surfacer, and wet sanding before painting. *(Courtesy of America Sikkens, Inc.)*

it is tough and durable. The topcoat thickness on a new car when it comes from the factory is only about 2.5 mils.

Refinishing materials and methods are constantly changing. The refinish technician or painter must respond to these changes to provide the perfect matches demanded by the customer on a refinishing job. Keeping up-to-date on the changes is crucial.

19.2 TYPES OF REFINISHING REPAIR

The conditions determining the type of topcoat paint to be used are: the extent of the area to be covered, the extent of paint film deterioration, and whether or not the vehicle has been repainted previously. The type of paint used and how it is applied are very important factors governing work efficiency and speed.

There are three general types of refinishing repairs:

- Spot refinishing repairs (Figure 19–3)
- Panel refinishing repairs (Figure 19–4)
- Overall repainting of the entire vehicle (Figure 19–5)

SPOT REPAIR

Spot repair involves painting an area smaller than a panel. The paint must be blended out to match the existing finish.

Spot repair generally involves the following:

1. Minor body repair
2. Metal conditioning
3. Application of undercoat system
4. Application of topcoat to blend into the old finish surrounding the repair

Spot repairs are recommended where a complete panel repair is either uneconomical or impractical. This might be due to the tiny size of the damage, its hidden location, or another factor. It is also commonly used when two large panels have no break line. For example, when painting a quarter panel, the paint is blended into the sail panel where the roof joins the quarter panel.

Solid Color Spot Repair

You must use experience and common sense as to the range of blending needed for spot repairs with

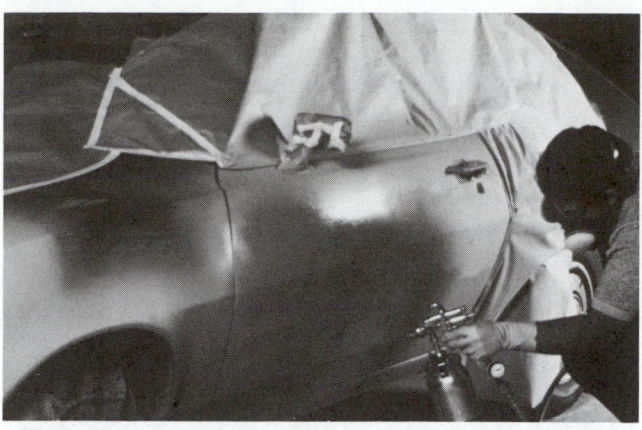

FIGURE 19-4 Panel or block refinishing repair, basically the same as spot repair, covers an entire panel or panels of the car (door, hood, and so on).

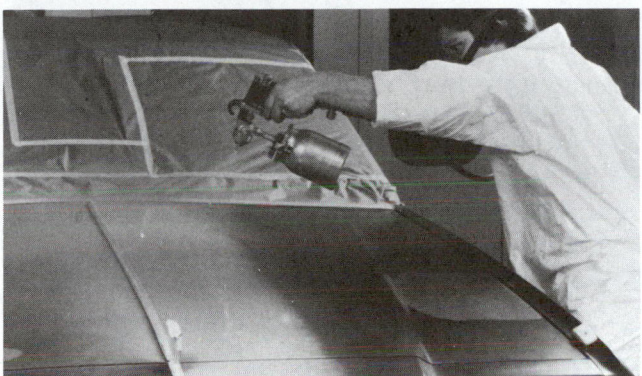

FIGURE 19-5 Overall repainting can be dictated by the extent of the repair, the condition of the finish, or by the owner's preference. The whole vehicle must be prepared, masked, and refinished. *(Courtesy of Maaco Enterprises, Inc.)*

solid colors. Blending with a solid color of paint is used for situations such as light damage to the fender edge.

In this case, there are two methods of blending: at the hood-to-fender gap and at the body line. When blending at a body line, you do not have to paint the upper portion of the fender where paint fading shows more. This can help avoid problems with color and texture differences.

When a spot repair is made with a solid color, the blend area should be treated with rubbing compound or sanded with #1000 grit sandpaper before refinishing.

Metallic Color Spot Repair

Matching a metallic color is complicated because you have to match the color and the density of the metallic flakes in the paint. If a spot repair is done with a metallic color, skill is required in matching the color and the metallic flakes through proper distribution of the paint. The blend area will be less noticeable if it is angled away from the body line.

PANEL REPAIR

Panel repair involves painting a complete body part separated by a definite boundary, such as a door or fender. You do not always have to blend the paint with panel repairs. Blending is needed when it is difficult to match the paint, as with a metallic or pearl color. However, blending is routinely done for such areas as between the quarter panel and roof panel.

Solid Color Panel Repair

Panel repair with a solid color covers an entire panel (door, hood, etc.). The paint match is made at the panel joints. Panel repair is a very common repair. It usually results in a better-looking repair and can be just as fast as spot repair on most panels.

For a complete panel repair, mask off the area not to be painted. If a panel has damage at two different locations, the whole panel should be repaired. Blending can be done to the molding or extended below the molding. Blending is also needed with hard-to-match colors.

When blending new paint into old paint, you must use common sense to determine where to blend the paint. Try to find a body section that gives a natural break line or small area for blending.

Metallic Color Panel Repair

New and old paint differences tend to be very noticeable with bright metallic colors. It is almost impossible to match the new metallic finish exactly with the previous one. You must extend the blending over a wide area to help hide the differences. This will make the repair less visible. If a panel has damage at both ends or if the whole panel is to be refinished, the blending might have to extend onto adjacent panels.

OVERALL REPAINTING

Overall refinishing or repainting is just what it says, the whole vehicle is painted. Some reasons for refinishing the entire car include the following:

- Size and/or number of spots to be repaired
- Dull, cracked, or worn finish
- Color change desired by car owner

There are specialty paint shops that do mostly overall painting as well as custom shops that develop glamour finishes for custom, antique, and classic cars.

19.3 APPLYING UNDERCOATS

Discussed in previous chapters, a *primer* (or *"prep coat"* as it is sometimes called) is generally the first coat in any finishing system. It is designed to prepare the bare substrate and to accept and hold the color topcoat. Primers should provide maximum adhesion to the surface and produce a corrosion-resistant foundation.

Straight primers generally do not fill surface imperfections and therefore often do not require sanding. Straight primers are predominantly used by original equipment manufacturers rather than paint and body repair shops. Primers are usually enamel-type products because they provide better adhesion and corrosion resistance than lacquers. (Where the original surfaces are plastic or fiberglass, some lacquer primers are still used.)

There are several special primers available to the shop refinisher. For example, the two-part, self-etching epoxy primers are probably the most versatile and valuable primer products on the market today.

An *epoxy primer* system consists of the primer itself and an activator (or *catalyst*). When mixed together (known as *catalyzing*), an induction time is often required. **Induction time** is the period of time that the components must be allowed to sit in a container together to thoroughly react with each other before spraying. It must be remembered that these two-part mixtures also have a "pot life," or a limited time before they become unusable.

Epoxy primers are a good choice where a great deal of fill is required. On blasted, coarse surfaces, for example, a high-build epoxy primer adheres to the bare substrate and provides the fill necessary for a smooth finish.

Chip-resistant primers are specially formulated for use on lower body sections that are prone to gravel and stone impact damage. They give improved resistance against corrosion and help reduce drumming noise.

Adhesion to plastic panels that have come from the factory unprimed can be improved by the use of a special plastic primer. Flexible plastics in particular require *plastic primer*. Many rigid plastics also benefit from priming with a plastic primer. Plastic primers promote bonding and eliminate peeling and other adhesion problems.

Another ready-to-spray primer product designed specifically for use with basecoat/clearcoat OEM finishes is the *adhesion promoter primer*, also called *midcoat primer*. This product is a water-clear primer with good durability and excellent adhesion to very hard clearcoats. It is recommended that it be applied beyond the repaired area on a spot repair before any other primer is used. Its purpose is to provide a surface to which a blend edge can adhere.

APPLYING PRIMER-SURFACERS

Primer-surfacers are the most popular of all the undercoats in refinish applications. They are used to build up featheredged areas for rough surfaces and to provide a smooth base for topcoats.

There are five types of primer-surfacers:

- Acrylic lacquer
- Self-etching
- Acrylic urethane
- Alkyd or synthetic enamel
- Nitrocellulose lacquer (older, phased-out product)

All primer-surfacers should be thinned as per label directions and applied in medium coats. Allow label recommended flash time between coats (typically about 15 to 20 minutes). Applying heavy coats or not allowing enough flash time will result in a thick coat of surfacer that can gum up the sandpaper, featheredge poorly, and cause loss of gloss or peeling of the topcoat. Excessive thickness of a primer-surfacer coat under a topcoat could lead to premature crazing and/or cracking conditions.

While a primer-surfacer will cover small imperfections, it will not handle deep scratches, gouges, or other similar defects. When these conditions exist, the painter has two other options:

- Where the scratches are a bit deeper, it is possible to apply putty over the primer-surfacer.
- When the scratches are too severe for putty, the painter has to use body filler on the metal and then prime it (just as it is used in the metalworking area to repair larger dents).

APPLYING PRIMER-SEALERS

A primer-sealer, like a primer, is used to prime bare metal to resist rust and corrosion and to provide topcoat adhesion. In addition, primer-sealers can be used to seal aged painted surfaces that have been sanded. In contrast to primer-surfacers, primer-sealers do not fill and do not have to be sanded.

Primer-sealers should be thinned or reduced as per label directions and applied in medium coats. A primer-sealer is generally an enamel-based product that must be reduced with an enamel reducer. The main reason for using this type of primer is to ensure that the ready-to-paint surface is consistent in color (to prevent primer spots from showing through the finish) and porosity (to prevent topcoat solvents from soaking into the primer spots, causing dull spots or a featheredge ring). See Figure 19–6.

APPLYING SEALERS

Sealers differ from primer-sealers in that they cannot be used as a primer. Sealers are sprayed over a primer or primer-surfacer or a sanded old finish (Figure 19–7). Sealers can be clear or colored. They are used in automotive refinishing for five specific purposes:

1. To provide better adhesion between the paint material to be applied and the repair surface.
2. To act as a barrier-type material that prevents or retards the mass penetration of refinish solvents into the color and/or undercoat being repaired.
3. To provide uniform hold out. If the old finish is good and hard and if a primer-surfacer with good hold out is used for spot repairing, a sealer is not mandatory. Obviously, if only one or nei-

ther of these conditions is present, a sealer is recommended.

4. To prevent show-through or sand-through. If sand scratches are present in the undercoat, particularly if noticeable to the eye, they will show through the topcoat. The safest procedure is to apply a coat of sealer, especially on large areas of sand scratches. Small areas, such as sand scratches around a featheredge, can be removed by compounding.
5. To provide a solid color over which to apply the colorcoat. If there have been previous repairs with different primers, the sealer will make the whole repair area the same color. This will reduce the amount of topcoat needed, especially with transparent colorcoats.

Sealers also provide the following desirable characteristics:

- Improve adhesion of the repair color to very hard undercoats and enamel surfaces
- Improve gloss
- Prevent bleeding (when designed for this purpose)
- Can be used on small, clean, bare metal surfaces

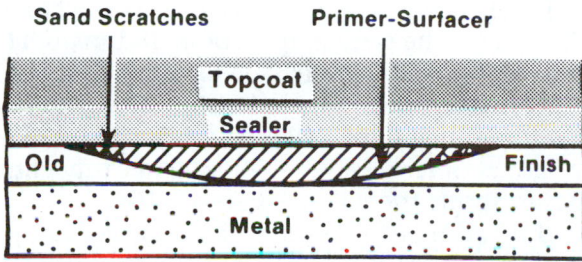

FIGURE 19-6 (A) If a primer-surfacer has been used, there might be a difference in the hold out between the two types of finishes. (B) If so, a sealer will solve the problem, and might prevent show through of sand scratches. *(Courtesy of DuPont Automotive Products)*

Always read the label directions on material before using it (Figure 19-8). The directions will be specific to the product. They will include detailed application, mixing, and safety information.

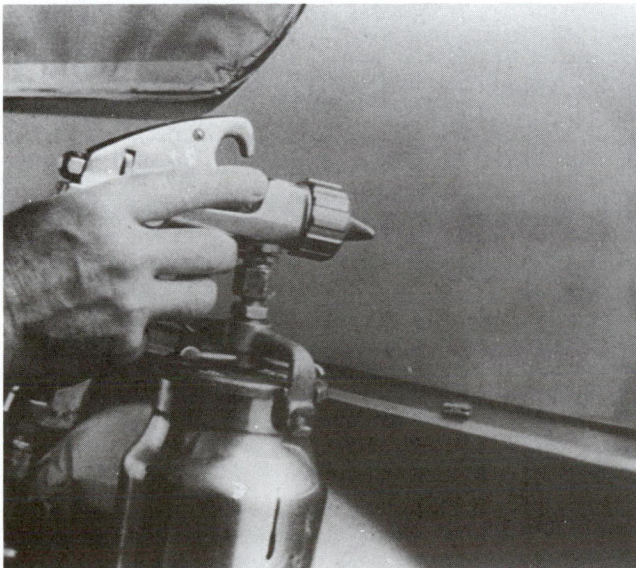

FIGURE 19-7 Sealer should be used when making panel refinishing repairs or when doing a complete refinishing job.

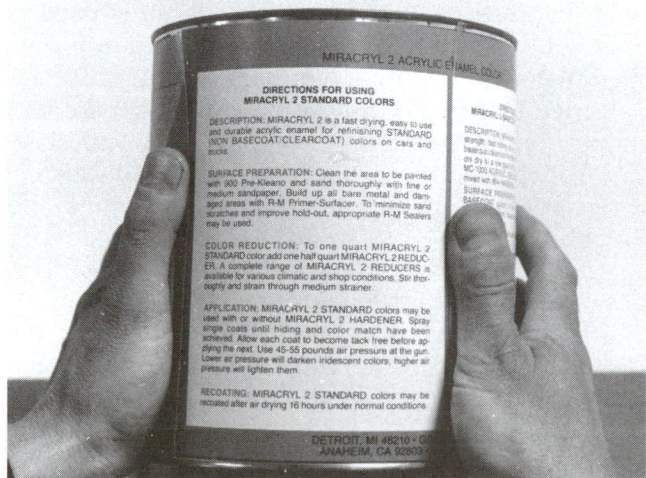

FIGURE 19-8 Carefully read the manufacturer's directions that appear on the paint container. They give instructions for safely mixing and spraying the specific type of primer or paint.

Universal sealers have been developed that can be used under the majority of topcoat systems. For example, *barrier coat sealers* are designed for extremely sensitive substrates. Old finishes that are on the verge of cracking or that have been exposed to too much sun can require the use of a barrier coat sealer. These sealers eliminate lift when an enamel topcoat is going to be sprayed over a previous enamel paint job that has been buffed through to an OEM lacquer.

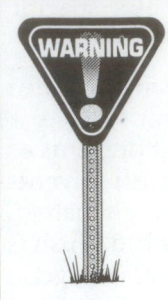

Avoid spraying lacquer paint over the top of enamel. The enamel can lift and cause problems. You can spray enamel over an old lacquer finish, however.

19.4 TOPCOATS

The expert refinisher takes special pride in producing a beautiful finish that matches both the color (or color effect) and the texture of the original finish. Therefore, it is of great importance to fully understand the topcoat materials and how they are applied.

There are several types of paint:

1. Lacquer paints
2. Enamel paints
3. Waterbase paints
4. Basecoat/clearcoat paints
5. Polyurethane single-stage paints
6. Multicoat paints

There are variations within these categories, and it is important that you know what type of finishes manufacturers use because slightly different methods are required for refinishing them.

Table 19–1 shows the relative durability of the various types of topcoat systems previously discussed. It also gives a comparison to original equipment (OE) high-baked thermosetting acrylic enamels.

Costs can be deceiving, particularly with the polyurethane and acrylic urethane colors, both of which are substantially more expensive in their unreduced prices per gallon. When costs are viewed in terms of life cycle—for example, the length of time the finish will continue to present an acceptable appearance before requiring refinishing—acrylic urethane enamels prove to be the most economical.

The type of paint used for the topcoat ultimately determines the attractiveness of the color, gloss, and finish.

Table 19–2 is a general summary of the properties of paint used for repainting.

Lacquer is an older paint that dries quickly because of solvent evaporation. Lacquers have been phased out for more durable enamel and waterbased paints by both OEMs and body shops.

Lacquer topcoats usually must be compounded or rubbed with a compound or polish to bring out their gloss. The acrylic lacquer basecoat/clearcoat finish can allow a lacquer basecoat to be clearcoated with an enamel topcoat. This helps eliminate the need for compounding or polishing to bring out the gloss.

Acrylic enamel and *acrylic urethane enamel* are two specific types of enamel paint. Both are commonly used in the industry. Acrylic urethanes are slightly harder than plain acrylics. Each is available in a variety of colors.

Medium-sized reflective pigment particles, such as mica, are added to **pearl paints** to give the paint a luster or shine that tends to change color with viewing angle (Figure 19–9). As you will learn later, pearlescent paints are now common and are the most difficult to match when repainting.

Large reflective pigment flakes are added to metallic paints. The size, shape, color, and material in the flakes can vary. Often called **metal flakes**, the flakes can be made of tiny but visible bits of metal or polyester. When light strikes the flakes, they reflect the light at different angles, looking like tiny glittering stars inside the paint (Figure 19–10).

If this is new to you, start looking at paint jobs more closely. See if you can tell the difference between a solid color, a pearl, and a metallic.

NOTE: For more information on metallic and pearl paints, refer to the next chapter, which explains color matching in detail.

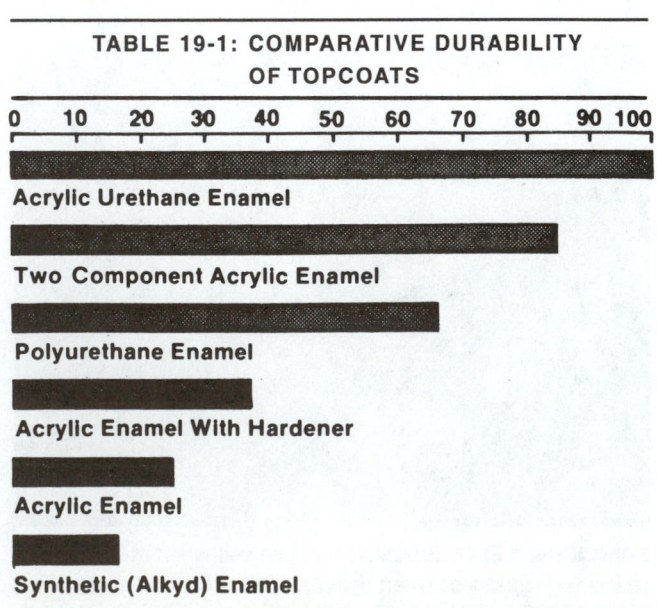

TABLE 19-1: COMPARATIVE DURABILITY OF TOPCOATS

| 0 | 10 | 20 | 30 | 40 | 50 | 60 | 70 | 80 | 90 | 100 |

Acrylic Urethane Enamel

Two Component Acrylic Enamel

Polyurethane Enamel

Acrylic Enamel With Hardener

Acrylic Enamel

Synthetic (Alkyd) Enamel

TABLE 19-2: SUMMARY OF TOPCOAT PAINT FEATURES

Nomenclature		One-Component Type			Two-Component Type	
		Alkyd Enamels	Acrylic Lacquer	Acrylic Enamel	Polyurethane	Acrylic Urethane Enamel
Spray characteristics		Excellent	Excellent	Good	Good	Good
Possible thickness per application		Fair	Fair	Good	Excellent	Excellent
Gloss	without polishing	Fair	Good	Good	Excellent	Excellent
	after polishing	Good	Good	Good	—	Good
Hardness		Good	Good	Good	Excellent	Excellent
Weather resistance (frosting, yellowing)		Fair	Fair	Good	Excellent	Excellent
Gasoline resistance		Fair	Fair	Fair	Excellent	Good
Adhesion		Good	Good	Fair	Excellent	Excellent
Pollutant resistance		Fair	Fair	Fair	Excellent	Excellent
Drying time	to touch	68°F 5–10 minutes	68°F 10 minutes	68°F 10 minutes	68°F 20–30 minutes	68°F 10–20 minutes
	for surface repair	68°F 6 hours 140°F 40 minutes	68°F 8 hours 158°F 30 minutes	68°F 8 hours 158°F 30 minutes	—	68°F 4 hours 158°F 15 minutes
	to let stand outside	68°F 24 hours 140°F 40 minutes	68°F 24 hours 158°F 40 minutes	68°F 24 hours 158°F 40 minutes	68°F 48 hours 158°F 1 hour	68°F 16 hours 158°F 30 minutes

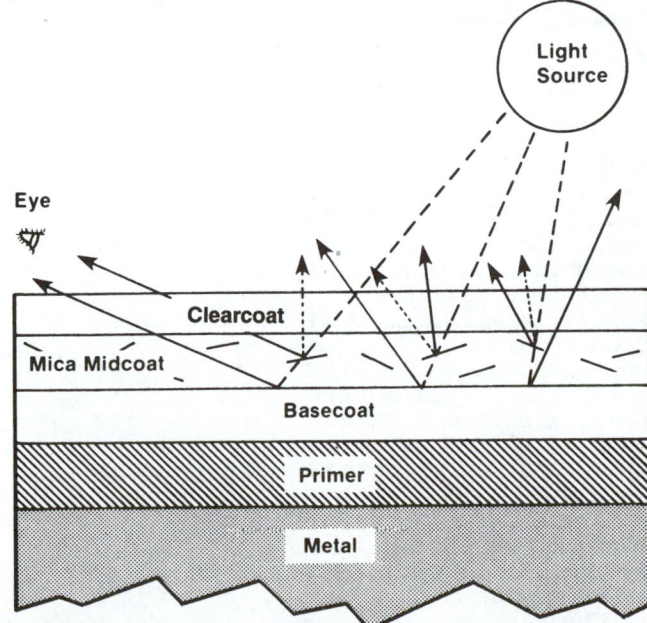

FIGURE 19-9 As this example shows, tri-coat finishes will alter the hue (color) of the finish as the angle of view changes. This makes them more difficult to match, as detailed in the next chapter.

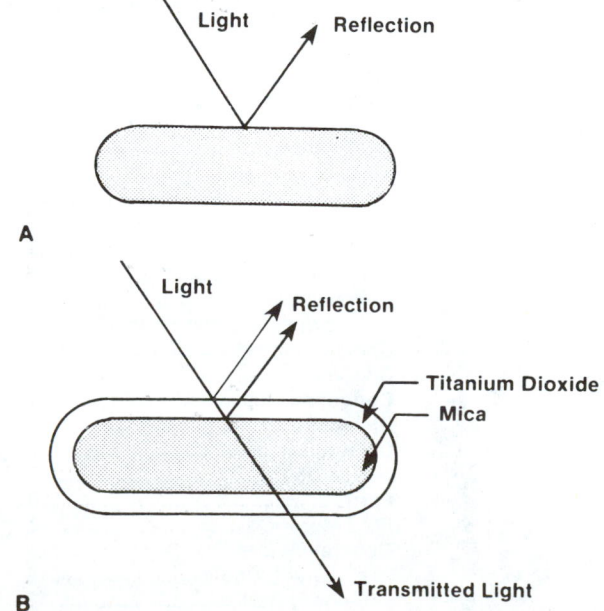

FIGURE 19-10 As opposed to (A) an aluminum flake, which only reflects light off its surface, (B) colored mica particles can be designed to reflect, absorb, and refract differing amounts of light striking them, thereby changing the color.

Waterbase paint, as implied, uses water to carry the pigment. It dries through evaporation of the water. Some manufacturers are starting to use waterbase paints on new vehicles. This is to help satisfy stricter emission regulations in some geographic areas. The basecoat of color is water-based. Then an enamel topcoat is applied over the waterbase paint to protect it from the environment.

Waterbase primers have been used for years as a fix for lifting problems and serve as excellent barrier coats when there are paint incompatibility problems.

OEM FINISHES

Today's passenger car and light truck **OEM finishes** (factory paint jobs) are either "thermo-setting" acrylic enamels (paint is furnace hardened at the factory),

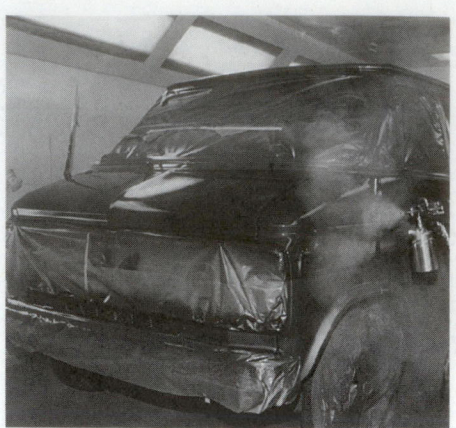

FIGURE 19–11 Before repainting, make sure that the paint on the vehicle is not too thick. If you apply more paint over a repainted vehicle, problems will usually result. *(Courtesy of PPG Industries, Inc.)*

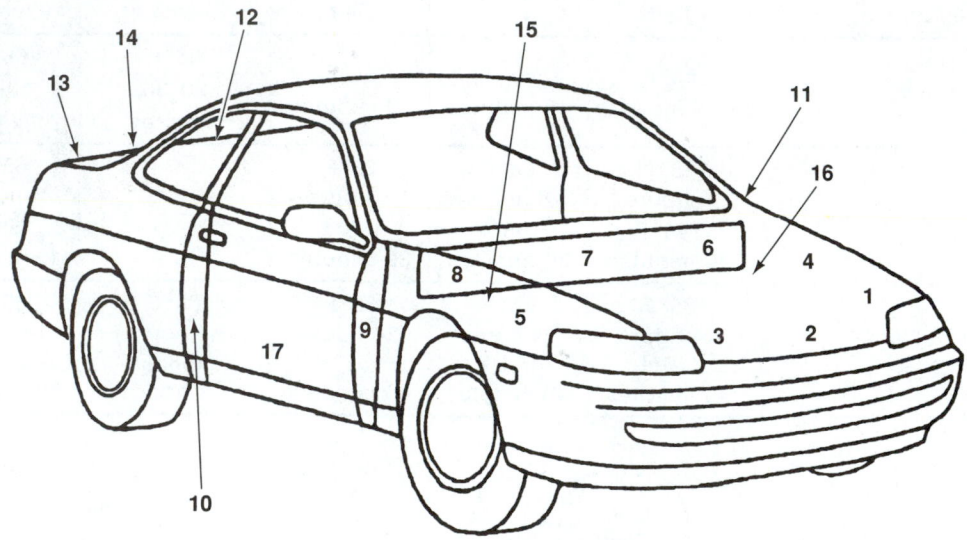

Model	Position	Model	Position
Acura	9	Honda	8,10
Alfa Romeo	4,13	Hyundai	6,7
AMC	9,10	Isuzu	2,10
Audi	12,13	Lexus	7,8
Austin Rover	17	Mazda	1,2,3,4,6,8
BMW	4,5	Mercedes	2,7,9
Chrysler	3,5,16	Mitsubishi	7
Chrysler Corp	3,5	Montero / Pickup	3
Caravan / Voyager / Ram Van	6	Cordia / Tredia	4
Chrysler Imports	1,2,4	Others	1,2,3
Colt Vista	16	Nissan	1,3,4,6,8,15,*
Conquest	7	Peugeot	2,3,4,5,8
Diahatsu	1,6,7	Porsche	9
Datsun	2	Renault	1,3,4,5,8
Dodge D50	3	Rover	1,3,4,5
Ford	10	Saab	5,6,8
Ford Motor Co.	10	Subaru	2
General Motors		Suzuki	7,11
A, J and L Bodies	14	Toyota Passenger	7,8,14
E and K Bodies	12	Truck	4
B,C,H and N Bodies	13	Volkswagen	2,11
GM Imports	2,12,13,14	Volvo	6,7,8
		Yugo	12

FIGURE 19–12 This chart can be used to locate the service label containing paint codes for most makes of vehicles.

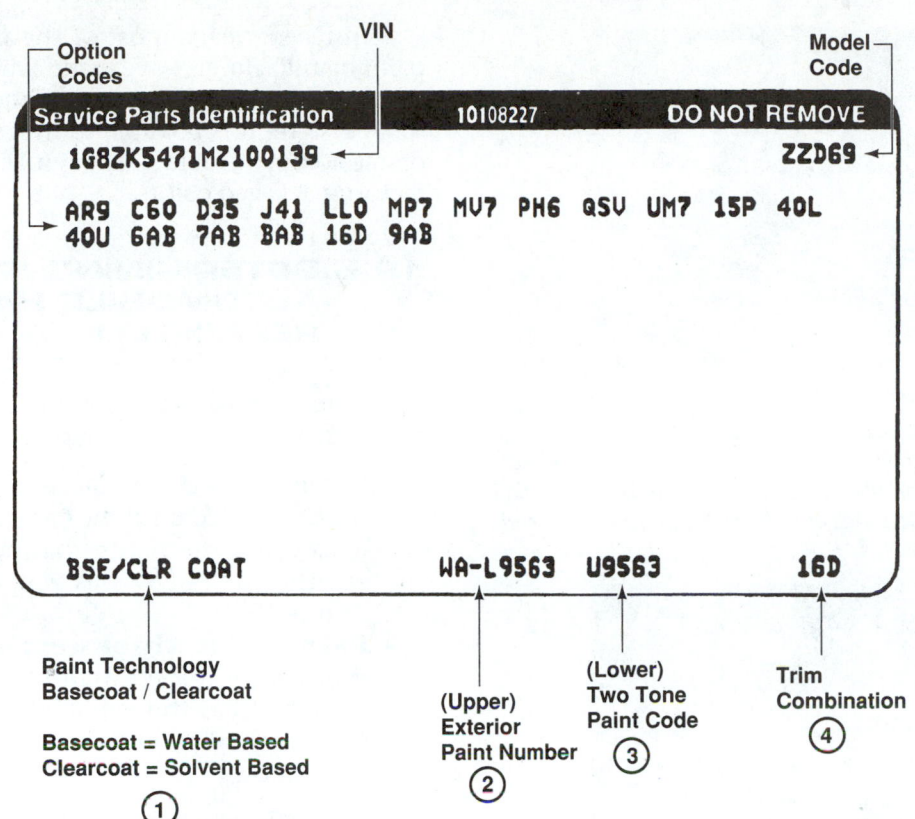

Option Codes — VIN — Model Code

| Service Parts Identification | 10108227 | DO NOT REMOVE |

1G8ZK5471MZ100139 ←

ZZD69 ←

AR9 C60 D35 J41 LL0 MP7 MU7 PH6 QSU UM7 15P 40L
40U 6AB 7AB 8AB 16D 9AB

BSE/CLR COAT   WA-L9563 U9563   16D

Paint Technology
Basecoat / Clearcoat

Basecoat = Water Based
Clearcoat = Solvent Based
①

(Upper) Exterior Paint Number ②

(Lower) Two Tone Paint Code ③

Trim Combination ④

FIGURE 19-13 Body ID plate example gives information on paint on this specific vehicle. *(Courtesy of General Motors Corp.)*

new high solids basecoat/clearcoat enamels, or sometimes waterbase, low emission paints. Common enamel finishes are baked in huge ovens to shorten the drying times and cure the paint. This is done before installing the interior and other nonmetal parts.

Vehicle manufacturers use several different types of finish materials and coating and application processes. Each type of finish requires different planning and repair steps.

The most common types of OEM coating processes include the following:

1. Single-stage
2. Two-stage (basecoat/clearcoat)
3. Three-stage paint (tri-coat)
4. Multistage

These will be explained fully in the next chapter of this book.

COLOR SELECTION AND IDENTIFICATION

There are two important steps that you must take before proceeding with any refinishing job:

1. Decide what type of repair is called for: spot, panel, or overall repainting (Figure 19–11).
2. Order all materials that are needed to complete the repair, particularly the topcoat color to be used.

If spot or panel repair is planned, it is important to purchase the topcoat color that will accurately match the old color. To match the existing color exactly, tinting and blending might be required. When planning an overall repainting, the customer might wish to match an old finish or might want a completely new one.

To order a matching topcoat color, first locate the vehicle identification plate (VIP). Write down the car manufacturer's paint code shown on the plate. Figure 19–12 gives a chart that will help you locate paint code numbers on most vehicles.

A service parts label for one make of vehicle is given in Figure 19–13. Note the breakdown of information on the paint.

Most auto refinishing shops have a color book (Figure 19–14). The **color book** contains color chips and color information for almost all makes and models worldwide. If the vehicle still has the original paint on it and the paint has not faded, use of paint mixed to match the paint code color will normally provide good results. However, keep in mind that tinting and blending might be needed for a perfect color match.

As a double-check, it is wise to compare the color chip with the actual vehicle color (Figure 19–15). There is always the chance that the car has been repainted with a different color. Place the chip on the paint and compare them. Remember that paint chips

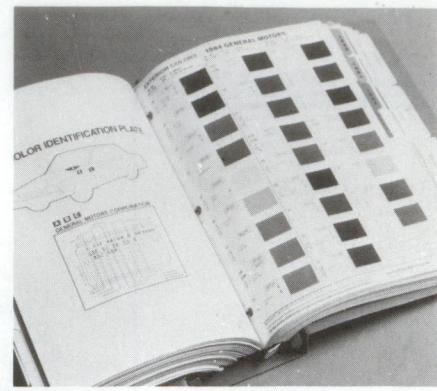

FIGURE 19–14 Refer to the shop's color book containing color chips provided for each make and model year. This will give a general indicator of the color on the vehicle and the paint required. *(Courtesy of DuPont Automotive Products)*

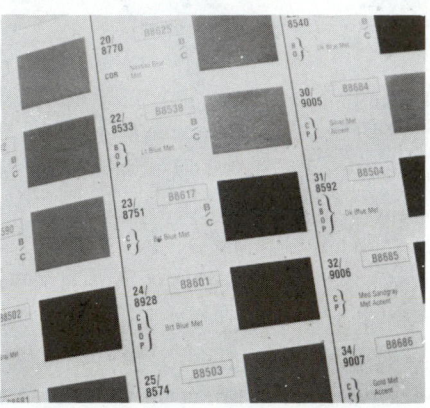

FIGURE 19–15 Locate the manufacturer's paint code number in the book. The identifying color chip is located next to it. Place the chip over the paint on the vehicle to check the general color match. The chip may vary from the actual color on the vehicle, however. *(Courtesy of DuPont Automotive Products)*

are not perfectly accurate because they are placed on paper and not over an undercoat.

If the color match is correct, order the topcoat from a local supplier by color stock number. Refinish suppliers supply topcoat colors in two ways:

- If it is a recent model or a popular color, chances are they will have it ready-mixed in pint, quart, and gallon cans. These ready-mixed colors are called *factory packaged*.
- If it is an older or less popular color, they might have to mix it in pint, quart, or gallon quantities. Paint manufacturers work extensively to develop OEM matches with mixing color formulas. *Custom-mixed colors* are those colors that are mixed to order at the paint supply distributor. Custom-mixed color can always be identified easily because the contents of the container must be written on the label by the paint distributor who mixed the paint.

In recent years, most of the major automobile paint manufacturers have made available to refinisher shops a color mixing system. Under such an "intermix" system, it is possible to mix thousands of colors at a considerable cost savings over the cost of factory-packaged colors.

19.5 DETERMINING IF THE AUTOMOBILE HAS BEEN REPAINTED

There are three ways to determine if the automobile has been repainted in the past. They are:

- **Sanding method.** Sand an edge on the area to be repainted until the bare metal appears. The makeup of the paint coating will determine whether or not it was repainted previously (Figure 19–16).
- **Paint film thickness measurement method.** A thicker than normal paint coating indicates that the vehicle has been repainted. The standard paint film thicknesses of new vehicles are

Domestic vehicles	3 to 5 mils
European cars	5 to 8 mils
Japanese vehicles	3 to 5 mils

Normally, an electromagnetic thickness gauge or a mechanical thickness gauge, mentioned in previous chapters, is used to measure the paint film thickness. If your measurements show coating about twice as thick (6 to 10 mils), the vehicle has probably been repainted. Reduction of mil thickness by paint removal would be needed before painting again.

- **Inspection method**. With the inspection method, inspect closely for signs of repainting

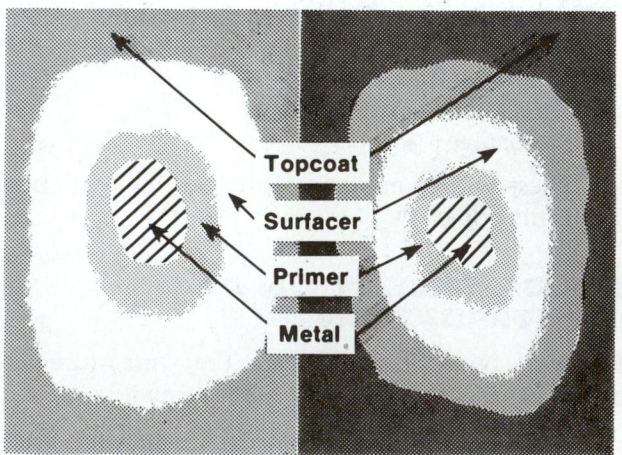

FIGURE 19–16 Sanding through a small area to bare metal will show what is under the paint. You can tell if the vehicle has been repainted or whether it has a factory paint job.

FIGURE 19–17 Visual inspection will help you tell if the vehicle might have been repainted or repaired. Look for paint lines from masking tape, overspray on trim pieces, different paint textures, etc., which indicate repainting.

FIGURE 19–18 To determine if a finish is lacquer, wipe a small area with thinner. If the rag comes away clean, you probably have enamel.

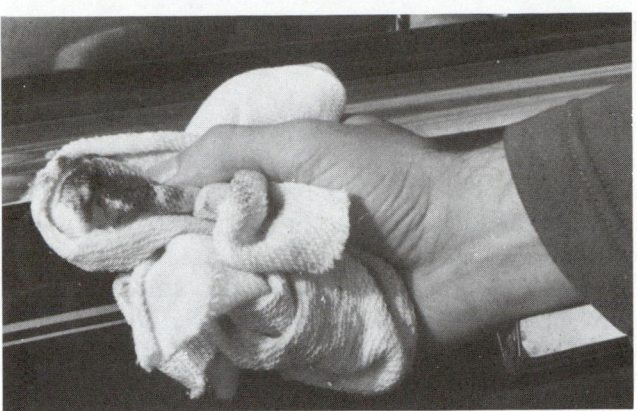

FIGURE 19–19 If a lacquer-thinner-soaked rag has dissolved paint on it, you probably have lacquer paint on the vehicle.

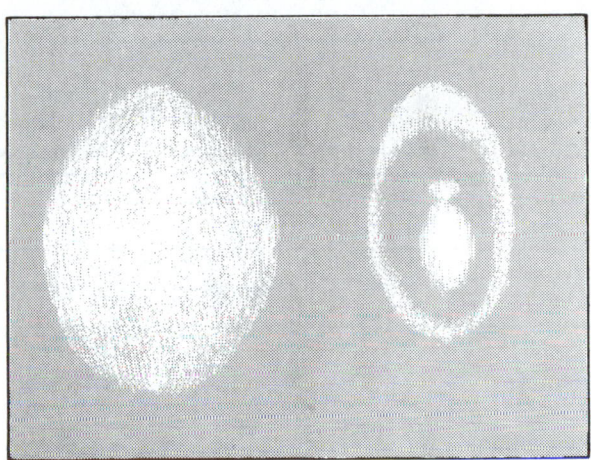

FIGURE 19–20 Wet sand a hidden area to dull the paint film. Then heat the area with an infrared lamp. If a gloss returns to the dulled spot, the paint is acrylic lacquer.

(Figure 19–17). Look for masking tape–created paint lines, overspray, and other signs of repairing. With a good paint job, this can sometimes be difficult because all signs of repainting will be hidden.

19.6 DETERMINING TYPE OF PAINT ON VEHICLE

Before planning any refinishing job, you must find out what type of paint is on the vehicle. The vehicle might have its original paint or it could have been repainted with a different type of paint.

Methods for finding out the type of paint on a vehicle include the following:

1. With the *solvent application method*, rub the paint with a white cloth soaked in lacquer thinner to see how easily the paint will dissolve (Figure 19–18). If the paint film dissolves and leaves a mark or color stain on the rag, it is some type of air-dried paint. If it does not dissolve, it is either an oven-dried or a two-part reaction-type paint. An acrylic urethane paint film will not dissolve as easily as an air-dried paint, but sometimes the thinner will penetrate sufficiently to blur the paint gloss (Figure 19–19).

2. With the *heat application method*, wet sand an area with #1000 grit sandpaper to dull the paint film. Then heat the area with an infrared lamp. If a gloss returns to the dulled appearance, the paint is acrylic lacquer (Figure 19–20).

3. With the *hardness method*, you must check the general hardness of the paint. Paints do not dry to the same hardness. Generally, two-part reaction and oven-dried paints dry to a harder film than air-dried paint.

Identifying cars that are clearcoated is easy. Looking at the vehicle identification code and the color chip book is a quick way to find out if the car has the basecoat/clearcoat system. If the code has been removed or destroyed, sanding a small spot in a concealed area of the vehicle to be finished, using a fine sandpaper, can help determine the type of finish. If the dust is white, the car has a basecoat/clearcoat finish. If the dust is the color of the car, it is a solid color.

Table 19–3 classifies standard paints for determining previous painted coating. Table 19–4 lists the types of previously applied paints and those topcoats that can be applied over them.

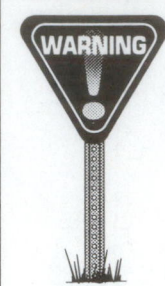 Always make sure you are using the correct solvent. With old containers of solvent in the shop, it is very easy to pick up the wrong can. This can make your paint job a nightmare!

TABLE 19-3: CLASSIFICATION STANDARD FOR PREVIOUS PAINT COATINGS

| Previous Paint Coating | Classification Method | | |
	Visual Inspection	Solvent Method	Heat Application Method
Alkyd enamel	Caulking surface	Does not dissolve	Some softening
Acrylic lacquer	—	Dissolve	Softens
Acrylic enamel	—	—	Some softening
Polyurethane	Polished skin	—	—
Acrylic urethane lacquer	Polished skin	Difficult to dissolve	Some softening
Acrylic urethane enamel	Gloss with some orange peel	—	—

19.7 SELECTING SOLVENTS (REDUCERS AND THINNERS)

There are two vital variables that affect the spraying of materials: *temperature* and *humidity*. Unless a shop has year-round temperature control, these variables must be carefully observed and compensated for with use of the proper solvent. Of the two variables, temperature is the most critical.

Here is how temperature and humidity affect sprayed material:

- Hot, dry weather produces a faster dry time.
- Hot, humid, or warm, dry weather produces a fast dry time, but is slower than hot, dry weather. (High humidity can cause problems.)
- Normal weather—70 degrees Fahrenheit (21 degrees Celsius) with 45 to 55 percent relative humidity—produces a normal dry time.
- Cold, dry weather produces a slower dry time than normal.
- Cold, wet, or humid weather produces the slowest dry time.

Thus, to do quality refinishing, many auto paint shops use up to four different types of solvents for

TABLE 19-4: APPLICATION CHART—PREVIOUSLY APPLIED PAINT AND REPAINTING CHART

| Topcoat | Previously Applied Paint | | | | |
	Alkyd Enamel	Acrylic Lacquer	Acrylic Enamel	Polyurethane Enamel	Acrylic Urethane Enamel
Alkyd enamel	A	B	A	A	A
Acrylic lacquer	A	A	B	B	B
Acrylic enamel	A	A	A	A	A
Polyurethane enamel	A	A	A	A	A
Acrylic urethane enamel	A	A	A	A	A

A Okay to repaint with
B Okay if primer-surfacer or sealer specified by paint manufacturer is used

each paint system employed during the course of a year:

- **Slow-drying solvent.** The flash time evaporation rate for slow lacquer thinners ranges from 3½ to 5 minutes at about 75 degrees Fahrenheit (24 degrees Celsius) when applied wet. For a slow-drying enamel reducer, the flash time is slightly longer.
- **Medium-drying solvent.** The average flash time evaporation rate for medium thinner is about 2 minutes when applied wet. Flash time for a medium reducer is slightly longer.
- **Fast-drying solvent.** The flash time evaporation rate for fast thinners ranges from 15 to 20 seconds when applied wet at about 75 degrees Fahrenheit (24 degrees Celsius). The flash time for a fast reducer is slightly longer.
- **Retarder.** This is a very slow-drying solvent. The flash time evaporation rate of retarder is about 30 minutes when applied wet at about 75 degrees Fahrenheit (24 degrees Celsius).

 A good general rule to follow when selecting the proper solvent is: The faster the shop drying conditions, the slower drying the solvent you should use. In hot, dry weather, use a slow-drying solvent. In cold, wet weather, use a fast-drying solvent.

READING LABEL DIRECTIONS

Before applying any topcoat finish, remember that it is necessary to very carefully read the paint manufacturer's directions that appear on the paint container. While different types of paints may have the same general characteristics, each manufacturer has its own formulations for its products. For this reason, the best source of data on how to apply a specific brand of paint is the container label. Another source of good information can be found in the manufacturer's literature.

When thinning or reducing, refer to the product label for the proper solvent percentage (Figure 19–21) so there will be no danger of over- or under-reducing.

Some of the more important label and literature data that should be checked include the following:

- Proper viscosity, using either Ford or Zahn cup
- Spray gun pressure
- Use of additives, reducers, thinners, and activators when necessary
- Application techniques
- Number of paint coats required for different refinishing jobs
- Blending and mist coat procedures, if necessary
- Cleanup procedures

Mixing instructions are normally given on the materials label. This might be a percentage or parts of one ingredient compared to the other. A simple example is given in Figure 19–22.

A **percentage reduction** means that each material must be added in certain proportions or parts. For example, if a paint requires 50 percent reduction, this means that one part reducer or thinner (solvent) must be mixed with two parts paint.

Mixing by parts means that for a specific volume of paint or other material, a specific amount of another material must be added. If you are mixing a gallon of paint in a spray gun pressure tank, for example, and directions call for 25 percent reduction, you would add one quart of reducer. There are four quarts in a gallon and you want one part or 25 percent reducer for each four parts of paint.

Proportional numbers denote the amount of each material needed. The first number is usually the parts of paint needed. The second number is usually the solvent (thinner or reducer). A third number might be used to denote the amount of hardener or other additives required.

DIRECTIONS FOR USING MIRACRYL 2 STANDARD COLORS

DESCRIPTION: MIRACRYL 2 is a fast drying, easy to use and durable acrylic enamel for refinishing STANDARD (NON BASECOAT/CLEARCOAT) colors on cars and trucks.

SURFACE PREPARATION: Clean the area to be painted with 900 Pre-Kleano and sand thoroughly with fine or medium sandpaper. Build up all bare metal and damaged areas with R-M Primer-Surfacer. To minimize sand scratches and improve hold-out, appropriate R-M Sealers may be used.

COLOR REDUCTION: To one quart MIRACRYL 2 STANDARD color add one half quart MIRACRYL 2 REDUCER. A complete range of MIRACRYL 2 REDUCERS is available for various climatic and shop conditions. Stir thoroughly and strain through medium strainer.

APPLICATION: MIRACRYL 2 STANDARD colors may be used with or without MIRACRYL 2 HARDENER. Spray single coats until hiding and color match have been achieved. Allow each coat to become tack free before applying the next. Use 45-55 pounds air pressure at the gun. Lower air pressure will darken iridescent colors, higher air pressure will lighten them.

RECOATING: MIRACRYL 2 STANDARD colors may be recoated after air drying 16 hours under normal conditions.

FIGURE 19–21 Before thinning or reducing a paint material, check the label for proper percentage. It will give instructions that must be followed.

For example, the number 2:1:1 means for two parts of paint, add one part solvent and one part hardener. For a gallon of paint, you would add a quart of solvent and a quart of catalyst. This can vary, so always refer to the exact directions on the materials.

A **mixing chart** converts a percentage into how many parts of each material must be mixed. One is given in Figure 19–23. Study the percentages and parts of each material that must be mixed.

USING PAINT MIXING STICKS

Mentioned in Chapter 18, **paint mixing sticks** have graduated scales that allow you to easily con-

vert ingredient percentages into part proportions. They are used by painters to help mix paints, solvents, catalysts (hardeners), and other additives right before spraying. They are often provided by the paint system manufacturer. Several are needed to provide mixing guides for each type of paint (Figure 19–24).

Each mixing stick will have a ratio or percentage printed at the top. Paint measurement marks and numbers are placed along the side of the tool for pouring out the correct quantity of each material quickly. Select the mixing stick with the correct ratio and type of paint to be used (Figure 19–25).

Use a can, pail, or container with straight sides. Tapered sides on the container would upset your measurements (Figure 19–26). The container should be big enough to hold all the paint, hardener, and solvent needed for the job. A gallon or several liter container saves mixing time for an overall or com-

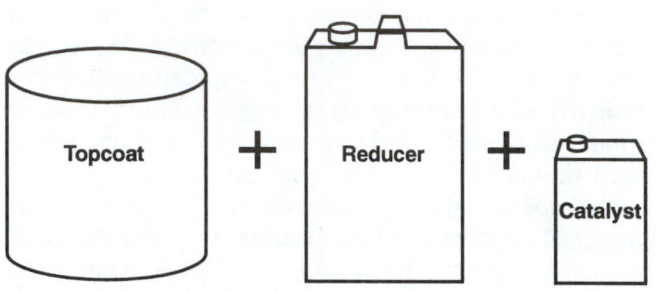

FIGURE 19-22 Today's refinish materials work best when mixed according to manufacturers directions. *(Courtesy of PPG Industries, Inc.)*

Paint mixing sticks should not be confused with **paint stirring sticks** (wooden sticks used for mixing the contents after they are poured into the spray gun cup or a container).

Reduction / Thinning Percentage		Reduction / Thinning Proportions	Paint	Solvent
10%	=	5 parts paint / 1 part solvent	10%	
25%	=	4 parts paint / 1 part solvent	25%	
33%	=	3 parts paint / 1 part solvent	33%	
50%	=	2 parts paint / 1 part solvent	50%	
75%	=	4 parts paint / 3 parts solvent	75%	
100%	=	1 part paint / 1 part solvent	100%	
125%	=	4 parts paint / 5 parts solvent	125%	
150%	=	2 parts paint / 3 parts solvent	150%	
200%	=	1 part paint / 2 parts solvent	200%	
250%	=	2 parts paint / 5 parts solvent	250%	

FIGURE 19-23 The chart shows conversions for different reduction/thinning percentages.

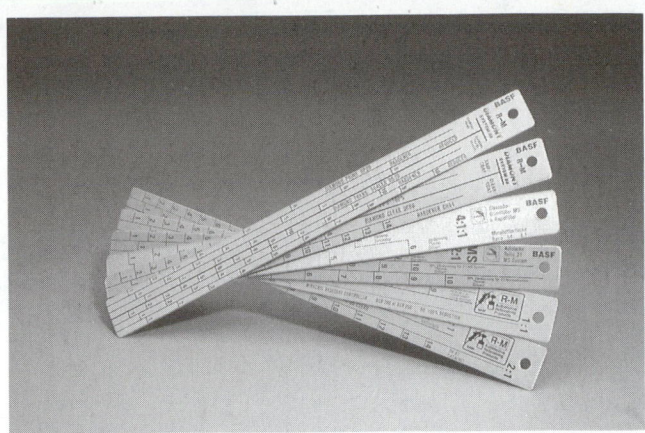

FIGURE 19-24 Paint mixing sticks provide a fast, easy way to mix paint, solvent, and hardener properly. *[Courtesy of BASF]*

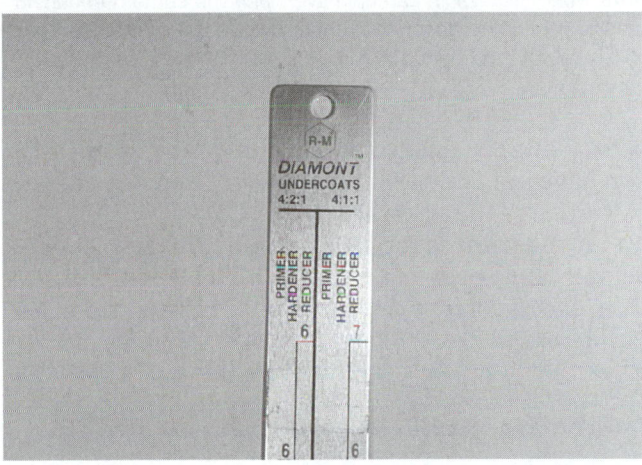

A

B

FIGURE 19-25 (A) Top of mixing stick will give mixing ratio. Find stick that matches instructions on paint can label. (B) Scales on side of stick show how much to pour into container for paint, solvent, and hardener.

plete paint job. If you are only painting a small area and one cup will do, mix the materials in the spray gun cup. If you are using a pressure tank spray gun, measure and mix the materials in the tank.

To give you an example, let us say that the paint mixing stick is for a mixing ratio of 4:1:1. This means that you need four parts color, one part hardener, and one part solvent.

1. Place the correct paint mixing stick for the type of paint into the container (Figure 19–27).
2. Filter the amount of paint or primer you plan to use into the bucket. Stop pouring when the paint is even with any of the numbers on the left of the stick. The material must be even with any number on the left column. This might be 1 through 7, depending on the quantity of paint needed.

   For more paint, you might fill to lines 6 or 7. For a spot repair, you may only need to pour material even with lines 2 or 3, for example. Make sure the paint is perfectly even with any of the numbers on the mixing stick. If color is listed on the left, the paint should be even with any number in that column.
3. Pour in reducer until even with the same number (from Step 2) in the next column on the stick. If the paint already in the container aligned with a 3, pour in reducer until it aligns with 3 but in the second column.
4. Pour in the final ingredient (usually hardener) until it aligns with the same number in the last column on the mixing stick. If the paint and reducer were aligned with a 3, pour hardener in until it aligns with the 3 in its mixing stick column.

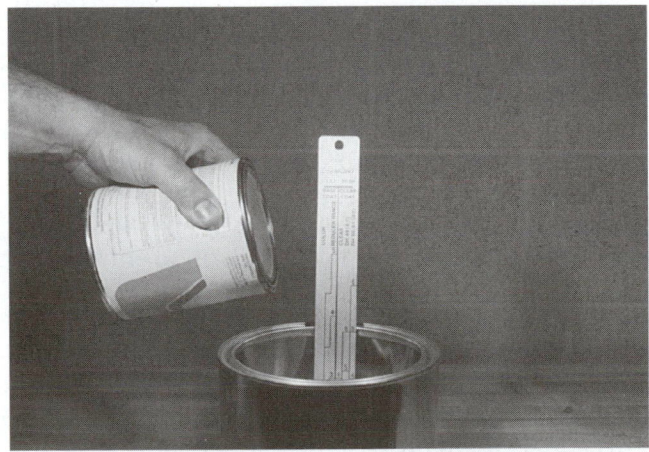

FIGURE 19-26 Pour amount of paint needed into a clean container until it is even with one of the number scales on the stick for paint. A straight-sided container is needed for accurate measurements.

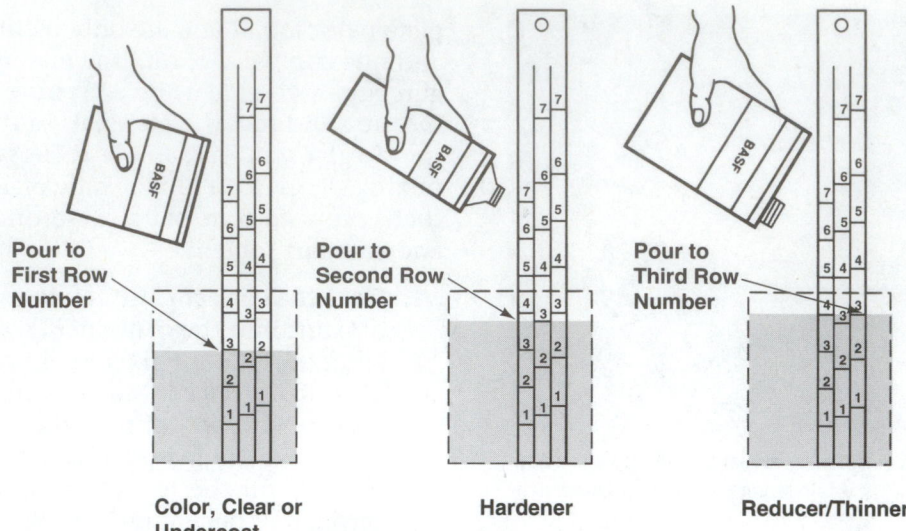

Pour to First Row Number

Pour to Second Row Number

Pour to Third Row Number

Color, Clear or Undercoat

Hardener

Reducer/Thinner

FIGURE 19-27 Study basic steps for using mixing stick. (BASF) (A) Depending upon information on top of the mixing stick, you must usually pour in paint or primer first. Pour in the amount of material needed for the job. Stop when you reach any of the numbers on the mixing stick. (B) If used, pour hardener or catalyst in next. Pour material in can until even with the same number on the mixing stick but in the next column. If the paint was even with number 5, pour in hardener to number five on the next or center column on the stick. (C) Pour in solvent until the liquid is even with 5 on the stick in the right column. Pour all materials slowly so you do not add too much. Stir materials with metal stick.

After adding the correct amount of each material, thoroughly mix with the metal mixing stick. You can then fill your spray paint gun with properly mixed paint.

When filling your spray gun, always use a paint strainer over the cup. The **paint strainer** is a paper funnel/mesh strainer that keeps debris out of the spray gun. It should be used when anything is poured into the spray gun cup. Contaminants can accidentally get into new materials. Dirt and dust can also fall off the top of containers when you are pouring!

USING A VISCOSITY CUP

Discussed in Chapter 18, a *viscosity cup* can also be used to measure the thickness or fluidity of the mixed materials, usually paint. It is a small, stainless steel cup attached to a handle.

To use a viscosity cup, dip it into the mixed paint until submerged. Lift the cup out and hold it over the paint container. As soon as the cup is lifted out, start timing how long it takes the cup to empty. The paint will leak out of a small specific size **orifice** (hole) in the bottom of the cup.

Use the second hand on your watch or a stopwatch to time draining. When the paint stream breaks into drops, note how much time has elapsed. This equals the paint viscosity in seconds.

The paint manufacturer will give a recommended viscosity value in *viscosity cup seconds*. It will vary between 17 and 30 seconds depending upon the type of paint and type of cup used. Refer to the paint specifications for an exact value.

If the paint drains too quickly out of the viscosity cup, you have added too much solvent. More

paint would be needed. If the cup drains too slowly, you have not added enough solvent. Remix until the paint passes the viscosity cup test.

If the paint is too thick, your paint job will develop orange peel or a rough film. If the paint is too thin, excess solvent can cause poor hiding and other problems.

PRIMER-SURFACER SOLVENTS

When thinning or reducing primer-surfacers, select a solvent that is slow enough for shop conditions. A slow solvent will help provide a relatively smooth surface, requiring a minimum amount of sanding. Generally, faster-drying thinners and reducers will cause the film to be rough, thus increasing sanding time.

The use of an excessively slow-drying solvent should also be avoided. There is a tendency to spray heavy coats of primer-surfacer due to its high-solids content. When slow-drying solvents are used, longer flash times and dry times are required. If the refinisher fails to allow sufficient flash and dry time, the primer-surfacer might shrink and crack at the featheredge.

19.8 PREPAINTING REVIEW

Before explaining how to apply the topcoats or finish coats, you should have a quick review of the major steps for preparing the vehicle for spraying:

1. Thoroughly wash the car with soap and water.
2. Chemically clean the car with a wax and grease remover to remove wax buildup, tar, and other nonwater-soluble grime.

3. Wet sand the repair area by hand with a block with #400 grit sandpaper or dry or wet sand by machine with #320 grit sandpaper. Note that the grit recommended can vary with the type of paint.

4. Reclean the area with the wax and grease remover.

5. If there is any surface rust, blast off the rust and treat the area with a rust converter to eliminate any hidden corrosion.

6. Spot-prime any bare metal areas with a self-etch primer, epoxy primer, or suitable product.

7. Apply primer-surfacer as needed to fill low areas and eliminate sand scratches.

8. Wet sand the primer-surfacer level to the surrounding area. Flush off the sanding grit with water. When dry, wipe and blow off to remove dust. Typically use #400 or finer grit for hand sanding and #320 grit if machine sanding (Figure 19–28) for most finishes.

9. After properly masking the vehicle and preparing the surface for repainting as just described, once again blow away any remaining dust with an air gun. Give a final touch-up cleaning with the wax and grease remover.

10. Check all masking tape and paper one last time. Make sure none of the tape has pulled up and the paper has not been torn. Inspect all edges and the paper closely for openings that could allow overspray leaks.

11. Confirm that the air circulation system in the spray booth is turned on and working properly.

12. While wearing protective gear, blow off any remaining dust with an air gun. As you blow off surfaces, wipe the vehicle down with a **tack cloth** (rag with coating that holds dust and debris). This is shown in Figure 19–29.

13. After wiping with a tack cloth, be careful not to touch the surface being refinished or stir up dust in the booth. Also, do not open the spray booth doors.

SPRAYING VARIABLES

Before the topcoat material can be applied by spraying, also double-check the following:

- The paint is properly stirred or mixed.
- It has been thinned or reduced with the correct solvent to the desired viscosity.
- The surface temperature has been checked.

Stir Paint Right Before Use

Failure to properly stir all the settled pigment into the liquid is a principal cause of paint problems. Stirring or mixing can be done by hand or by machine.

The part of paint that settles is the pigment, which gives the paint its color, opacity, and specific performance properties. The weight of pigments vary greatly. Some of the commonly used pigments are seven to eight times as heavy as the liquid part of the paint. Because of their weight, the heavy pigments slowly settle. Some of the pigments are light and fluffy and have very little tendency to settle. The commonly used pigments that settle quite rapidly are the whites, chrome yellows, chrome oranges, chrome greens, and red and yellow iron oxides.

The consistency or viscosity of the liquid part of the paint has much to do with the rate of settling.

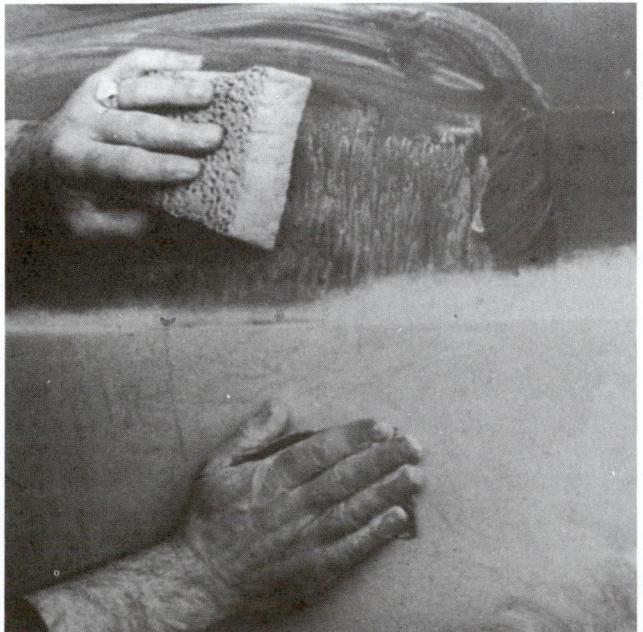

FIGURE 19–28 Wet sanding is commonly done before applying the topcoat. This will smooth, clean, and texture old paint and the repaired area before painting.

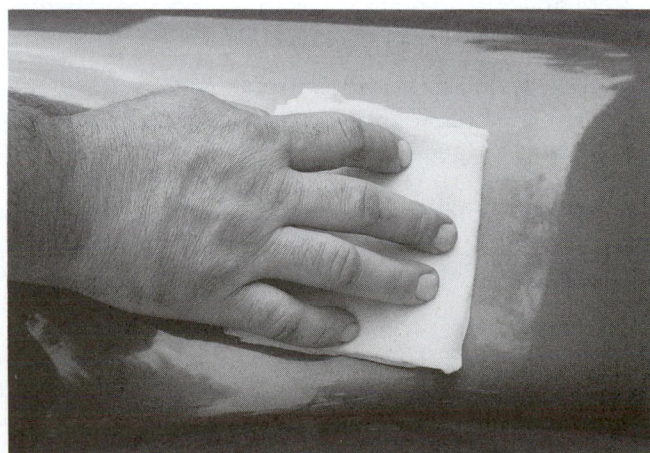

FIGURE 19–29 Once in the spray booth, use a tack cloth on all surfaces to be painted to remove dust and other small debris.

SHOP TALK

Do not use sharp sticks or screwdrivers for stirring. At least a 1-inch (25 mm)-wide flat-bottomed, clean stirring paddle or steel spatula should be used.

The heavier the consistency, the slower the settling. Heavy pigments will settle out of a straight thinner in a few minutes, whereas in a paint vehicle it will take weeks or months. Careful judgment in thinning only sufficient material to do the job and discarding the small amount that is not used is the mark of an experienced painter.

If a color that contains one or more of the heavy pigments is thinned or reduced to spraying consistency and allowed to stand 10 to 15 minutes without being stirred, it will have settled enough in that time to be off-color when sprayed.

After a can of paint has been thoroughly agitated, empty out the contents of the can into another container or the gun cup, wash the can clean with a little solvent, and add this to the paint. Keep in mind that this will affect your mixing ratio.

If a paint residue has become very hard, the liquid part should be poured off and the remaining part well broken up. The liquid should then be slowly poured back and mixed in by vigorous stirring.

When using an intermix system, agitate all base mixing colors for a minimum of 15 minutes. Before putting the can on the scale, put in just enough universal retarder to cover the bottom. This will prevent small color additions from drying out and not mixing in with the rest of the colors.

There are several important points to remember when mixing a basecoat color:

- Always read the label directions first.
- Use only the manufacturer's recommended hardeners and reducers (or basecoat stabilizers, if recommended).
- Use the proper reducer for the shop conditions and size of the job.
- Use only the proper mixing ratios.

Mentioned earlier, the choice of a reducer is an important one. The refinisher must be careful to choose the product to complement the basecoat (and later the clearcoat) in relation to the shop's temperature and humidity. Most major paint manufacturers offer a choice of reducers to offset atmospheric conditions that may cause color shifting. Problems such as soak-in (too slow a reducer) or dry overspray (too fast a reducer) can also occur if the incorrect reducer is used.

In the case of some new systems, a basecoat stabilizer or additive replaces the standard reducer. The **stabilizer** contains a basecoat resin designed to give it a faster recoat time and allow better metallic control. This is especially important if any blending is desired, because it prevents wash-out or a halo-like effect at the edge of the repair.

Check Paint Viscosity

Topcoat paint materials are usually shipped with as high a viscosity as practical to help in slowing down the rate of settling. In order to apply these paint materials, they must be reduced or thinned to a viscosity that can be properly atomized by the spray gun.

Compared with a two-component reaction, the air-dry type of topcoat has a higher resin viscosity so vaporization is not as rapid. Because the initial drying time is faster, more solvent is required to improve the finish, and it is also necessary to increase the spraying air pressure. However, increasing the air pressure will cause the solvent to evaporate faster, resulting in an even faster initial drying time and possibly a defective finish. For preventive measures, consider decreasing the distance between the gun and painting surface, increasing the discharge volume, increasing the number of coats, speeding up the painting operation, and so on.

Metallic colors with a lower viscosity than solid colors are used to prevent unevenness. The distance between the gun and the painted surface is greater and there is also more pattern overlap. Application is similar to that for a dry coat.

Clearcoat is applied in a method similar to a solid color but care must be taken to avoid a heavy coat. This can cause the paint film to be rough (orange peel) or other defects. A small spray gun nozzle and adequate gun pressure are needed to get good paint atomization and *paint laydown* (smooth, shiny paint film).

The recommended viscosities for spot repairs, body repairs, and overall paint, as well as spray gun pressure, are given later in this chapter. All spray gun air pressure recommendations in this reading are at the gun.

Check Temperatures

The temperatures of concern when refinishing are the room temperature, the temperature of the surface of the car, and the temperature of the paint.

Temperature is especially important in downdraft spray booths. With all the air rushing by, the temperature of the paint applied is raised 6 or 7 degrees almost immediately. If the paint dries too fast, solvents will be trapped in the basecoat. This can cause problems when applying the clearcoat.

Also, be careful when bringing in a car that has been stored outside in the cold. If the surface is too cold, the solvents will not evaporate as quickly as they should. Color matching and paint curing problems can result. A thermometer should be placed on the surface of the car (Figure 19–30). Make sure the surface is the temperature suggested by the manufacturer before spraying.

TYPES OF SPRAY COATS

There are varying degrees of thickness for a sprayed coat. Generally, they are referred to as light, medium, or heavy. The easiest way to control this degree of thickness is by the speed with which the gun is moved. That is, the slower the speed, the heavier the coat.

Generally, you want the thinnest coat possible that will produce complete coverage. Thinner paint will be more durable and less prone to problems than thicker coats.

There are also several other terms to describe spray coats:

- Tack
- Full wet coat
- Mist
- Dust
- Shading or blending
- Banding

Many refinishers apply each coat the way they want the finished job to look. Some use a first light tack coat, then a wet coat, and finally a mist coat. Either method will help you produce a smooth, shiny paint job.

The first **tack coat** allows the application of heavier wet coats without sagging or runs. This is a light covering coat applied to the surface first and then allowed to *flash* until it is just tacky, which usually takes only a few minutes. The finish wet coats are then sprayed over the tack coat. A tack coat is usually used with enamel paint.

A **medium wet coat** is not as heavy a paint application as a full wet coat. It is done by holding the gun a little farther from the surface or by moving the gun a little faster. It is often used to produce a smooth paint surface while avoiding paint runs and sags.

A **full wet coat** is a heavy, glossy coat that is applied in a thickness almost heavy enough to run. It requires skill and practice to spray such a coat. You must use good lighting and carefully watch the spray go onto the vehicle to prevent runs.

The **mist coat** is an application of slower-drying thinners over a colorcoat. It helps to level the final coat, melt in the overspray, and control mot-

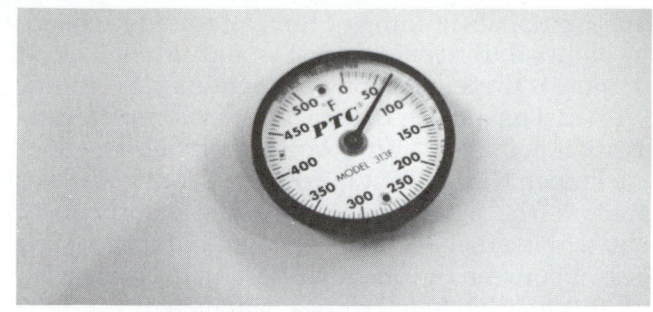

FIGURE 19–30 A surface thermometer is handy for making sure body temperature will not cause problems when the paint hits the surface.

tling of metallics. Uniforming finishes or blenders are applied as mist coats in spot and panel repairing. It is often the last coat when refinishing.

A **dust coat** is a light, dry coat of finish. It is accomplished by holding the gun a little farther from the surface being sprayed. It can be used to help even out coverage. It is often the last coat used when spraying the base or colorcoat of a metallic basecoat/clearcoat system. The dryer dust coat will help the metallic color look the same in all areas.

19.9 REPAINTING SPRAY METHODS

As mentioned in earlier chapters, repainting spray methods are classified according to the condition of the previous (original) paint coat, the size of the area to be repainted, and the location.

SPOT AND PANEL SPRAYING

Spot repainting repairs are recommended where a complete panel repair is unjustified, being either uneconomical (size of repair or amount of masking involved) or impractical (difficulty of rendering the repair invisible, particularly in the case of metallic finishes). Blending is necessary so that slight differences in color or texture are not detectable.

Blend coats are progressive applications of paint on the boundary of spot repair areas so that a color difference is not noticeable. They are applied in two or more coats. The second and third coats are thinner and sprayed over a wider area than the first.

Banding is a single coat applied in a small spray pattern to the frame in an area to be sprayed. This technique assures the painter of coverage at the edges without spraying beyond the spray area and reduces overspray. Banding is often used in spraying panel repairs with a primer-surfacer.

Sometimes a banding coat is thinned more than the normal application that follows. This is especially true when the paint to be sprayed is of high viscosity.

The additional thinning of the paint used for banding allows it to fully enter cracks and seams. A good example is the application of a textured vinyl finish.

Panel repainting is done to repair complete panels separated by a definite boundary, such as a door or a fender. Normally, it is not necessary to shade or graduate the paint unless the paint is difficult to match or is a metallic color. Shading is done for such areas as between the quarter panel and the roof panel, but this is still referred to as a panel repair.

Paint blending involves tapering the new paint gradually into the old paint. This makes any difference in the new paint less noticeable. It helps hide any slight differences in color or texture. Blending is the key to a successful spot repair.

To blend paint, apply the topcoat with a fanning motion, working from the center outward. This allows each coat to blend out a bit farther than the previous one.

An alternative method is to apply the finish in short strokes from the center outward. Again, extend each coat so that it blends out farther than the previous one.

With either method, the spray pattern should be narrowed and the fluid delivery reduced. To minimize overspray, the air pressure may also have to be reduced, depending on the material being sprayed.

Spot Repainting with Solid Colors

Definite rules cannot be laid down as to when and where to do spot repair. The spot repair method should be used for light damage in areas that will not show the repair easily. With spot repairs, you will have to blend the painted repair area out and over the unpainted, unrepaired area. As shown in Figure 19–31, each coat of paint should blend out over more of the spot repair area. The last or third coat will be thin near its edges to blend into the existing paint with less difference.

Shown in Figure 19–32, when spot repainting with a solid color, the shaded area should be treated with compound or sanded using a #1000 grit sandpaper before repainting.

Apply the topcoat with a spiraling circular motion, working the spray gun from the center outward. This allows each coat to slightly overlap the previous one. An alternative method is to apply the finish in short arcing strokes from the center outward (Figure 19–33). Again extend each coat so that it slightly overlaps the previous one. But in either method, the spray pattern should be narrowed and the fluid delivery reduced by adjusting the spray controls. To minimize overspray, the air pressure should be reduced, depending on the material to be sprayed. The spray pattern should never be reduced to a completely round jet; otherwise, both paint control and overlapping of passes become difficult.

Panel Repairs with Solid Colors

For complete panel or block repair (Figure 19–34), it is wise to properly mask off the area not to be painted (Figure 19–35).

If a panel has damage at two different locations, as shown in Figure 19–36, repairs should be done in

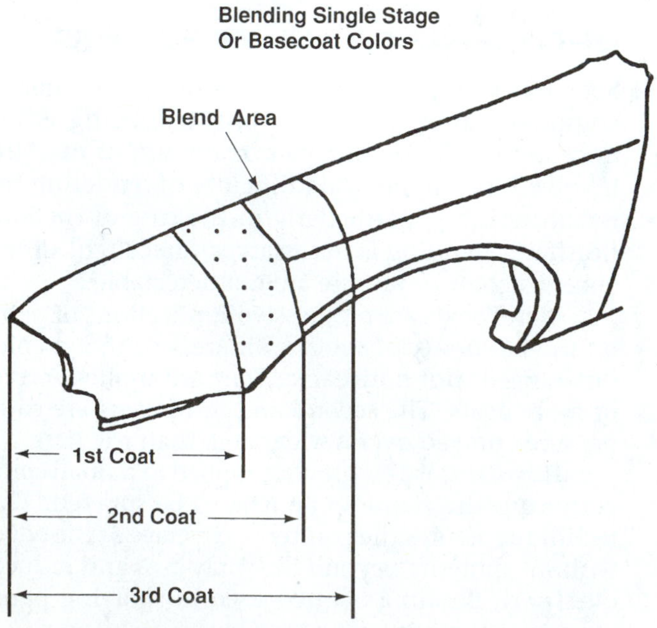

**Blending Single Stage
Or Basecoat Colors**

Blend Area

1st Coat

2nd Coat

3rd Coat

FIGURE 19-31 With a spot repair, each coat of paint should blend out over more of the repair area. The last or third coat will be thin near its edges to blend into the existing paint with less difference. *(Courtesy of I-CAR)*

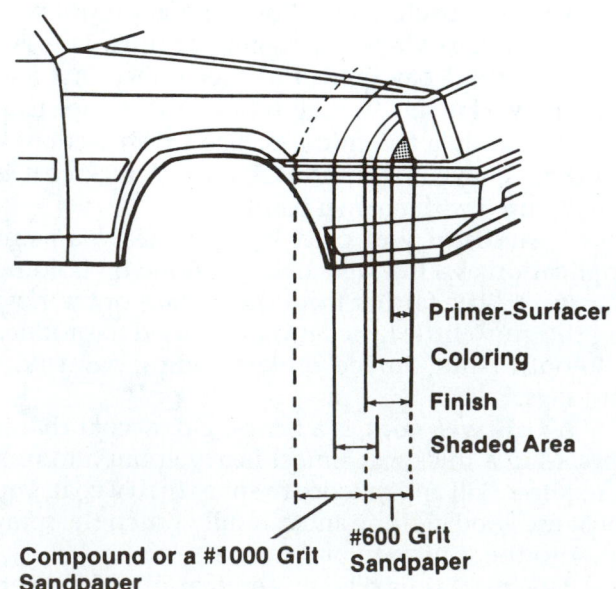

Primer-Surfacer

Coloring

Finish

Shaded Area

#600 Grit
Sandpaper

Compound or a #1000 Grit
Sandpaper

FIGURE 19-32 Note the major steps to prepare for spot repainting.

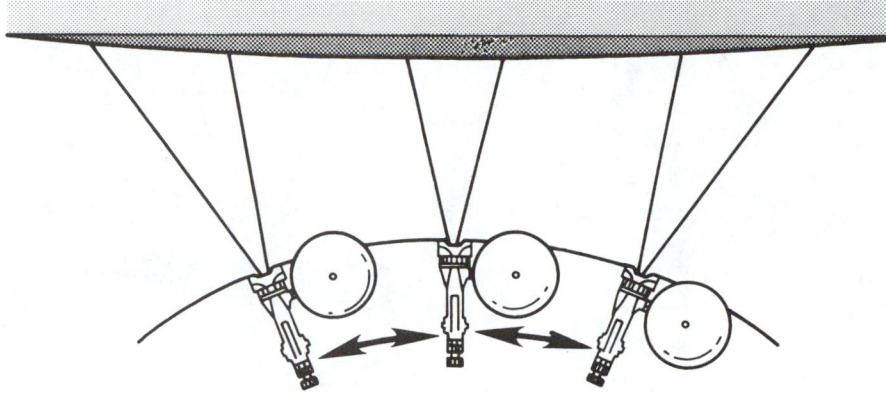

FIGURE 19-33 When spraying a shade or blend area, it might be necessary to break the spray gun rule against gun arcing, but the arcing should be kept to a minimum. It will help thin coats at outer edges of spot repair so new and old paints blend together without being noticed.

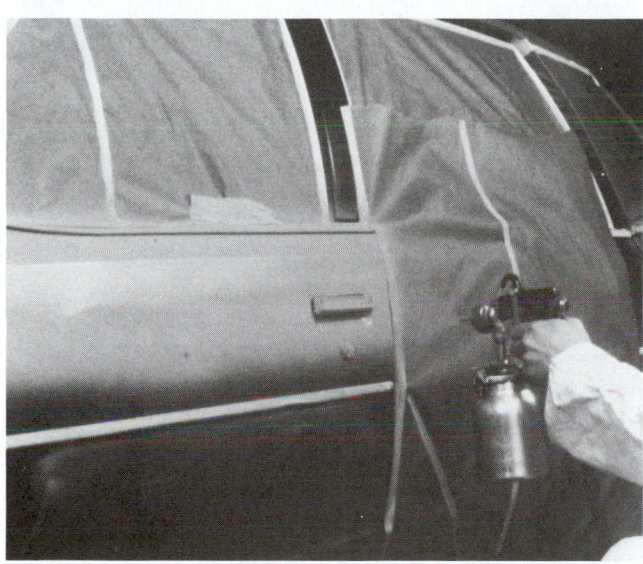

FIGURE 19-34 With panel repairs, paint the complete panel. With proper color selection, mixing, and application, old and new paints should match when unmasked.

FIGURE 19-35 If only a small area has been primed, apply the first colorcoat over this area only. Then, paint the whole panel to help cover the primer area.

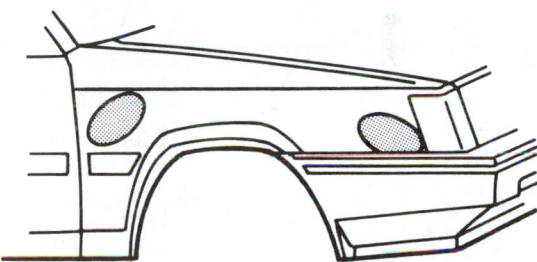

FIGURE 19-36 If you must repaint two areas on the same panel, painting the panel would be easier than spot repairing both.

a panel style. As shown, shading can be done at the molding area or extended below the molding.

For the quarter panel, it is generally necessary to graduate the shade area into the quarter pillar. If the vehicle has a ventilation louver on the quarter pillar, as shown in Figure 19–37, shading at that area will be less noticeable.

Spot Repainting with Metallic Colors

If spot repainting with a metallic color at locations shown in Figure 19–38, skill is required in matching the color tone and bringing out the metallic image through proper distribution of the paint. The shaded area will be less noticeable if it is angled away from the press line (Figure 19–39).

Panel Repairs with Metallic Colors

Unevenness tends to be very noticeable with clear and bright metallic colors. Therefore, if it is impossible to match the paint exactly with the previous coat, extend the shade area over a wide range, as shown in Figure 19–40, to make it less distinguishable.

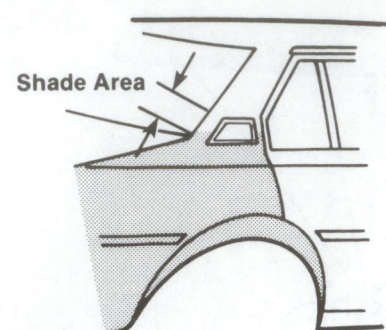

FIGURE 19-37 When repainting a panel without a definite break line, blend or shade the paint in a small or hidden area. Then, any difference in the blend or shaded area will be less noticeable.

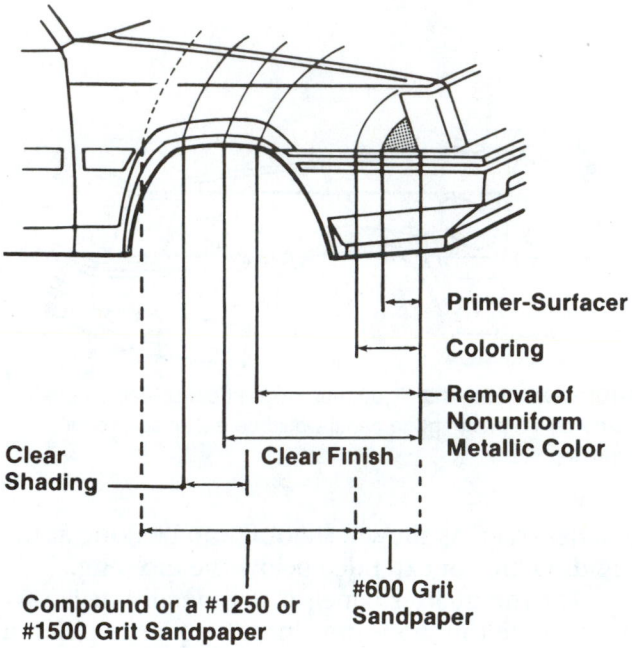

FIGURE 19-38 Spot repainting of metallic colors can be more challenging than with solid colors. Note major operations.

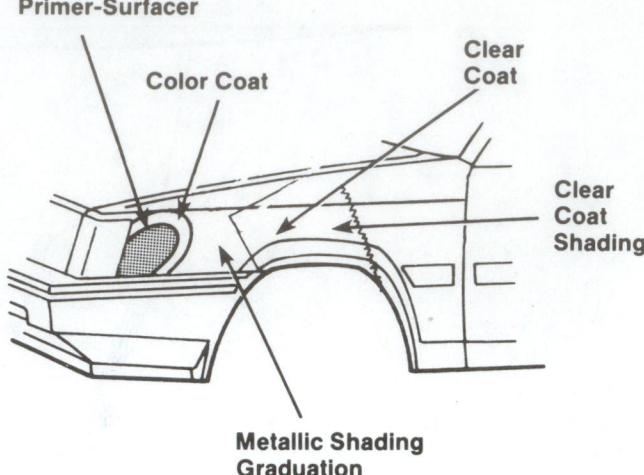

FIGURE 19-39 Here is another method of making a shaded or blended area less noticeable. Only colorcoat the small area that is being repaired. Then, clearcoat most or all of the panel.

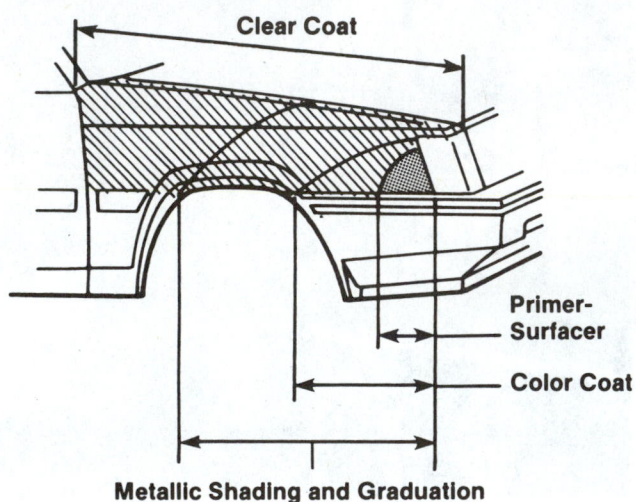

FIGURE 19-40 Panel repainting of metallic colors is best done by colorcoating the smaller area. Then, clearcoat the whole fender area in break lines.

If one panel has damage at both ends or if the whole panel is to be repainted, shading of metallic paint must extend onto the adjacent panels. As shown in Figure 19–41, the clearcoat extends beyond the fender onto one of the adjacent door panels and to the second press line of the hood.

NOTE: The next chapter explains color matching, which relates to the information just given.

OVERALL SPRAYING

For overall spraying, keep a wet edge while maintaining minimal overspray on the horizontal surfaces. This prevents spray from settling onto areas that have already dried, which would cause a gritty surface.

Avoid sags in the overlap line by changing the point of overlapping (Figure 19–42).

Although there is not a single perfect procedure for repainting a car overall, most refinishers will agree that the diagrams in Figure 19–43 illustrate the best patterns. With a conventional booth, by starting with the top of the car and proceeding to the trunk deck lid, side, and so on, the painter can best keep a wet edge while maintaining minimum overspray on the horizontal surfaces. This prevents spray dust from settling onto areas that have already dried, causing a gritty or contaminated surface.

If possible, it is better for two painters to work in a downdraft booth. The spray pattern is different from that of the conventional booth because of the direction of the airflow (top to bottom). Following

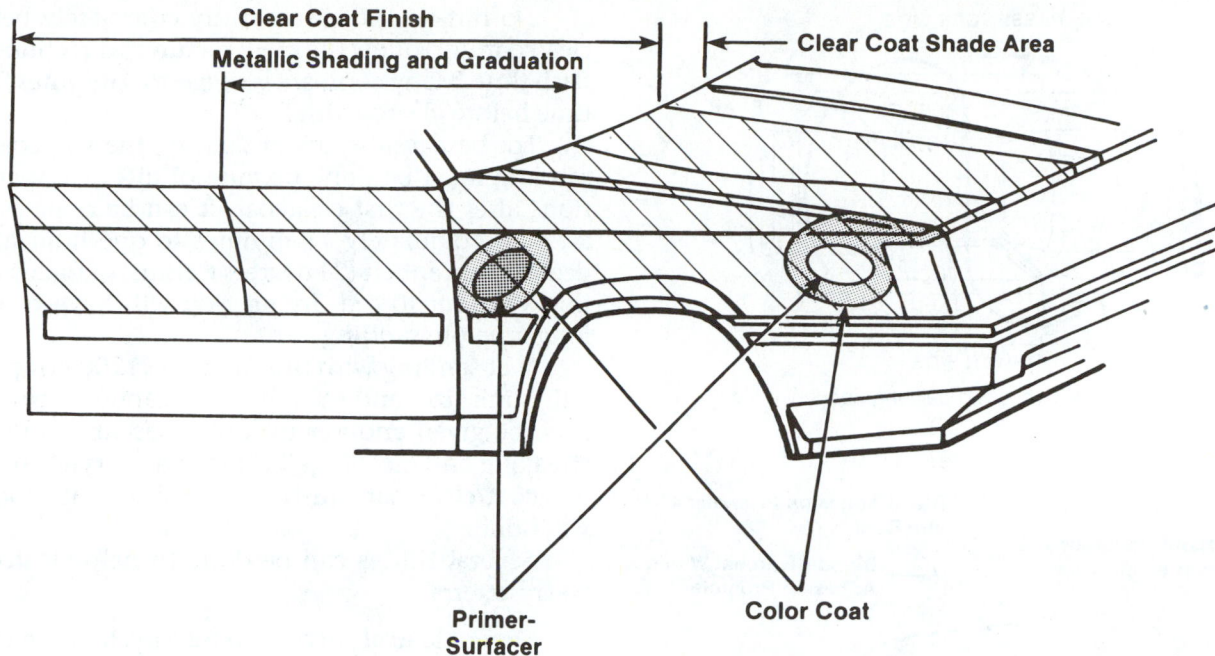

FIGURE 19-41 Panel repairs of metallic colors with damage at more than one area may require blending or shading to other panels to make the repair less noticeable.

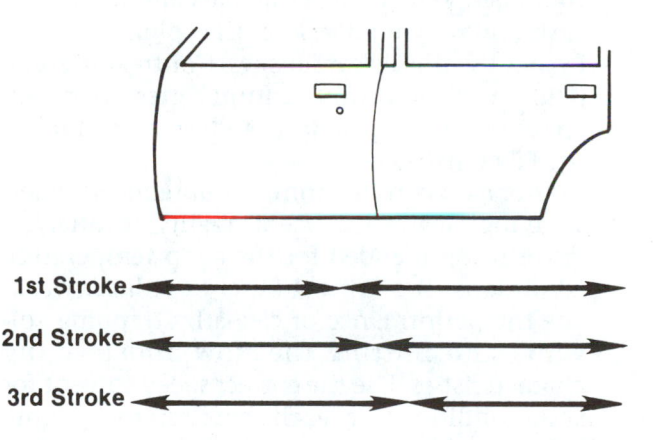

FIGURE 19-42 Changing overlapping point will help hide paint repair.

the pattern shown in Figure 19–44 allows the three main horizontal surfaces to remain as wet as possible while maintaining minimum overspray. These procedures also allow the painter(s) to continue to apply additional coats as needed without a significant loss of time due to flash off between coats.

APPLYING BASECOAT/ CLEARCOATS

Basecoat/clearcoat systems are more difficult to use than single-stage paints. To help match and repair basecoat/clearcoat systems, paint manufacturers offer special clearcoats for use over color basecoats. Since more and more cars and trucks have basecoat/

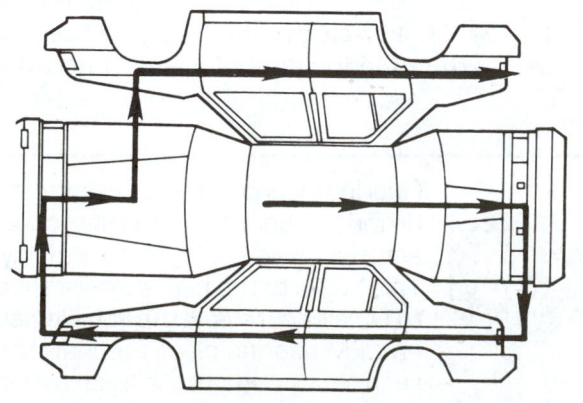

PAINTING ORDER FOR 1 PERSON

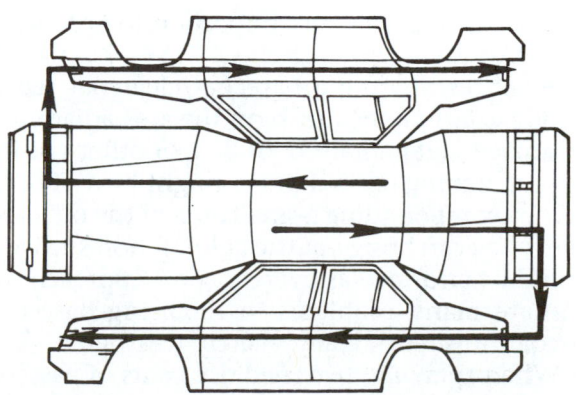

PAINTING ORDER FOR 2 PERSONS

FIGURE 19-43 Study the sequence for overall painting procedures for one and two people.

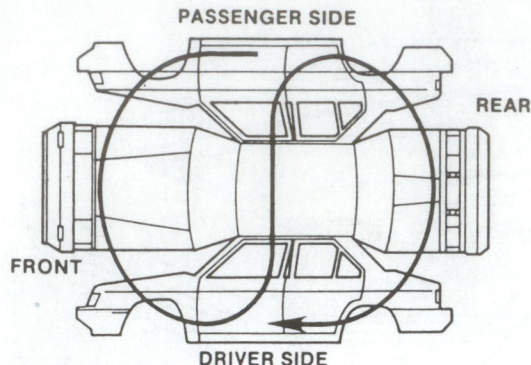

FIGURE 19-44 This is the painting sequence for a downdraft booth.

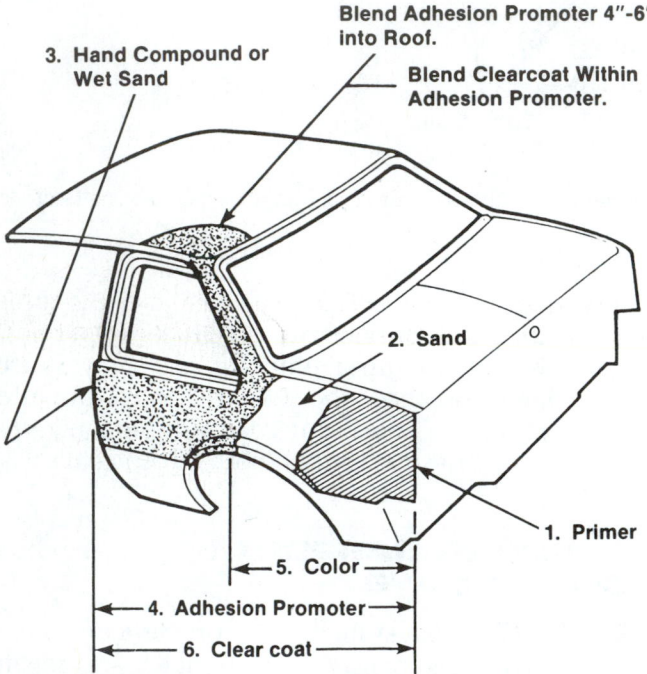

3. Hand Compound or Wet Sand

Blend Adhesion Promoter 4"-6" into Roof.

Blend Clearcoat Within Adhesion Promoter.

2. Sand

1. Primer

5. Color

4. Adhesion Promoter

6. Clear coat

FIGURE 19-45 Study major steps for panel repair using adhesion promoter.

clearcoat finishes, it is very important to become familiar with them.

When estimating a basecoat/clearcoat repair, carefully examine the finish on the area adjacent to the damage. If it is chalked, dulled, or otherwise impaired, matching the old finish might prove impossible. Try compounding a small area of the old paint to see if you can bring out the color. If not, such jobs should be performed as overalls. This approach will eliminate many problems in repairing basecoat/clearcoat finishes that are severely weathered.

When spraying, two medium coats of basecoat should be applied. The basecoat does not need to be glossy, and only enough should be used so as to achieve hiding. Two or three medium wet coats of clear should be applied next, with at least 15 minutes flash time between coats.

Do not let the basecoat dry completely before clearcoating. Follow the manufacturer's directions for flash time. Many companies suggest 30 minutes flash time before clearcoating.

For best results, avoid sanding the basecoat. If sanding must be done because of dirt or imperfections after the first clearcoat, it can be done safely after approximately 15 minutes to one hour at 70 degrees Fahrenheit (21 degrees Celsius) of forced drying. Without forced drying, you will have to wait longer before sanding.

Wet sanding with fine #600 to #1200 grit paper will minimize sand scratches. The sanded area must then be given another mist of basecoat to prevent streaking and mottling. Buffing of an acrylic enamel basecoat/clearcoat finish is needed only if sanding was done.

Several things can be done to help clearcoats wear better:

- Do not load clearcoats on heavily. Because these finishes are clear, refinishers have a tendency to use too much in an attempt to increase the desired glamour effect. As a result, they "bury" the colorcoat. Remember, clearcoats are not always clear and they tend to alter the color.
- Do not use thick clear finishes. Contrary to some opinions, clears do not perform better when they are underreduced. Thin or reduce according to the label instructions.
- Do not use economy thinners or reducers when spraying clearcoats. Use a quality thinner/reducer recommended for the shop temperature conditions. Too fast a thinner or reducer weakens the performance of clears by trapping solvents and hurting the flow and leveling characteristics. Use the correct speed solvent for shop conditions. Let each coat flash thoroughly before applying the next one.
- If necessary, apply an adhesion promoter over the entire panel to be refinished, or at least past where any color or clear will be applied (Figure 19–45). Apply a clearcoat over the entire panel within the adhesion promoter area. If necessary,

SHOP TALK

Fluorine clearcoats are designed to provide superior weathering characteristics and paint film durability. This is due to the higher resistance to ultraviolet rays. If the vehicle has an OEM fluorine clear topcoat, the refinish system must also use a fluorine clearcoat. Read label directions. Some might require slightly different application and blending procedures.

blend the clear as described earlier. Remember to step out the coats of clear.

APPLYING TRI-COAT FINISHES

Although tri-coats require somewhat different refinish procedures and techniques, they are essentially the same as the repairs done on basecoat/clearcoat finishes. Following are some key points to keep in mind when performing a tri-coat repair:

- Follow the recommendations for this type of repair furnished by the paint manufacturer.
- Pay close attention when the instructions call for the use of adhesion promoters, antistatic materials, and so on.
- Make a test or let-down panel, as described in the next chapter. A mismatch in the basecoat, mica coats, or clearcoat can affect the overall finish match.
- Keep the repair area as small as possible.
- Avoid a halo effect by applying the first coat of mica to the basecoat only.
- The more intermediate coats that are applied, the darker the finish will appear.
- Allow a larger area in which to blend the mica intermediate coats. They require more room to blend than a standard basecoat.
- Do not try to substitute another type of paint for the recommended basecoat.
- Always check the basecoat color against the OEM basecoat. To do this, find an uncleared mica-free area of the vehicle. Some car companies leave an exposed portion of basecoat beneath the right and left sill plates that is perfect for this.

FORCE DRYING ENAMEL TOPCOATS

Force drying enamels by means of heat convection ovens or infrared lights will greatly reduce the drying period. However, care must be exercised to avoid wrinkling, blistering, pinholing, or discoloration. It is generally better to force dry at lower temperatures for longer periods than to run high temperatures for shorter periods.

Usually, acrylic enamel will dry in 15 to 20 minutes at 175 degrees Fahrenheit (79 degrees Celsius). While enamel can be force dried up to 200 degrees Fahrenheit (93 degrees Celsius), maximum care should be taken to allow sufficient flash time for solvents to escape before force drying, or blistering is likely to result. Generally speaking, pastel colors are heat sensitive, and extreme caution must be used in force drying them to avoid discoloration.

It is especially necessary to avoid heavy coats of enamel in hot weather or when force drying. The temperature at which the enamel is sprayed, rather

When force drying, be careful to measure the surface temperature of the vehicle, not air temperature. Also, do not bake at too high of a temperature (above 160 degrees Fahrenheit or 71 degrees Celsius) because part damage can occur.

than the temperature at which it is dried, determines which reducer to select.

APPLYING WATERBASE MATERIALS

Waterbase paint, as implied, uses water to carry the pigment. It dries through evaporation of the water. Some manufacturers are starting to use waterbase paints on new vehicles to help satisfy stricter emission regulations in some geographic areas. The basecoat of color is water-based. Then an enamel topcoat is applied over the waterbase paint to protect it from the environment.

Waterbase primers have been used for years as a fix for lifting problems. Waterborne primers serve as an excellent barrier coat when there are paint incompatibility problems.

Waterbased paints come *premixed* (ready to spray) and they are not normally reduced. In an emergency, distilled water can be added to make a thinner, more liquid solution.

When using waterborne materials, the water used for equipment cleaning

1. Contains hazardous materials and should be disposed of as hazardous waste.
2. Cannot be poured down the drain for disposal.
3. Must not be combined with other waste solvents such as reducers or thinners and must be kept in a separate container for storing water wastes.

Due to clean air regulations, some solvents are no longer being used. To meet clean air regulations, traditional solvents are being replaced by water or other solvents. Check local ordinances.

When using a waterbase primer, do not wet sand it since it is water soluble. Dry sand it only. Waterbase primers and sealers are sometimes used as barrier coats. They will help seal a paint problem, such as lifting when trying to repaint over a featheredged area.

REMOVAL OF MASKING MATERIALS

If the finish has been force dried, remove the masking tape while the finish is still warm. If the finish is

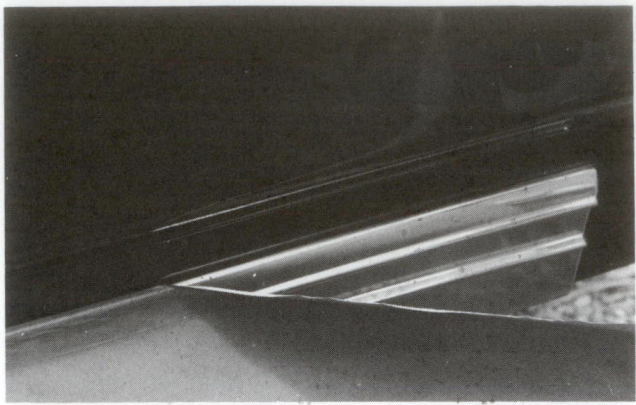

FIGURE 19-46 When removing masking tape, do not touch the paint with fingers, tape, or the paper or plastic material.

allowed to cool, the tape will be difficult to remove and can leave adhesive behind.

Pull the tape slowly so that it comes off evenly (Figure 19–46). Take care not to touch any painted areas because the paint might not be completely dry. Fingerprints or tape marks could result.

If you use liquid masking material, wash it off with soap and water. Do not wash the freshly painted surfaces until they are fully dry.

Never allow lacquer paint to dry thoroughly before removing the tape because the paint could peel off along with the tape. With fast-drying lacquer, it is best to remove the tape right after refinishing.

Never mask a vehicle and let it sit for a prolonged period. Do not let the masking paper and tape get wet. Either will cause problems. The tape edge can roll up and allow paint to spray onto parts not to be painted. Also, the tape can stick and be difficult to remove. You might have to carefully wash off the adhesive.

NOTE: Information relating to painting a vehicle is given in several chapters of this text. Refer to the index for added information if needed.

19.10 SPATTER FINISHES

The interior of luggage compartments of some vehicles—the side walls and floor—may be repainted with an aftermarket spatter finish. The material is

often water-reducible and can be applied in one heavy or two medium coats. Hand paddle mixing is usually sufficient. Do not shake on a paint shaker. When ordering spatter paint from a paint jobber, mention the make of the vehicle and the model year. The application procedure is as follows:

1. After all metal repair work and priming have been completed, clean the luggage compartment surfaces with a solvent.
2. Mask off the compartment area, as required.
3. Read the label directions carefully and follow them to the letter. As a rule, open the spray fan nozzle to give only $3/4$ of the full pattern. The fluid feed should be wide open. Use the lowest air pressure that causes the desired spray pattern:
   - For smaller spatters, increase the air pressure.
   - For larger spatters, reduce the air pressure.
4. Apply the coating. If two coats are needed, allow several minutes of flash time. Allow the surface to dry completely before putting the vehicle back into service.

19.11 TOPCOATS FOR PLASTIC AUTOMOTIVE PARTS

After plastic parts have been repaired as described in Chapter 14 and surface prepped as detailed in Chapter 17, the final color can be applied. Automotive plastics can generally be topcoated using conventional paint systems.

Follow the manufacturer's recommendations to determine if a particular paint system can be used on a specific type of plastic, or if a special plastic primer or flex agent is required.

Table 19–5 lists the more popular automotive plastics and suggested finishing systems.

Most rigid (hard) plastics generally require no primers. The paint will adhere properly to the plastic. Semirigid (flexible) plastics might require the addition of a flex agent to the paint system. The additive is needed because semirigid plastics expand,

Plastic parts are normally painted before they are installed. However, if painting is done on the car, it is important that the surfaces are properly masked off.

TABLE 19-5: FINISHING SYSTEMS FOR POPULAR PLASTICS

KEY		Standard Lacquer System	Flexible Lacquer/ Enamel System	Polypropylene System	Vinyl System	Urethane System
I Interior **E** Exterior **P** Primer **NP** No primer **SP** Special primer/adhesion promoter **NA** None approved ***** Flexible primer and/or additive recommended						
ABS	Acrylonitrile-Butadiene-Styrene	I/NP E/NP				
ABS/PVC	ABS/Vinyl (Soft)		I/NP E/NP		I/NP	
EP I, EP II, or TPO	Ethylene Propylene			E/SP*		
PA	Nylon	E/P				
PC	Lexan	I/NP				
PE	Polyethylene	NA	NA	NA	NA	NA
PP	Polypropylene			I/SP		
PPO	Noryl	I/NP				
PS	Polystyrene	NA	NA	NA	NA	NA
PUR, RIM, or RRIM	Thermoset Polyurethane		E*			E
PVC	Polyvinyl Chloride (Vinyl)		E/NP I/NP		E/NP	E I/NP
SAN	Styrene Acrylonitrile	I/NP				
SMC	Sheet Molded Compound (Polyester)	E/P				
UP	Polyester (Fiberglass)	E/P				
TPUR	Thermoplastic Polyurethane		E*			E
TPR	Thermoplastic Rubber		E*			E

contract, and bend more easily than other substrates. The **paint flex agent** will keep the paint film flexible so it can accommodate the movement of the substrate without cracking.

Some product manufacturers require that different flexible additives be used in the various paint systems. Others offer a *universal flexible additive* that can be used in a variety of paint systems. These products eliminate the need to stock several flexible additives and help keep costs down.

As always, it is best not to mix manufacturers' products. The flexible additive, the topcoat, the undercoat products, and the reducer or thinner used should all be provided by the same manufacturer. Mixing labels or using different manufacturers' products on the same job can result in poor performance.

PAINTING RIGID INTERIOR PLASTIC PARTS

As mentioned, rigid or hard ABS plastic parts generally require no primer or primer-sealer. Interior colors are color-keyed to trim combination numbers located on the body number plate (see Chapter 14).

Each major paint supplier provides an interior color chart that identifies the stock number, color name, gloss factor, and trim combination number for each conventional interior color.

When painting rigid interior surfaces, proceed as follows:

1. Wash the part with a cleaning liquid or solvent.
2. Apply interior acrylic lacquer color or special vinyl paint according to trim combination (see

paint supplier color chart for trim and color code). Apply only enough color for proper hiding to avoid washout of grain effect.

3. Allow to dry, following label directions, and then install the part.

PAINTING RIGID EXTERIOR PLASTIC PARTS

Painting of rigid exterior plastic parts is basically the same as for rigid interior plastic parts (Figure 19–47). While most rigid exterior plastics do not require a primer, some paint manufacturers recommend giving ABS exterior parts a primer coat before the colorcoat. When applying a coat to rigid (hard) plastic parts, proceed as follows:

1. Wash the part thoroughly with a cleaning solvent.
2. Colorcoat the part using the appropriate color of acrylic lacquer, acrylic enamel, urethane, or basecoat/clearcoat systems.
3. Allow the colorcoat to dry, and reinstall the part.

In finishing fiberglass after the primer-sealer has been applied, the color or topcoat is applied following the basic procedures as for body steel.

When refinishing a previously painted sheet molded compound (SMC) with either a blend or full panel paint procedure, it is necessary to apply a coat of an adhesion promoter. This must be applied 6 to 8 inches (152 to 203 mm) beyond the blend area, when performing a spot repair. In the event of refinishing a full panel, the entire part must be coated. A flash time of at least 30 minutes is required before applying the base color. This will ensure adequate adhesion of the topcoat.

Typically, a spot repair can be accomplished in the following manner. The area that will receive the basecoat color should be sanded with #400 grit wet or dry paper. The blend area that will be clearcoated

SHOP TALK The sandpaper grit used will depend upon the type and brand of refinishing (paint) system used. Check the manufacturer's instructions to select the correct grit sandpaper and obtain the best results.

should be sanded with #600 grit or finer wet or dry paper (Figure 19–48). It is important that the adhesion promoter extend beyond the blend area. The application of paint to new parts does not require an adhesion promoter prior to applying the topcoat.

When refinishing rigid plastic parts, the label directions for the product selected should be followed accordingly. Specific reduction ratios supplied by the paint manufacturer should be followed. Only enough film thickness to achieve full hiding is necessary; usually two or three medium wet coats are sufficient. The basecoat should be allowed to dry at least 20 minutes before the clearcoat is applied. The clearcoat can be either lacquer or enamel.

PAINTING INTERIOR/EXTERIOR FLEXIBLE PLASTIC PARTS

Most flexible or semirigid plastics require an additive to the paint to allow the paint to flex without cracking. There are several flexible topcoat systems available for the painter's selection.

Basecoat/clearcoat material can be either enamel- or lacquer-based. The trend is toward enamel systems. Some manufacturers do not recommend the use of flex additives in their base color material when using a clear topcoat, but do recommend its use for their clearcoats. Refer to label directions if in doubt.

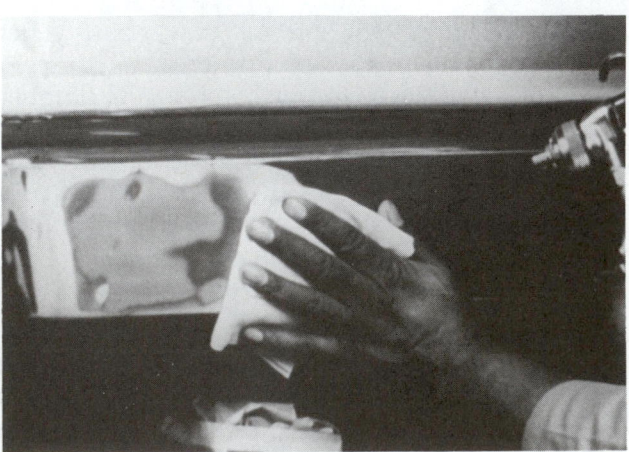

FIGURE 19–47 Use a tack cloth on a rigid exterior plastic part before spraying. *(Courtesy of 3M)*

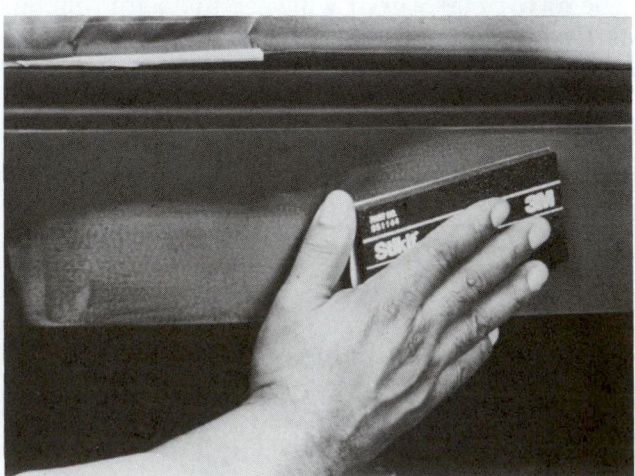

FIGURE 19–48 After spraying, wet sand surface with a #600 grit or finer if needed. *(Courtesy of 3M)*

FIGURE 19-49 When refinishing plastic parts, mix the basecoat according to the manufacturer's instructions.

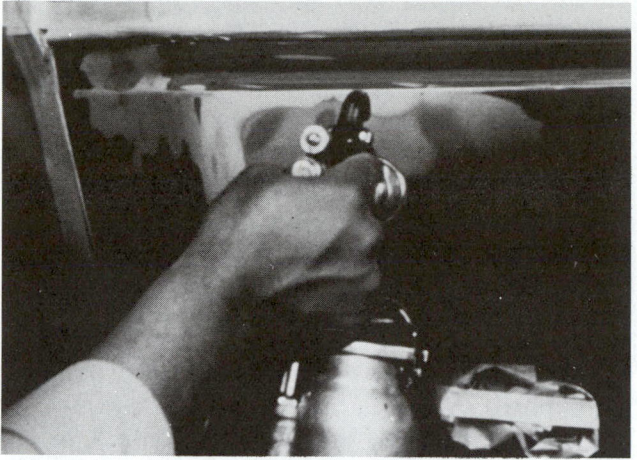

FIGURE 19-50 Apply enough colorcoats to achieve the proper color match on plastic parts.

To apply a flexible (elastomeric) finish, proceed as follows:

1. Thoroughly sand the entire part with the recommended grit abrasive paper (typcially #400 grit). Clean the surface with a cleaning solvent.
2. Following the manufacturer's instructions, mix the base color, the flex additive, and the recommended solvent (Figure 19–49). Mix the base color and flex additive thoroughly before adding the amount of solvent best suited for the shop temperature.

   Remember to mix only the amount of flex agent that is going to be used, since the reduced material cannot be stored.
3. Using the recommended air pressure at the gun, apply a sufficient number of wet double coats to achieve complete hiding and the proper color match (Figure 19–50). Wet double coats are applied as follows: Spray the first pass left to right. Spray the second pass right to left, directly over the first pass. Drop the nozzle so that 50 percent of the pattern overlaps the bottom half of the initial double coat. Continue the pattern until complete. Be sure to allow flash time between coats.

4. Allow the basecoat adequate drying time before applying the clearcoat (typically 30 to 60 minutes). Do not sand the basecoat before applying the clearcoat. When not applying a topcoat, air dry for approximately 4 hours before installing or putting the part into service.
5. If sanding of the basecoat is necessary to remove imperfections, such as dirt or sags, sand with the correct grit sandpaper (about #400 grit or finer) and reclean the area(s). Apply one additional coat of base material and let dry.

Apply the clearcoat, if desired, in the following manner:

1. Mix and reduce the material (paint, solvent, and hardener, if needed) as per the label instructions. Use flex additive when spraying flexible parts.
2. Strain the mixture into the gun. Check that you have the recommended air pressure at the gun. Apply 2 to 3 coats of paint.
3. Allow each coat to flash completely before applying the next one. Refer to label directions for flash times at specific booth temperatures (an average flash time for catalyzed enamel is 20 to 30 minutes).
4. Generally, allow at least 4 hours air dry time or force dry for 30 minutes with a heat light at 180 degrees Fahrenheit before putting the part back into service.

Flexible replacement panels are factory primed with a flexible enamel-based primer. The only preparation required prior to topcoating is cleaning with solvent, sanding with the proper grit paper (approximately #400 grit), and a second cleaning after the sanding is completed. In the event the OEM primer is scratched and has left the plastic substrate exposed, or the part has been repaired with a flexible filler material, it is necessary to cover the exposed area

SHOP TALK

Compounding is not necessary when a flexible additive is used in the topcoat; the mixture will dry with acceptable gloss. Compounding dulls the gloss of elastomeric finishes, causing a flat appearance. In this case, the finish cannot be brought back to the same gloss level without applying more paint. Make sure you produce a good finish when spraying flexible finishes.

SHOP TALK Spot repairs on OEM-finished flexible panels and parts are not recommended because of the failure of flexible color to flow or "wet out" properly at the blend area.

with a flexible primer-surfacer prior to topcoating. If the exposed surface is not primed, the area will be highlighted after the topcoats are applied. A fast-evaporating solvent should be used to reduce the primer-surfacer and prevent swelling of the base material by absorption.

Keep flex paint material off regular vehicle finishes. If applied to them, there could be a problem color matching the gloss differences. If retopcoated with lacquer, the finish could lift or wrinkle, requiring removal of the affected area. Keep conventional acrylic lacquers and enamels off all flexible exterior parts. If these finishes are applied to flexible parts, the finish will crack as the parts are flexed and will spoil the appearance of the car.

PAINTING INTERIOR/EXTERIOR POLYPROPYLENE PARTS

The system for painting polypropylene (PP) parts involves the use of a special primer. Since this plastic is hard, it can be colorcoated after priming with conventional interior acrylic lacquer.

The most common exterior use of polypropylene plastic parts is for bumpers, which come in two types:

- One with a tinted base material (black, gray, or dark gray)
- One that is partially painted, also called a colored bumper (Figure 19–51)

The quality of paint used for PP bumpers is different from that for metal surfaces; adhesive and

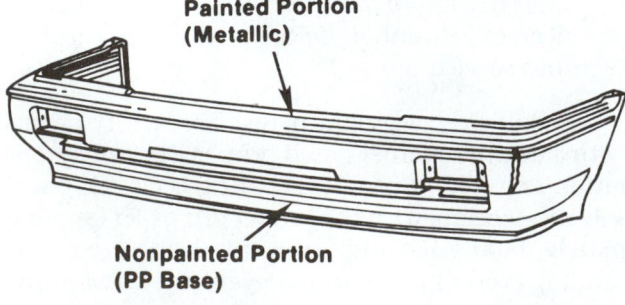

Painted Portion
(Metallic)

Nonpainted Portion
(PP Base)

FIGURE 19-51 Refer to service manual when needed to refinish colored bumper.

softening agents are required. Therefore, a special PP primer must be used for the undercoat and a flexing agent added to the topcoat. If not, peeling will result.

If a PP bumper has major structural damage, it must be replaced. Replacement bumpers of this type are usually primed and ready to be painted. If they are not primed, a special PP primer must be applied over the entire bumper. Before starting the painting, be sure to wash the surface with solvent. When applying only a regular colorcoat, proceed as follows:

1. Apply properly thinned/reduced, proportioned, and mixed PP primer and flexible additive as directed by the manufacturer. Allow recommended drying time (1 to 2 hours typically) before applying any colorcoats.
2. Apply proportioned and mixed refinish material and hardener additive. Flexible additive should not be used in the topcoat.
3. Allow 8 hours (overnight if possible) drying time to assure paint hardness.

If a basecoat/clearcoat is being used, read the container labels and proceed as follows:

1. Apply properly thinned/reduced and agitated PP primer. Allow recommended drying time ($\frac{1}{2}$ to 1 hour typically) before the application of colorcoat. Flexible additive should not be used with basecoat/clearcoat finishes.
2. Allow specified drying time before applying the clearcoat.
3. Apply properly thinned/reduced, proportioned, and mixed clearcoat. Add hardener if an enamel clearcoat. Allow 8 hours to overnight drying time to assure finish coat hardness.

Minor surface scratches can usually be repaired by following the same procedures used for finishing replacement PP bumpers, with the following changes:

1. If the scratches do not penetrate the substrate, follow the entire procedure but do not apply primer.
2. If scratches penetrate the substrate, use a recommended plastic body filler and prime the repair area only.

A summary of repainting procedures of a PP bumper can be found in Figure 19–52 and Table 19–6.

Repainting of Urethane Bumpers

Urethane bumpers include the colored type that have been painted and the tinted black type. Although both are made of urethane, the black type has been made with an additive that helps prevent deterioration due to sunlight and rain. If painted, a black bumper would change color due to the additive. Light

TABLE 19-6: SUMMARY OF THE REPAINTING STAGES OF A PP BUMPER

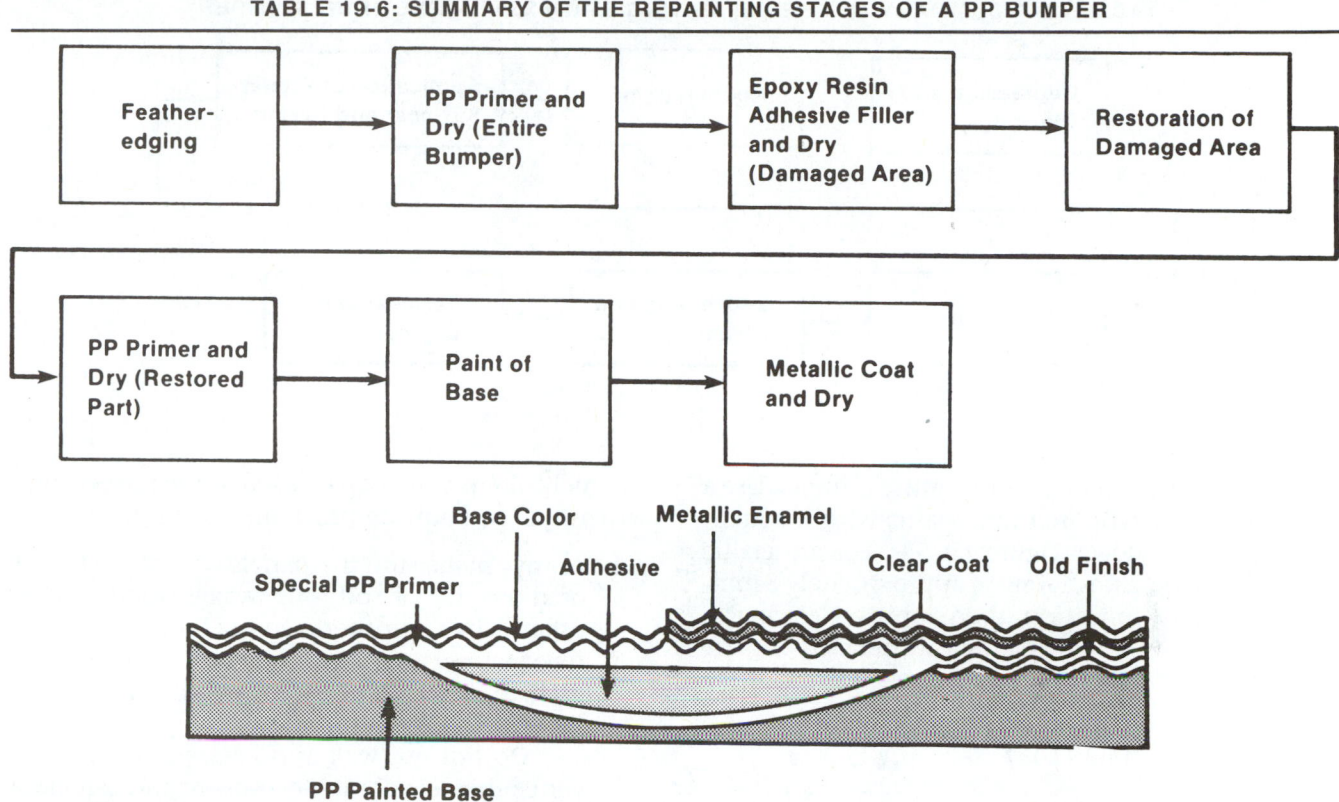

FIGURE 19-52 Note materials used when repainting a damaged PP bumper. *(Courtesy of Toyota Motor Corp.)*

FIGURE 19-53 If needed, apply a coat of special plastic primer to the surface to be finished. *(Courtesy of Urethane Supply Co.)*

FIGURE 19-54 After the topcoat has been applied, the bumper should match the body color. *(Courtesy of Urethane Supply Co.)*

colors such as white would cause a noticeable change. Therefore, black bumpers cannot be painted.

Described here is the procedure for painting a colored urethane bumper:

1. Mask off the area to be repainted and clean with a silicone removing solvent. Keep in mind that insufficient cleaning will result in peeling or blistering.

2. Apply a coat of primer-surfacer over the entire surface (Figure 19–53). Repair any scratches with a brush.

3. It is extremely difficult to match the paint for spot repainting, so the entire bumper should be repainted. Prepare the entire surface by wet sanding with a #600 grit abrasive paper.

4. Clean the topcoat surface again.

TABLE 19-7: SUMMARY OF THE REPAINTING STAGES OF A URETHANE BUMPER

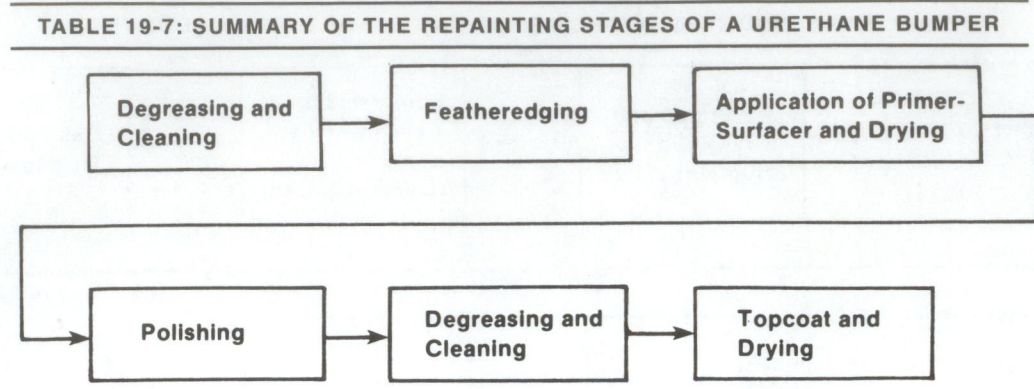

5. Apply the topcoat over the entire bumper. Use a two-part acrylic urethane paint with a softening agent added (Figure 19–54). For a metallic color, allow a flash time of approximately 5 minutes after application, then apply a clearcoat.
6. Follow the dry time recommended by the manufacturer.

A summary of repainting procedures of urethane can be found in Table 19–7.

PAINTING INTERIOR VINYL AND SOFT ABS PLASTIC PARTS

The outer cover material of flexible instrument panel cover assemblies is made mostly of ABS plastic modified with PVC or vinyl. The same is true of many padded door trim assemblies. The soft cushion padding under ABS covers is urethane foam plastic.

The most widely used flexible vinyls (polyvinyl chloride) are coated fabrics as used in seat trim, some door trim assemblies, headlinings, and sun visors. Examples of hard vinyls are door and front seat back assist handles, coat hooks, and exterior molding inserts.

The paint system for vinyl as well as for interior ABS plastic involves the use of vinyl lacquer or an enamel. Originally, this heavy-bodied finish was used over painted steel tops to simulate vinyl fabric tops. By changing reductions and air pressures, the vinyl color will dry to a leatherlike texture similar in appearance to a fabric textured vinyl top. This product is frequently used to restore faded vinyl tops.

Vinyl color has been used as a flat black topcoat to produce accent stripes and nonglare hood trim. Vinyl paint is also suggested as a base coat for duplicating the OEM chip-resistant coating on rocker panels. Once dry, most vinyl colors can be recoated with acrylic lacquer or acrylic enamel to match the car color. Vinyl system finishes are usually available in a wide array of colors.

No primer or other undercoat is required. Also, no thinning is necessary since vinyl lacquer or

enamel color is usually packaged at the proper spray viscosity. The painting procedure is as follows:

1. Always make sure the panels or parts to be colored are free of soil, oils, waxes, food, and all other debris. Synthetic enamel reducer or a vinyl cleaning and preparation solvent should be used to clean vinyl. Isopropyl alcohol will remove ballpoint pen ink.

   Do not use wax and grease removing solvents; they evaporate too slowly and can cause poor adhesion and cracking of coatings. If an extremely soiled condition exists, detergent and water can be used for a first washing before the solvents are used. Be sure all moisture has completely evaporated before any coatings are applied. Infrared or quartz radiation is the most effective method of evaporation.

2. As soon as the surface dries, apply interior vinyl color in wet coats. Allow flash time between coats according to label directions. Use the proper vinyl color shown by the interior trim code combination. Apply only enough color for proper hiding to avoid washout of the grain effect. Use the air pressure recommended by the finish manufacturer.

3. Before color flashes completely, apply one wet double coat of vinyl clear topcoat. Use a topcoat with the appropriate gloss level to match adjacent similar components. The clearcoat is necessary to control the gloss requirement and to prevent crocking (rubbing off) of the colorcoat after drying. Remember that instrument panel covers require a nonglare final topcoat.

4. Allow to dry according to label directions before installing the part or putting the vehicle back in service.

Leather interior parts can be refinished in much the same manner as vinyl plastic. It must be remembered that vinyl is not dyed but colored with pigment and coatings. The same is true of leathers used for upholstering vehicles. Leathers are coated in Europe

SHOP TALK

American-made cars with leather seats use a vinyl impregnated leather. These can be coated with vinyl color. Do not use on leather generally without testing for scratch-off on a test piece after 24 hours curing time.

with nitrocellulose lacquers and urethanes. In the United States, leathers are coated with acrylics and urethanes. Vinyl colors are usually used to repaint leather.

When painting leather, some interesting applications can be achieved. For example, leather can be given a dual tone accent. This is accomplished by using a basecoat to cover the panel and supply the primary color; then a darker color is applied over the base color in a shadowy manner. Some interior colors have metallic flakes to add sparkle, while others have a pearlescent pigment.

PAINTING EXTERIOR VINYL ROOFS

To paint exterior vinyl roofs, proceed as follows:

1. Wash the old top with a bleach-type detergent, a brush, and plenty of water. Rinse the top and entire car thoroughly with clean water.
2. Clean the top thoroughly with paint finish cleaning solvent or with vinyl prep conditioner.
3. Blow out all gap spacing and crevices around the top and tack-wipe the top as required.
4. Be sure masking is carefully done and cover the entire hood and deck lid. The adhesion property of the vinyl system will make overspray difficult to remove.
5. Using the manufacturer's recommended paint system, start the color application with a banding coat at low air pressure and a narrow fan. Spray into drip rails and cracks where the windshield and back window molding meet the roof.
6. Increase the air pressure as directed by the manufacturer and open the fan to a normal spray pattern. Apply vinyl color, working toward the center.
7. On the opposite side of the car, start at the center and maintain wet application of the near side. Keep the application wet with a full and uniform 50 to 75 percent stroke overlap. Keep the spray gun as perpendicular to the surface as possible. Control the hose by positioning it over the shoulders and back.

8. Apply a second full wet coat for complete hiding and uniformity of wetness. Adjust gun distance and speed for the desired texture. If streaks or dry spots are present, apply a good thinner/reducer through the spray gun to wet out the dry spots and even out the spray pattern.
9. For the final coat, many manufacturers recommend an application of one wet coat of highly thinned vinyl paint over the entire vinyl roof area to obtain a uniform appearance. Refer to the paint label for directions.
10. After 1 hour of drying, remove the masking. Allow to dry a minimum of 4 hours before putting the car into service.

Vinyl roof repairs are described in Chapter 14 and should be completed before applying a new finish.

Vinyl Preserver

A clear protective dressing is available for use on vinyl roof tops, upholstery, and other areas covered with vinyl color—floor mats, tires, wires, hoses, and batteries. Its water- and dirt-repellent film withstands sun, salt, and snow. It is also ideal for spray applications to preclean the engine compartment.

When using over vinyls, apply a thin, even coat of preserver with a cellulose sponge or clean soft cloth (Figure 19–55). It dries to the touch in 10 to 20 minutes, is water repellent in 1 hour, and detergent resistant in 1 day. Do not thin the preserver; use it at can consistency.

RETEXTURING INTERIOR PLASTIC PARTS

Many different textures or grains are found in the average automobile interior. When retexturing a repaired

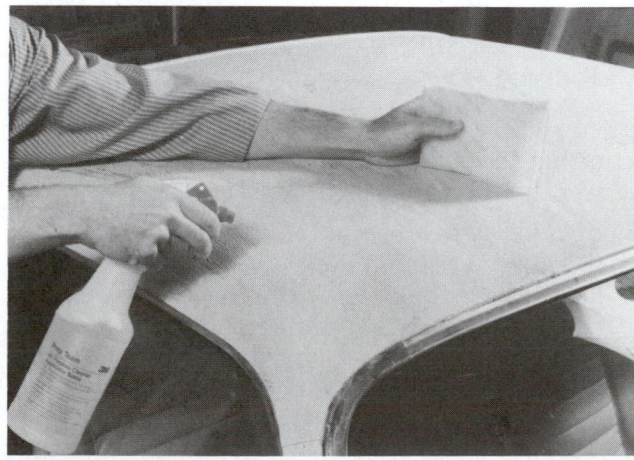

FIGURE 19-55 Apply vinyl gloss and preserver to protect material and to bring out gloss.

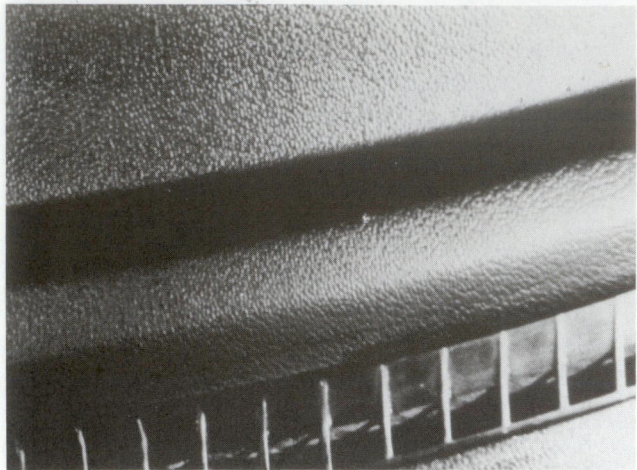

FIGURE 19-56 This is a retextured plastic surface. Repairs like this are only feasible when hidden from plain view, as they are difficult to match.

FIGURE 19-57 An aerosol sprayer can be used to apply a new finish to a repaired area.

part, it is important to keep in mind that the existing texture does not have to be duplicated. There is no need to spend time and effort trying to get the retextured area to look exactly like the rest of the piece; a variation in the grain is meaningless. Only the coarseness of the grain must be duplicated in order to achieve professional results (Figure 19–56).

Retexturing can be done one of two ways:

- By blending the new texture out into the old
- By retexturing to a natural break line on the panel

Use a refillable aerosol sprayer; the lower pressure will prevent the material from atomizing, which means a faster texture buildup. To achieve a coarse texture, use the material unthinned. For a finer texture, use a small amount of thinner or reducer. A typical retexturing procedure is as follows:

1. Mix the texture material as per the manufacturer's instructions. Direct the first coat toward the repaired area only. Hold the sprayer 18 to 24 inches (457 to 609 mm) from the surface (Figure 19–57), and always dry spray the material. Spraying it on wet will destroy the grain effect.
2. Allow flash time between coats. As many as eight to ten light coats may be needed to achieve the required buildup. Remember that this is by no means a one-shot application.
3. When buildup has been achieved, begin blending the texture out and away from the repaired area, similar to a colorcoat. Force drying between coats speeds up the process.
4. When the texture material has dried, nib sand with #220 or finer grit paper. This will blend the newly textured area into the original texture of the panel.
5. If not satisfied with the texture, apply more light coats of material and repeat the sanding.

After retexturing, the part should be blown dust free in preparation for the refinishing. Do not use any type of cleaner on a newly retextured area.

SUMMARY

There are several rules you should remember when refinishing a vehicle. Ask yourself these questions and answer them before painting:

- Are the vehicle surfaces straight and ready for painting? A common mistake is to overlook a surface problem before painting. Since the surface is usually sanded dull, imperfections are easy to overlook. Double-check metal straightening work, plastic filler, and primer before proceeding. Overlooking even a small paint chip can ruin the job.
- Are all surfaces perfectly clean and scuffed? Paint will not adhere to a dirty or glossy smooth surface. If the paint peels, you will have hours of rework. Blow off and tack rag the vehicle before painting.
- Are you painting in an ideal environment? Are the spray booth and the vehicle the right temperature? Are the booth filters and blower working properly? Do not try to paint a vehicle in an open shop. Dirt will almost always settle in the paint. Paint only in a spray booth!
- Are the paint type and mixing correct? Use the right kind of paint and mix it following label directions. Are all additives mixed into the paint?

If you forget to add hardener, you will not be able to wet sand or buff the paint for weeks.

- Is the paint spray gun working properly? Test and adjust the spray pattern on a sheet of paper or old part. If the gun is spitting or not working properly, you do not want to try to paint the vehicle. This would result in hours of rework.

- Does the spray gun cup leak? A leaking cup can drip and ruin the paint job. You must tilt the gun down when spraying the roof, hood, and trunk lid. If the cup is leaking, paint might drip out and onto these surfaces.

- Are you allowing enough flash time between coats? Flash time is the time needed for a fresh coat of paint to partially dry. Flash time is needed to prevent the paint from sagging or running. Always refer to the material manufacturer's guidelines.

- Are you applying the paint properly? Hold the gun the right distance from the surface, typically 8 to 10 inches (203 to 254 mm). Aim the gun directly at the surface while moving it at the correct speed. Do not fan the gun unless you are blending the paint.

- Is the paint going on the surface properly? Closely watch the paint as it deposits on the surface being painted. Check for application problems, such as dry coat, excessive wet coat, and improper spray overlap. Constant inspection of the wet paint as it hits the surface is critical to doing good paint work. Is the lighting good enough to see the paint go onto all surfaces?

- Are you applying the correct film thickness? A common mistake is to apply too much paint. Today, thin is in. You only want to apply enough paint to provide good color coverage and gloss. Film thicknesses are much less than in the past. Use the number of coats recommended by the paint manufacturer.

- Could you accidentally touch the paint? Keep air hoses and yourself a safe distance from the wet paint. It is easy to brush up against the wet paint and damage it.

- Use proper methods and materials when refinishing plastics. They must be treated differently than metal.

ASE-STYLE REVIEW QUESTIONS

1. Which of the following statements is incorrect?
   A. The objective of blending is to create the illusion that only one color is beeing seen.
   B. It is not practical to try to get a perfect color match when blending.
   C. When preparing the rest of the car, always prepare the adjacent panels for blending.
   D. None of the above.

2. When spot repainting with solid colors, Technician A applies the topcoat with a circular motion from the center outward. Technician B treats the blend area with #1000 grit sandpaper before topcoating. Who is correct?
   A. Technician A
   B. Technician B
   C. Both A and B
   D. Neither A nor B

3. Which of the following statements concerning basecoat/clearcoat applications is incorrect?
   A. Sanding of the basecoat should be avoided.
   B. Do not thin or reduce clearcoats before applying them.
   C. Two medium coats of basecoat are needed.
   D. All of the above.

4. When applying a flexible finish, _____.
   A. Thoroughly sand the entire part with #400 grit adhesive paper
   B. Mix the base color and solvent thoroughly before adding the flex additive
   C. Store any unused elastomeric material in a covered container
   D. Both A and B

5. When painting a polypropylene part, _____.
   A. A flexing agent must be added to the topcoat
   B. A special PP primer is required for the undercoat
   C. Both A and B
   D. Neither A nor B

6. When spraying a metallic or pearl luster paint, Technician A uses an agitator cup. Technician B tested the color match with the aid of a spray-out card. Who is correct?
   A. Technician A
   B. Technician B
   C. Both A and B
   D. Neither A or B

ESSAY QUESTIONS

1. Summarize the four very important functions of a topcoat.

2. Describe some reasons for overall refinishing.

3. Summarize some characteristics of sealers.

4. What are some methods for finding out the type of paint on a vehicle?

5. Summarize some of the more important paint label and literature data that should be checked.

6. In your own words, how do you use paint mixing sticks?

CRITICAL THINKING PROBLEMS

1. If you mix the paint too thin or too thick, what can happen?

2. If the first coat you spray on the vehicle is a very wet coat, what can happen after the next two coats?

MATH PROBLEMS

1. If a spec temperature is given as 75 degrees Fahrenheit, what is its equivalent in Celsius? (Use formula: degrees Fahrenheit minus 32 times .556 equals degrees Celsius.)

2. If directions say to reduce a paint 33 percent and you need one gallon of paint, what parts of each do you need?

CHAPTER 20

Color Matching and Custom Painting

INTRODUCTION

Color matching involves the steps needed to make the new finish match the existing finish on the vehicle. Even if you use the body color code numbers and correct paint formula, the new paint may not be the exact same color. With today's multistage finishes and factory robotic painting, it can be very difficult to match colors when making spot and panel repairs.

Multistage paints like metallic pearl paints, for example, are very difficult to match. Even if you have a new vehicle with a fresh OEM finish, it can be a challenge to make your repair look like the paint already on the vehicle. Today, there are many variables that affect the color and appearance of your paint work (Figure 20–1).

FIGURE 20–1 Finish on today's car is much more difficult to match than in the past. Multistage metallic and pearl paints can be a challenge to match. *(Courtesy of Chrysler Corporation)*

OBJECTIVES

After studying this chapter, you should be able to:

✔ Describe color theory and how it relates to refinishing.

✔ Define the terms relating to color.

✔ Describe the use of a computerized color-matching system.

✔ Make let-down and spray-out test panels.

✔ Explain how to tint solid and metallic colors.

✔ Summarize the repair procedures for multi-stage finishes.

✔ Describe basic methods of doing custom paint work.

✔ Answer ASE test questions about color matching.

KEY TERMS

basecoat patch
chroma
color
color blindness
color directory
color matching
color spectrum
color tree
computerized paint matching systems
custom painting
custom-mixed colors
draw-down bars
factory-packaged colors
flip-flop
fluorescent light
halo effect
head-on view
hue
incandescent light
intermix system
let-down panels
light temperature
lumen rating
metamerism
plotting color
side-tone view
spectrophotometer
spray-out panels
sunlight
tinting
value
variance chips
white light
zone concept

ASE TASK LIST

Job Skills covered in this chapter include:

PAINTING AND REFINISHING TEST (B2) TASK LIST

A. Surface Preparation

17. Prepare the repaired and adjacent areas for blending.

C. Paint Mixing, Matching, and Applying

1. Determine type and color of paint already on vehicle.
3. Use appropriate spray technique (gun arc, gun angle, gun distance, gun speed, and spray pattern overlap) for finish being applied.
4. Apply selected product on test panel or let-down panel in accordance with manufacturer's recommendations.

7. Check for color matching of all applied finishes.
12. Apply multistage (mica, pearl, etc.) coats for spot repair, panel blending, and overall refinishing.
13. Identify paint color formula and proper usage of mixing equipment and materials.

D. Solving Paint Application Problems

14. Identify that color is off-shade or does not match; determine the cause(s), and correct the condition.

This chapter will help you develop the skills needed to match any type of paint. It will summarize color theory, color evaluation, color matching, computer analysis of paint, tinting, and other factors.

20.1 COLOR THEORY

Color is caused by how objects reflect light at different frequencies or wavelengths into our eyes. The color seen depends on the kind and amount of light waves the surface reflects. When these light waves strike the retina in your eye, they are converted into electrical impulses that the brain sees as color.

When the eye sees a colored object, that object is absorbing all of the light except for the color that it appears to be. A red ball appears red because the ball absorbs all of the colors in the light shining on it except for the reds. In contrast, a black object absorbs almost all light, while chrome absorbs none.

White light is actually a mixture of various colors of light. By passing light through a prism, light is broken down into its separate colors, called the **color spectrum**. This is the same principle of how rain breaks up light to form a rainbow.

The colors in the spectrum are easily remembered with the phrase "Roy G. Biv," which stands for the colors:

Red
Orange
Yellow
Green
Blue
Indigo
Violet

LIGHTING

Sunlight contains the entire visible spectrum of light. It is the standard by which other light sources are measured. Since the vehicle will be seen in sunlight, you should always use sunlight, or daylight-corrected lighting, when making color evaluations.

Each light source has a different mixture of colored light (Figure 20–2). When light sources are plotted on a graph, a difference can be seen in the amount of colored light each light source contains. The same color of paint will look very different under different kinds of light. That is why it is so important to check the color match in daylight or under a balanced artificial light.

Compared to daylight, **incandescent light** has more yellows, oranges, and reds. **Fluorescent light** has more violets and reds. They should not be used when analyzing a color.

The index for measuring how close a lamp in indoor lighting is to actual daylight is the *Color Rendering Index,* abbreviated *CRI.* A CRI of 100 duplicates daylight. A range of 85 to 100 is preferred for spray booth lighting.

Lamps may also have a **lumen rating** for brightness. Lamps are normally between 1,000 to 2,000 lumens, with a higher lumen rating producing a brighter light.

Lamps may also be rated for **light temperature** in "Kelvin." Daylight is 6,200 Kelvin. For painting, a lamp rating of 6,000 to 7,000 Kelvin is recommended.

What the eye sees as color is really light reflected from an object. The eye may see different shades of a color depending on the type of light source used.

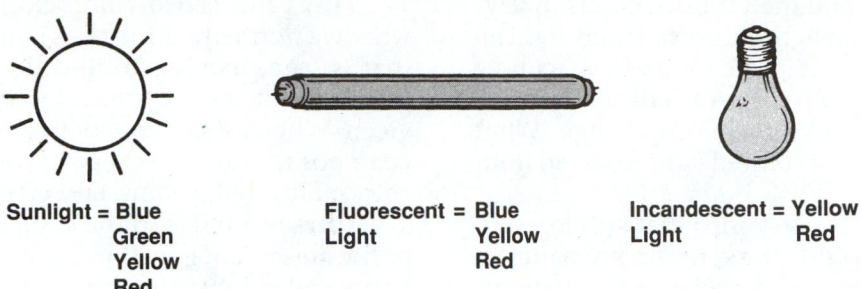

Sunlight = Blue
           Green
           Yellow
           Red

Fluorescent = Blue
Light   **Yellow**
                 Red

Incandescent = Yellow
Light   **Red**

FIGURE 20-2 When evaluating paint color, use sunlight or color-corrected lighting. Colors will look different under fluorescent or incandescent lighting.

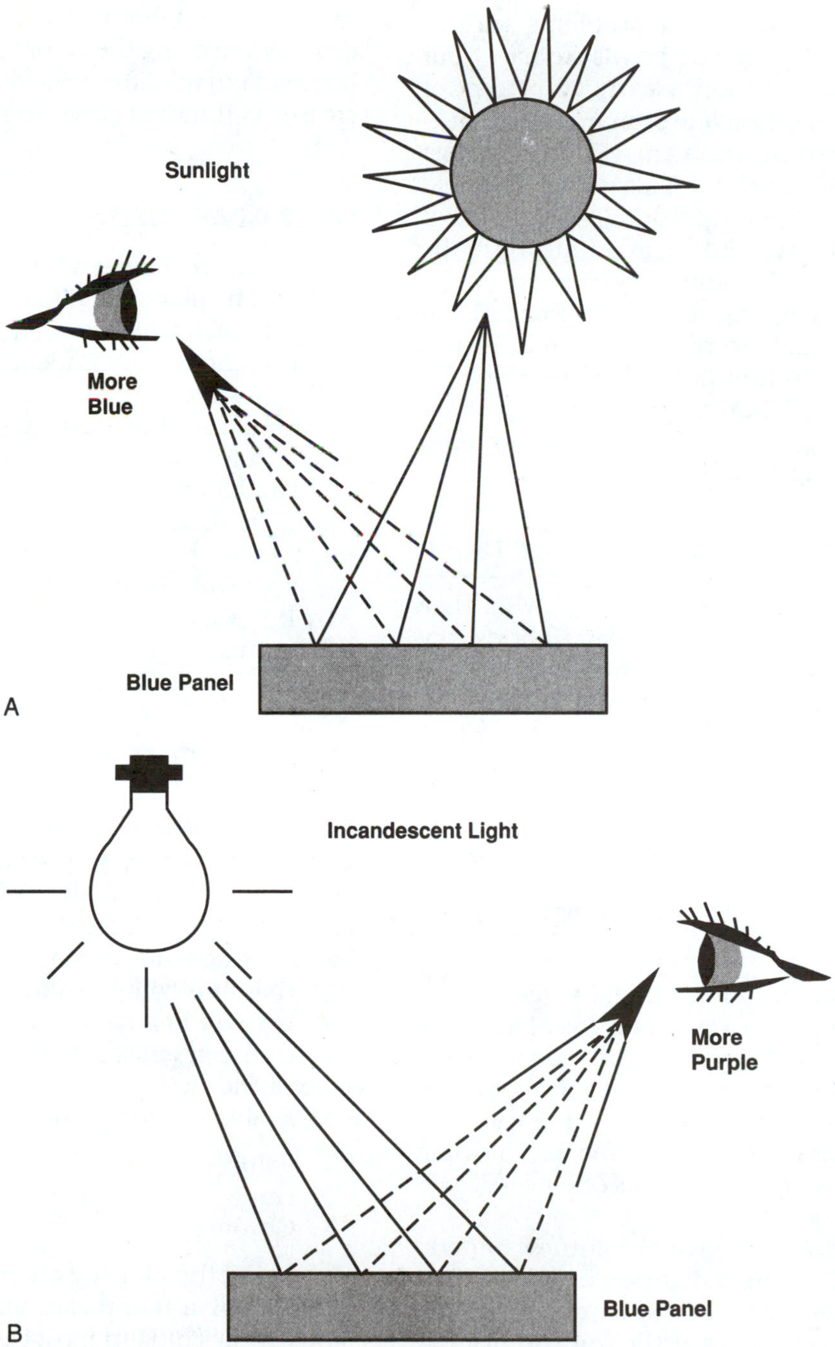

Sunlight

**More
Blue**

Blue Panel

A

Incandescent Light

**More
Purple**

Blue Panel

B

FIGURE 20-3 Sunlight is the best lighting to use when evaluating color. *(Courtesy of I-CAR)*

Always evaluate and match paint colors in daylight or while using daylight-corrected lighting. The characteristics of the light in the shop will affect how you perceive the color of paint on a vehicle. The color of the paint on the panel does not change. What changes is the amount of colored light reflected from the panel.

For this reason, choose lamps that are closest to simulating actual sunlight. The spray booth manufacturer or representative may have recommendations.

COLOR BLINDNESS

Color blindness makes it difficult for a person's eye to see colors accurately. If problems arise in matching certain colors, it may be wise to have your eyes checked for color vision. Nearly 10 percent of all men have trouble seeing one or more colors. Blue-greens are most often the problem. However, almost no women have this difficulty. This is one of the reasons women do the touch-up work in auto plants. If you have a color vision problem, ask someone in the shop to help you match colors.

To do finish matching, the painter or refinish technician must be able to recognize colors as they actually are. It is important not only to see the color that is to be worked on, but also the overtones within that color, including the shades of darkness or lightness and the richness or fullness of the color.

20.2 DIMENSIONS OF COLOR

Many people describe color in terms of what they see. You might have heard these descriptions: sky blue, ruby red, grass green, or midnight blue. These terms cause confusion when describing colors.

To minimize the confusion when painting, color should be based upon three *dimensions of color,* which are the following:

1. Value (lightness or darkness), (Figure 20–3).
2. Hue (color, cast, or tint).
3. Chroma (saturation, richness, intensity, muddiness).

These three dimensions are used to organize colors into a logical sequence on a color tree. The **color tree** is used to locate colors three-dimensionally when matching colors. Colors move around the color tree in a specific sequence—from blue to red to yellow to green. This sequence is easier to remember if you think of "BRYG." These are the first letters of blue, red, yellow, and green.

Value refers to the degree of lightness or darkness of the color. It is one dimension of color. When using the color tree, the value scale runs vertically through the tree. It is white at the top and black at the bottom. It is neutral gray at the center.

Hue, also called color, cast, or tint, describes what we normally think of as color. Hue is the color that is seen, moving around the outer edge of the color tree. It moves from blue to red to yellow to green. When using the color tree, the hue scale shows color position around the color tree. It uses four main colors: blue, red, yellow, and green.

Chroma refers to the color's level of intensity, or the amount of gray (black and white) in a color. It is also called saturation, richness, intensity, or muddiness. It moves along the spokes that radiate outward from the central gray axis of the color tree. Weak, washed-out colors with the least chroma are at the core of the color tree. Highly chromatic colors that are rich, vibrant, and intense are at the outer edge. When using the color tree, chroma increases as it moves outward from the neutral gray center. It decreases as it moves closer to the neutral gray center.

METAMERISM

Metamerism happens when different light sources affect paint pigments differently. A paint may have some blue in it which is not noticeable in daylight, but which becomes very evident under street lights. The problem is that the refinish and OEM paint formulas are not made of the same pigments. This causes the pigments to look different under different light sources.

Paint manufacturers formulate refinish paints to minimize the effect of metamerism. Metamerism is most often a problem when a painter leaves or varies the paint formula during tinting operations. That is why it is important to use only the tints called for in the formula.

20.3 PLOTTING COLOR

Plotting color is the process of identifying paint color in a graphic way based on value, hue, and chroma. This knowledge will help you become familiar with the refinishing evaluation process.

Plotting is not an exact science. It only has meaning when it is compared to the color on the vehicle. Plotting will help you recognize the differences between changes in value, hue, and chroma and adjust color as needed.

Evaluate colors in this order:

1. Value
2. Hue
3. Chroma

Using the plotting chart (Figure 20–4), develop a tinting plan that makes sense for all three dimensions of the color. In most cases only one dimension of a color will require adjustment.

PLOTTING CHART

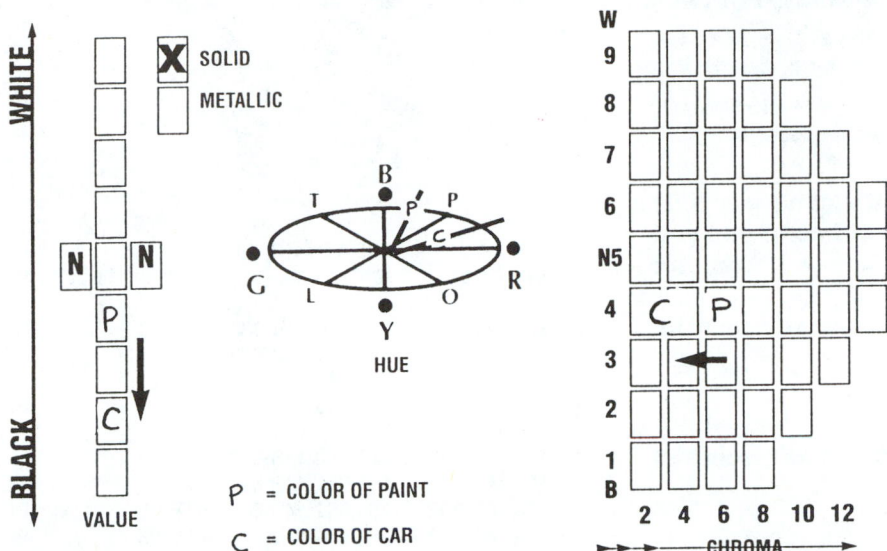

FIGURE 20-4 When plotting color, letter "C" stands for the car or vehicle and "P" stands for paint. Draw an arrow from P to C to show proper direction of the tint. First, you would tint paint to adjust value. You would adjust hue second, by adding color to move hue around the wheel. Chroma is rarely adjusted. *(Courtesy of PPG Industries, Inc.)*

When plotting a color:

1. Adjust value first. Add white to increase value. Add black or color pigments to decrease value. Remember that changing value affects chroma and changing hue affects value.
2. Adjust hue second. Adding color pigments moves the directed color around the color wheel. Adding color pigments affects chroma.
3. Chroma is seldom adjusted. Increasing chroma decreases value. Use gray to decrease with minimum effect on value.

When working with colors that are between the four main colors on the color wheel, a painter should remember the following:

1. Orange, bronze, and gold colors can move toward red or yellow.
2. Maroon and purple can move toward blue or red.
3. Lime can move toward green or yellow.
4. Aqua and turquoise can move toward green or blue.

When adjusting hue, a paint can be moved to either side of the dominant color or toward one of two dominant colors if the paint is between major colors. Always use bluer, redder, yellower, or greener to describe how the hue is to be moved.

20.4 PAINT COLOR MATCHING

Because a variety of factors can cause paint mismatch problems, it is very important to follow a step-by-step process on every job. There are acceptable variances in color that occur at both the refinishing and OEM levels.

In other words, paint can fall to either side of the color standard and still be acceptable. Thus, there can be a difference in the two finishes even though they are officially the same color. This can pose problems when painting.

Remember that whatever the reason for the color variance, you must match the only color standard that really matters—the vehicle itself. For better or worse, the vehicle is the standard!

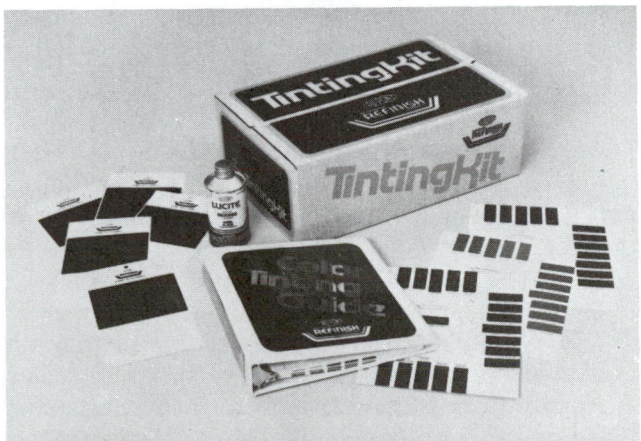

FIGURE 20-5 Color manuals give instructions, color chip samples, and charts for matching colors. Typical tinting guide and kit is designed for the refinisher to provide a color-matching tool that will help to visualize color changes and develop the experience and skill to accomplish successful color tinting. *(Courtesy of DuPont Automotive Products)*

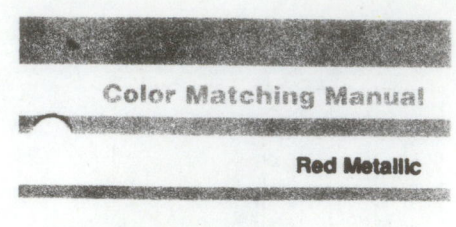

Color Matching Manual

Red Metallic

If you have a RED METALLIC color and you want to:	ALPHA-CRYL		MIRATHANE & MIRACRYL		SUPER-MAX		DIAMONT	
Make it more BLUE	AT-120	AT-155	MB-214	MB-633	TE-20		BC-300	BC-833
	AT-151		MB-255	MB-643			BC-402	BC-838
	AT-153		MB-557				BC-405	BC-840
	AT-154		MB-630				BC-470	BC-880
Make it more RED	AT-150	AT-162	MB-550	MB-630	TE-50	TE-79	BC-810	BC-840
	AT-151	AT-163	MB-551	MB-633	TE-51		BC-820	BC-880
	AT-152	AT-165	MB-552	MB-643	TE-65		BC-825	
	AT-153	AT-170	MB-554	MB-731	TE-66		BC-830	
	AT-154	AT-179	MB-557		TE-67		BC-833	
	AT-155	AT-1613	MB-559		TE-78		BC-838	
Make it more YELLOW	AT-174		MB-560	MB-731	TE-72	TE-86	BC-600	BC-805
	AT-176		MB-566	MB-733	TE-74	TE-87	BC-605	BC-810
	AT-180		MB-568	MB-763	TE-78		BC-610	BC-850
	AT-184		MB-711	MB-764	TE-80		BC-620	
	AT-186		MB-716	MB-765	TE-84		BC-710	
	AT-187		MB-722		TE-85		BC-800	
Make it LIGHTER	AT-111	AT-118	MB-073	MB-084	TE-13		BC-110	BC-170
	AT-112	AT-1101	MB-074	MB-085	TE-14		BC-113	BC-180
	AT-113	AT-1102	MB-075	MB-086	TE-17		BC-115	
	AT-114	AT-1173	MB-081				BC-120	
	AT-115	AT-1174	MB-082				BC-140	
	AT-116	AT-1175	MB-083				BC-145	
Make it DARKER	AT-141		MB-431		TE-41		BC-200	
	AT-142		MB-451		TE-42		BC-209	
	AT-143		MB-455		TE-43		BC-250	
Make it BRIGHTER & More SPARKLE	AT-112		MB-084		TE-14		BC-170	
	AT-115		MB-085		TE-17		BC-180	
Make it GRAYER	AT-141		MB-431		TE-31	TE-42	BC-200	
	AT-142		MB-451		TE-32	TE-43	BC-209	
	AT-143		MB-455		TE-41		BC-250	
Make the PITCH LIGHTER (Basecoat only)	AT-104		MB-040				BC-101	
	# 850		# 850					

FIGURE 20-6 Study typical page from a color manual. It gives formulas for making a paint match the vehicle more closely. *(Courtesy of BASF)*

COLOR DIRECTORY

A **color directory** is a publication containing color chips and other paint-related information for most makes and models of vehicles. Most collision repair and refinishing shops have an aftermarket refinishing color directory (Figure 20–5).

To use a color directory, locate the vehicle manufacturer's paint code. Then you can identify the color chip next to it. It is wise to compare the color chip with the vehicle color. There is always the chance that the vehicle has been repainted with a different color.

Figure 20–6 shows an actual page from a color-matching manual. Note how it gives recommendations for changing or tinting a specific color (metallic red in this example).

The OEM has a group of sample panels, called *color standards,* for each vehicle color. These color

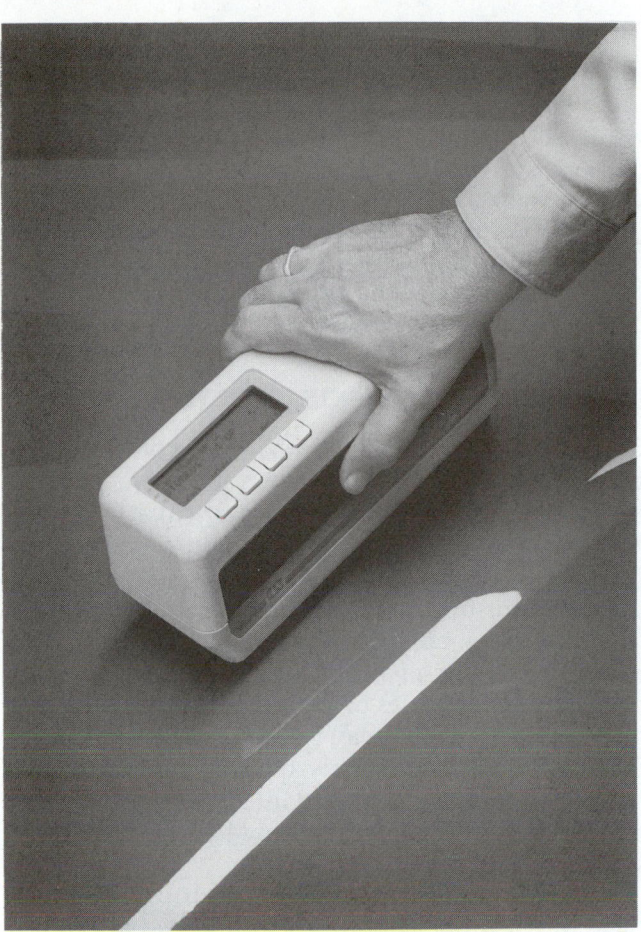

FIGURE 20-7 A spectrophotometer or electronic color analyzer use electronic technology to read the actual color of the vehicle. It can then communicate with the computer system and paint formula software to mix the correct paint color. *(Courtesy of PPG Industries, Inc.)*

standard panels are also provided to the aftermarket paint manufacturers for formulation of refinish colors. The OEM uses these color standard panels to identify acceptable color variances. For example, a blue color can vary slightly to either the green or the red side of the standard and still be acceptable.

Remember that the vehicle must be the standard because it may already have been repaired, possibly several times in different areas.

SPECTROPHOTOMETER

A **spectrophotometer** is an electronic device for analyzing the color of the paint on a vehicle. It electronically reads the color frequencies in the finish to quickly and accurately find the correct paint formula for a color match (Figure 20–7).

The spectrophotometer wand or box is placed on the surface (either the vehicle or test panel) to be checked. Most systems require that a test panel be

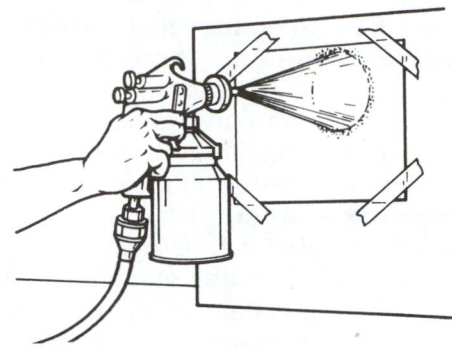

FIGURE 20-8 To get an idea of how the solid color will look on the vehicle, you might want to make a test card by spray painting a piece of cardboard or masking paper.

sprayed for a comparison (Figure 20–8). The multiangle spectrophotometers get a reading at 25, 45, and 75 degrees. Each angle is read for several variables.

Depending upon the type of system, they will read the following:

1. *LAB* or lightness, red to green hue, and yellow to blue hue
2. *LCH* or lightness, chroma, and hue

Most systems compare the vehicle and test panel to one another. The refinish technician will get a reading on the relative lightness/darkness, hue, and chroma of the vehicle to the panel checked. It is still up to the painter to decide how to move the paint closer to that of the vehicle. Decisions on which tint and how much to add must still be made using human judgment.

PAINT FORMULA

The *paint formula* gives the percentage of each ingredient that is needed to match an OEM color. The formula will be available from the mixing system or from the local paint jobber.

A **basecoat patch** is a small area on the vehicle's surface without clearcoat to enable the technician to check for color match with tri-coat colors. The manufacturer masks the patch before clearcoating to help you match the color more easily. The basecoat patch is sometimes located on the driver's rocker panel. Other common locations for the exposed basecoat are under the deck lid or hood somewhere. Refer to the vehicle's service manual to find the exact location of the basecoat patch.

If the color match is correct, order the topcoat by color stock number. Refinish suppliers supply topcoat colors in two ways—factory colors and custom colors. If it is a recent model or a popular color, chances are they will have the paint ready-mixed. These ready-mixed paints are called **factory-packaged colors**.

If it is an older color, they may have to mix it. **Custom-mixed colors** are those colors that are mixed to order at the paint supply distributor. Custom-mixed colors can always be identified easily. The contents of the can must be written or typed on the label by the person who mixed the paint.

An **intermix system** is a full set of paint pigments and solvents that can be mixed at the body shop. Most automobile paint manufacturers have made a color mixing system available to painters. With such an intermix system, it is possible to mix thousands of colors at a savings over the cost of factory-packaged colors or jobber-mixed colors.

An in-shop intermix system also allows you to remix the paint to better match the color of the vehicle more easily. You do not have to go back to the paint supplier for a remix.

COMPUTERIZED PAINT MATCHING SYSTEMS

Computerized paint matching systems use data from the spectrophotometer to help match the paint color. Many spectrophotometer systems can input their color data into a computer. The computer can then use its stored data to help determine how to mix or tint the paint.

Depending upon the sophistication of the system, a computerized paint matching system may be able to do the folowing:

1. Compare the actual color of the vehicle to a computer-stored set of color formulations.
2. Make a recommendation on which tint in the formula will move the sample panel closer to the color of the vehicle.
3. Automatically keep a record of the mixing or tinting procedure. This will let you quickly match the paint if the vehicle returns for another repair.
4. Give a list of tints by number and name.
5. Provide notes on tint strength or hiding characteristics.
6. Summarize how each tint affects value, hue, and chroma.
7. Give cautionary notes, if needed, for using each tint.

TABLE 20-1: HOW COLORS ARE DESCROBED	
Lighter—Darker (called depth)	1. Direct look (panel to panel) 2. Side angle look (panel to panel)
· Cast differences	1. Redder 2. Bluer 3. Greener 4. Yellower
Cleanliness	1. Grayer (dirtier or more muddy) 2. Brighter (cleaner appearance)

Some computerized paint systems provide color variance information. Before making the decision to tint, determine if a color variance chip or formula is available. This may provide a blendable match and reduce or eliminate the need to tint. Computerized paint systems also provide tinting information.

This computer matching information can be printed out with a copy of the paint formula when the paint is mixed. The printout can be referred to during the tinting operation.

COLOR VARIANCE PROGRAMS

Color variance programs compare the color of refinish paints to OEM color standards and actual painted parts obtained from shops. If a particular OEM finish variation is noted often enough, a paint company may develop a *color variation formula* to match the OEM finish.

Variance chips are several samples of color used to help match paint color variations. One paint code may have a series of variance chips on each side of a typical color. They can be used to adjust the paint formulation more precisely than by just using one chip. The chips are organized into books by the paint manufacturer. Many paint manufacturers have color variance programs.

To use variance chips, lay the chips on the vehicle under proper lighting. Find the variance chip that best matches the color on the vehicle. Using the number for this chip, the computer will then give instructions for mixing the paint to match the variance chip and the vehicle. Computerized formulas will give variance to change the color as needed.

20.5 ANALYZING COLOR

Through various application techniques, lightness/darkness, cast, and brightness can be adjusted so that the paint technician can achieve a good match (Table 20–1). To determine what must be done, the color must first be analyzed to determine whether it is too light or too dark, looking at the finish from an angle and head-on. Then the refinisher must check the cast to see if the sprayed color is redder, bluer, greener, or yellower than the original finish. Before adjusting begins, the color must be checked to see if the finish just sprayed is brighter or grayer than the original. The sprayed portion must always be allowed to dry before any adjustments are made.

Adjustments for lightness or darkness rely primarily on shop conditions, spraying techniques, and solvent usage (Table 20–2). Other variables include the amount of paint applied, the air pressure at the spray gun, and the amount of color added to the mix.

Once the lightness or darkness has been adjusted, tinting might be required to get the right cast. Each color can only vary in cast in two directions.

TABLE 20-2: ADJUSTING LIGHTNESS/DARKNESS

	To Make Colors	
Variable	**Lighter**	**Darker**
Shop Condition 1. Temperature 2. Humidity 3. Ventilation	1. Increase 2. Decrease 3. Increase	1. Decrease 2. Increase 3. Decrease
Spraying Techniques 1. Gun distance 2. Gun speed 3. Flash time between coats 4. Mist coat	1. Increase distance 2. Increase speed 3. Allow more flash time 4. Will not lighten color	1. Decrease distance 2. Decrease speed 3. Allow less flash time 4. Wetter mist coat
Solvent Usage 1. Type solvent 2. Reduction of color 3. Use of retarder	1. Use faster evaporator solvent 2. Increase amount of solvent 3. Do not use retarder	1. Use slower evaporator solvent 2. Decrease amount of solvent 3. Add retarder to solvent

- Colors that are either greener or redder in cast include the following:

Blues	Purples
Yellows	Beiges
Golds	Browns

- Colors yellower or bluer in cast are as follows:

Greens	Blacks
Maroons	Grays or Silvers
Whites	

- Colors yellower or redder in cast are as follows:

Bronzes	Reds
Oranges	

- Colors bluer or greener in cast include the following:

Aqua	Turquoise

Charts and manuals available from manufacturers can help you decide on what tint color to use for the appropriate system. Once the color necessary to correctly adjust the cast is determined (Table 20–3), the amount must then be calculated, utilizing the least amount necessary to effectively change the color. The color must be thoroughly mixed; the gun triggered to clear the chamber; and then a small panel can be sprayed, allowed to dry, and checked against the original color.

After the color is correct in lightness/darkness and cast, the color may be made grayer or dirtier. Attempting to make a color brighter at this point will throw off the previous two corrections. To gray the finish, a wet coat must be sprayed, followed by a coat sprayed at half trigger at a slightly greater distance and with a small amount of white mixed with a very small amount of black.

TABLE 20-3: METHOD OF CHANGING CASTS

Color	Add		Cast
Blue	Green	to kill	Red
Blue	Red	to kill	Green
Green	Yellow	to kill	Blue
Green	Blue	to kill	Yellow
Red	Yellow	to kill	Blue
Red	Blue	to kill	Yellow
Gold	Yellow	to kill	Red
Gold	Red	to kill	Yellow
Maroon	Yellow	to kill	Blue
Maroon	Blue	to kill	Yellow
Bronze	Yellow	to kill	Red
Bronze	Red	to kill	Yellow
Orange	Yellow	to kill	Red
Orange	Red	to kill	Yellow
Yellow	Green	to kill	Red
Yellow	Red	to kill	Green
White	White	to kill	Blue
White	White	to kill	Yellow
Beige	Green	to kill	Red
Beige	Red	to kill	Green
Purple	Blue	to kill	Red
Purple	Red	to kill	Blue
Aqua	Blue	to kill	Green
Aqua	Green	to kill	Blue

Use three angles to determine whether or not a color adjustment is necessary:

- **Head-on.** Viewing the repaired area from an angle that is perpendicular to the vehicle.
- *Near specular.* Viewing the repaired area from an angle just past the reflection of the light source.

- **Side tone.** Viewing the repaired area at an angle of less than 45 degrees.

The color of the repaired area should be the same as the rest of the vehicle. If not, correct it until it is the same when viewed from all three angles.

20.6 TINTING

Tinting involves altering the paint color slightly to better match the new finish with the old finish. Tinting may be one of the least understood tools of finish matching. There are three basic reasons for tinting:

1. To adjust color variations in shades to match the color from the manufacturer
2. To adjust color on an aged or weathered finish
3. To make a color for which there is no formula or for which there are no paint codes available

Although a paint technician cannot control the variables that can affect colors at the manufacturing plant, there are variables in the shop that can be controlled. For example, a painter can control the following:

1. Agitation of the paint
2. Application techniques
3. Amount of material applied
4. Spray gun set-up
5. Atomizing air pressure
6. Type and amount of solvent used for reduction
7. Identification of the correct paint code and mixing formula

All of these will help you make the new paint match the existing finish.

Varying the spraying technique can affect color. The three shades shown in Figure 20–9 represent paint that was sprayed out of the same gun. In other words, the application technique can cause the color to vary.

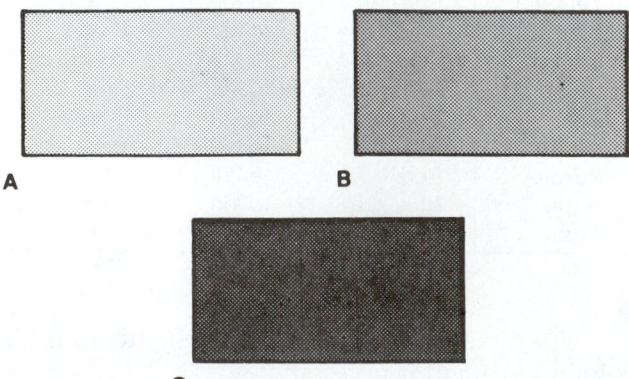

FIGURE 20-9 A spray gun can be used to alter colors slightly. (A) Dry spray typically lightens color. (B) Normal spray will duplicate color best. (C) Wet spray typically darkens color.

Painters who spray wet, end up with a darker color than those who spray drier, especially with metallics.

Color matching is probably the single most recurring problem in the automotive refinishing industry (Figure 20–10). Most of the color matching problems are experienced when attempting to match metallic colors. Although some problems are encountered with solid colors, they cause the fewest problems for the average technician.

The first step in the color matching procedure is to identify the original color from the manufacturer's paint code in the vehicle paint code plate. Keep in mind that this might not be exactly the right color because automotive finishes gradually change color when exposed to light. Each finish also weathers differently depending on its pigment composition. For these reasons, the ability to tint becomes very important.

TINTING COLORS FOR A PERFECT MATCH

There are three basic reasons for tinting colors:

- To adjust color variations in shades for cars of the same color as they come from the manufacturer.
- To adjust color because of aged or weathered finish.
- To make a color for which there is no formula. There are cars painted with bench colors (a color that was never formulated or a color that has no color codes available).

To do color matching, the refinisher must be able to recognize colors as they actually are. It is important not only to see the color that is to be worked on, but also the overtones within that color, including the shades of darkness or lightness and the richness or fullness of the color.

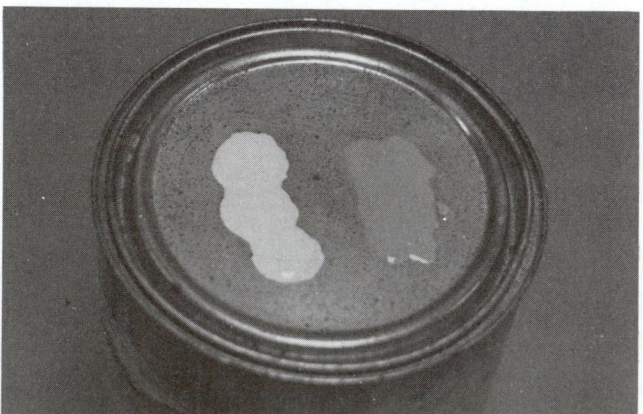

FIGURE 20-10 Paint must be sprayed while duplicating the actual painting process to check color match. Paint puddles on the paint can lid will not reflect the exact color from spraying. They will help determine what the overcast of the tinting color is.

Tinting should only be used as a last resort. If the color of the refinish paint varies from the original finish, check the following possible reasons for the mismatch before deciding to tint the paint:

- The original may have faded. Check the paint on unexposed areas such as door jambs or under the trunk or hood to determine if the finish has faded. If this is the case, restore the paint's luster by compounding the old finish well beyond the repair area.
- Was the wrong color used? Check the auto manufacturer's code and the paint company's stock number for the color being used to make sure that it is the right one. It may be necessary to know the VIN as well as the paint code in order to check the manufacturer's code.
- The pigment and/or flakes may not have been mixed thoroughly. Leaving pigment, flake, or pearl in the bottom of the can could cause a mismatch, so agitate thoroughly.
- Has the amount of thinner or reducer been measured carefully? Overthinning will lighten or desaturate a color. Remember that it is easy to add more thinner, but it cannot be taken out.
- Clean and compound the old finish to remove all chalking and oxidation before making a color comparison.
- When using a test panel, allow the paint enough time to dry. Allow proper flash and dry times for each coat because paint usually gets darker as it dries. If using a clearcoat, remember that compounding the clearcoat will make the paint appear darker. If testing for a base/clear finish, color judgment cannot be made until the clearcoat is applied to the basecoat. Further information on making a test panel is given later in this chapter.
- Vary the spraying technique. Here is a list of shading adjustments:

  Darker
  1. Open fluid valve more.
  2. Reduce size of fan pattern.
  3. Decrease gun distance.
  4. Slow down stroke.
  5. Allow less flash time.

  Lighter
  1. Close fluid valve slightly.
  2. Increase size of fan pattern.
  3. Increase gun distance.
  4. Speed up stroke.
  5. Increase flash time.

Oddly enough, a mismatch in a panel repair will usually show up more than a mismatch in a spot repair—even though the spot repair is smaller—because a panel, such as a car door, has a distinct edge, and the repair, obviously, cuts off at that edge. Any mismatch—as in the case of front and rear doors—will be right next to the adjoining panel and will show a sharp contrast.

A spot repair, on the other hand, is performed by blending the repair into the surrounding area. In spot repairing, the first coat is applied to the immediate area being repaired. Subsequent coats extend beyond this area gradually. Finally, a blend coat extends beyond the colorcoats. Thus, if there is a slight mismatch, the blend coat and the last colorcoat will allow enough show-through of the old finish to make the color difference a gradual one.

HINTS ON COLOR TINTING

Here are some additional tips that might prove helpful when tinting a color:

- Check the color in daylight as well as artificial light. It might not look the same in both lights. When a refinish color matches in one light but not another, it often indicates that the same pigments were not used in the refinish material as in the original finish. The original equipment supplier is usually very careful to use the same pigments in the refinish material as are used in the original finish. To control the uniformity of colors, every batch is checked for exact color match in three different lights—yellow, blue, and daylight.
- Be sure the panel to be matched is thoroughly cleaned and compounded so the true color can be clearly seen.
- Determine what the color problem is and select the proper tinting colors. Do not use mixed colors from the bench for this because they probably have overcasts of the wrong shades. Adjust the color to make the hue redder, greener, bluer, or yellower.
- To understand what the overcasts are to a tinting color, put a few drops on a quart lid with a few drops of white; then intermix these two and make a finger smear on the lid. This will allow the refinisher to determine what the overcast of that tinting color is.
- Do all tinting systematically.
  1. Use a measuring device such as described in Chapters 18 and 19.
  2. Keep a list of the tinting colors used.
  3. Keep a record of the amount used.
- A formula of the color is a help in tinting because it shows the original base colors and indicates which color has faded out and has to be toned down in the refinish material.
- Be sure to mix all tinting colors thoroughly before using; also thoroughly mix the tinting color every time any color is added.
- Add tinting colors in small amounts because it is very easy to overtint. Keep in mind that more color can be added but it cannot be taken out.

- Do not tint the whole can of paint at one time. Make progressive tryouts with small samples until a color match is achieved.
- Be conservative when tinting near the limits of the color range. Correct the most noticeable color differences first.
- Use caution when adding white to metallics or pearls and always use low-strength whites.
- Stay with the same pearls and metallic flakes used in the formula.
- Do not use reduced material when using the draw-down bar.
- An agitator cup should be used when spraying metallics or pearls.
- Allow the color to dry before attempting to adjust it. To shorten dry time, use heat lamps, heat guns, or other drying methods. Be sure the method chosen has been approved for use in the paint/body shop.
- To check the true color, spray out a small panel and allow it to dry. Compare it to the panel to be matched. When it is possible, an old panel from the car to be matched is good to use because it can be masked in the center for an excellent comparison.
- Keep the tint on the light side until the final match is determined. Do not make a final judgment of the color match while the color is wet or still damp because it will keep changing until it is completely dry.
- Once the color is tinted "close enough," complete the repair. Many times that last "just a little bit closer" is the thing that ruins a successful tinting.

Tinting can be divided into two categories: major color tinting and minor color tinting.

Major color tinting consists of making up a color for which there is no color mixing formula available. Find a color chip as close as possible to the desired color and look at that formula. Break the formula down into percentages. For example, gold metallic is 45 percent coarse metallic, 20 percent sparkle metallic, 15 percent gold toner, 10 percent yellow gold, 3 percent soft white, and 2 percent soft black. Using these percentages, make only half of a can; stir and tint to match the desired color.

Minor color tinting is used to adjust a color in a given repair situation to achieve an acceptable color match. To achieve this, each major paint supplier has a basic color tinting kit, a set of instructions, and a tinting guide that is available to the paint jobber and the paint shop technician. Any painter, once familiar with the tinting information and kits that are available, should be able to do minor color tinting to achieve top quality color matches. For more information on the tinting colors and/or for a copy of a company's tinting guide with color chips, contact the shop's local paint jobber or a paint manufacturer's sales and service representative.

20.7 MATCHING SOLID COLORS

For many years all vehicles were solid colors, such as black, white, tan, blue, green, maroon, and so on. Solid colors reflect light in only one direction. Solid colors are still used on vehicles, but to a lesser degree when compared with a few years ago.

Matching solid colors is easier than matching metallic or mica paints. You only need to match the color pigment and not the metal or mica flakes suspended in the paint.

20.8 MATCHING METALLIC FINISHES

In most cases, solid-colored finishes—when properly prepared, thinned, and sprayed—will provide a good color match. The matching of metallic (polychrome) finishes, however, is probably the most skillful operation the refinisher has to perform. There are more cars on the road with metallic rather than solid-colored finishes; that means there are more metallic repairs to be made.

The reason much difficulty is experienced in matching metallic color is that metallic colors are made with a pigment and aluminum flake in the binder that allows light to penetrate beyond the surface of the paint film. When viewing a metallic color at right angles or perpendicular to the surface, you see the face of the color. When viewing it at a 45-degree angle or less, you see the pitch or side tone of a color.

As the position of metallic and/or pigment particles changes in a color film, the color shade of the metallic finish changes accordingly. Each metallic particle is like a tiny mirror. That is what changes the appearance of metallic colors when viewed from different angles. Metallic color also appears to be different when viewed under different kinds of light, such as daylight, shade, sun, or artificial light.

Metallic paints contain small flakes of aluminum suspended in liquid (Figure 20–11). The position of the flakes and the thickness of the paint affect the overall color (Figure 20–12). The flakes reflect light while the color absorbs a higher amount of the light. The thicker the layer of paint, the greater the light absorption.

A *dry metallic paint spray* makes the paint appear lighter and more silvery (Figure 20–13). The aluminum flakes are trapped at various angles near the

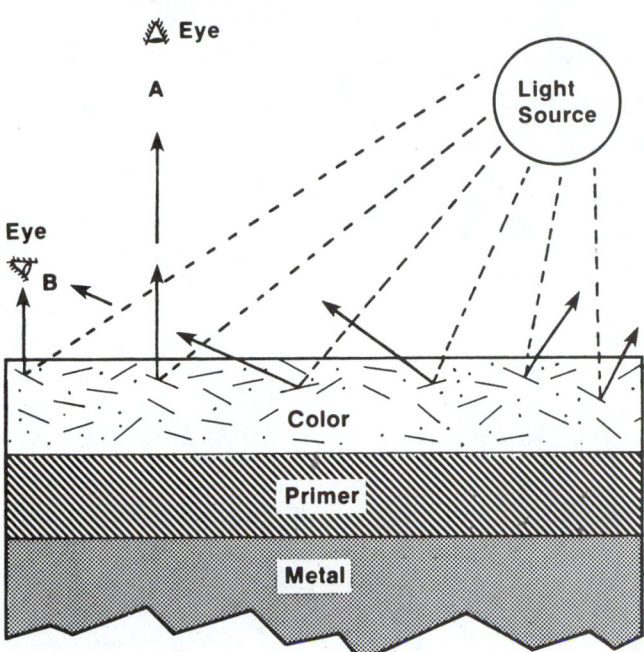

FIGURE 20-11 Metallic color construction showing face and side-tone appearance of color. *(Courtesy of General Motors Corp.)*

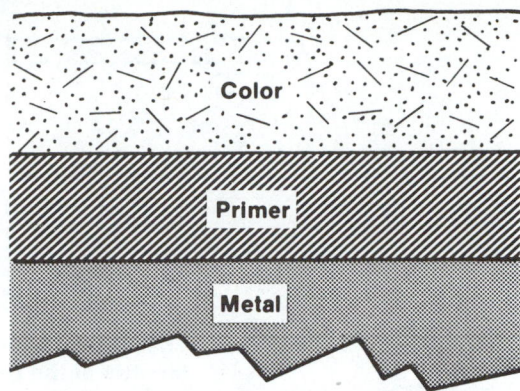

FIGURE 20-12 Note flake pattern in standard shade of metallic color. *(Courtesy of General Motors Corp.)*

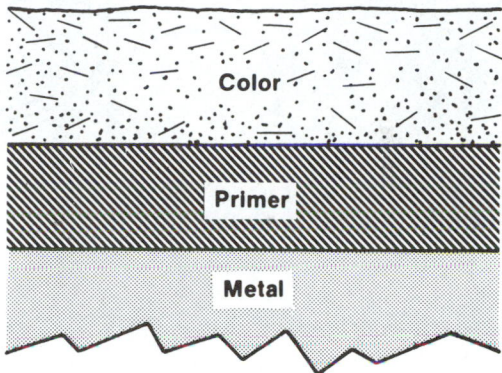

FIGURE 20-13 Note pattern in light shade of metallic color. *(Courtesy of General Motors Corp.)*

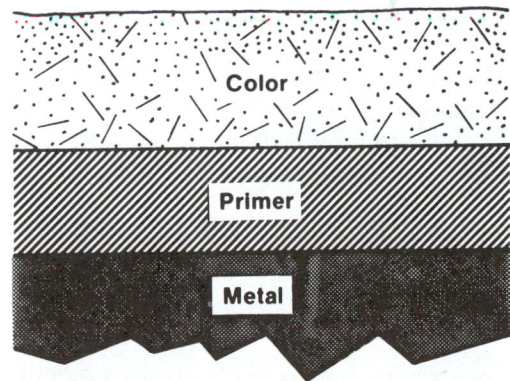

FIGURE 20-14 Note dark shade of metallic color. *(Courtesy of General Motors Corp.)*

surface of the paint film. Light reflection is not uniform. The light has less paint film to travel through; little of it is absorbed. The result is nonuniform light reflection and minimum light absorption.

A *wet metallic paint spray* makes the color appear darker and less silvery (Figure 20–14). The flakes have sufficient time to settle in the wet paint. The flakes lie parallel to and deeper within the paint film. Light reflection is uniform and, because the light has to go farther into the paint film, light absorption is greater. The result is a painted surface that appears deeper and darker in color.

NOTE: Metallic colors must be stirred and mixed thoroughly before use. The pigment quickly settles below the binder. Also, the aluminum flakes settle below the pigment. If flakes stay at the bottom of the can, the paint will not match the color on the vehicle being refinished.

An *agitator paint cup* will help keep the metallic flakes mixed and evenly distributed in the cup. A small air motor in the gun moves a mixing paddle up and down. This helps keep the paint stirred.

In summary, the shades of metallic colors are controlled by the following:

1. Choice of solvents
2. Color reduction
3. Air pressure
4. Wetness of application
5. Spraying techniques

A good paint technician must know how to handle metallic colors. They are very sensitive to the solvents with which they are reduced and the air pressure with which they are applied.

Metallic colors are also affected by a number of variables. These include spraying conditions such as temperature, humidity, and ventilation. They also involve elements of the spraying process such as amount of reduction, evaporation, speed of solvents, air pressure, and type of equipment.

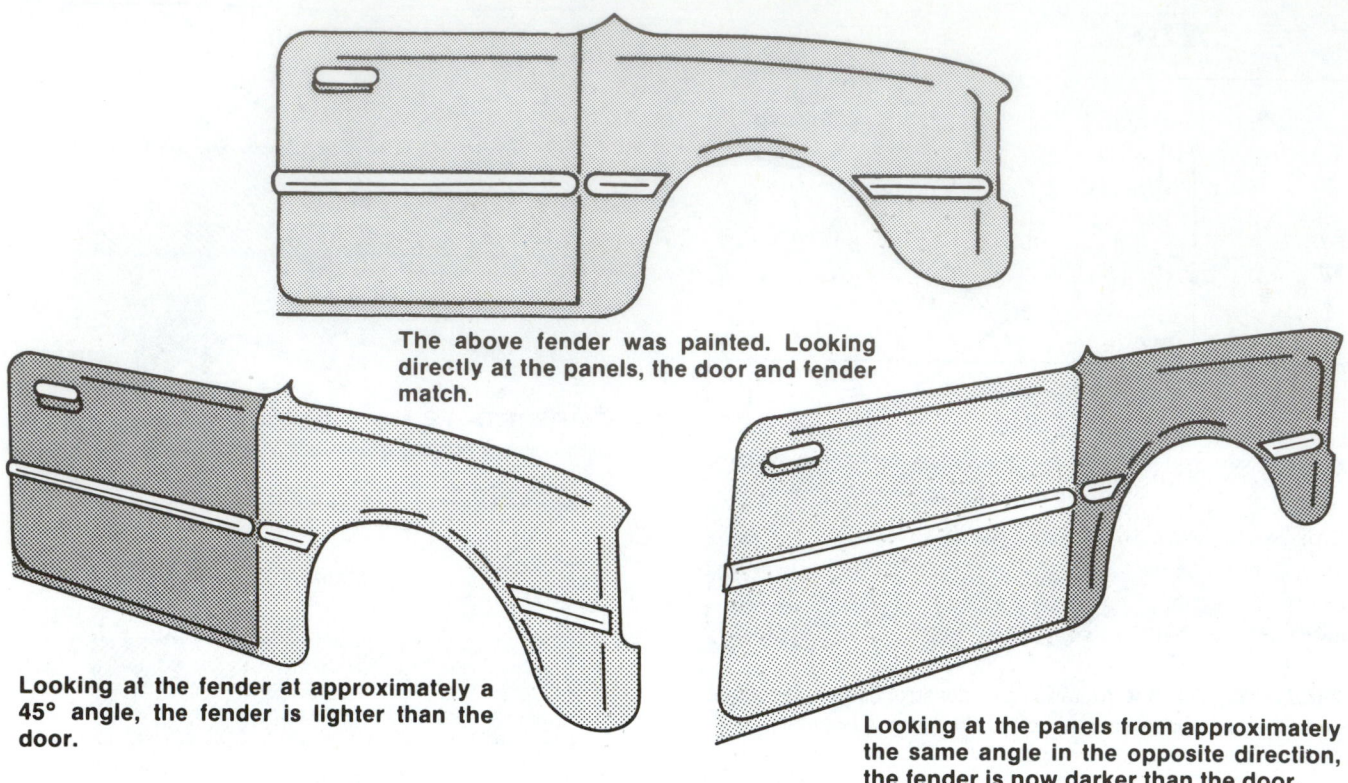

The above fender was painted. Looking directly at the panels, the door and fender match.

Looking at the fender at approximately a 45° angle, the fender is lighter than the door.

Looking at the panels from approximately the same angle in the opposite direction, the fender is now darker than the door.

FIGURE 20-15 Study the cause of flip-flop. *(Courtesy of Martin-Senour Automotive Finishes)*

To darken a metallic color, do the following:

1. Use a larger fluid needle and fluid nozzle.
2. Increase fluid flow.
3. Decrease fan width.
4. Decrease air pressure.
5. Decrease travel speed.
6. Use a slower evaporating solvent.

To lighten a metallic color, do just the opposite.

FLIP-FLOP OF COLOR

Flip-flop is a condition that occurs in metallics, involving the positioning of the aluminum particles and the manner in which light is reflected to the observer (Figure 20–15). The cause of this effect results from the percentage of aluminum particles oriented in a specific direction and their depth in the paint film. The direction and intensity of the light being reflected back through the paint film cause the flip-flop phenomenon that is observed.

The first approach to correct the problem is to adjust your spraying technique to compensate for this effect. Spraying the fender a little wetter will slightly darken the appearance when looking directly into the panel. When viewed from an angle, the resulting appearance is lighter. This occurs be-

cause the aluminum particles are positioned flatter and deeper in the paint film.

Spraying the panel slightly drier reverses the effect, giving a light appearance when looking directly at the panel because the aluminum particles are closer to the surface. The result is a darker appearance viewed at the angle, as light becomes trapped. Both of these techniques are a compromise and should be used to correct minor conditions of flip-flop because the match in one direction can be changed too severely to be acceptable.

If spray techniques cannot correct this condition, the addition of a small amount of white toner will eliminate the sharp contrast from light to dark when the surface is viewed at various angles. The white acts to dull the transparency, giving a more uniform, subdued reflection through the paint film. Care should be taken when adding white since the change occurs quickly. Once too much white is added, recovering the color match becomes virtually impossible.

When confronted with an extremely difficult flip-flop condition, the best method involves adding white, plus blending the color into the adjacent panels. When blending, extend the color in stages. In a basecoat/clearcoat system, for example, spray the blend into the adjacent panel as needed. Then apply the clear to both panels.

20.9 MATCHING BASECOAT/ CLEARCOAT FINISHES

In the past few years, OEMs around the world have increasingly adopted basecoat/clearcoat systems as the finish of choice for new cars rolling off assembly lines. The technology for basecoat/clearcoat finishes was developed in Europe. The durability and popularity of these finishes prompted Japanese and American automobile manufacturers to begin offering them, too. In fact, most automotive finishing experts agree that the basecoat/clearcoat system will be used on the vast majority of refinished vehicles before the turn of the century.

More recently, refinishers have been switching to the new acrylic urethane basecoat/clearcoat system. Many refinish paint manufacturers now offer this excellent two-stage system, which has quickly achieved a strong following. The reason for its popularity stems from not only meeting the practical needs of the painter, but also because of its advanced technology. Acrylic urethanes are fast, offer the best color matches, and provide better coverage and hiding. Thus, they increase productivity and improve customer satisfaction.

With good surface preparation and a clean shop, however, body shop paint technicians will find that the new acrylic enamel basecoat/clearcoats are very easy to apply and result in better-looking, more durable finishes. For example, painters no longer have to balance flow for metallic control in a metallic color, and streaking and mottling are eliminated. Spot repair is made easy because basecoat/clearcoat finishes make it simple to blend in an edge. And the need for buffing is just about eliminated. But most important to body shop paint technicians is the outstanding color match that state-of-the-art basecoat/clearcoat systems provide.

One challenge the body shop painter faces is becoming well educated about the many different systems available in order to select the best one. No customer likes a streaked finish or color drift. No body shop can afford to lose potential income because shop time is tied up due to the difficult application procedures and long dry times. When choosing a refinishing system, the key factors involved are probably appearance and ease of application.

It is a good idea to look for a system that provides fast dry time and locks in the metallic flakes for consistent color match. Also, the amount of pigment in the basecoat should be considered. A good basecoat will contain enough pigment to achieve hiding in 1 to $1\frac{1}{2}$ mils (two coats basecoat).

Identifying cars that are clearcoated is easy. Looking at the vehicle identification code and the color chip book is a quick way to find out if the car has the basecoat/clearcoat system. If the code has

been removed or destroyed, sanding a small spot in a concealed area of the vehicle to be finished, using a fine sandpaper, can help determine the type of finish. If the dust is white, the car has a basecoat/ clearcoat finish. If the dust is the color of the car, it does not.

Clearcoats may be blended by using additives provided by the manufacturer or by painting technique. When working with basecoat/clearcoat finishes, remember the following:

1. Clearcoats are not all perfectly clear. They may change the appearance of a color.
2. Blend the basecoat and clearcoat the entire panel.
3. You may have to step-out the clearcoat if it must be blended.
4. You should clearcoat the entire surface of horizontal panels.
5. Blend into the smallest area possible to help hide the repair.

FLUORINE CLEARCOAT REPAIRS

The basic steps for spot repair with a fluorine clearcoat system include the following:

1. Compound or sand with #1200- to #1500-grit sandpaper for better adhesion (Figure 20–16).
2. Apply first coat of basecoat.
3. Apply second coat of basecoat.
4. Apply third coat of basecoat or until hiding is obtained.
5. Apply color blender if necessary. Dry at 140° Fahrenheit (60° Celsius) for 20 minutes.

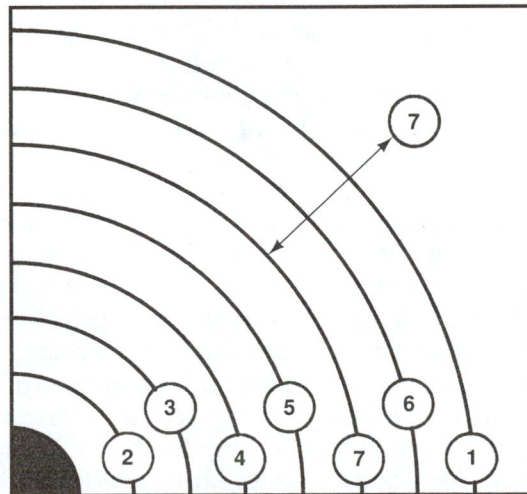

FIGURE 20–16 Compare numbers on the illustration to the following: 1: Compound or sand with #1200- to #1500- grit sandpaper; 2: Apply first coat of basecoat; 3: Apply second coat of basecoat; 4: Apply third coat of basecoat or until hiding is obtained; 5: Apply color blender if necessary. Dry at 140° F (60° C) for 20 minutes. 6: Apply 3 to 4 coats of fluorine clearcoat. Dry properly between coats. 7: Polish with fine compound. *(Courtesy of Nissan Motor Co.)*

6. Apply 3 to 4 coats of fluorine clearcoat. The area between steps 5 and 6 is faded out or blended if required. Dry at 60° to 70° Fahrenheit (16° to 21° Celsius) for 10 minutes between coats. After applying final coat, force dry at 170° Fahrenheit (77° Celsius) for 45 minutes.
7. Polish with fine compounds.

When working with tri-coat and multistage finishes, a technician must match the basecoat prior to applying the intermediate and clearcoats. Some vehicle manufacturers leave an area of basecoat that is not coated with mica or clear for comparing the basecoat spray-out with the vehicle color.

20.10 MATCHING PEARL LUSTER FINISHES

Most recently, in their effort to attract buyers in a market where cars are starting to look alike, car manufacturers are offering highly iridescent colors applied in three layers. The first stage is a mica or pearl coat, and the final stage is the clear topcoat.

In 1960, a synthetic pearl luster pigment using a mica particle covered with thin layers of titanium dioxide was developed. The mica particles made very good carriers for the titanium dioxide because of their highly transparent qualities. The titanium dioxide layers provide the rainbow or pearl effect as light reflects and passes through them.

This transparent quality of the pearl luster pigments is what allows a much higher reflective brilliance that cannot be obtained with aluminum flakes. Aluminum flakes act as miniature mirrors that reflect light. However, they will not let light pass through to the color. If too many aluminum flakes are added to a brilliant color, the color will be washed out.

Pearl luster pigments of titanium dioxide–covered mica flakes reflect light while also allowing some to pass through to other mica flakes and colored pigments below. The brilliance and high iridescent effect of the finish are created this way. Because the pearl coat stage of this system is translucent and reflective, the amount of mica is critical for matching.

It must be remembered that the color code might not be exactly the right color because all automotive finishes gradually change color when exposed to light. Some colors fade lighter; others go darker. Yellow, for example, fades fairly rapidly. If the yellow fades from a cream, the color will usually go lighter and whiter. If the yellow fades from a green composed of blue and yellow, the color will go bluer and usually darker.

Every color weathers a little differently from any other color, depending on its pigment composition. Other factors that affect weathering are the care and the part of the country in which the vehicle is driven.

In general, cars that are in a garage a good share of the time change less. Those that are rubbed and polished a lot change more. Those parked under trees change depending on the type of tree spray or drippings. Those in the South change more rapidly due to increased ultraviolet radiation from the sun. Those in industrial areas or in areas of the country where there are natural chemicals in the air, such as alkali flats, change depending on the chemical to which they are exposed. If a refinish material made with the same pigmentation of the original equipment material is used, the weathering will be the same (unless the type of care or exposure has changed) and in a few months, the refinished area will change to the same weathered color as the original finish. Weathering is fastest in the first few months, and then it slows down. For that reason, a fresh touch-up spot will catch up in time.

In order to give the customer the best match at the time the vehicle is delivered, most shops tint colors to match the weathered color on the car. This matching process is made much simpler through the use of test cards. First, the original finish must be compounded to bring it up to its original shine. Then thin cardboard (not corrugated) can be taped to a piece of paper or another protective backing and propped up so that it hangs vertically. The cardboard is spray painted and held against the car to see if it matches. A number of adjustments can be made to make slight changes in the paint's color. Several test cards may be needed in order to achieve a good match. The ability to match is mostly a matter of experience and can be honed with practice.

The repaint formulas available for the pearl colors on new cars usually have colored pigments and mica pigments combined. These formulas are applied like any other two-stage paint, except when they are to be sprayed over an area that has been primed. Several additional coats with the proper flash times between them will be necessary because of the transparent quality of the paint. Allow more spraying time when scheduling jobs with mica finishes.

Several things should be kept in mind when working with pearl luster paints:

- Mica flakes are heavy. Keep the paint agitated to ensure even distribution.
- Spray test panels. Do not test on the car.
- Continually blend.
- Do not rush. Allow enough flash time between coats.
- Spray in a well-lighted booth.
- Ultraviolet light can help in checking the pearlescent effect.
- Direct sunlight is the best source of light for evaluating touch-ups.
- Check work from three angles in direct sunlight: straight in, from a 45-degree angle to the surface

with the light behind the observer, and from the opposite 45-degree angle with the light ahead of the observer.

Some experimenting with tinting of the base colors, tinting of the pearl coat with colored pigments or pearl pigments, or a combination of both may be necessary to accomplish a good match.

BLENDING MICA COATS

This is a typical mica intermediate coat blending procedure:

1. Apply mica intermediate coat to the area covered by the basecoat.
2. Apply a second mica intermediate coat well beyond the edge of the first coat.
3. Apply a third mica intermediate coat so it extends just beyond the edge of the first coat but within the second coat.
4. Apply a fourth mica intermediate coat to just beyond the edge of the second coat.

Spot repair recommendations for tri-coats are now available. However, the zone repair is still a workable repair option and may be required on certain vehicles.

MATCHING MICA PAINTS

Many basecoat/clearcoat finishes contain mica pigments. Some of these finishes have proven to be especially challenging to match.

Because the finish may not provide full hiding, the color of the primer may show through enough to change the look of the paint. If the painter applies the refinish paint to full hiding, there may be a color mismatch because the primer color is no longer visible through the topcoat.

A color effect test panel is required for basecoat/clearcoat finishes that contain mica. To make this test panel, do the following:

1. Apply a primer that matches the primer color on the vehicle. Allow to dry.
2. Mask the panel into thirds.
3. Apply one coat of basecoat to the exposed section. Allow to dry.
4. Unmask the next section, and apply another coat of basecoat.
5. Unmask the last section of panel, and apply an additional two coats of basecoat to the entire panel. Allow to dry.
6. Apply two coats of clear to the panel. Allow to dry.
7. Compare the test panel against the vehicle to determine the number of basecoat coats that provide the best match.

Applying an extra coat of refinish basecoat to the vehicle may change the color to the point of not matching. Reds or other colors that traditionally provide poor hiding work best for making this panel.

When working with multistage finishes, do not tint the mica intermediate coat. Adjust the number of intermediate coats that are applied. Micas used in basecoat/clearcoat finishes are treated the same as metallic colors. They require special attention to side-tone angles. Mica will make the paint appear lighter and less chromatic on the head-on angle.

Mica will also reduce the metallic effect. For example, if the vehicle seems darker and more chromatic on the side-tone, add mica. If the test panel is darker and more chromatic on the side-tone, there is too much mica pigment.

Mica cannot be removed from the paint; however, this condition can be corrected. To correct the paint if there is too much mica, add small amounts of both black and white. You can also add small amounts of the coarsest metallic in the formula.

Keeping good records during and after the tinting operation is important. Use the plotting chart or a separate record sheet.

20.11 MATCHING TRI-COAT FINISHES

Tri-coating, a three-stage basecoat/clearcoat technique, has been used for 20 years in glamour coating custom cars and for other special applications. It has now found its way into the production line of several manufacturers' deluxe models.

As the name implies, tri-coat finishes consist of three distinct layers that produce a pearlescent appearance: a basecoat, a mid-coat or interference coat, and a clearcoat. Unlike other coating systems such as metallics and some micas, which change the value of a color (lightness/darkness) when viewed from different angles, tri-coats actually change the hue (color) as the angle of view changes. Thus a three-coat finish might look red viewed from straight on and blue when seen from the side. The effect is similar to that of a thin layer of oil floating on water.

In a three-coat finish, the mid-coat is the layer contributing the most to the final appearance of the color. Particles making up the mid-coat can be designed to reflect, absorb, and refract differing amounts of the light striking them, as in pearl luster finishes. Changing the amount, or color, of the mica flakes' coating drastically alters the color of the finish when viewed from straight on or from an angle. While most of the three-coat finishes currently used by auto makers are pastels, darker shades are also feasible. Three-coat finishes have given automotive stylists an exciting new palette of available colors.

Anywhere from one to five coats of pearl luster mid-coat need to be applied to achieve the desired effect.

TRI-COAT SPOT REPAIR

A **halo effect** is an unwanted shiny ring or halo that appears around a pearl or mica paint repair. It is caused by the paint being wetter in the middle and drier near the outer edges of the repair.

Avoid a halo effect by applying the first coat of mica to the basecoat only. The more intermediate mica coats that are applied, the darker the finish will appear. Allow a larger area in which to blend the mica intermediate coats. They require more room to blend than a standard basecoat. Keep the tri-coat repair area as small as possible.

This is one manufacturer's method for a spot or partial repair on a tri-coat system:

1. Apply adhesion promoter to all unsealed panels. Adhesion promoter should extend beyond the repair area.
2. Apply primer to the area over the body filler.
3. Apply two or more coats of basecoat to areas for full hiding. Extend each coat slightly beyond the previous one, allowing dry time between coats (Figure 20–17).
4. Check the let-down panel for the total number of mica intermediate coats needed to match the OEM finish. Apply the intermediate coat to the repair area, extending each coat beyond the last. Allow adequate flash time between coats.
5. Apply two coats of clear over the entire panel. The clear may have to be blended into the sail panel.

TRI-COAT PANEL REPAIR

These are typical steps for a panel repair with a multistage finish:

1. Apply adhesion promoter to both doors.
2. Apply primer to the area over the body filler.
3. Apply two or more coats of basecoat to areas for full hiding. Extend each coat slightly beyond the previous one, allowing to dry between coats (Figure 20–18).
4. Check the let-down panel for the total number of mica intermediate coats needed to match the OEM finish to areas. Extend each coat beyond the previous one, with only the last coat extending into the adjacent panel. Allow adequate flash time between coats.
5. Apply two coats of clear to both doors.

PAINTING VARIABLES

A *variable* is a painting condition (temperature, humidity, and ventilation) or a painting process (reduction, evaporation, speed of solvents, air pressure, and type of equipment). If a paint technician is to get good color matches, it is important to understand how certain paint variables affect the shades of metallic colors.

Variables are divided into two categories: positive and negative. *Positive variables* are those things that a painter does to duplicate the original finish, which in turn results in a good color match. They are:

- Slowness of solvent evaporation. This allows the refinisher to reproduce the factory finish.
- Wetness of color application.

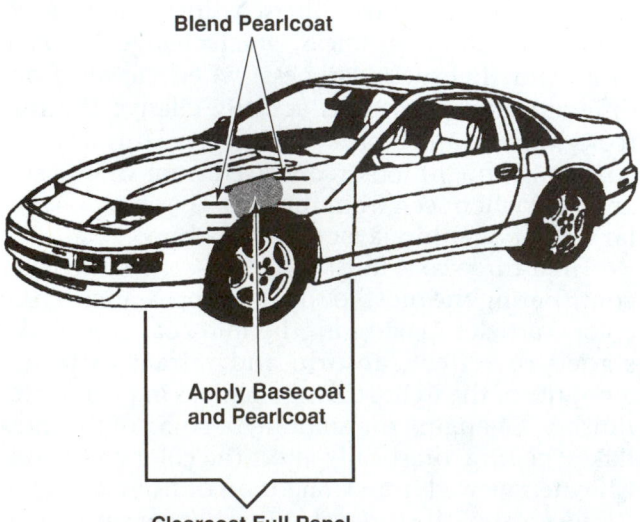

FIGURE 20-17 Note spot repairing a multistage finish. You must blend pearl coat over area larger than repair. Then, clear the whole panel. *(Courtesy of Nissan Motor Co.)*

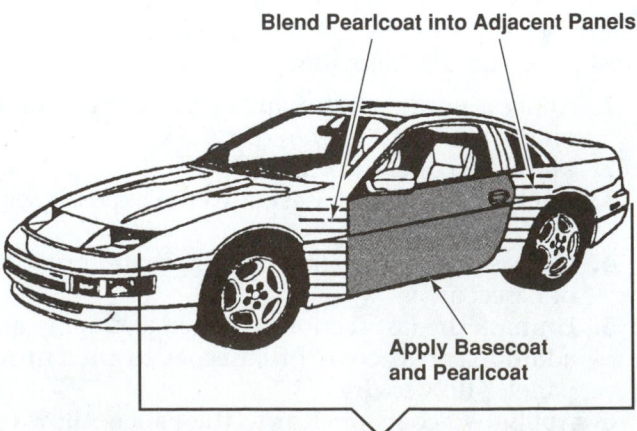

FIGURE 20-18 Note panel repair of multistage finish. You will usually have to blend pearl into adjacent panels. Then, clear the whole side of the vehicle. *(Courtesy of Nissan Motor Co.)*

• Proper spraying technique and the correct air pressure.

Negative variables are those that cause the shades of colors to be off standard. Most common are:

• Improper reduction
• Improper agitation
• Improper application; primarily too high or too low air pressure

In summary, the shades of metallic colors are controlled by:

• Choice of solvents
• Color reduction
• Air pressure
• Wetness of application
• Spraying techniques

20.12 TEST PANELS

While a test spray-out panel is merely a recommendation in many refinish applications, with pearl luster and three-coat finishes it is vital. Test panels for three coats are needed to determine the correct amount of mid-coat color. The mid-coat color is the most critical portion of the three-coat repair. Gun pressure, reduction, and spray techniques affect the amount of color. The extra time spent spraying one or more test panels will be repaid with a satisfied customer who does not come back later with a complaint.

There are three ways to check color match:

1. Spray-out panels (used with conventional paints)
2. Let-down panels (used with multistage paints)
3. Draw-down panels (used to check show-through)

SPRAY-OUT PANEL

The **spray-out panel** checks the paint color and also shows the effects of a painter's technique on a test piece. Spray-out panels are prepared by applying the paint as near to actual spraying conditions as possible. When done properly, a spray-out panel shows the paint exactly as it will look when sprayed on a vehicle.

Before making a spray-out panel, double-check the paint code. Reduce the paint correctly. Set up the spray gun for the material that will be sprayed. Adjust the air pressure at the spray gun for the material to be applied.

When making a spray-out panel, apply a primer that matches the primer on the vehicle. Apply basecoat to full hiding. Allow proper flash time between coats. Apply clearcoat to half the panel. The panel should be fully dry or cured prior to evaluating a color match.

When evaluating a color match on basecoat/clearcoat systems, clear only half of the panel. The uncleared section can be used to check the color match of the basecoat prior to applying the clearcoat. You can refer to any noncleared patch on the vehicle.

NOTE: On basecoat/clearcoat finishes, the basecoat can also be applied to a piece of clear plastic. After the basecoat dries, the sheet is turned over to give an idea of what the basecoat will look like with clearcoat applied.

LET-DOWN PANEL

A **let-down panel** is used to evaluate the color match on tri-coat and multistage paint systems (Figure 20–19). Directions for making a let-down panel are available from the manufacturer of the paint system used. Each coat of the mica intermediate coat will darken the appearance of the finish.

Here is how to make a let-down panel for a tri-coat finish:

1. Prepare the let-down panel with the same primers being used on the vehicle. If a primer-sealer is going to be used, apply the primer-sealer to the let-down panel also. Generally, a light color

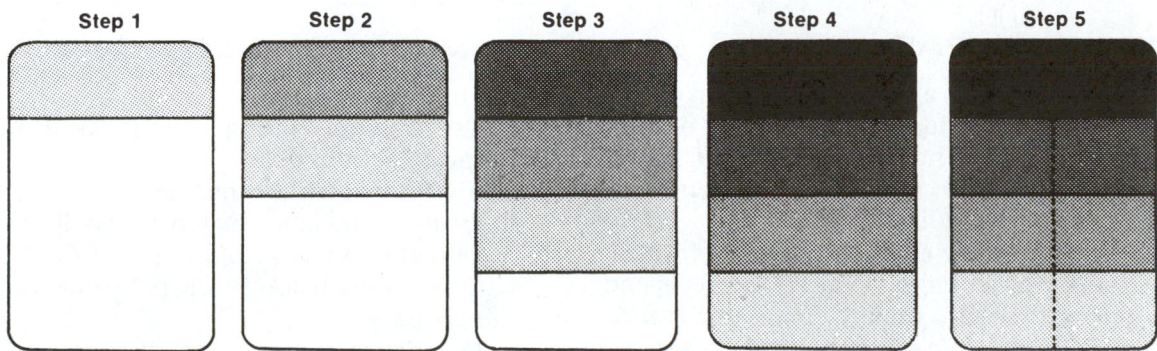

FIGURE 20–19 This example shows the step-by-step creation of a three-coat test panel. A test panel is always recommended when spraying paint. However, when spraying three-stage paint, it is necessary in order to achieve a proper color match.

primer-surfacer (or primer-sealer) is preferred for tri-coat finishes.

2. Apply the basecoat color to hide the test sheet, using the same air pressure and spray pattern that will be used on the vehicle. Duplicating the actual spray techniques when preparing the let-down panel is an important point. Make sure not to vary your procedures.

3. After the panel has dried, divide it into four equal sections. Next, mask off the lower three quarters of the panel, exposing the top quarter.

4. Apply one coat of mica mid-coat color over the top quarter of the panel.

5. After the mica coat has flashed, remove the masking paper and move it down to the middle of the panel, exposing the top half.

6. Apply another coat of mica mid-coat color over the exposed top half of the panel.

7. After this second coat has flashed, remove the masking paper and move it down to expose three quarters of the panel.

8. Apply another coat of mica mid-coat color over the exposed three quarters of the panel.

9. After flashing, remove the masking paper entirely.

10. Apply a fourth coat of mica mid-coat color. As always, spray the coating in the same way as would be done on the vehicle.

11. After the entire let-down panel has dried, mask off the panel lengthwise this time.

12. Apply the manufacturer's recommended number of clearcoats to the exposed half.

13. Compare the different shades on the let-down panel with the paint on the vehicle. Use the same number of coats used on the matching section of the let-down panel to achieve the correct paint match.

Once made, the let-down panel can be kept and used on vehicles with the same color code. On the back of the panel note the color code, gun settings, and paint technician's name. A panel must be made for each different tri-coat color and each painter.

DRAW-DOWN BAR

The **draw-down bar** is a machined bar that evenly distributes 1 mil of unreduced paint to a card or sheet of plastic. It is used to check how much hiding the paint will provide. More than one coat may be required to achieve full hiding.

The draw-down bar cannot show the effects of reduction and application technique. Do not depend on the draw-down bar to make a final color evaluation. A refinish technician must always make a spray-out panel for final color evaluation. Draw-downs cannot be used for tri-coat or multistage finishes.

ZONE CONCEPT

The **zone concept** divides the horizontal surfaces of the vehicle into zones defined by character lines and moldings. It requires refinishing of an entire zone or zones with basecoat, mica intermediate coats, and clearcoats.

20.13 WHY A COLOR MISMATCH?

Tinting should only be used as a last resort. When the paint does not match the vehicle, you must determine if the:

1. Paint code was properly identified.
2. Proper paint code was used.
3. Spray-out test panel was made properly.

Was the spray-out test panel checked:

1. Against a clean vehicle?
2. In the proper light?
3. On both face and side-tone views?

Finally, check the paint company variance chips for a formula that may obtain a better match. If there is not a variance formula, determine if the color is close enough to blend. If not, the paint must be tinted to have a blendable match.

A paint mismatch does not automatically mean tinting. There are a number of things to be checked first. Tinting should only be done to move the paint close enough for blending. Do not try to tint to a perfect match. Blending the paint out over a larger area takes care of any final color variations.

It must be remembered that the color might not be exactly right because all automotive finishes gradually change color over time. Some colors fade lighter; others go darker. Yellow, for instance, fades fairly rapidly. If the yellow fades to a cream, the color will usually go lighter and whiter.

20.14 TINTING SUMMARY

As a brief review of tinting, remember the following:

1. Tint only to blend.
2. Use only one tinting base at a time and check the color after every "hit."
3. Always tint within the formula.
4. Evaluate the color match in daylight or under daylight-corrected lighting.
5. Do not use a black tinting base unless absolutely necessary.
6. Do not use a white tinting base for metallic colors, unless absolutely necessary.
7. Evaluate the color match against a clean vehicle.

TABLE 20-4: GENERAL RULES FOR TINTING (See the last page of the color insert.)

Base Color	General Usage	Base Color	General Usage
1. Yellow Gold	Use for reddish-yellow tint in solids and metallics. Lightens flop in metallics.	15. Indo Orange	Use as orange tint in pastel solids and metallics.
2. Lt. Chrome Yellow	Use in substantial amounts to give bright greenish-yellow hue. Not used in metallics.	16. Moly Orange (Red Shade)	Use as reddish-orange tint in solids only.
3. Oxide Yellow	Use for reddish-yellow tint in solids, has yellow tint with light flop in metallics.	17. Red Oxide	Use for clean red tone in beige solids, not commonly used in metallics.
4. Indo Yellow	Use for greenish-yellow tint in all colors. Lightens flop in metallics.	18. Transparent Red Oxide	Use to give red-gold tint in metallics. Also beige tint in solids.
5. Transparent Yellow Oxide	Use for yellow tint in solids and metallics. Lightens flop in metallics.	19. Deep Violet	Use as purple tint in pastel solids and metallics. Also use in blues and grays for violet tones.
6. Rich Brown	Use to give clean golden tint to metallics and clean beige to solid colors.	20. Quindo Violet	Use as blue-red tint in pastel solids and metallics.
7. Black	Use where small amounts of black are needed. Has brown or yellow undertone.	21. Magenta Maroon	Use as blue-red tint in pastel solids and metallics.
8. Strong Black	Use where a large amount of black is needed. Has brown or yellow undertone.	22. Phthalo Green (Yellow Shade)	Use as green tint in pastel solids and metallic. Has yellow-green tint.
9. Organic Orange (Light)	Use as reddish-orange tint in pastel solid and metallic colors.	23. Phthalo Green	Use as green tint in all colors.
10. Oxide Red	Use for clean red tint in beige colors. Not commonly used in metallics.	24. Scarlet Red	Use in bright red solid colors.
11. Permanent Red	Use for blue-red tint in pastel solids and metallics. Lightens flop in metallics.	25. Perrindo Maroon	Use as rich brown-maroon tint in solids and metallics.
12. Organic Scarlet	Use as red in pastel solids and metallics.	26. Phthalo Blue (Medium)	Use as blue tint in pastels and metallics. Has a clean red-blue tint.
13. Phthalo Blue (Green Shade)	Use as blue tint in pastel and metallics. Has a very green-blue tint.	27. Phthalo	Use as green tint in all colors.
14. Permanent Blue	Use as blue tint in pastel solids and metallics. Has a very red-blue tint.	28. Phthalo Green (Yellow)	Use as yellow-green tint in pastel solids and metallics.

1. Base colors have two tones: mass tone (as appears in can) and tint tone (small amount mixed with white or aluminum).
2. Colors darken as they dry. Always match on the light side.
3. The same color arrived at with two different formulations (using different pigments) might vary in color under different lights and might weather differently.
4. Metallic colors have varying degrees of flop (oblique angle view). Colors with a deep rich flop contain coarse aluminum and tinting colors with greater transparency or depth.
   a. To maintain a rich flop, use coarse aluminum, and if required, tinting colors with greater transparency.
   b. When the flop requires a grayer appearance, use finer aluminum.

This table and the last color page are strictly for tinting existing colors and not original formulation work. When tinting colors, it is not recommended to add more than 50 parts per pint, 100 parts per quart or 400 parts per gallon.

(Courtesy of PPG Industries, Inc.)

8. Do not to use the "dab method" of color matching. A dab of paint placed on a surface and compared to existing paint will not give accurate results.
9. Adding small amounts of coarse metallic tinting base can darken side-tone.
10. Adding small amounts of metallic tinting base can lighten a metallic color.

Evaluating and tinting metallic colors is more difficult than evaluating and tinting solid colors. Evaluation of both metallic and solid colors should be done in this order:

1. Verify formula.
2. Plot the color on the plotting chart.
3. Compare the test panel to the vehicle.

4. Check and adjust value.
5. Check and adjust hue.
6. Check and adjust chroma.

Using kill charts and adding tints not in the formula will also move the color, but not around the outside of the color wheel. Adding tints suggested by a *kill chart* will move the paint more directly toward the gray center of the color wheel, affecting both hue and chroma.

There may be situations where this may be desirable, but as a rule of thumb, do not use kill charts to change hue. The proper tint for changing hue is already in the formula. Adding tints which are not in the formula may also cause problems with metamerism.

Adding paint instead of tints will add the dominate tint and all the other tints in that paint formula. Again, use only tint bases and do not add paint.

Some paint companies produce metallic tinting bases designed to correct a specific problem, usually having to do with changes in side-tone. Use these products according to the manufacturer's recommendations. Using the paint formula and tinting guide, select the tinting base that will move the paint in the right direction.

When tinting, a painter should do the following:

1. Use only half the can of paint.
2. Use only tint bases. Never add paint.
3. Use only tints that are in the paint formula.
4. Add tint bases in small amounts, and check the paint following each addition.
5. Keep records of each tint base and the amount added. This will be useful if more paint is needed.

A chart summarizing the general tinting rules is given in Table 20–4. Read through it carefully.

20.15 CUSTOM PAINTING

Custom painting involves using multiple colors, metal flake paints, multilayer masking, and special spraying techniques to produce a personalized paint job. Multicolor stripes, flames, murals, landscapes, names, and other artwork can be added to the finish. Complex images require you to have special artistic talent. If you cannot paint an attractive image on paper, you will not be able to do it on a vehicle (Figure 20–20).

Custom painting requires considerable talent, skill, and knowledge. You need to plan the custom job carefully to determine how to mask and spray or apply each color (Figure 20–21). Custom painters are good at using airbrushes, striping tools, and masking materials.

Before custom painting, make sure the base finish is in good condition. You do not want to waste your time trying to paint over a weathered or problem finish. Wet sand and clean the area to be custom painted. Use surface preparation methods detailed in other chapters.

Custom masks can be made by drawing designs on thin posterboard and then cutting them out (Figure 20–22). The posterboard designs are then taped onto the vehicle and spray painted. By using an airbrush and translucent paint, various attractive effects can be produced.

Card masking involves using a simple masking pattern to produce a custom paint effect. Usually, an airbrush is used to mist the paint over the edge of the masking card. The card can be moved to repeat the pattern to produce a wide range of paint effects (Figure 20–23).

Lace painting involves spraying through lace fabric to produce a custom pattern in the paint. Various lace designs can be purchased at fabric stores. The cloth pattern will allow the paint to pass through

FIGURE 20–20 Experience and artistic abilities are needed to do custom paint work. It is best to learn on small projects first, like motorcycle gas tanks. Then, with experience, you can do custom paint work on cars and trucks. *(Courtesy of Badger Airbrush)*

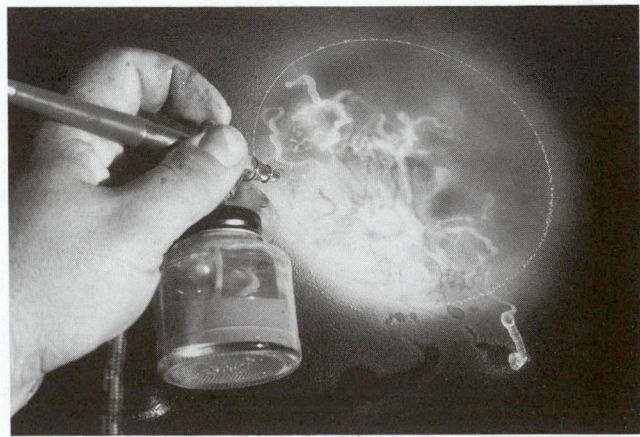

FIGURE 20–21 Planning is vital to a successful custom paint job. Multiple layers of masking and spraying require complex planning for attractive results. *(Courtesy of Badger Airbrush)*

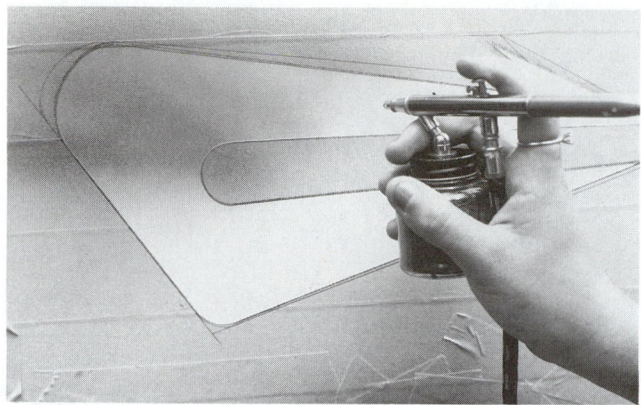

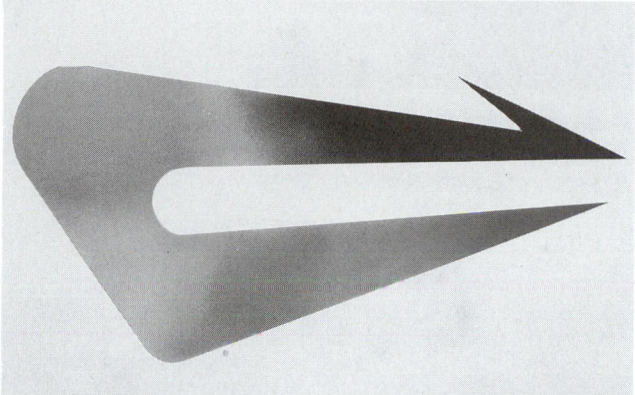

FIGURE 20-22 Custom masks are often made of posterboard. The design is taped onto the vehicle and sprayed. Using an airbrush and translucent paint, various attractive effects can result. *(Courtesy of Badger Airbrush)*

FIGURE 20-23 An airbrush is used to mist paint over edge of masking card. Card is then moved to repeat the pattern for custom paint effect. *(Courtesy of Badger Airbrush)*

the holes in the lace but mask it in other areas (Figure 20–24).

A *marble effect* can be made by forcing crumpled plastic against a freshly painted stripe or area. You might want to spray the area with two colors to reveal the desired color under the wet coat. When the plastic is lifted off, it will remove the top layer of paint in random areas, leaving a marble-type effect (Figure 20–25).

FIGURE 20-24 Lace painting involves spraying through lace fabric to produce a custom pattern in paint. Various lace designs can be purchased at fabric stores. *(Courtesy of Badger Airbrush)*

Spider-webbing is done by forcing paint through the airbrush in a very thin, fibrous-type spray. Air pressure from the gun can also be used to spread and smear the wet paint to produce a varying effect (Figure 20–26).

Painted flames are a custom painting technique often used on hot rods or older street rods. Fine-line masking tape is used to form the outline of the flames. Then the area around this shape is covered with masking paper or plastic.

First the base color for the flames is applied. Then a second translucent color is blended inside the flame area (Figure 20–27A). A third color may be used to darken the outer edges and center area of the flames. The flames are finally wet sanded and polished when dry (Figure 20–27B).

Painted lettering involves masking off letters over the finish and spraying or brushing them on with a different color. This can be time-consuming but is sometimes requested by customers.

When doing custom paint work, do not "bite off more than you can chew." Start out simple with minor complexity. As you learn to successfully do custom work, you can progress to more complex paint work. Experience is the best teacher with custom paint work.

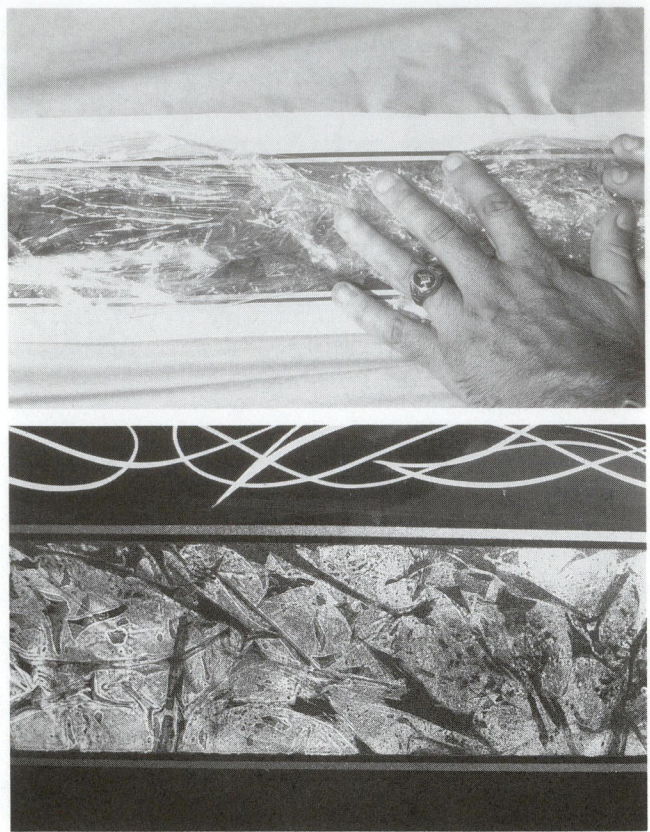

FIGURE 20-25 A marble effect can be made by forcing crumpled plastic against a freshly painted stripe or area. *(Courtesy of Badger Airbrush)*

A good idea is to practice techniques on old parts using leftover paint. Then you can learn from your mistakes and successes without working on a customer's vehicle.

SUMMARY

- Color matching involves the steps needed to make the new finish match the existing finish on the vehicle.

- Color is caused by objects reflecting light at different frequencies or wavelengths into our eyes.

- Since the vehicle will be seen in sunlight, you should always use sunlight, or daylight-corrected lighting, when making color evaluations.

- To minimize the confusion when painting, color should be based upon the three dimensions of color, which are the following:
  1. Value (lightness or darkness)
  2. Hue (color, cast, or tint)
  3. Chroma (saturation, richness, intensity, muddiness)

- A color directory is a publication containing color chips and other paint-related information for most makes and models of vehicles.

FIGURE 20-26 Spider-webbing is done by forcing paint through the airbrush in a very thin, fibrous-type spray. *(Courtesy of Badger Airbrush)*

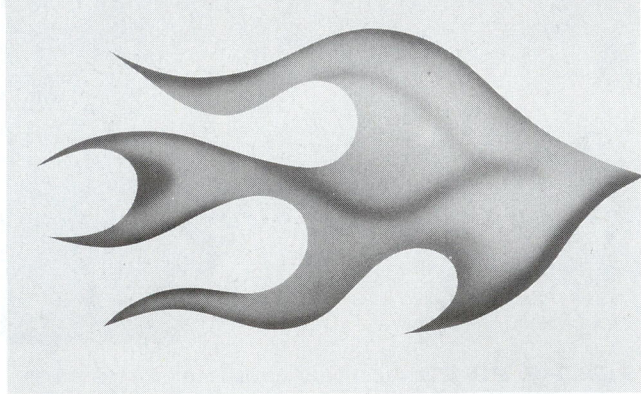

FIGURE 20-27 Painted flames are a custom painting technique often used on hot rods or street rods. *(Courtesy of Badger Airbrush)*

- A spectrophotometer is an electronic device for analyzing the color of the paint on a vehicle.

- Computerized paint matching systems use data from the spectrophotometer to match the paint color.

- Tinting involves altering the paint color slightly to better match the new finish with the old finish.

- A dry metallic paint spray makes the paint appear lighter and more silver.

- A wet metallic paint spray makes the color appear darker and less silver.

- Flip-flop is a condition that occurs in metallics involving the positioning of the aluminum particles and the manner in which light is reflected to the observer.

- Clearcoats are not all perfectly clear. They may change the appearance of a color.

- A halo effect is an unwanted shiny ring or halo that appears around a pearl or mica paint repair.

- A variable is a painting condition (temperature, humidity, and ventilation) or a painting process (reduction, evaporation, speed of solvents, air pressure, and type of equipment).

- The spray-out panel checks the paint color and also shows the effects of the paint technician's technique on a test piece.

- Custom painting involves using multiple colors, metal flake paints, multilayer masking, and special spraying techniques to produce a personalized paint job.

ASE-STYLE REVIEW QUESTIONS

1. When evaluating the color of a finish, it should be viewed under
   A. Sunlight
   B. Incandescent light
   C. Fluorescent light
   D. Drop light

2. Technician A says that a spectrophotometer can be used to help match a color. Technician B says that a computerized paint matching system can help determine how to mix or tint the paint. Who is correct?
   A. Technician A
   B. Technician B
   C. Both A and B
   D. Neither A or B

3. This refers to the degree of lightness or darkness of the color.
   A. Value
   B. Hue
   C. Tint
   D. Shade

4. This refers to the color's level of intensity, or the amount of gray (black and white) in a color.
   A. Hue
   B. Chroma
   C. Tint
   D. Value

5. The basic reasons for tinting are
   A. To adjust color variations in shades to match the color from the manufacturer.
   B. To adjust color on an aged or weathered finish.
   C. To make a color for which there is no formula or for which there are no paint codes available.
   D. All of the above
   E. None of the above

6. This is a full set of paint pigments and solvents that can be mixed at the body shop.
   A. Factory-packaged colors
   B. Custom-mixed colors
   C. Intermix system
   D. Manufacturer system

ESSAY QUESTIONS

1. What is a color directory?

2. What is a spectrophotometer?

3. List three tasks that a computerized paint matching system may be able to do.

4. Explain variance chips in detail.

5. How do we see the color of a finish or any object?

6. List six ways to darken a metallic color.

7. Define the term "flip-flop."

8. Summarize the basic steps for spot repair with a fluorine clearcoat system.

CRITICAL THINKING PROBLEMS

1. If the color of the refinish job varies from the original, what are some possible reasons for the mismatch?

2. Why should you view a color from different angles when checking a color match?

MATH PROBLEMS

1. A paint technician uses a mil gauge to measure the paint thickness on a vehicle. The gauge shows a thickness that varies from 8 mils to 10 mils. If the new paint must be applied with a thickness of 2 to 4 mils, how thick will the paint become?

CHAPTER 21

Paint Problems and Final Detailing

INTRODUCTION

Paint problems include a wide range of defects that can be found before or after painting. You must be able to analyze and correct paint problems efficiently. If you fail to find a problem before painting, you will have extra work to do sanding off the new paint and fixing the original problem. You must also be able to find and solve paint problems that are found after you paint a vehicle.

After doing the body work and painting the vehicle, you must check your work quality. Ideally, the vehicle can be released to the customer after a minor cleanup. Sometimes however, you will find small imperfections in the paint film that must be corrected. On rare occasions, you may have to solve major paint problems on existing or freshly painted surfaces.

Final detailing locates and corrects any paint defect which may cause customer complaints. It also involves cleaning of the car before delivery.

This chapter will provide you with the information needed to analyze and correct paint problems. It will also outline how to final detail a vehicle.

OBJECTIVES

After studying this chapter, you should be able to:

✔ Define the most common paint problems.

✔ Explain the causes of paint problems.

✔ Repair common paint problems.

✔ Use a dirt-nib file.

✔ Complete minor paint touch-up work.

✔ Wet sand to remove some minor paint problems.

✔ Hand and machine compound a finish.

✔ Final detail a vehicle properly.

✔ Answer ASE-style review questions.

KEY TERMS

alkali spotting	mottling
bleeding	orange peel
blistering	orbital action polisher
blushing	paint problems
burn-through	paint surface chips
chalking	paint surface
chipping	protrusion
color sanding	peeling
cracking	pinholing
dirt-nib files	plastic filler bleed-
dulled finish	through
edge masking	polishing
featheredge splitting	rubbing compounds
final detailing	runs
finesse sanding	sag
fisheyes	sandscratch swelling
hand compounds	solvent popping
lifting	swirl marks
line checking	water spotting
machine compounds	wet spots
microchecking	wrinkling

 **ASE TASK LIST**

Job Skills covered in this chapter include:

PAINTING AND REFINISHING TEST (B2) TASK LIST

A. Surface Preparation

3. Inspect and identify substrate, type of finish, and surface condition; develop a plan for refinishing.
5. Dry or wet sand areas to be refinished.
9. Mix primer, primer-surfacer, or primer-sealer; spray onto surface of repaired area.
11. Dry or wet sand area to which primer-surfacer and/or two-component putty have been applied.
12. Remove dust from area to be refinished, including cracks or moldings of adjacent areas.
13. Clean area to be refinished using a proper cleaning solution.

C. Paint Mixing, Matching, and Applying

7. Check for color matching of all applied finishes.
8. Sand, buff, and polish finishes where necessary.
12. Apply multistage (mica, pearl, etc.) coats for spot repair, panel blending, and overall refinishing.

D. Solving Paint Application Problems

1. Identify blistering (raising of the paint surface); determine the cause(s), and correct the condition.
2. Identify blushing (milky or hazy formation); determine the cause(s), and correct the condition.
3. Identify contaminants in the painted surface; determine the source(s), and correct the condition.
4. Identify a dry spray appearance in the paint surface; determine the cause(s), and correct the condition.
5. Identify the presence of fisheyes (craterlike openings) in the finish after it has been applied; determine the cause(s), and correct the condition.
6. Identify lifting (surface distortion or shriveling) while the topcoat is being applied; determine the cause(s), and correct the condition.
7. Identify mottling or streaking in metallic and mica paint finishes; determine the cause(s), and correct the condition.
8. Identify orange peel appearance of the refinished surface; determine the cause(s), and correct the condition.
9. Identify an overspray condition; determine the cause(s), and correct the condition.
10. Identify solvent popping (pinholing) in the freshly painted surface; determine the cause(s), and correct the condition.

11. Identify sags and runs in the paint surface; determine the cause(s), and correct the condition.
12. Identify sandscratch swelling; determine the cause(s), and correct the condition.
13. Identify shrinking or splitting while finish is drying; determine the cause(s), and correct the condition.
14. Identify that color is off-shade or does not match; determine the cause(s), and correct the condition.
15. Identify tape tracking; determine the cause(s), and correct the condition.
16. Identify loss of gloss in the paint surface; determine the cause(s), and correct the condition.

E. Finish Defects, Causes, and Cures

1. Identify poor adhesion; determine the cause(s), and correct the condition.
2. Identify paint cracking (crowsfeet or line-checking, microchecking, etc.); determine the cause(s), and correct the condition.
3. Check for rust spots (corrosion); determine the cause(s), and correct the condition.
4. Identify blistering in the paint surface; determine the cause(s), and correct the condition.
5. Identify water spotting on paint surface; correct the condition.
6. Identify finish damage caused by bird droppings, tree sap, and other natural causes; correct the condition.
7. Identify finish damage caused by airborne contaminants (acids, soot, rail dust, and other industrial-related causes); correct the condition.
8. Identify die-back conditions (dulling of the paint film showing haziness, or film distortion showing shrinkage); correct the condition.
9. Identify chalking (oxidation); correct the condition.
10. Identify body filler bleed-through or staining; correct the condition.
11. Identify solvent popping (pinholing); correct the condition.
12. Identify damage caused by buffing painted surfaces; correct the condition.

F. Safety Precautions and Miscellaneous

5. Apply/remove decals, transfers, tapes, woodgrains, pinstripes (painted and taped), etc.

21.1 PAINTING PROBLEMS

Most refinishing problems can usually be repaired, but this "rework" requires time and money. Therefore, it is wise to prevent common paint problems before they occur. Unfortunately, there is a variety of causes for defects in a paint finish. They usually originate in the preparation of the base metal, the painting procedure, the paint ingredients, the environment, and external influences.

If you see paint defects while spraying, you must decide whether to stop work immediately or wait until the painting is finished to correct them. This depends on the type and extent of the problem. For example, if there is poor, wavy bodywork, stop right away. If the problem is a small piece of dust in the paint, keep spraying. You can normally fix this surface imperfection after the paint cures.

One of the best ways to reduce the likelihood of defects occurring is to closely follow the proper fundamental painting procedures outlined in the paint manufacturer's instructions.

NOTE: Refer to the color insert in the middle of this textbook for color examples of paint problems.

ACID AND ALKALI SPOTTING

Acid and **alkali spotting** cause an obvious discoloration of the surface. Various paint pigments react differently when in contact with acids or alkalies (Figure 21–1).

Acid and Alkali Spotting Causes

The cause of acid and alkali spotting is a chemical change of pigments. This chemical change results from atmospheric contamination in the presence of moisture. This problem is found on older finishes that have been exposed to industrial pollution.

Acid and Alkali Spotting Prevention

1. Keep finish away from contaminated atmosphere.
2. Immediately following contamination, the surface should be vigorously flushed with cool water and detergent.

Acid and Alkali Spotting Solution

1. Wash with detergent water and follow with a vinegar bath.
2. Sand and refinish. You might try wet sanding and compounding if there is only minor spotting.
3. If contamination has reached the metal or substrate, the spot must be sanded down to the metal before refinishing.

BLEEDING

Bleeding is the original finish discoloring—or seeping through—the new topcoat color (Figure 21–2).

Bleeding Causes

Bleeding is caused by contamination, usually in the form of soluble dyes or pigments on the older finish before it was repainted. (This is especially true with older shades of red.)

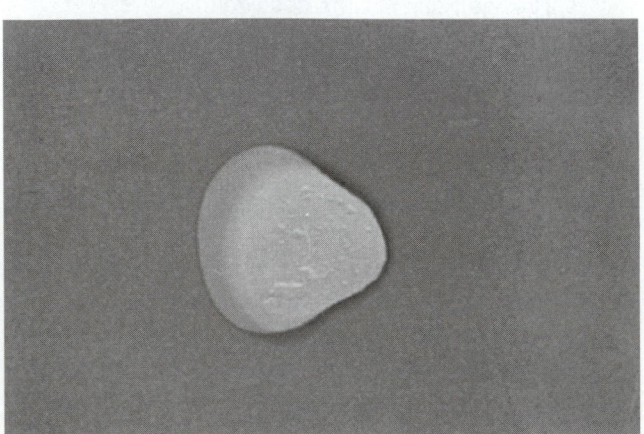

FIGURE 21-1 This is a close-up of acid or alkali spotting of paint. *(Courtesy of PPG Industries, Inc.)*

FIGURE 21-2 This paint has suffered from bleeding. *(Courtesy of PPG Industries, Inc.)*

Bleeding Prevention

Thoroughly clean areas to be painted before sanding, especially when applying lighter colors over darker colors. (Avoid using lighter colors over older shades of red without sealing first.)

Bleeding Solution

Apply two medium coats of bleeder sealer. Seal in accordance with label instructions. Then reapply colorcoat.

BLISTERING

Blistering shows up as small, swelled areas on the finish, like water blisters on human skin. There will be a lack of gloss if the blisters are small. You will find broken-edged craters if the blisters have burst (Figure 21–3).

Blistering Causes

1. Improper surface cleaning or preparation. Tiny specks of dirt left on the surface can act as sponges and hold moisture. When the finish is exposed to the sun (or abrupt changes in atmospheric pressure), moisture expands and builds up pressure. If the pressure is great enough, blisters form.
2. Wrong thinner or reducer. Use of a fast-dry thinner or reducer, especially when the material is sprayed too dry or at an excessive pressure. Air or moisture can be trapped in the film.
3. Excessive film thickness. Insufficient drying time between coats or too heavy application of the undercoats can trap solvents that escape later and blister the colorcoat.

4. Contamination of compressed air lines. Oil, water, or dirt in lines.

Blistering Prevention

1. Thoroughly clean areas to be painted before sanding. Be sure surface is completely dry before applying either undercoats or topcoats. Do not touch a cleaned area because the oils in the hands will contaminate the surface.
2. Select the thinner or reducer most suitable for existing shop conditions.
3. Allow proper drying time for undercoats and topcoats. Be sure to let each coat flash before applying the next.
4. Drain and clean air pressure regulator daily to remove trapped moisture and dirt. Air compressor tank should also be drained daily.

Blistering Solution

If damage is extensive and severe, paint must be removed down to undercoat or metal, depending on the depth of the blisters. Then refinish. In less severe cases, blisters can be sanded out, resurfaced, and retopcoated.

BLUSHING

Blushing is a problem that makes the finish turn "milky looking" (Figure 21–4).

Blushing Causes

1. In hot humid weather, moisture droplets become trapped in the wet paint film. Air currents from the spray gun and the evaporation

FIGURE 21-3 Here is a blistering paint problem. *(Courtesy of PPG Industries, Inc.)*

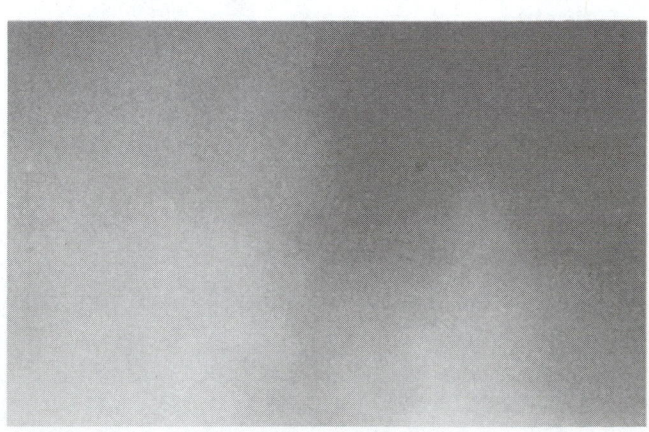

FIGURE 21-4 Blushing makes the paint look milky. *(Courtesy of PPG Industries, Inc.)*

of the thinner or reducer tend to make the surface being sprayed lower in temperature than the surrounding atmosphere. This causes moisture in the air to condense on the wet paint film.

2. Excessive air pressure.
3. Using too fast a thinner or reducer.

Blushing Prevention

1. In hot humid weather try to schedule painting early in the morning when temperature and humidity conditions are more suitable.
2. Use proper gun adjustments and techniques.
3. Select the thinner or reducer that is suitable for existing shop conditions.

Blushing Solution

Add retarder to the thinned or reduced color and apply additional coats.

CHALKING

Chalking is a problem that causes a lack of gloss on the paint surface. Extreme cases will show up as a powdery surface.

Chalking Causes (Other Than Normal Exposure)

1. Wrong thinner or reducer, which can harm topcoat durability.
2. Materials not uniformly mixed.
3. Starved paint film.
4. Excessive mist coats when finishing a metallic color application.

Chalking Prevention

1. Select the thinner or reducer that is best suited for existing shop conditions.
2. Stir all pigmented undercoats and topcoats thoroughly.
3. Meet or slightly exceed minimum film thickness.
4. Apply metallic color as evenly as possible so that misting is not required. When mist coats are necessary to even out flake, avoid using straight reducer.

Chalking Solution

Remove surface in affected area by sanding, then clean and refinish.

CHIPPING

Chipping is a condition where small chips of a finish lose adhesion to the substrate, usually caused by the impact of stones or hard objects. While refinishers have no control over local road conditions—and thus cannot prevent such occurrences—they can take steps to minimize the effects if they know beforehand that these conditions will exist. (For details on the causes, prevention, and solution for chipping, see Peeling.)

CRACKING

Cracking is a series of deep cracks resembling mud cracks in a dry pond. Often in the form of three-legged stars and in no definite pattern, they usually go through the colorcoat and sometimes the undercoat as well.

Cracking Causes

1. Excessive film thickness. Excessively thick topcoats magnify normal stresses and strains that can result in cracking even under normal conditions.
2. Materials not uniformly mixed.
3. Insufficient flash time.
4. Incorrect use of additive.

Cracking Prevention

1. Do not pile on topcoats. Allow sufficient flash and dry time between coats. Do not dry by gun fanning.
2. Stir all pigmented undercoats and topcoats thoroughly. Strain and add fisheye eliminator to topcoats where necessary.
3. Repeat.
4. Read and carefully follow label instructions. Additives not specifically designed for a colorcoat can weaken the final paint film and make it more susceptible to cracking.

Cracking Solution

The affected areas must be sanded to a smooth finish or, in extreme cases, removed down to the bare metal and refinished.

LINE CHECKING

Line checking is similar to cracking, except that the lines or cracks are more parallel and range from very short up to about 18 inches (Figure 21–5).

Line Checking Causes

1. Excessive film thickness.
2. Improper surface preparation. Oftentimes caused by the application of a new finish over an old film that had cracked and was not completely removed.

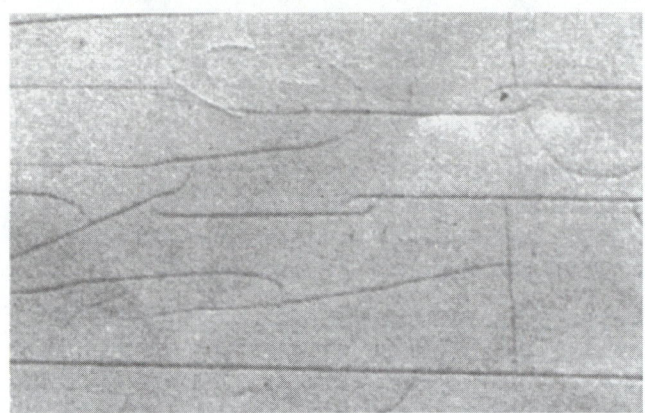

FIGURE 21-5 Note this line checking problem. *(Courtesy of PPG Industries, Inc.)*

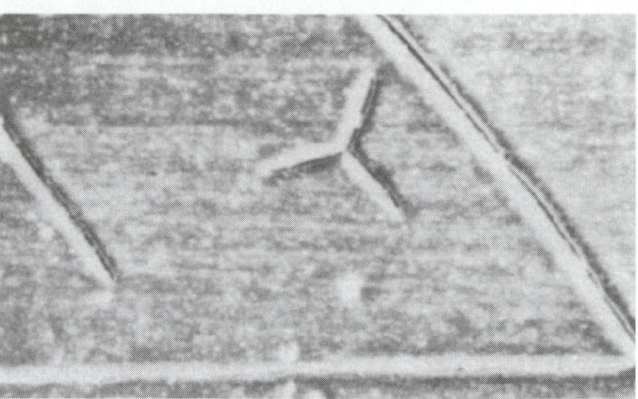

FIGURE 21-6 Crazing looks something like bird tracks. *(Courtesy of PPG Industries, Inc.)*

Line Checking Prevention

1. Do not pile on topcoats. Allow sufficient flash and dry time. Do not dry by gun fanning.
2. Thoroughly clean areas to be painted before sanding. Be sure the surface is completely dry before applying any undercoats or topcoats.

Line Checking Solution

Remove the colorcoat down to the primer and apply a new colorcoat.

MICROCHECKING

Microchecking appears as severe dulling of the film, but when examined with a magnifying glass, it contains many small, microscopic cracks.

Microchecking Causes

Microchecking is the beginning of film breakdown and might be an indication that film failures such as cracking or crazing will develop.

Microchecking Solution

Sand off the colorcoat to remove the cracks; then recoat as required.

CRAZING

Crazing results in fine splits or small cracks—often called crows-feet—that completely checker an area in an irregular manner (Figure 21–6). This problem was common with older lacquer finishes.

Crazing Causes

1. Shop temperature is too cold.
2. Surface tension of original material is under stress

and literally shatters under the softening action of the solvents being applied.
3. OEM lacquer crazes due to age and temperature extremes.

Crazing Prevention

1. Select the thinner or reducer that is suitable for existing shop conditions.
2. Schedule painting to avoid temperature and humidity extremes in the shop or between the temperature of the shop and the job.
3. Bring the vehicle to room temperature before refinishing.

Crazing Solution

1. Continue to apply wet coats of topcoat to melt the crazing and flow pattern together, using the wettest/slowest possible solvent that shop conditions will allow.
2. Use a fast-flashing thinner, which will allow a bridging of subsequent topcoats over the crazing area. (This is one case where bridging is a cure and not a cause for trouble.)
3. If OEM lacquer, remove the old, crazed finish. Then repaint the vehicle.

DIRT IN FINISH

Dirt in finish simply means foreign particles dried in the paint film.

Dirt in Finish Causes

1. Improper solvent cleaning, blowing off, and tack rag wiping of the surface to be painted.
2. Defective air regulator cleaning filter.
3. Dirty working area.
4. Defective or dirty air inlet filters.

5. Dirty spray gun.
6. Technician not wearing proper clothing.

Dirt in Finish Prevention

1. Blow out all cracks and body joints.
2. Solvent clean and tack rag surface thoroughly.
3. Be sure equipment is clean.
4. Work in clean spray area.
5. Replace inlet air filters if dirty or defective.
6. Strain out foreign matter from paint.
7. Keep all containers closed when not in use to prevent contamination.
8. Wear a paint suit and head/hair covering.

Dirt in Finish Solution

1. Rub out finish with rubbing compounds.
2. If dirt is deep in finish, sand and compound to restore gloss. Metallic finishes might show mottling with this treatment and will then require additional colorcoats.

DULLED FINISH

Dulled finish means the gloss retards as the film dries.

DULLED FINISH CAUSES

1. Compounding before thinner evaporates.
2. Using poorly balanced thinner or reducer.
3. Poorly cleaned surface.
4. Topcoats put on wet subcoats.
5. Washing with caustic cleaners.
6. Inferior polishes.

Dulled Finish Prevention

1. Clean work area and surface thoroughly (Figure 21–7).

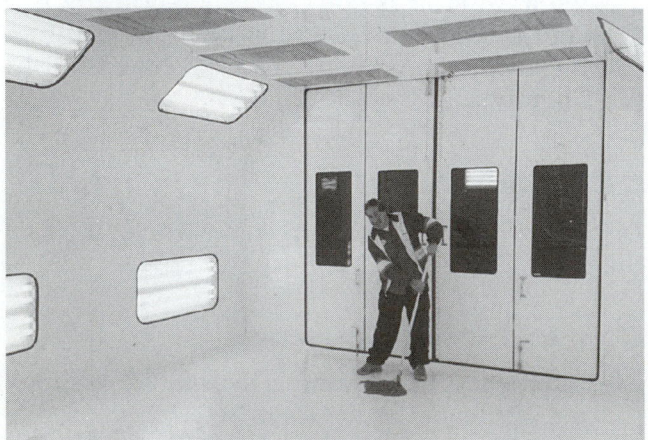

FIGURE 21-7 Cleanliness in the paint booth and proper surface preparation are critical to prevention of paint problems. (*Courtesy of Binks Manufacturing Co.*)

2. Use recommended materials.
3. Allow all coatings sufficient drying time.

Dulled Finish Solution

Allow finish to dry hard and rub with a mild rubbing compound.

FEATHEREDGE SPLITTING

Featheredge splitting appears as stretch marks (or cracking) along the featheredge. It occurs during or shortly after the topcoat is applied over lacquer primer-surfacer.

Featheredge Splitting Causes

1. Piling on the undercoat in heavy, wet coats. The solvent is trapped in undercoat layers that have not had sufficient time to set up.
2. Material not uniformly mixed. Because of the high pigment content of primer-surfacers, it is possible for settling to occur after it has been thinned. Delayed use of this material without restirring results in applying a film with loosely held pigment containing voids and crevices throughout, causing the film to act like a sponge.
3. Wrong thinner.
4. Improper surface cleaning or preparation. When not properly cleaned, primer-surfacer coats can crawl or draw away from the edge because of poor wetting and adhesion.
5. Improper drying. Fanning with a spray gun after the primer-surfacer is applied will result in drying the surface before solvent or air from the lower layers is released.
6. Excessive use (and film build) of putty.

Featheredge Splitting Prevention

1. Apply properly reduced primer-surfacer in thin to medium coats with enough time between coats to allow solvents and air to escape.
2. Stir all pigmented undercoats and topcoats thoroughly. Select thinner that is suitable for existing shop conditions.
3. Select only thinners that are recommended for existing shop conditions.
4. Thoroughly clean areas to be painted before sanding.
5. Apply primer-surfacer in thin to medium coats with enough time between coats to allow solvents and air to escape.
6. Spot putty should be limited to filling minor imperfections. Putty applied too heavily will eventually shrink, causing featheredge splitting.

Featheredge Splitting Solution

Remove finish from the affected areas and refinish.

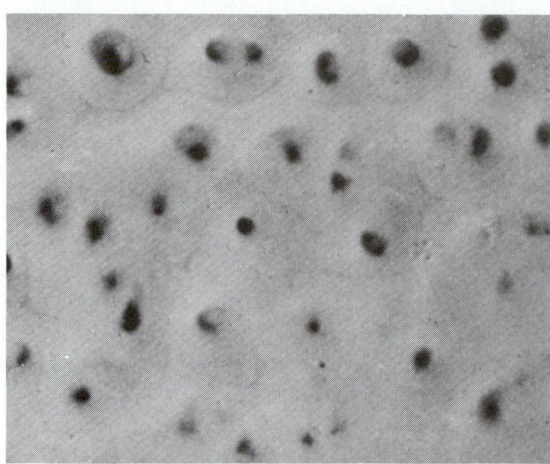

FIGURE 21-8 Fisheyes are small indentations in the surface film that look something like the eyes of a fish. *(Courtesy of PPG Industries, Inc.)*

FIGURE 21-9 With lifting, the paint tends to raise up and swell. *(Courtesy of PPG Industries, Inc.)*

FISHEYES

Fisheyes are small, craterlike openings that appear in the finish after it has been applied. Sometimes the previous finish can be seen in these spots or craters under the new paint (Figure 21–8).

Fisheye Causes

1. Improper surface cleaning or preparation. Many waxes and polishes contain silicone, the most common cause of fisheyes. Silicones adhere firmly to the paint film and require extra effort for their removal. Even small quantities in sanding dust, rags, or from cars being polished nearby can cause this failure.
2. Effects of the old finish or previous repair. Old finish or previous repair can contain excessive amounts of silicone from additives used during their application. Usually, solvent wiping will not remove embedded silicone.
3. Contamination of air lines.

Fisheye Prevention

1. Precautions should be taken to remove all traces of silicone by thoroughly cleaning with wax and grease solvent. The use of fisheye eliminator is in no way a replacement for good surface preparation.
2. Add fisheye eliminator.
3. Drain and clean air pressure regulator daily to remove trapped moisture and dirt. Air compressor tank should also be drained daily.

Fisheye Solution

After the affected coat shows a fisheye problem, apply another double coat of color containing the recommended amount of fisheye eliminator. In severe cases, affected areas should be sanded down and refinished.

LIFTING

Lifting is a condition that causes surface distortion or shriveling, while the topcoat is being applied or is drying (Figure 21–9).

Lifting Causes

1. Use of incompatible materials. Solvents in a new topcoat attack the old surface, which results in a distorted or wrinkled effect.
2. Insufficient flash time. Lifting will occur when the paint film is an alkyd enamel and is only partially cured. The solvents from the coat being applied cause localized swelling or partial dissolving that later distorts the final surface.
3. Improper drying. When synthetic enamel-type undercoats are not thoroughly dry, topcoating with lacquer can result in lifting.
4. Effect of old finish or previous repair. Lacquer applied over a fresh air-dry enamel finish will cause lifting.
5. Improper surface cleaning or preparation. Use of an enamel-type primer or sealer over an original lacquer finish, which is to be topcoated with a lacquer, will result in lifting due to a sandwich effect.
6. Wrong thinner or reducer. The use of lacquer thinners in enamel increases the amount of substrate swelling and distortion, which can lead to lifting, particularly when two-toning or recoating.

Lifting Prevention

1. Avoid incompatible materials, such as thinners with enamel products or incompatible sealers and primers.

2. Do not pile on topcoats. Allow sufficient flash and dry time. The final topcoat should be applied when the previous coat is still soluble or after it has completely dried and is impervious to topcoat solvents.

3. Select the thinner or reducer that is correct for the finish applied and suitable for existing shop conditions.

Lifting Solution

Remove finish from affected areas and refinish.

MOTTLING

Mottling occurs only in metallics when the flakes float together to form a spotty or striped appearance (Figure 21–10).

Mottling Causes

1. Wrong thinner or reducer.
2. Materials not uniformly mixed.
3. Spraying too wet.
4. Holding spray gun too close to work.
5. Uneven spray pattern.
6. Low shop temperature.

Mottling Prevention

1. Select the thinner or reducer that is suitable for existing shop conditions and mix properly. In cold, damp weather use a faster-drying solvent.
2. Stir all pigmented topcoats—especially metallics—thoroughly.
3. Use proper gun adjustments, techniques, and air pressure.
4. Keep your spray gun clean (especially the needle fluid tip and air cap) and in good working condition.

Mottling Solution

Allow colorcoat to set up and apply a drier double coat or two single coats, depending upon which topcoat is to be applied.

ORANGE PEEL

Orange peel is an uneven surface formation—much like that of the skin of an orange. It is caused by poor fusion of atomized paint droplets. Paint droplets dry out before they can flow and level out smoothly (Figure 21–11).

Orange Peel Causes

1. Improper gun adjustment and techniques. Too little air pressure, wide fan patterns, or spraying at excessive gun distances cause droplets to become too dry during their travel time to the work surface, and they remain as formed by the gun nozzle.
2. Extreme shop temperature. When air temperature is too high, droplets lose more solvent and dry out before they can flow and level properly.
3. Improper drying. Gun fanning before paint droplets have a chance to flow together will cause orange peel.
4. Improper flash or recoat time between coats. If first coats of enamel are allowed to become too dry, solvent in the paint droplets of following coats will be absorbed into the first coat before proper flow is achieved.
5. Wrong thinner or reducer. Underdiluted paint or paint thinned with fast evaporating thinners or reducers causes the atomized droplets to become too dry before reaching the surface.

FIGURE 21-10 Mottling is a streaking problem common to metallic paints. [Courtesy of PPG Industries, Inc.]

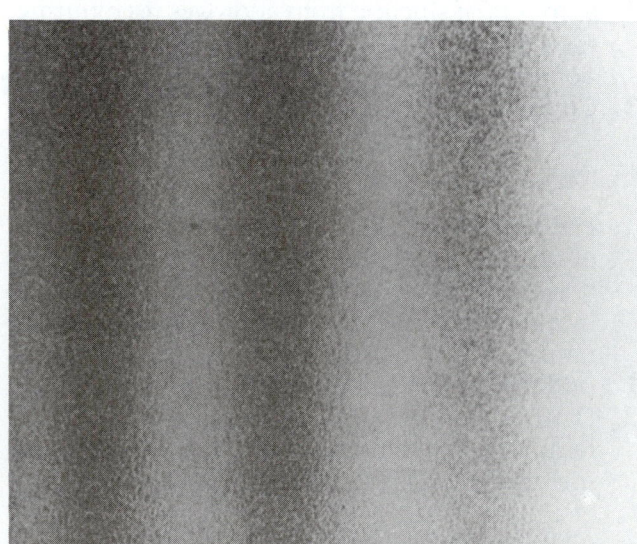

FIGURE 21-11 Orange peel is a rough, bumpy paint film. [Courtesy of PPG Industries, Inc.]

6. Too little thinner or reducer.
7. Materials not uniformly mixed. Many finishes are formulated with components that aid fusion. If these are not properly mixed, orange peel will result.

Orange Peel Prevention

1. Use proper gun adjustments, techniques, and air pressure.
2. Schedule painting to avoid temperature and humidity extremes. Select the thinner or reducer that is suitable for existing conditions. The use of a slower-evaporating thinner or reducer will overcome this.
3. Allow sufficient flash and dry time. Do not dry by fanning.
4. Allow proper drying time for undercoats and topcoats (not too long or not too short).
5. Select the thinner or reducer that is most suitable for existing shop conditions to provide good flow and leveling of the topcoat.
6. Reduce to the recommended viscosity with the proper thinner/reducer.
7. Stir all pigmented undercoats and topcoats thoroughly.

Orange Peel Solution

Compounding might help; use a mild polishing compound for minor orange peel and a rubbing compound for rougher orange peel. In extreme cases, wet sand the area before compounding. If you burn through the topcoat, you will have to refinish the area, using a slower-evaporating thinner or reducer at the correct air pressure.

PEELING

Peeling is caused by a loss of adhesion between paint and substrate (topcoat to primer and/or old finish, or primer to metal) (Figure 21–12).

Peeling Causes

1. Improper cleaning or preparation. Failure to remove sanding dust and other surface contaminants will keep the finish coat from coming into proper contact with the substrate.
2. Improper metal treatment.
3. Materials not uniformly mixed.
4. Failure to use proper sealer.

Peeling Prevention

1. Thoroughly clean areas to be painted. It is always good shop practice to wash the sanding

dust off the area to be refinished with cleanup solvent.
2. Use the correct metal conditioner and conversion coating.
3. Stir all pigmented undercoats and topcoats thoroughly.
4. In general, sealers are recommended to improve adhesion of topcoats.

Peeling Solution

Remove finish from an area slightly larger than the affected area and refinish.

PINHOLING

Pinholing is tiny holes or groups of holes in the finish, or in putty or body filler, usually the result of trapped solvents, air, or moisture.

Pinholing Causes

1. Improper surface cleaning or preparation. Moisture left on primer-surfacers will pass through the wet topcoat to cause pinholing.
2. Contamination of air lines. Moisture or oil in air lines will enter paint while it is being applied and cause pinholes when released during the drying stage.
3. Wrong gun adjustment or technique. If adjustments or techniques result in an application that is too wet, or if the gun is held too close to the surface, pinholes will occur when the air or excessive solvent is released during drying.
4. Wrong thinner or reducer. The use of a solvent that is too fast for the shop temperature tends to make the refinisher spray too close to the surface in order to get an adequate flow. When the

FIGURE 21-12 With peeling, a noticeable area of paint lifts. [Courtesy of PPG Industries, Inc.]

solvent is too slow, it is trapped by subsequent topcoats.

5. Improper drying. Fanning a newly applied finish can drive air into the surface or cause a dry skin—both of which result in pinholing when solvents retained in lower layers come to the surface.

Pinholing Prevention

1. Thoroughly clean all areas to be painted. Be sure the surface is completely dry before applying undercoats or topcoats.
2. Drain and clean air pressure regulator daily to remove trapped moisture and dirt. Air compressor tank should also be drained daily.
3. Use proper gun adjustments, techniques, and air pressure.
4. Select the thinner or reducer that is suitable for existing shop conditions.
5. Allow sufficient flash and dry time. Do not dry by fanning.

Pinholing Solution

Sand affected area smooth and refinish.

PLASTIC FILLER BLEED-THROUGH

Plastic filler bleed-through results in discoloration (normally yellowing) of the topcoat color (Figure 21–13).

Plastic Filler Bleed-through Causes

1. Too much hardener.
2. Applying topcoat before plastic filler is cured.
3. Improper filling.

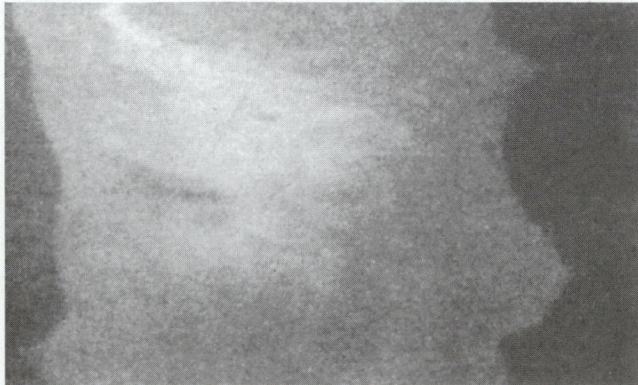

FIGURE 21-13 With a bleed-through problem, colors under the paint are visible. *[Courtesy of PPG Industries, Inc.]*

Plastic Filler Bleed-through Prevention

1. Use correct amount of hardener.
2. Allow adequate cure time before refinishing.
3. Use proper filler (stain free).

Plastic Filler Bleed-through Solution

1. Remove filler patch.
2. Cure topcoat, sand, and refinish.

PLASTIC FILLER NOT DRYING

Plastic filler not drying causes the filler to remain soft after applying.

Plastic Filler Not Drying Causes

1. Insufficient amount of hardener.
2. Hardener exposed to sunlight.

Plastic Filler Not Drying Prevention

1. Add recommended amount of hardener.
2. Be sure hardener is fresh and avoid exposure to sunlight.

Plastic Filler Not Drying Solution

Scrape off plastic filler and reapply.

RUNS OR SAGS

Runs occur when gravity produces a mass slippage of an overwet, thick paint film. The weight of the film will cause it to slide or roll down the surface. A large area of paint will flow down and form large globules of paint (Figure 21–14).

A **sag** is a partial slipping down of the paint created by a film that is too heavy to support itself. It appears something like a curtain.

Runs or Sags Causes

1. Too much thinner or reducer.
2. Wrong thinner or reducer.
3. Excessive film thickness without allowing proper dry time.
4. Low air pressure (causing lack of atomization), holding gun too close, or making too slow a gun pass.
5. Shop or surface too cold.

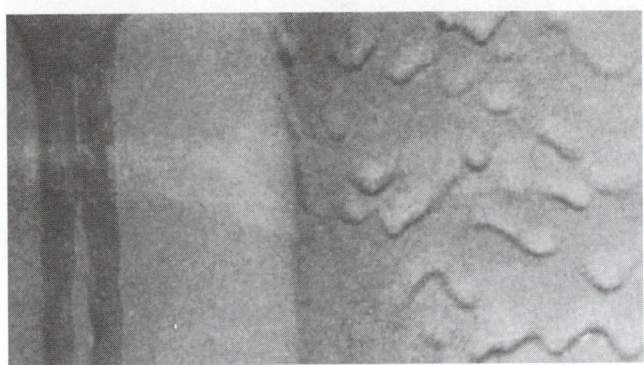

FIGURE 21-14 Runs and sags are due to too much wet paint in one place. *(Courtesy of PPG Industries, Inc.)*

FIGURE 21-15 Rust under a finish is due to improper surface preparation. *(Courtesy of PPG Industries, Inc.)*

Runs or Sags Prevention

1. Read and carefully follow the instructions on the label.
2. Select the proper thinner/reducer.
3. Do not pile on finishes. Allow sufficient flash and dry time in between coats.
4. Use proper gun adjustment, techniques, and air pressure.
5. Allow vehicle surface to warm up to at least room temperature before attempting to refinish. Try to maintain an appropriate shop temperature for paint areas.

Runs or Sags Solution

Wash the affected area and let dry until it can be sanded to a smooth surface and refinish. Sand runs until smooth. Polish to bring back the gloss. If you sand or burn through, refinish the panel.

RUST UNDER FINISH

Rust under finish will show up as raised surface spots, peeling, or blistering (Figure 21–15).

Rust under Finish Causes

1. Improper metal preparation.
2. Broken paint film allows moisture to creep under surrounding finish.
3. Water in air lines.
4. Fingerprints (moisture on skin).

Rust under Finish Prevention

1. Locate source of moisture and seal.
2. When replacing ornaments or molding, be careful not to break paint film and allow dissimilar metals to come in contact. This contact can produce electrolysis that might cause a tearing away or loss of good bond with the film.
3. Wear plastic or rubber gloves.

Rust under Finish Solution

1. Seal entrance of moisture from inner part of panels.
2. Sand down to bare metal, prepare it, and treat with phosphate before refinishing.

SANDSCRATCH SWELLING

Sandscratch swelling is enlarged sand scratches caused by the swelling action of topcoat solvents.

Sandscratch Swelling Causes

1. Improper surface cleaning or preparation. Use of too coarse sandpaper or omitting a sealer in panel repairs greatly exaggerates swelling caused by thinner penetration.
2. Improper thinner or reducer, especially a slow-dry thinner or reducer when sealer has been omitted.
3. Underreduced or wrong thinner (too fast) used in primer-surfacer causes "bridging" of scratches.

Sandscratch Swelling Prevention

1. Use appropriate grits of sanding materials for the topcoats being used.
2. Seal to eliminate sandscratch swelling. Select thinner or reducer suitable for existing shop conditions.
3. Use proper thinner and reducer for primer-surfacer.

Sandscratch Swelling Solution

Sand affected area down to a smooth surface and apply the appropriate sealer before refinishing.

SOLVENT POPPING

Solvent popping is blisters on the paint surface caused by trapped solvents in the topcoats or primer-surfacer—a situation that is further aggravated by force drying or uneven heating.

Solvent Popping Causes

1. Improper surface cleaning or preparation.
2. Wrong thinner or reducer. Use of fast-dry thinner or reducer, especially when the material is sprayed too dry or at excessive pressure, can cause solvent popping by trapping air in the film.
3. Excessive film thickness. Insufficient drying time between coats and too heavy application of the undercoats can trap solvents causing popping of the colorcoat as they later escape.

Solvent Popping Prevention

1. Thoroughly clean areas to be painted.
2. Select the thinner or reducer suitable for existing shop conditions.
3. Do not pile on undercoats or topcoats. Allow sufficient flash and dry time. Allow proper drying time for undercoats and topcoats. Allow each coat of primer-surfacer to flash naturally—do not fan.

Solvent Popping Solution

If damage is extensive and severe, paint must be removed down to the undercoat or metal, depending on the depth of the blisters; then refinish. In less severe cases, sand out, resurface, and retopcoat.

UNDERCOAT SHOW-THROUGH

Undercoat show-through is a variation in surface color (Figure 21–16).

Undercoat Show-through Causes

1. Insufficient colorcoats.
2. Repeated compounding.

Undercoat Show-through Prevention

1. Apply good coverage of color.
2. Avoid excessive compounding or polishing.

Undercoat Show-through Solution

Sand and refinish.

WATER SPOTTING

Water spotting is the general dulling of gloss in spots or masses of spots (Figure 21–17).

Water Spotting Causes

1. Water evaporating on finish before it is thoroughly dry.
2. Washing finish in bright sunlight.

Water Spotting Prevention

1. Do not apply water to a fresh paint job and try to keep a newly finished car out of the rain.

FIGURE 21-16 This is an undercoat show-through problem. *[Courtesy of PPG Industries, Inc.]*

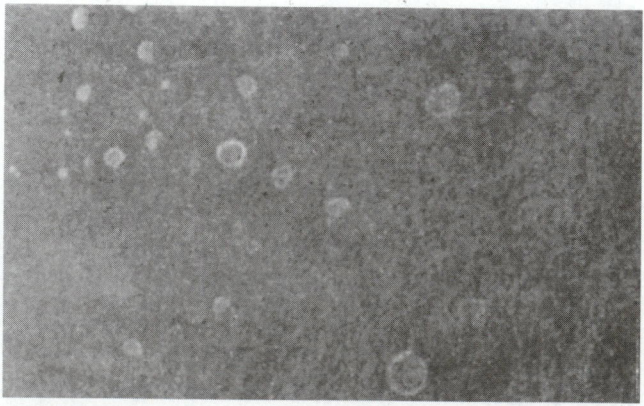

FIGURE 21-17 Water spotting dulls paint gloss. *[Courtesy of PPG Industries, Inc.]*

Allow sufficient dry time before delivering the car to the customer.

2. Wash car in the shade and wipe completely dry.

Water Spotting Solution

Compound or polish with rubbing or polishing compound. In severe cases, sand affected areas and refinish.

WET SPOTS

Wet spots is discoloration and/or the slow drying of various areas.

Wet Spots Causes

1. Improper cleaning and preparation.
2. Improper drying of excessive undercoat film build.
3. Sanding with contaminated solvent.

Wet Spots Prevention

1. Thoroughly clean all areas to be painted.
2. Allow proper drying time for undercoats.
3. Wet sand with clean water.

Wet Spots Solution

Wash or sand all affected areas thoroughly and then refinish.

WRINKLING

Wrinkling is a severe puckering of the paint film that appears like the skin of a prune (fruit) and is more common with enamel paints. There is a loss of gloss as paint dries. Minute wrinkling may not be visible to the naked eye (Figure 21–18).

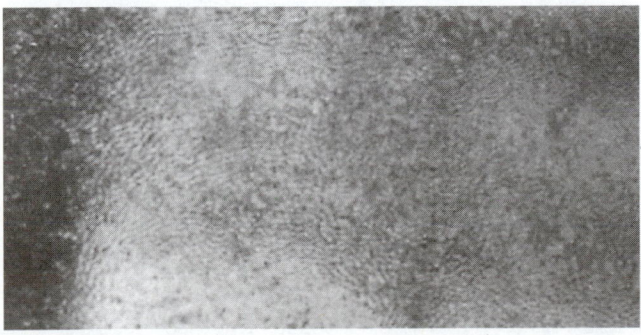

FIGURE 21-18 Wrinkling gives the paint a "prune look." *(Courtesy of PPG Industries, Inc.)*

Wrinkling Causes

1. Improper dry. When a freshly applied topcoat is baked or force dried too soon, softening of the undercoats can occur. This increases topcoat solvent penetration and swelling. In addition, baking or force drying causes surface layers to dry too soon. The combination of these forces causes wrinkling.
2. Piling on heavy or wet coats. When enamel coats are too thick, the lower wet coats are not able to release their solvents and set up at the same rate as the surface layer, which results in wrinkling.
3. Improper reducer or incompatible materials. A fast-dry reducer or the use of a lacquer thinner in enamel can cause wrinkling.
4. Improper or rapid change in shop temperature. Drafts of warm air cause enamel surfaces to set up and shrink before sublayers have released their solvents, which results in localized skinning in uneven patterns.

Wrinkling Prevention

1. Allow proper drying time for undercoats and topcoats. When force drying alkyd enamel, baking additive is required to retard surface setup until lower layers harden. Lesser amounts can be used in hot weather. Read and carefully follow label instructions.
2. Do not pile on topcoats. Allow sufficient flash and dry time.
3. Select proper reducer and avoid using incompatible materials such as a reducer with lacquer products or thinner with enamel products.
4. Schedule painting to avoid temperature extremes or rapid changes.

Wrinkling Solution

Remove wrinkled enamel and refinish.

21.2 REMOVING MASKING MATERIALS

After the topcoat has dried, the masking paper and tape must be removed. If the finish has been force dried, the masking should be removed while the paint finish is still warm. If the finish is allowed to cool, the tape is more difficult to remove and leaves adhesive particles on the areas that were masked.

The tape should be removed slowly so that it comes off evenly. Pull the tape away from the paint edge—never across it. If the vehicle was not force dried, take care not to touch any painted areas because the paint might not be completely dry. Fingerprints or tape marks could result if the surface is touched. Also, be careful of loose-fitting clothing or belt buckles that could accidentally rub against the paint.

21.3 FINAL DETAILING

The objective in **final detailing** is to locate and correct any defect that may cause customer complaints. Corrective steps for final detailing are:

1. Sanding and filing
2. Compounding
3. Machine glazing
4. Hand glazing

Each of these steps has its own requirements. As a general rule, finer and finer grades of products will be used for all of these steps. Use progressively finer wet sandpaper and compounds. Also, a single product line should be used throughout the repair and the manufacturer's recommendations should always be followed.

PAINT SURFACE CHIPS

Paint surface chips result from mechanical impact damage to the paint film: door dings, damage from road debris, and so on. If the whole vehicle is not refinished, you should take the time to touch up chips in the old paint. Use the paint mixed for the repair. It will usually have hardener in it to speed curing.

Degrease the area with wax and grease remover. If you use a small paint brush, slowly move the touch-

up paint straight into each chip (Figure 21–19). On smaller chips, a toothpick will reach into the chip more efficiently (Figure 21–20). If you are using a solid color, use a thicker viscosity touch-up paint to fill the chip in one application. If you have metallic paint, use thinner touch-up paint and several coats to help match the color.

A

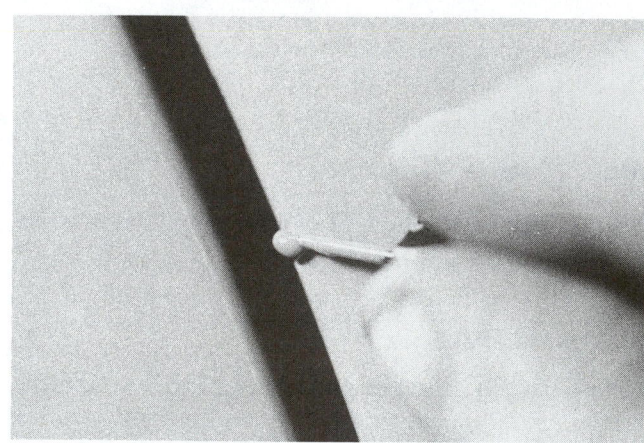

B

C

FIGURE 21-20 Basic steps for chip touch-up. (A) Small chip on edge of panel. (B) Use toothpick to deposit catalyzed paint into chip. (C) Finished chip repair.

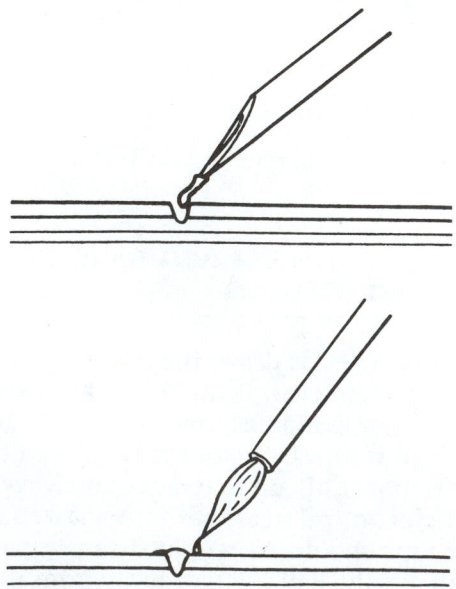

FIGURE 21-19 A small paint brush can be used to fill larger paint chips. Make sure the enamel has hardener in it so you can wet sand and compound the repair.

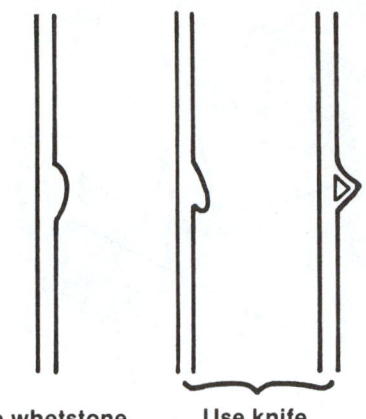

Use whetstone.   Use knife.

FIGURE 21–21 Note types of paint protrusions and their repair method.

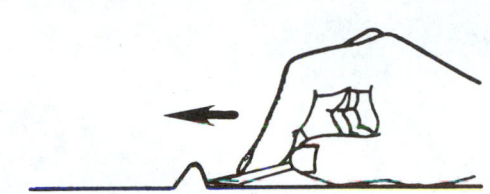

FIGURE 21–22 Large protrusions in paint can often be cut off with a knife or razor blade. Be careful not to damage the surrounding surface.

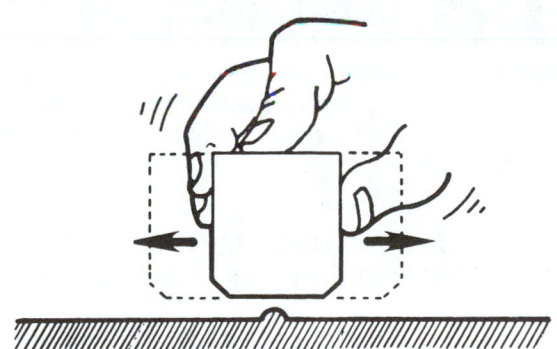

FIGURE 21–23 Whetstone will cut off smaller protrusions in paint.

Allow the paint to cure sufficiently before wet sanding and polishing the chip repairs level.

PAINT SURFACE PROTRUSIONS

A **paint surface protrusion** is a particle of paint or other debris sticking out of the paint film after refinishing (Figure 21–21). This problem results from a lack of cleanliness. The spray gun, paint materials, or booth were contaminated.

If the protrusion is large, repair it with a knife or single-edge razor blade as follows:

1. Being careful not to take off more finish than is necessary, cut off the protrusion (Figure 21–22).

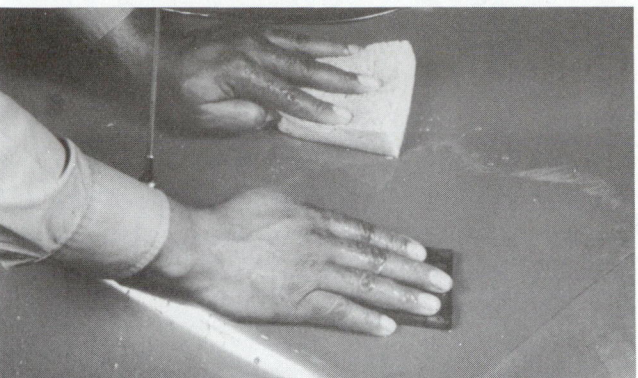

FIGURE 21–24 After finesse sanding, wet sand with a block to further smooth the surface.

The tip of the knife or razor blade should be pointed slightly upward.
2. Smooth the area with a finesse sanding block (Figure 21–23) or #3500-grit sandpaper (Figure 21–24).
3. Blow off any particles. Finish with an extra-fine rubbing compound.

DIRT-NIB FILING

Dirt-nib files can be used to remove any defect which is on or above the surface of the paint. This includes runs, sags, and dirt. A dirt-nib file will remove the protrusion with minimum damage to the surrounding paint film. Dirt-nib files are available commercially, or you can make your own.

To make a dirt-nib file, cut off a short piece of a vixen body file (Figure 21–25). After dressing down the sharp corners on a grinder, place a piece of #400 grit sandpaper on a flat surface and draw the piece of file back and forth over the sandpaper until the teeth become smooth. A light machine oil will speed up the operation.

Do not try to break off a piece of the vixen body file. It can shatter!

To use a dirt-nib file, place the file lightly on the paint film. Use short, straight strokes in one direction only. Two or three light passes will remove most defects. After filing, the area must be sanded to remove the file marks.

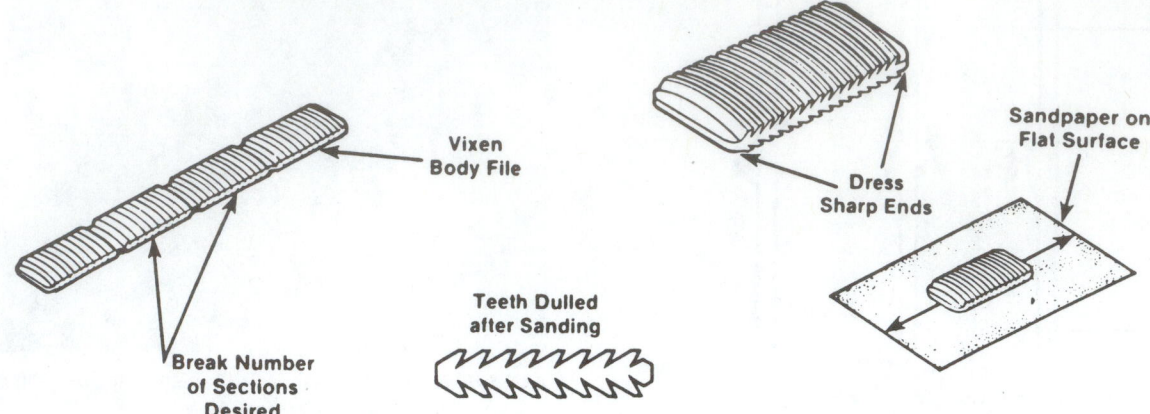

FIGURE 21-25 Surface file can be purchased or made from an old file. Be sure to wear eye protection.

FINESSE SANDING

After dirt-nib filing, **finesse sanding** can be used to level the protrusion with the surrounding paint film. Use a finesse sanding block as follows:

1. Dress the surface of the block with wet #220 grit sandpaper to make it smooth and flat.
2. Thoroughly soak the block in clear water.
3. Place the block over the protrusion and move it back and forth. If necessary, use a little water to help make the movement smoother.
4. When the protrusion has all but disappeared, blow off the loose particles and finish the job with rubbing, then polishing compound.

Table 21–1 summarizes finesse finishing procedures.

 **WARNING** If you fail to catalyze enamel paints with a hardener, you may not be able to file, wet sand, and repair minor paint problems right away. If slight defects in the finished surface should occur, they should not be compounded until the paint has had a chance to cure. This could involve a period of several days if not catalyzed.

COLOR SANDING

Color sanding can be done to smooth the paint surface on larger areas, as when removing orange peel (Figure 21–26).

Color sanding should normally be done with a backing pad or rubber sanding block to avoid crowning of the paint surface. A pad or block will help keep large, relatively flat surfaces level and uncrowned. On restricted and curved surfaces, you can use only your hand to color sand.

Sanding blocks and sandpapers are available in a variety of grit sizes. For surface repairs, use coarser

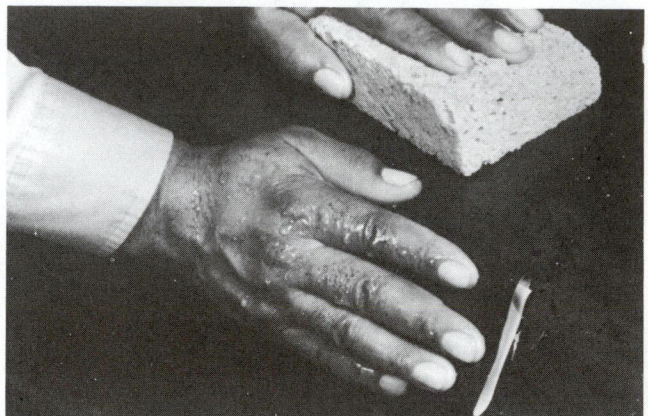

FIGURE 21-26 Wet sanding is a common method of repairing minor paint surface problems. You must be careful not to cut through the colorcoat or refinishing will be needed. *(Courtesy of 3M)*

wet sandpapers, #400 to #600. For finesse finishing, #1000, #1200, and finer grits of wet sandpaper are typically used.

Wet sand in a back-and-forth or small circular motion. Use plenty of water to flush away paint debris. Dip the block in a bucket of water. You can also use a sponge, garden hose, or spray bottle to flow water over the area. Some air sanders can also be equipped with a wet sanding attachment. It uses a small plastic hose to feed flushing water up to the sanding pad.

Check the defect often when using a sanding block. You do not want to cut too deeply into the finish. If you cut through the clearcoat or color, repainting will be necessary. Wash surfaces thoroughly with clean water and a sponge after color sanding.

RUBBING COMPOUND

Rubbing compounds generally contain coarser grit abrasives than polishes. They are used to more rapidly cut the surface film by hand or machine. Rubbing compounds produce a low surface gloss.

TABLE 21-1: FINESSE FINISHING PROCEDURES

Paint Type	Paint Condition	Procedure			
		Wet Sanding	Compounding	Machine Glazing	Hand Glazing
Refinish paints: cured enamels/urethanes* (air-dried more than 48 hours or baked)	1. Minor dust nibs or mismatched orange peel (light sanding) 2. Heavy orange peel, dust nibs, paint, runs or sags	1. Fine 1500 2. Fine 1200	1. — 2. Microfinishing compound	1. Finishing material 2. Finishing material	1. Hand glaze 2. Hand glaze
Refinish paints: fresh enamels/urethanes* (air-dried 24 to 48 hours)	1. Minor dust nibs or mismatched orange peel (light sanding) 2. Heavy orange peel, dust nibs, paint runs or sags	1. Fine 1500 2. Fine 1200	1. Microfinishing compound 2. Microfinishing compound	1. Microfinishing glaze 2. Microfinishing glaze	1. Hand glaze 2. Hand galze
Refinish paints: acrylic lacquer	1. Low gloss or overspray 2. Low gloss, minor orange peel, or overspray 3. Low gloss, moderate orange peel, or dust nibs 4. Low gloss, heavy orange peel, paint runs or sags	1. — 2. — 3. Fine 1200 4. Fine 1000	1. — 2. Paste or rubbing compound (heavy cut) 3. Microfinishing compound (medium cut) 4. Paste or rubbing compound (heavy cut)	1. Machine glaze 2. Machine glaze 3. Machine glaze 4. Machine glaze	1. Hand glaze 2. Hand glaze 3. Hand glaze 4. Hand glaze
All factory applied (OEM)	1. New car prep or fine wheel marks 2. Coarse swirl marks, chemical spotting or light oxidation 3. Overspray or medium oxidation 4. Heavy oxidation or minor acid rain pitting 5. Dust nibs, minor scratches, or major acid rain pitting 6. Orange peel, paint runs or sags	1. — 2. — 3. — 4. — 5. Fine 1500 6. Fine 1200 or 1500	1. — 2. — 3. Microfinishing compound (medium cut) 4. Rubbing compound (heavy cut) 5. — 6. Microfinishing compound (medium cut)	1. — 2. Finishing material 3. Finishing material 4. Finishing material 5. Finishing material 6. Finishing material	1. Hand glaze liquid polish 2. Hand glaze 3. Hand glaze 4. Hand glaze 5. Hand glaze 6. Hand glaze

*Enamels/urethanes—as referred to in this chart—are catalyzed paint systems (including acrylic enamel, urethane, acrylic urethanes, acrylic urethane enamels, polyurethane enamels, and polyurethane acrylic enamels) and nonisocyanate-activated paint systems used in color or clear coats.

Rubbing compounds are available in various cutting strengths for both hand and machine compounding. **Hand compounds** are oil-based to provide lubrication. Small areas or blended areas are best done by hand compounding (Figure 21–27). On large surfaces, machine compounding is recommended.

Rubbing compounds are hand compounds used to do the following:

1. Eliminate fine sand scratches around a repair area
2. Correct a gritty surface

FIGURE 21-27 Here is a basic procedure for removing a minor flaw that is not all the way through the colorcoat. (A) Minor damage to paint film. (B) Place rubbing compound on rag. (C) Use rubbing compound to cut and level area around flaw. (D) Use star wheel to clean polishing pad. (E) Machine polish area after using rubbing compound. (F) Finished repair looks like surrounding paint.

3. Smooth and bring out some of the gloss of lacquer topcoats
4. Remove light scratches and small dirt particles before painting as a final smoothing step.

HAND COMPOUNDING

Fold a soft, lint-free flannel cloth into a thick pad or roll it into a ball and apply a small amount of hand compound to it. Use straight, back-and-forth strokes and medium-to-hard pressure until the desired smoothness is achieved.

Hand compounding takes a lot of elbow grease and is time consuming. To keep the compounding of topcoats to a minimum, it is important to apply the finish as wet as possible (without sags or runs) by using the proper thinner for the shop temperature.

When using hand polishes or glazes, apply the glaze to the surface using a clean dry cloth. Rub the glaze thoroughly into the surface. Then wipe dry. Glazes can fill and cover up some scratches which should be buffed out. These kinds of scratches will reappear after the vehicle has been washed a few times.

Table 21–2 shows some applications for different rubbing and polishing compounds.

MACHINE COMPOUNDING

Machine compounds are water-based to disperse the abrasive while using a power buffer. A buffing pad is rotated by an electric or air polisher to force the compound over the paint surface. If done properly, this will quickly bring it to a medium gloss (Figure 21–28).

Swirl marks are patterns of very fine scratches produced by power buffing or compounding. They are caused by a dirty, worn buffing pad, too much pressure on the buffer, or too coarse a compound. Alway keep your buffing pad clean and replace it when worn.

Never lay the face of a buffing pad on a workbench or any surface that could contaminate the pad with dirt and debris. This could scratch the paint badly.

Edge masking involves taping over panel edges and body lines prior to machine buffing or polishing to protect the paint from burn-through. Masking tape is applied to these surfaces to protect them. Buff right up to the tape.

Burn-through occurs when the pad removes too much paint on an edge or lip of a body surface. Since machine buffing will cut more quickly on these areas, always protect them with masking tape or paper.

After the compounding is completed, remove the tape and compound the edge by hand—just enough to produce a smooth finish. Keep in mind that body

TABLE 21-2: POLISHING AND RUBBING COMPOUNDS

Grade	Liquid	Paste	Use and Application
Very fine	Machine or hand	—	Used to remove swirl marks on topcoat. Spread material evenly with buffing wheel pad before starting compounding.
Fine	Machine or hand	Hand (add water for machine use)	Used to level orange peel. Can also be used to clean, polish, and restore older finishes leaving no wheel marks or swirls.
Medium	Machine or hand	Paste (add water for machine use)	Used for quick-leveling orange peel. Can be used to repair other minor paint defects.
Coarse	Machine	Machine	Used for compounding before final topcoating (see Chapter 18)

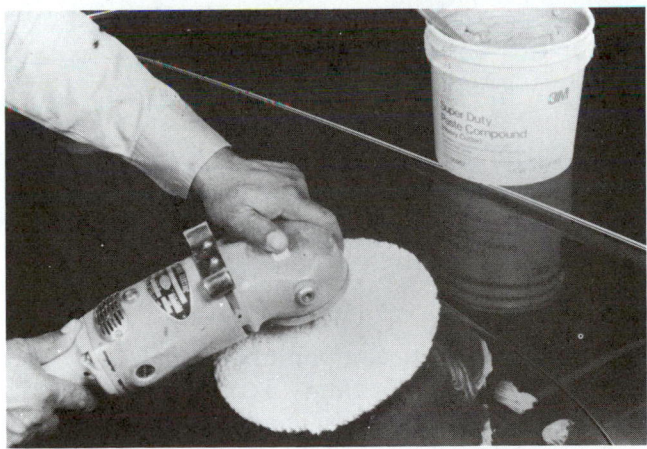

FIGURE 21-28 Rubbing compounds are coarser than polishing compounds. They can be used to quickly remove bad paint.

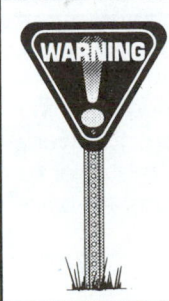

Never use a power buffer with a hand compound. This will cause deep scratches, swirl marks, and burn-through. Use only machine compounds when power buffing.

SHOP TALK

A common mistake for the beginner is to burn through new paint while machine compounding. In an effort to make the paint job really shine, he or she cuts right through the paint to the primer or basecoat. The result will usually be a time-consuming repaint of the panel.

FIGURE 21-29 Polishing compounds are for bringing out shine. *(Courtesy of PPG Industries, Inc.)*

lines usually retain less paint than flat surfaces and thus should get only minimum compounding.

Many finesse finishing systems recommend the use of specific buffing and polishing pads. Wool pads have been replaced with finer, more uniform synthetic pads.

Apply machine compound (Figure 21–29) over a small area using a medium- to coarse-bristled paintbrush or a squeeze bottle. Then compound using a slow-speed power buffer. It is important to use a slow-speed machine to avoid static buildup and high surface temperatures. Do not push down on the buffer. Let the weight of the machine do the work.

CAUTION

Wear eye protection when machine compounding or buffing. It is very easy for the pastelike material to fly into your face and eyes—a painful mistake!

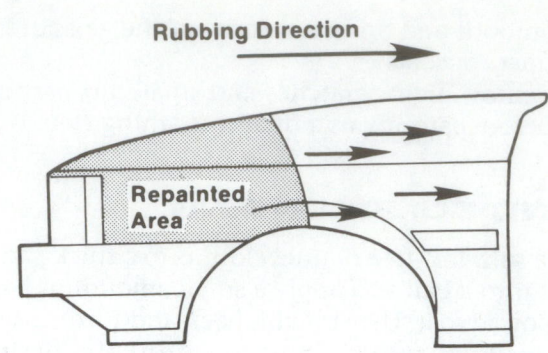

FIGURE 21-30 Note rubbing or polishing direction over repair area.

Because the compound has a tendency to dry out, do not try to do too large an area at one time. Always keep the machine moving to prevent cutting through or burning the topcoat. As the compound starts to dry out, lift up a little on the machine so pad speed increases. This will make the surface start to shine.

POLISHING

Polishing involves using a very fine compound to bring the paint surface up to full gloss. It is usually done after compounding. You can hand polish small or hard-to-reach areas. Machine polish larger areas to save time. Figure 21–30 shows a typical polishing procedure for a repainted area.

Slight defects in the topcoat can be repaired by polishing. The choice of compounds depends on the extent of the damage. Final polishing should always be done with an extra-fine polishing compound.

Using Buffers and Polishers

When using buffing and polishing pads, do the following:

1. Inspect, clean, or replace pads often to avoid residue buildup.
2. Use separate pads for different grades and types of products.
3. When applying the compound, apply an "X" of the product to the surface. Work it around the face of the pad before hitting the machine's trigger. This will help prevent compound from flying all over.

The buffer has an effect on the cutting action. For example, the higher the rpm, the higher the cutting rate and the lower the rpm, the lower the cutting rate. The faster the orbital buffer is moved across the panel, the slower the cutting rate. The slower the buffer is moved, the higher the cutting rate.

FIGURE 21-31 Dual-action buffer will help remove swirl marks. *(Courtesy of S. M. Arnold Corp.)*

Excessive buffing heat can cause swirl marks, warping, discoloring, and hazing, and the material can dry out too quickly. If the area is hot to the touch, there is too much heat. Cool it with water.

When using a buffer, a painter should do the following:

1. Keep the pad flat or at about a 5-degree angle to the surface.
2. Let the weight of the machine do the work.
3. Use care around panel edges and character lines to avoid burn-through.
4. Check the repair area often and apply more product as needed.
5. Compound until the product begins to dry. Do not keep polishing on a dry surface.

Static electricity when machine compounding causes the product to cling to the surface being repaired. Avoid static by grounding the vehicle. You might also want to add 25 percent rubbing alcohol to the water used to cool the surface.

When using rubbing compounds and machine glazes, do the following:

1. Use a single manufacturer's product line.
2. Follow the manufacturer's recommendations for use.
3. Use the materials sparingly.
4. Use the buffing wheel to distribute the material evenly over the area to be repaired.
5. Keep the pad flat and directly over the surface being repaired.
6. Use a slow, circular motion.
7. Use the finest product possible. Using a finer product may take a little longer initially, but will generally require less time to complete the repair.
8. Reduce swirl marks by avoiding coarse products and worn buffing pads.

Instead of a circular action buffer, you should use an orbital action machine for final polishing. An **orbital action polisher** will move the polishing compound in a random manner to prevent swirl marks left from machine compounding. Swirl marks are tiny lines in the paint film from the abrasive action of the coarser compound (Figure 21–31).

21.4 CARING FOR A NEW FINISH

A newly refinished vehicle must receive special care, as the paint can still be curing for several months. Each paint manufacturer will have specific recommendations for caring for a new finish. Explain all precautions to the vehicle owner.

To care for a new finish, you and the customer should do the following:

1. Avoid commercial car washes and harsh cleaners for 1 to 3 months.
2. Hand wash using only water and a soft sponge for the first month. Dry with cotton towels only. Do not use a chamois.
3. Avoid waxing and polishing for up to three months. After that time, use a wax designed for basecoat/clearcoat finishes, as they are the least aggressive.
4. Avoid scraping ice and snow near newly refinished surfaces.
5. Flush gas, oil, or fluid spills with water as soon as possible for the first month. Do not wipe off.

21.5 DECAL REPLACEMENT

Whenever overlays or decals are badly damaged, the only solution is replacement (Figure 21–32). To remove the decal, use a heat gun to soften the adhesive on the transfer (Figure 21–33). Start at one edge and slowly peel the decal back. Work the heat over the area until the sheet is completely off.

After the old decal is removed, repair any damaged metal and prime the repair. With either a new panel or a repaired one, sand the surface smooth (Figure 21–34). Then clean with wax and grease remover.

The first step in the reinstallation of the decal is to make a template of the area to be covered. As an example, we will describe installing a woodgrain transfer. Tack a piece of masking paper over the area for the transfer material. Align it along a body line.

With the template paper securely taped to the panel, mark the centerline of the panel on both the panel and the template. Smooth the paper flush against the panel and mark the front, rear, and bottom edges of the panel. If the woodgrain transfer on adjacent panels has a plank design, mark the top horizontal plank line on the front and rear edges of the panel.

FIGURE 21-32 Decals are commonly used today. You should know how they are removed and replaced.

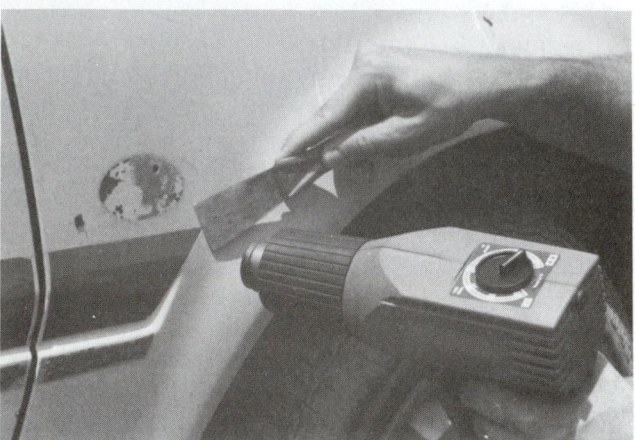

FIGURE 21-33 Heat gun will soften decals and overlays so they can be peeled off more easily. A dull scraper may help removal but do not gouge paint surface with tool.

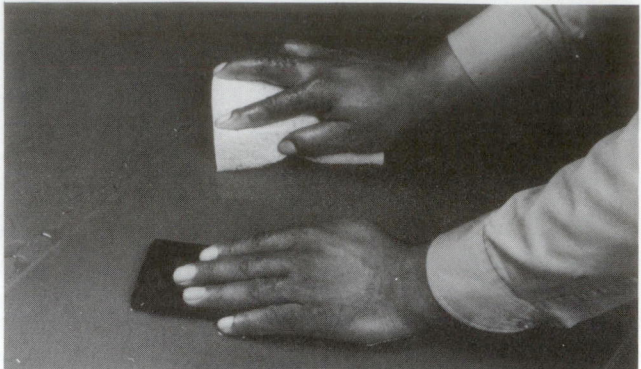

FIGURE 21-34 After removal of decal, wet sand area to smooth and texture it for new decal.

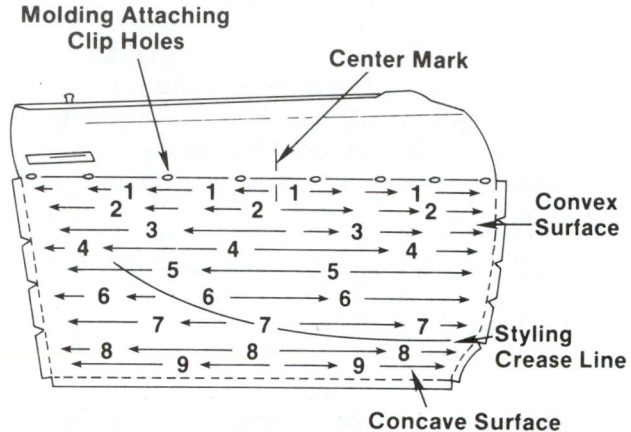

FIGURE 21-35 Study transfer installation sequence. This is for a right front door. *(Courtesy of General Motors Corp.)*

Remove the template from the panel and lay it out on a flat, clean work surface. Measure $^3/_4$ inches (19 mm) out from the panel outline and mark another perimeter line. Oversizing the template this way will allow room for fitting the transfer to the panel. With a pair of scissors, cut out the template along the perimeter line. Mark the front edge of the template on the backside of the paper.

Now, roll out a sufficient amount of overlay, and cut it to length. Lay the transfer face down on the work surface. Turn the template over and place it face down on the transfer. Make sure that the woodgrain is running left to right and that the horizontal planking lines on the template align with plank lines on the transfer (Figure 21-35). Trace the outline of the template on the transfer backing paper and cut the paper to shape with a pair of scissors. Align the woodgrain as close as possible.

Hold the transfer cutout against the panel again. Carefully position the top edge of the transfer with the centerline of the trim clip holes and mark the centerline of the transfer with the centerline of the panel.

Lay the transfer face down again on the work surface and peel off the adhesive backing paper. With a sponge and a solution of water and liquid detergent, wet the adhesive side of the woodgrain overlay and the panel.

Align the decal with the clip holes and panel centerline. Lightly press the top of the transfer to the panel, making sure to align any plank lines or grain. With the transfer aligned, squeegee the center 3 or 4 inches (76 or 101 mm) of the transfer. Use an upward motion with the squeegee, forcing the liquid solution out along the top. This anchors the transfer in position.

Raise one side of the decal and with short strokes gradually squeegee the top edge of the transfer in place. Make sure the transfer edge stays even with the centerline of the molding clip holes. Then, squeegee the top edge with a long horizontal stroke. Repeat this procedure for the other side top edge. Then raise the transfer and with the squeegee press down another 2 or 3 inches (50 or 76 mm) in the center.

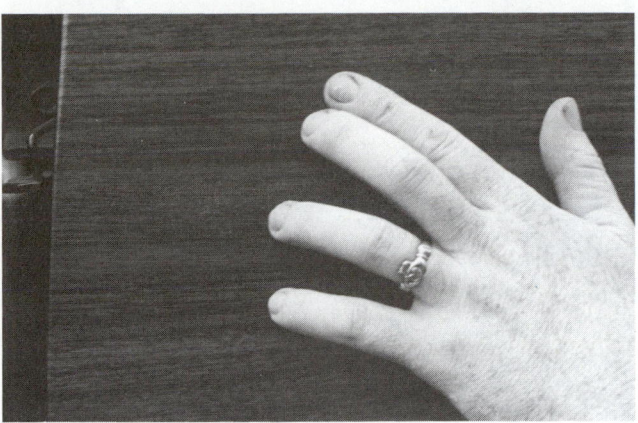

FIGURE 21-36 Inspect decal or transfer for air bubbles. Bubbles can be pierced with needle to remove trapped air.

Use overlapping horizontal strokes to bond another band of decal across the top of the panel. Progressively work down and across the panel in this manner. If the decal gets tacky and sticks to the panel before it is pressed in place, break the grab with a fast firm pull. Periodically rewet the panel to decrease the tack as well as to make the transfer easier to position.

When reaching the edges of the decal cut 90-degree notches in the corners and V-shaped notches along the edges where necessary to fit the transfer to the panel. Avoid excessive pulling and stretching, as the decal can tear.

Apply vinyl trim adhesive to door hem edges. Apply the adhesive sparingly to avoid a lumpy build-up under the transfer. Heat the edges of the door hem flanges and the transfer. Then wrap the transfer around the flange edge and firmly press it to the backside. Apply heat to any depressions or hole edges, and firmly press the transfer to ensure a good bond. Cut the excess decal away from panel edges and holes with a razor blade.

Inspect the application from an angle where light reflections will expose any irregularities. Pierce bubbles with a fine needle from an acute angle and press these down firmly (Figure 21-36). Reinstall all moldings and other hardware.

In addition to woodgrain, other decorative effects can be achieved by using decals. For example, many customers like to personalize their cars, trucks, or vans with stripe designs. These stripes and decals are easy to apply as long as the supplier's directions are followed.

21.6 FINAL DETAILING

Final detailing or *"get ready"* is the last, thorough cleanup before returning the vehicle to the customer. You must do all the little things that make a big

WARNING Avoid using strong cleaning agents on the plastic parts in the dash panel. Some will dissolve and damage plastic, a costly mistake. You should also avoid having anything with silicone in it in the body shop.

difference to customer satisfaction. The interior and exterior of the vehicle should be cleaner than when the customer brought it in.

Vacuum the interior of the vehicle carefully. Clean the seats, door panels, seat belts, and carpets. If dusty, clean and treat the vinyl surface with a conditioner. Be sure to remove all excess reconditioner from the seat crevices and folds. Stubborn stains should be cleaned with a recommended carpet cleaning solution.

Carefully remove any overspray that may have been left on windows or chrome. If done without dripping on the new finish, you may use paint solvent (thinner or reducer). Clean and polish chrome, moldings, and bumpers. Thoroughly clean all the glass, including windows, mirrors, and lights (Figure 21-37).

Use a brush with soap and water to clean the tires and wheels. Do not let dirty wheels spoil the appearance of an otherwise quality job. Coat them with a conditioner.

FIGURE 21-37 Before releasing vehicle, make sure everything is perfectly clean. Clean windows, interior, and lights to help satisfy customer. *(Courtesy of 3M)*

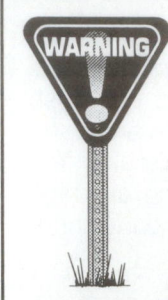

WARNING Steel wool should not be used to polish chrome because pieces of the wool can easily become embedded in the new finish. Instead, use a commercial chrome polisher polish.

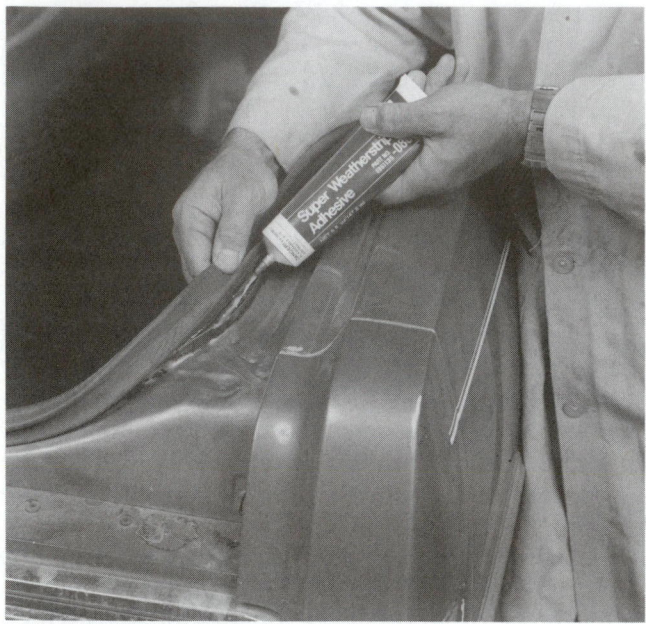

FIGURE 21-38 To prevent water and air leaks, make sure all weatherstripping is properly installed. Here the technician is using weatherstrip adhesive to secure trunk weatherstripping. *(Courtesy of 3M)*

Use chassis black to blacken wheel openings and any other exposed undercarriage parts, since overspray often gets on these areas. The customer generally will not notice this, but it certainly will be noticed if not done.

FIGURE 21-39 Always send your customers home satisfied and they will someday return with more work for your shop.

Replace wipers, moldings, and emblems that were removed before finishing. Take the time to clean off these items and be certain that everything is replaced. Make sure all weatherstipping is installed properly (Figure 21–38).

If the vehicle has a vinyl top, do not forget to wipe it with a damp cloth or a commercial vinyl cleaner.

As a finishing touch, clean the engine compartment. The easiest way to do this is to spray it with a heavy-duty engine cleaner. Then flush out the engine compartment with high-pressure water. A clean engine compartment usually makes a big impression on the customer.

Finally, inspect the vehicle with a careful eye for details. If a window is smeared, clean it again. If a piece of masking tape remains, remove it. If an emblem is missing, replace it before the customer asks where it is.

If the vehicle gets dirty while waiting to be picked up, wipe it down with a clean cloth. The number one objective should always be a satisfied customer (Figure 21–39).

SUMMARY

■ Paint problems include a wide range of troubles that can be found before or after painting. You must be able to analyze and correct paint problems efficiently.

■ If you see paint defects while spraying, you must decide whether to stop work immediately or wait until the painting is finished to correct the problem. This depends on the type and extent of the problem.

■ The objective in final detailing is to locate and correct any defect which may cause customer complaints.

■ Paint surface chips result from mechanical impact damage to the paint film: door dings, damage from road debris, and so on.

■ A paint surface protrusion is a particle of paint or other debris sticking out of the paint film after refinishing.

■ Dirt-nib files can be used to remove any defect which is on or above the surface of the paint.

■ After dirt-nib filing, finesse sanding can be used to level the protrusion with the surrounding paint film.

- Wet sanding can be done to smooth the paint surface on larger areas, as when removing orange peel.
- Rubbing compounds generally contain coarser grit abrasives than polishes.
- Machine compounds are water-based to disperse the abrasive while using a power buffer.
- Edge masking involves taping over panel edges and body lines prior to machine buffing or polishing to protect the paint from burn-through.
- Polishing involves using a very fine compound to bring the paint surface up to full gloss. A newly refinished vehicle must receive special care, as the paint can still be curing for several months.
- Final detailing or "get ready" is the last, thorough cleanup before returning the vehicle to the customer. You must do all the little things that make a big difference to customer satisfaction.

ASE-STYLE REVIEW QUESTIONS

1. Which type of compound features coarse particles?
   - A. Polishing
   - B. Lacquer
   - C. Epoxy
   - D. Rubbing

2. How long should a new topcoat be cured before automotive finish wax is applied?
   - A. 24 hours
   - B. 1 week
   - C. 30 to 45 days
   - D. 60 to 90 days

3. Technician A says to try color sanding and polishing if acid spotting is not too severe. Technician B says that sanding and refinishing might be needed. Who is correct?
   - A. Technician A
   - B. Technician B
   - C. Both A and B
   - D. Neither A nor B

4. Technician A says to pierce bubbles in a decal with a fine needle from an acute angle and press down firmly. Technician B says to apply heat to any depressions or holes and press firmly to ensure a good bond. Who is correct?
   - A. Technician A
   - B. Technician B
   - C. Both A and B
   - D. Neither A or B

5. A new paint job shows signs of blushing. Technician A says to lower air pressure at the gun. Technician B says to use a slower reducer. Who is correct?
   - A. Technician A
   - B. Technician B
   - C. Both A and B
   - D. Neither A nor B

ESSAY QUESTIONS

1. Explain how to prevent paint cracking.

2. Describe some causes for crazing.

3. How can you prevent orange peel?

4. Summarize final detailing.

CRITICAL THINKING PROBLEMS

1. Explain when a rubbing compound would be used versus a polishing compound.

2. How can you prevent burn-through on edges when machine polishing?

3. How do you care for a new finish?

MATH PROBLEMS

1. Instructions for an overlay state that the overlay should be trimmed to extend off each end of a panel by $3/4$ inch. The panel measures $21\frac{1}{4}$ inches long. How long should the overlay be cut?

CHAPTER 22

Estimating Repair Costs and Entrepreneurship

INTRODUCTION

An **estimate**, also called a *damage report* or *appraisal*, calculates the cost of parts, materials, and labor for repairing a collision-damaged vehicle. Developed by the estimator and the equipment available, it is a written or printed summary of the repairs needed. The estimate is used by the customer, the insurance company, shop management, and the technician.

Estimates must be accurately written. Repair costs are a major consideration for both the owner and the collision repair shop. Insurance companies

OBJECTIVES

After studying this chapter, you should be able to:

✔ Explain how damage repair estimates are determined.

✔ Describe the basic procedures of writing up an estimate for the body shop and customer or insurance company.

✔ Outline the sequence for estimating vehicle damage.

✔ Describe the method of determining the repairability of a damaged vehicle based on observation and by consulting appropriate sources.

✔ Begin making damage appraisal judgments about whether or not new parts or repair and straightening procedures are necessary.

✔ Explain the difference between flat-rate labor time and overlap labor time when estimating labor costs.

✔ Convert flat-rate labor time into dollars.

✔ Estimate roughly the time required for painting a given collision repair job.

✔ Estimate material costs based on a refinishing materials list.

✔ Explain the term "entrepreneur."

✔ Answer ASE-style test questions pertaining to the estimating process.

KEY TERMS

bar codes	job overlap
blue book	laptop computer
central processing unit	mainframe computer
collision-estimating guides	memory
command	menu
compact disc	modem
computer	monitor
computerized estimating	overhaul
cursor	PC
damage report	peripherals
database	printer
data scanner	productivity
deductible clause	professionalism
dial-up estimating system	program
digital images	reliability
direct repair programs	remove and install
downloading	remove and repair
drive	salvage parts
entrepreneur	salvage value
erase	save
estimate	social skills
estimating program	software
flat rate	system software
floppy disk	systematic approach
front clip assembly	terminals
hard disk	total loss
hardware	work order
in-shop estimating system	work traits

ASE TASK LIST

Job Skills covered in this chapter include:

NONSTRUCTURAL ANALYSIS AND DAMAGE REPAIR TEST (B3) TASK LIST

A. Preparation

1. Review damage report; analyze damage to determine appropriate methods for overall repair.

D. Moveable Glass and Hardware

2. Repair or replace electrically driven power sun roofs and related controls.

generally require at least two written estimates from two separate shops. Usually, the shop that submits the lowest estimate gets the work. Therefore, the profit margins for a collision repair shop depend heavily on the accuracy of estimates. Insurance companies can also write their own estimates.

This chapter will help you understand how both manual and electronic estimates are prepared (Figure 22–1).

22.1 THE ESTIMATE

Before any decision involving the repair of the damaged vehicle can be made, a detailed repair estimate, sometimes called a *damage appraisal* or **damage report,** must be made.

Before writing the formal damage report (Figure 22–2), fill in all preliminary information that identifies the owner of the vehicle and who will pay the bill. Record the make, year, body style or type, license plate number, mileage, and date. The written estimate should also give a detailed description of all the labor operations that must be performed and a complete listing of parts and materials needed to make the necessary repairs. All this information—pricing of labor, parts and materials, and their totals—can help prevent any misunderstanding between the shop and the vehicle owner.

At lease three copies of the written estimate should be made. One is kept by the shop; one is given to the insurance company; and the other is given to the customer. An estimate is an *approximate bid* for a given period of time—usually for 30 days. The reason for a given time period is obvious: part prices change and damaged parts can deteriorate.

Direct repair programs (DRPs) are made up of cooperating insurance companies and body shops. An insurance company will approve a body shop for making repairs on vehicles insured by their company. The insurance company will have the customer take the damaged vehicle to one of the approved body shops for repairs. This eliminates the time needed to get and approve one of several estimates.

FIGURE 22-1 Although damage to a vehicle may appear minimal, repair costs may be high due to modern construction and crush zones. Computer-based estimating will help speed the process and result in more accurate estimates. *(Courtesy of Mitchell Manuals, Inc.)*

The estimate can be considered the authorization to complete the repair work as listed, but only when it is agreed upon and signed by the owner or by the owner and insurance company appraiser. The estimate explains the legal conditions under which the repair work is accepted by the collision repair shop. It protects the shop against the possibility of undetected damage that might be revealed later as repairs progress.

Many insurance policies contain a **deductible clause**, which means that the owner is responsible for the first given amount of the estimate (usually

B & J
Collision Estimating Services

20 WEST BROAD ST.
AUBURN, PA 17922
TELEPHONE: (717) 555-7764

ESTIMATE OF REPAIRS
Nº 002128

SHEET NO._____OF_____SHEETS

NAME	ADDRESS	PHONE	DATE
KAREN Miller	1143 Railroad St. Cressona, PA. 17929	HOME 395-2719 BUS. 623 7347	12-17-91

YEAR	MAKE	MODEL	LICENSE NO.	MILEAGE	SERIAL/ V.I. NO.
1990	Ford	P.V.	MAE 917	14864	1FTDF15YSGNA69994

INSURANCE COMPANY	TYPE OF INSURANCE	ADJUSTER	PHONE	CAR LOCATED AT
Amerisure	COL			mf6. 4/91

PARTS NECESSARY AND ESTIMATE OF LABOR REQUIRED	PAINT COST ESTIMATE		PARTS COST ESTIMATE		LABOR COST ESTIMATE
① FRONT FACE BAR Chrome NO gds or PAds			205	82	.5
① " STONE deflector	1	0	40	50	.5
① Left HeadLamp door (with argent Grill)			52	32	2
① " " Shield			3	20	
① " front fender	3	1	133	00	1 6
① " " " APRON			60	47	1 0
② Wheels 15" 55.60 ea.			111	20	6
② Stems and Balance			2	50	5
② HUB CAPS			43	46	—
Repair Radiator Support			—		2 0
ALign FRONT End			—		1 5
① Left door trim Panel			73	40	
STRIPE Left FRONT fender			15	00	.5
LABOR 25.0 HRS @ 23.00			575	00	
(note may Be FRONT SUSPENSION damage)					
PAINT mat.			89	10	
UNderCoaT			15	00	
TOTALS			1,529	97	

INSURED PAYS $_____INS. CO. PAYS $_____R.O. NO._____

INS. CHECK PAYABLE TO_____

The above is an estimate, based on our inspection, and does not cover additional parts or labor which may be required after the work has been opened up. Occasionally, after work has started, worn, broken or damaged parts are discovered which are not evident on first inspection. Quotations on parts and labor are current and subject to change. Not responsible for any delays caused by un-availability of parts or delays in parts shipment by supplier or transporter.

ESTIMATOR_____

AUTHORIZATION FOR REPAIRS. You are hereby authorized to make the above specified repairs to the car described herein.

SIGNED X_____   DATE_____ 19_____

GRAND TOTAL	1,529.97
TOWING & STORAGE	
TAX	91.80
TOTAL OF ESTIMATE	$ 1,621.77

FIGURE 22-2 This is an example of a hand-written damage estimate or damage appraisal for parts and labor.

REPAIR ORDER

B & J
Collision Estimating Services

20 WEST BROAD ST.
AUBURN, PA 17922
TELEPHONE: (717) 555-7764

REPAIR ORDER
NO. 004007

SHEET NO. _____ OF _____ SHEETS

CUSTOMER'S NAME	ADDRESS	PHONE	DATE
Karen Miller	1143 Railroad St. Cressona, PA. 17929	HOME 395-2719 BUS. 623-7347	12-17-91

YEAR	MAKE	MODEL	LICENSE NO.	MILEAGE	SERIAL/VIN
1990	Ford	PV	MAE 917	14,864	1FTDF15YSGNA69994

ORDER GENERATED BY	MECHANIC	PAINTER
Bob	JT	Bill

VEHICLE IN BODY SHOP	IN	DATE 6-29	HOUR 0800	VEHICLE IN PAINT SHOP	IN	DATE 7-1	HOUR 0800
	OUT	DATE 6-30	HOUR 1130		OUT	DATE 7-2	HOUR 1430

BODY SHOP

INSTRUCTIONS	NO. OF MECHANICS	PART NO.	PART IN HOUSE	ACTUAL TIME
Repr. Front Bumper	1	E1TZ17757A	6-26	.5
" Stone Deflector	"	EOTZ17779A	"	.6
" L. Headlight Door	"	E2TZ13064B	"	.6
" " " Shield	"	EOTZ13B042B	"	.2
" Frt Fender (Left)	"	E1TZ16006B	"	1.6
" " " Apron	"	E4TZ16055A	"	1.0
" 2 Frt Wheels	"	E7TZ1015C	"	.6
" 2 Hubcaps	"	E2TZ1130A	"	.5
Repair Rad. Support	"			2.5
Align Front End	"			1.5

PAINT SHOP

INSTRUCTIONS	NO. OF PAINTERS	MATERIALS AND PAINT TYPE AND COLOR	PAINT IN HOUSE	ACTUAL TIME
Paint Damaged & Replaced Items	1	W. White - Centari	6.26	2.0
Refinish & Repair	"	Paper, Putty, Tape		3.5
Replace Door Trim Panel	"	EO1Z1020897B		.5
Stripe Fender (Left f.)	"	Tape		.5
Undercoat	"	2 Qts		1.0

SHOP FOREMAN _____ DATE _____

FIGURE 22-3 Typical repair order is used by shop personnel.

$100 to $500). The remaining cost is paid by the insurance company. In such cases, both the customer and the insurance company should authorize the estimate.

Following are additional time factors to be considered when writing an estimate. Such added time is usually negotiated between the estimator and the insurance company or customer:

1. Time for the setup of the vehicle on a frame machine and damage diagnosis
2. Time for pushing, pulling, cutting, and so on, to remove collision-damaged parts, called *access time*
3. Time to straighten or align related parts
4. Time to remove undercoating, tar, grease, and similar materials
5. Time to repair rust or corrosion damage to adjacent parts
6. Time for the free-up of rusted or frozen parts
7. Time for drilling for ornamentation or mounting holes
8. Time for filling or plugging unneeded holes in new parts
9. Time to repair damaged replacement parts prior to installation
10. Time to check suspension and steering alignment/toe-in
11. Time for removal of shattered glass
12. Time to rebuild, recondition, and install aftermarket parts, not including refinishing time
13. Time for application of sound-deadening material, undercoating, caulking, and painting of the inner areas
14. Time to restore corrosion protection
15. Time to R & I main computer module when excessive temperatures (above 176 degrees Fahrenheit or 80 degrees Celsius) are necessary in repair or paint drying operations
16. Time to R & I wheel or hub cap locks
17. Time to replace accessories, such as trailer hitches, sun roofs, and fender flares

Until a few years ago, many body shops gave auto body damage reports or estimates free. Because the preparation of an estimate can involve a considerable amount of time and paperwork, this practice is seldom followed. Instead, most body shops charge a flat fee for the estimate or a percentage of the estimated repair costs. However, if the vehicle is repaired by that facility, this charge is usually deducted from the final bill.

In some parts of the country, so-called courtesy estimates or appraisals are made. These appraisals are generally made because a customer might need two or more estimates for the insurance company. The body shop will submit an appraisal for a small fee knowing quite well that the shop will not get the job. Also, courtesy estimates or appraisals are written for vehicles that might be totally wrecked. The estimator will write up to the total list range. If the vehicle is worth $4,000 and the actual cost of making the repairs would be $4,500, the shop will stop at a $4,000 estimate.

Another important function of the estimate is that it serves as a basis for writing the work order or operational plan. It is usually prepared from the damage appraisal of the estimator (using the written estimate) and a visual inspection by the shop supervisor and/or a technician (Figure 22–3).

The **work order** outlines the procedures that should be taken to put the vehicle back in preaccident condition. It is also a valuable tool to both the estimator and the shop foreman, since it lists the actual methods needed to do the job.

FIGURE 22-4 Crash-estimating guides for American-made and import vehicles give information needed to make an estimate of damage.

FIGURE 22-5 Estimating guide will give illustrations, part numbers, and repair time for each make and model vehicle.

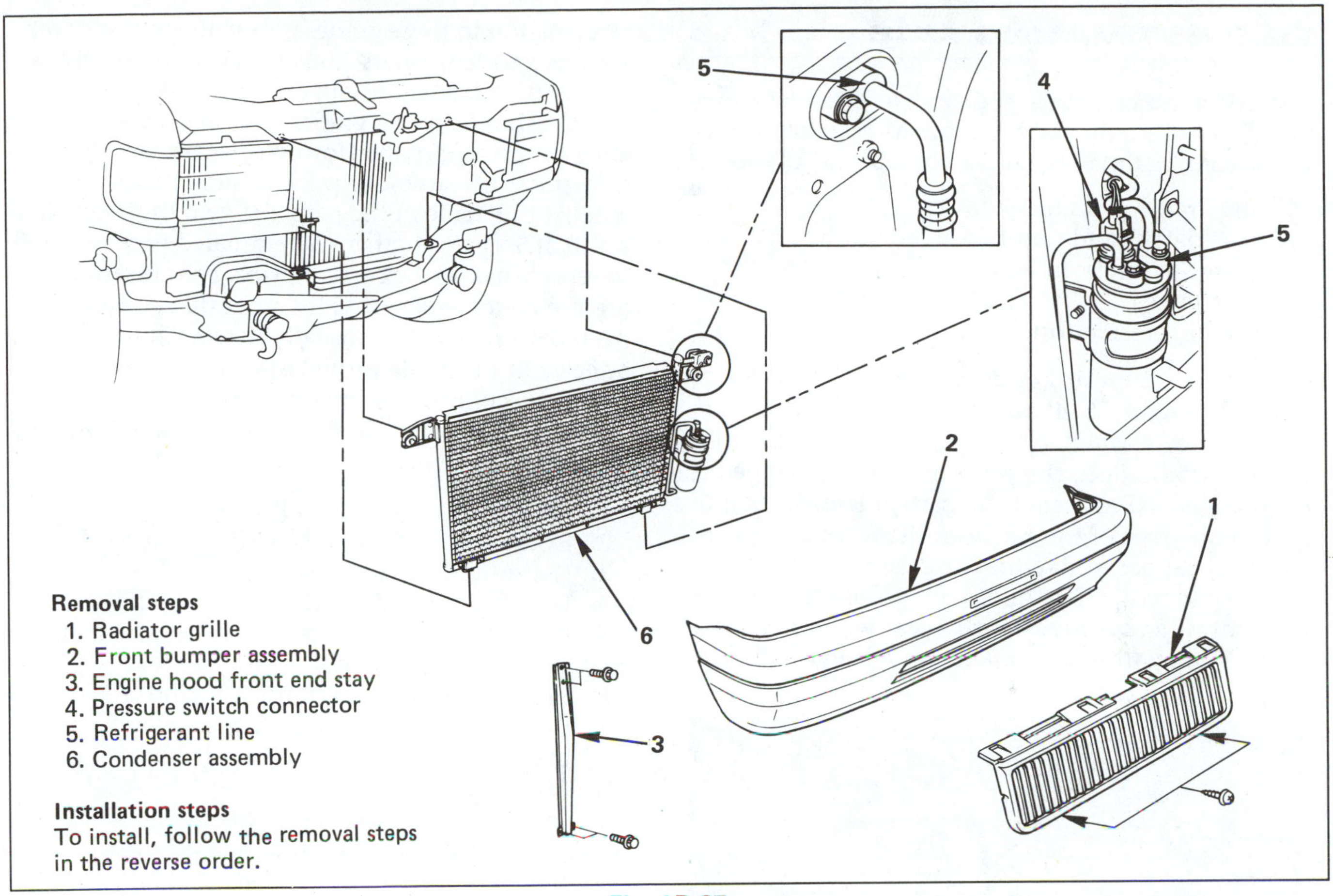

Removal steps
1. Radiator grille
2. Front bumper assembly
3. Engine hood front end stay
4. Pressure switch connector
5. Refrigerant line
6. Condenser assembly

Installation steps
To install, follow the removal steps in the reverse order.

Fig. 1B-37

↔ REMOVAL

Preparation:
Disconnect the battery ground cable
Discharge and recover refrigerant (Refer to "REFRIGERANT RECOVERY" in this section.)

1. **Radiator grille**

2. **Front bumper assembly**
   - Refer to Section 2B "BUMPERS".

3. **Engine hood front end stay**

4. **Pressure switch connector**

5. **Refrigerant line**
   - When removing the line connector, the connecting part should immediately be plugged or capped to prevent foreign matter from being mixed into the line.

6. **Condenser assembly**

→← INSTALLATION

6. **Condenser assembly**
   - If installing a new condenser, be sure to add 30cc (1.0 fl.oz.) of new compressor oil to a new one.

- Tighten the condenser fixing bolts to the specified torque.

	Torque N·m (lb·in)	6 (52)

5. **Refrigerant line**
   - Tighten the inlet line connector fixing bolt to the specified torque.

	Torque N·m (lb·ft)	15 (11)

   - Tighten the outlet line connector fixing bolt to the specified torque.

	Torque N·m (lb·in)	6 (52)

   - O-rings cannot be reused. Always replace with new ones.
   - Be sure to apply new compressor oil to the O-rings when connecting refrigerant line.

4. **Pressure switch connector**

3. **Engine hood front end stay**

2. **Front bumper assembly**

1. **Radiator grille**

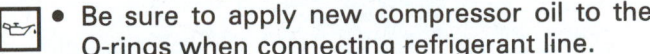

FIGURE 22-6 Exploded view in manual will help you find parts hidden under outer panels. They often require repair or replacement. *(Courtesy of Isuzu Motor Co.)*

22.2 ESTIMATING AIDS

Collision-estimating guides are essential when filling out the estimate. Whether in manual or electronic form, estimating guides contain the following:

1. Illustrated parts breakdowns
2. Part names and numbers
3. Flat-rate times
4. Part prices
5. Other information

Crash-estimating guides or collision-damage manuals (Figure 22–4) are essential tools of the estimator. They contain such items as vehicle identification information, the price of new parts, amount of time needed to install the parts, identification of almost any part of the car from the front bumper to the rear bumper, and refinishing data such as a paint code reference. These guides are published and updated at different times of the year when manufacturers change prices or make model revisions. The

FIGURE 22-7 Estimator must have a thorough knowledge of vehicle construction and body shop methods to develop an accurate estimate of repair costs. *(Courtesy of Tech-Cor)*

information in these guides is of value to body technicians and refinishers; thus they should be able to read and understand them (Figure 22–5).

Crash-estimating guides can be used as a reference for pricing parts. However, never use them to determine the absolute price. Usually these guides will list the name of the part, the year of the vehicle it will fit, its part number, the estimated time required to replace it, and the current price. The current prices are the factory-suggested list prices. Parts that have been discontinued are usually listed. The price that appears in the guide is the last available one.

Each collision-estimating guide will have procedure or "P" pages. These procedure pages provide important information such as:

1. Arrangement of material
2. Explanation of symbols used
3. Definitions of terms used
4. How to read and use the parts illustrations
5. Procedure explanations, including which operations are included and which are not
6. How discontinued parts information is displayed
7. How interchangeable part information is displayed
8. Additions to labor times
9. Labor times for overlap items
10. How to identify structural operations
11. How to identify mechanical operations

Service manuals and estimating manuals will also give exploded views of parts. You can use these illustrations to help determine which parts must be replaced during repairs (Figure 22–6).

ABBREVIATIONS USED IN ESTIMATING GUIDES

When using crash-estimating guides, the body technician and painter, as well as the estimator, must be familiar with the abbreviations that are used (Figure 22–7). The three most commonly used abbreviations are:

- **R & I: Remove and Install.** The item is removed as an assembly, set aside, and later reinstalled and aligned for a proper fit. This is generally done to gain access to another part. For example, R & I bumper would mean that the bumper assembly would have to be removed to install a new fender or quarter panel.
- **R & R: Remove and Replace.** Remove the old parts, transfer necessary items to new part, replace, and align.
- **O/H: Overhaul.** Remove an assembly from the vehicle, disassemble, clean, inspect, replace parts as needed, then reassemble, install, and adjust (except wheel and suspension alignment).

```
DAMAGE REPORT                                    SMITH
Date: 06/12/87                                   D.R.    25
B.A.R. #AA199902                                 Estimator: BOB

                TEAM MANAGEMENT NETWORK
                MANAGEMENT INFORMATION SYSTEMS
                   67 NORTH CATALINA AVENUE
                   PASADENA, CA  91106-
                      (818) 795-9406

VEHICLE OWNER: JENNIFER SMITH          DAY PHONE: (818)  795-9409-1009
                                     OTHER PHONE: (818)  793-0896-
ADDRESS: 974 REYNOLDS ROAD, SAN MARINO, CA 91106

YEAR: 86 96              MAKE: OLDSMOBILE      MODEL: CUTLASS
STYLE: 4 DOOR           COLOR: RED            LICENSE: 1QKG673
VIN: JSL39852UOTU2985223  PROD DATE: 08/85    MILEAGE: 12987
                                                    535-1212
INSURANCE CO.: ALL AMERICAN INSURANCE CO.    PHONE.:818-793-0897-1922
ADJUSTOR: BOB BRIDGES              CLAIM NO.: 975-2975-11   DED.: 250.00
```

```
     REPLACE                             PARTS   LBR   PAINT
NO.  REPAIR   DESCRIPTION OF DAMAGE  QTY  COST    HRS   HRS     MISC

 1   Replace  LEFT FRONT FENDER        1  238.55  4.2   3.6
 2   Replace  LEFT FRONT FENDER MLDG   1   23.28  0.5   0.0
 3   Replace  LEFT FRONT MARKER LIGHT  1   46.49  0.7   0.0
 4   Replace  LEFT FRONT WHEEL OPENING 1  127.50  3.0   2.0
 5   Repair   LEFT FRONT DOOR OUTER PANEL 1 0.00  5.0   3.2
 6            CLEAN UP BROKEN GLASS    1    0.00  0.5   0.0
 7   Replace  LEFT FRONT DOOR GLASS    1    0.00  0.0   0.0 X   200.00
 8   Replace  LEFT FRONT DOOR HANDLE   1   40.28  0.8   0.0
 9            AIM HEADLAMPS            1    0.00  1.0   0.0
10            SETUP / DIAGNOSIS        1    0.00  0.0   0.0 F   114.00
11   Repair   FRAME - 3 PULLS          1    0.00  0.0   0.0 F   152.00
12   Repair   LEFT CENTER POST         1    0.00  3.7   0.7
13            DETAIL                   1    0.00  0.0   0.0

          Subtotals  ===>             476.10 19.4  9.5        466.00
```

```
          Parts   (Subject to Invoice)       476.10
          Labor   19.4      27.00            523.80
          Shop Supplies                       26.10
          Paint    9.5      27.00            256.50
          Paint Supplies                     142.50
          Frame    7.0      38.00            266.00
          Sublet/Misc                        200.00
          Subtotal                    $     1891.00
          Tax on $   644.70   6.5000          41.91
          ------------------------------------------
          GRAND TOTAL                 $     1932.91

                  Page:   1
```

FIGURE 22-8 This is a computerized estimate printout. This streamlines the process over trying to do an estimate by hand.

Overhaul time should be used only if the time for the individual parts (less overlap) is more than the overhaul time.

In addition to these abbreviations, Appendix B in this textbook gives a listing of other terms that are accepted by most estimating guide publishers and estimators in general when filling out written forms.

22.3 COMPUTER ESTIMATING

Computerized estimating systems using a **PC** (personal computer) provide more accurate and consistent damage reports. The use of computers makes dealing with thousands of parts on hundreds of vehicles more manageable.

Computerized estimating systems store the collision-estimating guide information in a computer. This eliminates a lot of time in looking up parts and labor times, and manually entering and totaling them on a form. The computer prepares a damage report, while still allowing the possibility of a manual override when necessary. Computerized estimating and shop management systems are being used by an ever increasing number of body shops (Figure 22–8). Regardless of size, body shops face concerns that demand increased efficiency and speed. A dramatic increase in the number of vehicle makes and models has led to a tremendous increase in essential repair information. Part numbers, part prices, and repair times are constantly updated and changed. Calculating the needed costs for repairs is now more complicated than ever. Damage report writing, accounting, job costing, time and inventory control, and business analysis can all be performed quicker and easier using a computer.

COMPUTER COMPONENTS AND JARGON

A **computer** is an electronic device for storing and manipulating information. The computer is simply a machine that helps people use large databases of information (Figure 22–9).

Like the auto, the computer is made up of various systems and components (Figure 22–10). *Computer input devices* are items like the keyboard or bar code reader that allow you to put information into the computer. *Output devices* are items such as cathode ray tubes (CRTs) or monitors and printers that allow you to read information inside the computer.

An awareness of computer terminology is essential to understanding the computer. The following is a list of definitions to acquaint the damage report writer with computer terminology, increase an understanding of the computer, and serve as a reference guide.

Central processing unit (CPU) The portion of the CPU that directs the sequence of operations by electrical signals and governs the actions of the units that make up the computer. The central processing unit is the "computer brain" where calculations take place.

Command An instruction to the computer to perform a predefined operation. It can be given by typing on the keyboard or installing data from another device.

Compact disc (CD) Disc that stores optical data instead of magnetic data. It can hold several times the information as a floppy disk. A compact disc is a read-only disc. You cannot erase or alter data on a compact disc.

FIGURE 22-9 Computer estimating is much more efficient than trying to figure repair costs long hand. *(Courtesy of Mitchell Manuals)*

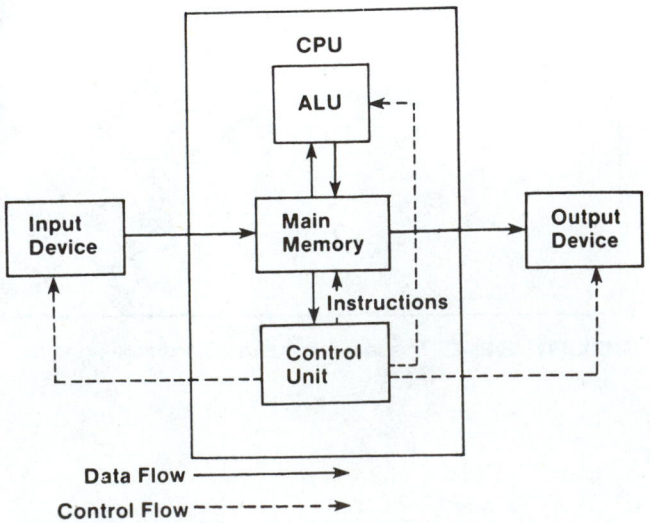

FIGURE 22-10 Note basic components of a personal or business computer system.

Cursor A movable, blinking marker or pointer on the terminal video screen that shows where the next point of entry or change will be made.

Drive A device that holds a disk, cassette tape, or CD-ROM so that the computer can read data from and write data onto it.

Erase To remove information or data from memory or from disk storage.

Floppy disk A thin magnetic disk (in protective paper or plastic jacket) that stores data or programs. It is erasable and reusable, and comes in several diameters. It is often used to backup data in case the information is lost on the main hard drive.

Hard disk A device, usually inside the computer, to store large amounts of data magnetically. Unlike floppies, it is rigid and not readily interchangeable but can store and retrieve data faster than a floppy.

Hardware The physical parts of a computer, such as keyboard, screen, disk drives, printer, and so on.

Laptop computer A small portable computer with a flat display screen for working in remote locations (Figure 22–11).

Mainframe computer A powerful computer capable of storing and processing extremely large amounts of data. Also called the "host" computer in some systems.

Memory Computer chips that hold information electronically. This information is lost when the computer is shut off. Memory chip information must be saved to a disc to make it permanent.

Menu A displayed list of options from which the user selects an action to be performed.

Modem Device for sending computer data over phone lines for use by another computer at a remote location.

Monitor Computer screen for viewing information.

Peripherals Hardware that can be added to a basic computer, such as a printer, disk drive, CD-ROM drive, or modem (Figure 22–12).

Program Disk information that loads into memory to allow the computer to do a specific task.

Save To write data from memory to disc for safekeeping.

Software Also known as programs; instructions written in a computer language that tell the computer what the user wants to do. Smaller programs

A

B

FIGURE 22–11 This laptop computer can be taken out to the vehicle during a damage analysis. It can also be used in the office to finalize the damage report. *(Courtesy of Mitchell Manuals)*

FIGURE 22–12 Peripherals are components added to the basic computer such as (A) printer and (B) disk drive to load floppy disks.

come on floppy disks. Larger, more complex programs come on CD-ROMs.

System software Disk information that allows the computer to start up and operate.

Terminal An input/output device used to enter data into a computer and record the output. Terminals are divided into two categories: hard copy (printers) and soft copy (video terminals).

COMPUTER SYSTEM TYPES

Computers are revolutionizing the art of estimating by making enormous amounts of data readily accessible. There are basically two types of computer estimating systems: dial-up systems and resident or in-shop systems.

The **dial-up estimating system** relies on data stored in a mainframe computer at a remote location. The data in the mainframe is usually accessed via a telephone modem (device for sending and receiving computer data over phone lines). The shop may be charged a fee whenever data from the mainframe is accessed.

An **in-shop estimating system** has all of the data needed at the shop office, usually on CD-ROMs (Figure 22–13). They do not require a fee for each estimate. However, the shop must purchase current CDs or subscribe and pay an annual fee to get updated CDs. Modern computer estimating systems store electronic databases containing up-to-date part prices and labor times that can be retrieved from the shop's computer system.

With the use of these sophisticated programs, estimators can now use the stored data to make virtually error-free estimates rapidly at a very low cost.

Data Wand and Bar Codes

A **data scanner** is a small hand-held electronic information storage tool. It has a keypad for inputting data and a read head for reading bar codes (Figure 22–14). Many computerized systems use a data scanner with a bar code wand.

SHOP TALK Even with the best computerization, you must still use a logical, sound inspection sequence to locate every damaged part. If you miss a part or two, you or the shop's profits will suffer.

FIGURE 22-14 Some computer estimating systems have a data wand. The keypad allows you to input vehicle information—VIN, make, model, year, etc. It can also be used in conjunction with a booklet that contains bar codes for each section and part of the vehicle. By running the wand over the bar code corresponding to parts, the data wand will store information in electronic memory. *[Courtesy of Mitchell Manuals]*

FIGURE 22-13 CD-ROM provides a vast database for the computer estimating system. One CD-ROM can hold a whole set of books in electronic form. *[Courtesy of Mitchell Manuals]*

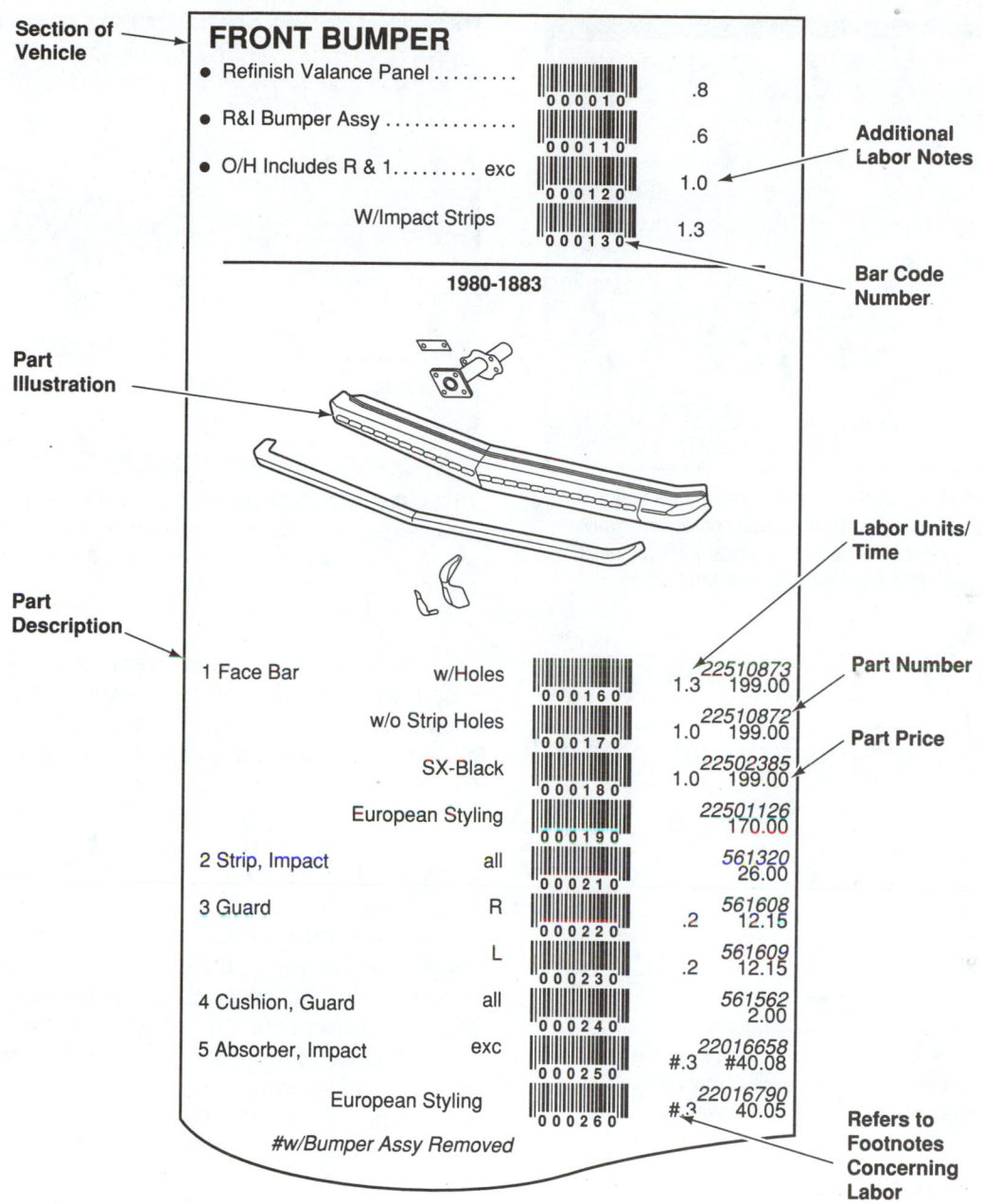

Section of Vehicle

FRONT BUMPER

- Refinish Valance Panel   000010   .8
- R&I Bumper Assy   000110   .6
- O/H Includes R & 1 . . . . exc   000120   1.0
     W/Impact Strips   000130   1.3

Additional Labor Notes

Bar Code Number

1980-1883

Part Illustration

Part Description

Labor Units/ Time

Part Number

Part Price

1 Face Bar	w/Holes	000160	1.3	22510873 199.00
	w/o Strip Holes	000170	1.0	22510872 199.00
	SX-Black	000180	1.0	22502385 199.00
	European Styling	000190		22501126 170.00
2 Strip, Impact	all	000210		561320 26.00
3 Guard	R	000220	.2	561608 12.15
	L	000230	.2	561609 12.15
4 Cushion, Guard	all	000240		561562 2.00
5 Absorber, Impact	exc	000250	#.3	22016658 #40.08
	European Styling	000260	#.3	22016790 40.05

#w/Bumper Assy Removed

Refers to Footnotes Concerning Labor

FIGURE 22-15 Note how the bar code guide has information that can be used with the data wand. *(Courtesy of Mitchell Manuals)*

Bar codes are a series of lines that represent other data, like a damage part. The data wand can be moved over bar codes printed in the special collision-estimating guide. For example, you can use the data wand keyboard to input the vehicle VIN, make, model, and other information. By pressing the keys on the data wand this information will be stored in its internal memory. If a fender is damaged, run the wand over the bar code next to that part. The unit will then store the data electronically for later use when in the office (Figure 22–15).

Downloading means to hardwire two information-holding devices (scanner, computer, laptop, etc.) so they can exchange digital data or information. The

bar code scanner is used to gather and store data for later downloading into the office's personal computer.

COMPUTER DATABASE

Once information has been gathered at the vehicle, the estimator takes this data into the office. The damage data is then electronically downloaded (entered) into the personal computer. This allows the operator to use the speed and convenience of a computer and estimating program to make up the estimate. There is no need to type in the names and numbers for damage parts.

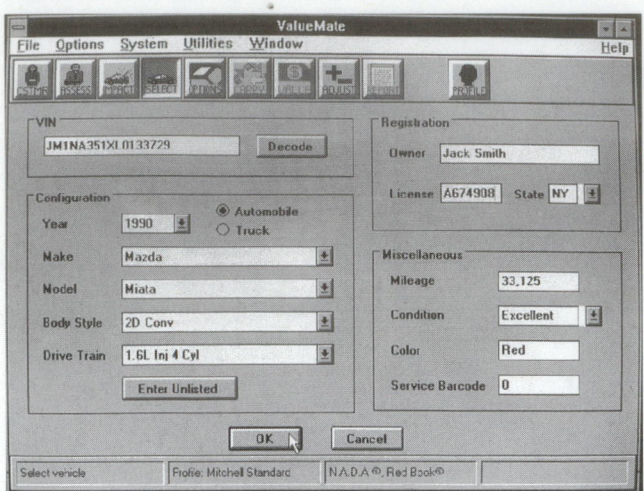

FIGURE 22-16 An estimating program is designed for helping the estimator quickly and accurately calculate repair costs. A data wand can download information for manipulation by personal computer. *(Courtesy of Mitchell Manuals)*

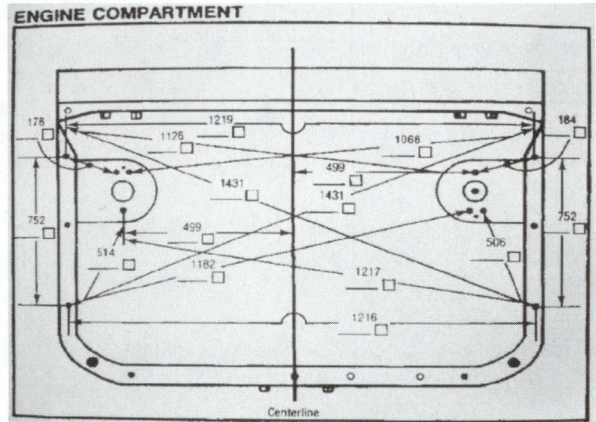

FIGURE 22-17 Computer estimating systems often have a digitized database of the dimension manual for faster retrieval of specifications. *(Courtesy of Mitchell Manuals)*

An **estimating program** is software (computer instructions) that will automatically help to find the parts needed, provide the labor rates, and calculate the total cost for the repairs (Figure 22–16). The estimating program is often on computer **floppy disks** (removable magnetic disks of data) or on a compact disc (optical data on disk).

To use the program, you must load the data into the PC's memory. This will then let you pull up images from an electronic dimension manual form (Figure 22–17). Modern systems will retrieve exploded views of assemblies in electronic form (Figure 22–18).

Repair overlap refers to two or more repair operations that are completed together in a larger operation. Many programs automatically deduct for overlap. The program notes operations as "Incl." if they are part of a larger operation. The program may also identify judgment times.

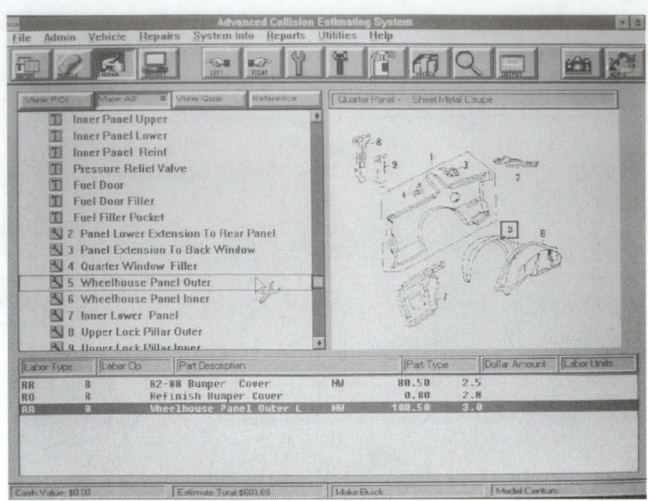

FIGURE 22-18 Here is a computer display of the exploded view of a fender assembly. The estimator can quickly see all parts and select which ones require replacement or repair. *(Courtesy of Mitchell Manuals)*

The estimating program can access a huge database (computer file) of information. The computer **database** includes part numbers, part illustrations, labor times, labor rates, and other data for filling out the estimate. Quick searches can be done to move through this database quickly and efficiently.

The most common and modern way of storing an estimating database is the CD-ROM. One CD-ROM or compact disc can hold all of the crash-estimating guides and dimension manuals as well as other information for every make and model vehicle. The compact disc database contains the same basic information that is in the collision-estimating guide.

A CD-ROM drive is wired to the office computer. The PC can then operate the drive and access this huge amount of data. CD-ROM data can be pulled into the PC's memory for manipulation.

Labor time information for a given operation is taken from the same sources as in the collision-estimating guides. A portable hand-held computer is used with a light wand to scan parts and labor operations from a reference book. This information is then fed into a personal computer for making calculations.

Advanced computer estimating systems will even allow you to input photos of the wrecked vehicle into the computer.

An *electronic camera* can be used to take photos of the vehicle's damage and store them as digital data (Figure 22–19). The camera can be connected to the computer to download these **digital images** (pictures stored as computer data) into memory or onto the computer's internal hard drive.

The estimator can look at these photos on the computer screen or monitor while finalizing the estimate. The electronic images can also be sent to the

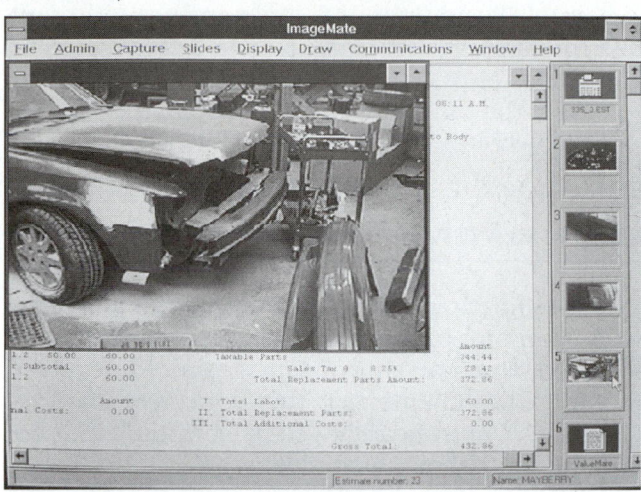

FIGURE 22-19 This advanced computer estimating system will let you download pictures taken with an electronic camera or a video camera. The estimator can refer to these images when working on an estimate. *(Courtesy of Mitchell Manuals)*

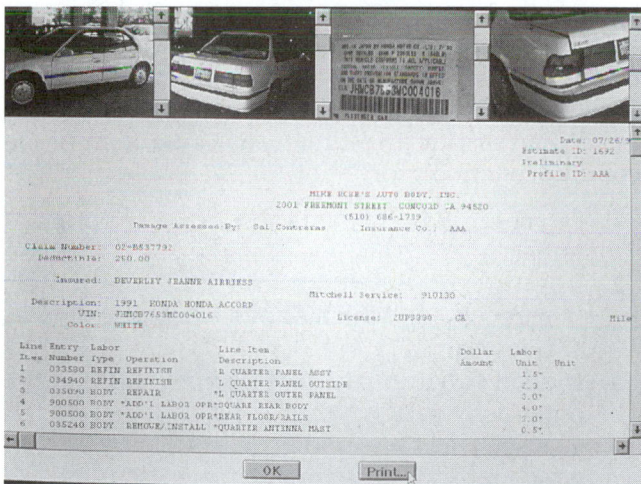

FIGURE 22-20 Pictures of the damaged vehicle can be pulled up and reexamined. They can also be sent to the insurance adjuster so he or she can view the vehicle damage without leaving the office. *(Courtesy of Mitchell Manuals)*

insurance adjuster so he or she can evaluate the vehicle's damage (Figure 22–20).

A **printer** can be used to make a *hard copy* (printed images on paper) for the customer, insurance adjusters, and shop personnel (Figure 22–21). You can quickly print vehicle dimension drawings, part views, or the estimate.

When working with any estimating system, either manual or computer, those working with the damage report must understand the following:

1. Where vehicle options are listed, so these can be double-checked for accuracy
2. Which labor rates are shown for a given operation and how to identify them

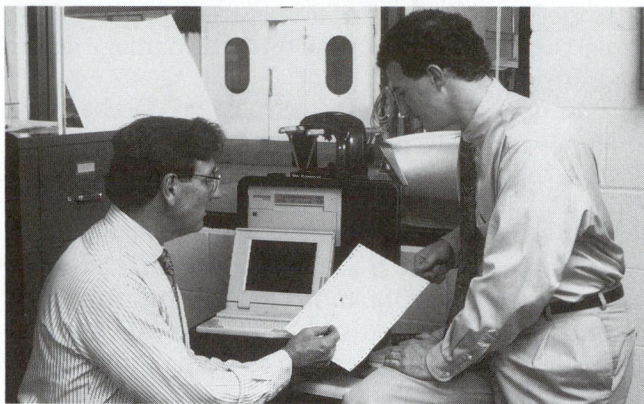

FIGURE 22-21 Personnel are using hardcopy printouts of estimate to review repair process. *(Courtesy of Mitchell Manuals)*

3. Which times and amounts are judgment items, and how these are generated
4. Which times and amounts have been overridden
5. Which times are included, and which are not

Only by understanding all of these can a judgment be made on whether the damage report is accurate. Each system will produce a unique look.

Manual Data Entry

Older manual data entry systems require the user to enter part prices, labor times, labor rates, and other pertinent information into the program by using the terminal keyboard. Information must be typed in line by line just as you would enter items on the lines of a handwritten damage report form. Data used with manual entry systems is found in the standard collision-estimating guides used to create handwritten reports. Once this data is entered into the program format, it can be manipulated and changed to create the damage report. The report can be stored on magnetic disk or tape for future reference or printed out in hard copy form.

The main disadvantage of manual entry is the time and effort needed to type in the data entries. With manual entry systems the accuracy of the data entered is also dependent on the accuracy of the typist or operator. Simple typed errors can generate wildly inaccurate results.

ON-SITE DATA BANKS

Mentioned briefly, with on-site data bank systems the secondary data storage becomes a component of the user's computer system; there is no need for a telephone modem or special hookup. Massive amounts of data can be accessed using several keystrokes. There is no need to type entire line entries. A single keystroke can enter a line of data onto the damage report program sheet.

Companies license their massive databases to independent computer software suppliers for inclusion in their estimating systems. Users subscribe to the data bank source in exactly the same way they subscribe to the estimating manuals. Updated computer data is provided at regular intervals on magnetic tape or CD-ROM and is loaded into the user's secondary storage system. Once the data bank is loaded, it can be used as often as needed to produce damage reports or appraisals.

Advantages of a "resident" data source include:

- No need to worry about transmission errors or breakdowns that can occur with telephone-based linkups. The data is also accessible at the user's convenience, day or night, peak working hours or off periods.
- The ability to factor in many features not user-controlled in central data bank systems. The user has the opportunity to view items on the screen before they are selected. There is access to footnotes, procedure pages, refinish times and notes, and lists of included operations.
- The entire estimate can be previewed and any errors or omissions can be corrected BEFORE printing. With central data bank systems the report is transmitted to and printed at the user location, then reviewed. If there are mistakes due to user or data entry personnel errors, the report must be revised and reprinted while time is wasted.
- The ability to transfer data directly from the database to the damage report. The system allows the user to add items and alter prices as necessary. The user controls all phases of what goes on the damage report without sacrificing the speed and convenience of a computerized database.
- With data banks based on printed estimating guides, the guides can serve as a verifiable source of the computer estimate entries. For example, Motor and Mitchell include a full subscription to their crash-estimating guides with their database. The user can refer to the printed guides to

check illustrations, industry nomenclature, parts interchange, and paint code references. The guides also serve as a portable reference away from the computer and a backup in the event of a power loss or system breakdown.

22.4 ESTIMATING SEQUENCE

When estimating any type of damage to a vehicle, whether minor or major, a logical sequence must be followed. The analyzing of damage procedures given in the earlier chapters of this book, especially in Chapter 10, must be followed by both the estimator and the shop's body technicians.

Before making a written estimate, the estimator should make a visual inspection of the entire vehicle, paying special attention to damaged subassemblies and parts that are mounted to (or part of) a damaged component. The estimator must consider basically the same points as the technician does before making any decision on the repair work. That is, most estimators start from the outside of the car and work inward, listing everything—by car section—that is found bent, broken, crushed, or missing.

When writing an estimate, these are some of the conditions to look for:

1. Direct outer body damage starting at points of impact
2. Improper fit of doors and movable panels (Figure 22–22), which shows major movement of body panels
3. Hidden structural damage (Figure 22–23) under outer body panels
4. Cracked sealer and undercoating (Figure 22–24), which indicate structural damage

FIGURE 22-22 When doing an estimate, look for uneven gaps around doors and panels. They indicate major movement of body parts from collision impact.

> **SHOP TALK**
> The estimator should develop a basic pattern or system for analyzing damage. This pattern should be used on every vehicle to prevent missed parts or overlooked problems.

FIGURE 22-23 Damage to hidden or partially hidden inner panels can be more difficult to repair than outer body panels. Keep this in mind when doing an estimate. *(Courtesy of Tech-Cor)*

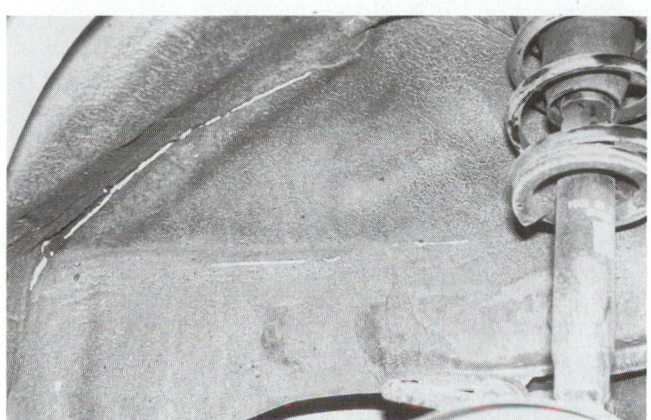

FIGURE 22-24 Cracked sealer and undercoating is an indicator of damage. Measurements would be needed to evaluate the extent of the damage. *(Courtesy of Tech-Cor)*

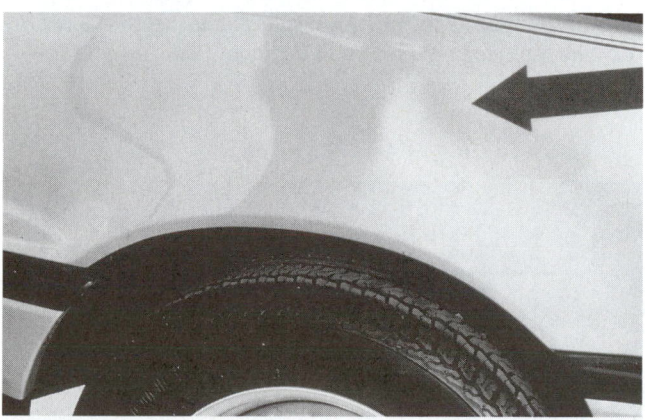

FIGURE 22-25 Indirect damage is common with today's unibody vehicles. An impact force wave can travel through the body structure and buckle panels at the other end of the vehicle. *(Courtesy of Tech-Cor)*

5. Indirect damage from shock wave travel through body parts (Figure 22–25)
6. Mechanical damage to engine parts (Figure 22–26), steering and suspension parts (Figure 22–27), and the drivetrain

FIGURE 22-26 Engine parts, like pulleys, motor mounts, manifolds, etc., are often damaged by a hard hit. *(Courtesy of Tech-Cor)*

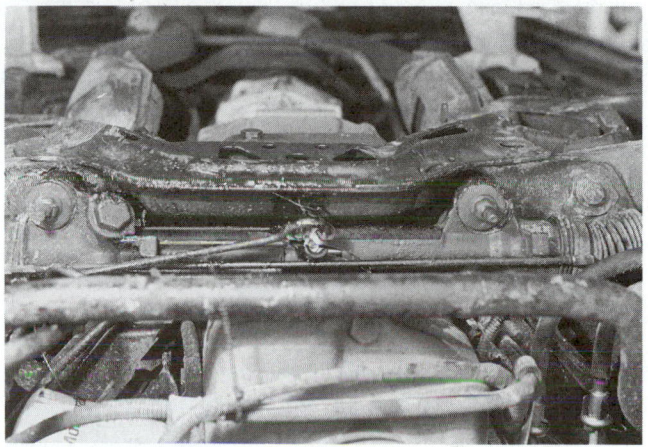

FIGURE 22-27 When inspecting under vehicle, check steering and suspension for damage. This rack-and-pinion assembly was broken during a wreck and would require expensive replacement.

7. Damage to interior of vehicle (Figure 22–28) such as air bag deployment or dash damage
8. Bent wheels and damaged tires (Figure 22–29)
9. Broken glass and mirrors (Figure 22–30)

An example, given here, lists the repairs or replacements needed if the front grille and some of the related parts are damaged.

Front Grille	Replace
Front Grille	Refinish
Opening Panel	Replace
Deflector (or Valance Panel)	Replace
Headlamp Door	Replace
Grille Opening Panel	Refinish

Notice that the parts to be repaired, straightened, replaced, or refinished are listed in a definite sequence according to factory disassembly operations or exploded views as provided in shop manuals or crash-estimating guides.

The estimator must be on constant guard against missing related and hidden damage. As described in

FIGURE 22-28 The interior is another area that can be costly to fix after a bad accident. Air bags, the dash panel, seats, glass, and carpet can all be damaged by flying bodies.

FIGURE 22-30 Inspect all glass and mirrors for cracks. They can be small and easy to overlook when writing an estimate.

FIGURE 22-29 Inspect wheels and tires closely for damage. Aluminum wheels are very susceptible to damage. Check the inside and outside of the wheel lips and tire sidewalls. *(Courtesy of Tech-Cor)*

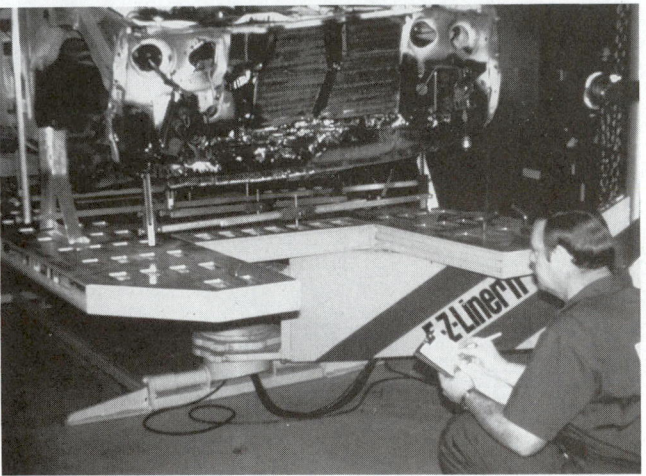

FIGURE 22-31 You may need to measure the vehicle to get a better idea of the extent of damage. *(Courtesy of Chief Automotive)*

Chapter 10, damage that happens to a vehicle during the moment of impact or immediately after impact is referred to as *related* or *indirect damage*.

Often, in the case of unibody cars, related damage is not near the area of impact but some distance away.

Hidden or secondary damage, as stressed many times in this book, can occur almost anywhere or to any part or component on a car that has been involved in a collision. Naturally, if the damage is only minor—creased fender, gouge in door, headlamp assembly smashed, and so forth—the chances of hidden damage are somewhat remote. However, due to the tremendous forces involved in a major, severe collision, the estimator must always suspect some form of hidden damage. For example, the snout of an engine crankshaft can snap from frontal impact forces. Badly damaged fan blades and pulleys can clue you to a damaged engine water pump.

Front motor mounts can be sheared and yet the engine resettles back into position, hiding this type of damage. Castings of the engine, transmission, or

FIGURE 22-32 Inspection is often done on lift or bench. Then you can easily measure damage and inspect under vehicle. *(Courtesy of Car-O-Liner Co.)*

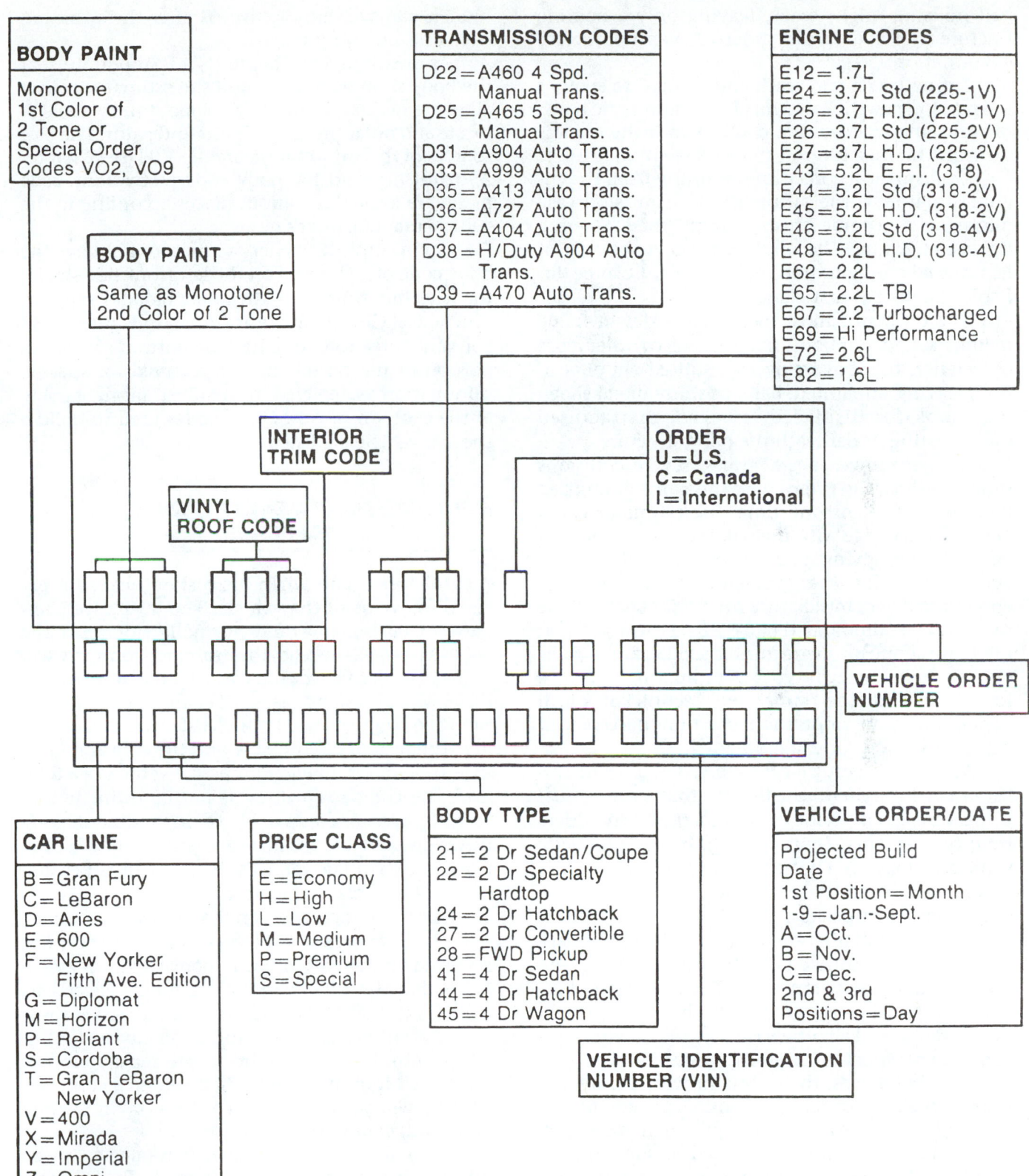

BODY PAINT

Monotone
1st Color of
2 Tone or
Special Order
Codes VO2, VO9

BODY PAINT

Same as Monotone/
2nd Color of 2 Tone

TRANSMISSION CODES

D22 = A460 4 Spd.
     Manual Trans.
D25 = A465 5 Spd.
     Manual Trans.
D31 = A904 Auto Trans.
D33 = A999 Auto Trans.
D35 = A413 Auto Trans.
D36 = A727 Auto Trans.
D37 = A404 Auto Trans.
D38 = H/Duty A904 Auto
     Trans.
D39 = A470 Auto Trans.

ENGINE CODES

E12 = 1.7L
E24 = 3.7L Std (225-1V)
E25 = 3.7L H.D. (225-1V)
E26 = 3.7L Std (225-2V)
E27 = 3.7L H.D. (225-2V)
E43 = 5.2L E.F.I. (318)
E44 = 5.2L Std (318-2V)
E45 = 5.2L H.D. (318-2V)
E46 = 5.2L Std (318-4V)
E48 = 5.2L H.D. (318-4V)
E62 = 2.2L
E65 = 2.2L TBI
E67 = 2.2 Turbocharged
E69 = Hi Performance
E72 = 2.6L
E82 = 1.6L

**INTERIOR
TRIM CODE**

**VINYL
ROOF CODE**

ORDER
U = U.S.
C = Canada
I = International

**VEHICLE ORDER
NUMBER**

CAR LINE

B = Gran Fury
C = LeBaron
D = Aries
E = 600
F = New Yorker
    Fifth Ave. Edition
G = Diplomat
M = Horizon
P = Reliant
S = Cordoba
T = Gran LeBaron
    New Yorker
V = 400
X = Mirada
Y = Imperial
Z = Omni

PRICE CLASS

E = Economy
H = High
L = Low
M = Medium
P = Premium
S = Special

BODY TYPE

21 = 2 Dr Sedan/Coupe
22 = 2 Dr Specialty
    Hardtop
24 = 2 Dr Hatchback
27 = 2 Dr Convertible
28 = FWD Pickup
41 = 4 Dr Sedan
44 = 4 Dr Hatchback
45 = 4 Dr Wagon

VEHICLE ORDER/DATE

Projected Build
Date
1st Position = Month
1-9 = Jan.-Sept.
A = Oct.
B = Nov.
C = Dec.
2nd & 3rd
Positions = Day

**VEHICLE IDENTIFICATION
NUMBER (VIN)**

FIGURE 22-33 Typical Chrysler body code plate interpretation. Other vehicle manufacturers follow the same basic system.

bell housing might crack, leaving only a hairline fracture that is difficult to detect even upon close examination.

Spring leaves can crack and be hard to spot, or the parking pawl, located inside the automatic transmission, might have snapped in two if the car was hit with the shift lever in the park position.

Transmission cooler lines running from the engine radiator to the automatic transmission case might have become crushed or pinched. Steering linkage might have become bent and yet because of its unusual configuration, misalignment can be difficult to detect at first glance.

The estimator must look for buckles in frame members. Look for bolts, flat washers, or other types of fasteners that have moved or shifted out of position, leaving unpainted, bare, or shiny metal showing. Look for displaced, cracked, or fractured undercoating underneath the body structure.

As mentioned several times, severe collisions from any direction often cause the frame or unitized body to distort. At one time, the estimator could check this damage with the naked eye and would be fairly close in giving a repair cost. But with today's auto construction, the "eyeballing" technique is not enough to detect misalignment. It is far better to use measuring equipment (Figure 22–31) to check for misalignment. However, unless the estimator is thoroughly skilled in such types of estimating, it is wise to consult with the frame/body technician for an appraisal to determine the labor time necessary to get it back into proper alignment.

When estimating a car involved in a severe collision, raise the car off the floor so that a good visual inspection can be made of all underbody and drivetrain components (Figure 22–32). In some unibody vehicles, it may even be necessary to remove the drivetrain and suspension components to make a thorough damage inspection.

Doors out of alignment and cracked stationary glass are often a solid clue to hidden damage to the frame or underbody structure. The operation of instrument panel gauges and lights, heater, air conditioner, radio, and all other comfort and convenience components must be carefully checked.

In other words, the estimator with the help of the technician must give the entire car—top to bottom and front to rear—a thorough, intensive inspection and take nothing for granted. If something is missed on the original estimate, the shop must contact the insurance company for further negotiations. Some estimates do include a so-called hidden damage clause, called an estimate supplement, that permits added charges to the original estimate when

hidden damages are discovered after the work has been opened up for repairs.

As mentioned in Chapter 1, it is important that the body shop technician and the estimator collect all of the necessary information about the vehicle to locate and order the correct parts and paint. The best sources of this information are the VIN (see Chapter 2), paint tag, and the body code plate. Body code plates are located in various places according to the vehicle manufacturer's desires.

For example, Chrysler products might have the body code plate located on the left front side shield, the wheelhousing, or on the upper radiator support. The typical Chrysler plate as shown in Figure 22–33 contains three rows of data. The bottom row and a portion of the center row are reserved for specific information as depicted in the illustration. Starting at the center row, other sales codes used to build a specific vehicle are listed.

22.5 REPAIRABILITY OF THE VEHICLE

A **total loss** occurs when the cost of the repairs exceeds the value of the vehicle. The insurance company or customer would normally not want the vehicle repaired. Instead, the insurance company will write a check to cover the cost of a replacement vehicle. An equivalent year, make, and model vehicle can then be purchased by the customer.

The insurance company and the collision facility estimator usually determine if a vehicle is a total loss. The company will evaluate the estimate and market prices for comparable vehicles when making this decision. Older vehicles are written up as a total loss more than late-model vehicles. This is because of their low replacement cost.

The first critical decision that the estimator must make is the repairability of the vehicle. If the car was involved in a severe collision, there is a strong possibility that the total repair costs might exceed its market value. For example, a car with a market value of $3,150 (if undamaged), and an estimate that totals $3,710, is one situation where repairs are not practical. More than likely the insurance company adjuster will agree with such a decision and would authorize "totaling" the vehicle.

Naturally, when a car that is totaled is towed into a body shop (for example, frame badly distorted, body crushed, engine and transmission severely damaged), there is usually never any doubt that writing a complete estimate is not necessary. All that is usually needed is a courtesy estimate that lists most major

86 FORD 1986-85

Av'g. Trd-In	BODY TYPE	Model	M.S.R.P.	Wgt.	Av'g. Loan	Av'g. Retail
1986 LTD CROWN VICTORIA-AT-PS-AC-Continued						
200	Add Power Windows (Std. 2D & LX)		$282		200	250
200	Add Power Seats				200	250
125	Add Rear Window Defroster		145		125	150
100	Add Luggage Rack (S/W)		110		100	125
150	Add Cruise Control		176		150	200
150	Add Tilt Steering Wheel		115		150	200
150	Add Wire Wheel Covers		205		150	200
275	Add Leather Seats		418		250	350
225	Add Brougham Roof		793		225	300
150	Add Custom or 2-Tone Paint		117		150	200
850	Deduct W/out Air Conditioning				850	850
1986 THUNDERBIRD-AT-PS-AC					**Start Oct. 1985**	
THUNDERBIRD-V6	Veh. Ident. 1FA( )( )P46( )( )( )G( )100001 Up					
8100	Coupe 2D	46	$11020	3089	7300	9300
9300	Coupe 2D elan	46	12554	3145	8375	10600
THUNDERBIRD-V8	Veh. Ident. 1FA( )( )P46( )( )( )G( )100001 Up					
8300	Coupe 2D	46	$11568	3275	7475	9525
9500	Coupe 2D elan	46	13102	3331	8550	10825
THUNDERBIRD-4 Cyl.-AT/5 Spd.	Veh. Ident. 1FA( )( )P46( )( )( )G( )100001 Up					
8950	Turbo Coupe 2D	46	$14143	3172	8075	10200
400	Add Power Sunroof		$701		375	475
150	Add AM/FM Stereo/Tape		127		150	200
75	Add Power Door Locks		220		75	100
100	Add Power Windows (Std. elan & Turbo)		207		100	125
100	Add Power Seats				100	125
75	Add Rear Window Defroster		145		75	100
75	Add Cruise Control		176		75	100
75	Add Tilt Steering Wheel		115		75	100
100	Add Wire Wheel Covers		212		100	125
200	Add Leather Seats		415		200	250
100	Add Custom or 2-Tone Paint (Std. Turbo)				100	125
650	Deduct W/out Air Conditioning				650	650
1985 ESCORT/EXP-AT-PS-FWD						
	Start Sept. 1984, 1985.5 March 1985					
ESCORT-4 Cyl.	Veh. Ident. 1FA( )P( )(Model)( )( )F( )( )100001 Up					
3300	Hatchback 2D	04	$5620 *	1990	2975	4125
3400	Hatchback 4D	13	5827 *	2055	3075	4225
3500	Hatchback 2D L	04	5876 *	1979	3150	4325
3600	Hatchback 4D L	13	6091 *	2044	3250	4450
3750	Station Wagon 4D L	09	6305 *	2071	3375	4600
3775	Hatchback 2D GL	05	6374 *	2047	3400	4650
3875	Hatchback 4D GL	14	6588 *	2114	3500	4750
4025	Station Wagon 4D GL	10	6765 *	2139	3625	4900
4025	Hatchback 4D LX	15	7840 *	2175	3625	4900
4175	Station Wagon 4D LX	11	7931 *	2198	3775	5050
4175	Hatchback 2D GT	07	7585 *	2140	3775	5050
4250	Hatchback 2D Turbo GT	07	8680 *	2172	3825	5150
1985.5 ESCORT-4 Cyl.	Veh. Ident. 1FA( )P( )(Model)( )( )F( )( )100001 Up					
3350	Hatchback 2D	31	$5856 *	2142	3025	4175
3550	Hatchback 2D L	31	6127 *		3200	4400
3650	Hatchback 4D L	36	6341 *	2195	3300	4500
3800	Station Wagon 4D L	34	6622 *	2223	3425	4675
3825	Hatchback 2D GL	32	6641 *		3450	4700

DEDUCT FOR RECONDITIONING
1987 APRIL 1987

A

FIGURE 22-34 Page from the auto industry blue book lists the market value of most vehicles. *(Courtesy of National Automobile Dealers Used Car Guide Co.)*

FIGURE 22-36 Salvage yards are a good source of parts. Recycled parts may save repair time and improve repair quality because they have factory corrosion protection. New parts might require more time to clean, treat, and refinish. *(Courtesy of Tech-Cor)*

parts that need repair or replacement. This would be enough to show that the vehicle is a total loss.

The auto industry **blue book** (Figure 22–34) or equivalent lists the market value (not the selling price in a used- or new-car lot) of various model years for all cars. Refer to this publication when such decisions of "repair or scrap" must be made for a badly damaged car. Of course, cars that have been wrapped around trees, smashed almost flat, or sliced in half by extreme collision forces are other examples of cars considered totaled (Figure 22–35).

Some totaled vehicles have a **salvage value**. That is, parts of a wrecked car or truck, such as the engine, transmission, drive shaft, and body parts might be usable and saleable. The totaled vehicle will usually be auctioned or sold to a salvage yard or recycler. The recycler will then disassemble the vehicle and sell its parts for a profit (Figure 22–36).

22.6 PART PRICES

Once the vehicle is judged repairable, the next decision is whether or not the collision damage requires new parts or simply repair and straightening. Naturally, it would be simple to list only new parts and units to restore the damaged vehicle to its precollision condition. But as mentioned, this is sometimes impractical. The estimator must have the ability to compare the cost of repairs against the cost of new parts or units.

As a general rule, repair costs should never exceed replacement costs. If there is some doubt that repairs and straightening will not produce a quality job, then use new parts. Remember, sheet metal parts

FIGURE 22-35 Example of a totaled vehicle. At first glance, vehicle may seem repairable at reasonable cost. Hidden damage and indirect damage would make repairs too costly. *(Courtesy of Tech-Cor)*

SHOP TALK Remember that salvage parts could have been previously damaged and repaired. For this reason, carefully inspect all salvage parts before they are installed to be sure that they are structurally sound.

usually offer the most opportunities for repair and straightening. As a result, sheet metal repairs, replacement, and refinishing of panels generally account for the largest number of estimate dollars.

To reduce parts costs, many insurance appraisers and some customers might want the body shop to use salvage parts. **Salvage parts** are parts in good condition that were removed from totally wrecked vehicles by salvage yards.

Some customers might object to used parts in their repaired car. Inform them that used parts might be preferable to new parts. New parts do not always have factory corrosion protection. Factory rust protection is difficult and time consuming to match in the shop. Used parts might be the only option on older vehicles because manufacturers stop making parts after a period of time.

Many salvage yard dealers offer a free computer parts location service. They can send out a request by shortwave radio or computer modem for your needed parts to hundreds of other salvage yards all over the country.

Each damaged car poses different problems that must be answered to arrive at a repair versus replacement decision. The most difficult and largest number of questions arise with major collision wreckage. For example, let us say that you are estimating a car with a "hard hit" in the right front. You might have to decide whether to install a partial or complete right, front frame rail. Do you want to section the damaged rail and splice on a partial front-half section? Or should you remove and replace the complete rail? Which would save time and money while still producing a solid structural repair? These kinds of questions take time and thought to answer! Therefore, the estimator must know correct repair procedures to write an esimate properly.

Estimating damage to the body shell is an even more difficult job. Knowing the corrective forces that would be needed to restore the body shell to its original dimensions requires on-the-job work experience. If body shell distortion is suspected, carefully

measure body dimensions to help determine whether it is more efficient to straighten or replace the panels.

Generally, clues to body distortion (twist, sag, sidesway) are apparent if any cracking of stationary glass (windshields/back window) is noted. *Stress cracks* indicate minor panel damage and movement. Always use a shop light to look around affected areas to find stress cracks between panel joints. Sealer and undercoating often crack to show you hints of damage.

In some situations, such as a severely damaged front end, an estimator might want to consider a complete front clip. The **front clip assembly** generally includes all body parts from the front bumper to the rear of the fenders. This would include the front bumper and supports, grille, radiator and support, hood and its hinges, front fenders and skirts, front lamps, wiring, and related parts. Often, this method of repair to the front end will decrease the total time the car is in the shop and possibly save the vehicle from being totaled.

Always inspect the inside and outside lips of wheels for cracks and damage. New wheels should be added to the estimate if needed. Tires should also be given a close inspection both on the outside and inside of the sidewall. If the estimator finds extensive or deep cuts, the tire is unsafe. A new tire must be listed in the estimate.

Sometimes, wheel damage occurs and yet the tire looks undamaged. Keep in mind that an impact that damaged a wheel will often cut cords inside the tire. This will cause tire runout and vibration when returned to service. In such instances, have someone dismount the tire and make an inspection of the inner surfaces.

If a tire must be replaced, the estimator must adjust for tread wear. From this information, the owner of the car should then be charged a "betterment" cost for the installation of a new tire. In other words, the owner should be charged for the amount of tread wear on the original tire, thus sharing in the purchase price of the new tire(s).

Collision or estimating guides can be used as a reference for pricing parts. However, never use the guides to determine the absolute price. Check with local part suppliers to get your final estimate part prices.

Collision guides will list part names, part numbers, the estimated time to repair or replace each part, and the current part price (Figure 22–37). These current prices are the factory-suggested list prices. Parts that are no longer available or discontinued by the vehicle manufacturer are usually listed; the price that appears in the guide (preceded by "D") is the last available one. The prefix "A" in a guide price list indicates

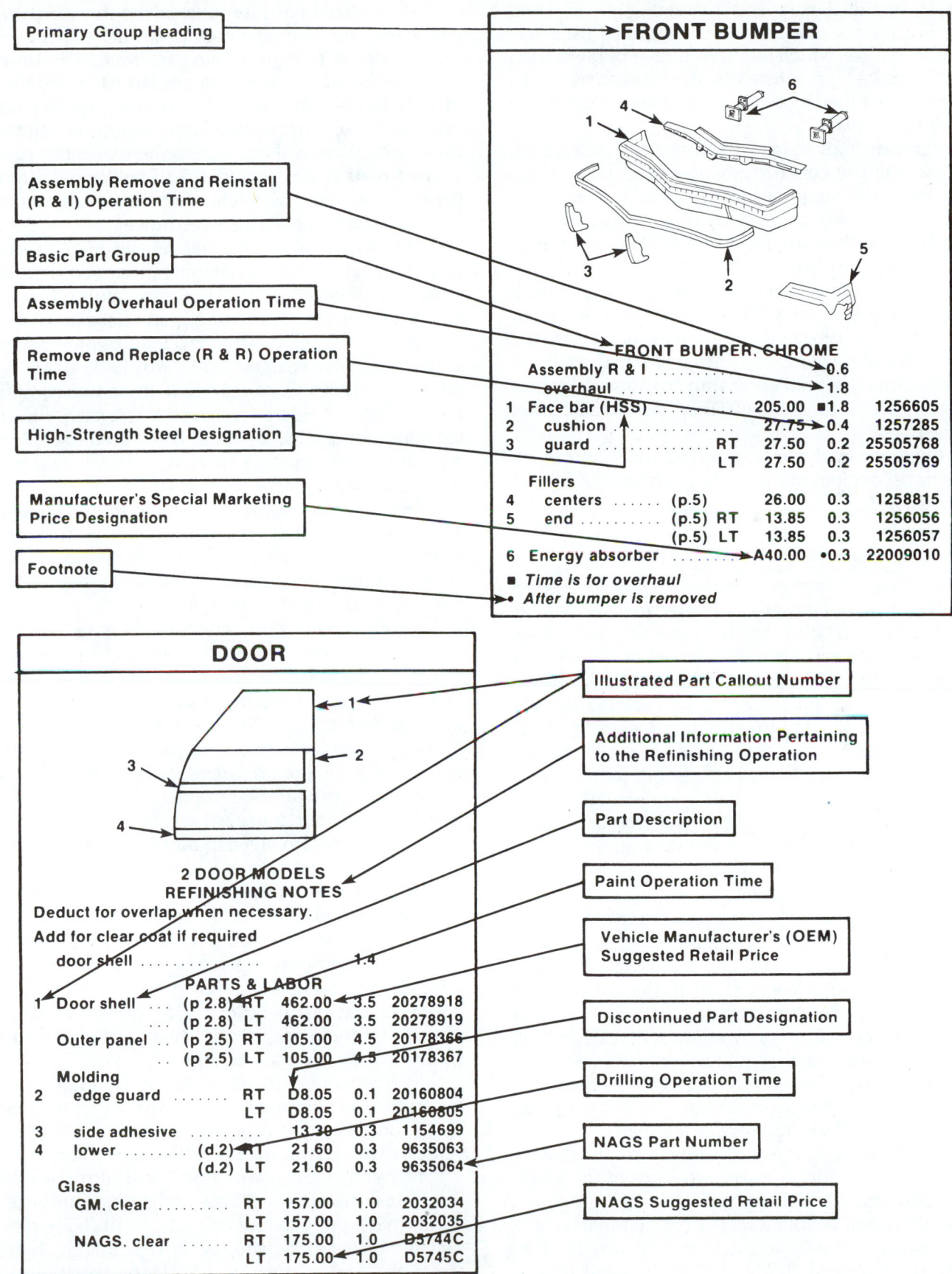

Primary Group Heading

Assembly Remove and Reinstall (R & I) Operation Time

Basic Part Group

Assembly Overhaul Operation Time

Remove and Replace (R & R) Operation Time

High-Strength Steel Designation

Manufacturer's Special Marketing Price Designation

Footnote

FRONT BUMPER

FRONT BUMPER, CHROME

Assembly R & I				0.6	
overhaul				1.8	
1	Face bar (HSS)		205.00	■1.8	1256605
2	cushion		27.75	0.4	1257285
3	guard	RT	27.50	0.2	25505768
		LT	27.50	0.2	25505769
	Fillers				
4	centers (p.5)		26.00	0.3	1258815
5	end (p.5)	RT	13.85	0.3	1256056
	(p.5)	LT	13.85	0.3	1256057
6	Energy absorber		A40.00	•0.3	22009010

■ *Time is for overhaul*
• *After bumper is removed*

DOOR

Illustrated Part Callout Number

Additional Information Pertaining to the Refinishing Operation

Part Description

Paint Operation Time

Vehicle Manufacturer's (OEM) Suggested Retail Price

Discontinued Part Designation

Drilling Operation Time

NAGS Part Number

NAGS Suggested Retail Price

2 DOOR MODELS
REFINISHING NOTES
Deduct for overlap when necessary.
Add for clear coat if required
door shell 1.4

PARTS & LABOR

1	Door shell (p 2.8)	RT	462.00	3.5	20278918
	(p 2.8)	LT	462.00	3.5	20278919
	Outer panel (p 2.5)	RT	105.00	4.5	20178366
	(p 2.5)	LT	105.00	4.5	20178367
	Molding				
2	edge guard	RT	D8.05	0.1	20160804
		LT	D8.05	0.1	20160805
3	side adhesive		13.30	0.3	1154699
4	lower (d.2)	RT	21.60	0.3	9635063
	(d.2)	LT	21.60	0.3	9635064
	Glass				
	GM. clear	RT	157.00	1.0	2032034
		LT	157.00	1.0	2032035
	NAGS. clear	RT	175.00	1.0	D5744C
		LT	175.00	1.0	D5745C

FIGURE 22-37 Study how guide gives part pricing information.

that it is included in a special marketing program and does not have a manufacturer's suggested list price. The actual price, which might be higher or lower than the "D" or "A" price listed in the estimating guide, can be determined by contacting a parts dealer in the shop's local area.

Part prices given in estimating guides usually do not include the cost of state and local taxes, shipping from the supplier, bolts, rivets, screws, nuts, washers, clips, fasteners, body repair materials, and refinishing, unless otherwise noted. These costs must be added to the estimate.

It is important that estimators take into consideration shop efficiency. The facilities of a particular body shop, the type of equipment available, and the experience of the body technicians to do the job in a specific time limit, all enter into the final "repair or replace" decision. This will affect the total cost of the estimate.

Once the part prices, material costs, and repair time have been determined, they can be entered into the estimate.

22.7 LABOR COSTS

The **flat rate** is a preset amount of time and money charged for a specific repair operation. Estimating guides provide an explanation of what the flat-rate labor time includes and does not include. For example, replacing a panel or fender includes transfer of the part attached to the panel. It does not include the installation of moldings, antennas, refinishing, pin striping, decals, or other accessory parts. Also, it does not consider rusted bolts, undercoating, and alignment or straightening of damaged adjacent parts or bolts. You should add a nominal amount of time to cover these types of unwritten repair operations (access time, for example).

The flat-rate labor time reported in collision damage manuals is to be used as a guide only. These times are principally based on data reported by vehicle manufacturers who have arrived at them by repeated performance of each operation a sufficient number of times under normal shop conditions. An explanation is listed in the estimating guides (Figure 22–38) of the established requirements for the average mechanic, working under average conditions and following procedures outlined in their service manuals.

The times reported apply only to standard stock models listed in identification sections of the guide. The labor times do not apply to cars with equipment other than that supplied by the car manufacturer as standard or regular production options. If other equipment is used (body spoilers, ground effects, etc.), the time must be adjusted higher to compensate for these added variables.

All reported flat-rate times given in estimating guides include time necessary to ensure proper fit of the individual new part being replaced. The times listed are based on new, undamaged parts installed as an individual operation. Additional time has not been added to compensate for collision damage to the vehicle. Removal and replacement of exchanged or used parts is not considered. If additional aligning or repair must be made, such factors should be considered when making an estimate.

Job overlap means that replacement of one part duplicates some labor operations required to replace an adjacent or attached part. With job overlap, reductions in estimating guide flat-rate operation times must be considered. For example, when replacing a quarter panel and a rear body panel on the same vehicle, that area where these two components join is considered overlap. Where a labor overlap condition exists, less time is required to replace adjoining components collectively than is required when they are replaced individually.

Overlap labor estimating guide information is generally included at the beginning of each group. In those instances where overlap information is not given, appropriate allowances should be negotiated after an on-the-spot evaluation.

Another labor cost reduction is noted as "included operations." These are jobs that can be performed individually but are also part of another operation. As an example, when replacing a door, the suggested time for the door replacement would include the replacement of all parts attached to the door, except for the ornamentation. Unless using a salvage door, it would be impossible to replace the new door without transferring these parts. Consequently, the time involved in transferring these parts would be included operations and should be disregarded because the times for the individual items are already included in the door replacement time.

An experienced estimator will not overlook the removal of exterior trim and body sheet metal hardware. These parts must often be removed prior to repair or painting operations. They must then be replaced after these operations. This is a labor cost that should not be overlooked (Figure 22–39).

Also remember to add tasks like headlight aiming and wheel alignment. As an estimator, you must charge out every job function needed to restore the vehicle to its precollision operation, safety, and value.

In recent years, estimators must also consider the special materials—plastic and aluminum—used in cars that require special handling by the body technician. For example, the use of lightweight, high-strength steel in various locations throughout vehicles, as noted in earlier chapters of this book, requires specific repair procedures.

1. BUMPER

FRONT OR REAR

NOTE: Disconnect at energy absorber (mounting bracket) or frame mounting.

ASSEMBLY REMOVE AND REINSTALL (R & I)

Included:
- R & I unit as assembly
- Alignment to vehicle

ASSEMBLY OVERHAUL

Included:
- R & I assembly
- Disassemble
- Replace damaged parts
- Reassemble unit
- Align to vehicle

FACE BAR REMOVE AND REPLACE (R & R)

Included:
- R & R face bar
- R & I guards and face bar cushions (unless otherwise noted in text)

ALL BUMPER OPERATIONS

Does Not Include:
- Additional time for frozen or broken fasteners
- Refinishing
- Optional moldings, name plates, emblems, and ornamentation; air bags and lamps
- Stripe tape, decals, overlays
- Removal of hydraulic energy absorbers
- Aim headlamps

2. FRAME

UNITIZED

Included:
- Welding as necessary and electrical wiring
- Floor mats, insulation, and trim (if required)

Does Not Include:
- Setup on frame machine and damage diagnosis
- R & I of all bolted-on parts and body sheet metal
- Wheel alignment
- Refinishing, undercoating, sound deadening material, and anticorrosion protection
- Removal of adjacent panels
- Time for pulling

CONVENTIONAL

Included:
- R & I front sheet metal and body assembly
- Front and rear suspension parts
- Steering parts and powertrain as assembly
- Brake line disconnect and bleed
- R & I fuel tank and bumper assembly
- Control linkage and electrical wiring

Does Not Include:
- Setup on frame machine and damage diagnosis
- Wheel alignment
- Refinishing, undercoating, sound deadening material, and anticorrosion protection
- Time for pulling

FRONT OR REAR SUSPENSION CROSS MEMBERS

Included:
- Welding as necessary

Does Not Include:
- R & I of all bolted on parts
- Wheel alignment

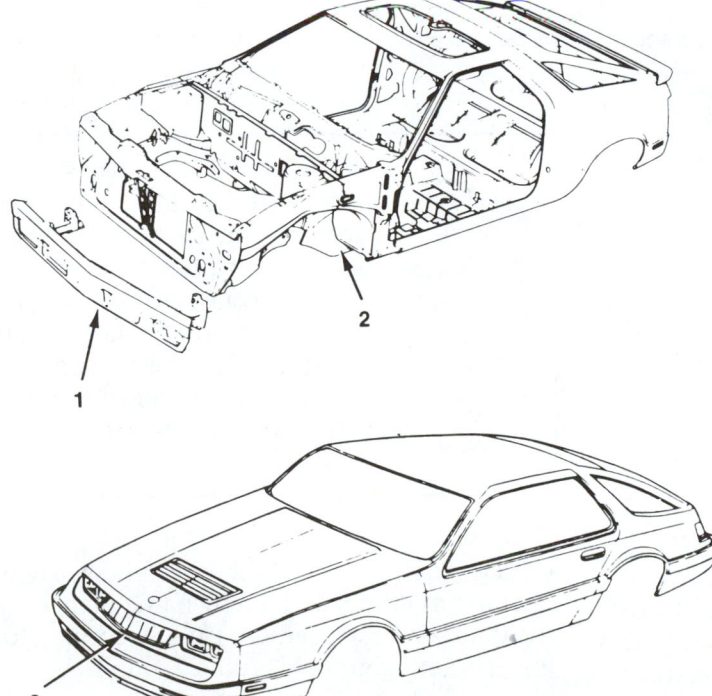

3. GRILLE

Included:
- Grille remove and reinstall
- Lamps (when mounted in grille)
- Standard equipment molding, name plates, and ornamentation

Does Not Include:
- Stripe tape, decals
- Optional molding, name plates, and ornamentation
- Refinishing
- Optional lamps
- Aim headlamps

GRILLE/HEADER PANEL/FASCIA

Included: (unless otherwise noted)
- Grille remove and reinstall
- Lamps
- Alignment to vehicle
- Fillers and extensions

Does Not Include:
- Bumper assembly remove and reinstall
- Refinishing
- Stripe tape, decal, overlays
- Molding, name plates, and ornamentation
- Drill time
- Aim headlamps

FIGURE 22-38 Read through guide and note included and not-included operations for repairs.

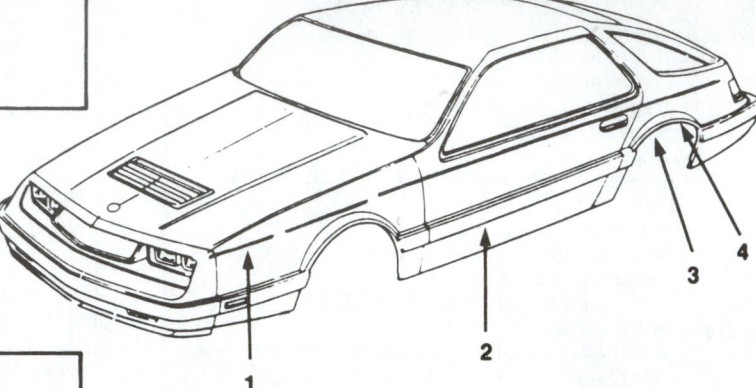

1. DECALS, STRIPE TAPE, AND OVERLAYS

Included:
- Installation of material only

Does Not Include:
- Removal of old material (estimate accordingly)
- R & R or R & I of moldings, name plates, and ornamentation
- Straightening or repairing damaged panels
- R & I outside door hardware
- Refinishing prior to application

2. ADHESIVE TYPE MOLDING

Panel replacement time does not include time for installation of molding, name plates, or ornamentation.

If new parts are used on a new panel, add one half the replacement time. If old parts are used on a new panel, add entire replacement time.

3. BOLT OR CLIP-ON MOLDING

Panel replacement time does not include time for installation of molding, name plates, ornamentation, or for the drilling of mounting holes.

If new parts are used on a new panel, add drilling time and one half the replacement time. If old parts are used on a new panel, add drilling time and entire replacement time.

4. DRILLING OPERATIONS

Time to drill mounting holes is listed in parentheses next to part description; (d.2) indicates .2 for drilling. Times shown are for round holes only. If holes must be other than round, estimate time accordingly.

If new parts are used on a new panel, add drilling time and one half the replacement time. If old parts are used on a new panel, add drilling time and entire replacement time.

FIGURE 22-39 Note typical example of repair operations for molding, decal, and overlay installation.

TOTAL ESTIMATED LABOR COSTS

Once all the repairs and labor times have been entered on the estimate, the computer will total your figures for you. If you do not have computerized estimating, it is necessary to refer to Table 22–1. This conversion table can be used to convert flat-rate labor time into dollars to fit local labor or operating rates per hour. With a computer estimating system, this form is built into the software and is calculated automatically.

When establishing flat labor rates, the shop overhead (including such items as rent, management and supervision, supplies, and depreciation on equipment) must be determined. Then the actual labor cost of all employees (including office help) and the profit required to keep the business operating must be added to the shop overhead to obtain a dollar flat-rate for repairs. This flat-rate cost is usually figured on an hourly basis.

Labor times shown in all collision-estimating guides are listed in hours and tenths of an hour. If a

SHOP TALK When using a conversion table such as Table 22–1, read across from the labor time column to the appropriate labor rate column. For time or dollar rates not listed, use a combination of columns given in the table. For dollar rates ending with 50 cents, add 50 from the cent column to the appropriate rate column.

vehicle requires a new right front fender, a new wheel opening molding, and a new name plate ornament, the total labor replacement time, according to a leading estimating guide (Figure 22–40), is

Front right fender	3.0 hours
Wheel opening molding	0.2 hours
Installation of nameplate	0.2 hours
Total labor	3.4 hours

TABLE 22-1: TIME/DOLLAR CONVERSION TABLE*

Dollar Per Hour Rates

Time	$.50	$1.00	$10.00	$15.00	$20.00	$25.00	$30.00	$35.00	$40.00	$45.00	$50.00
0.6	.30	.60	6.00	9.00	12.00	15.00	18.00	21.00	24.00	27.00	30.00
0.7	.35	.70	7.00	10.50	14.00	17.50	21.00	24.50	28.00	31.50	35.00
0.8	.40	.80	8.00	12.00	16.00	20.00	24.00	28.00	32.00	36.00	40.00
0.9	.45	.90	9.00	13.50	18.00	22.50	27.00	31.50	36.00	40.50	45.00
1.0	.50	1.00	10.00	15.00	20.00	25.00	30.00	35.00	40.00	45.00	50.00
1.1	.55	1.10	11.00	16.50	22.00	27.50	33.00	38.50	44.00	49.50	55.00
1.2	.60	1.20	12.00	18.00	24.00	30.00	36.00	42.00	48.00	54.00	60.00
1.3	.65	1.30	13.00	19.50	26.00	32.50	39.00	45.50	52.00	58.50	65.00
1.4	.70	1.40	14.00	21.00	28.00	35.00	42.00	49.00	56.00	63.00	70.00
1.5	.75	1.50	15.00	22.50	30.00	37.50	45.00	52.50	60.00	67.50	75.00
1.6	.80	1.60	16.00	24.00	32.00	40.00	48.00	56.00	64.00	72.00	80.00
1.7	.85	1.70	17.00	25.50	34.00	42.50	51.00	59.50	68.00	76.50	85.00
1.8	.90	1.80	18.00	27.00	36.00	45.00	54.00	63.00	72.00	81.00	90.00
1.9	.95	1.90	19.00	28.50	38.00	47.50	57.00	66.50	76.00	85.50	95.00
2.0	1.00	2.00	20.00	30.00	40.00	50.00	60.00	70.00	80.00	90.00	100.00
2.1	1.05	2.10	21.00	31.50	42.00	52.50	63.00	73.50	84.00	94.50	105.00
2.2	1.10	2.20	22.00	33.00	44.00	55.00	66.00	77.00	88.00	99.00	110.00
2.3	1.15	2.30	23.00	34.50	46.00	57.50	69.00	80.50	92.00	103.50	115.00
2.4	1.20	2.40	24.00	36.00	48.00	60.00	72.00	84.00	96.00	108.00	120.00
2.5	1.25	2.50	25.00	37.50	50.00	62.50	75.00	87.50	100.00	112.50	125.00
2.6	1.30	2.60	26.00	39.00	52.00	65.00	78.00	91.00	104.00	117.00	130.00
2.7	1.35	2.70	27.00	40.50	54.00	67.50	81.00	94.50	108.00	121.50	135.00
2.8	1.40	2.80	28.00	42.00	56.00	70.00	84.00	98.00	112.00	126.00	140.00
2.9	1.45	2.90	29.00	43.50	58.00	72.50	87.00	101.50	116.00	130.50	145.00
3.0	1.50	3.00	30.00	45.00	60.00	75.00	90.00	105.00	120.00	135.00	150.00
3.1	1.55	3.10	31.00	46.50	62.00	77.50	93.00	108.50	124.00	139.50	155.00
3.2	1.60	3.20	32.00	48.00	64.00	80.00	96.00	112.00	128.00	144.00	160.00
3.3	1.65	3.30	33.00	49.50	66.00	82.50	99.00	115.50	132.00	148.50	165.00
3.4	1.70	3.40	34.00	51.00	68.00	85.00	102.00	119.00	136.00	153.00	170.00
3.5	1.75	3.50	35.00	52.50	70.00	87.50	105.00	122.50	140.00	157.50	175.00
3.6	1.80	3.60	36.00	54.00	72.00	90.00	108.00	126.00	144.00	162.00	180.00
3.7	1.85	3.70	37.00	55.50	74.00	92.50	111.00	129.50	148.00	166.50	185.00
3.8	1.90	3.80	38.00	57.00	76.00	95.00	114.00	133.00	152.00	171.00	190.00
3.9	1.95	3.90	39.00	58.50	78.00	97.50	117.00	136.50	156.00	175.50	195.00
4.0	2.00	4.00	40.00	60.00	80.00	100.00	120.00	140.00	160.00	180.00	200.00
4.1	2.05	4.10	41.00	61.50	82.00	102.50	123.00	143.50	164.00	184.50	205.00
4.2	2.10	4.20	42.00	63.00	84.00	105.00	126.00	147.00	168.00	189.00	210.00
4.3	2.15	4.30	43.00	64.50	86.00	107.50	129.00	150.50	172.00	193.50	215.00
4.4	2.20	4.40	44.00	66.00	88.00	110.00	132.00	154.00	176.00	198.00	220.00
4.5	2.25	4.50	45.00	67.50	90.00	112.50	135.00	157.50	180.00	202.50	225.00
4.6	2.30	4.60	46.00	69.00	92.00	115.00	138.00	161.00	184.00	207.00	230.00
4.7	2.35	4.70	47.00	70.50	94.00	117.50	141.00	164.50	188.00	211.50	235.00
4.8	2.40	4.80	48.00	72.00	96.00	120.00	144.00	168.00	192.00	216.00	240.00
4.9	2.45	4.90	49.00	73.50	98.00	122.50	147.00	171.50	196.00	220.50	245.00
5.0	2.50	5.00	50.00	75.00	100.00	125.00	150.00	175.00	200.00	225.00	250.00
5.1	2.55	5.10	51.00	76.50	102.00	127.50	153.00	178.50	204.00	229.50	255.00
5.2	2.60	5.20	52.00	78.00	104.00	130.00	156.00	182.00	208.00	234.00	260.00
5.3	2.65	5.30	53.00	79.50	106.00	132.50	159.00	185.50	212.00	238.50	265.00
5.4	2.70	5.40	54.00	81.00	108.00	135.00	162.00	189.00	216.00	243.00	270.00
5.5	2.75	5.50	55.00	82.50	110.00	137.50	165.00	192.50	220.00	247.50	275.00
5.6	2.80	5.60	56.00	84.00	112.00	140.00	168.00	196.00	224.00	252.00	280.00
5.7	2.85	5.70	57.00	85.50	114.00	142.50	171.00	199.50	228.00	256.50	285.00
5.8	2.90	5.80	58.00	87.00	116.00	145.00	174.00	203.00	232.00	261.00	290.00
5.9	2.95	5.90	59.00	88.50	118.00	147.50	177.00	206.50	236.00	265.50	295.00
6.0	3.00	6.00	60.00	90.00	120.00	150.00	180.00	210.00	240.00	270.00	300.00

*Complete time/dollar conversion can be found in most crash guides.

But remember, there is a drilling time of 0.1 hours (d.2) involved when installing the new molding on the new fender; thus, the total labor time is 3.5 hours. This does not include refinishing or materials.

Referring to Table 22–1 and knowing that the shop's dollar per hour operating rate is (for this example) $25.00, the total estimated labor cost will be $87.50. When the dollar labor costs are determined, they are written on the estimate form.

On the whole, flat-rated operations comprise all work in which components must be removed and replaced. Estimated work is work done to straighten and/or repair body members. However, it is not always that simple. Sometimes there are combinations. A good example of this is when body members must be straightened first before cutting them off to make sure the new member will line up before welding. The procedure for this operation, even though it is a flat-rated R & R, must have some extra time (that must be estimated) added on to the total time for the procedure.

22.8 REFINISHING TIME

Making a correct estimate of the amount of labor time required to refinish panels, doors, hoods, and so forth is a vital part of an estimator's job function. Although the wide range of materials and conditions sometimes makes it difficult to arrive at a precise refinishing cost, there are a number of generally approved concepts that will help you arrive at a fair judgment of the amount of materials that will be needed for the job.

Flat-rate manuals published by vehicle manufacturers list a labor time plus a materials allowance (in dollars) for a multitude of individual paint operations. However, independently published estimating guides and crash books list paint labor times but not the dollar value of the materials required.

In some estimating guides, the time required for painting is shown in the parentheses adjacent to the part name. For the fender shown in Figure 22–40, the p 2.2, for example, indicates that 2.2 hours are needed to paint it. The basic colorcoat application generally includes the following:

- Clean panel, light sanding
- Mask adjacent panels
- Prime, scuff sand
- Final sand, clean
- Mix paint, load sprayer
- Apply colorcoat
- Remove masking
- Buff or compound (if required)
- Clean equipment

FENDER

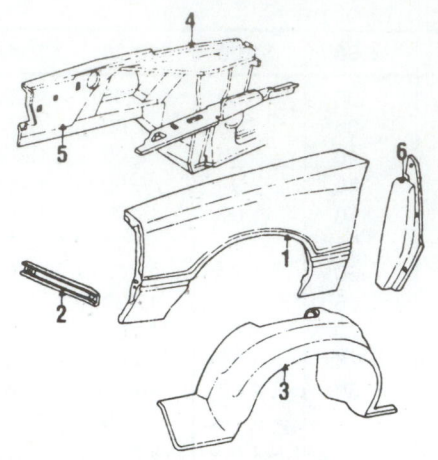

FENDER
REFINISHING NOTES

Deduct for overlap when necessary. See GUIDE to ESTIMATING pages.

Add for clear coat
fender. 1.1

PARTS & LABOR

1 Fender	(p2.2)	RT	242.43 #	3.0	E4ZZ16005A
	(p2.2)	LT	242.43 #	3.0	E4ZZ16006A
Transfer					
silver			8.17 †	0.3	E4ZZ16720B
black			8.17 †	0.3	E4ZZ16720A
Mouldings					
side, rear	84	RT	8.45	0.2	E4ZZ16A038B
	84	LT	8.45	0.2	E4ZZ16A039B
	85-86	RT	8.45	0.2	E5ZZ16A038CP
	85-86	LT	8.45	0.2	E5ZZ16A039CP
2 Brace			3.86		D9ZZ16A023A
Rear mount bracket			2.51		D8BZ16C078A
3 Splash shield		RT	51.70	0.5	E6ZZ16102A
		LT	51.70	0.5	E6ZZ16103A
4 Wheelhouse w/siderail					
	(p1.0) 84-85	RT	217.15 §	6.5	E7ZZ16054A
	(p1.0) 84-85	LT	217.15 §	6.5	E6ZZ16055A
	(p1.0) 86	RT	217.15 §	6.5	E7ZZ16054A
	(p1.0) 86	LT	217.15 §	6.5	E6ZZ16055A
5 front extension	84-85	RT	29.68 ‡	2.0	E4LY16054B
	84-85	LT	29.68 ‡	2.0	E7ZZ16055B
	86	RT	29.68 ‡	2.0	E7ZZ16054B
	86	LT	29.68 ‡	2.0	E7ZZ16055B
reinforcement					
upper		RT	12.23		E6ZZ16154A
		LT	11.71		D8BZ16155A
lower		RT	11.59		D9ZZ16060A
		LT	11.43		E6ZZ16060A
6 Sound absorber		RT	7.92		D9ZZ16071A
		LT	7.92		D9ZZ16072A

ANTENNA

Antenna assembly		15.00	0.5	E3AZ18813A

\# With antenna add .5.
§ After fender, radiator support, & front suspension crossmember are detached.
† Time does not include removal of old transfer.
‡ After fender and necessary parts are removed.

FIGURE 22–40 Note method of figuring refinishing and labor costs.

The basic colorcoat application generally does not include the following:

- Cost of paints or materials
- Matching and/or tinting color

- Grinding, filling, and smoothing welded seams
- Blending into adjacent panels
- Removal of protective coatings
- Spatter paint
- Clearcoat
- Custom painting
- Undercoating
- Anticorrosion materials
- Sound deadening
- Edging panel
- Underside of hood or trunk lids
- Covering entire vehicle prior to refinishing if necessary
- Protective coatings
- Additional time to produce custom, non-OEM finishes

Painting times given in most estimates are for one color on new replacement parts, OUTER SURFACES ONLY. Additions to paint times are usually made for the following operations:

- Underside of hood   Add 0.6
- Underside of trunk lid   Add 0.6
- Edging new part
  - First panel   Add 0.5
  - Each additional   Add 0.3
- Anticorrosion coating   Add 0.3
- Two-tone operations
  (unless otherwise specified in text)
  - First panel   Add 0.6
  - Each adjacent   Add 0.4
- Stone chip (protective material)
  - First panel   Add 0.5
  - Each additional   Add 0.3
- Clearcoat (basecoat/clearcoat)
  after deduction for overlap
  - First major panel   Add 50% of color-coat time
  - Each additional   Add 25% of color-coat time

Reductions in paint times can be considered when there is overlap:

- Overlap-Adjacent parts
  - First major panel   Full time
  - Each additional
    (except extensions)   Deduct 0.4
  - Extensions   Deduct 0.2
- Overlap-Additional parts (nonadjacent)
  - First major panel   Full time
  - Each additional   Deduct 0.2

The labor allowance given in a collision manual does not include any material costs. These must be estimated using a refinishing materials list (locally compiled) or one that is accepted on a national or

TABLE 22-2: REFINISH MATERIALS COST GUIDE (TYPICAL)

Hours	Amount of Material in Quarts	Cost of Materials†
0.5 to 1.0	1	$20.00
1.1 to 1.5	1-1/4	25.00
1.6 to 2.0	1-1/2	31.00
2.1 to 2.5	1-3/4	34.00
2.6 to 3.0	2	39.40
3.1 to 3.5	2-1/4	43.80
3.6 to 4.0	2-1/2	45.60
4.1 to 4.5	2-3/4	48.00
4.6 to 5.0	3	58.60
5.1 to 5.5	3-1/4	60.00
5.6 to 6.0	3-1/2	66.00
6.1 to 6.5	3-3/4	68.00
6.6 to 7.0	4	76.00
7.1 to 7.5	4-1/4	86.40
7.6 to 8.0	4-1/2	90.00
8.1 to 8.5	4-3/4	92.20
8.6 to 9.0	5	100.00
9.1 to 9.5	5-1/4	104.00
9.6 to 10.0	5-1/2	110.00
10.1 to 10.5	5-3/4	116.00
10.6 to 11.0	6	118.00
11.1 to 11.5	6-1/4	120.00
11.6 to 12.0	6-1/2	124.00
12.1 to 12.5	6-3/4	126.00
12.6 to 13.0	7	132.00
13.1 to 13.5	7-1/4	136.00
13.6 to 14.0	7-1/2	140.00
14.1 to 14.5	7-3/4	144.00
14.6 to 15.0	8	148.00
15.1 to 15.5	8-1/2	150.00
15.6 to 16.0		152.00

†Cost of materials varies across United States. Some colors, including metallics, are often priced higher.

regional basis. The use of a guide (Table 22–2) permits shop owners and estimators to place a fair evaluation on the refinish materials actually used.

To see how a guide such as this is used, assume that the car must have the entire trunk lid, the left rear quarter panel, and the rocker panel refinished. Looking up the labor times in one of the published guides reveals that the trunk lid requires 2.5 hours, the left rear quarter panel states 3.0 hours, while the rocker panel lists only 1.0 hour. Adding them up gives the total as 6.5 hours. Looking at Table 22–2, note that the total falls within the figures 6.1 through 6.5, which indicates that $2^3/_4$ quarts of refinish materials are needed to complete the job.

If the total amount of labor figures out to 14.0 hours on a two-door sedan or 16.0 hours on a four-door sedan, then a complete refinishing job should be considered. One final item that must not be overlooked in the refinish estimate has to do with preparation materials, such as plastic body filler, solvents, sanding discs, file sandpaper, forming blades, sealers, undercoat, tape, masking paper, welding supplies, as well as rivets, nuts, bolts, and so forth. Many body shop estimators recommend a charge of $3.00 to $5.00 per refinisher's hour in order to cover such material costs. The paint/material estimate might be listed on the damage report as individual operations or as a subtotal of the entire refinishing job.

22.9 ESTIMATE TOTAL

Once all cost information is entered (parts, labor, and materials), it can be added together for a subtotal. To this figure, add any extra charges such as wrecker and towing charges, storage fees, and state and local taxes. These figures, plus the subtotal figure, will give the grand total estimate of the repair.

While more and more larger body shops are doing repair jobs such as wheel alignment, rust-proofing, and tire replacement, smaller shops still "farm out" or "sublet" these jobs to others. When this is done, the specialty shop bills the body shop for the work. Generally, this is done at a rate less than the normal retail labor cost for work. In this way, the body shop can charge the normal retail cost and still make a small profit on this area of repair. Shops that farm out work usually have a column headed "Sublet" where the retail labor cost is marked. This sublet figure is added to others to obtain the grand total of the estimate.

If the owner wants to have extra work performed (on damage that occurred prior to the collision), this should be noted as "customer requested" (C/R) repairs. Insurance companies will not pay for customer requested repairs not a result of the wreck. A separate estimate for these extra repairs must be made for the customer.

22.10 ENTREPRENEURSHIP

An **entrepreneur** is a person who starts his or her own business. This business might be a body shop, materials supply house, parts warehouse, or similar endeavor. You might want to consider being an entrepreneur someday (Figure 22–41).

If you start your own business, you need to understand personnel management, bookkeeping, payroll, and state and local laws controlling the industry.

After mastering the knowledge needed to do collision repair work, you will be more marketable. These skills of a body technician can help when applying for other jobs or starting a business.

Try to learn something new that is job-related every day. Always learn from you mistakes. Read technical magazines and other publications. While working, try to think of more efficient ways to do things. Consider new tools and techniques. Participate in trade

FIGURE 22-41 If you prove yourself a dependable collision repair shop worker, it will help you advance into other positions. This is a vehicle assembly plant with numerous types of job openings. *(Courtesy of Chrysler Corp.)*

associations, such as I-CAR, VICA, NACE, ASE, etc. This attitude will help you become more productive.

Always try to develop a more systematic approach for doing your job. A **systematic approach** involves organizing and using a logical sequence of steps to accomplish a task or job. A systematic approach will result in your selecting more efficient ways of working.

Ask yourself these kinds of questions:

1. Did I take all needed tools to the vehicle?
2. Is there a better tool for a specific task?
3. Have I been reading the manufacturer's instructions (paints, plastic filler, equipment, etc.)?
4. Is my body in the right position to protect myself and feel comfortable while working?

Work traits are the little things, besides skills, that make you a good or bad employee. Work traits often determine whether you keep a good job or get fired. The most important work traits are listed below:

1. **Reliability** means being at work on time, being at work every day, and doing the job right. This is the most important job trait. Without it, you will have a difficult time keeping a job.

   For example, if you always miss work, you will affect everyone in the shop. If the vehicle has been promised on Friday and you are not there, either someone else has to do your work or the vehicle may not get done. Then the customer will be unhappy and will never return to your shop again. If this continues, word of mouth will ruin the reputation of the shop and profits will decline.

2. **Social skills** are important so that other workers and customers like you. Many times you will need the help of another worker to complete a difficult, two-person task. If you are unliked by others, you will find many tasks almost impossible.

A good body shop will have a team of workers who help each other succeed and prosper. They will exchange information, help each other with small tasks, and enjoy working together. You spend much of your life on the job; so why not enjoy it?

3. **Productivity** is a measure of how much work you get done. A highly productive technician will turn out a large amount of work. This will result in higher pay for the technician and more profits for the shop.

You want to balance productivity and quality. If you try to cut corners and get too much done too quickly, quality will often suffer. You must use proper work habits, common sense, and hard work to be productive.

4. **Professionalism** is a broad trait that includes everything from being able to follow orders to pride in workmanship. A professional does ev-

erything "by the book." He or she never cuts corners (leaves out the hard-to-reach bolt) trying to get the repair finished. Always think, dress, and act like a professional.

The following define a professional:

1. Customer oriented.
2. Up to date on vehicle developments.
3. Knows that vehicle safety and integrity depend on repair quality.
4. Pays attention to detail.
5. Ensures that his or her work is up to specs.
6. Participates in trade associations.
7. ASE or I-CAR certified.
8. Always tries to learn something new to improve skills.
9. Keeps his or her tools clean and organized.
10. Helps other technicians when they need it.

SUMMARY

- An estimate, also called a damage report or appraisal, calculates the cost of parts, materials, and labor for repairing a collision-damaged vehicle.

- An estimate is a firm bid for a given period of time—usually 30 days.

- The work order outlines the procedures that should be taken to put the vehicle back in top condition. Estimating guides contain the following:
  1. Illustrated parts breakdowns
  2. Part names and numbers
  3. Flat-rate times
  4. Part prices
  5. Other miscellaneous information

- Computerized estimating systems using a PC (personal computer) provide more accurate and consistent damage reports.

- The dial-up estimating system relies on data stored in a mainframe computer at a remote location.

- An in-shop estimating system has all of the data needed at the shop office, usually on CD-ROMs.

- A data scanner is a small hand-held electronic information storage tool.

- A total loss occurs when the cost of the repairs exceeds the value of the vehicle.

- The flat-rate is a preset amount of time and money charged for a specific repair operation.

- An entrepreneur is a person who starts his or her own business.

ASE-STYLE REVIEW QUESTIONS

1. Technician A uses flat rate labor times reported in collision damage manuals only as a guide for estimating. Technician B uses these guides as a reference for pricing parts. Who is correct?

   A. Technician A
   B. Technician B
   C. Both A and B
   D. Neither A or B

2. How long does an estimate usually remain a firm bid?

   A. 1 year
   B. 6 months
   C. 90 days
   D. 30 days

3. Which of the following abbreviations found in crash-estimating guides means that the item in question should be removed as an assembly, set aside, and later reinstalled?

   A. R & R
   B. R & I
   C. O & H
   D. O/H

4. Which of the following conditions is a good sign of minor body distortion?

   A. Stress cracking
   B. Front clip
   C. Side clip
   D. All of the above

5. What is the name given to jobs that can be performed individually but are also part of another operation?

   A. R & R
   B. Overlapping procedures
   C. Included operations
   D. Flat-rate

6. What is the smallest increment in which labor times are listed in crash-estimating guides?

   A. Hours
   B. Half hours
   C. Quarter hours
   D. Tenths of an hour

7. Technician A says a totaled vehicle has no salvage value. Technician B says that the customer will usually determine if a vehicle is a total loss. Who is correct?

   A. Technician A
   B. Technician B
   C. Both A and B
   D. Neither A or B

8. Which of the following is not included in the refinishing times for painting a new panel?

   A. Removing moldings
   B. Featheredging body putty
   C. Masking handles
   D. All of the above

9. Which of the following operations is included in the refinishing times listed in the crash manuals?

   A. Scuff sanding the repair area
   B. Masking the complete vehicle to prevent overspray damage
   C. Cleaning the equipment
   D. All of the above

10. Printed estimating guides produced from the same data base as the computer program _____.

    A. Serve as a verification check for data used to generate the computer estimate
    B. Serve as a backup system in the event of breakdown or power loss
    C. Serve as a portable reference guide and a source of additional useful information
    D. All of the above

11. A computer program or instructions written in a computer language that tell the computer what to do is also known as _____.

    A. Hardware
    B. Jargon
    C. Software
    D. Menu

ESSAY QUESTIONS

1. List the advantages of a resident data source.

2. When writing an estimate, what are some of the problems to look for?

3. Doors out of alignment and cracked stationary glass are often a solid clue to what?

4. What does "salvage value" mean?

5. How does shop efficiency affect estimating?

6. List six traits of a professional technician.

CRITICAL THINKING PROBLEMS

1. If you make your estimate too low or too high, how could each affect the shop operation?

2. A car was driven over a cement median curb at high speed by a drunken driver. You find cracked sealer around the front frame rails. What could this tell you about vehicle damage?

MATH PROBLEMS

1. A repair will require 3.0 hours at $40 per hour. How much will the total labor be for this job?

2. A gallon of paint is $95 and you only need a quart for the repair. How much should you charge the customer for the paint?

ESTIMATING

P22-1 The estimator must understand all aspects of collision repair to do accurate calculations of repair costs.

P22-2 The estimator must inspect for obvious signs of damage to the outer body structure. He or she must also be knowledgeable enough to find hidden damage.

P22-3 Today's estimators use PCs and specialized software to automate the estimating process.

Auto Body Shop Terms
Términos del Taller de Carrocería

Abrasive Material such as sand, crushed steel grit, aluminum oxide, silicon carbide, or crushed slag used for cleaning or surface roughening.

Abrasivo Una material tal como la arena, la granalla de acero para bruñir, el carburo de silicio, o la escoria machacada que sirve para limpiar o deslustrar una superficie.

Access time Time required to remove extensively damaged collision parts by cutting, pushing, pulling, and so on.

Tiempo de acceso El tiempo requerido para remover las partes dañadas extensivamente en una colisión por medio de cortar, empujar, jalar y etcétera.

Accessible area An area that can be reached without parts being removed from vehicle.

Area de acceso Una área que se puede alcanzar sin remover partes del vehículo.

Activator An additive needed to cure a two or multi-package enamel (see hardener).

Activador Un aditamento requerido para que un esmalte de dos o múltiples etapas se cura (vea endurecedor).

Actual cash value (ACV) Current market value of a standard production vehicle and its accessory options as determined by used car guidebook listings or car dealer assessments.

Valor actual El valor de mercado actual de un vehículo de producción regular y sus accesorios opcionales que se determina por la guía de coches usados o por los precios de los comerciantes.

Additives Chemical substances added to a finish in relatively small amounts to impart or improve desirable properties.

Aditamentos Las substancias químicas agregadas a un acabado en cantidades relativamente pequeñas para conferir o mejorar las propriedades deseadas.

Adhesion Ability of one substance to stick to another.

Adhesión La habilidad de una substancia de pegarse a otra.

Adhesion promoter A water-white, ready-to-spray lacquer material that provides a chemical etch to OEM finishes.

Promovedor de adhesión Una material de laca blanca soluble en agua, listo para atomizar que provee un grabado químico en los acabados OEM.

Adjuster An insurance company representative who is responsible for approving the collision repair bid and satisfying a customer's vehicle damage claim; same as appraiser.

Ajustador Un representante de una compañía de aseguranzas responsable de aprobar el precio de reparación de una colisión y de satisfacer la reclamación de los daños al vehículo; igual que un avaluador.

Aerodynamic Shape with low wind resistance.

Aerodinámico Una forma de poca resistencia al viento.

Aging Allowing a material to stand for some time.

Envejecimiento Dejando que una material se queda por un tiempo.

Agitator A paint mixer of any type.

Agitador Una mezcladora de pinturas de cualquier tipo.

Air Usually used under pressure as a propellant. Trapped air can cause bubbling, popping, foam, and so on.

Aire Suele usarse bajo presión como propulsor. El aire atrapado puede causar las burbújas, los estallidos, la espuma y todo lo demás.

Air bag system System that uses impact sensors, vehicle's on-board computer, an inflation module, and a nylon bag in the steering column and dash to protect the driver and passenger during a head-on collision.

Sistema de bolsa de aire Un sistema que usa los sensores, la computador a bordo del vehículo, un módulo inflador, y las bolsas de nylon en la columna de dirección y en el tablero para proteger al conductor y al pasajero en caso de una colisión frontal.

Air dry Ability of a coating to dry at room temperature.

Secado al aire La habilidad de una capa a secarse en temperaturas ambientes.

Air drying Allowing paint to dry at ambient (surrounding) temperatures without aid of an external heat source.

Secar al aire Permitir que la pintura se seca en las temperaturas del ambiente (actuales) sin aplicar el calor externo.

Air spray A system of applying paint in the form of tiny droplets in air. Paint is broken into droplets (atomized) by a spray gun as a result of being forced into a high velocity air stream. Shape and paint density of the resulting droplet cloud can be controlled by air pressure, paint viscosity, and gun tip geometry.

Atomización con aire Un sistema de aplicar la pintura en forma de diminutas gotitas suspendidas en el aire. La pintura se fracciona en gotitas (se atomiza) en una pistola rociadora por medio de la fuerza de un chorro de aire de alta velocidad. La forma y la densidad de la niebla de gotitas de pintura se controla por medio de la presión y la viscosidad de la pintura y por la geometría de la boquilla de la pistola.

Air supply system A breathing air pumping system supplying fresh air to a painter.

Sistema de suministro de aire Un sistema de respiración con una bomba de aire que provee el aire fresco al pintor.

Airless spraying A method of spray application in which atomization is affected by forcing paint under high-pressure through a very small orifice in a spray gun cap. On emerging, the paint instantly expands and breaks into very fine particles.

Atomización sin aire Un método de aplicar un rocío en el cual la atomización se efectúa forzando la pintura bajo alta presión por un orificio muy pequeño en la tapa de la pistola de atomización. Saliendo, la pintura se expanda al instante y se separa en partículos muy finos.

Alcohol A class of chemical compounds used as diluents, solvents, or cosolvents.

Alcohol Una clase de compuestos químicos que sirven de diluyentes, disolventes o codisolventes.

Align To adjust to a line or predetermined relative position.

Alinear Ajustar en una linea o a una posición relativa predeterminada.

Alkyd Chemical combination of an alcohol, an acid and an oil. Widely useful vehicle for paint. Properties depend on ingredients. Mainly used in fleet operations.

Preparación alquídica Una combinación química de un alcohol, un ácido y un aceite. Es un medio sumamente útil para la pintura. Sus propriedades dependen de las ingredientes. Se usa primariamente en las operaciones de flete.

Aluminum Metal useful as a substrate or pigment. When used as substrate requires paint to prevent corrosion.

Aluminio Un metal que sirve de substrato o de pigmento. De substrato, requiere la pintura para prevenir la corrosión.

Anchor To hold in place.

Anclar Sostener fijo.

Anodizing An electrolytic surface treatment for aluminum which builds up an aluminum oxide coating.

Anódico Un tratamiento electrolítico de la superficie del aluminio que recarga una capa del óxido de aluminio.

Antifouling A paint which contains toxic substances to inhibit growth of certain organisms on ship bottoms.

Preservativo Una pintura que contiene substancias tóxicas que impiden el crecimiento de ciertos organismos el la parte inferior de los barcos.

Appraiser An insurance company representative who is responsible for approving a collision repair bid and satisfying a customer's vehicle damage claim; same as an adjuster.

Avaluador Un representante de una compañía de aseguranzas responsable de aprobar el precio de reparación de una colisión y de satisfacer la reclamación de los daños al vehículo; igual que un ajustador.

Asbestos dust Cancer-causing agent used in the manufacture of older brake and clutch assemblies.

Polvo de amianto Un agente carcígeno usado en la fabricación de las asambleas antiguas de frenos y embragues.

ASE certification Testing program to help prove that you are a knowledgeable collision repair technician.

Certificación ASE Una programa de examenes para certificar que es Ud. un técnico de reparación de colisión experimentado.

Asphyxiation Refers to anything that prevents normal breathing. There are many mists, gases, and fumes in a body shop that can damage your lungs and affect your ability to breathe.

Asfixiación Se refiere a cualquier cosa que impide la respiración normal. Hay muchas brumas, gases y humos en un taller de carrocería que pueden dañar sus pulmones y afectar su habilidad de respirar.

Assembly A number of parts that are either bolted or welded together to form a single unit.

Asamblea Una cantidad de partes que se empernan o se sueldan juntos para formar una sola unedad.

Atomize To break a liquid into a fine mist of droplets.

Atomizar Fraccionar un líquido a una niebla fina de gotitas.

Autoignition temperature Approximate lowest temperature at which a flammable gas or vapor-air mixture will spontaneously ignite without spark or flame.

Temperatura de autoencendido La temperatura aproximativa más baja en la cual un gas inflamable o una mezcla de aire y vapor se encenderán sin la presencia de una chispa o una llama.

Automatic welding Welding with equipment that performs the entire welding operation without constant observation and adjustment of controls by an operator. Equipment may or may not perform loading and unloading of the work.

Soldadura automático La soldadura por medio de un aparato que ejecuta la operación de soldar total sin requerir un operador para observar o ajustar los controles constantemente. El aparato puede ser capaz de emplazar o desplazar el trabajo o no.

Back light Vehicle window located behind the passengers.
Ventanilla trasera La ventana ubicada detrás de los pasajeros.

Baffle Panel used to direct air to the radiator.
Deflector Un panel para dirigir el aire al radiador.

Banding A single coat applied in a small spray pattern to frame-in an area to be sprayed.
Rociar un cuadro Una capa única aplicado con un patrón de rociado muy denso para encuadrar una area que se debe rociar.

Bar gauges Gauges used to accurately measure and diagnose body and frame collision damages for all conventional and unitized vehicles.
Bancada Los calibradores para medir y diagnosticar precisamente los daños de colisión a la carrocería y al bastidor en todos los vehículos convencionales y de monocasco.

Basecoat/clearcoat A paint system in which the color effect is given by a highly pigmented basecoat. Gloss and durability are given by a subsequent clearcoat. The basecoat can be either a solid color or metallic.
Capa de fondo/capa transparente Un sistema de pintar en el que el efecto del color se logra por medio de una capa de fondo con mucho pigmento. Una capa transparente subsiguiente produce la brillantez y la durabilidad. La capa de fondo puede ser de un color sólido o metálico.

Basecoat Coat of paint on which the final coats are applied.
Capa de fondo Una capa de pintura sobre la cual las capas finales se aplican.

Belt Line Horizontal molding or crown along the side of the vehicle at the bottom of the glass.
Perímetro (ceja) La moldadura o cresta horizontal por el largo del lado del vehículo en la parte inferior del vidrio.

Bench A vehicle underbody anchoring device. The vehicle is anchored to the bench to check its frame and suspension dimensions for damage and to allow for straightening procedures.
Banco Un dispositivo para sujetar la plataforma inferior del vehículo. El vehículo se fija al banco para determinar los daños a las dimensiones del bastidor y suspensión y permitir que se efectúan los procedimientos del enderezado.

Binder Ingredient in a paint that holds pigment particles together.
Aglomerante Un ingrediente en la pintura que mantiene unidos los partículos de pigmento.

Bleeding Original color showing through after a new topcoat has been applied.
Color sangrado El color original se puede ver al través de la capa superior que se ha depositado.

Blistering Formation of hollow bubbles or water droplets in a paint film. Blistering is usually caused by expansion of air or moisture trapped beneath the paint film. Blisters can form around salt crystals trapped under a film because salt attracts moisture and is dissolved.
Formar ampollas La formación de las burbujas o las gotitas de agua dentro de una película de pintura. Las ampollas suelen ser causadas por una expansión del aire o la humedad que se ha atrapado debajo de la película de pintura. Las ampollas pueden formarse alrededor de los cristales de sal atrapados debajo de la película porque el sal atrae la humedad y se disuelve.

Blushing Hazing or whitening of a film caused by absorption and retention of moisture in the drying paint film.
Aspecto lechoso Una película que tiene un haz o se blanquea por causa de la absorción y retención de la humedad en una película de pintura que se esta secando.

Body Apparent viscosity of a paint as assessed when stirring it; consistency of a liquid.
Cuerpo La viscosidad aparente de una pintura que se nota al mezclarlo; la consistencia de un líquido.

Body filler A heavy-bodied plastic material that cures very hard and is used to fill small dents in metal.
Masilla Una material espesa de plástico que se endurece mucho al curarse y sirve para rellenar las abolladuras pequeñas en el metal.

Body panels Sheets of shaped steel, aluminum, or plastic that form the car body.
Paneles de la carrocería Las chapas de acero, aluminio o plástico que comprenden la carrocería de un coche.

Body trim Upholstery or body ornamentation.
Molduras Las piezas embellecedoras del interior o de la carrocería.

Body-over-frame A vehicle that has separate body and chassis parts bolted to the frame.
Carrocería sobre bastidor Un vehículo que tiene partes independientes de plancha y chasis empernados al bastidor.

Bounce-back Atomized particles of paint that rebound from the surface being sprayed and contribute to overspray.
Resalto Partículos atomizados de esmalte que brincan de la superficie que se esta rociando y contribuyan al desparrame.

Braze welding A method of welding using a filler metal. Unlike brazing, the filler metal is not distributed in the joint by capillary action.
Soldadura con bronce Un método de soldar usando un metal de relleno. Es distincto de la soldadura fuerte porque el metal de relleno no se distribuye por acción capilario.

Bridging A characteristic of undercoat performance that occurs when a scratch or surface imperfection is not completely filled. Generally caused by under-reducing primers or using too fast a solvent.
Puente superficial Una característica de la calidad de una capa de apresto que ocurre cuando una raya o una imperfección de la superficie no se rellena completamente. Generalmente se causa por una imprimación de baja reducción o por el uso de un solvente de evaporación demasiado rápida.

Brittle Lack of flexibility; usually combined with a lack of toughness.

Frágil Una carencia de la flexibilidad; suele combinarse con una carencia de tenacidad.

Bronzing Formation of a metallic appearing haze on a paint film.

Broncear La formación de un haz con una aparencia metálica en la película de la pintura.

Brush A method of applying paint or an applicator for applying paint.

Brocha Un metodo de aplicar la pintura o un aplicador para aplicar la pintura.

Brushing Act of applying paint by a brush or ability of a paint to be applied by a brush.

Con brocha El acto de aplicar la pintura con una brocha o la habilidad de una pintura de aplicarse con una brocha.

Build Amount of paint film deposited (measured in mils).

Recarga La cantidad de pintura en la película que se ha depositado (medida en mils).

Bulge High crown or area of stretched metal.

Protuberancia Una cresta alta o una área de metal estirado.

Burnishing Polishing or buffing a finish by hand or machine using a compound or liquid manufactured for this purpose.

Bruñir Pulimentar un acabado por mano o con máquina usando un compuesto o un líquido que se ha fabricado para este uso.

Butt weld Weld made along a line where two pieces are put edge to edge.

Soldadura a tope Una soldadura hecha por una linea en la que dos piezas se colocan borde a borde.

Butyl acetate A solvent for paint. Commonly used in lacquers. Usually the base for comparing evaporation rates of organic liquids.

Acetato butílico Un solvente de la pintura. Se usa generalmente con las lacas. Suele ser la base en la comparación de las relaciones de evaporación de los líquidos orgánicos.

Calcium A metal component of dryers and pigments.

Calcio Un componente metálico de los secadores y pigmentos.

Camber Inward or outward tilt of wheel at top from true vertical. It is the tire wearing angle in degrees.

Camber (comba) La inclinación fuera de vertical perfecto hacia adentro o hacia afuera en la parte superior de una rueda. Es el ángulo en grados que desgasta las llantas.

Caster Backward or forward tilt of kingpin or spindle support arm at top from true vertical. It is the directional control angle measured in degrees.

Angulo de caster La inclinación fuera de vertical perfecto hacia atrás o hacia afrente en la parte superior de un pivote o el brazo de soporte del husillo. Es el ángulo de control direccional que se mide en grados.

Catalyst A substance that causes or speeds up a chemical reaction when mixed with another substance but does not change by itself.

Catalizador Una substancia que causa o acelera una reacción química al ser mezclada con otra substancia pero que no cambia en si misma.

Centering gauge Used in sets of four, this frame gauge locates horizontal datum planes and centerlines on a vehicle.

Calibrador para centrar Una bancada que se usa en juegos de cuatro para localisar los planos de referencia horizontales y el eje mediano de un vehículo.

Chalking A result of weathering of a paint film, characterized by loose pigment particles on the surface of the paint. May be beneficial or harmful.

Formación de una capa de tisa El resultado del desgaste climático en una película de pintura, caracterizado por los partículos de pigmento sueltos en la superficie de la pintura. Puede ser beneficioso o dañoso.

Chassis Assembly of mechanisms that makes up major operating systems of vehicle; basically includes everything under the body—suspension system, brake system, wheels and tires, and steering system.

Chasis Una asamblea de los mecanismos que forman los sistemas principales de operación de un vehículo; incluye esencialmente todo lo que se encuentra abajo de la carrocería—el sistema de suspensión, el sistema de frenos, las ruedas y las llantas, y el sistema de dirección.

Check A small crack.

Grieta Una pequeña grieta.

Checking A type of failure in which cracks in the paint film begin at the surface and progress downward. The result is usually a straight V-shaped crack, which is narrower at the bottom than the top.

Agrietamiento Un tipo de fallo en el que las grietas en la película de pintura comienzan en la superficie y viajan hacia abajo. El resultado suele ser una grieta recta en forma de V, más delgada en la parte inferior que en la parte superior.

Chemical burns Injuries that result when a corrosive chemical burns the skin or eyes. May be caused by paint removers, cleaners, or other chemicals found in a body shop.

Quemaduras químicas Los daños que resultan cuando un producto químico corrosivo quema al piel o los ojos. Puede causarse por los decapadores de pintura, las limpiadores químicas u otros productos químicos de un taller de carrocería.

Chemical staining Spotty discoloration of the topcoat caused by atmospheric conditions (usually occurs near industrial activity).

Manchas químicas Las manchas descoloridas en la capa superior causadas por las condiciones atmosféricas (suele ocurrir cerca de una área industrial).

Chipping Breaking away of a small portion of paint film due to its inability to flex under impact or with the thermal expansion and contraction of the substrate.

Desconchar Se cae una pequeña parte de la película de pintura debido a su inabilidad de doblarse bajo impacto o por la dilatación y contracción térmica del substrato.

Chroma Strength or intensity of a color; the amount a color differs from the white, gray, or black of the neutral axis of the color tree. Often referred to as saturation/desaturation.

Saturación cromático La fuerza o intensidad de un color; la cantidad que difere un color del blanco, gris o negro en el eje neutro de una escala de colorimetría. Esta intensidad suele referirse como la saturación/desaturación.

Chronic effect An adverse effect on a human or animal. Symptoms develop slowly over a long period of time or recur frequently.

Efecto crónico Un efecto desfavorable padecido por un ser humano o un animal. Las símptomas se desarollan poco a poco durante un largo período de tiempo o se reaparecen frecuentemente.

Clean A bright clear color. Also clean as after washing or cleaning.

Limpio Un color vívido y claro. Limpio tambien se refiere al estado de limpieza.

Cleaner Material used to clean a substrate.

Limpiadora Una material que sirve para limpiar un substrato.

Clear A paint containing no pigment or only transparent pigments.

Transparente Una pintura que no tiene pigmento o que sólo contiene los pigmentos transparentes.

Clearance Amount of space between adjacent panels.

Holgura La cantidad del espacio entre dos paneles contiguas.

Clearcoat A transparent top coating on a painted surface so that the color coat beneath it is visible.

Capa transparente Una capa superficial transparente puesta sobre una superficie pintada para que la capa de color queda visible.

Closed structural members Boxed-in sections, typically accessible only from the outside, such as rails, pillars, and so forth.

Miembros estructurales cerrados Las secciones encerradas, a las que típicamente sólo se puede llegar desde afuera, tal como los largueros, los montantes y lo demás.

Clouding Formation or presence of a haze in a liquid or in a film.

Velarse La formación o presencia de un haz en un líquido o una película.

Coat (double) Two single coats, one followed by the other with little or no flash time.

Capa (doble) Dos capas sencillas, una seguida por la otra con poco o no tiempo de vaporización instantánea.

Coat (single) A coat produced by two passes of a spray gun, one pass overlapping the other in half steps.

Capa (sencilla) Una capa producida por dos desplazamientos de la pistola de atomización, un desplazamiento solapando al otro en etapas medias.

Coating Act of applying paint or the actual film left on a substrate by a paint.

Aplicar una capa El acto de depositar la pintura o la película de pintura que queda sobre un substrato.

Coatings Covering materials used to protect an area.

Capas Las materiales de revestimiento que sirven para proteger una área.

Cobwebbing Tendency of sprayed paint to form strings or strands rather than droplets as it leaves the gun.

Formación de filamentos La tendencia de la pintura atomizada de salir en hilos en vez de en gotitas al salir de la pistola.

Collision Commonly called a "crash" or "wreck", is accidental damage caused by an impact on the vehicle body and chassis.

Colisión Comunmente llamado un "choque" o un "accidente", se refiere a los daños accidentales causados por un impacto en la carrocería o el chasis del vehículo.

Color Visual appearance of a material—red, blue, green, etc. Colors are seen differently by different people.

Color La aparencia visual de una material—el rojo, el azul, el verde, etc. Los colores varían en su aspecto según quien los vea.

Color holdout A generic term used to describe the ability of a primer-sealer to allow the finish coat to maintain its high degree of gloss.

Retención de brillo Un término genérico que describe la habilidad de una preparación imprimación-sellador en permitir que la capa de acabado mantenga un alto nivel de brillantez.

Color retention Permanence of a color under a set of conditions.

Retención de color La permanencia de un color bajo ciertas condiciones.

Color sanding Process of wet sanding to remove surface imperfections from the topcoat. Must be polished after color sanding to return the gloss.

Lijado en color El proceso de lijar en húmedo para remover las imperfecciónes de la capa superior. Se debe pulimentar despues del lijado en húmedo para restaurar la brillantez.

Compatibility Ability of two or more materials to be used together successfully.

Compatibilidad La habilidad de dos materias o más de combinarse con buenos resultados.

Compounding Action of using an abrasive material—either by hand or machine—to smooth and bring out gloss of the applied topcoat.

Pulimentación El uso de una material abrasiva—aplicada sea por mano o por máquina—para que la capa superficial queda lisa y brillante.

Computer Electronic device for storing and manipulating information.

Computadora Un dispositivo electrónico que sirve para almacenar y manipular la información.

Concentration Amount of any substance in a solution.

Concentración La cantidad de cualquier substancia en una solución.

Condensation A change from a vapor to a liquid on a cold surface (commonly moisture). A type of polymerization characterized by the reaction of two or more monomers to form a polymer plus some other product, usually water.

Condensación Un cambio de un vapor a un líquido sobre una superficie fría (comunmente la humidez). Un tipo de polimerización caracterizado por la reacción de dos monómeros o más al formar un polímero y algún otro producto, comunmente el agua.

Consistency Fluidity of a system.
Consistencia La fluidez de un sistema.

Contaminants Foreign substances on a surface to be painted (or in paint) that adversely affect the finish.
Contaminantes Las materias extrañas en la superficia de acabado (o en la pintura) que pueden afectar negativamente al acabado.

Control points Points on a vehicle, including holes, flats, or other identifying areas, used to position panels and rails during manufacturing of the vehicle.
Puntos para comprobación Los puntos de un vehículo, incluyendo los agujeros, las areas planas, u otras áreas de identificación que sirven para posicionar los paneles y los largueros durante la fabricación del vehículo.

Conventional frame Vehicle construction type in which the engine and body are bolted to a separate frame.
Chasis universal Un tipo de construcción del vehículo en el que el motor y la carrocería se empernan en un bastidor portante.

Conventional points Points on a unibody used as references to make a repair.
Puntos convencionales Los puntos de referencia en un monocasco que sirven para efectuar un reparación.

Conversion coating A chemical treatment used on galvanized steel, uncoated steel, and aluminum to prevent rust.
Revestimiento por conversión Un tratamiento químico empleado en el acero galvanizado, el acero no tratado y el aluminio para prevenir la corrosión.

Copper A difficult metal substrate to paint. Used in the manufacture of pigments and dryers.
Cobre Una materia metálica de substrato que es muy difícil de pintar. Se usa en la fabricación de los pigmentos y los secantes.

Corrosion Chemical reaction of air, moisture, or corrosive materials on a metal surface; usually referred to as rusting or oxidation.
Corrosión La reacción química del aire, la humedad, o las materias corroídas en una superficie metálica; normalmente se refiere como comido por la herrumbre o la oxidación.

Corrosion protection Involves using various methods to protect steel body parts from rusting.
Protección de la corrosión Involucra el uso de varios métodos para proteger las partes de acero de la carrocería de la oxidación.

Coverage Area a given amount of paint will cover. The ability of a topcoat to hide undercoats.
Poder cubriente El área que puede cubrir una cantidad de pintura. La habilidad de la capa superior de encubrir las capas de apresto.

Cowl panels Panels forward of the passenger compartment to which the fenders, hood, and dashboard are bolted.
Cubretableros Los paneles delante del habitáculo de pasajeros al cual se empernan las aletas, el capót, y el panel de instrumentos.

Cracking Splitting of a paint film. Usually occurs as straight lines which penetrate the entire film thickness. Results when overbaked or excessively thick films are low temperature cycled.
Agrietura Hendidura de una película de pintura. Suele aparecer como lineas rectas que penetran la totalidad del espesor de la película. Resulta cuando las películas que se han cocido demasiado o que son de un espesor excesivo se ciclan en bajas temperaturas.

Cratering Formation of holes in the paint film where paint fails to cover due to surface contamination.
Formación de cráteres La formación de los agujeros en una película de pintura donde no cubre la pintura debido a la contaminación de la superficie.

Crawling A wet film defect that results in a paint film pulling away from certain areas or not wetting certain areas, leaving those areas uncoated.
Risado Un defecto de una película húmeda que hace retroceder una película de pintura de ciertas áreas o que no humedece algunas áreas, dejándolas descubiertas.

Crazing A film failure that results in surface distortion or fine cracking.
Agrietada Un fallo de la película que produce una distorsión en la superficie o las grietas muy finas.

Crush zones Sections built into the frame or body designed to collapse and absorb some of the energy of a collision.
Zonas de impacto Las secciones incorporadas en el diseño del bastidor o la carocerría que se hunden así absorbiendo algo de la energía producida por una colisión.

Cure Process of drying or hardening of a paint film.
Curar El proceso de secar o endurecer de una película de pintura.

Curing Chemical reaction in drying of paints which dry by chemical change.
Endurecimiento Una reacción química de secado de las pinturas que se secan por un cambio químico.

Datum line An imaginary line that appears on frame blueprints or charts to help determine the correct height.
Linea de nivel Una linea imaginaria que aparece en los dibujos o diagramas del bastidor para determinar la altura correcta.

Dedicated fixture measuring system A bench with fixtures that are set in specific points for body measurement.
Compás de puntas para medición Un banco que tiene montajes fijos en puntos específicos para la medición de la carrocería.

Degradation Gradual or rapid disintegration of a paint film.

Degradación La disintegración gradual o rápida de una película de pintura.

Degreasing Cleaning a substrate (usually metal) by removing greases, oils, and other surface contaminants.

Decapar La limpieza de un subestrato (comunmente de metal) removiendo las grasas, el aceite, y otros contaminantes de la superficie.

Dehydration Removal of water.

Deshidratación La remoción del agua.

Density Weight of any material per unit of volume. Commonly grams per cubic centimeter. Water is minus 1.

Densidad El peso de cualquier material por unedad de su volumen. Suele medirse en gramos por centímetro cúbico. La densidad del agua es de uno negativo.

Detailing Final cleanup and touch-up on a vehicle.

Detallar La limpieza y los últimos retoques en un vehículo.

Dew point Temperature at which water vapor condenses from the air.

Punto de rocío La temperatura en que el vapor de agua condensa del aire.

Diamond A damage condition where one side of the vehicle has been moved to rear or front causing the frame/body to be out of square.

Deformación romboidal Una condición de daños en que un lado del vehículo se ha movido hacia atrás o hacia afrente resultando en que la carrocería/el bastidor estén fuera de escuadra.

Diluent A liquid, not a true solvent, used to lower the cost of a paint thinner system.

Diluyente Un líquido, no un solvente puro, que sirve para rebajar los gastos de un sistema de diluir las pinturas.

Dilution ratio Amount of a diluent which can be added to any true solvent when the mixture is used to dissolve a certain weight of polymer. Limits are determined by dissolving the polymer in solvent and then adding diluent until gelation or kick-out occurs.

Relación de dilución La cantidad del diluyente que se puede agregar a cualquier solvente puro cuando esta mezcla se usará para disolver un polímero de cierto peso. Se determinan los límites disolviendo el polímero en el solvente y luego añadiendo el diluyente hasta que ocurre la gelificación o el desnate.

Dipping Applying paint, primer, or sealer to an article by immersing the article in a container of paint and then withdrawing it and allowing the excess paint to drain from the part.

Inmersión Aplicar el esmalte, la imprimación o el sellante a un artículo sumergiéndolo en un envase de pintura y luego sacándolo y dejándo que la pintura excesa se escurre de la parte.

Direct damage Damage that occurs to an area that is in direct contact with the damaging force or impact.

Daño directo El daño que ocurre en una área que esta en contacto directo con la fuerza dañosa o el impacto.

Dirty A color that is not bright and clean. A color that appears grayish. Also a condition requiring cleaning.

Sucio Un color que no es vívido ni claro. Un color que aparece agrisado. Tambien refiere al estado que requiere la limpieza.

Dispersion Act of distributing solid particles uniformly throughout a liquid, commonly dispersion of pigment in a vehicle. Does not imply fracturing of individual particles, only their distribution.

Dispersión El acto de distribuir en una manera uniforme los partículos sólidos en un líquido, suele referirse como la dispersión del pigmento en un medio. No implica la fracción de los partículos individuales, sólo su distribución.

Dispersion coatings Types of paint in which binder molecules are present as colloidal particles instead of solutions.

Capas de dispersión Los tipos de pintura en los cuales las moléculas aglomerantes se presentan en forma de partículos coloidales en vez de soluciones.

Distillation range Boiling range of a liquid.

Nivel de destilación La gama de ebullición de un líquido.

Dog tracking Off-center tracking of the rear wheels as related to the front wheels.

Posición oblicua El rastreo fuera de central de las ruedas traseras en relación a las ruedas delanteras.

Doghouse Front clip or front body section. The front section, also called nose section, includes everything between the front bumper and the firewall.

Doghouse (perrera) El sujetador delantero o la sección delantera de la carrocería. La parte delantera, se puede denominar también el "nariz", incluye todo entre el parachoques delantero y el tabique.

DOL U.S. Department of Labor.

DOL El departamento de trabajo de los E.U.

Double coat Spray first pass left to right; spray second pass right to left directly over the first pattern.

Capa cruzada La primera mano se deposita de izquierda a derecha; la segunda mano de derecha a izquierda directamente sobre el primer patrón.

Drivetrain Engine, transmission, and drive shafts/axles.

Tren de propulsión El motor, la transmisión, y las flechas/los ejes de propulsión.

Dry Change from a liquid to a solid, which takes place after a paint is deposited on a surface. Included is evaporation of solvents and any chemical changes that occur. The result of drying is a useful film.

Secar El cambio del líquido al sólido que ocurre después de que se haya depositado la pintura sobre la superficie. Incluye la evaporación de los solventes y cualquier cambios químicos que ocurren. El resultado del secado es una película útil.

Dry spray An imperfect coat, usually caused by spraying too far from the surface being painted or on too hot a surface.

Rociado en seco Una capa defectuosa, normalmente resulta del rociado desde una distancia excesiva de la superfice o de una superficie demasiada caliente.

Dryer A catalyst added to a paint to speed up cure or dry.
Secador Un catalizador agregado a una pintura que acelera el endurecemiento o el secado.

Drying Process of changing a coat of paint from a liquid to a solid state due to evaporation of the solvent, a chemical reaction of the binding medium, or a combination of these causes.
Secado El proceso de cambiar una capa de pintura de un líquido a un estado sólido debido a la evaporación del solvente, una reacción química del medio aglomerante, o una combinación de éstas causas.

Durability Length of life. Usually applies to a paint used for exterior purposes.
Durabilidad Tiempo de duración. Suele aplicarse a una pintura de so exterior.

Edge joint A joint between edges of two or more parallel or nearly parallel members.
Junta de orilla Una junta entre las bordes de dos o más miembros que son paralelos o casi paralelos.

Elastomer A man-made compound with flexible and elastic properties. The resin system of elastomeric enamels and lacquers is made of these elastomeric compounds.
Elastómer Una compuesta sintética de propriedades flexibles e elásticas. Las resinas de los esmaltes y las lacas elastoméricas se fabrican de estas compuestas elastoméricas.

Electrocution Electricity passes through a human body. Severe injury or death can result.
Electrocución La electricidad pasa por el cuerpo humano. Puede causar los daños severos o la muerte.

Electrostatic spraying Application of paint by high-voltage atomization.
Rociado electrostático La aplicación de la pintura por medio de la atomización en alta voltaje.

Emulsion A suspension of fine polymer particles in a liquid, usually water; dispersed particles may be binder, pigments, or other ingredients. Emulsions can be made by certain polymerization techniques or by certain processes.
Emulsión Una suspensión de partículos finos de un polímero en un líquido, normalmente el agua; los partículos dispersados pueden ser un aglomerante, los pigmentos u otros ingredientes. Las emulsiones se pueden preparar por medio de técnicas específicas o por otros procesos de polimerización.

Enamel A type of paint that dries in two stages: first by evaporation of the solvent and then by oxidation of the binder.
Esmalte Un tipo de pintura que se seca en dos etapas: primero por evaporación del solvente y luego por la oxidación del aglomerante.

Epoxy A class of resins characterized by good chemical resistance.
Resina epósica Una clase de resinas caracterizadas por su buena resistencia química.

Estimating Analyzing damage and calculating how much it will cost to repair a vehicle.

Dar presupuesto estimativo Analizar los daños y calcular los costos de efectuar las reparaciones del vehículo.

Etching Chemically removing a layer of base metal to prepare a surface for painting or bonding.
Decapado con ácido Removiendo una capa de la chapa metálica con una solución química para prepararla para la pintura o para la aglomeración.

Evaporation Change from liquid to a gas.
Evaporación Cambiar de un líquido a un gas.

Evaporation rate Speed with which any liquid evaporates. Generally the rate is expressed as a number related to the evaporation rate of butyl acetate.
Relación de evaporación La velocidad con que se evapora cualquier líquido. Generalmente se describe como un número relativo a la relación de evaporación del acetato butílico.

Exposure limit Limit set to minimize occupational exposure to a hazardous substance.
Límite de exposición El límite establecido para minimizar la exposición debido al trabajo efectuado con una substancia tóxica.

Extender pigment An inert, usually colorless and semitransparent pigment used in paints to fortify and lower the price of pigment systems.
Carga Un pigmento inerte, normalmente sin color y semitransparente que se usa en las pinturas para fortalecer y rebajar los precios de los sistemas de pigmentos.

Exterior Outside. Usually refers to an area not protected from weather or not in an enclosed area such as a building.
Exterior De afuera. Suele indicar una área no protegida del intemperie o que no esta encerrada tal como un edificio.

Fading Loss of color.
Descolorarse La pérdida del color.

Fan Spray pattern of a spray gun.
Cónico Un patrón de rociado de una pistola atomizador.

Fanning Use of pressurized air through a spray gun to speed up the drying time of a finish; fanning is not recommended!
Ventilación El uso del aire bajo presión de una pistola atomizador para acelerar el secado de un acabado; no se recomienda!

Fatigue failure A type of metal failure resulting from repeated stress that finally alters the character of metal so that it cracks.
Rotura por fatiga Un tipo de fallo del metal que resulta por la tensión repetida que con tiempo modifica la característa del metal y causa la agrietura.

Featheredge Tapering edges of the damaged area with sandpaper or special solvents.
Difuminar Adelgazar gradualmente el espesor de los bordes del área dañada con el papel de lija o con los solventes especiales.

Featheredge splitting Stretchmarks or cracks along the featheredge, which occur during drying or shortly after the topcoat has been applied over a primer-surfacer.

Agrieturas en el enlace difuminado La irregularidades de estirón o las agrietas por el enlace difuminado, las cuales ocurren durante el secado o muy poco después de que se haya aplicado la capa superior sobre una imprimación-preparación.

Filler Any material used to fill (level) a damaged area.

Relleno Cualquier material que sirve para rellenar (aplanar) una área dañada.

Film A very thin continuous sheet of material. Paint forms a film on the surface to which it is applied.

Película Una hoja de material contínua muy delgada. La pintura forma una película sobre la superficie a que se ha aplicada.

Finish A protective coating of paint; to apply a paint or paint system.

Acabado Una capa protectiva de la pintura; aplicar la pintura o un sistema de aplicar la pintura.

Fisheye A surface depression or crater in a wet paint film, usually caused by silicone comtamination of the paint.

Cráter Una depresión o un agujero en una película de pintura húmeda, suele ser causada por una contaminación de silicio en la pintura.

Fixtures A set of accessories for a dedicated measuring system designed to attach to a bench to fit reference points for a specific family of body styles.

Bancadas universales Un juego de accesorios de un compás de puntas para medición diseñado para conectarse a un banco y quedarse en puntos fijos de referencia para una plataforma específica de modelos de carrocería.

Flaking A paint failure characterized by large pieces of paint falling off a substrate.

Despegue Un fallo de la pintura caracterizado por la caída de trozos grandes de pintura de un substrato.

Flash First stage of drying where some of the solvents evaporate, which dulls the surface from an exceedingly high gloss to a normal gloss.

Vaporización instantánea La primera etapa del secado en la cual algo de los solventes se evapora, lo que deslustra la superficie de una brillantez fuertísima a una brillantez normal.

Flash point Temperature at which the vapor of a liquid will ignite when a spark is struck.

Temperatura de inflamabilidad La temperatura en la que el vapor de un líquido se encenderá al encenderse una chispa.

Flash time Time between coats or paint application and/or baking.

Tiempo de vaporización instantánea El tiempo entre las capas o aplicaciones de pintura y/o el horneo.

Flat Lacking in gloss or shine.

Mate Que carece en brillo o lustre.

Flat rate A preset amount of time required and money charged to perform a specific repair operation.

Tarifa fija Una cantidad establecida del tiempo requerido y del dinero cobrado para efectuar una reparación específica.

Flexibility Ability of a paint film to withstand dimensional changes.

Flexibilidad La habilidad de una película de pintura en resistir los cambios de dimensión.

Flow Leveling characteristics of a wet paint film. Also, ability of a liquid to run evenly from a surface and to leave a smooth film behind.

Flujo Las características aplanantes de una película de pintura húmeda. Tambien refiere a la habilidad de un líquido de cubrir una superficie de una manera uniforme dejando una película lisa.

Fog coat Following a wet coat in which mottling or streaking occur, move the gun back two or three times the normal distance and apply a fog coat with a continuing fluid flow and circular motion until the condition is corrected. Then move to another area.

Capa de niebla Siguiendo una capa húmeda en la que ocurre los tachones o las estriaciones, remueve la pistola a una distancia de dos o tres veces lo normal y aplique una capa de niebla con un flujo flúido y contínuo en una dirección circular hasta que se corrije la condición. Luego continúa en otra área.

Frame Heavy metal structure that supports the auto body and other external components.

Bastidor Una estructura pesada de metal que sostiene la carrocería del auto y otros componentes externos.

Frame gauges Gauges that can be hung from the car frame to check its alignment.

Calibradores del bastidor Los calibres que se pueden colgar del bastidor del coche para cerciorar su alineamiento.

Frame straightener A pneumatic- or hydraulic-powered machine used to align and straighten a distorted frame or body.

Enderezadora del chasis Una máquina de fuerza neumática o hidráulica que sirve en alinear y enderezar las deformaciones de un bastidor o una carrocería.

Front-wheel drive A vehicle that has its drive wheels located on the front axle.

Tracción de las ruedas delanteras Un vehículo cuyos ruedas de propulsión se encuentran en el eje delantero.

Frosting Formation of a surface haze or defects in a drying paint film.

Escarchada La formación de un haz en la superficie o los defectos en una película de pintura.

Full body section Simultaneous section repairs to both rocker panels, windshield pillars, and floor pan; required to join undamaged front half of one vehicle to undamaged rear half of another vehicle.

Elementos totales de la carrocería Las reparaciones simultáneas a ambos estribos, montantes de la parabrisa y el panel del piso; requeridas para unir la mitad delantera no dañada de un vehículo a la mitad trasera no dañada de un otro vehículo.

Full frame Strong, thick steel structure that extends from the front to the rear of vehicle. It is used on many full size cars and trucks.

Bastidor larga Una estructura fuerte, de acero espeso que extienda desde la parte delantera hasta la parte trasera de un vehículo. Se usa mucho en los coches y camionetas de tamaños grandes.

Galvanized Metal coated with zinc.

Galvanizado El metal cubierto del zinc.

Gap Distance between two points.

Holgura La distancia entre dos puntos.

Gas metal arc cutting An arc cutting process used to sever metals by melting them with the heat of an arc between a continuous metal (consumable) electrode and the work. Shielding is obtained entirely from an externally supplied gas or gas mixture.

Cortes de metal con arco protegido de gas Un proceso de cortar con arco que sirve para remover los metales derritiéndolos con el calor del arco entre un electrodo (consumible) alámbrico y el trabajo. La protección se obtiene completamente por un gas o una mezcla de gases de suministro externo.

Gas metal arc welding Also called MIG welding, an arc welding process that produces coalescence of metals by heating them with an arc between a continuous filler metal (consumable) electrode and the work. Shielding is obtained entirely from an externally supplied gas or gas mixture.

Soldadura de metal con arco protegido por gas Se puede denominar también la soldadura MIG, es un proceso de soldadura de arco que produce una unión de metales al calentarlos con un arco entre un electrodo (consumible) alámbrico de relleno metálico y el trabajo. La protección se obtiene completamente por un gas o una mezcla de gases de suministro externo.

Gauge A measure of thickness of sheet metal.

Espesor La medición del espesor de la chapa de metal.

Gloss Ability of a surface to reflect light. Measured by determining the percentage of light reflected from a surface at certain angles.

Brillantez La habilidad de una superficie en reflejar la luz. Se mide determinando el porcentaje de la luz reflejada de una superficie en ciertos ángulos.

Goggles A device with colored lenses or clear safety glass that protects the eyes from harmful radiation during welding and cutting operations.

Gafas protectivas Un dispositivo con lentes obscurecidos o de vidrio inastillable transparentes para proteger los ojos de la radiación dañosa al efectuar la soldadura o los cortes.

Grille Decorative panel in front of the radiator.

Rejilla del radiador Un panel embellecedor en frente del radiador.

Grit A measure of size of particles on sandpaper or discs.

Grano Una medida del tamaño de los partículos en el papel o los discos de lijar.

Guide coat A reference coat of a different color often applied to a primer-surfacer. It is sanded off to visually determine if the panel is straight.

Capa de guía Una capa de referencia de un color distincto que se aplica muchas veces en la imprimación-preparación. Se quita completamente por lijado para determinar si esta plana el panel.

Hardener A curing agent used in plastics.

Endurecedor Un agente para curar que se usa en los plásticos.

Hardness That quality of a dry paint film which gives film resistance to surface damage or deformation. Hard to measure accurately or to separate from such other properties as adhesion, toughness, mar resistance, etc.

Dureza La calidad de una película de pintura seca que la hace resistente a los daños o las deformaciones de la superficie. Es difícil juzgar esta calidad con certeza o separarla de otras propriedades tal como la adhesión, la tenacidad, la resistencia a daños, etc.

Hardware Computer, printers, hard drives, CD-ROM drives, and other computer equipment.

"Hardware" (equipo) La computadora, la imprimidora, los discos duros, la unidad de CD-ROM, y otros equipos de computadora.

Haze Development of a cloud in a film or in a clear liquid.

Haz El desarrollo de un velo en una película o en un líquido transparente.

Header bar Framework or inner construction that joins the upper sections of the windshield or the backglass and pillars; forms the upper portion of the windshield or the backglass opening and reinforces the top panel.

Transversal cabezal Una parte del bastidor o de la construcción interior que une las partes superiores de la parabrisas o la ventanilla trasera y los montantes; forma la parte superior de los marcos de la parabrisa o la ventanilla trasera y refuerza el panel superior.

Headliner Cloth or plastic material covering the roof area inside the passenger compartment.

Tapizado del techo Material de tela o de plástico que cubre el área del techo dentro del habitáculo de los pasajeros.

Hiding Degree to which a paint obscures the surface to which it is applied.

Poder cubriente El grado al que una pintura enmascara la superficie a la que se aplica.

High-strength steel A low-alloy steel that is much stronger than hot-rolled or cold-rolled sheet steels; normally used in the manufacture of vehicle structural parts.

Acero de gran resistencia Un acero de bajo aleación que es mucho más fuerte que las chapas de acero laminadas en caliente o en frío; normalmente se utiliza en la fabricación de las partes estructurales del vehículo.

Hinge pillar Framework or inner construction to which door hinges fasten.

Montante del bisagra Una parte del bastidor o de la construcción interior a la cual se conectan las bisagras de las puertas.

Hold out Ability of a surface to keep the topcoat from sinking in.

Integridad del soporte La habilidad de una superficie en prevenir que la capa superior se penetre demasiado.

Hood panel Large metal panel that generally fills the space between the two fenders and closes off the engine compartment so that rain cannot fall on the engine.

Panel del capót Un panel grande que generalmente llena el espacio entre las dos aletas e encierra el compartimento del motor para que no caiga la lluvia sobre el motor.

Hot spot An unprotected area subject to corrosion.

Punto descubierto Una área no protegida que se puede corroer.

Hot spray Technique of applying hot paint. The higher temperature reduces viscosity so that higher percent solids can be used.

Rociado en caliente Una tecnica para aplicar la pintura caliente. Las tempertaturas más altas reducen la viscosidad para que se puede usar un porcentaje más alto de sólidos.

Hot-melt Generally an adhesive. A polymer applied in its molten state to a substrate. Dries by cooling to solid.

Derritido en caliente Generalmente un adhesivo. Un polímero en su forma fundido aplicado en un substrato. Se seca enfriándose a un sólido.

HSLA High-strength low-alloy steel used in unibody construction.

HSLA El acero de gran resistencia y baja aleación utilizado en la construcción de los monocuerpos.

HSS High tensile strength steel whose strength is derived from heat treatment. This steel will tear or fracture if collision stresses exceed its tensile strength.

HSS El acero de gran resistencia a la rotura cuyo fuerza se deriva del tratamiento térmico. Este acero se puede romper o quebrar si las tensiones de colsión exceden su resistencia de a la rotura.

Hue Characteristic by which one color differs from another, such as red, blue, green, etc.

Matiz La característica por la cual un color es distincto de otro, tal como el rojo, el azul, el verde, etc.

Humidity Amount of water vapor in the atmosphere.

Humedad La cantidad del vapor de agua en la atmósfera.

Hydrocarbon A compound which contains only carbon and hydrogen.

Hidrocarburo Una compuesta que contiene sólo el carbono y el hidrógeno.

Hydrometer An instrument to measure specific gravity (density of a liquid). It is used for testing batteries.

Hidrómetro Un instrumento para medir la gravedad específica (densidad de un líquido). Sirve para comprobar las baterias.

I-CAR Inter-Industry Conference on Auto Collision Repair, and advanced training organization dedicated to promoting high-quality practices in the collision repair industry.

I-CAR La conferencia industrial para las reparaciones de colisiones vehículares, y la organización de entrenamiento avanzado dedicada a la promoción de las practicas de alta calidad en la industra de reparaciones de colisión.

Independent front suspension Conventional front suspension system in which each front wheel moves independently of the other.

Suspensión delantera independiente El sistema convencional de la suspensión delantera en el que cada rueda delantera mueva de una manera independiente.

Independent rear suspension Rear suspension that has no cross axle shaft. Each wheel and related suspension is allowed to act individually.

Suspensión trasera independiente La suspensión trasera que no tiene una flecha transaxial. Permite que cada rueda con su propria suspensión actua individualmente.

Induction heating Development of heat in a substrate by application of an electromagnetic field to that substrate.

Calentamiento por inducción El desarrollo del calor en un substrato por medio de la aplicación de un campo electromagnético en el substrato.

Infrared baking Drying a paint film by heat developed by an infrared source.

Horneado infrarrojo El secado de una película de pintura que se efectúa por medio de un fuente infrarrojo.

Infrared light That portion of the spectrum responsible for most of the heating effects of the sun's light, 8,000 to 100,000 Angstrom units.

Luz infrarrojo Esa porción del espectro que es responsable por la mayoría de los efectos del calentamiento del luz del sol, las unedades Angstrom 8,000 a 100,000.

Inhibitor An additive to a paint which slows some processes, e.g., gelling, skinning, or yellowing.

Inhibidor Un aditamento a la pintura que decelera algunos procesos, por ejemplo, la coagulación, la formación de una costra, o el amarilleo.

Insurance adjuster Reviews estimates and determines which one best reflects how the vehicle should be repaired.

Ajustador de aseguranza Repasa los presupuestos estimativos y determina cual refleja mejor la manera en que se debe efectuar las reparaciones del vehículo.

Interior trim All upholstery and moldings on the inside of the vehicle.

Moldura interior Todas las guarniciones y molduras que se hayan en el interior de un vehículo.

Iron A metallic substrate which requires painting to prevent corrosion.

Hierro Un substrato metálico que requiere la pintura para prevenir la corrosión.

Isocyanate resin A principal ingredient in urethane hardeners. Because of its toxic effects, the painter is always advised to wear a respirator that is approved by NIOSH.

Resina isocianúrica Un ingrediente principal en los endurecedores uretanos. Como tiene efectos tóxicos, siempre se recomienda que el pintor usa un respirador que haya sido aprobador por el NIOSH.

Jack stands Used to support the vehicle when working under it.
Torres Sirven para apoyar el vehículo al trabajar abajo de él.

Jig Mechanical device for holding work in its exact position while it is being welded.
Montaje Un dispositivo mecánico para anclar el trabajo en una posición exacta mientras que se efectúa la soldadura.

Joint Point or line at which two pieces are connected.
Junta Una punta o una linea en la que se conectan dos piezas.

Jounce Compression of the springs, or an upward movement of the wheel and a downward movement of the frame.
Sacudo Una compresión de los resortes, o un movimiento hacia arriba de la rueda y un movimiento hacia abajo del bastidor.

Kick-out Precipitation of dissolved binder from a solution as a result of solvent incompatibility.
Desnate La precipitación de un aglomerante disuelto en la solución como resultado de una incompatiblidad del solvente.

Lacquer Type of paint that dries by solvent evaporation; can be rubbed to improve appearance.
Laca Un tipo de pintura que se seca por medio de la evaporación del solvente; se puede mejorar su aparencia con un lustrado.

Lap weld A weld made along the edge of an overlapping piece.
Soldado de recubrimiento Una soldadura hecha en el borde de una pieza solapada.

Laser system Type of universal measuring system that uses laser optics in partial or total vehicle dimensioning.
Sistema de laser Un sistema de medición universal que utiliza un laser óptico para dimensionar una parte o todo el vehículo.

Lifting Attack on an undercoat by the solvent in a top coat which results in distortion or wrinkling of the undercoat.
Despegue El ataque en una capa de apresto por el solvente de una capa superior que resulta en la distorsión o las arrugas en la capa de apresto.

Lightness Whiteness of a paint measured by the amount of light reflected by a surface. A perfect white is one which reflects one hundred percent of the light in the visible spectrum.
Claridad La blancura de una pintura medida por la cantidad de reflección de luz en una superficie. Un blanco perfecto refleja el cien porciento de la luz del espectro visible.

Manual welding Welding in which the entire operation is performed and controlled by hand.
Soldadura por mano La soldadura en que la operación completa se efectúa y se controla por mano.

Mash Vehicle body damage condition in which the length of any section or frame member is shorter than factory specifications.
Chato La condición de daños de la carrocería del vehículo en que la longitud de cualquier sección o miembro del bastidor es más corta de la especificada por la fábrica.

Masking Paper or plastic that protects surfaces and parts from paint overspray.
Protección El papel o plástico que protege las superficies y las partes del desparrame de pintura.

Material safety data sheets (MSDS) Available from all product manufacturers; they detail chemical composition and precautionary information for all products that can present a health or safety hazard.
Hojas de Dato de Seguridad de los Materiales (MSDS) Disponibles de todos los fabricantes de productos; detallan la composición química y la información de precaución para cada producto que puede causar daños al salud o peligro a la seguridad.

Measurement systems Allow you to check for frame or body misalignment resulting from a major collision.
Sistemas de medición Le permite cerciorar la averías de alineación causadas por una colisión grave.

Mechanical parts Power train, accessories, and suspension system.
Partes mecánicas El tren de potencia, los accesorios, y la sistema de suspensión.

Member Any essential part of a machine or assembly.
Miembro Cualquier parte esencial de una máquina o una asamblea.

Metal conditioner Chemical cleaner that removes rust and corrosion from bare metal and helps prevent further rusting.
Acondicionador de metal Una limpiador química que remueva la oxidacion y la corrosión del metal descubierto y previene que se oxide más.

Metallic paint finish Paint colors that contain metallic flakes in addition to pigment.
Acabado con pintura metálica Los colores que contienen escamas metálicas además del pigmento.

Metallics Finishes which include metal flakes in addition to pigment.
Metálicos Los acabados que incluyen las escamas de metal además del pigmento.

Metamerism A term used to describe two or more colors that match when viewed under one light source, but do not match when viewed under a second light source.
Metámero Un término para describir dos o más colores que parecen iguales al verlos bajo una fuente luminosa, pero que no parecen iguales cuando se ven bajo otra fuente luminosa.

Mica A color pigment or particles found in pearl paints.
Mica Un pigmento de color o los partículos que se encuentran en las pinturas anacaradas.

MIG (Metal Inert Gas) Gas-shielded metal arc welding.
MIG (Gas inerte de metal) La soldadura de metal con arco protegido por gas.

Mil A measure of paint film thickness equal to one one-thousandth of an inch.

Mil Una medida del espesor de una película de pintura que iguala a un milésima de una pulgada.

Mildew Fungus growth which appears on substrates in warm, humid areas.

Moho Un crecimiento de hongo que aparece en los substratos en las áreas tibias y húmedas.

Misaligned Having uneven spacing, as between body panels.

No alineado Que tiene el espaciamiento desigual, tal como entre los paneles de la carrocería.

Mist Liquid droplets suspended in the air; generated by condensation from the gaseous to the liquid state, or by breaking up a liquid into a dispersed state by splashing, foaming, or atomizing.

Neblina Las diminutas gotitas suspendidas en el aire; generada por la condensación del estado gaseoso al estado líquido, o por salpicar, espumar o atomizar un líquido a un estado disperso.

Mist coat A light spray coat of high volume solvent for blending and/or gloss enhancement.

Capa de neblina Una capa de rocío ligera de un volumen muy alto de solvente para casar colores o mejorar la brillantez.

Model year Production period for new model vehicles or new engines.

Año de modelo El período de producción de los nuevos modelos de vehículos o los motores nuevos.

Molecule Smallest possible unit or amount of any substance which retains characteristics of that substance.

Molécula La unedad o cantidad más pequeña de cualquiera substancia que retiene las características de esa substancia.

Monocoque Unibody vehicle construction in which the sheet metal of the body provides most of the structural strength of the vehicle.

Monocasco La construcción de un vehículo de carrocería autoportante en que las chapas de metal de la carrocería proveen la mayoría de la fuerza estructural del vehículo.

Mottling A film defect appearing as blotches or surface imperfections (metallics).

Abigarrado Un defecto de la película que aparece como tachas o imperfeciones en la superficie (metálicas).

Multiple-pull systems Equipment that pulls in two or more directions to correct damage.

Sistemas de poleas múltiples El equipo que jale en dos direcciones o más para correjir los daños.

Nonferrous Metals that contain no iron. Aluminum, brass, bronze, copper, lead, nickel, and titanium are nonferrous.

No férreo Los metales que no contienen el hierro. El aluminio, el latón, el bronce, el cobre, el plomo, el níquel, y el titanio son no férreos.

OEM Original Equipment Manufacturer.

OEM Fabricante de equipo original.

Oils Obtained from various natural sources, commonly vegetable oils. A relatively viscous liquid which has an oily or slippery feel.

Aceites Derivados de varios origenes naturales, comúnmente los aceites vegetales. Un líquido relativamente viscoso que tiene una sensación grasoso o resbaladizo.

Opaque Impervious to light or not transparent. Light cannot be seen through it.

Opaco Impervio a la luz o no transparente. La luz no puede pasar por él.

Open structural members Flat panels, typically accessible from both sides, such as floor panels and trunk floors.

Miembros estructurales abiertos Los paneles planos, típicamente accesible de ambos lados, tal como los paneles del piso y los pisos de la cajuela.

Orange peel An irregularity in the surface of a paint film resulting from the inability of a wet film to "level out" after being applied. Orange peel characteristically appears to the eye as an uneven or dimpled surface, but usually feels smooth to the touch.

Piel de naranja Una irregularidad en la superficie de una película de pintura causada por la inabilidad de una película húmeda en "aplanarse" después de su aplicación. Típicamente, el piel de naranja tiene un aspecto de una superficie desigual o con hoyitos, pero suele tener una sensación lisa al tacto.

Outer panels Sheet metal sections that, when attached to inner panels, form the exterior of a vehicle.

Paneles exteriores Las secciones de chapa de metal que, al conectarse a los paneles interiores, forman el exterior del vehículo.

Oven A piece of equipment used to bake finishes.

Horno Una pieza de equipo que sirve para hornear los acabados.

Overall repainting A type of refinish repair in which the vehicle is completely repainted.

Reacabado total Un tipo de reparación de acabado en que se vuelve a pintar completamente al vehículo.

Overlap Amount of the spray pattern that covers the previous spray stroke. In estimating, when two operations or tasks share common steps or procedures.

Solape La cantidad del patrón de rociar que cubre la pasada previa. En fijar los presupuestos, cuando dos operaciones o trabajos comparten los mismos pasos o procedimientos.

Overpulling Stretching metal too much, resulting in a need to replace the part.

Sobrejalar Estirarse demasiado al acero, lo que resulta en una necesidad de reemplazar un elemento.

Overspray An overlap of dry spray gun particles on areas that were not meant to be painted. Paint that falls on the area next to the one being painted.

Desparrame Un recubrimiento de partículas secas de la pistola de atomización en las áreas que no se deseaba pintar. La pintura que cae en el área contigua a la que se esta pintando.

Oxidation Chemical combination of oxygen and the vehicle of a paint which leads to drying. Also, the destructive combination of oxygen and metal, for example rusting.

Oxidación La combinación química del oxígeno y el medio de una pintura que resulta en el secado. Tambien, la combinación destructiva del oxígeno y el metal, por ejemplo la corrosión.

Oxyacetylene welding A welding process that produces joining of metals by heating them with a gas flame obtained from the combustion of acetylene with oxygen. Not commonly used in collision repair.

Soldadura por oxiacetileno Un proceso de la soldadura que produce una fusión de los metales al calentarlas con una llama de gas obtenido por la combustión del acetileno y el oxígeno. No suele emplearse en las reparaciones de colisiones.

Ozone A portion of the atmosphere that protects earth from ultraviolet radiation from the sun.

Ozono La porción de la atmósfera que protege la tierra de la radiación ultravioleta del sol.

Paint film Actual thickness of paint on a surface.

Película de pintura El espesor actual de la pintura sobre una superficie.

Panel repair A type of refinish repair job in which a complete section (door, hood, deck lid, and so on) is repainted.

Reparación del panel Un tipo de reparación del acabado en que una sección completa (puerta, capót delantera y trasera y etc.) se vuelve a pintar.

Panel spotting Painting only a portion of a panel.

Pintura localizada Pintar sólo una porción de un panel.

Parallelogram steering System in which a short idler arm is mounted on the right side so that it is parallel to the pitman arm.

Dirección de paralelogramo Un sistema en que un brazo loco corte se monta en el lado derecho para que queda paralelo al brazo pitman.

Peeling Loss of adhesion of a paint film which results in large pieces of the film splitting away from the surface.

Despegue La pérdida de la adhesión de una película de pintura que resulta en que grandes escamas de una película se caigan de la superficie.

Penetration Depth of fusion into metal being welded.

Penetración La profundidad de fusión en el metal que se esta soldando.

Percent solids Percent mass of a paint due to its non-liquid components.

Porcentaje de sólidos El porcentaje de masa de una pintura debido a sus componentes que no son líquidos.

pH A measure of the acidity of a substance in an aqueous solution. 7 is neutral, below 7 is acid, and above 7 is alkaline.

pH Una medida del acidéz de una substancia en una solución acuosa. El 7 es neutro, menos del 7 es ácido y más del 7 es alcalino.

Phosphate coating A chemical coating on a steel surface that provides the best adhesion for undercoats, also called conversion coating.

Imprimación fosfatante Una capa química en una superficie de acero que provee la mejor adhesión para las capas de apresto, se puede denominar también revestimiento por conversión.

Phosphoric acid Acid used in metal cleaner solutions or as a catalyst in certain paint resin systems.

Acido fosfórico Un ácido que se usa en las soluciones para limpiar los metales o para un catalizador en ciertos sistemas de résinas de pintura.

Pigment Material in the form of fine powders used to impart color, opacity, and other effects to paint.

Pigmento Una materia en forma de los polvos finos que sirve para conferir el color, la opacidad u otros efectos a la pintura.

Pillars Vertical body members that hold the roof panel in place and provide protection in case of a rollover accident.

Montantes Los miembros verticales de la carrocería que mantienen al panel del techo en su lugar y proveen la protección en caso de una colisión que hace voltear el coche.

Pinholing Holes left in paint film or plastic filler by unwanted air bubbles.

Picaduras Los agujeros residuos en la película o en la masilla causadas por las burbújas de aire no deseadas.

Pitting Appearance of holes or pits in a paint film while it is wet. Related to crawling and poor wetting.

Cráteres La aparencia de los agujeros o los cráteres en una película de pintura mientras que este húmeda. Asociado con el rizado y la humectación incompleta.

Plasma A gas that has been heated to at least a partially ionized condition, which enables it to conduct an electric current.

Plasma Un gas que se ha calentado a una condición al menos parcialmente ionizada, lo que permite que conduzca un corriente eléctrico.

Plasma arc cutting Cutting process in which metal is severed by melting a localized area with a constructed arc and removing molten material with a high velocity jet of hot, ionized gas issuing from an orifice.

Corte con arco de plasma Un proceso de cortar en que el metal se quita al derritir una área localizada con un arco construido y se remueva al material derritido con un chorro caliente de alta velocidad del gas ionizado que sale de un orificio.

Plastic filler A compound of resin and fiberglass used to fill dents and level surfaces.

Masilla Una compuesta de resina y la fibra de vidrio que sirve para rellenar las abolladuras y aplanar las superficies.

Plasticizer A material added to a paint system to make film more flexible.

Plastificante Una material añadida a un sistema de pintura para hacer más flexible una película.

Plug weld Adding metal into a hole and fusing all metal.

Soldadura de tapón Colocando el metal en un agujero y uniendo todo el metal.

Polyurethane A chemical compound used in the production of resins for enamel finishes.
Poliuretano Una compuesta química usada en la producción de las resinas de los acabados de esmalte.

Porosity Gas pockets or voids in metal.
Porosidad Los baches de gas o las cavidades en un metal.

Pot life Amount of time a painter has to apply a plastic or paint finish to which a catalyst or hardener has been added.
Vida de la mezcla La cantidad del tiempo disponible al pintor para aplicar un plástico o un acabado de pintura al cual se ha agregado un catalizado o un endurecedor.

PPM Parts per million; a unit for measuring the concentration of a gas vapor in contaminated air. Also used to indicate the concentration of a particular substance in a liquid or solid.
PPM Las partes por milión; una unedad para medir la concentración de un vapor de gas en el aire contaminado. Tambien indica la concentración de una substancia particular en un líquido o un sólido.

Primary damage Damage that occurs at the point of impact on vehicle.
Daños primarios Los daños que ocurren en el punto de impacto del vehículo.

Prime coat First coat in a paint system. Its main purpose is to improve adhesion and provide corrosion protection.
Capa de imprimación La primera capa en un sistema de pintura. Su propósito principal es de mejorar la adhesión y proveer la protección contra la corrosión.

Primer A type of paint applied to a surface to increase its compatibility with topcoat or to improve the adhesion or corrosion resistance of substrate.
Imprimación Un tipo de pintura aplicada a una superficie para mejorar su compatabilidad con una capa superior o para mejorar la adhesión o la resistencia a la corrosión de un substrato.

Primer-sealer An undercoat that improves the adhesion of a topcoat and seals old painted surfaces that have been sanded.
Imprimación-sellante Una capa de apresto que mejora la adhesión de una capa superior y sella las superficias pintadas antiguas que se han deslustrado.

Primer-surfacer A high-solids primer that fills small imperfections in the substrate and usually must be sanded.
Imprimación-preparación Una imprimación de muchos sólidos que rellena las imperfecciones pequeñas en un subestrato y que se tiene que lijar.

Priming Process to help smooth a body surface and help the topcoats of paint adhere or bond to body.
Imprimación Un proceso de alisar la superficie de la carrocería y así facilitar la adhesión o la aglomeración de las capas superiores en la carrocería.

Pulling Applying force to a part to make a change.
Jalando Aplicando la fuerza en una parte para efectuar un cambio.

PVC Pigment volume content; percent by volume of pigment in total volume of solid material in a paint.

PVC El contenido de pigmento por volumen; un porcentaje del pigmento por volumen del volumen total de la material sólida en la pintura.

Quarter panel Side panel, which extends from the rear door to the end of a car.
Aleta trasera El panel del lado, que extienda de la puerta trasera al posterior del vehículo.

Rack-and-pinion steering Steering gear in which a pinion gear on the end of the steering shaft meshes with a rack gear on the steering linkage.
Dirección de piñón y cremallera El engrenaje de dirección en que un engrenaje de piñón en la extremidad de la flecha de dirección se dienta con un engrenaje de cremallera en la biela de dirección.

Reduce To lower the viscosity of a paint by the addition of solvent or thinner.
Reducir Disminuir la viscosidad de una pintura al añadir un solvente o un adelgazador.

Reducer Solvent combination used to thin enamel is usually referred to as a reducer.
Reductor Se refiere a una combinación de solventes que sirven para adelgazar el esmalte.

Reference points Points on a vehicle, including holes, flats, or other identifying areas, used to position parts during repairs.
Puntas de comprobación Las puntas en un vehículo, incluyendo los agujeros, los planos u otras áreas de identificación, que sirven para colocar las partes durante la reparación.

Refinish To remove or seal an old finish and apply a new topcoat; to repaint.
Reacabar Remover o sellar un acabado antiguo y aplicar una capa superior nueva; volver a pintar.

Reflow A heat process by which lacquers are melted to produce better flow "leveling".
Derritir Un proceso de calentamiento por el cual las lacas se derriten para producir un "aplanamiento" del flujo.

Resistance welding Welds made by passing an electric current through metal between the electrodes of a welding gun.
Soldadura por resistencia Las soldaduras que se efectúan pasando un corriente eléctrico por el metal entre los electrodos de una pistola de soldar.

Respirator A device worn over the mouth and nose to filter particles and fumes out of the air being breathed.
Respirador Un dispositivo que se coloca sobre la boca y la nariz para filtrar las partículas y los humos del aire que se respira.

Retarder A slow evaporating thinner or additive used to retard drying.
Retardador Un adelgazador o aditamento de lenta evaporación que sirve para retrasar el secado.

Right-to-know laws They give essential information and stipulations for safely working with hazardous materials.
Leyes de derecho a la información Proveen la información esencial y las estipulaciones para los proce-

dimientos de seguridad para trabajar con las materias tóxicas.

Rocker panel Narrow, outer panel attached below the car door. Door sills or strong beams that fit at the bottom of door openings.

Estribo Un panel estrecho exterior conectado debajo de la puerta del coche. Los umbrales de las puertas o las vigas fuertes que quedan en la parte inferior de los marcos de las puertas.

Roof rails Framework or inner construction that reinforces and supports the sides of the roof panel.

Largueros el techo El armazón o la construcción interior que refuerza y soporta los lados del panel del techo.

Rubbing (or polishing) compounds Abrasives that smooth and polish paint film.

Compuestas de pulimento Los abrasivos para alisar y pulir una película de pintura.

Runs and sags Heavy applications of sprayed material that fail to adhere uniformly to surface (most common of application problems). See sagging.

Goteos y rizaduras Los deplazamientos pesados de la material atomizada que falla en adherir de una forma uniforme en la superficie (el problema más comun en las aplicaciones). Vea rizadura.

Rust Corrosion product which forms on iron or steel when exposed to moisture.

Oxidación Un producto de la corrosión que forma en el hierro o en el acero cuando se exponen a la humedad.

SAE Society of Automotive Engineers.

SAE La Asociación de Ingenieros automotivos.

Sag A type of frame damage in which one or both side rails bend and sag at the cowl, causing buckles to be formed on top of the side rails.

Arrufo Un tipo de daño al bastidor en que uno o ambos largueros laterales se doblan y curvan en el sobretablero, formando los alabeos en las partes superiores de los largueros laterales.

Sagging Excessive flow on a vertical surface resulting in drips and other imperfections on a painted surface. Occurs not only when paint is wet, but also during baking in certain types of paints.

Rizaduras Un flujo excesivo en una superficie vertical que resulte en los goteos u otras imperfecciones en una superficie pintada. Ocurre no sólo estando húmeda la pintura, sino tambien durante el horneo en ciertos tipos de pintura.

Sandblasting A method of cleaning metal, usually steel, by means of a blast with an abrasive, using sand.

Decapado con arena Un método de limpiar al metal, generalmente al acero, por medio de un chorro de un abrasivo, utilizando la arena.

Sander (polisher) A power tool designed to speed the rate of sanding (or polishing) surfaces.

Lijador (pulidor) Una hierramenta de motor diseñada para acelerar el proceso de lijar (o pulir) las superficies.

Sanding Using an abrasive coated paper or plastic backing to level and smooth a body surface being repaired.

Lijar Usar un papel cubierto de un abrasivo o de un forro de plástico para aplanar y lijar la superficie de la carrocería que se esta reparando.

Sanding block A hard, flexible block to provide a smooth, consistent backing for hand sanding.

Bloque para lijar Un bloque duro y flexible que provee un respaldo liso y consistente para el lijado a mano.

Sandscratch swelling A swelling of sandscratches in an old surface caused by solvents in the topcoat.

Hinchazón de las rayas de limado Un hinchazón en las rayas de limado en una superficie antigua causado por los solventes en la capa superior.

Scanner Electronic device for reading and storing data or computer information.

Scanner (detector) Un dispositivo electrónico para leer y almacenar la información o los datos de computadora.

Scuff To rough up a surface by rubbing lightly with sandpaper to provide a suitable surface for painting.

Deslustrar Deslucir una superficie frotandola ligeramente con el papel de lija para proveer una superficie adecuada para pintarse.

Sealer An intercoat between the topcoat and the primer or old finish, giving better adhesion.

Sellante Una capa intermedia entre la capa superior y la imprimación o el acabado antiguo, dándole mejor adhesión.

Secondary damage Indirect damage that occurs due to misplaced energy that causes stresses in suspension and/or body dimensions at areas other than the primary impact zone.

Daños secundarios Los daños indirectos que ocurren debido a la energía mal dirigido causando las tensiones en la suspensión y/ en las dimensiones de la carrocería en las áreas que no son de la zona del primer impacto.

Sectioning A means of replacing partial areas of a vehicle.

Reemplazar por elementos Un metodo de remplazar las áreas parciales de un vehículo.

Seeding Development of tiny insoluble particles in a container of paint which results in a rough or gritty film.

Granular El desarrollo de minísculas partículas insolubles dentro del envase de pintura que produce una película áspera o granosa.

Semi-gloss An intermediate gloss level between high and low gloss.

Semi-brillante Un nivel de brillantez intermediano entre mucho o poco brillo.

Service manual A book published annually by each vehicle manufacturer that lists specifications and service procedures for each make and model of vehicle; also called a shop manual.

Manual de servicio Un libro publicado cada año por cada fabricante de vehículos que registra las especificaciones y los procedimientos de reparación para cada marca y modelo de vehículo; se puede denominar también un manual de taller.

Settling Gravity separation of one or more components from a paint, resulting in a layer of separated material at the bottom of a container.

Sedimentación La separación por gravedad de un componente o más de una pintura, causando una capa de material separada en el fondo del envase.

Sheen Gloss or flatness of a film when viewed at a low angle.

Brillo La brillantez o el tono mate de una película visto desde un ángulo bajo.

Shim Small metal piece used behind panels to bring them into alignment.

Cuña Un pedacito de metal colocado detrás de los paneles para emparejarlas.

Shock towers (strut towers) Reinforced body areas for holding the upper parts of the suspension system; coil springs and struts or shock absorbers fit into the shock towers.

Torres apoyaderos (torres de los postes) Las áreas de la carrocería reinforzadas que sirven para apoyar las partes superiores del sistema de suspensión: los muelles de embrague y los postes o los amortiguadores quedan dentro de los torres apoyaderos.

Show through Sand scratches in undercoat that are visible through topcoat.

Imperfecciones visibles Las rayas de limado de la capa de apresto son visibles en la capa superior.

Silicone An ingredient in waxes and polishes which makes them feel smooth. If silicone is not removed from the surface before painting, fisheyes will form in the finish.

Silicona Una ingrediente en las ceras y los pulimentos que les hace sentir lisas. Si no se remueva la silicona de la superficie antes de pintar, se formarán cráteres en el acabado.

Single pull system Straightening system capable of pulling only one direction at a time.

Sistema de una dirección Un sistema de enderezado capaz de jalar solamente en una direccíon a la vez.

Skinning Formation of a thin tough film on the surface of a liquid paint film. Usually due to a reaction with the air or rapid solvent loss.

Formar una costra La formación de una película delgada pero fuerte en la superficie de una película de pintura líquida. Suele causarse por una reacción con la atmósfera o una pérdida rápida del solvente.

Software Stored computer information in floppy disks, computer programs,and CDs.

"Software" (programa) La información de la computadora almacenada en los discos flexibles, las programas de computadora y los CDs.

Solids Percentage, on a weight basis, of solid material in a paint after solvents have evaporated.

Sólidos El porcentaje, por peso, de la materia sólida en la pintura después de evaporarse los solventes.

Solvency Ability of a pure or mixed liquid to dissolve resin.

Solvencia La habilidad de un líquido puro o mezclado de disolver la resina.

Solvent A liquid which will dissolve something.

Solvente Un líquido que puede disolver algo.

Solvent popping Blisters that form on a paint film, which are caused by trapped solvents.

Granos de solvente Las ampollas que forman en una película de pintura, causadas por los solventes atrapados.

Specific gravity Ratio of the weight of a specific volume of a substance in the air compared to the weight of an equal volume of water. Similar to density for practical purposes.

Gravedad específica La relación del peso de un volúmen específico de una substancia en la atmósfera comparado al peso de un volumen correspondiente del agua. Parecido a la densidad para los usos ordinarios.

Specifications Data as supplied by the manufacturer covering all measurements and areas of the vehicle.

Especificaciones La información proveido por el fabricante que toma en cuenta todas las medidas y las áreas de un vehículo.

Spectro-photometer An instrument to measure color. Compares reflectance of a test sample to reflectance of a standard at all points of a visible spectrum.

Espectrofómetro Un instrumento para medir el color. Compara lo reflectivo de un modelo en todos puntos de un espectro visible.

Spontaneous combustion Fire that starts by itself.

Combustión espontánea Un incendio que se encienda si mismo.

Spot putty A material made for filling small holes or sand scratches.

Masilla de retoque Una materia hecha para rellenar los agujeros chicos o las rayas del limado.

Spot repair Type of refinish repair job in which a section of a car smaller than a panel is refinished (often called ding and dent work).

Reparación localizada Un tipo de reparación del acabado en que una seccion de un coche que es más pequeña del panel se reacaba (muchas veces se refiere como trabajo de repiques y abolladuras).

Spot weld Weld in which an arc is directed to penetrate both pieces of metal, while triggering a timed impulse of wire feed.

Soldadura por puntos Una soldadura en que un arco dirigido penetra ambos pedazos de metal, mientras que distribuya una alimentación de alambre sincronizado.

Spray gun A painting tool powered by air pressure that atomizes liquids. Spray paint is atomized in a spray gun, and a stream of atomized paint is directed at the part to be painted.

Pistola de atomización Una herramienta de pintar neumática que atomiza los líquidos. La pintura rociada se atomiza en la pistola, y un chorro de pintura atomizada se dirige hacia la parte para pintar.

Steel A ferrous metal commonly used as a substrate for paint, which must be painted to prevent corrosion. Most common material used in the construction of motor vehicles.

Acero Un metal férreo que comúnmente sirve de substrato para pintar, que debe pintarse para prevenir la corrosión. La material más común en la construcción de los vehículos motorizados.

Steering system Mechanism that enables the driver to turn the wheels to change the direction of a vehicle's movement.

Sistem de dirección Un mecanismo que permite que el conductor gira las ruedas así cambiando la dirección del movimiento del vehículo.

Stitch welding Using intermittent welds to join two or more parts.

Soldadura intermitente Usando las soldaduras intermitentes para unir dos partes o más.

Straight-in damage Damage resulting from a direct impact rather than from a glancing hit.

Daño directo Los daños incurridos de un impacto directo en vez de un choque desviado.

Strength Opacity and/or tinting power of a pigment. Measure of the ability of a pigment to hide color.

Intensidad La opacidad o la capacidad del tinte de un pigmento. La medición de la habilidad de un pigmento de encubrir un color.

Stress Relieving or taking tension off a part.

Estabilizar Aliviar o quitar la tensión de una parte.

Stripping Removing old paint by applying a chemical paint remover which softens and lifts it, or by using air-powered blasting equipment or by power sanding.

Decapar Quitar la pintura antigua por medio de la aplicación de un removedor químico de pintura que la suaviza para despegarla, o por el uso del equipo neumático de chorro o por el lijado motorizado.

Structural member Any primary load-bearing portion of the body structure that affects its over-the-road performance or crashworthiness.

Miembro estructural Cualquier porción primaria que apoya un cargo de la carrocería que afecta su comportamiento durante la marcha o su mérito de resistir los daños de choques.

Structural panels Panels used in a unibody that become part of the whole unit and are vital to the strength of the body.

Paneles estructurales Los paneles de un monocasco que son partes íntegros de la unedad y que son esenciales para la fuerza de la carrocería.

Strut suspension Suspension system that attaches to spring tower and lower control arm.

Suspensión de poste Un sistema de suspensión que conecta al torre del muelle y el brazo inferior de control.

Stud welding General term for joining of a metal stud or similar part to a workpiece. Welding can be accomplished by arc, resistance, friction, or other suitable process with or without external gas shielding.

Soldadura de espiga Un término general indicando la unión de un espiga o una parte parecida a la pieza de trabajo. La soldadura se puede efectuar por arco, por resistencia, por fricción u otro proceso conforme con o sín que sea protegida por gas.

Sub frame Unitized body frame with only the front and rear stub sections of frame rails and no side rail portions.

Bastidor auxiliar Un bastidor de carrocería autoportante que sólo tiene las manguetas de las secciones delanteras y traseras de los largueros del bastidor sin las porciones laterales de los largueros.

Subassembly Several parts put together before the whole is attached.

Elementos prefabricados Varias partes que se han armado antes de que se conecta el conjunto.

Substrate Surface that is to be finished (painted). It can be anything from an old finish or primer to an unpainted surface.

Substrato La superficie que se va acabar (pintar). Puede consister de un acabado antiguo o una imprimación o una superficie sin pintura.

Surface preparation Also called "surface prep." It involves inspection and treatment of the old surface to prepare it for refinishing or painting.

Preparación de la superficie Se puede denominar también "preparación." Involucra la inspección y el tratamiento de la superficie antigua para preparla para el reacabado o la pintura.

Surfacer A heavily-pigmented paint designed to be applied to a substrate for the purpose of smoothing or uniforming the surface for subsequent coats of paint.

Preparación Una pintura con mucho pigmento diseñado a aplicarse al substrato con el propósito de alisar o hacer uniforme a la superficie para las capas de pintura subsequentes.

Suspension system Springs and other parts that support the upper part of a vehicle on its axles and wheels.

Sistema de suspensión Los muelles u otras partes que soportan la parte superior del vehículo en sus ejes y ruedas.

Symmetrical design Design in which both sides of a unibody are identical in structure and measurement.

Diseño simétrico Un diseño de un monocasco en que ambos lados son indénticos en su estructura y medida.

Tack Stickiness of a paint film or adhesive.

Viscosidad Lo pegajoso de una película de pintura o de un adhesivo.

Tack cloth A cheesecloth that has been treated with nondrying varnish to make it tacky. Used to pick up dust and lint from the surface to be painted.

Trapo de limpieza Una estopilla impregnada con una laca que no se seca para hacerla muy pegajosa. Sirve para recojer el polvo y la borra de la superficie que se tiene que pintar.

Tensile strength Resistance of a film to distortion.

Resistencia a la tensión La resistencia de una película a distorcionarse.

Thermoplastic A polymer or other solid material that becomes soft or fluid when it is heated and hardens again when it is cooled.

Termoplásticos Un polímero u otra material sólida que se ablanda o se pone flúido al ser calentado y se endurece de nuevo al enfriarse.

Thermosetting A polymer or other solid which will not soften when it is heated. Thermosetting implies cross-linking and reaction on drying.
Termoendurecible Un polímero u otro sólido que no se ablanda al ser calentado. El termoendurecible indica la reticulación y la reacción en el secado.

Thinner Solvent combination used to thin lacquers and acrylics to spraying viscosity.
Adelgazador Una combinación de solventes que sirve para diluir las lacas y los acrílicos a una viscosidad adecuada para rociar.

Three-stage Consists of three distinct paint layers that produce a pearlescent appearance: basecoat, midcoat, and clearcoat.
De tres etapas Consiste de tres capas distinctas de pintura que produce una aparencia anacarada: capa de fondo, capa intremedio y capa transparente.

TIG (Tungsten inert gas) Gas-shielded tungsten electrode arc welding.
TIG (Tungsten con gas inerte) La soldadura de arco con un electrodo de tungsten protegido por gas.

Tint A very light color, usually a pastel. To add color to white or another color.
Coloración Un color muy claro, normalmente de un matiz suave. Añadir el color al blanco o a otro color.

Tinting strength Ability of a pigment to change the color of a paint to which it is added. A small amount of pigment with high tinting strength makes a big change in the color.
Intensidad del coloración La habilidad de un pigmento en cambia el color de la pintura al que ha sido añadido. Una pequeña cantidad de pigmento con gran intensidad de tinte cambia mucho el color.

Toe Position of the front of the wheels compared to the rear of wheels.
Convergencia La posición de la parte delantera de las ruedas en relación a la parte trasera de las ruedas.

Toe-in The front edges of the wheels are closer together than the rear edges of the wheels.
Angulo de convergencia Los bordes delanteros de las ruedas son más cercanos que los bordes traseros de las ruedas.

Toe-out The rear edges of the wheels are closer together than the front edges of the wheels.
Angulo de divergencia Los bordes traseros de las ruedas son más cercanos que los bordes delanteros de las ruedas.

Tolerances Acceptable alteration limit of vehicle dimensions as provided by the manufacturer (for example, plus or minus inch or 3 mm is typical).
Tolerancias El límite aceptable según el fabricante de las alteraciones hechas a los dimensiones del vehículo (por ejemplo, el más o menos de pulgada o 3 mm es una tolerancia típica).

Topcoat Final layer of paint applied to a substrate. Several coats of topcoat may be applied in some cases.
Capa superior La última capa que se aplica en un substrato. Varias capas superiores pueden aplicarse en algunos casos.

Toxicity Basic biological property of a material reflecting its inherent capacity to produce injury; adverse effects resulting from over-exposure to a material, generally via mouth, skin, eyes or respiratory tract.
Toxicidad Una propriedad básica biológica de una materia que refleja su capacidad inherente de causar daños; los efectos adversos que resultan de exponerse demasiado a una materia, generalmente por la boca, el piel, los ojos o el sistema respiratorio.

Trim Decorative metal pieces on a car body.
Moldura Las piezas metálicas embellecedoras de una carrocería.

Tunnel Formed in floor panel for transmission and drive shaft.
Nervio Formado en el panel del piso para la transmisión y el eje propulsor.

Turning radius A tire wearing angle measured in degrees; the amount one front wheel turns sharper than the other.
Radio de viraje Un ángulo del desgaste de las llantas que se mide en grados; la cantidad que vira más una rueda delantera que la otra.

Turpentine A solvent obtained from distillate of pine trees. Largely replaced by petroleum derivatives now.
Aguarrás Un solvente que se obtiene como un distilato de los árboles de pino. Por la mayor parte se ha reemplazado por los derivados petróleos.

Two-part A paint or lacquer supplied in two parts which must be mixed together in correct proportions before use. The mixture remains usable for a limited time.
De dos partes Una pintura o laca proporcionada en dos partes que se debe mezclar juntas en las proporciones correctas antes de usarse. La mezcla solamente se puede usar por un tiempo limitado.

Two-stage Consists of two distinct layers of paint: basecoat and clearcoat.
De dos etapas Consiste de dos capas distinctas de pintura: la capa de fondo y la capa transparente.

Two-tone Two different colors on a single paint job; to apply two colors on same paint job.
De dos colores Una obra de pintura que involucra dos colores distinctos; aplicar dos colores en la misma obra.

Ultrasonic plastic welding Method of repairing auto plastics in which welding time is controlled by the power supply. This method is best suited to rigid plastics.
Soldadura ultrasónica de plástico Un método de reparación de los plásticos de auto en que el tiempo de la soldadura se controla por la toma del corriente. Este método se adapta mejor en los plásticos rígidos.

Ultrasonic stud welding A variation of shear joint used to join plastic parts in which a weld is made along the circumference of a stud.
Soldadura ultrasónica de espiga Una variación de una junta de resistencia que sirve en la unión de las partes de plástico en que una soldadura se efectúa por la circunferencia de un espiga.

Ultraviolet light That portion of the spectrum which is largely responsible for degradation of paints. Invisible to the eye. "Black light." 2000 to 4000 Angstrom units.

Luz ultravioleta Esa porción del espectro que es responsable en gran parte por la degradación de las pinturas. Es invisible a la simple vista. La "luz negra." Entre el 2000 a 4000 unedades Angstrom.

Undercoat A first coat; primer, sealer, or surfacer.

Capa de apresto Una primera capa; la imprimación, el sellante o la preparación.

Undercut A groove melted into the base metal adjacent to the toe or root of a weld and left unfilled by weld metal. When sanding, the filler is sanded below the height of the surrounding panel.

Socavación Una ranura derritida en el metal base contigua al pie o al raíz de una soldadura que no se rellena de metal. Al lijar, el relleno queda en una depresión relativa a la altura del panel contíguo.

Unibody Vehicle in which parts of the body structure serve as support for overall vehicle parts.

Monocasco Un vehículo en el cual las partes de la estructura de la carrocería sirve de soporte para todas las partes del vehículo.

Universal measuring system A measuring system that has devices mounted on a frame which can be adjusted for various vehicle bodies. This system can utilize laser systems, universal benches, or mechanical gauges.

Sistema de medición universal Un sistema de medición que tiene los dispositivos montados sobre un armazón que se puede ajustar para varias carrocerías de vehículo. Este sistema puede utilisar los sistemas de laser, los bancos universales o los calibres mecánicos.

Urethane A type of paint or polymer characterized by the presence of chemical linkages. Urethanes are noted for their toughness and abrasion resistance.

Uretano Un tipo de pintura o un polímero caracterizado por la presencia de las uniones químicas. Los uretanos se notan por su tenacidad y resistencia a la abrasión.

UV stabilizers Chemicals added to paint to absorb the ultraviolet radiation present in sunlight.

Estabilizador de UV Las químicas agregadas a la pintura para absorber la radiación ultravioleta presentes en el luz del sol.

Value Lightness or darkness of a color.

Valor Lo claro o obscuro de un color.

Vehicle All of a paint except the pigment. This includes solvents, diluents, resins, gums, dryers, etc. The liquid portion of a paint. Also, a car or truck.

Medio La totalidad de la pintura menos el pigmento. Esto incluye los solventes, los diluyentes, las resinas, las gomas, los secadores, etc. La porción líquida de una pintura.

Vehicle identification number (VIN) Number assigned to each vehicle by its manufacturer, primarily for registration and identification purposes.

Número de identificacíon del vehículo El número asignado a cada vehículo por su fabricante, primariamente con el propósito de la registración y la identificación.

Veiling Formation of a web or strings in a paint as it emerges from a spray gun.

Membrana La formación de una tela o los hilos en una pintura al salir de una pistola atomizador.

Vinyl A class of monomers which can be combined to form vinyl polymers. Vinyl polymers are addition polymers, widely used to make chemical resistant finishes, tough plastic articles, phonograph records, and floor tiles.

Vinilo Un clase de los monómeros que se pueden combinar para formar los polímeros vinílicos. Los polímeros vinílicos son polímeros de agregación, se usan extensamente para crear los acabados de resistencia química, los artículos de plástico durables, los disco fonográficos, y las losas para el piso.

Viscosity Consistency or body of a paint; thickness or thinness of a liquid.

Viscosidad La consistencia o cuerpo de una pintura; lo espeso o líquido de un líquido.

VOC Volatile organic compound; found in most paint solvents.

VOC Compuesta orgánica volátil: presente en la mayoría de los solventes de pintura.

Volatility Tendency of a liquid to evaporate. Liquids with high boiling points have low volatility, and vice versa. Volatility affects flash-off time and fire hazard considerations.

Volatilidad La tendencia de un líquido a evaporarse. Los líquidos con un punto de ebullición alto tienen una volatilidad baja y vice versa. La volatilidad afecta el tiempo de vaporización instantánea y las consideraciones de peligros de encendios.

Water spotting A condition caused by water evaporating on a finish before it is thoroughly dry which results in a dulling of the gloss in spots.

Aguas Una condición causada por la evaporación del agua en un acabado antes de que esté completamente seco que resulta en las manchas deslustradas en el brillo.

Wax A slippery solid sometimes added to paints to add some property. Also, specially prepared material designed to shine or improve a surface.

Cera Un sólido resbaloso que a veces se agrega a las pinturas para darlas una propriedad. Tambien, una material de preparación especial diseñado para relumbrar o mejorar una superficie.

Weathering Change in a paint film caused by natural forces such as sunlight, rain, dust, wind, etc.

Daños del intemperie El cambio en una película de pintura causado por las fuerzas naturales tal como el luz del sol, la lluvia, el polvo, el viento, etc.

Weld To join two metal or plastic pieces together by bringing them to their melting points, often involving the use of a welding rod to add metal or plastic to a joint.

Soldar Unir dos pedazos de metal o de plástico calentándolos a su punto de fusión, muchas veces involucra el uso de un electrodo de soldadura para añadir el metal o el plástico a una junta.

Wet sanding Using a water resistant, ultra-fine sandpaper and water to level paint.

Lijado en húmedo Usando un papel de lija impermeable de grano ultra fino y el agua para aplanar la pintura.

Wet spots Discoloration caused where paint fails to dry and adhere uniformly (usually caused by grease or finger marks).

Manchas húmedas La discoloración causado por el fallo de la pintura en secar y adherir en una forma uniforma (generalmente se causa por la grasa o las huellas digitales).

Wet-on-wet finish Technique of applying a fresh coat of paint over an earlier coat which has been allowed to "flash" but not cure.

Acabado en húmedo Una técnica de aplicar una capa nueva de pintura sobre una capa anterior cuyo solvente ha evaporado pero que no ha endurecido.

Wheel alignment Positioning suspension and steering components to assure a vehicle's proper handling and optimum tire wear.

Alineamiento de las ruedas Posicionar la suspensión y los componentes de dirección para asegurar el manejo satisfactorio del vehículo y el rendimiento óptimo de las llantas.

Wheel balancing Proper distribution of weight around a tire and wheel assembly to counteract centrifugal forces acting on heavy areas to maintain a true-running wheel perpendicular to its rotating axis.

Balanceo de la ruedas La distribución correcta del peso alrededor de una llanta y la asamblea de la rueda para neutralizar las fuerzas centrífugas que afectan las áreas más pesadas así manteniendo una rueda que gira exactamente perpendicular a su eje giratorio.

Wheelbase Distance between front and rear axles.

Batalla (empate) La distancia entre los ejes delanteros y traseros.

Wheelhouses Deep curved panels that form compartments in which the wheels rotate. They are generally bolted to front fenders and spot welded to rear quarter panels on frame type vehicles. On unibody vehicles, they are welded in place.

Pasos de las ruedas Los paneles curvados que forman los compartimentos en los cuales giran las ruedas. Normalmente se empernan a las aletas delanteras y se soldan por punto a los paneles costales traseros en los vehículos de tipo bastidor. En los vehículos monocasco, se sueldan en posición.

Wrinkle Pattern formed on surface of a paint film by improperly formulated or by specially formulated coatings. Appearance of tiny ridges or folds in film.

Arruga Un diseño que forma en la superficie de una película de pintura debido a las capas formuladas mal preparadas o de preparación especial. Aparecen diminutas pliegues en la película.

Wrinkling Surface distortion (shriveling or skinning) that occurs in a thick coat of enamel before the under layer has properly dried.

Formar arrugas Una distorsión en la superficie (se arruga o forma una costra) que ocurre en una capa espesa de esmalte antes de que la capa inferior haya secado correctamente.

Zinc Metal coating often used to prevent corrosion.

Zinc Un tratamiento del metal usado comunmente para prevenir la corrosión.

Zoning Method of systematically observing a damaged vehicle. It includes checking primary damage, secondary damage, suspension, steering and mechanics, interior parts, and finally, the exterior paint and trim of vehicle.

Comprobación por zonas Un método de inspeccionar sistemáticamente un vehículo dañado. Incluye verificando los daños primarios, los daños secundarios, la suspensión, la dirección y funciones mecánicos, las partes interiores, y finalmente, la pintura exterior y la moldura del vehículo.

Abbreviations Used by Body Technicians and Estimators

It is important that the estimator and body shop technician be able to communicate verbally as well as in writing. Both in estimates and work procedure reports most estimators use abbreviations. Generally, these abbreviations are the same as those used in the estimating crash guides. Some abbreviations are even used verbally. For example, three of the most commonly used abbreviations in a body shop are:

- **R & I: Remove and Reinstall.** The item is removed as an assembly, set aside, and later reinstalled and aligned for a proper fit. This is generally done to gain access to another part. For example, R & I bumper would mean that the bumper assembly would have to be removed to install a new fender or quarter panel.
- **R & R: Remove and Repair.** Remove the old parts, transfer necessary items to new part, replace, and align.
- **O/H: Overhaul.** Remove an assembly from the vehicle, disassemble, clean, inspect, replace parts as needed, then reassemble, install and adjust (except wheel and suspension alignment).

In addition to these abbreviations, the following terms are those accepted by most estimating guides, shop manuals, and estimators. They are the ones used in most written forms.

A . . . . . . Manufacturer has no list price for the part.
AC . . . . . Air Conditioner
ACRS . . . Air Cushion Restraint System
adj . . . . . adjuster or adjustable
AIR . . . . Air Injector Reactor
alt . . . . . alternator
alum . . . aluminum
amp . . . . ampere
assy . . . . assembly
AT . . . . . Automatic Trans
auto . . . . automatic
aux . . . . auxiliary
bbl . . . . . barrel
bk . . . . . back
blwr . . . . blower
bmpr . . . bumper
brg . . . . bearing
brkt . . . . bracket
Bro . . . . . Brougham
btry . . . . battery
btwn . . . between
B-U . . . . Back-Up
bush . . . bushing
Calif . . . California
chnl . . . . channel

c/mbr . . cross member
cntr . . . . center
col . . . . . column
comp . . . compressor
compt . . compartment
cond . . . conditioning or conditioner
cont . . . . control
conv . . . converter or convertible
cor . . . . . corner
cov . . . . cover
Cpe . . . . Coupe
C/R . . . . Customer Request
crossmbr cross member
c/shaft . . crankshaft
ctl . . . . . control
Ctry . . . . Country
Cust . . . . Custom
cyl . . . . . cylinder
D . . . . . . Discontinued part
d . . . . . . drilling operational time
dbl . . . . . double
def . . . . . deflector
dehyd . . dehydrator
desc . . . . description

dia . . . . . diameter
diag . . . . diagonal
dist . . . . distributor
div . . . . . division
Dlxe . . . . DeLuxe
dr . . . . . . door
ea . . . . . . each
elec . . . . electric
emiss . . . emission
eng . . . . engine
EP . . . . . Exhaust Purging
equip . . . equipment
evap . . . . evaporator
exc . . . . . except
exh . . . . exhaust
extn . . . . extension
flr . . . . . . floor
Fndr . . . . Fender
Fr & Rr . . Front & Rear
frm . . . . from
fr or rr . . front or rear
ft . . . . . . foot
gal . . . . . gallon
gen . . . . generator
grds . . . . guards
grv . . . . . groove
H'back . . Hatchback

HD . . . . . Heavy Duty
H/L . . . . Headlights
HDC . . . Heavy Duty Cooling
hdr . . . . header
HEI . . . . High Energy Ignition
Hi Per . . High Performance
horiz . . . horizontal
H.P. . . . . High Performance
hsg. . . . . housing
HSLA . . . High Strength Alloy
            Steel
HSS . . . . High Strength Steel
HT . . . . . Hard Top
H'Top . . Hard Top
hyd . . . . hydraulic
Hydra . . Hydramatic
ign . . . . . ignition
in . . . . . . inch
incl . . . . includes
inr . . . . . inner
inst . . . . instrument
inter . . . intermediate
L . . . . . . Left
lic . . . . . license
lp . . . . . . lamp
lwr . . . . . lower
max . . . . maximum
mdl . . . . model
mldg . . . molding
MT . . . . . Manual Transmission
mtd . . . . mounted
mtg . . . . mounting
muff . . . muffler
NAGS . . National Auto Glass
            Specification
neg . . . . negative
OD . . . . . Outside diameter

OEM . . . Original Equipment
            Manufacturer
OH . . . . . Overhaul
opng . . . opening
orna . . . . ornament
otr . . . . . outer
p . . . . . . paint operational
            time
pass . . . . passenger
pkg . . . . package
plr . . . . . pillar
pnl . . . . . panel
pos . . . . . positive
PS . . . . . Power Steering
Pwr . . . . Power
qtr . . . . . quarter
R . . . . . . Right
rad . . . . . radiator
R & R . . . Remove & Reinstall
R-L . . . . . Right or Left
rec . . . . . receiver
refl . . . . . reflector
reg . . . . . regulator
reinf . . . reinforcement
reson . . . resonator
Rr . . . . . . Rear
Sed . . . . . Sedan
ser . . . . . serial or series
shld . . . . shield
sidembr . side member
Sig . . . . . single
s/m . . . . side marker light
spd . . . . . speed
spec . . . . special
Sta . . . . . Station
stab . . . . stabilizer
stat . . . . stationary

Std . . . . . Standard
stl . . . . . steel
strg . . . . steering
Sub . . . . Suburban
sup . . . . super
supt . . . . support
surr . . . . surround
susp . . . . suspension
SW . . . . . Station Wagon
tach . . . . tachometer
t & t . . . . tilt & telescope or tilt
            & travel
TE . . . . . Thermactor Emission
tel . . . . . telescope
t/l . . . . . taillight
trans . . . transmission
t/s . . . . . turn signal
upr . . . . . upper
vent . . . . ventilator
vert . . . . vertical
vib . . . . . vibration
VIR . . . . Valve-In-Receiver
w/ . . . . . with
WB . . . . Wheelbase
WD . . . . Wheel Drive
Wgn . . . Wagon
w'house . wheelhouse
whl . . . . wheel
whlse . . . wheelhouse
wndo . . . window
wo/ . . . . without
w/o . . . . wheel opening
wshd . . . windshield
w'strip . . weatherstrip
xmember crossmember

Decimal and Metric Equivalents

		DECIMAL AND METRIC EQUIVALENTS			
Fractions	**Decimal (in.)**	**Metric (mm)**	**Fractions**	**Decimal (in.)**	**Metric (mm)**
1/64	.015625	.397	33/64	.515625	13.097
1/32	.03125	.794	17/32	.53125	13.494
3/64	.046875	1.191	35/64	.546875	13.891
1/16	.0625	1.588	9/16	.5625	14.288
5/64	.078125	1.984	36/64	.578125	14.684
3/32	.09375	2.381	19/32	.59375	15.081
7/64	.109375	2.778	39/64	.609375	15.478
1/8	.125	3.175	5/8	.625	15.875
9/64	.140625	3.572	41/64	.640625	16.272
5/32	.15625	3.969	21/32	.65625	16.669
11/64	.171875	4.366	43/64	.671875	17.066
3/16	.1875	4.763	11/16	.6875	17.463
13/64	.203125	5.159	45/64	.703125	17.859
7/32	.21875	5.556	23/32	.71875	18.256
15/64	.234275	5.953	47/64	.734375	18.653
1/4	.250	6.35	3/4	.750	19.05
17/64	.265625	6.747	49/64	.765625	19.447
9/32	.28125	7.144	25/32	.78125	19.844
19/64	.296875	7.54	51/64	.796875	20.241
5/16	.3125	7.938	13/16	.8125	20.638
21/64	.328125	8.334	53/64	.828125	21.034
11/32	.34375	8.731	27/32	.84375	21.431
23/64	.359375	9.128	55/64	.859375	21.828
3/8	.375	9.525	7/8	.875	22.225
25/64	.390625	9.922	57/64	.890625	22.622
13/32	.40625	10.319	29/32	.90625	23.019
27/64	.421875	10.716	59/64	.921875	23.416
7/16	.4375	11.113	15/16	.9375	23.813
29/64	.453125	11.509	61/64	.953125	24.209
15/32	.46875	11.906	31/32	.96875	24.606
31/64	.484375	12.303	63/64	.984375	25.003
1/2	.500	12.7	1	1.00	25.4

Fluid and Air Nozzle Selection

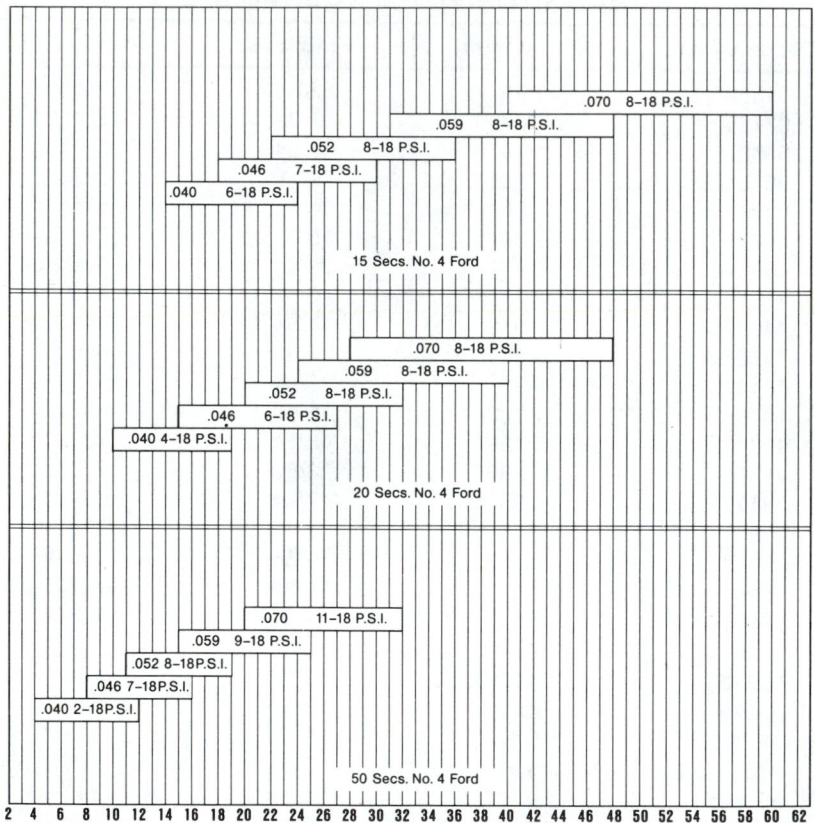

Relationship of Orifice Size, Viscosity, Pressure

FLOW RATE IN OUNCES PER MINUTE

VISCOSITY CONVERSION CHART
FOR MATERIALS AT 77 DEG. F. WITHOUT SPECIAL THIXOTROPIC CHARACTERISTICS

LIGHT CONSISTENCY — Watery or light oil type materials with translucent or very fine grind color.

MEDIUM CONSISTENCY — Light creamy or thin syrup type materials with medium to fine color or filler grind.

HEAVY CONSISTENCY — Fluffy cream or slow pouring syrup type materials with medium to coarse grind color or filler and highly filled materials.

Poise	.1	.2	.3	.4	.5	.6	.7	.8	.9	1	1.2	1.5	1.7	2.0	2.5	3.0	3.5	4.0	4.5	5	10	25	50	75	100	150
Centipoise (CPS)	10	20	30	40	50	60	70	80	90	100	120	150	170	200	250	300	350	400	450	500	1000	2500	5000	7500	10,000	15,000
Brookfield (CPS)	10	20	30	40	50	60	70	80	90	100	120	150	170	200	250	300	350	400	450	500	1000	2500	5000	7500	10,000	15,000
Fisher #1 (sec.)	20	30	39	50																						
Fisher #2 (sec.)		15	18	21	24	29	33	39	44	50	62															
Ford #3 (sec.)		12	19	25	29	33	36	41	45	50	58	70														
Ford #4 (sec.)	5	10	14	18	22	25	28	31	32	34	41	47	52	58	67	74										
Gardner-Holdt Bubble Units (sec.)	A-4	A-3	A-1	A		B		C		D	E	F	G	H	J	L	N	P	Q	S	W	Z-1	Z-3	Z-4	Z-5	Z-6
Krebs Unit (sec.)					30	33	35	37	38	40	43	47	49	52	57	60	62	64	66	68	85	114	140			
Parlin #7 (sec.)	27	32	43	50	57	64	68																			
Parlin #10 (sec.)	11	13	15	16	17	18	20	22	23	25	30	35	40	45												
Parlin #15 (sec.)														10		15		20		25	47	135	232	348	465	697
Parlin #20 (sec.)																				8	17	55	83	125	167	250
Parlin #30 (sec.)																							19	29	38	58
Saybolt (Universal) (SSU) (sec.)	60	100	160	210	260	320	370	430	480	530	580	740	845	1000	1240	1330	1475	1950	2215	2480	4600	11,600	23,500	35,000	46,500	69,500
Stormer (150 Gr.) (sec.)		10	12	14	16	18	20	22	25	27	32	38	44	49						114	223	450	1090	1635	2180	
Zahn #1 (sec.)	30	37	44	52	60	68																				
Zahn #2 (sec.)	16	18	20	22	24	27	30	34	37	41	49	62	70	82												
Zahn #3 (sec.)									10		14	17	19	23	29	34	40	46	51	57						
Zahn #4 (sec.)										10	11	13	15	17	21	24	27	30	34	37						
Zahn #5 (sec.)														10	13	15	18	20	22	25	49					
Sears Craftsman Cup (sec.)				19	20	21	23	24	26	27	31	36	39	44												
Dupont M-50 (sec.)	16	18	20	22	24	27	30	34	37	41	49	62	70	82												

CAUTION

Your viscosity cup is a precision instrument requiring careful handling, cleaning, and storage. Improper care will adversely affect its accuracy.

TABLE I
MAXIMUM ALLOWABLE CONCENTRATIONS OF SUBSTANCE VAPORS

Chemical Substance	TLV*
Acetone	750
Alcohols	
Ethyl (Ethanol)	1000
Isopropyl	400
Benzene (Benzol)	1
Carbon Tetrachloride (Skin)	5
Chlorobenze	10
Chloroprene	10
Chlorotoluene	50
Methyl Cellosolve Acetate (Skin)	25
Cyclohexane	300
Cyclohexanone	25
Cyclohexene	300

Chemical Substance	TLV*
Ethylene Dichloride	10
Methyl Isocyanate (Skin)	0.02
Methylene Chloride	50
Methyl Ethyl Ketone (MEK)	200
Methyl Ethyl Ketone Peroxide (MEKP)	0.2
Methyl Isobutyl Ketone (MIBK)	50
Perchloroethylene	25
Toluene (Toluol)	50
Toluene Diisocyanate	0.005
Trichloroethylene	50
Turpentine	100
Xylene (Xylol)	100

*Threshold Limit Value - Time Weighted Average (TLV - TWA) The time-weighted average concentration for a normal 8-hour workday and a 40-hour work week, to which nearly all workers may be repeatedly exposed, day after day, without adverse effect. Courtesy: American Conference of Governmental Industrial Hygienists "Threshold Limit Value" 1994

TABLE II
TOXIC EFFECTS FROM OVEREXPOSURE TO CHEMICAL SUBSTANCES

Chemical Substance	Effect of Inhalation
Acetone	Irritating to mucous membranes: choking sensation
Alcohol, Ethyl	Intoxication
Alcohol, Isopropyl	Intoxication: headache
Benzene (Benzol)	Injury to blood-forming organs, and to heart, liver, kidneys, etc.
Carbon Tetrachloride	Nausea, headache, vomiting; injury to liver (nephritis)
Ethylene Dichloride	Irritating to nose; retching; unconsciousness
Methyl Ethyl Ketone (MEK)	Irritating to nasal passages; choking sensation
Methyl Isobutyl Ketone (MIBK)	Irritating to mucous membranes; choking sensation
Naphtha, V.M.+P.	Headache; vomiting; muscular twitching
Toluene (Toluol)	Same as for benzene except little damage to blood-forming organs
Trichlorethylene	Similar to carbon tetrachloride; disturbed heart action
Turpentine	Irritating to nose and throat; headache; vomiting; stomach pains
Xylene (Xylol)	Same as for benzene except little damage to blood-forming organs

TABLE III
FLASH POINTS AND FLAMMABLE LIMITS OF CHEMICAL SUBSTANCES

Chemical Substances	Approximate Boiling Point Range, Deg. F	Average Flash Point Open Cup, Deg. F	Explosive Limits, % by Vol. LEL (Lower)	UEL (Upper)
Acetone	132-134	0	2.6	12.8
Alcohol, Butyl	240-245	115	1.5	11.2
Alcohol, Ethyl	173-176	55	3.3	19.0
Alcohol, Isopropyl	178-180	53	2.0	12.0
Benzene (Benzol)	174-176	12	1.3	7.1
Carbon Tetrachloride	170-172	None	Non-flammable	
Cellosolve, Acetate	293-313	117	1.7	—
Cyclohexane	179-182	32	1.3	8.0
Cyclohexanone	303-313	111	—	—
Cyclohexene	180-183	22	—	—
Ethylene Dichloride	179-186	56	6.2	15.9
Methyl Ethyl Ketone (MEK)	174-176	21	1.8	10.0
Methyl Isobutyl Ketone (MIBK)	237-246	73	1.4	7.5
Methylene Chloride	104-105	None	Non-flammable	
Mineral Spirits	307-389	104	6.9	7.5
Naphtha, V.M.+P.	212-320	45	0.9	6.0
Toluene (Toluol)	230-232	40	1.2	7.1
1,1,1, Trichlorethane	165-194	None	Non-flammable	
Trichlorethane	188-190	None	8.0-10.5	—
Turpentine	307-347	95	0.8	None
Xylene (Xylol)	281-291	81-115	1.1	7.0

Acid neutralization, 272
Acid rain, 654–55
Acid spotting, 826
Adhesion promoters, 546, 548, 549, 766
Adhesive door glass method, 429
Adhesives, 57, 162–63, 171, 582–84, 587–88
Adhesive-type windshield installation method, 580
Air bag systems, 530, 531, 609–13
Airbrushes, 740–42
Air caps, 102, 704
Air chisels, 110–12, 402, 568, 571, 573
Air compressors, 16, 132, 133–40, 144–45, 146–47
Air condensers, 141, 142, 143, 146
Air conditioning systems, 459, 500, 503–8, 578
Air distribution systems, 133, 141
Air drills, 105–7, 115
Air dryers, 143–44
Air files, 108
Air filters, 141, 142, 143, 146
Air filtration systems, 745–49
Air grinders, 108–9, 286, 288, 289, 291, 292
Air hacksaws, 113
Air hammers, 110–12, 115
Air impact wrenches, 102–4, 105, 117, 166, 462
Air injection emission control systems, 508
Airless plastic welding, 553, 554–55, 559
Airless spraying equipment, 734–35
Air makeup systems, 743–45
Air plasma arc cutters, 118, 229–33, 400, 401
Air polishers, 109–10
Air-powered tools, 22, 99–117, 145
Air pressure regulators, 23, 103, 141–42, 143
Air punches, 112
Air purification equipment, 143–44
Air ratchet wrenches, 104–5, 115, 462
Air sanders, 107–8, 120, 671
Air screwdrivers, 107
Air shears, 112
Air springs, 469–70
Air suspension systems, 471
Air wrenches, 102–5, 116–17, 166
Aligning punches, 86
Alignment, 354–91
Alignment gauges, 486
Alkali spotting, 826
Aluminum, 182, 192, 203–5, 224–25, 261–62, 685–86
Aluminum fillers, 270–71, 273
Aluminum tape, 546–47
Ammeters, 520, 525, 527
Analog displays, 529
Analog multimeters, 518, 520
Anchorage inspection, 608
Anchor chains, 359
Anchoring clamps, 358–59, 373, 377
Anchoring equipment, 357
Anchor points, 411

Anchor rails, 358
Antichip coating, 156, 157, 766
Anticorrosion materials, 635, 640–41
Antifreeze, 496–97, 498
Antilacerative glass, 578
Antilock brake systems, 422, 492, 493, 494–95, 531
Antispatter compounds, 190–91, 203, 204
A-pillars, 36, 411, 415
Arc brazing, 226–27, 402–3
Arming sensors, 610
Awls, 87, 88

Backbone frames, 41
Backhand welding, 188, 194, 195
Back masking, 696
Backup lights, 600–601
Ball joints, 466, 468
Ball joint separators, 112
Basecoat/clearcoat paint systems, 9, 152, 271, 773–74, 785–87, 813
Batteries, 22, 23, 459, 515–16, 525, 527, 533–34
Battery chargers, 120, 121
Beam and knee system, 382
Bench-rack system, 362
Bench straightening systems, 361–62
Bench vises, 78
Blade fuses, 521
Blasting, 680–82
Bleeding (air removal), 474, 492–93, 497
Bleeding (paint discoloration), 155, 271, 272, 826–27, 834
Blocking devices, 388
Body clips, 170
Body code plates, 10
Body construction types, 32–63
Body dimensions, 300, 303, 304, 315–50, 405–7
Body dimensions charts, 315–16
Body dimensions manuals, 14, 325–26
Body files, 91, 260, 261, 283
Body hammers, 81, 85, 96, 249–50
Body picks, 84–85, 246, 254
Body solder, 261, 269, 281–83
Body spoons, 83–84, 250, 251
Bodyworking tools, 79–90, 96–97
Bolts, 164–67
Bolt-through door glass method, 429
Bonding door glass method, 429
Bonnets (polishing/buffing pads), 109–10
Bottle jacks, 123, 124
Box sections, 243, 244, 388–89, 413
B-pillars, 36, 411, 412, 413, 415–16
Brake calipers, 461, 488, 490, 493
Brake lights, 523, 600–601
Brake systems, 488–95
Brass brazing, 176, 226
Braze welding, 176, 180, 223, 225–29, 402–3
Breaker bars, 72
Buckles (clasps), 607–8
Buckles (deformations), 242–45, 247
Buffing, 118, 248, 249, 844–45
"Bullseyes," 675
Bumper adjustment/replacement, 602–4

Bumper assemblies, 36, 57
Bumper painting, 792–94
Bumper reinforcements, 237, 238
Bumper stops, 438
Bumping files, 84, 250
Bumping hammers, 81–83
Burn-through, 177, 195, 203, 205, 843
Butt joints, 411, 413, 415, 416
Butt welding, 194–99, 203, 414, 415, 416, 554, 555

Camber, 478, 479, 480, 483, 485, 486
Camber gauges, 483, 487
Camshafts, 454
Carburizing flames, 222, 223, 228, 282
Card masking, 820
Carpeting, 627
Caster, 478–79
Caster gauges, 487
Catalysts, 154, 766
Catalytic converters, 507, 508
Centering gauges, 317, 326–28, 338, 405–6, 407
Chainless anchoring systems, 359–61
Chain tighteners, 359
Channel-locks, 76
Chipping (paint), 828
Chisels, 86–87
Chromium plating preparation, 686
Churning knives, 755
Circuit breakers, 521–22
Circuit numbering, 527
Circuit protection devices, 520
Circuits, 513–14
Clamping pliers, 106
Clamping tools, 192, 193, 231
Claw rippers, 112
Clutches, 454
Coil spring suspension systems, 466, 468, 470
Color analysis, 806–8
Color blindness, 802
Color codes, 527, 553
Color dimensions, 802
Color directories, 804–5
Color flip-flop, 812
Color matching, 799, 803–5, 808–9, 810–12, 818
Color matching scales, 755
Color Rendering Index, 800
Color sanding, 672, 840
Color selection/identification, 771–72
Color standards, 804–5, 806
Color theory, 800–802
Color tinting, 808–10, 818–20
Composite plastics, 540
Composite unibodies, 57
Compounding, 162, 689, 791
Compressed air, 19, 22, 229, 230, 733
Compressed air hoses, 144–45
Compressed air supply equipment, 132–49
Computer-assisted steering systems, 474
Computer estimating, 3, 858–64
Computerized automotive systems, 422, 452, 529–34

Computerized paint matching systems, 806
Computerized straightening equipment, 366, 368
Computerized suspension systems, 468–71
Computerized wheel alignment, 487, 488
Computer-simulated crash testing, 58
Computer-type color matching scales, 755
Constant velocity joints, 454, 455, 464
Continuous welding, 184, 193, 194, 202, 402, 555
Conversion coatings, 157, 684
Convertible body shapes, 34–35
Cooling systems, 496–502
Cornering angle, 482, 484, 485
Corporate Average Fuel Economy, 38
Corrosion, 157, 412, 633
Corrosion protection, 6, 37, 45, 389, 632–57
    antifreeze and, 497
    dent puller/pull rod holes and, 85
    epoxy primers and, 156
    fiberglass fillers for, 267
    in panel replacement, 423–24
    for space frame designs, 57
    in spot welding, 215
Couplings, 145
Cowls, 36, 334, 335, 379, 437
Crash testing, 58, 60
Cream hardeners, 158, 269, 274–76
Crosscutting method, 248, 249
Crossflow radiators, 500, 502
Crossflow spray booths, 745
Cross members, 309–10, 330, 383, 384, 406, 463
Crush zones, 58, 61, 310–11, 419
Custom body panels, 443–44
Custom masks, 820
Custom-mixed colors, 772, 806
Custom painting, 820–23
Cylinders, 453

Damage analysis forms, 338, 343
Damage appraisals, 5, 300–352, 369–70, 851
Damage classification, 241–47
Damage estimating, 3, 5–6, 13, 393–94, 395, 542, 850–78
Dashboard gauges, 529
Dash pads, 549–50, 560
Dash panels, 36, 36–37, 44, 379, 622–25
Databases, 861–62, 863–64
Data measurement charts, 319, 320
Data scanners, 860
Datum plane, 328, 329
Decal replacement, 845–47
Dedicated bench measuring system, 346–50, 361–62
Deductible clauses, 851, 854
Dent pullers, 85–86, 254, 255
Dent removal, 219, 220, 249–57, 414, 560
Destructive testing, 216, 217
Detailing, 10, 824, 838–45, 847–48
Diagnostic charts, 14, 527, 532, 533
Diagnostic leads, 532
Diagnostic tools, 516
Diagonal line measurement method, 323, 325
Dial-up estimating systems, 860
Diamond damage, 308–9, 333, 334, 384
Die grinders, 109
Differential assemblies, 454
Digital displays, 529
Digital images, 862–63
Digital multimeters, 518, 519

Digital signals, 530
Dinged doors, 2, 54, 246
Dinging hammers, 82, 249–50, 253
Dinging spoons, 84, 250
Ding repair, 286–87
Direct current, 519
Direct current reverse polarity, 184, 207
Direct repair programs, 851
Dirt in finish, 829–30
Dirt-nib files, 839–40
Disc adhesive, 108
Disc brakes, 490
Disc grinders, 108, 109, 425
Disc sanders, 107, 204, 248–49, 546, 547
"Dog tracking," 468, 482
Dollies, 83, 246, 250–52, 260
Door adjustment, 428, 430–33
Door aligners, 364, 365
Door alignment, 308, 315
Door components, 36, 55–56, 57
Door dings, 2, 54, 246
Door guard beams, 237, 238
Door handle tools, 89, 90
Door hinges, 36, 57, 423, 424, 430–33
Door jamb guns, 740–42
Door latches, 57
Door locks, 592, 593–95
Door panels, 395, 396, 424–27
Door skins, 424, 511, 571
Door spraying, 647–48
Door trim, 424
Door windows, 428–30, 433–34, 591–93
Downdraft spray booths, 743, 745
Downflow radiators, 500, 502
Drain valves, 138, 141
Draw-down bars, 818
Drill bits, 105
Drill pads, 567–68
Drill presses, 118
Drill-type spot cutters, 106, 399, 400
Drive shafts, 454, 463, 484, 485
Drive sockets, 71–72
Drivetrains, 453, 454, 459, 462
Drum brakes, 489–90
Drums (brake rotors), 488
Drying processes, 154–55, 267–68
Drying rooms, 9–10, 750–53, 754
Dual-diagonal split hydraulic system, 488, 489
Dual-phase steel (ultrahigh-strength steel), 237, 238, 399
Dulled finish, 830
Dust coats, 781
Dustless sanding systems, 20, 120
Dust particle masks, 19
Dust respirators, 15

Edge distance (welding), 215, 216
Edge masking, 843
Edging tools, 112
Elastic deformation, 239
Elastic stress, 240
Elastomers, 156, 789
Electrical circuits, 513–14
Electrical conductors, 513
Electrical cooling fans, 500, 601
Electrical current, 513, 520
Electrical diagnostic equipment, 516
Electrical drills, 105, 123
Electrical harnesses, 27, 511, 517, 523
Electrical insulation, 93, 94, 95–96, 513
Electrical polarity, 184, 207
Electrical resistance, 513, 518–19
Electrical systems, 447, 511–29
Electrical tools, 22, 117–23
Electrical welding, 117

Electrical window regulators, 434–35, 522
Electronic air suspension systems, 471
Electronic color analyzers, 805, 806
Electronic displays, 529–30, 534
Electronic leak detectors, 506
Electronic measuring equipment, 368
Electronic mil gauges, 680
Electronic power steering systems, 474
Electronic stethoscopes, 602
Electrostatic spraying, 735–37
Emission control systems, 508–10
Enamel paints, 152, 153, 154, 727, 768, 787
Energy reserve modules, 611
Engine holders, 364, 485
Engines
    cooling systems for, 496–502
    parts of, 452, 453–54
    removal of, 364, 365, 369, 457–64
    self-diagnosis of, 531
    starting problems of, 524–25
Entrepreneurship, 878–79
Epoxy adhesives, 163, 171, 429, 548–49
Epoxy fillers, 269
Epoxy primers, 156, 261, 641, 644, 766
Exhaust systems, 418, 462, 506–8, 510

Fabricated panels, 397
Far drying equipment, 753
Fasteners, 164–71, 398, 444
Fault codes, 531–33
Featheredge splitting, 830–31
Featheredging, 271, 285, 291, 566, 570, 672–75
Fender adjustments, 436–37
Fender aprons/skirts, 36, 42, 192, 378, 406, 465
Fender buckling, 244
Fender covers, 97, 437
Fender dollies, 83
Fiberglass cores, 572
Fiberglass fillers, 158, 269–70, 273, 289–90, 291, 292, 293–94, 296
Fiberglass repair materials, 158, 565, 566, 569, 570
Fiber-reinforced plastics, 540, 562, 564–69
Fillers, 91, 92, 156, 157–58, 259–60, 265–81
Final detailing, 824, 838–45, 847–48
Finesse work, 840, 841
Fisheye eliminator, 156
Fisheyes (paint), 831
Fixture measuring system, 346–50, 362
Flame abnormalities, 225
Flame paintings, 821, 822
Flame treatment, 546, 550
Flame types, 222, 223
Flange welding, 199, 218
Flash stage/time, 154, 687
Flatrate charges, 872
Flatteners (paint agents), 156
Flat welding, 192, 193, 198
Flex agents, 156, 789
Flexibility tests, 542
Flexible additives, 789
Flexible part repair, 548–49
Flexible plastic, 689–90, 790–92
Flexible putties, 550
Flexible sanders, 92
Flexockets, 73
Flip-flop (color), 812
"Flooding the pattern," 719
Fluid control valves, 101, 706
Fluid hoses, 144, 145
Fluid needle valves, 101, 706
Fluid tips, 706

Fluorescent light, 800
Fluoride compounds, 207
Fluorine clearcoats, 786, 813–14
Flux-cored arc welding, 205–8
Fluxes, 227, 228–29
Flywheels, 454
Foam fillers, 416–17, 560
Force drying, 787
Ford cups, 711–12
Four-channel anti-lock braking systems, 495
Four-point anchoring, 373
Four-wheel alignment, 342, 345, 487, 488
Four-wheel drive vehicles, 455, 456, 457, 500
Four-wheel steering systems, 476–77
Frame gauges, 336
Frame/panel straighteners (frame racks), 7, 124, 354, 460, 461
    accessories for, 363–65
    complexities of, 360
    pin pulling with, 256
    portable, 38, 125, 361, 372
    power train removal for, 457
Frame rails, 36, 411, 419–23
Frontal collisions, 304, 305, 306, 311, 498, 501
Front-engine front-drive bodies, 49–51, 380
Front-engine rear-drive bodies, 45–48, 50, 451, 453, 454, 462
Front/rear split hydraulic systems, 488–89
Front suspensions, 50, 466–68
Front-wheel drive vehicles, 454–55, 470, 481, 488
Fuel economy, 38
Fuel evaporative systems, 510
Fuel lines, 460, 461, 512
Fuel tanks, 22
Full cutout windshield installation method, 580, 582, 584–88
Fuses, 521
Fusion welding, 176

Galvanic corrosion, 639
Galvanizing, 203, 634, 685
Garnish moldings, 578, 580, 582
Gaskets, 168, 428, 429
Gasket windshield installation method, 578, 580, 582–84
Glass runs (channels), 433, 434
Glazing putties, 158–59, 266, 271–72
    application of, 279–80, 284, 285, 688–89
    flexible, 550
Grain mold dies, 619
Grain structures, 239
Gravel guard, 156, 157, 766
Gravity-feed spray guns, 138, 705, 707–8, 709–10, 738
Grinding, 262, 272–73, 401–2, 403, 404, 558, 680
Grinding discs, 161, 273, 664–66
Grinding wheels, 118
Grit ratings, 160–61, 248, 662–63, 675
Ground cables (ground clamps), 186, 230
Grounded outlets, 147
Grounding points, 513, 524
Grounding straps, 534
Grounding wires, 120–21, 518, 737
G-sensors, 471
Guide coats, 672, 688

Halo effect, 816
Hammering, 273, 274, 377, 386
Hammer-off-dolly method, 252–54, 259, 261

Hammer-on-dolly method, 251–52, 254, 261
Hammers, 80–83, 94–95, 96–97, 192, 246
Hand compounds, 841, 843
Hand glazes, 162
Hand-rubbing compounds, 162, 689
Hand sanding, 666–69
Hand tools, 22, 64–98
Hard brazing, 176, 226
Hardeners, 267–68, 269, 544, 572
Hardness paint-type identification method, 773
Hardtop doors, 428, 430, 431
Hardtop vehicles, 34
Hat channels, 413
Hatchback body shapes, 34, 35, 48, 441
Headlight aimers, 522–23
Headlights, 522, 598, 600
Headliners, 628
Head-on viewing angle, 807
Headrest guides, 629
Headrests, 628
Heat application paint-type identification method, 773
Heat crayons, 192, 224, 238, 262, 387–88
Heater hoses, 460
Heater systems, 496, 502–3
Heat guns, 120, 121, 561
Heat lamps, 277, 566
Heat shrinking, 256, 257–60, 262
Heat sink compound, 192, 296
Heel dollies, 83
Heeling, 723
Hemming, 425, 426
High-speed plastic welding, 551, 552
High-strength low-alloy steels, 178, 181, 210, 237, 238
High-strength steels, 178, 205, 210, 223–24, 237, 399
High-tensile-strength steel, 237–38
High-volume low-pressure spraying systems (HVLP), 102, 138, 139, 718, 737–40
Hinge buckles, 242, 243–44
Hinged parts, 57, 442
Hinge locks (strikers), 57, 432, 442, 443
Hitch pins, 169
Hog ring output shafts, 103
Hog rings, 629
Hole saw-type spot cutters, 106, 399, 400
Hood bumper stops, 438
Hood hinges, 36, 438
Hood latches, 438
Hoods, 36, 407–8, 437–40
Hood safety catches, 438
Horn switches, 524
Hose clamps, 171, 511
Hose routing, 474
Hoses, 510–11
Hot-air plastic welding, 118, 550–53, 554
Hot-rolled sheet metal, 235
Hot wires, 513, 514
Humidity, 277, 638, 774
Hydraulic actuators, 470–71
Hydraulic circuits, 514
Hydraulic equipment, 123, 126–27
Hydraulic jacks, 20, 126
Hydraulic lifts, 23, 127–30, 756
Hydraulic presses, 22–23
Hydraulic rams, 126, 357, 374–77
Hydraulic spraying equipment, 734–35
Hydrometers, 497, 515
Hydroscopic fillers, 266

Identification numbers, 11–12, 58
Idler arms, 471
Ignition switches, 524
Impact damage, 310–13

Impact forces, 239
Impact points, 253, 305–6, 314
Impact sensors, 610
Impact sockets, 70, 102
Impact wrenches, 102–4, 105, 117, 166, 462
Impedance, 518
Impellers, 499
Incandescent light, 800
"Included angle," 480
Independent rear suspension, 455, 470, 481
Industrial fallout, 655
Inertia damage, 315
Inertia sensors, 610
Inertia switches, 520
Infrared drying equipment, 752–53
Injuries, 15
In-line oilers, 114
Instrument clusters, 624–25
Insulation (electricity), 93, 94, 95–96, 513
Insurance companies, 2, 122, 394, 395, 851, 854, 868
Insurance rating system, 58
Insurance representatives, 3, 5, 422
Intake filters, 132, 140
Intermittent shorts, 514
Intermittent welding, 200, 555
Intermix color systems, 772, 806
International symbols, 542
International System of Units, 165
Interval timers, 202

Jacks, 20, 124, 125–26
Jack stands, 20, 123, 461
Jigs, 422, 423
Job overlap, 872
Joint brazing, 227, 228
Joint filling, 281, 282
Joint fit-up, 177
Joint preparation, 412
Joint sealing, 648–53
Joint sectioning, 411–12
Joint welding, 177, 228
Jounce-rebound checks, 483–85, 485
Jumper cables/wires, 515, 516, 517–18

Keyless entry systems, 594–95
Kick panels, 36
Kick-ups, 47
Kill charts, 820

Labor costs, 872–76
Lace painting, 820–21
Lacquer-based primers, 645
Lacquer paints, 152, 153, 768
Ladder frames, 41
Laminated plate glass, 576, 578
Lap belts, 606–8
Lap joints, 218, 411
Lap welding, 194, 199, 201, 202, 218, 415, 416, 419
Laser measuring systems, 338, 342, 344–46
Lead fillers, 261, 269, 281–83
Leaf spring suspension systems, 466, 468, 470
Leaks, 595–98
Let-down panels, 817–18
Liftback body shapes, 34, 35, 48, 441
Lifters (engine parts), 454
Lifting (human action), 19
Lifting (paint), 831–32
Lifts (apparatuses), 23, 127–30, 756
Light bulbs, 520, 522, 523
Light checks, 597–98
Light-emitting diodes, 530
Lighting, 800–802

Lighting systems, 522–24, 525
Line checking (paint), 828–29
Liquid crystal diodes, 530
Liquid masking material, 697–98
Load (electricity), 513
Loading (part loading), 238
Locking pliers, 77, 218, 219
Locking strip gaskets, 579, 580, 583
Lock strikers, 57, 432, 442, 443
Lubrication, 113–14, 727

Machine compounds, 162, 689, 843
MacPherson fixtures, 348
MacPherson strut centerline gauges, 332–33, 345
MacPherson strut suspension systems, 466–67, 468, 485
MacPherson strut towers, 364, 384
Magnetic gauges, 530
Magnet screwdrivers, 74
Mallets, 80–81
Marble paint effects, 821, 822
Martensitic steel, 237, 238
Masking areas, 8–9
Masking materials, 159–60, 691–98, 787–88, 837–38
Master cylinders, 488, 490–91, 492, 493, 494
Measurement gauges, 304, 319, 333–36, 350
Mechanical automobile systems, 447, 450, 453
Mechanical joining methods, 174, 176
Mechanical measuring systems, 336–38, 339
Mechanical weight-type color matching scales, 755
Melt-flow plastic welding, 559
Metal activity, 639
Metal alloys, 227
Metal combustibles, 24
Metal conditioners, 19, 157, 289, 292–93, 642–43, 684
Metal cutting shears, 88, 113
Metal files, 91–92
Metal fillers, 261, 269, 270–71, 273, 281–83
Metal flakes, 153, 768
Metallic paints, 153, 765, 783–84, 810–12
Metal patches, 293, 294–95, 389
Metal powders, 578
Metal preparation, 683–86
Metal stress relieving, 385–88
Metalworking techniques, 235–63, 273–74
Metalworking tools, 79–90, 96–97
Metamerism, 802
Metric measures, 165, 321
Mica paints, 815
Microchecking (paint), 829
Midcoat primers (adhesion promoters), 546, 548, 549, 766
Mid-engine rear-drive bodies, 51–53
MIG welding, 181–205
Mild steels, 237, 239
Mil gauges, 680, 713
Mill pads, 567–68
Minivan body shapes, 35
Mirrored targets, 338, 344
Mist coats, 781
Mixing boards, 157, 274, 275
Mixing charts, 776
Mixing instructions, 775
Mixing sticks, 712, 776–78
Modems, 859
Modular rail straightening system, 358
Moisture separators/regulators, 141, 146
Moldings, 578, 582, 613–16

Mold rivet welding, 220
Motor mounts, 457, 462, 463
Motors, 520
Motor starters, 139
Mottling, 832
Mounting tabs, 561–62
Mounts ("biscuits"), 39, 41, 307, 315
Mufflers, 507, 508
Multimeters, 518, 519, 520, 521
Multipass butt welding, 198
Multiple-pull systems, 356, 371, 385
Multipurpose vehicle body shapes, 35

Near drying equipment, 753
Near specular viewing angle, 807
Needle nose pliers, 76–77, 93
Needle scalers, 112
Neutral flames, 222, 223
Nibblers, 112, 113
Noises, 595–96, 597, 601–2
Nondestructive testing, 217–18
Nuts (fasteners), 165, 166, 167–68

Offset bumping hammers, 82
Offset butt joints, 411, 416, 417
Oil dipsticks, 454
Oilless compressors, 16, 135–36
Open-end wrenches, 66–67
Open hat channels, 413
Open time (work life), 564
Orange peel (paint defect), 728, 832–33
Orbital sanders, 107–8, 120, 161, 670–71
Output shafts, 103
Overall repainting, 765, 784–85
Overhead welding, 189, 192, 193, 202–3
Overlap joints, 414, 416, 417
Overload protection, 139–40
Overspray, 159, 695, 724, 729, 731
Oxidizing flames, 222, 223
Oxyacetylene torches, 220–25
Oxygen sensors, 534

Paddle agitators, 709, 710, 754
Paint additives, 154
Paint binders, 153
Paint blending, 782
Paint damage, 497, 679–80
Paint edges, 159, 695
Paint film thickness measurement method, 772
Paint formulas, 805–6
Paint hangers, 753–54
Paint hardeners, 154–55
Painting equipment, 702–61
Painting problems, 824–38
Painting variables, 816–17
Paint laydown, 780
Paintless dent removal, 85, 254
Paint mixing sticks, 712, 776–78
Paint paddles, 757–58
Paint pigments, 153
Paint reference charts, 14
Paint shakers, 754
Paint solvents, 153–54, 271, 272
Paint strainers, 757, 758, 778
Paint strippers, 157, 682–83
Paint surface chips, 838
Paint surface protrusions, 839
Paint systems, 156
Paint types, 152
Panel crimpers, 112
Panel cutters, 88
Panel drying ovens, 754
Panels, 36, 387, 427–28
    customized, 443–44
    paint repair of, 765, 782–84, 816
    removal of, 393–94

replacement of, 5, 393, 403–10, 426–27, 567–69
Panel saws, 113
Panel spotters, 218–20
Paraffins, 268–69
Parallel arm suspension systems, 467
Parallel circuits, 513
Parallelogram steering systems, 471, 472
Parking brakes, 495
Partial cutout windshield installation method, 580, 582, 588–90
Part loading, 238
Part location diagrams, 528, 529
Part prices, 869–72
Parts, 57
Parts managers, 11–12
Passive restraint systems, 604
Patching, 293, 294–95, 389
Pattern control valves, 706
Pearlescent paints, 153, 768, 814–15
Peel (part loading), 238
Peeling (paint), 833
Pencil mil gauges, 680
Perimeter frames, 37, 41–42, 43, 307
Permanent plastic (permanent stress), 240
Picking hammers, 81, 82, 261–62
Picks, 84–85, 246, 254
Pillars, 36, 327, 328
Pilot arcs, 229, 230, 233
Pinholing (paint), 833–34
Piston compressors, 132, 134–36, 137
Pistons, 453
Pitman arms, 471
Pivot measure system, 338, 341
Plasma arc cutters, 118, 229–33, 400, 401
Plastic accessory panels, 444
Plastic adhesives, 162–63, 544, 547–48, 564
Plastic deformation, 239–40
Plastic fillers, 157–58, 256, 265, 267–69, 270, 289–90, 565, 834
Plastic gaskets, 429
Plastic hose clamps, 171
Plastic materials, 537–75
Plastic media blasting, 110, 157, 288, 294
Plastic memory, 561
Plastic parts, 537, 539, 540, 561–62, 689–91, 788–96
Plastic primers, 766
Plastic retainers, 170
Plastic welding, 118, 550–60, 573
Platform frames, 44
Pliers, 76–77, 93
Plotting color, 802–3
Plug welding, 192, 194, 199–201, 238, 415, 416, 418
Polarity (electricity), 184, 207
Polishing, 844–45
Polishing compounds, 162, 843
Polishing pads, 109–10
Polyester fillers, 267, 269, 273
Polyester glazing putties, 271–72, 273, 279
Polyester primer-fillers, 272, 273, 284
Polypropylene plastic parts, 690, 792
Polyurethane accessory panels, 444
Polyurethane enamels, 154
Positive crankcase ventilation systems, 508, 510
Power brakes, 488, 490, 494
Power grinding, 670–72
Power handles, 72
Power jacks, 123–27, 255–56
Power knives, 584, 585
Power riveters, 113
Power sanding, 19, 21, 669–72

Power saws, 413, 415
Power steering, 460, 473–75
Power tools, 22–23, 99–131
Power towers (power posts), 359, 382
Power trains, 452–53, 456–57, 459–66
Power washers, 119–20
Power winches, 361
Power window regulators, 434–35, 522
Prep procedures, 8–9, 778–79
Prep solvents (wax/grease removers), 156, 272, 677–78
Pressure blasters, 681
Pressure bleeding, 493
Pressure caps, 496, 497, 498
Pressure cup spray guns, 708
Pressure-feed spray guns, 138, 705, 716–17
Pressure switches, 139
Pressure tank spray guns, 708
Pressure welding, 176
Primer-fillers, 155, 272, 273, 284
Primers, 155–56, 278, 279, 285, 586, 643–48, 686–88, 765–66
Primer-sealers, 155, 766
Primer-surfacers, 155–56, 279, 284, 766, 778
Propellant charges, 610
Propeller shafts, 455
Pulling towers (pulling posts), 38, 39, 357, 362, 365
Pull pin spot welders, 255–56
Pull rods, 85–86
Pulse welding, 184, 193, 194, 202, 402, 555
Punches, 86–87, 112
Push rods, 454

Quarter glass, 433–34
Quarter panels, 37, 380–81, 396, 408–9, 648
Quenching, 259, 262, 283

Rack-and-pinion steering systems, 465, 471, 472, 473, 474, 484
Rack systems, 361, 368
Radiators, 496, 497–98, 500–502
    core supports of, 36, 42, 378, 406–7
    drain cocks of, 459
    hoses of, 499
Rails, 36, 411, 419–23
Rail straightening system, 358
Ratchet wrenches, 104–5, 115, 462
Rear suspension, 455, 465, 468, 470, 481
Rear-wheel drive vehicles, 455, 463, 464, 481, 500
Recycled components, 223–24, 412, 421, 870
Re-dressing, 95, 97
Reducers, 153, 154, 678, 775–76
Reduction flames, 222, 223, 228, 282
Refinishing, 150, 152–64, 677, 702–98, 876–78
Refrigerants, 504–5, 506
Reinforced plastics, 540, 562, 564–69, 571, 573
Relays (electricity), 520
Resin, 154, 158, 265, 271, 544, 566, 572
Respirators, 15–17, 154, 203, 756
Restraint bars, 364, 365
Restraint systems, 604–14
Retarders, 154, 775
Retexturing, 795–96
Retractors, 608
Reveal moldings, 578, 581, 582
Reverse welding, 188, 194, 195
Ribbon sealers, 164, 587
Rivet guns, 88–89

Rocker arms, 454
Rocker panels, 36, 397, 411, 413–15, 648
Roof panels, 36, 239
Roof rails, 375
Roof structures, 333
Rotors (brake drums), 488
Rotors (disk brake components), 490
Rough-out operations, 248
Rubbing compounds, 162, 283, 840–44
Rust converters, 641
Rust deactivators, 291, 292–93
Rustouts, 269, 287, 290–96, 507, 508
Rust under finish, 835

Safety catches, 438
Safety practices, 14–26
Safety stands, 20, 123, 461
Safety valves, 146
Safing sensors, 610
Sag damage, 307–8, 334–35, 383
Salvage value, 869
Sandblasters, 110, 111, 288
Sander pads, 108, 248
Sanding guides, 280
Sanding methods, 666–77
Sandpapers, 161–62, 661–64
Sand scratches, 676–77
Sandscratch swelling, 835–36
Sash channel door glass method, 429, 430
Saws, 113, 413, 415
Scaling, 387
Scan tools, 532–33
Scissor lifts, 128, 363, 430
Scratch awls, 87, 88
Screwdrivers, 73–76, 95–96, 107
Screws, 168, 169
Scrub radius, 480
Scuff sanding, 661, 676
Sealers, 155, 163–64, 640, 766–68
Seam sealers, 164, 648–53
Seam welding, 184
Seats, 628–29, 795
Secondary damage, 310, 312, 313, 315
Sectioning, 397, 410–12, 413–16, 567
Sedan body shapes, 34, 48, 428, 430, 433
Self-centering gauges, 317, 326–28, 338, 405–6, 407
Self-diagnosis, 530, 531–33
Self-etching primers, 155, 643, 644–45
Semiunitized frames, 44
Sensor arrows, 612
Sensors, 531
Sensor wiring, 534
Series circuits, 513
Service parts identification labels, 59
Sheet metal brakes, 90
Sheet metal patches, 567
Sheet molded compounds, 540, 563, 564, 567, 568, 571
Shielded metal arc welding, 178
Shift linkage cables, 460
Shims, 384, 427, 436, 437, 440, 479
Shock absorber chisels, 112
Shock absorbers, 466
Shock hazards, 14, 24–25
Shock towers, 36, 333, 342, 346, 347, 364, 384, 465
Short circuit arc method, 182
Short circuits, 514, 517, 520
Shoulder belts, 606–8
Shrinking hammers, 82, 83, 261
Shrinking metal, 256, 257–60, 262
Side draft spray booths, 745
Sidesway damage, 307, 335–36
Side tone viewing angle, 808
Silencer pads, 425

Single-pull systems, 356, 370, 375–76
Siphon blasters, 681
"Skinning" (film formation), 758
Skip welding, 191–92, 204
Slapping spoons, 250
Sledgehammers, 81
Slide bars, 72–73
Slide hammers, 84, 219, 220, 254, 256
Slip-joint pliers, 76
Smoothing hammers, 112
Snap ring pliers, 170
Snips, 88, 96
Socket wrenches, 70–73
Soda blasters, 110, 157
Sodium quartz drying equipment, 752–53
Soft brazing (soldering), 176, 226, 227, 229
Solder, 261, 269, 281–83
Solenoids, 520, 524, 531
Solenoid valves, 495
Solvent popping, 836
Solvents, 8, 144, 146, 268–69, 543–44, 773, 774–75, 778
Sonic measuring systems, 346
Space frame construction, 34, 54, 57
Spade fuses, 521
Spark lighters, 222
Spark plugs, 524–25
Spatter finishes, 788
Spectrophotometers, 805, 806
Spectrum, 800
Speed welding, 551, 552, 558–59
Spider-webbing paint effects, 821, 822
Splash shields, 398, 436
Spoilers, 443, 444
Spoons (body tools), 83–84, 250, 251
Sport vehicle body shapes, 35
Spot-off timers, 202
Spot paint repair, 764–65, 781–82, 792, 809, 816
Spot putties, 159, 271–72, 688–89
Spot weld cutters, 106, 399
Spot welding, 117, 176
    advantages of, 208, 210
    comparative applications of, 180, 194
    components in, 211–12
    consumable, 184
    cutting around, 425
    inspection of, 216–18
    plug welding and, 199
    of quarter panels, 409, 410
    separation of, 399–402, 420, 421
    special applications of, 218–20
    techniques of, 201
Spray booths, 742–50
Spray gun cleaning tanks, 726–27
Spray guns, 100–102, 145, 703–34
Spraying, 9, 19, 20, 22, 714–61, 779–81, 808
Spray-out panels, 817
Spreader rams, 126
Spreader valves, 101
Spring compressors, 467
Spring hammering, 250, 262, 387
Squeeze-type spot welders, 179, 204, 208, 212–16, 219–20
Stabilizers, 780
Staggered butt joints, 411, 416, 417
Starting systems, 524–27
Static electricity, 19, 534, 624
Station wagon body shapes, 35, 48, 441–43
Steering axis inclination, 479–80
Steering knuckles, 466
Steering systems, 464–65, 468, 471–77
Stick-electrode arc welding, 178

Stitch-tamp plastic welding, 559
Stitch welding, 191–92, 194, 197, 202–3, 204, 205
Strain relievers, 122
Stressed hull structure, 44
Stress relieving, 385–88
Stretched metal, 257–59
Strikers (components), 57, 432, 442, 443
Strippers, 157, 682–83
Structural adhesives, 57, 426–27, 546–47, 569
Structural panels, 397–99, 467
Structural plastic fillers, 565
Strut bar brackets, 406
Strut centerline gauges, 332–33, 345
Strut fixtures, 348
Strut rear suspension, 470
Strut suspension systems, 466, 479
Stud welders, 219, 220, 254–55
Suction cups, 86, 254
Suction-feed spray guns, 138, 705, 706, 710, 738
Surface preparation, 272–74, 283–84, 286, 288–89, 291, 642–43, 658–700
Surface rust, 287, 290–91
Surfacing tools, 90–92
Surform files, 91–92, 277–78, 292
Surplus flames, 222, 223, 228, 282
Suspension systems, 464–71
Swirl marks, 109, 843
Switches, 520

Tabulation charts, 319, 320
Tachometers, 529
Tack welding, 193, 194, 195, 197, 296, 555
Tailgates, 441–43, 591, 593
Taillights, 519–10, 523, 600–601
Tail pipe cutters, 112
Tail pipes, 507
Tap and die sets, 79
Tape measures, 77–78, 321, 322
Telescoping trams, 340
Temperature, 277,.712, 774, 780–81
Temperature sensors, 529–30
Tempered glass, 576–78
Tensile strength (tensile stress), 166, 237, 238
Test lights, 516–17, 518
Test panels, 817–18
Test welds, 404, 412
T-handles, 72–73
Thermal conditioning equipment, 143–44
Thermal crayons, 192, 224, 238, 262, 387–88
Thermal paint, 192, 224, 238
Thermoplastics, 538
"Thermo-setting" acrylic enamels, 152–53, 770
Thermostats, 496, 498–99
Thinners, 153, 154, 775–76
Three-coat finishes, 787, 815–16, 817–18
Three-dimensional measurements, 331, 344
Three-way catalytic converters, 508
Three-way pulling, 381
Three-way tailgates, 441–43
Thrust line alignment, 482
Tinsley gauges, 713
Tin snips, 88
Tinted glass, 578

Tinting, 808–10, 818–20
Tire covers, 692
Toe adjustment, 480–81, 482, 485, 486
Toe dollies, 83
Topcoats, 9, 152, 763–64, 768–70, 787, 788–89
Torque boxes, 43, 45, 50, 328–29, 347
Torque converters, 454
Torque pattern, 166, 167
Torque rods, 440
Torque wrenches, 104, 105, 475–76
Torsion bar suspension system, 466, 468
Tracking, 482
Tracking gauges, 319
Traction, 356
Trade associations, 28
Tram gauges, 317, 319–23, 337, 338, 384, 407
Transaxles, 369, 454, 455, 460
Transfer cases, 455
Transmission cooler lines, 497
Transmission fluid, 496, 497
Transmission jacks, 125
Transmissions, 369, 454, 463, 464
Transversely mounted FF engine supports, 49, 50
Tri-coat finishes, 787, 815–16, 817–18
Trunk floors, 37, 164, 411, 417–18
Trunk hinges, 439, 440
Trunk lids, 37, 325, 440
Trunk locks, 440, 593–95
Trunk spraying, 647
Tungsten inert gas welding, 208
Turbine generators, 139, 739
Turnbuckles, 91
Turning radius angle, 482, 484, 485
Twin I-beam suspension system, 466, 467
Twin lead testers, 517
Twist damage, 309, 312–13, 333, 334, 383–84

Ultrahigh-strength steel, 237, 238, 399
Ultrasonic welding, 553, 562
Undercoats, 9, 152, 763, 765–66, 836
Undercutting (welding), 189
Universal dollies, 83
Universal flexible additives, 789
Universal joints, 73, 110, 454
Universal measuring systems, 336–46, 362
Universal sealers, 768
Upholstery tools, 89, 90
Urethane adhesives, 584, 587
Urethane basecoat/clearcoat systems, 813
Urethane bumpers, 792–94
Urethane enamels, 154, 768
Urethane foam, 417, 560

Vacuum boosters, 494
Vacuum cleaners, 21, 119, 120
Vacuum fluorescent diodes, 530
Vacuum hoses, 459
Vacuum leak detectors, 598
Vacuum suction cups, 254
Vacuum-type bleeders, 493
Variance chips, 806
Vector pulling system, 374–75
Ventilation, 20, 22, 207, 508
Vinyl adhesives, 163
Vinyl coating, 156, 157, 766

Vinyl interior parts, 794–95
Vinyl layers, 576, 578
Vinyl paints, 561
Vinyl prep, 561
Vinyl preserver, 795
Vinyl repairs, 560–61
Vinyl roofs, 616–22, 795
Viscometers, 710
Viscosity, 153, 710–11, 712, 780
Viscosity cups, 711–12, 778
Vise grips, 77, 218, 219
Vises, 78
Voltage, 513, 515, 516, 519–20
Voltmeters, 516, 524

Washers (fasteners), 168, 169
Wash filtration systems, 746–48
Wash-primers, 685
Waterbase paints, 152, 153–54, 770, 787
Water jackets, 496
Water leaks, 595, 596–97
Water pumps, 496, 499–500
Water spotting, 836–37
Wax/grease removers, 156, 272, 677–78
Weatherstrip adhesives, 162
Weatherstripping, 57, 440, 579, 598
Webbing, 607
Weld-bond adhesives, 426
Welding, 57, 174–234, 246–47, 258
Welding cables, 534
Welding gear, 17–18
Welding vise grips, 77
Weld rod adhesion tests, 542
Weld-through primers, 404, 640, 644
Wet/dry vacuum cleaners, 119
Wet filtration systems, 746–48
Wet metallic paint sprays, 811
Wet sanding stands, 753
Wet sandpapers, 161, 284, 285, 668–69
Wet spots (paint), 837
Wheel alignment, 5, 50, 354, 478–88
Wheel cylinders, 488
Wheel masks, 159
Wheel pullers, 477
Wheels, 466, 468, 474, 475–76
Wheel sensors, 495
Window glass, 428–30, 433–34, 576–90, 591–93
Window regulators, 36, 57, 434–36, 522, 591
Window tools, 90
Windshields, 576, 577, 578–90, 598
Windshield wipers, 524, 590
Wiring diagrams, 527
Wiring harnesses, 27, 511, 517, 523
Work hardening, 240–41, 242
Wrenches, 64–73, 78–79, 93–94

X-frames, 41, 42

Yield points, 239
Yield strength (yield stress), 237, 238

Zahn cups, 712
Zero planes, 330
Zirconia alumina grain, 662
Zone concept, 818
Zone-tempered glass, 577